PHYSICS

For Scientists and Engineers
with Modern Physics

Fifth Edition

Solar System Data

Body	Mass (kg)	Mean Radius (m)	Period (s)	Distance from the Sun (m)
Mercury	3.18×10^{23}	2.43×10^6	7.60×10^6	5.79×10^{10}
Venus	4.88×10^{24}	6.06×10^6	1.94×10^7	1.08×10^{11}
Earth	5.98×10^{24}	6.37×10^6	3.156×10^7	1.496×10^{11}
Mars	6.42×10^{23}	3.37×10^6	5.94×10^7	2.28×10^{11}
Jupiter	1.90×10^{27}	6.99×10^7	3.74×10^8	7.78×10^{11}
Saturn	5.68×10^{26}	5.85×10^7	9.35×10^8	1.43×10^{12}
Uranus	8.68×10^{25}	2.33×10^7	2.64×10^9	2.87×10^{12}
Neptune	1.03×10^{26}	2.21×10^7	5.22×10^9	4.50×10^{12}
Pluto	$\approx 1.4 \times 10^{22}$	$\approx 1.5 \times 10^6$	7.82×10^9	5.91×10^{12}
Moon	7.36×10^{22}	1.74×10^6	—	—
Sun	1.991×10^{30}	6.96×10^8	—	—

Physical Data Often Used[a]

Average Earth–Moon distance	3.84×10^8 m
Average Earth–Sun distance	1.496×10^{11} m
Average radius of the Earth	6.37×10^6 m
Density of air (0°C and 1 atm)	1.29 kg/m^3
Density of water (20°C and 1 atm)	1.00×10^3 kg/m^3
Free-fall acceleration	9.80 m/s^2
Mass of the Earth	5.98×10^{24} kg
Mass of the Moon	7.36×10^{22} kg
Mass of the Sun	1.99×10^{30} kg
Standard atmospheric pressure	1.013×10^5 Pa

[a] These are the values of the constants as used in the text.

Some Prefixes for Powers of Ten

Power	Prefix	Abbreviation	Power	Prefix	Abbreviation
10^{-24}	yocto	y	10^1	deka	da
10^{-21}	zepto	z	10^2	hecto	h
10^{-18}	atto	a	10^3	kilo	k
10^{-15}	femto	f	10^6	mega	M
10^{-12}	pico	p	10^9	giga	G
10^{-9}	nano	n	10^{12}	tera	T
10^{-6}	micro	μ	10^{15}	peta	P
10^{-3}	milli	m	10^{18}	exa	E
10^{-2}	centi	c	10^{21}	zetta	Z
10^{-1}	deci	d	10^{24}	yotta	Y

PHYSICS

For Scientists and Engineers
with Modern Physics

Fifth Edition

Raymond A. Serway

Robert J. Beichner
North Carolina State University

John W. Jewett, Jr., Contributing Author
California State Polytechnic University—Pomona

SAUNDERS COLLEGE PUBLISHING

A Division of Harcourt College Publishers

FORT WORTH   PHILADELPHIA   SAN DIEGO   NEW YORK   ORLANDO   AUSTIN   SAN ANTONIO   TORONTO   MONTREAL   LONDON   SYDNEY   TOKYO

Publisher: Emily Barrosse
Publisher: John Vondeling
Marketing Manager: Pauline Mula
Developmental Editor: Susan Dust Pashos
Project Editor: Frank Messina
Production Manager: Charlene Catlett Squibb
Manager of Art and Design: Carol Clarkson Bleistine
Text and Cover Designer: Ruth A. Hoover

Cover Image and Credit: Victoria Falls, Zimbabwe, at sunset. (© *Schafer & Hill/Tony Stone Images*)
Frontmatter Images and Credits: Title page: ripples on water. (© *Yagi Studio/Superstock, Inc.*); water droplets on flower (© *Richard H. Johnson/FPG International Corp.*); p. vii: Speed skaters (*Bill Bachman/Photo Researchers, Inc.*); p. viii: Sky surfer (*Jump Run Productions/The Image Bank*); p. ix: jogger (*Jim Cummins/FPG International*); p. x: Resistors on circuit board (*Superstock*); p. xi: Fuel element of a nuclear reactor (*Courtesy of U.S. Department of Energy/Photo Researchers, Inc.*); p. xiii: Long-jumper (*Chuck Muhlstock/FPG International*); p.xiv: Penny-farthing bicycle race (© *Steve Lovegrove/Tasmanian Photo Library*); p. xv: "Corkscrew" roller coaster (*Robin Smith/Tony Stone Images*); p.xvi: Cyclists pedaling uphill (*David Madison/Tony Stone Images*); p. xvii: Twin Falls on the island of Kauai, Hawaii (*Bruce Byers/FPG*); p. xviii: Bowling ball striking a pin (*Ben Rose/The Image Bank*); p. xix: Sprinters at staggered starting positions (© *Gerard Vandystadt/Photo Researchers, Inc.*); p. xx: U.S. Air Force F-117A stealth fighter in flight (*Courtesy of U.S. Air Force/Langley Air Force Base*); p. xxi: Cheering crowd (*Gregg Adams/Tony Stone Images*); p. xxiv: Welder (© *The Telegraph Colour Library/FPG*); p. xxv: Basketball player dunking ball (*Ron Chapple/FPG International*); p. xxvi: Athlete throwing discus (*Bruce Ayres/Tony Stone Images*); p. xxvii: Chum salmon (*Daniel J. Cox/Tony Stone Images*); p. xxix: Bottle-nosed dolphin (*Stuart Westmorland/Tony Stone Images*).

PHYSICS FOR SCIENTISTS AND ENGINEERS WITH MODERN PHYSICS, Fifth Edition
ISBN 0-03-022657-0
Library of Congress Catalog Card Number: 99-61821

Address for domestic orders:
Saunders College Publishing, 6277 Sea Harbor Drive, Orlando, FL 32887-6777
1-800-782-4479
e-mail collegesales@harcourt.com

Address for international orders:
International Customer Service, Harcourt, Inc.
6277 Sea Harbor Drive, Orlando FL 32887-6777
(407) 345-3800
Fax (407) 345-4060
e-mail hbintl@harcourt.com

Address for editorial correspondence:
Saunders College Publishing, Public Ledger Building, Suite 1250
150 S. Independence Mall West, Philadelphia, PA 19106-3412

Web Site Address
http://www.harcourtcollege.com

Printed in the United States of America
9012345678 032 10 987654321

SAUNDERS COLLEGE PUBLISHING

soon to become

A Harcourt Higher Learning Company

Soon you will find Saunders College Publishing's distinguished innovation, leadership, and support under a different name . . . a new brand that continues our unsurpassed quality, service, and commitment to education.

We are combining the strengths of our college imprints into one worldwide brand: ◐Harcourt Our mission is to make learning accessible to anyone, anywhere, anytime—reinforcing our commitment to lifelong learning.

We'll soon be Harcourt College Publishers. Ask for us by name.

One Company
"Where Learning Comes to Life."

Contents Overview

Table of Contents

part IV Electricity and Magnetism 707

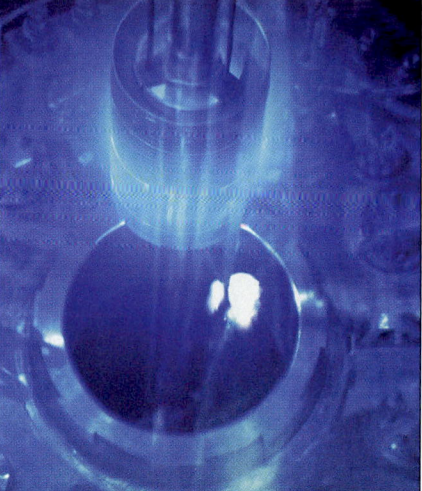

Preface

*I*n writing this fifth edition of *Physics for Scientists and Engineers,* we have made a major effort to improve the clarity of presentation and to include new pedagogical features that help support the learning and teaching processes. Drawing on positive feedback from users of the fourth edition and reviewers' suggestions, we have made refinements in order to better meet the needs of students and teachers. We have also streamlined the supplements package, which now includes a CD-ROM containing student tutorials and interactive problem-solving software, as well as offerings on the World Wide Web.

This textbook is intended for a course in introductory physics for students majoring in science or engineering. The entire contents of the text could be covered in a three-semester course, but it is possible to use the material in shorter sequences with the omission of selected chapters and sections. The mathematical background of the student taking this course should ideally include one semester of calculus. If that is not possible, the student should be enrolled in a concurrent course in introductory calculus.

OBJECTIVES

This introductory physics textbook has two main objectives: to provide the student with a clear and logical presentation of the basic concepts and principles of physics, and to strengthen an understanding of the concepts and principles through a broad range of interesting applications to the real world. To meet these objectives, we have placed emphasis on sound physical arguments and problem-solving methodology. At the same time, we have attempted to motivate the student through practical examples that demonstrate the role of physics in other disciplines, including engineering, chemistry, and medicine.

CHANGES IN THE FIFTH EDITION

A large number of changes and improvements have been made in preparing the fifth edition of this text. Some of the new features are based on our experiences and on current trends in science education. Other changes have been incorporated in response to comments and suggestions offered by users of the fourth edition and by reviewers of the manuscript. The following represent the major changes in the fifth edition:

Improved Illustrations

- **Time-sequenced events** are represented by circled letters in selected mechanics illustrations. For example, Figure 2.1b (see page 25) shows such letters at the appropriate places on a position–time graph. This construction helps students "translate" the observed motion into its graphical representation.

- **Motion diagrams** are used early in the text to illustrate the difference between velocity and acceleration, concepts easily confused by beginning students. (For example, see Figure 2.9 on page 34, Figure 4.5 on page 81, and Figure 4.8 on page 84.) Students will benefit greatly from sketching their own motion diagrams, as they are asked to do in the Quick Quizzes found in Chapter 4.

- **Greater realism** is achieved with the superimposition of photographs and line art in selected figures (see pages 96 and 97). Also, the three-dimensional appearance of "blocks" in figures accompanying examples and problems in mechanics has been improved (see pages 142 and 143).

More Realistic Worked Examples   Readers familiar with the fourth edition may recall that Example 12.5 involved raising a cylinder onto a step of height *h*. In this idealized example, we calculated the minimum force **F** necessary to raise the cylinder, as well as the reaction force exerted by the step on the cylinder. In the fifth edition, we are pleased to present the revised Example 12.5 (see page 370), in which we calculate the force that a person must apply to a wheelchair's main wheel to roll it up over an uncut sidewalk curb. Although the revised Example 12.5 involves essentially the same calculation as its fourth-edition predecessor (with some changes in notation), we think that the increased realism makes the example more interesting and provides new motivation for studying physics.

Puzzlers   Every chapter begins with an interesting photograph and a caption that includes a Puzzler. Each Puzzler poses a thought-provoking question that is intended to motivate students' curiosity and enhance their interest in the chapter's subject matter. Part or all of the answer to each Puzzler is contained within the chapter text and is indicated by the ⚙ icon.

Chapter Outlines   The opening page of each chapter now includes an outline of the chapter's major headings. This outline gives students and instructors a preview of the chapter's content.

QuickLabs   This new feature encourages students to perform simple experiments on their own, thereby engaging them actively in the learning process. Most QuickLab experiments can be performed with low-cost items such as string, rubber bands, tape, a ruler, drinking straws, and balloons. In most cases, students are asked to observe the outcome of an experiment and to explain their results in terms of what they have learned in the chapter. When appropriate, students are asked to record data and to graph their results.

Quick Quizzes   Several Quick Quiz questions are included in each chapter to provide students with opportunities to test their understanding of the physical concepts presented. Many questions are written in multiple-choice format and require students to make decisions and defend them on the basis of sound reasoning. Some of them have been written to help students overcome common misconceptions. (Instructors should look to the Instructor's Notes in the margins of the Instructor's Annotated Edition for tips regarding certain Quick Quizzes.) Answers to all Quick Quiz questions are found at the end of each chapter.

Marginal Comments and Icons   To provide students with further guidance, common misconceptions and pitfalls are pointed out in comments in the margin of the text. Often, references to the *Saunders Core Concepts in Physics CD-ROM* and useful World Wide Web site addresses are given in these comments to encourage students to expand their understanding of physical concepts. The 🪐 icon in the text margin refers students to the specific module and screen number(s) of the *Saunders Core*

Concepts in Physics CD-ROM that deals with the topic under discussion. A text illustration, example, Quick Quiz, or problem marked with the  icon indicates that it is accompanied by an Interactive Physics™ simulation that can be found on the *Student Tools CD-ROM*. See the Student Ancillaries section (page xviii) for descriptions of these two electronic learning packages.

Applications   Some chapters include Applications, which are about the same length as or slightly longer than worked examples. The Applications demonstrate to students how the physical principles covered in a chapter apply to practical problems of everyday life or engineering. For instance, Applications discuss antilock brakes within the context of static and kinetic friction (see Chapter 5); analyze the tension and compressional forces on the structural components of a truss bridge (see Chapter 12); explore the power delivered in automobile and diesel engine cycles (see Chapter 22); and discuss the construction and circuit wiring of holiday lightbulb strings (see Chapter 28).

Problems   A substantial revision of the end-of-chapter problems was made in an effort to improve their clarity and quality. Approximately 20 percent of the problems (about 650) are new, and most of these new problems are at the intermediate level (as identified by blue problem numbers). Many of the new problems require students to make order-of-magnitude calculations. All problems have been carefully edited and reworded when necessary. Solutions to approximately 20 percent of the end-of-chapter problems are included in the *Student Solutions Manual and Study Guide*. These problems are identified in the text by boxes around their numbers. A smaller subset of solutions are posted on the World Wide Web (**http://www.saunderscollege.com/physics/**) and are accessible to students and instructors using *Physics for Scientists and Engineers*. These problems are identified in the text by the **WEB** icon. See the next section for a complete description of other features of the problem set.

Line-by-Line Revision   The entire text has been carefully edited to improve clarity of presentation and precision of language. We believe that the result is a book that is both accurate and enjoyable to read.

Typographical and Notation Changes   The Text Features section (see page xvi) mentions the use of **boldface** type and screens for emphasizing important statements and definitions. Boldfaced passages in the text of the fifth edition replace the less legible passages appearing in italics in the fourth edition. Similarly, the symbols for vectors stand out very clearly from the surrounding text owing to the strong boldface type used in the fifth edition. As a step toward making equations more transparent and therefore more easily understood, the use of the subscripts "i" and "f" for initial and final values replaces the fourth edition's older notation, which makes use of subscript 0 (usually pronounced "naught") for an initial value and no subscript for a final value. In equations describing motion or direction, variables carry the subscripts x, y, or z whenever added clarity is needed.

Content Changes   Examination of the full Table of Contents might lead one to the impression that the content and organization of the textbook are essentially the same as those of the fourth edition. However, a number of subtle yet significant improvements in content have been made. Following are some examples:

• Section 16.8 contains a more complete and careful derivation of the power or rate of energy transfer for sinusoidal waves on strings. A similar development occurs in Section 17.3, which deals with the intensity of periodic sound waves.

- Section 18.2 contains an improved discussion of the envelope function of a standing wave.

- Chapter 20 contains an updated discussion of the distinction between heat and internal energy. Both heat and work are described and clarified as means of changing the energy of a system.

- Chapter 22 contains a new description of microstates and macrostates of a system, beginning with Section 22.6 on entropy and continuing through the end of the chapter.

- Section 24.3 contains a new list of guidelines for choosing a gaussian surface, allowing the student to take advantage of the symmetry of a charge distribution when determining the electric field.

- Chapter 25 contains new two- and three-dimensional graphs of the electric potential near a point charge and an electric dipole.

- In Chapter 27 and in following chapters, we use "Ohm's law" to refer only to the direct proportionality between current density and electric field seen in some (but not all) materials. See Section 27.2 and the corresponding Instructor's Note for a full explanation.

- Section 29.3 now makes explicit comparison between the potential energy of an electric dipole in an electric field and that of a magnetic dipole in a magnetic field. The section also contains new examples on satellite attitude control and the d'Arsonval galvanometer.

- Chapter 33 contains new information on rectifier circuits, including diodes. The material on rectifiers and filter circuits is now included in Optional Section 33.9, which follows the section on transformers and power transmission.

- In Chapter 35, reflection and refraction are now covered in separate sections, and discussion of Huygens's principle now precedes the section on dispersion and prisms. New Figure 35.8 illustrates retroreflection, which has many practical applications.

- Section 38.2 contains a new subsection considering two-slit diffraction patterns, in which the effects of diffraction and interference are combined.

- Within Section 39.4, new subsections cover space–time graphs and the relativistic Doppler effect. References to the concept of "rest mass" have been deleted.

Many sections in these and other chapters have been streamlined, deleted, or combined with other sections to allow for a more balanced presentation. In this extended version of the text, the former Chapter 44 on superconductivity in the fourth edition of *Physics for Scientists and Engineers with Modern Physics* has been deleted, and an abridged section on this topic has been added to Chapter 43. Some of the sections deleted from the fourth edition may be found on the textbook's Web sites for both instructors and students.

Instructor's Notes   For the first time, tips and comments are offered to instructors in blue marginal Instructor's Notes, which appear only in the Instructor's Annotated Edition. These annotations expand on common student misconceptions; call attention to certain worked examples, QuickLabs, and Quick Quizzes; or cite key physics education research literature that bears on the topic at hand. In some chapters, Instructor's Notes appear as footnotes in the end-of-chapter problem sets; these notes point out related groups of problems found in other chapters of the textbook. The Instructor's Annotated Edition includes Chapters 1 to 39.

CONTENT

The material in this book covers fundamental topics in classical physics and provides an introduction to modern physics. The book is divided into six parts. Part 1 (Chapters 1 to 15) deals with the fundamentals of Newtonian mechanics and the physics of fluids, Part 2 (Chapters 16 to 18) covers wave motion and sound, Part 3 (Chapters 19 to 22) addresses heat and thermodynamics, Part 4 (Chapters 23 to 34) treats electricity and magnetism, Part 5 (Chapters 35 to 38) covers light and optics, and Part 6 (Chapters 39 to 46) deals with relativity and modern physics. Each part opener includes an overview of the subject matter covered in that part, as well as some historical perspectives.

TEXT FEATURES

Most instructors would agree that the textbook selected for a course should be the student's primary guide for understanding and learning the subject matter. Furthermore, the textbook should be easily accessible and should be styled and written to facilitate instruction and learning. With these points in mind, we have included many pedagogical features in the textbook that are intended to enhance its usefulness to both students and instructors. These features are as follows:

Previews   Most chapters begin with a brief preview that includes a discussion of the chapters' objectives and content.

Important Statements and Equations   Most important statements and definitions are set in boldface type or are highlighted with a tan background screen for added emphasis and ease of review. Similarly, important equations are highlighted with a tan background screen to facilitate location.

Problem-Solving Hints   We have included general strategies for solving the types of problems featured both in the examples and in the end-of-chapter problems. This feature helps students to identify necessary steps in problem-solving and to eliminate any uncertainty they might have. Problem-Solving Hints are highlighted with a light blue-gray screen for emphasis and ease of location.

Marginal Notes   Comments and notes appearing in the margin can be used to locate important statements, equations, and concepts in the text.

Illustrations   The three-dimensional appearance of many illustrations has been improved in this fifth edition.

Mathematical Level   We have introduced calculus gradually, keeping in mind that students often take introductory courses in calculus and physics concurrently. Most steps are shown when basic equations are developed, and reference is often made to mathematical appendices at the end of the textbook. Vector products are introduced later in the text, where they are needed in physical applications. The dot product is introduced in Chapter 7, which addresses work and energy; the cross product is introduced in Chapter 11, which deals with rotational dynamics.

Worked Examples   A large number of worked examples of varying difficulty are presented to promote students' understanding of concepts. In many cases, the examples serve as models for solving the end-of-chapter problems. The examples are set off in boxes, and the answers to examples with numerical solutions are highlighted with a light blue-gray screen.

Worked Example Exercises   Many of the worked examples are followed immediately by exercises with answers. These exercises are intended to promote interactivity between the student and the textbook and to immediately reinforce the student's understanding of concepts and problem-solving techniques. The exercises represent extensions of the worked examples.

Conceptual Examples   As in the fourth edition, we have made a concerted effort to emphasize critical thinking and the teaching of physical concepts. We have accomplished this by including Conceptual Examples (for instance, see page 41). These examples provide students with a means of reviewing and applying the concepts presented in a section. Some Conceptual Examples demonstrate the connection between concepts presented in a chapter and other disciplines. The Conceptual Examples can serve as models for students when they are asked to respond to end-of-chapter questions, which are largely conceptual in nature.

Questions   Questions requiring verbal responses are provided at the end of each chapter. Over 1,000 questions are included in this edition. Some questions provide students with a means of testing their mastery of the concepts presented in the chapter. Others could serve as a basis for initiating classroom discussions. Answers to selected questions are included in the *Student Solutions Manual and Study Guide.*

Significant Figures   Significant figures in both worked examples and end-of-chapter problems have been handled with care. Most numerical examples and problems are worked out to either two or three significant figures, depending on the accuracy of the data provided.

Problems   An extensive set of problems is included at the end of each chapter; in all, over 3,000 problems are given throughout the text. Answers to odd-numbered problems are provided at the end of the book in a section whose pages have colored edges for ease of location. For the convenience of both the student and the instructor, about two thirds of the problems are keyed to specific sections of the chapter. The remaining problems, labeled "Additional Problems," are not keyed to specific sections.

Usually, the problems within a given section are presented so that the straightforward problems (those with black problem numbers) appear first; these straightforward problems are followed by problems of increasing difficulty. For ease of identification, the numbers of intermediate-level problems are printed in blue, and those of a small number of challenging problems are printed in magenta.

Review Problems   Many chapters include review problems that require the student to draw on numerous concepts covered in the chapter, as well as on those discussed in previous chapters. These problems could be used by students in preparing for tests and by instructors for special assignments and classroom discussions.

Paired Problems   Some end-of-chapter numerical problems are paired with the same problems in symbolic form. Two paired problems are identified by a common tan background screen.

Computer- and Calculator-Based Problems   Most chapters include one or more problems whose solution requires the use of a computer or graphing calculator. These problems are identified by the ▣ icon. Modeling of physical phenomena enables students to obtain graphical representations of variables and to perform numerical analyses.

Units   The international system of units (SI) is used throughout the text. The British engineering system of units (conventional system) is used only to a limited extent in the chapters on mechanics, heat, and thermodynamics.

Summaries   Each chapter contains a summary that reviews the important concepts and equations discussed in that chapter.

Appendices and Endpapers   Several appendices are provided at the end of the textbook. Most of the appendix material represents a review of mathematical concepts and techniques used in the text, including scientific notation, algebra, geometry, trigonometry, differential calculus, and integral calculus. Reference to these appendices is made throughout the text. Most mathematical review sections in the appendices include worked examples and exercises with answers. In addition to the mathematical reviews, the appendices contain tables of physical data, conversion factors, atomic masses, and the SI units of physical quantities, as well as a periodic table of the elements. Other useful information, including fundamental constants and physical data, planetary data, a list of standard prefixes, mathematical symbols, the Greek alphabet, and standard abbreviations of units of measure, appears on the endpapers.

ANCILLARIES

The ancillary package has been updated substantially and streamlined in response to suggestions from users of the fourth edition. The most essential changes in the student package are a *Student Solutions Manual and Study Guide* with a tighter focus on problem-solving, the *Student Tools CD-ROM,* and the *Saunders Core Concepts in Physics CD-ROM* developed by Archipelago Productions. Instructors will find increased support for their teaching efforts with new electronic materials.

Student Ancillaries

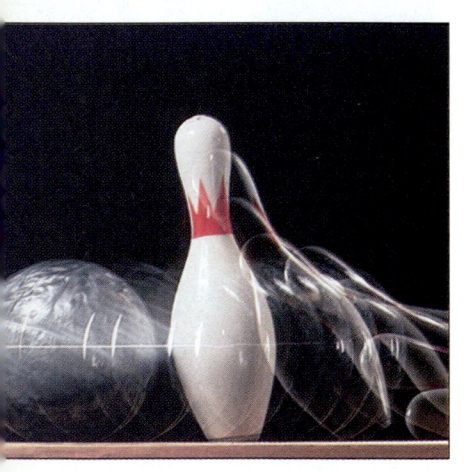

Student Solutions Manual and Study Guide by John R. Gordon, Ralph McGrew, and Raymond A. Serway, with contributions by Duane Deardorff. This two-volume manual features detailed solutions to 20 percent of the end-of-chapter problems from the textbook. Problems in the textbook whose complete solutions are found in the manual are identified by boxes around their numbers. The solutions to many problems follow the **GOAL** protocol described in the textbook (see page 47). The manual also features a list of important equations and concepts, as well as answers to selected end-of-chapter questions.

Pocket Guide by V. Gordon Lind. This 5″ × 7″ paperback is a section-by-section capsule of the textbook and serves as a handy guide for looking up important concepts, formulas, and problem-solving hints.

Student Tools CD-ROM   This CD-ROM contains tools that are designed to enhance the learning of physical concepts and train students to become better problem-solvers. It includes a textbook version of the highly acclaimed Interactive Physics™ software by MSC Working Knowledge and more than 100 Interactive Physics™ simulations keyed to appropriate figures, worked examples, Quick Quizzes, and selected end-of-chapter problems (as identified by the  icon).

Saunders Core Concepts in Physics CD-ROM   This CD-ROM package developed by Archipelago Productions applies the power of multimedia to the introductory physics course, offering full-motion animation and video, engaging interactive graphics, clear and concise text, and guiding narration. *Saunders Core Concepts in Physics CD-ROM* focuses on those concepts students usually find most difficult in the course, drawing from topics in mechanics, thermodynamics, electric fields, magnetic fields, and optics. The animations and graphics are presented to aid the student in developing accurate conceptual models of difficult topics—topics often too complex to be explained in words or chalkboard illustrations. The CD-ROM also presents step-by-step explorations of problem-solving strategies and provides animations of problems in order to promote conceptual understanding and sharpen problem-solving skills. Topics in the textbook that are further explored on the CD-ROM are identified by marginal 💿 icons that give the appropriate module and screen number(s). Students should look to the CD-ROM to aid in their understanding of these topics.

Student Web Site   Students will have access to an abundance of material at **http://www.saunderscollege.com/physics/**. The Web Site features special topic essays by guest authors, practice problems with answers, and optional topics that accompany selected chapters of the textbook. Also included are selected solutions from the *Student Solutions Manual and Study Guide,* a sampling of the *Pocket Guide,* and a glossary that includes more than 300 physics terms. Students also can take practice quizzes in our Practice Exercises and Testing area.

Physics Laboratory Manual, Second Edition by David Loyd. Updated and redesigned, this manual supplements the learning of basic physical principles while introducing laboratory procedures and equipment. Each chapter includes a pre-laboratory assignment, objectives, an equipment list, the theory behind the experiment, step-by-step experimental procedures, and questions. A laboratory report form is provided for each experiment so that students can record data and make calculations. Students are encouraged to apply statistical analysis to their data so that they can develop their ability to judge the validity of their results.

So You Want to Take Physics: A Preparatory Course with Calculus by Rodney Cole. This introductory-level book is useful to those students who need additional preparation before or during a calculus-based course in physics. The friendly, straightforward style makes it easier to understand how mathematics is used in the context of physics.

Life Science Applications for Physics by Jerry Faughn. This book provides examples, readings, and problems from the biological sciences as they relate to physics. Topics include "Friction in Human Joints," "Physics of the Human Circulatory System," "Physics of the Nervous System," and "Ultrasound and Its Applications." This supplement is useful in those courses taken by a significant number of pre-med students.

Instructor's Ancillaries

Instructor's Manual with Solutions by Ralph McGrew, Jeff Saul, and Charles Teague, with contributions by Duane Deardorff and Rhett Allain. This manual contains chapter summaries, answers to even-numbered problems, and complete worked solutions to all the problems in the textbook. The solutions to problems new to the fifth edition are marked for easy identification by the instructor. New to this edition of the manual are suggestions on how to teach difficult topics and how to help students overcome common misconceptions. These suggestions are based on recent research in physics education.

Instructor's Web Site The instructor's area at **http://www.saunderscollege.com/physics/** includes a listing of overhead transparencies; a guide to relevant experiments in David Loyd's *Physics Laboratory Manual, Second Edition*; a correlation guide between sections in *Physics for Scientists and Engineers* and modules in the *Saunders Core Concepts in Physics CD-ROM*; supplemental problems with answers; optional topics to accompany selected chapters of the textbook; and a problems correlation guide.

Instructor's Resource CD-ROM This CD-ROM accompanying the fifth edition of *Physics for Scientists and Engineers* has been created to provide instructors with an exciting new tool for classroom presentation. The CD-ROM contains a collection of graphics files of line art from the textbook. These files can be opened directly, can be imported into a variety of presentation packages, or can be used in the presentation package included on the CD-ROM. The labels for each piece of art have been enlarged and boldfaced to facilitate classroom viewing. The CD-ROM contains electronic files of the *Instructor's Manual, Test Bank,* and *Practice Problems with Solutions.*

CAPA: A Computer-Assisted Personalized Approach CAPA is a network system for learning, teaching, assessment, and administration. It provides students with personalized problem sets, quizzes, and examinations consisting of qualitative conceptual problems and quantitative problems, including problems from *Physics for Scientists and Engineers.* CAPA was developed through a collaborative effort of the Physics–Astronomy, Computer Science, and Chemistry Departments at Michigan State University. Students are given instant feedback and relevant hints via the Internet and may correct errors without penalty before an assignment's due date. The system records each student's participation and performance on assignments, quizzes, and examinations; and records are available on-line to both the individual student and to his or her instructor. For more information, visit the CAPA Web site at **http://www.pa.msu.edu/educ/CAPA/**

WebAssign: A Web-Based Homework System WebAssign is a Web-based homework delivery, collection, grading, and recording service developed at North Carolina State University. Instructors who sign up for WebAssign can assign homework to their students, using questions and problems taken directly from *Physics for Scientists and Engineers.* WebAssign gives students immediate feedback on their homework that helps them to master information and skills, leading to greater competence and better grades. WebAssign can free instructors from the drudgery of grading homework and recording scores, allowing them to devote more time to meeting with students and preparing classroom presentations. Details about and a demonstration of WebAssign are available at **http://webassign.net/info**. For more information about ordering this service, contact WebAssign at **webassign@ncsu.edu**

Homework Service With this service, instructors can reduce their grading workload by assigning thought-provoking homework problems using the World Wide Web. Instructors browse problem banks that include problems from *Physics for Scientists and Engineers,* select those they wish to assign to their students, and then let the Homework Service take over the delivery and grading. This system was developed and is maintained by Fred Moore at the University of Texas (**moore@physics.utexas.edu**). Students download their unique problems, submit their answers, and obtain immediate feedback; if students' answers are incorrect, they can resubmit them. This rapid grading feature facilitates effective learning. After the due date of their assignments, students can obtain the solutions to their problems. Minimal on-line connect time is required. The Homework Service uses algorithm-based problems: This means that each student solves sets of problems that are different from those given to other students. Details about and a demonstration of this service are available at **http://hw10.ph.utexas.edu/instInst.html**

Printed Test Bank by Edward Adelson. The *Printed Test Bank* contains approximately 2,300 multiple-choice questions. It is provided for the instructor who does not have access to a computer. About 20% of the old test items have been replaced with new, concept-based, thought-provoking questions.

Computerized Test Bank Available in Windows™ and Macintosh® formats, the *Computerized Test Bank* contains more than 2,300 multiple-choice questions, representing every chapter of the text. The *Computerized Test Bank* enables the instructor to create many unique tests by allowing the editing of questions and the addition of new questions. The software program solves all problems and prints each answer on a separate grading key. All questions have been reviewed for accuracy.

Overhead Transparency Acetates This collection of transparencies consists of 300 full-color figures from the text and features large print for easy viewing in the classroom.

Instructor's Manual for Physics Laboratory Manual by David Loyd. Each chapter contains a discussion of the experiment, teaching hints, answers to selected questions, and a post-laboratory quiz with short-answer and essay questions. It also includes a list of the suppliers of scientific equipment and a summary of the equipment needed for each of the laboratory experiments in the manual.

Saunders College Publishing, a division of Harcourt College Publishers, may provide complementary instructional aids and supplements or supplement packages to those adopters qualified under our adoption policy. Please contact your sales representative for more information. If as an adopter or potential user you receive supplements you do not need, please return them to your sales representative or send them to

Attn: Returns Department
Troy Warehouse
465 South Lincoln Drive
Troy, MO 63379

TEACHING OPTIONS

The topics in this textbook are presented in the following sequence: classical mechanics, mechanical waves, and heat and thermodynamics, followed by electricity and magnetism, electromagnetic waves, optics, and relativity. This presentation represents a more traditional sequence, with the subject of mechanical waves being presented before the topics of electricity and magnetism. Some instructors may prefer to cover this material after completing electricity and magnetism (i.e., after Chapter 34). The chapter on relativity was placed near the end of the text because this topic often is treated as an introduction to the era of "modern physics." If time permits, instructors may choose to cover Chapter 39 in the first semester after completing Chapter 14, as it concludes the material on Newtonian mechanics.

Instructors teaching a two-semester sequence can delete sections and chapters without any loss of continuity. We have labeled these as "Optional" in the Table of Contents and in the appropriate sections of the text. For student enrichment, instructors can assign some of these sections or chapters as extra reading.

ACKNOWLEDGMENTS

The fifth edition of this textbook was prepared with the guidance and assistance of many professors who reviewed part or all of the manuscript, the pre-revision text, or both. We wish to acknowledge the following scholars and express our sincere appreciation for their suggestions, criticisms, and encouragement:

Edward Adelson, *Ohio State University*
Roger Bengtson, *University of Texas at Austin*
Joseph Biegen, *Broome Community College*
Ronald J. Bieniek, *University of Missouri at Rolla*
Ronald Brown, *California Polytechnic State University—San Luis Obispo*
Michael E. Browne, *University of Idaho*
Tim Burns, *Leeward Community College*
Randall Caton, *Christopher Newport University*
Sekhar Chivukula, *Boston University*
Alfonso Díaz-Jiménez, *ADJOIN Research Center*
N. John DiNardo, *Drexel University*
F. Eugene Dunnum, *University of Florida*
William Ellis, *Cornell University*
F. Paul Esposito, *University of Cincinnati*
Paul Fahey, *University of Scranton*

Arnold Feldman, *University of Hawaii at Manoa*
Alexander Firestone, *Iowa State University*
Robert Forsythe, *Broome Community College*
Philip Fraundorf, *University of Missouri at St. Louis*
John Gerty, *Broome Community College*
John B. Gruber, *San Jose State University*
John Hubisz, *North Carolina State University*
Joey Huston, *Michigan State University*
Calvin S. Kalman, *Concordia University*
Natalie Kerr, M.D., *University of Tennessee, Memphis*
Peter Killen, *University of Queensland (Australia)*
Earl Koller, *Stevens Institute of Technology*
David LaGraffe, *U.S. Military Academy*
Ying-Cheng Lai, *University of Kansas*

Donald Larson, *Drexel University*
Robert Lieberman, *Cornell University*
Ralph McGrew, *Broome Community College*
David Mills, *Monash University (Australia)*
Clement J. Moses, *Utica College*
Peter Parker, *Yale University*
John Parsons, *Columbia University*
Arnold Perlmutter, *University of Miami*
Henry Schriemer, *Queen's University (Canada)*
Paul Snow, *University of Bath (U.K.)*
Edward W. Thomas, *Georgia Institute of Technology*
Charles C. Vuille, *Embry-Riddle Aeronautical University*
Xiaojun Wang, *Georgia Southern University*
Gail Welsh, *Salisbury State University*

This book was carefully checked for accuracy by James H. Smith (*University of Illinois at Urbana-Champaign*), Gregory Snow (*University of Nebraska—Lincoln*), Edward Gibson (*California State University—Sacramento*), Ronald Jodoin (*Rochester Institute of Technology*), Arnold Perlmutter (*University of Miami*), Michael Paesler (*North Carolina State University*), and Clement J. Moses (*Utica College*).

We thank the following people for their suggestions and assistance during the preparation of earlier editions of this textbook:

George Alexandrakis, *University of Miami*
Elmer E. Anderson, *University of Alabama*
Wallace Arthur, *Fairleigh Dickinson University*
Duane Aston, *California State University— Sacramento*
Stephen Baker, *Rice University*
Richard Barnes, *Iowa State University*
Stanley Bashkin, *University of Arizona*
Robert Bauman, *University of Alabama*
Marvin Blecher, *Virginia Polytechnic Institute and State University*
Jeffrey J. Braun, *University of Evansville*
Kenneth Brownstein, *University of Maine*
William A. Butler, *Eastern Illinois University*
Louis H. Cadwell, *Providence College*

Ron Canterna, *University of Wyoming*
Bo Casserberg, *University of Minnesota*
Soumya Chakravarti, *California Polytechnic State University*
C. H. Chan, *The University of Alabama at Huntsville*
Edward Chang, *University of Massachusetts at Amherst*
Don Chodrow, *James Madison University*
Clifton Bob Clark, *University of North Carolina at Greensboro*
Walter C. Connolly, *Appalachian State University*
Hans Courant, *University of Minnesota*
Lance E. De Long, *University of Kentucky*
James L. DuBard, *Binghamton-Southern College*

F. Paul Esposito, *University of Cincinnati*
Jerry S. Faughn, *Eastern Kentucky University*
Paul Feldker, *Florissant Valley Community College*
Joe L. Ferguson, *Mississippi State University*
R. H. Garstang, *University of Colorado at Boulder*
James B. Gerhart, *University of Washington*
John R. Gordon, *James Madison University*
Clark D. Hamilton, *National Bureau of Standards*
Mark Heald, *Swarthmore College*
Herb Helbig, *Rome Air Development Center*
Howard Herzog, *Broome Community College*
Paul Holoday, *Henry Ford Community College*

Jerome W. Hosken, *City College of San Francisco*

Larry Hmurcik, *University of Bridgeport*

William Ingham, *James Madison University*

Mario Iona, *University of Denver*

Karen L. Johnston, *North Carolina State University*

Brij M. Khorana, *Rose-Hulman Institute of Technology*

Larry Kirkpatrick, *Montana State University*

Carl Kocher, *Oregon State University*

Robert E. Kribel, *Jacksonville State University*

Barry Kunz, *Michigan Technological University*

Douglas A. Kurtze, *Clarkson University*

Fred Lipschultz, *University of Connecticut*

Francis A. Liuima, *Boston College*

Robert Long, *Worcester Polytechnic Institute*

Roger Ludin, *California Polytechnic State University*

Nolen G. Massey, *University of Texas at Arlington*

Charles E. McFarland, *University of Missouri at Rolla*

Ralph V. McGrew, *Broome Community College*

James Monroe, *The Pennsylvania State University, Beaver Campus*

Bruce Morgan, *United States Naval Academy*

Clement J. Moses, *Utica College*

Curt Moyer, *Clarkson University*

David Murdock, *Tennessee Technological University*

A. Wilson Nolle, *University of Texas at Austin*

Thomas L. O'Kuma, *San Jacinto College North*

Fred A. Otter, *University of Connecticut*

George Parker, *North Carolina State University*

William F. Parks, *University of Missouri at Rolla*

Philip B. Peters, *Virginia Military Institute*

Eric Peterson, *Highland Community College*

Richard Reimann, *Boise State University*

Joseph W. Rudmin, *James Madison University*

Jill Rugare, *DeVry Institute of Technology*

Charles Scherr, *University of Texas at Austin*

Eric Sheldon, *University of Massachusetts—Lowell*

John Shelton, *College of Lake County*

Stan Shepard, *The Pennsylvania State University*

James H. Smith, *University of Illinois at Urbana-Champaign*

Richard R. Sommerfield, *Foothill College*

Kervork Spartalian, *University of Vermont*

Robert W. Stewart, *University of Victoria*

James Stith, *American Institute of Physics*

Charles D. Teague, *Eastern Kentucky University*

Edward W. Thomas, *Georgia Institute of Technology*

Carl T. Tomizuka, *University of Arizona*

Herman Trivilino, *San Jacinto College North*

Som Tyagi, *Drexel University*

Steve Van Wyk, *Chapman College*

Joseph Veit, *Western Washington University*

T. S. Venkataraman, *Drexel University*

Noboru Wada, *Colorado School of Mines*

James Walker, *Washington State University*

Gary Williams, *University of California, Los Angeles*

George Williams, *University of Utah*

Edward Zimmerman, *University of Nebraska, Lincoln*

Earl Zwicker, *Illinois Institute of Technology*

We are grateful to Ralph McGrew for organizing the end-of-chapter problems, writing many new problems, and his suggestions for improving the content of the textbook. The new end-of-chapter problems were written by Rich Cohen, John DiNardo, Robert Forsythe, Ralph McGrew, and Ronald Bieniek, with suggestions by Liz McGrew, Alexandra Héder, and Richard McGrew. We thank Laurent Hodges for permission to use selected end-of-chapter problems. We are grateful to John R. Gordon, Ralph McGrew, and Duane Deardorff for writing the *Student Solutions Manual and Study Guide,* and we thank Michael Rudmin for its attractive layout. Ralph McGrew, Jeff Saul, and Charles Teague have prepared an excellent *Instructor's Manual,* and we thank them. We thank Gloria Langer, Linda Miller, and Jennifer Serway for their excellent work in preparing the *Instructor's Manual* and the supplemental materials that appear on our Web site.

Special thanks and recognition go to the professional staff at Saunders College Publishing—in particular, Susan Pashos, Sally Kusch, Carol Bleistine, Frank Messina, Suzanne Hakanen, Ruth Hoover, Alexandra Buczek, Pauline Mula, Walter Neary, and John Vondeling—for their fine work during the development and production of this textbook. We are most appreciative of the intelligent line editing by Irene Nunes, the final copy editing by Sue Nelson and Mary Patton, the excellent artwork produced by Rolin Graphics, and the dedicated photo research efforts of Dena Digilio Betz.

Finally, we are deeply indebted to our wives and children for their love, support, and long-term sacrifices.

Raymond A. Serway
Chapel Hill, North Carolina

Robert J. Beichner
Raleigh, North Carolina

John W. Jewett, Jr.
Pomona, California

To the Student

It is appropriate to offer some words of advice that should be of benefit to you, the student. Before doing so, we assume that you have read the Preface, which describes the various features of the text that will help you through the course.

HOW TO STUDY

Very often instructors are asked, "How should I study physics and prepare for examinations?" There is no simple answer to this question, but we would like to offer some suggestions that are based on our own experiences in learning and teaching over the years.

First and foremost, maintain a positive attitude toward the subject matter, keeping in mind that physics is the most fundamental of all natural sciences. Other science courses that follow will use the same physical principles, so it is important that you understand and are able to apply the various concepts and theories discussed in the text.

CONCEPTS AND PRINCIPLES

It is essential that you understand the basic concepts and principles before attempting to solve assigned problems. You can best accomplish this goal by carefully reading the textbook before you attend your lecture on the covered material. When reading the text, you should jot down those points that are not clear to you. We've purposely left wide margins in the text to give you space for doing this. Also be sure to make a diligent attempt at answering the questions in the Quick Quizzes as you come to them in your reading. We have worked hard to prepare questions that help you judge for yourself how well you understand the material. The QuickLabs provide an occasional break from your reading and will help you to experience some of the new concepts you are trying to learn. During class, take careful notes and ask questions about those ideas that are unclear to you. Keep in mind that few people are able to absorb the full meaning of scientific material after only one reading. Several readings of the text and your notes may be necessary. Your lectures and laboratory work supplement reading of the textbook and should clarify some of the more difficult material. You should minimize your memorization of material. Successful memorization of passages from the text, equations, and derivations does not necessarily indicate that you understand the material. Your understanding of the material will be enhanced through a combination of efficient study habits, discussions with other students and with instructors, and your ability to solve the problems presented in the textbook. Ask questions whenever you feel clarification of a concept is necessary.

STUDY SCHEDULE

It is important that you set up a regular study schedule, preferably one that is daily. Make sure that you to read the syllabus for the course and adhere to the schedule set by your instructor. The lectures will be much more meaningful if you read the corresponding textual material before attending them. As a general rule, you should devote about two hours of study time for every hour you are in class. If you are having trouble with the course, seek the advice of the instructor or other students who have taken the course. You may find it necessary to seek further instruction from experienced students. Very often, instructors offer review sessions in addition to regular class periods. It is important that you avoid the practice of delaying study until a day or two before an exam. More often than not, this approach has disastrous results. Rather than undertake an all-night study session, briefly review the basic concepts and equations and get a good night's rest. If you feel you need additional help in understanding the concepts, in preparing for exams, or in problem-solving, we suggest that you acquire a copy of the *Student Solutions Manual and Study Guide* that accompanies this textbook; this manual should be available at your college bookstore.

USE THE FEATURES

You should make full use of the various features of the text discussed in the preface. For example, marginal notes are useful for locating and describing important equations and concepts, and **boldfaced** type indicates important statements and definitions. Many useful tables are contained in the Appendices, but most are incorporated in the text where they are most often referenced. Appendix B is a convenient review of mathematical techniques.

Answers to odd-numbered problems are given at the end of the textbook, answers to Quick Quizzes are located at the end of each chapter, and answers to selected end-of-chapter questions are provided in the *Student Solutions Manual and Study Guide*. The exercises (with answers) that follow some worked examples represent extensions of those examples; in most of these exercises, you are expected to perform a simple calculation (see Example 4.7 on page 90). Their purpose is to test your problem-solving skills as you read through the text. Problem-Solving Hints are included in selected chapters throughout the text and give you additional information about how you should solve problems. The Table of Contents provides an overview of the entire text, while the Index enables you to locate specific material quickly. Footnotes sometimes are used to supplement the text or to cite other references on the subject discussed.

After reading a chapter, you should be able to define any new quantities introduced in that chapter and to discuss the principles and assumptions that were used to arrive at certain key relations. The chapter summaries and the review sections of the *Student Solutions Manual and Study Guide* should help you in this regard. In some cases, it may be necessary for you to refer to the index of the text to locate certain topics. You should be able to correctly associate with each physical quantity the symbol used to represent that quantity and the unit in which the quantity is specified. Furthermore, you should be able to express each important relation in a concise and accurate prose statement.

PROBLEM-SOLVING

R. P. Feynman, Nobel laureate in physics, once said, "You do not know anything until you have practiced." In keeping with this statement, we strongly advise that you develop the skills necessary to solve a wide range of problems. Your ability to solve problems will be one of the main tests of your knowledge of physics, and therefore you should try to solve as many problems as possible. It is good practice to try to find alternate solutions to the same problem. For example, you can solve problems in mechanics using Newton's laws, but very often an alternative method that draws on energy considerations is more direct. You should not deceive yourself into thinking that you understand a problem merely because you have seen it solved in class. You must be able to solve the problem and similar problems on your own.

The approach to solving problems should be carefully planned. A systematic plan is especially important when a problem involves several concepts. First, read the problem several times until you are confident you understand what is being asked. Look for any key words that will help you interpret the problem and perhaps allow you to make certain assumptions. Your ability to interpret a question properly is an integral part of problem-solving. Second, you should acquire the habit of writing down the information given in a problem and those quantities that need to be found; for example, you might construct a table listing both the quantities given and the quantities to be found. This procedure is sometimes used in the worked examples of the textbook. Finally, after you have decided on the method you feel is appropriate for a given problem, proceed with your solution. General problem-solving strategies of this type are included in the text and are highlighted with a light blue-gray screen. We have also developed the **GOAL** protocol (see page 47) to help guide you through complex problems. If you follow the steps of this procedure (**G**ather information, **O**rganize your approach, carry out your **A**nalysis, and finally **L**earn from your work), you will not only find it easier to come up with a solution, but you will also gain more from your efforts.

Often, students fail to recognize the limitations of certain formulas or physical laws in a particular situation. It is very important that you understand and remember the assumptions that underlie a particular theory or formalism. For example, certain equations in kinematics apply only to a particle moving with constant acceleration. These equations are not valid for describing motion whose acceleration is not constant, such as the motion of an object connected to a spring or the motion of an object through a fluid.

General Problem-Solving Strategy

Most courses in general physics require the student to learn the skills of problem-solving, and exams are largely composed of problems that test such skills. This brief section describes some useful ideas that will enable you to increase your accuracy in solving problems, enhance your understanding of physical concepts, eliminate initial panic or lack of direction in approaching a problem, and organize your work. One way to help accomplish these goals is to adopt a problem-solving strategy. Many chapters in this text include Problem-Solving Hints that should help you through the "rough spots."

In developing problem-solving strategies, five basic steps are commonly followed:

- Draw a suitable diagram with appropriate labels and coordinate axes (if needed).

- As you examine what is being asked in the problem, identify the basic physical principle (or principles) that are involved, listing the knowns and the unknowns.

- Select a basic relationship or derive an equation that can be used to find the unknown, and then solve the equation for the unknown symbolically.

- Substitute the given values along with the appropriate units into the equation.

- Obtain a numerical value for the unknown. The problem is verified and receives a check mark if the following questions can be properly answered: Do the units match? Is the answer reasonable? Is the plus or minus sign proper or meaningful?

One of the purposes of this strategy is to promote accuracy. Properly drawn diagrams can eliminate many sign errors. Diagrams also help to isolate the physical principles of the problem. Obtaining symbolic solutions and carefully labeling knowns and unknowns will help eliminate other careless errors. The use of symbolic solutions should help you think in terms of the physics of the problem. A check of units at the end of the problem can indicate a possible algebraic error. The physical layout and organization of your problem will make the final product more understandable and easier to follow. Once you have developed an organized system for examining problems and extracting relevant information, you will become a more confident problem-solver.

EXPERIMENTS

Physics is a science based on experimental observations. In view of this fact, we recommend that you try to supplement your reading of the text by performing various types of "hands-on" experiments, either at home or in the laboratory. Most chapters include one or two QuickLabs that describe simple experiments you can do on your own. These can be used to test ideas and models discussed in class or in the textbook. For example, the common Slinky™ toy is an excellent tool for studying traveling waves; a ball swinging on the end of a long string can be used to investigate pendulum motion; various masses attached to the end of a vertical spring or rubber band can be used to determine their elastic nature; an old pair of Polaroid sunglasses, some discarded lenses, and a magnifying glass are the components of various experiments in optics; and the approximate measure of the acceleration of gravity can be determined simply by measuring with a stopwatch the time it takes for a ball to drop from a known height. The list of such experiments is endless. When physical models are not available, be imaginative and try to develop models of your own.

NEW MEDIA

We strongly encourage you to use one or more of the following multimedia products that accompany this textbook. It is far easier to understand physics if you see it in action, and these new materials will enable you to become a part of that action.

Student Tools CD-ROM   The dual-platform (Windows™- and Macintosh®-compatible) *Student Tools CD-ROM* is available with each new copy of the textbook. This CD-ROM contains a textbook version of the Interactive Physics™ program by MSC Working Knowledge. Interactive Physics™ simulations are keyed to the following figures, worked examples, Quick Quizzes, and end-of-chapter problems (identified in the text with the 🖳 icon).

Saunders Core Concepts in Physics CD-ROM   In addition, you may purchase the *Saunders Core Concepts in Physics CD-ROM* developed by Archipelago Productions. This CD-ROM provides a complete multimedia presentation of selected topics in mechanics, thermodynamics, electromagnetism, and optics. It contains more than 350 movies—both animated and live video—that bring to life laboratory demonstrations, "real-world" examples, graphic models, and step-by-step explanations of essential mathematics. Those CD-ROM modules that supplement the material in *Physics for Scientists and Engineers* are identified in the margin of the text by the  icon.

AN INVITATION TO PHYSICS

It is our sincere hope that you too will find physics an exciting and enjoyable experience and that you will profit from this experience, regardless of your chosen profession. Welcome to the exciting world of physics!

The scientist does not study nature because it is useful; he studies it because he delights in it, and he delights in it because it is beautiful. If nature were not beautiful, it would not be worth knowing, and if nature were not worth knowing, life would not be worth living.

—Henri Poincaré

About the Authors

Raymond A. Serway received his doctorate at Illinois Institute of Technology and is Professor Emeritus at James Madison University. In 1990, he received the Madison Scholar award at James Madison University, where he taught for 17 years. Dr. Serway began his teaching career at Clarkson University, where he conducted research and taught from 1967 to 1980. He was the recipient of the Distinguished Teaching Award at Clarkson University in 1977 and of the Alumni Achievement Award from Utica College in 1985. As Guest Scientist at the IBM Research Laboratory in Zurich, Switzerland, he worked with K. Alex Müller, 1987 Nobel Prize recipient. Dr. Serway also was a visiting scientist at Argonne National Laboratory, where he collaborated with his mentor and friend, Sam Marshall. In addition to earlier editions of this textbook, Dr. Serway is the author of *Principles of Physics, Second Edition,* and co-author of *College Physics, Fifth Edition,* and *Modern Physics, Second Edition*; he also is the author of the high-school textbook *Physics,* published by Holt, Rinehart, & Winston. In addition, Dr. Serway has published more than 40 research papers in the field of condensed matter physics and has given more than 60 presentations at professional meetings. Dr. Serway and his wife Elizabeth enjoy traveling, golfing, and spending quality time with their four children and four grandchildren.

Robert J. Beichner received his doctorate at the State University of New York at Buffalo. Currently, he is Associate Professor of Physics at North Carolina State University, where he directs the Physics Education Research and Development Group. He has more than 20 years of teaching experience at the community-college, four year–college, and university levels. His research interests are centered on improving physics instruction: In his work, he has published studies of video-based laboratories, collaborative learning, technology-supplemented learning environments, and the assessment of student understanding of various physics topics. Dr. Beichner has held several leadership roles in the field of physics education research and has given numerous talks and colloquia on his work. In addition to being an author of this textbook, he is the co-author of two CD-ROMs, several commercially available software packages, and two books for preservice elementary school teachers. Dr. Beichner is an avid sea kayaker and enjoys spending time with his wife Mary and their two daughters, Sarah and Julie.

John W. Jewett, Jr. earned his doctorate at Ohio State University, specializing in optical and magnetic properties of condensed matter. He is currently Professor of Physics at California State Polytechnic University—Pomona. Throughout his teaching career, Dr. Jewett has been active in promoting science education. In addition to receiving four National Science Foundation grants, he helped found and direct the Southern California Area Modern Physics Institute (SCAMPI). He also is the director of Science IMPACT (Institute for Modern Pedagogy and Creative Teaching), which works with teachers and schools to develop effective science curricula. Both organizations operate in the United States and abroad. Dr. Jewett's honors include four Meritorious Performance and Professional Promise awards, selection as Outstanding Professor at California State Polytechnic University for 1991–1992, and the Excellence in Undergraduate Physics Teaching Award from the American Association of Physics Teachers (AAPT) in 1998. He has given many presentations both domestically and abroad, including multiple presentations at national meetings of the AAPT. He will co-author the third edition of *Principles of Physics* with Dr. Serway. Dr. Jewett enjoys playing piano, traveling, and collecting antique quack medical devices, as well as spending time with his wife Lisa and their children.

Pedagogical Color Chart

Part 1 (Chapters 1–15) : Mechanics

Displacement and position vectors

Linear (**p**) and angular (**L**) momentum vectors

Linear (**v**) and angular ($\boldsymbol{\omega}$) velocity vectors

Torque vectors ($\boldsymbol{\tau}$)

Velocity component vectors

Linear or rotational motion directions

Force vectors (**F**)

Force component vectors

Springs

Acceleration vectors (**a**)

Pulleys

Acceleration component vectors

Part 4 (Chapters 23–34) : Electricity and Magnetism

Electric fields

Capacitors

Magnetic fields

Inductors (coils)

Positive charges

Voltmeters

Negative charges

Ammeters

Resistors

Galvanometers

Batteries and other dc power supplies

ac Generators

Switches

Ground symbol

Part 5 (Chapters 35–38) : Light and Optics

Light rays

Objects

Lenses and prisms

Images

Mirrors

Help your students get to the "Core" of physics with *Saunders Core Concepts in Physics CD-ROM!*

Available with Serway & Beichner's *Physics for Scientists and Engineers, Fifth Edition*

Core Concepts in Physics CD-ROM covers the core principles of calculus-based physics through the use of on-screen simulations, animations, and videos. The combination of a conceptual presentation of physical principles and problem-solving helps students "see" material often too difficult for instructors to represent with chalkboard diagrams or with written and verbal descriptions.

Core Concepts in Physics CD-ROM contains more than 350 animated and live videos. "Real world" examples, graphics, models, and step-by-step explanations of mathematics help to spark students' interest and facilitate their exploration of the fascinating world of physics!

SERWAY · BEICHNER
PHYSICS
For Scientists and Engineers
FIFTH EDITION

PEDAGOGICAL COLOR CHART

Part 1 (Chapters 1–15)
Mechanics

Displacement and position vectors

Linear ($\mathbf{v}$) and angular ($\boldsymbol{\omega}$) velocity vectors

Velocity component vectors

Force vectors ($\mathbf{F}$)

Force component vectors

Acceleration vectors ($\mathbf{a}$)

Acceleration component vectors

Linear ($\mathbf{p}$) and angular ($\mathbf{L}$) momentum vectors

Torque vectors ($\boldsymbol{\tau}$)

Linear or rotational motion directions

Springs

Pulleys

Part 4 (Chapters 23–34)
Electricity and Magnetism

Electric fields

Magnetic fields

Positive charges

(continued on back)

SAUNDERS COLLEGE PUBLISHING

A Division of Harcourt College Publishers

SERWAY · BEICHNER
PHYSICS
For Scientists and Engineers
FIFTH EDITION

PEDAGOGICAL COLOR CHART

Part 4 (Chapters 23–34)
Electricity and Magnetism
(*continued*)

Negative charges

Batteries and other
dc power supplies

Switches

Resistors

Capacitors

Inductors (coils)

Voltmeters

Ammeters

Galvanometers

ac Generators

Ground symbol

Part 5 (Chapters 35–38)
Light and Optics

Light rays

Lenses and
prisms

Mirrors

Objects

Images

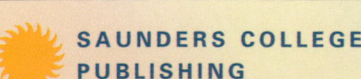
SAUNDERS COLLEGE
PUBLISHING

A Division of Harcourt College Publishers

**Application,
Visualization, and Practice.
Student Tools CD–ROM helps
students understand the
natural forces and principles
of physics.**

This CD-ROM contains tools that are designed to enhance the learning of physical concepts and to assist your students as they become better problem-solvers. It includes a textbook version of the highly acclaimed *Interactive Physics*™ software by Knowledge Revolution and more than 100 *Interactive Physics*™ simulations keyed to appropriate worked examples, figures, Quick Quizzes, and selected end-of-chapter problems. *The Student Tools CD-ROM* also features a graphing calculator and support for working the end-of-chapter problems that require a computer.

Mechanics

Physics, the most fundamental physical science, is concerned with the basic principles of the Universe. It is the foundation upon which the other sciences—astronomy, biology, chemistry, and geology—are based. The beauty of physics lies in the simplicity of the fundamental physical theories and in the manner in which just a small number of fundamental concepts, equations, and assumptions can alter and expand our view of the world around us.

All of physics can be divided into five main areas:

1. Classical mechanics, which is concerned with the motion of objects that are large relative to atoms and move at speeds much slower than the speed of light
2. Relativity, which is a theory describing objects moving at any speed, even speeds approaching the speed of light
3. Thermodynamics, which deals with heat, work, temperature, and the statistical behavior of a large number of particles
4. Electromagnetism, which is concerned with electricity, magnetism, and electromagnetic fields
5. Quantum mechanics, a collection of theories dealing with the behavior of matter at both the submicroscopic and macroscopic levels

The first part of this textbook deals with classical mechanics, sometimes referred to as *Newtonian mechanics* or simply *mechanics.* This is an appropriate place to begin an introductory text because many of the basic principles used to understand mechanical systems can later be used to describe such natural phenomena as waves and energy transfer. Furthermore, the laws of conservation of energy and momentum introduced in mechanics retain their importance in the fundamental theories of other areas of physics.

Today, classical mechanics is of vital importance to students from all disciplines. It is highly successful in describing the motions of different bodies, such as planets, rockets, and baseballs. In the first part of the text, we shall describe the laws of classical mechanics and examine a wide range of phenomena that can be understood with these fundamental ideas. We shall treat the various disciplines of classical physics in separate sections; however, we shall see that the disciplines of mechanics and electromagnetism are basic to all other branches of classical and modern physics.

c h a p t e r

Physics and Measurement

Like all other sciences, physics is based on experimental observations and quantitative measurements. The main objective of physics is to find the limited number of fundamental laws that govern natural phenomena and to use them to develop theories that can predict the results of future experiments. The fundamental laws used in developing theories are expressed in the language of mathematics, the tool that provides a bridge between theory and experiment

When a discrepancy between theory and experiment arises, new theories must be formulated to remove the discrepancy. Many times a theory is satisfactory only under limited conditions; a more general theory might be satisfactory without such limitations. For example, the laws of motion discovered by Isaac Newton (1642–1727) in the 17th century accurately describe the motion of bodies at normal speeds but do not apply to objects moving at speeds comparable with the speed of light. In contrast, the special theory of relativity developed by Albert Einstein (1879–1955) in the early 1900s gives the same results as Newton's laws at low speeds but also correctly describes motion at speeds approaching the speed of light. Hence, Einstein's is a more general theory of motion.

Classical physics, which means all of the physics developed before 1900, includes the theories, concepts, laws, and experiments in classical mechanics, thermodynamics, and electromagnetism.

Important contributions to classical physics were provided by Newton, who developed classical mechanics as a systematic theory and was one of the originators of calculus as a mathematical tool. Major developments in mechanics continued in the 18th century, but the fields of thermodynamics and electricity and magnetism were not developed until the latter part of the 19th century, principally because before that time the apparatus for controlled experiments was either too crude or unavailable.

A new era in physics, usually referred to as *modern physics,* began near the end of the 19th century. Modern physics developed mainly because of the discovery that many physical phenomena could not be explained by classical physics. The two most important developments in modern physics were the theories of relativity and quantum mechanics. Einstein's theory of relativity revolutionized the traditional concepts of space, time, and energy; quantum mechanics, which applies to both the microscopic and macroscopic worlds, was originally formulated by a number of distinguished scientists to provide descriptions of physical phenomena at the atomic level.

Scientists constantly work at improving our understanding of phenomena and fundamental laws, and new discoveries are made every day. In many research areas, a great deal of overlap exists between physics, chemistry, geology, and biology, as well as engineering. Some of the most notable developments are (1) numerous space missions and the landing of astronauts on the Moon, (2) microcircuitry and high-speed computers, and (3) sophisticated imaging techniques used in scientific research and medicine. The impact such developments and discoveries have had on our society has indeed been great, and it is very likely that future discoveries and developments will be just as exciting and challenging and of great benefit to humanity.

1.1 ▶ STANDARDS OF LENGTH, MASS, AND TIME

The laws of physics are expressed in terms of basic quantities that require a clear definition. In mechanics, the three basic quantities are length (L), mass (M), and time (T). All other quantities in mechanics can be expressed in terms of these three.

If we are to report the results of a measurement to someone who wishes to reproduce this measurement, a *standard* must be defined. It would be meaningless if a visitor from another planet were to talk to us about a length of 8 "glitches" if we do not know the meaning of the unit glitch. On the other hand, if someone familiar with our system of measurement reports that a wall is 2 meters high and our unit of length is defined to be 1 meter, we know that the height of the wall is twice our basic length unit. Likewise, if we are told that a person has a mass of 75 kilograms and our unit of mass is defined to be 1 kilogram, then that person is 75 times as massive as our basic unit.[1] Whatever is chosen as a standard must be readily accessible and possess some property that can be measured reliably—measurements taken by different people in different places must yield the same result.

In 1960, an international committee established a set of standards for length, mass, and other basic quantities. The system established is an adaptation of the metric system, and it is called the **SI system** of units. (The abbreviation SI comes from the system's French name "Système International.") In this system, the units of length, mass, and time are the meter, kilogram, and second, respectively. Other SI standards established by the committee are those for temperature (the *kelvin*), electric current (the *ampere*), luminous intensity (the *candela*), and the amount of substance (the *mole*). In our study of mechanics we shall be concerned only with the units of length, mass, and time.

Length

In A.D. 1120 the king of England decreed that the standard of length in his country would be named the *yard* and would be precisely equal to the distance from the tip of his nose to the end of his outstretched arm. Similarly, the original standard for the foot adopted by the French was the length of the royal foot of King Louis XIV. This standard prevailed until 1799, when the legal standard of length in France became the *meter*, defined as one ten-millionth the distance from the equator to the North Pole along one particular longitudinal line that passes through Paris.

Many other systems for measuring length have been developed over the years, but the advantages of the French system have caused it to prevail in almost all countries and in scientific circles everywhere. As recently as 1960, the length of the meter was defined as the distance between two lines on a specific platinum–iridium bar stored under controlled conditions in France. This standard was abandoned for several reasons, a principal one being that the limited accuracy with which the separation between the lines on the bar can be determined does not meet the current requirements of science and technology. In the 1960s and 1970s, the meter was defined as 1 650 763.73 wavelengths of orange-red light emitted from a krypton-86 lamp. However, in October 1983, the **meter (m) was redefined as the distance traveled by light in vacuum during a time of 1/299 792 458 second.** In effect, this latest definition establishes that the speed of light in vacuum is precisely 299 792 458 m per second.

Table 1.1 lists approximate values of some measured lengths.

[1] The need for assigning numerical values to various measured physical quantities was expressed by Lord Kelvin (William Thomson) as follows: "I often say that when you can measure what you are speaking about, and express it in numbers, you should know something about it, but when you cannot express it in numbers, your knowledge is of a meagre and unsatisfactory kind. It may be the beginning of knowledge but you have scarcely in your thoughts advanced to the state of science."

TABLE 1.1   **Approximate Values of Some Measured Lengths**

	Length (m)
Distance from the Earth to most remote known quasar	1.4×10^{26}
Distance from the Earth to most remote known normal galaxies	9×10^{25}
Distance from the Earth to nearest large galaxy (M 31, the Andromeda galaxy)	2×10^{22}
Distance from the Sun to nearest star (Proxima Centauri)	4×10^{16}
One lightyear	9.46×10^{15}
Mean orbit radius of the Earth about the Sun	1.50×10^{11}
Mean distance from the Earth to the Moon	3.84×10^{8}
Distance from the equator to the North Pole	1.00×10^{7}
Mean radius of the Earth	6.37×10^{6}
Typical altitude (above the surface) of a satellite orbiting the Earth	2×10^{5}
Length of a football field	9.1×10^{1}
Length of a housefly	5×10^{-3}
Size of smallest dust particles	$\sim 10^{-4}$
Size of cells of most living organisms	$\sim 10^{-5}$
Diameter of a hydrogen atom	$\sim 10^{-10}$
Diameter of an atomic nucleus	$\sim 10^{-14}$
Diameter of a proton	$\sim 10^{-15}$

Mass

The basic SI unit of mass, **the kilogram (kg), is defined as the mass of a specific platinum–iridium alloy cylinder kept at the International Bureau of Weights and Measures at Sèvres, France.** This mass standard was established in 1887 and has not been changed since that time because platinum–iridium is an unusually stable alloy (Fig. 1.1a). A duplicate of the Sèvres cylinder is kept at the National Institute of Standards and Technology (NIST) in Gaithersburg, Maryland.

Table 1.2 lists approximate values of the masses of various objects.

web

Visit the Bureau at **www.bipm.fr** or the National Institute of Standards at **www.NIST.gov**

Time

Before 1960, the standard of time was defined in terms of the *mean solar day* for the year 1900.[2] The *mean solar second* was originally defined as $\left(\frac{1}{60}\right)\left(\frac{1}{60}\right)\left(\frac{1}{24}\right)$ of a mean solar day. The rotation of the Earth is now known to vary slightly with time, however, and therefore this motion is not a good one to use for defining a standard.

In 1967, consequently, the second was redefined to take advantage of the high precision obtainable in a device known as an *atomic clock* (Fig. 1.1b). In this device, the frequencies associated with certain atomic transitions can be measured to a precision of one part in 10^{12}. This is equivalent to an uncertainty of less than one second every 30 000 years. Thus, in 1967 the SI unit of time, the *second,* was redefined using the characteristic frequency of a particular kind of cesium atom as the "reference clock." The basic SI unit of time, **the second (s), is defined as 9 192 631 770 times the period of vibration of radiation from the cesium-133 atom.**[3] To keep these atomic clocks—and therefore all common clocks and

TABLE 1.2
Masses of Various Bodies (Approximate Values)

Body	Mass (kg)
Visible Universe	$\sim 10^{52}$
Milky Way galaxy	7×10^{41}
Sun	1.99×10^{30}
Earth	5.98×10^{24}
Moon	7.36×10^{22}
Horse	$\sim 10^{3}$
Human	$\sim 10^{2}$
Frog	$\sim 10^{-1}$
Mosquito	$\sim 10^{-5}$
Bacterium	$\sim 10^{-15}$
Hydrogen atom	1.67×10^{-27}
Electron	9.11×10^{-31}

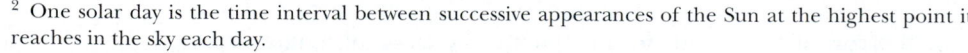

[2] One solar day is the time interval between successive appearances of the Sun at the highest point it reaches in the sky each day.

[3] *Period* is defined as the time interval needed for one complete vibration.

Figure 1.1   *(Top)* The National Standard Kilogram No. 20, an accurate copy of the International Standard Kilogram kept at Sèvres, France, is housed under a double bell jar in a vault at the National Institute of Standards and Technology (NIST). *(Bottom)* The primary frequency standard (an atomic clock) at the NIST. This device keeps time with an accuracy of about 3 millionths of a second per year. *(Courtesy of National Institute of Standards and Technology, U.S. Department of Commerce)*

watches that are set to them—synchronized, it has sometimes been necessary to add leap seconds to our clocks. This is not a new idea. In 46 B.C. Julius Caesar began the practice of adding extra days to the calendar during leap years so that the seasons occurred at about the same date each year.

Since Einstein's discovery of the linkage between space and time, precise measurement of time intervals requires that we know both the state of motion of the clock used to measure the interval and, in some cases, the location of the clock as well. Otherwise, for example, global positioning system satellites might be unable to pinpoint your location with sufficient accuracy, should you need rescuing.

Approximate values of time intervals are presented in Table 1.3.

In addition to SI, another system of units, the *British engineering system* (sometimes called the *conventional system*), is still used in the United States despite acceptance of SI by the rest of the world. In this system, the units of length, mass, and

TABLE 1.3   Approximate Values of Some Time Intervals

	Interval (s)
Age of the Universe	5×10^{17}
Age of the Earth	1.3×10^{17}
Average age of a college student	6.3×10^8
One year	3.16×10^7
One day (time for one rotation of the Earth about its axis)	8.64×10^4
Time between normal heartbeats	8×10^{-1}
Period of audible sound waves	$\sim 10^{-3}$
Period of typical radio waves	$\sim 10^{-6}$
Period of vibration of an atom in a solid	$\sim 10^{-13}$
Period of visible light waves	$\sim 10^{-15}$
Duration of a nuclear collision	$\sim 10^{-22}$
Time for light to cross a proton	$\sim 10^{-24}$

time are the foot (ft), slug, and second, respectively. In this text we shall use SI units because they are almost universally accepted in science and industry. We shall make some limited use of British engineering units in the study of classical mechanics.

In addition to the basic SI units of meter, kilogram, and second, we can also use other units, such as millimeters and nanoseconds, where the prefixes *milli-* and *nano-* denote various powers of ten. Some of the most frequently used prefixes for the various powers of ten and their abbreviations are listed in Table 1.4. For

TABLE 1.4   Prefixes for SI Units

Power	Prefix	Abbreviation
10^{-24}	yocto	y
10^{-21}	zepto	z
10^{-18}	atto	a
10^{-15}	femto	f
10^{-12}	pico	p
10^{-9}	nano	n
10^{-6}	micro	μ
10^{-3}	milli	m
10^{-2}	centi	c
10^{-1}	deci	d
10^1	deka	da
10^3	kilo	k
10^6	mega	M
10^9	giga	G
10^{12}	tera	T
10^{15}	peta	P
10^{18}	exa	E
10^{21}	zetta	Z
10^{24}	yotta	Y

example, 10^{-3} m is equivalent to 1 millimeter (mm), and 10^3 m corresponds to 1 kilometer (km). Likewise, 1 kg is 10^3 grams (g), and 1 megavolt (MV) is 10^6 volts (V).

1.2   THE BUILDING BLOCKS OF MATTER

A 1-kg cube of solid gold has a length of 3.73 cm on a side. Is this cube nothing but wall-to-wall gold, with no empty space? If the cube is cut in half, the two pieces still retain their chemical identity as solid gold. But what if the pieces are cut again and again, indefinitely? Will the smaller and smaller pieces always be gold? Questions such as these can be traced back to early Greek philosophers. Two of them— Leucippus and his student Democritus—could not accept the idea that such cuttings could go on forever. They speculated that the process ultimately must end when it produces a particle that can no longer be cut. In Greek, *atomos* means "not sliceable." From this comes our English word *atom*.

Let us review briefly what is known about the structure of matter. All ordinary matter consists of atoms, and each atom is made up of electrons surrounding a central nucleus. Following the discovery of the nucleus in 1911, the question arose: Does it have structure? That is, is the nucleus a single particle or a collection of particles? The exact composition of the nucleus is not known completely even today, but by the early 1930s a model evolved that helped us understand how the nucleus behaves. Specifically, scientists determined that occupying the nucleus are two basic entities, protons and neutrons. The *proton* carries a positive charge, and a specific element is identified by the number of protons in its nucleus. This number is called the **atomic number** of the element. For instance, the nucleus of a hydrogen atom contains one proton (and so the atomic number of hydrogen is 1), the nucleus of a helium atom contains two protons (atomic number 2), and the nucleus of a uranium atom contains 92 protons (atomic number 92). In addition to atomic number, there is a second number characterizing atoms—**mass number,** defined as the number of protons plus neutrons in a nucleus. As we shall see, the atomic number of an element never varies (i.e., the number of protons does not vary) but the mass number can vary (i.e., the number of neutrons varies). Two or more atoms of the same element having different mass numbers are **isotopes** of one another.

The existence of neutrons was verified conclusively in 1932. A *neutron* has no charge and a mass that is about equal to that of a proton. One of its primary purposes is to act as a "glue" that holds the nucleus together. If neutrons were not present in the nucleus, the repulsive force between the positively charged particles would cause the nucleus to come apart.

But is this where the breaking down stops? Protons, neutrons, and a host of other exotic particles are now known to be composed of six different varieties of particles called **quarks,** which have been given the names of *up, down, strange, charm, bottom,* and *top.* The up, charm, and top quarks have charges of $+\frac{2}{3}$ that of the proton, whereas the down, strange, and bottom quarks have charges of $-\frac{1}{3}$ that of the proton. The proton consists of two up quarks and one down quark (Fig. 1.2), which you can easily show leads to the correct charge for the proton. Likewise, the neutron consists of two down quarks and one up quark, giving a net charge of zero.

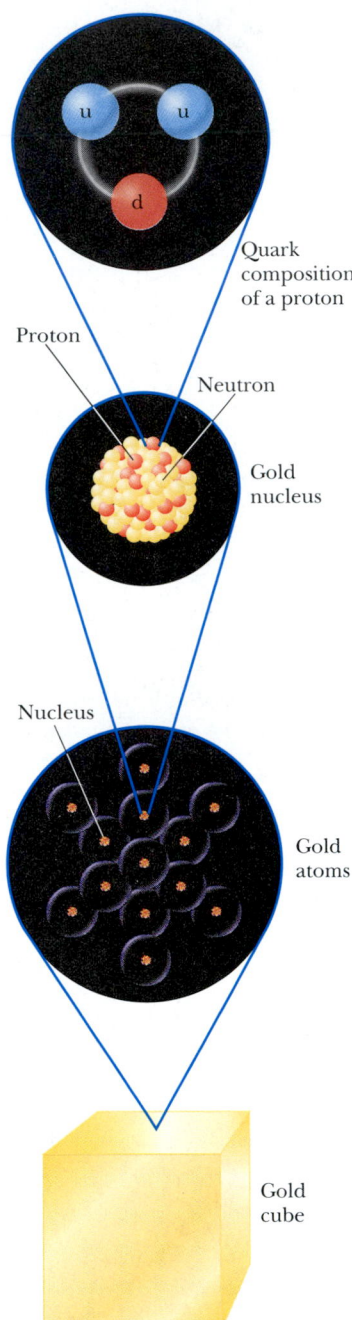

Quark composition of a proton

Proton

Neutron

Gold nucleus

Nucleus

Gold atoms

Gold cube

Figure 1.2   Levels of organization in matter. Ordinary matter consists of atoms, and at the center of each atom is a compact nucleus consisting of protons and neutrons. Protons and neutrons are composed of quarks. The quark composition of a proton is shown.

1.3 DENSITY

A property of any substance is its **density** ρ (Greek letter rho), defined as the amount of mass contained in a unit volume, which we usually express as *mass per unit volume:*

A table of the letters in the Greek alphabet is provided on the back endsheet of this textbook.

$$\rho \equiv \frac{m}{V}$$

(1.1)

For example, aluminum has a density of 2.70 g/cm^3, and lead has a density of 11.3 g/cm^3. Therefore, a piece of aluminum of volume 10.0 cm^3 has a mass of 27.0 g, whereas an equivalent volume of lead has a mass of 113 g. A list of densities for various substances is given Table 1.5.

The difference in density between aluminum and lead is due, in part, to their different *atomic masses*. The **atomic mass** of an element is the average mass of one atom in a sample of the element that contains all the element's isotopes, where the relative amounts of isotopes are the same as the relative amounts found in nature. The unit for atomic mass is the *atomic mass unit* (u), where 1 u = 1.660 540 2 $\times$ 10^{-27} kg. The atomic mass of lead is 207 u, and that of aluminum is 27.0 u. However, the ratio of atomic masses, 207 u/27.0 u = 7.67, does not correspond to the ratio of densities, (11.3 g/cm^3)/(2.70 g/cm^3) = 4.19. The discrepancy is due to the difference in atomic separations and atomic arrangements in the crystal structure of these two substances.

The mass of a nucleus is measured relative to the mass of the nucleus of the carbon-12 isotope, often written as ^{12}C. (This isotope of carbon has six protons and six neutrons. Other carbon isotopes have six protons but different numbers of neutrons.) Practically all of the mass of an atom is contained within the nucleus. Because the atomic mass of ^{12}C is defined to be exactly 12 u, the proton and neutron each have a mass of about 1 u.

One mole (mol) of a substance is that amount of the substance that contains as many particles (atoms, molecules, or other particles) as there are atoms in 12 g of the carbon-12 isotope. One mole of substance A contains the same number of particles as there are in 1 mol of any other substance B. For example, 1 mol of aluminum contains the same number of atoms as 1 mol of lead.

TABLE 1.5	Densities of Various Substances
Substance	**Density ρ (10^3 kg/m^3)**
Gold	19.3
Uranium	18.7
Lead	11.3
Copper	8.92
Iron	7.86
Aluminum	2.70
Magnesium	1.75
Water	1.00
Air	0.0012

Experiments have shown that this number, known as Avogadro's number, N_A, is

$$N_A = 6.022\ 137 \times 10^{23} \text{ particles/mol}$$

Avogadro's number is defined so that 1 mol of carbon-12 atoms has a mass of exactly 12 g. In general, the mass in 1 mol of any element is the element's atomic mass expressed in grams. For example, 1 mol of iron (atomic mass = 55.85 u) has a mass of 55.85 g (we say its *molar mass* is 55.85 g/mol), and 1 mol of lead (atomic mass = 207 u) has a mass of 207 g (its molar mass is 207 g/mol). Because there are 6.02×10^{23} particles in 1 mol of *any* element, the mass per atom for a given element is

$$m_{\text{atom}} = \frac{\text{molar mass}}{N_A} \tag{1.2}$$

For example, the mass of an iron atom is

$$m_{\text{Fe}} = \frac{55.85 \text{ g/mol}}{6.02 \times 10^{23} \text{ atoms/mol}} = 9.28 \times 10^{-23} \text{ g/atom}$$

EXAMPLE 1.1 ▸ **How Many Atoms in the Cube?**

A solid cube of aluminum (density 2.7 g/cm^3) has a volume of 0.20 cm^3. How many aluminum atoms are contained in the cube?

Solution   Since density equals mass per unit volume, the mass m of the cube is

$$m = \rho V = (2.7 \text{ g/cm}^3)(0.20 \text{ cm}^3) = 0.54 \text{ g}$$

To find the number of atoms N in this mass of aluminum, we can set up a proportion using the fact that one mole of alu-

minum (27 g) contains 6.02×10^{23} atoms:

$$\frac{N_A}{27 \text{ g}} = \frac{N}{0.54 \text{ g}}$$

$$\frac{6.02 \times 10^{23} \text{ atoms}}{27 \text{ g}} = \frac{N}{0.54 \text{ g}}$$

$$N = \frac{(0.54 \text{ g})(6.02 \times 10^{23} \text{ atoms})}{27 \text{ g}} = \boxed{1.2 \times 10^{22} \text{ atoms}}$$

1.4 ▸ DIMENSIONAL ANALYSIS

The word *dimension* has a special meaning in physics. It usually denotes the physical nature of a quantity. Whether a distance is measured in the length unit feet or the length unit meters, it is still a distance. We say the dimension—the physical nature—of distance is *length*.

The symbols we use in this book to specify length, mass, and time are L, M, and T, respectively. We shall often use brackets [ ] to denote the dimensions of a physical quantity. For example, the symbol we use for speed in this book is v, and in our notation the dimensions of speed are written $[v] = \text{L/T}$. As another example, the dimensions of area, for which we use the symbol A, are $[A] = \text{L}^2$. The dimensions of area, volume, speed, and acceleration are listed in Table 1.6.

In solving problems in physics, there is a useful and powerful procedure called *dimensional analysis*. This procedure, which should always be used, will help minimize the need for rote memorization of equations. Dimensional analysis makes use of the fact that **dimensions can be treated as algebraic quantities.** That is, quantities can be added or subtracted only if they have the same dimensions. Furthermore, the terms on both sides of an equation must have the same dimensions.

System	**Area** (L^2)	**Volume** (L^3)	**Speed** (L/T)	**Acceleration** (L/T^2)
SI	m^2	m^3	m/s	m/s^2
British engineering	ft^2	ft^3	ft/s	ft/s^2

TABLE 1.6   Dimensions and Common Units of Area, Volume, Speed, and Acceleration

By following these simple rules, you can use dimensional analysis to help determine whether an expression has the correct form. The relationship can be correct only if the dimensions are the same on both sides of the equation.

To illustrate this procedure, suppose you wish to derive a formula for the distance x traveled by a car in a time t if the car starts from rest and moves with constant acceleration a. In Chapter 2, we shall find that the correct expression is $x = \frac{1}{2}at^2$. Let us use dimensional analysis to check the validity of this expression. The quantity x on the left side has the dimension of length. For the equation to be dimensionally correct, the quantity on the right side must also have the dimension of length. We can perform a dimensional check by substituting the dimensions for acceleration, L/T^2, and time, T, into the equation. That is, the dimensional form of the equation $x = \frac{1}{2}at^2$ is

$$L = \frac{L}{T^2} \cdot T^2 = L$$

The units of time squared cancel as shown, leaving the unit of length.

A more general procedure using dimensional analysis is to set up an expression of the form

$$x \propto a^n t^m$$

where n and m are exponents that must be determined and the symbol $\propto$ indicates a proportionality. This relationship is correct only if the dimensions of both sides are the same. Because the dimension of the left side is length, the dimension of the right side must also be length. That is,

$$[a^n t^m] = L = LT^0$$

Because the dimensions of acceleration are L/T^2 and the dimension of time is T, we have

$$\left(\frac{L}{T^2}\right)^n T^m = L^1$$

$$L^n T^{m-2n} = L^1$$

Because the exponents of L and T must be the same on both sides, the dimensional equation is balanced under the conditions $m - 2n = 0$, $n = 1$, and $m = 2$. Returning to our original expression $x \propto a^n t^m$, we conclude that $x \propto at^2$. This result differs by a factor of 2 from the correct expression, which is $x = \frac{1}{2}at^2$. Because the factor $\frac{1}{2}$ is dimensionless, there is no way of determining it using dimensional analysis.

Quick Quiz 1.1

True or False: Dimensional analysis can give you the numerical value of constants of proportionality that may appear in an algebraic expression.

EXAMPLE 1.2   Analysis of an Equation

Show that the expression $v = at$ is dimensionally correct, where v represents speed, a acceleration, and t a time interval.

Solution   For the speed term, we have from Table 1.6

$$[v] = \frac{L}{T}$$

The same table gives us L/T^2 for the dimensions of acceleration, and so the dimensions of at are

$$[at] = \left(\frac{L}{T^2}\right)(T) = \frac{L}{T}$$

Therefore, the expression is dimensionally correct. (If the expression were given as $v = at^2$, it would be dimensionally *incorrect*. Try it and see!)

EXAMPLE 1.3   Analysis of a Power Law

Suppose we are told that the acceleration a of a particle moving with uniform speed v in a circle of radius r is proportional to some power of r, say r^n, and some power of v, say v^m. How can we determine the values of n and m?

Solution   Let us take a to be

$$a = kr^n v^m$$

where k is a dimensionless constant of proportionality. Knowing the dimensions of a, r, and v, we see that the dimensional equation must be

$$L/T^2 = L^n(L/T)^m = L^{n+m}/T^m$$

This dimensional equation is balanced under the conditions

$$n + m = 1 \quad \text{and} \quad m = 2$$

Therefore $n = -1$, and we can write the acceleration expression as

$$a = kr^{-1}v^2 = k\frac{v^2}{r}$$

When we discuss uniform circular motion later, we shall see that $k = 1$ if a consistent set of units is used. The constant k would not equal 1 if, for example, v were in km/h and you wanted a in m/s^2.

QuickLab

Estimate the weight (in pounds) of two large bottles of soda pop. Note that 1 L of water has a mass of about 1 kg. Use the fact that an object weighing 2.2 lb has a mass of 1 kg. Find some bathroom scales and check your estimate.

1.5   CONVERSION OF UNITS

Sometimes it is necessary to convert units from one system to another. Conversion factors between the SI units and conventional units of length are as follows:

$$1 \text{ mi} = 1\,609 \text{ m} = 1.609 \text{ km} \qquad 1 \text{ ft} = 0.304\,8 \text{ m} = 30.48 \text{ cm}$$

$$1 \text{ m} = 39.37 \text{ in.} = 3.281 \text{ ft} \qquad 1 \text{ in.} \equiv 0.025\,4 \text{ m} = 2.54 \text{ cm (exactly)}$$

A more complete list of conversion factors can be found in Appendix A.

Units can be treated as algebraic quantities that can cancel each other. For example, suppose we wish to convert 15.0 in. to centimeters. Because 1 in. is defined as exactly 2.54 cm, we find that

$$15.0 \text{ in.} = (15.0 \text{ in.})(2.54 \text{ cm/in.}) = 38.1 \text{ cm}$$

This works because multiplying by $\left(\frac{2.54 \text{ cm}}{1 \text{ in.}}\right)$ is the same as multiplying by 1, because the numerator and denominator describe identical things.

(Left) This road sign near Raleigh, North Carolina, shows distances in miles and kilometers. How accurate are the conversions? *(Billy E. Barnes/Stock Boston).*

(Right) This vehicle's speedometer gives speed readings in miles per hour and in kilometers per hour. Try confirming the conversion between the two sets of units for a few readings of the dial. *(Paul Silverman/Fundamental Photographs)*

EXAMPLE 1.4 ▶ The Density of a Cube

The mass of a solid cube is 856 g, and each edge has a length of 5.35 cm. Determine the density ρ of the cube in basic SI units.

Solution Because $1 \text{ g} = 10^{-3} \text{ kg}$ and $1 \text{ cm} = 10^{-2} \text{ m}$, the mass m and volume V in basic SI units are

$$m = 856 \text{ g} \times 10^{-3} \text{ kg/g} = 0.856 \text{ kg}$$

$$V = L^3 = (5.35 \text{ cm} \times 10^{-2} \text{ m/cm})^3$$
$$= (5.35)^3 \times 10^{-6} \text{ m}^3 = 1.53 \times 10^{-4} \text{ m}^3$$

Therefore,

$$\rho = \frac{m}{V} = \frac{0.856 \text{ kg}}{1.53 \times 10^{-4} \text{ m}^3} = \boxed{5.59 \times 10^3 \text{ kg/m}^3}$$

1.6 ESTIMATES AND ORDER-OF-MAGNITUDE CALCULATIONS

It is often useful to compute an approximate answer to a physical problem even where little information is available. Such an approximate answer can then be used to determine whether a more accurate calculation is necessary. Approximations are usually based on certain assumptions, which must be modified if greater accuracy is needed. Thus, we shall sometimes refer to the order of magnitude of a certain quantity as the power of ten of the number that describes that quantity. If, for example, we say that a quantity increases in value by three orders of magnitude, this means that its value is increased by a factor of $10^3 = 1000$. Also, if a quantity is given as 3×10^3, we say that the order of magnitude of that quantity is 10^3 (or in symbolic form, $3 \times 10^3 \sim 10^3$). Likewise, the quantity $8 \times 10^7 \sim 10^8$.

The spirit of order-of-magnitude calculations, sometimes referred to as "guesstimates" or "ball-park figures," is given in the following quotation: "Make an estimate before every calculation, try a simple physical argument . . . before every derivation, guess the answer to every puzzle. Courage: no one else needs to know what the guess is." [4] Inaccuracies caused by guessing too low for one number are often canceled out by other guesses that are too high. You will find that with practice your guesstimates get better and better. Estimation problems can be fun to work as you freely drop digits, venture reasonable approximations for unknown numbers, make simplifying assumptions, and turn the question around into something you can answer in your head.

EXAMPLE 1.5   Breaths in a Lifetime

Estimate the number of breaths taken during an average life span.

Solution   We shall start by guessing that the typical life span is about 70 years. The only other estimate we must make in this example is the average number of breaths that a person takes in 1 min. This number varies, depending on whether the person is exercising, sleeping, angry, serene, and so forth. To the nearest order of magnitude, we shall choose 10 breaths per minute as our estimate of the average. (This is certainly closer to the true value than 1 breath per minute or 100 breaths per minute.) The number of minutes in a year is approximately

$$1 \text{ yr} \times 400 \frac{\text{days}}{\text{yr}} \times 25 \frac{\text{h}}{\text{day}} \times 60 \frac{\text{min}}{\text{h}} = 6 \times 10^5 \text{ min}$$

Notice how much simpler it is to multiply 400×25 than it is to work with the more accurate 365×24. These approximate values for the number of days in a year and the number of hours in a day are close enough for our purposes. Thus, in 70 years there will be $(70 \text{ yr})(6 \times 10^5 \text{ min/yr}) = 4 \times 10^7$ min. At a rate of 10 breaths/min, an individual would take

$$4 \times 10^8 \text{ breaths in a lifetime.}$$

EXAMPLE 1.6   It's a Long Way to San Jose

Estimate the number of steps a person would take walking from New York to Los Angeles.

Solution   Without looking up the distance between these two cities, you might remember from a geography class that they are about 3 000 mi apart. The next approximation we must make is the length of one step. Of course, this length depends on the person doing the walking, but we can estimate that each step covers about 2 ft. With our estimated step size, we can determine the number of steps in 1 mi. Because this is a rough calculation, we round 5 280 ft/mi to 5 000 ft/mi. (What percentage error does this introduce?) This conversion factor gives us

$$\frac{5\ 000 \text{ ft/mi}}{2 \text{ ft/step}} = 2\ 500 \text{ steps/mi}$$

Now we switch to scientific notation so that we can do the calculation mentally:

$$(3 \times 10^3 \text{ mi})(2.5 \times 10^3 \text{ steps/mi}) = 7.5 \times 10^6 \text{ steps}$$

$$\sim \boxed{10^7 \text{ steps}}$$

So if we intend to walk across the United States, it will take us on the order of ten million steps. This estimate is almost certainly too small because we have not accounted for curving roads and going up and down hills and mountains. Nonetheless, it is probably within an order of magnitude of the correct answer.

EXAMPLE 1.7   How Much Gas Do We Use?

Estimate the number of gallons of gasoline used each year by all the cars in the United States.

Solution   There are about 270 million people in the United States, and so we estimate that the number of cars in the country is 100 million (guessing that there are between two and three people per car). We also estimate that the average distance each car travels per year is 10 000 mi. If we assume a gasoline consumption of 20 mi/gal or 0.05 gal/mi, then each car uses about 500 gal/yr. Multiplying this by the total number of cars in the United States gives an estimated total consumption of 5×10^{10} gal $\sim$ $\boxed{10^{11} \text{ gal.}}$

---

[4] E. Taylor and J. A. Wheeler, *Spacetime Physics,* San Francisco, W. H. Freeman & Company, Publishers, 1966, p. 60.

1.7 SIGNIFICANT FIGURES

When physical quantities are measured, the measured values are known only to within the limits of the experimental uncertainty. The value of this uncertainty can depend on various factors, such as the quality of the apparatus, the skill of the experimenter, and the number of measurements performed.

Suppose that we are asked to measure the area of a computer disk label using a meter stick as a measuring instrument. Let us assume that the accuracy to which we can measure with this stick is ± 0.1 cm. If the length of the label is measured to be 5.5 cm, we can claim only that its length lies somewhere between 5.4 cm and 5.6 cm. In this case, we say that the measured value has two significant figures. Likewise, if the label's width is measured to be 6.4 cm, the actual value lies between 6.3 cm and 6.5 cm. Note that the significant figures include the first estimated digit. Thus we could write the measured values as (5.5 ± 0.1) cm and (6.4 ± 0.1) cm.

Now suppose we want to find the area of the label by multiplying the two measured values. If we were to claim the area is (5.5 cm)(6.4 cm) = 35.2 cm², our answer would be unjustifiable because it contains three significant figures, which is greater than the number of significant figures in either of the measured lengths. A good rule of thumb to use in determining the number of significant figures that can be claimed is as follows:

> When multiplying several quantities, the number of significant figures in the final answer is the same as the number of significant figures in the *least* accurate of the quantities being multiplied, where "least accurate" means "having the lowest number of significant figures." The same rule applies to division.

Applying this rule to the multiplication example above, we see that the answer for the area can have only two significant figures because our measured lengths have only two significant figures. Thus, all we can claim is that the area is 35 cm², realizing that the value can range between (5.4 cm)(6.3 cm) = 34 cm² and (5.6 cm)(6.5 cm) = 36 cm².

Zeros may or may not be significant figures. Those used to position the decimal point in such numbers as 0.03 and 0.007 5 are not significant. Thus, there are one and two significant figures, respectively, in these two values. When the zeros come after other digits, however, there is the possibility of misinterpretation. For example, suppose the mass of an object is given as 1 500 g. This value is ambiguous because we do not know whether the last two zeros are being used to locate the decimal point or whether they represent significant figures in the measurement. To remove this ambiguity, it is common to use scientific notation to indicate the number of significant figures. In this case, we would express the mass as 1.5×10^3 g if there are two significant figures in the measured value, 1.50×10^3 g if there are three significant figures, and 1.500×10^3 g if there are four. The same rule holds when the number is less than 1, so that 2.3×10^{-4} has two significant figures (and so could be written 0.000 23) and 2.30×10^{-4} has three significant figures (also written 0.000 230). In general, **a significant figure is a reliably known digit** (other than a zero used to locate the decimal point).

For addition and subtraction, you must consider the number of decimal places when you are determining how many significant figures to report.

QuickLab

Determine the thickness of a page from this book. (Note that numbers that have no measurement errors—like the count of a number of pages—do not affect the significant figures in a calculation.) In terms of significant figures, why is it better to measure the thickness of as many pages as possible and then divide by the number of sheets?

When numbers are added or subtracted, the number of decimal places in the result should equal the smallest number of decimal places of any term in the sum.

For example, if we wish to compute $123 + 5.35$, the answer given to the correct number of significant figures is 128 and not 128.35. If we compute the sum $1.000\ 1 + 0.000\ 3 = 1.000\ 4$, the result has five significant figures, even though one of the terms in the sum, $0.000\ 3$, has only one significant figure. Likewise, if we perform the subtraction $1.002 - 0.998 = 0.004$, the result has only one significant figure even though one term has four significant figures and the other has three. In this book, **most of the numerical examples and end-of-chapter problems will yield answers having three significant figures.** When carrying out estimates we shall typically work with a single significant figure.

Quick Quiz 1.2

Suppose you measure the position of a chair with a meter stick and record that the center of the seat is $1.043\ 860\ 564\ 2$ m from a wall. What would a reader conclude from this recorded measurement?

EXAMPLE 1.8   The Area of a Rectangle

A rectangular plate has a length of (21.3 ± 0.2) cm and a width of (9.80 ± 0.1) cm. Find the area of the plate and the uncertainty in the calculated area.

Solution

$$\text{Area} = \ell w = (21.3 \pm 0.2\ \text{cm}) \times (9.80 \pm 0.1\ \text{cm})$$

$$\approx (21.3 \times 9.80 \pm 21.3 \times 0.1 \pm 0.2 \times 9.80)\ \text{cm}^2$$

$$\approx \boxed{(209 \pm 4)\ \text{cm}^2}$$

Because the input data were given to only three significant figures, we cannot claim any more in our result. Do you see why we did not need to multiply the uncertainties 0.2 cm and 0.1 cm?

EXAMPLE 1.9   Installing a Carpet

A carpet is to be installed in a room whose length is measured to be 12.71 m and whose width is measured to be 3.46 m. Find the area of the room.

Solution   If you multiply 12.71 m by 3.46 m on your calculator, you will get an answer of $43.976\ 6$ m^2. How many of these numbers should you claim? Our rule of thumb for multiplication tells us that you can claim only the number of significant figures in the least accurate of the quantities being measured. In this example, we have only three significant figures in our least accurate measurement, so we should express our final answer as $\boxed{44.0\ \text{m}^2}$.

Note that in reducing $43.976\ 6$ to three significant figures for our answer, we used a general rule for rounding off numbers that states that the last digit retained (the 9 in this example) is increased by 1 if the first digit dropped (here, the 7) is 5 or greater. (A technique for avoiding error accumulation is to delay rounding of numbers in a long calculation until you have the final result. Wait until you are ready to copy the answer from your calculator before rounding to the correct number of significant figures.)

SUMMARY

The three fundamental physical quantities of mechanics are length, mass, and time, which in the SI system have the units meters (m), kilograms (kg), and seconds (s), respectively. Prefixes indicating various powers of ten are used with these three basic units. The **density** of a substance is defined as its *mass per unit volume*. Different substances have different densities mainly because of differences in their atomic masses and atomic arrangements.

The number of particles in one mole of any element or compound, called **Avogadro's number,** N_A, is 6.02×10^{23}.

The method of *dimensional analysis* is very powerful in solving physics problems. Dimensions can be treated as algebraic quantities. By making estimates and making order-of-magnitude calculations, you should be able to approximate the answer to a problem when there is not enough information available to completely specify an exact solution.

When you compute a result from several measured numbers, each of which has a certain accuracy, you should give the result with the correct number of significant figures.

QUESTIONS

1. In this chapter we described how the Earth's daily rotation on its axis was once used to define the standard unit of time. What other types of natural phenomena could serve as alternative time standards?

2. Suppose that the three fundamental standards of the metric system were length, density, and time rather than length, mass, and time. The standard of density in this system is to be defined as that of water. What considerations about water would you need to address to make sure that the standard of density is as accurate as possible?

3. A hand is defined as 4 in.; a foot is defined as 12 in. Why should the hand be any less acceptable as a unit than the foot, which we use all the time?

4. Express the following quantities using the prefixes given in Table 1.4: (a) 3×10^{-4} m (b) 5×10^{-5} s (c) 72×10^2 g.

5. Suppose that two quantities A and B have different dimensions. Determine which of the following arithmetic operations *could* be physically meaningful: (a) $A + B$ (b) A/B (c) $B - A$ (d) AB.

6. What level of accuracy is implied in an order-of-magnitude calculation?

7. Do an order-of-magnitude calculation for an everyday situation you might encounter. For example, how far do you walk or drive each day?

8. Estimate your age in seconds.

9. Estimate the mass of this textbook in kilograms. If a scale is available, check your estimate.

PROBLEMS

1, 2, 3 = straightforward, intermediate, challenging   ☐ = full solution available in the *Student Solutions Manual and Study Guide*
WEB = solution posted at **http://www.saunderscollege.com/physics/**   🖥 = Computer useful in solving problem   🔲 = Interactive Physics
☐ = paired numerical/symbolic problems

Section 1.3   Density

1. The standard kilogram is a platinum–iridium cylinder 39.0 mm in height and 39.0 mm in diameter. What is the density of the material?

2. The mass of the planet Saturn (Fig. P1.2) is 5.64×10^{26} kg, and its radius is 6.00×10^7 m. Calculate its density.

3. How many grams of copper are required to make a hollow spherical shell having an inner radius of 5.70 cm and an outer radius of 5.75 cm? The density of copper is 8.92 g/cm^3.

4. What mass of a material with density ρ is required to make a hollow spherical shell having inner radius r_1 and outer radius r_2?

5. Iron has molar mass 55.8 g/mol. (a) Find the volume of 1 mol of iron. (b) Use the value found in (a) to determine the volume of one iron atom. (c) Calculate the cube root of the atomic volume, to have an estimate for the distance between atoms in the solid. (d) Repeat the calculations for uranium, finding its molar mass in the periodic table of the elements in Appendix C.

Figure P1.2   A view of Saturn from *Voyager 2*. *(Courtesy of NASA)*

6. Two spheres are cut from a certain uniform rock. One has radius 4.50 cm. The mass of the other is five times greater. Find its radius.

WEB 7. Calculate the mass of an atom of (a) helium, (b) iron, and (c) lead. Give your answers in atomic mass units and in grams. The molar masses are 4.00, 55.9, and 207 g/mol, respectively, for the atoms given.

8. On your wedding day your lover gives you a gold ring of mass 3.80 g. Fifty years later its mass is 3.35 g. As an average, how many atoms were abraded from the ring during each second of your marriage? The molar mass of gold is 197 g/mol.

9. A small cube of iron is observed under a microscope. The edge of the cube is 5.00×10^{-6} cm long. Find (a) the mass of the cube and (b) the number of iron atoms in the cube. The molar mass of iron is 55.9 g/mol, and its density is 7.86 g/cm^3.

10. A structural I-beam is made of steel. A view of its cross-section and its dimensions are shown in Figure P1.10.

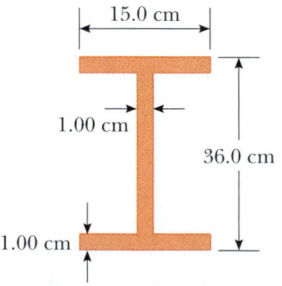

15.0 cm

1.00 cm

36.0 cm

1.00 cm

Figure P1.10

(a) What is the mass of a section 1.50 m long? (b) How many atoms are there in this section? The density of steel is 7.56×10^3 kg/m^3.

11. A child at the beach digs a hole in the sand and, using a pail, fills it with water having a mass of 1.20 kg. The molar mass of water is 18.0 g/mol. (a) Find the number of water molecules in this pail of water. (b) Suppose the quantity of water on the Earth is 1.32×10^{21} kg and remains constant. How many of the water molecules in this pail of water were likely to have been in an equal quantity of water that once filled a particular claw print left by a dinosaur?

Section 1.4   Dimensional Analysis

12. The radius r of a circle inscribed in any triangle whose sides are a, b, and c is given by

$$r = [(s - a)(s - b)(s - c)/s]^{1/2}$$

where s is an abbreviation for $(a + b + c)/2$. Check this formula for dimensional consistency.

13. The displacement of a particle moving under uniform acceleration is some function of the elapsed time and the acceleration. Suppose we write this displacement $s = ka^m t^n$, where k is a dimensionless constant. Show by dimensional analysis that this expression is satisfied if $m = 1$ and $n = 2$. Can this analysis give the value of k?

14. The period T of a simple pendulum is measured in time units and is described by

$$T = 2\pi \sqrt{\frac{\ell}{g}}$$

where ℓ is the length of the pendulum and g is the free-fall acceleration in units of length divided by the square of time. Show that this equation is dimensionally correct.

15. Which of the equations below are dimensionally correct?
    (a) $v = v_0 + ax$
    (b) $y = (2 \text{ m}) \cos(kx)$, where $k = 2 \text{ m}^{-1}$

16. Newton's law of universal gravitation is represented by

$$F = \frac{GMm}{r^2}$$

Here F is the gravitational force, M and m are masses, and r is a length. Force has the SI units kg·m/s^2. What are the SI units of the proportionality constant G?

WEB 17. The consumption of natural gas by a company satisfies the empirical equation $V = 1.50t + 0.008\,00t^2$, where V is the volume in millions of cubic feet and t the time in months. Express this equation in units of cubic feet and seconds. Put the proper units on the coefficients. Assume a month is 30.0 days.

Section 1.5   Conversion of Units

18. Suppose your hair grows at the rate 1/32 in. per day. Find the rate at which it grows in nanometers per second. Since the distance between atoms in a molecule is

on the order of 0.1 nm, your answer suggests how rapidly layers of atoms are assembled in this protein synthesis.

19. A rectangular building lot is 100 ft by 150 ft. Determine the area of this lot in m^2.

20. An auditorium measures 40.0 m $\times$ 20.0 m $\times$ 12.0 m. The density of air is 1.20 kg/m^3. What are (a) the volume of the room in cubic feet and (b) the weight of air in the room in pounds?

21. Assume that it takes 7.00 min to fill a 30.0-gal gasoline tank. (a) Calculate the rate at which the tank is filled in gallons per second. (b) Calculate the rate at which the tank is filled in cubic meters per second. (c) Determine the time, in hours, required to fill a 1-cubic-meter volume at the same rate. (1 U.S. gal = 231 in.^{3})

22. A creature moves at a speed of 5.00 furlongs per fortnight (not a very common unit of speed). Given that 1 furlong = 220 yards and 1 fortnight = 14 days, determine the speed of the creature in meters per second. What kind of creature do you think it might be?

23. A section of land has an area of 1 mi^2 and contains 640 acres. Determine the number of square meters in 1 acre.

24. A quart container of ice cream is to be made in the form of a cube. What should be the length of each edge in centimeters? (Use the conversion 1 gal = 3.786 L.)

25. A solid piece of lead has a mass of 23.94 g and a volume of 2.10 cm^3. From these data, calculate the density of lead in SI units (kg/m^3).

26. An astronomical unit (AU) is defined as the average distance between the Earth and the Sun. (a) How many astronomical units are there in one lightyear? (b) Determine the distance from the Earth to the Andromeda galaxy in astronomical units.

27. The mass of the Sun is 1.99×10^{30} kg, and the mass of an atom of hydrogen, of which the Sun is mostly composed, is 1.67×10^{-27} kg. How many atoms are there in the Sun?

28. (a) Find a conversion factor to convert from miles per hour to kilometers per hour. (b) In the past, a federal law mandated that highway speed limits would be 55 mi/h. Use the conversion factor of part (a) to find this speed in kilometers per hour. (c) The maximum highway speed is now 65 mi/h in some places. In kilometers per hour, how much of an increase is this over the 55-mi/h limit?

29. At the time of this book's printing, the U. S. national debt is about $6 trillion. (a) If payments were made at the rate of $1 000/s, how many years would it take to pay off a $6-trillion debt, assuming no interest were charged? (b) A dollar bill is about 15.5 cm long. If six trillion dollar bills were laid end to end around the Earth's equator, how many times would they encircle the Earth? Take the radius of the Earth at the equator to be 6 378 km. (*Note:* Before doing any of these calculations, try to guess at the answers. You may be very surprised.)

30. (a) How many seconds are there in a year? (b) If one micrometeorite (a sphere with a diameter of 1.00×10^{-6} m) strikes each square meter of the Moon each second, how many years will it take to cover the Moon to a depth of 1.00 m? (*Hint:* Consider a cubic box on the Moon 1.00 m on a side, and find how long it will take to fill the box.)

WEB **31.** One gallon of paint (volume = 3.78×10^{-3} m^3) covers an area of 25.0 m^2. What is the thickness of the paint on the wall?

32. A pyramid has a height of 481 ft, and its base covers an area of 13.0 acres (Fig. P1.32). If the volume of a pyramid is given by the expression $V = \frac{1}{3}Bh$, where B is the area of the base and h is the height, find the volume of this pyramid in cubic meters. (1 acre = 43 560 ft^2)

Figure P1.32   Problems 32 and 33.

33. The pyramid described in Problem 32 contains approximately two million stone blocks that average 2.50 tons each. Find the weight of this pyramid in pounds.

34. Assuming that 70% of the Earth's surface is covered with water at an average depth of 2.3 mi, estimate the mass of the water on the Earth in kilograms.

35. The amount of water in reservoirs is often measured in acre-feet. One acre-foot is a volume that covers an area of 1 acre to a depth of 1 ft. An acre is an area of 43 560 ft^2. Find the volume in SI units of a reservoir containing 25.0 acre-ft of water.

36. A hydrogen atom has a diameter of approximately 1.06×10^{-10} m, as defined by the diameter of the spherical electron cloud around the nucleus. The hydrogen nucleus has a diameter of approximately 2.40×10^{-15} m. (a) For a scale model, represent the diameter of the hydrogen atom by the length of an American football field (100 yards = 300 ft), and determine the diameter of the nucleus in millimeters. (b) The atom is how many times larger in volume than its nucleus?

37. The diameter of our disk-shaped galaxy, the Milky Way, is about 1.0×10^5 lightyears. The distance to Messier 31—which is Andromeda, the spiral galaxy nearest to the Milky Way—is about 2.0 million lightyears. If a scale model represents the Milky Way and Andromeda galaxies as dinner plates 25 cm in diameter, determine the distance between the two plates.

38. The mean radius of the Earth is 6.37×10^6 m, and that of the Moon is 1.74×10^8 cm. From these data calculate (a) the ratio of the Earth's surface area to that of the Moon and (b) the ratio of the Earth's volume to that of the Moon. Recall that the surface area of a sphere is $4\pi r^2$ and that the volume of a sphere is $\frac{4}{3}\pi r^3$.

WEB **39.** One cubic meter (1.00 m^3) of aluminum has a mass of 2.70×10^3 kg, and 1.00 m^3 of iron has a mass of 7.86×10^3 kg. Find the radius of a solid aluminum sphere that balances a solid iron sphere of radius 2.00 cm on an equal-arm balance.

40. Let ρ_{Al} represent the density of aluminum and ρ_{Fe} that of iron. Find the radius of a solid aluminum sphere that balances a solid iron sphere of radius r_{Fe} on an equal-arm balance.

Section 1.6 Estimates and Order-of-Magnitude Calculations

WEB **41.** Estimate the number of Ping-Pong balls that would fit into an average-size room (without being crushed). In your solution state the quantities you measure or estimate and the values you take for them.

42. McDonald's sells about 250 million packages of French fries per year. If these fries were placed end to end, estimate how far they would reach.

43. An automobile tire is rated to last for 50 000 miles. Estimate the number of revolutions the tire will make in its lifetime.

44. Approximately how many raindrops fall on a 1.0-acre lot during a 1.0-in. rainfall?

45. Grass grows densely everywhere on a quarter-acre plot of land. What is the order of magnitude of the number of blades of grass on this plot of land? Explain your reasoning. (1 acre = 43 560 ft^2.)

46. Suppose that someone offers to give you $1 billion if you can finish counting it out using only one-dollar bills. Should you accept this offer? Assume you can count one bill every second, and be sure to note that you need about 8 hours a day for sleeping and eating and that right now you are probably at least 18 years old.

47. Compute the order of magnitude of the mass of a bathtub half full of water and of the mass of a bathtub half full of pennies. In your solution, list the quantities you take as data and the value you measure or estimate for each.

48. Soft drinks are commonly sold in aluminum containers. Estimate the number of such containers thrown away or recycled each year by U.S. consumers. Approximately how many tons of aluminum does this represent?

49. To an order of magnitude, how many piano tuners are there in New York City? The physicist Enrico Fermi was famous for asking questions like this on oral Ph.D. qualifying examinations and for his own facility in making order-of-magnitude calculations.

Section 1.7 Significant Figures

50. Determine the number of significant figures in the following measured values: (a) 23 cm (b) 3.589 s (c) 4.67×10^3 m/s (d) 0.003 2 m.

51. The radius of a circle is measured to be 10.5 ± 0.2 m. Calculate the (a) area and (b) circumference of the circle and give the uncertainty in each value.

52. Carry out the following arithmetic operations: (a) the sum of the measured values 756, 37.2, 0.83, and 2.5; (b) the product $0.003\ 2 \times 356.3$; (c) the product $5.620 \times \pi$.

53. The radius of a solid sphere is measured to be (6.50 ± 0.20) cm, and its mass is measured to be (1.85 ± 0.02) kg. Determine the density of the sphere in kilograms per cubic meter and the uncertainty in the density.

54. How many significant figures are in the following numbers: (a) 78.9 ± 0.2, (b) 3.788×10^9, (c) 2.46×10^{-6}, and (d) 0.005 3?

55. A farmer measures the distance around a rectangular field. The length of the long sides of the rectangle is found to be 38.44 m, and the length of the short sides is found to be 19.5 m. What is the total distance around the field?

56. A sidewalk is to be constructed around a swimming pool that measures (10.0 ± 0.1) m by (17.0 ± 0.1) m. If the sidewalk is to measure (1.00 ± 0.01) m wide by (9.0 ± 0.1) cm thick, what volume of concrete is needed, and what is the approximate uncertainty of this volume?

ADDITIONAL PROBLEMS

57. In a situation where data are known to three significant digits, we write 6.379 m = 6.38 m and 6.374 m = 6.37 m. When a number ends in 5, we arbitrarily choose to write 6.375 m = 6.38 m. We could equally well write 6.375 m = 6.37 m, "rounding down" instead of "rounding up," since we would change the number 6.375 by equal increments in both cases. Now consider an order-of-magnitude estimate, in which we consider factors rather than increments. We write 500 m $\sim 10^3$ m because 500 differs from 100 by a factor of 5 whereas it differs from 1000 by only a factor of 2. We write 437 m $\sim 10^3$ m and 305 m $\sim 10^2$ m. What distance differs from 100 m and from 1000 m by equal factors, so that we could equally well choose to represent its order of magnitude either as $\sim 10^2$ m or as $\sim 10^3$ m?

58. When a droplet of oil spreads out on a smooth water surface, the resulting "oil slick" is approximately one molecule thick. An oil droplet of mass 9.00×10^{-7} kg and density 918 kg/m^3 spreads out into a circle of radius 41.8 cm on the water surface. What is the diameter of an oil molecule?

59. The basic function of the carburetor of an automobile is to "atomize" the gasoline and mix it with air to promote rapid combustion. As an example, assume that 30.0 cm³ of gasoline is atomized into N spherical droplets, each with a radius of 2.00×10^{-5} m. What is the total surface area of these N spherical droplets?

60. In physics it is important to use mathematical approximations. Demonstrate for yourself that for small angles ($< 20°$)

$$\tan \alpha \approx \sin \alpha \approx \alpha = \pi \alpha' / 180°$$

where α is in radians and α' is in degrees. Use a calculator to find the largest angle for which $\tan \alpha$ may be approximated by $\sin \alpha$ if the error is to be less than 10.0%.

61. A high fountain of water is located at the center of a circular pool as in Figure P1.61. Not wishing to get his feet wet, a student walks around the pool and measures its circumference to be 15.0 m. Next, the student stands at the edge of the pool and uses a protractor to gauge the angle of elevation of the top of the fountain to be 55.0°. How high is the fountain?

55.0°

Figure P1.61

62. Assume that a shape covers an area A and has a uniform height h. If its cross-sectional area is uniform over its height, then its volume is given by $V = Ah$. (a) Show that $V = Ah$ is dimensionally correct. (b) Show that the volumes of a cylinder and of a rectangular box can be written in the form $V = Ah$, identifying A in each case. (Note that A, sometimes called the "footprint" of the object, can have any shape and that the height can be replaced by average thickness in general.)

63. A useful fact is that there are about $\pi \times 10^7$ s in one year. Find the percentage error in this approximation, where "percentage error" is defined as

$$\frac{|\text{Assumed value} - \text{true value}|}{\text{True value}} \times 100\%$$

64. A crystalline solid consists of atoms stacked up in a repeating lattice structure. Consider a crystal as shown in Figure P1.64a. The atoms reside at the corners of cubes of side $L = 0.200$ nm. One piece of evidence for the regular arrangement of atoms comes from the flat surfaces along which a crystal separates, or "cleaves," when it is broken. Suppose this crystal cleaves along a face diagonal, as shown in Figure P1.64b. Calculate the spacing d between two adjacent atomic planes that separate when the crystal cleaves.

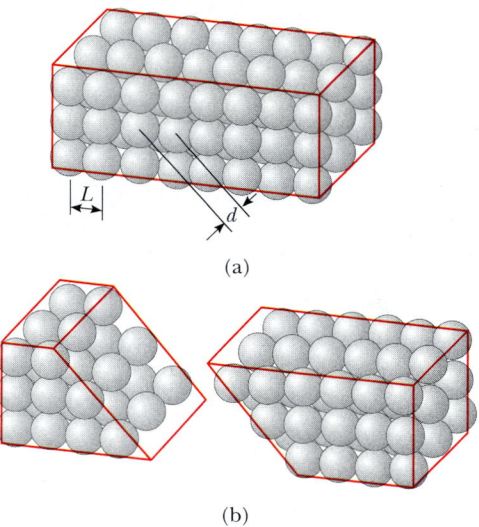

L

d

(a)

(b)

Figure P1.64

65. A child loves to watch as you fill a transparent plastic bottle with shampoo. Every horizontal cross-section of the bottle is a circle, but the diameters of the circles all have different values, so that the bottle is much wider in some places than in others. You pour in bright green shampoo with constant volume flow rate 16.5 cm³/s. At what rate is its level in the bottle rising (a) at a point where the diameter of the bottle is 6.30 cm and (b) at a point where the diameter is 1.35 cm?

66. As a child, the educator and national leader Booker T. Washington was given a spoonful (about 12.0 cm³) of molasses as a treat. He pretended that the quantity increased when he spread it out to cover uniformly all of a tin plate (with a diameter of about 23.0 cm). How thick a layer did it make?

67. Assume there are 100 million passenger cars in the United States and that the average fuel consumption is 20 mi/gal of gasoline. If the average distance traveled by each car is 10 000 mi/yr, how much gasoline would be saved per year if average fuel consumption could be increased to 25 mi/gal?

68. One cubic centimeter of water has a mass of 1.00×10^{-3} kg. (a) Determine the mass of 1.00 m³ of water. (b) Assuming biological substances are 98% water, estimate the mass of a cell that has a diameter of 1.0 μm, a human kidney, and a fly. Assume that a kidney is roughly a sphere with a radius of 4.0 cm and that a fly is roughly a cylinder 4.0 mm long and 2.0 mm in diameter.

69. The distance from the Sun to the nearest star is 4×10^{16} m. The Milky Way galaxy is roughly a disk of diameter $\sim 10^{21}$ m and thickness $\sim 10^{19}$ m. Find the order of magnitude of the number of stars in the Milky Way. Assume the 4×10^{16}-m distance between the Sun and the nearest star is typical.

70. The data in the following table represent measurements of the masses and dimensions of solid cylinders of alu-

minum, copper, brass, tin, and iron. Use these data to calculate the densities of these substances. Compare your results for aluminum, copper, and iron with those given in Table 1.5.

Substance	Mass (g)	Diameter (cm)	Length (cm)
Aluminum	51.5	2.52	3.75
Copper	56.3	1.23	5.06
Brass	94.4	1.54	5.69
Tin	69.1	1.75	3.74
Iron	216.1	1.89	9.77

ANSWERS TO QUICK QUIZZES

1.1 False. Dimensional analysis gives the units of the proportionality constant but provides no information about its numerical value. For example, experiments show that doubling the radius of a solid sphere increases its mass 8-fold, and tripling the radius increases the mass 27-fold. Therefore, its mass is proportional to the cube of its radius. Because $m \propto r^3$, we can write $m = kr^3$. Dimensional analysis shows that the proportionality constant k must have units kg/m^3, but to determine its numerical value requires either experimental data or geometrical reasoning.

1.2 Reporting all these digits implies you have determined the location of the center of the chair's seat to the nearest $\pm$ 0.000 000 000 1 m. This roughly corresponds to being able to count the atoms in your meter stick because each of them is about that size! It would probably be better to record the measurement as 1.044 m: this indicates that you know the position to the nearest millimeter, assuming the meter stick has millimeter markings on its scale.

THE WIZARD OF ID   **By Parker and Hart**

By permission of John Hart and Field Enterprises, Inc.

Motion in One Dimension

chapter

2

Chapter Outline

As a first step in studying classical mechanics, we describe motion in terms of space and time while ignoring the agents that caused that motion. This portion of classical mechanics is called *kinematics*. (The word *kinematics* has the same root as *cinema*. Can you see why?) In this chapter we consider only motion in one dimension. We first define displacement, velocity, and acceleration. Then, using these concepts, we study the motion of objects traveling in one dimension with a constant acceleration.

From everyday experience we recognize that motion represents a continuous change in the position of an object. In physics we are concerned with three types of motion: translational, rotational, and vibrational. A car moving down a highway is an example of translational motion, the Earth's spin on its axis is an example of rotational motion, and the back-and-forth movement of a pendulum is an example of vibrational motion. In this and the next few chapters, we are concerned only with translational motion. (Later in the book we shall discuss rotational and vibrational motions.)

In our study of translational motion, we describe the moving object as a *particle* regardless of its size. In general, **a particle is a point-like mass having infinitesimal size.** For example, if we wish to describe the motion of the Earth around the Sun, we can treat the Earth as a particle and obtain reasonably accurate data about its orbit. This approximation is justified because the radius of the Earth's orbit is large compared with the dimensions of the Earth and the Sun. As an example on a much smaller scale, it is possible to explain the pressure exerted by a gas on the walls of a container by treating the gas molecules as particles.

2.1   DISPLACEMENT, VELOCITY, AND SPEED

The motion of a particle is completely known if the particle's position in space is known at all times. Consider a car moving back and forth along the x axis, as shown in Figure 2.1a. When we begin collecting position data, the car is 30 m to the right of a road sign. (Let us assume that all data in this example are known to two significant figures. To convey this information, we should report the initial position as 3.0×10^1 m. We have written this value in this simpler form to make the discussion easier to follow.) We start our clock and once every 10 s note the car's location relative to the sign. As you can see from Table 2.1, the car is moving to the right (which we have defined as the positive direction) during the first 10 s of motion, from position Ⓐ to position Ⓑ. The position values now begin to decrease, however, because the car is backing up from position Ⓑ through position Ⓕ. In fact, at Ⓓ, 30 s after we start measuring, the car is alongside the sign we are using as our origin of coordinates. It continues moving to the left and is more than 50 m to the left of the sign when we stop recording information after our sixth data point. A graph of this information is presented in Figure 2.1b. Such a plot is called a *position–time graph*.

If a particle is moving, we can easily determine its change in position. **The displacement of a particle is defined as its change in position.** As it moves from an initial position x_i to a final position x_f, its displacement is given by $x_f - x_i$. We use the Greek letter delta (Δ) to denote the *change* in a quantity. Therefore, we write the displacement, or change in position, of the particle as

$$\Delta x \equiv x_f - x_i \qquad \qquad \text{(2.1)}$$

From this definition we see that Δx is positive if x_f is greater than x_i and negative if x_f is less than x_i.

TABLE 2.1
Position of the Car at Various Times

Position	t(s)	x(m)
Ⓐ	0	30
Ⓑ	10	52
Ⓒ	20	38
Ⓓ	30	0
Ⓔ	40	−37
Ⓕ	50	−53

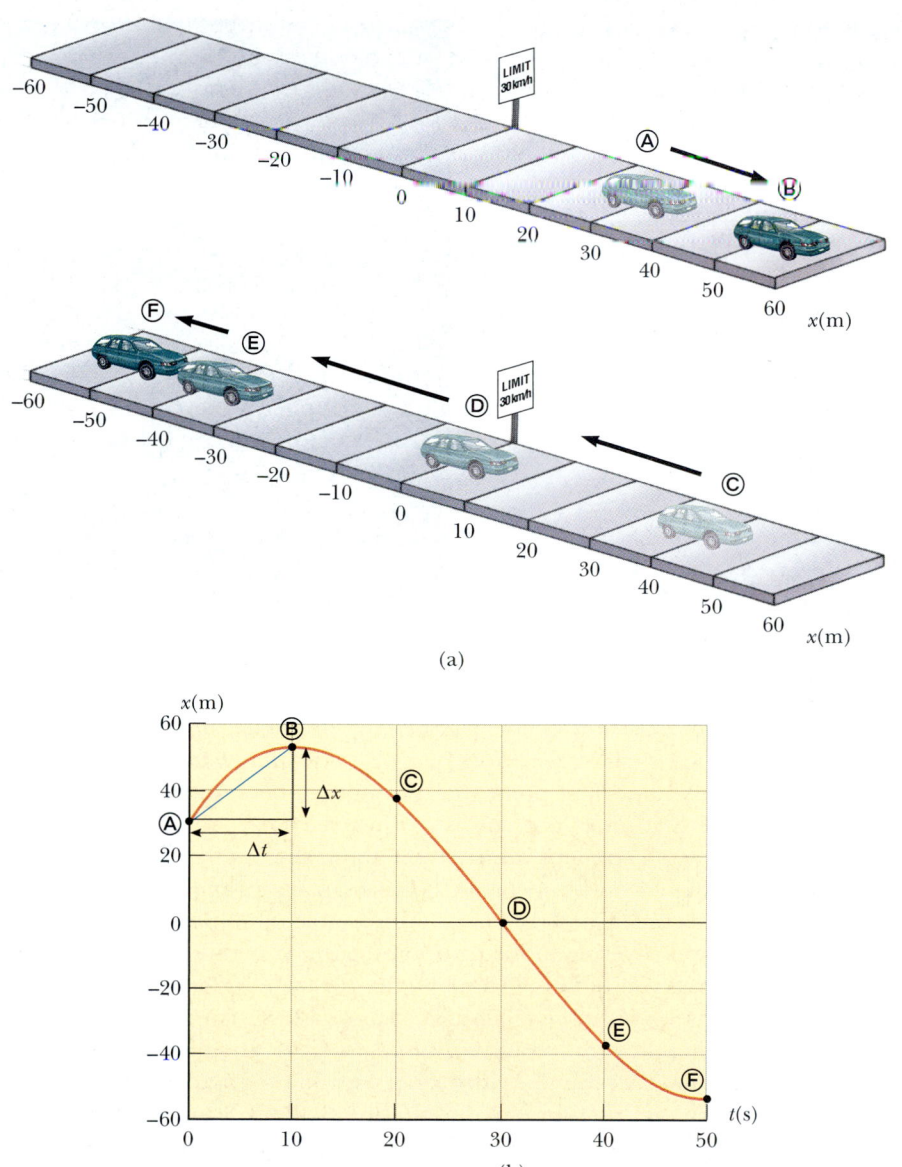

Figure 2.1 (a) A car moves back and forth along a straight line taken to be the *x* axis. Because we are interested only in the car's translational motion, we can treat it as a particle. (b) Position–time graph for the motion of the "particle."

A very easy mistake to make is not to recognize the difference between displacement and distance traveled (Fig. 2.2). A baseball player hitting a home run travels a distance of 360 ft in the trip around the bases. However, the player's displacement is zero because his final and initial positions are identical.

Displacement is an example of a vector quantity. Many other physical quantities, including velocity and acceleration, also are vectors. In general, **a vector is a physical quantity that requires the specification of both direction and magnitude.** By contrast, **a scalar is a quantity that has magnitude and no direction.** In this chapter, we use plus and minus signs to indicate vector direction. We can do this because the chapter deals with one-dimensional motion only; this means that any object we study can be moving only along a straight line. For example, for horizontal motion, let us arbitrarily specify to the right as being the positive direction. It follows that any object always moving to the right undergoes a

[handwritten margin notes:] Distance always positive. Displacement depends on where you start & when you stop & what direction. Can be pos, neg or zero.

Scalar = distance

Figure 2.2 Bird's-eye view of a baseball diamond. A batter who hits a home run travels 360 ft as he rounds the bases, but his displacement for the round trip is zero. *(Mark C. Burnett/Photo Researchers, Inc.)*

positive displacement $+\Delta x$, and any object moving to the left undergoes a negative displacement $-\Delta x$. We shall treat vectors in greater detail in Chapter 3.

There is one very important point that has not yet been mentioned. Note that the graph in Figure 2.1b does not consist of just six data points but is actually a smooth curve. The graph contains information about the entire 50-s interval during which we watched the car move. It is much easier to see changes in position from the graph than from a verbal description or even a table of numbers. For example, it is clear that the car was covering more ground during the middle of the 50-s interval than at the end. Between positions Ⓒ and Ⓓ, the car traveled almost 40 m, but during the last 10 s, between positions Ⓔ and Ⓕ, it moved less than half that far. A common way of comparing these different motions is to divide the displacement Δx that occurs between two clock readings by the length of that particular time interval Δt. This turns out to be a very useful ratio, one that we shall use many times. For convenience, the ratio has been given a special name — *average velocity*. **The average velocity $\bar{v}_x$ of a particle is defined as the particle's displacement Δx divided by the time interval Δt during which that displacement occurred:**

Average velocity

$$\bar{v}_x \equiv \frac{\Delta x}{\Delta t} \tag{2.2}$$

where the subscript x indicates motion along the x axis. From this definition we see that average velocity has dimensions of length divided by time (L/T) — meters per second in SI units.

Although the distance traveled for any motion is always positive, the average velocity of a particle moving in one dimension can be positive or negative, depending on the sign of the displacement. (The time interval Δt is always positive.) If the coordinate of the particle increases in time (that is, if $x_f > x_i$), then Δx is positive and $\bar{v}_x = \Delta x/\Delta t$ is positive. This case corresponds to motion in the positive x direction. If the coordinate decreases in time (that is, if $x_f < x_i$), then Δx is negative and hence $\bar{v}_x$ is negative. This case corresponds to motion in the negative x direction.

We can interpret average velocity geometrically by drawing a straight line between any two points on the position–time graph in Figure 2.1b. This line forms the hypotenuse of a right triangle of height Δx and base Δt. The slope of this line is the ratio $\Delta x/\Delta t$. For example, the line between positions Ⓐ and Ⓑ has a slope equal to the average velocity of the car between those two times, $(52 \text{ m} - 30 \text{ m})/(10 \text{ s} - 0) = 2.2 \text{ m/s}$.

In everyday usage, the terms *speed* and *velocity* are interchangeable. In physics, however, there is a clear distinction between these two quantities. Consider a marathon runner who runs more than 40 km, yet ends up at his starting point. His average velocity is zero! Nonetheless, we need to be able to quantify how fast he was running. A slightly different ratio accomplishes this for us. **The average speed of a particle, a scalar quantity, is defined as the total distance traveled divided by the total time it takes to travel that distance:**

$$\text{Average speed} = \frac{\text{total distance}}{\text{total time}}$$

Average speed

The SI unit of average speed is the same as the unit of average velocity: meters per second. However, unlike average velocity, average speed has no direction and hence carries no algebraic sign.

Knowledge of the average speed of a particle tells us nothing about the details of the trip. For example, suppose it takes you 8.0 h to travel 280 km in your car. The average speed for your trip is 35 km/h. However, you most likely traveled at various speeds during the trip, and the average speed of 35 km/h could result from an infinite number of possible speed values.

ЕXAMPLE 2.1 ▸ Calculating the Variables of Motion

Find the displacement, average velocity, and average speed of the car in Figure 2.1a between positions Ⓐ and Ⓕ.

Solution The units of displacement must be meters, and the numerical result should be of the same order of magnitude as the given position data (which means probably not 10 or 100 times bigger or smaller). From the position–time graph given in Figure 2.1b, note that $x_A = 30$ m at $t_A = 0$ s and that $x_F = -53$ m at $t_F = 50$ s. Using these values along with the definition of displacement, Equation 2.1, we find that

$$\Delta x = x_F - x_A = -53 \text{ m} - 30 \text{ m} = \boxed{-83 \text{ m}}$$

This result means that the car ends up 83 m in the negative direction (to the left, in this case) from where it started. This number has the correct units and is of the same order of

magnitude as the supplied data. A quick look at Figure 2.1a indicates that this is the correct answer.

It is difficult to estimate the average velocity without completing the calculation, but we expect the units to be meters per second. Because the car ends up to the left of where we started taking data, we know the average velocity must be negative. From Equation 2.2,

$$\bar{v}_x = \frac{\Delta x}{\Delta t} = \frac{x_f - x_i}{t_f - t_i} = \frac{x_F - x_A}{t_F - t_A}$$

$$= \frac{-53 \text{ m} - 30 \text{ m}}{50 \text{ s} - 0 \text{ s}} = \frac{-83 \text{ m}}{50 \text{ s}} = \boxed{-1.7 \text{ m/s}}$$

We find the car's average speed for this trip by adding the distances traveled and dividing by the total time:

$$\text{Average speed} = \frac{22 \text{ m} + 52 \text{ m} + 53 \text{ m}}{50 \text{ s}} = \boxed{2.5 \text{ m/s}}$$

2.2 ▸ INSTANTANEOUS VELOCITY AND SPEED

Often we need to know the velocity of a particle at a particular instant in time, rather than over a finite time interval. For example, even though you might want to calculate your average velocity during a long automobile trip, you would be especially interested in knowing your velocity at the *instant* you noticed the police

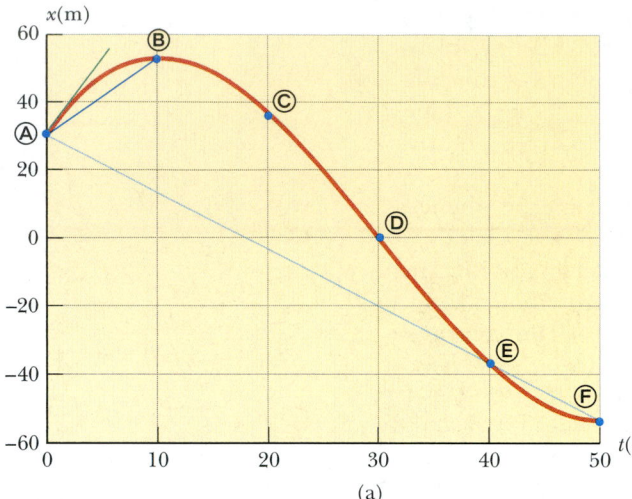

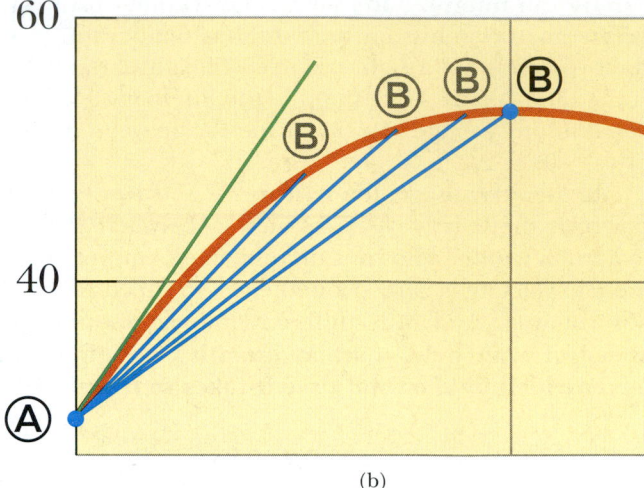

(a)   (b)

Figure 2.3   (a) Graph representing the motion of the car in Figure 2.1. (b) An enlargement of the upper left-hand corner of the graph shows how the blue line between positions Ⓐ and Ⓑ approaches the green tangent line as point Ⓑ gets closer to point Ⓐ.

car parked alongside the road in front of you. In other words, you would like to be able to specify your velocity just as precisely as you can specify your position by noting what is happening at a specific clock reading—that is, at some specific instant. It may not be immediately obvious how to do this. What does it mean to talk about how fast something is moving if we "freeze time" and talk only about an individual instant? This is a subtle point not thoroughly understood until the late 1600s. At that time, with the invention of calculus, scientists began to understand how to describe an object's motion at any moment in time.

To see how this is done, consider Figure 2.3a. We have already discussed the average velocity for the interval during which the car moved from position Ⓐ to position Ⓑ (given by the slope of the dark blue line) and for the interval during which it moved from Ⓐ to Ⓕ (represented by the slope of the light blue line). Which of these two lines do you think is a closer approximation of the initial velocity of the car? The car starts out by moving to the right, which we defined to be the positive direction. Therefore, being positive, the value of the average velocity during the Ⓐ to Ⓑ interval is probably closer to the initial value than is the value of the average velocity during the Ⓐ to Ⓕ interval, which we determined to be negative in Example 2.1. Now imagine that we start with the dark blue line and slide point Ⓑ to the left along the curve, toward point Ⓐ, as in Figure 2.3b. The line between the points becomes steeper and steeper, and as the two points get extremely close together, the line becomes a tangent line to the curve, indicated by the green line on the graph. The slope of this tangent line represents the velocity of the car at the moment we started taking data, at point Ⓐ. What we have done is determine the *instantaneous velocity* at that moment. In other words, the **instantaneous velocity v_x equals the limiting value of the ratio $\Delta x/\Delta t$ as Δt approaches zero:**[1]

Definition of instantaneous velocity

3.3

$$v_x \equiv \lim_{\Delta t \to 0} \frac{\Delta x}{\Delta t} \qquad \textbf{(2.3)}$$

[1] Note that the displacement Δx also approaches zero as Δt approaches zero. As Δx and Δt become smaller and smaller, the ratio $\Delta x/\Delta t$ approaches a value equal to the slope of the line tangent to the *x*-versus-*t* curve.

In calculus notation, this limit is called the *derivative* of x with respect to t, written dx/dt:

$$v_x \equiv \lim_{\Delta t \to 0} \frac{\Delta x}{\Delta t} = \frac{dx}{dt} \qquad\qquad \textbf{(2.4)}$$

The instantaneous velocity can be positive, negative, or zero. When the slope of the position–time graph is positive, such as at any time during the first 10 s in Figure 2.3, v_x is positive. After point Ⓑ, v_x is negative because the slope is negative. At the peak, the slope and the instantaneous velocity are zero.

From here on, we use the word *velocity* to designate instantaneous velocity. When it is *average velocity* we are interested in, we always use the adjective *average*.

The instantaneous speed of a particle is defined as the magnitude of its velocity. As with average speed, instantaneous speed has no direction associated with it and hence carries no algebraic sign. For example, if one particle has a velocity of $+25$ m/s along a given line and another particle has a velocity of -25 m/s along the same line, both have a speed[2] of 25 m/s.

EXAMPLE 2.2   Average and Instantaneous Velocity

A particle moves along the x axis. Its x coordinate varies with time according to the expression $x = -4t + 2t^2$, where x is in meters and t is in seconds.[3] The position–time graph for this motion is shown in Figure 2.4. Note that the particle moves in the negative x direction for the first second of motion, is at rest at the moment $t = 1$ s, and moves in the positive x direction for $t > 1$ s. (a) Determine the displacement of the particle in the time intervals $t = 0$ to $t = 1$ s and $t = 1$ s to $t = 3$ s.

Solution   During the first time interval, we have a negative slope and hence a negative velocity. Thus, we know that the displacement between Ⓐ and Ⓑ must be a negative number having units of meters. Similarly, we expect the displacement between Ⓑ and Ⓓ to be positive.

In the first time interval, we set $t_i = t_A = 0$ and $t_f = t_B = 1$ s. Using Equation 2.1, with $x = -4t + 2t^2$, we obtain for the first displacement

$$\Delta x_{A \to B} = x_f - x_i = x_B - x_A$$
$$= [-4(1) + 2(1)^2] - [-4(0) + 2(0)^2]$$
$$= \boxed{-2 \text{ m}}$$

To calculate the displacement during the second time interval, we set $t_i = t_B = 1$ s and $t_f = t_D = 3$ s:

$$\Delta x_{B \to D} = x_f - x_i = x_D - x_B$$

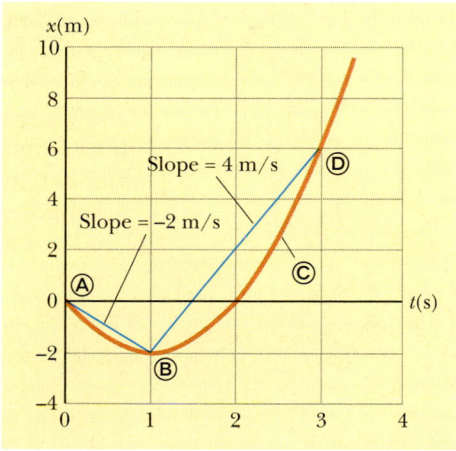

Figure 2.4   Position–time graph for a particle having an x coordinate that varies in time according to the expression $x = -4t + 2t^2$.

$$= [-4(3) + 2(3)^2] - [-4(1) + 2(1)^2]$$
$$= \boxed{+8 \text{ m}}$$

These displacements can also be read directly from the position–time graph.

---

[2] As with velocity, we drop the adjective for instantaneous speed: "Speed" means instantaneous speed.

[3] Simply to make it easier to read, we write the empirical equation as $x = -4t + 2t^2$ rather than as $x = (-4.00 \text{ m/s})t + (2.00 \text{ m/s}^2)t^{2.00}$. When an equation summarizes measurements, consider its coefficients to have as many significant digits as other data quoted in a problem. Consider its coefficients to have the units required for dimensional consistency. When we start our clocks at $t = 0$ s, we usually do not mean to limit the precision to a single digit. Consider any zero value in this book to have as many significant figures as you need.

(b) Calculate the average velocity during these two time intervals.

Solution   In the first time interval, $\Delta t = t_f - t_i = t_B - t_A = 1$ s. Therefore, using Equation 2.2 and the displacement calculated in (a), we find that

$$\bar{v}_{x(A \to B)} = \frac{\Delta x_{A \to B}}{\Delta t} = \frac{-2 \text{ m}}{1 \text{ s}} = \boxed{-2 \text{ m/s}}$$

In the second time interval, $\Delta t = 2$ s; therefore,

$$\bar{v}_{x(B \to D)} = \frac{\Delta x_{B \to D}}{\Delta t} = \frac{8 \text{ m}}{2 \text{ s}} = \boxed{+4 \text{ m/s}}$$

These values agree with the slopes of the lines joining these points in Figure 2.4.

(c) Find the instantaneous velocity of the particle at $t = 2.5$ s.

Solution   Certainly we can guess that this instantaneous velocity must be of the same order of magnitude as our previous results, that is, around 4 m/s. Examining the graph, we see that the slope of the tangent at position Ⓒ is greater than the slope of the blue line connecting points Ⓑ and Ⓓ. Thus, we expect the answer to be greater than 4 m/s. By measuring the slope of the position–time graph at $t = 2.5$ s, we find that

$$v_x = \boxed{+6 \text{ m/s}}$$

2.3 ▶ ACCELERATION

In the last example, we worked with a situation in which the velocity of a particle changed while the particle was moving. This is an extremely common occurrence. (How constant is your velocity as you ride a city bus?) It is easy to quantify changes in velocity as a function of time in exactly the same way we quantify changes in position as a function of time. When the velocity of a particle changes with time, the particle is said to be *accelerating*. For example, the velocity of a car increases when you step on the gas and decreases when you apply the brakes. However, we need a better definition of acceleration than this.

Suppose a particle moving along the x axis has a velocity v_{xi} at time t_i and a velocity v_{xf} at time t_f, as in Figure 2.5a.

> The average acceleration of the particle is defined as the *change* in velocity Δv_x divided by the time interval Δt during which that change occurred:

Average acceleration

$$\bar{a}_x \equiv \frac{\Delta v_x}{\Delta t} = \frac{v_{xf} - v_{xi}}{t_f - t_i} \tag{2.5}$$

As with velocity, when the motion being analyzed is one-dimensional, we can use positive and negative signs to indicate the direction of the acceleration. Because the dimensions of velocity are L/T and the dimension of time is T, accelera-

Figure 2.5   (a) A "particle" moving along the x axis from Ⓐ to Ⓑ has velocity v_{xi} at $t = t_i$ and velocity v_{xf} at $t = t_f$. (b) Velocity–time graph for the particle moving in a straight line. The slope of the blue straight line connecting Ⓐ and Ⓑ is the average acceleration in the time interval $\Delta t = t_f - t_i$.

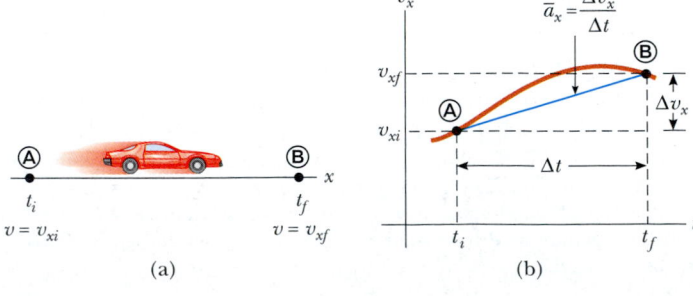

tion has dimensions of length divided by time squared, or L/T^2. The SI unit of acceleration is meters per second squared (m/s^2). It might be easier to interpret these units if you think of them as meters per second per second. For example, suppose an object has an acceleration of 2 m/s^2. You should form a mental image of the object having a velocity that is along a straight line and is increasing by 2 m/s during every 1-s interval. If the object starts from rest, you should be able to picture it moving at a velocity of + 2 m/s after 1 s, at + 4 m/s after 2 s, and so on.

In some situations, the value of the average acceleration may be different over different time intervals. It is therefore useful to define the *instantaneous acceleration* as the limit of the average acceleration as Δt approaches zero. This concept is analogous to the definition of instantaneous velocity discussed in the previous section. If we imagine that point Ⓑ is brought closer and closer to point Ⓐ in Figure 2.5a and take the limit of $\Delta v_x/\Delta t$ as Δt approaches zero, we obtain the instantaneous acceleration:

$$a_x \equiv \lim_{\Delta t \to 0} \frac{\Delta v_x}{\Delta t} = \frac{dv_x}{dt} \qquad \textbf{(2.6)}$$

Instantaneous acceleration

That is, **the instantaneous acceleration equals the derivative of the velocity with respect to time,** which by definition is the slope of the velocity–time graph (Fig. 2.5b). Thus, we see that just as the velocity of a moving particle is the slope of the particle's *x-t* graph, the acceleration of a particle is the slope of the particle's v_x-*t* graph. One can interpret the derivative of the velocity with respect to time as the time rate of change of velocity. If a_x is positive, then the acceleration is in the positive *x* direction; if a_x is negative, then the acceleration is in the negative *x* direction.

From now on we shall use the term *acceleration* to mean instantaneous acceleration. When we mean average acceleration, we shall always use the adjective *average*.

Because $v_x = dx/dt$, the acceleration can also be written

$$a_x = \frac{dv_x}{dt} = \frac{d}{dt}\left(\frac{dx}{dt}\right) = \frac{d^2x}{dt^2} \qquad \textbf{(2.7)}$$

That is, in one-dimensional motion, the acceleration equals the *second derivative* of *x* with respect to time.

Figure 2.6 illustrates how an acceleration–time graph is related to a velocity–time graph. The acceleration at any time is the slope of the velocity–time graph at that time. Positive values of acceleration correspond to those points in Figure 2.6a where the velocity is increasing in the positive *x* direction. The acceler-

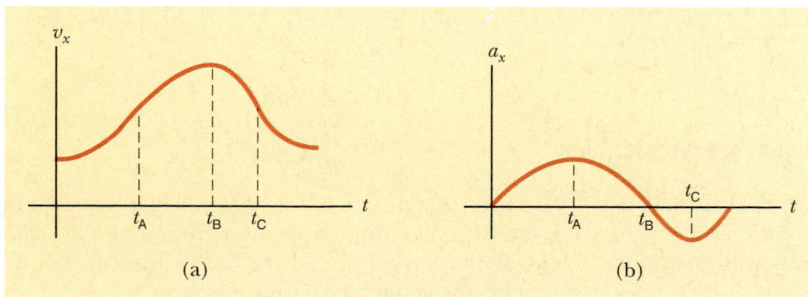

(a)

(b)

Figure 2.6 Instantaneous acceleration can be obtained from the v_x-*t* graph. (a) The velocity–time graph for some motion. (b) The acceleration–time graph for the same motion. The acceleration given by the a_x-*t* graph for any value of *t* equals the slope of the line tangent to the v_x-*t* graph at the same value of *t*.

ation reaches a maximum at time t_A, when the slope of the velocity–time graph is a maximum. The acceleration then goes to zero at time t_B, when the velocity is a maximum (that is, when the slope of the v_x-t graph is zero). The acceleration is negative when the velocity is decreasing in the positive x direction, and it reaches its most negative value at time t_C.

CONCEPTUAL EXAMPLE 2.3   Graphical Relationships Between x, v_x, and a_x

The position of an object moving along the x axis varies with time as in Figure 2.7a. Graph the velocity versus time and the acceleration versus time for the object.

Solution   The velocity at any instant is the slope of the tangent to the x-t graph at that instant. Between $t = 0$ and $t = t_A$, the slope of the x-t graph increases uniformly, and so the velocity increases linearly, as shown in Figure 2.7b. Between t_A and t_B, the slope of the x-t graph is constant, and so the velocity remains constant. At t_D, the slope of the x-t graph is zero, so the velocity is zero at that instant. Between t_D and t_E, the slope of the x-t graph and thus the velocity are negative and decrease uniformly in this interval. In the interval t_E to t_F, the slope of the x-t graph is still negative, and at t_F it goes to zero. Finally, after t_F, the slope of the x-t graph is zero, meaning that the object is at rest for $t > t_F$.

The acceleration at any instant is the slope of the tangent to the v_x-t graph at that instant. The graph of acceleration versus time for this object is shown in Figure 2.7c. The acceleration is constant and positive between 0 and t_A, where the slope of the v_x-t graph is positive. It is zero between t_A and t_B and for $t > t_F$ because the slope of the v_x-t graph is zero at these times. It is negative between t_B and t_E because the slope of the v_x-t graph is negative during this interval.

Figure 2.7   (a) Position–time graph for an object moving along the x axis. (b) The velocity–time graph for the object is obtained by measuring the slope of the position–time graph at each instant. (c) The acceleration–time graph for the object is obtained by measuring the slope of the velocity–time graph at each instant.

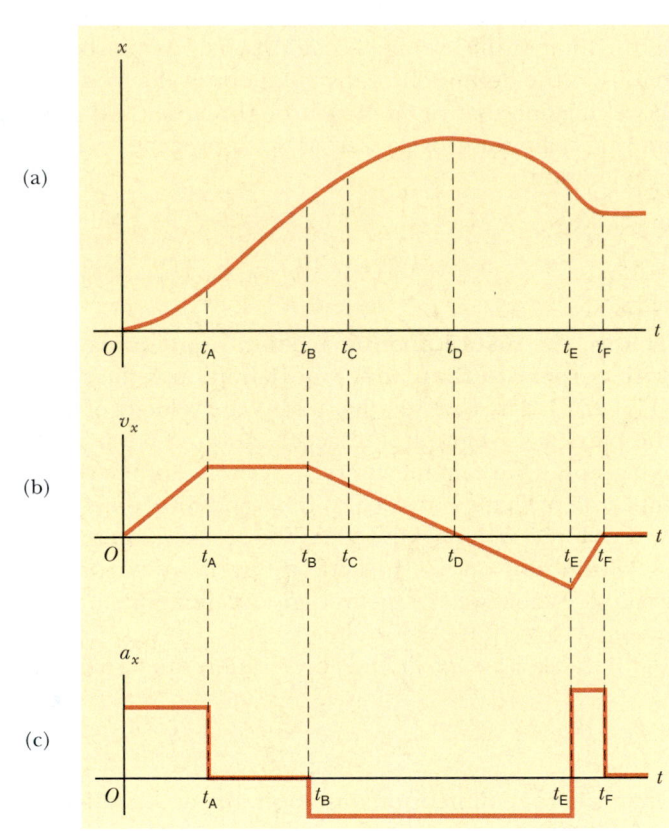

Quick Quiz 2.1

Make a velocity–time graph for the car in Figure 2.1a and use your graph to determine whether the car ever exceeds the speed limit posted on the road sign (30 km/h).

EXAMPLE 2.4   Average and Instantaneous Acceleration

The velocity of a particle moving along the x axis varies in time according to the expression $v_x = (40 - 5t^2)$ m/s, where t is in seconds. (a) Find the average acceleration in the time interval $t = 0$ to $t = 2.0$ s.

Solution   Figure 2.8 is a v_x-t graph that was created from the velocity versus time expression given in the problem statement. Because the slope of the entire v_x-t curve is negative, we expect the acceleration to be negative.

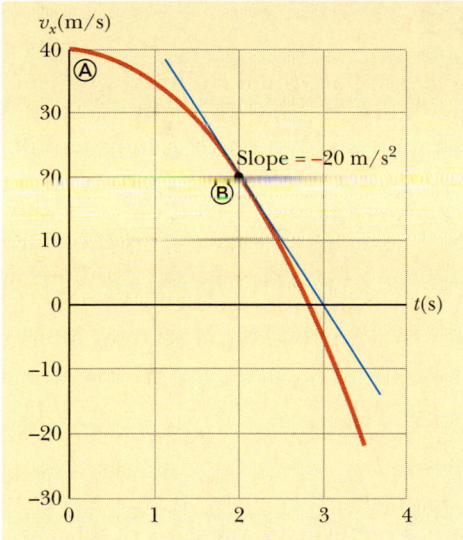

Figure 2.8 The velocity–time graph for a particle moving along the x axis according to the expression $v_x = (40 - 5t^2)$ m/s. The acceleration at $t = 2$ s is equal to the slope of the blue tangent line at that time.

$$\bar{a}_x = \frac{v_{xf} - v_{xi}}{t_f - t_i} = \frac{v_{xB} - v_{xA}}{t_B - t_A} = \frac{(20 - 40)\ \text{m/s}}{(2.0 - 0)\ \text{s}}$$

$$= \boxed{-10\ \text{m/s}^2}$$

The negative sign is consistent with our expectations—namely, that the average acceleration, which is represented by the slope of the line (not shown) joining the initial and final points on the velocity–time graph, is negative.

(b) Determine the acceleration at $t = 2.0$ s.

Solution The velocity at any time t is $v_{xi} = (40 - 5t^2)$ m/s, and the velocity at any later time $t + \Delta t$ is

$$v_{xf} = 40 - 5(t + \Delta t)^2 = 40 - 5t^2 - 10t\,\Delta t - 5(\Delta t)^2$$

Therefore, the change in velocity over the time interval Δt is

$$\Delta v_x = v_{xf} - v_{xi} = [-10t\,\Delta t - 5(\Delta t)^2]\ \text{m/s}$$

Dividing this expression by Δt and taking the limit of the result as Δt approaches zero gives the acceleration at *any* time t:

$$a_x = \lim_{\Delta t \to 0} \frac{\Delta v_x}{\Delta t} = \lim_{\Delta t \to 0} (-10t - 5\Delta t) = -10t\ \text{m/s}^2$$

Therefore, at $t = 2.0$ s,

$$a_x = (-10)(2.0)\ \text{m/s}^2 = \boxed{-20\ \text{m/s}^2}$$

What we have done by comparing the average acceleration during the interval between Ⓐ and Ⓑ $(-10\ \text{m/s}^2)$ with the instantaneous value at Ⓑ $(-20\ \text{m/s}^2)$ is compare the slope of the line (not shown) joining Ⓐ and Ⓑ with the slope of the tangent at Ⓑ.

Note that the acceleration is not constant in this example. Situations involving constant acceleration are treated in Section 2.5.

We find the velocities at $t_i = t_A = 0$ and $t_f = t_B = 2.0$ s by substituting these values of t into the expression for the velocity:

$$v_{xA} = (40 - 5t_A{}^2)\ \text{m/s} = [40 - 5(0)^2]\ \text{m/s} = +40\ \text{m/s}$$

$$v_{xB} = (40 - 5t_B{}^2)\ \text{m/s} = [40 - 5(2.0)^2]\ \text{m/s} = +20\ \text{m/s}$$

Therefore, the average acceleration in the specified time interval $\Delta t = t_B - t_A = 2.0$ s is

So far we have evaluated the derivatives of a function by starting with the definition of the function and then taking the limit of a specific ratio. Those of you familiar with calculus should recognize that there are specific rules for taking derivatives. These rules, which are listed in Appendix B.6, enable us to evaluate derivatives quickly. For instance, one rule tells us that the derivative of any constant is zero. As another example, suppose x is proportional to some power of t, such as in the expression

$$x = At^n$$

where A and n are constants. (This is a very common functional form.) The derivative of x with respect to t is

$$\frac{dx}{dt} = nAt^{n-1}$$

Applying this rule to Example 2.4, in which $v_x = 40 - 5t^2$, we find that $a_x = dv_x/dt = -10t$.

2.4 MOTION DIAGRAMS

The concepts of velocity and acceleration are often confused with each other, but in fact they are quite different quantities. It is instructive to use motion diagrams to describe the velocity and acceleration while an object is in motion. In order not to confuse these two vector quantities, for which both magnitude and direction are important, we use red for velocity vectors and violet for acceleration vectors, as shown in Figure 2.9. The vectors are sketched at several instants during the motion of the object, and the time intervals between adjacent positions are assumed to be equal. This illustration represents three sets of strobe photographs of a car moving from left to right along a straight roadway. The time intervals between flashes are equal in each diagram.

In Figure 2.9a, the images of the car are equally spaced, showing us that the car moves the same distance in each time interval. Thus, the car moves with *constant positive velocity* and has *zero acceleration.*

In Figure 2.9b, the images become farther apart as time progresses. In this case, the velocity vector increases in time because the car's displacement between adjacent positions increases in time. The car is moving with a *positive velocity* and a *positive acceleration.*

In Figure 2.9c, we can tell that the car slows as it moves to the right because its displacement between adjacent images decreases with time. In this case, the car moves to the right with a constant negative acceleration. The velocity vector decreases in time and eventually reaches zero. From this diagram we see that the acceleration and velocity vectors are *not* in the same direction. The car is moving with a *positive velocity* but with a *negative acceleration.*

You should be able to construct motion diagrams for a car that moves initially to the left with a constant positive or negative acceleration.

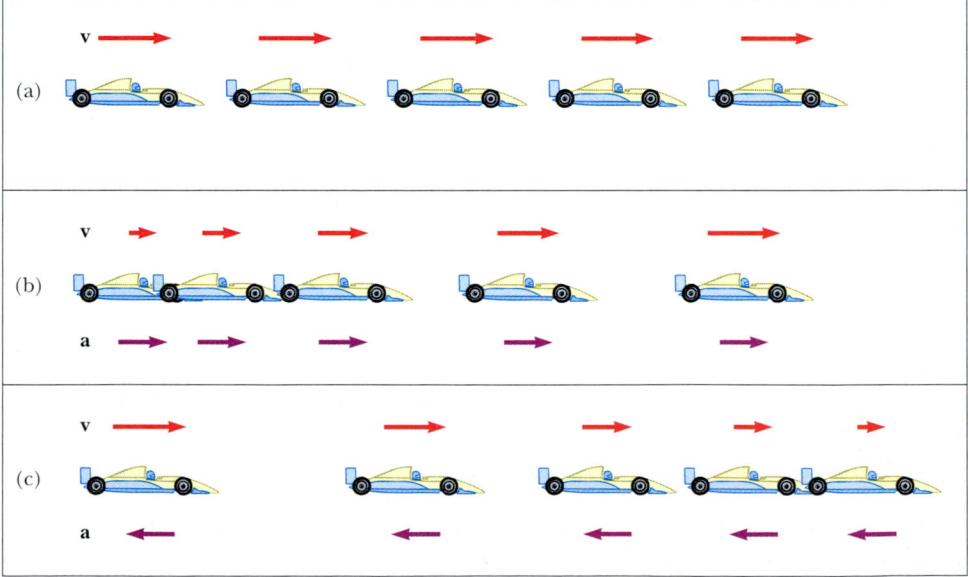

Figure 2.9   (a) Motion diagram for a car moving at constant velocity (zero acceleration). (b) Motion diagram for a car whose constant acceleration is in the direction of its velocity. The velocity vector at each instant is indicated by a red arrow, and the constant acceleration by a violet arrow. (c) Motion diagram for a car whose constant acceleration is in the direction *opposite* the velocity at each instant.

(a) If a car is traveling eastward, can its acceleration be westward? (b) If a car is slowing down, can its acceleration be positive?

2.5 ONE-DIMENSIONAL MOTION WITH CONSTANT ACCELERATION

If the acceleration of a particle varies in time, its motion can be complex and difficult to analyze. However, a very common and simple type of one-dimensional motion is that in which the acceleration is constant. When this is the case, the average acceleration over any time interval equals the instantaneous acceleration at any instant within the interval, and the velocity changes at the same rate throughout the motion.

If we replace $\bar{a}_x$ by a_x in Equation 2.5 and take $t_i = 0$ and t_f to be any later time t, we find that

$$a_x = \frac{v_{xf} - v_{xi}}{t}$$

or

$$v_{xf} = v_{xi} + a_x t \qquad \text{(for constant } a_x\text{)} \qquad \textbf{(2.8)}$$

Velocity as a function of time

This powerful expression enables us to determine an object's velocity at *any* time t if we know the object's initial velocity and its (constant) acceleration. A velocity–time graph for this constant-acceleration motion is shown in Figure 2.10a. The graph is a straight line, the (constant) slope of which is the acceleration a_x; this is consistent with the fact that $a_x = dv_x/dt$ is a constant. Note that the slope is positive; this indicates a positive acceleration. If the acceleration were negative, then the slope of the line in Figure 2.10a would be negative.

When the acceleration is constant, the graph of acceleration versus time (Fig. 2.10b) is a straight line having a slope of zero.

Describe the meaning of each term in Equation 2.8.

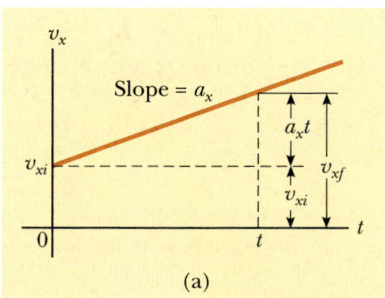
(a)

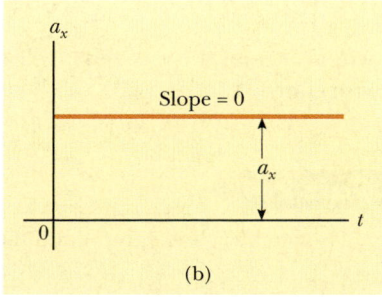
(b)

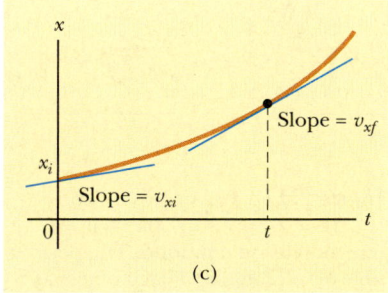
(c)

Figure 2.10 An object moving along the x axis with constant acceleration a_x. (a) The velocity–time graph. (b) The acceleration–time graph. (c) The position–time graph.

Because velocity at constant acceleration varies linearly in time according to Equation 2.8, we can express the average velocity in any time interval as the arithmetic mean of the initial velocity v_{xi} and the final velocity v_{xf}:

$$\bar{v}_x = \frac{v_{xi} + v_{xf}}{2} \qquad \text{(for constant } a_x\text{)} \qquad \textbf{(2.9)}$$

Note that this expression for average velocity applies *only* in situations in which the acceleration is constant.

We can now use Equations 2.1, 2.2, and 2.9 to obtain the displacement of any object as a function of time. Recalling that Δx in Equation 2.2 represents $x_f - x_i$, and now using t in place of Δt (because we take $t_i = 0$), we can say

Displacement as a function of velocity and time

$$x_f - x_i = \bar{v}_x t = \tfrac{1}{2}(v_{xi} + v_{xf})t \qquad \text{(for constant } a_x\text{)} \qquad \textbf{(2.10)}$$

We can obtain another useful expression for displacement at constant acceleration by substituting Equation 2.8 into Equation 2.10:

$$x_f - x_i = \tfrac{1}{2}(v_{xi} + v_{xi} + a_x t)t$$

$$x_f - x_i = v_{xi}t + \tfrac{1}{2}a_x t^2 \qquad \textbf{(2.11)}$$

The position–time graph for motion at constant (positive) acceleration shown in Figure 2.10c is obtained from Equation 2.11. Note that the curve is a parabola. The slope of the tangent line to this curve at $t = t_i = 0$ equals the initial velocity v_{xi}, and the slope of the tangent line at any later time t equals the velocity at that time, v_{xf}.

We can check the validity of Equation 2.11 by moving the x_i term to the right-hand side of the equation and differentiating the equation with respect to time:

$$v_{xf} = \frac{dx_f}{dt} = \frac{d}{dt}\left(x_i + v_{xi}t + \frac{1}{2}a_x t^2\right) = v_{xi} + a_x t$$

Finally, we can obtain an expression for the final velocity that does not contain a time interval by substituting the value of t from Equation 2.8 into Equation 2.10:

$$x_f - x_i = \frac{1}{2}(v_{xi} + v_{xf})\left(\frac{v_{xf} - v_{xi}}{a_x}\right) = \frac{v_{xf}^2 - v_{xi}^2}{2a_x}$$

$$v_{xf}^2 = v_{xi}^2 + 2a_x(x_f - x_i) \qquad \text{(for constant } a_x\text{)} \qquad \textbf{(2.12)}$$

For motion at *zero* acceleration, we see from Equations 2.8 and 2.11 that

$$\left.\begin{array}{l} v_{xf} = v_{xi} = v_x \\ x_f - x_i = v_x t \end{array}\right\} \qquad \text{when } a_x = 0$$

That is, when acceleration is zero, velocity is constant and displacement changes linearly with time.

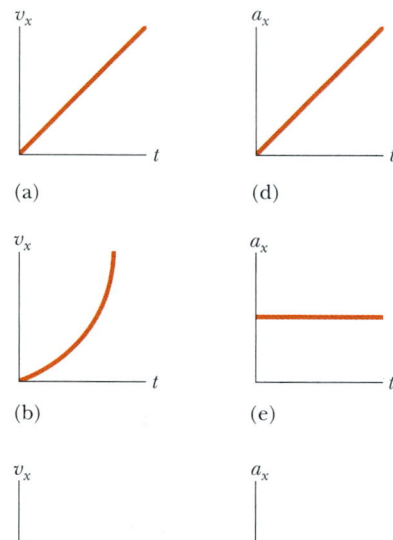

(a)   (b)   (c)   (d)   (e)   (f)

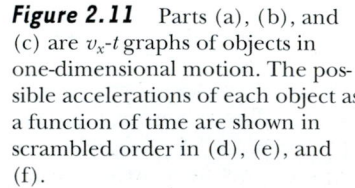

Figure 2.11 Parts (a), (b), and (c) are v_x-t graphs of objects in one-dimensional motion. The possible accelerations of each object as a function of time are shown in scrambled order in (d), (e), and (f).

Quick Quiz 2.4

In Figure 2.11, match each v_x-t graph with the a_x-t graph that best describes the motion.

Equations 2.8 through 2.12 are **kinematic expressions that may be used to solve any problem involving one-dimensional motion at constant accelera-**

TABLE 2.2	Kinematic Equations for Motion in a Straight Line Under Constant Acceleration
Equation	**Information Given by Equation**
$v_{xf} = v_{xi} + a_x t$	Velocity as a function of time
$x_f - x_i = \frac{1}{2}(v_{xi} + v_{xf})t$	Displacement as a function of velocity and time
$x_f - x_i = v_{xi}t + \frac{1}{2}a_x t^2$	Displacement as a function of time
$v_{xf}^2 = v_{xi}^2 + 2a_x(x_f - x_i)$	Velocity as a function of displacement

Note: Motion is along the x axis.

tion. Keep in mind that these relationships were derived from the definitions of velocity and acceleration, together with some simple algebraic manipulations and the requirement that the acceleration be constant.

The four kinematic equations used most often are listed in Table 2.2 for convenience. The choice of which equation you use in a given situation depends on what you know beforehand. Sometimes it is necessary to use two of these equations to solve for two unknowns. For example, suppose initial velocity v_{xi} and acceleration a_x are given. You can then find (1) the velocity after an interval t has elapsed, using $v_{xf} = v_{xi} + a_x t$, and (2) the displacement after an interval t has elapsed, using $x_f - x_i = v_{xi}t + \frac{1}{2}a_x t^2$. You should recognize that the quantities that vary during the motion are velocity, displacement, and time.

You will get a great deal of practice in the use of these equations by solving a number of exercises and problems. Many times you will discover that more than one method can be used to obtain a solution. Remember that these equations of kinematics *cannot* be used in a situation in which the acceleration varies with time. They can be used only when the acceleration is constant.

CONCEPTUAL EXAMPLE 2.5 ▶ The Velocity of Different Objects

Consider the following one-dimensional motions: (a) A ball thrown directly upward rises to a highest point and falls back into the thrower's hand. (b) A race car starts from rest and speeds up to 100 m/s. (c) A spacecraft drifts through space at constant velocity. Are there any points in the motion of these objects at which the instantaneous velocity is the same as the average velocity over the entire motion? If so, identify the point(s).

Solution (a) The average velocity for the thrown ball is zero because the ball returns to the starting point; thus its displacement is zero. (Remember that average velocity is de-

fined as $\Delta x/\Delta t$.) There is one point at which the instantaneous velocity is zero—at the top of the motion.

(b) The car's average velocity cannot be evaluated unambiguously with the information given, but it must be some value between 0 and 100 m/s. Because the car will have every instantaneous velocity between 0 and 100 m/s at some time during the interval, there must be some instant at which the instantaneous velocity is equal to the average velocity.

(c) Because the spacecraft's instantaneous velocity is constant, its instantaneous velocity at *any* time and its average velocity over *any* time interval are the same.

EXAMPLE 2.6 ▶ Entering the Traffic Flow

(a) Estimate your average acceleration as you drive up the entrance ramp to an interstate highway.

Solution This problem involves more than our usual amount of estimating! We are trying to come up with a value

of a_x, but that value is hard to guess directly. The other three variables involved in kinematics are position, velocity, and time. Velocity is probably the easiest one to approximate. Let us assume a final velocity of 100 km/h, so that you can merge with traffic. We multiply this value by 1 000 to convert kilometers to meters and then divide by 3 600 to convert hours to seconds. These two calculations together are roughly equivalent to dividing by 3. In fact, let us just say that the final velocity is $v_{xf} \approx 30$ m/s. (Remember, you can get away with this type of approximation and with dropping digits when performing mental calculations. If you were starting with British units, you could approximate 1 mi/h as roughly 0.5 m/s and continue from there.)

Now we assume that you started up the ramp at about one-third your final velocity, so that $v_{xi} \approx 10$ m/s. Finally, we assume that it takes about 10 s to get from v_{xi} to v_{xf}, basing this guess on our previous experience in automobiles. We can then find the acceleration, using Equation 2.8:

$$a_x = \frac{v_{xf} - v_{xi}}{t} \approx \frac{30 \text{ m/s} - 10 \text{ m/s}}{10 \text{ s}} = \boxed{2 \text{ m/s}^2}$$

Granted, we made many approximations along the way, but this type of mental effort can be surprisingly useful and often

yields results that are not too different from those derived from careful measurements.

(b) How far did you go during the first half of the time interval during which you accelerated?

Solution   We can calculate the distance traveled during the first 5 s from Equation 2.11:

$$x_f - x_i = v_{xi}t + \tfrac{1}{2}a_x t^2 \approx (10 \text{ m/s})(5 \text{ s}) + \tfrac{1}{2}(2 \text{ m/s}^2)(5 \text{ s})^2$$

$$= 50 \text{ m} + 25 \text{ m} = \boxed{75 \text{ m}}$$

This result indicates that if you had not accelerated, your initial velocity of 10 m/s would have resulted in a 50-m movement up the ramp during the first 5 s. The additional 25 m is the result of your increasing velocity during that interval.

Do not be afraid to attempt making educated guesses and doing some fairly drastic number rounding to simplify mental calculations. Physicists engage in this type of thought analysis all the time.

EXAMPLE 2.7   Carrier Landing

A jet lands on an aircraft carrier at 140 mi/h (≈ 63 m/s). (a) What is its acceleration if it stops in 2.0 s?

Solution   We define our x axis as the direction of motion of the jet. A careful reading of the problem reveals that in addition to being given the initial speed of 63 m/s, we also know that the final speed is zero. We also note that we are not given the displacement of the jet while it is slowing down. Equation 2.8 is the only equation in Table 2.2 that does not involve displacement, and so we use it to find the acceleration:

$$a_x = \frac{v_{xf} - v_{xi}}{t} \approx \frac{0 - 63 \text{ m/s}}{2.0 \text{ s}} = \boxed{-31 \text{ m/s}^2}$$

(b) What is the displacement of the plane while it is stopping?

Solution   We can now use any of the other three equations in Table 2.2 to solve for the displacement. Let us choose Equation 2.10:

$$x_f - x_i = \tfrac{1}{2}(v_{xi} + v_{xf})t = \tfrac{1}{2}(63 \text{ m/s} + 0)(2.0 \text{ s}) = \boxed{63 \text{ m}}$$

If the plane travels much farther than this, it might fall into the ocean. Although the idea of using arresting cables to enable planes to land safely on ships originated at about the time of the First World War, the cables are still a vital part of the operation of modern aircraft carriers.

EXAMPLE 2.8   Watch Out for the Speed Limit!

A car traveling at a constant speed of 45.0 m/s passes a trooper hidden behind a billboard. One second after the speeding car passes the billboard, the trooper sets out from the billboard to catch it, accelerating at a constant rate of 3.00 m/s². How long does it take her to overtake the car?

Solution   A careful reading lets us categorize this as a constant-acceleration problem. We know that after the 1-s delay in starting, it will take the trooper 15 additional seconds to accelerate up to 45.0 m/s. Of course, she then has to continue to pick up speed (at a rate of 3.00 m/s per second) to

catch up to the car. While all this is going on, the car continues to move. We should therefore expect our result to be well over 15 s. A sketch (Fig. 2.12) helps clarify the sequence of events.

First, we write expressions for the position of each vehicle as a function of time. It is convenient to choose the position of the billboard as the origin and to set $t_B \equiv 0$ as the time the trooper begins moving. At that instant, the car has already traveled a distance of 45.0 m because it has traveled at a constant speed of $v_x = 45.0$ m/s for 1 s. Thus, the initial position of the speeding car is $x_B = 45.0$ m.

Because the car moves with constant speed, its accelera-

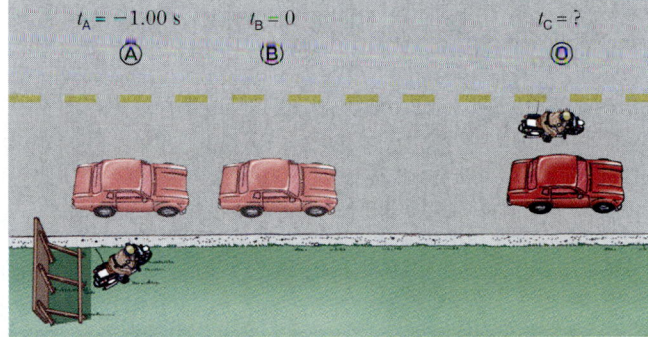

$v_{x\,car} = 45.0$ m/s
$a_{x\,car} = 0$
$a_{x\,trooper} = 3.00$ m/s^2

$t_A = -1.00$ s $t_B = 0$ $t_C = ?$
Ⓐ   Ⓑ   Ⓒ

Figure 2.12   A speeding car passes a hidden police officer.

tion is zero, and applying Equation 2.11 (with $a_x = 0$) gives for the car's position at any time t:

$$x_{car} = x_B + v_{x\,car}t = 45.0 \text{ m} + (45.0 \text{ m/s})t$$

A quick check shows that at $t = 0$, this expression gives the car's correct initial position when the trooper begins to move: $x_{car} = x_B = 45.0$ m. Looking at limiting cases to see whether they yield expected values is a very useful way to make sure that you are obtaining reasonable results.

The trooper starts from rest at $t = 0$ and accelerates at 3.00 m/s^2 away from the origin. Hence, her position after any time interval t can be found from Equation 2.11:

$$x_f = x_i + v_{xi}t + \tfrac{1}{2}a_x t^2$$

$$x_{trooper} = 0 + 0t + \tfrac{1}{2}a_x t^2 = \tfrac{1}{2}(3.00 \text{ m/s}^2)t^2$$

The trooper overtakes the car at the instant her position matches that of the car, which is position Ⓒ:

$$x_{trooper} = x_{car}$$

$$\tfrac{1}{2}(3.00 \text{ m/s}^2)t^2 = 45.0 \text{ m} + (45.0 \text{ m/s})t$$

This gives the quadratic equation

$$1.50t^2 - 45.0t - 45.0 = 0$$

The positive solution of this equation is $t = \boxed{31.0 \text{ s}}$.

(For help in solving quadratic equations, see Appendix B.2.) Note that in this 31.0-s time interval, the trooper travels a distance of about 1440 m. [This distance can be calculated from the car's constant speed: $(45.0 \text{ m/s})(31 + 1) \text{ s} = 1\,440$ m.]

Exercise   This problem can be solved graphically. On the same graph, plot position versus time for each vehicle, and from the intersection of the two curves determine the time at which the trooper overtakes the car.

2.6 ▷ FREELY FALLING OBJECTS

It is now well known that, in the absence of air resistance, all objects dropped near the Earth's surface fall toward the Earth with the same constant acceleration under the influence of the Earth's gravity. It was not until about 1600 that this conclusion was accepted. Before that time, the teachings of the great philosopher Aristotle (384–322 B.C.) had held that heavier objects fall faster than lighter ones.

It was the Italian Galileo Galilei (1564–1642) who originated our present-day ideas concerning falling objects. There is a legend that he demonstrated the law of falling objects by observing that two different weights dropped simultaneously from the Leaning Tower of Pisa hit the ground at approximately the same time. Although there is some doubt that he carried out this particular experiment, it is well established that Galileo performed many experiments on objects moving on inclined planes. In his experiments he rolled balls down a slight incline and measured the distances they covered in successive time intervals. The purpose of the incline was to reduce the acceleration; with the acceleration reduced, Galileo was able to make accurate measurements of the time intervals. By gradually increasing the slope of the incline, he was finally able to draw conclusions about freely falling objects because a freely falling ball is equivalent to a ball moving down a vertical incline.

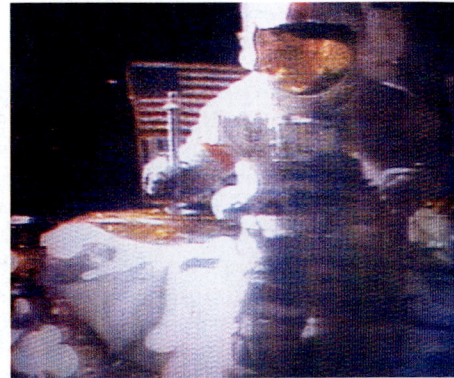

Astronaut David Scott released a hammer and a feather simultaneously, and they fell in unison to the lunar surface. *(Courtesy of NASA)*

Use a pencil to poke a hole in the bottom of a paper or polystyrene cup. Cover the hole with your finger and fill the cup with water. Hold the cup up in front of you and release it. Does water come out of the hole while the cup is falling? Why or why not?

Definition of free fall

Free-fall acceleration
$g = 9.80 \text{ m/s}^2$

You might want to try the following experiment. Simultaneously drop a coin and a crumpled-up piece of paper from the same height. If the effects of air resistance are negligible, both will have the same motion and will hit the floor at the same time. In the idealized case, in which air resistance is absent, such motion is referred to as *free fall*. If this same experiment could be conducted in a vacuum, in which air resistance is truly negligible, the paper and coin would fall with the same acceleration even when the paper is not crumpled. On August 2, 1971, such a demonstration was conducted on the Moon by astronaut David Scott. He simultaneously released a hammer and a feather, and in unison they fell to the lunar surface. This demonstration surely would have pleased Galileo!

When we use the expression *freely falling object,* we do not necessarily refer to an object dropped from rest. **A freely falling object is any object moving freely under the influence of gravity alone, regardless of its initial motion. Objects thrown upward or downward and those released from rest are all falling freely once they are released. Any freely falling object experiences an acceleration directed *downward*, regardless of its initial motion.**

We shall denote the magnitude of the *free-fall acceleration* by the symbol g. The value of g near the Earth's surface decreases with increasing altitude. Furthermore, slight variations in g occur with changes in latitude. It is common to define "up" as the $+y$ direction and to use y as the position variable in the kinematic equations. At the Earth's surface, the value of g is approximately 9.80 m/s^2. Unless stated otherwise, we shall use this value for g when performing calculations. For making quick estimates, use $g = 10 \text{ m/s}^2$.

If we neglect air resistance and assume that the free-fall acceleration does not vary with altitude over short vertical distances, then the motion of a freely falling object moving vertically is equivalent to motion in one dimension under constant acceleration. Therefore, the equations developed in Section 2.5 for objects moving with constant acceleration can be applied. The only modification that we need to make in these equations for freely falling objects is to note that the motion is in the vertical direction (the y direction) rather than in the horizontal (x) direction and that the acceleration is downward and has a magnitude of 9.80 m/s^2. Thus, we always take $a_y = -g = -9.80 \text{ m/s}^2$, where the minus sign means that the acceleration of a freely falling object is downward. In Chapter 14 we shall study how to deal with variations in g with altitude.

CONCEPTUAL EXAMPLE 2.9   The Daring Sky Divers

A sky diver jumps out of a hovering helicopter. A few seconds later, another sky diver jumps out, and they both fall along the same vertical line. Ignore air resistance, so that both sky divers fall with the same acceleration. Does the difference in their speeds stay the same throughout the fall? Does the vertical distance between them stay the same throughout the fall? If the two divers were connected by a long bungee cord, would the tension in the cord increase, lessen, or stay the same during the fall?

Solution   At any given instant, the speeds of the divers are different because one had a head start. In any time interval

Δt after this instant, however, the two divers increase their speeds by the same amount because they have the same acceleration. Thus, the difference in their speeds remains the same throughout the fall.

The first jumper always has a greater speed than the second. Thus, in a given time interval, the first diver covers a greater distance than the second. Thus, the separation distance between them increases.

Once the distance between the divers reaches the length of the bungee cord, the tension in the cord begins to increase. As the tension increases, the distance between the divers becomes greater and greater.

EXAMPLE 2.10 ▶ Describing the Motion of a Tossed Ball

A ball is tossed straight up at 25 m/s. Estimate its velocity at 1-s intervals.

Solution Let us choose the upward direction to be positive. Regardless of whether the ball is moving upward or downward, its vertical velocity changes by approximately −10 m/s for every second it remains in the air. It starts out at 25 m/s. After 1 s has elapsed, it is still moving upward but at 15 m/s because its acceleration is downward (downward acceleration causes its velocity to decrease). After another second, its upward velocity has dropped to 5 m/s. Now comes the tricky part—after another half second, its velocity is zero.

The ball has gone as high as it will go. After the last half of this 1-s interval, the ball is moving at −5 m/s. (The minus sign tells us that the ball is now moving in the negative direction, that is, downward. Its velocity has changed from +5 m/s to −5 m/s during that 1-s interval. The change in velocity is still $-5 - [+5] = -10$ m/s in that second.) It continues downward, and after another 1 s has elapsed, it is falling at a velocity of −15 m/s. Finally, after another 1 s, it has reached its original starting point and is moving downward at −25 m/s. If the ball had been tossed vertically off a cliff so that it could continue downward, its velocity would continue to change by about −10 m/s every second.

CONCEPTUAL EXAMPLE 2.11 ▶ Follow the Bouncing Ball

A tennis ball is dropped from shoulder height (about 1.5 m) and bounces three times before it is caught. Sketch graphs of its position, velocity, and acceleration as functions of time, with the $+y$ direction defined as upward.

Solution For our sketch let us stretch things out horizontally so that we can see what is going on. (Even if the ball were moving horizontally, this motion would not affect its vertical motion.)

From Figure 2.13 we see that the ball is in contact with the floor at points Ⓑ, Ⓓ, and Ⓕ. Because the velocity of the ball changes from negative to positive three times during these bounces, the slope of the position–time graph must change in the same way. Note that the time interval between bounces decreases. Why is that?

During the rest of the ball's motion, the slope of the velocity–time graph should be -9.80 m/s². The acceleration–time graph is a horizontal line at these times because the acceleration does not change when the ball is in free fall. When the ball is in contact with the floor, the velocity

changes substantially during a very short time interval, and so the acceleration must be quite great. This corresponds to the very steep upward lines on the velocity–time graph and to the spikes on the acceleration–time graph.

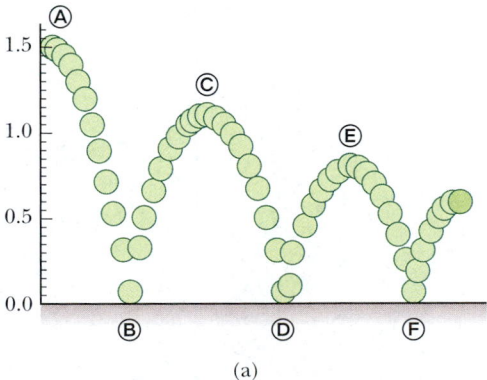

(a)

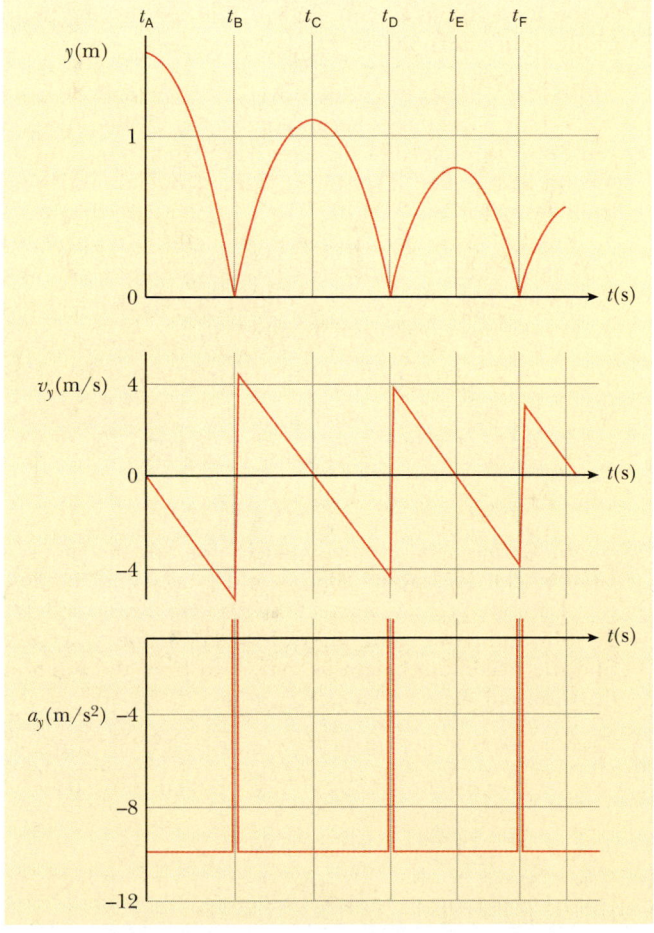

Figure 2.13 (a) A ball is dropped from a height of 1.5 m and bounces from the floor. (The horizontal motion is not considered here because it does not affect the vertical motion.) (b) Graphs of position, velocity, and acceleration versus time.

(b)

Quick Quiz 2.5

Which values represent the ball's velocity and acceleration at points Ⓐ, Ⓒ, and Ⓔ in Figure 2.13?

(a) $v_y = 0$, $a_y = 0$
(b) $v_y = 0$, $a_y = 9.80 \text{ m/s}^2$
(c) $v_y = 0$, $a_y = -9.80 \text{ m/s}^2$
(d) $v_y = -9.80 \text{ m/s}$, $a_y = 0$

EXAMPLE 2.12   Not a Bad Throw for a Rookie!

A stone thrown from the top of a building is given an initial velocity of 20.0 m/s straight upward. The building is 50.0 m high, and the stone just misses the edge of the roof on its way down, as shown in Figure 2.14. Using $t_A = 0$ as the time the stone leaves the thrower's hand at position Ⓐ, determine (a) the time at which the stone reaches its maximum height, (b) the maximum height, (c) the time at which the stone returns to the height from which it was thrown, (d) the velocity of the stone at this instant, and (e) the velocity and position of the stone at $t = 5.00$ s.

Solution   (a) As the stone travels from Ⓐ to Ⓑ, its velocity must change by 20 m/s because it stops at Ⓑ. Because gravity causes vertical velocities to change by about 10 m/s for every second of free fall, it should take the stone about 2 s to go from Ⓐ to Ⓑ in our drawing. (In a problem like this, a sketch definitely helps you organize your thoughts.) To calculate the time t_B at which the stone reaches maximum height, we use Equation 2.8, $v_{yB} = v_{yA} + a_y t$, noting that $v_{yB} = 0$ and setting the start of our clock readings at $t_A \equiv 0$:

$$20.0 \text{ m/s} + (-9.80 \text{ m/s}^2)t = 0$$

$$t = t_B = \frac{20.0 \text{ m/s}}{9.80 \text{ m/s}^2} = \boxed{2.04 \text{ s}}$$

Our estimate was pretty close.

(b) Because the average velocity for this first interval is 10 m/s (the average of 20 m/s and 0 m/s) and because it travels for about 2 s, we expect the stone to travel about 20 m. By substituting our time interval into Equation 2.11, we can find the maximum height as measured from the position of the thrower, where we set $y_i = y_A = 0$:

$$y_{max} = y_B = v_{yA} t + \tfrac{1}{2} a_y t^2$$

$$y_B = (20.0 \text{ m/s})(2.04 \text{ s}) + \tfrac{1}{2}(-9.80 \text{ m/s}^2)(2.04 \text{ s})^2$$

$$= \boxed{20.4 \text{ m}}$$

Our free-fall estimates are very accurate.

(c) There is no reason to believe that the stone's motion from Ⓑ to Ⓒ is anything other than the reverse of its motion

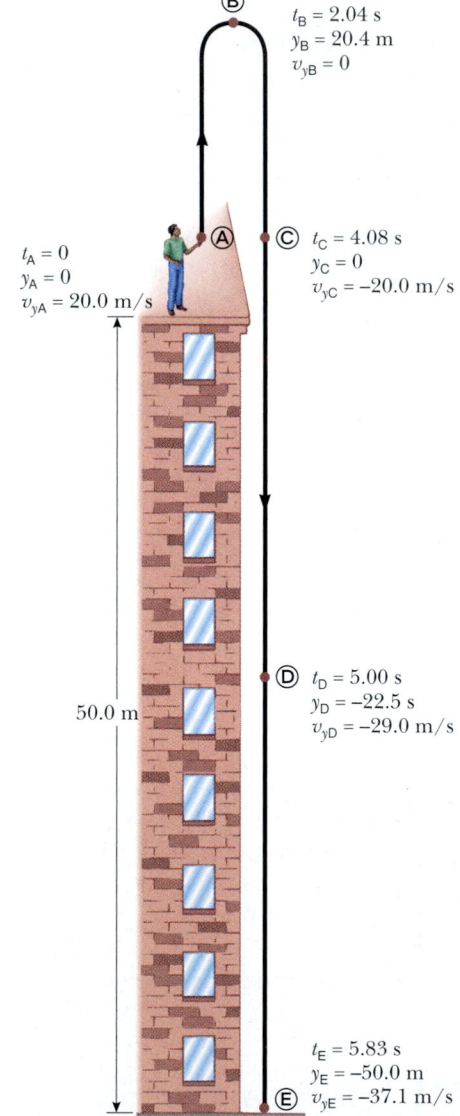

Figure 2.14   Position and velocity versus time for a freely falling stone thrown initially upward with a velocity $v_{yi} = 20.0$ m/s.

Ⓑ
$t_B = 2.04$ s
$y_B = 20.4$ m
$v_{yB} = 0$

$t_A = 0$
$y_A = 0$
$v_{yA} = 20.0$ m/s

Ⓐ Ⓒ $t_C = 4.08$ s
$y_C = 0$
$v_{yC} = -20.0$ m/s

Ⓓ $t_D = 5.00$ s
$y_D = -22.5$ s
$v_{yD} = -29.0$ m/s

50.0 m

$t_E = 5.83$ s
$y_E = -50.0$ m
Ⓔ $v_{yE} = -37.1$ m/s

from Ⓐ to Ⓑ. Thus, the time needed for it to go from Ⓐ to Ⓒ should be twice the time needed for it to go from Ⓐ to Ⓑ. When the stone is back at the height from which it was thrown (position Ⓒ), the y coordinate is again zero. Using Equation 2.11, with $y_f = y_C = 0$ and $y_i = y_A = 0$, we obtain

$$y_C - y_A = v_{yA} t + \tfrac{1}{2}a_y t^2$$

$$0 = 20.0t - 4.90t^2$$

This is a quadratic equation and so has two solutions for $t = t_C$. The equation can be factored to give

$$t(20.0 - 4.90t) = 0$$

One solution is $t = 0$, corresponding to the time the stone starts its motion. The other solution is $t = 4.08$ s, which is the solution we are after. Notice that it is double the value we calculated for t_B.

(d) Again, we expect everything at Ⓒ to be the same as it is at Ⓐ, except that the velocity is now in the opposite direction. The value for t found in (c) can be inserted into Equation 2.8 to give

$$v_{yC} = v_{yA} + a_y t = 20.0 \text{ m/s} + (-9.80 \text{ m/s}^2)(4.08 \text{ s})$$

$$= -20.0 \text{ m/s}$$

The velocity of the stone when it arrives back at its original height is equal in magnitude to its initial velocity but opposite in direction. This indicates that the motion is symmetric.

(e) For this part we consider what happens as the stone falls from position Ⓑ, where it had zero vertical velocity, to position Ⓓ. Because the elapsed time for this part of the motion is about 3 s, we estimate that the acceleration due to gravity will have changed the speed by about 30 m/s. We can calculate this from Equation 2.8, where we take $t = t_D - t_B$:

$$v_{yD} = v_{yB} + a_y t = 0 \text{ m/s} + (-9.80 \text{ m/s}^2)(5.00 \text{ s} - 2.04 \text{ s})$$

$$= -29.0 \text{ m/s}$$

We could just as easily have made our calculation between positions Ⓐ and Ⓓ by making sure we use the correct time interval, $t = t_D - t_A = 5.00$ s:

$$v_{yD} = v_{yA} + a_y t = 20.0 \text{ m/s} + (-9.80 \text{ m/s}^2)(5.00 \text{ s})$$

$$= -29.0 \text{ m/s}$$

To demonstrate the power of our kinematic equations, we can use Equation 2.11 to find the position of the stone at $t_D = 5.00$ s by considering the change in position between a different pair of positions, Ⓒ and Ⓓ. In this case, the time is $t_D - t_C$:

$$y_D = y_C + v_{yC}t + \tfrac{1}{2}a_y t^2$$

$$= 0 \text{ m} + (-20.0 \text{ m/s})(5.00 \text{ s} - 4.08 \text{ s})$$

$$+ \tfrac{1}{2}(-9.80 \text{ m/s}^2)(5.00 \text{ s} - 4.08 \text{ s})^2$$

$$= -22.5 \text{ m}$$

Exercise Find (a) the velocity of the stone just before it hits the ground at Ⓔ and (b) the total time the stone is in the air.

Answer (a) -37.1 m/s (b) 5.83 s

Optional Section

2.7 KINEMATIC EQUATIONS DERIVED FROM CALCULUS

This is an optional section that assumes the reader is familiar with the techniques of integral calculus. If you have not yet studied integration in your calculus course, you should skip this section or cover it after you become familiar with integration.

The velocity of a particle moving in a straight line can be obtained if its position as a function of time is known. Mathematically, the velocity equals the derivative of the position coordinate with respect to time. It is also possible to find the displacement of a particle if its velocity is known as a function of time. In calculus, the procedure used to perform this task is referred to either as *integration* or as finding the *antiderivative*. Graphically, it is equivalent to finding the area under a curve.

Suppose the v_x-t graph for a particle moving along the x axis is as shown in Figure 2.15. Let us divide the time interval $t_f - t_i$ into many small intervals, each of duration Δt_n. From the definition of average velocity we see that the displacement during any small interval, such as the one shaded in Figure 2.15, is given by $\Delta x_n = \bar{v}_{xn} \Delta t_n$, where $\bar{v}_{xn}$ is the average velocity in that interval. Therefore, the displacement during this small interval is simply the area of the shaded rectangle.

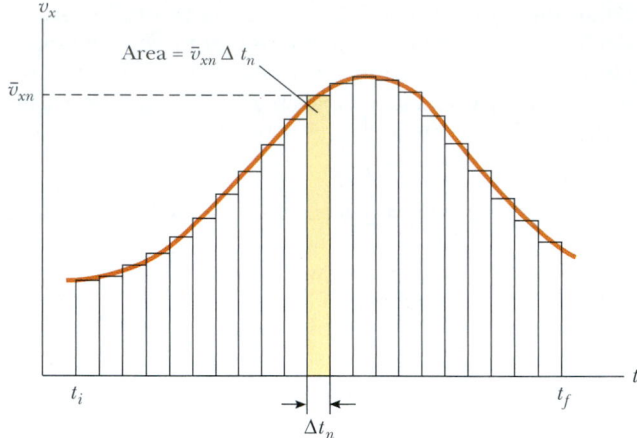

Figure 2.15 Velocity versus time for a particle moving along the *x* axis. The area of the shaded rectangle is equal to the displacement Δx in the time interval Δt_n, while the total area under the curve is the total displacement of the particle.

The total displacement for the interval $t_f - t_i$ is the sum of the areas of all the rectangles:

$$\Delta x = \sum_n \bar{v}_{xn}\, \Delta t_n$$

where the symbol Σ (upper case Greek sigma) signifies a sum over all terms. In this case, the sum is taken over all the rectangles from t_i to t_f. Now, as the intervals are made smaller and smaller, the number of terms in the sum increases and the sum approaches a value equal to the area under the velocity–time graph. Therefore, in the limit $n \to \infty$, or $\Delta t_n \to 0$, the displacement is

$$\Delta x = \lim_{\Delta t_n \to 0} \sum_n v_{xn}\, \Delta t_n \tag{2.13}$$

or

$$\text{Displacement} = \text{area under the } v_x\text{-}t \text{ graph}$$

Note that we have replaced the average velocity $\bar{v}_{xn}$ with the instantaneous velocity v_{xn} in the sum. As you can see from Figure 2.15, this approximation is clearly valid in the limit of very small intervals. We conclude that if we know the v_x-t graph for motion along a straight line, we can obtain the displacement during any time interval by measuring the area under the curve corresponding to that time interval.

The limit of the sum shown in Equation 2.13 is called a **definite integral** and is written

Definite integral

$$\lim_{\Delta t_n \to 0} \sum_n v_{xn}\Delta t_n = \int_{t_i}^{t_f} v_x(t)\; dt \tag{2.14}$$

where $v_x(t)$ denotes the velocity at any time t. If the explicit functional form of $v_x(t)$ is known and the limits are given, then the integral can be evaluated.

Sometimes the v_x-t graph for a moving particle has a shape much simpler than that shown in Figure 2.15. For example, suppose a particle moves at a constant ve-

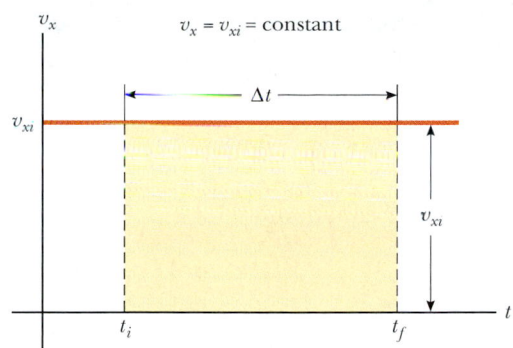

Figure 2.16 The velocity–time curve for a particle moving with constant velocity v_{xi}. The displacement of the particle during the time interval $t_f - t_i$ is equal to the area of the shaded rectangle.

locity v_{xi}. In this case, the v_x-t graph is a horizontal line, as shown in Figure 2.16, and its displacement during the time interval Δt is simply the area of the shaded rectangle:

$$\Delta x = v_{xi}\Delta t \qquad \text{(when } v_{xf} = v_{xi} = \text{constant)}$$

As another example, consider a particle moving with a velocity that is proportional to t, as shown in Figure 2.17. Taking $v_x = a_x t$, where a_x is the constant of proportionality (the acceleration), we find that the displacement of the particle during the time interval $t = 0$ to $t = t_A$ is equal to the area of the shaded triangle in Figure 2.17:

$$\Delta x = \tfrac{1}{2}(t_A)(a_x t_A) = \tfrac{1}{2}a_x t_A^{2}$$

Kinematic Equations

We now use the defining equations for acceleration and velocity to derive two of our kinematic equations, Equations 2.8 and 2.11.

The defining equation for acceleration (Eq. 2.6),

$$a_x = \frac{dv_x}{dt}$$

may be written as $dv_x = a_x dt$ or, in terms of an integral (or antiderivative), as

$$v_x = \int a_x \, dt + C_1$$

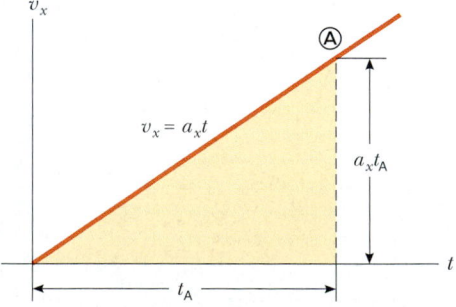

Figure 2.17 The velocity–time curve for a particle moving with a velocity that is proportional to the time.

where C_1 is a constant of integration. For the special case in which the acceleration is constant, the a_x can be removed from the integral to give

$$v_x = a_x \int dt + C_1 = a_x t + C_1 \qquad \textbf{(2.15)}$$

The value of C_1 depends on the initial conditions of the motion. If we take $v_x = v_{xi}$ when $t = 0$ and substitute these values into the last equation, we have

$$v_{xi} = a_x(0) + C_1$$

$$C_1 = v_{xi}$$

Calling $v_x = v_{xf}$ the velocity after the time interval t has passed and substituting this and the value just found for C_1 into Equation 2.15, we obtain kinematic Equation 2.8:

$$v_{xf} = v_{xi} + a_x t \qquad \text{(for constant } a_x)$$

Now let us consider the defining equation for velocity (Eq. 2.4):

$$v_x = \frac{dx}{dt}$$

We can write this as $dx = v_x dt$ or in integral form as

$$x = \int v_x \, dt + C_2$$

where C_2 is another constant of integration. Because $v_x = v_{xf} = v_{xi} + a_x t$, this expression becomes

$$x = \int (v_{xi} + a_x t) \, dt + C_2$$

$$x = \int v_{xi} \, dt + a_x \int t \, dt + C_2$$

$$x = v_{xi} t + \tfrac{1}{2} a_x t^2 + C_2$$

To find C_2, we make use of the initial condition that $x = x_i$ when $t = 0$. This gives $C_2 = x_i$. Therefore, after substituting x_f for x, we have

$$x_f = x_i + v_{xi} t + \tfrac{1}{2} a_x t^2 \qquad \text{(for constant } a_x)$$

Once we move x_i to the left side of the equation, we have kinematic Equation 2.11. Recall that $x_f - x_i$ is equal to the displacement of the object, where x_i is its initial position.

Besides what you might expect to learn about physics concepts, a very valuable skill you should hope to take away from your physics course is the ability to solve complicated problems. The way physicists approach complex situations and break them down into manageable pieces is extremely useful. We have developed a memory aid to help you easily recall the steps required for successful problem solving. When working on problems, the secret is to keep your GOAL in mind!

GOAL PROBLEM-SOLVING STEPS

Gather information

The first thing to do when approaching a problem is to understand the situation. Carefully read the problem statement, looking for key phrases like "at rest" or "freely falls." What information is given? Exactly what is the question asking? Don't forget to gather information from your own experiences and common sense. What should a reasonable answer look like? You wouldn't expect to calculate the speed of an automobile to be 5×10^6 m/s. Do you know what units to expect? Are there any limiting cases you can consider? What happens when an angle approaches 0° or 90° or when a mass becomes huge or goes to zero? Also make sure you carefully study any drawings that accompany the problem.

Organize your approach

Once you have a really good idea of what the problem is about, you need to think about what to do next. Have you seen this type of question before? Being able to classify a problem can make it much easier to lay out a plan to solve it. You should almost always make a quick drawing of the situation. Label important events with circled letters. Indicate any known values, perhaps in a table or directly on your sketch.

Analyze the problem

Because you have already categorized the problem, it should not be too difficult to select relevant equations that apply to this type of situation. Use algebra (and calculus, if necessary) to solve for the unknown variable in terms of what is given. Substitute in the appropriate numbers, calculate the result, and round it to the proper number of significant figures.

Learn from your efforts

This is the most important part. Examine your numerical answer. Does it meet your expectations from the first step? What about the algebraic form of the result—before you plugged in numbers? Does it make sense? (Try looking at the variables in it to see whether the answer would change in a physically meaningful way if they were drastically increased or decreased or even became zero.) Think about how this problem compares with others you have done. How was it similar? In what critical ways did it differ? Why was this problem assigned? You should have learned something by doing it. Can you figure out what?

When solving complex problems, you may need to identify a series of subproblems and apply the GOAL process to each. For very simple problems, you probably don't need GOAL at all. But when you are looking at a problem and you don't know what to do next, remember what the letters in GOAL stand for and use that as a guide.

SUMMARY

After a particle moves along the x axis from some initial position x_i to some final position x_f, its **displacement** is

$$\Delta x \equiv x_f - x_i \tag{2.1}$$

The **average velocity** of a particle during some time interval is the displacement Δx divided by the time interval Δt during which that displacement occurred:

$$\overline{v}_x \equiv \frac{\Delta x}{\Delta t} \tag{2.2}$$

The **average speed** of a particle is equal to the ratio of the total distance it travels to the total time it takes to travel that distance.

The **instantaneous velocity** of a particle is defined as the limit of the ratio $\Delta x / \Delta t$ as Δt approaches zero. By definition, this limit equals the derivative of x with respect to t, or the time rate of change of the position:

$$v_x \equiv \lim_{\Delta t \to 0} \frac{\Delta x}{\Delta t} = \frac{dx}{dt} \tag{2.4}$$

The **instantaneous speed** of a particle is equal to the magnitude of its velocity.

The **average acceleration** of a particle is defined as the ratio of the change in its velocity Δv_x divided by the time interval Δt during which that change occurred:

$$\overline{a}_x \equiv \frac{\Delta v_x}{\Delta t} = \frac{v_{xf} - v_{xi}}{t_f - t_i} \tag{2.5}$$

The **instantaneous acceleration** is equal to the limit of the ratio $\Delta v_x / \Delta t$ as Δt approaches 0. By definition, this limit equals the derivative of v_x with respect to t, or the time rate of change of the velocity:

$$a_x \equiv \lim_{\Delta t \to 0} \frac{\Delta v_x}{\Delta t} = \frac{dv_x}{dt} \tag{2.6}$$

The **equations of kinematics** for a particle moving along the x axis with uniform acceleration a_x (constant in magnitude and direction) are

$$v_{xf} = v_{xi} + a_x t \tag{2.8}$$

$$x_f - x_i = \overline{v}_x t = \tfrac{1}{2}(v_{xi} + v_{xf})t \tag{2.10}$$

$$x_f - x_i = v_{xi}t + \tfrac{1}{2}a_x t^2 \tag{2.11}$$

$$v_{xf}^2 = v_{xi}^2 + 2a_x(x_f - x_i) \tag{2.12}$$

You should be able to use these equations and the definitions in this chapter to analyze the motion of any object moving with constant acceleration.

An object falling freely in the presence of the Earth's gravity experiences a free-fall acceleration directed toward the center of the Earth. If air resistance is neglected, if the motion occurs near the surface of the Earth, and if the range of the motion is small compared with the Earth's radius, then the free-fall acceleration g is constant over the range of motion, where g is equal to 9.80 m/s^2.

Complicated problems are best approached in an organized manner. You should be able to recall and apply the steps of the GOAL strategy when you need them.

QUESTIONS

1. Average velocity and instantaneous velocity are generally different quantities. Can they ever be equal for a specific type of motion? Explain.

2. If the average velocity is nonzero for some time interval, does this mean that the instantaneous velocity is never zero during this interval? Explain.

3. If the average velocity equals zero for some time interval Δt and if $v_x(t)$ is a continuous function, show that the instantaneous velocity must go to zero at some time in this interval. (A sketch of x versus t might be useful in your proof.)

4. Is it possible to have a situation in which the velocity and acceleration have opposite signs? If so, sketch a velocity–time graph to prove your point.

5. If the velocity of a particle is nonzero, can its acceleration be zero? Explain.

6. If the velocity of a particle is zero, can its acceleration be nonzero? Explain.

7. Can an object having constant acceleration ever stop and stay stopped?

8. A stone is thrown vertically upward from the top of a building. Does the stone's displacement depend on the location of the origin of the coordinate system? Does the stone's velocity depend on the origin? (Assume that the coordinate system is stationary with respect to the building.) Explain.

9. A student at the top of a building of height h throws one ball upward with an initial speed v_{yi} and then throws a second ball downward with the same initial speed. How do the final speeds of the balls compare when they reach the ground?

10. Can the magnitude of the instantaneous velocity of an object ever be greater than the magnitude of its average velocity? Can it ever be less?

11. If the average velocity of an object is zero in some time interval, what can you say about the displacement of the object for that interval?

12. A rapidly growing plant doubles in height each week. At the end of the 25th day, the plant reaches the height of a building. At what time was the plant one-fourth the height of the building?

13. Two cars are moving in the same direction in parallel lanes along a highway. At some instant, the velocity of car A exceeds the velocity of car B. Does this mean that the acceleration of car A is greater than that of car B? Explain.

14. An apple is dropped from some height above the Earth's surface. Neglecting air resistance, how much does the apple's speed increase each second during its descent?

15. Consider the following combinations of signs and values for velocity and acceleration of a particle with respect to a one-dimensional x axis:

	Velocity	Acceleration
a.	Positive	Positive
b.	Positive	Negative
c.	Positive	Zero
d.	Negative	Positive
e.	Negative	Negative
f.	Negative	Zero
g.	Zero	Positive
h.	Zero	Negative

Describe what the particle is doing in each case, and give a real-life example for an automobile on an east-west one-dimensional axis, with east considered to be the positive direction.

16. A pebble is dropped into a water well, and the splash is heard 16 s later, as illustrated in Figure Q2.16. Estimate the distance from the rim of the well to the water's surface.

17. Average velocity is an entirely contrived quantity, and other combinations of data may prove useful in other contexts. For example, the ratio $\Delta t / \Delta x$, called the "slowness" of a moving object, is used by geophysicists when discussing the motion of continental plates. Explain what this quantity means.

Figure Q2.16

PROBLEMS

1, 2, 3 = straightforward, intermediate, challenging   ☐ = full solution available in the *Student Solutions Manual and Study Guide*
WEB = solution posted at **http://www.saunderscollege.com/physics/**   💻 = Computer useful in solving problem   🔧 = Interactive Physics
☐ = paired numerical/symbolic problems

Section 2.1   Displacement, Velocity, and Speed

1. The position of a pinewood derby car was observed at various times; the results are summarized in the table below. Find the average velocity of the car for (a) the first second, (b) the last 3 s, and (c) the entire period of observation.

x (m)	0	2.3	9.2	20.7	36.8	57.5
t (s)	0	1.0	2.0	3.0	4.0	5.0

2. A motorist drives north for 35.0 min at 85.0 km/h and then stops for 15.0 min. He then continues north, traveling 130 km in 2.00 h. (a) What is his total displacement? (b) What is his average velocity?

3. The displacement versus time for a certain particle moving along the x axis is shown in Figure P2.3. Find the average velocity in the time intervals (a) 0 to 2 s, (b) 0 to 4 s, (c) 2 s to 4 s, (d) 4 s to 7 s, (e) 0 to 8 s.

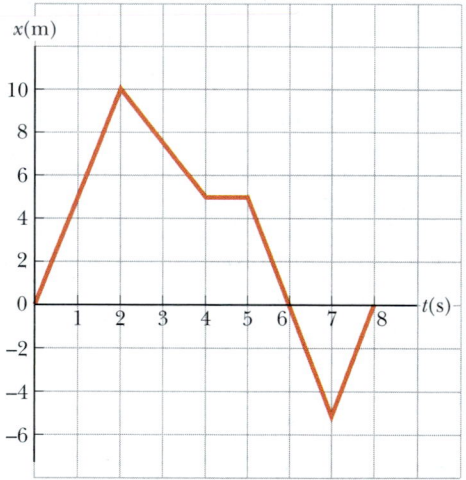

Figure P2.3   Problems 3 and 11.

4. A particle moves according to the equation $x = 10t^2$, where x is in meters and t is in seconds. (a) Find the average velocity for the time interval from 2.0 s to 3.0 s. (b) Find the average velocity for the time interval from 2.0 s to 2.1 s.

5. A person walks first at a constant speed of 5.00 m/s along a straight line from point A to point B and then back along the line from B to A at a constant speed of 3.00 m/s. What are (a) her average speed over the entire trip and (b) her average velocity over the entire trip?

6. A person first walks at a constant speed v_1 along a straight line from A to B and then back along the line from B to A at a constant speed v_2. What are (a) her average speed over the entire trip and (b) her average velocity over the entire trip?

Section 2.2   Instantaneous Velocity and Speed

7. At $t = 1.00$ s, a particle moving with constant velocity is located at $x = -3.00$ m, and at $t = 6.00$ s the particle is located at $x = 5.00$ m. (a) From this information, plot the position as a function of time. (b) Determine the velocity of the particle from the slope of this graph.

8. The position of a particle moving along the x axis varies in time according to the expression $x = 3t^2$, where x is in meters and t is in seconds. Evaluate its position (a) at $t = 3.00$ s and (b) at 3.00 s $+ \Delta t$. (c) Evaluate the limit of $\Delta x/\Delta t$ as Δt approaches zero to find the velocity at $t = 3.00$ s.

WEB 9. A position–time graph for a particle moving along the x axis is shown in Figure P2.9. (a) Find the average velocity in the time interval $t = 1.5$ s to $t = 4.0$ s. (b) Determine the instantaneous velocity at $t = 2.0$ s by measuring the slope of the tangent line shown in the graph. (c) At what value of t is the velocity zero?

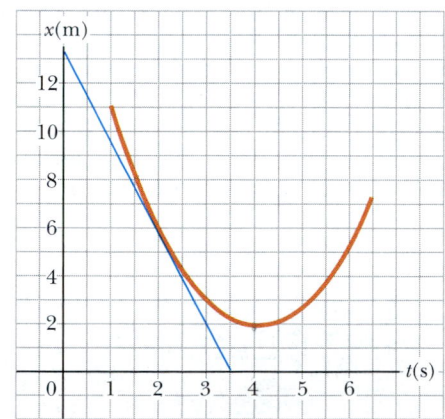

Figure P2.9

10. (a) Use the data in Problem 1 to construct a smooth graph of position versus time. (b) By constructing tangents to the $x(t)$ curve, find the instantaneous velocity of the car at several instants. (c) Plot the instantaneous velocity versus time and, from this, determine the average acceleration of the car. (d) What was the initial velocity of the car?

11. Find the instantaneous velocity of the particle described in Figure P2.3 at the following times: (a) $t = 1.0$ s, (b) $t = 3.0$ s, (c) $t = 4.5$ s, and (d) $t = 7.5$ s.

Section 2.3   Acceleration

12. A particle is moving with a velocity of 60.0 m/s in the positive x direction at $t = 0$. Between $t = 0$ and $t = 15.0$ s, the velocity decreases uniformly to zero. What was the acceleration during this 15.0-s interval? What is the significance of the sign of your answer?

13. A 50.0-g superball traveling at 25.0 m/s bounces off a brick wall and rebounds at 22.0 m/s. A high-speed camera records this event. If the ball is in contact with the wall for 3.50 ms, what is the magnitude of the average acceleration of the ball during this time interval? (*Note:* 1 ms $= 10^{-3}$ s.)

14. A particle starts from rest and accelerates as shown in Figure P2.14. Determine: (a) the particle's speed at $t = 10$ s and at $t = 20$ s, and (b) the distance traveled in the first 20 s.

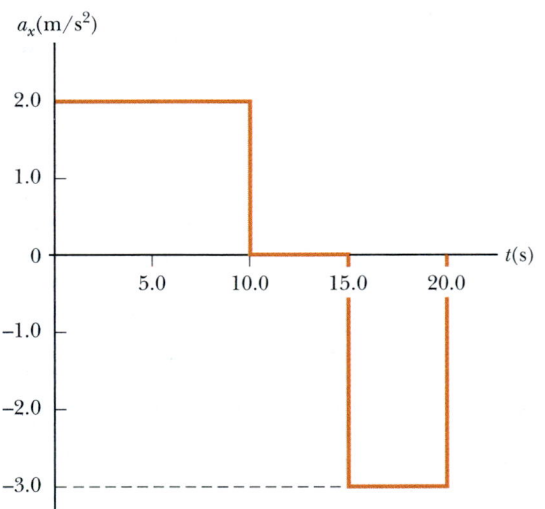

Figure P2.14

15. A velocity–time graph for an object moving along the x axis is shown in Figure P2.15. (a) Plot a graph of the acceleration versus time. (b) Determine the average acceleration of the object in the time intervals $t = 5.00$ s to $t = 15.0$ s and $t = 0$ to $t = 20.0$ s.

16. A student drives a moped along a straight road as described by the velocity–time graph in Figure P2.16. Sketch this graph in the middle of a sheet of graph paper. (a) Directly above your graph, sketch a graph of the position versus time, aligning the time coordinates of the two graphs. (b) Sketch a graph of the acceleration versus time directly below the v_x-t graph, again aligning the time coordinates. On each graph, show the

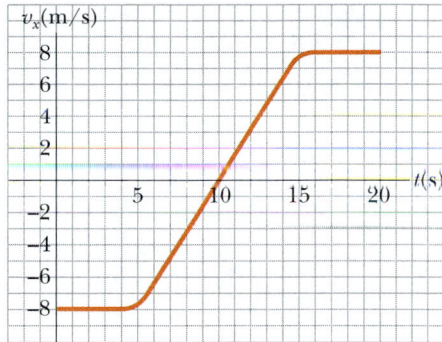

Figure P2.15

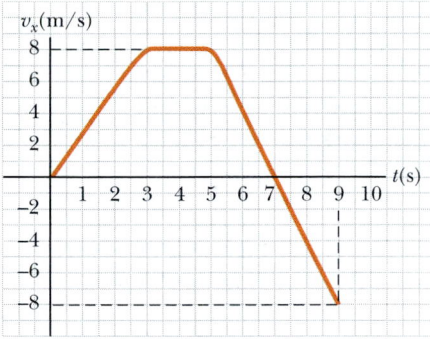

Figure P2.16

numerical values of x and a_x for all points of inflection. (c) What is the acceleration at $t = 6$ s? (d) Find the position (relative to the starting point) at $t = 6$ s. (e) What is the moped's final position at $t = 9$ s?

WEB 17. A particle moves along the x axis according to the equation $x = 2.00 + 3.00t - t^2$, where x is in meters and t is in seconds. At $t = 3.00$ s, find (a) the position of the particle, (b) its velocity, and (c) its acceleration.

18. An object moves along the x axis according to the equation $x = (3.00t^2 - 2.00t + 3.00)$ m. Determine (a) the average speed between $t = 2.00$ s and $t = 3.00$ s, (b) the instantaneous speed at $t = 2.00$ s and at $t = 3.00$ s, (c) the average acceleration between $t = 2.00$ s and $t = 3.00$ s, and (d) the instantaneous acceleration at $t = 2.00$ s and $t = 3.00$ s.

19. Figure P2.19 shows a graph of v_x versus t for the motion of a motorcyclist as he starts from rest and moves along the road in a straight line. (a) Find the average acceleration for the time interval $t = 0$ to $t = 6.00$ s. (b) Estimate the time at which the acceleration has its greatest positive value and the value of the acceleration at that instant. (c) When is the acceleration zero? (d) Estimate the maximum negative value of the acceleration and the time at which it occurs.

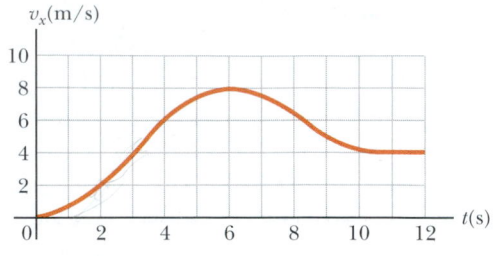

Figure P2.19

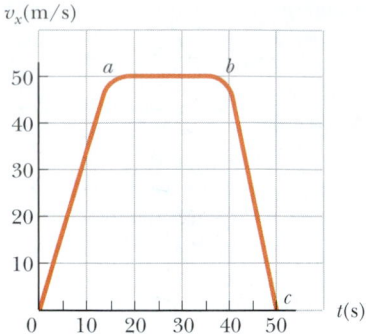

Figure P2.26

Section 2.4   Motion Diagrams

20. Draw motion diagrams for (a) an object moving to the right at constant speed, (b) an object moving to the right and speeding up at a constant rate, (c) an object moving to the right and slowing down at a constant rate, (d) an object moving to the left and speeding up at a constant rate, and (e) an object moving to the left and slowing down at a constant rate. (f) How would your drawings change if the changes in speed were not uniform; that is, if the speed were not changing at a constant rate?

Section 2.5   One-Dimensional Motion with Constant Acceleration

21. Jules Verne in 1865 proposed sending people to the Moon by firing a space capsule from a 220-m-long cannon with a final velocity of 10.97 km/s. What would have been the unrealistically large acceleration experienced by the space travelers during launch? Compare your answer with the free-fall acceleration, 9.80 m/s^2.

22. A certain automobile manufacturer claims that its super-deluxe sports car will accelerate from rest to a speed of 42.0 m/s in 8.00 s. Under the (improbable) assumption that the acceleration is constant, (a) determine the acceleration of the car. (b) Find the distance the car travels in the first 8.00 s. (c) What is the speed of the car 10.0 s after it begins its motion, assuming it continues to move with the same acceleration?

23. A truck covers 40.0 m in 8.50 s while smoothly slowing down to a final speed of 2.80 m/s. (a) Find its original speed. (b) Find its acceleration.

24. The minimum distance required to stop a car moving at 35.0 mi/h is 40.0 ft. What is the minimum stopping distance for the same car moving at 70.0 mi/h, assuming the same rate of acceleration?

WEB 25. A body moving with uniform acceleration has a velocity of 12.0 cm/s in the positive x direction when its x coordinate is 3.00 cm. If its x coordinate 2.00 s later is -5.00 cm, what is the magnitude of its acceleration?

26. Figure P2.26 represents part of the performance data of a car owned by a proud physics student. (a) Calculate from the graph the total distance traveled. (b) What distance does the car travel between the times $t = 10$ s and $t = 40$ s? (c) Draw a graph of its ac-

celeration versus time between $t = 0$ and $t = 50$ s. (d) Write an equation for x as a function of time for each phase of the motion, represented by (i) $0a$, (ii) ab, (iii) bc. (e) What is the average velocity of the car between $t = 0$ and $t = 50$ s?

27. A particle moves along the x axis. Its position is given by the equation $x = 2.00 + 3.00t - 4.00t^2$ with x in meters and t in seconds. Determine (a) its position at the instant it changes direction and (b) its velocity when it returns to the position it had at $t = 0$.

28. The initial speed of a body is 5.20 m/s. What is its speed after 2.50 s (a) if it accelerates uniformly at 3.00 m/s^2 and (b) if it accelerates uniformly at -3.00 m/s^2?

29. A drag racer starts her car from rest and accelerates at 10.0 m/s^2 for the entire distance of 400 m ($\frac{1}{4}$ mi). (a) How long did it take the race car to travel this distance? (b) What is the speed of the race car at the end of the run?

30. A car is approaching a hill at 30.0 m/s when its engine suddenly fails, just at the bottom of the hill. The car moves with a constant acceleration of -2.00 m/s^2 while coasting up the hill. (a) Write equations for the position along the slope and for the velocity as functions of time, taking $x = 0$ at the bottom of the hill, where $v_i = 30.0$ m/s. (b) Determine the maximum distance the car travels up the hill.

31. A jet plane lands with a speed of 100 m/s and can accelerate at a maximum rate of -5.00 m/s^2 as it comes to rest. (a) From the instant the plane touches the runway, what is the minimum time it needs before it can come to rest? (b) Can this plane land at a small tropical island airport where the runway is 0.800 km long?

32. The driver of a car slams on the brakes when he sees a tree blocking the road. The car slows uniformly with an acceleration of -5.60 m/s^2 for 4.20 s, making straight skid marks 62.4 m long ending at the tree. With what speed does the car then strike the tree?

33. *Help! One of our equations is missing!* We describe constant-acceleration motion with the variables and parameters v_{xi}, v_{xf}, a_x, t, and $x_f - x_i$. Of the equations in Table 2.2, the first does not involve $x_f - x_i$. The second does not contain a_x, the third omits v_{xf}, and the last

Figure P2.37   *(Left)* Col. John Stapp on rocket sled. *(Courtesy of the U.S. Air Force)*
(Right) Col. Stapp's face is contorted by the stress of rapid negative acceleration. *(Photri, Inc.)*

leaves out t. So to complete the set there should be an equation *not* involving v_{xi}. Derive it from the others. Use it to solve Problem 32 in one step.

34. An indestructible bullet 2.00 cm long is fired straight through a board that is 10.0 cm thick. The bullet strikes the board with a speed of 420 m/s and emerges with a speed of 280 m/s. (a) What is the average acceleration of the bullet as it passes through the board? (b) What is the total time that the bullet is in contact with the board? (c) What thickness of board (calculated to 0.1 cm) would it take to stop the bullet, assuming the bullet's acceleration through all parts of the board is the same?

35. A truck on a straight road starts from rest, accelerating at 2.00 m/s² until it reaches a speed of 20.0 m/s. Then the truck travels for 20.0 s at constant speed until the brakes are applied, stopping the truck in a uniform manner in an additional 5.00 s. (a) How long is the truck in motion? (b) What is the average velocity of the truck for the motion described?

36. A train is traveling down a straight track at 20.0 m/s when the engineer applies the brakes. This results in an acceleration of -1.00 m/s² as long as the train is in motion. How far does the train move during a 40.0-s time interval starting at the instant the brakes are applied?

37. For many years the world's land speed record was held by Colonel John P. Stapp, USAF (Fig. P2.37). On March 19, 1954, he rode a rocket-propelled sled that moved down the track at 632 mi/h. He and the sled were safely brought to rest in 1.40 s. Determine (a) the negative acceleration he experienced and (b) the distance he traveled during this negative acceleration.

38. An electron in a cathode-ray tube (CRT) accelerates uniformly from 2.00×10^4 m/s to 6.00×10^6 m/s over 1.50 cm. (a) How long does the electron take to travel this 1.50 cm? (b) What is its acceleration?

39. A ball starts from rest and accelerates at 0.500 m/s² while moving down an inclined plane 9.00 m long. When it reaches the bottom, the ball rolls up another plane, where, after moving 15.0 m, it comes to rest.

(a) What is the speed of the ball at the bottom of the first plane? (b) How long does it take to roll down the first plane? (c) What is the acceleration along the second plane? (d) What is the ball's speed 8.00 m along the second plane?

40. Speedy Sue, driving at 30.0 m/s, enters a one-lane tunnel. She then observes a slow-moving van 155 m ahead traveling at 5.00 m/s. Sue applies her brakes but can accelerate only at -2.00 m/s² because the road is wet. Will there be a collision? If so, determine how far into the tunnel and at what time the collision occurs. If not, determine the distance of closest approach between Sue's car and the van.

Section 2.6   Freely Falling Objects

Note: In all problems in this section, ignore the effects of air resistance.

41. A golf ball is released from rest from the top of a very tall building. Calculate (a) the position and (b) the velocity of the ball after 1.00 s, 2.00 s, and 3.00 s.

42. *Every morning at seven o'clock*
There's twenty terriers drilling on the rock.
The boss comes around and he says, "Keep still
And bear down heavy on the cast-iron drill
And drill, ye terriers, drill." And drill, ye terriers, drill.
It's work all day for sugar in your tea . . .
And drill, ye terriers, drill.

One day a premature blast went off
And a mile in the air went big Jim Goff. And drill . . .

Then when next payday came around
Jim Goff a dollar short was found.
When he asked what for, came this reply:
"You were docked for the time you were up in the sky." And
drill . . .

—American folksong

What was Goff's hourly wage? State the assumptions you make in computing it.

WEB **43.** A student throws a set of keys vertically upward to her sorority sister, who is in a window 4.00 m above. The keys are caught 1.50 s later by the sister's outstretched hand. (a) With what initial velocity were the keys thrown? (b) What was the velocity of the keys just before they were caught?

44. A ball is thrown directly downward with an initial speed of 8.00 m/s from a height of 30.0 m. How many seconds later does the ball strike the ground?

45. Emily challenges her friend David to catch a dollar bill as follows: She holds the bill vertically, as in Figure P2.45, with the center of the bill between David's index finger and thumb. David must catch the bill after Emily releases it without moving his hand downward. If his reaction time is 0.2 s, will he succeed? Explain your reasoning.

Figure P2.45   *(George Semple)*

46. A ball is dropped from rest from a height h above the ground. Another ball is thrown vertically upward from the ground at the instant the first ball is released. Determine the speed of the second ball if the two balls are to meet at a height $h/2$ above the ground.

47. A baseball is hit so that it travels straight upward after being struck by the bat. A fan observes that it takes 3.00 s for the ball to reach its maximum height. Find (a) its initial velocity and (b) the maximum height it reaches.

48. A woman is reported to have fallen 144 ft from the 17th floor of a building, landing on a metal ventilator box, which she crushed to a depth of 18.0 in. She suffered only minor injuries. Calculate (a) the speed of the woman just before she collided with the ventilator box, (b) her average acceleration while in contact with the box, and (c) the time it took to crush the box.

WEB **49.** A daring ranch hand sitting on a tree limb wishes to drop vertically onto a horse galloping under the tree. The speed of the horse is 10.0 m/s, and the distance from the limb to the saddle is 3.00 m. (a) What must be the horizontal distance between the saddle and limb when the ranch hand makes his move? (b) How long is he in the air?

50. A ball thrown vertically upward is caught by the thrower after 20.0 s. Find (a) the initial velocity of the ball and (b) the maximum height it reaches.

51. A ball is thrown vertically upward from the ground with an initial speed of 15.0 m/s. (a) How long does it take the ball to reach its maximum altitude? (b) What is its maximum altitude? (c) Determine the velocity and acceleration of the ball at $t = 2.00$ s.

52. The height of a helicopter above the ground is given by $h = 3.00t^3$, where h is in meters and t is in seconds. After 2.00 s, the helicopter releases a small mailbag. How long after its release does the mailbag reach the ground?

(Optional)
2.7   Kinematic Equations Derived from Calculus

53. Automotive engineers refer to the time rate of change of acceleration as the "jerk." If an object moves in one dimension such that its jerk J is constant, (a) determine expressions for its acceleration a_x, velocity v_x, and position x, given that its initial acceleration, speed, and position are a_{xi}, v_{xi}, and x_i, respectively. (b) Show that $a_x^2 = a_{xi}^2 + 2J(v_x - v_{xi})$.

54. The speed of a bullet as it travels down the barrel of a rifle toward the opening is given by the expression $v = (-5.0 \times 10^7)t^2 + (3.0 \times 10^5)t$, where v is in meters per second and t is in seconds. The acceleration of the bullet just as it leaves the barrel is zero. (a) Determine the acceleration and position of the bullet as a function of time when the bullet is in the barrel. (b) Determine the length of time the bullet is accelerated. (c) Find the speed at which the bullet leaves the barrel. (d) What is the length of the barrel?

55. The acceleration of a marble in a certain fluid is proportional to the speed of the marble squared and is given (in SI units) by $a = -3.00v^2$ for $v > 0$. If the marble enters this fluid with a speed of 1.50 m/s, how long will it take before the marble's speed is reduced to half of its initial value?

ADDITIONAL PROBLEMS

56. A motorist is traveling at 18.0 m/s when he sees a deer in the road 38.0 m ahead. (a) If the maximum negative acceleration of the vehicle is -4.50 m/s², what is the maximum reaction time Δt of the motorist that will allow him to avoid hitting the deer? (b) If his reaction time is actually 0.300 s, how fast will he be traveling when he hits the deer?

57. Another scheme to catch the roadrunner has failed. A safe falls from rest from the top of a 25.0-m-high cliff toward Wile E. Coyote, who is standing at the base. Wile first notices the safe after it has fallen 15.0 m. How long does he have to get out of the way?

58. A dog's hair has been cut and is now getting longer by 1.04 mm each day. With winter coming on, this rate of hair growth is steadily increasing by 0.132 mm/day every week. By how much will the dog's hair grow during five weeks?

59. A test rocket is fired vertically upward from a well. A catapult gives it an initial velocity of 80.0 m/s at ground level. Subsequently, its engines fire and it accelerates upward at 4.00 m/s^2 until it reaches an altitude of 1000 m. At that point its engines fail, and the rocket goes into free fall, with an acceleration of -9.80 m/s^2. (a) How long is the rocket in motion above the ground? (b) What is its maximum altitude? (c) What is its velocity just before it collides with the Earth? (*Hint:* Consider the motion while the engine is operating separate from the free-fall motion.)

60. A motorist drives along a straight road at a constant speed of 15.0 m/s. Just as she passes a parked motorcycle police officer, the officer starts to accelerate at 2.00 m/s^2 to overtake her. Assuming the officer maintains this acceleration, (a) determine the time it takes the police officer to reach the motorist. Also find (b) the speed and (c) the total displacement of the officer as he overtakes the motorist.

61. In Figure 2.10a, the area under the velocity–time curve between the vertical axis and time t (vertical dashed line) represents the displacement. As shown, this area consists of a rectangle and a triangle. Compute their areas and compare the sum of the two areas with the expression on the righthand side of Equation 2.11.

62. A commuter train travels between two downtown stations. Because the stations are only 1.00 km apart, the train never reaches its maximum possible cruising speed. The engineer minimizes the time t between the two stations by accelerating at a rate $a_1 = 0.100$ m/s^2 for a time t_1 and then by braking with acceleration $a_2 = -0.500$ m/s^2 for a time t_2. Find the minimum time of travel t and the time t_1.

63. In a 100-m race, Maggie and Judy cross the finish line in a dead heat, both taking 10.2 s. Accelerating uniformly, Maggie took 2.00 s and Judy 3.00 s to attain maximum speed, which they maintained for the rest of the race. (a) What was the acceleration of each sprinter? (b) What were their respective maximum speeds? (c) Which sprinter was ahead at the 6.00-s mark, and by how much?

64. A hard rubber ball, released at chest height, falls to the pavement and bounces back to nearly the same height. When it is in contact with the pavement, the lower side of the ball is temporarily flattened. Suppose the maximum depth of the dent is on the order of 1 cm. Compute an order-of-magnitude estimate for the maximum acceleration of the ball while it is in contact with the pavement. State your assumptions, the quantities you estimate, and the values you estimate for them.

65. A teenager has a car that speeds up at 3.00 m/s^2 and slows down at 4.50 m/s^2. On a trip to the store, he accelerates from rest to 12.0 m/s, drives at a constant speed for 5.00 s, and then comes to a momentary stop at an intersection. He then accelerates to 18.0 m/s, drives at a constant speed for 20.0 s, slows down for 2.67 s, continues for 4.00 s at this speed, and then comes to a stop. (a) How long does the trip take? (b) How far has he traveled? (c) What is his average speed for the trip? (d) How long would it take to walk to the store and back if he walks at 1.50 m/s?

66. A rock is dropped from rest into a well. (a) If the sound of the splash is heard 2.40 s later, how far below the top of the well is the surface of the water? The speed of sound in air (at the ambient temperature) is 336 m/s. (b) If the travel time for the sound is neglected, what percentage error is introduced when the depth of the well is calculated?

67. An inquisitive physics student and mountain climber climbs a 50.0-m cliff that overhangs a calm pool of water. He throws two stones vertically downward, 1.00 s apart, and observes that they cause a single splash. The first stone has an initial speed of 2.00 m/s. (a) How long after release of the first stone do the two stones hit the water? (b) What was the initial velocity of the second stone? (c) What is the velocity of each stone at the instant the two hit the water?

68. A car and train move together along parallel paths at 25.0 m/s, with the car adjacent to the rear of the train. Then, because of a red light, the car undergoes a uniform acceleration of -2.50 m/s^2 and comes to rest. It remains at rest for 45.0 s and then accelerates back to a speed of 25.0 m/s at a rate of 2.50 m/s^2. How far behind the rear of the train is the car when it reaches the speed of 25.0 m/s, assuming that the speed of the train has remained 25.0 m/s?

69. Kathy Kool buys a sports car that can accelerate at the rate of 4.90 m/s^2. She decides to test the car by racing with another speedster, Stan Speedy. Both start from rest, but experienced Stan leaves the starting line 1.00 s before Kathy. If Stan moves with a constant acceleration of 3.50 m/s^2 and Kathy maintains an acceleration of 4.90 m/s^2, find (a) the time it takes Kathy to overtake Stan, (b) the distance she travels before she catches up with him, and (c) the speeds of both cars at the instant she overtakes him.

70. To protect his food from hungry bears, a boy scout raises his food pack with a rope that is thrown over a tree limb at height h above his hands. He walks away from the vertical rope with constant velocity v_{boy}, holding the free end of the rope in his hands (Fig. P2.70).

Figure P2.70

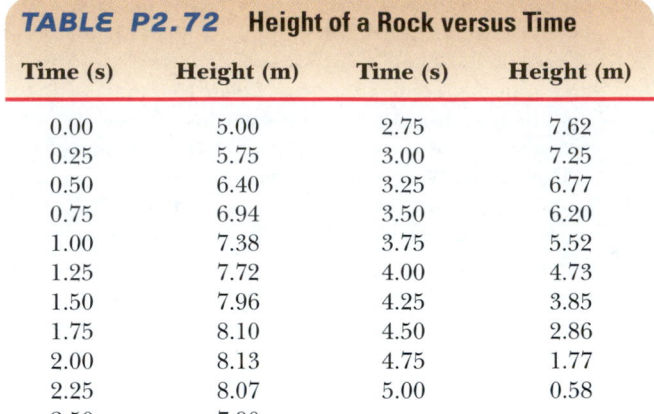

TABLE P2.72	Height of a Rock versus Time		
Time (s)	Height (m)	Time (s)	Height (m)
0.00	5.00	2.75	7.62
0.25	5.75	3.00	7.25
0.50	6.40	3.25	6.77
0.75	6.94	3.50	6.20
1.00	7.38	3.75	5.52
1.25	7.72	4.00	4.73
1.50	7.96	4.25	3.85
1.75	8.10	4.50	2.86
2.00	8.13	4.75	1.77
2.25	8.07	5.00	0.58
2.50	7.90		

(a) Show that the speed v of the food pack is $x(x^2 + h^2)^{-1/2} v_{\text{boy}}$, where x is the distance he has walked away from the vertical rope. (b) Show that the acceleration a of the food pack is $h^2(x^2 + h^2)^{-3/2} v_{\text{boy}}^2$. (c) What values do the acceleration and velocity have shortly after he leaves the point under the pack ($x = 0$)? (d) What values do the pack's velocity and acceleration approach as the distance x continues to increase?

 71. In Problem 70, let the height h equal 6.00 m and the speed v_{boy} equal 2.00 m/s. Assume that the food pack starts from rest. (a) Tabulate and graph the speed–time graph. (b) Tabulate and graph the acceleration–time graph. (Let the range of time be from 0 to 5.00 s and the time intervals be 0.500 s.)

 72. Astronauts on a distant planet toss a rock into the air. With the aid of a camera that takes pictures at a steady rate, they record the height of the rock as a function of time as given in Table P2.72. (a) Find the average velocity of the rock in the time interval between each measurement and the next. (b) Using these average veloci-ties to approximate instantaneous velocities at the midpoints of the time intervals, make a graph of velocity as a function of time. Does the rock move with constant acceleration? If so, plot a straight line of best fit on the graph and calculate its slope to find the acceleration.

73. Two objects, A and B, are connected by a rigid rod that has a length L. The objects slide along perpendicular guide rails, as shown in Figure P2.73. If A slides to the left with a constant speed v, find the speed of B when $\alpha = 60.0°$.

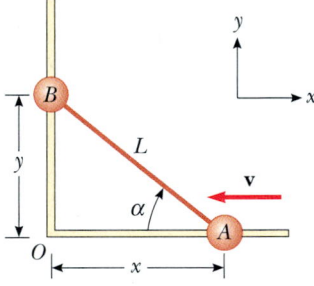

Figure P2.73

Answers to Quick Quizzes

2.1 Your graph should look something like the one in (a). This v_x-t graph shows that the maximum speed is about 5.0 m/s, which is 18 km/h (= 11 mi/h), and so the driver was not speeding. Can you derive the acceleration–time graph from the velocity–time graph? It should look something like the one in (b).

2.2 (a) Yes. This occurs when the car is slowing down, so that the direction of its acceleration is opposite the direction of its motion. (b) Yes. If the motion is in the direction chosen as negative, a positive acceleration causes a decrease in speed.

2.3 The left side represents the final velocity of an object. The first term on the right side is the velocity that the object had initially when we started watching it. The second term is the change in that initial velocity that is caused by the acceleration. If this second term is positive, then the initial velocity has increased ($v_{xf} > v_{xi}$). If this term is negative, then the initial velocity has decreased ($v_{xf} < v_{xi}$).

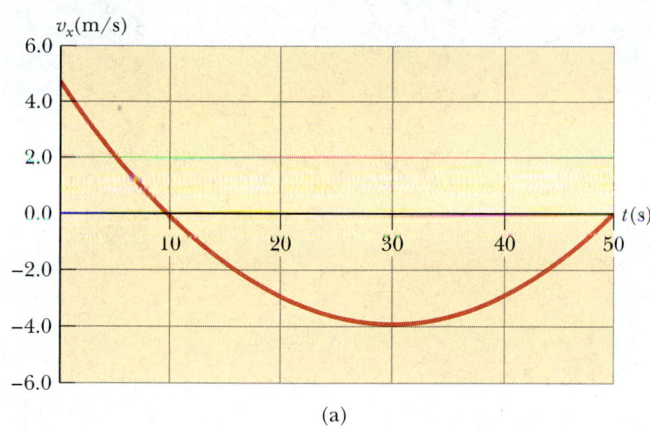
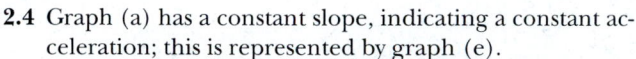

(a)

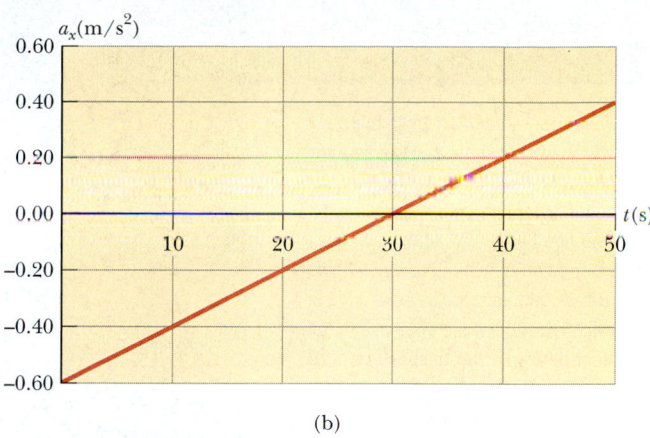

(b)

2.4 Graph (a) has a constant slope, indicating a constant acceleration; this is represented by graph (e).

Graph (b) represents a speed that is increasing constantly but not at a uniform rate. Thus, the acceleration must be increasing, and the graph that best indicates this is (d).

Graph (c) depicts a velocity that first increases at a constant rate, indicating constant acceleration. Then the velocity stops increasing and becomes constant, indicating zero acceleration. The best match to this situation is graph (f).

2.5 (c). As can be seen from Figure 2.13b, the ball is at rest for an infinitesimally short time at these three points. Nonetheless, gravity continues to act even though the ball is instantaneously not moving.

PUZZLER

When this honeybee gets back to its hive, it will tell the other bees how to return to the food it has found. By moving in a special, very precisely defined pattern, the bee conveys to other workers the information they need to find a flower bed. Bees communicate by "speaking in vectors." What does the bee have to tell the other bees in order to specify where the flower bed is located relative to the hive? *(E. Webber/Visuals Unlimited)*

chapter

3

Vectors

e often need to work with physical quantities that have both numerical and directional properties. As noted in Section 2.1, quantities of this nature are represented by vectors. This chapter is primarily concerned with vector algebra and with some general properties of vector quantities. We discuss the addition and subtraction of vector quantities, together with some common applications to physical situations.

Vector quantities are used throughout this text, and it is therefore imperative that you master both their graphical and their algebraic properties.

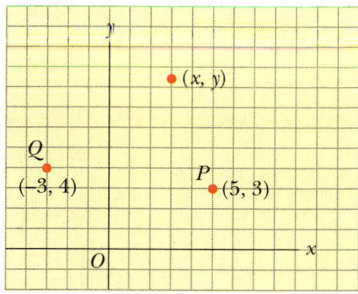

Figure 3.1 Designation of points in a cartesian coordinate system. Every point is labeled with coordinates (x, y).

3.1 COORDINATE SYSTEMS

Many aspects of physics deal in some form or other with locations in space. In Chapter 2, for example, we saw that the mathematical description of an object's motion requires a method for describing the object's position at various times. This description is accomplished with the use of coordinates, and in Chapter 2 we used the cartesian coordinate system, in which horizontal and vertical axes intersect at a point taken to be the origin (Fig. 3.1). Cartesian coordinates are also called *rectangular coordinates*.

Sometimes it is more convenient to represent a point in a plane by its *plane polar coordinates* (r, θ), as shown in Figure 3.2a. In this *polar coordinate system*, r is the distance from the origin to the point having cartesian coordinates (x, y), and θ is the angle between r and a fixed axis. This fixed axis is usually the positive x axis, and θ is usually measured counterclockwise from it. From the right triangle in Figure 3.2b, we find that $\sin \theta = y/r$ and that $\cos \theta = x/r$. (A review of trigonometric functions is given in Appendix B.4.) Therefore, starting with the plane polar coordinates of any point, we can obtain the cartesian coordinates, using the equations

$$x = r \cos \theta \tag{3.1}$$

$$y = r \sin \theta \tag{3.2}$$

Furthermore, the definitions of trigonometry tell us that

$$\tan \theta = \frac{y}{x} \tag{3.3}$$

$$r = \sqrt{x^2 + y^2} \tag{3.4}$$

These four expressions relating the coordinates (x, y) to the coordinates (r, θ) apply only when θ is defined, as shown in Figure 3.2a—in other words, when positive θ is an angle measured *counterclockwise* from the positive x axis. (Some scientific calculators perform conversions between cartesian and polar coordinates based on these standard conventions.) If the reference axis for the polar angle θ is chosen to be one other than the positive x axis or if the sense of increasing θ is chosen differently, then the expressions relating the two sets of coordinates will change.

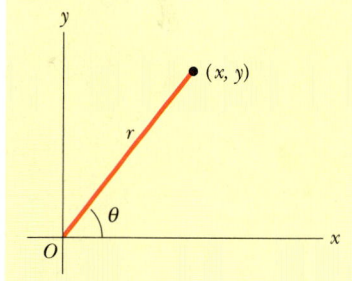

$$\sin \theta = \frac{y}{r}$$
$$\cos \theta = \frac{x}{r}$$
$$\tan \theta = \frac{y}{x}$$

(a)

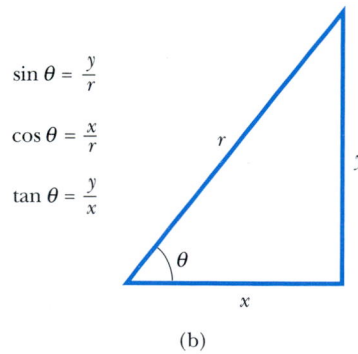

(b)

Figure 3.2 (a) The plane polar coordinates of a point are represented by the distance r and the angle θ, where θ is measured counterclockwise from the positive x axis. (b) The right triangle used to relate (x, y) to (r, θ).

Quick Quiz 3.1

Would the honeybee at the beginning of the chapter use cartesian or polar coordinates when specifying the location of the flower? Why? What is the honeybee using as an origin of coordinates?

You may want to read *Talking Apes and Dancing Bees* (1997) by Betsy Wyckoff.

EXAMPLE 3.1   Polar Coordinates

The cartesian coordinates of a point in the xy plane are $(x, y) = (-3.50, -2.50)$ m, as shown in Figure 3.3. Find the polar coordinates of this point.

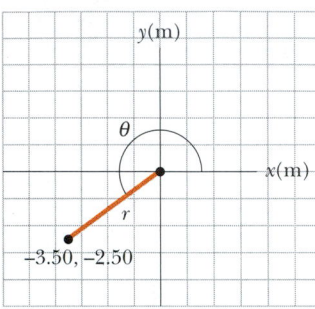

Figure 3.3   Finding polar coordinates when cartesian coordinates are given.

Solution

$$r = \sqrt{x^2 + y^2} = \sqrt{(-3.50 \text{ m})^2 + (-2.50 \text{ m})^2} = \boxed{4.30 \text{ m}}$$

$$\tan \theta = \frac{y}{x} = \frac{-2.50 \text{ m}}{-3.50 \text{ m}} = 0.714$$

$$\theta = \boxed{216°}$$

Note that you must use the signs of x and y to find that the point lies in the third quadrant of the coordinate system. That is, $\theta = 216°$ and not $35.5°$.

3.2   VECTOR AND SCALAR QUANTITIES

As noted in Chapter 2, some physical quantities are scalar quantities whereas others are vector quantities. When you want to know the temperature outside so that you will know how to dress, the only information you need is a number and the unit "degrees C" or "degrees F." Temperature is therefore an example of a **scalar quantity,** which is defined as a quantity that is completely specified by a number and appropriate units. That is,

> A scalar quantity is specified by a single value with an appropriate unit and has no direction.

Other examples of scalar quantities are volume, mass, and time intervals. The rules of ordinary arithmetic are used to manipulate scalar quantities.

If you are getting ready to pilot a small plane and need to know the wind velocity, you must know both the speed of the wind and its direction. Because direction is part of the information it gives, velocity is a **vector quantity,** which is defined as a physical quantity that is completely specified by a number and appropriate units plus a direction. That is,

> A vector quantity has both magnitude and direction.

Another example of a vector quantity is displacement, as you know from Chapter 2. Suppose a particle moves from some point Ⓐ to some point Ⓑ along a straight path, as shown in Figure 3.4. We represent this displacement by drawing an arrow from Ⓐ to Ⓑ, with the tip of the arrow pointing away from the starting point. The direction of the arrowhead represents the direction of the displacement, and the length of the arrow represents the magnitude of the displacement. If the particle travels along some other path from Ⓐ to Ⓑ, such as the broken line in Figure 3.4, its displacement is still the arrow drawn from Ⓐ to Ⓑ.

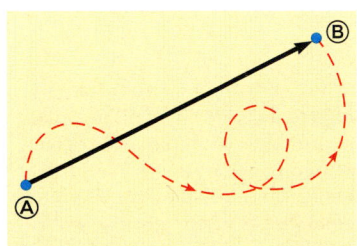

Figure 3.4   As a particle moves from Ⓐ to Ⓑ along an arbitrary path represented by the broken line, its displacement is a vector quantity shown by the arrow drawn from Ⓐ to Ⓑ.

(a)

(b)

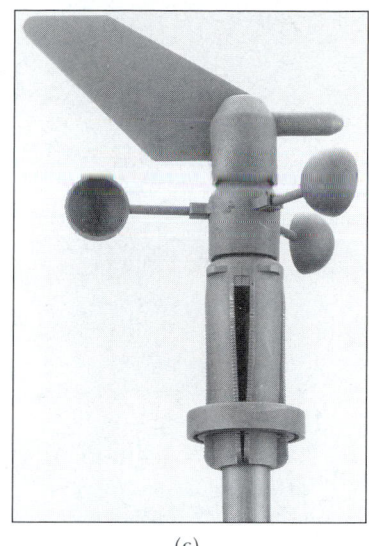

(c)

(a) The number of apples in the basket is one example of a scalar quantity. Can you think of other examples? *(Superstock)* (b) Jennifer pointing to the right. A vector quantity is one that must be specified by both magnitude and direction. *(Photo by Ray Serway)* (c) An anemometer is a device meteorologists use in weather forecasting. The cups spin around and reveal the magnitude of the wind velocity. The pointer indicates the direction. *(Courtesy of Peet Bros.Company, 1308 Doris Avenue, Ocean, NJ 07712)*

In this text, we use a boldface letter, such as **A**, to represent a vector quantity. Another common method for vector notation that you should be aware of is the use of an arrow over a letter, such as $\vec{A}$. The magnitude of the vector **A** is written either A or $|\mathbf{A}|$. The magnitude of a vector has physical units, such as meters for displacement or meters per second for velocity.

3.3 SOME PROPERTIES OF VECTORS

Equality of Two Vectors

For many purposes, two vectors **A** and **B** may be defined to be equal if they have the same magnitude and point in the same direction. That is, **A** = **B** only if $A = B$ *and* if **A** and **B** point in the same direction along parallel lines. For example, all the vectors in Figure 3.5 are equal even though they have different starting points. This property allows us to move a vector to a position parallel to itself in a diagram without affecting the vector.

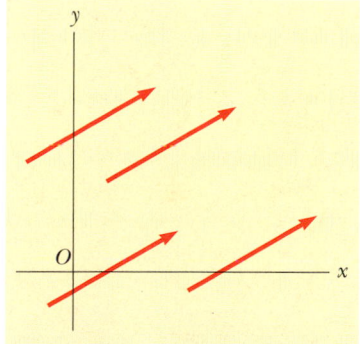

Figure 3.5 These four vectors are equal because they have equal lengths and point in the same direction.

Adding Vectors

The rules for adding vectors are conveniently described by geometric methods. To add vector **B** to vector **A**, first draw vector **A**, with its magnitude represented by a convenient scale, on graph paper and then draw vector **B** to the same scale with its tail starting from the tip of **A**, as shown in Figure 3.6. The **resultant vector R = A + B** is the vector drawn from the tail of **A** to the tip of **B**. This procedure is known as the **triangle method of addition.**

For example, if you walked 3.0 m toward the east and then 4.0 m toward the north, as shown in Figure 3.7, you would find yourself 5.0 m from where you

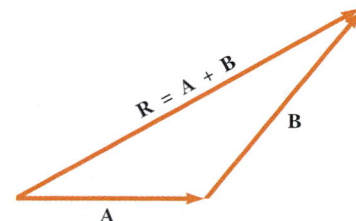

Figure 3.6 When vector **B** is added to vector **A**, the resultant **R** is the vector that runs from the tail of **A** to the tip of **B**.

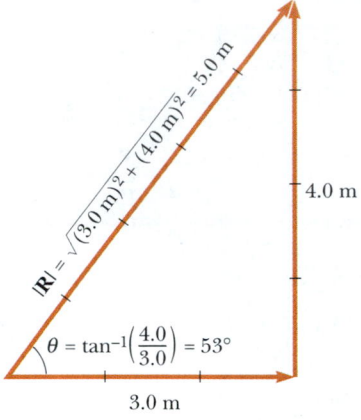

Figure 3.7 Vector addition. Walking first 3.0 m due east and then 4.0 m due north leaves you $|\mathbf{R}| =$ 5.0 m from your starting point.

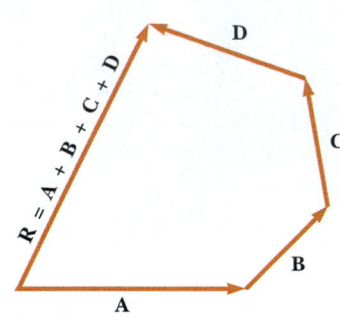

Figure 3.8 Geometric construction for summing four vectors. The resultant vector $\mathbf{R}$ is by definition the one that completes the polygon.

started, measured at an angle of 53° north of east. Your total displacement is the vector sum of the individual displacements.

A geometric construction can also be used to add more than two vectors. This is shown in Figure 3.8 for the case of four vectors. The resultant vector $\mathbf{R} = \mathbf{A} + \mathbf{B} + \mathbf{C} + \mathbf{D}$ is the vector that completes the polygon. In other words, **R is the vector drawn from the tail of the first vector to the tip of the last vector.**

An alternative graphical procedure for adding two vectors, known as the **parallelogram rule of addition,** is shown in Figure 3.9a. In this construction, the tails of the two vectors $\mathbf{A}$ and $\mathbf{B}$ are joined together and the resultant vector $\mathbf{R}$ is the diagonal of a parallelogram formed with $\mathbf{A}$ and $\mathbf{B}$ as two of its four sides.

When two vectors are added, the sum is independent of the order of the addition. (This fact may seem trivial, but as you will see in Chapter 11, the order is important when vectors are multiplied). This can be seen from the geometric construction in Figure 3.9b and is known as the **commutative law of addition:**

> Commutative law

$$\mathbf{A} + \mathbf{B} = \mathbf{B} + \mathbf{A} \qquad (3.5)$$

When three or more vectors are added, their sum is independent of the way in which the individual vectors are grouped together. A geometric proof of this rule

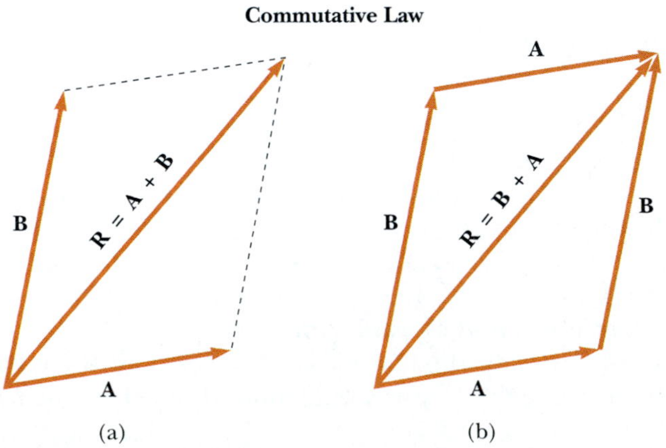

Commutative Law

Figure 3.9 (a) In this construction, the resultant $\mathbf{R}$ is the diagonal of a parallelogram having sides $\mathbf{A}$ and $\mathbf{B}$. (b) This construction shows that $\mathbf{A} + \mathbf{B} = \mathbf{B} + \mathbf{A}$—in other words, that vector addition is commutative.

(a)   (b)

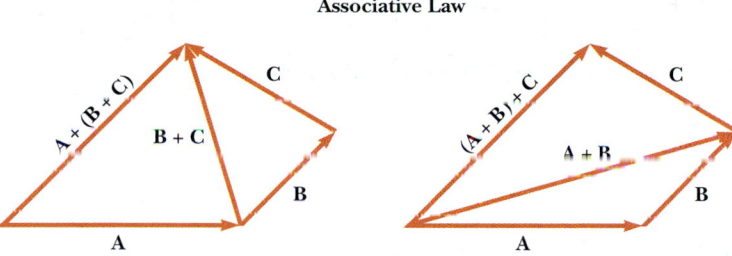

Associative Law

Figure 3.10 Geometric constructions for verifying the associative law of addition.

for three vectors is given in Figure 3.10. This is called the **associative law of addition:**

$$\mathbf{A} + (\mathbf{B} + \mathbf{C}) = (\mathbf{A} + \mathbf{B}) + \mathbf{C} \tag{3.6}$$

Associative law

In summary, **a vector quantity has both magnitude and direction and also obeys the laws of vector addition** as described in Figures 3.6 to 3.10. When two or more vectors are added together, *all* of them *must* have the same units. It would be meaningless to add a velocity vector (for example, 60 km/h to the east) to a displacement vector (for example, 200 km to the north) because they represent different physical quantities. The same rule also applies to scalars. For example, it would be meaningless to add time intervals to temperatures.

Negative of a Vector

The negative of the vector **A** is defined as the vector that when added to **A** gives zero for the vector sum. That is, $\mathbf{A} + (-\mathbf{A}) = 0$. The vectors **A** and $-\mathbf{A}$ have the same magnitude but point in opposite directions.

Subtracting Vectors

The operation of vector subtraction makes use of the definition of the negative of a vector. We define the operation $\mathbf{A} - \mathbf{B}$ as vector $-\mathbf{B}$ added to vector **A**:

$$\mathbf{A} - \mathbf{B} = \mathbf{A} + (-\mathbf{B}) \tag{3.7}$$

The geometric construction for subtracting two vectors in this way is illustrated in Figure 3.11a.

Another way of looking at vector subtraction is to note that the difference $\mathbf{A} - \mathbf{B}$ between two vectors **A** and **B** is what you have to add to the second vector to obtain the first. In this case, the vector $\mathbf{A} - \mathbf{B}$ points from the tip of the second vector to the tip of the first, as Figure 3.11b shows.

Vector Subtraction

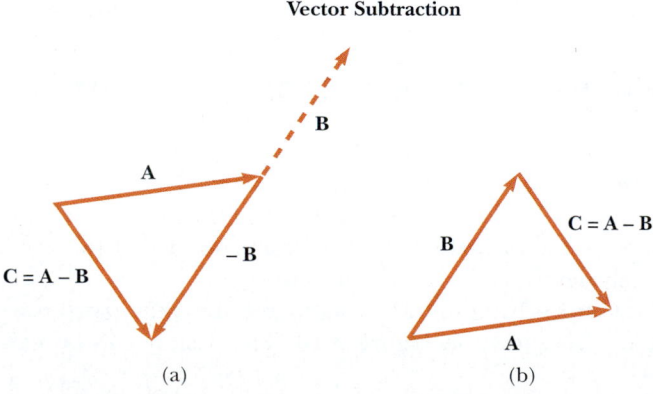

(a)   (b)

Figure 3.11 (a) This construction shows how to subtract vector **B** from vector **A**. The vector $-\mathbf{B}$ is equal in magnitude to vector **B** and points in the opposite direction. To subtract **B** from **A**, apply the rule of vector addition to the combination of **A** and $-\mathbf{B}$: Draw **A** along some convenient axis, place the tail of $-\mathbf{B}$ at the tip of **A**, and **C** is the difference $\mathbf{A} - \mathbf{B}$. (b) A second way of looking at vector subtraction. The difference vector $\mathbf{C} = \mathbf{A} - \mathbf{B}$ is the vector that we must add to **B** to obtain **A**.

EXAMPLE 3.2   A Vacation Trip

A car travels 20.0 km due north and then 35.0 km in a direction 60.0° west of north, as shown in Figure 3.12. Find the magnitude and direction of the car's resultant displacement.

Solution   In this example, we show two ways to find the resultant of two vectors. We can solve the problem geometrically, using graph paper and a protractor, as shown in Figure 3.12. (In fact, even when you know you are going to be carry-ing out a calculation, you should sketch the vectors to check your results.) The displacement **R** is the resultant when the two individual displacements **A** and **B** are added.

To solve the problem algebraically, we note that the magnitude of **R** can be obtained from the law of cosines as applied to the triangle (see Appendix B.4). With $\theta = 180° - 60° = 120°$ and $R^2 = A^2 + B^2 - 2AB \cos \theta$, we find that

$$R = \sqrt{A^2 + B^2 - 2AB \cos\theta}$$

$$= \sqrt{(20.0 \text{ km})^2 + (35.0 \text{ km})^2 - 2(20.0 \text{ km})(35.0 \text{ km})\cos 120°}$$

$$= \boxed{48.2 \text{ km}}$$

The direction of **R** measured from the northerly direction can be obtained from the law of sines (Appendix B.4):

$$\frac{\sin \beta}{B} = \frac{\sin \theta}{R}$$

$$\sin \beta = \frac{B}{R} \sin \theta = \frac{35.0 \text{ km}}{48.2 \text{ km}} \sin 120° = 0.629$$

$$\beta = \boxed{38.9°}$$

The resultant displacement of the car is 48.2 km in a direction 38.9° west of north. This result matches what we found graphically.

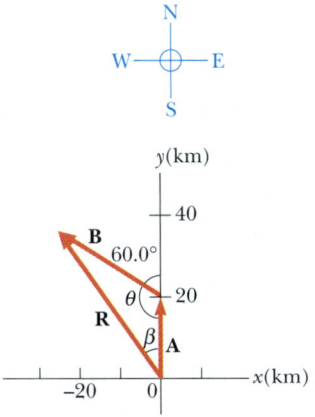

Figure 3.12   Graphical method for finding the resultant displacement vector **R** = **A** + **B**.

Multiplying a Vector by a Scalar

If vector **A** is multiplied by a positive scalar quantity m, then the product m**A** is a vector that has the same direction as **A** and magnitude mA. If vector **A** is multiplied by a negative scalar quantity $-m$, then the product $-m$**A** is directed opposite **A**. For example, the vector 5**A** is five times as long as **A** and points in the same direction as **A**; the vector $-\frac{1}{3}$**A** is one-third the length of **A** and points in the direction opposite **A**.

Quick Quiz 3.2

If vector **B** is added to vector **A**, under what condition does the resultant vector **A** + **B** have magnitude $A + B$? Under what conditions is the resultant vector equal to zero?

3.4   COMPONENTS OF A VECTOR AND UNIT VECTORS

The geometric method of adding vectors is not recommended whenever great accuracy is required or in three-dimensional problems. In this section, we describe a method of adding vectors that makes use of the *projections* of vectors along coordinate axes. These projections are called the **components** of the vector. Any vector can be completely described by its components.

Consider a vector **A** lying in the xy plane and making an arbitrary angle θ with the positive x axis, as shown in Figure 3.13. This vector can be expressed as the

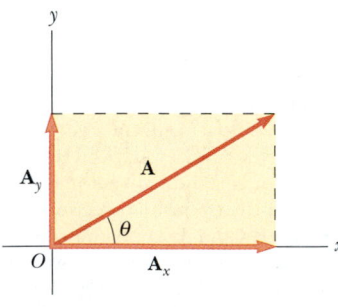

Figure 3.13   Any vector **A** lying in the xy plane can be represented by a vector $\mathbf{A}_x$ lying along the x axis and by a vector $\mathbf{A}_y$ lying along the y axis, where $\mathbf{A} = \mathbf{A}_x + \mathbf{A}_y$.

sum of two other vectors $\mathbf{A}_x$ and $\mathbf{A}_y$. From Figure 3.13, we see that the three vectors form a right triangle and that $\mathbf{A} = \mathbf{A}_x + \mathbf{A}_y$. (If you cannot see why this equality holds, go back to Figure 3.9 and review the parallelogram rule.) We shall often refer to the "components of a vector $\mathbf{A}$," written A_x and A_y (without the boldface notation). The component A_x represents the projection of $\mathbf{A}$ along the x axis, and the component A_y represents the projection of $\mathbf{A}$ along the y axis. These components can be positive or negative. The component A_x is positive if $\mathbf{A}_x$ points in the positive x direction and is negative if $\mathbf{A}_x$ points in the negative x direction. The same is true for the component A_y.

From Figure 3.13 and the definition of sine and cosine, we see that $\cos\theta = A_x/A$ and that $\sin\theta = A_y/A$. Hence, the components of $\mathbf{A}$ are

$$A_x = A\cos\theta \tag{3.8}$$

$$A_y = A\sin\theta \tag{3.9}$$

Components of the vector $\mathbf{A}$

These components form two sides of a right triangle with a hypotenuse of length A. Thus, it follows that the magnitude and direction of $\mathbf{A}$ are related to its components through the expressions

$$A = \sqrt{A_x^{\,2} + A_y^{\,2}} \tag{3.10}$$

Magnitude of $\mathbf{A}$

$$\theta = \tan^{-1}\left(\frac{A_y}{A_x}\right) \tag{3.11}$$

Direction of $\mathbf{A}$

Note that **the signs of the components A_x and A_y depend on the angle θ.** For example, if $\theta = 120°$, then A_x is negative and A_y is positive. If $\theta = 225°$, then both A_x and A_y are negative. Figure 3.14 summarizes the signs of the components when $\mathbf{A}$ lies in the various quadrants.

When solving problems, you can specify a vector $\mathbf{A}$ either with its components A_x and A_y or with its magnitude and direction A and θ.

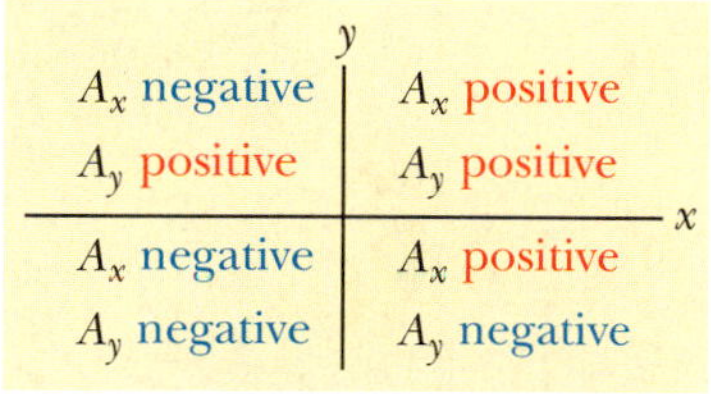

	y	
A_x negative		A_x positive
A_y positive		A_y positive
A_x negative		A_x positive x
A_y negative		A_y negative

Quick Quiz 3.3

Can the component of a vector ever be greater than the magnitude of the vector?

Figure 3.14 The signs of the components of a vector $\mathbf{A}$ depend on the quadrant in which the vector is located.

Suppose you are working a physics problem that requires resolving a vector into its components. In many applications it is convenient to express the components in a coordinate system having axes that are not horizontal and vertical but are still perpendicular to each other. If you choose reference axes or an angle other than the axes and angle shown in Figure 3.13, the components must be modified accordingly. Suppose a vector $\mathbf{B}$ makes an angle θ' with the x' axis defined in Figure 3.15. The components of $\mathbf{B}$ along the x' and y' axes are $B_{x'} = B\cos\theta'$ and $B_{y'} = B\sin\theta'$, as specified by Equations 3.8 and 3.9. The magnitude and direction of $\mathbf{B}$ are obtained from expressions equivalent to Equations 3.10 and 3.11. Thus, we can express the components of a vector in *any* coordinate system that is convenient for a particular situation.

Unit Vectors

Vector quantities often are expressed in terms of unit vectors. **A unit vector is a dimensionless vector having a magnitude of exactly 1.** Unit vectors are used to specify a given direction and have no other physical significance. They are used solely as a convenience in describing a direction in space. We shall use the symbols

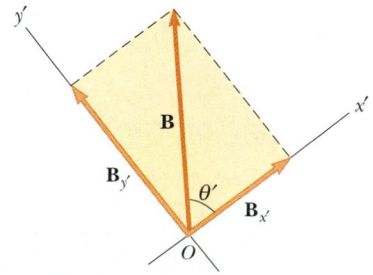

Figure 3.15 The component vectors of $\mathbf{B}$ in a coordinate system that is tilted.

i, **j**, and **k** to represent unit vectors pointing in the positive *x*, *y*, and *z* directions, respectively. The unit vectors **i**, **j**, and **k** form a set of mutually perpendicular vectors in a right-handed coordinate system, as shown in Figure 3.16a. The magnitude of each unit vector equals 1; that is, $|\mathbf{i}| = |\mathbf{j}| = |\mathbf{k}| = 1$.

Consider a vector **A** lying in the *xy* plane, as shown in Figure 3.16b. The product of the component A_x and the unit vector **i** is the vector $A_x\mathbf{i}$, which lies on the *x* axis and has magnitude $|A_x|$. (The vector $A_x\mathbf{i}$ is an alternative representation of vector $\mathbf{A}_x$.) Likewise, $A_y\mathbf{j}$ is a vector of magnitude $|A_y|$ lying on the *y* axis. (Again, vector $A_y\mathbf{j}$ is an alternative representation of vector $\mathbf{A}_y$.) Thus, the unit–vector notation for the vector **A** is

$$\mathbf{A} = A_x\mathbf{i} + A_y\mathbf{j} \tag{3.12}$$

For example, consider a point lying in the *xy* plane and having cartesian coordinates (x, y), as in Figure 3.17. The point can be specified by the **position vector r,** which in unit–vector form is given by

$$\mathbf{r} = x\mathbf{i} + y\mathbf{j} \tag{3.13}$$

This notation tells us that the components of **r** are the lengths *x* and *y*.

Now let us see how to use components to add vectors when the geometric method is not sufficiently accurate. Suppose we wish to add vector **B** to vector **A**, where vector **B** has components B_x and B_y. All we do is add the *x* and *y* components separately. The resultant vector $\mathbf{R} = \mathbf{A} + \mathbf{B}$ is therefore

$$\mathbf{R} = (A_x\mathbf{i} + A_y\mathbf{j}) + (B_x\mathbf{i} + B_y\mathbf{j})$$

or

$$\mathbf{R} = (A_x + B_x)\mathbf{i} + (A_y + B_y)\mathbf{j} \tag{3.14}$$

Because $\mathbf{R} = R_x\mathbf{i} + R_y\mathbf{j}$, we see that the components of the resultant vector are

$$R_x = A_x + B_x$$
$$R_y = A_y + B_y \tag{3.15}$$

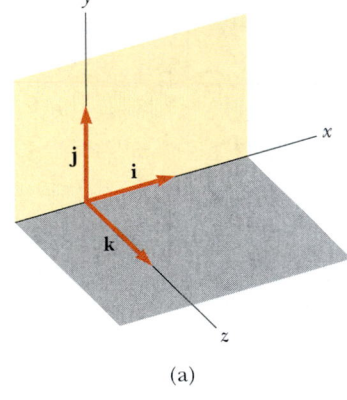

(a)

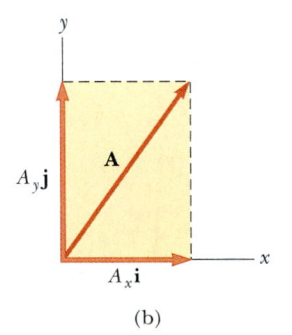

(b)

Figure 3.16   (a) The unit vectors **i**, **j**, and **k** are directed along the *x*, *y*, and *z* axes, respectively. (b) Vector $\mathbf{A} = A_x\mathbf{i} + A_y\mathbf{j}$ lying in the *xy* plane has components A_x and A_y.

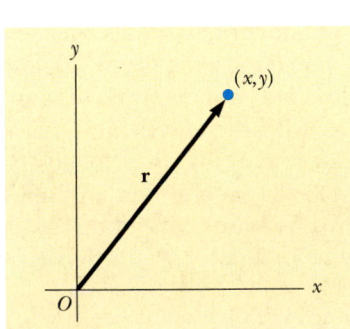

Figure 3.17   The point whose cartesian coordinates are (x, y) can be represented by the position vector $\mathbf{r} = x\mathbf{i} + y\mathbf{j}$.

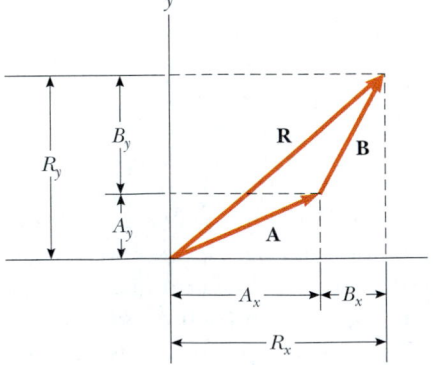

Figure 3.18   This geometric construction for the sum of two vectors shows the relationship between the components of the resultant **R** and the components of the individual vectors.

We obtain the magnitude of **R** and the angle it makes with the x axis from its components, using the relationships

$$R = \sqrt{R_x{}^2 + R_y{}^2} = \sqrt{(A_x + B_x)^2 + (A_y + B_y)^2} \qquad \textbf{(3.16)}$$

$$\tan \theta = \frac{R_y}{R_x} = \frac{A_y + B_y}{A_x + B_x} \qquad \textbf{(3.17)}$$

We can check this addition by components with a geometric construction, as shown in Figure 3.18. Remember that you must note the *signs* of the components when using either the algebraic or the geometric method.

At times, we need to consider situations involving motion in three component directions. The extension of our methods to three-dimensional vectors is straightforward. If **A** and **B** both have x, y, and z components, we express them in the form

$$\mathbf{A} = A_x\mathbf{i} + A_y\mathbf{j} + A_z\mathbf{k} \qquad \textbf{(3.18)}$$

$$\mathbf{B} = B_x\mathbf{i} + B_y\mathbf{j} + B_z\mathbf{k} \qquad \textbf{(3.19)}$$

The sum of **A** and **B** is

$$\mathbf{R} = (A_x + B_x)\mathbf{i} + (A_y + B_y)\mathbf{j} + (A_z + B_z)\mathbf{k} \qquad \textbf{(3.20)}$$

Note that Equation 3.20 differs from Equation 3.14: in Equation 3.20, the resultant vector also has a z component $R_z = A_z + B_z$.

QuickLab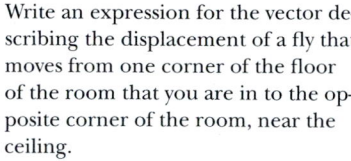

Write an expression for the vector describing the displacement of a fly that moves from one corner of the floor of the room that you are in to the opposite corner of the room, near the ceiling.

Quick Quiz 3.4

If one component of a vector is not zero, can the magnitude of the vector be zero? Explain.

Quick Quiz 3.5

If **A** + **B** = 0, what can you say about the components of the two vectors?

> ## Problem-Solving Hints
>
> **Adding Vectors**
>
> When you need to add two or more vectors, use this step-by-step procedure:
>
> - Select a coordinate system that is convenient. (Try to reduce the number of components you need to find by choosing axes that line up with as many vectors as possible.)
> - Draw a labeled sketch of the vectors described in the problem.
> - Find the x and y components of all vectors and the resultant components (the algebraic sum of the components) in the x and y directions.
> - If necessary, use the Pythagorean theorem to find the magnitude of the resultant vector and select a suitable trigonometric function to find the angle that the resultant vector makes with the x axis.

EXAMPLE 3.3 ▶ The Sum of Two Vectors

Find the sum of two vectors **A** and **B** lying in the *xy* plane and given by

$$\mathbf{A} = (2.0\mathbf{i} + 2.0\mathbf{j}) \text{ m} \qquad \text{and} \qquad \mathbf{B} = (2.0\mathbf{i} - 4.0\mathbf{j}) \text{ m}$$

Solution Comparing this expression for **A** with the general expression $\mathbf{A} = A_x\mathbf{i} + A_y\mathbf{j}$, we see that $A_x = 2.0$ m and that $A_y = 2.0$ m. Likewise, $B_x = 2.0$ m and $B_y = -4.0$ m. We obtain the resultant vector **R**, using Equation 3.14:

$$\mathbf{R} = \mathbf{A} + \mathbf{B} = (2.0 + 2.0)\mathbf{i} \text{ m} + (2.0 - 4.0)\mathbf{j} \text{ m}$$
$$= (4.0\mathbf{i} - 2.0\mathbf{j}) \text{ m}$$

or

$$R_x = 4.0 \text{ m} \qquad R_y = -2.0 \text{ m}$$

The magnitude of **R** is given by Equation 3.16:

$$R = \sqrt{R_x^2 + R_y^2} = \sqrt{(4.0 \text{ m})^2 + (-2.0 \text{ m})^2} = \sqrt{20} \text{ m}$$
$$= \boxed{4.5 \text{ m}}$$

We can find the direction of **R** from Equation 3.17:

$$\tan\theta = \frac{R_y}{R_x} = \frac{-2.0 \text{ m}}{4.0 \text{ m}} = -0.50$$

Your calculator likely gives the answer $-27°$ for $\theta = \tan^{-1}(-0.50)$. This answer is correct if we interpret it to mean 27° clockwise from the *x* axis. Our standard form has been to quote the angles measured counterclockwise from the $+x$ axis, and that angle for this vector is $\theta = \boxed{333°}$.

EXAMPLE 3.4 ▶ The Resultant Displacement

A particle undergoes three consecutive displacements: $\mathbf{d}_1 = (15\mathbf{i} + 30\mathbf{j} + 12\mathbf{k})$ cm, $\mathbf{d}_2 = (23\mathbf{i} - 14\mathbf{j} - 5.0\mathbf{k})$ cm,   and $\mathbf{d}_3 = (-13\mathbf{i} + 15\mathbf{j})$ cm. Find the components of the resultant displacement and its magnitude.

Solution Rather than looking at a sketch on flat paper, visualize the problem as follows: Start with your fingertip at the front left corner of your horizontal desktop. Move your fingertip 15 cm to the right, then 30 cm toward the far side of the desk, then 12 cm vertically upward, then 23 cm to the right, then 14 cm horizontally toward the front edge of the desk, then 5.0 cm vertically toward the desk, then 13 cm to the left, and (finally!) 15 cm toward the back of the desk. The

mathematical calculation keeps track of this motion along the three perpendicular axes:

$$\mathbf{R} = \mathbf{d}_1 + \mathbf{d}_2 + \mathbf{d}_3$$
$$= (15 + 23 - 13)\mathbf{i} \text{ cm} + (30 - 14 + 15)\mathbf{j} \text{ cm}$$
$$+ (12 - 5.0 + 0)\mathbf{k} \text{ cm}$$
$$= (25\mathbf{i} + 31\mathbf{j} + 7.0\mathbf{k}) \text{ cm}$$

The resultant displacement has components $R_x = 25$ cm, $R_y = 31$ cm, and $R_z = 7.0$ cm. Its magnitude is

$$R = \sqrt{R_x^2 + R_y^2 + R_z^2}$$
$$= \sqrt{(25 \text{ cm})^2 + (31 \text{ cm})^2 + (7.0 \text{ cm})^2} = \boxed{40 \text{ cm}}$$

EXAMPLE 3.5 ▶ Taking a Hike

A hiker begins a trip by first walking 25.0 km southeast from her car. She stops and sets up her tent for the night. On the second day, she walks 40.0 km in a direction 60.0° north of east, at which point she discovers a forest ranger's tower. (a) Determine the components of the hiker's displacement for each day.

Solution If we denote the displacement vectors on the first and second days by **A** and **B**, respectively, and use the car as the origin of coordinates, we obtain the vectors shown in Figure 3.19. Displacement **A** has a magnitude of 25.0 km and is directed 45.0° below the positive *x* axis. From Equations 3.8 and 3.9, its components are

$$A_x = A\cos(-45.0°) = (25.0 \text{ km})(0.707) = \boxed{17.7 \text{ km}}$$

$$A_y = A\sin(-45.0°) = -(25.0 \text{ km})(0.707) = \boxed{-17.7 \text{ km}}$$

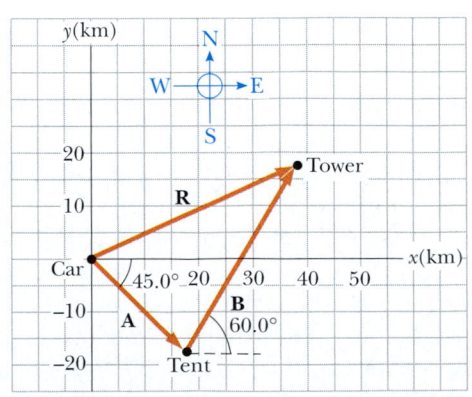

Figure 3.19   The total displacement of the hiker is the vector $\mathbf{R} = \mathbf{A} + \mathbf{B}$.

The negative value of A_y indicates that the hiker walks in the negative y direction on the first day. The signs of A_x and A_y also are evident from Figure 3.19.

The second displacement **B** has a magnitude of 40.0 km and is 60.0° north of east. Its components are

$$B_x = B \cos 60.0° = (40.0 \text{ km})(0.500) = \boxed{20.0 \text{ km}}$$

$$B_y = B \sin 60.0° = (40.0 \text{ km})(0.866) = \boxed{34.6 \text{ km}}$$

(b) Determine the components of the hiker's resultant displacement **R** for the trip. Find an expression for **R** in terms of unit vectors.

Solution   The resultant displacement for the trip **R** = **A** + **B** has components given by Equation 3.15:

$$R_x = A_x + B_x = 17.7 \text{ km} + 20.0 \text{ km} = \boxed{37.7 \text{ km}}$$

$$R_y = A_y + B_y = -17.7 \text{ km} + 34.6 \text{ km} = \boxed{16.9 \text{ km}}$$

In unit vector form, we can write the total displacement as

$$\boxed{\mathbf{R} = (37.7\mathbf{i} + 16.9\mathbf{j}) \text{ km}}$$

Exercise   Determine the magnitude and direction of the total displacement.

Answer   41.3 km, 24.1° north of east from the car.

EXAMPLE 3.6 ▶   Let's Fly Away!

A commuter airplane takes the route shown in Figure 3.20. First, it flies from the origin of the coordinate system shown to city A, located 175 km in a direction 30.0° north of east. Next, it flies 153 km 20.0° west of north to city B. Finally, it flies 195 km due west to city C. Find the location of city C relative to the origin.

Solution   It is convenient to choose the coordinate system shown in Figure 3.20, where the x axis points to the east and the y axis points to the north. Let us denote the three consecutive displacements by the vectors **a**, **b**, and **c**. Displacement **a** has a magnitude of 175 km and the components

$$a_x = a \cos(30.0°) = (175 \text{ km})(0.866) = 152 \text{ km}$$

$$a_y = a \sin(30.0°) = (175 \text{ km})(0.500) = 87.5 \text{ km}$$

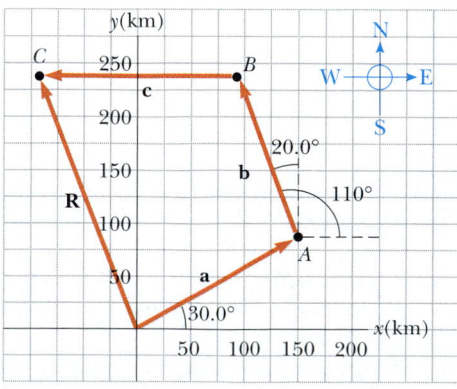

Figure 3.20   The airplane starts at the origin, flies first to city A, then to city B, and finally to city C.

Displacement **b**, whose magnitude is 153 km, has the components

$$b_x = b \cos(110°) = (153 \text{ km})(-0.342) = -52.3 \text{ km}$$

$$b_y = b \sin(110°) = (153 \text{ km})(0.940) = 144 \text{ km}$$

Finally, displacement **c**, whose magnitude is 195 km, has the components

$$c_x = c \cos(180°) = (195 \text{ km})(-1) = -195 \text{ km}$$

$$c_y = c \sin(180°) = 0$$

Therefore, the components of the position vector **R** from the starting point to city C are

$$R_x = a_x + b_x + c_x = 152 \text{ km} - 52.3 \text{ km} - 195 \text{ km}$$

$$= \boxed{-95.3 \text{ km}}$$

$$R_y = a_y + b_y + c_y = 87.5 \text{ km} + 144 \text{ km} + 0$$

$$= \boxed{232 \text{ km}}$$

In unit–vector notation, $\boxed{\mathbf{R} = (-95.3\mathbf{i} + 232\mathbf{j}) \text{ km.}}$ That is, the airplane can reach city C from the starting point by first traveling 95.3 km due west and then by traveling 232 km due north.

Exercise   Find the magnitude and direction of **R**.

Answer   251 km, 22.3° west of north.

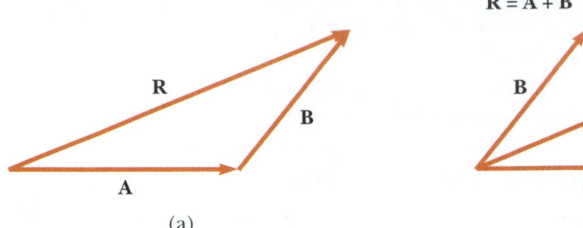

(a) (b)

Figure 3.21 (a) Vector addition by the triangle method. (b) Vector addition by the parallelogram rule.

SUMMARY

Scalar quantities are those that have only magnitude and no associated direction. **Vector quantities** have both magnitude and direction and obey the laws of vector addition.

We can add two vectors **A** and **B** graphically, using either the triangle method or the parallelogram rule. In the triangle method (Fig. 3.21a), the resultant vector **R** = **A** + **B** runs from the tail of **A** to the tip of **B**. In the parallelogram method (Fig. 3.21b), **R** is the diagonal of a parallelogram having **A** and **B** as two of its sides. You should be able to add or subtract vectors, using these graphical methods.

The x component A_x of the vector **A** is equal to the projection of **A** along the x axis of a coordinate system, as shown in Figure 3.22, where $A_x = A \cos \theta$. The y component A_y of **A** is the projection of **A** along the y axis, where $A_y = A \sin \theta$. Be sure you can determine which trigonometric functions you should use in all situations, especially when θ is defined as something other than the counterclockwise angle from the positive x axis.

If a vector **A** has an x component A_x and a y component A_y, the vector can be expressed in unit–vector form as **A** = A_x**i** + A_y**j**. In this notation, **i** is a unit vector pointing in the positive x direction, and **j** is a unit vector pointing in the positive y direction. Because **i** and **j** are unit vectors, $|\mathbf{i}| = |\mathbf{j}| = 1$.

We can find the resultant of two or more vectors by resolving all vectors into their x and y components, adding their resultant x and y components, and then using the Pythagorean theorem to find the magnitude of the resultant vector. We can find the angle that the resultant vector makes with respect to the x axis by using a suitable trigonometric function.

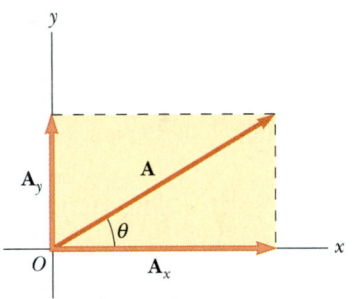

Figure 3.22 The addition of the two vectors A_x and A_y gives vector **A**. Note that $\mathbf{A}_x = A_x\mathbf{i}$ and $\mathbf{A}_y = A_y\mathbf{j}$, where A_x and A_y are the *components* of vector **A**.

QUESTIONS

1. Two vectors have unequal magnitudes. Can their sum be zero? Explain.
2. Can the magnitude of a particle's displacement be greater than the distance traveled? Explain.
3. The magnitudes of two vectors **A** and **B** are $A = 5$ units and $B = 2$ units. Find the largest and smallest values possible for the resultant vector **R** = **A** + **B**.
4. Vector **A** lies in the xy plane. For what orientations of vector **A** will both of its components be negative? For what orientations will its components have opposite signs?
5. If the component of vector **A** along the direction of vector

B is zero, what can you conclude about these two vectors?
6. Can the magnitude of a vector have a negative value? Explain.
7. Which of the following are vectors and which are not: force, temperature, volume, ratings of a television show, height, velocity, age?
8. Under what circumstances would a nonzero vector lying in the xy plane ever have components that are equal in magnitude?
9. Is it possible to add a vector quantity to a scalar quantity? Explain.

PROBLEMS

1, 2, 3 = straightforward, intermediate, challenging   □ = full solution available in the *Student Solutions Manual and Study Guide*
WEB = solution posted at **http://www.saunderscollege.com/physics/**   💻 = Computer useful in solving problem   📱 − Interactive Physics
□ = paired numerical/symbolic problems

Section 3.1 Coordinate Systems

WEB 1. The polar coordinates of a point are $r = 5.50$ m and $\theta = 240°$. What are the cartesian coordinates of this point?

2. Two points in the xy plane have cartesian coordinates $(2.00, -4.00)$ m and $(-3.00, 3.00)$ m. Determine (a) the distance between these points and (b) their polar coordinates.

3. If the cartesian coordinates of a point are given by $(2, y)$ and its polar coordinates are $(r, 30°)$, determine y and r.

4. Two points in a plane have polar coordinates $(2.50$ m, $30.0°)$ and $(3.80$ m, $120.0°)$. Determine (a) the cartesian coordinates of these points and (b) the distance between them.

5. A fly lands on one wall of a room. The lower left-hand corner of the wall is selected as the origin of a two-dimensional cartesian coordinate system. If the fly is located at the point having coordinates $(2.00, 1.00)$ m, (a) how far is it from the corner of the room? (b) what is its location in polar coordinates?

6. If the polar coordinates of the point (x, y) are (r, θ), determine the polar coordinates for the points (a) $(-x, y)$, (b) $(-2x, -2y)$, and (c) $(3x, -3y)$.

Section 3.2 Vector and Scalar Quantities

Section 3.3 Some Properties of Vectors

7. An airplane flies 200 km due west from city A to city B and then 300 km in the direction 30.0° north of west from city B to city C. (a) In straight-line distance, how far is city C from city A? (b) Relative to city A, in what direction is city C?

8. A pedestrian moves 6.00 km east and then 13.0 km north. Using the graphical method, find the magnitude and direction of the resultant displacement vector.

9. A surveyor measures the distance across a straight river by the following method: Starting directly across from a tree on the opposite bank, she walks 100 m along the riverbank to establish a baseline. Then she sights across to the tree. The angle from her baseline to the tree is 35.0°. How wide is the river?

10. A plane flies from base camp to lake A, a distance of 280 km at a direction 20.0° north of east. After dropping off supplies, it flies to lake B, which is 190 km and 30.0° west of north from lake A. Graphically determine the distance and direction from lake B to the base camp.

11. Vector **A** has a magnitude of 8.00 units and makes an angle of 45.0° with the positive x axis. Vector **B** also has a magnitude of 8.00 units and is directed along the neg-

ative x axis. Using graphical methods, find (a) the vector sum **A** + **B** and (b) the vector difference **A** − **B**.

12. A force **F**$_1$ of magnitude 6.00 units acts at the origin in a direction 30.0° above the positive x axis. A second force **F**$_2$ of magnitude 5.00 units acts at the origin in the direction of the positive y axis. Find graphically the magnitude and direction of the resultant force **F**$_1$ + **F**$_2$.

WEB 13. A person walks along a circular path of radius 5.00 m. If the person walks around one half of the circle, find (a) the magnitude of the displacement vector and (b) how far the person walked. (c) What is the magnitude of the displacement if the person walks all the way around the circle?

14. A dog searching for a bone walks 3.50 m south, then 8.20 m at an angle 30.0° north of east, and finally 15.0 m west. Using graphical techniques, find the dog's resultant displacement vector.

WEB 15. Each of the displacement vectors **A** and **B** shown in Figure P3.15 has a magnitude of 3.00 m. Find graphically (a) **A** + **B**, (b) **A** − **B**, (c) **B** − **A**, (d) **A** − 2**B**. Report all angles counterclockwise from the positive x axis.

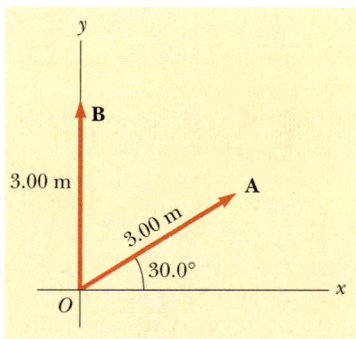

Figure P3.15   Problems 15 and 39.

16. Arbitrarily define the "instantaneous vector height" of a person as the displacement vector from the point halfway between the feet to the top of the head. Make an order-of-magnitude estimate of the total vector height of all the people in a city of population 100 000 (a) at 10 a.m. on a Tuesday and (b) at 5 a.m. on a Saturday. Explain your reasoning.

17. A roller coaster moves 200 ft horizontally and then rises 135 ft at an angle of 30.0° above the horizontal. It then travels 135 ft at an angle of 40.0° downward. What is its displacement from its starting point? Use graphical techniques.

18. The driver of a car drives 3.00 km north, 2.00 km northeast (45.0° east of north), 4.00 km west, and then

3.00 km southeast (45.0° east of south). Where does he end up relative to his starting point? Work out your answer graphically. Check by using components. (The car is not near the North Pole or the South Pole.)

19. Fox Mulder is trapped in a maze. To find his way out, he walks 10.0 m, makes a 90.0° right turn, walks 5.00 m, makes another 90.0° right turn, and walks 7.00 m. What is his displacement from his initial position?

Section 3.4 Components of a Vector and Unit Vectors

20. Find the horizontal and vertical components of the 100-m displacement of a superhero who flies from the top of a tall building following the path shown in Figure P3.20.

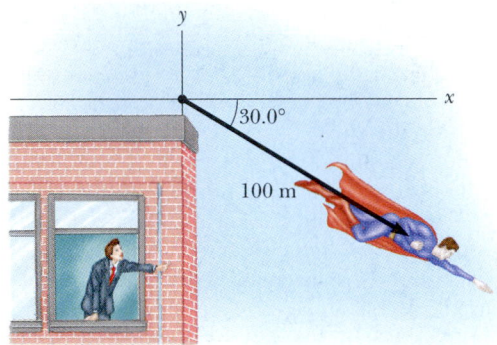

Figure P3.20

21. A person walks 25.0° north of east for 3.10 km. How far would she have to walk due north and due east to arrive at the same location?

22. While exploring a cave, a spelunker starts at the entrance and moves the following distances: She goes 75.0 m north, 250 m east, 125 m at an angle 30.0° north of east, and 150 m south. Find the resultant displacement from the cave entrance.

23. In the assembly operation illustrated in Figure P3.23, a robot first lifts an object upward along an arc that forms one quarter of a circle having a radius of 4.80 cm and

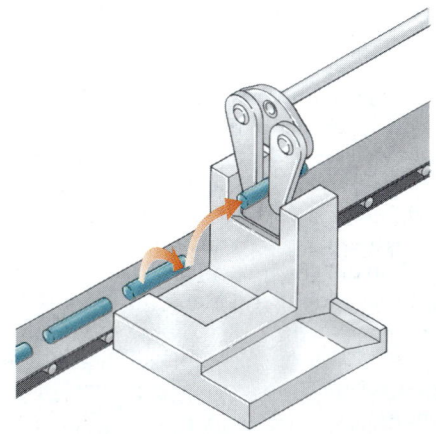

Figure P3.23

lying in an east–west vertical plane. The robot then moves the object upward along a second arc that forms one quarter of a circle having a radius of 3.70 cm and lying in a north–south vertical plane. Find (a) the magnitude of the total displacement of the object and (b) the angle the total displacement makes with the vertical.

24. Vector $\mathbf{B}$ has x, y, and z components of 4.00, 6.00, and 3.00 units, respectively. Calculate the magnitude of $\mathbf{B}$ and the angles that $\mathbf{B}$ makes with the coordinate axes.

WEB 25. A vector has an x component of -25.0 units and a y component of 40.0 units. Find the magnitude and direction of this vector.

26. A map suggests that Atlanta is 730 mi in a direction 5.00° north of east from Dallas. The same map shows that Chicago is 560 mi in a direction 21.0° west of north from Atlanta. Assuming that the Earth is flat, use this information to find the displacement from Dallas to Chicago.

27. A displacement vector lying in the xy plane has a magnitude of 50.0 m and is directed at an angle of 120° to the positive x axis. Find the x and y components of this vector and express the vector in unit–vector notation.

28. If $\mathbf{A} = 2.00\mathbf{i} + 6.00\mathbf{j}$ and $\mathbf{B} = 3.00\mathbf{i} - 2.00\mathbf{j}$, (a) sketch the vector sum $\mathbf{C} = \mathbf{A} + \mathbf{B}$ and the vector difference $\mathbf{D} = \mathbf{A} - \mathbf{B}$. (b) Find solutions for $\mathbf{C}$ and $\mathbf{D}$, first in terms of unit vectors and then in terms of polar coordinates, with angles measured with respect to the $+x$ axis.

29. Find the magnitude and direction of the resultant of three displacements having x and y components (3.00, 2.00) m, $(-5.00, 3.00)$ m, and (6.00, 1.00) m.

30. Vector $\mathbf{A}$ has x and y components of -8.70 cm and 15.0 cm, respectively; vector $\mathbf{B}$ has x and y components of 13.2 cm and -6.60 cm, respectively. If $\mathbf{A} - \mathbf{B} + 3\mathbf{C} = 0$, what are the components of $\mathbf{C}$?

31. Consider two vectors $\mathbf{A} = 3\mathbf{i} - 2\mathbf{j}$ and $\mathbf{B} = -\mathbf{i} - 4\mathbf{j}$. Calculate (a) $\mathbf{A} + \mathbf{B}$, (b) $\mathbf{A} - \mathbf{B}$, (c) $|\mathbf{A} + \mathbf{B}|$, (d) $|\mathbf{A} - \mathbf{B}|$, (e) the directions of $\mathbf{A} + \mathbf{B}$ and $\mathbf{A} - \mathbf{B}$.

32. A boy runs 3.00 blocks north, 4.00 blocks northeast, and 5.00 blocks west. Determine the length and direction of the displacement vector that goes from the starting point to his final position.

33. Obtain expressions in component form for the position vectors having polar coordinates (a) 12.8 m, 150°; (b) 3.30 cm, 60.0°; (c) 22.0 in., 215°.

34. Consider the displacement vectors $\mathbf{A} = (3\mathbf{i} + 3\mathbf{j})$ m, $\mathbf{B} = (\mathbf{i} - 4\mathbf{j})$ m, and $\mathbf{C} = (-2\mathbf{i} + 5\mathbf{j})$ m. Use the component method to determine (a) the magnitude and direction of the vector $\mathbf{D} = \mathbf{A} + \mathbf{B} + \mathbf{C}$ and (b) the magnitude and direction of $\mathbf{E} = -\mathbf{A} - \mathbf{B} + \mathbf{C}$.

35. A particle undergoes the following consecutive displacements: 3.50 m south, 8.20 m northeast, and 15.0 m west. What is the resultant displacement?

36. In a game of American football, a quarterback takes the ball from the line of scrimmage, runs backward for 10.0 yards, and then sideways parallel to the line of scrimmage for 15.0 yards. At this point, he throws a forward

pass 50.0 yards straight downfield perpendicular to the line of scrimmage. What is the magnitude of the football's resultant displacement?

37. The helicopter view in Figure P3.37 shows two people pulling on a stubborn mule. Find (a) the single force that is equivalent to the two forces shown and (b) the force that a third person would have to exert on the mule to make the resultant force equal to zero. The forces are measured in units of newtons.

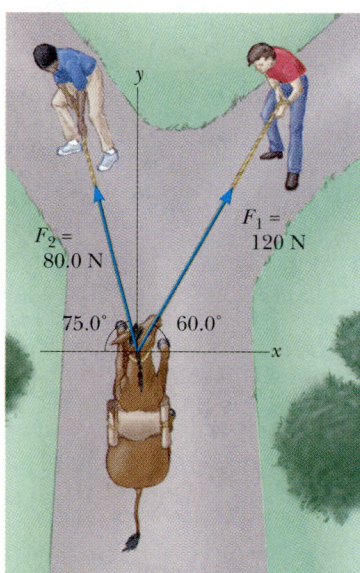

Figure P3.37

38. A novice golfer on the green takes three strokes to sink the ball. The successive displacements are 4.00 m to the north, 2.00 m northeast, and 1.00 m 30.0° west of south. Starting at the same initial point, an expert golfer could make the hole in what single displacement?

39. Find the x and y components of the vectors $\mathbf{A}$ and $\mathbf{B}$ shown in Figure P3.15; then derive an expression for the resultant vector $\mathbf{A} + \mathbf{B}$ in unit–vector notation.

40. You are standing on the ground at the origin of a coordinate system. An airplane flies over you with constant velocity parallel to the x axis and at a constant height of 7.60×10^3 m. At $t = 0$, the airplane is directly above you, so that the vector from you to it is given by $\mathbf{P}_0 = (7.60 \times 10^3 \text{ m})\mathbf{j}$. At $t = 30.0$ s, the position vector leading from you to the airplane is $\mathbf{P}_{30} = (8.04 \times 10^3 \text{ m})\mathbf{i} + (7.60 \times 10^3 \text{ m})\mathbf{j}$. Determine the magnitude and orientation of the airplane's position vector at $t = 45.0$ s.

41. A particle undergoes two displacements. The first has a magnitude of 150 cm and makes an angle of 120° with the positive x axis. The *resultant* displacement has a magnitude of 140 cm and is directed at an angle of 35.0° to the positive x axis. Find the magnitude and direction of the second displacement.

42. Vectors $\mathbf{A}$ and $\mathbf{B}$ have equal magnitudes of 5.00. If the sum of $\mathbf{A}$ and $\mathbf{B}$ is the vector $6.00\,\mathbf{j}$, determine the angle between $\mathbf{A}$ and $\mathbf{B}$.

43. The vector $\mathbf{A}$ has x, y, and z components of 8.00, 12.0, and -4.00 units, respectively. (a) Write a vector expression for $\mathbf{A}$ in unit–vector notation. (b) Obtain a unit–vector expression for a vector $\mathbf{B}$ one-fourth the length of $\mathbf{A}$ pointing in the same direction as $\mathbf{A}$. (c) Obtain a unit–vector expression for a vector $\mathbf{C}$ three times the length of $\mathbf{A}$ pointing in the direction opposite the direction of $\mathbf{A}$.

44. Instructions for finding a buried treasure include the following: Go 75.0 paces at 240°, turn to 135° and walk 125 paces, then travel 100 paces at 160°. The angles are measured counterclockwise from an axis pointing to the east, the $+ x$ direction. Determine the resultant displacement from the starting point.

45. Given the displacement vectors $\mathbf{A} = (3\mathbf{i} - 4\mathbf{j} + 4\mathbf{k})$ m and $\mathbf{B} = (2\mathbf{i} + 3\mathbf{j} - 7\mathbf{k})$ m, find the magnitudes of the vectors (a) $\mathbf{C} = \mathbf{A} + \mathbf{B}$ and (b) $\mathbf{D} = 2\mathbf{A} - \mathbf{B}$, also expressing each in terms of its x, y, and z components.

46. A radar station locates a sinking ship at range 17.3 km and bearing 136° clockwise from north. From the same station a rescue plane is at horizontal range 19.6 km, 153° clockwise from north, with elevation 2.20 km. (a) Write the vector displacement from plane to ship, letting $\mathbf{i}$ represent east, $\mathbf{j}$ north, and $\mathbf{k}$ up. (b) How far apart are the plane and ship?

47. As it passes over Grand Bahama Island, the eye of a hurricane is moving in a direction 60.0° north of west with a speed of 41.0 km/h. Three hours later, the course of the hurricane suddenly shifts due north and its speed slows to 25.0 km/h. How far from Grand Bahama is the eye 4.50 h after it passes over the island?

48. (a) Vector $\mathbf{E}$ has magnitude 17.0 cm and is directed 27.0° counterclockwise from the $+ x$ axis. Express it in unit–vector notation. (b) Vector $\mathbf{F}$ has magnitude 17.0 cm and is directed 27.0° counterclockwise from the $+ y$ axis. Express it in unit–vector notation. (c) Vector $\mathbf{G}$ has magnitude 17.0 cm and is directed 27.0° clockwise from the $+ y$ axis. Express it in unit–vector notation.

49. Vector $\mathbf{A}$ has a negative x component 3.00 units in length and a positive y component 2.00 units in length. (a) Determine an expression for $\mathbf{A}$ in unit–vector notation. (b) Determine the magnitude and direction of $\mathbf{A}$. (c) What vector $\mathbf{B}$, when added to vector $\mathbf{A}$, gives a resultant vector with no x component and a negative y component 4.00 units in length?

50. An airplane starting from airport A flies 300 km east, then 350 km at 30.0° west of north, and then 150 km north to arrive finally at airport B. (a) The next day, another plane flies directly from airport A to airport B in a straight line. In what direction should the pilot travel in this direct flight? (b) How far will the pilot travel in this direct flight? Assume there is no wind during these flights.

WEB **51.** Three vectors are oriented as shown in Figure P3.51, where $|\mathbf{A}| = 20.0$ units, $|\mathbf{B}| = 40.0$ units, and $|\mathbf{C}| = 30.0$ units. Find (a) the x and y components of the resultant vector (expressed in unit–vector notation) and (b) the magnitude and direction of the resultant vector.

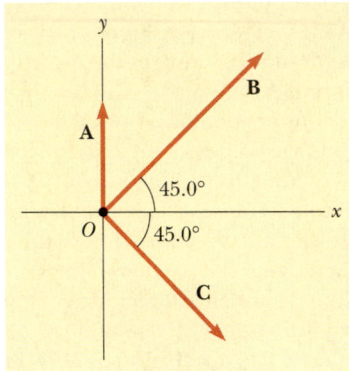

Figure P3.51

52. If $\mathbf{A} = (6.00\mathbf{i} - 8.00\mathbf{j})$ units, $\mathbf{B} = (-8.00\mathbf{i} + 3.00\mathbf{j})$ units, and $\mathbf{C} = (26.0\mathbf{i} + 19.0\mathbf{j})$ units, determine a and b such that $a\mathbf{A} + b\mathbf{B} + \mathbf{C} = 0$.

ADDITIONAL PROBLEMS

53. Two vectors $\mathbf{A}$ and $\mathbf{B}$ have precisely equal magnitudes. For the magnitude of $\mathbf{A} + \mathbf{B}$ to be 100 times greater than the magnitude of $\mathbf{A} - \mathbf{B}$, what must be the angle between them?

54. Two vectors $\mathbf{A}$ and $\mathbf{B}$ have precisely equal magnitudes. For the magnitude of $\mathbf{A} + \mathbf{B}$ to be greater than the magnitude of $\mathbf{A} - \mathbf{B}$ by the factor n, what must be the angle between them?

55. A vector is given by $\mathbf{R} = 2.00\mathbf{i} + 1.00\mathbf{j} + 3.00\mathbf{k}$. Find (a) the magnitudes of the x, y, and z components, (b) the magnitude of $\mathbf{R}$, and (c) the angles between $\mathbf{R}$ and the x, y, and z axes.

56. Find the sum of these four vector forces: 12.0 N to the right at 35.0° above the horizontal, 31.0 N to the left at 55.0° above the horizontal, 8.40 N to the left at 35.0° below the horizontal, and 24.0 N to the right at 55.0° below the horizontal. (*Hint:* Make a drawing of this situation and select the best axes for x and y so that you have the least number of components. Then add the vectors, using the component method.)

57. A person going for a walk follows the path shown in Figure P3.57. The total trip consists of four straight-line paths. At the end of the walk, what is the person's resultant displacement measured from the starting point?

58. In general, the instantaneous position of an object is specified by its position vector $\mathbf{P}$ leading from a fixed

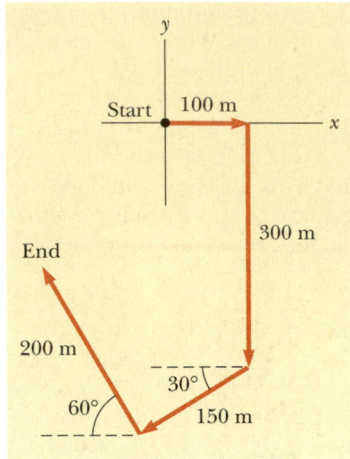

Figure P3.57

origin to the location of the object. Suppose that for a certain object the position vector is a function of time, given by $\mathbf{P} = 4\mathbf{i} + 3\mathbf{j} - 2t\mathbf{j}$, where P is in meters and t is in seconds. Evaluate $d\mathbf{P}/dt$. What does this derivative represent about the object?

59. A jet airliner, moving initially at 300 mi/h to the east, suddenly enters a region where the wind is blowing at 100 mi/h in a direction 30.0° north of east. What are the new speed and direction of the aircraft relative to the ground?

60. A pirate has buried his treasure on an island with five trees located at the following points: $A(30.0$ m, -20.0 m$)$, $B(60.0$ m, 80.0 m$)$, $C(-10.0$ m, -10.0 m$)$, $D(40.0$ m, -30.0 m$)$, and $E(-70.0$ m, 60.0 m$)$. All points are measured relative to some origin, as in Figure P3.60. Instructions on the map tell you to start at A and move toward B, but to cover only one-half the distance between A and B. Then, move toward C, covering one-third the distance between your current location and C. Next, move toward D, covering one-fourth the distance between where you are and D. Finally, move toward E, covering one-fifth the distance between you and E, stop, and dig. (a) What are the coordinates of the point where the pirate's treasure is buried? (b) Re-

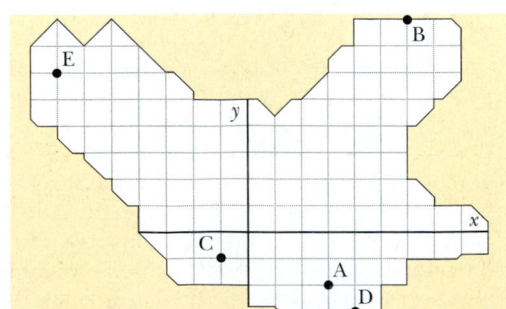

Figure P3.60

arrange the order of the trees, (for instance, B(30.0 m, − 20.0 m), A(60.0 m, 80.0 m), E(− 10.0 m, − 10.0 m), C(40.0 m, − 30.0 m), and D(− 70.0 m, 60.0 m), and repeat the calculation to show that the answer does not depend on the order of the trees.

61. A rectangular parallelepiped has dimensions a, b, and c, as in Figure P3.61. (a) Obtain a vector expression for the face diagonal vector $\mathbf{R}_1$. What is the magnitude of this vector? (b) Obtain a vector expression for the body diagonal vector $\mathbf{R}_2$. Note that $\mathbf{R}_1$, $c\mathbf{k}$, and $\mathbf{R}_2$ make a right triangle, and prove that the magnitude of $\mathbf{R}_2$ is $\sqrt{a^2 + b^2 + c^2}$.

62. A point lying in the xy plane and having coordinates (x, y) can be described by the position vector given by $\mathbf{r} = x\mathbf{i} + y\mathbf{j}$. (a) Show that the displacement vector for a particle moving from (x_1, y_1) to (x_2, y_2) is given by $\mathbf{d} = (x_2 − x_1)\mathbf{i} + (y_2 − y_1)\mathbf{j}$. (b) Plot the position vectors $\mathbf{r}_1$ and $\mathbf{r}_2$ and the displacement vector $\mathbf{d}$, and verify by the graphical method that $\mathbf{d} = \mathbf{r}_2 − \mathbf{r}_1$.

63. A point P is described by the coordinates (x, y) with respect to the normal cartesian coordinate system shown in Figure P3.63. Show that (x', y'), the coordinates of this point in the rotated coordinate system, are related to (x, y) and the rotation angle α by the expressions

$$x' = x \cos \alpha + y \sin \alpha$$

$$y' = − x \sin \alpha + y \cos \alpha$$

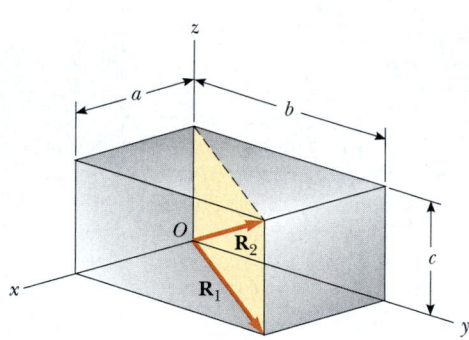

Figure P3.61

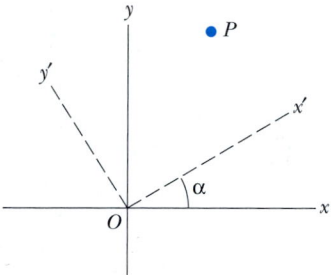

Figure P3.63

ANSWERS TO QUICK QUIZZES

3.1 The honeybee needs to communicate to the other honeybees how far it is to the flower and in what direction they must fly. This is exactly the kind of information that polar coordinates convey, as long as the origin of the coordinates is the beehive.

3.2 The resultant has magnitude $A + B$ when vector $\mathbf{A}$ is oriented in the same direction as vector $\mathbf{B}$. The resultant vector is $\mathbf{A} + \mathbf{B} = 0$ when vector $\mathbf{A}$ is oriented in the direction opposite vector $\mathbf{B}$ and $A = B$.

3.3 No. In two dimensions, a vector and its components form a right triangle. The vector is the hypotenuse and must be longer than either side. Problem 61 extends this concept to three dimensions.

3.4 No. The magnitude of a vector $\mathbf{A}$ is equal to $\sqrt{A_x^2 + A_y^2 + A_z^2}$. Therefore, if any component is nonzero, A cannot be zero. This generalization of the Pythagorean theorem is left for you to prove in Problem 61.

3.5 The fact that $\mathbf{A} + \mathbf{B} = 0$ tells you that $\mathbf{A} = − \mathbf{B}$. Therefore, the components of the two vectors must have opposite signs and equal magnitudes: $A_x = − B_x$, $A_y = − B_y$, and $A_z = − B_z$.

PUZZLER

This airplane is used by NASA for astronaut training. When it flies along a certain curved path, anything inside the plane that is not strapped down begins to float. What causes this strange effect? *(NASA)*

web

For more information about how this plane is used, visit

http://imocc.imoc.com/~acft-ops /rgpindex.htm

c h a p t e r

4

Motion in Two Dimensions

Chapter Outline

4.1 The Displacement, Velocity, and Acceleration Vectors

4.2 Two-Dimensional Motion with Constant Acceleration

4.3 Projectile Motion

4.4 Uniform Circular Motion

4.5 Tangential and Radial Acceleration

4.6 Relative Velocity and Relative Acceleration

In this chapter we deal with the kinematics of a particle moving in two dimensions. Knowing the basics of two-dimensional motion will allow us to examine—in future chapters—a wide variety of motions, ranging from the motion of satellites in orbit to the motion of electrons in a uniform electric field. We begin by studying in greater detail the vector nature of displacement, velocity, and acceleration. As in the case of one-dimensional motion, we derive the kinematic equations for two-dimensional motion from the fundamental definitions of these three quantities. We then treat projectile motion and uniform circular motion as special cases of motion in two dimensions. We also discuss the concept of relative motion, which shows why observers in different frames of reference may measure different displacements, velocities, and accelerations for a given particle.

4.1 ▷ THE DISPLACEMENT, VELOCITY, AND ACCELERATION VECTORS

In Chapter 2 we found that the motion of a particle moving along a straight line is completely known if its position is known as a function of time. Now let us extend this idea to motion in the xy plane. We begin by describing the position of a particle by its position vector **r**, drawn from the origin of some coordinate system to the particle located in the xy plane, as in Figure 4.1. At time t_i the particle is at point Ⓐ, and at some later time t_f it is at point Ⓑ. The path from Ⓐ to Ⓑ is not necessarily a straight line. As the particle moves from Ⓐ to Ⓑ in the time interval $\Delta t = t_f - t_i$, its position vector changes from $\mathbf{r}_i$ to $\mathbf{r}_f$. As we learned in Chapter 2, displacement is a vector, and the displacement of the particle is the difference between its final position and its initial position. We now formally define the **displacement vector** $\Delta \mathbf{r}$ for the particle of Figure 4.1 as being the difference between its final position vector and its initial position vector:

$$\Delta \mathbf{r} \equiv \mathbf{r}_f - \mathbf{r}_i \tag{4.1}$$

The direction of $\Delta \mathbf{r}$ is indicated in Figure 4.1. As we see from the figure, the magnitude of $\Delta \mathbf{r}$ is *less* than the distance traveled along the curved path followed by the particle.

As we saw in Chapter 2, it is often useful to quantify motion by looking at the ratio of a displacement divided by the time interval during which that displacement occurred. In two-dimensional (or three-dimensional) kinematics, everything is the same as in one-dimensional kinematics except that we must now use vectors rather than plus and minus signs to indicate the direction of motion.

We define the **average velocity** of a particle during the time interval Δt as the displacement of the particle divided by that time interval:

$$\overline{\mathbf{v}} \equiv \frac{\Delta \mathbf{r}}{\Delta t} \tag{4.2}$$

Multiplying or dividing a vector quantity by a scalar quantity changes only the magnitude of the vector, not its direction. Because displacement is a vector quantity and the time interval is a scalar quantity, we conclude that the average velocity is a vector quantity directed along $\Delta \mathbf{r}$.

Note that the average velocity between points is *independent of the path* taken. This is because average velocity is proportional to displacement, which depends

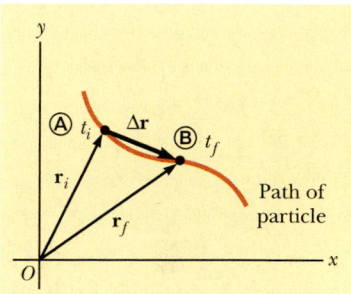

Figure 4.1 A particle moving in the xy plane is located with the position vector **r** drawn from the origin to the particle. The displacement of the particle as it moves from Ⓐ to Ⓑ in the time interval $\Delta t = t_f - t_i$ is equal to the vector $\Delta \mathbf{r} = \mathbf{r}_f - \mathbf{r}_i$.

Displacement vector

Average velocity

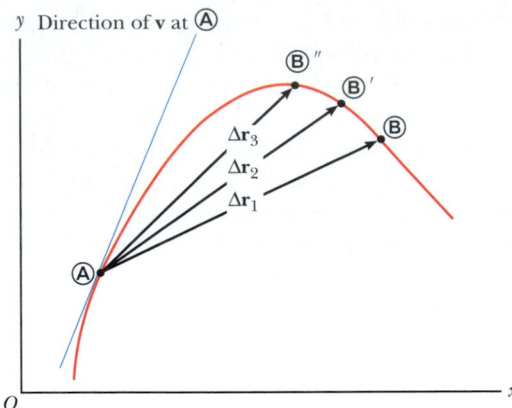

Figure 4.2   As a particle moves between two points, its average velocity is in the direction of the displacement vector $\Delta\mathbf{r}$. As the end point of the path is moved from Ⓑ to Ⓑ′ to Ⓑ″, the respective displacements and corresponding time intervals become smaller and smaller. In the limit that the end point approaches Ⓐ, Δt approaches zero, and the direction of $\Delta\mathbf{r}$ approaches that of the line tangent to the curve at Ⓐ. By definition, the instantaneous velocity at Ⓐ is in the direction of this tangent line.

only on the initial and final position vectors and not on the path taken. As we did with one-dimensional motion, we conclude that if a particle starts its motion at some point and returns to this point via any path, its average velocity is zero for this trip because its displacement is zero.

Consider again the motion of a particle between two points in the *xy* plane, as shown in Figure 4.2. As the time interval over which we observe the motion becomes smaller and smaller, the direction of the displacement approaches that of the line tangent to the path at Ⓐ.

> The **instantaneous velocity v** is defined as the limit of the average velocity $\Delta\mathbf{r}/\Delta t$ as Δt approaches zero:
>
> $$\mathbf{v} \equiv \lim_{\Delta t \to 0} \frac{\Delta\mathbf{r}}{\Delta t} = \frac{d\mathbf{r}}{dt} \tag{4.3}$$

Instantaneous velocity

That is, the instantaneous velocity equals the derivative of the position vector with respect to time. The direction of the instantaneous velocity vector at any point in a particle's path is along a line tangent to the path at that point and in the direction of motion (Fig. 4.3).

The magnitude of the instantaneous velocity vector $v = |\mathbf{v}|$ is called the *speed*, which, as you should remember, is a scalar quantity.

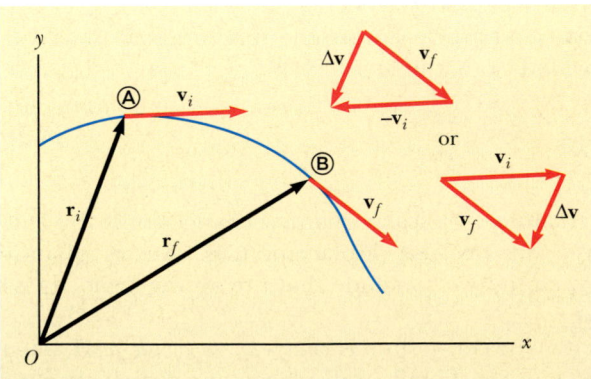

Figure 4.3   A particle moves from position Ⓐ to position Ⓑ. Its velocity vector changes from $\mathbf{v}_i$ to $\mathbf{v}_f$. The vector diagrams at the upper right show two ways of determining the vector $\Delta\mathbf{v}$ from the initial and final velocities.

As a particle moves from one point to another along some path, its instantaneous velocity vector changes from $\mathbf{v}_i$ at time t_i to $\mathbf{v}_f$ at time t_f. Knowing the velocity at these points allows us to determine the average acceleration of the particle:

The **average acceleration** of a particle as it moves from one position to another is defined as the change in the instantaneous velocity vector $\Delta\mathbf{v}$ divided by the time Δt during which that change occurred:

$$\bar{\mathbf{a}} \equiv \frac{\mathbf{v}_f - \mathbf{v}_i}{t_f - t_i} = \frac{\Delta\mathbf{v}}{\Delta t} \tag{4.4}$$

Average acceleration

Because it is the ratio of a vector quantity $\Delta\mathbf{v}$ and a scalar quantity Δt, we conclude that average acceleration $\bar{\mathbf{a}}$ is a vector quantity directed along $\Delta\mathbf{v}$. As indicated in Figure 4.3, the direction of $\Delta\mathbf{v}$ is found by adding the vector $-\mathbf{v}_i$ (the negative of $\mathbf{v}_i$) to the vector $\mathbf{v}_f$, because by definition $\Delta\mathbf{v} = \mathbf{v}_f - \mathbf{v}_i$.

When the average acceleration of a particle changes during different time intervals, it is useful to define its instantaneous acceleration $\mathbf{a}$:

The **instantaneous acceleration a** is defined as the limiting value of the ratio $\Delta\mathbf{v}/\Delta t$ as Δt approaches zero:

$$\mathbf{a} \equiv \lim_{\Delta t \to 0} \frac{\Delta\mathbf{v}}{\Delta t} = \frac{d\mathbf{v}}{dt} \tag{4.5}$$

Instantaneous acceleration

In other words, the instantaneous acceleration equals the derivative of the velocity vector with respect to time.

It is important to recognize that various changes can occur when a particle accelerates. First, the magnitude of the velocity vector (the speed) may change with time as in straight-line (one-dimensional) motion. Second, the direction of the velocity vector may change with time even if its magnitude (speed) remains constant, as in curved-path (two-dimensional) motion. Finally, both the magnitude and the direction of the velocity vector may change simultaneously.

Quick Quiz 4.1

The gas pedal in an automobile is called the *accelerator.* (a) Are there any other controls in an automobile that can be considered accelerators? (b) When is the gas pedal not an accelerator?

4.2 TWO-DIMENSIONAL MOTION WITH CONSTANT ACCELERATION

Let us consider two-dimensional motion during which the acceleration remains constant in both magnitude and direction.

The position vector for a particle moving in the xy plane can be written

$$\mathbf{r} = x\mathbf{i} + y\mathbf{j} \tag{4.6}$$

where x, y, and $\mathbf{r}$ change with time as the particle moves while $\mathbf{i}$ and $\mathbf{j}$ remain constant. If the position vector is known, the velocity of the particle can be obtained from Equations 4.3 and 4.6, which give

$$\mathbf{v} = v_x\mathbf{i} + v_y\mathbf{j} \tag{4.7}$$

Because **a** is assumed constant, its components a_x and a_y also are constants. Therefore, we can apply the equations of kinematics to the x and y components of the velocity vector. Substituting $v_{xf} = v_{xi} + a_x t$ and $v_{yf} = v_{yi} + a_y t$ into Equation 4.7 to determine the final velocity at any time t, we obtain

$$\mathbf{v}_f = (v_{xi} + a_x t)\mathbf{i} + (v_{yi} + a_y t)\mathbf{j}$$
$$= (v_{xi}\mathbf{i} + v_{yi}\mathbf{j}) + (a_x\mathbf{i} + a_y\mathbf{j})t$$

Velocity vector as a function of time

$$\mathbf{v}_f = \mathbf{v}_i + \mathbf{a}t \qquad (4.8)$$

This result states that the velocity of a particle at some time t equals the vector sum of its initial velocity $\mathbf{v}_i$ and the additional velocity $\mathbf{a}t$ acquired in the time t as a result of constant acceleration.

Similarly, from Equation 2.11 we know that the x and y coordinates of a particle moving with constant acceleration are

$$x_f = x_i + v_{xi}t + \tfrac{1}{2}a_x t^2 \qquad y_f = y_i + v_{yi}t + \tfrac{1}{2}a_y t^2$$

Substituting these expressions into Equation 4.6 (and labeling the final position vector $\mathbf{r}_f$) gives

$$\mathbf{r}_f = (x_i + v_{xi}t + \tfrac{1}{2}a_x t^2)\mathbf{i} + (y_i + v_{yi}t + \tfrac{1}{2}a_y t^2)\mathbf{j}$$
$$= (x_i\mathbf{i} + y_i\mathbf{j}) + (v_{xi}\mathbf{i} + v_{yi}\mathbf{j})t + \tfrac{1}{2}(a_x\mathbf{i} + a_y\mathbf{j})t^2$$

Position vector as a function of time

$$\mathbf{r}_f = \mathbf{r}_i + \mathbf{v}_i t + \tfrac{1}{2}\mathbf{a}t^2 \qquad (4.9)$$

This equation tells us that the displacement vector $\Delta\mathbf{r} = \mathbf{r}_f - \mathbf{r}_i$ is the vector sum of a displacement $\mathbf{v}_i t$ arising from the initial velocity of the particle and a displacement $\tfrac{1}{2}\mathbf{a}t^2$ resulting from the uniform acceleration of the particle.

Graphical representations of Equations 4.8 and 4.9 are shown in Figure 4.4. For simplicity in drawing the figure, we have taken $\mathbf{r}_i = 0$ in Figure 4.4a. That is, we assume the particle is at the origin at $t = t_i = 0$. Note from Figure 4.4a that $\mathbf{r}_f$ is generally not along the direction of either $\mathbf{v}_i$ or $\mathbf{a}$ because the relationship between these quantities is a vector expression. For the same reason, from Figure 4.4b we see that $\mathbf{v}_f$ is generally not along the direction of $\mathbf{v}_i$ or $\mathbf{a}$. Finally, note that $\mathbf{v}_f$ and $\mathbf{r}_f$ are generally not in the same direction.

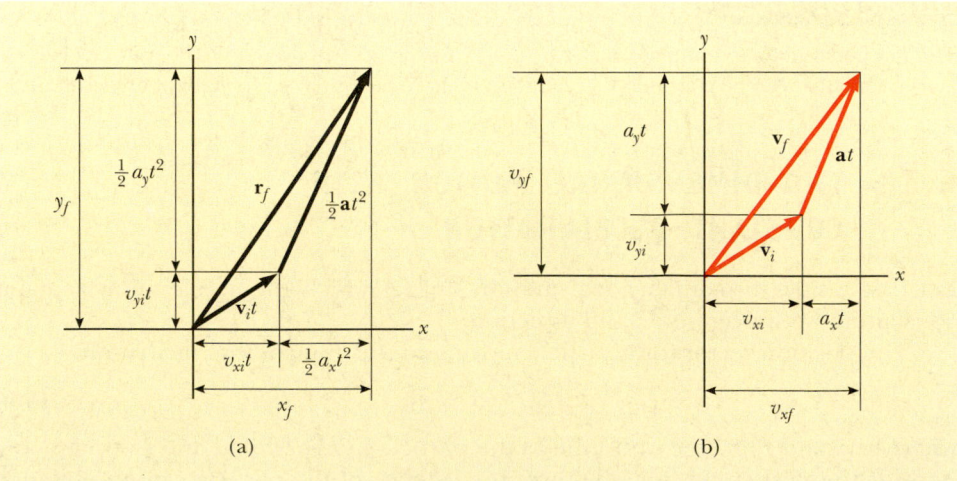

Figure 4.4 Vector representations and components of (a) the displacement and (b) the velocity of a particle moving with a uniform acceleration **a**. To simplify the drawing, we have set $\mathbf{r}_i = 0$.

Because Equations 4.8 and 4.9 are vector expressions, we may write them in component form:

$$\mathbf{v}_f = \mathbf{v}_i + \mathbf{a}t$$

$$\begin{cases} v_{xf} = v_{xi} + a_x t \\ v_{yf} = v_{yi} + a_y t \end{cases} \quad \text{(4.8a)}$$

$$\mathbf{r}_f = \mathbf{r}_i + \mathbf{v}_i t + \tfrac{1}{2}\mathbf{a}t^2$$

$$\begin{cases} x_f = x_i + v_{xi}t + \tfrac{1}{2}a_x t^2 \\ y_f = y_i + v_{yi}t + \tfrac{1}{2}a_y t^2 \end{cases} \quad \text{(1.0a)}$$

These components are illustrated in Figure 4.4. The component form of the equations for $\mathbf{v}_f$ and $\mathbf{r}_f$ show us that two-dimensional motion at constant acceleration is equivalent to two *independent* motions—one in the x direction and one in the y direction—having constant accelerations a_x and a_y.

 ## EXAMPLE 4.1 ▶ Motion in a Plane

A particle starts from the origin at $t = 0$ with an initial velocity having an x component of 20 m/s and a y component of -15 m/s. The particle moves in the xy plane with an x component of acceleration only, given by $a_x = 4.0$ m/s^2. (a) Determine the components of the velocity vector at any time and the total velocity vector at any time.

Solution After carefully reading the problem, we realize we can set $v_{xi} = 20$ m/s, $v_{yi} = -15$ m/s, $a_x = 4.0$ m/s^2, and $a_y = 0$. This allows us to sketch a rough motion diagram of the situation. The x component of velocity starts at 20 m/s and increases by 4.0 m/s every second. The y component of velocity never changes from its initial value of -15 m/s. From this information we sketch some velocity vectors as shown in Figure 4.5. Note that the spacing between successive images increases as time goes on because the velocity is increasing.

The equations of kinematics give

$$v_{xf} = v_{xi} + a_x t = (20 + 4.0t) \text{ m/s}$$

$$v_{yf} = v_{yi} + a_y t = -15 \text{ m/s} + 0 = -15 \text{ m/s}$$

Therefore,

$$\mathbf{v}_f = v_{xf}\mathbf{i} + v_{yf}\mathbf{j} = \boxed{[(20 + 4.0t)\mathbf{i} - 15\mathbf{j}] \text{ m/s}}$$

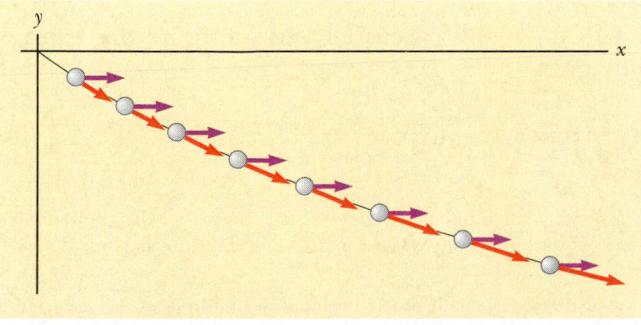

Figure 4.5 Motion diagram for the particle.

We could also obtain this result using Equation 4.8 directly, noting that $\mathbf{a} = 4.0\mathbf{i}$ m/s^2 and $\mathbf{v}_i = (20\mathbf{i} - 15\mathbf{j})$ m/s. According to this result, the x component of velocity increases while the y component remains constant; this is consistent with what we predicted. After a long time, the x component will be so great that the y component will be negligible. If we were to extend the object's path in Figure 4.5, eventually it would become nearly parallel to the x axis. It is always helpful to make comparisons between final answers and initial stated conditions.

(b) Calculate the velocity and speed of the particle at $t = 5.0$ s.

Solution With $t = 5.0$ s, the result from part (a) gives

$$\mathbf{v}_f = \{[20 + 4.0(5.0)]\mathbf{i} - 15\mathbf{j}\} \text{ m/s} = \boxed{(40\mathbf{i} - 15\mathbf{j}) \text{ m/s}}$$

This result tells us that at $t = 5.0$ s, $v_{xf} = 40$ m/s and $v_{yf} = -15$ m/s. Knowing these two components for this two-dimensional motion, we can find both the direction and the magnitude of the velocity vector. To determine the angle θ that $\mathbf{v}$ makes with the x axis at $t = 5.0$ s, we use the fact that $\tan \theta = v_{yf}/v_{xf}$:

$$\theta = \tan^{-1}\left(\frac{v_{yf}}{v_{xf}}\right) = \tan^{-1}\left(\frac{-15 \text{ m/s}}{40 \text{ m/s}}\right) = \boxed{-21°}$$

where the minus sign indicates an angle of 21° below the positive x axis. The speed is the magnitude of $\mathbf{v}_f$:

$$v_f = |\mathbf{v}_f| = \sqrt{v_{xf}^2 + v_{yf}^2} = \sqrt{(40)^2 + (-15)^2} \text{ m/s} = \boxed{43 \text{ m/s}}$$

In looking over our result, we notice that if we calculate v_i from the x and y components of $\mathbf{v}_i$, we find that $v_f > v_i$. Does this make sense?

(c) Determine the x and y coordinates of the particle at any time t and the position vector at this time.

Solution   Because $x_i = y_i = 0$ at $t = 0$, Equation 2.11 gives

$$x_f = v_{xi}t + \tfrac{1}{2}a_x t^2 = \boxed{(20t + 2.0t^2)\ \text{m}}$$

$$y_f = v_{yi}t = \boxed{(-15t)\ \text{m}}$$

Therefore, the position vector at any time t is

$$\mathbf{r}_f = x_f\mathbf{i} + y_f\mathbf{j} = \boxed{[(20t + 2.0t^2)\mathbf{i} - 15t\mathbf{j}]\ \text{m}}$$

(Alternatively, we could obtain $\mathbf{r}_f$ by applying Equation 4.9 directly, with $\mathbf{v}_i = (20\mathbf{i} - 15\mathbf{j})$ m/s and $\mathbf{a} = 4.0\mathbf{i}$ m/s^2. Try it!) Thus, for example, at $t = 5.0$ s, $x = 150$ m, $y = -75$ m, and $\mathbf{r}_f = (150\mathbf{i} - 75\mathbf{j})$ m. The magnitude of the displacement of the particle from the origin at $t = 5.0$ s is the magnitude of $\mathbf{r}_f$ at this time:

$$r_f = |\mathbf{r}_f| = \sqrt{(150)^2 + (-75)^2}\ \text{m} = 170\ \text{m}$$

Note that this is *not* the distance that the particle travels in this time! Can you determine this distance from the available data?

4.3 PROJECTILE MOTION

Anyone who has observed a baseball in motion (or, for that matter, any other object thrown into the air) has observed projectile motion. The ball moves in a curved path, and its motion is simple to analyze if we make two assumptions:

Assumptions of projectile motion

(1) the free-fall acceleration $\mathbf{g}$ is constant over the range of motion and is directed downward,[1] and (2) the effect of air resistance is negligible.[2] With these assumptions, we find that the path of a projectile, which we call its *trajectory*, is *always* a parabola. **We use these assumptions throughout this chapter.**

To show that the trajectory of a projectile is a parabola, let us choose our reference frame such that the y direction is vertical and positive is upward. Because air resistance is neglected, we know that $a_y = -g$ (as in one-dimensional free fall) and that $a_x = 0$. Furthermore, let us assume that at $t = 0$, the projectile leaves the origin ($x_i = y_i = 0$) with speed v_i, as shown in Figure 4.6. The vector $\mathbf{v}_i$ makes an angle θ_i with the horizontal, where θ_i is the angle at which the projectile leaves the origin. From the definitions of the cosine and sine functions we have

$$\cos\theta_i = v_{xi}/v_i \qquad \sin\theta_i = v_{yi}/v_i$$

Therefore, the initial x and y components of velocity are

$$v_{xi} = v_i\cos\theta_i \qquad v_{yi} = v_i\sin\theta_i$$

Substituting the x component into Equation 4.9a with $x_i = 0$ and $a_x = 0$, we find that

Horizontal position component

$$x_f = v_{xi}t = (v_i\cos\theta_i)\,t \qquad\qquad \textbf{(4.10)}$$

Repeating with the y component and using $y_i = 0$ and $a_y = -g$, we obtain

Vertical position component

$$y_f = v_{yi}t + \tfrac{1}{2}a_y t^2 = (v_i\sin\theta_i)\,t - \tfrac{1}{2}gt^2 \qquad\qquad \textbf{(4.11)}$$

Next, we solve Equation 4.10 for $t = x_f/(v_i\cos\theta_i)$ and substitute this expression for t into Equation 4.11; this gives

$$y = (\tan\theta_i)x - \left(\frac{g}{2v_i^{\,2}\cos^2\theta_i}\right)x^2 \qquad\qquad \textbf{(4.12)}$$

[1] This assumption is reasonable as long as the range of motion is small compared with the radius of the Earth (6.4×10^6 m). In effect, this assumption is equivalent to assuming that the Earth is flat over the range of motion considered.

[2] This assumption is generally *not* justified, especially at high velocities. In addition, any spin imparted to a projectile, such as that applied when a pitcher throws a curve ball, can give rise to some very interesting effects associated with aerodynamic forces, which will be discussed in Chapter 15.

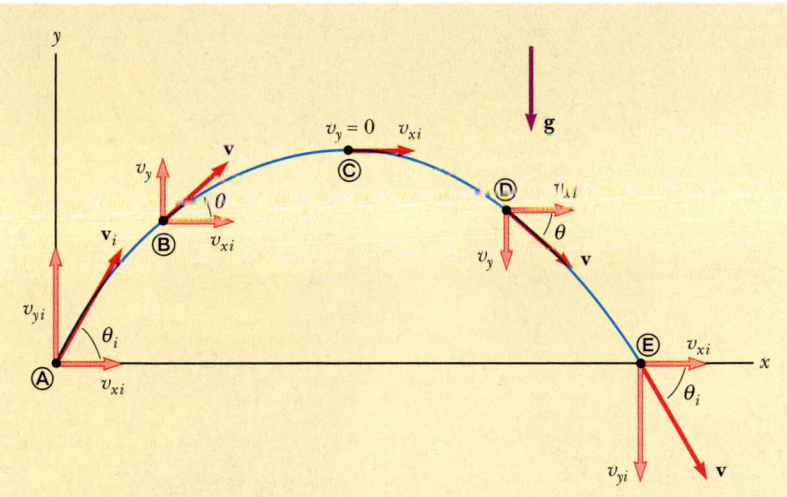

Figure 4.6 The parabolic path of a projectile that leaves the origin with a velocity $\mathbf{v}_i$. The velocity vector $\mathbf{v}$ changes with time in both magnitude and direction. This change is the result of acceleration in the negative y direction. The x component of velocity remains constant in time because there is no acceleration along the horizontal direction. The y component of velocity is zero at the peak of the path.

A welder cuts holes through a heavy metal construction beam with a hot torch. The sparks generated in the process follow parabolic paths. *(©The Telegraph Colour Library/FPG)*

This equation is valid for launch angles in the range $0 < \theta_i < \pi/2$. We have left the subscripts off the x and y because the equation is valid for any point (x, y) along the path of the projectile. The equation is of the form $y = ax - bx^2$, which is the equation of a parabola that passes through the origin. Thus, we have shown that the trajectory of a projectile is a parabola. Note that the trajectory is completely specified if both the initial speed v_i and the launch angle θ_i are known.

The vector expression for the position vector of the projectile as a function of time follows directly from Equation 4.9, with $\mathbf{r}_i = 0$ and $\mathbf{a} = \mathbf{g}$:

$$\mathbf{r} = \mathbf{v}_i t + \tfrac{1}{2}\mathbf{g}t^2$$

This expression is plotted in Figure 4.7.

QuickLab

Place two tennis balls at the edge of a tabletop. Sharply snap one ball horizontally off the table with one hand while gently tapping the second ball off with your other hand. Compare how long it takes the two to reach the floor. Explain your results.

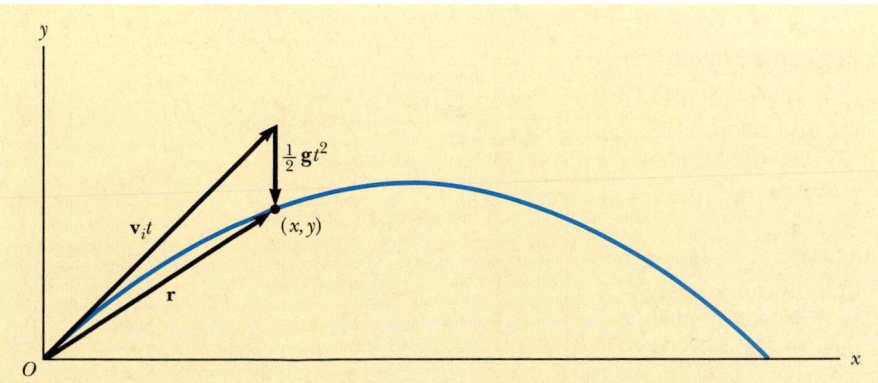

Figure 4.7 The position vector $\mathbf{r}$ of a projectile whose initial velocity at the origin is $\mathbf{v}_i$. The vector $\mathbf{v}_i t$ would be the displacement of the projectile if gravity were absent, and the vector $\tfrac{1}{2}\mathbf{g}t^2$ is its vertical displacement due to its downward gravitational acceleration.

Multiflash exposure of a tennis player executing a forehand swing. Note that the ball follows a parabolic path characteristic of a projectile. Such photographs can be used to study the quality of sports equipment and the performance of an athlete. *(© Zimmerman, FPG International)*

It is interesting to realize that the motion of a particle can be considered the superposition of the term $\mathbf{v}_i t$, the displacement if no acceleration were present, and the term $\frac{1}{2}\mathbf{g}t^2$, which arises from the acceleration due to gravity. In other words, if there were no gravitational acceleration, the particle would continue to move along a straight path in the direction of $\mathbf{v}_i$. Therefore, the vertical distance $\frac{1}{2}\mathbf{g}t^2$ through which the particle "falls" off the straight-line path is the same distance that a freely falling body would fall during the same time interval. We conclude that **projectile motion is the superposition of two motions: (1) constant-velocity motion in the horizontal direction** and (2) **free-fall motion in the vertical direction.** Except for t, the time of flight, the horizontal and vertical components of a projectile's motion are completely independent of each other.

EXAMPLE 4.2   **Approximating Projectile Motion**

A ball is thrown in such a way that its initial vertical and horizontal components of velocity are 40 m/s and 20 m/s, respectively. Estimate the total time of flight and the distance the ball is from its starting point when it lands.

Solution   We start by remembering that the two velocity components are independent of each other. By considering the vertical motion first, we can determine how long the ball remains in the air. Then, we can use the time of flight to estimate the horizontal distance covered.

A motion diagram like Figure 4.8 helps us organize what we know about the problem. The acceleration vectors are all the same, pointing downward with a magnitude of nearly 10 m/s². The velocity vectors change direction. Their hori-

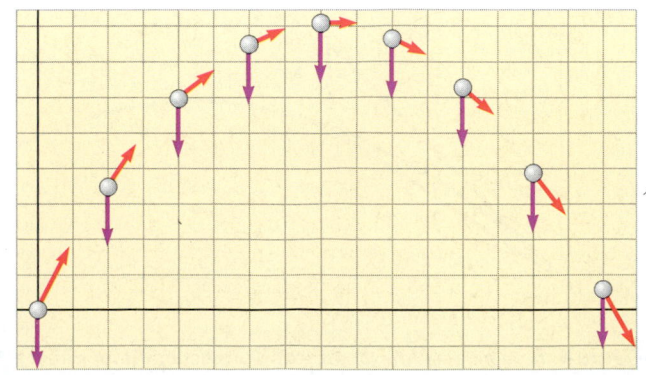

Figure 4.8   Motion diagram for a projectile.

zontal components are all the same: 20 m/s. Because the vertical motion is free fall, the vertical components of the velocity vectors change, second by second, from 40 m/s to roughly 30, 20, and 10 m/s in the upward direction, and then to 0 m/s. Subsequently, its velocity becomes 10, 20, 30, and 40 m/s in the downward direction. Thus it takes the ball about 4 s to go up and another 4 s to come back down, for a total time of flight of approximately 8 s. Because the horizontal component of velocity is 20 m/s, and because the ball travels at this speed for 8 s, it ends up approximately 160 m from its starting point.

Horizontal Range and Maximum Height of a Projectile

Let us assume that a projectile is fired from the origin at $t_i = 0$ with a positive v_{yi} component, as shown in Figure 4.9. Two points are especially interesting to analyze: the peak point Ⓐ, which has cartesian coordinates $(R/2, h)$, and the point Ⓑ, which has coordinates $(R, 0)$. The distance R is called the *horizontal range* of the projectile, and the distance h is its *maximum height*. Let us find h and R in terms of v_i, θ_i, and g.

We can determine h by noting that at the peak, $v_{yA} = 0$. Therefore, we can use Equation 4.8a to determine the time t_A it takes the projectile to reach the peak:

$$v_{yf} = v_{yi} + a_y t$$

$$0 = v_i \sin \theta_i - g t_A$$

$$t_A = \frac{v_i \sin \theta_i}{g}$$

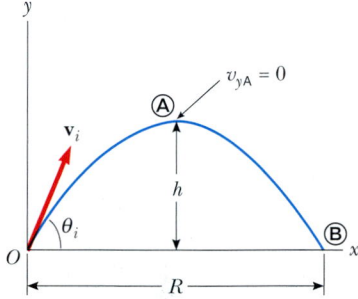

Figure 4.9 A projectile fired from the origin at $t_i = 0$ with an initial velocity $\mathbf{v}_i$. The maximum height of the projectile is h, and the horizontal range is R. At Ⓐ, the peak of the trajectory, the particle has coordinates $(R/2, h)$.

Substituting this expression for t_A into the y part of Equation 4.9a and replacing $y_f = y_A$ with h, we obtain an expression for h in terms of the magnitude and direction of the initial velocity vector:

$$h = (v_i \sin \theta_i) \frac{v_i \sin \theta_i}{g} - \frac{1}{2}g \left(\frac{v_i \sin \theta_i}{g} \right)^2$$

$$h = \frac{v_i^2 \sin^2 \theta_i}{2g} \qquad (4.13)$$

◀ Maximum height of projectile

The range R is the horizontal distance that the projectile travels in twice the time it takes to reach its peak, that is, in a time $t_B = 2t_A$. Using the x part of Equation 4.9a, noting that $v_{xi} = v_{xB} = v_i \cos \theta_i$, and setting $R \equiv x_B$ at $t = 2t_A$, we find that

$$R = v_{xi}t_B = (v_i \cos \theta_i)2t_A$$

$$= (v_i \cos \theta_i) \frac{2v_i \sin \theta_i}{g} = \frac{2v_i^2 \sin \theta_i \cos \theta_i}{g}$$

Using the identity $\sin 2\theta = 2 \sin \theta \cos \theta$ (see Appendix B.4), we write R in the more compact form

$$R = \frac{v_i^2 \sin 2\theta_i}{g} \qquad (4.14)$$

◀ Range of projectile

Keep in mind that Equations 4.13 and 4.14 are useful for calculating h and R only if v_i and θ_i are known (which means that only $\mathbf{v}_i$ has to be specified) and if the projectile lands at the same height from which it started, as it does in Figure 4.9.

The maximum value of R from Equation 4.14 is $R_{max} = v_i^2/g$. This result follows from the fact that the maximum value of $\sin 2\theta_i$ is 1, which occurs when $2\theta_i = 90°$. Therefore, R is a maximum when $\theta_i = 45°$.

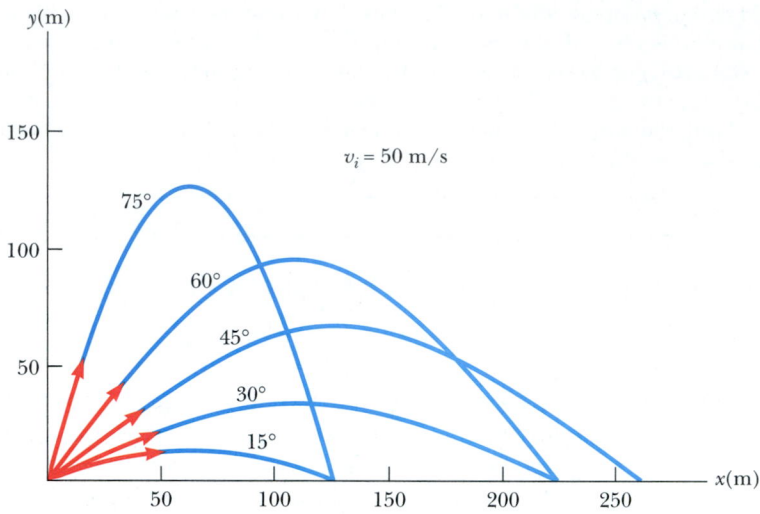

Figure 4.10   A projectile fired from the origin with an initial speed of 50 m/s at various angles of projection. Note that complementary values of θ_i result in the same value of x (range of the projectile).

QuickLab

To carry out this investigation, you need to be outdoors with a small ball, such as a tennis ball, as well as a wrist-watch. Throw the ball straight up as hard as you can and determine the initial speed of your throw and the approximate maximum height of the ball, using only your watch. What happens when you throw the ball at some angle $\theta \neq 90°$? Does this change the time of flight (perhaps because it is easier to throw)? Can you still determine the maximum height and initial speed?

Figure 4.10 illustrates various trajectories for a projectile having a given initial speed but launched at different angles. As you can see, the range is a maximum for $\theta_i = 45°$. In addition, for any θ_i other than 45°, a point having cartesian coordinates $(R, 0)$ can be reached by using either one of two complementary values of θ_i, such as 75° and 15°. Of course, the maximum height and time of flight for one of these values of θ_i are different from the maximum height and time of flight for the complementary value.

Quick Quiz 4.2

As a projectile moves in its parabolic path, is there any point along the path where the velocity and acceleration vectors are (a) perpendicular to each other? (b) parallel to each other? (c) Rank the five paths in Figure 4.10 with respect to time of flight, from the shortest to the longest.

Problem-Solving Hints

Projectile Motion

We suggest that you use the following approach to solving projectile motion problems:

- Select a coordinate system and resolve the initial velocity vector into x and y components.
- Follow the techniques for solving constant-velocity problems to analyze the horizontal motion. Follow the techniques for solving constant-acceleration problems to analyze the vertical motion. The x and y motions share the same time of flight t.

EXAMPLE 4.3   The Long-Jump

A long-jumper leaves the ground at an angle of 20.0° above the horizontal and at a speed of 11.0 m/s. (a) How far does he jump in the horizontal direction? (Assume his motion is equivalent to that of a particle.)

Solution   Because the initial speed and launch angle are given, the most direct way of solving this problem is to use the range formula given by Equation 4.14. However, it is more instructive to take a more general approach and use Figure 4.9. As before, we set our origin of coordinates at the

In a long-jump event, 1993 United States champion Mike Powell can leap horizontal distances of at least 8 m. *(Chuck Muhlstock/FPG International)*

takeoff point and label the peak as Ⓐ and the landing point as Ⓑ. The horizontal motion is described by Equation 4.10:

$$x_f = x_B = (v_i \cos \theta_i) t_B = (11.0 \text{ m/s})(\cos 20.0°) t_B$$

The value of x_B can be found if the total time of the jump is known. We are able to find t_B by remembering that $a_y = -g$ and by using the y part of Equation 4.8a. We also note that at the top of the jump the vertical component of velocity v_{yA} is zero:

$$v_{yf} = v_{yA} = v_i \sin \theta_i - gt_A$$

$$0 = (11.0 \text{ m/s}) \sin 20.0° - (9.80 \text{ m/s}^2) t_A$$

$$t_A = 0.384 \text{ s}$$

This is the time needed to reach the *top* of the jump. Because of the symmetry of the vertical motion, an identical time interval passes before the jumper returns to the ground. Therefore, the *total time* in the air is $t_B = 2t_A = 0.768$ s. Substituting this value into the above expression for x_f gives

$$x_f = x_B = (11.0 \text{ m/s})(\cos 20.0°)(0.768 \text{ s}) = \boxed{7.94 \text{ m}}$$

This is a reasonable distance for a world-class athlete.

(b) What is the maximum height reached?

Solution   We find the maximum height reached by using Equation 4.11:

$$
\begin{aligned}
y_{max} = y_A &= (v_i \sin \theta_i) t_A - \tfrac{1}{2} g t_A^2 \\
&= (11.0 \text{ m/s})(\sin 20.0°)(0.384 \text{ s}) \\
&\quad - \tfrac{1}{2}(9.80 \text{ m/s}^2)(0.384 \text{ s})^2 \\
&= \boxed{0.722 \text{ m}}
\end{aligned}
$$

Treating the long-jumper as a particle is an oversimplification. Nevertheless, the values obtained are reasonable.

Exercise   To check these calculations, use Equations 4.13 and 4.14 to find the maximum height and horizontal range.

EXAMPLE 4.4   A Bull's-Eye Every Time

In a popular lecture demonstration, a projectile is fired at a target in such a way that the projectile leaves the gun at the same time the target is dropped from rest, as shown in Figure 4.11. Show that if the gun is initially aimed at the stationary target, the projectile hits the target.

Solution   We can argue that a collision results under the conditions stated by noting that, as soon as they are released, the projectile and the target experience the same accelera-

tion $a_y = -g$. First, note from Figure 4.11b that the initial y coordinate of the target is $x_T \tan \theta_i$ and that it falls through a distance $\tfrac{1}{2} g t^2$ in a time t. Therefore, the y coordinate of the target at any moment after release is

$$y_T = x_T \tan \theta_i - \tfrac{1}{2} g t^2$$

Now if we use Equation 4.9a to write an expression for the y coordinate of the projectile at any moment, we obtain

$$y_P = x_P \tan \theta_i - \tfrac{1}{2} g t^2$$

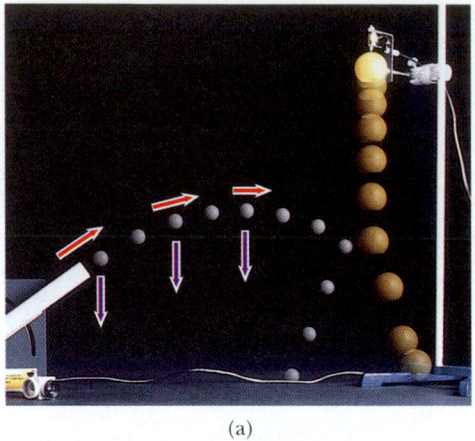

(a)

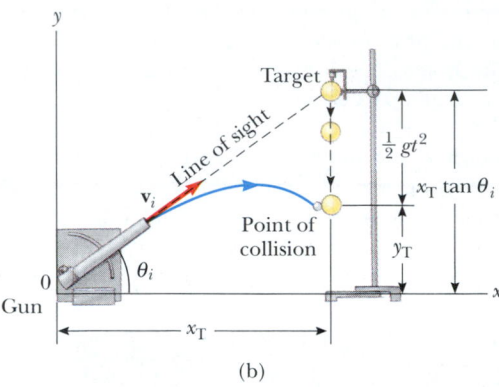

(b)

Figure 4.11   (a) Multiflash photograph of projectile–target demonstration. If the gun is aimed directly at the target and is fired at the same instant the target begins to fall, the projectile will hit the target. Note that the velocity of the projectile (red arrows) changes in direction and magnitude, while the downward acceleration (violet arrows) remains constant. *(Central Scientific Company.)* (b) Schematic diagram of the projectile–target demonstration. Both projectile and target fall through the same vertical distance in a time *t* because both experience the same acceleration $a_y = -g$.

Thus, by comparing the two previous equations, we see that when the *y* coordinates of the projectile and target are the same, their *x* coordinates are the same and a collision results. That is, when $y_P = y_T$, $x_P = x_T$. You can obtain the same result, using expressions for the position vectors for the projectile and target.

Note that a collision will *not* always take place owing to a further restriction: A collision can result only when $v_i \sin \theta_i \geq \sqrt{gd/2}$, where *d* is the initial elevation of the target above the *floor*. If $v_i \sin \theta_i$ is less than this value, the projectile will strike the floor before reaching the target.

EXAMPLE 4.5 ▶ That's Quite an Arm!

A stone is thrown from the top of a building upward at an angle of 30.0° to the horizontal and with an initial speed of 20.0 m/s, as shown in Figure 4.12. If the height of the building is 45.0 m, (a) how long is it before the stone hits the ground?

Solution   We have indicated the various parameters in Figure 4.12. When working problems on your own, you should always make a sketch such as this and label it.

The initial *x* and *y* components of the stone's velocity are

$$v_{xi} = v_i \cos \theta_i = (20.0 \text{ m/s})(\cos 30.0°) = 17.3 \text{ m/s}$$

$$v_{yi} = v_i \sin \theta_i = (20.0 \text{ m/s})(\sin 30.0°) = 10.0 \text{ m/s}$$

To find *t*, we can use $y_f = v_{yi}t + \frac{1}{2}a_yt^2$ (Eq. 4.9a) with $y_f = -45.0$ m, $a_y = -g$, and $v_{yi} = 10.0$ m/s (there is a minus sign on the numerical value of y_f because we have chosen the top of the building as the origin):

$$-45.0 \text{ m} = (10.0 \text{ m/s})t - \frac{1}{2}(9.80 \text{ m/s}^2)t^2$$

Solving the quadratic equation for *t* gives, for the positive root, $t = $ 4.22 s. Does the negative root have any physical

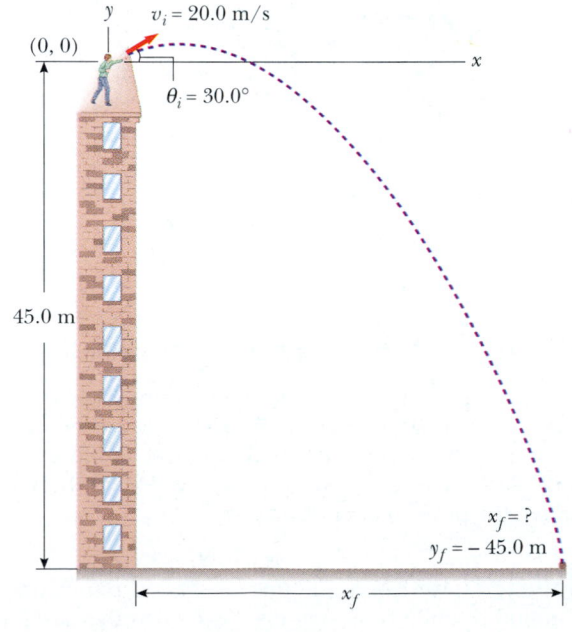

Figure 4.12

meaning? (Can you think of another way of finding t from the information given?)

(b) What is the speed of the stone just before it strikes the ground?

Solution   We can use Equation 4.8a, $v_{yf} = v_{yi} + a_y t$, with $t = 4.22$ s to obtain the y component of the velocity just before the stone strikes the ground:

$$v_{yf} = 10.0 \text{ m/s} - (9.80 \text{ m/s}^2)(4.22 \text{ s}) = -31.4 \text{ m/s}$$

The negative sign indicates that the stone is moving downward. Because $v_{xf} = v_{xi} = 17.3$ m/s, the required speed is

$$v_f = \sqrt{v_{xf}^2 + v_{yf}^2} = \sqrt{(17.3)^2 + (-31.4)^2} \text{ m/s} = \boxed{35.0 \text{ m/s}}$$

Exercise   Where does the stone strike the ground?

Answer   73.0 m from the base of the building.

EXAMPLE 4.6 ▸   The Stranded Explorers

An Alaskan rescue plane drops a package of emergency rations to a stranded party of explorers, as shown in Figure 4.13. If the plane is traveling horizontally at 40.0 m/s and is 100 m above the ground, where does the package strike the ground relative to the point at which it was released?

Solution   For this problem we choose the coordinate system shown in Figure 4.13, in which the origin is at the point of release of the package. Consider first the horizontal motion of the package. The only equation available to us for finding the distance traveled along the horizontal direction is $x_f = v_{xi}t$ (Eq. 4.9a). The initial x component of the package

velocity is the same as that of the plane when the package is released: 40.0 m/s. Thus, we have

$$x_f = (40.0 \text{ m/s})t$$

If we know t, the length of time the package is in the air, then we can determine x_f, the distance the package travels in the horizontal direction. To find t, we use the equations that describe the vertical motion of the package. We know that at the instant the package hits the ground, its y coordinate is $y_f = -100$ m. We also know that the initial vertical component of the package velocity v_{yi} is zero because at the moment of release, the package had only a horizontal component of velocity.

From Equation 4.9a, we have

$$y_f = -\tfrac{1}{2}gt^2$$

$$-100 \text{ m} = -\tfrac{1}{2}(9.80 \text{ m/s}^2)t^2$$

$$t = 4.52 \text{ s}$$

Substitution of this value for the time of flight into the equation for the x coordinate gives

$$x_f = (40.0 \text{ m/s})(4.52 \text{ s}) = \boxed{181 \text{ m}}$$

The package hits the ground 181 m to the right of the drop point.

Exercise   What are the horizontal and vertical components of the velocity of the package just before it hits the ground?

Answer $v_{xf} = 40.0$ m/s; $v_{yf} = -44.3$ m/s.

Exercise   Where is the plane when the package hits the ground? (Assume that the plane does not change its speed or course.)

Answer   Directly over the package.

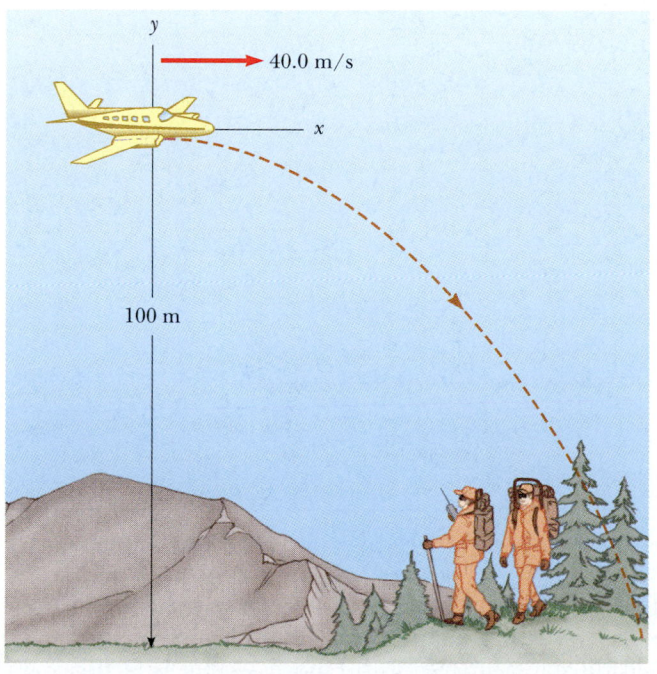

Figure 4.13

EXAMPLE 4.7   The End of the Ski Jump

A ski jumper leaves the ski track moving in the horizontal direction with a speed of 25.0 m/s, as shown in Figure 4.14. The landing incline below him falls off with a slope of 35.0°. Where does he land on the incline?

Solution   It is reasonable to expect the skier to be airborne for less than 10 s, and so he will not go farther than 250 m horizontally. We should expect the value of d, the distance traveled along the incline, to be of the same order of magnitude. It is convenient to select the beginning of the jump as the origin ($x_i = 0$, $y_i = 0$). Because $v_{xi} = 25.0$ m/s and $v_{yi} = 0$, the x and y component forms of Equation 4.9a are

$$(1) \qquad x_f = v_{xi}t = (25.0 \text{ m/s})t$$

$$(2) \qquad y_f = \tfrac{1}{2}a_y t^2 = -\tfrac{1}{2}(9.80 \text{ m/s}^2)t^2$$

From the right triangle in Figure 4.14, we see that the jumper's x and y coordinates at the landing point are $x_f = d \cos 35.0°$ and $y_f = -d \sin 35.0°$. Substituting these relationships into (1) and (2), we obtain

$$(3) \qquad d \cos 35.0° = (25.0 \text{ m/s})t$$

$$(4) \qquad -d \sin 35.0° = -\tfrac{1}{2}(9.80 \text{ m/s}^2)t^2$$

Solving (3) for t and substituting the result into (4), we find that $d = 109$ m. Hence, the x and y coordinates of the point at which he lands are

$$x_f = d \cos 35.0° = (109 \text{ m}) \cos 35.0° = \boxed{89.3 \text{ m}}$$

$$y_f = -d \sin 35.0° = -(109 \text{ m}) \sin 35.0° = \boxed{-62.5 \text{ m}}$$

Exercise   Determine how long the jumper is airborne and his vertical component of velocity just before he lands.

Answer   3.57 s; -35.0 m/s.

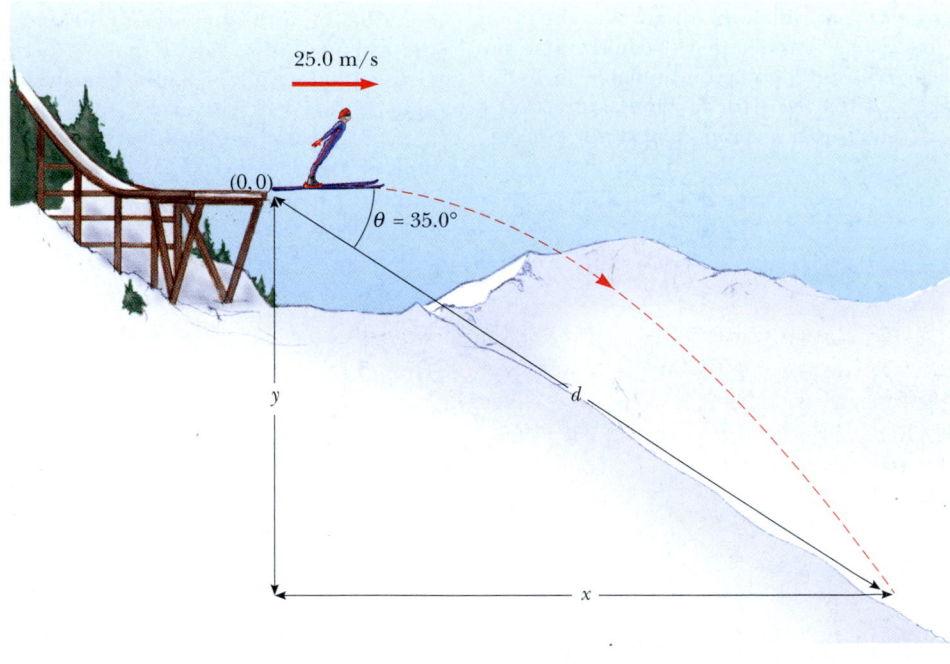

25.0 m/s

(0, 0)

$\theta = 35.0°$

y

d

x

Figure 4.14

    What would have occurred if the skier in the last example happened to be carrying a stone and let go of it while in midair? Because the stone has the same initial velocity as the skier, it will stay near him as he moves—that is, it floats alongside him. This is a technique that NASA uses to train astronauts. The plane pictured at the beginning of the chapter flies in the same type of projectile path that the skier and stone follow. The passengers and cargo in the plane fall along-

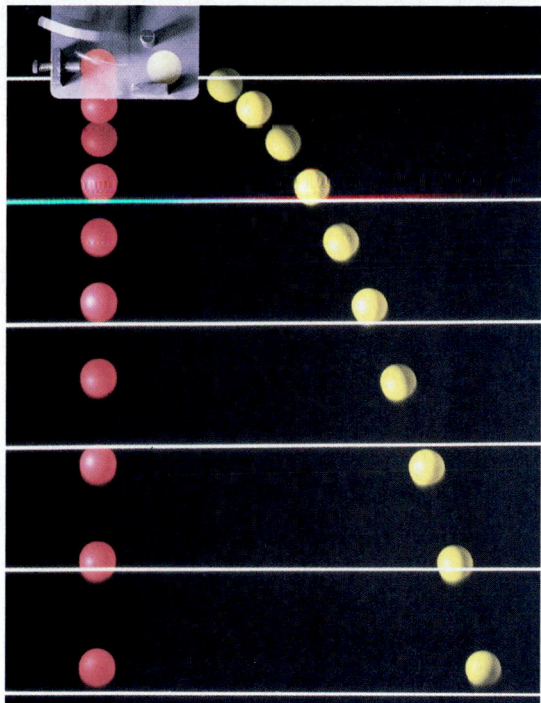

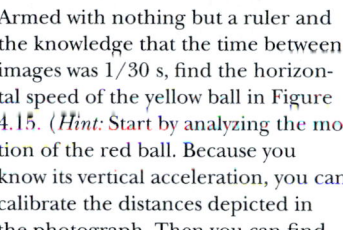

Figure 4.15 This multiflash photograph of two balls released simultaneously illustrates both free fall (red ball) and projectile motion (yellow ball). The yellow ball was projected horizontally, while the red ball was released from rest. *(Richard Megna/Fundamental Photographs)*

side each other; that is, they have the same trajectory. An astronaut can release a piece of equipment and it will float freely alongside her hand. The same thing happens in the space shuttle. The craft and everything in it are falling as they orbit the Earth.

4.4 UNIFORM CIRCULAR MOTION

Figure 4.16a shows a car moving in a circular path with constant linear speed v. Such motion is called **uniform circular motion.** Because the car's direction of motion changes, the car has an acceleration, as we learned in Section 4.1. For any motion, the velocity vector is tangent to the path. Consequently, when an object moves in a circular path, its velocity vector is perpendicular to the radius of the circle.

We now show that the acceleration vector in uniform circular motion is always perpendicular to the path and always points toward the center of the circle. An ac-

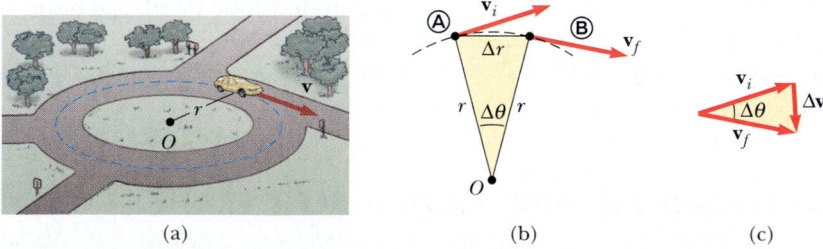

Figure 4.16 (a) A car moving along a circular path at constant speed experiences uniform circular motion. (b) As a particle moves from Ⓐ to Ⓑ, its velocity vector changes from $\mathbf{v}_i$ to $\mathbf{v}_f$. (c) The construction for determining the direction of the change in velocity $\Delta\mathbf{v}$, which is toward the center of the circle for small $\Delta\mathbf{r}$.

celeration of this nature is called a **centripetal** (center-seeking) acceleration, and its magnitude is

$$a_r = \frac{v^2}{r}$$

(4.15)

where r is the radius of the circle and the notation a_r is used to indicate that the centripetal acceleration is along the radial direction.

To derive Equation 4.15, consider Figure 4.16b, which shows a particle first at point Ⓐ and then at point Ⓑ. The particle is at Ⓐ at time t_i, and its velocity at that time is $\mathbf{v}_i$. It is at Ⓑ at some later time t_f, and its velocity at that time is $\mathbf{v}_f$. Let us assume here that $\mathbf{v}_i$ and $\mathbf{v}_f$ differ only in direction; their magnitudes (speeds) are the same (that is, $v_i = v_f = v$). To calculate the acceleration of the particle, let us begin with the defining equation for average acceleration (Eq. 4.4):

$$\overline{\mathbf{a}} = \frac{\mathbf{v}_f - \mathbf{v}_i}{t_f - t_i} = \frac{\Delta \mathbf{v}}{\Delta t}$$

This equation indicates that we must subtract $\mathbf{v}_i$ from $\mathbf{v}_f$, being sure to treat them as vectors, where $\Delta \mathbf{v} = \mathbf{v}_f - \mathbf{v}_i$ is the change in the velocity. Because $\mathbf{v}_i + \Delta \mathbf{v} = \mathbf{v}_f$, we can find the vector $\Delta \mathbf{v}$, using the vector triangle in Figure 4.16c.

Now consider the triangle in Figure 4.16b, which has sides Δr and r. This triangle and the one in Figure 4.16c, which has sides Δv and v, are similar. This fact enables us to write a relationship between the lengths of the sides:

$$\frac{\Delta v}{v} = \frac{\Delta r}{r}$$

This equation can be solved for Δv and the expression so obtained substituted into $\overline{a} = \Delta v / \Delta t$ (Eq. 4.4) to give

$$\overline{a} = \frac{v \, \Delta r}{r \, \Delta t}$$

Now imagine that points Ⓐ and Ⓑ in Figure 4.16b are extremely close together. In this case $\Delta \mathbf{v}$ points toward the center of the circular path, and because the acceleration is in the direction of $\Delta \mathbf{v}$, it too points toward the center. Furthermore, as Ⓐ and Ⓑ approach each other, Δt approaches zero, and the ratio $\Delta r / \Delta t$ approaches the speed v. Hence, in the limit $\Delta t \to 0$, the magnitude of the acceleration is

$$a_r = \frac{v^2}{r}$$

Thus, we conclude that in uniform circular motion, the acceleration is directed toward the center of the circle and has a magnitude given by v^2/r, where v is the speed of the particle and r is the radius of the circle. You should be able to show that the dimensions of a_r are L/T^2. We shall return to the discussion of circular motion in Section 6.1.

4.5   TANGENTIAL AND RADIAL ACCELERATION

Now let us consider a particle moving along a curved path where the velocity changes both in direction and in magnitude, as shown in Figure 4.17. As is always the case, the velocity vector is tangent to the path, but now the direction of the ac-

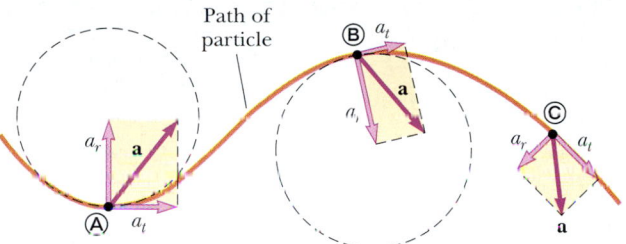

Figure 4.17   The motion of a particle along an arbitrary curved path lying in the xy plane. If the velocity vector **v** (always tangent to the path) changes in direction and magnitude, the component vectors of the acceleration **a** are a tangential component a_t and a radial component a_r.

celeration vector **a** changes from point to point. This vector can be resolved into two component vectors: a radial component vector $\mathbf{a}_r$ and a tangential component vector $\mathbf{a}_t$. Thus, **a** can be written as the vector sum of these component vectors:

$$\mathbf{a} = \mathbf{a}_r + \mathbf{a}_t \tag{4.16}$$

Total acceleration

The tangential acceleration causes the change in the speed of the particle. It is parallel to the instantaneous velocity, and its magnitude is

$$a_t = \frac{d|\mathbf{v}|}{dt} \tag{4.17}$$

Tangential acceleration

The radial acceleration arises from the change in direction of the velocity vector as described earlier and has an absolute magnitude given by

$$a_r = \frac{v^2}{r} \tag{4.18}$$

Radial acceleration

where r is the radius of curvature of the path at the point in question. Because $\mathbf{a}_r$ and $\mathbf{a}_t$ are mutually perpendicular component vectors of **a**, it follows that $a = \sqrt{a_r^2 + a_t^2}$. As in the case of uniform circular motion, $\mathbf{a}_r$ in nonuniform circular motion always points toward the center of curvature, as shown in Figure 4.17. Also, at a given speed, a_r is large when the radius of curvature is small (as at points Ⓐ and Ⓑ in Figure 4.17) and small when r is large (such as at point Ⓒ). The direction of $\mathbf{a}_t$ is either in the same direction as **v** (if v is increasing) or opposite **v** (if v is decreasing).

In uniform circular motion, where v is constant, $a_t = 0$ and the acceleration is always completely radial, as we described in Section 4.4. (*Note:* Eq. 4.18 is identical to Eq. 4.15.) In other words, uniform circular motion is a special case of motion along a curved path. Furthermore, if the direction of **v** does not change, then there is no radial acceleration and the motion is one-dimensional (in this case, $a_r = 0$, but a_t may not be zero).

Quick Quiz 4.3

(a) Draw a motion diagram showing velocity and acceleration vectors for an object moving with constant speed counterclockwise around a circle. Draw similar diagrams for an object moving counterclockwise around a circle but (b) slowing down at constant tangential acceleration and (c) speeding up at constant tangential acceleration.

It is convenient to write the acceleration of a particle moving in a circular path in terms of unit vectors. We do this by defining the unit vectors $\hat{\mathbf{r}}$ and $\hat{\boldsymbol{\theta}}$ shown in

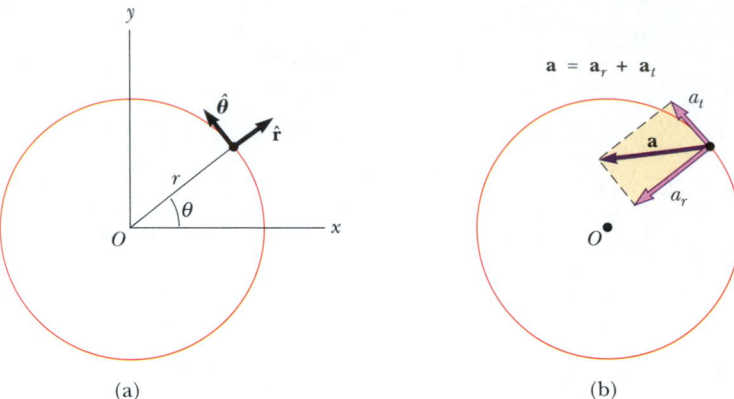

Figure 4.18   (a) Descriptions of the unit vectors $\hat{\mathbf{r}}$ and $\hat{\boldsymbol{\theta}}$. (b) The total acceleration **a** of a particle moving along a curved path (which at any instant is part of a circle of radius r) is the sum of radial and tangential components. The radial component is directed toward the center of curvature. If the tangential component of acceleration becomes zero, the particle follows uniform circular motion.

Figure 4.18a, where $\hat{\mathbf{r}}$ is a unit vector lying along the radius vector and directed radially outward from the center of the circle and $\hat{\boldsymbol{\theta}}$ is a unit vector tangent to the circle. The direction of $\hat{\boldsymbol{\theta}}$ is in the direction of increasing θ, where θ is measured counterclockwise from the positive x axis. Note that both $\hat{\mathbf{r}}$ and $\hat{\boldsymbol{\theta}}$ "move along with the particle" and so vary in time. Using this notation, we can express the total acceleration as

$$\mathbf{a} = \mathbf{a}_t + \mathbf{a}_r = \frac{d|\mathbf{v}|}{dt}\,\hat{\boldsymbol{\theta}} - \frac{v^2}{r}\,\hat{\mathbf{r}} \qquad \textbf{(4.19)}$$

These vectors are described in Figure 4.18b. The negative sign on the v^2/r term in Equation 4.19 indicates that the radial acceleration is always directed radially inward, *opposite* $\hat{\mathbf{r}}$.

Quick Quiz 4.4

Based on your experience, draw a motion diagram showing the position, velocity, and acceleration vectors for a pendulum that, from an initial position 45° to the right of a central vertical line, swings in an arc that carries it to a final position 45° to the left of the central vertical line. The arc is part of a circle, and you should use the center of this circle as the origin for the position vectors.

EXAMPLE 4.8   The Swinging Ball

A ball tied to the end of a string 0.50 m in length swings in a vertical circle under the influence of gravity, as shown in Figure 4.19. When the string makes an angle $\theta = 20°$ with the vertical, the ball has a speed of 1.5 m/s. (a) Find the magnitude of the radial component of acceleration at this instant.

Solution   The diagram from the answer to Quick Quiz 4.4 (p. 109) applies to this situation, and so we have a good idea of how the acceleration vector varies during the motion. Fig-

ure 4.19 lets us take a closer look at the situation. The radial acceleration is given by Equation 4.18. With $v = 1.5$ m/s and $r = 0.50$ m, we find that

$$a_r = \frac{v^2}{r} = \frac{(1.5 \text{ m/s})^2}{0.50 \text{ m}} = \boxed{4.5 \text{ m/s}^2}$$

(b) What is the magnitude of the tangential acceleration when $\theta = 20°$?

Solution   When the ball is at an angle θ to the vertical, it has a tangential acceleration of magnitude $g \sin \theta$ (the component of **g** tangent to the circle). Therefore, at $\theta = 20°$,

$$a_t = g \sin 20° = \boxed{3.4 \text{ m/s}^2}.$$

(c) Find the magnitude and direction of the total acceleration **a** at $\theta = 20°$.

Solution   Because $\mathbf{a} = \mathbf{a}_r + \mathbf{a}_t$, the magnitude of **a** at $\theta = 20°$ is

$$a = \sqrt{a_r^{\,2} + a_t^{\,2}} = \sqrt{(4.5)^2 + (3.4)^2} \text{ m/s}^2 = \boxed{5.6 \text{ m/s}^2}$$

If ϕ is the angle between **a** and the string, then

$$\phi = \tan^{-1} \frac{a_t}{a_r} = \tan^{-1}\left(\frac{3.4 \text{ m/s}^2}{4.5 \text{ m/s}^2}\right) = \boxed{37°}$$

Note that **a**, $\mathbf{a}_t$, and $\mathbf{a}_r$ all change in direction *and* magnitude as the ball swings through the circle. When the ball is at its lowest elevation ($\theta = 0$), $a_t = 0$ because there is no tangential component of **g** at this angle; also, a_r is a *maximum* because v is a maximum. If the ball has enough speed to reach its highest position ($\theta = 180°$), then a_t is again zero but a_r is a minimum because v is now a minimum. Finally, in the two horizontal positions ($\theta = 90°$ and $270°$), $|\mathbf{a}_t| = g$ and a_r has a value between its minimum and maximum values.

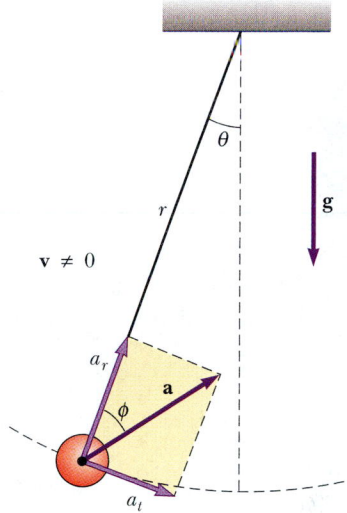

Figure 4.19   Motion of a ball suspended by a string of length r. The ball swings with nonuniform circular motion in a vertical plane, and its acceleration **a** has a radial component a_r and a tangential component a_t.

4.6 RELATIVE VELOCITY AND RELATIVE ACCELERATION

In this section, we describe how observations made by different observers in different frames of reference are related to each other. We find that observers in different frames of reference may measure different displacements, velocities, and accelerations for a given particle. That is, two observers moving relative to each other generally do not agree on the outcome of a measurement.

For example, suppose two cars are moving in the same direction with speeds of 50 mi/h and 60 mi/h. To a passenger in the slower car, the speed of the faster car is 10 mi/h. Of course, a stationary observer will measure the speed of the faster car to be 60 mi/h, not 10 mi/h. Which observer is correct? They both are! This simple example demonstrates that the velocity of an object depends on the frame of reference in which it is measured.

Suppose a person riding on a skateboard (observer A) throws a ball in such a way that it appears in this person's frame of reference to move first straight upward and then straight downward along the same vertical line, as shown in Figure 4.20a. A stationary observer B sees the path of the ball as a parabola, as illustrated in Figure 4.20b. Relative to observer B, the ball has a vertical component of velocity (resulting from the initial upward velocity and the downward acceleration of gravity) *and* a horizontal component.

Another example of this concept that of is a package dropped from an airplane flying with a constant velocity; this is the situation we studied in Example 4.6. An observer on the airplane sees the motion of the package as a straight line toward the Earth. The stranded explorer on the ground, however, sees the trajectory of the package as a parabola. If, once it drops the package, the airplane con-

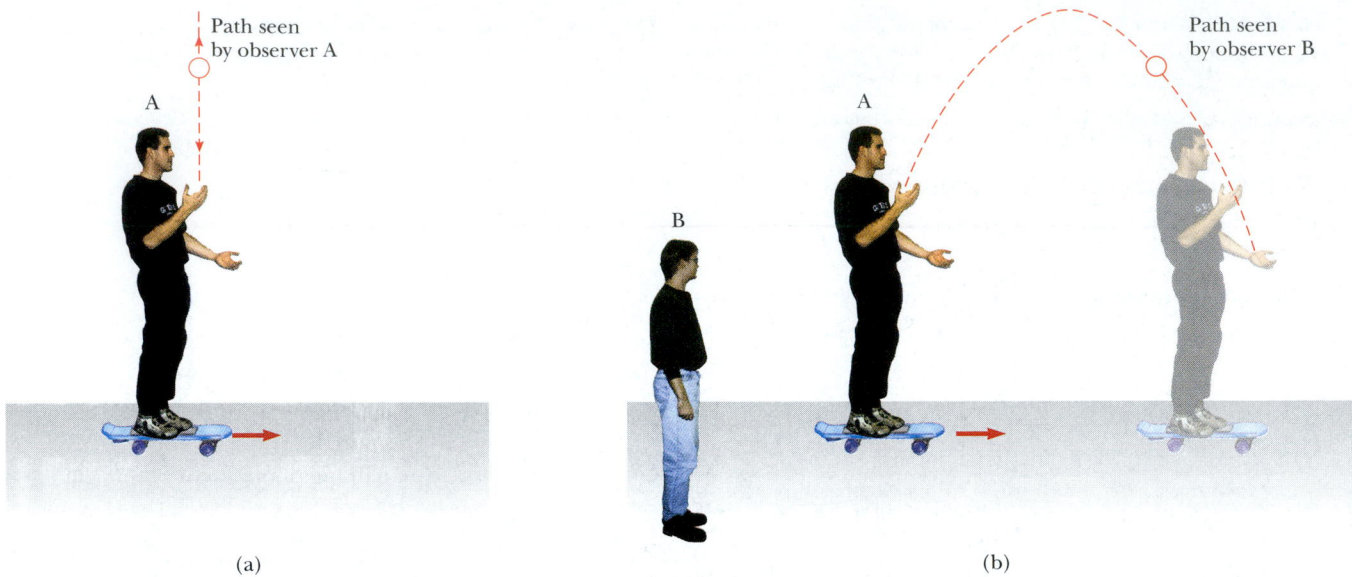

Figure 4.20 (a) Observer A on a moving vehicle throws a ball upward and sees it rise and fall in a straight-line path. (b) Stationary observer B sees a parabolic path for the same ball.

tinues to move horizontally with the same velocity, then the package hits the ground directly beneath the airplane (if we assume that air resistance is neglected)!

In a more general situation, consider a particle located at point Ⓐ in Figure 4.21. Imagine that the motion of this particle is being described by two observers, one in reference frame *S*, fixed relative to the Earth, and another in reference frame *S′*, moving to the right relative to *S* (and therefore relative to the Earth) with a constant velocity $\mathbf{v}_0$. (Relative to an observer in *S′*, *S* moves to the left with a velocity $-\mathbf{v}_0$.) Where an observer stands in a reference frame is irrelevant in this discussion, but for purposes of this discussion let us place each observer at her or his respective origin.

We label the position of the particle relative to the *S* frame with the position vector $\mathbf{r}$ and that relative to the *S′* frame with the position vector $\mathbf{r}′$, both after some time *t*. The vectors $\mathbf{r}$ and $\mathbf{r}′$ are related to each other through the expression $\mathbf{r} = \mathbf{r}′ + \mathbf{v}_0 t$, or

$$\mathbf{r}′ = \mathbf{r} - \mathbf{v}_0 t \qquad (4.20)$$

Galilean coordinate transformation

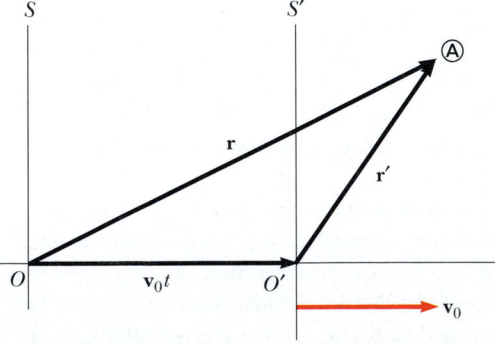

Figure 4.21 A particle located at Ⓐ is described by two observers, one in the fixed frame of reference *S*, and the other in the frame *S′*, which moves to the right with a constant velocity $\mathbf{v}_0$. The vector $\mathbf{r}$ is the particle's position vector relative to *S*, and $\mathbf{r}′$ is its position vector relative to *S′*.

The woman standing on the beltway sees the walking man pass by at a slower speed than the woman standing on the stationary floor does.

That is, after a time t, the S' frame is displaced to the right of the S frame by an amount $\mathbf{v}_0 t$.

If we differentiate Equation 4.20 with respect to time and note that $\mathbf{v}_0$ is constant, we obtain

$$\frac{d\mathbf{r}'}{dt} = \frac{d\mathbf{r}}{dt} - \mathbf{v}_0$$

$$\mathbf{v}' = \mathbf{v} - \mathbf{v}_0 \qquad \textbf{(4.21)}$$

◄ Galilean velocity transformation

where $\mathbf{v}'$ is the velocity of the particle observed in the S' frame and $\mathbf{v}$ is its velocity observed in the S frame. Equations 4.20 and 4.21 are known as **Galilean transformation equations.** They relate the coordinates and velocity of a particle as measured in a frame fixed relative to the Earth to those measured in a frame moving with uniform motion relative to the Earth.

Although observers in two frames measure different velocities for the particle, they measure the *same acceleration* when $\mathbf{v}_0$ is constant. We can verify this by taking the time derivative of Equation 4.21:

$$\frac{d\mathbf{v}'}{dt} = \frac{d\mathbf{v}}{dt} - \frac{d\mathbf{v}_0}{dt}$$

Because $\mathbf{v}_0$ is constant, $d\mathbf{v}_0/dt = 0$. Therefore, we conclude that $\mathbf{a}' = \mathbf{a}$ because $\mathbf{a}' = d\mathbf{v}'/dt$ and $\mathbf{a} = d\mathbf{v}/dt$. That is, **the acceleration of the particle measured by an observer in the Earth's frame of reference is the same as that measured by any other observer moving with constant velocity relative to the Earth's frame.**

Quick Quiz 4.5

A passenger in a car traveling at 60 mi/h pours a cup of coffee for the tired driver. Describe the path of the coffee as it moves from a Thermos bottle into a cup as seen by (a) the passenger and (b) someone standing beside the road and looking in the window of the car as it drives past. (c) What happens if the car accelerates while the coffee is being poured?

EXAMPLE 4.9   A Boat Crossing a River

A boat heading due north crosses a wide river with a speed of 10.0 km/h relative to the water. The water in the river has a uniform speed of 5.00 km/h due east relative to the Earth. Determine the velocity of the boat relative to an observer standing on either bank.

Solution   We know $\mathbf{v}_{br}$, the velocity of the *boat* relative to the *river*, and $\mathbf{v}_{rE}$, the velocity of the *river* relative to the *Earth*. What we need to find is $\mathbf{v}_{bE}$, the velocity of the *boat* relative to the *Earth*. The relationship between these three quantities is

$$\mathbf{v}_{bE} = \mathbf{v}_{br} + \mathbf{v}_{rE}$$

The terms in the equation must be manipulated as vector quantities; the vectors are shown in Figure 4.22. The quantity $\mathbf{v}_{br}$ is due north, $\mathbf{v}_{rE}$ is due east, and the vector sum of the two, $\mathbf{v}_{bE}$, is at an angle θ, as defined in Figure 4.22. Thus, we can find the speed v_{bE} of the boat relative to the Earth by using the Pythagorean theorem:

$$v_{bE} = \sqrt{v_{br}^2 + v_{rE}^2} = \sqrt{(10.0)^2 + (5.00)^2} \text{ km/h}$$

$$= \boxed{11.2 \text{ km/h}}$$

The direction of $\mathbf{v}_{bE}$ is

$$\theta = \tan^{-1}\left(\frac{v_{rE}}{v_{br}}\right) = \tan^{-1}\left(\frac{5.00}{10.0}\right) = 26.6°$$

The boat is moving at a speed of 11.2 km/h in the direction 26.6° east of north relative to the Earth.

Exercise   If the width of the river is 3.0 km, find the time it takes the boat to cross it.

Answer   18 min.

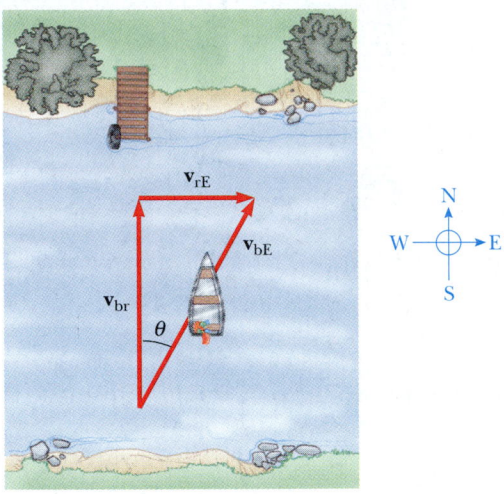

Figure 4.22

EXAMPLE 4.10   Which Way Should We Head?

If the boat of the preceding example travels with the same speed of 10.0 km/h relative to the river and is to travel due north, as shown in Figure 4.23, what should its heading be?

Solution   As in the previous example, we know $\mathbf{v}_{rE}$ and the magnitude of the vector $\mathbf{v}_{br}$, and we want $\mathbf{v}_{bE}$ to be directed across the river. Figure 4.23 shows that the boat must head upstream in order to travel directly northward across the river. Note the difference between the triangle in Figure 4.22 and the one in Figure 4.23—specifically, that the hypotenuse in Figure 4.23 is no longer $\mathbf{v}_{bE}$. Therefore, when we use the Pythagorean theorem to find $\mathbf{v}_{bE}$ this time, we obtain

$$v_{bE} = \sqrt{v_{br}^2 - v_{rE}^2} = \sqrt{(10.0)^2 - (5.00)^2} \text{ km/h} = 8.66 \text{ km/h}$$

Now that we know the magnitude of $\mathbf{v}_{bE}$, we can find the direction in which the boat is heading:

$$\theta = \tan^{-1}\left(\frac{v_{rE}}{v_{bE}}\right) = \tan^{-1}\left(\frac{5.00}{8.66}\right) = \boxed{30.0°}$$

The boat must steer a course 30.0° west of north.

Exercise   If the width of the river is 3.0 km, find the time it takes the boat to cross it.

Answer   21 min.

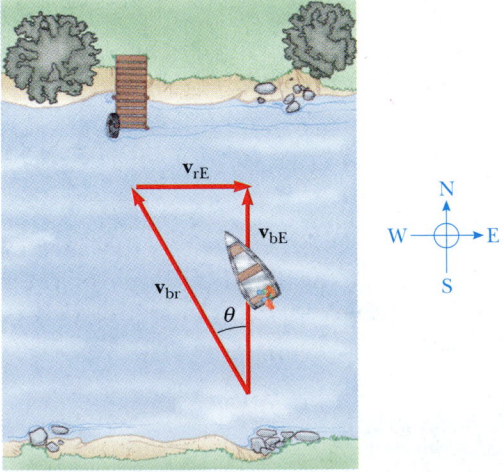

Figure 4.23

SUMMARY

If a particle moves with *constant* acceleration $\mathbf{a}$ and has velocity $\mathbf{v}_i$ and position $\mathbf{r}_i$ at $t = 0$, its velocity and position vectors at some later time t are

$$\mathbf{v}_f = \mathbf{v}_i + \mathbf{a}t \tag{4.8}$$

$$\mathbf{r}_f = \mathbf{r}_i + \mathbf{v}_i t + \tfrac{1}{2}\mathbf{a}t^2 \tag{4.9}$$

For two-dimensional motion in the xy plane under constant acceleration, each of these vector expressions is equivalent to two component expressions—one for the motion in the x direction and one for the motion in the y direction. You should be able to break the two-dimensional motion of any object into these two components.

Projectile motion is one type of two-dimensional motion under constant acceleration, where $a_x = 0$ and $a_y = -g$. It is useful to think of projectile motion as the superposition of two motions: (1) constant-velocity motion in the x direction and (2) free-fall motion in the vertical direction subject to a constant downward acceleration of magnitude $g = 9.80 \text{ m/s}^2$. You should be able to analyze the motion in terms of separate horizontal and vertical components of velocity, as shown in Figure 4.24.

A particle moving in a circle of radius r with constant speed v is in **uniform circular motion**. It undergoes a centripetal (or radial) acceleration $\mathbf{a}_r$ because the direction of $\mathbf{v}$ changes in time. The magnitude of $\mathbf{a}_r$ is

$$a_r = \frac{v^2}{r} \tag{4.18}$$

and its direction is always toward the center of the circle.

If a particle moves along a curved path in such a way that both the magnitude and the direction of $\mathbf{v}$ change in time, then the particle has an acceleration vector that can be described by two component vectors: (1) a radial component vector $\mathbf{a}_r$ that causes the change in direction of $\mathbf{v}$ and (2) a tangential component vector $\mathbf{a}_t$ that causes the change in magnitude of $\mathbf{v}$. The magnitude of $\mathbf{a}_r$ is v^2/r, and the magnitude of $\mathbf{a}_t$ is $d|\mathbf{v}|/dt$. You should be able to sketch motion diagrams for an object following a curved path and show how the velocity and acceleration vectors change as the object's motion varies.

The velocity $\mathbf{v}$ of a particle measured in a fixed frame of reference S can be related to the velocity $\mathbf{v}'$ of the same particle measured in a moving frame of reference S' by

$$\mathbf{v}' = \mathbf{v} - \mathbf{v}_0 \tag{4.21}$$

where $\mathbf{v}_0$ is the velocity of S' relative to S. You should be able to translate back and forth between different frames of reference.

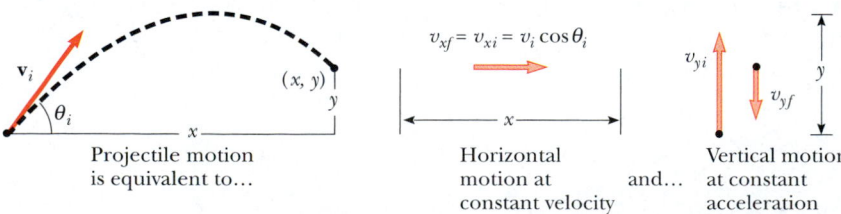

Figure 4.24   Analyzing projectile motion in terms of horizontal and vertical components.

QUESTIONS

1. Can an object accelerate if its speed is constant? Can an object accelerate if its velocity is constant?

2. If the average velocity of a particle is zero in some time interval, what can you say about the displacement of the particle for that interval?

3. If you know the position vectors of a particle at two points along its path and also know the time it took to get from one point to the other, can you determine the particle's instantaneous velocity? Its average velocity? Explain.

4. Describe a situation in which the velocity of a particle is always perpendicular to the position vector.

5. Explain whether or not the following particles have an acceleration: (a) a particle moving in a straight line with constant speed and (b) a particle moving around a curve with constant speed.

6. Correct the following statement: "The racing car rounds the turn at a constant velocity of 90 mi/h."

7. Determine which of the following moving objects have an approximately parabolic trajectory: (a) a ball thrown in an arbitrary direction, (b) a jet airplane, (c) a rocket leaving the launching pad, (d) a rocket whose engines fail a few minutes after launch, (e) a tossed stone moving to the bottom of a pond.

8. A rock is dropped at the same instant that a ball at the same initial elevation is thrown horizontally. Which will have the greater speed when it reaches ground level?

9. A spacecraft drifts through space at a constant velocity. Suddenly, a gas leak in the side of the spacecraft causes a constant acceleration of the spacecraft in a direction perpendicular to the initial velocity. The orientation of the spacecraft does not change, and so the acceleration remains perpendicular to the original direction of the velocity. What is the shape of the path followed by the spacecraft in this situation?

10. A ball is projected horizontally from the top of a building. One second later another ball is projected horizontally from the same point with the same velocity. At what point in the motion will the balls be closest to each other? Will the first ball always be traveling faster than the second ball? How much time passes between the moment the first ball hits the ground and the moment the second one hits the ground? Can the horizontal projection velocity of the second ball be changed so that the balls arrive at the ground at the same time?

11. A student argues that as a satellite orbits the Earth in a circular path, the satellite moves with a constant velocity and therefore has no acceleration. The professor claims that the student is wrong because the satellite must have a centripetal acceleration as it moves in its circular orbit. What is wrong with the student's argument?

12. What is the fundamental difference between the unit vectors $\hat{\mathbf{r}}$ and $\hat{\boldsymbol{\theta}}$ and the unit vectors $\mathbf{i}$ and $\mathbf{j}$?

13. At the end of its arc, the velocity of a pendulum is zero. Is its acceleration also zero at this point?

14. If a rock is dropped from the top of a sailboat's mast, will it hit the deck at the same point regardless of whether the boat is at rest or in motion at constant velocity?

15. A stone is thrown upward from the top of a building. Does the stone's displacement depend on the location of the origin of the coordinate system? Does the stone's velocity depend on the location of the origin?

16. Is it possible for a vehicle to travel around a curve without accelerating? Explain.

17. A baseball is thrown with an initial velocity of $(10\mathbf{i} + 15\mathbf{j})$ m/s. When it reaches the top of its trajectory, what are (a) its velocity and (b) its acceleration? Neglect the effect of air resistance.

18. An object moves in a circular path with constant speed v. (a) Is the velocity of the object constant? (b) Is its acceleration constant? Explain.

19. A projectile is fired at some angle to the horizontal with some initial speed v_i, and air resistance is neglected. Is the projectile a freely falling body? What is its acceleration in the vertical direction? What is its acceleration in the horizontal direction?

20. A projectile is fired at an angle of $30°$ from the horizontal with some initial speed. Firing at what other projectile angle results in the same range if the initial speed is the same in both cases? Neglect air resistance.

21. A projectile is fired on the Earth with some initial velocity. Another projectile is fired on the Moon with the same initial velocity. If air resistance is neglected, which projectile has the greater range? Which reaches the greater altitude? (Note that the free-fall acceleration on the Moon is about 1.6 m/s².)

22. As a projectile moves through its parabolic trajectory, which of these quantities, if any, remain constant: (a) speed, (b) acceleration, (c) horizontal component of velocity, (d) vertical component of velocity?

23. A passenger on a train that is moving with constant velocity drops a spoon. What is the acceleration of the spoon relative to (a) the train and (b) the Earth?

PROBLEMS

1, 2, 3 = straightforward, intermediate, challenging ☐ = full solution available in the *Student Solutions Manual and Study Guide*
WEB = solution posted at **http://www.saunderscollege.com/physics/** 💻 – Computer useful in solving problem 📕 = Interactive Physics
☐ = paired numerical/symbolic problems

Section 4.1 The Displacement, Velocity, and Acceleration Vectors

WEB **1.** A motorist drives south at 20.0 m/s for 3.00 min, then turns west and travels at 25.0 m/s for 2.00 min, and finally travels northwest at 30.0 m/s for 1.00 min. For this 6.00-min trip, find (a) the total vector displacement, (b) the average speed, and (c) the average velocity. Use a coordinate system in which east is the positive x axis.

2. Suppose that the position vector for a particle is given as $\mathbf{r} = x\mathbf{i} + y\mathbf{j}$, with $x = at + b$ and $y = ct^2 + d$, where $a = 1.00$ m/s, $b = 1.00$ m, $c = 0.125$ m/s^2, and $d = 1.00$ m. (a) Calculate the average velocity during the time interval from $t = 2.00$ s to $t = 4.00$ s. (b) Determine the velocity and the speed at $t = 2.00$ s.

3. A golf ball is hit off a tee at the edge of a cliff. Its x and y coordinates versus time are given by the following expressions:

$$x = (18.0 \text{ m/s})t$$

and

$$y = (4.00 \text{ m/s})t - (4.90 \text{ m/s}^2)t^2$$

(a) Write a vector expression for the ball's position as a function of time, using the unit vectors $\mathbf{i}$ and $\mathbf{j}$. By taking derivatives of your results, write expressions for (b) the velocity vector as a function of time and (c) the acceleration vector as a function of time. Now use unit vector notation to write expressions for (d) the position, (e) the velocity, and (f) the acceleration of the ball, all at $t = 3.00$ s.

4. The coordinates of an object moving in the xy plane vary with time according to the equations

$$x = -(5.00 \text{ m})\sin \omega t$$

and

$$y = (4.00 \text{ m}) - (5.00 \text{ m})\cos \omega t$$

where t is in seconds and ω has units of seconds^{-1}. (a) Determine the components of velocity and components of acceleration at $t = 0$. (b) Write expressions for the position vector, the velocity vector, and the acceleration vector at any time $t > 0$. (c) Describe the path of the object on an xy graph.

Section 4.2 Two-Dimensional Motion with Constant Acceleration

5. At $t = 0$, a particle moving in the xy plane with constant acceleration has a velocity of $\mathbf{v}_i = (3.00\mathbf{i} - 2.00\mathbf{j})$ m/s when it is at the origin. At $t = 3.00$ s, the particle's velocity is $\mathbf{v} = (9.00\mathbf{i} + 7.00\mathbf{j})$ m/s. Find (a) the acceleration of the particle and (b) its coordinates at any time t.

6. The vector position of a particle varies in time according to the expression $\mathbf{r} = (3.00\mathbf{i} - 6.00t^2\mathbf{j})$ m. (a) Find expressions for the velocity and acceleration as functions of time. (b) Determine the particle's position and velocity at $t = 1.00$ s.

7. A fish swimming in a horizontal plane has velocity $\mathbf{v}_i = (4.00\mathbf{i} + 1.00\mathbf{j})$ m/s at a point in the ocean whose displacement from a certain rock is $\mathbf{r}_i = (10.0\mathbf{i} - 4.00\mathbf{j})$ m. After the fish swims with constant acceleration for 20.0 s, its velocity is $\mathbf{v} = (20.0\mathbf{i} - 5.00\mathbf{j})$ m/s. (a) What are the components of the acceleration? (b) What is the direction of the acceleration with respect to the unit vector $\mathbf{i}$? (c) Where is the fish at $t = 25.0$ s if it maintains its original acceleration and in what direction is it moving?

8. A particle initially located at the origin has an acceleration of $\mathbf{a} = 3.00\mathbf{j}$ m/s^2 and an initial velocity of $\mathbf{v}_i = 5.00\mathbf{i}$ m/s. Find (a) the vector position and velocity at any time t and (b) the coordinates and speed of the particle at $t = 2.00$ s.

Section 4.3 Projectile Motion

(Neglect air resistance in all problems and take $g = 9.80$ m/s^2.)

WEB **9.** In a local bar, a customer slides an empty beer mug down the counter for a refill. The bartender is momentarily distracted and does not see the mug, which slides off the counter and strikes the floor 1.40 m from the base of the counter. If the height of the counter is 0.860 m, (a) with what velocity did the mug leave the counter and (b) what was the direction of the mug's velocity just before it hit the floor?

10. In a local bar, a customer slides an empty beer mug down the counter for a refill. The bartender is momentarily distracted and does not see the mug, which slides off the counter and strikes the floor at distance d from the base of the counter. If the height of the counter is h, (a) with what velocity did the mug leave the counter and (b) what was the direction of the mug's velocity just before it hit the floor?

11. One strategy in a snowball fight is to throw a first snowball at a high angle over level ground. While your opponent is watching the first one, you throw a second one at a low angle and timed to arrive at your opponent before or at the same time as the first one. Assume both snowballs are thrown with a speed of 25.0 m/s. The first one is thrown at an angle of 70.0° with respect to the horizontal. (a) At what angle should the second (low-angle) snowball be thrown if it is to land at the same point as the first? (b) How many seconds later should the second snowball be thrown if it is to land at the same time as the first?

12. A tennis player standing 12.6 m from the net hits the ball at 3.00° above the horizontal. To clear the net, the ball must rise at least 0.330 m. If the ball just clears the net at the apex of its trajectory, how fast was the ball moving when it left the racket?

13. An artillery shell is fired with an initial velocity of 300 m/s at 55.0° above the horizontal. It explodes on a mountainside 42.0 s after firing. What are the x and y coordinates of the shell where it explodes, relative to its firing point?

14. An astronaut on a strange planet finds that she can jump a maximum horizontal distance of 15.0 m if her initial speed is 3.00 m/s. What is the free-fall acceleration on the planet?

15. A projectile is fired in such a way that its horizontal range is equal to three times its maximum height. What is the angle of projection? Give your answer to three significant figures.

16. A ball is tossed from an upper-story window of a building. The ball is given an initial velocity of 8.00 m/s at an angle of 20.0° below the horizontal. It strikes the ground 3.00 s later. (a) How far horizontally from the base of the building does the ball strike the ground? (b) Find the height from which the ball was thrown. (c) How long does it take the ball to reach a point 10.0 m below the level of launching?

17. A cannon with a muzzle speed of 1 000 m/s is used to start an avalanche on a mountain slope. The target is 2 000 m from the cannon horizontally and 800 m above the cannon. At what angle, above the horizontal, should the cannon be fired?

18. Consider a projectile that is launched from the origin of an xy coordinate system with speed v_i at initial angle θ_i above the horizontal. Note that at the apex of its trajectory the projectile is moving horizontally, so that the slope of its path is zero. Use the expression for the trajectory given in Equation 4.12 to find the x coordinate that corresponds to the maximum height. Use this x coordinate and the symmetry of the trajectory to determine the horizontal range of the projectile.

 19. A placekicker must kick a football from a point 36.0 m (about 40 yards) from the goal, and half the crowd hopes the ball will clear the crossbar, which is 3.05 m high. When kicked, the ball leaves the ground with a speed of 20.0 m/s at an angle of 53.0° to the horizontal. (a) By how much does the ball clear or fall short of clearing the crossbar? (b) Does the ball approach the crossbar while still rising or while falling?

20. A firefighter 50.0 m away from a burning building directs a stream of water from a fire hose at an angle of 30.0° above the horizontal, as in Figure P4.20. If the speed of the stream is 40.0 m/s, at what height will the water strike the building?

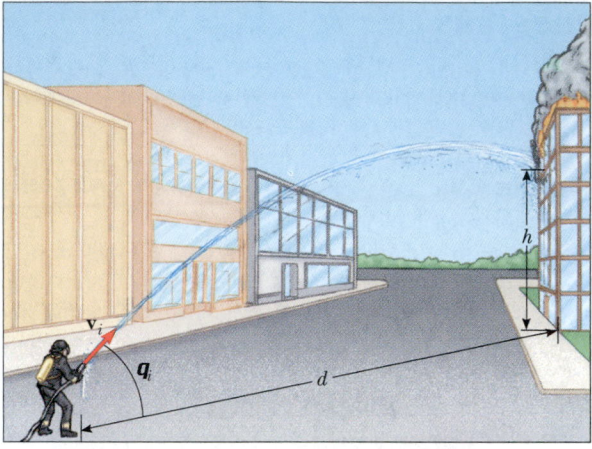

Figure P4.20   Problems 20 and 21. *(Frederick McKinney/FPG International)*

21. A firefighter a distance d from a burning building directs a stream of water from a fire hose at angle θ_i above the horizontal as in Figure P4.20. If the initial speed of the stream is v_i, at what height h does the water strike the building?

22. A soccer player kicks a rock horizontally off a cliff 40.0 m high into a pool of water. If the player hears the sound of the splash 3.00 s later, what was the initial speed given to the rock? Assume the speed of sound in air to be 343 m/s.

23. A basketball star covers 2.80 m horizontally in a jump to dunk the ball (Fig. P4.23). His motion through space can be modeled as that of a particle at a point called his center of mass (which we shall define in Chapter 9). His center of mass is at elevation 1.02 m when he leaves the floor. It reaches a maximum height of 1.85 m above the floor and is at elevation 0.900 m when he touches down again. Determine (a) his time of flight (his "hang time"), (b) his horizontal and (c) vertical velocity components at the instant of takeoff, and (d) his takeoff angle. (e) For comparison, determine the hang time of a whitetail deer making a jump with center-of-mass elevations $y_i = 1.20$ m, $y_{max} = 2.50$ m, $y_f = 0.700$ m.

Figure P4.23 (Top, Ron Chapple/FPG International; bottom, Bill Lea/Dembinsky Photo Associates)

Section 4.4   Uniform Circular Motion

24. The orbit of the Moon about the Earth is approximately circular, with a mean radius of 3.84×10^8 m. It takes 27.3 days for the Moon to complete one revolution about the Earth. Find (a) the mean orbital speed of the Moon and (b) its centripetal acceleration.

WEB **25.** The athlete shown in Figure P4.25 rotates a 1.00-kg discus along a circular path of radius 1.06 m. The maximum speed of the discus is 20.0 m/s. Determine the magnitude of the maximum radial acceleration of the discus.

Figure P4.25 (Sam Sargent/Liaison International)

26. From information on the endsheets of this book, compute, for a point located on the surface of the Earth at the equator, the radial acceleration due to the rotation of the Earth about its axis.

27. A tire 0.500 m in radius rotates at a constant rate of 200 rev/min. Find the speed and acceleration of a small stone lodged in the tread of the tire (on its outer edge). (*Hint:* In one revolution, the stone travels a distance equal to the circumference of its path, $2\pi r$.)

28. During liftoff, Space Shuttle astronauts typically feel accelerations up to 1.4g, where $g = 9.80$ m/s². In their training, astronauts ride in a device where they experience such an acceleration as a centripetal acceleration. Specifically, the astronaut is fastened securely at the end of a mechanical arm that then turns at constant speed in a horizontal circle. Determine the rotation rate, in revolutions per second, required to give an astronaut a centripetal acceleration of 1.40g while the astronaut moves in a circle of radius 10.0 m.

29. Young David who slew Goliath experimented with slings before tackling the giant. He found that he could revolve a sling of length 0.600 m at the rate of 8.00 rev/s. If he increased the length to 0.900 m, he could revolve the sling only 6.00 times per second. (a) Which rate of rotation gives the greater speed for the stone at the end of the sling? (b) What is the centripetal acceleration of the stone at 8.00 rev/s? (c) What is the centripetal acceleration at 6.00 rev/s?

30. The astronaut orbiting the Earth in Figure P4.30 is preparing to dock with a Westar VI satellite. The satellite is in a circular orbit 600 km above the Earth's surface, where the free-fall acceleration is 8.21 m/s². The radius of the Earth is 6 400 km. Determine the speed of the satellite and the time required to complete one orbit around the Earth.

Figure P4.30   (Courtesy of NASA)

Section 4.5   Tangential and Radial Acceleration

31. A train slows down as it rounds a sharp horizontal curve, slowing from 90.0 km/h to 50.0 km/h in the 15.0 s that it takes to round the curve. The radius of the curve is 150 m. Compute the acceleration at the moment the train speed reaches 50.0 km/h. Assume that the train slows down at a uniform rate during the 15.0-s interval.

32. An automobile whose speed is increasing at a rate of 0.600 m/s² travels along a circular road of radius 20.0 m. When the instantaneous speed of the automobile is 4.00 m/s, find (a) the tangential acceleration component, (b) the radial acceleration component, and (c) the magnitude and direction of the total acceleration.

33. Figure P4.33 shows the total acceleration and velocity of a particle moving clockwise in a circle of radius 2.50 m

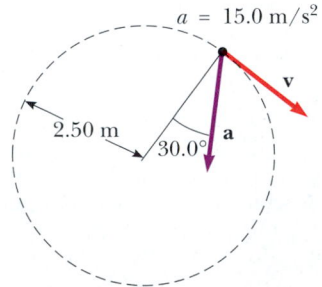

Figure P4.33

at a given instant of time. At this instant, find (a) the radial acceleration, (b) the speed of the particle, and (c) its tangential acceleration.

34. A student attaches a ball to the end of a string 0.600 m in length and then swings the ball in a vertical circle. The speed of the ball is 4.30 m/s at its highest point and 6.50 m/s at its lowest point. Find the acceleration of the ball when the string is vertical and the ball is at (a) its highest point and (b) its lowest point.

35. A ball swings in a vertical circle at the end of a rope 1.50 m long. When the ball is 36.9° past the lowest point and on its way up, its total acceleration is $(-22.5\mathbf{i} + 20.2\mathbf{j})$ m/s². At that instant, (a) sketch a vector diagram showing the components of this acceleration, (b) determine the magnitude of its radial acceleration, and (c) determine the speed and velocity of the ball.

Section 4.6   Relative Velocity and Relative Acceleration

36. Heather in her Corvette accelerates at the rate of $(3.00\mathbf{i} - 2.00\mathbf{j})$ m/s², while Jill in her Jaguar accelerates at $(1.00\mathbf{i} + 3.00\mathbf{j})$ m/s². They both start from rest at the origin of an xy coordinate system. After 5.00 s, (a) what is Heather's speed with respect to Jill, (b) how far apart are they, and (c) what is Heather's acceleration relative to Jill?

37. A river has a steady speed of 0.500 m/s. A student swims upstream a distance of 1.00 km and swims back to the starting point. If the student can swim at a speed of 1.20 m/s in still water, how long does the trip take? Compare this with the time the trip would take if the water were still.

38. How long does it take an automobile traveling in the left lane at 60.0 km/h to pull alongside a car traveling in the right lane at 40.0 km/h if the cars' front bumpers are initially 100 m apart?

39. The pilot of an airplane notes that the compass indicates a heading due west. The airplane's speed relative to the air is 150 km/h. If there is a wind of 30.0 km/h toward the north, find the velocity of the airplane relative to the ground.

40. Two swimmers, Alan and Beth, start at the same point in a stream that flows with a speed v. Both move at the same speed c (c > v) relative to the stream. Alan swims downstream a distance L and then upstream the same distance. Beth swims such that her motion relative to the ground is perpendicular to the banks of the stream. She swims a distance L in this direction and then back. The result of the motions of Alan and Beth is that they both return to the starting point. Which swimmer returns first? (*Note:* First guess at the answer.)

41. A child in danger of drowning in a river is being carried downstream by a current that has a speed of 2.50 km/h. The child is 0.600 km from shore and 0.800 km upstream of a boat landing when a rescue boat sets out. (a) If the boat proceeds at its maximum speed of 20.0 km/h relative to the water, what heading relative to the shore should the pilot take? (b) What angle does

the boat velocity make with the shore? (c) How long does it take the boat to reach the child?

42. A bolt drops from the ceiling of a train car that is accelerating northward at a rate of 2.50 m/s^2. What is the acceleration of the bolt relative to (a) the train car and (b) the Earth?

43. A science student is riding on a flatcar of a train traveling along a straight horizontal track at a constant speed of 10.0 m/s. The student throws a ball into the air along a path that he judges to make an initial angle of 60.0° with the horizontal and to be in line with the track. The student's professor, who is standing on the ground nearby, observes the ball to rise vertically. How high does she see the ball rise?

ADDITIONAL PROBLEMS

44. A ball is thrown with an initial speed v_i at an angle θ_i with the horizontal. The horizontal range of the ball is R, and the ball reaches a maximum height $R/6$. In terms of R and g, find (a) the time the ball is in motion, (b) the ball's speed at the peak of its path, (c) the initial vertical component of its velocity, (d) its initial speed, and (e) the angle θ_i. (f) Suppose the ball is thrown at the same initial speed found in part (d) but at the angle appropriate for reaching the maximum height. Find this height. (g) Suppose the ball is thrown at the same initial speed but at the angle necessary for maximum range. Find this range.

45. As some molten metal splashes, one droplet flies off to the east with initial speed v_i at angle θ_i above the horizontal, and another droplet flies off to the west with the same speed at the same angle above the horizontal, as in Figure P4.45. In terms of v_i and θ_i, find the distance between the droplets as a function of time.

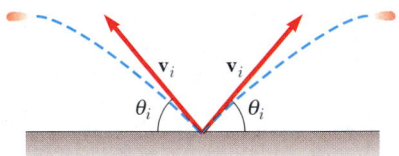

Figure P4.45

46. A ball on the end of a string is whirled around in a horizontal circle of radius 0.300 m. The plane of the circle is 1.20 m above the ground. The string breaks and the ball lands 2.00 m (horizontally) away from the point on the ground directly beneath the ball's location when the string breaks. Find the radial acceleration of the ball during its circular motion.

47. A projectile is fired up an incline (incline angle ϕ) with an initial speed v_i at an angle θ_i with respect to the horizontal ($\theta_i > \phi$), as shown in Figure P4.47. (a) Show that the projectile travels a distance d up the incline, where

$$d = \frac{2v_i^2 \cos \theta_i \sin(\theta_i - \phi)}{g \cos^2 \phi}$$

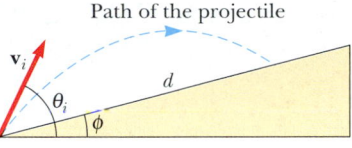

Figure P4.47

(b) For what value of θ_i is d a maximum, and what is that maximum value of d?

48. A student decides to measure the muzzle velocity of the pellets from his BB gun. He points the gun horizontally. On a vertical wall a distance x away from the gun, a target is placed. The shots hit the target a vertical distance y below the gun. (a) Show that the vertical displacement component of the pellets when traveling through the air is given by $y = Ax^2$, where A is a constant. (b) Express the constant A in terms of the initial velocity and the free-fall acceleration. (c) If $x = 3.00$ m and $y = 0.210$ m, what is the initial speed of the pellets?

49. A home run is hit in such a way that the baseball just clears a wall 21.0 m high, located 130 m from home plate. The ball is hit at an angle of 35.0° to the horizontal, and air resistance is negligible. Find (a) the initial speed of the ball, (b) the time it takes the ball to reach the wall, and (c) the velocity components and the speed of the ball when it reaches the wall. (Assume the ball is hit at a height of 1.00 m above the ground.)

50. An astronaut standing on the Moon fires a gun so that the bullet leaves the barrel initially moving in a horizontal direction. (a) What must be the muzzle speed of the bullet so that it travels completely around the Moon and returns to its original location? (b) How long does this trip around the Moon take? Assume that the free-fall acceleration on the Moon is one-sixth that on the Earth.

51. A pendulum of length 1.00 m swings in a vertical plane (Fig. 4.19). When the pendulum is in the two horizontal positions $\theta = 90°$ and $\theta = 270°$, its speed is 5.00 m/s. (a) Find the magnitude of the radial acceleration and tangential acceleration for these positions. (b) Draw a vector diagram to determine the direction of the total acceleration for these two positions. (c) Calculate the magnitude and direction of the total acceleration.

52. A basketball player who is 2.00 m tall is standing on the floor 10.0 m from the basket, as in Figure P4.52. If he shoots the ball at a 40.0° angle with the horizontal, at what initial speed must he throw so that it goes through the hoop without striking the backboard? The basket height is 3.05 m.

53. A particle has velocity components

$$v_x = +4 \text{ m/s} \qquad v_y = -(6 \text{ m/s}^2)t + 4 \text{ m/s}$$

Calculate the speed of the particle and the direction $\theta = \tan^{-1}(v_y/v_x)$ of the velocity vector at $t = 2.00$ s.

54. When baseball players throw the ball in from the outfield, they usually allow it to take one bounce before it reaches the infielder on the theory that the ball arrives

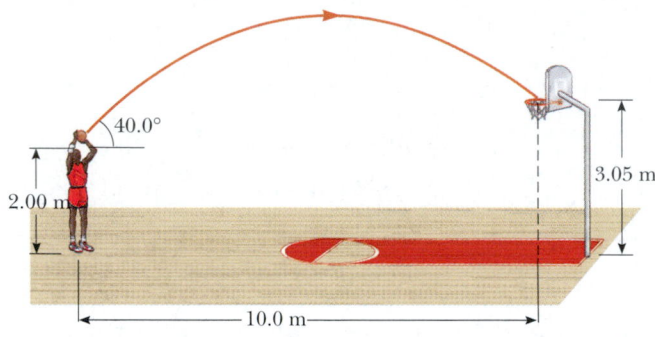

Figure P4.52

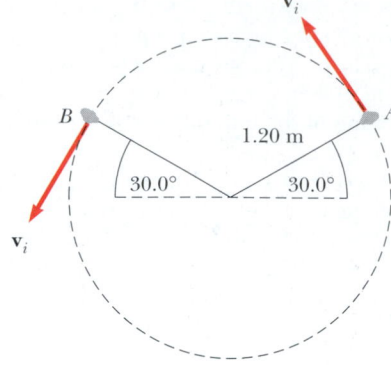

Figure P4.57

sooner that way. Suppose that the angle at which a bounced ball leaves the ground is the same as the angle at which the outfielder launched it, as in Figure P4.54, but that the ball's speed after the bounce is one half of what it was before the bounce. (a) Assuming the ball is always thrown with the same initial speed, at what angle θ should the ball be thrown in order to go the same distance D with one bounce (blue path) as a ball thrown upward at 45.0° with no bounce (green path)? (b) Determine the ratio of the times for the one-bounce and no-bounce throws.

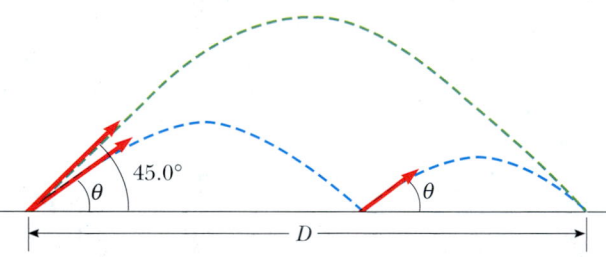

Figure P4.54

55. A boy can throw a ball a maximum horizontal distance of 40.0 m on a level field. How far can he throw the same ball vertically upward? Assume that his muscles give the ball the same speed in each case.

56. A boy can throw a ball a maximum horizontal distance of R on a level field. How far can he throw the same ball vertically upward? Assume that his muscles give the ball the same speed in each case.

57. A stone at the end of a sling is whirled in a vertical circle of radius 1.20 m at a constant speed $v_i = 1.50$ m/s as in Figure P4.57. The center of the string is 1.50 m above the ground. What is the range of the stone if it is released when the sling is inclined at 30.0° with the horizontal (a) at A? (b) at B? What is the acceleration of the stone (c) just before it is released at A? (d) just after it is released at A?

58. A quarterback throws a football straight toward a receiver with an initial speed of 20.0 m/s, at an angle of 30.0° above the horizontal. At that instant, the receiver is 20.0 m from the quarterback. In what direction and with what constant speed should the receiver run to catch the football at the level at which it was thrown?

59. A bomber is flying horizontally over level terrain, with a speed of 275 m/s relative to the ground, at an altitude of 3 000 m. Neglect the effects of air resistance. (a) How far will a bomb travel horizontally between its release from the plane and its impact on the ground? (b) If the plane maintains its original course and speed, where will it be when the bomb hits the ground? (c) At what angle from the vertical should the telescopic bombsight be set so that the bomb will hit the target seen in the sight at the time of release?

60. A person standing at the top of a hemispherical rock of radius R kicks a ball (initially at rest on the top of the rock) to give it horizontal velocity $\mathbf{v}_i$ as in Figure P4.60. (a) What must be its minimum initial speed if the ball is never to hit the rock after it is kicked? (b) With this initial speed, how far from the base of the rock does the ball hit the ground?

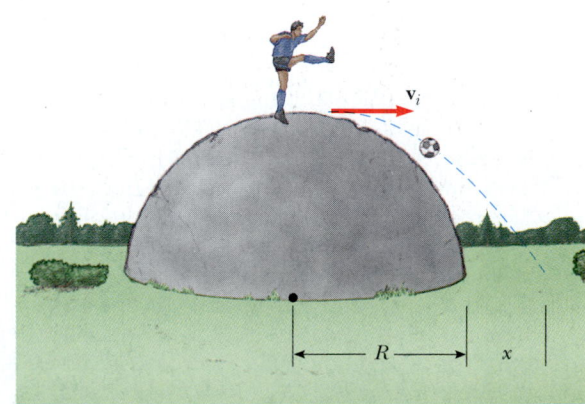

Figure P4.60

61. A hawk is flying horizontally at 10.0 m/s in a straight line, 200 m above the ground. A mouse it has been carrying struggles free from its grasp. The hawk continues on its path at the same speed for 2.00 s before attempting to retrieve its prey. To accomplish the retrieval, it dives in a straight line at constant speed and recaptures the mouse 3.00 m above the ground. (a) Assuming no air resistance, find the diving speed of the hawk. (b) What angle did the hawk make with the horizontal during its descent? (c) For how long did the mouse "enjoy" free fall?

62. A truck loaded with cannonball watermelons stops suddenly to avoid running over the edge of a washed-out bridge (Fig. P4.62). The quick stop causes a number of melons to fly off the truck. One melon rolls over the edge with an initial speed $v_i = 10.0$ m/s in the horizontal direction. A cross-section of the bank has the shape of the bottom half of a parabola with its vertex at the edge of the road, and with the equation $y^2 = 16x$, where x and y are measured in meters. What are the x and y coordinates of the melon when it splatters on the bank?

$v_i = 10$ m/s

Figure P4.62

63. A catapult launches a rocket at an angle of 53.0° above the horizontal with an initial speed of 100 m/s. The rocket engine immediately starts a burn, and for 3.00 s the rocket moves along its initial line of motion with an acceleration of 30.0 m/s². Then its engine fails, and the rocket proceeds to move in free fall. Find (a) the maximum altitude reached by the rocket, (b) its total time of flight, and (c) its horizontal range.

64. A river flows with a uniform velocity **v**. A person in a motorboat travels 1.00 km upstream, at which time she passes a log floating by. Always with the same throttle setting, the boater continues to travel upstream for another 60.0 min and then returns downstream to her starting point, which she reaches just as the same log does. Find the velocity of the river. (*Hint:* The time of travel of the boat after it meets the log equals the time of travel of the log.)

WEB **65.** A car is parked on a steep incline overlooking the ocean, where the incline makes an angle of 37.0° below the horizontal. The negligent driver leaves the car in neutral, and the parking brakes are defective. The car rolls from rest down the incline with a constant acceleration of 4.00 m/s², traveling 50.0 m to the edge of a vertical cliff. The cliff is 30.0 m above the ocean. Find (a) the speed of the car when it reaches the edge of the cliff and the time it takes to get there, (b) the velocity of the car when it lands in the ocean, (c) the total time the car is in motion, and (d) the position of the car when it lands in the ocean, relative to the base of the cliff.

 66. The determined coyote is out once more to try to capture the elusive roadrunner. The coyote wears a pair of Acme jet-powered roller skates, which provide a constant horizontal acceleration of 15.0 m/s² (Fig. P4.66). The coyote starts off at rest 70.0 m from the edge of a cliff at the instant the roadrunner zips past him in the direction of the cliff. (a) If the roadrunner moves with constant speed, determine the minimum speed he must have to reach the cliff before the coyote. At the brink of the cliff, the roadrunner escapes by making a sudden turn, while the coyote continues straight ahead. (b) If the cliff is 100 m above the floor of a canyon, determine where the coyote lands in the canyon (assume his skates remain horizontal and continue to operate when he is in "flight"). (c) Determine the components of the coyote's impact velocity.

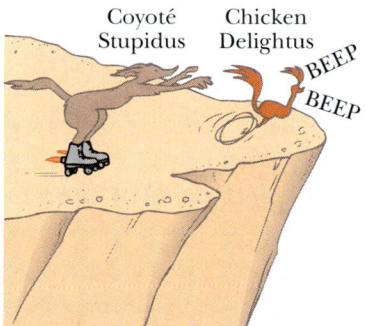

Coyoté   Chicken
Stupidus   Delightus
                          BEEP
                      BEEP

Figure P4.66

67. A skier leaves the ramp of a ski jump with a velocity of 10.0 m/s, 15.0° above the horizontal, as in Figure P4.67. The slope is inclined at 50.0°, and air resistance is negligible. Find (a) the distance from the ramp to where the jumper lands and (b) the velocity components just before the landing. (How do you think the results might be affected if air resistance were included? Note that jumpers lean forward in the shape of an airfoil, with their hands at their sides, to increase their distance. Why does this work?)

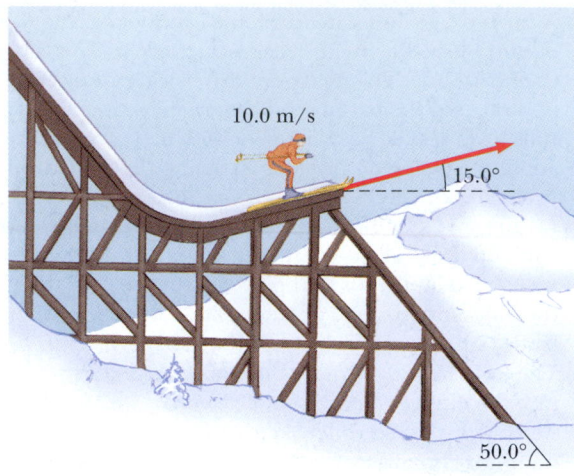

Figure P4.67

68. Two soccer players, Mary and Jane, begin running from nearly the same point at the same time. Mary runs in an easterly direction at 4.00 m/s, while Jane takes off in a direction 60.0° north of east at 5.40 m/s. (a) How long is it before they are 25.0 m apart? (b) What is the velocity of Jane relative to Mary? (c) How far apart are they after 4.00 s?

69. Do not hurt yourself; do not strike your hand against anything. Within these limitations, describe what you do to give your hand a large acceleration. Compute an order-of-magnitude estimate of this acceleration, stating the quantities you measure or estimate and their values.

70. An enemy ship is on the western side of a mountain island, as shown in Figure P4.70. The enemy ship can maneuver to within 2 500 m of the 1 800-m-high mountain peak and can shoot projectiles with an initial speed of 250 m/s. If the eastern shoreline is horizontally 300 m from the peak, what are the distances from the eastern shore at which a ship can be safe from the bombardment of the enemy ship?

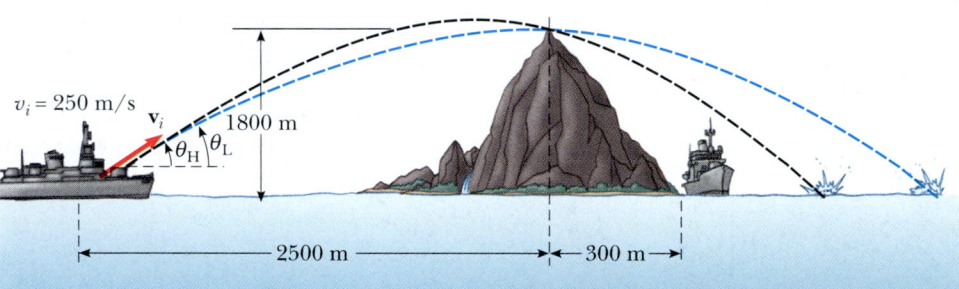

Figure P4.70

ANSWERS TO QUICK QUIZZES

4.1 (a) Because acceleration occurs whenever the velocity changes in any way—with an increase or decrease in speed, a change in direction, or both—the brake pedal can also be considered an accelerator because it causes the car to slow down. The steering wheel is also an accelerator because it changes the direction of the velocity vector. (b) When the car is moving with constant speed, the gas pedal is not causing an acceleration; it is an accelerator only when it causes a change in the speedometer reading.

4.2 (a) At only one point—the peak of the trajectory—are the velocity and acceleration vectors perpendicular to each other. (b) If the object is thrown straight up or down, **v** and **a** are parallel to each other throughout the downward motion. Otherwise, the velocity and acceleration vectors are never parallel to each other. (c) The greater the maximum height, the longer it takes the projectile to reach that altitude and then fall back down from

it. So, as the angle increases from 0° to 90°, the time of flight increases. Therefore, the 15° angle gives the shortest time of flight, and the 75° angle gives the longest.

4.3 (a) Because the object is moving with a constant speed, the velocity vector is always the same length; because the motion is circular, this vector is always tangent to the circle. The only acceleration is that which changes the direction of the velocity vector; it points radially inward.

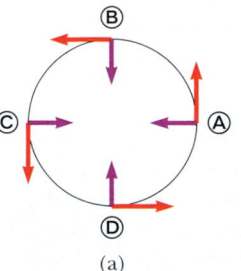

(a)

(b) Now there is a component of the acceleration vector that is tangent to the circle and points in the direction opposite the velocity. As a result, the acceleration vector does not point toward the center. The object is slowing down, and so the velocity vectors become shorter and shorter.

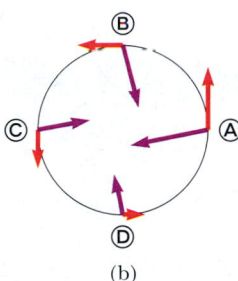

(b)

(c) Now the tangential component of the acceleration points in the same direction as the velocity. The object is speeding up, and so the velocity vectors become longer and longer. Because the speed changes rapidly here, but gradually in part (b), the acceleration vectors are longer here than in part (b).

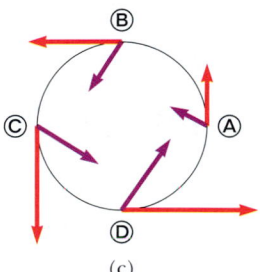

(c)

4.4 The motion diagram is as shown below. Note that each position vector points from the pivot point at the center of the circle to the position of the ball.

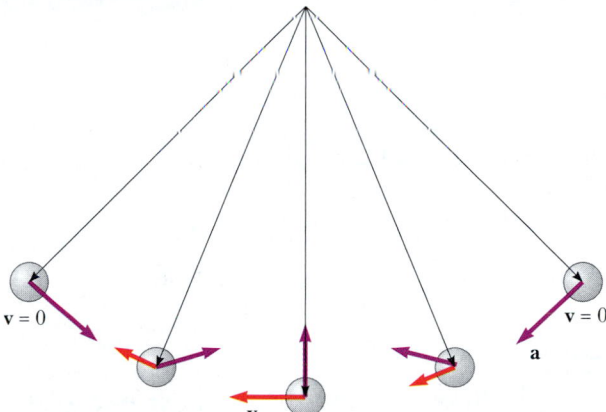

4.5 (a) The passenger sees the coffee pouring nearly vertically into the cup, just as if she were standing on the ground pouring it. (b) The stationary observer sees the coffee moving in a parabolic path with a constant horizontal velocity of 60 mi/h ($= 88$ ft/s) and a downward acceleration of $- g$. If it takes the coffee 0.10 s to reach the cup, the stationary observer sees the coffee moving 8.8 ft horizontally before it hits the cup! (c) If the car slows suddenly, the coffee reaches the place where the cup *would have been* had there been no change in velocity and continues falling because the cup has not yet reached that location. If the car rapidly speeds up, the coffee falls behind the cup. If the car accelerates sideways, the coffee again ends up somewhere other than the cup.

c h a p t e r

5

The Laws of Motion

Chapter Outline

In Chapters 2 and 4, we described motion in terms of displacement, velocity, and acceleration without considering what might cause that motion. What might cause one particle to remain at rest and another particle to accelerate? In this chapter, we investigate what causes changes in motion. The two main factors we need to consider are the forces acting on an object and the mass of the object. We discuss the three basic laws of motion, which deal with forces and masses and were formulated more than three centuries ago by Isaac Newton. Once we understand these laws, we can answer such questions as "What mechanism changes motion?" and "Why do some objects accelerate more than others?"

5.1 THE CONCEPT OF FORCE

Everyone has a basic understanding of the concept of force from everyday experience. When you push your empty dinner plate away, you exert a force on it. Similarly, you exert a force on a ball when you throw or kick it. In these examples, the word *force* is associated with muscular activity and some change in the velocity of an object. Forces do not always cause motion, however. For example, as you sit reading this book, the force of gravity acts on your body and yet you remain stationary. As a second example, you can push (in other words, exert a force) on a large boulder and not be able to move it.

What force (if any) causes the Moon to orbit the Earth? Newton answered this and related questions by stating that forces are what cause any change in the velocity of an object. Therefore, if an object moves with uniform motion (constant velocity), no force is required for the motion to be maintained. The Moon's velocity is not constant because it moves in a nearly circular orbit around the Earth. We now know that this change in velocity is caused by the force exerted on the Moon by the Earth. Because only a force can cause a change in velocity, we can think of force as *that which causes a body to accelerate*. In this chapter, we are concerned with the relationship between the force exerted on an object and the acceleration of that object.

A body accelerates because of an external force

What happens when several forces act simultaneously on an object? In this case, the object accelerates only if the net force acting on it is not equal to zero. The **net force** acting on an object is defined as the vector sum of all forces acting on the object. (We sometimes refer to the net force as the *total force*, the *resultant force*, or the *unbalanced force*.) **If the net force exerted on an object is zero, then the acceleration of the object is zero and its velocity remains constant.** That is, if the net force acting on the object is zero, then the object either remains at rest or continues to move with constant velocity. When the velocity of an object is constant (including the case in which the object remains at rest), the object is said to be in **equilibrium.**

Definition of equilibrium

When a coiled spring is pulled, as in Figure 5.1a, the spring stretches. When a stationary cart is pulled sufficently hard that friction is overcome, as in Figure 5.1b, the cart moves. When a football is kicked, as in Figure 5.1c, it is both deformed and set in motion. These situations are all examples of a class of forces called *contact forces*. That is, they involve physical contact between two objects. Other examples of contact forces are the force exerted by gas molecules on the walls of a container and the force exerted by your feet on the floor.

Another class of forces, known as *field forces,* do not involve physical contact between two objects but instead act through empty space. The force of gravitational attraction between two objects, illustrated in Figure 5.1d, is an example of this class of force. This gravitational force keeps objects bound to the Earth. The plan-

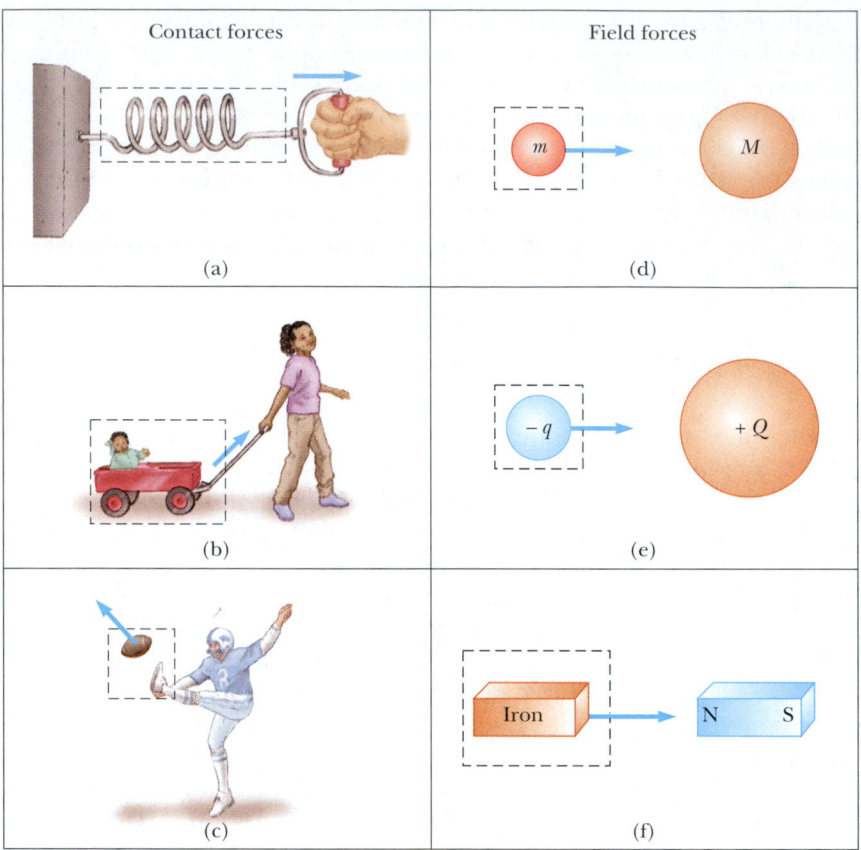

Figure 5.1 Some examples of applied forces. In each case a force is exerted on the object within the boxed area. Some agent in the environment external to the boxed area exerts a force on the object.

ets of our Solar System are bound to the Sun by the action of gravitational forces. Another common example of a field force is the electric force that one electric charge exerts on another, as shown in Figure 5.1e. These charges might be those of the electron and proton that form a hydrogen atom. A third example of a field force is the force a bar magnet exerts on a piece of iron, as shown in Figure 5.1f. The forces holding an atomic nucleus together also are field forces but are very short in range. They are the dominating interaction for particle separations of the order of 10^{-15} m.

Early scientists, including Newton, were uneasy with the idea that a force can act between two disconnected objects. To overcome this conceptual problem, Michael Faraday (1791–1867) introduced the concept of a *field*. According to this approach, when object 1 is placed at some point P near object 2, we say that object 1 interacts with object 2 by virtue of the gravitational field that exists at P. The gravitational field at P is created by object 2. Likewise, a gravitational field created by object 1 exists at the position of object 2. In fact, all objects create a gravitational field in the space around themselves.

The distinction between contact forces and field forces is not as sharp as you may have been led to believe by the previous discussion. When examined at the atomic level, all the forces we classify as contact forces turn out to be caused by electric (field) forces of the type illustrated in Figure 5.1e. Nevertheless, in developing models for macroscopic phenomena, it is convenient to use both classifications of forces. The only known *fundamental* forces in nature are all field forces: (1) gravitational forces between objects, (2) electromagnetic forces between electric charges, (3) strong nuclear forces between subatomic particles, and (4) weak nuclear forces that arise in certain radioactive decay processes. In classical physics, we are concerned only with gravitational and electromagnetic forces.

Measuring the Strength of a Force

It is convenient to use the deformation of a spring to measure force. Suppose we apply a vertical force to a spring scale that has a fixed upper end, as shown in Figure 5.2a. The spring elongates when the force is applied, and a pointer on the scale reads the value of the applied force. We can calibrate the spring by defining the unit force $\mathbf{F}_1$ as the force that produces a pointer reading of 1.00 cm. (Because force is a vector quantity, we use the bold-faced symbol $\mathbf{F}$.) If we now apply a different downward force $\mathbf{F}_2$ whose magnitude is 2 units, as seen in Figure 5.2b, the pointer moves to 2.00 cm. Figure 5.2c shows that the combined effect of the two collinear forces is the sum of the effects of the individual forces.

Now suppose the two forces are applied simultaneously with $\mathbf{F}_1$ downward and $\mathbf{F}_2$ horizontal, as illustrated in Figure 5.2d. In this case, the pointer reads $\sqrt{5}\ \mathrm{cm}^2 = 2.24$ cm. The single force $\mathbf{F}$ that would produce this same reading is the sum of the two vectors $\mathbf{F}_1$ and $\mathbf{F}_2$, as described in Figure 5.2d. That is, $|\mathbf{F}| = \sqrt{F_1{}^2 + F_2{}^2} = 2.24$ units, and its direction is $\theta = \tan^{-1}(-0.500) = -26.6°$. **Because forces are vector quantities, you must use the rules of vector addition to obtain the net force acting on an object.**

QuickLab

Find a tennis ball, two drinking straws, and a friend. Place the ball on a table. You and your friend can each apply a force to the ball by blowing through the straws (held horizontally a few centimeters above the table) so that the air rushing out strikes the ball. Try a variety of configurations: Blow in opposite directions against the ball, blow in the same direction, blow at right angles to each other, and so forth. Can you verify the vector nature of the forces?

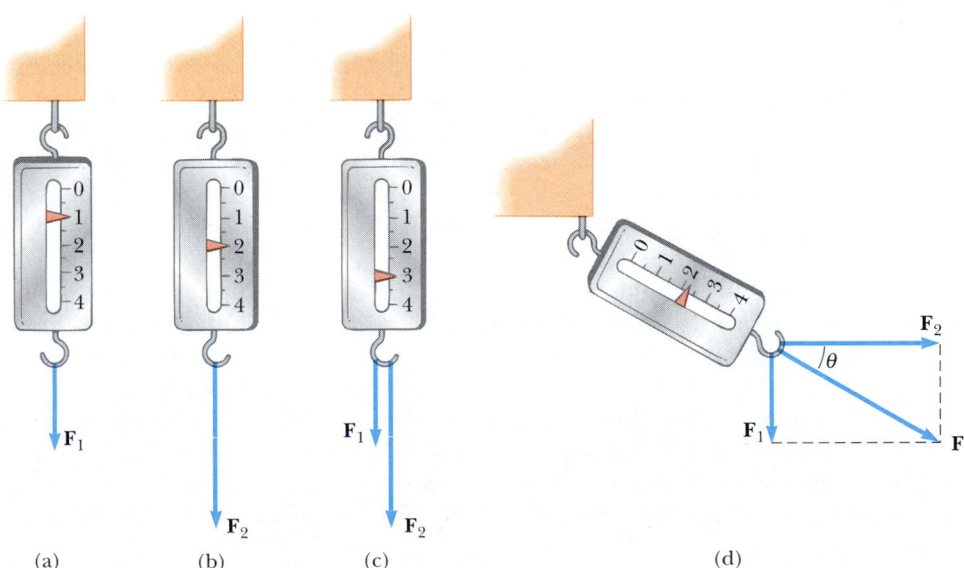

(a)   (b)   (c)   (d)

Figure 5.2   The vector nature of a force is tested with a spring scale. (a) A downward force $\mathbf{F}_1$ elongates the spring 1 cm. (b) A downward force $\mathbf{F}_2$ elongates the spring 2 cm. (c) When $\mathbf{F}_1$ and $\mathbf{F}_2$ are applied simultaneously, the spring elongates by 3 cm. (d) When $\mathbf{F}_1$ is downward and $\mathbf{F}_2$ is horizontal, the combination of the two forces elongates the spring $\sqrt{1^2 + 2^2}$ cm $= \sqrt{5}$ cm.

5.2   NEWTON'S FIRST LAW AND INERTIAL FRAMES

Before we state Newton's first law, consider the following simple experiment. Suppose a book is lying on a table. Obviously, the book remains at rest. Now imagine that you push the book with a horizontal force great enough to overcome the force of friction between book and table. (This force you exert, the force of friction, and any other forces exerted on the book by other objects are referred to as *external forces*.) You can keep the book in motion with constant velocity by applying a force that is just equal in magnitude to the force of friction and acts in the opposite direction. If you then push harder so that the magnitude of your applied force exceeds the magnitude of the force of friction, the book accelerates. If you stop pushing, the book stops after moving a short distance because the force of friction retards its motion. Suppose you now push the book across a smooth, highly waxed floor. The book again comes to rest after you stop pushing but not as quickly as before. Now imagine a floor so highly polished that friction is absent; in this case, the book, once set in motion, moves until it hits a wall.

Before about 1600, scientists felt that the natural state of matter was the state of rest. Galileo was the first to take a different approach to motion and the natural state of matter. He devised thought experiments, such as the one we just discussed for a book on a frictionless surface, and concluded that it is not the nature of an object to stop once set in motion: rather, it is its nature to *resist changes in its motion*. In his words, "Any velocity once imparted to a moving body will be rigidly maintained as long as the external causes of retardation are removed."

This new approach to motion was later formalized by Newton in a form that has come to be known as **Newton's first law of motion:**

> In the absence of external forces, an object at rest remains at rest and an object in motion continues in motion with a constant velocity (that is, with a constant speed in a straight line).

In simpler terms, we can say that **when no force acts on an object, the acceleration of the object is zero.** If nothing acts to change the object's motion, then its velocity does not change. From the first law, we conclude that any *isolated object* (one that does not interact with its environment) is either at rest or moving with constant velocity. The tendency of an object to resist any attempt to change its velocity is called the **inertia** of the object. Figure 5.3 shows one dramatic example of a consequence of Newton's first law.

Another example of uniform (constant-velocity) motion on a nearly frictionless surface is the motion of a light disk on a film of air (the lubricant), as shown in Figure 5.4. If the disk is given an initial velocity, it coasts a great distance before stopping.

Finally, consider a spaceship traveling in space and far removed from any planets or other matter. The spaceship requires some propulsion system to change its velocity. However, if the propulsion system is turned off when the spaceship reaches a velocity **v**, the ship coasts at that constant velocity and the astronauts get a free ride (that is, no propulsion system is required to keep them moving at the velocity **v**).

Inertial Frames

As we saw in Section 4.6, a moving object can be observed from any number of reference frames. Newton's first law, sometimes called the *law of inertia*, defines a special set of reference frames called *inertial frames*. **An inertial frame of reference**

QuickLab

Use a drinking straw to impart a strong, short-duration burst of air against a tennis ball as it rolls along a tabletop. Make the force perpendicular to the ball's path. What happens to the ball's motion? What is different if you apply a continuous force (constant magnitude and direction) that is directed along the direction of motion?

Newton's first law

Definition of inertia

Definition of inertial frame

Figure 5.3 Unless a net external force acts on it, an object at rest remains at rest and an object in motion continues in motion with constant velocity. In this case, the wall of the building did not exert a force on the moving train that was large enough to stop it. (*Roger Viollet, Mill Valley, CA, University Science Books, 1982*)

Isaac Newton **English physicist and mathematician (1642–1727)** Isaac Newton was one of the most brilliant scientists in history. Before the age of 30, he formulated the basic concepts and laws of mechanics, discovered the law of universal gravitation, and invented the mathematical methods of calculus. As a consequence of his theories, Newton was able to explain the motions of the planets, the ebb and flow of the tides, and many special features of the motions of the Moon and the Earth. He also interpreted many fundamental observations concerning the nature of light. His contributions to physical theories dominated scientific thought for two centuries and remain important today. (*Giraudon/Art Resource*)

is one that is not accelerating. Because Newton's first law deals only with objects that are not accelerating, it holds only in inertial frames. Any reference frame that moves with constant velocity relative to an inertial frame is itself an inertial frame. (The Galilean transformations given by Equations 4.20 and 4.21 relate positions and velocities between two inertial frames.)

A reference frame that moves with constant velocity relative to the distant stars is the best approximation of an inertial frame, and for our purposes we can consider planet Earth as being such a frame. The Earth is not really an inertial frame because of its orbital motion around the Sun and its rotational motion about its own axis. As the Earth travels in its nearly circular orbit around the Sun, it experiences an acceleration of about 4.4×10^{-3} m/s^2 directed toward the Sun. In addition, because the Earth rotates about its own axis once every 24 h, a point on the equator experiences an additional acceleration of 3.37×10^{-2} m/s^2 directed toward the center of the Earth. However, these accelerations are small compared with g and can often be neglected. For this reason, we assume that the Earth is an inertial frame, as is any other frame attached to it.

If an object is moving with constant velocity, an observer in one inertial frame (say, one at rest relative to the object) claims that the acceleration of the object and the resultant force acting on it are zero. An observer in *any other* inertial frame also finds that $\mathbf{a} = 0$ and $\Sigma\mathbf{F} = 0$ for the object. According to the first law, a body at rest and one moving with constant velocity are equivalent. A passenger in a car moving along a straight road at a constant speed of 100 km/h can easily pour coffee into a cup. But if the driver steps on the gas or brake pedal or turns the steering wheel while the coffee is being poured, the car accelerates and it is no longer an inertial frame. The laws of motion do not work as expected, and the coffee ends up in the passenger's lap!

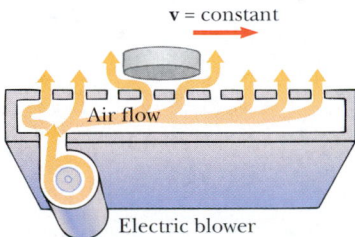

Figure 5.4 Air hockey takes advantage of Newton's first law to make the game more exciting.

Quick Quiz 5.1

True or false: (a) It is possible to have motion in the absence of a force. (b) It is possible to have force in the absence of motion.

5.3 MASS

Imagine playing catch with either a basketball or a bowling ball. Which ball is more likely to keep moving when you try to catch it? Which ball has the greater tendency to remain motionless when you try to throw it? Because the bowling ball is more resistant to changes in its velocity, we say it has greater inertia than the basketball. As noted in the preceding section, inertia is a measure of how an object responds to an external force.

Definition of mass

Mass is that property of an object that specifies how much inertia the object has, and as we learned in Section 1.1, the SI unit of mass is the kilogram. The greater the mass of an object, the less that object accelerates under the action of an applied force. For example, if a given force acting on a 3-kg mass produces an acceleration of 4 m/s^2, then the same force applied to a 6-kg mass produces an acceleration of 2 m/s^2.

To describe mass quantitatively, we begin by comparing the accelerations a given force produces on different objects. Suppose a force acting on an object of mass m_1 produces an acceleration $\mathbf{a}_1$, and the *same force* acting on an object of mass m_2 produces an acceleration $\mathbf{a}_2$. The ratio of the two masses is defined as the *inverse* ratio of the magnitudes of the accelerations produced by the force:

$$\frac{m_1}{m_2} \equiv \frac{a_2}{a_1} \tag{5.1}$$

If one object has a known mass, the mass of the other object can be obtained from acceleration measurements.

Mass is an inherent property of an object and is independent of the object's surroundings and of the method used to measure it. Also, **mass is a scalar quantity** and thus obeys the rules of ordinary arithmetic. That is, several masses can be combined in simple numerical fashion. For example, if you combine a 3-kg mass with a 5-kg mass, their total mass is 8 kg. We can verify this result experimentally by comparing the acceleration that a known force gives to several objects separately with the acceleration that the same force gives to the same objects combined as one unit.

Mass and weight are different quantities

Mass should not be confused with weight. **Mass and weight are two different quantities.** As we see later in this chapter, the weight of an object is equal to the magnitude of the gravitational force exerted on the object and varies with location. For example, a person who weighs 180 lb on the Earth weighs only about 30 lb on the Moon. On the other hand, the mass of a body is the same everywhere: an object having a mass of 2 kg on the Earth also has a mass of 2 kg on the Moon.

5.4 NEWTON'S SECOND LAW

Newton's first law explains what happens to an object when no forces act on it. It either remains at rest or moves in a straight line with constant speed. Newton's second law answers the question of what happens to an object that has a nonzero resultant force acting on it.

Imagine pushing a block of ice across a frictionless horizontal surface. When you exert some horizontal force **F**, the block moves with some acceleration **a**. If you apply a force twice as great, the acceleration doubles. If you increase the applied force to 3**F**, the acceleration triples, and so on. From such observations, we conclude that **the acceleration of an object is directly proportional to the resultant force acting on it.**

The acceleration of an object also depends on its mass, as stated in the preceding section. We can understand this by considering the following experiment. If you apply a force **F** to a block of ice on a frictionless surface, then the block undergoes some acceleration **a**. If the mass of the block is doubled, then the same applied force produces an acceleration **a**/2. If the mass is tripled, then the same applied force produces an acceleration **a**/3, and so on. According to this observation, we conclude that **the magnitude of the acceleration of an object is inversely proportional to its mass.**

These observations are summarized in **Newton's second law:**

> The acceleration of an object is directly proportional to the net force acting on it and inversely proportional to its mass.

Newton's second law

Thus, we can relate mass and force through the following mathematical statement of Newton's second law:[1]

$$\sum \mathbf{F} = m\mathbf{a} \tag{5.2}$$

Note that this equation is a vector expression and hence is equivalent to three component equations:

$$\sum F_x = ma_x \qquad \sum F_y = ma_y \qquad \sum F_z = ma_z \tag{5.3}$$

Newton's second law— component form

Quick Quiz 5.2

Is there any relationship between the net force acting on an object and the direction in which the object moves?

Unit of Force

The SI unit of force is the **newton,** which is defined as the force that, when acting on a 1-kg mass, produces an acceleration of 1 m/s^2. From this definition and Newton's second law, we see that the newton can be expressed in terms of the following fundamental units of mass, length, and time:

$$1 \text{ N} \equiv 1 \text{ kg} \cdot \text{m/s}^2 \tag{5.4}$$

Definition of newton

In the British engineering system, the unit of force is the **pound,** which is defined as the force that, when acting on a 1-slug mass,[2] produces an acceleration of 1 ft/s^2:

$$1 \text{ lb} \equiv 1 \text{ slug} \cdot \text{ft/s}^2 \tag{5.5}$$

A convenient approximation is that $1 \text{ N} \approx \frac{1}{4} \text{ lb}$.

---

[1] Equation 5.2 is valid only when the speed of the object is much less than the speed of light. We treat the relativistic situation in Chapter 39.

[2] The *slug* is the unit of mass in the British engineering system and is that system's counterpart of the SI unit the *kilogram*. Because most of the calculations in our study of classical mechanics are in SI units, the slug is seldom used in this text.

TABLE 5.1   Units of Force, Mass, and Acceleration[a]

System of Units	Mass	Acceleration	Force
SI	kg	m/s^2	$N = kg \cdot m/s^2$
British engineering	slug	ft/s^2	$lb = slug \cdot ft/s^2$

[a] 1 N = 0.225 lb.

The units of force, mass, and acceleration are summarized in Table 5.1.

 We can now understand how a single person can hold up an airship but is not able to change its motion abruptly, as stated at the beginning of the chapter. The mass of the blimp is greater than 6 800 kg. In order to make this large mass accelerate appreciably, a very large force is required—certainly one much greater than a human can provide.

EXAMPLE 5.1   An Accelerating Hockey Puck

A hockey puck having a mass of 0.30 kg slides on the horizontal, frictionless surface of an ice rink. Two forces act on the puck, as shown in Figure 5.5. The force $\mathbf{F}_1$ has a magnitude of 5.0 N, and the force $\mathbf{F}_2$ has a magnitude of 8.0 N. Determine both the magnitude and the direction of the puck's acceleration.

Solution   The resultant force in the x direction is

$$\sum F_x = F_{1x} + F_{2x} = F_1 \cos(-20°) + F_2 \cos 60°$$
$$= (5.0\ N)(0.940) + (8.0\ N)(0.500) = 8.7\ N$$

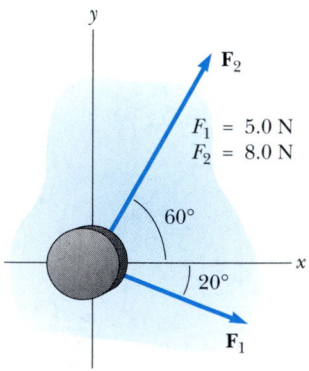

Figure 5.5   A hockey puck moving on a frictionless surface accelerates in the direction of the resultant force $\mathbf{F}_1 + \mathbf{F}_2$.

The resultant force in the y direction is

$$\sum F_y = F_{1y} + F_{2y} = F_1 \sin(-20°) + F_2 \sin 60°$$
$$= (5.0\ N)(-0.342) + (8.0\ N)(0.866) = 5.2\ N$$

Now we use Newton's second law in component form to find the x and y components of acceleration:

$$a_x = \frac{\sum F_x}{m} = \frac{8.7\ N}{0.30\ kg} = 29\ m/s^2$$
$$a_y = \frac{\sum F_y}{m} = \frac{5.2\ N}{0.30\ kg} = 17\ m/s^2$$

The acceleration has a magnitude of

$$a = \sqrt{(29)^2 + (17)^2}\ m/s^2 = \boxed{34\ m/s^2}$$

and its direction relative to the positive x axis is

$$\theta = \tan^{-1}\left(\frac{a_y}{a_x}\right) = \tan^{-1}\left(\frac{17}{29}\right) = \boxed{30°}$$

We can graphically add the vectors in Figure 5.5 to check the reasonableness of our answer. Because the acceleration vector is along the direction of the resultant force, a drawing showing the resultant force helps us check the validity of the answer.

Exercise   Determine the components of a third force that, when applied to the puck, causes it to have zero acceleration.

Answer $F_{3x} = -8.7\ N,\ F_{3y} = -5.2\ N$.

5.5 THE FORCE OF GRAVITY AND WEIGHT

We are well aware that all objects are attracted to the Earth. The attractive force exerted by the Earth on an object is called the **force of gravity $\mathbf{F}_g$**. This force is directed toward the center of the Earth,[3] and its magnitude is called the **weight** of the object.

We saw in Section 2.6 that a freely falling object experiences an acceleration $\mathbf{g}$ acting toward the center of the Earth. Applying Newton's second law $\Sigma \mathbf{F} = m\mathbf{a}$ to a freely falling object of mass m, with $\mathbf{a} = \mathbf{g}$ and $\Sigma \mathbf{F} = \mathbf{F}_g$, we obtain

$$\mathbf{F}_g = m\mathbf{g} \tag{5.6}$$

Thus, the weight of an object, being defined as the magnitude of $\mathbf{F}_g$, is mg. (You should not confuse the italicized symbol g for gravitational acceleration with the nonitalicized symbol g used as the abbreviation for "gram.")

Because it depends on g, weight varies with geographic location. Hence, weight, unlike mass, is not an inherent property of an object. Because g decreases with increasing distance from the center of the Earth, bodies weigh less at higher altitudes than at sea level. For example, a 1 000-kg palette of bricks used in the construction of the Empire State Building in New York City weighed about 1 N less by the time it was lifted from sidewalk level to the top of the building. As another example, suppose an object has a mass of 70.0 kg. Its weight in a location where $g = 9.80 \text{ m/s}^2$ is $F_g = mg = 686 \text{ N}$ (about 150 lb). At the top of a mountain, however, where $g = 9.77 \text{ m/s}^2$, its weight is only 684 N. Therefore, if you want to lose weight without going on a diet, climb a mountain or weigh yourself at 30 000 ft during an airplane flight!

Because weight $= F_g = mg$, we can compare the masses of two objects by measuring their weights on a spring scale. At a given location, the ratio of the weights of two objects equals the ratio of their masses.

QuickLab

Drop a pen and your textbook simultaneously from the same height and watch as they fall. How can they have the same acceleration when their weights are so different?

The life-support unit strapped to the back of astronaut Edwin Aldrin weighed 300 lb on the Earth. During his training, a 50-lb mock-up was used. Although this effectively simulated the reduced weight the unit would have on the Moon, it did not correctly mimic the unchanging mass. It was just as difficult to accelerate the unit (perhaps by jumping or twisting suddenly) on the Moon as on the Earth. *(Courtesy of NASA)*

---

[3] This statement ignores the fact that the mass distribution of the Earth is not perfectly spherical.

CONCEPTUAL EXAMPLE 5.2   How Much Do You Weigh in an Elevator?

You have most likely had the experience of standing in an elevator that accelerates upward as it moves toward a higher floor. In this case, you feel heavier. In fact, if you are standing on a bathroom scale at the time, the scale measures a force magnitude that is greater than your weight. Thus, you have tactile and measured evidence that leads you to believe you are heavier in this situation. *Are* you heavier?

Solution   No, your weight is unchanged. To provide the acceleration upward, the floor or scale must exert on your feet an upward force that is greater in magnitude than your weight. It is this greater force that you feel, which you interpret as feeling heavier. The scale reads this upward force, not your weight, and so its reading increases.

Quick Quiz 5.3

A baseball of mass m is thrown upward with some initial speed. If air resistance is neglected, what forces are acting on the ball when it reaches (a) half its maximum height and (b) its maximum height?

5.6   NEWTON'S THIRD LAW

4.5   If you press against a corner of this textbook with your fingertip, the book pushes back and makes a small dent in your skin. If you push harder, the book does the same and the dent in your skin gets a little larger. This simple experiment illustrates a general principle of critical importance known as **Newton's third law:**

Newton's third law

> If two objects interact, the force $\mathbf{F}_{12}$ exerted by object 1 on object 2 is equal in magnitude to and opposite in direction to the force $\mathbf{F}_{21}$ exerted by object 2 on object 1:
>
> $$\mathbf{F}_{12} = -\mathbf{F}_{21} \qquad \text{(5.7)}$$

This law, which is illustrated in Figure 5.6a, states that a force that affects the motion of an object must come from a second, *external,* object. The external object, in turn, is subject to an equal-magnitude but oppositely directed force exerted on it.

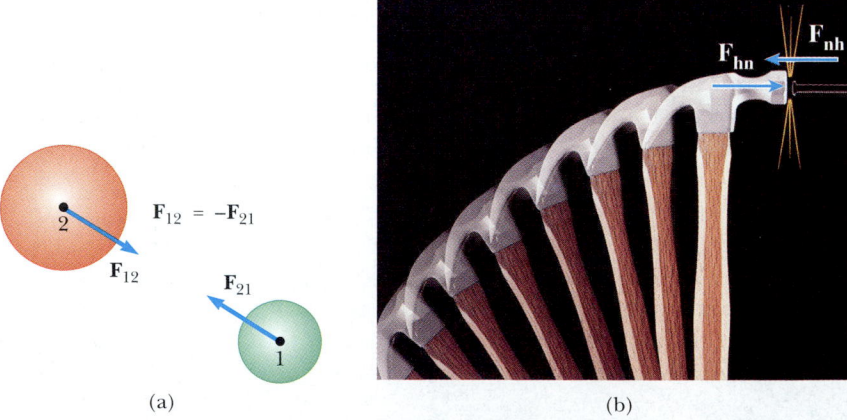

Figure 5.6   Newton's third law. (a) The force $\mathbf{F}_{12}$ exerted by object 1 on object 2 is equal in magnitude to and opposite in direction to the force $\mathbf{F}_{21}$ exerted by object 2 on object 1. (b) The force $\mathbf{F}_{hn}$ exerted by the hammer on the nail is equal to and opposite the force $\mathbf{F}_{nh}$ exerted by the nail on the hammer. *(John Gillmoure/The Stock Market)*

This is equivalent to stating that **a single isolated force cannot exist.** The force that object 1 exerts on object 2 is sometimes called the *action force*, while the force object 2 exerts on object 1 is called the *reaction force*. In reality, either force can be labeled the action or the reaction force. **The action force is equal in magnitude to the reaction force and opposite in direction. In all cases, the action and reaction forces act on different objects.** For example, the force acting on a freely falling projectile is $\mathbf{F}_g = m\mathbf{g}$, which is the force of gravity exerted by the Earth on the projectile. The reaction to this force is the force exerted by the projectile on the Earth, $\mathbf{F}'_g = -\mathbf{F}_g$. The reaction force $\mathbf{F}'_g$ accelerates the Earth toward the projectile just as the action force $\mathbf{F}_g$ accelerates the projectile toward the Earth. However, because the Earth has such a great mass, its acceleration due to this reaction force is negligibly small.

Another example of Newton's third law is shown in Figure 5.6b. The force exerted by the hammer on the nail (the action force $\mathbf{F}_{hn}$) is equal in magnitude and opposite in direction to the force exerted by the nail on the hammer (the reaction force $\mathbf{F}_{nh}$). It is this latter force that causes the hammer to stop its rapid forward motion when it strikes the nail.

You experience Newton's third law directly whenever you slam your fist against a wall or kick a football. You should be able to identify the action and reaction forces in these cases.

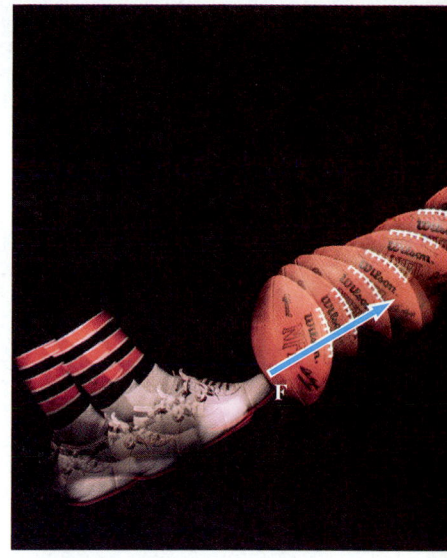

Compression of a football as the force exerted by a player's foot sets the ball in motion. *(Ralph Cowan/ Tony Stone Images)*

Quick Quiz 5.4

A person steps from a boat toward a dock. Unfortunately, he forgot to tie the boat to the dock, and the boat scoots away as he steps from it. Analyze this situation in terms of Newton's third law.

The force of gravity $\mathbf{F}_g$ was defined as the attractive force the Earth exerts on an object. If the object is a TV at rest on a table, as shown in Figure 5.7a, why does the TV not accelerate in the direction of $\mathbf{F}_g$? The TV does not accelerate because the table holds it up. What is happening is that the table exerts on the TV an upward force $\mathbf{n}$ called the **normal force.**[4] The normal force is a contact force that prevents the TV from falling through the table and can have any magnitude needed to balance the downward force $\mathbf{F}_g$, up to the point of breaking the table. If someone stacks books on the TV, the normal force exerted by the table on the TV increases. If someone lifts up on the TV, the normal force exerted by the table on the TV decreases. (The normal force becomes zero if the TV is raised off the table.)

The two forces in an action–reaction pair **always act on different objects.** For the hammer-and-nail situation shown in Figure 5.6b, one force of the pair acts on the hammer and the other acts on the nail. For the unfortunate person stepping out of the boat in Quick Quiz 5.4, one force of the pair acts on the person, and the other acts on the boat.

For the TV in Figure 5.7, the force of gravity $\mathbf{F}_g$ and the normal force $\mathbf{n}$ are *not* an action–reaction pair because they act on the same body—the TV. The two reaction forces in this situation—$\mathbf{F}'_g$ and $\mathbf{n}'$—are exerted on objects other than the TV. Because the reaction to $\mathbf{F}_g$ is the force $\mathbf{F}'_g$ exerted by the TV on the Earth and the reaction to $\mathbf{n}$ is the force $\mathbf{n}'$ exerted by the TV on the table, we conclude that

$$\mathbf{F}_g = -\mathbf{F}'_g \qquad \text{and} \qquad \mathbf{n} = -\mathbf{n}'$$

Definition of normal force

---

[4] *Normal* in this context means *perpendicular.*

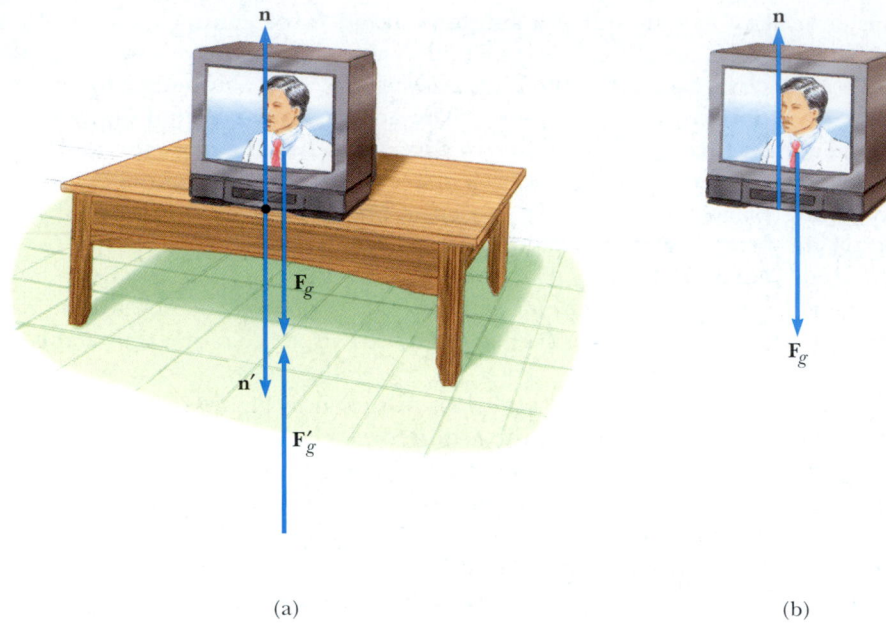

(a)   (b)

Figure 5.7   When a TV is at rest on a table, the forces acting on the TV are the normal force **n** and the force of gravity $\mathbf{F}_g$, as illustrated in part (b). The reaction to **n** is the force **n**′ exerted by the TV on the table. The reaction to $\mathbf{F}_g$ is the force $\mathbf{F}_g'$ exerted by the TV on the Earth.

The forces **n** and **n**′ have the same magnitude, which is the same as that of $\mathbf{F}_g$ until the table breaks. From the second law, we see that, because the TV is in equilibrium (**a** = 0), it follows[5] that $F_g = n = mg$.

▶ **Quick Quiz 5.5**

If a fly collides with the windshield of a fast-moving bus, (a) which experiences the greater impact force: the fly or the bus, or is the same force experienced by both? (b) Which experiences the greater acceleration: the fly or the bus, or is the same acceleration experienced by both?

CONCEPTUAL EXAMPLE 5.3 ▶ **You Push Me and I'll Push You**

A large man and a small boy stand facing each other on frictionless ice. They put their hands together and push against each other so that they move apart. (a) Who moves away with the higher speed?

Solution   This situation is similar to what we saw in Quick Quiz 5.5. According to Newton's third law, the force exerted by the man on the boy and the force exerted by the boy on the man are an action–reaction pair, and so they must be equal in magnitude. (A bathroom scale placed between their hands would read the same, regardless of which way it faced.)

Therefore, the boy, having the lesser mass, experiences the greater acceleration. Both individuals accelerate for the same amount of time, but the greater acceleration of the boy over this time interval results in his moving away from the interaction with the higher speed.

(b) Who moves farther while their hands are in contact?

Solution   Because the boy has the greater acceleration, he moves farther during the interval in which the hands are in contact.

---

[5] Technically, we should write this equation in the component form $F_{gy} = n_y = mg_y$. This component notation is cumbersome, however, and so in situations in which a vector is parallel to a coordinate axis, we usually do not include the subscript for that axis because there is no other component.

5.7 SOME APPLICATIONS OF NEWTON'S LAWS

In this section we apply Newton's laws to objects that are either in equilibrium (**a** = 0) or accelerating along a straight line under the action of constant external forces. We assume that the objects behave as particles so that we need not worry about rotational motion. We also neglect the effects of friction in those problems involving motion; this is equivalent to stating that the surfaces are *frictionless*. Finally, we usually neglect the mass of any ropes involved. In this approximation, the magnitude of the force exerted at any point along a rope is the same at all points along the rope. In problem statements, the synonymous terms *light, lightweight,* and *of negligible mass* are used to indicate that a mass is to be ignored when you work the problems.

When we apply Newton's laws to an object, we are interested only in external forces that act on the object. For example, in Figure 5.7 the only external forces acting on the TV are **n** and **F**$_g$. The reactions to these forces, **n**′ and **F**$_g'$, act on the table and on the Earth, respectively, and therefore do not appear in Newton's second law applied to the TV.

When a rope attached to an object is pulling on the object, the rope exerts a force **T** on the object, and the magnitude of that force is called the **tension** in the rope. Because it is the magnitude of a vector quantity, tension is a scalar quantity.

Consider a crate being pulled to the right on a frictionless, horizontal surface, as shown in Figure 5.8a. Suppose you are asked to find the acceleration of the crate and the force the floor exerts on it. First, note that the horizontal force being applied to the crate acts through the rope. Use the symbol **T** to denote the force exerted by the rope on the crate. The magnitude of **T** is equal to the tension in the rope. A dotted circle is drawn around the crate in Figure 5.8a to remind you that you are interested only in the forces acting on the crate. These are illustrated in Figure 5.8b. In addition to the force **T**, this force diagram for the crate includes the force of gravity **F**$_g$ and the normal force **n** exerted by the floor on the crate. Such a force diagram, referred to as a **free-body diagram,** shows all external forces acting on the object. The construction of a correct free-body diagram is an important step in applying Newton's laws. The *reactions* to the forces we have listed—namely, the force exerted by the crate on the rope, the force exerted by the crate on the Earth, and the force exerted by the crate on the floor—are *not* included in the free-body diagram because they act on *other* bodies and not on the crate.

We can now apply Newton's second law in component form to the crate. The only force acting in the *x* direction is **T**. Applying $\Sigma F_x = ma_x$ to the horizontal motion gives

$$\sum F_x = T = ma_x \qquad \text{or} \qquad a_x = \frac{T}{m}$$

No acceleration occurs in the *y* direction. Applying $\Sigma F_y = ma_y$ with $a_y = 0$ yields

$$n + (-F_g) = 0 \qquad \text{or} \qquad n = F_g$$

That is, the normal force has the same magnitude as the force of gravity but is in the opposite direction.

If **T** is a constant force, then the acceleration $a_x = T/m$ also is constant. Hence, the constant-acceleration equations of kinematics from Chapter 2 can be used to obtain the crate's displacement Δx and velocity v_x as functions of time. Be-

Tension

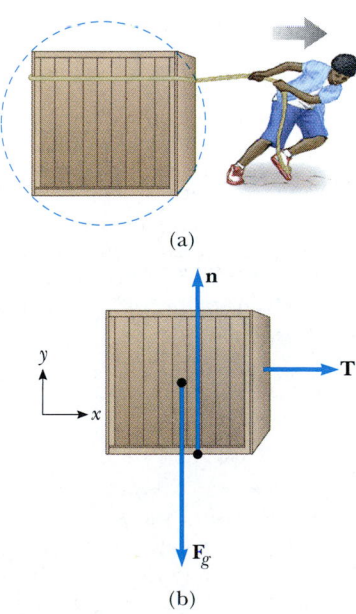

(a)

(b)

Figure 5.8 (a) A crate being pulled to the right on a frictionless surface. (b) The free-body diagram representing the external forces acting on the crate.

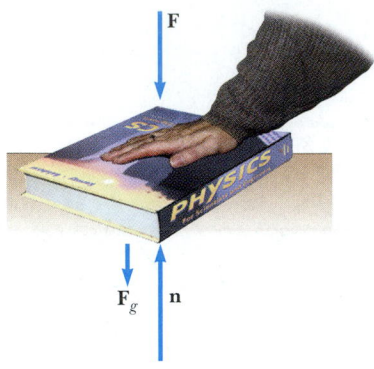

Figure 5.9 When one object pushes downward on another object with a force **F**, the normal force **n** is greater than the force of gravity: $n = F_g + F$.

cause $a_x = T/m =$ constant, Equations 2.8 and 2.11 can be written as

$$v_{xf} = v_{xi} + \left(\frac{T}{m}\right)t$$

$$\Delta x = v_{xi}t + \tfrac{1}{2}\left(\frac{T}{m}\right)t^2$$

In the situation just described, the magnitude of the normal force **n** is equal to the magnitude of $\mathbf{F}_g$, but this is not always the case. For example, suppose a book is lying on a table and you push down on the book with a force **F**, as shown in Figure 5.9. Because the book is at rest and therefore not accelerating, $\Sigma F_y = 0$, which gives $n - F_g - F = 0$, or $n = F_g + F$. Other examples in which $n \neq F_g$ are presented later.

Consider a lamp suspended from a light chain fastened to the ceiling, as in Figure 5.10a. The free-body diagram for the lamp (Figure 5.10b) shows that the forces acting on the lamp are the downward force of gravity $\mathbf{F}_g$ and the upward force **T** exerted by the chain. If we apply the second law to the lamp, noting that $\mathbf{a} = 0$, we see that because there are no forces in the x direction, $\Sigma F_x = 0$ provides no helpful information. The condition $\Sigma F_y = ma_y = 0$ gives

$$\sum F_y = T - F_g = 0 \qquad \text{or} \qquad T = F_g$$

Again, note that **T** and $\mathbf{F}_g$ are *not* an action–reaction pair because they act on the same object—the lamp. The reaction force to **T** is **T**′, the downward force exerted by the lamp on the chain, as shown in Figure 5.10c. The ceiling exerts on the chain a force **T**″ that is equal in magnitude to the magnitude of **T**′ and points in the opposite direction.

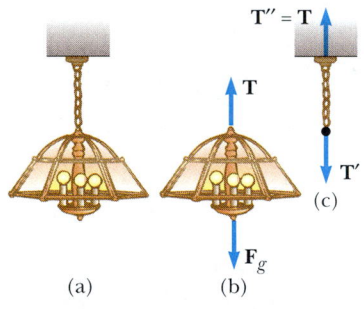

Figure 5.10 (a) A lamp suspended from a ceiling by a chain of negligible mass. (b) The forces acting on the lamp are the force of gravity $\mathbf{F}_g$ and the force exerted by the chain **T**. (c) The forces acting on the chain are the force exerted by the lamp **T**′ and the force exerted by the ceiling **T**″.

Problem-Solving Hints

Applying Newton's Laws

The following procedure is recommended when dealing with problems involving Newton's laws:

- Draw a simple, neat diagram of the system.
- Isolate the object whose motion is being analyzed. Draw a free-body diagram for this object. For systems containing more than one object, draw *separate* free-body diagrams for each object. *Do not* include in the free-body diagram forces exerted by the object on its surroundings. Establish convenient coordinate axes for each object and find the components of the forces along these axes.
- Apply Newton's second law, $\Sigma \mathbf{F} = m\mathbf{a}$, in component form. Check your dimensions to make sure that all terms have units of force.
- Solve the component equations for the unknowns. Remember that you must have as many independent equations as you have unknowns to obtain a complete solution.
- Make sure your results are consistent with the free-body diagram. Also check the predictions of your solutions for extreme values of the variables. By doing so, you can often detect errors in your results.

EXAMPLE 5.4 A Traffic Light at Rest

A traffic light weighing 125 N hangs from a cable tied to two other cables fastened to a support. The upper cables make angles of 37.0° and 53.0° with the horizontal. Find the tension in the three cables.

Solution Figure 5.11a shows the type of drawing we might make of this situation. We then construct two free-body diagrams—one for the traffic light, shown in Figure 5.11b, and one for the knot that holds the three cables together, as seen in Figure 5.11c. This knot is a convenient object to choose because all the forces we are interested in act through it. Because the acceleration of the system is zero, we know that the net force on the light and the net force on the knot are both zero.

In Figure 5.11b the force $\mathbf{T}_3$ exerted by the vertical cable supports the light, and so $T_3 = F_g =$ 125 N. Next, we choose the coordinate axes shown in Figure 5.11c and resolve the forces acting on the knot into their components:

Force	x Component	y Component
$\mathbf{T}_1$	$-T_1 \cos 37.0°$	$T_1 \sin 37.0°$
$\mathbf{T}_2$	$T_2 \cos 53.0°$	$T_2 \sin 53.0°$
$\mathbf{T}_3$	0	-125 N

Knowing that the knot is in equilibrium ($\mathbf{a} = 0$) allows us to write

$$(1) \qquad \sum F_x = -T_1 \cos 37.0° + T_2 \cos 53.0° = 0$$

$$(2) \qquad \sum F_y = T_1 \sin 37.0° + T_2 \sin 53.0°$$
$$+ (-125 \text{ N}) = 0$$

From (1) we see that the horizontal components of $\mathbf{T}_1$ and $\mathbf{T}_2$ must be equal in magnitude, and from (2) we see that the sum of the vertical components of $\mathbf{T}_1$ and $\mathbf{T}_2$ must balance the weight of the light. We solve (1) for T_2 in terms of T_1 to obtain

$$T_2 = T_1 \left(\frac{\cos 37.0°}{\cos 53.0°} \right) = 1.33 T_1$$

This value for T_2 is substituted into (2) to yield

$$T_1 \sin 37.0° + (1.33 T_1)(\sin 53.0°) - 125 \text{ N} = 0$$

$$T_1 = \boxed{75.1 \text{ N}}$$

$$T_2 = 1.33 T_1 = \boxed{99.9 \text{ N}}$$

This problem is important because it combines what we have learned about vectors with the new topic of forces. The general approach taken here is very powerful, and we will repeat it many times.

Exercise In what situation does $T_1 = T_2$?

Answer When the two cables attached to the support make equal angles with the horizontal.

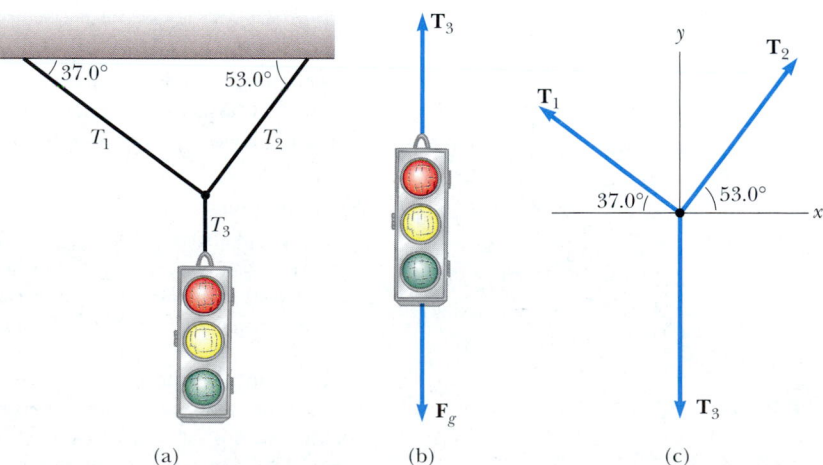

(a) (b) (c)

Figure 5.11 (a) A traffic light suspended by cables. (b) Free-body diagram for the traffic light. (c) Free-body diagram for the knot where the three cables are joined.

CONCEPTUAL EXAMPLE 5.5   Forces Between Cars in a Train

In a train, the cars are connected by *couplers,* which are under tension as the locomotive pulls the train. As you move down the train from locomotive to caboose, does the tension in the couplers increase, decrease, or stay the same as the train speeds up? When the engineer applies the brakes, the couplers are under compression. How does this compression force vary from locomotive to caboose? (Assume that only the brakes on the wheels of the engine are applied.)

Solution   As the train speeds up, the tension decreases from the front of the train to the back. The coupler between

the locomotive and the first car must apply enough force to accelerate all of the remaining cars. As you move back along the train, each coupler is accelerating less mass behind it. The last coupler has to accelerate only the caboose, and so it is under the least tension.

When the brakes are applied, the force again decreases from front to back. The coupler connecting the locomotive to the first car must apply a large force to slow down all the remaining cars. The final coupler must apply a force large enough to slow down only the caboose.

EXAMPLE 5.6   Crate on a Frictionless Incline

A crate of mass m is placed on a frictionless inclined plane of angle θ. (a) Determine the acceleration of the crate after it is released.

Solution   Because we know the forces acting on the crate, we can use Newton's second law to determine its acceleration. (In other words, we have classified the problem; this gives us a hint as to the approach to take.) We make a sketch as in Figure 5.12a and then construct the free-body diagram for the crate, as shown in Figure 5.12b. The only forces acting on the crate are the normal force **n** exerted by the inclined plane, which acts perpendicular to the plane, and the force of gravity $\mathbf{F}_g = m\mathbf{g}$, which acts vertically downward. For problems involving inclined planes, it is convenient to choose the coordinate axes with x downward along the incline and y perpendicular to it, as shown in Figure 5.12b. (It is possible to solve the problem with "standard" horizontal and vertical axes. You may want to try this, just for practice.) Then, we re-

place the force of gravity by a component of magnitude $mg \sin \theta$ along the positive x axis and by one of magnitude $mg \cos \theta$ along the negative y axis.

Now we apply Newton's second law in component form, noting that $a_y = 0$:

$$(1) \qquad \sum F_x = mg \sin \theta = ma_x$$

$$(2) \qquad \sum F_y = n - mg \cos \theta = 0$$

Solving (1) for a_x, we see that the acceleration along the incline is caused by the component of $\mathbf{F}_g$ directed down the incline:

$$(3) \qquad a_x = g \sin \theta$$

Note that this acceleration component is independent of the mass of the crate! It depends only on the angle of inclination and on g.

From (2) we conclude that the component of $\mathbf{F}_g$ perpendicular to the incline is balanced by the normal force; that is, $n = mg \cos \theta$. This is one example of a situation in which the normal force is *not* equal in magnitude to the weight of the object.

Special Cases   Looking over our results, we see that in the extreme case of $\theta = 90°$, $a_x = g$ and $n = 0$. This condition corresponds to the crate's being in free fall. When $\theta = 0$, $a_x = 0$ and $n = mg$ (its maximum value); in this case, the crate is sitting on a horizontal surface.

(b) Suppose the crate is released from rest at the top of the incline, and the distance from the front edge of the crate to the bottom is d. How long does it take the front edge to reach the bottom, and what is its speed just as it gets there?

Solution   Because $a_x =$ constant, we can apply Equation 2.11, $x_f - x_i = v_{xi}t + \frac{1}{2}a_x t^2$, to analyze the crate's motion.

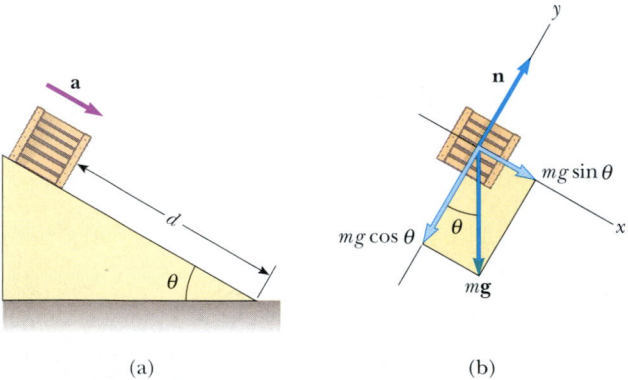

(a)   (b)

Figure 5.12   (a) A crate of mass m sliding down a frictionless incline. (b) The free-body diagram for the crate. Note that its acceleration along the incline is $g \sin \theta$.

With the displacement $x_f - x_i = d$ and $v_{xi} = 0$, we obtain

$$d = \tfrac{1}{2} a_x t^2$$

(4) $\qquad t = \sqrt{\dfrac{2d}{a_x}} = \sqrt{\dfrac{2d}{g \sin \theta}}$

Using Equation 2.12, $v_{xf}{}^2 = v_{xi}{}^2 + 2a_x(x_f - x_i)$, with $v_{xi} = 0$, we find that

$$v_{xf}{}^2 = 2a_x d$$

(5) $\qquad v_{xf} = \sqrt{2a_x d} = \sqrt{2gd \sin \theta}$

We see from equations (4) and (5) that the time t needed to reach the bottom and the speed v_{xf}, like acceleration, are independent of the crate's mass. This suggests a simple method you can use to measure g, using an inclined air track: Measure the angle of inclination, some distance traveled by a cart along the incline, and the time needed to travel that distance. The value of g can then be calculated from (4).

EXAMPLE 5.7 ▸ One Block Pushes Another

Two blocks of masses m_1 and m_2 are placed in contact with each other on a frictionless horizontal surface. A constant horizontal force **F** is applied to the block of mass m_1. (a) Determine the magnitude of the acceleration of the two-block system.

Solution Common sense tells us that both blocks must experience the same acceleration because they remain in contact with each other. Just as in the preceding example, we make a labeled sketch and free-body diagrams, which are shown in Figure 5.13. In Figure 5.13a the dashed line indicates that we treat the two blocks together as a system. Because **F** is the only external horizontal force acting on the system (the two blocks), we have

$$\sum F_x(\text{system}) = F = (m_1 + m_2)a_x$$

(1) $\qquad a_x = \dfrac{F}{m_1 + m_2}$

Treating the two blocks together as a system simplifies the solution but does not provide information about internal forces.

(b) Determine the magnitude of the contact force between the two blocks.

Solution To solve this part of the problem, we must treat each block separately with its own free-body diagram, as in Figures 5.13b and 5.13c. We denote the contact force by **P**. From Figure 5.13c, we see that the only horizontal force acting on block 2 is the contact force **P** (the force exerted by block 1 on block 2), which is directed to the right. Applying Newton's second law to block 2 gives

(2) $\qquad \sum F_x = P = m_2 a_x$

Substituting into (2) the value of a_x given by (1), we obtain

(3) $\qquad P = m_2 a_x = \left(\dfrac{m_2}{m_1 + m_2}\right)F$

From this result, we see that the contact force **P** exerted by block 1 on block 2 is *less* than the applied force **F**. This is consistent with the fact that the force required to accelerate block 2 alone must be less than the force required to produce the same acceleration for the two-block system.

It is instructive to check this expression for P by considering the forces acting on block 1, shown in Figure 5.13b. The horizontal forces acting on this block are the applied force **F** to the right and the contact force **P′** to the left (the force exerted by block 2 on block 1). From Newton's third law, **P′** is the reaction to **P**, so that $|\mathbf{P'}| = |\mathbf{P}|$. Applying Newton's second law to block 1 produces

(4) $\qquad \sum F_x = F - P' = F - P = m_1 a_x$

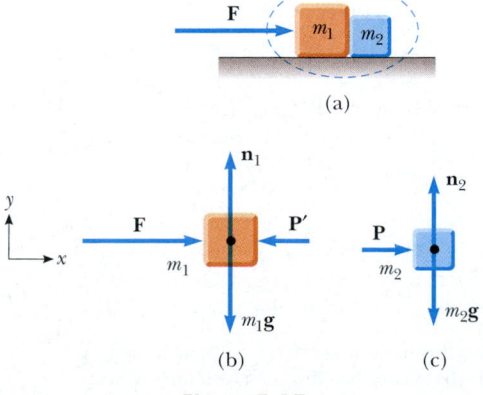

(a)

(b)   (c)

Figure 5.13

Substituting into (4) the value of a_x from (1), we obtain

$$P = F - m_1 a_x = F - \frac{m_1 F}{m_1 + m_2} = \left(\frac{m_2}{m_1 + m_2}\right)F$$

This agrees with (3), as it must.

Exercise   If $m_1 = 4.00$ kg, $m_2 = 3.00$ kg, and $F = 9.00$ N, find the magnitude of the acceleration of the system and the magnitude of the contact force.

Answer $a_x = 1.29$ m/s^2; $P = 3.86$ N.

EXAMPLE 5.8   Weighing a Fish in an Elevator

A person weighs a fish of mass m on a spring scale attached to the ceiling of an elevator, as illustrated in Figure 5.14. Show that if the elevator accelerates either upward or downward, the spring scale gives a reading that is different from the weight of the fish.

Solution   The external forces acting on the fish are the downward force of gravity $\mathbf{F}_g = m\mathbf{g}$ and the force $\mathbf{T}$ exerted by the scale. By Newton's third law, the tension T is also the reading of the scale. If the elevator is either at rest or moving at constant velocity, the fish is not accelerating, and so $\Sigma F_y = T - mg = 0$ or $T = mg$ (remember that the scalar mg is the weight of the fish).

If the elevator moves upward with an acceleration $\mathbf{a}$ relative to an observer standing outside the elevator in an inertial frame (see Fig. 5.14a), Newton's second law applied to the fish gives the net force on the fish:

$$(1) \qquad \sum F_y = T - mg = ma_y$$

where we have chosen upward as the positive direction. Thus, we conclude from (1) that the scale reading T is greater than the weight mg if $\mathbf{a}$ is upward, so that a_y is positive, and that the reading is less than mg if $\mathbf{a}$ is downward, so that a_y is negative.

For example, if the weight of the fish is 40.0 N and $\mathbf{a}$ is upward, so that $a_y = +2.00$ m/s^2, the scale reading from (1) is

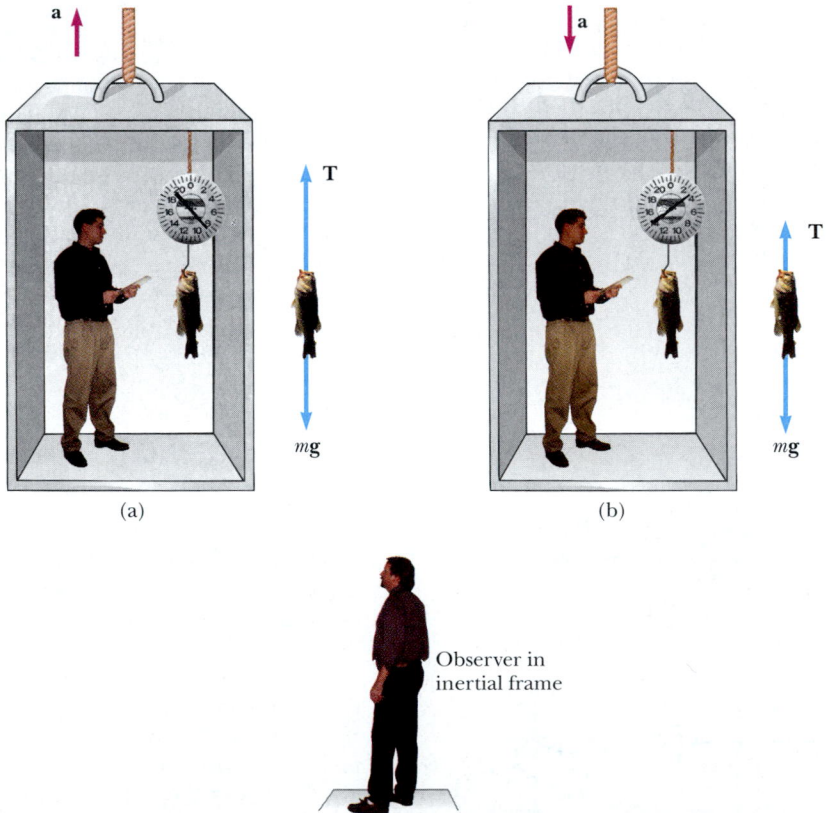

(a)   *(b)*

Observer in inertial frame

Figure 5.14   Apparent weight versus true weight. (a) When the elevator accelerates upward, the spring scale reads a value greater than the weight of the fish. (b) When the elevator accelerates downward, the spring scale reads a value less than the weight of the fish.

$$(2) \quad T = ma_y + mg = mg\left(\frac{a_y}{g} + 1\right)$$

$$= (40.0 \text{ N})\left(\frac{2.00 \text{ m/s}^2}{9.80 \text{ m/s}^2} + 1\right)$$

$$= \boxed{48.2 \text{ N}}$$

If **a** is downward so that $a_y = -2.00 \text{ m/s}^2$, then (2) gives us

$$T = mg\left(\frac{a_y}{g} + 1\right) = (40.0 \text{ N})\left(\frac{-2.00 \text{ m/s}^2}{9.80 \text{ m/s}^2} + 1\right)$$

$$= \boxed{31.8 \text{ N}}$$

Hence, if you buy a fish by weight in an elevator, make sure the fish is weighed while the elevator is either at rest or accelerating downward! Furthermore, note that from the information given here one cannot determine the direction of motion of the elevator.

Special Cases If the elevator cable breaks, the elevator falls freely and $a_y = -g$. We see from (2) that the scale reading T is zero in this case; that is, the fish appears to be weightless. If the elevator accelerates downward with an acceleration greater than g, the fish (along with the person in the elevator) eventually hits the ceiling because the acceleration of fish and person is still that of a freely falling object relative to an outside observer.

EXAMPLE 5.9 ▶ Atwood's Machine

When two objects of unequal mass are hung vertically over a frictionless pulley of negligible mass, as shown in Figure 5.15a, the arrangement is called an *Atwood machine*. The device is sometimes used in the laboratory to measure the free-fall acceleration. Determine the magnitude of the acceleration of the two objects and the tension in the lightweight cord.

Solution If we were to define our system as being made up of both objects, as we did in Example 5.7, we would have to determine an *internal* force (tension in the cord). We must define two systems here—one for each object—and apply Newton's second law to each. The free-body diagrams for the two objects are shown in Figure 5.15b. Two forces act on each object: the upward force **T** exerted by the cord and the downward force of gravity.

We need to be very careful with signs in problems such as this, in which a string or rope passes over a pulley or some other structure that causes the string or rope to bend. In Figure 5.15a, notice that if object 1 accelerates upward, then object 2 accelerates downward. Thus, for consistency with signs, if we define the upward direction as positive for object 1, we must define the downward direction as positive for object 2. With this sign convention, both objects accelerate in the same direction as defined by the choice of sign. With this sign convention applied to the forces, the y component of the net force exerted on object 1 is $T - m_1g$, and the y component of the net force exerted on object 2 is $m_2g - T$. Because the objects are connected by a cord, their accelerations must be equal in magnitude. (Otherwise the cord would stretch or break as the distance between the objects increased.) If we assume $m_2 > m_1$, then object 1 must accelerate upward and object 2 downward.

When Newton's second law is applied to object 1, we obtain

$$(1) \quad \sum F_y = T - m_1g = m_1 a_y$$

Similarly, for object 2 we find

$$(2) \quad \sum F_y = m_2g - T = m_2 a_y$$

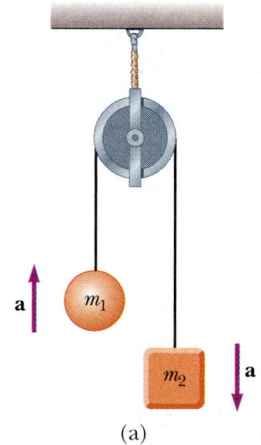

(a)

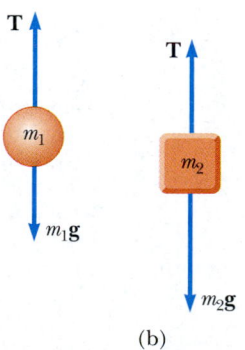

(b)

Figure 5.15 Atwood's machine. (a) Two objects ($m_2 > m_1$) connected by a cord of negligible mass strung over a frictionless pulley. (b) Free-body diagrams for the two objects.

When (2) is added to (1), T drops out and we get

$$-m_1g + m_2g = m_1a_y + m_2a_y$$

(3) $$a_y = \left(\frac{m_2 - m_1}{m_1 + m_2}\right)g$$

When (3) is substituted into (1), we obtain

(4) $$T = \left(\frac{2m_1m_2}{m_1 + m_2}\right)g$$

The result for the acceleration in (3) can be interpreted as the ratio of the unbalanced force on the system $(m_2g - m_1g)$ to the total mass of the system $(m_1 + m_2)$, as expected from Newton's second law.

Special Cases   When $m_1 = m_2$, then $a_y = 0$ and $T = m_1g$, as we would expect for this balanced case. If $m_2 \gg m_1$, then $a_y \approx g$ (a freely falling body) and $T \approx 2m_1g$.

Exercise   Find the magnitude of the acceleration and the string tension for an Atwood machine in which $m_1 = 2.00$ kg and $m_2 = 4.00$ kg.

Answer $a_y = 3.27$ m/s², $T = 26.1$ N.

EXAMPLE 5.10   Acceleration of Two Objects Connected by a Cord

A ball of mass m_1 and a block of mass m_2 are attached by a lightweight cord that passes over a frictionless pulley of negligible mass, as shown in Figure 5.16a. The block lies on a frictionless incline of angle θ. Find the magnitude of the acceleration of the two objects and the tension in the cord.

Solution   Because the objects are connected by a cord (which we assume does not stretch), their accelerations have the same magnitude. The free-body diagrams are shown in Figures 5.16b and 5.16c. Applying Newton's second law in component form to the ball, with the choice of the upward direction as positive, yields

(1) $$\sum F_x = 0$$
(2) $$\sum F_y = T - m_1g = m_1a_y = m_1a$$

Note that in order for the ball to accelerate upward, it is necessary that $T > m_1g$. In (2) we have replaced a_y with a because the acceleration has only a y component.

For the block it is convenient to choose the positive x' axis along the incline, as shown in Figure 5.16c. Here we choose the positive direction to be down the incline, in the $+x'$ di-

rection. Applying Newton's second law in component form to the block gives

(3) $$\sum F_{x'} = m_2g\sin\theta - T = m_2a_{x'} = m_2a$$
(4) $$\sum F_{y'} = n - m_2g\cos\theta = 0$$

In (3) we have replaced $a_{x'}$ with a because that is the acceleration's only component. In other words, the two objects have accelerations of the same magnitude a, which is what we are trying to find. Equations (1) and (4) provide no information regarding the acceleration. However, if we solve (2) for T and then substitute this value for T into (3) and solve for a, we obtain

(5) $$a = \frac{m_2g\sin\theta - m_1g}{m_1 + m_2}$$

When this value for a is substituted into (2), we find

(6) $$T = \frac{m_1m_2g(\sin\theta + 1)}{m_1 + m_2}$$

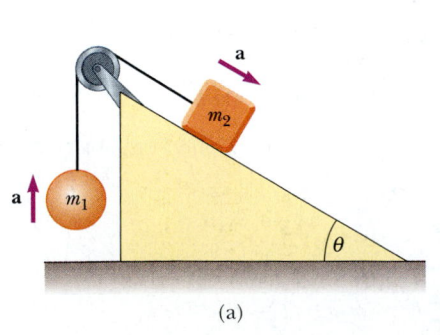

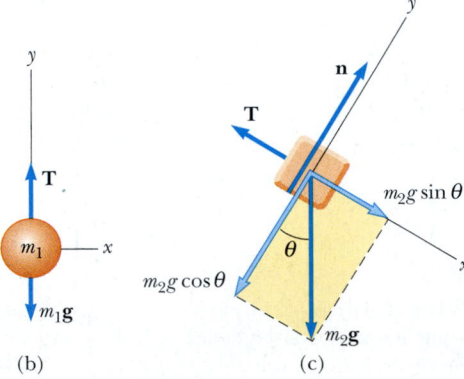

Figure 5.16   (a) Two objects connected by a lightweight cord strung over a frictionless pulley. (b) Free-body diagram for the ball. (c) Free-body diagram for the block. (The incline is frictionless.)

(a)   (b)   (c)

Note that the block accelerates down the incline only if $m_2 \sin \theta > m_1$ (that is, if **a** is in the direction we assumed). If $m_1 > m_2 \sin \theta$, then the acceleration is up the incline for the block and downward for the ball. Also note that the result for the acceleration (5) can be interpreted as the resultant force acting on the system divided by the total mass of the system; this is consistent with Newton's second law. Finally, if $\theta = 90°$, then the results for a and T are identical to those of Example 5.9.

Exercise If $m_1 = 10.0$ kg, $m_2 = 5.00$ kg, and $\theta = 45.0°$, find the acceleration of each object.

Answer $a = -4.22$ m/s^2, where the negative sign indicates that the block accelerates up the incline and the ball accelerates downward.

5.8 ▶ FORCES OF FRICTION

When a body is in motion either on a surface or in a viscous medium such as air or water, there is resistance to the motion because the body interacts with its surroundings. We call such resistance a **force of friction**. Forces of friction are very important in our everyday lives. They allow us to walk or run and are necessary for the motion of wheeled vehicles.

Have you ever tried to move a heavy desk across a rough floor? You push harder and harder until all of a sudden the desk seems to "break free" and subsequently moves relatively easily. It takes a greater force to start the desk moving than it does to keep it going once it has started sliding. To understand why this happens, consider a book on a table, as shown in Figure 5.17a. If we apply an external horizontal force **F** to the book, acting to the right, the book remains stationary if **F** is not too great. The force that counteracts **F** and keeps the book from moving acts to the left and is called the **frictional force f**.

As long as the book is not moving, $f = F$. Because the book is stationary, we call this frictional force the **force of static friction f$_s$**. Experiments show that this force arises from contacting points that protrude beyond the general level of the surfaces in contact, even for surfaces that are apparently very smooth, as shown in the magnified view in Figure 5.17a. (If the surfaces are clean and smooth at the atomic level, they are likely to weld together when contact is made.) The frictional force arises in part from one peak's physically blocking the motion of a peak from the opposing surface, and in part from chemical bonding of opposing points as they come into contact. If the surfaces are rough, bouncing is likely to occur, further complicating the analysis. Although the details of friction are quite complex at the atomic level, this force ultimately involves an electrical interaction between atoms or molecules.

If we increase the magnitude of **F**, as shown in Figure 5.17b, the magnitude of **f$_s$** increases along with it, keeping the book in place. The force **f$_s$** cannot increase indefinitely, however. Eventually the surfaces in contact can no longer supply sufficient frictional force to counteract **F**, and the book accelerates. When it is on the verge of moving, f_s is a maximum, as shown in Figure 5.17c. When F exceeds $f_{s,\text{max}}$, the book accelerates to the right. Once the book is in motion, the retarding frictional force becomes less than $f_{s,\text{max}}$ (see Fig. 5.17c). When the book is in motion, we call the retarding force the **force of kinetic friction f$_k$**. If $F = f_k$, then the book moves to the right with constant speed. If $F > f_k$, then there is an unbalanced force $F - f_k$ in the positive x direction, and this force accelerates the book to the right. If the applied force **F** is removed, then the frictional force **f$_k$** acting to the left accelerates the book in the negative x direction and eventually brings it to rest.

Experimentally, we find that, to a good approximation, both $f_{s,\text{max}}$ and f_k are proportional to the normal force acting on the book. The following empirical laws of friction summarize the experimental observations:

Force of static friction

Force of kinetic friction

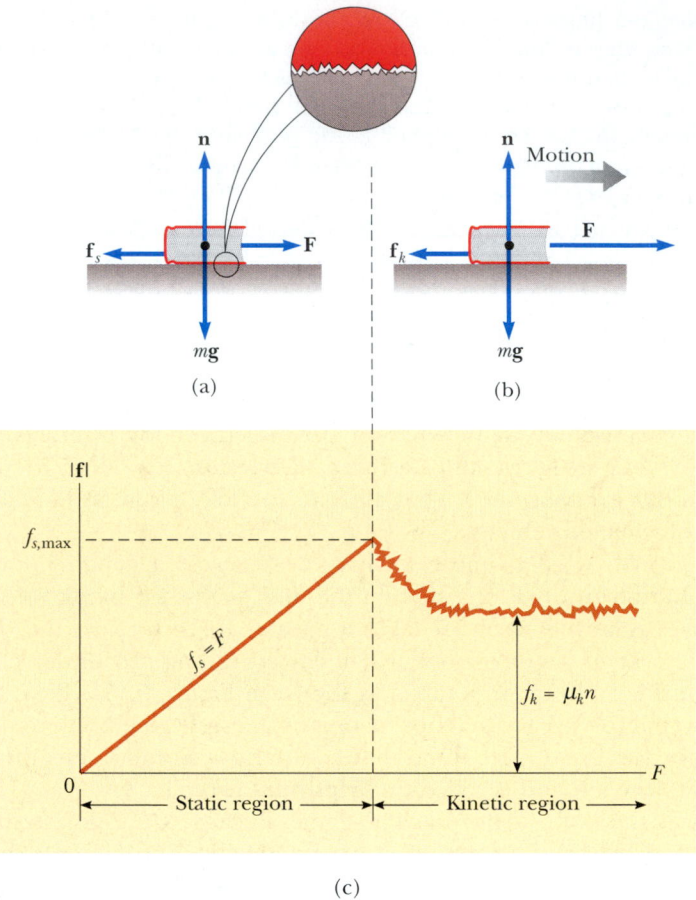

Figure 5.17   The direction of the force of friction **f** between a book and a rough surface is opposite the direction of the applied force **F**. Because the two surfaces are both rough, contact is made only at a few points, as illustrated in the "magnified" view. (a) The magnitude of the force of static friction equals the magnitude of the applied force. (b) When the magnitude of the applied force exceeds the magnitude of the force of kinetic friction, the book accelerates to the right. (c) A graph of frictional force versus applied force. Note that $f_{s,max} > f_k$.

- The direction of the force of static friction between any two surfaces in contact with each other is opposite the direction of relative motion and can have values

$$f_s \leq \mu_s n \qquad (5.8)$$

where the dimensionless constant μ_s is called the **coefficient of static friction** and n is the magnitude of the normal force. The equality in Equation 5.8 holds when one object is on the verge of moving, that is, when $f_s = f_{s,max} = \mu_s n$. The inequality holds when the applied force is less than $\mu_s n$.
- The direction of the force of kinetic friction acting on an object is opposite the direction of the object's sliding motion relative to the surface applying the frictional force and is given by

$$f_k = \mu_k n \qquad (5.9)$$

where μ_k is the **coefficient of kinetic friction.**
- The values of μ_k and μ_s depend on the nature of the surfaces, but μ_k is generally less than μ_s. Typical values range from around 0.03 to 1.0. Table 5.2 lists some reported values.

TABLE 5.2 **Coefficients of Friction**[a]		
	μ_s	μ_k
Steel on steel	0.74	0.57
Aluminum on steel	0.61	0.47
Copper on steel	0.53	0.36
Rubber on concrete	1.0	0.8
Wood on wood	0.25–0.5	0.2
Glass on glass	0.94	0.4
Waxed wood on wet snow	0.14	0.1
Waxed wood on dry snow	—	0.04
Metal on metal (lubricated)	0.15	0.06
Ice on ice	0.1	0.03
Teflon on Teflon	0.04	0.04
Synovial joints in humans	0.01	0.003

[a] All values are approximate. In some cases, the coefficient of friction can exceed 1.0.

- The coefficients of friction are nearly independent of the area of contact between the surfaces. To understand why, we must examine the difference between the *apparent contact area,* which is the area we see with our eyes, and the *real contact area,* represented by two irregular surfaces touching, as shown in the magnified view in Figure 5.17a. It seems that increasing the apparent contact area does not increase the real contact area. When we increase the apparent area (without changing anything else), there is less force per unit area driving the jagged points together. This decrease in force counteracts the effect of having more points involved.

> If you would like to learn more about this subject, read the article "Friction at the Atomic Scale" by J. Krim in the October 1996 issue of *Scientific American.*

Although the coefficient of kinetic friction can vary with speed, we shall usually neglect any such variations in this text. We can easily demonstrate the approximate nature of the equations by trying to get a block to slip down an incline at constant speed. Especially at low speeds, the motion is likely to be characterized by alternate episodes of sticking and movement.

QuickLab

Can you apply the ideas of Example 5.12 to determine the coefficients of static and kinetic friction between the cover of your book and a quarter? What should happen to those coefficients if you make the measurements between your book and *two* quarters taped one on top of the other?

Quick Quiz 5.6

A crate is sitting in the center of a flatbed truck. The truck accelerates to the right, and the crate moves with it, not sliding at all. What is the direction of the frictional force exerted by the truck on the crate? (a) To the left. (b) To the right. (c) No frictional force because the crate is not sliding.

CONCEPTUAL EXAMPLE 5.11   Why Does the Sled Accelerate?

A horse pulls a sled along a level, snow-covered road, causing the sled to accelerate, as shown in Figure 5.18a. Newton's third law states that the sled exerts an equal and opposite force on the horse. In view of this, how can the sled accelerate? Under what condition does the system (horse plus sled) move with constant velocity?

Solution   It is important to remember that the forces described in Newton's third law act on different objects—the horse exerts a force on the sled, and the sled exerts an equal-magnitude and oppositely directed force on the horse. Because we are interested only in the motion of the sled, we do not consider the forces it exerts on the horse. When deter-

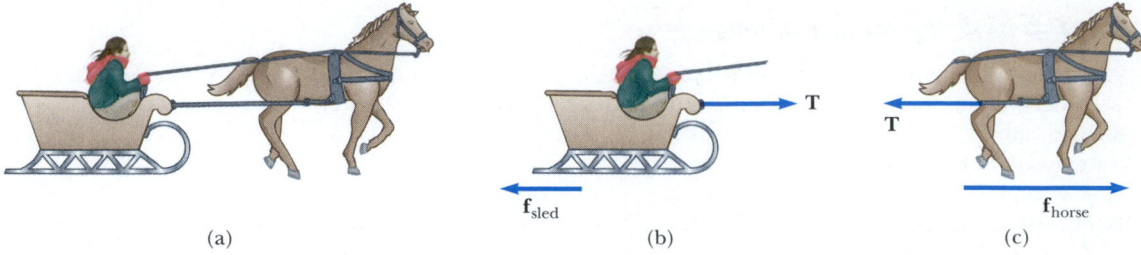

(a)   (b)   (c)

Figure 5.18

mining the motion of an object, you must add only the forces on that object. The horizontal forces exerted on the sled are the forward force **T** exerted by the horse and the backward force of friction $\mathbf{f}_{\text{sled}}$ between sled and snow (see Fig. 5.18b). When the forward force exceeds the backward force, the sled accelerates to the right.

The force that accelerates the system (horse plus sled) is the frictional force $\mathbf{f}_{\text{horse}}$ exerted by the Earth on the horse's feet. The horizontal forces exerted on the horse are the forward force $\mathbf{f}_{\text{horse}}$ exerted by the Earth and the backward tension force **T** exerted by the sled (Fig. 5.18c). The resultant of these two forces causes the horse to accelerate. When $\mathbf{f}_{\text{horse}}$ balances $\mathbf{f}_{\text{sled}}$, the system moves with constant velocity.

Exercise   Are the normal force exerted by the snow on the horse and the gravitational force exerted by the Earth on the horse a third-law pair?

Answer   No, because they act on the same object. Third-law force pairs are equal in magnitude and opposite in direction, and the forces act on *different* objects.

EXAMPLE 5.12   Experimental Determination of μ_s and μ_k

The following is a simple method of measuring coefficients of friction: Suppose a block is placed on a rough surface inclined relative to the horizontal, as shown in Figure 5.19. The incline angle is increased until the block starts to move. Let us show that by measuring the critical angle θ_c at which this slipping just occurs, we can obtain μ_s.

Solution   The only forces acting on the block are the force of gravity $m\mathbf{g}$, the normal force **n**, and the force of static friction $\mathbf{f}_s$. These forces balance when the block is on the verge of slipping but has not yet moved. When we take x to be parallel to the plane and y perpendicular to it, Newton's second law applied to the block for this balanced situation gives

Static case: (1) $\sum F_x = mg \sin \theta - f_s = ma_x = 0$

            (2) $\sum F_y = n - mg \cos \theta = ma_y = 0$

We can eliminate mg by substituting $mg = n/\cos \theta$ from (2) into (1) to get

(3) $f_s = mg \sin \theta = \left(\dfrac{n}{\cos \theta}\right) \sin \theta = n \tan \theta$

When the incline is at the critical angle θ_c, we know that $f_s = f_{s,\text{max}} = \mu_s n$, and so at this angle, (3) becomes

$$\mu_s n = n \tan \theta_c$$

Static case: $\mu_s = \tan \theta_c$

For example, if the block just slips at $\theta_c = 20°$, then we find that $\mu_s = \tan 20° = 0.364$.

Once the block starts to move at $\theta \geq \theta_c$, it accelerates down the incline and the force of friction is $f_k = \mu_k n$. However, if θ is reduced to a value less than θ_c, it may be possible to find an angle θ_c' such that the block moves down the incline with constant speed ($a_x = 0$). In this case, using (1) and (2) with f_s replaced by f_k gives

Kinetic case: $\mu_k = \tan \theta_c'$

where $\theta_c' < \theta_c$.

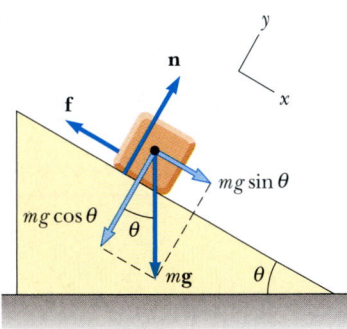

Figure 5.19   The external forces exerted on a block lying on a rough incline are the force of gravity $m\mathbf{g}$, the normal force **n**, and the force of friction **f**. For convenience, the force of gravity is resolved into a component along the incline $mg \sin \theta$ and a component perpendicular to the incline $mg \cos \theta$.

EXAMPLE 5.13   The Sliding Hockey Puck

A hockey puck on a frozen pond is given an initial speed of 20.0 m/s. If the puck always remains on the ice and slides 115 m before coming to rest, determine the coefficient of kinetic friction between the puck and ice.

Solution The forces acting on the puck after it is in motion are shown in Figure 5.20. If we assume that the force of kinetic friction f_k remains constant, then this force produces a uniform acceleration of the puck in the direction opposite its velocity, causing the puck to slow down. First, we find this acceleration in terms of the coefficient of kinetic friction, using Newton's second law. Knowing the acceleration of the puck and the distance it travels, we can then use the equations of kinematics to find the coefficient of kinetic friction.

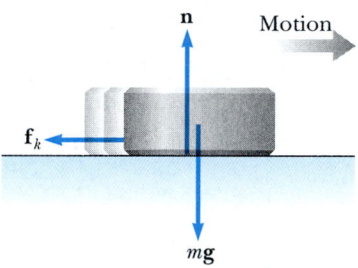

Figure 5.20 *After* the puck is given an initial velocity to the right, the only external forces acting on it are the force of gravity $m\mathbf{g}$, the normal force $\mathbf{n}$, and the force of kinetic friction $\mathbf{f}_k$.

Defining rightward and upward as our positive directions, we apply Newton's second law in component form to the puck and obtain

$$(1) \qquad \sum F_x = -f_k = ma_x$$

$$(2) \qquad \sum F_y = n - mg = 0 \qquad (a_y = 0)$$

But $f_k = \mu_k n$, and from (2) we see that $n = mg$. Therefore, (1) becomes

$$-\mu_k n = -\mu_k mg = ma_x$$

$$a_x = -\mu_k g$$

The negative sign means the acceleration is to the left; this means that the puck is slowing down. The acceleration is independent of the mass of the puck and is constant because we assume that μ_k remains constant.

Because the acceleration is constant, we can use Equation 2.12, $v_{xf}^2 = v_{xi}^2 + 2a_x(x_f - x_i)$, with $x_i = 0$ and $v_{xf} = 0$:

$$v_{xi}^2 + 2ax_f = v_{xi}^2 - 2\mu_k g x_f = 0$$

$$\mu_k = \frac{v_{xi}^2}{2g x_f}$$

$$\mu_k = \frac{(20.0 \text{ m/s})^2}{2(9.80 \text{ m/s}^2)(115 \text{ m})} = \boxed{0.177}$$

Note that μ_k is dimensionless.

EXAMPLE 5.14   Acceleration of Two Connected Objects When Friction Is Present

A block of mass m_1 on a rough, horizontal surface is connected to a ball of mass m_2 by a lightweight cord over a lightweight, frictionless pulley, as shown in Figure 5.21a. A force of magnitude F at an angle θ with the horizontal is applied to the block as shown. The coefficient of kinetic friction between the block and surface is μ_k. Determine the magnitude of the acceleration of the two objects.

Solution We start by drawing free-body diagrams for the two objects, as shown in Figures 5.21b and 5.21c. (Are you beginning to see the similarities in all these examples?) Next, we apply Newton's second law in component form to each object and use Equation 5.9, $f_k = \mu_k n$. Then we can solve for the acceleration in terms of the parameters given.

The applied force $\mathbf{F}$ has x and y components $F \cos \theta$ and $F \sin \theta$, respectively. Applying Newton's second law to both objects and assuming the motion of the block is to the right, we obtain

Motion of block: $(1) \qquad \sum F_x = F \cos \theta - f_k - T = m_1 a_x$
$$= m_1 a$$

$$(2) \qquad \sum F_y = n + F \sin \theta - m_1 g$$
$$= m_1 a_y = 0$$

Motion of ball: $\qquad \sum F_x = m_2 a_x = 0$

$$(3) \qquad \sum F_y = T - m_2 g = m_2 a_y = m_2 a$$

Note that because the two objects are connected, we can equate the magnitudes of the x component of the acceleration of the block and the y component of the acceleration of the ball. From Equation 5.9 we know that $f_k = \mu_k n$, and from (2) we know that $n = m_1 g - F \sin \theta$ (note that in this case n is *not* equal to $m_1 g$); therefore,

$$(4) \qquad f_k = \mu_k(m_1 g - F \sin \theta)$$

That is, the frictional force is reduced because of the positive *y* component of **F**. Substituting (4) and the value of *T* from (3) into (1) gives

$$F \cos \theta - \mu_k(m_1 g - F \sin \theta) - m_2(a + g) = m_1 a$$

Solving for *a*, we obtain

$$(5) \qquad a = \frac{F(\cos \theta + \mu_k \sin \theta) - g(m_2 + \mu_k m_1)}{m_1 + m_2}$$

Note that the acceleration of the block can be either to the right or to the left,[6] depending on the sign of the numerator in (5). If the motion is to the left, then we must reverse the sign of f_k in (1) because the force of kinetic friction must oppose the motion. In this case, the value of *a* is the same as in (5), with μ_k replaced by $-\mu_k$.

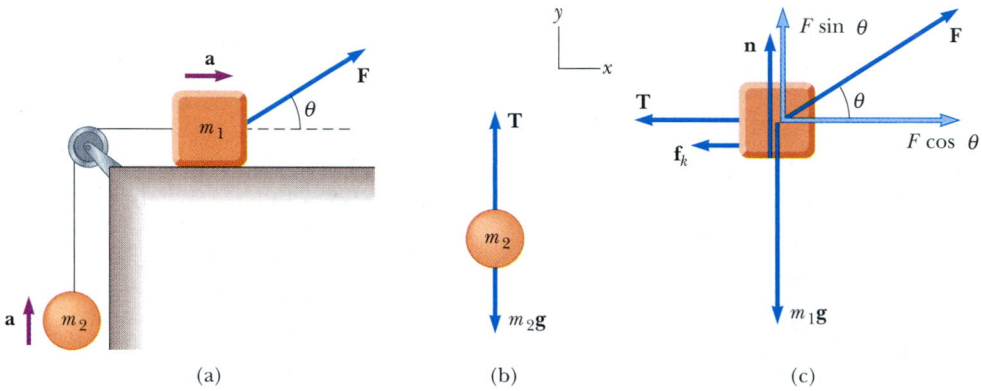

(a)   (b)   (c)

Figure 5.21   (a) The external force **F** applied as shown can cause the block to accelerate to the right. (b) and (c) The free-body diagrams, under the assumption that the block accelerates to the right and the ball accelerates upward. The magnitude of the force of kinetic friction in this case is given by $f_k = \mu_k n = \mu_k(m_1 g - F \sin \theta)$.

APPLICATION ▸   **Automobile Antilock Braking Systems (ABS)**

If an automobile tire is rolling and not slipping on a road surface, then the maximum frictional force that the road can exert on the tire is the force of static friction $\mu_s n$. One must use static friction in this situation because at the point of contact between the tire and the road, no sliding of one surface over the other occurs if the tire is not skidding. However, if the tire starts to skid, the frictional force exerted on it is reduced to the force of kinetic friction $\mu_k n$. Thus, to maximize the frictional force and minimize stopping distance, the wheels must maintain pure rolling motion and not skid. An additional benefit of maintaining wheel rotation is that directional control is not lost as it is in skidding.

Unfortunately, in emergency situations drivers typically press down as hard as they can on the brake pedal, "locking the brakes." This stops the wheels from rotating, ensuring a skid and reducing the frictional force from the static to the kinetic case. To address this problem, automotive engineers

have developed antilock braking systems (ABS) that very briefly release the brakes when a wheel is just about to stop turning. This maintains rolling contact between the tire and the pavement. When the brakes are released momentarily, the stopping distance is greater than it would be if the brakes were being applied continuously. However, through the use of computer control, the "brake-off" time is kept to a minimum. As a result, the stopping distance is much less than what it would be if the wheels were to skid.

Let us model the stopping of a car by examining real data. In a recent issue of *AutoWeek*,[7] the braking performance for a Toyota Corolla was measured. These data correspond to the braking force acquired by a highly trained, professional driver. We begin by assuming constant acceleration. (Why do we need to make this assumption?) The magazine provided the initial speed and stopping distance in non-SI units. After converting these values to SI we use $v_{xf}^2 = v_{xi}^2 + 2a_x x$ to deter-

---

[6] Equation 5 shows that when $\mu_k m_1 > m_2$, there is a range of values of *F* for which no motion occurs at a given angle θ.

[7] *AutoWeek* magazine, 48:22–23, 1998.

mine the acceleration at different speeds. These do not vary greatly, and so our assumption of constant acceleration is reasonable.

Initial Speed		Stopping Distance		Acceleration
(mi/h)	(m/s)	(ft)	(m)	(m/s²)
30	13.4	34	10.4	− 8.67
60	26.8	143	43.6	− 8.25
80	35.8	251	76.5	− 8.36

We take an average value of acceleration of -8.4 m/s², which is approximately $0.86g$. We then calculate the coefficient of friction from $\Sigma F = \mu_s mg = ma$; this gives $\mu_s = 0.86$ for the Toyota. This is lower than the rubber-to-concrete value given in Table 5.2. Can you think of any reasons for this?

Let us now estimate the stopping distance of the car if the wheels were skidding. Examining Table 5.2 again, we see that the difference between the coefficients of static and kinetic friction for rubber against concrete is about 0.2. Let us therefore assume that our coefficients differ by the same amount, so that $\mu_k \approx 0.66$. This allows us to calculate estimated stopping distances for the case in which the wheels are locked and the car skids across the pavement. The results illustrate the advantage of not allowing the wheels to skid.

Initial Speed (mi/h)	Stopping Distance no skid (m)	Stopping distance skidding (m)
30	10.4	13.9
60	43.6	55.5
80	76.5	98.9

An ABS keeps the wheels rotating, with the result that the higher coefficient of static friction is maintained between the tires and road. This approximates the technique of a professional driver who is able to maintain the wheels at the point of maximum frictional force. Let us estimate the ABS performance by assuming that the magnitude of the acceleration is not quite as good as that achieved by the professional driver but instead is reduced by 5%.

We now plot in Figure 5.22 vehicle speed versus distance from where the brakes were applied (at an initial speed of 80 mi/h = 37.5 m/s) for the three cases of amateur driver, professional driver, and estimated ABS performance (amateur driver). We find that a markedly shorter distance is necessary for stopping without locking the wheels and skidding and a satisfactory value of stopping distance when the ABS computer maintains tire rotation.

The purpose of the ABS is to help typical drivers (whose tendency is to lock the wheels in an emergency) to better maintain control of their automobiles and minimize stopping distance.

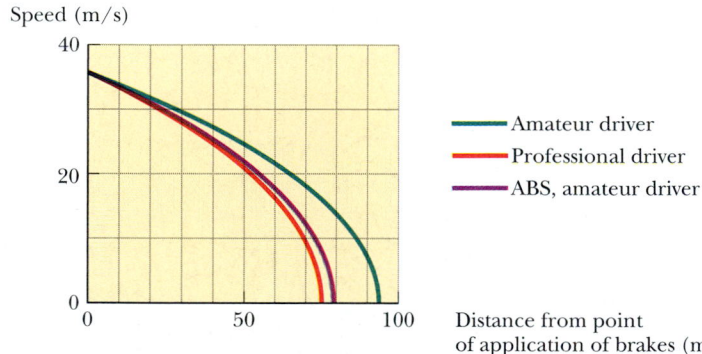

Speed (m/s)

Distance from point of application of brakes (m)

— Amateur driver
— Professional driver
— ABS, amateur driver

Figure 5.22   This plot of vehicle speed versus distance from where the brakes were applied shows that an antilock braking system (ABS) approaches the performance of a trained professional driver.

SUMMARY

Newton's first law states that, in the absence of an external force, a body at rest remains at rest and a body in uniform motion in a straight line maintains that motion. An **inertial frame** is one that is not accelerating.

Newton's second law states that the acceleration of an object is directly proportional to the net force acting on it and inversely proportional to its mass. The net force acting on an object equals the product of its mass and its acceleration: $\Sigma \mathbf{F} = m\mathbf{a}$. You should be able to apply the x and y component forms of this equation to describe the acceleration of any object acting under the influence of speci-

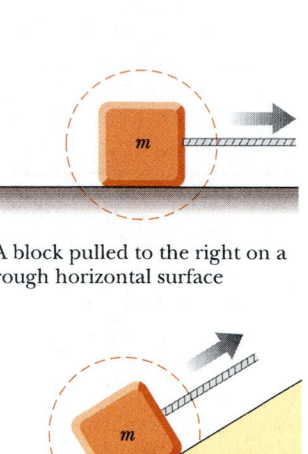

A block pulled to the right on a rough horizontal surface

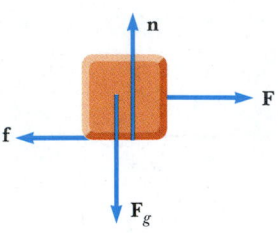

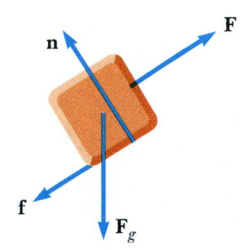

A block pulled up a rough incline

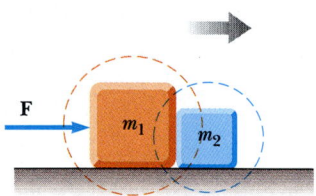

F

Two blocks in contact, pushed to the right on a frictionless surface

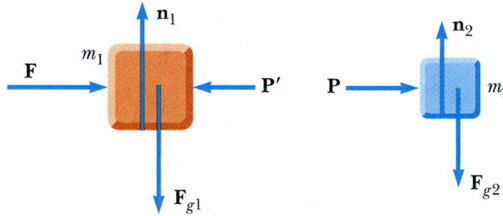

Note: **P** = − **P′** because they are an action–reaction pair.

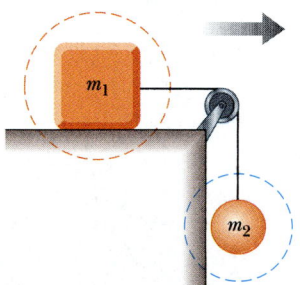

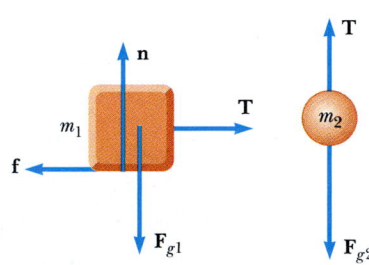

Two masses connected by a light cord. The surface is rough, and the pulley is frictionless.

Figure 5.23 Various systems (*left*) and the corresponding free-body diagrams (*right*).

fied forces. If the object is either stationary or moving with constant velocity, then the forces must vectorially cancel each other.

The **force of gravity** exerted on an object is equal to the product of its mass (a scalar quantity) and the free-fall acceleration: $\mathbf{F}_g = m\mathbf{g}$. The **weight** of an object is the magnitude of the force of gravity acting on the object.

Newton's third law states that if two objects interact, then the force exerted by object 1 on object 2 is equal in magnitude and opposite in direction to the force exerted by object 2 on object 1. Thus, an isolated force cannot exist in nature. Make sure you can identify third-law pairs and the two objects upon which they act.

The **maximum force of static friction** $\mathbf{f}_{s,\max}$ between an object and a surface is proportional to the normal force acting on the object. In general, $f_s \leq \mu_s n$, where μ_s is the **coefficient of static friction** and n is the magnitude of the normal force. When an object slides over a surface, the direction of the **force of kinetic friction** $\mathbf{f}_k$ is opposite the direction of sliding motion and is also proportional to the magnitude of the normal force. The magnitude of this force is given by $f_k = \mu_k n$, where μ_k is the **coefficient of kinetic friction.**

More on Free-Body Diagrams

To be successful in applying Newton's second law to a system, you must be able to recognize all the forces acting on the system. That is, you must be able to construct the correct free-body diagram. The importance of constructing the free-body diagram cannot be overemphasized. In Figure 5.23 a number of systems are presented together with their free-body diagrams. You should examine these carefully and then construct free-body diagrams for other systems described in the end-of-chapter problems. When a system contains more than one element, it is important that you construct a separate free-body diagram for *each* element.

As usual, $\mathbf{F}$ denotes some applied force, $\mathbf{F}_g = m\mathbf{g}$ is the force of gravity, $\mathbf{n}$ denotes a normal force, $\mathbf{f}$ is the force of friction, and $\mathbf{T}$ is the force whose magnitude is the tension exerted on an object.

QUESTIONS

1. A passenger sitting in the rear of a bus claims that he was injured when the driver slammed on the brakes, causing a suitcase to come flying toward the passenger from the front of the bus. If you were the judge in this case, what disposition would you make? Why?

2. A space explorer is in a spaceship moving through space far from any planet or star. She notices a large rock, taken as a specimen from an alien planet, floating around the cabin of the spaceship. Should she push it gently toward a storage compartment or kick it toward the compartment? Why?

3. A massive metal object on a rough metal surface may undergo contact welding to that surface. Discuss how this affects the frictional force between object and surface.

4. The observer in the elevator of Example 5.8 would claim that the weight of the fish is T, the scale reading. This claim is obviously wrong. Why does this observation differ from that of a person in an inertial frame outside the elevator?

5. Identify the action–reaction pairs in the following situations: a man takes a step; a snowball hits a woman in the back; a baseball player catches a ball; a gust of wind strikes a window.

6. A ball is held in a person's hand. (a) Identify all the external forces acting on the ball and the reaction to each. (b) If the ball is dropped, what force is exerted on it while it is falling? Identify the reaction force in this case. (Neglect air resistance.)

7. If a car is traveling westward with a constant speed of 20 m/s, what is the resultant force acting on it?

8. "When the locomotive in Figure 5.3 broke through the wall of the train station, the force exerted by the locomotive on the wall was greater than the force the wall could exert on the locomotive." Is this statement true or in need of correction? Explain your answer.

9. A rubber ball is dropped onto the floor. What force causes the ball to bounce?

10. What is wrong with the statement, "Because the car is at rest, no forces are acting on it"? How would you correct this statement?

11. Suppose you are driving a car along a highway at a high speed. Why should you avoid slamming on your brakes if you want to stop in the shortest distance? That is, why should you keep the wheels turning as you brake?

12. If you have ever taken a ride in an elevator of a high-rise building, you may have experienced a nauseating sensation of "heaviness" and "lightness" depending on the direction of the acceleration. Explain these sensations. Are we truly weightless in free-fall?

13. The driver of a speeding empty truck slams on the brakes and skids to a stop through a distance d. (a) If the truck carried a heavy load such that its mass were doubled, what would be its skidding distance? (b) If the initial speed of the truck is halved, what would be its skidding distance?

14. In an attempt to define Newton's third law, a student states that the action and reaction forces are equal in magnitude and opposite in direction to each other. If this is the case, how can there ever be a net force on an object?

15. What force causes (a) a propeller-driven airplane to move? (b) a rocket? (c) a person walking?

16. Suppose a large and spirited Freshman team is beating the Sophomores in a tug-of-war contest. The center of the rope being tugged is gradually accelerating toward the Freshman team. State the relationship between the strengths of these two forces: First, the force the Freshmen exert on the Sophomores; and second, the force the Sophomores exert on the Freshmen.

17. If you push on a heavy box that is at rest, you must exert some force to start its motion. However, once the box is sliding, you can apply a smaller force to maintain that motion. Why?

18. A weight lifter stands on a bathroom scale. He pumps a barbell up and down. What happens to the reading on the scale as this is done? Suppose he is strong enough to actually *throw* the barbell upward. How does the reading on the scale vary now?

19. As a rocket is fired from a launching pad, its speed *and* acceleration increase with time as its engines continue to operate. Explain why this occurs even though the force of the engines exerted on the rocket remains constant.

20. In the motion picture *It Happened One Night* (Columbia Pictures, 1934), Clark Gable is standing inside a stationary bus in front of Claudette Colbert, who is seated. The bus suddenly starts moving forward, and Clark falls into Claudette's lap. Why did this happen?

PROBLEMS

1, 2, 3 = straightforward, intermediate, challenging   ☐ = full solution available in the *Student Solutions Manual and Study Guide*
WEB = solution posted at **http://www.saunderscollege.com/physics/**   💻 = Computer useful in solving problem   🔧 = Interactive Physics
☐ = paired numerical/symbolic problems

Sections 5.1 through 5.6

1. A force **F** applied to an object of mass m_1 produces an acceleration of 3.00 m/s². The same force applied to a second object of mass m_2 produces an acceleration of 1.00 m/s². (a) What is the value of the ratio m_1/m_2? (b) If m_1 and m_2 are combined, find their acceleration under the action of the force **F**.

2. A force of 10.0 N acts on a body of mass 2.00 kg. What are (a) the body's acceleration, (b) its weight in newtons, and (c) its acceleration if the force is doubled?

3. A 3.00-kg mass undergoes an acceleration given by **a** = $(2.00\mathbf{i} + 5.00\mathbf{j})$ m/s². Find the resultant force $\Sigma\mathbf{F}$ and its magnitude.

4. A heavy freight train has a mass of 15 000 metric tons. If the locomotive can pull with a force of 750 000 N, how long does it take to increase the speed from 0 to 80.0 km/h?

5. A 5.00-g bullet leaves the muzzle of a rifle with a speed of 320 m/s. The expanding gases behind it exert what force on the bullet while it is traveling down the barrel of the rifle, 0.820 m long? Assume constant acceleration and negligible friction.

6. After uniformly accelerating his arm for 0.090 0 s, a pitcher releases a baseball of weight 1.40 N with a velocity of 32.0 m/s horizontally forward. If the ball starts from rest, (a) through what distance does the ball accelerate before its release? (b) What force does the pitcher exert on the ball?

7. After uniformly accelerating his arm for a time t, a pitcher releases a baseball of weight $-F_g\mathbf{j}$ with a velocity $v\mathbf{i}$. If the ball starts from rest, (a) through what distance does the ball accelerate before its release? (b) What force does the pitcher exert on the ball?

8. Define one pound as the weight of an object of mass 0.453 592 37 kg at a location where the acceleration due to gravity is 32.174 0 ft/s². Express the pound as one quantity with one SI unit.

WEB 9. A 4.00-kg object has a velocity of $3.00\mathbf{i}$ m/s at one instant. Eight seconds later, its velocity has increased to $(8.00\mathbf{i} + 10.0\mathbf{j})$ m/s. Assuming the object was subject to a constant total force, find (a) the components of the force and (b) its magnitude.

10. The average speed of a nitrogen molecule in air is about 6.70×10^2 m/s, and its mass is 4.68×10^{-26} kg. (a) If it takes 3.00×10^{-13} s for a nitrogen molecule to hit a wall and rebound with the same speed but moving in the opposite direction, what is the average acceleration of the molecule during this time interval? (b) What average force does the molecule exert on the wall?

11. An electron of mass 9.11×10^{-31} kg has an initial speed of 3.00×10^5 m/s. It travels in a straight line, and its speed increases to 7.00×10^5 m/s in a distance of 5.00 cm. Assuming its acceleration is constant, (a) determine the force exerted on the electron and (b) compare this force with the weight of the electron, which we neglected.

12. A woman weighs 120 lb. Determine (a) her weight in newtons and (b) her mass in kilograms.

13. If a man weighs 900 N on the Earth, what would he weigh on Jupiter, where the acceleration due to gravity is 25.9 m/s²?

14. The distinction between mass and weight was discovered after Jean Richer transported pendulum clocks from Paris to French Guiana in 1671. He found that they ran slower there quite systematically. The effect was reversed when the clocks returned to Paris. How much weight would you personally lose in traveling from Paris, where $g = 9.809\ 5$ m/s², to Cayenne, where $g = 9.780\ 8$ m/s²? (We shall consider how the free-fall acceleration influences the period of a pendulum in Section 13.4.)

15. Two forces $\mathbf{F}_1$ and $\mathbf{F}_2$ act on a 5.00-kg mass. If $F_1 = 20.0$ N and $F_2 = 15.0$ N, find the accelerations in (a) and (b) of Figure P5.15.

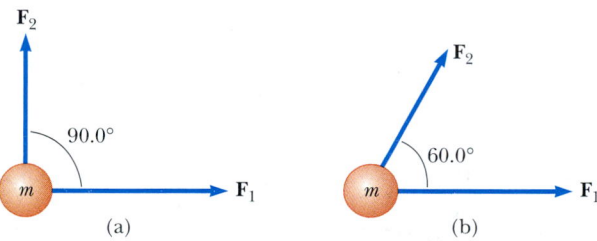

(a)   **(b)**

Figure P5.15

16. Besides its weight, a 2.80-kg object is subjected to one other constant force. The object starts from rest and in 1.20 s experiences a displacement of $(4.20\ \text{m})\mathbf{i} - (3.30\ \text{m})\mathbf{j}$, where the direction of $\mathbf{j}$ is the upward vertical direction. Determine the other force.

17. You stand on the seat of a chair and then hop off. (a) During the time you are in flight down to the floor, the Earth is lurching up toward you with an acceleration of what order of magnitude? In your solution explain your logic. Visualize the Earth as a perfectly solid object. (b) The Earth moves up through a distance of what order of magnitude?

18. Forces of 10.0 N north, 20.0 N east, and 15.0 N south are simultaneously applied to a 4.00-kg mass as it rests on an air table. Obtain the object's acceleration.

19. A boat moves through the water with two horizontal forces acting on it. One is a 2000-N forward push caused by the motor; the other is a constant 1800-N resistive force caused by the water. (a) What is the acceler-

ation of the 1 000-kg boat? (b) If it starts from rest, how far will it move in 10.0 s? (c) What will be its speed at the end of this time?

20. Three forces, given by $\mathbf{F}_1 = (-2.00\mathbf{i} + 2.00\mathbf{j})$ N, $\mathbf{F}_2 = (5.00\mathbf{i} - 3.00\mathbf{j})$ N, and $\mathbf{F}_3 = (-45.0\mathbf{i})$ N, act on an object to give it an acceleration of magnitude 3.75 m/s² (a) What is the direction of the acceleration? (b) What is the mass of the object? (c) If the object is initially at rest, what is its speed after 10.0 s? (d) What are the velocity components of the object after 10.0 s?

21. A 15.0-lb block rests on the floor. (a) What force does the floor exert on the block? (b) If a rope is tied to the block and run vertically over a pulley, and the other end is attached to a free-hanging 10.0-lb weight, what is the force exerted by the floor on the 15.0-lb block? (c) If we replace the 10.0-lb weight in part (b) with a 20.0-lb weight, what is the force exerted by the floor on the 15.0-lb block?

Section 5.7   Some Applications of Newton's Laws

22. A 3.00-kg mass is moving in a plane, with its x and y coordinates given by $x = 5t^2 - 1$ and $y = 3t^3 + 2$, where x and y are in meters and t is in seconds. Find the magnitude of the net force acting on this mass at $t = 2.00$ s.

23. The distance between two telephone poles is 50.0 m. When a 1.00-kg bird lands on the telephone wire midway between the poles, the wire sags 0.200 m. Draw a free-body diagram of the bird. How much tension does the bird produce in the wire? Ignore the weight of the wire.

24. A bag of cement of weight 325 N hangs from three wires as shown in Figure P5.24. Two of the wires make angles $\theta_1 = 60.0°$ and $\theta_2 = 25.0°$ with the horizontal. If the system is in equilibrium, find the tensions T_1, T_2, and T_3 in the wires.

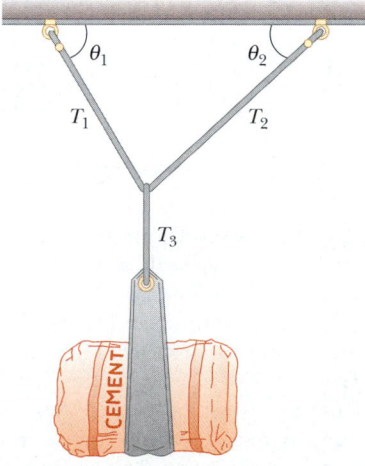

Figure P5.24   Problems 24 and 25.

25. A bag of cement of weight F_g hangs from three wires as shown in Figure P5.24. Two of the wires make angles θ_1 and θ_2 with the horizontal. If the system is in equilibrium, show that the tension in the left-hand wire is

$$T_1 = F_g \cos \theta_2 / \sin(\theta_1 + \theta_2)$$

26. You are a judge in a children's kite-flying contest, and two children will win prizes for the kites that pull most strongly and least strongly on their strings. To measure string tensions, you borrow a weight hanger, some slotted weights, and a protractor from your physics teacher and use the following protocol, illustrated in Figure P5.26: Wait for a child to get her kite well-controlled, hook the hanger onto the kite string about 30 cm from her hand, pile on weights until that section of string is horizontal, record the mass required, and record the angle between the horizontal and the string running up to the kite. (a) Explain how this method works. As you construct your explanation, imagine that the children's parents ask you about your method, that they might make false assumptions about your ability without concrete evidence, and that your explanation is an opportunity to give them confidence in your evaluation technique. (b) Find the string tension if the mass required to make the string horizontal is 132 g and the angle of the kite string is 46.3°.

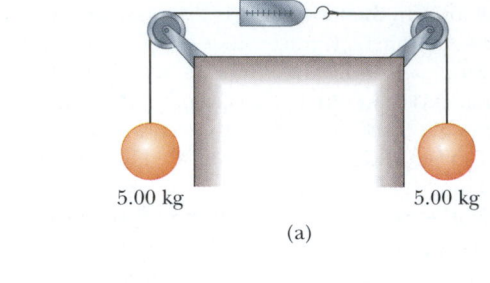

5.00 kg   5.00 kg

(a)

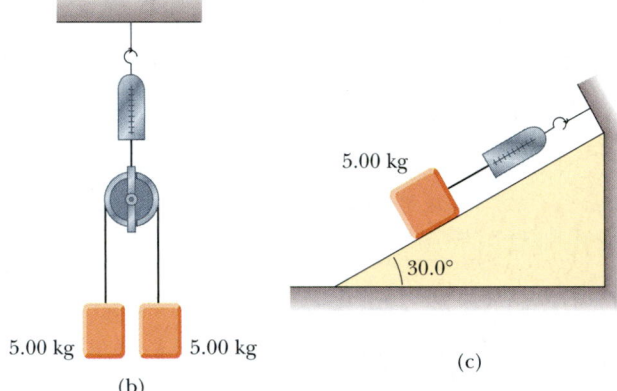

5.00 kg

30.0°

5.00 kg   5.00 kg   (c)

(b)

Figure P5.27

Figure P5.26

27. The systems shown in Figure P5.27 are in equilibrium. If the spring scales are calibrated in newtons, what do they read? (Neglect the masses of the pulleys and strings, and assume the incline is frictionless.)

28. A fire helicopter carries a 620-kg bucket of water at the end of a cable 20.0 m long. As the aircraft flies back from a fire at a constant speed of 40.0 m/s, the cable makes an angle of 40.0° with respect to the vertical. (a) Determine the force of air resistance on the bucket. (b) After filling the bucket with sea water, the pilot re-

turns to the fire at the same speed with the bucket now making an angle of 7.00° with the vertical. What is the mass of the water in the bucket?

WEB **29.** A 1.00-kg mass is observed to accelerate at 10.0 m/s² in a direction 30.0° north of east (Fig. P5.29). The force $\mathbf{F}_2$ acting on the mass has a magnitude of 5.00 N and is directed north. Determine the magnitude and direction of the force $\mathbf{F}_1$ acting on the mass.

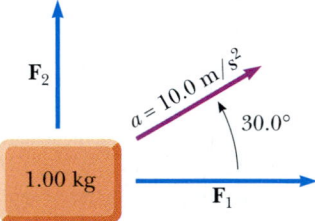

$\mathbf{F}_2$

$a = 10.0 \text{ m/s}^2$

30.0°

1.00 kg

$\mathbf{F}_1$

Figure P5.29

30. A simple accelerometer is constructed by suspending a mass m from a string of length L that is tied to the top of a cart. As the cart is accelerated the string-mass system makes a constant angle θ with the vertical. (a) Assuming that the string mass is negligible compared with m, derive an expression for the cart's acceleration in terms of θ and show that it is independent of the mass m and the length L. (b) Determine the acceleration of the cart when $\theta = 23.0°$.

31. Two people pull as hard as they can on ropes attached to a boat that has a mass of 200 kg. If they pull in the same direction, the boat has an acceleration of 1.52 m/s² to the right. If they pull in opposite directions, the boat has an acceleration of 0.518 m/s² to the left. What is the force exerted by each person on the boat? (Disregard any other forces on the boat.)

32. Draw a free-body diagram for a block that slides down a frictionless plane having an inclination of $\theta = 15.0°$ (Fig. P5.32). If the block starts from rest at the top and the length of the incline is 2.00 m, find (a) the acceleration of the block and (b) its speed when it reaches the bottom of the incline.

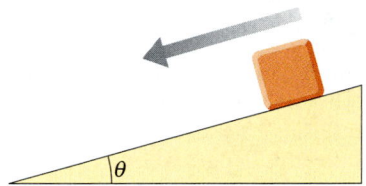

Figure P5.32

WEB 33. A block is given an initial velocity of 5.00 m/s up a frictionless 20.0° incline. How far up the incline does the block slide before coming to rest?

34. Two masses are connected by a light string that passes over a frictionless pulley, as in Figure P5.34. If the incline is frictionless and if $m_1 = 2.00$ kg, $m_2 = 6.00$ kg, and $\theta = 55.0°$, find (a) the accelerations of the masses, (b) the tension in the string, and (c) the speed of each mass 2.00 s after being released from rest.

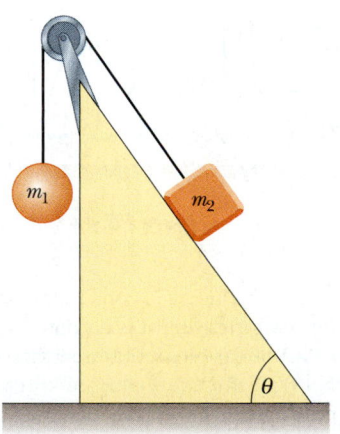

Figure P5.34

35. Two masses m_1 and m_2 situated on a frictionless, horizontal surface are connected by a light string. A force **F** is exerted on one of the masses to the right (Fig. P5.35). Determine the acceleration of the system and the tension T in the string.

Figure P5.35   Problems 35 and 51.

36. Two masses of 3.00 kg and 5.00 kg are connected by a light string that passes over a frictionless pulley, as was shown in Figure 5.15a. Determine (a) the tension in the string, (b) the acceleration of each mass, and (c) the distance each mass will move in the first second of motion if they start from rest.

37. In the system shown in Figure P5.37, a horizontal force F_x acts on the 8.00-kg mass. The horizontal surface is frictionless. (a) For what values of F_x does the 2.00-kg mass accelerate upward? (b) For what values of F_x is the tension in the cord zero? (c) Plot the acceleration of the 8.00-kg mass versus F_x. Include values of F_x from -100 N to $+100$ N.

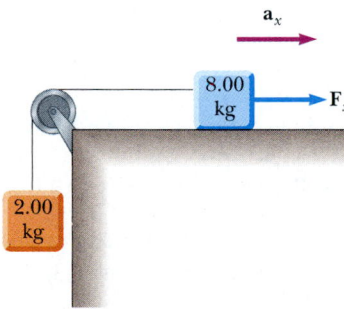

Figure P5.37

38. Mass m_1 on a frictionless horizontal table is connected to mass m_2 by means of a very light pulley P_1 and a light fixed pulley P_2 as shown in Figure P5.38. (a) If a_1 and a_2

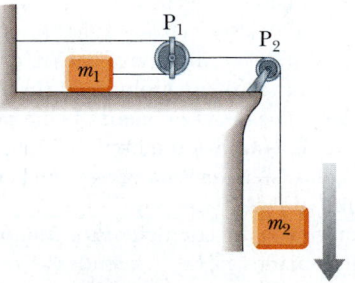

Figure P5.38

are the accelerations of m_1 and m_2, respectively, what is the relationship between these accelerations? Express (b) the tensions in the strings and (c) the accelerations a_1 and a_2 in terms of the masses m_1 and m_2 and g.

39. A 72.0-kg man stands on a spring scale in an elevator. Starting from rest, the elevator ascends, attaining its maximum speed of 1.20 m/s in 0.800 s. It travels with this constant speed for the next 5.00 s. The elevator then undergoes a uniform acceleration in the negative y direction for 1.50 s and comes to rest. What does the spring scale register (a) before the elevator starts to move? (b) during the first 0.800 s? (c) while the elevator is traveling at constant speed? (d) during the time it is slowing down?

Section 5.8   Forces of Friction

40. The coefficient of static friction is 0.800 between the soles of a sprinter's running shoes and the level track surface on which she is running. Determine the maximum acceleration she can achieve. Do you need to know that her mass is 60.0 kg?

41. A 25.0-kg block is initially at rest on a horizontal surface. A horizontal force of 75.0 N is required to set the block in motion. After it is in motion, a horizontal force of 60.0 N is required to keep the block moving with constant speed. Find the coefficients of static and kinetic friction from this information.

42. A racing car accelerates uniformly from 0 to 80.0 mi/h in 8.00 s. The external force that accelerates the car is the frictional force between the tires and the road. If the tires do not slip, determine the minimum coefficient of friction between the tires and the road.

43. A car is traveling at 50.0 mi/h on a horizontal highway. (a) If the coefficient of friction between road and tires on a rainy day is 0.100, what is the minimum distance in which the car will stop? (b) What is the stopping distance when the surface is dry and $\mu_s = 0.600$?

44. A woman at an airport is towing her 20.0-kg suitcase at constant speed by pulling on a strap at an angle of θ above the horizontal (Fig. P5.44). She pulls on the strap with a 35.0-N force, and the frictional force on the suitcase is 20.0 N. Draw a free-body diagram for the suitcase. (a) What angle does the strap make with the horizontal? (b) What normal force does the ground exert on the suitcase?

WEB 45. A 3.00-kg block starts from rest at the top of a 30.0° incline and slides a distance of 2.00 m down the incline in 1.50 s. Find (a) the magnitude of the acceleration of the block, (b) the coefficient of kinetic friction between block and plane, (c) the frictional force acting on the block, and (d) the speed of the block after it has slid 2.00 m.

46. To determine the coefficients of friction between rubber and various surfaces, a student uses a rubber eraser and an incline. In one experiment the eraser begins to slip down the incline when the angle of inclination is

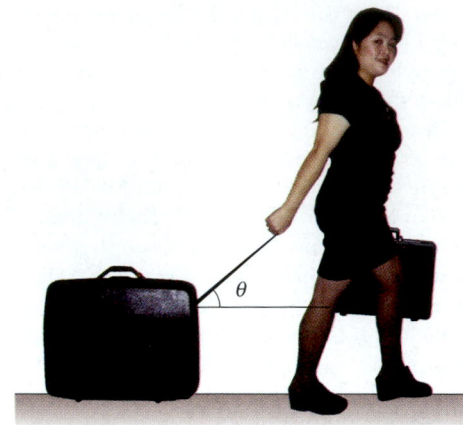

Figure P5.44

36.0° and then moves down the incline with constant speed when the angle is reduced to 30.0°. From these data, determine the coefficients of static and kinetic friction for this experiment.

47. A boy drags his 60.0-N sled at constant speed up a 15.0° hill. He does so by pulling with a 25.0-N force on a rope attached to the sled. If the rope is inclined at 35.0° to the horizontal, (a) what is the coefficient of kinetic friction between sled and snow? (b) At the top of the hill, he jumps on the sled and slides down the hill. What is the magnitude of his acceleration down the slope?

48. Determine the stopping distance for a skier moving down a slope with friction with an initial speed of 20.0 m/s (Fig. P5.48). Assume $\mu_k = 0.180$ and $\theta = 5.00°$.

Figure P5.48

49. A 9.00-kg hanging weight is connected by a string over a pulley to a 5.00-kg block that is sliding on a flat table (Fig. P5.49). If the coefficient of kinetic friction is 0.200, find the tension in the string.

50. Three blocks are connected on a table as shown in Figure P5.50. The table is rough and has a coefficient of ki-

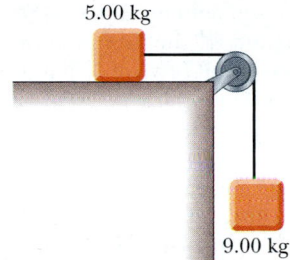

Figure P5.49

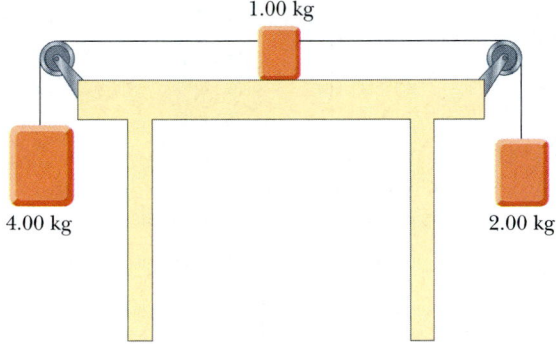

Figure P5.50

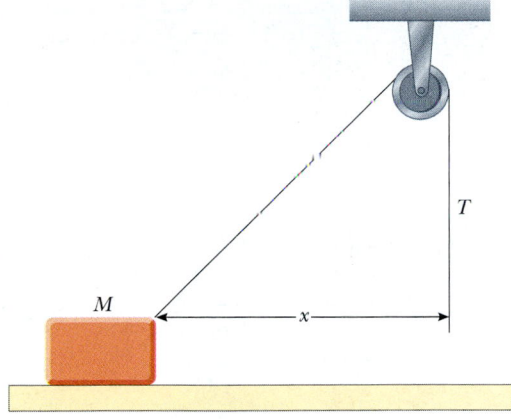

Figure P5.52

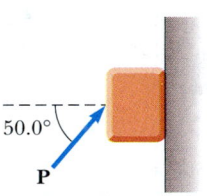

Figure P5.53

netic friction of 0.350. The three masses are 4.00 kg, 1.00 kg, and 2.00 kg, and the pulleys are frictionless. Draw a free-body diagram for each block. (a) Determine the magnitude and direction of the acceleration of each block. (b) Determine the tensions in the two cords.

 51. Two blocks connected by a rope of negligible mass are being dragged by a horizontal force **F** (see Fig. P5.35). Suppose that $F = 68.0$ N, $m_1 = 12.0$ kg, $m_2 = 18.0$ kg, and the coefficient of kinetic friction between each block and the surface is 0.100. (a) Draw a free-body diagram for each block. (b) Determine the tension T and the magnitude of the acceleration of the system.

52. A block of mass 2.20 kg is accelerated across a rough surface by a rope passing over a pulley, as shown in Figure P5.52. The tension in the rope is 10.0 N, and the pulley is 10.0 cm above the top of the block. The coefficient of kinetic friction is 0.400. (a) Determine the acceleration of the block when $x = 0.400$ m. (b) Find the value of x at which the acceleration becomes zero.

53. A block of mass 3.00 kg is pushed up against a wall by a force **P** that makes a 50.0° angle with the horizontal as shown in Figure P5.53. The coefficient of static friction between the block and the wall is 0.250. Determine the possible values for the magnitude of **P** that allow the block to remain stationary.

ADDITIONAL PROBLEMS

54. A time-dependent force $\mathbf{F} = (8.00\mathbf{i} - 4.00t\mathbf{j})$ N (where t is in seconds) is applied to a 2.00-kg object initially at rest. (a) At what time will the object be moving with a speed of 15.0 m/s? (b) How far is the object from its initial position when its speed is 15.0 m/s? (c) What is the object's displacement at the time calculated in (a)?

55. An inventive child named Pat wants to reach an apple in a tree without climbing the tree. Sitting in a chair connected to a rope that passes over a frictionless pulley (Fig. P5.55), Pat pulls on the loose end of the rope with such a force that the spring scale reads 250 N. Pat's weight is 320 N, and the chair weighs 160 N. (a) Draw free-body diagrams for Pat and the chair considered as separate systems, and draw another diagram for Pat and the chair considered as one system. (b) Show that the acceleration of the system is *upward* and find its magnitude. (c) Find the force Pat exerts on the chair.

56. Three blocks are in contact with each other on a frictionless, horizontal surface, as in Figure P5.56. A horizontal force **F** is applied to m_1. If $m_1 = 2.00$ kg, $m_2 = 3.00$ kg, $m_3 = 4.00$ kg, and $F = 18.0$ N, draw a separate free-body diagram for each block and find (a) the acceleration of the blocks, (b) the *resultant* force on each block, and (c) the magnitudes of the contact forces between the blocks.

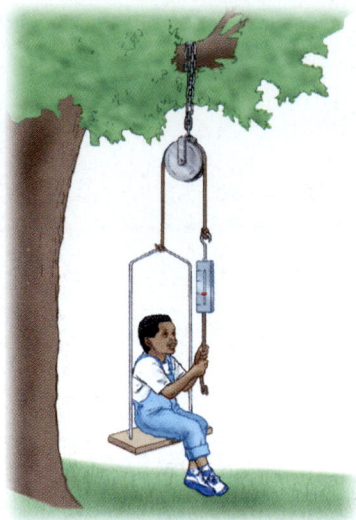

Figure P5.55

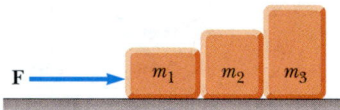

Figure P5.56

57. A high diver of mass 70.0 kg jumps off a board 10.0 m above the water. If his downward motion is stopped 2.00 s after he enters the water, what average upward force did the water exert on him?

58. Consider the three connected objects shown in Figure P5.58. If the inclined plane is frictionless and the system is in equilibrium, find (in terms of m, g, and θ) (a) the mass M and (b) the tensions T_1 and T_2. If the value of M is double the value found in part (a), find (c) the acceleration of each object, and (d) the tensions T_1 and T_2. If the coefficient of static friction between m and $2m$ and the inclined plane is μ_s, and

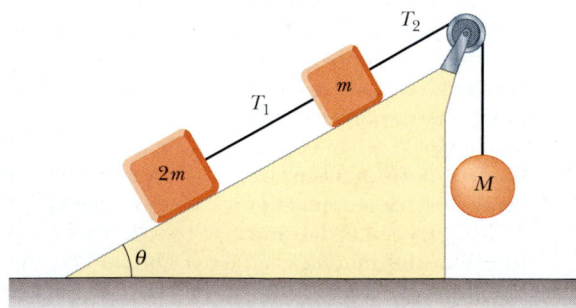

Figure P5.58

the system is in equilibrium, find (e) the minimum value of M and (f) the maximum value of M. (g) Compare the values of T_2 when M has its minimum and maximum values.

WEB 59. A mass M is held in place by an applied force $\mathbf{F}$ and a pulley system as shown in Figure P5.59. The pulleys are massless and frictionless. Find (a) the tension in each section of rope, T_1, T_2, T_3, T_4, and T_5 and (b) the magnitude of $\mathbf{F}$. (*Hint:* Draw a free-body diagram for each pulley.)

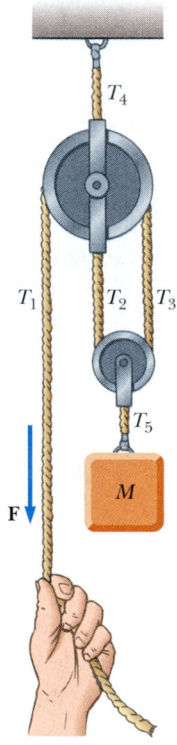

Figure P5.59

60. Two forces, given by $\mathbf{F}_1 = (-6.00\mathbf{i} - 4.00\mathbf{j})$ N and $\mathbf{F}_2 = (-3.00\mathbf{i} + 7.00\mathbf{j})$ N, act on a particle of mass 2.00 kg that is initially at rest at coordinates $(-2.00$ m, $+4.00$ m). (a) What are the components of the particle's velocity at $t = 10.0$ s? (b) In what direction is the particle moving at $t = 10.0$ s? (c) What displacement does the particle undergo during the first 10.0 s? (d) What are the coordinates of the particle at $t = 10.0$ s?

61. A crate of weight $\mathbf{F}_g$ is pushed by a force $\mathbf{P}$ on a horizontal floor. (a) If the coefficient of static friction is μ_s and $\mathbf{P}$ is directed at an angle θ below the horizontal, show that the minimum value of P that will move the crate is given by

$$P = \mu_s F_g \sec \theta (1 - \mu_s \tan \theta)^{-1}$$

(b) Find the minimum value of P that can produce motion when $\mu_s = 0.400$, $F_g = 100$ N, and $\theta = 0°$, 15.0°, 30.0°, 45.0°, and 60.0°.

62. **Review Problem.** A block of mass $m = 2.00$ kg is released from rest $h = 0.500$ m from the surface of a table, at the top of a $\theta = 30.0°$ incline as shown in Figure P5.62. The frictionless incline is fixed on a table of height $H = 2.00$ m. (a) Determine the acceleration of the block as it slides down the incline. (b) What is the velocity of the block as it leaves the incline? (c) How far from the table will the block hit the floor? (d) How much time has elapsed between when the block is released and when it hits the floor? (e) Does the mass of the block affect any of the above calculations?

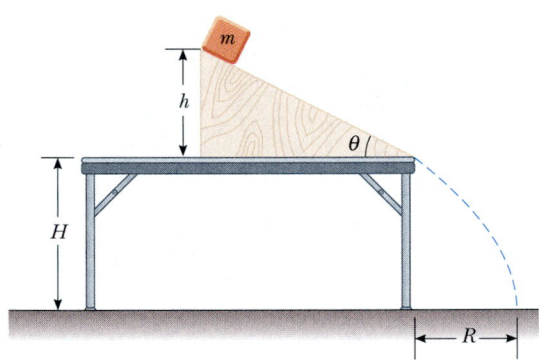

Figure P5.62

63. A 1.30-kg toaster is not plugged in. The coefficient of static friction between the toaster and a horizontal countertop is 0.350. To make the toaster start moving, you carelessly pull on its electric cord. (a) For the cord tension to be as small as possible, you should pull at what angle above the horizontal? (b) With this angle, how large must the tension be?

64. A 2.00-kg aluminum block and a 6.00-kg copper block are connected by a light string over a frictionless pulley. They sit on a steel surface, as shown in Figure P5.64, and $\theta = 30.0°$. Do they start to move once any holding mechanism is released? If so, determine (a) their acceleration and (b) the tension in the string. If not, determine the sum of the magnitudes of the forces of friction acting on the blocks.

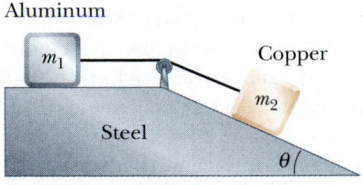

Figure P5.64

65. A block of mass $m = 2.00$ kg rests on the left edge of a block of larger mass $M = 8.00$ kg. The coefficient of kinetic friction between the two blocks is 0.300, and the surface on which the 8.00-kg block rests is frictionless. A constant horizontal force of magnitude $F = 10.0$ N is applied to the 2.00-kg block, setting it in motion as shown in Figure P5.65a. If the length L that the leading edge of the smaller block travels on the larger block is 3.00 m, (a) how long will it take before this block makes it to the right side of the 8.00-kg block, as shown in Figure P5.65b? (*Note:* Both blocks are set in motion when **F** is applied.) (b) How far does the 8.00-kg block move in the process?

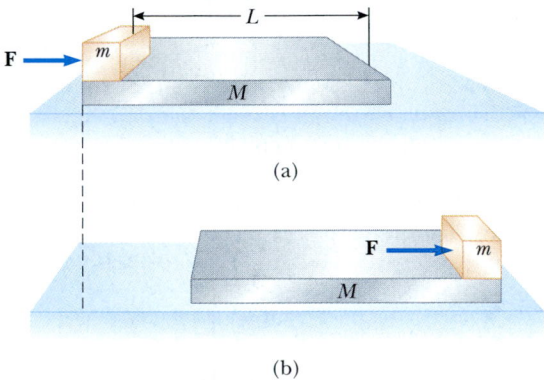

Figure P5.65

66. A student is asked to measure the acceleration of a cart on a "frictionless" inclined plane as seen in Figure P5.32, using an air track, a stopwatch, and a meter stick. The height of the incline is measured to be 1.774 cm, and the total length of the incline is measured to be $d = 127.1$ cm. Hence, the angle of inclination θ is determined from the relation $\sin \theta = 1.774/127.1$. The cart is released from rest at the top of the incline, and its displacement x along the incline is measured versus time, where $x = 0$ refers to the initial position of the cart. For x values of 10.0 cm, 20.0 cm, 35.0 cm, 50.0 cm, 75.0 cm, and 100 cm, the measured times to undergo these displacements (averaged over five runs) are 1.02 s, 1.53 s, 2.01 s, 2.64 s, 3.30 s, and 3.75 s, respectively. Construct a graph of x versus t^2, and perform a linear least-squares fit to the data. Determine the acceleration of the cart from the slope of this graph, and compare it with the value you would get using $a' = g \sin \theta$, where $g = 9.80$ m/s².

67. A 2.00-kg block is placed on top of a 5.00-kg block as in Figure P5.67. The coefficient of kinetic friction between the 5.00-kg block and the surface is 0.200. A horizontal force **F** is applied to the 5.00-kg block. (a) Draw a free-body diagram for each block. What force accelerates the 2.00-kg block? (b) Calculate the magnitude of the force necessary to pull both blocks to the right with an

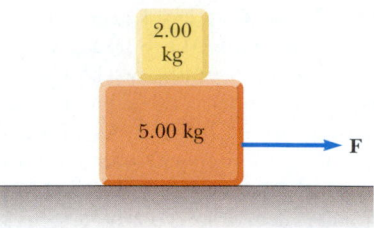

Figure P5.67

acceleration of 3.00 m/s². (c) Find the minimum coefficient of static friction between the blocks such that the 2.00-kg block does not slip under an acceleration of 3.00 m/s².

68. A 5.00-kg block is placed on top of a 10.0-kg block (Fig. P5.68). A horizontal force of 45.0 N is applied to the 10.0-kg block, and the 5.00-kg block is tied to the wall. The coefficient of kinetic friction between all surfaces is 0.200. (a) Draw a free-body diagram for each block and identify the action–reaction forces between the blocks. (b) Determine the tension in the string and the magnitude of the acceleration of the 10.0-kg block.

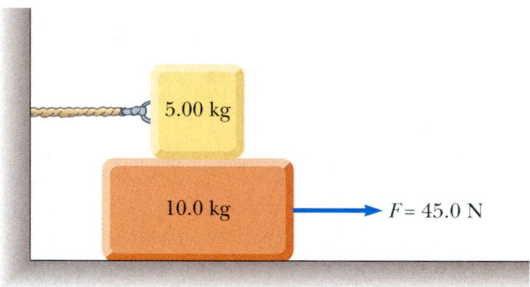

Figure P5.68

69. What horizontal force must be applied to the cart shown in Figure P5.69 so that the blocks remain stationary relative to the cart? Assume all surfaces, wheels, and pulley are frictionless. (*Hint:* Note that the force exerted by the string accelerates m_1.)

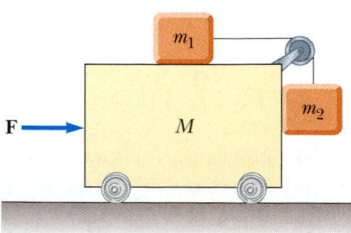

Figure P5.69   Problems 69 and 70.

70. Initially the system of masses shown in Figure P5.69 is held motionless. All surfaces, pulley, and wheels are frictionless. Let the force **F** be zero and assume that m_2 can move only vertically. At the instant after the system of masses is released, find (a) the tension T in the string, (b) the acceleration of m_2, (c) the acceleration of M, and (d) the acceleration of m_1. (*Note:* The pulley accelerates along with the cart.)

71. A block of mass 5.00 kg sits on top of a second block of mass 15.0 kg, which in turn sits on a horizontal table. The coefficients of friction between the two blocks are $\mu_s = 0.300$ and $\mu_k = 0.100$. The coefficients of friction between the lower block and the rough table are $\mu_s = 0.500$ and $\mu_k = 0.400$. You apply a constant horizontal force to the lower block, just large enough to make this block start sliding out from between the upper block and the table. (a) Draw a free-body diagram of each block, naming the forces acting on each. (b) Determine the magnitude of each force on each block at the instant when you have started pushing but motion has not yet started. (c) Determine the acceleration you measure for each block.

72. Two blocks of mass 3.50 kg and 8.00 kg are connected by a string of negligible mass that passes over a frictionless pulley (Fig. P5.72). The inclines are frictionless. Find (a) the magnitude of the acceleration of each block and (b) the tension in the string.

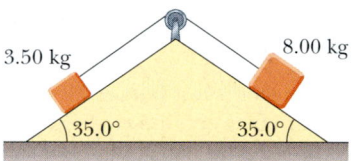

Figure P5.72   Problems 72 and 73.

73. The system shown in Figure P5.72 has an acceleration of magnitude 1.50 m/s². Assume the coefficients of kinetic friction between block and incline are the same for both inclines. Find (a) the coefficient of kinetic friction and (b) the tension in the string.

74. In Figure P5.74, a 500-kg horse pulls a sledge of mass 100 kg. The system (horse plus sledge) has a forward acceleration of 1.00 m/s² when the frictional force exerted on the sledge is 500 N. Find (a) the tension in the connecting rope and (b) the magnitude and direction of the force of friction exerted on the horse. (c) Verify that the total forces of friction the ground exerts on the system will give the system an acceleration of 1.00 m/s².

75. A van accelerates down a hill (Fig. P5.75), going from rest to 30.0 m/s in 6.00 s. During the acceleration, a toy ($m = 0.100$ kg) hangs by a string from the van's ceiling. The acceleration is such that the string remains perpendicular to the ceiling. Determine (a) the angle θ and (b) the tension in the string.

Figure P5.74

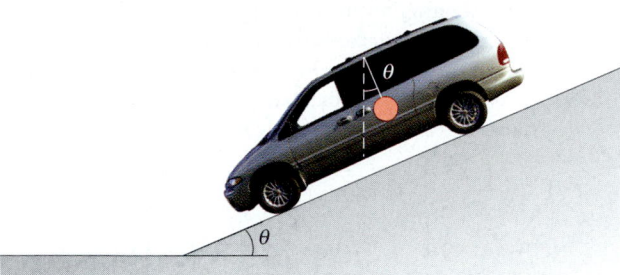

Figure P5.75

76. A mobile is formed by supporting four metal butterflies of equal mass m from a string of length L. The points of support are evenly spaced a distance ℓ apart as shown in Figure P5.76. The string forms an angle θ_1 with the ceiling at each end point. The center section of string is horizontal. (a) Find the tension in each section of string in terms of θ_1, m, and g. (b) Find the angle θ_2, in terms of θ_1, that the sections of string between the outside butterflies and the inside butterflies form with the horizontal. (c) Show that the distance D between the end points of the string is

$$D = \frac{L}{5}\left\{2\cos\theta_1 + 2\cos\left[\tan^{-1}\left(\frac{1}{2}\tan\theta_1\right)\right] + 1\right\}$$

77. Before 1960 it was believed that the maximum attainable coefficient of static friction for an automobile tire was less than 1. Then about 1962, three companies independently developed racing tires with coefficients of 1.6. Since then, tires have improved, as illustrated in this problem. According to the 1990 *Guinness Book of Records*, the fastest time in which a piston-engine car initially at rest has covered a distance of one-quarter mile is 4.96 s. This record was set by Shirley Muldowney in September 1989 (Fig. P5.77). (a) Assuming that the rear wheels nearly lifted the front wheels off the pavement, what minimum value of μ_s is necessary to achieve the record time? (b) Suppose Muldowney were able to double her engine power, keeping other things equal. How would this change affect the elapsed time?

Figure P5.77   *(Mike Powell/AllSport USA)*

78. An 8.40-kg mass slides down a fixed, frictionless inclined plane. Use a computer to determine and tabulate the normal force exerted on the mass and its acceleration for a series of incline angles (measured from the horizontal) ranging from 0 to 90° in 5° increments. Plot a graph of the normal force and the acceleration as functions of the incline angle. In the limiting cases of 0 and 90°, are your results consistent with the known behavior?

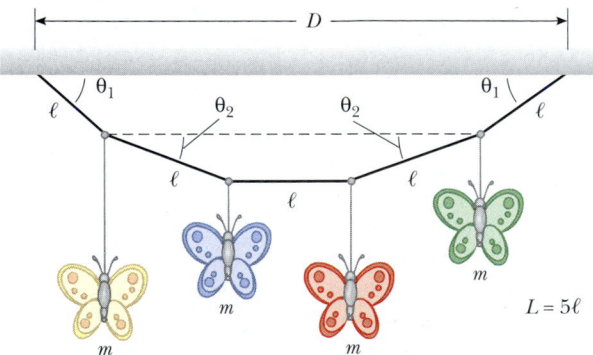

$L = 5\ell$

Figure P5.76

ANSWERS TO QUICK QUIZZES

5.1 (a) True. Newton's first law tells us that motion requires no force: An object in motion continues to move at constant velocity in the absence of external forces. (b) True. A stationary object can have several forces acting on it, but if the vector sum of all these external forces is zero, there is no net force and the object remains stationary. It also is possible to have a net force and no motion, but only for an instant. A ball tossed vertically upward stops at the peak of its path for an infinitesimally short time, but the force of gravity is still acting on it. Thus, although $\mathbf{v} = 0$ at the peak, the net force acting on the ball is *not* zero.

5.2 No. Direction of motion is part of an object's *velocity*, and force determines the direction of acceleration, not that of velocity.

5.3 (a) Force of gravity. (b) Force of gravity. The only external force acting on the ball at *all* points in its trajectory is the downward force of gravity.

5.4 As the person steps out of the boat, he pushes against it with his foot, expecting the boat to push back on him so that he accelerates toward the dock. However, because the boat is untied, the force exerted by the foot causes the boat to scoot away from the dock. As a result, the person is not able to exert a very large force on the boat before it moves out of reach. Therefore, the boat does not exert a very large reaction force on him, and he ends up not being accelerated sufficiently to make it to the dock. Consequently, he falls into the water instead. If a small dog were to jump from the untied boat toward the dock, the force exerted by the boat on the dog would probably be enough to ensure the dog's successful landing because of the dog's small mass.

5.5 (a) The same force is experienced by both. The fly and bus experience forces that are equal in magnitude but opposite in direction. (b) The fly. Because the fly has such a small mass, it undergoes a very large acceleration. The huge mass of the bus means that it more effectively resists any change in its motion.

5.6 (b) The crate accelerates to the right. Because the only horizontal force acting on it is the force of static friction between its bottom surface and the truck bed, that force must also be directed to the right.

Calvin and Hobbes   by Bill Watterson

(CALVIN AND HOBBES ©1995 Watterson. Reprinted with permission of UNIVERSAL PRESS SYNDICATE. All rights reserved.)

This sky diver is falling at more than 50 m/s (120 mi/h), but once her parachute opens, her downward velocity will be greatly reduced. Why does she slow down rapidly when her chute opens, enabling her to fall safely to the ground? If the chute does not function properly, the sky diver will almost certainly be seriously injured. What force exerted on her limits her maximum speed? *(Guy Savage/Photo Researchers, Inc.)*

c h a p t e r

Circular Motion and Other Applications of Newton's Laws

151

In the preceding chapter we introduced Newton's laws of motion and applied them to situations involving linear motion. Now we discuss motion that is slightly more complicated. For example, we shall apply Newton's laws to objects traveling in circular paths. Also, we shall discuss motion observed from an accelerating frame of reference and motion in a viscous medium. For the most part, this chapter is a series of examples selected to illustrate the application of Newton's laws to a wide variety of circumstances.

6.1 ▸ NEWTON'S SECOND LAW APPLIED TO UNIFORM CIRCULAR MOTION

In Section 4.4 we found that a particle moving with uniform speed v in a circular path of radius r experiences an acceleration $\mathbf{a}_r$ that has a magnitude

$$a_r = \frac{v^2}{r}$$

The acceleration is called the *centripetal acceleration* because $\mathbf{a}_r$ is directed toward the center of the circle. Furthermore, $\mathbf{a}_r$ is *always* perpendicular to $\mathbf{v}$. (If there were a component of acceleration parallel to $\mathbf{v}$, the particle's speed would be changing.)

Consider a ball of mass m that is tied to a string of length r and is being whirled at constant speed in a horizontal circular path, as illustrated in Figure 6.1. Its weight is supported by a low-friction table. Why does the ball move in a circle? Because of its inertia, the tendency of the ball is to move in a straight line; however, the string prevents motion along a straight line by exerting on the ball a force that makes it follow the circular path. This force is directed along the string toward the center of the circle, as shown in Figure 6.1. This force can be any one of our familiar forces causing an object to follow a circular path.

If we apply Newton's second law along the radial direction, we find that the value of the net force causing the centripetal acceleration can be evaluated:

Force causing centripetal acceleration

$$\sum F_r = ma_r = m\frac{v^2}{r} \tag{6.1}$$

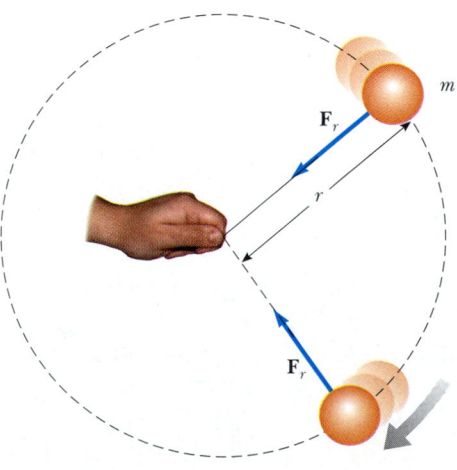

Figure 6.1 Overhead view of a ball moving in a circular path in a horizontal plane. A force $\mathbf{F}_r$ directed toward the center of the circle keeps the ball moving in its circular path.

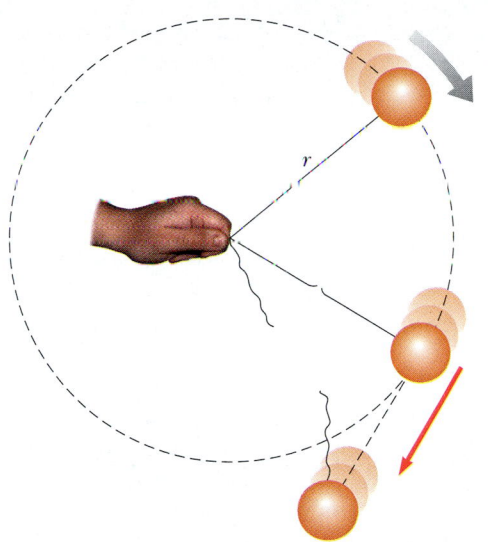

Figure 6.2 When the string breaks, the ball moves in the direction tangent to the circle.

An athlete in the process of throwing the hammer at the 1996 Olympic Games in Atlanta, Georgia. The force exerted by the chain is the force causing the circular motion. Only when the athlete releases the hammer will it move along a straight-line path tangent to the circle. *(Mike Powell/AllSport USA)*

A force causing a centripetal acceleration acts toward the center of the circular path and causes a change in the direction of the velocity vector. If that force should vanish, the object would no longer move in its circular path; instead, it would move along a straight-line path tangent to the circle. This idea is illustrated in Figure 6.2 for the ball whirling at the end of a string. If the string breaks at some instant, the ball moves along the straight-line path tangent to the circle at the point where the string broke.

Quick Quiz 6.1

Is it possible for a car to move in a circular path in such a way that it has a tangential acceleration but no centripetal acceleration?

CONCEPTUAL EXAMPLE 6.1 **Forces That Cause Centripetal Acceleration**

The force causing centripetal acceleration is sometimes called a *centripetal force*. We are familiar with a variety of forces in nature—friction, gravity, normal forces, tension, and so forth. Should we add *centripetal* force to this list?

Solution No; centripetal force *should not* be added to this list. This is a pitfall for many students. Giving the force causing circular motion a name—centripetal force—leads many students to consider it a new kind of force rather than a new *role* for force. A common mistake in force diagrams is to draw all the usual forces and then to add another vector for the centripetal force. But it is not a separate force—it is simply one of our familiar forces *acting in the role of a force that causes a circular motion.*

Consider some examples. For the motion of the Earth around the Sun, the centripetal force is *gravity*. For an object sitting on a rotating turntable, the centripetal force is *friction*. For a rock whirled on the end of a string, the centripetal force is the force of *tension* in the string. For an amusement-park patron pressed against the inner wall of a rapidly rotating circular room, the centripetal force is the *normal force* exerted by the wall. What's more, the centripetal force could be a combination of two or more forces. For example, as a Ferris-wheel rider passes through the lowest point, the centripetal force on her is the difference between the normal force exerted by the seat and her weight.

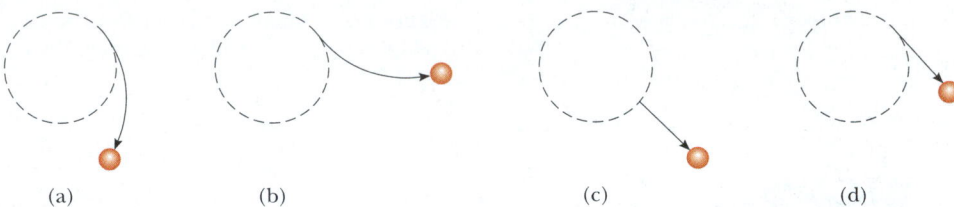

(a)   (b)   (c)   (d)

Figure 6.3   A ball that had been moving in a circular path is acted on by various external forces that change its path.

Quick Quiz 6.2

QuickLab

Tie a string to a tennis ball, swing it in a circle, and then, while it is swinging, let go of the string to verify your answer to the last part of Quick Quiz 6.2.

A ball is following the dotted circular path shown in Figure 6.3 under the influence of a force. At a certain instant of time, the force on the ball changes abruptly to a new force, and the ball follows the paths indicated by the solid line with an arrowhead in each of the four parts of the figure. For each part of the figure, describe the magnitude and direction of the force required to make the ball move in the solid path. If the dotted line represents the path of a ball being whirled on the end of a string, which path does the ball follow if the string breaks?

Let us consider some examples of uniform circular motion. In each case, be sure to recognize the external force (or forces) that causes the body to move in its circular path.

EXAMPLE 6.2   How Fast Can It Spin?

A ball of mass 0.500 kg is attached to the end of a cord 1.50 m long. The ball is whirled in a horizontal circle as was shown in Figure 6.1. If the cord can withstand a maximum tension of 50.0 N, what is the maximum speed the ball can attain before the cord breaks? Assume that the string remains horizontal during the motion.

Solution   It is difficult to know what might be a reasonable value for the answer. Nonetheless, we know that it cannot be too large, say 100 m/s, because a person cannot make a ball move so quickly. It makes sense that the stronger the cord, the faster the ball can twirl before the cord breaks. Also, we expect a more massive ball to break the cord at a lower speed. (Imagine whirling a bowling ball!)

Because the force causing the centripetal acceleration in this case is the force **T** exerted by the cord on the ball, Equation 6.1 yields for $\Sigma F_r = ma_r$

$$T = m\frac{v^2}{r}$$

Solving for v, we have

$$v = \sqrt{\frac{Tr}{m}}$$

This shows that v increases with T and decreases with larger m, as we expect to see—for a given v, a large mass requires a large tension and a small mass needs only a small tension. The maximum speed the ball can have corresponds to the maximum tension. Hence, we find

$$v_{max} = \sqrt{\frac{T_{max}r}{m}} = \sqrt{\frac{(50.0\ \text{N})(1.50\ \text{m})}{0.500\ \text{kg}}}$$

$$= \boxed{12.2\ \text{m/s}}$$

Exercise   Calculate the tension in the cord if the speed of the ball is 5.00 m/s.

Answer   8.33 N.

EXAMPLE 6.3   The Conical Pendulum

A small object of mass m is suspended from a string of length L. The object revolves with constant speed v in a horizontal circle of radius r, as shown in Figure 6.4. (Because the string sweeps out the surface of a cone, the system is known as a *conical pendulum*.) Find an expression for v.

Solution   Let us choose θ to represent the angle between string and vertical. In the free-body diagram shown in Figure 6.4, the force **T** exerted by the string is resolved into a vertical component $T\cos\theta$ and a horizontal component $T\sin\theta$ acting toward the center of revolution. Because the object does not accelerate in the vertical direction, $\Sigma F_y = ma_y = 0$, and the upward vertical component of **T** must balance the downward force of gravity. Therefore,

$$(1) \qquad T \cos \theta = mg$$

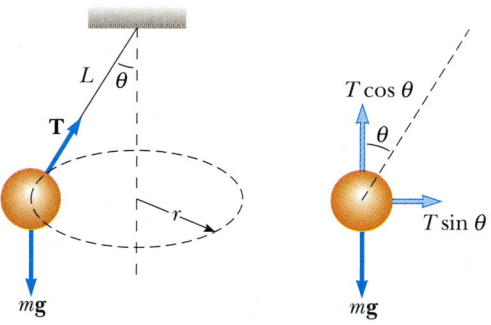

Figure 6.4 The conical pendulum and its free-body diagram.

Because the force providing the centripetal acceleration in this example is the component $T \sin \theta$, we can use Newton's second law and Equation 6.1 to obtain

$$(2) \qquad \sum F_r = T \sin \theta = ma_r = \frac{mv^2}{r}$$

Dividing (2) by (1) and remembering that $\sin \theta / \cos \theta = \tan \theta$, we eliminate T and find that

$$\tan \theta = \frac{v^2}{rg}$$

$$v = \sqrt{rg \tan \theta}$$

From the geometry in Figure 6.4, we note that $r = L \sin \theta$; therefore,

$$v = \boxed{\sqrt{Lg \sin \theta \tan \theta}}$$

Note that the speed is independent of the mass of the object.

EXAMPLE 6.4 ▶ What Is the Maximum Speed of the Car?

A 1 500-kg car moving on a flat, horizontal road negotiates a curve, as illustrated in Figure 6.5. If the radius of the curve is 35.0 m and the coefficient of static friction between the tires and dry pavement is 0.500, find the maximum speed the car can have and still make the turn successfully.

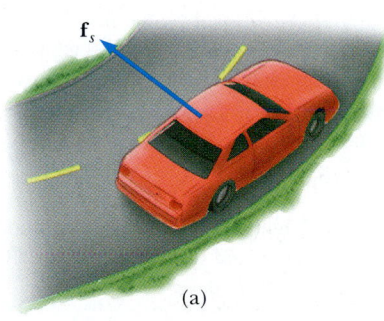

(a)

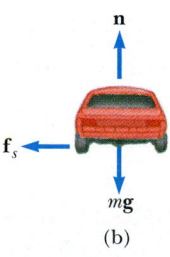

(b)

Figure 6.5 (a) The force of static friction directed toward the center of the curve keeps the car moving in a circular path. (b) The free-body diagram for the car.

Solution From experience, we should expect a maximum speed less than 50 m/s. (A convenient mental conversion is that 1 m/s is roughly 2 mi/h.) In this case, the force that enables the car to remain in its circular path is the force of static friction. (Because no slipping occurs at the point of contact between road and tires, the acting force is a force of static friction directed toward the center of the curve. If this force of static friction were zero—for example, if the car were on an icy road—the car would continue in a straight line and slide off the road.) Hence, from Equation 6.1 we have

$$(1) \qquad f_s = m\frac{v^2}{r}$$

The maximum speed the car can have around the curve is the speed at which it is on the verge of skidding outward. At this point, the friction force has its maximum value $f_{s,\text{max}} = \mu_s n$. Because the car is on a horizontal road, the magnitude of the normal force equals the weight ($n = mg$) and thus $f_{s,\text{max}} = \mu_s mg$. Substituting this value for f_s into (1), we find that the maximum speed is

$$v_{\text{max}} = \sqrt{\frac{f_{s,\text{max}} r}{m}} = \sqrt{\frac{\mu_s mgr}{m}} = \sqrt{\mu_s gr}$$

$$= \sqrt{(0.500)(9.80 \text{ m/s}^2)(35.0 \text{ m})} = \boxed{13.1 \text{ m/s}}$$

Note that the maximum speed does not depend on the mass of the car. That is why curved highways do not need multiple speed limit signs to cover the various masses of vehicles using the road.

Exercise   On a wet day, the car begins to skid on the curve when its speed reaches 8.00 m/s. What is the coefficient of static friction in this case?

Answer   0.187.

EXAMPLE 6.5 ▸ The Banked Exit Ramp

A civil engineer wishes to design a curved exit ramp for a highway in such a way that a car will not have to rely on friction to round the curve without skidding. In other words, a car moving at the designated speed can negotiate the curve even when the road is covered with ice. Such a ramp is usually *banked;* this means the roadway is tilted toward the inside of the curve. Suppose the designated speed for the ramp is to be 13.4 m/s (30.0 mi/h) and the radius of the curve is 50.0 m. At what angle should the curve be banked?

Solution   On a level (unbanked) road, the force that causes the centripetal acceleration is the force of static friction between car and road, as we saw in the previous example. However, if the road is banked at an angle θ, as shown in Figure 6.6, the normal force **n** has a horizontal component

$n \sin \theta$ pointing toward the center of the curve. Because the ramp is to be designed so that the force of static friction is zero, only the component $n \sin \theta$ causes the centripetal acceleration. Hence, Newton's second law written for the radial direction gives

$$(1) \qquad \sum F_r = n \sin \theta = \frac{mv^2}{r}$$

The car is in equilibrium in the vertical direction. Thus, from $\sum F_y = 0$, we have

$$(2) \qquad n \cos \theta = mg$$

Dividing (1) by (2) gives

$$\tan \theta = \frac{v^2}{rg}$$

$$\theta = \tan^{-1}\left[\frac{(13.4 \text{ m/s})^2}{(50.0 \text{ m})(9.80 \text{ m/s}^2)}\right] = \boxed{20.1°}$$

If a car rounds the curve at a speed less than 13.4 m/s, friction is needed to keep it from sliding down the bank (to the left in Fig. 6.6). A driver who attempts to negotiate the curve at a speed greater than 13.4 m/s has to depend on friction to keep from sliding up the bank (to the right in Fig. 6.6). The banking angle is independent of the mass of the vehicle negotiating the curve.

Exercise   Write Newton's second law applied to the radial direction when a frictional force **f**$_s$ is directed down the bank, toward the center of the curve.

Answer $n \sin \theta + f_s \cos \theta = \dfrac{mv^2}{r}$

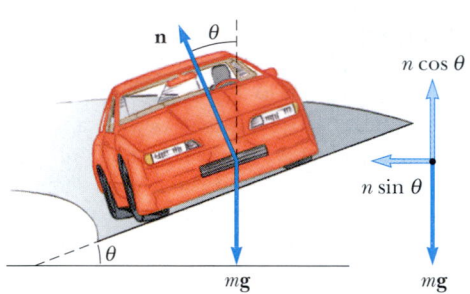

Figure 6.6   Car rounding a curve on a road banked at an angle θ to the horizontal. When friction is neglected, the force that causes the centripetal acceleration and keeps the car moving in its circular path is the horizontal component of the normal force. Note that **n** is the *sum* of the forces exerted by the road on the wheels.

EXAMPLE 6.6 ▸ Satellite Motion

This example treats a satellite moving in a circular orbit around the Earth. To understand this situation, you must know that the gravitational force between spherical objects and small objects that can be modeled as particles having

masses m_1 and m_2 and separated by a distance r is attractive and has a magnitude

$$F_g = G\frac{m_1 m_2}{r^2}$$

where $G = 6.673 \times 10^{-11}$ N·m^2/kg^2. This is Newton's law of gravitation, which we study in Chapter 14.

Consider a satellite of mass m moving in a circular orbit around the Earth at a constant speed v and at an altitude h above the Earth's surface, as illustrated in Figure 6.7. Determine the speed of the satellite in terms of G, h, R_E (the radius of the Earth), and M_E (the mass of the Earth).

Solution The only external force acting on the satellite is the force of gravity, which acts toward the center of the Earth

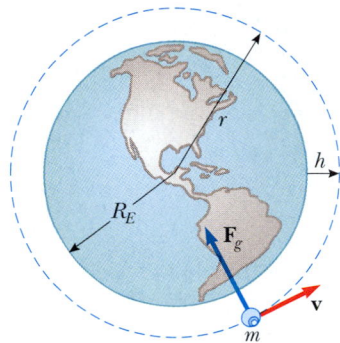

Figure 6.7 A satellite of mass m moving around the Earth at a constant speed v in a circular orbit of radius $r = R_E + h$. The force $\mathbf{F}_g$ acting on the satellite that causes the centripetal acceleration is the gravitational force exerted by the Earth on the satellite.

and keeps the satellite in its circular orbit. Therefore,

$$F_r = F_g = G\frac{M_E m}{r^2}$$

From Newton's second law and Equation 6.1 we obtain

$$G\frac{M_E m}{r^2} = m\frac{v^2}{r}$$

Solving for v and remembering that the distance r from the center of the Earth to the satellite is $r = R_E + h$, we obtain

$$(1) \qquad v = \sqrt{\frac{GM_E}{r}} = \boxed{\sqrt{\frac{GM_E}{R_E + h}}}$$

If the satellite were orbiting a different planet, its velocity would increase with the mass of the planet and decrease as the satellite's distance from the center of the planet increased.

Exercise A satellite is in a circular orbit around the Earth at an altitude of 1 000 km. The radius of the Earth is equal to 6.37×10^6 m, and its mass is 5.98×10^{24} kg. Find the speed of the satellite, and then find the *period*, which is the time it needs to make one complete revolution.

Answer 7.36×10^3 m/s; 6.29×10^3 s = 105 min.

EXAMPLE 6.7 ▶ Let's Go Loop-the-Loop!

A pilot of mass m in a jet aircraft executes a loop-the-loop, as shown in Figure 6.8a. In this maneuver, the aircraft moves in a vertical circle of radius 2.70 km at a constant speed of 225 m/s. Determine the force exerted by the seat on the pilot (a) at the bottom of the loop and (b) at the top of the loop. Express your answers in terms of the weight of the pilot mg.

Solution We expect the answer for (a) to be greater than that for (b) because at the bottom of the loop the normal and gravitational forces act in opposite directions, whereas at the top of the loop these two forces act in the same direction. It is the vector sum of these two forces that gives the force of constant magnitude that keeps the pilot moving in a circular path. To yield net force vectors with the same magnitude, the normal force at the bottom (where the normal and gravitational forces are in opposite directions) must be greater than that at the top (where the normal and gravitational forces are in the same direction). (a) The free-body diagram for the pilot at the bottom of the loop is shown in Figure 6.8b. The only forces acting on him are the downward force of gravity $\mathbf{F}_g = m\mathbf{g}$ and the upward force $\mathbf{n}_{bot}$ exerted by the seat. Because the net upward force that provides the centripetal ac-

celeration has a magnitude $n_{bot} - mg$, Newton's second law for the radial direction combined with Equation 6.1 gives

$$\sum F_r = n_{bot} - mg = m\frac{v^2}{r}$$

$$n_{bot} = mg + m\frac{v^2}{r} = mg\left(1 + \frac{v^2}{rg}\right)$$

Substituting the values given for the speed and radius gives

$$n_{bot} = mg\left[1 + \frac{(225 \text{ m/s})^2}{(2.70 \times 10^3 \text{ m})(9.80 \text{ m/s}^2)}\right] = \boxed{2.91\,mg}$$

Hence, the magnitude of the force $\mathbf{n}_{bot}$ exerted by the seat on the pilot is *greater* than the weight of the pilot by a factor of 2.91. This means that the pilot experiences an apparent weight that is greater than his true weight by a factor of 2.91.

(b) The free-body diagram for the pilot at the top of the loop is shown in Figure 6.8c. As we noted earlier, both the gravitational force exerted by the Earth and the force $\mathbf{n}_{top}$ exerted by the seat on the pilot act downward, and so the net downward force that provides the centripetal acceleration has

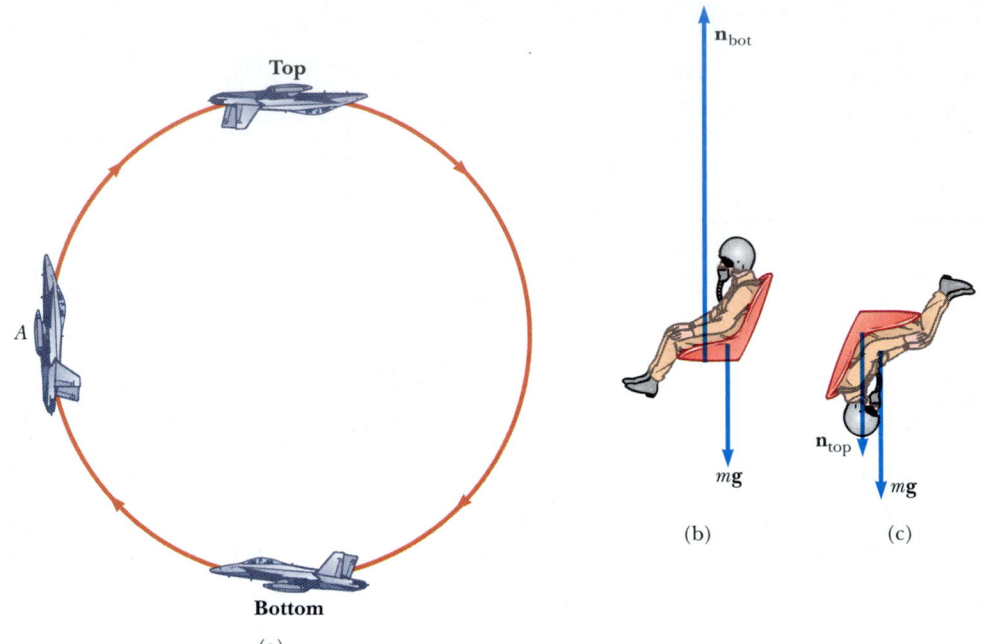

Figure 6.8   (a) An aircraft executes a loop-the-loop maneuver as it moves in a vertical circle at constant speed. (b) Free-body diagram for the pilot at the bottom of the loop. In this position the pilot experiences an apparent weight greater than his true weight. (c) Free-body diagram for the pilot at the top of the loop.

a magnitude $n_{top} + mg$. Applying Newton's second law yields

$$\sum F_r = n_{top} + mg = m\frac{v^2}{r}$$

$$n_{top} = m\frac{v^2}{r} - mg = mg\left(\frac{v^2}{rg} - 1\right)$$

$$n_{top} = mg\left[\frac{(225 \text{ m/s})^2}{(2.70 \times 10^3 \text{ m})(9.80 \text{ m/s}^2)} - 1\right] = \boxed{0.913\,mg}$$

In this case, the magnitude of the force exerted by the seat on the pilot is *less* than his true weight by a factor of 0.913, and the pilot feels lighter.

Exercise   Determine the magnitude of the radially directed force exerted on the pilot by the seat when the aircraft is at point A in Figure 6.8a, midway up the loop.

Answer $n_A = 1.913\,mg$ directed to the right.

Quick Quiz 6.3

A bead slides freely along a curved wire at constant speed, as shown in the overhead view of Figure 6.9. At each of the points Ⓐ, Ⓑ, and Ⓒ, draw the vector representing the force that the wire exerts on the bead in order to cause it to follow the path of the wire at that point.

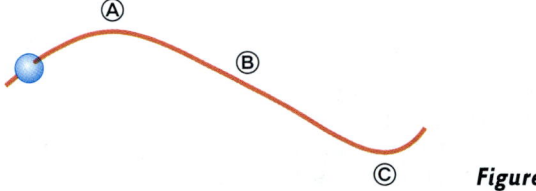

Figure 6.9

6.2   NONUNIFORM CIRCULAR MOTION

In Chapter 4 we found that if a particle moves with varying speed in a circular path, there is, in addition to the centripetal (radial) component of acceleration, a tangential component having magnitude dv/dt. Therefore, the force acting on the

Some examples of forces acting during circular motion. *(Left)* As these speed skaters round a curve, the force exerted by the ice on their skates provides the centripetal acceleration. *(Right)* Passengers on a "corkscrew" roller coaster. What are the origins of the forces in this example? *(Left: Bill Bachman/Photo Researchers, Inc.; Right: Robin Smith/Tony Stone Images)*

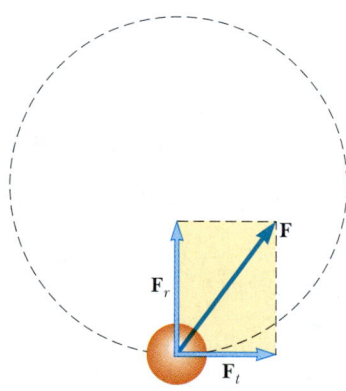

Figure 6.10 When the force acting on a particle moving in a circular path has a tangential component F_t, the particle's speed changes. The total force exerted on the particle in this case is the vector sum of the radial force and the tangential force. That is, $\mathbf{F} = \mathbf{F}_r + \mathbf{F}_t$.

particle must also have a tangential and a radial component. Because the total acceleration is $\mathbf{a} = \mathbf{a}_r + \mathbf{a}_t$, the total force exerted on the particle is $\mathbf{F} = \mathbf{F}_r + \mathbf{F}_t$, as shown in Figure 6.10. The vector $\mathbf{F}_r$ is directed toward the center of the circle and is responsible for the centripetal acceleration. The vector $\mathbf{F}_t$ tangent to the circle is responsible for the tangential acceleration, which represents a change in the speed of the particle with time. The following example demonstrates this type of motion.

EXAMPLE 6.8 ▶ Keep Your Eye on the Ball

A small sphere of mass m is attached to the end of a cord of length R and whirls in a *vertical* circle about a fixed point O, as illustrated in Figure 6.11a. Determine the tension in the cord at any instant when the speed of the sphere is v and the cord makes an angle θ with the vertical.

Solution Unlike the situation in Example 6.7, the speed is *not* uniform in this example because, at most points along the path, a tangential component of acceleration arises from the gravitational force exerted on the sphere. From the free-body diagram in Figure 6.11b, we see that the only forces acting on

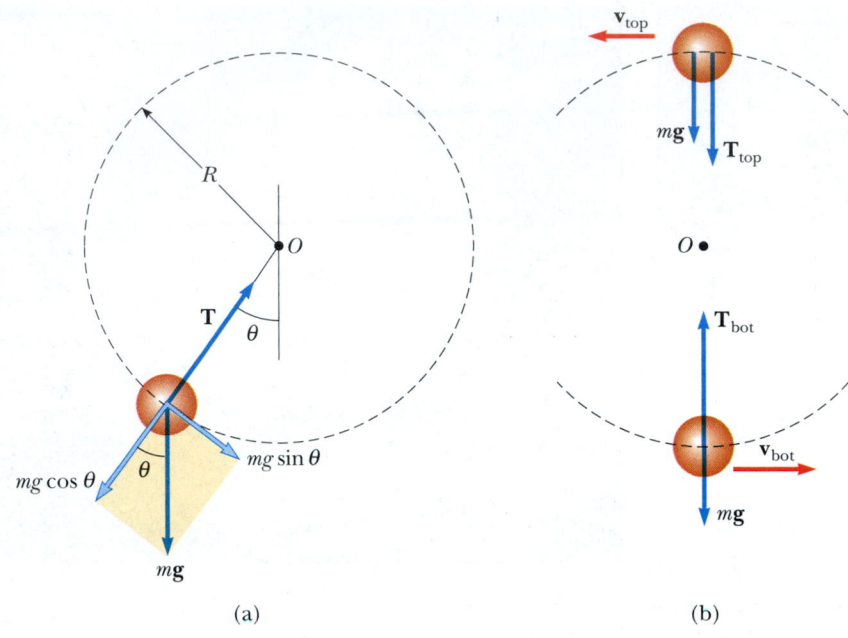

(a)

(b)

Figure 6.11 (a) Forces acting on a sphere of mass m connected to a cord of length R and rotating in a vertical circle centered at O. (b) Forces acting on the sphere at the top and bottom of the circle. The tension is a maximum at the bottom and a minimum at the top.

the sphere are the gravitational force $\mathbf{F}_g = m\mathbf{g}$ exerted by the Earth and the force $\mathbf{T}$ exerted by the cord. Now we resolve $\mathbf{F}_g$ into a tangential component $mg \sin\theta$ and a radial component $mg \cos\theta$. Applying Newton's second law to the forces acting on the sphere in the tangential direction yields

$$\sum F_t = mg \sin\theta = ma_t$$

$$a_t = g \sin\theta$$

This tangential component of the acceleration causes v to change in time because $a_t = dv/dt$.

Applying Newton's second law to the forces acting on the sphere in the radial direction and noting that both $\mathbf{T}$ and $\mathbf{a}_r$ are directed toward O, we obtain

$$\sum F_r = T - mg \cos\theta = \frac{mv^2}{R}$$

$$T = m\left(\frac{v^2}{R} + g \cos\theta\right)$$

Special Cases At the top of the path, where $\theta = 180°$, we have $\cos 180° = -1$, and the tension equation becomes

$$T_{top} = m\left(\frac{v_{top}^2}{R} - g\right)$$

This is the minimum value of T. Note that at this point $a_t = 0$ and therefore the acceleration is purely radial and directed downward.

At the bottom of the path, where $\theta = 0$, we see that, because $\cos 0 = 1$,

$$T_{bot} = m\left(\frac{v_{bot}^2}{R} + g\right)$$

This is the maximum value of T. At this point, a_t is again 0 and the acceleration is now purely radial and directed upward.

Exercise At what position of the sphere would the cord most likely break if the average speed were to increase?

Answer At the bottom, where T has its maximum value.

Optional Section

6.3 ▶ MOTION IN ACCELERATED FRAMES

When Newton's laws of motion were introduced in Chapter 5, we emphasized that they are valid only when observations are made in an inertial frame of reference. In this section, we analyze how an observer in a noninertial frame of reference (one that is accelerating) applies Newton's second law.

To understand the motion of a system that is noninertial because an object is moving along a curved path, consider a car traveling along a highway at a high speed and approaching a curved exit ramp, as shown in Figure 6.12a. As the car takes the sharp left turn onto the ramp, a person sitting in the passenger seat slides to the right and hits the door. At that point, the force exerted on her by the door keeps her from being ejected from the car. What causes her to move toward the door? A popular, but improper, explanation is that some mysterious force acting from left to right pushes her outward. (This is often called the "centrifugal" force, but we shall not use this term because it often creates confusion.) The passenger invents this **fictitious force** to explain what is going on in her accelerated frame of reference, as shown in Figure 6.12b. (The driver also experiences this effect but holds on to the steering wheel to keep from sliding to the right.)

The phenomenon is correctly explained as follows. Before the car enters the ramp, the passenger is moving in a straight-line path. As the car enters the ramp and travels a curved path, the passenger tends to move along the original straight-line path. This is in accordance with Newton's first law: The natural tendency of a body is to continue moving in a straight line. However, if a sufficiently large force (toward the center of curvature) acts on the passenger, as in Figure 6.12c, she will move in a curved path along with the car. The origin of this force is the force of friction between her and the car seat. If this frictional force is not large enough, she will slide to the right as the car turns to the left under her. Eventually, she encounters the door, which provides a force large enough to enable her to follow the same curved path as the car. She slides toward the door not because of some mysterious outward force but because **the force of friction is not sufficiently great to allow her to travel along the circular path followed by the car.**

In general, if a particle moves with an acceleration **a** relative to an observer in an inertial frame, that observer may use Newton's second law and correctly claim that $\Sigma \mathbf{F} = m\mathbf{a}$. If another observer in an accelerated frame tries to apply Newton's second law to the motion of the particle, the person must introduce fictitious forces to make Newton's second law work. These forces "invented" by the observer in the accelerating frame appear to be real. However, we emphasize that **these fictitious forces do not exist when the motion is observed in an inertial frame.** Fictitious forces are used only in an accelerating frame and do not represent "real" forces acting on the particle. (By real forces, we mean the interaction of the particle with its environment.) If the fictitious forces are properly defined in the accelerating frame, the description of motion in this frame is equivalent to the description given by an inertial observer who considers only real forces. Usually, we analyze motions using inertial reference frames, but there are cases in which it is more convenient to use an accelerating frame.

4.8

QuickLab

Use a string, a small weight, and a protractor to measure your acceleration as you start sprinting from a standing position.

Fictitious forces

(a)

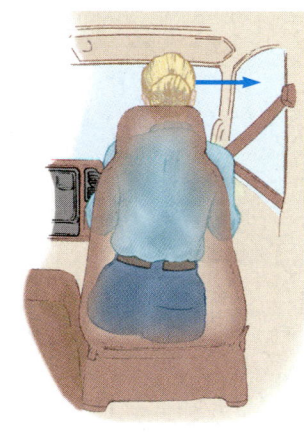

(b)

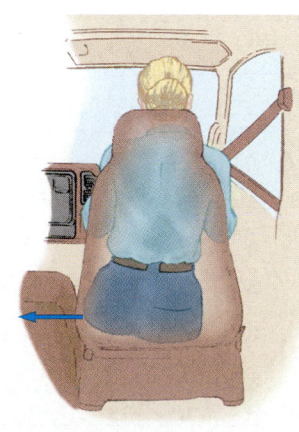

(c)

Figure 6.12 (a) A car approaching a curved exit ramp. What causes a front-seat passenger to move toward the right-hand door? (b) From the frame of reference of the passenger, a (fictitious) force pushes her toward the right door. (c) Relative to the reference frame of the Earth, the car seat applies a leftward force to the passenger, causing her to change direction along with the rest of the car.

EXAMPLE 6.9   **Fictitious Forces in Linear Motion**

A small sphere of mass m is hung by a cord from the ceiling of a boxcar that is accelerating to the right, as shown in Figure 6.13. According to the inertial observer at rest (Fig. 6.13a), the forces on the sphere are the force **T** exerted by the cord and the force of gravity. The inertial observer concludes that the acceleration of the sphere is the same as that of the boxcar and that this acceleration is provided by the horizontal component of **T**. Also, the vertical component of **T** balances the force of gravity. Therefore, she writes Newton's second law as $\Sigma\mathbf{F} = \mathbf{T} + m\mathbf{g} = m\mathbf{a}$, which in component form becomes

$$\text{Inertial observer}\quad\begin{cases}(1) & \sum F_x = T \sin\theta = ma \\ (2) & \sum F_y = T\cos\theta - mg = 0\end{cases}$$

Thus, by solving (1) and (2) simultaneously for a, the inertial observer can determine the magnitude of the car's acceleration through the relationship

$$a = g \tan\theta$$

Because the deflection of the cord from the vertical serves as a measure of acceleration, *a simple pendulum can be used as an accelerometer.*

According to the noninertial observer riding in the car (Fig. 6.13b), the cord still makes an angle θ with the vertical; however, to her the sphere is at rest and so its acceleration is zero. Therefore, she introduces a fictitious force to balance the horizontal component of **T** and claims that the net force on the sphere is *zero!* In this noninertial frame of reference, Newton's second law in component form yields

$$\text{Noninertial observer}\quad\begin{cases}\sum F_x' = T\sin\theta - F_{\text{fictitious}} = 0 \\ \sum F_y' = T\cos\theta - mg = 0\end{cases}$$

If we recognize that $F_{\text{fictitious}} = ma_{\text{inertial}} = ma$, then these expressions are equivalent to (1) and (2); therefore, the noninertial observer obtains the same mathematical results as the inertial observer does. However, the physical interpretation of the deflection of the cord differs in the two frames of reference.

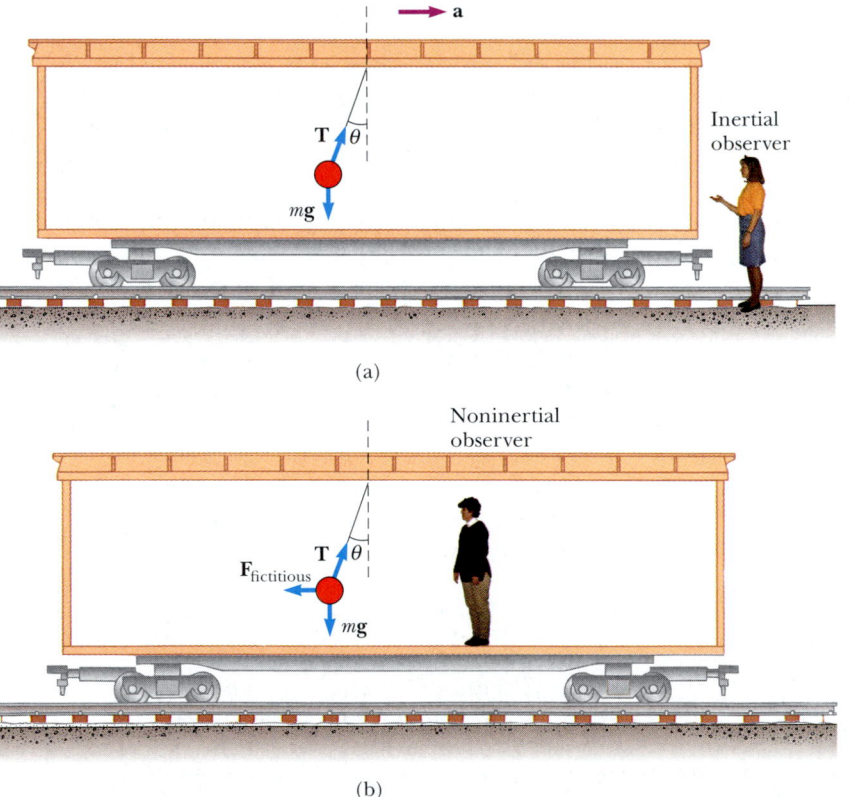

(a)

(b)

Figure 6.13   A small sphere suspended from the ceiling of a boxcar accelerating to the right is deflected as shown. (a) An inertial observer at rest outside the car claims that the acceleration of the sphere is provided by the horizontal component of **T**. (b) A noninertial observer riding in the car says that the net force on the sphere is zero and that the deflection of the cord from the vertical is due to a fictitious force $\mathbf{F}_{\text{fictitious}}$ that balances the horizontal component of **T**.

EXAMPLE 6.10   Fictitious Force in a Rotating System

Suppose a block of mass m lying on a horizontal, frictionless turntable is connected to a string attached to the center of the turntable, as shown in Figure 6.14. According to an inertial observer, if the block rotates uniformly, it undergoes an acceleration of magnitude v^2/r, where v is its linear speed. The inertial observer concludes that this centripetal acceleration is provided by the force **T** exerted by the string and writes Newton's second law as $T = mv^2/r$.

According to a noninertial observer attached to the turntable, the block is at rest and its acceleration is zero. Therefore, she must introduce a fictitious outward force of magnitude mv^2/r to balance the inward force exerted by the string. According to her, the net force on the block is zero, and she writes Newton's second law as $T - mv^2/r = 0$.

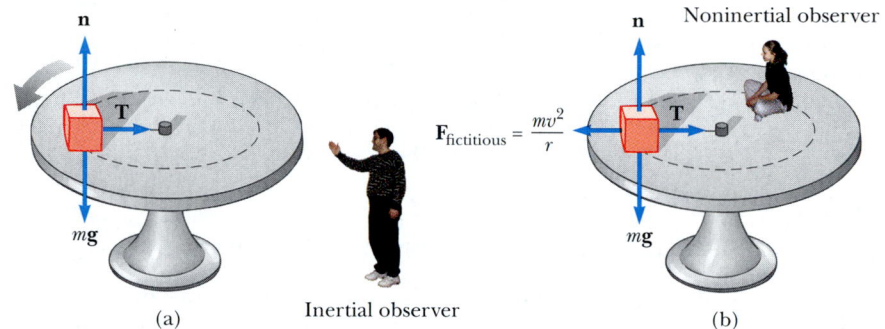

Figure 6.14   A block of mass m connected to a string tied to the center of a rotating turntable. (a) The inertial observer claims that the force causing the circular motion is provided by the force **T** exerted by the string on the block. (b) The noninertial observer claims that the block is not accelerating, and therefore she introduces a fictitious force of magnitude mv^2/r that acts outward and balances the force **T**.

Optional Section

6.4 ▸ MOTION IN THE PRESENCE OF RESISTIVE FORCES

In the preceding chapter we described the force of kinetic friction exerted on an object moving on some surface. We completely ignored any interaction between the object and the medium through which it moves. Now let us consider the effect of that medium, which can be either a liquid or a gas. The medium exerts a **resistive force R** on the object moving through it. Some examples are the air resistance associated with moving vehicles (sometimes called *air drag*) and the viscous forces that act on objects moving through a liquid. The magnitude of **R** depends on such factors as the speed of the object, and the direction of **R** is always opposite the direction of motion of the object relative to the medium. The magnitude of **R** nearly always increases with increasing speed.

The magnitude of the resistive force can depend on speed in a complex way, and here we consider only two situations. In the first situation, we assume the resistive force is proportional to the speed of the moving object; this assumption is valid for objects falling slowly through a liquid and for very small objects, such as dust particles, moving through air. In the second situation, we assume a resistive force that is proportional to the square of the speed of the moving object; large objects, such as a skydiver moving through air in free fall, experience such a force.

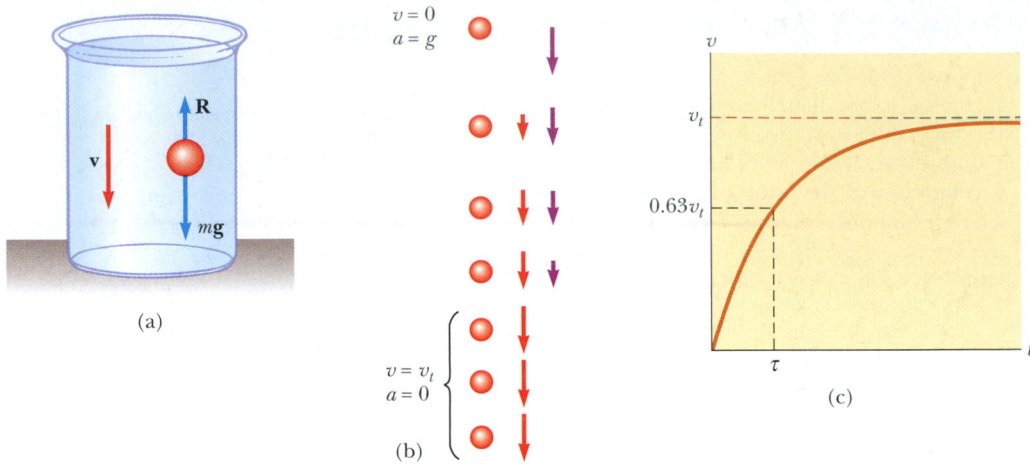

Figure 6.15   (a) A small sphere falling through a liquid. (b) Motion diagram of the sphere as it falls. (c) Speed–time graph for the sphere. The sphere reaches a maximum, or terminal, speed v_t, and the time constant τ is the time it takes to reach $0.63v_t$.

Resistive Force Proportional to Object Speed

If we assume that the resistive force acting on an object moving through a liquid or gas is proportional to the object's speed, then the magnitude of the resistive force can be expressed as

$$R = bv \tag{6.2}$$

where v is the speed of the object and b is a constant whose value depends on the properties of the medium and on the shape and dimensions of the object. If the object is a sphere of radius r, then b is proportional to r.

Consider a small sphere of mass m released from rest in a liquid, as in Figure 6.15a. Assuming that the only forces acting on the sphere are the resistive force bv and the force of gravity F_g, let us describe its motion.[1] Applying Newton's second law to the vertical motion, choosing the downward direction to be positive, and noting that $\Sigma F_y = mg - bv$, we obtain

$$mg - bv = ma = m\frac{dv}{dt} \tag{6.3}$$

where the acceleration dv/dt is downward. Solving this expression for the acceleration gives

$$\frac{dv}{dt} = g - \frac{b}{m}v \tag{6.4}$$

This equation is called a *differential equation,* and the methods of solving it may not be familiar to you as yet. However, note that initially, when $v = 0$, the resistive force $- bv$ is also zero and the acceleration dv/dt is simply g. As t increases, the resistive force increases and the acceleration decreases. Eventually, the acceleration becomes zero when the magnitude of the resistive force equals the sphere's weight. At this point, the sphere reaches its **terminal speed** v_t, and from then on

Terminal speed

---

[1] There is also a *buoyant force* acting on the submerged object. This force is constant, and its magnitude is equal to the weight of the displaced liquid. This force changes the apparent weight of the sphere by a constant factor, so we will ignore the force here. We discuss buoyant forces in Chapter 15.

it continues to move at this speed with zero acceleration, as shown in Figure 6.15b. We can obtain the terminal speed from Equation 6.3 by setting $a = dv/dt = 0$. This gives

$$mg - bv_t = 0 \qquad \text{or} \qquad v_t = mg/b$$

The expression for v that satisfies Equation 6.4 with $v = 0$ at $t = 0$ is

$$v = \frac{mg}{b}(1 - e^{-bt/m}) - v_t(1 - e^{-t/\tau}) \qquad \textbf{(6.5)}$$

This function is plotted in Figure 6.15c. The **time constant** $\tau = m/b$ (Greek letter tau) is the time it takes the sphere to reach 63.2% ($= 1 - 1/e$) of its terminal speed. This can be seen by noting that when $t = \tau$, Equation 6.5 yields $v = 0.632v_t$.

We can check that Equation 6.5 is a solution to Equation 6.4 by direct differentiation:

$$\frac{dv}{dt} = \frac{d}{dt}\left(\frac{mg}{b} - \frac{mg}{b}e^{-bt/m}\right) = -\frac{mg}{b}\frac{d}{dt}e^{-bt/m} = ge^{-bt/m}$$

(See Appendix Table B.4 for the derivative of e raised to some power.) Substituting into Equation 6.4 both this expression for dv/dt and the expression for v given by Equation 6.5 shows that our solution satisfies the differential equation.

Aerodynamic car. A streamlined body reduces air drag and increases fuel efficiency. *(Courtesy of Ford Motor Company)*

EXAMPLE 6.11   Sphere Falling in Oil

A small sphere of mass 2.00 g is released from rest in a large vessel filled with oil, where it experiences a resistive force proportional to its speed. The sphere reaches a terminal speed of 5.00 cm/s. Determine the time constant τ and the time it takes the sphere to reach 90% of its terminal speed.

Solution Because the terminal speed is given by $v_t = mg/b$, the coefficient b is

$$b = \frac{mg}{v_t} = \frac{(2.00 \text{ g})(980 \text{ cm/s}^2)}{5.00 \text{ cm/s}} = 392 \text{ g/s}$$

Therefore, the time constant τ is

$$\tau = \frac{m}{b} = \frac{2.00 \text{ g}}{392 \text{ g/s}} = \boxed{5.10 \times 10^{-3} \text{ s}}$$

The speed of the sphere as a function of time is given by Equation 6.5. To find the time t it takes the sphere to reach a speed of $0.900v_t$, we set $v = 0.900v_t$ in Equation 6.5 and solve for t:

$$0.900v_t = v_t(1 - e^{-t/\tau})$$

$$1 - e^{-t/\tau} = 0.900$$

$$e^{-t/\tau} = 0.100$$

$$-\frac{t}{\tau} = \ln(0.100) = -2.30$$

$$t = 2.30\tau = 2.30(5.10 \times 10^{-3} \text{ s}) = 11.7 \times 10^{-3} \text{ s}$$

$$= \boxed{11.7 \text{ ms}}$$

Thus, the sphere reaches 90% of its terminal (maximum) speed in a very short time.

Exercise What is the sphere's speed through the oil at $t = 11.7$ ms? Compare this value with the speed the sphere would have if it were falling in a vacuum and so were influenced only by gravity.

Answer 4.50 cm/s in oil versus 11.5 cm/s in free fall.

Air Drag at High Speeds

For objects moving at high speeds through air, such as airplanes, sky divers, cars, and baseballs, the resistive force is approximately proportional to the square of the speed. In these situations, the magnitude of the resistive force can be expressed as

$$R = \tfrac{1}{2}D\rho Av^2 \qquad \textbf{(6.6)}$$

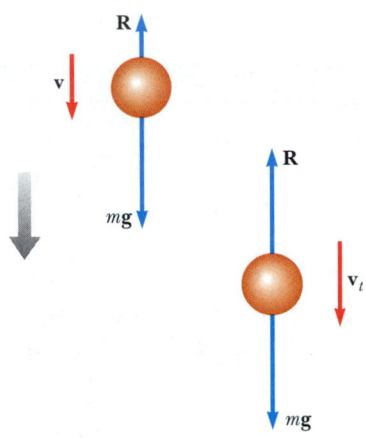

Figure 6.16 An object falling through air experiences a resistive force **R** and a gravitational force $\mathbf{F}_g = m\mathbf{g}$. The object reaches terminal speed (on the right) when the net force acting on it is zero, that is, when $\mathbf{R} = -\mathbf{F}_g$ or $R = mg$. Before this occurs, the acceleration varies with speed according to Equation 6.8.

where ρ is the density of air, A is the cross-sectional area of the falling object measured in a plane perpendicular to its motion, and D is a dimensionless empirical quantity called the *drag coefficient*. The drag coefficient has a value of about 0.5 for spherical objects but can have a value as great as 2 for irregularly shaped objects.

Let us analyze the motion of an object in free fall subject to an upward air resistive force of magnitude $R = \frac{1}{2} D\rho A v^2$. Suppose an object of mass m is released from rest. As Figure 6.16 shows, the object experiences two external forces: the downward force of gravity $\mathbf{F}_g = m\mathbf{g}$ and the upward resistive force **R**. (There is also an upward buoyant force that we neglect.) Hence, the magnitude of the net force is

$$\sum F = mg - \tfrac{1}{2}D\rho A v^2 \tag{6.7}$$

where we have taken downward to be the positive vertical direction. Substituting $\Sigma F = ma$ into Equation 6.7, we find that the object has a downward acceleration of magnitude

$$a = g - \left(\frac{D\rho A}{2m}\right)v^2 \tag{6.8}$$

We can calculate the terminal speed v_t by using the fact that when the force of gravity is balanced by the resistive force, the net force on the object is zero and therefore its acceleration is zero. Setting $a = 0$ in Equation 6.8 gives

$$g - \left(\frac{D\rho A}{2m}\right)v_t^2 = 0$$

$$v_t = \sqrt{\frac{2mg}{D\rho A}} \tag{6.9}$$

Using this expression, we can determine how the terminal speed depends on the dimensions of the object. Suppose the object is a sphere of radius r. In this case, $A \propto r^2$ (from $A = \pi r^2$) and $m \propto r^3$ (because the mass is proportional to the volume of the sphere, which is $V = \frac{4}{3}\pi r^3$). Therefore, $v_t \propto \sqrt{r}$.

Table 6.1 lists the terminal speeds for several objects falling through air.

The high cost of fuel has prompted many truck owners to install wind deflectors on their cabs to reduce drag.

TABLE 6.1   Terminal Speed for Various Objects Falling Through Air

Object	Mass (kg)	Cross-Sectional Area (m²)	v_t (m/s)
Sky diver	75	0.70	60
Baseball (radius 3.7 cm)	0.145	4.2×10^{-3}	43
Golf ball (radius 2.1 cm)	0.046	1.4×10^{-3}	44
Hailstone (radius 0.50 cm)	4.8×10^{-4}	7.9×10^{-5}	14
Raindrop (radius 0.20 cm)	3.4×10^{-5}	1.3×10^{-5}	9.0

web

For more information about the sport of sky surfing, visit **http://espn.sportszone.com /editors/xgames/surf/index.html**

CONCEPTUAL EXAMPLE 6.12

Consider a sky surfer who jumps from a plane with her feet attached firmly to her surfboard, does some tricks, and then opens her parachute. Describe the forces acting on her during these maneuvers.

Solution When the surfer first steps out of the plane, she has no vertical velocity. The downward force of gravity causes her to accelerate toward the ground. As her downward speed increases, so does the upward resistive force exerted by the air on her body and the board. This upward force reduces their acceleration, and so their speed increases more slowly. Eventually, they are going so fast that the upward resistive force matches the downward force of gravity. Now the net force is zero and they no longer accelerate, but reach their terminal speed. At some point after reaching terminal speed, she opens her parachute, resulting in a drastic increase in the upward resistive force. The net force (and thus the acceleration) is now upward, in the direction opposite the direction of the velocity. This causes the downward velocity to decrease rapidly; this means the resistive force on the chute also decreases. Eventually the upward resistive force and the downward force of gravity balance each other and a much smaller terminal speed is reached, permitting a safe landing.

(Contrary to popular belief, the velocity vector of a sky diver never points upward. You may have seen a videotape in which a sky diver appeared to "rocket" upward once the chute opened. In fact, what happened is that the diver slowed down while the person holding the camera continued falling at high speed.)

A sky surfer takes advantage of the upward force of the air on her board. (*Jump Run Productions/Image Bank*)

EXAMPLE 6.13   Falling Coffee Filters

The dependence of resistive force on speed is an empirical relationship. In other words, it is based on observation rather than on a theoretical model. A series of stacked filters is dropped, and the terminal speeds are measured. Table 6.2 presents data for these coffee filters as they fall through the air. The time constant τ is small, so that a dropped filter quickly reaches terminal speed. Each filter has a mass of 1.64 g. When the filters are nested together, they stack in such a way that the front-facing surface area does not increase. Determine the relationship between the resistive force exerted by the air and the speed of the falling filters.

Solution   At terminal speed, the upward resistive force balances the downward force of gravity. So, a single filter falling at its terminal speed experiences a resistive force of

$$R = mg = \left(\frac{1.64 \text{ g}}{1000 \text{ g/kg}} \right) (9.80 \text{ m/s}^2) = 0.016 \text{ 1 N}$$

Two filters nested together experience 0.032 2 N of resistive force, and so forth. A graph of the resistive force on the filters as a function of terminal speed is shown in Figure 6.17a. A straight line would not be a good fit, indicating that the resistive force is not proportional to the speed. The curved line is for a second-order polynomial, indicating a proportionality of the resistive force to the square of the speed. This proportionality is more clearly seen in Figure 6.17b, in which the resistive force is plotted as a function of the square of the terminal speed.

TABLE 6.2
Terminal Speed for Stacked Coffee Filters

Number of Filters	v_t (m/s)[a]
1	1.01
2	1.40
3	1.63
4	2.00
5	2.25
6	2.40
7	2.57
8	2.80
9	3.05
10	3.22

[a] All values of v_t are approximate.

Pleated coffee filters can be nested together so that the force of air resistance can be studied. (*Charles D. Winters*)

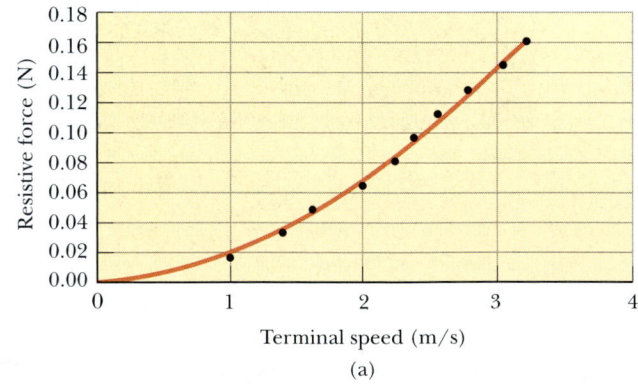

(a)

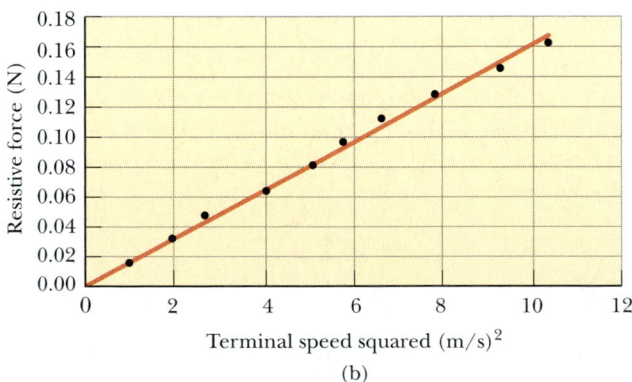

(b)

Figure 6.17   (a) Relationship between the resistive force acting on falling coffee filters and their terminal speed. The curved line is a second-order polynomial fit. (b) Graph relating the resistive force to the square of the terminal speed. The fit of the straight line to the data points indicates that the resistive force is proportional to the terminal speed squared. Can you find the proportionality constant?

EXAMPLE 6.14 ▶ Resistive Force Exerted on a Baseball

A pitcher hurls a 0.145-kg baseball past a batter at 40.2 m/s (=90 mi/h). Find the resistive force acting on the ball at this speed.

Solution   We do not expect the air to exert a huge force on the ball, and so the resistive force we calculate from Equation 6.6 should not be more than a few newtons. First, we must determine the drag coefficient D. We do this by imagining that we drop the baseball and allow it to reach terminal speed. We solve Equation 6.9 for D and substitute the appropriate values for m, v_t, and A from Table 6.1. Taking the density of air as 1.29 kg/m^3, we obtain

$$D = \frac{2\,mg}{v_t^2\,\rho A} = \frac{2(0.145\text{ kg})(9.80\text{ m/s}^2)}{(43\text{ m/s})^2\,(1.29\text{ kg/m}^3)(4.2\times 10^{-3}\text{ m}^2)}$$

$$= 0.284$$

This number has no dimensions. We have kept an extra digit beyond the two that are significant and will drop it at the end of our calculation.

We can now use this value for D in Equation 6.6 to find the magnitude of the resistive force:

$$R = \tfrac{1}{2}\,D\rho A v^2$$

$$= \tfrac{1}{2}(0.284)(1.29\text{ kg/m}^3)(4.2\times 10^{-3}\text{ m}^2)(40.2\text{ m/s})^2$$

$$= \boxed{1.2\text{ N}}$$

Optional Section

6.5 ▶ NUMERICAL MODELING IN PARTICLE DYNAMICS[2]

As we have seen in this and the preceding chapter, the study of the dynamics of a particle focuses on describing the position, velocity, and acceleration as functions of time. Cause-and-effect relationships exist among these quantities: Velocity causes position to change, and acceleration causes velocity to change. Because acceleration is the direct result of applied forces, any analysis of the dynamics of a particle usually begins with an evaluation of the net force being exerted on the particle.

Up till now, we have used what is called the *analytical method* to investigate the position, velocity, and acceleration of a moving particle. Let us review this method briefly before learning about a second way of approaching problems in dynamics. (Because we confine our discussion to one-dimensional motion in this section, boldface notation will not be used for vector quantities.)

If a particle of mass m moves under the influence of a net force ΣF, Newton's second law tells us that the acceleration of the particle is $a = \Sigma F/m$. In general, we apply the analytical method to a dynamics problem using the following procedure:

1. Sum all the forces acting on the particle to get the net force ΣF.
2. Use this net force to determine the acceleration from the relationship $a = \Sigma F/m$.
3. Use this acceleration to determine the velocity from the relationship $dv/dt = a$.
4. Use this velocity to determine the position from the relationship $dx/dt = v$.

The following straightforward example illustrates this method.

EXAMPLE 6.15 ▶ An Object Falling in a Vacuum—Analytical Method

Consider a particle falling in a vacuum under the influence of the force of gravity, as shown in Figure 6.18. Use the analytical method to find the acceleration, velocity, and position of the particle.

Solution   The only force acting on the particle is the downward force of gravity of magnitude F_g, which is also the net force. Applying Newton's second law, we set the net force acting on the particle equal to the mass of the particle times

---

[2] The authors are most grateful to Colonel James Head of the U.S. Air Force Academy for preparing this section. See the *Student Tools CD-ROM* for some assistance with numerical modeling.

its acceleration (taking upward to be the positive y direction):

$$F_g = ma_y = -mg$$

Thus, $a_y = -g$, which means the acceleration is constant. Because $dv_y/dt = a_y$, we see that $dv_y/dt = -g$, which may be integrated to yield

$$v_y(t) = v_{yi} - gt$$

Then, because $v_y = dy/dt$, the position of the particle is obtained from another integration, which yields the well-known result

$$y(t) = y_i + v_{yi}t - \tfrac{1}{2}gt^2$$

In these expressions, y_i and v_{yi} represent the position and speed of the particle at $t_i = 0$.

Figure 6.18   An object falling in vacuum under the influence of gravity.

The analytical method is straightforward for many physical situations. In the "real world," however, complications often arise that make analytical solutions difficult and perhaps beyond the mathematical abilities of most students taking introductory physics. For example, the net force acting on a particle may depend on the particle's position, as in cases where the gravitational acceleration varies with height. Or the force may vary with velocity, as in cases of resistive forces caused by motion through a liquid or gas.

Another complication arises because the expressions relating acceleration, velocity, position, and time are differential equations rather than algebraic ones. Differential equations are usually solved using integral calculus and other special techniques that introductory students may not have mastered.

When such situations arise, scientists often use a procedure called *numerical modeling* to study motion. The simplest numerical model is called the Euler method, after the Swiss mathematician Leonhard Euler (1707–1783).

The Euler Method

In the **Euler method** for solving differential equations, derivatives are approximated as ratios of finite differences. Considering a small increment of time Δt, we can approximate the relationship between a particle's speed and the magnitude of its acceleration as

$$a(t) \approx \frac{\Delta v}{\Delta t} = \frac{v(t + \Delta t) - v(t)}{\Delta t}$$

Then the speed $v(t + \Delta t)$ of the particle at the end of the time interval Δt is approximately equal to the speed $v(t)$ at the beginning of the time interval plus the magnitude of the acceleration during the interval multiplied by Δt:

$$v(t + \Delta t) \approx v(t) + a(t)\Delta t \tag{6.10}$$

Because the acceleration is a function of time, this estimate of $v(t + \Delta t)$ is accurate only if the time interval Δt is short enough that the change in acceleration during it is very small (as is discussed later). Of course, Equation 6.10 is exact if the acceleration is constant.

The position $x(t + \Delta t)$ of the particle at the end of the interval Δt can be found in the same manner:

$$v(t) \approx \frac{\Delta x}{\Delta t} = \frac{x(t + \Delta t) - x(t)}{\Delta t}$$

$$x(t + \Delta t) \approx x(t) + v(t)\Delta t \qquad \textbf{(6.11)}$$

You may be tempted to add the term $\frac{1}{2}a(\Delta t)^2$ to this result to make it look like the familiar kinematics equation, but this term is not included in the Euler method because Δt is assumed to be so small that Δt^2 is nearly zero.

If the acceleration at any instant t is known, the particle's velocity and position at a time $t + \Delta t$ can be calculated from Equations 6.10 and 6.11. The calculation then proceeds in a series of finite steps to determine the velocity and position at any later time. The acceleration is determined from the net force acting on the particle, and this force may depend on position, velocity, or time:

$$a(x, v, t) = \frac{\sum F(x, v, t)}{m} \qquad \textbf{(6.12)}$$

It is convenient to set up the numerical solution to this kind of problem by numbering the steps and entering the calculations in a table, a procedure that is illustrated in Table 6.3.

The equations in the table can be entered into a spreadsheet and the calculations performed row by row to determine the velocity, position, and acceleration as functions of time. The calculations can also be carried out by using a program written in either BASIC, C++, or FORTRAN or by using commercially available mathematics packages for personal computers. Many small increments can be taken, and accurate results can usually be obtained with the help of a computer. Graphs of velocity versus time or position versus time can be displayed to help you visualize the motion.

One advantage of the Euler method is that the dynamics is not obscured—the fundamental relationships between acceleration and force, velocity and acceleration, and position and velocity are clearly evident. Indeed, these relationships form the heart of the calculations. There is no need to use advanced mathematics, and the basic physics governs the dynamics.

The Euler method is completely reliable for infinitesimally small time increments, but for practical reasons a finite increment size must be chosen. For the finite difference approximation of Equation 6.10 to be valid, the time increment must be small enough that the acceleration can be approximated as being constant during the increment. We can determine an appropriate size for the time in-

See the spreadsheet file "Baseball with Drag" in the *Student Tools CD-ROM* for an example of how this technique can be applied to find the initial speed of the baseball described in Example 6.14. We cannot use our regular approach because our kinematics equations assume constant acceleration. Euler's method provides a way to circumvent this difficulty.

A detailed solution to Problem 41 involving iterative integration appears in the *Student Solutions Manual and Study Guide* and is posted on the Web at **http:/ www.saunderscollege.com/physics**

TABLE 6.3	**The Euler Method for Solving Dynamics Problems**			
Step	**Time**	**Position**	**Velocity**	**Acceleration**
0	t_0	x_0	v_0	$a_0 = F(x_0, v_0, t_0)/m$
1	$t_1 = t_0 + \Delta t$	$x_1 = x_0 + v_0 \Delta t$	$v_1 = v_0 + a_0 \Delta t$	$a_1 = F(x_1, v_1, t_1)/m$
2	$t_2 = t_1 + \Delta t$	$x_2 = x_1 + v_1 \Delta t$	$v_2 = v_1 + a_1 \Delta t$	$a_2 = F(x_2, v_2, t_2)/m$
3	$t_3 = t_2 + \Delta t$	$x_3 = x_2 + v_2 \Delta t$	$v_3 = v_2 + a_2 \Delta t$	$a_3 = F(x_3, v_3, t_3)/m$
$\vdots$	$\vdots$	$\vdots$	$\vdots$	$\vdots$
n	t_n	x_n	v_n	a_n

crement by examining the particular problem being investigated. The criterion for the size of the time increment may need to be changed during the course of the motion. In practice, however, we usually choose a time increment appropriate to the initial conditions and use the same value throughout the calculations.

The size of the time increment influences the accuracy of the result, but unfortunately it is not easy to determine the accuracy of a Euler-method solution without a knowledge of the correct analytical solution. One method of determining the accuracy of the numerical solution is to repeat the calculations with a smaller time increment and compare results. If the two calculations agree to a certain number of significant figures, you can assume that the results are correct to that precision.

SUMMARY

Newton's second law applied to a particle moving in uniform circular motion states that the net force causing the particle to undergo a centripetal acceleration is

$$\sum F_r = ma_r = \frac{mv^2}{r} \tag{6.1}$$

You should be able to use this formula in situations where the force providing the centripetal acceleration could be the force of gravity, a force of friction, a force of string tension, or a normal force.

A particle moving in nonuniform circular motion has both a centripetal component of acceleration and a nonzero tangential component of acceleration. In the case of a particle rotating in a vertical circle, the force of gravity provides the tangential component of acceleration and part or all of the centripetal component of acceleration. Be sure you understand the directions and magnitudes of the velocity and acceleration vectors for nonuniform circular motion.

An observer in a noninertial (accelerating) frame of reference must introduce **fictitious forces** when applying Newton's second law in that frame. If these fictitious forces are properly defined, the description of motion in the noninertial frame is equivalent to that made by an observer in an inertial frame. However, the observers in the two frames do not agree on the causes of the motion. You should be able to distinguish between inertial and noninertial frames and identify the fictitious forces acting in a noninertial frame.

A body moving through a liquid or gas experiences a **resistive force** that is speed-dependent. This resistive force, which opposes the motion, generally increases with speed. The magnitude of the resistive force depends on the shape of the body and on the properties of the medium through which the body is moving. In the limiting case for a falling body, when the magnitude of the resistive force equals the body's weight, the body reaches its **terminal speed.** You should be able to apply Newton's laws to analyze the motion of objects moving under the influence of resistive forces. You may need to apply **Euler's method** if the force depends on velocity, as it does for air drag.

QUESTIONS

1. Because the Earth rotates about its axis and revolves around the Sun, it is a noninertial frame of reference. Assuming the Earth is a uniform sphere, why would the apparent weight of an object be greater at the poles than at the equator?

2. Explain why the Earth bulges at the equator.

3. Why is it that an astronaut in a space capsule orbiting the Earth experiences a feeling of weightlessness?
4. Why does mud fly off a rapidly turning automobile tire?
5. Imagine that you attach a heavy object to one end of a spring and then whirl the spring and object in a horizontal circle (by holding the free end of the spring). Does the spring stretch? If so, why? Discuss this in terms of the force causing the circular motion.
6. It has been suggested that rotating cylinders about 10 mi in length and 5 mi in diameter be placed in space and used as colonies. The purpose of the rotation is to simulate gravity for the inhabitants. Explain this concept for producing an effective gravity.
7. Why does a pilot tend to black out when pulling out of a steep dive?

8. Describe a situation in which a car driver can have a centripetal acceleration but no tangential acceleration.
9. Describe the path of a moving object if its acceleration is constant in magnitude at all times and (a) perpendicular to the velocity; (b) parallel to the velocity.
10. Analyze the motion of a rock falling through water in terms of its speed and acceleration as it falls. Assume that the resistive force acting on the rock increases as the speed increases.
11. Consider a small raindrop and a large raindrop falling through the atmosphere. Compare their terminal speeds. What are their accelerations when they reach terminal speed?

PROBLEMS

1, **2**, **3** = straightforward, intermediate, challenging ☐ = full solution available in the *Student Solutions Manual and Study Guide*
WEB = solution posted at **http://www.saunderscollege.com/physics/** 🖥 = Computer useful in solving problem 🌐 = Interactive Physics
☐ = paired numerical/symbolic problems

Section 6.1   Newton's Second Law
Applied to Uniform Circular Motion

1. A toy car moving at constant speed completes one lap around a circular track (a distance of 200 m) in 25.0 s. (a) What is its average speed? (b) If the mass of the car is 1.50 kg, what is the magnitude of the force that keeps it in a circle?
2. A 55.0-kg ice skater is moving at 4.00 m/s when she grabs the loose end of a rope, the opposite end of which is tied to a pole. She then moves in a circle of radius 0.800 m around the pole. (a) Determine the force exerted by the rope on her arms. (b) Compare this force with her weight.
3. A light string can support a stationary hanging load of 25.0 kg before breaking. A 3.00-kg mass attached to the string rotates on a horizontal, frictionless table in a circle of radius 0.800 m. What range of speeds can the mass have before the string breaks?
4. In the Bohr model of the hydrogen atom, the speed of the electron is approximately 2.20×10^6 m/s. Find (a) the force acting on the electron as it revolves in a circular orbit of radius 0.530×10^{-10} m and (b) the centripetal acceleration of the electron.
5. In a cyclotron (one type of particle accelerator), a deuteron (of atomic mass 2.00 u) reaches a final speed of 10.0% of the speed of light while moving in a circular path of radius 0.480 m. The deuteron is maintained in the circular path by a magnetic force. What magnitude of force is required?
6. A satellite of mass 300 kg is in a circular orbit around the Earth at an altitude equal to the Earth's mean radius (see Example 6.6). Find (a) the satellite's orbital

speed, (b) the period of its revolution, and (c) the gravitational force acting on it.
7. Whenever two Apollo astronauts were on the surface of the Moon, a third astronaut orbited the Moon. Assume the orbit to be circular and 100 km above the surface of the Moon. If the mass of the Moon is 7.40×10^{22} kg and its radius is 1.70×10^6 m, determine (a) the orbiting astronaut's acceleration, (b) his orbital speed, and (c) the period of the orbit.
8. The speed of the tip of the minute hand on a town clock is 1.75×10^{-3} m/s. (a) What is the speed of the tip of the second hand of the same length? (b) What is the centripetal acceleration of the tip of the second hand?
9. A coin placed 30.0 cm from the center of a rotating, horizontal turntable slips when its speed is 50.0 cm/s. (a) What provides the force in the radial direction when the coin is stationary relative to the turntable? (b) What is the coefficient of static friction between coin and turntable?
10. The cornering performance of an automobile is evaluated on a skid pad, where the maximum speed that a car can maintain around a circular path on a dry, flat surface is measured. The centripetal acceleration, also called the lateral acceleration, is then calculated as a multiple of the free-fall acceleration g. The main factors affecting the performance are the tire characteristics and the suspension system of the car. A Dodge Viper GTS can negotiate a skid pad of radius 61.0 m at 86.5 km/h. Calculate its maximum lateral acceleration.
11. A crate of eggs is located in the middle of the flatbed of a pickup truck as the truck negotiates an unbanked

curve in the road. The curve may be regarded as an arc of a circle of radius 35.0 m. If the coefficient of static friction between crate and truck is 0.600, how fast can the truck be moving without the crate sliding?

12. A car initially traveling eastward turns north by traveling in a circular path at uniform speed as in Figure P6.12. The length of the arc *ABC* is 235 m, and the car completes the turn in 36.0 s. (a) What is the acceleration when the car is at *B* located at an angle of 35.0°? Express your answer in terms of the unit vectors **i** and **j**. Determine (b) the car's average speed and (c) its average acceleration during the 36.0-s interval.

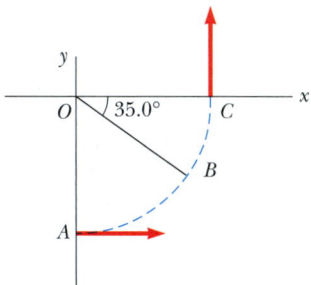

Figure P6.12

13. Consider a conical pendulum with an 80.0-kg bob on a 10.0-m wire making an angle of $\theta = 5.00°$ with the vertical (Fig. P6.13). Determine (a) the horizontal and vertical components of the force exerted by the wire on the pendulum and (b) the radial acceleration of the bob.

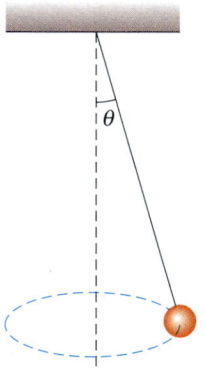

Figure P6.13

Section 6.2   Nonuniform Circular Motion

14. A car traveling on a straight road at 9.00 m/s goes over a hump in the road. The hump may be regarded as an arc of a circle of radius 11.0 m. (a) What is the apparent weight of a 600-N woman in the car as she rides over the

hump? (b) What must be the speed of the car over the hump if she is to experience weightlessness? (That is, if her apparent weight is zero.)

WEB 15. Tarzan ($m = 85.0$ kg) tries to cross a river by swinging from a vine. The vine is 10.0 m long, and his speed at the bottom of the swing (as he just clears the water) is 8.00 m/s. Tarzan doesn't know that the vine has a breaking strength of 1 000 N. Does he make it safely across the river?

16. A hawk flies in a horizontal arc of radius 12.0 m at a constant speed of 4.00 m/s. (a) Find its centripetal acceleration. (b) It continues to fly along the same horizontal arc but steadily increases its speed at the rate of 1.20 m/s². Find the acceleration (magnitude and direction) under these conditions.

17. A 40.0-kg child sits in a swing supported by two chains, each 3.00 m long. If the tension in each chain at the lowest point is 350 N, find (a) the child's speed at the lowest point and (b) the force exerted by the seat on the child at the lowest point. (Neglect the mass of the seat.)

18. A child of mass *m* sits in a swing supported by two chains, each of length *R*. If the tension in each chain at the lowest point is *T*, find (a) the child's speed at the lowest point and (b) the force exerted by the seat on the child at the lowest point. (Neglect the mass of the seat.)

WEB 19. A pail of water is rotated in a vertical circle of radius 1.00 m. What must be the minimum speed of the pail at the top of the circle if no water is to spill out?

20. A 0.400-kg object is swung in a vertical circular path on a string 0.500 m long. If its speed is 4.00 m/s at the top of the circle, what is the tension in the string there?

21. A roller-coaster car has a mass of 500 kg when fully loaded with passengers (Fig. P6.21). (a) If the car has a speed of 20.0 m/s at point *A*, what is the force exerted by the track on the car at this point? (b) What is the maximum speed the car can have at *B* and still remain on the track?

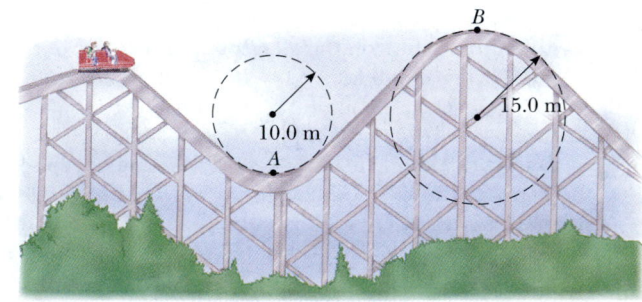

Figure P6.21

22. A roller coaster at the Six Flags Great America amusement park in Gurnee, Illinois, incorporates some of the latest design technology and some basic physics. Each vertical loop, instead of being circular, is shaped like a teardrop (Fig. P6.22). The cars ride on the inside of the loop at the top, and the speeds are high enough to ensure that the cars remain on the track. The biggest loop is 40.0 m high, with a maximum speed of 31.0 m/s (nearly 70 mi/h) at the bottom. Suppose the speed at the top is 13.0 m/s and the corresponding centripetal acceleration is 2g. (a) What is the radius of the arc of the teardrop at the top? (b) If the total mass of the cars plus people is M, what force does the rail exert on this total mass at the top? (c) Suppose the roller coaster had a loop of radius 20.0 m. If the cars have the same speed, 13.0 m/s at the top, what is the centripetal acceleration at the top? Comment on the normal force at the top in this situation.

Figure P6.22   (*Frank Cezus/FPG International*)

(*Optional*)
Section 6.3   Motion in Accelerated Frames

23. A merry-go-round makes one complete revolution in 12.0 s. If a 45.0-kg child sits on the horizontal floor of the merry-go-round 3.00 m from the center, find (a) the child's acceleration and (b) the horizontal force of friction that acts on the child. (c) What minimum coefficient of static friction is necessary to keep the child from slipping?

24. A 5.00-kg mass attached to a spring scale rests on a frictionless, horizontal surface as in Figure P6.24. The spring scale, attached to the front end of a boxcar, reads 18.0 N when the car is in motion. (a) If the spring scale reads zero when the car is at rest, determine the acceleration of the car. (b) What will the spring scale read if the car moves with constant velocity? (c) Describe the forces acting on the mass as observed by someone in the car and by someone at rest outside the car.

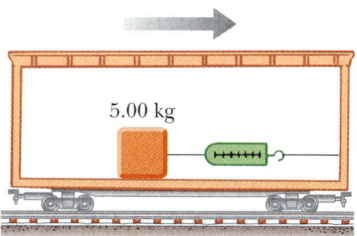

5.00 kg

Figure P6.24

25. A 0.500-kg object is suspended from the ceiling of an accelerating boxcar as was seen in Figure 6.13. If $a = 3.00$ m/s^2, find (a) the angle that the string makes with the vertical and (b) the tension in the string.

26. The Earth rotates about its axis with a period of 24.0 h. Imagine that the rotational speed can be increased. If an object at the equator is to have zero apparent weight, (a) what must the new period be? (b) By what factor would the speed of the object be increased when the planet is rotating at the higher speed? (*Hint:* See Problem 53 and note that the apparent weight of the object becomes zero when the normal force exerted on it is zero. Also, the distance traveled during one period is $2\pi R$, where R is the Earth's radius.)

27. A person stands on a scale in an elevator. As the elevator starts, the scale has a constant reading of 591 N. As the elevator later stops, the scale reading is 391 N. Assume the magnitude of the acceleration is the same during starting and stopping, and determine (a) the weight of the person, (b) the person's mass, and (c) the acceleration of the elevator.

28. A child on vacation wakes up. She is lying on her back. The tension in the muscles on both sides of her neck is 55.0 N as she raises her head to look past her toes and out the motel window. Finally, it is not raining! Ten minutes later she is screaming and sliding feet first down a water slide at a constant speed of 5.70 m/s, riding high on the outside wall of a horizontal curve of radius 2.40 m (Fig. P6.28). She raises her head to look forward past her toes; find the tension in the muscles on both sides of her neck.

Figure P6.28

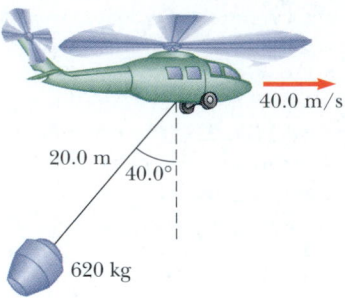

Figure P6.34

29. A plumb bob does not hang exactly along a line directed to the center of the Earth, because of the Earth's rotation. How much does the plumb bob deviate from a radial line at 35.0° north latitude? Assume that the Earth is spherical.

(Optional)
Section 6.4   Motion in the Presence of Resistive Forces

30. A sky diver of mass 80.0 kg jumps from a slow-moving aircraft and reaches a terminal speed of 50.0 m/s. (a) What is the acceleration of the sky diver when her speed is 30.0 m/s? What is the drag force exerted on the diver when her speed is (b) 50.0 m/s? (c) 30.0 m/s?

31. A small piece of Styrofoam packing material is dropped from a height of 2.00 m above the ground. Until it reaches terminal speed, the magnitude of its acceleration is given by $a = g - bv$. After falling 0.500 m, the Styrofoam effectively reaches its terminal speed, and then takes 5.00 s more to reach the ground. (a) What is the value of the constant b? (b) What is the acceleration at $t = 0$? (c) What is the acceleration when the speed is 0.150 m/s?

32. (a) Estimate the terminal speed of a wooden sphere (density 0.830 g/cm³) falling through the air if its radius is 8.00 cm. (b) From what height would a freely falling object reach this speed in the absence of air resistance?

33. Calculate the force required to pull a copper ball of radius 2.00 cm upward through a fluid at the constant speed 9.00 cm/s. Take the drag force to be proportional to the speed, with proportionality constant 0.950 kg/s. Ignore the buoyant force.

34. A fire helicopter carries a 620-kg bucket at the end of a cable 20.0 m long as in Figure P6.34. As the helicopter flies to a fire at a constant speed of 40.0 m/s, the cable makes an angle of 40.0° with respect to the vertical. The bucket presents a cross-sectional area of 3.80 m² in a plane perpendicular to the air moving past it. Determine the drag coefficient assuming that the resistive force is proportional to the square of the bucket's speed.

35. A small, spherical bead of mass 3.00 g is released from rest at $t = 0$ in a bottle of liquid shampoo. The terminal speed is observed to be $v_t = 2.00$ cm/s. Find (a) the value of the constant b in Equation 6.4, (b) the time τ the bead takes to reach $0.632v_t$, and (c) the value of the resistive force when the bead reaches terminal speed.

36. The mass of a sports car is 1 200 kg. The shape of the car is such that the aerodynamic drag coefficient is 0.250 and the frontal area is 2.20 m². Neglecting all other sources of friction, calculate the initial acceleration of the car if, after traveling at 100 km/h, it is shifted into neutral and is allowed to coast.

WEB 37. A motorboat cuts its engine when its speed is 10.0 m/s and coasts to rest. The equation governing the motion of the motorboat during this period is $v = v_i e^{-ct}$, where v is the speed at time t, v_i is the initial speed, and c is a constant. At $t = 20.0$ s, the speed is 5.00 m/s. (a) Find the constant c. (b) What is the speed at $t = 40.0$ s? (c) Differentiate the expression for $v(t)$ and thus show that the acceleration of the boat is proportional to the speed at any time.

38. Assume that the resistive force acting on a speed skater is $f = -kmv^2$, where k is a constant and m is the skater's mass. The skater crosses the finish line of a straight-line race with speed v_f and then slows down by coasting on his skates. Show that the skater's speed at any time t after crossing the finish line is $v(t) = v_f/(1 + ktv_f)$.

39. You can feel a force of air drag on your hand if you stretch your arm out of the open window of a speeding car. (*Note:* Do not get hurt.) What is the order of magnitude of this force? In your solution, state the quantities you measure or estimate and their values.

(Optional)
6.5   Numerical Modeling in Particle Dynamics

40. A 3.00-g leaf is dropped from a height of 2.00 m above the ground. Assume the net downward force exerted on the leaf is $F = mg - bv$, where the drag factor is $b = 0.030\ 0$ kg/s. (a) Calculate the terminal speed of the leaf. (b) Use Euler's method of numerical analysis to find the speed and position of the leaf as functions of time, from the instant it is released until 99% of terminal speed is reached. (*Hint:* Try $\Delta t = 0.005$ s.)

WEB 41. A hailstone of mass 4.80×10^{-4} kg falls through the air and experiences a net force given by

$$F = -mg + Cv^2$$

where $C = 2.50 \times 10^{-5}$ kg/m. (a) Calculate the terminal speed of the hailstone. (b) Use Euler's method of numerical analysis to find the speed and position of the hailstone at 0.2-s intervals, taking the initial speed to be zero. Continue the calculation until the hailstone reaches 99% of terminal speed.

42. A 0.142-kg baseball has a terminal speed of 42.5 m/s (95 mi/h). (a) If a baseball experiences a drag force of magnitude $R = Cv^2$, what is the value of the constant C? (b) What is the magnitude of the drag force when the speed of the baseball is 36.0 m/s? (c) Use a computer to determine the motion of a baseball thrown vertically upward at an initial speed of 36.0 m/s. What maximum height does the ball reach? How long is it in the air? What is its speed just before it hits the ground?

43. A 50.0-kg parachutist jumps from an airplane and falls with a drag force proportional to the square of the speed $R = Cv^2$. Take $C = 0.200$ kg/m with the parachute closed and $C = 20.0$ kg/m with the chute open. (a) Determine the terminal speed of the parachutist in both configurations, before and after the chute is opened. (b) Set up a numerical analysis of the motion and compute the speed and position as functions of time, assuming the jumper begins the descent at 1 000 m above the ground and is in free fall for 10.0 s before opening the parachute. (*Hint:* When the parachute opens, a sudden large acceleration takes place; a smaller time step may be necessary in this region.)

44. Consider a 10.0-kg projectile launched with an initial speed of 100 m/s, at an angle of 35.0° elevation. The resistive force is $\mathbf{R} = -b\mathbf{v}$, where $b = 10.0$ kg/s. (a) Use a numerical method to determine the horizontal and vertical positions of the projectile as functions of time. (b) What is the range of this projectile? (c) Determine the elevation angle that gives the maximum range for the projectile. (*Hint:* Adjust the elevation angle by trial and error to find the greatest range.)

45. A professional golfer hits a golf ball of mass 46.0 g with her 5-iron, and the ball first strikes the ground 155 m (170 yards) away. The ball experiences a drag force of magnitude $R = Cv^2$ and has a terminal speed of 44.0 m/s. (a) Calculate the drag constant C for the golf ball. (b) Use a numerical method to analyze the trajectory of this shot. If the initial velocity of the ball makes an angle of 31.0° (the loft angle) with the horizontal, what initial speed must the ball have to reach the 155-m distance? (c) If the same golfer hits the ball with her 9-iron (47.0° loft) and it first strikes the ground 119 m away, what is the initial speed of the ball? Discuss the differences in trajectories between the two shots.

ADDITIONAL PROBLEMS

46. An 1 800-kg car passes over a bump in a road that follows the arc of a circle of radius 42.0 m as in Figure P6.46. (a) What force does the road exert on the car as the car passes the highest point of the bump if the car travels at 16.0 m/s? (b) What is the maximum speed the car can have as it passes this highest point before losing contact with the road?

47. A car of mass m passes over a bump in a road that follows the arc of a circle of radius R as in Figure P6.46. (a) What force does the road exert on the car as the car passes the highest point of the bump if the car travels at a speed v? (b) What is the maximum speed the car can have as it passes this highest point before losing contact with the road?

Figure P6.46 Problems 46 and 47.

48. In one model of a hydrogen atom, the electron in orbit around the proton experiences an attractive force of about 8.20×10^{-8} N. If the radius of the orbit is 5.30×10^{-11} m, how many revolutions does the electron make each second? (This number of revolutions per unit time is called the *frequency* of the motion.) See the inside front cover for additional data.

49. A student builds and calibrates an accelerometer, which she uses to determine the speed of her car around a certain unbanked highway curve. The accelerometer is a plumb bob with a protractor that she attaches to the roof of her car. A friend riding in the car with her observes that the plumb bob hangs at an angle of 15.0° from the vertical when the car has a speed of 23.0 m/s. (a) What is the centripetal acceleration of the car rounding the curve? (b) What is the radius of the curve? (c) What is the speed of the car if the plumb bob deflection is 9.00° while the car is rounding the same curve?

50. Suppose the boxcar shown in Figure 6.13 is moving with constant acceleration a up a hill that makes an angle ϕ with the horizontal. If the hanging pendulum makes a constant angle θ with the perpendicular to the ceiling, what is a?

51. An air puck of mass 0.250 kg is tied to a string and allowed to revolve in a circle of radius 1.00 m on a fric-

tionless horizontal table. The other end of the string passes through a hole in the center of the table, and a mass of 1.00 kg is tied to it (Fig. P6.51). The suspended mass remains in equilibrium while the puck on the tabletop revolves. What are (a) the tension in the string, (b) the force exerted by the string on the puck, and (c) the speed of the puck?

52. An air puck of mass m_1 is tied to a string and allowed to revolve in a circle of radius R on a frictionless horizontal table. The other end of the string passes through a hole in the center of the table, and a mass m_2 is tied to it (Fig. P6.51). The suspended mass remains in equilibrium while the puck on the tabletop revolves. What are (a) the tension in the string? (b) the central force exerted on the puck? (c) the speed of the puck?

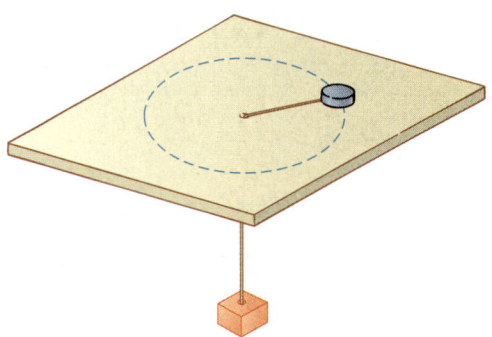

Figure P6.51   Problems 51 and 52.

WEB 53. Because the Earth rotates about its axis, a point on the equator experiences a centripetal acceleration of 0.033 7 m/s², while a point at one of the poles experiences no centripetal acceleration. (a) Show that at the equator the gravitational force acting on an object (the true weight) must exceed the object's apparent weight. (b) What is the apparent weight at the equator and at the poles of a person having a mass of 75.0 kg? (Assume the Earth is a uniform sphere and take $g = 9.800$ m/s².)

54. A string under a tension of 50.0 N is used to whirl a rock in a horizontal circle of radius 2.50 m at a speed of 20.4 m/s. The string is pulled in and the speed of the rock increases. When the string is 1.00 m long and the speed of the rock is 51.0 m/s, the string breaks. What is the breaking strength (in newtons) of the string?

55. A child's toy consists of a small wedge that has an acute angle θ (Fig. P6.55). The sloping side of the wedge is frictionless, and a mass m on it remains at constant height if the wedge is spun at a certain constant speed. The wedge is spun by rotating a vertical rod that is firmly attached to the wedge at the bottom end. Show

that, when the mass sits a distance L up along the sloping side, the speed of the mass must be

$$v = (g\,L \sin\,\theta)^{1/2}$$

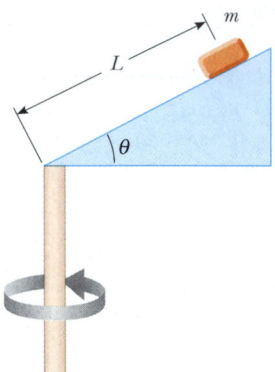

Figure P6.55

56. The pilot of an airplane executes a constant-speed loop-the-loop maneuver. His path is a vertical circle. The speed of the airplane is 300 mi/h, and the radius of the circle is 1 200 ft. (a) What is the pilot's apparent weight at the lowest point if his true weight is 160 lb? (b) What is his apparent weight at the highest point? (c) Describe how the pilot could experience apparent weightlessness if both the radius and the speed can be varied. (*Note:* His apparent weight is equal to the force that the seat exerts on his body.)

57. For a satellite to move in a stable circular orbit at a constant speed, its centripetal acceleration must be inversely proportional to the square of the radius r of the orbit. (a) Show that the tangential speed of a satellite is proportional to $r^{-1/2}$. (b) Show that the time required to complete one orbit is proportional to $r^{3/2}$.

58. A penny of mass 3.10 g rests on a small 20.0-g block supported by a spinning disk (Fig. P6.58). If the coeffi-

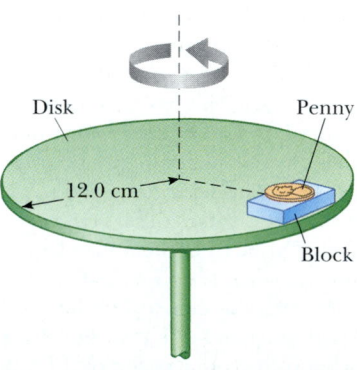

Figure P6.58

cients of friction between block and disk are 0.750 (static) and 0.640 (kinetic) while those for the penny and block are 0.450 (kinetic) and 0.520 (static), what is the maximum rate of rotation (in revolutions per minute) that the disk can have before either the block or the penny starts to slip?

59. Figure P6.59 shows a Ferris wheel that rotates four times each minute and has a diameter of 18.0 m. (a) What is the centripetal acceleration of a rider? What force does the seat exert on a 40.0-kg rider (b) at the lowest point of the ride and (c) at the highest point of the ride? (d) What force (magnitude and direction) does the seat exert on a rider when the rider is halfway between top and bottom?

Figure P6.59   *(Color Box/FPG)*

60. A space station, in the form of a large wheel 120 m in diameter, rotates to provide an "artificial gravity" of 3.00 m/s² for persons situated at the outer rim. Find the rotational frequency of the wheel (in revolutions per minute) that will produce this effect.

61. An amusement park ride consists of a rotating circular platform 8.00 m in diameter from which 10.0-kg seats are suspended at the end of 2.50-m massless chains (Fig. P6.61). When the system rotates, the chains make an angle $\theta = 28.0°$ with the vertical. (a) What is the speed of each seat? (b) Draw a free-body diagram of a 40.0-kg child riding in a seat and find the tension in the chain.

62. A piece of putty is initially located at point A on the rim of a grinding wheel rotating about a horizontal axis. The putty is dislodged from point A when the diameter through A is horizontal. The putty then rises vertically and returns to A the instant the wheel completes one revolution. (a) Find the speed of a point on the rim of the wheel in terms of the acceleration due to gravity and the radius R of the wheel. (b) If the mass of the putty is m, what is the magnitude of the force that held it to the wheel?

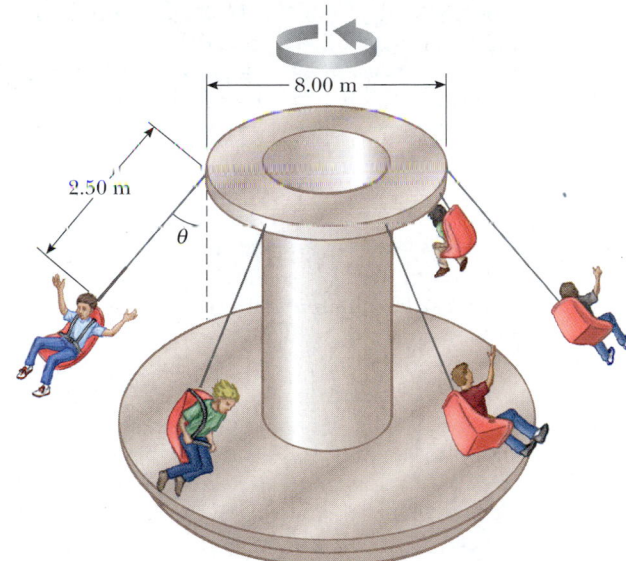

Figure P6.61

63. An amusement park ride consists of a large vertical cylinder that spins about its axis fast enough that any person inside is held up against the wall when the floor drops away (Fig. P6.63). The coefficient of static friction between person and wall is μ_s, and the radius of the cylinder is R. (a) Show that the maximum period of revolution necessary to keep the person from falling is $T = (4\pi^2 R\mu_s/g)^{1/2}$. (b) Obtain a numerical value for T

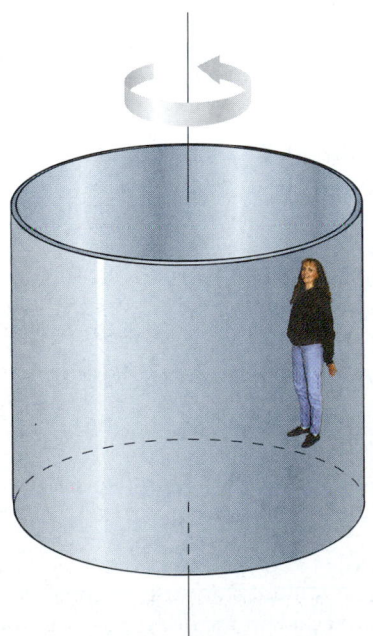

Figure P6.63

if $R = 4.00$ m and $\mu_s = 0.400$. How many revolutions per minute does the cylinder make?

64. *An example of the Coriolis effect.* Suppose air resistance is negligible for a golf ball. A golfer tees off from a location precisely at $\phi_i = 35.0°$ north latitude. He hits the ball due south, with range 285 m. The ball's initial velocity is at 48.0° above the horizontal. (a) For what length of time is the ball in flight? The cup is due south of the golfer's location, and he would have a hole-in-one if the Earth were not rotating. As shown in Figure P6.64, the Earth's rotation makes the tee move in a circle of radius $R_E \cos \phi_i = (6.37 \times 10^6$ m$) \cos 35.0°$, completing one revolution each day. (b) Find the eastward speed of the tee, relative to the stars. The hole is also moving eastward, but it is 285 m farther south and thus at a slightly lower latitude ϕ_f. Because the hole moves eastward in a slightly larger circle, its speed must be greater than that of the tee. (c) By how much does the hole's speed exceed that of the tee? During the time the ball is in flight, it moves both upward and downward, as well as southward with the projectile motion you studied in Chapter 4, but it also moves eastward with the speed you found in part (b). The hole moves to the east at a faster speed, however, pulling ahead of the ball with the relative speed you found in part (c). (d) How far to the west of the hole does the ball land?

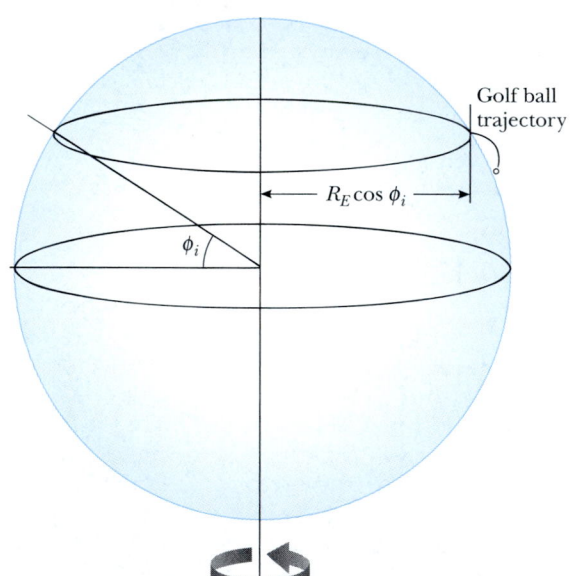

Figure P6.64

65. A curve in a road forms part of a horizontal circle. As a car goes around it at constant speed 14.0 m/s, the total force exerted on the driver has magnitude 130 N. What are the magnitude and direction of the total force exerted on the driver if the speed is 18.0 m/s instead?

66. A car rounds a banked curve as shown in Figure 6.6. The radius of curvature of the road is R, the banking angle is θ, and the coefficient of static friction is μ_s. (a) Determine the range of speeds the car can have without slipping up or down the banked surface. (b) Find the minimum value for μ_s such that the minimum speed is zero. (c) What is the range of speeds possible if $R = 100$ m, $\theta = 10.0°$, and $\mu_s = 0.100$ (slippery conditions)?

67. A single bead can slide with negligible friction on a wire that is bent into a circle of radius 15.0 cm, as in Figure P6.67. The circle is always in a vertical plane and rotates steadily about its vertical diameter with a period of 0.450 s. The position of the bead is described by the angle θ that the radial line from the center of the loop to the bead makes with the vertical. (a) At what angle up from the lowest point can the bead stay motionless relative to the turning circle? (b) Repeat the problem if the period of the circle's rotation is 0.850 s.

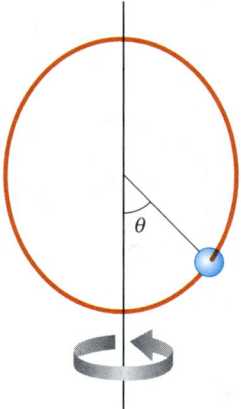

Figure P6.67

68. The expression $F = arv + br^2v^2$ gives the magnitude of the resistive force (in newtons) exerted on a sphere of radius r (in meters) by a stream of air moving at speed v (in meters per second), where a and b are constants with appropriate SI units. Their numerical values are $a = 3.10 \times 10^{-4}$ and $b = 0.870$. Using this formula, find the terminal speed for water droplets falling under their own weight in air, taking the following values for the drop radii: (a) 10.0 μm, (b) 100 μm, (c) 1.00 mm. Note that for (a) and (c) you can obtain accurate answers without solving a quadratic equation, by considering which of the two contributions to the air resistance is dominant and ignoring the lesser contribution.

69. A model airplane of mass 0.750 kg flies in a horizontal circle at the end of a 60.0-m control wire, with a speed of 35.0 m/s. Compute the tension in the wire if it makes a constant angle of 20.0° with the horizontal. The forces exerted on the airplane are the pull of the control wire,

its own weight, and aerodynamic lift, which acts at 20.0° inward from the vertical as shown in Figure P6.69.

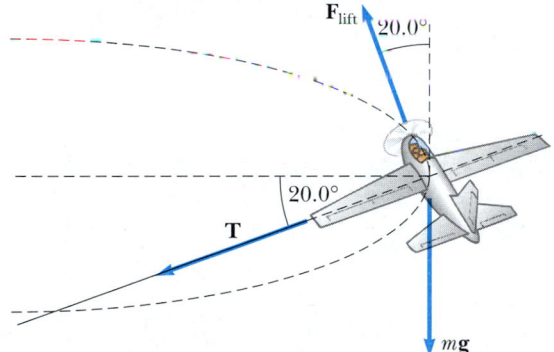

Figure P6.69

70. A 9.00-kg object starting from rest falls through a viscous medium and experiences a resistive force $\mathbf{R} = -b\mathbf{v}$, where $\mathbf{v}$ is the velocity of the object. If the object's speed reaches one-half its terminal speed in 5.54 s, (a) determine the terminal speed. (b) At what time is the speed of the object three-fourths the terminal speed? (c) How far has the object traveled in the first 5.54 s of motion?

71. Members of a skydiving club were given the following data to use in planning their jumps. In the table, d is the distance fallen from rest by a sky diver in a "free-fall stable spread position" versus the time of fall t. (a) Convert the distances in feet into meters. (b) Graph d (in meters) versus t. (c) Determine the value of the terminal speed v_t by finding the slope of the straight portion of the curve. Use a least-squares fit to determine this slope.

t (s)	d (ft)
1	16
2	62
3	138
4	242
5	366
6	504
7	652
8	808
9	971
10	1 138
11	1 309
12	1 483
13	1 657
14	1 831
15	2 005
16	2 179
17	2 353
18	2 527
19	2 701
20	2 875

ANSWERS TO QUICK QUIZZES

6.1 No. The tangential acceleration changes just the speed part of the velocity vector. For the car to move in a circle, the *direction* of its velocity vector must change, and the only way this can happen is for there to be a centripetal acceleration.

6.2 (a) The ball travels in a circular path that has a larger radius than the original circular path, and so there must be some external force causing the change in the velocity vector's direction. The external force must not be as strong as the original tension in the string because if it were, the ball would follow the original path. (b) The ball again travels in an arc, implying some kind of external force. As in part (a), the external force is directed toward the center of the new arc and not toward the center of the original circular path. (c) The ball undergoes an abrupt change in velocity—from tangent to the circle to perpendicular to it—and so must have experienced a large force that had one component opposite the ball's velocity (tangent to the circle) and another component radially outward. (d) The ball travels in a straight line tangent to the original path. If there is an external force, it cannot have a component perpendicular to this line because if it did, the path would curve. In

fact, if the string breaks and there is no other force acting on the ball, Newton's first law says the ball will travel along such a tangent line at constant speed.

6.3 At Ⓐ the path is along the circumference of the larger circle. Therefore, the wire must be exerting a force on the bead directed toward the center of the circle. Because the speed is constant, there is no tangential force component. At Ⓑ the path is not curved, and so the wire exerts no force on the bead. At Ⓒ the path is again curved, and so the wire is again exerting a force on the bead. This time the force is directed toward the center of the smaller circle. Because the radius of this circle is smaller, the magnitude of the force exerted on the bead is larger here than at Ⓐ.

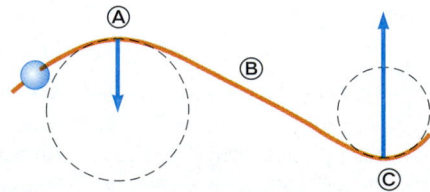

 **PUZZLER**

Chum salmon "climbing a ladder" in the McNeil River in Alaska. Why are fish ladders like this often built around dams? Do the ladders reduce the amount of work that the fish must do to get past the dam? *(Daniel J. Cox/Tony Stone Images)*

c h a p t e r

7

Work and Kinetic Energy

The concept of energy is one of the most important topics in science and engineering. In everyday life, we think of energy in terms of fuel for transportation and heating, electricity for lights and appliances, and foods for consumption. However, these ideas do not really define energy. They merely tell us that fuels are needed to do a job and that those fuels provide us with something we call *energy*.

In this chapter, we first introduce the concept of work. Work is done by a force acting on an object when the point of application of that force moves through some distance and the force has a component along the line of motion. Next, we define kinetic energy, which is energy an object possesses because of its motion. In general, we can think of *energy* as the capacity that an object has for performing work. We shall see that the concepts of work and kinetic energy can be applied to the dynamics of a mechanical system without resorting to Newton's laws. In a complex situation, in fact, the "energy approach" can often allow a much simpler analysis than the direct application of Newton's second law. However, it is important to note that the work–energy concepts are based on Newton's laws and therefore allow us to make predictions that are always in agreement with these laws.

This alternative method of describing motion is especially useful when the force acting on a particle varies with the position of the particle. In this case, the acceleration is not constant, and we cannot apply the kinematic equations developed in Chapter 2. Often, a particle in nature is subject to a force that varies with the position of the particle. Such forces include the gravitational force and the force exerted on an object attached to a spring. Although we could analyze situations like these by applying numerical methods such as those discussed in Section 6.5, utilizing the ideas of work and energy is often much simpler. We describe techniques for treating complicated systems with the help of an extremely important theorem called the *work–kinetic energy theorem,* which is the central topic of this chapter.

7.1 WORK DONE BY A CONSTANT FORCE

Almost all the terms we have used thus far—velocity, acceleration, force, and so on—convey nearly the same meaning in physics as they do in everyday life. Now, however, we encounter a term whose meaning in physics is distinctly different from its everyday meaning. That new term is *work*.

To understand what *work* means to the physicist, consider the situation illustrated in Figure 7.1. A force is applied to a chalkboard eraser, and the eraser slides along the tray. If we are interested in how effective the force is in moving the

(a)

(b)

(c)

Figure 7.1   An eraser being pushed along a chalkboard tray.   *(Charles D. Winters)*

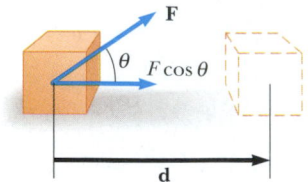

Figure 7.2 If an object undergoes a displacement **d** under the action of a constant force **F**, the work done by the force is $(F \cos \theta) d$.

eraser, we need to consider not only the magnitude of the force but also its direction. If we assume that the magnitude of the applied force is the same in all three photographs, it is clear that the push applied in Figure 7.1b does more to move the eraser than the push in Figure 7.1a. On the other hand, Figure 7.1c shows a situation in which the applied force does not move the eraser at all, regardless of how hard it is pushed. (Unless, of course, we apply a force so great that we break something.) So, in analyzing forces to determine the work they do, we must consider the vector nature of forces. We also need to know how far the eraser moves along the tray if we want to determine the work required to cause that motion. Moving the eraser 3 m requires more work than moving it 2 cm.

Let us examine the situation in Figure 7.2, where an object undergoes a displacement **d** along a straight line while acted on by a constant force **F** that makes an angle θ with **d**.

Work done by a constant force

> The **work** W done on an object by an agent exerting a constant force on the object is the product of the component of the force in the direction of the displacement and the magnitude of the displacement:
>
> $$W = Fd \cos \theta \qquad \textbf{(7.1)}$$

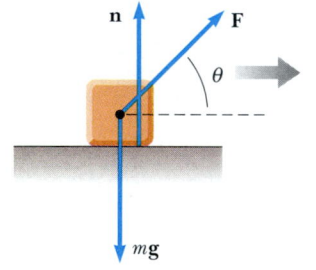

Figure 7.3 When an object is displaced on a frictionless, horizontal, surface, the normal force **n** and the force of gravity $m\mathbf{g}$ do no work on the object. In the situation shown here, **F** is the only force doing work on the object.

As an example of the distinction between this definition of work and our everyday understanding of the word, consider holding a heavy chair at arm's length for 3 min. At the end of this time interval, your tired arms may lead you to think that you have done a considerable amount of work on the chair. According to our definition, however, you have done no work on it whatsoever.[1] You exert a force to support the chair, but you do not move it. A force does no work on an object if the object does not move. This can be seen by noting that if $d = 0$, Equation 7.1 gives $W = 0$—the situation depicted in Figure 7.1c.

Also note from Equation 7.1 that the work done by a force on a moving object is zero when the force applied is perpendicular to the object's displacement. That is, if $\theta = 90°$, then $W = 0$ because $\cos 90° = 0$. For example, in Figure 7.3, the work done by the normal force on the object and the work done by the force of gravity on the object are both zero because both forces are perpendicular to the displacement and have zero components in the direction of **d**.

The sign of the work also depends on the direction of **F** relative to **d**. The work done by the applied force is positive when the vector associated with the component $F \cos \theta$ is in the same direction as the displacement. For example, when an object is lifted, the work done by the applied force is positive because the direction of that force is upward, that is, in the same direction as the displacement. When the vector associated with the component $F \cos \theta$ is in the direction opposite the displacement, W is negative. For example, as an object is lifted, the work done by the gravitational force on the object is negative. The factor $\cos \theta$ in the definition of W (Eq. 7.1) automatically takes care of the sign. It is important to note that **work is an energy transfer;** if energy is transferred *to* the system (object), W is positive; if energy is transferred *from* the system, W is negative.

5.3

---

[1] Actually, you do work while holding the chair at arm's length because your muscles are continuously contracting and relaxing; this means that they are exerting internal forces on your arm. Thus, work is being done by your body—but internally on itself rather than on the chair.

If an applied force **F** acts along the direction of the displacement, then $\theta = 0$ and $\cos 0 = 1$. In this case, Equation 7.1 gives

$$W = Fd$$

Work is a scalar quantity, and its units are force multiplied by length. Therefore, the SI unit of work is the **newton·meter** (N·m). This combination of units is used so frequently that it has been given a name of its own: the **joule** (J).

Can the component of a force that gives an object a centripetal acceleration do any work on the object? (One such force is that exerted by the Sun on the Earth that holds the Earth in a circular orbit around the Sun.)

In general, a particle may be moving with either a constant or a varying velocity under the influence of several forces. In these cases, because work is a scalar quantity, the total work done as the particle undergoes some displacement is the algebraic sum of the amounts of work done by all the forces.

EXAMPLE 7.1 **Mr. Clean**

A man cleaning a floor pulls a vacuum cleaner with a force of magnitude $F = 50.0$ N at an angle of 30.0° with the horizontal (Fig. 7.4a). Calculate the work done by the force on the vacuum cleaner as the vacuum cleaner is displaced 3.00 m to the right.

Solution Because they aid us in clarifying which forces are acting on the object being considered, drawings like Figure 7.4b are helpful when we are gathering information and organizing a solution. For our analysis, we use the definition of work (Eq. 7.1):

$$W = (F \cos \theta) d$$
$$= (50.0 \text{ N})(\cos 30.0°)(3.00 \text{ m}) = 130 \text{ N·m}$$
$$= \boxed{130 \text{ J}}$$

One thing we should learn from this problem is that the normal force **n**, the force of gravity $\mathbf{F}_g = m\mathbf{g}$, and the upward component of the applied force $(50.0 \text{ N})(\sin 30.0°)$ do no work on the vacuum cleaner because these forces are perpendicular to its displacement.

Exercise Find the work done by the man on the vacuum cleaner if he pulls it 3.0 m with a horizontal force of 32 N.

Answer 96 J.

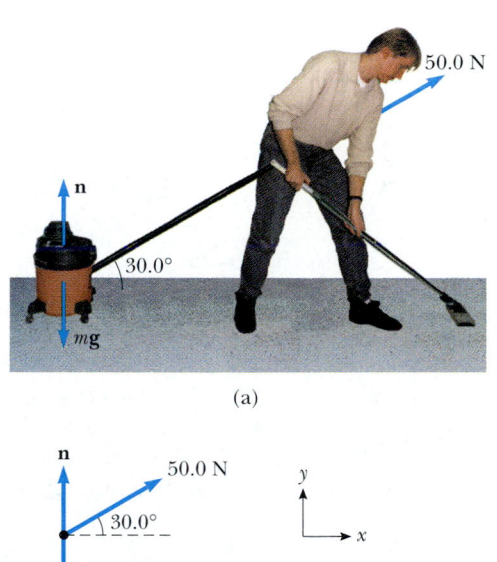

Figure 7.4 (a) A vacuum cleaner being pulled at an angle of 30.0° with the horizontal. (b) Free-body diagram of the forces acting on the vacuum cleaner.

The weightlifter does no work on the weights as he holds them on his shoulders. (If he could rest the bar on his shoulders and lock his knees, he would be able to support the weights for quite some time.) Did he do any work when he raised the weights to this height? *(Gerard Vandystadt/Photo Researchers, Inc.)*

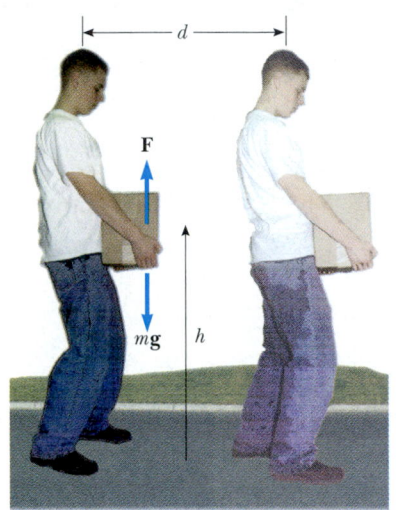

Figure 7.5 A person lifts a box of mass *m* a vertical distance *h* and then walks horizontally a distance *d*.

Quick Quiz 7.2

A person lifts a heavy box of mass *m* a vertical distance *h* and then walks horizontally a distance *d* while holding the box, as shown in Figure 7.5. Determine (a) the work he does on the box and (b) the work done on the box by the force of gravity.

7.2   THE SCALAR PRODUCT OF TWO VECTORS

Because of the way the force and displacement vectors are combined in Equation 7.1, it is helpful to use a convenient mathematical tool called the **scalar product.** This tool allows us to indicate how **F** and **d** interact in a way that depends on how close to parallel they happen to be. We write this scalar product **F·d.** (Because of the dot symbol, the scalar product is often called the **dot product.**) Thus, we can express Equation 7.1 as a scalar product:

Work expressed as a dot product

$$W = \mathbf{F} \cdot \mathbf{d} = Fd \cos \theta \qquad (7.2)$$

In other words, **F·d** (read "F dot d") is a shorthand notation for $Fd \cos \theta$.

Scalar product of any two vectors **A** and **B**

> In general, the scalar product of any two vectors **A** and **B** is a scalar quantity equal to the product of the magnitudes of the two vectors and the cosine of the angle θ between them:
>
> $$\mathbf{A} \cdot \mathbf{B} \equiv AB \cos \theta \qquad (7.3)$$

This relationship is shown in Figure 7.6. Note that **A** and **B** need not have the same units.

In Figure 7.6, $B\cos\theta$ is the projection of **B** onto **A**. Therefore, Equation 7.3 says that $\mathbf{A}\cdot\mathbf{B}$ is the product of the magnitude of **A** and the projection of **B** onto **A**.[2]

From the right-hand side of Equation 7.3 we also see that the scalar product is **commutative**.[3] That is,

$$\mathbf{A}\cdot\mathbf{B} = \mathbf{B}\cdot\mathbf{A}$$

Finally, the scalar product obeys the **distributive law of multiplication,** so that

$$\mathbf{A}\cdot(\mathbf{B} + \mathbf{C}) = \mathbf{A}\cdot\mathbf{B} + \mathbf{A}\cdot\mathbf{C}$$

The dot product is simple to evaluate from Equation 7.3 when **A** is either perpendicular or parallel to **B**. If **A** is perpendicular to **B** ($\theta = 90°$), then $\mathbf{A}\cdot\mathbf{B} = 0$. (The equality $\mathbf{A}\cdot\mathbf{B} = 0$ also holds in the more trivial case when either **A** or **B** is zero.) If vector **A** is parallel to vector **B** and the two point in the same direction ($\theta = 0$), then $\mathbf{A}\cdot\mathbf{B} = AB$. If vector **A** is parallel to vector **B** but the two point in opposite directions ($\theta = 180°$), then $\mathbf{A}\cdot\mathbf{B} = -AB$. The scalar product is negative when $90° < \theta < 180°$.

The unit vectors **i**, **j**, and **k**, which were defined in Chapter 3, lie in the positive x, y, and z directions, respectively, of a right-handed coordinate system. Therefore, it follows from the definition of $\mathbf{A}\cdot\mathbf{B}$ that the scalar products of these unit vectors are

$$\mathbf{i}\cdot\mathbf{i} = \mathbf{j}\cdot\mathbf{j} = \mathbf{k}\cdot\mathbf{k} = 1 \tag{7.4}$$

$$\mathbf{i}\cdot\mathbf{j} = \mathbf{i}\cdot\mathbf{k} = \mathbf{j}\cdot\mathbf{k} = 0 \tag{7.5}$$

Equations 3.18 and 3.19 state that two vectors **A** and **B** can be expressed in component vector form as

$$\mathbf{A} = A_x\mathbf{i} + A_y\mathbf{j} + A_z\mathbf{k}$$

$$\mathbf{B} = B_x\mathbf{i} + B_y\mathbf{j} + B_z\mathbf{k}$$

Using the information given in Equations 7.4 and 7.5 shows that the scalar product of **A** and **B** reduces to

$$\mathbf{A}\cdot\mathbf{B} = A_xB_x + A_yB_y + A_zB_z \tag{7.6}$$

(Details of the derivation are left for you in Problem 7.10.) In the special case in which $\mathbf{A} = \mathbf{B}$, we see that

$$\mathbf{A}\cdot\mathbf{A} = A_x{}^2 + A_y{}^2 + A_z{}^2 = A^2$$

> The order of the dot product can be reversed

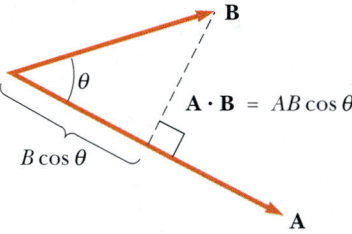

$$\mathbf{A}\cdot\mathbf{B} = AB\cos\theta$$

Figure 7.6 The scalar product $\mathbf{A}\cdot\mathbf{B}$ equals the magnitude of **A** multiplied by $B\cos\theta$, which is the projection of **B** onto **A**.

> Dot products of unit vectors

Quick Quiz 7.3

If the dot product of two vectors is positive, must the vectors have positive rectangular components?

---

[2] This is equivalent to stating that $\mathbf{A}\cdot\mathbf{B}$ equals the product of the magnitude of **B** and the projection of **A** onto **B**.

[3] This may seem obvious, but in Chapter 11 you will see another way of combining vectors that proves useful in physics and is not commutative.

EXAMPLE 7.2   **The Scalar Product**

The vectors **A** and **B** are given by $\mathbf{A} = 2\mathbf{i} + 3\mathbf{j}$ and $\mathbf{B} = -\mathbf{i} + 2\mathbf{j}$. (a) Determine the scalar product $\mathbf{A} \cdot \mathbf{B}$.

Solution

$$\mathbf{A} \cdot \mathbf{B} = (2\mathbf{i} + 3\mathbf{j}) \cdot (-\mathbf{i} + 2\mathbf{j})$$
$$= -2\mathbf{i} \cdot \mathbf{i} + 2\mathbf{i} \cdot 2\mathbf{j} - 3\mathbf{j} \cdot \mathbf{i} + 3\mathbf{j} \cdot 2\mathbf{j}$$
$$= -2(1) + 4(0) - 3(0) + 6(1)$$
$$= -2 + 6 = \boxed{4}$$

where we have used the facts that $\mathbf{i} \cdot \mathbf{i} = \mathbf{j} \cdot \mathbf{j} = 1$ and $\mathbf{i} \cdot \mathbf{j} = \mathbf{j} \cdot \mathbf{i} = 0$. The same result is obtained when we use Equation 7.6 directly, where $A_x = 2$, $A_y = 3$, $B_x = -1$, and $B_y = 2$.

(b) Find the angle θ between **A** and **B**.

Solution   The magnitudes of **A** and **B** are

$$A = \sqrt{A_x^2 + A_y^2} = \sqrt{(2)^2 + (3)^2} = \sqrt{13}$$
$$B = \sqrt{B_x^2 + B_y^2} = \sqrt{(-1)^2 + (2)^2} = \sqrt{5}$$

Using Equation 7.3 and the result from part (a) we find that

$$\cos \theta = \frac{\mathbf{A} \cdot \mathbf{B}}{AB} = \frac{4}{\sqrt{13}\sqrt{5}} = \frac{4}{\sqrt{65}}$$

$$\theta = \cos^{-1} \frac{4}{8.06} = \boxed{60.2°}$$

EXAMPLE 7.3   **Work Done by a Constant Force**

A particle moving in the xy plane undergoes a displacement $\mathbf{d} = (2.0\mathbf{i} + 3.0\mathbf{j})$ m as a constant force $\mathbf{F} = (5.0\mathbf{i} + 2.0\mathbf{j})$ N acts on the particle. (a) Calculate the magnitude of the displacement and that of the force.

Solution

$$d = \sqrt{x^2 + y^2} = \sqrt{(2.0)^2 + (3.0)^2} = \boxed{3.6 \text{ m}}$$

$$F = \sqrt{F_x^2 + F_y^2} = \sqrt{(5.0)^2 + (2.0)^2} = \boxed{5.4 \text{ N}}$$

(b) Calculate the work done by **F**.

Solution   Substituting the expressions for **F** and **d** into Equations 7.4 and 7.5, we obtain

$$W = \mathbf{F} \cdot \mathbf{d} = (5.0\mathbf{i} + 2.0\mathbf{j}) \cdot (2.0\mathbf{i} + 3.0\mathbf{j}) \text{ N} \cdot \text{m}$$
$$= 5.0\mathbf{i} \cdot 2.0\mathbf{i} + 5.0\mathbf{i} \cdot 3.0\mathbf{j} + 2.0\mathbf{j} \cdot 2.0\mathbf{i} + 2.0\mathbf{j} \cdot 3.0\mathbf{j}$$
$$= 10 + 0 + 0 + 6 = 16 \text{ N} \cdot \text{m} = \boxed{16 \text{ J}}$$

Exercise   Calculate the angle between **F** and **d**.

Answer   35°.

7.3   WORK DONE BY A VARYING FORCE

Consider a particle being displaced along the x axis under the action of a varying force. The particle is displaced in the direction of increasing x from $x = x_i$ to $x = x_f$. In such a situation, we cannot use $W = (F \cos \theta) d$ to calculate the work done by the force because this relationship applies only when **F** is constant in magnitude and direction. However, if we imagine that the particle undergoes a very small displacement Δx, shown in Figure 7.7a, then the x component of the force F_x is approximately constant over this interval; for this small displacement, we can express the work done by the force as

$$\Delta W = F_x \Delta x$$

This is just the area of the shaded rectangle in Figure 7.7a. If we imagine that the F_x versus x curve is divided into a large number of such intervals, then the total work done for the displacement from x_i to x_f is approximately equal to the sum of a large number of such terms:

$$W \approx \sum_{x_i}^{x_f} F_x \Delta x$$

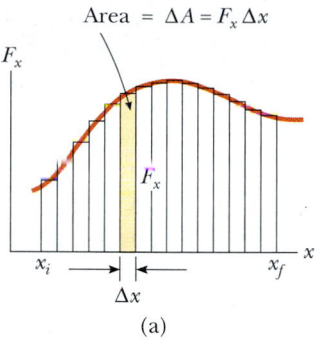

Area $= \Delta A = F_x \Delta x$

(a)

(b)

Figure 7.7   (a) The work done by the force component F_x for the small displacement Δx is $F_x \Delta x$, which equals the area of the shaded rectangle. The total work done for the displacement from x_i to x_f is approximately equal to the sum of the areas of all the rectangles. (b) The work done by the component F_x of the varying force as the particle moves from x_i to x_f is exactly equal to the area under this curve.

If the displacements are allowed to approach zero, then the number of terms in the sum increases without limit but the value of the sum approaches a definite value equal to the area bounded by the F_x curve and the x axis:

$$\lim_{\Delta x \to 0} \sum_{x_i}^{x_f} F_x \Delta x = \int_{x_i}^{x_f} F_x \, dx$$

This definite integral is numerically equal to the area under the F_x-versus-x curve between x_i and x_f. Therefore, we can express the work done by F_x as the particle moves from x_i to x_f as

$$W = \int_{x_i}^{x_f} F_x \, dx \tag{7.7}$$

Work done by a varying force

This equation reduces to Equation 7.1 when the component $F_x = F \cos \theta$ is constant.

If more than one force acts on a particle, the total work done is just the work done by the resultant force. If we express the resultant force in the x direction as $\sum F_x$, then the total work, or *net work*, done as the particle moves from x_i to x_f is

$$\sum W = W_{\text{net}} = \int_{x_i}^{x_f} \left(\sum F_x \right) dx \tag{7.8}$$

EXAMPLE 7.4   **Calculating Total Work Done from a Graph**

A force acting on a particle varies with x, as shown in Figure 7.8. Calculate the work done by the force as the particle moves from $x = 0$ to $x = 6.0$ m.

Solution   The work done by the force is equal to the area under the curve from $x_A = 0$ to $x_C = 6.0$ m. This area is equal to the area of the rectangular section from Ⓐ to Ⓑ plus

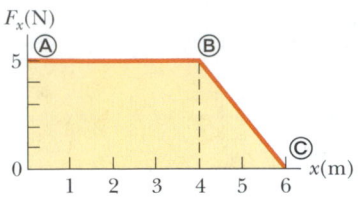

Figure 7.8 The force acting on a particle is constant for the first 4.0 m of motion and then decreases linearly with x from $x_B = 4.0$ m to $x_C = 6.0$ m. The net work done by this force is the area under the curve.

the area of the triangular section from Ⓑ to Ⓒ. The area of the rectangle is $(4.0)(5.0)$ N·m = 20 J, and the area of the triangle is $\frac{1}{2}(2.0)(5.0)$ N·m = 5.0 J. Therefore, the total work done is 25 J.

EXAMPLE 7.5 Work Done by the Sun on a Probe

The interplanetary probe shown in Figure 7.9a is attracted to the Sun by a force of magnitude

$$F = -1.3 \times 10^{22}/x^2$$

where x is the distance measured outward from the Sun to the probe. Graphically and analytically determine how much work is done by the Sun on the probe as the probe–Sun separation changes from 1.5×10^{11} m to 2.3×10^{11} m.

Graphical Solution The minus sign in the formula for the force indicates that the probe is attracted to the Sun. Because the probe is moving away from the Sun, we expect to calculate a negative value for the work done on it.

A spreadsheet or other numerical means can be used to generate a graph like that in Figure 7.9b. Each small square of the grid corresponds to an area $(0.05$ N$)(0.1 \times 10^{11}$ m$) = 5 \times 10^8$ N·m. The work done is equal to the shaded area in Figure 7.9b. Because there are approximately 60 squares shaded, the total area (which is negative because it is below the x axis) is about -3×10^{10} N·m. This is the work done by the Sun on the probe.

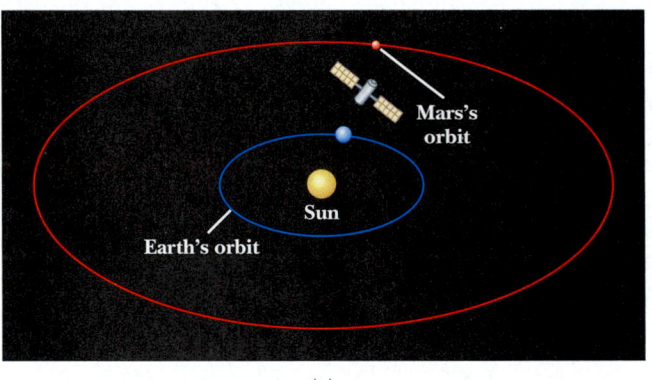

(a)

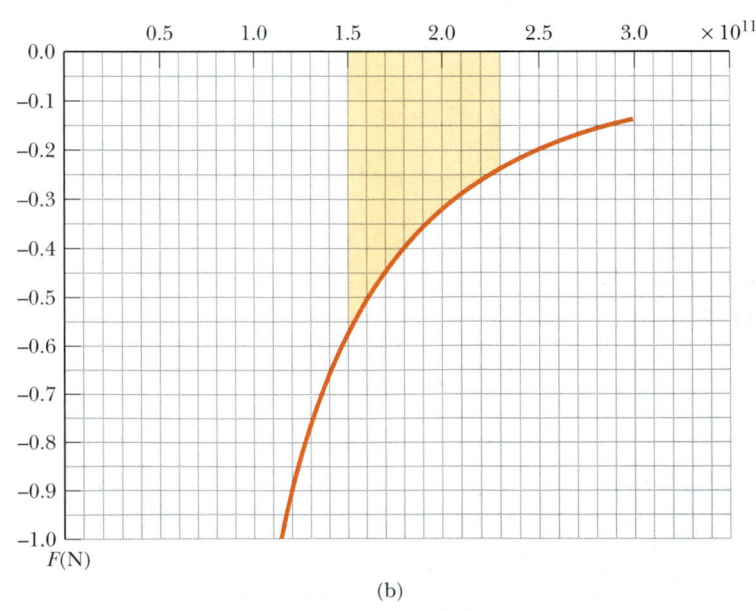

(b)

Figure 7.9 (a) An interplanetary probe moves from a position near the Earth's orbit radially outward from the Sun, ending up near the orbit of Mars. (b) Attractive force versus distance for the interplanetary probe.

Analytical Solution   We can use Equation 7.7 to calculate a more precise value for the work done on the probe by the Sun. To solve this integral, we use the first formula of Table B.5 in Appendix B with $n = -2$:

$$W = \int_{1.5 \times 10^{11}}^{2.3 \times 10^{11}} \left(\frac{-1.3 \times 10^{22}}{x^2} \right) dx$$

$$= (-1.3 \times 10^{22}) \int_{1.5 \times 10^{11}}^{2.3 \times 10^{11}} x^{-2} \, dx$$

$$= (-1.3 \times 10^{22})(-x^{-1}) \Big|_{1.5 \times 10^{11}}^{2.3 \times 10^{11}}$$

$$= (-1.3 \times 10^{22}) \left(\frac{-1}{2.3 \times 10^{11}} - \frac{-1}{1.5 \times 10^{11}} \right)$$

$$= -3.0 \times 10^{10} \, \text{J}$$

Exercise   Does it matter whether the path of the probe is not directed along a radial line away from the Sun?

Answer   No; the value of W depends only on the initial and final positions, not on the path taken between these points.

Work Done by a Spring

A common physical system for which the force varies with position is shown in Figure 7.10. A block on a horizontal, frictionless surface is connected to a spring. If the spring is either stretched or compressed a small distance from its unstretched (equilibrium) configuration, it exerts on the block a force of magnitude

$$F_s = -kx \qquad (7.9)$$

Spring force

where x is the displacement of the block from its unstretched ($x = 0$) position and k is a positive constant called the **force constant** of the spring. In other words, the force required to stretch or compress a spring is proportional to the amount of stretch or compression x. This force law for springs, known as **Hooke's law**, is valid only in the limiting case of small displacements. The value of k is a measure of the *stiffness* of the spring. Stiff springs have large k values, and soft springs have small k values.

Quick Quiz 7.4

What are the units for k, the force constant in Hooke's law?

The negative sign in Equation 7.9 signifies that the force exerted by the spring is always directed *opposite* the displacement. When $x > 0$ as in Figure 7.10a, the spring force is directed to the left, in the negative x direction. When $x < 0$ as in Figure 7.10c, the spring force is directed to the right, in the positive x direction. When $x = 0$ as in Figure 7.10b, the spring is unstretched and $F_s = 0$. Because the spring force always acts toward the equilibrium position ($x = 0$), it sometimes is called a *restoring force*. If the spring is compressed until the block is at the point $-x_{max}$ and is then released, the block moves from $-x_{max}$ through zero to $+x_{max}$. If the spring is instead stretched until the block is at the point x_{max} and is then released, the block moves from $+x_{max}$ through zero to $-x_{max}$. It then reverses direction, returns to $+x_{max}$, and continues oscillating back and forth.

Suppose the block has been pushed to the left a distance x_{max} from equilibrium and is then released. Let us calculate the work W_s done by the spring force as the block moves from $x_i = -x_{max}$ to $x_f = 0$. Applying Equation 7.7 and assuming the block may be treated as a particle, we obtain

$$W_s = \int_{x_i}^{x_f} F_s \, dx = \int_{-x_{max}}^{0} (-kx) \, dx = \tfrac{1}{2} k x_{max}^2 \qquad (7.10)$$

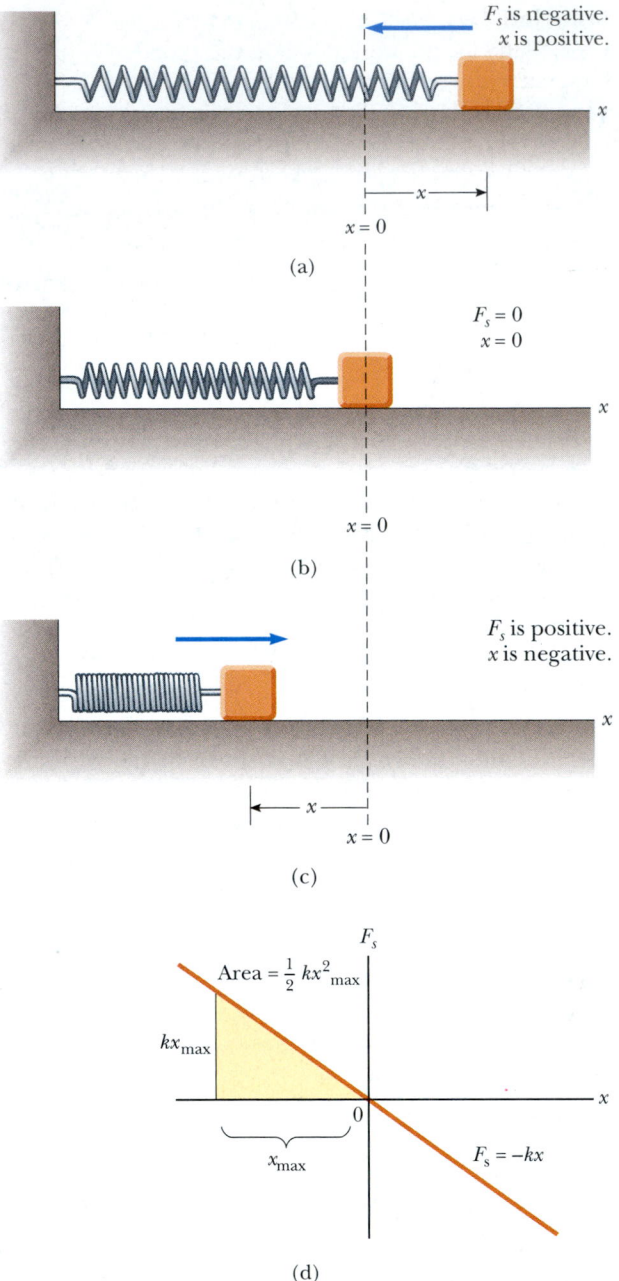

Figure 7.10 The force exerted by a spring on a block varies with the block's displacement x from the equilibrium position $x = 0$. (a) When x is positive (stretched spring), the spring force is directed to the left. (b) When x is zero (natural length of the spring), the spring force is zero. (c) When x is negative (compressed spring), the spring force is directed to the right. (d) Graph of F_s versus x for the block–spring system. The work done by the spring force as the block moves from $-x_{max}$ to 0 is the area of the shaded triangle, $\frac{1}{2}kx_{max}^2$.

where we have used the indefinite integral $\int x^n dx = x^{n+1}/(n + 1)$ with $n = 1$. The work done by the spring force is positive because the force is in the same direction as the displacement (both are to the right). When we consider the work done by the spring force as the block moves from $x_i = 0$ to $x_f = x_{max}$, we find that

$W_s = -\frac{1}{2}kx_{max}^2$ because for this part of the motion the displacement is to the right and the spring force is to the left. Therefore, the *net* work done by the spring force as the block moves from $x_i = -x_{max}$ to $x_f = x_{max}$ is *zero*.

Figure 7.10d is a plot of F_s versus x. The work calculated in Equation 7.10 is the area of the shaded triangle, corresponding to the displacement from $-x_{max}$ to 0. Because the triangle has base x_{max} and height kx_{max}, its area is $\frac{1}{2}kx_{max}^2$, the work done by the spring as given by Equation 7.10.

If the block undergoes an arbitrary displacement from $x = x_i$ to $x = x_f$, the work done by the spring force is

$$W_s = \int_{x_i}^{x_f} (-kx)\,dx = \frac{1}{2}kx_i^2 - \frac{1}{2}kx_f^2 \qquad (7.11)$$

Work done by a spring

For example, if the spring has a force constant of 80 N/m and is compressed 3.0 cm from equilibrium, the work done by the spring force as the block moves from $x_i = -3.0$ cm to its unstretched position $x_f = 0$ is 3.6×10^{-2} J. From Equation 7.11 we also see that the work done by the spring force is zero for any motion that ends where it began ($x_i = x_f$). We shall make use of this important result in Chapter 8, in which we describe the motion of this system in greater detail.

Equations 7.10 and 7.11 describe the work done by the spring on the block. Now let us consider the work done on the spring by an *external agent* that stretches the spring very slowly from $x_i = 0$ to $x_f = x_{max}$, as in Figure 7.11. We can calculate this work by noting that at any value of the displacement, the *applied force* $\mathbf{F}_{app}$ is equal to and opposite the spring force $\mathbf{F}_s$, so that $F_{app} = -(-kx) = kx$. Therefore, the work done by this applied force (the external agent) is

$$W_{F_{app}} = \int_0^{x_{max}} F_{app}\,dx = \int_0^{x_{max}} kx\,dx = \frac{1}{2}kx_{max}^2$$

This work is equal to the negative of the work done by the spring force for this displacement.

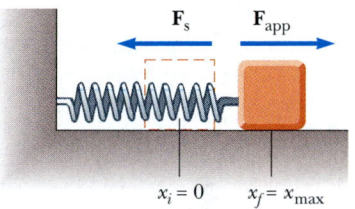

Figure 7.11 A block being pulled from $x_i = 0$ to $x_f = x_{max}$ on a frictionless surface by a force $\mathbf{F}_{app}$. If the process is carried out very slowly, the applied force is equal to and opposite the spring force at all times.

EXAMPLE 7.6   Measuring *k* for a Spring

A common technique used to measure the force constant of a spring is described in Figure 7.12. The spring is hung vertically, and an object of mass *m* is attached to its lower end. Under the action of the "load" *mg*, the spring stretches a distance *d* from its equilibrium position. Because the spring force is upward (opposite the displacement), it must balance the downward force of gravity *m***g** when the system is at rest. In this case, we can apply Hooke's law to give $|\mathbf{F}_s| = kd = mg$, or

$$k = \frac{mg}{d}$$

For example, if a spring is stretched 2.0 cm by a suspended object having a mass of 0.55 kg, then the force constant is

$$k = \frac{mg}{d} = \frac{(0.55\ \text{kg})(9.80\ \text{m/s}^2)}{2.0 \times 10^{-2}\ \text{m}} = \boxed{2.7 \times 10^2\ \text{N/m}}$$

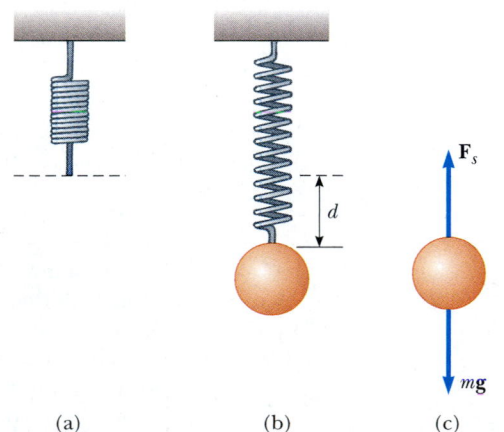

(a)   (b)   (c)

Figure 7.12 Determining the force constant *k* of a spring. The elongation *d* is caused by the attached object, which has a weight *mg*. Because the spring force balances the force of gravity, it follows that $k = mg/d$.

7.4   KINETIC ENERGY AND THE WORK–KINETIC ENERGY THEOREM

It can be difficult to use Newton's second law to solve motion problems involving complex forces. An alternative approach is to relate the speed of a moving particle to its displacement under the influence of some net force. If the work done by the net force on a particle can be calculated for a given displacement, then the change in the particle's speed can be easily evaluated.

Figure 7.13 shows a particle of mass m moving to the right under the action of a constant net force $\Sigma\mathbf{F}$. Because the force is constant, we know from Newton's second law that the particle moves with a constant acceleration $\mathbf{a}$. If the particle is displaced a distance d, the net work done by the total force $\Sigma\mathbf{F}$ is

$$\sum W = \left(\sum F\right)d = (ma)\,d \tag{7.12}$$

In Chapter 2 we found that the following relationships are valid when a particle undergoes constant acceleration:

$$d = \tfrac{1}{2}(v_i + v_f)t \qquad a = \frac{v_f - v_i}{t}$$

where v_i is the speed at $t = 0$ and v_f is the speed at time t. Substituting these expressions into Equation 7.12 gives

$$\sum W = m\left(\frac{v_f - v_i}{t}\right)\tfrac{1}{2}(v_i + v_f)t$$

$$\sum W = \tfrac{1}{2}mv_f^2 - \tfrac{1}{2}mv_i^2 \tag{7.13}$$

The quantity $\tfrac{1}{2}mv^2$ represents the energy associated with the motion of the particle. This quantity is so important that it has been given a special name—**kinetic energy.** The net work done on a particle by a constant net force $\Sigma\mathbf{F}$ acting on it equals the change in kinetic energy of the particle.

In general, the kinetic energy K of a particle of mass m moving with a speed v is defined as

Kinetic energy is energy associated with the motion of a body

$$K \equiv \tfrac{1}{2}mv^2 \tag{7.14}$$

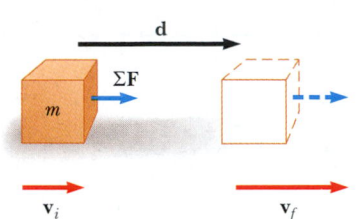

Figure 7.13 A particle undergoing a displacement **d** and a change in velocity under the action of a constant net force $\Sigma\mathbf{F}$.

TABLE 7.1	Kinetic Energies for Various Objects		
Object	**Mass (kg)**	**Speed (m/s)**	**Kinetic Energy (J)**
Earth orbiting the Sun	5.98×10^{24}	2.98×10^4	2.65×10^{33}
Moon orbiting the Earth	7.35×10^{22}	1.02×10^3	3.82×10^{28}
Rocket moving at escape speed[a]	500	1.12×10^4	3.14×10^{10}
Automobile at 55 mi/h	2 000	25	6.3×10^5
Running athlete	70	10	3.5×10^3
Stone dropped from 10 m	1.0	14	9.8×10^1
Golf ball at terminal speed	0.046	44	4.5×10^1
Raindrop at terminal speed	3.5×10^{-5}	9.0	1.4×10^{-3}
Oxygen molecule in air	5.3×10^{-26}	500	6.6×10^{-21}

[a] *Escape speed* is the minimum speed an object must attain near the Earth's surface if it is to escape the Earth's gravitational force.

Kinetic energy is a scalar quantity and has the same units as work. For example, a 2.0-kg object moving with a speed of 4.0 m/s has a kinetic energy of 16 J. Table 7.1 lists the kinetic energies for various objects.

It is often convenient to write Equation 7.13 in the form

$$\sum W = K_f - K_i = \Delta K \qquad \text{(7.15)}$$

That is, $K_i + \sum W = K_f$.

Work–kinetic energy theorem

Equation 7.15 is an important result known as the **work–kinetic energy theorem.** It is important to note that when we use this theorem, we must include *all* of the forces that do work on the particle in the calculation of the net work done. From this theorem, we see that the speed of a particle increases if the net work done on it is positive because the final kinetic energy is greater than the initial kinetic energy. The particle's speed decreases if the net work done is negative because the final kinetic energy is less than the initial kinetic energy.

The work–kinetic energy theorem as expressed by Equation 7.15 allows us to think of kinetic energy as the work a particle can do in coming to rest, or the amount of energy stored in the particle. For example, suppose a hammer (our particle) is on the verge of striking a nail, as shown in Figure 7.14. The moving hammer has kinetic energy and so can do work on the nail. The work done on the nail is equal to *Fd*, where *F* is the average force exerted on the nail by the hammer and *d* is the distance the nail is driven into the wall.[4]

We derived the work–kinetic energy theorem under the assumption of a constant net force, but it also is valid when the force varies. To see this, suppose the net force acting on a particle in the *x* direction is $\sum F_x$. We can apply Newton's second law, $\sum F_x = ma_x$, and use Equation 7.8 to express the net work done as

$$\sum W = \int_{x_i}^{x_f} \left(\sum F_x \right) dx = \int_{x_i}^{x_f} ma_x \, dx$$

If the resultant force varies with *x*, the acceleration and speed also depend on *x*. Because we normally consider acceleration as a function of *t*, we now use the following chain rule to express *a* in a slightly different way:

$$a = \frac{dv}{dt} = \frac{dv}{dx}\frac{dx}{dt} = v\frac{dv}{dx}$$

Substituting this expression for *a* into the above equation for $\sum W$ gives

$$\sum W = \int_{x_i}^{x_f} mv\frac{dv}{dx}\, dx = \int_{v_i}^{v_f} mv\, dv$$

$$\sum W = \tfrac{1}{2}mv_f^2 - \tfrac{1}{2}mv_i^2 \qquad \text{(7.16)}$$

Figure 7.14 The moving hammer has kinetic energy and thus can do work on the nail, driving it into the wall.

The net work done on a particle equals the change in its kinetic energy

The limits of the integration were changed from *x* values to *v* values because the variable was changed from *x* to *v*. Thus, we conclude that the net work done on a particle by the net force acting on it is equal to the change in the kinetic energy of the particle. This is true whether or not the net force is constant.

---

[4] Note that because the nail and the hammer are *systems* of particles rather than single particles, part of the hammer's kinetic energy goes into warming the hammer and the nail upon impact. Also, as the nail moves into the wall in response to the impact, the large frictional force between the nail and the wood results in the continuous transformation of the kinetic energy of the nail into further temperature increases in the nail and the wood, as well as in deformation of the wall. Energy associated with temperature changes is called *internal energy* and will be studied in detail in Chapter 20.

Situations Involving Kinetic Friction

One way to include frictional forces in analyzing the motion of an object sliding on a *horizontal* surface is to describe the kinetic energy lost because of friction. Suppose a book moving on a horizontal surface is given an initial horizontal velocity $\mathbf{v}_i$ and slides a distance d before reaching a final velocity $\mathbf{v}_f$ as shown in Figure 7.15. The external force that causes the book to undergo an acceleration in the negative x direction is the force of kinetic friction $\mathbf{f}_k$ acting to the left, opposite the motion. The initial kinetic energy of the book is $\frac{1}{2}mv_i^2$, and its final kinetic energy is $\frac{1}{2}mv_f^2$. Applying Newton's second law to the book can show this. Because the only force acting on the book in the x direction is the friction force, Newton's second law gives $-f_k = ma_x$. Multiplying both sides of this expression by d and using Equation 2.12 in the form $v_{xf}^2 - v_{xi}^2 = 2a_x d$ for motion under constant acceleration give $-f_k d = (ma_x)d = \frac{1}{2}mv_{xf}^2 - \frac{1}{2}mv_{xi}^2$ or

$$\Delta K_{\text{friction}} = -f_k d \qquad \textbf{(7.17a)}$$

This result specifies that the amount by which the force of kinetic friction changes the kinetic energy of the book is equal to $-f_k d$. Part of this lost kinetic energy goes into warming up the book, and the rest goes into warming up the surface over which the book slides. In effect, the quantity $-f_k d$ is equal to the work done by kinetic friction on the book *plus* the work done by kinetic friction on the surface. (We shall study the relationship between temperature and energy in Part III of this text.) When friction—as well as other forces—acts on an object, the work–kinetic energy theorem reads

$$K_i + \sum W_{\text{other}} - f_k d = K_f \qquad \textbf{(7.17b)}$$

Here, $\sum W_{\text{other}}$ represents the sum of the amounts of work done on the object by forces other than kinetic friction.

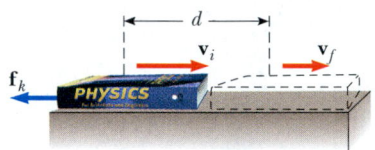

Loss in kinetic energy due to friction

Figure 7.15   A book sliding to the right on a horizontal surface slows down in the presence of a force of kinetic friction acting to the left. The initial velocity of the book is $\mathbf{v}_i$, and its final velocity is $\mathbf{v}_f$. The normal force and the force of gravity are not included in the diagram because they are perpendicular to the direction of motion and therefore do not influence the book's velocity.

> **Quick Quiz 7.5**
>
> Can frictional forces ever *increase* an object's kinetic energy?

> **EXAMPLE 7.7**   **A Block Pulled on a Frictionless Surface**

A 6.0-kg block initially at rest is pulled to the right along a horizontal, frictionless surface by a constant horizontal force of 12 N. Find the speed of the block after it has moved 3.0 m.

Solution   We have made a drawing of this situation in Figure 7.16a. We could apply the equations of kinematics to determine the answer, but let us use the energy approach for

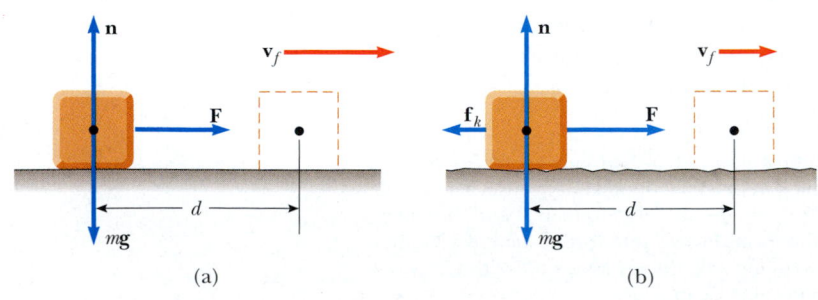

Figure 7.16   A block pulled to the right by a constant horizontal force. (a) Frictionless surface. (b) Rough surface.

practice. The normal force balances the force of gravity on the block, and neither of these vertically acting forces does work on the block because the displacement is horizontal. Because there is no friction, the net external force acting on the block is the 12-N force. The work done by this force is

$$W = Fd = (12 \text{ N})(3.0 \text{ m}) = 36 \text{ N} \cdot \text{m} = 36 \text{ J}$$

Using the work–kinetic energy theorem and noting that the initial kinetic energy is zero, we obtain

$$W = K_f - K_i = \tfrac{1}{2}mv_f^2 - 0$$

$$v_f^2 = \frac{2W}{m} = \frac{2(36 \text{ J})}{6.0 \text{ kg}} = 12 \text{ m}^2/\text{s}^2$$

$$v_f = \boxed{3.5 \text{ m/s}}$$

Exercise Find the acceleration of the block and determine its final speed, using the kinematics equation $v_{xf}^2 = v_{xi}^2 + 2a_x d$.

Answer $a_x = 2.0 \text{ m/s}^2$; $v_f = 3.5 \text{ m/s}$.

EXAMPLE 7.8   A Block Pulled on a Rough Surface

Find the final speed of the block described in Example 7.7 if the surface is not frictionless but instead has a coefficient of kinetic friction of 0.15.

Solution The applied force does work just as in Example 7.7:

$$W = Fd = (12 \text{ N})(3.0 \text{ m}) = 36 \text{ J}$$

In this case we must use Equation 7.17a to calculate the kinetic energy lost to friction $\Delta K_{\text{friction}}$. The magnitude of the frictional force is

$$f_k = \mu_k n = \mu_k mg = (0.15)(6.0 \text{ kg})(9.80 \text{ m/s}^2) = 8.82 \text{ N}$$

The change in kinetic energy due to friction is

$$\Delta K_{\text{friction}} = -f_k d = -(8.82 \text{ N})(3.0 \text{ m}) = -26.5 \text{ J}$$

The final speed of the block follows from Equation 7.17b:

$$\tfrac{1}{2}mv_i^2 + \sum W_{\text{other}} - f_k d = \tfrac{1}{2}mv_f^2$$

$$0 + 36 \text{ J} - 26.5 \text{ J} = \tfrac{1}{2}(6.0 \text{ kg})v_f^2$$

$$v_f^2 = 2(9.5 \text{ J})/(6.0 \text{ kg}) = 3.18 \text{ m}^2/\text{s}^2$$

$$v_f = \boxed{1.8 \text{ m/s}}$$

After sliding the 3-m distance on the rough surface, the block is moving at a speed of 1.8 m/s; in contrast, after covering the same distance on a frictionless surface (see Example 7.7), its speed was 3.5 m/s.

Exercise Find the acceleration of the block from Newton's second law and determine its final speed, using equations of kinematics.

Answer $a_x = 0.53 \text{ m/s}^2$; $v_f = 1.8 \text{ m/s}$.

CONCEPTUAL EXAMPLE 7.9   Does the Ramp Lessen the Work Required?

A man wishes to load a refrigerator onto a truck using a ramp, as shown in Figure 7.17. He claims that less work would be required to load the truck if the length L of the ramp were increased. Is his statement valid?

Solution No. Although less force is required with a longer ramp, that force must act over a greater distance if the same amount of work is to be done. Suppose the refrigerator is wheeled on a dolly up the ramp at constant speed. The

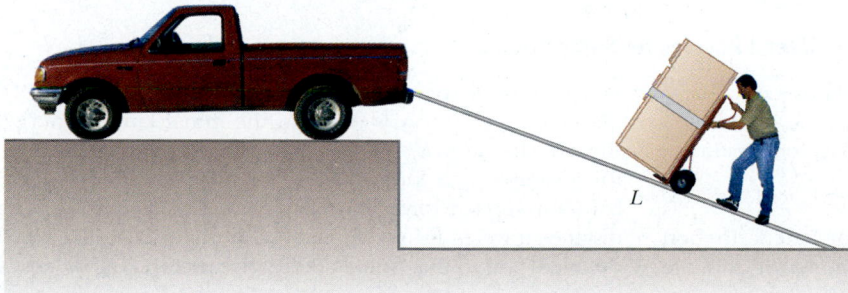

Figure 7.17 A refrigerator attached to a frictionless wheeled dolly is moved up a ramp at constant speed.

normal force exerted by the ramp on the refrigerator is directed 90° to the motion and so does no work on the refrigerator. Because $\Delta K = 0$, the work–kinetic energy theorem gives

$$\sum W = W_{\text{by man}} + W_{\text{by gravity}} = 0$$

The work done by the force of gravity equals the weight of the refrigerator mg times the vertical height h through which it is displaced times cos 180°, or $W_{\text{by gravity}} = - mgh$. (The minus sign arises because the downward force of gravity is opposite the displacement.) Thus, the man must do work mgh on the refrigerator, regardless of the length of the ramp.

QuickLab

Attach two paperclips to a ruler so that one of the clips is twice the distance from the end as the other. Place the ruler on a table with two small wads of paper against the clips, which act as stops. Sharply swing the ruler through a small angle, stopping it abruptly with your finger. The outer paper wad will have twice the speed of the inner paper wad as the two slide on the table away from the ruler. Compare how far the two wads slide. How does this relate to the results of Conceptual Example 7.10?

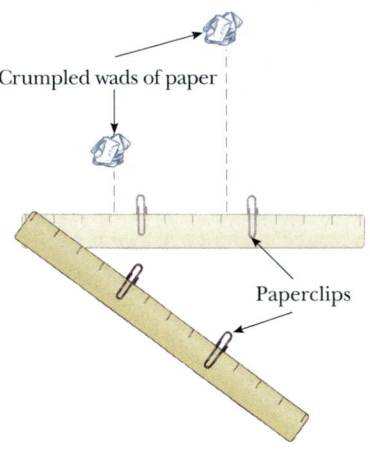

Crumpled wads of paper

Paperclips

Consider the chum salmon attempting to swim upstream in the photograph at the beginning of this chapter. The "steps" of a fish ladder built around a dam do not change the total amount of work that must be done by the salmon as they leap through some vertical distance. However, the ladder allows the fish to perform that work in a series of smaller jumps, and the net effect is to raise the vertical position of the fish by the height of the dam.

These cyclists are working hard and expending energy as they pedal uphill in Marin County, CA.
(David Madison/Tony Stone Images)

CONCEPTUAL EXAMPLE 7.10   Useful Physics for Safer Driving

A certain car traveling at an initial speed v slides a distance d to a halt after its brakes lock. Assuming that the car's initial speed is instead $2v$ at the moment the brakes lock, estimate the distance it slides.

Solution Let us assume that the force of kinetic friction between the car and the road surface is constant and the same for both speeds. The net force multiplied by the displacement of the car is equal to the initial kinetic energy of the car (because $K_f = 0$). If the speed is doubled, as it is in this example, the kinetic energy is quadrupled. For a given constant applied force (in this case, the frictional force), the distance traveled is four times as great when the initial speed is doubled, and so the estimated distance that the car slides is $4d$.

EXAMPLE 7.11   A Block–Spring System

A block of mass 1.6 kg is attached to a horizontal spring that has a force constant of 1.0×10^3 N/m, as shown in Figure 7.10. The spring is compressed 2.0 cm and is then released from rest. (a) Calculate the speed of the block as it passes through the equilibrium position $x = 0$ if the surface is frictionless.

Solution   In this situation, the block starts with $v_i = 0$ at $x_i = -2.0$ cm, and we want to find v_f at $x_f = 0$. We use Equation 7.10 to find the work done by the spring with $x_{max} = x_i = -2.0$ cm $= -2.0 \times 10^{-2}$ m:

$$W_s = \tfrac{1}{2}kx_{max}^2 = \tfrac{1}{2}(1.0 \times 10^3 \text{ N/m})(-2.0 \times 10^{-2}\text{ m})^2 = 0.20\text{ J}$$

Using the work–kinetic energy theorem with $v_i = 0$, we obtain the change in kinetic energy of the block due to the work done on it by the spring:

$$W_s = \tfrac{1}{2}mv_f^2 - \tfrac{1}{2}mv_i^2$$

$$0.20\text{ J} = \tfrac{1}{2}(1.6\text{ kg})v_f^2 - 0$$

$$v_f^2 = \frac{0.40\text{ J}}{1.6\text{ kg}} = 0.25\text{ m}^2/\text{s}^2$$

$$v_f = \boxed{0.50\text{ m/s}}$$

(b) Calculate the speed of the block as it passes through the equilibrium position if a constant frictional force of 4.0 N retards its motion from the moment it is released.

Solution   Certainly, the answer has to be less than what we found in part (a) because the frictional force retards the motion. We use Equation 7.17 to calculate the kinetic energy lost because of friction and add this negative value to the kinetic energy found in the absence of friction. The kinetic energy lost due to friction is

$$\Delta K = -f_k d = -(4.0\text{ N})(2.0 \times 10^{-2}\text{ m}) = -0.080\text{ J}$$

In part (a), the final kinetic energy without this loss was found to be 0.20 J. Therefore, the final kinetic energy in the presence of friction is

$$K_f = 0.20\text{ J} - 0.080\text{ J} = 0.12\text{ J} = \tfrac{1}{2}mv_f^2$$

$$\tfrac{1}{2}(1.6\text{ kg})v_f^2 = 0.12\text{ J}$$

$$v_f^2 = \frac{0.24\text{ J}}{1.6\text{ kg}} = 0.15\text{ m}^2/\text{s}^2$$

$$v_f = \boxed{0.39\text{ m/s}}$$

As expected, this value is somewhat less than the 0.50 m/s we found in part (a). If the frictional force were greater, then the value we obtained as our answer would have been even smaller.

7.5   POWER

Imagine two identical models of an automobile: one with a base-priced four-cylinder engine; and the other with the highest-priced optional engine, a mighty eight-cylinder powerplant. Despite the differences in engines, the two cars have the same mass. Both cars climb a roadway up a hill, but the car with the optional engine takes much less time to reach the top. Both cars have done the same amount of work against gravity, but in different time periods. From a practical viewpoint, it is interesting to know not only the work done by the vehicles but also the *rate* at which it is done. In taking the ratio of the amount of work done to the time taken to do it, we have a way of quantifying this concept. The time rate of doing work is called **power.**

If an external force is applied to an object (which we assume acts as a particle), and if the work done by this force in the time interval Δt is W, then the **average power** expended during this interval is defined as

$$\overline{\mathcal{P}} \equiv \frac{W}{\Delta t}$$

Average power

The work done on the object contributes to the increase in the energy of the object. Therefore, a more general definition of power is the *time rate of energy transfer.* In a manner similar to how we approached the definition of velocity and acceleration, we can define the **instantaneous power** $\mathcal{P}$ as the limiting value of the average power as Δt approaches zero:

$$\mathcal{P} \equiv \lim_{\Delta t \to 0} \frac{W}{\Delta t} = \frac{dW}{dt}$$

where we have represented the increment of work done by dW. We find from Equation 7.2, letting the displacement be expressed as $d\mathbf{s}$, that $dW = \mathbf{F} \cdot d\mathbf{s}$. Therefore, the instantaneous power can be written

Instantaneous power

$$\mathcal{P} = \frac{dW}{dt} = \mathbf{F} \cdot \frac{d\mathbf{s}}{dt} = \mathbf{F} \cdot \mathbf{v} \qquad (7.18)$$

where we use the fact that $\mathbf{v} = d\mathbf{s}/dt$.

The SI unit of power is joules per second (J/s), also called the *watt* (W) (after James Watt, the inventor of the steam engine):

The watt

$$1 \text{ W} = 1 \text{ J/s} = 1 \text{ kg·m}^2/\text{s}^3$$

The symbol W (not italic) for watt should not be confused with the symbol W (italic) for work.

A unit of power in the British engineering system is the horsepower (hp):

$$1 \text{ hp} = 746 \text{ W}$$

A unit of energy (or work) can now be defined in terms of the unit of power. One **kilowatt hour** (kWh) is the energy converted or consumed in 1 h at the constant rate of 1 kW = 1 000 J/s. The numerical value of 1 kWh is

The kilowatt hour is a unit of energy

$$1 \text{ kWh} = (10^3 \text{ W})(3\,600 \text{ s}) = 3.60 \times 10^6 \text{ J}$$

It is important to realize that a kilowatt hour is a unit of energy, not power. When you pay your electric bill, you pay the power company for the total electrical energy you used during the billing period. This energy is the power used multiplied by the time during which it was used. For example, a 300-W lightbulb run for 12 h would convert $(0.300 \text{ kW})(12 \text{ h}) = 3.6$ kWh of electrical energy.

Quick Quiz 7.6

Suppose that an old truck and a sports car do the same amount of work as they climb a hill but that the truck takes much longer to accomplish this work. How would graphs of $\mathcal{P}$ versus t compare for the two vehicles?

EXAMPLE 7.12   Power Delivered by an Elevator Motor

An elevator car has a mass of 1 000 kg and is carrying passengers having a combined mass of 800 kg. A constant frictional force of 4 000 N retards its motion upward, as shown in Figure 7.18a. (a) What must be the minimum power delivered by the motor to lift the elevator car at a constant speed of 3.00 m/s?

Solution The motor must supply the force of magnitude T that pulls the elevator car upward. Reading that the speed is constant provides the hint that $a = 0$, and therefore we know from Newton's second law that $\Sigma F_y = 0$. We have drawn

a free-body diagram in Figure 7.18b and have arbitrarily specified that the upward direction is positive. From Newton's second law we obtain

$$\sum F_y = T - f - Mg = 0$$

where M is the *total* mass of the system (car plus passengers), equal to 1 800 kg. Therefore,

$$\begin{aligned}
T &= f + Mg \\
&= 4.00 \times 10^3 \text{ N} + (1.80 \times 10^3 \text{ kg})(9.80 \text{ m/s}^2) \\
&= 2.16 \times 10^4 \text{ N}
\end{aligned}$$

Using Equation 7.18 and the fact that **T** is in the same direction as **v**, we find that

$$\mathscr{P} = \mathbf{T \cdot v} = Tv$$

$$= (2.16 \times 10^4 \text{ N})(3.00 \text{ m/s}) = \boxed{6.48 \times 10^4 \text{ W}}$$

(b) What power must the motor deliver at the instant its speed is v if it is designed to provide an upward acceleration of 1.00 m/s^2?

Solution   Now we expect to obtain a value greater than we did in part (a), where the speed was constant, because the motor must now perform the additional task of accelerating the car. The only change in the setup of the problem is that now $a > 0$. Applying Newton's second law to the car gives

$$\sum F_y = T - f - Mg = Ma$$
$$T = M(a + g) + f$$
$$= (1.80 \times 10^3 \text{ kg})(1.00 + 9.80)\text{m/s}^2 + 4.00 \times 10^3 \text{ N}$$
$$= 2.34 \times 10^4 \text{ N}$$

Therefore, using Equation 7.18, we obtain for the required power

$$\mathscr{P} = Tv = \boxed{(2.34 \times 10^4 v) \text{ W}}$$

where v is the instantaneous speed of the car in meters per second. The power is less than that obtained in part (a) as

long as the speed is less than $\mathscr{P}/T = 2.77 \text{ m/s}$, but it is greater when the elevator's speed exceeds this value.

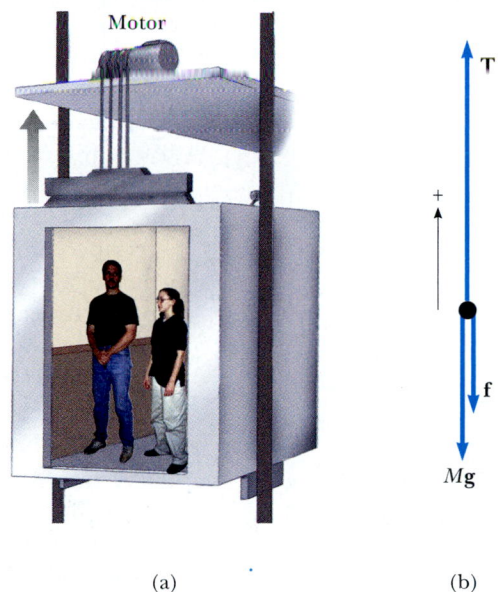

(a)   (b)

Figure 7.18   (a) The motor exerts an upward force **T** on the elevator car. The magnitude of this force is the tension T in the cable connecting the car and motor. The downward forces acting on the car are a frictional force **f** and the force of gravity $\mathbf{F}_g = M\mathbf{g}$. (b) The free-body diagram for the elevator car.

CONCEPTUAL EXAMPLE 7.13

In part (a) of the preceding example, the motor delivers power to lift the car, and yet the car moves at constant speed. A student analyzing this situation notes that the kinetic energy of the car does not change because its speed does not change. This student then reasons that, according to the work–kinetic energy theorem, $W = \Delta K = 0$. Knowing that $\mathscr{P} = W/t$, the student concludes that the power delivered by the motor also must be zero. How would you explain this apparent paradox?

Solution   The work–kinetic energy theorem tells us that the *net* force acting on the system multiplied by the displacement is equal to the change in the kinetic energy of the system. In our elevator case, the net force is indeed zero (that is, $T - Mg - f = 0$), and so $W = (\sum F_y)d = 0$. However, the power from the motor is calculated not from the *net* force but rather from the force exerted by the motor acting in the direction of motion, which in this case is T and not zero.

Optional Section

7.6 ENERGY AND THE AUTOMOBILE

Automobiles powered by gasoline engines are very inefficient machines. Even under ideal conditions, less than 15% of the chemical energy in the fuel is used to power the vehicle. The situation is much worse under stop-and-go driving conditions in a city. In this section, we use the concepts of energy, power, and friction to analyze automobile fuel consumption.

Many mechanisms contribute to energy loss in an automobile. About 67% of the energy available from the fuel is lost in the engine. This energy ends up in the atmosphere, partly via the exhaust system and partly via the cooling system. (As we shall see in Chapter 22, the great energy loss from the exhaust and cooling systems is required by a fundamental law of thermodynamics.) Approximately 10% of the available energy is lost to friction in the transmission, drive shaft, wheel and axle bearings, and differential. Friction in other moving parts dissipates approximately 6% of the energy, and 4% of the energy is used to operate fuel and oil pumps and such accessories as power steering and air conditioning. This leaves a mere 13% of the available energy to propel the automobile! This energy is used mainly to balance the energy loss due to flexing of the tires and the friction caused by the air, which is more commonly referred to as *air resistance.*

Let us examine the power required to provide a force in the forward direction that balances the combination of the two frictional forces. The coefficient of rolling friction μ between the tires and the road is about 0.016. For a 1 450-kg car, the weight is 14 200 N and the force of rolling friction has a magnitude of $\mu n = \mu mg = 227$ N. As the speed of the car increases, a small reduction in the normal force occurs as a result of a decrease in atmospheric pressure as air flows over the top of the car. (This phenomenon is discussed in Chapter 15.) This reduction in the normal force causes a slight reduction in the force of rolling friction f_r with increasing speed, as the data in Table 7.2 indicate.

Now let us consider the effect of the resistive force that results from the movement of air past the car. For large objects, the resistive force f_a associated with air friction is proportional to the square of the speed (in meters per second; see Section 6.4) and is given by Equation 6.6:

$$f_a = \tfrac{1}{2} D\rho A v^2$$

where D is the drag coefficient, ρ is the density of air, and A is the cross-sectional area of the moving object. We can use this expression to calculate the f_a values in Table 7.2, using $D = 0.50$, $\rho = 1.293$ kg/m^3, and $A \approx 2$ m^2.

The magnitude of the total frictional force f_t is the sum of the rolling frictional force and the air resistive force:

$$f_t = f_r + f_a$$

At low speeds, road friction is the predominant resistive force, but at high speeds air drag predominates, as shown in Table 7.2. Road friction can be decreased by a reduction in tire flexing (for example, by an increase in the air pres-

TABLE 7.2   Frictional Forces and Power Requirements for a Typical Car[a]

v (m/s)	n (N)	f_r (N)	f_a (N)	f_t (N)	$\mathscr{P} = f_t v$ (kW)
0	14 200	227	0	227	0
8.9	14 100	226	51	277	2.5
17.8	13 900	222	204	426	7.6
26.8	13 600	218	465	683	18.3
35.9	13 200	211	830	1 041	37.3
44.8	12 600	202	1 293	1 495	67.0

[a] In this table, n is the normal force, f_r is road friction, f_a is air friction, f_t is total friction, and $\mathscr{P}$ is the power delivered to the wheels.

sure slightly above recommended values) and by the use of radial tires. Air drag can be reduced through the use of a smaller cross-sectional area and by streamlining the car. Although driving a car with the windows open increases air drag and thus results in a 3% decrease in mileage, driving with the windows closed and the air conditioner running results in a 12% decrease in mileage.

The total power needed to maintain a constant speed v is $f_t v$, and it is this power that must be delivered to the wheels. For example, from Table 7.2 we see that at $v = 26.8$ m/s (60 mi/h) the required power is

$$\mathcal{P} = f_t v = (683\ \text{N})\left(26.8\frac{\text{m}}{\text{s}}\right) = 18.3\ \text{kW}$$

This power can be broken down into two parts: (1) the power $f_r v$ needed to compensate for road friction, and (2) the power $f_a v$ needed to compensate for air drag. At $v = 26.8$ m/s, we obtain the values

$$\mathcal{P}_r = f_r v = (218\ \text{N})\left(26.8\frac{\text{m}}{\text{s}}\right) = 5.84\ \text{kW}$$

$$\mathcal{P}_a = f_a v = (465\ \text{N})\left(26.8\frac{\text{m}}{\text{s}}\right) = 12.5\ \text{kW}$$

Note that $\mathcal{P} = \mathcal{P}_r + \mathcal{P}_a$.

On the other hand, at $v = 44.8$ m/s (100 mi/h), $\mathcal{P}_r = 9.05$ kW, $\mathcal{P}_a = 57.9$ kW, and $\mathcal{P} = 67.0$ kW. This shows the importance of air drag at high speeds.

EXAMPLE 7.14 ▶ **Gas Consumed by a Compact Car**

A compact car has a mass of 800 kg, and its efficiency is rated at 18%. (That is, 18% of the available fuel energy is delivered to the wheels.) Find the amount of gasoline used to accelerate the car from rest to 27 m/s (60 mi/h). Use the fact that the energy equivalent of 1 gal of gasoline is 1.3×10^8 J.

Solution The energy required to accelerate the car from rest to a speed v is its final kinetic energy $\frac{1}{2}mv^2$:

$$K = \tfrac{1}{2}mv^2 = \tfrac{1}{2}(800\ \text{kg})(27\ \text{m/s})^2 = 2.9 \times 10^5\ \text{J}$$

If the engine were 100% efficient, each gallon of gasoline

would supply 1.3×10^8 J of energy. Because the engine is only 18% efficient, each gallon delivers only $(0.18)(1.3 \times 10^8\ \text{J}) = 2.3 \times 10^7$ J. Hence, the number of gallons used to accelerate the car is

$$\text{Number of gallons} = \frac{2.9 \times 10^5\ \text{J}}{2.3 \times 10^7\ \text{J/gal}} = \boxed{0.013\ \text{gal}}$$

At cruising speed, this much gasoline is sufficient to propel the car nearly 0.5 mi. This demonstrates the extreme energy requirements of stop-and-start driving.

EXAMPLE 7.15 ▶ **Power Delivered to Wheels**

Suppose the compact car in Example 7.14 gets 35 mi/gal at 60 mi/h. How much power is delivered to the wheels?

Solution By simply canceling units, we determine that the car consumes 60 mi/h ÷ 35 mi/gal = 1.7 gal/h. Using the fact that each gallon is equivalent to 1.3×10^8 J, we find that the total power used is

$$\mathcal{P} = \frac{(1.7\ \text{gal/h})(1.3 \times 10^8\ \text{J/gal})}{3.6 \times 10^3\ \text{s/h}}$$

$$= \frac{2.2 \times 10^8\ \text{J}}{3.6 \times 10^3\ \text{s}} = 62\ \text{kW}$$

Because 18% of the available power is used to propel the car, the power delivered to the wheels is $(0.18)(62\ \text{kW}) = $ 11 kW. This is 40% less than the 18.3-kW value obtained for the 1 450-kg car discussed in the text. Vehicle mass is clearly an important factor in power-loss mechanisms.

EXAMPLE 7.16 **Car Accelerating Up a Hill**

Consider a car of mass m that is accelerating up a hill, as shown in Figure 7.19. An automotive engineer has measured the magnitude of the total resistive force to be

$$f_t = (218 + 0.70v^2) \text{ N}$$

where v is the speed in meters per second. Determine the power the engine must deliver to the wheels as a function of speed.

Solution The forces on the car are shown in Figure 7.19, in which $\mathbf{F}$ is the force of friction from the road that propels the car; the remaining forces have their usual meaning. Applying Newton's second law to the motion along the road surface, we find that

$$\sum F_x = F - f_t - mg \sin \theta = ma$$

$$F = ma + mg \sin \theta + f_t$$

$$= ma + mg \sin \theta + (218 + 0.70v^2)$$

Therefore, the power required to move the car forward is

$$\mathcal{P} = Fv = mva + mvg \sin \theta + 218v + 0.70v^3$$

The term mva represents the power that the engine must deliver to accelerate the car. If the car moves at constant speed, this term is zero and the total power requirement is reduced. The term $mvg \sin \theta$ is the power required to provide a force to balance a component of the force of gravity as the car moves up the incline. This term would be zero for motion on a horizontal surface. The term $218v$ is the power required to provide a force to balance road friction, and the term $0.70v^3$ is the power needed to do work on the air.

If we take $m = 1\,450$ kg, $v = 27$ m/s ($= 60$ mi/h), $a =$

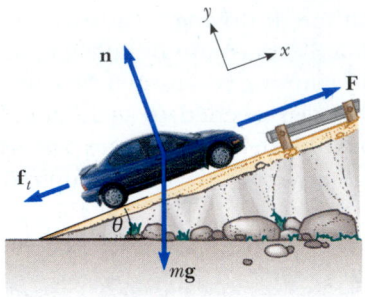

Figure 7.19

1.0 m/s², and $\theta = 10°$, then the various terms in $\mathcal{P}$ are calculated to be

$$mva = (1\,450 \text{ kg})(27 \text{ m/s})(1.0 \text{ m/s}^2)$$

$$= 39 \text{ kW} = 52 \text{ hp}$$

$$mvg \sin \theta = (1\,450 \text{ kg})(27 \text{ m/s})(9.80 \text{ m/s}^2)(\sin 10°)$$

$$= 67 \text{ kW} = 89 \text{ hp}$$

$$218v = 218(27 \text{ m/s}) = 5.9 \text{ kW} = 7.9 \text{ hp}$$

$$0.70v^3 = 0.70(27 \text{ m/s})^3 = 14 \text{ kW} = 19 \text{ hp}$$

Hence, the total power required is 126 kW, or 168 hp.

Note that the power requirements for traveling at constant speed on a horizontal surface are only 20 kW, or 27 hp (the sum of the last two terms). Furthermore, if the mass were halved (as in the case of a compact car), then the power required also is reduced by almost the same factor.

Optional Section

7.7 **KINETIC ENERGY AT HIGH SPEEDS**

The laws of Newtonian mechanics are valid only for describing the motion of particles moving at speeds that are small compared with the speed of light in a vacuum c ($= 3.00 \times 10^8$ m/s). When speeds are comparable to c, the equations of Newtonian mechanics must be replaced by the more general equations predicted by the theory of relativity. One consequence of the theory of relativity is that the kinetic energy of a particle of mass m moving with a speed v is no longer given by $K = mv^2/2$. Instead, one must use the relativistic form of the kinetic energy:

Relativistic kinetic energy

$$K = mc^2\left(\frac{1}{\sqrt{1 - (v/c)^2}} - 1\right) \qquad \textbf{(7.19)}$$

According to this expression, speeds greater than c are not allowed because, as v approaches c, K approaches ∞. This limitation is consistent with experimental observations on subatomic particles, which have shown that no particles travel at speeds greater than c. (In other words, c is the ultimate speed.) From this relativistic point of view, the work–kinetic energy theorem says that v can only approach c because it would take an infinite amount of work to attain the speed $v = c$.

All formulas in the theory of relativity must reduce to those in Newtonian mechanics at low particle speeds. It is instructive to show that this is the case for the kinetic energy relationship by analyzing Equation 7.19 when v is small compared with c. In this case, we expect K to reduce to the Newtonian expression. We can check this by using the binomial expansion (Appendix B.5) applied to the quantity $[1 - (v/c)^2]^{-1/2}$, with $v/c \ll 1$. If we let $x = (v/c)^2$, the expansion gives

$$\frac{1}{(1 - x)^{1/2}} = 1 + \frac{x}{2} + \frac{3}{8}x^2 + \cdots$$

Making use of this expansion in Equation 7.19 gives

$$K = mc^2\left(1 + \frac{v^2}{2c^2} + \frac{3}{8}\frac{v^4}{c^4} + \cdots - 1\right)$$

$$= \frac{1}{2}mv^2 + \frac{3}{8}m\frac{v^4}{c^2} + \cdots$$

$$= \frac{1}{2}mv^2 \quad \text{for} \quad \frac{v}{c} \ll 1$$

Thus, we see that the relativistic kinetic energy expression does indeed reduce to the Newtonian expression for speeds that are small compared with c. We shall return to the subject of relativity in Chapter 39.

SUMMARY

The work done by a constant force $\mathbf{F}$ acting on a particle is defined as the product of the component of the force in the direction of the particle's displacement and the magnitude of the displacement. Given a force $\mathbf{F}$ that makes an angle θ with the displacement vector $\mathbf{d}$ of a particle acted on by the force, you should be able to determine the work done by $\mathbf{F}$ using the equation

$$W \equiv Fd \cos \theta \qquad \textbf{(7.1)}$$

The **scalar product** (dot product) of two vectors $\mathbf{A}$ and $\mathbf{B}$ is defined by the relationship

$$\mathbf{A \cdot B} \equiv AB \cos \theta \qquad \textbf{(7.3)}$$

where the result is a scalar quantity and θ is the angle between the two vectors. The scalar product obeys the commutative and distributive laws.

If a varying force does work on a particle as the particle moves along the x axis from x_i to x_f, you must use the expression

$$W \equiv \int_{x_i}^{x_f} F_x \, dx \qquad \textbf{(7.7)}$$

where F_x is the component of force in the x direction. If several forces are acting on the particle, the net work done by all of the forces is the sum of the amounts of work done by all of the forces.

The **kinetic energy** of a particle of mass m moving with a speed v (where v is small compared with the speed of light) is

$$K \equiv \tfrac{1}{2}mv^2 \tag{7.14}$$

The **work–kinetic energy theorem** states that the net work done on a particle by external forces equals the change in kinetic energy of the particle:

$$\sum W = K_f - K_i = \tfrac{1}{2}mv_f^2 - \tfrac{1}{2}mv_i^2 \tag{7.16}$$

If a frictional force acts, then the work–kinetic energy theorem can be modified to give

$$K_i + \sum W_{\text{other}} - f_k d = K_f \tag{7.17b}$$

The **instantaneous power** $\mathcal{P}$ is defined as the time rate of energy transfer. If an agent applies a force $\mathbf{F}$ to an object moving with a velocity $\mathbf{v}$, the power delivered by that agent is

$$\mathcal{P} \equiv \frac{dW}{dt} = \mathbf{F} \cdot \mathbf{v} \tag{7.18}$$

QUESTIONS

1. Consider a tug-of-war in which two teams pulling on a rope are evenly matched so that no motion takes place. Assume that the rope does not stretch. Is work done on the rope? On the pullers? On the ground? Is work done on anything?

2. For what values of θ is the scalar product (a) positive and (b) negative?

3. As the load on a spring hung vertically is increased, one would not expect the F_s-versus-x curve to always remain linear, as shown in Figure 7.10d. Explain qualitatively what you would expect for this curve as m is increased.

4. Can the kinetic energy of an object be negative? Explain.

5. (a) If the speed of a particle is doubled, what happens to its kinetic energy? (b) If the net work done on a particle is zero, what can be said about the speed?

6. In Example 7.16, does the required power increase or decrease as the force of friction is reduced?

7. An automobile sales representative claims that a "souped-up" 300-hp engine is a necessary option in a compact car (instead of a conventional 130-hp engine). Suppose you intend to drive the car within speed limits (≤ 55 mi/h) and on flat terrain. How would you counter this sales pitch?

8. One bullet has twice the mass of another bullet. If both bullets are fired so that they have the same speed, which has the greater kinetic energy? What is the ratio of the kinetic energies of the two bullets?

9. When a punter kicks a football, is he doing any work on the ball while his toe is in contact with it? Is he doing any work on the ball after it loses contact with his toe? Are any forces doing work on the ball while it is in flight?

10. Discuss the work done by a pitcher throwing a baseball. What is the approximate distance through which the force acts as the ball is thrown?

11. Two sharpshooters fire 0.30-caliber rifles using identical shells. The barrel of rifle A is 2.00 cm longer than that of rifle B. Which rifle will have the higher muzzle speed? (*Hint:* The force of the expanding gases in the barrel accelerates the bullets.)

12. As a simple pendulum swings back and forth, the forces acting on the suspended mass are the force of gravity, the tension in the supporting cord, and air resistance. (a) Which of these forces, if any, does no work on the pendulum? (b) Which of these forces does negative work at all times during its motion? (c) Describe the work done by the force of gravity while the pendulum is swinging.

13. The kinetic energy of an object depends on the frame of reference in which its motion is measured. Give an example to illustrate this point.

14. An older model car accelerates from 0 to a speed v in 10 s. A newer, more powerful sports car accelerates from 0 to $2v$ in the same time period. What is the ratio of powers expended by the two cars? Consider the energy coming from the engines to appear only as kinetic energy of the cars.

PROBLEMS

1, **2**, **3** = straightforward, intermediate, challenging ☐ = full solution available in the *Student Solutions Manual and Study Guide*
WEB = solution posted at **http://www.saunderscollege.com/physics/** 🖥 = Computer useful in solving problem 🕹 = Interactive Physics
☐ = paired numerical/symbolic problems

Section 7.1   Work Done by a Constant Force

1. A tugboat exerts a constant force of 5 000 N on a ship moving at constant speed through a harbor. How much work does the tugboat do on the ship in a distance of 3.00 km?

2. A shopper in a supermarket pushes a cart with a force of 35.0 N directed at an angle of 25.0° downward from the horizontal. Find the work done by the shopper as she moves down an aisle 50.0 m in length.

3. A raindrop ($m = 3.35 \times 10^{-5}$ kg) falls vertically at constant speed under the influence of gravity and air resistance. After the drop has fallen 100 m, what is the work done (a) by gravity and (b) by air resistance?

4. A sledge loaded with bricks has a total mass of 18.0 kg and is pulled at constant speed by a rope. The rope is inclined at 20.0° above the horizontal, and the sledge moves a distance of 20.0 m on a horizontal surface. The coefficient of kinetic friction between the sledge and the surface is 0.500. (a) What is the tension of the rope? (b) How much work is done on the sledge by the rope? (c) What is the energy lost due to friction?

5. A block of mass 2.50 kg is pushed 2.20 m along a frictionless horizontal table by a constant 16.0-N force directed 25.0° below the horizontal. Determine the work done by (a) the applied force, (b) the normal force exerted by the table, and (c) the force of gravity. (d) Determine the total work done on the block.

6. A 15.0-kg block is dragged over a rough, horizontal surface by a 70.0-N force acting at 20.0° above the horizontal. The block is displaced 5.00 m, and the coefficient of kinetic friction is 0.300. Find the work done by (a) the 70-N force, (b) the normal force, and (c) the force of gravity. (d) What is the energy loss due to friction? (e) Find the total change in the block's kinetic energy.

WEB 7. Batman, whose mass is 80.0 kg, is holding onto the free end of a 12.0-m rope, the other end of which is fixed to a tree limb above. He is able to get the rope in motion as only Batman knows how, eventually getting it to swing enough so that he can reach a ledge when the rope makes a 60.0° angle with the vertical. How much work was done against the force of gravity in this maneuver?

Section 7.2   The Scalar Product of Two Vectors

In Problems 8 to 14, calculate all numerical answers to three significant figures.

8. Vector **A** has a magnitude of 5.00 units, and vector **B** has a magnitude of 9.00 units. The two vectors make an angle of 50.0° with each other. Find **A · B**.

9. Vector **A** extends from the origin to a point having polar coordinates (7, 70°), and vector **B** extends from the origin to a point having polar coordinates (4, 130°). Find **A · B**.

10. Given two arbitrary vectors **A** and **B**, show that **A·B** = $A_x B_x + A_y B_y + A_z B_z$. (*Hint:* Write **A** and **B** in unit vector form and use Equations 7.4 and 7.5.)

WEB 11. A force **F** = $(6\mathbf{i} - 2\mathbf{j})$ N acts on a particle that undergoes a displacement **d** = $(3\mathbf{i} + \mathbf{j})$ m. Find (a) the work done by the force on the particle and (b) the angle between **F** and **d**.

12. For **A** = $3\mathbf{i} + \mathbf{j} - \mathbf{k}$, **B** = $-\mathbf{i} + 2\mathbf{j} + 5\mathbf{k}$, and **C** = $2\mathbf{j} - 3\mathbf{k}$, find **C·(A − B)**.

13. Using the definition of the scalar product, find the angles between (a) **A** = $3\mathbf{i} - 2\mathbf{j}$ and **B** = $4\mathbf{i} - 4\mathbf{j}$; (b) **A** = $-2\mathbf{i} + 4\mathbf{j}$ and **B** = $3\mathbf{i} - 4\mathbf{j} + 2\mathbf{k}$; (c) **A** = $\mathbf{i} - 2\mathbf{j} + 2\mathbf{k}$ and **B** = $3\mathbf{j} + 4\mathbf{k}$.

14. Find the scalar product of the vectors in Figure P7.14.

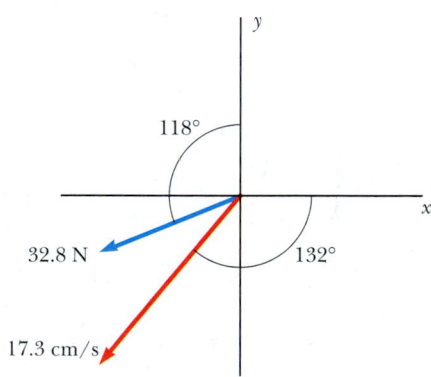

Figure P7.14

Section 7.3   Work Done by a Varying Force

15. The force acting on a particle varies as shown in Figure P7.15. Find the work done by the force as the particle moves (a) from $x = 0$ to $x = 8.00$ m, (b) from $x = 8.00$ m to $x = 10.0$ m, and (c) from $x = 0$ to $x = 10.0$ m.

16. The force acting on a particle is $F_x = (8x - 16)$ N, where x is in meters. (a) Make a plot of this force versus x from $x = 0$ to $x = 3.00$ m. (b) From your graph, find the net work done by this force as the particle moves from $x = 0$ to $x = 3.00$ m.

WEB 17. A particle is subject to a force F_x that varies with position as in Figure P7.17. Find the work done by the force on the body as it moves (a) from $x = 0$ to $x = 5.00$ m,

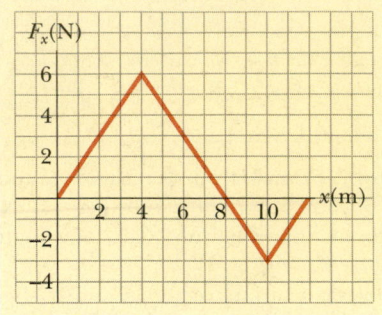

Figure P7.15

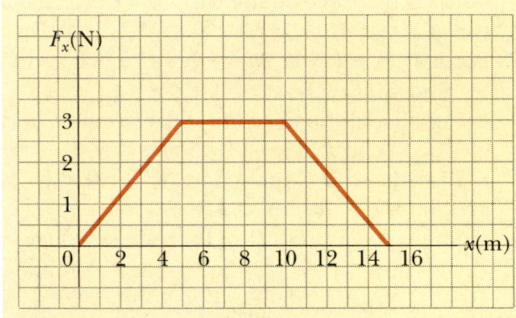

Figure P7.17   Problems 17 and 32.

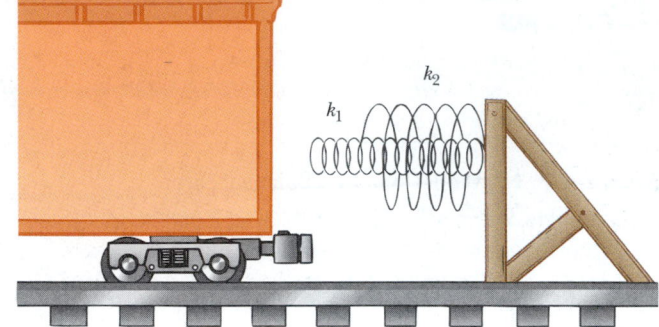

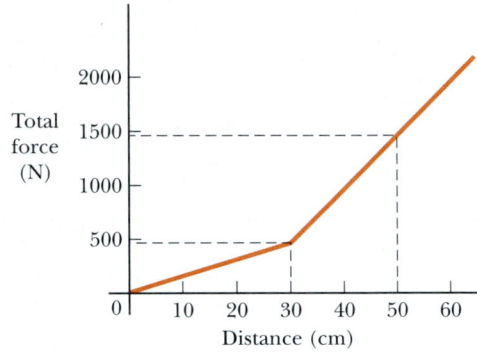

Figure P7.21

(b) from $x = 5.00$ m to $x = 10.0$ m, and (c) from $x = 10.0$ m to $x = 15.0$ m. (d) What is the total work done by the force over the distance $x = 0$ to $x = 15.0$ m?

18. A force $\mathbf{F} = (4x\mathbf{i} + 3y\mathbf{j})$ N acts on an object as it moves in the x direction from the origin to $x = 5.00$ m. Find the work done on the object by the force.

19. When a 4.00-kg mass is hung vertically on a certain light spring that obeys Hooke's law, the spring stretches 2.50 cm. If the 4.00-kg mass is removed, (a) how far will the spring stretch if a 1.50-kg mass is hung on it and (b) how much work must an external agent do to stretch the same spring 4.00 cm from its unstretched position?

20. An archer pulls her bow string back 0.400 m by exerting a force that increases uniformly from zero to 230 N. (a) What is the equivalent spring constant of the bow? (b) How much work is done by the archer in pulling the bow?

21. A 6 000-kg freight car rolls along rails with negligible friction. The car is brought to rest by a combination of two coiled springs, as illustrated in Figure P7.21. Both springs obey Hooke's law with $k_1 = 1\ 600$ N/m and $k_2 = 3\ 400$ N/m. After the first spring compresses a distance of 30.0 cm, the second spring (acting with the first) increases the force so that additional compression occurs, as shown in the graph. If the car is brought to

rest 50.0 cm after first contacting the two-spring system, find the car's initial speed.

22. A 100-g bullet is fired from a rifle having a barrel 0.600 m long. Assuming the origin is placed where the bullet begins to move, the force (in newtons) exerted on the bullet by the expanding gas is 15 000 + 10 000x − 25 000x^2, where x is in meters. (a) Determine the work done by the gas on the bullet as the bullet travels the length of the barrel. (b) If the barrel is 1.00 m long, how much work is done and how does this value compare with the work calculated in part (a)?

23. If it takes 4.00 J of work to stretch a Hooke's-law spring 10.0 cm from its unstressed length, determine the extra work required to stretch it an additional 10.0 cm.

24. If it takes work W to stretch a Hooke's-law spring a distance d from its unstressed length, determine the extra work required to stretch it an additional distance d.

25. A small mass m is pulled to the top of a frictionless half-cylinder (of radius R) by a cord that passes over the top of the cylinder, as illustrated in Figure P7.25. (a) If the mass moves at a constant speed, show that $F = mg\cos\theta$. (Hint: If the mass moves at a constant speed, the component of its acceleration tangent to the cylinder must be zero at all times.) (b) By directly integrating $W = \int \mathbf{F} \cdot d\mathbf{s}$, find the work done in moving the mass at constant speed from the bottom to the top of the half-

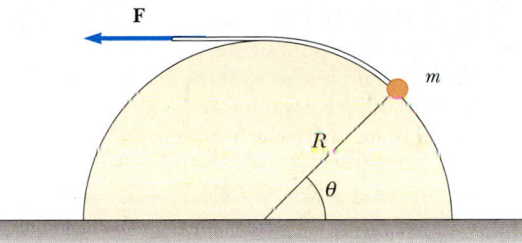

Figure P7.25

cylinder. Here $d\mathbf{s}$ represents an incremental displacement of the small mass.

26. Express the unit of the force constant of a spring in terms of the basic units meter, kilogram, and second.

Section 7.4 Kinetic Energy and the Work–Kinetic Energy Theorem

27. A 0.600-kg particle has a speed of 2.00 m/s at point A and kinetic energy of 7.50 J at point B. What is (a) its kinetic energy at A? (b) its speed at B? (c) the total work done on the particle as it moves from A to B?

28. A 0.300-kg ball has a speed of 15.0 m/s. (a) What is its kinetic energy? (b) If its speed were doubled, what would be its kinetic energy?

29. A 3.00-kg mass has an initial velocity $\mathbf{v}_i = (6.00\mathbf{i} - 2.00\mathbf{j})$ m/s. (a) What is its kinetic energy at this time? (b) Find the total work done on the object if its velocity changes to $(8.00\mathbf{i} + 4.00\mathbf{j})$ m/s. (*Hint:* Remember that $v^2 = \mathbf{v}\cdot\mathbf{v}$.)

30. A mechanic pushes a 2 500-kg car, moving it from rest and making it accelerate from rest to a speed v. He does 5 000 J of work in the process. During this time, the car moves 25.0 m. If friction between the car and the road is negligible, (a) what is the final speed v of the car? (b) What constant horizontal force did he exert on the car?

31. A mechanic pushes a car of mass m, doing work W in making it accelerate from rest. If friction between the car and the road is negligible, (a) what is the final speed of the car? During the time the mechanic pushes the car, the car moves a distance d. (b) What constant horizontal force did the mechanic exert on the car?

32. A 4.00-kg particle is subject to a total force that varies with position, as shown in Figure P7.17. The particle starts from rest at $x = 0$. What is its speed at (a) $x = 5.00$ m, (b) $x = 10.0$ m, (c) $x = 15.0$ m?

33. A 40.0-kg box initially at rest is pushed 5.00 m along a rough, horizontal floor with a constant applied horizontal force of 130 N. If the coefficient of friction between the box and the floor is 0.300, find (a) the work done by the applied force, (b) the energy loss due to friction, (c) the work done by the normal force, (d) the work done by gravity, (e) the change in kinetic energy of the box, and (f) the final speed of the box.

34. You can think of the work–kinetic energy theorem as a second theory of motion, parallel to Newton's laws in describing how outside influences affect the motion of an object. In this problem, work out parts (a) and (b) separately from parts (c) and (d) to compare the predictions of the two theories. In a rifle barrel, a 15.0-g bullet is accelerated from rest to a speed of 780 m/s. (a) Find the work that is done on the bullet. (b) If the rifle barrel is 72.0 cm long, find the magnitude of the average total force that acted on it, as $F = W/(d\cos\theta)$. (c) Find the constant acceleration of a bullet that starts from rest and gains a speed of 780 m/s over a distance of 72.0 cm. (d) Find the total force that acted on it as $\Sigma F = ma$.

35. A crate of mass 10.0 kg is pulled up a rough incline with an initial speed of 1.50 m/s. The pulling force is 100 N parallel to the incline, which makes an angle of 20.0° with the horizontal. The coefficient of kinetic friction is 0.400, and the crate is pulled 5.00 m. (a) How much work is done by gravity? (b) How much energy is lost because of friction? (c) How much work is done by the 100-N force? (d) What is the change in kinetic energy of the crate? (e) What is the speed of the crate after it has been pulled 5.00 m?

36. A block of mass 12.0 kg slides from rest down a frictionless 35.0° incline and is stopped by a strong spring with $k = 3.00 \times 10^4$ N/m. The block slides 3.00 m from the point of release to the point where it comes to rest against the spring. When the block comes to rest, how far has the spring been compressed?

WEB 37. A sled of mass m is given a kick on a frozen pond. The kick imparts to it an initial speed $v_i = 2.00$ m/s. The coefficient of kinetic friction between the sled and the ice is $\mu_k = 0.100$. Utilizing energy considerations, find the distance the sled moves before it stops.

38. A picture tube in a certain television set is 36.0 cm long. The electrical force accelerates an electron in the tube from rest to 1.00% of the speed of light over this distance. Determine (a) the kinetic energy of the electron as it strikes the screen at the end of the tube, (b) the magnitude of the average electrical force acting on the electron over this distance, (c) the magnitude of the average acceleration of the electron over this distance, and (d) the time of flight.

39. A bullet with a mass of 5.00 g and a speed of 600 m/s penetrates a tree to a depth of 4.00 cm. (a) Use work and energy considerations to find the average frictional force that stops the bullet. (b) Assuming that the frictional force is constant, determine how much time elapsed between the moment the bullet entered the tree and the moment it stopped.

40. An Atwood's machine (see Fig. 5.15) supports masses of 0.200 kg and 0.300 kg. The masses are held at rest beside each other and then released. Neglecting friction, what is the speed of each mass the instant it has moved 0.400 m?

41. A 2.00-kg block is attached to a spring of force constant 500 N/m, as shown in Figure 7.10. The block is pulled 5.00 cm to the right of equilibrium and is then released from rest. Find the speed of the block as it passes through equilibrium if (a) the horizontal surface is frictionless and (b) the coefficient of friction between the block and the surface is 0.350.

Section 7.5   Power

42. Make an order-of-magnitude estimate of the power a car engine contributes to speeding up the car to highway speed. For concreteness, consider your own car (if you use one). In your solution, state the physical quantities you take as data and the values you measure or estimate for them. The mass of the vehicle is given in the owner's manual. If you do not wish to consider a car, think about a bus or truck for which you specify the necessary physical quantities.

WEB **43.** A 700-N Marine in basic training climbs a 10.0-m vertical rope at a constant speed in 8.00 s. What is his power output?

44. If a certain horse can maintain 1.00 hp of output for 2.00 h, how many 70.0-kg bundles of shingles can the horse hoist (using some pulley arrangement) to the roof of a house 8.00 m tall, assuming 70.0% efficiency?

45. A certain automobile engine delivers 2.24×10^4 W (30.0 hp) to its wheels when moving at a constant speed of 27.0 m/s ($\approx$ 60 mi/h). What is the resistive force acting on the automobile at that speed?

46. A skier of mass 70.0 kg is pulled up a slope by a motor-driven cable. (a) How much work is required for him to be pulled a distance of 60.0 m up a 30.0° slope (assumed frictionless) at a constant speed of 2.00 m/s? (b) A motor of what power is required to perform this task?

47. A 650-kg elevator starts from rest. It moves upward for 3.00 s with constant acceleration until it reaches its cruising speed of 1.75 m/s. (a) What is the average power of the elevator motor during this period? (b) How does this power compare with its power when it moves at its cruising speed?

48. An energy-efficient lightbulb, taking in 28.0 W of power, can produce the same level of brightness as a conventional bulb operating at 100-W power. The lifetime of the energy-efficient bulb is 10 000 h and its purchase price is $17.0, whereas the conventional bulb has a lifetime of 750 h and costs $0.420 per bulb. Determine the total savings obtained through the use of one energy-efficient bulb over its lifetime as opposed to the use of conventional bulbs over the same time period. Assume an energy cost of $0.080 0 per kilowatt hour.

(Optional)
Section 7.6   Energy and the Automobile

49. A compact car of mass 900 kg has an overall motor efficiency of 15.0%. (That is, 15.0% of the energy supplied by the fuel is delivered to the wheels of the car.) (a) If

burning 1 gal of gasoline supplies 1.34×10^8 J of energy, find the amount of gasoline used by the car in accelerating from rest to 55.0 mi/h. Here you may ignore the effects of air resistance and rolling resistance. (b) How many such accelerations will 1 gal provide? (c) The mileage claimed for the car is 38.0 mi/gal at 55 mi/h. What power is delivered to the wheels (to overcome frictional effects) when the car is driven at this speed?

50. Suppose the empty car described in Table 7.2 has a fuel economy of 6.40 km/L (15 mi/gal) when traveling at 26.8 m/s (60 mi/h). Assuming constant efficiency, determine the fuel economy of the car if the total mass of the passengers and the driver is 350 kg.

51. When an air conditioner is added to the car described in Problem 50, the additional output power required to operate the air conditioner is 1.54 kW. If the fuel economy of the car is 6.40 km/L without the air conditioner, what is it when the air conditioner is operating?

(Optional)
Section 7.7   Kinetic Energy at High Speeds

52. An electron moves with a speed of $0.995c$. (a) What is its kinetic energy? (b) If you use the classical expression to calculate its kinetic energy, what percentage error results?

53. A proton in a high-energy accelerator moves with a speed of $c/2$. Using the work–kinetic energy theorem, find the work required to increase its speed to (a) $0.750c$ and (b) $0.995c$.

54. Find the kinetic energy of a 78.0-kg spacecraft launched out of the Solar System with a speed of 106 km/s using (a) the classical equation $K = \frac{1}{2}mv^2$ and (b) the relativistic equation.

ADDITIONAL PROBLEMS

55. A baseball outfielder throws a 0.150-kg baseball at a speed of 40.0 m/s and an initial angle of 30.0°. What is the kinetic energy of the baseball at the highest point of the trajectory?

56. While running, a person dissipates about 0.600 J of mechanical energy per step per kilogram of body mass. If a 60.0-kg runner dissipates a power of 70.0 W during a race, how fast is the person running? Assume a running step is 1.50 m in length.

57. A particle of mass m moves with a constant acceleration **a**. If the initial position vector and velocity of the particle are $\mathbf{r}_i$ and $\mathbf{v}_i$, respectively, use energy arguments to show that its speed v_f at any time satisfies the equation

$$v_f{}^2 = v_i{}^2 + 2\mathbf{a} \cdot (\mathbf{r}_f - \mathbf{r}_i)$$

where $\mathbf{r}_f$ is the position vector of the particle at that same time.

58. The direction of an arbitrary vector **A** can be completely specified with the angles α, β, and γ that the vector makes with the x, y, and z axes, respectively. If $\mathbf{A} = A_x\mathbf{i} + A_y\mathbf{j} + A_z\mathbf{k}$, (a) find expressions for $\cos\alpha$, $\cos\beta$, and $\cos\gamma$ (known as *direction cosines*) and (b) show that these angles satisfy the relation $\cos^2\alpha + \cos^2\beta + \cos^2\gamma = 1$. (*Hint:* Take the scalar product of $\mathbf{A}$ with $\mathbf{i}$, $\mathbf{j}$, and $\mathbf{k}$ separately.)

59. A 4.00-kg particle moves along the x axis. Its position varies with time according to $x = t + 2.0t^3$, where x is in meters and t is in seconds. Find (a) the kinetic energy at any time t, (b) the acceleration of the particle and the force acting on it at time t, (c) the power being delivered to the particle at time t, and (d) the work done on the particle in the interval $t = 0$ to $t = 2.00$ s.

60. A traveler at an airport takes an escalator up one floor (Fig. P7.60). The moving staircase would itself carry him upward with vertical velocity component v between entry and exit points separated by height h. However, while the escalator is moving, the hurried traveler climbs the steps of the escalator at a rate of n steps/s. Assume that the height of each step is h_s. (a) Determine the amount of work done by the traveler during his escalator ride, given that his mass is m. (b) Determine the work the escalator motor does on this person.

Figure P7.60 (©*Ron Chapple/FPG*)

61. When a certain spring is stretched beyond its proportional limit, the restoring force satisfies the equation $F = -kx + \beta x^3$. If $k = 10.0$ N/m and $\beta = 100$ N/m^3,

calculate the work done by this force when the spring is stretched 0.100 m.

62. In a control system, an accelerometer consists of a 4.70-g mass sliding on a low-friction horizontal rail. A low-mass spring attaches the mass to a flange at one end of the rail. When subject to a steady acceleration of $0.800g$, the mass is to assume a location 0.500 cm away from its equilibrium position. Find the stiffness constant required for the spring.

63. A 2 100-kg pile driver is used to drive a steel I-beam into the ground. The pile driver falls 5.00 m before coming into contact with the beam, and it drives the beam 12.0 cm into the ground before coming to rest. Using energy considerations, calculate the average force the beam exerts on the pile driver while the pile driver is brought to rest.

64. A cyclist and her bicycle have a combined mass of 75.0 kg. She coasts down a road inclined at 2.00° with the horizontal at 4.00 m/s and down a road inclined at 4.00° at 8.00 m/s. She then holds on to a moving vehicle and coasts on a level road. What power must the vehicle expend to maintain her speed at 3.00 m/s? Assume that the force of air resistance is proportional to her speed and that other frictional forces remain constant. (*Warning:* You must *not* attempt this dangerous maneuver.)

65. A single constant force $\mathbf{F}$ acts on a particle of mass m. The particle starts at rest at $t = 0$. (a) Show that the instantaneous power delivered by the force at any time t is $(F^2/m)t$. (b) If $F = 20.0$ N and $m = 5.00$ kg, what is the power delivered at $t = 3.00$ s?

66. A particle is attached between two identical springs on a horizontal frictionless table. Both springs have spring constant k and are initially unstressed. (a) If the particle is pulled a distance x along a direction perpendicular to the initial configuration of the springs, as in Figure P7.66, show that the force exerted on the particle by the springs is

$$\mathbf{F} = -2kx\left(1 - \frac{L}{\sqrt{x^2 + L^2}}\right)\mathbf{i}$$

(b) Determine the amount of work done by this force in moving the particle from $x = A$ to $x = 0$.

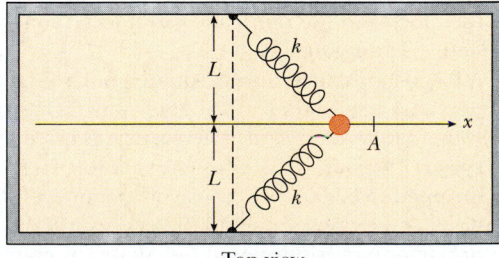

Top view

Figure P7.66

67. Review Problem. Two constant forces act on a 5.00-kg object moving in the *xy* plane, as shown in Figure P7.67. Force $\mathbf{F}_1$ is 25.0 N at 35.0°, while $\mathbf{F}_2$ = 42.0 N at 150°. At time *t* = 0, the object is at the origin and has velocity (4.0**i** + 2.5**j**) m/s. (a) Express the two forces in unit–vector notation. Use unit–vector notation for your other answers. (b) Find the total force on the object. (c) Find the object's acceleration. Now, considering the instant *t* = 3.00 s, (d) find the object's velocity, (e) its location, (f) its kinetic energy from $\frac{1}{2}mv_f^2$, and (g) its kinetic energy from $\frac{1}{2}mv_i^2 + \Sigma\mathbf{F}\cdot\mathbf{d}$.

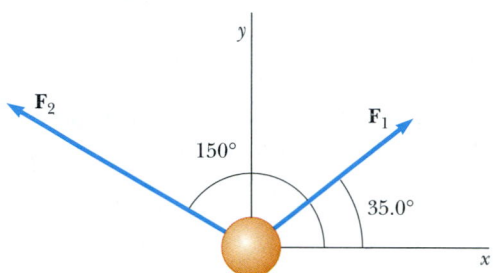

Figure P7.67

68. When different weights are hung on a spring, the spring stretches to different lengths as shown in the following table. (a) Make a graph of the applied force versus the extension of the spring. By least-squares fitting, determine the straight line that best fits the data. (You may not want to use all the data points.) (b) From the slope of the best-fit line, find the spring constant *k*. (c) If the spring is extended to 105 mm, what force does it exert on the suspended weight?

F (N)	2.0	4.0	6.0	8.0	10	12	14	16	18
L (mm)	15	32	49	64	79	98	112	126	149

69. A 200-g block is pressed against a spring of force constant 1.40 kN/m until the block compresses the spring 10.0 cm. The spring rests at the bottom of a ramp inclined at 60.0° to the horizontal. Using energy considerations, determine how far up the incline the block moves before it stops (a) if there is no friction between the block and the ramp and (b) if the coefficient of kinetic friction is 0.400.

70. A 0.400-kg particle slides around a horizontal track. The track has a smooth, vertical outer wall forming a circle with a radius of 1.50 m. The particle is given an initial speed of 8.00 m/s. After one revolution, its speed has dropped to 6.00 m/s because of friction with the rough floor of the track. (a) Find the energy loss due to friction in one revolution. (b) Calculate the coefficient of kinetic friction. (c) What is the total number of revolutions the particle makes before stopping?

Figure P7.71

WEB

71. The ball launcher in a pinball machine has a spring that has a force constant of 1.20 N/cm (Fig. P7.71). The surface on which the ball moves is inclined 10.0° with respect to the horizontal. If the spring is initially compressed 5.00 cm, find the launching speed of a 100-g ball when the plunger is released. Friction and the mass of the plunger are negligible.

72. In diatomic molecules, the constituent atoms exert attractive forces on each other at great distances and repulsive forces at short distances. For many molecules, the Lennard–Jones law is a good approximation to the magnitude of these forces:

$$F = F_0\left[2\left(\frac{\sigma}{r}\right)^{13} - \left(\frac{\sigma}{r}\right)^7\right]$$

where *r* is the center-to-center distance between the atoms in the molecule, σ is a length parameter, and F_0 is the force when *r* = σ. For an oxygen molecule, F_0 = 9.60 × 10⁻¹¹ N and σ = 3.50 × 10⁻¹⁰ m. Determine the work done by this force if the atoms are pulled apart from *r* = 4.00 × 10⁻¹⁰ m to *r* = 9.00 × 10⁻¹⁰ m.

73. A horizontal string is attached to a 0.250-kg mass lying on a rough, horizontal table. The string passes over a light, frictionless pulley, and a 0.400-kg mass is then attached to its free end. The coefficient of sliding friction between the 0.250-kg mass and the table is 0.200. Using the work–kinetic energy theorem, determine (a) the speed of the masses after each has moved 20.0 m from rest and (b) the mass that must be added to the 0.250-kg mass so that, given an initial velocity, the masses continue to move at a constant speed. (c) What mass must be removed from the 0.400-kg mass so that the same outcome as in part (b) is achieved?

74. Suppose a car is modeled as a cylinder moving with a speed *v*, as in Figure P7.74. In a time Δt, a column of air

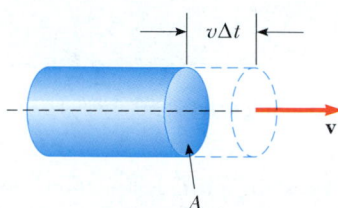

Figure P7.74

of mass Δm must be moved a distance $v\,\Delta t$ and, hence, must be given a kinetic energy $\frac{1}{2}(\Delta m)v^2$. Using this model, show that the power loss due to air resistance is $\frac{1}{2}\rho Av^3$ and that the resistive force is $\frac{1}{2}\rho Av^2$, where ρ is the density of air.

75. A particle moves along the x axis from $x = 12.8$ m to $x = 23.7$ m under the influence of a force

$$F = \frac{375}{x^3 + 3.75\,x}$$

where F is in newtons and x is in meters. Using numerical integration, determine the total work done by this force during this displacement. Your result should be accurate to within 2%.

76. More than 2 300 years ago the Greek teacher Aristotle wrote the first book called *Physics*. The following passage, rephrased with more precise terminology, is from the end of the book's Section Eta:

Let $\mathscr{P}$ be the power of an agent causing motion; w, the thing moved; d, the distance covered; and t, the time taken. Then (1) a power equal to $\mathscr{P}$ will in a period of time equal to t move $w/2$ a distance $2d$; or (2) it will move $w/2$ the given distance d in time $t/2$. Also, if (3) the given power $\mathscr{P}$ moves the given object w a distance $d/2$ in time $t/2$, then (4) $\mathscr{P}/2$ will move $w/2$ the given distance d in the given time t.

(a) Show that Aristotle's proportions are included in the equation $\mathscr{P}t = bwd$, where b is a proportionality constant. (b) Show that our theory of motion includes this part of Aristotle's theory as one special case. In particular, describe a situation in which it is true, derive the equation representing Aristotle's proportions, and determine the proportionality constant.

ANSWERS TO QUICK QUIZZES

7.1 No. The force does no work on the object because the force is pointed toward the center of the circle and is therefore perpendicular to the motion.

7.2 (a) Assuming the person lifts with a force of magnitude mg, the weight of the box, the work he does during the vertical displacement is mgh because the force is in the direction of the displacement. The work he does during the horizontal displacement is zero because now the force he exerts on the box is perpendicular to the displacement. The net work he does is $mgh + 0 = mgh$. (b) The work done by the gravitational force on the box as the box is displaced vertically is $-mgh$ because the direction of this force is opposite the direction of the displacement. The work done by the gravitational force is zero during the horizontal displacement because now the direction of this force is perpendicular to the direction of the displacement. The net work done by the gravitational force $-mgh + 0 = -mgh$. The total work done on the box is $+mgh - mgh = 0$.

7.3 No. For example, consider the two vectors $\mathbf{A} = 3\mathbf{i} - 2\mathbf{j}$ and $\mathbf{B} = 2\mathbf{i} - \mathbf{j}$. Their dot product is $\mathbf{A}\cdot\mathbf{B} = 8$, yet both vectors have negative y components.

7.4 Force divided by displacement, which in SI units is newtons per meter (N/m).

7.5 Yes, whenever the frictional force has a component along the direction of motion. Consider a crate sitting on the bed of a truck as the truck accelerates to the east. The static friction force exerted on the crate by the truck acts to the east to give the crate the same acceleration as the truck (assuming that the crate does not slip). Because the crate accelerates, its kinetic energy must increase.

7.6 Because the two vehicles perform the same amount of work, the areas under the two graphs are equal. However, the graph for the low-power truck extends over a longer time interval and does not extend as high on the $\mathscr{P}$ axis as the graph for the sports car does.

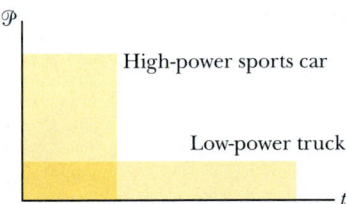

PUZZLER

A common scene at a carnival is the Ring-the-Bell attraction, in which the player swings a heavy hammer downward in an attempt to project a mass upward to ring a bell. What is the best strategy to win the game and impress your friends? *(Robert E. Daemmrich/Tony Stone Images)*

c h a p t e r

Potential Energy and Conservation of Energy

214

n Chapter 7 we introduced the concept of kinetic energy, which is the energy associated with the motion of an object. In this chapter we introduce another form of energy—*potential energy,* which is the energy associated with the arrangement of a system of objects that exert forces on each other. Potential energy can be thought of as stored energy that can either do work or be converted to kinetic energy.

The potential energy concept can be used only when dealing with a special class of forces called *conservative forces.* When only conservative forces act within an isolated system, the kinetic energy gained (or lost) by the system as its members change their relative positions is balanced by an equal loss (or gain) in potential energy. This balancing of the two forms of energy is known as the *principle of conservation of mechanical energy.*

Energy is present in the Universe in various forms, including mechanical, electromagnetic, chemical, and nuclear. Furthermore, one form of energy can be converted to another. For example, when an electric motor is connected to a battery, the chemical energy in the battery is converted to electrical energy in the motor, which in turn is converted to mechanical energy as the motor turns some device. The transformation of energy from one form to another is an essential part of the study of physics, engineering, chemistry, biology, geology, and astronomy.

When energy is changed from one form to another, the total amount present does not change. Conservation of energy means that although the form of energy may change, if an object (or system) loses energy, that same amount of energy appears in another object or in the object's surroundings.

8.1 ▶ POTENTIAL ENERGY

An object that possesses kinetic energy can do work on another object—for example, a moving hammer driving a nail into a wall. Now we shall introduce another form of energy. This energy, called **potential energy U,** is the energy associated with a system of objects.

Before we describe specific forms of potential energy, we must first define a *system,* which consists of two or more objects that exert forces on one another. **If the arrangement of the system changes, then the potential energy of the system changes.** If the system consists of only two particle-like objects that exert forces on each other, then the work done by the force acting on one of the objects causes a transformation of energy between the object's kinetic energy and other forms of the system's energy.

Gravitational Potential Energy

As an object falls toward the Earth, the Earth exerts a gravitational force $m\mathbf{g}$ on the object, with the direction of the force being the same as the direction of the object's motion. The gravitational force does work on the object and thereby increases the object's kinetic energy. Imagine that a brick is dropped from rest directly above a nail in a board lying on the ground. When the brick is released, it falls toward the ground, gaining speed and therefore gaining kinetic energy. The brick–Earth system has potential energy when the brick is at any distance above the ground (that is, it has the *potential* to do work), and this potential energy is converted to kinetic energy as the brick falls. The conversion from potential energy to kinetic energy occurs continuously over the entire fall. When the brick reaches the nail and the board lying on the ground, it does work on the nail, driving it into the board. What determines how much work the brick is able to do on the nail? It is easy to see that the heavier the brick, the farther in it drives the nail; also the higher the brick is before it is released, the more work it does when it strikes the nail.

The product of the magnitude of the gravitational force mg acting on an object and the height y of the object is so important in physics that we give it a name: the **gravitational potential energy.** The symbol for gravitational potential energy is U_g, and so the defining equation for gravitational potential energy is

$$U_g \equiv mgy \qquad \qquad \textbf{(8.1)}$$

Gravitational potential energy is the potential energy of the object–Earth system. This potential energy is transformed into kinetic energy of the system by the gravitational force. In this type of system, in which one of the members (the Earth) is much more massive than the other (the object), the massive object can be modeled as stationary, and the kinetic energy of the system can be represented entirely by the kinetic energy of the lighter object. Thus, the kinetic energy of the system is represented by that of the object falling toward the Earth. Also note that Equation 8.1 is valid only for objects near the surface of the Earth, where $\mathbf{g}$ is approximately constant.[1]

Let us now directly relate the work done on an object by the gravitational force to the gravitational potential energy of the object–Earth system. To do this, let us consider a brick of mass m at an initial height y_i above the ground, as shown in Figure 8.1. If we neglect air resistance, then the only force that does work on the brick as it falls is the gravitational force exerted on the brick $m\mathbf{g}$. The work W_g done by the gravitational force as the brick undergoes a downward displacement $\mathbf{d}$ is

$$W_g = (m\mathbf{g}) \cdot \mathbf{d} = (-mg\mathbf{j}) \cdot (y_f - y_i)\,\mathbf{j} = mgy_i - mgy_f$$

where we have used the fact that $\mathbf{j} \cdot \mathbf{j} = 1$ (Eq. 7.4). If an object undergoes both a horizontal and a vertical displacement, so that $\mathbf{d} = (x_f - x_i)\mathbf{i} + (y_f - y_i)\mathbf{j}$, then the work done by the gravitational force is still $mgy_i - mgy_f$ because $-mg\mathbf{j} \cdot (x_f - x_i)\mathbf{i} = 0$. Thus, the work done by the gravitational force depends only on the change in y and not on any change in the horizontal position x.

We just learned that the quantity mgy is the gravitational potential energy of the system U_g, and so we have

$$W_g = U_i - U_f = -(U_f - U_i) = -\Delta U_g \qquad \qquad \textbf{(8.2)}$$

From this result, we see that the work done on any object by the gravitational force is equal to the negative of the change in the system's gravitational potential energy. Also, this result demonstrates that it is only the *difference* in the gravitational potential energy at the initial and final locations that matters. This means that we are free to place the origin of coordinates in any convenient location. Finally, the work done by the gravitational force on an object as the object falls to the Earth is the same as the work done were the object to start at the same point and slide down an incline to the Earth. Horizontal motion does not affect the value of W_g.

The unit of gravitational potential energy is the same as that of work—the joule. Potential energy, like work and kinetic energy, is a scalar quantity.

---

[1] The assumption that the force of gravity is constant is a good one as long as the vertical displacement is small compared with the Earth's radius.

Gravitational potential energy

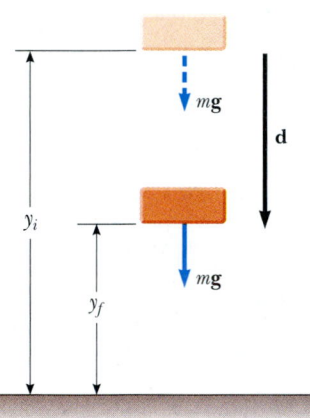

Figure 8.1   The work done on the brick by the gravitational force as the brick falls from a height y_i to a height y_f is equal to $mgy_i - mgy_f$.

Quick Quiz 8.1

Can the gravitational potential energy of a system ever be negative?

EXAMPLE 8.1 ▸ The Bowler and the Sore Toe

A bowling ball held by a careless bowler slips from the bowler's hands and drops on the bowler's toe. Choosing floor level as the $y = 0$ point of your coordinate system, estimate the total work done on the ball by the force of gravity as the ball falls. Repeat the calculation, using the top of the bowler's head as the origin of coordinates.

Solution First, we need to estimate a few values. A bowling ball has a mass of approximately 7 kg, and the top of a person's toe is about 0.03 m above the floor. Also, we shall assume the ball falls from a height of 0.5 m. Holding nonsignificant digits until we finish the problem, we calculate the gravitational potential energy of the ball–Earth system just before the ball is released to be $U_i = mgy_i = (7\ \text{kg})$ $(9.80\ \text{m/s}^2)(0.5\ \text{m}) = 34.3\ \text{J}$. A similar calculation for when

the ball reaches his toe gives $U_f = mgy_f = (7\ \text{kg})$ $(9.80\ \text{m/s}^2)(0.03\ \text{m}) = 2.06\ \text{J}$. So, the work done by the gravitational force is $W_g = U_i - U_f = 32.24\ \text{J}$. We should probably keep only one digit because of the roughness of our estimates; thus, we estimate that the gravitational force does 30 J of work on the bowling ball as it falls. The system had 30 J of gravitational potential energy relative to the top of the toe before the ball began its fall.

When we use the bowler's head (which we estimate to be 1.50 m above the floor) as our origin of coordinates, we find that $U_i = mgy_i = (7\ \text{kg})(9.80\ \text{m/s}^2)(-1\ \text{m}) = -68.6\ \text{J}$ and that $U_f = mgy_f = (7\ \text{kg})(9.80\ \text{m/s}^2)(-1.47\ \text{m}) = -100.8\ \text{J}$. The work being done by the gravitational force is still

$$W_g = U_i - U_f = 32.24\ \text{J} \approx \boxed{30\ \text{J}.}$$

Elastic Potential Energy

Now consider a system consisting of a block plus a spring, as shown in Figure 8.2. The force that the spring exerts on the block is given by $F_s = -kx$. In the previous chapter, we learned that the work done by the spring force on a block connected to the spring is given by Equation 7.11:

$$W_s = \tfrac{1}{2}kx_i^2 - \tfrac{1}{2}kx_f^2 \tag{8.3}$$

In this situation, the initial and final x coordinates of the block are measured from its equilibrium position, $x = 0$. Again we see that W_s depends only on the initial and final x coordinates of the object and is zero for any closed path. The **elastic potential energy** function associated with the system is defined by

$$U_s \equiv \tfrac{1}{2}kx^2 \tag{8.4}$$

Elastic potential energy stored in a spring

The elastic potential energy of the system can be thought of as the energy stored in the deformed spring (one that is either compressed or stretched from its equilibrium position). To visualize this, consider Figure 8.2, which shows a spring on a frictionless, horizontal surface. When a block is pushed against the spring (Fig. 8.2b) and the spring is compressed a distance x, the elastic potential energy stored in the spring is $kx^2/2$. When the block is released from rest, the spring snaps back to its original length and the stored elastic potential energy is transformed into kinetic energy of the block (Fig. 8.2c). The elastic potential energy stored in the spring is zero whenever the spring is undeformed ($x = 0$). Energy is stored in the spring only when the spring is either stretched or compressed. Furthermore, the elastic potential energy is a maximum when the spring has reached its maximum compression or extension (that is, when $|x|$ is a maximum). Finally, because the elastic potential energy is proportional to x^2, we see that U_s is always positive in a deformed spring.

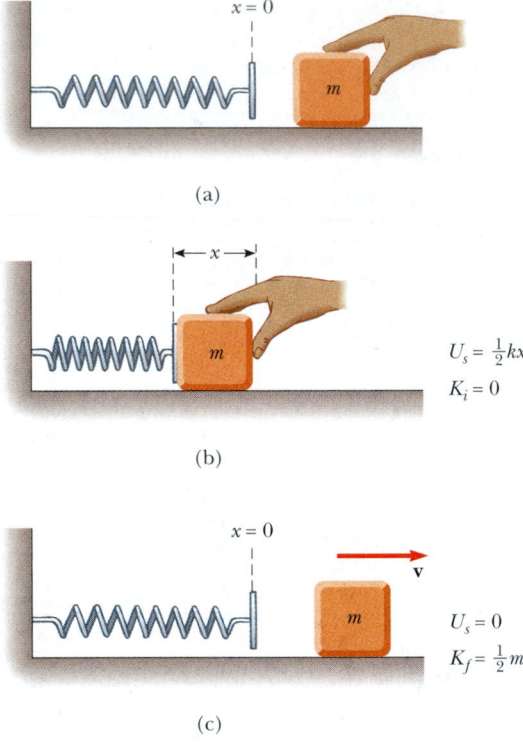

(a)

(b)

$$U_s = \frac{1}{2}kx^2$$
$$K_i = 0$$

(c)

$$U_s = 0$$
$$K_f = \frac{1}{2}mv^2$$

Figure 8.2 (a) An undeformed spring on a frictionless horizontal surface. (b) A block of mass m is pushed against the spring, compressing it a distance x. (c) When the block is released from rest, the elastic potential energy stored in the spring is transferred to the block in the form of kinetic energy.

8.2 CONSERVATIVE AND NONCONSERVATIVE FORCES

The work done by the gravitational force does not depend on whether an object falls vertically or slides down a sloping incline. All that matters is the change in the object's elevation. On the other hand, the energy loss due to friction on that incline depends on the distance the object slides. In other words, the path makes no difference when we consider the work done by the gravitational force, but it does make a difference when we consider the energy loss due to frictional forces. We can use this varying dependence on path to classify forces as either conservative or nonconservative.

Of the two forces just mentioned, the gravitational force is conservative and the frictional force is nonconservative.

Conservative Forces

Properties of a conservative force

Conservative forces have two important properties:

1. A force is conservative if the work it does on a particle moving between any two points is independent of the path taken by the particle.
2. The work done by a conservative force on a particle moving through any closed path is zero. (A closed path is one in which the beginning and end points are identical.)

The gravitational force is one example of a conservative force, and the force that a spring exerts on any object attached to the spring is another. As we learned in the preceding section, the work done by the gravitational force on an object moving between any two points near the Earth's surface is $W_g = mgy_i - mgy_f$. From this equation we see that W_g depends only on the initial and final y coordinates of the object and hence is independent of the path. Furthermore, W_g is zero when the object moves over any closed path (where $y_i = y_f$).

For the case of the object–spring system, the work W_s done by the spring force is given by $W_s = \frac{1}{2}kx_i^2 - \frac{1}{2}kx_f^2$ (Eq. 8.3). Again, we see that the spring force is conservative because W_s depends only on the initial and final x coordinates of the object and is zero for any closed path.

We can associate a potential energy with any conservative force and can do this *only* for conservative forces. In the previous section, the potential energy associated with the gravitational force was defined as $U_g \equiv mgy$. In general, the work W_c done on an object by a conservative force is equal to the initial value of the potential energy associated with the object minus the final value:

$$W_c = U_i - U_f = -\Delta U \qquad \textbf{(8.5)}$$

This equation should look familiar to you. It is the general form of the equation for work done by the gravitational force (Eq. 8.2) and that for the work done by the spring force (Eq. 8.3).

Work done by a conservative force

Nonconservative Forces

A force is nonconservative if it causes a change in mechanical energy E, which we define as the sum of kinetic and potential energies. For example, if a book is sent sliding on a horizontal surface that is not frictionless, the force of kinetic friction reduces the book's kinetic energy. As the book slows down, its kinetic energy decreases. As a result of the frictional force, the temperatures of the book and surface increase. The type of energy associated with temperature is *internal energy*, which we will study in detail in Chapter 20. Experience tells us that this internal energy cannot be transferred back to the kinetic energy of the book. In other words, the energy transformation is not reversible. Because the force of kinetic friction changes the mechanical energy of a system, it is a nonconservative force.

Properties of a nonconservative force

From the work–kinetic energy theorem, we see that the work done by a conservative force on an object causes a change in the kinetic energy of the object. The change in kinetic energy depends only on the initial and final positions of the object, and not on the path connecting these points. Let us compare this to the sliding book example, in which the nonconservative force of friction is acting between the book and the surface. According to Equation 7.17a, the change in kinetic energy of the book due to friction is $\Delta K_{\text{friction}} = -f_k d$, where d is the length of the path over which the friction force acts. Imagine that the book slides from A to B over the straight-line path of length d in Figure 8.3. The change in kinetic energy is $-f_k d$. Now, suppose the book slides over the semicircular path from A to B. In this case, the path is longer and, as a result, the change in kinetic energy is greater in magnitude than that in the straight-line case. For this particular path, the change in kinetic energy is $-f_k \pi d/2$, since d is the diameter of the semicircle. Thus, we see that for a nonconservative force, the change in kinetic energy depends on the path followed between the initial and final points. If a potential energy is involved, then the change in the total mechanical energy depends on the path followed. We shall return to this point in Section 8.5.

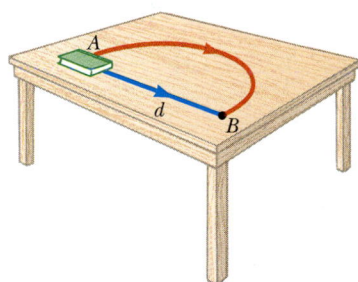

Figure 8.3 The loss in mechanical energy due to the force of kinetic friction depends on the path taken as the book is moved from A to B. The loss in mechanical energy is greater along the red path than along the blue path.

8.3 CONSERVATIVE FORCES AND POTENTIAL ENERGY

In the preceding section we found that the work done on a particle by a conservative force does not depend on the path taken by the particle. The work depends only on the particle's initial and final coordinates. As a consequence, we can define a **potential energy function** U such that the work done by a conservative force equals the decrease in the potential energy of the system. The work done by a conservative force $\mathbf{F}$ as a particle moves along the x axis is[2]

$$W_c = \int_{x_i}^{x_f} F_x\,dx = -\Delta U \tag{8.6}$$

where F_x is the component of $\mathbf{F}$ in the direction of the displacement. That is, **the work done by a conservative force equals the negative of the change in the potential energy associated with that force,** where the change in the potential energy is defined as $\Delta U = U_f - U_i$.

We can also express Equation 8.6 as

$$\Delta U = U_f - U_i = -\int_{x_i}^{x_f} F_x\,dx \tag{8.7}$$

Therefore, ΔU is negative when F_x and dx are in the same direction, as when an object is lowered in a gravitational field or when a spring pushes an object toward equilibrium.

The term *potential energy* implies that the object has the potential, or capability, of either gaining kinetic energy or doing work when it is released from some point under the influence of a conservative force exerted on the object by some other member of the system. It is often convenient to establish some particular location x_i as a reference point and measure all potential energy differences with respect to it. We can then define the potential energy function as

$$U_f(x) = -\int_{x_i}^{x_f} F_x\,dx + U_i \tag{8.8}$$

The value of U_i is often taken to be zero at the reference point. It really does not matter what value we assign to U_i, because any nonzero value merely shifts $U_f(x)$ by a constant amount, and only the *change* in potential energy is physically meaningful.

If the conservative force is known as a function of position, we can use Equation 8.8 to calculate the change in potential energy of a system as an object within the system moves from x_i to x_f. It is interesting to note that in the case of one-dimensional displacement, a force is always conservative if it is a function of position only. This is not necessarily the case for motion involving two- or three-dimensional displacements.

8.4 CONSERVATION OF MECHANICAL ENERGY

An object held at some height h above the floor has no kinetic energy. However, as we learned earlier, the gravitational potential energy of the object–Earth system is equal to mgh. If the object is dropped, it falls to the floor; as it falls, its speed and thus its kinetic energy increase, while the potential energy of the system decreases. If factors such as air resistance are ignored, whatever potential energy the system loses as the object moves downward appears as kinetic energy of the object. In other words, the sum of the kinetic and potential energies—the total mechanical energy E—remains constant. This is an example of the principle of **conservation**

---

[2] For a general displacement, the work done in two or three dimensions also equals $U_i - U_f$, where $U = U(x, y, z)$. We write this formally as $W = \int_i^f \mathbf{F} \cdot d\mathbf{s} = U_i - U_f$.

of mechanical energy. For the case of an object in free fall, this principle tells us that any increase (or decrease) in potential energy is accompanied by an equal decrease (or increase) in kinetic energy. Note that **the total mechanical energy of a system remains constant in any isolated system of objects that interact only through conservative forces.**

Because the total mechanical energy E of a system is defined as the sum of the kinetic and potential energies, we can write

$$E \equiv K + U \qquad\qquad (8.9)$$

| Total mechanical energy |

We can state the principle of conservation of energy as $E_i = E_f$, and so we have

$$K_i + U_i = K_f + U_f \qquad\qquad (8.10)$$

| The mechanical energy of an isolated system remains constant |

It is important to note that Equation 8.10 is valid only when no energy is added to or removed from the system. Furthermore, there must be no nonconservative forces doing work within the system.

Consider the carnival Ring-the-Bell event illustrated at the beginning of the chapter. The participant is trying to convert the initial kinetic energy of the hammer into gravitational potential energy associated with a weight that slides on a vertical track. If the hammer has sufficient kinetic energy, the weight is lifted high enough to reach the bell at the top of the track. To maximize the hammer's kinetic energy, the player must swing the heavy hammer as rapidly as possible. The fast-moving hammer does work on the pivoted target, which in turn does work on the weight. Of course, greasing the track (so as to minimize energy loss due to friction) would also help but is probably not allowed!

If more than one conservative force acts on an object within a system, a potential energy function is associated with each force. In such a case, we can apply the principle of conservation of mechanical energy for the system as

$$K_i + \sum U_i = K_f + \sum U_f \qquad\qquad (8.11)$$

where the number of terms in the sums equals the number of conservative forces present. For example, if an object connected to a spring oscillates vertically, two conservative forces act on the object: the spring force and the gravitational force.

QuickLab

Dangle a shoe from its lace and use it as a pendulum. Hold it to the side, release it, and note how high it swings at the end of its arc. How does this height compare with its initial height? You may want to check Question 8.3 as part of your investigation.

Twin Falls on the Island of Kauai, Hawaii. The gravitational potential energy of the water–Earth system when the water is at the top of the falls is converted to kinetic energy once that water begins falling. How did the water get to the top of the cliff? In other words, what was the original source of the gravitational potential energy when the water was at the top? (*Hint:* This same source powers nearly everything on the planet.) (*Bruce Byers/FPG*)

Figure 8.4 A ball connected to a massless spring suspended vertically. What forms of potential energy are associated with the ball–spring–Earth system when the ball is displaced downward?

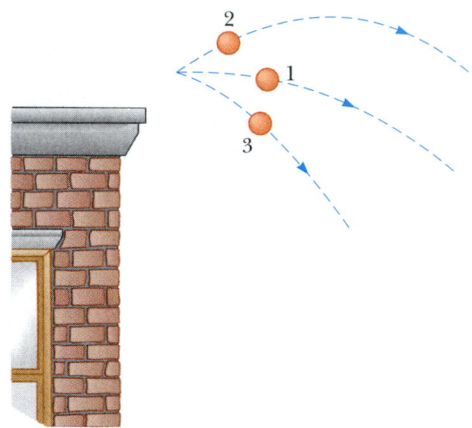

Quick Quiz 8.2

A ball is connected to a light spring suspended vertically, as shown in Figure 8.4. When displaced downward from its equilibrium position and released, the ball oscillates up and down. If air resistance is neglected, is the total mechanical energy of the system (ball plus spring plus Earth) conserved? How many forms of potential energy are there for this situation?

Quick Quiz 8.3

Three identical balls are thrown from the top of a building, all with the same initial speed. The first is thrown horizontally, the second at some angle above the horizontal, and the third at some angle below the horizontal, as shown in Figure 8.5. Neglecting air resistance, rank the speeds of the balls at the instant each hits the ground.

Figure 8.5 Three identical balls are thrown with the same initial speed from the top of a building.

EXAMPLE 8.2   Ball in Free Fall

A ball of mass m is dropped from a height h above the ground, as shown in Figure 8.6. (a) Neglecting air resistance, determine the speed of the ball when it is at a height y above the ground.

Solution   Because the ball is in free fall, the only force acting on it is the gravitational force. Therefore, we apply the principle of conservation of mechanical energy to the ball–Earth system. Initially, the system has potential energy but no kinetic energy. As the ball falls, the total mechanical energy remains constant and equal to the initial potential energy of the system.

At the instant the ball is released, its kinetic energy is $K_i = 0$ and the potential energy of the system is $U_i = mgh$. When the ball is at a distance y above the ground, its kinetic energy is $K_f = \frac{1}{2}mv_f^2$ and the potential energy relative to the ground is $U_f = mgy$. Applying Equation 8.10, we obtain

$$K_i + U_i = K_f + U_f$$

$$0 + mgh = \tfrac{1}{2}mv_f^2 + mgy$$

$$v_f^2 = 2g(h - y)$$

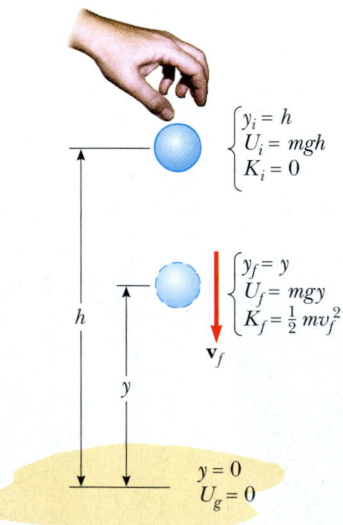

Figure 8.6 A ball is dropped from a height h above the ground. Initially, the total energy of the ball–Earth system is potential energy, equal to mgh relative to the ground. At the elevation y, the total energy is the sum of the kinetic and potential energies.

$$v_f = \sqrt{2g(h - y)}$$

The speed is always positive. If we had been asked to find the ball's velocity, we would use the negative value of the square root as the y component to indicate the downward motion.

(b) Determine the speed of the ball at y if at the instant of release it already has an initial speed v_i at the initial altitude h.

Solution   In this case, the initial energy includes kinetic energy equal to $\frac{1}{2}mv_i^2$, and Equation 8.10 gives

$$\tfrac{1}{2}mv_i^2 + mgh = \tfrac{1}{2}mv_f^2 + mgy$$

$$v_f^2 = v_i^2 + 2g(h - y)$$

$$v_f = \sqrt{v_i^2 + 2g(h - y)}$$

This result is consistent with the expression $v_{yf}^2 = v_{yi}^2 - 2g(y_f - y_i)$ from kinematics, where $y_i = h$. Furthermore, this result is valid even if the initial velocity is at an angle to the horizontal (the projectile situation) for two reasons: (1) energy is a scalar, and the kinetic energy depends only on the magnitude of the velocity; and (2) the change in the gravitational potential energy depends only on the change in position in the vertical direction.

EXAMPLE 8.3   The Pendulum

A pendulum consists of a sphere of mass m attached to a light cord of length L, as shown in Figure 8.7. The sphere is released from rest when the cord makes an angle θ_A with the vertical, and the pivot at P is frictionless. (a) Find the speed of the sphere when it is at the lowest point Ⓑ.

Solution   The only force that does work on the sphere is the gravitational force. (The force of tension is always perpendicular to each element of the displacement and hence does no work.) Because the gravitational force is conservative, the total mechanical energy of the pendulum–Earth system is constant. (In other words, we can classify this as an "energy conservation" problem.) As the pendulum swings, continuous transformation between potential and kinetic energy occurs. At the instant the pendulum is released, the energy of the system is entirely potential energy. At point Ⓑ the pendulum has kinetic energy, but the system has lost some potential energy. At Ⓒ the system has regained its initial potential energy, and the kinetic energy of the pendulum is again zero.

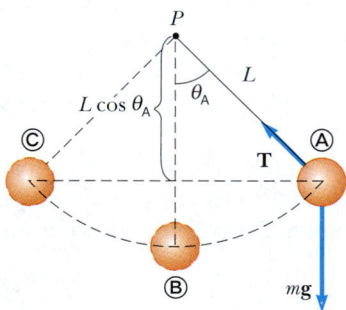

Figure 8.7 If the sphere is released from rest at the angle θ_A it will never swing above this position during its motion. At the start of the motion, position Ⓐ, the energy is entirely potential. This initial potential energy is all transformed into kinetic energy at the lowest elevation Ⓑ. As the sphere continues to move along the arc, the energy again becomes entirely potential energy at Ⓒ.

If we measure the y coordinates of the sphere from the center of rotation, then $y_A = -L\cos\theta_A$ and $y_B = -L$. Therefore, $U_A = -mgL\cos\theta_A$ and $U_B = -mgL$. Applying the principle of conservation of mechanical energy to the system gives

$$K_A + U_A = K_B + U_B$$

$$0 - mgL\cos\theta_A = \tfrac{1}{2}mv_B^2 - mgL$$

$$(1)\qquad v_B = \sqrt{2\,gL(1 - \cos\theta_A)}$$

(b) What is the tension T_B in the cord at Ⓑ?

Solution   Because the force of tension does no work, we cannot determine the tension using the energy method. To find T_B, we can apply Newton's second law to the radial direction. First, recall that the centripetal acceleration of a particle moving in a circle is equal to v^2/r directed toward the center of rotation. Because $r = L$ in this example, we obtain

$$(2)\qquad \sum F_r = T_B - mg = ma_r = m\frac{v_B^2}{L}$$

Substituting (1) into (2) gives the tension at point Ⓑ:

$$(3)\qquad T_B = mg + 2\,mg(1 - \cos\theta_A)$$

$$= mg(3 - 2\cos\theta_A)$$

From (2) we see that the tension at Ⓑ is greater than the weight of the sphere. Furthermore, (3) gives the expected result that $T_B = mg$ when the initial angle $\theta_A = 0$.

Exercise   A pendulum of length 2.00 m and mass 0.500 kg is released from rest when the cord makes an angle of 30.0° with the vertical. Find the speed of the sphere and the tension in the cord when the sphere is at its lowest point.

Answer   2.29 m/s; 6.21 N.

8.5   WORK DONE BY NONCONSERVATIVE FORCES

As we have seen, if the forces acting on objects within a system are conservative, then the mechanical energy of the system remains constant. However, if some of the forces acting on objects within the system are not conservative, then the mechanical energy of the system does not remain constant. Let us examine two types of nonconservative forces: an applied force and the force of kinetic friction.

Work Done by an Applied Force

When you lift a book through some distance by applying a force to it, the force you apply does work W_{app} on the book, while the gravitational force does work W_g on the book. If we treat the book as a particle, then the net work done on the book is related to the change in its kinetic energy as described by the work–kinetic energy theorem given by Equation 7.15:

$$W_{app} + W_g = \Delta K \tag{8.12}$$

Because the gravitational force is conservative, we can use Equation 8.2 to express the work done by the gravitational force in terms of the change in potential energy, or $W_g = -\Delta U$. Substituting this into Equation 8.12 gives

$$W_{app} = \Delta K + \Delta U \tag{8.13}$$

Note that the right side of this equation represents the change in the mechanical energy of the book–Earth system. This result indicates that your applied force transfers energy to the system in the form of kinetic energy of the book and gravitational potential energy of the book–Earth system. Thus, we conclude that if an object is part of a system, then **an applied force can transfer energy into or out of the system.**

Situations Involving Kinetic Friction

Kinetic friction is an example of a nonconservative force. If a book is given some initial velocity on a horizontal surface that is not frictionless, then the force of kinetic friction acting on the book opposes its motion and the book slows down and eventually stops. The force of kinetic friction reduces the kinetic energy of the book by transforming kinetic energy to internal energy of the book and part of the horizontal surface. Only part of the book's kinetic energy is transformed to internal energy in the book. The rest appears as internal energy in the surface. (When you trip and fall while running across a gymnasium floor, not only does the skin on your knees warm up but so does the floor!)

As the book moves through a distance d, the only force that does work is the force of kinetic friction. This force causes a decrease in the kinetic energy of the book. This decrease was calculated in Chapter 7, leading to Equation 7.17a, which we repeat here:

$$\Delta K_{friction} = -f_k d \tag{8.14}$$

If the book moves on an incline that is not frictionless, a change in the gravitational potential energy of the book–Earth system also occurs, and $-f_k d$ is the amount by which the mechanical energy of the system changes because of the force of kinetic friction. In such cases,

$$\Delta E = \Delta K + \Delta U = -f_k d \tag{8.15}$$

where $E_i + \Delta E = E_f$.

QuickLab

Find a friend and play a game of racquetball. After a long volley, feel the ball and note that it is warm. Why is that?

Quick Quiz 8.4

Write down the work–kinetic energy theorem for the general case of two objects that are connected by a spring and acted upon by gravity and some other external applied force. Include the effects of friction as $\Delta E_{\text{friction}}$.

Problem-Solving Hints

Conservation of Energy

We can solve many problems in physics using the principle of conservation of energy. You should incorporate the following procedure when you apply this principle:

- Define your system, which may include two or more interacting particles, as well as springs or other systems in which elastic potential energy can be stored. Choose the initial and final points.
- Identify zero points for potential energy (both gravitational and spring). If there is more than one conservative force, write an expression for the potential energy associated with each force.
- Determine whether any nonconservative forces are present. Remember that if friction or air resistance is present, mechanical energy *is not conserved*.
- If mechanical energy is *conserved*, you can write the total initial energy $E_i = K_i + U_i$ at some point. Then, write an expression for the total final energy $E_f = K_f + U_f$ at the final point that is of interest. Because mechanical energy is *conserved*, you can equate the two total energies and solve for the quantity that is unknown.
- If frictional forces are present (and thus mechanical energy is *not conserved*), first write expressions for the total initial and total final energies. In this case, the difference between the total final mechanical energy and the total initial mechanical energy equals the change in mechanical energy in the system due to friction.

EXAMPLE 8.4   Crate Sliding Down a Ramp

A 3.00-kg crate slides down a ramp. The ramp is 1.00 m in length and inclined at an angle of 30.0°, as shown in Figure 8.8. The crate starts from rest at the top, experiences a constant frictional force of magnitude 5.00 N, and continues to move a short distance on the flat floor after it leaves the ramp. Use energy methods to determine the speed of the crate at the bottom of the ramp.

Solution   Because $v_i = 0$, the initial kinetic energy at the top of the ramp is zero. If the y coordinate is measured from the bottom of the ramp (the final position where the potential energy is zero) with the upward direction being positive, then $y_i = 0.500$ m. Therefore, the total mechanical energy of the crate–Earth system at the top is all potential energy:

$$E_i = K_i + U_i = 0 + U_i = mgy_i$$
$$= (3.00 \text{ kg})(9.80 \text{ m/s}^2)(0.500 \text{ m}) = 14.7 \text{ J}$$

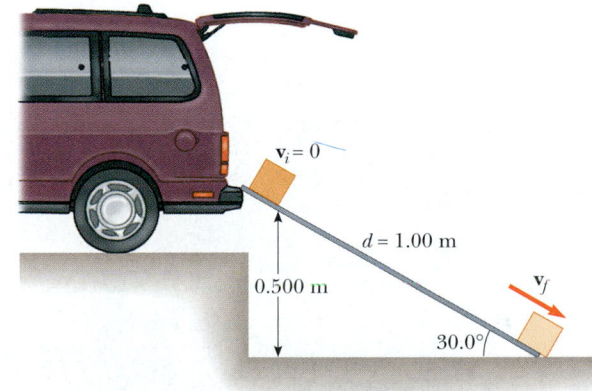

Figure 8.8   A crate slides down a ramp under the influence of gravity. The potential energy decreases while the kinetic energy increases.

When the crate reaches the bottom of the ramp, the potential energy of the system is *zero* because the elevation of the crate is $y_f = 0$. Therefore, the total mechanical energy of the system when the crate reaches the bottom is all kinetic energy:

$$E_f = K_f + U_f = \tfrac{1}{2}mv_f^2 + 0$$

We cannot say that $E_i = E_f$ because a nonconservative force reduces the mechanical energy of the system: the force of kinetic friction acting on the crate. In this case, Equation 8.15 gives $\Delta E = -f_k d$, where d is the displacement along the ramp. (Remember that the forces normal to the ramp do no work on the crate because they are perpendicular to the displacement.) With $f_k = 5.00$ N and $d = 1.00$ m, we have

$$\Delta E = -f_k d = -(5.00 \text{ N})(1.00 \text{ m}) = -5.00 \text{ J}$$

This result indicates that the system loses some mechanical energy because of the presence of the nonconservative frictional force. Applying Equation 8.15 gives

$$E_f - E_i = \tfrac{1}{2}mv_f^2 - mgy_i = -f_k d$$

$$\tfrac{1}{2}mv_f^2 = 14.7 \text{ J} - 5.00 \text{ J} = 9.70 \text{ J}$$

$$v_f^2 = \frac{19.4 \text{ J}}{3.00 \text{ kg}} = 6.47 \text{ m}^2/\text{s}^2$$

$$v_f = \boxed{2.54 \text{ m/s}}$$

Exercise Use Newton's second law to find the acceleration of the crate along the ramp, and use the equations of kinematics to determine the final speed of the crate.

Answer 3.23 m/s^2; 2.54 m/s.

Exercise Assuming the ramp to be frictionless, find the final speed of the crate and its acceleration along the ramp.

Answer 3.13 m/s; 4.90 m/s^2.

EXAMPLE 8.5   Motion on a Curved Track

A child of mass m rides on an irregularly curved slide of height $h = 2.00$ m, as shown in Figure 8.9. The child starts from rest at the top. (a) Determine his speed at the bottom, assuming no friction is present.

Solution The normal force **n** does no work on the child because this force is always perpendicular to each element of the displacement. Because there is no friction, the mechanical energy of the child–Earth system is conserved. If we measure the y coordinate in the upward direction from the bottom of the slide, then $y_i = h$, $y_f = 0$, and we obtain

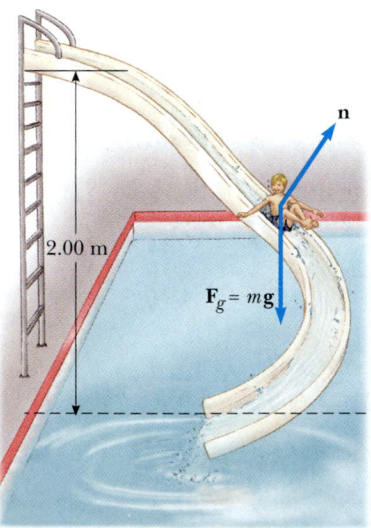

Figure 8.9 If the slide is frictionless, the speed of the child at the bottom depends only on the height of the slide.

$$K_i + U_i = K_f + U_f$$

$$0 + mgh = \tfrac{1}{2}mv_f^2 + 0$$

$$v_f = \sqrt{2gh}$$

Note that the result is the same as it would be had the child fallen vertically through a distance h! In this example, $h = 2.00$ m, giving

$$v_f = \sqrt{2gh} = \sqrt{2(9.80 \text{ m/s}^2)(2.00 \text{ m})} = \boxed{6.26 \text{ m/s}}$$

(b) If a force of kinetic friction acts on the child, how much mechanical energy does the system lose? Assume that $v_f = 3.00$ m/s and $m = 20.0$ kg.

Solution In this case, mechanical energy is *not* conserved, and so we must use Equation 8.15 to find the loss of mechanical energy due to friction:

$$\Delta E = E_f - E_i = (K_f + U_f) - (K_i + U_i)$$

$$= (\tfrac{1}{2}mv_f^2 + 0) - (0 + mgh) = \tfrac{1}{2}mv_f^2 - mgh$$

$$= \tfrac{1}{2}(20.0 \text{ kg})(3.00 \text{ m/s})^2 - (20.0 \text{ kg})(9.80 \text{ m/s}^2)(2.00 \text{ m})$$

$$= \boxed{-302 \text{ J}}$$

Again, ΔE is negative because friction is reducing mechanical energy of the system (the final mechanical energy is less than the initial mechanical energy). Because the slide is curved, the normal force changes in magnitude and direction during the motion. Therefore, the frictional force, which is proportional to n, also changes during the motion. Given this changing frictional force, do you think it is possible to determine μ_k from these data?

EXAMPLE 8.6 Let's Go Skiing!

A skier starts from rest at the top of a frictionless incline of height 20.0 m, as shown in Figure 8.10. At the bottom of the incline, she encounters a horizontal surface where the coefficient of kinetic friction between the skis and the snow is 0.210. How far does she travel on the horizontal surface before coming to rest?

Solution   First, let us calculate her speed at the bottom of the incline, which we choose as our zero point of potential energy. Because the incline is frictionless, the mechanical energy of the skier–Earth system remains constant, and we find, as we did in the previous example, that

$$v_B = \sqrt{2gh} = \sqrt{2(9.80 \text{ m/s}^2)(20.0 \text{ m})} = 19.8 \text{ m/s}$$

Now we apply Equation 8.15 as the skier moves along the rough horizontal surface from Ⓑ to Ⓒ. The change in mechanical energy along the horizontal is $\Delta E = -f_k d$, where d is the horizontal displacement.

To find the distance the skier travels before coming to rest, we take $K_C = 0$. With $v_B = 19.8$ m/s and the frictional force given by $f_k = \mu_k n = \mu_k mg$, we obtain

$$\Delta E = E_C - E_B = -\mu_k mgd$$

$$(K_C + U_C) - (K_B + U_B) = (0 + 0) - (\tfrac{1}{2}mv_B^2 + 0)$$

$$= -\mu_k mgd$$

$$d = \frac{v_B^2}{2\mu_k g} = \frac{(19.8 \text{ m/s})^2}{2(0.210)(9.80 \text{ m/s}^2)}$$

$$= \boxed{95.2 \text{ m}}$$

Exercise   Find the horizontal distance the skier travels before coming to rest if the incline also has a coefficient of kinetic friction equal to 0.210.

Answer   40.3 m.

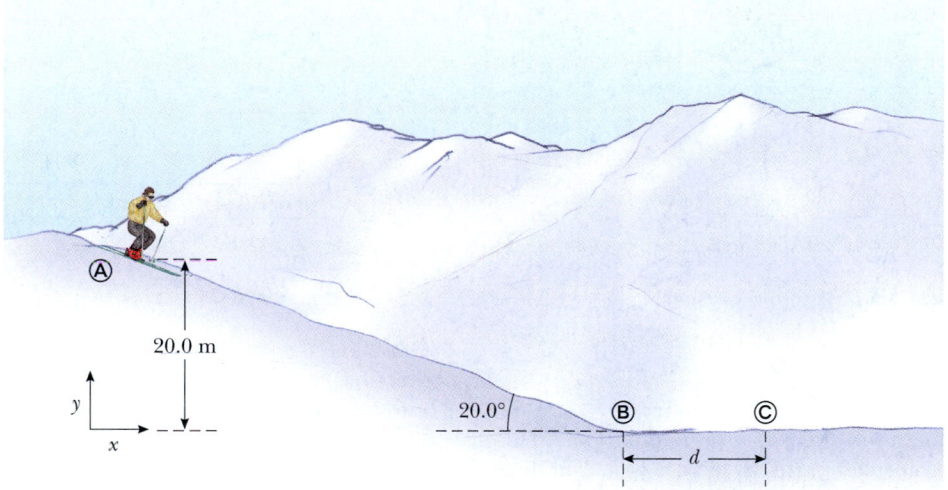

Figure 8.10   The skier slides down the slope and onto a level surface, stopping after a distance d from the bottom of the hill.

EXAMPLE 8.7 The Spring-Loaded Popgun

The launching mechanism of a toy gun consists of a spring of unknown spring constant (Fig. 8.11a). When the spring is compressed 0.120 m, the gun, when fired vertically, is able to launch a 35.0-g projectile to a maximum height of 20.0 m above the position of the projectile before firing. (a) Neglecting all resistive forces, determine the spring constant.

Solution   Because the projectile starts from rest, the initial kinetic energy is zero. If we take the zero point for the gravitational

tional potential energy of the projectile–Earth system to be at the lowest position of the projectile x_A, then the initial gravitational potential energy also is zero. The mechanical energy of this system is constant because no nonconservative forces are present.

Initially, the only mechanical energy in the system is the elastic potential energy stored in the spring of the gun, $U_{sA} = kx^2/2$, where the compression of the spring is $x = 0.120$ m. The projectile rises to a maximum height

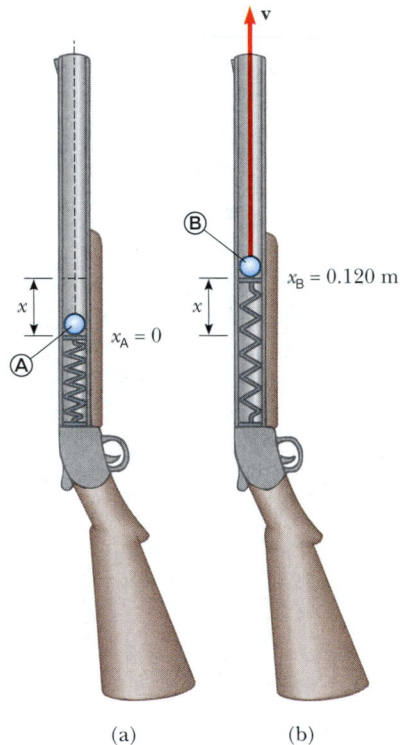

ⓒ ○ $x_C = 20.0$ m

v

Ⓑ

$x_B = 0.120$ m

x

x

$x_A = 0$

Ⓐ

(a)   (b)

Figure 8.11   A spring-loaded popgun.

$x_C = h = 20.0$ m, and so the final gravitational potential energy when the projectile reaches its peak is mgh. The final kinetic energy of the projectile is zero, and the final elastic potential energy stored in the spring is zero. Because the mechanical energy of the system is constant, we find that

$$E_A = E_C$$

$$K_A + U_{gA} + U_{sA} = K_C + U_{gC} + U_{sC}$$

$$0 + 0 + \tfrac{1}{2}kx^2 = 0 + mgh + 0$$

$$\tfrac{1}{2}k(0.120 \text{ m})^2 = (0.0350 \text{ kg})(9.80 \text{ m/s}^2)(20.0 \text{ m})$$

$$k = \boxed{953 \text{ N/m}}$$

(b) Find the speed of the projectile as it moves through the equilibrium position of the spring (where $x_B = 0.120$ m) as shown in Figure 8.11b.

Solution   As already noted, the only mechanical energy in the system at Ⓐ is the elastic potential energy $kx^2/2$. The total energy of the system as the projectile moves through the equilibrium position of the spring comprises the kinetic energy of the projectile $mv_B^2/2$, and the gravitational potential energy mgx_B. Hence, the principle of the conservation of mechanical energy in this case gives

$$E_A = E_B$$

$$K_A + U_{gA} + U_{sA} = K_B + U_{gB} + U_{sB}$$

$$0 + 0 + \tfrac{1}{2}kx^2 = \tfrac{1}{2}mv_B^2 + mgx_B + 0$$

Solving for v_B gives

$$v_B = \sqrt{\frac{kx^2}{m} - 2gx_B}$$

$$= \sqrt{\frac{(953 \text{ N/m})(0.120 \text{ m})^2}{0.0350 \text{ kg}} - 2(9.80 \text{ m/s}^2)(0.120 \text{ m})}$$

$$= \boxed{19.7 \text{ m/s}}$$

You should compare the different examples we have presented so far in this chapter. Note how breaking the problem into a sequence of labeled events helps in the analysis.

Exercise   What is the speed of the projectile when it is at a height of 10.0 m?

Answer   14.0 m/s.

EXAMPLE 8.8   Block–Spring Collision

A block having a mass of 0.80 kg is given an initial velocity $v_A = 1.2$ m/s to the right and collides with a spring of negligible mass and force constant $k = 50$ N/m, as shown in Figure 8.12. (a) Assuming the surface to be frictionless, calculate the maximum compression of the spring after the collision.

Solution   Our system in this example consists of the block and spring. Before the collision, at Ⓐ, the block has kinetic energy and the spring is uncompressed, so that the elastic potential energy stored in the spring is zero. Thus, the total mechanical energy of the system before the collision is just $\tfrac{1}{2}mv_A^2$. After the collision, at Ⓒ, the spring is fully compressed; now the block is at rest and so has zero kinetic energy, while the energy stored in the spring has its maximum value $\tfrac{1}{2}kx^2 = \tfrac{1}{2}kx_m^2$, where the origin of coordinates $x = 0$ is chosen to be the equilibrium position of the spring and x_m is

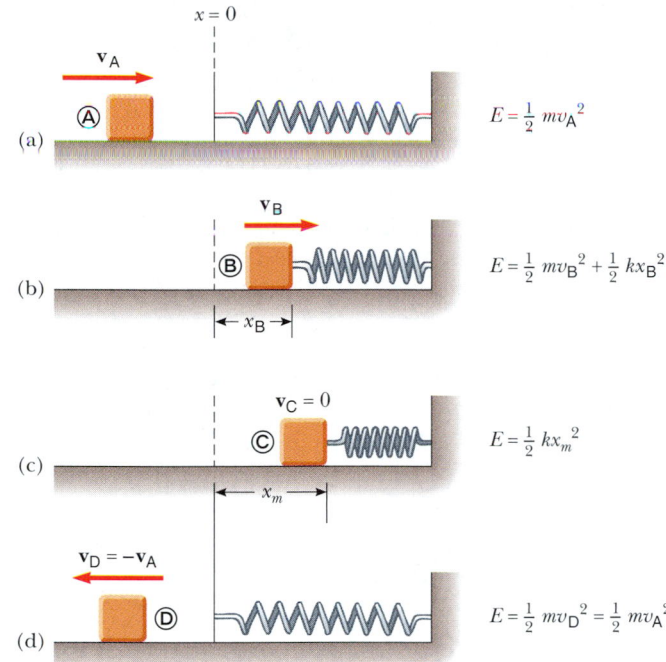

(a) $E = \frac{1}{2} mv_A{}^2$

(b) $E = \frac{1}{2} mv_B{}^2 + \frac{1}{2} kx_B{}^2$

(c) $E = \frac{1}{2} kx_m{}^2$

(d) $E = \frac{1}{2} mv_D{}^2 = \frac{1}{2} mv_A{}^2$

Figure 8.12   A block sliding on a smooth, horizontal surface collides with a light spring. (a) Initially the mechanical energy is all kinetic energy. (b) The mechanical energy is the sum of the kinetic energy of the block and the elastic potential energy in the spring. (c) The energy is entirely potential energy. (d) The energy is transformed back to the kinetic energy of the block. The total energy remains constant throughout the motion.

Multiflash photograph of a pole vault event. How many forms of energy can you identify in this picture? *(© Harold E. Edgerton/Courtesy of Palm Press, Inc.)*

the maximum compression of the spring, which in this case happens to be x_C. The total mechanical energy of the system is conserved because no nonconservative forces act on objects within the system.

Because mechanical energy is conserved, the kinetic energy of the block before the collision must equal the maximum potential energy stored in the fully compressed spring:

$$E_A = E_C$$

$$K_A + U_{sA} = K_C + U_{sC}$$

$$\tfrac{1}{2} mv_A{}^2 + 0 = 0 + \tfrac{1}{2} kx_m{}^2$$

$$x_m = \sqrt{\frac{m}{k}} \, v_A = \sqrt{\frac{0.80 \text{ kg}}{50 \text{ N/m}}} \, (1.2 \text{ m/s})$$

$$= \boxed{0.15 \text{ m}}$$

Note that we have not included U_g terms because no change in vertical position occurred.

(b) Suppose a constant force of kinetic friction acts between the block and the surface, with $\mu_k = 0.50$. If the speed

of the block at the moment it collides with the spring is $v_A = 1.2$ m/s, what is the maximum compression in the spring?

Solution   In this case, mechanical energy is *not* conserved because a frictional force acts on the block. The magnitude of the frictional force is

$$f_k = \mu_k n = \mu_k mg = 0.50(0.80 \text{ kg})(9.80 \text{ m/s}^2) = 3.92 \text{ N}$$

Therefore, the change in the block's mechanical energy due to friction as the block is displaced from the equilibrium position of the spring (where we have set our origin) to x_B is

$$\Delta E = -f_k x_B = -3.92 x_B$$

Substituting this into Equation 8.15 gives

$$\Delta E = E_f - E_i = (0 + \tfrac{1}{2} kx_B{}^2) - (\tfrac{1}{2} mv_A{}^2 + 0) = -f_k x_B$$

$$\tfrac{1}{2}(50) x_B{}^2 - \tfrac{1}{2}(0.80)(1.2)^2 = -3.92 x_B$$

$$25 x_B{}^2 + 3.92 x_B - 0.576 = 0$$

Solving the quadratic equation for x_B gives $x_B = 0.092$ m and $x_B = -0.25$ m. The   physically   meaningful   root   is $x_B =$

0.092 m.   The negative root does not apply to this situation

because the block must be to the right of the origin (positive value of x) when it comes to rest. Note that 0.092 m is less than the distance obtained in the frictionless case of part (a). This result is what we expect because friction retards the motion of the system.

EXAMPLE 8.9 ▶ Connected Blocks in Motion

Two blocks are connected by a light string that passes over a frictionless pulley, as shown in Figure 8.13. The block of mass m_1 lies on a horizontal surface and is connected to a spring of force constant k. The system is released from rest when the spring is unstretched. If the hanging block of mass m_2 falls a distance h before coming to rest, calculate the coefficient of kinetic friction between the block of mass m_1 and the surface.

Solution   The key word *rest* appears twice in the problem statement, telling us that the initial and final velocities and kinetic energies are zero. (Also note that because we are concerned only with the beginning and ending points of the motion, we do not need to label events with circled letters as we did in the previous two examples. Simply using i and f is sufficient to keep track of the situation.) In this situation, the system consists of the two blocks, the spring, and the Earth. We need to consider two forms of potential energy: gravitational and elastic. Because the initial and final kinetic energies of the system are zero, $\Delta K = 0$, and we can write

$$(1) \qquad \Delta E = \Delta U_g + \Delta U_s$$

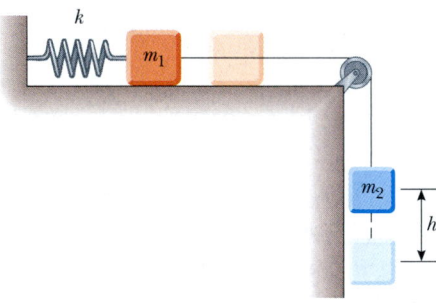

Figure 8.13   As the hanging block moves from its highest elevation to its lowest, the system loses gravitational potential energy but gains elastic potential energy in the spring. Some mechanical energy is lost because of friction between the sliding block and the surface.

where $\Delta U_g = U_{gf} - U_{gi}$ is the change in the system's gravitational potential energy and $\Delta U_s = U_{sf} - U_{si}$ is the change in the system's elastic potential energy. As the hanging block falls a distance h, the horizontally moving block moves the same distance h to the right. Therefore, using Equation 8.15, we find that the loss in energy due to friction between the horizontally sliding block and the surface is

$$(2) \qquad \Delta E = -f_k h = -\mu_k m_1 g h$$

The change in the gravitational potential energy of the system is associated with only the falling block because the vertical coordinate of the horizontally sliding block does not change. Therefore, we obtain

$$(3) \qquad \Delta U_g = U_{gf} - U_{gi} = 0 - m_2 g h$$

where the coordinates have been measured from the lowest position of the falling block.

The change in the elastic potential energy stored in the spring is

$$(4) \qquad \Delta U_s = U_{sf} - U_{si} = \tfrac{1}{2} k h^2 - 0$$

Substituting Equations (2), (3), and (4) into Equation (1) gives

$$-\mu_k m_1 g h = -m_2 g h + \tfrac{1}{2} k h^2$$

$$\mu_k = \frac{m_2 g - \tfrac{1}{2} k h}{m_1 g}$$

This setup represents a way of measuring the coefficient of kinetic friction between an object and some surface. As you can see from the problem, sometimes it is easier to work with the changes in the various types of energy rather than the actual values. For example, if we wanted to calculate the numerical value of the gravitational potential energy associated with the horizontally sliding block, we would need to specify the height of the horizontal surface relative to the lowest position of the falling block. Fortunately, this is not necessary because the gravitational potential energy associated with the first block does not change.

EXAMPLE 8.10 ▶ A Grand Entrance

You are designing apparatus to support an actor of mass 65 kg who is to "fly" down to the stage during the performance of a play. You decide to attach the actor's harness to a 130-kg sandbag by means of a lightweight steel cable running smoothly over two frictionless pulleys, as shown in Figure 8.14a. You need 3.0 m of cable between the harness and the nearest pulley so that the pulley can be hidden behind a curtain. For the apparatus to work successfully, the sandbag must never lift above the floor as the actor swings from above the

stage to the floor. Let us call the angle that the actor's cable makes with the vertical θ. What is the maximum value θ can have before the sandbag lifts off the floor?

Solution   We need to draw on several concepts to solve this problem. First, we use the principle of the conservation of mechanical energy to find the actor's speed as he hits the floor as a function of θ and the radius R of the circular path through which he swings. Next, we apply Newton's second law to the actor at the bottom of his path to find the cable tension as a function of the given parameters. Finally, we note that the sandbag lifts off the floor when the upward force exerted on it by the cable exceeds the gravitational force acting on it; the normal force is zero when this happens.

Applying conservation of energy to the actor–Earth system gives

$$K_i + U_i = K_f + U_f$$

$$(1) \qquad 0 + m_{actor} g y_i = \tfrac{1}{2} m_{actor} v_f^2 + 0$$

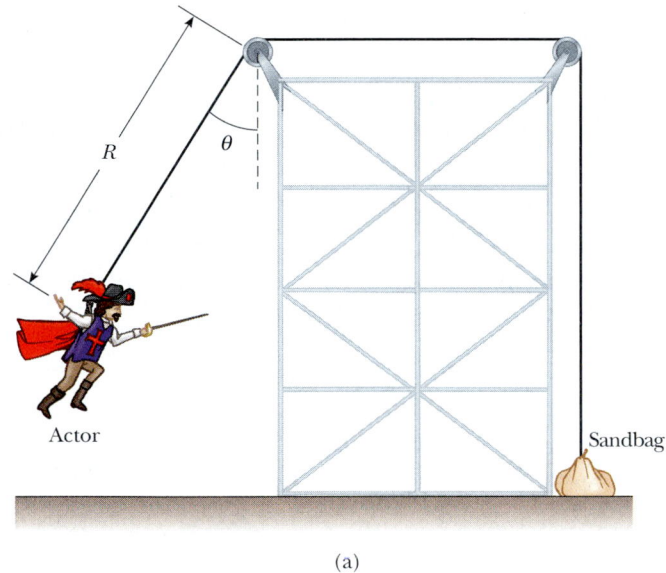

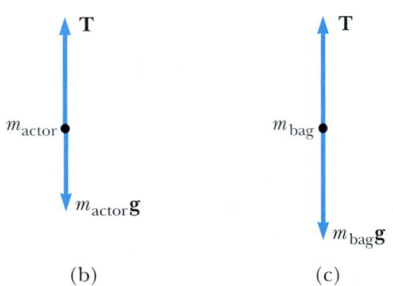

(b)   (c)

Figure 8.14   (a) An actor uses some clever staging to make his entrance. (b) Free-body diagram for actor at the bottom of the circular path. (c) Free-body diagram for sandbag.

where y_i is the initial height of the actor above the floor and v_f is the speed of the actor at the instant before he lands. (Note that $K_i = 0$ because he starts from rest and that $U_f = 0$ because we set the level of the actor's harness when he is standing on the floor as the zero level of potential energy.) From the geometry in Figure 8.14a, we see that $y_i = R - R \cos \theta = R(1 - \cos \theta)$. Using this relationship in Equation (1), we obtain

$$(2) \qquad v_f^2 = 2gR(1 - \cos \theta)$$

Now we apply Newton's second law to the actor when he is at the bottom of the circular path, using the free-body diagram in Figure 8.14b as a guide:

$$\sum F_y = T - m_{actor} g = m_{actor} \frac{v_f^2}{R}$$

$$(3) \qquad T = m_{actor} g + m_{actor} \frac{v_f^2}{R}$$

A force of the same magnitude as T is transmitted to the sandbag. If it is to be just lifted off the floor, the normal force on it becomes zero, and we require that $T = m_{bag} g$, as shown in Figure 8.14c. Using this condition together with Equations (2) and (3), we find that

$$m_{bag} g = m_{actor} g + m_{actor} \frac{2gR(1 - \cos \theta)}{R}$$

Solving for θ and substituting in the given parameters, we obtain

$$\cos \theta = \frac{3 m_{actor} - m_{bag}}{2 m_{actor}} = \frac{3(65 \text{ kg}) - 130 \text{ kg}}{2(65 \text{ kg})} = \frac{1}{2}$$

$$\theta = \boxed{60°}$$

Notice that we did not need to be concerned with the length R of the cable from the actor's harness to the leftmost pulley. The important point to be made from this problem is that it is sometimes necessary to combine energy considerations with Newton's laws of motion.

Exercise   If the initial angle $\theta = 40°$, find the speed of the actor and the tension in the cable just before he reaches the floor. (*Hint:* You cannot ignore the length $R = 3.0$ m in this calculation.)

Answer   3.7 m/s; 940 N.

8.6 ## RELATIONSHIP BETWEEN CONSERVATIVE FORCES AND POTENTIAL ENERGY

Once again let us consider a particle that is part of a system. Suppose that the particle moves along the x axis, and assume that a conservative force with an x component F_x acts on the particle. Earlier in this chapter, we showed how to determine the change in potential energy of a system when we are given the conservative force. We now show how to find F_x if the potential energy of the system is known.

In Section 8.2 we learned that the work done by the conservative force as its point of application undergoes a displacement Δx equals the negative of the change in the potential energy associated with that force; that is, $W = F_x \Delta x = -\Delta U$. If the point of application of the force undergoes an infinitesimal displacement dx, we can express the infinitesimal change in the potential energy of the system dU as

$$dU = -F_x \, dx$$

Therefore, the conservative force is related to the potential energy function through the relationship[3]

Relationship between force and potential energy

$$F_x = -\frac{dU}{dx} \qquad\qquad \textbf{(8.16)}$$

That is, **any conservative force acting on an object within a system equals the negative derivative of the potential energy of the system with respect to x.**

We can easily check this relationship for the two examples already discussed. In the case of the deformed spring, $U_s = \frac{1}{2}kx^2$, and therefore

$$F_s = -\frac{dU_s}{dx} = -\frac{d}{dx}(\tfrac{1}{2}kx^2) = -kx$$

which corresponds to the restoring force in the spring. Because the gravitational potential energy function is $U_g = mgy$, it follows from Equation 8.16 that $F_g = -mg$ when we differentiate U_g with respect to y instead of x.

We now see that U is an important function because a conservative force can be derived from it. Furthermore, Equation 8.16 should clarify the fact that adding a constant to the potential energy is unimportant because the derivative of a constant is zero.

Quick Quiz 8.5

What does the slope of a graph of $U(x)$ versus x represent?

Optional Section

8.7 ENERGY DIAGRAMS AND THE EQUILIBRIUM OF A SYSTEM

The motion of a system can often be understood qualitatively through a graph of its potential energy versus the separation distance between the objects in the system. Consider the potential energy function for a block–spring system, given by $U_s = \frac{1}{2}kx^2$. This function is plotted versus x in Figure 8.15a. (A common mistake is to think that potential energy on the graph represents height. This is clearly not

---

[3] In three dimensions, the expression is $\mathbf{F} = -\mathbf{i}\dfrac{\partial U}{\partial x} - \mathbf{j}\dfrac{\partial U}{\partial y} - \mathbf{k}\dfrac{\partial U}{\partial z}$, where $\dfrac{\partial U}{\partial x}$, and so forth, are partial derivatives. In the language of vector calculus, $\mathbf{F}$ equals the negative of the gradient of the scalar quantity $U(x, y, z)$.

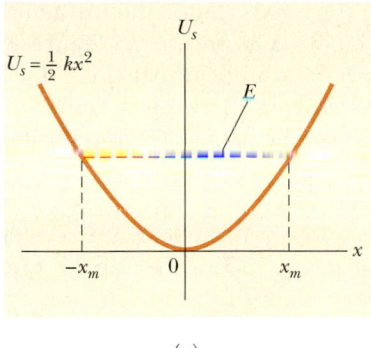

(a)

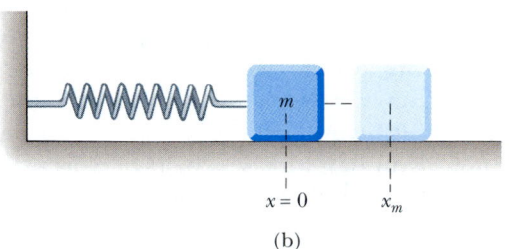

(b)

Figure 8.15 (a) Potential energy as a function of x for the block–spring system shown in (b). The block oscillates between the turning points, which have the coordinates $x = \pm x_m$. Note that the restoring force exerted by the spring always acts toward $x = 0$, the position of stable equilibrium.

the case here, where the block is only moving horizontally.) The force F_s exerted by the spring on the block is related to U_s through Equation 8.16:

$$F_s = -\frac{dU_s}{dx} = -kx$$

As we saw in Quick Quiz 8.5, the force is equal to the negative of the slope of the U versus x curve. When the block is placed at rest at the equilibrium position of the spring ($x = 0$), where $F_s = 0$, it will remain there unless some external force F_{ext} acts on it. If this external force stretches the spring from equilibrium, x is positive and the slope dU/dx is positive; therefore, the force F_s exerted by the spring is negative, and the block accelerates back toward $x = 0$ when released. If the external force compresses the spring, then x is negative and the slope is negative; therefore, F_s is positive, and again the mass accelerates toward $x = 0$ upon release.

From this analysis, we conclude that the $x = 0$ position for a block–spring system is one of **stable equilibrium.** That is, any movement away from this position results in a force directed back toward $x = 0$. In general, **positions of stable equilibrium correspond to points for which $U(x)$ is a minimum.**

From Figure 8.15 we see that if the block is given an initial displacement x_m and is released from rest, its total energy initially is the potential energy stored in the spring $\frac{1}{2}kx_m{}^2$. As the block starts to move, the system acquires kinetic energy and loses an equal amount of potential energy. Because the total energy must remain constant, the block oscillates (moves back and forth) between the two points $x = -x_m$ and $x = +x_m$, called the *turning points*. In fact, because no energy is lost (no friction), the block will oscillate between $-x_m$ and $+x_m$ forever. (We discuss these oscillations further in Chapter 13.) From an energy viewpoint, the energy of the system cannot exceed $\frac{1}{2}kx_m{}^2$; therefore, the block must stop at these points and, because of the spring force, must accelerate toward $x = 0$.

Another simple mechanical system that has a position of stable equilibrium is a ball rolling about in the bottom of a bowl. Anytime the ball is displaced from its lowest position, it tends to return to that position when released.

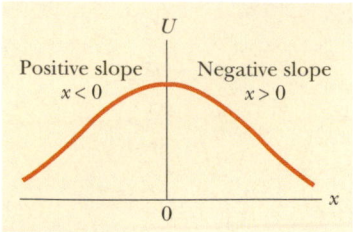

Figure 8.16 A plot of *U* versus *x* for a particle that has a position of unstable equilibrium located at *x* = 0. For any finite displacement of the particle, the force on the particle is directed away from *x* = 0.

Now consider a particle moving along the *x* axis under the influence of a conservative force F_x, where the *U* versus *x* curve is as shown in Figure 8.16. Once again, $F_x = 0$ at *x* = 0, and so the particle is in equilibrium at this point. However, this is a position of **unstable equilibrium** for the following reason: Suppose that the particle is displaced to the right (*x* > 0). Because the slope is negative for *x* > 0, $F_x = -dU/dx$ is positive and the particle accelerates *away from x* = 0. If instead the particle is at *x* = 0 and is displaced to the left (*x* < 0), the force is negative because the slope is positive for *x* < 0, and the particle again accelerates *away from* the equilibrium position. The position *x* = 0 in this situation is one of unstable equilibrium because for any displacement from this point, the force pushes the particle *farther away from* equilibrium. The force pushes the particle toward a position of lower potential energy. A pencil balanced on its point is in a position of unstable equilibrium. If the pencil is displaced slightly from its absolutely vertical position and is then released, it will surely fall over. In general, **positions of unstable equilibrium correspond to points for which U(x) is a maximum.**

Finally, a situation may arise where *U* is constant over some region and hence $F_x = 0$. This is called a position of **neutral equilibrium.** Small displacements from this position produce neither restoring nor disrupting forces. A ball lying on a flat horizontal surface is an example of an object in neutral equilibrium.

EXAMPLE 8.11   Force and Energy on an Atomic Scale

The potential energy associated with the force between two neutral atoms in a molecule can be modeled by the Lennard–Jones potential energy function:

$$U(x) = 4\epsilon\left[\left(\frac{\sigma}{x}\right)^{12} - \left(\frac{\sigma}{x}\right)^{6}\right]$$

where *x* is the separation of the atoms. The function *U(x)* contains two parameters σ and ϵ that are determined from experiments. Sample values for the interaction between two atoms in a molecule are $\sigma = 0.263$ nm and $\epsilon = 1.51 \times 10^{-22}$ J. (a) Using a spreadsheet or similar tool, graph this function and find the most likely distance between the two atoms.

Solution   We expect to find stable equilibrium when the two atoms are separated by some equilibrium distance and the potential energy of the system of two atoms (the molecule) is a minimum. One can minimize the function *U(x)* by taking its derivative and setting it equal to zero:

$$\frac{dU(x)}{dx} = 4\epsilon\frac{d}{dx}\left[\left(\frac{\sigma}{x}\right)^{12} - \left(\frac{\sigma}{x}\right)^{6}\right] = 0$$

$$= 4\epsilon\left[\frac{-12\sigma^{12}}{x^{13}} - \frac{-6\sigma^{6}}{x^{7}}\right] = 0$$

Solving for *x*—the equilibrium separation of the two atoms in the molecule—and inserting the given information yield

$$x = \boxed{2.95 \times 10^{-10}\text{ m.}}$$

We graph the Lennard–Jones function on both sides of this critical value to create our energy diagram, as shown in Figure 8.17a. Notice how *U(x)* is extremely large when the atoms are very close together, is a minimum when the atoms

are at their critical separation, and then increases again as the atoms move apart. When *U(x)* is a minimum, the atoms are in stable equilbrium; this indicates that this is the most likely separation between them.

(b) Determine $F_x(x)$—the force that one atom exerts on the other in the molecule as a function of separation—and argue that the way this force behaves is physically plausible when the atoms are close together and far apart.

Solution   Because the atoms combine to form a molecule, we reason that the force must be attractive when the atoms are far apart. On the other hand, the force must be repulsive when the two atoms get very close together. Otherwise, the molecule would collapse in on itself. Thus, the force must change sign at the critical separation, similar to the way spring forces switch sign in the change from extension to compression. Applying Equation 8.16 to the Lennard–Jones potential energy function gives

$$F_x = -\frac{dU(x)}{dx} = -4\epsilon\frac{d}{dx}\left[\left(\frac{\sigma}{x}\right)^{12} - \left(\frac{\sigma}{x}\right)^{6}\right]$$

$$= \boxed{4\epsilon\left[\frac{12\sigma^{12}}{x^{13}} - \frac{6\sigma^{6}}{x^{7}}\right]}$$

This result is graphed in Figure 8.17b. As expected, the force is positive (repulsive) at small atomic separations, zero when the atoms are at the position of stable equilibrium [recall how we found the minimum of *U(x)*], and negative (attractive) at greater separations. Note that the force approaches zero as the separation between the atoms becomes very great.

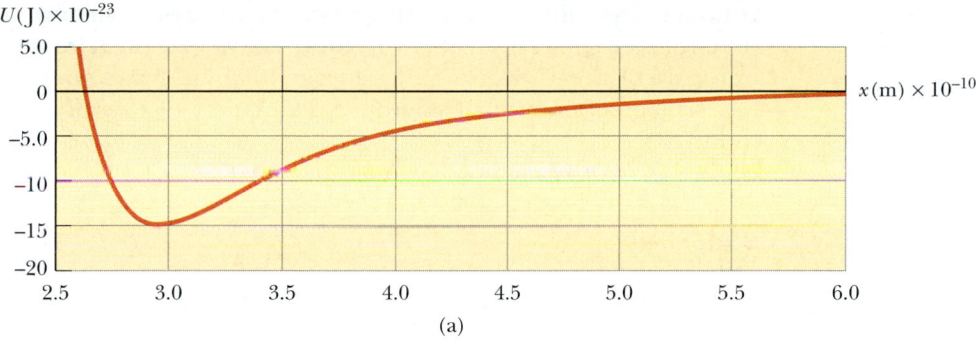

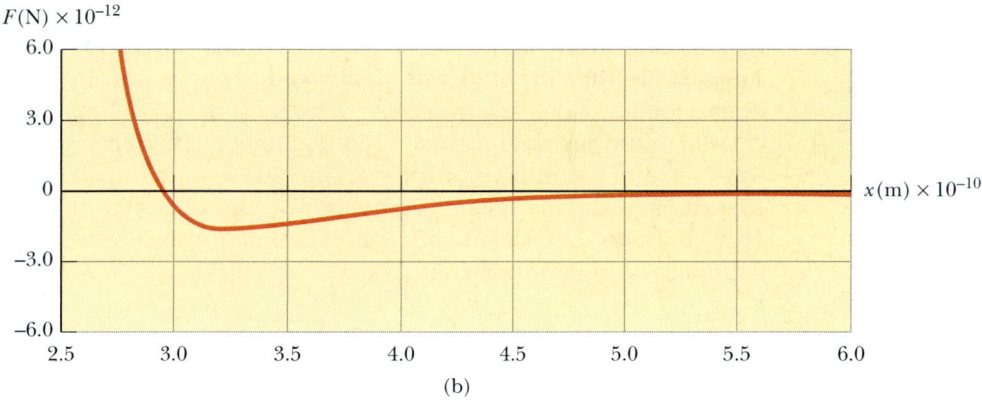

Figure 8.17 (a) Potential energy curve associated with a molecule. The distance x is the separation between the two atoms making up the molecule. (b) Force exerted on one atom by the other.

8.8 ▸ CONSERVATION OF ENERGY IN GENERAL

We have seen that the total mechanical energy of a system is constant when only conservative forces act within the system. Furthermore, we can associate a potential energy function with each conservative force. On the other hand, as we saw in Section 8.5, mechanical energy is lost when nonconservative forces such as friction are present.

In our study of thermodynamics later in this course, we shall find that mechanical energy can be transformed into energy stored *inside* the various objects that make up the system. This form of energy is called *internal energy*. For example, when a block slides over a rough surface, the mechanical energy lost because of friction is transformed into internal energy that is stored temporarily inside the block and inside the surface, as evidenced by a measurable increase in the temperature of both block and surface. We shall see that on a submicroscopic scale, this internal energy is associated with the vibration of atoms about their equilibrium positions. Such internal atomic motion involves both kinetic and potential energy. Therefore, if we include in our energy expression this increase in the internal energy of the objects that make up the system, then energy is conserved.

This is just one example of how you can analyze an isolated system and always find that the total amount of energy it contains does not change, as long as you account for all forms of energy. That is, **energy can never be created or destroyed. Energy may be transformed from one form to another, but the**

Total energy is always conserved

total energy of an isolated system is always constant. From a universal point of view, we can say that the **total energy of the Universe is constant.** If one part of the Universe gains energy in some form, then another part must lose an equal amount of energy. No violation of this principle has ever been found.

Optional Section

8.9 ▶ MASS–ENERGY EQUIVALENCE

This chapter has been concerned with the important principle of energy conservation and its application to various physical phenomena. Another important principle, **conservation of mass,** states that **in any physical or chemical process, mass is neither created nor destroyed.** That is, the mass before the process equals the mass after the process.

For centuries, scientists believed that energy and mass were two quantities that were separately conserved. However, in 1905 Einstein made the brilliant discovery that the mass of any system is a measure of the energy of that system. Hence, energy and mass are related concepts. The relationship between the two is given by Einstein's most famous formula:

$$E_R = mc^2 \qquad (8.17)$$

where c is the speed of light and E_R is the energy equivalent of a mass m. The subscript R on the energy refers to the **rest energy** of an object of mass m—that is, the energy of the object when its speed is $v = 0$.

The rest energy associated with even a small amount of matter is enormous. For example, the rest energy of 1 kg of any substance is

$$E_R = mc^2 = (1 \text{ kg})(3 \times 10^8 \text{ m/s})^2 = 9 \times 10^{16} \text{ J}$$

This is equivalent to the energy content of about 15 million barrels of crude oil—about one day's consumption in the United States! If this energy could easily be released as useful work, our energy resources would be unlimited.

In reality, only a small fraction of the energy contained in a material sample can be released through chemical or nuclear processes. The effects are greatest in nuclear reactions, in which fractional changes in energy, and hence mass, of approximately 10^{-3} are routinely observed. A good example is the enormous amount of energy released when the uranium-235 nucleus splits into two smaller nuclei. This happens because the sum of the masses of the product nuclei is slightly less than the mass of the original ^{235}U nucleus. The awesome nature of the energy released in such reactions is vividly demonstrated in the explosion of a nuclear weapon.

Equation 8.17 indicates that *energy has mass.* Whenever the energy of an object changes in any way, its mass changes as well. If ΔE is the change in energy of an object, then its change in mass is

$$\Delta m = \frac{\Delta E}{c^2} \qquad (8.18)$$

Anytime energy ΔE in any form is supplied to an object, the change in the mass of the object is $\Delta m = \Delta E/c^2$. However, because c^2 is so large, the changes in mass in any ordinary mechanical experiment or chemical reaction are too small to be detected.

EXAMPLE 8.12 ▶   Here Comes the Sun

The Sun converts an enormous amount of matter to energy. Each second, 4.19×10^9 kg—approximately the capacity of 400 average-sized cargo ships—is changed to energy. What is the power output of the Sun?

Solution We find the energy liberated per second by means of a straightforward conversion:

$$E_R = (4.19 \times 10^9 \text{ kg})(3.00 \times 10^8 \text{ m/s})^2 = 3.77 \times 10^{26} \text{ J}$$

We then apply the definition of power:

$$\mathscr{P} = \frac{3.77 \times 10^{26} \text{ J}}{1.00 \text{ s}} = \boxed{3.77 \times 10^{26} \text{ W}}$$

The Sun radiates uniformly in all directions, and so only a very tiny fraction of its total output is collected by the Earth. Nonetheless this amount is sufficient to supply energy to nearly everything on the Earth. (Nuclear and geothermal energy are the only alternatives.) Plants absorb solar energy and convert it to chemical potential energy (energy stored in the plant's molecules). When an animal eats the plant, this chemical potential energy can be turned into kinetic and other forms of energy. You are reading this book with solar-powered eyes!

Optional Section

8.10 ▶ QUANTIZATION OF ENERGY

Certain physical quantities such as electric charge are *quantized;* that is, the quantities have discrete values rather than continuous values. The quantized nature of energy is especially important in the atomic and subatomic world. As an example, let us consider the energy levels of the hydrogen atom (which consists of an electron orbiting around a proton). The atom can occupy only certain energy levels, called *quantum states,* as shown in Figure 8.18a. The atom cannot have any energy values lying between these quantum states. The lowest energy level E_1 is called the

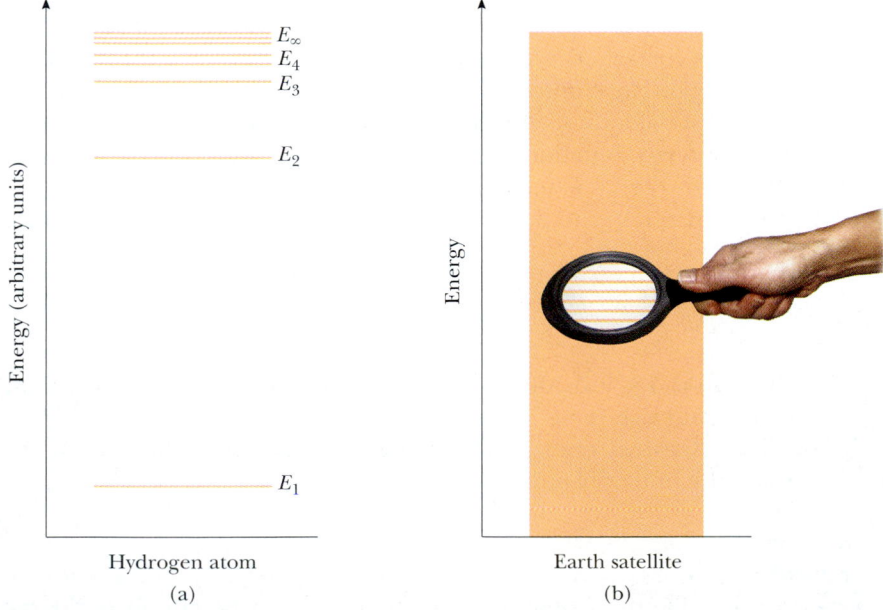

Figure 8.18 Energy-level diagrams: (a) Quantum states of the hydrogen atom. The lowest state E_1 is the ground state. (b) The energy levels of an Earth satellite are also quantized but are so close together that they cannot be distinguished from one another.

ground state of the atom. The ground state corresponds to the state that an isolated atom usually occupies. The atom can move to higher energy states by absorbing energy from some external source or by colliding with other atoms. The highest energy on the scale shown in Figure 8.18a, E_∞, corresponds to the energy of the atom when the electron is completely removed from the proton. The energy difference $E_\infty - E_1$ is called the **ionization energy.** Note that the energy levels get closer together at the high end of the scale.

Next, consider a satellite in orbit about the Earth. If you were asked to describe the possible energies that the satellite could have, it would be reasonable (but incorrect) to say that it could have any arbitrary energy value. Just like that of the hydrogen atom, however, **the energy of the satellite is quantized.** If you were to construct an energy level diagram for the satellite showing its allowed energies, the levels would be so close to one another, as shown in Figure 8.18b, that it would be difficult to discern that they were not continuous. In other words, we have no way of experiencing quantization of energy in the macroscopic world; hence, we can ignore it in describing everyday experiences.

SUMMARY

If a particle of mass m is at a distance y above the Earth's surface, the **gravitational potential energy** of the particle–Earth system is

$$U_g = mgy \tag{8.1}$$

The **elastic potential energy** stored in a spring of force constant k is

$$U_s \equiv \tfrac{1}{2}kx^2 \tag{8.4}$$

You should be able to apply these two equations in a variety of situations to determine the potential an object has to perform work.

A force is **conservative** if the work it does on a particle moving between two points is independent of the path the particle takes between the two points. Furthermore, a force is conservative if the work it does on a particle is zero when the particle moves through an arbitrary closed path and returns to its initial position. A force that does not meet these criteria is said to be **nonconservative.**

A **potential energy** function U can be associated only with a conservative force. If a conservative force $\mathbf{F}$ acts on a particle that moves along the x axis from x_i to x_f, then the change in the potential energy of the system equals the negative of the work done by that force:

$$U_f - U_i = -\int_{x_i}^{x_f} F_x \, dx \tag{8.7}$$

You should be able to use calculus to find the potential energy associated with a conservative force and vice versa.

The **total mechanical energy of a system** is defined as the sum of the kinetic energy and the potential energy:

$$E \equiv K + U \tag{8.9}$$

If no external forces do work on a system and if no nonconservative forces are acting on objects inside the system, then the total mechanical energy of the system is constant:

$$K_i + U_i = K_f + U_f \tag{8.10}$$

If nonconservative forces (such as friction) act on objects inside a system, then mechanical energy is not conserved. In these situations, the difference between the total final mechanical energy and the total initial mechanical energy of the system equals the energy transferred to or from the system by the nonconservative forces.

QUESTIONS

1. Many mountain roads are constructed so that they spiral around a mountain rather than go straight up the slope. Discuss this design from the viewpoint of energy and power.

2. A ball is thrown straight up into the air. At what position is its kinetic energy a maximum? At what position is the gravitational potential energy a maximum?

3. A bowling ball is suspended from the ceiling of a lecture hall by a strong cord. The bowling ball is drawn away from its equilibrium position and released from rest at the tip of the student's nose as in Figure Q8.3. If the student remains stationary, explain why she will not be struck by the ball on its return swing. Would the student be safe if she pushed the ball as she released it?

4. One person drops a ball from the top of a building, while another person at the bottom observes its motion. Will these two people agree on the value of the potential energy of the ball–Earth system? on its change in potential energy? on the kinetic energy of the ball?

5. When a person runs in a track event at constant velocity, is any work done? (*Note:* Although the runner moves with constant velocity, the legs and arms accelerate.) How does air resistance enter into the picture? Does the center of mass of the runner move horizontally?

6. Our body muscles exert forces when we lift, push, run, jump, and so forth. Are these forces conservative?

7. If three conservative forces and one nonconservative force act on a system, how many potential energy terms appear in the equation that describes this system?

8. Consider a ball fixed to one end of a rigid rod whose other end pivots on a horizontal axis so that the rod can rotate in a vertical plane. What are the positions of stable and unstable equilibrium?

9. Is it physically possible to have a situation where $E - U < 0$?

10. What would the curve of U versus x look like if a particle were in a region of neutral equilibrium?

11. Explain the energy transformations that occur during (a) the pole vault, (b) the shot put, (c) the high jump. What is the source of energy in each case?

12. Discuss some of the energy transformations that occur during the operation of an automobile.

13. If only one external force acts on a particle, does it necessarily change the particle's (a) kinetic energy? (b) velocity?

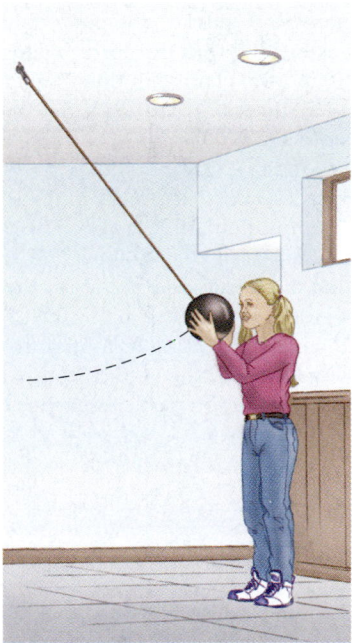

Figure Q8.3

PROBLEMS

1, 2, 3 = straightforward, intermediate, challenging ☐ = full solution available in the *Student Solutions Manual and Study Guide*
WEB = solution posted at **http://www.saunderscollege.com/physics/** 🖥 = Computer useful in solving problem 🔧 = Interactive Physics
☐ = paired numerical/symbolic problems

Section 8.1   Potential Energy

Section 8.2   Conservative and Nonconservative Forces

1. A 1 000-kg roller coaster is initially at the top of a rise, at point *A*. It then moves 135 ft, at an angle of 40.0° below the horizontal, to a lower point *B*. (a) Choose point *B* to be the zero level for gravitational potential energy. Find the potential energy of the roller coaster–Earth system at points *A* and *B* and the change in its potential energy as the coaster moves. (b) Repeat part (a), setting the zero reference level at point *A*.

2. A 40.0-N child is in a swing that is attached to ropes 2.00 m long. Find the gravitational potential energy of the child–Earth system relative to the child's lowest position when (a) the ropes are horizontal, (b) the ropes make a 30.0° angle with the vertical, and (c) the child is at the bottom of the circular arc.

3. A 4.00-kg particle moves from the origin to position C, which has coordinates $x = 5.00$ m and $y = 5.00$ m (Fig. P8.3). One force on it is the force of gravity acting in the negative y direction. Using Equation 7.2, calculate the work done by gravity as the particle moves from O to C along (a) OAC, (b) OBC, and (c) OC. Your results should all be identical. Why?

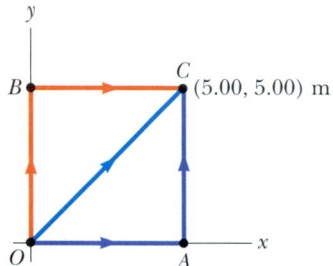

Figure P8.3   Problems 3, 4, and 5.

4. (a) Suppose that a constant force acts on an object. The force does not vary with time, nor with the position or velocity of the object. Start with the general definition for work done by a force

$$W = \int_i^f \mathbf{F} \cdot d\mathbf{s}$$

and show that the force is conservative. (b) As a special case, suppose that the force $\mathbf{F} = (3\mathbf{i} + 4\mathbf{j})$ N acts on a particle that moves from O to C in Figure P8.3. Calculate the work done by $\mathbf{F}$ if the particle moves along each one of the three paths OAC, OBC, and OC. (Your three answers should be identical.)

5. A force acting on a particle moving in the xy plane is given by $\mathbf{F} = (2\,y\mathbf{i} + x^2\,\mathbf{j})$ N, where x and y are in meters. The particle moves from the origin to a final position having coordinates $x = 5.00$ m and $y = 5.00$ m, as in Figure P8.3. Calculate the work done by $\mathbf{F}$ along (a) OAC, (b) OBC, (c) OC. (d) Is $\mathbf{F}$ conservative or nonconservative? Explain.

Section 8.3   Conservative Forces and Potential Energy
Section 8.4   Conservation of Mechanical Energy

6. At time t_i, the kinetic energy of a particle in a system is 30.0 J and the potential energy of the system is 10.0 J. At some later time t_f, the kinetic energy of the particle is 18.0 J. (a) If only conservative forces act on the particle, what are the potential energy and the total energy at

time t_f? (b) If the potential energy of the system at time t_f is 5.00 J, are any nonconservative forces acting on the particle? Explain.

7. WEB A single conservative force acts on a 5.00-kg particle. The equation $F_x = (2x + 4)$ N, where x is in meters, describes this force. As the particle moves along the x axis from $x = 1.00$ m to $x = 5.00$ m, calculate (a) the work done by this force, (b) the change in the potential energy of the system, and (c) the kinetic energy of the particle at $x = 5.00$ m if its speed at $x = 1.00$ m is 3.00 m/s.

8. A single constant force $\mathbf{F} = (3\mathbf{i} + 5\mathbf{j})$ N acts on a 4.00-kg particle. (a) Calculate the work done by this force if the particle moves from the origin to the point having the vector position $\mathbf{r} = (2\mathbf{i} - 3\mathbf{j})$ m. Does this result depend on the path? Explain. (b) What is the speed of the particle at $\mathbf{r}$ if its speed at the origin is 4.00 m/s? (c) What is the change in the potential energy of the system?

9. A single conservative force acting on a particle varies as $\mathbf{F} = (-Ax + Bx^2)\mathbf{i}$ N, where A and B are constants and x is in meters. (a) Calculate the potential energy function $U(x)$ associated with this force, taking $U = 0$ at $x = 0$. (b) Find the change in potential energy and change in kinetic energy as the particle moves from $x = 2.00$ m to $x = 3.00$ m.

10. A particle of mass 0.500 kg is shot from P as shown in Figure P8.10. The particle has an initial velocity $\mathbf{v}_i$ with a horizontal component of 30.0 m/s. The particle rises to a maximum height of 20.0 m above P. Using the law of conservation of energy, determine (a) the vertical component of $\mathbf{v}_i$, (b) the work done by the gravitational force on the particle during its motion from P to B, and (c) the horizontal and the vertical components of the velocity vector when the particle reaches B.

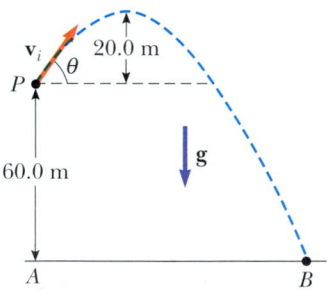

Figure P8.10

11. A 3.00-kg mass starts from rest and slides a distance d down a frictionless 30.0° incline. While sliding, it comes into contact with an unstressed spring of negligible mass, as shown in Figure P8.11. The mass slides an additional 0.200 m as it is brought momentarily to rest by compression of the spring ($k = 400$ N/m). Find the initial separation d between the mass and the spring.

12. A mass *m* starts from rest and slides a distance *d* down a frictionless incline of angle *θ*. While sliding, it contacts an unstressed spring of negligible mass, as shown in Figure P8.11. The mass slides an additional distance *x* as it is brought momentarily to rest by compression of the spring (of force constant *k*). Find the initial separation *d* between the mass and the spring.

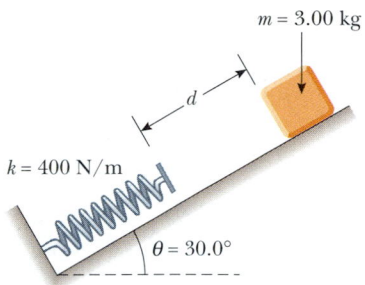

Figure P8.11 Problems 11 and 12.

13. A particle of mass $m = 5.00$ kg is released from point Ⓐ and slides on the frictionless track shown in Figure P8.13. Determine (a) the particle's speed at points Ⓑ and Ⓒ and (b) the net work done by the force of gravity in moving the particle from Ⓐ to Ⓒ.

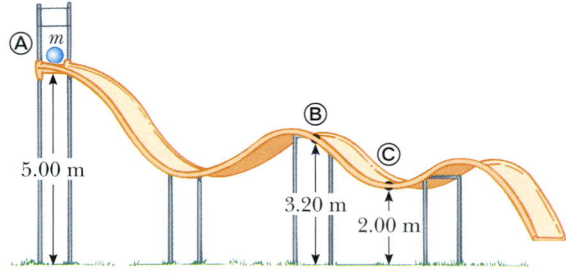

Figure P8.13

14. A simple, 2.00-m-long pendulum is released from rest when the support string is at an angle of 25.0° from the vertical. What is the speed of the suspended mass at the bottom of the swing?

15. A bead slides without friction around a loop-the-loop (Fig. P8.15). If the bead is released from a height $h = 3.50R$, what is its speed at point *A*? How great is the normal force on it if its mass is 5.00 g?

16. A 120-g mass is attached to the bottom end of an unstressed spring. The spring is hanging vertically and has a spring constant of 40.0 N/m. The mass is dropped. (a) What is its maximum speed? (b) How far does it drop before coming to rest momentarily?

17. A block of mass 0.250 kg is placed on top of a light verti-

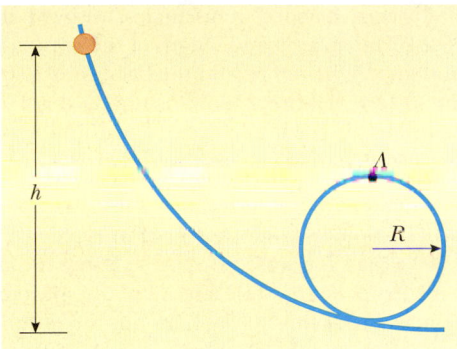

Figure P8.15

cal spring of constant $k = 5\,000$ N/m and is pushed downward so that the spring is compressed 0.100 m. After the block is released, it travels upward and then leaves the spring. To what maximum height above the point of release does it rise?

18. Dave Johnson, the bronze medalist at the 1992 Olympic decathlon in Barcelona, leaves the ground for his high jump with a vertical velocity component of 6.00 m/s. How far up does his center of gravity move as he makes the jump?

19. A 0.400-kg ball is thrown straight up into the air and reaches a maximum altitude of 20.0 m. Taking its initial position as the point of zero potential energy and using energy methods, find (a) its initial speed, (b) its total mechanical energy, and (c) the ratio of its kinetic energy to the potential energy of the ball–Earth system when the ball is at an altitude of 10.0 m.

20. In the dangerous "sport" of bungee-jumping, a daring student jumps from a balloon with a specially designed

Figure P8.20 Bungee-jumping. *(Gamma)*

elastic cord attached to his ankles, as shown in Figure P8.20. The unstretched length of the cord is 25.0 m, the student weighs 700 N, and the balloon is 36.0 m above the surface of a river below. Assuming that Hooke's law describes the cord, calculate the required force constant if the student is to stop safely 4.00 m above the river.

21. Two masses are connected by a light string passing over a light frictionless pulley, as shown in Figure P8.21. The 5.00-kg mass is released from rest. Using the law of conservation of energy, (a) determine the speed of the 3.00-kg mass just as the 5.00-kg mass hits the ground and (b) find the maximum height to which the 3.00-kg mass rises.

22. Two masses are connected by a light string passing over a light frictionless pulley, as shown in Figure P8.21. The mass m_1 (which is greater than m_2) is released from rest. Using the law of conservation of energy, (a) determine the speed of m_2 just as m_1 hits the ground in terms of m_1, m_2, and h, and (b) find the maximum height to which m_2 rises.

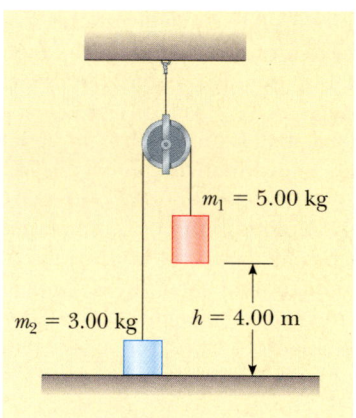

Figure P8.21   Problems 21 and 22.

23. A 20.0-kg cannon ball is fired from a cannon with a muzzle speed of 1 000 m/s at an angle of 37.0° with the horizontal. A second ball is fired at an angle of 90.0°. Use the law of conservation of mechanical energy to find (a) the maximum height reached by each ball and (b) the total mechanical energy at the maximum height for each ball. Let $y = 0$ at the cannon.

24. A 2.00-kg ball is attached to the bottom end of a length of 10-lb (44.5-N) fishing line. The top end of the fishing line is held stationary. The ball is released from rest while the line is taut and horizontal ($\theta = 90.0°$). At what angle θ (measured from the vertical) will the fishing line break?

25. The circus apparatus known as the *trapeze* consists of a bar suspended by two parallel ropes, each of length ℓ. The trapeze allows circus performers to swing in a verti-

cal circular arc (Fig. P8.25). Suppose a performer with mass m and holding the bar steps off an elevated platform, starting from rest with the ropes at an angle of θ_i with respect to the vertical. Suppose the size of the performer's body is small compared with the length ℓ, that she does not pump the trapeze to swing higher, and that air resistance is negligible. (a) Show that when the ropes make an angle of θ with respect to the vertical, the performer must exert a force

$$F = mg\,(3\cos\theta - 2\cos\theta_i)$$

in order to hang on. (b) Determine the angle θ_i at which the force required to hang on at the bottom of the swing is twice the performer's weight.

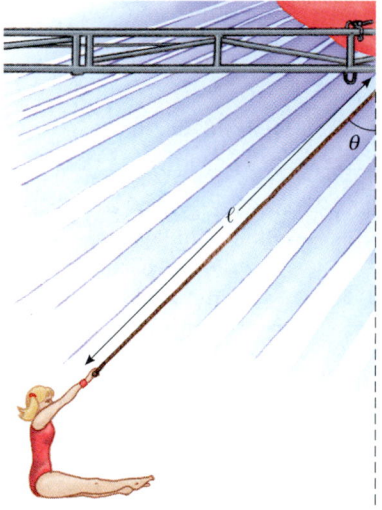

Figure P8.25

26. After its release at the top of the first rise, a roller-coaster car moves freely with negligible friction. The roller coaster shown in Figure P8.26 has a circular loop of radius 20.0 m. The car barely makes it around the loop: At the top of the loop, the riders are upside down and feel weightless. (a) Find the speed of the roller coaster car at the top of the loop (position 3). Find the speed of the roller coaster car (b) at position 1 and (c) at position 2. (d) Find the difference in height between positions 1 and 4 if the speed at position 4 is 10.0 m/s.

27. A light rigid rod is 77.0 cm long. Its top end is pivoted on a low-friction horizontal axle. The rod hangs straight down at rest, with a small massive ball attached to its bottom end. You strike the ball, suddenly giving it a horizontal velocity so that it swings around in a full circle. What minimum speed at the bottom is required to make the ball go over the top of the circle?

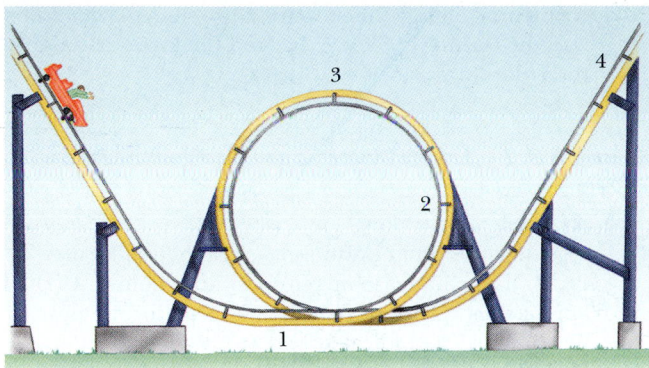

Figure P8.26

Section 8.5 Work Done by Nonconservative Forces

28. A 70.0-kg diver steps off a 10.0-m tower and drops straight down into the water. If he comes to rest 5.00 m beneath the surface of the water, determine the average resistance force that the water exerts on the diver.

29. A force F_x, shown as a function of distance in Figure P8.29, acts on a 5.00-kg mass. If the particle starts from rest at $x = 0$ m, determine the speed of the particle at $x = 2.00$, 4.00, and 6.00 m.

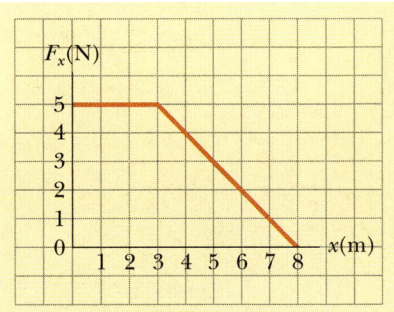

Figure P8.29

30. A softball pitcher swings a ball of mass 0.250 kg around a vertical circular path of radius 60.0 cm before releasing it from her hand. The pitcher maintains a component of force on the ball of constant magnitude 30.0 N in the direction of motion around the complete path. The speed of the ball at the top of the circle is 15.0 m/s. If the ball is released at the bottom of the circle, what is its speed upon release?

WEB 31. The coefficient of friction between the 3.00-kg block and the surface in Figure P8.31 is 0.400. The system starts from rest. What is the speed of the 5.00-kg ball when it has fallen 1.50 m?

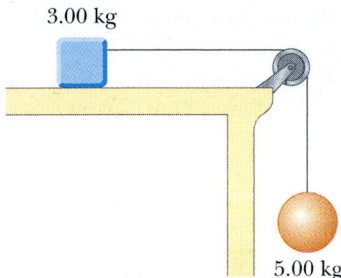

Figure P8.31

32. A 2 000-kg car starts from rest and coasts down from the top of a 5.00-m-long driveway that is sloped at an angle of 20.0° with the horizontal. If an average friction force of 4 000 N impedes the motion of the car, find the speed of the car at the bottom of the driveway.

33. A 5.00-kg block is set into motion up an inclined plane with an initial speed of 8.00 m/s (Fig. P8.33). The block comes to rest after traveling 3.00 m along the plane, which is inclined at an angle of 30.0° to the horizontal. For this motion determine (a) the change in the block's kinetic energy, (b) the change in the potential energy, and (c) the frictional force exerted on it (assumed to be constant). (d) What is the coefficient of kinetic friction?

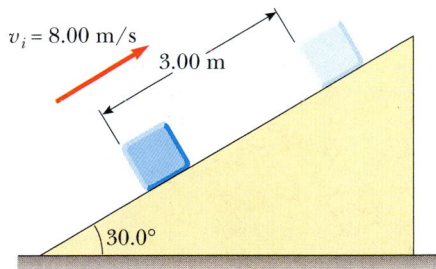

Figure P8.33

34. A boy in a wheelchair (total mass, 47.0 kg) wins a race with a skateboarder. He has a speed of 1.40 m/s at the crest of a slope 2.60 m high and 12.4 m long. At the bottom of the slope, his speed is 6.20 m/s. If air resistance and rolling resistance can be modeled as a constant frictional force of 41.0 N, find the work he did in pushing forward on his wheels during the downhill ride.

35. A parachutist of mass 50.0 kg jumps out of a balloon at a height of 1 000 m and lands on the ground with a speed of 5.00 m/s. How much energy was lost to air friction during this jump?

36. An 80.0-kg sky diver jumps out of a balloon at an altitude of 1 000 m and opens the parachute at an altitude of 200.0 m. (a) Assuming that the total retarding force

on the diver is constant at 50.0 N with the parachute closed and constant at 3 600 N with the parachute open, what is the speed of the diver when he lands on the ground? (b) Do you think the sky diver will get hurt? Explain. (c) At what height should the parachute be opened so that the final speed of the sky diver when he hits the ground is 5.00 m/s? (d) How realistic is the assumption that the total retarding force is constant? Explain.

37. A toy cannon uses a spring to project a 5.30-g soft rubber ball. The spring is originally compressed by 5.00 cm and has a stiffness constant of 8.00 N/m. When it is fired, the ball moves 15.0 cm through the barrel of the cannon, and there is a constant frictional force of 0.032 0 N between the barrel and the ball. (a) With what speed does the projectile leave the barrel of the cannon? (b) At what point does the ball have maximum speed? (c) What is this maximum speed?

38. A 1.50-kg mass is held 1.20 m above a relaxed, massless vertical spring with a spring constant of 320 N/m. The mass is dropped onto the spring. (a) How far does it compress the spring? (b) How far would it compress the spring if the same experiment were performed on the Moon, where $g = 1.63$ m/s^2? (c) Repeat part (a), but this time assume that a constant air-resistance force of 0.700 N acts on the mass during its motion.

39. A 3.00-kg block starts at a height $h = 60.0$ cm on a plane that has an inclination angle of 30.0°, as shown in Figure P8.39. Upon reaching the bottom, the block slides along a horizontal surface. If the coefficient of friction on both surfaces is $\mu_k = 0.200$, how far does the block slide on the horizontal surface before coming to rest? (*Hint:* Divide the path into two straight-line parts.)

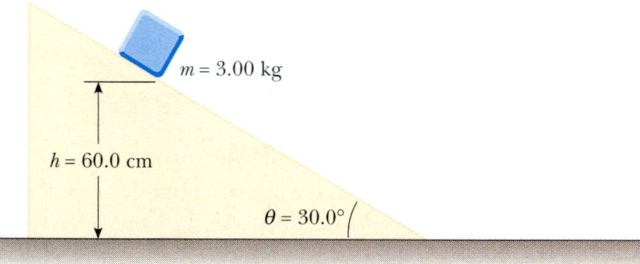

m = 3.00 kg

h = 60.0 cm

θ = 30.0°

Figure P8.39

40. A 75.0-kg sky diver is falling with a terminal speed of 60.0 m/s. Determine the rate at which he is losing mechanical energy.

Section 8.6   Relationship Between Conservative Forces and Potential Energy

WEB **41.** The potential energy of a two-particle system separated by a distance *r* is given by $U(r) = A/r$, where *A* is a constant. Find the radial force $\mathbf{F}_r$ that each particle exerts on the other.

42. A potential energy function for a two-dimensional force is of the form $U = 3x^3y - 7x$. Find the force that acts at the point (x, y).

(Optional)
Section 8.7   Energy Diagrams and the Equilibrium of a System

43. A particle moves along a line where the potential energy depends on its position *r*, as graphed in Figure P8.43. In the limit as *r* increases without bound, $U(r)$ approaches $+1$ J. (a) Identify each equilibrium position for this particle. Indicate whether each is a point of stable, unstable, or neutral equilibrium. (b) The particle will be bound if its total energy is in what range? Now suppose the particle has energy -3 J. Determine (c) the range of positions where it can be found, (d) its maximum kinetic energy, (e) the location at which it has maximum kinetic energy, and (f) its *binding energy*—that is, the additional energy that it would have to be given in order for it to move out to $r \rightarrow \infty$.

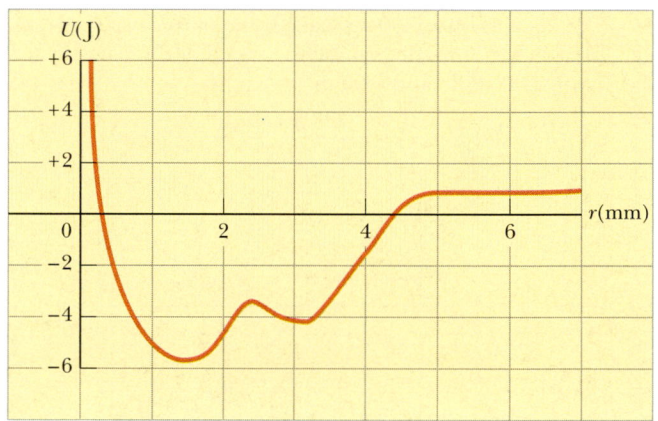

Figure P8.43

44. A right circular cone can be balanced on a horizontal surface in three different ways. Sketch these three equilibrium configurations and identify them as positions of stable, unstable, or neutral equilibrium.

45. For the potential energy curve shown in Figure P8.45, (a) determine whether the force F_x is positive, negative, or zero at the five points indicated. (b) Indicate points of stable, unstable, and neutral equilibrium. (c) Sketch the curve for F_x versus *x* from $x = 0$ to $x = 9.5$ m.

46. A hollow pipe has one or two weights attached to its inner surface, as shown in Figure P8.46. Characterize each configuration as being stable, unstable, or neutral equilibrium and explain each of your choices ("CM" indicates center of mass).

47. A particle of mass *m* is attached between two identical springs on a horizontal frictionless tabletop. The

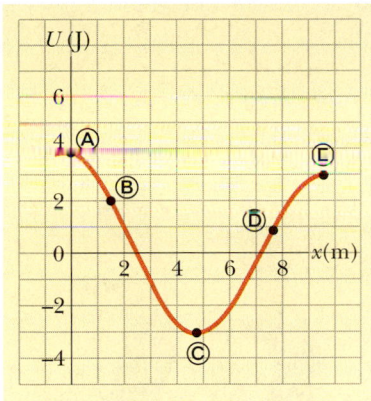

Figure P8.45

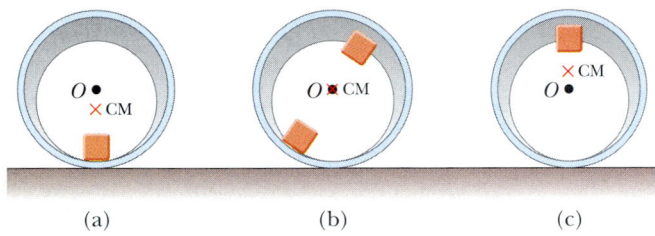

Figure P8.46

springs have spring constant k, and each is initially un-stressed. (a) If the mass is pulled a distance x along a direction perpendicular to the initial configuration of the springs, as in Figure P8.47, show that the potential energy of the system is

$$U(x) = kx^2 + 2kL(L - \sqrt{x^2 + L^2})$$

(*Hint:* See Problem 66 in Chapter 7.) (b) Make a plot of $U(x)$ versus x and identify all equilibrium points. Assume that $L = 1.20$ m and $k = 40.0$ N/m. (c) If the mass is pulled 0.500 m to the right and then released, what is its speed when it reaches the equilibrium point $x = 0$?

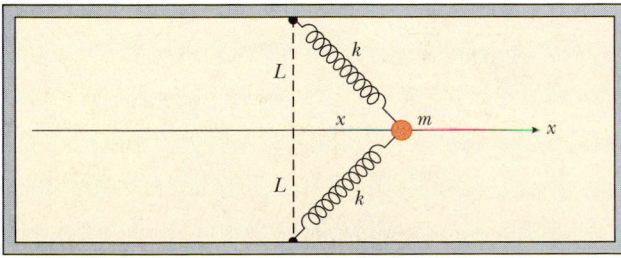

Top View

Figure P8.47

(Optional)
Section 8.9   Mass–Energy Equivalence

48. Find the energy equivalents of (a) an electron of mass 9.11×10^{-31} kg, (b) a uranium atom with a mass of 4.00×10^{-25} kg, (c) a paper clip of mass 2.00 g, and (d) the Earth (of mass 5.99×10^{24} kg).

49. The expression for the kinetic energy of a particle moving with speed v is given by Equation 7.19, which can be written as $K = \gamma mc^2 - mc^2$, where $\gamma = [1 - (v/c)^2]^{-1/2}$. The term γmc^2 is the total energy of the particle, and the term mc^2 is its rest energy. A proton moves with a speed of $0.990c$, where c is the speed of light. Find (a) its rest energy, (b) its total energy, and (c) its kinetic energy.

ADDITIONAL PROBLEMS

50. A block slides down a curved frictionless track and then up an inclined plane as in Figure P8.50. The coefficient of kinetic friction between the block and the incline is μ_k. Use energy methods to show that the maximum height reached by the block is

$$y_{max} = \frac{h}{1 + \mu_k \cot \theta}$$

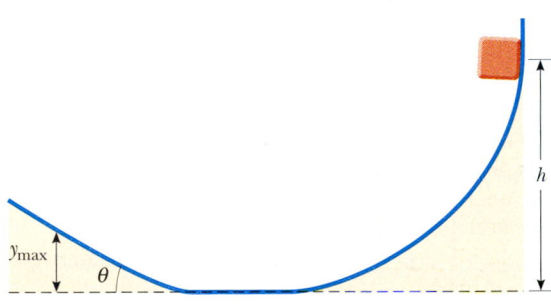

Figure P8.50

51. Close to the center of a campus is a tall silo topped with a hemispherical cap. The cap is frictionless when wet. Someone has somehow balanced a pumpkin at the highest point. The line from the center of curvature of the cap to the pumpkin makes an angle $\theta_i = 0°$ with the vertical. On a rainy night, a breath of wind makes the pumpkin start sliding downward from rest. It loses contact with the cap when the line from the center of the hemisphere to the pumpkin makes a certain angle with the vertical; what is this angle?

52. A 200-g particle is released from rest at point Ⓐ along the horizontal diameter on the inside of a frictionless, hemispherical bowl of radius $R = 30.0$ cm (Fig. P8.52). Calculate (a) the gravitational potential energy when the particle is at point Ⓐ relative to point Ⓑ, (b) the kinetic energy of the particle at point Ⓑ, (c) its speed at point Ⓑ, and (d) its kinetic energy and the potential energy at point Ⓒ.

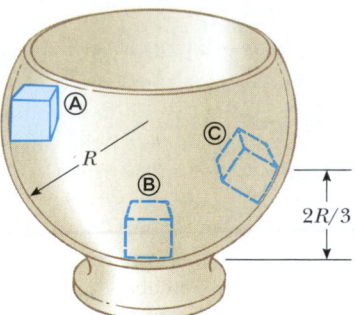

Figure P8.52   Problems 52 and 53.

Figure P8.56

WEB **53.** The particle described in Problem 52 (Fig. P8.52) is released from rest at Ⓐ, and the surface of the bowl is rough. The speed of the particle at Ⓑ is 1.50 m/s. (a) What is its kinetic energy at Ⓑ? (b) How much energy is lost owing to friction as the particle moves from Ⓐ to Ⓑ? (c) Is it possible to determine μ from these results in any simple manner? Explain.

54. **Review Problem.**   The mass of a car is 1 500 kg. The shape of the body is such that its aerodynamic drag coefficient is $D = 0.330$ and the frontal area is 2.50 m². Assuming that the drag force is proportional to v^2 and neglecting other sources of friction, calculate the power the car requires to maintain a speed of 100 km/h as it climbs a long hill sloping at 3.20°.

55. Make an order-of-magnitude estimate of your power output as you climb stairs. In your solution, state the physical quantities you take as data and the values you measure or estimate for them. Do you consider your peak power or your sustainable power?

56. A child's pogo stick (Fig. P8.56) stores energy in a spring ($k = 2.50 \times 10^4$ N/m). At position Ⓐ ($x_A = -0.100$ m), the spring compression is a maximum and the child is momentarily at rest. At position Ⓑ ($x_B = 0$), the spring is relaxed and the child is moving upward. At position Ⓒ, the child is again momentarily at rest at the top of the jump. Assuming that the combined mass of the child and the pogo stick is 25.0 kg, (a) calculate the total energy of the system if both potential energies are zero at $x = 0$, (b) determine x_C, (c) calculate the speed of the child at $x = 0$, (d) determine the value of x for which the kinetic energy of the system is a maximum, and (e) calculate the child's maximum upward speed.

57. A 10.0-kg block is released from point Ⓐ in Figure P8.57. The track is frictionless except for the portion between Ⓑ and Ⓒ, which has a length of 6.00 m. The block travels down the track, hits a spring of force constant $k = 2\ 250$ N/m, and compresses the spring 0.300 m from its equilibrium position before coming to rest momentarily. Determine the coefficient of kinetic friction between the block and the rough surface between Ⓑ and Ⓒ.

58. A 2.00-kg block situated on a rough incline is connected to a spring of negligible mass having a spring constant of 100 N/m (Fig. P8.58). The pulley is frictionless. The block is released from rest when the spring is unstretched. The block moves 20.0 cm down the incline before coming to rest. Find the coefficient of kinetic friction between block and incline.

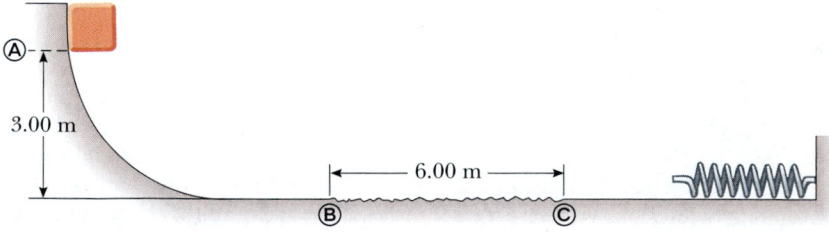

Figure P8.57

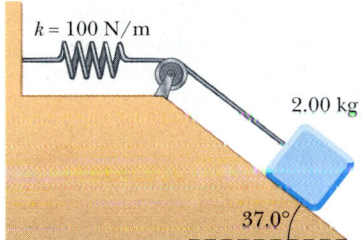

Figure P8.58   Problems 58 and 59.

59. **Review Problem.**   Suppose the incline is frictionless for the system described in Problem 58 (see Fig. P8.58). The block is released from rest with the spring initially unstretched. (a) How far does it move down the incline before coming to rest? (b) What is its acceleration at its lowest point? Is the acceleration constant? (c) Describe the energy transformations that occur during the descent.

60. The potential energy function for a system is given by $U(x) = -x^3 + 2x^2 + 3x$. (a) Determine the force F_x as a function of x. (b) For what values of x is the force equal to zero? (c) Plot $U(x)$ versus x and F_x versus x, and indicate points of stable and unstable equilibrium.

61. A 20.0-kg block is connected to a 30.0-kg block by a string that passes over a frictionless pulley. The 30.0-kg block is connected to a spring that has negligible mass and a force constant of 250 N/m, as shown in Figure P8.61. The spring is unstretched when the system is as shown in the figure, and the incline is frictionless. The 20.0-kg block is pulled 20.0 cm down the incline (so that the 30.0-kg block is 40.0 cm above the floor) and is released from rest. Find the speed of each block when the 30.0-kg block is 20.0 cm above the floor (that is, when the spring is unstretched).

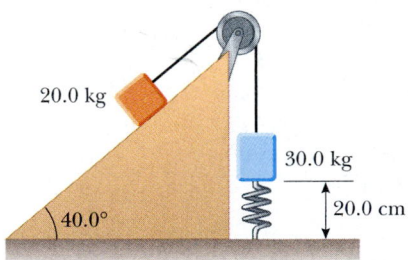

Figure P8.61

62. A 1.00-kg mass slides to the right on a surface having a coefficient of friction $\mu = 0.250$ (Fig. P8.62). The mass has a speed of $v_i = 3.00$ m/s when it makes contact with a light spring that has a spring constant $k = 50.0$ N/m. The mass comes to rest after the spring has been compressed a distance d. The mass is then forced toward the

left by the spring and continues to move in that direction beyond the spring's unstretched position. Finally, the mass comes to rest at a distance D to the left of the unstretched spring. Find (a) the distance of compression d, (b) the speed v of the mass at the unstretched position when the mass is moving to the left, and (c) the distance D between the unstretched spring and the point at which the mass comes to rest.

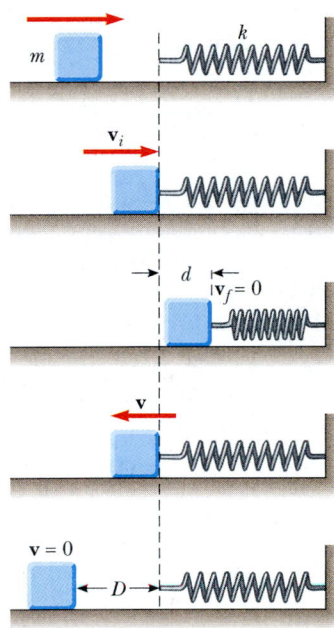

Figure P8.62

63. A block of mass 0.500 kg is pushed against a horizontal spring of negligible mass until the spring is compressed a distance Δx (Fig. P8.63). The spring constant is 450 N/m. When it is released, the block travels along a frictionless, horizontal surface to point B, at the bottom of a vertical circular track of radius $R = 1.00$ m, and continues to move up the track. The speed of the block at the bottom of the track is $v_B = 12.0$ m/s, and the block experiences an average frictional force of 7.00 N while sliding up the track. (a) What is Δx? (b) What speed do you predict for the block at the top of the track? (c) Does the block actually reach the top of the track, or does it fall off before reaching the top?

64. A uniform chain of length 8.00 m initially lies stretched out on a horizontal table. (a) If the coefficient of static friction between the chain and the table is 0.600, show that the chain will begin to slide off the table if at least 3.00 m of it hangs over the edge of the table. (b) Determine the speed of the chain as all of it leaves the table, given that the coefficient of kinetic friction between the chain and the table is 0.400.

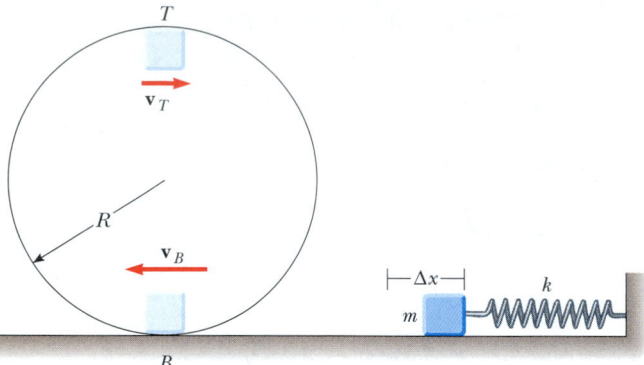

Figure P8.63

65. An object of mass m is suspended from a post on top of a cart by a string of length L as in Figure P8.65a. The cart and object are initially moving to the right at constant speed v_i. The cart comes to rest after colliding and sticking to a bumper as in Figure P8.65b, and the suspended object swings through an angle θ. (a) Show that the speed is $v_i = \sqrt{2gL(1 - \cos \theta)}$. (b) If $L = 1.20$ m and $\theta = 35.0°$, find the initial speed of the cart. (*Hint:* The force exerted by the string on the object does no work on the object.)

Figure P8.65

66. A child slides without friction from a height h along a curved water slide (Fig. P8.66). She is launched from a height $h/5$ into the pool. Determine her maximum airborne height y in terms of h and θ.

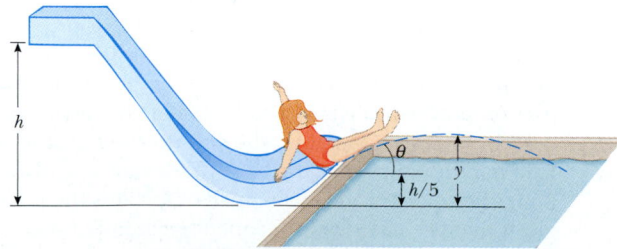

Figure P8.66

67. A ball having mass m is connected by a strong string of length L to a pivot point and held in place in a vertical position. A wind exerting constant force of magnitude F is blowing from left to right as in Figure P8.67a. (a) If the ball is released from rest, show that the maximum height H it reaches, as measured from its initial height, is

$$H = \frac{2L}{1 + (mg/F)^2}$$

Check that the above formula is valid both when $0 \le H \le L$ and when $L \le H \le 2L$. (*Hint:* First determine the potential energy associated with the *constant* wind force.) (b) Compute the value of H using the values $m = 2.00$ kg, $L = 2.00$ m, and $F = 14.7$ N. (c) Using these same values, determine the *equilibrium* height of the ball. (d) Could the equilibrium height ever be greater than L? Explain.

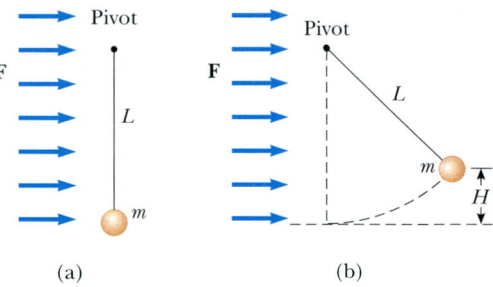

Figure P8.67

68. A ball is tied to one end of a string. The other end of the string is fixed. The ball is set in motion around a vertical circle without friction. At the top of the circle, the ball has a speed of $v_i = \sqrt{Rg}$, as shown in Figure P8.68. At what angle θ should the string be cut so that the ball will travel through the center of the circle?

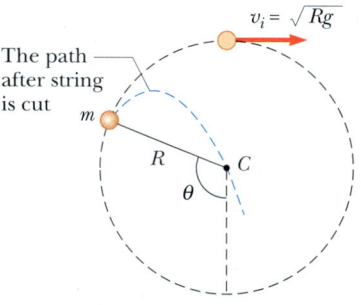

Figure P8.68

69. A ball at the end of a string whirls around in a vertical circle. If the ball's total energy remains constant, show that the tension in the string at the bottom is greater than the tension at the top by a value six times the weight of the ball.

70. A pendulum comprising a string of length L and a sphere swings in the vertical plane. The string hits a peg located a distance d below the point of suspension (Fig. P8.70). (a) Show that if the sphere is released from a height below that of the peg, it will return to this height after striking the peg. (b) Show that if the pendulum is released from the horizontal position ($\theta = 90°$) and is to swing in a complete circle centered on the peg, then the minimum value of d must be $3L/5$.

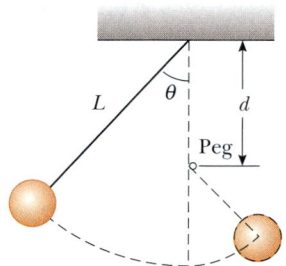

Figure P8.70

71. Jane, whose mass is 50.0 kg, needs to swing across a river (having width D) filled with man-eating crocodiles to save Tarzan from danger. However, she must swing into a wind exerting constant horizontal force **F** on a vine having length L and initially making an angle θ with the vertical (Fig. P8.71). Taking $D = 50.0$ m, $F = 110$ N, $L = 40.0$ m, and $\theta = 50.0°$, (a) with what minimum speed must Jane begin her swing to just make it to

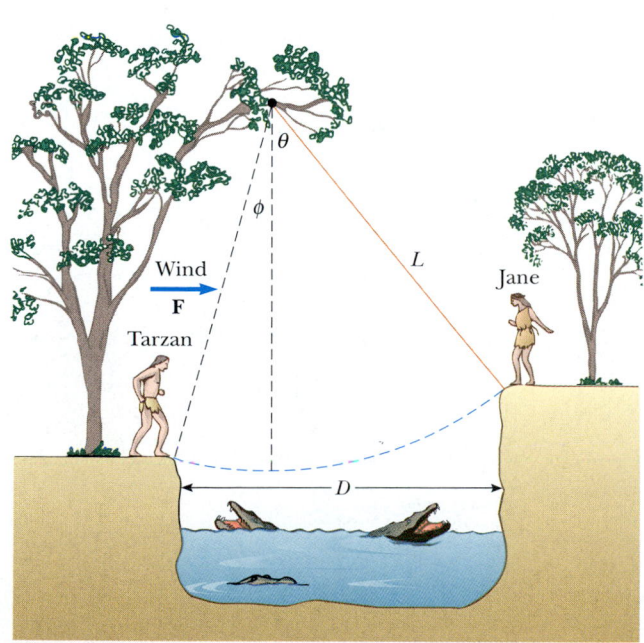

Figure P8.71

the other side? (*Hint:* First determine the potential energy associated with the wind force.) (b) Once the rescue is complete, Tarzan and Jane must swing back across the river. With what minimum speed must they begin their swing? Assume that Tarzan has a mass of 80.0 kg.

72. A child starts from rest and slides down the frictionless slide shown in Figure P8.72. In terms of R and H, at what height h will he lose contact with the section of radius R?

Figure P8.72

73. A 5.00-kg block free to move on a horizontal, frictionless surface is attached to one end of a light horizontal spring. The other end of the spring is fixed. The spring is compressed 0.100 m from equilibrium and is then released. The speed of the block is 1.20 m/s when it passes the equilibrium position of the spring. The same experiment is now repeated with the frictionless surface replaced by a surface for which $\mu_k = 0.300$. Determine the speed of the block at the equilibrium position of the spring.

74. A 50.0-kg block and a 100-kg block are connected by a string as in Figure P8.74. The pulley is frictionless and of negligible mass. The coefficient of kinetic friction between the 50.0-kg block and the incline is $\mu_k = 0.250$. Determine the change in the kinetic energy of the 50.0-kg block as it moves from Ⓐ to Ⓑ, a distance of 20.0 m.

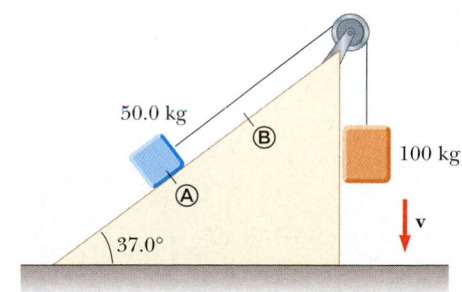

Figure P8.74

ANSWERS TO QUICK QUIZZES

8.1 Yes, because we are free to choose any point whatsoever as our origin of coordinates, which is the $U_g = 0$ point. If the object is below the origin of coordinates that we choose, then $U_g < 0$ for the object–Earth system.

8.2 Yes, the total mechanical energy of the system is conserved because the only forces acting are conservative: the force of gravity and the spring force. There are two forms of potential energy: (1) gravitational potential energy and (2) elastic potential energy stored in the spring.

8.3 The first and third balls speed up after they are thrown, while the second ball initially slows down but then speeds up after reaching its peak. The paths of all three balls are parabolas, and the balls take different times to reach the ground because they have different initial velocities. However, all three balls have the same speed at the moment they hit the ground because all start with the same kinetic energy and undergo the same change in gravitational potential energy. In other words, $E_{\text{total}} = \frac{1}{2}mv^2 + mgh$ is the same for all three balls at the start of the motion.

8.4 Designate one object as No. 1 and the other as No. 2. The external force does work W_{app} on the system. If $W_{\text{app}} > 0$, then the system energy increases. If $W_{\text{app}} < 0$, then the system energy decreases. The effect of friction is to decrease the total system energy. Equation 8.15 then becomes

$$\Delta E = W_{\text{app}} - \Delta E_{\text{friction}}$$
$$= \Delta K + \Delta U$$
$$= [K_{1f} + K_{2f}) - (K_{1i} + K_{2i})]$$
$$+ [(U_{g1f} + U_{g2f} + U_{sf}) - (U_{g1i} + U_{g2i} + U_{si})]$$

You may find it easier to think of this equation with its terms in a different order, saying

total initial energy + net change = total final energy

$$K_{1i} + K_{2i} + U_{g1i} + U_{g2i} + U_{si} + W_{\text{app}} - f_k d =$$
$$K_{1f} + K_{2f} + U_{g1f} + U_{g2f} + U_{sf}$$

8.5 The slope of a $U(x)$-versus-x graph is by definition $dU(x)/dx$. From Equation 8.16, we see that this expression is equal to the negative of the x component of the conservative force acting on an object that is part of the system.

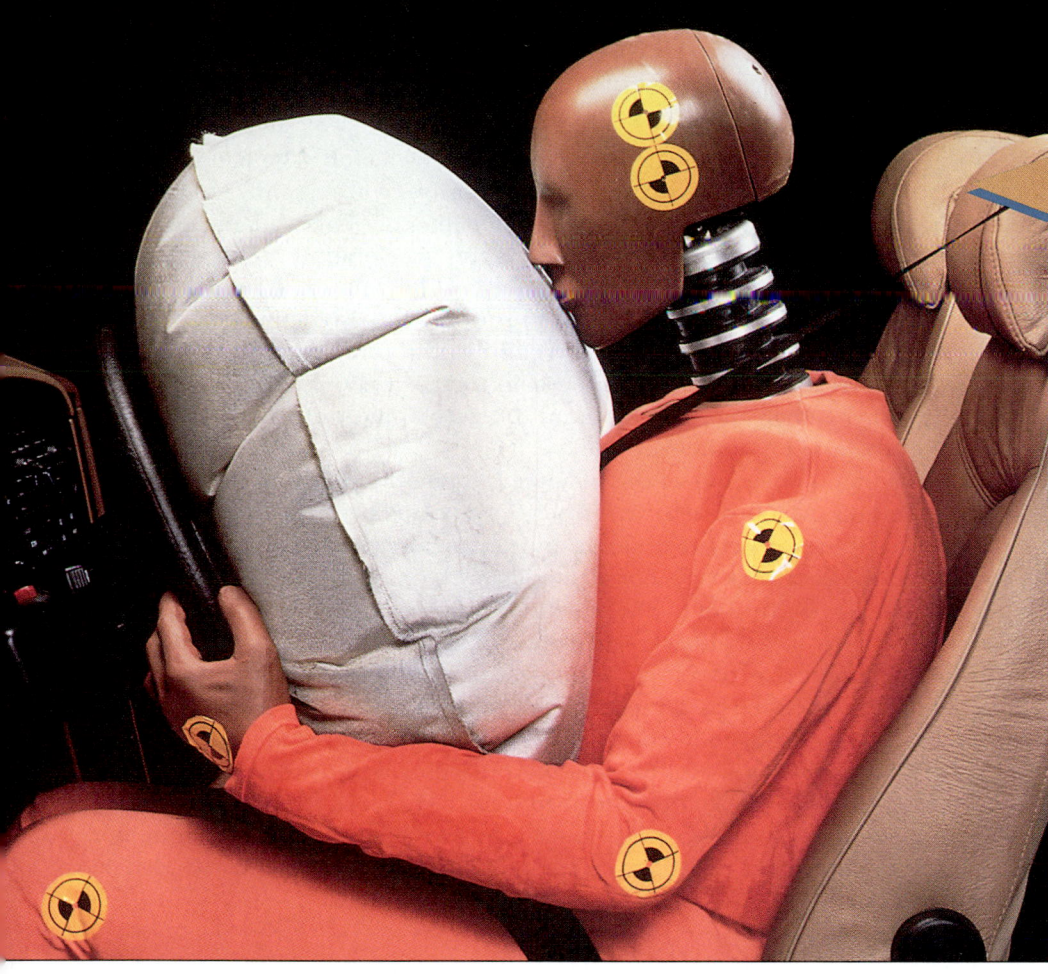

PUZZLER

Airbags have saved countless lives by reducing the forces exerted on vehicle occupants during collisions. How can airbags change the force needed to bring a person from a high speed to a complete stop? Why are they usually safer than seat belts alone? *(Courtesy of Saab)*

c h a p t e r

9

Linear Momentum and Collisions

Consider what happens when a golf ball is struck by a club. The ball is given a very large initial velocity as a result of the collision; consequently, it is able to travel more than 100 m through the air. The ball experiences a large acceleration. Furthermore, because the ball experiences this acceleration over a very short time interval, the average force exerted on it during the collision is very great. According to Newton's third law, the ball exerts on the club a reaction force that is equal in magnitude to and opposite in direction to the force exerted by the club on the ball. This reaction force causes the club to accelerate. Because the club is much more massive than the ball, however, the acceleration of the club is much less than the acceleration of the ball.

One of the main objectives of this chapter is to enable you to understand and analyze such events. As a first step, we introduce the concept of *momentum,* which is useful for describing objects in motion and as an alternate and more general means of applying Newton's laws. For example, a very massive football player is often said to have a great deal of momentum as he runs down the field. A much less massive player, such as a halfback, can have equal or greater momentum if his speed is greater than that of the more massive player. This follows from the fact that momentum is defined as the product of mass and velocity. The concept of momentum leads us to a second conservation law, that of conservation of momentum. This law is especially useful for treating problems that involve collisions between objects and for analyzing rocket propulsion. The concept of the center of mass of a system of particles also is introduced, and we shall see that the motion of a system of particles can be described by the motion of one representative particle located at the center of mass.

9.1 LINEAR MOMENTUM AND ITS CONSERVATION

In the preceding two chapters we studied situations too complex to analyze easily with Newton's laws. In fact, Newton himself used a form of his second law slightly different from $\Sigma \mathbf{F} = m\mathbf{a}$ (Eq. 5.2)—a form that is considerably easier to apply in complicated circumstances. Physicists use this form to study everything from subatomic particles to rocket propulsion. In studying situations such as these, it is often useful to know both something about the object and something about its motion. We start by defining a new term that incorporates this information:

> The **linear momentum** of a particle of mass m moving with a velocity $\mathbf{v}$ is defined to be the product of the mass and velocity:
>
> $$\mathbf{p} \equiv m\mathbf{v} \qquad (9.1)$$

Definition of linear momentum of a particle

Linear momentum is a vector quantity because it equals the product of a scalar quantity m and a vector quantity $\mathbf{v}$. Its direction is along $\mathbf{v}$, it has dimensions ML/T, and its SI unit is kg·m/s.

If a particle is moving in an arbitrary direction, $\mathbf{p}$ must have three components, and Equation 9.1 is equivalent to the component equations

$$p_x = mv_x \qquad p_y = mv_y \qquad p_z = mv_z \qquad (9.2)$$

As you can see from its definition, the concept of momentum provides a quantitative distinction between heavy and light particles moving at the same velocity. For example, the momentum of a bowling ball moving at 10 m/s is much greater than that of a tennis ball moving at the same speed. Newton called the product $m\mathbf{v}$

quantity of motion; this is perhaps a more graphic description than our present-day word *momentum,* which comes from the Latin word for movement.

Two objects have equal kinetic energies. How do the magnitudes of their momenta compare? (a) $p_1 < p_2$, (b) $p_1 = p_2$, (c) $p_1 > p_2$, (d) not enough information to tell.

Using Newton's second law of motion, we can relate the linear momentum of a particle to the resultant force acting on the particle: **The time rate of change of the linear momentum of a particle is equal to the net force acting on the particle:**

$$\sum \mathbf{F} = \frac{d\mathbf{p}}{dt} = \frac{d(m\mathbf{v})}{dt} \qquad (9.3)$$

Newton's second law for a particle

In addition to situations in which the velocity vector varies with time, we can use Equation 9.3 to study phenomena in which the mass changes. The real value of Equation 9.3 as a tool for analysis, however, stems from the fact that when the net force acting on a particle is zero, the time derivative of the momentum of the particle is zero, and therefore its linear momentum[1] is constant. Of course, if the particle is *isolated,* then by necessity $\Sigma \mathbf{F} = 0$ and $\mathbf{p}$ remains unchanged. This means that $\mathbf{p}$ is conserved. Just as the law of conservation of energy is useful in solving complex motion problems, the law of conservation of momentum can greatly simplify the analysis of other types of complicated motion.

Conservation of Momentum for a Two-Particle System

Consider two particles 1 and 2 that can interact with each other but are isolated from their surroundings (Fig. 9.1). That is, the particles may exert a force on each other, but no external forces are present. It is important to note the impact of Newton's third law on this analysis. If an internal force from particle 1 (for example, a gravitational force) acts on particle 2, then there must be a second internal force—equal in magnitude but opposite in direction—that particle 2 exerts on particle 1.

Suppose that at some instant, the momentum of particle 1 is $\mathbf{p}_1$ and that of particle 2 is $\mathbf{p}_2$. Applying Newton's second law to each particle, we can write

$$\mathbf{F}_{21} = \frac{d\mathbf{p}_1}{dt} \qquad \text{and} \qquad \mathbf{F}_{12} = \frac{d\mathbf{p}_2}{dt}$$

where $\mathbf{F}_{21}$ is the force exerted by particle 2 on particle 1 and $\mathbf{F}_{12}$ is the force exerted by particle 1 on particle 2. Newton's third law tells us that $\mathbf{F}_{12}$ and $\mathbf{F}_{21}$ are equal in magnitude and opposite in direction. That is, they form an action–reaction pair $\mathbf{F}_{12} = -\mathbf{F}_{21}$. We can express this condition as

$$\mathbf{F}_{21} + \mathbf{F}_{12} = 0$$

or as

$$\frac{d\mathbf{p}_1}{dt} + \frac{d\mathbf{p}_2}{dt} = \frac{d}{dt}(\mathbf{p}_1 + \mathbf{p}_2) = 0$$

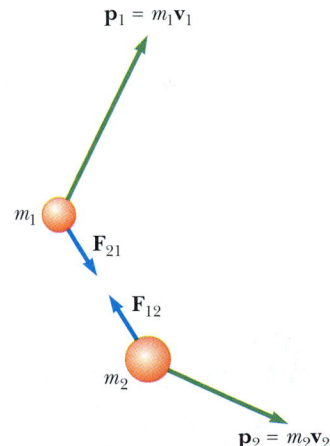

Figure 9.1 At some instant, the momentum of particle 1 is $\mathbf{p}_1 = m_1\mathbf{v}_1$ and the momentum of particle 2 is $\mathbf{p}_2 = m_2\mathbf{v}_2$. Note that $\mathbf{F}_{12} = -\mathbf{F}_{21}$. The total momentum of the system $\mathbf{p}_{\text{tot}}$ is equal to the vector sum $\mathbf{p}_1 + \mathbf{p}_2$.

[1] In this chapter, the terms *momentum* and *linear momentum* have the same meaning. Later, in Chapter 11, we shall use the term *angular momentum* when dealing with rotational motion.

Because the time derivative of the total momentum $\mathbf{p}_{tot} = \mathbf{p}_1 + \mathbf{p}_2$ is *zero*, we conclude that the *total* momentum of the system must remain constant:

$$\mathbf{p}_{tot} = \sum_{system} \mathbf{p} = \mathbf{p}_1 + \mathbf{p}_2 = \text{constant} \qquad (9.4)$$

or, equivalently,

$$\mathbf{p}_{1i} + \mathbf{p}_{2i} = \mathbf{p}_{1f} + \mathbf{p}_{2f} \qquad (9.5)$$

where $\mathbf{p}_{1i}$ and $\mathbf{p}_{2i}$ are the initial values and $\mathbf{p}_{1f}$ and $\mathbf{p}_{2f}$ the final values of the momentum during the time interval dt over which the reaction pair interacts. Equation 9.5 in component form demonstrates that the total momenta in the x, y, and z directions are all independently conserved:

$$\sum_{system} p_{ix} = \sum_{system} p_{fx} \qquad \sum_{system} p_{iy} = \sum_{system} p_{fy} \qquad \sum_{system} p_{iz} = \sum_{system} p_{fz} \qquad (9.6)$$

This result, known as the **law of conservation of linear momentum,** can be extended to any number of particles in an isolated system. It is considered one of the most important laws of mechanics. We can state it as follows:

Conservation of momentum

> Whenever two or more particles in an isolated system interact, the total momentum of the system remains constant.

This law tells us that **the total momentum of an isolated system at all times equals its initial momentum.**

Notice that we have made no statement concerning the nature of the forces acting on the particles of the system. The only requirement is that the forces must be *internal* to the system.

Quick Quiz 9.2

Your physical education teacher throws a baseball to you at a certain speed, and you catch it. The teacher is next going to throw you a medicine ball whose mass is ten times the mass of the baseball. You are given the following choices: You can have the medicine ball thrown with (a) the same speed as the baseball, (b) the same momentum, or (c) the same kinetic energy. Rank these choices from easiest to hardest to catch.

EXAMPLE 9.1   The Floating Astronaut

A SkyLab astronaut discovered that while concentrating on writing some notes, he had gradually floated to the middle of an open area in the spacecraft. Not wanting to wait until he floated to the opposite side, he asked his colleagues for a push. Laughing at his predicament, they decided not to help, and so he had to take off his uniform and throw it in one direction so that he would be propelled in the opposite direction. Estimate his resulting velocity.

Figure 9.2   A hapless astronaut has discarded his uniform to get somewhere.

Solution   We begin by making some reasonable guesses of relevant data. Let us assume we have a 70-kg astronaut who threw his 1-kg uniform at a speed of 20 m/s. For convenience, we set the positive direction of the x axis to be the direction of the throw (Fig. 9.2). Let us also assume that the x axis is tangent to the circular path of the spacecraft.

We take the system to consist of the astronaut and the uniform. Because of the gravitational force (which keeps the astronaut, his uniform, and the entire spacecraft in orbit), the system is not really isolated. However, this force is directed perpendicular to the motion of the system. Therefore, momentum is constant in the x direction because there are no external forces in this direction.

The total momentum of the system before the throw is zero ($m_1\mathbf{v}_{1i} + m_2\mathbf{v}_{2i} = 0$). Therefore, the total momentum after the throw must be zero; that is,

$$m_1\mathbf{v}_{1f} + m_2\mathbf{v}_{2f} = 0$$

With $m_1 = 70$ kg, $\mathbf{v}_{2f} = 20\mathbf{i}$ m/s, and $m_2 = 1$ kg, solving for $\mathbf{v}_{1f}$, we find the recoil velocity of the astronaut to be

$$\mathbf{v}_{1f} = -\frac{m_2}{m_1}\mathbf{v}_{2f} = -\left(\frac{1\text{ kg}}{70\text{ kg}}\right)(20\mathbf{i}\text{ m/s}) = \boxed{-0.3\mathbf{i}\text{ m/s}}$$

The negative sign for $\mathbf{v}_{1f}$ indicates that the astronaut is moving to the left after the throw, in the direction opposite the direction of motion of the uniform, in accordance with Newton's third law. Because the astronaut is much more massive than his uniform, his acceleration and consequent velocity are much smaller than the acceleration and velocity of the uniform.

EXAMPLE 9.2 ▶ Breakup of a Kaon at Rest

One type of nuclear particle, called the *neutral kaon* (K^0), breaks up into a pair of other particles called *pions* (π^+ and π^-) that are oppositely charged but equal in mass, as illustrated in Figure 9.3. Assuming the kaon is initially at rest, prove that the two pions must have momenta that are equal in magnitude and opposite in direction.

Solution   The breakup of the kaon can be written

$$K^0 \longrightarrow \pi^+ + \pi^-$$

If we let $\mathbf{p}^+$ be the momentum of the positive pion and $\mathbf{p}^-$ the momentum of the negative pion, the final momentum of the system consisting of the two pions can be written

$$\mathbf{p}_f = \mathbf{p}^+ + \mathbf{p}^-$$

Because the kaon is at rest before the breakup, we know that $\mathbf{p}_i = 0$. Because momentum is conserved, $\mathbf{p}_i = \mathbf{p}_f = 0$, so that $\mathbf{p}^+ + \mathbf{p}^- = 0$, or

$$\mathbf{p}^+ = -\mathbf{p}^-$$

The important point behind this problem is that even though it deals with objects that are very different from those in the preceding example, the physics is identical: Linear momentum is conserved in an isolated system.

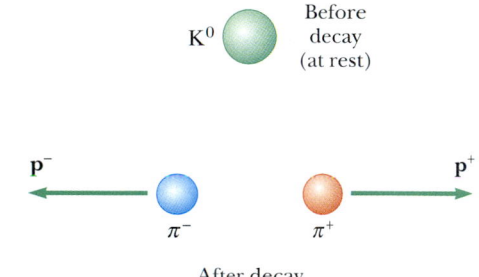

Figure 9.3   A kaon at rest breaks up spontaneously into a pair of oppositely charged pions. The pions move apart with momenta that are equal in magnitude but opposite in direction.

9.2 ▶ IMPULSE AND MOMENTUM

As we have seen, the momentum of a particle changes if a net force acts on the particle. Knowing the change in momentum caused by a force is useful in solving some types of problems. To begin building a better understanding of this important concept, let us assume that a single force $\mathbf{F}$ acts on a particle and that this force may vary with time. According to Newton's second law, $\mathbf{F} = d\mathbf{p}/dt$, or

$$d\mathbf{p} = \mathbf{F}\,dt \tag{9.7}$$

We can integrate[2] this expression to find the change in the momentum of a particle when the force acts over some time interval. If the momentum of the particle

---

[2]Note that here we are integrating force with respect to time. Compare this with our efforts in Chapter 7, where we integrated force with respect to position to express the work done by the force.

changes from $\mathbf{p}_i$ at time t_i to $\mathbf{p}_f$ at time t_f, integrating Equation 9.7 gives

$$\Delta \mathbf{p} = \mathbf{p}_f - \mathbf{p}_i = \int_{t_i}^{t_f} \mathbf{F} \, dt \tag{9.8}$$

To evaluate the integral, we need to know how the force varies with time. The quantity on the right side of this equation is called the **impulse** of the force $\mathbf{F}$ acting on a particle over the time interval $\Delta t = t_f - t_i$. Impulse is a vector defined by

Impulse of a force

$$\mathbf{I} \equiv \int_{t_i}^{t_f} \mathbf{F} \, dt = \Delta \mathbf{p} \tag{9.9}$$

Impulse–momentum theorem

The impulse of the force $\mathbf{F}$ acting on a particle equals the change in the momentum of the particle caused by that force.

This statement, known as the **impulse–momentum theorem,**[3] is equivalent to Newton's second law. From this definition, we see that impulse is a vector quantity having a magnitude equal to the area under the force–time curve, as described in Figure 9.4a. In this figure, it is assumed that the force varies in time in the general manner shown and is nonzero in the time interval $\Delta t = t_f - t_i$. The direction of the impulse vector is the same as the direction of the change in momentum. Impulse has the dimensions of momentum—that is, ML/T. Note that impulse is *not* a property of a particle; rather, it is a measure of the degree to which an external force changes the momentum of the particle. Therefore, when we say that an impulse is given to a particle, we mean that momentum is transferred from an external agent to that particle.

Because the force imparting an impulse can generally vary in time, it is convenient to define a time-averaged force

$$\overline{\mathbf{F}} \equiv \frac{1}{\Delta t} \int_{t_i}^{t_f} \mathbf{F} \, dt \tag{9.10}$$

where $\Delta t = t_f - t_i$. (This is an application of the mean value theorem of calculus.) Therefore, we can express Equation 9.9 as

$$\mathbf{I} \equiv \overline{\mathbf{F}} \, \Delta t \tag{9.11}$$

This time-averaged force, described in Figure 9.4b, can be thought of as the constant force that would give to the particle in the time interval Δt the same impulse that the time-varying force gives over this same interval.

In principle, if $\mathbf{F}$ is known as a function of time, the impulse can be calculated from Equation 9.9. The calculation becomes especially simple if the force acting on the particle is constant. In this case, $\overline{\mathbf{F}} = \mathbf{F}$ and Equation 9.11 becomes

$$\mathbf{I} = \mathbf{F} \, \Delta t \tag{9.12}$$

In many physical situations, we shall use what is called the **impulse approximation, in which we assume that one of the forces exerted on a particle acts for a short time but is much greater than any other force present.** This approximation is especially useful in treating collisions in which the duration of the

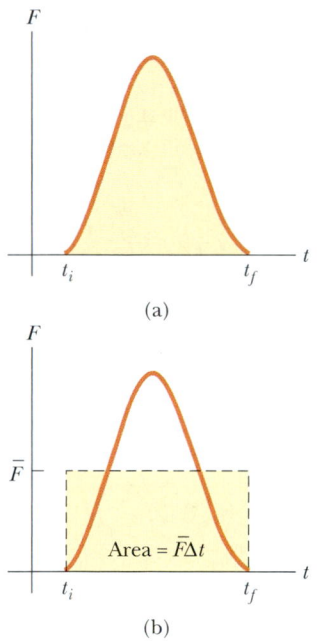

Figure 9.4 (a) A force acting on a particle may vary in time. The impulse imparted to the particle by the force is the area under the force versus time curve. (b) In the time interval Δt, the time-averaged force (horizontal dashed line) gives the same impulse to a particle as does the time-varying force described in part (a).

---

[3]Although we assumed that only a single force acts on the particle, the impulse–momentum theorem is valid when several forces act; in this case, we replace $\mathbf{F}$ in Equation 9.9 with $\Sigma \mathbf{F}$.

During the brief time the club is in contact with the ball, the ball gains momentum as a result of the collision, and the club loses the same amount of momentum. *(Courtesy of Michel Hans/Photo Researchers, Inc.)*

collision is very short. When this approximation is made, we refer to the force as an *impulsive force*. For example, when a baseball is struck with a bat, the time of the collision is about 0.01 s and the average force that the bat exerts on the ball in this time is typically several thousand newtons. Because this is much greater than the magnitude of the gravitational force, the impulse approximation justifies our ignoring the weight of the ball and bat. When we use this approximation, it is important to remember that $\mathbf{p}_i$ and $\mathbf{p}_f$ represent the momenta *immediately* before and after the collision, respectively. Therefore, in any situation in which it is proper to use the impulse approximation, the particle moves very little during the collision.

QuickLab

If you can find someone willing, play catch with an egg. What is the best way to move your hands so that the egg does not break when you change its momentum to zero?

Quick Quiz 9.3

Two objects are at rest on a frictionless surface. Object 1 has a greater mass than object 2. When a force is applied to object 1, it accelerates through a distance d. The force is removed from object 1 and is applied to object 2. At the moment when object 2 has accelerated through the same distance d, which statements are true? (a) $p_1 < p_2$, (b) $p_1 = p_2$, (c) $p_1 > p_2$, (d) $K_1 < K_2$, (e) $K_1 = K_2$, (f) $K_1 > K_2$.

EXAMPLE 9.3   Teeing Off

A golf ball of mass 50 g is struck with a club (Fig. 9.5). The force exerted on the ball by the club varies from zero, at the instant before contact, up to some maximum value (at which the ball is deformed) and then back to zero when the ball leaves the club. Thus, the force–time curve is qualitatively described by Figure 9.4. Assuming that the ball travels 200 m, estimate the magnitude of the impulse caused by the collision.

Solution   Let us use Ⓐ to denote the moment when the club first contacts the ball, Ⓑ to denote the moment when

the club loses contact with the ball as the ball starts on its trajectory, and Ⓒ to denote its landing. Neglecting air resistance, we can use Equation 4.14 for the range of a projectile:

$$R = x_C = \frac{v_B{}^2}{g} \sin 2\theta_B$$

Let us assume that the launch angle θ_B is 45°, the angle that provides the maximum range for any given launch velocity. This assumption gives $\sin 2\theta_B = 1$, and the launch velocity of

the ball is

$$v_B = \sqrt{x_C g} = \sqrt{(200 \text{ m})(9.80 \text{ m/s}^2)} = 44 \text{ m/s}$$

Considering the time interval for the collision, $v_i = v_A = 0$ and $v_f = v_B$ for the ball. Hence, the magnitude of the impulse imparted to the ball is

$$I = \Delta p = mv_B - mv_A = (50 \times 10^{-3} \text{ kg})(44 \text{ m/s}) - 0$$

$$= \boxed{2.2 \text{ kg} \cdot \text{m/s}}$$

Exercise   If the club is in contact with the ball for a time of 4.5×10^{-4} s, estimate the magnitude of the average force exerted by the club on the ball.

Answer 4.9×10^3 N, a value that is extremely large when compared with the weight of the ball, 0.49 N.

Figure 9.5   A golf ball being struck by a club. (*© Harold E. Edgerton/Courtesy of Palm Press, Inc.*)

EXAMPLE 9.4 ▶ How Good Are the Bumpers?

In a particular crash test, an automobile of mass 1 500 kg collides with a wall, as shown in Figure 9.6. The initial and final velocities of the automobile are $\mathbf{v}_i = -15.0\mathbf{i}$ m/s and $\mathbf{v}_f = 2.60\mathbf{i}$ m/s, respectively. If the collision lasts for 0.150 s, find the impulse caused by the collision and the average force exerted on the automobile.

Solution   Let us assume that the force exerted on the car by the wall is large compared with other forces on the car so that we can apply the impulse approximation. Furthermore, we note that the force of gravity and the normal force exerted by the road on the car are perpendicular to the motion and therefore do not affect the horizontal momentum.

The initial and final momenta of the automobile are

$$\mathbf{p}_i = m\mathbf{v}_i = (1\,500 \text{ kg})(-15.0\mathbf{i} \text{ m/s}) = -2.25 \times 10^4 \mathbf{i} \text{ kg} \cdot \text{m/s}$$

$$\mathbf{p}_f = m\mathbf{v}_f = (1\,500 \text{ kg})(2.60\mathbf{i} \text{ m/s}) = 0.39 \times 10^4 \mathbf{i} \text{ kg} \cdot \text{m/s}$$

Hence, the impulse is

$$\mathbf{I} = \Delta\mathbf{p} = \mathbf{p}_f - \mathbf{p}_i = 0.39 \times 10^4 \mathbf{i} \text{ kg} \cdot \text{m/s} - (-2.25 \times 10^4 \text{ kg} \cdot \text{m/s})$$

$$\mathbf{I} = \boxed{2.64 \times 10^4 \mathbf{i} \text{ kg} \cdot \text{m/s}}$$

The average force exerted on the automobile is

$$\overline{\mathbf{F}} = \frac{\Delta\mathbf{p}}{\Delta t} = \frac{2.64 \times 10^4 \mathbf{i} \text{ kg} \cdot \text{m/s}}{0.150 \text{ s}} = \boxed{1.76 \times 10^5 \mathbf{i} \text{ N}}$$

Before

−15.0 m/s

After

2.60 m/s

Figure 9.6   (a) This car's momentum changes as a result of its collision with the wall. (b) In a crash test, much of the car's initial kinetic energy is transformed into energy used to damage the car. (*b, courtesy of General Motors*)

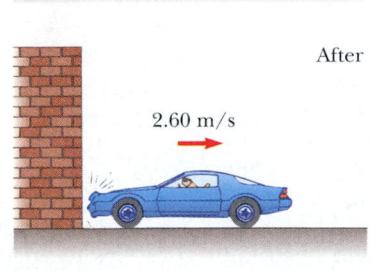

(a)

(b)

Note that the magnitude of this force is large compared with the weight of the car ($mg = 1.47 \times 10^4$ N), which justifies our initial assumption. Of note in this problem is how the signs of the velocities indicated the reversal of directions. What would the mathematics be describing if both the initial and final velocities had the same sign?

 Quick Quiz 9.4

Rank an automobile dashboard, seatbelt, and airbag in terms of (a) the impulse and (b) the average force they deliver to a front-seat passenger during a collision.

9.3 COLLISIONS

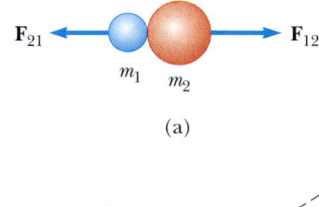

In this section we use the law of conservation of linear momentum to describe what happens when two particles collide. We use the term **collision** to represent the event of two particles' coming together for a short time and thereby producing impulsive forces on each other. **These forces are assumed to be much greater than any external forces present.**

A collision may entail physical contact between two macroscopic objects, as described in Figure 9.7a, but the notion of what we mean by collision must be generalized because "physical contact" on a submicroscopic scale is ill-defined and hence meaningless. To understand this, consider a collision on an atomic scale (Fig. 9.7b), such as the collision of a proton with an alpha particle (the nucleus of a helium atom). Because the particles are both positively charged, they never come into physical contact with each other; instead, they repel each other because of the strong electrostatic force between them at close separations. When two particles 1 and 2 of masses m_1 and m_2 collide as shown in Figure 9.7, the impulsive forces may vary in time in complicated ways, one of which is described in Figure 9.8. If $\mathbf{F}_{21}$ is the force exerted by particle 2 on particle 1, and if we assume that no external forces act on the particles, then the change in momentum of particle 1 due to the collision is given by Equation 9.8:

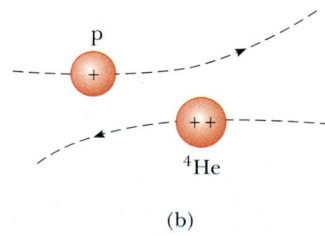

Figure 9.7 (a) The collision between two objects as the result of direct contact. (b) The "collision" between two charged particles.

$$\Delta\mathbf{p}_1 = \int_{t_i}^{t_f} \mathbf{F}_{21} \, dt$$

Likewise, if $\mathbf{F}_{12}$ is the force exerted by particle 1 on particle 2, then the change in momentum of particle 2 is

$$\Delta\mathbf{p}_2 = \int_{t_i}^{t_f} \mathbf{F}_{12} \, dt$$

From Newton's third law, we conclude that

$$\Delta\mathbf{p}_1 = -\Delta\mathbf{p}_2$$

$$\Delta\mathbf{p}_1 + \Delta\mathbf{p}_2 = 0$$

Because the total momentum of the system is $\mathbf{p}_{system} = \mathbf{p}_1 + \mathbf{p}_2$, we conclude that the *change* in the momentum of the system due to the collision is zero:

$$\mathbf{p}_{system} = \mathbf{p}_1 + \mathbf{p}_2 = constant$$

This is precisely what we expect because no external forces are acting on the system (see Section 9.2). Because the impulsive forces are internal, they do not change the total momentum of the system (only external forces can do that).

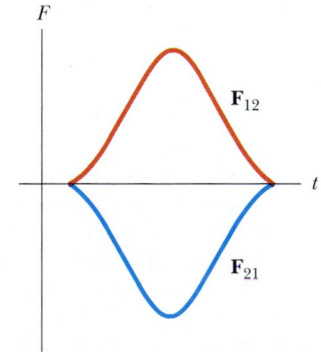

Figure 9.8 The impulse force as a function of time for the two colliding particles described in Figure 9.7a. Note that $\mathbf{F}_{12} = -\mathbf{F}_{21}$.

Momentum is conserved for any collision

Therefore, we conclude that **the total momentum of an isolated system just before a collision equals the total momentum of the system just after the collision.**

EXAMPLE 9.5   Carry Collision Insurance!

A car of mass 1800 kg stopped at a traffic light is struck from the rear by a 900-kg car, and the two become entangled. If the smaller car was moving at 20.0 m/s before the collision, what is the velocity of the entangled cars after the collision?

Solution   We can guess that the final speed is less than 20.0 m/s, the initial speed of the smaller car. The total momentum of the system (the two cars) before the collision must equal the total momentum immediately after the collision because momentum is conserved in any type of collision. The magnitude of the total momentum before the collision is equal to that of the smaller car because the larger car is initially at rest:

$$p_i = m_1 v_{1i} = (900 \text{ kg})(20.0 \text{ m/s}) = 1.80 \times 10^4 \text{ kg·m/s}$$

After the collision, the magnitude of the momentum of

the entangled cars is

$$p_f = (m_1 + m_2) v_f = (2\ 700 \text{ kg}) v_f$$

Equating the momentum before to the momentum after and solving for v_f, the final velocity of the entangled cars, we have

$$v_f = \frac{p_i}{m_1 + m_2} = \frac{1.80 \times 10^4 \text{ kg·m/s}}{2\ 700 \text{ kg}} = \boxed{6.67 \text{ m/s}}$$

The direction of the final velocity is the same as the velocity of the initially moving car.

Exercise   What would be the final speed if the two cars each had a mass of 900 kg?

Answer   10.0 m/s.

When the bowling ball and pin collide, part of the ball's momentum is transferred to the pin. Consequently, the pin acquires momentum and kinetic energy, and the ball loses momentum and kinetic energy. However, the total momentum of the system (ball and pin) remains constant. *(Ben Rose/The Image Bank)*

Elastic collision

Quick Quiz 9.5

As a ball falls toward the Earth, the ball's momentum increases because its speed increases. Does this mean that momentum is not conserved in this situation?

Quick Quiz 9.6

A skater is using very low-friction rollerblades. A friend throws a Frisbee straight at her. In which case does the Frisbee impart the greatest impulse to the skater: (a) she catches the Frisbee and holds it, (b) she catches it momentarily but drops it, (c) she catches it and at once throws it back to her friend?

9.4   ELASTIC AND INELASTIC COLLISIONS IN ONE DIMENSION

As we have seen, momentum is conserved in any collision in which external forces are negligible. In contrast, kinetic energy may or may not be constant, depending on the type of collision. In fact, whether or not kinetic energy is the same before and after the collision is used to classify collisions as being either elastic or inelastic.

An **elastic collision** between two objects is one in which *total kinetic energy (as well as total momentum) is the same before and after the collision.* Billiard-ball collisions and the collisions of air molecules with the walls of a container at ordinary temperatures are approximately elastic. Truly elastic collisions do occur, however, between atomic and subatomic particles. Collisions between certain objects in the macroscopic world, such as billiard-ball collisions, are only approximately elastic because some deformation and loss of kinetic energy take place.

An **inelastic collision** is one in which *total kinetic energy is not the same before and after the collision (even though momentum is constant).* Inelastic collisions are of two types. When the colliding objects stick together after the collision, as happens when a meteorite collides with the Earth, the collision is called **perfectly inelastic.** When the colliding objects do not stick together, but some kinetic energy is lost, as in the case of a rubber ball colliding with a hard surface, the collision is called **inelastic** (with no modifying adverb). For example, when a rubber ball collides with a hard surface, the collision is inelastic because some of the kinetic energy of the ball is lost when the ball is deformed while it is in contact with the surface.

In most collisions, kinetic energy is *not* the same before and after the collision because some of it is converted to internal energy, to elastic potential energy when the objects are deformed, and to rotational energy. Elastic and perfectly inelastic collisions are limiting cases; most collisions fall somewhere between them.

In the remainder of this section, we treat collisions in one dimension and consider the two extreme cases—perfectly inelastic and elastic collisions. The important distinction between these two types of collisions is that **momentum is constant in all collisions, but kinetic energy is constant only in elastic collisions.**

Perfectly Inelastic Collisions

Consider two particles of masses m_1 and m_2 moving with initial velocities $\mathbf{v}_{1i}$ and $\mathbf{v}_{2i}$ along a straight line, as shown in Figure 9.9. The two particles collide head-on, stick together, and then move with some common velocity $\mathbf{v}_f$ after the collision. Because momentum is conserved in any collision, we can say that the total momentum before the collision equals the total momentum of the composite system after the collision:

$$m_1\mathbf{v}_{1i} + m_2\mathbf{v}_{2i} = (m_1 + m_2)\mathbf{v}_f \tag{9.13}$$

$$\mathbf{v}_f = \frac{m_1\mathbf{v}_{1i} + m_2\mathbf{v}_{2i}}{m_1 + m_2} \tag{9.14}$$

Quick Quiz 9.7

Which is worse, crashing into a brick wall at 40 mi/h or crashing head-on into an oncoming car that is identical to yours and also moving at 40 mi/h?

Elastic Collisions

Now consider two particles that undergo an elastic head-on collision (Fig. 9.10). In this case, both momentum and kinetic energy are conserved; therefore, we have

$$m_1 v_{1i} + m_2 v_{2i} = m_1 v_{1f} + m_2 v_{2f} \tag{9.15}$$

$$\tfrac{1}{2} m_1 v_{1i}^2 + \tfrac{1}{2} m_2 v_{2i}^2 = \tfrac{1}{2} m_1 v_{1f}^2 + \tfrac{1}{2} m_2 v_{2f}^2 \tag{9.16}$$

Because all velocities in Figure 9.10 are either to the left or the right, they can be represented by the corresponding speeds along with algebraic signs indicating direction. We shall indicate v as positive if a particle moves to the right and negative

Inelastic collision

QuickLab

Hold a Ping-Pong ball or tennis ball on top of a basketball. Drop them both at the same time so that the basketball hits the floor, bounces up, and hits the smaller falling ball. What happens and why?

Before collision

After collision

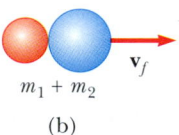

Figure 9.9 Schematic representation of a perfectly inelastic head-on collision between two particles: (a) before collision and (b) after collision.

Before collision

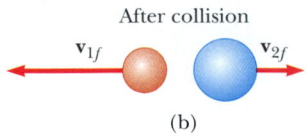

(a)

After collision

$\mathbf{v}_{1f}$ $\mathbf{v}_{2f}$

(b)

Figure 9.10 Schematic representation of an elastic head-on collision between two particles: (a) before collision and (b) after collision.

if it moves to the left. As has been seen in earlier chapters, it is common practice to call these values "speed" even though this term technically refers to the magnitude of the velocity vector, which does not have an algebraic sign.

In a typical problem involving elastic collisions, there are two unknown quantities, and Equations 9.15 and 9.16 can be solved simultaneously to find these. An alternative approach, however—one that involves a little mathematical manipulation of Equation 9.16—often simplifies this process. To see how, let us cancel the factor $\frac{1}{2}$ in Equation 9.16 and rewrite it as

$$m_1(v_{1i}{}^2 - v_{1f}{}^2) = m_2(v_{2f}{}^2 - v_{2i}{}^2)$$

and then factor both sides:

$$m_1(v_{1i} - v_{1f})(v_{1i} + v_{1f}) = m_2(v_{2f} - v_{2i})(v_{2f} + v_{2i}) \qquad \text{(9.17)}$$

Next, let us separate the terms containing m_1 and m_2 in Equation 9.15 to get

$$m_1(v_{1i} - v_{1f}) = m_2(v_{2f} - v_{2i}) \qquad \text{(9.18)}$$

To obtain our final result, we divide Equation 9.17 by Equation 9.18 and get

$$v_{1i} + v_{1f} = v_{2f} + v_{2i}$$

$$v_{1i} - v_{2i} = -(v_{1f} - v_{2f}) \qquad \text{(9.19)}$$

This equation, in combination with Equation 9.15, can be used to solve problems dealing with elastic collisions. According to Equation 9.19, the relative speed of the two particles before the collision $v_{1i} - v_{2i}$ equals the negative of their relative speed after the collision, $-(v_{1f} - v_{2f})$.

Suppose that the masses and initial velocities of both particles are known. Equations 9.15 and 9.19 can be solved for the final speeds in terms of the initial speeds because there are two equations and two unknowns:

$$v_{1f} = \left(\frac{m_1 - m_2}{m_1 + m_2}\right)v_{1i} + \left(\frac{2m_2}{m_1 + m_2}\right)v_{2i} \qquad \text{(9.20)}$$

Elastic collision: relationships between final and initial velocities

$$v_{2f} = \left(\frac{2m_1}{m_1 + m_2}\right)v_{1i} + \left(\frac{m_2 - m_1}{m_1 + m_2}\right)v_{2i} \qquad \text{(9.21)}$$

It is important to remember that the appropriate signs for v_{1i} and v_{2i} must be included in Equations 9.20 and 9.21. For example, if particle 2 is moving to the left initially, then v_{2i} is negative.

Let us consider some special cases: If $m_1 = m_2$, then $v_{1f} = v_{2i}$ and $v_{2f} = v_{1i}$. That is, the particles exchange speeds if they have equal masses. This is approximately what one observes in head-on billiard ball collisions—the cue ball stops, and the struck ball moves away from the collision with the same speed that the cue ball had.

If particle 2 is initially at rest, then $v_{2i} = 0$ and Equations 9.20 and 9.21 become

$$v_{1f} = \left(\frac{m_1 - m_2}{m_1 + m_2}\right)v_{1i} \qquad \text{(9.22)}$$

Elastic collision: particle 2 initially at rest

$$v_{2f} = \left(\frac{2m_1}{m_1 + m_2}\right)v_{1i} \qquad \text{(9.23)}$$

If m_1 is much greater than m_2 and $v_{2i} = 0$, we see from Equations 9.22 and 9.23 that $v_{1f} \approx v_{1i}$ and $v_{2f} \approx 2v_{1i}$. That is, when a very heavy particle collides head-on with a very light one that is initially at rest, the heavy particle continues its motion unaltered after the collision, and the light particle rebounds with a speed equal to about twice the initial speed of the heavy particle. An example of such a collision would be that of a moving heavy atom, such as uranium, with a light atom, such as hydrogen.

If m_2 is much greater than m_1 and particle 2 is initially at rest, then $v_{1f} \approx -v_{1i}$ and $v_{2f} \approx v_{2i} = 0$. That is, when a very light particle collides head-on with a very heavy particle that is initially at rest, the light particle has its velocity reversed and the heavy one remains approximately at rest.

EXAMPLE 9.6   The Ballistic Pendulum

The ballistic pendulum (Fig. 9.11) is a system used to measure the speed of a fast-moving projectile, such as a bullet. The bullet is fired into a large block of wood suspended from some light wires. The bullet embeds in the block, and the entire system swings through a height h. The collision is perfectly inelastic, and because momentum is conserved, Equation 9.14 gives the speed of the system right after the collision, when we assume the impulse approximation. If we call the bullet particle 1 and the block particle 2, the total kinetic energy right after the collision is

$$(1) \qquad K_f = \tfrac{1}{2}(m_1 + m_2)\,v_f^2$$

With $v_{2i} = 0$, Equation 9.14 becomes

$$(2) \qquad v_f = \frac{m_1 v_{1i}}{m_1 + m_2}$$

Substituting this value of v_f into (1) gives

$$K_f = \frac{m_1^2 v_{1i}^2}{2(m_1 + m_2)}$$

Note that this kinetic energy immediately after the collision is less than the initial kinetic energy of the bullet. In all the energy changes that take place *after* the collision, however, the total amount of mechanical energy remains constant; thus, we can say that after the collision, the kinetic energy of the block and bullet at the bottom is transformed to potential energy at the height h:

$$\frac{m_1^2 v_{1i}^2}{2(m_1 + m_2)} = (m_1 + m_2)gh$$

Solving for v_{1i}, we obtain

$$v_{1i} = \left(\frac{m_1 + m_2}{m_1}\right)\sqrt{2gh}$$

This expression tells us that it is possible to obtain the initial speed of the bullet by measuring h and the two masses.

Because the collision is perfectly inelastic, some mechanical energy is converted to internal energy and it would be *incorrect* to equate the initial kinetic energy of the incoming bullet to the final gravitational potential energy of the bullet–block combination.

Exercise   In a ballistic pendulum experiment, suppose that $h = 5.00$ cm, $m_1 = 5.00$ g, and $m_2 = 1.00$ kg. Find (a) the initial speed of the bullet and (b) the loss in mechanical energy due to the collision.

Answer   199 m/s; 98.5 J.

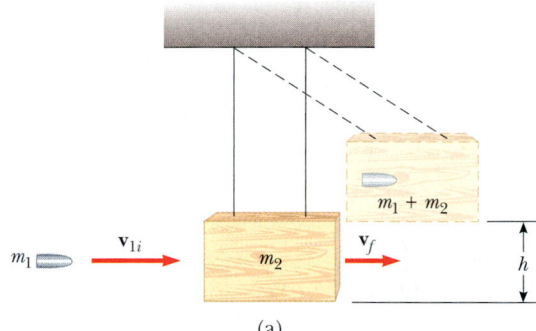

(a)

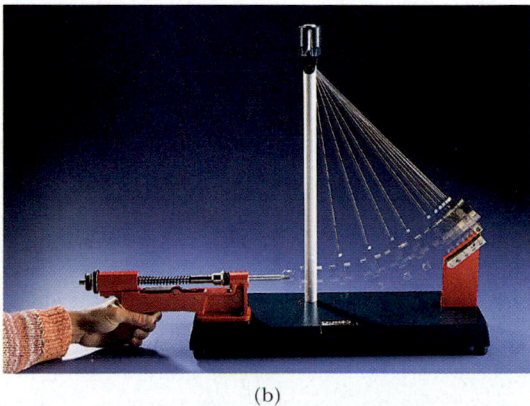

(b)

Figure 9.11   (a) Diagram of a ballistic pendulum. Note that $\mathbf{v}_{1i}$ is the velocity of the bullet just before the collision and $\mathbf{v}_f = \mathbf{v}_{1f} = \mathbf{v}_{2f}$ is the velocity of the bullet + block system just after the perfectly inelastic collision. (b) Multiflash photograph of a ballistic pendulum used in the laboratory. *(Courtesy of Central Scientific Company.)*

EXAMPLE 9.7 ▶ A Two-Body Collision with a Spring

A block of mass $m_1 = 1.60$ kg initially moving to the right with a speed of 4.00 m/s on a frictionless horizontal track collides with a spring attached to a second block of mass $m_2 = 2.10$ kg initially moving to the left with a speed of 2.50 m/s, as shown in Figure 9.12a. The spring constant is 600 N/m. (a) At the instant block 1 is moving to the right with a speed of 3.00 m/s, as in Figure 9.12b, determine the velocity of block 2.

Solution First, note that the initial velocity of block 2 is -2.50 m/s because its direction is to the left. Because momentum is conserved for the system of two blocks, we have

$$m_1 v_{1i} + m_2 v_{2i} = m_1 v_{1f} + m_2 v_{2f}$$

$$(1.60 \text{ kg})(4.00 \text{ m/s}) + (2.10 \text{ kg})(-2.50 \text{ m/s})$$
$$= (1.60 \text{ kg})(3.00 \text{ m/s}) + (2.10 \text{ kg}) v_{2f}$$

$$v_{2f} = \boxed{-1.74 \text{ m/s}}$$

The negative value for v_{2f} means that block 2 is still moving to the left at the instant we are considering.

(b) Determine the distance the spring is compressed at that instant.

Solution To determine the distance that the spring is compressed, shown as x in Figure 9.12b, we can use the concept of conservation of mechanical energy because no friction or other nonconservative forces are acting on the system. Thus, we have

$$\tfrac{1}{2}m_1 v_{1i}^2 + \tfrac{1}{2}m_2 v_{2i}^2 = \tfrac{1}{2}m_1 v_{1f}^2 + \tfrac{1}{2}m_2 v_{2f}^2 + \tfrac{1}{2}kx^2$$

Substituting the given values and the result to part (a) into this expression gives

$$x = \boxed{0.173 \text{ m}}$$

It is important to note that we needed to use the principles of both conservation of momentum and conservation of mechanical energy to solve the two parts of this problem.

Exercise Find the velocity of block 1 and the compression in the spring at the instant that block 2 is at rest.

Answer 0.719 m/s to the right; 0.251 m.

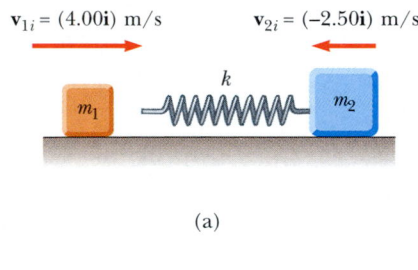

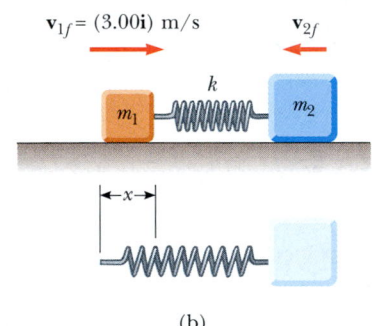

Figure 9.12

EXAMPLE 9.8 ▶ Slowing Down Neutrons by Collisions

In a nuclear reactor, neutrons are produced when a $^{235}_{92}$U atom splits in a process called *fission*. These neutrons are moving at about 10^7 m/s and must be slowed down to about 10^3 m/s before they take part in another fission event. They are slowed down by being passed through a solid or liquid material called a *moderator*. The slowing-down process involves elastic collisions. Let us show that a neutron can lose most of its kinetic energy if it collides elastically with a moderator containing light nuclei, such as deuterium (in "heavy water," D_2O) or carbon (in graphite).

Solution Let us assume that the moderator nucleus of mass m_m is at rest initially and that a neutron of mass m_n and initial speed v_{ni} collides with it head-on.

Because these are elastic collisions, the first thing we do is recognize that both momentum and kinetic energy are constant. Therefore, Equations 9.22 and 9.23 can be applied to the head-on collision of a neutron with a moderator nucleus. We can represent this process by a drawing such as Figure 9.10.

The initial kinetic energy of the neutron is

$$K_{ni} = \tfrac{1}{2} m_n v_{ni}^2$$

After the collision, the neutron has kinetic energy $\tfrac{1}{2} m_n v_{nf}^2$, and we can substitute into this the value for v_{nf} given by Equation 9.22:

$$K'_{nf} = \tfrac{1}{2} m_n v_{nf}^2 = \frac{m_n}{2}\left(\frac{m_n - m_m}{m_n + m_m}\right)^2 v_{ni}^2$$

Therefore, the fraction f_n of the initial kinetic energy possessed by the neutron after the collision is

$$(1) \qquad f_n = \frac{K_{nf}}{K_{ni}} = \left(\frac{m_n - m_m}{m_n + m_m}\right)^2$$

From this result, we see that the final kinetic energy of the neutron is small when m_m is close to m_n and zero when $m_n = m_m$.

We can use Equation 9.23, which gives the final speed of the particle that was initially at rest, to calculate the kinetic energy of the moderator nucleus after the collision:

$$K_{mf} = \tfrac{1}{2} m_m v_{mf}^2 = \frac{2 m_n^2 m_m}{(m_n + m_m)^2} v_{ni}^2$$

Hence, the fraction f_m of the initial kinetic energy transferred to the moderator nucleus is

$$(2) \qquad f_m = \frac{K_{mf}}{K_{ni}} = \frac{4 m_n m_m}{(m_n + m_m)^2}$$

Because the total kinetic energy of the system is conserved, (2) can also be obtained from (1) with the condition that $f_n + f_m = 1$, so that $f_m = 1 - f_n$.

Suppose that heavy water is used for the moderator. For collisions of the neutrons with deuterium nuclei in D_2O ($m_m = 2m_n$), $f_n = 1/9$ and $f_m = 8/9$. That is, 89% of the neutron's kinetic energy is transferred to the deuterium nucleus. In practice, the moderator efficiency is reduced because head-on collisions are very unlikely.

How do the results differ when graphite (^{12}C, as found in pencil lead) is used as the moderator?

Quick Quiz 9.8

An ingenious device that illustrates conservation of momentum and kinetic energy is shown in Figure 9.13a. It consists of five identical hard balls supported by strings of equal lengths. When ball 1 is pulled out and released, after the almost-elastic collision between it and ball 2, ball 5 moves out, as shown in Figure 9.13b. If balls 1 and 2 are pulled out and released, balls 4 and 5 swing out, and so forth. Is it ever possible that, when ball 1 is released, balls 4 and 5 will swing out on the opposite side and travel with half the speed of ball 1, as in Figure 9.13c?

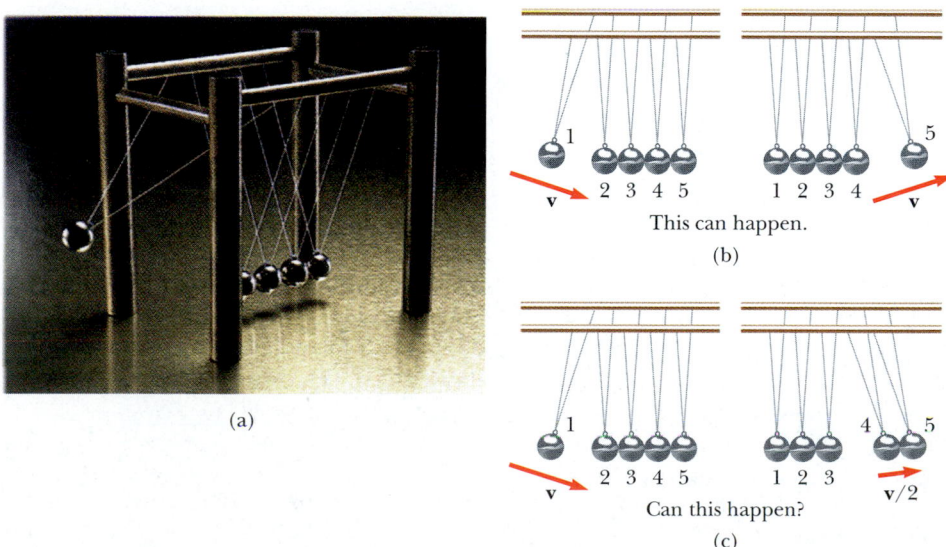

(a)

This can happen.
(b)

Can this happen?
(c)

Figure 9.13 An executive stress reliever.

9.5 ▷ TWO-DIMENSIONAL COLLISIONS

In Sections 9.1 and 9.3, we showed that the momentum of a system of two particles is constant when the system is isolated. For any collision of two particles, this result implies that the momentum in each of the directions x, y, and z is constant. However, an important subset of collisions takes place in a plane. The game of billiards is a familiar example involving multiple collisions of objects moving on a two-dimensional surface. For such two-dimensional collisions, we obtain two component equations for conservation of momentum:

$$m_1 v_{1ix} + m_2 v_{2ix} = m_1 v_{1fx} + m_2 v_{2fx}$$

$$m_1 v_{1iy} + m_2 v_{2iy} = m_1 v_{1fy} + m_2 v_{2fy}$$

Let us consider a two-dimensional problem in which particle 1 of mass m_1 collides with particle 2 of mass m_2, where particle 2 is initially at rest, as shown in Figure 9.14. After the collision, particle 1 moves at an angle θ with respect to the horizontal and particle 2 moves at an angle ϕ with respect to the horizontal. This is called a *glancing* collision. Applying the law of conservation of momentum in component form, and noting that the initial y component of the momentum of the two-particle system is zero, we obtain

$$m_1 v_{1i} = m_1 v_{1f} \cos\theta + m_2 v_{2f} \cos\phi \tag{9.24}$$

$$0 = m_1 v_{1f} \sin\theta - m_2 v_{2f} \sin\phi \tag{9.25}$$

where the minus sign in Equation 9.25 comes from the fact that after the collision, particle 2 has a y component of velocity that is downward. We now have two independent equations. As long as no more than two of the seven quantities in Equations 9.24 and 9.25 are unknown, we can solve the problem.

If the collision is elastic, we can also use Equation 9.16 (conservation of kinetic energy), with $v_{2i} = 0$, to give

$$\tfrac{1}{2} m_1 v_{1i}^2 = \tfrac{1}{2} m_1 v_{1f}^2 + \tfrac{1}{2} m_2 v_{2f}^2 \tag{9.26}$$

Knowing the initial speed of particle 1 and both masses, we are left with four unknowns (v_{1f}, v_{2f}, θ, ϕ). Because we have only three equations, one of the four remaining quantities must be given if we are to determine the motion after the collision from conservation principles alone.

If the collision is inelastic, kinetic energy is *not* conserved and Equation 9.26 does *not* apply.

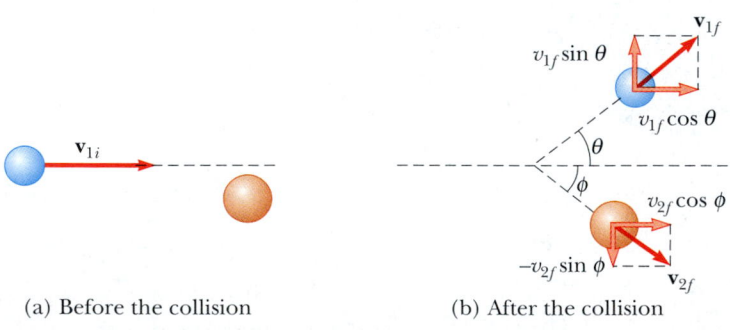

(a) Before the collision   (b) After the collision

Figure 9.14   An elastic glancing collision between two particles.

Problem-Solving Hints

Collisions

The following procedure is recommended when dealing with problems involving collisions between two objects:

- Set up a coordinate system and define your velocities with respect to that system. It is usually convenient to have the x axis coincide with one of the initial velocities.
- In your sketch of the coordinate system, draw and label all velocity vectors and include all the given information.
- Write expressions for the x and y components of the momentum of each object before and after the collision. Remember to include the appropriate signs for the components of the velocity vectors.
- Write expressions for the total momentum in the x direction before and after the collision and equate the two. Repeat this procedure for the total momentum in the y direction. These steps follow from the fact that, because the momentum of the *system* is conserved in any collision, the total momentum along any direction must also be constant. Remember, it is the momentum of the *system* that is constant, not the momenta of the individual objects.
- If the collision is inelastic, kinetic energy is not conserved, and additional information is probably required. If the collision is perfectly inelastic, the final velocities of the two objects are equal. Solve the momentum equations for the unknown quantities.
- If the collision is elastic, kinetic energy is conserved, and you can equate the total kinetic energy before the collision to the total kinetic energy after the collision to get an additional relationship between the velocities.

EXAMPLE 9.9 Collision at an Intersection

A 1 500-kg car traveling east with a speed of 25.0 m/s collides at an intersection with a 2 500-kg van traveling north at a speed of 20.0 m/s, as shown in Figure 9.15. Find the direction and magnitude of the velocity of the wreckage after the collision, assuming that the vehicles undergo a perfectly inelastic collision (that is, they stick together).

Solution Let us choose east to be along the positive x direction and north to be along the positive y direction. Before the collision, the only object having momentum in the x direction is the car. Thus, the magnitude of the total initial momentum of the system (car plus van) in the x direction is

$$\sum p_{xi} = (1\,500 \text{ kg})(25.0 \text{ m/s}) = 3.75 \times 10^4 \text{ kg·m/s}$$

Let us assume that the wreckage moves at an angle θ and speed v_f after the collision. The magnitude of the total momentum in the x direction after the collision is

$$\sum p_{xf} = (4\,000 \text{ kg})v_f \cos \theta$$

Because the total momentum in the x direction is constant, we can equate these two equations to obtain

$$(1) \qquad 3.75 \times 10^4 \text{ kg·m/s} = (4\,000 \text{ kg})v_f \cos \theta$$

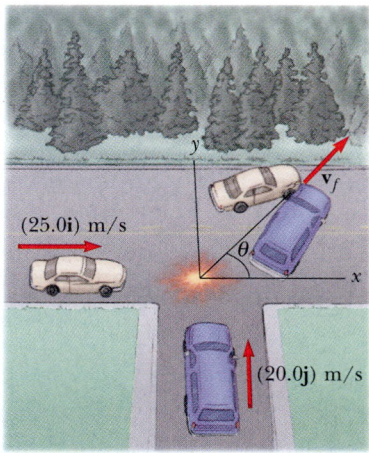

Figure 9.15 An eastbound car colliding with a northbound van.

Similarly, the total initial momentum of the system in the y direction is that of the van, and the magnitude of this momentum is $(2\,500 \text{ kg})(20.0 \text{ m/s})$. Applying conservation of

momentum to the y direction, we have

$$\sum p_{yi} = \sum p_{yf}$$

$$(2\ 500\ \text{kg})(20.0\ \text{m/s}) = (4\ 000\ \text{kg})v_f \sin \theta$$

(2) $$5.00 \times 10^4\ \text{kg} \cdot \text{m/s} = (4\ 000\ \text{kg})v_f \sin \theta$$

If we divide (2) by (1), we get

$$\frac{\sin \theta}{\cos \theta} = \tan \theta = \frac{5.00 \times 10^4}{3.75 \times 10^4} = 1.33$$

$$\theta = \boxed{53.1°}$$

When this angle is substituted into (2), the value of v_f is

$$v_f = \frac{5.00 \times 10^4\ \text{kg} \cdot \text{m/s}}{(4\ 000\ \text{kg})\sin 53.1°} = \boxed{15.6\ \text{m/s}}$$

It might be instructive for you to draw the momentum vectors of each vehicle before the collision and the two vehicles together after the collision.

EXAMPLE 9.10   Proton–Proton Collision

Proton 1 collides elastically with proton 2 that is initially at rest. Proton 1 has an initial speed of 3.50×10^5 m/s and makes a glancing collision with proton 2, as was shown in Figure 9.14. After the collision, proton 1 moves at an angle of $37.0°$ to the horizontal axis, and proton 2 deflects at an angle ϕ to the same axis. Find the final speeds of the two protons and the angle ϕ.

Solution   Because both particles are protons, we know that $m_1 = m_2$. We also know that $\theta = 37.0°$ and $v_{1i} = 3.50 \times 10^5$ m/s. Equations 9.24, 9.25, and 9.26 become

$$v_{1f}\cos 37.0° + v_{2f}\cos \phi = 3.50 \times 10^5\ \text{m/s}$$

$$v_{1f}\sin 37.0° - v_{2f}\sin \phi = 0$$

$$v_{1f}{}^2 + v_{2f}{}^2 = (3.50 \times 10^5\ \text{m/s})^2$$

Solving these three equations with three unknowns simultaneously gives

$$v_{1f} = \boxed{2.80 \times 10^5\ \text{m/s}}$$

$$v_{2f} = \boxed{2.11 \times 10^5\ \text{m/s}}$$

$$\phi = \boxed{53.0°}$$

Note that $\theta + \phi = 90°$. This result is not accidental. **Whenever two equal masses collide elastically in a glancing collision and one of them is initially at rest, their final velocities are always at right angles to each other.** The next example illustrates this point in more detail.

EXAMPLE 9.11   Billiard Ball Collision

In a game of billiards, a player wishes to sink a target ball 2 in the corner pocket, as shown in Figure 9.16. If the angle to the corner pocket is 35°, at what angle θ is the cue ball 1 deflected? Assume that friction and rotational motion are unimportant and that the collision is elastic.

Solution   Because the target ball is initially at rest, conservation of energy (Eq. 9.16) gives

$$\tfrac{1}{2}m_1 v_{1i}{}^2 = \tfrac{1}{2}m_1 v_{1f}{}^2 + \tfrac{1}{2}m_2 v_{2f}{}^2$$

But $m_1 = m_2$, so that

(1) $$v_{1i}{}^2 = v_{1f}{}^2 + v_{2f}{}^2$$

Applying conservation of momentum to the two-dimensional collision gives

(2) $$\mathbf{v}_{1i} = \mathbf{v}_{1f} + \mathbf{v}_{2f}$$

Note that because $m_1 = m_2$, the masses also cancel in (2). If we square both sides of (2) and use the definition of the dot

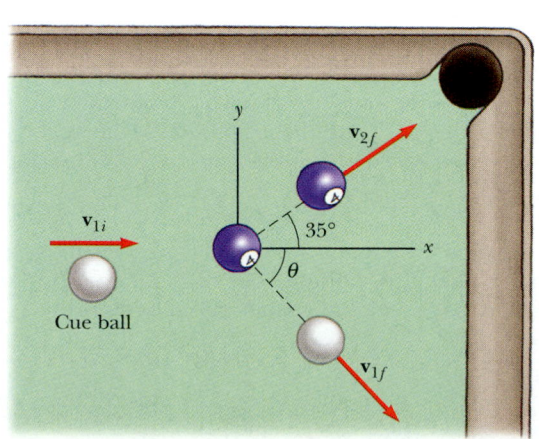

Figure 9.16

product of two vectors from Section 7.2, we get

$$v_{1i}{}^2 = (\mathbf{v}_{1f} + \mathbf{v}_{2f}) \cdot (\mathbf{v}_{1f} + \mathbf{v}_{2f}) = v_{1f}{}^2 + v_{2f}{}^2 + 2\mathbf{v}_{1f} \cdot \mathbf{v}_{2f}$$

Because the angle between $\mathbf{v}_{1f}$ and $\mathbf{v}_{2f}$ is $\theta + 35°$, $\mathbf{v}_{1f} \cdot \mathbf{v}_{2f} = v_{1f}v_{2f}\cos(\theta + 35°)$, and so

$$(3) \qquad v_{1i}{}^2 = v_{1f}{}^2 + v_{2f}{}^2 + 2v_{1f}v_{2f}\cos(\theta + 35°)$$

Subtracting (1) from (3) gives

$$0 = 2v_{1f}v_{2f}\cos(\theta + 35°)$$

$$0 = \cos(\theta + 35°)$$

$$\theta + 35° = 90° \qquad \text{or} \qquad \theta = \boxed{55°}$$

This result shows that whenever two equal masses undergo a glancing elastic collision and one of them is initially at rest, they move at right angles to each other after the collision. The same physics describes two very different situations, protons in Example 9.10 and billiard balls in this example.

9.6 THE CENTER OF MASS

In this section we describe the overall motion of a mechanical system in terms of a special point called the **center of mass** of the system. The mechanical system can be either a system of particles, such as a collection of atoms in a container, or an extended object, such as a gymnast leaping through the air. We shall see that the center of mass of the system moves as if all the mass of the system were concentrated at that point. Furthermore, if the resultant external force on the system is $\Sigma\mathbf{F}_{\text{ext}}$ and the total mass of the system is M, the center of mass moves with an acceleration given by $\mathbf{a} = \Sigma\mathbf{F}_{\text{ext}}/M$. That is, the system moves as if the resultant external force were applied to a single particle of mass M located at the center of mass. This behavior is independent of other motion, such as rotation or vibration of the system. This result was implicitly assumed in earlier chapters because many examples referred to the motion of extended objects that were treated as particles.

Consider a mechanical system consisting of a pair of particles that have different masses and are connected by a light, rigid rod (Fig. 9.17). One can describe the position of the center of mass of a system as being the *average position* of the system's mass. The center of mass of the system is located somewhere on the line joining the

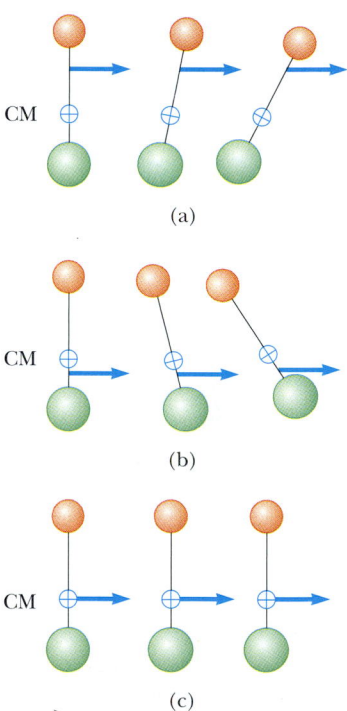

Figure 9.17 Two particles of unequal mass are connected by a light, rigid rod. (a) The system rotates clockwise when a force is applied between the less massive particle and the center of mass. (b) The system rotates counterclockwise when a force is applied between the more massive particle and the center of mass. (c) The system moves in the direction of the force without rotating when a force is applied at the center of mass.

This multiflash photograph shows that as the acrobat executes a somersault, his center of mass follows a parabolic path, the same path that a particle would follow. *(Globus Bros. Studios/The Stock Market)*

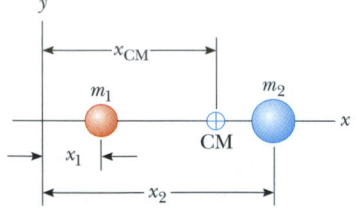

Figure 9.18 The center of mass of two particles of unequal mass on the x axis is located at x_{CM}, a point between the particles, closer to the one having the larger mass.

particles and is closer to the particle having the larger mass. If a single force is applied at some point on the rod somewhere between the center of mass and the less massive particle, the system rotates clockwise (see Fig. 9.17a). If the force is applied at a point on the rod somewhere between the center of mass and the more massive particle, the system rotates counterclockwise (see Fig. 9.17b). If the force is applied at the center of mass, the system moves in the direction of **F** without rotating (see Fig. 9.17c). Thus, the center of mass can be easily located.

The center of mass of the pair of particles described in Figure 9.18 is located on the x axis and lies somewhere between the particles. Its x coordinate is

$$x_{CM} \equiv \frac{m_1 x_1 + m_2 x_2}{m_1 + m_2} \tag{9.27}$$

For example, if $x_1 = 0$, $x_2 = d$, and $m_2 = 2m_1$, we find that $x_{CM} = \frac{2}{3}d$. That is, the center of mass lies closer to the more massive particle. If the two masses are equal, the center of mass lies midway between the particles.

We can extend this concept to a system of many particles in three dimensions. The x coordinate of the center of mass of n particles is defined to be

$$x_{CM} \equiv \frac{m_1 x_1 + m_2 x_2 + m_3 x_3 + \cdots + m_n x_n}{m_1 + m_2 + m_3 + \cdots + m_n} = \frac{\sum\limits_i m_i x_i}{\sum\limits_i m_i} \tag{9.28}$$

where x_i is the x coordinate of the ith particle. For convenience, we express the total mass as $M \equiv \sum\limits_i m_i$, where the sum runs over all n particles. The y and z coordinates of the center of mass are similarly defined by the equations

$$y_{CM} \equiv \frac{\sum\limits_i m_i y_i}{M} \quad \text{and} \quad z_{CM} \equiv \frac{\sum\limits_i m_i z_i}{M} \tag{9.29}$$

The center of mass can also be located by its position vector, $\mathbf{r}_{CM}$. The cartesian coordinates of this vector are x_{CM}, y_{CM}, and z_{CM}, defined in Equations 9.28 and 9.29. Therefore,

$$\mathbf{r}_{CM} = x_{CM}\mathbf{i} + y_{CM}\mathbf{j} + z_{CM}\mathbf{k}$$
$$= \frac{\sum\limits_i m_i x_i \mathbf{i} + \sum\limits_i m_i y_i \mathbf{j} + \sum\limits_i m_i z_i \mathbf{k}}{M}$$

Vector position of the center of mass for a system of particles

$$\mathbf{r}_{CM} \equiv \frac{\sum\limits_i m_i \mathbf{r}_i}{M} \tag{9.30}$$

where $\mathbf{r}_i$ is the position vector of the ith particle, defined by

$$\mathbf{r}_i \equiv x_i\mathbf{i} + y_i\mathbf{j} + z_i\mathbf{k}$$

Although locating the center of mass for an extended object is somewhat more cumbersome than locating the center of mass of a system of particles, the basic ideas we have discussed still apply. We can think of an extended object as a system containing a large number of particles (Fig. 9.19). The particle separation is very small, and so the object can be considered to have a continuous mass distribution. By dividing the object into elements of mass Δm_i, with coordinates x_i, y_i, z_i, we see that the x coordinate of the center of mass is approximately

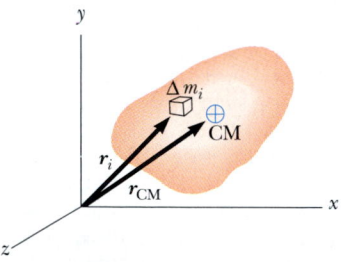

Figure 9.19 An extended object can be considered a distribution of small elements of mass Δm_i. The center of mass is located at the vector position $\mathbf{r}_{CM}$, which has coordinates x_{CM}, y_{CM}, and z_{CM}.

$$x_{CM} \approx \frac{\sum\limits_i x_i \Delta m_i}{M}$$

with similar expressions for y_{CM} and z_{CM}. If we let the number of elements n approach infinity, then x_{CM} is given precisely. In this limit, we replace the sum by an integral and Δm_i by the differential element dm:

$$x_{\text{CM}} = \lim_{\Delta m_i \to 0} \frac{\sum_i x_i \, \Delta m_i}{M} = \frac{1}{M} \int x \, dm \tag{9.31}$$

Likewise, for y_{CM} and z_{CM} we obtain

$$y_{\text{CM}} = \frac{1}{M} \int y \, dm \qquad \text{and} \qquad z_{\text{CM}} = \frac{1}{M} \int z \, dm \tag{9.32}$$

We can express the vector position of the center of mass of an extended object in the form

$$\mathbf{r}_{\text{CM}} = \frac{1}{M} \int \mathbf{r} \, dm \tag{9.33}$$

which is equivalent to the three expressions given by Equations 9.31 and 9.32.

The center of mass of any symmetric object lies on an axis of symmetry and on any plane of symmetry.[4] For example, the center of mass of a rod lies in the rod, midway between its ends. The center of mass of a sphere or a cube lies at its geometric center.

One can determine the center of mass of an irregularly shaped object by suspending the object first from one point and then from another. In Figure 9.20, a wrench is hung from point A, and a vertical line AB (which can be established with a plumb bob) is drawn when the wrench has stopped swinging. The wrench is then hung from point C, and a second vertical line CD is drawn. The center of mass is halfway through the thickness of the wrench, under the intersection of these two lines. In general, if the wrench is hung freely from any point, the vertical line through this point must pass through the center of mass.

Because an extended object is a continuous distribution of mass, each small mass element is acted upon by the force of gravity. The net effect of all these forces is equivalent to the effect of a single force, $M\mathbf{g}$, acting through a special point, called the **center of gravity.** If $\mathbf{g}$ is constant over the mass distribution, then the center of gravity coincides with the center of mass. If an extended object is pivoted at its center of gravity, it balances in any orientation.

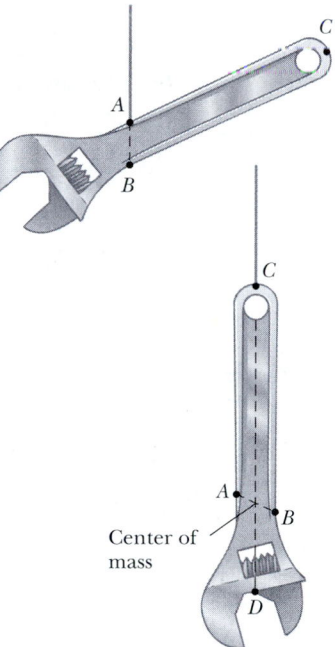

Figure 9.20 An experimental technique for determining the center of mass of a wrench. The wrench is hung freely first from point A and then from point C. The intersection of the two lines AB and CD locates the center of mass.

If a baseball bat is cut at the location of its center of mass as shown in Figure 9.21, do the two pieces have the same mass?

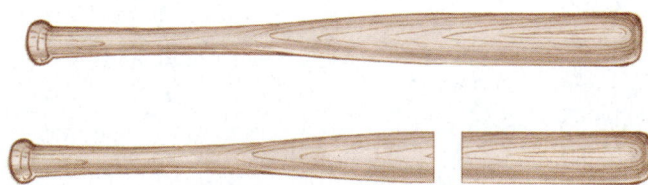

Figure 9.21 A baseball bat cut at the location of its center of mass.

QuickLab

Cut a triangle from a piece of cardboard and draw a set of adjacent strips inside it, parallel to one of the sides. Put a dot at the approximate location of the center of mass of each strip and then draw a straight line through the dots and into the angle opposite your starting side. The center of mass for the triangle must lie on this bisector of the angle. Repeat these steps for the other two sides. The three angle bisectors you have drawn will intersect at the center of mass of the triangle. If you poke a hole anywhere in the triangle and hang the cardboard from a string attached at that hole, the center of mass will be vertically aligned with the hole.

---

[4]This statement is valid only for objects that have a uniform mass per unit volume.

EXAMPLE 9.12   **The Center of Mass of Three Particles**

A system consists of three particles located as shown in Figure 9.22a. Find the center of mass of the system.

Solution   We set up the problem by labeling the masses of the particles as shown in the figure, with $m_1 = m_2 = 1.0$ kg and $m_3 = 2.0$ kg. Using the basic defining equations for the coordinates of the center of mass and noting that $z_{CM} = 0$, we obtain

$$x_{CM} = \frac{\sum_i m_i x_i}{M} = \frac{m_1 x_1 + m_2 x_2 + m_3 x_3}{m_1 + m_2 + m_3}$$

$$= \frac{(1.0 \text{ kg})(1.0 \text{ m}) + (1.0 \text{ kg})(2.0 \text{ m}) + (2.0 \text{ kg})(0 \text{ m})}{1.0 \text{ kg} + 1.0 \text{ kg} + 2.0 \text{ kg}}$$

$$= \frac{3.0 \text{ kg} \cdot \text{m}}{4.0 \text{ kg}} = 0.75 \text{ m}$$

$$y_{CM} = \frac{\sum_i m_i y_i}{M} = \frac{m_1 y_1 + m_2 y_2 + m_3 y_3}{m_1 + m_2 + m_3}$$

$$= \frac{(1.0 \text{ kg})(0) + (1.0 \text{ kg})(0) + (2.0 \text{ kg})(2.0 \text{ m})}{4.0 \text{ kg}}$$

$$= \frac{4.0 \text{ kg} \cdot \text{m}}{4.0 \text{ kg}} = 1.0 \text{ m}$$

The position vector to the center of mass measured from the origin is therefore

$$\mathbf{r}_{CM} = x_{CM}\mathbf{i} + y_{CM}\mathbf{j} = \boxed{0.75\mathbf{i} \text{ m} + 1.0\mathbf{j} \text{ m}}$$

We can verify this result graphically by adding together $m_1\mathbf{r}_1 + m_2\mathbf{r}_2 + m_3\mathbf{r}_3$ and dividing the vector sum by M, the total mass. This is shown in Figure 9.22b.

Figure 9.22   (a) Two 1-kg masses and a single 2-kg mass are located as shown. The vector indicates the location of the system's center of mass. (b) The vector sum of $m_i\mathbf{r}_i$.

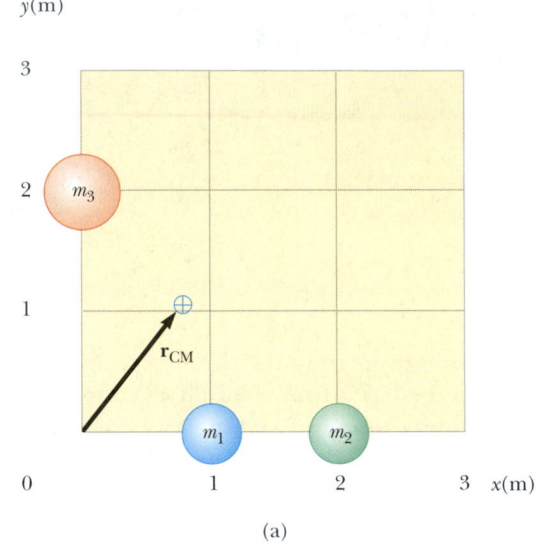

(a)

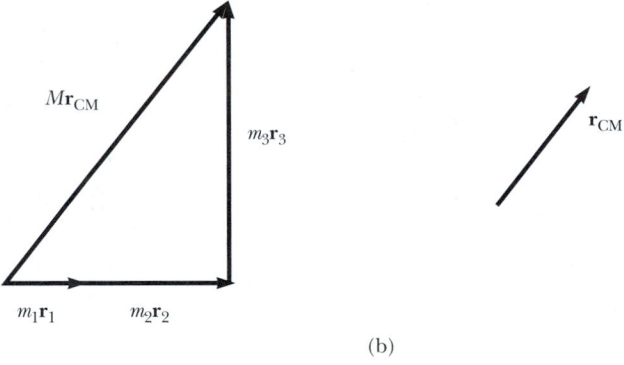

(b)

EXAMPLE 9.13   **The Center of Mass of a Rod**

(a) Show that the center of mass of a rod of mass M and length L lies midway between its ends, assuming the rod has a uniform mass per unit length.

Solution   The rod is shown aligned along the x axis in Figure 9.23, so that $y_{CM} = z_{CM} = 0$. Furthermore, if we call the mass per unit length λ (this quantity is called the *linear mass density*), then $\lambda = M/L$ for the uniform rod we assume here. If we divide the rod into elements of length dx, then the mass of each element is $dm = \lambda \, dx$. For an arbitrary element located a distance x from the origin, Equation 9.31 gives

$$x_{CM} = \frac{1}{M} \int x \, dm = \frac{1}{M} \int_0^L x\lambda \, dx = \frac{\lambda}{M} \frac{x^2}{2} \Big|_0^L = \frac{\lambda L^2}{2M}$$

Because $\lambda = M/L$, this reduces to

$$x_{CM} = \frac{L^2}{2M}\left(\frac{M}{L}\right) = \boxed{\frac{L}{2}}$$

One can also use symmetry arguments to obtain the same result.

(b) Suppose a rod is *nonuniform* such that its mass per unit length varies linearly with x according to the expression $\lambda = \alpha x$, where α is a constant. Find the x coordinate of the center of mass as a fraction of L.

Solution   In this case, we replace dm by λdx where λ is not constant. Therefore, x_{CM} is

$$x_{CM} = \frac{1}{M} \int x \, dm = \frac{1}{M} \int_0^L x\lambda \, dx = \frac{1}{M} \int_0^L x\alpha x \, dx$$

$$= \frac{\alpha}{M} \int_0^L x^2 \, dx = \frac{\alpha L^3}{3M}$$

We can eliminate α by noting that the total mass of the rod is related to α through the relationship

$$M = \int dm = \int_0^L \lambda \, dx = \int_0^L \alpha x \, dx = \frac{\alpha L^2}{2}$$

Substituting this into the expression for x_{CM} gives

$$x_{CM} = \frac{\alpha L^3}{3\alpha L^2 / 2} = \boxed{\frac{2}{3} L}$$

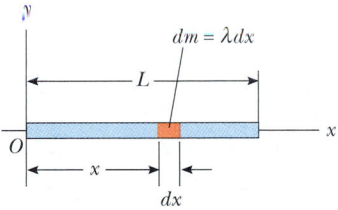

Figure 9.23 The center of mass of a uniform rod of length L is located at $x_{CM} = L/2$.

EXAMPLE 9.14 ▶ The Center of Mass of a Right Triangle

An object of mass M is in the shape of a right triangle whose dimensions are shown in Figure 9.24. Locate the coordinates of the center of mass, assuming the object has a uniform mass per unit area.

Solution By inspection we can estimate that the x coordinate of the center of mass must be past the center of the base, that is, greater than $a/2$, because the largest part of the triangle lies beyond that point. A similar argument indicates that its y coordinate must be less than $b/2$. To evaluate the x coordinate, we divide the triangle into narrow strips of width dx and height y as in Figure 9.24. The mass dm of each strip is

$$dm = \frac{\text{total mass of object}}{\text{total area of object}} \times \text{area of strip}$$

$$= \frac{M}{1/2 \, ab}(y \, dx) = \left(\frac{2M}{ab}\right) y \, dx$$

Therefore, the x coordinate of the center of mass is

$$x_{CM} = \frac{1}{M} \int x \, dm = \frac{1}{M} \int_0^a x \left(\frac{2M}{ab}\right) y \, dx = \frac{2}{ab} \int_0^a xy \, dx$$

To evaluate this integral, we must express y in terms of x. From similar triangles in Figure 9.24, we see that

$$\frac{y}{x} = \frac{b}{a} \qquad \text{or} \qquad y = \frac{b}{a} x$$

With this substitution, x_{CM} becomes

$$x_{CM} = \frac{2}{ab} \int_0^a x \left(\frac{b}{a} x\right) dx = \frac{2}{a^2} \int_0^a x^2 \, dx = \frac{2}{a^2} \left[\frac{x^3}{3}\right]_0^a$$

$$= \boxed{\frac{2}{3} a}$$

By a similar calculation, we get for the y coordinate of the center of mass

$$y_{CM} = \boxed{\frac{1}{3} b}$$

These values fit our original estimates.

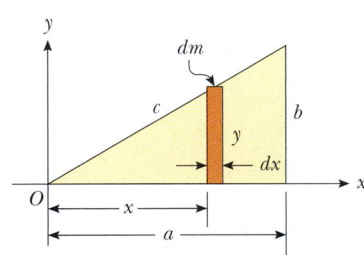

Figure 9.24

9.7 ▶ MOTION OF A SYSTEM OF PARTICLES

We can begin to understand the physical significance and utility of the center of mass concept by taking the time derivative of the position vector given by Equation 9.30. From Section 4.1 we know that the time derivative of a position vector is by definition a velocity. Assuming M remains constant for a system of particles, that is, no particles enter or leave the system, we get the following expression for the **velocity of the center of mass** of the system:

Velocity of the center of mass

$$\mathbf{v}_{CM} = \frac{d\mathbf{r}_{CM}}{dt} = \frac{1}{M}\sum_i m_i \frac{d\mathbf{r}_i}{dt} = \frac{\sum_i m_i \mathbf{v}_i}{M} \tag{9.34}$$

where $\mathbf{v}_i$ is the velocity of the ith particle. Rearranging Equation 9.34 gives

Total momentum of a system of particles

$$M\mathbf{v}_{CM} = \sum_i m_i \mathbf{v}_i = \sum_i \mathbf{p}_i = \mathbf{p}_{tot} \tag{9.35}$$

Therefore, we conclude that the **total linear momentum of the system** equals the total mass multiplied by the velocity of the center of mass. In other words, the total linear momentum of the system is equal to that of a single particle of mass M moving with a velocity $\mathbf{v}_{CM}$.

If we now differentiate Equation 9.34 with respect to time, we get the **acceleration of the center of mass** of the system:

Acceleration of the center of mass

$$\mathbf{a}_{CM} = \frac{d\mathbf{v}_{CM}}{dt} = \frac{1}{M}\sum_i m_i \frac{d\mathbf{v}_i}{dt} = \frac{1}{M}\sum_i m_i \mathbf{a}_i \tag{9.36}$$

Rearranging this expression and using Newton's second law, we obtain

$$M\mathbf{a}_{CM} = \sum_i m_i \mathbf{a}_i = \sum_i \mathbf{F}_i \tag{9.37}$$

where $\mathbf{F}_i$ is the net force on particle i.

The forces on any particle in the system may include both external forces (from outside the system) and internal forces (from within the system). However, by Newton's third law, the internal force exerted by particle 1 on particle 2, for example, is equal in magnitude and opposite in direction to the internal force exerted by particle 2 on particle 1. Thus, when we sum over all internal forces in Equation 9.37, they cancel in pairs and the net force on the system is caused *only* by external forces. Thus, we can write Equation 9.37 in the form

Newton's second law for a system of particles

$$\sum \mathbf{F}_{ext} = M\mathbf{a}_{CM} = \frac{d\mathbf{p}_{tot}}{dt} \tag{9.38}$$

That is, the resultant external force on a system of particles equals the total mass of the system multiplied by the acceleration of the center of mass. If we compare this with Newton's second law for a single particle, we see that

The center of mass of a system of particles of combined mass M moves like an equivalent particle of mass M would move under the influence of the resultant external force on the system.

Finally, we see that if the resultant external force is zero, then from Equation 9.38 it follows that

$$\frac{d\mathbf{p}_{tot}}{dt} = M\mathbf{a}_{CM} = 0$$

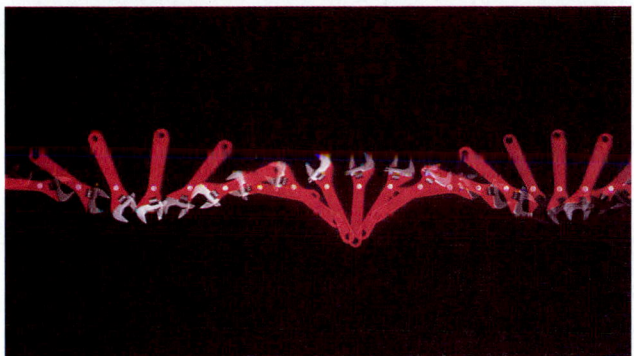

Figure 9.25 Multiflash photograph showing an overhead view of a wrench moving on a horizontal surface. The center of mass of the wrench moves in a straight line as the wrench rotates about this point, shown by the white dots. *(Richard Megna/ Fundamental Photographs)*

so that

$$\mathbf{p}_{\text{tot}} = M\mathbf{v}_{\text{CM}} = \text{constant} \qquad (\text{when } \Sigma \mathbf{F}_{\text{ext}} = 0) \qquad \textbf{(9.39)}$$

That is, the total linear momentum of a system of particles is conserved if no net external force is acting on the system. It follows that for an isolated system of particles, both the total momentum and the velocity of the center of mass are constant in time, as shown in Figure 9.25. This is a generalization to a many-particle system of the law of conservation of momentum discussed in Section 9.1 for a two-particle system.

Suppose an isolated system consisting of two or more members is at rest. The center of mass of such a system remains at rest unless acted upon by an external force. For example, consider a system made up of a swimmer standing on a raft, with the system initially at rest. When the swimmer dives horizontally off the raft, the center of mass of the system remains at rest (if we neglect friction between raft and water). Furthermore, the linear momentum of the diver is equal in magnitude to that of the raft but opposite in direction.

As another example, suppose an unstable atom initially at rest suddenly breaks up into two fragments of masses M_A and M_B, with velocities $\mathbf{v}_A$ and $\mathbf{v}_B$, respectively. Because the total momentum of the system before the breakup is zero, the total momentum of the system after the breakup must also be zero. Therefore, $M_A\mathbf{v}_A + M_B\mathbf{v}_B = 0$. If the velocity of one of the fragments is known, the recoil velocity of the other fragment can be calculated.

EXAMPLE 9.15 The Sliding Bear

Suppose you tranquilize a polar bear on a smooth glacier as part of a research effort. How might you estimate the bear's mass using a measuring tape, a rope, and knowledge of your own mass?

Solution Tie one end of the rope around the bear, and then lay out the tape measure on the ice with one end at the bear's original position, as shown in Figure 9.26. Grab hold of the free end of the rope and position yourself as shown, noting your location. Take off your spiked shoes and pull on the rope hand over hand. Both you and the bear will slide over the ice until you meet. From the tape, observe how far you have slid, x_p, and how far the bear has slid, x_b. The point where you meet the bear is the constant location of the center of mass of the system (bear plus you), and so you can determine the mass of the bear from $m_b x_b = m_p x_p$. (Unfortunately, you cannot get back to your spiked shoes and so are in big trouble if the bear wakes up!)

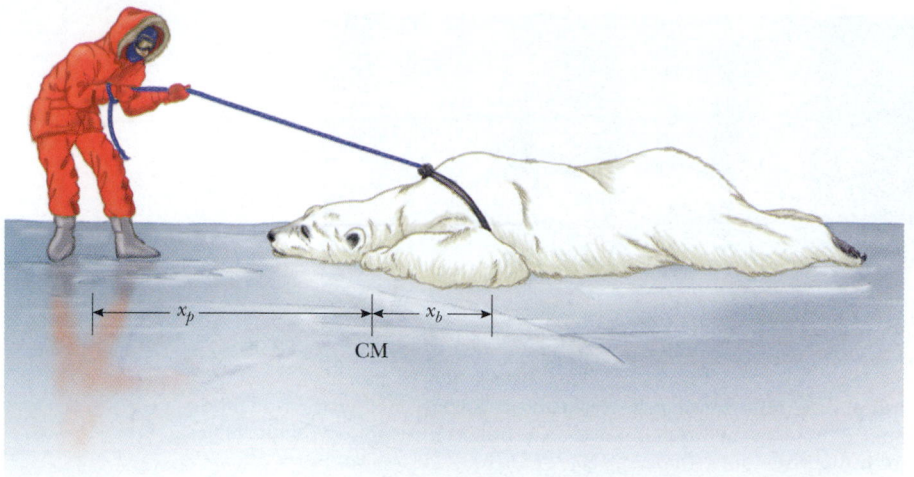

Figure 9.26   The center of mass of an isolated system remains at rest unless acted on by an external force. How can you determine the mass of the polar bear?

CONCEPTUAL EXAMPLE 9.16   Exploding Projectile

A projectile fired into the air suddenly explodes into several fragments (Fig. 9.27). What can be said about the motion of the center of mass of the system made up of all the fragments after the explosion?

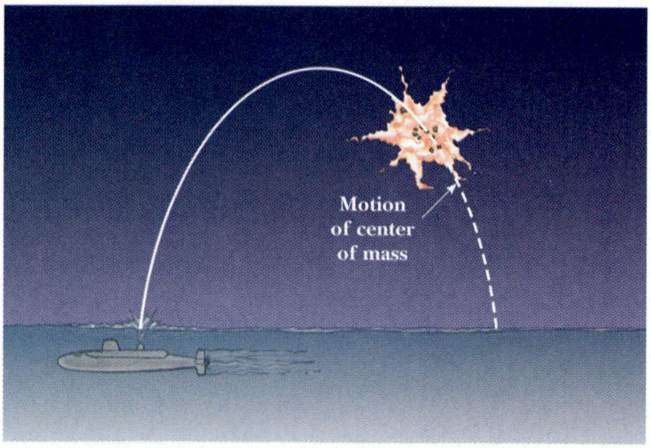

Solution   Neglecting air resistance, the only external force on the projectile is the gravitational force. Thus, if the projectile did not explode, it would continue to move along the parabolic path indicated by the broken line in Figure 9.27. Because the forces caused by the explosion are internal, they do not affect the motion of the center of mass. Thus, after the explosion the center of mass of the system (the fragments) follows the same parabolic path the projectile would have followed if there had been no explosion.

Motion
of center
of mass

Figure 9.27   When a projectile explodes into several fragments, the center of mass of the system made up of all the fragments follows the same parabolic path the projectile would have taken had there been no explosion.

EXAMPLE 9.17   The Exploding Rocket

A rocket is fired vertically upward. At the instant it reaches an altitude of 1 000 m and a speed of 300 m/s, it explodes into three equal fragments. One fragment continues to move upward with a speed of 450 m/s following the explosion. The second fragment has a speed of 240 m/s and is moving east right after the explosion. What is the velocity of the third fragment right after the explosion?

Solution   Let us call the total mass of the rocket M; hence, the mass of each fragment is $M/3$. Because the forces of the explosion are internal to the system and cannot affect its total momentum, the total momentum $\mathbf{p}_i$ of the rocket just before the explosion must equal the total momentum $\mathbf{p}_f$ of the fragments right after the explosion.

Before the explosion:

$$\mathbf{p}_i = M\mathbf{v}_i = M(300\mathbf{j}) \text{ m/s}$$

After the explosion:

$$\mathbf{p}_f = \frac{M}{3}(240\mathbf{i}) \text{ m/s} + \frac{M}{3}(450\mathbf{j}) \text{ m/s} + \frac{M}{3}\mathbf{v}_f$$

where $\mathbf{v}_f$ is the unknown velocity of the third fragment. Equating these two expressions (because $\mathbf{p}_i = \mathbf{p}_f$) gives

$$\frac{M}{3}\mathbf{v}_f + M(80\mathbf{i}) \text{ m/s} + M(150\mathbf{j}) \text{ m/s} = M(300\mathbf{j}) \text{ m/s}$$

$$\mathbf{v}_f = (-240\mathbf{i} + 450\mathbf{j}) \text{ m/s}$$

What does the sum of the momentum vectors for all the fragments look like?

Exercise Find the position of the center of mass of the system of fragments relative to the ground 3.00 s after the explosion. Assume the rocket engine is nonoperative after the explosion.

Answer The x coordinate does not change; $y_{\text{CM}} = 1.86$ km.

Optional Section

9.8 ▶ ROCKET PROPULSION

When ordinary vehicles, such as automobiles and locomotives, are propelled, the driving force for the motion is friction. In the case of the automobile, the driving force is the force exerted by the road on the car. A locomotive "pushes" against the tracks; hence, the driving force is the force exerted by the tracks on the locomotive. However, a rocket moving in space has no road or tracks to push against. Therefore, the source of the propulsion of a rocket must be something other than friction. Figure 9.28 is a dramatic photograph of a spacecraft at liftoff. **The operation of a rocket depends upon the law of conservation of linear momentum as applied to a system of particles, where the system is the rocket plus its ejected fuel.**

Rocket propulsion can be understood by first considering the mechanical system consisting of a machine gun mounted on a cart on wheels. As the gun is fired,

Figure 9.28 Liftoff of the space shuttle *Columbia*. Enormous thrust is generated by the shuttle's liquid-fuel engines, aided by the two solid-fuel boosters. Many physical principles from mechanics, thermodynamics, and electricity and magnetism are involved in such a launch. *(Courtesy of NASA)*

The force from a nitrogen-propelled, hand-controlled device allows an astronaut to move about freely in space without restrictive tethers. *(Courtesy of NASA)*

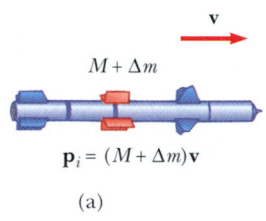

$$\mathbf{p}_i = (M + \Delta m)\mathbf{v}$$

(a)

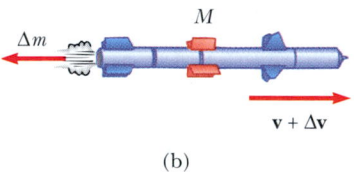

(b)

Figure 9.29   Rocket propulsion. (a) The initial mass of the rocket plus all its fuel is $M + \Delta m$ at a time t, and its speed is v. (b) At a time $t + \Delta t$, the rocket's mass has been reduced to M and an amount of fuel Δm has been ejected. The rocket's speed increases by an amount Δv.

Expression for rocket propulsion

each bullet receives a momentum $m\mathbf{v}$ in some direction, where $\mathbf{v}$ is measured with respect to a stationary Earth frame. The momentum of the system made up of cart, gun, and bullets must be conserved. Hence, for each bullet fired, the gun and cart must receive a compensating momentum in the opposite direction. That is, the reaction force exerted by the bullet on the gun accelerates the cart and gun, and the cart moves in the direction opposite that of the bullets. If n is the number of bullets fired each second, then the average force exerted on the gun is $\mathbf{F}_{av} = nm\mathbf{v}$.

In a similar manner, as a rocket moves in free space, its linear momentum changes when some of its mass is released in the form of ejected gases. **Because the gases are given momentum when they are ejected out of the engine, the rocket receives a compensating momentum in the opposite direction.** Therefore, the rocket is accelerated as a result of the "push," or thrust, from the exhaust gases. In free space, the center of mass of the system (rocket plus expelled gases) moves uniformly, independent of the propulsion process.[5]

Suppose that at some time t, the magnitude of the momentum of a rocket plus its fuel is $(M + \Delta m)v$, where v is the speed of the rocket relative to the Earth (Fig. 9.29a). Over a short time interval Δt, the rocket ejects fuel of mass Δm, and so at the end of this interval the rocket's speed is $v + \Delta v$, where Δv is the change in speed of the rocket (Fig. 9.29b). If the fuel is ejected with a speed v_e relative to the rocket (the subscript "e" stands for *exhaust*, and v_e is usually called the *exhaust speed*), the velocity of the fuel relative to a stationary frame of reference is $v - v_e$. Thus, if we equate the total initial momentum of the system to the total final momentum, we obtain

$$(M + \Delta m)v = M(v + \Delta v) + \Delta m(v - v_e)$$

where M represents the mass of the rocket and its remaining fuel after an amount of fuel having mass Δm has been ejected. Simplifying this expression gives

$$M \Delta v = v_e \Delta m$$

We also could have arrived at this result by considering the system in the center-of-mass frame of reference, which is a frame having the same velocity as the center of mass of the system. In this frame, the total momentum of the system is zero; therefore, if the rocket gains a momentum $M \Delta v$ by ejecting some fuel, the exhausted fuel obtains a momentum $v_e \Delta m$ in the *opposite* direction, so that $M \Delta v - v_e \Delta m = 0$. If we now take the limit as Δt goes to zero, we get $\Delta v \to dv$ and $\Delta m \to dm$. Futhermore, the increase in the exhaust mass dm corresponds to an equal decrease in the rocket mass, so that $dm = -dM$. Note that dM is given a negative sign because it represents a decrease in mass. Using this fact, we obtain

$$M \, dv = v_e \, dm = -v_e \, dM \qquad \textbf{(9.40)}$$

Integrating this equation and taking the initial mass of the rocket plus fuel to be M_i and the final mass of the rocket plus its remaining fuel to be M_f, we obtain

$$\int_{v_i}^{v_f} dv = -v_e \int_{M_i}^{M_f} \frac{dM}{M}$$

$$v_f - v_i = v_e \ln\!\left(\frac{M_i}{M_f}\right) \qquad \textbf{(9.41)}$$

---

[5]It is interesting to note that the rocket and machine gun represent cases of the reverse of a perfectly inelastic collision: Momentum is conserved, but the kinetic energy of the system increases (at the expense of chemical potential energy in the fuel).

This is the basic expression of rocket propulsion. First, it tells us that the increase in rocket speed is proportional to the exhaust speed of the ejected gases, v_e. Therefore, the exhaust speed should be very high. Second, the increase in rocket speed is proportional to the natural logarithm of the ratio M_i/M_f. Therefore, this ratio should be as large as possible, which means that the mass of the rocket without its fuel should be as small as possible and the rocket should carry as much fuel as possible.

The **thrust** on the rocket is the force exerted on it by the ejected exhaust gases. We can obtain an expression for the thrust from Equation 9.40:

$$\text{Thrust} = M\frac{dv}{dt} = \left| v_e \frac{dM}{dt} \right| \qquad\qquad \textbf{(9.42)}$$

This expression shows us that the thrust increases as the exhaust speed increases and as the rate of change of mass (called the *burn rate*) increases.

EXAMPLE 9.18   **A Rocket in Space**

A rocket moving in free space has a speed of 3.0×10^3 m/s relative to the Earth. Its engines are turned on, and fuel is ejected in a direction opposite the rocket's motion at a speed of 5.0×10^3 m/s relative to the rocket. (a) What is the speed of the rocket relative to the Earth once the rocket's mass is reduced to one-half its mass before ignition?

Solution   We can guess that the speed we are looking for must be greater than the original speed because the rocket is accelerating. Applying Equation 9.41, we obtain

$$v_f = v_i + v_e \ln\left(\frac{M_i}{M_f}\right)$$

$$= 3.0 \times 10^3 \text{ m/s} + (5.0 \times 10^3 \text{ m/s})\ln\left(\frac{M_i}{0.5\ M_i}\right)$$

$$= \boxed{6.5 \times 10^3 \text{ m/s}}$$

(b) What is the thrust on the rocket if it burns fuel at the rate of 50 kg/s?

Solution

$$\text{Thrust} = \left| v_e \frac{dM}{dt} \right| = (5.0 \times 10^3 \text{ m/s})(50 \text{ kg/s})$$

$$= \boxed{2.5 \times 10^5 \text{ N}}$$

EXAMPLE 9.19   **Fighting a Fire**

Two firefighters must apply a total force of 600 N to steady a hose that is discharging water at 3 600 L/min. Estimate the speed of the water as it exits the nozzle.

Solution   The water is exiting at 3 600 L/min, which is 60 L/s. Knowing that 1 L of water has a mass of 1 kg, we can say that about 60 kg of water leaves the nozzle every second. As the water leaves the hose, it exerts on the hose a thrust that must be counteracted by the 600-N force exerted on the hose by the firefighters. So, applying Equation 9.42 gives

$$\text{Thrust} = \left| v_e \frac{dM}{dt} \right|$$

$$600 \text{ N} = |v_e(60 \text{ kg/s})|$$

$$v_e = \boxed{10 \text{ m/s}}$$

Firefighting is dangerous work. If the nozzle should slip from

their hands, the movement of the hose due to the thrust it receives from the rapidly exiting water could injure the firefighters.

Firefighters attack a burning house with a hose line. (© *Bill Stormont/The Stock Market*)

SUMMARY

The **linear momentum p** of a particle of mass m moving with a velocity **v** is

$$\mathbf{p} \equiv m\mathbf{v} \tag{9.1}$$

The law of **conservation of linear momentum** indicates that the total momentum of an isolated system is conserved. If two particles form an isolated system, their total momentum is conserved regardless of the nature of the force between them. Therefore, the total momentum of the system at all times equals its initial total momentum, or

$$\mathbf{p}_{1i} + \mathbf{p}_{2i} = \mathbf{p}_{1f} + \mathbf{p}_{2f} \tag{9.5}$$

The **impulse** imparted to a particle by a force **F** is equal to the change in the momentum of the particle:

$$\mathbf{I} \equiv \int_{t_i}^{t_f} \mathbf{F} \, dt = \Delta\mathbf{p} \tag{9.9}$$

This is known as the **impulse–momentum theorem.**

Impulsive forces are often very strong compared with other forces on the system and usually act for a very short time, as in the case of collisions.

When two particles collide, the total momentum of the system before the collision always equals the total momentum after the collision, regardless of the nature of the collision. An **inelastic collision** is one for which the total kinetic energy is not conserved. A **perfectly inelastic collision** is one in which the colliding bodies stick together after the collision. An **elastic collision** is one in which kinetic energy is constant.

In a two- or three-dimensional collision, the components of momentum in each of the three directions (x, y, and z) are conserved independently.

The position vector of the center of mass of a system of particles is defined as

$$\mathbf{r}_{CM} \equiv \frac{\sum_i m_i \mathbf{r}_i}{M} \tag{9.30}$$

where $M = \sum_i m_i$ is the total mass of the system and $\mathbf{r}_i$ is the position vector of the ith particle.

The position vector of the center of mass of a rigid body can be obtained from the integral expression

$$\mathbf{r}_{CM} = \frac{1}{M} \int \mathbf{r} \, dm \tag{9.33}$$

The velocity of the center of mass for a system of particles is

$$\mathbf{v}_{CM} = \frac{\sum_i m_i \mathbf{v}_i}{M} \tag{9.34}$$

The total momentum of a system of particles equals the total mass multiplied by the velocity of the center of mass.

Newton's second law applied to a system of particles is

$$\sum \mathbf{F}_{ext} = M\mathbf{a}_{CM} = \frac{d\mathbf{p}_{tot}}{dt} \tag{9.38}$$

where $\mathbf{a}_{CM}$ is the acceleration of the center of mass and the sum is over all external forces. The center of mass moves like an imaginary particle of mass M under the influence of the resultant external force on the system. It follows from Equation 9.38 that the total momentum of the system is conserved if there are no external forces acting on it.

QUESTIONS

1. If the kinetic energy of a particle is zero, what is its linear momentum?

2. If the speed of a particle is doubled, by what factor is its momentum changed? By what factor is its kinetic energy changed?

3. If two particles have equal kinetic energies, are their momenta necessarily equal? Explain.

4. If two particles have equal momenta, are their kinetic energies necessarily equal? Explain.

5. An isolated system is initially at rest. Is it possible for parts of the system to be in motion at some later time? If so, explain how this might occur.

6. If two objects collide and one is initially at rest, is it possible for both to be at rest after the collision? Is it possible for one to be at rest after the collision? Explain.

7. Explain how linear momentum is conserved when a ball bounces from a floor.

8. Is it possible to have a collision in which all of the kinetic energy is lost? If so, cite an example.

9. In a perfectly elastic collision between two particles, does the kinetic energy of each particle change as a result of the collision?

10. When a ball rolls down an incline, its linear momentum increases. Does this imply that momentum is not conserved? Explain.

11. Consider a perfectly inelastic collision between a car and a large truck. Which vehicle loses more kinetic energy as a result of the collision?

12. Can the center of mass of a body lie outside the body? If so, give examples.

13. Three balls are thrown into the air simultaneously. What is the acceleration of their center of mass while they are in motion?

14. A meter stick is balanced in a horizontal position with the index fingers of the right and left hands. If the two fingers are slowly brought together, the stick remains balanced and the two fingers always meet at the 50-cm mark regardless of their original positions (try it!). Explain.

15. A sharpshooter fires a rifle while standing with the butt of the gun against his shoulder. If the forward momentum of a bullet is the same as the backward momentum of the gun, why is it not as dangerous to be hit by the gun as by the bullet?

16. A piece of mud is thrown against a brick wall and sticks to the wall. What happens to the momentum of the mud? Is momentum conserved? Explain.

17. Early in this century, Robert Goddard proposed sending a rocket to the Moon. Critics took the position that in a vacuum, such as exists between the Earth and the Moon, the gases emitted by the rocket would have nothing to push against to propel the rocket. According to *Scientific American* (January 1975), Goddard placed a gun in a vacuum and fired a blank cartridge from it. (A blank cartridge fires only the wadding and hot gases of the burning gunpowder.) What happened when the gun was fired?

18. A pole-vaulter falls from a height of 6.0 m onto a foam rubber pad. Can you calculate his speed just before he reaches the pad? Can you estimate the force exerted on him due to the collision? Explain.

19. Explain how you would use a balloon to demonstrate the mechanism responsible for rocket propulsion.

20. Does the center of mass of a rocket in free space accelerate? Explain. Can the speed of a rocket exceed the exhaust speed of the fuel? Explain.

21. A ball is dropped from a tall building. Identify the system for which linear momentum is conserved.

22. A bomb, initially at rest, explodes into several pieces. (a) Is linear momentum conserved? (b) Is kinetic energy conserved? Explain.

23. NASA often uses the gravity of a planet to "slingshot" a probe on its way to a more distant planet. This is actually a collision where the two objects do not touch. How can the probe have its speed increased in this manner?

24. The Moon revolves around the Earth. Is the Moon's linear momentum conserved? Is its kinetic energy conserved? Assume that the Moon's orbit is circular.

25. A raw egg dropped to the floor breaks apart upon impact. However, a raw egg dropped onto a thick foam rubber cushion from a height of about 1 m rebounds without breaking. Why is this possible? (If you try this experiment, be sure to catch the egg after the first bounce.)

26. On the subject of the following positions, state your own view and argue to support it: (a) The best theory of motion is that force causes acceleration. (b) The true measure of a force's effectiveness is the work it does, and the best theory of motion is that work on an object changes its energy. (c) The true measure of a force's effect is impulse, and the best theory of motion is that impulse injected into an object changes its momentum.

PROBLEMS

1, **2**, } = straightforward, intermediate, challenging   ☐ = full solution available in the *Student Solutions Manual and Study Guide*
WEB = solution posted at **http://www.saunderscollege.com/physics/**   🖳 = Computer useful in solving problem   🎯 = Interactive Physics
☐ = paired numerical/symbolic problems

Section 9.1   Linear Momentum and Its Conservation

1. A 3.00-kg particle has a velocity of $(3.00\mathbf{i} - 4.00\mathbf{j})$ m/s.
   (a) Find its x and y components of momentum.
   (b) Find the magnitude and direction of its momentum.

2. A 0.100-kg ball is thrown straight up into the air with an initial speed of 15.0 m/s. Find the momentum of the ball (a) at its maximum height and (b) halfway up to its maximum height.

3. A 40.0-kg child standing on a frozen pond throws a 0.500-kg stone to the east with a speed of 5.00 m/s. Neglecting friction between child and ice, find the recoil velocity of the child.

4. A pitcher claims he can throw a baseball with as much momentum as a 3.00-g bullet moving with a speed of 1 500 m/s. A baseball has a mass of 0.145 kg. What must be its speed if the pitcher's claim is valid?

5. How fast can you set the Earth moving? In particular, when you jump straight up as high as you can, you give the Earth a maximum recoil speed of what order of magnitude? Model the Earth as a perfectly solid object. In your solution, state the physical quantities you take as data and the values you measure or estimate for them.

6. Two blocks of masses M and $3M$ are placed on a horizontal, frictionless surface. A light spring is attached to one of them, and the blocks are pushed together with the spring between them (Fig. P9.6). A cord initially holding the blocks together is burned; after this, the block of mass $3M$ moves to the right with a speed of 2.00 m/s. (a) What is the speed of the block of mass M? (b) Find the original elastic energy in the spring if $M = 0.350$ kg.

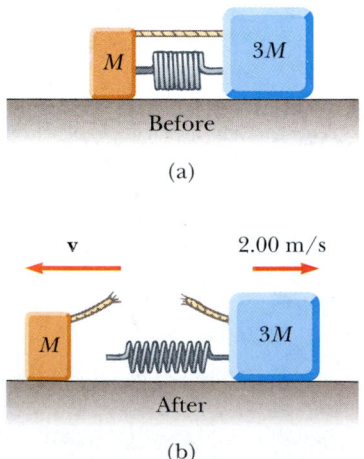

Figure P9.6

7. (a) A particle of mass m moves with momentum p. Show that the kinetic energy of the particle is given by $K = p^2/2m$. (b) Express the magnitude of the particle's momentum in terms of its kinetic energy and mass.

Section 9.2   Impulse and Momentum

8. A car is stopped for a traffic signal. When the light turns green, the car accelerates, increasing its speed from zero to 5.20 m/s in 0.832 s. What linear impulse and average force does a 70.0-kg passenger in the car experience?

9. An estimated force–time curve for a baseball struck by a bat is shown in Figure P9.9. From this curve, determine (a) the impulse delivered to the ball, (b) the average force exerted on the ball, and (c) the peak force exerted on the ball.

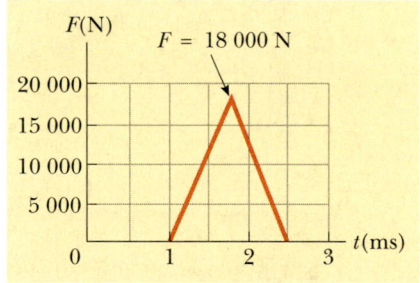

Figure P9.9

10. A tennis player receives a shot with the ball (0.060 0 kg) traveling horizontally at 50.0 m/s and returns the shot with the ball traveling horizontally at 40.0 m/s in the opposite direction. (a) What is the impulse delivered to the ball by the racket? (b) What work does the racket do on the ball?

WEB 11. A 3.00-kg steel ball strikes a wall with a speed of 10.0 m/s at an angle of 60.0° with the surface. It bounces off with the same speed and angle (Fig. P9.11). If the ball is in contact with the wall for 0.200 s, what is the average force exerted on the ball by the wall?

12. In a slow-pitch softball game, a 0.200-kg softball crossed the plate at 15.0 m/s at an angle of 45.0° below the horizontal. The ball was hit at 40.0 m/s, 30.0° above the horizontal. (a) Determine the impulse delivered to the ball. (b) If the force on the ball increased linearly for 4.00 ms, held constant for 20.0 ms, and then decreased to zero linearly in another 4.00 ms, what was the maximum force on the ball?

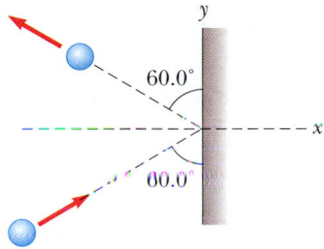

Figure P9.11

13. A garden hose is held in the manner shown in Figure P9.13. The hose is initially full of motionless water. What additional force is necessary to hold the nozzle stationary after the water is turned on if the discharge rate is 0.600 kg/s with a speed of 25.0 m/s?

Figure P9.13

14. A professional diver performs a dive from a platform 10 m above the water surface. Estimate the order of magnitude of the average impact force she experiences in her collision with the water. State the quantities you take as data and their values.

Section 9.3 Collisions

Section 9.4 Elastic and Inelastic Collisions in One Dimension

15. High-speed stroboscopic photographs show that the head of a golf club of mass 200 g is traveling at 55.0 m/s just before it strikes a 46.0-g golf ball at rest on a tee. After the collision, the club head travels (in the same direction) at 40.0 m/s. Find the speed of the golf ball just after impact.

16. A 75.0-kg ice skater, moving at 10.0 m/s, crashes into a stationary skater of equal mass. After the collision, the two skaters move as a unit at 5.00 m/s. Suppose the average force a skater can experience without breaking a bone is 4 500 N. If the impact time is 0.100 s, does a bone break?

17. A 10.0-g bullet is fired into a stationary block of wood ($m = 5.00$ kg). The relative motion of the bullet stops

inside the block. The speed of the bullet-plus-wood combination immediately after the collision is measured as 0.600 m/s. What was the original speed of the bullet?

18. As shown in Figure P9.18, a bullet of mass m and speed v passes completely through a pendulum bob of mass M. The bullet emerges with a speed of $v/2$. The pendulum bob is suspended by a stiff rod of length ℓ and negligible mass. What is the minimum value of v such that the pendulum bob will barely swing through a complete vertical circle?

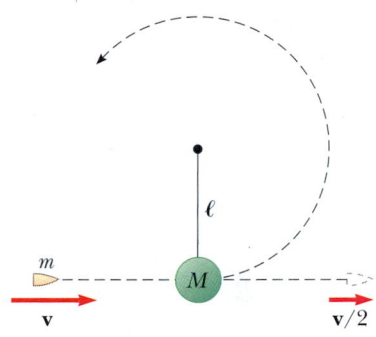

Figure P9.18

19. A 45.0-kg girl is standing on a plank that has a mass of 150 kg. The plank, originally at rest, is free to slide on a frozen lake, which is a flat, frictionless supporting surface. The girl begins to walk along the plank at a constant speed of 1.50 m/s relative to the plank. (a) What is her speed relative to the ice surface? (b) What is the speed of the plank relative to the ice surface?

20. Gayle runs at a speed of 4.00 m/s and dives on a sled, which is initially at rest on the top of a frictionless snow-covered hill. After she has descended a vertical distance of 5.00 m, her brother, who is initially at rest, hops on her back and together they continue down the hill. What is their speed at the bottom of the hill if the total vertical drop is 15.0 m? Gayle's mass is 50.0 kg, the sled has a mass of 5.00 kg and her brother has a mass of 30.0 kg.

21. A 1 200-kg car traveling initially with a speed of 25.0 m/s in an easterly direction crashes into the rear end of a 9 000-kg truck moving in the same direction at 20.0 m/s (Fig. P9.21). The velocity of the car right after the collision is 18.0 m/s to the east. (a) What is the velocity of the truck right after the collision? (b) How much mechanical energy is lost in the collision? Account for this loss in energy.

22. A railroad car of mass 2.50×10^4 kg is moving with a speed of 4.00 m/s. It collides and couples with three other coupled railroad cars, each of the same mass as the single car and moving in the same direction with an initial speed of 2.00 m/s. (a) What is the speed of the four cars after the collision? (b) How much energy is lost in the collision?

Figure P9.21

23. Four railroad cars, each of mass 2.50×10^4 kg, are coupled together and coasting along horizontal tracks at a speed of v_i toward the south. A very strong but foolish movie actor, riding on the second car, uncouples the front car and gives it a big push, increasing its speed to 4.00 m/s southward. The remaining three cars continue moving toward the south, now at 2.00 m/s. (a) Find the initial speed of the cars. (b) How much work did the actor do? (c) State the relationship between the process described here and the process in Problem 22.

24. A 7.00-kg bowling ball collides head-on with a 2.00-kg bowling pin. The pin flies forward with a speed of 3.00 m/s. If the ball continues forward with a speed of 1.80 m/s, what was the initial speed of the ball? Ignore rotation of the ball.

WEB **25.** A neutron in a reactor makes an elastic head-on collision with the nucleus of a carbon atom initially at rest. (a) What fraction of the neutron's kinetic energy is transferred to the carbon nucleus? (b) If the initial kinetic energy of the neutron is 1.60×10^{-13} J, find its final kinetic energy and the kinetic energy of the carbon nucleus after the collision. (The mass of the carbon nucleus is about 12.0 times greater than the mass of the neutron.)

26. Consider a frictionless track *ABC* as shown in Figure P9.26. A block of mass $m_1 = 5.00$ kg is released from *A*. It makes a head-on elastic collision at *B* with a block of mass $m_2 = 10.0$ kg that is initially at rest. Calculate the maximum height to which m_1 rises after the collision.

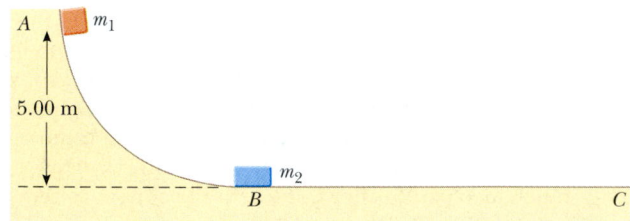

Figure P9.26

WEB **27.** A 12.0-g bullet is fired into a 100-g wooden block initially at rest on a horizontal surface. After impact, the block slides 7.50 m before coming to rest. If the coefficient of friction between the block and the surface is

0.650, what was the speed of the bullet immediately before impact?

28. A 7.00-g bullet, when fired from a gun into a 1.00-kg block of wood held in a vise, would penetrate the block to a depth of 8.00 cm. This block of wood is placed on a frictionless horizontal surface, and a 7.00-g bullet is fired from the gun into the block. To what depth will the bullet penetrate the block in this case?

Section 9.5   Two-Dimensional Collisions

29. A 90.0-kg fullback running east with a speed of 5.00 m/s is tackled by a 95.0-kg opponent running north with a speed of 3.00 m/s. If the collision is perfectly inelastic, (a) calculate the speed and direction of the players just after the tackle and (b) determine the energy lost as a result of the collision. Account for the missing energy.

30. The mass of the blue puck in Figure P9.30 is 20.0% greater than the mass of the green one. Before colliding, the pucks approach each other with equal and opposite momenta, and the green puck has an initial speed of 10.0 m/s. Find the speeds of the pucks after the collision if half the kinetic energy is lost during the collision.

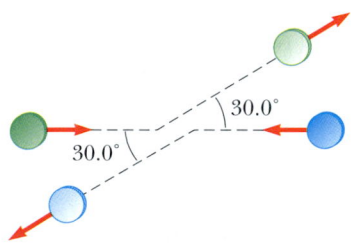

Figure P9.30

31. Two automobiles of equal mass approach an intersection. One vehicle is traveling with velocity 13.0 m/s toward the east and the other is traveling north with a speed of v_{2i}. Neither driver sees the other. The vehicles collide in the intersection and stick together, leaving parallel skid marks at an angle of 55.0° north of east. The speed limit for both roads is 35 mi/h, and the driver of the northward-moving vehicle claims he was within the speed limit when the collision occurred. Is he telling the truth?

32. A proton, moving with a velocity of $v_i\mathbf{i}$, collides elastically with another proton that is initially at rest. If the two protons have equal speeds after the collision, find (a) the speed of each proton after the collision in terms of v_i and (b) the direction of the velocity vectors after the collision.

33. A billiard ball moving at 5.00 m/s strikes a stationary ball of the same mass. After the collision, the first ball moves at 4.33 m/s and at an angle of 30.0° with respect to the original line of motion. Assuming an elastic collision (and ignoring friction and rotational motion), find the struck ball's velocity.

34. A 0.300-kg puck, initially at rest on a horizontal, frictionless surface, is struck by a 0.200-kg puck moving initially along the x axis with a speed of 2.00 m/s. After the collision, the 0.200-kg puck has a speed of 1.00 m/s at an angle of $\theta = 53.0°$ to the positive x axis (see Fig. 9.14). (a) Determine the velocity of the 0.300-kg puck after the collision. (b) Find the fraction of kinetic energy lost in the collision.

35. A 3.00-kg mass with an initial velocity of $5.00\mathbf{i}$ m/s collides with and sticks to a 2.00-kg mass with an initial velocity of $-3.00\mathbf{j}$ m/s. Find the final velocity of the composite mass.

36. Two shuffleboard disks of equal mass, one orange and the other yellow, are involved in an elastic, glancing collision. The yellow disk is initially at rest and is struck by the orange disk moving with a speed of 5.00 m/s. After the collision, the orange disk moves along a direction that makes an angle of 37.0° with its initial direction of motion, and the velocity of the yellow disk is perpendicular to that of the orange disk (after the collision). Determine the final speed of each disk.

37. Two shuffleboard disks of equal mass, one orange and the other yellow, are involved in an elastic, glancing collision. The yellow disk is initially at rest and is struck by the orange disk moving with a speed v_i. After the collision, the orange disk moves along a direction that makes an angle θ with its initial direction of motion, and the velocity of the yellow disk is perpendicular to that of the orange disk (after the collision). Determine the final speed of each disk.

38. During the battle of Gettysburg, the gunfire was so intense that several bullets collided in midair and fused together. Assume a 5.00-g Union musket ball was moving to the right at a speed of 250 m/s, 20.0° above the horizontal, and that a 3.00-g Confederate ball was moving to the left at a speed of 280 m/s, 15.0° above the horizontal. Immediately after they fuse together, what is their velocity?

WEB **39.** An unstable nucleus of mass 17.0×10^{-27} kg initially at rest disintegrates into three particles. One of the particles, of mass 5.00×10^{-27} kg, moves along the y axis with a velocity of 6.00×10^6 m/s. Another particle, of mass 8.40×10^{-27} kg, moves along the x axis with a speed of 4.00×10^6 m/s. Find (a) the velocity of the third particle and (b) the total kinetic energy increase in the process.

Section 9.6   The Center of Mass

40. Four objects are situated along the y axis as follows: A 2.00-kg object is at + 3.00 m, a 3.00-kg object is at + 2.50 m, a 2.50-kg object is at the origin, and a 4.00-kg object is at − 0.500 m. Where is the center of mass of these objects?

41. A uniform piece of sheet steel is shaped as shown in Figure P9.41. Compute the x and y coordinates of the center of mass of the piece.

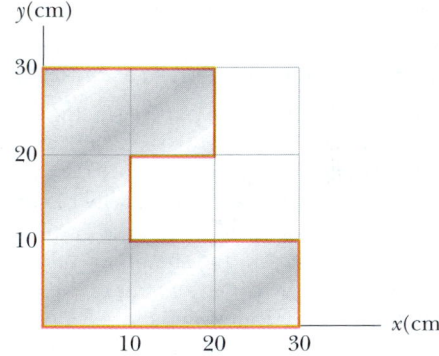

Figure P9.41

42. The mass of the Earth is 5.98×10^{24} kg, and the mass of the Moon is 7.36×10^{22} kg. The distance of separation, measured between their centers, is 3.84×10^8 m. Locate the center of mass of the Earth−Moon system as measured from the center of the Earth.

43. A water molecule consists of an oxygen atom with two hydrogen atoms bound to it (Fig. P9.43). The angle between the two bonds is 106°. If the bonds are 0.100 nm long, where is the center of mass of the molecule?

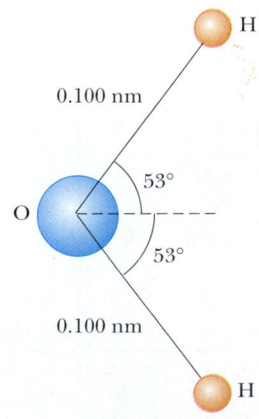

Figure P9.43

44. A 0.400-kg mass m_1 has position $\mathbf{r}_1 = 12.0\mathbf{j}$ cm. A 0.800-kg mass m_2 has position $\mathbf{r}_2 = -12.0\mathbf{i}$ cm. Another 0.800-kg mass m_3 has position $\mathbf{r}_3 = (12.0\mathbf{i} - 12.0\mathbf{j})$ cm. Make a drawing of the masses. Start from the origin and, to the scale 1 cm = 1 kg·cm, construct the vector $m_1\mathbf{r}_1$, then the vector $m_1\mathbf{r}_1 + m_2\mathbf{r}_2$, then the vector $m_1\mathbf{r}_1 + m_2\mathbf{r}_2 + m_3\mathbf{r}_3$, and at last $\mathbf{r}_{CM} = (m_1\mathbf{r}_1 + m_2\mathbf{r}_2 + m_3\mathbf{r}_3)/(m_1 + m_2 + m_3)$. Observe that the head of the vector $\mathbf{r}_{CM}$ indicates the position of the center of mass.

45. A rod of length 30.0 cm has linear density (mass-per-length) given by

$$\lambda = 50.0 \text{ g/m} + 20.0x \text{ g/m}^2$$

where x is the distance from one end, measured in meters. (a) What is the mass of the rod? (b) How far from the $x = 0$ end is its center of mass?

Section 9.7   Motion of a System of Particles

46. Consider a system of two particles in the xy plane: $m_1 = 2.00$ kg is at $\mathbf{r}_1 = (1.00\mathbf{i} + 2.00\mathbf{j})$ m and has velocity $(3.00\mathbf{i} + 0.500\mathbf{j})$ m/s; $m_2 = 3.00$ kg is at $\mathbf{r}_2 = (-4.00\mathbf{i} - 3.00\mathbf{j})$ m and has velocity $(3.00\mathbf{i} - 2.00\mathbf{j})$ m/s. (a) Plot these particles on a grid or graph paper. Draw their position vectors and show their velocities. (b) Find the position of the center of mass of the system and mark it on the grid. (c) Determine the velocity of the center of mass and also show it on the diagram. (d) What is the total linear momentum of the system?

47. Romeo (77.0 kg) entertains Juliet (55.0 kg) by playing his guitar from the rear of their boat at rest in still water, 2.70 m away from Juliet who is in the front of the boat. After the serenade, Juliet carefully moves to the rear of the boat (away from shore) to plant a kiss on Romeo's cheek. How far does the 80.0-kg boat move toward the shore it is facing?

48. Two masses, 0.600 kg and 0.300 kg, begin uniform motion at the same speed, 0.800 m/s, from the origin at $t = 0$ and travel in the directions shown in Figure P9.48. (a) Find the velocity of the center of mass in unit–vector notation. (b) Find the magnitude and direction

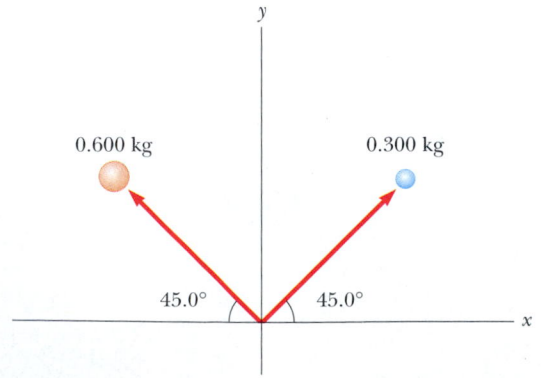

Figure P9.48

of the velocity of the center of mass. (c) Write the position vector of the center of mass as a function of time.

49. A 2.00-kg particle has a velocity of $(2.00\mathbf{i} - 3.00\mathbf{j})$ m/s, and a 3.00-kg particle has a velocity of $(1.00\mathbf{i} + 6.00\mathbf{j})$ m/s. Find (a) the velocity of the center of mass and (b) the total momentum of the system.

50. A ball of mass 0.200 kg has a velocity of $1.50\mathbf{i}$ m/s; a ball of mass 0.300 kg has a velocity of $-0.400\mathbf{i}$ m/s. They meet in a head-on elastic collision. (a) Find their velocities after the collision. (b) Find the velocity of their center of mass before and after the collision.

(Optional)
Section 9.8   Rocket Propulsion

WEB 51. The first stage of a Saturn V space vehicle consumes fuel and oxidizer at the rate of 1.50×10^4 kg/s, with an exhaust speed of 2.60×10^3 m/s. (a) Calculate the thrust produced by these engines. (b) Find the initial acceleration of the vehicle on the launch pad if its initial mass is 3.00×10^6 kg. [*Hint:* You must include the force of gravity to solve part (b).]

52. A large rocket with an exhaust speed of $v_e = 3\,000$ m/s develops a thrust of 24.0 million newtons. (a) How much mass is being blasted out of the rocket exhaust per second? (b) What is the maximum speed the rocket can attain if it starts from rest in a force-free environment with $v_e = 3.00$ km/s and if 90.0% of its initial mass is fuel and oxidizer?

53. A rocket for use in deep space is to have the capability of boosting a total load (payload plus rocket frame and engine) of 3.00 metric tons to a speed of 10 000 m/s. (a) It has an engine and fuel designed to produce an exhaust speed of 2 000 m/s. How much fuel plus oxidizer is required? (b) If a different fuel and engine design could give an exhaust speed of 5 000 m/s, what amount of fuel and oxidizer would be required for the same task?

54. A rocket car has a mass of 2 000 kg unfueled and a mass of 5 000 kg when completely fueled. The exhaust velocity is 2 500 m/s. (a) Calculate the amount of fuel used to accelerate the completely fueled car from rest to 225 m/s (about 500 mi/h). (b) If the burn rate is constant at 30.0 kg/s, calculate the time it takes the car to reach this speed. Neglect friction and air resistance.

ADDITIONAL PROBLEMS

55. **Review Problem.** A 60.0-kg person running at an initial speed of 4.00 m/s jumps onto a 120-kg cart initially at rest (Fig. P9.55). The person slides on the cart's top surface and finally comes to rest relative to the cart. The coefficient of kinetic friction between the person and the cart is 0.400. Friction between the cart and ground can be neglected. (a) Find the final velocity of the person and cart relative to the ground. (b) Find the frictional force acting on the person while he is sliding

across the top surface of the cart. (c) How long does the frictional force act on the person? (d) Find the change in momentum of the person and the change in momentum of the cart. (e) Determine the displacement of the person relative to the ground while he is sliding on the cart. (f) Determine the displacement of the cart relative to the ground while the person is sliding. (g) Find the change in kinetic energy of the person. (h) Find the change in kinetic energy of the cart. (i) Explain why the answers to parts (g) and (h) differ. (What kind of collision is this, and what accounts for the loss of mechanical energy?)

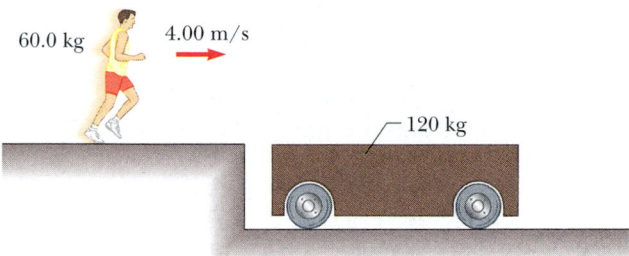

60.0 kg   4.00 m/s

120 kg

Figure P9.55

56. A golf ball ($m = 46.0$ g) is struck a blow that makes an angle of 45.0° with the horizontal. The ball lands 200 m away on a flat fairway. If the golf club and ball are in contact for 7.00 ms, what is the average force of impact? (Neglect air resistance.)

57. An 8.00-g bullet is fired into a 2.50-kg block that is initially at rest at the edge of a frictionless table of height 1.00 m (Fig. P9.57). The bullet remains in the block, and after impact the block lands 2.00 m from the bottom of the table. Determine the initial speed of the bullet.

58. A bullet of mass m is fired into a block of mass M that is initially at rest at the edge of a frictionless table of height h (see Fig. P9.57). The bullet remains in the block, and after impact the block lands a distance d from the bottom of the table. Determine the initial speed of the bullet.

59. An 80.0-kg astronaut is working on the engines of his ship, which is drifting through space with a constant velocity. The astronaut, wishing to get a better view of the Universe, pushes against the ship and much later finds himself 30.0 m behind the ship and at rest with respect to it. Without a thruster, the only way to return to the ship is to throw his 0.500-kg wrench directly away from the ship. If he throws the wrench with a speed of 20.0 m/s relative to the ship, how long does it take the astronaut to reach the ship?

60. A small block of mass $m_1 = 0.500$ kg is released from rest at the top of a curve-shaped frictionless wedge of mass $m_2 = 3.00$ kg, which sits on a frictionless horizontal surface, as shown in Figure P9.60a. When the block leaves the wedge, its velocity is measured to be 4.00 m/s to the right, as in Figure P9.60b. (a) What is the velocity of the wedge after the block reaches the horizontal surface? (b) What is the height h of the wedge?

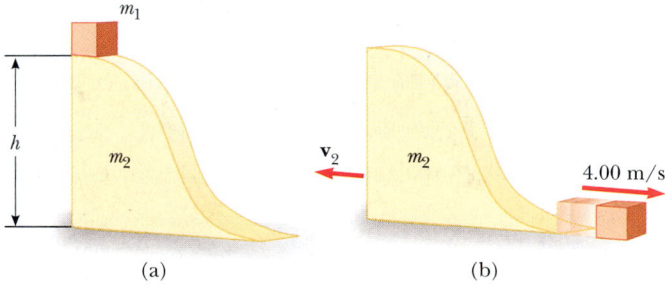

(a)   (b)

Figure P9.60

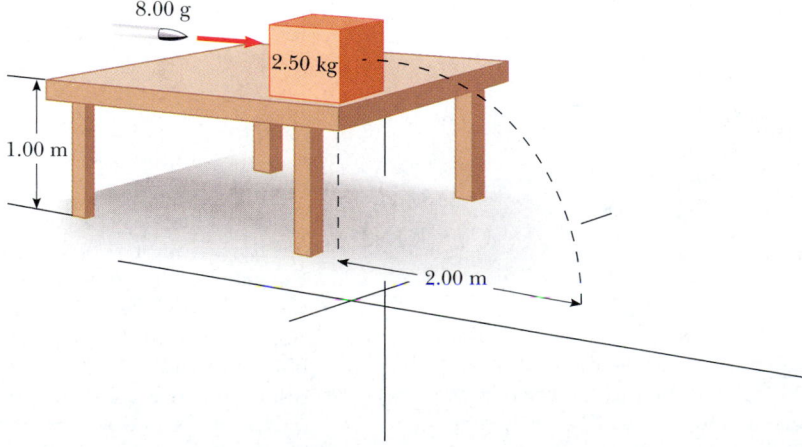

8.00 g

2.50 kg

1.00 m

2.00 m

Figure P9.57 Problems 57 and 58.

61. Tarzan, whose mass is 80.0 kg, swings from a 3.00-m vine that is horizontal when he starts. At the bottom of his arc, he picks up 60.0-kg Jane in a perfectly inelastic collision. What is the height of the highest tree limb they can reach on their upward swing?

62. A jet aircraft is traveling at 500 mi/h (223 m/s) in horizontal flight. The engine takes in air at a rate of 80.0 kg/s and burns fuel at a rate of 3.00 kg/s. If the exhaust gases are ejected at 600 m/s relative to the aircraft, find the thrust of the jet engine and the delivered horsepower.

63. A 75.0-kg firefighter slides down a pole while a constant frictional force of 300 N retards her motion. A horizontal 20.0-kg platform is supported by a spring at the bottom of the pole to cushion the fall. The firefighter starts from rest 4.00 m above the platform, and the spring constant is 4 000 N/m. Find (a) the firefighter's speed just before she collides with the platform and (b) the maximum distance the spring is compressed. (Assume the frictional force acts during the entire motion.)

64. A cannon is rigidly attached to a carriage, which can move along horizontal rails but is connected to a post by a large spring, initially unstretched and with force constant $k = 2.00 \times 10^4$ N/m, as shown in Figure P9.64. The cannon fires a 200-kg projectile at a velocity of 125 m/s directed 45.0° above the horizontal. (a) If the mass of the cannon and its carriage is 5 000 kg, find the recoil speed of the cannon. (b) Determine the maximum extension of the spring. (c) Find the maximum force the spring exerts on the carriage. (d) Consider the system consisting of the cannon, carriage, and shell. Is the momentum of this system conserved during the firing? Why or why not?

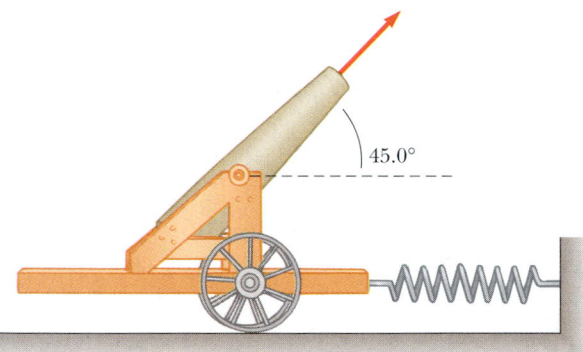

Figure P9.64

65. A chain of length L and total mass M is released from rest with its lower end just touching the top of a table, as shown in Figure P9.65a. Find the force exerted by the table on the chain after the chain has fallen through a distance x, as shown in Figure P9.65b. (Assume each link comes to rest the instant it reaches the table.)

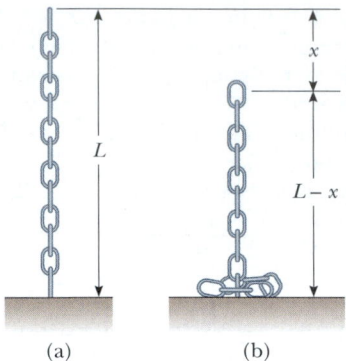

(a)   (b)

Figure P9.65

66. Two gliders are set in motion on an air track. A spring of force constant k is attached to the near side of one glider. The first glider of mass m_1 has a velocity of $\mathbf{v}_1$, and the second glider of mass m_2 has a velocity of $\mathbf{v}_2$, as shown in Figure P9.66 ($v_1 > v_2$). When m_1 collides with the spring attached to m_2 and compresses the spring to its maximum compression x_m, the velocity of the gliders is $\mathbf{v}$. In terms of $\mathbf{v}_1$, $\mathbf{v}_2$, m_1, m_2, and k, find (a) the velocity $\mathbf{v}$ at maximum compression, (b) the maximum compression x_m, and (c) the velocities of each glider after m_1 has lost contact with the spring.

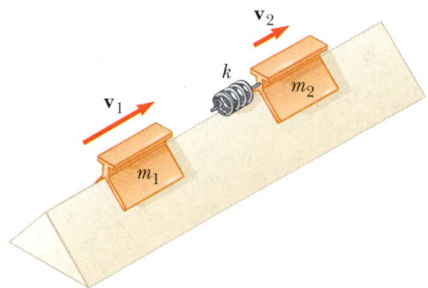

Figure P9.66

67. Sand from a stationary hopper falls onto a moving conveyor belt at the rate of 5.00 kg/s, as shown in Figure P9.67. The conveyor belt is supported by frictionless rollers and moves at a constant speed of 0.750 m/s under the action of a constant horizontal external force $\mathbf{F}_{ext}$ supplied by the motor that drives the belt. Find (a) the sand's rate of change of momentum in the horizontal direction, (b) the force of friction exerted by the belt on the sand, (c) the external force $\mathbf{F}_{ext}$, (d) the work done by $\mathbf{F}_{ext}$ in 1 s, and (e) the kinetic energy acquired by the falling sand each second due to the change in its horizontal motion. (f) Why are the answers to parts (d) and (e) different?

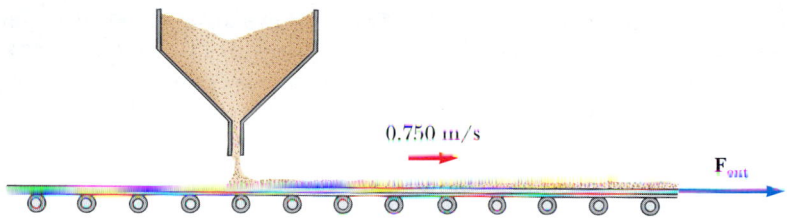

0.750 m/s

F_{out}

Figure P9.67

68. A rocket has total mass $M_i = 360$ kg, including 330 kg of fuel and oxidizer. In interstellar space it starts from rest, turns on its engine at time $t = 0$, and puts out exhaust with a relative speed of $v_e = 1\,500$ m/s at the constant rate $k = 2.50$ kg/s. Although the fuel will last for an actual burn time of 330 kg/(2.5 kg/s) = 132 s, define a "projected depletion time" as $T_p = M_i/k = 360$ kg/(2.5 kg/s) = 144 s. (This would be the burn time if the rocket could use its payload, fuel tanks, and even the walls of the combustion chamber as fuel.) (a) Show that during the burn the velocity of the rocket is given as a function of time by

$$v(t) = -v_e \ln(1 - t/T_p)$$

(b) Make a graph of the velocity of the rocket as a function of time for times running from 0 to 132 s. (c) Show that the acceleration of the rocket is

$$a(t) = v_e/(T_p - t)$$

(d) Graph the acceleration as a function of time. (e) Show that the displacement of the rocket from its initial position at $t = 0$ is

$$x(t) = v_e(T_p - t)\ln(1 - t/T_p) + v_e t$$

(f) Graph the displacement during the burn.

69. A 40.0-kg child stands at one end of a 70.0-kg boat that is 4.00 m in length (Fig. P9.69). The boat is initially 3.00 m from the pier. The child notices a turtle on a rock near the far end of the boat and proceeds to walk to that end to catch the turtle. Neglecting friction be-

tween the boat and the water, (a) describe the subsequent motion of the system (child plus boat). (b) Where is the child *relative to the pier* when he reaches the far end of the boat? (c) Will he catch the turtle? (Assume he can reach out 1.00 m from the end of the boat.)

70. A student performs a ballistic pendulum experiment, using an apparatus similar to that shown in Figure 9.11b. She obtains the following average data: $h = 8.68$ cm, $m_1 = 68.8$ g, and $m_2 = 263$ g. The symbols refer to the quantities in Figure 9.11a. (a) Determine the initial speed v_{1i} of the projectile. (b) In the second part of her experiment she is to obtain v_{1i} by firing the same projectile horizontally (with the pendulum removed from the path) and measuring its horizontal displacement x and vertical displacement y (Fig. P9.70). Show that the initial speed of the projectile is related to x and y through the relationship

$$v_{1i} = \frac{x}{\sqrt{2y/g}}$$

What numerical value does she obtain for v_{1i} on the basis of her measured values of $x = 257$ cm and $y = 85.3$ cm? What factors might account for the difference in this value compared with that obtained in part (a)?

Figure P9.69

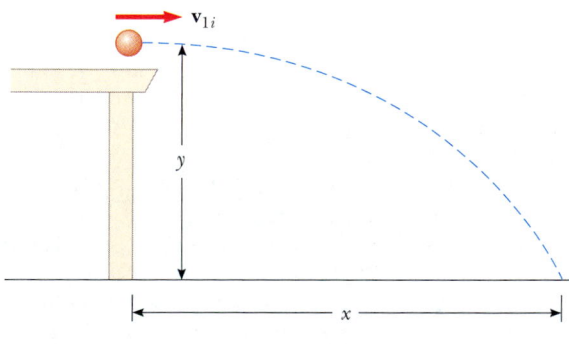

v_{1i}

y

x

Figure P9.70

71. A 5.00-g bullet moving with an initial speed of 400 m/s is fired into and passes through a 1.00-kg block, as shown in Figure P9.71. The block, initially at rest on a

frictionless, horizontal surface, is connected to a spring of force constant 900 N/m. If the block moves 5.00 cm to the right after impact, find (a) the speed at which the bullet emerges from the block and (b) the energy lost in the collision.

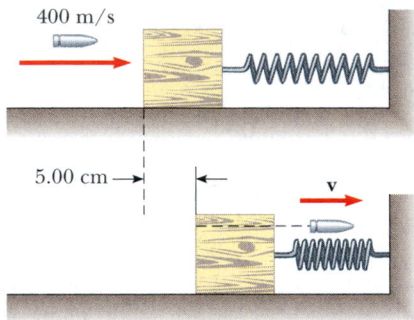

400 m/s

5.00 cm →

v

Figure P9.71

72. Two masses m and $3m$ are moving toward each other along the x axis with the same initial speeds v_i. Mass m is traveling to the left, while mass $3m$ is traveling to the right. They undergo a head-on elastic collision and each rebounds along the same line as it approached. Find the final speeds of the masses.

73. Two masses m and $3m$ are moving toward each other along the x axis with the same initial speeds v_i. Mass m is traveling to the left, while mass $3m$ is traveling to the right. They undergo an elastic glancing collision such

that mass m is moving downward after the collision at right angles from its initial direction. (a) Find the final speeds of the two masses. (b) What is the angle θ at which the mass $3m$ is scattered?

74. **Review Problem.**   There are (one can say) three co-equal theories of motion: Newton's second law, stating that the total force on an object causes its acceleration; the work–kinetic energy theorem, stating that the total work on an object causes its change in kinetic energy; and the impulse–momentum theorem, stating that the total impulse on an object causes its change in momentum. In this problem, you compare predictions of the three theories in one particular case. A 3.00-kg object has a velocity of $7.00\mathbf{j}$ m/s. Then, a total force $12.0\mathbf{i}$ N acts on the object for 5.00 s. (a) Calculate the object's final velocity, using the impulse–momentum theorem. (b) Calculate its acceleration from $\mathbf{a} = (\mathbf{v}_f - \mathbf{v}_i)/t$. (c) Calculate its acceleration from $\mathbf{a} = \Sigma\mathbf{F}/m$. (d) Find the object's vector displacement from $\mathbf{r} = \mathbf{v}_i t + \frac{1}{2}\mathbf{a}\,t^2$. (e) Find the work done on the object from $W = \mathbf{F}\cdot\mathbf{r}$. (f) Find the final kinetic energy from $\frac{1}{2}mv_f^2 = \frac{1}{2}m\mathbf{v}_f\cdot\mathbf{v}_f$. (g) Find the final kinetic energy from $\frac{1}{2}mv_i^2 + W$.

75. A rocket has a total mass of $M_i = 360$ kg, including 330 kg of fuel and oxidizer. In interstellar space it starts from rest. Its engine is turned on at time $t = 0$, and it puts out exhaust with a relative speed of $v_e = 1\,500$ m/s at the constant rate 2.50 kg/s. The burn lasts until the fuel runs out at time 330 kg/(2.5 kg/s) = 132 s. Set up and carry out a computer analysis of the motion according to Euler's method. Find (a) the final velocity of the rocket and (b) the distance it travels during the burn.

ANSWERS TO QUICK QUIZZES

9.1 (d). Two identical objects ($m_1 = m_2$) traveling in the same direction at the same speed ($v_1 = v_2$) have the same kinetic energies and the same momenta. However, this is not true if the two objects are moving at the same speed but in different directions. In the latter case, $K_1 = K_2$, but the differing velocity directions indicate that $\mathbf{p}_1 \neq \mathbf{p}_2$ because momentum is a vector quantity.

It also is possible for particular combinations of masses and velocities to satisfy $K_1 = K_2$ but not $p_1 = p_2$. For example, a 1-kg object moving at 2 m/s has the same kinetic energy as a 4-kg object moving at 1 m/s, but the two clearly do not have the same momenta.

9.2 (b), (c), (a). The slower the ball, the easier it is to catch. If the momentum of the medicine ball is the same as the momentum of the baseball, the speed of the medicine ball must be 1/10 the speed of the baseball because the medicine ball has 10 times the mass. If the kinetic energies are the same, the speed of the medicine ball must be $1/\sqrt{10}$ the speed of the baseball because of the squared speed term in the formula for K. The medicine

ball is hardest to catch when it has the same speed as the baseball.

9.3 (c) and (e). Object 2 has a greater acceleration because of its smaller mass. Therefore, it takes less time to travel the distance d. Thus, even though the force applied to objects 1 and 2 is the same, the change in momentum is less for object 2 because Δt is smaller. Therefore, because the initial momenta were the same (both zero), $p_1 > p_2$. The work $W = Fd$ done on both objects is the same because both F and d are the same in the two cases. Therefore, $K_1 = K_2$.

9.4 Because the passenger is brought from the car's initial speed to a full stop, the change in momentum (the impulse) is the same regardless of whether the passenger is stopped by dashboard, seatbelt, or airbag. However, the dashboard stops the passenger very quickly, the seatbelt takes a little more time, and the airbag takes the most time. Therefore, the dashboard applies the greatest force, the seatbelt an intermediate force, and the airbag the least force. Airbags are designed to work in conjunction with seatbelts. The airbag keeps your head from snapping forward. Make sure you wear your seatbelt at all times while in a moving vehicle. Make sure you wear your seatbelt at all times while in a moving vehicle.

9.5 If we define the ball as our system, momentum is not conserved. The ball's speed—and hence its momentum—continually increase. This is consistent with the fact that the gravitational force is external to this chosen system. However, if we define our system as the ball and the Earth, momentum is conserved, for the Earth also has momentum because the ball exerts a gravitational force on it. As the ball falls, the Earth moves up to meet it (although the Earth's speed is on the order of 10^{25} times less than that of the ball!). This upward movement changes the Earth's momentum. The change in the Earth's momentum is numerically equal to the change in the ball's momentum but is in the opposite direction. Therefore, the total momentum of the Earth–ball system is conserved. Because the Earth's mass is so great, its upward motion is negligibly small.

9.6 (c). The greatest impulse (greatest change in momentum) is imparted to the Frisbee when the skater reverses its momentum vector by catching it and throwing it back. Since this is when the skater imparts the greatest impulse to the Frisbee, then this also is when the Frisbee imparts the greatest impulse to her.

9.7 Both are equally bad. Imagine watching the collision from a safer location alongside the road. As the "crush zones" of the two cars are compressed, you will see that the actual point of contact is stationary. You would see the same thing if your car were to collide with a solid wall.

9.8 No, such movement can never occur if we assume the collisions are elastic. The momentum of the system before the collision is mv, where m is the mass of ball 1 and v is its speed just before the collision. After the collision, we would have two balls, each of mass m and moving with a speed of $v/2$. Thus, the total momentum of the system after the collision would be $m(v/2) + m(v/2) = mv$. Thus, momentum is conserved. However, the kinetic energy just before the collision is $K_i = \frac{1}{2}mv^2$, and that after the collision is $K_f = \frac{1}{2}m(v/2)^2 + \frac{1}{2}m(v/2)^2 = \frac{1}{4}mv^2$. Thus, kinetic energy is *not* conserved. Both momentum and kinetic energy are conserved only when one ball moves out when one ball is released, two balls move out when two are released, and so on.

9.9 No they will not! The piece with the handle will have less mass than the piece made up of the end of the bat. To see why this is so, take the origin of coordinates as the center of mass before the bat was cut. Replace each cut piece by a small sphere located at the center of mass for each piece. The sphere representing the handle piece is farther from the origin, but the product of lesser mass and greater distance balances the product of greater mass and lesser distance for the end piece:

PUZZLER

Did you know that the CD inside this player spins at different speeds, depending on which song is playing? Why would such a strange characteristic be incorporated into the design of every CD player? *(George Semple)*

c h a p t e r

10

Rotation of a Rigid Object About a Fixed Axis

When an extended object, such as a wheel, rotates about its axis, the motion cannot be analyzed by treating the object as a particle because at any given time different parts of the object have different linear velocities and linear accelerations. For this reason, it is convenient to consider an extended object as a large number of particles, each of which has its own linear velocity and linear acceleration.

In dealing with a rotating object, analysis is greatly simplified by assuming that the object is rigid. A **rigid object** is one that is nondeformable—that is, it is an object in which the separations between all pairs of particles remain constant. All real bodies are deformable to some extent; however, our rigid-object model is useful in many situations in which deformation is negligible.

Rigid object

In this chapter, we treat the rotation of a rigid object about a fixed axis, which is commonly referred to as *pure rotational motion*.

10.1 ANGULAR DISPLACEMENT, VELOCITY, AND ACCELERATION

Figure 10.1 illustrates a planar (flat), rigid object of arbitrary shape confined to the xy plane and rotating about a fixed axis through O. The axis is perpendicular to the plane of the figure, and O is the origin of an xy coordinate system. Let us look at the motion of only one of the millions of "particles" making up this object. A particle at P is at a fixed distance r from the origin and rotates about it in a circle of radius r. (In fact, *every* particle on the object undergoes circular motion about O.) It is convenient to represent the position of P with its polar coordinates (r, θ), where r is the distance from the origin to P and θ is measured *counterclockwise* from some preferred direction—in this case, the positive x axis. In this representation, the only coordinate that changes in time is the angle θ; r remains constant. (In cartesian coordinates, both x and y vary in time.) As the particle moves along the circle from the positive x axis ($\theta = 0$) to P, it moves through an arc of length s, which is related to the angular position θ through the relationship

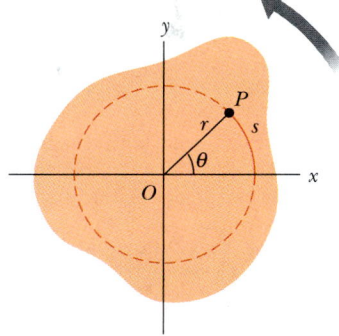

Figure 10.1 A rigid object rotating about a fixed axis through O perpendicular to the plane of the figure. (In other words, the axis of rotation is the z axis.) A particle at P rotates in a circle of radius r centered at O.

$$s = r\theta \tag{10.1a}$$

$$\theta = \frac{s}{r} \tag{10.1b}$$

It is important to note the units of θ in Equation 10.1b. Because θ is the ratio of an arc length and the radius of the circle, it is a pure number. However, we commonly give θ the artificial unit **radian** (rad), where

one radian is the angle subtended by an arc length equal to the radius of the arc.

Radian

Because the circumference of a circle is $2\pi r$, it follows from Equation 10.1b that $360°$ corresponds to an angle of $2\pi r/r$ rad $= 2\pi$ rad (one revolution). Hence, 1 rad $= 360°/2\pi \approx 57.3°$. To convert an angle in degrees to an angle in radians, we use the fact that 2π rad $= 360°$:

$$\theta \, (\text{rad}) = \frac{\pi}{180°} \, \theta \, (\text{deg})$$

For example, $60°$ equals $\pi/3$ rad, and $45°$ equals $\pi/4$ rad.

In a short track event, such as a 200-m or 400-m sprint, the runners begin from staggered positions on the track. Why don't they all begin from the same line? *(© Gerard Vandystadt/Photo Researchers, Inc.)*

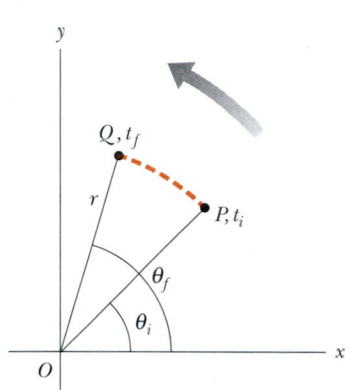

Figure 10.2 A particle on a rotating rigid object moves from P to Q along the arc of a circle. In the time interval $\Delta t = t_f - t_i$, the radius vector sweeps out an angle $\Delta \theta = \theta_f - \theta_i$.

As the particle in question on our rigid object travels from position P to position Q in a time Δt as shown in Figure 10.2, the radius vector sweeps out an angle $\Delta \theta = \theta_f - \theta_i$. This quantity $\Delta \theta$ is defined as the **angular displacement** of the particle:

$$\Delta \theta = \theta_f - \theta_i \qquad \textbf{(10.2)}$$

We define the **average angular speed** $\overline{\omega}$ (omega) as the ratio of this angular displacement to the time interval Δt:

Average angular speed

$$\overline{\omega} \equiv \frac{\theta_f - \theta_i}{t_f - t_i} = \frac{\Delta \theta}{\Delta t} \qquad \textbf{(10.3)}$$

In analogy to linear speed, the **instantaneous angular speed** ω is defined as the limit of the ratio $\Delta \theta / \Delta t$ as Δt approaches zero:

Instantaneous angular speed

$$\omega \equiv \lim_{\Delta t \to 0} \frac{\Delta \theta}{\Delta t} = \frac{d\theta}{dt} \qquad \textbf{(10.4)}$$

Angular speed has units of radians per second (rad/s), or rather second^{-1} (s^{-1}) because radians are not dimensional. We take ω to be positive when θ is increasing (counterclockwise motion) and negative when θ is decreasing (clockwise motion).

If the instantaneous angular speed of an object changes from ω_i to ω_f in the time interval Δt, the object has an angular acceleration. The **average angular acceleration** $\overline{\alpha}$ (alpha) of a rotating object is defined as the ratio of the change in the angular speed to the time interval Δt:

Average angular acceleration

$$\overline{\alpha} \equiv \frac{\omega_f - \omega_i}{t_f - t_i} = \frac{\Delta \omega}{\Delta t} \qquad \textbf{(10.5)}$$

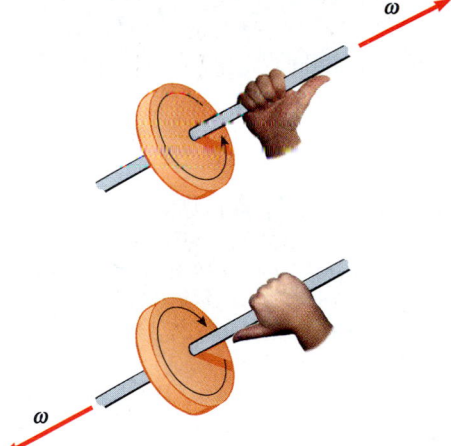

Figure 10.3 The right-hand rule for determining the direction of the angular velocity vector.

In analogy to linear acceleration, the **instantaneous angular acceleration** is defined as the limit of the ratio $\Delta\omega/\Delta t$ as Δt approaches zero:

$$\alpha \equiv \lim_{\Delta t \to 0} \frac{\Delta\omega}{\Delta t} = \frac{d\omega}{dt} \tag{10.6}$$

Instantaneous angular acceleration

Angular acceleration has units of radians per second squared (rad/s^2), or just second^{-2} (s^{-2}). Note that α is positive when the rate of counterclockwise rotation is increasing or when the rate of clockwise rotation is decreasing.

When rotating about a fixed axis, every particle on a rigid object rotates through the same angle and has the same angular speed and the same angular acceleration. That is, the quantities θ, ω, and α characterize the rotational motion of the entire rigid object. Using these quantities, we can greatly simplify the analysis of rigid-body rotation.

Angular position (θ), angular speed (ω), and angular acceleration (α) are analogous to linear position (x), linear speed (v), and linear acceleration (a). The variables θ, ω, and α differ dimensionally from the variables x, v, and a only by a factor having the unit of length.

We have not specified any direction for ω and α. Strictly speaking, these variables are the magnitudes of the angular velocity and the angular acceleration vectors $\boldsymbol{\omega}$ and $\boldsymbol{\alpha}$, respectively, and they should always be positive. Because we are considering rotation about a fixed axis, however, we can indicate the directions of the vectors by assigning a positive or negative sign to ω and α, as discussed earlier with regard to Equations 10.4 and 10.6. For rotation about a fixed axis, the only direction that uniquely specifies the rotational motion is the direction along the axis of rotation. Therefore, the directions of $\boldsymbol{\omega}$ and $\boldsymbol{\alpha}$ are along this axis. If an object rotates in the xy plane as in Figure 10.1, the direction of $\boldsymbol{\omega}$ is out of the plane of the diagram when the rotation is counterclockwise and into the plane of the diagram when the rotation is clockwise. To illustrate this convention, it is convenient to use the *right-hand rule* demonstrated in Figure 10.3. When the four fingers of the right hand are wrapped in the direction of rotation, the extended right thumb points in the direction of $\boldsymbol{\omega}$. The direction of $\boldsymbol{\alpha}$ follows from its definition $d\boldsymbol{\omega}/dt$. It is the same as the direction of $\boldsymbol{\omega}$ if the angular speed is increasing in time, and it is antiparallel to $\boldsymbol{\omega}$ if the angular speed is decreasing in time.

Quick Quiz 10.1

Describe a situation in which $\omega < 0$ and $\boldsymbol{\omega}$ and $\boldsymbol{\alpha}$ are antiparallel.

10.2 ROTATIONAL KINEMATICS: ROTATIONAL MOTION WITH CONSTANT ANGULAR ACCELERATION

In our study of linear motion, we found that the simplest form of accelerated motion to analyze is motion under constant linear acceleration. Likewise, for rotational motion about a fixed axis, the simplest accelerated motion to analyze is motion under constant angular acceleration. Therefore, we next develop kinematic relationships for this type of motion. If we write Equation 10.6 in the form $d\omega = \alpha \, dt$, and let $t_i = 0$ and $t_f = t$, we can integrate this expression directly:

$$\omega_f = \omega_i + \alpha t \qquad \text{(for constant } \alpha) \tag{10.7}$$

Substituting Equation 10.7 into Equation 10.4 and integrating once more we obtain

$$\theta_f = \theta_i + \omega_i t + \tfrac{1}{2}\alpha t^2 \qquad \text{(for constant } \alpha) \tag{10.8}$$

Rotational kinematic equations

If we eliminate t from Equations 10.7 and 10.8, we obtain

$$\omega_f^2 = \omega_i^2 + 2\alpha(\theta_f - \theta_i) \qquad \text{(for constant } \alpha) \tag{10.9}$$

Notice that these kinematic expressions for rotational motion under constant angular acceleration are of the same form as those for linear motion under constant linear acceleration with the substitutions $x \rightarrow \theta$, $v \rightarrow \omega$, and $a \rightarrow \alpha$. Table 10.1 compares the kinematic equations for rotational and linear motion.

EXAMPLE 10.1   Rotating Wheel

A wheel rotates with a constant angular acceleration of 3.50 rad/s². If the angular speed of the wheel is 2.00 rad/s at $t_i = 0$, (a) through what angle does the wheel rotate in 2.00 s?

Solution We can use Figure 10.2 to represent the wheel, and so we do not need a new drawing. This is a straightforward application of an equation from Table 10.1:

$$\theta_f - \theta_i = \omega_i t + \tfrac{1}{2}\alpha t^2 = (2.00 \text{ rad/s})(2.00 \text{ s})$$
$$+ \tfrac{1}{2} (3.50 \text{ rad/s}^2)(2.00 \text{ s})^2$$

$$= 11.0 \text{ rad} = (11.0 \text{ rad})(57.3°/\text{rad}) = \boxed{630°}$$

$$= \frac{630°}{360°/\text{rev}} = \boxed{1.75 \text{ rev}}$$

(b) What is the angular speed at $t = 2.00$ s?

Solution Because the angular acceleration and the angular speed are both positive, we can be sure our answer must be greater than 2.00 rad/s.

$$\omega_f = \omega_i + \alpha t = 2.00 \text{ rad/s} + (3.50 \text{ rad/s}^2)(2.00 \text{ s})$$
$$= \boxed{9.00 \text{ rad/s}}$$

We could also obtain this result using Equation 10.9 and the results of part (a). Try it! You also may want to see if you can formulate the linear motion analog to this problem.

Exercise Find the angle through which the wheel rotates between $t = 2.00$ s and $t = 3.00$ s.

Answer 10.8 rad.

TABLE 10.1	Kinematic Equations for Rotational and Linear Motion Under Constant Acceleration
Rotational Motion About a Fixed Axis	**Linear Motion**
$\omega_f = \omega_i + \alpha t$	$v_f = v_i + at$
$\theta_f = \theta_i + \omega_i t + \frac{1}{2}\alpha t^2$	$x_f = x_i + v_i t + \frac{1}{2}at^2$
$\omega_f{}^2 = \omega_i{}^2 + 2\alpha(\theta_f - \theta_i)$	$v_f{}^2 = v_i{}^2 + 2a(x_f - x_i)$

10.3 ANGULAR AND LINEAR QUANTITIES

In this section we derive some useful relationships between the angular speed and acceleration of a rotating rigid object and the linear speed and acceleration of an arbitrary point in the object. To do so, we must keep in mind that when a rigid object rotates about a fixed axis, as in Figure 10.4, every particle of the object moves in a circle whose center is the axis of rotation.

We can relate the angular speed of the rotating object to the tangential speed of a point P on the object. Because point P moves in a circle, the linear velocity vector $\mathbf{v}$ is always tangent to the circular path and hence is called *tangential velocity*. The magnitude of the tangential velocity of the point P is by definition the tangential speed $v = ds/dt$, where s is the distance traveled by this point measured along the circular path. Recalling that $s = r\theta$ (Eq. 10.1a) and noting that r is constant, we obtain

$$v = \frac{ds}{dt} = r\frac{d\theta}{dt}$$

Because $d\theta/dt = \omega$ (see Eq. 10.4), we can say

$$v = r\omega \qquad (10.10)$$

That is, the tangential speed of a point on a rotating rigid object equals the perpendicular distance of that point from the axis of rotation multiplied by the angular speed. Therefore, although every point on the rigid object has the same *angular* speed, not every point has the same *linear* speed because r is not the same for all points on the object. Equation 10.10 shows that the linear speed of a point on the rotating object increases as one moves outward from the center of rotation, as we would intuitively expect. The outer end of a swinging baseball bat moves much faster than the handle.

We can relate the angular acceleration of the rotating rigid object to the tangential acceleration of the point P by taking the time derivative of v:

$$a_t = \frac{dv}{dt} = r\frac{d\omega}{dt}$$

$$a_t = r\alpha \qquad (10.11)$$

That is, the tangential component of the linear acceleration of a point on a rotating rigid object equals the point's distance from the axis of rotation multiplied by the angular acceleration.

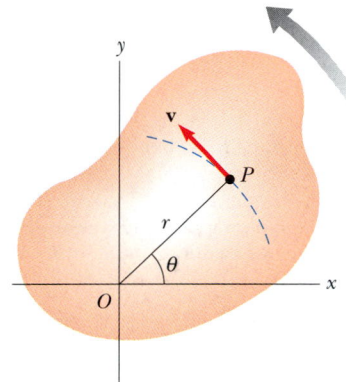

Figure 10.4 As a rigid object rotates about the fixed axis through O, the point P has a linear velocity $\mathbf{v}$ that is always tangent to the circular path of radius r.

Relationship between linear and angular speed

QuickLab

Spin a tennis ball or basketball and watch it gradually slow down and stop. Estimate α and a_t as accurately as you can.

Relationship between linear and angular acceleration

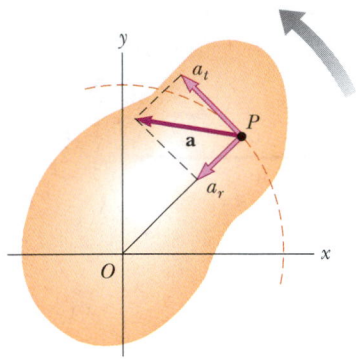

Figure 10.5 As a rigid object rotates about a fixed axis through O, the point P experiences a tangential component of linear acceleration a_t and a radial component of linear acceleration a_r. The total linear acceleration of this point is $\mathbf{a} = \mathbf{a}_t + \mathbf{a}_r$.

In Section 4.4 we found that a point rotating in a circular path undergoes a centripetal, or radial, acceleration $\mathbf{a}_r$ of magnitude v^2/r directed toward the center of rotation (Fig. 10.5). Because $v = r\omega$ for a point P on a rotating object, we can express the radial acceleration of that point as

$$a_r = \frac{v^2}{r} = r\omega^2 \qquad (10.12)$$

The total linear acceleration vector of the point is $\mathbf{a} = \mathbf{a}_t + \mathbf{a}_r$. ($\mathbf{a}_t$ describes the change in how fast the point is moving, and $\mathbf{a}_r$ represents the change in its direction of travel.) Because $\mathbf{a}$ is a vector having a radial and a tangential component, the magnitude of $\mathbf{a}$ for the point P on the rotating rigid object is

$$a = \sqrt{a_t{}^2 + a_r{}^2} = \sqrt{r^2\alpha^2 + r^2\omega^4} = r\sqrt{\alpha^2 + \omega^4} \qquad (10.13)$$

Quick Quiz 10.2

When a wheel of radius R rotates about a fixed axis, do all points on the wheel have (a) the same angular speed and (b) the same linear speed? If the angular speed is constant and equal to ω, describe the linear speeds and linear accelerations of the points located at (c) $r = 0$, (d) $r = R/2$, and (e) $r = R$, all measured from the center of the wheel.

EXAMPLE 10.2   CD Player

On a compact disc, audio information is stored in a series of pits and flat areas on the surface of the disc. The information is stored digitally, and the alternations between pits and flat areas on the surface represent binary ones and zeroes to be read by the compact disc player and converted back to sound waves. The pits and flat areas are detected by a system consisting of a laser and lenses. The length of a certain number of ones and zeroes is the same everywhere on the disc, whether the information is near the center of the disc or near its outer edge. In order that this length of ones and zeroes always passes by the laser–lens system in the same time period, the linear speed of the disc surface at the location of the lens must be constant. This requires, according to Equation 10.10, that the angular speed vary as the laser–lens system moves radially along the disc. In a typical compact disc player, the disc spins counterclockwise (Fig. 10.6), and the constant speed of the surface at the point of the laser–lens system is 1.3 m/s. (a) Find the angular speed of the disc in revolutions per minute when information is being read from the innermost first track ($r = 23$ mm) and the outermost final track ($r = 58$ mm).

Solution Using Equation 10.10, we can find the angular speed; this will give us the required linear speed at the position of the inner track,

$$\omega_i = \frac{v}{r_i} = \frac{1.3 \text{ m/s}}{2.3 \times 10^{-2}\text{ m}} = 56.5 \text{ rad/s}$$

$$= (56.5 \text{ rad/s})\left(\frac{1}{2\pi} \text{ rev/rad}\right)(60 \text{ s/min})$$

$$= 5.4 \times 10^2 \text{ rev/min}$$

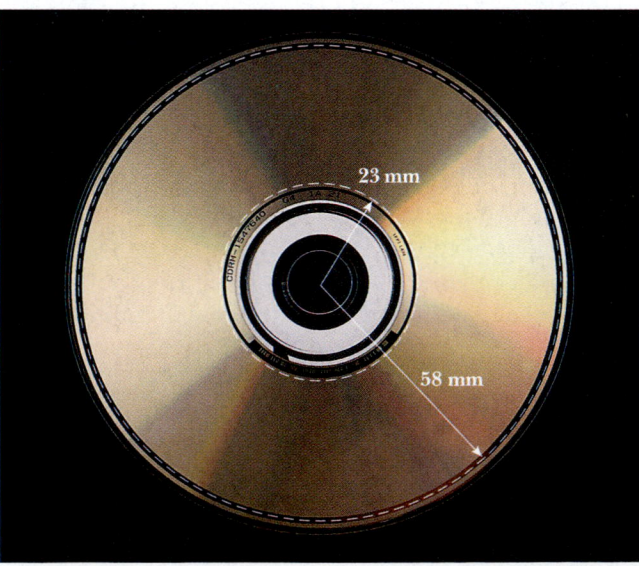

Figure 10.6 A compact disc. (*George Semple*)

For the outer track,

$$\omega_f = \frac{v}{r_f} = \frac{1.3 \text{ m/s}}{5.8 \times 10^{-2} \text{ m}} = 22.4 \text{ rad/s}$$

$$= 2.1 \times 10^2 \text{ rev/min}$$

The player adjusts the angular speed ω of the disc within this range so that information moves past the objective lens at a constant rate. These angular velocity values are positive because the direction of rotation is counterclockwise.

(b) The maximum playing time of a standard music CD is 74 minutes and 33 seconds. How many revolutions does the disc make during that time?

Solution We know that the angular speed is always decreasing, and we assume that it is decreasing steadily, with α constant. The time interval t is (74 min)(60 s/min) + 33 s = 4 473 s. We are looking for the angular position θ_f, where we set the initial angular position $\theta_i = 0$. We can use Equation 10.3, replacing the average angular speed $\bar\omega$ with its mathematical equivalent $(\omega_i + \omega_f)/2$:

$$\theta_f = \theta_i + \tfrac{1}{2}(\omega_i + \omega_f)t$$

$$= 0 + \tfrac{1}{2}(540 \text{ rev/min} + 210 \text{ rev/min})$$

$$(1 \text{ min}/60 \text{ s})(4 473 \text{ s})$$

$$= 2.8 \times 10^4 \text{ rev}$$

(c) What total length of track moves past the objective lens during this time?

Solution Because we know the (constant) linear velocity and the time interval, this is a straightforward calculation:

$$x_f = v_i t = (1.3 \text{ m/s})(4 473 \text{ s}) = 5.8 \times 10^3 \text{ m}$$

More than 3.6 miles of track spins past the objective lens!

(d) What is the angular acceleration of the CD over the 4 473-s time interval? Assume that α is constant.

Solution We have several choices for approaching this problem. Let us use the most direct approach by utilizing Equation 10.5, which is based on the definition of the term we are seeking. We should obtain a negative number for the angular acceleration because the disc spins more and more slowly in the positive direction as time goes on. Our answer should also be fairly small because it takes such a long time — more than an hour — for the change in angular speed to be accomplished:

$$\alpha = \frac{\omega_f - \omega_i}{t} = \frac{22.4 \text{ rad/s} - 56.5 \text{ rad/s}}{4 473 \text{ s}}$$

$$= -7.6 \times 10^{-3} \text{ rad/s}^2$$

The disc experiences a very gradual decrease in its rotation rate, as expected.

10.4 ROTATIONAL ENERGY

Let us now look at the kinetic energy of a rotating rigid object, considering the object as a collection of particles and assuming it rotates about a fixed z axis with an angular speed ω (Fig. 10.7). Each particle has kinetic energy determined by its mass and linear speed. If the mass of the ith particle is m_i and its linear speed is v_i, its kinetic energy is

$$K_i = \tfrac{1}{2}m_i v_i^2$$

To proceed further, we must recall that although every particle in the rigid object has the same angular speed ω, the individual linear speeds depend on the distance r_i from the axis of rotation according to the expression $v_i = r_i\omega$ (see Eq. 10.10). The *total* kinetic energy of the rotating rigid object is the sum of the kinetic energies of the individual particles:

$$K_R = \sum_i K_i = \sum_i \tfrac{1}{2}m_i v_i^2 = \tfrac{1}{2}\sum_i m_i r_i^2 \omega^2$$

We can write this expression in the form

$$K_R = \tfrac{1}{2}\left(\sum_i m_i r_i^2\right)\omega^2 \qquad \textbf{(10.14)}$$

where we have factored ω^2 from the sum because it is common to every particle.

web

If you want to learn more about the physics of CD players, visit the Special Interest Group on CD Applications and Technology at **www.sigcat.org**

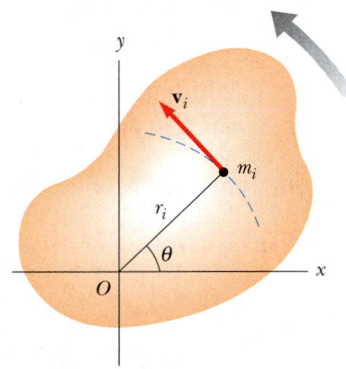

Figure 10.7 A rigid object rotating about a z axis with angular speed ω. The kinetic energy of the particle of mass m_i is $\tfrac{1}{2}m_i v_i^2$. The total kinetic energy of the object is called its rotational kinetic energy.

We simplify this expression by defining the quantity in parentheses as the **moment of inertia I**:

Moment of inertia

$$I \equiv \sum_i m_i r_i{}^2 \tag{10.15}$$

From the definition of moment of inertia, we see that it has dimensions of ML^2 (kg·m^2 in SI units).[1] With this notation, Equation 10.14 becomes

Rotational kinetic energy

$$K_R = \tfrac{1}{2} I \omega^2 \tag{10.16}$$

Although we commonly refer to the quantity $\tfrac{1}{2} I \omega^2$ as **rotational kinetic energy,** it is not a new form of energy. It is ordinary kinetic energy because it is derived from a sum over individual kinetic energies of the particles contained in the rigid object. However, the mathematical form of the kinetic energy given by Equation 10.16 is a convenient one when we are dealing with rotational motion, provided we know how to calculate I.

It is important that you recognize the analogy between kinetic energy associated with linear motion $\tfrac{1}{2} m v^2$ and rotational kinetic energy $\tfrac{1}{2} I \omega^2$. The quantities I and ω in rotational motion are analogous to m and v in linear motion, respectively. (In fact, I takes the place of m every time we compare a linear-motion equation with its rotational counterpart.) The moment of inertia is a measure of the resistance of an object to changes in its rotational motion, just as mass is a measure of the tendency of an object to resist changes in its linear motion. Note, however, that mass is an intrinsic property of an object, whereas I depends on the physical arrangement of that mass. Can you think of a situation in which an object's moment of inertia changes even though its mass does not?

EXAMPLE 10.3   **The Oxygen Molecule**

Consider an oxygen molecule (O_2) rotating in the xy plane about the z axis. The axis passes through the center of the molecule, perpendicular to its length. The mass of each oxygen atom is 2.66×10^{-26} kg, and at room temperature the average separation between the two atoms is $d = 1.21 \times 10^{-10}$ m (the atoms are treated as point masses). (a) Calculate the moment of inertia of the molecule about the z axis.

Solution   This is a straightforward application of the definition of I. Because each atom is a distance $d/2$ from the z axis, the moment of inertia about the axis is

$$I = \sum_i m_i r_i{}^2 = m\left(\frac{d}{2}\right)^2 + m\left(\frac{d}{2}\right)^2 = \tfrac{1}{2} m d^2$$

$$= \tfrac{1}{2}(2.66 \times 10^{-26} \text{ kg})(1.21 \times 10^{-10} \text{ m})^2$$

$$= 1.95 \times 10^{-46} \text{ kg·m}^2$$

This is a very small number, consistent with the minuscule masses and distances involved.

(b) If the angular speed of the molecule about the z axis is 4.60×10^{12} rad/s, what is its rotational kinetic energy?

Solution   We apply the result we just calculated for the moment of inertia in the formula for K_R:

$$K_R = \tfrac{1}{2} I \omega^2$$

$$= \tfrac{1}{2}(1.95 \times 10^{-46} \text{ kg·m}^2)(4.60 \times 10^{12} \text{ rad/s})^2$$

$$= 2.06 \times 10^{-21} \text{ J}$$

---

[1] Civil engineers use moment of inertia to characterize the elastic properties (rigidity) of such structures as loaded beams. Hence, it is often useful even in a nonrotational context.

EXAMPLE 10.4 ▸ Four Rotating Masses

Four tiny spheres are fastened to the corners of a frame of negligible mass lying in the xy plane (Fig. 10.8). We shall assume that the spheres' radii are small compared with the dimensions of the frame. (a) If the system rotates about the y axis with an angular speed ω, find the moment of inertia and the rotational kinetic energy about this axis.

Solution First, note that the two spheres of mass m, which lie on the y axis, do not contribute to I_y (that is, $r_i = 0$ for these spheres about this axis). Applying Equation 10.15, we obtain

$$I_y = \sum_i m_i r_i^2 = Ma^2 + Ma^2 = \boxed{2Ma^2}$$

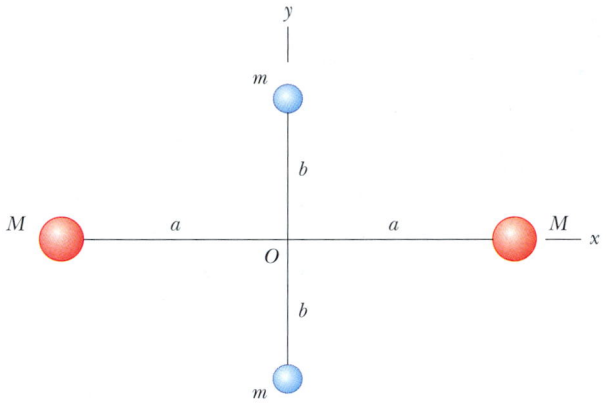

Figure 10.8 The four spheres are at a fixed separation as shown. The moment of inertia of the system depends on the axis about which it is evaluated.

Therefore, the rotational kinetic energy about the y axis is

$$K_R = \tfrac{1}{2}I_y\omega^2 = \tfrac{1}{2}(2Ma^2)\omega^2 = \boxed{Ma^2\omega^2}$$

The fact that the two spheres of mass m do not enter into this result makes sense because they have no motion about the axis of rotation; hence, they have no rotational kinetic energy. By similar logic, we expect the moment of inertia about the x axis to be $I_x = 2mb^2$ with a rotational kinetic energy about that axis of $K_R = mb^2\omega^2$.

(b) Suppose the system rotates in the xy plane about an axis through O (the z axis). Calculate the moment of inertia and rotational kinetic energy about this axis.

Solution Because r_i in Equation 10.15 is the *perpendicular* distance to the axis of rotation, we obtain

$$I_z = \sum_i m_i r_i^2 = Ma^2 + Ma^2 + mb^2 + mb^2 = \boxed{2Ma^2 + 2mb^2}$$

$$K_R = \tfrac{1}{2}I_z\omega^2 = \tfrac{1}{2}(2Ma^2 + 2mb^2)\omega^2 = \boxed{(Ma^2 + mb^2)\omega^2}$$

Comparing the results for parts (a) and (b), we conclude that the moment of inertia and therefore the rotational kinetic energy associated with a given angular speed depend on the axis of rotation. In part (b), we expect the result to include all four spheres and distances because all four spheres are rotating in the xy plane. Furthermore, the fact that the rotational kinetic energy in part (a) is smaller than that in part (b) indicates that it would take less effort (work) to set the system into rotation about the y axis than about the z axis.

10.5 ▸ CALCULATION OF MOMENTS OF INERTIA

We can evaluate the moment of inertia of an extended rigid object by imagining the object divided into many small volume elements, each of which has mass Δm. We use the definition $I = \sum_i r_i^2 \Delta m_i$ and take the limit of this sum as $\Delta m \to 0$. In this limit, the sum becomes an integral over the whole object:

$$I = \lim_{\Delta m_i \to 0} \sum_i r_i^2 \Delta m_i = \int r^2\, dm \qquad \textbf{(10.17)}$$

It is usually easier to calculate moments of inertia in terms of the volume of the elements rather than their mass, and we can easily make that change by using Equation 1.1, $\rho = m/V$, where ρ is the density of the object and V is its volume. We want this expression in its differential form $\rho = dm/dV$ because the volumes we are dealing with are very small. Solving for $dm = \rho\, dV$ and substituting the result into Equation 10.17 gives

$$I = \int \rho r^2 \, dV$$

If the object is homogeneous, then ρ is constant and the integral can be evaluated for a known geometry. If ρ is not constant, then its variation with position must be known to complete the integration.

The density given by $\rho = m/V$ sometimes is referred to as *volume density* for the obvious reason that it relates to volume. Often we use other ways of expressing density. For instance, when dealing with a sheet of uniform thickness t, we can define a *surface density* $\sigma = \rho t$, which signifies *mass per unit area*. Finally, when mass is distributed along a uniform rod of cross-sectional area A, we sometimes use *linear density* $\lambda = M/L = \rho A$, which is the *mass per unit length*.

EXAMPLE 10.5   Uniform Hoop

Find the moment of inertia of a uniform hoop of mass M and radius R about an axis perpendicular to the plane of the hoop and passing through its center (Fig. 10.9).

Solution   All mass elements dm are the same distance $r = R$ from the axis, and so, applying Equation 10.17, we obtain for the moment of inertia about the z axis through O:

$$I_z = \int r^2 \, dm = R^2 \int dm = \boxed{MR^2}$$

Note that this moment of inertia is the same as that of a single particle of mass M located a distance R from the axis of rotation.

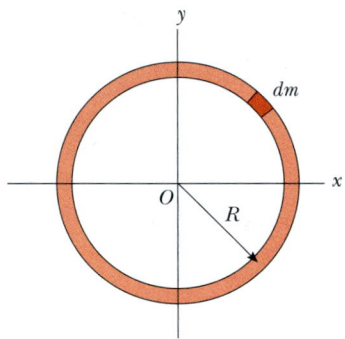

Figure 10.9   The mass elements dm of a uniform hoop are all the same distance from O.

Quick Quiz 10.3

(a) Based on what you have learned from Example 10.5, what do you expect to find for the moment of inertia of two particles, each of mass $M/2$, located anywhere on a circle of radius R around the axis of rotation? (b) How about the moment of inertia of four particles, each of mass $M/4$, again located a distance R from the rotation axis?

EXAMPLE 10.6   Uniform Rigid Rod

Calculate the moment of inertia of a uniform rigid rod of length L and mass M (Fig. 10.10) about an axis perpendicular to the rod (the y axis) and passing through its center of mass.

Solution   The shaded length element dx has a mass dm equal to the mass per unit length λ multiplied by dx:

$$dm = \lambda \, dx = \frac{M}{L} \, dx$$

Substituting this expression for dm into Equation 10.17, with $r = x$, we obtain

$$I_y = \int r^2 \, dm = \int_{-L/2}^{L/2} x^2 \frac{M}{L} \, dx = \frac{M}{L} \int_{-L/2}^{L/2} x^2 \, dx$$

$$= \frac{M}{L} \left[\frac{x^3}{3} \right]_{-L/2}^{L/2} = \boxed{\tfrac{1}{12} ML^2}$$

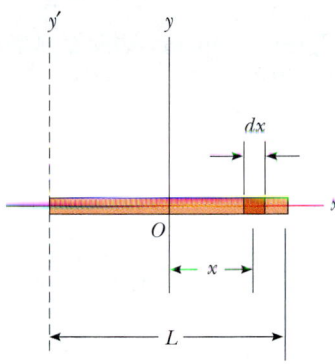

Figure 10.10 A uniform rigid rod of length L. The moment of inertia about the y axis is less than that about the y' axis. The latter axis is examined in Example 10.8.

EXAMPLE 10.7   Uniform Solid Cylinder

A uniform solid cylinder has a radius R, mass M, and length L. Calculate its moment of inertia about its central axis (the z axis in Fig. 10.11).

Solution   It is convenient to divide the cylinder into many

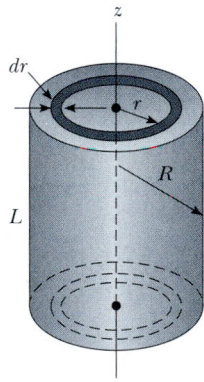

Figure 10.11 Calculating I about the z axis for a uniform solid cylinder.

cylindrical shells, each of which has radius r, thickness dr, and length L, as shown in Figure 10.11. The volume dV of each shell is its cross-sectional area multiplied by its length: $dV = dA \cdot L = (2\pi r \, dr) L$. If the mass per unit volume is ρ, then the mass of this differential volume element is $dm = \rho \, dV = \rho 2\pi rL \, dr$. Substituting this expression for dm into Equation 10.17, we obtain

$$I_z = \int r^2 \, dm = 2\pi\rho L \int_0^R r^3 \, dr = \tfrac{1}{2}\pi\rho L R^4$$

Because the total volume of the cylinder is $\pi R^2 L$, we see that $\rho = M/V = M/\pi R^2 L$. Substituting this value for ρ into the above result gives

$$(1) \qquad I_z = \boxed{\tfrac{1}{2}MR^2}$$

Note that this result does not depend on L, the length of the cylinder. In other words, it applies equally well to a long cylinder and a flat disc. Also note that this is exactly half the value we would expect were all the mass concentrated at the outer edge of the cylinder or disc. (See Example 10.5.)

Table 10.2 gives the moments of inertia for a number of bodies about specific axes. The moments of inertia of rigid bodies with simple geometry (high symmetry) are relatively easy to calculate provided the rotation axis coincides with an axis of symmetry. The calculation of moments of inertia about an arbitrary axis can be cumbersome, however, even for a highly symmetric object. Fortunately, use of an important theorem, called the **parallel-axis theorem,** often simplifies the calculation. Suppose the moment of inertia about an axis through the center of mass of an object is I_{CM}. The parallel-axis theorem states that the moment of inertia about any axis parallel to and a distance D away from this axis is

$$I = I_{CM} + MD^2 \qquad\qquad (10.18)$$

Parallel-axis theorem

TABLE 10.2   Moments of Inertia of Homogeneous Rigid Bodies with Different Geometries

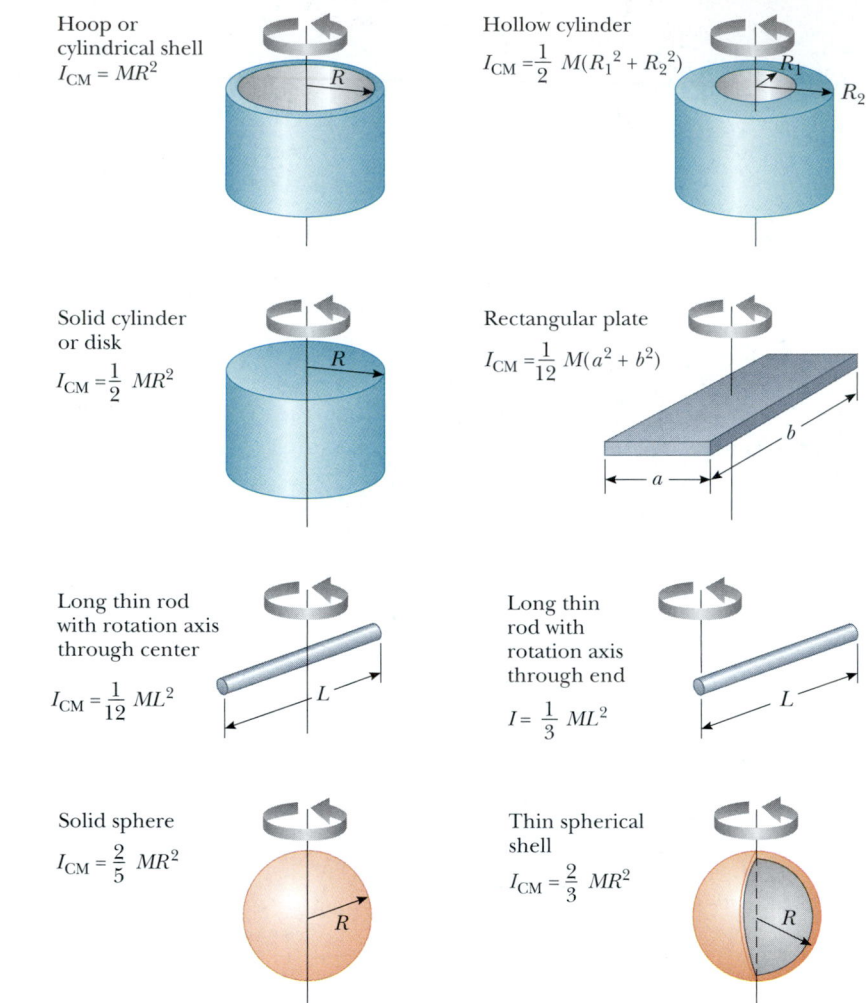

Hoop or cylindrical shell
$I_{CM} = MR^2$

Hollow cylinder
$I_{CM} = \frac{1}{2} M(R_1^2 + R_2^2)$

Solid cylinder or disk
$I_{CM} = \frac{1}{2} MR^2$

Rectangular plate
$I_{CM} = \frac{1}{12} M(a^2 + b^2)$

Long thin rod with rotation axis through center
$I_{CM} = \frac{1}{12} ML^2$

Long thin rod with rotation axis through end
$I = \frac{1}{3} ML^2$

Solid sphere
$I_{CM} = \frac{2}{5} MR^2$

Thin spherical shell
$I_{CM} = \frac{2}{3} MR^2$

Proof of the Parallel-Axis Theorem (Optional).   Suppose that an object rotates in the *xy* plane about the *z* axis, as shown in Figure 10.12, and that the coordinates of the center of mass are x_{CM}, y_{CM}. Let the mass element dm have coordinates x, y. Because this element is a distance $r = \sqrt{x^2 + y^2}$ from the *z* axis, the moment of inertia about the *z* axis is

$$I = \int r^2 \, dm = \int (x^2 + y^2) \, dm$$

However, we can relate the coordinates x, y of the mass element dm to the coordinates of this same element located in a coordinate system having the object's center of mass as its origin. If the coordinates of the center of mass are x_{CM}, y_{CM} in the original coordinate system centered on *O*, then from Figure 10.12a we see that the relationships between the unprimed and primed coordinates are $x = x' + x_{CM}$

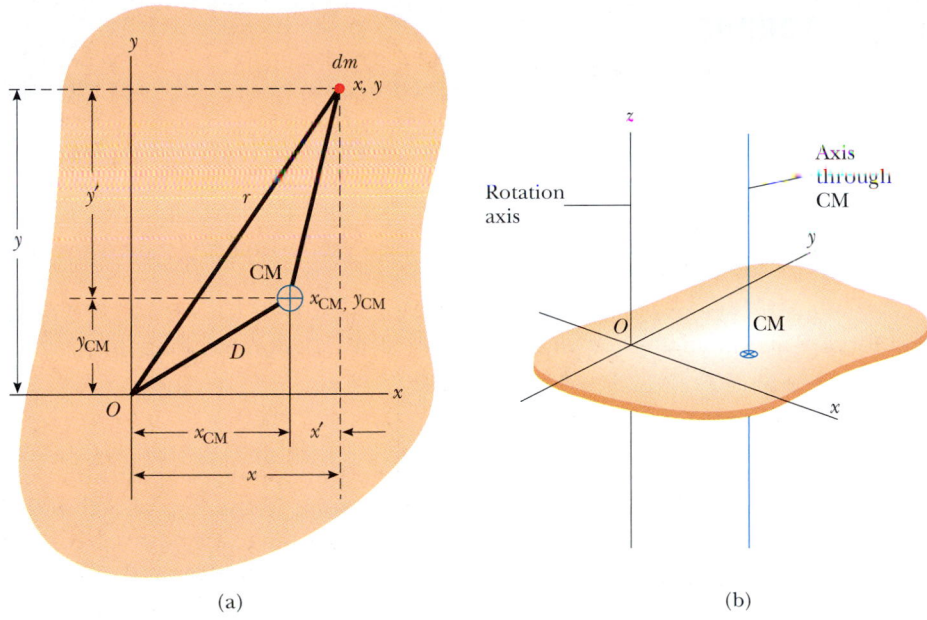

Figure 10.12 (a) The parallel-axis theorem: If the moment of inertia about an axis perpendicular to the figure through the center of mass is I_{CM}, then the moment of inertia about the z axis is $I_z = I_{CM} + MD^2$. (b) Perspective drawing showing the z axis (the axis of rotation) and the parallel axis through the CM.

and $y = y' + y_{CM}$. Therefore,

$$I = \int [(x' + x_{CM})^2 + (y' + y_{CM})^2] \, dm$$

$$= \int [(x')^2 + (y')^2] \, dm + 2x_{CM} \int x' \, dm + 2y_{CM} \int y' \, dm + (x_{CM}{}^2 + y_{CM}{}^2) \int dm$$

The first integral is, by definition, the moment of inertia about an axis that is parallel to the z axis and passes through the center of mass. The second two integrals are zero because, by definition of the center of mass, $\int x' \, dm = \int y' \, dm = 0$. The last integral is simply MD^2 because $\int dm = M$ and $D^2 = x_{CM}{}^2 + y_{CM}{}^2$. Therefore, we conclude that

$$I = I_{CM} + MD^2$$

EXAMPLE 10.8 ▶ **Applying the Parallel-Axis Theorem**

Consider once again the uniform rigid rod of mass M and length L shown in Figure 10.10. Find the moment of inertia of the rod about an axis perpendicular to the rod through one end (the y' axis in Fig. 10.10).

Solution Intuitively, we expect the moment of inertia to be greater than $I_{CM} = \frac{1}{12}ML^2$ because it should be more difficult to change the rotational motion of a rod spinning about an axis at one end than one that is spinning about its center. Because the distance between the center-of-mass axis and the y' axis is $D = L/2$, the parallel-axis theorem gives

$$I = I_{CM} + MD^2 = \tfrac{1}{12} ML^2 + M\left(\frac{L}{2}\right)^2 = \boxed{\tfrac{1}{3} ML^2}$$

So, it is four times more difficult to change the rotation of a rod spinning about its end than it is to change the motion of one spinning about its center.

Exercise Calculate the moment of inertia of the rod about a perpendicular axis through the point $x = L/4$.

Answer $I = \frac{7}{48} ML^2$.

10.6 ▷ TORQUE

Why are a door's doorknob and hinges placed near opposite edges of the door? This question actually has an answer based on common sense ideas. The harder we push against the door and the farther we are from the hinges, the more likely we are to open or close the door. When a force is exerted on a rigid object pivoted about an axis, the object tends to rotate about that axis. The tendency of a force to rotate an object about some axis is measured by a vector quantity called **torque** τ (tau).

Consider the wrench pivoted on the axis through O in Figure 10.13. The applied force $\mathbf{F}$ acts at an angle ϕ to the horizontal. We define the magnitude of the torque associated with the force $\mathbf{F}$ by the expression

Definition of torque

$$\tau \equiv rF \sin \phi = Fd \qquad (10.19)$$

where r is the distance between the pivot point and the point of application of $\mathbf{F}$ and d is the perpendicular distance from the pivot point to the line of action of $\mathbf{F}$. (The *line of action* of a force is an imaginary line extending out both ends of the vector representing the force. The dashed line extending from the tail of $\mathbf{F}$ in Figure 10.13 is part of the line of action of $\mathbf{F}$.) From the right triangle in Figure 10.13 that has the wrench as its hypotenuse, we see that $d = r \sin \phi$. This quantity d is

Moment arm

called the **moment arm** (or *lever arm*) of $\mathbf{F}$.

It is very important that you recognize that *torque is defined only when a reference axis is specified.* Torque is the product of a force and the moment arm of that force, and moment arm is defined only in terms of an axis of rotation.

In Figure 10.13, the only component of $\mathbf{F}$ that tends to cause rotation is $F \sin \phi$, the component perpendicular to r. The horizontal component $F \cos \phi$, because it passes through O, has no tendency to produce rotation. From the definition of torque, we see that the rotating tendency increases as $\mathbf{F}$ increases and as d increases. This explains the observation that it is easier to close a door if we push at the doorknob rather than at a point close to the hinge. We also want to apply our push as close to perpendicular to the door as we can. Pushing sideways on the doorknob will not cause the door to rotate.

If two or more forces are acting on a rigid object, as shown in Figure 10.14, each tends to produce rotation about the pivot at O. In this example, $\mathbf{F}_2$ tends to

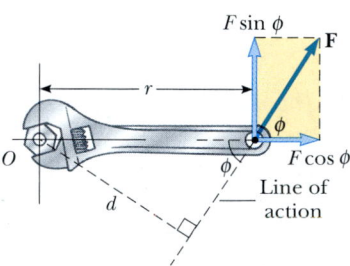

Figure 10.13 The force $\mathbf{F}$ has a greater rotating tendency about O as F increases and as the moment arm d increases. It is the component $F \sin \phi$ that tends to rotate the wrench about O.

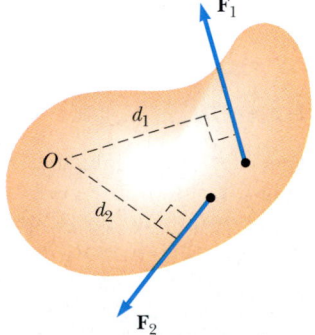

Figure 10.14 The force $\mathbf{F}_1$ tends to rotate the object counterclockwise about O, and $\mathbf{F}_2$ tends to rotate it clockwise.

rotate the object clockwise, and $\mathbf{F}_1$ tends to rotate it counterclockwise. We use the convention that the sign of the torque resulting from a force is positive if the turning tendency of the force is counterclockwise and is negative if the turning tendency is clockwise. For example, in Figure 10.14, the torque resulting from $\mathbf{F}_1$, which has a moment arm d_1, is positive and equal to $+F_1 d_1$; the torque from $\mathbf{F}_2$ is negative and equal to $-F_2 d_2$. Hence, the net torque about O is

$$\sum \tau = \tau_1 + \tau_2 = F_1 d_1 - F_2 d_2$$

Torque should not be confused with force. Forces can cause a change in linear motion, as described by Newton's second law. Forces can also cause a change in rotational motion, but the effectiveness of the forces in causing this change depends on both the forces and the moment arms of the forces, in the combination that we call *torque*. Torque has units of force times length—newton·meters in SI units—and should be reported in these units. Do not confuse torque and work, which have the same units but are very different concepts.

EXAMPLE 10.9 ▶ The Net Torque on a Cylinder

A one-piece cylinder is shaped as shown in Figure 10.15, with a core section protruding from the larger drum. The cylinder is free to rotate around the central axis shown in the drawing. A rope wrapped around the drum, which has radius R_1, exerts a force $\mathbf{F}_1$ to the right on the cylinder. A rope wrapped around the core, which has radius R_2, exerts a force $\mathbf{F}_2$ downward on the cylinder. (a) What is the net torque acting on the cylinder about the rotation axis (which is the z axis in Figure 10.15)?

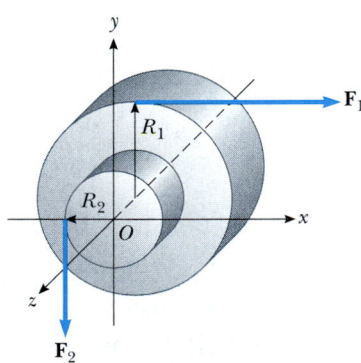

Figure 10.15   A solid cylinder pivoted about the z axis through O. The moment arm of $\mathbf{F}_1$ is R_1, and the moment arm of $\mathbf{F}_2$ is R_2.

Solution   The torque due to $\mathbf{F}_1$ is $-R_1 F_1$ (the sign is negative because the torque tends to produce clockwise rotation). The torque due to $\mathbf{F}_2$ is $+R_2 F_2$ (the sign is positive because the torque tends to produce counterclockwise rotation). Therefore, the net torque about the rotation axis is

$$\sum \tau = \tau_1 + \tau_2 = \boxed{-R_1 F_1 + R_2 F_2}$$

We can make a quick check by noting that if the two forces are of equal magnitude, the net torque is negative because $R_1 > R_2$. Starting from rest with both forces acting on it, the cylinder would rotate clockwise because $\mathbf{F}_1$ would be more effective at turning it than would $\mathbf{F}_2$.

(b) Suppose $F_1 = 5.0$ N, $R_1 = 1.0$ m, $F_2 = 15.0$ N, and $R_2 = 0.50$ m. What is the net torque about the rotation axis, and which way does the cylinder rotate from rest?

$$\sum \tau = -(5.0 \text{ N})(1.0 \text{ m}) + (15.0 \text{ N})(0.50 \text{ m}) = \boxed{2.5 \text{ N} \cdot \text{m}}$$

Because the net torque is positive, if the cylinder starts from rest, it will commence rotating counterclockwise with increasing angular velocity. (If the cylinder's initial rotation is clockwise, it will slow to a stop and then rotate counterclockwise with increasing angular speed.)

10.7 ▶ RELATIONSHIP BETWEEN TORQUE AND ANGULAR ACCELERATION

In this section we show that the angular acceleration of a rigid object rotating about a fixed axis is proportional to the net torque acting about that axis. Before discussing the more complex case of rigid-body rotation, however, it is instructive

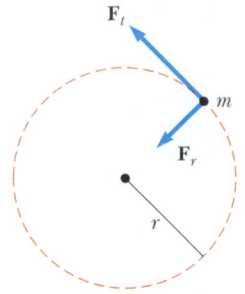

Figure 10.16 A particle rotating in a circle under the influence of a tangential force $\mathbf{F}_t$. A force $\mathbf{F}_r$ in the radial direction also must be present to maintain the circular motion.

Relationship between torque and angular acceleration

first to discuss the case of a particle rotating about some fixed point under the influence of an external force.

Consider a particle of mass m rotating in a circle of radius r under the influence of a tangential force $\mathbf{F}_t$ and a radial force $\mathbf{F}_r$, as shown in Figure 10.16. (As we learned in Chapter 6, the radial force must be present to keep the particle moving in its circular path.) The tangential force provides a tangential acceleration $\mathbf{a}_t$, and

$$F_t = ma_t$$

The torque about the center of the circle due to $\mathbf{F}_t$ is

$$\tau = F_t r = (ma_t)r$$

Because the tangential acceleration is related to the angular acceleration through the relationship $a_t = r\alpha$ (see Eq. 10.11), the torque can be expressed as

$$\tau = (mr\alpha)r = (mr^2)\alpha$$

Recall from Equation 10.15 that mr^2 is the moment of inertia of the rotating particle about the z axis passing through the origin, so that

$$\tau = I\alpha \qquad \textbf{(10.20)}$$

That is, **the torque acting on the particle is proportional to its angular acceleration,** and the proportionality constant is the moment of inertia. It is important to note that $\tau = I\alpha$ is the rotational analog of Newton's second law of motion, $F = ma$.

Now let us extend this discussion to a rigid object of arbitrary shape rotating about a fixed axis, as shown in Figure 10.17. The object can be regarded as an infinite number of mass elements dm of infinitesimal size. If we impose a cartesian coordinate system on the object, then each mass element rotates in a circle about the origin, and each has a tangential acceleration $\mathbf{a}_t$ produced by an external tangential force $d\mathbf{F}_t$. For any given element, we know from Newton's second law that

$$dF_t = (dm)a_t$$

The torque $d\tau$ associated with the force $d\mathbf{F}_t$ acts about the origin and is given by

$$d\tau = r\,dF_t = (r\,dm)a_t$$

Because $a_t = r\alpha$, the expression for $d\tau$ becomes

$$d\tau = (r\,dm)r\alpha = (r^2\,dm)\alpha$$

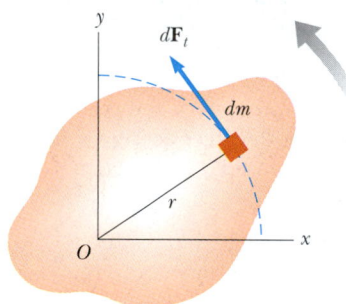

Figure 10.17 A rigid object rotating about an axis through O. Each mass element dm rotates about O with the same angular acceleration α, and the net torque on the object is proportional to α.

It is important to recognize that although each mass element of the rigid object may have a different linear acceleration $\mathbf{a}_t$, they all have the *same* angular acceleration α. With this in mind, we can integrate the above expression to obtain the net torque about O due to the external forces:

$$\sum \tau = \int (r^2\,dm)\alpha = \alpha \int r^2\,dm$$

where α can be taken outside the integral because it is common to all mass elements. From Equation 10.17, we know that $\int r^2\,dm$ is the moment of inertia of the object about the rotation axis through O, and so the expression for $\sum \tau$ becomes

Torque is proportional to angular acceleration

$$\sum \tau = I\alpha \qquad \textbf{(10.21)}$$

Note that this is the same relationship we found for a particle rotating in a circle (see Eq. 10.20). So, again we see that the net torque about the rotation axis is proportional to the angular acceleration of the object, with the proportionality factor being I, a quantity that depends upon the axis of rotation and upon the size and shape of the object. In view of the complex nature of the system, it is interesting to note that the relationship $\Sigma\tau = I\alpha$ is strikingly simple and in complete agreement with experimental observations. The simplicity of the result lies in the manner in which the motion is described.

Although each point on a rigid object rotating about a fixed axis may not experience the same force, linear acceleration, or linear speed, each point experiences the same angular acceleration and angular speed at any instant. Therefore, at any instant the rotating rigid object as a whole is characterized by specific values for angular acceleration, net torque, and angular speed.

Every point has the same ω and α

QuickLab

Tip over a child's tall tower of blocks. Try this several times. Does the tower "break" at the same place each time? What affects where the tower comes apart as it tips? If the tower were made of toy bricks that snap together, what would happen? (Refer to Conceptual Example 10.11.)

Finally, note that the result $\Sigma\tau = I\alpha$ also applies when the forces acting on the mass elements have radial components as well as tangential components. This is because the line of action of all radial components must pass through the axis of rotation, and hence all radial components produce zero torque about that axis.

EXAMPLE 10.10   Rotating Rod

A uniform rod of length L and mass M is attached at one end to a frictionless pivot and is free to rotate about the pivot in the vertical plane, as shown in Figure 10.18. The rod is released from rest in the horizontal position. What is the initial angular acceleration of the rod and the initial linear acceleration of its right end?

Solution   We cannot use our kinematic equations to find α or a because the torque exerted on the rod varies with its position, and so neither acceleration is constant. We have enough information to find the torque, however, which we can then use in the torque–angular acceleration relationship (Eq. 10.21) to find α and then a.

The only force contributing to torque about an axis through the pivot is the gravitational force $M\mathbf{g}$ exerted on the rod. (The force exerted by the pivot on the rod has zero torque about the pivot because its moment arm is zero.) To compute the torque on the rod, we can assume that the gravitational force acts at the center of mass of the rod, as shown in Figure 10.18. The torque due to this force about an axis through the pivot is

$$\tau = Mg\left(\frac{L}{2}\right)$$

With $\Sigma\tau = I\alpha$, and $I = \frac{1}{3}ML^2$ for this axis of rotation (see Table 10.2), we obtain

$$\alpha = \frac{\tau}{I} = \frac{Mg(L/2)}{1/3\,ML^2} = \boxed{\frac{3g}{2L}}$$

All points on the rod have this angular acceleration.

To find the linear acceleration of the right end of the rod, we use the relationship $a_t = r\alpha$ (Eq. 10.11), with $r = L$:

$$a_t = L\alpha = \boxed{\tfrac{3}{2}g}$$

This result—that $a_t > g$ for the free end of the rod—is rather interesting. It means that if we place a coin at the tip of the rod, hold the rod in the horizontal position, and then release the rod, the tip of the rod falls faster than the coin does!

Other points on the rod have a linear acceleration that is less than $\tfrac{3}{2}g$. For example, the middle of the rod has an acceleration of $\tfrac{3}{4}g$.

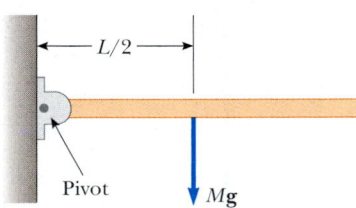

Figure 10.18   The uniform rod is pivoted at the left end.

CONCEPTUAL EXAMPLE 10.11   **Falling Smokestacks and Tumbling Blocks**

When a tall smokestack falls over, it often breaks somewhere along its length before it hits the ground, as shown in Figure 10.19. The same thing happens with a tall tower of children's toy blocks. Why does this happen?

Solution   As the smokestack rotates around its base, each higher portion of the smokestack falls with an increasing tangential acceleration. (The tangential acceleration of a given point on the smokestack is proportional to the distance of that portion from the base.) As the acceleration increases, higher portions of the smokestack experience an acceleration greater than that which could result from gravity alone; this is similar to the situation described in Example 10.10. This can happen only if these portions are being pulled downward by a force in addition to the gravitational force. The force that causes this to occur is the shear force from lower portions of the smokestack. Eventually the shear force that provides this acceleration is greater than the smokestack can withstand, and the smokestack breaks.

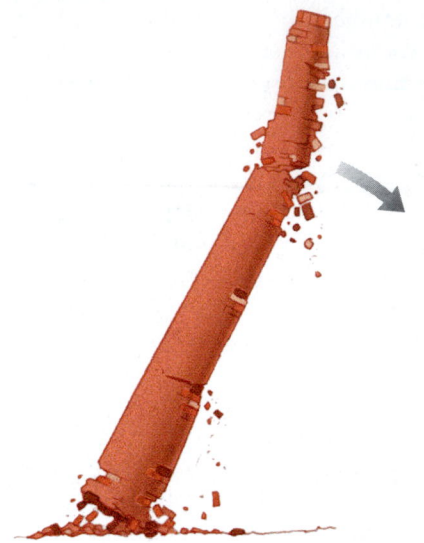

Figure 10.19   A falling smokestack.

EXAMPLE 10.12   **Angular Acceleration of a Wheel**

A wheel of radius R, mass M, and moment of inertia I is mounted on a frictionless, horizontal axle, as shown in Figure 10.20. A light cord wrapped around the wheel supports an object of mass m. Calculate the angular acceleration of the wheel, the linear acceleration of the object, and the tension in the cord.

Solution   The torque acting on the wheel about its axis of rotation is $\tau = TR$, where T is the force exerted by the cord on the rim of the wheel. (The gravitational force exerted by the Earth on the wheel and the normal force exerted by the axle on the wheel both pass through the axis of rotation and thus produce no torque.) Because $\Sigma \tau = I\alpha$, we obtain

$$\sum \tau = I\alpha = TR$$

$$(1) \qquad \alpha = \frac{TR}{I}$$

Now let us apply Newton's second law to the motion of the object, taking the downward direction to be positive:

$$\sum F_y = mg - T = ma$$

$$(2) \qquad a = \frac{mg - T}{m}$$

Equations (1) and (2) have three unknowns, α, a, and T. Because the object and wheel are connected by a string that does not slip, the linear acceleration of the suspended object is equal to the linear acceleration of a point on the rim of the

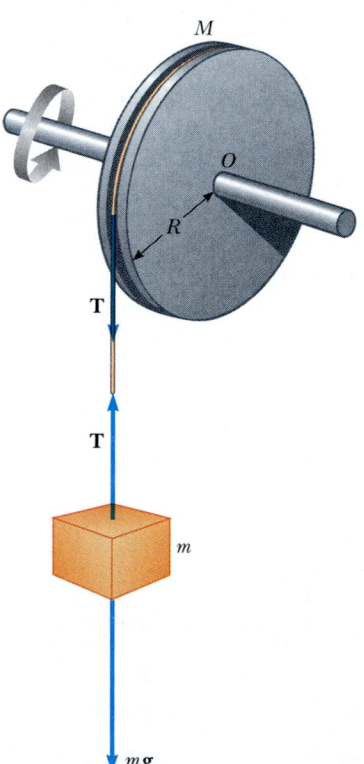

Figure 10.20   The tension in the cord produces a torque about the axle passing through O.

wheel. Therefore, the angular acceleration of the wheel and this linear acceleration are related by $a = R\alpha$. Using this fact together with Equations (1) and (2), we obtain

$$(3) \qquad a = R\alpha = \frac{TR^2}{I} = \frac{\mu\gamma - T}{\mu}$$

$$(4) \qquad T = \frac{mg}{1 + \frac{mR^2}{I}}$$

Substituting Equation (4) into Equation (2), and solving for a and α, we find that

$$a = \frac{g}{1 + I/mR^2}$$

$$\alpha = \frac{a}{R} = \frac{g}{R + I/mR}$$

Exercise The wheel in Figure 10.20 is a solid disc of $M = 2.00$ kg, $R = 30.0$ cm, and $I = 0.090\ 0$ kg·m^2. The suspended object has a mass of $m = 0.500$ kg. Find the tension in the cord and the angular acceleration of the wheel.

Answer 3.27 N; 10.9 rad/s^2.

EXAMPLE 10.13 ▶ Atwood's Machine Revisited

Two blocks having masses m_1 and m_2 are connected to each other by a light cord that passes over two identical, frictionless pulleys, each having a moment of inertia I and radius R, as shown in Figure 10.21a. Find the acceleration of each block and the tensions T_1, T_2, and T_3 in the cord. (Assume no slipping between cord and pulleys.)

Solution We shall define the downward direction as positive for m_1 and upward as the positive direction for m_2. This allows us to represent the acceleration of both masses by a single variable a and also enables us to relate a positive a to a positive (counterclockwise) angular acceleration α. Let us write Newton's second law of motion for each block, using the free-body diagrams for the two blocks as shown in Figure 10.21b:

$$(1) \qquad m_1 g - T_1 = m_1 a$$

$$(2) \qquad T_3 - m_2 g = m_2 a$$

Next, we must include the effect of the pulleys on the motion. Free-body diagrams for the pulleys are shown in Figure 10.21c. The net torque about the axle for the pulley on the left is $(T_1 - T_2)R$, while the net torque for the pulley on the right is $(T_2 - T_3)R$. Using the relation $\Sigma\tau = I\alpha$ for each pulley and noting that each pulley has the same angular acceleration α, we obtain

$$(3) \qquad (T_1 - T_2)R = I\alpha$$

$$(4) \qquad (T_2 - T_3)R = I\alpha$$

We now have four equations with four unknowns: a, T_1, T_2, and T_3. These can be solved simultaneously. Adding Equations (3) and (4) gives

$$(5) \qquad (T_1 - T_3)R = 2I\alpha$$

Adding Equations (1) and (2) gives

$$T_3 - T_1 + m_1 g - m_2 g = (m_1 + m_2)a$$

$$(6) \qquad T_1 - T_3 = (m_1 - m_2)g - (m_1 + m_2)a$$

Substituting Equation (6) into Equation (5), we have

$$[(m_1 - m_2)g - (m_1 + m_2)a]R = 2I\alpha$$

Because $\alpha = a/R$, this expression can be simplified to

$$(m_1 - m_2)g - (m_1 + m_2)a = 2I\frac{a}{R^2}$$

$$(7) \qquad a = \frac{(m_1 - m_2)g}{m_1 + m_2 + 2\dfrac{I}{R^2}}$$

This value of a can then be substituted into Equations (1)

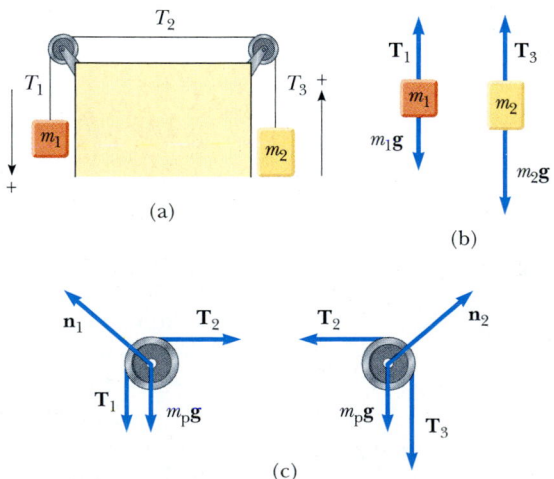

Figure 10.21 (a) Another look at Atwood's machine. (b) Free-body diagrams for the blocks. (c) Free-body diagrams for the pulleys, where $m_p\mathbf{g}$ represents the force of gravity acting on each pulley.

and (2) to give T_1 and T_3. Finally, T_2 can be found from Equation (3) or Equation (4). Note that if $m_1 > m_2$, the acceleration is positive; this means that the left block accelerates downward, the right block accelerates upward, and both pulleys accelerate counterclockwise. If $m_1 < m_2$, then all the values are negative and the motions are reversed. If $m_1 = m_2$, then no acceleration occurs at all. You should compare these results with those found in Example 5.9 on page 129.

10.8 ▶ WORK, POWER, AND ENERGY IN ROTATIONAL MOTION

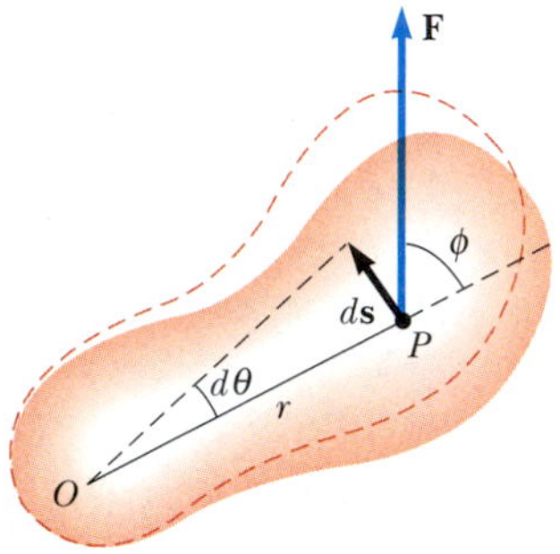

Figure 10.22 A rigid object rotates about an axis through O under the action of an external force **F** applied at P.

In this section, we consider the relationship between the torque acting on a rigid object and its resulting rotational motion in order to generate expressions for the power and a rotational analog to the work–kinetic energy theorem. Consider the rigid object pivoted at O in Figure 10.22. Suppose a single external force **F** is applied at P, where **F** lies in the plane of the page. The work done by **F** as the object rotates through an infinitesimal distance $ds = r\,d\theta$ in a time dt is

$$dW = \mathbf{F} \cdot d\mathbf{s} = (F \sin \phi) r\, d\theta$$

where $F \sin \phi$ is the tangential component of **F**, or, in other words, the component of the force along the displacement. Note that *the radial component of* **F** *does no work because it is perpendicular to the displacement.*

Because the magnitude of the torque due to **F** about O is defined as $rF \sin \phi$ by Equation 10.19, we can write the work done for the infinitesimal rotation as

$$dW = \tau\, d\theta \tag{10.22}$$

The rate at which work is being done by **F** as the object rotates about the fixed axis is

$$\frac{dW}{dt} = \tau \frac{d\theta}{dt}$$

Because dW/dt is the instantaneous power $\mathcal{P}$ (see Section 7.5) delivered by the force, and because $d\theta/dt = \omega$, this expression reduces to

$$\mathcal{P} = \frac{dW}{dt} = \tau\omega \tag{10.23}$$

This expression is analogous to $\mathcal{P} = Fv$ in the case of linear motion, and the expression $dW = \tau\, d\theta$ is analogous to $dW = F_x\, dx$.

Work and Energy in Rotational Motion

In studying linear motion, we found the energy concept—and, in particular, the work–kinetic energy theorem—extremely useful in describing the motion of a system. The energy concept can be equally useful in describing rotational motion. From what we learned of linear motion, we expect that when a symmetric object rotates about a fixed axis, the work done by external forces equals the change in the rotational energy.

To show that this is in fact the case, let us begin with $\Sigma \tau = I\alpha$. Using the chain rule from the calculus, we can express the resultant torque as

$$\sum \tau = I\alpha = I\frac{d\omega}{dt} = I\frac{d\omega}{d\theta}\frac{d\theta}{dt} = I\frac{d\omega}{d\theta}\omega$$

TABLE 10.3 Useful Equations in Rotational and Linear Motion

Rotational Motion About a Fixed Axis	Linear Motion
Angular speed $\omega = d\theta/dt$	Linear speed $v = dx/dt$
Angular acceleration $\alpha = d\omega/dt$	Linear acceleration $a = dv/dt$
Resultant torque $\Sigma\tau = I\alpha$	Resultant force $\Sigma F = ma$
If $\alpha =$ constant $\begin{cases} \omega_f = \omega_i + \alpha t \\ \theta_f - \theta_i = \omega_i t + \frac{1}{2}\alpha t^2 \\ \omega_f{}^2 = \omega_i{}^2 + 2\alpha(\theta_f - \theta_i) \end{cases}$	If $a =$ constant $\begin{cases} v_f = v_i + at \\ x_f - x_i = v_i t + \frac{1}{2}at^2 \\ v_f{}^2 = v_i{}^2 + 2a(x_f - x_i) \end{cases}$
Work $W = \int_{\theta_i}^{\theta_f} \tau \, d\theta$	Work $W = \int_{x_i}^{x_f} F_x \, dx$
Rotational kinetic energy $K_R = \frac{1}{2}I\omega^2$	Kinetic energy $K = \frac{1}{2}mv^2$
Power $\mathcal{P} = \tau\omega$	Power $\mathcal{P} = Fv$
Angular momentum $L = I\omega$	Linear momentum $p = mv$
Resultant torque $\Sigma\tau = dL/dt$	Resultant force $\Sigma F = dp/dt$

Rearranging this expression and noting that $\Sigma\tau \, d\theta = dW$, we obtain

$$\sum \tau \, d\theta = dW = I\omega \, d\omega$$

Integrating this expression, we get for the total work done by the net external force acting on a rotating system

$$\sum W = \int_{\theta_i}^{\theta_f} \sum \tau \, d\theta = \int_{\omega_i}^{\omega_f} I\omega \, d\omega = \tfrac{1}{2}I\omega_f{}^2 - \tfrac{1}{2}I\omega_i{}^2 \qquad \textbf{(10.24)}$$

Work–kinetic energy theorem for rotational motion

where the angular speed changes from ω_i to ω_f as the angular position changes from θ_i to θ_f. That is,

the net work done by external forces in rotating a symmetric rigid object about a fixed axis equals the change in the object's rotational energy.

Table 10.3 lists the various equations we have discussed pertaining to rotational motion, together with the analogous expressions for linear motion. The last two equations in Table 10.3, involving angular momentum L, are discussed in Chapter 11 and are included here only for the sake of completeness.

Quick Quiz 10.4

For a hoop lying in the *xy* plane, which of the following requires that more work be done by an external agent to accelerate the hoop from rest to an angular speed ω: (a) rotation about the *z* axis through the center of the hoop, or (b) rotation about an axis parallel to *z* passing through a point *P* on the hoop rim?

EXAMPLE 10.14 Rotating Rod Revisited

A uniform rod of length L and mass M is free to rotate on a frictionless pin passing through one end (Fig 10.23). The rod is released from rest in the horizontal position. (a) What is its angular speed when it reaches its lowest position?

Solution The question can be answered by considering the mechanical energy of the system. When the rod is horizontal, it has no rotational energy. The potential energy relative to the lowest position of the center of mass of the rod (O') is $MgL/2$. When the rod reaches its lowest position, the

energy is entirely rotational energy, $\frac{1}{2}I\omega^2$, where I is the moment of inertia about the pivot. Because $I = \frac{1}{3}ML^2$ (see Table 10.2) and because mechanical energy is constant, we have $E_i = E_f$ or

$$\tfrac{1}{2}MgL = \tfrac{1}{2}I\omega^2 = \tfrac{1}{2}(\tfrac{1}{3}ML^2)\omega^2$$

$$\omega = \sqrt{\frac{3g}{L}}$$

(b) Determine the linear speed of the center of mass and the linear speed of the lowest point on the rod when it is in the vertical position.

Solution These two values can be determined from the relationship between linear and angular speeds. We know ω from part (a), and so the linear speed of the center of mass is

$$v_{CM} = r\omega = \frac{L}{2}\omega = \tfrac{1}{2}\sqrt{3gL}$$

Because r for the lowest point on the rod is twice what it is for the center of mass, the lowest point has a linear speed equal to

$$2v_{CM} = \sqrt{3gL}$$

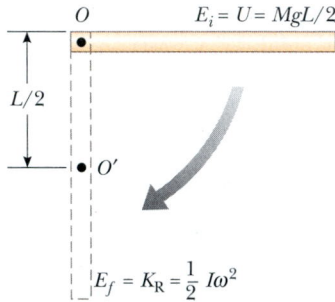

Figure 10.23 A uniform rigid rod pivoted at O rotates in a vertical plane under the action of gravity.

EXAMPLE 10.15 Connected Cylinders

Consider two cylinders having masses m_1 and m_2, where $m_1 \neq m_2$, connected by a string passing over a pulley, as shown in Figure 10.24. The pulley has a radius R and moment of

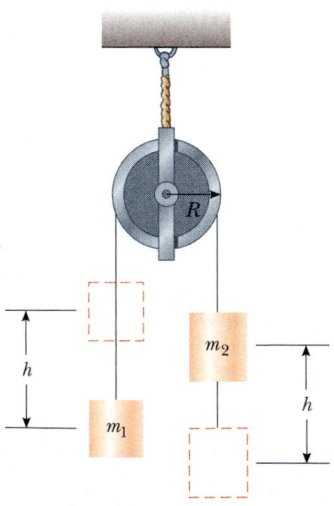

Figure 10.24

inertia I about its axis of rotation. The string does not slip on the pulley, and the system is released from rest. Find the linear speeds of the cylinders after cylinder 2 descends through a distance h, and the angular speed of the pulley at this time.

Solution We are now able to account for the effect of a massive pulley. Because the string does not slip, the pulley rotates. We neglect friction in the axle about which the pulley rotates for the following reason: Because the axle's radius is small relative to that of the pulley, the frictional torque is much smaller than the torque applied by the two cylinders, provided that their masses are quite different. Mechanical energy is constant; hence, the increase in the system's kinetic energy (the system being the two cylinders, the pulley, and the Earth) equals the decrease in its potential energy. Because $K_i = 0$ (the system is initially at rest), we have

$$\Delta K = K_f - K_i = (\tfrac{1}{2}m_1v_f^2 + \tfrac{1}{2}m_2v_f^2 + \tfrac{1}{2}I\omega_f^2) - 0$$

where v_f is the same for both blocks. Because $v_f = R\omega_f$, this expression becomes

$$\Delta K = \tfrac{1}{2}\left(m_1 + m_2 + \frac{I}{R^2}\right)v_f^2$$

From Figure 10.24, we see that the system loses potential energy as cylinder 2 descends and gains potential energy as cylinder 1 rises. That is, $\Delta U_2 = -m_2 gh$ and $\Delta U_1 = m_1 gh$. Applying the principle of conservation of energy in the form $\Delta K + \Delta U_1 + \Delta U_2 = 0$ gives

$$\frac{1}{2}\left(m_1 + m_2 + \frac{I}{R^2}\right)v_f^2 + m_1 gh - m_2 gh = 0$$

$$v_f = \left[\frac{2(m_2 - m_1)gh}{\left(m_1 + m_2 + \dfrac{I}{R^2}\right)}\right]^{1/2}$$

Because $v_f = R\omega_f$, the angular speed of the pulley at this instant is

$$\omega_f = \frac{v_f}{R} = \frac{1}{R}\left[\frac{2(m_2 - m_1)gh}{\left(m_1 + m_2 + \dfrac{I}{R^2}\right)}\right]^{1/2}$$

Exercise Repeat the calculation of v_f, using $\Sigma\tau = I\alpha$ applied to the pulley and Newton's second law applied to the two cylinders. Use the procedures presented in Examples 10.12 and 10.13.

SUMMARY

If a particle rotates in a circle of radius r through an angle θ (measured in radians), the arc length it moves through is $s = r\theta$.

The **angular displacement** of a particle rotating in a circle or of a rigid object rotating about a fixed axis is

$$\Delta\theta = \theta_f - \theta_i \qquad \text{(10.2)}$$

The **instantaneous angular speed** of a particle rotating in a circle or of a rigid object rotating about a fixed axis is

$$\omega = \frac{d\theta}{dt} \qquad \text{(10.4)}$$

The **instantaneous angular acceleration** of a rotating object is

$$\alpha = \frac{d\omega}{dt} \qquad \text{(10.6)}$$

When a rigid object rotates about a fixed axis, every part of the object has the same angular speed and the same angular acceleration.

If a particle or object rotates about a fixed axis under constant angular acceleration, one can apply equations of kinematics that are analogous to those for linear motion under constant linear acceleration:

$$\omega_f = \omega_i + \alpha t \qquad \text{(10.7)}$$

$$\theta_f = \theta_i + \omega_i t + \tfrac{1}{2}\alpha t^2 \qquad \text{(10.8)}$$

$$\omega_f^2 = \omega_i^2 + 2\alpha(\theta_f - \theta_i) \qquad \text{(10.9)}$$

A useful technique in solving problems dealing with rotation is to visualize a linear version of the same problem.

When a rigid object rotates about a fixed axis, the angular position, angular speed, and angular acceleration are related to the linear position, linear speed, and linear acceleration through the relationships

$$s = r\theta \qquad \text{(10.1a)}$$

$$v = r\omega \qquad \text{(10.10)}$$

$$a_t = r\alpha \qquad \text{(10.11)}$$

You must be able to easily alternate between the linear and rotational variables that describe a given situation.

The **moment of inertia of a system of particles** is

$$I \equiv \sum_i m_i r_i^2 \tag{10.15}$$

If a rigid object rotates about a fixed axis with angular speed ω, its **rotational energy** can be written

$$K_R = \tfrac{1}{2} I \omega^2 \tag{10.16}$$

where I is the moment of inertia about the axis of rotation.

The **moment of inertia of a rigid object** is

$$I = \int r^2 \, dm \tag{10.17}$$

where r is the distance from the mass element dm to the axis of rotation.

The magnitude of the **torque** associated with a force $\mathbf{F}$ acting on an object is

$$\tau = Fd \tag{10.19}$$

where d is the moment arm of the force, which is the perpendicular distance from some origin to the line of action of the force. Torque is a measure of the tendency of the force to change the rotation of the object about some axis.

If a rigid object free to rotate about a fixed axis has a **net external torque** acting on it, the object undergoes an angular acceleration α, where

$$\sum \tau = I\alpha \tag{10.21}$$

The rate at which work is done by an external force in rotating a rigid object about a fixed axis, or the **power** delivered, is

$$\mathcal{P} = \tau\omega \tag{10.23}$$

The net work done by external forces in rotating a rigid object about a fixed axis equals the change in the rotational kinetic energy of the object:

$$\sum W = \tfrac{1}{2} I \omega_f^2 - \tfrac{1}{2} I \omega_i^2 \tag{10.24}$$

QUESTIONS

1. What is the angular speed of the second hand of a clock? What is the direction of $\boldsymbol{\omega}$ as you view a clock hanging vertically? What is the magnitude of the angular acceleration vector $\boldsymbol{\alpha}$ of the second hand?

2. A wheel rotates counterclockwise in the xy plane. What is the direction of $\boldsymbol{\omega}$? What is the direction of $\boldsymbol{\alpha}$ if the angular velocity is decreasing in time?

3. Are the kinematic expressions for θ, ω, and α valid when the angular displacement is measured in degrees instead of in radians?

4. A turntable rotates at a constant rate of 45 rev/min. What is its angular speed in radians per second? What is the magnitude of its angular acceleration?

5. Suppose $a = b$ and $M > m$ for the system of particles described in Figure 10.8. About what axis (x, y, or z) does the moment of inertia have the smallest value? the largest value?

6. Suppose the rod in Figure 10.10 has a nonuniform mass distribution. In general, would the moment of inertia about the y axis still equal $ML^2/12$? If not, could the moment of inertia be calculated without knowledge of the manner in which the mass is distributed?

7. Suppose that only two external forces act on a rigid body, and the two forces are equal in magnitude but opposite in direction. Under what condition does the body rotate?

8. Explain how you might use the apparatus described in Example 10.12 to determine the moment of inertia of the wheel. (If the wheel does not have a uniform mass density, the moment of inertia is not necessarily equal to $\tfrac{1}{2}MR^2$.)

9. Using the results from Example 10.12, how would you calculate the angular speed of the wheel and the linear speed of the suspended mass at $t = 2$ s, if the system is released from rest at $t = 0$? Is the expression $v = R\omega$ valid in this situation?

10. If a small sphere of mass M were placed at the end of the rod in Figure 10.23, would the result for ω be greater than, less than, or equal to the value obtained in Example 10.14?

11. Explain why changing the axis of rotation of an object changes its moment of inertia.

12. Is it possible to change the translational kinetic energy of an object without changing its rotational energy?

13. Two cylinders having the same dimensions are set into rotation about their long axes with the same angular speed.

One is hollow, and the other is filled with water. Which cylinder will be easier to stop rotating? Explain your answer.

14. Must an object be rotating to have a nonzero moment of inertia?

15. If you see an object rotating, is there necessarily a net torque acting on it?

16. Can a (momentarily) stationary object have a nonzero angular acceleration?

17. The polar diameter of the Earth is slightly less than the equatorial diameter. How would the moment of inertia of the Earth change if some mass from near the equator were removed and transferred to the polar regions to make the Earth a perfect sphere?

PROBLEMS

1, 2, 3 = straightforward, intermediate, challenging ☐ = full solution available in the *Student Solutions Manual and Study Guide*
WEB = solution posted at **http://www.saunderscollege.com/physics/** 🖥 = Computer useful in solving problem 🖼 = Interactive Physics
☐ = paired numerical/symbolic problems

Section 10.2   Rotational Kinematics: Rotational Motion with Constant Angular Acceleration

1. A wheel starts from rest and rotates with constant angular acceleration and reaches an angular speed of 12.0 rad/s in 3.00 s. Find (a) the magnitude of the angular acceleration of the wheel and (b) the angle (in radians) through which it rotates in this time.

2. What is the angular speed in radians per second of (a) the Earth in its orbit about the Sun and (b) the Moon in its orbit about the Earth?

3. An airliner arrives at the terminal, and its engines are shut off. The rotor of one of the engines has an initial clockwise angular speed of 2 000 rad/s. The engine's rotation slows with an angular acceleration of magnitude 80.0 rad/s^2. (a) Determine the angular speed after 10.0 s. (b) How long does it take for the rotor to come to rest?

4. (a) The positions of the hour and minute hand on a clock face coincide at 12 o'clock. Determine all other times (up to the second) at which the positions of the hands coincide. (b) If the clock also has a second hand, determine all times at which the positions of all three hands coincide, given that they all coincide at 12 o'clock.

WEB 5. An electric motor rotating a grinding wheel at 100 rev/min is switched off. Assuming constant negative acceleration of magnitude 2.00 rad/s^2, (a) how long does it take the wheel to stop? (b) Through how many radians does it turn during the time found in part (a)?

6. A centrifuge in a medical laboratory rotates at a rotational speed of 3 600 rev/min. When switched off, it rotates 50.0 times before coming to rest. Find the constant angular acceleration of the centrifuge.

7. The angular position of a swinging door is described by $\theta = 5.00 + 10.0t + 2.00t^2$ rad. Determine the angular position, angular speed, and angular acceleration of the door (a) at $t = 0$ and (b) at $t = 3.00$ s.

8. The tub of a washer goes into its spin cycle, starting from rest and gaining angular speed steadily for 8.00 s, when it is turning at 5.00 rev/s. At this point the person doing the laundry opens the lid, and a safety switch turns off the washer. The tub smoothly slows to rest in 12.0 s. Through how many revolutions does the tub turn while it is in motion?

9. A rotating wheel requires 3.00 s to complete 37.0 revolutions. Its angular speed at the end of the 3.00-s interval is 98.0 rad/s. What is the constant angular acceleration of the wheel?

10. (a) Find the angular speed of the Earth's rotation on its axis. As the Earth turns toward the east, we see the sky turning toward the west at this same rate.
(b) *The rainy Pleiads wester*
        And seek beyond the sea
    The head that I shall dream of
        That shall not dream of me.

        A. E. Housman   *(© Robert E. Symons)*

Cambridge, England, is at longitude 0°, and Saskatoon, Saskatchewan, is at longitude 107° west. How much time elapses after the Pleiades set in Cambridge until these stars fall below the western horizon in Saskatoon?

Section 10.3   Angular and Linear Quantities

11. Make an order-of-magnitude estimate of the number of revolutions through which a typical automobile tire

turns in 1 yr. State the quantities you measure or estimate and their values.

12. The diameters of the main rotor and tail rotor of a single-engine helicopter are 7.60 m and 1.02 m, respectively. The respective rotational speeds are 450 rev/min and 4 138 rev/min. Calculate the speeds of the tips of both rotors. Compare these speeds with the speed of sound, 343 m/s.

Figure P10.12   *(Ross Harrrison Koty/Tony Stone Images)*

13. A racing car travels on a circular track with a radius of 250 m. If the car moves with a constant linear speed of 45.0 m/s, find (a) its angular speed and (b) the magnitude and direction of its acceleration.

14. A car is traveling at 36.0 km/h on a straight road. The radius of its tires is 25.0 cm. Find the angular speed of one of the tires, with its axle taken as the axis of rotation.

15. A wheel 2.00 m in diameter lies in a vertical plane and rotates with a constant angular acceleration of 4.00 rad/s². The wheel starts at rest at $t = 0$, and the radius vector of point P on the rim makes an angle of 57.3° with the horizontal at this time. At $t = 2.00$ s, find (a) the angular speed of the wheel, (b) the linear speed and acceleration of the point P, and (c) the angular position of the point P.

16. A discus thrower accelerates a discus from rest to a speed of 25.0 m/s by whirling it through 1.25 rev. As-

Figure P10.16   *(Bruce Ayers/Tony Stone Images)*

sume the discus moves on the arc of a circle 1.00 m in radius. (a) Calculate the final angular speed of the discus. (b) Determine the magnitude of the angular acceleration of the discus, assuming it to be constant. (c) Calculate the acceleration time.

17. A car accelerates uniformly from rest and reaches a speed of 22.0 m/s in 9.00 s. If the diameter of a tire is 58.0 cm, find (a) the number of revolutions the tire makes during this motion, assuming that no slipping occurs. (b) What is the final rotational speed of a tire in revolutions per second?

18. A 6.00-kg block is released from A on the frictionless track shown in Figure P10.18. Determine the radial and tangential components of acceleration for the block at P.

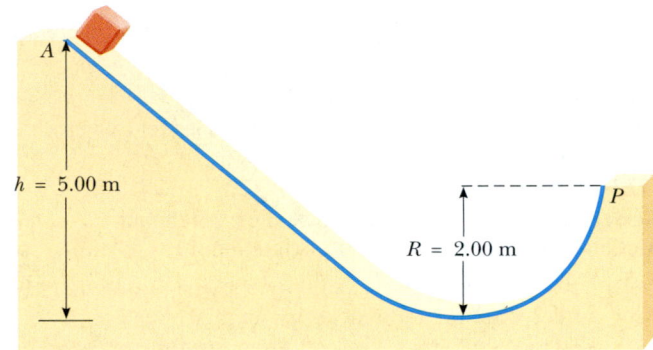

Figure P10.18

WEB 19. A disc 8.00 cm in radius rotates at a constant rate of 1 200 rev/min about its central axis. Determine (a) its angular speed, (b) the linear speed at a point 3.00 cm from its center, (c) the radial acceleration of a point on the rim, and (d) the total distance a point on the rim moves in 2.00 s.

20. A car traveling on a flat (unbanked) circular track accelerates uniformly from rest with a tangential acceleration of 1.70 m/s². The car makes it one quarter of the way around the circle before it skids off the track. Determine the coefficient of static friction between the car and track from these data.

21. A small object with mass 4.00 kg moves counterclockwise with constant speed 4.50 m/s in a circle of radius 3.00 m centered at the origin. (a) It started at the point with cartesian coordinates (3 m, 0). When its angular displacement is 9.00 rad, what is its position vector, in cartesian unit-vector notation? (b) In what quadrant is the particle located, and what angle does its position vector make with the positive x axis? (c) What is its velocity vector, in unit–vector notation? (d) In what direction is it moving? Make a sketch of the position and velocity vectors. (e) What is its acceleration, expressed in unit–vector notation? (f) What total force acts on the object? (Express your answer in unit vector notation.)

22. A standard cassette tape is placed in a standard cassette player. Each side plays for 30 min. The two tape wheels of the cassette fit onto two spindles in the player. Suppose that a motor drives one spindle at a constant angular speed of ~ 1 rad/s and that the other spindle is free to rotate at any angular speed. Estimate the order of magnitude of the thickness of the tape.

Section 10.4 Rotational Energy

23. Three small particles are connected by rigid rods of negligible mass lying along the y axis (Fig. P10.23). If the system rotates about the x axis with an angular speed of 2.00 rad/s, find (a) the moment of inertia about the x axis and the total rotational kinetic energy evaluated from $\frac{1}{2}I\omega^2$ and (b) the linear speed of each particle and the total kinetic energy evaluated from $\Sigma \frac{1}{2} m_i v_i^2$.

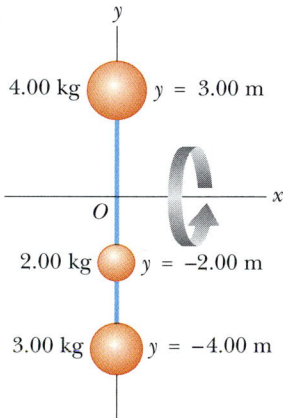

Figure P10.23

24. The center of mass of a pitched baseball (3.80-cm radius) moves at 38.0 m/s. The ball spins about an axis through its center of mass with an angular speed of 125 rad/s. Calculate the ratio of the rotational energy to the translational kinetic energy. Treat the ball as a uniform sphere.

WEB **25.** The four particles in Figure P10.25 are connected by rigid rods of negligible mass. The origin is at the center of the rectangle. If the system rotates in the xy plane about the z axis with an angular speed of 6.00 rad/s, calculate (a) the moment of inertia of the system about the z axis and (b) the rotational energy of the system.

26. The hour hand and the minute hand of Big Ben, the famous Parliament tower clock in London, are 2.70 m long and 4.50 m long and have masses of 60.0 kg and 100 kg, respectively. Calculate the total rotational kinetic energy of the two hands about the axis of rotation. (You may model the hands as long thin rods.)

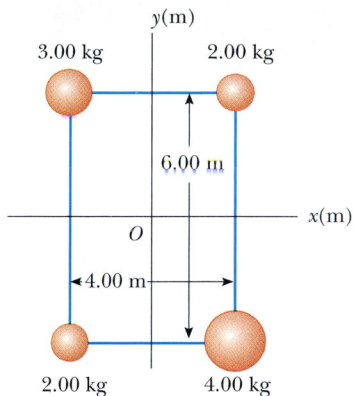

Figure P10.25

Figure P10.26 Problems 26 and 74. (*John Lawrence/Tony Stone Images*)

27. Two masses M and m are connected by a rigid rod of length L and of negligible mass, as shown in Figure P10.27. For an axis perpendicular to the rod, show that the system has the minimum moment of inertia when the axis passes through the center of mass. Show that this moment of inertia is $I = \mu L^2$, where $\mu = mM/(m + M)$.

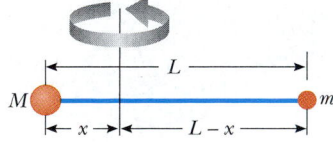

Figure P10.27

Section 10.5 Calculation of Moments of Inertia

28. Three identical thin rods, each of length L and mass m, are welded perpendicular to each other, as shown in Figure P10.28. The entire setup is rotated about an axis

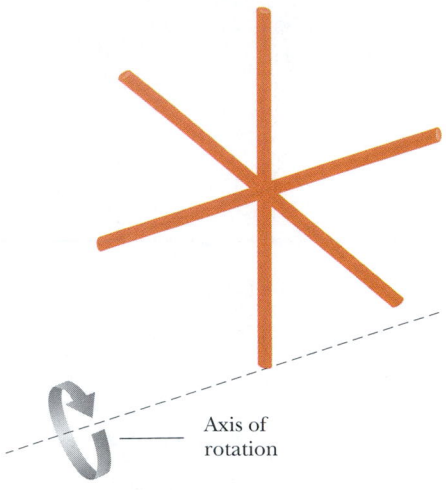

Figure P10.28

that passes through the end of one rod and is parallel to another. Determine the moment of inertia of this arrangement.

29. Figure P10.29 shows a side view of a car tire and its radial dimensions. The rubber tire has two sidewalls of uniform thickness 0.635 cm and a tread wall of uniform thickness 2.50 cm and width 20.0 cm. Suppose its density is uniform, with the value $1.10 \times 10^3 \, kg/m^3$. Find its moment of inertia about an axis through its center perpendicular to the plane of the sidewalls.

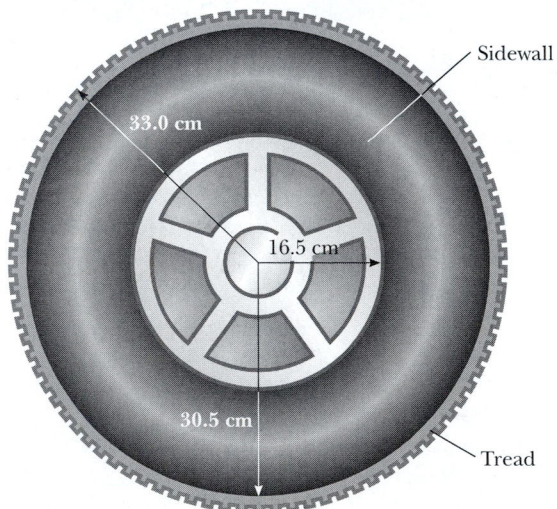

Figure P10.29

30. Use the parallel-axis theorem and Table 10.2 to find the moments of inertia of (a) a solid cylinder about an axis parallel to the center-of-mass axis and passing through the edge of the cylinder and (b) a solid sphere about an axis tangent to its surface.

31. *Attention! About face!* Compute an order-of-magnitude estimate for the moment of inertia of your body as you stand tall and turn around a vertical axis passing through the top of your head and the point halfway between your ankles. In your solution state the quantities you measure or estimate and their values.

Section 10.6   Torque

32. Find the mass m needed to balance the 1 500-kg truck on the incline shown in Figure P10.32. Assume all pulleys are frictionless and massless.

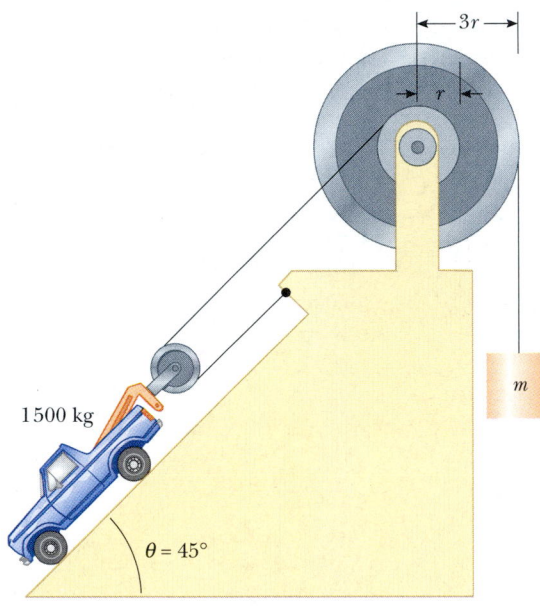

Figure P10.32

33. Find the net torque on the wheel in Figure P10.33 about the axle through O if $a = 10.0$ cm and $b = 25.0$ cm.

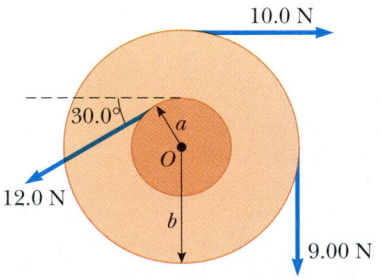

Figure P10.33

34. The fishing pole in Figure P10.34 makes an angle of 20.0° with the horizontal. What is the torque exerted by

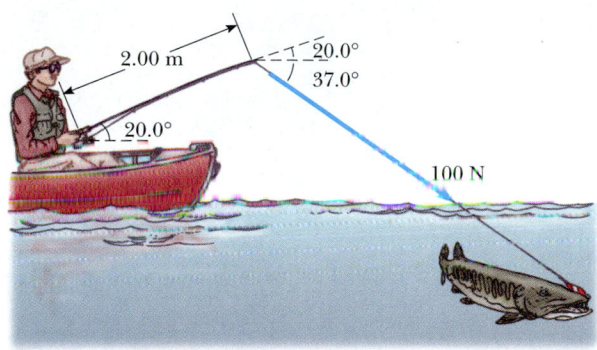

Figure P10.34

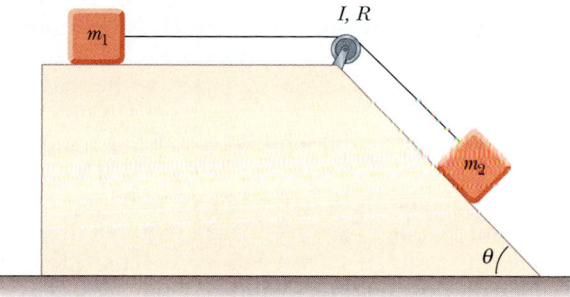

Figure P10.39

the fish about an axis perpendicular to the page and passing through the fisher's hand?

35. The tires of a 1 500-kg car are 0.600 m in diameter, and the coefficients of friction with the road surface are $\mu_s = 0.800$ and $\mu_k = 0.600$. Assuming that the weight is evenly distributed on the four wheels, calculate the maximum torque that can be exerted by the engine on a driving wheel such that the wheel does not spin. If you wish, you may suppose that the car is at rest.

36. Suppose that the car in Problem 35 has a disk brake system. Each wheel is slowed by the frictional force between a single brake pad and the disk-shaped rotor. On this particular car, the brake pad comes into contact with the rotor at an average distance of 22.0 cm from the axis. The coefficients of friction between the brake pad and the disk are $\mu_s = 0.600$ and $\mu_k = 0.500$. Calculate the normal force that must be applied to the rotor such that the car slows as quickly as possible.

Section 10.7 Relationship Between Torque and Angular Acceleration

WEB 37. A model airplane having a mass of 0.750 kg is tethered by a wire so that it flies in a circle 30.0 m in radius. The airplane engine provides a net thrust of 0.800 N perpendicular to the tethering wire. (a) Find the torque the net thrust produces about the center of the circle. (b) Find the angular acceleration of the airplane when it is in level flight. (c) Find the linear acceleration of the airplane tangent to its flight path.

38. The combination of an applied force and a frictional force produces a constant total torque of 36.0 N·m on a wheel rotating about a fixed axis. The applied force acts for 6.00 s; during this time the angular speed of the wheel increases from 0 to 10.0 rad/s. The applied force is then removed, and the wheel comes to rest in 60.0 s. Find (a) the moment of inertia of the wheel, (b) the magnitude of the frictional torque, and (c) the total number of revolutions of the wheel.

39. A block of mass $m_1 = 2.00$ kg and a block of mass $m_2 = 6.00$ kg are connected by a massless string over a pulley

in the shape of a disk having radius $R = 0.250$ m and mass $M = 10.0$ kg. These blocks are allowed to move on a fixed block–wedge of angle $\theta = 30.0°$, as shown in Figure P10.39. The coefficient of kinetic friction for both blocks is 0.360. Draw free-body diagrams of both blocks and of the pulley. Determine (a) the acceleration of the two blocks and (b) the tensions in the string on both sides of the pulley.

40. A potter's wheel—a thick stone disk with a radius of 0.500 m and a mass of 100 kg—is freely rotating at 50.0 rev/min. The potter can stop the wheel in 6.00 s by pressing a wet rag against the rim and exerting a radially inward force of 70.0 N. Find the effective coefficient of kinetic friction between the wheel and the rag.

41. A bicycle wheel has a diameter of 64.0 cm and a mass of 1.80 kg. Assume that the wheel is a hoop with all of its mass concentrated on the outside radius. The bicycle is placed on a stationary stand on rollers, and a resistive force of 120 N is applied tangent to the rim of the tire. (a) What force must be applied by a chain passing over a 9.00-cm-diameter sprocket if the wheel is to attain an acceleration of 4.50 rad/s²? (b) What force is required if the chain shifts to a 5.60-cm-diameter sprocket?

Section 10.8 Work , Power, and Energy in Rotational Motion

42. A cylindrical rod 24.0 cm long with a mass of 1.20 kg and a radius of 1.50 cm has a ball with a diameter of 8.00 cm and a mass of 2.00 kg attached to one end. The arrangement is originally vertical and stationary, with the ball at the top. The apparatus is free to pivot about the bottom end of the rod. (a) After it falls through 90°, what is its rotational kinetic energy? (b) What is the angular speed of the rod and ball? (c) What is the linear speed of the ball? (d) How does this compare with the speed if the ball had fallen freely through the same distance of 28 cm?

43. A 15.0-kg mass and a 10.0-kg mass are suspended by a pulley that has a radius of 10.0 cm and a mass of 3.00 kg (Fig. P10.43). The cord has a negligible mass and causes the pulley to rotate without slipping. The pulley

rotates without friction. The masses start from rest 3.00 m apart. Treating the pulley as a uniform disk, determine the speeds of the two masses as they pass each other.

44. A mass m_1 and a mass m_2 are suspended by a pulley that has a radius R and a mass M (see Fig. P10.43). The cord has a negligible mass and causes the pulley to rotate without slipping. The pulley rotates without friction. The masses start from rest a distance d apart. Treating the pulley as a uniform disk, determine the speeds of the two masses as they pass each other.

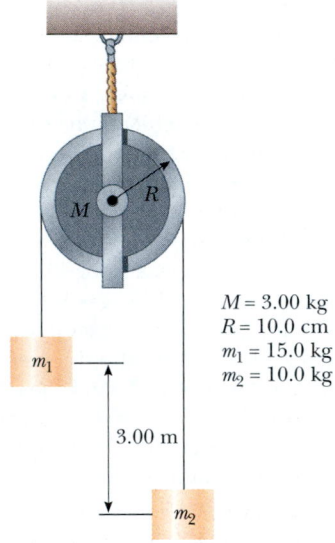

$$M = 3.00 \text{ kg}$$
$$R = 10.0 \text{ cm}$$
$$m_1 = 15.0 \text{ kg}$$
$$m_2 = 10.0 \text{ kg}$$

3.00 m

Figure P10.43   Problems 43 and 44.

45. A weight of 50.0 N is attached to the free end of a light string wrapped around a reel with a radius of 0.250 m and a mass of 3.00 kg. The reel is a solid disk, free to rotate in a vertical plane about the horizontal axis passing through its center. The weight is released 6.00 m above the floor. (a) Determine the tension in the string, the acceleration of the mass, and the speed with which the weight hits the floor. (b) Find the speed calculated in part (a), using the principle of conservation of energy.

46. A constant torque of 25.0 N·m is applied to a grindstone whose moment of inertia is 0.130 kg·m². Using energy principles, find the angular speed after the grindstone has made 15.0 revolutions. (Neglect friction.)

47. This problem describes one experimental method of determining the moment of inertia of an irregularly shaped object such as the payload for a satellite. Figure P10.47 shows a mass m suspended by a cord wound around a spool of radius r, forming part of a turntable supporting the object. When the mass is released from rest, it descends through a distance h, acquiring a speed

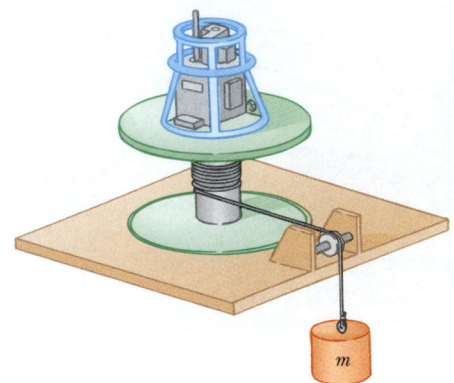

Figure P10.47

v. Show that the moment of inertia I of the equipment (including the turntable) is $mr^2(2gh/v^2 - 1)$.

48. A bus is designed to draw its power from a rotating flywheel that is brought up to its maximum rate of rotation (3 000 rev/min) by an electric motor. The flywheel is a solid cylinder with a mass of 1 000 kg and a diameter of 1.00 m. If the bus requires an average power of 10.0 kW, how long does the flywheel rotate?

49. (a) A uniform, solid disk of radius R and mass M is free to rotate on a frictionless pivot through a point on its rim (Fig. P10.49). If the disk is released from rest in the position shown by the blue circle, what is the speed of its center of mass when the disk reaches the position indicated by the dashed circle? (b) What is the speed of the lowest point on the disk in the dashed position? (c) Repeat part (a), using a uniform hoop.

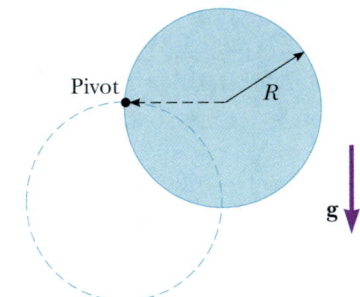

Figure P10.49

50. A horizontal 800-N merry-go-round is a solid disk of radius 1.50 m and is started from rest by a constant horizontal force of 50.0 N applied tangentially to the cylinder. Find the kinetic energy of the solid cylinder after 3.00 s.

ADDITIONAL PROBLEMS

51. Toppling chimneys often break apart in mid-fall (Fig. P10.51) because the mortar between the bricks cannot

withstand much shear stress. As the chimney begins to fall, shear forces must act on the topmost sections to accelerate them tangentially so that they can keep up with the rotation of the lower part of the stack. For simplicity, let us model the chimney as a uniform rod of length ℓ pivoted at the lower end. The rod starts at rest in a vertical position (with the frictionless pivot at the bottom) and falls over under the influence of gravity. What fraction of the length of the rod has a tangential acceleration greater than $g \sin \theta$, where θ is the angle the chimney makes with the vertical?

Figure P10.51   A building demolition site in Baltimore, MD. At the left is a chimney, mostly concealed by the building, that has broken apart on its way down. Compare with Figure 10.19. (*Jerry Wachter/Photo Researchers, Inc.*)

52. Review Problem.   A mixing beater consists of three thin rods: Each is 10.0 cm long, diverges from a central hub, and is separated from the others by 120°. All turn in the same plane. A ball is attached to the end of each rod. Each ball has a cross-sectional area of 4.00 cm^2 and is shaped so that it has a drag coefficient of 0.600. Calculate the power input required to spin the beater at 1 000 rev/min (a) in air and (b) in water.

53.   A grinding wheel is in the form of a uniform solid disk having a radius of 7.00 cm and a mass of 2.00 kg. It starts from rest and accelerates uniformly under the action of the constant torque of 0.600 N·m that the motor exerts on the wheel. (a) How long does the wheel take to reach its final rotational speed of 1 200 rev/min? (b) Through how many revolutions does it turn while accelerating?

54.   The density of the Earth, at any distance r from its center, is approximately

$$\rho = [14.2 - 11.6 \, r/R] \times 10^3 \, \text{kg/m}^3$$

where R is the radius of the Earth. Show that this density leads to a moment of inertia $I = 0.330MR^2$ about an axis through the center, where M is the mass of the Earth.

55.   A 4.00-m length of light nylon cord is wound around a uniform cylindrical spool of radius 0.500 m and mass 1.00 kg. The spool is mounted on a frictionless axle and is initially at rest. The cord is pulled from the spool with a constant acceleration of magnitude 2.50 m/s^2. (a) How much work has been done on the spool when it reaches an angular speed of 8.00 rad/s? (b) Assuming that there is enough cord on the spool, how long does it take the spool to reach this angular speed? (c) Is there enough cord on the spool?

56.   A flywheel in the form of a heavy circular disk of diameter 0.600 m and mass 200 kg is mounted on a frictionless bearing. A motor connected to the flywheel accelerates it from rest to 1 000 rev/min. (a) What is the moment of inertia of the flywheel? (b) How much work is done on it during this acceleration? (c) When the angular speed reaches 1 000 rev/min, the motor is disengaged. A friction brake is used to slow the rotational rate to 500 rev/min. How much energy is dissipated as internal energy in the friction brake?

57.   A shaft is turning at 65.0 rad/s at time zero. Thereafter, its angular acceleration is given by

$$\alpha = -10 \, \text{rad/s}^2 - 5t \, \text{rad/s}^3$$

where t is the elapsed time. (a) Find its angular speed at $t = 3.00$ s. (b) How far does it turn in these 3 s?

58.   For any given rotational axis, the *radius of gyration K* of a rigid body is defined by the expression $K^2 = I/M$, where M is the total mass of the body and I is its moment of inertia. Thus, the radius of gyration is equal to the distance between an imaginary point mass M and the axis of rotation such that I for the point mass about that axis is the same as that for the rigid body. Find the radius of gyration of (a) a solid disk of radius R, (b) a uniform rod of length L, and (c) a solid sphere of radius R, all three of which are rotating about a central axis.

59.   A long, uniform rod of length L and mass M is pivoted about a horizontal, frictionless pin passing through one end. The rod is released from rest in a vertical position, as shown in Figure P10.59. At the instant the rod is horizontal, find (a) its angular speed, (b) the magnitude of its angular acceleration, (c) the x and y components of the acceleration of its center of mass, and (d) the components of the reaction force at the pivot.

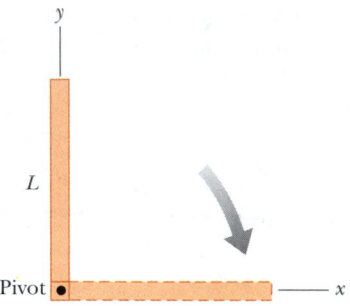

Figure P10.59

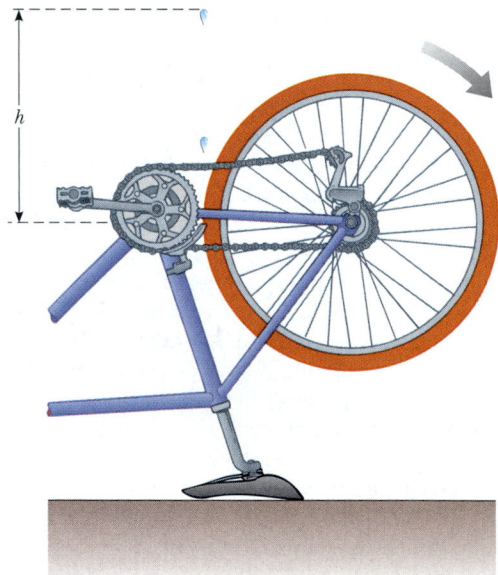

Figure P10.60   Problems 60 and 61.

A drop that breaks loose on the next turn rises a distance $h_2 < h_1$ above the tangent point. The height to which the drops rise decreases because the angular speed of the wheel decreases. From this information, determine the magnitude of the average angular acceleration of the wheel.

62. The top shown in Figure P10.62 has a moment of inertia of 4.00×10^{-4} kg·m^2 and is initially at rest. It is free to rotate about the stationary axis AA'. A string, wrapped around a peg along the axis of the top, is pulled in such a manner that a constant tension of 5.57 N is maintained. If the string does not slip while it is unwound from the peg, what is the angular speed of the top after 80.0 cm of string has been pulled off the peg?

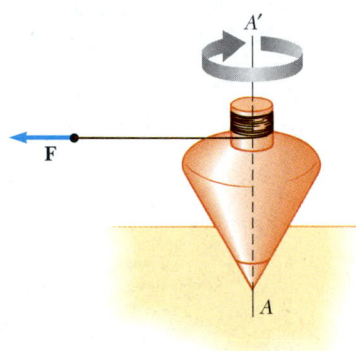

Figure P10.62

63. A cord is wrapped around a pulley of mass m and of radius r. The free end of the cord is connected to a block of mass M. The block starts from rest and then slides down an incline that makes an angle θ with the horizontal. The coefficient of kinetic friction between block and incline is μ. (a) Use energy methods to show that the block's speed as a function of displacement d down the incline is

$$v = [4gdM(m + 2M)^{-1}(\sin \theta - \mu \cos \theta)]^{1/2}$$

(b) Find the magnitude of the acceleration of the block in terms of μ, m, M, g, and θ.

64. (a) What is the rotational energy of the Earth about its spin axis? The radius of the Earth is 6 370 km, and its mass is 5.98×10^{24} kg. Treat the Earth as a sphere of moment of inertia $\frac{2}{5}MR^2$. (b) The rotational energy of the Earth is decreasing steadily because of tidal friction. Estimate the change in one day, given that the rotational period increases by about 10 μs each year.

65. The speed of a moving bullet can be determined by allowing the bullet to pass through two rotating paper disks mounted a distance d apart on the same axle (Fig. P10.65). From the angular displacement $\Delta \theta$ of the two

60. A bicycle is turned upside down while its owner repairs a flat tire. A friend spins the other wheel, of radius 0.381 m, and observes that drops of water fly off tangentially. She measures the height reached by drops moving vertically (Fig. P10.60). A drop that breaks loose from the tire on one turn rises $h = 54.0$ cm above the tangent point. A drop that breaks loose on the next turn rises 51.0 cm above the tangent point. The height to which the drops rise decreases because the angular speed of the wheel decreases. From this information, determine the magnitude of the average angular acceleration of the wheel.

61. A bicycle is turned upside down while its owner repairs a flat tire. A friend spins the other wheel of radius R and observes that drops of water fly off tangentially. She measures the height reached by drops moving vertically (see Fig. P10.60). A drop that breaks loose from the tire on one turn rises a distance h_1 above the tangent point.

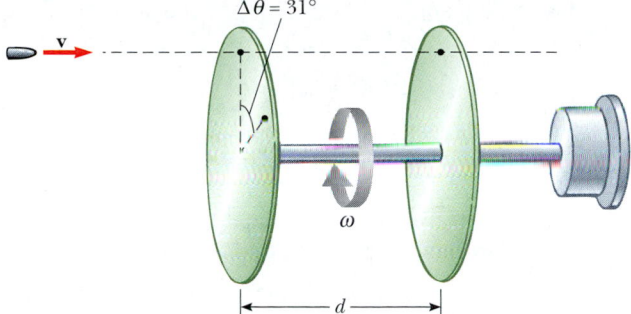

Figure P10.65

bullet holes in the disks and the rotational speed of the disks, we can determine the speed v of the bullet. Find the bullet speed for the following data: $d = 80$ cm, $\omega = 900$ rev/min, and $\Delta\theta = 31.0°$.

66. A wheel is formed from a hoop and n equally spaced spokes extending from the center of the hoop to its rim. The mass of the hoop is M, and the radius of the hoop (and hence the length of each spoke) is R. The mass of each spoke is m. Determine (a) the moment of inertia of the wheel about an axis through its center and perpendicular to the plane of the wheel and (b) the moment of inertia of the wheel about an axis through its rim and perpendicular to the plane of the wheel.

67. A uniform, thin, solid door has a height of 2.20 m, a width of 0.870 m, and a mass of 23.0 kg. Find its moment of inertia for rotation on its hinges. Are any of the data unnecessary?

68. A uniform, hollow, cylindrical spool has inside radius $R/2$, outside radius R, and mass M (Fig. P10.68). It is mounted so that it rotates on a massless horizontal axle. A mass m is connected to the end of a string wound around the spool. The mass m falls from rest through a distance y in time t. Show that the torque due to the frictional forces between spool and axle is
$$\tau_f = R[m(g - 2y/t^2) - M(5y/4t^2)]$$

69. An electric motor can accelerate a Ferris wheel of moment of inertia $I = 20\ 000$ kg·m² from rest to

10.0 rev/min in 12.0 s. When the motor is turned off, friction causes the wheel to slow down from 10.0 to 8.00 rev/min in 10.0 s. Determine (a) the torque generated by the motor to bring the wheel to 10.0 rev/min and (b) the power that would be needed to maintain this rotational speed.

70. The pulley shown in Figure P10.70 has radius R and moment of inertia I. One end of the mass m is connected to a spring of force constant k, and the other end is fastened to a cord wrapped around the pulley. The pulley axle and the incline are frictionless. If the pulley is wound counterclockwise so that the spring is stretched a distance d from its unstretched position and is then released from rest, find (a) the angular speed of the pulley when the spring is again unstretched and (b) a numerical value for the angular speed at this point if $I = 1.00$ kg·m², $R = 0.300$ m, $k = 50.0$ N/m, $m = 0.500$ kg, $d = 0.200$ m, and $\theta = 37.0°$.

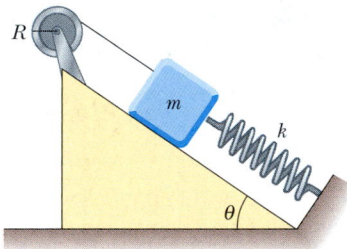

Figure P10.70

71. Two blocks, as shown in Figure P10.71, are connected by a string of negligible mass passing over a pulley of radius 0.250 m and moment of inertia I. The block on the frictionless incline is moving upward with a constant acceleration of 2.00 m/s². (a) Determine T_1 and T_2, the tensions in the two parts of the string. (b) Find the moment of inertia of the pulley.

72. A common demonstration, illustrated in Figure P10.72, consists of a ball resting at one end of a uniform board

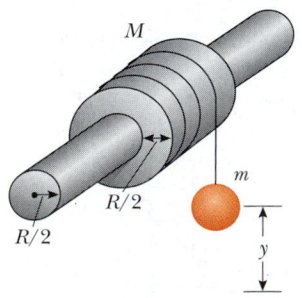

Figure P10.68

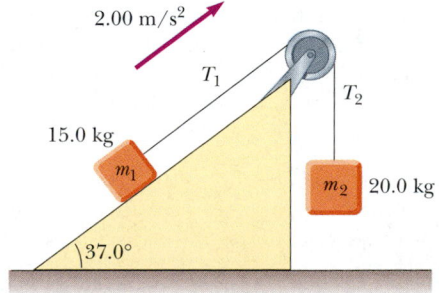

Figure P10.71

of length ℓ, hinged at the other end, and elevated at an angle θ. A light cup is attached to the board at r_c so that it will catch the ball when the support stick is suddenly

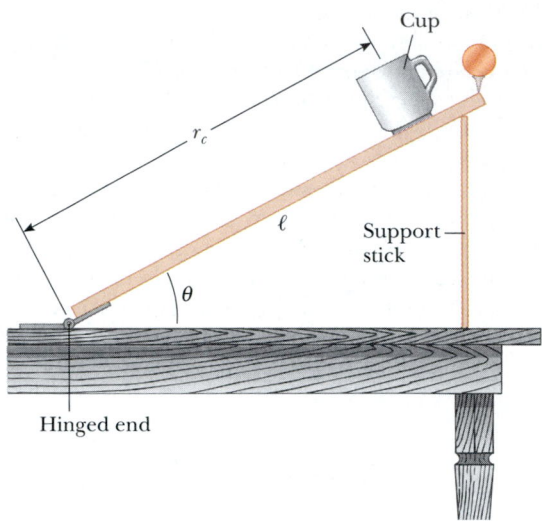

Figure P10.72

removed. (a) Show that the ball will lag behind the falling board when θ is less than $35.3°$; and that (b) the ball will fall into the cup when the board is supported at

this limiting angle and the cup is placed at

$$r_c = \frac{2\,\ell}{3\,\cos\theta}$$

(c) If a ball is at the end of a 1.00-m stick at this critical angle, show that the cup must be 18.4 cm from the moving end.

73. As a result of friction, the angular speed of a wheel changes with time according to the relationship

$$d\theta/dt = \omega_0 e^{-\sigma t}$$

where ω_0 and σ are constants. The angular speed changes from 3.50 rad/s at $t = 0$ to 2.00 rad/s at $t = 9.30$ s. Use this information to determine σ and ω_0. Then, determine (a) the magnitude of the angular acceleration at $t = 3.00$ s, (b) the number of revolutions the wheel makes in the first 2.50 s, and (c) the number of revolutions it makes before coming to rest.

74. The hour hand and the minute hand of Big Ben, the famous Parliament tower clock in London, are 2.70 m long and 4.50 m long and have masses of 60.0 kg and 100 kg, respectively (see Fig. P10.26). (a) Determine the total torque due to the weight of these hands about the axis of rotation when the time reads (i) 3:00, (ii) 5:15, (iii) 6:00, (iv) 8:20, and (v) 9:45. (You may model the hands as long thin rods.) (b) Determine all times at which the total torque about the axis of rotation is zero. Determine the times to the nearest second, solving a transcendental equation numerically.

ANSWERS TO QUICK QUIZZES

10.1 The fact that ω is negative indicates that we are dealing with an object that is rotating in the clockwise direction. We also know that when $\boldsymbol{\omega}$ and $\boldsymbol{\alpha}$ are antiparallel, ω must be decreasing—the object is slowing down. Therefore, the object is spinning more and more slowly (with less and less angular speed) in the clockwise, or negative, direction. This has a linear analogy to a sky diver opening her parachute. The velocity is negative—downward. When the sky diver opens the parachute, a large upward force causes an upward acceleration. As a result, the acceleration and velocity vectors are in opposite directions. Consequently, the parachutist slows down.

10.2 (a) Yes, all points on the wheel have the same angular speed. This is why we use angular quantities to describe

rotational motion. (b) No, not all points on the wheel have the same linear speed. (c) $v = 0$, $a = 0$. (d) $v = R\omega/2$, $a = a_r = v^2/(R/2) = R\omega^2/2$ (a_t is zero at all points because ω is constant). (e) $v = R\omega$, $a = R\omega^2$.

10.3 (a) $I = MR^2$. (b) $I = MR^2$. The moment of inertia of a system of masses equidistant from an axis of rotation is always the sum of the masses multiplied by the square of the distance from the axis.

10.4 (b) Rotation about the axis through point P requires more work. The moment of inertia of the hoop about the center axis is $I_{CM} = MR^2$, whereas, by the parallel-axis theorem, the moment of inertia about the axis through point P is $I_P = I_{CM} + MR^2 = MR^2 + MR^2 = 2MR^2$.

Rolling Motion and Angular Momentum

c h a p t e r

11

327

n the preceding chapter we learned how to treat a rigid body rotating about a fixed axis; in the present chapter, we move on to the more general case in which the axis of rotation is not fixed in space. We begin by describing such motion, which is called *rolling motion*. The central topic of this chapter is, however, angular momentum, a quantity that plays a key role in rotational dynamics. In analogy to the conservation of linear momentum, we find that the angular momentum of a rigid object is always conserved if no external torques act on the object. Like the law of conservation of linear momentum, the law of conservation of angular momentum is a fundamental law of physics, equally valid for relativistic and quantum systems.

11.1 ▶ ROLLING MOTION OF A RIGID OBJECT

In this section we treat the motion of a rigid object rotating about a moving axis. In general, such motion is very complex. However, we can simplify matters by restricting our discussion to a homogeneous rigid object having a high degree of symmetry, such as a cylinder, sphere, or hoop. Furthermore, we assume that the object undergoes rolling motion along a flat surface. We shall see that if an object such as a cylinder rolls without slipping on the surface (we call this *pure rolling motion*), a simple relationship exists between its rotational and translational motions.

Suppose a cylinder is rolling on a straight path. As Figure 11.1 shows, the center of mass moves in a straight line, but a point on the rim moves in a more complex path called a *cycloid*. This means that the axis of rotation remains parallel to its initial orientation in space. Consider a uniform cylinder of radius R rolling without slipping on a horizontal surface (Fig. 11.2). As the cylinder rotates through an angle θ, its center of mass moves a linear distance $s = R\theta$ (see Eq. 10.1a). Therefore, the linear speed of the center of mass for pure rolling motion is given by

$$v_{CM} = \frac{ds}{dt} = R\frac{d\theta}{dt} = R\omega \tag{11.1}$$

where ω is the angular velocity of the cylinder. Equation 11.1 holds whenever a cylinder or sphere rolls without slipping and is the **condition for pure rolling**

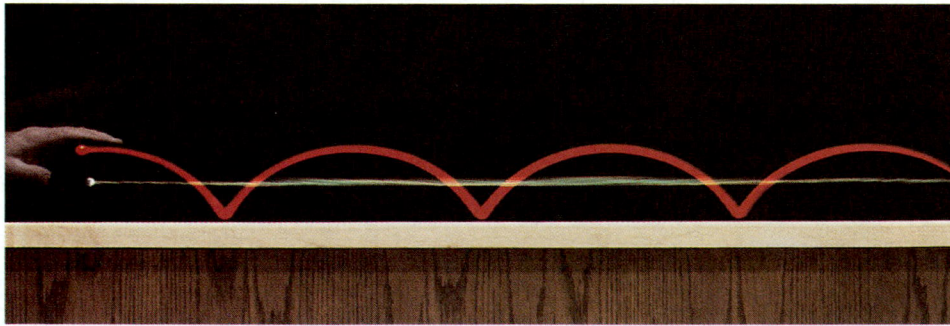

Figure 11.1   One light source at the center of a rolling cylinder and another at one point on the rim illustrate the different paths these two points take. The center moves in a straight line (green line), whereas the point on the rim moves in the path called a *cycloid* (red curve). *(Henry Leap and Jim Lehman)*

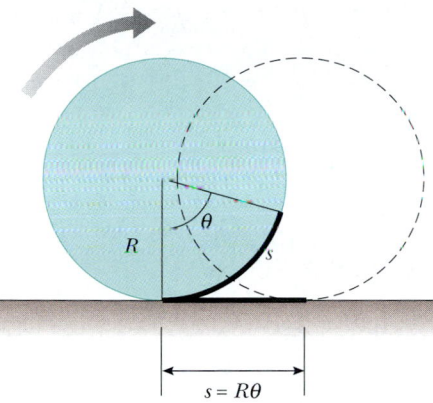

Figure 11.2 In pure rolling motion, as the cylinder rotates through an angle θ, its center of mass moves a linear distance $s = R\theta$.

motion. The magnitude of the linear acceleration of the center of mass for pure rolling motion is

$$a_{\text{CM}} = \frac{dv_{\text{CM}}}{dt} = R\frac{d\omega}{dt} = R\alpha \tag{11.2}$$

where α is the angular acceleration of the cylinder.

The linear velocities of the center of mass and of various points on and within the cylinder are illustrated in Figure 11.3. A short time after the moment shown in the drawing, the rim point labeled P will have rotated from the six o'clock position to, say, the seven o'clock position, the point Q will have rotated from the ten o'clock position to the eleven o'clock position, and so on. Note that the linear velocity of any point is in a direction perpendicular to the line from that point to the contact point P. At any instant, the part of the rim that is at point P is at rest relative to the surface because slipping does not occur.

All points on the cylinder have the same angular speed. Therefore, because the distance from P' to P is twice the distance from P to the center of mass, P' has a speed $2v_{\text{CM}} = 2R\omega$. To see why this is so, let us model the rolling motion of the cylinder in Figure 11.4 as a combination of translational (linear) motion and rotational motion. For the pure translational motion shown in Figure 11.4a, imagine that the cylinder does not rotate, so that each point on it moves to the right with speed v_{CM}. For the pure rotational motion shown in Figure 11.4b, imagine that a rotation axis through the center of mass is stationary, so that each point on the cylinder has the same rotational speed ω. The combination of these two motions represents the rolling motion shown in Figure 11.4c. Note in Figure 11.4c that the top of the cylinder has linear speed $v_{\text{CM}} + R\omega = v_{\text{CM}} + v_{\text{CM}} = 2v_{\text{CM}}$, which is greater than the linear speed of any other point on the cylinder. As noted earlier, the center of mass moves with linear speed v_{CM} while the contact point between the surface and cylinder has a linear speed of zero.

We can express the total kinetic energy of the rolling cylinder as

$$K = \tfrac{1}{2}I_P\omega^2 \tag{11.3}$$

where I_P is the moment of inertia about a rotation axis through P. Applying the parallel-axis theorem, we can substitute $I_P = I_{\text{CM}} + MR^2$ into Equation 11.3 to obtain

$$K = \tfrac{1}{2}I_{\text{CM}}\omega^2 + \tfrac{1}{2}MR^2\omega^2$$

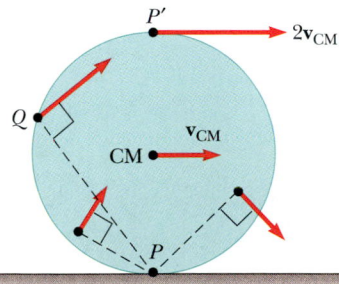

Figure 11.3 All points on a rolling object move in a direction perpendicular to an axis through the instantaneous point of contact P. In other words, all points rotate about P. The center of mass of the object moves with a velocity $\mathbf{v}_{\text{CM}}$, and the point P' moves with a velocity $2\mathbf{v}_{\text{CM}}$.

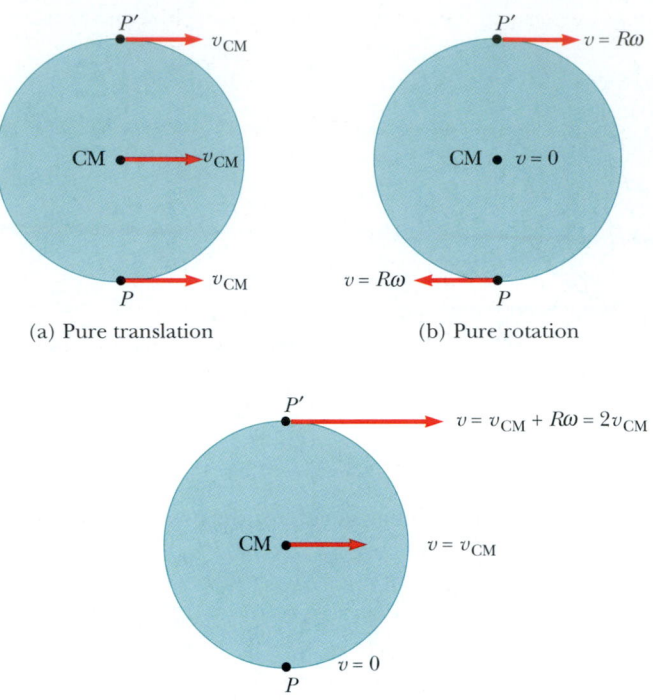

(a) Pure translation

(b) Pure rotation

(c) Combination of translation and rotation

Figure 11.4   The motion of a rolling object can be modeled as a combination of pure translation and pure rotation.

or, because $v_{CM} = R\omega$,

$$K = \tfrac{1}{2}I_{CM}\omega^2 + \tfrac{1}{2}Mv_{CM}{}^2 \qquad \textbf{(11.4)}$$

Total kinetic energy of a rolling body

The term $\tfrac{1}{2}I_{CM}\omega^2$ represents the rotational kinetic energy of the cylinder about its center of mass, and the term $\tfrac{1}{2}Mv_{CM}{}^2$ represents the kinetic energy the cylinder would have if it were just translating through space without rotating. Thus, we can say that the **total kinetic energy of a rolling object is the sum of the rotational kinetic energy about the center of mass and the translational kinetic energy of the center of mass.**

We can use energy methods to treat a class of problems concerning the rolling motion of a sphere down a rough incline. (The analysis that follows also applies to the rolling motion of a cylinder or hoop.) We assume that the sphere in Figure 11.5 rolls without slipping and is released from rest at the top of the incline. Note that accelerated rolling motion is possible only if a frictional force is present between the sphere and the incline to produce a net torque about the center of mass. Despite the presence of friction, no loss of mechanical energy occurs because the contact point is at rest relative to the surface at any instant. On the other hand, if the sphere were to slip, mechanical energy would be lost as motion progressed.

Using the fact that $v_{CM} = R\omega$ for pure rolling motion, we can express Equation 11.4 as

$$K = \tfrac{1}{2}I_{CM}\left(\frac{v_{CM}}{R}\right)^2 + \tfrac{1}{2}Mv_{CM}{}^2$$

$$K = \tfrac{1}{2}\left(\frac{I_{CM}}{R^2} + M\right)v_{CM}{}^2 \qquad \textbf{(11.5)}$$

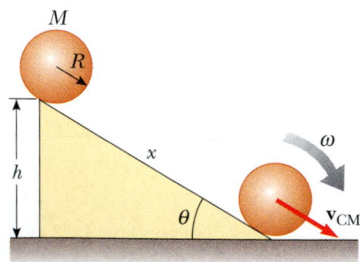

Figure 11.5   A sphere rolling down an incline. Mechanical energy is conserved if no slipping occurs.

By the time the sphere reaches the bottom of the incline, work equal to Mgh has been done on it by the gravitational field, where h is the height of the incline. Because the sphere starts from rest at the top, its kinetic energy at the bottom, given by Equation 11.5, must equal this work done. Therefore, the speed of the center of mass at the bottom can be obtained by equating these two quantities:

$$\frac{1}{2}\left(\frac{I_{\rm CM}}{R^2} + M\right)v_{\rm CM}{}^2 = Mgh$$

$$v_{\rm CM} = \left(\frac{2gh}{1 + I_{\rm CM}/MR^2}\right)^{1/2} \qquad \textbf{(11.6)}$$

Quick Quiz 11.1

Imagine that you slide your textbook across a gymnasium floor with a certain initial speed. It quickly stops moving because of friction between it and the floor. Yet, if you were to start a basketball rolling with the same initial speed, it would probably keep rolling from one end of the gym to the other. Why does a basketball roll so far? Doesn't friction affect its motion?

EXAMPLE 11.1   Sphere Rolling Down an Incline

For the solid sphere shown in Figure 11.5, calculate the linear speed of the center of mass at the bottom of the incline and the magnitude of the linear acceleration of the center of mass.

Solution   The sphere starts from the top of the incline with potential energy $U_g = Mgh$ and kinetic energy $K = 0$. As we have seen before, if it fell vertically from that height, it would have a linear speed of $\sqrt{2gh}$ at the moment before it hit the floor. After rolling down the incline, the linear speed of the center of mass must be less than this value because some of the initial potential energy is diverted into rotational kinetic energy rather than all being converted into translational kinetic energy. For a uniform solid sphere, $I_{\rm CM} = \frac{2}{5}MR^2$ (see Table 10.2), and therefore Equation 11.6 gives

$$v_{\rm CM} = \left(\frac{2gh}{1 + \dfrac{2/5MR^2}{MR^2}}\right)^{1/2} = \left(\frac{10}{7}gh\right)^{1/2}$$

which is less than $\sqrt{2gh}$.

To calculate the linear acceleration of the center of mass, we note that the vertical displacement is related to the displacement x along the incline through the relationship $h = $ $x\sin\theta$. Hence, after squaring both sides, we can express the equation above as

$$v_{\rm CM}{}^2 = \tfrac{10}{7}\,gx\sin\theta$$

Comparing this with the expression from kinematics, $v_{\rm CM}{}^2 = 2a_{\rm CM}x$ (see Eq. 2.12), we see that the acceleration of the center of mass is

$$a_{\rm CM} = \tfrac{5}{7}g\sin\theta$$

These results are quite interesting in that both the speed and the acceleration of the center of mass are *independent* of the mass and the radius of the sphere! That is, **all homogeneous solid spheres experience the same speed and acceleration on a given incline.**

If we repeated the calculations for a hollow sphere, a solid cylinder, or a hoop, we would obtain similar results in which only the factor in front of $g\sin\theta$ would differ. The constant factors that appear in the expressions for $v_{\rm CM}$ and $a_{\rm CM}$ depend only on the moment of inertia about the center of mass for the specific body. In all cases, the acceleration of the center of mass is *less* than $g\sin\theta$, the value the acceleration would have if the incline were frictionless and no rolling occurred.

EXAMPLE 11.2   Another Look at the Rolling Sphere

In this example, let us use dynamic methods to verify the results of Example 11.1. The free-body diagram for the sphere is illustrated in Figure 11.6.

Solution   Newton's second law applied to the center of mass gives

$$(1) \qquad \Sigma F_x = Mg\sin\theta - f = Ma_{\rm CM}$$

$$\Sigma F_y = n - Mg\cos\theta = 0$$

where x is measured along the slanted surface of the incline.

Now let us write an expression for the torque acting on the sphere. A convenient axis to choose is the one that passes

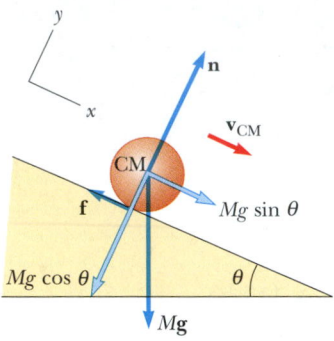

Figure 11.6 Free-body diagram for a solid sphere rolling down an incline.

clockwise direction,

$$\tau_{CM} = fR = I_{CM}\alpha$$

Because $I_{CM} = \frac{2}{5}MR^2$ and $\alpha = a_{CM}/R$, we obtain

$$(2) \qquad f = \frac{I_{CM}\alpha}{R} = \left(\frac{\frac{2}{5}MR^2}{R}\right)\frac{a_{CM}}{R} = \frac{2}{5}Ma_{CM}$$

Substituting Equation (2) into Equation (1) gives

$$a_{CM} = \boxed{\frac{5}{7}g\sin\theta}$$

which agrees with the result of Example 11.1.

Note that $\Sigma\mathbf{F} = m\mathbf{a}$ applies only if $\Sigma\mathbf{F}$ is the net external force exerted on the sphere and $\mathbf{a}$ is the acceleration of its center of mass. In the case of our sphere rolling down an incline, even though the frictional force does not change the total kinetic energy of the sphere, it does contribute to $\Sigma\mathbf{F}$ and thus decreases the acceleration of the center of mass. As a result, the final translational kinetic energy is less than it would be in the absence of friction. As mentioned in Example 11.1, some of the initial potential energy is converted to rotational kinetic energy.

through the center of the sphere and is perpendicular to the plane of the figure.[1] Because $\mathbf{n}$ and $M\mathbf{g}$ go through the center of mass, they have zero moment arm about this axis and thus do not contribute to the torque. However, the force of static friction produces a torque about this axis equal to fR in the clockwise direction; therefore, because τ is also in the

QuickLab

Hold a basketball and a tennis ball side by side at the top of a ramp and release them at the same time. Which reaches the bottom first? Does the outcome depend on the angle of the ramp? What if the angle were 90° (that is, if the balls were in free fall)?

Quick Quiz 11.2

Which gets to the bottom first: a ball rolling without sliding down incline A or a box sliding down a frictionless incline B having the same dimensions as incline A?

11.2   THE VECTOR PRODUCT AND TORQUE

Consider a force $\mathbf{F}$ acting on a rigid body at the vector position $\mathbf{r}$ (Fig. 11.7). **The origin O is assumed to be in an inertial frame, so Newton's first law is valid in this case.** As we saw in Section 10.6, the *magnitude* of the torque due to this force relative to the origin is, by definition, $rF\sin\phi$, where ϕ is the angle between $\mathbf{r}$ and $\mathbf{F}$. The axis about which $\mathbf{F}$ tends to produce rotation is perpendicular to the plane formed by $\mathbf{r}$ and $\mathbf{F}$. If the force lies in the xy plane, as it does in Figure 11.7, the torque τ is represented by a vector parallel to the z axis. The force in Figure 11.7 creates a torque that tends to rotate the body counterclockwise about the z axis; thus the direction of τ is toward increasing z, and τ is therefore in the positive z direction. If we reversed the direction of $\mathbf{F}$ in Figure 11.7, then τ would be in the negative z direction.

The torque τ involves the two vectors $\mathbf{r}$ and $\mathbf{F}$, and its direction is perpendicular to the plane of $\mathbf{r}$ and $\mathbf{F}$. We can establish a mathematical relationship between τ, $\mathbf{r}$, and $\mathbf{F}$, using a new mathematical operation called the **vector product,** or **cross product:**

Torque

$$\tau \equiv \mathbf{r} \times \mathbf{F} \qquad\qquad (11.7)$$

[1] Although a coordinate system whose origin is at the center of mass of a rolling object is not an inertial frame, the expression $\tau_{CM} = I\alpha$ still applies in the center-of-mass frame.

We now give a formal definition of the vector product. Given any two vectors **A** and **B**, the **vector product A** × **B** is defined as a third vector **C**, the magnitude of which is $AB \sin \theta$, where θ is the angle between **A** and **B**. That is, if **C** is given by

$$\mathbf{C} = \mathbf{A} \times \mathbf{B} \tag{11.8}$$

then its magnitude is

$$C \equiv AB \sin \theta \tag{11.9}$$

The quantity $AB \sin \theta$ is equal to the area of the parallelogram formed by **A** and **B**, as shown in Figure 11.8. The *direction* of **C** is perpendicular to the plane formed by **A** and **B**, and the best way to determine this direction is to use the right-hand rule illustrated in Figure 11.8. The four fingers of the right hand are pointed along **A** and then "wrapped" into **B** through the angle θ. The direction of the erect right thumb is the direction of **A** × **B** = **C.** Because of the notation, **A** × **B** is often read "**A** cross **B**"; hence, the term *cross product*.

Some properties of the vector product that follow from its definition are as follows:

1. Unlike the scalar product, the vector product is *not* commutative. Instead, the order in which the two vectors are multiplied in a cross product is important:

$$\mathbf{A} \times \mathbf{B} = -\mathbf{B} \times \mathbf{A} \tag{11.10}$$

   Therefore, if you change the order of the vectors in a cross product, you must change the sign. You could easily verify this relationship with the right-hand rule.
2. If **A** is parallel to **B** ($\theta = 0°$ or $180°$), then **A** × **B** = 0; therefore, it follows that **A** × **A** = 0.
3. If **A** is perpendicular to **B**, then $|\mathbf{A} \times \mathbf{B}| = AB$.
4. The vector product obeys the distributive law:

$$\mathbf{A} \times (\mathbf{B} + \mathbf{C}) = \mathbf{A} \times \mathbf{B} + \mathbf{A} \times \mathbf{C} \tag{11.11}$$

5. The derivative of the cross product with respect to some variable such as t is

$$\frac{d}{dt}(\mathbf{A} \times \mathbf{B}) = \mathbf{A} \times \frac{d\mathbf{B}}{dt} + \frac{d\mathbf{A}}{dt} \times \mathbf{B} \tag{11.12}$$

where it is important to preserve the multiplicative order of **A** and **B**, in view of Equation 11.10.

It is left as an exercise to show from Equations 11.9 and 11.10 and from the definition of unit vectors that the cross products of the rectangular unit vectors **i**,

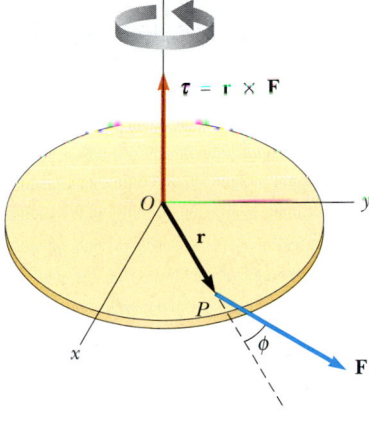

Figure 11.7 The torque vector τ lies in a direction perpendicular to the plane formed by the position vector **r** and the applied force vector **F**.

▶ Properties of the vector product

Right-hand rule

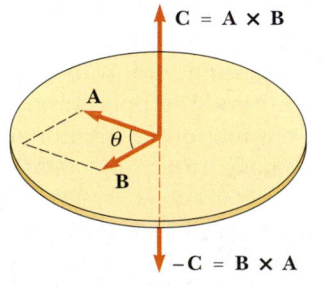

Figure 11.8 The vector product **A** × **B** is a third vector **C** having a magnitude $AB \sin \theta$ equal to the area of the parallelogram shown. The direction of **C** is perpendicular to the plane formed by **A** and **B**, and this direction is determined by the right-hand rule.

Cross products of unit vectors

j, and **k** obey the following rules:

$$\mathbf{i} \times \mathbf{i} = \mathbf{j} \times \mathbf{j} = \mathbf{k} \times \mathbf{k} = 0 \qquad \text{(11.13a)}$$

$$\mathbf{i} \times \mathbf{j} = -\mathbf{j} \times \mathbf{i} = \mathbf{k} \qquad \text{(11.13b)}$$

$$\mathbf{j} \times \mathbf{k} = -\mathbf{k} \times \mathbf{j} = \mathbf{i} \qquad \text{(11.13c)}$$

$$\mathbf{k} \times \mathbf{i} = -\mathbf{i} \times \mathbf{k} = \mathbf{j} \qquad \text{(11.13d)}$$

Signs are interchangeable in cross products. For example, $\mathbf{A} \times (-\mathbf{B}) = -\mathbf{A} \times \mathbf{B}$ and $\mathbf{i} \times (-\mathbf{j}) = -\mathbf{i} \times \mathbf{j}$.

The cross product of any two vectors **A** and **B** can be expressed in the following determinant form:

$$\mathbf{A} \times \mathbf{B} = \begin{vmatrix} \mathbf{i} & \mathbf{j} & \mathbf{k} \\ A_x & A_y & A_z \\ B_x & B_y & B_z \end{vmatrix} = \mathbf{i} \begin{vmatrix} A_y & A_z \\ B_y & B_z \end{vmatrix} - \mathbf{j} \begin{vmatrix} A_x & A_z \\ B_x & B_z \end{vmatrix} + \mathbf{k} \begin{vmatrix} A_x & A_y \\ B_x & B_y \end{vmatrix}$$

Expanding these determinants gives the result

$$\mathbf{A} \times \mathbf{B} = (A_y B_z - A_z B_y)\mathbf{i} - (A_x B_z - A_z B_x)\mathbf{j} + (A_x B_y - A_y B_x)\mathbf{k} \qquad \text{(11.14)}$$

EXAMPLE 11.3 ▸ The Cross Product

Two vectors lying in the xy plane are given by the equations $\mathbf{A} = 2\mathbf{i} + 3\mathbf{j}$ and $\mathbf{B} = -\mathbf{i} + 2\mathbf{j}$. Find $\mathbf{A} \times \mathbf{B}$ and verify that $\mathbf{A} \times \mathbf{B} = -\mathbf{B} \times \mathbf{A}$.

Solution Using Equations 11.13a through 11.13d, we obtain

$$\mathbf{A} \times \mathbf{B} = (2\mathbf{i} + 3\mathbf{j}) \times (-\mathbf{i} + 2\mathbf{j})$$

$$= 2\mathbf{i} \times 2\mathbf{j} + 3\mathbf{j} \times (-\mathbf{i}) = 4\mathbf{k} + 3\mathbf{k} = \boxed{7\mathbf{k}}$$

(We have omitted the terms containing $\mathbf{i} \times \mathbf{i}$ and $\mathbf{j} \times \mathbf{j}$ because, as Equation 11.13a shows, they are equal to zero.)

We can show that $\mathbf{A} \times \mathbf{B} = -\mathbf{B} \times \mathbf{A}$, since

$$\mathbf{B} \times \mathbf{A} = (-\mathbf{i} + 2\mathbf{j}) \times (2\mathbf{i} + 3\mathbf{j})$$

$$= -\mathbf{i} \times 3\mathbf{j} + 2\mathbf{j} \times 2\mathbf{i} = -3\mathbf{k} - 4\mathbf{k} = \boxed{-7\mathbf{k}}$$

Therefore, $\mathbf{A} \times \mathbf{B} = -\mathbf{B} \times \mathbf{A}$.

As an alternative method for finding $\mathbf{A} \times \mathbf{B}$, we could use Equation 11.14, with $A_x = 2$, $A_y = 3$, $A_z = 0$ and $B_x = -1$, $B_y = 2$, $B_z = 0$:

$$\mathbf{A} \times \mathbf{B} = (0)\mathbf{i} - (0)\mathbf{j} + [(2)(2) - (3)(-1)]\mathbf{k} = 7\mathbf{k}$$

Exercise Use the results to this example and Equation 11.9 to find the angle between **A** and **B**.

Answer 60.3°

11.3 ▸ ANGULAR MOMENTUM OF A PARTICLE

Imagine a rigid pole sticking up through the ice on a frozen pond (Fig. 11.9). A skater glides rapidly toward the pole, aiming a little to the side so that she does not hit it. As she approaches a point beside the pole, she reaches out and grabs the pole, an action that whips her rapidly into a circular path around the pole. Just as the idea of linear momentum helps us analyze translational motion, a rotational analog—*angular momentum*—helps us describe this skater and other objects undergoing rotational motion.

To analyze the motion of the skater, we need to know her mass and her velocity, as well as her position relative to the pole. In more general terms, consider a particle of mass m located at the vector position $\mathbf{r}$ and moving with linear velocity $\mathbf{v}$ (Fig. 11.10).

The instantaneous angular momentum $\mathbf{L}$ of the particle relative to the origin O is defined as the cross product of the particle's instantaneous position vector $\mathbf{r}$ and its instantaneous linear momentum $\mathbf{p}$:

$$\mathbf{L} \equiv \mathbf{r} \times \mathbf{p} \qquad (11.15)$$

Angular momentum of a particle

The SI unit of angular momentum is kg·m²/s. It is important to note that both the magnitude and the direction of $\mathbf{L}$ depend on the choice of origin. Following the right-hand rule, note that the direction of $\mathbf{L}$ is perpendicular to the plane formed by $\mathbf{r}$ and $\mathbf{p}$. In Figure 11.10, $\mathbf{r}$ and $\mathbf{p}$ are in the xy plane, and so $\mathbf{L}$ points in the z direction. Because $\mathbf{p} = m\mathbf{v}$, the magnitude of $\mathbf{L}$ is

$$L = mvr \sin \phi \qquad (11.16)$$

where ϕ is the angle between $\mathbf{r}$ and $\mathbf{p}$. It follows that L is zero when $\mathbf{r}$ is parallel to $\mathbf{p}$ ($\phi = 0$ or $180°$). In other words, when the linear velocity of the particle is along a line that passes through the origin, the particle has zero angular momentum with respect to the origin. On the other hand, if $\mathbf{r}$ is perpendicular to $\mathbf{p}$ ($\phi = 90°$), then $L = mvr$. At that instant, the particle moves exactly as if it were on the rim of a wheel rotating about the origin in a plane defined by $\mathbf{r}$ and $\mathbf{p}$.

Quick Quiz 11.3

Recall the skater described at the beginning of this section. What would be her angular momentum relative to the pole if she were skating directly toward it?

In describing linear motion, we found that the net force on a particle equals the time rate of change of its linear momentum, $\Sigma \mathbf{F} = d\mathbf{p}/dt$ (see Eq. 9.3). We now show that the net torque acting on a particle equals the time rate of change of its angular momentum. Let us start by writing the net torque on the particle in the form

$$\sum \boldsymbol{\tau} = \mathbf{r} \times \sum \mathbf{F} = \mathbf{r} \times \frac{d\mathbf{p}}{dt} \qquad (11.17)$$

Now let us differentiate Equation 11.15 with respect to time, using the rule given by Equation 11.12:

$$\frac{d\mathbf{L}}{dt} = \frac{d}{dt}(\mathbf{r} \times \mathbf{p}) = \mathbf{r} \times \frac{d\mathbf{p}}{dt} + \frac{d\mathbf{r}}{dt} \times \mathbf{p}$$

Remember, it is important to adhere to the order of terms because $\mathbf{A} \times \mathbf{B} = -\mathbf{B} \times \mathbf{A}$. The last term on the right in the above equation is zero because $\mathbf{v} = d\mathbf{r}/dt$ is parallel to $\mathbf{p} = m\mathbf{v}$ (property 2 of the vector product). Therefore,

$$\frac{d\mathbf{L}}{dt} = \mathbf{r} \times \frac{d\mathbf{p}}{dt} \qquad (11.18)$$

Comparing Equations 11.17 and 11.18, we see that

$$\sum \boldsymbol{\tau} = \frac{d\mathbf{L}}{dt} \qquad (11.19)$$

Figure 11.9 As the skater passes the pole, she grabs hold of it. This causes her to swing around the pole rapidly in a circular path.

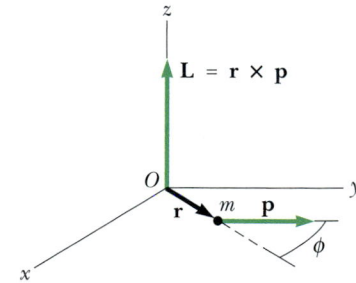

Figure 11.10 The angular momentum $\mathbf{L}$ of a particle of mass m and linear momentum $\mathbf{p}$ located at the vector position $\mathbf{r}$ is a vector given by $\mathbf{L} = \mathbf{r} \times \mathbf{p}$. The value of $\mathbf{L}$ depends on the origin about which it is measured and is a vector perpendicular to both $\mathbf{r}$ and $\mathbf{p}$.

The net torque equals time rate of change of angular momentum which is the rotational analog of Newton's second law, $\Sigma \mathbf{F} = d\mathbf{p}/dt$. Note that torque causes the angular momentum $\mathbf{L}$ to change just as force causes linear momentum $\mathbf{p}$ to change. This rotational result, Equation 11.19, states that

> the net torque acting on a particle is equal to the time rate of change of the particle's angular momentum.

It is important to note that Equation 11.19 is valid only if $\Sigma \boldsymbol{\tau}$ and $\mathbf{L}$ are measured about the same origin. (Of course, the same origin must be used in calculating all of the torques.) Furthermore, **the expression is valid for any origin fixed in an inertial frame.**

Angular Momentum of a System of Particles

The total angular momentum of a system of particles about some point is defined as the vector sum of the angular momenta of the individual particles:

$$\mathbf{L} = \mathbf{L}_1 + \mathbf{L}_2 + \cdots + \mathbf{L}_n = \sum_i \mathbf{L}_i$$

where the vector sum is over all n particles in the system.

Because individual angular momenta can change with time, so can the total angular momentum. In fact, from Equations 11.18 and 11.19, we find that the time rate of change of the total angular momentum equals the vector sum of all torques acting on the system, both those associated with internal forces between particles and those associated with external forces. However, the net torque associated with all internal forces is zero. To understand this, recall that Newton's third law tells us that internal forces between particles of the system are equal in magnitude and opposite in direction. If we assume that these forces lie along the line of separation of each pair of particles, then the torque due to each action–reaction force pair is zero. That is, the moment arm d from O to the line of action of the forces is equal for both particles. In the summation, therefore, we see that the net internal torque vanishes. We conclude that the total angular momentum of a system can vary with time only if a net external torque is acting on the system, so that we have

$$\sum \boldsymbol{\tau}_{\text{ext}} = \sum_i \frac{d\mathbf{L}_i}{dt} = \frac{d}{dt} \sum_i \mathbf{L}_i = \frac{d\mathbf{L}}{dt} \qquad \textbf{(11.20)}$$

That is,

> the time rate of change of the total angular momentum of a system about some origin in an inertial frame equals the net external torque acting on the system about that origin.

Note that Equation 11.20 is the rotational analog of Equation 9.38, $\Sigma \mathbf{F}_{\text{ext}} = d\mathbf{p}/dt$, for a system of particles.

EXAMPLE 11.4   **Circular Motion**

A particle moves in the xy plane in a circular path of radius r, as shown in Figure 11.11. (a) Find the magnitude and direction of its angular momentum relative to O when its linear velocity is **v**.

Solution   You might guess that because the linear momentum of the particle is always changing (in direction, not magnitude), the direction of the angular momentum must also change. In this example, however, this is not the case. The magnitude of **L** is given by

$$L = mvr \sin 90° = \boxed{mvr} \qquad \text{(for } \mathbf{r} \text{ perpendicular to } \mathbf{v}\text{)}$$

This value of L is constant because all three factors on the right are constant. The direction of **L** also is constant, even

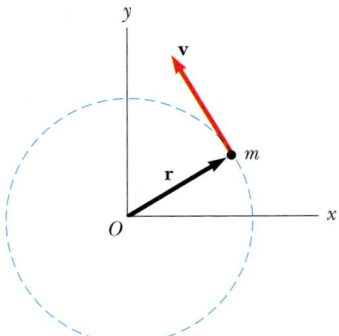

Figure 11.11   A particle moving in a circle of radius r has an angular momentum about O that has magnitude mvr. The vector $\mathbf{L} = \mathbf{r} \times \mathbf{p}$ points *out* of the diagram.

though the direction of $\mathbf{p} = m\mathbf{v}$ keeps changing. You can visualize this by sliding the vector **v** in Figure 11.11 parallel to itself until its tail meets the tail of **r** and by then applying the right-hand rule. (You can use **v** to determine the direction of $\mathbf{L} = \mathbf{r} \times \mathbf{p}$ because the direction of **p** is the same as the direction of **v**.) Line up your fingers so that they point along **r** and wrap your fingers into the vector **v**. Your thumb points upward and away from the page; this is the direction of **L**. Hence, we can write the vector expression $\mathbf{L} = (mvr)\mathbf{k}$. If the particle were to move clockwise, **L** would point downward and into the page.

(b) Find the magnitude and direction of **L** in terms of the particle's angular speed ω.

Solution   Because $v = r\omega$ for a particle rotating in a circle, we can express L as

$$L = mvr = mr^2\omega = \boxed{I\omega}$$

where I is the moment of inertia of the particle about the z axis through O. Because the rotation is counterclockwise, the direction of $\boldsymbol{\omega}$ is along the z axis (see Section 10.1). The direction of **L** is the same as that of $\boldsymbol{\omega}$, and so we can write the angular momentum as $\mathbf{L} = I\boldsymbol{\omega} = I\omega\mathbf{k}$.

Exercise   A car of mass 1 500 kg moves with a linear speed of 40 m/s on a circular race track of radius 50 m. What is the magnitude of its angular momentum relative to the center of the track?

Answer 3.0×10^6 kg·m²/s

11.4 ANGULAR MOMENTUM OF A ROTATING RIGID OBJECT

Consider a rigid object rotating about a fixed axis that coincides with the z axis of a coordinate system, as shown in Figure 11.12. Let us determine the angular momentum of this object. Each particle of the object rotates in the xy plane about the z axis with an angular speed ω. The magnitude of the angular momentum of a particle of mass m_i about the origin O is $m_i v_i r_i$. Because $v_i = r_i\omega$, we can express the magnitude of the angular momentum of this particle as

$$L_i = m_i r_i^2 \omega$$

The vector $\mathbf{L}_i$ is directed along the z axis, as is the vector $\boldsymbol{\omega}$.

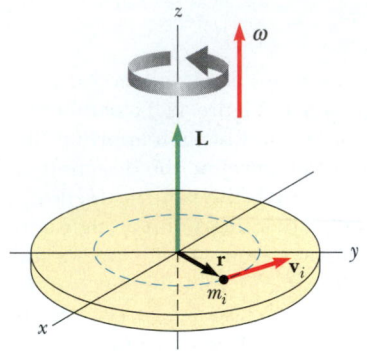

Figure 11.12 When a rigid body rotates about an axis, the angular momentum **L** is in the same direction as the angular velocity $\boldsymbol{\omega}$, according to the expression **L** = $I\boldsymbol{\omega}$.

We can now find the angular momentum (which in this situation has only a z component) of the whole object by taking the sum of L_i over all particles:

$$L_z = \sum_i m_i r_i^2 \omega = \left(\sum_i m_i r_i^2 \right) \omega$$

$$L_z = I\omega \qquad \qquad \textbf{(11.21)}$$

where I is the moment of inertia of the object about the z axis.

Now let us differentiate Equation 11.21 with respect to time, noting that I is constant for a rigid body:

$$\frac{dL_z}{dt} = I\frac{d\omega}{dt} = I\alpha \qquad \qquad \textbf{(11.22)}$$

where α is the angular acceleration relative to the axis of rotation. Because dL_z/dt is equal to the net external torque (see Eq. 11.20), we can express Equation 11.22 as

$$\sum \tau_{\text{ext}} = \frac{dL_z}{dt} = I\alpha \qquad \qquad \textbf{(11.23)}$$

That is, the net external torque acting on a rigid object rotating about a fixed axis equals the moment of inertia about the rotation axis multiplied by the object's angular acceleration relative to that axis.

Equation 11.23 also is valid for a rigid object rotating about a moving axis provided the moving axis (1) passes through the center of mass and (2) is a symmetry axis.

You should note that if a symmetrical object rotates about a fixed axis passing through its center of mass, you can write Equation 11.21 in vector form as **L** = $I\boldsymbol{\omega}$, where **L** is the total angular momentum of the object measured with respect to the axis of rotation. Furthermore, the expression is valid for any object, regardless of its symmetry, if **L** stands for the component of angular momentum along the axis of rotation.[2]

EXAMPLE 11.5   Bowling Ball

Estimate the magnitude of the angular momentum of a bowling ball spinning at 10 rev/s, as shown in Figure 11.13.

Solution We start by making some estimates of the relevant physical parameters and model the ball as a uniform solid sphere. A typical bowling ball might have a mass of 6 kg and a radius of 12 cm. The moment of inertia of a solid sphere about an axis through its center is, from Table 10.2,

$$I = \tfrac{2}{5}MR^2 = \tfrac{2}{5}(6 \text{ kg})(0.12 \text{ m})^2 = 0.035 \text{ kg} \cdot \text{m}^2$$

Therefore, the magnitude of the angular momentum is

---

[2] In general, the expression **L** = $I\boldsymbol{\omega}$ is not always valid. If a rigid object rotates about an arbitrary axis, **L** and $\boldsymbol{\omega}$ may point in different directions. In this case, the moment of inertia cannot be treated as a scalar. Strictly speaking, **L** = $I\boldsymbol{\omega}$ applies only to rigid objects of any shape that rotate about one of three mutually perpendicular axes (called *principal axes*) through the center of mass. This is discussed in more advanced texts on mechanics.

$$L = I\omega = (0.035 \text{ kg} \cdot \text{m}^2)(10 \text{ rev/s})(2\pi \text{ rad/rev})$$
$$= 2.2 \text{ kg} \cdot \text{m}^2/\text{s}$$

Because of the roughness of our estimates, we probably want to keep only one significant figure, and so $L \approx$ 2 kg · m²/s.

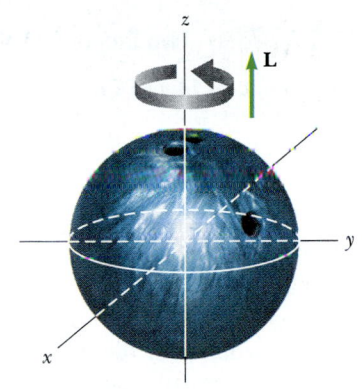

Figure 11.13 A bowling ball that rotates about the z axis in the direction shown has an angular momentum **L** in the positive z direction. If the direction of rotation is reversed, **L** points in the negative z direction.

EXAMPLE 11.6 Rotating Rod

A rigid rod of mass M and length ℓ is pivoted without friction at its center (Fig. 11.14). Two particles of masses m_1 and m_2 are connected to its ends. The combination rotates in a vertical plane with an angular speed ω. (a) Find an expression for the magnitude of the angular momentum of the system.

Solution This is different from the last example in that we now must account for the motion of more than one object. The moment of inertia of the system equals the sum of the moments of inertia of the three components: the rod and the two particles. Referring to Table 10.2 to obtain the expression for the moment of inertia of the rod, and using the expression $I = mr^2$ for each particle, we find that the total moment of inertia about the z axis through O is

$$I = \frac{1}{12}M\ell^2 + m_1\left(\frac{\ell}{2}\right)^2 + m_2\left(\frac{\ell}{2}\right)^2$$
$$= \frac{\ell^2}{4}\left(\frac{M}{3} + m_1 + m_2\right)$$

Therefore, the magnitude of the angular momentum is

$$L = I\omega = \frac{\ell^2}{4}\left(\frac{M}{3} + m_1 + m_2\right)\omega$$

(b) Find an expression for the magnitude of the angular acceleration of the system when the rod makes an angle θ with the horizontal.

Solution If the masses of the two particles are equal, then the system has no angular acceleration because the net torque on the system is zero when $m_1 = m_2$. If the initial angle θ is exactly $\pi/2$ or $-\pi/2$ (vertical position), then the rod will be in equilibrium. To find the angular acceleration of the system at any angle θ, we first calculate the net torque on the system and then use $\Sigma\tau_{ext} = I\alpha$ to obtain an expression for α.

The torque due to the force m_1g about the pivot is

$$\tau_1 = m_1 g\frac{\ell}{2}\cos\theta \qquad (\tau_1 \text{ out of page})$$

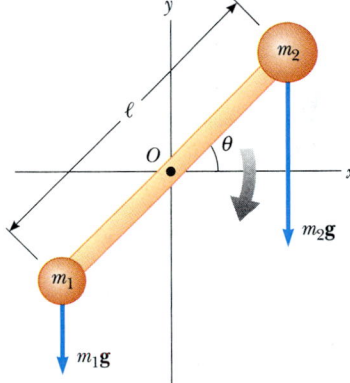

Figure 11.14 Because gravitational forces act on the rotating rod, there is in general a net nonzero torque about O when $m_1 \neq m_2$. This net torque produces an angular acceleration given by $\alpha = \Sigma\tau_{ext}/I$.

The torque due to the force m_2g about the pivot is

$$\tau_2 = -m_2 g\frac{\ell}{2}\cos\theta \qquad (\tau_2 \text{ into page})$$

Hence, the net torque exerted on the system about O is

$$\Sigma\tau_{ext} = \tau_1 + \tau_2 = \tfrac{1}{2}(m_1 - m_2)g\ell\cos\theta$$

The direction of $\Sigma\tau_{ext}$ is out of the page if $m_1 > m_2$ and is into the page if $m_2 > m_1$.

To find α, we use $\Sigma\tau_{ext} = I\alpha$, where I was obtained in part (a):

$$\alpha = \frac{\Sigma\tau_{ext}}{I} = \frac{2(m_1 - m_2)g\cos\theta}{\ell(M/3 + m_1 + m_2)}$$

Note that α is zero when θ is $\pi/2$ or $-\pi/2$ (vertical position) and is a maximum when θ is 0 or π (horizontal position).

Exercise If $m_2 > m_1$, at what value of θ is ω a maximum?

Answer $\theta = -\pi/2$.

EXAMPLE 11.7   **Two Connected Masses**

A sphere of mass m_1 and a block of mass m_2 are connected by a light cord that passes over a pulley, as shown in Figure 11.15. The radius of the pulley is R, and the moment of inertia about its axle is I. The block slides on a frictionless, horizontal surface. Find an expression for the linear acceleration of the two objects, using the concepts of angular momentum and torque.

Solution   We need to determine the angular momentum of the system, which consists of the two objects and the pulley. Let us calculate the angular momentum about an axis that coincides with the axle of the pulley.

At the instant the sphere and block have a common speed v, the angular momentum of the sphere is $m_1 vR$, and that of the block is $m_2 vR$. At the same instant, the angular momentum of the pulley is $I\omega = Iv/R$. Hence, the total angular momentum of the system is

$$(1) \qquad L = m_1 vR + m_2 vR + I\frac{v}{R}$$

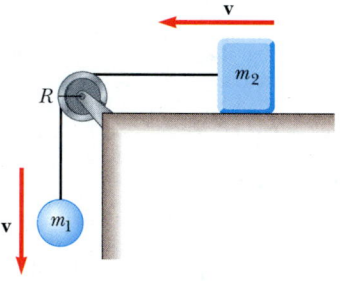

Figure 11.15

Now let us evaluate the total external torque acting on the system about the pulley axle. Because it has a moment arm of zero, the force exerted by the axle on the pulley does not contribute to the torque. Furthermore, the normal force acting on the block is balanced by the force of gravity $m_2\mathbf{g}$, and so these forces do not contribute to the torque. The force of gravity $m_1\mathbf{g}$ acting on the sphere produces a torque about the axle equal in magnitude to $m_1 gR$, where R is the moment arm of the force about the axle. (Note that in this situation, the tension is *not* equal to $m_1 g$.) This is the total external torque about the pulley axle; that is, $\Sigma\tau_{\text{ext}} = m_1 gR$. Using this result, together with Equation (1) and Equation 11.23, we find

$$\sum \tau_{\text{ext}} = \frac{dL}{dt}$$

$$m_1 gR = \frac{d}{dt}\left[(m_1 + m_2)Rv + I\frac{v}{R}\right]$$

$$(2) \qquad m_1 gR = (m_1 + m_2)R\frac{dv}{dt} + \frac{I}{R}\frac{dv}{dt}$$

Because $dv/dt = a$, we can solve this for a to obtain

$$a = \frac{m_1 g}{(m_1 + m_2) + I/R^2}$$

You may wonder why we did not include the forces that the cord exerts on the objects in evaluating the net torque about the axle. The reason is that these forces are internal to the system under consideration, and we analyzed the system as a whole. Only external torques contribute to the change in the system's angular momentum.

11.5   CONSERVATION OF ANGULAR MOMENTUM

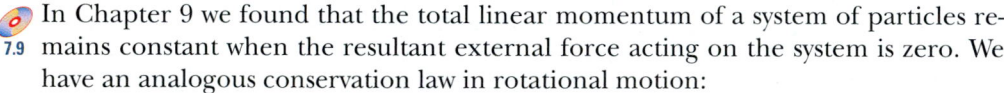

In Chapter 9 we found that the total linear momentum of a system of particles remains constant when the resultant external force acting on the system is zero. We have an analogous conservation law in rotational motion:

> **The total angular momentum of a system is constant in both magnitude and direction if the resultant external torque acting on the system is zero.**

Conservation of angular momentum

This follows directly from Equation 11.20, which indicates that if

$$\sum \boldsymbol{\tau}_{\text{ext}} = \frac{d\mathbf{L}}{dt} = 0 \tag{11.24}$$

then

$$\mathbf{L} = \text{constant} \tag{11.25}$$

For a system of particles, we write this conservation law as $\Sigma \mathbf{L}_n = \text{constant}$, where the index n denotes the nth particle in the system.

If the mass of an object undergoes redistribution in some way, then the object's moment of inertia changes; hence, its angular speed must change because $L = I\omega$. In this case we express the conservation of angular momentum in the form

$$\mathbf{L}_i = \mathbf{L}_f = \text{constant} \qquad \textbf{(11.26)}$$

If the system is an object rotating about a *fixed* axis, such as the z axis, we can write $L_z = I\omega$, where L_z is the component of $\mathbf{L}$ along the axis of rotation and I is the moment of inertia about this axis. In this case, we can express the conservation of angular momentum as

$$I_i\omega_i = I_f\omega_f = \text{constant} \qquad \textbf{(11.27)}$$

This expression is valid both for rotation about a fixed axis and for rotation about an axis through the center of mass of a moving system as long as that axis remains parallel to itself. We require only that the net external torque be zero.

Although we do not prove it here, there is an important theorem concerning the angular momentum of an object relative to the object's center of mass:

> The resultant torque acting on an object about an axis through the center of mass equals the time rate of change of angular momentum regardless of the motion of the center of mass.

This theorem applies even if the center of mass is accelerating, provided τ and $\mathbf{L}$ are evaluated relative to the center of mass.

In Equation 11.26 we have a third conservation law to add to our list. We can now state that the energy, linear momentum, and angular momentum of an isolated system all remain constant:

$$\left.\begin{array}{c} K_i + U_i = K_f + U_f \\[4pt] \mathbf{p}_i = \mathbf{p}_f \\[4pt] \mathbf{L}_i = \mathbf{L}_f \end{array}\right\} \quad \text{For an isolated system}$$

There are many examples that demonstrate conservation of angular momentum. You may have observed a figure skater spinning in the finale of a program. The angular speed of the skater increases when the skater pulls his hands and feet close to his body, thereby decreasing I. Neglecting friction between skates and ice, no external torques act on the skater. The change in angular speed is due to the fact that, because angular momentum is conserved, the product $I\omega$ remains constant, and a decrease in the moment of inertia of the skater causes an increase in the angular speed. Similarly, when divers or acrobats wish to make several somersaults, they pull their hands and feet close to their bodies to rotate at a higher rate. In these cases, the external force due to gravity acts through the center of mass and hence exerts no torque about this point. Therefore, the angular momentum about the center of mass must be conserved—that is, $I_i\omega_i = I_f\omega_f$. For example, when divers wish to double their angular speed, they must reduce their moment of inertia to one-half its initial value.

Angular momentum is conserved as figure skater Todd Eldredge pulls his arms toward his body.
(© 1998 David Madison)

Quick Quiz 11.4

A particle moves in a straight line, and you are told that the net torque acting on it is zero about some unspecified point. Decide whether the following statements are true or false: (a) The net force on the particle must be zero. (b) The particle's velocity must be constant.

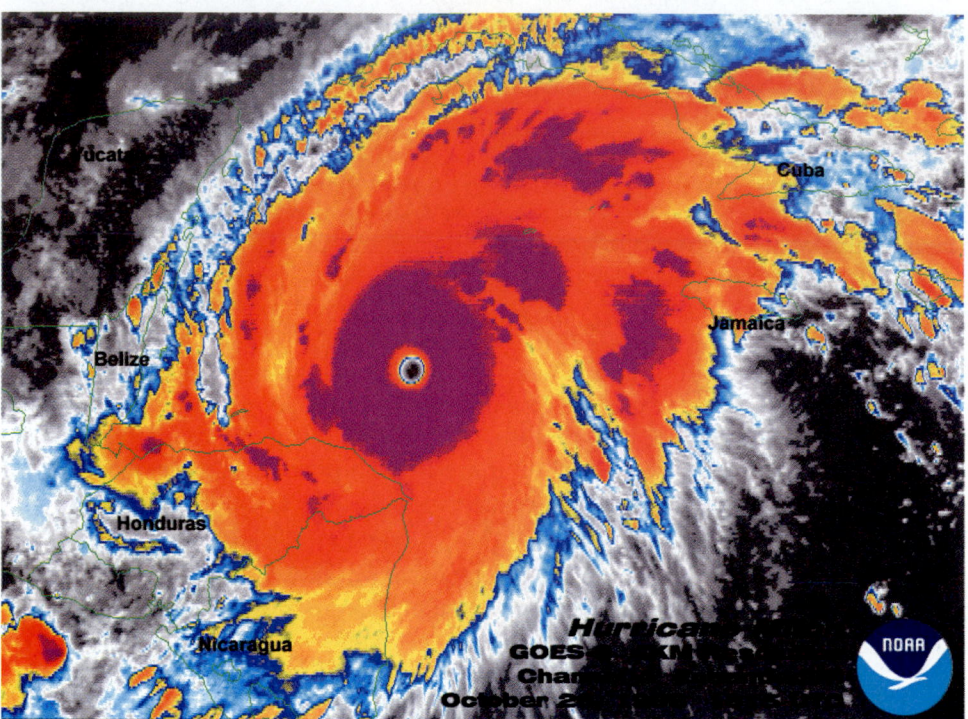

A color-enhanced, infrared image of Hurricane Mitch, which devastated large areas of Honduras and Nicaragua in October 1998. The spiral, nonrigid mass of air undergoes rotation and has angular momentum. *(Courtesy of NOAA)*

EXAMPLE 11.8   Formation of a Neutron Star

A star rotates with a period of 30 days about an axis through its center. After the star undergoes a supernova explosion, the stellar core, which had a radius of 1.0×10^4 km, collapses into a neutron star of radius 3.0 km. Determine the period of rotation of the neutron star.

Solution The same physics that makes a skater spin faster with his arms pulled in describes the motion of the neutron star. Let us assume that during the collapse of the stellar core, (1) no torque acts on it, (2) it remains spherical, and (3) its mass remains constant. Also, let us use the symbol T for the period, with T_i being the initial period of the star and T_f being the period of the neutron star. The period is the length of time a point on the star's equator takes to make one complete circle around the axis of rotation. The angular speed of a star is given by $\omega = 2\pi/T$. Therefore, because I is proportional to r^2, Equation 11.27 gives

$$T_f = T_i\left(\frac{r_f}{r_i}\right)^2 = (30 \text{ days})\left(\frac{3.0 \text{ km}}{1.0 \times 10^4 \text{ km}}\right)^2$$

$$= 2.7 \times 10^{-6} \text{ days} = \boxed{0.23 \text{ s}}$$

Thus, the neutron star rotates about four times each second; this result is approximately the same as that for a spinning figure skater.

EXAMPLE 11.9   The Merry-Go-Round

A horizontal platform in the shape of a circular disk rotates in a horizontal plane about a frictionless vertical axle (Fig. 11.16). The platform has a mass $M = 100$ kg and a radius $R = 2.0$ m. A student whose mass is $m = 60$ kg walks slowly from the rim of the disk toward its center. If the angular speed of the system is 2.0 rad/s when the student is at the rim, what is the angular speed when he has reached a point $r = 0.50$ m from the center?

Solution The speed change here is similar to the increase in angular speed of the spinning skater when he pulls his arms inward. Let us denote the moment of inertia of the platform as I_p and that of the student as I_s. Treating the student as a point mass, we can write the initial moment of inertia I_i of the system (student plus platform) about the axis of rotation:

$$I_i = I_{pi} + I_{si} = \tfrac{1}{2}MR^2 + mR^2$$

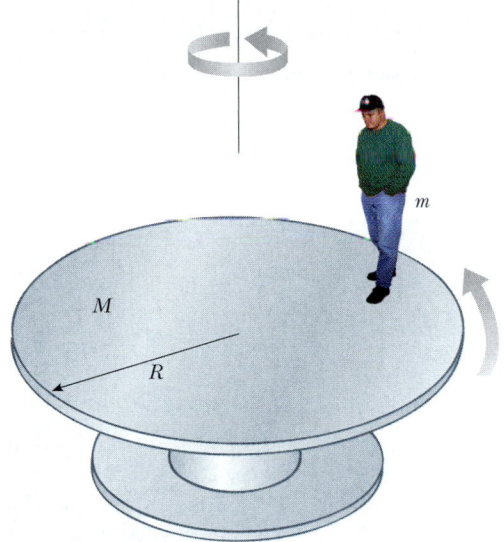

Figure 11.16 As the student walks toward the center of the rotating platform, the angular speed of the system increases because the angular momentum must remain constant.

When the student has walked to the position $r < R$, the moment of inertia of the system reduces to

$$I_f = I_{pf} + I_{sf} = \tfrac{1}{2}MR^2 + mr^2$$

Note that we still use the greater radius R when calculating I_{pf} because the radius of the platform has not changed. Because no external torques act on the system about the axis of rotation, we can apply the law of conservation of angular momentum:

$$I_i\omega_i = I_f\omega_f$$

$$\left(\tfrac{1}{2}MR^2 + mR^2\right)\omega_i = \left(\tfrac{1}{2}MR^2 + mr^2\right)\omega_f$$

$$\omega_f = \left(\frac{\tfrac{1}{2}MR^2 + mR^2}{\tfrac{1}{2}MR^2 + mr^2}\right)\omega_i$$

$$\omega_f = \left(\frac{200 + 240}{200 + 15}\right)(2.0\,\text{rad/s}) = \boxed{4.1\,\text{rad/s}}$$

As expected, the angular speed has increased.

Exercise Calculate the initial and final rotational energies of the system.

Answer $K_i = 880\,\text{J}; K_f = 1.8 \times 10^3\,\text{J}.$

Quick Quiz 11.5

Note that the rotational energy of the system described in Example 11.9 increases. What accounts for this increase in energy?

EXAMPLE 11.10   The Spinning Bicycle Wheel

In a favorite classroom demonstration, a student holds the axle of a spinning bicycle wheel while seated on a stool that is free to rotate (Fig. 11.17). The student and stool are initially at rest while the wheel is spinning in a horizontal plane with an initial angular momentum $\mathbf{L}_i$ that points upward. When the wheel is inverted about its center by 180°, the student and

stool start rotating. In terms of $\mathbf{L}_i$, what are the magnitude and the direction of $\mathbf{L}$ for the student plus stool?

Solution The system consists of the student, the wheel, and the stool. Initially, the total angular momentum of the system $\mathbf{L}_i$ comes entirely from the spinning wheel. As the wheel is inverted, the student applies a torque to the wheel, but this torque is internal to the system. No external torque is acting on the system about the vertical axis. Therefore, the angular momentum of the system is conserved. Initially, we have

$$\mathbf{L}_{\text{system}} = \mathbf{L}_i = \mathbf{L}_{\text{wheel}} \qquad \text{(upward)}$$

After the wheel is inverted, we have $\mathbf{L}_{\text{inverted wheel}} = -\mathbf{L}_i$. For angular momentum to be conserved, some other part of the system has to start rotating so that the total angular momentum remains the initial angular momentum $\mathbf{L}_i$. That other part of the system is the student plus the stool she is sitting on. So, we can now state that

$$\mathbf{L}_f = \mathbf{L}_i = \mathbf{L}_{\text{student+stool}} - \mathbf{L}_i$$

$$\mathbf{L}_{\text{student+stool}} = \boxed{2\mathbf{L}_i}$$

Figure 11.17 The wheel is initially spinning when the student is at rest. What happens when the wheel is inverted?

EXAMPLE 11.11 Disk and Stick

A 2.0-kg disk traveling at 3.0 m/s strikes a 1.0-kg stick that is lying flat on nearly frictionless ice, as shown in Figure 11.18. Assume that the collision is elastic. Find the translational speed of the disk, the translational speed of the stick, and the rotational speed of the stick after the collision. The moment of inertia of the stick about its center of mass is 1.33 kg·m².

Solution Because the disk and stick form an isolated system, we can assume that total energy, linear momentum, and angular momentum are all conserved. We have three unknowns, and so we need three equations to solve simultaneously. The first comes from the law of the conservation of linear momentum:

$$p_i = p_f$$

$$m_d v_{di} = m_d v_{df} + m_s v_s$$

$$(2.0 \text{ kg})(3.0 \text{ m/s}) = (2.0 \text{ kg}) v_{df} + (1.0 \text{ kg}) v_s$$

(1) $$6.0 \text{ kg·m/s} - (2.0 \text{ kg}) v_{df} = (1.0 \text{ kg}) v_s$$

Now we apply the law of conservation of angular momentum, using the initial position of the center of the stick as our reference point. We know that the component of angular momentum of the disk along the axis perpendicular to the plane of the ice is negative (the right-hand rule shows that $\mathbf{L}_d$ points into the ice).

$$L_i = L_f$$

$$-r m_d v_{di} = -r m_d v_{df} + I\omega$$

$$-(2.0 \text{ m})(2.0 \text{ kg})(3.0 \text{ m/s}) = -(2.0 \text{ m})(2.0 \text{ kg}) v_{df} + (1.33 \text{ kg·m}^2)\omega$$

$$-12 \text{ kg·m}^2/\text{s} = -(4.0 \text{ kg·m}) v_{df} + (1.33 \text{ kg·m}^2)\omega$$

(2) $$-9.0 \text{ rad/s} + (3.0 \text{ rad/m}) v_{df} = \omega$$

We used the fact that radians are dimensionless to ensure consistent units for each term.

Finally, the elastic nature of the collision reminds us that kinetic energy is conserved; in this case, the kinetic energy consists of translational and rotational forms:

$$K_i = K_f$$

$$\tfrac{1}{2} m_d v_{di}{}^2 = \tfrac{1}{2} m_d v_{df}{}^2 + \tfrac{1}{2} m_s v_s{}^2 + \tfrac{1}{2} I\omega^2$$

$$\tfrac{1}{2}(2.0 \text{ kg})(3.0 \text{ m/s})^2 = \tfrac{1}{2}(2.0 \text{ kg}) v_{df}{}^2 + \tfrac{1}{2}(1.0 \text{ kg}) v_s{}^2 + \tfrac{1}{2}(1.33 \text{ kg·m}^2/\text{s})\omega^2$$

(3) $$54 \text{ m}^2/\text{s}^2 = 6.0 v_{df}{}^2 + 3.0 v_s{}^2 + (4.0 \text{ m}^2)\omega^2$$

In solving Equations (1), (2), and (3) simultaneously, we find that $v_{df} = 2.3$ m/s, $v_s = 1.3$ m/s, and $\omega = -2.0$ rad/s. These values seem reasonable. The disk is moving more slowly than it was before the collision, and the stick has a small translational speed. Table 11.1 summarizes the initial and final values of variables for the disk and the stick and verifies the conservation of linear momentum, angular momentum, and kinetic energy.

Exercise Verify the values in Table 11.1.

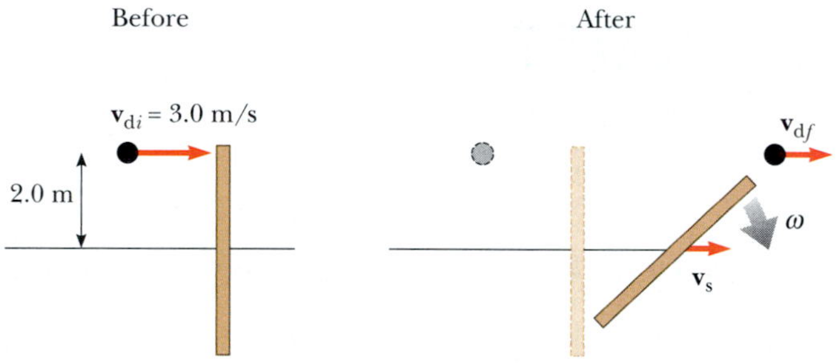

Before   After

Figure 11.18 Overhead view of a disk striking a stick in an elastic collision, which causes the stick to rotate.

TABLE 11.1	Comparison of Values in Example 11.11 Before and After the Collision[a]					
	v (m/s)	ω (rad/s)	p (kg·m/s)	L (kg·m²/s)	K_{trans} (J)	K_{rot} (J)
Before						
Disk	3.0	—	6.0	−12	9.0	—
Stick	0	0	0	0	0	0
Total	—	—	6.0	−12	9.0	0
After						
Disk	2.3	—	4.7	−9.3	5.4	—
Stick	1.3	−2.0	1.3	−2.7	0.9	2.7
Total	—	—	6.0	−12	6.3	2.7

[a]Notice that linear momentum, angular momentum, and total kinetic energy are conserved.

Optional Section

11.6 THE MOTION OF GYROSCOPES AND TOPS

A very unusual and fascinating type of motion you probably have observed is that of a top spinning about its axis of symmetry, as shown in Figure 11.19a. If the top spins very rapidly, the axis rotates about the *z* axis, sweeping out a cone (see Fig. 11.19b). The motion of the axis about the vertical—known as **precessional motion**—is usually slow relative to the spin motion of the top.

It is quite natural to wonder why the top does not fall over. Because the center of mass is not directly above the pivot point *O*, a net torque is clearly acting on the top about *O*—a torque resulting from the force of gravity *M***g**. The top would certainly fall over if it were not spinning. Because it is spinning, however, it has an angular momentum **L** directed along its symmetry axis. As we shall show, the motion of this symmetry axis about the *z* axis (the precessional motion) occurs because the torque produces a change in the *direction* of the symmetry axis. This is an excellent example of the importance of the directional nature of angular momentum.

The two forces acting on the top are the downward force of gravity *M***g** and the normal force **n** acting upward at the pivot point *O*. The normal force produces no torque about the pivot because its moment arm through that point is zero. However, the force of gravity produces a torque $\boldsymbol{\tau} = \mathbf{r} \times M\mathbf{g}$ about *O*, where the direction of $\boldsymbol{\tau}$ is perpendicular to the plane formed by **r** and *M***g**. By necessity, the vector $\boldsymbol{\tau}$ lies in a horizontal *xy* plane perpendicular to the angular momentum vector. The net torque and angular momentum of the top are related through Equation 11.19:

$$\boldsymbol{\tau} = \frac{d\mathbf{L}}{dt}$$

From this expression, we see that the nonzero torque produces a change in angular momentum $d\mathbf{L}$—a change that is in the same direction as $\boldsymbol{\tau}$. Therefore, like the torque vector, $d\mathbf{L}$ must also be at right angles to **L**. Figure 11.19b illustrates the resulting precessional motion of the symmetry axis of the top. In a time Δt, the change in angular momentum is $\Delta \mathbf{L} = \mathbf{L}_f - \mathbf{L}_i = \boldsymbol{\tau} \, \Delta t$. Because $\Delta \mathbf{L}$ is perpendicular to **L**, the magnitude of **L** does not change ($|\mathbf{L}_i| = |\mathbf{L}_f|$). Rather, what is changing is the *direction* of **L**. Because the change in angular momentum $\Delta \mathbf{L}$ is in the direction of $\boldsymbol{\tau}$, which lies in the *xy* plane, the top undergoes precessional motion.

The essential features of precessional motion can be illustrated by considering the simple gyroscope shown in Figure 11.20a. This device consists of a wheel free to spin about an axle that is pivoted at a distance *h* from the center of mass of the wheel. When given an angular velocity $\boldsymbol{\omega}$ about the axle, the wheel has an angular momentum $\mathbf{L} = I\boldsymbol{\omega}$ directed along the axle as shown. Let us consider the torque acting on the wheel about the pivot *O*. Again, the force **n** exerted by the support on the axle produces no torque about *O*, and the force of gravity *M***g** produces a torque of magnitude *Mgh* about *O*, where the axle is perpendicular to the support. The direction of this torque is perpendicular to the axle (and perpendicular to **L**), as shown in Figure 11.20a. This torque causes the angular momentum to change in the direction perpendicular to the axle. Hence, the axle moves in the direction of the torque—that is, in the horizontal plane.

To simplify the description of the system, we must make an assumption: The total angular momentum of the precessing wheel is the sum of the angular momentum $I\boldsymbol{\omega}$ due to the spinning and the angular momentum due to the motion of

Precessional motion

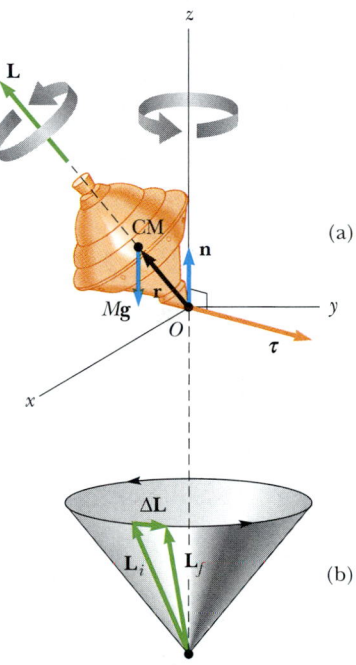

Figure 11.19 Precessional motion of a top spinning about its symmetry axis. (a) The only external forces acting on the top are the normal force **n** and the force of gravity *M***g**. The direction of the angular momentum **L** is along the axis of symmetry. The right-hand rule indicates that $\boldsymbol{\tau} = \mathbf{r} \times \mathbf{F} = \mathbf{r} \times M\mathbf{g}$ is in the *xy* plane. (b). The direction of $\Delta \mathbf{L}$ is parallel to that of $\boldsymbol{\tau}$ in part (a). The fact that $\mathbf{L}_f = \Delta \mathbf{L} + \mathbf{L}_i$ indicates that the top precesses about the *z* axis.

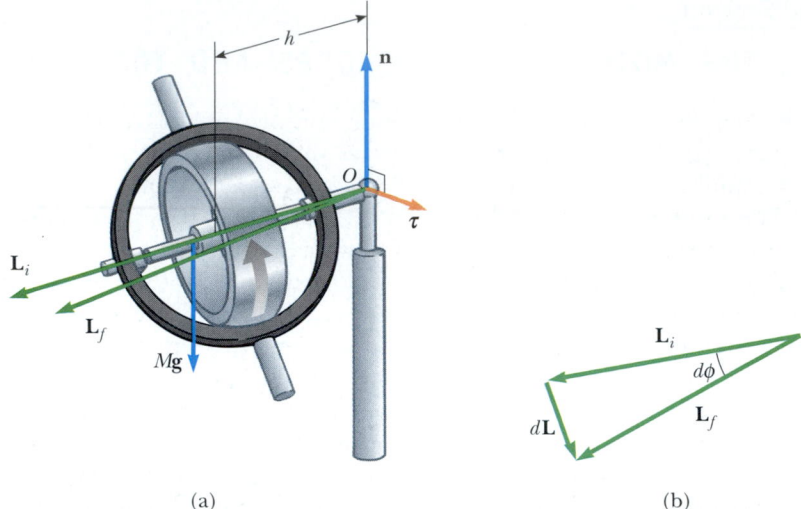

(a)   (b)

Figure 11.20   (a) The motion of a simple gyroscope pivoted a distance *h* from its center of mass. The force of gravity *M***g** produces a torque about the pivot, and this torque is perpendicular to the axle. (b) This torque results in a change in angular momentum *d***L** in a direction perpendicular to the axle. The axle sweeps out an angle *dφ* in a time *dt*.

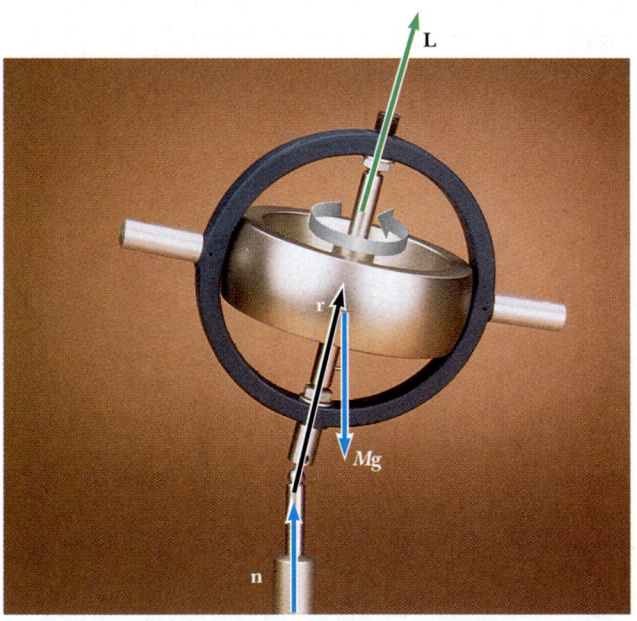

This toy gyroscope undergoes precessional motion about the vertical axis as it spins about its axis of symmetry. The only forces acting on it are the force of gravity *M***g** and the upward force of the pivot **n**. The direction of its angular momentum **L** is along the axis of symmetry. The torque and Δ**L** are directed into the page. *(Courtesy of Central Scientific Company)*

the center of mass about the pivot. In our treatment, we shall neglect the contribution from the center-of-mass motion and take the total angular momentum to be just *I***ω**. In practice, this is a good approximation if **ω** is made very large.

In a time *dt*, the torque due to the gravitational force changes the angular momentum of the system by *d***L** = **τ** *dt*. When added vectorially to the original total angular momentum $I\omega$, this additional angular momentum causes a shift in the direction of the total angular momentum.

The vector diagram in Figure 11.20b shows that in the time dt, the angular momentum vector rotates through an angle $d\phi$, which is also the angle through which the axle rotates. From the vector triangle formed by the vectors $\mathbf{L}_i$, $\mathbf{L}_f$, and $d\mathbf{L}$, we see that

$$\sin(d\phi) \approx d\phi = \frac{dL}{L} = \frac{(Mgh)\,dt}{L}$$

where we have used the fact that, for small values of any angle θ, $\sin\theta \approx \theta$. Dividing through by dt and using the relationship $L = I\omega$, we find that the rate at which the axle rotates about the vertical axis is

$$\omega_p = \frac{d\phi}{dt} = \frac{Mgh}{I\omega} \qquad\qquad (11.28)$$

The angular speed ω_p is called the **precessional frequency.** This result is valid only when $\omega_p \ll \omega$. Otherwise, a much more complicated motion is involved. As you can see from Equation 11.28, the condition $\omega_p \ll \omega$ is met when $I\omega$ is great compared with Mgh. Furthermore, note that the precessional frequency decreases as ω increases—that is, as the wheel spins faster about its axis of symmetry.

> Precessional frequency

Quick Quiz 11.6

How much work is done by the force of gravity when a top precesses through one complete circle?

Optional Section

11.7   ANGULAR MOMENTUM AS A FUNDAMENTAL QUANTITY

We have seen that the concept of angular momentum is very useful for describing the motion of macroscopic systems. However, the concept also is valid on a submicroscopic scale and has been used extensively in the development of modern theories of atomic, molecular, and nuclear physics. In these developments, it was found that the angular momentum of a system is a fundamental quantity. The word *fundamental* in this context implies that angular momentum is an intrinsic property of atoms, molecules, and their constituents, a property that is a part of their very nature.

To explain the results of a variety of experiments on atomic and molecular systems, we rely on the fact that the angular momentum has discrete values. These discrete values are multiples of the fundamental unit of angular momentum $\hbar = h/2\pi$, where h is called Planck's constant:

Fundamental unit of angular momentum $= \hbar = 1.054 \times 10^{-34}$ kg·m²/s

Let us accept this postulate without proof for the time being and show how it can be used to estimate the angular speed of a diatomic molecule. Consider the O_2 molecule as a rigid rotor, that is, two atoms separated by a fixed distance d and rotating about the center of mass (Fig. 11.21). Equating the angular momentum to the fundamental unit $\hbar$, we can estimate the lowest angular speed:

$$I_{CM}\omega \approx \hbar \qquad \text{or} \qquad \omega \approx \frac{\hbar}{I_{CM}}$$

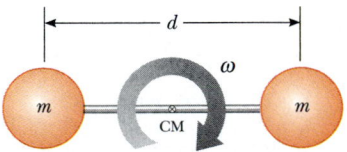

Figure 11.21   The rigid-rotor model of a diatomic molecule. The rotation occurs about the center of mass in the plane of the page.

In Example 10.3, we found that the moment of inertia of the O_2 molecule about this axis of rotation is $1.95 \times 10^{-46} \text{ kg} \cdot \text{m}^2$. Therefore,

$$\omega \approx \frac{\hbar}{I_{CM}} = \frac{1.054 \times 10^{-34} \text{ kg} \cdot \text{m}^2/\text{s}}{1.95 \times 10^{-46} \text{ kg} \cdot \text{m}^2} = 5.41 \times 10^{11} \text{ rad/s}$$

Actual angular speeds are multiples of this smallest possible value.

This simple example shows that certain classical concepts and models, when properly modified, might be useful in describing some features of atomic and molecular systems. A wide variety of phenomena on the submicroscopic scale can be explained only if we assume discrete values of the angular momentum associated with a particular type of motion.

The Danish physicist Niels Bohr (1885–1962) accepted and adopted this radical idea of discrete angular momentum values in developing his theory of the hydrogen atom. Strictly classical models were unsuccessful in describing many properties of the hydrogen atom. Bohr postulated that the electron could occupy only those circular orbits about the proton for which the orbital angular momentum was equal to $n\hbar$, where n is an integer. That is, he made the bold assumption that orbital angular momentum is quantized. From this simple model, the rotational frequencies of the electron in the various orbits can be estimated (see Problem 43).

SUMMARY

The **total kinetic energy** of a rigid object rolling on a rough surface without slipping equals the rotational kinetic energy about its center of mass, $\frac{1}{2}I_{CM}\omega^2$, plus the translational kinetic energy of the center of mass, $\frac{1}{2}Mv_{CM}{}^2$:

$$K = \tfrac{1}{2}I_{CM}\omega^2 + \tfrac{1}{2}Mv_{CM}{}^2 \tag{11.4}$$

The **torque** $\boldsymbol{\tau}$ due to a force $\mathbf{F}$ about an origin in an inertial frame is defined to be

$$\boldsymbol{\tau} \equiv \mathbf{r} \times \mathbf{F} \tag{11.7}$$

Given two vectors $\mathbf{A}$ and $\mathbf{B}$, the **cross product** $\mathbf{A} \times \mathbf{B}$ is a vector $\mathbf{C}$ having a magnitude

$$C \equiv AB \sin\theta \tag{11.9}$$

where θ is the angle between $\mathbf{A}$ and $\mathbf{B}$. The direction of the vector $\mathbf{C} = \mathbf{A} \times \mathbf{B}$ is perpendicular to the plane formed by $\mathbf{A}$ and $\mathbf{B}$, and this direction is determined by the right-hand rule.

The **angular momentum** $\mathbf{L}$ of a particle having linear momentum $\mathbf{p} = m\mathbf{v}$ is

$$\mathbf{L} \equiv \mathbf{r} \times \mathbf{p} \tag{11.15}$$

where $\mathbf{r}$ is the vector position of the particle relative to an origin in an inertial frame.

The **net external torque** acting on a particle or rigid object is equal to the time rate of change of its angular momentum:

$$\sum \boldsymbol{\tau}_{\text{ext}} = \frac{d\mathbf{L}}{dt} \tag{11.20}$$

The z *component* of **angular momentum** of a rigid object rotating about a fixed z axis is

$$L_z = I\omega \tag{11.21}$$

where I is the moment of inertia of the object about the axis of rotation and ω is its angular speed.

The **net external torque** acting on a rigid object equals the product of its moment of inertia about the axis of rotation and its angular acceleration:

$$\sum \tau_{\text{ext}} = I\alpha \qquad \qquad \textbf{(11.23)}$$

If the net external torque acting on a system is zero, then the total angular momentum of the system is constant. Applying this **law of conservation of angular momentum** to a system whose moment of inertia changes gives

$$I_i\omega_i = I_f\omega_f = \text{constant} \qquad \qquad \textbf{(11.27)}$$

QUESTIONS

1. Is it possible to calculate the torque acting on a rigid body without specifying a center of rotation? Is the torque independent of the location of the center of rotation?
2. Is the triple product defined by $\mathbf{A} \cdot (\mathbf{B} \times \mathbf{C})$ a scalar or a vector quantity? Explain why the operation $(\mathbf{A} \cdot \mathbf{B}) \times \mathbf{C}$ has no meaning.
3. In some motorcycle races, the riders drive over small hills, and the motorcycles become airborne for a short time. If a motorcycle racer keeps the throttle open while leaving the hill and going into the air, the motorcycle tends to nose upward. Why does this happen?
4. If the torque acting on a particle about a certain origin is zero, what can you say about its angular momentum about that origin?
5. Suppose that the velocity vector of a particle is completely specified. What can you conclude about the direction of its angular momentum vector with respect to the direction of motion?
6. If a single force acts on an object, and the torque caused by that force is nonzero about some point, is there any other point about which the torque is zero?
7. If a system of particles is in motion, is it possible for the total angular momentum to be zero about some origin? Explain.
8. A ball is thrown in such a way that it does not spin about its own axis. Does this mean that the angular momentum is zero about an arbitrary origin? Explain.
9. In a tape recorder, the tape is pulled past the read-and-write heads at a constant speed by the drive mechanism. Consider the reel from which the tape is pulled—as the tape is pulled off it, the radius of the roll of remaining tape decreases. How does the torque on the reel change with time? How does the angular speed of the reel change with time? If the tape mechanism is suddenly turned on so that the tape is quickly pulled with a great force, is the tape more likely to break when pulled from a nearly full reel or a nearly empty reel?
10. A scientist at a hotel sought assistance from a bellhop to carry a mysterious suitcase. When the unaware bellhop rounded a corner carrying the suitcase, it suddenly

moved away from him for some unknown reason. At this point, the alarmed bellhop dropped the suitcase and ran off. What do you suppose might have been in the suitcase?

11. When a cylinder rolls on a horizontal surface as in Figure 11.3, do any points on the cylinder have only a vertical component of velocity at some instant? If so, where are they?
12. Three objects of uniform density—a solid sphere, a solid cylinder, and a hollow cylinder—are placed at the top of an incline (Fig. Q11.12). If they all are released from rest at the same elevation and roll without slipping, which object reaches the bottom first? Which reaches it last? You should try this at home and note that the result is independent of the masses and the radii of the objects.

Figure Q11.12 Which object wins the race?

13. A mouse is initially at rest on a horizontal turntable mounted on a frictionless vertical axle. If the mouse begins to walk around the perimeter, what happens to the turntable? Explain.
14. Stars originate as large bodies of slowly rotating gas. Because of gravity, these regions of gas slowly decrease in size. What happens to the angular speed of a star as it shrinks? Explain.
15. Often, when a high diver wants to execute a flip in midair, she draws her legs up against her chest. Why does this make her rotate faster? What should she do when she wants to come out of her flip?
16. As a tether ball winds around a thin pole, what happens to its angular speed? Explain.

17. Two solid spheres—a large, massive sphere and a small sphere with low mass—are rolled down a hill. Which sphere reaches the bottom of the hill first? Next, a large, low-density sphere and a small, high-density sphere having the same mass are rolled down the hill. Which one reaches the bottom first in this case?

18. Suppose you are designing a car for a coasting race—the cars in this race have no engines; they simply coast down a hill. Do you want to use large wheels or small wheels? Do you want to use solid, disk-like wheels or hoop-like wheels? Should the wheels be heavy or light?

19. Why do tightrope walkers carry a long pole to help themselves keep their balance?

20. Two balls have the same size and mass. One is hollow, whereas the other is solid. How would you determine which is which without breaking them apart?

21. A particle is moving in a circle with constant speed. Locate one point about which the particle's angular momentum is constant and another about which it changes with time.

22. If global warming occurs over the next century, it is likely that some polar ice will melt and the water will be distributed closer to the equator. How would this change the moment of inertia of the Earth? Would the length of the day (one revolution) increase or decrease?

PROBLEMS

1, 2, 3 = straightforward, intermediate, challenging   ☐ = full solution available in the *Student Solutions Manual and Study Guide*
WEB = solution posted at **http://www.saunderscollege.com/physics/**   🖥 = Computer useful in solving problem   🖼 = Interactive Physics
☐ = paired numerical/symbolic problems

Section 11.1   Rolling Motion of a Rigid Object

WEB 1. A cylinder of mass 10.0 kg rolls without slipping on a horizontal surface. At the instant its center of mass has a speed of 10.0 m/s, determine (a) the translational kinetic energy of its center of mass, (b) the rotational energy about its center of mass, and (c) its total energy.

2. A bowling ball has a mass of 4.00 kg, a moment of inertia of 1.60×10^{-2} kg·m², and a radius of 0.100 m. If it rolls down the lane without slipping at a linear speed of 4.00 m/s, what is its total energy?

3. A bowling ball has a mass M, a radius R, and a moment of inertia $\frac{2}{5}MR^2$. If it starts from rest, how much work must be done on it to set it rolling without slipping at a linear speed v? Express the work in terms of M and v.

4. A uniform solid disk and a uniform hoop are placed side by side at the top of an incline of height h. If they are released from rest and roll without slipping, determine their speeds when they reach the bottom. Which object reaches the bottom first?

5. (a) Determine the acceleration of the center of mass of a uniform solid disk rolling down an incline making an angle θ with the horizontal. Compare this acceleration with that of a uniform hoop. (b) What is the minimum coefficient of friction required to maintain pure rolling motion for the disk?

6. A ring of mass 2.40 kg, inner radius 6.00 cm, and outer radius 8.00 cm rolls (without slipping) up an inclined plane that makes an angle of $\theta = 36.9°$ (Fig. P11.6). At the moment the ring is at position $x = 2.00$ m up the plane, its speed is 2.80 m/s. The ring continues up the plane for some additional distance and then rolls back down. It does not roll off the top end. How far up the plane does it go?

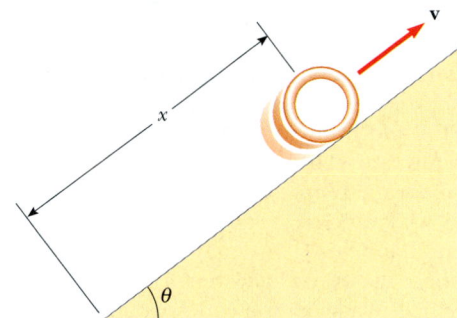

Figure P11.6

7. A metal can containing condensed mushroom soup has a mass of 215 g, a height of 10.8 cm, and a diameter of 6.38 cm. It is placed at rest on its side at the top of a 3.00-m-long incline that is at an angle of 25.0° to the horizontal and is then released to roll straight down. Assuming energy conservation, calculate the moment of inertia of the can if it takes 1.50 s to reach the bottom of the incline. Which pieces of data, if any, are unnecessary for calculating the solution?

8. A tennis ball is a hollow sphere with a thin wall. It is set rolling without slipping at 4.03 m/s on the horizontal section of a track, as shown in Figure P11.8. It rolls around the inside of a vertical circular loop 90.0 cm in diameter and finally leaves the track at a point 20.0 cm below the horizontal section. (a) Find the speed of the ball at the top of the loop. Demonstrate that it will not fall from the track. (b) Find its speed as it leaves the track. (c) Suppose that static friction between the ball and the track was negligible, so that the ball slid instead of rolling. Would its speed

Figure P11.8

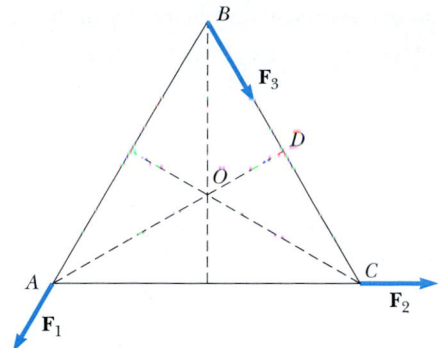

Figure P11.18

then be higher, lower, or the same at the top of the loop? Explain.

Section 11.2 The Vector Product and Torque

9. Given $\mathbf{M} = 6\mathbf{i} + 2\mathbf{j} - \mathbf{k}$ and $\mathbf{N} = 2\mathbf{i} - \mathbf{j} - 3\mathbf{k}$, calculate the vector product $\mathbf{M} \times \mathbf{N}$.

10. The vectors 42.0 cm at 15.0° and 23.0 cm at 65.0° both start from the origin. Both angles are measured counterclockwise from the x axis. The vectors form two sides of a parallelogram. (a) Find the area of the parallelogram. (b) Find the length of its longer diagonal.

WEB 11. Two vectors are given by $\mathbf{A} = -3\mathbf{i} + 4\mathbf{j}$ and $\mathbf{B} = 2\mathbf{i} + 3\mathbf{j}$. Find (a) $\mathbf{A} \times \mathbf{B}$ and (b) the angle between $\mathbf{A}$ and $\mathbf{B}$.

12. For the vectors $\mathbf{A} = -3\mathbf{i} + 7\mathbf{j} - 4\mathbf{k}$ and $\mathbf{B} = 6\mathbf{i} - 10\mathbf{j} + 9\mathbf{k}$, evaluate the expressions (a) $\cos^{-1} (\mathbf{A} \cdot \mathbf{B}/AB)$ and (b) $\sin^{-1} (|\mathbf{A} \times \mathbf{B}|/AB)$. (c) Which give(s) the angle between the vectors?

13. A force of $\mathbf{F} = 2.00\mathbf{i} + 3.00\mathbf{j}$ N is applied to an object that is pivoted about a fixed axis aligned along the z coordinate axis. If the force is applied at the point $\mathbf{r} = (4.00\mathbf{i} + 5.00\mathbf{j} + 0\mathbf{k})$ m, find (a) the magnitude of the net torque about the z axis and (b) the direction of the torque vector $\boldsymbol{\tau}$.

14. A student claims that she has found a vector $\mathbf{A}$ such that $(2\mathbf{i} - 3\mathbf{j} + 4\mathbf{k}) \times \mathbf{A} = (4\mathbf{i} + 3\mathbf{j} - \mathbf{k})$. Do you believe this claim? Explain.

15. Vector $\mathbf{A}$ is in the negative y direction, and vector $\mathbf{B}$ is in the negative x direction. What are the directions of (a) $\mathbf{A} \times \mathbf{B}$ and (b) $\mathbf{B} \times \mathbf{A}$?

16. A particle is located at the vector position $\mathbf{r} = (\mathbf{i} + 3\mathbf{j})$ m, and the force acting on it is $\mathbf{F} = (3\mathbf{i} + 2\mathbf{j})$ N. What is the torque about (a) the origin and (b) the point having coordinates (0, 6) m?

17. If $|\mathbf{A} \times \mathbf{B}| = \mathbf{A} \cdot \mathbf{B}$, what is the angle between $\mathbf{A}$ and $\mathbf{B}$?

18. Two forces $\mathbf{F}_1$ and $\mathbf{F}_2$ act along the two sides of an equilateral triangle, as shown in Figure P11.18. Point O is the intersection of the altitudes of the triangle. Find a third force $\mathbf{F}_3$ to be applied at B and along BC that will make the total torque about the point O be zero. Will the total torque change if $\mathbf{F}_3$ is applied not at B, but rather at any other point along BC?

Section 11.3 Angular Momentum of a Particle

19. A light, rigid rod 1.00 m in length joins two particles—with masses 4.00 kg and 3.00 kg—at its ends. The combination rotates in the xy plane about a pivot through the center of the rod (Fig. P11.19). Determine the angular momentum of the system about the origin when the speed of each particle is 5.00 m/s.

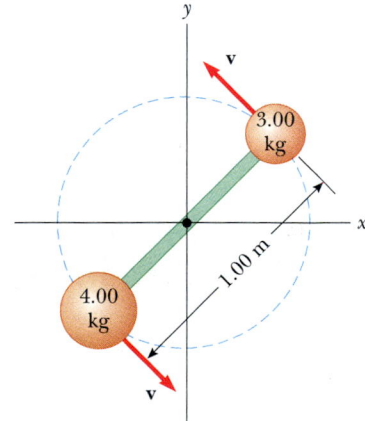

Figure P11.19

20. A 1.50-kg particle moves in the xy plane with a velocity of $\mathbf{v} = (4.20\mathbf{i} - 3.60\mathbf{j})$ m/s. Determine the particle's angular momentum when its position vector is $\mathbf{r} = (1.50\mathbf{i} + 2.20\mathbf{j})$ m.

WEB 21. The position vector of a particle of mass 2.00 kg is given as a function of time by $\mathbf{r} = (6.00\mathbf{i} + 5.00t\mathbf{j})$ m. Determine the angular momentum of the particle about the origin as a function of time.

22. A conical pendulum consists of a bob of mass m in motion in a circular path in a horizontal plane, as shown in Figure P11.22. During the motion, the supporting wire of length ℓ maintains the constant angle θ with the vertical. Show that the magnitude of the angular momen-

tum of the mass about the center of the circle is

$$L = (m^2 g \ell^3 \sin^4 \theta / \cos \theta)^{1/2}$$

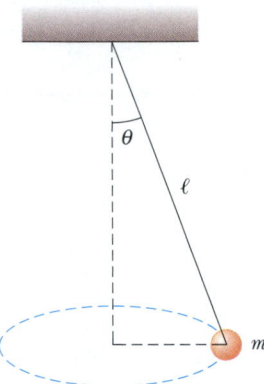

Figure P11.22

23. A particle of mass m moves in a circle of radius R at a constant speed v, as shown in Figure P11.23. If the motion begins at point Q, determine the angular momentum of the particle about point P as a function of time.

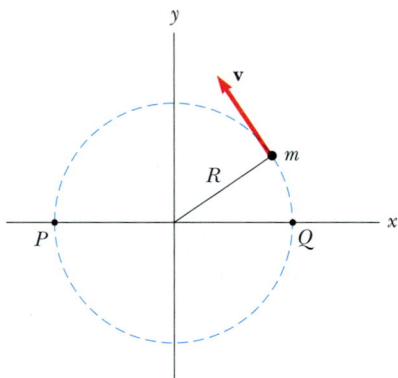

Figure P11.23

24. A 4.00-kg mass is attached to a light cord that is wound around a pulley (see Fig. 10.20). The pulley is a uniform solid cylinder with a radius of 8.00 cm and a mass of 2.00 kg. (a) What is the net torque on the system about the point O? (b) When the mass has a speed v, the pulley has an angular speed $\omega = v/R$. Determine the total angular momentum of the system about O. (c) Using the fact that $\tau = d\mathbf{L}/dt$ and your result from part (b), calculate the acceleration of the mass.

25. A particle of mass m is shot with an initial velocity $\mathbf{v}_i$ and makes an angle θ with the horizontal, as shown in Figure P11.25. The particle moves in the gravitational field of the Earth. Find the angular momentum of the parti-

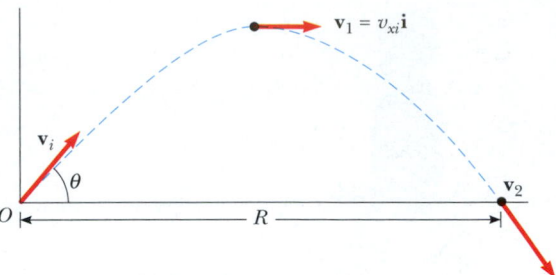

Figure P11.25

cle about the origin when the particle is (a) at the origin, (b) at the highest point of its trajectory, and (c) just about to hit the ground. (d) What torque causes its angular momentum to change?

26. Heading straight toward the summit of Pike's Peak, an airplane of mass 12 000 kg flies over the plains of Kansas at a nearly constant altitude of 4.30 km and with a constant velocity of 175 m/s westward. (a) What is the airplane's vector angular momentum relative to a wheat farmer on the ground directly below the airplane? (b) Does this value change as the airplane continues its motion along a straight line? (c) What is its angular momentum relative to the summit of Pike's Peak?

27. A ball of mass m is fastened at the end of a flagpole connected to the side of a tall building at point P, as shown in Figure P11.27. The length of the flagpole is ℓ, and θ is the angle the flagpole makes with the horizontal. Suppose that the ball becomes loose and starts to fall. Determine the angular momentum (as a function of time) of the ball about point P. Neglect air resistance.

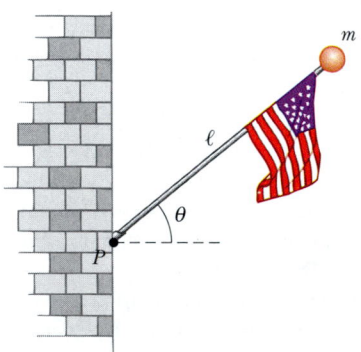

Figure P11.27

28. A fireman clings to a vertical ladder and directs the nozzle of a hose horizontally toward a burning building. The rate of water flow is 6.31 kg/s, and the nozzle speed is 12.5 m/s. The hose passes between the fireman's feet, which are 1.30 m vertically below the nozzle. Choose the origin to be inside the hose between the fireman's

feet. What torque must the fireman exert on the hose? That is, what is the rate of change of angular momentum of the water?

Section 11.4 Angular Momentum of a Rotating Rigid Object

29. A uniform solid sphere with a radius of 0.500 m and a mass of 15.0 kg turns counterclockwise about a vertical axis through its center. Find its vector angular momentum when its angular speed is 3.00 rad/s.

30. A uniform solid disk with a mass of 3.00 kg and a radius of 0.200 m rotates about a fixed axis perpendicular to its face. If the angular speed is 6.00 rad/s, calculate the angular momentum of the disk when the axis of rotation (a) passes through its center of mass and (b) passes through a point midway between the center and the rim.

31. A particle with a mass of 0.400 kg is attached to the 100-cm mark of a meter stick with a mass of 0.100 kg. The meter stick rotates on a horizontal, frictionless table with an angular speed of 4.00 rad/s. Calculate the angular momentum of the system when the stick is pivoted about an axis (a) perpendicular to the table through the 50.0-cm mark and (b) perpendicular to the table through the 0-cm mark.

32. The hour and minute hands of Big Ben, the famous Parliament Building tower clock in London, are 2.70 m and 4.50 m long and have masses of 60.0 kg and 100 kg, respectively. Calculate the total angular momentum of these hands about the center point. Treat the hands as long thin rods.

Section 11.5 Conservation of Angular Momentum

33. A cylinder with a moment of inertia of I_1 rotates about a vertical, frictionless axle with angular velocity ω_i. A second cylinder that has a moment of inertia of I_2 and initially is not rotating drops onto the first cylinder (Fig. P11.33). Because of friction between the surfaces, the two eventually reach the same angular speed ω_f. (a) Calculate ω_f. (b) Show that the kinetic energy of the system decreases in this interaction and calculate the ratio of the final rotational energy to the initial rotational energy.

34. A playground merry-go-round of radius $R = 2.00$ m has a moment of inertia of $I = 250$ kg·m^2 and is rotating at 10.0 rev/min about a frictionless vertical axle. Facing the axle, a 25.0-kg child hops onto the merry-go-round and manages to sit down on its edge. What is the new angular speed of the merry-go round?

35. A student sits on a freely rotating stool holding two weights, each of which has a mass of 3.00 kg. When his arms are extended horizontally, the weights are 1.00 m from the axis of rotation and he rotates with an angular speed of 0.750 rad/s. The moment of inertia of the student plus stool is 3.00 kg·m^2 and is assumed to be constant. The student pulls the weights inward horizontally to a position 0.300 m from the rotation axis. (a) Find the new angular speed of the student. (b) Find the kinetic energy of the student before and after he pulls the weights inward.

36. A uniform rod with a mass of 100 g and a length of 50.0 cm rotates in a horizontal plane about a fixed, vertical, frictionless pin passing through its center. Two small beads, each having a mass 30.0 g, are mounted on the rod so that they are able to slide without friction along its length. Initially, the beads are held by catches at positions 10.0 cm on each side of center; at this time, the system rotates at an angular speed of 20.0 rad/s. Suddenly, the catches are released, and the small beads slide outward along the rod. Find (a) the angular speed of the system at the instant the beads reach the ends of the rod and (b) the angular speed of the rod after the beads fly off the rod's ends.

WEB 37. A 60.0-kg woman stands at the rim of a horizontal turntable having a moment of inertia of 500 kg·m^2 and a radius of 2.00 m. The turntable is initially at rest and is free to rotate about a frictionless, vertical axle through its center. The woman then starts walking around the rim clockwise (as viewed from above the system) at a constant speed of 1.50 m/s relative to the Earth. (a) In what direction and with what angular speed does the turntable rotate? (b) How much work does the woman do to set herself and the turntable into motion?

38. A puck with a mass of 80.0 g and a radius of 4.00 cm slides along an air table at a speed of 1.50 m/s, as shown in Figure P11.38a. It makes a glancing collision

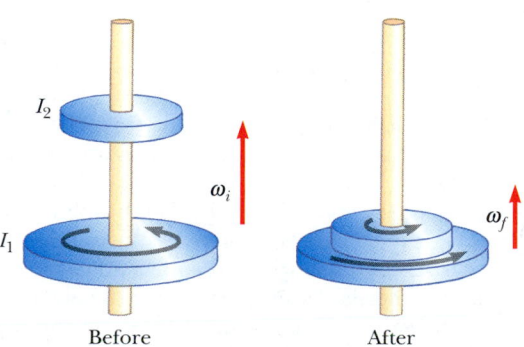

Before   After

Figure P11.33

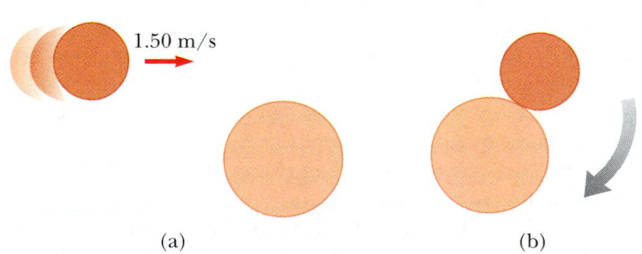

1.50 m/s

(a)   (b)

Figure P11.38

with a second puck having a radius of 6.00 cm and a mass of 120 g (initially at rest) such that their rims just touch. Because their rims are coated with instant-acting glue, the pucks stick together and spin after the collision (Fig. P11.38b). (a) What is the angular momentum of the system relative to the center of mass? (b) What is the angular speed about the center of mass?

39. A wooden block of mass M resting on a frictionless horizontal surface is attached to a rigid rod of length ℓ and of negligible mass (Fig. P11.39). The rod is pivoted at the other end. A bullet of mass m traveling parallel to the horizontal surface and normal to the rod with speed v hits the block and becomes embedded in it. (a) What is the angular momentum of the bullet–block system? (b) What fraction of the original kinetic energy is lost in the collision?

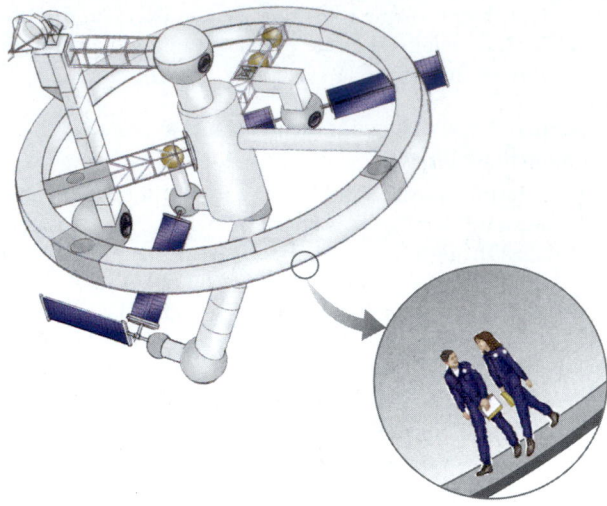

Figure P11.40

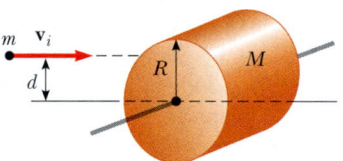

Figure P11.41

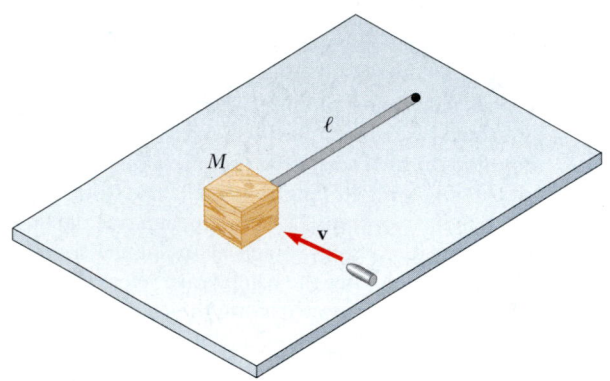

Figure P11.39

40. A space station shaped like a giant wheel has a radius of 100 m and a moment of inertia of 5.00×10^8 kg·m². A crew of 150 are living on the rim, and the station's rotation causes the crew to experience an acceleration of $1g$ (Fig. P11.40). When 100 people move to the center of the station for a union meeting, the angular speed changes. What acceleration is experienced by the managers remaining at the rim? Assume that the average mass of each inhabitant is 65.0 kg.

41. A wad of sticky clay of mass m and velocity $\mathbf{v}_i$ is fired at a solid cylinder of mass M and radius R (Fig. P11.41). The cylinder is initially at rest and is mounted on a fixed horizontal axle that runs through the center of mass. The line of motion of the projectile is perpendicular to the axle and at a distance d, less than R, from the center. (a) Find the angular speed of the system just after the clay strikes and sticks to the surface of the cylinder. (b) Is mechanical energy conserved in this process? Explain your answer.

42. Suppose a meteor with a mass of 3.00×10^{13} kg is moving at 30.0 km/s relative to the center of the Earth and strikes the Earth. What is the order of magnitude of the

maximum possible decrease in the angular speed of the Earth due to this collision? Explain your answer.

(Optional)
Section 11.7   Angular Momentum as a Fundamental Quantity

43. In the Bohr model of the hydrogen atom, the electron moves in a circular orbit of radius 0.529×10^{-10} m around the proton. Assuming that the orbital angular momentum of the electron is equal to $h/2\pi$, calculate (a) the orbital speed of the electron, (b) the kinetic energy of the electron, and (c) the angular speed of the electron's motion.

ADDITIONAL PROBLEMS

44. Review Problem.   A rigid, massless rod has three equal masses attached to it, as shown in Figure P11.44. The rod is free to rotate in a vertical plane about a frictionless axle perpendicular to the rod through the point P, and it is released from rest in the horizontal position at $t = 0$. Assuming m and d are known, find (a) the moment of inertia of the system about the pivot, (b) the torque acting on the system at $t = 0$, (c) the angular acceleration of the system at $t = 0$, (d) the linear acceleration of the mass labeled "3" at $t = 0$, (e) the maximum kinetic energy of the system, (f) the maximum angular speed attained by the rod, (g) the maximum angular momentum of the system, and (h) the maximum speed attained by the mass labeled "2."

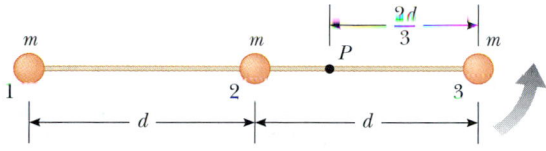

Figure P11.44

45. A uniform solid sphere of radius r is placed on the inside surface of a hemispherical bowl having a much greater radius R. The sphere is released from rest at an angle θ to the vertical and rolls without slipping (Fig. P11.45). Determine the angular speed of the sphere when it reaches the bottom of the bowl.

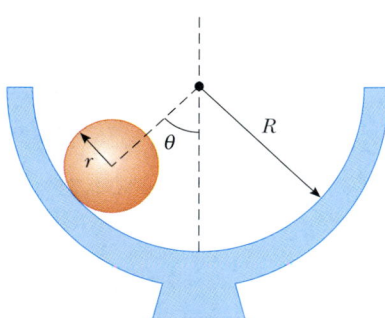

Figure P11.45

46. A 100-kg uniform horizontal disk of radius 5.50 m turns without friction at 2.50 rev/s on a vertical axis through its center, as shown in Figure P11.46. A feedback mechanism senses the angular speed of the disk, and a drive motor at A ensures that the angular speed remains constant. While the disk turns, a 1.20-kg mass on top of the disk slides outward in a radial slot. The 1.20-kg mass starts at the center of the disk at time $t = 0$ and moves outward with a constant speed of 1.25 cm/s relative to the disk until it reaches the edge at $t = 440$ s. The sliding mass experiences no friction. Its motion is constrained by a brake at B so that its radial speed remains constant. The constraint produces tension in a light string tied to the mass. (a) Find the torque as a function of time that the drive motor must provide while the mass is sliding. (b) Find the value of this torque at $t = 440$ s, just before the sliding mass finishes its motion. (c) Find the power that the drive motor must deliver as a function of time. (d) Find the value of the power when the sliding mass is just reaching the end of the slot. (e) Find the string tension as a function of

time. (f) Find the work done by the drive motor during the 440-s motion. (g) Find the work done by the string brake on the sliding mass. (h) Find the total work done on the system consisting of the disk and the sliding mass.

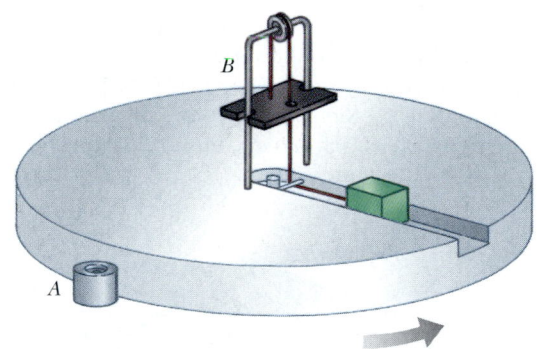

Figure P11.46

47. A string is wound around a uniform disk of radius R and mass M. The disk is released from rest when the string is vertical and its top end is tied to a fixed bar (Fig. P11.47). Show that (a) the tension in the string is one-third the weight of the disk, (b) the magnitude of the acceleration of the center of mass is $2g/3$, and (c) the speed of the center of mass is $(4gh/3)^{1/2}$ as the disk descends. Verify your answer to part (c) using the energy approach.

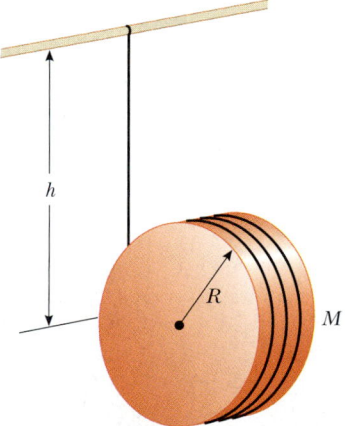

Figure P11.47

48. Comet Halley moves about the Sun in an elliptical orbit, with its closest approach to the Sun being about 0.590 AU and its greatest distance from the Sun being 35.0 AU (1 AU = the average Earth–Sun distance). If the comet's speed at its closest approach is 54.0 km/s,

what is its speed when it is farthest from the Sun? The angular momentum of the comet about the Sun is conserved because no torque acts on the comet. The gravitational force exerted by the Sun on the comet has a moment arm of zero.

49. A constant horizontal force **F** is applied to a lawn roller having the form of a uniform solid cylinder of radius R and mass M (Fig. P11.49). If the roller rolls without slipping on the horizontal surface, show that (a) the acceleration of the center of mass is $2\mathbf{F}/3M$ and that (b) the minimum coefficient of friction necessary to prevent slipping is $F/3Mg$. (*Hint:* Consider the torque with respect to the center of mass.)

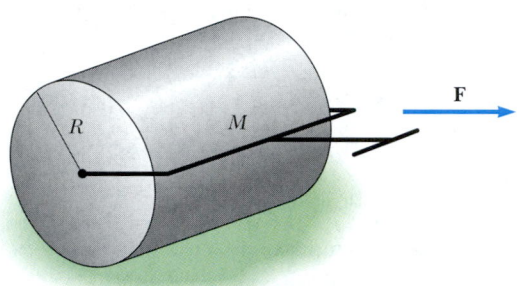

Figure P11.49

50. A light rope passes over a light, frictionless pulley. A bunch of bananas of mass M is fastened at one end, and a monkey of mass M clings to the other (Fig. P11.50).

Figure P11.50

The monkey climbs the rope in an attempt to reach the bananas. (a) Treating the system as consisting of the monkey, bananas, rope, and pulley, evaluate the net torque about the pulley axis. (b) Using the results to part (a), determine the total angular momentum about the pulley axis and describe the motion of the system. Will the monkey reach the bananas?

51. A solid sphere of mass m and radius r rolls without slipping along the track shown in Figure P11.51. The sphere starts from rest with its lowest point at height h above the bottom of a loop of radius R, which is much larger than r. (a) What is the minimum value that h can have (in terms of R) if the sphere is to complete the loop? (b) What are the force components on the sphere at point P if $h = 3R$?

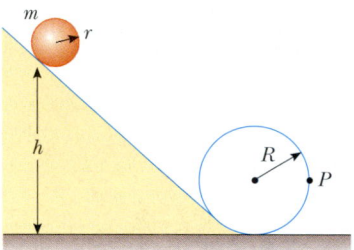

Figure P11.51

52. A thin rod with a mass of 0.630 kg and a length of 1.24 m is at rest, hanging vertically from a strong fixed hinge at its top end. Suddenly, a horizontal impulsive force $(14.7\mathbf{i})$ N is applied to it. (a) Suppose that the force acts at the bottom end of the rod. Find the acceleration of the rod's center of mass and the horizontal force that the hinge exerts. (b) Suppose that the force acts at the midpoint of the rod. Find the acceleration of this point and the horizontal hinge reaction. (c) Where can the impulse be applied so that the hinge exerts no horizontal force? (This point is called the *center of percussion*.)

53. At one moment, a bowling ball is both sliding and spinning on a horizontal surface such that its rotational kinetic energy equals its translational kinetic energy. Let v_{CM} represent the ball's center-of-mass speed relative to the surface. Let v_r represent the speed of the topmost point on the ball's surface relative to the center of mass. Find the ratio v_{CM}/v_r.

54. A projectile of mass m moves to the right with speed v_i (Fig. P11.54a). The projectile strikes and sticks to the end of a stationary rod of mass M and length d that is pivoted about a frictionless axle through its center (Fig. P11.54b). (a) Find the angular speed of the system right after the collision. (b) Determine the fractional loss in mechanical energy due to the collision.

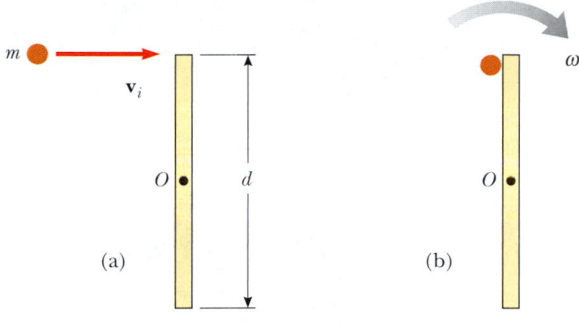

Figure P11.54

55. A mass m is attached to a cord passing through a small hole in a frictionless, horizontal surface (Fig. P11.55). The mass is initially orbiting with speed v_i in a circle of radius r_i. The cord is then slowly pulled from below, and the radius of the circle decreases to r. (a) What is the speed of the mass when the radius is r? (b) Find the tension in the cord as a function of r? (c) How much work W is done in moving m from r_i to r? (*Note:* The tension depends on r.) (d) Obtain numerical values for v, T, and W when $r = 0.100$ m, $m = 50.0$ g, $r_i = 0.300$ m, and $v_i = 1.50$ m/s.

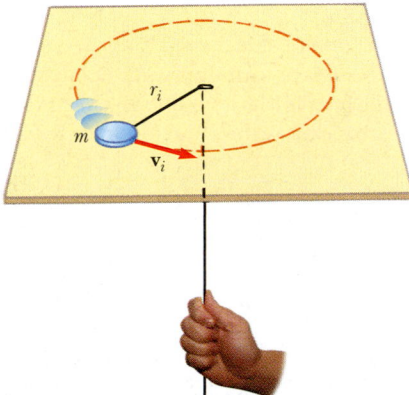

Figure P11.55

56. A bowler releases a bowling ball with no spin, sending it sliding straight down the alley toward the pins. The ball continues to slide for some distance before its motion becomes rolling without slipping; of what order of magnitude is this distance? State the quantities you take as data, the values you measure or estimate for them, and your reasoning.

57. Following Thanksgiving dinner, your uncle falls into a deep sleep while sitting straight up and facing the television set. A naughty grandchild balances a small spheri-

cal grape at the top of his bald head, which itself has the shape of a sphere. After all of the children have had time to giggle, the grape starts from rest and rolls down your uncle's head without slipping. It loses contact with your uncle's scalp when the radial line joining it to the center of curvature makes an angle θ with the vertical. What is the measure of angle θ?

58. A thin rod of length h and mass M is held vertically with its lower end resting on a frictionless horizontal surface. The rod is then released to fall freely. (a) Determine the speed of its center of mass just before it hits the horizontal surface. (b) Now suppose that the rod has a fixed pivot at its lower end. Determine the speed of the rod's center of mass just before it hits the surface.

WEB **59.** Two astronauts (Fig. P11.59), each having a mass of 75.0 kg, are connected by a 10.0-m rope of negligible mass. They are isolated in space, orbiting their center of mass at speeds of 5.00 m/s. (a) Treating the astronauts as particles, calculate the magnitude of the angular momentum and (b) the rotational energy of the system. By pulling on the rope, one of the astronauts shortens the distance between them to 5.00 m. (c) What is the new angular momentum of the system? (d) What are the astronauts' new speeds? (e) What is the new rotational energy of the system? (f) How much work is done by the astronaut in shortening the rope?

60. Two astronauts (see Fig. P11.59), each having a mass M, are connected by a rope of length d having negligible mass. They are isolated in space, orbiting their center of mass at speeds v. Treating the astronauts as particles, calculate (a) the magnitude of the angular momentum and (b) the rotational energy of the system. By pulling on the rope, one of the astronauts shortens the distance between them to $d/2$. (c) What is the new angular momentum of the system? (d) What are the astronauts' new speeds? (e) What is the new rotational energy of the system? (f) How much work is done by the astronaut in shortening the rope?

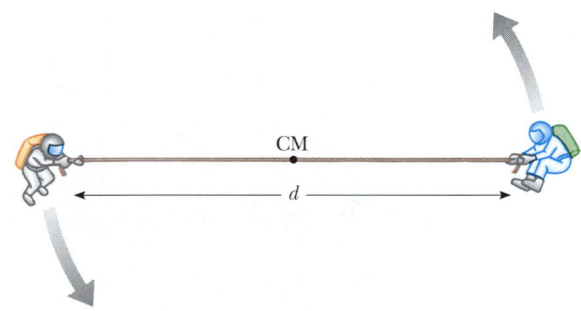

Figure P11.59   Problems 59 and 60.

61. A solid cube of wood of side $2a$ and mass M is resting on a horizontal surface. The cube is constrained to rotate about an axis AB (Fig. P11.61). A bullet of mass m and speed v is shot at the face opposite $ABCD$ at a height of $4a/3$. The bullet becomes embedded in the cube. Find the minimum value of v required to tip the cube so that it falls on face $ABCD$. Assume $m \ll M$.

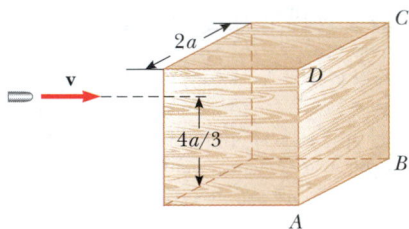

Figure P11.61

62. A large, cylindrical roll of paper of initial radius R lies on a long, horizontal surface with the open end of the paper nailed to the surface. The roll is given a slight shove ($v_i \approx 0$) and begins to unroll. (a) Determine the speed of the center of mass of the roll when its radius has diminished to r. (b) Calculate a numerical value for this speed at $r = 1.00$ mm, assuming $R = 6.00$ m. (c) What happens to the energy of the system when the paper is completely unrolled? (*Hint:* Assume that the roll has a uniform density and apply energy methods.)

63. A spool of wire of mass M and radius R is unwound under a constant force $\mathbf{F}$ (Fig. P11.63). Assuming that the spool is a uniform solid cylinder that does not slip, show that (a) the acceleration of the center of mass is $4\mathbf{F}/3M$ and that (b) the force of friction is to the *right* and is equal in magnitude to $F/3$. (c) If the cylinder starts from rest and rolls without slipping, what is the speed of its center of mass after it has rolled through a distance d?

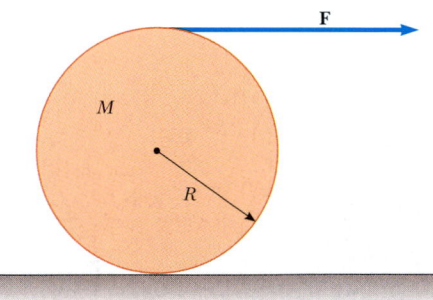

Figure P11.63

64. A uniform solid disk is set into rotation with an angular speed ω_i about an axis through its center. While still rotating at this speed, the disk is placed into contact with

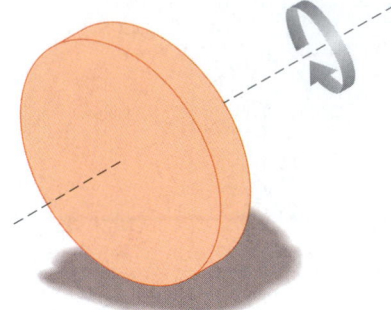

Figure P11.64   Problems 64 and 65.

a horizontal surface and released, as shown in Figure P11.64. (a) What is the angular speed of the disk once pure rolling takes place? (b) Find the fractional loss in kinetic energy from the time the disk is released until the time pure rolling occurs. (*Hint:* Consider torques about the center of mass.)

65. Suppose a solid disk of radius R is given an angular speed ω_i about an axis through its center and is then lowered to a horizontal surface and released, as shown in Problem 64 (see Fig. P11.64). Furthermore, assume that the coefficient of friction between the disk and the surface is μ. (a) Show that the time it takes for pure rolling motion to occur is $R\omega_i/3\mu g$. (b) Show that the distance the disk travels before pure rolling occurs is $R^2\omega_i^2/18\mu g$.

66. A solid cube of side $2a$ and mass M is sliding on a frictionless surface with uniform velocity $\mathbf{v}$, as shown in Figure P11.66a. It hits a small obstacle at the end of the table; this causes the cube to tilt, as shown in Figure

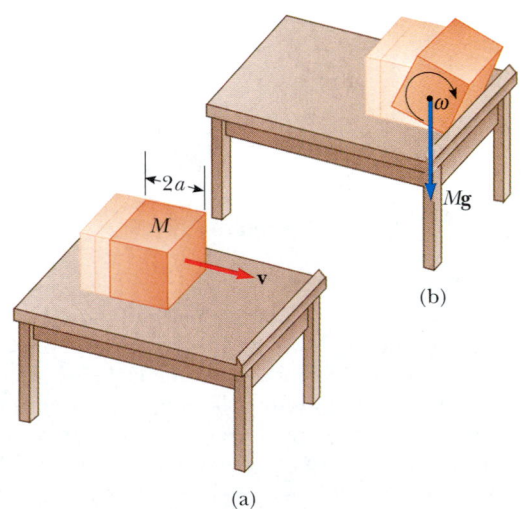

(a)

Figure P11.66

P11.66b. Find the minimum value of **v** such that the cube falls off the table. Note that the moment of inertia of the cube about an axis along one of its edges is $8Ma^2/3$. (*Hint:* The cube undergoes an inelastic collision at the edge.)

67. A plank with a mass $M = 6.00$ kg rides on top of two identical solid cylindrical rollers that have $R = 5.00$ cm and $m = 2.00$ kg (Fig. P11.67). The plank is pulled by a constant horizontal force of magnitude $F = 6.00$ N applied to the end of the plank and perpendicular to the axes of the cylinders (which are parallel). The cylinders roll without slipping on a flat surface. Also, no slipping occurs between the cylinders and the plank. (a) Find the acceleration of the plank and that of the rollers. (b) What frictional forces are acting?

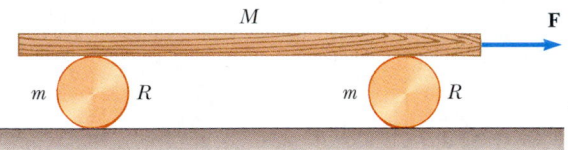

Figure P11.67

68. A spool of wire rests on a horizontal surface as in Figure P11.68. As the wire is pulled, the spool does not slip at the contact point P. On separate trials, each one of the forces $\mathbf{F}_1$, $\mathbf{F}_2$, $\mathbf{F}_3$, and $\mathbf{F}_4$ is applied to the spool. For each one of these forces, determine the direction in which the spool will roll. Note that the line of action of $\mathbf{F}_2$ passes through P.

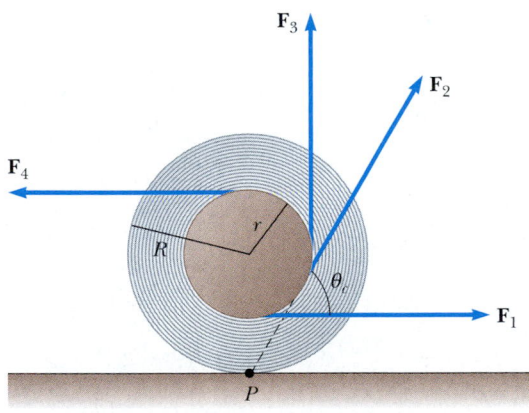

Figure P11.68   Problems 68 and 69.

69. The spool of wire shown in Figure P11.68 has an inner radius r and an outer radius R. The angle θ between the applied force and the horizontal can be varied. Show

that the critical angle for which the spool does not slip and remains stationary is

$$\cos \theta_c = \frac{r}{R}$$

(*Hint:* At the critical angle, the line of action of the applied force passes through the contact point.)

70. In a demonstration that employs a ballistics cart, a ball is projected vertically upward from a cart moving with constant velocity along the horizontal direction. The ball lands in the catching cup of the cart because both the cart and the ball have the same horizontal component of velocity. Now consider a ballistics cart on an incline making an angle θ with the horizontal, as shown in Figure P11.70. The cart (including its wheels) has a mass M, and the moment of inertia of each of the two wheels is $mR^2/2$. (a) Using conservation of energy considerations (assuming that there is no friction between the cart and the axles) and assuming pure rolling motion (that is, no slipping), show that the acceleration of the cart along the incline is

$$a_x = \left(\frac{M}{M + 2m}\right) g \sin \theta$$

(b) Note that the x component of acceleration of the ball released by the cart is $g \sin \theta$. Thus, the x component of the cart's acceleration is *smaller* than that of the ball by the factor $M/(M + 2m)$. Use this fact and kinematic equations to show that the ball overshoots the cart by an amount Δx, where

$$\Delta x = \left(\frac{4m}{M + 2m}\right)\left(\frac{\sin \theta}{\cos^2 \theta}\right) \frac{v_{yi}^2}{g}$$

and v_{yi} is the initial speed of the ball imparted to it by the spring in the cart. (c) Show that the distance d that the ball travels measured along the incline is

$$d = \frac{2v_{yi}^2}{g} \frac{\sin \theta}{\cos^2 \theta}$$

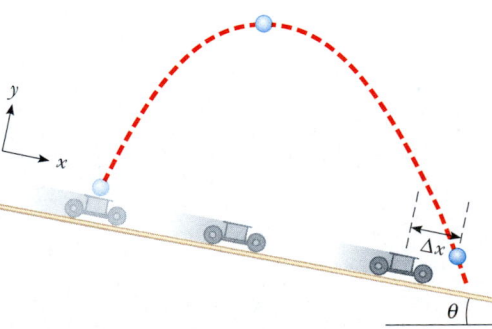

Figure P11.70

ANSWERS TO QUICK QUIZZES

11.1 There is very little resistance to motion that can reduce the kinetic energy of the rolling ball. Even though there is friction between the ball and the floor (if there were not, then no rotation would occur, and the ball would slide), there is no relative motion of the two surfaces (by the definition of "rolling"), and so kinetic friction cannot reduce K. (Air drag and friction associated with deformation of the ball eventually stop the ball.)

11.2 The box. Because none of the box's initial potential energy is converted to rotational kinetic energy, at any time $t > 0$ its translational kinetic energy is greater than that of the rolling ball.

11.3 Zero. If she were moving directly toward the pole, **r** and **p** would be antiparallel to each other, and the sine of the angle between them is zero; therefore, $L = 0$.

11.4 Both (a) and (b) are false. The net force is not necessarily zero. If the line of action of the net force passes through the point, then the net torque about an axis passing through that point is zero even though the net force is not zero. Because the net force is not necessarily zero, you cannot conclude that the particle's velocity is constant.

11.5 The student does work as he walks from the rim of the platform toward its center.

11.6 Because it is perpendicular to the precessional motion of the top, the force of gravity does no work. This is a situation in which a force causes motion but does no work.

P U Z Z L E R

This one-bottle wine holder is an interesting example of a mechanical system that seems to defy gravity. The system (holder plus bottle) is balanced when its center of gravity is directly over the lowest support point. What two conditions are necessary for an object to exhibit this kind of stability? *(Charles D. Winters)*

c h a p t e r

Static Equilibrium and Elasticity

In Chapters 10 and 11 we studied the dynamics of rigid objects—that is, objects whose parts remain at a fixed separation with respect to each other when subjected to external forces. Part of this chapter addresses the conditions under which a rigid object is in equilibrium. The term *equilibrium* implies either that the object is at rest or that its center of mass moves with constant velocity. We deal here only with the former case, in which the object is described as being in *static equilibrium*. Static equilibrium represents a common situation in engineering practice, and the principles it involves are of special interest to civil engineers, architects, and mechanical engineers. If you are an engineering student you will undoubtedly take an advanced course in statics in the future.

The last section of this chapter deals with how objects deform under load conditions. Such deformations are usually elastic and do not affect the conditions for equilibrium. An *elastic* object returns to its original shape when the deforming forces are removed. Several elastic constants are defined, each corresponding to a different type of deformation.

12.1 ▶ THE CONDITIONS FOR EQUILIBRIUM

In Chapter 5 we stated that one necessary condition for equilibrium is that the net force acting on an object be zero. If the object is treated as a particle, then this is the only condition that must be satisfied for equilibrium. The situation with real (extended) objects is more complex, however, because these objects cannot be treated as particles. For an extended object to be in static equilibrium, a second condition must be satisfied. This second condition involves the net torque acting on the extended object. Note that equilibrium does not require the absence of motion. For example, a rotating object can have constant angular velocity and still be in equilibrium.

Consider a single force **F** acting on a rigid object, as shown in Figure 12.1. The effect of the force depends on its point of application *P*. If **r** is the position vector of this point relative to *O*, the torque associated with the force **F** about *O* is given by Equation 11.7:

$$\boldsymbol{\tau} = \mathbf{r} \times \mathbf{F}$$

Recall from the discussion of the vector product in Section 11.2 that the vector **τ** is perpendicular to the plane formed by **r** and **F**. You can use the right-hand rule to determine the direction of **τ**: Curl the fingers of your right hand in the direction of rotation that **F** tends to cause about an axis through *O*: your thumb then points in the direction of **τ**. Hence, in Figure 12.1 **τ** is directed toward you out of the page.

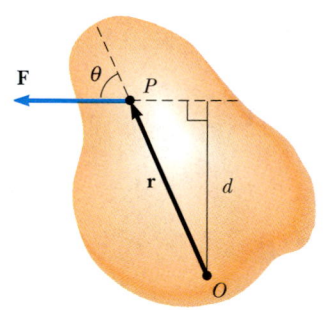

Figure 12.1 A single force **F** acts on a rigid object at the point *P*.

As you can see from Figure 12.1, the tendency of **F** to rotate the object about an axis through *O* depends on the moment arm *d*, as well as on the magnitude of **F**. Recall that the magnitude of **τ** is *Fd* (see Eq. 10.19). Now suppose a rigid object is acted on first by force $\mathbf{F}_1$ and later by force $\mathbf{F}_2$. If the two forces have the same magnitude, they will produce the same effect on the object only if they have the same direction and the same line of action. In other words,

Equivalent forces

> two forces $\mathbf{F}_1$ and $\mathbf{F}_2$ are **equivalent** if and only if $F_1 = F_2$ and if and only if the two produce the same torque about any axis.

The two forces shown in Figure 12.2 are equal in magnitude and opposite in direction. They are *not* equivalent. The force directed to the right tends to rotate the object clockwise about an axis perpendicular to the diagram through *O*, whereas the force directed to the left tends to rotate it counterclockwise about that axis.

Suppose an object is pivoted about an axis through its center of mass, as shown in Figure 12.3. Two forces of equal magnitude act in opposite directions along parallel lines of action. A pair of forces acting in this manner form what is called a **couple.** (The two forces shown in Figure 12.2 also form a couple.) Do not make the mistake of thinking that the forces in a couple are a result of Newton's third law. They cannot be third-law forces because they act on the same object. Third-law force pairs act on different objects. Because each force produces the same torque *Fd*, the net torque has a magnitude of 2*Fd*. Clearly, the object rotates clockwise and undergoes an angular acceleration about the axis. With respect to rotational motion, this is a nonequilibrium situation. The net torque on the object gives rise to an angular acceleration α according to the relationship $\Sigma\tau = 2Fd = I\alpha$ (see Eq. 10.21).

In general, an object is in rotational equilibrium only if its angular acceleration $\alpha = 0$. Because $\Sigma\tau = I\alpha$ for rotation about a fixed axis, our second necessary condition for equilibrium is that **the net torque about any axis must be zero.** We now have two necessary conditions for equilibrium of an object:

1. The resultant external force must equal zero. $\sum \mathbf{F} = 0$ **(12.1)**

2. The resultant external torque about *any* axis must be zero. $\sum \tau = 0$ **(12.2)**

The first condition is a statement of translational equilibrium; it tells us that the linear acceleration of the center of mass of the object must be zero when viewed from an inertial reference frame. The second condition is a statement of rotational equilibrium and tells us that the angular acceleration about any axis must be zero. In the special case of **static equilibrium,** which is the main subject of this chapter, the object is at rest and so has no linear or angular speed (that is, $v_{CM} = 0$ and $\omega = 0$).

> ### Quick Quiz 12.1
>
> (a) Is it possible for a situation to exist in which Equation 12.1 is satisfied while Equation 12.2 is not? (b) Can Equation 12.2 be satisfied while Equation 12.1 is not?

The two vector expressions given by Equations 12.1 and 12.2 are equivalent, in general, to six scalar equations: three from the first condition for equilibrium, and three from the second (corresponding to *x*, *y*, and *z* components). Hence, in a complex system involving several forces acting in various directions, you could be faced with solving a set of equations with many unknowns. Here, we restrict our discussion to situations in which all the forces lie in the *xy* plane. (Forces whose vector representations are in the same plane are said to be *coplanar.*) With this restriction, we must deal with only three scalar equations. Two of these come from balancing the forces in the *x* and *y* directions. The third comes from the torque equation—namely, that the net torque about *any* point in the *xy* plane must be zero. Hence, the two conditions of equilibrium provide the equations

$$\sum F_x = 0 \qquad \sum F_y = 0 \qquad \sum \tau_z = 0 \qquad \textbf{(12.3)}$$

where the axis of the torque equation is arbitrary, as we now show.

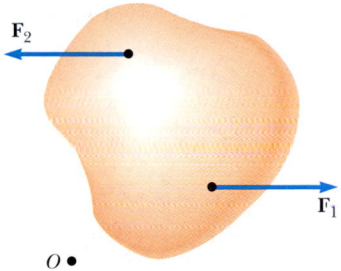

Figure 12.2 The forces $\mathbf{F}_1$ and $\mathbf{F}_2$ are not equivalent because they do not produce the same torque about some axis, even though they are equal in magnitude and opposite in direction.

Conditions for equilibrium

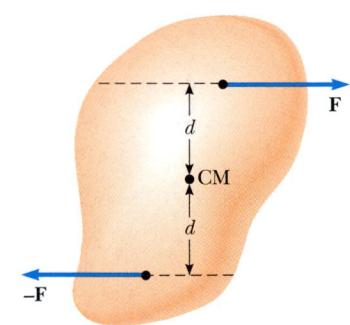

Figure 12.3 Two forces of equal magnitude form a couple if their lines of action are different parallel lines. In this case, the object rotates clockwise. The net torque about any axis is 2*Fd*.

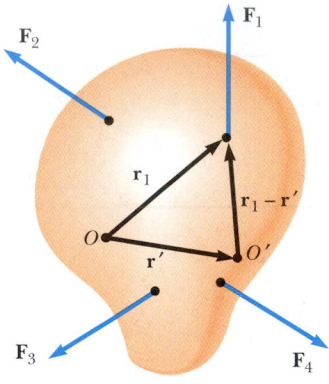

Figure 12.4 Construction showing that if the net torque is zero about origin *O*, it is also zero about any other origin, such as *O'*.

Regardless of the number of forces that are acting, if an object is in translational equilibrium and if the net torque is zero about one axis, then the net torque must also be zero about any other axis. The point can be inside or outside the boundaries of the object. Consider an object being acted on by several forces such that the resultant force $\sum \mathbf{F} = \mathbf{F}_1 + \mathbf{F}_2 + \mathbf{F}_3 + \cdots = 0$. Figure 12.4 describes this situation (for clarity, only four forces are shown). The point of application of $\mathbf{F}_1$ relative to *O* is specified by the position vector $\mathbf{r}_1$. Similarly, the points of application of $\mathbf{F}_2$, $\mathbf{F}_3$, . . . are specified by $\mathbf{r}_2$, $\mathbf{r}_3$, . . . (not shown). The net torque about an axis through *O* is

$$\sum \boldsymbol{\tau}_O = \mathbf{r}_1 \times \mathbf{F}_1 + \mathbf{r}_2 \times \mathbf{F}_2 + \mathbf{r}_3 \times \mathbf{F}_3 + \cdots$$

Now consider another arbitrary point *O'* having a position vector $\mathbf{r}'$ relative to *O*. The point of application of $\mathbf{F}_1$ relative to *O'* is identified by the vector $\mathbf{r}_1 - \mathbf{r}'$. Likewise, the point of application of $\mathbf{F}_2$ relative to *O'* is $\mathbf{r}_2 - \mathbf{r}'$, and so forth. Therefore, the torque about an axis through *O'* is

$$\sum \boldsymbol{\tau}_{O'} = (\mathbf{r}_1 - \mathbf{r}') \times \mathbf{F}_1 + (\mathbf{r}_2 - \mathbf{r}') \times \mathbf{F}_2 + (\mathbf{r}_3 - \mathbf{r}') \times \mathbf{F}_3 + \cdots$$

$$= \mathbf{r}_1 \times \mathbf{F}_1 + \mathbf{r}_2 \times \mathbf{F}_2 + \mathbf{r}_3 \times \mathbf{F}_3 + \cdots - \mathbf{r}' \times (\mathbf{F}_1 + \mathbf{F}_2 + \mathbf{F}_3 + \cdots)$$

Because the net force is assumed to be zero (given that the object is in translational equilibrium), the last term vanishes, and we see that the torque about *O'* is equal to the torque about *O*. Hence, **if an object is in translational equilibrium and the net torque is zero about one point, then the net torque must be zero about any other point.**

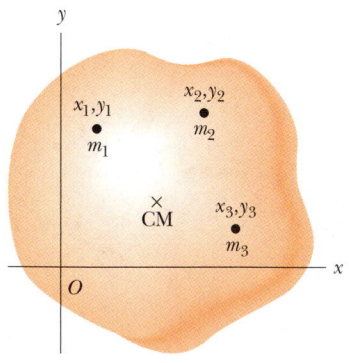

Figure 12.5 An object can be divided into many small particles each having a specific mass and specific coordinates. These particles can be used to locate the center of mass.

12.2 MORE ON THE CENTER OF GRAVITY

We have seen that the point at which a force is applied can be critical in determining how an object responds to that force. For example, two equal-magnitude but oppositely directed forces result in equilibrium if they are applied at the same point on an object. However, if the point of application of one of the forces is moved, so that the two forces no longer act along the same line of action, then a force couple results and the object undergoes an angular acceleration. (This is the situation shown in Figure 12.3.)

Whenever we deal with a rigid object, one of the forces we must consider is the force of gravity acting on it, and we must know the point of application of this force. As we learned in Section 9.6, on every object is a special point called its center of gravity. All the various gravitational forces acting on all the various mass elements of the object are equivalent to a single gravitational force acting through this point. Thus, to compute the torque due to the gravitational force on an object of mass *M*, we need only consider the force *M*$\mathbf{g}$ acting at the center of gravity of the object.

How do we find this special point? As we mentioned in Section 9.6, if we assume that $\mathbf{g}$ is uniform over the object, then the center of gravity of the object coincides with its center of mass. To see that this is so, consider an object of arbitrary shape lying in the *xy* plane, as illustrated in Figure 12.5. Suppose the object is divided into a large number of particles of masses m_1, m_2, m_3, . . . having coordinates (x_1, y_1), (x_2, y_2), (x_3, y_3), ... In

Equation 9.28 we defined the x coordinate of the center of mass of such an object to be

$$x_{CM} = \frac{m_1 x_1 + m_2 x_2 + m_3 x_3 + \cdots}{m_1 + m_2 + m_3 + \cdots} = \frac{\sum_i m_i x_i}{\sum_i m_i}$$

We use a similar equation to define the y coordinate of the center of mass, replacing each x with its y counterpart.

Let us now examine the situation from another point of view by considering the force of gravity exerted on each particle, as shown in Figure 12.6. Each particle contributes a torque about the origin equal in magnitude to the particle's weight mg multiplied by its moment arm. For example, the torque due to the force $m_1 \mathbf{g}_1$ is $m_1 g_1 x_1$, where g_1 is the magnitude of the gravitational field at the position of the particle of mass m_1. We wish to locate the center of gravity, the point at which application of the single gravitational force $M\mathbf{g}$ (where $M = m_1 + m_2 + m_3 + \cdots$ is the total mass of the object) has the same effect on rotation as does the combined effect of all the individual gravitational forces $m_i \mathbf{g}_i$. Equating the torque resulting from $M\mathbf{g}$ acting at the center of gravity to the sum of the torques acting on the individual particles gives

$$(m_1 g_1 + m_2 g_2 + m_3 g_3 + \cdots)x_{CG} = m_1 g_1 x_1 + m_2 g_2 x_2 + m_3 g_3 x_3 + \cdots$$

This expression accounts for the fact that the gravitational field strength g can in general vary over the object. If we assume uniform g over the object (as is usually the case), then the g terms cancel and we obtain

$$x_{CG} = \frac{m_1 x_1 + m_2 x_2 + m_3 x_3 + \cdots}{m_1 + m_2 + m_3 + \cdots} \qquad \textbf{(12.4)}$$

Comparing this result with Equation 9.28, we see that **the center of gravity is located at the center of mass as long as the object is in a uniform gravitational field.**

In several examples presented in the next section, we are concerned with homogeneous, symmetric objects. The center of gravity for any such object coincides with its geometric center.

12.3 EXAMPLES OF RIGID OBJECTS IN STATIC EQUILIBRIUM

The photograph of the one-bottle wine holder on the first page of this chapter shows one example of a balanced mechanical system that seems to defy gravity. For the system (wine holder plus bottle) to be in equilibrium, the net external force must be zero (see Eq. 12.1) and the net external torque must be zero (see Eq. 12.2). The second condition can be satisfied only when the center of gravity of the system is directly over the support point.

In working static equilibrium problems, it is important to recognize all the external forces acting on the object. Failure to do so results in an incorrect analysis. When analyzing an object in equilibrium under the action of several external forces, use the following procedure.

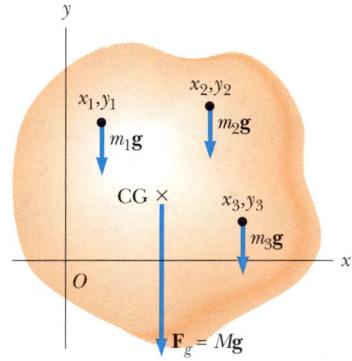

Figure 12.6 The center of gravity of an object is located at the center of mass if **g** is constant over the object.

A large balanced rock at the Garden of the Gods in Colorado Springs, Colorado—an example of stable equilibrium. *(David Serway)*

> **Problem-Solving Hints**
> **Objects in Static Equilibrium**
>
> - Draw a simple, neat diagram of the system.
> - Isolate the object being analyzed. Draw a free-body diagram and then show and label all external forces acting on the object, indicating where those forces are applied. Do not include forces exerted by the object on its surroundings. (For systems that contain more than one object, draw a *separate* free-body diagram for each one.) Try to guess the correct direction for each force. If the direction you select leads to a negative force, do not be alarmed; this merely means that the direction of the force is the opposite of what you guessed.
> - Establish a convenient coordinate system for the object and find the components of the forces along the two axes. Then apply the first condition for equilibrium. Remember to keep track of the signs of all force components.
> - Choose a convenient axis for calculating the net torque on the object. Remember that the choice of origin for the torque equation is arbitrary; therefore, choose an origin that simplifies your calculation as much as possible. Note that a force that acts along a line passing through the point chosen as the origin gives zero contribution to the torque and thus can be ignored.

The first and second conditions for equilibrium give a set of linear equations containing several unknowns, and these equations can be solved simultaneously.

EXAMPLE 12.1   ▶   **The Seesaw**

A uniform 40.0-N board supports a father and daughter weighing 800 N and 350 N, respectively, as shown in Figure 12.7. If the support (called the *fulcrum*) is under the center of gravity of the board and if the father is 1.00 m from the center, (a) determine the magnitude of the upward force **n** exerted on the board by the support.

Solution   First note that, in addition to **n**, the external forces acting on the board are the downward forces exerted by each person and the force of gravity acting on the board. We know that the board's center of gravity is at its geometric center because we were told the board is uniform. Because the system is in static equilibrium, the upward force **n** must balance all the downward forces. From $\Sigma F_y = 0$, we have, once we define upward as the positive y direction,

$$n - 800 \text{ N} - 350 \text{ N} - 40.0 \text{ N} = 0$$

$$n = \boxed{1\ 190 \text{ N}}$$

(The equation $\Sigma F_x = 0$ also applies, but we do not need to consider it because no forces act horizontally on the board.)

(b) Determine where the child should sit to balance the system.

Solution   To find this position, we must invoke the second condition for equilibrium. Taking an axis perpendicular to the page through the center of gravity of the board as the axis for our torque equation (this means that the torques

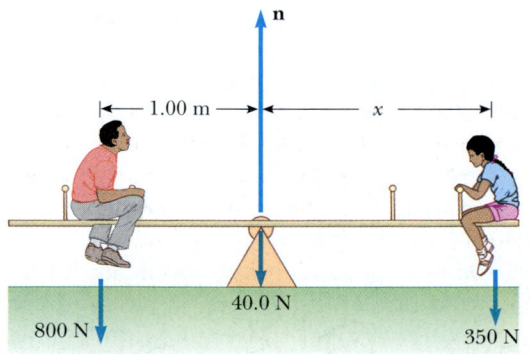

Figure 12.7   A balanced system.

produced by **n** and the force of gravity acting on the board about this axis are zero), we see from $\Sigma \tau = 0$ that

$$(800 \text{ N})(1.00 \text{ m}) - (350 \text{ N})x = 0$$

$$x = \quad 2.29 \text{ m}$$

(c) Repeat part (b) for another axis.

Solution   To illustrate that the choice of axis is arbitrary, let us choose an axis perpendicular to the page and passing through the location of the father. Recall that the sign of the torque associated with a force is positive if that force tends to rotate the system counterclockwise, while the sign of the torque is negative if the force tends to rotate the system clockwise. In this case, $\Sigma \tau = 0$ yields

$$n(1.00 \text{ m}) - (40.0 \text{ N})(1.00 \text{ m}) - (350 \text{ N})(1.00 \text{ m} + x) = 0$$

From part (a) we know that $n = 1\ 190$ N. Thus, we can solve for x to find $x = 2.29 \text{ m}$.   This result is in agreement with the one we obtained in part (b).

Quick Quiz 12.2

In Example 12.1, if the fulcrum did not lie under the board's center of gravity, what other information would you need to solve the problem?

EXAMPLE 12.2   A Weighted Hand

A person holds a 50.0-N sphere in his hand. The forearm is horizontal, as shown in Figure 12.8a. The biceps muscle is attached 3.00 cm from the joint, and the sphere is 35.0 cm from the joint. Find the upward force exerted by the biceps on the forearm and the downward force exerted by the upper arm on the forearm and acting at the joint. Neglect the weight of the forearm.

Solution   We simplify the situation by modeling the forearm as a bar as shown in Figure 12.8b, where **F** is the upward force exerted by the biceps and **R** is the downward force exerted by the upper arm at the joint. From the first condition for equilibrium, we have, with upward as the positive y direction,

$$(1) \qquad \sum F_y = F - R - 50.0 \text{ N} = 0$$

From the second condition for equilibrium, we know that the sum of the torques about any point must be zero. With the joint O as the axis, we have

$$Fd - mg\ell = 0$$

$$F(3.00 \text{ cm}) - (50.0 \text{ N})(35.0 \text{ cm}) = 0$$

$$F = \quad 583 \text{ N}$$

This value for F can be substituted into Equation (1) to give $R = 533$ N. As this example shows, the forces at joints and in muscles can be extremely large.

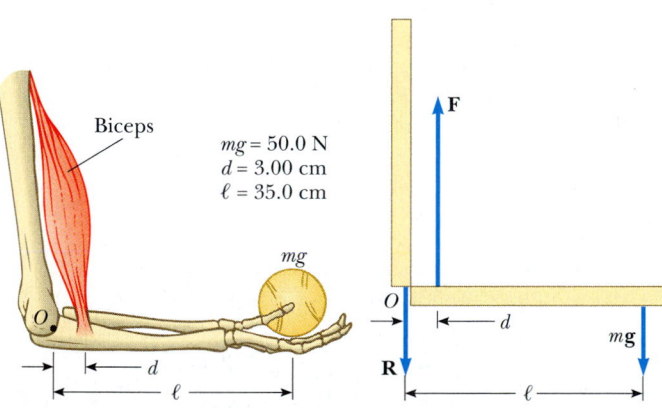

$mg = 50.0$ N
$d = 3.00$ cm
$\ell = 35.0$ cm

Figure 12.8   (a) The biceps muscle pulls upward with a force **F** that is essentially at right angles to the forearm. (b) The mechanical model for the system described in part (a).

Exercise   In reality, the biceps makes an angle of 15.0° with the vertical; thus, **F** has both a vertical and a horizontal component. Find the magnitude of **F** and the components of **R** when you include this fact in your analysis.

Answer $F = 604$ N, $R_x = 156$ N, $R_y = 533$ N.

EXAMPLE 12.3   Standing on a Horizontal Beam

A uniform horizontal beam with a length of 8.00 m and a weight of 200 N is attached to a wall by a pin connection. Its far end is supported by a cable that makes an angle of 53.0° with the horizontal (Fig. 12.9a). If a 600-N person stands 2.00 m from the wall, find the tension in the cable, as well as the magnitude and direction of the force exerted by the wall on the beam.

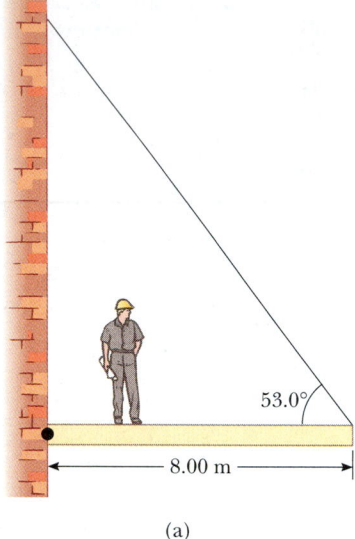

(a)

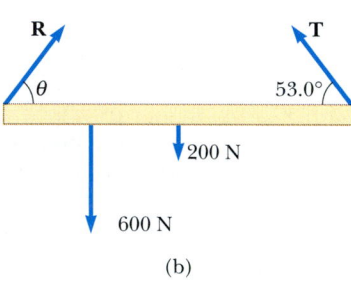

(b)

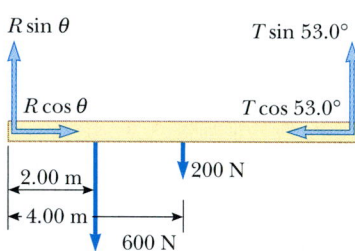

Figure 12.9   (a) A uniform beam supported by a cable. (b) The free-body diagram for the beam. (c) The free-body diagram for the beam showing the components of **R** and **T**.

Solution   First we must identify all the external forces acting on the beam: They are the 200-N force of gravity, the force **T** exerted by the cable, the force **R** exerted by the wall at the pivot, and the 600-N force that the person exerts on the beam. These forces are all indicated in the free-body diagram for the beam shown in Figure 12.9b. When we consider directions for forces, it sometimes is helpful if we imagine what would happen if a force were suddenly removed. For example, if the wall were to vanish suddenly,

the left end of the beam would probably move to the left as it begins to fall. This tells us that the wall is not only holding the beam up but is also pressing outward against it. Thus, we draw the vector **R** as shown in Figure 12.9b. If we resolve **T** and **R** into horizontal and vertical components, as shown in Figure 12.9c, and apply the first condition for equilibrium, we obtain

(1) $\sum F_x = R \cos \theta - T \cos 53.0° = 0$

(2) $\sum F_y = R \sin \theta + T \sin 53.0°$
$$- 600\text{ N} - 200\text{ N} = 0$$

where we have chosen rightward and upward as our positive directions. Because R, T, and θ are all unknown, we cannot obtain a solution from these expressions alone. (The number of simultaneous equations must equal the number of unknowns for us to be able to solve for the unknowns.)

Now let us invoke the condition for rotational equilibrium. A convenient axis to choose for our torque equation is the one that passes through the pin connection. The feature that makes this point so convenient is that the force **R** and the horizontal component of **T** both have a moment arm of zero; hence, these forces provide no torque about this point. Recalling our counterclockwise-equals-positive convention for the sign of the torque about an axis and noting that the moment arms of the 600-N, 200-N, and $T \sin 53.0°$ forces are 2.00 m, 4.00 m, and 8.00 m, respectively, we obtain

$$\sum \tau = (T \sin 53.0°)(8.00\text{ m})$$
$$- (600\text{ N})(2.00\text{ m}) - (200\text{ N})(4.00\text{ m}) = 0$$

$$T = \boxed{313\text{ N}}$$

Thus, the torque equation with this axis gives us one of the unknowns directly! We now substitute this value into Equations (1) and (2) and find that

$$R \cos \theta = 188\text{ N}$$

$$R \sin \theta = 550\text{ N}$$

We divide the second equation by the first and, recalling the trigonometric identity $\sin \theta / \cos \theta = \tan \theta$, we obtain

$$\tan \theta = \frac{550\text{ N}}{188\text{ N}} = 2.93$$

$$\theta = \boxed{71.1°}$$

This positive value indicates that our estimate of the direction of **R** was accurate.

Finally,

$$R = \frac{188\text{ N}}{\cos \theta} = \frac{188\text{ N}}{\cos 71.1°} = \boxed{580\text{ N}}$$

If we had selected some other axis for the torque equation, the solution would have been the same. For example, if we had chosen an axis through the center of gravity of the beam, the torque equation would involve both T and R. However, this equation, coupled with Equations (1) and (2), could still be solved for the unknowns. Try it!

When many forces are involved in a problem of this nature, it is convenient to set up a table. For instance, for the example just given, we could construct the following table. Setting the sum of the terms in the last column equal to zero represents the condition of rotational equilibrium.

Force Component	Moment Arm Relative to O (m)	Torque About O (N·m)
$T \sin 53.0°$	8.00	$(8.00)\,T \sin 53.0°$
$T \cos 53.0°$	0	0
200 N	4.00	$-(4.00)(200)$
600 N	2.00	$-(2.00)(600)$
$R \sin \theta$	0	0
$R \cos \theta$	0	0

EXAMPLE 12.4   The Leaning Ladder

A uniform ladder of length ℓ and weight $mg = 50$ N rests against a smooth, vertical wall (Fig. 12.10a). If the coefficient of static friction between the ladder and the ground is $\mu_s = 0.40$, find the minimum angle θ_{min} at which the ladder does not slip.

Solution The free-body diagram showing all the external forces acting on the ladder is illustrated in Figure 12.10b. The reaction force $\mathbf{R}$ exerted by the ground on the ladder is the vector sum of a normal force $\mathbf{n}$ and the force of static friction $\mathbf{f}_s$. The reaction force $\mathbf{P}$ exerted by the wall on the ladder is horizontal because the wall is frictionless. Notice how we have included only forces that act on the ladder. For example, the forces exerted by the ladder on the ground and on the wall are not part of the problem and thus do not appear in the free-body diagram. Applying the first condition

for equilibrium to the ladder, we have

$$\sum F_x = f - P = 0$$
$$\sum F_y = n - mg = 0$$

From the second equation we see that $n = mg = 50$ N. Furthermore, when the ladder is on the verge of slipping, the force of friction must be a maximum, which is given by $f_{s,max} = \mu_s n = 0.40(50 \text{ N}) = 20$ N. (Recall Eq. 5.8: $f_s \le \mu_s n$.) Thus, at this angle, $P = 20$ N.

To find θ_{min}, we must use the second condition for equilibrium. When we take the torques about an axis through the origin O at the bottom of the ladder, we have

$$\sum \tau_O = P\ell \sin \theta - mg \frac{\ell}{2} \cos \theta = 0$$

Because $P = 20$ N when the ladder is about to slip, and because $mg = 50$ N, this expression gives

$$\tan \theta_{min} = \frac{mg}{2P} = \frac{50 \text{ N}}{40 \text{ N}} = 1.25$$

$$\theta_{min} = \boxed{51°}$$

An alternative approach is to consider the intersection O' of the lines of action of forces $m\mathbf{g}$ and $\mathbf{P}$. Because the torque about any origin must be zero, the torque about O' must be zero. This requires that the line of action of $\mathbf{R}$ (the resultant of $\mathbf{n}$ and $\mathbf{f}$) pass through O'. In other words, because the ladder is stationary, the three forces acting on it must all pass through some common point. (We say that such forces are *concurrent*.) With this condition, you could then obtain the angle ϕ that $\mathbf{R}$ makes with the horizontal (where ϕ is greater than θ). Because this approach depends on the length of the ladder, you would have to know the value of ℓ to obtain a value for θ_{min}.

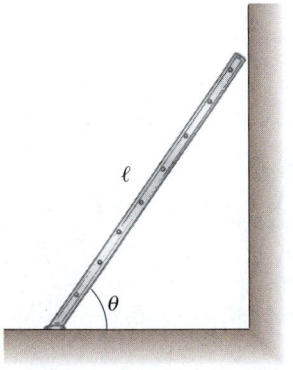

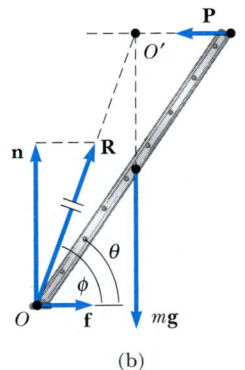

(a)　　　　　　　　(b)

Figure 12.10   (a) A uniform ladder at rest, leaning against a smooth wall. The ground is rough. (b) The free-body diagram for the ladder. Note that the forces $\mathbf{R}$, $m\mathbf{g}$, and $\mathbf{P}$ pass through a common point O'.

Exercise For the angles labeled in Figure 12.10, show that $\tan \phi = 2 \tan \theta$.

EXAMPLE 12.5   Negotiating a Curb

(a) Estimate the magnitude of the force **F** a person must apply to a wheelchair's main wheel to roll up over a sidewalk curb (Fig. 12.11a). This main wheel, which is the one that comes in contact with the curb, has a radius r, and the height of the curb is h.

Solution   Normally, the person's hands supply the required force to a slightly smaller wheel that is concentric with the main wheel. We assume that the radius of the smaller wheel is the same as the radius of the main wheel, and so we can use r for our radius. Let us estimate a combined weight of $mg = 1\ 400$ N for the person and the wheelchair and choose a wheel radius of $r = 30$ cm, as shown in Figure 12.11b. We also pick a curb height of $h = 10$ cm. We assume that the wheelchair and occupant are symmetric, and that each wheel supports a weight of 700 N. We then proceed to analyze only one of the wheels.

When the wheel is just about to be raised from the street, the reaction force exerted by the ground on the wheel at point Q goes to zero. Hence, at this time only three forces act on the wheel, as shown in Figure 12.11c. However, the force **R**, which is the force exerted on the wheel by the curb, acts at point P, and so if we choose to have our axis of rotation pass through point P, we do not need to include **R** in our torque equation. From the triangle OPQ shown in Figure 12.11b, we see that the moment arm d of the gravitational force $m\mathbf{g}$ acting on the wheel relative to point P is

$$d = \sqrt{r^2 - (r - h)^2} = \sqrt{2rh - h^2}$$

The moment arm of **F** relative to point P is $2r - h$. Therefore, the net torque acting on the wheel about point P is

$$mgd - F(2r - h) = 0$$

$$mg\sqrt{2rh - h^2} - F(2r - h) = 0$$

$$F = \frac{mg\sqrt{2rh - h^2}}{2r - h}$$

$$F = \frac{(700\ \text{N})\sqrt{2(0.3\ \text{m})(0.1\ \text{m}) - (0.1\ \text{m})^2}}{2(0.3\ \text{m}) - 0.1\ \text{m}} = \boxed{300\ \text{N}}$$

(Notice that we have kept only one digit as significant.) This result indicates that the force that must be applied to each wheel is substantial. You may want to estimate the force required to roll a wheelchair up a typical sidewalk accessibility ramp for comparison.

(b) Determine the magnitude and direction of **R**.

Solution   We use the first condition for equilibrium to determine the direction:

$$\sum F_x = F - R\cos\theta = 0$$

$$\sum F_y = R\sin\theta - mg = 0$$

Dividing the second equation by the first gives

$$\tan\theta = \frac{mg}{F} = \frac{700\ \text{N}}{300\ \text{N}}; \quad \theta = \boxed{70°}$$

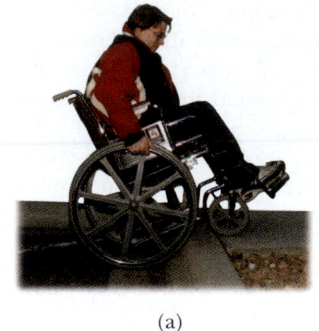

(a)

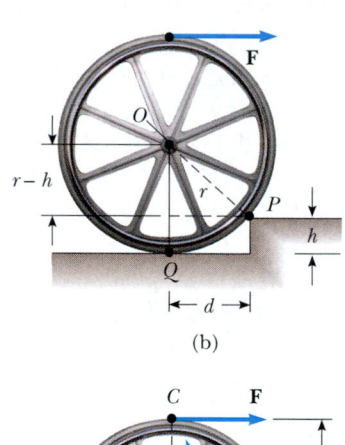

(b)

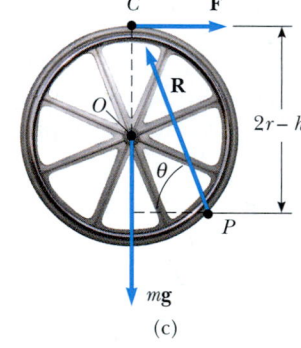

(c)

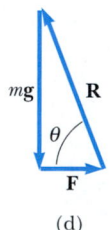

(d)

Figure 12.11   (a) A wheelchair and person of total weight mg being raised over a curb by a force **F**. (b) Details of the wheel and curb. (c) The free-body diagram for the wheel when it is just about to be raised. Three forces act on the wheel at this instant: **F**, which is exerted by the hand; **R**, which is exerted by the curb; and the gravitational force $m\mathbf{g}$. (d) The vector sum of the three external forces acting on the wheel is zero.

We can use the right triangle shown in Figure 12.11d to obtain n:

$$n = \sqrt{(mg)^2 + F^2} = \sqrt{(700 \text{ N})^2 + (300 \text{ N})^2} = \boxed{800 \text{ N}}$$

Exercise Solve this problem by noting that the three forces acting on the wheel are concurrent (that is, that all three pass through the point C). The three forces form the sides of the triangle shown in Figure 12.11d.

APPLICATION ▸ Analysis of a Truss

Roofs, bridges, and other structures that must be both strong and lightweight often are made of trusses similar to the one shown in Figure 12.12a. Imagine that this truss structure represents part of a bridge. To approach this problem, we assume that the structural components are connected by pin joints. We also assume that the entire structure is free to slide horizontally because it sits on "rockers" on each end, which allow it to move back and forth as it undergoes thermal expansion and contraction. Assuming the mass of the bridge structure is negligible compared with the load, let us calculate the forces of tension or compression in all the structural components when it is supporting a 7 200-N load at the center (see Problem 58).

The force notation that we use here is not of our usual format. Until now, we have used the notation F_{AB} to mean "the force exerted by A on B." For this application, however, all double-letter subscripts on F indicate only the body exerting the force. The body on which a given force acts is not named in the subscript. For example, in Figure 12.12, F_{AB} is the force exerted by strut AB on the pin at A.

First, we apply Newton's second law to the truss as a whole in the vertical direction. Internal forces do not enter into this accounting. We balance the weight of the load with the normal forces exerted at the two ends by the supports on which the bridge rests:

$$\sum F_y = n_A + n_E - F_g = 0$$
$$n_A + n_E = 7\ 200 \text{ N}$$

Next, we calculate the torque about A, noting that the overall length of the bridge structure is $L = 50$ m:

$$\sum \tau = Ln_E - (L/2)F_g = 0$$
$$n_E = F_g/2 = 3\ 600 \text{ N}$$

Although we could repeat the torque calculation for the right end (point E), it should be clear from symmetry arguments that $n_A = 3\ 600$ N.

Now let us balance the vertical forces acting on the pin at point A. If we assume that strut AB is in compression, then the force F_{AB} that the strut exerts on the pin at point A has a negative y component. (If the strut is actually in tension, our calculations will result in a negative value for the magnitude of the force, still of the correct size):

$$\sum F_y = n_A - F_{AB} \sin 30° = 0$$
$$F_{AB} = 7\ 200 \text{ N}$$

The positive result shows that our assumption of compression was correct.

We can now find the forces acting on the beam between A and C by considering the horizontal forces acting on the pin at point A. Because point A is not accelerating, we can safely assume that F_{AC} must point toward the right (Fig. 12.12b); this indicates that the beam between points A and C is under tension:

$$\sum F_x = F_{AC} - F_{AB} \cos 30° = 0$$
$$F_{AC} = (7\ 200 \text{ N}) \cos 30° = 6\ 200 \text{ N}$$

Now let us consider the vertical forces acting on the pin at point C. We shall assume that strut BC is in tension. (Imagine the subsequent motion of the pin at point C if strut BC were to break suddenly.) On the basis of symmetry, we assert that $F_{BC} = F_{DC}$ and that $F_{AC} = F_{EC}$:

$$\sum F_y = 2 F_{BC} \sin 30° - 7\ 200 \text{ N} = 0$$
$$F_{BC} = 7\ 200 \text{ N}$$

Finally, we balance the horizontal forces on B, assuming that strut BD is in compression:

$$\sum F_x = F_{AB} \cos 30° + F_{BC} \cos 30° - F_{BD} = 0$$
$$(7\ 200 \text{ N}) \cos 30° + (7\ 200 \text{ N}) \cos 30° - F_{BD} = 0$$
$$F_{BD} = 12\ 000 \text{ N}$$

Thus, the top beam in a bridge of this design must be very strong.

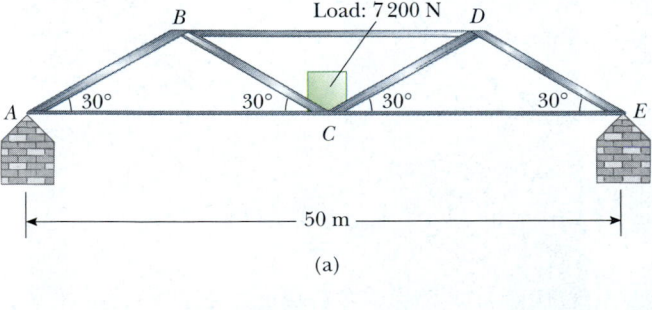

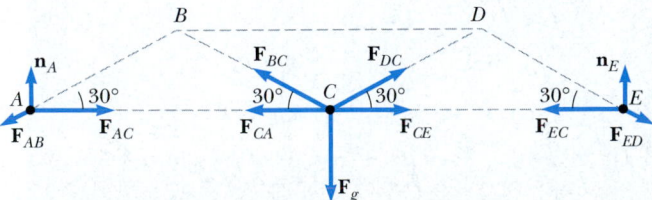

Figure 12.12 (a) Truss structure from a bridge. (b) The forces acting on the pins at points A, C, and E. As an exercise, you should diagram the forces acting on the pin at point B.

12.4 ▶ ELASTIC PROPERTIES OF SOLIDS

In our study of mechanics thus far, we have assumed that objects remain undeformed when external forces act on them. In reality, all objects are deformable. That is, it is possible to change the shape or the size of an object (or both) by applying external forces. As these changes take place, however, internal forces in the object resist the deformation.

We shall discuss the deformation of solids in terms of the concepts of stress and strain. **Stress** is a quantity that is proportional to the force causing a deformation; more specifically, stress is the external force acting on an object per unit cross-sectional area. **Strain** is a measure of the degree of deformation. It is found that, for sufficiently small stresses, **strain is proportional to stress;** the constant of proportionality depends on the material being deformed and on the nature of the deformation. We call this proportionality constant the **elastic modulus.** The elastic modulus is therefore the ratio of the stress to the resulting strain:

$$\text{Elastic modulus} \equiv \frac{\text{stress}}{\text{strain}} \qquad \textbf{(12.5)}$$

In a very real sense it is a comparison of what is done to a solid object (a force is applied) and how that object responds (it deforms to some extent).

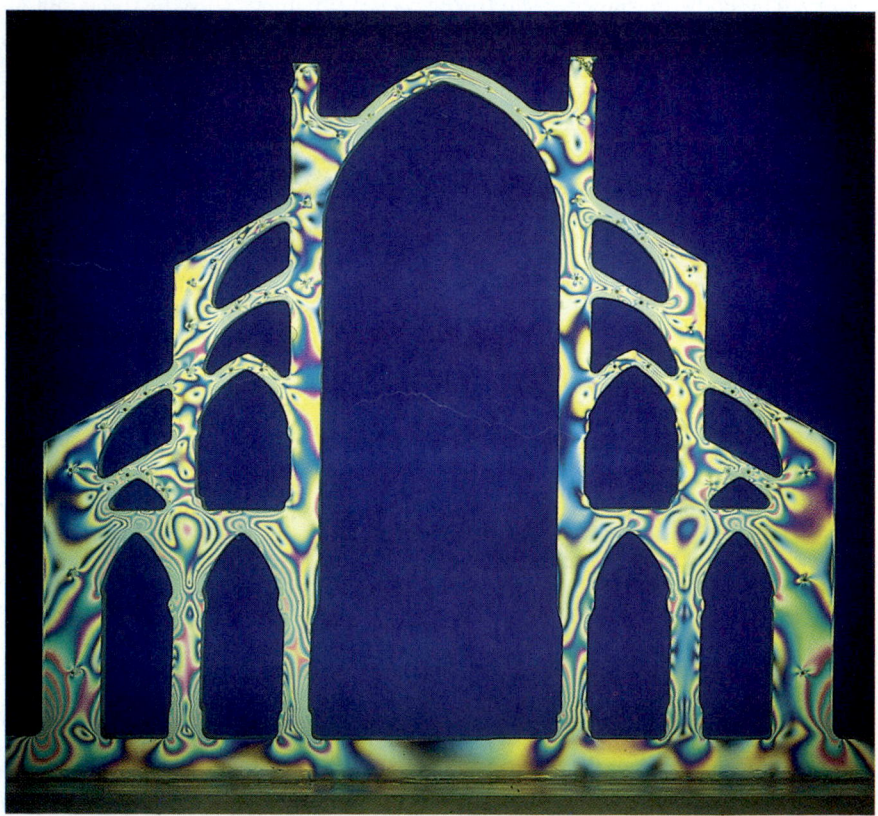

A plastic model of an arch structure under load conditions. The wavy lines indicate regions where the stresses are greatest. Such models are useful in designing architectural components. *(Peter Aprahamian/Sharples Stress Engineers Ltd./Science Photo Library)*

We consider three types of deformation and define an elastic modulus for each:

1. **Young's modulus,** which measures the resistance of a solid to a change in its length
2. **Shear modulus,** which measures the resistance to motion of the planes of a solid sliding past each other
3. **Bulk modulus,** which measures the resistance of solids or liquids to changes in their volume

Young's Modulus: Elasticity in Length

Consider a long bar of cross-sectional area A and initial length L_i that is clamped at one end, as in Figure 12.13. When an external force is applied perpendicular to the cross section, internal forces in the bar resist distortion ("stretching"), but the bar attains an equilibrium in which its length L_f is greater than L_i and in which the external force is exactly balanced by internal forces. In such a situation, the bar is said to be stressed. We define the **tensile stress** as the ratio of the magnitude of the external force F to the cross-sectional area A. The **tensile strain** in this case is defined as the ratio of the change in length ΔL to the original length L_i. We define **Young's modulus** by a combination of these two ratios:

$$Y = \frac{\text{tensile stress}}{\text{tensile strain}} = \frac{F/A}{\Delta L/L_i} \qquad (12.6)$$

Young's modulus is typically used to characterize a rod or wire stressed under either tension or compression. Note that because strain is a dimensionless quantity, Y has units of force per unit area. Typical values are given in Table 12.1. Experiments show (a) that for a fixed applied force, the change in length is proportional to the original length and (b) that the force necessary to produce a given strain is proportional to the cross-sectional area. Both of these observations are in accord with Equation 12.6.

The **elastic limit** of a substance is defined as the maximum stress that can be applied to the substance before it becomes permanently deformed. It is possible to exceed the elastic limit of a substance by applying a sufficiently large stress, as seen in Figure 12.14. Initially, a stress–strain curve is a straight line. As the stress increases, however, the curve is no longer straight. When the stress exceeds the elas-

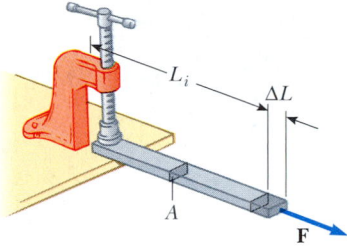

Figure 12.13 A long bar clamped at one end is stretched by an amount ΔL under the action of a force **F**.

Young's modulus

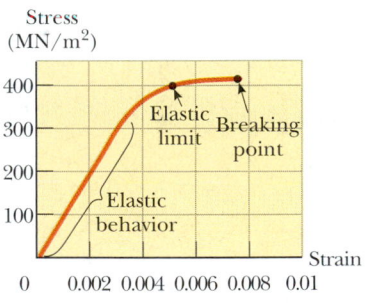

Figure 12.14 Stress-versus-strain curve for an elastic solid.

TABLE 12.1	Typical Values for Elastic Modulus		
Substance	Young's Modulus (N/m^2)	Shear Modulus (N/m^2)	Bulk Modulus (N/m^2)
Tungsten	35×10^{10}	14×10^{10}	20×10^{10}
Steel	20×10^{10}	8.4×10^{10}	6×10^{10}
Copper	11×10^{10}	4.2×10^{10}	14×10^{10}
Brass	9.1×10^{10}	3.5×10^{10}	6.1×10^{10}
Aluminum	7.0×10^{10}	2.5×10^{10}	7.0×10^{10}
Glass	$6.5–7.8 \times 10^{10}$	$2.6–3.2 \times 10^{10}$	$5.0–5.5 \times 10^{10}$
Quartz	5.6×10^{10}	2.6×10^{10}	2.7×10^{10}
Water	—	—	0.21×10^{10}
Mercury	—	—	2.8×10^{10}

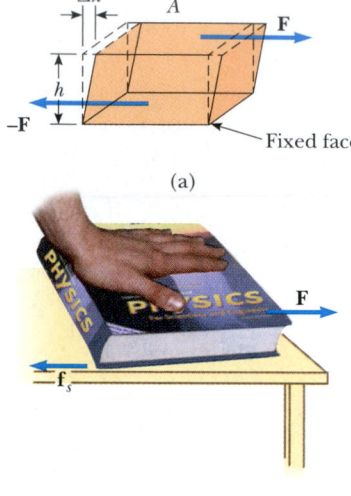

(a)

(b)

Figure 12.15   (a) A shear deformation in which a rectangular block is distorted by two forces of equal magnitude but opposite directions applied to two parallel faces. (b) A book under shear stress.

Shear modulus

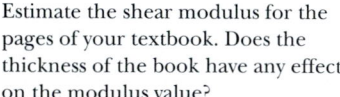

Estimate the shear modulus for the pages of your textbook. Does the thickness of the book have any effect on the modulus value?

Bulk modulus

tic limit, the object is permanently distorted and does not return to its original shape after the stress is removed. Hence, the shape of the object is permanently changed. As the stress is increased even further, the material ultimately breaks.

Quick Quiz 12.3

What is Young's modulus for the elastic solid whose stress–strain curve is depicted in Figure 12.14?

Quick Quiz 12.4

A material is said to be *ductile* if it can be stressed well beyond its elastic limit without breaking. A *brittle* material is one that breaks soon after the elastic limit is reached. How would you classify the material in Figure 12.14?

Shear Modulus: Elasticity of Shape

Another type of deformation occurs when an object is subjected to a force tangential to one of its faces while the opposite face is held fixed by another force (Fig. 12.15a). The stress in this case is called a shear stress. If the object is originally a rectangular block, a shear stress results in a shape whose cross-section is a parallelogram. A book pushed sideways, as shown in Figure 12.15b, is an example of an object subjected to a shear stress. To a first approximation (for small distortions), no change in volume occurs with this deformation.

We define the **shear stress** as F/A, the ratio of the tangential force to the area A of the face being sheared. The **shear strain** is defined as the ratio $\Delta x/h$, where Δx is the horizontal distance that the sheared face moves and h is the height of the object. In terms of these quantities, the **shear modulus** is

$$S = \frac{\text{shear stress}}{\text{shear strain}} = \frac{F/A}{\Delta x/h} \qquad (12.7)$$

Values of the shear modulus for some representative materials are given in Table 12.1. The unit of shear modulus is force per unit area.

Bulk Modulus: Volume Elasticity

Bulk modulus characterizes the response of a substance to uniform squeezing or to a reduction in pressure when the object is placed in a partial vacuum. Suppose that the external forces acting on an object are at right angles to all its faces, as shown in Figure 12.16, and that they are distributed uniformly over all the faces. As we shall see in Chapter 15, such a uniform distribution of forces occurs when an object is immersed in a fluid. An object subject to this type of deformation undergoes a change in volume but no change in shape. The **volume stress** is defined as the ratio of the magnitude of the normal force F to the area A. The quantity $P = F/A$ is called the **pressure.** If the pressure on an object changes by an amount $\Delta P = \Delta F/A$, then the object will experience a volume change ΔV. The **volume strain** is equal to the change in volume ΔV divided by the initial volume V_i. Thus, from Equation 12.5, we can characterize a volume ("bulk") compression in terms of the **bulk modulus,** which is defined as

$$B \equiv \frac{\text{volume stress}}{\text{volume strain}} = -\frac{\Delta F/A}{\Delta V/V_i} = -\frac{\Delta P}{\Delta V/V_i} \qquad (12.8)$$

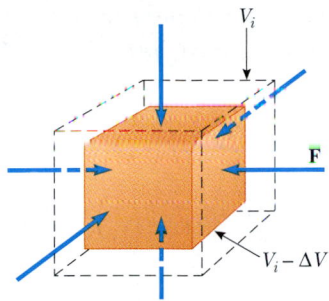

Figure 12.16 When a solid is under uniform pressure, it undergoes a change in volume but no change in shape. This cube is compressed on all sides by forces normal to its six faces.

A negative sign is inserted in this defining equation so that B is a positive number. This maneuver is necessary because an increase in pressure (positive ΔP) causes a decrease in volume (negative ΔV) and vice versa.

Table 12.1 lists bulk moduli for some materials. If you look up such values in a different source, you often find that the reciprocal of the bulk modulus is listed. The reciprocal of the bulk modulus is called the **compressibility** of the material.

Note from Table 12.1 that both solids and liquids have a bulk modulus. However, no shear modulus and no Young's modulus are given for liquids because a liquid does not sustain a shearing stress or a tensile stress (it flows instead).

Prestressed Concrete

If the stress on a solid object exceeds a certain value, the object fractures. The maximum stress that can be applied before fracture occurs depends on the nature of the material and on the type of applied stress. For example, concrete has a tensile strength of about 2×10^6 N/m^2, a compressive strength of 20×10^6 N/m^2, and a shear strength of 2×10^6 N/m^2. If the applied stress exceeds these values, the concrete fractures. It is common practice to use large safety factors to prevent failure in concrete structures.

Concrete is normally very brittle when it is cast in thin sections. Thus, concrete slabs tend to sag and crack at unsupported areas, as shown in Figure 12.17a. The slab can be strengthened by the use of steel rods to reinforce the concrete, as illustrated in Figure 12.17b. Because concrete is much stronger under compression (squeezing) than under tension (stretching) or shear, vertical columns of concrete can support very heavy loads, whereas horizontal beams of concrete tend to sag and crack. However, a significant increase in shear strength is achieved if the reinforced concrete is prestressed, as shown in Figure 12.17c. As the concrete is being poured, the steel rods are held under tension by external forces. The external

QuickLab

Support a new flat eraser (art gum or Pink Pearl will do) on two parallel pencils at least 3 cm apart. Press down on the middle of the top surface just enough to make the top face of the eraser curve a bit. Is the top face under tension or compression? How about the bottom? Why does a flat slab of concrete supported at the ends tend to crack on the bottom face and not the top?

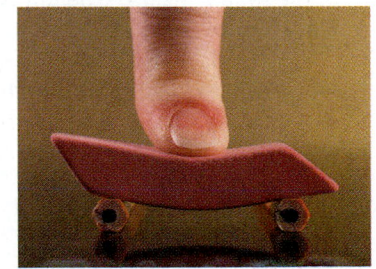

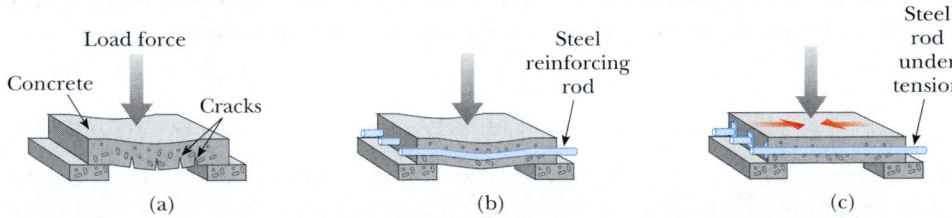

Figure 12.17 (a) A concrete slab with no reinforcement tends to crack under a heavy load. (b) The strength of the concrete is increased by using steel reinforcement rods. (c) The concrete is further strengthened by prestressing it with steel rods under tension.

forces are released after the concrete cures; this results in a permanent tension in the steel and hence a compressive stress on the concrete. This enables the concrete slab to support a much heavier load.

EXAMPLE 12.6   Stage Design

Recall Example 8.10, in which we analyzed a cable used to support an actor as he swung onto the stage. The tension in the cable was 940 N. What diameter should a 10-m-long steel wire have if we do not want it to stretch more than 0.5 cm under these conditions?

Solution   From the definition of Young's modulus, we can solve for the required cross-sectional area. Assuming that the cross section is circular, we can determine the diameter of the wire. From Equation 12.6, we have

$$Y = \frac{F/A}{\Delta L/L_i}$$

$$A = \frac{FL_i}{Y\Delta L} = \frac{(940 \text{ N})(10 \text{ m})}{(20 \times 10^{10} \text{ N/m}^2)(0.005 \text{ m})} = 9.4 \times 10^{-6} \text{ m}^2$$

The radius of the wire can be found from $A = \pi r^2$:

$$r = \sqrt{\frac{A}{\pi}} = \sqrt{\frac{9.4 \times 10^{-6} \text{ m}^2}{\pi}} = 1.7 \times 10^{-3} \text{ m} = 1.7 \text{ mm}$$

$$d = 2r = 2(1.7 \text{ mm}) = \boxed{3.4 \text{ mm}}$$

To provide a large margin of safety, we would probably use a flexible cable made up of many smaller wires having a total cross-sectional area substantially greater than our calculated value.

EXAMPLE 12.7   Squeezing a Brass Sphere

A solid brass sphere is initially surrounded by air, and the air pressure exerted on it is 1.0×10^5 N/m² (normal atmospheric pressure). The sphere is lowered into the ocean to a depth at which the pressure is 2.0×10^7 N/m². The volume of the sphere in air is 0.50 m³. By how much does this volume change once the sphere is submerged?

Solution   From the definition of bulk modulus, we have

$$B = -\frac{\Delta P}{\Delta V/V_i}$$

$$\Delta V = -\frac{V_i \Delta P}{B}$$

Because the final pressure is so much greater than the initial pressure, we can neglect the initial pressure and state that $\Delta P = P_f - P_i \approx P_f = 2.0 \times 10^7$ N/m². Therefore,

$$\Delta V = -\frac{(0.50 \text{ m}^3)(2.0 \times 10^7 \text{ N/m}^2)}{6.1 \times 10^{10} \text{ N/m}^2} = \boxed{-1.6 \times 10^{-4} \text{ m}^3}$$

The negative sign indicates a decrease in volume.

SUMMARY

A rigid object is in **equilibrium** if and only if **the resultant external force acting on it is zero and the resultant external torque on it is zero about any axis:**

$$\sum \mathbf{F} = 0 \tag{12.1}$$

$$\sum \boldsymbol{\tau} = 0 \tag{12.2}$$

The first condition is the condition for translational equilibrium, and the second is the condition for rotational equilibrium. These two equations allow you to analyze a great variety of problems. Make sure you can identify forces unambiguously, create a free-body diagram, and then apply Equations 12.1 and 12.2 and solve for the unknowns.

The force of gravity exerted on an object can be considered as acting at a single point called the **center of gravity.** The center of gravity of an object coincides with its center of mass if the object is in a uniform gravitational field.

We can describe the elastic properties of a substance using the concepts of stress and strain. **Stress** is a quantity proportional to the force producing a deformation; **strain** is a measure of the degree of deformation. Strain is proportional to stress, and the constant of proportionality is the **elastic modulus:**

$$\text{Elastic modulus} \equiv \frac{\text{stress}}{\text{strain}} \qquad \textbf{(12.5)}$$

Three common types of deformation are (1) the resistance of a solid to elongation under a load, characterized by **Young's modulus** Y; (2) the resistance of a solid to the motion of internal planes sliding past each other, characterized by the **shear modulus** S; and (3) the resistance of a solid or fluid to a volume change, characterized by the **bulk modulus** B.

QUESTIONS

1. Can a body be in equilibrium if only one external force acts on it? Explain.
2. Can a body be in equilibrium if it is in motion? Explain.
3. Locate the center of gravity for the following uniform objects: (a) sphere, (b) cube, (c) right circular cylinder.
4. The center of gravity of an object may be located outside the object. Give a few examples for which this is the case.
5. You are given an arbitrarily shaped piece of plywood, together with a hammer, nail, and plumb bob. How could you use these items to locate the center of gravity of the plywood? (*Hint:* Use the nail to suspend the plywood.)
6. For a chair to be balanced on one leg, where must the center of gravity of the chair be located?
7. Can an object be in equilibrium if the only torques acting on it produce clockwise rotation?
8. A tall crate and a short crate of equal mass are placed side by side on an incline (without touching each other). As the incline angle is increased, which crate will topple first? Explain.
9. When lifting a heavy object, why is it recommended to keep the back as vertical as possible, lifting from the knees, rather than bending over and lifting from the waist?
10. Give a few examples in which several forces are acting on a system in such a way that their sum is zero but the system is not in equilibrium.
11. If you measure the net torque and the net force on a system to be zero, (a) could the system still be rotating with respect to you? (b) Could it be translating with respect to you?
12. A ladder is resting inclined against a wall. Would you feel safer climbing up the ladder if you were told that the ground is frictionless but the wall is rough or that the wall is frictionless but the ground is rough? Justify your answer.
13. What kind of deformation does a cube of Jell-O exhibit when it "jiggles"?
14. Ruins of ancient Greek temples often have intact vertical columns, but few horizontal slabs of stone are still in place. Can you think of a reason why this is so?

PROBLEMS

1, 2, 3 = straightforward, intermediate, challenging   ☐ = full solution available in the *Student Solutions Manual and Study Guide*
WEB = solution posted at **http://www.saunderscollege.com/physics/**   🖥 = Computer useful in solving problem   🎮 = Interactive Physics
☐ = paired numerical/symbolic problems

Section 12.1   The Conditions for Equilibrium

1. A baseball player holds a 36-oz bat (weight = 10.0 N) with one hand at the point O (Fig. P12.1). The bat is in equilibrium. The weight of the bat acts along a line 60.0 cm to the right of O. Determine the force and the torque exerted on the bat by the player.

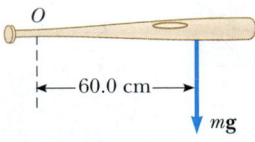

Figure P12.1

2. Write the necessary conditions of equilibrium for the body shown in Figure P12.2. Take the origin of the torque equation at the point O.

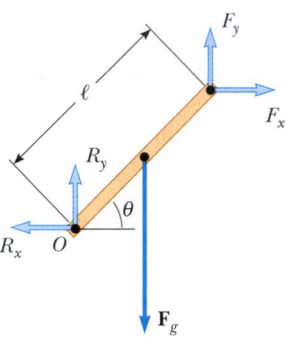

Figure P12.2

WEB **3.** A uniform beam of mass m_b and length ℓ supports blocks of masses m_1 and m_2 at two positions, as shown in Figure P12.3. The beam rests on two points. For what value of x will the beam be balanced at P such that the normal force at O is zero?

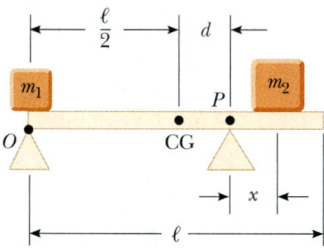

Figure P12.3

4. A student gets his car stuck in a snow drift. Not at a loss, having studied physics, he attaches one end of a stout rope to the vehicle and the other end to the trunk of a nearby tree, allowing for a very small amount of slack. The student then exerts a force **F** on the center of the rope in the direction perpendicular to the car–tree line, as shown in Figure P12.4. If the rope is inextensible and if the magnitude of the applied force is 500 N, what is the force on the car? (Assume equilibrium conditions.)

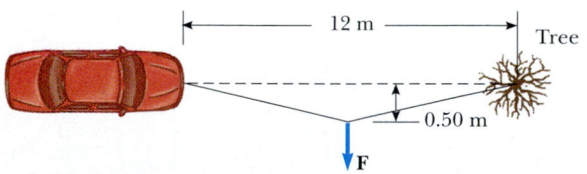

Figure P12.4

Section 12.2   More on the Center of Gravity

5. A 3.00-kg particle is located on the x axis at $x = -5.00$ m, and a 4.00-kg particle is located on the x axis at $x = 3.00$ m. Find the center of gravity of this two-particle system.

6. A circular pizza of radius R has a circular piece of radius $R/2$ removed from one side, as shown in Figure P12.6. Clearly, the center of gravity has moved from C to C' along the x axis. Show that the distance from C to C' is $R/6$. (Assume that the thickness and density of the pizza are uniform throughout.)

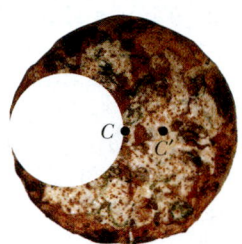

Figure P12.6

7. A carpenter's square has the shape of an L, as shown in Figure P12.7. Locate its center of gravity.

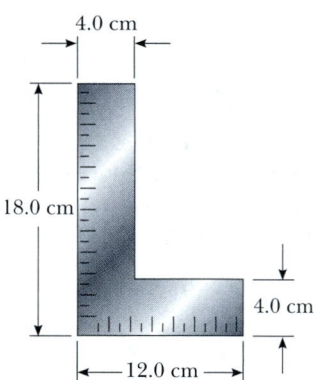

Figure P12.7

8. Pat builds a track for his model car out of wood, as illustrated in Figure P12.8. The track is 5.00 cm wide, 1.00 m high, and 3.00 m long, and it is solid. The runway is cut so that it forms a parabola described by the equation $y = (x - 3)^2/9$. Locate the horizontal position of the center of gravity of this track.

WEB **9.** Consider the following mass distribution: 5.00 kg at $(0, 0)$ m, 3.00 kg at $(0, 4.00)$ m, and 4.00 kg at $(3.00, 0)$ m. Where should a fourth mass of 8.00 kg be placed so that the center of gravity of the four-mass arrangement will be at $(0, 0)$?

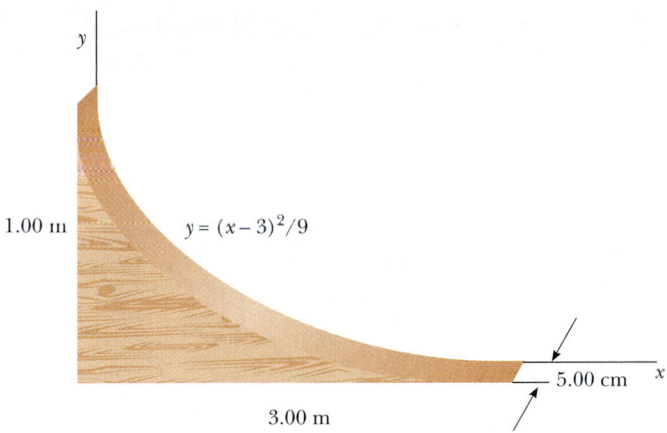

$y = (x-3)^2/9$

1.00 m

5.00 cm x

3.00 m

Figure P12.8

10. Figure P12.10 shows three uniform objects: a rod, a right triangle, and a square. Their masses in kilograms and their coordinates in meters are given. Determine the center of gravity for the three-object system.

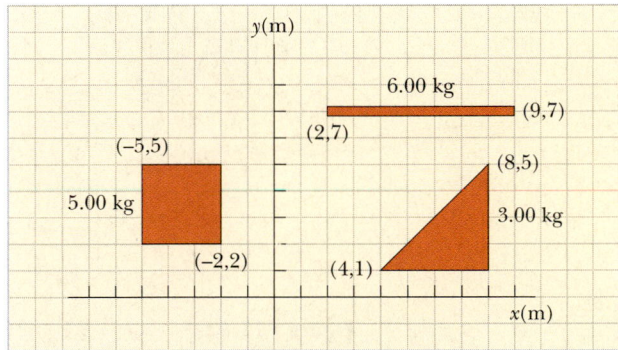

y(m)

6.00 kg

(9,7)

(2,7)

(−5,5)

(8,5)

5.00 kg

3.00 kg

(−2,2)

(4,1)

x(m)

Figure P12.10

Section 12.3   Examples of Rigid Objects in Static Equilibrium

11. Stephen is pushing his sister Joyce in a wheelbarrow when it is stopped by a brick 8.00 cm high (Fig. P12.11). The handles make an angle of 15.0° from the horizontal. A downward force of 400 N is exerted on the wheel, which has a radius of 20.0 cm. (a) What force must Stephen apply along the handles to just start the wheel over the brick? (b) What is the force (magnitude and direction) that the brick exerts on the wheel just as the wheel begins to lift over the brick? Assume in both parts (a) and (b) that the brick remains fixed and does not slide along the ground.

12. Two pans of a balance are 50.0 cm apart. The fulcrum of the balance has been shifted 1.00 cm away from the center by a dishonest shopkeeper. By what percentage is the true weight of the goods being marked up by the shopkeeper? (Assume that the balance has negligible mass.)

15.0°

Figure P12.11

13. A 15.0-m uniform ladder weighing 500 N rests against a frictionless wall. The ladder makes a 60.0° angle with the horizontal. (a) Find the horizontal and vertical forces that the ground exerts on the base of the ladder when an 800-N firefighter is 4.00 m from the bottom. (b) If the ladder is just on the verge of slipping when the firefighter is 9.00 m up, what is the coefficient of static friction between the ladder and the ground?

14. A uniform ladder of length L and mass m_1 rests against a frictionless wall. The ladder makes an angle θ with the horizontal. (a) Find the horizontal and vertical forces that the ground exerts on the base of the ladder when a firefighter of mass m_2 is a distance x from the bottom. (b) If the ladder is just on the verge of slipping when the firefighter is a distance d from the bottom, what is the coefficient of static friction between the ladder and the ground?

15. Figure P12.15 shows a claw hammer as it is being used to pull a nail out of a horizontal surface. If a force of magnitude 150 N is exerted horizontally as shown, find

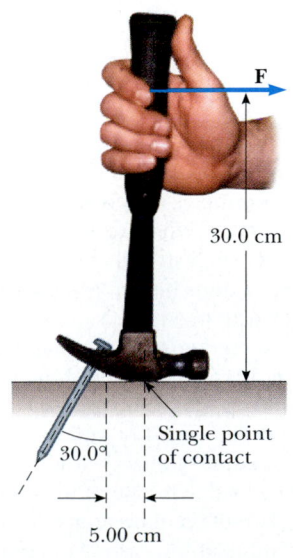

F

30.0 cm

Single point of contact

30.0°

5.00 cm

Figure P12.15

(a) the force exerted by the hammer claws on the nail and (b) the force exerted by the surface on the point of contact with the hammer head. Assume that the force the hammer exerts on the nail is parallel to the nail.

16. A uniform plank with a length of 6.00 m and a mass of 30.0 kg rests horizontally across two horizontal bars of a scaffold. The bars are 4.50 m apart, and 1.50 m of the plank hangs over one side of the scaffold. Draw a free-body diagram for the plank. How far can a painter with a mass of 70.0 kg walk on the overhanging part of the plank before it tips?

17. A 1 500-kg automobile has a wheel base (the distance between the axles) of 3.00 m. The center of mass of the automobile is on the center line at a point 1.20 m behind the front axle. Find the force exerted by the ground on each wheel.

18. A vertical post with a square cross section is 10.0 m tall. Its bottom end is encased in a base 1.50 m tall that is precisely square but slightly loose. A force of 5.50 N to the right acts on the top of the post. The base maintains the post in equilibrium. Find the force that the top of the right sidewall of the base exerts on the post. Find the force that the bottom of the left sidewall of the base exerts on the post.

19. A flexible chain weighing 40.0 N hangs between two hooks located at the same height (Fig. P12.19). At each hook, the tangent to the chain makes an angle $\theta = 42.0°$ with the horizontal. Find (a) the magnitude of the force each hook exerts on the chain and (b) the tension in the chain at its midpoint. (*Hint:* For part (b), make a free-body diagram for half the chain.)

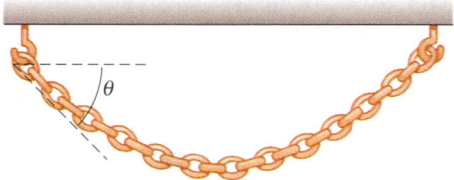

Figure P12.19

20. A hemispherical sign 1.00 m in diameter and of uniform mass density is supported by two strings, as shown in Figure P12.20. What fraction of the sign's weight is supported by each string?

21. Sir Lost-a-Lot dons his armor and sets out from the castle on his trusty steed in his quest to improve communication between damsels and dragons (Fig. P12.21). Unfortunately, his squire lowered the draw bridge too far and finally stopped lowering it when it was 20.0° below the horizontal. Lost-a-Lot and his horse stop when their combined center of mass is 1.00 m from the end of the bridge. The bridge is 8.00 m long and has a mass of 2 000 kg. The lift cable is attached to the bridge 5.00 m from the hinge at the castle end and to a point on the castle wall 12.0 m above the bridge. Lost-a-Lot's mass

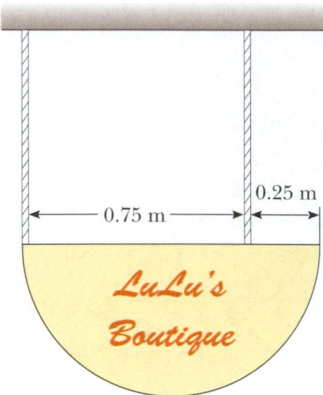

Figure P12.20

Figure P12.21

combined with that of his armor and steed is 1 000 kg. Determine (a) the tension in the cable, as well as (b) the horizontal and (c) the vertical force components acting on the bridge at the hinge.

22. Two identical, uniform bricks of length L are placed in a stack over the edge of a horizontal surface such that the maximum possible overhang without falling is achieved, as shown in Figure P12.22. Find the distance x.

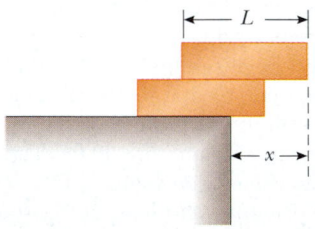

Figure P12.22

23. A vaulter holds a 29.4-N pole in equilibrium by exerting an upward force **U** with her leading hand and a downward force **D** with her trailing hand, as shown in Figure P12.23. Point C is the center of gravity of the pole. What are the magnitudes of **U** and **D**?

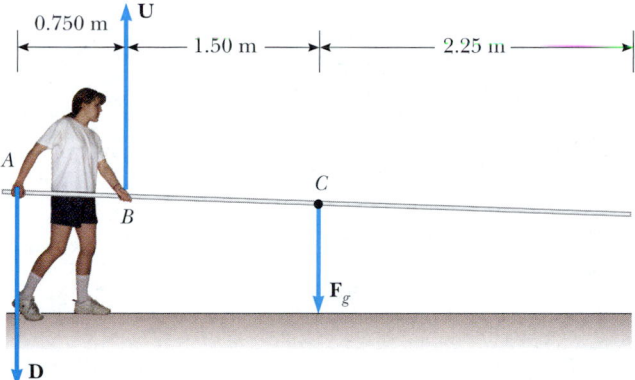

Figure P12.23

Section 12.4 Elastic Properties of Solids

24. Assume that Young's modulus for bone is 1.50×10^{10} N/m² and that a bone will fracture if more than 1.50×10^8 N/m² is exerted. (a) What is the maximum force that can be exerted on the femur bone in the leg if it has a minimum effective diameter of 2.50 cm? (b) If a force of this magnitude is applied compressively, by how much does the 25.0-cm-long bone shorten?

25. A 200-kg load is hung on a wire with a length of 4.00 m, a cross-sectional area of 0.200×10^{-4} m², and a Young's modulus of 8.00×10^{10} N/m². What is its increase in length?

26. A steel wire 1 mm in diameter can support a tension of 0.2 kN. Suppose you need a cable made of these wires to support a tension of 20 kN. The cable's diameter should be of what order of magnitude?

27. A child slides across a floor in a pair of rubber-soled shoes. The frictional force acting on each foot is 20.0 N. The footprint area of each shoe's sole is 14.0 cm², and the thickness of each sole is 5.00 mm. Find the horizontal distance by which the upper and lower surfaces of each sole are offset. The shear modulus of the rubber is 3.00×10^6 N/m².

28. Review Problem. A 30.0-kg hammer strikes a steel spike 2.30 cm in diameter while moving with a speed of 20.0 m/s. The hammer rebounds with a speed of 10.0 m/s after 0.110 s. What is the average strain in the spike during the impact?

29. If the elastic limit of copper is 1.50×10^8 N/m², determine the minimum diameter a copper wire can have under a load of 10.0 kg if its elastic limit is not to be exceeded.

30. Review Problem. A 2.00-m-long cylindrical steel wire with a cross-sectional diameter of 4.00 mm is placed over a light frictionless pulley, with one end of the wire connected to a 5.00-kg mass and the other end connected to a 3.00-kg mass. By how much does the wire stretch while the masses are in motion?

31. Review Problem. A cylindrical steel wire of length L_i with a cross-sectional diameter d is placed over a light frictionless pulley, with one end of the wire connected to a mass m_1 and the other end connected to a mass m_2. By how much does the wire stretch while the masses are in motion?

32. Calculate the density of sea water at a depth of 1 000 m, where the water pressure is about 1.00×10^7 N/m². (The density of sea water is 1.030×10^3 kg/m³ at the surface.)

WEB 33. If the shear stress exceeds about 4.00×10^8 N/m², steel ruptures. Determine the shearing force necessary (a) to shear a steel bolt 1.00 cm in diameter and (b) to punch a 1.00-cm-diameter hole in a steel plate 0.500 cm thick.

34. (a) Find the minimum diameter of a steel wire 18.0 m long that elongates no more than 9.00 mm when a load of 380 kg is hung on its lower end. (b) If the elastic limit for this steel is 3.00×10^8 N/m², does permanent deformation occur with this load?

35. When water freezes, it expands by about 9.00%. What would be the pressure increase inside your automobile's engine block if the water in it froze? (The bulk modulus of ice is 2.00×10^9 N/m².)

36. For safety in climbing, a mountaineer uses a 50.0-m nylon rope that is 10.0 mm in diameter. When supporting the 90.0-kg climber on one end, the rope elongates by 1.60 m. Find Young's modulus for the rope material.

ADDITIONAL PROBLEMS

37. A bridge with a length of 50.0 m and a mass of 8.00×10^4 kg is supported on a smooth pier at each end, as illustrated in Figure P12.37. A truck of mass 3.00×10^4 kg

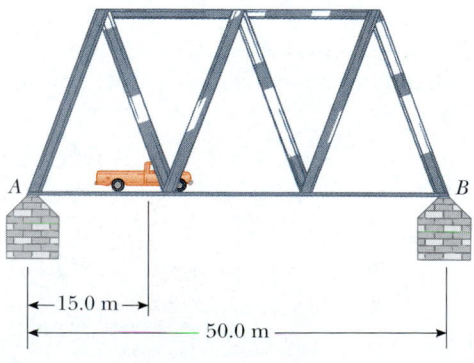

Figure P12.37

is located 15.0 m from one end. What are the forces on the bridge at the points of support?

38. A frame in the shape of the letter A is formed from two uniform pieces of metal, each of weight 26.0 N and length 1.00 m. They are hinged at the top and held together by a horizontal wire 1.20 m in length (Fig. P12.38). The structure rests on a frictionless surface. If the wire is connected at points a distance of 0.650 m from the top of the frame, determine the tension in the wire.

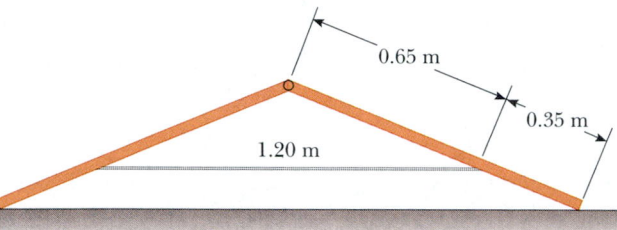

Figure P12.38

39. Refer to Figure 12.17c. A lintel of prestressed reinforced concrete is 1.50 m long. The cross-sectional area of the concrete is 50.0 cm². The concrete encloses one steel reinforcing rod with a cross-sectional area of 1.50 cm². The rod joins two strong end plates. Young's modulus for the concrete is 30.0×10^9 N/m². After the concrete cures and the original tension T_1 in the rod is released, the concrete will be under a compressive stress of 8.00×10^6 N/m². (a) By what distance will the rod compress the concrete when the original tension in the rod is released? (b) Under what tension T_2 will the rod still be? (c) How much longer than its unstressed length will the rod then be? (d) When the concrete was poured, the rod should have been stretched by what extension distance from its unstressed length? (e) Find the required original tension T_1 in the rod.

40. A solid sphere of radius R and mass M is placed in a trough, as shown in Figure P12.40. The inner surfaces of the trough are frictionless. Determine the forces exerted by the trough on the sphere at the two contact points.

41. A 10.0-kg monkey climbs up a 120-N uniform ladder of length L, as shown in Figure P12.41. The upper and

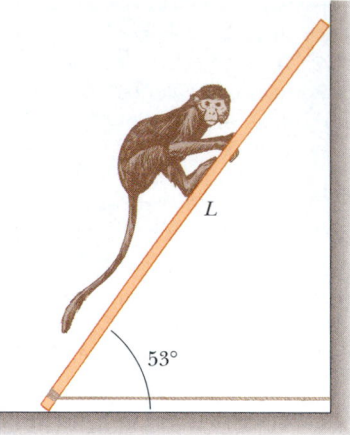

Figure P12.41

lower ends of the ladder rest on frictionless surfaces. The lower end is fastened to the wall by a horizontal rope that can support a maximum tension of 110 N. (a) Draw a free-body diagram for the ladder. (b) Find the tension in the rope when the monkey is one third the way up the ladder. (c) Find the maximum distance d that the monkey can climb up the ladder before the rope breaks. Express your answer as a fraction of L.

42. A hungry bear weighing 700 N walks out on a beam in an attempt to retrieve a basket of food hanging at the end of the beam (Fig. P12.42). The beam is uniform, weighs 200 N, and is 6.00 m long; the basket weighs 80.0 N. (a) Draw a free-body diagram for the beam. (b) When the bear is at $x = 1.00$ m, find the tension in the wire and the components of the force exerted by the wall on the left end of the beam. (c) If the wire can withstand a maximum tension of 900 N, what is the maximum distance that the bear can walk before the wire breaks?

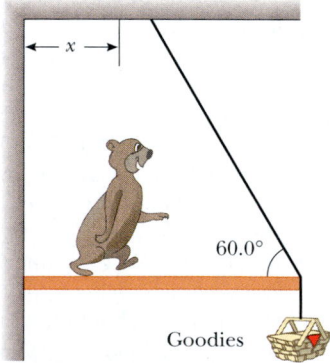

Figure P12.42

43. Old MacDonald had a farm, and on that farm he had a gate (Fig. P12.43). The gate is 3.00 m wide and 1.80 m

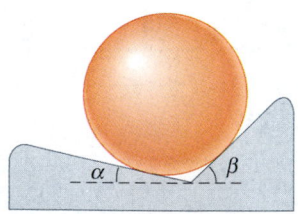

Figure P12.40

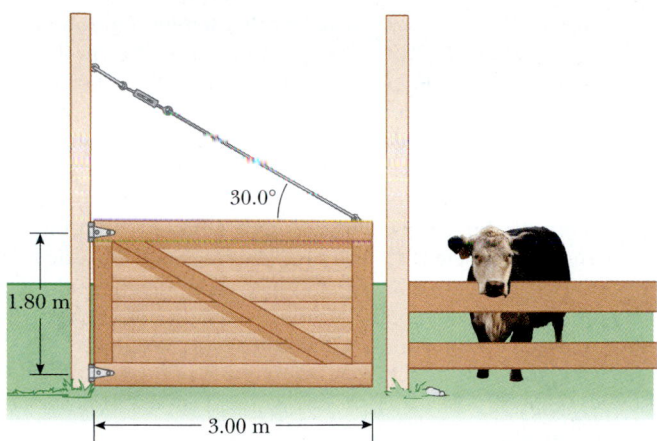

Figure P12.43

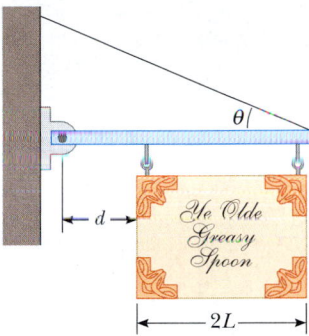

Figure P12.45

high, with hinges attached to the top and bottom. The guy wire makes an angle of 30.0° with the top of the gate and is tightened by a turn buckle to a tension of 200 N. The mass of the gate is 40.0 kg. (a) Determine the horizontal force exerted on the gate by the bottom hinge. (b) Find the horizontal force exerted by the upper hinge. (c) Determine the combined vertical force exerted by both hinges. (d) What must the tension in the guy wire be so that the horizontal force exerted by the upper hinge is zero?

44. A 1 200-N uniform boom is supported by a cable, as illustrated in Figure P12.44. The boom is pivoted at the bottom, and a 2 000-N object hangs from its top. Find the tension in the cable and the components of the reaction force exerted on the boom by the floor.

46. A crane of mass 3 000 kg supports a load of 10 000 kg as illustrated in Figure P12.46. The crane is pivoted with a frictionless pin at *A* and rests against a smooth support at *B*. Find the reaction forces at *A* and *B*.

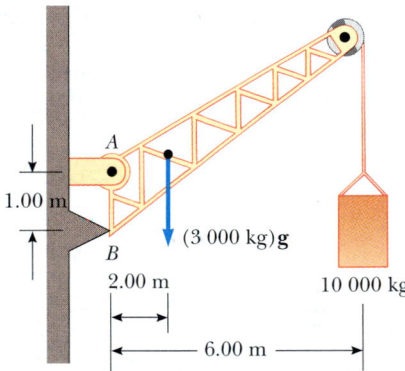

Figure P12.46

47. A ladder having a uniform density and a mass *m* rests against a frictionless vertical wall, making an angle 60.0° with the horizontal. The lower end rests on a flat surface, where the coefficient of static friction is $\mu_s =$ 0.400. A window cleaner having a mass $M = 2\,m$ attempts to climb the ladder. What fraction of the length *L* of the ladder will the worker have reached when the ladder begins to slip?

48. A uniform ladder weighing 200 N is leaning against a wall (see Fig. 12.10). The ladder slips when $\theta = 60.0°$. Assuming that the coefficients of static friction at the wall and the ground are the same, obtain a value for μ_s.

49. A 10 000-N shark is supported by a cable attached to a 4.00-m rod that can pivot at its base. Calculate the tension in the tie-rope between the wall and the rod if it is holding the system in the position shown in Figure P12.49. Find the horizontal and vertical forces exerted on the base of the rod. (Neglect the weight of the rod.)

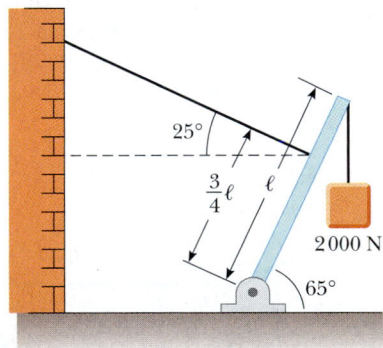

Figure P12.44

WEB **45.** A uniform sign of weight F_g and width $2L$ hangs from a light, horizontal beam hinged at the wall and supported by a cable (Fig. P12.45). Determine (a) the tension in the cable and (b) the components of the reaction force exerted by the wall on the beam in terms of F_g, *d*, *L*, and θ.

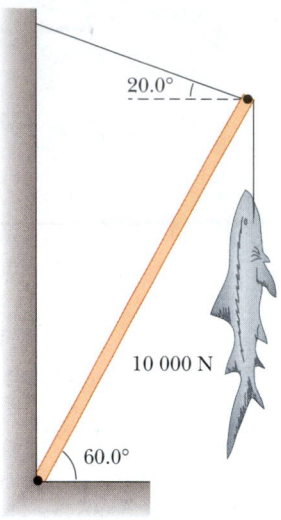

Figure P12.49

10 000 N

20.0°

60.0°

50. When a person stands on tiptoe (a strenuous position), the position of the foot is as shown in Figure P12.50a. The total weight of the body $\mathbf{F}_g$ is supported by the force $\mathbf{n}$ exerted by the floor on the toe. A mechanical model for the situation is shown in Figure P12.50b,

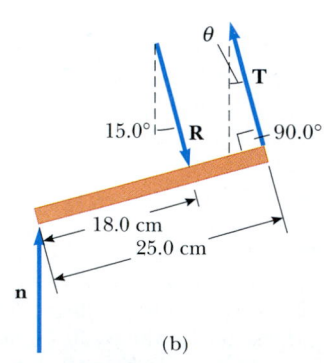

Achilles tendon

Tibia

(a)

θ

$\mathbf{T}$

15.0° $\mathbf{R}$ 90.0°

18.0 cm

25.0 cm

$\mathbf{n}$

(b)

Figure P12.50

where $\mathbf{T}$ is the force exerted by the Achilles tendon on the foot and $\mathbf{R}$ is the force exerted by the tibia on the foot. Find the values of T, R, and θ when $F_g = 700$ N.

51. A person bends over and lifts a 200-N object as shown in Figure P12.51a, with his back in a horizontal position (a terrible way to lift an object). The back muscle attached at a point two thirds the way up the spine maintains the position of the back, and the angle between the spine and this muscle is 12.0°. Using the mechanical model shown in Figure P12.51b and taking the weight of the upper body to be 350 N, find the tension in the back muscle and the compressional force in the spine.

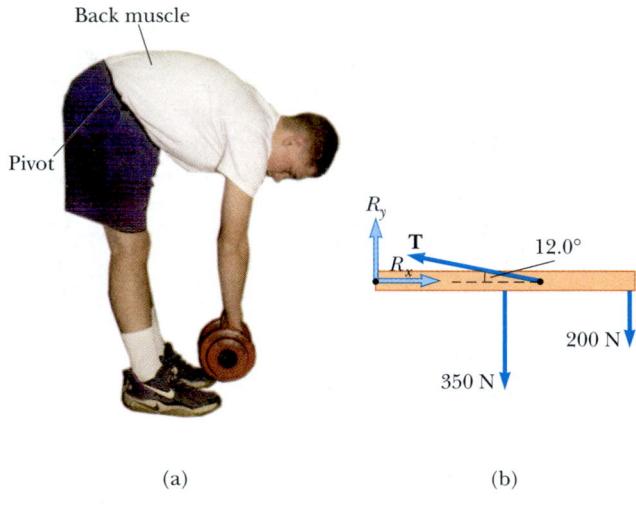

Back muscle

Pivot

R_y

$\mathbf{T}$ 12.0°

R_x

200 N

350 N

(a)   (b)

Figure P12.51

52. Two 200-N traffic lights are suspended from a single cable, as shown in Figure 12.52. Neglecting the cable's weight, (a) prove that if $\theta_1 = \theta_2$, then $T_1 = T_2$. (b) Determine the three tensions T_1, T_2, and T_3 if $\theta_1 = \theta_2 = 8.00°$.

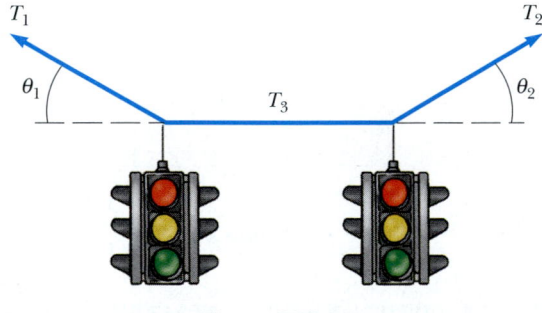

T_1 T_2

θ_1 T_3 θ_2

Figure P12.52

53. A force acts on a rectangular cabinet weighing 400 N, as illustrated in Figure P12.53. (a) If the cabinet slides with constant speed when $F = 200$ N and $h = 0.400$ m, find the coefficient of kinetic friction and the position of the resultant normal force. (b) If $F = 300$ N, find the value of h for which the cabinet just begins to tip.

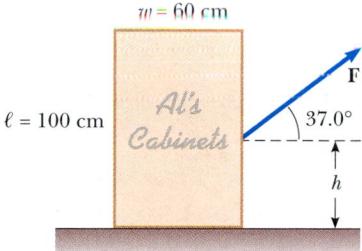

Figure P12.53 Problems 53 and 54.

54. Consider the rectangular cabinet of Problem 53, but with a force **F** applied horizontally at its upper edge. (a) What is the minimum force that must be applied for the cabinet to start tipping? (b) What is the minimum coefficient of static friction required to prevent the cabinet from sliding with the application of a force of this magnitude? (c) Find the magnitude and direction of the minimum force required to tip the cabinet if the point of application can be chosen anywhere on it.

55. A uniform rod of weight F_g and length L is supported at its ends by a frictionless trough, as shown in Figure P12.55. (a) Show that the center of gravity of the rod is directly over point O when the rod is in equilibrium. (b) Determine the equilibrium value of the angle θ.

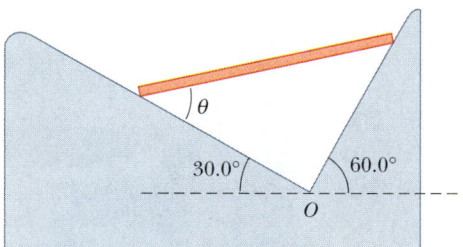

Figure P12.55

56. **Review Problem.** A cue stick strikes a cue ball and delivers a horizontal impulse in such a way that the ball rolls without slipping as it starts to move. At what height above the ball's center (in terms of the radius of the ball) was the blow struck?

57. A uniform beam of mass m is inclined at an angle θ to the horizontal. Its upper end produces a 90° bend in a very rough rope tied to a wall, and its lower end rests on a rough floor (Fig. P12.57). (a) If the coefficient of static friction between the beam and the floor is μ_s, determine an expression for the maximum mass M that can

be suspended from the top before the beam slips. (b) Determine the magnitude of the reaction force at the floor and the magnitude of the force exerted by the beam on the rope at P in terms of m, M, and μ_s.

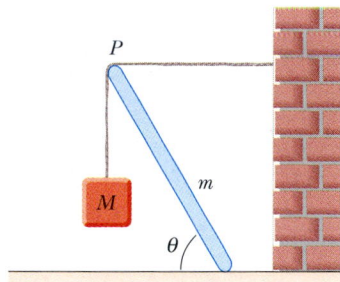

Figure P12.57

58. Figure P12.58 shows a truss that supports a downward force of 1 000 N applied at the point B. The truss has negligible weight. The piers at A and C are smooth. (a) Apply the conditions of equilibrium to prove that $n_A = 366$ N and that $n_C = 634$ N. (b) Show that, because forces act on the light truss only at the hinge joints, each bar of the truss must exert on each hinge pin only a force along the length of that bar—a force of tension or compression. (c) Find the force of tension or compression in each of the three bars.

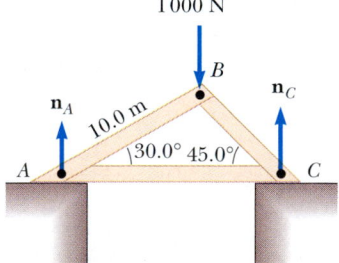

Figure P12.58

 59. A stepladder of negligible weight is constructed as shown in Figure P12.59. A painter with a mass of 70.0 kg stands on the ladder 3.00 m from the bottom. Assuming that the floor is frictionless, find (a) the tension in the horizontal bar connecting the two halves of the ladder, (b) the normal forces at A and B, and (c) the components of the reaction force at the single hinge C that the left half of the ladder exerts on the right half. (*Hint:* Treat each half of the ladder separately.)

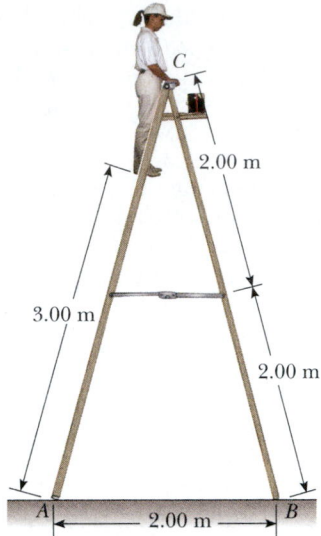

Figure P12.59

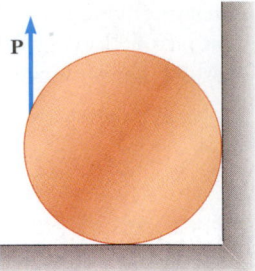

Figure P12.62

static friction between the cylinder and all surfaces is 0.500. In terms of F_g, find the maximum force **P** that can be applied that does not cause the cylinder to rotate. (*Hint:* When the cylinder is on the verge of slipping, both friction forces are at their maximum values. Why?)

WEB **63.** **Review Problem.**   A wire of length L_i, Young's modulus Y, and cross-sectional area A is stretched elastically by an amount ΔL. According to Hooke's law, the restoring force is $-k \Delta L$. (a) Show that $k = YA/L_i$. (b) Show that the work done in stretching the wire by an amount ΔL is $W = YA(\Delta L)^2/2L_i$.

64. Two racquetballs are placed in a glass jar, as shown in Figure P12.64. Their centers and the point A lie on a straight line. (a) Assuming that the walls are frictionless, determine P_1, P_2, and P_3. (b) Determine the magnitude of the force exerted on the right ball by the left ball. Assume each ball has a mass of 170 g.

60. A flat dance floor of dimensions 20.0 m by 20.0 m has a mass of 1 000 kg. Three dance couples, each of mass 125 kg, start in the top left, top right, and bottom left corners. (a) Where is the initial center of gravity? (b) The couple in the bottom left corner moves 10.0 m to the right. Where is the new center of gravity? (c) What was the average velocity of the center of gravity if it took that couple 8.00 s to change position?

61. A shelf bracket is mounted on a vertical wall by a single screw, as shown in Figure P12.61. Neglecting the weight of the bracket, find the horizontal component of the force that the screw exerts on the bracket when an 80.0-N vertical force is applied as shown. (*Hint:* Imagine that the bracket is slightly loose.)

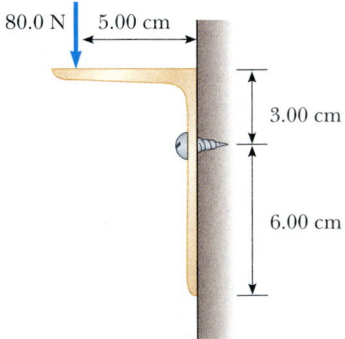

Figure P12.61

62. Figure P12.62 shows a vertical force applied tangentially to a uniform cylinder of weight F_g. The coefficient of

Figure P12.64

65. In Figure P12.65, the scales read $F_{g1} = 380$ N and $F_{g2} = 320$ N. Neglecting the weight of the supporting plank,

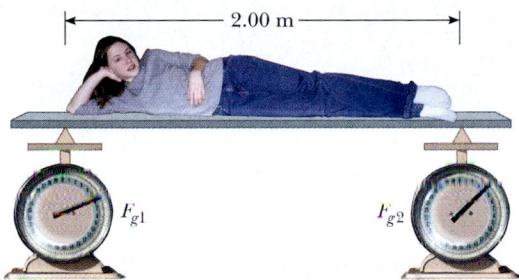

Figure P12.65

how far from the woman's feet is her center of mass, given that her height is 2.00 m?

66. A steel cable 3.00 cm^2 in cross-sectional area has a mass of 2.40 kg per meter of length. If 500 m of the cable is hung over a vertical cliff, how much does the cable stretch under its own weight? (For Young's modulus for steel, refer to Table 12.1.)

67. (a) Estimate the force with which a karate master strikes a board if the hand's speed at time of impact is 10.0 m/s and decreases to 1.00 m/s during a 0.002 00-s time-of-contact with the board. The mass of coordinated hand-and-arm is 1.00 kg. (b) Estimate the shear stress if this force is exerted on a 1.00-cm-thick pine board that is 10.0 cm wide. (c) If the maximum shear stress a pine board can receive before breaking is 3.60×10^6 N/m^2, will the board break?

68. A bucket is made from thin sheet metal. The bottom and top of the bucket have radii of 25.0 cm and 35.0 cm, respectively. The bucket is 30.0 cm high and filled with water. Where is the center of gravity? (Ignore the weight of the bucket itself.)

69. **Review Problem.** A trailer with a loaded weight of F_g is being pulled by a vehicle with a force **P**, as illustrated in Figure P12.69. The trailer is loaded such that its center of mass is located as shown. Neglect the force of rolling friction and let a represent the x component of the acceleration of the trailer. (a) Find the vertical component of **P** in terms of the given parameters. (b) If $a =$ 2.00 m/s^2 and $h = 1.50$ m, what must be the value of d

so that $P_y = 0$ (that is, no vertical load on the vehicle)? (c) Find the values of P_x and P_y given that $F_g = 1\,500$ N, $d = 0.800$ m, $L = 3.00$ m, $h = 1.50$ m, and $a = -2.00$ m/s^2.

70. **Review Problem.** An aluminum wire is 0.850 m long and has a circular cross section of diameter 0.780 mm. Fixed at the top end, the wire supports a 1.20-kg mass that swings in a horizontal circle. Determine the angular velocity required to produce strain 1.00×10^{-3}.

71. A 200-m-long bridge truss extends across a river (Fig. P12.71). Calculate the force of tension or compression in each structural component when a 1 360-kg car is at the center of the bridge. Assume that the structure is free to slide horizontally to permit thermal expansion and contraction, that the structural components are connected by pin joints, and that the masses of the structural components are small compared with the mass of the car.

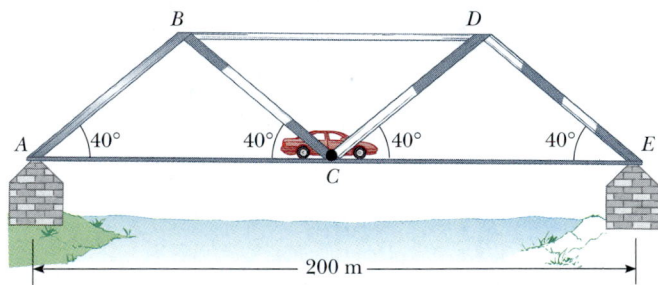

Figure P12.71

72. A 100-m-long bridge truss is supported at its ends so that it can slide freely (Fig. P12.72). A 1 500-kg car is halfway between points A and C. Show that the weight of the car is evenly distributed between points A and C, and calculate the force in each structural component. Specify whether each structural component is under tension or compression. Assume that the structural components are connected by pin joints and that the masses of the components are small compared with the mass of the car.

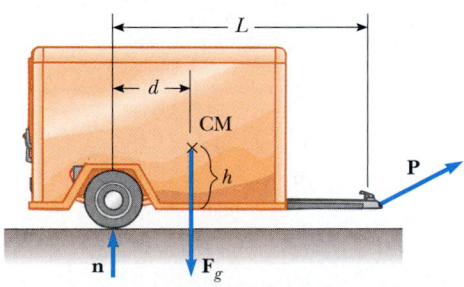

Figure P12.69

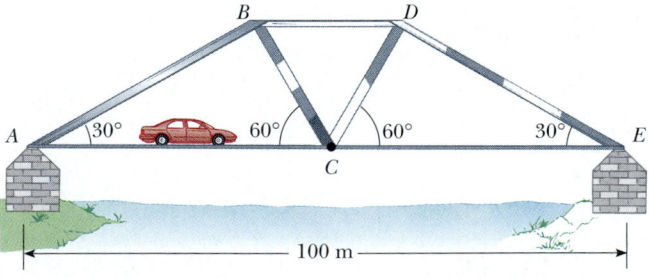

Figure P12.72

ANSWERS TO QUICK QUIZZES

12.1 (a) Yes, as Figure 12.3 shows. The unbalanced torques cause an angular acceleration even though the linear acceleration is zero. (b) Yes, again. This happens when the lines of action of all the forces intersect at a common point. If a net force acts on the object, then the object has a translational acceleration. However, because there is no net torque on the object, the object has no angular acceleration. There are other instances in which torques cancel but the forces do not. You should be able to draw at least two.

12.2 The location of the board's center of gravity relative to the fulcrum.

12.3 Young's modulus is given by the ratio of stress to strain, which is the slope of the elastic behavior section of the graph in Figure 12.14. Reading from the graph, we note that a stress of approximately $3 \times 10^8 \, \text{N/m}^2$ results in a strain of 0.003. The slope, and hence Young's modulus, are therefore $10 \times 10^{10} \, \text{N/m}^2$.

12.4 A substantial part of the graph extends beyond the elastic limit, indicating permanent deformation. Thus, the material is ductile.

PUZZLER

Inside the pocket watch is a small disk (called a torsional pendulum) that oscillates back and forth at a very precise rate and controls the watch gears. A grandfather clock keeps accurate time because of its pendulum. The tall wooden case provides the space needed by the long pendulum as it advances the clock gears with each swing. In both of these timepieces, the vibration of a carefully shaped component is critical to accurate operation. What properties of oscillating objects make them so useful in timing devices? *(Photograph of pocket watch, George Semple; photograph of grandfather clock, Charles D. Winters)*

chapter

13

Oscillatory Motion

A very special kind of motion occurs when the force acting on a body is proportional to the displacement of the body from some equilibrium position. If this force is always directed toward the equilibrium position, repetitive back-and-forth motion occurs about this position. Such motion is called *periodic motion, harmonic motion, oscillation,* or *vibration* (the four terms are completely equivalent).

You are most likely familiar with several examples of periodic motion, such as the oscillations of a block attached to a spring, the swinging of a child on a playground swing, the motion of a pendulum, and the vibrations of a stringed musical instrument. In addition to these everyday examples, numerous other systems exhibit periodic motion. For example, the molecules in a solid oscillate about their equilibrium positions; electromagnetic waves, such as light waves, radar, and radio waves, are characterized by oscillating electric and magnetic field vectors; and in alternating-current electrical circuits, voltage, current, and electrical charge vary periodically with time.

Most of the material in this chapter deals with *simple harmonic motion,* in which an object oscillates such that its position is specified by a sinusoidal function of time with no loss in mechanical energy. In real mechanical systems, damping (frictional) forces are often present. These forces are considered in optional Section 13.6 at the end of this chapter.

13.1 ▶ SIMPLE HARMONIC MOTION

Consider a physical system that consists of a block of mass m attached to the end of a spring, with the block free to move on a horizontal, frictionless surface (Fig. 13.1). When the spring is neither stretched nor compressed, the block is at the position $x = 0$, called the *equilibrium position* of the system. We know from experience that such a system oscillates back and forth if disturbed from its equilibrium position.

We can understand the motion in Figure 13.1 qualitatively by first recalling that when the block is displaced a small distance x from equilibrium, the spring exerts on the block a force that is proportional to the displacement and given by Hooke's law (see Section 7.3):

$$F_s = -kx \tag{13.1}$$

We call this a **restoring force** because it is is always directed toward the equilibrium position and therefore *opposite* the displacement. That is, when the block is displaced to the right of $x = 0$ in Figure 13.1, then the displacement is positive and the restoring force is directed to the left. When the block is displaced to the left of $x = 0$, then the displacement is negative and the restoring force is directed to the right.

Applying Newton's second law to the motion of the block, together with Equation 13.1, we obtain

$$F_s = -kx = ma$$

$$a = -\frac{k}{m}x \tag{13.2}$$

That is, the acceleration is proportional to the displacement of the block, and its direction is opposite the direction of the displacement. Systems that behave in this way are said to exhibit **simple harmonic motion. An object moves with simple harmonic motion whenever its acceleration is proportional to its displacement from some equilibrium position and is oppositely directed.**

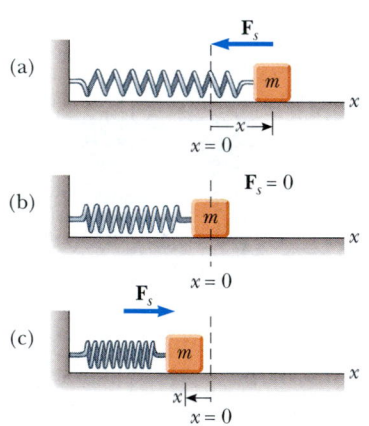

Figure 13.1 A block attached to a spring moving on a frictionless surface. (a) When the block is displaced to the right of equilibrium ($x > 0$), the force exerted by the spring acts to the left. (b) When the block is at its equilibrium position ($x = 0$), the force exerted by the spring is zero. (c) When the block is displaced to the left of equilibrium ($x < 0$), the force exerted by the spring acts to the right.

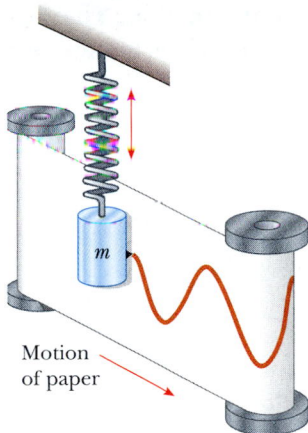

Motion
of paper

Figure 13.2 An experimental apparatus for demonstrating simple harmonic motion. A pen attached to the oscillating mass traces out a wavelike pattern on the moving chart paper.

8.2
&
8.3

An experimental arrangement that exhibits simple harmonic motion is illustrated in Figure 13.2. A mass oscillating vertically on a spring has a pen attached to it. While the mass is oscillating, a sheet of paper is moved perpendicular to the direction of motion of the spring, and the pen traces out a wavelike pattern.

In general, a particle moving along the x axis exhibits simple harmonic motion when x, the particle's displacement from equilibrium, varies in time according to the relationship

$$x = A \cos(\omega t + \phi)$$ **(13.3)**

where A, ω, and ϕ are constants. To give physical significance to these constants, we have labeled a plot of x as a function of t in Figure 13.3a. This is just the pattern that is observed with the experimental apparatus shown in Figure 13.2. The **amplitude** A of the motion is the maximum displacement of the particle in either the positive or negative x direction. The constant ω is called the **angular frequency** of the motion and has units of radians per second. (We shall discuss the geometric significance of ω in Section 13.2.) The constant angle ϕ, called the **phase constant** (or phase angle), is determined by the initial displacement and velocity of the particle. If the particle is at its maximum position $x = A$ at $t = 0$, then $\phi = 0$ and the curve of x versus t is as shown in Figure 13.3b. If the particle is at some other position at $t = 0$, the constants ϕ and A tell us what the position was at time $t = 0$. The quantity $(\omega t + \phi)$ is called the **phase** of the motion and is useful in comparing the motions of two oscillators.

Note from Equation 13.3 that the trigonometric function x is *periodic* and repeats itself every time ωt increases by 2π rad. **The period T of the motion is the time it takes for the particle to go through one full cycle.** We say that the particle has made *one oscillation*. This definition of T tells us that the value of x at time t equals the value of x at time $t + T$. We can show that $T = 2\pi/\omega$ by using the preceding observation that the phase $(\omega t + \phi)$ increases by 2π rad in a time T:

$$\omega t + \phi + 2\pi = \omega(t + T) + \phi$$

Hence, $\omega T = 2\pi$, or

$$T = \frac{2\pi}{\omega}$$ **(13.4)**

Displacement versus time for simple harmonic motion

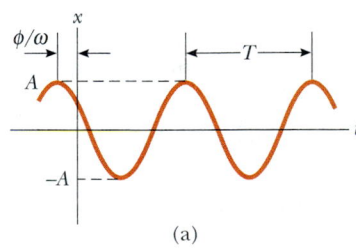

(a)

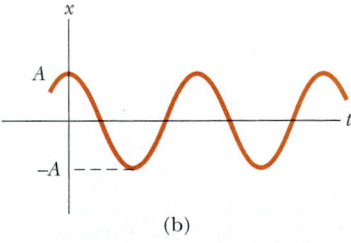

(b)

Figure 13.3 (a) An x–t curve for a particle undergoing simple harmonic motion. The amplitude of the motion is A, the period is T, and the phase constant is ϕ. (b) The x–t curve in the special case in which $x = A$ at $t = 0$ and hence $\phi = 0$.

The inverse of the period is called the **frequency** f of the motion. **The frequency represents the number of oscillations that the particle makes per unit time:**

Frequency

$$f = \frac{1}{T} = \frac{\omega}{2\pi} \qquad (13.5)$$

The units of f are cycles per second $= s^{-1}$, or **hertz** (Hz).

Rearranging Equation 13.5, we obtain the angular frequency:

Angular frequency

$$\omega = 2\pi f = \frac{2\pi}{T} \qquad (13.6)$$

Quick Quiz 13.1

What would the phase constant ϕ have to be in Equation 13.3 if we were describing an oscillating object that happened to be at the origin at $t = 0$?

Quick Quiz 13.2

An object undergoes simple harmonic motion of amplitude A. Through what total distance does the object move during one complete cycle of its motion? (a) $A/2$. (b) A. (c) $2A$. (d) $4A$.

We can obtain the linear velocity of a particle undergoing simple harmonic motion by differentiating Equation 13.3 with respect to time:

Velocity in simple harmonic motion

$$v = \frac{dx}{dt} = -\omega A \sin(\omega t + \phi) \qquad (13.7)$$

The acceleration of the particle is

Acceleration in simple harmonic motion

$$a = \frac{dv}{dt} = -\omega^2 A \cos(\omega t + \phi) \qquad (13.8)$$

Because $x = A \cos(\omega t + \phi)$, we can express Equation 13.8 in the form

$$a = -\omega^2 x \qquad (13.9)$$

From Equation 13.7 we see that, because the sine function oscillates between ± 1, the extreme values of v are $\pm \omega A$. Because the cosine function also oscillates between ± 1, Equation 13.8 tells us that the extreme values of a are $\pm \omega^2 A$. Therefore, the maximum speed and the magnitude of the maximum acceleration of a particle moving in simple harmonic motion are

Maximum values of speed and acceleration in simple harmonic motion

$$v_{\text{max}} = \omega A \qquad (13.10)$$

$$a_{\text{max}} = \omega^2 A \qquad (13.11)$$

Figure 13.4a represents the displacement versus time for an arbitrary value of the phase constant. The velocity and acceleration curves are illustrated in Figure 13.4b and c. These curves show that the phase of the velocity differs from the phase of the displacement by $\pi/2$ rad, or $90°$. That is, when x is a maximum or a minimum, the velocity is zero. Likewise, when x is zero, the speed is a maximum.

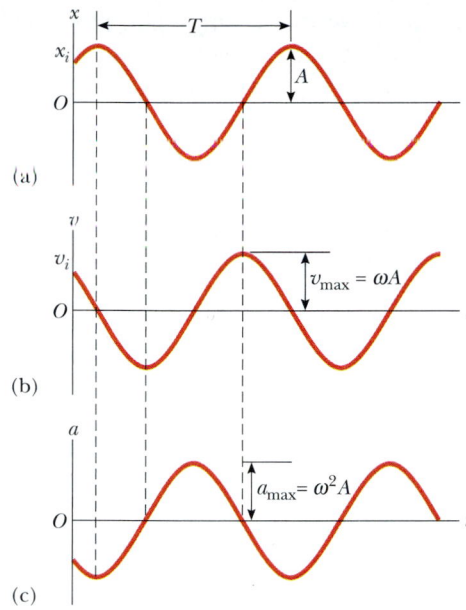

Figure 13.4 Graphical representation of simple harmonic motion. (a) Displacement versus time. (b) Velocity versus time. (c) Acceleration versus time. Note that at any specified time the velocity is 90° out of phase with the displacement and the acceleration is 180° out of phase with the displacement.

Furthermore, note that the phase of the acceleration differs from the phase of the displacement by π rad, or 180°. That is, when x is a maximum, a is a maximum in the opposite direction.

The phase constant ϕ is important when we compare the motion of two or more oscillating objects. Imagine two identical pendulum bobs swinging side by side in simple harmonic motion, with one having been released later than the other. The pendulum bobs have different phase constants. Let us show how the phase constant and the amplitude of any particle moving in simple harmonic motion can be determined if we know the particle's initial speed and position and the angular frequency of its motion.

Suppose that at $t = 0$ the initial position of a single oscillator is $x = x_i$ and its initial speed is $v = v_i$. Under these conditions, Equations 13.3 and 13.7 give

$$x_i = A \cos \phi \tag{13.12}$$

$$v_i = -\omega A \sin \phi \tag{13.13}$$

Dividing Equation 13.13 by Equation 13.12 eliminates A, giving $v_i/x_i = -\omega \tan \phi$, or

$$\tan \phi = -\frac{v_i}{\omega x_i} \tag{13.14}$$

Furthermore, if we square Equations 13.12 and 13.13, divide the velocity equation by ω^2, and then add terms, we obtain

$$x_i{}^2 + \left(\frac{v_i}{\omega}\right)^2 = A^2 \cos^2 \phi + A^2 \sin^2 \phi$$

Using the identity $\sin^2 \phi + \cos^2 \phi = 1$, we can solve for A:

$$A = \sqrt{x_i{}^2 + \left(\frac{v_i}{\omega}\right)^2} \tag{13.15}$$

The following properties of a particle moving in simple harmonic motion are important:

Properties of simple harmonic motion

- The acceleration of the particle is proportional to the displacement but is in the opposite direction. This is the *necessary and sufficient condition for simple harmonic motion,* as opposed to all other kinds of vibration.
- The displacement from the equilibrium position, velocity, and acceleration all vary sinusoidally with time but are not in phase, as shown in Figure 13.4.
- The frequency and the period of the motion are independent of the amplitude. (We show this explicitly in the next section.)

Quick Quiz 13.3

Can we use Equations 2.8, 2.10, 2.11, and 2.12 (see pages 35 and 36) to describe the motion of a simple harmonic oscillator?

EXAMPLE 13.1 ▶ An Oscillating Object

An object oscillates with simple harmonic motion along the x axis. Its displacement from the origin varies with time according to the equation

$$x = (4.00 \text{ m}) \cos\left(\pi t + \frac{\pi}{4} \right)$$

where t is in seconds and the angles in the parentheses are in radians. (a) Determine the amplitude, frequency, and period of the motion.

Solution By comparing this equation with Equation 13.3, the general equation for simple harmonic motion— $x = A \cos(\omega t + \phi)$ —we see that $A = 4.00$ m and $\omega = \pi$ rad/s. Therefore, $f = \omega/2\pi = \pi/2\pi = 0.500$ Hz and $T = 1/f = 2.00$ s.

(b) Calculate the velocity and acceleration of the object at any time t.

Solution

$$v = \frac{dx}{dt} = -(4.00 \text{ m}) \sin\left(\pi t + \frac{\pi}{4} \right) \frac{d}{dt} (\pi t)$$

$$= \boxed{-(4.00\pi \text{ m/s}) \sin\left(\pi t + \frac{\pi}{4} \right)}$$

$$a = \frac{dv}{dt} = -(4.00\pi \text{ m/s}) \cos\left(\pi t + \frac{\pi}{4} \right) \frac{d}{dt} (\pi t)$$

$$= \boxed{-(4.00\pi^2 \text{ m/s}^2) \cos\left(\pi t + \frac{\pi}{4} \right)}$$

(c) Using the results of part (b), determine the position, velocity, and acceleration of the object at $t = 1.00$ s.

Solution Noting that the angles in the trigonometric functions are in radians, we obtain, at $t = 1.00$ s,

$$x = (4.00 \text{ m}) \cos\left(\pi + \frac{\pi}{4} \right) = (4.00 \text{ m}) \cos\left(\frac{5\pi}{4} \right)$$

$$= (4.00 \text{ m})(-0.707) = \boxed{-2.83 \text{ m}}$$

$$v = -(4.00\pi \text{ m/s}) \sin\left(\frac{5\pi}{4} \right) = -(4.00\pi \text{ m/s})(-0.707)$$

$$= \boxed{8.89 \text{ m/s}}$$

$$a = -(4.00\pi^2 \text{ m/s}^2) \cos\left(\frac{5\pi}{4} \right)$$

$$= -(4.00\pi^2 \text{ m/s}^2)(-0.707) = \boxed{27.9 \text{ m/s}^2}$$

(d) Determine the maximum speed and maximum acceleration of the object.

Solution In the general expressions for v and a found in part (b), we use the fact that the maximum values of the sine and cosine functions are unity. Therefore, v varies between $\pm 4.00\pi$ m/s, and a varies between $\pm 4.00\pi^2$ m/s². Thus,

$$v_{\text{max}} = 4.00\pi \text{ m/s} = \boxed{12.6 \text{ m/s}}$$

$$a_{\text{max}} = 4.00\pi^2 \text{ m/s}^2 = \boxed{39.5 \text{ m/s}^2}$$

We obtain the same results using $v_{\text{max}} = \omega A$ and $a_{\text{max}} = \omega^2 A$, where $A = 4.00$ m and $\omega = \pi$ rad/s.

(e) Find the displacement of the object between $t = 0$ and $t = 1.00$ s.

Solution   The x coordinate at $t = 0$ is

$$x_i = (4.00 \text{ m}) \cos\left(0 + \frac{\pi}{4}\right) = (4.00 \text{ m})(0.707) = 2.83 \text{ m}$$

In part (c), we found that the x coordinate at $t = 1.00$ s is -2.83 m; therefore, the displacement between $t = 0$ and $t = 1.00$ s is

$$\Delta x = x_f - x_i = -2.83 \text{ m} - 2.83 \text{ m} = \boxed{-5.66 \text{ m}}$$

Because the object's velocity changes sign during the first second, the magnitude of Δx is not the same as the distance traveled in the first second. (By the time the first second is over, the object has been through the point $x = -2.83$ m once, traveled to $x = -4.00$ m, and come back to $x = 2.83$ m.)

Exercise   What is the phase of the motion at $t = 2.00$ s?

Answer $9\pi/4$ rad.

13.2   THE BLOCK–SPRING SYSTEM REVISITED

Let us return to the block–spring system (Fig. 13.5). Again we assume that the surface is frictionless; hence, when the block is displaced from equilibrium, the only force acting on it is the restoring force of the spring. As we saw in Equation 13.2, when the block is displaced a distance x from equilibrium, it experiences an acceleration $a = -(k/m)x$. If the block is displaced a maximum distance $x = A$ at some initial time and then released from rest, its initial acceleration at that instant is $-kA/m$ (its extreme negative value). When the block passes through the equilibrium position $x = 0$, its acceleration is zero. At this instant, its speed is a maximum. The block then continues to travel to the left of equilibrium and finally reaches $x = -A$, at which time its acceleration is kA/m (maximum positive) and its speed is again zero. Thus, we see that the block oscillates between the turning points $x = \pm A$.

Let us now describe the oscillating motion in a quantitative fashion. Recall that $a = dv/dt = d^2x/dt^2$, and so we can express Equation 13.2 as

$$\frac{d^2x}{dt^2} = -\frac{k}{m}x \tag{13.16}$$

If we denote the ratio k/m with the symbol ω^2, this equation becomes

$$\frac{d^2x}{dt^2} = -\omega^2 x \tag{13.17}$$

Now we require a solution to Equation 13.17—that is, a function $x(t)$ that satisfies this second-order differential equation. Because Equations 13.17 and 13.9 are equivalent, each solution must be that of simple harmonic motion:

$$x = A \cos(\omega t + \phi)$$

To see this explicitly, assume that $x = A \cos(\omega t + \phi)$. Then

$$\frac{dx}{dt} = A \frac{d}{dt} \cos(\omega t + \phi) = -\omega A \sin(\omega t + \phi)$$

$$\frac{d^2x}{dt^2} = -\omega A \frac{d}{dt} \sin(\omega t + \phi) = -\omega^2 A \cos(\omega t + \phi)$$

Comparing the expressions for x and d^2x/dt^2, we see that $d^2x/dt^2 = -\omega^2 x$, and Equation 13.17 is satisfied. We conclude that **whenever the force acting on a particle is linearly proportional to the displacement from some equilibrium**

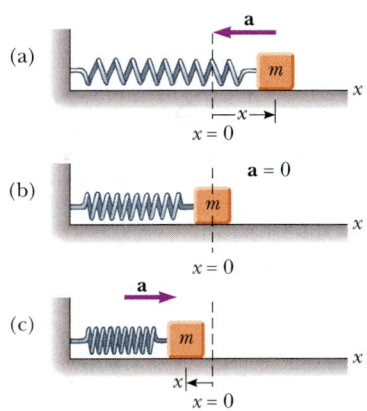

Figure 13.5   A block of mass m attached to a spring on a frictionless surface undergoes simple harmonic motion. (a) When the block is displaced to the right of equilibrium, the displacement is positive and the acceleration is negative. (b) At the equilibrium position, $x = 0$, the acceleration is zero and the speed is a maximum. (c) When the block is displaced to the left of equilibrium, the displacement is negative and the acceleration is positive.

position and in the opposite direction ($F = -kx$), the particle moves in simple harmonic motion.

Recall that the period of any simple harmonic oscillator is $T = 2\pi/\omega$ (Eq. 13.4) and that the frequency is the inverse of the period. We know from Equations 13.16 and 13.17 that $\omega = \sqrt{k/m}$, so we can express the period and frequency of the block–spring system as

Period and frequency for a block–spring system

$$T = \frac{2\pi}{\omega} = 2\pi\sqrt{\frac{m}{k}} \qquad (13.18)$$

$$f = \frac{1}{T} = \frac{1}{2\pi}\sqrt{\frac{k}{m}} \qquad (13.19)$$

That is, **the frequency and period depend only on the mass of the block and on the force constant of the spring.** Furthermore, the frequency and period are independent of the amplitude of the motion. As we might expect, the frequency is greater for a stiffer spring (the stiffer the spring, the greater the value of k) and decreases with increasing mass.

QuickLab

Hang an object from a rubber band and start it oscillating. Measure T. Now tie four identical rubber bands together, end to end. How should k for this longer band compare with k for the single band? Again, time the oscillations with the same object. Can you verify Equation 13.19?

Special Case 1.   Let us consider a special case to better understand the physical significance of Equation 13.3, the defining expression for simple harmonic motion. We shall use this equation to describe the motion of an oscillating block–spring system. Suppose we pull the block a distance A from equilibrium and then release it from rest at this stretched position, as shown in Figure 13.6. Our solution for x must obey the initial conditions that $x_i = A$ and $v_i = 0$ at $t = 0$. It does if we choose $\phi = 0$, which gives $x = A\cos\omega t$ as the solution. To check this solution, we note that it satisfies the condition that $x_i = A$ at $t = 0$ because $\cos 0 = 1$. Thus, we see that A and ϕ contain the information on initial conditions.

Now let us investigate the behavior of the velocity and acceleration for this special case. Because $x = A\cos\omega t$,

$$v = \frac{dx}{dt} = -\omega A\sin\omega t$$

$$a = \frac{dv}{dt} = -\omega^2 A\cos\omega t$$

From the velocity expression we see that, because $\sin 0 = 0$, $v_i = 0$ at $t = 0$, as we require. The expression for the acceleration tells us that $a = -\omega^2 A$ at $t = 0$. Physically, this negative acceleration makes sense because the force acting on the block is directed to the left when the displacement is positive. In fact, at the extreme po-

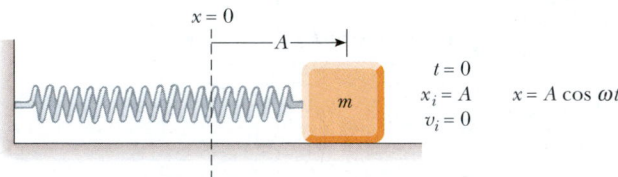

Figure 13.6   A block–spring system that starts from rest at $x_i = A$. In this case, $\phi = 0$ and thus $x = A\cos\omega t$.

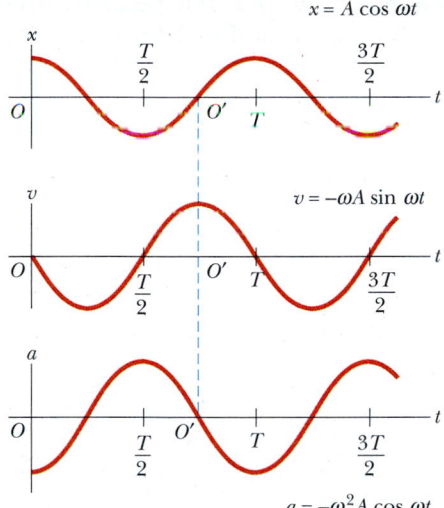

$$x = A \cos \omega t$$

$$v = -\omega A \sin \omega t$$

$$a = -\omega^2 A \cos \omega t$$

Figure 13.7 Displacement, velocity, and acceleration versus time for a block–spring system like the one shown in Figure 13.6, undergoing simple harmonic motion under the initial conditions that at $t = 0$, $x_i = A$ and $v_i = 0$ (Special Case 1). The origins at O' correspond to Special Case 2, the block–spring system under the initial conditions shown in Figure 13.8.

sition shown in Figure 13.6, $F_s = -kA$ (to the left) and the initial acceleration is $-\omega^2 A = -kA/m$.

Another approach to showing that $x = A \cos \omega t$ is the correct solution involves using the relationship $\tan \phi = -v_i/\omega x_i$ (Eq. 13.14). Because $v_i = 0$ at $t = 0$, $\tan \phi = 0$ and thus $\phi = 0$. (The tangent of π also equals zero, but $\phi = \pi$ gives the wrong value for x_i.)

Figure 13.7 is a plot of displacement, velocity, and acceleration versus time for this special case. Note that the acceleration reaches extreme values of $\pm \omega^2 A$ while the displacement has extreme values of $\pm A$ because the force is maximal at those positions. Furthermore, the velocity has extreme values of $\pm \omega A$, which both occur at $x = 0$. Hence, the quantitative solution agrees with our qualitative description of this system.

Special Case 2. Now suppose that the block is given an initial velocity $\mathbf{v}_i$ to the right at the instant it is at the equilibrium position, so that $x_i = 0$ and $v = v_i$ at $t = 0$ (Fig. 13.8). The expression for x must now satisfy these initial conditions. Because the block is moving in the positive x direction at $t = 0$ and because $x_i = 0$ at $t = 0$, the expression for x must have the form $x = A \sin \omega t$.

Applying Equation 13.14 and the initial condition that $x_i = 0$ at $t = 0$, we find that $\tan \phi = -\infty$ and $\phi = -\pi/2$. Hence, Equation 13.3 becomes $x = A \cos (\omega t - \pi/2)$, which can be written $x = A \sin \omega t$. Furthermore, from Equation 13.15 we see that $A = v_i/\omega$; therefore, we can express x as

$$x = \frac{v_i}{\omega} \sin \omega t$$

The velocity and acceleration in this case are

$$v = \frac{dx}{dt} = v_i \cos \omega t$$

$$a = \frac{dv}{dt} = -\omega v_i \sin \omega t$$

These results are consistent with the facts that (1) the block always has a maximum

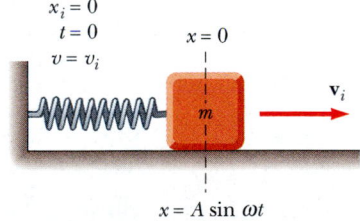

Figure 13.8 The block–spring system starts its motion at the equilibrium position at $t = 0$. If its initial velocity is v_i to the right, the block's x coordinate varies as $x = (v_i/\omega) \sin \omega t$.

speed at $x = 0$ and (2) the force and acceleration are zero at this position. The graphs of these functions versus time in Figure 13.7 correspond to the origin at O'.

Quick Quiz 13.4

What is the solution for x if the block is initially moving to the left in Figure 13.8?

EXAMPLE 13.2   Watch Out for Potholes!

A car with a mass of 1 300 kg is constructed so that its frame is supported by four springs. Each spring has a force constant of 20 000 N/m. If two people riding in the car have a combined mass of 160 kg, find the frequency of vibration of the car after it is driven over a pothole in the road.

Solution   We assume that the mass is evenly distributed. Thus, each spring supports one fourth of the load. The total mass is 1 460 kg, and therefore each spring supports 365 kg.

Hence, the frequency of vibration is, from Equation 13.19,

$$f = \frac{1}{2\pi}\sqrt{\frac{k}{m}} = \frac{1}{2\pi}\sqrt{\frac{20\,000\text{ N/m}}{365\text{ kg}}} = \boxed{1.18\text{ Hz}}$$

Exercise   How long does it take the car to execute two complete vibrations?

Answer   1.70 s.

EXAMPLE 13.3   A Block–Spring System

A block with a mass of 200 g is connected to a light spring for which the force constant is 5.00 N/m and is free to oscillate on a horizontal, frictionless surface. The block is displaced 5.00 cm from equilibrium and released from rest, as shown in Figure 13.6. (a) Find the period of its motion.

Solution   From Equations 13.16 and 13.17, we know that the angular frequency of any block–spring system is

$$\omega = \sqrt{\frac{k}{m}} = \sqrt{\frac{5.00\text{ N/m}}{200 \times 10^{-3}\text{ kg}}} = 5.00\text{ rad/s}$$

and the period is

$$T = \frac{2\pi}{\omega} = \frac{2\pi}{5.00\text{ rad/s}} = \boxed{1.26\text{ s}}$$

(b) Determine the maximum speed of the block.

Solution   We use Equation 13.10:

$$v_{\max} = \omega A = (5.00\text{ rad/s})(5.00 \times 10^{-2}\text{ m}) = \boxed{0.250\text{ m/s}}$$

(c) What is the maximum acceleration of the block?

Solution   We use Equation 13.11:

$$a_{\max} = \omega^2 A = (5.00\text{ rad/s})^2(5.00 \times 10^{-2}\text{ m}) = \boxed{1.25\text{ m/s}^2}$$

(d) Express the displacement, speed, and acceleration as functions of time.

Solution   This situation corresponds to Special Case 1, where our solution is $x = A\cos \omega t$. Using this expression and the results from (a), (b), and (c), we find that

$$x = A\cos \omega t = \boxed{(0.050\text{ m})\cos 5.00t}$$

$$v = \omega A\sin \omega t = \boxed{-(0.250\text{ m/s})\sin 5.00t}$$

$$a = \omega^2 A\cos \omega t = \boxed{-(1.25\text{ m/s}^2)\cos 5.00t}$$

13.3   ENERGY OF THE SIMPLE HARMONIC OSCILLATOR

Let us examine the mechanical energy of the block–spring system illustrated in Figure 13.6. Because the surface is frictionless, we expect the total mechanical energy to be constant, as was shown in Chapter 8. We can use Equation 13.7 to express the kinetic energy as

$$K = \tfrac{1}{2}\,mv^2 = \tfrac{1}{2}\,m\omega^2 A^2 \sin^2(\omega t + \phi) \qquad \textbf{(13.20)}$$

The elastic potential energy stored in the spring for any elongation x is given by $\tfrac{1}{2}\,kx^2$ (see Eq. 8.4). Using Equation 13.3, we obtain

$$U = \tfrac{1}{2}\,kx^2 = \tfrac{1}{2}\,kA^2 \cos^2(\omega t + \phi) \qquad \textbf{(13.21)}$$

We see that K and U are *always* positive quantities. Because $\omega^2 = k/m$, we can express the total mechanical energy of the simple harmonic oscillator as

$$E = K + U = \tfrac{1}{2}\,kA^2[\sin^2(\omega t + \phi) + \cos^2(\omega t + \phi)]$$

From the identity $\sin^2\theta + \cos^2\theta = 1$, we see that the quantity in square brackets is unity. Therefore, this equation reduces to

$$E = \tfrac{1}{2}\,kA^2 \qquad \textbf{(13.22)}$$

That is, **the total mechanical energy of a simple harmonic oscillator is a constant of the motion and is proportional to the square of the amplitude.** Note that U is small when K is large, and vice versa, because the sum must be constant. In fact, the total mechanical energy is equal to the maximum potential energy stored in the spring when $x = \pm A$ because $v = 0$ at these points and thus there is no kinetic energy. At the equilibrium position, where $U = 0$ because $x = 0$, the total energy, all in the form of kinetic energy, is again $\tfrac{1}{2}\,kA^2$. That is,

$$E = \tfrac{1}{2}\,mv^2_{\text{max}} = \tfrac{1}{2}\,m\omega^2 A^2 = \tfrac{1}{2}\,m\,\frac{k}{m}\,A^2 = \tfrac{1}{2}\,kA^2 \qquad (\text{at } x = 0)$$

Plots of the kinetic and potential energies versus time appear in Figure 13.9a, where we have taken $\phi = 0$. As already mentioned, both K and U are always positive, and at all times their sum is a constant equal to $\tfrac{1}{2}\,kA^2$, the total energy of the system. The variations of K and U with the displacement x of the block are plotted

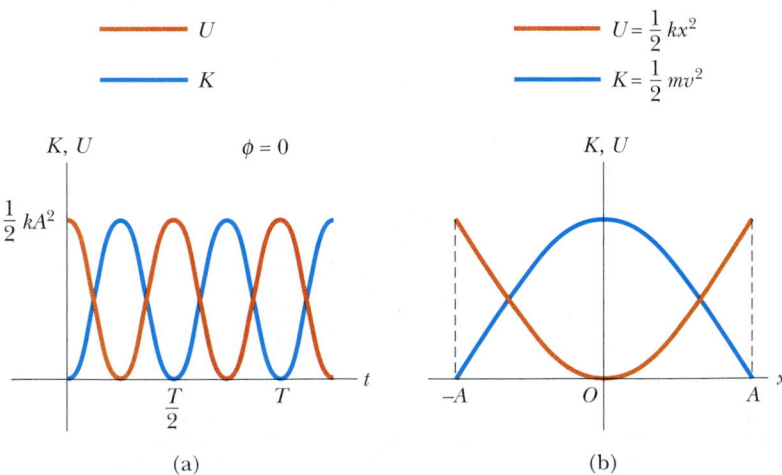

Figure 13.9 (a) Kinetic energy and potential energy versus time for a simple harmonic oscillator with $\phi = 0$. (b) Kinetic energy and potential energy versus displacement for a simple harmonic oscillator. In either plot, note that $K + U = $ constant.

in Figure 13.9b. Energy is continuously being transformed between potential energy stored in the spring and kinetic energy of the block.

Figure 13.10 illustrates the position, velocity, acceleration, kinetic energy, and potential energy of the block–spring system for one full period of the motion. Most of the ideas discussed so far are incorporated in this important figure. Study it carefully.

Finally, we can use the principle of conservation of energy to obtain the velocity for an arbitrary displacement by expressing the total energy at some arbitrary position x as

$$E = K + U = \tfrac{1}{2} mv^2 + \tfrac{1}{2} kx^2 = \tfrac{1}{2} kA^2$$

Velocity as a function of position for a simple harmonic oscillator

$$v = \pm \sqrt{\frac{k}{m} (A^2 - x^2)} = \pm \omega \sqrt{A^2 - x^2} \qquad \textbf{(13.23)}$$

When we check Equation 13.23 to see whether it agrees with known cases, we find that it substantiates the fact that the speed is a maximum at $x = 0$ and is zero at the turning points $x = \pm A$.

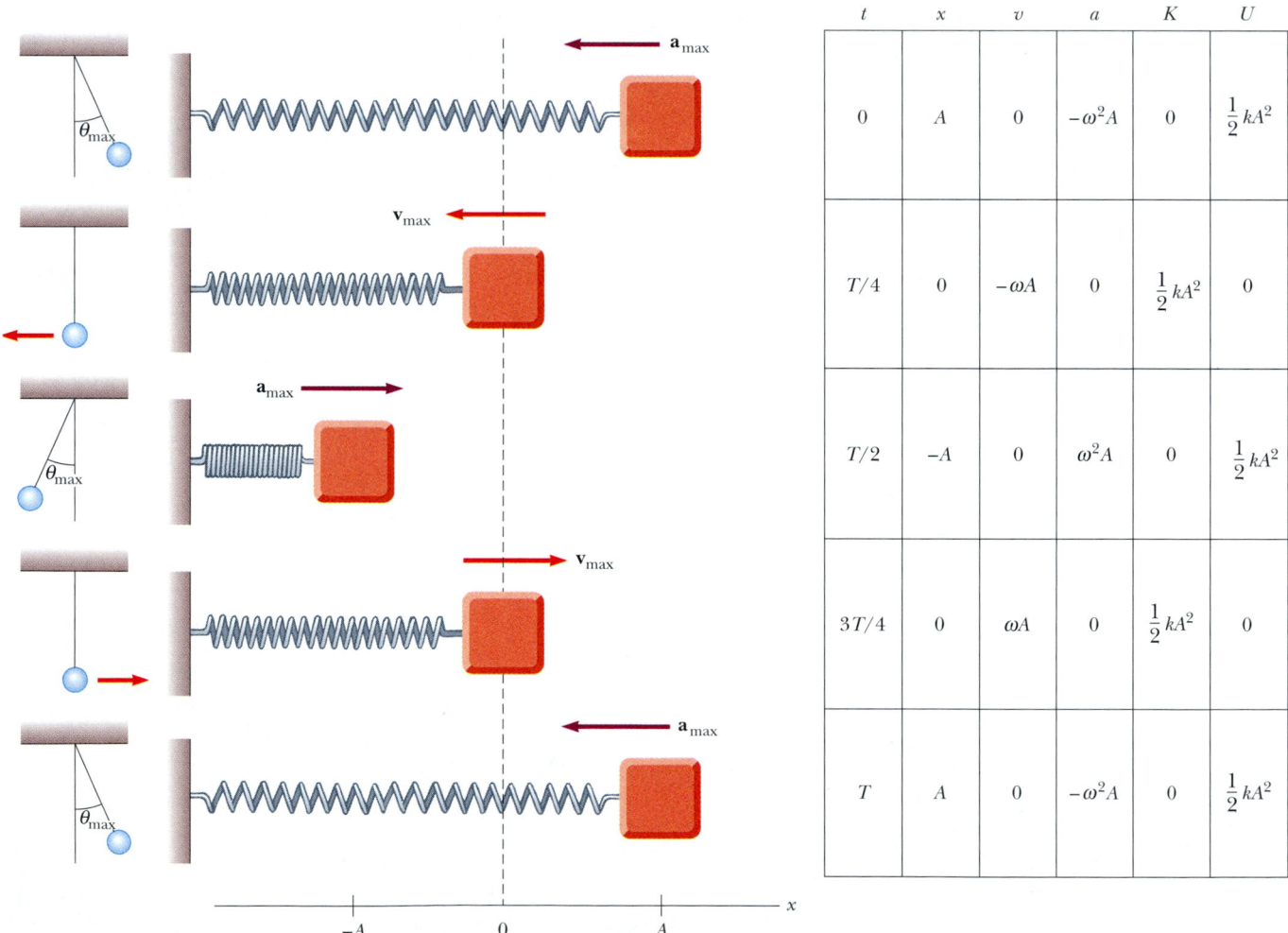

t	x	v	a	K	U
0	A	0	$-\omega^2 A$	0	$\tfrac{1}{2} kA^2$
$T/4$	0	$-\omega A$	0	$\tfrac{1}{2} kA^2$	0
$T/2$	$-A$	0	$\omega^2 A$	0	$\tfrac{1}{2} kA^2$
$3T/4$	0	ωA	0	$\tfrac{1}{2} kA^2$	0
T	A	0	$-\omega^2 A$	0	$\tfrac{1}{2} kA^2$

Figure 13.10 Simple harmonic motion for a block–spring system and its relationship to the motion of a simple pendulum. The parameters in the table refer to the block–spring system, assuming that $x = A$ at $t = 0$; thus, $x = A \cos \omega t$ (see Special Case 1).

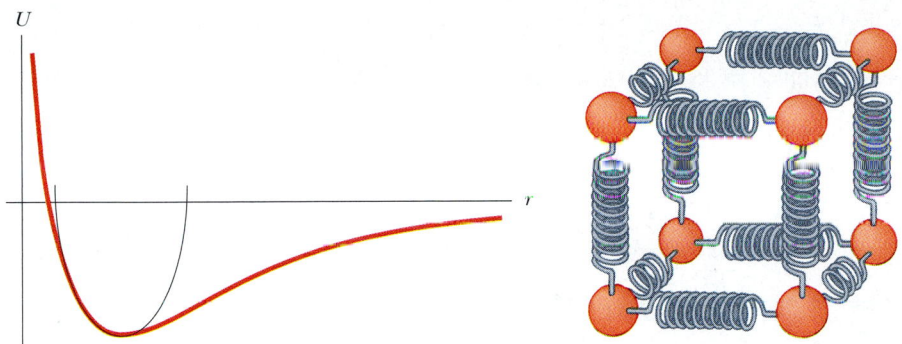

Figure 13.11 (a) If the atoms in a molecule do not move too far from their equilibrium positions, a graph of potential energy versus separation distance between atoms is similar to the graph of potential energy versus position for a simple harmonic oscillator. (b) Tiny springs approximate the forces holding atoms together.

You may wonder why we are spending so much time studying simple harmonic oscillators. We do so because they are good models of a wide variety of physical phenomena. For example, recall the Lennard–Jones potential discussed in Example 8.11. This complicated function describes the forces holding atoms together. Figure 13.11a shows that, for small displacements from the equilibrium position, the potential energy curve for this function approximates a parabola, which represents the potential energy function for a simple harmonic oscillator. Thus, we can approximate the complex atomic binding forces as tiny springs, as depicted in Figure 13.11b.

The ideas presented in this chapter apply not only to block–spring systems and atoms, but also to a wide range of situations that include bungee jumping, tuning in a television station, and viewing the light emitted by a laser. You will see more examples of simple harmonic oscillators as you work through this book.

EXAMPLE 13.4 **Oscillations on a Horizontal Surface**

A 0.500-kg cube connected to a light spring for which the force constant is 20.0 N/m oscillates on a horizontal, frictionless track. (a) Calculate the total energy of the system and the maximum speed of the cube if the amplitude of the motion is 3.00 cm.

Solution Using Equation 13.22, we obtain

$$E = K + U = \tfrac{1}{2} kA^2 = \tfrac{1}{2}(20.0\ \text{N/m})(3.00 \times 10^{-2}\ \text{m})^2$$

$$= \boxed{9.00 \times 10^{-3}\,\text{J}}$$

When the cube is at $x = 0$, we know that $U = 0$ and $E = \tfrac{1}{2} mv_{\text{max}}^2$; therefore,

$$\tfrac{1}{2} mv_{\text{max}}^2 = 9.00 \times 10^{-3}\,\text{J}$$

$$v_{\text{max}} = \sqrt{\frac{18.0 \times 10^{-3}\,\text{J}}{0.500\ \text{kg}}} = \boxed{0.190\ \text{m/s}}$$

(b) What is the velocity of the cube when the displacement is 2.00 cm?

Solution We can apply Equation 13.23 directly:

$$v = \pm\sqrt{\frac{k}{m}(A^2 - x^2)}$$

$$= \pm\sqrt{\frac{20.0\ \text{N/m}}{0.500\ \text{kg}}[(0.030\,0\ \text{m})^2 - (0.020\,0\ \text{m})^2]}$$

$$= \boxed{\pm 0.141\ \text{m/s}}$$

The positive and negative signs indicate that the cube could be moving to either the right or the left at this instant.

(c) Compute the kinetic and potential energies of the system when the displacement is 2.00 cm.

Solution   Using the result of (b), we find that

$$K = \tfrac{1}{2} mv^2 = \tfrac{1}{2} (0.500 \text{ kg})(0.141 \text{ m/s})^2 = \boxed{5.00 \times 10^{-3} \text{ J}}$$

$$U = \tfrac{1}{2} kx^2 = \tfrac{1}{2} (20.0 \text{ N/m})(0.020\,0 \text{ m})^2 = \boxed{4.00 \times 10^{-3} \text{ J}}$$

Note that $K + U = E$.

Exercise   For what values of x is the speed of the cube 0.100 m/s?

Answer ± 2.55 cm.

13.4 ▶ THE PENDULUM

8.11
&
8.12

The **simple pendulum** is another mechanical system that exhibits periodic motion. It consists of a particle-like bob of mass m suspended by a light string of length L that is fixed at the upper end, as shown in Figure 13.12. The motion occurs in the vertical plane and is driven by the force of gravity. We shall show that, provided the angle θ is small (less than about 10°), the motion is that of a simple harmonic oscillator.

The forces acting on the bob are the force **T** exerted by the string and the gravitational force $m\mathbf{g}$. The tangential component of the gravitational force, $mg \sin \theta$, always acts toward $\theta = 0$, opposite the displacement. Therefore, the tangential force is a restoring force, and we can apply Newton's second law for motion in the tangential direction:

$$\sum F_t = -mg \sin \theta = m \frac{d^2 s}{dt^2}$$

where s is the bob's displacement measured along the arc and the minus sign indicates that the tangential force acts toward the equilibrium (vertical) position. Because $s = L\theta$ (Eq. 10.1a) and L is constant, this equation reduces to

$$\frac{d^2 \theta}{dt^2} = -\frac{g}{L} \sin \theta$$

The right side is proportional to $\sin \theta$ rather than to θ; hence, with $\sin \theta$ present, we would not expect simple harmonic motion because this expression is not of the form of Equation 13.17. However, if we assume that θ is small, we can use the approximation $\sin \theta \approx \theta$; thus the equation of motion for the simple pen-

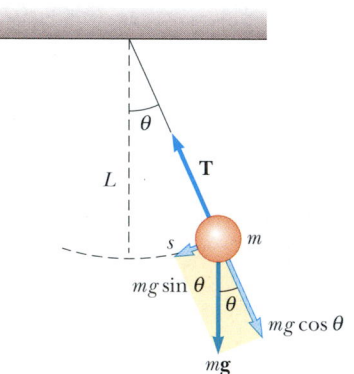

Figure 13.12   When θ is small, a simple pendulum oscillates in simple harmonic motion about the equilibrium position $\theta = 0$. The restoring force is $mg \sin \theta$, the component of the gravitational force tangent to the arc.

 The motion of a simple pendulum, captured with multiflash photography. Is the oscillating motion simple harmonic in this case?   *(Richard Megna/Fundamental Photographs)*

dulum becomes

$$\frac{d^2\theta}{dt^2} = -\frac{g}{L}\theta \tag{13.24}$$

Equation of motion for a simple pendulum (small θ)

Now we have an expression of the same form as Equation 13.17, and we conclude that the motion for small amplitudes of oscillation is simple harmonic motion. Therefore, θ can be written as $\theta = \theta_{max}\cos(\omega t + \phi)$, where θ_{max} is the *maximum angular displacement* and the angular frequency ω is

$$\omega = \sqrt{\frac{g}{L}} \tag{13.25}$$

Angular frequency of motion for a simple pendulum

The Foucault pendulum at the Franklin Institute in Philadelphia. This type of pendulum was first used by the French physicist Jean Foucault to verify the Earth's rotation experimentally. As the pendulum swings, the vertical plane in which it oscillates appears to rotate as the bob successively knocks over the indicators arranged in a circle on the floor. In reality, the plane of oscillation is fixed in space, and the Earth rotating beneath the swinging pendulum moves the indicators into position to be knocked down, one after the other. *(© Bob Emott, Photographer)*

The period of the motion is

Period of motion for a simple pendulum

$$T = \frac{2\pi}{\omega} = 2\pi\sqrt{\frac{L}{g}}$$ **(13.26)**

In other words, **the period and frequency of a simple pendulum depend only on the length of the string and the acceleration due to gravity.** Because the period is independent of the mass, we conclude that all simple pendulums that are of equal length and are at the same location (so that g is constant) oscillate with the same period. The analogy between the motion of a simple pendulum and that of a block–spring system is illustrated in Figure 13.10.

 The simple pendulum can be used as a timekeeper because its period depends only on its length and the local value of g. It is also a convenient device for making precise measurements of the free-fall acceleration. Such measurements are important because variations in local values of g can provide information on the location of oil and of other valuable underground resources.

Quick Quiz 13.5

A block of mass m is first allowed to hang from a spring in static equilibrium. It stretches the spring a distance L beyond the spring's unstressed length. The block and spring are then set into oscillation. Is the period of this system less than, equal to, or greater than the period of a simple pendulum having a length L and a bob mass m?

EXAMPLE 13.5   A Connection Between Length and Time

Christian Huygens (1629–1695), the greatest clockmaker in history, suggested that an international unit of length could be defined as the length of a simple pendulum having a period of exactly 1 s. How much shorter would our length unit be had his suggestion been followed?

Solution   Solving Equation 13.26 for the length gives

$$L = \frac{T^2 g}{4\pi^2} = \frac{(1\ \text{s})^2(9.80\ \text{m/s}^2)}{4\pi^2} = \boxed{0.248\ \text{m}}$$

Thus, the meter's length would be slightly less than one-fourth its current length. Note that the number of significant digits depends only on how precisely we know g because the time has been defined to be exactly 1 s.

Physical Pendulum

Firmly hold a ruler so that about half of it is over the edge of your desk. With your other hand, pull down and then release the free end, watching how it vibrates. Now slide the ruler so that only about a quarter of it is free to vibrate. This time when you release it, how does the vibrational period compare with its earlier value? Why?

Suppose you balance a wire coat hanger so that the hook is supported by your extended index finger. When you give the hanger a small displacement (with your other hand) and then release it, it oscillates. If a hanging object oscillates about a fixed axis that does not pass through its center of mass and the object cannot be approximated as a point mass, we cannot treat the system as a simple pendulum. In this case the system is called a **physical pendulum.**

Consider a rigid body pivoted at a point O that is a distance d from the center of mass (Fig. 13.13). The force of gravity provides a torque about an axis through O, and the magnitude of that torque is $mgd \sin\theta$, where θ is as shown in Figure 13.13. Using the law of motion $\Sigma\tau = I\alpha$, where I is the moment of inertia about

the axis through O, we obtain

$$- mgd \sin \theta = I \frac{d^2\theta}{dt^2}$$

The minus sign indicates that the torque about O tends to decrease θ. That is, the force of gravity produces a restoring torque. Because this equation gives us the angular acceleration $d^2\theta/dt^2$ of the pivoted body, we can consider it the equation of motion for the system. If we again assume that θ is small, the approximation $\sin \theta \approx \theta$ is valid, and the equation of motion reduces to

$$\frac{d^2\theta}{dt^2} = -\left(\frac{mgd}{I}\right)\theta = -\omega^2\theta \qquad \textbf{(13.27)}$$

Because this equation is of the same form as Equation 13.17, the motion is simple harmonic motion. That is, the solution of Equation 13.27 is $\theta = \theta_{max} \cos(\omega t + \phi)$, where θ_{max} is the maximum angular displacement and

$$\omega = \sqrt{\frac{mgd}{I}}$$

The period is

$$T = \frac{2\pi}{\omega} = 2\pi\sqrt{\frac{I}{mgd}} \qquad \textbf{(13.28)}$$

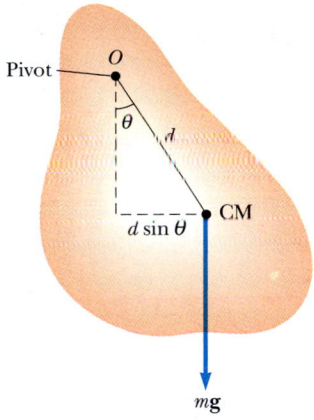

Figure 13.13   A physical pendulum.

Period of motion for a physical pendulum

One can use this result to measure the moment of inertia of a flat rigid body. If the location of the center of mass—and hence the value of d—are known, the moment of inertia can be obtained by measuring the period. Finally, note that Equation 13.28 reduces to the period of a simple pendulum (Eq. 13.26) when $I = md^2$—that is, when all the mass is concentrated at the center of mass.

 EXAMPLE 13.6 ▶ **A Swinging Rod**

A uniform rod of mass M and length L is pivoted about one end and oscillates in a vertical plane (Fig. 13.14). Find the period of oscillation if the amplitude of the motion is small.

Solution   In Chapter 10 we found that the moment of inertia of a uniform rod about an axis through one end is $\frac{1}{3}ML^2$. The distance d from the pivot to the center of mass is $L/2$. Substituting these quantities into Equation 13.28 gives

$$T = 2\pi\sqrt{\frac{\frac{1}{3}ML^2}{Mg\frac{L}{2}}} = 2\pi\sqrt{\frac{2L}{3g}}$$

Comment   In one of the Moon landings, an astronaut walking on the Moon's surface had a belt hanging from his space suit, and the belt oscillated as a physical pendulum. A scientist on the Earth observed this motion on television and used it to estimate the free-fall acceleration on the Moon. How did the scientist make this calculation?

Exercise   Calculate the period of a meter stick that is pivoted about one end and is oscillating in a vertical plane.

Answer   1.64 s.

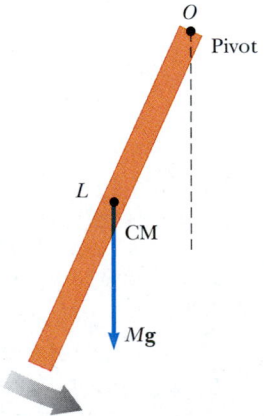

Figure 13.14   A rigid rod oscillating about a pivot through one end is a physical pendulum with $d = L/2$ and, from Table 10.2, $I = \frac{1}{3}ML^2$.

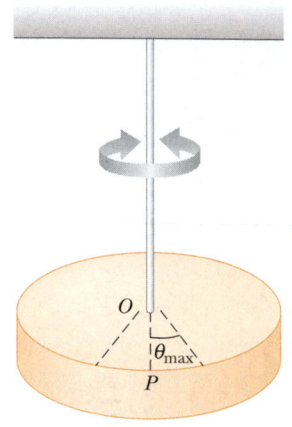

Figure 13.15 A torsional pendulum consists of a rigid body suspended by a wire attached to a rigid support. The body oscillates about the line *OP* with an amplitude θ_{max}.

Figure 13.16 The balance wheel of this antique pocket watch is a torsional pendulum and regulates the time-keeping mechanism. *(George Semple)*

Torsional Pendulum

Figure 13.15 shows a rigid body suspended by a wire attached at the top to a fixed support. When the body is twisted through some small angle θ, the twisted wire exerts on the body a restoring torque that is proportional to the angular displacement. That is,

$$\tau = -\kappa\theta$$

where κ (kappa) is called the *torsion constant* of the support wire. The value of κ can be obtained by applying a known torque to twist the wire through a measurable angle θ. Applying Newton's second law for rotational motion, we find

$$\tau = -\kappa\theta = I\frac{d^2\theta}{dt^2}$$

$$\frac{d^2\theta}{dt^2} = -\frac{\kappa}{I}\theta \qquad \text{(13.29)}$$

Again, this is the equation of motion for a simple harmonic oscillator, with $\omega = \sqrt{\kappa/I}$ and a period

Period of motion for a torsional pendulum

$$T = 2\pi\sqrt{\frac{I}{\kappa}} \qquad \text{(13.30)}$$

This system is called a *torsional pendulum*. There is no small-angle restriction in this situation as long as the elastic limit of the wire is not exceeded. Figure 13.16 shows the balance wheel of a watch oscillating as a torsional pendulum, energized by the mainspring.

 13.5 ▷ **COMPARING SIMPLE HARMONIC MOTION WITH UNIFORM CIRCULAR MOTION**

8.8

We can better understand and visualize many aspects of simple harmonic motion by studying its relationship to uniform circular motion. Figure 13.17 is an overhead view of an experimental arrangement that shows this relationship. A ball is attached to the rim of a turntable of radius *A*, which is illuminated from the side by a lamp. The ball casts a shadow on a screen. We find that as the turntable rotates with constant angular speed, the shadow of the ball moves back and forth in simple harmonic motion.

Consider a particle located at point P on the circumference of a circle of radius A, as shown in Figure 13.18a, with the line OP making an angle ϕ with the x axis at $t = 0$. We call this circle a *reference circle* for comparing simple harmonic motion and uniform circular motion, and we take the position of P at $t = 0$ as our reference position. If the particle moves along the circle with constant angular speed ω until OP makes an angle θ with the x axis, as illustrated in Figure 13.18b, then at some time $t > 0$, the angle between OP and the x axis is $\theta = \omega t + \phi$. As the particle moves along the circle, the projection of P on the x axis, labeled point Q, moves back and forth along the x axis, between the limits $x = \pm A$.

Note that points P and Q always have the same x coordinate. From the right triangle OPQ, we see that this x coordinate is

$$x = A \cos(\omega t + \phi) \tag{13.31}$$

This expression shows that the point Q moves with simple harmonic motion along the x axis. Therefore, we conclude that

simple harmonic motion along a straight line can be represented by the projection of uniform circular motion along a diameter of a reference circle.

We can make a similar argument by noting from Figure 13.18b that the projection of P along the y axis also exhibits simple harmonic motion. Therefore, **uniform circular motion can be considered a combination of two simple harmonic motions,** one along the x axis and one along the y axis, with the two differing in phase by $90°$.

This geometric interpretation shows that the time for one complete revolution of the point P on the reference circle is equal to the period of motion T for simple harmonic motion between $x = \pm A$. That is, the angular speed ω of P is the same as the angular frequency ω of simple harmonic motion along the x axis (this is why we use the same symbol). The phase constant ϕ for simple harmonic motion corresponds to the initial angle that OP makes with the x axis. The radius A of the reference circle equals the amplitude of the simple harmonic motion.

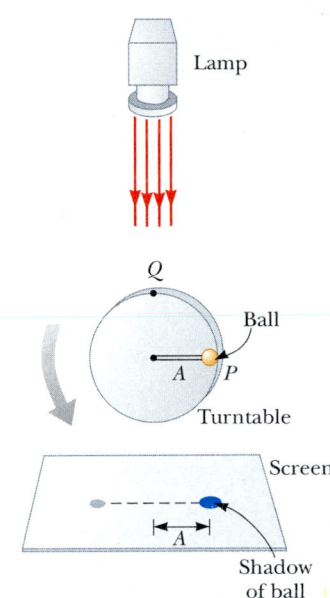

Figure 13.17 An experimental setup for demonstrating the connection between simple harmonic motion and uniform circular motion. As the ball rotates on the turntable with constant angular speed, its shadow on the screen moves back and forth in simple harmonic motion.

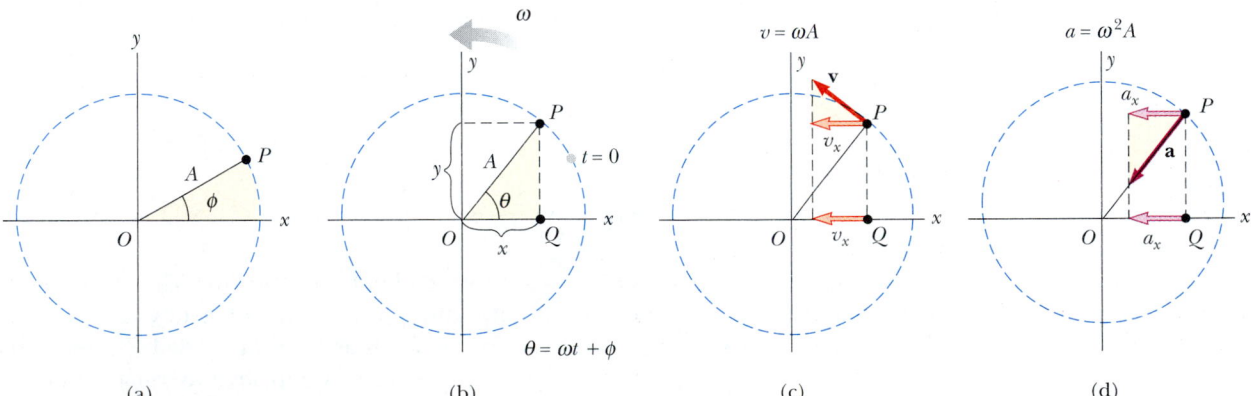

Figure 13.18 Relationship between the uniform circular motion of a point P and the simple harmonic motion of a point Q. A particle at P moves in a circle of radius A with constant angular speed ω. (a) A reference circle showing the position of P at $t = 0$. (b) The x coordinates of points P and Q are equal and vary in time as $x = A \cos(\omega t + \phi)$. (c) The x component of the velocity of P equals the velocity of Q. (d) The x component of the acceleration of P equals the acceleration of Q.

Because the relationship between linear and angular speed for circular motion is $v = r\omega$ (see Eq. 10.10), the particle moving on the reference circle of radius A has a velocity of magnitude ωA. From the geometry in Figure 13.18c, we see that the x component of this velocity is $-\omega A \sin(\omega t + \phi)$. By definition, the point Q has a velocity given by dx/dt. Differentiating Equation 13.31 with respect to time, we find that the velocity of Q is the same as the x component of the velocity of P.

The acceleration of P on the reference circle is directed radially inward toward O and has a magnitude $v^2/A = \omega^2 A$. From the geometry in Figure 13.18d, we see that the x component of this acceleration is $-\omega^2 A \cos(\omega t + \phi)$. This value is also the acceleration of the projected point Q along the x axis, as you can verify by taking the second derivative of Equation 13.31.

EXAMPLE 13.7   Circular Motion with Constant Angular Speed

A particle rotates counterclockwise in a circle of radius 3.00 m with a constant angular speed of 8.00 rad/s. At $t = 0$, the particle has an x coordinate of 2.00 m and is moving to the right. (a) Determine the x coordinate as a function of time.

Solution   Because the amplitude of the particle's motion equals the radius of the circle and $\omega = 8.00$ rad/s, we have

$$x = A \cos(\omega t + \phi) = (3.00 \text{ m}) \cos(8.00t + \phi)$$

We can evaluate ϕ by using the initial condition that $x = 2.00$ m at $t = 0$:

$$2.00 \text{ m} = (3.00 \text{ m}) \cos(0 + \phi)$$

$$\phi = \cos^{-1}\left(\frac{2.00 \text{ m}}{3.00 \text{ m}}\right)$$

If we were to take our answer as $\phi = 48.2°$, then the coordinate $x = (3.00 \text{ m}) \cos(8.00t + 48.2°)$ would be decreasing at time $t = 0$ (that is, moving to the left). Because our particle is first moving to the right, we must choose $\phi = -48.2° = -0.841$ rad. The x coordinate as a function of time is then

$$x = (3.00 \text{ m}) \cos(8.00t - 0.841)$$

Note that ϕ in the cosine function must be in radians.

(b) Find the x components of the particle's velocity and acceleration at any time t.

Solution

$$v_x = \frac{dx}{dt} = (-3.00 \text{ m})(8.00 \text{ rad/s}) \sin(8.00t - 0.841)$$

$$= -(24.0 \text{ m/s}) \sin(8.00t - 0.841)$$

$$a_x = \frac{dv_x}{dt} = (-24.0 \text{ m/s})(8.00 \text{ rad/s}) \cos(8.00t - 0.841)$$

$$= -(192 \text{ m/s}^2) \cos(8.00t - 0.841)$$

From these results, we conclude that $v_{max} = 24.0$ m/s and that $a_{max} = 192$ m/s^2. Note that these values also equal the tangential speed ωA and the centripetal acceleration $\omega^2 A$.

Optional Section

13.6   DAMPED OSCILLATIONS

The oscillatory motions we have considered so far have been for ideal systems—that is, systems that oscillate indefinitely under the action of a linear restoring force. In many real systems, dissipative forces, such as friction, retard the motion. Consequently, the mechanical energy of the system diminishes in time, and the motion is said to be *damped*.

One common type of retarding force is the one discussed in Section 6.4, where the force is proportional to the speed of the moving object and acts in the direction opposite the motion. This retarding force is often observed when an object moves through air, for instance. Because the retarding force can be expressed as $\mathbf{R} = -b\mathbf{v}$ (where b is a constant called the *damping coefficient*) and the restoring force of the system is $-kx$, we can write Newton's second law as

$$\sum F_x = -kx - bv = ma_x$$

$$-kx - b\frac{dx}{dt} = m\frac{d^2x}{dt^2} \qquad \textbf{(13.32)}$$

The solution of this equation requires mathematics that may not be familiar to you yet; we simply state it here without proof. When the retarding force is small compared with the maximum restoring force—that is, when b is small—the solution to Equation 13.32 is

$$x = Ae^{-\frac{b}{2m}t}\cos(\omega t + \phi) \qquad \textbf{(13.33)}$$

where the angular frequency of oscillation is

$$\omega = \sqrt{\frac{k}{m} - \left(\frac{b}{2m}\right)^2} \qquad \textbf{(13.34)}$$

This result can be verified by substituting Equation 13.33 into Equation 13.32.

Figure 13.19a shows the displacement as a function of time for an object oscillating in the presence of a retarding force, and Figure 13.19b depicts one such system: a block attached to a spring and submersed in a viscous liquid. We see that **when the retarding force is much smaller than the restoring force, the oscillatory character of the motion is preserved but the amplitude decreases in time, with the result that the motion ultimately ceases.** Any system that behaves in this way is known as a **damped oscillator.** The dashed blue lines in Figure 13.19a, which define the *envelope* of the oscillatory curve, represent the exponential factor in Equation 13.33. This envelope shows that **the amplitude decays exponentially with time.** For motion with a given spring constant and block mass, the oscillations dampen more rapidly as the maximum value of the retarding force approaches the maximum value of the restoring force.

It is convenient to express the angular frequency of a damped oscillator in the form

$$\omega = \sqrt{\omega_0^2 - \left(\frac{b}{2m}\right)^2}$$

where $\omega_0 = \sqrt{k/m}$ represents the angular frequency in the absence of a retarding force (the undamped oscillator) and is called the **natural frequency** of the system. When the magnitude of the maximum retarding force $R_{\max} = bv_{\max} < kA$, the system is said to be **underdamped.** As the value of R approaches kA, the amplitudes of the oscillations decrease more and more rapidly. This motion is represented by the blue curve in Figure 13.20. When b reaches a critical value b_c such that $b_c/2m = \omega_0$, the system does not oscillate and is said to be **critically damped.** In this case the system, once released from rest at some nonequilibrium position, returns to equilibrium and then stays there. The graph of displacement versus time for this case is the red curve in Figure 13.20.

If the medium is so viscous that the retarding force is greater than the restoring force—that is, if $R_{\max} = bv_{\max} > kA$ and $b/2m > \omega_0$—the system is **overdamped.** Again, the displaced system, when free to move, does not oscillate but simply returns to its equilibrium position. As the damping increases, the time it takes the system to approach equilibrium also increases, as indicated by the black curve in Figure 13.20.

In any case in which friction is present, whether the system is overdamped or underdamped, the energy of the oscillator eventually falls to zero. The lost mechanical energy dissipates into internal energy in the retarding medium.

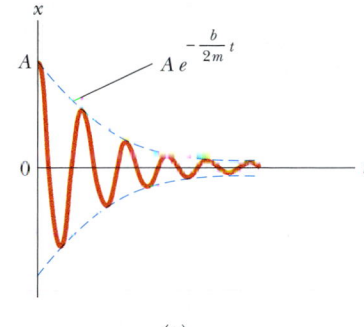

(a)

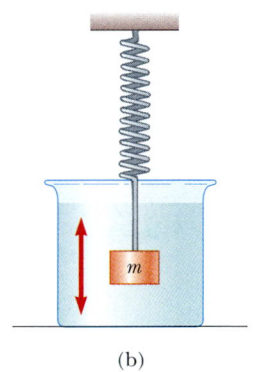

(b)

Figure 13.19 (a) Graph of displacement versus time for a damped oscillator. Note the decrease in amplitude with time. (b) One example of a damped oscillator is a mass attached to a spring and submersed in a viscous liquid.

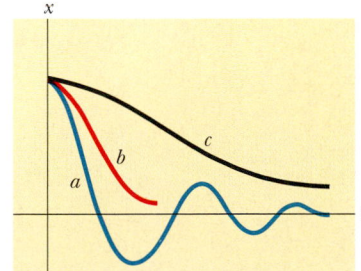

Figure 13.20 Graphs of displacement versus time for (a) an underdamped oscillator, (b) a critically damped oscillator, and (c) an overdamped oscillator.

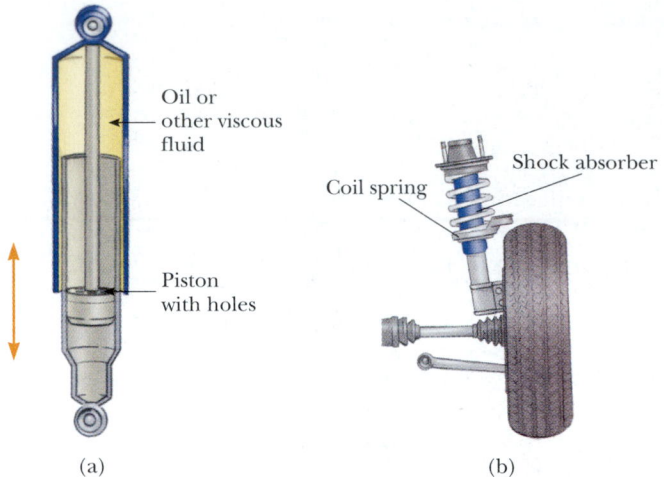

(a)   (b)

Figure 13.21   (a) A shock absorber consists of a piston oscillating in a chamber filled with oil. As the piston oscillates, the oil is squeezed through holes between the piston and the chamber, causing a damping of the piston's oscillations. (b) One type of automotive suspension system, in which a shock absorber is placed inside a coil spring at each wheel.

web

To learn more about shock absorbers, visit **http://www.monroe.com**

Quick Quiz 13.6

An automotive suspension system consists of a combination of springs and shock absorbers, as shown in Figure 13.21. If you were an automotive engineer, would you design a suspension system that was underdamped, critically damped, or overdamped? Discuss each case.

Optional Section

13.7 ▸ FORCED OSCILLATIONS

It is possible to compensate for energy loss in a damped system by applying an external force that does positive work on the system. At any instant, energy can be put into the system by an applied force that acts in the direction of motion of the oscillator. For example, a child on a swing can be kept in motion by appropriately timed pushes. The amplitude of motion remains constant if the energy input per cycle exactly equals the energy lost as a result of damping. Any motion of this type is called **forced oscillation.**

A common example of a forced oscillator is a damped oscillator driven by an external force that varies periodically, such as $F = F_{ext} \cos \omega t$, where ω is the angular frequency of the periodic force and F_{ext} is a constant. Adding this driving force to the left side of Equation 13.32 gives

$$F_{ext} \cos \omega t - kx - b \frac{dx}{dt} = m \frac{d^2x}{dt^2} \qquad \textbf{(13.35)}$$

(As earlier, we present the solution of this equation without proof.) After a sufficiently long period of time, when the energy input per cycle equals the energy lost per cycle, a steady-state condition is reached in which the oscillations proceed with constant amplitude. At this time, when the system is in a steady state, the solution of Equation 13.35 is

$$x = A \cos(\omega t + \phi) \qquad \textbf{(13.36)}$$

where

$$A = \frac{F_{\text{ext}}/m}{\sqrt{(\omega^2 - \omega_0^2)^2 + \left(\dfrac{b\omega}{m}\right)^2}}$$

(13.37)

and where $\omega_0 = \sqrt{k/m}$ is the angular frequency of the undamped oscillator ($b = 0$). One could argue that in steady state the oscillator must physically have the same frequency as the driving force, and thus the solution given by Equation 13.36 is expected. In fact, when this solution is substituted into Equation 13.35, one finds that it is indeed a solution, provided the amplitude is given by Equation 13.37.

Equation 13.37 shows that, because an external force is driving it, the motion of the forced oscillator is not damped. The external agent provides the necessary energy to overcome the losses due to the retarding force. Note that the system oscillates at the angular frequency ω of the driving force. For small damping, the amplitude becomes very large when the frequency of the driving force is near the natural frequency of oscillation. The dramatic increase in amplitude near the natural frequency ω_0 is called **resonance,** and for this reason ω_0 is sometimes called the **resonance frequency** of the system.

The reason for large-amplitude oscillations at the resonance frequency is that energy is being transferred to the system under the most favorable conditions. We can better understand this by taking the first time derivative of x in Equation 13.36, which gives an expression for the velocity of the oscillator. We find that v is proportional to $\sin(\omega t + \phi)$. When the applied force **F** is in phase with the velocity, the rate at which work is done on the oscillator by **F** equals the dot product **F · v**. Remember that "rate at which work is done" is the definition of power. Because the product **F · v** is a maximum when **F** and **v** are in phase, we conclude that **at resonance the applied force is in phase with the velocity and that the power transferred to the oscillator is a maximum.**

Figure 13.22 is a graph of amplitude as a function of frequency for a forced oscillator with and without damping. Note that the amplitude increases with decreasing damping ($b \to 0$) and that the resonance curve broadens as the damping increases. Under steady-state conditions and at any driving frequency, the energy transferred into the system equals the energy lost because of the damping force; hence, the average total energy of the oscillator remains constant. In the absence of a damping force ($b = 0$), we see from Equation 13.37 that the steady-state amplitude approaches infinity as $\omega \to \omega_0$. In other words, if there are no losses in the system and if we continue to drive an initially motionless oscillator with a periodic force that is in phase with the velocity, the amplitude of motion builds without limit (see the red curve in Fig. 13.22). This limitless building does not occur in practice because some damping is always present.

The behavior of a driven oscillating system after the driving force is removed depends on b and on how close ω was to ω_0. This behavior is sometimes quantified by a parameter called the *quality factor* Q. The closer a system is to being undamped, the greater its Q. The amplitude of oscillation drops by a factor of e ($= 2.718 \ldots$) in Q/π cycles.

Later in this book we shall see that resonance appears in other areas of physics. For example, certain electrical circuits have natural frequencies. A bridge has natural frequencies that can be set into resonance by an appropriate driving force. A dramatic example of such resonance occurred in 1940, when the Tacoma Narrows Bridge in the state of Washington was destroyed by resonant vibrations. Although the winds were not particularly strong on that occasion, the bridge ultimately collapsed (Fig. 13.23) because the bridge design had no built-in safety features.

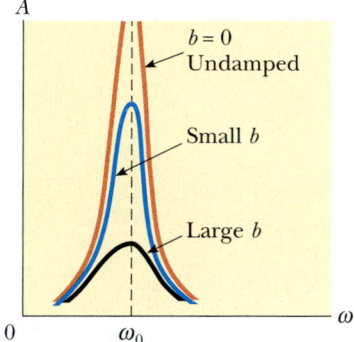

Figure 13.22 Graph of amplitude versus frequency for a damped oscillator when a periodic driving force is present. When the frequency of the driving force equals the natural frequency ω_0, resonance occurs. Note that the shape of the resonance curve depends on the size of the damping coefficient b.

QuickLab

Tie several objects to strings and suspend them from a horizontal string, as illustrated in the figure. Make two of the hanging strings approximately the same length. If one of this pair, such as P, is set into sideways motion, all the others begin to oscillate. But Q, whose length is the same as that of P, oscillates with the greatest amplitude. Must all the masses have the same value?

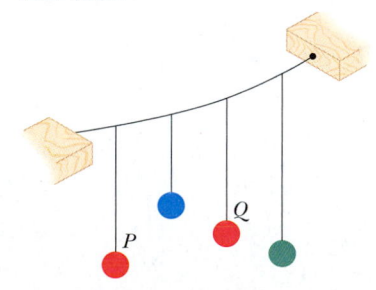

(a)   (b)

Figure 13.23   (a) In 1940 turbulent winds set up torsional vibrations in the Tacoma Narrows Bridge, causing it to oscillate at a frequency near one of the natural frequencies of the bridge structure. (b) Once established, this resonance condition led to the bridge's collapse. *(UPI/Bettmann Newsphotos)*

Many other examples of resonant vibrations can be cited. A resonant vibration that you may have experienced is the "singing" of telephone wires in the wind. Machines often break if one vibrating part is at resonance with some other moving part. Soldiers marching in cadence across a bridge have been known to set up resonant vibrations in the structure and thereby cause it to collapse. Whenever any real physical system is driven near its resonance frequency, you can expect oscillations of very large amplitudes.

SUMMARY

When the acceleration of an object is proportional to its displacement from some equilibrium position and is in the direction opposite the displacement, the object moves with simple harmonic motion. The position x of a simple harmonic oscillator varies periodically in time according to the expression

$$x = A \cos(\omega t + \phi) \tag{13.3}$$

where A is the **amplitude** of the motion, ω is the **angular frequency,** and ϕ is the **phase constant.** The value of ϕ depends on the initial position and initial velocity of the oscillator. You should be able to use this formula to describe the motion of an object undergoing simple harmonic motion.

The time T needed for one complete oscillation is defined as the **period** of the motion:

$$T = \frac{2\pi}{\omega} \tag{13.4}$$

The inverse of the period is the **frequency** of the motion, which equals the number of oscillations per second.

The velocity and acceleration of a simple harmonic oscillator are

$$v = \frac{dx}{dt} = -\omega A \sin(\omega t + \phi) \tag{13.7}$$

$$a = \frac{dv}{dt} = -\omega^2 A \cos(\omega t + \phi) \tag{13.8}$$

$$v = \pm\omega\sqrt{A^2 - x^2} \tag{13.23}$$

Thus, the maximum speed is ωA, and the maximum acceleration is $\omega^2 A$. The speed is zero when the oscillator is at its turning points, $x = \pm A$, and is a maximum when the oscillator is at the equilibrium position $x = 0$. The magnitude of the acceleration is a maximum at the turning points and zero at the equilibrium position. You should be able to find the velocity and acceleration of an oscillating object at any time if you know the amplitude, angular frequency, and phase constant.

A block–spring system moves in simple harmonic motion on a frictionless surface, with a period

$$T = \frac{2\pi}{\omega} = 2\pi\sqrt{\frac{m}{k}} \qquad \textbf{(13.18)}$$

The kinetic energy and potential energy for a simple harmonic oscillator vary with time and are given by

$$K = \tfrac{1}{2}mv^2 = \tfrac{1}{2}m\omega^2 A^2 \sin^2(\omega t + \phi) \qquad \textbf{(13.20)}$$

$$U = \tfrac{1}{2}kx^2 = \tfrac{1}{2}kA^2 \cos^2(\omega t + \phi) \qquad \textbf{(13.21)}$$

These three formulas allow you to analyze a wide variety of situations involving oscillations. Be sure you recognize how the mass of the block and the spring constant of the spring enter into the calculations.

The total energy of a simple harmonic oscillator is a constant of the motion and is given by

$$E = \tfrac{1}{2}kA^2 \qquad \textbf{(13.22)}$$

The potential energy of the oscillator is a maximum when the oscillator is at its turning points and is zero when the oscillator is at the equilibrium position. The kinetic energy is zero at the turning points and a maximum at the equilibrium position. You should be able to determine the division of energy between potential and kinetic forms at any time t.

A **simple pendulum** of length L moves in simple harmonic motion. For small angular displacements from the vertical, its period is

$$T = 2\pi\sqrt{\frac{L}{g}} \qquad \textbf{(13.26)}$$

For small angular displacements from the vertical, a **physical pendulum** moves in simple harmonic motion about a pivot that does not go through the center of mass. The period of this motion is

$$T = 2\pi\sqrt{\frac{I}{mgd}} \qquad \textbf{(13.28)}$$

where I is the moment of inertia about an axis through the pivot and d is the distance from the pivot to the center of mass. You should be able to distinguish when to use the simple-pendulum formula and when the system must be considered a physical pendulum.

Uniform circular motion can be considered a combination of two simple harmonic motions, one along the x axis and the other along the y axis, with the two differing in phase by $90°$.

QUESTIONS

1. Is a bouncing ball an example of simple harmonic motion? Is the daily movement of a student from home to school and back simple harmonic motion? Why or why not?

2. If the coordinate of a particle varies as $x = -A\cos\omega t$, what is the phase constant in Equation 13.3? At what position does the particle begin its motion?

3. Does the displacement of an oscillating particle between $t = 0$ and a later time t necessarily equal the position of the particle at time t? Explain.

4. Determine whether the following quantities can be in the same direction for a simple harmonic oscillator: (a) displacement and velocity, (b) velocity and acceleration, (c) displacement and acceleration.

5. Can the amplitude A and the phase constant ϕ be determined for an oscillator if only the position is specified at $t = 0$? Explain.

6. Describe qualitatively the motion of a mass–spring system when the mass of the spring is not neglected.

7. Make a graph showing the potential energy of a stationary block hanging from a spring, $U = \frac{1}{2}ky^2 + mgy$. Why is the lowest part of the graph offset from the origin?

8. A block–spring system undergoes simple harmonic motion with an amplitude A. Does the total energy change if the mass is doubled but the amplitude is not changed? Do the kinetic and potential energies depend on the mass? Explain.

9. What happens to the period of a simple pendulum if the pendulum's length is doubled? What happens to the period if the mass of the suspended bob is doubled?

10. A simple pendulum is suspended from the ceiling of a stationary elevator, and the period is determined. Describe the changes, if any, in the period when the elevator

(a) accelerates upward, (b) accelerates downward, and (c) moves with constant velocity.

11. A simple pendulum undergoes simple harmonic motion when θ is small. Is the motion periodic when θ is large? How does the period of motion change as θ increases?

12. Will damped oscillations occur for any values of b and k? Explain.

13. As it possible to have damped oscillations when a system is at resonance? Explain.

14. At resonance, what does the phase constant ϕ equal in Equation 13.36? (*Hint:* Compare this equation with the expression for the driving force, which must be in phase with the velocity at resonance.)

15. Some parachutes have holes in them to allow air to move smoothly through them. Without such holes, sometimes the air that has gathered beneath the chute as a parachutist falls is released from under its edges alternately and periodically, at one side and then at the other. Why might this periodic release of air cause a problem?

16. If a grandfather clock were running slowly, how could we adjust the length of the pendulum to correct the time?

17. A pendulum bob is made from a sphere filled with water. What would happen to the frequency of vibration of this pendulum if the sphere had a hole in it that allowed the water to leak out slowly?

PROBLEMS

1, 2, 3 = straightforward, intermediate, challenging   ☐ = full solution available in the *Student Solutions Manual and Study Guide*
WEB = solution posted at **http://www.saunderscollege.com/physics/**   🖥 = Computer useful in solving problem   🕹 = Interactive Physics
☐ = paired numerical/symbolic problems

Section 13.1   Simple Harmonic Motion

1. The displacement of a particle at $t = 0.250$ s is given by the expression $x = (4.00 \text{ m}) \cos(3.00\pi t + \pi)$, where x is in meters and t is in seconds. Determine (a) the frequency and period of the motion, (b) the amplitude of the motion, (c) the phase constant, and (d) the displacement of the particle at $t = 0.250$ s.

2. A ball dropped from a height of 4.00 m makes a perfectly elastic collision with the ground. Assuming that no energy is lost due to air resistance, (a) show that the motion is periodic and (b) determine the period of the motion. (c) Is the motion simple harmonic? Explain.

3. A particle moves in simple harmonic motion with a frequency of 3.00 oscillations/s and an amplitude of 5.00 cm. (a) Through what total distance does the particle move during one cycle of its motion? (b) What is its maximum speed? Where does this occur? (c) Find the maximum acceleration of the particle. Where in the motion does the maximum acceleration occur?

4. In an engine, a piston oscillates with simple harmonic motion so that its displacement varies according to the expression

$$x = (5.00 \text{ cm}) \cos(2t + \pi/6)$$

where x is in centimeters and t is in seconds. At $t = 0$,

find (a) the displacement of the particle, (b) its velocity, and (c) its acceleration. (d) Find the period and amplitude of the motion.

WEB 5. A particle moving along the x axis in simple harmonic motion starts from its equilibrium position, the origin, at $t = 0$ and moves to the right. The amplitude of its motion is 2.00 cm, and the frequency is 1.50 Hz. (a) Show that the displacement of the particle is given by $x = (2.00 \text{ cm}) \sin(3.00\pi t)$. Determine (b) the maximum speed and the earliest time ($t > 0$) at which the particle has this speed, (c) the maximum acceleration and the earliest time ($t > 0$) at which the particle has this acceleration, and (d) the total distance traveled between $t = 0$ and $t = 1.00$ s.

6. The initial position and initial velocity of an object moving in simple harmonic motion are x_i and v_i; the angular frequency of oscillation is ω. (a) Show that the position and velocity of the object for all time can be written as

$$x(t) = x_i \cos \omega t + \left(\frac{v_i}{\omega}\right) \sin \omega t$$

$$v(t) = -x_i\omega \sin \omega t + v_i \cos \omega t$$

(b) If the amplitude of the motion is A, show that

$$v^2 - ax = v_i^2 - a_i x_i = \omega^2 A^2$$

Section 13.2 The Block–Spring System Revisited

Note: Neglect the mass of the spring in all problems in this section.

7. A spring stretches by 3.90 cm when a 10.0-g mass is hung from it. If a 25.0 g mass attached to this spring oscillates in simple harmonic motion, calculate the period of the motion.

8. A simple harmonic oscillator takes 12.0 s to undergo five complete vibrations. Find (a) the period of its motion, (b) the frequency in hertz, and (c) the angular frequency in radians per second.

9. A 0.500-kg mass attached to a spring with a force constant of 8.00 N/m vibrates in simple harmonic motion with an amplitude of 10.0 cm. Calculate (a) the maximum value of its speed and acceleration, (b) the speed and acceleration when the mass is 6.00 cm from the equilibrium position, and (c) the time it takes the mass to move from $x = 0$ to $x = 8.00$ cm.

10. A 1.00-kg mass attached to a spring with a force constant of 25.0 N/m oscillates on a horizontal, frictionless track. At $t = 0$, the mass is released from rest at $x = -3.00$ cm. (That is, the spring is compressed by 3.00 cm.) Find (a) the period of its motion; (b) the maximum values of its speed and acceleration; and (c) the displacement, velocity, and acceleration as functions of time.

11. A 7.00-kg mass is hung from the bottom end of a vertical spring fastened to an overhead beam. The mass is set into vertical oscillations with a period of 2.60 s. Find the force constant of the spring.

12. A block of unknown mass is attached to a spring with a spring constant of 6.50 N/m and undergoes simple harmonic motion with an amplitude of 10.0 cm. When the mass is halfway between its equilibrium position and the end point, its speed is measured to be $+ 30.0$ cm/s. Calculate (a) the mass of the block, (b) the period of the motion, and (c) the maximum acceleration of the block.

13. A particle that hangs from a spring oscillates with an angular frequency of 2.00 rad/s. The spring–particle system is suspended from the ceiling of an elevator car and hangs motionless (relative to the elevator car) as the car descends at a constant speed of 1.50 m/s. The car then stops suddenly. (a) With what amplitude does the particle oscillate? (b) What is the equation of motion for the particle? (Choose upward as the positive direction.)

14. A particle that hangs from a spring oscillates with an angular frequency ω. The spring–particle system is suspended from the ceiling of an elevator car and hangs motionless (relative to the elevator car) as the car descends at a constant speed v. The car then stops suddenly. (a) With what amplitude does the particle oscillate? (b) What is the equation of motion for the particle? (Choose upward as the positive direction.)

15. A 1.00-kg mass is attached to a horizontal spring. The spring is initially stretched by 0.100 m, and the mass is

released from rest there. It proceeds to move without friction. After 0.500 s, the speed of the mass is zero. What is the maximum speed of the mass?

Section 13.3 Energy of the Simple Harmonic Oscillator

Note: Neglect the mass of the spring in all problems in this section.

16. A 200-g mass is attached to a spring and undergoes simple harmonic motion with a period of 0.250 s. If the total energy of the system is 2.00 J, find (a) the force constant of the spring and (b) the amplitude of the motion.

17. An automobile having a mass of 1 000 kg is driven into a brick wall in a safety test. The bumper behaves as a spring of constant 5.00×10^6 N/m and compresses 3.16 cm as the car is brought to rest. What was the speed of the car before impact, assuming that no energy is lost during impact with the wall?

18. A mass–spring system oscillates with an amplitude of 3.50 cm. If the spring constant is 250 N/m and the mass is 0.500 kg, determine (a) the mechanical energy of the system, (b) the maximum speed of the mass, and (c) the maximum acceleration.

19. A 50.0-g mass connected to a spring with a force constant of 35.0 N/m oscillates on a horizontal, frictionless surface with an amplitude of 4.00 cm. Find (a) the total energy of the system and (b) the speed of the mass when the displacement is 1.00 cm. Find (c) the kinetic energy and (d) the potential energy when the displacement is 3.00 cm.

20. A 2.00-kg mass is attached to a spring and placed on a horizontal, smooth surface. A horizontal force of 20.0 N is required to hold the mass at rest when it is pulled 0.200 m from its equilibrium position (the origin of the *x* axis). The mass is now released from rest with an initial displacement of $x_i = 0.200$ m, and it subsequently undergoes simple harmonic oscillations. Find (a) the force constant of the spring, (b) the frequency of the oscillations, and (c) the maximum speed of the mass. Where does this maximum speed occur? (d) Find the maximum acceleration of the mass. Where does it occur? (e) Find the total energy of the oscillating system. Find (f) the speed and (g) the acceleration when the displacement equals one third of the maximum value.

21. A 1.50-kg block at rest on a tabletop is attached to a horizontal spring having force constant of 19.6 N/m. The spring is initially unstretched. A constant 20.0-N horizontal force is applied to the object, causing the spring to stretch. (a) Determine the speed of the block after it has moved 0.300 m from equilibrium, assuming that the surface between the block and the tabletop is frictionless. (b) Answer part (a) for a coefficient of kinetic friction of 0.200 between the block and the tabletop.

22. The amplitude of a system moving in simple harmonic motion is doubled. Determine the change in (a) the total energy, (b) the maximum speed, (c) the maximum acceleration, and (d) the period.

23. A particle executes simple harmonic motion with an amplitude of 3.00 cm. At what displacement from the midpoint of its motion does its speed equal one half of its maximum speed?

24. A mass on a spring with a constant of 3.24 N/m vibrates, with its position given by the equation $x = (5.00 \text{ cm}) \cos(3.60t \text{ rad/s})$. (a) During the first cycle, for $0 < t < 1.75$ s, when is the potential energy of the system changing most rapidly into kinetic energy? (b) What is the maximum rate of energy transformation?

Section 13.4   The Pendulum

25. A man enters a tall tower, needing to know its height. He notes that a long pendulum extends from the ceiling almost to the floor and that its period is 12.0 s. (a) How tall is the tower? (b) If this pendulum is taken to the Moon, where the free-fall acceleration is 1.67 m/s^2, what is its period there?

26. A "seconds" pendulum is one that moves through its equilibrium position once each second. (The period of the pendulum is 2.000 s.) The length of a seconds pendulum is 0.992 7 m at Tokyo and 0.994 2 m at Cambridge, England. What is the ratio of the free-fall accelerations at these two locations?

27. A rigid steel frame above a street intersection supports standard traffic lights, each of which is hinged to hang immediately below the frame. A gust of wind sets a light swinging in a vertical plane. Find the order of magnitude of its period. State the quantities you take as data and their values.

28. The angular displacement of a pendulum is represented by the equation $\theta = (0.320 \text{ rad}) \cos \omega t$, where θ is in radians and $\omega = 4.43 \text{ rad/s}$. Determine the period and length of the pendulum.

WEB 29. A simple pendulum has a mass of 0.250 kg and a length of 1.00 m. It is displaced through an angle of 15.0° and then released. What are (a) the maximum speed, (b) the maximum angular acceleration, and (c) the maximum restoring force?

30. A simple pendulum is 5.00 m long. (a) What is the period of simple harmonic motion for this pendulum if it is hanging in an elevator that is accelerating upward at 5.00 m/s^2? (b) What is its period if the elevator is accelerating downward at 5.00 m/s^2? (c) What is the period of simple harmonic motion for this pendulum if it is placed in a truck that is accelerating horizontally at 5.00 m/s^2?

31. A particle of mass m slides without friction inside a hemispherical bowl of radius R. Show that, if it starts from rest with a small displacement from equilibrium, the particle moves in simple harmonic motion with an angular frequency equal to that of a simple pendulum of length R. That is, $\omega = \sqrt{g/R}$.

32. A mass is attached to the end of a string to form a simple pendulum. The period of its harmonic motion is measured for small angular displacements and three lengths; in each case, the motion is clocked with a stopwatch for 50 oscillations. For lengths of 1.000 m, 0.750 m, and 0.500 m, total times of 99.8 s, 86.6 s, and 71.1 s, respectively, are measured for the 50 oscillations. (a) Determine the period of motion for each length. (b) Determine the mean value of g obtained from these three independent measurements, and compare it with the accepted value. (c) Plot T^2 versus L, and obtain a value for g from the slope of your best-fit straight-line graph. Compare this value with that obtained in part (b).

33. A physical pendulum in the form of a planar body moves in simple harmonic motion with a frequency of 0.450 Hz. If the pendulum has a mass of 2.20 kg and the pivot is located 0.350 m from the center of mass, determine the moment of inertia of the pendulum.

34. A very light, rigid rod with a length of 0.500 m extends straight out from one end of a meter stick. The stick is suspended from a pivot at the far end of the rod and is set into oscillation. (a) Determine the period of oscillation. (b) By what percentage does this differ from a 1.00-m-long simple pendulum?

35. Consider the physical pendulum of Figure 13.13. (a) If I_{CM} is its moment of inertia about an axis that passes through its center of mass and is parallel to the axis that passes through its pivot point, show that its period is

$$T = 2\pi \sqrt{\frac{I_{\text{CM}} + md^2}{mgd}}$$

where d is the distance between the pivot point and the center of mass. (b) Show that the period has a minimum value when d satisfies $md^2 = I_{\text{CM}}$.

36. A torsional pendulum is formed by attaching a wire to the center of a meter stick with a mass of 2.00 kg. If the resulting period is 3.00 min, what is the torsion constant for the wire?

37. A clock balance wheel has a period of oscillation of 0.250 s. The wheel is constructed so that 20.0 g of mass is concentrated around a rim of radius 0.500 cm. What are (a) the wheel's moment of inertia and (b) the torsion constant of the attached spring?

Section 13.5   Comparing Simple Harmonic Motion with Uniform Circular Motion

38. While riding behind a car that is traveling at 3.00 m/s, you notice that one of the car's tires has a small hemispherical boss on its rim, as shown in Figure P13.38. (a) Explain why the boss, from your viewpoint behind the car, executes simple harmonic motion. (b) If the radius of the car's tires is 0.300 m, what is the boss's period of oscillation?

39. Consider the simplified single-piston engine shown in Figure P13.39. If the wheel rotates with constant angular speed, explain why the piston rod oscillates in simple harmonic motion.

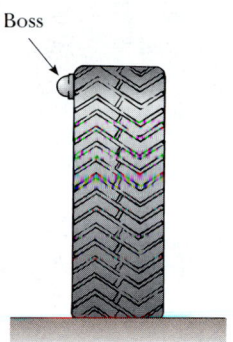

Figure P13.38

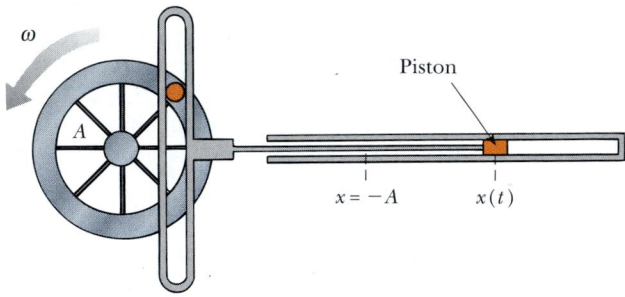

Figure P13.39

(Optional)

Section 13.6   Damped Oscillations

40. Show that the time rate of change of mechanical energy for a damped, undriven oscillator is given by $dE/dt = -bv^2$ and hence is always negative. (*Hint:* Differentiate the expression for the mechanical energy of an oscillator, $E = \frac{1}{2}mv^2 + \frac{1}{2}kx^2$, and use Eq. 13.32.)

41. A pendulum with a length of 1.00 m is released from an initial angle of 15.0°. After 1 000 s, its amplitude is reduced by friction to 5.50°. What is the value of $b/2m$?

42. Show that Equation 13.33 is a solution of Equation 13.32 provided that $b^2 < 4mk$.

(Optional)

Section 13.7   Forced Oscillations

43. A baby rejoices in the day by crowing and jumping up and down in her crib. Her mass is 12.5 kg, and the crib mattress can be modeled as a light spring with a force constant of 4.30 kN/m. (a) The baby soon learns to bounce with maximum amplitude and minimum effort by bending her knees at what frequency? (b) She learns to use the mattress as a trampoline—losing contact with it for part of each cycle—when her amplitude exceeds what value?

44. A 2.00-kg mass attached to a spring is driven by an external force $F = (3.00 \text{ N})\cos(2\pi t)$. If the force constant of the spring is 20.0 N/m, determine (a) the pe-

riod and (b) the amplitude of the motion. (*Hint:* Assume that there is no damping—that is, that $b = 0$—and use Eq. 13.37.)

45. Considering an *undamped*, forced oscillator ($b = 0$), show that Equation 13.36 is a solution of Equation 13.35, with an amplitude given by Equation 13.37.

46. A weight of 40.0 N is suspended from a spring that has a force constant of 200 N/m. The system is undamped and is subjected to a harmonic force with a frequency of 10.0 Hz, which results in a forced-motion amplitude of 2.00 cm. Determine the maximum value of the force.

47. Damping is negligible for a 0.150-kg mass hanging from a light 6.30-N/m spring. The system is driven by a force oscillating with an amplitude of 1.70 N. At what frequency will the force make the mass vibrate with an amplitude of 0.440 m?

48. You are a research biologist. Before dining at a fine restaurant, you set your pager to vibrate instead of beep, and you place it in the side pocket of your suit coat. The arm of your chair presses the light cloth against your body at one spot. Fabric with a length of 8.21 cm hangs freely below that spot, with the pager at the bottom. A co-worker telephones you. The motion of the vibrating pager makes the hanging part of your coat swing back and forth with remarkably large amplitude. The waiter, maître d', wine steward, and nearby diners notice immediately and fall silent. Your daughter pipes up and says, "Daddy, look! Your cockroaches must have gotten out again!" Find the frequency at which your pager vibrates.

ADDITIONAL PROBLEMS

49. A car with bad shock absorbers bounces up and down with a period of 1.50 s after hitting a bump. The car has a mass of 1 500 kg and is supported by four springs of equal force constant k. Determine the value of k.

50. A large passenger with a mass of 150 kg sits in the middle of the car described in Problem 49. What is the new period of oscillation?

51. A compact mass M is attached to the end of a uniform rod, of equal mass M and length L, that is pivoted at the top (Fig. P13.51). (a) Determine the tensions in the rod

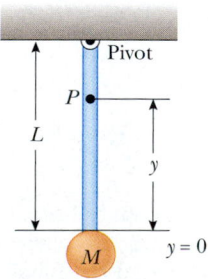

Figure P13.51

at the pivot and at the point P when the system is stationary. (b) Calculate the period of oscillation for small displacements from equilibrium, and determine this period for $L = 2.00$ m. (*Hint:* Assume that the mass at the end of the rod is a point mass, and use Eq. 13.28.)

52. A mass, $m_1 = 9.00$ kg, is in equilibrium while connected to a light spring of constant $k = 100$ N/m that is fastened to a wall, as shown in Figure P13.52a. A second mass, $m_2 = 7.00$ kg, is slowly pushed up against mass m_1, compressing the spring by the amount $A = 0.200$ m (see Fig. P13.52b). The system is then released, and both masses start moving to the right on the frictionless surface. (a) When m_1 reaches the equilibrium point, m_2 loses contact with m_1 (see Fig. P13.52c) and moves to the right with speed v. Determine the value of v. (b) How far apart are the masses when the spring is fully stretched for the first time (D in Fig. P13.52d)? (*Hint:* First determine the period of oscillation and the amplitude of the m_1–spring system after m_2 loses contact with m_1.)

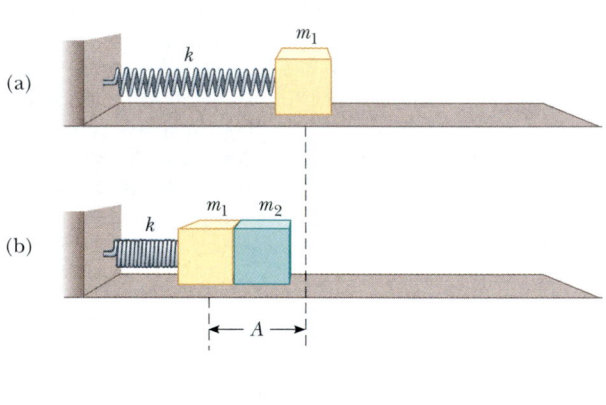

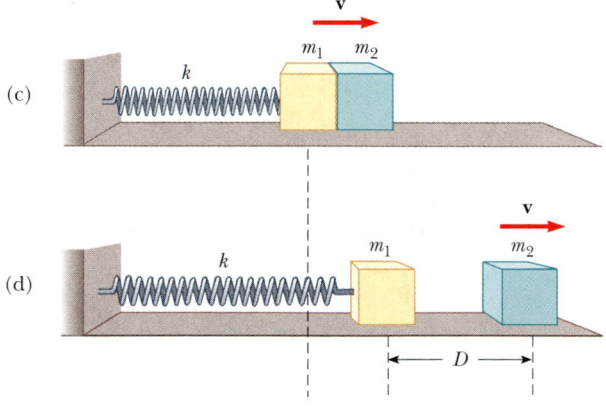

Figure P13.52

WEB **53.** A large block P executes horizontal simple harmonic motion as it slides across a frictionless surface with a frequency of $f = 1.50$ Hz. Block B rests on it, as shown

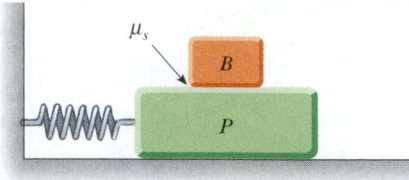

Figure P13.53   Problems 53 and 54.

in Figure P13.53, and the coefficient of static friction between the two is $\mu_s = 0.600$. What maximum amplitude of oscillation can the system have if block B is not to slip?

54. A large block P executes horizontal simple harmonic motion as it slides across a frictionless surface with a frequency f. Block B rests on it, as shown in Figure P13.53, and the coefficient of static friction between the two is μ_s. What maximum amplitude of oscillation can the system have if the upper block is not to slip?

55. The mass of the deuterium molecule (D_2) is twice that of the hydrogen molecule (H_2). If the vibrational frequency of H_2 is 1.30×10^{14} Hz, what is the vibrational frequency of D_2? Assume that the "spring constant" of attracting forces is the same for the two molecules.

56. A solid sphere (radius $= R$) rolls without slipping in a cylindrical trough (radius $= 5R$), as shown in Figure P13.56. Show that, for small displacements from equilibrium perpendicular to the length of the trough, the sphere executes simple harmonic motion with a period $T = 2\pi\sqrt{28R/5g}$.

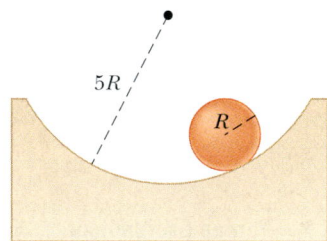

Figure P13.56

57. A light cubical container of volume a^3 is initially filled with a liquid of mass density ρ. The container is initially supported by a light string to form a pendulum of length L_i, measured from the center of mass of the filled container. The liquid is allowed to flow from the bottom of the container at a constant rate (dM/dt). At any time t, the level of the liquid in the container is h

and the length of the pendulum is L (measured relative to the instantaneous center of mass). (a) Sketch the apparatus and label the dimensions a, h, L_i, and L. (b) Find the time rate of change of the period as a function of time t. (c) Find the period as a function of time.

58. After a thrilling plunge, bungee-jumpers bounce freely on the bungee cords through many cycles. Your little brother can make a pest of himself by figuring out the mass of each person, using a proportion he set up by solving this problem: A mass m is oscillating freely on a vertical spring with a period T (Fig. P13.58a). An unknown mass m' on the same spring oscillates with a period T'. Determine (a) the spring constant k and (b) the unknown mass m'.

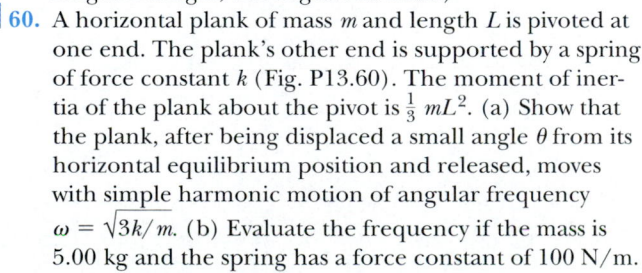

(a)   (b)

Figure P13.58   (a) Mass–spring system for Problems 58 and 68. (b) Bungee-jumping from a bridge. *(Telegraph Colour Library/ FPG International)*

59. A pendulum of length L and mass M has a spring of force constant k connected to it at a distance h below its point of suspension (Fig. P13.59). Find the frequency of vibration of the system for small values of the amplitude (small θ). (Assume that the vertical suspension of length L is rigid, but neglect its mass.)

60. A horizontal plank of mass m and length L is pivoted at one end. The plank's other end is supported by a spring of force constant k (Fig. P13.60). The moment of inertia of the plank about the pivot is $\frac{1}{3}mL^2$. (a) Show that the plank, after being displaced a small angle θ from its horizontal equilibrium position and released, moves with simple harmonic motion of angular frequency $\omega = \sqrt{3k/m}$. (b) Evaluate the frequency if the mass is 5.00 kg and the spring has a force constant of 100 N/m.

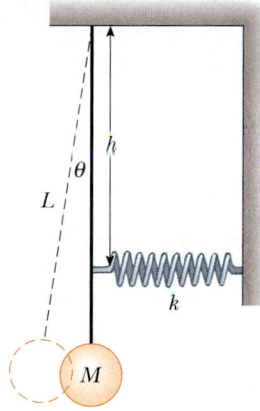

Figure P13.59

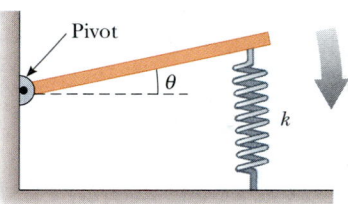

Figure P13.60

61. One end of a light spring with a force constant of 100 N/m is attached to a vertical wall. A light string is tied to the other end of the horizontal spring. The string changes from horizontal to vertical as it passes over a 4.00-cm-diameter solid pulley that is free to turn on a fixed smooth axle. The vertical section of the string supports a 200-g mass. The string does not slip at its contact with the pulley. Find the frequency of oscillation of the mass if the mass of the pulley is (a) negligible, (b) 250 g, and (c) 750 g.

62. A 2.00-kg block hangs without vibrating at the end of a spring ($k = 500$ N/m) that is attached to the ceiling of an elevator car. The car is rising with an upward acceleration of $g/3$ when the acceleration suddenly ceases (at $t = 0$). (a) What is the angular frequency of oscillation of the block after the acceleration ceases? (b) By what amount is the spring stretched during the acceleration of the elevator car? (c) What are the amplitude of the oscillation and the initial phase angle observed by a rider in the car? Take the upward direction to be positive.

63. A simple pendulum with a length of 2.23 m and a mass of 6.74 kg is given an initial speed of 2.06 m/s at its equilibrium position. Assume that it undergoes simple harmonic motion, and determine its (a) period, (b) total energy, and (c) maximum angular displacement.

64. People who ride motorcycles and bicycles learn to look out for bumps in the road and especially for *washboarding*, which is a condition of many equally spaced ridges worn into the road. What is so bad about washboarding? A motorcycle has several springs and shock absorbers in its suspension, but you can model it as a single spring supporting a mass. You can estimate the spring constant by thinking about how far the spring compresses when a big biker sits down on the seat. A motorcyclist traveling at highway speed must be particularly careful of washboard bumps that are a certain distance apart. What is the order of magnitude of their separation distance? State the quantities you take as data and the values you estimate or measure for them.

65. A wire is bent into the shape of one cycle of a cosine curve. It is held in a vertical plane so that the height y of the wire at any horizontal distance x from the center is given by $y = 20.0$ cm$[1 - \cos(0.160x$ rad/m$)]$. A bead can slide without friction on the stationary wire. Show that if its excursion away from $x = 0$ is never large, the bead moves with simple harmonic motion. Determine its angular frequency. (*Hint:* $\cos\theta \cong 1 - \theta^2/2$ for small θ.)

66. A block of mass M is connected to a spring of mass m and oscillates in simple harmonic motion on a horizontal, frictionless track (Fig. P13.66). The force constant of the spring is k, and the equilibrium length is ℓ. Find (a) the kinetic energy of the system when the block has a speed v, and (b) the period of oscillation. (*Hint:* Assume that all portions of the spring oscillate in phase and that the velocity of a segment dx is proportional to the distance x from the fixed end; that is, $v_x = [x/\ell]v$. Also, note that the mass of a segment of the spring is $dm = [m/\ell]dx$.)

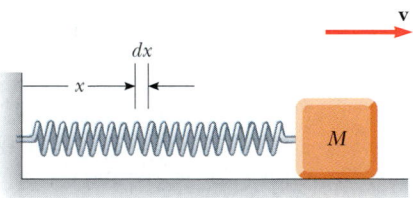

Figure P13.66

WEB 67. A ball of mass m is connected to two rubber bands of length L, each under tension T, as in Figure P13.67. The ball is displaced by a small distance y perpendicular to the length of the rubber bands. Assuming that the tension does not change, show that (a) the restoring force is $-(2T/L)y$ and (b) the system exhibits simple harmonic motion with an angular frequency $\omega = \sqrt{2T/mL}$.

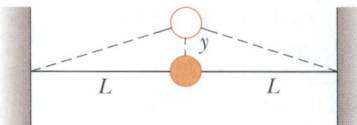

Figure P13.67

68. When a mass M, connected to the end of a spring of mass $m_s = 7.40$ g and force constant k, is set into simple harmonic motion, the period of its motion is

$$T = 2\pi\sqrt{\frac{M + (m_s/3)}{k}}$$

A two-part experiment is conducted with the use of various masses suspended vertically from the spring, as shown in Figure P13.58a. (a) Static extensions of 17.0, 29.3, 35.3, 41.3, 47.1, and 49.3 cm are measured for M values of 20.0, 40.0, 50.0, 60.0, 70.0, and 80.0 g, respectively. Construct a graph of Mg versus x, and perform a linear least-squares fit to the data. From the slope of your graph, determine a value for k for this spring. (b) The system is now set into simple harmonic motion, and periods are measured with a stopwatch. With $M = 80.0$ g, the total time for 10 oscillations is measured to be 13.41 s. The experiment is repeated with M values of 70.0, 60.0, 50.0, 40.0, and 20.0 g, with corresponding times for 10 oscillations of 12.52, 11.67, 10.67, 9.62, and 7.03 s. Compute the experimental value for T for each of these measurements. Plot a graph of T^2 versus M, and determine a value for k from the slope of the linear least-squares fit through the data points. Compare this value of k with that obtained in part (a). (c) Obtain a value for m_s from your graph, and compare it with the given value of 7.40 g.

69. A small, thin disk of radius r and mass m is attached rigidly to the face of a second thin disk of radius R and mass M, as shown in Figure P13.69. The center of the small disk is located at the edge of the large disk. The large disk is mounted at its center on a frictionless axle. The assembly is rotated through a small angle θ from its equilibrium position and released. (a) Show that the

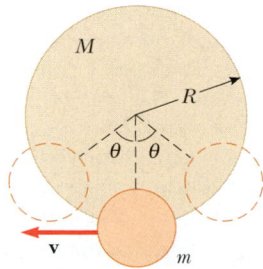

Figure P13.69

speed of the center of the small disk as it passes through the equilibrium position is

$$v = 2\left[\frac{Rg(1 - \cos\theta)}{(M/m) + (r/R)^2 + 2}\right]^{1/2}$$

(b) Show that the period of the motion is

$$T = 2\pi\left[\frac{(M + 2m)R^2 + mr^2}{2mgR}\right]^{1/2}$$

70. Consider the damped oscillator illustrated in Figure 13.19. Assume that the mass is 375 g, the spring constant is 100 N/m, and $b = 0.100$ kg/s. (a) How long does it takes for the amplitude to drop to half its initial value? (b) How long does it take for the mechanical energy to drop to half its initial value? (c) Show that, in general, the fractional rate at which the amplitude decreases in a damped harmonic oscillator is one-half the fractional rate at which the mechanical energy decreases.

71. A mass m is connected to two springs of force constants k_1 and k_2, as shown in Figure P13.71a and b. In each case, the mass moves on a frictionless table and is displaced from equilibrium and then released. Show that in the two cases the mass exhibits simple harmonic motion with periods

(a) $$T = 2\pi\sqrt{\frac{m(k_1 + k_2)}{k_1 k_2}}$$

(b) $$T = 2\pi\sqrt{\frac{m}{k_1 + k_2}}$$

72. Consider a simple pendulum of length $L = 1.20$ m that is displaced from the vertical by an angle $\theta_{\max}$ and then released. You are to predict the subsequent angular displacements when $\theta_{\max}$ is small and also when it is large. Set up and carry out a numerical method to integrate

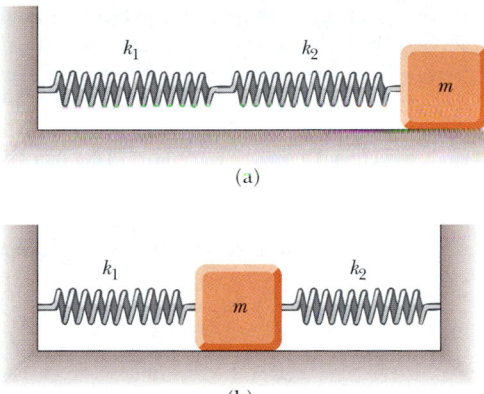

(a)

(b)

Figure P13.71

the equation of motion for the simple pendulum:

$$\frac{d^2\theta}{dt^2} = -\frac{g}{L}\sin\theta$$

Take the initial conditions to be $\theta = \theta_{\max}$ and $d\theta/dt = 0$ at $t = 0$. On one trial choose $\theta_{\max} = 5.00°$, and on another trial take $\theta_{\max} = 100°$. In each case, find the displacement θ as a function of time. Using the same values for $\theta_{\max}$, compare your results for θ with those obtained from $\theta_{\max}\cos\omega t$. How does the period for the large value of $\theta_{\max}$ compare with that for the small value of $\theta_{\max}$? *Note:* Using the Euler method to solve this differential equation, you may find that the amplitude tends to increase with time. The fourth-order Runge–Kutta method would be a better choice to solve the differential equation. However, if you choose Δt small enough, the solution that you obtain using Euler's method can still be good.

ANSWERS TO QUICK QUIZZES

13.1 Because A can never be zero, ϕ must be any value that results in the cosine function's being zero at $t = 0$. In other words, $\phi = \cos^{-1}(0)$. This is true at $\phi = \pi/2$, $3\pi/2$ or, more generally, $\phi = \pm n\pi/2$, where n is any nonzero odd integer. If we want to restrict our choices of ϕ to values between 0 and 2π, we need to know whether the object was moving to the right or to the left at $t = 0$. If it was moving with a positive velocity, then $\phi = 3\pi/2$. If $v_i < 0$, then $\phi = \pi/2$.

13.2 (d) $4A$. From its maximum positive position to the equilibrium position, it travels a distance A, by definition of *amplitude*. It then goes an equal distance past the equilibrium position to its maximum negative position. It then repeats these two motions in the reverse direction to return to its original position and complete one cycle.

13.3 No, because in simple harmonic motion, the acceleration is not constant.

13.4 $x = -A\sin\omega t$, where $A = v_i/\omega$.

13.5 From Hooke's law, the spring constant must be $k = mg/L$. If we substitute this value for k into Equation 13.18, we find that

$$T = 2\pi\sqrt{\frac{m}{k}} = 2\pi\sqrt{\frac{m}{mg/L}} = 2\pi\sqrt{\frac{L}{g}}$$

This is the same as Equation 13.26, which gives the period of a simple pendulum. Thus, when an object stretches a vertically hung spring, the period of the system is the same as that of a simple pendulum having a length equal to the amount of static extension of the spring.

13.6 If your goal is simply to stop the bounce from an absorbed shock as rapidly as possible, you should critically damp the suspension. Unfortunately, the stiffness of this design makes for an uncomfortable ride. If you underdamp the suspension, the ride is more comfortable but the car bounces. If you overdamp the suspension, the wheel is displaced from its equilibrium position longer than it should be. (For example, after hitting a bump, the spring stays compressed for a short time and the wheel does not quickly drop back down into contact with the road after the wheel is past the bump—a dangerous situation.) Because of all these considerations, automotive engineers usually design suspensions to be slightly underdamped. This allows the suspension to absorb a shock rapidly (minimizing the roughness of the ride) and then return to equilibrium after only one or two noticeable oscillations.

PUZZLER

More than 300 years ago, Isaac Newton realized that the same gravitational force that causes apples to fall to the Earth also holds the Moon in its orbit. In recent years, scientists have used the Hubble Space Telescope to collect evidence of the gravitational force acting even farther away, such as at this protoplanetary disk in the constellation Taurus. What properties of an object such as a protoplanet or the Moon determine the strength of its gravitational attraction to another object? *(Left, Larry West/FPG International; right, Courtesy of NASA)*

web

For more information about the Hubble, visit the Space Telescope Science Institute at **http://www.stsci.edu/**

c h a p t e r

The Law of Gravity

423

Before 1687, a large amount of data had been collected on the motions of the Moon and the planets, but a clear understanding of the forces causing these motions was not available. In that year, Isaac Newton provided the key that unlocked the secrets of the heavens. He knew, from his first law, that a net force had to be acting on the Moon because without such a force the Moon would move in a straight-line path rather than in its almost circular orbit. Newton reasoned that this force was the gravitational attraction exerted by the Earth on the Moon. He realized that the forces involved in the Earth–Moon attraction and in the Sun–planet attraction were not something special to those systems, but rather were particular cases of a general and universal attraction between objects. In other words, Newton saw that the same force of attraction that causes the Moon to follow its path around the Earth also causes an apple to fall from a tree. As he put it, "I deduced that the forces which keep the planets in their orbs must be reciprocally as the squares of their distances from the centers about which they revolve; and thereby compared the force requisite to keep the Moon in her orb with the force of gravity at the surface of the Earth; and found them answer pretty nearly."

In this chapter we study the law of gravity. We place emphasis on describing the motion of the planets because astronomical data provide an important test of the validity of the law of gravity. We show that the laws of planetary motion developed by Johannes Kepler follow from the law of gravity and the concept of conservation of angular momentum. We then derive a general expression for gravitational potential energy and examine the energetics of planetary and satellite motion. We close by showing how the law of gravity is also used to determine the force between a particle and an extended object.

14.1 NEWTON'S LAW OF UNIVERSAL GRAVITATION

You may have heard the legend that Newton was struck on the head by a falling apple while napping under a tree. This alleged accident supposedly prompted him to imagine that perhaps all bodies in the Universe were attracted to each other in the same way the apple was attracted to the Earth. Newton analyzed astronomical data on the motion of the Moon around the Earth. From that analysis, he made the bold assertion that the force law governing the motion of planets was the *same* as the force law that attracted a falling apple to the Earth. This was the first time that "earthly" and "heavenly" motions were unified. We shall look at the mathematical details of Newton's analysis in Section 14.5.

In 1687 Newton published his work on the law of gravity in his treatise *Mathematical Principles of Natural Philosophy*. **Newton's law of universal gravitation** states that

 every particle in the Universe attracts every other particle with a force that is directly proportional to the product of their masses and inversely proportional to the square of the distance between them.

If the particles have masses m_1 and m_2 and are separated by a distance r, the magnitude of this gravitational force is

The law of gravity

$$F_g = G \frac{m_1 m_2}{r^2}$$

(14.1)

where G is a constant, called the *universal gravitational constant,* that has been measured experimentally. As noted in Example 6.6, its value in SI units is

$$G = 6.673 \times 10^{-11} \, \text{N} \cdot \text{m}^2/\text{kg}^2 \tag{14.2}$$

The form of the force law given by Equation 14.1 is often referred to as an **inverse square law** because the magnitude of the force varies as the inverse square of the separation of the particles.[1] We shall see other examples of this type of force law in subsequent chapters. We can express this force in vector form by defining a unit vector $\hat{\mathbf{r}}_{12}$ (Fig. 14.1). Because this unit vector is directed from particle 1 to particle 2, the force exerted by particle 1 on particle 2 is

$$\mathbf{F}_{12} = -G\frac{m_1 m_2}{r^2}\hat{\mathbf{r}}_{12} \tag{14.3}$$

where the minus sign indicates that particle 2 is attracted to particle 1, and hence the force must be directed toward particle 1. By Newton's third law, the force exerted by particle 2 on particle 1, designated $\mathbf{F}_{21}$, is equal in magnitude to $\mathbf{F}_{12}$ and in the opposite direction. That is, these forces form an action–reaction pair, and $\mathbf{F}_{21} = -\mathbf{F}_{12}$.

Several features of Equation 14.3 deserve mention. The gravitational force is a field force that always exists between two particles, regardless of the medium that separates them. Because the force varies as the inverse square of the distance between the particles, it decreases rapidly with increasing separation. We can relate this fact to the geometry of the situation by noting that the intensity of light emanating from a point source drops off in the same $1/r^2$ manner, as shown in Figure 14.2.

Another important point about Equation 14.3 is that **the gravitational force exerted by a finite-size, spherically symmetric mass distribution on a particle outside the distribution is the same as if the entire mass of the distribution were concentrated at the center.** For example, the force exerted by the

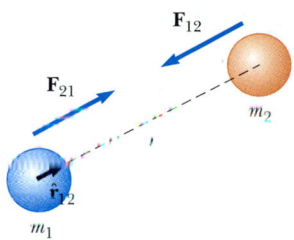

Figure 14.1 The gravitational force between two particles is attractive. The unit vector $\hat{\mathbf{r}}_{12}$ is directed from particle 1 to particle 2. Note that $\mathbf{F}_{21} = -\mathbf{F}_{12}$.

Properties of the gravitational force

QuickLab

Inflate a balloon just enough to form a small sphere. Measure its diameter. Use a marker to color in a 1-cm square on its surface. Now continue inflating the balloon until it reaches twice the original diameter. Measure the size of the square you have drawn. Also note how the color of the marked area has changed. Have you verified what is shown in Figure 14.2?

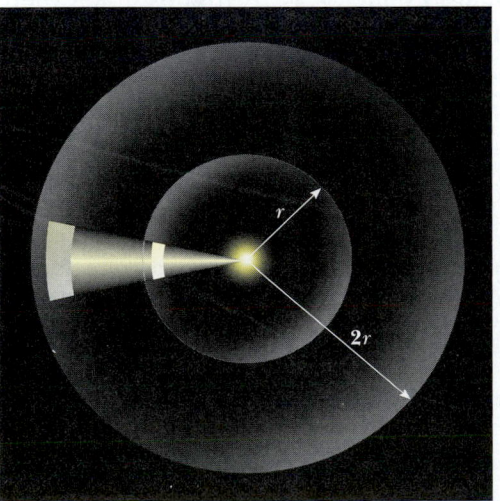

Figure 14.2 Light radiating from a point source drops off as $1/r^2$, a relationship that matches the way the gravitational force depends on distance. When the distance from the light source is doubled, the light has to cover four times the area and thus is one fourth as bright.

[1] An inverse relationship between two quantities x and y is one in which $y = k/x$, where k is a constant. A direct proportion between x and y exists when $y = kx$.

Earth on a particle of mass m near the Earth's surface has the magnitude

$$F_g = G\frac{M_E m}{R_E^2} \qquad \textbf{(14.4)}$$

where M_E is the Earth's mass and R_E its radius. This force is directed toward the center of the Earth.

We have evidence of the fact that the gravitational force acting on an object is directly proportional to its mass from our observations of falling objects, discussed in Chapter 2. All objects, regardless of mass, fall in the absence of air resistance at the same acceleration g near the surface of the Earth. According to Newton's second law, this acceleration is given by $g = F_g/m$, where m is the mass of the falling object. If this ratio is to be the same for all falling objects, then F_g must be directly proportional to m, so that the mass cancels in the ratio. If we consider the more general situation of a gravitational force between any two objects with mass, such as two planets, this same argument can be applied to show that the gravitational force is proportional to one of the masses. We can choose *either* of the masses in the argument, however; thus, the gravitational force must be directly proportional to *both* masses, as can be seen in Equation 14.3.

14.2 MEASURING THE GRAVITATIONAL CONSTANT

The universal gravitational constant G was measured in an important experiment by Henry Cavendish (1731–1810) in 1798. The Cavendish apparatus consists of two small spheres, each of mass m, fixed to the ends of a light horizontal rod suspended by a fine fiber or thin metal wire, as illustrated in Figure 14.3. When two large spheres, each of mass M, are placed near the smaller ones, the attractive force between smaller and larger spheres causes the rod to rotate and twist the wire suspension to a new equilibrium orientation. The angle of rotation is measured by the deflection of a light beam reflected from a mirror attached to the vertical suspension. The deflection of the light is an effective technique for amplifying the motion. The experiment is carefully repeated with different masses at various separations. In addition to providing a value for G, the results show experimentally that the force is attractive, proportional to the product mM, and inversely proportional to the square of the distance r.

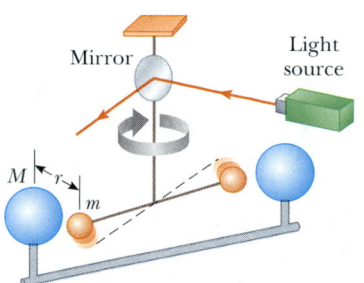

Figure 14.3 Schematic diagram of the Cavendish apparatus for measuring G. As the small spheres of mass m are attracted to the large spheres of mass M, the rod between the two small spheres rotates through a small angle. A light beam reflected from a mirror on the rotating apparatus measures the angle of rotation. The dashed line represents the original position of the rod.

EXAMPLE 14.1   Billiards, Anyone?

Three 0.300-kg billiard balls are placed on a table at the corners of a right triangle, as shown in Figure 14.4. Calculate the gravitational force on the cue ball (designated m_1) resulting from the other two balls.

Solution   First we calculate separately the individual forces on the cue ball due to the other two balls, and then we find the vector sum to get the resultant force. We can see graphically that this force should point upward and toward the right. We locate our coordinate axes as shown in Figure 14.4, placing our origin at the position of the cue ball.

The force exerted by m_2 on the cue ball is directed upward and is given by

$$\mathbf{F}_{21} = G\frac{m_2 m_1}{r_{21}^2}\mathbf{j}$$

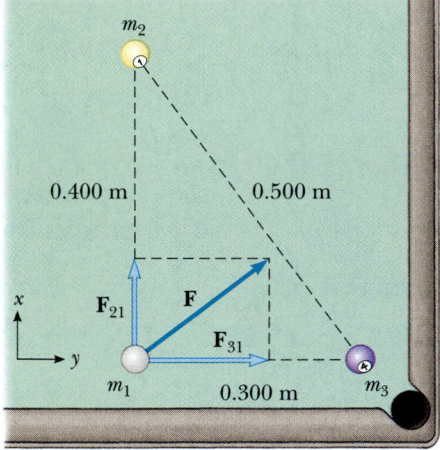

Figure 14.4 The resultant gravitational force acting on the cue ball is the vector sum $\mathbf{F}_{21} + \mathbf{F}_{31}$.

$$= \left(6.67 \times 10^{-11}\ \frac{\mathrm{N \cdot m^2}}{\mathrm{kg^2}}\right)\frac{(0.300\ \mathrm{kg})(0.300\ \mathrm{kg})}{(0.400\ \mathrm{m})^2}\mathbf{j}$$

$$= 3.75 \times 10^{-11}\mathbf{j}\ \mathrm{N}$$

This result shows that the gravitational forces between everyday objects have extremely small magnitudes. The force exerted by m_3 on the cue ball is directed to the right:

$$\mathbf{F}_{31} = G\frac{m_3 m_1}{r_{31}^2}\mathbf{i}$$

$$= \left(6.67 \times 10^{-11}\ \frac{\mathrm{N \cdot m^2}}{\mathrm{kg^2}}\right)\frac{(0.300\ \mathrm{kg})(0.300\ \mathrm{kg})}{(0.300\ \mathrm{m})^2}\mathbf{i}$$

$$= 6.67 \times 10^{-11}\mathbf{i}\ \mathrm{N}$$

Therefore, the resultant force on the cue ball is

$$\mathbf{F} = \mathbf{F}_{21} + \mathbf{F}_{31} = \boxed{(3.75\mathbf{j} + 6.67\mathbf{i}) \times 10^{-11}\ \mathrm{N}}$$

and the magnitude of this force is

$$F = \sqrt{F_{21}^2 + F_{31}^2} = \sqrt{(3.75)^2 + (6.67)^2} \times 10^{-11}$$

$$= 7.65 \times 10^{-11}\ \mathrm{N}$$

Exercise Find the direction of **F**.

Answer 29.3° counterclockwise from the positive *x* axis.

14.3 FREE-FALL ACCELERATION AND THE GRAVITATIONAL FORCE

In Chapter 5, when defining mg as the weight of an object of mass m, we referred to g as the magnitude of the free-fall acceleration. Now we are in a position to obtain a more fundamental description of g. Because the force acting on a freely falling object of mass m near the Earth's surface is given by Equation 14.4, we can equate mg to this force to obtain

$$mg = G\frac{M_E m}{R_E^2}$$

$$g = G\frac{M_E}{R_E^2} \tag{14.5}$$

Free-fall acceleration near the Earth's surface

Now consider an object of mass m located a distance h above the Earth's surface or a distance r from the Earth's center, where $r = R_E + h$. The magnitude of the gravitational force acting on this object is

$$F_g = G\frac{M_E m}{r^2} = G\frac{M_E m}{(R_E + h)^2}$$

The gravitational force acting on the object at this position is also $F_g = mg'$, where g' is the value of the free-fall acceleration at the altitude h. Substituting this expression for F_g into the last equation shows that g' is

Variation of g with altitude

$$g' = \frac{GM_E}{r^2} = \frac{GM_E}{(R_E + h)^2}$$ (14.6)

Thus, it follows that g' *decreases* with *increasing altitude*. Because the weight of a body is mg', we see that as $r \to \infty$, its weight approaches zero.

EXAMPLE 14.2   **Variation of g with Altitude h**

The International Space Station is designed to operate at an altitude of 350 km. When completed, it will have a weight (measured at the Earth's surface) of 4.22×10^6 N. What is its weight when in orbit?

Solution   Because the station is above the surface of the Earth, we expect its weight in orbit to be less than its weight on Earth, 4.22×10^6 N. Using Equation 14.6 with $h = 350$ km, we obtain

$$g' = \frac{GM_E}{(R_E + h)^2}$$

$$= \frac{(6.67 \times 10^{-11} \text{ N·m}^2/\text{kg}^2)(5.98 \times 10^{24} \text{ kg})}{(6.37 \times 10^6 \text{ m} + 0.350 \times 10^6 \text{ m})^2}$$

$$= 8.83 \text{ m/s}^2$$

Because $g'/g = 8.83/9.80 = 0.901$, we conclude that the weight of the station at an altitude of 350 km is 90.1% of the value at the Earth's surface. So the station's weight in orbit is

$$(0.901)(4.22 \times 10^6 \text{ N}) = \boxed{3.80 \times 10^6 \text{ N}}$$

Values of g' at other altitudes are listed in Table 14.1.

TABLE 14.1   **Free-Fall Acceleration g' at Various Altitudes Above the Earth's Surface**

Altitude h (km)	g' (m/s^2)
1 000	7.33
2 000	5.68
3 000	4.53
4 000	3.70
5 000	3.08
6 000	2.60
7 000	2.23
8 000	1.93
9 000	1.69
10 000	1.49
50 000	0.13
∞	0

web
The official web site for the International Space Station is **www.station.nasa.gov**

EXAMPLE 14.3   **The Density of the Earth**

Using the fact that $g = 9.80$ m/s^2 at the Earth's surface, find the average density of the Earth.

Solution   Using $g = 9.80$ m/s^2 and $R_E = 6.37 \times 10^6$ m, we find from Equation 14.5 that $M_E = 5.96 \times 10^{24}$ kg. From this result, and using the definition of density from Chapter 1, we obtain

$$\rho_E = \frac{M_E}{V_E} = \frac{M_E}{\frac{4}{3}\pi R_E^3} = \frac{5.96 \times 10^{24} \text{ kg}}{\frac{4}{3}\pi(6.37 \times 10^6 \text{ m})^3}$$

$$= \boxed{5.50 \times 10^3 \text{ kg/m}^3}$$

Because this value is about twice the density of most rocks at the Earth's surface, we conclude that the inner core of the Earth has a density much higher than the average value. It is most amazing that the Cavendish experiment, which determines G (and can be done on a tabletop), combined with simple free-fall measurements of g, provides information about the core of the Earth.

Astronauts F. Story Musgrave and Jeffrey A. Hoffman, along with the Hubble Space Telescope and the space shuttle *Endeavor,* are all falling around the Earth. *(Courtesy of NASA)*

14.4 ▶ KEPLER'S LAWS

People have observed the movements of the planets, stars, and other celestial bodies for thousands of years. In early history, scientists regarded the Earth as the center of the Universe. This so-called geocentric model was elaborated and formalized by the Greek astronomer Claudius Ptolemy (c. 100–c. 170) in the second century A.D. and was accepted for the next 1 400 years. In 1543 the Polish astronomer Nicolaus Copernicus (1473–1543) suggested that the Earth and the other planets revolved in circular orbits around the Sun (the heliocentric model).

The Danish astronomer Tycho Brahe (1546–1601) wanted to determine how the heavens were constructed, and thus he developed a program to determine the positions of both stars and planets. It is interesting to note that those observations of the planets and 777 stars visible to the naked eye were carried out with only a large sextant and a compass. (The telescope had not yet been invented.)

The German astronomer Johannes Kepler was Brahe's assistant for a short while before Brahe's death, whereupon he acquired his mentor's astronomical data and spent 16 years trying to deduce a mathematical model for the motion of the planets. Such data are difficult to sort out because the Earth is also in motion around the Sun. After many laborious calculations, Kepler found that Brahe's data on the revolution of Mars around the Sun provided the answer.

Johannes Kepler **German astronomer (1571–1630)** The German astronomer Johannes Kepler is best known for developing the laws of planetary motion based on the careful observations of Tycho Brahe. *(Art Resource)*

For more information about Johannes Kepler, visit our Web site at **www.saunderscollege.com/physics/**

Kepler's analysis first showed that the concept of circular orbits around the Sun had to be abandoned. He eventually discovered that the orbit of Mars could be accurately described by an **ellipse.** Figure 14.5 shows the geometric description of an ellipse. The longest dimension is called the major axis and is of length $2a$, where a is the **semimajor axis.** The shortest dimension is the minor axis, of length $2b$, where b is the **semiminor axis.** On either side of the center is a **focal point,** a distance c from the center, where $a^2 = b^2 + c^2$. The Sun is located at one of the focal points of Mars's orbit. Kepler generalized his analysis to include the motions of all planets. The complete analysis is summarized in three statements known as **Kepler's laws:**

Kepler's laws

1. All planets move in elliptical orbits with the Sun at one focal point.
2. The radius vector drawn from the Sun to a planet sweeps out equal areas in equal time intervals.
3. The square of the orbital period of any planet is proportional to the cube of the semimajor axis of the elliptical orbit.

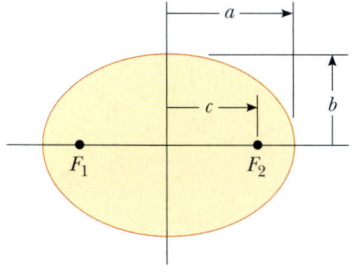

Figure 14.5   Plot of an ellipse. The semimajor axis has a length a, and the semiminor axis has a length b. The focal points are located at a distance c from the center, where $a^2 = b^2 + c^2$.

Most of the planetary orbits are close to circular in shape; for example, the semimajor and semiminor axes of the orbit of Mars differ by only 0.4%. Mercury and Pluto have the most elliptical orbits of the nine planets. In addition to the planets, there are many asteroids and comets orbiting the Sun that obey Kepler's laws. Comet Halley is such an object; it becomes visible when it is close to the Sun every 76 years. Its orbit is very elliptical, with a semiminor axis 76% smaller than its semimajor axis.

Although we do not prove it here, Kepler's first law is a direct consequence of the fact that the gravitational force varies as $1/r^2$. That is, under an inverse-square gravitational-force law, the orbit of a planet can be shown mathematically to be an ellipse with the Sun at one focal point. Indeed, half a century after Kepler developed his laws, Newton demonstrated that these laws are a consequence of the gravitational force that exists between any two masses. Newton's law of universal gravitation, together with his development of the laws of motion, provides the basis for a full mathematical solution to the motion of planets and satellites.

14.5 ▶ THE LAW OF GRAVITY AND THE MOTION OF PLANETS

In formulating his law of gravity, Newton used the following reasoning, which supports the assumption that the gravitational force is proportional to the inverse square of the separation between the two interacting bodies. He compared the acceleration of the Moon in its orbit with the acceleration of an object falling near the Earth's surface, such as the legendary apple (Fig. 14.6). Assuming that both accelerations had the same cause—namely, the gravitational attraction of the Earth—Newton used the inverse-square law to reason that the acceleration of the Moon toward the Earth (centripetal acceleration) should be proportional to $1/r_M^2$, where r_M is the distance between the centers of the Earth and the Moon. Furthermore, the acceleration of the apple toward the Earth should be proportional to $1/R_E^2$, where R_E is the radius of the Earth, or the distance between the centers of the Earth and the apple. Using the values $r_M = 3.84 \times 10^8$ m and

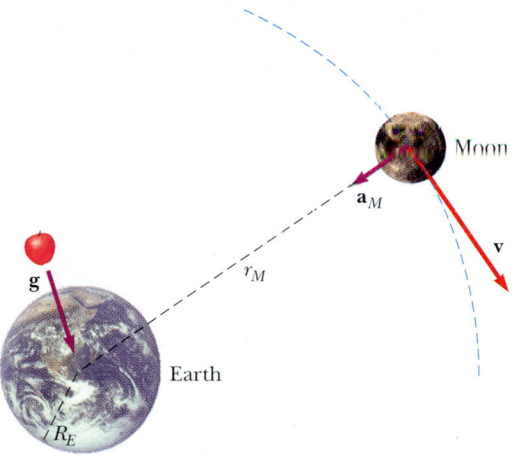

Figure 14.6   As it revolves around the Earth, the Moon experiences a centripetal acceleration $\mathbf{a}_M$ directed toward the Earth. An object near the Earth's surface, such as the apple shown here, experiences an acceleration $\mathbf{g}$. (Dimensions are not to scale.)

$R_E = 6.37 \times 10^6$ m, Newton predicted that the ratio of the Moon's acceleration a_M to the apple's acceleration g would be

$$\frac{a_M}{g} = \frac{(1/r_M)^2}{(1/R_E)^2} = \left(\frac{R_E}{r_M}\right)^2 = \left(\frac{6.37 \times 10^6 \text{ m}}{3.84 \times 10^8 \text{ m}}\right)^2 = 2.75 \times 10^{-4}$$

Therefore, the centripetal acceleration of the Moon is

$$a_M = (2.75 \times 10^{-4})(9.80 \text{ m/s}^2) = 2.70 \times 10^{-3} \text{ m/s}^2$$

Acceleration of the Moon

Newton also calculated the centripetal acceleration of the Moon from a knowledge of its mean distance from the Earth and its orbital period, $T = 27.32$ days $= 2.36 \times 10^6$ s. In a time T, the Moon travels a distance $2\pi r_M$, which equals the circumference of its orbit. Therefore, its orbital speed is $2\pi r_M/T$ and its centripetal acceleration is

$$a_M = \frac{v^2}{r_M} = \frac{(2\pi r_M/T)^2}{r_M} = \frac{4\pi^2 r_M}{T^2} = \frac{4\pi^2(3.84 \times 10^8 \text{ m})}{(2.36 \times 10^6 \text{ s})^2}$$

$$= 2.72 \times 10^{-3} \text{ m/s}^2 \approx \frac{9.80 \text{ m/s}^2}{60^2}$$

In other words, because the Moon is roughly 60 Earth radii away, the gravitational acceleration at that distance should be about $1/60^2$ of its value at the Earth's surface. This is just the acceleration needed to account for the circular motion of the Moon around the Earth. The nearly perfect agreement between this value and the value Newton obtained using g provides strong evidence of the inverse-square nature of the gravitational force law.

Although these results must have been very encouraging to Newton, he was deeply troubled by an assumption he made in the analysis. To evaluate the acceleration of an object at the Earth's surface, Newton treated the Earth as if its mass were all concentrated at its center. That is, he assumed that the Earth acted as a particle as far as its influence on an exterior object was concerned. Several years later, in 1687, on the basis of his pioneering work in the development of calculus, Newton proved that this assumption was valid and was a natural consequence of the law of universal gravitation.

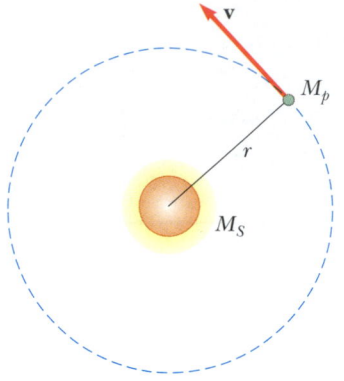

Figure 14.7 A planet of mass M_p moving in a circular orbit around the Sun. The orbits of all planets except Mercury and Pluto are nearly circular.

Kepler's third law

Kepler's Third Law

It is informative to show that Kepler's third law can be predicted from the inverse-square law for circular orbits.[2] Consider a planet of mass M_p moving around the Sun of mass M_S in a circular orbit, as shown in Figure 14.7. Because the gravitational force exerted by the Sun on the planet is a radially directed force that keeps the planet moving in a circle, we can apply Newton's second law ($\Sigma F = ma$) to the planet:

$$\frac{GM_S M_p}{r^2} = \frac{M_p v^2}{r}$$

Because the orbital speed v of the planet is simply $2\pi r/T$, where T is its period of revolution, the preceding expression becomes

$$\frac{GM_S}{r^2} = \frac{(2\pi r/T)^2}{r}$$

$$T^2 = \left(\frac{4\pi^2}{GM_S}\right) r^3 = K_S r^3 \qquad \textbf{(14.7)}$$

where K_S is a constant given by

$$K_S = \frac{4\pi^2}{GM_S} = 2.97 \times 10^{-19} \text{ s}^2/\text{m}^3$$

Equation 14.7 is Kepler's third law. It can be shown that the law is also valid for elliptical orbits if we replace r with the length of the semimajor axis a. Note that the constant of proportionality K_S is independent of the mass of the planet. Therefore, Equation 14.7 is valid for *any* planet.[3] Table 14.2 contains a collection of useful planetary data. The last column verifies that T^2/r^3 is a constant. The small variations in the values in this column reflect uncertainties in the measured values of the periods and semimajor axes of the planets.

If we were to consider the orbit around the Earth of a satellite such as the Moon, then the proportionality constant would have a different value, with the Sun's mass replaced by the Earth's mass.

EXAMPLE 14.4   **The Mass of the Sun**

Calculate the mass of the Sun using the fact that the period of the Earth's orbit around the Sun is 3.156×10^7 s and its distance from the Sun is 1.496×10^{11} m.

Solution   Using Equation 14.7, we find that

$$M_S = \frac{4\pi^2 r^3}{GT^2} = \frac{4\pi^2 (1.496 \times 10^{11} \text{ m})^3}{(6.67 \times 10^{-11} \text{ N}\cdot\text{m}^2/\text{kg}^2)(3.156 \times 10^7 \text{ s})^2}$$

$$= \boxed{1.99 \times 10^{30} \text{ kg}}$$

In Example 14.3, an understanding of gravitational forces enabled us to find out something about the density of the Earth's core, and now we have used this understanding to determine the mass of the Sun.

---

[2] The orbits of all planets except Mercury and Pluto are very close to being circular; hence, we do not introduce much error with this assumption. For example, the ratio of the semiminor axis to the semimajor axis for the Earth's orbit is $b/a = 0.999\ 86$.

[3] Equation 14.7 is indeed a proportion because the ratio of the two quantities T^2 and r^3 is a constant. The variables in a proportion are not required to be limited to the first power only.

TABLE 14.2 Useful Planetary Data

Body	Mass (kg)	Mean Radius (m)	Period of Revolution (s)	Mean Distance from Sun (m)	$\dfrac{T^2}{r^3}$ (s^2/m^3)
Mercury	3.18×10^{23}	2.43×10^6	7.60×10^6	5.79×10^{10}	2.97×10^{-19}
Venus	4.88×10^{24}	6.06×10^6	1.94×10^7	1.08×10^{11}	2.99×10^{-19}
Earth	5.98×10^{24}	6.37×10^6	3.156×10^7	1.496×10^{11}	2.97×10^{-19}
Mars	6.42×10^{23}	3.37×10^6	5.94×10^7	2.28×10^{11}	2.98×10^{-19}
Jupiter	1.90×10^{27}	6.99×10^7	3.74×10^8	7.78×10^{11}	2.97×10^{-19}
Saturn	5.68×10^{26}	5.85×10^7	9.35×10^8	1.43×10^{12}	2.99×10^{-19}
Uranus	8.68×10^{25}	2.33×10^7	2.64×10^9	2.87×10^{12}	2.95×10^{-19}
Neptune	1.03×10^{26}	2.21×10^7	5.22×10^9	4.50×10^{12}	2.99×10^{-19}
Pluto	$\approx 1.4 \times 10^{22}$	$\approx 1.5 \times 10^6$	7.82×10^9	5.91×10^{12}	2.96×10^{-19}
Moon	7.36×10^{22}	1.74×10^6	—	—	—
Sun	1.991×10^{30}	6.96×10^8	—	—	—

Kepler's Second Law and Conservation of Angular Momentum

Consider a planet of mass M_p moving around the Sun in an elliptical orbit (Fig. 14.8). The gravitational force acting on the planet is always along the radius vector, directed toward the Sun, as shown in Figure 14.9a. When a force is directed toward or away from a fixed point and is a function of r only, it is called a **central force.** The torque acting on the planet due to this force is clearly zero; that is, because **F** is parallel to **r**,

$$\boldsymbol{\tau} = \mathbf{r} \times \mathbf{F} = \mathbf{r} \times F\hat{\mathbf{r}} = 0$$

(You may want to revisit Section 11.2 to refresh your memory on the vector product.) Recall from Equation 11.19, however, that torque equals the time rate of change of angular momentum: $\boldsymbol{\tau} = d\mathbf{L}/dt$. Therefore, **because the gravitational**

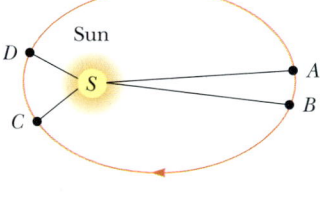

Figure 14.8 Kepler's second law is called the law of equal areas. When the time interval required for a planet to travel from A to B is equal to the time interval required for it to go from C to D, the two areas swept out by the planet's radius vector are equal. Note that in order for this to be true, the planet must be moving faster between C and D than between A and B.

Separate views of Jupiter and of Periodic Comet Shoemaker–Levy 9—both taken with the Hubble Space Telescope about two months before Jupiter and the comet collided in July 1994—were put together with the use of a computer. Their relative sizes and distances were altered. The black spot on Jupiter is the shadow of its moon Io. *(Courtesy of NASA)*

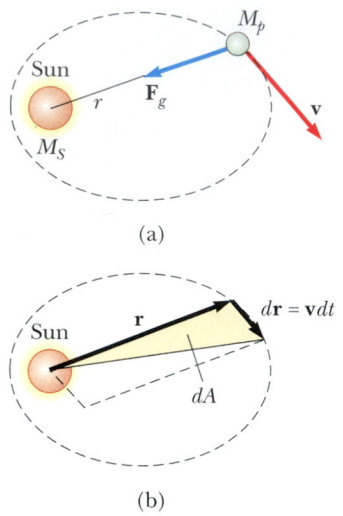

Figure 14.9 (a) The gravitational force acting on a planet is directed toward the Sun, along the radius vector. (b) As a planet orbits the Sun, the area swept out by the radius vector in a time *dt* is equal to one-half the area of the parallelogram formed by the vectors **r** and *d***r** = **v***dt*.

force exerted by the Sun on a planet results in no torque on the planet, the angular momentum **L** of the planet is constant:

$$\mathbf{L} = \mathbf{r} \times \mathbf{p} = \mathbf{r} \times M_p\mathbf{v} = M_p\mathbf{r} \times \mathbf{v} = \text{constant} \qquad (14.8)$$

Because **L** remains constant, the planet's motion at any instant is restricted to the plane formed by **r** and **v**.

We can relate this result to the following geometric consideration. The radius vector **r** in Figure 14.9b sweeps out an area *dA* in a time *dt*. This area equals one-half the area $|\mathbf{r} \times d\mathbf{r}|$ of the parallelogram formed by the vectors **r** and *d***r** (see Section 11.2). Because the displacement of the planet in a time *dt* is *d***r** = **v***dt*, we can say that

$$dA = \tfrac{1}{2}|\mathbf{r} \times d\mathbf{r}| = \tfrac{1}{2}|\mathbf{r} \times \mathbf{v}\,dt| = \frac{L}{2M_p}\,dt$$

$$\boxed{\frac{dA}{dt} = \frac{L}{2M_p} = \text{constant}} \qquad (14.9)$$

where *L* and M_p are both constants. Thus, we conclude that

> the radius vector from the Sun to a planet sweeps out equal areas in equal time intervals.

It is important to recognize that this result, which is Kepler's second law, is a consequence of the fact that the force of gravity is a central force, which in turn implies that angular momentum is constant. Therefore, Kepler's second law applies to *any* situation involving a central force, whether inverse-square or not.

EXAMPLE 14.5   Motion in an Elliptical Orbit

A satellite of mass *m* moves in an elliptical orbit around the Earth (Fig. 14.10). The minimum distance of the satellite from the Earth is called the *perigee* (indicated by *p* in Fig. 14.10), and the maximum distance is called the *apogee* (indicated by *a*). If the speed of the satellite at *p* is v_p, what is its speed at *a*?

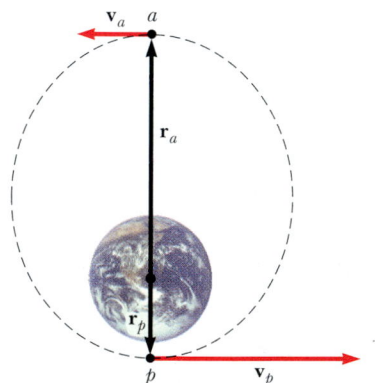

Figure 14.10 As a satellite moves around the Earth in an elliptical orbit, its angular momentum is constant. Therefore, $mv_a r_a = mv_p r_p$, where the subscripts *a* and *p* represent apogee and perigee, respectively.

Solution As the satellite moves from perigee toward apogee, it is moving farther from the Earth. Thus, a component of the gravitational force exerted by the Earth on the satellite is opposite the velocity vector. Negative work is done on the satellite, which causes it to slow down, according to the work–kinetic energy theorem. As a result, we expect the speed at apogee to be lower than the speed at perigee.

The angular momentum of the satellite relative to the Earth is $\mathbf{r} \times m\mathbf{v} = m\mathbf{r} \times \mathbf{v}$. At the points *a* and *p*, **v** is perpendicular to **r**. Therefore, the magnitude of the angular momentum at these positions is $L_a = mv_a r_a$ and $L_p = mv_p r_p$. Because angular momentum is constant, we see that

$$mv_a r_a = mv_p r_p$$

$$v_a = \frac{r_p}{r_a}\,v_p$$

How would you explain the fact that Saturn and Jupiter have periods much greater than one year?

14.6 THE GRAVITATIONAL FIELD

When Newton published his theory of universal gravitation, it was considered a success because it satisfactorily explained the motion of the planets. Since 1687 the same theory has been used to account for the motions of comets, the deflection of a Cavendish balance, the orbits of binary stars, and the rotation of galaxies. Nevertheless, both Newton's contemporaries and his successors found it difficult to accept the concept of a force that acts through a distance, as mentioned in Section 5.1. They asked how it was possible for two objects to interact when they were not in contact with each other. Newton himself could not answer that question.

An approach to describing interactions between objects that are not in contact came well after Newton's death, and it enables us to look at the gravitational interaction in a different way. As described in Section 5.1, this alternative approach uses the concept of a **gravitational field** that exists at every point in space. When a particle of mass m is placed at a point where the gravitational field is $\mathbf{g}$, the particle experiences a force $\mathbf{F}_g = m\mathbf{g}$. In other words, the field exerts a force on the particle. Hence, the gravitational field $\mathbf{g}$ is defined as

$$\mathbf{g} \equiv \frac{\mathbf{F}_g}{m} \tag{14.10}$$

Gravitational field

That is, the gravitational field at a point in space equals the gravitational force experienced by a *test particle* placed at that point divided by the mass of the test particle. Notice that the presence of the test particle is not necessary for the field to exist—the Earth creates the gravitational field. We call the object creating the field the *source particle* (although the Earth is clearly not a particle; we shall discuss shortly the fact that we can approximate the Earth as a particle for the purpose of finding the gravitational field that it creates). We can detect the presence of the field and measure its strength by placing a test particle in the field and noting the force exerted on it.

Although the gravitational force is inherently an interaction between two objects, the concept of a gravitational field allows us to "factor out" the mass of one of the objects. In essence, we are describing the "effect" that any object (in this case, the Earth) has on the empty space around itself in terms of the force that *would* be present *if* a second object were somewhere in that space.[4]

As an example of how the field concept works, consider an object of mass m near the Earth's surface. Because the gravitational force acting on the object has a magnitude $GM_E m/r^2$ (see Eq. 14.4), the field $\mathbf{g}$ at a distance r from the center of the Earth is

$$\mathbf{g} = \frac{\mathbf{F}_g}{m} = -\frac{GM_E}{r^2}\,\hat{\mathbf{r}} \tag{14.11}$$

where $\hat{\mathbf{r}}$ is a unit vector pointing radially outward from the Earth and the minus

---

[4] We shall return to this idea of mass affecting the space around it when we discuss Einstein's theory of gravitation in Chapter 39.

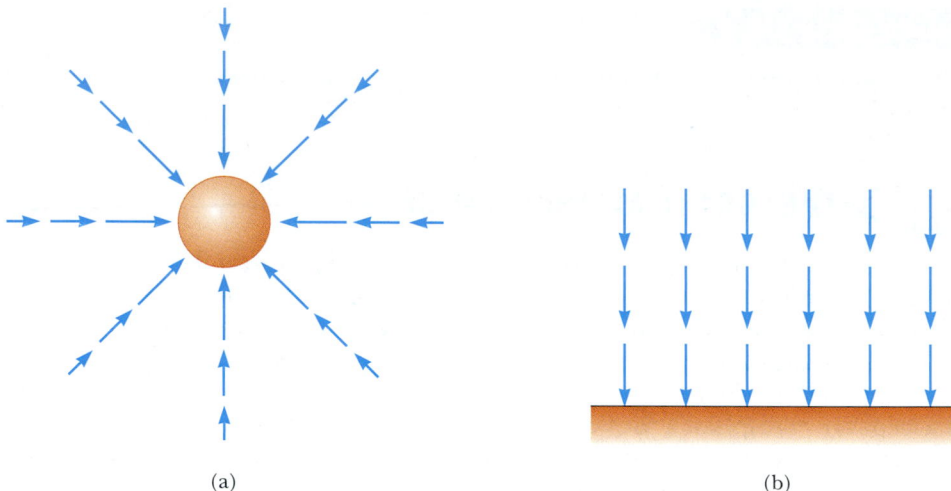

(a)   (b)

Figure 14.11   (a) The gravitational field vectors in the vicinity of a uniform spherical mass such as the Earth vary in both direction and magnitude. The vectors point in the direction of the acceleration a particle would experience if it were placed in the field. The magnitude of the field vector at any location is the magnitude of the free-fall acceleration at that location. (b) The gravitational field vectors in a small region near the Earth's surface are uniform in both direction and magnitude.

sign indicates that the field points toward the center of the Earth, as illustrated in Figure 14.11a. Note that the field vectors at different points surrounding the Earth vary in both direction and magnitude. In a small region near the Earth's surface, the downward field **g** is approximately constant and uniform, as indicated in Figure 14.11b. Equation 14.11 is valid at all points *outside* the Earth's surface, assuming that the Earth is spherical. At the Earth's surface, where $r = R_E$, **g** has a magnitude of 9.80 N/kg.

14.7 ▶ GRAVITATIONAL POTENTIAL ENERGY

In Chapter 8 we introduced the concept of gravitational potential energy, which is the energy associated with the position of a particle. We emphasized that the gravitational potential energy function $U = mgy$ is valid only when the particle is near the Earth's surface, where the gravitational force is constant. Because the gravitational force between two particles varies as $1/r^2$, we expect that a more general potential energy function—one that is valid without the restriction of having to be near the Earth's surface—will be significantly different from $U = mgy$.

Before we calculate this general form for the gravitational potential energy function, let us first verify that *the gravitational force is conservative*. (Recall from Section 8.2 that a force is conservative if the work it does on an object moving between any two points is independent of the path taken by the object.) To do this, we first note that the gravitational force is a central force. By definition, a central force is any force that is directed along a radial line to a fixed center and has a magnitude that depends only on the radial coordinate r. Hence, a central force can be represented by $F(r)\hat{\mathbf{r}}$, where $\hat{\mathbf{r}}$ is a unit vector directed from the origin to the particle, as shown in Figure 14.12.

Consider a central force acting on a particle moving along the general path P to Q in Figure 14.12. The path from P to Q can be approximated by a series of steps according to the following procedure. In Figure 14.12, we draw several thin wedges, which are shown as dashed lines. The outer boundary of our set of wedges is a path consisting of short radial line segments and arcs (gray in the figure). We select the length of the radial dimension of each wedge such that the short arc at the wedge's wide end intersects the actual path of the particle. Then we can approximate the actual path with a series of zigzag movements that alternate between moving along an arc and moving along a radial line.

By definition, a central force is always directed along one of the radial segments; therefore, the work done by **F** along any radial segment is

$$dW = \mathbf{F} \cdot d\mathbf{r} = F(r)\ dr$$

You should recall that, by definition, the work done by a force that is perpendicular to the displacement is zero. Hence, the work done in moving along any arc is zero because **F** is perpendicular to the displacement along these segments. Therefore, the total work done by **F** is the sum of the contributions along the radial segments:

$$W = \int_{r_i}^{r_f} F(r)\ dr$$

where the subscripts i and f refer to the initial and final positions. Because the integrand is a function only of the radial position, this integral depends only on the initial and final values of r. Thus, the work done is the same over *any* path from P to Q. Because the work done is independent of the path and depends only on the end points, we conclude that *any central force is conservative*. We are now assured that a potential energy function can be obtained once the form of the central force is specified.

Recall from Equation 8.2 that the change in the gravitational potential energy associated with a given displacement is defined as the negative of the work done by the gravitational force during that displacement:

$$\Delta U = U_f - U_i = -\int_{r_i}^{r_f} F(r)\ dr \qquad \textbf{(14.12)}$$

We can use this result to evaluate the gravitational potential energy function. Consider a particle of mass m moving between two points P and Q above the Earth's surface (Fig. 14.13). The particle is subject to the gravitational force given by Equation 14.1. We can express this force as

$$F(r) = -\frac{GM_E m}{r^2}$$

where the negative sign indicates that the force is attractive. Substituting this expression for $F(r)$ into Equation 14.12, we can compute the change in the gravita-

Work done by a central force

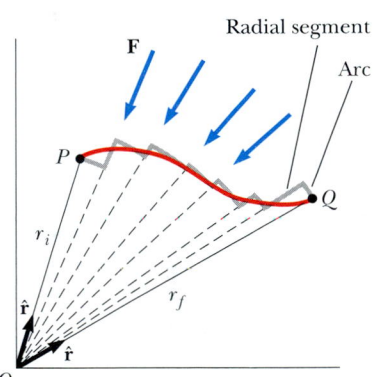

Figure 14.12 A particle moves from P to Q while acted on by a central force **F**, which is directed radially. The path is broken into a series of radial segments and arcs. Because the work done along the arcs is zero, the work done is independent of the path and depends only on r_f and r_i.

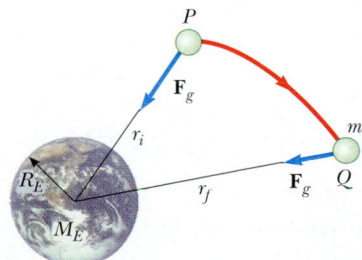

Figure 14.13 As a particle of mass m moves from P to Q above the Earth's surface, the gravitational potential energy changes according to Equation 14.12.

tional potential energy function:

$$U_f - U_i = GM_E m \int_{r_i}^{r_f} \frac{dr}{r^2} = GM_E m \left[-\frac{1}{r} \right]_{r_i}^{r_f}$$

$$U_f - U_i = -GM_E m \left(\frac{1}{r_f} - \frac{1}{r_i} \right) \qquad (14.13)$$

Change in gravitational potential energy

As always, the choice of a reference point for the potential energy is completely arbitrary. It is customary to choose the reference point where the force is zero. Taking $U_i = 0$ at $r_i = \infty$, we obtain the important result

$$U = -\frac{GM_E m}{r} \qquad (14.14)$$

Gravitational potential energy of the Earth–particle system for $r \geq R_E$

This expression applies to the Earth–particle system where the two masses are separated by a distance r, provided that $r \geq R_E$. The result is not valid for particles inside the Earth, where $r < R_E$. (The situation in which $r < R_E$ is treated in Section 14.10.) Because of our choice of U_i, the function U is always negative (Fig. 14.14).

Although Equation 14.14 was derived for the particle–Earth system, it can be applied to any two particles. That is, the gravitational potential energy associated with any pair of particles of masses m_1 and m_2 separated by a distance r is

$$U = -\frac{Gm_1 m_2}{r} \qquad (14.15)$$

This expression shows that the gravitational potential energy for any pair of particles varies as $1/r$, whereas the force between them varies as $1/r^2$. Furthermore, the potential energy is negative because the force is attractive and we have taken the potential energy as zero when the particle separation is infinite. Because the force between the particles is attractive, we know that an external agent must do positive work to increase the separation between them. The work done by the external agent produces an increase in the potential energy as the two particles are separated. That is, U becomes less negative as r increases.

When two particles are at rest and separated by a distance r, an external agent has to supply an energy at least equal to $+ Gm_1 m_2 / r$ in order to separate the particles to an infinite distance. It is therefore convenient to think of the absolute value of the potential energy as the *binding energy* of the system. If the external agent supplies an energy greater than the binding energy, the excess energy of the system will be in the form of kinetic energy when the particles are at an infinite separation.

We can extend this concept to three or more particles. In this case, the total potential energy of the system is the sum over all pairs of particles.[5] Each pair contributes a term of the form given by Equation 14.15. For example, if the system contains three particles, as in Figure 14.15, we find that

$$U_{\text{total}} = U_{12} + U_{13} + U_{23} = -G \left(\frac{m_1 m_2}{r_{12}} + \frac{m_1 m_3}{r_{13}} + \frac{m_2 m_3}{r_{23}} \right) \qquad (14.16)$$

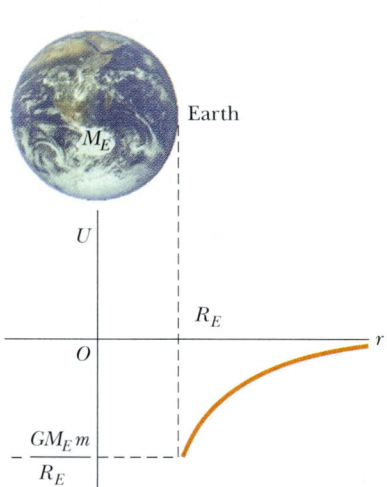

Figure 14.14 Graph of the gravitational potential energy U versus r for a particle above the Earth's surface. The potential energy goes to zero as r approaches infinity.

The absolute value of U_{total} represents the work needed to separate the particles by an infinite distance.

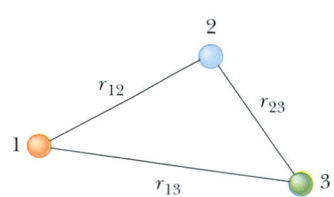

Figure 14.15 Three interacting particles.

[5] The fact that potential energy terms can be added for all pairs of particles stems from the experimental fact that gravitational forces obey the superposition principle.

> **EXAMPLE 14.6** **The Change in Potential Energy**
>
> A particle of mass m is displaced through a small vertical distance Δy near the Earth's surface. Show that in this situation the general expression for the change in gravitational potential energy given by Equation 14.13 reduces to the familiar relationship $\Delta U = mg\,\Delta y$.
>
> **Solution** We can express Equation 14.13 in the form
>
> $$\Delta U = -GM_E m \left(\frac{1}{r_f} - \frac{1}{r_i} \right) = GM_E m \left(\frac{r_f - r_i}{r_i r_f} \right)$$
>
> If both the initial and final positions of the particle are close to the Earth's surface, then $r_f - r_i = \Delta y$ and $r_i r_f \approx R_E^2$. (Recall that r is measured from the center of the Earth.) Therefore, the change in potential energy becomes
>
> $$\Delta U \approx \frac{GM_E m}{R_E^2} \Delta y = mg\,\Delta y$$
>
> where we have used the fact that $g = GM_E/R_E^2$ (Eq. 14.5). Keep in mind that the reference point is arbitrary because it is the *change* in potential energy that is meaningful.

14.8 ENERGY CONSIDERATIONS IN PLANETARY AND SATELLITE MOTION

Consider a body of mass m moving with a speed v in the vicinity of a massive body of mass M, where $M \gg m$. The system might be a planet moving around the Sun, a satellite in orbit around the Earth, or a comet making a one-time flyby of the Sun. If we assume that the body of mass M is at rest in an inertial reference frame, then the total mechanical energy E of the two-body system when the bodies are separated by a distance r is the sum of the kinetic energy of the body of mass m and the potential energy of the system, given by Equation 14.15:[6]

$$E = K + U$$

$$E = \tfrac{1}{2}mv^2 - \frac{GMm}{r} \tag{14.17}$$

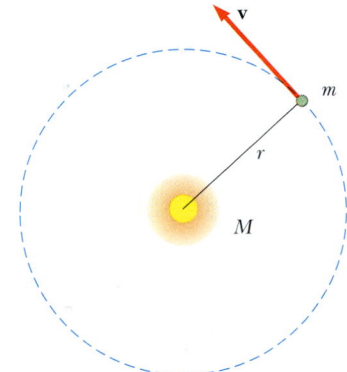

Figure 14.16 A body of mass m moving in a circular orbit about a much larger body of mass M.

This equation shows that E may be positive, negative, or zero, depending on the value of v. However, for a bound system,[7] such as the Earth–Sun system, E is necessarily *less than zero* because we have chosen the convention that $U \to 0$ as $r \to \infty$.

We can easily establish that $E < 0$ for the system consisting of a body of mass m moving in a circular orbit about a body of mass $M \gg m$ (Fig. 14.16). Newton's second law applied to the body of mass m gives

$$\frac{GMm}{r^2} = ma = \frac{mv^2}{r}$$

---

[6] You might recognize that we have ignored the acceleration and kinetic energy of the larger body. To see that this simplification is reasonable, consider an object of mass m falling toward the Earth. Because the center of mass of the object–Earth system is effectively stationary, it follows that $mv = M_E v_E$. Thus, the Earth acquires a kinetic energy equal to

$$\tfrac{1}{2}M_E v_E^2 = \tfrac{1}{2}\frac{m^2}{M_E}v^2 = \frac{m}{M_E}K$$

where K is the kinetic energy of the object. Because $M_E \gg m$, this result shows that the kinetic energy of the Earth is negligible.

[7] Of the three examples provided at the beginning of this section, the planet moving around the Sun and a satellite in orbit around the Earth are bound systems—the Earth will always stay near the Sun, and the satellite will always stay near the Earth. The one-time comet flyby represents an unbound system—the comet interacts once with the Sun but is not bound to it. Thus, in theory the comet can move infinitely far away from the Sun.

Multiplying both sides by r and dividing by 2 gives

$$\tfrac{1}{2}mv^2 = \frac{GMm}{2r} \tag{14.18}$$

Substituting this into Equation 14.17, we obtain

$$E = \frac{GMm}{2r} - \frac{GMm}{r}$$

Total energy for circular orbits

$$E = -\frac{GMm}{2r} \tag{14.19}$$

This result clearly shows that **the total mechanical energy is negative in the case of circular orbits.** Note that **the kinetic energy is positive and equal to one-half the absolute value of the potential energy.** The absolute value of E is also equal to the binding energy of the system, because this amount of energy must be provided to the system to move the two masses infinitely far apart.

The total mechanical energy is also negative in the case of elliptical orbits. The expression for E for elliptical orbits is the same as Equation 14.19 with r replaced by the semimajor axis length a. Furthermore, the total energy is constant if we assume that the system is isolated. Therefore, as the body of mass m moves from P to Q in Figure 14.13, the total energy remains constant and Equation 14.17 gives

$$E = \tfrac{1}{2}mv_i^2 - \frac{GMm}{r_i} = \tfrac{1}{2}mv_f^2 - \frac{GMm}{r_f} \tag{14.20}$$

Combining this statement of energy conservation with our earlier discussion of conservation of angular momentum, we see that **both the total energy and the total angular momentum of a gravitationally bound, two-body system are constants of the motion.**

EXAMPLE 14.7   Changing the Orbit of a Satellite

The space shuttle releases a 470-kg communications satellite while in an orbit that is 280 km above the surface of the Earth. A rocket engine on the satellite boosts it into a geosynchronous orbit, which is an orbit in which the satellite stays directly over a single location on the Earth. How much energy did the engine have to provide?

Solution   First we must determine the radius of a geosynchronous orbit. Then we can calculate the change in energy needed to boost the satellite into orbit.

The period of the orbit T must be one day (86 400 s), so that the satellite travels once around the Earth in the same time that the Earth spins once on its axis. Knowing the period, we can then apply Kepler's third law (Eq. 14.7) to find the radius, once we replace K_S with $K_E = 4\pi^2/GM_E = 9.89 \times 10^{-14}$ s²/m³:

$$T^2 = K_E r^3$$

$$r = \sqrt[3]{\frac{T^2}{K_E}} = \sqrt[3]{\frac{(86\ 400\ \text{s})^2}{9.89 \times 10^{-14}\ \text{s}^2/\text{m}^3}} = 4.23 \times 10^7\ \text{m} = R_f$$

This is a little more than 26 000 mi above the Earth's surface.

We must also determine the initial radius (not the altitude above the Earth's surface) of the satellite's orbit when it was still in the shuttle's cargo bay. This is simply

$$R_E + 280\ \text{km} = 6.65 \times 10^6\ \text{m} = R_i$$

Now, applying Equation 14.19, we obtain, for the total initial and final energies,

$$E_i = -\frac{GM_E m}{2R_i} \qquad E_f = -\frac{GM_E m}{2R_f}$$

The energy required from the engine to boost the satellite is

$$E_{\text{engine}} = E_f - E_i = -\frac{GM_E m}{2}\left(\frac{1}{R_f} - \frac{1}{R_i}\right)$$

$$= -\frac{(6.67 \times 10^{-11}\ \text{N} \cdot \text{m}^2/\text{kg}^2)(5.98 \times 10^{24}\ \text{kg})(470\ \text{kg})}{2}$$

$$\times \left(\frac{1}{4.23 \times 10^7\ \text{m}} - \frac{1}{6.65 \times 10^6\ \text{m}}\right)$$

$$= 1.19 \times 10^{10}\ \text{J}$$

This is the energy equivalent of 89 gal of gasoline. NASA engineers must account for the changing mass of the spacecraft as it ejects burned fuel, something we have not done here. Would you expect the calculation that includes the effect of this changing mass to yield a greater or lesser amount of energy required from the engine?

If we wish to determine how the energy is distributed after the engine is fired, we find from Equation 14.18 that the change in kinetic energy is $\Delta K = (GM_E m/2)(1/R_f - 1/R_i) = -1.19 \times 10^{10}$ J (a decrease),

and the corresponding change in potential energy is $\Delta U = -GM_E m(1/R_f - 1/R_i) = 2.38 \times 10^{10}$ J (an increase). Thus, the change in mechanical energy of the system is $\Delta E = \Delta K + \Delta U = 1.19 \times 10^{10}$ J, as we already calculated. The firing of the engine results in an increase in the total mechanical energy of the system. Because an increase in potential energy is accompanied by a decrease in kinetic energy, we conclude that the speed of an orbiting satellite decreases as its altitude increases.

Escape Speed

Suppose an object of mass m is projected vertically upward from the Earth's surface with an initial speed v_i, as illustrated in Figure 14.17. We can use energy considerations to find the minimum value of the initial speed needed to allow the object to escape the Earth's gravitational field. Equation 14.17 gives the total energy of the object at any point. At the surface of the Earth, $v = v_i$ and $r = r_i = R_E$. When the object reaches its maximum altitude, $v = v_f = 0$ and $r = r_f = r_{max}$. Because the total energy of the system is constant, substituting these conditions into Equation 14.20 gives

$$\tfrac{1}{2}mv_i^2 - \frac{GM_E m}{R_E} = -\frac{GM_E m}{r_{max}}$$

Solving for v_i^2 gives

$$v_i^2 = 2GM_E\left(\frac{1}{R_E} - \frac{1}{r_{max}}\right) \qquad \textbf{(14.21)}$$

Therefore, if the initial speed is known, this expression can be used to calculate the maximum altitude h because we know that

$$h = r_{max} - R_E$$

We are now in a position to calculate **escape speed,** which is the minimum speed the object must have at the Earth's surface in order to escape from the influence of the Earth's gravitational field. Traveling at this minimum speed, the object continues to move farther and farther away from the Earth as its speed asymptotically approaches zero. Letting $r_{max} \to \infty$ in Equation 14.21 and taking $v_i = v_{esc}$, we obtain

$$v_{esc} = \sqrt{\frac{2GM_E}{R_E}} \qquad \textbf{(14.22)}$$

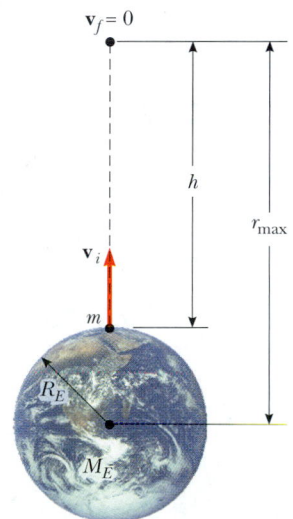

Figure 14.17 An object of mass m projected upward from the Earth's surface with an initial speed v_i reaches a maximum altitude h.

Escape speed

Note that this expression for v_{esc} is independent of the mass of the object. In other words, a spacecraft has the same escape speed as a molecule. Furthermore, the result is independent of the direction of the velocity and ignores air resistance.

If the object is given an initial speed equal to v_{esc}, its total energy is equal to zero. This can be seen by noting that when $r \to \infty$, the object's kinetic energy and its potential energy are both zero. If v_i is greater than v_{esc}, the total energy is greater than zero and the object has some residual kinetic energy as $r \to \infty$.

EXAMPLE 14.8   **Escape Speed of a Rocket**

Calculate the escape speed from the Earth for a 5 000-kg spacecraft, and determine the kinetic energy it must have at the Earth's surface in order to escape the Earth's gravitational field.

Solution   Using Equation 14.22 gives

$$v_{esc} = \sqrt{\frac{2GM_E}{R_E}}$$

$$= \sqrt{\frac{2(6.67 \times 10^{-11}\,\text{N} \cdot \text{m}^2/\text{kg}^2)(5.98 \times 10^{24}\,\text{kg})}{6.37 \times 10^6\,\text{m}}}$$

$$= \boxed{1.12 \times 10^4\,\text{m/s}}$$

This corresponds to about 25 000 mi/h.

The kinetic energy of the spacecraft is

$$K = \tfrac{1}{2}mv_{esc}^2 = \tfrac{1}{2}(5.00 \times 10^3\,\text{kg})(1.12 \times 10^4\,\text{m/s})^2$$

$$= \boxed{3.14 \times 10^{11}\,\text{J}}$$

This is equivalent to about 2 300 gal of gasoline.

Equations 14.21 and 14.22 can be applied to objects projected from any planet. That is, in general, the escape speed from the surface of any planet of mass M and radius R is

$$v_{esc} = \sqrt{\frac{2GM}{R}}$$

Escape speeds for the planets, the Moon, and the Sun are provided in Table 14.3. Note that the values vary from 1.1 km/s for Pluto to about 618 km/s for the Sun. These results, together with some ideas from the kinetic theory of gases (see Chapter 21), explain why some planets have atmospheres and others do not. As we shall see later, a gas molecule has an average kinetic energy that depends on the temperature of the gas. Hence, lighter molecules, such as hydrogen and helium, have a higher average speed than heavier species at the same temperature. When the average speed of the lighter molecules is not much less than the escape speed of a planet, a significant fraction of them have a chance to escape from the planet.

This mechanism also explains why the Earth does not retain hydrogen molecules and helium atoms in its atmosphere but does retain heavier molecules, such as oxygen and nitrogen. On the other hand, the very large escape speed for Jupiter enables that planet to retain hydrogen, the primary constituent of its atmosphere.

TABLE 14.3

Escape Speeds from the Surfaces of the Planets, Moon, and Sun

Body	v_{esc} (km/s)
Mercury	4.3
Venus	10.3
Earth	11.2
Moon	2.3
Mars	5.0
Jupiter	60
Saturn	36
Uranus	22
Neptune	24
Pluto	1.1
Sun	618

Quick Quiz 14.2

If you were a space prospector and discovered gold on an asteroid, it probably would not be a good idea to jump up and down in excitement over your find. Why?

Quick Quiz 14.3

Figure 14.18 is a drawing by Newton showing the path of a stone thrown from a mountaintop. He shows the stone landing farther and farther away when thrown at higher and higher speeds (at points *D, E, F,* and *G*), until finally it is thrown all the way around the Earth. Why didn't Newton show the stone landing at *B* and *A* before it was going fast enough to complete an orbit?

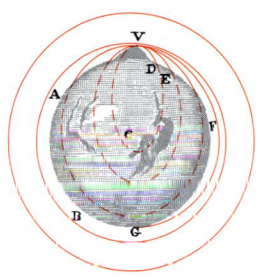

Figure 14.18   "The greater the velocity . . . with which [a stone] is projected, the farther it goes before it falls to the Earth. We may therefore suppose the velocity to be so increased, that it would describe an arc of 1, 2, 5, 10, 100, 1000 miles before it arrived at the Earth, till at last, exceeding the limits of the Earth, it should pass into space without touching." Sir Isaac Newton, *System of the World.*

Optional Section

14.9   THE GRAVITATIONAL FORCE BETWEEN AN EXTENDED OBJECT AND A PARTICLE

We have emphasized that the law of universal gravitation given by Equation 14.3 is valid only if the interacting objects are treated as particles. In view of this, how can we calculate the force between a particle and an object having finite dimensions? This is accomplished by treating the extended object as a collection of particles and making use of integral calculus. We first evaluate the potential energy function, and then calculate the gravitational force from that function.

We obtain the potential energy associated with a system consisting of a particle of mass m and an extended object of mass M by dividing the object into many elements, each having a mass ΔM_i (Fig. 14.19). The potential energy associated with the system consisting of any one element and the particle is $U = -Gm\,\Delta M_i/r_i$, where r_i is the distance from the particle to the element ΔM_i. The total potential energy of the overall system is obtained by taking the sum over all elements as $\Delta M_i \rightarrow 0$. In this limit, we can express U in integral form as

$$U = -Gm \int \frac{dM}{r} \tag{14.23}$$

Once U has been evaluated, we obtain the force exerted by the extended object on the particle by taking the negative derivative of this scalar function (see Section 8.6). If the extended object has spherical symmetry, the function U depends only on r, and the force is given by $-dU/dr$. We treat this situation in Section 14.10. In principle, one can evaluate U for any geometry; however, the integration can be cumbersome.

An alternative approach to evaluating the gravitational force between a particle and an extended object is to perform a vector sum over all mass elements of the object. Using the procedure outlined in evaluating U and the law of universal gravitation in the form shown in Equation 14.3, we obtain, for the total force exerted on the particle

$$\mathbf{F}_g = -Gm \int \frac{dM}{r^2}\,\hat{\mathbf{r}} \tag{14.24}$$

where $\hat{\mathbf{r}}$ is a unit vector directed from the element dM toward the particle (see Fig. 14.19) and the minus sign indicates that the direction of the force is opposite that of $\hat{\mathbf{r}}$. This procedure is not always recommended because working with a vector function is more difficult than working with the scalar potential energy function. However, if the geometry is simple, as in the following example, the evaluation of $\mathbf{F}$ can be straightforward.

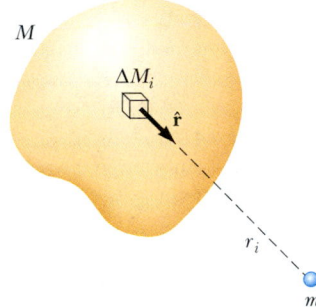

Figure 14.19   A particle of mass m interacting with an extended object of mass M. The total gravitational force exerted by the object on the particle can be obtained by dividing the object into numerous elements, each having a mass ΔM_i, and then taking a vector sum over the forces exerted by all elements.

Total force exerted on a particle by an extended object

EXAMPLE 14.9   Gravitational Force Between a Particle and a Bar

The left end of a homogeneous bar of length L and mass M is at a distance h from a particle of mass m (Fig. 14.20). Calculate the total gravitational force exerted by the bar on the particle.

Solution   The arbitrary segment of the bar of length dx has a mass dM. Because the mass per unit length is constant, it follows that the ratio of masses dM/M is equal to the ratio

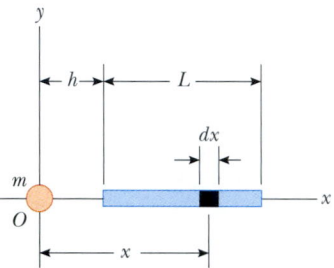

Figure 14.20   The gravitational force exerted by the bar on the particle is directed to the right. Note that the bar is *not* equivalent to a particle of mass M located at the center of mass of the bar.

of lengths dx/L, and so $dM = (M/L)\,dx$. In this problem, the variable r in Equation 14.24 is the distance x shown in Figure 14.20, the unit vector $\hat{\mathbf{r}}$ is $\hat{\mathbf{r}} = -\mathbf{i}$, and the force acting on the particle is to the right; therefore, Equation 14.24 gives us

$$\mathbf{F}_g = -Gm \int_h^{h+L} \frac{M\,dx}{L} \frac{1}{x^2}(-\mathbf{i}) = Gm\frac{M}{L} \int_h^{h+L} \frac{dx}{x^2}\mathbf{i}$$

$$\mathbf{F}_g = \frac{GmM}{L}\left[-\frac{1}{x}\right]_h^{h+L}\mathbf{i} = \frac{GmM}{h(h+L)}\mathbf{i}$$

We see that the force exerted on the particle is in the positive x direction, which is what we expect because the gravitational force is attractive.

Note that in the limit $L \to 0$, the force varies as $1/h^2$, which is what we expect for the force between two point masses. Furthermore, if $h \gg L$, the force also varies as $1/h^2$. This can be seen by noting that the denominator of the expression for $\mathbf{F}_g$ can be expressed in the form $h^2(1 + L/h)$, which is approximately equal to h^2 when $h \gg L$. Thus, when bodies are separated by distances that are great relative to their characteristic dimensions, they behave like particles.

Optional Section

14.10   THE GRAVITATIONAL FORCE BETWEEN A PARTICLE AND A SPHERICAL MASS

We have already stated that a large sphere attracts a particle outside it as if the total mass of the sphere were concentrated at its center. We now describe the force acting on a particle when the extended object is either a spherical shell or a solid sphere, and then apply these facts to some interesting systems.

Spherical Shell

Case 1.   If a particle of mass m is located outside a spherical shell of mass M at, for instance, point P in Figure 14.21a, the shell attracts the particle as though the mass of the shell were concentrated at its center. We can show this, as Newton did, with integral calculus. Thus, as far as the gravitational force acting on a particle outside the shell is concerned, a spherical shell acts no differently from the solid spherical distributions of mass we have seen.

Case 2.   If the particle is located inside the shell (at point P in Fig. 14.21b), the gravitational force acting on it can be shown to be zero.

We can express these two important results in the following way:

> Force on a particle due to a spherical shell

$$\mathbf{F}_g = -\frac{GMm}{r^2}\hat{\mathbf{r}} \qquad \text{for } r \geq R \qquad \textbf{(14.25a)}$$

$$\mathbf{F}_g = 0 \qquad \text{for } r < R \qquad \textbf{(14.25b)}$$

The gravitational force as a function of the distance r is plotted in Figure 14.21c.

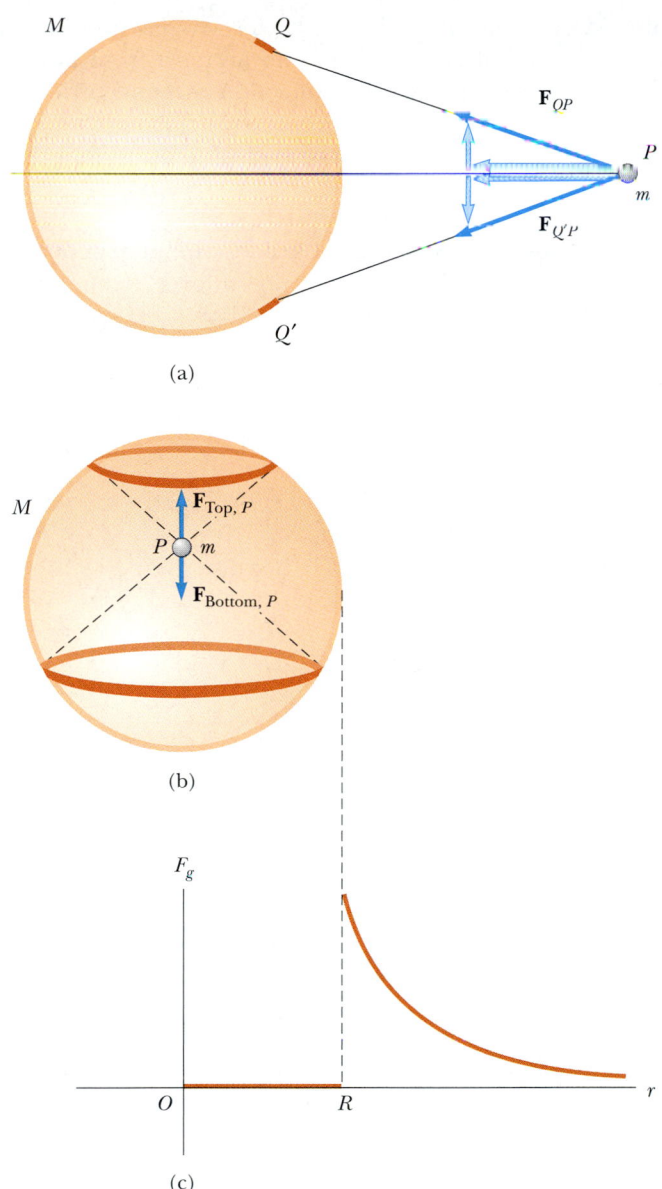

(a)

(b)

(c)

Figure 14.21 (a) The nonradial components of the gravitational forces exerted on a particle of mass m located at point P outside a spherical shell of mass M cancel out. (b) The spherical shell can be broken into rings. Even though point P is closer to the top ring than to the bottom ring, the bottom ring is larger, and the gravitational forces exerted on the particle at P by the matter in the two rings cancel each other. Thus, for a particle located at any point P inside the shell, there is no gravitational force exerted on the particle by the mass M of the shell. (c) The magnitude of the gravitational force versus the radial distance r from the center of the shell.

The shell does not act as a gravitational shield, which means that a particle inside a shell may experience forces exerted by bodies outside the shell.

Solid Sphere

Case 1. If a particle of mass m is located outside a homogeneous solid sphere of mass M (at point P in Fig. 14.22), the sphere attracts the particle as though the mass of the sphere were concentrated at its center. We have used this notion at several places in this chapter already, and we can argue it from Equation 14.25a. A solid sphere can be considered to be a collection of concentric spherical shells. The masses of all of the shells can be interpreted as being concentrated at their common center, and the gravitational force is equivalent to that due to a particle of mass M located at that center.

Case 2.   If a particle of mass m is located inside a homogeneous solid sphere of mass M (at point Q in Fig. 14.22), the gravitational force acting on it is due *only* to the mass M' contained within the sphere of radius $r < R$, shown in Figure 14.22. In other words,

> Force on a particle due to a solid sphere

$$\mathbf{F}_g = -\frac{GmM}{r^2}\,\hat{\mathbf{r}} \qquad \text{for } r \geq R \tag{14.26a}$$

$$\mathbf{F}_g = -\frac{GmM'}{r^2}\,\hat{\mathbf{r}} \qquad \text{for } r < R \tag{14.26b}$$

This also follows from spherical-shell Case 1 because the part of the sphere that is

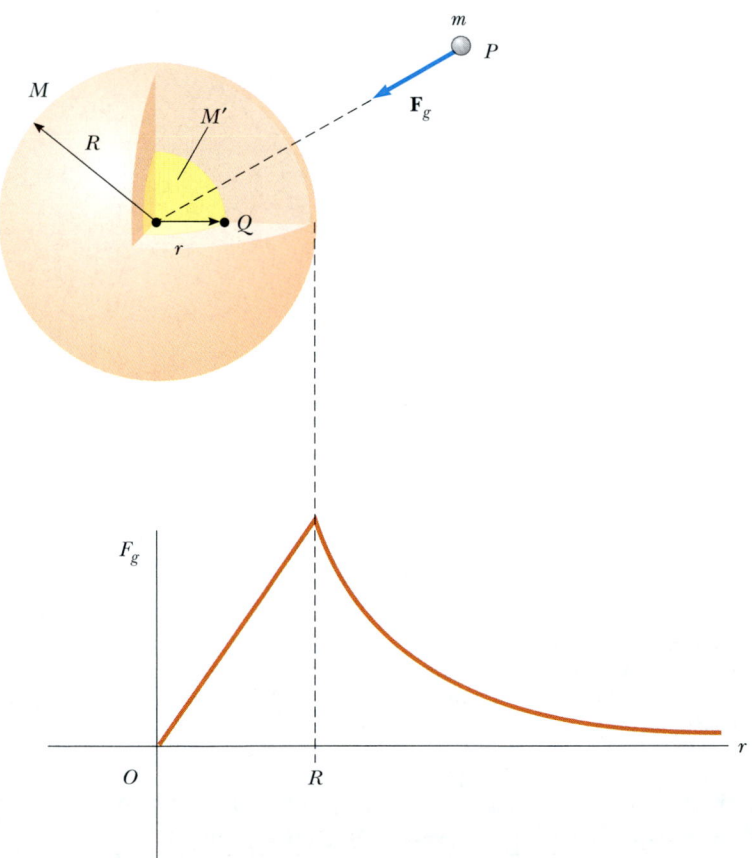

Figure 14.22   The gravitational force acting on a particle when it is outside a uniform solid sphere is GMm/r^2 and is directed toward the center of the sphere. The gravitational force acting on the particle when it is inside such a sphere is proportional to r and goes to zero at the center.

farther from the center than Q can be treated as a series of concentric spherical shells that do not exert a net force on the particle because the particle is inside them. Because the sphere is assumed to have a uniform density, it follows that the ratio of masses M'/M is equal to the ratio of volumes V'/V, where V is the total volume of the sphere and V' is the volume within the sphere of radius r only:

$$\frac{M'}{M} = \frac{V'}{V} = \frac{\frac{4}{3}\pi r^3}{\frac{4}{3}\pi R^3} = \frac{r^3}{R^3}$$

Solving this equation for M' and substituting the value obtained into Equation 14.26b, we have

$$\mathbf{F}_g = -\frac{GmM}{R^3}\, r\hat{\mathbf{r}} \qquad \text{for } r < R \qquad\qquad \textbf{(14.27)}$$

This equation tells us that at the center of the solid sphere, where $r = 0$, the gravitational force goes to zero, as we intuitively expect. The force as a function of r is plotted in Figure 14.22.

Case 3.   If a particle is located inside a solid sphere having a density ρ that is spherically symmetric but not uniform, then M' in Equation 14.26b is given by an integral of the form $M' = \int \rho\, dV$, where the integration is taken over the volume contained within the sphere of radius r in Figure 14.22. We can evaluate this integral if the radial variation of ρ is given. In this case, we take the volume element dV as the volume of a spherical shell of radius r and thickness dr, and thus $dV = 4\pi r^2\, dr$. For example, if $\rho = Ar$, where A is a constant, it is left to a problem (Problem 63) to show that $M' = \pi A r^4$. Hence, we see from Equation 14.26b that F is proportional to r^2 in this case and is zero at the center.

Quick Quiz 14.4

A particle is projected through a small hole into the interior of a spherical shell. Describe the motion of the particle inside the shell.

EXAMPLE 14.10   **A Free Ride, Thanks to Gravity**

An object of mass m moves in a smooth, straight tunnel dug between two points on the Earth's surface (Fig. 14.23). Show that the object moves with simple harmonic motion, and find the period of its motion. Assume that the Earth's density is uniform.

Solution   The gravitational force exerted on the object acts toward the Earth's center and is given by Equation 14.27:

$$\mathbf{F}_g = -\frac{GmM}{R^3}\, r\hat{\mathbf{r}}$$

We receive our first indication that this force should result in simple harmonic motion by comparing it to Hooke's law, first seen in Section 7.3. Because the gravitational force on the object is linearly proportional to the displacement, the object experiences a Hooke's law force.

The y component of the gravitational force on the object is balanced by the normal force exerted by the tunnel wall, and the x component is

$$F_x = -\frac{GmM_E}{R_E^{\,3}}\, r\cos\theta$$

Because the x coordinate of the object is $x = r\cos\theta$, we can write

$$F_x = -\frac{GmM_E}{R_E^{\,3}}\, x$$

Applying Newton's second law to the motion along the x direction gives

$$F_x = -\frac{GmM_E}{R_E^{\,3}}\, x = ma_x$$

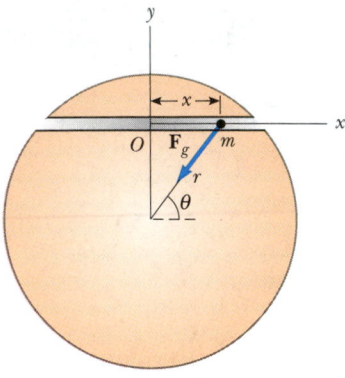

Figure 14.23   An object moves along a tunnel dug through the Earth. The component of the gravitational force $\mathbf{F}_g$ along the x axis is the driving force for the motion. Note that this component always acts toward O.

Solving for a_x, we obtain

$$a_x = -\frac{GM_E}{R_E^{3}}\,x$$

If we use the symbol ω^2 for the coefficient of x — $GM_E/R_E^{3} = \omega^2$ — we see that

$$(1) \qquad a_x = -\omega^2 x$$

an expression that matches the mathematical form of Equation 13.9, which gives the acceleration of a particle in simple harmonic motion: $a_x = -\omega^2 x$. Therefore, Equation (1),

which we have derived for the acceleration of our object in the tunnel, is the acceleration equation for simple harmonic motion at angular speed ω with

$$\omega = \sqrt{\frac{GM_E}{R_E^{3}}}$$

Thus, the object in the tunnel moves in the same way as a block hanging from a spring! The period of oscillation is

$$T = \frac{2\pi}{\omega} = 2\pi\sqrt{\frac{R_E^{3}}{GM_E}}$$

$$= 2\pi\sqrt{\frac{(6.37 \times 10^{6}\ \text{m})^{3}}{(6.67 \times 10^{-11}\ \text{N·m}^2/\text{kg}^2)(5.98 \times 10^{24}\ \text{kg})}}$$

$$= 5.06 \times 10^{3}\ \text{s} = \boxed{84.3\ \text{min}}$$

This period is the same as that of a satellite traveling in a circular orbit just above the Earth's surface (ignoring any trees, buildings, or other objects in the way). Note that the result is independent of the length of the tunnel.

A proposal has been made to operate a mass-transit system between any two cities, using the principle described in this example. A one-way trip would take about 42 min. A more precise calculation of the motion must account for the fact that the Earth's density is not uniform. More important, there are many practical problems to consider. For instance, it would be impossible to achieve a frictionless tunnel, and so some auxiliary power source would be required. Can you think of other problems?

SUMMARY

Newton's law of universal gravitation states that the gravitational force of attraction between any two particles of masses m_1 and m_2 separated by a distance r has the magnitude

$$F_g = G\frac{m_1 m_2}{r^2} \qquad \textbf{(14.1)}$$

where $G = 6.673 \times 10^{-11}\ \text{N·m}^2/\text{kg}^2$ is the universal gravitational constant. This equation enables us to calculate the force of attraction between masses under a wide variety of circumstances.

An object at a distance h above the Earth's surface experiences a gravitational force of magnitude mg', where g' is the free-fall acceleration at that elevation:

$$g' = \frac{GM_E}{r^2} = \frac{GM_E}{(R_E + h)^2} \qquad \textbf{(14.6)}$$

In this expression, M_E is the mass of the Earth and R_E is its radius. Thus, the weight of an object decreases as the object moves away from the Earth's surface.

Kepler's laws of planetary motion state that

1. All planets move in elliptical orbits with the Sun at one focal point.
2. The radius vector drawn from the Sun to a planet sweeps out equal areas in equal time intervals.
3. The square of the orbital period of any planet is proportional to the cube of the semimajor axis of the elliptical orbit.

Kepler's third law can be expressed as

$$T^2 = \left(\frac{4\pi^2}{GM_S}\right)r^3 \tag{14.7}$$

where M_S is the mass of the Sun and r is the orbital radius. For elliptical orbits, Equation 14.7 is valid if r is replaced by the semimajor axis a. Most planets have nearly circular orbits around the Sun.

The **gravitational field** at a point in space equals the gravitational force experienced by any test particle located at that point divided by the mass of the test particle:

$$\mathbf{g} = \frac{\mathbf{F}_g}{m} \tag{14.10}$$

The gravitational force is conservative, and therefore a potential energy function can be defined. The **gravitational potential energy** associated with two particles separated by a distance r is

$$U = -\frac{Gm_1m_2}{r} \tag{14.15}$$

where U is taken to be zero as $r \rightarrow \infty$. The total potential energy for a system of particles is the sum of energies for all pairs of particles, with each pair represented by a term of the form given by Equation 14.15.

If an isolated system consists of a particle of mass m moving with a speed v in the vicinity of a massive body of mass M, the total energy E of the system is the sum of the kinetic and potential energies:

$$E = \tfrac{1}{2}mv^2 - \frac{GMm}{r} \tag{14.17}$$

The total energy is a constant of the motion. If the particle moves in a circular orbit of radius r around the massive body and if $M \gg m$, the total energy of the system is

$$E = -\frac{GMm}{2r} \tag{14.19}$$

The total energy is negative for any bound system.

The **escape speed** for an object projected from the surface of the Earth is

$$v_{esc} = \sqrt{\frac{2GM_E}{R_E}} \tag{14.22}$$

QUESTIONS

1. Use Kepler's second law to convince yourself that the Earth must move faster in its orbit during December, when it is closest to the Sun, than during June, when it is farthest from the Sun.

2. The gravitational force that the Sun exerts on the Moon is about twice as great as the gravitational force that the Earth exerts on the Moon. Why doesn't the Sun pull the Moon away from the Earth during a total eclipse of the Sun?

3. If a system consists of five particles, how many terms appear in the expression for the total potential energy? How many terms appear if the system consists of N particles?

4. Is it possible to calculate the potential energy function associated with a particle and an extended body without knowing the geometry or mass distribution of the extended body?

5. Does the escape speed of a rocket depend on its mass? Explain.

6. Compare the energies required to reach the Moon for a 10^5-kg spacecraft and a 10^3-kg satellite.

7. Explain why it takes more fuel for a spacecraft to travel from the Earth to the Moon than for the return trip. Estimate the difference.

8. Why don't we put a geosynchronous weather satellite in orbit around the 45th parallel? Wouldn't this be more useful for the United States than such a satellite in orbit around the equator?

9. Is the potential energy associated with the Earth–Moon system greater than, less than, or equal to the kinetic energy of the Moon relative to the Earth?

10. Explain why no work is done on a planet as it moves in a circular orbit around the Sun, even though a gravita-tional force is acting on the planet. What is the net work done on a planet during each revolution as it moves around the Sun in an elliptical orbit?

11. Explain why the force exerted on a particle by a uniform sphere must be directed toward the center of the sphere. Would this be the case if the mass distribution of the sphere were not spherically symmetric?

12. Neglecting the density variation of the Earth, what would be the period of a particle moving in a smooth hole dug between opposite points on the Earth's surface, passing through its center?

13. At what position in its elliptical orbit is the speed of a planet a maximum? At what position is the speed a minimum?

14. If you were given the mass and radius of planet X, how would you calculate the free-fall acceleration on the surface of this planet?

15. If a hole could be dug to the center of the Earth, do you think that the force on a mass m would still obey Equation 14.1 there? What do you think the force on m would be at the center of the Earth?

16. In his 1798 experiment, Cavendish was said to have "weighed the Earth." Explain this statement.

17. The gravitational force exerted on the *Voyager* spacecraft by Jupiter accelerated it toward escape speed from the Sun. How is this possible?

18. How would you find the mass of the Moon?

19. The *Apollo 13* spaceship developed trouble in the oxygen system about halfway to the Moon. Why did the spaceship continue on around the Moon and then return home, rather than immediately turn back to Earth?

PROBLEMS

1, 2, 3 = straightforward, intermediate, challenging   ☐ = full solution available in the *Student Solutions Manual and Study Guide*
WEB = solution posted at **http://www.saunderscollege.com/physics/**   🖥 = Computer useful in solving problem   🌐 = Interactive Physics
☐ = paired numerical/symbolic problems

Section 14.1   Newton's Law of Universal Gravitation
Section 14.2   Measuring the Gravitational Constant
Section 14.3   Free-Fall Acceleration and the Gravitational Force

1. Determine the order of magnitude of the gravitational force that you exert on another person 2 m away. In your solution, state the quantities that you measure or estimate and their values.

2. A 200-kg mass and a 500-kg mass are separated by 0.400 m. (a) Find the net gravitational force exerted by these masses on a 50.0-kg mass placed midway between them. (b) At what position (other than infinitely re-mote ones) can the 50.0-kg mass be placed so as to ex-perience a net force of zero?

3. Three equal masses are located at three corners of a square of edge length ℓ, as shown in Figure P14.3. Find the gravitational field **g** at the fourth corner due to these masses.

4. Two objects attract each other with a gravitational force of magnitude 1.00×10^{-8} N when separated by 20.0 cm. If the total mass of the two objects is 5.00 kg, what is the mass of each?

5. Three uniform spheres of masses 2.00 kg, 4.00 kg, and 6.00 kg are placed at the corners of a right triangle, as illustrated in Figure P14.5. Calculate the resultant gravi-

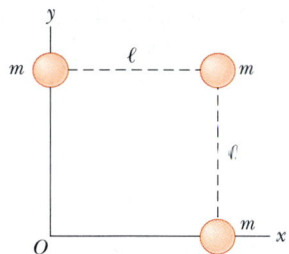

Figure P14.3

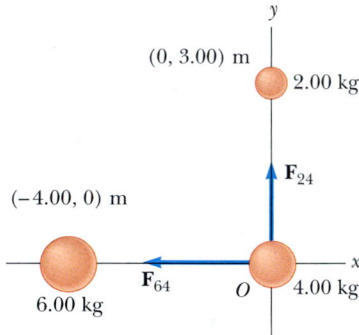

Figure P14.5

tational force on the 4.00-kg mass, assuming that the spheres are isolated from the rest of the Universe.

6. The free-fall acceleration on the surface of the Moon is about one-sixth that on the surface of the Earth. If the radius of the Moon is about $0.250R_E$, find the ratio of their average densities, ρ_{Moon}/ρ_{Earth}.

7. During a solar eclipse, the Moon, Earth, and Sun all lie on the same line, with the Moon between the Earth and the Sun. (a) What force is exerted by the Sun on the Moon? (b) What force is exerted by the Earth on the Moon? (c) What force is exerted by the Sun on the Earth?

8. The center-to-center distance between the Earth and the Moon is 384 400 km. The Moon completes an orbit in 27.3 days. (a) Determine the Moon's orbital speed. (b) If gravity were switched off, the Moon would move along a straight line tangent to its orbit, as described by Newton's first law. In its actual orbit in 1.00 s, how far does the Moon fall below the tangent line and toward the Earth?

WEB 9. When a falling meteoroid is at a distance above the Earth's surface of 3.00 times the Earth's radius, what is its acceleration due to the Earth's gravity?

10. Two ocean liners, each with a mass of 40 000 metric tons, are moving on parallel courses, 100 m apart. What is the magnitude of the acceleration of one of the liners toward the other due to their mutual gravitational attraction? (Treat the ships as point masses.)

11. A student proposes to measure the gravitational constant G by suspending two spherical masses from the ceiling of a tall cathedral and measuring the deflection of the cables from the vertical. Draw a free-body diagram of one of the masses. If two 100.0-kg masses are suspended at the end of 45.00-m-long cables, and the cables are attached to the ceiling 1.000 m apart, what is the separation of the masses?

12. On the way to the Moon, the Apollo astronauts reached a point where the Moon's gravitational pull became stronger than the Earth's. (a) Determine the distance of this point from the center of the Earth. (b) What is the acceleration due to the Earth's gravity at this point?

Section 14.4   Kepler's Laws

Section 14.5   The Law of Gravity and the Motion of Planets

13. A particle of mass m moves along a straight line with constant speed in the x direction, a distance b from the x axis (Fig. P14.13). Show that Kepler's second law is satisfied by demonstrating that the two shaded triangles in the figure have the same area when $t_4 - t_3 = t_2 - t_1$.

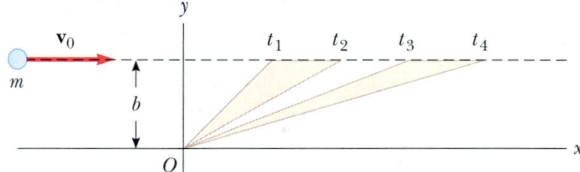

Figure P14.13

14. A communications satellite in geosynchronous orbit remains above a single point on the Earth's equator as the planet rotates on its axis. (a) Calculate the radius of its orbit. (b) The satellite relays a radio signal from a transmitter near the north pole to a receiver, also near the north pole. Traveling at the speed of light, how long is the radio wave in transit?

15. Plaskett's binary system consists of two stars that revolve in a circular orbit about a center of mass midway between them. This means that the masses of the two stars are equal (Fig. P14.15). If the orbital velocity of each star is 220 km/s and the orbital period of each is 14.4 days, find the mass M of each star. (For comparison, the mass of our Sun is 1.99×10^{30} kg.)

16. Plaskett's binary system consists of two stars that revolve in a circular orbit about a center of gravity midway between them. This means that the masses of the two stars are equal (see Fig. P14.15). If the orbital speed of each star is v and the orbital period of each is T, find the mass M of each star.

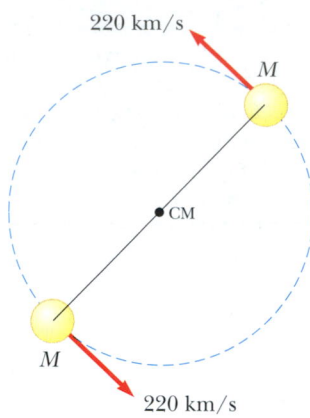

Figure P14.15   Problems 15 and 16.

17. The *Explorer VIII* satellite, placed into orbit November 3, 1960, to investigate the ionosphere, had the following orbit parameters: perigee, 459 km; apogee, 2 289 km (both distances above the Earth's surface); and period, 112.7 min. Find the ratio v_p/v_a of the speed at perigee to that at apogee.

18. Comet Halley (Fig. P14.18) approaches the Sun to within 0.570 AU, and its orbital period is 75.6 years (AU is the symbol for astronomical unit, where 1 AU = 1.50×10^{11} m is the mean Earth–Sun distance). How far from the Sun will Halley's comet travel before it starts its return journey?

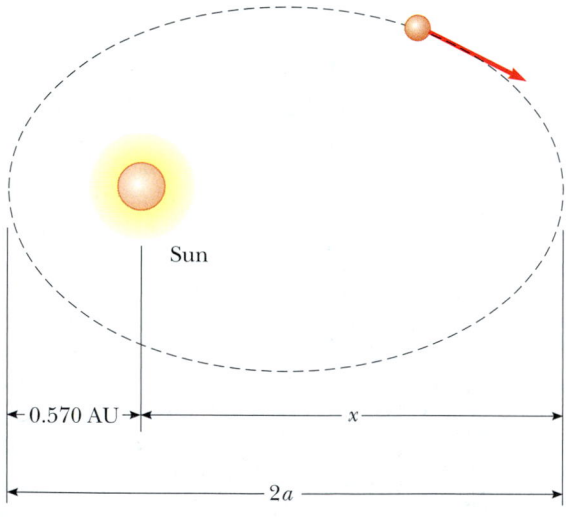

Figure P14.18

WEB 19. Io, a satellite of Jupiter, has an orbital period of 1.77 days and an orbital radius of 4.22×10^5 km. From these data, determine the mass of Jupiter.

20. Two planets, X and Y, travel counterclockwise in circular orbits about a star, as shown in Figure P14.20. The radii of their orbits are in the ratio 3:1. At some time, they are aligned as in Figure P14.20a, making a straight line with the star. During the next five years, the angular displacement of planet X is 90.0°, as shown in Figure P14.20b. Where is planet Y at this time?

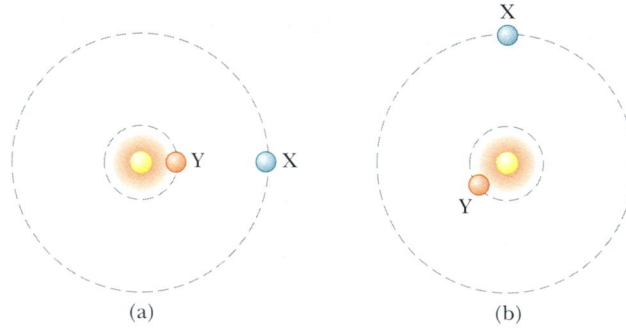

(a)   (b)

Figure P14.20

21. A synchronous satellite, which always remains above the same point on a planet's equator, is put in orbit around Jupiter so that scientists can study the famous red spot. Jupiter rotates once every 9.84 h. Use the data in Table 14.2 to find the altitude of the satellite.

22. Neutron stars are extremely dense objects that are formed from the remnants of supernova explosions. Many rotate very rapidly. Suppose that the mass of a certain spherical neutron star is twice the mass of the Sun and that its radius is 10.0 km. Determine the greatest possible angular speed it can have for the matter at the surface of the star on its equator to be just held in orbit by the gravitational force.

23. The Solar and Heliospheric Observatory (SOHO) spacecraft has a special orbit, chosen so that its view of the Sun is never eclipsed and it is always close enough to the Earth to transmit data easily. It moves in a near-circle around the Sun that is smaller than the Earth's circular orbit. Its period, however, is not less than 1 yr but is just equal to 1 yr. It is always located between the Earth and the Sun along the line joining them. Both objects exert gravitational forces on the observatory. Show that the spacecraft's distance from the Earth must be between 1.47×10^9 m and 1.48×10^9 m. In 1772 Joseph Louis Lagrange determined theoretically the special location that allows this orbit. The SOHO spacecraft took this position on February 14, 1996. (*Hint:* Use data that are precise to four digits. The mass of the Earth is 5.983×10^{24} kg.)

Section 14.6   The Gravitational Field

24. A spacecraft in the shape of a long cylinder has a length of 100 m, and its mass with occupants is 1 000 kg. It has

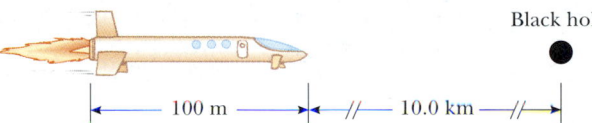

Black hole

100 m — // — 10.0 km — //

Figure P14.24

strayed too close to a 1.0-m-radius black hole having a mass 100 times that of the Sun (Fig. P14.24). The nose of the spacecraft is pointing toward the center of the black hole, and the distance between the nose and the black hole is 10.0 km. (a) Determine the total force on the spacecraft. (b) What is the difference in the gravitational fields acting on the occupants in the nose of the ship and on those in the rear of the ship, farthest from the black hole?

25. Compute the magnitude and direction of the gravitational field at a point P on the perpendicular bisector of two equal masses separated by a distance $2a$, as shown in Figure P14.25.

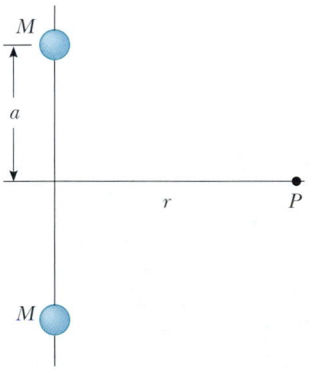

Figure P14.25

26. Find the gravitational field at a distance r along the axis of a thin ring of mass M and radius a.

Section 14.7   Gravitational Potential Energy

Note: Assume that $U = 0$ as $r \rightarrow \infty$.

27. A satellite of the Earth has a mass of 100 kg and is at an altitude of 2.00×10^6 m. (a) What is the potential energy of the satellite–Earth system? (b) What is the magnitude of the gravitational force exerted by the Earth on the satellite? (c) What force does the satellite exert on the Earth?

28. How much energy is required to move a 1 000-kg mass from the Earth's surface to an altitude twice the Earth's radius?

29. After our Sun exhausts its nuclear fuel, its ultimate fate may be to collapse to a *white-dwarf* state, in which it has approximately the same mass it has now but a radius

equal to the radius of the Earth. Calculate (a) the average density of the white dwarf, (b) the acceleration due to gravity at its surface, and (c) the gravitational potential energy associated with a 1.00-kg object at its surface.

30. At the Earth's surface a projectile is launched straight up at a speed of 10.0 km/s. To what height will it rise? Ignore air resistance.

31. A system consists of three particles, each of mass 5.00 g, located at the corners of an equilateral triangle with sides of 30.0 cm. (a) Calculate the potential energy of the system. (b) If the particles are released simultaneously, where will they collide?

32. How much work is done by the Moon's gravitational field as a 1 000-kg meteor comes in from outer space and impacts the Moon's surface?

Section 14.8   Energy Considerations in Planetary and Satellite Motion

33. A 500-kg satellite is in a circular orbit at an altitude of 500 km above the Earth's surface. Because of air friction, the satellite is eventually brought to the Earth's surface, and it hits the Earth with a speed of 2.00 km/s. How much energy was transformed to internal energy by means of friction?

34. (a) What is the minimum speed, relative to the Sun, that is necessary for a spacecraft to escape the Solar System if it starts at the Earth's orbit? (b) *Voyager 1* achieved a maximum speed of 125 000 km/h on its way to photograph Jupiter. Beyond what distance from the Sun is this speed sufficient for a spacecraft to escape the Solar System?

35. A satellite with a mass of 200 kg is placed in Earth orbit at a height of 200 km above the surface. (a) Assuming a circular orbit, how long does the satellite take to complete one orbit? (b) What is the satellite's speed? (c) What is the minimum energy necessary to place this satellite in orbit (assuming no air friction)?

36. A satellite of mass m is placed in Earth orbit at an altitude h. (a) Assuming a circular orbit, how long does the satellite take to complete one orbit? (b) What is the satellite's speed? (c) What is the minimum energy necessary to place this satellite in orbit (assuming no air friction)?

37. A spaceship is fired from the Earth's surface with an initial speed of 2.00×10^4 m/s. What will its speed be when it is very far from the Earth? (Neglect friction.)

38. A 1 000-kg satellite orbits the Earth at a constant altitude of 100 km. How much energy must be added to the system to move the satellite into a circular orbit at an altitude of 200 km?

39. A "treetop satellite" moves in a circular orbit just above the surface of a planet, which is assumed to offer no air resistance. Show that its orbital speed v and the escape speed from the planet are related by the expression $v_{esc} = \sqrt{2}v$.

40. The planet Uranus has a mass about 14 times the Earth's mass, and its radius is equal to about 3.7 Earth

radii. (a) By setting up ratios with the corresponding Earth values, find the acceleration due to gravity at the cloud tops of Uranus. (b) Ignoring the rotation of the planet, find the minimum escape speed from Uranus.

41. Determine the escape velocity for a rocket on the far side of Ganymede, the largest of Jupiter's moons. The radius of Ganymede is 2.64×10^6 m, and its mass is 1.495×10^{23} kg. The mass of Jupiter is 1.90×10^{27} kg, and the distance between Jupiter and Ganymede is 1.071×10^9 m. Be sure to include the gravitational effect due to Jupiter, but you may ignore the motions of Jupiter and Ganymede as they revolve about their center of mass (Fig. P14.41).

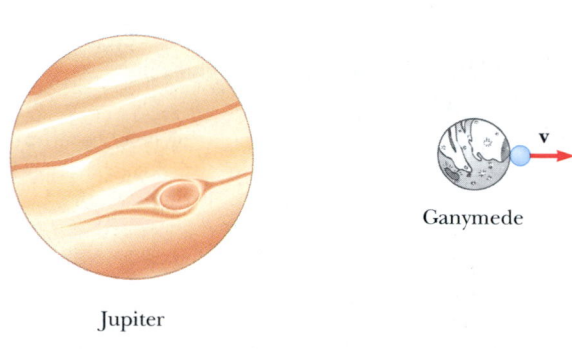

Jupiter   Ganymede

Figure P14.41

42. In Robert Heinlein's *The Moon is a Harsh Mistress,* the colonial inhabitants of the Moon threaten to launch rocks down onto the Earth if they are not given independence (or at least representation). Assuming that a rail gun could launch a rock of mass m at twice the lunar escape speed, calculate the speed of the rock as it enters the Earth's atmosphere. (By *lunar escape speed* we mean the speed required to escape entirely from a stationary Moon alone in the Universe.)

43. Derive an expression for the work required to move an Earth satellite of mass m from a circular orbit of radius $2R_E$ to one of radius $3R_E$.

(Optional)
Section 14.9   The Gravitational Force Between an Extended Object and a Particle

44. Consider two identical uniform rods of length L and mass m lying along the same line and having their closest points separated by a distance d (Fig. P14.44). Show that the mutual gravitational force between these rods has a magnitude

$$F = \frac{Gm^2}{L^2} \ln\left(\frac{(L+d)^2}{d(2L+d)}\right)$$

45. A uniform rod of mass M is in the shape of a semicircle of radius R (Fig. P14.45). Calculate the force on a point mass m placed at the center of the semicircle.

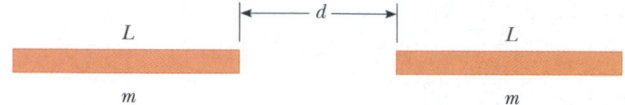

Figure P14.44

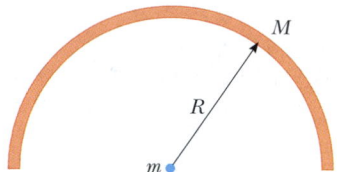

Figure P14.45

(Optional)
Section 14.10   The Gravitational Force Between a Particle and a Spherical Mass

46. (a) Show that the period calculated in Example 14.10 can be written as

$$T = 2\pi\sqrt{\frac{R_E}{g}}$$

where g is the free-fall acceleration on the surface of the Earth. (b) What would this period be if tunnels were made through the Moon? (c) What practical problem regarding these tunnels on Earth would be removed if they were built on the Moon?

47. A 500-kg uniform solid sphere has a radius of 0.400 m. Find the magnitude of the gravitational force exerted by the sphere on a 50.0-g particle located (a) 1.50 m from the center of the sphere, (b) at the surface of the sphere, and (c) 0.200 m from the center of the sphere.

48. A uniform solid sphere of mass m_1 and radius R_1 is inside and concentric with a spherical shell of mass m_2 and radius R_2 (Fig. P14.48). Find the gravitational force exerted by the spheres on a particle of mass m located at (a) $r = a$, (b) $r = b$, and (c) $r = c$, where r is measured from the center of the spheres.

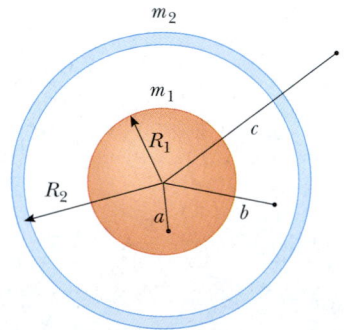

Figure P14.48

ADDITIONAL PROBLEMS

49. Let Δg_M represent the difference in the gravitational fields produced by the Moon at the points on the Earth's surface nearest to and farthest from the Moon. Find the fraction $\Delta g_M/g$, where g is the Earth's gravitational field. (This difference is responsible for the occurrence of the *lunar tides* on the Earth.)

50. Two spheres having masses M and $2M$ and radii R and $3R$, respectively, are released from rest when the distance between their centers is $12R$. How fast will each sphere be moving when they collide? Assume that the two spheres interact only with each other.

51. In Larry Niven's science-fiction novel *Ringworld*, a rigid ring of material rotates about a star (Fig. P14.51). The rotational speed of the ring is 1.25×10^6 m/s, and its radius is 1.53×10^{11} m. (a) Show that the centripetal acceleration of the inhabitants is 10.2 m/s². (b) The inhabitants of this ring world experience a normal contact force $\mathbf{n}$. Acting alone, this normal force would produce an inward acceleration of 9.90 m/s². Additionally, the star at the center of the ring exerts a gravitational force on the ring and its inhabitants. The difference between the total acceleration and the acceleration provided by the normal force is due to the gravitational attraction of the central star. Show that the mass of the star is approximately 10^{32} kg.

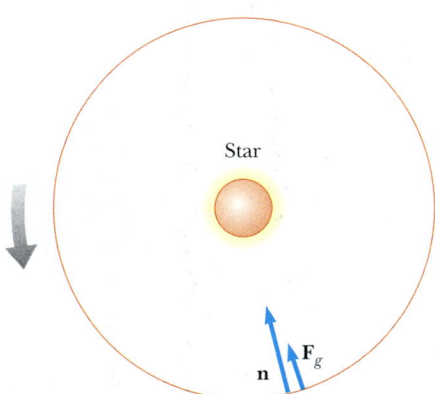

Figure P14.51

52. (a) Show that the rate of change of the free-fall acceleration with distance above the Earth's surface is

$$\frac{dg}{dr} = -\frac{2GM_E}{R_E{}^3}$$

This rate of change over distance is called a *gradient*. (b) If h is small compared to the radius of the Earth, show that the difference in free-fall acceleration between two points separated by vertical distance h is

$$|\Delta g| = \frac{2GM_E h}{R_E{}^3}$$

(c) Evaluate this difference for $h = 6.00$ m, a typical height for a two-story building.

53. A particle of mass m is located inside a uniform solid sphere of radius R and mass M, at a distance r from its center. (a) Show that the gravitational potential energy of the system is $U = (GmM/2R^3)r^2 - 3GmM/2R$. (b) Write an expression for the amount of work done by the gravitational force in bringing the particle from the surface of the sphere to its center.

54. *Voyagers 1* and *2* surveyed the surface of Jupiter's moon Io and photographed active volcanoes spewing liquid sulfur to heights of 70 km above the surface of this moon. Find the speed with which the liquid sulfur left the volcano. Io's mass is 8.9×10^{22} kg, and its radius is 1 820 km.

55. As an astronaut, you observe a small planet to be spherical. After landing on the planet, you set off, walking always straight ahead, and find yourself returning to your spacecraft from the opposite side after completing a lap of 25.0 km. You hold a hammer and a falcon feather at a height of 1.40 m, release them, and observe that they fall together to the surface in 29.2 s. Determine the mass of the planet.

56. A cylindrical habitat in space, 6.00 km in diameter and 30 km long, was proposed by G. K. O'Neill in 1974. Such a habitat would have cities, land, and lakes on the inside surface and air and clouds in the center. All of these would be held in place by the rotation of the cylinder about its long axis. How fast would the cylinder have to rotate to imitate the Earth's gravitational field at the walls of the cylinder?

WEB 57. In introductory physics laboratories, a typical Cavendish balance for measuring the gravitational constant G uses lead spheres with masses of 1.50 kg and 15.0 g whose centers are separated by about 4.50 cm. Calculate the gravitational force between these spheres, treating each as a point mass located at the center of the sphere.

58. Newton's law of universal gravitation is valid for distances covering an enormous range, but it is thought to fail for very small distances, where the structure of space itself is uncertain. The crossover distance, far less than the diameter of an atomic nucleus, is called the *Planck length*. It is determined by a combination of the constants G, c, and h, where c is the speed of light in vacuum and h is Planck's constant (introduced briefly in Chapter 11 and discussed in greater detail in Chapter 40) with units of angular momentum. (a) Use dimensional analysis to find a combination of these three universal constants that has units of length. (b) Determine the order of magnitude of the Planck length. (*Hint:* You will need to consider noninteger powers of the constants.)

59. Show that the escape speed from the surface of a planet of uniform density is directly proportional to the radius of the planet.

60. (a) Suppose that the Earth (or another object) has density $\rho(r)$, which can vary with radius but is spherically symmetric. Show that at any particular radius r inside the Earth, the gravitational field strength $g(r)$ will increase as r increases, if and only if the density there exceeds 2/3 the average density of the portion of the Earth inside the radius r. (b) The Earth as a whole has an average density of 5.5 g/cm^3, while the density at the surface is 1.0 g/cm^3 on the oceans and about 3 g/cm^3 on land. What can you infer from this?

WEB 61. Two hypothetical planets of masses m_1 and m_2 and radii r_1 and r_2, respectively, are nearly at rest when they are an infinite distance apart. Because of their gravitational attraction, they head toward each other on a collision course. (a) When their center-to-center separation is d, find expressions for the speed of each planet and their *relative* velocity. (b) Find the kinetic energy of each planet *just* before they collide, if $m_1 = 2.00 \times 10^{24}$ kg, $m_2 = 8.00 \times 10^{24}$ kg, $r_1 = 3.00 \times 10^6$ m, and $r_2 = 5.00 \times 10^6$ m. (*Hint:* Both energy and momentum are conserved.)

62. The maximum distance from the Earth to the Sun (at our aphelion) is 1.521×10^{11} m, and the distance of closest approach (at perihelion) is 1.471×10^{11} m. If the Earth's orbital speed at perihelion is 30.27 km/s, determine (a) the Earth's orbital speed at aphelion, (b) the kinetic and potential energies at perihelion, and (c) the kinetic and potential energies at aphelion. Is the total energy constant? (Neglect the effect of the Moon and other planets.)

63. A sphere of mass M and radius R has a nonuniform density that varies with r, the distance from its center, according to the expression $\rho = Ar$, for $0 \leq r \leq R$. (a) What is the constant A in terms of M and R? (b) Determine an expression for the force exerted on a particle of mass m placed outside the sphere. (c) Determine an expression for the force exerted on the particle if it is inside the sphere. (*Hint:* See Section 14.10 and note that the distribution is spherically symmetric.)

64. (a) Determine the amount of work (in joules) that must be done on a 100-kg payload to elevate it to a height of 1 000 km above the Earth's surface. (b) Determine the amount of additional work that is required to put the payload into circular orbit at this elevation.

65. X-ray pulses from Cygnus X-1, a celestial x-ray source, have been recorded during high-altitude rocket flights. The signals can be interpreted as originating when a blob of ionized matter orbits a black hole with a period of 5.0 ms. If the blob is in a circular orbit about a black hole whose mass is $20 M_{Sun}$, what is the orbital radius?

66. Studies of the relationship of the Sun to its galaxy—the Milky Way—have revealed that the Sun is located near the outer edge of the galactic disk, about 30 000 lightyears from the center. Furthermore, it has been found that the Sun has an orbital speed of approximately 250 km/s around the galactic center. (a) What is the period of the Sun's galactic motion? (b) What is the order of magnitude of the mass of the Milky Way galaxy? Suppose that the galaxy is made mostly of stars,

of which the Sun is typical. What is the order of magnitude of the number of stars in the Milky Way?

67. The oldest artificial satellite in orbit is *Vanguard I*, launched March 3, 1958. Its mass is 1.60 kg. In its initial orbit, its minimum distance from the center of the Earth was 7.02 Mm, and its speed at this perigee point was 8.23 km/s. (a) Find its total energy. (b) Find the magnitude of its angular momentum. (c) Find its speed at apogee and its maximum (apogee) distance from the center of the Earth. (d) Find the semimajor axis of its orbit. (e) Determine its period.

68. A rocket is given an initial speed vertically upward of $v_i = 2\sqrt{Rg}$ at the surface of the Earth, which has radius R and surface free-fall acceleration g. The rocket motors are quickly cut off, and thereafter the rocket coasts under the action of gravitational forces only. (Ignore atmospheric friction and the Earth's rotation.) Derive an expression for the subsequent speed v as a function of the distance r from the center of the Earth in terms of g, R, and r.

69. Two stars of masses M and m, separated by a distance d, revolve in circular orbits about their center of mass (Fig. P14.69). Show that each star has a period given by

$$T^2 = \frac{4\pi^2 d^3}{G(M + m)}$$

(*Hint:* Apply Newton's second law to each star, and note that the center-of-mass condition requires that $Mr_2 = mr_1$, where $r_1 + r_2 = d$.)

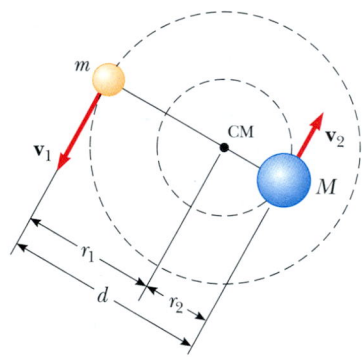

Figure P14.69

70. (a) A 5.00-kg mass is released 1.20×10^7 m from the center of the Earth. It moves with what acceleration relative to the Earth? (b) A 2.00×10^{24} kg mass is released 1.20×10^7 m from the center of the Earth. It moves with what acceleration relative to the Earth? Assume that the objects behave as pairs of particles, isolated from the rest of the Universe.

71. The acceleration of an object moving in the gravitational field of the Earth is

$$\mathbf{a} = -\frac{GM_E}{r^3}\mathbf{r}$$

where **r** is the position vector directed from the center of the Earth to the object. Choosing the origin at the center of the Earth and assuming that the small object is moving in the xy plane, we find that the rectangular (cartesian) components of its acceleration are

$$a_x = -\frac{GM_E x}{(x^2 + y^2)^{3/2}} \qquad a_y = -\frac{GM_E y}{(x^2 + y^2)^{3/2}}$$

Use a computer to set up and carry out a numerical pre-

diction of the motion of the object, according to Euler's method. Assume that the initial position of the object is $x = 0$ and $y = 2R_E$, where R_E is the radius of the Earth. Give the object an initial velocity of 5 000 m/s in the x direction. The time increment should be made as small as practical. Try 5 s. Plot the x and y coordinates of the object as time goes on. Does the object hit the Earth? Vary the initial velocity until you find a circular orbit.

ANSWERS TO QUICK QUIZZES

14.1 Kepler's third law (Eq. 14.7), which applies to all the planets, tells us that the period of a planet is proportional to $r^{3/2}$. Because Saturn and Jupiter are farther from the Sun than the Earth is, they have longer periods. The Sun's gravitational field is much weaker at Saturn and Jupiter than it is at the Earth. Thus, these planets experience much less centripetal acceleration than the Earth does, and they have correspondingly longer periods.

14.2 The mass of the asteroid might be so small that you would be able to exceed escape velocity by leg power alone. You would jump up, but you would never come back down!

14.3 Kepler's first law applies not only to planets orbiting the Sun but also to any relatively small object orbiting another under the influence of gravity. Any elliptical path that does not touch the Earth before reaching point G will continue around the other side to point V in a complete orbit (see figure in next column).

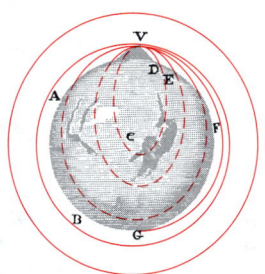

14.4 The gravitational force is zero inside the shell (Eq. 14.25b). Because the force on it is zero, the particle moves with constant velocity in the direction of its original motion outside the shell until it hits the wall opposite the entry hole. Its path thereafter depends on the nature of the collision and on the particle's original direction.

c h a p t e r

15

Fluid Mechanics

Matter is normally classified as being in one of three states: solid, liquid, or gas. From everyday experience, we know that a solid has a definite volume and shape. A brick maintains its familiar shape and size day in and day out. We also know that a liquid has a definite volume but no definite shape. Finally, we know that an unconfined gas has neither a definite volume nor a definite shape. These definitions help us picture the states of matter, but they are somewhat artificial. For example, asphalt and plastics are normally considered solids, but over long periods of time they tend to flow like liquids. Likewise, most substances can be a solid, a liquid, or a gas (or a combination of any of these), depending on the temperature and pressure. In general, the time it takes a particular substance to change its shape in response to an external force determines whether we treat the substance as a solid, as a liquid, or as a gas.

A **fluid** is a collection of molecules that are randomly arranged and held together by weak cohesive forces and by forces exerted by the walls of a container. Both liquids and gases are fluids.

In our treatment of the mechanics of fluids, we shall see that we do not need to learn any new physical principles to explain such effects as the buoyant force acting on a submerged object and the dynamic lift acting on an airplane wing. First, we consider the mechanics of a fluid at rest—that is, *fluid statics*—and derive an expression for the pressure exerted by a fluid as a function of its density and depth. We then treat the mechanics of fluids in motion—that is, *fluid dynamics*. We can describe a fluid in motion by using a model in which we make certain simplifying assumptions. We use this model to analyze some situations of practical importance. An analysis leading to *Bernoulli's equation* enables us to determine relationships between the pressure, density, and velocity at every point in a fluid.

15.1 ▶ PRESSURE

Fluids do not sustain shearing stresses or tensile stresses; thus, the only stress that can be exerted on an object submerged in a fluid is one that tends to compress the object. In other words, the force exerted by a fluid on an object is always perpendicular to the surfaces of the object, as shown in Figure 15.1.

The pressure in a fluid can be measured with the device pictured in Figure 15.2. The device consists of an evacuated cylinder that encloses a light piston connected to a spring. As the device is submerged in a fluid, the fluid presses on the top of the piston and compresses the spring until the inward force exerted by the fluid is balanced by the outward force exerted by the spring. The fluid pressure can be measured directly if the spring is calibrated in advance. If F is the magnitude of the force exerted on the piston and A is the surface area of the piston,

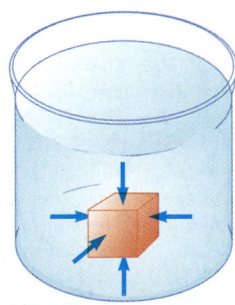

Figure 15.1 At any point on the surface of a submerged object, the force exerted by the fluid is perpendicular to the surface of the object. The force exerted by the fluid on the walls of the container is perpendicular to the walls at all points.

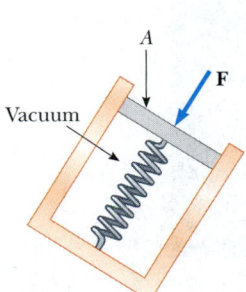

Figure 15.2 A simple device for measuring the pressure exerted by a fluid.

then the **pressure** P of the fluid at the level to which the device has been submerged is defined as the ratio F/A:

$$P \equiv \frac{F}{A} \qquad (15.1)$$

Note that pressure is a scalar quantity because it is proportional to the magnitude of the force on the piston.

To define the pressure at a specific point, we consider a fluid acting on the device shown in Figure 15.2. If the force exerted by the fluid over an infinitesimal surface element of area dA containing the point in question is dF, then the pressure at that point is

$$P = \frac{dF}{dA} \qquad (15.2)$$

As we shall see in the next section, the pressure exerted by a fluid varies with depth. Therefore, to calculate the total force exerted on a flat wall of a container, we must integrate Equation 15.2 over the surface area of the wall.

Because pressure is force per unit area, it has units of newtons per square meter (N/m^2) in the SI system. Another name for the SI unit of pressure is **pascal** (Pa):

$$1 \text{ Pa} \equiv 1 \text{ N/m}^2 \qquad (15.3)$$

| Definition of pressure |

Snowshoes keep you from sinking into soft snow because they spread the downward force you exert on the snow over a large area, reducing the pressure on the snow's surface. *(Earl Young/FPG)*

Quick Quiz 15.1

Suppose you are standing directly behind someone who steps back and accidentally stomps on your foot with the heel of one shoe. Would you be better off if that person were a professional basketball player wearing sneakers or a petite woman wearing spike-heeled shoes? Explain.

Quick Quiz 15.2

After a long lecture, the daring physics professor stretches out for a nap on a bed of nails, as shown in Figure 15.3. How is this possible?

QuickLab

Place a tack between your thumb and index finger, as shown in the figure. Now very gently squeeze the tack and note the sensation. The pointed end of the tack causes pain, and the blunt end does not. According to Newton's third law, the force exerted by the tack on the thumb is equal in magnitude and opposite in direction to the force exerted by the tack on the index finger. However, the pressure at the pointed end of the tack is much greater than the pressure at the blunt end. (Remember that pressure is force per unit area.)

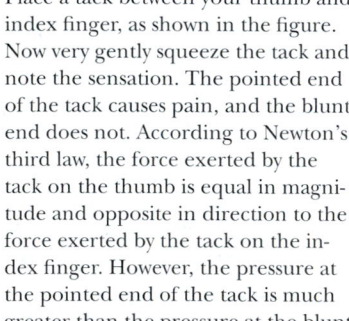

Tack

Figure 15.3 *(Jim Lehman)*

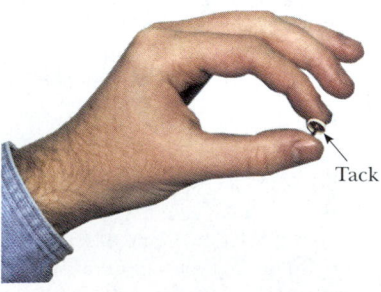

EXAMPLE 15.1   The Water Bed

The mattress of a water bed is 2.00 m long by 2.00 m wide and 30.0 cm deep. (a) Find the weight of the water in the mattress.

Solution   The density of water is 1 000 kg/m³ (Table 15.1), and so the mass of the water is

$$M = \rho V = (1\ 000\ \text{kg/m}^3)(1.20\ \text{m}^3) = 1.20 \times 10^3\ \text{kg}$$

and its weight is

$$Mg = (1.20 \times 10^3\ \text{kg})(9.80\ \text{m/s}^2) = \boxed{1.18 \times 10^4\ \text{N}}$$

This is approximately 2 650 lb. (A regular bed weighs approx-

imately 300 lb.) Because this load is so great, such a water bed is best placed in the basement or on a sturdy, well-supported floor.

(b) Find the pressure exerted by the water on the floor when the bed rests in its normal position. Assume that the entire lower surface of the bed makes contact with the floor.

Solution   When the bed is in its normal position, the cross-sectional area is 4.00 m²; thus, from Equation 15.1, we find that

$$P = \frac{1.18 \times 10^4\ \text{N}}{4.00\ \text{m}^2} = \boxed{2.95 \times 10^3\ \text{Pa}}$$

TABLE 15.1	**Densities of Some Common Substances at Standard Temperature (0°C) and Pressure (Atmospheric)**		
Substance	ρ **(kg/m³)**	**Substance**	ρ **(kg/m³)**
Air	1.29	Ice	0.917×10^3
Aluminum	2.70×10^3	Iron	7.86×10^3
Benzene	0.879×10^3	Lead	11.3×10^3
Copper	8.92×10^3	Mercury	13.6×10^3
Ethyl alcohol	0.806×10^3	Oak	0.710×10^3
Fresh water	1.00×10^3	Oxygen gas	1.43
Glycerine	1.26×10^3	Pine	0.373×10^3
Gold	19.3×10^3	Platinum	21.4×10^3
Helium gas	1.79×10^{-1}	Seawater	1.03×10^3
Hydrogen gas	8.99×10^{-2}	Silver	10.5×10^3

15.2 ▶ VARIATION OF PRESSURE WITH DEPTH

As divers well know, water pressure increases with depth. Likewise, atmospheric pressure decreases with increasing altitude; it is for this reason that aircraft flying at high altitudes must have pressurized cabins.

We now show how the pressure in a liquid increases linearly with depth. As Equation 1.1 describes, the *density* of a substance is defined as its mass per unit volume: $\rho \equiv m/V$. Table 15.1 lists the densities of various substances. These values vary slightly with temperature because the volume of a substance is temperature dependent (as we shall see in Chapter 19). Note that under standard conditions (at 0°C and at atmospheric pressure) the densities of gases are about 1/1 000 the densities of solids and liquids. This difference implies that the average molecular spacing in a gas under these conditions is about ten times greater than that in a solid or liquid.

Now let us consider a fluid of density ρ at rest and open to the atmosphere, as shown in Figure 15.4. We assume that ρ is constant; this means that the fluid is incompressible. Let us select a sample of the liquid contained within an imaginary cylinder of cross-sectional area A extending from the surface to a depth h. The

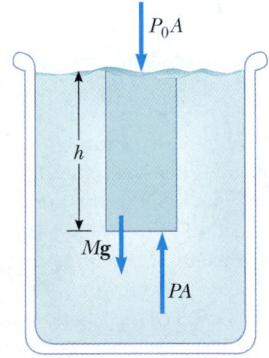

Figure 15.4   How pressure varies with depth in a fluid. The net force exerted on the volume of water within the darker region must be zero.

QuickLab

Poke two holes in the side of a paper or polystyrene cup—one near the top and the other near the bottom. Fill the cup with water and watch the water flow out of the holes. Why does water exit from the bottom hole at a higher speed than it does from the top hole?

pressure exerted by the outside liquid on the bottom face of the cylinder is P, and the pressure exerted on the top face of the cylinder is the atmospheric pressure P_0. Therefore, the upward force exerted by the outside fluid on the bottom of the cylinder is PA, and the downward force exerted by the atmosphere on the top is P_0A. The mass of liquid in the cylinder is $M = \rho V = \rho Ah$; therefore, the weight of the liquid in the cylinder is $Mg = \rho Ahg$. Because the cylinder is in equilibrium, the net force acting on it must be zero. Choosing upward to be the positive y direction, we see that

$$\sum F_y = PA - P_0A - Mg = 0$$

or

| Variation of pressure with depth |

$$PA - P_0A - \rho Ahg = 0$$

$$PA - P_0A = \rho Ahg$$

$$P = P_0 + \rho gh \qquad (15.4)$$

That is, **the pressure P at a depth h below the surface of a liquid open to the atmosphere is *greater* than atmospheric pressure by an amount ρgh.** In our calculations and working of end-of-chapter problems, we usually take atmospheric pressure to be

$$P_0 = 1.00 \text{ atm} = 1.013 \times 10^5 \text{ Pa}$$

Equation 15.4 implies that the pressure is the same at all points having the same depth, independent of the shape of the container.

Quick Quiz 15.3

In the derivation of Equation 15.4, why were we able to ignore the pressure that the liquid exerts on the sides of the cylinder?

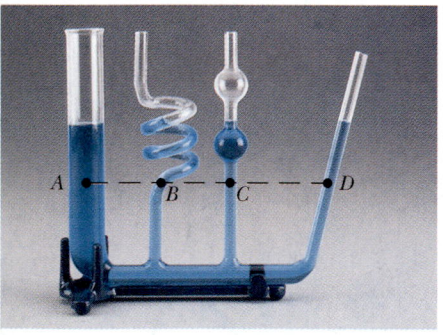

This arrangement of interconnected tubes demonstrates that the pressure in a liquid is the same at all points having the same elevation. For example, the pressure is the same at points *A*, *B*, *C*, and *D*. *(Courtesy of Central Scientific Company)*

In view of the fact that the pressure in a fluid depends on depth and on the value of P_0, any increase in pressure at the surface must be transmitted to every other point in the fluid. This concept was first recognized by the French scientist Blaise Pascal (1623–1662) and is called **Pascal's law: A change in the pressure applied to a fluid is transmitted undiminished to every point of the fluid and to the walls of the container.**

An important application of Pascal's law is the hydraulic press illustrated in Figure 15.5a. A force of magnitude F_1 is applied to a small piston of surface area A_1. The pressure is transmitted through a liquid to a larger piston of surface area A_2. Because the pressure must be the same on both sides, $P = F_1/A_1 = F_2/A_2$. Therefore, the force F_2 is greater than the force F_1 by a factor A_2/A_1, which is called the *force-multiplying factor.* Because liquid is neither added nor removed, the volume pushed down on the left as the piston moves down a distance d_1 equals the volume pushed up on the right as the right piston moves up a distance d_2. That is, $A_1d_1 = A_2d_2$; thus, the force-multiplying factor can also be written as d_1/d_2. Note that $F_1d_1 = F_2d_2$. Hydraulic brakes, car lifts, hydraulic jacks, and forklifts all make use of this principle (Fig. 15.5b).

Quick Quiz 15.4

A grain silo has many bands wrapped around its perimeter (Fig. 15.6). Why is the spacing between successive bands smaller at the lower portions of the silo, as shown in the photograph?

(a)   (b)

Figure 15.5   (a) Diagram of a hydraulic press. Because the increase in pressure is the same on the two sides, a small force $\mathbf{F}_1$ at the left produces a much greater force $\mathbf{F}_2$ at the right. (b) A vehicle undergoing repair is supported by a hydraulic lift in a garage. *(David Frazier)*

Figure 15.6   *(Henry Leap)*

EXAMPLE 15.2   **The Car Lift**

In a car lift used in a service station, compressed air exerts a force on a small piston that has a circular cross section and a radius of 5.00 cm. This pressure is transmitted by a liquid to a piston that has a radius of 15.0 cm. What force must the compressed air exert to lift a car weighing 13 300 N? What air pressure produces this force?

Solution   Because the pressure exerted by the compressed air is transmitted undiminished throughout the liquid, we have

$$F_1 = \left(\frac{A_1}{A_2}\right) F_2 = \frac{\pi(5.00 \times 10^{-2}\ \text{m})^2}{\pi(15.0 \times 10^{-2}\ \text{m})^2}\ (1.33 \times 10^4\ \text{N})$$

$$= 1.48 \times 10^3\ \text{N}$$

The air pressure that produces this force is

$$P = \frac{F_1}{A_1} = \frac{1.48 \times 10^3\ \text{N}}{\pi(5.00 \times 10^{-2}\ \text{m})^2} = \boxed{1.88 \times 10^5\ \text{Pa}}$$

This pressure is approximately twice atmospheric pressure.

The input work (the work done by $\mathbf{F}_1$) is equal to the output work (the work done by $\mathbf{F}_2$), in accordance with the principle of conservation of energy.

EXAMPLE 15.3   **A Pain in the Ear**

Estimate the force exerted on your eardrum due to the water above when you are swimming at the bottom of a pool that is 5.0 m deep.

Solution   First, we must find the unbalanced pressure on the eardrum; then, after estimating the eardrum's surface area, we can determine the force that the water exerts on it.

The air inside the middle ear is normally at atmospheric pressure P_0. Therefore, to find the net force on the eardrum, we must consider the difference between the total pressure at the bottom of the pool and atmospheric pressure:

$$P_{bot} - P_0 = \rho g h$$
$$= (1.00 \times 10^3 \text{ kg/m}^3)(9.80 \text{ m/s}^2)(5.0 \text{ m})$$
$$= 4.9 \times 10^4 \text{ Pa}$$

We estimate the surface area of the eardrum to be approximately $1 \text{ cm}^2 = 1 \times 10^{-4} \text{ m}^2$. This means that the force on it

is $F = (P_{bot} - P_0)A \approx 5$ N. Because a force on the eardrum of this magnitude is extremely uncomfortable, swimmers often "pop their ears" while under water, an action that pushes air from the lungs into the middle ear. Using this technique equalizes the pressure on the two sides of the eardrum and relieves the discomfort.

EXAMPLE 15.4   **The Force on a Dam**

Water is filled to a height H behind a dam of width w (Fig. 15.7). Determine the resultant force exerted by the water on the dam.

Solution Because pressure varies with depth, we cannot calculate the force simply by multiplying the area by the pressure. We can solve the problem by finding the force dF ex-

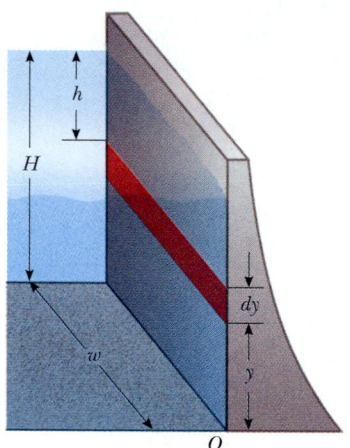

Figure 15.7 Because pressure varies with depth, the total force exerted on a dam must be obtained from the expression $F = \int P \, dA$, where dA is the area of the dark strip.

erted on a narrow horizontal strip at depth h and then integrating the expression to find the total force. Let us imagine a vertical y axis, with $y = 0$ at the bottom of the dam and our strip a distance y above the bottom.

We can use Equation 15.4 to calculate the pressure at the depth h; we omit atmospheric pressure because it acts on both sides of the dam:

$$P = \rho g h = \rho g (H - y)$$

Using Equation 15.2, we find that the force exerted on the shaded strip of area $dA = w \, dy$ is

$$dF = P \, dA = \rho g (H - y) w \, dy$$

Therefore, the total force on the dam is

$$F = \int P \, dA = \int_0^H \rho g (H - y) w \, dy = \boxed{\tfrac{1}{2}\rho g w H^2}$$

Note that the thickness of the dam shown in Figure 15.7 increases with depth. This design accounts for the greater and greater pressure that the water exerts on the dam at greater depths.

Exercise Find an expression for the average pressure on the dam from the total force exerted by the water on the dam.

Answer $\tfrac{1}{2}\rho g H$.

15.3 PRESSURE MEASUREMENTS

One simple device for measuring pressure is the open-tube manometer illustrated in Figure 15.8a. One end of a U-shaped tube containing a liquid is open to the atmosphere, and the other end is connected to a system of unknown pressure P. The difference in pressure $P - P_0$ is equal to $\rho g h$; hence, $P = P_0 + \rho g h$. The pressure P is called the **absolute pressure,** and the difference $P - P_0$ is called the **gauge pressure.** The latter is the value that normally appears on a pressure gauge. For example, the pressure you measure in your bicycle tire is the gauge pressure.

Another instrument used to measure pressure is the common *barometer,* which was invented by Evangelista Torricelli (1608–1647). The barometer consists of a

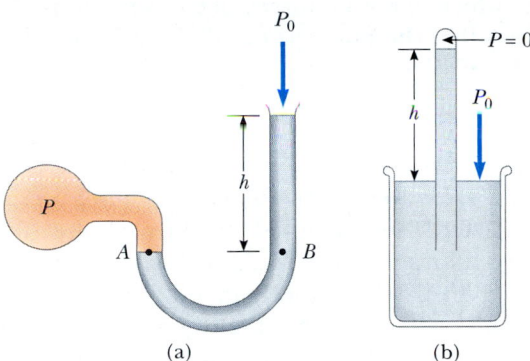

Figure 15.8 Two devices for measuring pressure: (a) an open-tube manometer and (b) a mercury barometer.

long, mercury-filled tube closed at one end and inverted into an open container of mercury (Fig. 15.8b). The closed end of the tube is nearly a vacuum, and so its pressure can be taken as zero. Therefore, it follows that $P_0 = \rho gh$, where h is the height of the mercury column.

One atmosphere ($P_0 = 1$ atm) of pressure is defined as the pressure that causes the column of mercury in a barometer tube to be exactly 0.760 0 m in height at 0°C, with $g = 9.806\ 65\ \text{m/s}^2$. At this temperature, mercury has a density of $13.595 \times 10^3\ \text{kg/m}^3$; therefore,

$$P_0 = \rho gh = (13.595 \times 10^3\ \text{kg/m}^3)(9.806\ 65\ \text{m/s}^2)(0.760\ 0\ \text{m})$$
$$= 1.013 \times 10^5\ \text{Pa} = 1\ \text{atm}$$

Quick Quiz 15.5

Other than the obvious problem that occurs with freezing, why don't we use water in a barometer in the place of mercury?

15.4 ▸ BUOYANT FORCES AND ARCHIMEDES'S PRINCIPLE

Have you ever tried to push a beach ball under water? This is extremely difficult to do because of the large upward force exerted by the water on the ball. The upward force exerted by water on any immersed object is called a **buoyant force.** We can determine the magnitude of a buoyant force by applying some logic and Newton's second law. Imagine that, instead of air, the beach ball is filled with water. If you were standing on land, it would be difficult to hold the water-filled ball in your arms. If you held the ball while standing neck deep in a pool, however, the force you would need to hold it would almost disappear. In fact, the required force would be zero if we were to ignore the thin layer of plastic of which the beach ball is made. Because the water-filled ball is in equilibrium while it is submerged, the magnitude of the upward buoyant force must equal its weight.

If the submerged ball were filled with air rather than water, then the upward buoyant force exerted by the surrounding water would still be present. However, because the weight of the water is now replaced by the much smaller weight of that volume of air, the net force is upward and quite great; as a result, the ball is pushed to the surface.

Archimedes (c. 287–212 B.C.)
Archimedes, a Greek mathematician, physicist, and engineer, was perhaps the greatest scientist of antiquity. He was the first to compute accurately the ratio of a circle's circumference to its diameter, and he showed how to calculate the volume and surface area of spheres, cylinders, and other geometric shapes. He is well known for discovering the nature of the buoyant force.

Archimedes was also a gifted inventor. One of his practical inventions, still in use today, is Archimedes's screw—an inclined, rotating, coiled tube originally used to lift water from the holds of ships. He also invented the catapult and devised systems of levers, pulleys, and weights for raising heavy loads. Such inventions were successfully used to defend his native city Syracuse during a two-year siege by the Romans.

The manner in which buoyant forces act is summarized by **Archimedes's principle,** which states that **the magnitude of the buoyant force always equals the weight of the fluid displaced by the object.** The buoyant force acts vertically upward through the point that was the center of gravity of the displaced fluid.

Note that Archimedes's principle does not refer to the makeup of the object experiencing the buoyant force. The object's composition is not a factor in the buoyant force. We can verify this in the following manner: Suppose we focus our attention on the indicated cube of liquid in the container illustrated in Figure 15.9. This cube is in equilibrium as it is acted on by two forces. One of these forces is the gravitational force $\mathbf{F}_g$. What cancels this downward force? Apparently, the rest of the liquid in the container is holding the cube in equilibrium. Thus, the magnitude of the buoyant force $\mathbf{B}$ exerted on the cube is exactly equal to the magnitude of $\mathbf{F}_g$, which is the weight of the liquid inside the cube:

$$B = F_g$$

Now imagine that the cube of liquid is replaced by a cube of steel of the same dimensions. What is the buoyant force acting on the steel? The liquid surrounding a cube behaves in the same way no matter what the cube is made of. Therefore, **the buoyant force acting on the steel cube is the same as the buoyant force acting on a cube of liquid of the same dimensions.** In other words, the magnitude of the buoyant force is the same as the weight of the *liquid* cube, not the steel cube. Although mathematically more complicated, this same principle applies to submerged objects of any shape, size, or density.

Although we have described the magnitude and direction of the buoyant force, we still do not know its origin. Why would a fluid exert such a strange force, almost as if the fluid were trying to expel a foreign body? To understand why, look again at Figure 15.9. The pressure at the bottom of the cube is greater than the pressure at the top by an amount ρgh, where h is the length of any side of the cube. The pressure difference ΔP between the bottom and top faces of the cube is equal to the buoyant force per unit area of those faces—that is, $\Delta P = B/A$. Therefore, $B = (\Delta P)A = (\rho gh)A = \rho gV$, where V is the volume of the cube. Because the mass of the fluid in the cube is $M = \rho V$, we see that

$$B = F_g = \rho Vg = Mg \tag{15.5}$$

where Mg is the weight of the fluid in the cube. Thus, the buoyant force is a result of the pressure differential on a submerged or partly submerged object.

Before we proceed with a few examples, it is instructive for us to compare the forces acting on a totally submerged object with those acting on a floating (partly submerged) object.

Case 1: Totally Submerged Object   When an object is totally submerged in a fluid of density ρ_f, the magnitude of the upward buoyant force is $B = \rho_f V_o g$, where V_o is the volume of the object. If the object has a mass M and density ρ_o, its weight is equal to $F_g = Mg = \rho_o V_o g$, and the net force on it is $B - F_g = (\rho_f - \rho_o) V_o g$. Hence, if the density of the object is less than the density of the fluid, then the downward force of gravity is less than the buoyant force, and the unconstrained object accelerates upward (Fig. 15.10a). If the density of the object is greater than the density of the fluid, then the upward buoyant force is less than the downward force of gravity, and the unsupported object sinks (Fig. 15.10b).

Case 2: Floating Object   Now consider an object of volume V_o in static equilibrium floating on a fluid—that is, an object that is only partially submerged. In this

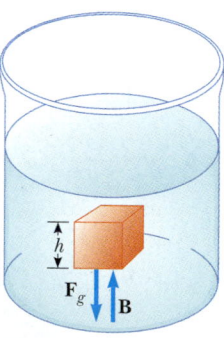

Figure 15.9   The external forces acting on the cube of liquid are the force of gravity $\mathbf{F}_g$ and the buoyant force $\mathbf{B}$. Under equilibrium conditions, $B = F_g$.

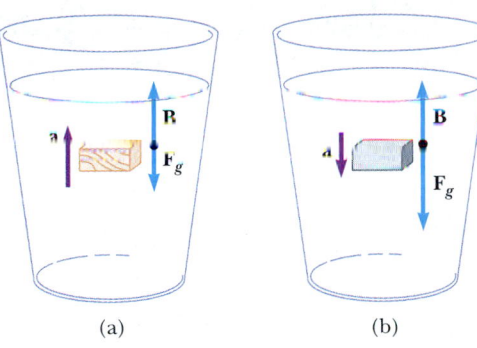

(a)   (b)

Figure 15.10 (a) A totally submerged object that is less dense than the fluid in which it is submerged experiences a net upward force. (b) A totally submerged object that is denser than the fluid sinks.

Hot-air balloons. Because hot air is less dense than cold air, a net upward force acts on the balloons. *(Richard Megna/Fundamental Photographs)*

case, the upward buoyant force is balanced by the downward gravitational force acting on the object. If V_f is the volume of the fluid displaced by the object (this volume is the same as the volume of that part of the object that is beneath the fluid level), the buoyant force has a magnitude $B = \rho_f V_f g$. Because the weight of the object is $F_g = Mg = \rho_o V_o g$, and because $F_g = B$, we see that $\rho_f V_f g = \rho_o V_o g$, or

$$\frac{\rho_o}{\rho_f} = \frac{V_f}{V_o} \qquad (15.6)$$

Under normal conditions, the average density of a fish is slightly greater than the density of water. It follows that the fish would sink if it did not have some mechanism for adjusting its density. The fish accomplishes this by internally regulating the size of its air-filled swim bladder to balance the change in the magnitude of the buoyant force acting on it. In this manner, fish are able to swim to various depths. Unlike a fish, a scuba diver cannot achieve neutral buoyancy (at which the buoyant force just balances the weight) by adjusting the magnitude of the buoyant force B. Instead, the diver adjusts F_g by manipulating lead weights.

Quick Quiz 15.6

Steel is much denser than water. In view of this fact, how do steel ships float?

Quick Quiz 15.7

A glass of water contains a single floating ice cube (Fig. 15.11). When the ice melts, does the water level go up, go down, or remain the same?

Quick Quiz 15.8

When a person in a rowboat in a small pond throws an anchor overboard, does the water level of the pond go up, go down, or remain the same?

Figure 15.11

EXAMPLE 15.5 **Eureka!**

Archimedes supposedly was asked to determine whether a crown made for the king consisted of pure gold. Legend has it that he solved this problem by weighing the crown first in air and then in water, as shown in Figure 15.12. Suppose the scale read 7.84 N in air and 6.86 N in water. What should Archimedes have told the king?

Solution When the crown is suspended in air, the scale

reads the true weight $T_1 = F_g$ (neglecting the buoyancy of air). When it is immersed in water, the buoyant force **B** reduces the scale reading to an apparent weight of $T_2 = F_g - B$. Hence, the buoyant force exerted on the crown is the difference between its weight in air and its weight in water:

$$B = F_g - T_2 = 7.84 \text{ N} - 6.86 \text{ N} = 0.98 \text{ N}$$

Because this buoyant force is equal in magnitude to the weight of the displaced water, we have $\rho_w g V_w = 0.98$ N, where V_w is the volume of the displaced water and ρ_w is its density. Also, the volume of the crown V_c is equal to the volume of the displaced water because the crown is completely submerged. Therefore,

$$V_c = V_w = \frac{0.98 \text{ N}}{g\rho_w} = \frac{0.98 \text{ N}}{(9.8 \text{ m/s}^2)(1\,000 \text{ kg/m}^3)}$$

$$= 1.0 \times 10^{-4} \text{ m}^3$$

Finally, the density of the crown is

$$\rho_c = \frac{m_c}{V_c} = \frac{m_c g}{V_c g} = \frac{7.84 \text{ N}}{(1.0 \times 10^{-4} \text{ m}^3)(9.8 \text{ m/s}^2)}$$

$$= 8.0 \times 10^3 \text{ kg/m}^3$$

From Table 15.1 we see that the density of gold is 19.3×10^3 kg/m³. Thus, Archimedes should have told the king that

he had been cheated. Either the crown was hollow, or it was not made of pure gold.

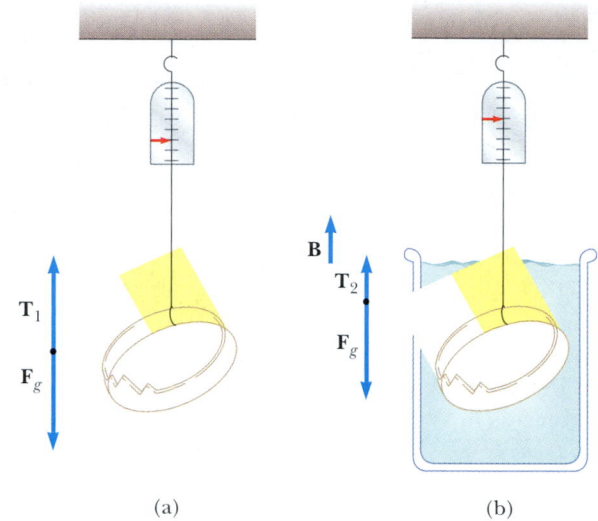

(a)   (b)

Figure 15.12   (a) When the crown is suspended in air, the scale reads its true weight $T_1 = F_g$ (the buoyancy of air is negligible). (b) When the crown is immersed in water, the buoyant force **B** reduces the scale reading to the apparent weight $T_2 = F_g - B$.

EXAMPLE 15.6   **A Titanic Surprise**

An iceberg floating in seawater, as shown in Figure 15.13a, is extremely dangerous because much of the ice is below the surface. This hidden ice can damage a ship that is still a considerable distance from the visible ice. What fraction of the iceberg lies below the water level?

Solution   This problem corresponds to Case 2. The weight of the iceberg is $F_{gi} = \rho_i V_i g$, where $\rho_i = 917$ kg/m³ and V_i is the volume of the whole iceberg. The magnitude of the up-

ward buoyant force equals the weight of the displaced water: $B = \rho_w V_w g$, where V_w, the volume of the displaced water, is equal to the volume of the ice beneath the water (the shaded region in Fig. 15.13b) and ρ_w is the density of seawater, $\rho_w = 1\,030$ kg/m³. Because $\rho_i V_i g = \rho_w V_w g$, the fraction of ice beneath the water's surface is

$$f = \frac{V_w}{V_i} = \frac{\rho_i}{\rho_w} = \frac{917 \text{ kg/m}^3}{1\,030 \text{ kg/m}^3} = 0.890 \quad \text{or} \quad \boxed{89.0\%}$$

(a)

(b)

Figure 15.13   (a) Much of the volume of this iceberg is beneath the water. *(Geraldine Prentice/Tony Stone Images)* (b) A ship can be damaged even when it is not near the exposed ice.

15.5 ▶ FLUID DYNAMICS

Thus far, our study of fluids has been restricted to fluids at rest. We now turn our attention to fluids in motion. Instead of trying to study the motion of each particle of the fluid as a function of time, we describe the properties of a moving fluid at each point as a function of time.

Flow Characteristics

When fluid is in motion, its flow can be characterized as being one of two main types. The flow is said to be **steady,** or **laminar,** if each particle of the fluid follows a smooth path, such that the paths of different particles never cross each other, as shown in Figure 15.14. In steady flow, the velocity of the fluid at any point remains constant in time.

Above a certain critical speed, fluid flow becomes **turbulent;** turbulent flow is irregular flow characterized by small whirlpool-like regions, as shown in Figure 15.15.

The term **viscosity** is commonly used in the description of fluid flow to characterize the degree of internal friction in the fluid. This internal friction, or *viscous force*, is associated with the resistance that two adjacent layers of fluid have to moving relative to each other. Viscosity causes part of the kinetic energy of a fluid to be converted to internal energy. This mechanism is similar to the one by which an object sliding on a rough horizontal surface loses kinetic energy.

Because the motion of real fluids is very complex and not fully understood, we make some simplifying assumptions in our approach. In our model of an **ideal fluid,** we make the following four assumptions:

1. **The fluid is nonviscous.** In a nonviscous fluid, internal friction is neglected. An object moving through the fluid experiences no viscous force.
2. **The flow is steady.** In steady (laminar) flow, the velocity of the fluid at each point remains constant.

Properties of an ideal fluid

Figure 15.14 Laminar flow around an automobile in a test wind tunnel. *(Andy Sacks/Tony Stone Images)*

Figure 15.15 Hot gases from a cigarette made visible by smoke particles. The smoke first moves in laminar flow at the bottom and then in turbulent flow above. *(Werner Wolff/Black Star)*

3. **The fluid is incompressible.** The density of an incompressible fluid is constant.
4. **The flow is irrotational.** In irrotational flow, the fluid has no angular momentum about any point. If a small paddle wheel placed anywhere in the fluid does not rotate about the wheel's center of mass, then the flow is irrotational.

15.6   STREAMLINES AND THE EQUATION OF CONTINUITY

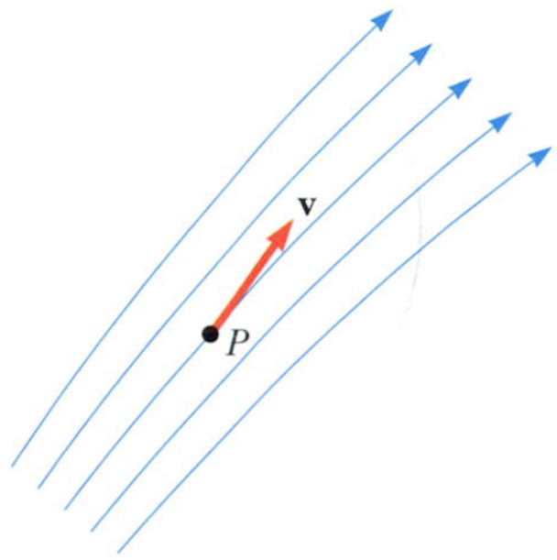

Figure 15.16   A particle in laminar flow follows a streamline, and at each point along its path the particle's velocity is tangent to the streamline.

The path taken by a fluid particle under steady flow is called a **streamline.** The velocity of the particle is always tangent to the streamline, as shown in Figure 15.16. A set of streamlines like the ones shown in Figure 15.16 form a *tube of flow.* Note that fluid particles cannot flow into or out of the sides of this tube; if they could, then the streamlines would cross each other.

Consider an ideal fluid flowing through a pipe of nonuniform size, as illustrated in Figure 15.17. The particles in the fluid move along streamlines in steady flow. In a time t, the fluid at the bottom end of the pipe moves a distance $\Delta x_1 = v_1 t$. If A_1 is the cross-sectional area in this region, then the mass of fluid contained in the left shaded region in Figure 15.17 is $m_1 = \rho A_1 \Delta x_1 = \rho A_1 v_1 t$, where ρ is the (nonchanging) density of the ideal fluid. Similarly, the fluid that moves through the upper end of the pipe in the time t has a mass $m_2 = \rho A_2 v_2 t$. However, because *mass is conserved* and because the flow is steady, the mass that crosses A_1 in a time t must equal the mass that crosses A_2 in the time t. That is, $m_1 = m_2$, or $\rho A_1 v_1 t = \rho A_2 v_2 t$; this means that

Equation of continuity

$$A_1 v_1 = A_2 v_2 = \text{constant} \qquad\qquad (15.7)$$

This expression is called the **equation of continuity.** It states that

> the product of the area and the fluid speed at all points along the pipe is a constant for an incompressible fluid.

This equation tells us that the speed is high where the tube is constricted (small A) and low where the tube is wide (large A). The product Av, which has the dimensions of volume per unit time, is called either the *volume flux* or the *flow rate.* The condition $Av = \text{constant}$ is equivalent to the statement that the volume of fluid that enters one end of a tube in a given time interval equals the volume leaving the other end of the tube in the same time interval if no leaks are present.

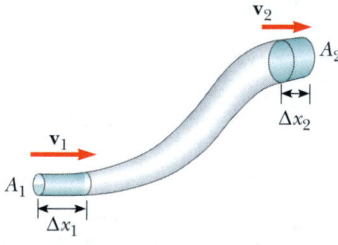

Figure 15.17   A fluid moving with steady flow through a pipe of varying cross-sectional area. The volume of fluid flowing through area A_1 in a time interval t must equal the volume flowing through area A_2 in the same time interval. Therefore, $A_1 v_1 = A_2 v_2$.

Figure 15.18   *(George Semple)*

Quick Quiz 15.9

As water flows from a faucet, as shown in Figure 15.18, why does the stream of water become narrower as it descends?

EXAMPLE 15.7 ▸ **Niagara Falls**

Each second, 5 525 m³ of water flows over the 670-m-wide cliff of the Horseshoe Falls portion of Niagara Falls. The water is approximately 2 m deep as it reaches the cliff. What is its speed at that instant?

Solution The cross-sectional area of the water as it reaches the edge of the cliff is $A = (670 \text{ m})(2 \text{ m}) = 1\,340 \text{ m}^2$. The flow rate of 5 525 m³/s is equal to Av. This gives

$$v = \frac{5\,525 \text{ m}^3/\text{s}}{A} = \frac{5\,525 \text{ m}^3/\text{s}}{1\,340 \text{ m}^2} = \boxed{4 \text{ m/s}}$$

Note that we have kept only one significant figure because our value for the depth has only one significant figure.

Exercise A barrel floating along in the river plunges over the Falls. How far from the base of the cliff is the barrel when it reaches the water 49 m below?

Answer $13 \text{ m} \approx 10 \text{ m}$.

15.7 ▸ BERNOULLI'S EQUATION

When you press your thumb over the end of a garden hose so that the opening becomes a small slit, the water comes out at high speed, as shown in Figure 15.19. Is the water under greater pressure when it is inside the hose or when it is out in the air? You can answer this question by noting how hard you have to push your thumb against the water inside the end of the hose. The pressure inside the hose is definitely greater than atmospheric pressure.

The relationship between fluid speed, pressure, and elevation was first derived in 1738 by the Swiss physicist Daniel Bernoulli. Consider the flow of an ideal fluid through a nonuniform pipe in a time t, as illustrated in Figure 15.20. Let us call the lower shaded part section 1 and the upper shaded part section 2. The force exerted by the fluid in section 1 has a magnitude P_1A_1. The work done by this force in a time t is $W_1 = F_1\Delta x_1 = P_1A_1\Delta x_1 = P_1V$, where V is the volume of section 1. In a similar manner, the work done by the fluid in section 2 in the same time t is $W_2 = -P_2A_2\Delta x_2 = -P_2V$. (The volume that passes through section 1 in a time t equals the volume that passes through section 2 in the same time.) This work is negative because the fluid force opposes the displacement. Thus, the net work done by these forces in the time t is

$$W = (P_1 - P_2)V$$

Daniel Bernoulli (1700–1782)
Daniel Bernoulli, a Swiss physicist and mathematician, made important discoveries in fluid dynamics. Born into a family of mathematicians, he was the only member of the family to make a mark in physics.
Bernoulli's most famous work, *Hydrodynamica*, was published in 1738; it is both a theoretical and a practical study of equilibrium, pressure, and speed in fluids. He showed that as the speed of a fluid increases, its pressure decreases.
In *Hydrodynamica* Bernoulli also attempted the first explanation of the behavior of gases with changing pressure and temperature; this was the beginning of the kinetic theory of gases, a topic we study in Chapter 21.
(Corbis–Bettmann)

Figure 15.19 The speed of water spraying from the end of a hose increases as the size of the opening is decreased with the thumb.
(George Semple)

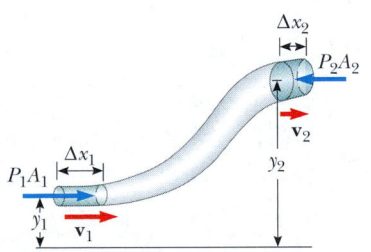

Figure 15.20 A fluid in laminar flow through a constricted pipe. The volume of the shaded section on the left is equal to the volume of the shaded section on the right.

QuickLab

Place two soda cans on their sides approximately 2 cm apart on a table. Align your mouth at table level and with the space between the cans. Blow a horizontal stream of air through this space. What do the cans do? Is this what you expected? Compare this with the force acting on a car parked close to the edge of a road when a big truck goes by. How does the outcome relate to Equation 15.9?

Part of this work goes into changing the kinetic energy of the fluid, and part goes into changing the gravitational potential energy. If m is the mass that enters one end and leaves the other in a time t, then the change in the kinetic energy of this mass is

$$\Delta K = \tfrac{1}{2}mv_2{}^2 - \tfrac{1}{2}mv_1{}^2$$

The change in gravitational potential energy is

$$\Delta U = mgy_2 - mgy_1$$

We can apply Equation 8.13, $W = \Delta K + \Delta U$, to this volume of fluid to obtain

$$(P_1 - P_2)V = \tfrac{1}{2}mv_2{}^2 - \tfrac{1}{2}mv_1{}^2 + mgy_2 - mgy_1$$

If we divide each term by V and recall that $\rho = m/V$, this expression reduces to

$$P_1 - P_2 = \tfrac{1}{2}\rho v_2{}^2 - \tfrac{1}{2}\rho v_1{}^2 + \rho g y_2 - \rho g y_1$$

Rearranging terms, we obtain

$$P_1 + \tfrac{1}{2}\rho v_1{}^2 + \rho g y_1 = P_2 + \tfrac{1}{2}\rho v_2{}^2 + \rho g y_2 \qquad \textbf{(15.8)}$$

This is **Bernoulli's equation** as applied to an ideal fluid. It is often expressed as

Bernoulli's equation

$$\boxed{P + \tfrac{1}{2}\rho v^2 + \rho g y = \text{constant}} \qquad \textbf{(15.9)}$$

This expression specifies that, in laminar flow, the sum of the pressure (P), kinetic energy per unit volume ($\tfrac{1}{2}\rho v^2$), and gravitational potential energy per unit volume ($\rho g y$) has the same value at all points along a streamline.

When the fluid is at rest, $v_1 = v_2 = 0$ and Equation 15.8 becomes

$$P_1 - P_2 = \rho g(y_2 - y_1) = \rho g h$$

This is in agreement with Equation 15.4.

EXAMPLE 15.8   **The Venturi Tube**

The horizontal constricted pipe illustrated in Figure 15.21, known as a *Venturi tube,* can be used to measure the flow speed of an incompressible fluid. Let us determine the flow speed at point 2 if the pressure difference $P_1 - P_2$ is known.

Solution   Because the pipe is horizontal, $y_1 = y_2$, and applying Equation 15.8 to points 1 and 2 gives

$$(1) \qquad P_1 + \tfrac{1}{2}\rho v_1{}^2 = P_2 + \tfrac{1}{2}\rho v_2{}^2$$

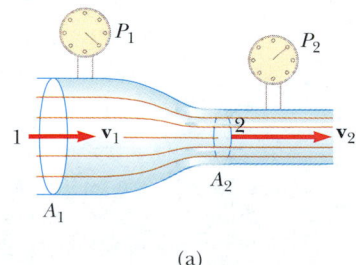

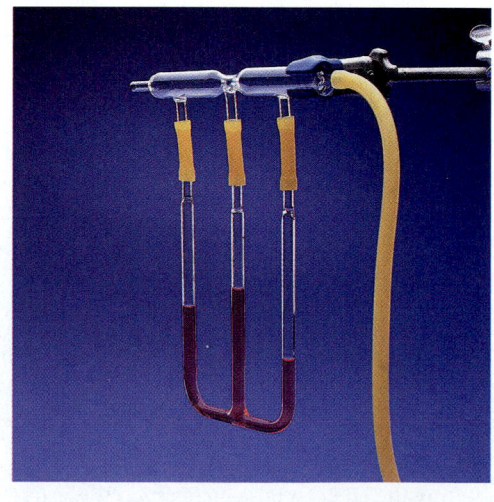

Figure 15.21   (a) Pressure P_1 is greater than pressure P_2 because $v_1 < v_2$. This device can be used to measure the speed of fluid flow. (b) A Venturi tube. *(Courtesy of Central Scientific Company)*

(a)   (b)

From the equation of continuity, $A_1 v_1 = A_2 v_2$, we find that

$$(2) \qquad v_1 = \frac{A_2}{A_1} v_2$$

Substituting this expression into equation (1) gives

$$P_1 + \tfrac{1}{2}\rho\left(\frac{A_2}{A_1}\right)^2 v_2{}^2 = P_2 + \tfrac{1}{2}\rho v_2{}^2$$

$$v_2 = A_1 \sqrt{\frac{2(P_1 - P_2)}{\rho(A_1{}^2 - A_2{}^2)}}$$

We can use this result and the continuity equation to obtain an expression for v_1. Because $A_2 < A_1$, Equation (2) shows us that $v_2 > v_1$. This result, together with equation (1), indicates that $P_1 > P_2$. In other words, the pressure is reduced in the constricted part of the pipe. This result is somewhat analogous to the following situation: Consider a very crowded room in which people are squeezed together. As soon as a door is opened and people begin to exit, the squeezing (pressure) is least near the door, where the motion (flow) is greatest.

EXAMPLE 15.9 ▸ A Good Trick

It is possible to blow a dime off a table and into a tumbler. Place the dime about 2 cm from the edge of the table. Place the tumbler on the table horizontally with its open edge about 2 cm from the dime, as shown in Figure 15.22a. If you blow forcefully across the top of the dime, it will rise, be caught in the airstream, and end up in the tumbler. The mass of a dime is $m = 2.24$ g, and its surface area is $A = 2.50 \times 10^{-4}$ m^2. How hard are you blowing when the dime rises and travels into the tumbler?

Solution Figure 15.22b indicates we must calculate the upward force acting on the dime. First, note that a thin stationary layer of air is present between the dime and the table. When you blow across the dime, it deflects most of the moving air from your breath across its top, so that the air above the dime has a greater speed than the air beneath it. This fact, together with Bernoulli's equation, demonstrates that the air moving across the top of the dime is at a lower pressure than the air beneath the dime. If we neglect the small thickness of the dime, we can apply Equation 15.8 to obtain

$$P_{\text{above}} + \tfrac{1}{2}\rho v_{\text{above}}^2 = P_{\text{beneath}} + \tfrac{1}{2}\rho v_{\text{beneath}}^2$$

Because the air beneath the dime is almost stationary, we can neglect the last term in this expression and write the difference as $P_{\text{beneath}} - P_{\text{above}} = \tfrac{1}{2}\rho v_{\text{above}}^2$. If we multiply this pressure difference by the surface area of the dime, we obtain the upward force acting on the dime. At the very least, this upward force must balance the gravitational force acting on the dime, and so, taking the density of air from Table 15.1, we can state that

$$F_g = mg = (P_{\text{beneath}} - P_{\text{above}})A = (\tfrac{1}{2}\rho v_{\text{above}}^2)A$$

$$v_{\text{above}} = \sqrt{\frac{2mg}{\rho A}} = \sqrt{\frac{2(2.24 \times 10^{-3}\ \text{kg})(9.80\ \text{m/s}^2)}{(1.29\ \text{kg/m}^3)(2.50 \times 10^{-4}\ \text{m}^2)}}$$

$$v_{\text{above}} = 11.7\ \text{m/s}$$

The air you blow must be moving faster than this if the upward force is to exceed the weight of the dime. Practice this trick a few times and then impress all your friends!

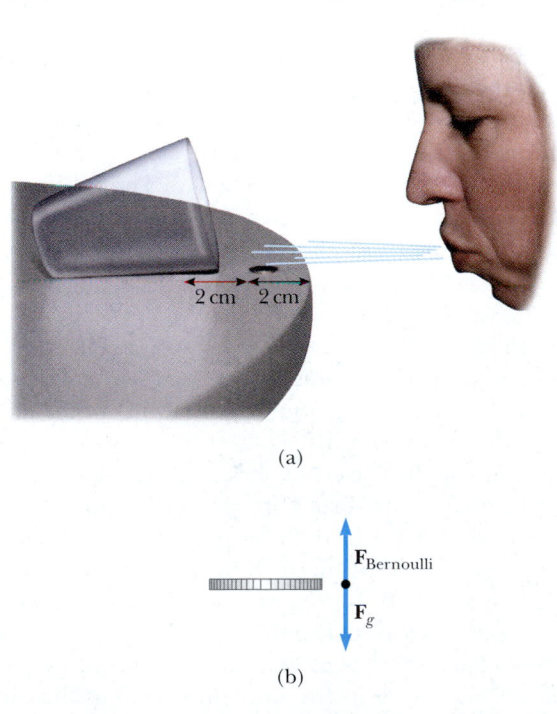

(a)

(b)

Figure 15.22

EXAMPLE 15.10   Torricelli's Law

An enclosed tank containing a liquid of density ρ has a hole in its side at a distance y_1 from the tank's bottom (Fig. 15.23). The hole is open to the atmosphere, and its diameter is much smaller than the diameter of the tank. The air above the liquid is maintained at a pressure P. Determine the speed at

which the liquid leaves the hole when the liquid's level is a distance h above the hole.

Solution   Because $A_2 \gg A_1$, the liquid is approximately at rest at the top of the tank, where the pressure is P. Applying Bernoulli's equation to points 1 and 2 and noting that at the hole P_1 is equal to atmospheric pressure P_0, we find that

$$P_0 + \tfrac{1}{2}\rho v_1{}^2 + \rho g y_1 = P + \rho g y_2$$

But $y_2 - y_1 = h$; thus, this expression reduces to

$$v_1 = \sqrt{\frac{2(P - P_0)}{\rho} + 2gh}$$

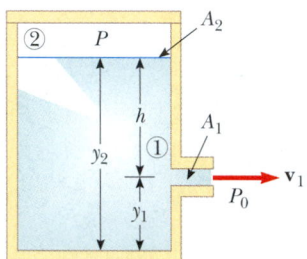

When P is much greater than P_0 (so that the term $2gh$ can be neglected), the exit speed of the water is mainly a function of P. If the tank is open to the atmosphere, then $P = P_0$ and $v_1 = \sqrt{2gh}$. In other words, for an open tank, the speed of liquid coming out through a hole a distance h below the surface is equal to that acquired by an object falling freely through a vertical distance h. This phenomenon is known as **Torricelli's law.**

Figure 15.23   When P is much larger than atmospheric pressure P_0, the liquid speed as the liquid passes through the hole in the side of the container is given approximately by $v_1 = \sqrt{2(P - P_0)/\rho}$.

Optional Section

15.8   OTHER APPLICATIONS OF BERNOULLI'S EQUATION

The lift on an aircraft wing can be explained, in part, by the Bernoulli effect. Airplane wings are designed so that the air speed above the wing is greater than that below the wing. As a result, the air pressure above the wing is less than the pressure below, and a net upward force on the wing, called *lift*, results.

Another factor influencing the lift on a wing is shown in Figure 15.24. The wing has a slight upward tilt that causes air molecules striking its bottom to be deflected downward. This deflection means that the wing is exerting a downward force on the air. According to Newton's third law, the air must exert an equal and opposite force on the wing.

Finally, turbulence also has an effect. If the wing is tilted too much, the flow of air across the upper surface becomes turbulent, and the pressure difference across the wing is not as great as that predicted by Bernoulli's equation. In an extreme case, this turbulence may cause the aircraft to stall.

In general, an object moving through a fluid experiences lift as the result of any effect that causes the fluid to change its direction as it flows past the object. Some factors that influence lift are the shape of the object, its orientation with respect to the fluid flow, any spinning motion it might have, and the texture of its surface. For example, a golf ball struck with a club is given a rapid backspin, as shown in Figure 15.25a. The dimples on the ball help "entrain" the air to follow the curvature of the ball's surface. This effect is most pronounced on the top half of the ball, where the ball's surface is moving in the same direction as the air flow. Figure 15.25b shows a thin layer of air wrapping part way around the ball and being deflected downward as a result. Because the ball pushes the air down, the air must push up on the ball. Without the dimples, the air is not as well entrained,

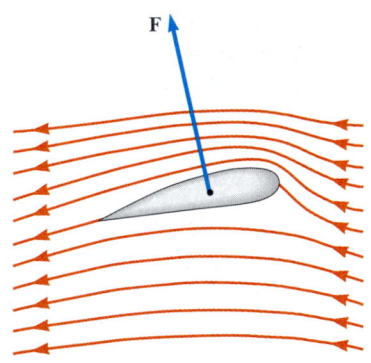

Figure 15.24   Streamline flow around an airplane wing. The pressure above the wing is less than the pressure below, and a dynamic lift upward results.

(a)

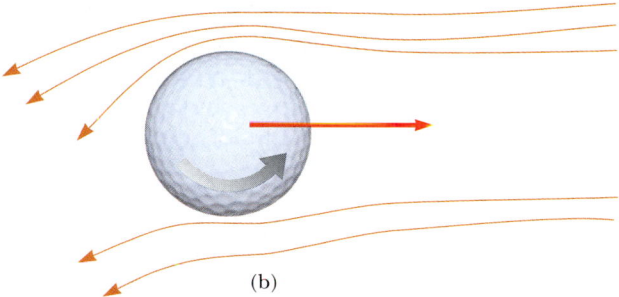

(b)

Figure 15.25 (a) A golf ball is made to spin when struck by the club. (*© Harold E. Edgerton/Courtesy of Palm Press, Inc.*) (b) The spinning ball experiences a lifting force that allows it to travel much farther than it would if it were not spinning.

and the golf ball does not travel as far. For the same reason, a tennis ball's fuzz helps the spinning ball "grab" the air rushing by and helps deflect it.

A number of devices operate by means of the pressure differentials that result from differences in a fluid's speed. For example, a stream of air passing over one end of an open tube, the other end of which is immersed in a liquid, reduces the pressure above the tube, as illustrated in Figure 15.26. This reduction in pressure causes the liquid to rise into the air stream. The liquid is then dispersed into a fine spray of droplets. You might recognize that this so-called atomizer is used in perfume bottles and paint sprayers. The same principle is used in the carburetor of a gasoline engine. In this case, the low-pressure region in the carburetor is produced by air drawn in by the piston through the air filter. The gasoline vaporizes in that region, mixes with the air, and enters the cylinder of the engine, where combustion occurs.

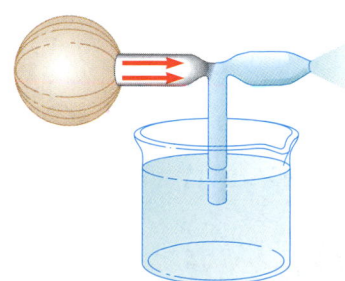

Figure 15.26 A stream of air passing over a tube dipped into a liquid causes the liquid to rise in the tube.

Quick Quiz 15.10

People in buildings threatened by a tornado are often told to open the windows to minimize damage. Why?

SUMMARY

The **pressure** P in a fluid is the force per unit area exerted by the fluid on a surface:

$$P \equiv \frac{F}{A} \tag{15.1}$$

In the SI system, pressure has units of newtons per square meter (N/m^2), and $1 \ N/m^2 = 1$ pascal (Pa).

The pressure in a fluid at rest varies with depth h in the fluid according to the expression

$$P = P_0 + \rho g h \tag{15.4}$$

where P_0 is atmospheric pressure ($= 1.013 \times 10^5 \ N/m^2$) and ρ is the density of the fluid, assumed uniform.

Pascal's law states that when pressure is applied to an enclosed fluid, the pressure is transmitted undiminished to every point in the fluid and to every point on the walls of the container.

When an object is partially or fully submerged in a fluid, the fluid exerts on the object an upward force called the **buoyant force.** According to **Archimedes's principle,** the magnitude of the buoyant force is equal to the weight of the fluid displaced by the object. Be sure you can apply this principle to a wide variety of situations, including sinking objects, floating ones, and neutrally buoyant ones.

You can understand various aspects of a fluid's dynamics by assuming that the fluid is nonviscous and incompressible and that the fluid's motion is a steady flow with no rotation.

Two important concepts regarding ideal fluid flow through a pipe of nonuniform size are as follows:

1. The flow rate (volume flux) through the pipe is constant; this is equivalent to stating that the product of the cross-sectional area A and the speed v at any point is a constant. This result is expressed in the **equation of continuity:**

$$A_1 v_1 = A_2 v_2 = \text{constant} \tag{15.7}$$

   You can use this expression to calculate how the velocity of a fluid changes as the fluid is constricted or as it flows into a more open area.
2. The sum of the pressure, kinetic energy per unit volume, and gravitational potential energy per unit volume has the same value at all points along a streamline. This result is summarized in **Bernoulli's equation:**

$$P + \tfrac{1}{2}\rho v^2 + \rho g y = \text{constant} \tag{15.9}$$

QUESTIONS

1. Two drinking glasses of the same weight but of different shape and different cross-sectional area are filled to the same level with water. According to the expression $P = P_0 + \rho g h$, the pressure at the bottom of both glasses is the same. In view of this, why does one glass weigh more than the other?

2. If the top of your head has a surface area of $100 \ cm^2$, what is the weight of the air above your head?

3. When you drink a liquid through a straw, you reduce the pressure in your mouth and let the atmosphere move the liquid. Explain why this is so. Can you use a straw to sip a drink on the Moon?

4. A helium-filled balloon rises until its density becomes the same as that of the surrounding air. If a sealed submarine begins to sink, will it go all the way to the bottom of the ocean or will it stop when its density becomes the same as that of the surrounding water?

5. A fish rests on the bottom of a bucket of water while the

bucket is being weighed. When the fish begins to swim around, does the weight change?

6. Does a ship ride higher in the water of an inland lake or in the ocean? Why?

7. Lead has a greater density than iron, and both metals are denser than water. Is the buoyant force on a lead object greater than, less than, or equal to the buoyant force on an iron object of the same volume?

8. The water supply for a city is often provided by reservoirs built on high ground. Water flows from the reservoir, through pipes, and into your home when you turn the tap on your faucet. Why is the flow of water more rapid out of a faucet on the first floor of a building than it is in an apartment on a higher floor?

9. Smoke rises in a chimney faster when a breeze is blowing than when there is no breeze at all. Use Bernoulli's equation to explain this phenomenon.

10. If a Ping–Pong ball is above a hair dryer, the ball can be suspended in the air column emitted by the dryer. Explain.

11. When ski jumpers are airborne (Fig. Q15.11), why do they bend their bodies forward and keep their hands at their sides?

18. Why do airplane pilots prefer to take off into the wind?

19. If you release a ball while inside a freely falling elevator, the ball remains in front of you rather than falling to the floor because the ball, the elevator, and you all experience the same downward acceleration **g**. What happens if you repeat this experiment with a helium-filled balloon? (This one is tricky.)

20. Two identical ships set out to sea. One is loaded with a cargo of Styrofoam, and the other is empty. Which ship is more submerged?

21. A small piece of steel is tied to a block of wood. When the wood is placed in a tub of water with the steel on top, half of the block is submerged. If the block is inverted so that the steel is underwater, does the amount of the block submerged increase, decrease, or remain the same? What happens to the water level in the tub when the block is inverted?

22. Prairie dogs (Fig. Q15.22) ventilate their burrows by building a mound over one entrance, which is open to a stream of air. A second entrance at ground level is open to almost stagnant air. How does this construction create an air flow through the burrow?

Figure Q15.11 (Galen Powell/Peter Arnold, Inc.)

Figure Q15.22 (Pamela Zilly/The Image Bank)

12. Explain why a sealed bottle partially filled with a liquid can float.

13. When is the buoyant force on a swimmer greater—after exhaling or after inhaling?

14. A piece of unpainted wood barely floats in a container partly filled with water. If the container is sealed and then pressurized above atmospheric pressure, does the wood rise, sink, or remain at the same level? (*Hint:* Wood is porous.)

15. A flat plate is immersed in a liquid at rest. For what orientation of the plate is the pressure on its flat surface uniform?

16. Because atmospheric pressure is about 10^5 N/m^2 and the area of a person's chest is about 0.13 m^2, the force of the atmosphere on one's chest is around 13 000 N. In view of this enormous force, why don't our bodies collapse?

17. How would you determine the density of an irregularly shaped rock?

23. An unopened can of diet cola floats when placed in a tank of water, whereas a can of regular cola of the same brand sinks in the tank. What do you suppose could explain this phenomenon?

24. Figure Q15.24 shows a glass cylinder containing four liquids of different densities. From top to bottom, the liquids are oil (orange), water (yellow), salt water (green), and mercury (silver). The cylinder also contains, from top to bottom, a Ping–Pong ball, a piece of wood, an egg, and a steel ball. (a) Which of these liquids has the lowest density, and which has the greatest? (b) What can you conclude about the density of each object?

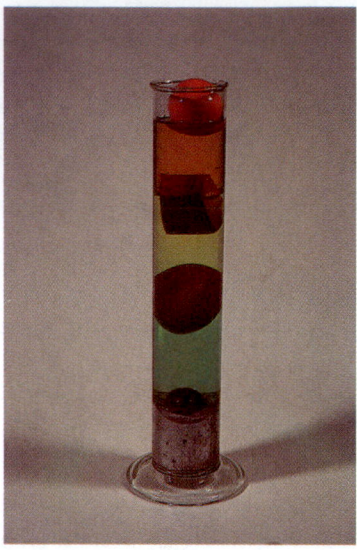

Figure Q15.24   *(Henry Leap and Jim Lehman)*

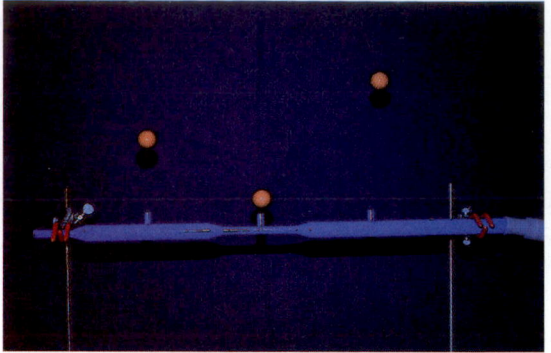

Figure Q15.25   *(Henry Leap and Jim Lehman)*

25. In Figure Q15.25, an air stream moves from right to left through a tube that is constricted at the middle. Three Ping–Pong balls are levitated in equilibrium above the vertical columns through which the air escapes. (a) Why is the ball at the right higher than the one in the middle?

(b) Why is the ball at the left lower than the ball at the right even though the horizontal tube has the same dimensions at these two points?

26. You are a passenger on a spacecraft. For your comfort, the interior contains air just like that at the surface of the Earth. The craft is coasting through a very empty region of space. That is, a nearly perfect vacuum exists just outside the wall. Suddenly a meteoroid pokes a hole, smaller than the palm of your hand, right through the wall next to your seat. What will happen? Is there anything you can or should do about it?

PROBLEMS

1, 2, 3 = straightforward, intermediate, challenging   ☐ = full solution available in the *Student Solutions Manual and Study Guide*
WEB = solution posted at **http://www.saunderscollege.com/physics/**   🖥 = Computer useful in solving problem   🔲 = Interactive Physics
☐ = paired numerical/symbolic problems

Section 15.1   Pressure

1. Calculate the mass of a solid iron sphere that has a diameter of 3.00 cm.

2. Find the order of magnitude of the density of the *nucleus* of an atom. What does this result suggest concerning the structure of matter? (Visualize a nucleus as protons and neutrons closely packed together. Each has mass 1.67×10^{-27} kg and radius on the order of 10^{-15} m.)

3. A 50.0-kg woman balances on one heel of a pair of high-heeled shoes. If the heel is circular and has a radius of 0.500 cm, what pressure does she exert on the floor?

4. The four tires of an automobile are inflated to a gauge pressure of 200 kPa. Each tire has an area of 0.024 0 m^2 in contact with the ground. Determine the weight of the automobile.

5. What is the total mass of the Earth's atmosphere? (The radius of the Earth is 6.37×10^6 m, and atmospheric pressure at the Earth's surface is 1.013×10^5 N/m^2.)

Section 15.2   Variation of Pressure with Depth

6. (a) Calculate the absolute pressure at an ocean depth of 1 000 m. Assume the density of seawater is 1 024 kg/m^3 and that the air above exerts a pressure of 101.3 kPa. (b) At this depth, what force must the frame around a circular submarine porthole having a diameter of 30.0 cm exert to counterbalance the force exerted by the water?

7. The spring of the pressure gauge shown in Figure 15.2 has a force constant of 1 000 N/m, and the piston has a diameter of 2.00 cm. When the gauge is lowered into water, at what depth does the piston move in by 0.500 cm?

8. The small piston of a hydraulic lift has a cross-sectional area of 3.00 cm^2, and its large piston has a cross-sectional area of 200 cm^2 (see Fig. 15.5a). What force must be applied to the small piston for it to raise a load of 15.0 kN? (In service stations, this force is usually generated with the use of compressed air.)

(a)

(b)

Figure P15.10

WEB **9.** What must be the contact area between a suction cup (completely exhausted) and a ceiling if the cup is to support the weight of an 80.0-kg student?

10. (a) A very powerful vacuum cleaner has a hose 2.86 cm in diameter. With no nozzle on the hose, what is the weight of the heaviest brick that the cleaner can lift (Fig. P15.10)? (b) A very powerful octopus uses one sucker of diameter 2.86 cm on each of the two shells of a clam in an attempt to pull the shells apart. Find the greatest force that the octopus can exert in salt water 32.3 m in depth. (*Caution:* Experimental verification can be interesting, but do not drop a brick on your foot. Do not overheat the motor of a vacuum cleaner. Do not get an octopus mad at you.)

11. For the cellar of a new house, a hole with vertical sides descending 2.40 m is dug in the ground. A concrete foundation wall is built all the way across the 9.60-m width of the excavation. This foundation wall is 0.183 m away from the front of the cellar hole. During a rainstorm, drainage from the street fills up the space in front of the concrete wall but not the cellar behind the wall. The water does not soak into the clay soil. Find the force that the water causes on the foundation wall. For comparison, the weight of the water is given by

$$2.40 \text{ m} \times 9.60 \text{ m} \times 0.183 \text{ m} \times 1\,000 \text{ kg/m}^3$$
$$\times\ 9.80 \text{ m/s}^2 = 41.3 \text{ kN}$$

12. A swimming pool has dimensions 30.0 m × 10.0 m and a flat bottom. When the pool is filled to a depth of 2.00 m with fresh water, what is the force caused by the water on the bottom? On each end? On each side?

13. A sealed spherical shell of diameter d is rigidly attached to a cart that is moving horizontally with an acceleration a, as shown in Figure P15.13. The sphere is nearly filled with a fluid having density ρ and also contains one small bubble of air at atmospheric pressure. Find an expression for the pressure P at the center of the sphere.

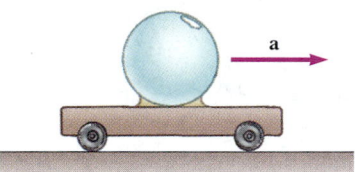

Figure P15.13

14. The tank shown in Figure P15.14 is filled with water to a depth of 2.00 m. At the bottom of one of the side walls is a rectangular hatch 1.00 m high and 2.00 m wide. The hatch is hinged at its top. (a) Determine the force that the water exerts on the hatch. (b) Find the torque exerted about the hinges.

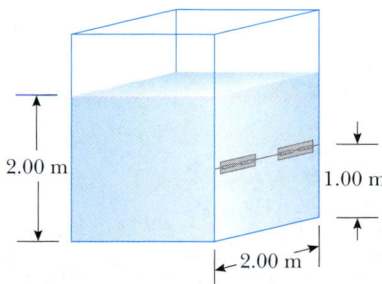

2.00 m

1.00 m

2.00 m

Figure P15.14

15. **Review Problem.** A solid copper ball with a diameter of 3.00 m at sea level is placed at the bottom of the ocean (at a depth of 10.0 km). If the density of seawater is 1 030 kg/m^3, by how much (approximately) does the diameter of the ball decrease when it reaches bottom? Take the bulk modulus of copper as $14.0 \times 10^{10} \text{ N/m}^2$.

Section 15.3   Pressure Measurements

16. Normal atmospheric pressure is 1.013×10^5 Pa. The approach of a storm causes the height of a mercury barometer to drop by 20.0 mm from the normal height. What is the atmospheric pressure? (The density of mercury is 13.59 g/cm³.)

WEB 17. Blaise Pascal duplicated Torricelli's barometer, using a red Bordeaux wine, of density 984 kg/m³, as the working liquid (Fig. P15.17). What was the height h of the wine column for normal atmospheric pressure? Would you expect the vacuum above the column to be as good as that for mercury?

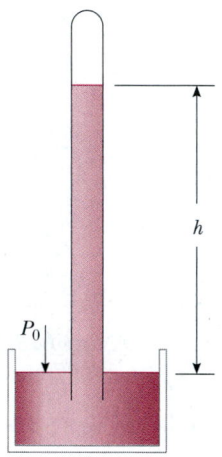

Figure P15.17

18. Mercury is poured into a U-tube, as shown in Figure P15.18a. The left arm of the tube has a cross-sectional area A_1 of 10.0 cm², and the right arm has a cross-sectional area A_2 of 5.00 cm². One-hundred grams of water are then poured into the right arm, as shown in Figure P15.18b. (a) Determine the length of the water column

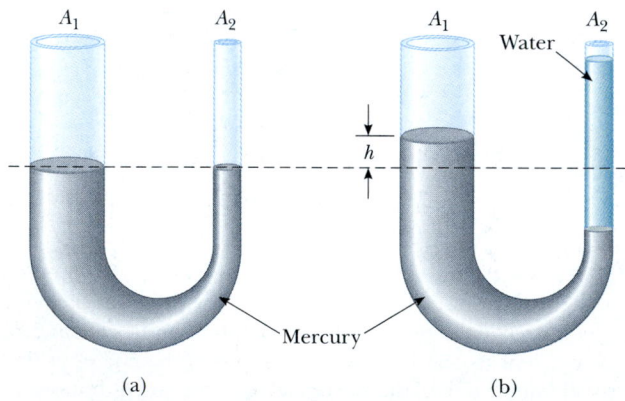

Figure P15.18

in the right arm of the U-tube. (b) Given that the density of mercury is 13.6 g/cm³, what distance h does the mercury rise in the left arm?

19. A U-tube of uniform cross-sectional area and open to the atmosphere is partially filled with mercury. Water is then poured into both arms. If the equilibrium configuration of the tube is as shown in Figure P15.19, with $h_2 = 1.00$ cm, determine the value of h_1.

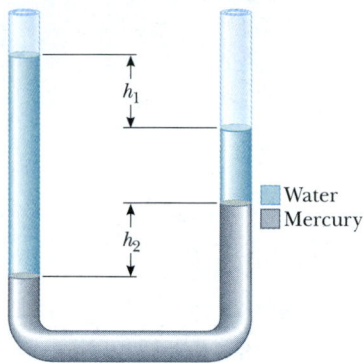

Figure P15.19

Section 15.4   Buoyant Forces and Archimedes's Principle

20. (a) A light balloon is filled with 400 m³ of helium. At 0°C, what is the mass of the payload that the balloon can lift? (b) In Table 15.1, note that the density of hydrogen is nearly one-half the density of helium. What load can the balloon lift if it is filled with hydrogen?

21. A Styrofoam slab has a thickness of 10.0 cm and a density of 300 kg/m³. When a 75.0-kg swimmer is resting on it, the slab floats in fresh water with its top at the same level as the water's surface. Find the area of the slab.

22. A Styrofoam slab has thickness h and density ρ_S. What is the area of the slab if it floats with its upper surface just awash in fresh water, when a swimmer of mass m is on top?

23. A piece of aluminum with mass 1.00 kg and density 2 700 kg/m³ is suspended from a string and then completely immersed in a container of water (Fig. P15.23). Calculate the tension in the string (a) before and (b) after the metal is immersed.

24. A 10.0-kg block of metal measuring 12.0 cm × 10.0 cm × 10.0 cm is suspended from a scale and immersed in water, as shown in Figure P15.23b. The 12.0-cm dimension is vertical, and the top of the block is 5.00 cm from the surface of the water. (a) What are the forces acting on the top and on the bottom of the block? (Take $P_0 = 1.013\ 0 \times 10^5$ N/m².) (b) What is the reading of the spring scale? (c) Show that the buoyant force equals the difference between the forces at the top and bottom of the block.

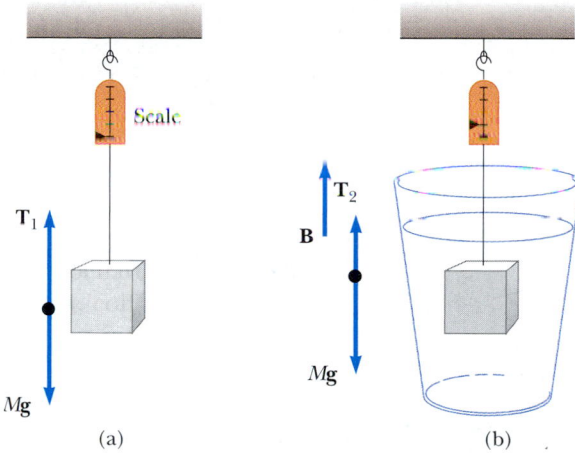

Figure P15.23   Problems 23 and 24.

WEB **25.** A cube of wood having a side dimension of 20.0 cm and a density of 650 kg/m^3 floats on water. (a) What is the distance from the horizontal top surface of the cube to the water level? (b) How much lead weight must be placed on top of the cube so that its top is just level with the water?

26. To an order of magnitude, how many helium-filled toy balloons would be required to lift you? Because helium is an irreplaceable resource, develop a theoretical answer rather than an experimental answer. In your solution, state what physical quantities you take as data and the values you measure or estimate for them.

27. A plastic sphere floats in water with 50.0% of its volume submerged. This same sphere floats in glycerin with 40.0% of its volume submerged. Determine the densities of the glycerin and the sphere.

28. A frog in a hemispherical pod finds that he just floats without sinking into a sea of blue-green ooze having a density of 1.35 g/cm^3 (Fig. P15.28). If the pod has a radius of 6.00 cm and a negligible mass, what is the mass of the frog?

Figure P15.28

29. How many cubic meters of helium are required to lift a balloon with a 400-kg payload to a height of 8 000 m? (Take ρ_{He} = 0.180 kg/m^3.) Assume that the balloon

maintains a constant volume and that the density of air decreases with the altitude z according to the expression $\rho_{air} = \rho_0 e^{-z/8\,000}$, where z is in meters and $\rho_0 = 1.25$ kg/m^3 is the density of air at sea level.

30. Review Problem. A long cylindrical tube of radius r is weighted on one end so that it floats upright in a fluid having a density ρ. It is pushed downward a distance x from its equilibrium position and then released. Show that the tube will execute simple harmonic motion if the resistive effects of the water are neglected, and determine the period of the oscillations.

31. A bathysphere used for deep-sea exploration has a radius of 1.50 m and a mass of 1.20×10^4 kg. To dive, this submarine takes on mass in the form of seawater. Determine the amount of mass that the submarine must take on if it is to descend at a constant speed of 1.20 m/s, when the resistive force on it is 1 100 N in the upward direction. Take 1.03×10^3 kg/m^3 as the density of seawater.

32. The United States possesses the eight largest warships in the world—aircraft carriers of the *Nimitz* class—and it is building one more. Suppose that one of the ships bobs up to float 11.0 cm higher in the water when 50 fighters take off from it at a location where g = 9.78 m/s^2. The planes have an average mass of 29 000 kg. Find the horizontal area enclosed by the waterline of the ship. (By comparison, its flight deck has an area of 18 000 m^2.)

Section 15.5   Fluid Dynamics

Section 15.6   Streamlines and the Equation of Continuity

Section 15.7   Bernoulli's Equation

33. (a) A water hose 2.00 cm in diameter is used to fill a 20.0-L bucket. If it takes 1.00 min to fill the bucket, what is the speed v at which water moves through the hose? (*Note:* 1 L = 1 000 cm^3.) (b) If the hose has a nozzle 1.00 cm in diameter, find the speed of the water at the nozzle.

34. A horizontal pipe 10.0 cm in diameter has a smooth reduction to a pipe 5.00 cm in diameter. If the pressure of the water in the larger pipe is 8.00×10^4 Pa and the pressure in the smaller pipe is 6.00×10^4 Pa, at what rate does water flow through the pipes?

35. A large storage tank, open at the top and filled with water, develops a small hole in its side at a point 16.0 m below the water level. If the rate of flow from the leak is 2.50×10^{-3} m^3/min, determine (a) the speed at which the water leaves the hole and (b) the diameter of the hole.

36. Through a pipe of diameter 15.0 cm, water is pumped from the Colorado River up to Grand Canyon Village, located on the rim of the canyon. The river is at an elevation of 564 m, and the village is at an elevation of 2 096 m. (a) What is the minimum pressure at which the water must be pumped if it is to arrive at the village?

(b) If 4 500 m³ are pumped per day, what is the speed of the water in the pipe? (c) What additional pressure is necessary to deliver this flow? (*Note:* You may assume that the acceleration due to gravity and the density of air are constant over this range of elevations.)

37. Water flows through a fire hose of diameter 6.35 cm at a rate of 0.012 0 m³/s. The fire hose ends in a nozzle with an inner diameter of 2.20 cm. What is the speed at which the water exits the nozzle?

38. Old Faithful Geyser in Yellowstone National Park erupts at approximately 1-h intervals, and the height of the water column reaches 40.0 m (Fig. P15.38). (a) Consider the rising stream as a series of separate drops. Analyze the free-fall motion of one of these drops to determine the speed at which the water leaves the ground. (b) Treating the rising stream as an ideal fluid in streamline flow, use Bernoulli's equation to determine the speed of the water as it leaves ground level. (c) What is the pressure (above atmospheric) in the heated underground chamber if its depth is 175 m? You may assume that the chamber is large compared with the geyser's vent.

Figure P15.38　(*Stan Osolinski/Dembinsky Photo Associates*)

(Optional)
Section 15.8　Other Applications of Bernoulli's Equation

39. An airplane has a mass of 1.60×10^4 kg, and each wing has an area of 40.0 m². During level flight, the pressure on the lower wing surface is 7.00×10^4 Pa. Determine the pressure on the upper wing surface.

40. A Venturi tube may be used as a fluid flow meter (see Fig. 15.21). If the difference in pressure is $P_1 - P_2 = 21.0$ kPa, find the fluid flow rate in cubic meters per second, given that the radius of the outlet tube is 1.00 cm, the radius of the inlet tube is 2.00 cm, and the fluid is gasoline ($\rho = 700$ kg/m³).

41. A Pitot tube can be used to determine the velocity of air flow by measuring the difference between the total pressure and the static pressure (Fig. P15.41). If the fluid in the tube is mercury, whose density is $\rho_{\text{Hg}} = 13\ 600$ kg/m³, and if $\Delta h = 5.00$ cm, find the speed of

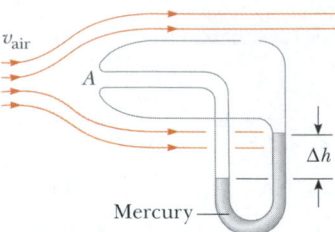

Figure P15.41

air flow. (Assume that the air is stagnant at point A, and take $\rho_{\text{air}} = 1.25$ kg/m³.)

42. An airplane is cruising at an altitude of 10 km. The pressure outside the craft is 0.287 atm; within the passenger compartment, the pressure is 1.00 atm and the temperature is 20°C. A small leak occurs in one of the window seals in the passenger compartment. Model the air as an ideal fluid to find the speed of the stream of air flowing through the leak.

43. A siphon is used to drain water from a tank, as illustrated in Figure P15.43. The siphon has a uniform diameter. Assume steady flow without friction. (a) If the distance $h = 1.00$ m, find the speed of outflow at the end of the siphon. (b) What is the limitation on the height of the top of the siphon above the water surface? (For the flow of liquid to be continuous, the pressure must not drop below the vapor pressure of the liquid.)

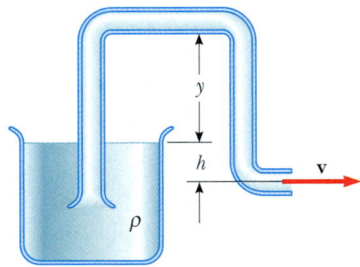

Figure P15.43

44. A hypodermic syringe contains a medicine with the density of water (Fig. P15.44). The barrel of the syringe has a cross-sectional area $A = 2.50 \times 10^{-5}$ m², and the needle has a cross-sectional area $a = 1.00 \times 10^{-8}$ m². In the absence of a force on the plunger, the pressure everywhere is 1 atm. A force **F** of magnitude 2.00 N acts on the plunger, making the medicine squirt horizontally from the needle. Determine the speed of the medicine as it leaves the needle's tip.

WEB 45. A large storage tank is filled to a height h_0. The tank is punctured at a height h above the bottom of the tank (Fig. P15.45). Find an expression for how far from the tank the exiting stream lands.

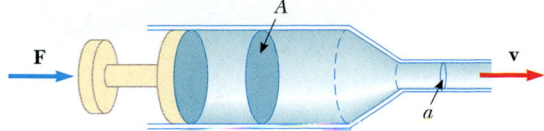

Figure P15.44

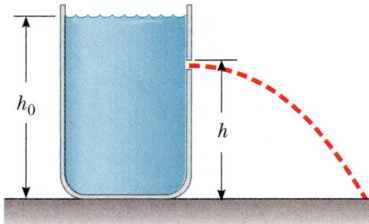

Figure P15.45 Problems 45 and 46.

46. A hole is punched at a height h in the side of a container of height h_0. The container is full of water, as shown in Figure P15.45. If the water is to shoot as far as possible horizontally, (a) how far from the bottom of the container should the hole be punched? (b) Neglecting frictional losses, how far (initially) from the side of the container will the water land?

ADDITIONAL PROBLEMS

47. A Ping–Pong ball has a diameter of 3.80 cm and an average density of 0.084 0 g/cm^3. What force would be required to hold it completely submerged under water?

48. Figure P15.48 shows a tank of water with a valve at the bottom. If this valve is opened, what is the maximum height attained by the water stream exiting the right side of the tank? Assume that $h = 10.0$ m, $L = 2.00$ m, and $\theta = 30.0°$, and that the cross-sectional area at point A is very large compared with that at point B.

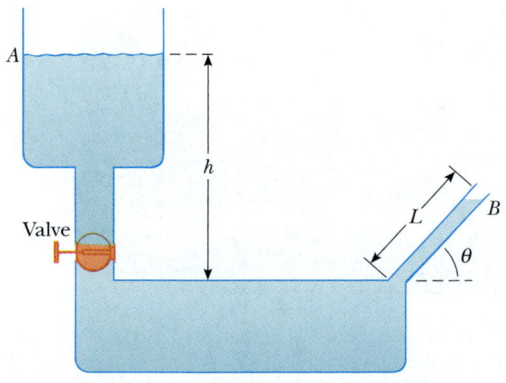

Figure P15.48

49. A helium-filled balloon is tied to a 2.00-m-long, 0.050 0-kg uniform string. The balloon is spherical with a radius of 0.400 m. When released, the balloon lifts a length h of string and then remains in equilibrium, as shown in Figure P15.49. Determine the value of h. The envelope of the balloon has a mass of 0.250 kg.

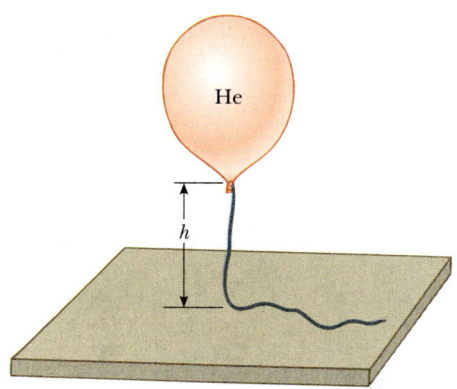

Figure P15.49

50. Water is forced out of a fire extinguisher by air pressure, as shown in Figure P15.50. How much gauge air pressure in the tank (above atmospheric) is required for the water jet to have a speed of 30.0 m/s when the water level is 0.500 m below the nozzle?

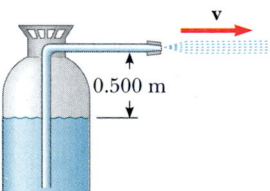

Figure P15.50

51. The true weight of an object is measured in a vacuum, where buoyant forces are absent. An object of volume V is weighed in air on a balance with the use of weights of density ρ. If the density of air is ρ_{air} and the balance reads F_g', show that the true weight F_g is

$$F_g = F_g' + \left(V - \frac{F_g'}{\rho g} \right) \rho_{air} g$$

52. Evangelista Torricelli was the first to realize that we live at the bottom of an ocean of air. He correctly surmised that the pressure of our atmosphere is attributable to the weight of the air. The density of air at 0°C at the Earth's surface is 1.29 kg/m^3. The density decreases with increasing altitude (as the atmosphere thins). On the other hand, if we assume that the density is constant

(1.29 kg/m^3) up to some altitude h, and zero above that altitude, then h would represent the thickness of our atmosphere. Use this model to determine the value of h that gives a pressure of 1.00 atm at the surface of the Earth. Would the peak of Mt. Everest rise above the surface of such an atmosphere?

53. A wooden dowel has a diameter of 1.20 cm. It floats in water with 0.400 cm of its diameter above water level (Fig. P15.53). Determine the density of the dowel.

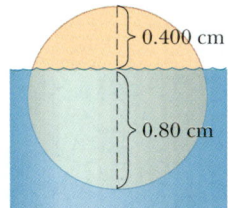

0.400 cm

0.80 cm

Figure P15.53

54. A light spring of constant $k = 90.0$ N/m rests vertically on a table (Fig. P15.54a). A 2.00-g balloon is filled with helium (density $= 0.180 \text{ kg/m}^3$) to a volume of 5.00 m^3 and is then connected to the spring, causing it to stretch as shown in Figure P15.54b. Determine the extension distance L when the balloon is in equilibrium.

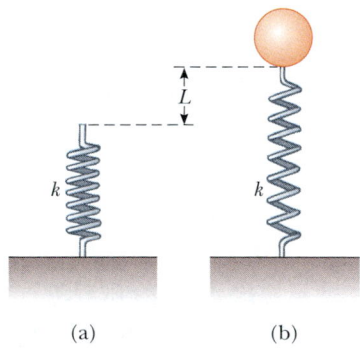

L

k k

(a)   (b)

Figure P15.54

55. A 1.00-kg beaker containing 2.00 kg of oil (density $= 916.0 \text{ kg/m}^3$) rests on a scale. A 2.00-kg block of iron is suspended from a spring scale and completely submerged in the oil, as shown in Figure P15.55. Determine the equilibrium readings of both scales.

56. A beaker of mass m_b containing oil of mass m_0 (density $= \rho_0$) rests on a scale. A block of iron of mass m_{Fe} is suspended from a spring scale and completely submerged in the oil, as shown in Figure P15.55. Determine the equilibrium readings of both scales.

WEB **57.** **Review Problem.** With reference to Figure 15.7, show that the total torque exerted by the water behind the

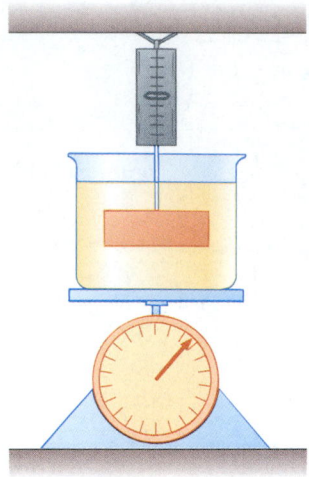

Figure P15.55   Problems 55 and 56.

dam about an axis through O is $\frac{1}{6}\rho gwH^3$. Show that the effective line of action of the total force exerted by the water is at a distance $\frac{1}{3}H$ above O.

58. In about 1657 Otto von Guericke, inventor of the air pump, evacuated a sphere made of two brass hemispheres. Two teams of eight horses each could pull the hemispheres apart only on some trials, and then "with greatest difficulty," with the resulting sound likened to a cannon firing (Fig. P15.58). (a) Show that the force F

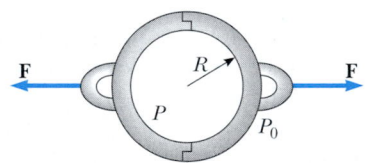

F   F

R

P

P_0

Figure P15.58   The colored engraving, dated 1672, illustrates Otto von Guericke's demonstration of the force due to air pressure as performed before Emperor Ferdinand III in 1657. (*The Granger Collection*)

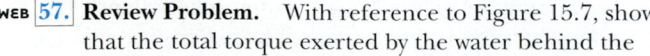

required to pull the evacuated hemispheres apart is $\pi R^2(P_0 - P)$, where R is the radius of the hemispheres and P is the pressure inside the hemispheres, which is much less than P_0. (b) Determine the force if $P = 0.100P_0$ and $R = 0.300$ m.

59. In 1983 the United States began coining the cent piece out of copper-clad zinc rather than pure copper. The mass of the old copper cent is 3.083 g, whereas that of the new cent is 2.517 g. Calculate the percent of zinc (by volume) in the new cent. The density of copper is 8.960 g/cm³, and that of zinc is 7.133 g/cm³. The new and old coins have the same volume.

60. A thin spherical shell with a mass of 4.00 kg and a diameter of 0.200 m is filled with helium (density = 0.180 kg/m³). It is then released from rest on the bottom of a pool of water that is 4.00 m deep. (a) Neglecting frictional effects, show that the shell rises with constant acceleration and determine the value of that acceleration. (b) How long does it take for the top of the shell to reach the water's surface?

61. An incompressible, nonviscous fluid initially rests in the vertical portion of the pipe shown in Figure P15.61a, where $L = 2.00$ m. When the valve is opened, the fluid flows into the horizontal section of the pipe. What is the speed of the fluid when all of it is in the horizontal section, as in Figure P15.61b? Assume that the cross-sectional area of the entire pipe is constant.

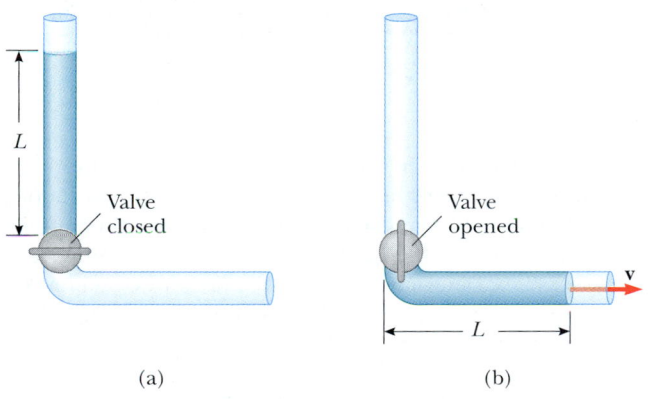

Figure P15.61

62. Review Problem.   A uniform disk with a mass of 10.0 kg and a radius of 0.250 m spins at 300 rev/min on a low-friction axle. It must be brought to a stop in 1.00 min by a brake pad that makes contact with the disk at an average distance of 0.220 m from the axis. The coefficient of friction between the pad and the disk is 0.500. A piston in a cylinder with a diameter of 5.00 cm presses the brake pad against the disk. Find the pressure that the brake fluid in the cylinder must have.

63. Figure P15.63 shows Superman attempting to drink water through a very long straw. With his great strength,

Figure P15.63

he achieves maximum possible suction. The walls of the tubular straw do not collapse. (a) Find the maximum height through which he can lift the water. (b) Still thirsty, the Man of Steel repeats his attempt on the Moon, which has no atmosphere. Find the difference between the water levels inside and outside the straw.

64. Show that the variation of atmospheric pressure with altitude is given by $P = P_0e^{-\alpha h}$, where $\alpha = \rho_0g/P_0$, P_0 is atmospheric pressure at some reference level $y = 0$, and ρ_0 is the atmospheric density at this level. Assume that the decrease in atmospheric pressure with increasing altitude is given by Equation 15.4, so that $dP/dy = -\rho g$, and assume that the density of air is proportional to the pressure.

65. A cube of ice whose edge measures 20.0 mm is floating in a glass of ice-cold water with one of its faces parallel to the water's surface. (a) How far below the water surface is the bottom face of the block? (b) Ice-cold ethyl alcohol is gently poured onto the water's surface to form a layer 5.00 mm thick above the water. The alcohol does not mix with the water. When the ice cube again attains hydrostatic equilibrium, what is the distance from the top of the water to the bottom face of the block? (c) Additional cold ethyl alcohol is poured onto the water's surface until the top surface of the alcohol coincides with the top surface of the ice cube (in hydrostatic equilibrium). How thick is the required layer of ethyl alcohol?

66. **Review Problem.**   A light balloon filled with helium with a density of 0.180 kg/m³ is tied to a light string of length $L = 3.00$ m. The string is tied to the ground, forming an "inverted" simple pendulum, as shown in Figure P15.66a. If the balloon is displaced slightly from its equilibrium position as shown in Figure P15.66b, (a) show that the ensuing motion is simple harmonic and (b) determine the period of the motion. Take the density of air to be 1.29 kg/m³ and ignore any energy loss due to air friction.

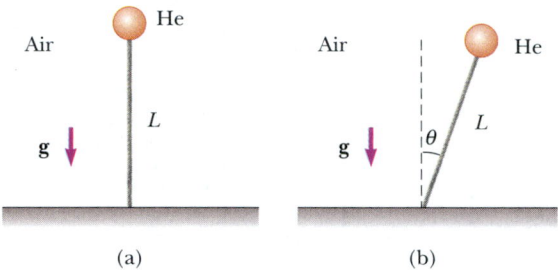

(a)   (b)

Figure P15.66

67. The water supply of a building is fed through a main 6.00-cm-diameter pipe. A 2.00-cm-diameter faucet tap located 2.00 m above the main pipe is observed to fill a 25.0-L container in 30.0 s. (a) What is the speed at which the water leaves the faucet? (b) What is the gauge pressure in the 6-cm main pipe? (Assume that the faucet is the only "leak" in the building.)

68. The *spirit-in-glass thermometer*, invented in Florence, Italy, around 1654, consists of a tube of liquid (the spirit) containing a number of submerged glass spheres with slightly different masses (Fig. P15.68). At sufficiently low temperatures, all the spheres float, but as the temperature rises, the spheres sink one after the other. The device is a crude but interesting tool for measuring temperature. Suppose that the tube is filled with ethyl alcohol, whose density is 0.789 45 g/cm³ at 20.0°C and decreases to 0.780 97 g/cm³ at 30.0°C. (a) If one of the spheres has a radius of 1.000 cm and is in equilibrium halfway up the tube at 20.0°C, determine its mass. (b) When the temperature increases to 30.0°C, what mass must a second sphere of the same radius have to be in equilibrium at the halfway point? (c) At 30.0°C the first sphere has fallen to the bottom of the tube. What upward force does the bottom of the tube exert on this sphere?

69. A U-tube open at both ends is partially filled with water (Fig. P15.69a). Oil having a density of 750 kg/m³ is then poured into the right arm and forms a column $L = 5.00$ cm in height (Fig. P15.69b). (a) Determine

Figure P15.68   *(Courtesy of Jeanne Maier)*

the difference h in the heights of the two liquid surfaces. (b) The right arm is shielded from any air motion while air is blown across the top of the left arm until the surfaces of the two liquids are at the same height (Fig. P15.69c). Determine the speed of the air being blown across the left arm. (Take the density of air as 1.29 kg/m³.)

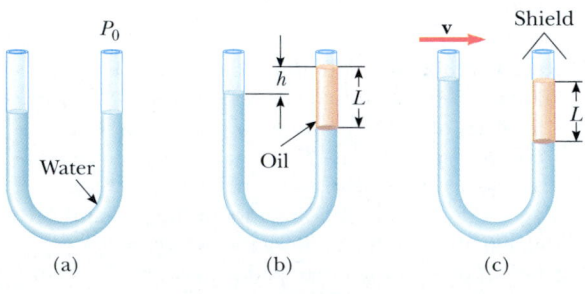

(a)   (b)   (c)

Figure P15.69

ANSWERS TO QUICK QUIZZES

15.1 You would be better off with the basketball player. Although weight is distributed over the larger surface area, equal to about half of the total surface area of the sneaker sole, the pressure (F/A) that he applies is relatively small. The woman's lesser weight is distributed over the very small cross-sectional area of the spiked heel. Some museums make women in high-heeled shoes wear slippers or special heel attachments so that they do not damage the wood floors.

15.2 If the professor were to try to support his entire weight on a single nail, the pressure exerted on his skin would be his entire weight divided by the very small surface area of the nail point. This extremely great pressure would cause the nail to puncture his skin. However, if the professor distributes his weight over several hundred nails, as shown in the photograph, the pressure exerted on his skin is considerably reduced because the surface area that supports his weight is now the total surface area of all the nail points. (Lying on the bed of nails is much more comfortable than sitting on the bed, and standing on the bed without shoes is definitely not recommended. Do not lie on a bed of nails unless you have been shown how to do so safely.)

15.3 Because the horizontal force exerted by the outside fluid on an element of the cylinder is equal and opposite the horizontal force exerted by the fluid on another element diametrically opposite the first, the net force on the cylinder in the horizontal direction is zero.

15.4 If you think of the grain stored in the silo as a fluid, then the pressure it exerts on the walls increases with increasing depth. The spacing between bands is smaller at the lower portions so that the greater outward forces acting on the walls can be overcome. The silo on the right shows another way of accomplishing the same thing: double banding at the bottom.

15.5 Because water is so much less dense than mercury, the column for a water barometer would have to be $h = P_0/\rho g = 10.3$ m high, and such a column is inconveniently tall.

15.6 The entire hull of a ship is full of air, and the density of air is about one-thousandth the density of water. Hence, the total weight of the ship equals the weight of the volume of water that is displaced by the portion of the ship that is below sea level.

15.7 Remains the same. In effect, the ice creates a "hole" in the water, and the weight of the water displaced from the hole is the same as all the weight of the cube. When the cube changes from ice to water, the water just fills the hole.

15.8 Goes down because the anchor displaces more water while in the boat than it does in the pond. While it is in the boat, the anchor can be thought of as a floating object that displaces a volume of water weighing as much as it does. When the anchor is thrown overboard, it sinks and displaces a volume of water equal to its own volume. Because the density of the anchor is greater than that of water, the volume of water that weighs the same as the anchor is greater than the volume of the anchor.

15.9 As the water falls, its speed increases. Because the flow rate Av must remain constant at all cross sections (see Eq. 15.7), the stream must become narrower as the speed increases.

15.10 The rapidly moving air characteristic of a tornado is at a pressure below atmospheric pressure. The stationary air inside the building remains at atmospheric pressure. The pressure difference results in an outward force on the roof and walls, and this force can be great enough to lift the roof off the building. Opening the windows helps to equalize the inside and outside pressures.

(CALVIN AND HOBBES ©1992 Watterson. Reprinted with permission of UNIVERSAL PRESS SYNDICATE. All rights reserved.)

part 2

Mechanical Waves

As we learned in Chapter 13, most elastic objects oscillate when a force is applied to them and then removed. That is, once such an object is distorted, its shape tends to be restored to some equilibrium configuration. Even the atoms in a solid oscillate about some equilibrium position, as if they were connected to their neighbors by imaginary springs.

Wave motion—the subject we study next—is closely related to the phenomenon of oscillation. Sound waves, earthquake waves, waves on stretched strings, and water waves are all produced by some source of oscillation. As a sound wave travels through the air, the air molecules oscillate back and forth; as a water wave travels across a pond, the water molecules oscillate up and down and backward and forward. In general, as waves travel through any medium, the particles of the medium move in repetitive cycles. Therefore, the motion of the particles bears a strong resemblance to the periodic motion of an oscillating pendulum or a mass attached to a spring.

To explain many other phenomena in nature, we must understand the concepts of oscillations and waves. For instance, although skyscrapers and bridges appear to be rigid, they actually oscillate, a fact that the architects and engineers who design and build them must take into account. To understand how radio and television work, we must understand the origin and nature of electromagnetic waves and how they propagate through space. Finally, much of what scientists have learned about atomic structure has come from information carried by waves. Therefore, we must first study oscillations and waves if we are to understand the concepts and theories of atomic physics.

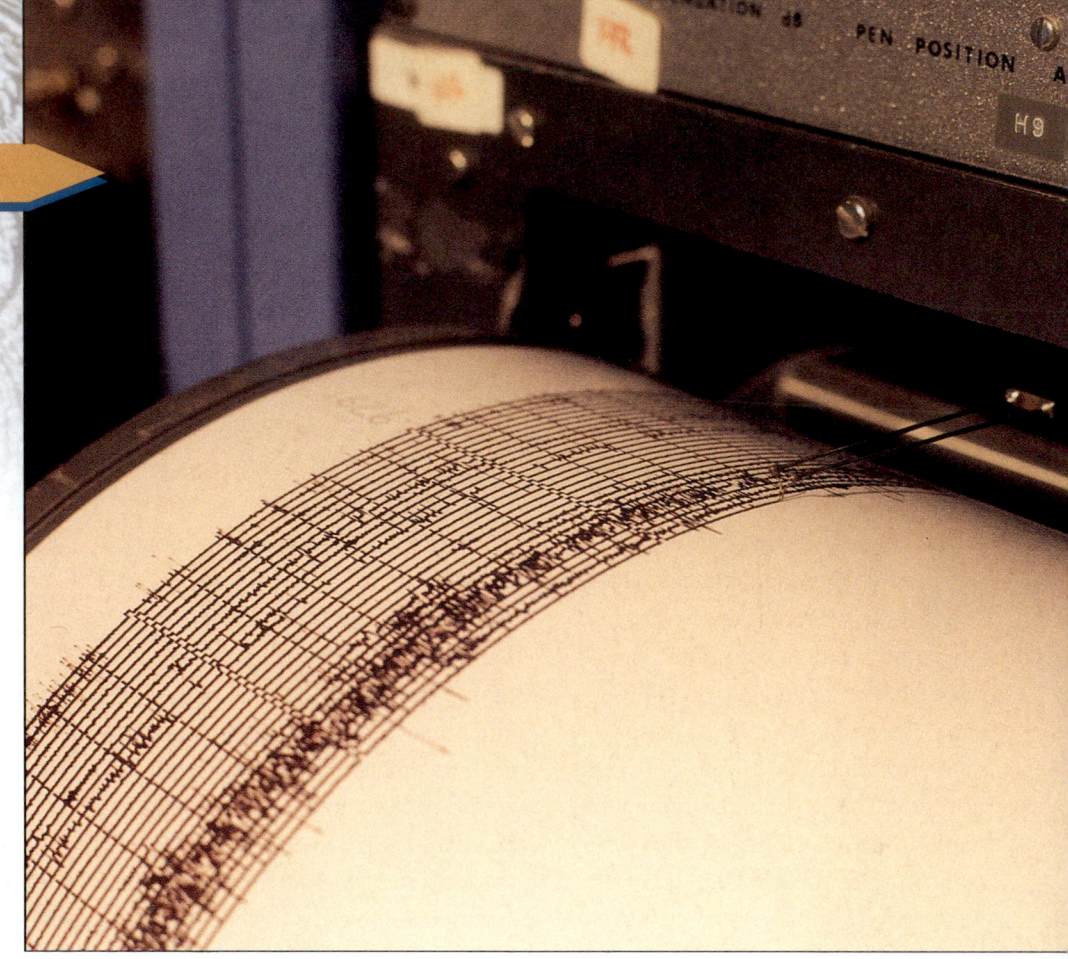

chapter

16

Wave Motion

Chapter Outline

Most of us experienced waves as children when we dropped a pebble into a pond. At the point where the pebble hits the water's surface, waves are created. These waves move outward from the creation point in expanding circles until they reach the shore. If you were to examine carefully the motion of a leaf floating on the disturbed water, you would see that the leaf moves up, down, and sideways about its original position but does not undergo any net displacement away from or toward the point where the pebble hit the water. The water molecules just beneath the leaf, as well as all the other water molecules on the pond's surface, behave in the same way. That is, the water *wave* moves from the point of origin to the shore, but the water is not carried with it.

An excerpt from a book by Einstein and Infeld gives the following remarks concerning wave phenomena:[1]

> A bit of gossip starting in Washington reaches New York [by word of mouth] very quickly, even though not a single individual who takes part in spreading it travels between these two cities. There are two quite different motions involved, that of the rumor, Washington to New York, and that of the persons who spread the rumor. The wind, passing over a field of grain, sets up a wave which spreads out across the whole field. Here again we must distinguish between the motion of the wave and the motion of the separate plants, which undergo only small oscillations... The particles constituting the medium perform only small vibrations, but the whole motion is that of a progressive wave. The essentially new thing here is that for the first time we consider the motion of something which is not matter, but energy propagated through matter.

The world is full of waves, the two main types being *mechanical* waves and *electromagnetic* waves. We have already mentioned examples of mechanical waves: sound waves, water waves, and "grain waves." In each case, some physical medium is being disturbed—in our three particular examples, air molecules, water molecules, and stalks of grain. Electromagnetic waves do not require a medium to propagate; some examples of electromagnetic waves are visible light, radio waves, television signals, and x-rays. Here, in Part 2 of this book, we study only mechanical waves.

The wave concept is abstract. When we observe what we call a water wave, what we see is a rearrangement of the water's surface. Without the water, there would be no wave. A wave traveling on a string would not exist without the string. Sound waves could not travel through air if there were no air molecules. With mechanical waves, what we interpret as a wave corresponds to the propagation of a disturbance through a medium.

Interference patterns produced by outward-spreading waves from many drops of liquid falling into a body of water. *(Martin Dohrn/ Science Photo Library/Photo Researchers, Inc.)*

[1] A. Einstein and L. Infeld, *The Evolution of Physics*, New York, Simon & Schuster, 1961. Excerpt from "What Is a Wave?"

The mechanical waves discussed in this chapter require (1) some source of disturbance, (2) a medium that can be disturbed, and (3) some physical connection through which adjacent portions of the medium can influence each other. We shall find that all waves carry energy. The amount of energy transmitted through a medium and the mechanism responsible for that transport of energy differ from case to case. For instance, the power of ocean waves during a storm is much greater than the power of sound waves generated by a single human voice.

16.1 BASIC VARIABLES OF WAVE MOTION

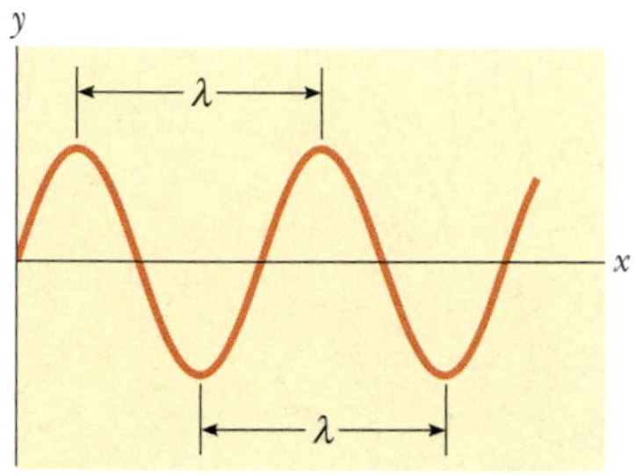

Figure 16.1 The wavelength λ of a wave is the distance between adjacent crests, adjacent troughs, or any other comparable adjacent identical points.

Imagine you are floating on a raft in a large lake. You slowly bob up and down as waves move past you. As you look out over the lake, you may be able to see the individual waves approaching. The point at which the displacement of the water from its normal level is highest is called the **crest** of the wave. The distance from one crest to the next is called the **wavelength** λ (Greek letter lambda). More generally, the wavelength is **the minimum distance between any two identical points (such as the crests) on adjacent waves,** as shown in Figure 16.1.

If you count the number of seconds between the arrivals of two adjacent waves, you are measuring the **period** T of the waves. In general, the period is **the time required for two identical points (such as the crests) of adjacent waves to pass by a point.**

The same information is more often given by the inverse of the period, which is called the **frequency** f. In general, the frequency of a periodic wave is **the number of crests (or troughs, or any other point on the wave) that pass a given point in a unit time interval.** The maximum displacement of a particle of the medium is called the **amplitude** A of the wave. For our water wave, this represents the highest distance of a water molecule above the undisturbed surface of the water as the wave passes by.

Waves travel with a specific speed, and this speed depends on the properties of the medium being disturbed. For instance, sound waves travel through room-temperature air with a speed of about 343 m/s (781 mi/h), whereas they travel through most solids with a speed greater than 343 m/s.

16.2 DIRECTION OF PARTICLE DISPLACEMENT

One way to demonstrate wave motion is to flick one end of a long rope that is under tension and has its opposite end fixed, as shown in Figure 16.2. In this manner, a single wave bump (called a *wave pulse*) is formed and travels along the rope with a definite speed. This type of disturbance is called a **traveling wave,** and Figure 16.2 represents four consecutive "snapshots" of the creation and propagation of the traveling wave. The rope is the medium through which the wave travels. Such a single pulse, in contrast to a train of pulses, has no frequency, no period, and no wavelength. However, the pulse does have definite amplitude and definite speed. As we shall see later, the properties of this particular medium that determine the speed of the wave are the tension in the rope and its mass per unit length. The shape of the wave pulse changes very little as it travels along the rope.[2]

As the wave pulse travels, each small segment of the rope, as it is disturbed, moves in a direction perpendicular to the wave motion. Figure 16.3 illustrates this

---

[2] Strictly speaking, the pulse changes shape and gradually spreads out during the motion. This effect is called *dispersion* and is common to many mechanical waves, as well as to electromagnetic waves. We do not consider dispersion in this chapter.

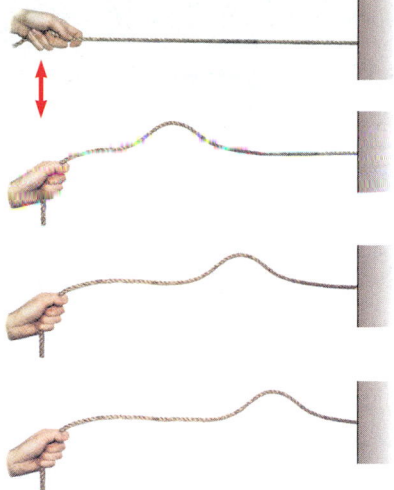

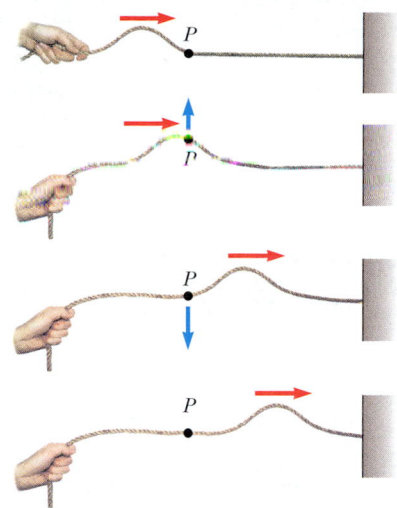

Figure 16.2   A wave pulse traveling down a stretched rope. The shape of the pulse is approximately unchanged as it travels along the rope.

Figure 16.3   A pulse traveling on a stretched rope is a transverse wave. The direction of motion of any element *P* of the rope (blue arrows) is perpendicular to the direction of wave motion (red arrows).

point for one particular segment, labeled *P*. Note that no part of the rope ever moves in the direction of the wave.

A traveling wave that causes the particles of the disturbed medium to move perpendicular to the wave motion is called a **transverse wave.**

Transverse wave

Compare this with another type of wave—one moving down a long, stretched spring, as shown in Figure 16.4. The left end of the spring is pushed briefly to the right and then pulled briefly to the left. This movement creates a sudden compression of a region of the coils. The compressed region travels along the spring (to the right in Figure 16.4). The compressed region is followed by a region where the coils are extended. Notice that the direction of the displacement of the coils is *parallel* to the direction of propagation of the compressed region.

A traveling wave that causes the particles of the medium to move parallel to the direction of wave motion is called a **longitudinal wave.**

Longitudinal wave

Sound waves, which we shall discuss in Chapter 17, are another example of longitudinal waves. The disturbance in a sound wave is a series of high-pressure and low-pressure regions that travel through air or any other material medium.

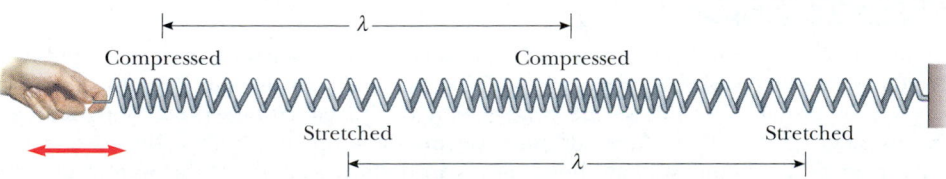

Figure 16.4   A longitudinal wave along a stretched spring. The displacement of the coils is in the direction of the wave motion. Each compressed region is followed by a stretched region.

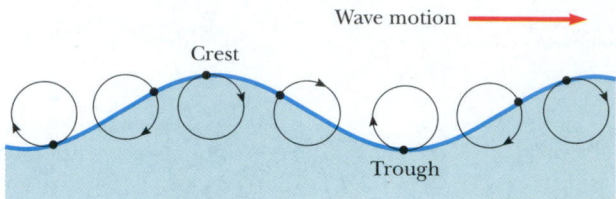

Figure 16.5   The motion of water molecules on the surface of deep water in which a wave is propagating is a combination of transverse and longitudinal displacements, with the result that molecules at the surface move in nearly circular paths. Each molecule is displaced both horizontally and vertically from its equilibrium position.

QuickLab

Make a "telephone" by poking a small hole in the bottom of two paper cups, threading a string through the holes, and tying knots in the ends of the string. If you speak into one cup while pulling the string taut, a friend can hear your voice in the other cup. What kind of wave is present in the string?

Some waves in nature exhibit a combination of transverse and longitudinal displacements. Surface water waves are a good example. When a water wave travels on the surface of deep water, water molecules at the surface move in nearly circular paths, as shown in Figure 16.5. Note that the disturbance has both transverse and longitudinal components. The transverse displacement is seen in Figure 16.5 as the variations in vertical position of the water molecules. The longitudinal displacement can be explained as follows: As the wave passes over the water's surface, water molecules at the crests move in the direction of propagation of the wave, whereas molecules at the troughs move in the direction opposite the propagation. Because the molecule at the labeled crest in Figure 16.5 will be at a trough after half a period, its movement in the direction of the propagation of the wave will be canceled by its movement in the opposite direction. This holds for every other water molecule disturbed by the wave. Thus, there is no net displacement of any water molecule during one complete cycle. Although the *molecules* experience no net displacement, the *wave* propagates along the surface of the water.

 The three-dimensional waves that travel out from the point under the Earth's surface at which an earthquake occurs are of both types—transverse and longitudinal. The longitudinal waves are the faster of the two, traveling at speeds in the range of 7 to 8 km/s near the surface. These are called **P waves,** with "P" standing for *primary* because they travel faster than the transverse waves and arrive at a seismograph first. The slower transverse waves, called **S waves** (with "S" standing for *secondary*), travel through the Earth at 4 to 5 km/s near the surface. By recording the time interval between the arrival of these two sets of waves at a seismograph, the distance from the seismograph to the point of origin of the waves can be determined. A single such measurement establishes an imaginary sphere centered on the seismograph, with the radius of the sphere determined by the difference in arrival times of the P and S waves. The origin of the waves is located somewhere on that sphere. The imaginary spheres from three or more monitoring stations located far apart from each other intersect at one region of the Earth, and this region is where the earthquake occurred.

Quick Quiz 16.1

(a) In a long line of people waiting to buy tickets, the first person leaves and a pulse of motion occurs as people step forward to fill the gap. As each person steps forward, the gap moves through the line. Is the propagation of this gap transverse or longitudinal? (b) Consider the "wave" at a baseball game: people stand up and shout as the wave arrives at their location, and the resultant pulse moves around the stadium. Is this wave transverse or longitudinal?

16.3 ▸ ONE-DIMENSIONAL TRAVELING WAVES

Consider a wave pulse traveling to the right with constant speed v on a long, taut string, as shown in Figure 16.6. The pulse moves along the x axis (the axis of the string), and the transverse (vertical) displacement of the string (the medium) is measured along the y axis. Figure 16.6a represents the shape and position of the pulse at time $t = 0$. At this time, the shape of the pulse, whatever it may be, can be represented as $y = f(x)$. That is, y, which is the vertical position of any point on the string, is some definite function of x. The displacement y, sometimes called the *wave function*, depends on both x and t. For this reason, it is often written $y(x, t)$, which is read "y as a function of x and t." Consider a particular point P on the string, identified by a specific value of its x coordinate. Before the pulse arrives at P, the y coordinate of this point is zero. As the wave passes P, the y coordinate of this point increases, reaches a maximum, and then decreases to zero. Therefore, **the wave function y represents the y coordinate of any point P of the medium at any time t.**

Because its speed is v, the wave pulse travels to the right a distance vt in a time t (see Fig. 16.6b). If the shape of the pulse does not change with time, we can represent the wave function y for all times after $t = 0$. Measured in a stationary reference frame having its origin at O, the wave function is

$$y = f(x - vt) \tag{16.1}$$

Wave traveling to the right

If the wave pulse travels to the left, the string displacement is

$$y = f(x + vt) \tag{16.2}$$

Wave traveling to the left

For any given time t, the wave function y as a function of x defines a curve representing the shape of the pulse at this time. This curve is equivalent to a "snapshot" of the wave at this time. For a pulse that moves without changing shape, the speed of the pulse is the same as that of any feature along the pulse, such as the crest shown in Figure 16.6. To find the speed of the pulse, we can calculate how far the crest moves in a short time and then divide this distance by the time interval. To follow the motion of the crest, we must substitute some particular value, say x_0, in Equation 16.1 for $x - vt$. Regardless of how x and t change individually, we must require that $x - vt = x_0$ in order to stay with the crest. This expression therefore represents the equation of motion of the crest. At $t = 0$, the crest is at $x = x_0$; at a

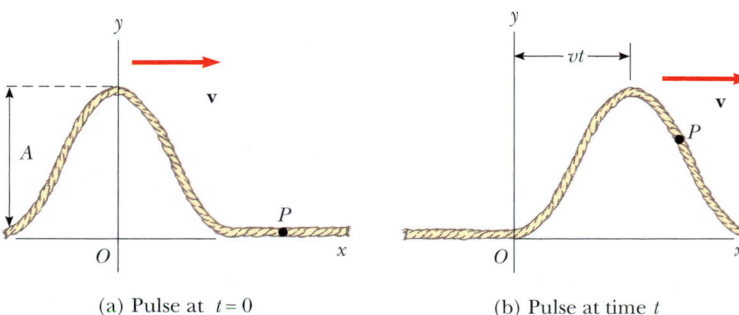

(a) Pulse at $t = 0$ (b) Pulse at time t

Figure 16.6   A one-dimensional wave pulse traveling to the right with a speed v. (a) At $t = 0$, the shape of the pulse is given by $y = f(x)$. (b) At some later time t, the shape remains unchanged and the vertical displacement of any point P of the medium is given by $y = f(x - vt)$.

time dt later, the crest is at $x = x_0 + v\,dt$. Therefore, in a time dt, the crest has moved a distance $dx = (x_0 + v\,dt) - x_0 = v\,dt$. Hence, the wave speed is

$$v = \frac{dx}{dt} \tag{16.3}$$

EXAMPLE 16.1   A Pulse Moving to the Right

A wave pulse moving to the right along the x axis is represented by the wave function

$$y(x, t) = \frac{2}{(x - 3.0t)^2 + 1}$$

where x and y are measured in centimeters and t is measured in seconds. Plot the wave function at $t = 0$, $t = 1.0$ s, and $t = 2.0$ s.

Solution First, note that this function is of the form $y = f(x - vt)$. By inspection, we see that the wave speed is $v = 3.0$ cm/s. Furthermore, the wave amplitude (the maximum value of y) is given by $A = 2.0$ cm. (We find the maximum value of the function representing y by letting $x - 3.0t = 0$.) The wave function expressions are

$$y(x, 0) = \frac{2}{x^2 + 1} \qquad \text{at } t = 0$$

$$y(x, 1.0) = \frac{2}{(x - 3.0)^2 + 1} \qquad \text{at } t = 1.0 \text{ s}$$

$$y(x, 2.0) = \frac{2}{(x - 6.0)^2 + 1} \qquad \text{at } t = 2.0 \text{ s}$$

We now use these expressions to plot the wave function versus x at these times. For example, let us evaluate $y(x, 0)$ at $x = 0.50$ cm:

$$y(0.50, 0) = \frac{2}{(0.50)^2 + 1} = 1.6 \text{ cm}$$

Likewise, at $x = 1.0$ cm, $y(1.0, 0) = 1.0$ cm, and at $x = 2.0$ cm, $y(2.0, 0) = 0.40$ cm. Continuing this procedure for other values of x yields the wave function shown in Figure 16.7a. In a similar manner, we obtain the graphs of $y(x, 1.0)$ and $y(x, 2.0)$, shown in Figure 16.7b and c, respectively. These snapshots show that the wave pulse moves to the right without changing its shape and that it has a constant speed of 3.0 cm/s.

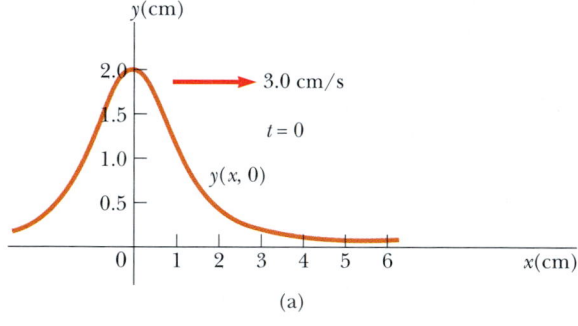

(a)

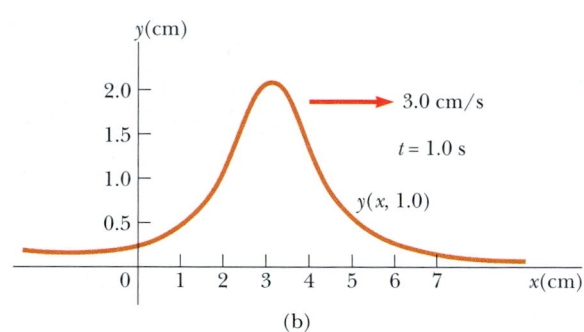

(b)

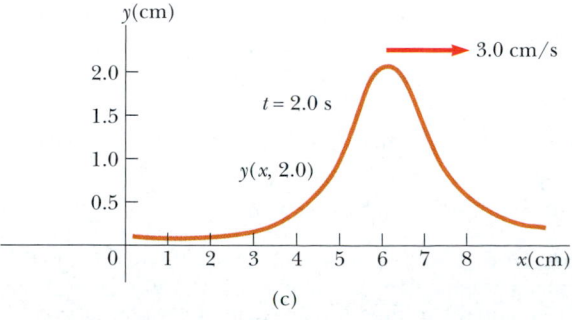

(c)

Figure 16.7 Graphs of the function $y(x, t) = 2/[(x - 3.0t)^2 + 1]$ at (a) $t = 0$, (b) $t = 1.0$ s, and (c) $t = 2.0$ s.

16.4 ▶ SUPERPOSITION AND INTERFERENCE

Many interesting wave phenomena in nature cannot be described by a single moving pulse. Instead, one must analyze complex waves in terms of a combination of many traveling waves. To analyze such wave combinations, one can make use of the **superposition principle:**

> If two or more traveling waves are moving through a medium, the resultant wave function at any point is the algebraic sum of the wave functions of the individual waves.

Linear waves obey the superposition principle

Waves that obey this principle are called *linear waves* and are generally characterized by small amplitudes. Waves that violate the superposition principle are called *nonlinear waves* and are often characterized by large amplitudes. In this book, we deal only with linear waves.

One consequence of the superposition principle is that **two traveling waves can pass through each other without being destroyed or even altered.** For instance, when two pebbles are thrown into a pond and hit the surface at different places, the expanding circular surface waves do not destroy each other but rather pass through each other. The complex pattern that is observed can be viewed as two independent sets of expanding circles. Likewise, when sound waves from two sources move through air, they pass through each other. The resulting sound that one hears at a given point is the resultant of the two disturbances.

Figure 16.8 is a pictorial representation of superposition. The wave function for the pulse moving to the right is y_1, and the wave function for the pulse moving

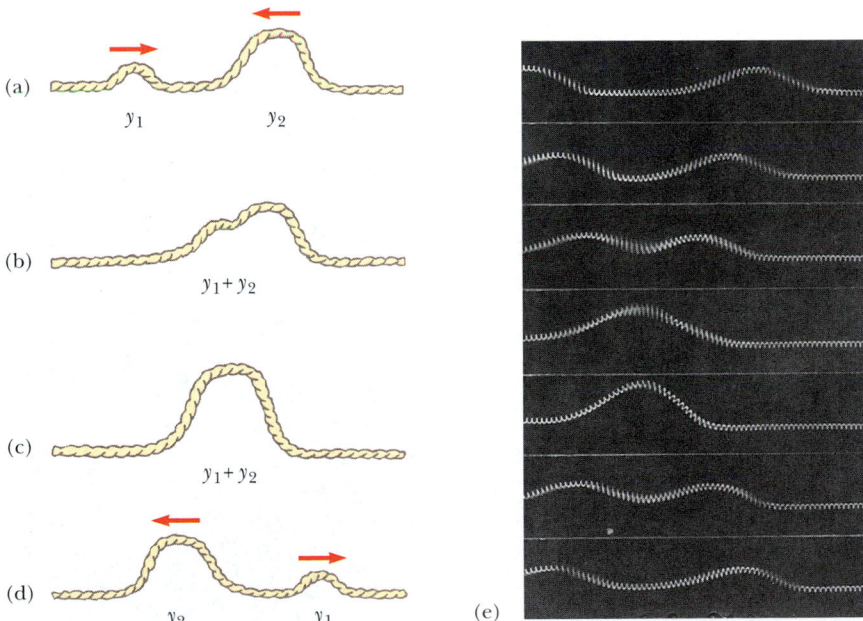

Figure 16.8 (a–d) Two wave pulses traveling on a stretched string in opposite directions pass through each other. When the pulses overlap, as shown in (b) and (c), the net displacement of the string equals the sum of the displacements produced by each pulse. Because each pulse displaces the string in the positive direction, we refer to the superposition of the two pulses as *constructive interference*. (e) Photograph of superposition of two equal, symmetric pulses traveling in opposite directions on a stretched spring. *(e, Education Development Center, Newton, MA)*

Interference of water waves produced in a ripple tank. The sources of the waves are two objects that oscillate perpendicular to the surface of the tank. *(Courtesy of Central Scientific Company)*

to the left is y_2. The pulses have the same speed but different shapes. Each pulse is assumed to be symmetric, and the displacement of the medium is in the positive y direction for both pulses. (Note, however, that the superposition principle applies even when the two pulses are not symmetric.) When the waves begin to overlap (Fig. 16.8b), the wave function for the resulting complex wave is given by $y_1 + y_2$.

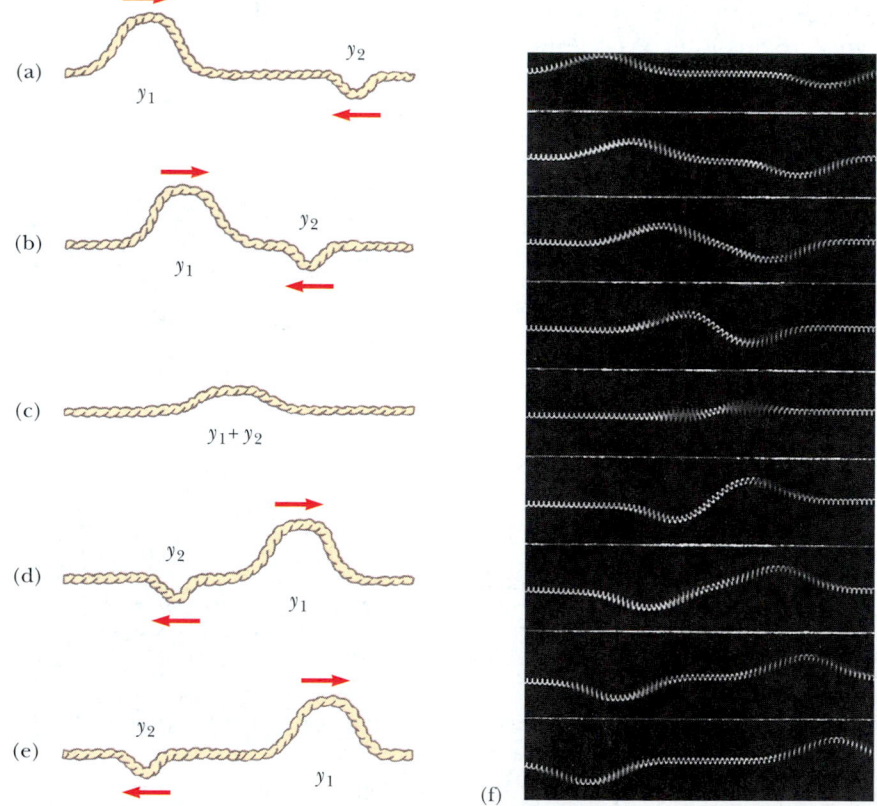

Figure 16.9   (a–e) Two wave pulses traveling in opposite directions and having displacements that are inverted relative to each other. When the two overlap in (c), their displacements partially cancel each other. (f) Photograph of superposition of two symmetric pulses traveling in opposite directions, where one pulse is inverted relative to the other. *(f, Education Development Center, Newton, MA)*

When the crests of the pulses coincide (Fig. 16.8c), the resulting wave given by $y_1 + y_2$ is symmetric. The two pulses finally separate and continue moving in their original directions (Fig. 16.8d). Note that the pulse shapes remain unchanged, as if the two pulses had never met!

The combination of separate waves in the same region of space to produce a resultant wave is called **interference.** For the two pulses shown in Figure 16.8, the displacement of the medium is in the positive y direction for both pulses, and the resultant wave (created when the pulses overlap) exhibits a displacement greater than that of either individual pulse. Because the displacements caused by the two pulses are in the same direction, we refer to their superposition as **constructive interference.**

Now consider two pulses traveling in opposite directions on a taut string where one pulse is inverted relative to the other, as illustrated in Figure 16.9. In this case, when the pulses begin to overlap, the resultant wave is given by $y_1 + y_2$, but the values of the function y_2 are negative. Again, the two pulses pass through each other; however, because the displacements caused by the two pulses are in opposite directions, we refer to their superposition as **destructive interference.**

Quick Quiz 16.2

Two pulses are traveling toward each other at 10 cm/s on a long string, as shown in Figure 16.10. Sketch the shape of the string at $t = 0.6$ s.

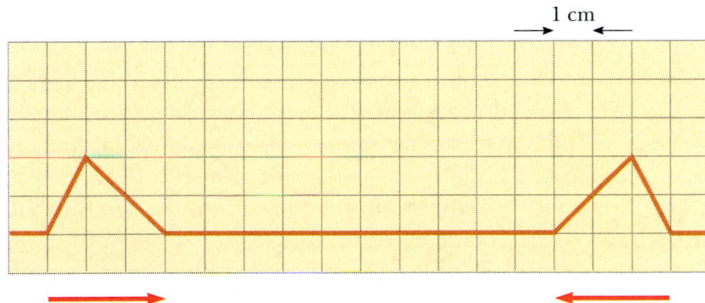

Figure 16.10 The pulses on this string are traveling at 10 cm/s.

16.5 THE SPEED OF WAVES ON STRINGS

In this section, we focus on determining the speed of a transverse pulse traveling on a taut string. Let us first conceptually argue the parameters that determine the speed. If a string under tension is pulled sideways and then released, the tension is responsible for accelerating a particular segment of the string back toward its equilibrium position. According to Newton's second law, the acceleration of the segment increases with increasing tension. If the segment returns to equilibrium more rapidly due to this increased acceleration, we would intuitively argue that the wave speed is greater. Thus, we expect the wave speed to increase with increasing tension.

Likewise, we can argue that the wave speed decreases if the mass per unit length of the string increases. This is because it is more difficult to accelerate a massive segment of the string than a light segment. If the tension in the string is T (not to be confused with the same symbol used for the period) and its mass per

The strings of this piano vary in both tension and mass per unit length. These differences in tension and density, in combination with the different lengths of the strings, allow the instrument to produce a wide range of sounds. *(Charles D. Winters)*

unit length is μ (Greek letter mu), then, as we shall show, the wave speed is

$$v = \sqrt{\frac{T}{\mu}}$$

(16.4)

First, let us verify that this expression is dimensionally correct. The dimensions of T are ML/T^2, and the dimensions of μ are M/L. Therefore, the dimensions of T/μ are L^2/T^2; hence, the dimensions of $\sqrt{T/\mu}$ are L/T—indeed, the dimensions of speed. No other combination of T and μ is dimensionally correct if we assume that they are the only variables relevant to the situation.

Now let us use a mechanical analysis to derive Equation 16.4. On our string under tension, consider a pulse moving to the right with a uniform speed v measured relative to a stationary frame of reference. Instead of staying in this reference frame, it is more convenient to choose as our reference frame one that moves along with the pulse with the same speed as the pulse, so that the pulse is at rest within the frame. This change of reference frame is permitted because Newton's laws are valid in either a stationary frame or one that moves with constant velocity. In our new reference frame, a given segment of the string initially to the right of the pulse moves to the left, rises up and follows the shape of the pulse, and then continues to move to the left. Figure 16.11a shows such a segment at the instant it is located at the top of the pulse.

The small segment of the string of length Δs shown in Figure 16.11a, and magnified in Figure 16.11b, forms an approximate arc of a circle of radius R. In our moving frame of reference (which is moving to the right at a speed v along with the pulse), the shaded segment is moving to the left with a speed v. This segment has a centripetal acceleration equal to v^2/R, which is supplied by components of the tension $\mathbf{T}$ in the string. The force $\mathbf{T}$ acts on either side of the segment and tangent to the arc, as shown in Figure 16.11b. The horizontal components of $\mathbf{T}$ cancel, and each vertical component $T \sin \theta$ acts radially toward the center of the arc. Hence, the total radial force is $2T \sin \theta$. Because the segment is small, θ is small, and we can use the small-angle approximation $\sin \theta \approx \theta$. Therefore, the total radial force is

$$\sum F_r = 2T \sin \theta \approx 2T\theta$$

The segment has a mass $m = \mu \Delta s$. Because the segment forms part of a circle and subtends an angle 2θ at the center, $\Delta s = R(2\theta)$, and hence

$$m = \mu \Delta s = 2\mu R\theta$$

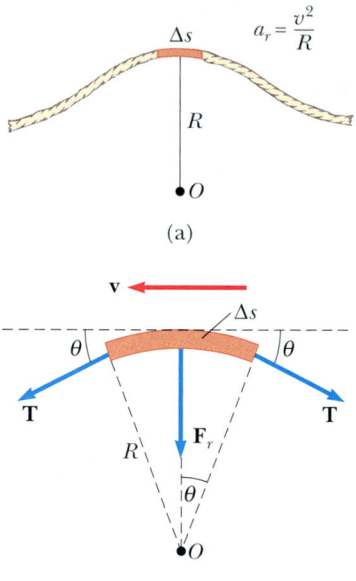

Figure 16.11 (a) To obtain the speed v of a wave on a stretched string, it is convenient to describe the motion of a small segment of the string in a moving frame of reference. (b) In the moving frame of reference, the small segment of length Δs moves to the left with speed v. The net force on the segment is in the radial direction because the horizontal components of the tension force cancel.

If we apply Newton's second law to this segment, the radial component of motion gives

$$\sum F_r = ma = \frac{mv^2}{R}$$

$$2T\theta = \frac{2\mu R\theta v^2}{R}$$

Solving for v gives Equation 16.4.

Notice that this derivation is based on the assumption that the pulse height is small relative to the length of the string. Using this assumption, we were able to use the approximation $\sin\theta \approx \theta$. Furthermore, the model assumes that the tension T is not affected by the presence of the pulse; thus, T is the same at all points on the string. Finally, this proof does *not* assume any particular shape for the pulse. Therefore, we conclude that a pulse of *any shape* travels along the string with speed $v = \sqrt{T/\mu}$ without any change in pulse shape.

EXAMPLE 16.2 The Speed of a Pulse on a Cord

A uniform cord has a mass of 0.300 kg and a length of 6.00 m (Fig. 16.12). The cord passes over a pulley and supports a 2.00-kg object. Find the speed of a pulse traveling along this cord.

Solution The tension T in the cord is equal to the weight of the suspended 2.00-kg mass:

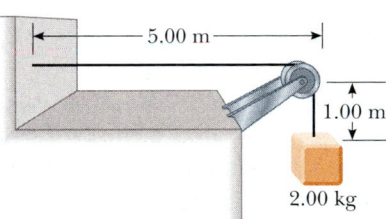

Figure 16.12 The tension T in the cord is maintained by the suspended object. The speed of any wave traveling along the cord is given by $v = \sqrt{T/\mu}$.

$$T = mg = (2.00\ \text{kg})(9.80\ \text{m/s}^2) = 19.6\ \text{N}$$

(This calculation of the tension neglects the small mass of the cord. Strictly speaking, the cord can never be exactly horizontal, and therefore the tension is not uniform.) The mass per unit length μ of the cord is

$$\mu = \frac{m}{\ell} = \frac{0.300\ \text{kg}}{6.00\ \text{m}} = 0.0500\ \text{kg/m}$$

Therefore, the wave speed is

$$v = \sqrt{\frac{T}{\mu}} = \sqrt{\frac{19.6\ \text{N}}{0.050\ 0\ \text{kg/m}}} = \boxed{19.8\ \text{m/s}}$$

Exercise Find the time it takes the pulse to travel from the wall to the pulley.

Answer 0.253 s.

Quick Quiz 16.3

Suppose you create a pulse by moving the free end of a taut string up and down once with your hand. The string is attached at its other end to a distant wall. The pulse reaches the wall in a time t. Which of the following actions, taken by itself, decreases the time it takes the pulse to reach the wall? More than one choice may be correct.
(a) Moving your hand more quickly, but still only up and down once by the same amount.
(b) Moving your hand more slowly, but still only up and down once by the same amount.
(c) Moving your hand a greater distance up and down in the same amount of time.
(d) Moving your hand a lesser distance up and down in the same amount of time.
(e) Using a heavier string of the same length and under the same tension.
(f) Using a lighter string of the same length and under the same tension.
(g) Using a string of the same linear mass density but under decreased tension.
(h) Using a string of the same linear mass density but under increased tension.

16.6 ▸ REFLECTION AND TRANSMISSION

We have discussed traveling waves moving through a uniform medium. We now consider how a traveling wave is affected when it encounters a change in the medium. For example, consider a pulse traveling on a string that is rigidly attached to a support at one end (Fig. 16.13). When the pulse reaches the support, a severe change in the medium occurs—the string ends. The result of this change is that the wave undergoes **reflection**—that is, the pulse moves back along the string in the opposite direction.

Note that the reflected pulse is inverted. This inversion can be explained as follows: When the pulse reaches the fixed end of the string, the string produces an upward force on the support. By Newton's third law, the support must exert an equal and opposite (downward) reaction force on the string. This downward force causes the pulse to invert upon reflection.

Now consider another case: this time, the pulse arrives at the end of a string that is free to move vertically, as shown in Figure 16.14. The tension at the free end is maintained because the string is tied to a ring of negligible mass that is free to slide vertically on a smooth post. Again, the pulse is reflected, but this time it is not inverted. When it reaches the post, the pulse exerts a force on the free end of the string, causing the ring to accelerate upward. The ring overshoots the height of the incoming pulse, and then the downward component of the tension force pulls the ring back down. This movement of the ring produces a reflected pulse that is not inverted and that has the same amplitude as the incoming pulse.

Finally, we may have a situation in which the boundary is intermediate between these two extremes. In this case, part of the incident pulse is reflected and part undergoes **transmission**—that is, some of the pulse passes through the boundary. For instance, suppose a light string is attached to a heavier string, as shown in Figure 16.15. When a pulse traveling on the light string reaches the boundary between the two, part of the pulse is reflected and inverted and part is transmitted to the heavier string. The reflected pulse is inverted for the same reasons described earlier in the case of the string rigidly attached to a support.

Note that the reflected pulse has a smaller amplitude than the incident pulse. In Section 16.8, we shall learn that the energy carried by a wave is related to its amplitude. Thus, according to the principle of the conservation of energy, when the pulse breaks up into a reflected pulse and a transmitted pulse at the boundary, the sum of the energies of these two pulses must equal the energy of the incident pulse. Because the reflected pulse contains only part of the energy of the incident pulse, its amplitude must be smaller.

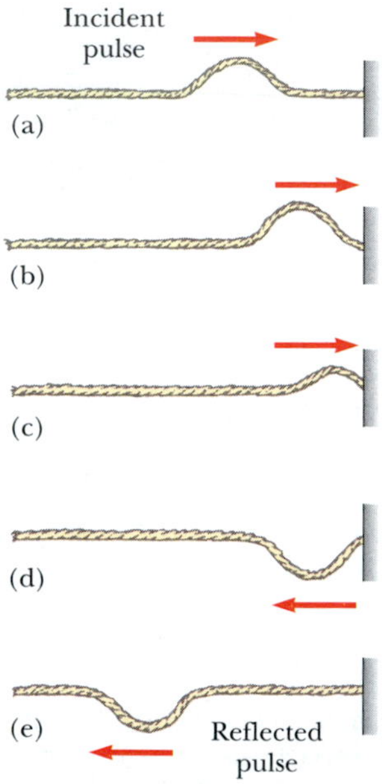

Figure 16.13 The reflection of a traveling wave pulse at the fixed end of a stretched string. The reflected pulse is inverted, but its shape is unchanged.

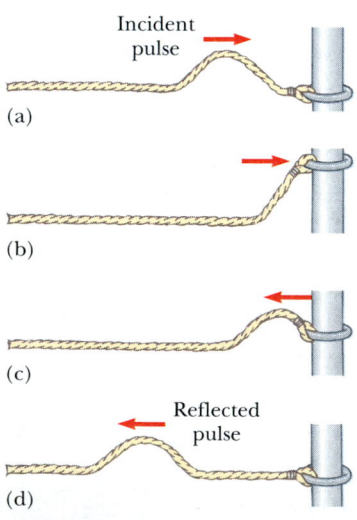

Figure 16.14 The reflection of a traveling wave pulse at the free end of a stretched string. The reflected pulse is not inverted.

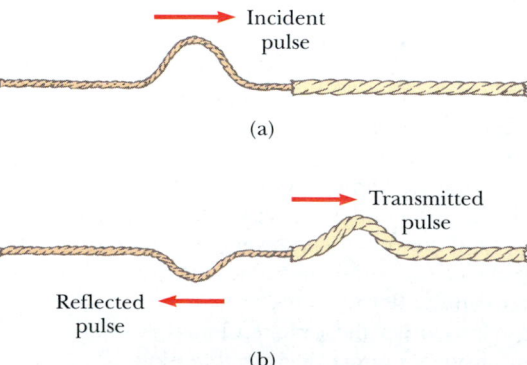

Figure 16.15 (a) A pulse traveling to the right on a light string attached to a heavier string. (b) Part of the incident pulse is reflected (and inverted), and part is transmitted to the heavier string.

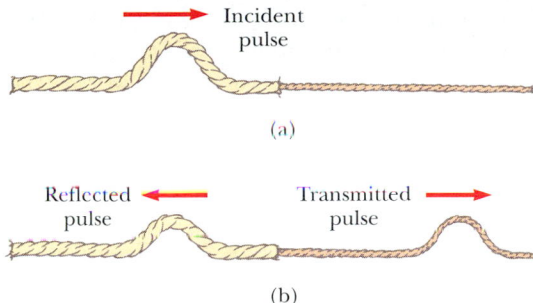

(a)

Reflected pulse ◄──   Transmitted pulse ──►

(b)

Figure 16.16 (a) A pulse traveling to the right on a heavy string attached to a lighter string. (b) The incident pulse is partially reflected and partially transmitted, and the reflected pulse is not inverted.

When a pulse traveling on a heavy string strikes the boundary between the heavy string and a lighter one, as shown in Figure 16.16, again part is reflected and part is transmitted. In this case, the reflected pulse is not inverted.

In either case, the relative heights of the reflected and transmitted pulses depend on the relative densities of the two strings. If the strings are identical, there is no discontinuity at the boundary and no reflection takes place.

According to Equation 16.4, the speed of a wave on a string increases as the mass per unit length of the string decreases. In other words, a pulse travels more slowly on a heavy string than on a light string if both are under the same tension. The following general rules apply to reflected waves: **When a wave pulse travels from medium A to medium B and $v_A > v_B$ (that is, when B is denser than A), the pulse is inverted upon reflection. When a wave pulse travels from medium A to medium B and $v_A < v_B$ (that is, when A is denser than B), the pulse is not inverted upon reflection.**

16.7 SINUSOIDAL WAVES

In this section, we introduce an important wave function whose shape is shown in Figure 16.17. The wave represented by this curve is called a **sinusoidal wave** because the curve is the same as that of the function $\sin \theta$ plotted against θ. The sinusoidal wave is the simplest example of a periodic continuous wave and can be used to build more complex waves, as we shall see in Section 18.8. The red curve represents a snapshot of a traveling sinusoidal wave at $t = 0$, and the blue curve represents a snapshot of the wave at some later time t. At $t = 0$, the function describing the positions of the particles of the medium through which the sinusoidal wave is traveling can be written

$$y = A \sin\left(\frac{2\pi}{\lambda} x\right) \tag{16.5}$$

where the constant A represents the wave amplitude and the constant λ is the wavelength. Thus, we see that the position of a particle of the medium is the same whenever x is increased by an integral multiple of λ. If the wave moves to the right with a speed v, then the wave function at some later time t is

$$y = A \sin\left[\frac{2\pi}{\lambda} (x - vt)\right] \tag{16.6}$$

That is, the traveling sinusoidal wave moves to the right a distance vt in the time t, as shown in Figure 16.17. Note that the wave function has the form $f(x - vt)$ and

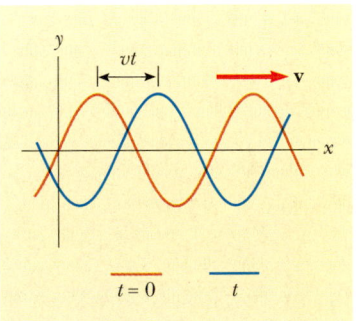

Figure 16.17 A one-dimensional sinusoidal wave traveling to the right with a speed v. The red curve represents a snapshot of the wave at $t = 0$, and the blue curve represents a snapshot at some later time t.

so represents a wave traveling to the right. If the wave were traveling to the left, the quantity $x - vt$ would be replaced by $x + vt$, as we learned when we developed Equations 16.1 and 16.2.

By definition, the wave travels a distance of one wavelength in one period T. Therefore, the wave speed, wavelength, and period are related by the expression

$$v = \frac{\lambda}{T} \tag{16.7}$$

Substituting this expression for v into Equation 16.6, we find that

$$y = A \sin\left[2\pi\left(\frac{x}{\lambda} - \frac{t}{T}\right)\right] \tag{16.8}$$

This form of the wave function clearly shows the *periodic* nature of y. At any given time t (a snapshot of the wave), y has the *same* value at the positions x, $x + \lambda$, $x + 2\lambda$, and so on. Furthermore, at any given position x, the value of y is the same at times t, $t + T$, $t + 2T$, and so on.

We can express the wave function in a convenient form by defining two other quantities, the **angular wave number** k and the **angular frequency** ω:

Angular wave number

$$k \equiv \frac{2\pi}{\lambda} \tag{16.9}$$

Angular frequency

$$\omega \equiv \frac{2\pi}{T} \tag{16.10}$$

Using these definitions, we see that Equation 16.8 can be written in the more compact form

Wave function for a sinusoidal wave

$$y = A \sin(kx - \omega t) \tag{16.11}$$

The frequency of a sinusoidal wave is related to the period by the expression

Frequency

$$f = \frac{1}{T} \tag{16.12}$$

The most common unit for frequency, as we learned in Chapter 13, is second^{-1}, or **hertz** (Hz). The corresponding unit for T is seconds.

Using Equations 16.9, 16.10, and 16.12, we can express the wave speed v originally given in Equation 16.7 in the alternative forms

$$v = \frac{\omega}{k} \tag{16.13}$$

Speed of a sinusoidal wave

$$v = \lambda f \tag{16.14}$$

The wave function given by Equation 16.11 assumes that the vertical displacement y is zero at $x = 0$ and $t = 0$. This need not be the case. If it is not, we generally express the wave function in the form

General expression for a sinusoidal wave

$$y = A \sin(kx - \omega t + \phi) \tag{16.15}$$

where ϕ is the **phase constant,** just as we learned in our study of periodic motion in Chapter 13. This constant can be determined from the initial conditions.

EXAMPLE 16.3 ▸ A Traveling Sinusoidal Wave

A sinusoidal wave traveling in the positive x direction has an amplitude of 15.0 cm, a wavelength of 40.0 cm, and a frequency of 8.00 Hz. The vertical displacement of the medium at $t = 0$ and $x = 0$ is also 15.0 cm, as shown in Figure 16.18. (a) Find the angular wave number k, period T, angular frequency ω, and speed v of the wave.

Solution (a) Using Equations 16.9, 16.10, 16.12, and 16.14, we find the following:

$$k = \frac{2\pi}{\lambda} = \frac{2\pi \text{ rad}}{40.0 \text{ cm}} = \boxed{0.157 \text{ rad/cm}}$$

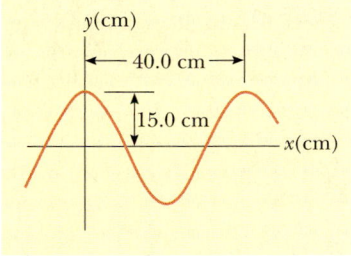

Figure 16.18 A sinusoidal wave of wavelength $\lambda = 40.0$ cm and amplitude $A = 15.0$ cm. The wave function can be written in the form $y = A \cos(kx - \omega t)$.

$$\omega = 2\pi f = 2\pi(8.00 \text{ s}^{-1}) = \boxed{50.3 \text{ rad/s}}$$

$$T = \frac{1}{f} = \frac{1}{8.00 \text{ s}^{-1}} = \boxed{0.125 \text{ s}}$$

$$v = \lambda f = (40.0 \text{ cm})(8.00 \text{ s}^{-1}) = \boxed{320 \text{ cm/s}}$$

(b) Determine the phase constant ϕ, and write a general expression for the wave function.

Solution Because $A = 15.0$ cm and because $y = 15.0$ cm at $x = 0$ and $t = 0$, substitution into Equation 16.15 gives

$$15.0 = (15.0) \sin \phi \qquad \text{or} \qquad \sin \phi = 1$$

We may take the principal value $\phi = \pi/2$ rad (or 90°). Hence, the wave function is of the form

$$y = A \sin\left(kx - \omega t + \frac{\pi}{2}\right) = A \cos(kx - \omega t)$$

By inspection, we can see that the wave function must have this form, noting that the cosine function has the same shape as the sine function displaced by 90°. Substituting the values for A, k, and ω into this expression, we obtain

$$y = (15.0 \text{ cm}) \cos(0.157x - 50.3t)$$

Sinusoidal Waves on Strings

In Figure 16.2, we demonstrated how to create a pulse by jerking a taut string up and down once. To create a train of such pulses, normally referred to as a *wave train,* or just plain *wave,* we can replace the hand with an oscillating blade. If the wave consists of a train of identical cycles, whatever their shape, the relationships $f = 1/T$ and $v = f\lambda$ among speed, frequency, period, and wavelength hold true. We can make more definite statements about the wave function if the source of the waves vibrates in simple harmonic motion. Figure 16.19 represents snapshots of the wave created in this way at intervals of $T/4$. Note that because the end of the blade oscillates in simple harmonic motion, **each particle of the string, such as that at P, also oscillates vertically with simple harmonic motion.** This must be the case because each particle follows the simple harmonic motion of the blade. Therefore, every segment of the string can be treated as a simple harmonic oscillator vibrating with a frequency equal to the frequency of oscillation of the blade.[3] Note that although each segment oscillates in the y direction, the wave travels in the x direction with a speed v. Of course, this is the definition of a transverse wave.

[3] In this arrangement, we are assuming that a string segment always oscillates in a vertical line. The tension in the string would vary if a segment were allowed to move sideways. Such motion would make the analysis very complex.

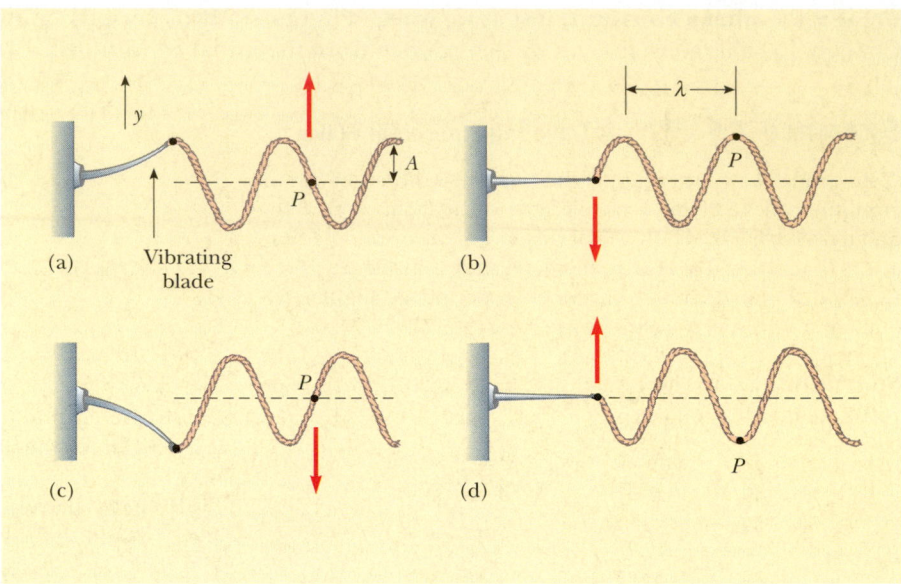

Figure 16.19   One method for producing a train of sinusoidal wave pulses on a string. The left end of the string is connected to a blade that is set into oscillation. Every segment of the string, such as the point P, oscillates with simple harmonic motion in the vertical direction.

If the wave at $t = 0$ is as described in Figure 16.19b, then the wave function can be written as

$$y = A \sin(kx - \omega t)$$

We can use this expression to describe the motion of any point on the string. The point P (or any other point on the string) moves only vertically, and so its x coordinate remains constant. Therefore, the **transverse speed v_y** (not to be confused with the wave speed v) and the **transverse acceleration a_y** are

$$v_y = \frac{dy}{dt}\bigg]_{x = \text{constant}} = \frac{\partial y}{\partial t} = -\omega A \cos(kx - \omega t) \qquad \textbf{(16.16)}$$

$$a_y = \frac{dv_y}{dt}\bigg]_{x = \text{constant}} = \frac{\partial v_y}{\partial t} = -\omega^2 A \sin(kx - \omega t) \qquad \textbf{(16.17)}$$

In these expressions, we must use partial derivatives (see Section 8.6) because y depends on both x and t. In the operation $\partial y/\partial t$, for example, we take a derivative with respect to t while holding x constant. The maximum values of the transverse speed and transverse acceleration are simply the absolute values of the coefficients of the cosine and sine functions:

$$v_{y,\,\text{max}} = \omega A \qquad \textbf{(16.18)}$$

$$a_{y,\,\text{max}} = \omega^2 A \qquad \textbf{(16.19)}$$

The transverse speed and transverse acceleration do not reach their maximum values simultaneously. The transverse speed reaches its maximum value (ωA) when $y = 0$, whereas the transverse acceleration reaches its maximum value ($\omega^2 A$) when $y = \pm A$. Finally, Equations 16.18 and 16.19 are identical in mathematical form to the corresponding equations for simple harmonic motion, Equations 13.10 and 13.11.

Quick Quiz 16.4

A sinusoidal wave is moving on a string. If you increase the frequency *f* of the wave, how do the transverse speed, wave speed, and wavelength change?

EXAMPLE 16.4   **A Sinusoidally Driven String**

The string shown in Figure 16.19 is driven at a frequency of 5.00 Hz. The amplitude of the motion is 12.0 cm, and the wave speed is 20.0 m/s. Determine the angular frequency ω and angular wave number k for this wave, and write an expression for the wave function.

Solution   Using Equations 16.10, 16.12, and 16.13, we find that

$$\omega = \frac{2\pi}{T} = 2\pi f = 2\pi(5.00 \text{ Hz}) = \boxed{31.4 \text{ rad/s}}$$

$$k = \frac{\omega}{v} = \frac{31.4 \text{ rad/s}}{20.0 \text{ m/s}} = \boxed{1.57 \text{ rad/m}}$$

Because $A = 12.0$ cm $= 0.120$ m, we have

$$y = A \sin(kx - \omega t) = (0.120 \text{ m}) \sin(1.57x - 31.4t)$$

Exercise   Calculate the maximum values for the transverse speed and transverse acceleration of any point on the string.

Answer   3.77 m/s; 118 m/s².

16.8   RATE OF ENERGY TRANSFER BY SINUSOIDAL WAVES ON STRINGS

As waves propagate through a medium, they transport energy. We can easily demonstrate this by hanging an object on a stretched string and then sending a pulse down the string, as shown in Figure 16.20. When the pulse meets the suspended object, the object is momentarily displaced, as illustrated in Figure 16.20b. In the process, energy is transferred to the object because work must be done for it to move upward. This section examines the rate at which energy is transported along a string. We shall assume a one-dimensional sinusoidal wave in the calculation of the energy transferred.

Consider a sinusoidal wave traveling on a string (Fig. 16.21). The source of the energy being transported by the wave is some external agent at the left end of the string; this agent does work in producing the oscillations. As the external agent performs work on the string, moving it up and down, energy enters the system of the string and propagates along its length. Let us focus our attention on a segment of the string of length Δx and mass Δm. Each such segment moves vertically with simple harmonic motion. Furthermore, all segments have the same angular frequency ω and the same amplitude A. As we found in Chapter 13, the elastic potential energy U associated with a particle in simple harmonic motion is $U = \frac{1}{2}ky^2$, where the simple harmonic motion is in the y direction. Using the relationship $\omega^2 = k/m$ developed in Equations 13.16 and 13.17, we can write this as

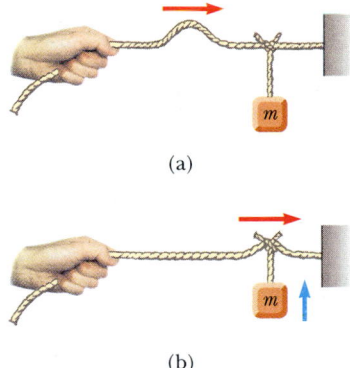

(a)

(b)

Figure 16.20   (a) A pulse traveling to the right on a stretched string on which an object has been suspended. (b) Energy is transmitted to the suspended object when the pulse arrives.

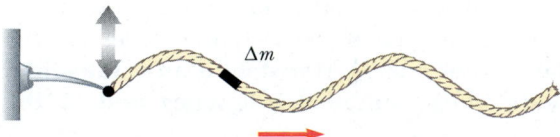

Figure 16.21   A sinusoidal wave traveling along the *x* axis on a stretched string. Every segment moves vertically, and every segment has the same total energy.

$U = \frac{1}{2}m\omega^2 y^2$. If we apply this equation to the segment of mass Δm, we see that the potential energy of this segment is

$$\Delta U = \frac{1}{2}(\Delta m)\,\omega^2 y^2$$

Because the mass per unit length of the string is $\mu = \Delta m/\Delta x$, we can express the potential energy of the segment as

$$\Delta U = \frac{1}{2}(\mu\Delta x)\,\omega^2 y^2$$

As the length of the segment shrinks to zero, $\Delta x \rightarrow dx$, and this expression becomes a differential relationship:

$$dU = \frac{1}{2}(\mu dx)\,\omega^2 y^2$$

We replace the general displacement y of the segment with the wave function for a sinusoidal wave:

$$dU = \frac{1}{2}\mu\omega^2[A\sin(kx - \omega t)]^2\,dx = \frac{1}{2}\mu\omega^2 A^2\sin^2(kx - \omega t)\,dx$$

If we take a snapshot of the wave at time $t = 0$, then the potential energy in a given segment is

$$dU = \frac{1}{2}\mu\omega^2 A^2\sin^2 kx\,dx$$

To obtain the total potential energy in one wavelength, we integrate this expression over all the string segments in one wavelength:

$$U_\lambda = \int dU = \int_0^\lambda \frac{1}{2}\mu\omega^2 A^2\sin^2 kx\,dx = \frac{1}{2}\mu\omega^2 A^2\int_0^\lambda \sin^2 kx\,dx$$

$$= \frac{1}{2}\mu\omega^2 A^2\left[\frac{1}{2}x - \frac{1}{4k}\sin 2\,kx\right]_0^\lambda = \frac{1}{2}\mu\omega^2 A^2(\tfrac{1}{2}\lambda) = \frac{1}{4}\mu\omega^2 A^2\lambda$$

Because it is in motion, each segment of the string also has kinetic energy. When we use this procedure to analyze the total kinetic energy in one wavelength of the string, we obtain the same result:

$$K_\lambda = \int dK = \frac{1}{4}\mu\omega^2 A^2\lambda$$

The total energy in one wavelength of the wave is the sum of the potential and kinetic energies:

$$E_\lambda = U_\lambda + K_\lambda = \frac{1}{2}\mu\omega^2 A^2\lambda \qquad \textbf{(16.20)}$$

As the wave moves along the string, this amount of energy passes by a given point on the string during one period of the oscillation. Thus, the power, or rate of energy transfer, associated with the wave is

$$\mathcal{P} = \frac{E_\lambda}{\Delta t} = \frac{\frac{1}{2}\mu\omega^2 A^2\lambda}{T} = \frac{1}{2}\mu\omega^2 A^2\left(\frac{\lambda}{T}\right)$$

Power of a wave

$$\mathcal{P} = \frac{1}{2}\mu\omega^2 A^2 v \qquad \textbf{(16.21)}$$

This shows that the rate of energy transfer by a sinusoidal wave on a string is proportional to (a) the wave speed, (b) the square of the frequency, and (c) the square of the amplitude. In fact: **the rate of energy transfer in any sinusoidal wave is proportional to the square of the angular frequency and to the square of the amplitude.**

EXAMPLE 16.5   Power Supplied to a Vibrating String

A taut string for which $\mu = 5.00 \times 10^{-2}$ kg/m is under a tension of 80.0 N. How much power must be supplied to the string to generate sinusoidal waves at a frequency of 60.0 Hz and an amplitude of 6.00 cm?

Solution   The wave speed on the string is, from Equation 16.4,

$$v = \sqrt{\frac{T}{\mu}} = \sqrt{\frac{80.0 \text{ N}}{5.00 \times 10^{-2} \text{ kg/m}}} = 40.0 \text{ m/s}$$

Because $f = 60.0$ Hz, the angular frequency ω of the sinus-

oidal waves on the string has the value

$$\omega = 2\pi f = 2\pi(60.0 \text{ Hz}) = 377 \text{ s}^{-1}$$

Using these values in Equation 16.21 for the power, with $A = 6.00 \times 10^{-2}$ m, we obtain

$$\mathcal{P} = \frac{1}{2}\mu\omega^2 A^2 v$$
$$= \frac{1}{2}(5.00 \times 10^{-2} \text{ kg/m})(377 \text{ s}^{-1})^2$$
$$\times (6.00 \times 10^{-2} \text{ m})^2(40.0 \text{ m/s})$$
$$= \boxed{512 \text{ W}}$$

Optional Section

16.9 ▶ THE LINEAR WAVE EQUATION

In Section 16.3 we introduced the concept of the wave function to represent waves traveling on a string. All wave functions $y(x, t)$ represent solutions of an equation called the *linear wave equation*. This equation gives a complete description of the wave motion, and from it one can derive an expression for the wave speed. Furthermore, the linear wave equation is basic to many forms of wave motion. In this section, we derive this equation as applied to waves on strings.

Suppose a traveling wave is propagating along a string that is under a tension T. Let us consider one small string segment of length Δx (Fig. 16.22). The ends of the segment make small angles θ_A and θ_B with the x axis. The net force acting on the segment in the vertical direction is

$$\sum F_y = T \sin \theta_B - T \sin \theta_A = T(\sin \theta_B - \sin \theta_A)$$

Because the angles are small, we can use the small-angle approximation $\sin \theta \approx \tan \theta$ to express the net force as

$$\sum F_y \approx T(\tan \theta_B - \tan \theta_A)$$

However, the tangents of the angles at A and B are defined as the slopes of the string segment at these points. Because the slope of a curve is given by $\partial y/\partial x$, we have

$$\sum F_y \approx T\left[\left(\frac{\partial y}{\partial x}\right)_B - \left(\frac{\partial y}{\partial x}\right)_A\right] \tag{16.22}$$

We now apply Newton's second law to the segment, with the mass of the segment given by $m = \mu\Delta x$:

$$\sum F_y = ma_y = \mu\Delta x\left(\frac{\partial^2 y}{\partial t^2}\right) \tag{16.23}$$

Combining Equation 16.22 with Equation 16.23, we obtain

$$\mu\Delta x\left(\frac{\partial^2 y}{\partial t^2}\right) = T\left[\left(\frac{\partial y}{\partial x}\right)_B - \left(\frac{\partial y}{\partial x}\right)_A\right]$$

$$\frac{\mu}{T}\frac{\partial^2 y}{\partial t^2} = \frac{(\partial y/\partial x)_B - (\partial y/\partial x)_A}{\Delta x} \tag{16.24}$$

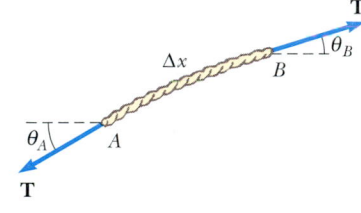

Figure 16.22   A segment of a string under tension T. The slopes at points A and B are given by $\tan \theta_A$ and $\tan \theta_B$, respectively.

The right side of this equation can be expressed in a different form if we note that the partial derivative of any function is defined as

$$\frac{\partial f}{\partial x} \equiv \lim_{\Delta x \to 0} \frac{f(x + \Delta x) - f(x)}{\Delta x}$$

If we associate $f(x + \Delta x)$ with $(\partial y/\partial x)_B$ and $f(x)$ with $(\partial y/\partial x)_A$, we see that, in the limit $\Delta x \to 0$, Equation 16.24 becomes

Linear wave equation

$$\frac{\mu}{T} \frac{\partial^2 y}{\partial t^2} = \frac{\partial^2 y}{\partial x^2} \tag{16.25}$$

This is the linear wave equation as it applies to waves on a string.

We now show that the sinusoidal wave function (Eq. 16.11) represents a solution of the linear wave equation. If we take the sinusoidal wave function to be of the form $y(x, t) = A \sin(kx - \omega t)$, then the appropriate derivatives are

$$\frac{\partial^2 y}{\partial t^2} = -\omega^2 A \sin(kx - \omega t)$$

$$\frac{\partial^2 y}{\partial x^2} = -k^2 A \sin(kx - \omega t)$$

Substituting these expressions into Equation 16.25, we obtain

$$-\frac{\mu \omega^2}{T} \sin(kx - \omega t) = -k^2 \sin(kx - \omega t)$$

This equation must be true for all values of the variables x and t in order for the sinusoidal wave function to be a solution of the wave equation. Both sides of the equation depend on x and t through the same function $\sin(kx - \omega t)$. Because this function divides out, we do indeed have an identity, provided that

$$k^2 = \frac{\mu \omega^2}{T}$$

Using the relationship $v = \omega/k$ (Eq. 16.13) in this expression, we see that

$$v^2 = \frac{\omega^2}{k^2} = \frac{T}{\mu}$$

$$v = \sqrt{\frac{T}{\mu}}$$

which is Equation 16.4. This derivation represents another proof of the expression for the wave speed on a taut string.

The linear wave equation (Eq. 16.25) is often written in the form

Linear wave equation in general

$$\frac{\partial^2 y}{\partial x^2} = \frac{1}{v^2} \frac{\partial^2 y}{\partial t^2} \tag{16.26}$$

This expression applies in general to various types of traveling waves. For waves on strings, y represents the vertical displacement of the string. For sound waves, y corresponds to displacement of air molecules from equilibrium or variations in either the pressure or the density of the gas through which the sound waves are propagating. In the case of electromagnetic waves, y corresponds to electric or magnetic field components.

We have shown that the sinusoidal wave function (Eq. 16.11) is one solution of the linear wave equation (Eq. 16.26). Although we do not prove it here, the linear wave equation is satisfied by *any* wave function having the form $y = f(x \pm vt)$. Furthermore, we have seen that the linear wave equation is a direct consequence of Newton's second law applied to any segment of the string.

SUMMARY

A **transverse wave** is one in which the particles of the medium move in a direction *perpendicular* to the direction of the wave velocity. An example is a wave on a taut string. A **longitudinal wave** is one in which the particles of the medium move in a direction *parallel* to the direction of the wave velocity. Sound waves in fluids are longitudinal. You should be able to identify examples of both types of waves.

Any one-dimensional wave traveling with a speed v in the x direction can be represented by a wave function of the form

$$y = f(x \pm vt) \tag{16.1, 16.2}$$

where the positive sign applies to a wave traveling in the negative x direction and the negative sign applies to a wave traveling in the positive x direction. The shape of the wave at any instant in time (a snapshot of the wave) is obtained by holding t constant.

The **superposition principle** specifies that when two or more waves move through a medium, the resultant wave function equals the algebraic sum of the individual wave functions. When two waves combine in space, they interfere to produce a resultant wave. The **interference** may be **constructive** (when the individual displacements are in the same direction) or **destructive** (when the displacements are in opposite directions).

The **speed** of a wave traveling on a taut string of mass per unit length μ and tension T is

$$v = \sqrt{\frac{T}{\mu}} \tag{16.4}$$

A wave is totally or partially reflected when it reaches the end of the medium in which it propagates or when it reaches a boundary where its speed changes discontinuously. If a wave pulse traveling on a string meets a fixed end, the pulse is reflected and inverted. If the pulse reaches a free end, it is reflected but not inverted.

The **wave function** for a one-dimensional sinusoidal wave traveling to the right can be expressed as

$$y = A \sin\left[\frac{2\pi}{\lambda}(x - vt)\right] = A \sin(kx - \omega t) \tag{16.6, 16.11}$$

where A is the **amplitude,** λ is the **wavelength,** k is the **angular wave number,** and ω is the **angular frequency.** If T is the **period** and f the **frequency,** v, k and ω can be written

$$v = \frac{\lambda}{T} = \lambda f \tag{16.7, 16.14}$$

$$k \equiv \frac{2\pi}{\lambda} \tag{16.9}$$

$$\omega \equiv \frac{2\pi}{T} = 2\pi f \tag{16.10, 16.12}$$

You should know how to find the equation describing the motion of particles in a wave from a given set of physical parameters.

The **power** transmitted by a sinusoidal wave on a stretched string is

$$\mathcal{P} = \tfrac{1}{2}\mu\omega^2 A^2 v \tag{16.21}$$

QUESTIONS

1. Why is a wave pulse traveling on a string considered a transverse wave?

2. How would you set up a longitudinal wave in a stretched spring? Would it be possible to set up a transverse wave in a spring?

3. By what factor would you have to increase the tension in a taut string to double the wave speed?

4. When traveling on a taut string, does a wave pulse always invert upon reflection? Explain.

5. Can two pulses traveling in opposite directions on the same string reflect from each other? Explain.

6. Does the vertical speed of a segment of a horizontal, taut string, through which a wave is traveling, depend on the wave speed?

7. If you were to shake one end of a taut rope periodically three times each second, what would be the period of the sinusoidal waves set up in the rope?

8. A vibrating source generates a sinusoidal wave on a string under constant tension. If the power delivered to the string is doubled, by what factor does the amplitude change? Does the wave speed change under these circumstances?

9. Consider a wave traveling on a taut rope. What is the difference, if any, between the speed of the wave and the speed of a small segment of the rope?

10. If a long rope is hung from a ceiling and waves are sent up the rope from its lower end, they do not ascend with constant speed. Explain.

11. What happens to the wavelength of a wave on a string when the frequency is doubled? Assume that the tension in the string remains the same.

12. What happens to the speed of a wave on a taut string when the frequency is doubled? Assume that the tension in the string remains the same.

13. How do transverse waves differ from longitudinal waves?

14. When all the strings on a guitar are stretched to the same tension, will the speed of a wave along the more massive bass strings be faster or slower than the speed of a wave on the lighter strings?

15. If you stretch a rubber hose and pluck it, you can observe a pulse traveling up and down the hose. What happens to the speed of the pulse if you stretch the hose more tightly? What happens to the speed if you fill the hose with water?

16. In a longitudinal wave in a spring, the coils move back and forth in the direction of wave motion. Does the speed of the wave depend on the maximum speed of each coil?

17. When two waves interfere, can the amplitude of the resultant wave be greater than either of the two original waves? Under what conditions?

18. A solid can transport both longitudinal waves and transverse waves, but a fluid can transport only longitudinal waves. Why?

PROBLEMS

1, **2**, 3 = straightforward, intermediate, challenging ☐ = full solution available in the *Student Solutions Manual and Study Guide*
WEB = solution posted at **http://www.saunderscollege.com/physics/** 🖥 = Computer useful in solving problem 🎯 = Interactive Physics
☐ = paired numerical/symbolic problems

Section 16.1   Basic Variables of Wave Motion

Section 16.2   Direction of Particle Displacement

Section 16.3   One-Dimensional Traveling Waves

1. At $t = 0$, a transverse wave pulse in a wire is described by the function

$$y = \frac{6}{x^2 + 3}$$

where x and y are in meters. Write the function $y(x, t)$ that describes this wave if it is traveling in the positive x direction with a speed of 4.50 m/s.

2. Two wave pulses A and B are moving in opposite directions along a taut string with a speed of 2.00 cm/s. The amplitude of A is twice the amplitude of B. The pulses are shown in Figure P16.2 at $t = 0$. Sketch the shape of the string at $t = 1, 1.5, 2, 2.5,$ and 3 s.

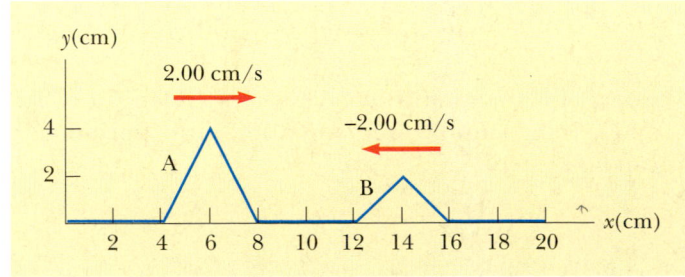

Figure P16.2

3. A wave moving along the x axis is described by

$$y(x, t) = 5.00e^{-(x+5.00t)^2}$$

where x is in meters and t is in seconds. Determine (a) the direction of the wave motion and (b) the speed of the wave.

4. Ocean waves with a crest-to-crest distance of 10.0 m can be described by the equation

$$y(x, t) = (0.800 \text{ m}) \sin[0.628(x - vt)]$$

where $v = 1.20$ m/s. (a) Sketch $y(x, t)$ at $t = 0$. (b) Sketch $y(x, t)$ at $t = 2.00$ s. Note how the entire wave form has shifted 2.40 m in the positive x direction in this time interval.

5. Two points, A and B, on the surface of the Earth are at the same longitude and 60.0° apart in latitude. Suppose that an earthquake at point A sends two waves toward point B. A transverse wave travels along the surface of the Earth at 4.50 km/s, and a longitudinal wave travels straight through the body of the Earth at 7.80 km/s. (a) Which wave arrives at point B first? (b) What is the time difference between the arrivals of the two waves at point B? Take the radius of the Earth to be 6 370 km.

6. A seismographic station receives S and P waves from an earthquake, 17.3 s apart. Suppose that the waves have traveled over the same path at speeds of 4.50 km/s and 7.80 km/s, respectively. Find the distance from the seismometer to the epicenter of the quake.

Section 16.4 Superposition and Interference

WEB 7. Two sinusoidal waves in a string are defined by the functions

$$y_1 = (2.00 \text{ cm}) \sin(20.0x - 32.0t)$$

and

$$y_2 = (2.00 \text{ cm}) \sin(25.0x - 40.0t)$$

where y and x are in centimeters and t is in seconds. (a) What is the phase difference between these two waves at the point $x = 5.00$ cm at $t = 2.00$ s? (b) What is the positive x value closest to the origin for which the two phases differ by $\pm \pi$ at $t = 2.00$ s? (This is where the sum of the two waves is zero.)

8. Two waves in one string are described by the wave functions

$$y_1 = 3.0 \cos(4.0x - 1.6t)$$

and

$$y_2 = 4.0 \sin(5.0x - 2.0t)$$

where y and x are in centimeters and t is in seconds. Find the superposition of the waves $y_1 + y_2$ at the points (a) $x = 1.00$, $t = 1.00$; (b) $x = 1.00$, $t = 0.500$; (c) $x = 0.500$, $t = 0$. (Remember that the arguments of the trigonometric functions are in radians.)

9. Two pulses traveling on the same string are described by the functions

$$y_1 = \frac{5}{(3x - 4t)^2 + 2}$$

and

$$y_2 = \frac{-5}{(3x + 4t - 6)^2 + 2}$$

(a) In which direction does each pulse travel? (b) At what time do the two cancel? (c) At what point do the two waves always cancel?

Section 16.5 The Speed of Waves on Strings

10. A phone cord is 4.00 m long. The cord has a mass of 0.200 kg. A transverse wave pulse is produced by plucking one end of the taut cord. The pulse makes four trips down and back along the cord in 0.800 s. What is the tension in the cord?

11. Transverse waves with a speed of 50.0 m/s are to be produced in a taut string. A 5.00-m length of string with a total mass of 0.060 0 kg is used. What is the required tension?

12. A piano string having a mass per unit length 5.00×10^{-3} kg/m is under a tension of 1 350 N. Find the speed with which a wave travels on this string.

13. An astronaut on the Moon wishes to measure the local value of g by timing pulses traveling down a wire that has a large mass suspended from it. Assume that the wire has a mass of 4.00 g and a length of 1.60 m, and that a 3.00-kg mass is suspended from it. A pulse requires 36.1 ms to traverse the length of the wire. Calculate g_{Moon} from these data. (You may neglect the mass of the wire when calculating the tension in it.)

14. Transverse pulses travel with a speed of 200 m/s along a taut copper wire whose diameter is 1.50 mm. What is the tension in the wire? (The density of copper is 8.92 g/cm^3.)

15. Transverse waves travel with a speed of 20.0 m/s in a string under a tension of 6.00 N. What tension is required to produce a wave speed of 30.0 m/s in the same string?

16. A simple pendulum consists of a ball of mass M hanging from a uniform string of mass m and length L, with $m \ll M$. If the period of oscillation for the pendulum is T, determine the speed of a transverse wave in the string when the pendulum hangs at rest.

17. The elastic limit of a piece of steel wire is 2.70×10^9 Pa. What is the maximum speed at which transverse wave pulses can propagate along this wire before this stress is exceeded? (The density of steel is 7.86×10^3 kg/m^3.)

18. **Review Problem.** A light string with a mass per unit length of 8.00 g/m has its ends tied to two walls separated by a distance equal to three-fourths the length of the string (Fig. P16.18). An object of mass m is sus-

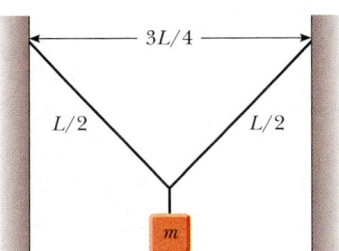

Figure P16.18

pended from the center of the string, putting a tension in the string. (a) Find an expression for the transverse wave speed in the string as a function of the hanging mass. (b) How much mass should be suspended from the string to produce a wave speed of 60.0 m/s?

19. **Review Problem.**   A light string with a mass of 10.0 g and a length $L = 3.00$ m has its ends tied to two walls that are separated by the distance $D = 2.00$ m. Two objects, each with a mass $M = 2.00$ kg, are suspended from the string, as shown in Figure P16.19. If a wave pulse is sent from point A, how long does it take for it to travel to point B?

20. **Review Problem.**   A light string of mass m and length L has its ends tied to two walls that are separated by the distance D. Two objects, each of mass M, are suspended from the string, as shown in Figure P16.19. If a wave pulse is sent from point A, how long does it take to travel to point B?

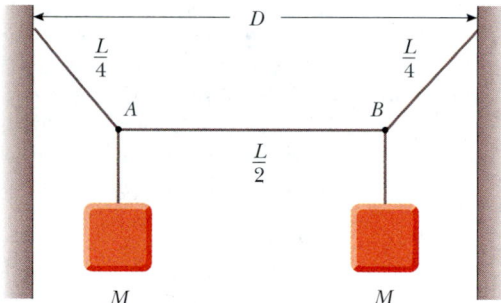

Figure P16.19   Problems 19 and 20.

WEB 21. A 30.0-m steel wire and a 20.0-m copper wire, both with 1.00-mm diameters, are connected end to end and are stretched to a tension of 150 N. How long does it take a transverse wave to travel the entire length of the two wires?

Section 16.6   Reflection and Transmission

22. A series of pulses, each of amplitude 0.150 m, are sent down a string that is attached to a post at one end. The pulses are reflected at the post and travel back along the string without loss of amplitude. What is the displacement at a point on the string where two pulses are crossing (a) if the string is rigidly attached to the post? (b) if the end at which reflection occurs is free to slide up and down?

Section 16.7   Sinusoidal Waves

23. (a) Plot y versus t at $x = 0$ for a sinusoidal wave of the form $y = (15.0 \text{ cm}) \cos(0.157x - 50.3t)$, where x and y are in centimeters and t is in seconds. (b) Determine

the period of vibration from this plot and compare your result with the value found in Example 16.3.

24. For a certain transverse wave, the distance between two successive crests is 1.20 m, and eight crests pass a given point along the direction of travel every 12.0 s. Calculate the wave speed.

25. A sinusoidal wave is traveling along a rope. The oscillator that generates the wave completes 40.0 vibrations in 30.0 s. Also, a given maximum travels 425 cm along the rope in 10.0 s. What is the wavelength?

26. Consider the sinusoidal wave of Example 16.3, with the wave function

$$y = (15.0 \text{ cm}) \cos(0.157x - 50.3t)$$

At a certain instant, let point A be at the origin and point B be the first point along the x axis where the wave is 60.0° out of phase with point A. What is the coordinate of point B?

27. When a particular wire is vibrating with a frequency of 4.00 Hz, a transverse wave of wavelength 60.0 cm is produced. Determine the speed of wave pulses along the wire.

28. A sinusoidal wave traveling in the $-x$ direction (to the left) has an amplitude of 20.0 cm, a wavelength of 35.0 cm, and a frequency of 12.0 Hz. The displacement of the wave at $t = 0$, $x = 0$ is $y = -3.00$ cm; at this same point, a particle of the medium has a positive velocity. (a) Sketch the wave at $t = 0$. (b) Find the angular wave number, period, angular frequency, and wave speed of the wave. (c) Write an expression for the wave function $y(x, t)$.

29. A sinusoidal wave train is described by the equation

$$y = (0.25 \text{ m}) \sin(0.30x - 40t)$$

where x and y are in meters and t is in seconds. Determine for this wave the (a) amplitude, (b) angular frequency, (c) angular wave number, (d) wavelength, (e) wave speed, and (f) direction of motion.

30. A transverse wave on a string is described by the expression

$$y = (0.120 \text{ m}) \sin(\pi x/8 + 4\pi t)$$

(a) Determine the transverse speed and acceleration of the string at $t = 0.200$ s for the point on the string located at $x = 1.60$ m. (b) What are the wavelength, period, and speed of propagation of this wave?

WEB 31. (a) Write the expression for y as a function of x and t for a sinusoidal wave traveling along a rope in the *negative x* direction with the following characteristics: $A = 8.00$ cm, $\lambda = 80.0$ cm, $f = 3.00$ Hz, and $y(0, t) = 0$ at $t = 0$. (b) Write the expression for y as a function of x and t for the wave in part (a), assuming that $y(x, 0) = 0$ at the point $x = 10.0$ cm.

32. A transverse sinusoidal wave on a string has a period $T = 25.0$ ms and travels in the negative x direction with a speed of 30.0 m/s. At $t = 0$, a particle on the string at

$x = 0$ has a displacement of 2.00 cm and travels downward with a speed of 2.00 m/s. (a) What is the amplitude of the wave? (b) What is the initial phase angle? (c) What is the maximum transverse speed of the string? (d) Write the wave function for the wave.

33. A sinusoidal wave of wavelength 2.00 m and amplitude 0.100 m travels on a string with a speed of 1.00 m/s to the right. Initially, the left end of the string is at the origin. Find (a) the frequency and angular frequency, (b) the angular wave number, and (c) the wave function for this wave. Determine the equation of motion for (d) the left end of the string and (e) the point on the string at $x = 1.50$ m to the right of the left end. (f) What is the maximum speed of any point on the string?

34. A sinusoidal wave on a string is described by the equation

$$y = (0.51 \text{ cm}) \sin(kx - \omega t)$$

where $k = 3.10$ rad/cm and $\omega = 9.30$ rad/s. How far does a wave crest move in 10.0 s? Does it move in the positive or negative x direction?

35. A wave is described by $y = (2.00 \text{ cm}) \sin(kx - \omega t)$, where $k = 2.11$ rad/m, $\omega = 3.62$ rad/s, x is in meters, and t is in seconds. Determine the amplitude, wavelength, frequency, and speed of the wave.

36. A transverse traveling wave on a taut wire has an amplitude of 0.200 mm and a frequency of 500 Hz. It travels with a speed of 196 m/s. (a) Write an equation in SI units of the form $y = A \sin(kx - \omega t)$ for this wave. (b) The mass per unit length of this wire is 4.10 g/m. Find the tension in the wire.

37. A wave on a string is described by the wave function

$$y = (0.100 \text{ m}) \sin(0.50x - 20t)$$

(a) Show that a particle in the string at $x = 2.00$ m executes simple harmonic motion. (b) Determine the frequency of oscillation of this particular point.

Section 16.8 Rate of Energy Transfer by Sinusoidal Waves on Strings

38. A taut rope has a mass of 0.180 kg and a length of 3.60 m. What power must be supplied to the rope to generate sinusoidal waves having an amplitude of 0.100 m and a wavelength of 0.500 m and traveling with a speed of 30.0 m/s?

39. A two-dimensional water wave spreads in circular wave fronts. Show that the amplitude A at a distance r from the initial disturbance is proportional to $1/\sqrt{r}$. (*Hint:* Consider the energy carried by one outward-moving ripple.)

40. Transverse waves are being generated on a rope under constant tension. By what factor is the required power increased or decreased if (a) the length of the rope is doubled and the angular frequency remains constant, (b) the amplitude is doubled and the angular frequency is halved, (c) both the wavelength and the amplitude are doubled, and (d) both the length of the rope and the wavelength are halved?

WEB 41. Sinusoidal waves 5.00 cm in amplitude are to be transmitted along a string that has a linear mass density of 4.00×10^{-2} kg/m. If the source can deliver a maximum power of 300 W and the string is under a tension of 100 N, what is the highest vibrational frequency at which the source can operate?

42. It is found that a 6.00-m segment of a long string contains four complete waves and has a mass of 180 g. The string is vibrating sinusoidally with a frequency of 50.0 Hz and a peak-to-valley displacement of 15.0 cm. (The "peak-to-valley" distance is the vertical distance from the farthest positive displacement to the farthest negative displacement.) (a) Write the function that describes this wave traveling in the positive x direction. (b) Determine the power being supplied to the string.

43. A sinusoidal wave on a string is described by the equation

$$y = (0.15 \text{ m}) \sin(0.80x - 50t)$$

where x and y are in meters and t is in seconds. If the mass per unit length of this string is 12.0 g/m, determine (a) the speed of the wave, (b) the wavelength, (c) the frequency, and (d) the power transmitted to the wave.

44. A horizontal string can transmit a maximum power of $\mathcal{P}$ (without breaking) if a wave with amplitude A and angular frequency ω is traveling along it. To increase this maximum power, a student folds the string and uses the "double string" as a transmitter. Determine the maximum power that can be transmitted along the "double string," supposing that the tension is constant.

(*Optional*)
Section 16.9 The Linear Wave Equation

45. (a) Evaluate A in the scalar equality $(7 + 3)4 = A$. (b) Evaluate A, B, and C in the vector equality $7.00\mathbf{i} + 3.00\mathbf{k} = A\mathbf{i} + B\mathbf{j} + C\mathbf{k}$. Explain how you arrive at your answers. (c) The functional equality or identity

$$A + B\cos(Cx + Dt + E) = (7.00 \text{ mm}) \cos(3x + 4t + 2)$$

is true for all values of the variables x and t, which are measured in meters and in seconds, respectively. Evaluate the constants A, B, C, D, and E. Explain how you arrive at your answers.

46. Show that the wave function $y = e^{b(x - vt)}$ is a solution of the wave equation (Eq. 16.26), where b is a constant.

47. Show that the wave function $y = \ln[b(x - vt)]$ is a solution to Equation 16.26, where b is a constant.

48. (a) Show that the function $y(x, t) = x^2 + v^2t^2$ is a solution to the wave equation. (b) Show that the function above can be written as $f(x + vt) + g(x - vt)$, and determine the functional forms for f and g. (c) Repeat parts (a) and (b) for the function $y(x, t) = \sin(x)\cos(vt)$.

ADDITIONAL PROBLEMS

49. The "wave" is a particular type of wave pulse that can sometimes be seen propagating through a large crowd gathered at a sporting arena to watch a soccer or American football match (Fig. P16.49). The particles of the medium are the spectators, with zero displacement corresponding to their being in the seated position and maximum displacement corresponding to their being in the standing position and raising their arms. When a large fraction of the spectators participate in the wave motion, a somewhat stable pulse shape can develop. The wave speed depends on people's reaction time, which is typically on the order of 0.1 s. Estimate the order of magnitude, in minutes, of the time required for such a wave pulse to make one circuit around a large sports stadium. State the quantities you measure or estimate and their values.

Figure P16.49   (Gregg Adams/Tony Stone Images)

50. A traveling wave propagates according to the expression $y = (4.0 \text{ cm}) \sin(2.0x - 3.0t)$, where x is in centimeters and t is in seconds. Determine (a) the amplitude, (b) the wavelength, (c) the frequency, (d) the period, and (e) the direction of travel of the wave.

WEB **51.** The wave function for a traveling wave on a taut string is (in SI units)

$$y(x, t) = (0.350 \text{ m}) \sin(10\pi t - 3\pi x + \pi/4)$$

(a) What are the speed and direction of travel of the wave? (b) What is the vertical displacement of the string at $t = 0$, $x = 0.100$ m? (c) What are the wavelength and frequency of the wave? (d) What is the maximum magnitude of the transverse speed of the string?

52. Motion picture film is projected at 24.0 frames per second. Each frame is a photograph 19.0 mm in height. At what constant speed does the film pass into the projector?

53. **Review Problem.**   A block of mass M, supported by a string, rests on an incline making an angle θ with the horizontal (Fig. P16.53). The string's length is L, and its mass is $m \ll M$. Derive an expression for the time it takes a transverse wave to travel from one end of the string to the other.

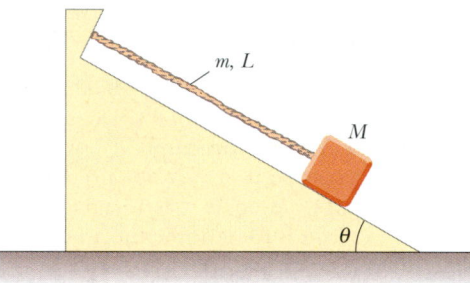

Figure P16.53

54. (a) Determine the speed of transverse waves on a string under a tension of 80.0 N if the string has a length of 2.00 m and a mass of 5.00 g. (b) Calculate the power required to generate these waves if they have a wavelength of 16.0 cm and an amplitude of 4.00 cm.

55. **Review Problem.**   A 2.00-kg block hangs from a rubber cord. The block is supported so that the cord is not stretched. The unstretched length of the cord is 0.500 m, and its mass is 5.00 g. The "spring constant" for the cord is 100 N/m. The block is released and stops at the lowest point. (a) Determine the tension in the cord when the block is at this lowest point. (b) What is the length of the cord in this "stretched" position? (c) Find the speed of a transverse wave in the cord if the block is held in this lowest position.

56. **Review Problem.**   A block of mass M hangs from a rubber cord. The block is supported so that the cord is not stretched. The unstretched length of the cord is L_0, and its mass is m, much less than M. The "spring constant" for the cord is k. The block is released and stops at the lowest point. (a) Determine the tension in the cord when the block is at this lowest point. (b) What is the length of the cord in this "stretched" position? (c) Find the speed of a transverse wave in the cord if the block is held in this lowest position.

57. A sinusoidal wave in a rope is described by the wave function

$$y = (0.20 \text{ m}) \sin(0.75\pi x + 18\pi t)$$

where x and y are in meters and t is in seconds. The rope has a linear mass density of 0.250 kg/m. If the tension in the rope is provided by an arrangement like the one illustrated in Figure 16.12, what is the value of the suspended mass?

58. A wire of density ρ is tapered so that its cross-sectional area varies with x, according to the equation

$$A = (1.0 \times 10^{-3}x + 0.010) \text{ cm}^2$$

(a) If the wire is subject to a tension T, derive a relationship for the speed of a wave as a function of position. (b) If the wire is aluminum and is subject to a tension of 24.0 N, determine the speed at the origin and at $x = 10.0$ m.

59. A rope of total mass m and length L is suspended vertically. Show that a transverse wave pulse travels the length of the rope in a time $t = 2\sqrt{L/g}$. (*Hint:* First find an expression for the wave speed at any point a distance x from the lower end by considering the tension in the rope as resulting from the weight of the segment below that point.)

60. If mass M is suspended from the bottom of the rope in Problem 59, (a) show that the time for a transverse wave to travel the length of the rope is

$$t = 2\sqrt{\frac{L}{mg}}\left[\sqrt{(M + m)} - \sqrt{M}\right]$$

(b) Show that this reduces to the result of Problem 59 when $M = 0$. (c) Show that for $m \ll M$, the expression in part (a) reduces to

$$t = \sqrt{\frac{mL}{Mg}}$$

61. It is stated in Problem 59 that a wave pulse travels from the bottom to the top of a rope of length L in a time $t = 2\sqrt{L/g}$. Use this result to answer the following questions. (It is *not* necessary to set up any new integrations.) (a) How long does it take for a wave pulse to travel halfway up the rope? (Give your answer as a fraction of the quantity $2\sqrt{L/g}$.) (b) A pulse starts traveling up the rope. How far has it traveled after a time $\sqrt{L/g}$?

62. Determine the speed and direction of propagation of each of the following sinusoidal waves, assuming that x is measured in meters and t in seconds:
(a) $y = 0.60 \cos(3.0x - 15t + 2)$
(b) $y = 0.40 \cos(3.0x + 15t - 2)$
(c) $y = 1.2 \sin(15t + 2.0x)$
(d) $y = 0.20 \sin(12t - x/2 + \pi)$

63. Review Problem. An aluminum wire under zero tension at room temperature is clamped at each end. The tension in the wire is increased by reducing the temperature, which results in a decrease in the wire's equilibrium length. What strain ($\Delta L/L$) results in a transverse wave speed of 100 m/s? Take the cross-sectional area of the wire to be 5.00×10^{-6} m^2, the density of the material to be 2.70×10^3 kg/m^3, and Young's modulus to be 7.00×10^{10} N/m^2.

64. (a) Show that the speed of longitudinal waves along a spring of force constant k is $v = \sqrt{kL/\mu}$, where L is the unstretched length of the spring and μ is the mass per unit length. (b) A spring with a mass of 0.400 kg has an unstretched length of 2.00 m and a force constant of 100 N/m. Using the result you obtained in (a), determine the speed of longitudinal waves along this spring.

65. A string of length L consists of two sections: The left half has mass per unit length $\mu = \mu_0/2$, whereas the right half has a mass per unit length $\mu' = 3\mu = 3\mu_0/2$. Tension in the string is T_0. Notice from the data given that this string has the same total mass as a uniform string of length L and of mass per unit length μ_0. (a) Find the speeds v and v' at which transverse wave pulses travel in the two sections. Express the speeds in terms of T_0 and μ_0, and also as multiples of the speed $v_0 = (T_0/\mu_0)^{1/2}$. (b) Find the time required for a pulse to travel from one end of the string to the other. Give your result as a multiple of $t_0 = L/v_0$.

66. A wave pulse traveling along a string of linear mass density μ is described by the relationship

$$y = [A_0 e^{-bx}] \sin(kx - \omega t)$$

where the factor in brackets before the sine function is said to be the amplitude. (a) What is the power $\mathcal{P}(x)$ carried by this wave at a point x? (b) What is the power carried by this wave at the origin? (c) Compute the ratio $\mathcal{P}(x)/\mathcal{P}(0)$.

67. An earthquake on the ocean floor in the Gulf of Alaska produces a *tsunami* (sometimes called a "tidal wave") that reaches Hilo, Hawaii, 4 450 km away, in a time of 9 h 30 min. Tsunamis have enormous wavelengths (100–200 km), and the propagation speed of these waves is $v \approx \sqrt{gd}$, where d is the average depth of the water. From the information given, find the average wave speed and the average ocean depth between Alaska and Hawaii. (This method was used in 1856 to estimate the average depth of the Pacific Ocean long before soundings were made to obtain direct measurements.)

ANSWERS TO QUICK QUIZZES

16.1 (a) It is longitudinal because the disturbance (the shift of position) is parallel to the direction in which the wave travels. (b) It is transverse because the people stand up and sit down (vertical motion), whereas the wave moves either to the left or to the right (motion perpendicular to the disturbance).

16.2

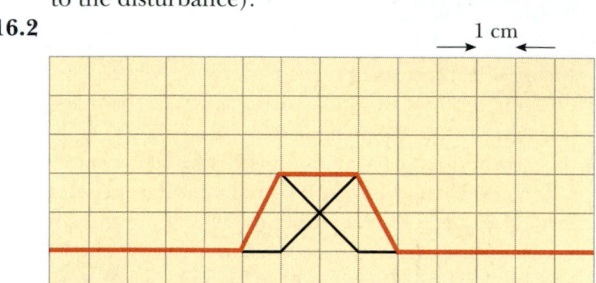

1 cm

16.3 Only answers (f) and (h) are correct. (a) and (b) affect the transverse speed of a particle of the string, but not the wave speed along the string. (c) and (d) change the amplitude. (e) and (g) increase the time by decreasing the wave speed.

16.4 The transverse speed increases because $v_{y, \text{max}} = \omega A = 2\pi f A$. The wave speed does not change because it depends only on the tension and mass per length of the string, neither of which has been modified. The wavelength must decrease because the wave speed $v = \lambda f$ remains constant.

PUZZLER

You can estimate the distance to an approaching storm by listening carefully to the sound of the thunder. How is this done? Why is the sound that follows a lightning strike sometimes a short, sharp thunderclap and other times a long-lasting rumble? *(Richard Kaylin/Tony Stone Images)*

chapter

17

Sound Waves

Sound waves are the most important example of longitudinal waves. They can travel through any material medium with a speed that depends on the properties of the medium. As the waves travel, the particles in the medium vibrate to produce changes in density and pressure along the direction of motion of the wave. These changes result in a series of high-pressure and low-pressure regions. If the source of the sound waves vibrates sinusoidally, the pressure variations are also sinusoidal. We shall find that the mathematical description of sinusoidal sound waves is identical to that of sinusoidal string waves, which was discussed in the previous chapter.

Sound waves are divided into three categories that cover different frequency ranges. (1) *Audible waves* are waves that lie within the range of sensitivity of the human ear. They can be generated in a variety of ways, such as by musical instruments, human vocal cords, and loudspeakers. (2) *Infrasonic waves* are waves having frequencies below the audible range. Elephants can use infrasonic waves to communicate with each other, even when separated by many kilometers. (3) *Ultrasonic waves* are waves having frequencies above the audible range. You may have used a "silent" whistle to retrieve your dog. The ultrasonic sound it emits is easily heard by dogs, although humans cannot detect it at all. Ultrasonic waves are also used in medical imaging.

We begin this chapter by discussing the speed of sound waves and then wave intensity, which is a function of wave amplitude. We then provide an alternative description of the intensity of sound waves that compresses the wide range of intensities to which the ear is sensitive to a smaller range. Finally, we treat effects of the motion of sources and/or listeners.

Undisturbed gas

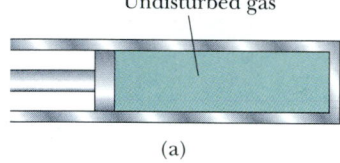

(a)

Compressed region

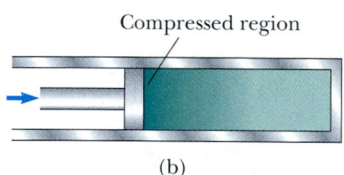

(b)

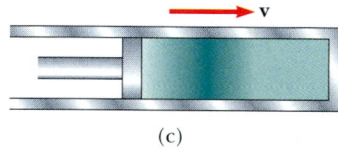

(c)

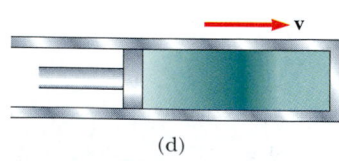

(d)

Figure 17.1   Motion of a longitudinal pulse through a compressible gas. The compression (darker region) is produced by the moving piston.

17.1 SPEED OF SOUND WAVES

Let us describe pictorially the motion of a one-dimensional longitudinal pulse moving through a long tube containing a compressible gas (Fig. 17.1). A piston at the left end can be moved to the right to compress the gas and create the pulse. Before the piston is moved, the gas is undisturbed and of uniform density, as represented by the uniformly shaded region in Figure 17.1a. When the piston is suddenly pushed to the right (Fig. 17.1b), the gas just in front of it is compressed (as represented by the more heavily shaded region); the pressure and density in this region are now higher than they were before the piston moved. When the piston comes to rest (Fig. 17.1c), the compressed region of the gas continues to move to the right, corresponding to a longitudinal pulse traveling through the tube with

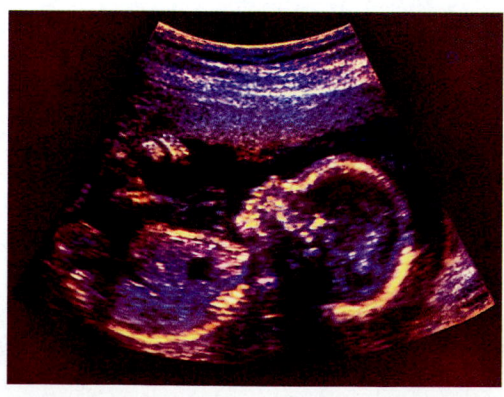

An ultrasound image of a human fetus in the womb after 20 weeks of development, showing the head, body, arms, and legs in profile. *(U.H.B. Trust/Tony Stone Images)*

speed v. Note that the piston speed does *not* equal v. Furthermore, the compressed region does not "stay with" the piston as the piston moves, because the speed of the wave may be greater than the speed of the piston.

The speed of sound waves depends on the compressibility and inertia of the medium. If the medium has a bulk modulus B (see Section 12.4) and density ρ, the speed of sound waves in that medium is

$$v = \sqrt{\frac{B}{\rho}}$$ (17.1)

Speed of sound

It is interesting to compare this expression with Equation 16.4 for the speed of transverse waves on a string, $v = \sqrt{T/\mu}$. In both cases, the wave speed depends on an elastic property of the medium—bulk modulus B or string tension T—and on an inertial property of the medium—ρ or μ. In fact, the speed of *all mechanical waves* follows an expression of the general form

$$v = \sqrt{\frac{\text{elastic property}}{\text{inertial property}}}$$

The speed of sound also depends on the temperature of the medium. For sound traveling through air, the relationship between wave speed and medium temperature is

$$v = (331 \text{ m/s}) \sqrt{1 + \frac{T_C}{273°C}}$$

where 331 m/s is the speed of sound in air at 0°C, and T_C is the temperature in degrees Celsius. Using this equation, one finds that at 20°C the speed of sound in air is approximately 343 m/s.

This information provides a convenient way to estimate the distance to a thunderstorm, as demonstrated in the QuickLab. During a lightning flash, the temperature of a long channel of air rises rapidly as the bolt passes through it. This temperature increase causes the air in the channel to expand rapidly, and this expansion creates a sound wave. The channel produces sound throughout its entire length at essentially the same instant. If the orientation of the channel is such that all of its parts are approximately the same distance from you, sounds from the different parts reach you at the same time, and you hear a short, intense thunderclap. However, if the distances between your ear and different portions of the channel vary, sounds from different portions arrive at your ears at different times. If the channel were a straight line, the resulting sound would be a steady roar, but the zigzag shape of the path produces variations in loudness.

QuickLab

The next time a thunderstorm approaches, count the seconds between a flash of lightning (which reaches you almost instantaneously) and the following thunderclap. Divide this time by 3 to determine the approximate number of kilometers (or by 5 to estimate the miles) to the storm.

To learn more about lightning, read E. Williams, "The Electrification of Thunderstorms" *Sci. Am.* 259(5):88–89, 1988.

Quick Quiz 17.1

The speed of sound in air is a function of (a) wavelength, (b) frequency, (c) temperature, (d) amplitude.

Quick Quiz 17.2

As a result of a distant explosion, an observer first senses a ground tremor and then hears the explosion later. Explain.

EXAMPLE 17.1 ▶ Speed of Sound in a Solid

If a solid bar is struck at one end with a hammer, a longitudinal pulse propagates down the bar with a speed $v = \sqrt{Y/\rho}$, where Y is the Young's modulus for the material (see Section 12.4). Find the speed of sound in an aluminum bar.

Solution From Table 12.1 we obtain $Y = 7.0 \times 10^{10}$ N/m² for aluminum, and from Table 1.5 we obtain $\rho = 2.70 \times 10^3$ kg/m³. Therefore,

$$v_{Al} = \sqrt{\frac{Y}{\rho}} = \sqrt{\frac{7.0 \times 10^{10} \text{ N/m}^2}{2.70 \times 10^3 \text{ kg/m}^3}} \approx \boxed{5.1 \text{ km/s}}$$

This typical value for the speed of sound in solids is much greater than the speed of sound in gases, as Table 17.1 shows. This difference in speeds makes sense because the molecules of a solid are bound together into a much more rigid structure than those in a gas and hence respond more rapidly to a disturbance.

EXAMPLE 17.2 ▶ Speed of Sound in a Liquid

(a) Find the speed of sound in water, which has a bulk modulus of 2.1×10^9 N/m² and a density of 1.00×10^3 kg/m³.

Solution Using Equation 17.1, we find that

$$v_{water} = \sqrt{\frac{B}{\rho}} = \sqrt{\frac{2.1 \times 10^9 \text{ N/m}^2}{1.00 \times 10^3 \text{ kg/m}^3}} = \boxed{1.4 \text{ km/s}}$$

In general, sound waves travel more slowly in liquids than in solids because liquids are more compressible than solids.

(b) Dolphins use sound waves to locate food. Experiments have shown that a dolphin can detect a 7.5-cm target 110 m away, even in murky water. For a bit of "dinner" at that distance, how much time passes between the moment the dolphin emits a sound pulse and the moment the dolphin hears its reflection and thereby detects the distant target?

Solution The total distance covered by the sound wave as it travels from dolphin to target and back is 2×110 m = 220 m. From Equation 2.2, we have

$$\Delta t = \frac{\Delta x}{v_x} = \frac{220 \text{ m}}{1\,400 \text{ m/s}} = \boxed{0.16 \text{ s}}$$

Bottle-nosed dolphin. *(Stuart Westmoreland/Tony Stone Images)*

17.2 ▶ PERIODIC SOUND WAVES

This section will help you better comprehend the nature of sound waves. You will learn that pressure variations control what we hear—an important fact for understanding how our ears work.

One can produce a one-dimensional periodic sound wave in a long, narrow tube containing a gas by means of an oscillating piston at one end, as shown in Figure 17.2. The darker parts of the colored areas in this figure represent regions where the gas is compressed and thus the density and pressure are above their equilibrium values. A compressed region is formed whenever the piston is pushed into the tube. This compressed region, called a **condensation,** moves through the tube as a pulse, continuously compressing the region just in front of itself. When the piston is pulled back, the gas in front of it expands, and the pressure and density in this region fall below their equilibrium values (represented by the lighter parts of the colored areas in Fig. 17.2). These low-pressure regions, called **rarefactions,** also propagate along the tube, following the condensations. Both regions move with a speed equal to the speed of sound in the medium.

As the piston oscillates sinusoidally, regions of condensation and rarefaction are continuously set up. The distance between two successive condensations (or two successive rarefactions) equals the wavelength λ. As these regions travel through the tube, any small volume of the medium moves with simple harmonic motion parallel to the direction of the wave. If $s(x, t)$ is the displacement of a small volume element from its equilibrium position, we can express this harmonic displacement function as

$$s(x, t) = s_{max} \cos(kx - \omega t) \tag{17.2}$$

where s_{max} **is the maximum displacement of the medium from equilibrium** (in other words, the **displacement amplitude** of the wave), k is the angular wavenumber, and ω is the angular frequency of the piston. Note that the displacement of the medium is along x, in the direction of motion of the sound wave, which means we are describing a longitudinal wave.

As we shall demonstrate shortly, the variation in the gas pressure ΔP, measured from the equilibrium value, is also periodic and for the displacement function in Equation 17.2 is given by

$$\Delta P = \Delta P_{max} \sin(kx - \omega t) \tag{17.3}$$

where **the pressure amplitude** ΔP_{max}—which is the **maximum change in pres-**

TABLE 17.1	
Speeds of Sound in Various Media	
Medium	v **(m/s)**
Gases	
Hydrogen (0°C)	1 286
Helium (0°C)	972
Air (20°C)	343
Air (0°C)	331
Oxygen (0°C)	317
Liquids at 25°C	
Glycerol	1 904
Sea water	1 533
Water	1 493
Mercury	1 450
Kerosene	1 324
Methyl alcohol	1 143
Carbon tetrachloride	926
Solids	
Diamond	12 000
Pyrex glass	5 640
Iron	5 130
Aluminum	5 100
Brass	4 700
Copper	3 560
Gold	3 240
Lucite	2 680
Lead	1 322
Rubber	1 600

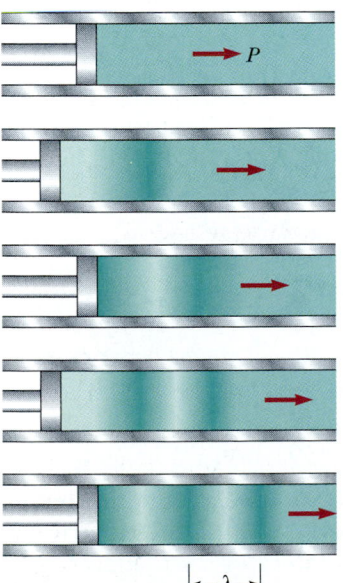

Figure 17.2   A sinusoidal longitudinal wave propagating through a gas-filled tube. The source of the wave is a sinusoidally oscillating piston at the left. The high-pressure and low-pressure regions are colored darkly and lightly, respectively.

sure from the equilibrium value—is given by

$$\Delta P_{max} = \rho v \omega s_{max}$$ (17.4)

Pressure amplitude

Thus, we see that a sound wave may be considered as either a displacement wave or a pressure wave. A comparison of Equations 17.2 and 17.3 shows that **the pressure wave is 90° out of phase with the displacement wave.** Graphs of these functions are shown in Figure 17.3. Note that the pressure variation is a maximum when the displacement is zero, and the displacement is a maximum when the pressure variation is zero.

Quick Quiz 17.3

If you blow across the top of an empty soft-drink bottle, a pulse of air travels down the bottle. At the moment the pulse reaches the bottom of the bottle, compare the displacement of air molecules with the pressure variation.

Derivation of Equation 17.3

From the definition of bulk modulus (see Eq. 12.8), the pressure variation in the gas is

$$\Delta P = -B \frac{\Delta V}{V_i}$$

The volume of gas that has a thickness Δx in the horizontal direction and a cross-sectional area A is $V_i = A \, \Delta x$. The change in volume ΔV accompanying the pressure change is equal to $A \, \Delta s$, where Δs is the difference between the value of s at $x + \Delta x$ and the value of s at x. Hence, we can express ΔP as

$$\Delta P = -B \frac{\Delta V}{V_i} = -B \frac{A \, \Delta s}{A \, \Delta x} = -B \frac{\Delta s}{\Delta x}$$

As Δx approaches zero, the ratio $\Delta s / \Delta x$ becomes $\partial s / \partial x$. (The partial derivative indicates that we are interested in the variation of s with position at a *fixed* time.) Therefore,

$$\Delta P = -B \frac{\partial s}{\partial x}$$

If the displacement is the simple sinusoidal function given by Equation 17.2, we find that

$$\Delta P = -B \frac{\partial}{\partial x} [s_{max} \cos(kx - \omega t)] = Bk s_{max} \sin(kx - \omega t)$$

Because the bulk modulus is given by $B = \rho v^2$ (see Eq. 17.1), the pressure variation reduces to

$$\Delta P = \rho v^2 k s_{max} \sin(kx - \omega t)$$

From Equation 16.13, we can write $k = \omega / v$; hence, ΔP can be expressed as

$$\Delta P = \rho v \omega s_{max} \sin(kx - \omega t)$$

Because the sine function has a maximum value of 1, we see that the maximum value of the pressure variation is $\Delta P_{max} = \rho v \omega s_{max}$ (see Eq. 17.4), and we arrive at Equation 17.3:

$$\Delta P = \Delta P_{max} \sin(kx - \omega t)$$

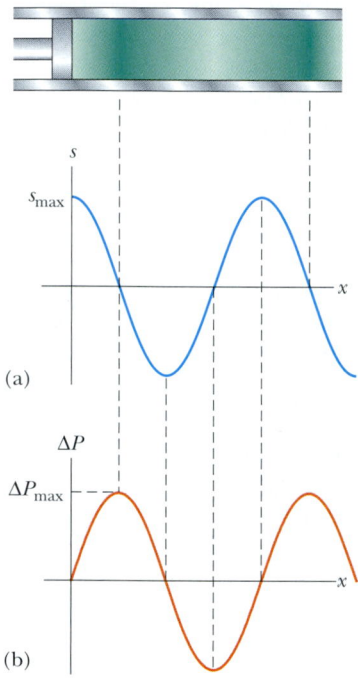

Figure 17.3   (a) Displacement amplitude versus position and (b) pressure amplitude versus position for a sinusoidal longitudinal wave. The displacement wave is 90° out of phase with the pressure wave.

17.3 ▶ INTENSITY OF PERIODIC SOUND WAVES

In the previous chapter, we showed that a wave traveling on a taut string transports energy. The same concept applies to sound waves. Consider a volume of air of mass Δm and width Δx in front of a piston oscillating with a frequency ω, as shown in Figure 17.4. The piston transmits energy to this volume of air in the tube, and the energy is propagated away from the piston by the sound wave.[1] To evaluate the rate of energy transfer for the sound wave, we shall evaluate the kinetic energy of this volume of air, which is undergoing simple harmonic motion. We shall follow a procedure similar to that in Section 16.8, in which we evaluated the rate of energy transfer for a wave on a string.

As the sound wave propagates away from the piston, the displacement of any volume of air in front of the piston is given by Equation 17.2. To evaluate the kinetic energy of this volume of air, we need to know its speed. We find the speed by taking the time derivative of Equation 17.2:

$$v(x, t) = \frac{\partial}{\partial t} s(x, t) = \frac{\partial}{\partial t} [s_{max} \cos(kx - \omega t)] = \omega s_{max} \sin(kx - \omega t)$$

Imagine that we take a "snapshot" of the wave at $t = 0$. The kinetic energy of a given volume of air at this time is

$$\Delta K = \tfrac{1}{2} \Delta m v^2 = \tfrac{1}{2} \Delta m (\omega s_{max} \sin kx)^2 = \tfrac{1}{2}\rho A \Delta x (\omega s_{max} \sin kx)^2$$

$$= \tfrac{1}{2}\rho A \Delta x (\omega s_{max})^2 \sin^2 kx$$

where A is the cross-sectional area of the moving air and $A\,\Delta x$ is its volume. Now, as in Section 16.8, we integrate this expression over a full wavelength to find the total kinetic energy in one wavelength. Letting the volume of air shrink to infinitesimal thickness, so that $\Delta x \rightarrow dx$, we have

$$K_\lambda = \int dK = \int_0^\lambda \tfrac{1}{2}\rho A (\omega s_{max})^2 \sin^2 kx\, dx = \tfrac{1}{2}\rho A (\omega s_{max})^2 \int_0^\lambda \sin^2 kx\, dx$$

$$= \tfrac{1}{2}\rho A (\omega s_{max})^2 \left(\tfrac{1}{2}\lambda\right) = \tfrac{1}{4}\rho A (\omega s_{max})^2 \lambda$$

As in the case of the string wave in Section 16.8, the total potential energy for one wavelength has the same value as the total kinetic energy; thus, the total mechani-

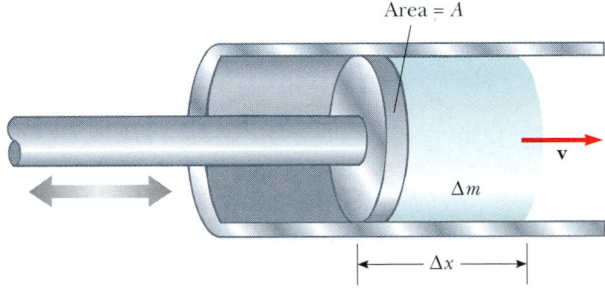

Figure 17.4 An oscillating piston transfers energy to the air in the tube, initially causing the volume of air of width Δx and mass Δm to oscillate with an amplitude s_{max}.

---

[1] Although it is not proved here, the work done by the piston equals the energy carried away by the wave. For a detailed mathematical treatment of this concept, see Chapter 4 in Frank S. Crawford, Jr., *Waves*, Berkeley Physics Course, vol. 3, New York, McGraw-Hill Book Company, 1968.

cal energy is

$$E_\lambda = K_\lambda + U_\lambda = \tfrac{1}{2}\rho A(\omega s_{max})^2 \lambda$$

As the sound wave moves through the air, this amount of energy passes by a given point during one period of oscillation. Hence, the rate of energy transfer is

$$\mathcal{P} = \frac{E_\lambda}{\Delta t} = \frac{\tfrac{1}{2}\rho A(\omega s_{max})^2 \lambda}{T} = \tfrac{1}{2}\rho A(\omega s_{max})^2 \left(\frac{\lambda}{T}\right) = \tfrac{1}{2}\rho A v(\omega s_{max})^2$$

where v is the speed of sound in air.

> We define the **intensity I** of a wave, or the power per unit area, to be the rate at which the energy being transported by the wave flows through a unit area A perpendicular to the direction of travel of the wave.

In the present case, therefore, the intensity is

Intensity of a sound wave

$$I = \frac{\mathcal{P}}{A} = \tfrac{1}{2}\rho v(\omega s_{max})^2 \tag{17.5}$$

Thus, we see that the intensity of a periodic sound wave is proportional to the square of the displacement amplitude and to the square of the angular frequency (as in the case of a periodic string wave). This can also be written in terms of the pressure amplitude ΔP_{max}; in this case, we use Equation 17.4 to obtain

$$I = \frac{\Delta P_{max}^2}{2\rho v} \tag{17.6}$$

EXAMPLE 17.3   ▶   Hearing Limits

The faintest sounds the human ear can detect at a frequency of 1 000 Hz correspond to an intensity of about 1.00×10^{-12} W/m² — the so-called *threshold of hearing*. The loudest sounds the ear can tolerate at this frequency correspond to an intensity of about 1.00 W/m² — the *threshold of pain*. Determine the pressure amplitude and displacement amplitude associated with these two limits.

Solution First, consider the faintest sounds. Using Equation 17.6 and taking $v = 343$ m/s as the speed of sound waves in air and $\rho = 1.20$ kg/m³ as the density of air, we obtain

$$\Delta P_{max} = \sqrt{2\rho v I}$$
$$= \sqrt{2(1.20 \text{ kg/m}^3)(343 \text{ m/s})(1.00 \times 10^{-12} \text{ W/m}^2)}$$
$$= 2.87 \times 10^{-5} \text{ N/m}^2$$

Because atmospheric pressure is about 10^5 N/m², this result

tells us that the ear can discern pressure fluctuations as small as 3 parts in 10^{10}!

We can calculate the corresponding displacement amplitude by using Equation 17.4, recalling that $\omega = 2\pi f$ (see Eqs. 16.10 and 16.12):

$$s_{max} = \frac{\Delta P_{max}}{\rho v \omega} = \frac{2.87 \times 10^{-5} \text{ N/m}^2}{(1.20 \text{ kg/m}^3)(343 \text{ m/s})(2\pi \times 1\,000 \text{ Hz})}$$
$$= 1.11 \times 10^{-11} \text{ m}$$

This is a remarkably small number! If we compare this result for s_{max} with the diameter of a molecule (about 10^{-10} m), we see that the ear is an extremely sensitive detector of sound waves.

In a similar manner, one finds that the loudest sounds the human ear can tolerate correspond to a pressure amplitude of 28.7 N/m² and a displacement amplitude equal to 1.11×10^{-5} m.

Sound Level in Decibels

The example we just worked illustrates the wide range of intensities the human ear can detect. Because this range is so wide, it is convenient to use a logarithmic scale, where the **sound level** β (Greek letter beta) is defined by the equation

$$\beta = 10 \log\left(\frac{I}{I_0}\right) \tag{17.7}$$

The constant I_0 is the *reference intensity*, taken to be at the threshold of hearing ($I_0 = 1.00 \times 10^{-12}$ W/m^2), and I is the intensity, in watts per square meter, at the sound level β, where β is measured in **decibels** (dB).[2] On this scale, the threshold of pain ($I = 1.00$ W/m^2) corresponds to a sound level of $\beta = 10 \log[(1 \text{ W/m}^2)/(10^{-12} \text{ W/m}^2)] = 10 \log(10^{12}) = 120$ dB, and the threshold of hearing corresponds to $\beta = 10 \log[(10^{-12} \text{ W/m}^2)/(10^{-12} \text{ W/m}^2)] = 0$ dB.

Prolonged exposure to high sound levels may seriously damage the ear. Ear plugs are recommended whenever sound levels exceed 90 dB. Recent evidence suggests that "noise pollution" may be a contributing factor to high blood pressure, anxiety, and nervousness. Table 17.2 gives some typical sound-level values.

TABLE 17.2
Sound Levels

Source of Sound	β (dB)
Nearby jet airplane	150
Jackhammer; machine gun	130
Siren; rock concert	120
Subway; power mower	100
Busy traffic	80
Vacuum cleaner	70
Normal conversation	50
Mosquito buzzing	40
Whisper	30
Rustling leaves	10
Threshold of hearing	0

EXAMPLE 17.4 ▸ Sound Levels

Two identical machines are positioned the same distance from a worker. The intensity of sound delivered by each machine at the location of the worker is 2.0×10^{-7} W/m^2. Find the sound level heard by the worker (a) when one machine is operating and (b) when both machines are operating.

Solution (a) The sound level at the location of the worker with one machine operating is calculated from Equation 17.7:

$$\beta_1 = 10 \log\left(\frac{2.0 \times 10^{-7} \text{ W/m}^2}{1.00 \times 10^{-12} \text{ W/m}^2}\right) = 10 \log(2.0 \times 10^5)$$

$$= \boxed{53 \text{ dB}}$$

(b) When both machines are operating, the intensity is doubled to 4.0×10^{-7} W/m^2; therefore, the sound level now is

$$\beta_2 = 10 \log\left(\frac{4.0 \times 10^{-7} \text{ W/m}^2}{1.00 \times 10^{-12} \text{ W/m}^2}\right) = 10 \log(4.0 \times 10^5)$$

$$= \boxed{56 \text{ dB}}$$

From these results, we see that when the intensity is doubled, the sound level increases by only 3 dB.

Quick Quiz 17.4

A violin plays a melody line and is then joined by nine other violins, all playing at the same intensity as the first violin, in a repeat of the same melody. (a) When all of the violins are playing together, by how many decibels does the sound level increase? (b) If ten more violins join in, how much has the sound level increased over that for the single violin?

---

[2] The unit *bel* is named after the inventor of the telephone, Alexander Graham Bell (1847–1922). The prefix *deci-* is the SI prefix that stands for 10^{-1}.

17.4 ▶ SPHERICAL AND PLANE WAVES

If a spherical body oscillates so that its radius varies sinusoidally with time, a spherical sound wave is produced (Fig. 17.5). The wave moves outward from the source at a constant speed if the medium is uniform.

Because of this uniformity, we conclude that the energy in a spherical wave propagates equally in all directions. That is, no one direction is preferred over any other. If $\mathscr{P}_{av}$ is the average power emitted by the source, then this power at any distance r from the source must be distributed over a spherical surface of area $4\pi r^2$. Hence, the wave intensity at a distance r from the source is

$$I = \frac{\mathscr{P}_{av}}{A} = \frac{\mathscr{P}_{av}}{4\pi r^2} \qquad (17.8)$$

Because $\mathscr{P}_{av}$ is the same for any spherical surface centered at the source, we see that the intensities at distances r_1 and r_2 are

$$I_1 = \frac{\mathscr{P}_{av}}{4\pi r_1{}^2} \qquad \text{and} \qquad I_2 = \frac{\mathscr{P}_{av}}{4\pi r_2{}^2}$$

Therefore, the ratio of intensities on these two spherical surfaces is

$$\frac{I_1}{I_2} = \frac{r_2{}^2}{r_1{}^2}$$

This inverse-square law states that the intensity decreases in proportion to the square of the distance from the source. Equation 17.5 tells us that the intensity is proportional to s_{max}^2. Setting the right side of Equation 17.5 equal to the right side

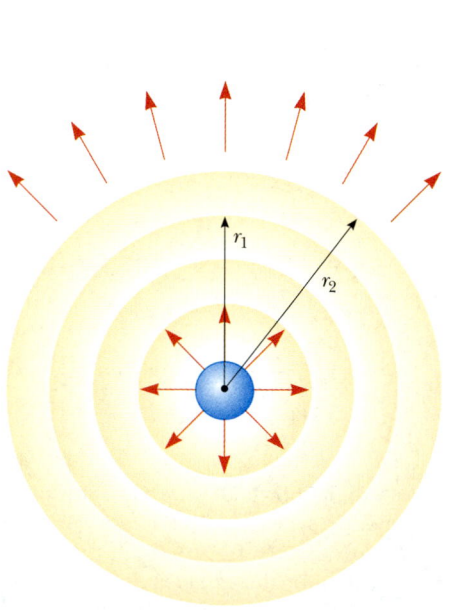

Figure 17.5 A spherical sound wave propagating radially outward from an oscillating spherical body. The intensity of the spherical wave varies as $1/r^2$.

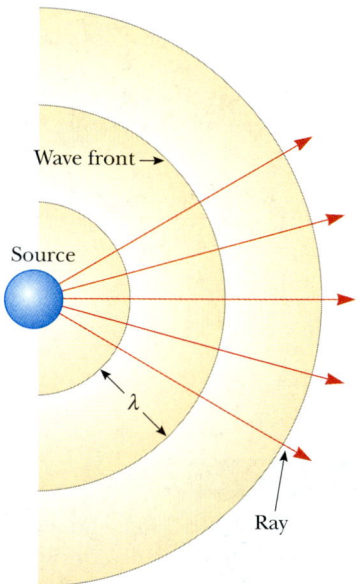

Figure 17.6 Spherical waves emitted by a point source. The circular arcs represent the spherical wave fronts that are concentric with the source. The rays are radial lines pointing outward from the source, perpendicular to the wave fronts.

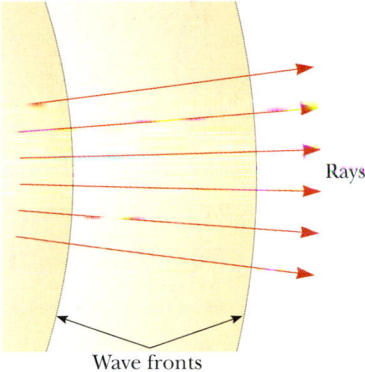

Figure 17.7 Far away from a point source, the wave fronts are nearly parallel planes, and the rays are nearly parallel lines perpendicular to the planes. Hence, a small segment of a spherical wave front is approximately a plane wave.

of Equation 17.8, we conclude that the displacement amplitude s_{max} of a spherical wave must vary as $1/r$. Therefore, we can write the wave function ψ (Greek letter psi) for an outgoing spherical wave in the form

$$\psi(r, t) = \frac{s_0}{r} \sin(kr - \omega t) \qquad \text{(17.9)}$$

where s_0, the displacement amplitude at unit distance from the source, is a constant parameter characterizing the whole wave.

It is useful to represent spherical waves with a series of circular arcs concentric with the source, as shown in Figure 17.6. Each arc represents a surface over which the phase of the wave is constant. We call such a surface of constant phase a **wave front.** The distance between adjacent wave fronts equals the wavelength λ. The radial lines pointing outward from the source are called **rays.**

Now consider a small portion of a wave front far from the source, as shown in Figure 17.7. In this case, the rays passing through the wave front are nearly parallel to one another, and the wave front is very close to being planar. Therefore, at distances from the source that are great compared with the wavelength, we can approximate a wave front with a plane. Any small portion of a spherical wave far from its source can be considered a **plane wave.**

Figure 17.8 illustrates a plane wave propagating along the x axis, which means that the wave fronts are parallel to the yz plane. In this case, the wave function depends only on x and t and has the form

$$\psi(x, t) = A \sin(kx - \omega t) \qquad \text{(17.10)}$$

That is, the wave function for a plane wave is identical in form to that for a one-dimensional traveling wave.

The intensity is the same at all points on a given wave front of a plane wave.

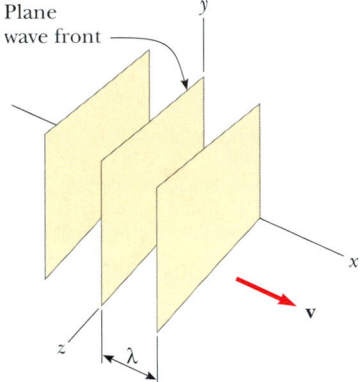

Figure 17.8 A representation of a plane wave moving in the positive x direction with a speed v. The wave fronts are planes parallel to the yz plane.

Representation of a plane wave

EXAMPLE 17.5 Intensity Variations of a Point Source

A point source emits sound waves with an average power output of 80.0 W. (a) Find the intensity 3.00 m from the source.

Solution A point source emits energy in the form of spherical waves (see Fig. 17.5). At a distance r from the source, the power is distributed over the surface area of a sphere, $4\pi r^2$. Therefore, the intensity at the distance r is given by Equation 17.8:

$$I = \frac{\mathscr{P}_{av}}{4\pi r^2} = \frac{80.0 \text{ W}}{4\pi(3.00 \text{ m})^2} = \boxed{0.707 \text{ W/m}^2}$$

an intensity that is close to the threshold of pain.

(b) Find the distance at which the sound level is 40 dB.

Solution We can find the intensity at the 40-dB sound level by using Equation 17.7 with $I_0 = 1.00 \times 10^{-12}$ W/m^2:

$$10 \log\left(\frac{I}{I_0}\right) = 40 \text{ dB}$$

$$\log I - \log I_0 = \frac{40}{10} = 4$$

$$\log I = 4 + \log 10^{-12}$$

$$\log I = -8$$

$$I = 1.00 \times 10^{-8} \text{ W/m}^2$$

Using this value for I in Equation 17.8 and solving for r, we obtain

$$r = \sqrt{\frac{\mathcal{P}_{\text{av}}}{4\pi I}} = \sqrt{\frac{80.0 \text{ W}}{4\pi \times 1.00 \times 10^{-8} \text{ W/m}^2}}$$

$$= 2.52 \times 10^4 \text{ m}$$

which equals about 16 miles!

17.5 THE DOPPLER EFFECT

Perhaps you have noticed how the sound of a vehicle's horn changes as the vehicle moves past you. The frequency of the sound you hear as the vehicle approaches you is higher than the frequency you hear as it moves away from you (see Quick-Lab). This is one example of the **Doppler effect.**[3]

To see what causes this apparent frequency change, imagine you are in a boat that is lying at anchor on a gentle sea where the waves have a period of $T = 3.0$ s. This means that every 3.0 s a crest hits your boat. Figure 17.9a shows this situation, with the water waves moving toward the left. If you set your watch to $t = 0$ just as one crest hits, the watch reads 3.0 s when the next crest hits, 6.0 s when the third crest hits, and so on. From these observations you conclude that the wave frequency is $f = 1/T = (1/3.0)$ Hz. Now suppose you start your motor and head directly into the oncoming waves, as shown in Figure 17.9b. Again you set your watch to $t = 0$ as a crest hits the front of your boat. Now, however, because you are moving toward the next wave crest as it moves toward you, it hits you less than 3.0 s after the first hit. In other words, the period you observe is shorter than the 3.0-s period you observed when you were stationary. Because $f = 1/T$, you observe a higher wave frequency than when you were at rest.

If you turn around and move in the same direction as the waves (see Fig. 17.9c), you observe the opposite effect. You set your watch to $t = 0$ as a crest hits the back of the boat. Because you are now moving away from the next crest, more than 3.0 s has elapsed on your watch by the time that crest catches you. Thus, you observe a lower frequency than when you were at rest.

These effects occur because the relative speed between your boat and the waves depends on the direction of travel and on the speed of your boat. When you are moving toward the right in Figure 17.9b, this relative speed is higher than that of the wave speed, which leads to the observation of an increased frequency. When you turn around and move to the left, the relative speed is lower, as is the observed frequency of the water waves.

Let us now examine an analogous situation with sound waves, in which the water waves become sound waves, the water becomes the air, and the person on the boat becomes an observer listening to the sound. In this case, an observer O is moving and a sound source S is stationary. For simplicity, we assume that the air is also stationary and that the observer moves directly toward the source. The observer moves with a speed v_O toward a stationary point source ($v_S = 0$) (Fig. 17.10). In general, *at rest* means at rest with respect to the medium, air.

[3] Named after the Austrian physicist Christian Johann Doppler (1803–1853), who discovered the effect for light waves.

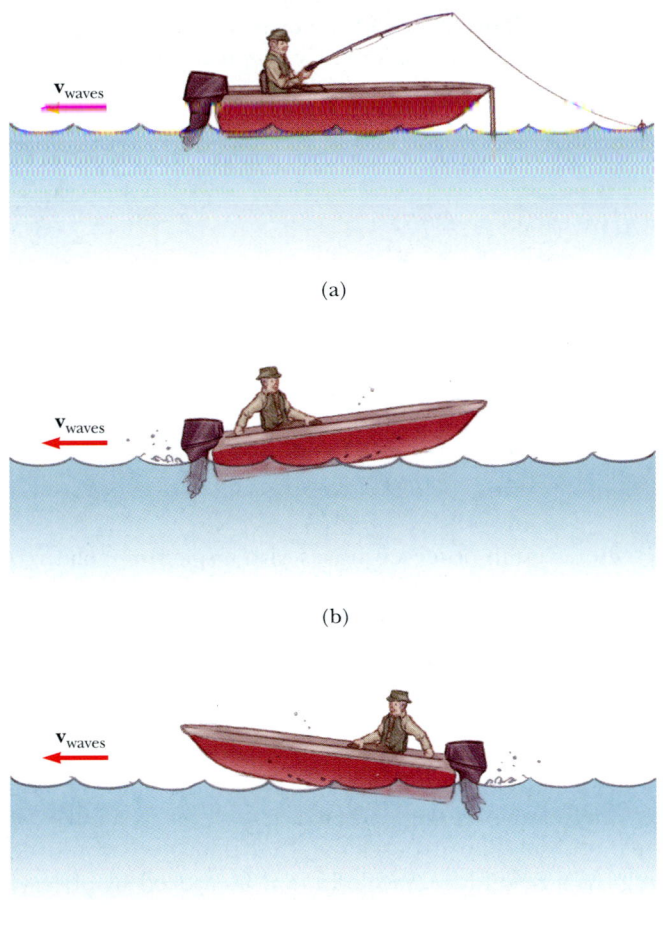

(a)

(b)

(c)

Figure 17.9 (a) Waves moving toward a stationary boat. The waves travel to the left, and their source is far to the right of the boat, out of the frame of the drawing. (b) The boat moving toward the wave source. (c) The boat moving away from the wave source.

We take the frequency of the source to be f, the wavelength to be λ, and the speed of sound to be v. If the observer were also stationary, he or she would detect f wave fronts per second. (That is, when $v_O = 0$ and $v_S = 0$, the observed frequency equals the source frequency.) When the observer moves toward the source,

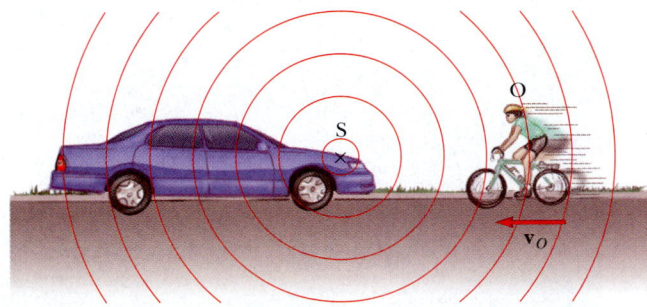

Figure 17.10 An observer O (the cyclist) moves with a speed v_O toward a stationary point source S, the horn of a parked car. The observer hears a frequency f' that is greater than the source frequency.

the speed of the waves relative to the observer is $v' = v + v_O$, as in the case of the boat, but the wavelength λ is unchanged. Hence, using Equation 16.14, $v = \lambda f$, we can say that the frequency heard by the observer is *increased* and is given by

$$f' = \frac{v'}{\lambda} = \frac{v + v_O}{\lambda}$$

Because $\lambda = v/f$, we can express f' as

$$f' = \left(1 + \frac{v_O}{v}\right)f \qquad \text{(observer moving toward source)} \qquad \textbf{(17.11)}$$

If the observer is moving away from the source, the speed of the wave relative to the observer is $v' = v - v_O$. The frequency heard by the observer in this case is *decreased* and is given by

$$f' = \left(1 - \frac{v_O}{v}\right)f \qquad \text{(observer moving away from source)} \qquad \textbf{(17.12)}$$

In general, whenever an observer moves with a speed v_O relative to a stationary source, the frequency heard by the observer is

Frequency heard with an observer in motion

$$f' = \left(1 \pm \frac{v_O}{v}\right)f \qquad \textbf{(17.13)}$$

where the positive sign is used when the observer moves toward the source and the negative sign is used when the observer moves away from the source.

Now consider the situation in which the source is in motion and the observer is at rest. If the source moves directly toward observer A in Figure 17.11a, the wave fronts heard by the observer are closer together than they would be if the source were not moving. As a result, the wavelength λ' measured by observer A is shorter than the wavelength λ of the source. During each vibration, which lasts for a time T (the period), the source moves a distance $v_S T = v_S / f$ and the wavelength is

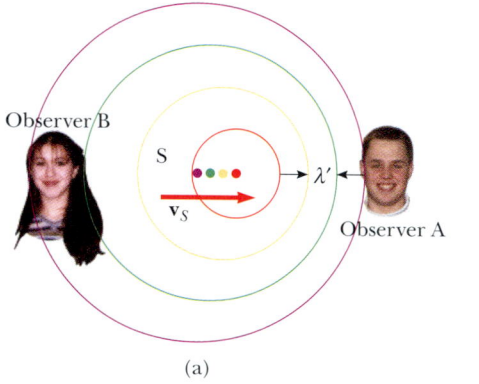

(a)

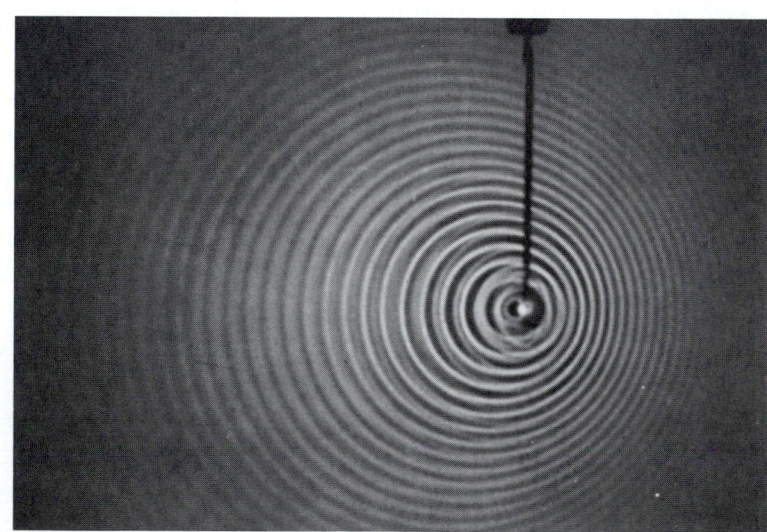

(b)

Figure 17.11   (a) A source S moving with a speed v_S toward a stationary observer A and away from a stationary observer B. Observer A hears an increased frequency, and observer B hears a decreased frequency. (b) The Doppler effect in water, observed in a ripple tank. A point source is moving to the right with speed v_S. *(Courtesy of the Educational Development Center, Newton, MA)*

shortened by this amount. Therefore, the observed wavelength λ' is

$$\lambda' = \lambda - \Delta\lambda = \lambda - \frac{v_S}{f}$$

Because $\lambda = v/f$, the frequency heard by observer A is

$$f' = \frac{v}{\lambda'} = \frac{v}{\lambda - \dfrac{v_S}{f}} = \frac{v}{\dfrac{v}{f} - \dfrac{v_S}{f}}$$

$$f' = \left(\frac{1}{1 - \dfrac{v_S}{v}}\right)f \qquad\qquad (17.14)$$

"I love hearing that lonesome wail of the train whistle as the magnitude of the frequency of the wave changes due to the Doppler effect." *(Sydney Harris)*

That is, the observed frequency is *increased* whenever the source is moving toward the observer.

When the source moves away from a stationary observer, as is the case for observer B in Figure 17.11a, the observer measures a wavelength λ' that is *greater* than λ and hears a *decreased* frequency:

$$f' = \left(\frac{1}{1 + \dfrac{v_S}{v}}\right)f \qquad\qquad (17.15)$$

Combining Equations 17.14 and 17.15, we can express the general relationship for the observed frequency when a source is moving and an observer is at rest as

$$f' = \left(\frac{1}{1 \mp \dfrac{v_S}{v}}\right)f \qquad\qquad (17.16)$$

Frequency heard with source in motion

Finally, if both source and observer are in motion, we find the following general relationship for the observed frequency:

$$f' = \left(\frac{v \pm v_O}{v \mp v_S}\right)f \qquad\qquad (17.17)$$

Frequency heard with observer and source in motion

In this expression, the upper signs $(+ v_O$ and $- v_S)$ refer to motion of one toward the other, and the lower signs $(- v_O$ and $+ v_S)$ refer to motion of one away from the other.

A convenient rule concerning signs for you to remember when working with all Doppler-effect problems is as follows:

The word *toward* is associated with an *increase* in observed frequency. The words *away from* are associated with a *decrease* in observed frequency.

Although the Doppler effect is most typically experienced with sound waves, it is a phenomenon that is common to all waves. For example, the relative motion of source and observer produces a frequency shift in light waves. The Doppler effect is used in police radar systems to measure the speeds of motor vehicles. Likewise, astronomers use the effect to determine the speeds of stars, galaxies, and other celestial objects relative to the Earth.

EXAMPLE 17.6 ▶ The Noisy Siren

As an ambulance travels east down a highway at a speed of 33.5 m/s (75 mi/h), its siren emits sound at a frequency of 400 Hz. What frequency is heard by a person in a car traveling west at 24.6 m/s (55 mi/h) (a) as the car approaches the ambulance and (b) as the car moves away from the ambulance?

Solution   (a) We can use Equation 17.17 in both cases, taking the speed of sound in air to be $v = 343$ m/s. As the ambulance and car approach each other, the person in the car hears the frequency

$$f' = \left(\frac{v + v_O}{v - v_S}\right)f = \left(\frac{343 \text{ m/s} + 24.6 \text{ m/s}}{343 \text{ m/s} - 33.5 \text{ m/s}}\right)(400 \text{ Hz})$$

$$= \boxed{475 \text{ Hz}}$$

(b) As the vehicles recede from each other, the person hears the frequency

$$f' = \left(\frac{v - v_O}{v + v_S}\right)f = \left(\frac{343 \text{ m/s} - 24.6 \text{ m/s}}{343 \text{ m/s} + 33.5 \text{ m/s}}\right)(400 \text{ Hz})$$

$$= \boxed{338 \text{ Hz}}$$

The *change* in frequency detected by the person in the car is $475 - 338 = 137$ Hz, which is more than 30% of the true frequency.

Exercise   Suppose the car is parked on the side of the highway as the ambulance speeds by. What frequency does the person in the car hear as the ambulance (a) approaches and (b) recedes?

Answer   (a) 443 Hz.   (b) 364 Hz.

Shock Waves

Now let us consider what happens when the speed v_S of a source *exceeds* the wave speed v. This situation is depicted graphically in Figure 17.12a. The circles represent spherical wave fronts emitted by the source at various times during its motion. At $t = 0$, the source is at S_0, and at a later time t, the source is at S_n. In the time t,

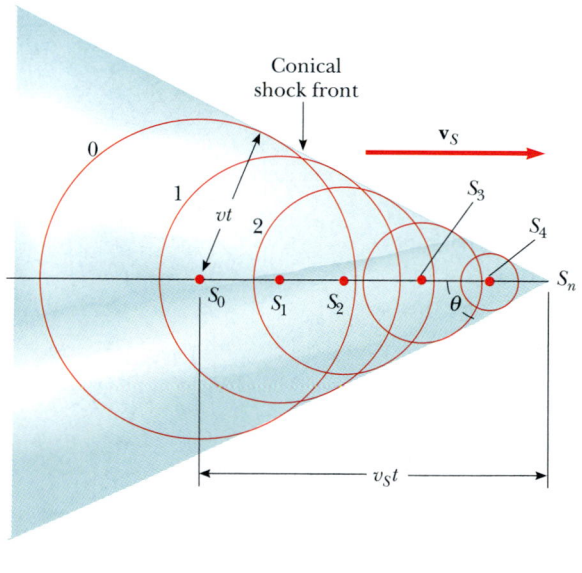

(a)

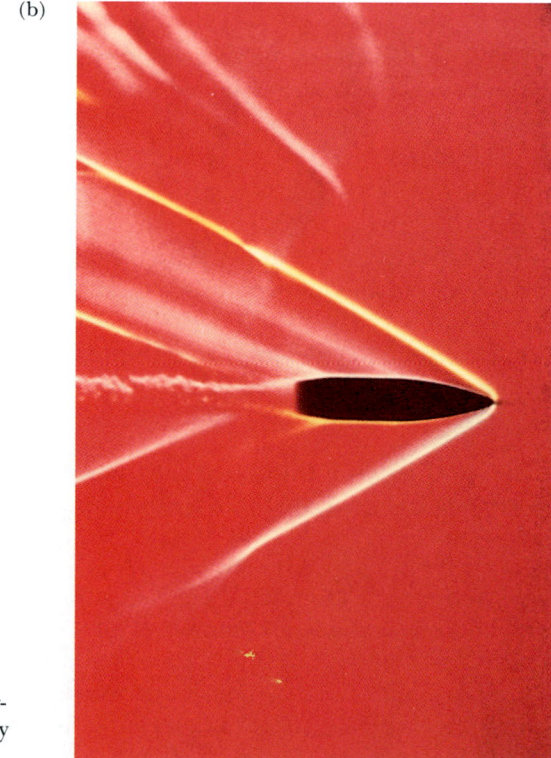

(b)

Figure 17.12   (a) A representation of a shock wave produced when a source moves from S_0 to S_n with a speed v_S, which is greater than the wave speed v in the medium. The envelope of the wave fronts forms a cone whose apex half-angle is given by $\sin \theta = v/v_S$. (b) A stroboscopic photograph of a bullet moving at supersonic speed through the hot air above a candle. Note the shock wave in the vicinity of the bullet. *(©1973 Kim Vandiver & Harold E. Edgerton/Courtesy of Palm Press, Inc.)*

Figure 17.13 The V-shaped bow wave of a boat is formed because the boat speed is greater than the speed of the water waves. A bow wave is analogous to a shock wave formed by an airplane traveling faster than sound. *(©1994 Comstock)*

the wave front centered at S_0 reaches a radius of vt. In this same amount of time, the source travels a distance v_St to S_n. At the instant the source is at S_n, waves are just beginning to be generated at this location, and hence the wave front has zero radius at this point. The tangent line drawn from S_n to the wave front centered on S_0 is tangent to all other wave fronts generated at intermediate times. Thus, we see that the envelope of these wave fronts is a cone whose apex half-angle θ is given by

$$\sin \theta = \frac{vt}{v_St} = \frac{v}{v_S}$$

The ratio v_S/v is referred to as the *Mach number,* and the conical wave front produced when $v_S > v$ (supersonic speeds) is known as a *shock wave.* An interesting analogy to shock waves is the V-shaped wave fronts produced by a boat (the *bow wave*) when the boat's speed exceeds the speed of the surface-water waves (Fig. 17.13).

Jet airplanes traveling at supersonic speeds produce shock waves, which are responsible for the loud "sonic boom" one hears. The shock wave carries a great deal of energy concentrated on the surface of the cone, with correspondingly great pressure variations. Such shock waves are unpleasant to hear and can cause damage to buildings when aircraft fly supersonically at low altitudes. In fact, an airplane flying at supersonic speeds produces a double boom because two shock fronts are formed, one from the nose of the plane and one from the tail (Fig. 17.14). People near the path of the space shuttle as it glides toward its landing point often report hearing what sounds like two very closely spaced cracks of thunder.

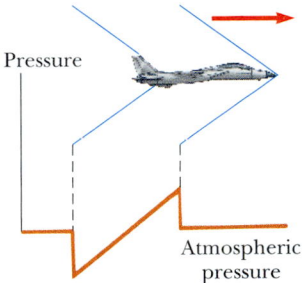

Figure 17.14 The two shock waves produced by the nose and tail of a jet airplane traveling at supersonic speeds.

Quick Quiz 17.5

An airplane flying with a constant velocity moves from a cold air mass into a warm air mass. Does the Mach number increase, decrease, or stay the same?

Quick Quiz 17.6

Suppose that an observer and a source of sound are both at rest and that a strong wind blows from the source toward the observer. Describe the effect of the wind (if any) on

(a) the observed frequency of the sound waves, (b) the observed wave speed, and (c) the observed wavelength.

SUMMARY

Sound waves are longitudinal and travel through a compressible medium with a speed that depends on the compressibility and inertia of that medium. The speed of sound in a medium having a bulk modulus B and density ρ is

$$v = \sqrt{\frac{B}{\rho}} \qquad (17.1)$$

With this formula you can determine the speed of a sound wave in many different materials.

For sinusoidal sound waves, the variation in the displacement is given by

$$s(x, t) = s_{\max} \cos(kx - \omega t) \qquad (17.2)$$

and the variation in pressure from the equilibrium value is

$$\Delta P = \Delta P_{\max} \sin(kx - \omega t) \qquad (17.3)$$

where $\Delta P_{\max}$ is the **pressure amplitude.** The pressure wave is 90° out of phase with the displacement wave. The relationship between $s_{\max}$ and $\Delta P_{\max}$ is given by

$$\Delta P_{\max} = \rho v \omega s_{\max} \qquad (17.4)$$

The intensity of a periodic sound wave, which is the power per unit area, is

$$I = \tfrac{1}{2}\rho v(\omega s_{\max})^2 = \frac{\Delta P_{\max}^2}{2\rho v} \qquad (17.5, 17.6)$$

The sound level of a sound wave, in decibels, is given by

$$\beta = 10 \log\left(\frac{I}{I_0}\right) \qquad (17.7)$$

The constant I_0 is a reference intensity, usually taken to be at the threshold of hearing (1.00×10^{-12} W/m^2), and I is the intensity of the sound wave in watts per square meter.

The intensity of a spherical wave produced by a point source is proportional to the average power emitted and inversely proportional to the square of the distance from the source:

$$I = \frac{\mathcal{P}_{av}}{4\pi r^2} \qquad (17.8)$$

The change in frequency heard by an observer whenever there is relative motion between a source of sound waves and the observer is called the **Doppler effect.** The observed frequency is

$$f' = \left(\frac{v \pm v_O}{v \mp v_S}\right)f \qquad (17.17)$$

The upper signs ($+ v_O$ and $- v_S$) are used with motion of one toward the other, and the lower signs ($- v_O$ and $+ v_S$) are used with motion of one away from the other. You can also use this formula when v_O or v_S is zero.

QUESTIONS

1. Why are sound waves characterized as longitudinal?
2. If an alarm clock is placed in a good vacuum and then activated, no sound is heard. Explain.
3. A sonic ranger is a device that determines the position of an object by sending out an ultrasonic sound pulse and measuring how long it takes for the sound wave to return after it reflects from the object. Typically, these devices cannot reliably detect an object that is less than half a meter from the sensor. Why is that?
4. In Example 17.5, we found that a point source with a power output of 80 W reduces to a sound level of 40 dB at a distance of about 16 miles. Why do you suppose you cannot normally hear a rock concert that is going on 16 miles away? (See Table 17.2.)
5. If the distance from a point source is tripled, by what factor does the intensity decrease?
6. Explain how the Doppler effect is used with microwaves to determine the speed of an automobile.
7. Explain what happens to the frequency of your echo as you move in a vehicle *toward* a canyon wall. What happens to the frequency as you move *away* from the wall?
8. Of the following sounds, which is most likely to have a sound level of 60 dB—a rock concert, the turning of a page in this text, normal conversation, or a cheering crowd at a football game?
9. Estimate the decibel level of each of the sounds in the previous question.
10. A binary star system consists of two stars revolving about their common center of mass. If we observe the light reaching us from one of these stars as it makes one complete revolution, what does the Doppler effect predict will happen to this light?
11. How can an object move with respect to an observer so that the sound from it is not shifted in frequency?
12. Why is it not possible to use sonar (sound waves) to determine the speed of an object traveling faster than the speed of sound in a given medium?
13. Why is it so quiet after a snowfall?
14. Why is the intensity of an echo less than that of the original sound?
15. If the wavelength of a sound source is reduced by a factor of 2, what happens to its frequency? Its speed?
16. In a recent discovery, a nearby star was found to have a large planet orbiting about it, although the planet could not be seen. In terms of the concept of a system rotating about its center of mass and the Doppler shift for light (which is in many ways similar to that for sound), explain how an astronomer could determine the presence of the invisible planet.
17. A friend sitting in her car far down the road waves to you and beeps her horn at the same time. How far away must her car be for you to measure the speed of sound to two significant figures by measuring the time it takes for the sound to reach you?

PROBLEMS

1, 2, 3 = straightforward, intermediate, challenging ☐ = full solution available in the *Student Solutions Manual and Study Guide*
WEB = solution posted at **http://www.saunderscollege.com/physics/** 🖥 = Computer useful in solving problem 🎮 = Interactive Physics
☐ = paired numerical/symbolic problems

Section 17.1 Speed of Sound Waves

1. Suppose that you hear a clap of thunder 16.2 s after seeing the associated lightning stroke. The speed of sound waves in air is 343 m/s, and the speed of light in air is 3.00×10^8 m/s. How far are you from the lightning stroke?
2. Find the speed of sound in mercury, which has a bulk modulus of approximately 2.80×10^{10} N/m² and a density of 13 600 kg/m³.
3. A flower pot is knocked off a balcony 20.0 m above the sidewalk and falls toward an unsuspecting 1.75-m-tall man who is standing below. How close to the sidewalk can the flower pot fall before it is too late for a shouted warning from the balcony to reach the man in time? Assume that the man below requires 0.300 s to respond to the warning.
4. You are watching a pier being constructed on the far shore of a saltwater inlet when some blasting occurs. You hear the sound in the water 4.50 s before it reaches you through the air. How wide is the inlet? (*Hint:* See Table 17.1. Assume that the air temperature is 20°C.)
5. Another approximation of the temperature dependence of the speed of sound in air (in meters per second) is given by the expression

$$v = 331.5 + 0.607 T_C$$

where T_C is the Celsius temperature. In dry air the temperature decreases about 1°C for every 150-m rise in altitude. (a) Assuming that this change is constant up to an altitude of 9 000 m, how long will it take the sound from an airplane flying at 9 000 m to reach the ground on a day when the ground temperature is 30°C? (b) Compare this to the time it would take if the air were at 30°C at all altitudes. Which interval is longer?
6. A bat can detect very small objects, such as an insect whose length is approximately equal to one wavelength

of the sound the bat makes. If bats emit a chirp at a frequency of 60.0 kHz, and if the speed of sound in air is 340 m/s, what is the smallest insect a bat can detect?

7. An airplane flies horizontally at a constant speed, piloted by rescuers who are searching for a disabled boat. When the plane is directly above the boat, the boat's crew blows a loud horn. By the time the plane's sound detector receives the horn's sound, the plane has traveled a distance equal to one-half its altitude above the ocean. If it takes the sound 2.00 s to reach the plane, determine (a) the speed of the plane and (b) its altitude. Take the speed of sound to be 343 m/s.

Section 17.2   Periodic Sound Waves

Note: In this section, use the following values as needed, unless otherwise specified. The equilibrium density of air is $\rho = 1.20$ kg/m^3; the speed of sound in air is $v = 343$ m/s. Pressure variations ΔP are measured relative to atmospheric pressure, 1.013×10^5 Pa.

8. A sound wave in air has a pressure amplitude equal to 4.00×10^{-3} Pa. Calculate the displacement amplitude of the wave at a frequency of 10.0 kHz.

9. A sinusoidal sound wave is described by the displacement

$$s(x, t) = (2.00 \ \mu\text{m}) \cos[(15.7 \ \text{m}^{-1})x - (858 \ \text{s}^{-1})t]$$

(a) Find the amplitude, wavelength, and speed of this wave. (b) Determine the instantaneous displacement of the molecules at the position $x = 0.050\ 0$ m at $t = 3.00$ ms. (c) Determine the maximum speed of a molecule's oscillatory motion.

10. As a sound wave travels through the air, it produces pressure variations (above and below atmospheric pressure) that are given by $\Delta P = 1.27 \sin(\pi x - 340\pi t)$ in SI units. Find (a) the amplitude of the pressure variations, (b) the frequency of the sound wave, (c) its wavelength in air, and (d) its speed.

11. Write an expression that describes the pressure variation as a function of position and time for a sinusoidal sound wave in air, if $\lambda = 0.100$ m and $\Delta P_{max} = 0.200$ Pa.

12. Write the function that describes the displacement wave corresponding to the pressure wave in Problem 11.

13. The tensile stress in a thick copper bar is 99.5% of its elastic breaking point of 13.0×10^{10} N/m^2. A 500-Hz sound wave is transmitted through the material. (a) What displacement amplitude will cause the bar to break? (b) What is the maximum speed of the particles at this moment?

14. Calculate the pressure amplitude of a 2.00-kHz sound wave in air if the displacement amplitude is equal to 2.00×10^{-8} m.

WEB 15. An experimenter wishes to generate in air a sound wave that has a displacement amplitude of 5.50×10^{-6} m. The pressure amplitude is to be limited to 8.40×10^{-1} Pa. What is the minimum wavelength the sound wave can have?

16. A sound wave in air has a pressure amplitude of 4.00 Pa and a frequency of 5.00 kHz. Take $\Delta P = 0$ at the point $x = 0$ when $t = 0$. (a) What is ΔP at $x = 0$ when $t = 2.00 \times 10^{-4}$ s? (b) What is ΔP at $x = 0.020\ 0$ m when $t = 0$?

Section 17.3   Intensity of Periodic Sound Waves

17. Calculate the sound level, in decibels, of a sound wave that has an intensity of 4.00 μW/m^2.

18. A vacuum cleaner has a measured sound level of 70.0 dB. (a) What is the intensity of this sound in watts per square meter? (b) What is the pressure amplitude of the sound?

19. The intensity of a sound wave at a fixed distance from a speaker vibrating at 1.00 kHz is 0.600 W/m^2. (a) Determine the intensity if the frequency is increased to 2.50 kHz while a constant displacement amplitude is maintained. (b) Calculate the intensity if the frequency is reduced to 0.500 kHz and the displacement amplitude is doubled.

20. The intensity of a sound wave at a fixed distance from a speaker vibrating at a frequency f is I. (a) Determine the intensity if the frequency is increased to f' while a constant displacement amplitude is maintained. (b) Calculate the intensity if the frequency is reduced to $f/2$ and the displacement amplitude is doubled.

WEB 21. A family ice show is held in an enclosed arena. The skaters perform to music with a sound level of 80.0 dB. This is too loud for your baby, who consequently yells at a level of 75.0 dB. (a) What total sound intensity engulfs you? (b) What is the combined sound level?

Section 17.4   Spherical and Plane Waves

22. For sound radiating from a point source, show that the difference in sound levels, β_1 and β_2, at two receivers is related to the ratio of the distances r_1 and r_2 from the source to the receivers by the expression

$$\beta_2 - \beta_1 = 20 \log\left(\frac{r_1}{r_2}\right)$$

23. A fireworks charge is detonated many meters above the ground. At a distance of 400 m from the explosion, the acoustic pressure reaches a maximum of 10.0 N/m^2. Assume that the speed of sound is constant at 343 m/s throughout the atmosphere over the region considered, that the ground absorbs all the sound falling on it, and that the air absorbs sound energy as described by the rate 7.00 dB/km. What is the sound level (in decibels) at 4.00 km from the explosion?

24. A loudspeaker is placed between two observers who are 110 m apart, along the line connecting them. If one observer records a sound level of 60.0 dB and the other records a sound level of 80.0 dB, how far is the speaker from each observer?

25. Two small speakers emit spherical sound waves of different frequencies. Speaker *A* has an output of 1.00 mW, and speaker *B* has an output of 1.50 mW. Determine the sound level (in decibels) at point *C* (Fig. P17.25) if (a) only speaker *A* emits sound, (b) only speaker *B* emits sound, (c) both speakers emit sound.

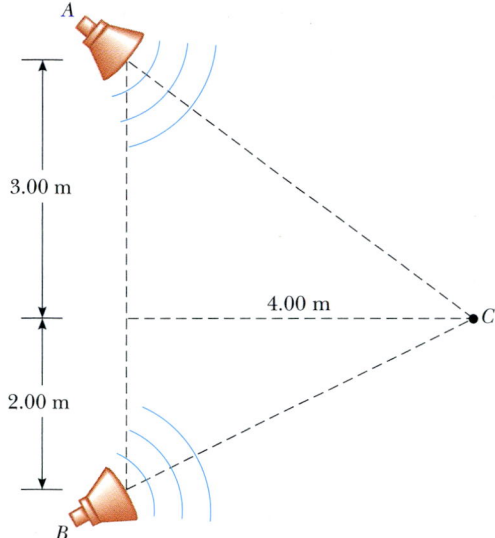

Figure P17.25

26. An experiment requires a sound intensity of 1.20 W/m^2 at a distance of 4.00 m from a speaker. What power output is required? Assume that the speakers radiate sound equally in all directions.

27. A source of sound (1 000 Hz) emits uniformly in all directions. An observer 3.00 m from the source measures a sound level of 40.0 dB. Calculate the average power output of the source.

28. A jackhammer, operated continuously at a construction site, behaves as a point source of spherical sound waves. A construction supervisor stands 50.0 m due north of this sound source and begins to walk due west. How far does she have to walk in order for the amplitude of the wave function to drop by a factor of 2.00?

29. The sound level at a distance of 3.00 m from a source is 120 dB. At what distances is the sound level (a) 100 dB and (b) 10.0 dB?

30. A fireworks rocket explodes 100 m above the ground. An observer directly under the explosion experiences an average sound intensity of 7.00×10^{-2} W/m^2 for 0.200 s. (a) What is the total sound energy of the explosion? (b) What sound level, in decibels, is heard by the observer?

31. As the people in a church sing on a summer morning, the sound level everywhere inside the church is 101 dB. The massive walls are opaque to sound, but all the windows and doors are open. Their total area is 22.0 m^2. (a) How much sound energy is radiated in 20.0 min? (b) Suppose the ground is a good reflector and sound

radiates uniformly in all horizontal and upward directions. Find the sound level 1.00 km away.

32. A spherical wave is radiating from a point source and is described by the wave function

$$\Delta P(r, t) = \left[\frac{25.0}{r} \right] \sin(1.25r - 1\,870t)$$

where ΔP is in pascals, *r* in meters, and *t* in seconds. (a) What is the pressure amplitude 4.00 m from the source? (b) Determine the speed of the wave and hence the material the wave might be traveling through. (c) Find the sound level of the wave, in decibels, 4.00 m from the source. (d) Find the instantaneous pressure 5.00 m from the source at 0.080 0 s.

Section 17.5 The Doppler Effect

33. A commuter train passes a passenger platform at a constant speed of 40.0 m/s. The train horn is sounded at its characteristic frequency of 320 Hz. (a) What change in frequency is detected by a person on the platform as the train passes? (b) What wavelength is detected by a person on the platform as the train approaches?

34. A driver travels northbound on a highway at a speed of 25.0 m/s. A police car, traveling southbound at a speed of 40.0 m/s, approaches with its siren sounding at a frequency of 2 500 Hz. (a) What frequency does the driver observe as the police car approaches? (b) What frequency does the driver detect after the police car passes him? (c) Repeat parts (a) and (b) for the case in which the police car is northbound.

35. Standing at a crosswalk, you hear a frequency of 560 Hz from the siren of an approaching police car. After the police car passes, the observed frequency of the siren is 480 Hz. Determine the car's speed from these observations.

36. Expectant parents are thrilled to hear their unborn baby's heartbeat, revealed by an ultrasonic motion detector. Suppose the fetus's ventricular wall moves in simple harmonic motion with an amplitude of 1.80 mm and a frequency of 115 per minute. (a) Find the maximum linear speed of the heart wall. Suppose the motion detector in contact with the mother's abdomen produces sound at 2 000 000.0 Hz, which travels through tissue at 1.50 km/s. (b) Find the maximum frequency at which sound arrives at the wall of the baby's heart. (c) Find the maximum frequency at which reflected sound is received by the motion detector. (By electronically "listening" for echoes at a frequency different from the broadcast frequency, the motion detector can produce beeps of audible sound in synchronization with the fetal heartbeat.)

37. A tuning fork vibrating at 512 Hz falls from rest and accelerates at 9.80 m/s^2. How far below the point of release is the tuning fork when waves with a frequency of 485 Hz reach the release point? Take the speed of sound in air to be 340 m/s.

38. A block with a speaker bolted to it is connected to a spring having spring constant $k = 20.0$ N/m, as shown in Figure P17.38. The total mass of the block and speaker is 5.00 kg, and the amplitude of this unit's motion is 0.500 m. (a) If the speaker emits sound waves of frequency 440 Hz, determine the highest and lowest frequencies heard by the person to the right of the speaker. (b) If the maximum sound level heard by the person is 60.0 dB when he is closest to the speaker, 1.00 m away, what is the minimum sound level heard by the observer? Assume that the speed of sound is 343 m/s.

Figure P17.38

39. A train is moving parallel to a highway with a constant speed of 20.0 m/s. A car is traveling in the same direction as the train with a speed of 40.0 m/s. The car horn sounds at a frequency of 510 Hz, and the train whistle sounds at a frequency of 320 Hz. (a) When the car is behind the train, what frequency does an occupant of the car observe for the train whistle? (b) When the car is in front of the train, what frequency does a train passenger observe for the car horn just after the car passes?

40. At the Winter Olympics, an athlete rides her luge down the track while a bell just above the wall of the chute rings continuously. When her sled passes the bell, she hears the frequency of the bell fall by the musical interval called a minor third. That is, the frequency she hears drops to five sixths of its original value. (a) Find the speed of sound in air at the ambient temperature $-10.0°$C. (b) Find the speed of the athlete.

41. A jet fighter plane travels in horizontal flight at Mach 1.20 (that is, 1.20 times the speed of sound in air). At the instant an observer on the ground hears the shock wave, what is the angle her line of sight makes with the horizontal as she looks at the plane?

42. When high-energy charged particles move through a transparent medium with a speed greater than the speed of light in that medium, a shock wave, or bow wave, of light is produced. This phenomenon is called the *Cerenkov effect* and can be observed in the vicinity of the core of a swimming-pool nuclear reactor due to

high-speed electrons moving through the water. In a particular case, the Cerenkov radiation produces a wave front with an apex half-angle of 53.0°. Calculate the speed of the electrons in the water. (The speed of light in water is 2.25×10^8 m/s.)

WEB **43.** A supersonic jet traveling at Mach 3.00 at an altitude of 20 000 m is directly over a person at time $t = 0$, as in Figure P17.43. (a) How long will it be before the person encounters the shock wave? (b) Where will the plane be when it is finally heard? (Assume that the speed of sound in air is 335 m/s.)

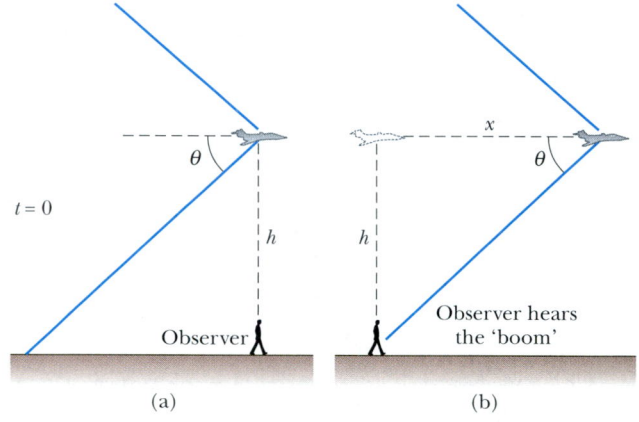

(a)   (b)

Figure P17.43

44. The tip of a circus ringmaster's whip travels at Mach 1.38 (that is, $v_S/v = 1.38$). What angle does the shock front make with the direction of the whip's motion?

ADDITIONAL PROBLEMS

45. A stone is dropped into a deep canyon and is heard to strike the bottom 10.2 s after release. The speed of sound waves in air is 343 m/s. How deep is the canyon? What would be the percentage error in the calculated depth if the time required for the sound to reach the canyon rim were ignored?

46. Unoccupied by spectators, a large set of football bleachers has solid seats and risers. You stand on the field in front of it and fire a starter's pistol or sharply clap two wooden boards together once. The sound pulse you produce has no frequency and no wavelength. You hear back from the bleachers a sound with definite pitch, which may remind you of a short toot on a trumpet, or of a buzzer or a kazoo. Account for this sound. Compute order-of-magnitude estimates for its frequency, wavelength, and duration on the basis of data that you specify.

47. Many artists sing very high notes in ornaments and cadenzas. The highest note written for a singer in a published score was F-sharp above high C, 1.480 kHz, sung by Zerbinetta in the original version of Richard Strauss's opera *Ariadne auf Naxos*. (a) Find the wavelength of this sound in air. (b) Suppose that the people in the fourth row of seats hear this note with a level of 81.0 dB. Find the displacement amplitude of the sound. (c) In response to complaints, Strauss later transposed the note down to F above high C, 1.397 kHz. By what increment did the wavelength change?

48. A sound wave in a cylinder is described by Equations 17.2 through 17.4. Show that $\Delta P = \pm \rho v \omega \sqrt{s_{max}^2 - s^2}$.

49. On a Saturday morning, pickup trucks carrying garbage to the town dump form a nearly steady procession on a country road, all traveling at 19.7 m/s. From this direction, two trucks arrive at the dump every three minutes. A bicyclist also is traveling toward the dump at 4.47 m/s. (a) With what frequency do the trucks pass him? (b) A hill does not slow the trucks but makes the out-of-shape cyclist's speed drop to 1.56 m/s. How often do the noisy trucks whiz past him now?

50. The ocean floor is underlain by a layer of basalt that constitutes the crust, or uppermost layer, of the Earth in that region. Below the crust is found denser peridotite rock, which forms the Earth's mantle. The boundary between these two layers is called the Mohorovicic discontinuity ("Moho" for short). If an explosive charge is set off at the surface of the basalt, it generates a seismic wave that is reflected back out at the Moho. If the speed of the wave in basalt is 6.50 km/s and the two-way travel time is 1.85 s, what is the thickness of this oceanic crust?

51. A worker strikes a steel pipeline with a hammer, generating both longitudinal and transverse waves. Reflected waves return 2.40 s apart. How far away is the reflection point? (For steel, $v_{long} = 6.20$ km/s and $v_{trans} = 3.20$ km/s.)

52. For a certain type of steel, stress is proportional to strain with Young's modulus as given in Table 12.1. The steel has the density listed for iron in Table 15.1. It bends permanently if subjected to compressive stress greater than its elastic limit, $\sigma = 400$ MPa, also called its *yield strength*. A rod 80.0 cm long, made of this steel, is projected at 12.0 m/s straight at a hard wall. (a) Find the speed of compressional waves moving along the rod. (b) After the front end of the rod hits the wall and stops, the back end of the rod keeps moving, as described by Newton's first law, until it is stopped by the excess pressure in a sound wave moving back through the rod. How much time elapses before the back end of the rod gets the message? (c) How far has the back end of the rod moved in this time? (d) Find the strain in the rod and (e) the stress. (f) If it is not to fail, show that the maximum impact speed a rod can have is given by the expression $\sigma/\sqrt{\rho Y}$.

53. To determine her own speed, a sky diver carries a buzzer that emits a steady tone at 1 800 Hz. A friend at the landing site on the ground directly below the sky diver listens to the amplified sound he receives from the buzzer. Assume that the air is calm and that the speed of sound is 343 m/s, independent of altitude. While the sky diver is falling at terminal speed, her friend on the ground receives waves with a frequency of 2 150 Hz. (a) What is the sky diver's speed of descent? (b) Suppose the sky diver is also carrying sound-receiving equipment that is sensitive enough to detect waves reflected from the ground. What frequency does she receive?

54. A train whistle ($f = 400$ Hz) sounds higher or lower in pitch depending on whether it is approaching or receding. (a) Prove that the difference in frequency between the approaching and receding train whistle is

$$\Delta f = \frac{2(u/v)}{1 - (u^2/v^2)} f$$

where u is the speed of the train and v is the speed of sound. (b) Calculate this difference for a train moving at a speed of 130 km/h. Take the speed of sound in air to be 340 m/s.

55. A bat, moving at 5.00 m/s, is chasing a flying insect. If the bat emits a 40.0-kHz chirp and receives back an echo at 40.4 kHz, at what relative speed is the bat moving toward or away from the insect? (Take the speed of sound in air to be $v = 340$ m/s.)

Figure P17.55   *(Joe McDonald/Visuals Unlimited)*

56. A supersonic aircraft is flying parallel to the ground. When the aircraft is directly overhead, an observer on the ground sees a rocket fired from the aircraft. Ten seconds later the observer hears the sonic boom, which is followed 2.80 s later by the sound of the rocket engine. What is the Mach number of the aircraft?

57. A police car is traveling east at 40.0 m/s along a straight road, overtaking a car that is moving east at 30.0 m/s. The police car has a malfunctioning siren that is stuck at 1 000 Hz. (a) Sketch the appearance of the wave fronts of the sound produced by the siren. Show the

wave fronts both to the east and to the west of the police car. (b) What would be the wavelength in air of the siren sound if the police car were at rest? (c) What is the wavelength in front of the car? (d) What is the wavelength behind the police car? (e) What frequency is heard by the driver being chased?

58. A copper bar is given a sharp compressional blow at one end. The sound of the blow, traveling through air at 0°C, reaches the opposite end of the bar 6.40 ms later than the sound transmitted through the metal of the bar. What is the length of the bar? (Refer to Table 17.1.)

59. The power output of a certain public address speaker is 6.00 W. Suppose it broadcasts equally in all directions. (a) Within what distance from the speaker would the sound be painful to the ear? (b) At what distance from the speaker would the sound be barely audible?

60. A jet flies toward higher altitude at a constant speed of 1 963 m/s in a direction that makes an angle θ with the horizontal (Fig. P17.60). An observer on the ground hears the jet for the first time when it is directly overhead. Determine the value of θ if the speed of sound in air is 340 m/s.

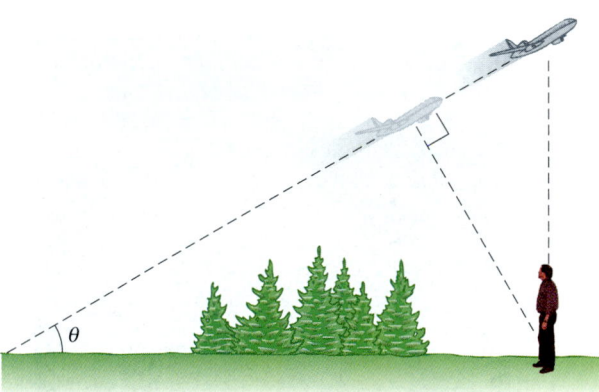

Figure P17.60

61. Two ships are moving along a line due east. The trailing vessel has a speed of 64.0 km/h relative to a land-based observation point, and the leading ship has a speed of 45.0 km/h relative to that point. The two ships are in a region of the ocean where the current is moving uniformly due west at 10.0 km/h. The trailing ship transmits a sonar signal at a frequency of 1 200.0 Hz. What frequency is monitored by the leading ship? (Use 1 520 m/s as the speed of sound in ocean water.)

62. A microwave oven generates a sound with intensity level 40.0 dB everywhere just outside it, when consuming 1.00 kW of power. Find the fraction of this power that is converted into the energy of sound waves. Assume the dimensions of the oven are 40.0 cm × 40.0 cm × 50.0 cm.

63. A meteoroid the size of a truck enters the Earth's atmosphere at a speed of 20.0 km/s and is not significantly slowed before entering the ocean. (a) What is the Mach angle of the shock wave from the meteoroid in the atmosphere? (Use 331 m/s as the sound speed.) (b) Assuming that the meteoroid survives the impact with the ocean surface, what is the (initial) Mach angle of the shock wave that the meteoroid produces in the water? (Use the wave speed for sea water given in Table 17.1.)

64. Consider a longitudinal (compressional) wave of wavelength λ traveling with speed v along the x direction through a medium of density ρ. The *displacement* of the molecules of the medium from their equilibrium position is

$$s = s_{max} \sin(kx - \omega t)$$

Show that the pressure variation in the medium is given by

$$\Delta P = -\left(\frac{2\pi\rho v^2}{\lambda} s_{max}\right)\cos(kx - \omega t)$$

WEB 65. By proper excitation, it is possible to produce both longitudinal and transverse waves in a long metal rod. A particular metal rod is 150 cm long and has a radius of 0.200 cm and a mass of 50.9 g. Young's modulus for the material is 6.80×10^{10} N/m². What must the tension in the rod be if the ratio of the speed of longitudinal waves to the speed of transverse waves is 8.00?

66. An interstate highway has been built through a neighborhood in a city. In the afternoon, the sound level in a rented room is 80.0 dB as 100 cars per minute pass outside the window. Late at night, the traffic flow on the freeway is only five cars per minute. What is the average late-night sound level in the room?

67. A siren creates a sound level of 60.0 dB at a location 500 m from the speaker. The siren is powered by a battery that delivers a total energy of 1.00 kJ. Assuming that the efficiency of the siren is 30.0% (that is, 30.0% of the supplied energy is transformed into sound energy), determine the total time the siren can sound.

68. A siren creates a sound level β at a distance d from the speaker. The siren is powered by a battery that delivers a total energy E. Assuming that the efficiency of the siren is e (that is, e is equal to the output sound energy divided by the supplied energy), determine the total time the siren can sound.

69. The Doppler equation presented in the text is valid when the motion between the observer and the source occurs on a straight line, so that the source and observer are moving either directly toward or directly away from each other. If this restriction is relaxed, one must use the more general Doppler equation

$$f' = \left(\frac{v + v_O \cos \theta_O}{v - v_S \cos \theta_S}\right)f$$

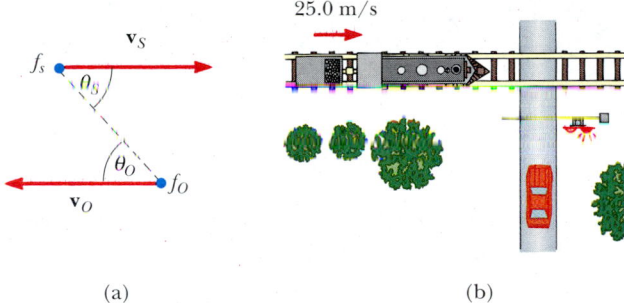

(a)   (b)

Figure P17.69

where θ_O and θ_S are defined in Figure P17.69a.
(a) Show that if the observer and source are moving away from each other, the preceding equation reduces to Equation 17.17 with lower signs. (b) Use the preceding equation to solve the following problem. A train moves at a constant speed of 25.0 m/s toward the intersection shown in Figure P17.69b. A car is stopped near the intersection, 30.0 m from the tracks. If the train's horn emits a frequency of 500 Hz, what frequency is heard by the passengers in the car when the train is 40.0 m from the intersection? Take the speed of sound to be 343 m/s.

70. Figure 17.5 illustrates that at distance r from a point source with power $\mathcal{P}_{av}$, the wave intensity is $I = \mathcal{P}_{av}/4\pi r^2$. Study Figure 17.11a and prove that at distance r straight in front of a point source with power $\mathcal{P}_{av}$, moving with constant speed v_S, the wave intensity is

$$I = \frac{\mathcal{P}_{av}}{4\pi r^2}\left(\frac{v - v_S}{v}\right)$$

71. Three metal rods are located relative to each other as shown in Figure P17.71, where $L_1 + L_2 = L_3$. The den-

sity values and Young's moduli for the three materials are $\rho_1 = 2.70 \times 10^3$ kg/m³, $Y_1 = 7.00 \times 10^{10}$ N/m²; $\rho_2 = 11.3 \times 10^3$ kg/m³, $Y_2 = 1.60 \times 10^{10}$ N/m²; $\rho_3 = 8.80 \times 10^3$ kg/m³, $Y_3 = 11.0 \times 10^{10}$ N/m². (a) If $L_3 = 1.50$ m, what must the ratio L_1/L_2 be if a sound wave is to travel the combined length of rods 1 and 2 in the same time it takes to travel the length of rod 3? (b) If the frequency of the source is 4.00 kHz, determine the phase difference between the wave traveling along rods 1 and 2 and the one traveling along rod 3.

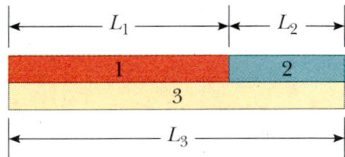

Figure P17.71

72. The volume knob on a radio has what is known as a "logarithmic taper." The electrical device connected to the knob (called a potentiometer) has a resistance R whose logarithm is proportional to the angular position of the knob: that is, $\log R \propto \theta$. If the intensity of the sound I (in watts per square meter) produced by the speaker is proportional to the resistance R, show that the sound level β (in decibels) is a linear function of θ.

73. The smallest wavelength possible for a sound wave in air is on the order of the separation distance between air molecules. Find the order of magnitude of the highest-frequency sound wave possible in air, assuming a wave speed of 343 m/s, a density of 1.20 kg/m³, and an average molecular mass of 4.82×10^{-26} kg.

ANSWERS TO QUICK QUIZZES

17.1 The only correct answer is (c). Although the speed of a wave is given by the product of its wavelength and frequency, it is not affected by changes in either one. For example, if the sound from a musical instrument increases in frequency, the wavelength decreases, and thus $v = \lambda f$ remains constant. The amplitude of a sound wave determines the size of the oscillations of air molecules but does not affect the speed of the wave through the air.

17.2 The ground tremor represents a sound wave moving through the Earth. Sound waves move faster through the Earth than through air because rock and other ground materials are much stiffer against compression. Therefore—the vibration through the ground and the sound in the air having started together—the vibration through the ground reaches the observer first.

17.3 Because the bottom of the bottle does not allow molecular motion, the displacement in this region is at its minimum value. Because the pressure variation is a maximum when the displacement is a minimum, the pressure variation at the bottom is a maximum.

17.4 (a) 10 dB. If we call the intensity of each violin I, the total intensity when all the violins are playing is $I + 9I = 10I$. Therefore, the addition of the nine violins increases the intensity of the sound over that of one violin by a factor of 10. From Equation 17.7 we see that an increase in intensity by a factor of 10 increases the sound level by 10 dB. (b) 13 dB. The intensity is now increased by a factor of 20 over that of a single violin.

17.5 The Mach number is the ratio of the plane's speed (which does not change) to the speed of sound, which is greater in the warm air than in the cold, as we learned in Section 17.1 (see Quick Quiz 17.1). The denominator of this fraction increases while the numerator stays constant. Therefore, the fraction as a whole—the Mach number—decreases.

17.6 (a) In the reference frame of the air, the observer is moving toward the source at the wind speed through stationary air, and the source is moving away from the observer with the same speed. In Equation 17.17, therefore, a plus sign is needed in both the numerator and the denominator:

$$f' = \left(\frac{v_{sound} + v_{wind}}{v_{sound} + v_{wind}} \right) f$$

meaning the observed frequency is the same as if no wind were blowing. (b) The observer "sees" the sound waves coming toward him at a higher speed $(v_{sound} + v_{wind})$. (c) At this higher speed, he attributes a greater wavelength $\lambda' = (v_{sound} + v_{wind})/f$ to the wave.

c h a p t e r

18

Superposition and Standing Waves

Chapter Outline

mportant in the study of waves is the combined effect of two or more waves traveling in the same medium. For instance, what happens to a string when a wave traveling along it hits a fixed end and is reflected back on itself? What is the air pressure variation at a particular seat in a theater when the instruments of an orchestra sound together?

When analyzing a linear medium—that is, one in which the restoring force acting on the particles of the medium is proportional to the displacement of the particles—we can apply the principle of superposition to determine the resultant disturbance. In Chapter 16 we discussed this principle as it applies to wave pulses. In this chapter we study the superposition principle as it applies to sinusoidal waves. If the sinusoidal waves that combine in a linear medium have the same frequency and wavelength, a stationary pattern—called a *standing wave*—can be produced at certain frequencies under certain circumstances. For example, a taut string fixed at both ends has a discrete set of oscillation patterns, called *modes of vibration,* that are related to the tension and linear mass density of the string. These modes of vibration are found in stringed musical instruments. Other musical instruments, such as the organ and the flute, make use of the natural frequencies of sound waves in hollow pipes. Such frequencies are related to the length and shape of the pipe and depend on whether the pipe is open at both ends or open at one end and closed at the other.

We also consider the superposition and interference of waves having different frequencies and wavelengths. When two sound waves having nearly the same frequency interfere, we hear variations in the loudness called *beats.* The beat frequency corresponds to the rate of alternation between constructive and destructive interference. Finally, we discuss how any non-sinusoidal periodic wave can be described as a sum of sine and cosine functions.

18.1 SUPERPOSITION AND INTERFERENCE OF SINUSOIDAL WAVES

Imagine that you are standing in a swimming pool and that a beach ball is floating a couple of meters away. You use your right hand to send a series of waves toward the beach ball, causing it to repeatedly move upward by 5 cm, return to its original position, and then move downward by 5 cm. After the water becomes still, you use your left hand to send an identical set of waves toward the beach ball and observe the same behavior. What happens if you use both hands at the same time to send two waves toward the beach ball? How the beach ball responds to the waves depends on whether the waves work together (that is, both waves make the beach ball go up at the same time and then down at the same time) or work against each other (that is, one wave tries to make the beach ball go up, while the other wave tries to make it go down). Because it is possible to have two or more waves in the same location at the same time, we have to consider how waves interact with each other and with their surroundings.

The superposition principle states that when two or more waves move in the same linear medium, the net displacement of the medium (that is, the resultant wave) at any point equals the algebraic sum of all the displacements caused by the individual waves. Let us apply this principle to two sinusoidal waves traveling in the same direction in a linear medium. If the two waves are traveling to the right and have the same frequency, wavelength, and amplitude but differ in phase, we can express their individual wave functions as

$$y_1 = A \sin(kx - \omega t) \qquad y_2 = A \sin(kx - \omega t + \phi)$$

where, as usual, $k = 2\pi/\lambda$, $\omega = 2\pi f$, and ϕ is the phase constant, which we introduced in the context of simple harmonic motion in Chapter 13. Hence, the resultant wave function y is

$$y = y_1 + y_2 = A[\sin(kx - \omega t) + \sin(kx - \omega t + \phi)]$$

To simplify this expression, we use the trigonometric identity

$$\sin a + \sin b = 2 \cos\left(\frac{a - b}{2}\right) \sin\left(\frac{a + b}{2}\right)$$

If we let $a = kx - \omega t$ and $b = kx - \omega t + \phi$, we find that the resultant wave function y reduces to

$$y = 2A \cos\left(\frac{\phi}{2}\right) \sin\left(kx - \omega t + \frac{\phi}{2}\right)$$

Resultant of two traveling sinusoidal waves

This result has several important features. The resultant wave function y also is sinusoidal and has the same frequency and wavelength as the individual waves, since the sine function incorporates the same values of k and ω that appear in the original wave functions. The amplitude of the resultant wave is $2A \cos(\phi/2)$, and its phase is $\phi/2$. If the phase constant ϕ equals 0, then $\cos(\phi/2) = \cos 0 = 1$, and the amplitude of the resultant wave is $2A$—twice the amplitude of either individual wave. In this case, in which $\phi = 0$, the waves are said to be everywhere *in phase* and thus **interfere constructively.** That is, the crests and troughs of the individual waves y_1 and y_2 occur at the same positions and combine to form the red curve y of amplitude $2A$ shown in Figure 18.1a. Because the individual waves are in phase, they are indistinguishable in Figure 18.1a, in which they appear as a single blue curve. In general, constructive interference occurs when $\cos(\phi/2) = \pm 1$. This is true, for example, when $\phi = 0, 2\pi, 4\pi, \ldots$ rad—that is, when ϕ is an *even* multiple of π.

Constructive interference

When ϕ is equal to π rad or to any *odd* multiple of π, then $\cos(\phi/2) = \cos(\pi/2) = 0$, and the crests of one wave occur at the same positions as the troughs of the second wave (Fig. 18.1b). Thus, the resultant wave has *zero* amplitude everywhere, as a consequence of **destructive interference.** Finally, when the phase constant has an arbitrary value other than 0 or other than an integer multiple of π rad (Fig. 18.1c), the resultant wave has an amplitude whose value is somewhere between 0 and $2A$.

Destructive interference

Interference of Sound Waves

One simple device for demonstrating interference of sound waves is illustrated in Figure 18.2. Sound from a loudspeaker S is sent into a tube at point P, where there is a T-shaped junction. Half of the sound power travels in one direction, and half travels in the opposite direction. Thus, the sound waves that reach the receiver R can travel along either of the two paths. The distance along any path from speaker to receiver is called the **path length r.** The lower path length r_1 is fixed, but the upper path length r_2 can be varied by sliding the U-shaped tube, which is similar to that on a slide trombone. When the difference in the path lengths $\Delta r = |r_2 - r_1|$ is either zero or some integer multiple of the wavelength λ (that is, $r = n\lambda$, where $n = 0, 1, 2, 3, \ldots$), the two waves reaching the receiver at any instant are in phase and interfere constructively, as shown in Figure 18.1a. For this case, a maximum in the sound intensity is detected at the receiver. If the path length r_2 is ad-

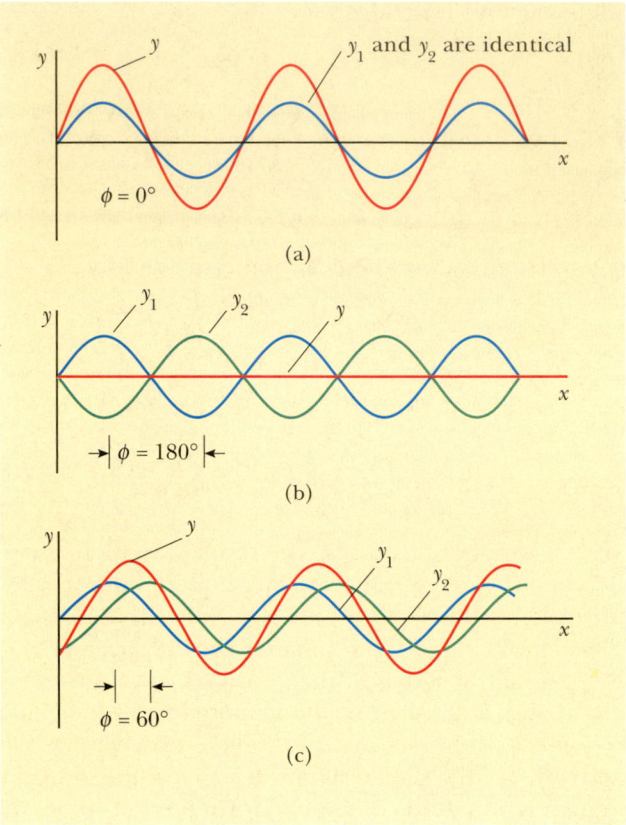

Figure 18.1 The superposition of two identical waves y_1 and y_2 (blue) to yield a resultant wave (red). (a) When y_1 and y_2 are in phase, the result is constructive interference. (b) When y_1 and y_2 are π rad out of phase, the result is destructive interference. (c) When the phase angle has a value other than 0 or π rad, the resultant wave y falls somewhere between the extremes shown in (a) and (b).

justed such that the path difference $\Delta r = \lambda/2, 3\lambda/2, \ldots, n\lambda/2$ (for n odd), the two waves are exactly π rad, or 180°, out of phase at the receiver and hence cancel each other. In this case of destructive interference, no sound is detected at the receiver. This simple experiment demonstrates that a phase difference may arise between two waves generated by the same source when they travel along paths of unequal lengths. This important phenomenon will be indispensable in our investigation of the interference of light waves in Chapter 37.

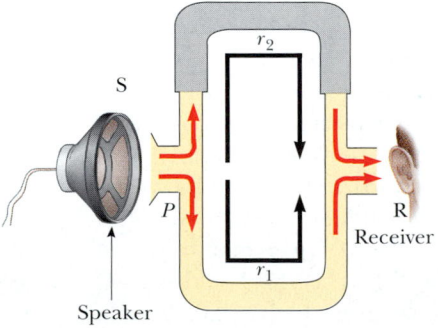

Figure 18.2 An acoustical system for demonstrating interference of sound waves. A sound wave from the speaker (S) propagates into the tube and splits into two parts at point P. The two waves, which superimpose at the opposite side, are detected at the receiver (R). The upper path length r_2 can be varied by sliding the upper section.

It is often useful to express the path difference in terms of the phase angle ϕ between the two waves. Because a path difference of one wavelength corresponds to a phase angle of 2π rad, we obtain the ratio $\phi/2\pi = \Delta r/\lambda$, or

$$\Delta r = \frac{\phi}{2\pi}\lambda \qquad (18.1)$$

Relationship between path difference and phase angle

Using the notion of path difference, we can express our conditions for constructive and destructive interference in a different way. If the path difference is any even multiple of $\lambda/2$, then the phase angle $\phi = 2n\pi$, where $n = 0, 1, 2, 3, \ldots$, and the interference is constructive. For path differences of odd multiples of $\lambda/2$, $\phi = (2n + 1)\pi$, where $n = 0, 1, 2, 3 \ldots$, and the interference is destructive. Thus, we have the conditions

$$\Delta r = (2n)\frac{\lambda}{2} \qquad \text{for constructive interference}$$

and $\qquad\qquad\qquad\qquad\qquad\qquad\qquad\qquad\qquad\qquad$ (18.2)

$$\Delta r = (2n + 1)\frac{\lambda}{2} \qquad \text{for destructive interference}$$

EXAMPLE 18.1 ▶ Two Speakers Driven by the Same Source

A pair of speakers placed 3.00 m apart are driven by the same oscillator (Fig. 18.3). A listener is originally at point O, which is located 8.00 m from the center of the line connecting the two speakers. The listener then walks to point P, which is a perpendicular distance 0.350 m from O, before reaching the *first minimum* in sound intensity. What is the frequency of the oscillator?

Solution To find the frequency, we need to know the wavelength of the sound coming from the speakers. With this information, combined with our knowledge of the speed of sound, we can calculate the frequency. We can determine the wavelength from the interference information given. The first minimum occurs when the two waves reaching the listener at point P are 180° out of phase—in other words, when their path difference Δr equals $\lambda/2$. To calculate the path difference, we must first find the path lengths r_1 and r_2.

Figure 18.3 shows the physical arrangement of the speakers, along with two shaded right triangles that can be drawn on the basis of the lengths described in the problem. From

these triangles, we find that the path lengths are

$$r_1 = \sqrt{(8.00 \text{ m})^2 + (1.15 \text{ m})^2} = 8.08 \text{ m}$$

and

$$r_2 = \sqrt{(8.00 \text{ m})^2 + (1.85 \text{ m})^2} = 8.21 \text{ m}$$

Hence, the path difference is $r_2 - r_1 = 0.13$ m. Because we require that this path difference be equal to $\lambda/2$ for the first minimum, we find that $\lambda = 0.26$ m.

To obtain the oscillator frequency, we use Equation 16.14, $v = \lambda f$, where v is the speed of sound in air, 343 m/s:

$$f = \frac{v}{\lambda} = \frac{343 \text{ m/s}}{0.26 \text{ m}} = \boxed{1.3 \text{ kHz}}$$

Exercise If the oscillator frequency is adjusted such that the first location at which a listener hears no sound is at a distance of 0.75 m from O, what is the new frequency?

Answer 0.63 kHz.

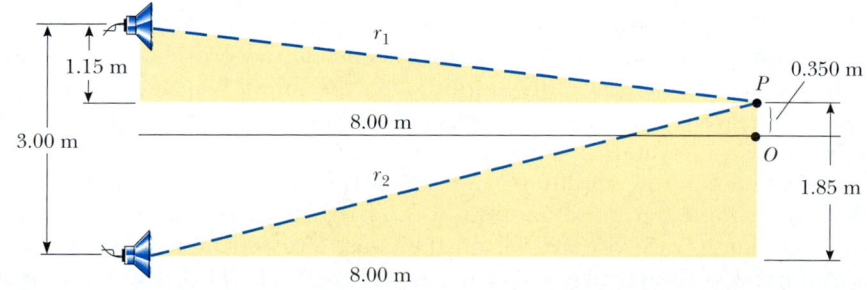

Figure 18.3

You can now understand why the speaker wires in a stereo system should be connected properly. When connected the wrong way—that is, when the positive (or red) wire is connected to the negative (or black) terminal—the speakers are said to be "out of phase" because the sound wave coming from one speaker destructively interferes with the wave coming from the other. In this situation, one speaker cone moves outward while the other moves inward. Along a line midway between the two, a rarefaction region from one speaker is superposed on a condensation region from the other speaker. Although the two sounds probably do not completely cancel each other (because the left and right stereo signals are usually not identical), a substantial loss of sound quality still occurs at points along this line.

18.2   STANDING WAVES

The sound waves from the speakers in Example 18.1 left the speakers in the forward direction, and we considered interference at a point in space in front of the speakers. Suppose that we turn the speakers so that they face each other and then have them emit sound of the same frequency and amplitude. We now have a situation in which two identical waves travel in opposite directions in the same medium. These waves combine in accordance with the superposition principle.

We can analyze such a situation by considering wave functions for two transverse sinusoidal waves having the same amplitude, frequency, and wavelength but traveling in opposite directions in the same medium:

$$y_1 = A \sin(kx - \omega t) \qquad y_2 = A \sin(kx + \omega t)$$

where y_1 represents a wave traveling to the right and y_2 represents one traveling to the left. Adding these two functions gives the resultant wave function y:

$$y = y_1 + y_2 = A \sin(kx - \omega t) + A \sin(kx + \omega t)$$

When we use the trigonometric identity $\sin(a \pm b) = \sin a \cos b \pm \cos a \sin b$, this expression reduces to

Wave function for a standing wave

$$y = (2A \sin kx) \cos \omega t \qquad \text{(18.3)}$$

which is the wave function of a standing wave. A **standing wave,** such as the one shown in Figure 18.4, is an oscillation pattern with a stationary outline that results from the superposition of two identical waves traveling in opposite directions.

Notice that Equation 18.3 does not contain a function of $kx \pm \omega t$. Thus, it is not an expression for a traveling wave. If we observe a standing wave, we have no sense of motion in the direction of propagation of either of the original waves. If we compare this equation with Equation 13.3, we see that Equation 18.3 describes a special kind of simple harmonic motion. Every particle of the medium oscillates in simple harmonic motion with the same frequency ω (according to the $\cos \omega t$ factor in the equation). However, the amplitude of the simple harmonic motion of a given particle (given by the factor $2A \sin kx$, the coefficient of the cosine function) depends on the location x of the particle in the medium. We need to distinguish carefully between the amplitude A of the individual waves and the amplitude $2A \sin kx$ of the simple harmonic motion of the particles of the medium. A given particle in a standing wave vibrates within the constraints of the *envelope* function $2A \sin kx$, where x is the particle's position in the medium. This is in contrast to the situation in a traveling sinusoidal wave, in which all particles oscillate with the

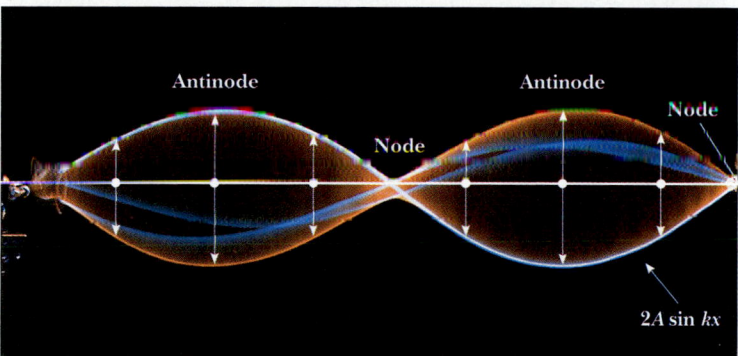

Figure 18.4   Multiflash photograph of a standing wave on a string. The time behavior of the vertical displacement from equilibrium of an individual particle of the string is given by $\cos \omega t$. That is, each particle vibrates at an angular frequency ω. The amplitude of the vertical oscillation of any particle on the string depends on the horizontal position of the particle. Each particle vibrates within the confines of the envelope function $2A \sin kx$. *(©1991 Richard Megna/Fundamental Photographs)*

same amplitude and the same frequency and in which the amplitude of the wave is the same as the amplitude of the simple harmonic motion of the particles.

The maximum displacement of a particle of the medium has a minimum value of zero when x satisfies the condition $\sin kx = 0$, that is, when

$$kx = \pi, 2\pi, 3\pi, \ldots$$

Because $k = 2\pi/\lambda$, these values for kx give

$$x = \frac{\lambda}{2}, \lambda, \frac{3\lambda}{2}, \ldots = \frac{n\lambda}{2} \qquad n = 0, 1, 2, 3, \ldots \qquad \textbf{(18.4)}$$

Position of nodes

These points of zero displacement are called **nodes.**

The particle with the greatest possible displacement from equilibrium has an amplitude of $2A$, and we define this as the amplitude of the standing wave. The positions in the medium at which this maximum displacement occurs are called **antinodes.** The antinodes are located at positions for which the coordinate x satisfies the condition $\sin kx = \pm 1$, that is, when

$$kx = \frac{\pi}{2}, \frac{3\pi}{2}, \frac{5\pi}{2}, \ldots$$

Thus, the positions of the antinodes are given by

$$x = \frac{\lambda}{4}, \frac{3\lambda}{4}, \frac{5\lambda}{4}, \ldots = \frac{n\lambda}{4} \qquad n = 1, 3, 5, \ldots \qquad \textbf{(18.5)}$$

Position of antinodes

In examining Equations 18.4 and 18.5, we note the following important features of the locations of nodes and antinodes:

The distance between adjacent antinodes is equal to $\lambda/2$.
The distance between adjacent nodes is equal to $\lambda/2$.
The distance between a node and an adjacent antinode is $\lambda/4$.

Displacement patterns of the particles of the medium produced at various times by two waves traveling in opposite directions are shown in Figure 18.5. The blue and green curves are the individual traveling waves, and the red curves are

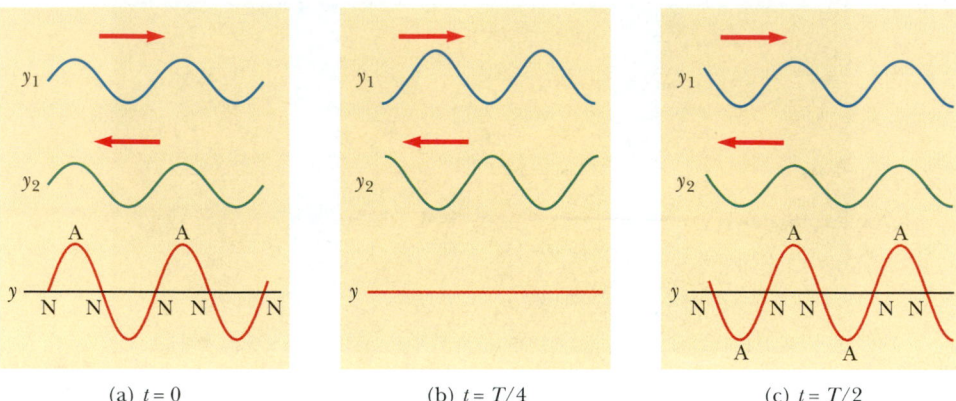

(a) $t = 0$ (b) $t = T/4$ (c) $t = T/2$

Figure 18.5   Standing-wave patterns produced at various times by two waves of equal amplitude traveling in opposite directions. For the resultant wave y, the nodes (N) are points of zero displacement, and the antinodes (A) are points of maximum displacement.

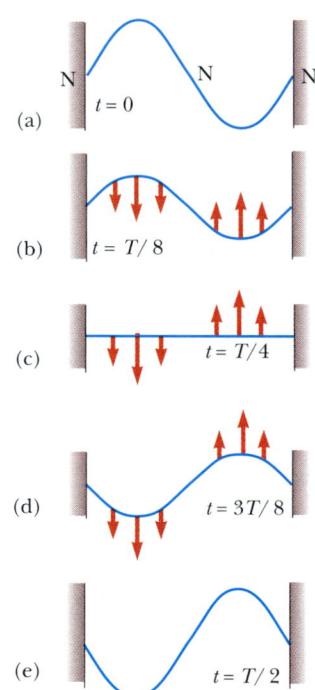

Figure 18.6   A standing-wave pattern in a taut string. The five "snapshots" were taken at half-cycle intervals. (a) At $t = 0$, the string is momentarily at rest; thus, $K = 0$, and all the energy is potential energy U associated with the vertical displacements of the string particles. (b) At $t = T/8$, the string is in motion, as indicated by the brown arrows, and the energy is half kinetic and half potential. (c) At $t = T/4$, the string is moving but horizontal (undeformed); thus, $U = 0$, and all the energy is kinetic. (d) The motion continues as indicated. (e) At $t = T/2$, the string is again momentarily at rest, but the crests and troughs of (a) are reversed. The cycle continues until ultimately, when a time interval equal to T has passed, the configuration shown in (a) is repeated.

the displacement patterns. At $t = 0$ (Fig. 18.5a), the two traveling waves are in phase, giving a displacement pattern in which each particle of the medium is experiencing its maximum displacement from equilibrium. One quarter of a period later, at $t = T/4$ (Fig. 18.5b), the traveling waves have moved one quarter of a wavelength (one to the right and the other to the left). At this time, the traveling waves are out of phase, and each particle of the medium is passing through the equilibrium position in its simple harmonic motion. The result is zero displacement for particles at all values of x—that is, the displacement pattern is a straight line. At $t = T/2$ (Fig. 18.5c), the traveling waves are again in phase, producing a displacement pattern that is inverted relative to the $t = 0$ pattern. In the standing wave, the particles of the medium alternate in time between the extremes shown in Figure 18.5a and c.

Energy in a Standing Wave

It is instructive to describe the energy associated with the particles of a medium in which a standing wave exists. Consider a standing wave formed on a taut string fixed at both ends, as shown in Figure 18.6. Except for the nodes, which are always stationary, all points on the string oscillate vertically with the same frequency but with different amplitudes of simple harmonic motion. Figure 18.6 represents snapshots of the standing wave at various times over one half of a period.

In a traveling wave, energy is transferred along with the wave, as we discussed in Chapter 16. We can imagine this transfer to be due to work done by one segment of the string on the next segment. As one segment moves upward, it exerts a force on the next segment, moving it through a displacement—that is, work is done. A particle of the string at a node, however, experiences no displacement. Thus, it cannot do work on the neighboring segment. As a result, no energy is transmitted along the string across a node, and energy does not propagate in a standing wave. For this reason, standing waves are often called **stationary waves.**

The energy of the oscillating string continuously alternates between elastic potential energy, when the string is momentarily stationary (see Fig. 18.6a), and kinetic energy, when the string is horizontal and the particles have their maximum speed (see Fig. 18.6c). At intermediate times (see Fig. 18.6b and d), the string particles have both potential energy and kinetic energy.

Quick Quiz 18.1

A standing wave described by Equation 18.3 is set up on a string. At what points on the string do the particles move the fastest?

EXAMPLE 18.2 ▸ Formation of a Standing Wave

Two waves traveling in opposite directions produce a standing wave. The individual wave functions $y = A \sin(kx - \omega t)$ are

$$y_1 = (4.0 \text{ cm}) \sin(3.0x - 2.0t)$$

and

$$y_2 = (4.0 \text{ cm}) \sin(3.0x + 2.0t)$$

where x and y are measured in centimeters. (a) Find the amplitude of the simple harmonic motion of the particle of the medium located at $x = 2.3$ cm.

Solution The standing wave is described by Equation 18.3; in this problem, we have $A = 4.0$ cm, $k = 3.0$ rad/cm, and $\omega = 2.0$ rad/s. Thus,

$$y = (2A \sin kx) \cos \omega t = [(8.0 \text{ cm}) \sin 3.0x] \cos 2.0t$$

Thus, we obtain the amplitude of the simple harmonic motion of the particle at the position $x = 2.3$ cm by evaluating the coefficient of the cosine function at this position:

$$y_{max} = (8.0 \text{ cm}) \sin 3.0x|_{x=2.3}$$

$$= (8.0 \text{ cm}) \sin(6.9 \text{ rad}) = \boxed{4.6 \text{ cm}}$$

(b) Find the positions of the nodes and antinodes.

Solution With $k = 2\pi/\lambda = 3.0$ rad/cm, we see that $\lambda = 2\pi/3$ cm. Therefore, from Equation 18.4 we find that the nodes are located at

$$x = n\frac{\lambda}{2} = \boxed{n\left(\frac{\pi}{3}\right) \text{cm}} \qquad n = 0, 1, 2, 3 \ldots$$

and from Equation 18.5 we find that the antinodes are located at

$$x = n\frac{\lambda}{4} = \boxed{n\left(\frac{\pi}{6}\right) \text{cm}} \qquad n = 1, 3, 5, \ldots$$

(c) What is the amplitude of the simple harmonic motion of a particle located at an antinode?

Solution According to Equation 18.3, the maximum displacement of a particle at an antinode is the amplitude of the standing wave, which is twice the amplitude of the individual traveling waves:

$$y_{max} = 2A = 2(4.0 \text{ cm}) = \boxed{8.0 \text{ cm}}$$

Let us check this result by evaluating the coefficient of our standing-wave function at the positions we found for the antinodes:

$$y_{max} = (8.0 \text{ cm}) \sin 3.0x|_{x=n(\pi/6)}$$

$$= (8.0 \text{ cm}) \sin\left[3.0n\left(\frac{\pi}{6}\right)\text{rad}\right]$$

$$= (8.0 \text{ cm}) \sin\left[n\left(\frac{\pi}{2}\right)\text{rad}\right] = 8.0 \text{ cm}$$

In evaluating this expression, we have used the fact that n is an odd integer; thus, the sine function is equal to unity.

18.3 ▸ STANDING WAVES IN A STRING FIXED AT BOTH ENDS

Consider a string of length L fixed at both ends, as shown in Figure 18.7. Standing waves are set up in the string by a continuous superposition of waves incident on and reflected from the ends. Note that the ends of the string, because they are fixed and must necessarily have zero displacement, are nodes by definition. The string has a number of natural patterns of oscillation, called **normal modes,** each of which has a characteristic frequency that is easily calculated.

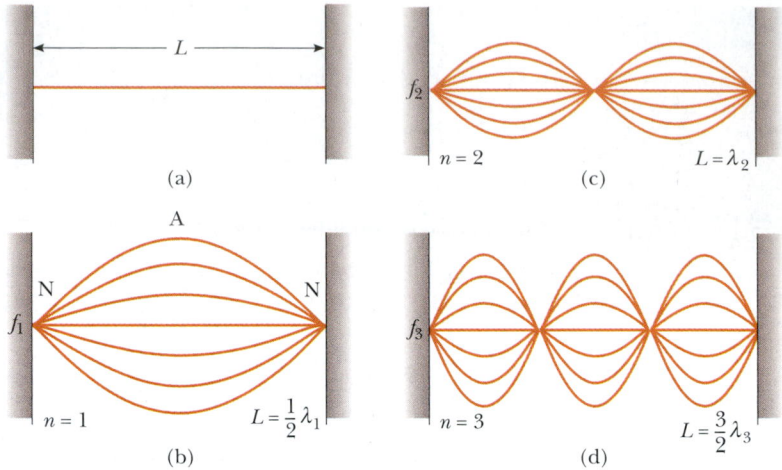

Figure 18.7   (a) A string of length L fixed at both ends. The normal modes of vibration form a harmonic series: (b) the fundamental, or first harmonic; (c) the second harmonic; (d) the third harmonic.

In general, the motion of an oscillating string fixed at both ends is described by the superposition of several normal modes. Exactly which normal modes are present depends on how the oscillation is started. For example, when a guitar string is plucked near its middle, the modes shown in Figure 18.7b and d, as well as other modes not shown, are excited.

In general, we can describe the normal modes of oscillation for the string by imposing the requirements that the ends be nodes and that the nodes and antinodes be separated by one fourth of a wavelength. The first normal mode, shown in Figure 18.7b, has nodes at its ends and one antinode in the middle. This is the longest-wavelength mode, and this is consistent with our requirements. This first normal mode occurs when the wavelength λ_1 is twice the length of the string, that is, $\lambda_1 = 2L$. The next normal mode, of wavelength λ_2 (see Fig. 18.7c), occurs when the wavelength equals the length of the string, that is, $\lambda_2 = L$. The third normal mode (see Fig. 18.7d) corresponds to the case in which $\lambda_3 = 2L/3$. In general, the wavelengths of the various normal modes for a string of length L fixed at both ends are

Wavelengths of normal modes

$$\lambda_n = \frac{2L}{n} \qquad n = 1, 2, 3, \ldots \tag{18.6}$$

where the index n refers to the nth normal mode of oscillation. These are the *possible* modes of oscillation for the string. The *actual* modes that are excited by a given pluck of the string are discussed below.

The natural frequencies associated with these modes are obtained from the relationship $f = v/\lambda$, where the wave speed v is the same for all frequencies. Using Equation 18.6, we find that the natural frequencies f_n of the normal modes are

Frequencies of normal modes as functions of wave speed and length of string

$$f_n = \frac{v}{\lambda_n} = n\frac{v}{2L} \qquad n = 1, 2, 3, \ldots \tag{18.7}$$

Because $v = \sqrt{T/\mu}$ (see Eq. 16.4), where T is the tension in the string and μ is its linear mass density, we can also express the natural frequencies of a taut string as

Frequencies of normal modes as functions of string tension and linear mass density

$$f_n = \frac{n}{2L}\sqrt{\frac{T}{\mu}} \qquad n = 1, 2, 3, \ldots \tag{18.8}$$

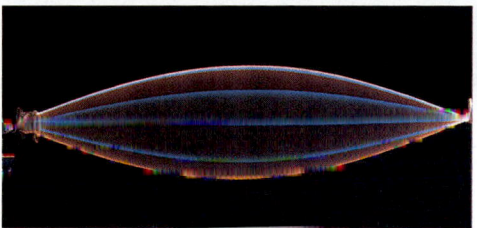 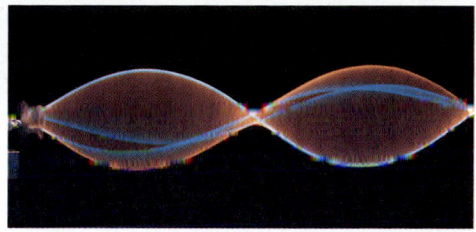 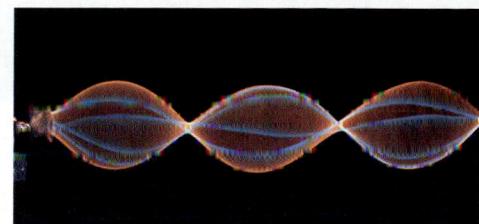

Multiflash photographs of standing-wave patterns in a cord driven by a vibrator at its left end. The single-loop pattern represents the first normal mode ($n = 1$). The double-loop pattern represents the second normal mode ($n = 2$), and the triple-loop pattern represents the third normal mode ($n = 3$). *(©1991 Richard Megna/Fundamental Photographs)*

The lowest frequency f_1, which corresponds to $n = 1$, is called either the **fundamental** or the **fundamental frequency** and is given by

$$f_1 = \frac{1}{2L}\sqrt{\frac{T}{\mu}} \tag{18.9}$$

Fundamental frequency of a taut string

The frequencies of the remaining normal modes are integer multiples of the fundamental frequency. Frequencies of normal modes that exhibit an integer-multiple relationship such as this form a **harmonic series,** and the normal modes are called **harmonics.** The fundamental frequency f_1 is the frequency of the first harmonic; the frequency $f_2 = 2f_1$ is the frequency of the second harmonic; and the frequency $f_n = nf_1$ is the frequency of the nth harmonic. Other oscillating systems, such as a drumhead, exhibit normal modes, but the frequencies are not related as integer multiples of a fundamental. Thus, we do not use the term *harmonic* in association with these types of systems.

In obtaining Equation 18.6, we used a technique based on the separation distance between nodes and antinodes. We can obtain this equation in an alternative manner. Because we require that the string be fixed at $x = 0$ and $x = L$, the wave function $y(x, t)$ given by Equation 18.3 must be zero at these points for all times. That is, the *boundary conditions* require that $y(0, t) = 0$ and that $y(L, t) = 0$ for all values of t. Because the standing wave is described by $y = (2A \sin kx) \cos \omega t$, the first boundary condition, $y(0, t) = 0$, is automatically satisfied because $\sin kx = 0$ at $x = 0$. To meet the second boundary condition, $y(L, t) = 0$, we require that $\sin kL = 0$. This condition is satisfied when the angle kL equals an integer multiple of π rad. Therefore, the allowed values of k are given by[1]

$$k_n L = n\pi \qquad n = 1, 2, 3, \ldots \tag{18.10}$$

Because $k_n = 2\pi/\lambda_n$, we find that

$$\left(\frac{2\pi}{\lambda_n}\right)L = n\pi \qquad \text{or} \qquad \lambda_n = \frac{2L}{n}$$

which is identical to Equation 18.6.

Let us now examine how these various harmonics are created in a string. If we wish to excite just a single harmonic, we need to distort the string in such a way that its distorted shape corresponded to that of the desired harmonic. After being released, the string vibrates at the frequency of that harmonic. This maneuver is difficult to perform, however, and it is not how we excite a string of a musical in-

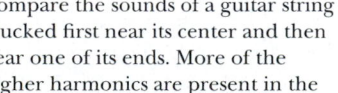

QuickLab

Compare the sounds of a guitar string plucked first near its center and then near one of its ends. More of the higher harmonics are present in the second situation. Can you hear the difference?

---

[1] We exclude $n = 0$ because this value corresponds to the trivial case in which no wave exists ($k = 0$).

strument. If the string is distorted such that its distorted shape is not that of just one harmonic, the resulting vibration includes various harmonics. Such a distortion occurs in musical instruments when the string is plucked (as in a guitar), bowed (as in a cello), or struck (as in a piano). When the string is distorted into a non-sinusoidal shape, only waves that satisfy the boundary conditions can persist on the string. These are the harmonics.

The frequency of a stringed instrument can be varied by changing either the tension or the string's length. For example, the tension in guitar and violin strings is varied by a screw adjustment mechanism or by tuning pegs located on the neck of the instrument. As the tension is increased, the frequency of the normal modes increases in accordance with Equation 18.8. Once the instrument is "tuned," players vary the frequency by moving their fingers along the neck, thereby changing the length of the oscillating portion of the string. As the length is shortened, the frequency increases because, as Equation 18.8 specifies, the normal-mode frequencies are inversely proportional to string length.

EXAMPLE 18.3   Give Me a C Note!

Middle C on a piano has a fundamental frequency of 262 Hz, and the first A above middle C has a fundamental frequency of 440 Hz. (a) Calculate the frequencies of the next two harmonics of the C string.

Solution   Knowing that the frequencies of higher harmonics are integer multiples of the fundamental frequency $f_1 = 262$ Hz, we find that

$$f_2 = 2f_1 = \boxed{524 \text{ Hz}}$$

$$f_3 = 3f_1 = \boxed{786 \text{ Hz}}$$

(b) If the A and C strings have the same linear mass density μ and length L, determine the ratio of tensions in the two strings.

Solution   Using Equation 18.8 for the two strings vibrating at their fundamental frequencies gives

$$f_{1A} = \frac{1}{2L}\sqrt{\frac{T_A}{\mu}} \quad \text{and} \quad f_{1C} = \frac{1}{2L}\sqrt{\frac{T_C}{\mu}}$$

Setting up the ratio of these frequencies, we find that

$$\frac{f_{1A}}{f_{1C}} = \sqrt{\frac{T_A}{T_C}}$$

$$\frac{T_A}{T_C} = \left(\frac{f_{1A}}{f_{1C}}\right)^2 = \left(\frac{440}{262}\right)^2 = \boxed{2.82}$$

(c) With respect to a real piano, the assumption we made in (b) is only partially true. The string densities are equal, but the length of the A string is only 64 percent of the length of the C string. What is the ratio of their tensions?

Solution   Using Equation 18.8 again, we set up the ratio of frequencies:

$$\frac{f_{1A}}{f_{1C}} = \frac{L_C}{L_A}\sqrt{\frac{T_A}{T_C}} = \left(\frac{100}{64}\right)\sqrt{\frac{T_A}{T_C}}$$

$$\frac{T_A}{T_C} = (0.64)^2\left(\frac{440}{262}\right)^2 = \boxed{1.16}$$

EXAMPLE 18.4   Guitar Basics

The high E string on a guitar measures 64.0 cm in length and has a fundamental frequency of 330 Hz. By pressing down on it at the first fret (Fig. 18.8), the string is shortened so that it plays an F note that has a frequency of 350 Hz. How far is the fret from the neck end of the string?

Solution   Equation 18.7 relates the string's length to the fundamental frequency. With $n = 1$, we can solve for the

speed of the wave on the string,

$$v = \frac{2L}{n}f_n = \frac{2(0.640 \text{ m})}{1}(330 \text{ Hz}) = 422 \text{ m/s}$$

Because we have not adjusted the tuning peg, the tension in the string, and hence the wave speed, remain constant. We can again use Equation 18.7, this time solving for L and sub-

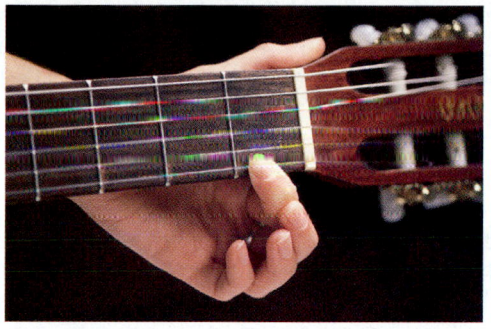

Figure 18.8 Playing an F note on a guitar. *(Charles D. Winters)*

stituting the new frequency to find the shortened string length:

$$L = n\frac{v}{2f_n} = (1)\frac{422 \text{ m/s}}{2(350 \text{ Hz})} = 0.603 \text{ m}$$

The difference between this length and the measured length of 64.0 cm is the distance from the fret to the neck end of the string, or 3.70 cm.

RESONANCE

We have seen that a system such as a taut string is capable of oscillating in one or more normal modes of oscillation. **If a periodic force is applied to such a system, the amplitude of the resulting motion is greater than normal when the frequency of the applied force is equal to or nearly equal to one of the natural frequencies of the system.** We discussed this phenomenon, known as *resonance*, briefly in Section 13.7. Although a block–spring system or a simple pendulum has only one natural frequency, standing-wave systems can have a whole set of natural frequencies. Because an oscillating system exhibits a large amplitude when driven at any of its natural frequencies, these frequencies are often referred to as **resonance frequencies.**

Figure 18.9 shows the response of an oscillating system to various driving frequencies, where one of the resonance frequencies of the system is denoted by f_0. Note that the amplitude of oscillation of the system is greatest when the frequency of the driving force equals the resonance frequency. The maximum amplitude is limited by friction in the system. If a driving force begins to work on an oscillating system initially at rest, the input energy is used both to increase the amplitude of the oscillation and to overcome the frictional force. Once maximum amplitude is reached, the work done by the driving force is used only to overcome friction.

A system is said to be *weakly damped* when the amount of friction to be overcome is small. Such a system has a large amplitude of motion when driven at one of its resonance frequencies, and the oscillations persist for a long time after the driving force is removed. A system in which considerable friction must be overcome is said to be *strongly damped*. For a given driving force applied at a resonance frequency, the maximum amplitude attained by a strongly damped oscillator is smaller than that attained by a comparable weakly damped oscillator. Once the driving force in a strongly damped oscillator is removed, the amplitude decreases rapidly with time.

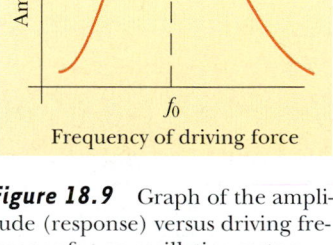

Figure 18.9 Graph of the amplitude (response) versus driving frequency for an oscillating system. The amplitude is a maximum at the resonance frequency f_0. Note that the curve is not symmetric.

Examples of Resonance

A playground swing is a pendulum having a natural frequency that depends on its length. Whenever we use a series of regular impulses to push a child in a swing, the swing goes higher if the frequency of the periodic force equals the natural fre-

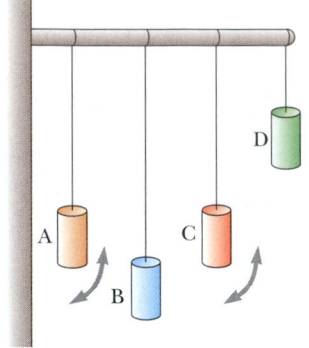

Figure 18.10 An example of resonance. If pendulum A is set into oscillation, only pendulum C, whose length matches that of A, eventually oscillates with large amplitude, or resonates. The arrows indicate motion perpendicular to the page.

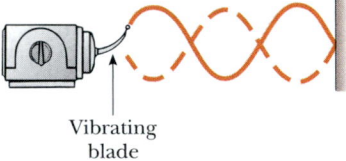

Vibrating
blade

Figure 18.11 Standing waves are set up in a string when one end is connected to a vibrating blade. When the blade vibrates at one of the natural frequencies of the string, large-amplitude standing waves are created.

quency of the swing. We can demonstrate a similar effect by suspending pendulums of different lengths from a horizontal support, as shown in Figure 18.10. If pendulum A is set into oscillation, the other pendulums begin to oscillate as a result of the longitudinal waves transmitted along the beam. However, pendulum C, the length of which is close to the length of A, oscillates with a much greater amplitude than pendulums B and D, the lengths of which are much different from that of pendulum A. Pendulum C moves the way it does because its natural frequency is nearly the same as the driving frequency associated with pendulum A.

Next, consider a taut string fixed at one end and connected at the opposite end to an oscillating blade, as illustrated in Figure 18.11. The fixed end is a node, and the end connected to the blade is very nearly a node because the amplitude of the blade's motion is small compared with that of the string. As the blade oscillates, transverse waves sent down the string are reflected from the fixed end. As we learned in Section 18.3, the string has natural frequencies that are determined by its length, tension, and linear mass density (see Eq. 18.8). When the frequency of the blade equals one of the natural frequencies of the string, standing waves are produced and the string oscillates with a large amplitude. In this resonance case, the wave generated by the oscillating blade is in phase with the reflected wave, and the string absorbs energy from the blade. If the string is driven at a frequency that is not one of its natural frequencies, then the oscillations are of low amplitude and exhibit no stable pattern.

Once the amplitude of the standing-wave oscillations is a maximum, the mechanical energy delivered by the blade and absorbed by the system is lost because of the damping forces caused by friction in the system. If the applied frequency differs from one of the natural frequencies, energy is transferred to the string at first, but later the phase of the wave becomes such that it forces the blade to receive energy from the string, thereby reducing the energy in the string.

Quick Quiz 18.2

Some singers can shatter a wine glass by maintaining a certain frequency of their voice for several seconds. Figure 18.12a shows a side view of a wine glass vibrating because of a sound wave. Sketch the standing-wave pattern in the rim of the glass as seen from above. If an inte-

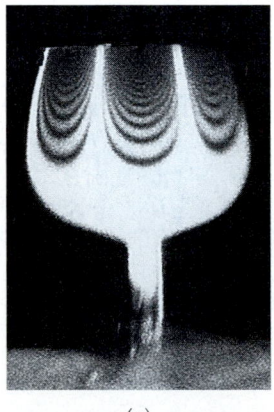

(a)

(b)

Figure 18.12 (a) Standing-wave pattern in a vibrating wine glass. The glass shatters if the amplitude of vibration becomes too great. *(Courtesy of Professor Thomas D. Rossing, Northern Illinois University)* (b) A wine glass shattered by the amplified sound of a human voice. *(©1992 Ben Rose/The Image Bank)*

gral number of waves "fit" around the circumference of the vibrating rim, how many wavelengths fit around the rim in Figure 18.12a?

Quick Quiz 18.3

"Rumble strips" (Fig. 18.13) are sometimes placed across a road to warn drivers that they are approaching a stop sign, or laid along the sides of the road to alert drivers when they are drifting out of their lane. Why are these sets of small bumps so effective at getting a driver's attention?

Figure 18.13 Rumble strips along the side of a highway. *(Charles D. Winters)*

18.5 STANDING WAVES IN AIR COLUMNS

Standing waves can be set up in a tube of air, such as that in an organ pipe, as the result of interference between longitudinal sound waves traveling in opposite directions. The phase relationship between the incident wave and the wave reflected from one end of the pipe depends on whether that end is open or closed. This relationship is analogous to the phase relationships between incident and reflected transverse waves at the end of a string when the end is either fixed or free to move (see Figs. 16.13 and 16.14).

In a pipe closed at one end, **the closed end is a displacement node because the wall at this end does not allow longitudinal motion of the air molecules.** As a result, at a closed end of a pipe, the reflected sound wave is 180° out of phase with the incident wave. Furthermore, because the pressure wave is 90° out of phase with the displacement wave (see Section 17.2), **the closed end of an air column corresponds to a pressure antinode** (that is, a point of maximum pressure variation).

The open end of an air column is approximately a displacement antinode[2] and a pressure node. We can understand why no pressure variation occurs at an open end by noting that the end of the air column is open to the atmosphere; thus, the pressure at this end must remain constant at atmospheric pressure.

---

[2] Strictly speaking, the open end of an air column is not exactly a displacement antinode. A condensation reaching an open end does not reflect until it passes beyond the end. For a thin-walled tube of circular cross section, this end correction is approximately $0.6R$, where R is the tube's radius. Hence, the effective length of the tube is longer than the true length L. We ignore this end correction in this discussion.

QuickLab

Snip off pieces at one end of a drinking straw so that the end tapers to a point. Chew on this end to flatten it, and you'll have created a double-reed instrument! Put your lips around the tapered end, press them tightly together, and blow through the straw. When you hear a steady tone, slowly snip off pieces of the straw from the other end. Be careful to maintain a constant pressure with your lips. How does the frequency change as the straw is shortened?

You may wonder how a sound wave can reflect from an open end, since there may not appear to be a change in the medium at this point. It is indeed true that the medium through which the sound wave moves is air both inside and outside the pipe. Remember that sound is a pressure wave, however, and a compression region of the sound wave is constrained by the sides of the pipe as long as the region is inside the pipe. As the compression region exits at the open end of the pipe, the constraint is removed and the compressed air is free to expand into the atmosphere. Thus, there is a change in the *character* of the medium between the inside of the pipe and the outside even though there is no change in the *material* of the medium. This change in character is sufficient to allow some reflection.

The first three normal modes of oscillation of a pipe open at both ends are shown in Figure 18.14a. When air is directed against an edge at the left, longitudinal standing waves are formed, and the pipe resonates at its natural frequencies. All normal modes are excited simultaneously (although not with the same amplitude). Note that both ends are displacement antinodes (approximately). In the first normal mode, the standing wave extends between two adjacent antinodes,

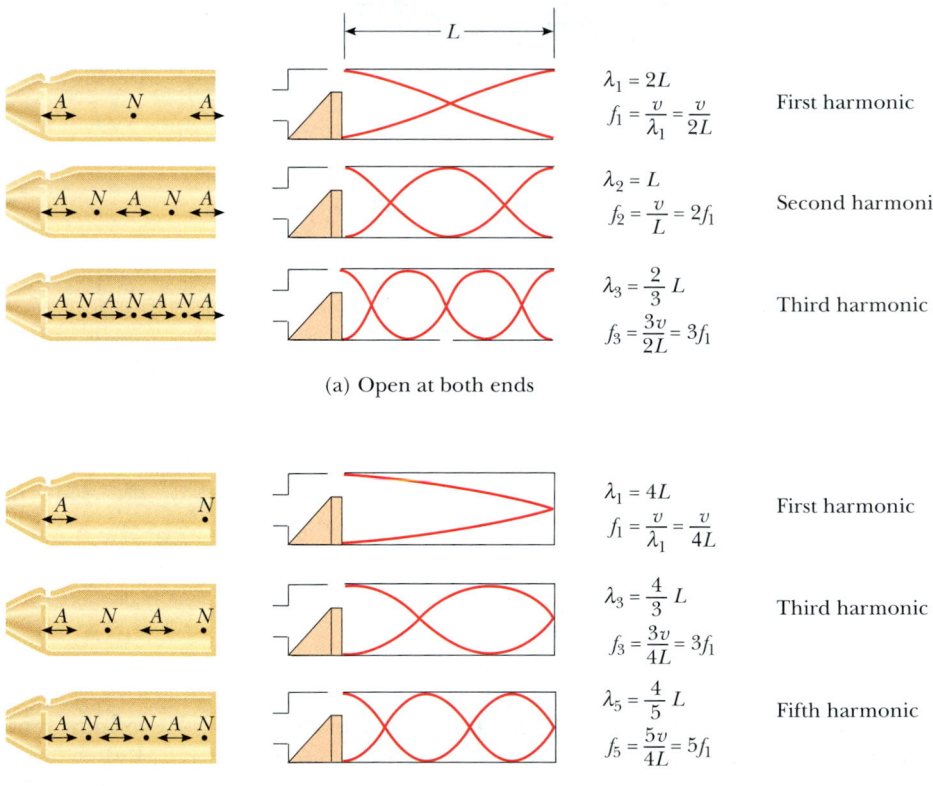

(a) Open at both ends

$\lambda_1 = 2L$
$f_1 = \dfrac{v}{\lambda_1} = \dfrac{v}{2L}$ First harmonic

$\lambda_2 = L$
$f_2 = \dfrac{v}{L} = 2f_1$ Second harmonic

$\lambda_3 = \dfrac{2}{3}L$
$f_3 = \dfrac{3v}{2L} = 3f_1$ Third harmonic

$\lambda_1 = 4L$
$f_1 = \dfrac{v}{\lambda_1} = \dfrac{v}{4L}$ First harmonic

$\lambda_3 = \dfrac{4}{3}L$
$f_3 = \dfrac{3v}{4L} = 3f_1$ Third harmonic

$\lambda_5 = \dfrac{4}{5}L$
$f_5 = \dfrac{5v}{4L} = 5f_1$ Fifth harmonic

(b) Closed at one end, open at the other

Figure 18.14 Motion of air molecules in standing longitudinal waves in a pipe, along with schematic representations of the waves. The graphs represent the displacement amplitudes, not the pressure amplitudes. (a) In a pipe open at both ends, the harmonic series created consists of all integer multiples of the fundamental frequency: $f_1, 2f_1, 3f_1, \ldots$. (b) In a pipe closed at one end and open at the other, the harmonic series created consists of only odd-integer multiples of the fundamental frequency: $f_1, 3f_1, 5f_1, \ldots$.

which is a distance of half a wavelength. Thus, the wavelength is twice the length of the pipe, and the fundamental frequency is $f_1 = v/2L$. As Figure 18.14a shows, the frequencies of the higher harmonics are $2f_1, 3f_1, \ldots$. Thus, we can say that

> in a pipe open at both ends, the natural frequencies of oscillation form a harmonic series that includes all integral multiples of the fundamental frequency.

Because all harmonics are present, and because the fundamental frequency is given by the same expression as that for a string (see Eq. 18.7), we can express the natural frequencies of oscillation as

$$f_n = n\frac{v}{2L} \qquad n = 1, 2, 3 \ldots \tag{18.11}$$

Natural frequencies of a pipe open at both ends

Despite the similarity between Equations 18.7 and 18.11, we must remember that v in Equation 18.7 is the speed of waves on the string, whereas v in Equation 18.11 is the speed of sound in air.

    If a pipe is closed at one end and open at the other, the closed end is a displacement node (see Fig. 18.14b). In this case, the standing wave for the fundamental mode extends from an antinode to the adjacent node, which is one fourth of a wavelength. Hence, the wavelength for the first normal mode is $4L$, and the fundamental frequency is $f_1 = v/4L$. As Figure 18.14b shows, the higher-frequency waves that satisfy our conditions are those that have a node at the closed end and an antinode at the open end; this means that the higher harmonics have frequencies $3f_1, 5f_1, \ldots$:

> In a pipe closed at one end and open at the other, the natural frequencies of oscillation form a harmonic series that includes only odd integer multiples of the fundamental frequency.

We express this result mathematically as

$$f_n = n\frac{v}{4L} \qquad n = 1, 3, 5, \ldots \tag{18.12}$$

Natural frequencies of a pipe closed at one end and open at the other

    It is interesting to investigate what happens to the frequencies of instruments based on air columns and strings during a concert as the temperature rises. The sound emitted by a flute, for example, becomes sharp (increases in frequency) as it warms up because the speed of sound increases in the increasingly warmer air inside the flute (consider Eq. 18.11). The sound produced by a violin becomes flat (decreases in frequency) as the strings expand thermally because the expansion causes their tension to decrease (see Eq. 18.8).

QuickLab

Blow across the top of an empty soda-pop bottle. From a measurement of the height of the bottle, estimate the frequency of the sound you hear. Note that the cross-sectional area of the bottle is not constant; thus, this is not a perfect model of a cylindrical air column.

Quick Quiz 18.4

A pipe open at both ends resonates at a fundamental frequency f_{open}. When one end is covered and the pipe is again made to resonate, the fundamental frequency is f_{closed}. Which of the following expressions describes how these two resonant frequencies compare? (a) $f_{\text{closed}} = f_{\text{open}}$ (b) $f_{\text{closed}} = \frac{1}{2}f_{\text{open}}$ (c) $f_{\text{closed}} = 2f_{\text{open}}$ (d) $f_{\text{closed}} = \frac{3}{2}f_{\text{open}}$

EXAMPLE 18.5 ▶ Wind in a Culvert

A section of drainage culvert 1.23 m in length makes a howling noise when the wind blows. (a) Determine the frequencies of the first three harmonics of the culvert if it is open at both ends. Take $v = 343$ m/s as the speed of sound in air.

Solution The frequency of the first harmonic of a pipe open at both ends is

$$f_1 = \frac{v}{2L} = \frac{343 \text{ m/s}}{2(1.23 \text{ m})} = \boxed{139 \text{ Hz}}$$

Because both ends are open, all harmonics are present; thus,

$f_2 = 2f_1 = \boxed{278 \text{ Hz}}$ and $f_3 = 3f_1 = \boxed{417 \text{ Hz}}$.

(b) What are the three lowest natural frequencies of the culvert if it is blocked at one end?

Solution The fundamental frequency of a pipe closed at one end is

$$f_1 = \frac{v}{4L} = \frac{343 \text{ m/s}}{4(1.23 \text{ m})} = \boxed{69.7 \text{ Hz}}$$

In this case, only odd harmonics are present; hence, the next two harmonics have frequencies $f_3 = 3f_1 = \boxed{209 \text{ Hz}}$ and $f_5 = 5f_1 = \boxed{349 \text{ Hz}}$.

(c) For the culvert open at both ends, how many of the harmonics present fall within the normal human hearing range (20 to 17 000 Hz)?

Solution Because all harmonics are present, we can express the frequency of the highest harmonic heard as $f_n = nf_1$, where n is the number of harmonics that we can hear. For $f_n = 17\,000$ Hz, we find that the number of harmonics present in the audible range is

$$n = \frac{17\,000 \text{ Hz}}{139 \text{ Hz}} = \boxed{122}$$

Only the first few harmonics are of sufficient amplitude to be heard.

EXAMPLE 18.6 ▶ Measuring the Frequency of a Tuning Fork

A simple apparatus for demonstrating resonance in an air column is depicted in Figure 18.15. A vertical pipe open at both ends is partially submerged in water, and a tuning fork vibrating at an unknown frequency is placed near the top of the pipe. The length L of the air column can be adjusted by moving the pipe vertically. The sound waves generated by the fork are reinforced when L corresponds to one of the resonance frequencies of the pipe.

For a certain tube, the smallest value of L for which a peak occurs in the sound intensity is 9.00 cm. What are (a) the frequency of the tuning fork and (b) the value of L for the next two resonance frequencies?

Solution (a) Although the pipe is open at its lower end to allow the water to enter, the water's surface acts like a wall at one end. Therefore, this setup represents a pipe closed at one end, and so the fundamental frequency is $f_1 = v/4L$. Taking $v = 343$ m/s for the speed of sound in air and $L = 0.090\,0$ m, we obtain

$$f_1 = \frac{v}{4L} = \frac{343 \text{ m/s}}{4(0.090\,0 \text{ m})} = \boxed{953 \text{ Hz}}$$

Because the tuning fork causes the air column to resonate at this frequency, this must be the frequency of the tuning fork.

(b) Because the pipe is closed at one end, we know from Figure 18.14b that the wavelength of the fundamental mode is $\lambda = 4L = 4(0.090\,0 \text{ m}) = 0.360$ m. Because the frequency of the tuning fork is constant, the next two normal modes (see Fig. 18.15b) correspond to lengths of $L = 3\lambda/4 = \boxed{0.270 \text{ m}}$ and $L = 5\lambda/4 = \boxed{0.450 \text{ m}}$.

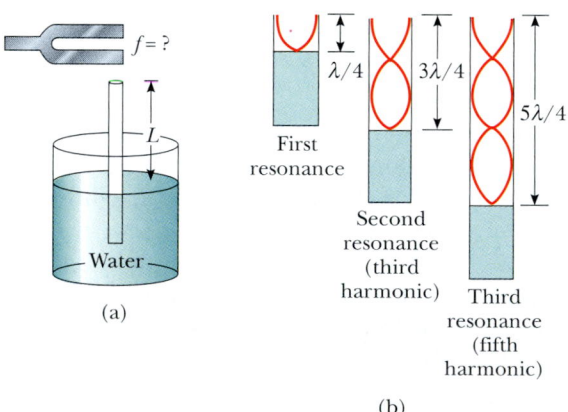

Figure 18.15 (a) Apparatus for demonstrating the resonance of sound waves in a tube closed at one end. The length L of the air column is varied by moving the tube vertically while it is partially submerged in water. (b) The first three normal modes of the system shown in part (a).

Optional Section

18.6 ▶ STANDING WAVES IN RODS AND PLATES

Standing waves can also be set up in rods and plates. A rod clamped in the middle and stroked at one end oscillates, as depicted in Figure 18.16a. The oscillations of the particles of the rod are longitudinal, and so the broken lines in Figure 18.16 represent *longitudinal* displacements of various parts of the rod. For clarity, we have drawn them in the transverse direction, just as we did for air columns. The midpoint is a displacement node because it is fixed by the clamp, whereas the ends are displacement antinodes because they are free to oscillate. The oscillations in this setup are analogous to those in a pipe open at both ends. The broken lines in Figure 18.16a represent the first normal mode, for which the wavelength is $2L$ and the frequency is $f = v/2L$, where v is the speed of longitudinal waves in the rod. Other normal modes may be excited by clamping the rod at different points. For example, the second normal mode (Fig. 18.16b) is excited by clamping the rod a distance $L/4$ away from one end.

Two-dimensional oscillations can be set up in a flexible membrane stretched over a circular hoop, such as that in a drumhead. As the membrane is struck at some point, wave pulses that arrive at the fixed boundary are reflected many times. The resulting sound is not harmonic because the oscillating drumhead and the drum's hollow interior together produce a set of standing waves having frequencies that are *not* related by integer multiples. Without this relationship, the sound may be more correctly described as *noise* than as music. This is in contrast to the situation in wind and stringed instruments, which produce sounds that we describe as musical.

Some possible normal modes of oscillation for a two-dimensional circular membrane are shown in Figure 18.17. The lowest normal mode, which has a frequency f_1, contains only one nodal curve; this curve runs around the outer edge of the membrane. The other possible normal modes show additional nodal curves that are circles and straight lines across the diameter of the membrane.

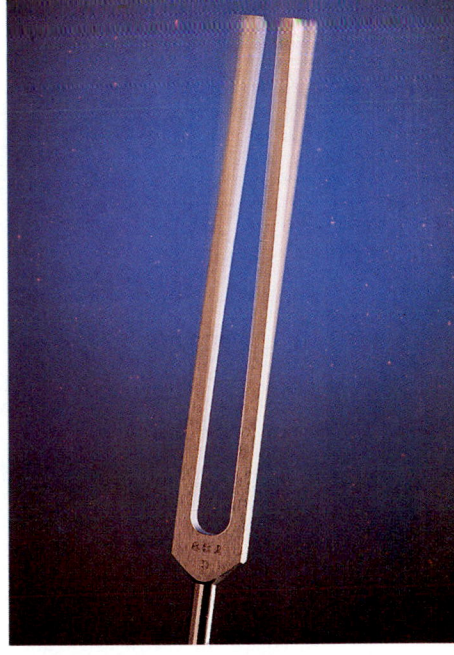

The sound from a tuning fork is produced by the vibrations of each of its prongs. *(Sam Dudgeon/Holt, Rinehart and Winston)*

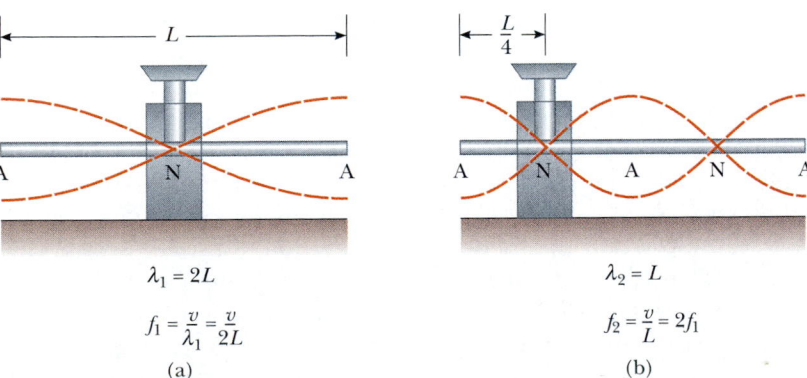

$\lambda_1 = 2L$

$f_1 = \dfrac{v}{\lambda_1} = \dfrac{v}{2L}$

(a)

$\lambda_2 = L$

$f_2 = \dfrac{v}{L} = 2f_1$

(b)

Figure 18.16 Normal-mode longitudinal vibrations of a rod of length L (a) clamped at the middle to produce the first normal mode and (b) clamped at a distance $L/4$ from one end to produce the second normal mode. Note that the dashed lines represent amplitudes parallel to the rod (longitudinal waves).

Wind chimes are usually designed so that the waves emanating from the vibrating rods blend into a harmonious sound. *(Joseph L. Fontenot/Visuals Unlimited)*

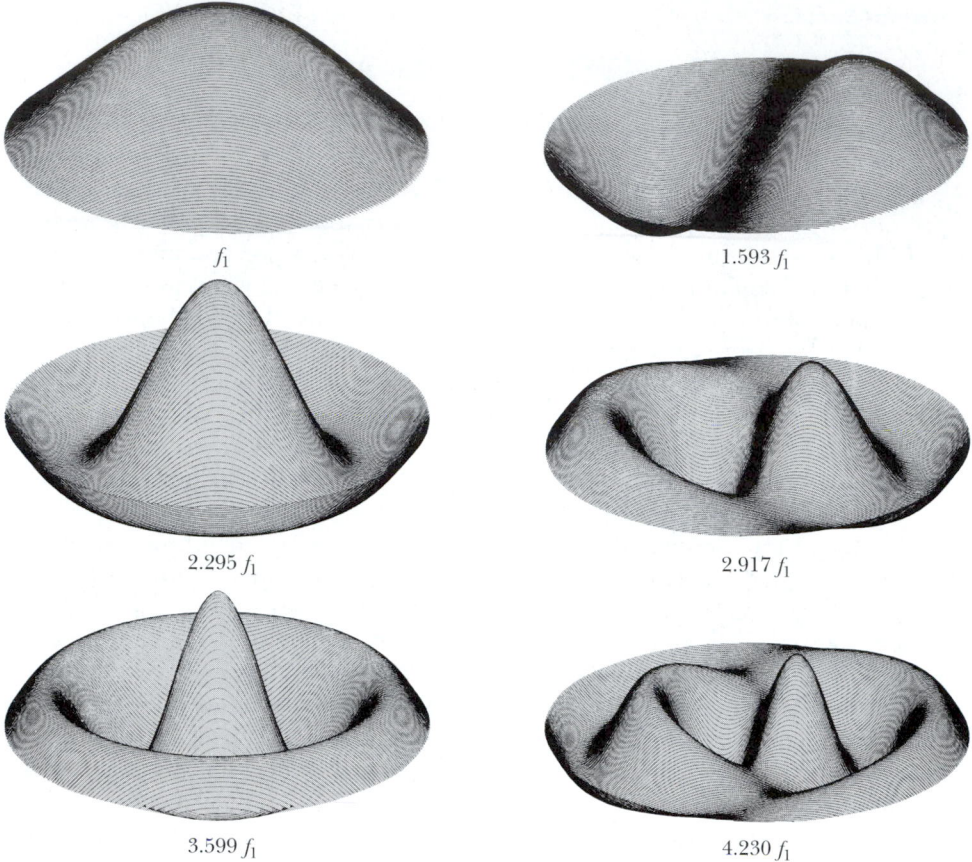

f_1

$1.593\,f_1$

$2.295\,f_1$

$2.917\,f_1$

$3.599\,f_1$

$4.230\,f_1$

Figure 18.17   Representation of some of the normal modes possible in a circular membrane fixed at its perimeter. The frequencies of oscillation do not form a harmonic series. *(From M. L. Warren,* Introductory Physics, *New York, W. H. Freeman & Co., Publishers, 1979, with permission.)*

18.7 ▶ BEATS: INTERFERENCE IN TIME

The interference phenomena with which we have been dealing so far involve the superposition of two or more waves having the same frequency. Because the resultant wave depends on the coordinates of the disturbed medium, we refer to the phenomenon as *spatial interference*. Standing waves in strings and pipes are common examples of spatial interference.

We now consider another type of interference, one that results from the superposition of two waves having slightly *different* frequencies. In this case, when the two waves are observed at the point of superposition, they are periodically in and out of phase. That is, there is a *temporal* (time) alternation between constructive and destructive interference. Thus, we refer to this phenomenon as *interference in time* or *temporal interference*. For example, if two tuning forks of slightly different frequencies are struck, one hears a sound of periodically varying intensity. This phenomenon is called **beating:**

Definition of beating

> Beating is the periodic variation in intensity at a given point due to the superposition of two waves having slightly different frequencies.

The number of intensity maxima one hears per second, or the *beat frequency*, equals the difference in frequency between the two sources, as we shall show below. The maximum beat frequency that the human ear can detect is about 20 beats/s. When the beat frequency exceeds this value, the beats blend indistinguishably with the compound sounds producing them.

A piano tuner can use beats to tune a stringed instrument by "beating" a note against a reference tone of known frequency. The tuner can then adjust the string tension until the frequency of the sound it emits equals the frequency of the reference tone. The tuner does this by tightening or loosening the string until the beats produced by it and the reference source become too infrequent to notice.

Consider two sound waves of equal amplitude traveling through a medium with slightly different frequencies f_1 and f_2. We use equations similar to Equation 16.11 to represent the wave functions for these two waves at a point that we choose as $x = 0$:

$$y_1 = A \cos \omega_1 t = A \cos 2\pi f_1 t$$

$$y_2 = A \cos \omega_2 t = A \cos 2\pi f_2 t$$

Using the superposition principle, we find that the resultant wave function at this point is

$$y = y_1 + y_2 = A(\cos 2\pi f_1 t + \cos 2\pi f_2 t)$$

The trigonometric identity

$$\cos a + \cos b = 2 \cos\left(\frac{a - b}{2}\right) \cos\left(\frac{a + b}{2}\right)$$

allows us to write this expression in the form

$$y = \left[2 A \cos 2\pi \left(\frac{f_1 - f_2}{2} \right) t \right] \cos 2\pi \left(\frac{f_1 + f_2}{2} \right) t \qquad \textbf{(18.13)}$$

Resultant of two waves of different frequencies but equal amplitude

Graphs of the individual waves and the resultant wave are shown in Figure 18.18. From the factors in Equation 18.13, we see that the resultant sound for a listener standing at any given point has an effective frequency equal to the average frequency $(f_1 + f_2)/2$ and an amplitude given by the expression in the square

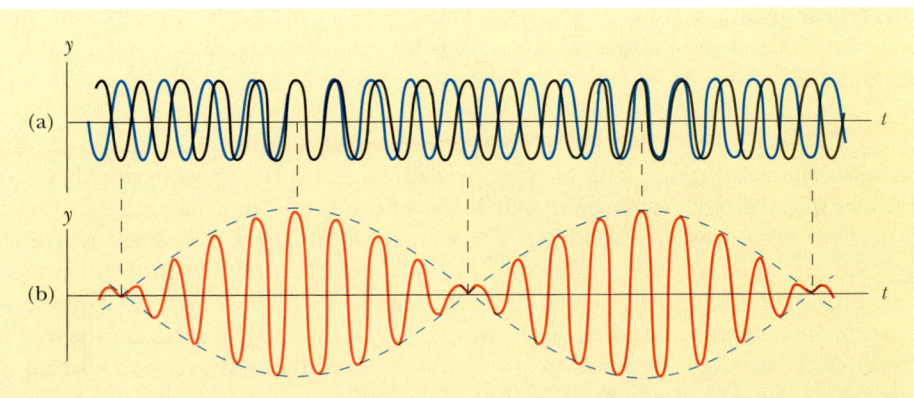

Figure 18.18 Beats are formed by the combination of two waves of slightly different frequencies. (a) The individual waves. (b) The combined wave has an amplitude (broken line) that oscillates in time.

brackets:

$$A_{\text{resultant}} = 2A \cos 2\pi \left(\frac{f_1 - f_2}{2} \right) t \qquad (18.14)$$

That is, the **amplitude and therefore the intensity of the resultant sound vary in time.** The broken blue line in Figure 18.18b is a graphical representation of Equation 18.14 and is a sine wave varying with frequency $(f_1 - f_2)/2$.

Note that a maximum in the amplitude of the resultant sound wave is detected whenever

$$\cos 2\pi \left(\frac{f_1 - f_2}{2} \right) t = \pm 1$$

This means there are *two* maxima in each period of the resultant wave. Because the amplitude varies with frequency as $(f_1 - f_2)/2$, the number of beats per second, or the beat frequency f_b, is twice this value. That is,

Beat frequency

$$\boxed{f_b = |f_1 - f_2|} \qquad (18.15)$$

For instance, if one tuning fork vibrates at 438 Hz and a second one vibrates at 442 Hz, the resultant sound wave of the combination has a frequency of 440 Hz (the musical note A) and a beat frequency of 4 Hz. A listener would hear a 440-Hz sound wave go through an intensity maximum four times every second.

Optional Section

18.8 ▶ NON-SINUSOIDAL WAVE PATTERNS

The sound-wave patterns produced by the majority of musical instruments are non-sinusoidal. Characteristic patterns produced by a tuning fork, a flute, and a clarinet, each playing the same note, are shown in Figure 18.19. Each instrument has its own characteristic pattern. Note, however, that despite the differences in the patterns, each pattern is periodic. This point is important for our analysis of these waves, which we now discuss.

We can distinguish the sounds coming from a trumpet and a saxophone even when they are both playing the same note. On the other hand, we may have difficulty distinguishing a note played on a clarinet from the same note played on an oboe. We can use the pattern of the sound waves from various sources to explain these effects.

The wave patterns produced by a musical instrument are the result of the superposition of various harmonics. This superposition results in the corresponding richness of musical tones. The human perceptive response associated with various mixtures of harmonics is the *quality* or *timbre* of the sound. For instance, the sound of the trumpet is perceived to have a "brassy" quality (that is, we have learned to associate the adjective *brassy* with that sound); this quality enables us to distinguish the sound of the trumpet from that of the saxophone, whose quality is perceived as "reedy." The clarinet and oboe, however, are both straight air columns excited by reeds; because of this similarity, it is more difficult for the ear to distinguish them on the basis of their sound quality.

The problem of analyzing non-sinusoidal wave patterns appears at first sight to be a formidable task. However, if the wave pattern is periodic, it can be represented as closely as desired by the combination of a sufficiently large number of si-

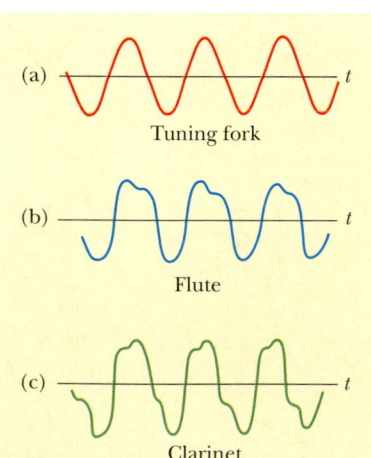

Figure 18.19 Sound wave patterns produced by (a) a tuning fork, (b) a flute, and (c) a clarinet, each at approximately the same frequency. *(Adapted from C. A. Culver, Musical Acoustics, 4th ed., New York, McGraw-Hill Book Company, 1956, p. 128.)*

(a) Tuning fork

(b) Flute

(c) Clarinet

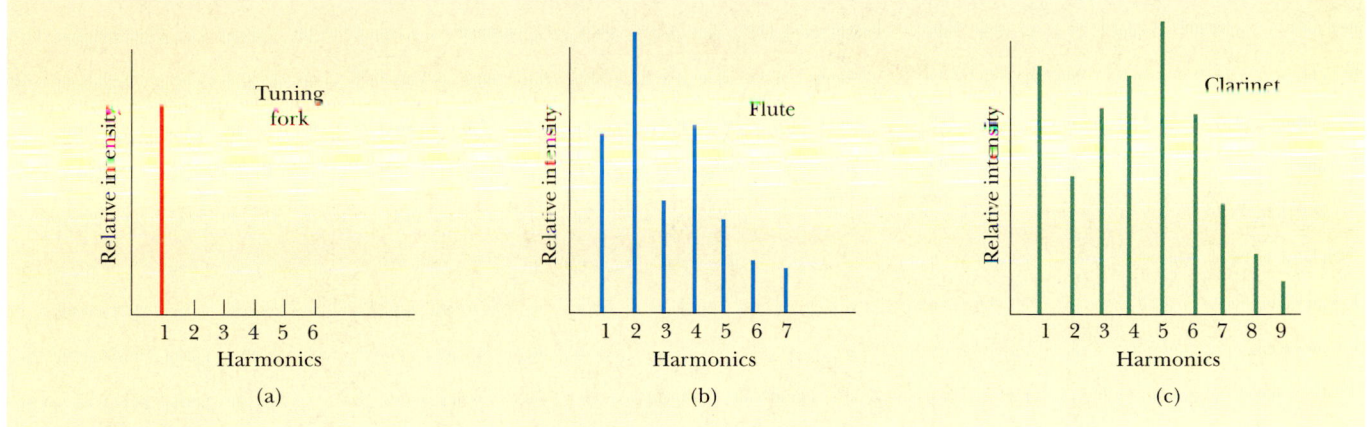

Figure 18.20 Harmonics of the wave patterns shown in Figure 18.19. Note the variations in intensity of the various harmonics. *(Adapted from C. A. Culver,* Musical Acoustics, 4th ed., *New York, McGraw-Hill Book Company, 1956.)*

nusoidal waves that form a harmonic series. In fact, we can represent any periodic function as a series of sine and cosine terms by using a mathematical technique based on **Fourier's theorem.**[3] The corresponding sum of terms that represents the periodic wave pattern is called a **Fourier series.**

Let $y(t)$ be any function that is periodic in time with period T, such that $y(t + T) = y(t)$. Fourier's theorem states that this function can be written as

$$y(t) = \sum_n (A_n \sin 2\pi f_n t + B_n \cos 2\pi f_n t) \tag{18.16}$$

Fourier's theorem

where the lowest frequency is $f_1 = 1/T$. The higher frequencies are integer multiples of the fundamental, $f_n = nf_1$, and the coefficients A_n and B_n represent the amplitudes of the various waves. Figure 18.20 represents a harmonic analysis of the wave patterns shown in Figure 18.19. Note that a struck tuning fork produces only one harmonic (the first), whereas the flute and clarinet produce the first and many higher ones.

Note the variation in relative intensity of the various harmonics for the flute and the clarinet. In general, any musical sound consists of a fundamental frequency f plus other frequencies that are integer multiples of f, all having different intensities.

We have discussed the *analysis* of a wave pattern using Fourier's theorem. The analysis involves determining the coefficients of the harmonics in Equation 18.16 from a knowledge of the wave pattern. The reverse process, called *Fourier synthesis,* can also be performed. In this process, the various harmonics are added together to form a resultant wave pattern. As an example of Fourier synthesis, consider the building of a square wave, as shown in Figure 18.21. The symmetry of the square wave results in only odd multiples of the fundamental frequency combining in its synthesis. In Figure 18.21a, the orange curve shows the combination of f and $3f$. In Figure 18.21b, we have added $5f$ to the combination and obtained the green curve. Notice how the general shape of the square wave is approximated, even though the upper and lower portions are not flat as they should be.

---

[3] Developed by Jean Baptiste Joseph Fourier (1786–1830).

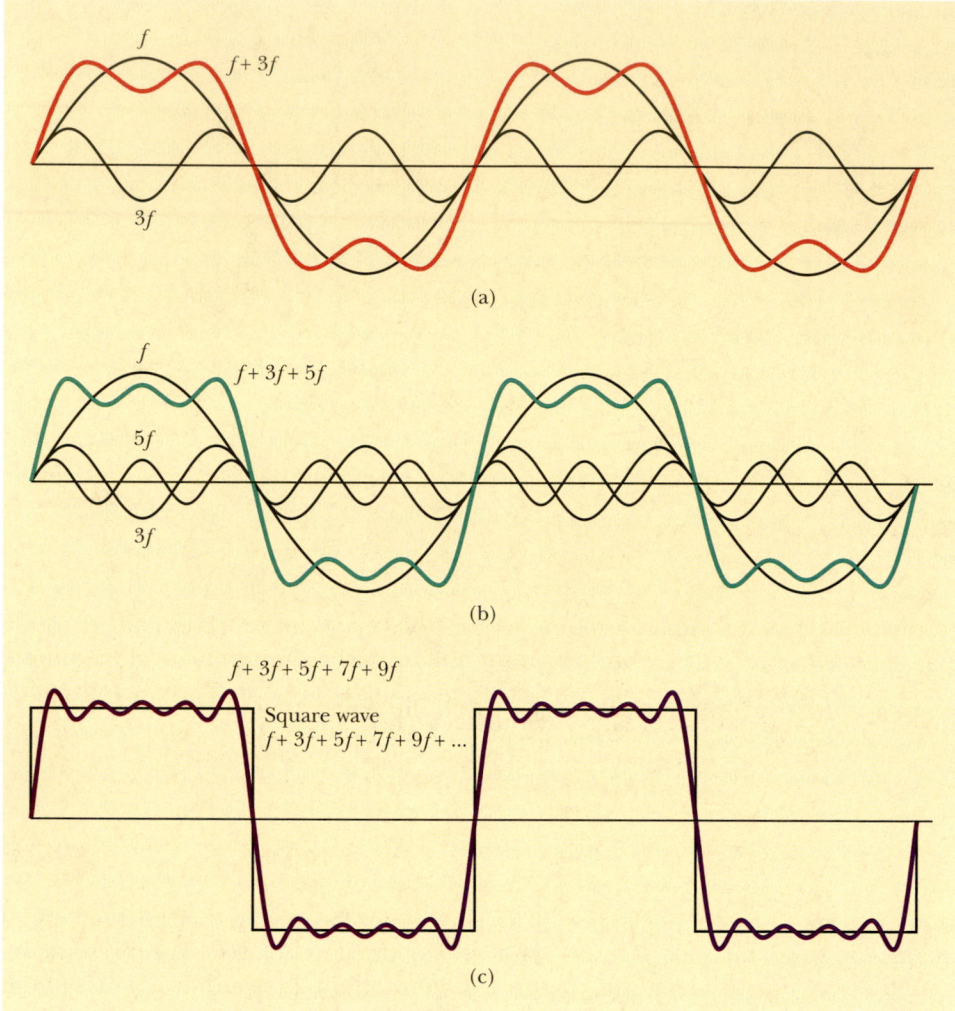

Figure 18.21 Fourier synthesis of a square wave, which is represented by the sum of odd multiples of the first harmonic, which has frequency f. (a) Waves of frequency f and $3f$ are added. (b) One more odd harmonic of frequency $5f$ is added. (c) The synthesis curve approaches the square wave when odd frequencies up to $9f$ are added.

This synthesizer can produce the characteristic sounds of different instruments by properly combining frequencies from electronic oscillators. *(Courtesy of Casio, Inc.)*

Figure 18.21c shows the result of adding odd frequencies up to $9f$. This approximation to the square wave (purple curve) is better than the approximations in parts a and b. To approximate the square wave as closely as possible, we would need to add all odd multiples of the fundamental frequency, up to infinite frequency.

Using modern technology, we can generate musical sounds electronically by mixing different amplitudes of any number of harmonics. These widely used electronic music synthesizers are capable of producing an infinite variety of musical tones.

SUMMARY

When two traveling waves having equal amplitudes and frequencies superimpose, the resultant wave has an amplitude that depends on the phase angle ϕ between the two waves. **Constructive interference** occurs when the two waves are in phase, corresponding to $\phi = 0$, 2π, 4π, . . . rad. **Destructive interference** occurs when the two waves are 180° out of phase, corresponding to $\phi - \pi$, 3π, 5π, . . . rad. Given two wave functions, you should be able to determine which if either of these two situations applies.

Standing waves are formed from the superposition of two sinusoidal waves having the same frequency, amplitude, and wavelength but traveling in opposite directions. The resultant standing wave is described by the wave function

$$y = (2A \sin kx) \cos \omega t \tag{18.3}$$

Hence, the amplitude of the standing wave is $2A$, and the amplitude of the simple harmonic motion of any particle of the medium varies according to its position as $2A \sin kx$. The points of zero amplitude (called **nodes**) occur at $x = n\lambda/2$ ($n = 0$, 1, 2, 3, . . .). The maximum amplitude points (called **antinodes**) occur at $x = n\lambda/4$ ($n = 1, 3, 5, . . .$). Adjacent antinodes are separated by a distance $\lambda/2$. Adjacent nodes also are separated by a distance $\lambda/2$. You should be able to sketch the standing-wave pattern resulting from the superposition of two traveling waves.

The natural frequencies of vibration of a taut string of length L and fixed at both ends are

$$f_n = \frac{n}{2L} \sqrt{\frac{T}{\mu}} \qquad n = 1, 2, 3, . . . \tag{18.8}$$

where T is the tension in the string and μ is its linear mass density. The natural frequencies of vibration f_1, $2f_1$, $3f_1$, . . . form a **harmonic series.**

An oscillating system is in **resonance** with some driving force whenever the frequency of the driving force matches one of the natural frequencies of the system. When the system is resonating, it responds by oscillating with a relatively large amplitude.

Standing waves can be produced in a column of air inside a pipe. If the pipe is open at both ends, all harmonics are present and the natural frequencies of oscillation are

$$f_n = n\frac{v}{2L} \qquad n = 1, 2, 3, . . . \tag{18.11}$$

If the pipe is open at one end and closed at the other, only the odd harmonics are present, and the natural frequencies of oscillation are

$$f_n = n\frac{v}{4L} \qquad n = 1, 3, 5, . . . \tag{18.12}$$

The phenomenon of **beating** is the periodic variation in intensity at a given point due to the superposition of two waves having slightly different frequencies.

QUESTIONS

1. For certain positions of the movable section shown in Figure 18.2, no sound is detected at the receiver—a situation corresponding to destructive interference. This suggests that perhaps energy is somehow lost! What happens to the energy transmitted by the speaker?

2. Does the phenomenon of wave interference apply only to sinusoidal waves?

3. When two waves interfere constructively or destructively, is there any gain or loss in energy? Explain.

4. A standing wave is set up on a string, as shown in Figure 18.6. Explain why no energy is transmitted along the string.

5. What is common to *all* points (other than the nodes) on a string supporting a standing wave?

6. What limits the amplitude of motion of a real vibrating system that is driven at one of its resonant frequencies?

7. In Balboa Park in San Diego, CA, there is a huge outdoor organ. Does the fundamental frequency of a particular

pipe of this organ change on hot and cold days? How about on days with high and low atmospheric pressure?

8. Explain why your voice seems to sound better than usual when you sing in the shower.

9. What is the purpose of the slide on a trombone or of the valves on a trumpet?

10. Explain why all harmonics are present in an organ pipe open at both ends, but only the odd harmonics are present in a pipe closed at one end.

11. Explain how a musical instrument such as a piano may be tuned by using the phenomenon of beats.

12. An airplane mechanic notices that the sound from a twin-engine aircraft rapidly varies in loudness when both engines are running. What could be causing this variation from loudness to softness?

13. Why does a vibrating guitar string sound louder when placed on the instrument than it would if it were allowed to vibrate in the air while off the instrument?

14. When the base of a vibrating tuning fork is placed against a chalkboard, the sound that it emits becomes louder. This is due to the fact that the vibrations of the tuning fork are transmitted to the chalkboard. Because it has a larger area than that of the tuning fork, the vibrating

chalkboard sets a larger number of air molecules into vibration. Thus, the chalkboard is a better radiator of sound than the tuning fork. How does this affect the length of time during which the fork vibrates? Does this agree with the principle of conservation of energy?

15. To keep animals away from their cars, some people mount short thin pipes on the front bumpers. The pipes produce a high-frequency wail when the cars are moving. How do they create this sound?

16. Guitarists sometimes play a "harmonic" by lightly touching a string at the exact center and plucking the string. The result is a clear note one octave higher than the fundamental frequency of the string, even though the string is not pressed to the fingerboard. Why does this happen?

17. If you wet your fingers and lightly run them around the rim of a fine wine glass, a high-frequency sound is heard. Why? How could you produce various musical notes with a set of wine glasses, each of which contains a different amount of water?

18. Despite a reasonably steady hand, one often spills coffee when carrying a cup of it from one place to another. Discuss resonance as a possible cause of this difficulty, and devise a means for solving the problem.

PROBLEMS

1, 2, 3 = straightforward, intermediate, challenging   ☐ = full solution available in the *Student Solutions Manual and Study Guide*
WEB = solution posted at **http://www.saunderscollege.com/physics/**   🖥 = Computer useful in solving problem   📟 = Interactive Physics
☐ = paired numerical/symbolic problems

Section 18.1 Superposition and Interference of Sinusoidal Waves

WEB **1.** Two sinusoidal waves are described by the equations

$$y_1 = (5.00 \text{ m}) \sin[\pi(4.00x - 1\,200t)]$$

and

$$y_2 = (5.00 \text{ m}) \sin[\pi(4.00x - 1\,200t - 0.250)]$$

where x, y_1, and y_2 are in meters and t is in seconds. (a) What is the amplitude of the resultant wave? (b) What is the frequency of the resultant wave?

2. A sinusoidal wave is described by the equation

$$y_1 = (0.080\,0 \text{ m}) \sin[2\pi(0.100x - 80.0t)]$$

where y_1 and x are in meters and t is in seconds. Write an expression for a wave that has the same frequency, amplitude, and wavelength as y_1 but which, when added to y_1, gives a resultant with an amplitude of $8\sqrt{3}$ cm.

3. Two waves are traveling in the same direction along a stretched string. The waves are 90.0° out of phase. Each wave has an amplitude of 4.00 cm. Find the amplitude of the resultant wave.

4. Two identical sinusoidal waves with wavelengths of 3.00 m travel in the same direction at a speed of 2.00 m/s. The second wave originates from the same

point as the first, but at a later time. Determine the minimum possible time interval between the starting moments of the two waves if the amplitude of the resultant wave is the same as that of each of the two initial waves.

5. A tuning fork generates sound waves with a frequency of 246 Hz. The waves travel in opposite directions along a hallway, are reflected by walls, and return. The hallway is 47.0 m in length, and the tuning fork is located 14.0 m from one end. What is the phase difference between the reflected waves when they meet? The speed of sound in air is 343 m/s.

6. Two identical speakers 10.0 m apart are driven by the same oscillator with a frequency of $f = 21.5$ Hz (Fig. P18.6). (a) Explain why a receiver at point A records a minimum in sound intensity from the two speakers. (b) If the receiver is moved in the plane of the speakers, what path should it take so that the intensity remains at a minimum? That is, determine the relationship between x and y (the coordinates of the receiver) that causes the receiver to record a minimum in sound intensity. Take the speed of sound to be 343 m/s.

7. Two speakers are driven by the same oscillator with frequency of 200 Hz. They are located 4.00 m apart on a

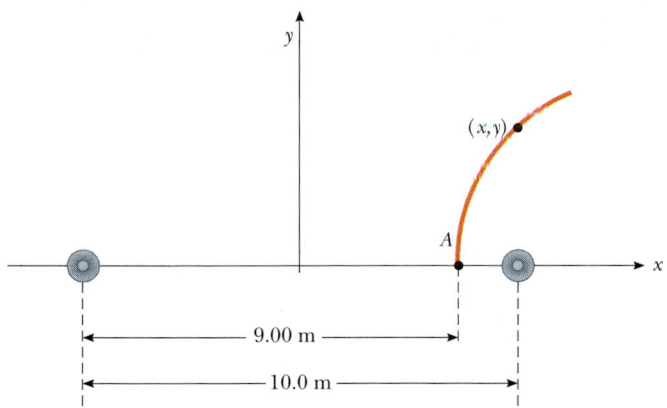

Figure P18.6

vertical pole. A man walks straight toward the lower speaker in a direction perpendicular to the pole, as shown in Figure P18.7. (a) How many times will he hear a minimum in sound intensity, and (b) how far is he from the pole at these moments? Take the speed of sound to be 330 m/s, and ignore any sound reflections coming off the ground.

8. Two speakers are driven by the same oscillator of frequency f. They are located a distance d from each other on a vertical pole. A man walks straight toward the lower speaker in a direction perpendicular to the pole, as shown in Figure P18.7. (a) How many times will he hear a minimum in sound intensity, and (b) how far is he from the pole at these moments? Take the speed of sound to be v, and ignore any sound reflections coming off the ground.

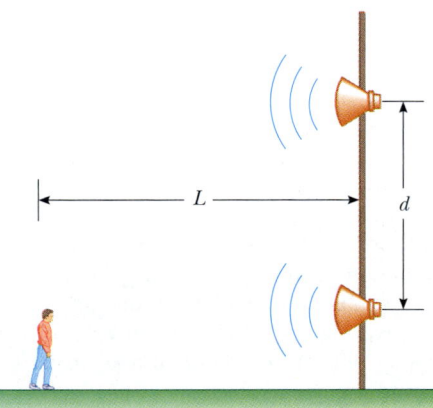

Figure P18.7 Problems 7 and 8.

Section 18.2 Standing Waves

9. Two sinusoidal waves traveling in opposite directions interfere to produce a standing wave described by the equation

$$y = (1.50 \text{ m}) \sin(0.400x) \cos(200t)$$

where x is in meters and t is in seconds. Determine the wavelength, frequency, and speed of the interfering waves.

10. Two waves in a long string are described by the equations

$$y_1 = (0.015 \ 0 \text{ m}) \cos\left(\frac{x}{2} - 40t\right)$$

and

$$y_2 = (0.015 \ 0 \text{ m}) \cos\left(\frac{x}{2} + 40t\right)$$

where y_1, y_2, and x are in meters and t is in seconds. (a) Determine the positions of the nodes of the resulting standing wave. (b) What is the maximum displacement at the position $x = 0.400$ m?

WEB 11. Two speakers are driven by a common oscillator at 800 Hz and face each other at a distance of 1.25 m. Locate the points along a line joining the two speakers where relative minima would be expected. (Use $v = 343$ m/s.)

12. Two waves that set up a standing wave in a long string are given by the expressions

$$y_1 = A \sin(kx - \omega t + \phi)$$

and

$$y_2 = A \sin(kx + \omega t)$$

Show (a) that the addition of the arbitrary phase angle changes only the position of the nodes, and (b) that the distance between the nodes remains constant in time.

13. Two sinusoidal waves combining in a medium are described by the equations

$$y_1 = (3.0 \text{ cm}) \sin \pi(x + 0.60t)$$

and

$$y_2 = (3.0 \text{ cm}) \sin \pi(x - 0.60t)$$

where x is in centimeters and t is in seconds. Determine the *maximum* displacement of the medium at (a) $x = 0.250$ cm, (b) $x = 0.500$ cm, and (c) $x = 1.50$ cm. (d) Find the three smallest values of x corresponding to antinodes.

14. A standing wave is formed by the interference of two traveling waves, each of which has an amplitude $A = \pi$ cm, angular wave number $k = (\pi/2)$ cm^{-1}, and angular frequency $\omega = 10\pi$ rad/s. (a) Calculate the distance between the first two antinodes. (b) What is the amplitude of the standing wave at $x = 0.250$ cm?

15. Verify by direct substitution that the wave function for a standing wave given in Equation 18.3, $y = 2A \sin kx \cos \omega t$, is a solution of the general linear

wave equation, Equation 16.26:

$$\frac{\partial^2 y}{\partial x^2} = \frac{1}{v^2}\frac{\partial^2 y}{\partial t^2}$$

Section 18.3   Standing Waves in a String Fixed at Both Ends

16. A 2.00-m-long wire having a mass of 0.100 kg is fixed at both ends. The tension in the wire is maintained at 20.0 N. What are the frequencies of the first three allowed modes of vibration? If a node is observed at a point 0.400 m from one end, in what mode and with what frequency is it vibrating?

17. Find the fundamental frequency and the next three frequencies that could cause a standing-wave pattern on a string that is 30.0 m long, has a mass per length of 9.00×10^{-3} kg/m, and is stretched to a tension of 20.0 N.

18. A standing wave is established in a 120-cm-long string fixed at both ends. The string vibrates in four segments when driven at 120 Hz. (a) Determine the wavelength. (b) What is the fundamental frequency of the string?

19. A cello A-string vibrates in its first normal mode with a frequency of 220 vibrations/s. The vibrating segment is 70.0 cm long and has a mass of 1.20 g. (a) Find the tension in the string. (b) Determine the frequency of vibration when the string vibrates in three segments.

20. A string of length L, mass per unit length μ, and tension T is vibrating at its fundamental frequency. Describe the effect that each of the following conditions has on the fundamental frequency: (a) The length of the string is doubled, but all other factors are held constant. (b) The mass per unit length is doubled, but all other factors are held constant. (c) The tension is doubled, but all other factors are held constant.

21. A 60.0-cm guitar string under a tension of 50.0 N has a mass per unit length of 0.100 g/cm. What is the highest resonance frequency of the string that can be heard by a person able to hear frequencies of up to 20 000 Hz?

22. A stretched wire vibrates in its first normal mode at a frequency of 400 Hz. What would be the fundamental frequency if the wire were half as long, its diameter were doubled, and its tension were increased four-fold?

23. A violin string has a length of 0.350 m and is tuned to concert G, with $f_G = 392$ Hz. Where must the violinist place her finger to play concert A, with $f_A = 440$ Hz? If this position is to remain correct to one-half the width of a finger (that is, to within 0.600 cm), what is the maximum allowable percentage change in the string's tension?

24. **Review Problem.**   A sphere of mass M is supported by a string that passes over a light horizontal rod of length L (Fig. P18.24). Given that the angle is θ and that the fundamental frequency of standing waves in the section of the string above the horizontal rod is f, determine the mass of this section of the string.

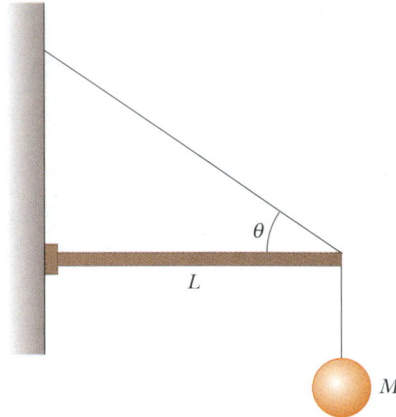

Figure P18.24

25. In the arrangement shown in Figure P18.25, a mass can be hung from a string (with a linear mass density of $\mu = 0.002\ 00$ kg/m) that passes over a light pulley. The string is connected to a vibrator (of constant frequency f), and the length of the string between point P and the pulley is $L = 2.00$ m. When the mass m is either 16.0 kg or 25.0 kg, standing waves are observed; however, no standing waves are observed with any mass between these values. (a) What is the frequency of the vibrator? (*Hint:* The greater the tension in the string, the smaller the number of nodes in the standing wave.) (b) What is the largest mass for which standing waves could be observed?

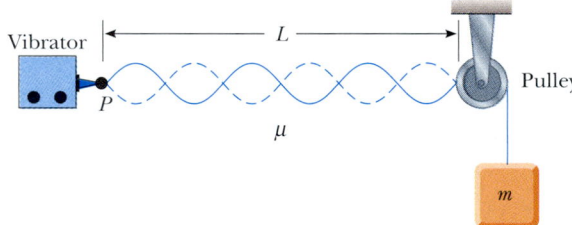

Figure P18.25

26. On a guitar, the fret closest to the bridge is a distance of 21.4 cm from it. The top string, pressed down at this last fret, produces the highest frequency that can be played on the guitar, 2 349 Hz. The next lower note has a frequency of 2 217 Hz. How far away from the last fret should the next fret be?

Section 18.4   Resonance

27. The chains suspending a child's swing are 2.00 m long. At what frequency should a big brother push to make the child swing with greatest amplitude?

28. Standing-wave vibrations are set up in a crystal goblet with four nodes and four antinodes equally spaced

around the 20.0-cm circumference of its rim. If transverse waves move around the glass at 900 m/s, an opera singer would have to produce a high harmonic with what frequency to shatter the glass with a resonant vibration?

29. An earthquake can produce a *seiche* (pronounced "saysh") in a lake, in which the water sloshes back and forth from end to end with a remarkably large amplitude and long period. Consider a seiche produced in a rectangular farm pond, as diagrammed in the cross-sectional view of Figure P18.29 (figure not drawn to scale). Suppose that the pond is 9.15 m long and of uniform depth. You measure that a wave pulse produced at one end reaches the other end in 2.50 s. (a) What is the wave speed? (b) To produce the seiche, you suggest that several people stand on the bank at one end and paddle together with snow shovels, moving them in simple harmonic motion. What must be the frequency of this motion?

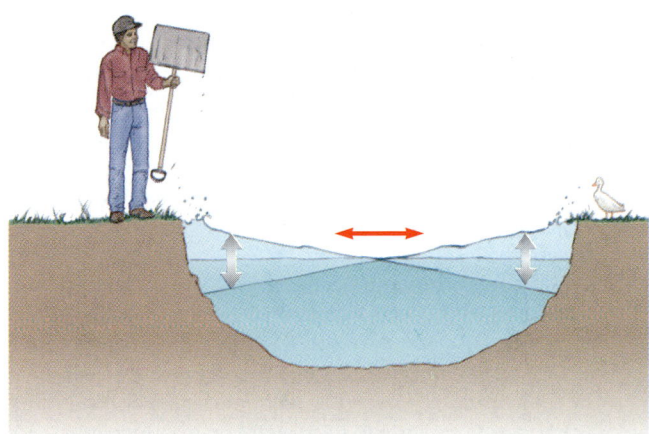

Figure P18.29

30. The Bay of Fundy, Nova Scotia, has the highest tides in the world. Assume that in mid-ocean and at the mouth of the bay, the Moon's gravity gradient and the Earth's rotation make the water surface oscillate with an amplitude of a few centimeters and a period of 12 h 24 min. At the head of the bay, the amplitude is several meters. Argue for or against the proposition that the tide is amplified by standing-wave resonance. Suppose that the bay has a length of 210 km and a depth everywhere of 36.1 m. The speed of long-wavelength water waves is given by $\sqrt{gd}$, where d is the water's depth.

Section 18.5   Standing Waves in Air Columns

Note: In this section, assume that the speed of sound in air is 343 m/s at 20°C and is described by the equation

$$v = (331 \text{ m/s})\sqrt{1 + \frac{T_C}{273°}}$$

at any Celsius temperature T_C.

31. Calculate the length of a pipe that has a fundamental frequency of 240 Hz if the pipe is (a) closed at one end and (b) open at both ends.

32. A glass tube (open at both ends) of length L is positioned near an audio speaker of frequency $f = 0.680$ kHz. For what values of L will the tube resonate with the speaker?

33. The overall length of a piccolo is 32.0 cm. The resonating air column vibrates as a pipe open at both ends. (a) Find the frequency of the lowest note that a piccolo can play, assuming that the speed of sound in air is 340 m/s. (b) Opening holes in the side effectively shortens the length of the resonant column. If the highest note that a piccolo can sound is 4 000 Hz, find the distance between adjacent antinodes for this mode of vibration.

34. The fundamental frequency of an open organ pipe corresponds to middle C (261.6 Hz on the chromatic musical scale). The third resonance of a closed organ pipe has the same frequency. What are the lengths of the two pipes?

35. Estimate the length of your ear canal, from its opening at the external ear to the eardrum. (Do not stick anything into your ear!) If you regard the canal as a tube that is open at one end and closed at the other, at approximately what fundamental frequency would you expect your hearing to be most sensitive? Explain why you can hear especially soft sounds just around this frequency.

36. An open pipe 0.400 m in length is placed vertically in a cylindrical bucket and nearly touches the bottom of the bucket, which has an area of 0.100 m². Water is slowly poured into the bucket until a sounding tuning fork of frequency 440 Hz, held over the pipe, produces resonance. Find the mass of water in the bucket at this moment.

WEB 37. A shower stall measures 86.0 cm × 86.0 cm × 210 cm. If you were singing in this shower, which frequencies would sound the richest (because of resonance)? Assume that the stall acts as a pipe closed at both ends, with nodes at opposite sides. Assume that the voices of various singers range from 130 Hz to 2 000 Hz. Let the speed of sound in the hot shower stall be 355 m/s.

38. When a metal pipe is cut into two pieces, the lowest resonance frequency in one piece is 256 Hz and that for the other is 440 Hz. (a) What resonant frequency would have been produced by the original length of pipe? (b) How long was the original pipe?

39. As shown in Figure P18.39, water is pumped into a long vertical cylinder at a rate of 18.0 cm³/s. The radius of the cylinder is 4.00 cm, and at the open top of the cylinder is a tuning fork vibrating with a frequency of 200 Hz. As the water rises, how much time elapses between successive resonances?

40. As shown in Figure P18.39, water is pumped into a long vertical cylinder at a volume flow rate R. The radius of

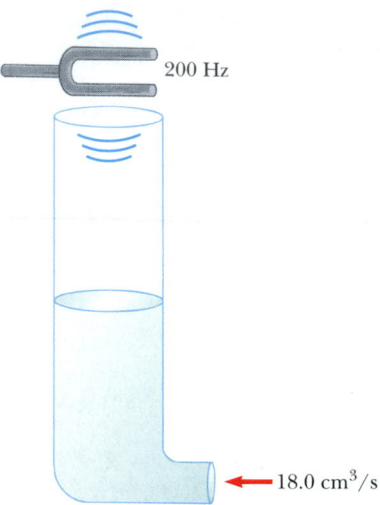

200 Hz

← 18.0 cm³/s

Figure P18.39   Problems 39 and 40.

the cylinder is r, and at the open top of the cylinder is a tuning fork vibrating with a frequency f. As the water rises, how much time elapses between successive resonances?

41. A tuning fork with a frequency of 512 Hz is placed near the top of the tube shown in Figure 18.15a. The water level is lowered so that the length L slowly increases from an initial value of 20.0 cm. Determine the next two values of L that correspond to resonant modes.

42. A student uses an audio oscillator of adjustable frequency to measure the depth of a water well. Two successive resonances are heard at 51.5 Hz and 60.0 Hz. How deep is the well?

43. A glass tube is open at one end and closed at the other by a movable piston. The tube is filled with air warmer than that at room temperature, and a 384-Hz tuning fork is held at the open end. Resonance is heard when the piston is 22.8 cm from the open end and again when it is 68.3 cm from the open end. (a) What speed of sound is implied by these data? (b) How far from the open end will the piston be when the next resonance is heard?

44. The longest pipe on an organ that has pedal stops is often 4.88 m. What is the fundamental frequency (at 0.00°C) if the nondriven end of the pipe is (a) closed and (b) open? (c) What are the frequencies at 20.0°C?

45. With a particular fingering, a flute sounds a note with a frequency of 880 Hz at 20.0°C. The flute is open at both ends. (a) Find the length of the air column. (b) Find the frequency it produces during the half-time performance at a late-season football game, when the ambient temperature is −5.00°C.

(Optional)
Section 18.6   Standing Waves in Rods and Plates

46. An aluminum rod is clamped one quarter of the way along its length and set into longitudinal vibration by a variable-frequency driving source. The lowest frequency that produces resonance is 4 400 Hz. The speed of sound in aluminum is 5 100 m/s. Determine the length of the rod.

47. An aluminum rod 1.60 m in length is held at its center. It is stroked with a rosin-coated cloth to set up a longitudinal vibration. (a) What is the fundamental frequency of the waves established in the rod? (b) What harmonics are set up in the rod held in this manner? (c) What would be the fundamental frequency if the rod were made of copper?

48. A 60.0-cm metal bar that is clamped at one end is struck with a hammer. If the speed of longitudinal (compressional) waves in the bar is 4 500 m/s, what is the lowest frequency with which the struck bar resonates?

Section 18.7   Beats: Interference in Time

WEB 49. In certain ranges of a piano keyboard, more than one string is tuned to the same note to provide extra loudness. For example, the note at 110 Hz has two strings that vibrate at this frequency. If one string slips from its normal tension of 600 N to 540 N, what beat frequency is heard when the hammer strikes the two strings simultaneously?

50. While attempting to tune the note C at 523 Hz, a piano tuner hears 2 beats/s between a reference oscillator and the string. (a) What are the possible frequencies of the string? (b) When she tightens the string slightly, she hears 3 beats/s. What is the frequency of the string now? (c) By what percentage should the piano tuner now change the tension in the string to bring it into tune?

51. A student holds a tuning fork oscillating at 256 Hz. He walks toward a wall at a constant speed of 1.33 m/s. (a) What beat frequency does he observe between the tuning fork and its echo? (b) How fast must he walk away from the wall to observe a beat frequency of 5.00 Hz?

(Optional)
Section 18.8   Non-Sinusoidal Wave Patterns

52. Suppose that a flutist plays a 523-Hz C note with first harmonic displacement amplitude $A_1 = 100$ nm. From Figure 18.20b, read, by proportion, the displacement amplitudes of harmonics 2 through 7. Take these as the values A_2 through A_7 in the Fourier analysis of the sound, and assume that $B_1 = B_2 = \ldots = B_7 = 0$. Construct a graph of the waveform of the sound. Your waveform will not look exactly like the flute waveform in Figure 18.19b because you simplify by ignoring cosine terms; nevertheless, it produces the same sensation to human hearing.

53. An A-major chord consists of the notes called A, C#, and E. It can be played on a piano by simultaneously striking strings that have fundamental frequencies of 440.00 Hz, 554.37 Hz, and 659.26 Hz. The rich consonance of the chord is associated with the near equality of the frequencies of some of the higher harmonics of the three tones. Consider the first five harmonics of each string and determine which harmonics show near equality.

ADDITIONAL PROBLEMS

54. Review Problem.   For the arrangement shown in Figure P18.54, $\theta = 30.0°$, the inclined plane and the small pulley are frictionless, the string supports the mass M at the bottom of the plane, and the string has a mass m that is small compared with M. The system is in equilibrium, and the vertical part of the string has a length h. Standing waves are set up in the vertical section of the string. Find (a) the tension in the string, (b) the whole length of the string (ignoring the radius of curvature of the pulley), (c) the mass per unit length of the string, (d) the speed of waves on the string, (e) the lowest-frequency standing wave, (f) the period of the standing wave having three nodes, (g) the wavelength of the standing wave having three nodes, and (h) the frequency of the beats resulting from the interference of the sound wave of lowest frequency generated by the string with another sound wave having a frequency that is 2.00% greater.

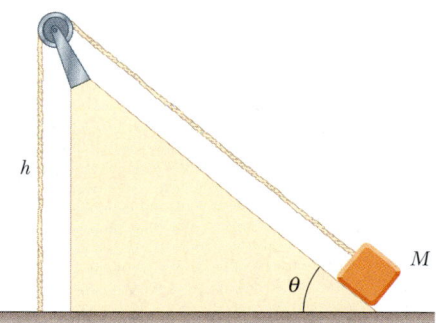

Figure P18.54

55. Two loudspeakers are placed on a wall 2.00 m apart. A listener stands 3.00 m from the wall directly in front of one of the speakers. The speakers are being driven by a single oscillator at a frequency of 300 Hz. (a) What is the phase difference between the two waves when they reach the observer? (b) What is the frequency closest to 300 Hz to which the oscillator may be adjusted such that the observer hears minimal sound?

56. On a marimba (Fig. P18.56), the wooden bar that sounds a tone when it is struck vibrates in a transverse standing wave having three antinodes and two nodes. The lowest-frequency note is 87.0 Hz; this note is produced by a bar 40.0 cm long. (a) Find the speed of transverse waves on the bar. (b) The loudness of the emitted sound is enhanced by a resonant pipe suspended vertically below the center of the bar. If the pipe is open at the top end only and the speed of sound in air is 340 m/s, what is the length of the pipe required to resonate with the bar in part (a)?

Figure P18.56   Marimba players in Mexico City. *(Murray Greenberg)*

57. Two train whistles have identical frequencies of 180 Hz. When one train is at rest in the station and is sounding its whistle, a beat frequency of 2.00 Hz is heard from a train moving nearby. What are the two possible speeds and directions that the moving train can have?

58. A speaker at the front of a room and an identical speaker at the rear of the room are being driven by the same oscillator at 456 Hz. A student walks at a uniform rate of 1.50 m/s along the length of the room. How many beats does the student hear per second?

59. While Jane waits on a railroad platform, she observes two trains approaching from the same direction at equal speeds of 8.00 m/s. Both trains are blowing their whistles (which have the same frequency), and one train is some distance behind the other. After the first train passes Jane, but before the second train passes her, she hears beats having a frequency of 4.00 Hz. What is the frequency of the trains' whistles?

60. A string fixed at both ends and having a mass of 4.80 g, a length of 2.00 m, and a tension of 48.0 N vibrates in its second ($n = 2$) natural mode. What is the wavelength in air of the sound emitted by this vibrating string?

61. A string 0.400 m in length has a mass per unit length of 9.00×10^{-3} kg/m. What must be the tension in the string if its second harmonic is to have the same frequency as the second resonance mode of a 1.75-m-long pipe open at one end?

62. In a major chord on the physical pitch musical scale, the frequencies are in the ratios $4:5:6:8$. A set of pipes, closed at one end, must be cut so that, when they are sounded in their first normal mode, they produce a major chord. (a) What is the ratio of the lengths of the pipes? (b) What are the lengths of the pipes needed if the lowest frequency of the chord is 256 Hz? (c) What are the frequencies of this chord?

63. Two wires are welded together. The wires are made of the same material, but the diameter of one wire is twice that of the other. They are subjected to a tension of 4.60 N. The thin wire has a length of 40.0 cm and a linear mass density of 2.00 g/m. The combination is fixed at both ends and vibrated in such a way that two antinodes are present, with the node between them being right at the weld. (a) What is the frequency of vibration? (b) How long is the thick wire?

64. Two identical strings, each fixed at both ends, are arranged near each other. If string A starts oscillating in its first normal mode, string B begins vibrating in its third ($n = 3$) natural mode. Determine the ratio of the tension of string B to the tension of string A.

65. A standing wave is set up in a string of variable length and tension by a vibrator of variable frequency. When the vibrator has a frequency f, in a string of length L and under a tension T, n antinodes are set up in the string. (a) If the length of the string is doubled, by what factor should the frequency be changed so that the same number of antinodes is produced? (b) If the frequency and length are held constant, what tension produces $n + 1$ antinodes? (c) If the frequency is tripled and the length of the string is halved, by what factor should the tension be changed so that twice as many antinodes are produced?

66. A 0.010 0-kg, 2.00-m-long wire is fixed at both ends and vibrates in its simplest mode under a tension of 200 N. When a tuning fork is placed near the wire, a beat frequency of 5.00 Hz is heard. (a) What could the frequency of the tuning fork be? (b) What should the tension in the wire be if the beats are to disappear?

WEB 67. If two adjacent natural frequencies of an organ pipe are determined to be 0.550 kHz and 0.650 kHz, calculate the fundamental frequency and length of the pipe. (Use $v = 340$ m/s.)

68. Two waves are described by the equations

$$y_1(x, t) = 5.0 \sin(2.0x - 10t)$$

and

$$y_2(x, t) = 10 \cos(2.0x - 10t)$$

where x is in meters and t is in seconds. Show that the resulting wave is sinusoidal, and determine the amplitude and phase of this sinusoidal wave.

69. The wave function for a standing wave is given in Equation 18.3 as $y = (2A \sin kx) \cos \omega t$. (a) Rewrite this wave function in terms of the wavelength λ and the wave speed v of the wave. (b) Write the wave function of the simplest standing-wave vibration of a stretched string of length L. (c) Write the wave function for the second harmonic. (d) Generalize these results, and write the wave function for the nth resonance vibration.

70. Review Problem. A 12.0-kg mass hangs in equilibrium from a string with a total length of $L = 5.00$ m and a linear mass density of $\mu = 0.001\ 00$ kg/m. The string is wrapped around two light, frictionless pulleys that are separated by a distance of $d = 2.00$ m (Fig. P18.70a). (a) Determine the tension in the string. (b) At what frequency must the string between the pulleys vibrate to form the standing-wave pattern shown in Figure P18.70b?

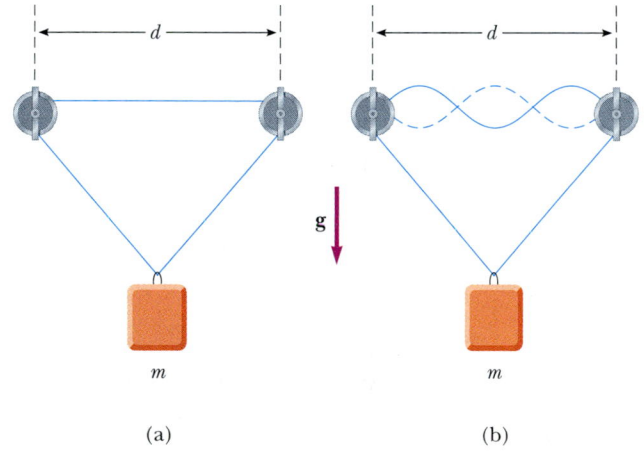

(a)   (b)

Figure P18.70

ANSWERS TO QUICK QUIZZES

18.1 At the antinodes. All particles have the same period $T = 2\pi/\omega$, but a particle at an antinode must travel through the greatest vertical distance in this amount of time and therefore must travel fastest.

18.2 For each natural frequency of the glass, the standing wave must "fit" exactly around the rim. In Figure 18.12a we see three antinodes on the near side of the glass, and thus there must be another three on the far side. This corresponds to three complete waves. In a top view, the wave pattern looks like this (although we have greatly exaggerated the amplitude):

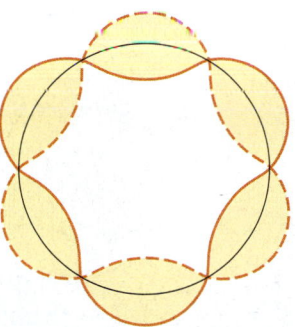

18.3 At highway speeds, a car crosses the ridges on the rumble strip at a rate that matches one of the car's natural frequencies of oscillation. This causes the car to oscillate substantially more than when it is traveling over the randomly spaced bumps of regular pavement. This sudden resonance oscillation alerts the driver that he or she must pay attention.

18.4 (b). With both ends open, the pipe has a fundamental frequency given by Equation 18.11: $f_{\text{open}} = v/2L$. With one end closed, the pipe has a fundamental frequency given by Equation 18.12:

$$f_{\text{closed}} = \frac{v}{4L} = \frac{1}{2}\frac{v}{2L} = \frac{1}{2}f_{\text{open}}$$

part 3

Thermodynamics

We now direct our attention to the study of thermodynamics, which deals with the concepts of heat and temperature. As we shall see, thermodynamics is very successful in explaining the bulk properties of matter and the correlation between these properties and the mechanics of both atoms and molecules.

Historically, the development of thermodynamics paralleled the development of the atomic theory of matter. By the 1820s, chemical experiments had provided solid evidence for the existence of atoms. At that time, scientists recognized that a connection between the theory of heat and temperature and the structure of matter must exist. In 1827, the botanist Robert Brown reported that grains of pollen suspended in a liquid move erratically from one place to another, as if under constant agitation. In 1905, Albert Einstein used kinetic theory to explain the cause of this erratic motion, which today is known as *Brownian motion.* Einstein explained this phenomenon by assuming that the grains are under constant bombardment by "invisible" molecules in the liquid, which themselves move erratically. This explanation gave scientists insight into molecular motion and gave credence to the idea that matter is made up of atoms. A connection was thus forged between the everyday world and the tiny, invisible building blocks that make up this world.

Thermodynamics also addresses more practical questions. Have you ever wondered how a refrigerator is able to cool its contents, what types of transformations occur in a power plant or in the engine of your automobile, or what happens to the kinetic energy of a moving object when the object comes to rest? The laws of thermodynamics and the concepts of heat and temperature provide explanations. In general, thermodynamics is concerned with transformations of matter in all of its states—solid, liquid, and gas.

◀ Richard A. Cooke III/Tony Stone Images.

PUZZLER

After this bottle of champagne was shaken, the cork was popped off and champagne spewed everywhere. Contrary to common belief, shaking a champagne bottle before opening it does not increase the pressure of the carbon dioxide (CO_2) inside. In fact, if you know the trick, you can open a thoroughly shaken bottle without spraying a drop. What's the secret? And why isn't the pressure inside the bottle greater after the bottle is shaken? *(Steve Niedorf/The Image Bank)*

c h a p t e r

Temperature

In our study of mechanics, we carefully defined such concepts as mass, force, and kinetic energy to facilitate our quantitative approach. Likewise, a quantitative description of thermal phenomena requires a careful definition of such important terms as *temperature, heat,* and *internal energy.* This chapter begins with a look at these three entities and with a description of one of the laws of thermodynamics (the poetically named "zeroth law"). We then discuss the three most common temperature scales—Celsius, Fahrenheit, and Kelvin.

Next, we consider why the composition of a body is an important factor when we are dealing with thermal phenomena. For example, gases expand appreciably when heated, whereas liquids and solids expand only slightly. If a gas is not free to expand as it is heated, its pressure increases. Certain substances may melt, boil, burn, or explode when they are heated, depending on their composition and structure.

This chapter concludes with a study of ideal gases on the macroscopic scale. Here, we are concerned with the relationships among such quantities as pressure, volume, and temperature. Later on, in Chapter 21, we shall examine gases on a microscopic scale, using a model that represents the components of a gas as small particles.

Molten lava flowing down a mountain in Kilauea, Hawaii. The temperature of the hot lava flowing from a central crater decreases until the lava is in thermal equilibrium with its surroundings. At that equilibrium temperature, the lava has solidified and formed the mountains. *(Ken Sakomoto/Black Star)*

19.1   TEMPERATURE AND THE ZEROTH LAW OF THERMODYNAMICS

We often associate the concept of temperature with how hot or cold an object feels when we touch it. Thus, our senses provide us with a qualitative indication of temperature. However, our senses are unreliable and often mislead us. For example, if we remove a metal ice tray and a cardboard box of frozen vegetables from the freezer, the ice tray feels colder than the box even though both are at the same temperature. The two objects feel different because metal is a better thermal conductor than cardboard is. What we need, therefore, is a reliable and reproducible method for establishing the relative hotness or coldness of bodies. Scientists have developed a variety of thermometers for making such quantitative measurements.

We are all familiar with the fact that two objects at different initial temperatures eventually reach some intermediate temperature when placed in contact with each other. For example, when a scoop of ice cream is placed in a room-temperature glass bowl, the ice cream melts and the temperature of the bowl decreases. Likewise, when an ice cube is dropped into a cup of hot coffee, it melts and the coffee's temperature decreases.

To understand the concept of temperature, it is useful to define two often-used phrases: *thermal contact* and *thermal equilibrium.* To grasp the meaning of thermal contact, let us imagine that two objects are placed in an insulated container such that they interact with each other but not with the rest of the world. If the objects are at different temperatures, energy is exchanged between them, even if they are initially not in physical contact with each other. **Heat is the transfer of energy from one object to another object as a result of a difference in temperature between the two.** We shall examine the concept of heat in greater detail in Chapter 20. For purposes of the current discussion, we assume that two objects are in **thermal contact** with each other if energy can be exchanged between them. **Thermal equilibrium** is a situation in which two objects in thermal contact with each other cease to exchange energy by the process of heat.

Let us consider two objects A and B, which are not in thermal contact, and a third object C, which is our thermometer. We wish to determine whether A and B

are in thermal equilibrium with each other. The thermometer (object C) is first placed in thermal contact with object A until thermal equilibrium is reached. From that moment on, the thermometer's reading remains constant, and we record this reading. The thermometer is then removed from object A and placed in thermal contact with object B. The reading is again recorded after thermal equilibrium is reached. If the two readings are the same, then object A and object B are in thermal equilibrium with each other.

We can summarize these results in a statement known as the **zeroth law of thermodynamics** (the law of equilibrium):

If objects A and B are separately in thermal equilibrium with a third object C, then objects A and B are in thermal equilibrium with each other.

Zeroth law of thermodynamics

This statement can easily be proved experimentally and is very important because it enables us to define temperature. We can think of **temperature** as the property that determines whether an object is in thermal equilibrium with other objects. **Two objects in thermal equilibrium with each other are at the same temperature.** Conversely, if two objects have different temperatures, then they are not in thermal equilibrium with each other.

19.2 THERMOMETERS AND THE CELSIUS TEMPERATURE SCALE

Thermometers are devices that are used to define and measure temperatures. All thermometers are based on the principle that some physical property of a system changes as the system's temperature changes. Some physical properties that change with temperature are (1) the volume of a liquid, (2) the length of a solid, (3) the pressure of a gas at constant volume, (4) the volume of a gas at constant pressure, (5) the electric resistance of a conductor, and (6) the color of an object. For a given substance and a given temperature range, a temperature scale can be established on the basis of any one of these physical properties.

A common thermometer in everyday use consists of a mass of liquid—usually mercury or alcohol—that expands into a glass capillary tube when heated (Fig. 19.1). In this case the physical property is the change in volume of a liquid. Any temperature change can be defined as being proportional to the change in length of the liquid column. The thermometer can be calibrated by placing it in thermal contact with some natural systems that remain at constant temperature. One such system is a mixture of water and ice in thermal equilibrium at atmospheric pressure. On the **Celsius temperature scale,** this mixture is defined to have a temperature of zero degrees Celsius, which is written as 0°C; this temperature is called the *ice point* of water. Another commonly used system is a mixture of water and steam in thermal equilibrium at atmospheric pressure; its temperature is 100°C, which is the *steam point* of water. Once the liquid levels in the thermometer have been established at these two points, the distance between the two points is divided into 100 equal segments to create the Celsius scale. Thus, each segment denotes a change in temperature of one Celsius degree. (This temperature scale used to be called the *centigrade scale* because there are 100 gradations between the ice and steam points of water.)

Thermometers calibrated in this way present problems when extremely accurate readings are needed. For instance, the readings given by an alcohol ther-

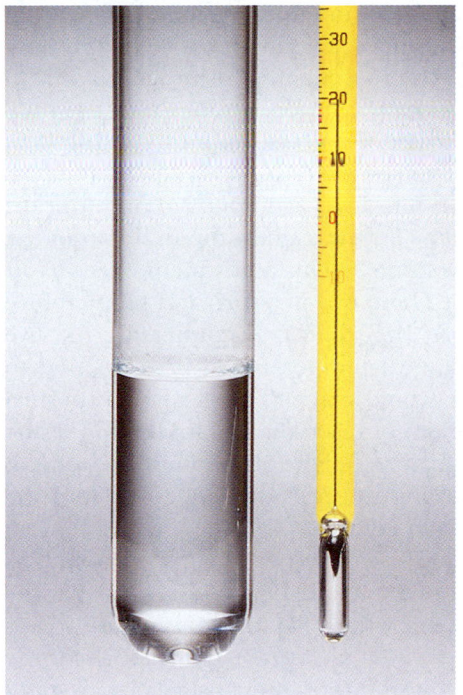
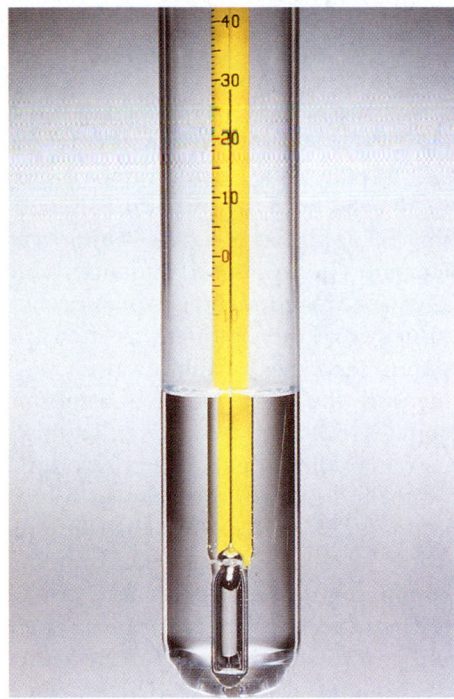

Figure 19.1   As a result of thermal expansion, the level of the mercury in the thermometer rises as the mercury is heated by water in the test tube. *(Charles D. Winters)*

mometer calibrated at the ice and steam points of water might agree with those given by a mercury thermometer only at the calibration points. Because mercury and alcohol have different thermal expansion properties, when one thermometer reads a temperature of, for example, 50°C, the other may indicate a slightly different value. The discrepancies between thermometers are especially large when the temperatures to be measured are far from the calibration points.[1]

An additional practical problem of any thermometer is the limited range of temperatures over which it can be used. A mercury thermometer, for example, cannot be used below the freezing point of mercury, which is − 39°C, and an alcohol thermometer is not useful for measuring temperatures above 85°C, the boiling point of alcohol. To surmount this problem, we need a universal thermometer whose readings are independent of the substance used in it. The gas thermometer, discussed in the next section, approaches this requirement.

19.3   THE CONSTANT-VOLUME GAS THERMOMETER AND THE ABSOLUTE TEMPERATURE SCALE

The temperature readings given by a gas thermometer are nearly independent of the substance used in the thermometer. One version is the constant-volume gas thermometer shown in Figure 19.2. The physical change exploited in this device is the variation of pressure of a fixed volume of gas with temperature. When the constant-volume gas thermometer was developed, it was calibrated by using the ice

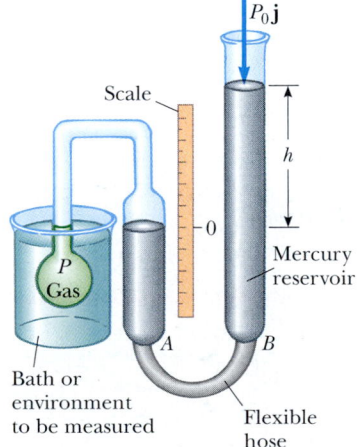

Figure 19.2   A constant-volume gas thermometer measures the pressure of the gas contained in the flask immersed in the bath. The volume of gas in the flask is kept constant by raising or lowering reservoir *B* to keep the mercury level in column *A* constant.

[1] Two thermometers that use the same liquid may also give different readings. This is due in part to difficulties in constructing uniform-bore glass capillary tubes.

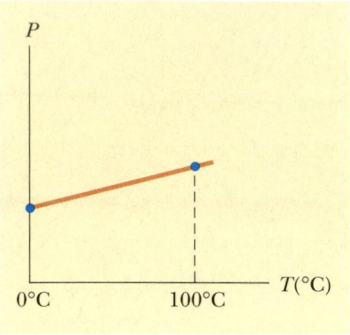

Figure 19.3 A typical graph of pressure versus temperature taken with a constant-volume gas thermometer. The two dots represent known reference temperatures (the ice and steam points of water).

10.3

web

For more information about the temperature standard, visit the National Institute of Standards and Technology at **http://www.nist.gov**

and steam points of water, as follows (a different calibration procedure, which we shall discuss shortly, is now used): The flask was immersed in an ice bath, and mercury reservoir B was raised or lowered until the top of the mercury in column A was at the zero point on the scale. The height h, the difference between the mercury levels in reservoir B and column A, indicated the pressure in the flask at 0°C.

The flask was then immersed in water at the steam point, and reservoir B was readjusted until the top of the mercury in column A was again at zero on the scale; this ensured that the gas's volume was the same as it was when the flask was in the ice bath (hence, the designation "constant volume"). This adjustment of reservoir B gave a value for the gas pressure at 100°C. These two pressure and temperature values were then plotted, as shown in Figure 19.3. The line connecting the two points serves as a calibration curve for unknown temperatures. If we wanted to measure the temperature of a substance, we would place the gas flask in thermal contact with the substance and adjust the height of reservoir B until the top of the mercury column in A was at zero on the scale. The height of the mercury column would indicate the pressure of the gas; knowing the pressure, we could find the temperature of the substance using the graph in Figure 19.3.

Now let us suppose that temperatures are measured with gas thermometers containing different gases at different initial pressures. Experiments show that the thermometer readings are nearly independent of the type of gas used, as long as the gas pressure is low and the temperature is well above the point at which the gas liquefies (Fig. 19.4). The agreement among thermometers using various gases improves as the pressure is reduced.

If you extend the curves shown in Figure 19.4 toward negative temperatures, you find, in every case, that the pressure is zero when the temperature is −273.15°C. This significant temperature is used as the basis for the **absolute temperature scale,** which sets −273.15°C as its zero point. This temperature is often referred to as **absolute zero.** The size of a degree on the absolute temperature scale is identical to the size of a degree on the Celsius scale. Thus, the conversion between these temperatures is

$$T_C = T - 273.15 \tag{19.1}$$

where T_C is the Celsius temperature and T is the absolute temperature.

Because the ice and steam points are experimentally difficult to duplicate, an absolute temperature scale based on a single fixed point was adopted in 1954 by the International Committee on Weights and Measures. From a list of fixed points associated with various substances (Table 19.1), the triple point of water was chosen as the reference temperature for this new scale. The **triple point of water** is the single combination of temperature and pressure at which liquid water, gaseous

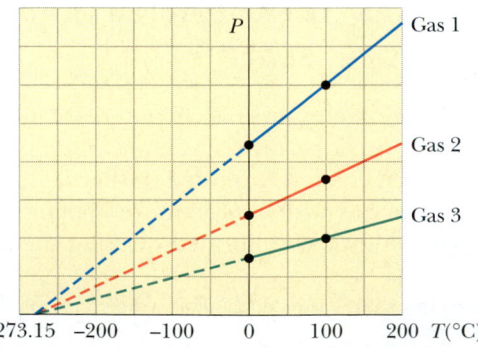

Figure 19.4 Pressure versus temperature for three dilute gases. Note that, for all gases, the pressure extrapolates to zero at the temperature −273.15°C.

TABLE 19.1 **Fixed-Point Temperatures[a]**		
Fixed Point	**Temperature (°C)**	**Temperature (K)**
Triple point of hydrogen	−259.34	13.81
Boiling point of helium	−268.93	4.215
Boiling point of hydrogen at 33.36 kPa pressure	−256.108	17.042
Boiling point of hydrogen	−252.87	20.28
Triple point of neon	−246.048	27.102
Triple point of oxygen	−218.789	54.361
Boiling point of oxygen	−182.962	90.188
Triple point of water	0.01	273.16
Boiling point of water	100.00	373.15
Freezing point of tin	231.968 1	505.118 1
Freezing point of zinc	419.58	692.73
Freezing point of silver	961.93	1 235.08
Freezing point of gold	1 064.43	1 337.58

[a] All values are from National Bureau of Standards Special Publication 420; U. S. Department of Commerce, May 1975. All values are at standard atmospheric pressure except for triple points and as noted.

water, and ice (solid water) coexist in equilibrium. This triple point occurs at a temperature of approximately 0.01°C and a pressure of 4.58 mm of mercury. On the new scale, which uses the unit *kelvin,* the temperature of water at the triple point was set at 273.16 kelvin, abbreviated 273.16 K. (*Note:* no degree sign "°" is used with the unit kelvin.) This choice was made so that the old absolute temperature scale based on the ice and steam points would agree closely with the new scale based on the triple point. This new absolute temperature scale (also called the **Kelvin scale**) employs the SI unit of absolute temperature, the **kelvin,** which is defined to be **1/273.16 of the difference between absolute zero and the temperature of the triple point of water.**

Figure 19.5 shows the absolute temperature for various physical processes and structures. The temperature of absolute zero (0 K) cannot be achieved, although laboratory experiments incorporating the laser cooling of atoms have come very close.

What would happen to a gas if its temperature could reach 0 K? As Figure 19.4 indicates, the pressure it exerts on the walls of its container would be zero. In Section 19.5 we shall show that the pressure of a gas is proportional to the average kinetic energy of its molecules. Thus, according to classical physics, the kinetic energy of the gas molecules would become zero at absolute zero, and molecular motion would cease; hence, the molecules would settle out on the bottom of the container. Quantum theory modifies this model and shows that some residual energy, called the *zero-point energy,* would remain at this low temperature.

The Celsius, Fahrenheit, and Kelvin Temperature Scales[2]

Equation 19.1 shows that the Celsius temperature T_C is shifted from the absolute (Kelvin) temperature T by 273.15°. Because the size of a degree is the same on the

[2] Named after Anders Celsius (1701–1744), Gabriel Fahrenheit (1686–1736), and William Thomson, Lord Kelvin (1824–1907), respectively.

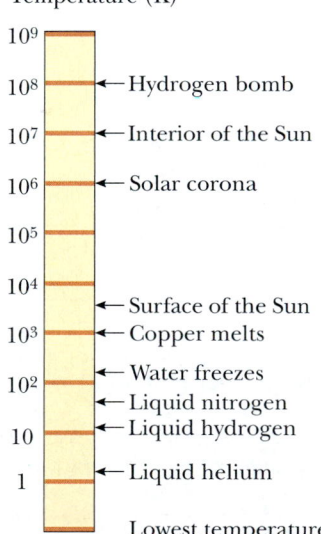

Figure 19.5 Absolute temperatures at which various physical processes occur. Note that the scale is logarithmic.

two scales, a temperature difference of 5°C is equal to a temperature difference of 5 K. The two scales differ only in the choice of the zero point. Thus, the ice-point temperature on the Kelvin scale, 273.15 K, corresponds to 0.00°C, and the Kelvin-scale steam point, 373.15 K, is equivalent to 100.00°C.

A common temperature scale in everyday use in the United States is the **Fahrenheit scale.** This scale sets the temperature of the ice point at 32°F and the temperature of the steam point at 212°F. The relationship between the Celsius and Fahrenheit temperature scales is

$$T_F = \tfrac{9}{5} T_C + 32°F \tag{19.2}$$

Quick Quiz 19.1

What is the physical significance of the factor $\tfrac{9}{5}$ in Equation 19.2? Why is this factor missing in Equation 19.1?

Extending the ideas considered in Quick Quiz 19.1, we use Equation 19.2 to find a relationship between changes in temperature on the Celsius, Kelvin, and Fahrenheit scales:

$$\Delta T_C = \Delta T = \tfrac{5}{9} \Delta T_F \tag{19.3}$$

EXAMPLE 19.1 Converting Temperatures

On a day when the temperature reaches 50°F, what is the temperature in degrees Celsius and in kelvins?

Solution Substituting $T_F = 50°F$ into Equation 19.2, we obtain

$$T_C = \tfrac{5}{9}(T_F - 32) = \tfrac{5}{9}(50 - 32) = \boxed{10°C}$$

From Equation 19.1, we find that

$$T = T_C + 273.15 = 10°C + 273.15 = \boxed{283 \text{ K}}$$

A convenient set of weather-related temperature equivalents to keep in mind is that 0°C is (literally) freezing at 32°F, 10°C is cool at 50°F, 30°C is warm at 86°F, and 40°C is a hot day at 104°F.

EXAMPLE 19.2 Heating a Pan of Water

A pan of water is heated from 25°C to 80°C. What is the change in its temperature on the Kelvin scale and on the Fahrenheit scale?

Solution From Equation 19.3, we see that the change in temperature on the Celsius scale equals the change on the Kelvin scale. Therefore,

$$\Delta T = \Delta T_C = 80°C - 25°C = 55°C = \boxed{55 \text{ K}}$$

From Equation 19.3, we also find that

$$\Delta T_F = \tfrac{9}{5}\Delta T_C = \tfrac{9}{5}(55°C) = \boxed{99°F}$$

19.4 THERMAL EXPANSION OF SOLIDS AND LIQUIDS

Our discussion of the liquid thermometer made use of one of the best-known changes in a substance: As its temperature increases, its volume almost always increases. (As we shall see shortly, in some substances the volume decreases when the temperature increases.) This phenomenon, known as **thermal expansion,** has

 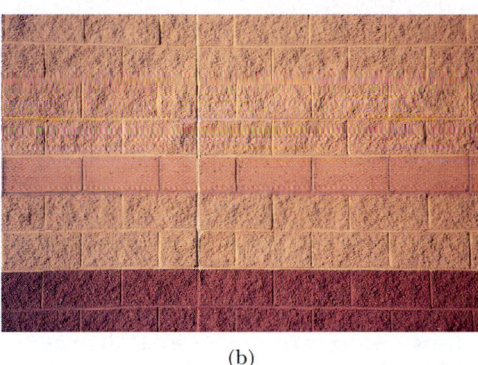

(a)   (b)

Figure 19.6 (a) Thermal-expansion joints are used to separate sections of roadways on bridges. Without these joints, the surfaces would buckle due to thermal expansion on very hot days or crack due to contraction on very cold days. (b) The long, vertical joint is filled with a soft material that allows the wall to expand and contract as the temperature of the bricks changes. *(a, Frank Siteman, Stock/Boston; b, George Semple)*

an important role in numerous engineering applications. For example, thermal-expansion joints, such as those shown in Figure 19.6, must be included in buildings, concrete highways, railroad tracks, brick walls, and bridges to compensate for dimensional changes that occur as the temperature changes.

Thermal expansion is a consequence of the change in the average separation between the constituent atoms in an object. To understand this, imagine that the atoms are connected by stiff springs, as shown in Figure 19.7. At ordinary temperatures, the atoms in a solid oscillate about their equilibrium positions with an amplitude of approximately 10^{-11} m and a frequency of approximately 10^{13} Hz. The average spacing between the atoms is about 10^{-10} m. As the temperature of the solid increases, the atoms oscillate with greater amplitudes; as a result, the average separation between them increases.[3] Consequently, the object expands.

If thermal expansion is sufficiently small relative to an object's initial dimensions, the change in any dimension is, to a good approximation, proportional to the first power of the temperature change. Suppose that an object has an initial length L_i along some direction at some temperature and that the length increases by an amount ΔL for a change in temperature ΔT. Because it is convenient to consider the fractional change in length per degree of temperature change, we define the **average coefficient of linear expansion** as

$$\alpha \equiv \frac{\Delta L / L_i}{\Delta T}$$

Experiments show that α is constant for small changes in temperature. For purposes of calculation, this equation is usually rewritten as

$$\Delta L = \alpha L_i \, \Delta T \qquad \textbf{(19.4)}$$

or as

$$L_f - L_i = \alpha L_i (T_f - T_i) \qquad \textbf{(19.5)}$$

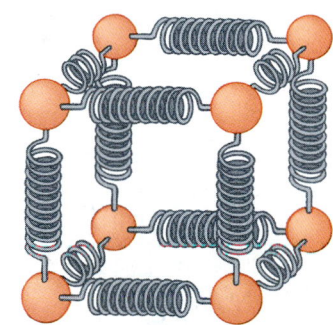

Figure 19.7 A mechanical model of the atomic configuration in a substance. The atoms (spheres) are imagined to be attached to each other by springs that reflect the elastic nature of the interatomic forces.

Average coefficient of linear expansion

The change in length of an object is proportional to the change in temperature

[3] More precisely, thermal expansion arises from the *asymmetrical* nature of the potential-energy curve for the atoms in a solid. If the oscillators were truly harmonic, the average atomic separations would not change regardless of the amplitude of vibration.

where L_f is the final length, T_i and T_f are the initial and final temperatures, and the proportionality constant α is the average coefficient of linear expansion for a given material and has units of $^\circ C^{-1}$.

It may be helpful to think of thermal expansion as an effective magnification or as a photographic enlargement of an object. For example, as a metal washer is heated (Fig. 19.8), all dimensions, including the radius of the hole, increase according to Equation 19.4.

Table 19.2 lists the average coefficient of linear expansion for various materials. Note that for these materials α is positive, indicating an increase in length with increasing temperature. This is not always the case. Some substances—calcite ($CaCO_3$) is one example—expand along one dimension (positive α) and contract along another (negative α) as their temperatures are increased.

Because the linear dimensions of an object change with temperature, it follows that surface area and volume change as well. The change in volume at constant pressure is proportional to the initial volume V_i and to the change in temperature according to the relationship

> The change in volume of a solid at constant pressure is proportional to the change in temperature

$$\Delta V = \beta V_i \Delta T \tag{19.6}$$

where β is the **average coefficient of volume expansion.** For a solid, the average coefficient of volume expansion is approximately three times the average linear expansion coefficient: $\beta = 3\alpha$. (This assumes that the average coefficient of linear expansion of the solid is the same in all directions.)

To see that $\beta = 3\alpha$ for a solid, consider a box of dimensions ℓ, w, and h. Its volume at some temperature T_i is $V_i = \ell w h$. If the temperature changes to $T_i + \Delta T$, its volume changes to $V_i + \Delta V$, where each dimension changes according to Equation 19.4. Therefore,

$$\begin{aligned}
V_i + \Delta V &= (\ell + \Delta\ell)(w + \Delta w)(h + \Delta h) \\
&= (\ell + \alpha\ell\,\Delta T)(w + \alpha w\,\Delta T)(h + \alpha h\,\Delta T) \\
&= \ell w h (1 + \alpha\,\Delta T)^3 \\
&= V_i[1 + 3\alpha\,\Delta T + 3(\alpha\,\Delta T)^2 + (\alpha\,\Delta T)^3]
\end{aligned}$$

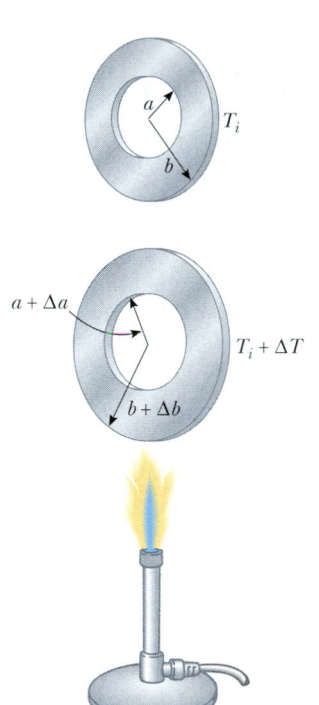

Figure 19.8 Thermal expansion of a homogeneous metal washer. As the washer is heated, all dimensions increase. (The expansion is exaggerated in this figure.)

TABLE 19.2	**Average Expansion Coefficients for Some Materials Near Room Temperature**		
Material	**Average Linear Expansion Coefficient (α) $(^\circ C)^{-1}$**	**Material**	**Average Volume Expansion Coefficient (β) $(^\circ C)^{-1}$**
Aluminum	24×10^{-6}	Alcohol, ethyl	1.12×10^{-4}
Brass and bronze	19×10^{-6}	Benzene	1.24×10^{-4}
Copper	17×10^{-6}	Acetone	1.5×10^{-4}
Glass (ordinary)	9×10^{-6}	Glycerin	4.85×10^{-4}
Glass (Pyrex)	3.2×10^{-6}	Mercury	1.82×10^{-4}
Lead	29×10^{-6}	Turpentine	9.0×10^{-4}
Steel	11×10^{-6}	Gasoline	9.6×10^{-4}
Invar (Ni–Fe alloy)	0.9×10^{-6}	Air at $0^\circ C$	3.67×10^{-3}
Concrete	12×10^{-6}	Helium	3.665×10^{-3}

If we now divide both sides by V_i and then isolate the term $\Delta V/V_i$, we obtain the fractional change in volume:

$$\frac{\Delta V}{V_i} = 3\alpha\,\Delta T + 3(\alpha\,\Delta T)^2 + (\alpha\,\Delta T)^3$$

Because $\alpha\,\Delta T \ll 1$ for typical values of ΔT ($< \sim 100°C$), we can neglect the terms $3(\alpha\,\Delta T)^2$ and $(\alpha\,\Delta T)^3$. Upon making this approximation, we see that

$$\frac{\Delta V}{V_i} = 3\alpha\,\Delta T$$

$$3\alpha = \frac{1}{V_i}\frac{\Delta V}{\Delta T}$$

Equation 19.6 shows that the right side of this expression is equal to β, and so we have $3\alpha = \beta$, the relationship we set out to prove. In a similar way, you can show that the change in area of a rectangular plate is given by $\Delta A = 2\alpha A_i\,\Delta T$ (see Problem 53).

As Table 19.2 indicates, each substance has its own characteristic average coefficient of expansion. For example, when the temperatures of a brass rod and a steel rod of equal length are raised by the same amount from some common initial value, the brass rod expands more than the steel rod does because brass has a greater average coefficient of expansion than steel does. A simple mechanism called a *bimetallic strip* utilizes this principle and is found in practical devices such as thermostats. It consists of two thin strips of dissimilar metals bonded together. As the temperature of the strip increases, the two metals expand by different amounts and the strip bends, as shown in Figure 19.9.

QuickLab

Tape two plastic straws tightly together along their entire length but with a 2 cm offset. Hold them in a stream of very hot water from a faucet so that water pours through one but not through the other. Quickly hold the straws up and sight along their length. You should be able to see a very slight curvature in the tape caused by the difference in expansion of the two straws. The effect is small, so look closely. Running cold water through the same straw and again sighting along the length will help you see the small change in shape more clearly.

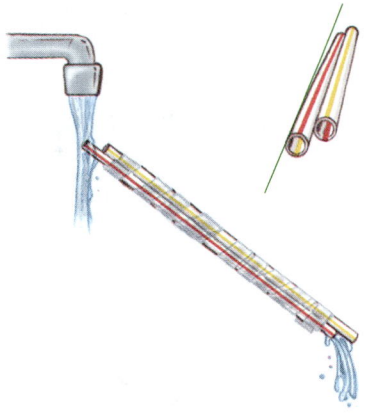

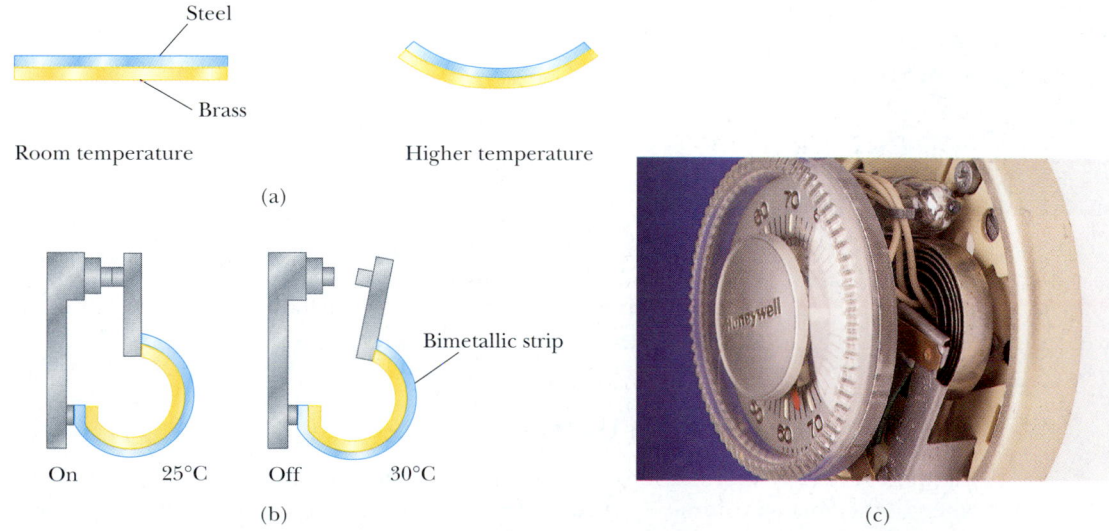

Figure 19.9 (a) A bimetallic strip bends as the temperature changes because the two metals have different expansion coefficients. (b) A bimetallic strip used in a thermostat to break or make electrical contact. (c) The interior of a thermostat, showing the coiled bimetallic strip. Why do you suppose the strip is coiled? *(c, Charles D. Winters)*

If you quickly plunge a room-temperature thermometer into very hot water, the mercury level will go *down* briefly before going up to a final reading. Why?

You are offered a prize for making the most sensitive glass thermometer using the materials in Table 19.2. Which glass and which working liquid would you choose?

EXAMPLE 19.3   **Expansion of a Railroad Track**

A steel railroad track has a length of 30.000 m when the temperature is 0.0°C. (a) What is its length when the temperature is 40.0°C?

Thermal expansion: The extreme temperature of a July day in Asbury Park, NJ, caused these railroad tracks to buckle and derail the train in the distance. *(AP/Wide World Photos)*

Solution Making use of Table 19.2 and noting that the change in temperature is 40.0°C, we find that the increase in length is

$$\Delta L = \alpha L_i \,\Delta T = [11 \times 10^{-6}(°\text{C})^{-1}](30.000 \text{ m})(40.0°\text{C})$$
$$= 0.013 \text{ m}$$

If the track is 30.000 m long at 0.0°C, its length at 40.0°C is

$$30.013 \text{ m.}$$

(b) Suppose that the ends of the rail are rigidly clamped at 0.0°C so that expansion is prevented. What is the thermal stress set up in the rail if its temperature is raised to 40.0°C?

Solution From the definition of Young's modulus for a solid (see Eq. 12.6), we have

$$\text{Tensile stress} = \frac{F}{A} = Y\frac{\Delta L}{L_i}$$

Because Y for steel is 20×10^{10} N/m² (see Table 12.1), we have

$$\frac{F}{A} = (20 \times 10^{10} \text{ N/m}^2)\left(\frac{0.013 \text{ m}}{30.000 \text{ m}}\right) = 8.7 \times 10^7 \text{ N/m}^2$$

Exercise If the rail has a cross-sectional area of 30.0 cm², what is the force of compression in the rail?

Answer 2.6×10^5 N = 58 000 lb!

The Unusual Behavior of Water

Liquids generally increase in volume with increasing temperature and have average coefficients of volume expansion about ten times greater than those of solids. Water is an exception to this rule, as we can see from its density-versus-temperature curve shown in Figure 19.10. As the temperature increases from 0°C to 4°C, water contracts and thus its density increases. Above 4°C, water expands with increasing temperature, and so its density decreases. In other words, the density of water reaches a maximum value of 1 000 kg/m³ at 4°C.

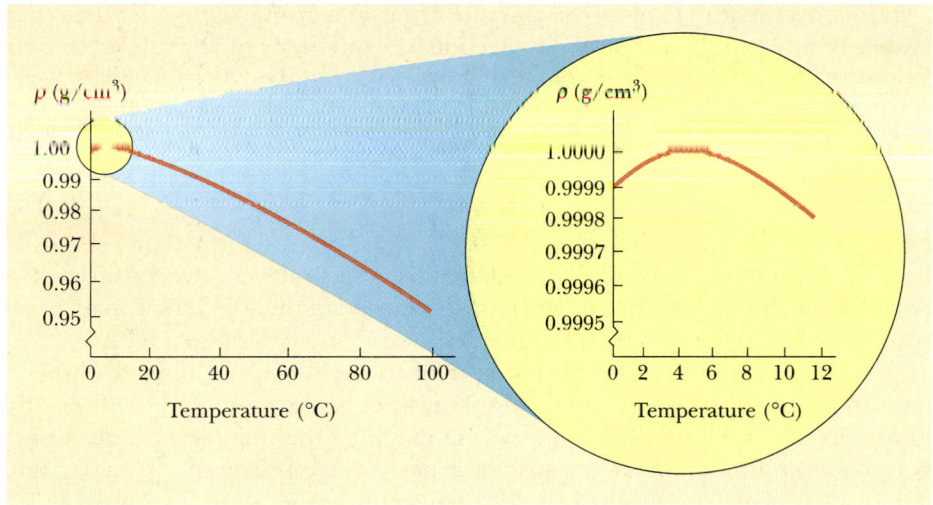

Figure 19.10 How the density of water at atmospheric pressure changes with temperature. The inset at the right shows that the maximum density of water occurs at 4°C.

We can use this unusual thermal-expansion behavior of water to explain why a pond begins freezing at the surface rather than at the bottom. When the atmospheric temperature drops from, for example, 7°C to 6°C, the surface water also cools and consequently decreases in volume. This means that the surface water is denser than the water below it, which has not cooled and decreased in volume. As a result, the surface water sinks, and warmer water from below is forced to the surface to be cooled. When the atmospheric temperature is between 4°C and 0°C, however, the surface water expands as it cools, becoming less dense than the water below it. The mixing process stops, and eventually the surface water freezes. As the water freezes, the ice remains on the surface because ice is less dense than water. The ice continues to build up at the surface, while water near the bottom remains at 4°C. If this were not the case, then fish and other forms of marine life would not survive.

19.5 MACROSCOPIC DESCRIPTION OF AN IDEAL GAS

In this section we examine the properties of a gas of mass m confined to a container of volume V at a pressure P and a temperature T. It is useful to know how these quantities are related. In general, the equation that interrelates these quantities, called the *equation of state,* is very complicated. However, if the gas is maintained at a very low pressure (or low density), the equation of state is quite simple and can be found experimentally. Such a low-density gas is commonly referred to as an *ideal gas.*[4]

[4] To be more specific, the assumption here is that the temperature of the gas must not be too low (the gas must not condense into a liquid) or too high, and that the pressure must be low. In reality, an ideal gas does not exist. However, the concept of an ideal gas is very useful in view of the fact that real gases at low pressures behave as ideal gases do. The concept of an ideal gas implies that the gas molecules do not interact except upon collision, and that the molecular volume is negligible compared with the volume of the container.

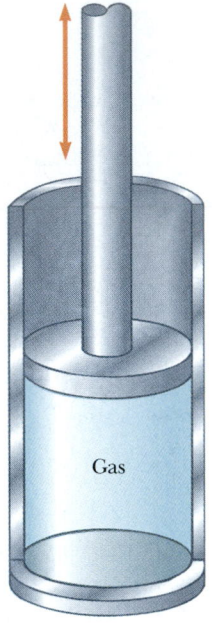

Figure 19.11 An ideal gas confined to a cylinder whose volume can be varied by means of a movable piston.

It is convenient to express the amount of gas in a given volume in terms of the number of moles n. As we learned in Section 1.3, one mole of any substance is that amount of the substance that contains Avogadro's number $N_A = 6.022 \times 10^{23}$ of constituent particles (atoms or molecules). The number of moles n of a substance is related to its mass m through the expression

$$n = \frac{m}{M} \tag{19.7}$$

where M is the molar mass of the substance (see Section 1.3), which is usually expressed in units of grams per mole (g/mol). For example, the molar mass of oxygen (O_2) is 32.0 g/mol. Therefore, the mass of one mole of oxygen is 32.0 g.

Now suppose that an ideal gas is confined to a cylindrical container whose volume can be varied by means of a movable piston, as shown in Figure 19.11. If we assume that the cylinder does not leak, the mass (or the number of moles) of the gas remains constant. For such a system, experiments provide the following information: First, when the gas is kept at a constant temperature, its pressure is inversely proportional to its volume (Boyle's law). Second, when the pressure of the gas is kept constant, its volume is directly proportional to its temperature (the law of Charles and Gay–Lussac). These observations are summarized by the **equation of state for an ideal gas:**

$$PV = nRT \tag{19.8}$$

In this expression, known as the **ideal gas law,** R is a universal constant that is the same for all gases and T is the absolute temperature in kelvins. Experiments on numerous gases show that as the pressure approaches zero, the quantity PV/nT approaches the same value R for all gases. For this reason, R is called the **universal gas constant.** In SI units, in which pressure is expressed in pascals (1 Pa = 1 N/m^2) and volume in cubic meters, the product PV has units of newton·meters, or joules, and R has the value

$$R = 8.315 \text{ J/mol·K} \tag{19.9}$$

If the pressure is expressed in atmospheres and the volume in liters (1 L = 10^3 cm^3 = 10^{-3} m^3), then R has the value

$$R = 0.082\ 14 \text{ L·atm/mol·K}$$

The universal gas constant

Using this value of R and Equation 19.8, we find that the volume occupied by 1 mol of any gas at atmospheric pressure and at 0°C (273 K) is 22.4 L.

Now that we have presented the equation of state, we are ready for a formal definition of an ideal gas: **An ideal gas is one for which PV/nT is constant at all pressures.**

The ideal gas law states that if the volume and temperature of a fixed amount of gas do not change, then the pressure also remains constant. Consider the bottle of champagne shown at the beginning of this chapter. Because the temperature of the bottle and its contents remains constant, so does the pressure, as can be shown by replacing the cork with a pressure gauge. Shaking the bottle displaces some carbon dioxide gas from the "head space" to form bubbles within the liquid, and these bubbles become attached to the inside of the bottle. (No new gas is generated by shaking.) When the bottle is opened, the pressure is reduced; this causes the volume of the bubbles to increase suddenly. If the bubbles are attached to the bottle (beneath the liquid surface), their rapid expansion expels liquid from the

QuickLab

Vigorously shake a can of soda pop and then thoroughly tap its bottom and sides to dislodge any bubbles trapped there. You should be able to open the can without spraying its contents all over.

bottle. If the sides and bottom of the bottle are first tapped until no bubbles remain beneath the surface, then when the champagne is opened, the drop in pressure will not force liquid from the bottle. Try the QuickLab, but practice before demonstrating to a friend!

The ideal gas law is often expressed in terms of the total number of molecules N. Because the total number of molecules equals the product of the number of moles n and Avogadro's number N_A, we can write Equation 19.8 as

$$PV = nRT = \frac{N}{N_A} RT$$

$$PV = Nk_B T \tag{19.10}$$

where k_B is **Boltzmann's constant,** which has the value

$$k_B = \frac{R}{N_A} = 1.38 \times 10^{-23} \, \text{J/K} \tag{19.11}$$

Boltzmann's constant

It is common to call quantities such as P, V, and T the **thermodynamic variables** of an ideal gas. If the equation of state is known, then one of the variables can always be expressed as some function of the other two.

EXAMPLE 19.4   How Many Gas Molecules in a Container?

An ideal gas occupies a volume of 100 cm³ at 20°C and 100 Pa. Find the number of moles of gas in the container.

$$n = \frac{PV}{RT} = \frac{(100 \, \text{Pa})(10^{-4} \, \text{m}^3)}{(8.315 \, \text{J/mol} \cdot \text{K})(293 \, \text{K})} = 4.10 \times 10^{-6} \, \text{mol}$$

Solution   The quantities given are volume, pressure, and temperature: $V = 100 \, \text{cm}^3 = 1.00 \times 10^{-4} \, \text{m}^3$, $P = 100 \, \text{Pa}$, and $T = 20°C = 293 \, \text{K}$. Using Equation 19.8, we find that

Exercise   How many molecules are in the container?

Answer 2.47×10^{18} molecules.

EXAMPLE 19.5   Filling a Scuba Tank

A certain scuba tank is designed to hold 66 ft³ of air when it is at atmospheric pressure at 22°C. When this volume of air is compressed to an absolute pressure of 3 000 lb/in.² and stored in a 10-L (0.35-ft³) tank, the air becomes so hot that the tank must be allowed to cool before it can be used. If the air does not cool, what is its temperature? (Assume that the air behaves like an ideal gas.)

Solution   If no air escapes from the tank during filling, then the number of moles n remains constant; therefore, using $PV = nRT$, and with n and R being constant, we obtain for the initial and final values:

$$\frac{P_i V_i}{T_i} = \frac{P_f V_f}{T_f}$$

The initial pressure of the air is 14.7 lb/in.², its final pressure is 3 000 lb/in.², and the air is compressed from an initial volume of 66 ft³ to a final volume of 0.35 ft³. The initial temperature, converted to SI units, is 295 K. Solving for T_f, we obtain

$$T_f = \left(\frac{P_f V_f}{P_i V_i} \right) T_i = \frac{(3\,000 \, \text{lb/in.}^2)(0.35 \, \text{ft}^3)}{(14.7 \, \text{lb/in.}^2)(66 \, \text{ft}^3)} (295 \, \text{K})$$

$$= 319 \, \text{K}$$

Exercise   What is the air temperature in degrees Celsius and in degrees Fahrenheit?

Answer   45.9°C; 115°F.

Quick Quiz 19.4

In the previous example we used SI units for the temperature in our calculation step but not for the pressures or volumes. When working with the ideal gas law, how do you decide when it is necessary to use SI units and when it is not?

EXAMPLE 19.6 Heating a Spray Can

A spray can containing a propellant gas at twice atmospheric pressure (202 kPa) and having a volume of 125 cm³ is at 22°C. It is then tossed into an open fire. When the temperature of the gas in the can reaches 195°C, what is the pressure inside the can? Assume any change in the volume of the can is negligible.

Solution We employ the same approach we used in Example 19.5, starting with the expression

$$\frac{P_i V_i}{T_i} = \frac{P_f V_f}{T_f}$$

Because the initial and final volumes of the gas are assumed to be equal, this expression reduces to

$$\frac{P_i}{T_i} = \frac{P_f}{T_f}$$

Solving for P_f gives

$$P_f = \left(\frac{T_f}{T_i}\right)(P_i) = \left(\frac{468 \text{ K}}{295 \text{ K}}\right)(202 \text{ kPa}) = \boxed{320 \text{ kPa}}$$

Obviously, the higher the temperature, the higher the pressure exerted by the trapped gas. Of course, if the pressure increases high enough, the can will explode. Because of this possibility, you should never dispose of spray cans in a fire.

SUMMARY

Two bodies are in **thermal equilibrium** with each other if they have the same temperature.

The **zeroth law of thermodynamics** states that if objects A and B are separately in thermal equilibrium with a third object C, then objects A and B are in thermal equilibrium with each other.

The SI unit of absolute temperature is the **kelvin,** which is defined to be the fraction 1/273.16 of the temperature of the triple point of water.

When the temperature of an object is changed by an amount ΔT, its length changes by an amount ΔL that is proportional to ΔT and to its initial length L_i:

$$\Delta L = \alpha L_i \Delta T \tag{19.4}$$

where the constant α is the **average coefficient of linear expansion.** The **average volume expansion coefficient** β for a solid is approximately equal to 3α.

An **ideal gas** is one for which PV/nT is constant at all pressures. An ideal gas is described by the **equation of state,**

$$PV = nRT \tag{19.8}$$

where n equals the number of moles of the gas, V is its volume, R is the universal gas constant (8.315 J/mol·K), and T is the absolute temperature. A real gas behaves approximately as an ideal gas if it is far from liquefaction.

QUESTIONS

1. Is it possible for two objects to be in thermal equilibrium if they are not in contact with each other? Explain.

2. A piece of copper is dropped into a beaker of water. If the water's temperature increases, what happens to the temperature of the copper? Under what conditions are the water and copper in thermal equilibrium?

3. In principle, any gas can be used in a constant-volume gas thermometer. Why is it not possible to use oxygen for temperatures as low as 15 K? What gas would you use? (Refer to the data in Table 19.1.)

4. Rubber has a negative average coefficient of linear expansion. What happens to the size of a piece of rubber as it is warmed?

5. Why should the amalgam used in dental fillings have the same average coefficient of expansion as a tooth? What would occur if they were mismatched?

6. Explain why the thermal expansion of a spherical shell made of a homogeneous solid is equivalent to that of a solid sphere of the same material.

7. A steel ring bearing has an inside diameter that is 0.1 mm smaller than the diameter of an axle. How can it be made to fit onto the axle without removing any material?

8. Markings to indicate length are placed on a steel tape in a room that has a temperature of 22°C. Are measurements made with the tape on a day when the temperature is 27°C greater than, less than, or the same length as the object's length? Defend your answer.

9. Determine the number of grams in 1 mol of each of the following gases: (a) hydrogen, (b) helium, and (c) carbon monoxide.

10. An inflated rubber balloon filled with air is immersed in a flask of liquid nitrogen that is at 77 K. Describe what happens to the balloon, assuming that it remains flexible while being cooled.

11. Two identical cylinders at the same temperature each contain the same kind of gas and the same number of moles of gas. If the volume of cylinder A is three times greater than the volume of cylinder B, what can you say about the relative pressures in the cylinders?

12. The pendulum of a certain pendulum clock is made of brass. When the temperature increases, does the clock run too fast, run too slowly, or remain unchanged? Explain.

13. An automobile radiator is filled to the brim with water while the engine is cool. What happens to the water when the engine is running and the water is heated? What do modern automobiles have in their cooling systems to prevent the loss of coolants?

14. Metal lids on glass jars can often be loosened by running them under hot water. How is this possible?

15. When the metal ring and metal sphere shown in Figure Q19.15 are both at room temperature, the sphere can just be passed through the ring. After the sphere is heated, it cannot be passed through the ring. Explain.

Figure Q19.15   *(Courtesy of Central Scientific Company)*

PROBLEMS

1, 2, 3 = straightforward, intermediate, challenging   ☐ = full solution available in the *Student Solutions Manual and Study Guide*
WEB = solution posted at **http://www.saunderscollege.com/physics/**   🖥 = Computer useful in solving problem 🔲 = Interactive Physics
☐ = paired numerical/symbolic problems

Section 19.1   Temperature and the Zeroth Law of Thermodynamics

Section 19.2   Thermometers and the Celsius Temperature Scale

Section 19.3   The Constant-Volume Gas Thermometer and the Absolute Temperature Scale

Note: A pressure of 1 atm = 1.013×10^5 Pa = 101.3 kPa.

1. Convert the following to equivalent temperatures on the Celsius and Kelvin scales: (a) the normal human body temperature, 98.6°F; (b) the air temperature on a cold day, − 5.00°F.

2. In a constant-volume gas thermometer, the pressure at 20.0°C is 0.980 atm. (a) What is the pressure at 45.0°C? (b) What is the temperature if the pressure is 0.500 atm?

WEB 3. A constant-volume gas thermometer is calibrated in dry ice (that is, carbon dioxide in the solid state, which has a temperature of − 80.0°C) and in boiling ethyl alcohol (78.0°C). The two pressures are 0.900 atm and

1.635 atm. (a) What Celsius value of absolute zero does the calibration yield? What is the pressure at (b) the freezing point of water and (c) the boiling point of water?

4. There is a temperature whose numerical value is the same on both the Celsius and Fahrenheit scales. What is this temperature?

5. Liquid nitrogen has a boiling point of $-195.81°C$ at atmospheric pressure. Express this temperature in (a) degrees Fahrenheit and (b) kelvins.

6. On a Strange temperature scale, the freezing point of water is $-15.0°S$ and the boiling point is $+60.0°S$. Develop a *linear* conversion equation between this temperature scale and the Celsius scale.

7. The temperature difference between the inside and the outside of an automobile engine is $450°C$. Express this temperature difference on the (a) Fahrenheit scale and (b) Kelvin scale.

8. The melting point of gold is $1\,064°C$, and the boiling point is $2\,660°C$. (a) Express these temperatures in kelvins. (b) Compute the difference between these temperatures in Celsius degrees and in kelvins.

Section 19.4   Thermal Expansion of Solids and Liquids

Note: When solving the problems in this section, use the data in Table 19.2.

9. A copper telephone wire has essentially no sag between poles 35.0 m apart on a winter day when the temperature is $-20.0°C$. How much longer is the wire on a summer day when $T_C = 35.0°C$?

10. The concrete sections of a certain superhighway are designed to have a length of 25.0 m. The sections are poured and cured at $10.0°C$. What minimum spacing should the engineer leave between the sections to eliminate buckling if the concrete is to reach a temperature of $50.0°C$?

11. An aluminum tube is 3.000 0 m long at $20.0°C$. What is its length at (a) $100.0°C$ and (b) $0.0°C$?

12. A brass ring with a diameter of 10.00 cm at $20.0°C$ is heated and slipped over an aluminum rod with a diameter of 10.01 cm at $20.0°C$. Assume that the average coefficients of linear expansion are constant. (a) To what temperature must this combination be cooled to separate them? Is this temperature attainable? (b) If the aluminum rod were 10.02 cm in diameter, what would be the required temperature?

13. A pair of eyeglass frames is made of epoxy plastic. At room temperature ($20.0°C$), the frames have circular lens holes 2.20 cm in radius. To what temperature must the frames be heated if lenses 2.21 cm in radius are to be inserted in them? The average coefficient of linear expansion for epoxy is 1.30×10^{-4} $(°C)^{-1}$.

14. The New River Gorge bridge in West Virginia is a steel arch bridge 518 m in length. How much does its length change between temperature extremes of $-20.0°C$ and $35.0°C$?

15. A square hole measuring 8.00 cm along each side is cut

in a sheet of copper. (a) Calculate the change in the area of this hole if the temperature of the sheet is increased by 50.0 K. (b) Does the result represent an increase or a decrease in the area of the hole?

16. The average coefficient of volume expansion for carbon tetrachloride is 5.81×10^{-4} $(°C)^{-1}$. If a 50.0-gal steel container is filled completely with carbon tetrachloride when the temperature is $10.0°C$, how much will spill over when the temperature rises to $30.0°C$?

WEB 17. The active element of a certain laser is a glass rod 30.0 cm long by 1.50 cm in diameter. If the temperature of the rod increases by $65.0°C$, what is the increase in (a) its length, (b) its diameter, and (c) its volume? (Assume that $\alpha = 9.00 \times 10^{-6}$ $(°C)^{-1}$.)

18. A volumetric glass flask made of Pyrex is calibrated at $20.0°C$. It is filled to the 100-mL mark with $35.0°C$ acetone with which it immediately comes to thermal equilibrium. (a) What is the volume of the acetone when it cools to $20.0°C$? (b) How significant is the change in volume of the flask?

19. A concrete walk is poured on a day when the temperature is $20.0°C$, in such a way that the ends are unable to move. (a) What is the stress in the cement on a hot day of $50.0°C$? (b) Does the concrete fracture? Take Young's modulus for concrete to be 7.00×10^9 N/m^2 and the tensile strength to be 2.00×10^9 N/m^2.

20. Figure P19.20 shows a circular steel casting with a gap. If the casting is heated, (a) does the width of the gap increase or decrease? (b) The gap width is 1.600 cm when the temperature is $30.0°C$. Determine the gap width when the temperature is $190°C$.

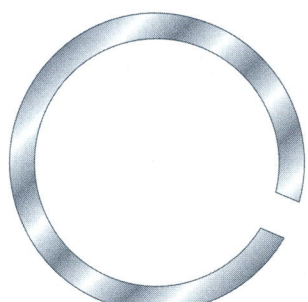

Figure P19.20

21. A steel rod undergoes a stretching force of 500 N. Its cross-sectional area is 2.00 cm^2. Find the change in temperature that would elongate the rod by the same amount that the 500-N force does. (*Hint:* Refer to Tables 12.1 and 19.2.)

22. A steel rod 4.00 cm in diameter is heated so that its temperature increases by $70.0°C$. It is then fastened between two rigid supports. The rod is allowed to cool to its original temperature. Assuming that Young's modulus for the steel is 20.6×10^{10} N/m^2 and that its average

coefficient of linear expansion is 11.0×10^{-6} (°C)$^{-1}$, calculate the tension in the rod.

23. A hollow aluminum cylinder 20.0 cm deep has an internal capacity of 2.000 L at 20.0°C. It is completely filled with turpentine and then warmed to 80.0°C. (a) How much turpentine overflows? (b) If the cylinder is then cooled back to 20.0°C, how far below the surface of the cylinder's rim does the turpentine's surface recede?

24. At 20.0°C, an aluminum ring has an inner diameter of 5.000 0 cm and a brass rod has a diameter of 5.050 0 cm. (a) To what temperature must the ring be heated so that it will just slip over the rod? (b) To what common temperature must the two be heated so that the ring just slips over the rod? Would this latter process work?

Section 19.5 Macroscopic Description of an Ideal Gas

25. Gas is contained in an 8.00-L vessel at a temperature of 20.0°C and a pressure of 9.00 atm. (a) Determine the number of moles of gas in the vessel. (b) How many molecules of gas are in the vessel?

26. A tank having a volume of 0.100 m^3 contains helium gas at 150 atm. How many balloons can the tank blow up if each filled balloon is a sphere 0.300 m in diameter at an absolute pressure of 1.20 atm?

27. An auditorium has dimensions 10.0 m $\times$ 20.0 m $\times$ 30.0 m. How many molecules of air fill the auditorium at 20.0°C and a pressure of 101 kPa?

28. Nine grams of water are placed in a 2.00-L pressure cooker and heated to 500°C. What is the pressure inside the container if no gas escapes?

WEB 29. The mass of a hot-air balloon and its cargo (not including the air inside) is 200 kg. The air outside is at 10.0°C and 101 kPa. The volume of the balloon is 400 m^3. To what temperature must the air in the balloon be heated before the balloon will lift off? (Air density at 10.0°C is 1.25 kg/m^3.)

30. One mole of oxygen gas is at a pressure of 6.00 atm and a temperature of 27.0°C. (a) If the gas is heated at constant volume until the pressure triples, what is the final temperature? (b) If the gas is heated until both the pressure and the volume are doubled, what is the final temperature?

31. (a) Find the number of moles in 1.00 m^3 of an ideal gas at 20.0°C and atmospheric pressure. (b) For air, Avogadro's number of molecules has a mass of 28.9 g. Calculate the mass of 1 m^3 of air. Compare the result with the tabulated density of air.

32. A cube 10.0 cm on each edge contains air (with equivalent molar mass 28.9 g/mol) at atmospheric pressure and temperature 300 K. Find (a) the mass of the gas, (b) its weight, and (c) the force it exerts on each face of the cube. (d) Comment on the underlying physical reason why such a small sample can exert such a great force.

33. An automobile tire is inflated with air originally at 10.0°C and normal atmospheric pressure. During the process, the air is compressed to 28.0% of its original volume and its temperature is increased to 40.0°C. (a) What is the tire pressure? (b) After the car is driven at high speed, the tire air temperature rises to 85.0°C and the interior volume of the tire increases by 2.00%. What is the new tire pressure (absolute) in pascals?

34. A spherical weather balloon is designed to expand to a maximum radius of 20.0 m when in flight at its working altitude, where the air pressure is 0.030 0 atm and the temperature is 200 K. If the balloon is filled at atmospheric pressure and 300 K, what is its radius at liftoff?

35. A room of volume 80.0 m^3 contains air having an equivalent molar mass of 28.9 g/mol. If the temperature of the room is raised from 18.0°C to 25.0°C, what mass of air (in kilograms) will leave the room? Assume that the air pressure in the room is maintained at 101 kPa.

36. A room of volume V contains air having equivalent molar mass M (in g/mol). If the temperature of the room is raised from T_1 to T_2, what mass of air will leave the room? Assume that the air pressure in the room is maintained at P_0.

37. At 25.0 m below the surface of the sea (density = 1 025 kg/m^3), where the temperature is 5.00°C, a diver exhales an air bubble having a volume of 1.00 cm^3. If the surface temperature of the sea is 20.0°C, what is the volume of the bubble right before it breaks the surface?

38. Estimate the mass of the air in your bedroom. State the quantities you take as data and the value you measure or estimate for each.

39. The pressure gauge on a tank registers the gauge pressure, which is the difference between the interior and exterior pressures. When the tank is full of oxygen (O_2), it contains 12.0 kg of the gas at a gauge pressure of 40.0 atm. Determine the mass of oxygen that has been withdrawn from the tank when the pressure reading is 25.0 atm. Assume that the temperature of the tank remains constant.

40. In state-of-the-art vacuum systems, pressures as low as 10^{-9} Pa are being attained. Calculate the number of molecules in a 1.00-m^3 vessel at this pressure if the temperature is 27°C.

41. Show that 1 mol of any gas (assumed to be ideal) at atmospheric pressure (101.3 kPa) and standard temperature (273 K) occupies a volume of 22.4 L.

42. A diving bell in the shape of a cylinder with a height of 2.50 m is closed at the upper end and open at the lower end. The bell is lowered from air into sea water (ρ = 1.025 g/cm^3). The air in the bell is initially at 20.0°C. The bell is lowered to a depth (measured to the bottom of the bell) of 45.0 fathoms, or 82.3 m. At this depth, the water temperature is 4.0°C, and the air in the bell is in thermal equilibrium with the water. (a) How high does sea water rise in the bell? (b) To what minimum pressure must the air in the bell be increased for the water that entered to be expelled?

ADDITIONAL PROBLEMS

43. A student measures the length of a brass rod with a steel tape at 20.0°C. The reading is 95.00 cm. What will the tape indicate for the length of the rod when the rod and the tape are at (a) − 15.0°C and (b) 55.0°C?

44. The density of gasoline is 730 kg/m³ at 0°C. Its average coefficient of volume expansion is 9.60×10^{-4} (°C)$^{-1}$. If 1.00 gal of gasoline occupies 0.003 80 m³, how many extra kilograms of gasoline would you get if you bought 10.0 gal of gasoline at 0°C rather than at 20.0°C from a pump that is not temperature compensated?

45. A steel ball bearing is 4.000 cm in diameter at 20.0°C. A bronze plate has a hole in it that is 3.994 cm in diameter at 20.0°C. What common temperature must they have so that the ball just squeezes through the hole?

46. Review Problem. An aluminum pipe 0.655 m long at 20.0°C and open at both ends is used as a flute. The pipe is cooled to a low temperature but is then filled with air at 20.0°C as soon as it is played. By how much does its fundamental frequency change as the temperature of the metal increases from 5.00°C to 20.0°C?

47. A mercury thermometer is constructed as shown in Figure P19.47. The capillary tube has a diameter of 0.004 00 cm, and the bulb has a diameter of 0.250 cm. Neglecting the expansion of the glass, find the change in height of the mercury column that occurs with a temperature change of 30.0°C.

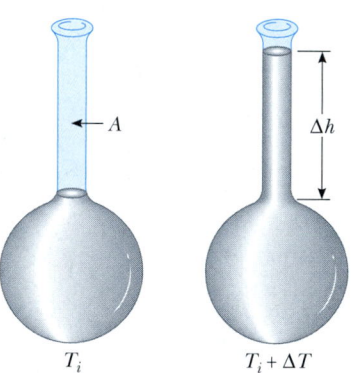

Figure P19.47   Problems 47 and 48.

48. A liquid with a coefficient of volume expansion β just fills a spherical shell of volume V_i at a temperature of T_i (see Fig. P19.47). The shell is made of a material that has an average coefficient of linear expansion α. The liquid is free to expand into an open capillary of area A projecting from the top of the sphere. (a) If the temperature increases by ΔT, show that the liquid rises in the capillary by the amount Δh given by the equation $\Delta h = (V_i/A)(\beta - 3\alpha)\,\Delta T$. (b) For a typical system, such as a mercury thermometer, why is it a good approximation to neglect the expansion of the shell?

WEB 49. A liquid has a density ρ. (a) Show that the fractional change in density for a change in temperature ΔT is $\Delta\rho/\rho = -\beta\,\Delta T$. What does the negative sign signify? (b) Fresh water has a maximum density of 1.000 0 g/cm³ at 4.0°C. At 10.0°C, its density is 0.999 7 g/cm³. What is β for water over this temperature interval?

50. A cylinder is closed by a piston connected to a spring of constant 2.00×10^3 N/m (Fig. P19.50). While the spring is relaxed, the cylinder is filled with 5.00 L of gas at a pressure of 1.00 atm and a temperature of 20.0°C. (a) If the piston has a cross-sectional area of 0.010 0 m² and a negligible mass, how high will it rise when the temperature is increased to 250°C? (b) What is the pressure of the gas at 250°C?

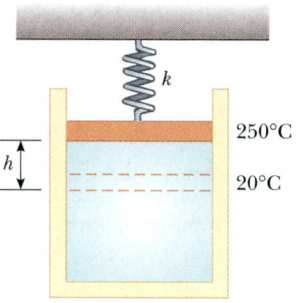

Figure P19.50

WEB 51. A vertical cylinder of cross-sectional area A is fitted with a tight-fitting, frictionless piston of mass m (Fig. P19.51). (a) If n moles of an ideal gas are in the cylinder at a temperature of T, what is the height h at which the piston is in equilibrium under its own weight? (b) What is the value for h if $n = 0.200$ mol, $T = 400$ K, $A = 0.008\ 00$ m², and $m = 20.0$ kg?

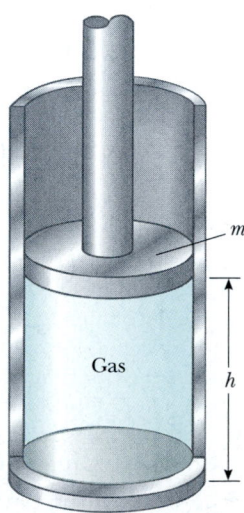

Figure P19.51

52. A bimetallic bar is made of two thin strips of dissimilar metals bonded together. As they are heated, the one with the greater average coefficient of expansion expands more than the other, forcing the bar into an arc, with the outer radius having a greater circumference (Fig. P19.52). (a) Derive an expression for the angle of bending θ as a function of the initial length of the strips, their average coefficients of linear expansion, the change in temperature, and the separation of the centers of the strips ($\Delta r = r_2 - r_1$). (b) Show that the angle of bending decreases to zero when ΔT decreases to zero or when the two average coefficients of expansion become equal. (c) What happens if the bar is cooled?

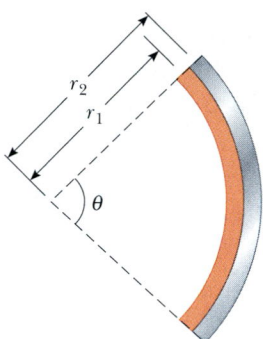

Figure P19.52

53. The rectangular plate shown in Figure P19.53 has an area A_i equal to ℓw. If the temperature increases by ΔT, show that the increase in area is $\Delta A = 2\alpha A_i \Delta T$, where α is the average coefficient of linear expansion. What approximation does this expression assume? (*Hint:* Note that each dimension increases according to the equation $\Delta L = \alpha L_i \Delta T$.)

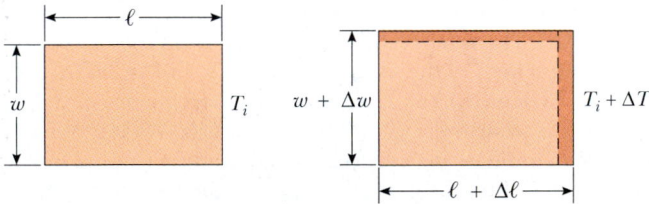

Figure P19.53

54. Precise temperature measurements are often made on the basis of the change in electrical resistance of a metal with temperature. The resistance varies approximately according to the expression $R = R_0(1 + AT_C)$, where R_0 and A are constants. A certain element has a resistance of 50.0 ohms (Ω) at 0°C and 71.5 Ω at the freezing point of tin (231.97°C). (a) Determine the constants A and R_0. (b) At what temperature is the resistance equal to 89.0 Ω?

55. Review Problem. A clock with a brass pendulum has a period of 1.000 s at 20.0°C. If the temperature increases to 30.0°C, (a) by how much does the period change, and (b) how much time does the clock gain or lose in one week?

56. Review Problem. Consider an object with any one of the shapes displayed in Table 10.2. What is the percentage increase in the moment of inertia of the object when it is heated from 0°C to 100°C if it is composed of (a) copper or (b) aluminum? (See Table 19.2. Assume that the average linear expansion coefficients do not vary between 0°C and 100°C.)

57. Review Problem. (a) Derive an expression for the buoyant force on a spherical balloon that is submerged in water as a function of the depth below the surface, the volume (V_i) of the balloon at the surface, the pressure (P_0) at the surface, and the density of the water. (Assume that water temperature does not change with depth.) (b) Does the buoyant force increase or decrease as the balloon is submerged? (c) At what depth is the buoyant force one-half the surface value?

58. (a) Show that the density of an ideal gas occupying a volume V is given by $\rho = PM/RT$, where M is the molar mass. (b) Determine the density of oxygen gas at atmospheric pressure and 20.0°C.

59. Starting with Equation 19.10, show that the total pressure P in a container filled with a mixture of several ideal gases is $P = P_1 + P_2 + P_3 + \ldots$, where P_1, P_2, $\ldots$ are the pressures that each gas would exert if it alone filled the container. (These individual pressures are called the *partial pressures* of the respective gases.) This is known as *Dalton's law of partial pressures*.

60. A sample of dry air that has a mass of 100.00 g, collected at sea level, is analyzed and found to consist of the following gases:

$$\text{nitrogen } (N_2) = 75.52 \text{ g}$$
$$\text{oxygen } (O_2) = 23.15 \text{ g}$$
$$\text{argon } (Ar) = 1.28 \text{ g}$$
$$\text{carbon dioxide } (CO_2) = 0.05 \text{ g}$$

as well as trace amounts of neon, helium, methane, and other gases. (a) Calculate the partial pressure (see Problem 59) of each gas when the pressure is 101.3 kPa. (b) Determine the volume occupied by the 100-g sample at a temperature of 15.00°C and a pressure of 1.013×10^5 Pa. What is the density of the air for these conditions? (c) What is the effective molar mass of the air sample?

61. Steel rails for an interurban rapid transit system form a continuous track that is held rigidly in place in concrete. (a) If the track was laid when the temperature was 0°C, what is the stress in the rails on a warm day when the temperature is 25.0°C? (b) What fraction of the yield strength of 52.2×10^7 N/m² does this stress represent?

62. (a) Use the equation of state for an ideal gas and the definition of the average coefficient of volume expansion, in the form $\beta = (1/V)\,dV/dT$, to show that the average coefficient of volume expansion for an ideal gas at constant pressure is given by $\beta = 1/T$, where T is the absolute temperature. (b) What value does this expression predict for β at 0°C? Compare this with the experimental values for helium and air in Table 19.2.

63. Two concrete spans of a 250-m-long bridge are placed end to end so that no room is allowed for expansion (Fig. P19.63a). If a temperature increase of 20.0°C occurs, what is the height y to which the spans rise when they buckle (Fig. P19.63b)?

64. Two concrete spans of a bridge of length L are placed end to end so that no room is allowed for expansion (see Fig. P19.63a). If a temperature increase of ΔT occurs, what is the height y to which the spans rise when they buckle (see Fig. P19.63b)?

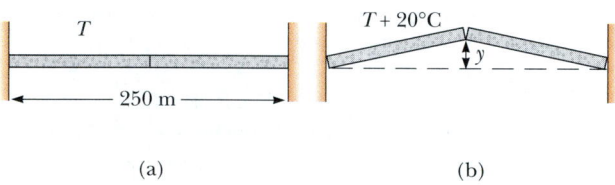

(a)   (b)

Figure P19.63   Problems 63 and 64.

65. A copper rod and a steel rod are heated. At 0°C the copper rod has length L_c, and the steel rod has length L_s. When the rods are being heated or cooled, the difference between their lengths stays constant at 5.00 cm. Determine the values of L_c and L_s.

66. A cylinder that has a 40.0-cm radius and is 50.0 cm deep is filled with air at 20.0°C and 1.00 atm (Fig. P19.66a). A 20.0-kg piston is now lowered into the cylinder, compressing the air trapped inside (Fig. P19.66b). Finally, a 75.0-kg man stands on the piston, further compressing the air, which remains at 20°C (Fig. P19.66c). (a) How far down (Δh) does the piston move when the man steps onto it? (b) To what temperature should the gas be heated to raise the piston and the man back to h_i?

67. The relationship $L_f = L_i(1 + \alpha\Delta T)$ is an approximation that works when the average coefficient of expansion is small. If α is large, one must integrate the relationship $dL/dT = \alpha L$ to determine the final length. (a) Assuming that the average coefficient of linear expansion is constant as L varies, determine a general expression for the final length. (b) Given a rod of length 1.00 m and a temperature change of 100.0 °C, determine the error caused by the approximation when $\alpha = 2.00 \times 10^{-5}$ (°C)$^{-1}$ (a typical value for a metal) and

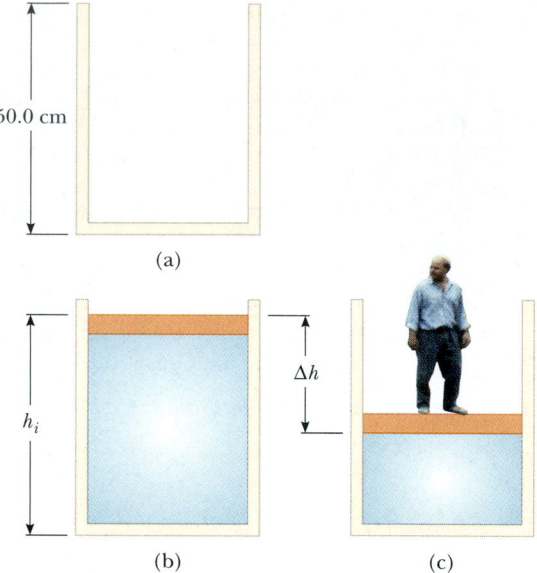

Figure P19.66

when $\alpha = 0.0200$ (°C)$^{-1}$ (an unrealistically large value for comparison).

68. A steel wire and a copper wire, each of diameter 2.000 mm, are joined end to end. At 40.0°C, each has an unstretched length of 2.000 m; they are connected between two fixed supports 4.000 m apart on a tabletop, so that the steel wire extends from $x = -2.000$ m to $x = 0$, the copper wire extends from $x = 0$ to $x = 2.000$ m, and the tension is negligible. The temperature is then lowered to 20.0°C. At this lower temperature, what are the tension in the wire and the x coordinate of the junction between the wires? (Refer to Tables 12.1 and 19.2.)

69. **Review Problem.**   A steel guitar string with a diameter of 1.00 mm is stretched between supports 80.0 cm apart. The temperature is 0.0°C. (a) Find the mass per unit length of this string. (Use 7.86×10^3 kg/m^3 as the mass density.) (b) The fundamental frequency of transverse oscillations of the string is 200 Hz. What is the tension in the string? (c) If the temperature is raised to 30.0°C, find the resulting values of the tension and the fundamental frequency. (Assume that both the Young's modulus [Table 12.1] and the average coefficient of linear expansion [Table 19.2] have constant values between 0.0°C and 30.0°C.)

70. A 1.00-km steel railroad rail is fastened securely at both ends when the temperature is 20.0°C. As the temperature increases, the rail begins to buckle. If its shape is an arc of a vertical circle, find the height h of the center of the buckle when the temperature is 25.0°C. (You will need to solve a transcendental equation.)

ANSWERS TO QUICK QUIZZES

19.1 The size of a degree on the Fahrenheit scale is $\frac{5}{9}$ the size of a degree on the Celsius scale. This is true because the Fahrenheit range of 32°F to 212°F is equivalent to the Celsius range of 0°C to 100°C. The factor $\frac{9}{5}$ in Equation 19.2 corrects for this difference. Equation 19.1 does not need this correction because the size of a Celsius degree is the same as the size of a kelvin.

19.2 The glass bulb containing most of the mercury warms up first because it is in direct thermal contact with the hot water. It expands slightly, and thus its volume increases. This causes the mercury level in the capillary tube to drop. As the mercury inside the bulb warms up, it also expands. Eventually, its increase in volume is sufficient to raise the mercury level in the capillary tube.

19.3 For the glass, choose Pyrex, which has a lower average coefficient of linear expansion than does ordinary glass. For the working liquid, choose gasoline, which has the largest average coefficient of volume expansion.

19.4 You do not have to convert the units for pressure and volume to SI units as long as the same units appear in the numerator and the denominator. This is not true for ratios of temperature units, as you can see by comparing the ratios 300 K/200 K and 26.85°C/(−73.15°C). You must always use absolute (kelvin) temperatures when applying the ideal gas law.

c h a p t e r

20

Heat and the First Law of Thermodynamics

Until about 1850, the fields of thermodynamics and mechanics were considered two distinct branches of science, and the law of conservation of energy seemed to describe only certain kinds of mechanical systems. However, mid–19th century experiments performed by the Englishman James Joule and others showed that energy may be added to (or removed from) a system either by heat or by doing work on the system (or having the system do work). Today we know that internal energy, which we formally define in this chapter, can be transformed to mechanical energy. Once the concept of energy was broadened to include internal energy, the law of conservation of energy emerged as a universal law of nature.

This chapter focuses on the concept of internal energy, the processes by which energy is transferred, the first law of thermodynamics, and some of the important applications of the first law. The first law of thermodynamics is the law of conservation of energy. It describes systems in which the only energy change is that of internal energy, which is due to transfers of energy by heat or work. Furthermore, the first law makes no distinction between the results of heat and the results of work. According to the first law, a system's internal energy can be changed either by an energy transfer by heat to or from the system or by work done on or by the system.

James Prescott Joule British physicist (1818–1889) Joule received some formal education in mathematics, philosophy, and chemistry but was in large part self-educated. His research led to the establishment of the principle of conservation of energy. His study of the quantitative relationship among electrical, mechanical, and chemical effects of heat culminated in his discovery in 1843 of the amount of work required to produce a unit of energy, called the mechanical equivalent of heat. *(By kind permission of the President and Council of the Royal Society)*

20.1 HEAT AND INTERNAL ENERGY

10.3

At the outset, it is important that we make a major distinction between internal energy and heat. **Internal energy is all the energy of a system that is associated with its microscopic components—atoms and molecules—when viewed from a reference frame at rest with respect to the object.** The last part of this sentence ensures that any bulk kinetic energy of the system due to its motion through space is not included in internal energy. Internal energy includes kinetic energy of translation, rotation, and vibration of molecules, potential energy within molecules, and potential energy between molecules. It is useful to relate internal energy to the temperature of an object, but this relationship is limited—we shall find in Section 20.3 that internal energy changes can also occur in the absence of temperature changes.

As we shall see in Chapter 21, the internal energy of a monatomic ideal gas is associated with the translational motion of its atoms. This is the only type of energy available for the microscopic components of this system. In this special case, the internal energy is simply the total kinetic energy of the atoms of the gas; the higher the temperature of the gas, the greater the average kinetic energy of the atoms and the greater the internal energy of the gas. More generally, in solids, liquids, and molecular gases, internal energy includes other forms of molecular energy. For example, a diatomic molecule can have rotational kinetic energy, as well as vibrational kinetic and potential energy.

Heat is defined as the transfer of energy across the boundary of a system due to a temperature difference between the system and its surroundings. When you *heat* a substance, you are transferring energy into it by placing it in contact with surroundings that have a higher temperature. This is the case, for example, when you place a pan of cold water on a stove burner—the burner is at a higher temperature than the water, and so the water gains energy. We shall also use the term *heat* to represent the amount of energy transferred by this method.

Scientists used to think of heat as a fluid called *caloric*, which they believed was transferred between objects; thus, they defined heat in terms of the temperature changes produced in an object during heating. Today we recognize the distinct difference between internal energy and heat. Nevertheless, we refer to quantities

Heat

using names that do not quite correctly define the quantities but which have become entrenched in physics tradition based on these early ideas. Examples of such quantities are *latent heat* and *heat capacity*.

As an analogy to the distinction between heat and internal energy, consider the distinction between work and mechanical energy discussed in Chapter 7. The work done on a system is a measure of the amount of energy transferred to the system from its surroundings, whereas the mechanical energy of the system (kinetic or potential, or both) is a consequence of the motion and relative positions of the members of the system. Thus, when a person does work on a system, energy is transferred from the person to the system. It makes no sense to talk about the work *of* a system—one can refer only to the work done *on* or *by* a system when some process has occurred in which energy has been transferred to or from the system. Likewise, it makes no sense to talk about the heat *of* a system—one can refer to *heat* only when energy has been transferred as a result of a temperature difference. Both heat and work are ways of changing the energy of a system.

It is also important to recognize that the internal energy of a system can be changed even when no energy is transferred by heat. For example, when a gas is compressed by a piston, the gas is warmed and its internal energy increases, but no transfer of energy by heat from the surroundings to the gas has occurred. If the gas then expands rapidly, it cools and its internal energy decreases, but no transfer of energy by heat from it to the surroundings has taken place. The temperature changes in the gas are due not to a difference in temperature between the gas and its surroundings but rather to the compression and the expansion. In each case, energy is transferred to or from the gas by *work*, and the energy change within the system is an increase or decrease of internal energy. The changes in internal energy in these examples are evidenced by corresponding changes in the temperature of the gas.

Units of Heat

As we have mentioned, early studies of heat focused on the resultant increase in temperature of a substance, which was often water. The early notions of heat based on caloric suggested that the flow of this fluid from one body to another caused changes in temperature. From the name of this mythical fluid, we have an energy unit related to thermal processes, the **calorie (cal),** which is defined as **the amount of energy transfer necessary to raise the temperature of 1 g of water from 14.5°C to 15.5°C.**[1] (Note that the "Calorie," written with a capital "C" and used in describing the energy content of foods, is actually a kilocalorie.) The unit of energy in the British system is the **British thermal unit (Btu),** which is defined as **the amount of energy transfer required to raise the temperature of 1 lb of water from 63°F to 64°F.**

Scientists are increasingly using the SI unit of energy, the *joule,* when describing thermal processes. In this textbook, heat and internal energy are usually measured in joules. (Note that both heat and work are measured in energy units. Do not confuse these two means of energy *transfer* with energy itself, which is also measured in joules.)

The calorie

---

[1] Originally, the calorie was defined as the "heat" necessary to raise the temperature of 1 g of water by 1°C. However, careful measurements showed that the amount of energy required to produce a 1°C change depends somewhat on the initial temperature; hence, a more precise definition evolved.

The Mechanical Equivalent of Heat

In Chapters 7 and 8, we found that whenever friction is present in a mechanical system, some mechanical energy is lost—in other words, mechanical energy is not conserved in the presence of nonconservative forces. Various experiments show that this lost mechanical energy does not simply disappear but is transformed into internal energy. We can perform such an experiment at home by simply hammering a nail into a scrap piece of wood. What happens to all the kinetic energy of the hammer once we have finished? Some of it is now in the nail as internal energy, as demonstrated by the fact that the nail is measurably warmer. Although this connection between mechanical and internal energy was first suggested by Benjamin Thompson, it was Joule who established the equivalence of these two forms of energy.

A schematic diagram of Joule's most famous experiment is shown in Figure 20.1. The system of interest is the water in a thermally insulated container. Work is done on the water by a rotating paddle wheel, which is driven by heavy blocks falling at a constant speed. The stirred water is warmed due to the friction between it and the paddles. If the energy lost in the bearings and through the walls is neglected, then the loss in potential energy associated with the blocks equals the work done by the paddle wheel on the water. If the two blocks fall through a distance h, the loss in potential energy is $2mgh$, where m is the mass of one block; it is this energy that causes the temperature of the water to increase. By varying the conditions of the experiment, Joule found that the loss in mechanical energy $2mgh$ is proportional to the increase in water temperature ΔT. The proportionality constant was found to be approximately $4.18 \, \text{J/g} \cdot {}^{\circ}\text{C}$. Hence, $4.18 \, \text{J}$ of mechanical energy raises the temperature of 1 g of water by $1{}^{\circ}\text{C}$. More precise measurements taken later demonstrated the proportionality to be $4.186 \, \text{J/g} \cdot {}^{\circ}\text{C}$ when the temperature of the water was raised from $14.5{}^{\circ}\text{C}$ to $15.5{}^{\circ}\text{C}$. We adopt this "15-degree calorie" value:

$$1 \, \text{cal} \equiv 4.186 \, \text{J} \qquad \textbf{(20.1)}$$

This equality is known, for purely historical reasons, as the **mechanical equivalent of heat.**

Benjamin Thompson (1753–1814). *(North Wind Picture Archives)*

Mechanical equivalent of heat

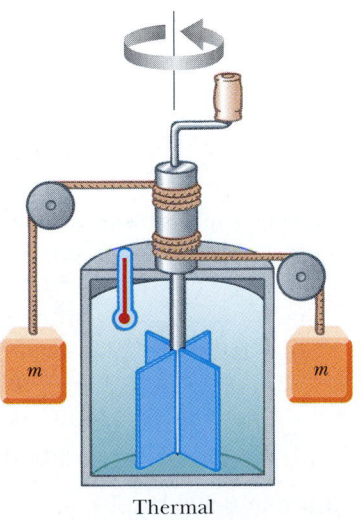

Thermal insulator

Figure 20.1   Joule's experiment for determining the mechanical equivalent of heat. The falling blocks rotate the paddles, causing the temperature of the water to increase.

EXAMPLE 20.1   **Losing Weight the Hard Way**

A student eats a dinner rated at 2 000 Calories. He wishes to do an equivalent amount of work in the gymnasium by lifting a 50.0-kg barbell. How many times must he raise the barbell to expend this much energy? Assume that he raises the barbell 2.00 m each time he lifts it and that he regains no energy when he drops the barbell to the floor.

Solution   Because 1 Calorie = 1.00×10^3 cal, the work required is 2.00×10^6 cal. Converting this value to joules, we have for the total work required:

$$W = (2.00 \times 10^6 \text{ cal})(4.186 \text{ J/cal}) = 8.37 \times 10^6 \text{ J}$$

The work done in lifting the barbell a distance h is equal to mgh, and the work done in lifting it n times is $nmgh$. We equate this to the total work required:

$$W = nmgh = 8.37 \times 10^6 \text{ J}$$

$$n = \frac{8.37 \times 10^6 \text{ J}}{(50.0 \text{ kg})(9.80 \text{ m/s}^2)(2.00 \text{ m})} = \boxed{8.54 \times 10^3 \text{ times}}$$

If the student is in good shape and lifts the barbell once every 5 s, it will take him about 12 h to perform this feat. Clearly, it is much easier for this student to lose weight by dieting.

20.2   HEAT CAPACITY AND SPECIFIC HEAT

10.3   When energy is added to a substance and no work is done, the temperature of the substance usually rises. (An exception to this statement is the case in which a substance undergoes a change of state—also called a *phase transition*—as discussed in the next section.) The quantity of energy required to raise the temperature of a given mass of a substance by some amount varies from one substance to another. For example, the quantity of energy required to raise the temperature of 1 kg of water by 1°C is 4 186 J, but the quantity of energy required to raise the temperature of 1 kg of copper by 1°C is only 387 J. In the discussion that follows, we shall use heat as our example of energy transfer, but we shall keep in mind that we could change the temperature of our system by doing work on it.

The **heat capacity** C of a particular sample of a substance is defined as the amount of energy needed to raise the temperature of that sample by 1°C. From this definition, we see that if heat Q produces a change ΔT in the temperature of a substance, then

Heat capacity

$$Q = C\Delta T \tag{20.2}$$

The **specific heat** c of a substance is the heat capacity per unit mass. Thus, if energy Q transferred by heat to mass m of a substance changes the temperature of the sample by ΔT, then the specific heat of the substance is

Specific heat

$$c \equiv \frac{Q}{m\Delta T} \tag{20.3}$$

Specific heat is essentially a measure of how thermally insensitive a substance is to the addition of energy. The greater a material's specific heat, the more energy must be added to a given mass of the material to cause a particular temperature change. Table 20.1 lists representative specific heats.

From this definition, we can express the energy Q transferred by heat between a sample of mass m of a material and its surroundings for a temperature change ΔT as

$$Q = mc\Delta T \tag{20.4}$$

For example, the energy required to raise the temperature of 0.500 kg of water by 3.00°C is $(0.500 \text{ kg})(4 \ 186 \text{ J/kg} \cdot °\text{C})(3.00°\text{C}) = 6.28 \times 10^3 \text{ J}$. Note that when the temperature increases, Q and ΔT are taken to be positive, and energy flows into

TABLE 20.1	**Specific Heats of Some Substances at 25°C and Atmospheric Pressure**	
	Specific Heat c	
Substance	**J/kg·°C**	**cal/g·°C**
Elemental Solids		
Aluminum	900	0.215
Beryllium	1 830	0.436
Cadmium	230	0.055
Copper	387	0.092 4
Germanium	322	0.077
Gold	129	0.030 8
Iron	448	0.107
Lead	128	0.030 5
Silicon	703	0.168
Silver	234	0.056
Other Solids		
Brass	380	0.092
Glass	837	0.200
Ice ($-5°C$)	2 090	0.50
Marble	860	0.21
Wood	1 700	0.41
Liquids		
Alcohol (ethyl)	2 400	0.58
Mercury	140	0.033
Water (15°C)	4 186	1.00
Gas		
Steam (100°C)	2 010	0.48

the system. When the temperature decreases, Q and ΔT are negative, and energy flows out of the system.

Specific heat varies with temperature. However, if temperature intervals are not too great, the temperature variation can be ignored and c can be treated as a constant.[2] For example, the specific heat of water varies by only about 1% from 0°C to 100°C at atmospheric pressure. Unless stated otherwise, we shall neglect such variations.

Measured values of specific heats are found to depend on the conditions of the experiment. In general, measurements made at constant pressure are different from those made at constant volume. For solids and liquids, the difference between the two values is usually no greater than a few percent and is often neglected. Most of the values given in Table 20.1 were measured at atmospheric pressure and room temperature. As we shall see in Chapter 21, the specific heats for

---

[2] The definition given by Equation 20.3 assumes that the specific heat does not vary with temperature over the interval $\Delta T = T_f - T_i$. In general, if c varies with temperature over the interval, then the correct expression for Q is

$$Q = m \int_{T_i}^{T_f} c \, dT$$

gases measured at constant pressure are quite different from values measured at constant volume.

Imagine you have 1 kg each of iron, glass, and water, and that all three samples are at 10°C. (a) Rank the samples from lowest to highest temperature after 100 J of energy is added to each. (b) Rank them from least to greatest amount of energy transferred by heat if each increases in temperature by 20°C.

It is interesting to note from Table 20.1 that water has the highest specific heat of common materials. This high specific heat is responsible, in part, for the moderate temperatures found near large bodies of water. As the temperature of a body of water decreases during the winter, energy is transferred from the cooling water to the air by heat, increasing the internal energy of the air. Because of the high specific heat of water, a relatively large amount of energy is transferred to the air for even modest temperature changes of the water. The air carries this internal energy landward when prevailing winds are favorable. For example, the prevailing winds on the West Coast of the United States are toward the land (eastward). Hence, the energy liberated by the Pacific Ocean as it cools keeps coastal areas much warmer than they would otherwise be. This explains why the western coastal states generally have more favorable winter weather than the eastern coastal states, where the prevailing winds do not tend to carry the energy toward land.

 A difference in specific heats causes the cheese topping on a slice of pizza to burn you more than a mouthful of crust at the same temperature. Both crust and cheese undergo the same change in temperature, starting at a high straight-from-the-oven value and ending at the temperature of the inside of your mouth, which is about 37°C. Because the cheese is much more likely to burn you, it must release much more energy as it cools than does the crust. If we assume roughly the same mass for both cheese and crust, then Equation 20.3 indicates that the specific heat of the cheese, which is mostly water, is greater than that of the crust, which is mostly air.

Conservation of Energy: Calorimetry

One technique for measuring specific heat involves heating a sample to some known temperature T_x, placing it in a vessel containing water of known mass and temperature $T_w < T_x$, and measuring the temperature of the water after equilibrium has been reached. Because a negligible amount of mechanical work is done in the process, the law of the conservation of energy requires that the amount of energy that leaves the sample (of unknown specific heat) equal the amount of energy that enters the water.[3] This technique is called **calorimetry,** and devices in which this energy transfer occurs are called **calorimeters.**

Conservation of energy allows us to write the equation

$$Q_{cold} = -Q_{hot} \qquad \qquad \textbf{(20.5)}$$

which simply states that the energy leaving the hot part of the system by heat is equal to that entering the cold part of the system. The negative sign in the equation is necessary to maintain consistency with our sign convention for heat. The

[3] For precise measurements, the water container should be included in our calculations because it also exchanges energy with the sample. However, doing so would require a knowledge of its mass and composition. If the mass of the water is much greater than that of the container, we can neglect the effects of the container.

heat Q_{hot} is negative because energy is leaving the hot sample. The negative sign in the equation ensures that the right-hand side is positive and thus consistent with the left-hand side, which is positive because energy is entering the cold water.

Suppose m_x is the mass of a sample of some substance whose specific heat we wish to determine. Let us call its specific heat c_x and its initial temperature T_x. Likewise, let m_w, c_w, and T_w represent corresponding values for the water. If T_f is the final equilibrium temperature after everything is mixed, then from Equation 20.4, we find that the energy transfer for the water is $m_w c_w (T_f - T_w)$, which is positive because $T_f > T_w$, and that the energy transfer for the sample of unknown specific heat is $m_x c_x (T_f - T_x)$, which is negative. Substituting these expressions into Equation 20.5 gives

$$m_w c_w (T_f - T_w) = - m_x c_x (T_f - T_x)$$

Solving for c_x gives

$$c_x = \frac{m_w c_w (T_f - T_w)}{m_x (T_x - T_f)}$$

EXAMPLE 20.2 ▶ Cooling a Hot Ingot

A 0.050 0-kg ingot of metal is heated to 200.0°C and then dropped into a beaker containing 0.400 kg of water initially at 20.0°C. If the final equilibrium temperature of the mixed system is 22.4°C, find the specific heat of the metal.

Solution   According to Equation 20.5, we can write

$$m_w c_w (T_f - T_w) = - m_x c_x (T_f - T_x)$$
$$(0.400 \text{ kg}) (4\ 186 \text{ J/kg} \cdot {}^{\circ}\text{C}) (22.4{}^{\circ}\text{C} - 20.0{}^{\circ}\text{C}) =$$
$$- (0.050\ 0 \text{ kg}) (c_x) (22.4{}^{\circ}\text{C} - 200.0{}^{\circ}\text{C})$$

From this we find that

$$c_x = \boxed{453 \text{ J/kg} \cdot {}^{\circ}\text{C}}$$

The ingot is most likely iron, as we can see by comparing this result with the data given in Table 20.1. Note that the temperature of the ingot is initially above the steam point. Thus, some of the water may vaporize when we drop the ingot into the water. We assume that we have a sealed system and thus that this steam cannot escape. Because the final equilibrium temperature is lower than the steam point, any steam that does result recondenses back into water.

Exercise   What is the amount of energy transferred to the water as the ingot is cooled?

Answer   4 020 J.

EXAMPLE 20.3 ▶ Fun Time for a Cowboy

A cowboy fires a silver bullet with a mass of 2.00 g and with a muzzle speed of 200 m/s into the pine wall of a saloon. Assume that all the internal energy generated by the impact remains with the bullet. What is the temperature change of the bullet?

Solution   The kinetic energy of the bullet is

$$\tfrac{1}{2} m v^2 = \tfrac{1}{2} (2.00 \times 10^{-3} \text{ kg}) (200 \text{ m/s})^2 = 40.0 \text{ J}$$

Because nothing in the environment is hotter than the bullet, the bullet gains no energy by heat. Its temperature increases because the 40.0 J of kinetic energy becomes 40.0 J of extra internal energy. The temperature change is the same as that which would take place if 40.0 J of energy were transferred by

heat from a stove to the bullet. If we imagine this latter process taking place, we can calculate ΔT from Equation 20.4. Using 234 J/kg · °C as the specific heat of silver (see Table 20.1), we obtain

$$\Delta T = \frac{Q}{mc} = \frac{40.0 \text{ J}}{(2.00 \times 10^{-3} \text{ kg}) (234 \text{ J/kg} \cdot {}^{\circ}\text{C})} = \boxed{85.5{}^{\circ}\text{C}}$$

Exercise   Suppose that the cowboy runs out of silver bullets and fires a lead bullet of the same mass and at the same speed into the wall. What is the temperature change of the bullet?

Answer   156°C.

20.3 ▸ LATENT HEAT

A substance often undergoes a change in temperature when energy is transferred between it and its surroundings. There are situations, however, in which the transfer of energy does not result in a change in temperature. This is the case whenever the physical characteristics of the substance change from one form to another; such a change is commonly referred to as a **phase change.** Two common phase changes are from solid to liquid (melting) and from liquid to gas (boiling); another is a change in the crystalline structure of a solid. All such phase changes involve a change in internal energy but no change in temperature. The increase in internal energy in boiling, for example, is represented by the breaking of bonds between molecules in the liquid state; this bond breaking allows the molecules to move farther apart in the gaseous state, with a corresponding increase in intermolecular potential energy.

As you might expect, different substances respond differently to the addition or removal of energy as they change phase because their internal molecular arrangements vary. Also, the amount of energy transferred during a phase change depends on the amount of substance involved. (It takes less energy to melt an ice cube than it does to thaw a frozen lake.) If a quantity Q of energy transfer is required to change the phase of a mass m of a substance, the ratio $L \equiv Q/m$ characterizes an important thermal property of that substance. Because this added or removed energy does not result in a temperature change, the quantity L is called the **latent heat** (literally, the "hidden" heat) of the substance. The value of L for a substance depends on the nature of the phase change, as well as on the properties of the substance.

From the definition of latent heat, and again choosing heat as our energy transfer mechanism, we find that the energy required to change the phase of a given mass m of a pure substance is

$$Q = mL \tag{20.6}$$

Latent heat of fusion L_f is the term used when the phase change is from solid to liquid (*to fuse* means "to combine by melting"), and **latent heat of vaporization**

	TABLE 20.2	**Latent Heats of Fusion and Vaporization**		
Substance	**Melting Point (°C)**	**Latent Heat of Fusion (J/kg)**	**Boiling Point (°C)**	**Latent Heat of Vaporization (J/kg)**
Helium	− 269.65	5.23×10^3	− 268.93	2.09×10^4
Nitrogen	− 209.97	2.55×10^4	− 195.81	2.01×10^5
Oxygen	− 218.79	1.38×10^4	− 182.97	2.13×10^5
Ethyl alcohol	− 114	1.04×10^5	78	8.54×10^5
Water	0.00	3.33×10^5	100.00	2.26×10^6
Sulfur	119	3.81×10^4	444.60	3.26×10^5
Lead	327.3	2.45×10^4	1 750	8.70×10^5
Aluminum	660	3.97×10^5	2 450	1.14×10^7
Silver	960.80	8.82×10^4	2 193	2.33×10^6
Gold	1 063.00	6.44×10^4	2 660	1.58×10^6
Copper	1 083	1.34×10^5	1 187	5.06×10^6

L_v is the term used when the phase change is from liquid to gas (the liquid "vaporizes").[4] The latent heats of various substances vary considerably, as data in Table 20.2 show.

Which is more likely to cause a serious burn, 100°C liquid water or an equal mass of 100°C steam?

To understand the role of latent heat in phase changes, consider the energy required to convert a 1.00-g block of ice at −30.0°C to steam at 120.0°C. Figure 20.2 indicates the experimental results obtained when energy is gradually added to the ice. Let us examine each portion of the red curve.

Part A.   On this portion of the curve, the temperature of the ice changes from −30.0°C to 0.0°C. Because the specific heat of ice is 2 090 J/kg · °C, we can calculate the amount of energy added by using Equation 20.4:

$$Q = m_i c_i \, \Delta T = (1.00 \times 10^{-3} \text{ kg})(2\,090 \text{ J/kg} \cdot °\text{C})(30.0°\text{C}) = 62.7 \text{ J}$$

Part B.   When the temperature of the ice reaches 0.0°C, the ice–water mixture remains at this temperature—even though energy is being added—until all the ice melts. The energy required to melt 1.00 g of ice at 0.0°C is, from Equation 20.6,

$$Q = mL_f = (1.00 \times 10^{-3} \text{ kg})(3.33 \times 10^5 \text{ J/kg}) = 333 \text{ J}$$

Thus, we have moved to the 396 J (= 62.7 J + 333 J) mark on the energy axis.

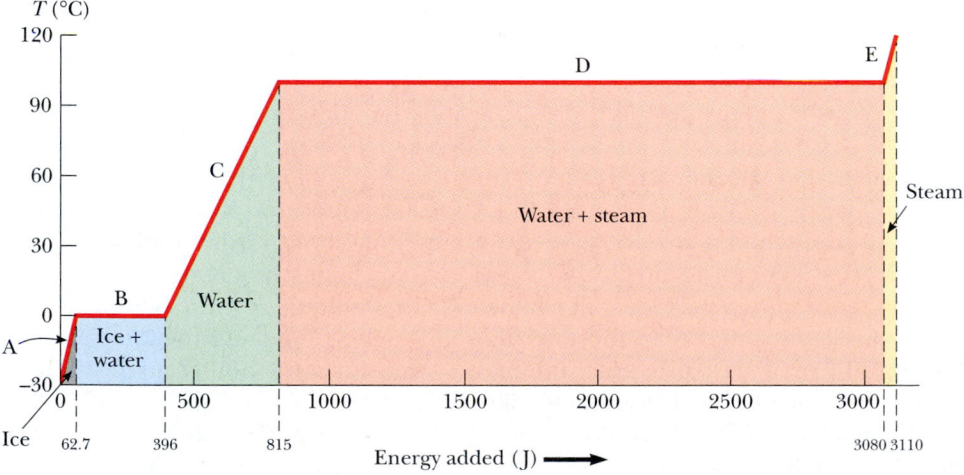

Figure 20.2   A plot of temperature versus energy added when 1.00 g of ice initially at −30.0°C is converted to steam at 120.0°C.

---

[4] When a gas cools, it eventually *condenses*—that is, it returns to the liquid phase. The energy given up per unit mass is called the *latent heat of condensation* and is numerically equal to the latent heat of vaporization. Likewise, when a liquid cools, it eventually solidifies, and the *latent heat of solidification* is numerically equal to the latent heat of fusion.

Part C.   Between 0.0°C and 100.0°C, nothing surprising happens. No phase change occurs, and so all energy added to the water is used to increase its temperature. The amount of energy necessary to increase the temperature from 0.0°C to 100.0°C is

$$Q = m_w c_w \, \Delta T = (1.00 \times 10^{-3} \, \text{kg})(4.19 \times 10^3 \, \text{J/kg} \cdot {}^\circ\text{C})(100.0{}^\circ\text{C}) = 419 \, \text{J}$$

Part D.   At 100.0°C, another phase change occurs as the water changes from water at 100.0°C to steam at 100.0°C. Similar to the ice–water mixture in part B, the water–steam mixture remains at 100.0°C—even though energy is being added—until all of the liquid has been converted to steam. The energy required to convert 1.00 g of water to steam at 100.0°C is

$$Q = mL_v = (1.00 \times 10^{-3} \, \text{kg})(2.26 \times 10^6 \, \text{J/kg}) = 2.26 \times 10^3 \, \text{J}$$

Part E.   On this portion of the curve, as in parts A and C, no phase change occurs; thus, all energy added is used to increase the temperature of the steam. The energy that must be added to raise the temperature of the steam from 100.0°C to 120.0°C is

$$Q = m_s c_s \, \Delta T = (1.00 \times 10^{-3} \, \text{kg})(2.01 \times 10^3 \, \text{J/kg} \cdot {}^\circ\text{C})(20.0{}^\circ\text{C}) = 40.2 \, \text{J}$$

The total amount of energy that must be added to change 1 g of ice at $-30.0{}^\circ\text{C}$ to steam at 120.0°C is the sum of the results from all five parts of the curve, which is $3.11 \times 10^3 \, \text{J}$. Conversely, to cool 1 g of steam at 120.0°C to ice at $-30.0{}^\circ\text{C}$, we must remove $3.11 \times 10^3 \, \text{J}$ of energy.

We can describe phase changes in terms of a rearrangement of molecules when energy is added to or removed from a substance. (For elemental substances in which the atoms do not combine to form molecules, the following discussion should be interpreted in terms of atoms. We use the general term *molecules* to refer to both molecular substances and elemental substances.) Consider first the liquid-to-gas phase change. The molecules in a liquid are close together, and the forces between them are stronger than those between the more widely separated molecules of a gas. Therefore, work must be done on the liquid against these attractive molecular forces if the molecules are to separate. The latent heat of vaporization is the amount of energy per unit mass that must be added to the liquid to accomplish this separation.

Similarly, for a solid, we imagine that the addition of energy causes the amplitude of vibration of the molecules about their equilibrium positions to become greater as the temperature increases. At the melting point of the solid, the amplitude is great enough to break the bonds between molecules and to allow molecules to move to new positions. The molecules in the liquid also are bound to each other, but less strongly than those in the solid phase. The latent heat of fusion is equal to the energy required per unit mass to transform the bonds among all molecules from the solid-type bond to the liquid-type bond.

As you can see from Table 20.2, the latent heat of vaporization for a given substance is usually somewhat higher than the latent heat of fusion. This is not surprising if we consider that the average distance between molecules in the gas phase is much greater than that in either the liquid or the solid phase. In the solid-to-liquid phase change, we transform solid-type bonds between molecules into liquid-type bonds between molecules, which are only slightly less strong. In the liquid-to-gas phase change, however, we break liquid-type bonds and create a situation in which the molecules of the gas essentially are not bonded to each other. Therefore, it is not surprising that more energy is required to vaporize a given mass of substance than is required to melt it.

Quick Quiz 20.3

Calculate the slopes for the A, C, and E portions of Figure 20.2. Rank the slopes from least to greatest and explain what this ordering means.

Problem-Solving Hints

Calorimetry Problems

If you are having difficulty in solving calorimetry problems, be sure to consider the following points:

- Units of measure must be consistent. For instance, if you are using specific heats measured in cal/g · °C, be sure that masses are in grams and temperatures are in Celsius degrees.
- Transfers of energy are given by the equation $Q = mc\Delta T$ only for those processes in which no phase changes occur. Use the equations $Q = mL_f$ and $Q = mL_v$ only when phase changes *are* taking place.
- Often, errors in sign are made when the equation $Q_{cold} = -Q_{hot}$ is used. Make sure that you use the negative sign in the equation, and remember that ΔT is always the final temperature minus the initial temperature.

EXAMPLE 20.4   Cooling the Steam

What mass of steam initially at 130°C is needed to warm 200 g of water in a 100-g glass container from 20.0°C to 50.0°C?

Solution   The steam loses energy in three stages. In the first stage, the steam is cooled to 100°C. The energy transfer in the process is

$$Q_1 = m_s c_s \,\Delta T = m_s(2.01 \times 10^3 \,\text{J/kg} \cdot \text{°C})(-30.0\text{°C})$$
$$= -m_s(6.03 \times 10^4 \,\text{J/kg})$$

where m_s is the unknown mass of the steam.

In the second stage, the steam is converted to water. To find the energy transfer during this phase change, we use $Q = -mL_v$, where the negative sign indicates that energy is leaving the steam:

$$Q_2 = -m_s(2.26 \times 10^6 \,\text{J/kg})$$

In the third stage, the temperature of the water created from the steam is reduced to 50.0°C. This change requires an energy transfer of

$$Q_3 = m_s c_w \,\Delta T = m_s(4.19 \times 10^3 \,\text{J/kg} \cdot \text{°C})(-50.0\text{°C})$$
$$= -m_s(2.09 \times 10^5 \,\text{J/kg})$$

Adding the energy transfers in these three stages, we obtain

$$Q_{hot} = Q_1 + Q_2 + Q_3$$
$$= -m_s(6.03 \times 10^4 \,\text{J/kg} + 2.26 \times 10^6 \,\text{J/kg}$$
$$+ 2.09 \times 10^5 \,\text{J/kg})$$
$$= -m_s(2.53 \times 10^6 \,\text{J/kg})$$

Now, we turn our attention to the temperature increase of the water and the glass. Using Equation 20.4, we find that

$$Q_{cold} = (0.200 \,\text{kg})(4.19 \times 10^3 \,\text{J/kg} \cdot \text{°C})(30.0\text{°C})$$
$$+ (0.100 \,\text{kg})(837 \,\text{J/kg} \cdot \text{°C})(30.0\text{°C})$$
$$= 2.77 \times 10^4 \,\text{J}$$

Using Equation 20.5, we can solve for the unknown mass:

$$Q_{cold} = -Q_{hot}$$
$$2.77 \times 10^4 \,\text{J} = -[-m_s(2.53 \times 10^6 \,\text{J/kg})]$$
$$m_s = \boxed{1.09 \times 10^{-2} \,\text{kg} = 10.9 \,\text{g}}$$

EXAMPLE 20.5   Boiling Liquid Helium

Liquid helium has a very low boiling point, 4.2 K, and a very low latent heat of vaporization, 2.09×10^4 J/kg. If energy is transferred to a container of boiling liquid helium from an immersed electric heater at a rate of 10.0 W, how long does it take to boil away 1.00 kg of the liquid?

Solution   Because $L_v = 2.09 \times 10^4$ J/kg, we must supply 2.09×10^4 J of energy to boil away 1.00 kg. Because 10.0 W = 10.0 J/s, 10.0 J of energy is transferred to the helium each second. Therefore, the time it takes to transfer 2.09×10^4 J

of energy is

$$t = \frac{2.09 \times 10^4 \, \text{J}}{10.0 \, \text{J/s}} = 2.09 \times 10^3 \, \text{s} \approx \boxed{35 \text{ min}}$$

Exercise   If 10.0 W of power is supplied to 1.00 kg of water at 100°C, how long does it take for the water to completely boil away?

Answer   62.8 h.

20.4   WORK AND HEAT IN THERMODYNAMIC PROCESSES

In the macroscopic approach to thermodynamics, we describe the *state* of a system using such variables as pressure, volume, temperature, and internal energy. The number of macroscopic variables needed to characterize a system depends on the nature of the system. For a homogeneous system, such as a gas containing only one type of molecule, usually only two variables are needed. However, it is important to note that a *macroscopic state* of an isolated system can be specified only if the system is in thermal equilibrium internally. In the case of a gas in a container, internal thermal equilibrium requires that every part of the gas be at the same pressure and temperature.

Consider a gas contained in a cylinder fitted with a movable piston (Fig. 20.3). At equilibrium, the gas occupies a volume V and exerts a uniform pressure P on the cylinder's walls and on the piston. If the piston has a cross-sectional area A, the

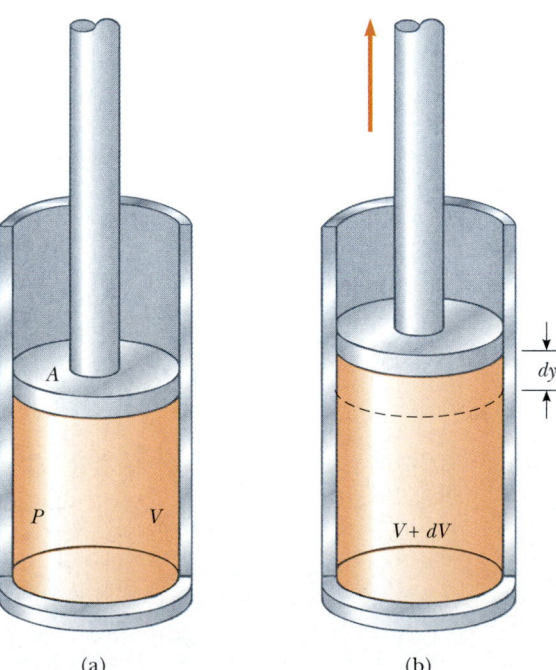

(a)   (b)

Figure 20.3   Gas contained in a cylinder at a pressure P does work on a moving piston as the system expands from a volume V to a volume $V + dV$.

force exerted by the gas on the piston is $F = PA$. Now let us assume that the gas expands **quasi-statically,** that is, slowly enough to allow the system to remain essentially in thermal equilibrium at all times. As the piston moves up a distance dy, the work done by the gas on the piston is

$$dW = F\, dy = PA\, dy$$

Because $A\, dy$ is the increase in volume of the gas dV, we can express the work done by the gas as

$$dW = P\, dV \qquad\qquad \textbf{(20.7)}$$

Because the gas expands, dV is positive, and so the work done by the gas is positive. If the gas were compressed, dV would be negative, indicating that the work done by the gas (which can be interpreted as work done *on* the gas) was negative.

In the thermodynamics problems that we shall solve, we shall identify the system of interest as a substance that is exchanging energy with the environment. In many problems, this will be a gas contained in a vessel; however, we will also consider problems involving liquids and solids. It is an unfortunate fact that, because of the separate historical development of thermodynamics and mechanics, positive work for a thermodynamic system is commonly defined as the work done *by* the system, rather than that done *on* the system. This is the reverse of the case for our study of work in mechanics. Thus, **in thermodynamics, positive work represents a transfer of energy out of the system.** We will use this convention to be consistent with common treatments of thermodynamics.

The total work done by the gas as its volume changes from V_i to V_f is given by the integral of Equation 20.7:

$$W = \int_{V_i}^{V_f} P\, dV \qquad\qquad \textbf{(20.8)}$$

To evaluate this integral, it is not enough that we know only the initial and final values of the pressure. We must also know the pressure at every instant during the expansion; we would know this if we had a functional dependence of P with respect to V. This important point is true for any process—the expansion we are discussing here, or any other. To fully specify a process, we must know the values of the thermodynamic variables at every state through which the system passes between the initial and final states. In the expansion we are considering here, we can plot the pressure and volume at each instant to create a *PV* diagram like the one shown in Figure 20.4. The value of the integral in Equation 20.8 is the area bounded by such a curve. Thus, we can say that

the work done by a gas in the expansion from an initial state to a final state is the area under the curve connecting the states in a *PV* diagram.

As Figure 20.4 shows, the work done in the expansion from the initial state i to the final state f depends on the path taken between these two states, where the *path* on a *PV* diagram is a description of the thermodynamic process through which the system is taken. To illustrate this important point, consider several paths connecting i and f (Fig. 20.5). In the process depicted in Figure 20.5a, the pressure of the gas is first reduced from P_i to P_f by cooling at constant volume V_i. The gas then expands from V_i to V_f at constant pressure P_f. The value of the work done along this path is equal to the area of the shaded rectangle, which is equal to

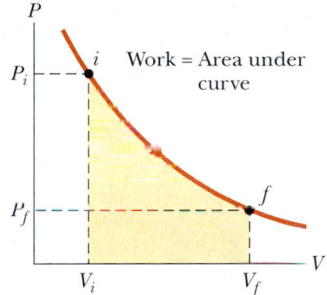

Figure 20.4 A gas expands quasi-statically (slowly) from state i to state f. The work done by the gas equals the area under the *PV* curve.

Work equals area under the curve in a *PV* diagram.

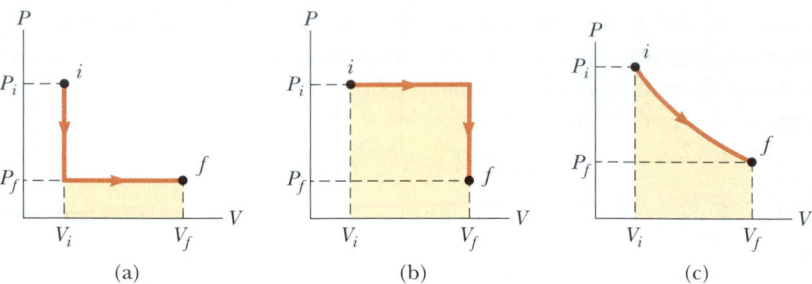

Figure 20.5   The work done by a gas as it is taken from an initial state to a final state depends on the path between these states.

$P_f(V_f - V_i)$. In Figure 20.5b, the gas first expands from V_i to V_f at constant pressure P_i. Then, its pressure is reduced to P_f at constant volume V_f. The value of the work done along this path is $P_i(V_f - V_i)$, which is greater than that for the process described in Figure 20.5a. Finally, for the process described in Figure 20.5c, where both P and V change continuously, the work done has some value intermediate between the values obtained in the first two processes. Therefore, we see that **the work done by a system depends on the initial and final states and on the path followed by the system between these states.**

The energy transfer by heat Q into or out of a system also depends on the process. Consider the situations depicted in Figure 20.6. In each case, the gas has the same initial volume, temperature, and pressure and is assumed to be ideal. In Figure 20.6a, the gas is thermally insulated from its surroundings except at the bottom of the gas-filled region, where it is in thermal contact with an energy reservoir. An *energy reservoir* is a source of energy that is considered to be so great that a finite transfer of energy from the reservoir does not change its temperature. The piston is held at its initial position by an external agent—a hand, for instance. When the force with which the piston is held is reduced slightly, the piston rises very slowly to its final position. Because the piston is moving upward, the gas is doing work on

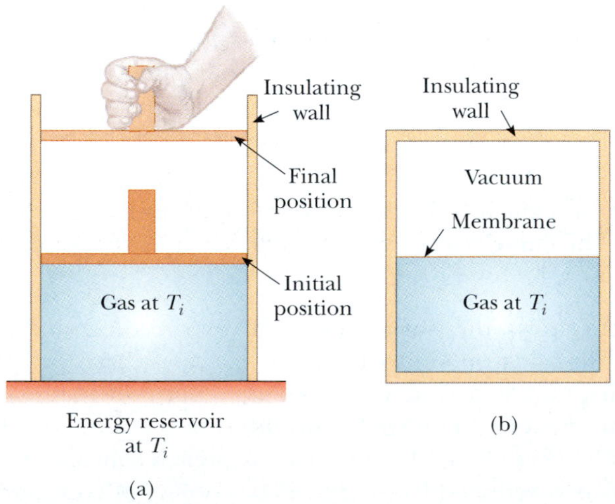

> Work done depends on the path between the initial and final states.

Figure 20.6   (a) A gas at temperature T_i expands slowly while absorbing energy from a reservoir in order to maintain a constant temperature. (b) A gas expands rapidly into an evacuated region after a membrane is broken.

the piston. During this expansion to the final volume V_f, just enough energy is transferred by heat from the reservoir to the gas to maintain a constant temperature T_i.

Now consider the completely thermally insulated system shown in Figure 20.6b. When the membrane is broken, the gas expands rapidly into the vacuum until it occupies a volume V_f and is at a pressure P_f. In this case, the gas does no work because there is no movable piston on which the gas applies a force. Furthermore, no energy is transferred by heat through the insulating wall.

The initial and final states of the ideal gas in Figure 20.6a are identical to the initial and final states in Figure 20.6b, but the paths are different. In the first case, the gas does work on the piston, and energy is transferred slowly to the gas. In the second case, no energy is transferred, and the value of the work done is zero. Therefore, we conclude that **energy transfer by heat, like work done, depends on the initial, final, and intermediate states of the system.** In other words, because heat and work depend on the path, neither quantity is determined solely by the end points of a thermodynamic process.

This device, called *Hero's engine*, was invented around 150 B.C. by Hero in Alexandria. When water is boiled in the flask, which is suspended by a cord, steam exits through two tubes at the sides (in opposite directions), creating a torque that rotates the flask. *(Courtesy of Central Scientific Company)*

20.5 ▷ THE FIRST LAW OF THERMODYNAMICS

When we introduced the law of conservation of mechanical energy in Chapter 8, we stated that the mechanical energy of a system is constant in the absence of nonconservative forces such as friction. That is, we did not include changes in the internal energy of the system in this mechanical model. The first law of thermodynamics is a generalization of the law of conservation of energy that encompasses changes in internal energy. It is a universally valid law that can be applied to many processes and provides a connection between the microscopic and macroscopic worlds.

We have discussed two ways in which energy can be transferred between a system and its surroundings. One is work done by the system, which requires that there be a macroscopic displacement of the point of application of a force (or pressure). The other is heat, which occurs through random collisions between the molecules of the system. Both mechanisms result in a change in the internal energy of the system and therefore usually result in measurable changes in the macroscopic variables of the system, such as the pressure, temperature, and volume of a gas.

To better understand these ideas on a quantitative basis, suppose that a system undergoes a change from an initial state to a final state. During this change, energy transfer by heat Q to the system occurs, and work W is done *by* the system. As an example, suppose that the system is a gas in which the pressure and volume change from P_i and V_i to P_f and V_f. If the quantity $Q - W$ is measured for various paths connecting the initial and final equilibrium states, we find that it is the same for all paths connecting the two states. We conclude that the quantity $Q - W$ is determined completely by the initial and final states of the system, and we call this quantity the **change in the internal energy** of the system. Although Q and W both depend on the path, **the quantity $Q - W$ is independent of the path.** If we use the symbol E_{int} to represent the internal energy, then the *change* in internal energy ΔE_{int} can be expressed as[5]

> $Q - W$ is the change in internal energy

$$\Delta E_{int} = Q - W \qquad (20.9)$$

> First-law equation

[5] It is an unfortunate accident of history that the traditional symbol for internal energy is U, which is also the traditional symbol for potential energy, as introduced in Chapter 8. To avoid confusion between potential energy and internal energy, we use the symbol E_{int} for internal energy in this book. If you take an advanced course in thermodynamics, however, be prepared to see U used as the symbol for internal energy.

where all quantities must have the same units of measure for energy.[6] Equation 20.9 is known as the **first-law equation** and is a key concept in many applications. As a reminder, we use the convention that Q is positive when energy enters the system and negative when energy leaves the system, and that W is positive when the system does work on the surroundings and negative when work is done on the system.

When a system undergoes an infinitesimal change in state in which a small amount of energy dQ is transferred by heat and a small amount of work dW is done, the internal energy changes by a small amount dE_{int}. Thus, for infinitesimal processes we can express the first-law equation as[7]

First-law equation for infinitesimal changes

$$dE_{int} = dQ - dW$$

The first-law equation is an energy conservation equation specifying that the only type of energy that changes in the system is the internal energy E_{int}. Let us look at some special cases in which this condition exists.

Isolated system

First, let us consider an *isolated system*—that is, one that does not interact with its surroundings. In this case, no energy transfer by heat takes place and the value of the work done by the system is zero; hence, the internal energy remains constant. That is, because $Q = W = 0$, it follows that $\Delta E_{int} = 0$, and thus $E_{int, i} = E_{int, f}$. We conclude that **the internal energy E_{int} of an isolated system remains constant.**

Next, we consider the case of a system (one not isolated from its surroundings) that is taken through a **cyclic process**—that is, a process that starts and ends at the same state. In this case, the change in the internal energy must again be zero, and therefore the energy Q added to the system must equal the work W done by the system during the cycle. That is, in a cyclic process,

Cyclic process

$$\Delta E_{int} = 0 \qquad \text{and} \qquad Q = W$$

On a PV diagram, a cyclic process appears as a closed curve. (The processes described in Figure 20.5 are represented by open curves because the initial and final states differ.) It can be shown that **in a cyclic process, the net work done by the system per cycle equals the area enclosed by the path representing the process on a PV diagram.**

If the value of the work done by the system during some process is zero, then the change in internal energy ΔE_{int} equals the energy transfer Q into or out of the system:

$$\Delta E_{int} = Q$$

If energy enters the system, then Q is positive and the internal energy increases. For a gas, we can associate this increase in internal energy with an increase in the kinetic energy of the molecules. Conversely, if no energy transfer occurs during some process but work is done by the system, then the change in internal energy equals the negative value of the work done by the system:

$$\Delta E_{int} = -W$$

---

[6] For the definition of work from our mechanics studies, the first law would be written as $\Delta E_{int} = Q + W$ because energy transfer into the system by either work or heat would increase the internal energy of the system. Because of the reversal of the definition of positive work discussed in Section 20.4, the first law appears as in Equation 20.9, with a minus sign.

[7] Note that dQ and dW are not true differential quantities; however, dE_{int} is. Because dQ and dW are *inexact differentials*, they are often represented by the symbols $đQ$ and $đW$. For further details on this point, see an advanced text on thermodynamics, such as R. P. Bauman, *Modern Thermodynamics and Statistical Mechanics*, New York, Macmillan Publishing Co., 1992.

For example, if a gas is compressed by a moving piston in an insulated cylinder, no energy is transferred by heat and the work done by the gas is negative; thus, the internal energy increases because kinetic energy is transferred from the moving piston to the gas molecules.

On a microscopic scale, no distinction exists between the result of heat and that of work. Both heat and work can produce a change in the internal energy of a system. Although the macroscopic quantities Q and W are *not* properties of a system, they are related to the change of the internal energy of a system through the first-law equation. Once we define a process, or path, we can either calculate or measure Q and W, and we can find the change in the system's internal energy using the first-law equation.

One of the important consequences of the first law of thermodynamics is that there exists a quantity known as internal energy whose value is determined by the state of the system. The internal energy function is therefore called a *state function*.

20.6 ▸ SOME APPLICATIONS OF THE FIRST LAW OF THERMODYNAMICS

Before we apply the first law of thermodynamics to specific systems, it is useful for us to first define some common thermodynamic processes. An **adiabatic process** is one during which no energy enters or leaves the system by heat—that is, $Q = 0$. An adiabatic process can be achieved either by thermally insulating the system from its surroundings (as shown in Fig. 20.6b) or by performing the process rapidly, so that there is little time for energy to transfer by heat. Applying the first law of thermodynamics to an adiabatic process, we see that

$$\Delta E_{int} = -W \qquad \text{(adiabatic process)} \qquad \textbf{(20.10)}$$

In an adiabatic process, $Q = 0$.

First-law equation for an adiabatic process

From this result, we see that if a gas expands adiabatically such that W is positive, then ΔE_{int} is negative and the temperature of the gas decreases. Conversely, the temperature of a gas increases when the gas is compressed adiabatically.

Adiabatic processes are very important in engineering practice. Some common examples are the expansion of hot gases in an internal combustion engine, the liquefaction of gases in a cooling system, and the compression stroke in a diesel engine.

The process described in Figure 20.6b, called an **adiabatic free expansion,** is unique. The process is adiabatic because it takes place in an insulated container. Because the gas expands into a vacuum, it does not apply a force on a piston as was depicted in Figure 20.6a, so no work is done on or by the gas. Thus, in this adiabatic process, both $Q = 0$ and $W = 0$. As a result, $\Delta E_{int} = 0$ for this process, as we can see from the first law. That is, **the initial and final internal energies of a gas are equal in an adiabatic free expansion.** As we shall see in the next chapter, the internal energy of an ideal gas depends only on its temperature. Thus, we expect no change in temperature during an adiabatic free expansion. This prediction is in accord with the results of experiments performed at low pressures. (Experiments performed at high pressures for real gases show a slight decrease or increase in temperature after the expansion. This change is due to intermolecular interactions, which represent a deviation from the model of an ideal gas.)

In an adiabatic free expansion, $\Delta E_{int} = 0$.

A process that occurs at constant pressure is called an **isobaric process.** In such a process, the values of the heat and the work are both usually nonzero. The

In an isobaric process, P remains constant.

work done by the gas is simply

$$W = P(V_f - V_i) \qquad \text{(isobaric process)} \qquad \textbf{(20.11)}$$

where P is the constant pressure.

A process that takes place at constant volume is called an **isovolumetric process.** In such a process, the value of the work done is clearly zero because the volume does not change. Hence, from the first law we see that in an isovolumetric process, because $W = 0$,

First-law equation for a constant-volume process

$$\Delta E_{\text{int}} = Q \qquad \text{(isovolumetric process)} \qquad \textbf{(20.12)}$$

This expression specifies that **if energy is added by heat to a system kept at constant volume, then all of the transferred energy remains in the system as an increase of the internal energy of the system.** For example, when a can of spray paint is thrown into a fire, energy enters the system (the gas in the can) by heat through the metal walls of the can. Consequently, the temperature, and thus the pressure, in the can increases until the can possibly explodes.

A process that occurs at constant temperature is called an **isothermal process.** A plot of P versus V at constant temperature for an ideal gas yields a hyperbolic curve called an *isotherm*. The internal energy of an ideal gas is a function of temperature only. Hence, in an isothermal process involving an ideal gas, $\Delta E_{\text{int}} = 0$. For an isothermal process, then, we conclude from the first law that the energy transfer Q must be equal to the work done by the gas—that is, $Q = W$. Any energy that enters the system by heat is transferred out of the system by work; as a result, no change of the internal energy of the system occurs.

In an isothermal process, T remains constant.

Quick Quiz 20.4

In the last three columns of the following table, fill in the boxes with $-$, $+$, or 0. For each situation, the system to be considered is identified.

Situation	System	Q	W	ΔE_{int}
(a) Rapidly pumping up a bicycle tire	Air in the pump			
(b) Pan of room-temperature water sitting on a hot stove	Water in the pan			
(c) Air quickly leaking out of a balloon	Air originally in balloon			

Isothermal Expansion of an Ideal Gas

Suppose that an ideal gas is allowed to expand quasi-statically at constant temperature, as described by the PV diagram shown in Figure 20.7. The curve is a hyperbola (see Appendix B, Eq. B.23), and the equation of state of an ideal gas with T constant indicates that the equation of this curve is $PV = $ constant. The isothermal expansion of the gas can be achieved by placing the gas in thermal contact with an energy reservoir at the same temperature, as shown in Figure 20.6a.

Let us calculate the work done by the gas in the expansion from state i to state f. The work done by the gas is given by Equation 20.8. Because the gas is ideal and the process is quasi-static, we can use the expression $PV = nRT$ for each point on

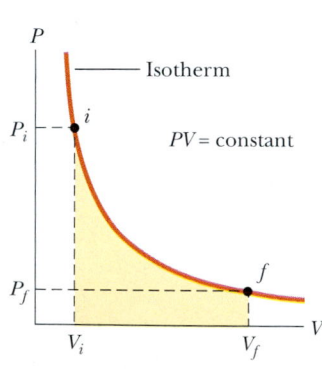

Figure 20.7 The PV diagram for an isothermal expansion of an ideal gas from an initial state to a final state. The curve is a hyperbola.

the path. Therefore, we have

$$W = \int_{V_i}^{V_f} P \, dV = \int_{V_i}^{V_f} \frac{nRT}{V} \, dV$$

Because T is constant in this case, it can be removed from the integral along with n and R:

$$W = nRT \int_{V_i}^{V_f} \frac{dV}{V} = nRT \ln V \Big|_{V_i}^{V_f}$$

To evaluate the integral, we used $\int (dx/x) = \ln x$. Evaluating this at the initial and final volumes, we have

$$W = nRT \ln\left(\frac{V_f}{V_i}\right) \qquad\qquad \textbf{(20.13)}$$

Work done by an ideal gas in an isothermal process

Numerically, this work W equals the shaded area under the PV curve shown in Figure 20.7. Because the gas expands, $V_f > V_i$, and the value for the work done by the gas is positive, as we expect. If the gas is compressed, then $V_f < V_i$, and the work done by the gas is negative.

Quick Quiz 20.5

Characterize the paths in Figure 20.8 as isobaric, isovolumetric, isothermal, or adiabatic. Note that $Q = 0$ for path B.

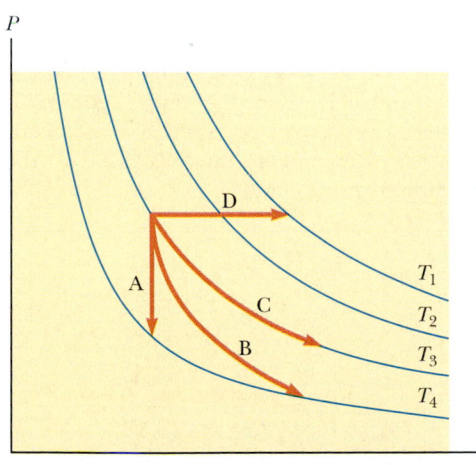

Figure 20.8 Identify the nature of paths A, B, C, and D.

EXAMPLE 20.6   An Isothermal Expansion

A 1.0-mol sample of an ideal gas is kept at 0.0°C during an expansion from 3.0 L to 10.0 L. (a) How much work is done by the gas during the expansion?

Solution Substituting the values into Equation 20.13, we have

$$W = nRT \ln\left(\frac{V_f}{V_i}\right)$$

$$W = (1.0 \text{ mol})(8.31 \text{ J/mol}\cdot\text{K})(273 \text{ K}) \ln\left(\frac{10.0}{3.0}\right)$$

$$= \boxed{2.7 \times 10^3 \text{ J}}$$

(b) How much energy transfer by heat occurs with the surroundings in this process?

Solution   From the first law, we find that

$$\Delta E_{int} = Q - W$$
$$0 = Q - W$$

$$Q = W = \boxed{2.7 \times 10^3 \text{ J}}$$

(c) If the gas is returned to the original volume by means of an isobaric process, how much work is done by the gas?

Solution   The work done in an isobaric process is given by Equation 20.11. We are not given the pressure, so we need to incorporate the ideal gas law:

$$W = P(V_f - V_i) = \frac{nRT_i}{V_i}(V_f - V_i)$$

$$= \frac{(1.0 \text{ mol})(8.31 \text{ J/mol·K})(273 \text{ K})}{10.0 \times 10^{-3} \text{ m}^3}$$

$$\times (3.0 \times 10^{-3} \text{ m}^3 - 10.0 \times 10^{-3} \text{ m}^3)$$

$$= \boxed{-1.6 \times 10^3 \text{ J}}$$

Notice that we use the initial temperature and volume to determine the value of the constant pressure because we do not know the final temperature. The work done by the gas is negative because the gas is being compressed.

EXAMPLE 20.7 ▸ Boiling Water

Suppose 1.00 g of water vaporizes isobarically at atmospheric pressure (1.013×10^5 Pa). Its volume in the liquid state is $V_i = V_{liquid} = 1.00 \text{ cm}^3$, and its volume in the vapor state is $V_f = V_{vapor} = 1\,671 \text{ cm}^3$. Find the work done in the expansion and the change in internal energy of the system. Ignore any mixing of the steam and the surrounding air—imagine that the steam simply pushes the surrounding air out of the way.

Solution   Because the expansion takes place at constant pressure, the work done by the system in pushing away the surrounding air is, from Equation 20.11,

$$W = P(V_f - V_i)$$
$$= (1.013 \times 10^5 \text{ Pa})(1\,671 \times 10^{-6} \text{ m}^3 - 1.00 \times 10^{-6} \text{ m}^3)$$

$$= \boxed{169 \text{ J}}$$

To determine the change in internal energy, we must know the energy transfer Q needed to vaporize the water. Using Equation 20.6 and the latent heat of vaporization for water, we have

$$Q = mL_v = (1.00 \times 10^{-3} \text{ kg})(2.26 \times 10^6 \text{ J/kg}) = 2\,260 \text{ J}$$

Hence, from the first law, the change in internal energy is

$$\Delta E_{int} = Q - W = 2\,260 \text{ J} - 169 \text{ J} = \boxed{2.09 \text{ kJ}}$$

The positive value for ΔE_{int} indicates that the internal energy of the system increases. We see that most (2 090 J/2 260 J = 93%) of the energy transferred to the liquid goes into increasing the internal energy of the system. Only 169 J/2 260 J = 7% leaves the system by work done by the steam on the surrounding atmosphere.

EXAMPLE 20.8 ▸ Heating a Solid

A 1.0-kg bar of copper is heated at atmospheric pressure. If its temperature increases from 20°C to 50°C, (a) what is the work done by the copper on the surrounding atmosphere?

Solution   Because the process is isobaric, we can find the work done by the copper using Equation 20.11, $W = P(V_f - V_i)$. We can calculate the change in volume of the copper using Equation 19.6. Using the average linear expansion coefficient for copper given in Table 19.2, and remembering that $\beta = 3\alpha$, we obtain

$$\Delta V = \beta V_i \Delta T$$
$$= [5.1 \times 10^{-5} (\text{°C})^{-1}](50\text{°C} - 20\text{°C}) V_i = 1.5 \times 10^{-3} V_i$$

The volume V_i is equal to m/ρ, and Table 15.1 indicates that the density of copper is $8.92 \times 10^3 \text{ kg/m}^3$. Hence,

$$\Delta V = (1.5 \times 10^{-3})\left(\frac{1.0 \text{ kg}}{8.92 \times 10^3 \text{ kg/m}^3}\right) = 1.7 \times 10^{-7} \text{ m}^3$$

The work done is

$$W = P\Delta V = (1.013 \times 10^5 \text{ N/m}^2)(1.7 \times 10^{-7} \text{ m}^3)$$

$$= \boxed{1.7 \times 10^{-2} \text{ J}}$$

(b) What quantity of energy is transferred to the copper by heat?

Solution   Taking the specific heat of copper from Table 20.1 and using Equation 20.4, we find that the energy transferred by heat is

$$Q = mc\Delta T = (1.0 \text{ kg})(387 \text{ J/kg} \cdot {}^{\circ}\text{C})(30^{\circ}\text{C}) = \boxed{1.2 \times 10^4 \text{ J}}$$

(c) What is the increase in internal energy of the copper?

Solution From the first law of thermodynamics, we have

$$\Delta E_{\text{int}} = Q - W = 1.2 \times 10^4 \text{ J} - 1.7 \times 10^{-2} \text{ J} = \boxed{1.2 \times 10^4 \text{ J}}$$

Note that almost all of the energy transferred into the system by heat goes into increasing the internal energy. The fraction of energy used to do work on the surrounding atmosphere is only about 10^{-6}! Hence, when analyzing the thermal expansion of a solid or a liquid, the small amount of work done by the system is usually ignored.

20.7 ENERGY TRANSFER MECHANISMS

It is important to understand the rate at which energy is transferred between a system and its surroundings and the mechanisms responsible for the transfer. Therefore, let us now look at three common energy transfer mechanisms that can result in a change in internal energy of a system.

Thermal Conduction

The energy transfer process that is most clearly associated with a temperature difference is **thermal conduction.** In this process, the transfer can be represented on an atomic scale as an exchange of kinetic energy between microscopic particles—molecules, atoms, and electrons—in which less energetic particles gain energy in collisions with more energetic particles. For example, if you hold one end of a long metal bar and insert the other end into a flame, you will find that the temperature of the metal in your hand soon increases. The energy reaches your hand by means of conduction. We can understand the process of conduction by examining what is happening to the microscopic particles in the metal. Initially, before the rod is inserted into the flame, the microscopic particles are vibrating about their equilibrium positions. As the flame heats the rod, those particles near the flame begin to vibrate with greater and greater amplitudes. These particles, in turn, collide with their neighbors and transfer some of their energy in the collisions. Slowly, the amplitudes of vibration of metal atoms and electrons farther and farther from the flame increase until, eventually, those in the metal near your hand are affected. This increased vibration represents an increase in the temperature of the metal and of your potentially burned hand.

The rate of thermal conduction depends on the properties of the substance being heated. For example, it is possible to hold a piece of asbestos in a flame indefinitely. This implies that very little energy is conducted through the asbestos. In general, metals are good thermal conductors, and materials such as asbestos, cork, paper, and fiberglass are poor conductors. Gases also are poor conductors because the separation distance between the particles is so great. Metals are good thermal conductors because they contain large numbers of electrons that are relatively free to move through the metal and so can transport energy over large distances. Thus, in a good conductor, such as copper, conduction takes place both by means of the vibration of atoms and by means of the motion of free electrons.

Conduction occurs only if there is a difference in temperature between two parts of the conducting medium. Consider a slab of material of thickness Δx and cross-sectional area A. One face of the slab is at a temperature T_1, and the other face is at a temperature $T_2 > T_1$ (Fig. 20.9). Experimentally, it is found that the

Melted snow pattern on a parking lot surface indicates the presence of underground hot water pipes used to aid snow removal. Energy from the water is conducted from the pipes to the pavement, where it causes the snow to melt. *(Courtesy of Dr. Albert A. Bartlett, University of Colorado, Boulder, CO)*

energy Q transferred in a time Δt flows from the hotter face to the colder one. The rate $Q/\Delta t$ at which this energy flows is found to be proportional to the cross-sectional area and the temperature difference $\Delta T = T_2 - T_1$, and inversely proportional to the thickness:

$$\frac{Q}{\Delta t} \propto A \frac{\Delta T}{\Delta x}$$

It is convenient to use the symbol for power $\mathcal{P}$ to represent the **rate of energy transfer**: $\mathcal{P} = Q/\Delta t$. Note that $\mathcal{P}$ has units of watts when Q is in joules and Δt is in seconds. For a slab of infinitesimal thickness dx and temperature difference dT, we can write the **law of thermal conduction** as

Law of thermal conduction

$$\mathcal{P} = kA \left| \frac{dT}{dx} \right| \qquad \text{(20.14)}$$

where the proportionality constant k is the **thermal conductivity** of the material and $|dT/dx|$ is the **temperature gradient** (the variation of temperature with position).

Suppose that a long, uniform rod of length L is thermally insulated so that energy cannot escape by heat from its surface except at the ends, as shown in Figure 20.10. One end is in thermal contact with an energy reservoir at temperature T_1, and the other end is in thermal contact with a reservoir at temperature $T_2 > T_1$. When a steady state has been reached, the temperature at each point along the rod is constant in time. In this case if we assume that k is not a function of temperature, the temperature gradient is the same everywhere along the rod and is

$$\left| \frac{dT}{dx} \right| = \frac{T_2 - T_1}{L}$$

Thus the rate of energy transfer by conduction through the rod is

$$\mathcal{P} = kA \frac{(T_2 - T_1)}{L} \qquad \text{(20.15)}$$

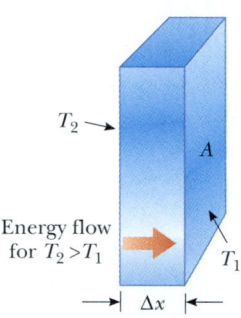

Figure 20.9 Energy transfer through a conducting slab with a cross-sectional area A and a thickness Δx. The opposite faces are at different temperatures T_1 and T_2.

Substances that are good thermal conductors have large thermal conductivity values, whereas good thermal insulators have low thermal conductivity values. Table 20.3 lists thermal conductivities for various substances. Note that metals are generally better thermal conductors than nonmetals are.

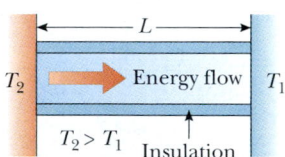

Figure 20.10 Conduction of energy through a uniform, insulated rod of length L. The opposite ends are in thermal contact with energy reservoirs at different temperatures.

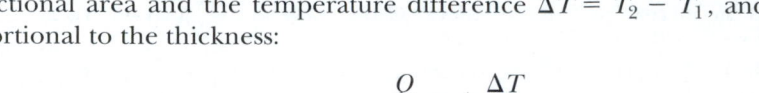

Quick Quiz 20.6

Will an ice cube wrapped in a wool blanket remain frozen for (a) a shorter length of time, (b) the same length of time, or (c) a longer length of time than an identical ice cube exposed to air at room temperature?

For a compound slab containing several materials of thicknesses L_1, L_2, . . . and thermal conductivities k_1, k_2, . . . , the rate of energy transfer through the slab at steady state is

$$\mathcal{P} = \frac{A(T_2 - T_1)}{\sum_i (L_i/k_i)} \qquad \text{(20.16)}$$

TABLE 20.3 **Thermal Conductivities**	
Substance	**Thermal Conductivity (W/m·°C)**
Metals (at 25°C)	
Aluminum	238
Copper	397
Gold	314
Iron	79.5
Lead	34.7
Silver	427
Nonmetals (approximate values)	
Asbestos	0.08
Concrete	0.8
Diamond	2 300
Glass	0.8
Ice	2
Rubber	0.2
Water	0.6
Wood	0.08
Gases (at 20°C)	
Air	0.023 4
Helium	0.138
Hydrogen	0.172
Nitrogen	0.023 4
Oxygen	0.023 8

where T_1 and T_2 are the temperatures of the outer surfaces (which are held constant) and the summation is over all slabs. The following example shows how this equation results from a consideration of two thicknesses of materials.

EXAMPLE 20.9   **Energy Transfer Through Two Slabs**

Two slabs of thickness L_1 and L_2 and thermal conductivities k_1 and k_2 are in thermal contact with each other, as shown in Figure 20.11. The temperatures of their outer surfaces are T_1 and T_2, respectively, and $T_2 > T_1$. Determine the temperature at the interface and the rate of energy transfer by conduction through the slabs in the steady-state condition.

Solution   If T is the temperature at the interface, then the rate at which energy is transferred through slab 1 is

$$(1) \qquad \mathcal{P}_1 = \frac{k_1 A(T - T_1)}{L_1}$$

The rate at which energy is transferred through slab 2 is

$$(2) \qquad \mathcal{P}_2 = \frac{k_2 A(T_2 - T)}{L_2}$$

When a steady state is reached, these two rates must be equal; hence,

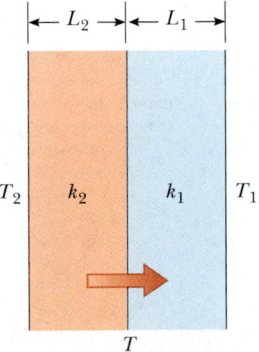

Figure 20.11   Energy transfer by conduction through two slabs in thermal contact with each other. At steady state, the rate of energy transfer through slab 1 equals the rate of energy transfer through slab 2.

$$\frac{k_1 A(T - T_1)}{L_1} = \frac{k_2 A(T_2 - T)}{L_2}$$

Solving for T gives

$$(3) \qquad T = \frac{k_1 L_2 T_1 + k_2 L_1 T_2}{k_1 L_2 + k_2 L_1}$$

Substituting (3) into either (1) or (2), we obtain

$$\mathcal{P} = \frac{A(T_2 - T_1)}{(L_1/k_1) + (L_2/k_2)}$$

Extension of this model to several slabs of materials leads to Equation 20.16.

Home Insulation

In engineering practice, the term L/k for a particular substance is referred to as the **R value** of the material. Thus, Equation 20.16 reduces to

$$\mathcal{P} = \frac{A(T_2 - T_1)}{\displaystyle\sum_i R_i} \qquad\qquad \textbf{(20.17)}$$

where $R_i = L_i/k_i$. The R values for a few common building materials are given in Table 20.4. In the United States, the insulating properties of materials used in buildings are usually expressed in engineering units, not SI units. Thus, in Table 20.4, measurements of R values are given as a combination of British thermal units, feet, hours, and degrees Fahrenheit.

At any vertical surface open to the air, a very thin stagnant layer of air adheres to the surface. One must consider this layer when determining the R value for a wall. The thickness of this stagnant layer on an outside wall depends on the speed of the wind. Energy loss from a house on a windy day is greater than the loss on a day when the air is calm. A representative R value for this stagnant layer of air is given in Table 20.4.

Energy is conducted from the inside to the exterior more rapidly on the part of the roof where the snow has melted. The dormer appears to have been added and insulated. The main roof does not appear to be well insulated. *(Courtesy of Dr. Albert A. Bartlett, University of Colorado, Boulder, CO)*

TABLE 20.4	*R* Values for Some Common Building Materials
Material	**R value (ft$^2 \cdot$ °F · h/Btu)**
Hardwood siding (1 in. thick)	0.91
Wood shingles (lapped)	0.87
Brick (4 in. thick)	4.00
Concrete block (filled cores)	1.93
Fiberglass batting (3.5 in. thick)	10.90
Fiberglass batting (6 in. thick)	18.80
Fiberglass board (1 in. thick)	4.35
Cellulose fiber (1 in. thick)	3.70
Flat glass (0.125 in. thick)	0.89
Insulating glass (0.25-in. space)	1.54
Air space (3.5 in. thick)	1.01
Stagnant air layer	0.17
Drywall (0.5 in. thick)	0.45
Sheathing (0.5 in. thick)	1.32

This thermogram of a home, made during cold weather, shows colors ranging from white and orange (areas of greatest energy loss) to blue and purple (areas of least energy loss). *(Daedalus Enterprises, Inc./Peter Arnold, Inc.)*

EXAMPLE 20.10   The *R* Value of a Typical Wall

Calculate the total R value for a wall constructed as shown in Figure 20.12a. Starting outside the house (toward the front in the figure) and moving inward, the wall consists of 4-in. brick, 0.5-in. sheathing, an air space 3.5 in. thick, and 0.5-in. drywall. Do not forget the stagnant air layers inside and outside the house.

Solution   Referring to Table 20.4, we find that

R_1 (outside stagnant air layer)	$= 0.17 \ \text{ft}^2 \cdot {}^\circ\text{F} \cdot \text{h}/\text{Btu}$
R_2 (brick)	$= 4.00 \ \text{ft}^2 \cdot {}^\circ\text{F} \cdot \text{h}/\text{Btu}$
R_3 (sheathing)	$= 1.32 \ \text{ft}^2 \cdot {}^\circ\text{F} \cdot \text{h}/\text{Btu}$
R_4 (air space)	$= 1.01 \ \text{ft}^2 \cdot {}^\circ\text{F} \cdot \text{h}/\text{Btu}$
R_5 (drywall)	$= 0.45 \ \text{ft}^2 \cdot {}^\circ\text{F} \cdot \text{h}/\text{Btu}$
R_6 (inside stagnant air layer)	$= 0.17 \ \text{ft}^2 \cdot {}^\circ\text{F} \cdot \text{h}/\text{Btu}$
R_{total}	$= 7.12 \ \text{ft}^2 \cdot {}^\circ\text{F} \cdot \text{h}/\text{Btu}$

Exercise   If a layer of fiberglass insulation 3.5 in. thick is placed inside the wall to replace the air space, as shown in Figure 20.12b, what is the new total R value? By what factor is the energy loss reduced?

Answer $R = 17 \ \text{ft}^2 \cdot {}^\circ\text{F} \cdot \text{h}/\text{Btu}$; 2.4.

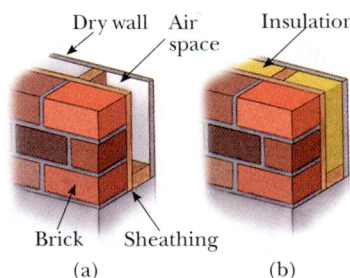

Figure 20.12   An exterior house wall containing (a) an air space and (b) insulation.

Convection

At one time or another, you probably have warmed your hands by holding them over an open flame. In this situation, the air directly above the flame is heated and expands. As a result, the density of this air decreases and the air rises. This warmed mass of air heats your hands as it flows by. **Energy transferred by the movement of a heated substance is said to have been transferred by convection.** When the movement results from differences in density, as with air around a fire, it is referred to as *natural convection*. Air flow at a beach is an example of natural convection, as is the mixing that occurs as surface water in a lake cools and sinks (see

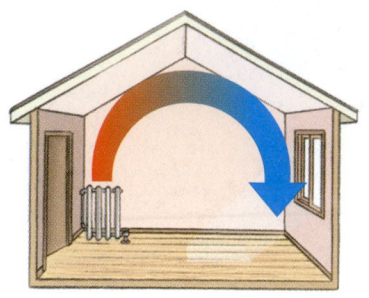

Figure 20.13 Convection currents are set up in a room heated by a radiator.

Chapter 19). When the heated substance is forced to move by a fan or pump, as in some hot-air and hot-water heating systems, the process is called *forced convection*.

If it were not for convection currents, it would be very difficult to boil water. As water is heated in a teakettle, the lower layers are warmed first. The heated water expands and rises to the top because its density is lowered. At the same time, the denser, cool water at the surface sinks to the bottom of the kettle and is heated.

The same process occurs when a room is heated by a radiator. The hot radiator warms the air in the lower regions of the room. The warm air expands and rises to the ceiling because of its lower density. The denser, cooler air from above sinks, and the continuous air current pattern shown in Figure 20.13 is established.

Radiation

The third means of energy transfer that we shall discuss is **radiation.** All objects radiate energy continuously in the form of electromagnetic waves (see Chapter 34) produced by thermal vibrations of the molecules. You are likely familiar with electromagnetic radiation in the form of the orange glow from an electric stove burner, an electric space heater, or the coils of a toaster.

The rate at which an object radiates energy is proportional to the fourth power of its absolute temperature. This is known as **Stefan's law** and is expressed in equation form as

$$\mathcal{P} = \sigma A e T^4 \tag{20.18}$$

where $\mathcal{P}$ is the power in watts radiated by the object, σ is a constant equal to $5.669\,6 \times 10^{-8}\ \mathrm{W/m^2 \cdot K^4}$, A is the surface area of the object in square meters, e is the **emissivity** constant, and T is the surface temperature in kelvins. The value of e can vary between zero and unity, depending on the properties of the surface of the object. The emissivity is equal to the fraction of the incoming radiation that the surface absorbs.

Approximately 1 340 J of electromagnetic radiation from the Sun passes perpendicularly through each 1 m² at the top of the Earth's atmosphere every second. This radiation is primarily visible and infrared light accompanied by a significant amount of ultraviolet radiation. We shall study these types of radiation in detail in Chapter 34. Some of this energy is reflected back into space, and some is absorbed by the atmosphere. However, enough energy arrives at the surface of the Earth each day to supply all our energy needs on this planet hundreds of times over—if only it could be captured and used efficiently. The growth in the number of solar energy–powered houses built in this country reflects the increasing efforts being made to use this abundant energy. Radiant energy from the Sun affects our day-to-day existence in a number of ways. For example, it influences the Earth's average temperature, ocean currents, agriculture, and rain patterns.

What happens to the atmospheric temperature at night is another example of the effects of energy transfer by radiation. If there is a cloud cover above the Earth, the water vapor in the clouds absorbs part of the infrared radiation emitted by the Earth and re-emits it back to the surface. Consequently, temperature levels at the surface remain moderate. In the absence of this cloud cover, there is nothing to prevent this radiation from escaping into space; thus the temperature decreases more on a clear night than on a cloudy one.

As an object radiates energy at a rate given by Equation 20.18, it also absorbs electromagnetic radiation. If the latter process did not occur, an object would eventually radiate all its energy, and its temperature would reach absolute zero. The energy an object absorbs comes from its surroundings, which consist of other objects that radiate energy. If an object is at a temperature T and its surroundings are at a temperature T_0, then the net energy gained or lost each second by the object as a result of radiation is

$$\mathcal{P}_{net} = \sigma A e (T^4 - T_0{}^4) \qquad\qquad (20.19)$$

When an object is in equilibrium with its surroundings, it radiates and absorbs energy at the same rate, and so its temperature remains constant. When an object is hotter than its surroundings, it radiates more energy than it absorbs, and its temperature decreases. An **ideal absorber** is defined as an object that absorbs all the energy incident on it, and for such a body, $e = 1$. Such an object is often referred to as a **black body.** An ideal absorber is also an ideal radiator of energy. In contrast, an object for which $e = 0$ absorbs none of the energy incident on it. Such an object reflects all the incident energy, and thus is an **ideal reflector.**

The Dewar Flask

The *Dewar flask*[8] is a container designed to minimize energy losses by conduction, convection, and radiation. Such a container is used to store either cold or hot liquids for long periods of time. (A Thermos bottle is a common household equivalent of a Dewar flask.) The standard construction (Fig. 20.14) consists of a double-walled Pyrex glass vessel with silvered walls. The space between the walls is evacuated to minimize energy transfer by conduction and convection. The silvered surfaces minimize energy transfer by radiation because silver is a very good reflector and has very low emissivity. A further reduction in energy loss is obtained by reducing the size of the neck. Dewar flasks are commonly used to store liquid nitrogen (boiling point: 77 K) and liquid oxygen (boiling point: 90 K).

To confine liquid helium (boiling point: 4.2 K), which has a very low heat of vaporization, it is often necessary to use a double Dewar system in which the Dewar flask containing the liquid is surrounded by a second Dewar flask. The space between the two flasks is filled with liquid nitrogen.

Newer designs of storage containers use "super insulation" that consists of many layers of reflecting material separated by fiberglass. All of this is in a vacuum, and no liquid nitrogen is needed with this design.

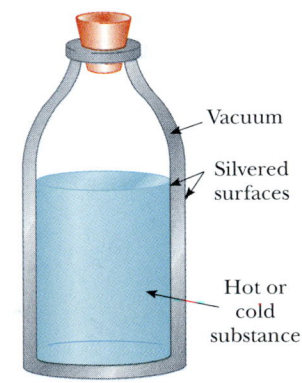

Figure 20.14   A cross-sectional view of a Dewar flask, which is used to store hot or cold substances.

Vacuum

Silvered surfaces

Hot or cold substance

EXAMPLE 20.11   **Who Turned Down the Thermostat?**

A student is trying to decide what to wear. The surroundings (his bedroom) are at 20.0°C. If the skin temperature of the unclothed student is 35°C, what is the net energy loss from his body in 10.0 min by radiation? Assume that the emissivity of skin is 0.900 and that the surface area of the student is 1.50 m².

Solution   Using Equation 20.19, we find that the net rate of energy loss from the skin is

$$\mathcal{P}_{net} = \sigma A e (T^4 - T_0{}^4)$$
$$= (5.67 \times 10^{-8}\ \mathrm{W/m^2 \cdot K^4})(1.50\ \mathrm{m^2})$$
$$\times\ (0.900)[(308\ \mathrm{K})^4 - (293\ \mathrm{K})^4] = 125\ \mathrm{W}$$

(Why is the temperature given in kelvins?) At this rate, the total energy lost by the skin in 10 min is

$$Q = \mathcal{P}_{net} \times \Delta t = (125\ \mathrm{W})(600\ \mathrm{s}) = \boxed{7.5 \times 10^4\ \mathrm{J}}$$

Note that the energy radiated by the student is roughly equivalent to that produced by two 60-W light bulbs!

---

[8] Invented by Sir James Dewar (1842–1923).

SUMMARY

Internal energy is all of a system's energy that is associated with the system's microscopic components. Internal energy includes kinetic energy of translation, rotation, and vibration of molecules, potential energy within molecules, and potential energy between molecules.

Heat is the transfer of energy across the boundary of a system resulting from a temperature difference between the system and its surroundings. We use the symbol Q for the amount of energy transferred by this process.

The **calorie** is the amount of energy necessary to raise the temperature of 1 g of water from $14.5°C$ to $15.5°C$. The **mechanical equivalent of heat** is 1 cal = 4.186 J.

The **heat capacity** C of any sample is the amount of energy needed to raise the temperature of the sample by $1°C$. The energy Q required to change the temperature of a mass m of a substance by an amount ΔT is

$$Q = mc\Delta T \tag{20.4}$$

where c is the **specific heat** of the substance.

The energy required to change the phase of a pure substance of mass m is

$$Q = mL \tag{20.6}$$

where L is the **latent heat** of the substance and depends on the nature of the phase change and the properties of the substance.

The **work done** by a gas as its volume changes from some initial value V_i to some final value V_f is

$$W = \int_{V_i}^{V_f} P \, dV \tag{20.8}$$

where P is the pressure, which may vary during the process. In order to evaluate W, the process must be fully specified—that is, P and V must be known during each step. In other words, the work done depends on the path taken between the initial and final states.

The **first law of thermodynamics** states that when a system undergoes a change from one state to another, the change in its internal energy is

$$\Delta E_{int} = Q - W \tag{20.9}$$

where Q is the energy transferred into the system by heat and W is the work done by the system. Although Q and W both depend on the path taken from the initial state to the final state, the quantity ΔE_{int} is path-independent. This central equation is a statement of conservation of energy that includes changes in internal energy.

In a **cyclic process** (one that originates and terminates at the same state), $\Delta E_{int} = 0$ and, therefore, $Q = W$. That is, the energy transferred into the system by heat equals the work done by the system during the process.

In an **adiabatic process,** no energy is transferred by heat between the system and its surroundings ($Q = 0$). In this case, the first law gives $\Delta E_{int} = -W$. That is, the internal energy changes as a consequence of work being done by the system. In the **adiabatic free expansion** of a gas, $Q = 0$ and $W = 0$; thus, $\Delta E_{int} = 0$. That is, the internal energy of the gas does not change in such a process.

An **isobaric process** is one that occurs at constant pressure. The work done in such a process is $W = P(V_f - V_i)$.

An **isovolumetric process** is one that occurs at constant volume. No work is done in such a process, so $\Delta E_{int} = Q$.

An **isothermal process** is one that occurs at constant temperature. The work done by an ideal gas during an isothermal process is

$$W = nRT \ln\left(\frac{V_f}{V_i}\right) \qquad \text{(20.13)}$$

Energy may be transferred by work, which we addressed in Chapter 7, and by conduction, convection, or radiation. **Conduction** can be viewed as an exchange of kinetic energy between colliding molecules or electrons. The rate at which energy flows by conduction through a slab of area A is

$$\mathcal{P} = kA\left|\frac{dT}{dx}\right| \qquad \text{(20.14)}$$

where k is the **thermal conductivity** of the material from which the slab is made and $|dT/dx|$ is the **temperature gradient.** This equation can be used in many situations in which the rate of transfer of energy through materials is important.

In **convection,** a heated substance moves from one place to another.

All bodies emit **radiation** in the form of electromagnetic waves at the rate

$$\mathcal{P} = \sigma AeT^4 \qquad \text{(20.18)}$$

A body that is hotter than its surroundings radiates more energy than it absorbs, whereas a body that is cooler than its surroundings absorbs more energy than it radiates.

QUESTIONS

1. The specific heat of water is about two times that of ethyl alcohol. Equal masses of alcohol and water are contained in separate beakers and are supplied with the same amount of energy. Compare the temperature increases of the two liquids.

2. Give one reason why coastal regions tend to have a more moderate climate than inland regions do.

3. A small metal crucible is taken from a 200°C oven and immersed in a tub full of water at room temperature (this process is often referred to as *quenching*). What is the approximate final equilibrium temperature?

4. What is the major problem that arises in measuring specific heats if a sample with a temperature greater than 100°C is placed in water?

5. In a daring lecture demonstration, an instructor dips his wetted fingers into molten lead (327°C) and withdraws them quickly, without getting burned. How is this possible? (This is a dangerous experiment that you should *not* attempt.)

6. The pioneers found that placing a large tub of water in a storage cellar would prevent their food from freezing on really cold nights. Explain why.

7. What is wrong with the statement, "Given any two bodies, the one with the higher temperature contains more heat."

8. Why is it possible for you to hold a lighted match, even when it is burned to within a few millimeters of your fingertips?

9. Why is it more comfortable to hold a cup of hot tea by the handle than by wrapping your hands around the cup itself?

10. Figure Q20.10 shows a pattern formed by snow on the roof of a barn. What causes the alternating pattern of snowcover and exposed roof?

Figure Q20.10 Alternating pattern on a snow-covered roof.
(Courtesy of Dr. Albert A. Bartlett, University of Colorado, Boulder, CO)

11. Why is a person able to remove a piece of dry aluminum foil from a hot oven with bare fingers but burns his or her fingers if there is moisture on the foil?

12. A tile floor in a bathroom may feel uncomfortably cold to your bare feet, but a carpeted floor in an adjoining room at the same temperature feels warm. Why?

13. Why can potatoes be baked more quickly when a metal skewer has been inserted through them?

14. Explain why a Thermos bottle has silvered walls and a vacuum jacket.

15. A piece of paper is wrapped around a rod made half of wood and half of copper. When held over a flame, the paper in contact with the wood burns but the paper in contact with the metal does not. Explain.

16. Why is it necessary to store liquid nitrogen or liquid oxygen in vessels equipped with either polystyrene insulation or a double-evacuated wall?

17. Why do heavy draperies hung over the windows help keep a home warm in the winter and cool in the summer?

18. If you wish to cook a piece of meat thoroughly on an open fire, why should you not use a high flame? (*Note:* Carbon is a good thermal insulator.)

19. When insulating a wood-frame house, is it better to place the insulation against the cooler, outside wall or against the warmer, inside wall? (In either case, an air barrier must be considered.)

20. In an experimental house, polystyrene beads were pumped into the air space between the panes of glass in double-pane windows at night in the winter, and they were pumped out to holding bins during the day. How would this procedure assist in conserving energy in the house?

21. Pioneers stored fruits and vegetables in underground cellars. Discuss the advantages of choosing this location as a storage site.

22. Concrete has a higher specific heat than soil does. Use this fact to explain (partially) why cities have a higher average night-time temperature than the surrounding countryside does. If a city is hotter than the surrounding countryside, would you expect breezes to blow from city to country or from country to city? Explain.

23. When camping in a canyon on a still night, a hiker notices that a breeze begins to stir as soon as the Sun strikes the surrounding peaks. What causes the breeze?

24. Updrafts of air are familiar to all pilots and are used to keep non-motorized gliders aloft. What causes these currents?

25. If water is a poor thermal conductor, why can it be heated quickly when placed over a flame?

26. The United States penny is now made of copper-coated zinc. Can a calorimetric experiment be devised to test for the metal content in a collection of pennies? If so, describe such a procedure.

27. If you hold water in a paper cup over a flame, you can bring the water to a boil without burning the cup. How is this possible?

28. When a sealed Thermos bottle full of hot coffee is shaken, what are the changes, if any, in (a) the temperature of the coffee and (b) the internal energy of the coffee?

29. Using the first law of thermodynamics, explain why the *total* energy of an isolated system is always constant.

30. Is it possible to convert internal energy into mechanical energy? Explain using examples.

31. Suppose that you pour hot coffee for your guests and one of them chooses to drink the coffee after it has been in the cup for several minutes. For the coffee to be warmest, should the person add the cream just after the coffee is poured or just before drinking it? Explain.

32. Suppose that you fill two identical cups both at room temperature with the same amount of hot coffee. One cup contains a metal spoon, while the other does not. If you wait for several minutes, which of the two contains the warmer coffee? Which energy transfer process accounts for this result?

33. A warning sign often seen on highways just before a bridge is "Caution—Bridge Surface Freezes Before Road Surface." Which of the three energy transfer processes is most important in causing a bridge surface to freeze before a road surface on very cold days?

PROBLEMS

1, 2, 3 = straightforward, intermediate, challenging   ☐ = full solution available in the *Student Solutions Manual and Study Guide*
WEB = solution posted at **http://www.saunderscollege.com/physics/**   🖥 = Computer useful in solving problem   🎯 = Interactive Physics
☐ = paired numerical/symbolic problems

Section 20.1   Heat and Internal Energy

1. Water at the top of Niagara Falls has a temperature of 10.0°C. It falls through a distance of 50.0 m. Assuming that all of its potential energy goes into warming of the water, calculate the temperature of the water at the bottom of the Falls.

2. Consider Joule's apparatus described in Figure 20.1. Each of the two masses is 1.50 kg, and the tank is filled with 200 g of water. What is the increase in the temperature of the water after the masses fall through a distance of 3.00 m?

Section 20.2   Heat Capacity and Specific Heat

3. The temperature of a silver bar rises by 10.0°C when it absorbs 1.23 kJ of energy by heat. The mass of the bar is 525 g. Determine the specific heat of silver.

4. A 50.0-g sample of copper is at 25.0°C. If 1 200 J of energy is added to it by heat, what is its final temperature?

WEB 5. A 1.50-kg iron horseshoe initially at 600°C is dropped into a bucket containing 20.0 kg of water at 25.0°C. What is the final temperature? (Neglect the heat capacity of the container and assume that a negligible amount of water boils away.)

6. An aluminum cup with a mass of 200 g contains 800 g of water in thermal equilibrium at 80.0°C. The combination of cup and water is cooled uniformly so that the temperature decreases at a rate of 1.50°C/min. At what rate is energy being removed by heat? Express your answer in watts.

7. An aluminum calorimeter with a mass of 100 g contains 250 g of water. The calorimeter and water are in thermal equilibrium at 10.0°C. Two metallic blocks are placed into the water. One is a 50.0-g piece of copper at 80.0°C; the other block has a mass of 70.0 g and is originally at a temperature of 100°C. The entire system stabilizes at a final temperature of 20.0°C. (a) Determine the specific heat of the unknown sample. (b) Guess the material of the unknown, using the data given in Table 20.1.

8. Lake Erie contains roughly 4.00×10^{11} m^3 of water. (a) How much energy is required to raise the temperature of this volume of water from 11.0°C to 12.0°C? (b) Approximately how many years would it take to supply this amount of energy with the use of a 1 000-MW wasted energy output of an electric power plant?

9. A 3.00-g copper penny at 25.0°C drops from a height of 50.0 m to the ground. (a) If 60.0% of the change in potential energy goes into increasing the internal energy, what is its final temperature? (b) Does the result you obtained in (a) depend on the mass of the penny? Explain.

10. If a mass m_h of water at T_h is poured into an aluminum cup of mass m_{Al} containing mass m_c of water at T_c, where $T_h > T_c$, what is the equilibrium temperature of the system?

11. A water heater is operated by solar power. If the solar collector has an area of 6.00 m^2 and the power delivered by sunlight is 550 W/m^2, how long does it take to increase the temperature of 1.00 m^3 of water from 20.0°C to 60.0°C?

Section 20.3 Latent Heat

12. How much energy is required to change a 40.0-g ice cube from ice at −10.0°C to steam at 110°C?

13. A 3.00-g lead bullet at 30.0°C is fired at a speed of 240 m/s into a large block of ice at 0°C, in which it becomes embedded. What quantity of ice melts?

14. Steam at 100°C is added to ice at 0°C. (a) Find the amount of ice melted and the final temperature when the mass of steam is 10.0 g and the mass of ice is 50.0 g. (b) Repeat this calculation, taking the mass of steam as 1.00 g and the mass of ice as 50.0 g.

15. A 1.00-kg block of copper at 20.0°C is dropped into a large vessel of liquid nitrogen at 77.3 K. How many kilograms of nitrogen boil away by the time the copper reaches 77.3 K? (The specific heat of copper is 0.092 0 cal/g · °C. The latent heat of vaporization of nitrogen is 48.0 cal/g.)

16. A 50.0-g copper calorimeter contains 250 g of water at 20.0°C. How much steam must be condensed into the water if the final temperature of the system is to reach 50.0°C?

17. In an insulated vessel, 250 g of ice at 0°C is added to 600 g of water at 18.0°C. (a) What is the final temperature of the system? (b) How much ice remains when the system reaches equilibrium?

18. Review Problem. Two speeding lead bullets, each having a mass of 5.00 g, a temperature of 20.0°C, and a speed of 500 m/s, collide head-on. Assuming a perfectly inelastic collision and no loss of energy to the atmosphere, describe the final state of the two-bullet system.

19. If 90.0 g of molten lead at 327.3°C is poured into a 300-g casting form made of iron and initially at 20.0°C, what is the final temperature of the system? (Assume that no energy loss to the environment occurs.)

Section 20.4 Work and Heat in Thermodynamic Processes

20. Gas in a container is at a pressure of 1.50 atm and a volume of 4.00 m^3. What is the work done by the gas (a) if it expands at constant pressure to twice its initial volume? (b) If it is compressed at constant pressure to one quarter of its initial volume?

21. A sample of ideal gas is expanded to twice its original volume of 1.00 m^3 in a quasi-static process for which $P = \alpha V^2$, with $\alpha = 5.00$ atm/m^6, as shown in Figure P20.21. How much work is done by the expanding gas?

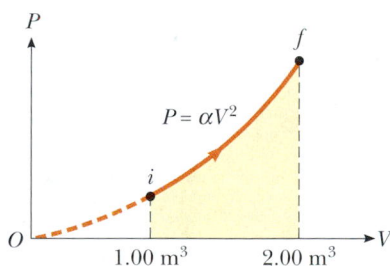

Figure P20.21

22. (a) Determine the work done by a fluid that expands from i to f as indicated in Figure P20.22. (b) How much

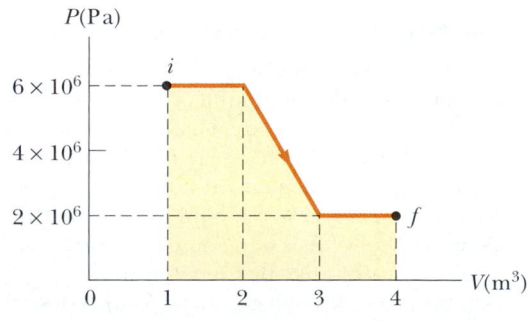

Figure P20.22

work is performed by the fluid if it is compressed from f to i along the same path?

23. One mole of an ideal gas is heated slowly so that it goes from PV state (P_i, V_i) to $(3P_i, 3V_i)$ in such a way that the pressure of the gas is directly proportional to the volume. (a) How much work is done in the process? (b) How is the temperature of the gas related to its volume during this process?

24. A sample of helium behaves as an ideal gas as energy is added by heat at constant pressure from 273 K to 373 K. If the gas does 20.0 J of work, what is the mass of helium present?

WEB 25. An ideal gas is enclosed in a cylinder with a movable piston on top. The piston has a mass of 8 000 g and an area of 5.00 cm² and is free to slide up and down, keeping the pressure of the gas constant. How much work is done as the temperature of 0.200 mol of the gas is raised from 20.0°C to 300°C?

26. An ideal gas is enclosed in a cylinder that has a movable piston on top. The piston has a mass m and an area A and is free to slide up and down, keeping the pressure of the gas constant. How much work is done as the temperature of n mol of the gas is raised from T_1 to T_2?

27. A gas expands from I to F along three possible paths, as indicated in Figure P20.27. Calculate the work in joules done by the gas along the paths IAF, IF, and IBF.

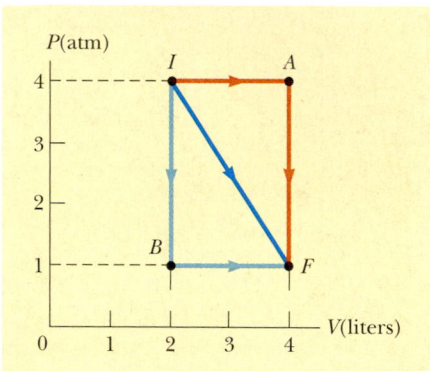

Figure P20.27

Section 20.5   The First Law of Thermodynamics

28. A gas is compressed from 9.00 L to 2.00 L at a constant pressure of 0.800 atm. In the process, 400 J of energy leaves the gas by heat. (a) What is the work done by the gas? (b) What is the change in its internal energy?

29. A thermodynamic system undergoes a process in which its internal energy decreases by 500 J. If, at the same time, 220 J of work is done on the system, what is the energy transferred to or from it by heat?

30. A gas is taken through the cyclic process described in Figure P20.30. (a) Find the net energy transferred to the system by heat during one complete cycle. (b) If the

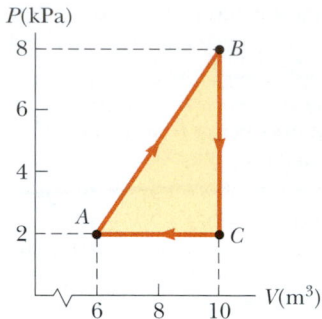

Figure P20.30   Problems 30 and 31.

cycle is reversed—that is, if the process follows the path $ACBA$—what is the net energy input per cycle by heat?

31. Consider the cyclic process depicted in Figure P20.30. If Q is negative for the process BC, and if ΔE_{int} is negative for the process CA, what are the signs of Q, W, and ΔE_{int} that are associated with each process?

32. A sample of an ideal gas goes through the process shown in Figure P20.32. From A to B, the process is adiabatic; from B to C, it is isobaric, with 100 kJ of energy flowing into the system by heat. From C to D, the process is isothermal; from D to A, it is isobaric, with 150 kJ of energy flowing out of the system by heat. Determine the difference in internal energy, $E_{int, B} - E_{int, A}$.

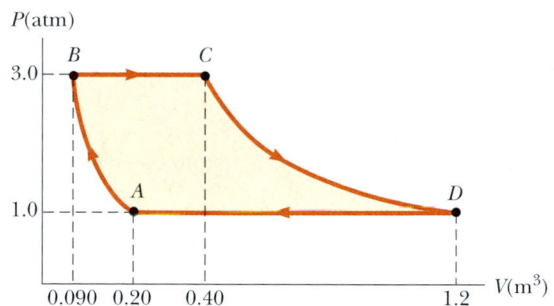

Figure P20.32

Section 20.6   Some Applications of the First Law of Thermodynamics

33. An ideal gas initially at 300 K undergoes an isobaric expansion at 2.50 kPa. If the volume increases from 1.00 m³ to 3.00 m³ and if 12.5 kJ of energy is transferred to the gas by heat, what are (a) the change in its internal energy and (b) its final temperature?

34. One mole of an ideal gas does 3 000 J of work on its surroundings as it expands isothermally to a final pressure of 1.00 atm and a volume of 25.0 L. Determine (a) the initial volume and (b) the temperature of the gas.

35. How much work is done by the steam when 1.00 mol of water at 100°C boils and becomes 1.00 mol of steam at

100°C and at 1.00 atm pressure? Assuming the steam to be an ideal gas, determine the change in internal energy of the steam as it vaporizes.

36. A 1.00-kg block of aluminum is heated at atmospheric pressure such that its temperature increases from 22.0°C to 40.0°C. Find (a) the work done by the aluminum, (b) the energy added to it by heat, and (c) the change in its internal energy.

37. A 2.00-mol sample of helium gas initially at 300 K and 0.400 atm is compressed isothermally to 1.20 atm. Assuming the behavior of helium to be that of an ideal gas, find (a) the final volume of the gas, (b) the work done by the gas, and (c) the energy transferred by heat.

38. One mole of water vapor at a temperature of 373 K cools down to 283 K. The energy given off from the cooling vapor by heat is absorbed by 10.0 mol of an ideal gas, causing it to expand at a constant temperature of 273 K. If the final volume of the ideal gas is 20.0 L, what is the initial volume of the ideal gas?

39. An ideal gas is carried through a thermodynamic cycle consisting of two isobaric and two isothermal processes, as shown in Figure P20.39. Show that the net work done in the entire cycle is given by the equation

$$W_{net} = P_1(V_2 - V_1) \ln \frac{P_2}{P_1}$$

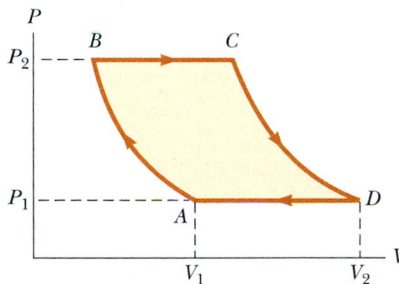

Figure P20.39

40. In Figure P20.40, the change in internal energy of a gas that is taken from A to C is $+800$ J. The work done along the path ABC is $+500$ J. (a) How much energy must be added to the system by heat as it goes from A through B and on to C? (b) If the pressure at point A is five times that at point C, what is the work done by the system in going from C to D? (c) What is the energy exchanged with the surroundings by heat as the gas is taken from C to A along the green path? (d) If the change in internal energy in going from point D to point A is $+500$ J, how much energy must be added to the system by heat as it goes from point C to point D?

Section 20.7 Energy Transfer Mechanisms

41. A steam pipe is covered with 1.50-cm-thick insulating material with a thermal conductivity of 0.200 cal/cm · °C · s.

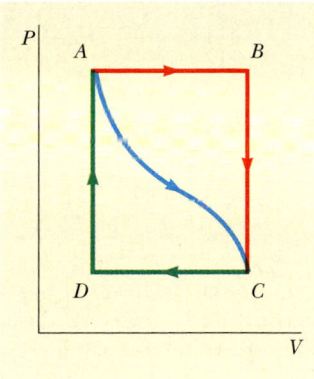

Figure P20.40

How much energy is lost every second by heat when the steam is at 200°C and the surrounding air is at 20.0°C? The pipe has a circumference of 20.0 cm and a length of 50.0 m. Neglect losses through the ends of the pipe.

42. A box with a total surface area of 1.20 m² and a wall thickness of 4.00 cm is made of an insulating material. A 10.0-W electric heater inside the box maintains the inside temperature at 15.0°C above the outside temperature. Find the thermal conductivity k of the insulating material.

43. A glass window pane has an area of 3.00 m² and a thickness of 0.600 cm. If the temperature difference between its surfaces is 25.0°C, what is the rate of energy transfer by conduction through the window?

44. A thermal window with an area of 6.00 m² is constructed of two layers of glass, each 4.00 mm thick and separated from each other by an air space of 5.00 mm. If the inside surface is at 20.0°C and the outside is at −30.0°C, what is the rate of energy transfer by conduction through the window?

45. A bar of gold is in thermal contact with a bar of silver of the same length and area (Fig. P20.45). One end of the compound bar is maintained at 80.0°C, while the opposite end is at 30.0°C. When the rate of energy transfer by conduction reaches steady state, what is the temperature at the junction?

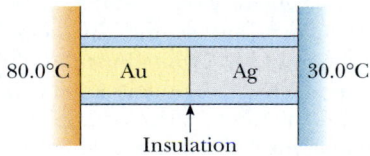

Figure P20.45

46. Two rods of the same length but made of different materials and having different cross-sectional areas are placed side by side, as shown in Figure P20.46. Deter-

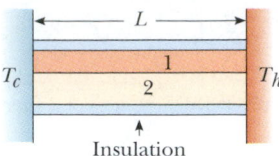

Figure P20.46

mine the rate of energy transfer by conduction in terms of the thermal conductivity and the area of each rod. Generalize your result to a system consisting of several rods.

47. Calculate the R value of (a) a window made of a single pane of flat glass $\frac{1}{8}$ in. thick; (b) a thermal window made of two single panes, each $\frac{1}{8}$ in. thick and separated by a $\frac{1}{4}$-in. air space. (c) By what factor is the thermal conduction reduced if the thermal window replaces the single-pane window?

48. The surface of the Sun has a temperature of about 5 800 K. The radius of the Sun is 6.96×10^8 m. Calculate the total energy radiated by the Sun each second. (Assume that $e = 0.965$.)

49. A large, hot pizza floats in outer space. What is the order of magnitude (a) of its rate of energy loss? (b) of its rate of temperature change? List the quantities you estimate and the value you estimate for each.

50. The tungsten filament of a certain 100-W light bulb radiates 2.00 W of light. (The other 98 W is carried away by convection and conduction.) The filament has a surface area of 0.250 mm^2 and an emissivity of 0.950. Find the filament's temperature. (The melting point of tungsten is 3 683 K.)

51. At high noon, the Sun delivers 1 000 W to each square meter of a blacktop road. If the hot asphalt loses energy only by radiation, what is its equilibrium temperature?

52. At our distance from the Sun, the intensity of solar radiation is 1 340 W/m^2. The temperature of the Earth is affected by the so-called "greenhouse effect" of the atmosphere. This effect makes our planet's emissivity for visible light higher than its emissivity for infrared light. For comparison, consider a spherical object with no atmosphere at the same distance from the Sun as the Earth. Assume that its emissivity is the same for all kinds of electromagnetic waves and that its temperature is uniform over its surface. Identify the projected area over which it absorbs sunlight and the surface area over which it radiates. Compute its equilibrium temperature. Chilly, isn't it? Your calculation applies to (a) the average temperature of the Moon, (b) astronauts in mortal danger aboard the crippled *Apollo 13* spacecraft, and (c) global catastrophe on the Earth if widespread fires caused a layer of soot to accumulate throughout the upper atmosphere so that most of the radiation from the Sun was absorbed there rather than at the surface below the atmosphere.

ADDITIONAL PROBLEMS

53. One hundred grams of liquid nitrogen at 77.3 K is stirred into a beaker containing 200 g of water at 5.00°C. If the nitrogen leaves the solution as soon as it turns to gas, how much water freezes? (The latent heat of vaporization of nitrogen is 48.0 cal/g, and the latent heat of fusion of water is 79.6 cal/g.)

54. A 75.0-kg cross-country skier moves across the snow (Fig. P20.54). The coefficient of friction between the skis and the snow is 0.200. Assume that all the snow beneath his skis is at 0°C and that all the internal energy generated by friction is added to the snow, which sticks to his skis until it melts. How far would he have to ski to melt 1.00 kg of snow?

Figure P20.54   A cross-country skier. *(Nathan Bilow/Leo de Wys, Inc.)*

55. An aluminum rod 0.500 m in length and with a cross-sectional area 2.50 cm^2 is inserted into a thermally insulated vessel containing liquid helium at 4.20 K. The rod is initially at 300 K. (a) If one half of the rod is inserted into the helium, how many liters of helium boil off by the time the inserted half cools to 4.20 K? (Assume that the upper half does not yet cool.) (b) If the upper end of the rod is maintained at 300 K, what is the approximate boil-off rate of liquid helium after the lower half has reached 4.20 K? (Aluminum has thermal conductivity of 31.0 J/s·cm·K at 4.2 K; ignore its temperature variation. Aluminum has a specific heat of 0.210 cal/g·°C and density of 2.70 g/cm^3. The density of liquid helium is 0.125 g/cm^3.)

56. On a cold winter day, you buy a hot dog from a street vendor. Into the pocket of your down parka you put the change he gives you: coins consisting of 9.00 g of copper at -12.0°C. Your pocket already contains 14.0 g of silver coins at 30.0°C. A short time later, the temperature of the copper coins is 4.00°C and is increasing at a rate of 0.500°C/s. At this time (a) what is the temperature of the silver coins, and (b) at what rate is it changing? (Neglect energy transferred to the surroundings.)

57. A *flow calorimeter* is an apparatus used to measure the specific heat of a liquid. The technique of flow calorimetry involves measuring the temperature difference between the input and output points of a flowing stream of the liquid while energy is added by heat at a known rate. In one particular experiment, a liquid with a density of 0.780 g/cm³ flows through the calorimeter at the rate of 4.00 cm³/s. At steady state, a temperature difference of 4.80°C is established between the input and output points when energy is supplied by heat at the rate of 30.0 J/s. What is the specific heat of the liquid?

58. A *flow calorimeter* is an apparatus used to measure the specific heat of a liquid. The technique of flow calorimetry involves measuring the temperature difference between the input and output points of a flowing stream of the liquid while energy is added by heat at a known rate. In one particular experiment, a liquid of density ρ flows through the calorimeter with volume flow rate R. At steady state, a temperature difference ΔT is established between the input and output points when energy is supplied at the rate $\mathcal{P}$. What is the specific heat of the liquid?

59. One mole of an ideal gas, initially at 300 K, is cooled at constant volume so that the final pressure is one-fourth the initial pressure. The gas then expands at constant pressure until it reaches the initial temperature. Determine the work done by the gas.

60. One mole of an ideal gas is contained in a cylinder with a movable piston. The initial pressure, temperature, and volume are P_i, V_i, and T_i, respectively. Find the work done by the gas for the following processes and show each process on a PV diagram: (a) An isobaric compression in which the final volume is one-half the initial volume. (b) An isothermal compression in which the final pressure is four times the initial pressure. (c) An isovolumetric process in which the final pressure is triple the initial pressure.

61. An ideal gas initially at P_i, V_i, and T_i is taken through a cycle as shown in Figure P20.61. (a) Find the net work done by the gas per cycle. (b) What is the net energy added by heat to the system per cycle? (c) Obtain a nu-merical value for the net work done per cycle for 1.00 mol of gas initially at 0°C.

62. Review Problem. An iron plate is held against an iron wheel so that a sliding frictional force of 50.0 N acts between the two pieces of metal. The relative speed at which the two surfaces slide over each other is 40.0 m/s. (a) Calculate the rate at which mechanical energy is converted to internal energy. (b) The plate and the wheel each have a mass of 5.00 kg, and each receives 50.0% of the internal energy. If the system is run as described for 10.0 s and each object is then allowed to reach a uniform internal temperature, what is the resultant temperature increase?

WEB 63. A "solar cooker" consists of a curved reflecting mirror that focuses sunlight onto the object to be warmed (Fig. P20.63). The solar power per unit area reaching the Earth at the location is 600 W/m², and the cooker has a diameter of 0.600 m. Assuming that 40.0% of the incident energy is transferred to the water, how long does it take to completely boil off 0.500 L of water initially at 20.0°C? (Neglect the heat capacity of the container.)

Figure P20.63

64. Water in an electric teakettle is boiling. The power absorbed by the water is 1.00 kW. Assuming that the pressure of the vapor in the kettle equals atmospheric pressure, determine the speed of effusion of vapor from the kettle's spout if the spout has a cross-sectional area of 2.00 cm².

65. Liquid water evaporates and even boils at temperatures other than 100°C, depending on the ambient pressure. Suppose that the latent heat of vaporization in Table 20.2 describes the liquid–vapor transition at all temperatures. A chamber contains 1.00 kg of water at 0°C under a piston, which just touches the water's surface. The piston is then raised quickly so that part of the water is vaporized and the other part is frozen (no liquid remains). Assuming that the temperature remains con-

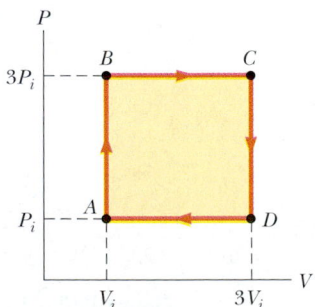

Figure P20.61

stant at 0°C, determine the mass of the ice that forms in the chamber.

66. A cooking vessel on a slow burner contains 10.0 kg of water and an unknown mass of ice in equilibrium at 0°C at time $t = 0$. The temperature of the mixture is measured at various times, and the result is plotted in Figure P20.66. During the first 50.0 min, the mixture remains at 0°C. From 50.0 min to 60.0 min, the temperature increases to 2.00°C. Neglecting the heat capacity of the vessel, determine the initial mass of the ice.

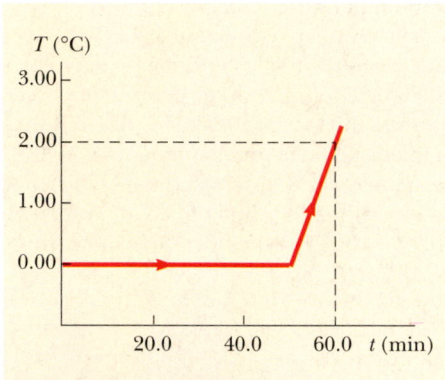

Figure P20.66

67. **Review Problem.**   (a) In air at 0°C, a 1.60-kg copper block at 0°C is set sliding at 2.50 m/s over a sheet of ice at 0°C. Friction brings the block to rest. Find the mass of the ice that melts. To describe the process of slowing down, identify the energy input Q, the work output W, the change in internal energy ΔE_{int}, and the change in mechanical energy ΔK for both the block and the ice. (b) A 1.60-kg block of ice at 0°C is set sliding at 2.50 m/s over a sheet of copper at 0°C. Friction brings the block to rest. Find the mass of the ice that melts. Identify Q, W, ΔE_{int}, and ΔK for the block and for the metal sheet during the process. (c) A thin 1.60-kg slab of copper at 20°C is set sliding at 2.50 m/s over an identical stationary slab at the same temperature. Friction quickly stops the motion. If no energy is lost to the environment by heat, find the change in temperature of both objects. Identify Q, W, ΔE_{int}, and ΔK for each object during the process.

68. The average thermal conductivity of the walls (including the windows) and roof of the house depicted in Figure P20.68 is 0.480 W/m·°C, and their average thickness is 21.0 cm. The house is heated with natural gas having a heat of combustion (that is, the energy provided per cubic meter of gas burned) of 9 300 kcal/m³. How many cubic meters of gas must be burned each day to maintain an inside temperature of 25.0°C if the out-

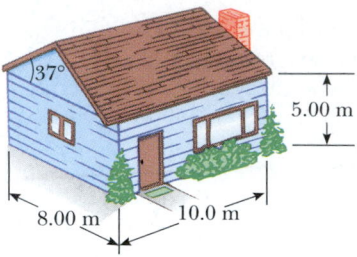

Figure P20.68

side temperature is 0.0°C? Disregard radiation and the energy lost by heat through the ground.

69. A pond of water at 0°C is covered with a layer of ice 4.00 cm thick. If the air temperature stays constant at −10.0°C, how long does it take the ice's thickness to increase to 8.00 cm? (*Hint:* To solve this problem, use Equation 20.14 in the form

$$\frac{dQ}{dt} = kA\frac{\Delta T}{x}$$

and note that the incremental energy dQ extracted from the water through the thickness x of ice is the amount required to freeze a thickness dx of ice. That is, $dQ = L\rho A\,dx$, where ρ is the density of the ice, A is the area, and L is the latent heat of fusion.)

70. The inside of a hollow cylinder is maintained at a temperature T_a while the outside is at a lower temperature T_b (Fig. P20.70). The wall of the cylinder has a thermal conductivity k. Neglecting end effects, show that the rate of energy conduction from the inner to the outer wall in the radial direction is

$$\frac{dQ}{dt} = 2\pi Lk\left[\frac{T_a - T_b}{\ln(b/a)}\right]$$

(*Hint:* The temperature gradient is dT/dr. Note that a radial flow of energy occurs through a concentric cylinder of area $2\pi rL$.)

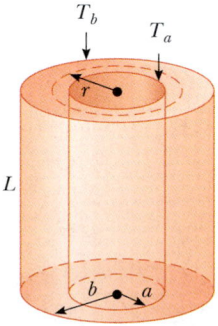

Figure P20.70

71. The passenger section of a jet airliner has the shape of a cylindrical tube with a length of 35.0 m and an inner radius of 2.50 m. Its walls are lined with an insulating material 6.00 cm in thickness and having a thermal conductivity of 4.00×10^{-5} cal/s·cm·°C. A heater must maintain the interior temperature at 25.0°C while the outside temperature is at -35.0°C. What power must be supplied to the heater if this temperature difference is to be maintained? (Use the result you obtained in Problem 70.)

72. A student obtains the following data in a calorimetry experiment designed to measure the specific heat of aluminum:

Initial temperature of water and calorimeter	70°C
Mass of water	0.400 kg
Mass of calorimeter	0.040 kg
Specific heat of calorimeter	0.63 kJ/kg·°C
Initial temperature of aluminum	27°C
Mass of aluminum	0.200 kg
Final temperature of mixture	66.3°C

Use these data to determine the specific heat of aluminum. Your result should be within 15% of the value listed in Table 20.1.

ANSWERS TO QUICK QUIZZES

20.1 (a) Water, glass, iron. Because water has the highest specific heat (4 186 J/kg·°C), it has the smallest change in temperature. Glass is next (837 J/kg·°C), and iron is last (448 J/kg·°C). (b) Iron, glass, water. For a given temperature increase, the energy transfer by heat is proportional to the specific heat.

20.2 Steam. According to Table 20.2, a kilogram of 100°C steam releases 2.26×10^6 J of energy as it condenses to 100°C water. After it releases this much energy into your skin, it is identical to 100°C water and will continue to burn you.

20.3 C, A, E. The slope is the ratio of the temperature change to the amount of energy input. Thus, the slope is proportional to the reciprocal of the specific heat. Water, which has the highest specific heat, has the least slope.

20.4

Situation	System	Q	W	ΔE_{int}
(a) Rapidly pumping up a bicycle tire	Air in the pump	0	−	+
(b) Pan of room-temperature water sitting on a hot stove	Water in the pan	+	0	+
(c) Air quickly leaking out of a balloon	Air originally in the balloon	0	+	−

(a) Because the pumping is rapid, no energy enters or leaves the system by heat; thus, $Q = 0$. Because work is done *on* the system, this work is negative. Thus, $\Delta E_{int} = Q - W$ must be positive. The air in the pump is warmer. (b) No work is done either by or on the system, but energy flows into the water by heat from the hot burner, making both Q and ΔE_{int} positive. (c) Because the leak is rapid, no energy flows into or out of the system by heat; hence, $Q = 0$. The air molecules escaping from the balloon do work on the surrounding air molecules as they push them out of the way. Thus, W is positive and ΔE_{int} is negative. The decrease in internal energy is evidenced by the fact that the escaping air becomes cooler.

20.5 A is isovolumetric, B is adiabatic, C is isothermal, and D is isobaric.

20.6 c. The blanket acts as a thermal insulator, slowing the transfer of energy by heat from the air into the cube.

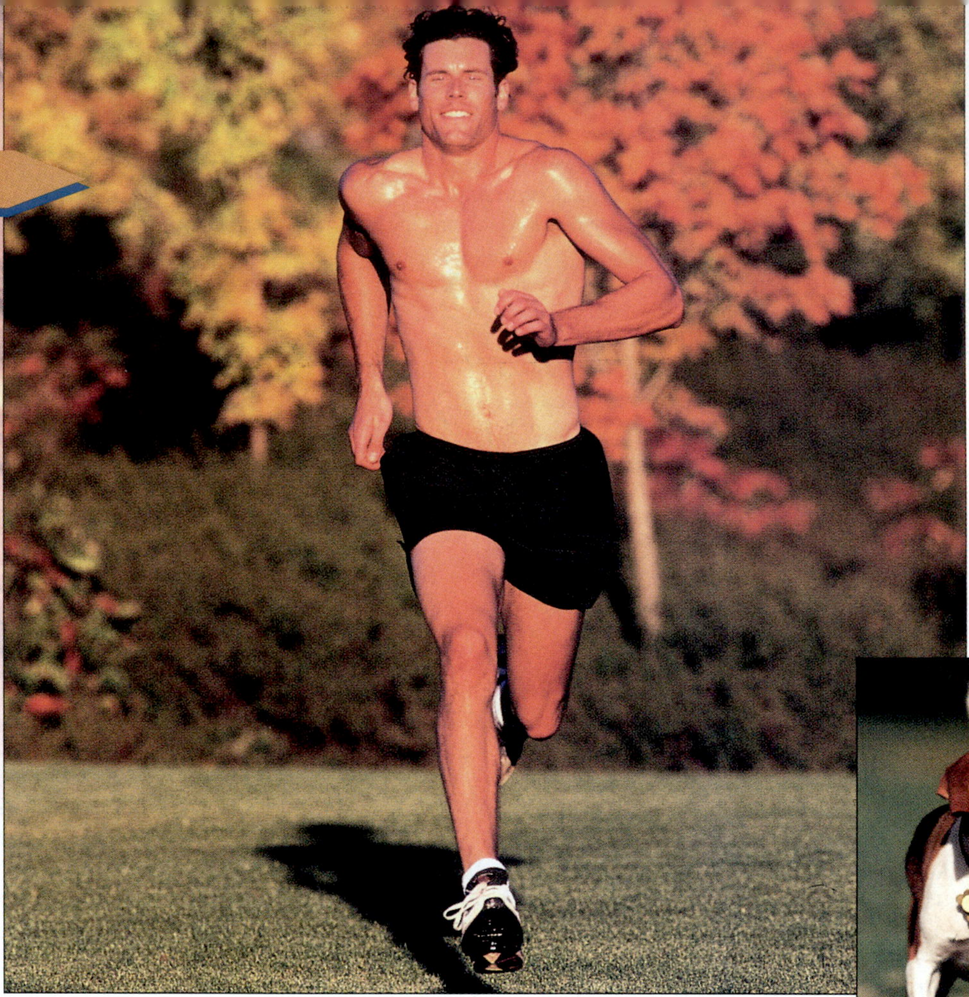

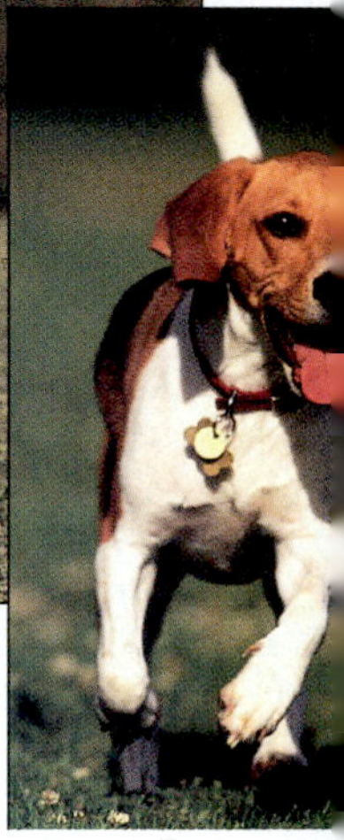

c h a p t e r

21

The Kinetic Theory of Gases

In Chapter 19 we discussed the properties of an ideal gas, using such macroscopic variables as pressure, volume, and temperature. We shall now show that such large-scale properties can be described on a microscopic scale, where matter is treated as a collection of molecules. Newton's laws of motion applied in a statistical manner to a collection of particles provide a reasonable description of thermodynamic processes. To keep the mathematics relatively simple, we shall consider molecular behavior of gases only, because in gases the interactions between molecules are much weaker than they are in liquids or solids. In the current view of gas behavior, called the *kinetic theory*, gas molecules move about in a random fashion, colliding with the walls of their container and with each other. Perhaps the most important feature of this theory is that it demonstrates that the kinetic energy of molecular motion and the internal energy of a gas system are equivalent. Furthermore, the kinetic theory provides us with a physical basis for our understanding of the concept of temperature.

In the simplest model of a gas, each molecule is considered to be a hard sphere that collides elastically with other molecules and with the container's walls. The hard-sphere model assumes that the molecules do not interact with each other except during collisions and that they are not deformed by collisions. This description is adequate only for monatomic gases, for which the energy is entirely translational kinetic energy. One must modify the theory for more complex molecules, such as oxygen (O_2) and carbon dioxide (CO_2), to include the internal energy associated with rotations and vibrations of the molecules.

21.1   MOLECULAR MODEL OF AN IDEAL GAS

We begin this chapter by developing a microscopic model of an ideal gas. The model shows that the pressure that a gas exerts on the walls of its container is a consequence of the collisions of the gas molecules with the walls. As we shall see, the model is consistent with the macroscopic description of Chapter 19. In developing this model, we make the following assumptions:

- The number of molecules is large, and the average separation between molecules is great compared with their dimensions. This means that the volume of the molecules is negligible when compared with the volume of the container.
- The molecules obey Newton's laws of motion, but as a whole they move randomly. By "randomly" we mean that any molecule can move in any direction with equal probability. We also assume that the distribution of speeds does not change in time, despite the collisions between molecules. That is, at any given moment, a certain percentage of molecules move at high speeds, a certain percentage move at low speeds, and a certain percentage move at speeds intermediate between high and low.
- The molecules undergo elastic collisions with each other and with the walls of the container. Thus, in the collisions, both kinetic energy and momentum are constant.
- The forces between molecules are negligible except during a collision. The forces between molecules are short-range, so the molecules interact with each other only during collisions.
- The gas under consideration is a pure substance. That is, all of its molecules are identical.

Although we often picture an ideal gas as consisting of single atoms, we can assume that the behavior of molecular gases approximates that of ideal gases rather

Assumptions of the molecular model of an ideal gas

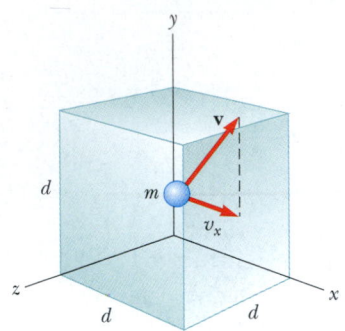

Figure 21.1 A cubical box with sides of length d containing an ideal gas. The molecule shown moves with velocity **v**.

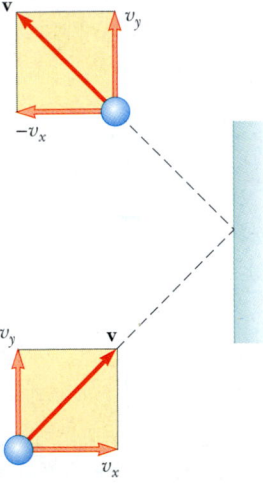

Figure 21.2 A molecule makes an elastic collision with the wall of the container. Its x component of momentum is reversed, while its y component remains unchanged. In this construction, we assume that the molecule moves in the xy plane.

well at low pressures. Molecular rotations or vibrations have no effect, on the average, on the motions that we considered here.

Now let us derive an expression for the pressure of an ideal gas consisting of N molecules in a container of volume V. The container is a cube with edges of length d (Fig. 21.1). Consider the collision of one molecule moving with a velocity **v** toward the right-hand face of the box. The molecule has velocity components v_x, v_y, and v_z. Previously, we used m to represent the mass of a sample, but throughout this chapter we shall use m to represent the mass of one molecule. As the molecule collides with the wall elastically, its x component of velocity is reversed, while its y and z components of velocity remain unaltered (Fig. 21.2). Because the x component of the momentum of the molecule is mv_x before the collision and $-mv_x$ after the collision, the change in momentum of the molecule is

$$\Delta p_x = -mv_x - (mv_x) = -2mv_x$$

Applying the impulse–momentum theorem (Eq. 9.9) to the molecule gives

$$F_1 \,\Delta t = \Delta p_x = -2mv_x$$

where F_1 is the magnitude of the average force exerted by the wall on the molecule in the time Δt. The subscript 1 indicates that we are currently considering only *one* molecule. For the molecule to collide twice with the same wall, it must travel a distance $2d$ in the x direction. Therefore, the time interval between two collisions with the same wall is $\Delta t = 2d/v_x$. Over a time interval that is long compared with Δt, the average force exerted on the molecule for each collision is

$$F_1 = \frac{-2mv_x}{\Delta t} = \frac{-2mv_x}{2d/v_x} = \frac{-mv_x{}^2}{d} \tag{21.1}$$

According to Newton's third law, the average force exerted by the molecule on the wall is equal in magnitude and opposite in direction to the force in Equation 21.1:

$$F_{1,\,\text{on wall}} = -F_1 = -\left(\frac{-mv_x{}^2}{d}\right) = \frac{mv_x{}^2}{d}$$

Each molecule of the gas exerts a force F_1 on the wall. We find the total force F exerted by all the molecules on the wall by adding the forces exerted by the individual molecules:

$$F = \frac{m}{d}\left(v_{x1}{}^2 + v_{x2}{}^2 + \cdots\right)$$

In this equation, v_{x1} is the x component of velocity of molecule 1, v_{x2} is the x component of velocity of molecule 2, and so on. The summation terminates when we reach N molecules because there are N molecules in the container.

To proceed further, we must note that the average value of the square of the velocity in the x direction for N molecules is

$$\overline{v_x{}^2} = \frac{v_{x1}{}^2 + v_{x2}{}^2 + \cdots + v_{xN}{}^2}{N}$$

Thus, the total force exerted on the wall can be written

$$F = \frac{Nm}{d}\,\overline{v_x{}^2}$$

Now let us focus on one molecule in the container whose velocity components are v_x, v_y, and v_z. The Pythagorean theorem relates the square of the speed of this molecule to the squares of these components:

$$v^2 = v_x{}^2 + v_y{}^2 + v_z{}^2$$

Hence, the average value of v^2 for all the molecules in the container is related to the average values of $v_x{}^2$, $v_y{}^2$, and $v_z{}^2$ according to the expression

$$\overline{v^2} = \overline{v_x{}^2} + \overline{v_y{}^2} + \overline{v_z{}^2}$$

Because the motion is completely random, the average values $\overline{v_x{}^2}$, $\overline{v_y{}^2}$, and $\overline{v_z{}^2}$ are equal to each other. Using this fact and the previous equation, we find that

$$\overline{v^2} = 3\overline{v_x{}^2}$$

Thus, the total force exerted on the wall is

$$F = \frac{N}{3}\left(\frac{m\overline{v^2}}{d}\right)$$

Using this expression, we can find the total pressure exerted on the wall:

$$P = \frac{F}{A} = \frac{F}{d^2} = \frac{1}{3}\left(\frac{N}{d^3}\ m\overline{v^2}\right) = \frac{1}{3}\left(\frac{N}{V}\right)m\overline{v^2}$$

$$P = \frac{2}{3}\left(\frac{N}{V}\right)\left(\frac{1}{2}\ m\overline{v^2}\right) \tag{21.2}$$

This result indicates that **the pressure is proportional to the number of molecules per unit volume and to the average translational kinetic energy of the molecules,** $\frac{1}{2}m\overline{v^2}$. In deriving this simplified model of an ideal gas, we obtain an important result that relates the large-scale quantity of pressure to an atomic quantity—the average value of the square of the molecular speed. Thus, we have established a key link between the atomic world and the large-scale world.

    You should note that Equation 21.2 verifies some features of pressure with which you are probably familiar. One way to increase the pressure inside a container is to increase the number of molecules per unit volume in the container. This is what you do when you add air to a tire. The pressure in the tire can also be increased by increasing the average translational kinetic energy of the air molecules in the tire. As we shall soon see, this can be accomplished by increasing the temperature of that air. It is for this reason that the pressure inside a tire increases as the tire warms up during long trips. The continuous flexing of the tire as it moves along the surface of a road results in work done as parts of the tire distort and in an increase in internal energy of the rubber. The increased temperature of the rubber results in the transfer of energy by heat into the air inside the tire. This transfer increases the air's temperature, and this increase in temperature in turn produces an increase in pressure.

Molecular Interpretation of Temperature

We can gain some insight into the meaning of temperature by first writing Equation 21.2 in the more familiar form

$$PV = \tfrac{2}{3}N\left(\tfrac{1}{2}m\overline{v^2}\right)$$

Let us now compare this with the equation of state for an ideal gas (Eq. 19.10):

$$PV = Nk_{\mathrm{B}}T$$

Ludwig Boltzmann Austrian theoretical physicist (1844–1906) Boltzmann made many important contributions to the development of the kinetic theory of gases, electromagnetism, and thermodynamics. His pioneering work in the field of kinetic theory led to the branch of physics known as *statistical mechanics*. *(Courtesy of AIP Niels Bohr Library, Lande Collection)*

Relationship between pressure and molecular kinetic energy

Recall that the equation of state is based on experimental facts concerning the macroscopic behavior of gases. Equating the right sides of these expressions, we find that

Temperature is proportional to average kinetic energy

$$T = \frac{2}{3k_B}\left(\frac{1}{2}m\overline{v^2}\right)$$ (21.3)

That is, **temperature is a direct measure of average molecular kinetic energy.**

By rearranging Equation 21.3, we can relate the translational molecular kinetic energy to the temperature:

Average kinetic energy per molecule

$$\frac{1}{2}m\overline{v^2} = \frac{3}{2}k_B T$$ (21.4)

That is, the average translational kinetic energy per molecule is $\frac{3}{2}k_B T$. Because $\overline{v_x^2} = \frac{1}{3}\overline{v^2}$, it follows that

$$\frac{1}{2}m\overline{v_x^2} = \frac{1}{2}k_B T$$ (21.5)

In a similar manner, it follows that the motions in the y and z directions give us

$$\frac{1}{2}m\overline{v_y^2} = \frac{1}{2}k_B T \qquad \text{and} \qquad \frac{1}{2}m\overline{v_z^2} = \frac{1}{2}k_B T$$

Thus, each translational degree of freedom contributes an equal amount of energy to the gas, namely, $\frac{1}{2}k_B T$. (In general, "degrees of freedom" refers to the number of independent means by which a molecule can possess energy.) A generalization of this result, known as the **theorem of equipartition of energy,** states that

Theorem of equipartition of energy

each degree of freedom contributes $\frac{1}{2}k_B T$ to the energy of a system.

The total translational kinetic energy of N molecules of gas is simply N times the average energy per molecule, which is given by Equation 21.4:

Total translational kinetic energy of N molecules

$$E_{trans} = N\left(\frac{1}{2}m\overline{v^2}\right) = \frac{3}{2}Nk_B T = \frac{3}{2}nRT$$ (21.6)

where we have used $k_B = R/N_A$ for Boltzmann's constant and $n = N/N_A$ for the number of moles of gas. If we consider a gas for which the only type of energy for the molecules is translational kinetic energy, we can use Equation 21.6 to express

TABLE 21.1   Some rms Speeds

Gas	Molar Mass (g/mol)	v_{rms} at 20°C (m/s)
H_2	2.02	1904
He	4.00	1352
H_2O	18.0	637
Ne	20.2	602
N_2 or CO	28.0	511
NO	30.0	494
CO_2	44.0	408
SO_2	64.1	338

the internal energy of the gas. This result implies that the internal energy of an ideal gas depends only on the temperature.

The square root of $\overline{v^2}$ is called the *root-mean-square* (rms) *speed* of the molecules. From Equation 21.4 we obtain, for the rms speed,

$$v_{\text{rms}} = \sqrt{\overline{v^2}} = \sqrt{\frac{3k_B T}{m}} = \sqrt{\frac{3RT}{M}} \tag{21.7}$$

Root-mean-square speed

where M is the molar mass in kilograms per mole. This expression shows that, at a given temperature, lighter molecules move faster, on the average, than do heavier molecules. For example, at a given temperature, hydrogen molecules, whose molar mass is 2×10^{-3} kg/mol, have an average speed four times that of oxygen molecules, whose molar mass is 32×10^{-3} kg/mol. Table 21.1 lists the rms speeds for various molecules at 20°C.

EXAMPLE 21.1   A Tank of Helium

A tank used for filling helium balloons has a volume of 0.300 m³ and contains 2.00 mol of helium gas at 20.0°C. Assuming that the helium behaves like an ideal gas, (a) what is the total translational kinetic energy of the molecules of the gas?

Solution   Using Equation 21.6 with $n = 2.00$ mol and $T = 293$ K, we find that

$$E_{\text{trans}} = \tfrac{3}{2}nRT = \tfrac{3}{2}(2.00 \text{ mol})(8.31 \text{ J/mol·K})(293 \text{ K})$$

$$= \boxed{7.30 \times 10^3 \text{ J}}$$

(b) What is the average kinetic energy per molecule?

Solution   Using Equation 21.4, we find that the average kinetic energy per molecule is

$$\tfrac{1}{2}m\overline{v^2} = \tfrac{3}{2}k_B T = \tfrac{3}{2}(1.38 \times 10^{-23} \text{ J/K})(293 \text{ K})$$

$$= \boxed{6.07 \times 10^{-21} \text{ J}}$$

Exercise   Using the fact that the molar mass of helium is 4.00×10^{-3} kg/mol, determine the rms speed of the atoms at 20.0°C.

Answer 1.35×10^3 m/s.

Quick Quiz 21.1

At room temperature, the average speed of an air molecule is several hundred meters per second. A molecule traveling at this speed should travel across a room in a small fraction of a second. In view of this, why does it take the odor of perfume (or other smells) several minutes to travel across the room?

21.2   MOLAR SPECIFIC HEAT OF AN IDEAL GAS

The energy required to raise the temperature of n moles of gas from T_i to T_f depends on the path taken between the initial and final states. To understand this, let us consider an ideal gas undergoing several processes such that the change in temperature is $\Delta T = T_f - T_i$ for all processes. The temperature change can be achieved by taking a variety of paths from one isotherm to another, as shown in Figure 21.3. Because ΔT is the same for each path, the change in internal energy ΔE_{int} is the same for all paths. However, we know from the first law, $Q = \Delta E_{\text{int}} + W$, that the heat Q is different for each path because W (the area under the curves) is different for each path. Thus, the heat associated with a given change in temperature does not have a unique value.

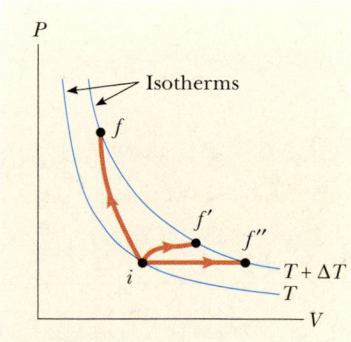

Figure 21.3   An ideal gas is taken from one isotherm at temperature T to another at temperature $T + \Delta T$ along three different paths.

We can address this difficulty by defining specific heats for two processes that frequently occur: changes at constant volume and changes at constant pressure. Because the number of moles is a convenient measure of the amount of gas, we define the **molar specific heats** associated with these processes with the following equations:

$$Q = nC_V\Delta T \qquad \text{(constant volume)} \tag{21.8}$$

$$Q = nC_P\Delta T \qquad \text{(constant pressure)} \tag{21.9}$$

where C_V is the **molar specific heat at constant volume** and C_P is the **molar specific heat at constant pressure.** When we heat a gas at constant pressure, not only does the internal energy of the gas increase, but the gas also does work because of the change in volume. Therefore, the heat $Q_{\text{constant } P}$ must account for both the increase in internal energy and the transfer of energy out of the system by work, and so $Q_{\text{constant } P}$ is greater than $Q_{\text{constant } V}$. Thus, C_P is greater than C_V.

In the previous section, we found that the temperature of a gas is a measure of the average translational kinetic energy of the gas molecules. This kinetic energy is associated with the motion of the center of mass of each molecule. It does not include the energy associated with the internal motion of the molecule—namely, vibrations and rotations about the center of mass. This should not be surprising because the simple kinetic theory model assumes a structureless molecule.

In view of this, let us first consider the simplest case of an ideal monatomic gas, that is, a gas containing one atom per molecule, such as helium, neon, or argon. When energy is added to a monatomic gas in a container of fixed volume (by heating, for example), all of the added energy goes into increasing the translational kinetic energy of the atoms. There is no other way to store the energy in a monatomic gas. Therefore, from Equation 21.6, we see that the total internal energy E_{int} of N molecules (or n mol) of an ideal monatomic gas is

> Internal energy of an ideal monatomic gas is proportional to its temperature

$$E_{\text{int}} = \tfrac{3}{2}Nk_BT = \tfrac{3}{2}nRT \tag{21.10}$$

Note that for a monatomic ideal gas, E_{int} is a function of T only, and the functional relationship is given by Equation 21.10. In general, the internal energy of an ideal gas is a function of T only, and the exact relationship depends on the type of gas, as we shall soon explore.

Quick Quiz 21.2

How does the internal energy of a gas change as its pressure is decreased while its volume is increased in such a way that the process follows the isotherm labeled T in Figure 21.4? (a) E_{int} increases. (b) E_{int} decreases. (c) E_{int} stays the same. (d) There is not enough information to determine ΔE_{int}.

If energy is transferred by heat to a system at *constant volume*, then no work is done by the system. That is, $W = \int P\,dV = 0$ for a constant-volume process. Hence, from the first law of thermodynamics, we see that

$$Q = \Delta E_{\text{int}} \tag{21.11}$$

In other words, all of the energy transferred by heat goes into increasing the internal energy (and temperature) of the system. A constant-volume process from i to f is described in Figure 21.4, where ΔT is the temperature difference between the two isotherms. Substituting the expression for Q given by Equation 21.8 into

Equation 21.11, we obtain

$$\Delta E_{int} = nC_V \Delta T \qquad \textbf{(21.12)}$$

If the molar specific heat is constant, we can express the internal energy of a gas as

$$E_{int} = nC_V T$$

This equation applies to all ideal gases—to gases having more than one atom per molecule, as well as to monatomic ideal gases.

In the limit of infinitesimal changes, we can use Equation 21.12 to express the molar specific heat at constant volume as

$$C_V = \frac{1}{n}\frac{dE_{int}}{dT} \qquad \textbf{(21.13)}$$

Let us now apply the results of this discussion to the monatomic gas that we have been studying. Substituting the internal energy from Equation 21.10 into Equation 21.13, we find that

$$C_V = \tfrac{3}{2}R \qquad \textbf{(21.14)}$$

This expression predicts a value of $C_V = \tfrac{3}{2}R = 12.5\ \text{J/mol·K}$ for all monatomic gases. This is in excellent agreement with measured values of molar specific heats for such gases as helium, neon, argon, and xenon over a wide range of temperatures (Table 21.2).

Now suppose that the gas is taken along the constant-pressure path $i \rightarrow f'$ shown in Figure 21.4. Along this path, the temperature again increases by ΔT. The energy that must be transferred by heat to the gas in this process is $Q = nC_P \Delta T$. Because the volume increases in this process, the work done by the gas is $W = P\Delta V$, where P is the constant pressure at which the process occurs. Applying

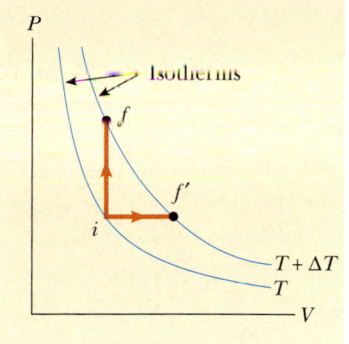

Figure 21.4 Energy is transferred by heat to an ideal gas in two ways. For the constant-volume path $i \rightarrow f$, all the energy goes into increasing the internal energy of the gas because no work is done. Along the constant-pressure path $i \rightarrow f'$, part of the energy transferred in by heat is transferred out by work done by the gas.

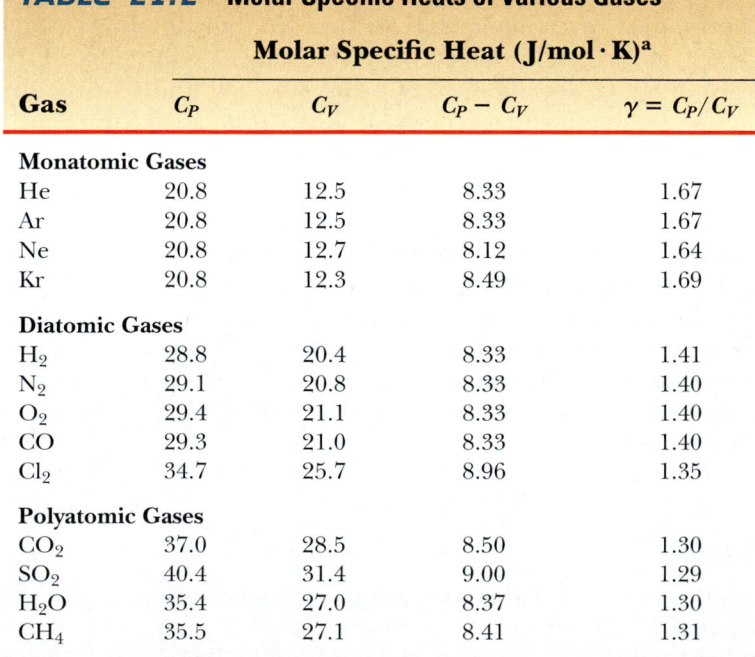

Gas	C_P	C_V	$C_P - C_V$	$\gamma = C_P/C_V$
TABLE 21.2 **Molar Specific Heats of Various Gases**				
Molar Specific Heat (J/mol·K)[a]				
Monatomic Gases				
He	20.8	12.5	8.33	1.67
Ar	20.8	12.5	8.33	1.67
Ne	20.8	12.7	8.12	1.64
Kr	20.8	12.3	8.49	1.69
Diatomic Gases				
H_2	28.8	20.4	8.33	1.41
N_2	29.1	20.8	8.33	1.40
O_2	29.4	21.1	8.33	1.40
CO	29.3	21.0	8.33	1.40
Cl_2	34.7	25.7	8.96	1.35
Polyatomic Gases				
CO_2	37.0	28.5	8.50	1.30
SO_2	40.4	31.4	9.00	1.29
H_2O	35.4	27.0	8.37	1.30
CH_4	35.5	27.1	8.41	1.31

[a] All values except that for water were obtained at 300 K.

the first law to this process, we have

$$\Delta E_{\text{int}} = Q - W = nC_P \Delta T - P\Delta V \qquad \textbf{(21.15)}$$

In this case, the energy added to the gas by heat is channeled as follows: Part of it does external work (that is, it goes into moving a piston), and the remainder increases the internal energy of the gas. But the change in internal energy for the process $i \rightarrow f'$ is equal to that for the process $i \rightarrow f$ because E_{int} depends only on temperature for an ideal gas and because ΔT is the same for both processes. In addition, because $PV = nRT$, we note that for a constant-pressure process, $P\Delta V = nR\Delta T$. Substituting this value for $P\Delta V$ into Equation 21.15 with $\Delta E_{\text{int}} = nC_V \Delta T$ (Eq. 21.12) gives

$$nC_V \Delta T = nC_P \Delta T - nR\Delta T$$

$$C_P - C_V = R \qquad \textbf{(21.16)}$$

This expression applies to *any* ideal gas. It predicts that the molar specific heat of an ideal gas at constant pressure is greater than the molar specific heat at constant volume by an amount R, the universal gas constant (which has the value 8.31 J/mol·K). This expression is applicable to real gases, as the data in Table 21.2 show.

Because $C_V = \frac{3}{2}R$ for a monatomic ideal gas, Equation 21.16 predicts a value $C_P = \frac{5}{2}R = 20.8$ J/mol·K for the molar specific heat of a monatomic gas at constant pressure. The ratio of these heat capacities is a dimensionless quantity γ (Greek letter gamma):

Ratio of molar specific heats for a monatomic ideal gas

$$\gamma = \frac{C_P}{C_V} = \frac{(5/2)R}{(3/2)R} = \frac{5}{3} = 1.67 \qquad \textbf{(21.17)}$$

Theoretical values of C_P and γ are in excellent agreement with experimental values obtained for monatomic gases, but they are in serious disagreement with the values for the more complex gases (see Table 21.2). This is not surprising because the value $C_V = \frac{3}{2}R$ was derived for a monatomic ideal gas, and we expect some additional contribution to the molar specific heat from the internal structure of the more complex molecules. In Section 21.4, we describe the effect of molecular structure on the molar specific heat of a gas. We shall find that the internal energy—and, hence, the molar specific heat—of a complex gas must include contributions from the rotational and the vibrational motions of the molecule.

We have seen that the molar specific heats of gases at constant pressure are greater than the molar specific heats at constant volume. This difference is a consequence of the fact that in a constant-volume process, no work is done and all of the energy transferred by heat goes into increasing the internal energy (and temperature) of the gas, whereas in a constant-pressure process, some of the energy transferred by heat is transferred out as work done by the gas as it expands. In the case of solids and liquids heated at constant pressure, very little work is done because the thermal expansion is small. Consequently, C_P and C_V are approximately equal for solids and liquids.

EXAMPLE 21.2 ▶ Heating a Cylinder of Helium

A cylinder contains 3.00 mol of helium gas at a temperature of 300 K. (a) If the gas is heated at constant volume, how much energy must be transferred by heat to the gas for its temperature to increase to 500 K?

Solution For the constant-volume process, we have

$$Q_1 = nC_V \Delta T$$

Because $C_V = 12.5$ J/mol·K for helium and $\Delta T = 200$ K, we obtain

$Q_1 = (3.00\ \text{mol})(12.5\ \text{J/mol} \cdot \text{K})(200\ \text{K}) = \boxed{7.50 \times 10^3\ \text{J}}$

$Q_2 = nC_P \Delta T = (3.00\ \text{mol})(20.8\ \text{J/mol} \cdot \text{K})(200\ \text{K})$

$= \boxed{12.5 \times 10^3\ \text{J}}$

(b) How much energy must be transferred by heat to the gas at constant pressure to raise the temperature to 500 K?

Exercise   What is the work done by the gas in this isobaric process?

Solution   Making use of Table 21.2, we obtain

Answer $W = Q_2 - Q_1 = 5.00 \times 10^3\ \text{J}$.

21.3   ADIABATIC PROCESSES FOR AN IDEAL GAS

As we noted in Section 20.6, an adiabatic process is one in which no energy is transferred by heat between a system and its surroundings. For example, if a gas is compressed (or expanded) very rapidly, very little energy is transferred out of (or into) the system by heat, and so the process is nearly adiabatic. (We must remember that the temperature of a system changes in an adiabatic process even though no energy is transferred by heat.) Such processes occur in the cycle of a gasoline engine, which we discuss in detail in the next chapter.

Another example of an adiabatic process is the very slow expansion of a gas that is thermally insulated from its surroundings. In general,

an **adiabatic process** is one in which no energy is exchanged by heat between a system and its surroundings.

Definition of an adiabatic process

Let us suppose that an ideal gas undergoes an adiabatic expansion. At any time during the process, we assume that the gas is in an equilibrium state, so that the equation of state $PV = nRT$ is valid. As we shall soon see, the pressure and volume at any time during an adiabatic process are related by the expression

$$PV^{\gamma} = \text{constant} \qquad\qquad (21.18)$$

Relationship between P and V for an adiabatic process involving an ideal gas

where $\gamma = C_P/C_V$ is assumed to be constant during the process. Thus, we see that all three variables in the ideal gas law—P, V, and T—change during an adiabatic process.

Proof That $PV^{\gamma} =$ constant for an Adiabatic Process

When a gas expands adiabatically in a thermally insulated cylinder, no energy is transferred by heat between the gas and its surroundings; thus, $Q = 0$. Let us take the infinitesimal change in volume to be dV and the infinitesimal change in temperature to be dT. The work done by the gas is $P\,dV$. Because the internal energy of an ideal gas depends only on temperature, the change in the internal energy in an adiabatic expansion is the same as that for an isovolumetric process between the same temperatures, $dE_{\text{int}} = nC_V\,dT$ (Eq. 21.12). Hence, the first law of thermodynamics, $\Delta E_{\text{int}} = Q - W$, with $Q = 0$, becomes

$$dE_{\text{int}} = nC_V\,dT = -P\,dV$$

Taking the total differential of the equation of state of an ideal gas, $PV = nRT$, we

QuickLab

Rapidly pump up a bicycle tire and then feel the coupling at the end of the hose. Why is the coupling warm?

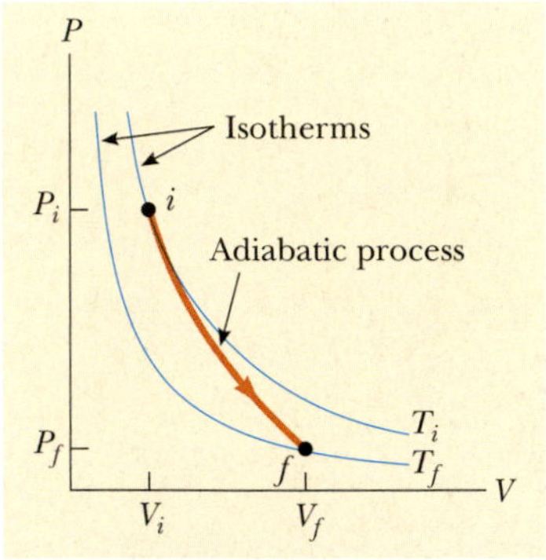

Figure 21.5 The *PV* diagram for an adiabatic expansion. Note that $T_f < T_i$ in this process.

Adiabatic process

see that

$$P\,dV + V\,dP = nR\,dT$$

Eliminating dT from these two equations, we find that

$$P\,dV + V\,dP = -\frac{R}{C_V}P\,dV$$

Substituting $R = C_P - C_V$ and dividing by PV, we obtain

$$\frac{dV}{V} + \frac{dP}{P} = -\left(\frac{C_P - C_V}{C_V}\right)\frac{dV}{V} = (1 - \gamma)\frac{dV}{V}$$

$$\frac{dP}{P} + \gamma\frac{dV}{V} = 0$$

Integrating this expression, we have

$$\ln P + \gamma \ln V = \text{constant}$$

which is equivalent to Equation 21.18:

$$PV^\gamma = \text{constant}$$

The *PV* diagram for an adiabatic expansion is shown in Figure 21.5. Because $\gamma > 1$, the *PV* curve is steeper than it would be for an isothermal expansion. By the definition of an adiabatic process, no energy is transferred by heat into or out of the system. Hence, from the first law, we see that ΔE_{int} is negative (the gas does work, so its internal energy decreases) and so ΔT also is negative. Thus, we see that the gas cools ($T_f < T_i$) during an adiabatic expansion. Conversely, the temperature increases if the gas is compressed adiabatically. Applying Equation 21.18 to the initial and final states, we see that

$$P_i V_i^\gamma = P_f V_f^\gamma \qquad (21.19)$$

Using the ideal gas law, we can express Equation 21.19 as

$$T_i V_i^{\gamma-1} = T_f V_f^{\gamma-1} \qquad (21.20)$$

EXAMPLE 21.3 **A Diesel Engine Cylinder**

Air at 20.0°C in the cylinder of a diesel engine is compressed from an initial pressure of 1.00 atm and volume of 800.0 cm³ to a volume of 60.0 cm³. Assume that air behaves as an ideal gas with $\gamma = 1.40$ and that the compression is adiabatic. Find the final pressure and temperature of the air.

Solution Using Equation 21.19, we find that

$$P_f = P_i\left(\frac{V_i}{V_f}\right)^\gamma = (1.00\ \text{atm})\left(\frac{800.0\ \text{cm}^3}{60.0\ \text{cm}^3}\right)^{1.40}$$

$$= \boxed{37.6\ \text{atm}}$$

Because $PV = nRT$ is valid during any process and because

no gas escapes from the cylinder,

$$\frac{P_i V_i}{T_i} = \frac{P_f V_f}{T_f}$$

$$T_f = \frac{P_f V_f}{P_i V_i}T_i = \frac{(37.6\ \text{atm})(60.0\ \text{cm}^3)}{(1.00\ \text{atm})(800.0\ \text{cm}^3)}(293\ \text{K})$$

$$= \boxed{826\ \text{K} = 553°\text{C}}$$

The high compression in a diesel engine raises the temperature of the fuel enough to cause its combustion without the use of spark plugs.

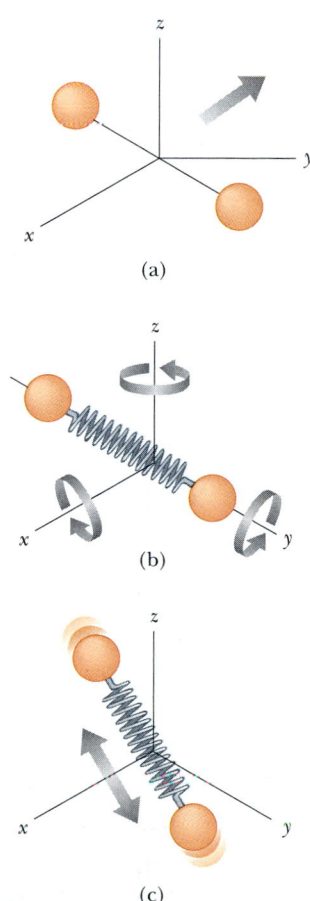

21.4 ▶ THE EQUIPARTITION OF ENERGY

We have found that model predictions based on molar specific heat agree quite well with the behavior of monatomic gases but not with the behavior of complex gases (see Table 21.2). Furthermore, the value predicted by the model for the quantity $C_P - C_V = R$ is the same for all gases. This is not surprising because this difference is the result of the work done by the gas, which is independent of its molecular structure.

To clarify the variations in C_V and C_P in gases more complex than monatomic gases, let us first explain the origin of molar specific heat. So far, we have assumed that the sole contribution to the internal energy of a gas is the translational kinetic energy of the molecules. However, the internal energy of a gas actually includes contributions from the translational, vibrational, and rotational motion of the molecules. The rotational and vibrational motions of molecules can be activated by collisions and therefore are "coupled" to the translational motion of the molecules. The branch of physics known as *statistical mechanics* has shown that, for a large number of particles obeying the laws of Newtonian mechanics, the available energy is, on the average, shared equally by each independent degree of freedom. Recall from Section 21.1 that the equipartition theorem states that, at equilibrium, each degree of freedom contributes $\frac{1}{2}k_B T$ of energy per molecule.

Let us consider a diatomic gas whose molecules have the shape of a dumbbell (Fig. 21.6). In this model, the center of mass of the molecule can translate in the x, y, and z directions (Fig. 21.6a). In addition, the molecule can rotate about three mutually perpendicular axes (Fig. 21.6b). We can neglect the rotation about the y axis because the moment of inertia I_y and the rotational energy $\frac{1}{2}I_y\omega^2$ about this axis are negligible compared with those associated with the x and z axes. (If the two atoms are taken to be point masses, then I_y is identically zero.) Thus, there are five degrees of freedom: three associated with the translational motion and two associated with the rotational motion. Because each degree of freedom contributes, on the average, $\frac{1}{2}k_B T$ of energy per molecule, the total internal energy for a system of N molecules is

$$E_{int} = 3N(\tfrac{1}{2}k_B T) + 2N(\tfrac{1}{2}k_B T) = \tfrac{5}{2}Nk_B T = \tfrac{5}{2}nRT$$

We can use this result and Equation 21.13 to find the molar specific heat at constant volume:

$$C_V = \frac{1}{n}\frac{dE_{int}}{dT} = \frac{1}{n}\frac{d}{dT}\left(\frac{5}{2}nRT\right) = \frac{5}{2}R$$

From Equations 21.16 and 21.17, we find that

$$C_P = C_V + R = \tfrac{7}{2}R$$

$$\gamma = \frac{C_P}{C_V} = \frac{\tfrac{7}{2}R}{\tfrac{5}{2}R} = \frac{7}{5} = 1.40$$

These results agree quite well with most of the data for diatomic molecules given in Table 21.2. This is rather surprising because we have not yet accounted for the possible vibrations of the molecule. In the vibratory model, the two atoms are joined by an imaginary spring (see Fig. 21.6c). The vibrational motion adds two more degrees of freedom, which correspond to the kinetic energy and the potential energy associated with vibrations along the length of the molecule. Hence, classical physics and the equipartition theorem predict an internal energy of

$$E_{int} = 3N(\tfrac{1}{2}k_B T) + 2N(\tfrac{1}{2}k_B T) + 2N(\tfrac{1}{2}k_B T) = \tfrac{7}{2}Nk_B T = \tfrac{7}{2}nRT$$

Figure 21.6 Possible motions of a diatomic molecule: (a) translational motion of the center of mass, (b) rotational motion about the various axes, and (c) vibrational motion along the molecular axis.

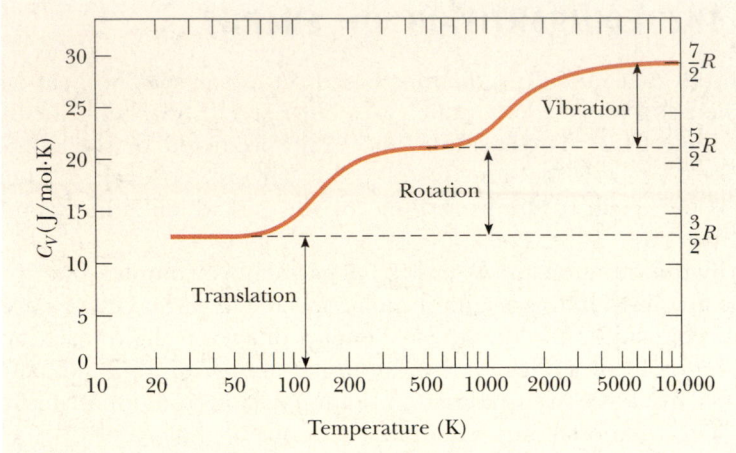

Figure 21.7   The molar specific heat of hydrogen as a function of temperature. The horizontal scale is logarithmic. Note that hydrogen liquefies at 20 K.

and a molar specific heat at constant volume of

$$ C_V = \frac{1}{n} \frac{dE_{\text{int}}}{dT} = \frac{1}{n} \frac{d}{dT} \left(\frac{7}{2} nRT \right) = \frac{7}{2} R $$

This value is inconsistent with experimental data for molecules such as H_2 and N_2 (see Table 21.2) and suggests a breakdown of our model based on classical physics.

For molecules consisting of more than two atoms, the number of degrees of freedom is even larger and the vibrations are more complex. This results in an even higher predicted molar specific heat, which is in qualitative agreement with experiment. The more degrees of freedom available to a molecule, the more "ways" it can store internal energy; this results in a higher molar specific heat.

We have seen that the equipartition theorem is successful in explaining some features of the molar specific heat of gas molecules with structure. However, the theorem does not account for the observed temperature variation in molar specific heats. As an example of such a temperature variation, C_V for H_2 is $\frac{5}{2}R$ from about 250 K to 750 K and then increases steadily to about $\frac{7}{2}R$ well above 750 K (Fig. 21.7). This suggests that much more significant vibrations occur at very high temperatures. At temperatures well below 250 K, C_V has a value of about $\frac{3}{2}R$, suggesting that the molecule has only translational energy at low temperatures.

A Hint of Energy Quantization

The failure of the equipartition theorem to explain such phenomena is due to the inadequacy of classical mechanics applied to molecular systems. For a more satisfactory description, it is necessary to use a quantum-mechanical model, in which the energy of an individual molecule is quantized. The energy separation between adjacent vibrational energy levels for a molecule such as H_2 is about ten times greater than the average kinetic energy of the molecule at room temperature. Consequently, collisions between molecules at low temperatures do not provide enough energy to change the vibrational state of the molecule. It is often stated that such degrees of freedom are "frozen out." This explains why the vibrational energy does not contribute to the molar specific heats of molecules at low temperatures.

The rotational energy levels also are quantized, but their spacing at ordinary temperatures is small compared with $k_B T$. Because the spacing between quantized energy levels is small compared with the available energy, the system behaves in accordance with classical mechanics. However, at sufficiently low temperatures (typically less than 50 K), where $k_B T$ is small compared with the spacing between rotational levels, intermolecular collisions may not be sufficiently energetic to alter the rotational states. This explains why C_V reduces to $\frac{3}{2}R$ for H_2 in the range from 20 K to approximately 100 K.

The Molar Specific Heat of Solids

The molar specific heats of solids also demonstrate a marked temperature dependence. Solids have molar specific heats that generally decrease in a nonlinear manner with decreasing temperature and approach zero as the temperature approaches absolute zero. At high temperatures (usually above 300 K), the molar specific heats approach the value of $3R \approx 25\,\text{J/mol}\cdot\text{K}$, a result known as the *DuLong–Petit law.* The typical data shown in Figure 21.8 demonstrate the temperature dependence of the molar specific heats for two semiconducting solids, silicon and germanium.

We can explain the molar specific heat of a solid at high temperatures using the equipartition theorem. For small displacements of an atom from its equilibrium position, each atom executes simple harmonic motion in the x, y, and z directions. The energy associated with vibrational motion in the x direction is

$$E = \tfrac{1}{2}mv_x^2 + \tfrac{1}{2}kx^2$$

The expressions for vibrational motions in the y and z directions are analogous. Therefore, each atom of the solid has six degrees of freedom. According to the equipartition theorem, this corresponds to an average vibrational energy of $6(\frac{1}{2}k_B T) = 3k_B T$ per atom. Therefore, the total internal energy of a solid consisting of N atoms is

$$E_{\text{int}} = 3Nk_B T = 3nRT \qquad \textbf{(21.21)}$$

From this result, we find that the molar specific heat of a solid at constant volume is

$$C_V = \frac{1}{n}\frac{dE_{\text{int}}}{dT} = 3R \qquad \textbf{(21.22)}$$

This result is in agreement with the empirical DuLong–Petit law. The discrepancies between this model and the experimental data at low temperatures are again due to the inadequacy of classical physics in describing the microscopic world.

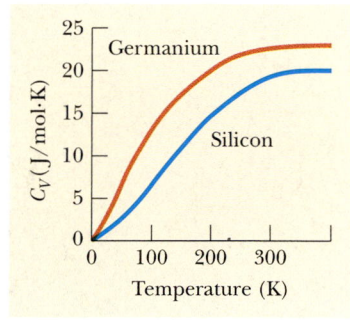

Figure 21.8 Molar specific heat of silicon and germanium. As *T* approaches zero, the molar specific heat also approaches zero. *(From C. Kittel,* Introduction to Solid State Physics, *New York, Wiley, 1971.)*

> Total internal energy of a solid

> Molar specific heat of a solid at constant volume

21.5 THE BOLTZMANN DISTRIBUTION LAW

Thus far we have neglected the fact that not all molecules in a gas have the same speed and energy. In reality, their motion is extremely chaotic. Any individual molecule is colliding with others at an enormous rate—typically, a billion times per second. Each collision results in a change in the speed and direction of motion of each of the participant molecules. From Equation 21.7, we see that average molecular speeds increase with increasing temperature. What we would like to know now is the relative number of molecules that possess some characteristic, such as a certain percentage of the total energy or speed. The ratio of the number of molecules that have the desired characteristic to the total number of molecules is the probability that a particular molecule has that characteristic.

The Exponential Atmosphere

We begin by considering the distribution of molecules in our atmosphere. Let us determine how the number of molecules per unit volume varies with altitude. Our model assumes that the atmosphere is at a constant temperature T. (This assumption is not entirely correct because the temperature of our atmosphere decreases by about 2°C for every 300-m increase in altitude. However, the model does illustrate the basic features of the distribution.)

According to the ideal gas law, a gas containing N molecules in thermal equilibrium obeys the relationship $PV = Nk_BT$. It is convenient to rewrite this equation in terms of the **number density** $n_V = N/V$, which represents the number of molecules per unit volume of gas. This quantity is important because it can vary from one point to another. In fact, our goal is to determine how n_V changes in our atmosphere. We can express the ideal gas law in terms of n_V as $P = n_Vk_BT$. Thus, if the number density n_V is known, we can find the pressure, and vice versa. The pressure in the atmosphere decreases with increasing altitude because a given layer of air must support the weight of all the atmosphere above it—that is, the greater the altitude, the less the weight of the air above that layer, and the lower the pressure.

To determine the variation in pressure with altitude, let us consider an atmospheric layer of thickness dy and cross-sectional area A, as shown in Figure 21.9. Because the air is in static equilibrium, the magnitude PA of the upward force exerted on the bottom of this layer must exceed the magnitude of the downward force on the top of the layer, $(P + dP)A$, by an amount equal to the weight of gas in this thin layer. If the mass of a gas molecule in the layer is m, and if a total of N molecules are in the layer, then the weight of the layer is given by $mgN = mgn_VV = mgn_VA\,dy$. Thus, we see that

$$PA - (P + dP)A = mgn_VA\,dy$$

This expression reduces to

$$dP = -mgn_V\,dy$$

Because $P = n_Vk_BT$ and T is assumed to remain constant, we see that $dP = k_BT\,dn_V$. Substituting this result into the previous expression for dP and rearranging terms, we have

$$\frac{dn_V}{n_V} = -\frac{mg}{k_BT}\,dy$$

Integrating this expression, we find that

$$n_V(y) = n_0e^{-mgy/k_BT} \qquad \textbf{(21.23)}$$

where the constant n_0 is the number density at $y = 0$. This result is known as the **law of atmospheres.**

According to Equation 21.23, the number density decreases exponentially with increasing altitude when the temperature is constant. The number density of our atmosphere at sea level is about $n_0 = 2.69 \times 10^{25}$ molecules/m³. Because the pressure is $P = n_Vk_BT$, we see from Equation 21.23 that the pressure of our atmosphere varies with altitude according to the expression

$$P = P_0e^{-mgy/k_BT} \qquad \textbf{(21.24)}$$

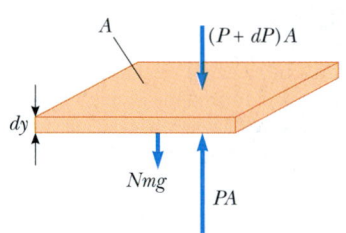

Figure 21.9 An atmospheric layer of gas in equilibrium.

Law of atmospheres

where $P_0 = n_0 k_B T$. A comparison of this model with the actual atmospheric pressure as a function of altitude shows that the exponential form is a reasonable approximation to the Earth's atmosphere.

EXAMPLE 21.4 ▸ **High-Flying Molecules**

What is the number density of air at an altitude of 11.0 km (the cruising altitude of a commercial jetliner) compared with its number density at sea level? Assume that the air temperature at this height is the same as that at the ground, 20°C.

Solution The number density of our atmosphere decreases exponentially with altitude according to the law of atmospheres, Equation 21.23. We assume an average molecular mass of 28.9 u = 4.80×10^{-26} kg. Taking $y = 11.0$ km, we calculate the power of the exponential in Equation 21.23 to be

$$\frac{mgy}{k_B T} = \frac{(4.80 \times 10^{-26} \text{ kg})(9.80 \text{ m/s}^2)(11\ 000 \text{ m})}{(1.38 \times 10^{-23} \text{ J/K})(293 \text{ K})} = 1.28$$

Thus, Equation 21.23 gives

$$n_V = n_0 e^{-mgy/k_B T} = n_0 e^{-1.28} = \boxed{0.278 n_0}$$

That is, the number density of air at an altitude of 11.0 km is only 27.8% of the number density at sea level, if we assume constant temperature. Because the temperature actually decreases with altitude, the number density of air is less than this in reality.

The pressure at this height is reduced in the same manner. For this reason, high-flying aircraft must have pressurized cabins to ensure passenger comfort and safety.

Computing Average Values

The exponential function $e^{-mgy/k_B T}$ that appears in Equation 21.23 can be interpreted as a probability distribution that gives the relative probability of finding a gas molecule at some height y. Thus, the probability distribution $p(y)$ is proportional to the number density distribution $n_V(y)$. This concept enables us to determine many properties of the atmosphere, such as the fraction of molecules below a certain height or the average potential energy of a molecule.

As an example, let us determine the average height $\bar{y}$ of a molecule in the atmosphere at temperature T. The expression for this average height is

$$\bar{y} = \frac{\displaystyle\int_0^\infty y n_V(y)\ dy}{\displaystyle\int_0^\infty n_V(y)\ dy} = \frac{\displaystyle\int_0^\infty y e^{-mgy/k_B T}\ dy}{\displaystyle\int_0^\infty e^{-mgy/k_B T}\ dy}$$

where the height of a molecule can range from 0 to ∞. The numerator in this expression represents the sum of the heights of the molecules times their number, while the denominator is the sum of the number of molecules. That is, the denominator is the total number of molecules. After performing the indicated integrations, we find that

$$\bar{y} = \frac{(k_B T/mg)^2}{k_B T/mg} = \frac{k_B T}{mg}$$

This expression states that the average height of a molecule increases as T increases, as expected.

We can use a similar procedure to determine the average potential energy of a gas molecule. Because the gravitational potential energy of a molecule at height y is $U = mgy$, the average potential energy is equal to $mg\bar{y}$. Because $\bar{y} = k_B T/mg$, we see that $\overline{U} = mg(k_B T/mg) = k_B T$. This important result indicates that the average gravitational potential energy of a molecule depends only on temperature, and not on m or g.

The Boltzmann Distribution

Because the gravitational potential energy of a molecule at height y is $U = mgy$, we can express the law of atmospheres (Eq. 21.23) as

$$n_V = n_0 e^{-U/k_B T}$$

This means that gas molecules in thermal equilibrium are distributed in space with a probability that depends on gravitational potential energy according to the exponential factor $e^{-U/k_B T}$.

This exponential expression describing the distribution of molecules in the atmosphere is powerful and applies to any type of energy. In general, the number density of molecules having energy E is

Boltzmann distribution law

$$n_V(E) = n_0 e^{-E/k_B T} \tag{21.25}$$

This equation is known as the **Boltzmann distribution law** and is important in describing the statistical mechanics of a large number of molecules. It states that **the probability of finding the molecules in a particular energy state varies exponentially as the negative of the energy divided by $k_B T$.** All the molecules would fall into the lowest energy level if the thermal agitation at a temperature T did not excite the molecules to higher energy levels.

EXAMPLE 21.5 Thermal Excitation of Atomic Energy Levels

As we discussed briefly in Section 8.10, atoms can occupy only certain discrete energy levels. Consider a gas at a temperature of 2 500 K whose atoms can occupy only two energy levels separated by 1.50 eV, where 1 eV (electron volt) is an energy unit equal to 1.6×10^{-19} J (Fig. 21.10). Determine the ratio of the number of atoms in the higher energy level to the number in the lower energy level.

Solution Equation 21.25 gives the relative number of atoms in a given energy level. In this case, the atom has two possible energies, E_1 and E_2, where E_1 is the lower energy level. Hence, the ratio of the number of atoms in the higher energy level to the number in the lower energy level is

$$\frac{n_V(E_2)}{n_V(E_1)} = \frac{n_0 e^{-E_2/k_B T}}{n_0 e^{-E_1/k_B T}} = e^{-(E_2 - E_1)/k_B T}$$

In this problem, $E_2 - E_1 = 1.50$ eV, and the denominator of the exponent is

$$k_B T = (1.38 \times 10^{-23}\,\text{J/K})(2\,500\,\text{K})/1.60 \times 10^{-19}\,\text{J/eV}$$
$$= 0.216\,\text{eV}$$

Therefore, the required ratio is

$$\frac{n(E_2)}{n(E_1)} = e^{-1.50\,\text{eV}/0.216\,\text{eV}} = e^{-6.94} = \boxed{9.64 \times 10^{-4}}$$

This result indicates that at $T = 2\,500$ K, only a small fraction of the atoms are in the higher energy level. In fact, for every atom in the higher energy level, there are about 1 000 atoms in the lower level. The number of atoms in the higher level increases at even higher temperatures, but the distribution law specifies that at equilibrium there are always more atoms in the lower level than in the higher level.

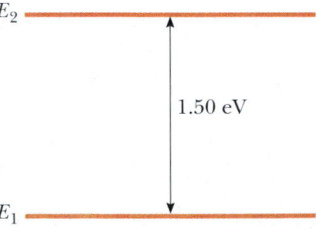

Figure 21.10 Energy level diagram for a gas whose atoms can occupy two energy levels.

21.6 DISTRIBUTION OF MOLECULAR SPEEDS

In 1860 James Clerk Maxwell (1831–1879) derived an expression that describes the distribution of molecular speeds in a very definite manner. His work and subsequent developments by other scientists were highly controversial because direct detection of molecules could not be achieved experimentally at that time. However, about 60 years later, experiments were devised that confirmed Maxwell's predictions.

Let us consider a container of gas whose molecules have some distribution of speeds. Suppose we want to determine how many gas molecules have a speed in the range from, for example, 400 to 410 m/s. Intuitively, we expect that the speed distribution depends on temperature. Furthermore, we expect that the distribution peaks in the vicinity of v_{rms}. That is, few molecules are expected to have speeds much less than or much greater than v_{rms} because these extreme speeds result only from an unlikely chain of collisions.

The observed speed distribution of gas molecules in thermal equilibrium is shown in Figure 21.11. The quantity N_v, called the **Maxwell–Boltzmann distribution function,** is defined as follows: If N is the total number of molecules, then the number of molecules with speeds between v and $v + dv$ is $dN = N_v dv$. This number is also equal to the area of the shaded rectangle in Figure 21.11. Furthermore, the fraction of molecules with speeds between v and $v + dv$ is $N_v dv / N$. This fraction is also equal to the probability that a molecule has a speed in the range v to $v + dv$.

The fundamental expression that describes the distribution of speeds of N gas molecules is

$$N_v = 4\pi N\left(\frac{m}{2\pi k_B T}\right)^{3/2} v^2 e^{-mv^2/2k_B T} \qquad \textbf{(21.26)}$$

where m is the mass of a gas molecule, k_B is Boltzmann's constant, and T is the absolute temperature.[1] Observe the appearance of the Boltzmann factor $e^{-E/k_B T}$ with $E = \frac{1}{2}mv^2$.

As indicated in Figure 21.11, the average speed $\bar{v}$ is somewhat lower than the rms speed. The *most probable speed* v_{mp} is the speed at which the distribution curve reaches a peak. Using Equation 21.26, one finds that

$$v_{rms} = \sqrt{\overline{v^2}} = \sqrt{3k_B T/m} = 1.73\sqrt{k_B T/m} \qquad \textbf{(21.27)}$$

$$\bar{v} = \sqrt{8k_B T/\pi m} = 1.60\sqrt{k_B T/m} \qquad \textbf{(21.28)}$$

$$v_{mp} = \sqrt{2k_B T/m} = 1.41\sqrt{k_B T/m} \qquad \textbf{(21.29)}$$

The details of these calculations are left for the student (see Problems 41 and 62). From these equations, we see that

$$v_{rms} > \bar{v} > v_{mp}$$

Figure 21.12 represents speed distribution curves for N_2. The curves were obtained by using Equation 21.26 to evaluate the distribution function at various speeds and at two temperatures. Note that the peak in the curve shifts to the right

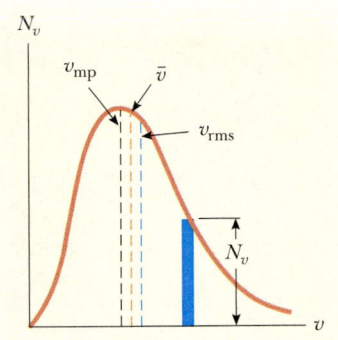

Figure 21.11 The speed distribution of gas molecules at some temperature. The number of molecules having speeds in the range dv is equal to the area of the shaded rectangle, $N_v dv$. The function N_v approaches zero as v approaches infinity.

Maxwell speed distribution function

rms speed

Average speed

Most probable speed

---

[1] For the derivation of this expression, see an advanced textbook on thermodynamics, such as that by R. P. Bauman, *Modern Thermodynamics with Statistical Mechanics*, New York, Macmillan Publishing Co., 1992.

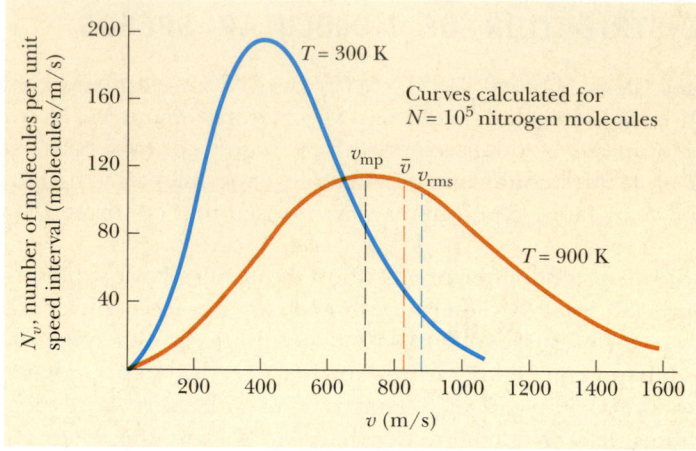

Figure 21.12   The speed distribution function for 10^5 nitrogen molecules at 300 K and 900 K. The total area under either curve is equal to the total number of molecules, which in this case equals 10^5. Note that $v_{rms} > \bar{v} > v_{mp}$.

as T increases, indicating that the average speed increases with increasing temperature, as expected. The asymmetric shape of the curves is due to the fact that the lowest speed possible is zero while the upper classical limit of the speed is infinity.

Quick Quiz 21.3

Consider the two curves in Figure 21.12. What is represented by the area under each of the curves between the 800-m/s and 1 000-m/s marks on the horizontal axis?

QuickLab

Fill one glass with very hot tap water and another with very cold water. Put a single drop of food coloring in each glass. Which drop disperses faster? Why?

Equation 21.26 shows that the distribution of molecular speeds in a gas depends both on mass and on temperature. At a given temperature, the fraction of molecules with speeds exceeding a fixed value increases as the mass decreases. This explains why lighter molecules, such as H_2 and He, escape more readily from the Earth's atmosphere than do heavier molecules, such as N_2 and O_2. (See the discussion of escape speed in Chapter 14. Gas molecules escape even more readily from the Moon's surface than from the Earth's because the escape speed on the Moon is lower than that on the Earth.)

 The speed distribution curves for molecules in a liquid are similar to those shown in Figure 21.12. We can understand the phenomenon of evaporation of a liquid from this distribution in speeds, using the fact that some molecules in the liquid are more energetic than others. Some of the faster-moving molecules in the liquid penetrate the surface and leave the liquid even at temperatures well below the boiling point. The molecules that escape the liquid by evaporation are those that have sufficient energy to overcome the attractive forces of the molecules in the liquid phase. Consequently, the molecules left behind in the liquid phase have a lower average kinetic energy; as a result, the temperature of the liquid decreases. Hence, evaporation is a cooling process. For example, an alcohol-soaked cloth often is placed on a feverish head to cool and comfort a patient.

The evaporation process

EXAMPLE 21.6 A System of Nine Particles

Nine particles have speeds of 5.00, 8.00, 12.0, 12.0, 12.0, 14.0, 14.0, 17.0, and 20.0 m/s. (a) Find the particles' average speed.

Solution The average speed is the sum of the speeds divided by the total number of particles:

$$\bar{v} = \frac{\begin{array}{c}(5.00 + 8.00 + 12.0 + 12.0 + 12.0 \\ + 14.0 + 14.0 + 17.0 + 20.0)\ \text{m/s}\end{array}}{9}$$

$$= \boxed{12.7\ \text{m/s}}$$

(b) What is the rms speed?

Solution The average value of the square of the speed is

$$\overline{v^2} = \frac{\begin{array}{c}(5.00^2 + 8.00^2 + 12.0^2 + 12.0^2 + 12.0^2 \\ + 14.0^2 + 14.0^2 + 17.0^2 + 20.0^2)\ \text{m}\end{array}}{9}$$

$$= 178\ \text{m}^2/\text{s}^2$$

Hence, the rms speed is

$$v_{\text{rms}} = \sqrt{\overline{v^2}} = \sqrt{178\ \text{m}^2/\text{s}^2} = \boxed{13.3\ \text{m/s}}$$

(c) What is the most probable speed of the particles?

Solution Three of the particles have a speed of 12 m/s, two have a speed of 14 m/s, and the remaining have different speeds. Hence, we see that the most probable speed v_{mp} is

$$\boxed{12\ \text{m/s.}}$$

Optional Section

21.7 MEAN FREE PATH

Most of us are familiar with the fact that the strong odor associated with a gas such as ammonia may take a fraction of a minute to diffuse throughout a room. However, because average molecular speeds are typically several hundred meters per second at room temperature, we might expect a diffusion time much less than 1 s. But, as we saw in Quick Quiz 21.1, molecules collide with one other because they are not geometrical points. Therefore, they do not travel from one side of a room to the other in a straight line. Between collisions, the molecules move with constant speed along straight lines. The average distance between collisions is called the **mean free path.** The path of an individual molecule is random and resembles that shown in Figure 21.13. As we would expect from this description, the mean free path is related to the diameter of the molecules and the density of the gas.

We now describe how to estimate the mean free path for a gas molecule. For this calculation, we assume that the molecules are spheres of diameter d. We see from Figure 21.14a that no two molecules collide unless their centers are less than a distance d apart as they approach each other. An equivalent way to describe the

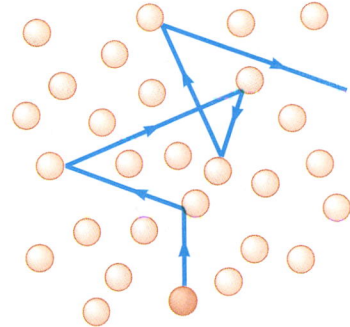

Figure 21.13 A molecule moving through a gas collides with other molecules in a random fashion. This behavior is sometimes referred to as a *random-walk process.* The mean free path increases as the number of molecules per unit volume decreases. Note that the motion is not limited to the plane of the paper.

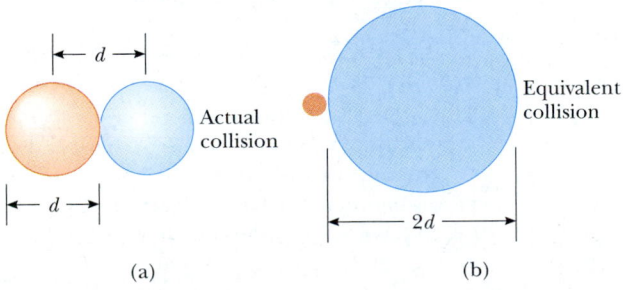

(a)   (b)

Figure 21.14 (a) Two spherical molecules, each of diameter d, collide if their centers are within a distance d of each other. (b) The collision between the two molecules is equivalent to a point molecule's colliding with a molecule having an effective diameter of $2d$.

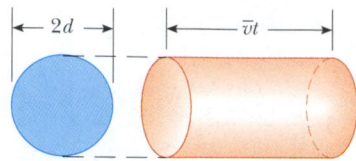

Figure 21.15 In a time t, a molecule of effective diameter $2d$ sweeps out a cylinder of length $\bar{v}t$, where $\bar{v}$ is its average speed. In this time, it collides with every point molecule within this cylinder.

collisions is to imagine that one of the molecules has a diameter $2d$ and that the rest are geometrical points (Fig. 21.14b). Let us choose the large molecule to be one moving with the average speed $\bar{v}$. In a time t, this molecule travels a distance $\bar{v}t$. In this time interval, the molecule sweeps out a cylinder having a cross-sectional area πd^2 and a length $\bar{v}t$ (Fig. 21.15). Hence, the volume of the cylinder is $\pi d^2 \bar{v}t$. If n_V is the number of molecules per unit volume, then the number of point-size molecules in the cylinder is $(\pi d^2 \bar{v}t)n_V$. The molecule of equivalent diameter $2d$ collides with every molecule in this cylinder in the time t. Hence, the number of collisions in the time t is equal to the number of molecules in the cylinder, $(\pi d^2 \bar{v}t)n_V$.

The mean free path ℓ equals the average distance $\bar{v}t$ traveled in a time t divided by the number of collisions that occur in that time:

$$\ell = \frac{\bar{v}t}{(\pi d^2 \bar{v}t)n_V} = \frac{1}{\pi d^2 n_V}$$

Because the number of collisions in a time t is $(\pi d^2 \bar{v}t)n_V$, the number of collisions per unit time, or **collision frequency f,** is

$$f = \pi d^2 \bar{v} n_V$$

The inverse of the collision frequency is the average time between collisions, known as the **mean free time.**

Our analysis has assumed that molecules in the cylinder are stationary. When the motion of these molecules is included in the calculation, the correct results are

Mean free path

$$\ell = \frac{1}{\sqrt{2}\,\pi d^2 n_V} \tag{21.30}$$

Collision frequency

$$f = \sqrt{2}\,\pi d^2 \bar{v} n_V = \frac{\bar{v}}{\ell} \tag{21.31}$$

EXAMPLE 21.7   Bouncing Around in the Air

Approximate the air around you as a collection of nitrogen molecules, each of which has a diameter of 2.00×10^{-10} m. (a) How far does a typical molecule move before it collides with another molecule?

Solution Assuming that the gas is ideal, we can use the equation $PV = Nk_BT$ to obtain the number of molecules per unit volume under typical room conditions:

$$n_V = \frac{N}{V} = \frac{P}{k_B T} = \frac{1.01 \times 10^5 \text{ N/m}^2}{(1.38 \times 10^{-23}\text{ J/K})(293\text{ K})}$$

$$= 2.50 \times 10^{25} \text{ molecules/m}^3$$

Hence, the mean free path is

$$\ell = \frac{1}{\sqrt{2}\,\pi d^2 n_V}$$

$$= \frac{1}{\sqrt{2}\,\pi(2.00 \times 10^{-10}\text{ m})^2(2.50 \times 10^{25}\text{ molecules/m}^3)}$$

$$= \boxed{2.25 \times 10^{-7}\text{ m}}$$

This value is about 10^3 times greater than the molecular diameter.

(b) On average, how frequently does one molecule collide with another?

Solution Because the rms speed of a nitrogen molecule at 20.0°C is 511 m/s (see Table 21.1), we know from Equations 21.27 and 21.28 that $\bar{v} = (1.60/1.73)(511$ m/s$) = 473$ m/s. Therefore, the collision frequency is

$$f = \frac{\bar{v}}{\ell} = \frac{473\text{ m/s}}{2.25 \times 10^{-7}\text{ m}} = \boxed{2.10 \times 10^9\text{/s}}$$

The molecule collides with other molecules at the average rate of about two billion times each second!

The mean free path ℓ is *not* the same as the average separation between particles. In fact, the average separation d between particles is approximately $n_V^{-1/3}$. In this example, the average molecular separation is

$$d = \frac{1}{n_V^{1/3}} = \frac{1}{(2.5 \times 10^{25})^{1/3}} = 3.4 \times 10^{-9}\text{ m}$$

SUMMARY

The pressure of N molecules of an ideal gas contained in a volume V is

$$P = \frac{2}{3}\frac{N}{V}\left(\frac{1}{2}m\overline{v^2}\right) \tag{21.2}$$

The average translational kinetic energy per molecule of a gas, $\frac{1}{2}m\overline{v^2}$, is related to the temperature T of the gas through the expression

$$\frac{1}{2}m\overline{v^2} = \frac{3}{2}k_B T \tag{21.4}$$

where k_B is Boltzmann's constant. Each translational degree of freedom $(x, y,$ or $z)$ has $\frac{1}{2}k_B T$ of energy associated with it.

The **theorem of equipartition of energy** states that the energy of a system in thermal equilibrium is equally divided among all degrees of freedom.

The total energy of N molecules (or n mol) of an ideal monatomic gas is

$$E_{int} = \frac{3}{2}Nk_B T = \frac{3}{2}nRT \tag{21.10}$$

The change in internal energy for n mol of any ideal gas that undergoes a change in temperature ΔT is

$$\Delta E_{int} = nC_V \Delta T \tag{21.12}$$

where C_V is the molar specific heat at constant volume.

The molar specific heat of an ideal monatomic gas at constant volume is $C_V = \frac{3}{2}R$; the molar specific heat at constant pressure is $C_P = \frac{5}{2}R$. The ratio of specific heats is $\gamma = C_P/C_V = \frac{5}{3}$.

If an ideal gas undergoes an adiabatic expansion or compression, the first law of thermodynamics, together with the equation of state, shows that

$$PV^\gamma = \text{constant} \tag{21.18}$$

The **Boltzmann distribution law** describes the distribution of particles among available energy states. The relative number of particles having energy E is

$$n_V(E) = n_0 e^{-E/k_B T} \tag{21.25}$$

The **Maxwell–Boltzmann distribution function** describes the distribution of speeds of molecules in a gas:

$$N_v = 4\pi N\left(\frac{m}{2\pi k_B T}\right)^{3/2} v^2 e^{-mv^2/2k_B T} \tag{21.26}$$

This expression enables us to calculate the **root-mean-square speed,** the **average speed,** and **the most probable speed:**

$$v_{rms} = \sqrt{\overline{v^2}} = \sqrt{3k_B T/m} = 1.73\sqrt{k_B T/m} \tag{21.27}$$

$$\overline{v} = \sqrt{8k_B T/\pi m} = 1.60\sqrt{k_B T/m} \tag{21.28}$$

$$v_{mp} = \sqrt{2k_B T/m} = 1.41\sqrt{k_B T/m} \tag{21.29}$$

QUESTIONS

1. Dalton's law of partial pressures states that the total pressure of a mixture of gases is equal to the sum of the partial pressures of gases making up the mixture. Give a convincing argument for this law on the basis of the kinetic theory of gases.
2. One container is filled with helium gas and another with argon gas. If both containers are at the same temperature, which gas molecules have the higher rms speed? Explain.
3. A gas consists of a mixture of He and N_2 molecules. Do the lighter He molecules travel faster than the N_2 molecules? Explain.
4. Although the average speed of gas molecules in thermal equilibrium at some temperature is greater than zero, the average velocity is zero. Explain why this statement must be true.
5. When alcohol is rubbed on your body, your body temperature decreases. Explain this effect.
6. A liquid partially fills a container. Explain why the temperature of the liquid decreases if the container is then partially evacuated. (Using this technique, one can freeze water at temperatures above 0°C.)
7. A vessel containing a fixed volume of gas is cooled. Does the mean free path of the gas molecules increase, decrease, or remain constant during the cooling process? What about the collision frequency?
8. A gas is compressed at a constant temperature. What happens to the mean free path of the molecules in the process?
9. If a helium-filled balloon initially at room temperature is placed in a freezer, will its volume increase, decrease, or remain the same?
10. What happens to a helium-filled balloon released into the air? Will it expand or contract? Will it stop rising at some height?
11. Which is heavier, dry air or air saturated with water vapor? Explain.
12. Why does a diatomic gas have a greater energy content per mole than a monatomic gas at the same temperature?
13. An ideal gas is contained in a vessel at 300 K. If the temperature is increased to 900 K, (a) by what factor does the rms speed of each molecule change? (b) By what factor does the pressure in the vessel change?
14. A vessel is filled with gas at some equilibrium pressure and temperature. Can all gas molecules in the vessel have the same speed?
15. In our model of the kinetic theory of gases, molecules were viewed as hard spheres colliding elastically with the walls of the container. Is this model realistic?
16. In view of the fact that hot air rises, why does it generally become cooler as you climb a mountain? (Note that air is a poor thermal conductor.)

PROBLEMS

1, 2, 3 = straightforward, intermediate, challenging   ☐ = full solution available in the *Student Solutions Manual and Study Guide*
WEB = solution posted at **http://www.saunderscollege.com/physics/**   🖥 = Computer useful in solving problem   🧊 = Interactive Physics
☐ = paired numerical/symbolic problems

Section 21.1   Molecular Model of an Ideal Gas

1. Use the definition of Avogadro's number to find the mass of a helium atom.
2. A sealed cubical container 20.0 cm on a side contains three times Avogadro's number of molecules at a temperature of 20.0°C. Find the force exerted by the gas on one of the walls of the container.
3. In a 30.0-s interval, 500 hailstones strike a glass window with an area of 0.600 m² at an angle of 45.0° to the window surface. Each hailstone has a mass of 5.00 g and a speed of 8.00 m/s. If the collisions are elastic, what are the average force and pressure on the window?
4. In a time t, N hailstones strike a glass window of area A at an angle θ to the window surface. Each hailstone has a mass m and a speed v. If the collisions are elastic, what are the average force and pressure on the window?
5. In a period of 1.00 s, 5.00×10^{23} nitrogen molecules strike a wall with an area of 8.00 cm². If the molecules move with a speed of 300 m/s and strike the wall head-on in perfectly elastic collisions, what is the pressure exerted on the wall? (The mass of one N_2 molecule is 4.68×10^{-26} kg.)
6. A 5.00-L vessel contains 2 mol of oxygen gas at a pressure of 8.00 atm. Find the average translational kinetic energy of an oxygen molecule under these conditions.
7. A spherical balloon with a volume of 4 000 cm³ contains helium at an (inside) pressure of 1.20×10^5 Pa. How many moles of helium are in the balloon if each helium atom has an average kinetic energy of 3.60×10^{-22} J?
8. The rms speed of a helium atom at a certain temperature is 1 350 m/s. Find by proportion the rms speed of an oxygen molecule at this temperature. (The molar mass of O_2 is 32.0 g/mol, and the molar mass of He is 4.00 g/mol.)
9. (a) How many atoms of helium gas fill a balloon of diameter 30.0 cm at 20.0°C and 1.00 atm? (b) What is the average kinetic energy of the helium atoms? (c) What is the root-mean-square speed of each helium atom?

10. A 5.00-liter vessel contains nitrogen gas at 27.0°C and 3.00 atm. Find (a) the total translational kinetic energy of the gas molecules and (b) the average kinetic energy per molecule.

WEB 11. A cylinder contains a mixture of helium and argon gas in equilibrium at 150°C. (a) What is the average kinetic energy for each type of gas molecule? (b) What is the root-mean-square speed for each type of molecule?

12. (a) Show that 1 Pa = 1 J/m³. (b) Show that the density in space of the translational kinetic energy of an ideal gas is $3P/2$.

Section 21.2 Molar Specific Heat of an Ideal Gas

Note: You may use the data given in Table 21.2.

13. Calculate the change in internal energy of 3.00 mol of helium gas when its temperature is increased by 2.00 K.

14. One mole of air ($C_V = 5R/2$) at 300 K and confined in a cylinder under a heavy piston occupies a volume of 5.00 L. Determine the new volume of the gas if 4.40 kJ of energy is transferred to the air by heat.

WEB 15. One mole of hydrogen gas is heated at constant pressure from 300 K to 420 K. Calculate (a) the energy transferred by heat to the gas, (b) the increase in its internal energy, and (c) the work done by the gas.

16. In a constant-volume process, 209 J of energy is transferred by heat to 1.00 mol of an ideal monatomic gas initially at 300 K. Find (a) the increase in internal energy of the gas, (b) the work it does, and (c) its final temperature.

17. A house has well-insulated walls. It contains a volume of 100 m³ of air at 300 K. (a) Calculate the energy required to increase the temperature of this air by 1.00°C. (b) If this energy could be used to lift an object of mass m through a height of 2.00 m, what is the value of m?

18. A vertical cylinder with a heavy piston contains air at 300 K. The initial pressure is 200 kPa, and the initial volume is 0.350 m³. Take the molar mass of air as 28.9 g/mol and assume that $C_V = 5R/2$. (a) Find the specific heat of air at constant volume in units of J/kg · °C. (b) Calculate the mass of the air in the cylinder. (c) Suppose the piston is held fixed. Find the energy input required to raise the temperature of the air to 700 K. (d) Assume again the conditions of the initial state and that the heavy piston is free to move. Find the energy input required to raise the temperature to 700 K.

19. A 1-L Thermos bottle is full of tea at 90°C. You pour out one cup and immediately screw the stopper back on. Make an order-of-magnitude estimate of the change in temperature of the tea remaining in the flask that results from the admission of air at room temperature. State the quantities you take as data and the values you measure or estimate for them.

20. For a diatomic ideal gas, $C_V = 5R/2$. One mole of this gas has pressure P and volume V. When the gas is heated, its pressure triples and its volume doubles. If this heating process includes two steps, the first at con-

stant pressure and the second at constant volume, determine the amount of energy transferred to the gas by heat.

21. One mole of an ideal monatomic gas is at an initial temperature of 300 K. The gas undergoes an isovolumetric process, acquiring 500 J of energy by heat. It then undergoes an isobaric process, losing this same amount of energy by heat. Determine (a) the new temperature of the gas and (b) the work done on the gas.

22. A container has a mixture of two gases: n_1 moles of gas 1, which has a molar specific heat C_1; and n_2 moles of gas 2, which has a molar specific heat C_2. (a) Find the molar specific heat of the mixture. (b) What is the molar specific heat if the mixture has m gases in the amounts $n_1, n_2, n_3, \ldots, n_m$, and molar specific heats $C_1, C_2, C_3, \ldots, C_m$, respectively?

23. One mole of an ideal diatomic gas with $C_V = 5R/2$ occupies a volume V_i at a pressure P_i. The gas undergoes a process in which the pressure is proportional to the volume. At the end of the process, it is found that the rms speed of the gas molecules has doubled from its initial value. Determine the amount of energy transferred to the gas by heat.

Section 21.3 Adiabatic Processes for an Ideal Gas

24. During the compression stroke of a certain gasoline engine, the pressure increases from 1.00 atm to 20.0 atm. Assuming that the process is adiabatic and that the gas is ideal, with $\gamma = 1.40$, (a) by what factor does the volume change and (b) by what factor does the temperature change? (c) If the compression starts with 0.016 0 mol of gas at 27.0°C, find the values of Q, W, and ΔE_{int} that characterize the process.

25. Two moles of an ideal gas ($\gamma = 1.40$) expands slowly and adiabatically from a pressure of 5.00 atm and a volume of 12.0 L to a final volume of 30.0 L. (a) What is the final pressure of the gas? (b) What are the initial and final temperatures? (c) Find Q, W, and ΔE_{int}.

26. Air ($\gamma = 1.40$) at 27.0°C and at atmospheric pressure is drawn into a bicycle pump that has a cylinder with an inner diameter of 2.50 cm and a length of 50.0 cm. The down stroke adiabatically compresses the air, which reaches a gauge pressure of 800 kPa before entering the tire. Determine (a) the volume of the compressed air and (b) the temperature of the compressed air. (c) The pump is made of steel and has an inner wall that is 2.00 mm thick. Assume that 4.00 cm of the cylinder's length is allowed to come to thermal equilibrium with the air. What will be the increase in wall temperature?

27. Air in a thundercloud expands as it rises. If its initial temperature was 300 K, and if no energy is lost by thermal conduction on expansion, what is its temperature when the initial volume has doubled?

28. How much work is required to compress 5.00 mol of air at 20.0°C and 1.00 atm to one tenth of the original vol-

ume by (a) an isothermal process and (b) an adiabatic process? (c) What is the final pressure in each of these two cases?

29. Four liters of a diatomic ideal gas ($\gamma = 1.40$) confined to a cylinder is subject to a closed cycle. Initially, the gas is at 1.00 atm and at 300 K. First, its pressure is tripled under constant volume. Then, it expands adiabatically to its original pressure. Finally, the gas is compressed isobarically to its original volume. (a) Draw a *PV* diagram of this cycle. (b) Determine the volume of the gas at the end of the adiabatic expansion. (c) Find the temperature of the gas at the start of the adiabatic expansion. (d) Find the temperature at the end of the cycle. (e) What was the net work done for this cycle?

30. A diatomic ideal gas ($\gamma = 1.40$) confined to a cylinder is subjected to a closed cycle. Initially, the gas is at P_i, V_i, and T_i. First, its pressure is tripled under constant volume. Then, it expands adiabatically to its original pressure. Finally, the gas is compressed isobarically to its original volume. (a) Draw a *PV* diagram of this cycle. (b) Determine the volume of the gas at the end of the adiabatic expansion. (c) Find the temperature of the gas at the start of the adiabatic expansion. (d) Find the temperature at the end of the cycle. (e) What was the net work done for this cycle?

31. During the power stroke in a four-stroke automobile engine, the piston is forced down as the mixture of gas and air undergoes an adiabatic expansion. Assume that (1) the engine is running at 2 500 rpm, (2) the gauge pressure right before the expansion is 20.0 atm, (3) the volumes of the mixture right before and after the expansion are 50.0 and 400 cm³, respectively (Fig.

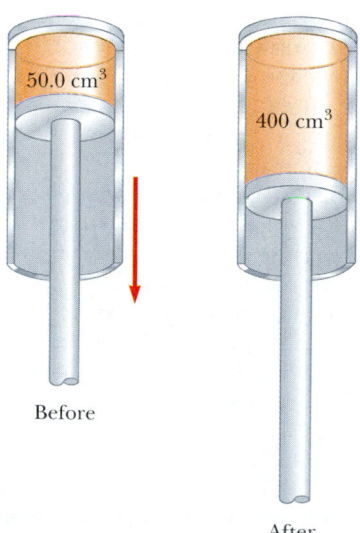

Before

After

Figure P21.31

P21.31), (4) the time involved in the expansion is one-fourth that of the total cycle, and (5) the mixture behaves like an ideal gas, with $\gamma = 1.40$. Find the average power generated during the expansion.

Section 21.4   The Equipartition of Energy

32. A certain molecule has *f* degrees of freedom. Show that a gas consisting of such molecules has the following properties: (1) its total internal energy is $fnRT/2$; (2) its molar specific heat at constant volume is $fR/2$; (3) its molar specific heat at constant pressure is $(f + 2)R/2$; (4) the ratio $\gamma = C_P/C_V = (f + 2)/f$.

WEB 33. Consider 2.00 mol of an ideal diatomic gas. Find the total heat capacity at constant volume and at constant pressure (a) if the molecules rotate but do not vibrate and (b) if the molecules both rotate and vibrate.

34. Inspecting the magnitudes of C_V and C_P for the diatomic and polyatomic gases in Table 21.2, we find that the values increase with increasing molecular mass. Give a qualitative explanation of this observation.

35. In a crude model (Fig. P21.35) of a rotating diatomic molecule of chlorine (Cl_2), the two Cl atoms are 2.00×10^{-10} m apart and rotate about their center of mass with angular speed $\omega = 2.00 \times 10^{12}$ rad/s. What is the rotational kinetic energy of one molecule of Cl_2, which has a molar mass of 70.0 g/mol?

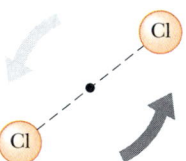

Figure P21.35

Section 21.5   The Boltzmann Distribution Law
Section 21.6   Distribution of Molecular Speeds

36. One cubic meter of atomic hydrogen at 0°C contains approximately 2.70×10^{25} atoms at atmospheric pressure. The first excited state of the hydrogen atom has an energy of 10.2 eV above the lowest energy level, which is called the *ground state*. Use the Boltzmann factor to find the number of atoms in the first excited state at 0°C and at 10 000°C.

37. If convection currents (weather) did not keep the Earth's lower atmosphere stirred up, its chemical composition would change somewhat with altitude because the various molecules have different masses. Use the law of atmospheres to determine how the equilibrium ratio of oxygen to nitrogen molecules changes between sea level and 10.0 km. Assume a uniform temperature of 300 K and take the masses to be 32.0 u for oxygen (O_2) and 28.0 u for nitrogen (N_2).

38. A mixture of two gases diffuses through a filter at rates proportional to the gases' rms speeds. (a) Find the ratio of speeds for the two isotopes of chlorine, ^{85}Cl and ^{37}Cl, as they diffuse through the air. (b) Which isotope moves faster?

39. Fifteen identical particles have various speeds: one has a speed of 2.00 m/s; two have a speed of 3.00 m/s; three have a speed of 5.00 m/s; four have a speed of 7.00 m/s; three have a speed of 9.00 m/s; and two have a speed of 12.0 m/s. Find (a) the average speed, (b) the rms speed, and (c) the most probable speed of these particles.

40. Gaseous helium is in thermal equilibrium with liquid helium at 4.20 K. Even though it is on the point of condensation, model the gas as ideal and determine the most probable speed of a helium atom (mass = 6.64×10^{-27} kg) in it.

41. From the Maxwell–Boltzmann speed distribution, show that the most probable speed of a gas molecule is given by Equation 21.29. Note that the most probable speed corresponds to the point at which the slope of the speed distribution curve, dN_v/dv, is zero.

42. **Review Problem.** At what temperature would the average speed of helium atoms equal (a) the escape speed from Earth, 1.12×10^4 m/s, and (b) the escape speed from the Moon, 2.37×10^3 m/s? (See Chapter 14 for a discussion of escape speed, and note that the mass of a helium atom is 6.64×10^{-27} kg.)

43. A gas is at 0°C. If we wish to double the rms speed of the gas's molecules, by how much must we raise its temperature?

44. The latent heat of vaporization for water at room temperature is 2 430 J/g. (a) How much kinetic energy does each water molecule that evaporates possess before it evaporates? (b) Find the pre-evaporation rms speed of a water molecule that is evaporating. (c) What is the effective temperature of these molecules (modeled as if they were already in a thin gas)? Why do these molecules not burn you?

(Optional)
Section 21.7 Mean Free Path

45. In an ultrahigh vacuum system, the pressure is measured to be 1.00×10^{-10} torr (where 1 torr = 133 Pa). Assume that the gas molecules have a molecular diameter of 3.00×10^{-10} m and that the temperature is 300 K. Find (a) the number of molecules in a volume of 1.00 m³, (b) the mean free path of the molecules, and (c) the collision frequency, assuming an average speed of 500 m/s.

46. In deep space it is reported that there is only one particle per cubic meter. Using the average temperature of 3.00 K and assuming that the particle is H$_2$ (with a diameter of 0.200 nm), (a) determine the mean free path of the particle and the average time between collisions.

(b) Repeat part (a), assuming that there is only one particle per cubic centimeter.

47. Show that the mean free path for the molecules of an ideal gas at temperature T and pressure P is

$$\ell = \frac{k_B T}{\sqrt{2}\,\pi d^2 P}$$

where d is the molecular diameter.

48. In a tank full of oxygen, how many molecular diameters d (on average) does an oxygen molecule travel (at 1.00 atm and 20.0°C) before colliding with another O$_2$ molecule? (The diameter of the O$_2$ molecule is approximately 3.60×10^{-10} m.)

49. Argon gas at atmospheric pressure and 20.0°C is confined in a 1.00-m³ vessel. The effective hard-sphere diameter of the argon atom is 3.10×10^{-10} m. (a) Determine the mean free path ℓ. (b) Find the pressure when the mean free path is $\ell = 1.00$ m. (c) Find the pressure when $\ell = 3.10 \times 10^{-10}$ m.

ADDITIONAL PROBLEMS

50. The dimensions of a room are 4.20 m × 3.00 m × 2.50 m. (a) Find the number of molecules of air in it at atmospheric pressure and 20.0°C. (b) Find the mass of this air, assuming that the air consists of diatomic molecules with a molar mass of 28.9 g/mol. (c) Find the average kinetic energy of a molecule. (d) Find the root-mean-square molecular speed. (e) On the assumption that the specific heat is a constant independent of temperature, we have $E_{int} = 5nRT/2$. Find the internal energy in the air. (f) Find the internal energy of the air in the room at 25.0°C.

51. The function $E_{int} = 3.50nRT$ describes the internal energy of a certain ideal gas. A sample comprising 2.00 mol of the gas always starts at pressure 100 kPa and temperature 300 K. For each one of the following processes, determine the final pressure, volume, and temperature; the change in internal energy of the gas; the energy added to the gas by heat; and the work done by the gas: (a) The gas is heated at constant pressure to 400 K. (b) The gas is heated at constant volume to 400 K. (c) The gas is compressed at constant temperature to 120 kPa. (d) The gas is compressed adiabatically to 120 kPa.

52. Twenty particles, each of mass m and confined to a volume V, have various speeds: two have speed v; three have speed $2v$; five have speed $3v$; four have speed $4v$; three have speed $5v$; two have speed $6v$; one has speed $7v$. Find (a) the average speed, (b) the rms speed, (c) the most probable speed, (d) the pressure that the particles exert on the walls of the vessel, and (e) the average kinetic energy per particle.

WEB 53. A cylinder contains n mol of an ideal gas that undergoes an adiabatic process. (a) Starting with the expression

$W = \int P\,dV$ and using the expression $PV^\gamma = $ constant, show that the work done is

$$W = \left(\frac{1}{\gamma - 1}\right)(P_iV_i - P_fV_f)$$

(b) Starting with the first law equation in differential form, prove that the work done also is equal to $nC_V(T_i - T_f)$. Show that this result is consistent with the equation given in part (a).

54. A vessel contains 1.00×10^4 oxygen molecules at 500 K. (a) Make an accurate graph of the Maxwell speed distribution function versus speed with points at speed intervals of 100 m/s. (b) Determine the most probable speed from this graph. (c) Calculate the average and rms speeds for the molecules and label these points on your graph. (d) From the graph, estimate the fraction of molecules having speeds in the range of 300 m/s to 600 m/s.

55. **Review Problem.**   Oxygen at pressures much greater than 1 atm is toxic to lung cells. By weight, what ratio of helium gas (He) to oxygen gas (O_2) must be used by a scuba diver who is to descend to an ocean depth of 50.0 m?

56. A cylinder with a piston contains 1.20 kg of air at 25.0°C and 200 kPa. Energy is transferred into the system by heat as it is allowed to expand, with the pressure rising to 400 kPa. Throughout the expansion, the relationship between pressure and volume is given by

$$P = CV^{1/2}$$

where C is a constant. (a) Find the initial volume. (b) Find the final volume. (c) Find the final temperature. (d) Find the work that the air does. (e) Find the energy transferred by heat. Take $M = 28.9$ g/mol.

WEB 57. The compressibility κ of a substance is defined as the fractional change in volume of that substance for a given change in pressure:

$$\kappa = -\frac{1}{V}\frac{dV}{dP}$$

(a) Explain why the negative sign in this expression ensures that κ is always positive. (b) Show that if an ideal gas is compressed isothermally, its compressibility is given by $\kappa_1 = 1/P$. (c) Show that if an ideal gas is compressed adiabatically, its compressibility is given by $\kappa_2 = 1/\gamma P$. (d) Determine values for κ_1 and κ_2 for a monatomic ideal gas at a pressure of 2.00 atm.

58. **Review Problem.**   (a) Show that the speed of sound in an ideal gas is

$$v = \sqrt{\frac{\gamma RT}{M}}$$

where M is the molar mass. Use the general expression for the speed of sound in a fluid from Section 17.1; the definition of the bulk modulus from Section 12.4; and the result of Problem 57 in this chapter. As a sound

wave passes through a gas, the compressions are either so rapid or so far apart that energy flow by heat is prevented by lack of time or by effective thickness of insulation. The compressions and rarefactions are adiabatic. (b) Compute the theoretical speed of sound in air at 20°C and compare it with the value given in Table 17.1. Take $M = 28.9$ g/mol. (c) Show that the speed of sound in an ideal gas is

$$v = \sqrt{\frac{\gamma k_B T}{m}}$$

where m is the mass of one molecule. Compare your result with the most probable, the average, and the rms molecular speeds.

59. For a Maxwellian gas, use a computer or programmable calculator to find the numerical value of the ratio $N_v(v)/N_v(v_{mp})$ for the following values of v: $v = (v_{mp}/50), (v_{mp}/10), (v_{mp}/2), v_{mp}, 2v_{mp}, 10v_{mp}, 50v_{mp}$. Give your results to three significant figures.

60. A pitcher throws a 0.142-kg baseball at 47.2 m/s (Fig. P21.60). As it travels 19.4 m, the ball slows to 42.5 m/s because of air resistance. Find the change in temperature of the air through which it passes. To find the greatest possible temperature change, you may make the following assumptions: Air has a molar heat capacity of $C_P = 7R/2$ and an equivalent molar mass of 28.9 g/mol. The process is so rapid that the cover of the baseball acts as thermal insulation, and the temperature of the ball itself does not change. A change in temperature happens initially only for the air in a cylinder 19.4 m in length and 3.70 cm in radius. This air is initially at 20.0°C.

Figure P21.60   Nolan Ryan hurls the baseball for his 5 000th strikeout. *(Joe Patronite/ALLSPORT)*

61. Consider the particles in a *gas centrifuge*, a device that separates particles of different mass by whirling them in a circular path of radius r at angular speed ω. Newton's second law applied to circular motion states that a force of magnitude equal to $m\omega^2 r$ acts on a particle. (a) Discuss how a gas centrifuge can be used to separate particles of different mass. (b) Show that the density of the particles as a function of r is

$$n(r) = n_0 e^{mr^2\omega^2/2k_B T}$$

62. Verify Equations 21.27 and 21.28 for the rms and average speeds of the molecules of a gas at a temperature T. Note that the average value of v^n is

$$\overline{v^n} = \frac{1}{N}\int_0^\infty v^n N_v\,dv$$

and make use of the definite integrals

$$\int_0^\infty x^3 e^{-ax^2}\,dx = \frac{1}{2a^2} \qquad \int_0^\infty x^4 e^{-ax^2}\,dx = \frac{3}{8a^2}\sqrt{\frac{\pi}{a}}$$

63. A sample of a monatomic ideal gas occupies 5.00 L at atmospheric pressure and 300 K (point A in Figure P21.63). It is heated at constant volume to 3.00 atm (point B). Then, it is allowed to expand isothermally to 1.00 atm (point C) and at last is compressed isobarically to its original state. (a) Find the number of moles in the sample. (b) Find the temperatures at points B and C and the volume at point C. (c) Assuming that the specific heat does not depend on temperature, so that $E_{int} = 3nRT/2$, find the internal energy at points A, B,

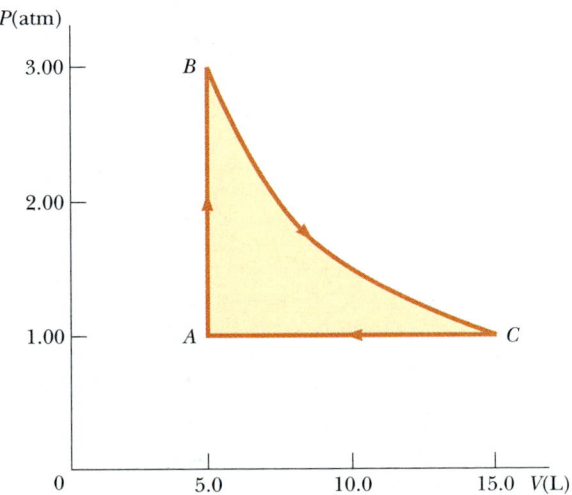

and C. (d) Tabulate P, V, T, and E_{int} at the states at points A, B, and C. (e) Now consider the processes $A \rightarrow B$, $B \rightarrow C$, and $C \rightarrow A$. Describe just how to carry out each process experimentally. (f) Find Q, W, and ΔE_{int} for each of the processes. (g) For the whole cycle $A \rightarrow B \rightarrow C \rightarrow A$, find Q, W, and ΔE_{int}.

64. *If you can't walk to outer space, can you walk at least half way?* (a) Show that the fraction of particles below an altitude h in the atmosphere is

$$f = 1 - e^{(-mgh/k_B T)}$$

(b) Use this result to show that half the particles are below the altitude $h' = k_B T \ln(2)/mg$. What is the value of h' for the Earth? (Assume a temperature of 270 K, and note that the average molar mass for air is 28.9 g/mol.)

65. This problem will help you to think about the size of molecules. In the city of Beijing, a restaurant keeps a pot of chicken broth simmering continuously. Every morning it is topped off to contain 10.0 L of water, along with a fresh chicken, vegetables, and spices. The soup is thoroughly stirred. The molar mass of water is 18.0 g/mol. (a) Find the number of molecules of water in the pot. (b) During a certain month, 90.0% of the broth was served each day to people who then emigrated immediately. Of the water molecules present in the pot on the first day of the month, when was the last one likely to have been ladled out of the pot? (c) The broth has been simmering for centuries, through wars, earthquakes, and stove repairs. Suppose that the water that was in the pot long ago has thoroughly mixed into the Earth's hydrosphere, of mass 1.32×10^{21} kg. How many of the water molecules originally in the pot are likely to be present in it again today?

66. **Review Problem.** (a) If it has enough kinetic energy, a molecule at the surface of the Earth can escape the Earth's gravitation. Using the principle of conservation of energy, show that the minimum kinetic energy needed for escape is mgR, where m is the mass of the molecule, g is the free-fall acceleration at the surface of the Earth, and R is the radius of the Earth. (b) Calculate the temperature for which the minimum escape kinetic energy is ten times the average kinetic energy of an oxygen molecule.

67. Using multiple laser beams, physicists have been able to cool and trap sodium atoms in a small region. In one experiment, the temperature of the atoms was reduced to 0.240 mK. (a) Determine the rms speed of the sodium atoms at this temperature. The atoms can be trapped for about 1.00 s. The trap has a linear dimension of roughly 1.00 cm. (b) Approximately how long would it take an atom to wander out of the trap region if there were no trapping action?

P(atm)

3.00 — B

2.00 —

1.00 — A C

0   5.0   10.0   15.0   V(L)

Figure P21.63

ANSWERS TO QUICK QUIZZES

21.1 Although a molecule moves very rapidly, it does not travel far before it collides with another molecule. The collision deflects the molecule from its original path. Eventually, a perfume molecule will make its way from one end of the room to the other, but the path it takes is much longer than the straight-line distance from the perfume bottle to your nose.

21.2 (c) E_{int} stays the same. According to Equation 21.10, E_{int} is a function of temperature only. Along an isotherm, T is constant by definition. Therefore, the internal energy of the gas does not change.

21.3 The area under each curve represents the number of molecules in that particular velocity range. The $T = 900$ K curve has many more molecules moving between 800 m/s and 1000 m/s than does the $T = 300$ K curve.

B.C.   **By John Hart**

By permission of John Hart and Field Enterprises, Inc.

chapter

22

Heat Engines, Entropy, and the Second Law of Thermodynamics

The first law of thermodynamics, which we studied in Chapter 20, is a state-ment of conservation of energy, generalized to include internal energy. This law states that a change in internal energy in a system can occur as a result of energy transfer by heat or by work, or by both. As was stated in Chapter 20, the law makes no distinction between the results of heat and the results of work—either heat or work can cause a change in internal energy. However, an important distinc-tion between the two is not evident from the first law. One manifestation of this distinction is that it is impossible to convert internal energy completely to mechan-ical energy by taking a substance through a thermodynamic cycle such as in a *heat engine,* a device we study in this chapter.

Although the first law of thermodynamics is very important, it makes no dis-tinction between processes that occur spontaneously and those that do not. How-ever, we find that only certain types of energy-conversion and energy-transfer processes actually take place. The *second law of thermodynamics,* which we study in this chapter, establishes which processes do and which do not occur in nature. The following are examples of processes that proceed in only one direction, governed by the second law:

- When two objects at different temperatures are placed in thermal contact with each other, energy always flows by heat from the warmer to the cooler, never from the cooler to the warmer.
- A rubber ball dropped to the ground bounces several times and eventually comes to rest, but a ball lying on the ground never begins bouncing on its own.
- An oscillating pendulum eventually comes to rest because of collisions with air molecules and friction at the point of suspension. The mechanical energy of the system is converted to internal energy in the air, the pendulum, and the suspen-sion; the reverse conversion of energy never occurs.

All these processes are *irreversible*—that is, they are processes that occur natu-rally in one direction only. No irreversible process has ever been observed to run backward—if it were to do so, it would violate the second law of thermodynamics.[1]

From an engineering standpoint, perhaps the most important implication of the second law is the limited efficiency of heat engines. The second law states that a machine capable of continuously converting internal energy completely to other forms of energy in a cyclic process cannot be constructed.

22.1 ▶ HEAT ENGINES AND THE SECOND LAW OF THERMODYNAMICS

A **heat engine** is a device that converts internal energy to mechanical energy. For instance, in a typical process by which a power plant produces electricity, coal or some other fuel is burned, and the high-temperature gases produced are used to convert liquid water to steam. This steam is directed at the blades of a turbine, set-ting it into rotation. The mechanical energy associated with this rotation is used to drive an electric generator. Another heat engine—the internal combustion en-gine in an automobile—uses energy from a burning fuel to perform work that re-sults in the motion of the automobile.

---

[1] Although we have never *observed* a process occurring in the time-reversed sense, it is *possible* for it to occur. As we shall see later in the chapter, however, such a process is highly improbable. From this view-point, we say that processes occur with a vastly greater probability in one direction than in the opposite direction.

Figure 22.1 This steam-driven locomotive runs from Durango to Silverton, Colorado. It obtains its energy by burning wood or coal. The generated energy vaporizes water into steam, which powers the locomotive. (This locomotive must take on water from tanks located along the route to replace steam lost through the funnel.) Modern locomotives use diesel fuel instead of wood or coal. Whether old-fashioned or modern, such locomotives are heat engines, which extract energy from a burning fuel and convert a fraction of it to mechanical energy. *(© Phil Degginger/Tony Stone Images)*

Lord Kelvin **British physicist and mathematician (1824–1907)** Born William Thomson in Belfast, Kelvin was the first to propose the use of an absolute scale of temperature. The Kelvin temperature scale is named in his honor. Kelvin's work in thermodynamics led to the idea that energy cannot pass spontaneously from a colder body to a hotter body. *(J. L. Charmet/SPL/Photo Researchers, Inc.)*

A heat engine carries some working substance through a cyclic process during which (1) the working substance absorbs energy from a high-temperature energy reservoir, (2) work is done by the engine, and (3) energy is expelled by the engine to a lower-temperature reservoir. As an example, consider the operation of a steam engine (Fig. 22.1), in which the working substance is water. The water in a boiler absorbs energy from burning fuel and evaporates to steam, which then does work by expanding against a piston. After the steam cools and condenses, the liquid water produced returns to the boiler and the cycle repeats.

It is useful to represent a heat engine schematically as in Figure 22.2. The engine absorbs a quantity of energy Q_h from the hot reservoir, does work W, and then gives up a quantity of energy Q_c to the cold reservoir. Because the working substance goes through a cycle, its initial and final internal energies are equal, and so $\Delta E_{int} = 0$. Hence, from the first law of thermodynamics, $\Delta E_{int} = Q - W$, and with no change in internal energy, **the net work W done by a heat engine is equal to the net energy Q_{net} flowing through it.** As we can see from Figure 22.2, $Q_{net} = Q_h - Q_c$; therefore,

$$W = Q_h - Q_c \qquad \textbf{(22.1)}$$

In this expression and in many others throughout this chapter, to be consistent with traditional treatments of heat engines, we take both Q_h and Q_c to be positive quantities, even though Q_c represents energy leaving the engine. In discussions of heat engines, we shall describe energy leaving a system with an explicit minus sign,

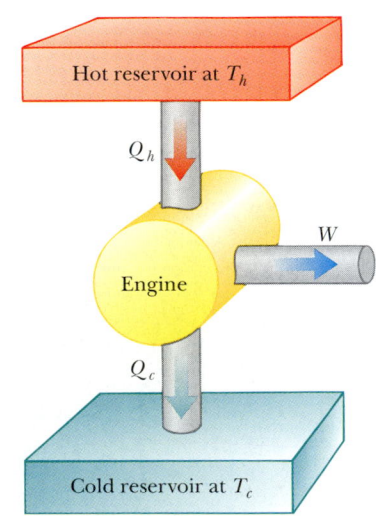

Figure 22.2 Schematic representation of a heat engine. The engine absorbs energy Q_h from the hot reservoir, expels energy Q_c to the cold reservoir, and does work W.

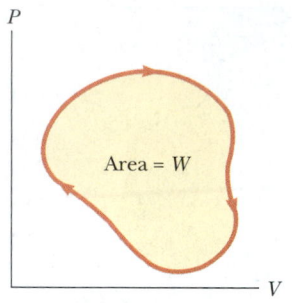

Figure 22.3 *PV* diagram for an arbitrary cyclic process. The value of the net work done equals the area enclosed by the curve.

as in Equation 22.1. Also note that we model the energy input and output for the heat engine as heat, as it often is; however, the energy transfer could occur by another mechanism.

The net work done in a cyclic process is the area enclosed by the curve representing the process on a *PV* diagram. This is shown for an arbitrary cyclic process in Figure 22.3.

The **thermal efficiency** *e* of a heat engine is defined as the ratio of the net work done by the engine during one cycle to the energy absorbed at the higher temperature during the cycle:

$$e = \frac{W}{Q_h} = \frac{Q_h - Q_c}{Q_h} = 1 - \frac{Q_c}{Q_h} \tag{22.2}$$

We can think of the efficiency as the ratio of what you get (mechanical work) to what you give (energy transfer at the higher temperature). In practice, we find that all heat engines expel only a fraction of the absorbed energy as mechanical work and that consequently the efficiency is less than 100%. For example, a good automobile engine has an efficiency of about 20%, and diesel engines have efficiencies ranging from 35% to 40%.

Equation 22.2 shows that a heat engine has 100% efficiency (*e* = 1) only if $Q_c = 0$—that is, if no energy is expelled to the cold reservoir. In other words, a heat engine with perfect efficiency would have to expel all of the absorbed energy as mechanical work. On the basis of the fact that efficiencies of real engines are well below 100%, the **Kelvin–Planck** form of the **second law of thermodynamics** states the following:

It is impossible to construct a heat engine that, operating in a cycle, produces no effect other than the absorption of energy from a reservoir and the performance of an equal amount of work.

This statement of the second law means that, during the operation of a heat engine, *W* can never be equal to Q_h, or, alternatively, that some energy Q_c must be

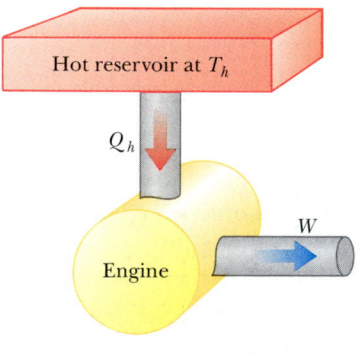

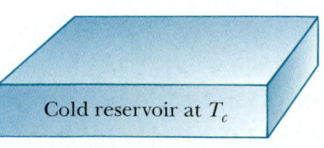

The impossible engine

Figure 22.4 Schematic diagram of a heat engine that absorbs energy Q_h from a hot reservoir and does an equivalent amount of work. It is impossible to construct such a perfect engine.

rejected to the environment. Figure 22.4 is a schematic diagram of the impossible "perfect" heat engine.

The first and second laws of thermodynamics can be summarized as follows: The first law specifies that **we cannot get more energy out of a cyclic process by work than the amount of energy we put in,** and the second law states that **we cannot break even because we must put more energy in, at the higher temperature, than the net amount of energy we get out by work.**

EXAMPLE 22.1 ▶ The Efficiency of an Engine

Find the efficiency of a heat engine that absorbs 2 000 J of energy from a hot reservoir and exhausts 1 500 J to a cold reservoir.

Solution To calculate the efficiency of the engine, we use

Equation 22.2:

$$e = 1 - \frac{Q_c}{Q_h} = 1 - \frac{1\,500\,\text{J}}{2\,000\,\text{J}} = 0.25, \text{ or } \boxed{25\%}$$

Refrigerators and Heat Pumps

Refrigerators and **heat pumps** are heat engines running in reverse. Here, we introduce them briefly for the purposes of developing an alternate statement of the second law; we shall discuss them more fully in Section 22.5.

In a refrigerator or heat pump, the engine absorbs energy Q_c from a cold reservoir and expels energy Q_h to a hot reservoir (Fig. 22.5). This can be accomplished only if work is done *on* the engine. From the first law, we know that the energy given up to the hot reservoir must equal the sum of the work done and the energy absorbed from the cold reservoir. Therefore, the refrigerator or heat pump transfers energy from a colder body (for example, the contents of a kitchen refrigerator or the winter air outside a building) to a hotter body (the air in the kitchen or a room in the building). In practice, it is desirable to carry out this process with a minimum of work. If it could be accomplished without doing any work, then the refrigerator or heat pump would be "perfect" (Fig. 22.6). Again, the existence of

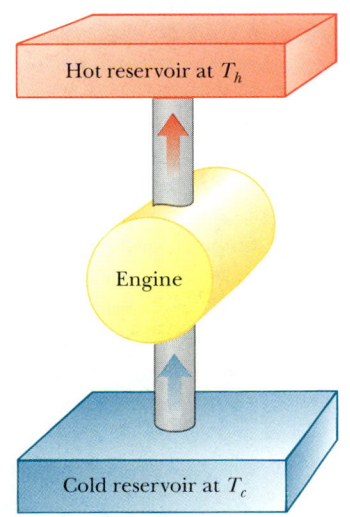

Impossible refrigerator

Figure 22.6 Schematic diagram of an impossible refrigerator or heat pump—that is, one that absorbs energy Q_c from a cold reservoir and expels an equivalent amount of energy to a hot reservoir with $W = 0$.

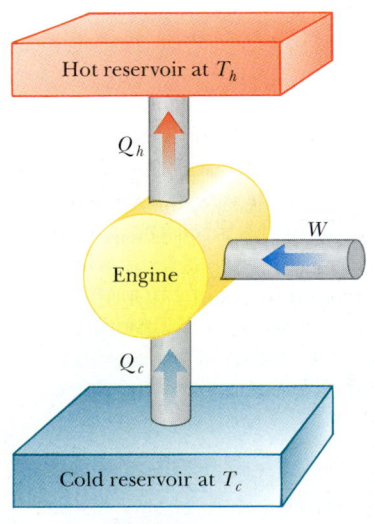

Refrigerator

Figure 22.5 Schematic diagram of a refrigerator, which absorbs energy Q_c from a cold reservoir and expels energy Q_h to a hot reservoir. Work W is done *on* the refrigerator. A heat pump, which can be used to heat or cool a building, works the same way.

such a device would be in violation of the second law of thermodynamics, which in the form of the **Clausius statement**[2] states:

> It is impossible to construct a cyclical machine whose sole effect is the continuous transfer of energy from one object to another object at a higher temperature without the input of energy by work.

In simpler terms, **energy does not flow spontaneously from a cold object to a hot object.** For example, we cool homes in summer using heat pumps called *air conditioners.* The air conditioner pumps energy from the cool room in the home to the warm air outside. This direction of energy transfer requires an input of energy to the air conditioner, which is supplied by the electric power company.

The Clausius and Kelvin–Planck statements of the second law of thermodynamics appear, at first sight, to be unrelated, but in fact they are equivalent in all respects. Although we do not prove so here, if either statement is false, then so is the other.[3]

22.2 REVERSIBLE AND IRREVERSIBLE PROCESSES

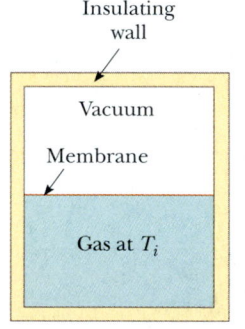

Insulating wall

Vacuum

Membrane

Gas at T_i

Figure 22.7 Adiabatic expansion of a gas.

In the next section we discuss a theoretical heat engine that is the most efficient possible. To understand its nature, we must first examine the meaning of reversible and irreversible processes. In a **reversible** process, the system undergoing the process can be returned to its initial conditions along the same path shown on a *PV* diagram, and every point along this path is an equilibrium state. A process that does not satisfy these requirements is **irreversible.**

All natural processes are known to be irreversible. From the endless number of examples that could be selected, let us examine the adiabatic free expansion of a gas, which was already discussed in Section 20.6, and show that it cannot be reversible. The system that we consider is a gas in a thermally insulated container, as shown in Figure 22.7. A membrane separates the gas from a vacuum. When the membrane is punctured, the gas expands freely into the vacuum. As a result of the puncture, the system has changed because it occupies a greater volume after the expansion. Because the gas does not exert a force through a distance on the surroundings, it does no work on the surroundings as it expands. In addition, no energy is transferred to or from the gas by heat because the container is insulated from its surroundings. Thus, in this adiabatic process, the system has changed but the surroundings have not.

For this process to be reversible, we need to be able to return the gas to its original volume and temperature without changing the surroundings. Imagine that we try to reverse the process by compressing the gas to its original volume. To do so, we fit the container with a piston and use an engine to force the piston inward. During this process, the surroundings change because work is being done by an outside agent on the system. In addition, the system changes because the compression increases the temperature of the gas. We can lower the temperature of the gas by allowing it to come into contact with an external energy reservoir. Although this step returns the gas to its original conditions, the surroundings are

[2] First expressed by Rudolf Clausius (1822–1888).

[3] See, for example, R. P. Bauman, *Modern Thermodynamics and Statistical Mechanics,* New York, Macmillan Publishing Co., 1992.

again affected because energy is being added to the surroundings from the gas. If this energy could somehow be used to drive the engine that we have used to compress the gas, then the net energy transfer to the surroundings would be zero. In this way, the system and its surroundings could be returned to their initial conditions, and we could identify the process as reversible. However, the Kelvin–Planck statement of the second law specifies that the energy removed from the gas to return the temperature to its original value cannot be completely converted to mechanical energy in the form of the work done by the engine in compressing the gas. Thus, we must conclude that the process is irreversible.

We could also argue that the adiabatic free expansion is irreversible by relying on the portion of the definition of a reversible process that refers to equilibrium states. For example, during the expansion, significant variations in pressure occur throughout the gas. Thus, there is no well-defined value of the pressure for the entire system at any time between the initial and final states. In fact, the process cannot even be represented as a path on a *PV* diagram. The *PV* diagram for an adiabatic free expansion would show the initial and final conditions as points, but these points would not be connected by a path. Thus, because the intermediate conditions between the initial and final states are not equilibrium states, the process is irreversible.

Although all real processes are always irreversible, some are almost reversible. If a real process occurs very slowly such that the system is always very nearly in an equilibrium state, then the process can be approximated as reversible. For example, let us imagine that we compress a gas very slowly by dropping some grains of sand onto a frictionless piston, as shown in Figure 22.8. We make the process isothermal by placing the gas in thermal contact with an energy reservoir, and we transfer just enough energy from the gas to the reservoir during the process to keep the temperature constant. The pressure, volume, and temperature of the gas are all well defined during the isothermal compression, so each state during the process is an equilibrium state. Each time we add a grain of sand to the piston, the volume of the gas decreases slightly while the pressure increases slightly. Each grain we add represents a change to a new equilibrium state. We can reverse the process by slowly removing grains from the piston.

A general characteristic of a reversible process is that no dissipative effects (such as turbulence or friction) that convert mechanical energy to internal energy can be present. Such effects can be impossible to eliminate completely. Hence, it is not surprising that real processes in nature are irreversible.

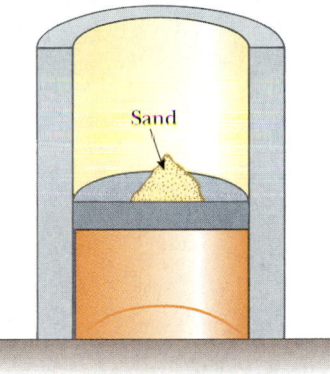

Figure 22.8 A gas in thermal contact with an energy reservoir is compressed slowly as individual grains of sand drop onto the piston. The compression is isothermal and reversible.

22.3 THE CARNOT ENGINE

In 1824 a French engineer named Sadi Carnot described a theoretical engine, now called a **Carnot engine,** that is of great importance from both practical and theoretical viewpoints. He showed that a heat engine operating in an ideal, reversible cycle—called a **Carnot cycle**—between two energy reservoirs is the most efficient engine possible. Such an ideal engine establishes an upper limit on the efficiencies of all other engines. That is, the net work done by a working substance taken through the Carnot cycle is the greatest amount of work possible for a given amount of energy supplied to the substance at the upper temperature. **Carnot's theorem** can be stated as follows:

No real heat engine operating between two energy reservoirs can be more efficient than a Carnot engine operating between the same two reservoirs.

Sadi Carnot French physicist **(1796–1832)** Carnot was the first to show the quantitative relationship between work and heat. In 1824 he published his only work—*Reflections on the Motive Power of Heat*—which reviewed the industrial, political, and economic importance of the steam engine. In it, he defined work as "weight lifted through a height." *(FPG)*

To argue the validity of this theorem, let us imagine two heat engines operating between the *same* energy reservoirs. One is a Carnot engine with efficiency e_C, and the other is an engine with efficiency e, which is greater than e_C. We use the more efficient engine to drive the Carnot engine as a Carnot refrigerator. Thus, the output by work of the more efficient engine is matched to the input by work of the

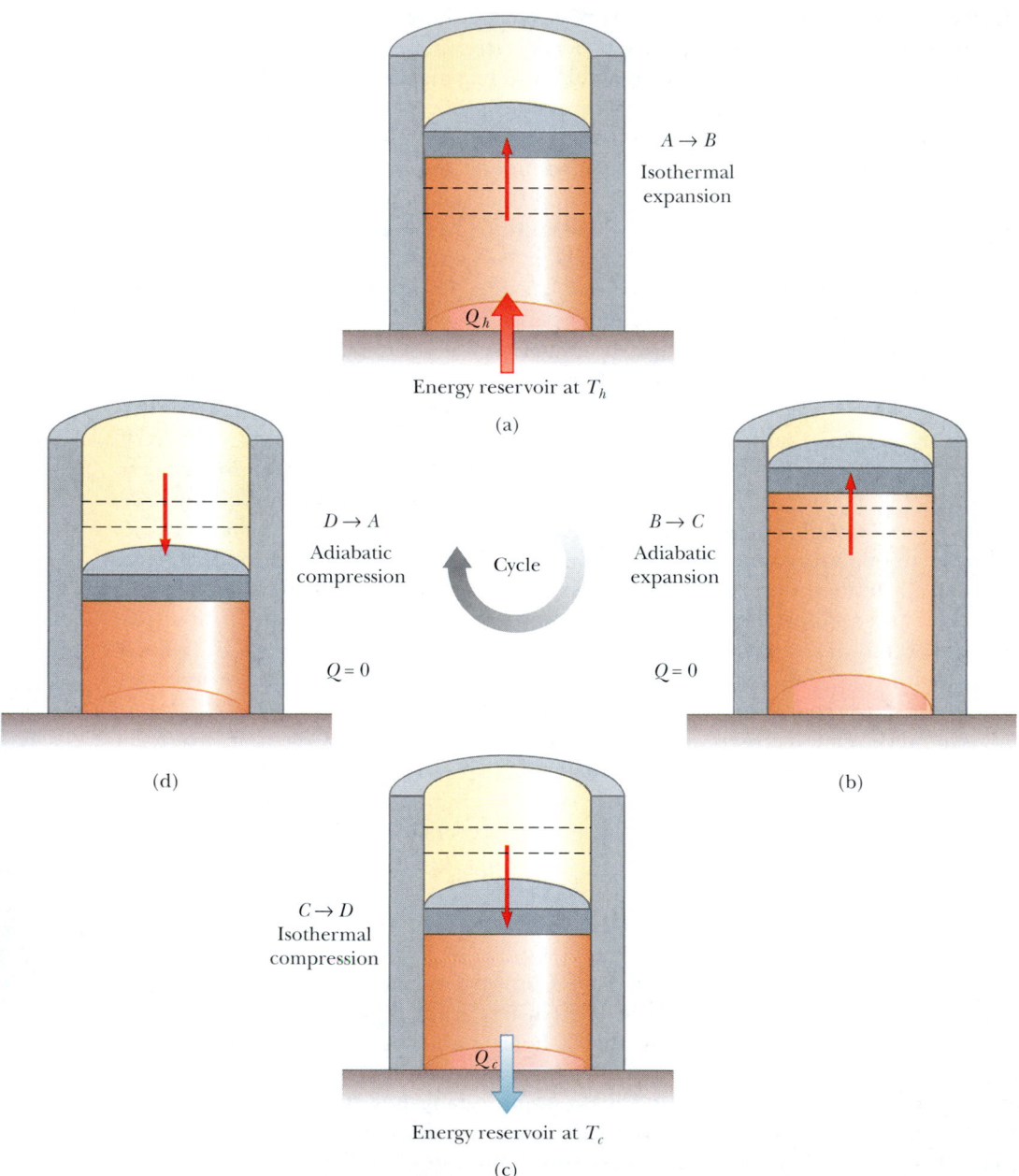

$A \rightarrow B$

Isothermal expansion

Q_h

Energy reservoir at T_h

(a)

$D \rightarrow A$

Adiabatic compression

Cycle

$Q = 0$

(d)

$B \rightarrow C$

Adiabatic expansion

$Q = 0$

(b)

$C \rightarrow D$

Isothermal compression

Q_c

Energy reservoir at T_c

(c)

Figure 22.9   The Carnot cycle. In process $A \rightarrow B$, the gas expands isothermally while in contact with a reservoir at T_h. In process $B \rightarrow C$, the gas expands adiabatically ($Q = 0$). In process $C \rightarrow D$, the gas is compressed isothermally while in contact with a reservoir at $T_c < T_h$. In process $D \rightarrow A$, the gas is compressed adiabatically. The upward arrows on the piston indicate that weights are being removed during the expansions, and the downward arrows indicate that weights are being added during the compressions.

Carnot refrigerator. For the *combination* of the engine and refrigerator, then, no exchange by work with the surroundings occurs. Because we have assumed that the engine is more efficient than the refrigerator, the net result of the combination is a transfer of energy from the cold to the hot reservoir without work being done on the combination. According to the Clausius statement of the second law, this is impossible. Hence, the assumption that $e > e_C$ must be false. **All real engines are less efficient than the Carnot engine because they do not operate through a reversible cycle.** The efficiency of a real engine is further reduced by such practical difficulties as friction and energy losses by conduction.

To describe the Carnot cycle taking place between temperatures T_c and T_h, we assume that the working substance is an ideal gas contained in a cylinder fitted with a movable piston at one end. The cylinder's walls and the piston are thermally nonconducting. Four stages of the Carnot cycle are shown in Figure 22.9, and the PV diagram for the cycle is shown in Figure 22.10. The Carnot cycle consists of two adiabatic processes and two isothermal processes, all reversible:

1. Process $A \rightarrow B$ (Fig. 22.9a) is an isothermal expansion at temperature T_h. The gas is placed in thermal contact with an energy reservoir at temperature T_h. During the expansion, the gas absorbs energy Q_h from the reservoir through the base of the cylinder and does work W_{AB} in raising the piston.
2. In process $B \rightarrow C$ (Fig. 22.9b), the base of the cylinder is replaced by a thermally nonconducting wall, and the gas expands adiabatically—that is, no energy enters or leaves the system. During the expansion, the temperature of the gas decreases from T_h to T_c and the gas does work W_{BC} in raising the piston.
3. In process $C \rightarrow D$ (Fig. 22.9c), the gas is placed in thermal contact with an energy reservoir at temperature T_c and is compressed isothermally at temperature T_c. During this time, the gas expels energy Q_c to the reservoir, and the work done by the piston on the gas is W_{CD}.
4. In the final process $D \rightarrow A$ (Fig. 22.9d), the base of the cylinder is replaced by a nonconducting wall, and the gas is compressed adiabatically. The temperature of the gas increases to T_h, and the work done by the piston on the gas is W_{DA}.

The net work done in this reversible, cyclic process is equal to the area enclosed by the path $ABCDA$ in Figure 22.10. As we demonstrated in Section 22.1, because the change in internal energy is zero, the net work W done in one cycle equals the net energy transferred into the system, $Q_h - Q_c$. The thermal efficiency of the engine is given by Equation 22.2:

$$e = \frac{W}{Q_h} = \frac{Q_h - Q_c}{Q_h} = 1 - \frac{Q_c}{Q_h}$$

In Example 22.2, we show that for a Carnot cycle

$$\frac{Q_c}{Q_h} = \frac{T_c}{T_h} \qquad \text{(22.3)}$$

Hence, the thermal efficiency of a Carnot engine is

$$e_C = 1 - \frac{T_c}{T_h} \qquad \text{(22.4)}$$

This result indicates that **all Carnot engines operating between the same two temperatures have the same efficiency.**

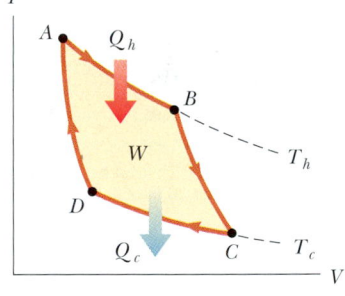

Figure 22.10 *PV* diagram for the Carnot cycle. The net work done, *W*, equals the net energy received in one cycle, $Q_h - Q_c$. Note that $\Delta E_{int} = 0$ for the cycle.

Ratio of energies for a Carnot cycle

Efficiency of a Carnot engine

Equation 22.4 can be applied to any working substance operating in a Carnot cycle between two energy reservoirs. According to this equation, the efficiency is zero if $T_c = T_h$, as one would expect. The efficiency increases as T_c is lowered and as T_h is raised. However, the efficiency can be unity (100%) only if $T_c = 0$ K. Such reservoirs are not available; thus, the maximum efficiency is always less than 100%. In most practical cases, T_c is near room temperature, which is about 300 K. Therefore, one usually strives to increase the efficiency by raising T_h.

EXAMPLE 22.2 Efficiency of the Carnot Engine

Show that the efficiency of a heat engine operating in a Carnot cycle using an ideal gas is given by Equation 22.4.

Solution During the isothermal expansion (process $A \rightarrow B$ in Figure 22.9), the temperature does not change. Thus, the internal energy remains constant. The work done by a gas during an isothermal expansion is given by Equation 20.13. According to the first law, this work is equal to Q_h, the energy absorbed, so that

$$Q_h = W_{AB} = nRT_h \ln \frac{V_B}{V_A}$$

In a similar manner, the energy transferred to the cold reservoir during the isothermal compression $C \rightarrow D$ is

$$Q_c = |W_{CD}| = nRT_c \ln \frac{V_C}{V_D}$$

We take the absolute value of the work because we are defining all values of Q for a heat engine as positive, as mentioned earlier. Dividing the second expression by the first, we find that

$$(1) \qquad \frac{Q_c}{Q_h} = \frac{T_c}{T_h} \frac{\ln(V_C/V_D)}{\ln(V_B/V_A)}$$

We now show that the ratio of the logarithmic quantities is unity by establishing a relationship between the ratio of volumes. For any quasi-static, adiabatic process, the pressure and volume are related by Equation 21.18:

$$(2) \qquad PV^\gamma = \text{constant}$$

During any reversible, quasi-static process, the ideal gas must also obey the equation of state, $PV = nRT$. Solving this ex-

pression for P and substituting into (2), we obtain

$$\frac{nRT}{V} V^\gamma = \text{constant}$$

which we can write as

$$TV^{\gamma-1} = \text{constant}$$

where we have absorbed nR into the constant right-hand side. Applying this result to the adiabatic processes $B \rightarrow C$ and $D \rightarrow A$, we obtain

$$T_h V_B^{\gamma-1} = T_c V_C^{\gamma-1}$$
$$T_h V_A^{\gamma-1} = T_c V_D^{\gamma-1}$$

Dividing the first equation by the second, we obtain

$$(V_B/V_A)^{\gamma-1} = (V_C/V_D)^{\gamma-1}$$

$$(3) \qquad \frac{V_B}{V_A} = \frac{V_C}{V_D}$$

Substituting (3) into (1), we find that the logarithmic terms cancel, and we obtain the relationship

$$\frac{Q_c}{Q_h} = \frac{T_c}{T_h}$$

Using this result and Equation 22.2, we see that the thermal efficiency of the Carnot engine is

$$e_C = 1 - \frac{Q_c}{Q_h} = 1 - \frac{T_c}{T_h}$$

which is Equation 22.4, the one we set out to prove.

EXAMPLE 22.3 The Steam Engine

A steam engine has a boiler that operates at 500 K. The energy from the burning fuel changes water to steam, and this steam then drives a piston. The cold reservoir's temperature is that of the outside air, approximately 300 K. What is the maximum thermal efficiency of this steam engine?

Solution Using Equation 22.4, we find that the maximum thermal efficiency for any engine operating between these temperatures is

$$e_C = 1 - \frac{T_c}{T_h} = 1 - \frac{300 \text{ K}}{500 \text{ K}} = 0.4, \text{ or } \boxed{40\%}$$

You should note that this is the highest *theoretical* efficiency of the engine. In practice, the efficiency is considerably lower.

Exercise Determine the maximum work that the engine can perform in each cycle if it absorbs 200 J of energy from the hot reservoir during each cycle.

Answer 80 J.

EXAMPLE 22.4 ▶ **The Carnot Efficiency**

The highest theoretical efficiency of a certain engine is 30%. If this engine uses the atmosphere, which has a temperature of 300 K, as its cold reservoir, what is the temperature of its hot reservoir?

Solution We use the Carnot efficiency to find T_h:

$$e_C = 1 - \frac{T_c}{T_h}$$

$$T_h = \frac{T_c}{1 - e_C} = \frac{300 \text{ K}}{1 - 0.30} = \boxed{430 \text{ K}}$$

22.4 ▶ GASOLINE AND DIESEL ENGINES

In a gasoline engine, six processes occur in each cycle; five of these are illustrated in Figure 22.11. In this discussion, we consider the interior of the cylinder above the piston to be the system that is taken through repeated cycles in the operation of the engine. For a given cycle, the piston moves up and down twice. This represents a four-stroke cycle consisting of two upstrokes and two downstrokes. The processes in the cycle can be approximated by the **Otto cycle,** a *PV* diagram of which is illustrated in Figure 22.12:

1. During the *intake stroke* $O \rightarrow A$ (Fig. 22.11a), the piston moves downward, and a gaseous mixture of air and fuel is drawn into the cylinder at atmospheric pressure. In this process, the volume increases from V_2 to V_1. This is the energy input part of the cycle, as energy enters the system (the interior of the cylinder) as internal energy stored in the fuel. This is energy transfer by *mass transfer—* that is, the energy is carried with a substance. It is similar to convection, which we studied in Chapter 20.

2. During the *compression stroke* $A \rightarrow B$ (Fig. 22.11b), the piston moves upward, the air–fuel mixture is compressed adiabatically from volume V_1 to volume V_2, and the temperature increases from T_A to T_B. The work done by the gas is negative, and its value is equal to the area under the curve AB in Figure 22.12.

3. In process $B \rightarrow C$, combustion occurs when the spark plug fires (Fig. 22.11c). This is not one of the strokes of the cycle because it occurs in a very short period of time while the piston is at its highest position. The combustion represents a rapid transformation from internal energy stored in chemical bonds in the fuel to internal energy associated with molecular motion, which is related to temperature. During this time, the pressure and temperature in the cylinder increase rapidly, with the temperature rising from T_B to T_C. The volume, however, remains approximately constant because of the short time interval. As a result, approximately no work is done by the gas. We can model this process in the *PV* diagram (Fig. 22.12) as that process in which the energy Q_h enters the system. However, in reality this process is a transformation of energy already in the cylinder (from process $O \rightarrow A$) rather than a transfer.

4. In the *power stroke* $C \rightarrow D$ (Fig. 22.11d), the gas expands adiabatically from V_2 to

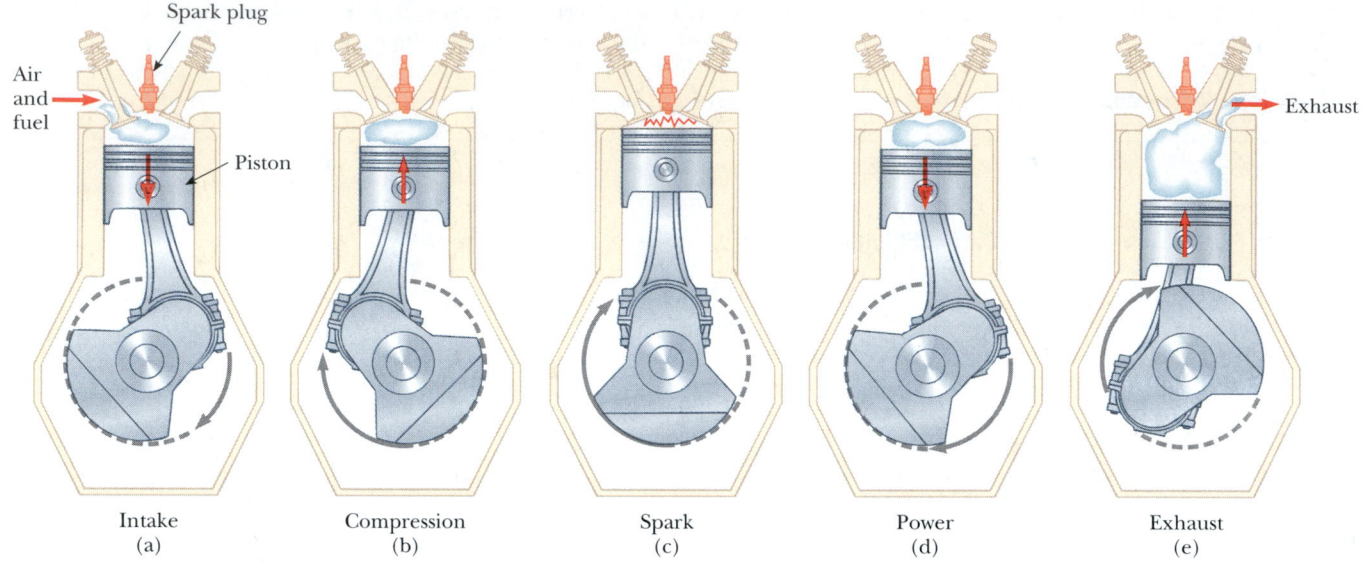

Figure 22.11   The four-stroke cycle of a conventional gasoline engine. (a) In the intake stroke, air is mixed with fuel. (b) The intake valve is then closed, and the air–fuel mixture is compressed by the piston. (c) The mixture is ignited by the spark plug, with the result that the temperature of the mixture increases. (d) In the power stroke, the gas expands against the piston. (e) Finally, the residual gases are expelled, and the cycle repeats.

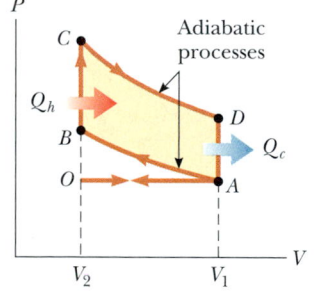

Figure 22.12   *PV* diagram for the Otto cycle, which approximately represents the processes occurring in an internal combustion engine.

V_1. This expansion causes the temperature to drop from T_C to T_D. Work is done by the gas in pushing the piston downward, and the value of this work is equal to the area under the curve *CD*.

5. In the process $D \rightarrow A$ (not shown in Fig. 22.11), an exhaust valve is opened as the piston reaches the bottom of its travel, and the pressure suddenly drops for a short time interval. During this interval, the piston is almost stationary and the volume is approximately constant. Energy is expelled from the interior of the cylinder and continues to be expelled during the next process.

6. In the final process, the *exhaust stroke* $A \rightarrow O$ (Fig. 22.11e), the piston moves upward while the exhaust valve remains open. Residual gases are exhausted at atmospheric pressure, and the volume decreases from V_1 to V_2. The cycle then repeats.

If the air–fuel mixture is assumed to be an ideal gas, then the efficiency of the Otto cycle is

Efficiency of the Otto cycle

$$e = 1 - \frac{1}{(V_1 / V_2)^{\gamma - 1}} \qquad \textbf{(22.5)}$$

where γ is the ratio of the molar specific heats C_P/C_V for the fuel–air mixture and V_1 / V_2 is the **compression ratio.** Equation 22.5, which we derive in Example 22.5, shows that the efficiency increases as the compression ratio increases. For a typical compression ratio of 8 and with $\gamma = 1.4$, we predict a theoretical efficiency of 56% for an engine operating in the idealized Otto cycle. This value is much greater than that achieved in real engines (15% to 20%) because of such effects as friction, energy transfer by conduction through the cylinder walls, and incomplete combustion of the air–fuel mixture.

Diesel engines operate on a cycle similar to the Otto cycle but do not employ a spark plug. The compression ratio for a diesel engine is much greater than that for a gasoline engine. Air in the cylinder is compressed to a very small volume, and, as a consequence, the cylinder temperature at the end of the compression stroke is very high. At this point, fuel is injected into the cylinder. The temperature is high enough for the fuel–air mixture to ignite without the assistance of a spark plug. Diesel engines are more efficient than gasoline engines because of their greater compression ratios and resulting higher combustion temperatures.

EXAMPLE 22.5   **Efficiency of the Otto Cycle**

Show that the thermal efficiency of an engine operating in an idealized Otto cycle (see Figs. 22.11 and 22.12) is given by Equation 22.5. Treat the working substance as an ideal gas.

Solution First, let us calculate the work done by the gas during each cycle. No work is done during processes $B \rightarrow C$ and $D \rightarrow A$. The work done by the gas during the adiabatic compression $A \rightarrow B$ is negative, and the work done by the gas during the adiabatic expansion $C \rightarrow D$ is positive. The value of the net work done equals the area of the shaded region bounded by the closed curve in Figure 22.12. Because the change in internal energy for one cycle is zero, we see from the first law that the net work done during one cycle equals the net energy flow through the system:

$$W = Q_h - Q_c$$

Because processes $B \rightarrow C$ and $D \rightarrow A$ take place at constant volume, and because the gas is ideal, we find from the definition of molar specific heat (Eq. 21.8) that

$$Q_h = nC_V(T_C - T_B) \quad \text{and} \quad Q_c = nC_V(T_D - T_A)$$

Using these expressions together with Equation 22.2, we obtain for the thermal efficiency

$$(1) \qquad e = \frac{W}{Q_h} = 1 - \frac{Q_c}{Q_h} = 1 - \frac{T_D - T_A}{T_C - T_B}$$

We can simplify this expression by noting that processes $A \rightarrow B$ and $C \rightarrow D$ are adiabatic and hence obey the relationship $TV^{\gamma-1} = \text{constant}$, which we obtained in Example 22.2. For the two adiabatic processes, then,

$$A \rightarrow B: \qquad T_A V_A{}^{\gamma-1} = T_B V_B{}^{\gamma-1}$$

$$C \rightarrow D: \qquad T_C V_C{}^{\gamma-1} = T_D V_D{}^{\gamma-1}$$

Using these equations and relying on the fact that $V_A = V_D = V_1$ and $V_B = V_C = V_2$, we find that

$$T_A V_1{}^{\gamma-1} = T_B V_2{}^{\gamma-1}$$

$$(2) \qquad T_A = T_B \left(\frac{V_2}{V_1} \right)^{\gamma-1}$$

$$T_D V_1{}^{\gamma-1} = T_C V_2{}^{\gamma-1}$$

$$(3) \qquad T_D = T_C \left(\frac{V_2}{V_1} \right)^{\gamma-1}$$

Subtracting (2) from (3) and rearranging, we find that

$$(4) \qquad \frac{T_D - T_A}{T_C - T_B} = \left(\frac{V_2}{V_1} \right)^{\gamma-1}$$

Substituting (4) into (1), we obtain for the thermal efficiency

$$(5) \qquad e = 1 - \frac{1}{(V_1 / V_2)^{\gamma-1}}$$

which is Equation 22.5.

We can also express this efficiency in terms of temperatures by noting from (2) and (3) that

$$\left(\frac{V_2}{V_1} \right)^{\gamma-1} = \frac{T_A}{T_B} = \frac{T_D}{T_C}$$

Therefore, (5) becomes

$$(6) \qquad e = 1 - \frac{T_A}{T_B} = 1 - \frac{T_D}{T_C}$$

During the Otto cycle, the lowest temperature is T_A and the highest temperature is T_C. Therefore, the efficiency of a Carnot engine operating between reservoirs at these two temperatures, which is given by the expression $e_C = 1 - (T_A/T_C)$, is *greater* than the efficiency of the Otto cycle given by (6), as expected.

APPLICATION   **Models of Gasoline and Diesel Engines**

We can use the thermodynamic principles discussed in this and earlier chapters to model the performance of gasoline and diesel engines. In both types of engine, a gas is first compressed in the cylinders of the engine and then the fuel–air mixture is ignited. Work is done on the gas during compression, but significantly more work is done on the piston by the mixture as the products of combustion expand in the cylinder. The power of the engine is transferred from the piston to the crankshaft by the connecting rod.

Two important quantities of either engine are the **displacement volume,** which is the volume displaced by the piston as it moves from the bottom to the top of the cylinder, and the com-

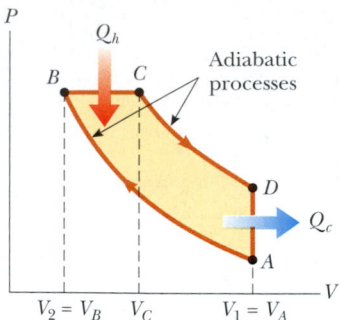

Figure 22.13   *PV* diagram for an ideal diesel engine.

pression ratio *r*, which is the ratio of the maximum and minimum volumes of the cylinder (see p. 680). In our notation, $r = V_A/V_B$, or V_1/V_2 in Eq. 22.5. Most gasoline and diesel engines operate with a four-cycle process (intake, compression, power, exhaust), in which the net work of the intake and exhaust cycles can be considered negligible. Therefore, power is developed only once for every two revolutions of the crankshaft.

In a diesel engine, only air (and no fuel) is present in the cylinder at the beginning of the compression. In the idealized diesel cycle of Figure 22.13, air in the cylinder undergoes an adiabatic compression from *A* to *B*. Starting at *B*, fuel is injected into the cylinder in such a way that the fuel–air mixture undergoes a constant-pressure expansion to an intermediate volume V_C $(B \rightarrow C)$. The high temperature of the mixture causes combustion, and the power stroke is an adiabatic expansion back to $V_D = V_A$ $(C \rightarrow D)$. The exhaust valve is opened, and a constant-volume output of energy occurs $(D \rightarrow A)$ as the cylinder empties.

To simplify our calculations, we assume that the mixture in the cylinder is air modeled as an ideal gas. We use specific heats *c* instead of molar specific heats *C* and assume constant values for air at 300 K. We express the specific heats and the universal gas constant in terms of unit masses rather than moles. Thus, $c_V = 0.718$ kJ/kg·K, $c_P = 1.005$ kJ/kg·K, $\gamma = c_P/c_V = 1.40$, and $R = c_P - c_V = 0.287$ kJ/kg·K = 0.287 kPa·m^3/kg·K.

A 3.00-L Gasoline Engine

Let us calculate the power delivered by a six-cylinder gasoline engine that has a displacement volume of 3.00 L operating at 4 000 rpm and having a compression ratio of $r = 9.50$. The air–fuel mixture enters a cylinder at atmospheric pressure and an ambient temperature of 27°C. During combustion, the mixture reaches a temperature of 1 350°C.

First, let us calculate the work done by an individual cylinder. Using the initial pressure $P_A = 100$ kPa and the initial temperature $T_A = 300$ K, we calculate the initial volume and the mass of the air–fuel mixture. We know that the ratio of the initial and final volumes is the compression ratio,

$$\frac{V_A}{V_B} = r = 9.50$$

We also know that the difference in volumes is the displacement volume. The 3.00-L rating of the engine is the total displacement volume for all six cylinders. Thus, for one cylinder,

$$V_A - V_B = \frac{3.00 \text{ L}}{6} = \frac{3.00 \times 10^{-3} \text{ m}^3}{6} = 0.500 \times 10^{-3} \text{ m}^3$$

Solving these two equations simultaneously, we find the initial and final volumes:

$$V_A = 0.559 \times 10^{-3} \text{ m}^3 \qquad V_B = 0.588 \times 10^{-4} \text{ m}^3$$

Using the ideal gas law (in the form $PV = mRT$, because we are using the universal gas constant in terms of mass rather than moles), we can find the mass of the air–fuel mixture:

$$m = \frac{P_A V_A}{R T_A} = \frac{(100 \text{ kPa})(0.559 \times 10^{-3} \text{ m}^3)}{(0.287 \text{ kPa} \cdot \text{m}^3/\text{kg} \cdot \text{K})(300 \text{ K})}$$
$$= 6.49 \times 10^{-4} \text{ kg}$$

Process $A \rightarrow B$ (see Fig. 22.12) is an adiabatic compression, and this means that $PV^\gamma = $ constant; hence,

$$P_B V_B{}^\gamma = P_A V_A{}^\gamma$$

$$P_B = P_A \left(\frac{V_A}{V_B}\right)^\gamma = P_A(r)^\gamma = (100 \text{ kPa})(9.50)^{1.40}$$
$$= 2.34 \times 10^3 \text{ kPa}$$

Using the ideal gas law, we find that the temperature after the compression is

$$T_B = \frac{P_B V_B}{mR} = \frac{(2.34 \times 10^3 \text{ kPa})(0.588 \times 10^{-4} \text{ m}^3)}{(6.49 \times 10^{-4} \text{ kg})(0.287 \text{ kPa} \cdot \text{m}^3/\text{kg} \cdot \text{K})}$$
$$= 739 \text{ K}$$

In process $B \rightarrow C$, the combustion that transforms the internal energy in chemical bonds into internal energy of molecular motion occurs at constant volume; thus, $V_C = V_B$. Combustion causes the temperature to increase to $T_C = 1 350$°C = 1 623 K. Using this value and the ideal gas law, we can calculate P_C:

$$P_C = \frac{mRT_C}{V_C}$$
$$= \frac{(6.49 \times 10^{-4} \text{ kg})(0.287 \text{ kPa} \cdot \text{m}^3/\text{kg} \cdot \text{K})(1 623 \text{ K})}{(0.588 \times 10^{-4} \text{ m}^3)}$$
$$= 5.14 \times 10^3 \text{ kPa}$$

Process $C \rightarrow D$ is an adiabatic expansion; the pressure after the expansion is

$$P_D = P_C \left(\frac{V_C}{V_D}\right)^\gamma = P_C \left(\frac{V_B}{V_A}\right)^\gamma = P_C \left(\frac{1}{r}\right)^\gamma$$
$$= (5.14 \times 10^3 \text{ kPa})\left(\frac{1}{9.50}\right)^{1.40} = 220 \text{ kPa}$$

Using the ideal gas law again, we find the final temperature:

$$T_D = \frac{P_D V_D}{mR} = \frac{(220 \text{ kPa})(0.559 \times 10^{-3} \text{ m}^3)}{(6.49 \times 10^{-4} \text{ kg})(0.287 \text{ kPa}\cdot\text{m}^3/\text{kg}\cdot\text{K})}$$

$$= 660 \text{ K}$$

Now that we have the temperatures at the beginning and end of each process of the cycle, we can calculate the net energy transfer and net work done by each cylinder every two cycles. From Equation 21.8, we can state

$$Q_h = Q_{in} = mc_V(T_C - T_B)$$
$$= (6.49 \times 10^{-4} \text{ kg})(0.718 \text{ kJ/kg}\cdot\text{K})(1\,623 \text{ K} - 739 \text{ K})$$
$$= 0.412 \text{ kJ}$$

$$Q_c = Q_{out} = mc_V(T_D - T_A)$$
$$= (6.49 \times 10^{-4} \text{ kg})(0.718 \text{ kJ/kg}\cdot\text{K})(660 \text{ K} - 300 \text{ K})$$
$$= 0.168 \text{ kJ}$$

$$W_{net} = Q_{in} - Q_{out} = 0.244 \text{ kJ}$$

From Equation 22.2, the efficiency is $e = W_{net}/Q_{in} = 59\%$. (We can also use Equation 22.5 to calculate the efficiency directly from the compression ratio.)

Recalling that power is delivered every other revolution of the crankshaft, we find that the net power for the six-cylinder engine operating at 4 000 rpm is

$$\mathcal{P}_{net} = 6\left(\frac{1}{2 \text{ rev}}\right)(4\,000 \text{ rev/min})(1 \text{ min}/60 \text{ s})(0.244 \text{ kJ})$$
$$= 49 \text{ kW} = 66 \text{ hp}$$

A 2.00-L Diesel Engine

Let us calculate the power delivered by a four-cylinder diesel engine that has a displacement volume of 2.00 L and is operating at 3 000 rpm. The compression ratio is $r = V_A/V_B = 22.0$, and the **cutoff ratio,** which is the ratio of the volume change during the constant-pressure process $B \rightarrow C$ in Figure 22.13, is $r_c = V_C/V_B = 2.00$. The air enters each cylinder at the beginning of the compression cycle at atmospheric pressure and at an ambient temperature of 27°C.

Our model of the diesel engine is similar to our model of the gasoline engine except that now the fuel is injected at point B and the mixture self-ignites near the end of the compression cycle $A \rightarrow B$, when the temperature reaches the ignition temperature. We assume that the energy input occurs in the constant-pressure process $B \rightarrow C$, and that the expansion process continues from C to D with no further energy transfer by heat.

Let us calculate the work done by an individual cylinder that has an initial volume of $V_A = (2.00 \times 10^{-3} \text{ m}^3)/4 = 0.500 \times 10^{-3} \text{ m}^3$. Because the compression ratio is quite high, we approximate the maximum cylinder volume to be the displacement volume. Using the initial pressure $P_A = 100$ kPa and initial temperature $T_A = 300$ K, we can calculate the mass of the air in the cylinder using the ideal gas law:

$$m = \frac{P_A V_A}{RT_A} = \frac{(100 \text{ kPa})(0.500 \times 10^{-3} \text{ m}^3)}{(0.287 \text{ kPa}\cdot\text{m}^3/\text{kg}\cdot\text{K})(300 \text{ K})} = 5.81 \times 10^{-4} \text{ kg}$$

Process $A \rightarrow B$ is an adiabatic compression, so $PV^\gamma = $ constant; thus,

$$P_B V_B^\gamma = P_A V_A^\gamma$$
$$P_B = P_A\left(\frac{V_A}{V_B}\right)^\gamma = (100 \text{ kPa})(22.0)^{1.40} = 7.57 \times 10^3 \text{ kPa}$$

Using the ideal gas law, we find that the temperature of the air after the compression is

$$T_B = \frac{P_B V_B}{mR} = \frac{(7.57 \times 10^3 \text{ kPa})(0.500 \times 10^{-3} \text{ m}^3)\left(\frac{1}{22.0}\right)}{(5.81 \times 10^{-4} \text{ kg})(0.287 \text{ kPa}\cdot\text{m}^3/\text{kg}\cdot\text{K})}$$
$$= 1.03 \times 10^3 \text{ K}$$

Process $B \rightarrow C$ is a constant-pressure expansion; thus, $P_C = P_B$. We know from the cutoff ratio of 2.00 that the volume doubles in this process. According to the ideal gas law, a doubling of volume in an isobaric process results in a doubling of the temperature, so

$$T_C = 2T_B = 2.06 \times 10^3 \text{ K}$$

Process $C \rightarrow D$ is an adiabatic expansion; therefore,

$$P_D = P_C\left(\frac{V_C}{V_D}\right)^\gamma = P_C\left(\frac{V_C}{V_B}\frac{V_B}{V_D}\right)^\gamma = P_C\left(r_c\frac{1}{r}\right)^\gamma$$
$$= (7.57 \times 10^3 \text{ kPa})\left(\frac{2.00}{22.0}\right)^{1.40} = 264 \text{ kPa}$$

We find the temperature at D from the ideal gas law:

$$T_D = \frac{P_D V_D}{mR} = \frac{(264 \text{ kPa})(0.500 \times 10^{-3} \text{ m}^3)}{(5.81 \times 10^{-4} \text{ kg})(0.287 \text{ kPa}\cdot\text{m}^3/\text{kg}\cdot\text{K})}$$
$$= 792 \text{ K}$$

Now that we have the temperatures at the beginning and the end of each process, we can calculate the net energy transfer by heat and the net work done by each cylinder every two cycles:

$$Q_h = Q_{in} = mc_P(T_C - T_B) = 0.601 \text{ kJ}$$
$$Q_c = Q_{out} = mc_V(T_D - T_A) = 0.205 \text{ kJ}$$
$$W_{net} = Q_{in} - Q_{out} = 0.396 \text{ kJ}$$

The efficiency is $e = W_{net}/Q_{in} = 66\%$.

The net power for the four-cylinder engine operating at 3 000 rpm is

$$\mathcal{P}_{net} = 4\left(\frac{1}{2 \text{ rev}}\right)(3\,000 \text{ rev/min})(1 \text{ min}/60 \text{ s})(0.396 \text{ kJ})$$
$$= 39.6 \text{ kW} = 53 \text{ hp}$$

Of course, modern engine design goes beyond this simple thermodynamic treatment, which uses idealized cycles.

22.5   HEAT PUMPS AND REFRIGERATORS

In Section 22.1 we introduced a heat pump as a mechanical device that moves energy from a region at lower temperature to a region at higher temperature. Heat pumps have long been used for cooling homes and buildings, and they are now becoming increasingly popular for heating them as well. The heat pump contains two sets of metal coils that can exchange energy by heat with the surroundings: one set on the outside of the building, in contact with the air or buried in the ground; and the other set in the interior of the building. In the heating mode, a circulating fluid flowing through the coils absorbs energy from the outside and releases it to the interior of the building from the interior coils. The fluid is cold and at low pressure when it is in the external coils, where it absorbs energy by heat from either the air or the ground. The resulting warm fluid is then compressed and enters the interior coils as a hot, high-pressure fluid, where it releases its stored energy to the interior air.

An air conditioner is simply a heat pump operating in the cooling mode, with its exterior and interior coils interchanged. Energy is absorbed into the circulating fluid in the interior coils; then, after the fluid is compressed, energy leaves the fluid through the external coils. The air conditioner must have a way to release energy to the outside. Otherwise, the work done on the air conditioner would represent energy added to the air inside the house, and the temperature would increase. In the same manner, a refrigerator cannot cool the kitchen if the refrigerator door is left open. The amount of energy leaving the external coils (Fig. 22.14) behind or underneath the refrigerator is greater than the amount of energy removed from the food or from the air in the kitchen if the door is left open. The difference between the energy out and the energy in is the work done by the electricity supplied to the refrigerator.

Figure 22.15 is a schematic representation of a heat pump. The cold temperature is T_c, the hot temperature is T_h, and the energy absorbed by the circulating fluid is Q_c. The heat pump does work W on the fluid, and the energy transferred from the pump to the building in the heating mode is Q_h.

The effectiveness of a heat pump is described in terms of a number called the **coefficient of performance** (COP). In the heating mode, the COP is defined as the ratio of the energy transferred to the hot reservoir to the work required to transfer that energy:

$$\text{COP (heating mode)} \equiv \frac{\text{Energy transferred at high temperature}}{\text{Work done by pump}} = \frac{Q_h}{W} \qquad \textbf{(22.6)}$$

Note that the COP is similar to the thermal efficiency for a heat engine in that it is a ratio of what you get (energy delivered to the interior of the building) to what you give (work input). Because Q_h is generally greater than W, typical values for the COP are greater than unity. It is desirable for the COP to be as high as possible, just as it is desirable for the thermal efficiency of an engine to be as high as possible.

If the outside temperature is 25°F or higher, then the COP for a heat pump is about 4. That is, the amount of energy transferred to the building is about four times greater than the work done by the motor in the heat pump. However, as the outside temperature decreases, it becomes more difficult for the heat pump to extract sufficient energy from the air, and so the COP decreases. In fact, the COP can fall below unity for temperatures below the midteens. Thus, the use of heat pumps that extract energy from the air, while satisfactory in moderate climates, is not appropriate in areas where winter temperatures are very low. It is possible to

Figure 22.14 The coils on the back of a refrigerator transfer energy by heat to the air. The second law of thermodynamics states that this amount of energy must be greater than the amount of energy removed from the contents of the refrigerator (or from the air in the kitchen, if the refrigerator door is left open). *(Charles D. Winters)*

use heat pumps in colder areas by burying the external coils deep in the ground. In this case, the energy is extracted from the ground, which tends to be warmer than the air in the winter.

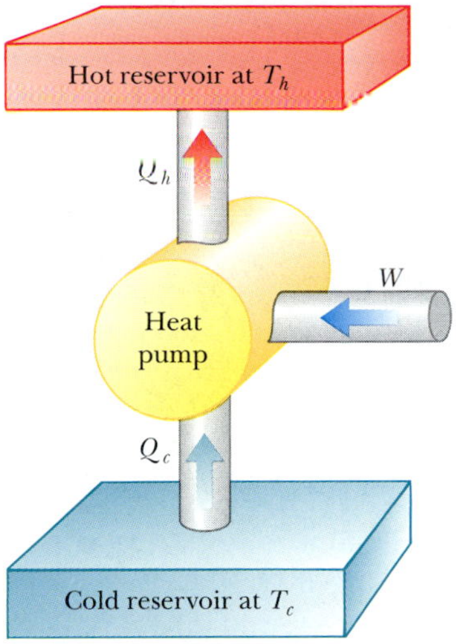

Figure 22.15 Schematic diagram of a heat pump, which absorbs energy Q_c from a cold reservoir and expels energy Q_h to a hot reservoir. Note that this diagram is the same as that for the refrigerator shown in Figure 22.5.

Quick Quiz 22.1

In an electric heater, electrical energy can be converted to internal energy with an efficiency of 100%. By what percentage does the cost of heating your home change when you replace your electric heating system with a heat pump that has a COP of 4? Assume that the motor running the heat pump is 100% efficient.

Theoretically, a Carnot-cycle heat engine run in reverse constitutes the most effective heat pump possible, and it determines the maximum COP for a given combination of hot and cold reservoir temperatures. Using Equations 22.1 and 22.3, we see that the maximum COP for a heat pump in its heating mode is

$$\text{COP}_\text{C}(\text{heating mode}) = \frac{Q_h}{W}$$

$$= \frac{Q_h}{Q_h - Q_c} = \frac{1}{1 - \dfrac{Q_c}{Q_h}} = \frac{1}{1 - \dfrac{T_c}{T_h}} = \frac{T_h}{T_h - T_c}$$

For a heat pump operating in the cooling mode, "what you get" is energy removed from the cold reservoir. The most effective refrigerator or air conditioner is one that removes the greatest amount of energy from the cold reservoir in exchange for the least amount of work. Thus, for these devices we define the COP in terms of Q_c:

$$\text{COP (cooling mode)} = \frac{Q_c}{W} \qquad \textbf{(22.7)}$$

A good refrigerator should have a high COP, typically 5 or 6.

The greatest possible COP for a heat pump in the cooling mode is that of a heat pump whose working substance is carried through a Carnot cycle in reverse:

$$\text{COP}_\text{C}\text{ (cooling mode)} = \frac{T_c}{T_h - T_c}$$

As the difference between the temperatures of the two reservoirs approaches zero in this expression, the theoretical COP approaches infinity. In practice, the low temperature of the cooling coils and the high temperature at the compressor limit the COP to values below 10.

QuickLab

Estimate the COP of your refrigerator by making rough temperature measurements of the stored food and of the exhaust coils (found either on the back of the unit or behind a panel on the bottom). Use just your hand if no thermometer is available.

22.6 ENTROPY

10.10 & 10.11

The zeroth law of thermodynamics involves the concept of temperature, and the first law involves the concept of internal energy. Temperature and internal energy are both state functions—that is, they can be used to describe the thermodynamic state of a system. Another state function—this one related to the second law of thermodynamics—is **entropy** S. In this section we define entropy on a macroscopic scale as it was first expressed by Clausius in 1865.

Consider any infinitesimal process in which a system changes from one equilibrium state to another. If dQ_r is the amount of energy transferred by heat when the system follows a reversible path between the states, then the change in entropy dS is equal to this amount of energy for the reversible process divided by the absolute temperature of the system:

Clausius definition of change in entropy

$$dS = \frac{dQ_r}{T} \qquad \textbf{(22.8)}$$

We have assumed that the temperature is constant because the process is infinitesimal. Since we have claimed that entropy is a state function, **the change in entropy during a process depends only on the end points and therefore is independent of the actual path followed.**

The subscript r on the quantity dQ_r is a reminder that the transferred energy is to be measured along a reversible path, even though the system may actually have followed some irreversible path. When energy is absorbed by the system, dQ_r is positive and the entropy of the system increases. When energy is expelled by the system, dQ_r is negative and the entropy of the system decreases. Note that Equation 22.8 defines not entropy but rather the *change* in entropy. Hence, the meaningful quantity in describing a process is the *change* in entropy.

Entropy was originally formulated as a useful concept in thermodynamics; however, its importance grew tremendously as the field of statistical mechanics developed because the analytical techniques of statistical mechanics provide an alternative means of interpreting entropy. In statistical mechanics, the behavior of a substance is described in terms of the statistical behavior of its atoms and molecules. One of the main results of this treatment is that **isolated systems tend toward disorder and that entropy is a measure of this disorder.** For example, consider the molecules of a gas in the air in your room. If half of the gas molecules had velocity vectors of equal magnitude directed toward the left and the other half had velocity vectors of the same magnitude directed toward the right, the situation would be very ordered. However, such a situation is extremely unlikely. If you could actually view the molecules, you would see that they move haphazardly in all directions, bumping into one another, changing speed upon collision, some going fast and others going slowly. This situation is highly disordered.

The cause of the tendency of an isolated system toward disorder is easily explained. To do so, we distinguish between *microstates* and *macrostates* of a system. A **microstate** is a particular description of the properties of the individual molecules of the system. For example, the description we just gave of the velocity vectors of the air molecules in your room being very ordered refers to a particular microstate, and the more likely likely haphazard motion is another microstate—one that represents disorder. A **macrostate** is a description of the conditions of the system from a macroscopic point of view and makes use of macroscopic variables such as pressure, density, and temperature. For example, in both of the microstates described for the air molecules in your room, the air molecules are distributed uniformly throughout the volume of the room; this uniform density distribution is a macrostate. We could not distinguish between our two microstates by making a macroscopic measurement—both microstates would appear to be the same macroscopically, and the two macrostates corresponding to these microstates are equivalent.

For any given macrostate of the system, a number of microstates are possible, or *accessible*. Among these microstates, it is assumed that all are equally probable. However, when all possible microstates are examined, it is found that far more of them are disordered than are ordered. Because all of the microstates are equally probable, it is highly likely that the actual macrostate is one resulting from one of the highly disordered microstates, simply because there are many more of them. Similarly, the probability of a macrostate's forming from disordered microstates is greater than the probability of a macrostate's forming from ordered microstates.

All physical processes that take place in a system tend to cause the system and its surroundings to move toward more probable macrostates. The more probable macrostate is always one of greater disorder. If we consider a system and its surroundings to include the entire Universe, then the Universe is always moving toward a macrostate corresponding to greater disorder. Because entropy is a measure of disorder, an alternative way of stating this is **the entropy of the Universe increases in all real processes.** This is yet another statement of the second law of thermodynamics that can be shown to be equivalent to the Kelvin–Planck and Clausius statements.

In real processes, the disorder of the Universe increases

To calculate the change in entropy for a finite process, we must recognize that T is generally not constant. If dQ_r is the energy transferred by heat when the system is at a temperature T, then the change in entropy in an arbitrary reversible process between an initial state and a final state is

$$\Delta S = \int_i^f dS = \int_i^f \frac{dQ_r}{T} \quad \text{(reversible path)} \tag{22.9}$$

Change in entropy for a finite process

As with an infinitesimal process, the change in entropy ΔS of a system going from one state to another has the same value for *all* paths connecting the two states. That is, the finite change in entropy ΔS of a system depends only on the properties of the initial and final equilibrium states. Thus, we are free to choose a particular reversible path over which to evaluate the entropy in place of the actual path, as long as the initial and final states are the same for both paths.

Quick Quiz 22.2

Which of the following is true for the entropy change of a system that undergoes a reversible, adiabatic process? (a) $\Delta S < 0$. (b) $\Delta S = 0$. (c) $\Delta S > 0$.

Let us consider the changes in entropy that occur in a Carnot heat engine operating between the temperatures T_c and T_h. In one cycle, the engine absorbs energy Q_h from the hot reservoir and expels energy Q_c to the cold reservoir. These energy transfers occur only during the isothermal portions of the Carnot cycle; thus, the constant temperature can be brought out in front of the integral sign in Equation 22.9. The integral then simply has the value of the total amount of energy transferred by heat. Thus, the total change in entropy for one cycle is

$$\Delta S = \frac{Q_h}{T_h} - \frac{Q_c}{T_c}$$

where the negative sign represents the fact that energy Q_c is expelled by the system, since we continue to define Q_c as a positive quantity when referring to heat engines. In Example 22.2 we showed that, for a Carnot engine,

$$\frac{Q_c}{Q_h} = \frac{T_c}{T_h}$$

Using this result in the previous expression for ΔS, we find that the total change in entropy for a Carnot engine operating in a cycle is *zero*:

The change in entropy for a Carnot cycle is zero

$$\Delta S = 0$$

Now let us consider a system taken through an arbitrary (non-Carnot) reversible cycle. Because entropy is a state function—and hence depends only on the properties of a given equilibrium state—we conclude that $\Delta S = 0$ for *any* reversible cycle. In general, we can write this condition in the mathematical form

$\Delta S = 0$ for any reversible cycle

$$\oint \frac{dQ_r}{T} = 0 \tag{22.10}$$

where the symbol $\oint$ indicates that the integration is over a closed path.

Quasi-Static, Reversible Process for an Ideal Gas

Let us suppose that an ideal gas undergoes a quasi-static, reversible process from an initial state having temperature T_i and volume V_i to a final state described by T_f and V_f. Let us calculate the change in entropy of the gas for this process.

Writing the first law of thermodynamics in differential form and rearranging the terms, we have $dQ_r = dE_{int} + dW$, where $dW = P\,dV$. For an ideal gas, recall that $dE_{int} = nC_V\,dT$ (Eq. 21.12), and from the ideal gas law, we have $P = nRT/V$. Therefore, we can express the energy transferred by heat in the process as

$$dQ_r = dE_{int} + P\,dV = nC_V\,dT + nRT\frac{dV}{V}$$

We cannot integrate this expression as it stands because the last term contains two variables, T and V. However, if we divide all terms by T, each of the terms on the right-hand side depends on only one variable:

$$\frac{dQ_r}{T} = nC_V\frac{dT}{T} + nR\frac{dV}{V} \tag{22.11}$$

Assuming that C_V is constant over the interval in question, and integrating Equation 22.11 from the initial state to the final state, we obtain

$$\Delta S = \int_i^f \frac{dQ_r}{T} = nC_V \ln\frac{T_f}{T_i} + nR \ln\frac{V_f}{V_i} \tag{22.12}$$

This expression demonstrates mathematically what we argued earlier—that ΔS depends only on the initial and final states and is independent of the path between the states. Also, note in Equation 22.12 that ΔS can be positive or negative, depending on the values of the initial and final volumes and temperatures. Finally, for a cyclic process ($T_i = T_f$ and $V_i = V_f$), we see from Equation 22.12 that $\Delta S = 0$. This is evidence that entropy is a state function.

EXAMPLE 22.6   **Change in Entropy—Melting**

A solid that has a latent heat of fusion L_f melts at a temperature T_m. (a) Calculate the change in entropy of this substance when a mass m of the substance melts.

Solution   Let us assume that the melting occurs so slowly that it can be considered a reversible process. In this case the temperature can be regarded as constant and equal to T_m.

Making use of Equations 22.9 and that for the latent heat of fusion $Q = mL_f$ (Eq. 20.6), we find that

$$\Delta S = \int \frac{dQ_r}{T} = \frac{1}{T_m}\int dQ = \frac{Q}{T_m} = \boxed{\frac{mL_f}{T_m}}$$

Note that we are able to remove T_m from the integral because the process is isothermal. Note also that ΔS is positive. This means that when a solid melts, its entropy increases because the molecules are much more disordered in the liquid state than they are in the solid state. The positive value for ΔS also means that the substance in its liquid state does not spontaneously transfer energy from itself to the surroundings and freeze because to do so would involve a spontaneous decrease in entropy.

(b) Estimate the value of the change in entropy of an ice cube when it melts.

Solution Let us assume an ice tray makes cubes that are about 3 cm on a side. The volume per cube is then (very roughly) 30 cm³. This much liquid water has a mass of 30 g. From Table 20.2 we find that the latent heat of fusion of ice is 3.33×10^5 J/kg. Substituting these values into our answer for part (a), we find that

$$\Delta S = \frac{mL_f}{T_m} = \frac{(0.03 \text{ kg})(3.33 \times 10^5 \text{ J/kg})}{273 \text{ K}} = \boxed{4 \times 10^1 \text{ J/K}}$$

We retain only one significant figure, in keeping with the nature of our estimations.

22.7 ENTROPY CHANGES IN IRREVERSIBLE PROCESSES

By definition, calculation of the change in entropy requires information about a reversible path connecting the initial and final equilibrium states. To calculate changes in entropy for real (irreversible) processes, we must remember that entropy (like internal energy) depends only on the *state* of the system. That is, entropy is a state function. Hence, the change in entropy when a system moves between any two equilibrium states depends only on the initial and final states. We can show that if this were not the case, the second law of thermodynamics would be violated.

We now calculate the entropy change in some irreversible process between two equilibrium states by devising a reversible process (or series of reversible processes) between the same two states and computing $\Delta S = \int dQ_r / T$ for the reversible process. In irreversible processes, it is critically important that we distinguish between Q, the actual energy transfer in the process, and Q_r, the energy that would have been transferred by heat along a reversible path. Only Q_r is the correct value to be used in calculating the entropy change.

As we shall see in the following examples, the change in entropy for a system and its surroundings is always positive for an irreversible process. In general, the total entropy—and therefore the disorder—always increase in an irreversible process. Keeping these considerations in mind, we can state the second law of thermodynamics as follows:

> The total entropy of an isolated system that undergoes a change can never decrease.

Furthermore, **if the process is irreversible, then the total entropy of an isolated system always increases. In a reversible process, the total entropy of an isolated system remains constant.**

When dealing with a system that is not isolated from its surroundings, remember that the increase in entropy described in the second law is that of the system *and* its surroundings. When a system and its surroundings interact in an irreversible process, the increase in entropy of one is greater than the decrease in entropy of the other. Hence, we conclude that **the change in entropy of the Universe must be greater than zero for an irreversible process and equal to zero for a reversible process.** Ultimately, the entropy of the Universe should reach a maximum value. At this value, the Universe will be in a state of uniform temperature and density. All physical, chemical, and biological processes will cease because a state of perfect disorder implies that no energy is available for doing work. This gloomy state of affairs is sometimes referred to as the heat death of the Universe.

Quick Quiz 22.3

In the presence of sunlight, a tree rearranges an unorganized collection of carbon dioxide and water molecules into the highly ordered collection of molecules we see as leaves and branches. True or false: This reduction of entropy in the tree is a violation of the second law of thermodynamics. Explain your response.

Entropy Change in Thermal Conduction

Let us now consider a system consisting of a hot reservoir and a cold reservoir in thermal contact with each other and isolated from the rest of the Universe. A process occurs during which energy Q is transferred by heat from the hot reservoir at temperature T_h to the cold reservoir at temperature T_c. Because the cold reservoir absorbs energy Q, its entropy increases by Q/T_c. At the same time, the hot reservoir loses energy Q, and so its entropy change is $-Q/T_h$. Because $T_h > T_c$, the increase in entropy of the cold reservoir is greater than the decrease in entropy of the hot reservoir. Therefore, the change in entropy of the system (and of the Universe) is greater than zero:

$$\Delta S_U = \frac{Q}{T_c} + \frac{-Q}{T_h} > 0$$

EXAMPLE 22.7   Which Way Does the Energy Flow?

A large, cold object is at 273 K, and a large, hot object is at 373 K. Show that it is impossible for a small amount of energy—for example, 8.00 J—to be transferred spontaneously from the cold object to the hot one without a decrease in the entropy of the Universe and therefore a violation of the second law.

Solution We assume that, during the energy transfer, the two objects do not undergo a temperature change. This is not a necessary assumption; we make it only to avoid using integral calculus in our calculations. The process as described is irreversible, and so we must find an equivalent reversible process. It is sufficient to assume that the objects are connected by a poor thermal conductor whose temperature spans the range from 273 K to 373 K. This conductor transfers energy slowly, and its state does not change during the process. Under this assumption, the energy transfer to or from each object is reversible, and we may set $Q = Q_r$. The entropy change of the hot object is

$$\Delta S_h = \frac{Q_r}{T_h} = \frac{8.00\text{ J}}{373\text{ K}} = 0.021\text{ }4\text{ J/K}$$

The cold object loses energy, and its entropy change is

$$\Delta S_c = \frac{Q_r}{T_c} = \frac{-8.00\text{ J}}{273\text{ K}} = -0.029\text{ }3\text{ J/K}$$

We consider the two objects to be isolated from the rest of the Universe. Thus, the entropy change of the Universe is just

that of our two-object system, which is

$$\Delta S_U = \Delta S_c + \Delta S_h = -0.007\text{ }9\text{ J/K}$$

This decrease in entropy of the Universe is in violation of the second law. That is, **the spontaneous transfer of energy from a cold to a hot object cannot occur.**

In terms of disorder, let us consider the violation of the second law if energy were to continue to transfer spontaneously from a cold object to a hot object. Before the transfer, a certain degree of order is associated with the different temperatures of the objects. The hot object's molecules have a higher average energy than the cold object's molecules. If energy spontaneously flows from the cold object to the hot object, then, over a period of time, the cold object will become colder and the hot object will become hotter. The difference in average molecular energy will become even greater; this would represent an increase in order for the system and a violation of the second law.

In comparison, the process that does occur naturally is the flow of energy from the hot object to the cold object. In this process, the difference in average molecular energy decreases; this represents a more random distribution of energy and an increase in disorder.

Exercise Suppose that 8.00 J of energy is transferred from a hot object to a cold one. What is the net entropy change of the Universe?

Answer $+0.007\text{ }9$ J/K.

Entropy Change in a Free Expansion

Let us again consider the adiabatic free expansion of a gas occupying an initial volume V_i (Fig. 22.16). A membrane separating the gas from an evacuated region is broken, and the gas expands (irreversibly) to a volume V_f. Let us find the changes in entropy of the gas and of the Universe during this process.

The process is clearly neither reversible nor quasi-static. The work done by the gas against the vacuum is zero, and because the walls are insulating, no energy is transferred by heat during the expansion. That is, $W = 0$ and $Q = 0$. Using the first law, we see that the change in internal energy is zero. Because the gas is ideal, E_{int} depends on temperature only, and we conclude that $\Delta T = 0$ or $T_i = T_f$.

To apply Equation 22.9, we cannot use $Q = 0$, the value for the irreversible process, but must instead find Q_r; that is, we must find an equivalent reversible path that shares the same initial and final states. A simple choice is an isothermal, reversible expansion in which the gas pushes slowly against a piston while energy enters the gas by heat from a reservoir to hold the temperature constant. Because T is constant in this process, Equation 22.9 gives

$$\Delta S = \int_i^f \frac{dQ_r}{T} = \frac{1}{T} \int_i^f dQ_r$$

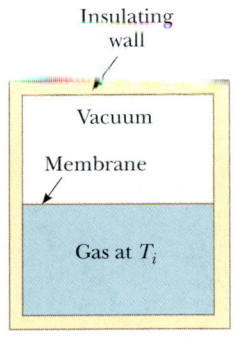

Figure 22.16 Adiabatic free expansion of a gas. When the membrane separating the gas from the evacuated region is ruptured, the gas expands freely and irreversibly. As a result, it occupies a greater final volume. The container is thermally insulated from its surroundings; thus, $Q = 0$.

For an isothermal process, the first law of thermodynamics specifies that $\int_i^f dQ_r$ is equal to the work done by the gas during the expansion from V_i to V_f, which is given by Equation 20.13. Using this result, we find that the entropy change for the gas is

$$\Delta S = nR \ln \frac{V_f}{V_i} \qquad \textbf{(22.13)}$$

Because $V_f > V_i$, we conclude that ΔS is positive. This positive result indicates that both the entropy and the disorder of the gas increase as a result of the irreversible, adiabatic expansion.

Because the free expansion takes place in an insulated container, no energy is transferred by heat from the surroundings. (Remember that the isothermal, reversible expansion is only a *replacement* process that we use to calculate the entropy change for the gas; it is not the *actual* process.) Thus, the free expansion has no effect on the surroundings, and the entropy change of the surroundings is zero. Thus, the entropy change for the Universe is positive; this is consistent with the second law.

EXAMPLE 22.8 ▶ Free Expansion of a Gas

Calculate the change in entropy for a process in which 2.00 mol of an ideal gas undergoes a free expansion to three times its initial volume.

Solution Using Equation 22.13 with $n = 2.00$ mol and $V_f = 3V_i$, we find that

$$\Delta S = nR \ln \frac{V_f}{V_i} = (2.00 \text{ mol})(8.31 \text{ J/mol} \cdot \text{K})(\ln 3)$$

$$= 18.3 \text{ J/K}$$

It is easy to see that the gas is more disordered after the expansion. Instead of being concentrated in a relatively small space, the molecules are scattered over a larger region.

Entropy Change in Calorimetric Processes

A substance of mass m_1, specific heat c_1, and initial temperature T_1 is placed in thermal contact with a second substance of mass m_2, specific heat c_2, and initial temperature $T_2 > T_1$. The two substances are contained in a calorimeter so that no energy is lost to the surroundings. The system of the two substances is allowed to reach thermal equilibrium. What is the total entropy change for the system?

First, let us calculate the final equilibrium temperature T_f. Using the techniques of Section 20.2—namely, Equation 20.5, $Q_{cold} = -Q_{hot}$, and Equation 20.4, $Q = mc \, \Delta T$, we obtain

$$m_1 c_1 \, \Delta T_1 = -m_2 c_2 \, \Delta T_2$$

$$m_1 c_1 (T_f - T_1) = -m_2 c_2 (T_f - T_2)$$

Solving for T_f, we have

$$T_f = \frac{m_1 c_1 T_1 + m_2 c_2 T_2}{m_1 c_1 + m_2 c_2} \tag{22.14}$$

The process is irreversible because the system goes through a series of non-equilibrium states. During such a transformation, the temperature of the system at any time is not well defined because different parts of the system have different temperatures. However, we can imagine that the hot substance at the initial temperature T_2 is slowly cooled to the temperature T_f as it comes into contact with a series of reservoirs differing infinitesimally in temperature, the first reservoir being at T_2 and the last being at T_f. Such a series of very small changes in temperature would approximate a reversible process. We imagine doing the same thing for the cold substance. Applying Equation 22.9 and noting that $dQ = mc \, dT$ for an infinitesimal change, we have

$$\Delta S = \int_1 \frac{dQ_{cold}}{T} + \int_2 \frac{dQ_{hot}}{T} = m_1 c_1 \int_{T_1}^{T_f} \frac{dT}{T} + m_2 c_2 \int_{T_2}^{T_f} \frac{dT}{T}$$

where we have assumed that the specific heats remain constant. Integrating, we find that

Change in entropy for a calorimetric process

$$\Delta S = m_1 c_1 \ln \frac{T_f}{T_1} + m_2 c_2 \ln \frac{T_f}{T_2} \tag{22.15}$$

where T_f is given by Equation 22.14. If Equation 22.14 is substituted into Equation 22.15, we can show that one of the terms in Equation 22.15 is always positive and the other is always negative. (You may want to verify this for yourself.) The positive term is always greater than the negative term, and this results in a positive value for ΔS. Thus, we conclude that the entropy of the Universe increases in this irreversible process.

Finally, you should note that Equation 22.15 is valid only when no mixing of different substances occurs, because a further entropy increase is associated with the increase in disorder during the mixing. If the substances are liquids or gases and mixing occurs, the result applies only if the two fluids are identical, as in the following example.

EXAMPLE 22.9   Calculating ΔS for a Calorimetric Process

Suppose that 1.00 kg of water at 0.00°C is mixed with an equal mass of water at 100°C. After equilibrium is reached, the mixture has a uniform temperature of 50.0°C. What is the change in entropy of the system?

Solution   We can calculate the change in entropy from Equation 22.15 using the values $m_1 = m_2 = 1.00$ kg, $c_1 = c_2 = 4\,186$ J/kg·K, $T_1 = 273$ K, $T_2 = 373$ K, and $T_f = 323$ K:

$$\Delta S = m_1 c_1 \ln \frac{T_f}{T_1} + m_2 c_2 \ln \frac{T_f}{T_2}$$

$$= (1.00 \text{ kg})(4\,186 \text{ J/kg} \cdot \text{K}) \ln \left(\frac{323 \text{ K}}{273 \text{ K}} \right)$$

$$+ (1.00 \text{ kg})(4\,186 \text{ J/kg} \cdot \text{K}) \ln \left(\frac{323 \text{ K}}{373 \text{ K}} \right)$$

$$= 704 \text{ J/K} - 602 \text{ J/K} = \boxed{102 \text{ J/K}}$$

That is, as a result of this irreversible process, the increase in entropy of the cold water is greater than the decrease in entropy of the warm water. Consequently, the increase in entropy of the system is 102 J/K.

Optional Section

22.8 ENTROPY ON A MICROSCOPIC SCALE[4]

As we have seen, we can approach entropy by relying on macroscopic concepts and using parameters such as pressure and temperature. We can also treat entropy from a microscopic viewpoint through statistical analysis of molecular motions. We now use a microscopic model to investigate once again the free expansion of an ideal gas, which was discussed from a macroscopic point of view in the preceding section.

In the kinetic theory of gases, gas molecules are represented as particles moving randomly. Let us suppose that the gas is initially confined to a volume V_i, as shown in Figure 22.17a. When the partition separating V_i from a larger container is removed, the molecules eventually are distributed throughout the greater volume V_f (Fig. 22.17b). For a given uniform distribution of gas in the volume, there are a large number of equivalent microstates, and we can relate the entropy of the gas to the number of microstates corresponding to a given macrostate.

We count the number of microstates by considering the variety of molecular locations involved in the free expansion. The instant after the partition is removed (and before the molecules have had a chance to rush into the other half of the container), all the molecules are in the initial volume. We assume that each molecule occupies some microscopic volume V_m. The total number of possible locations of a single molecule in a macroscopic initial volume V_i is the ratio $w_i = V_i/V_m$, which is a huge number. We use w_i here to represent the number of *ways* that the molecule can be placed in the volume, or the number of microstates, which is equivalent to the number of available locations. We assume that the molecule's occupying each of these locations is equally probable.

As more molecules are added to the system, the number of possible ways that the molecules can be positioned in the volume multiplies. For example, in considering two molecules, for every possible placement of the first, all possible placements of the second are available. Thus, there are w_1 ways of locating the first molecule, and for each of these, there are w_2 ways of locating the second molecule. The total number of ways of locating the two molecules is $w_1 w_2$.

Neglecting the very small probability of having two molecules occupy the same location, each molecule may go into any of the V_i/V_m locations, and so the number of ways of locating N molecules in the volume becomes $W_i = w_i^N = (V_i/V_m)^N$ (W_i is not to be confused with work.) Similarly, when the volume is increased to V_f, the number of ways of locating N molecules increases to $W_f = w_f^N = (V_f/V_m)^N$ The ratio of the number of ways of placing the molecules in the volume for the

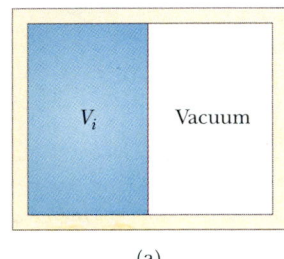

(a)

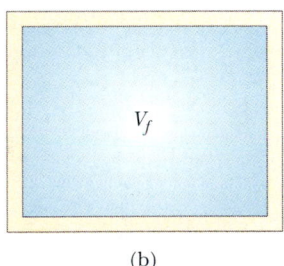

(b)

Figure 22.17 In a free expansion, the gas is allowed to expand into a region that was previously a vacuum.

[4] This section was adapted from A. Hudson and R. Nelson, *University Physics*, Philadelphia, Saunders College Publishing, 1990.

initial and final configurations is

$$\frac{W_f}{W_i} = \frac{(V_f/V_m)^N}{(V_i/V_m)^N} = \left(\frac{V_f}{V_i}\right)^N$$

If we now take the natural logarithm of this equation and multiply by Boltzmann's constant, we find that

$$k_B \ln\left(\frac{W_f}{W_i}\right) = nN_A k_B \ln\left(\frac{V_f}{V_i}\right)$$

where we have used the equality $N = nN_A$. We know from Equation 19.11 that $N_A k_B$ is the universal gas constant R; thus, we can write this equation as

$$k_B \ln W_f - k_B \ln W_i = nR \ln\left(\frac{V_f}{V_i}\right) \tag{22.16}$$

From Equation 22.13 we know that when n mol of a gas undergoes a free expansion from V_i to V_f, the change in entropy is

$$S_f - S_i = nR \ln\left(\frac{V_f}{V_i}\right) \tag{22.17}$$

Note that the right-hand sides of Equations 22.16 and 22.17 are identical. Thus, we make the following important connection between entropy and the number of microstates for a given macrostate:

Entropy (microscopic definition)

$$S \equiv k_B \ln W \tag{22.18}$$

The more microstates there are that correspond to a given macrostate, the greater is the entropy of that macrostate. As we have discussed previously, there are many more disordered microstates than ordered microstates. Thus, Equation 22.18 indicates mathematically that **entropy is a measure of microscopic disorder.** Although in our discussion we used the specific example of the free expansion of an ideal gas, a more rigorous development of the statistical interpretation of entropy would lead us to the same conclusion.

Imagine the container of gas depicted in Figure 22.18a as having all of its molecules traveling at speeds greater than the mean value on the left side and all of its molecules traveling at speeds less than the mean value on the right side (an ordered microstate). Compare this with the uniform mixture of fast- and slow-mov-

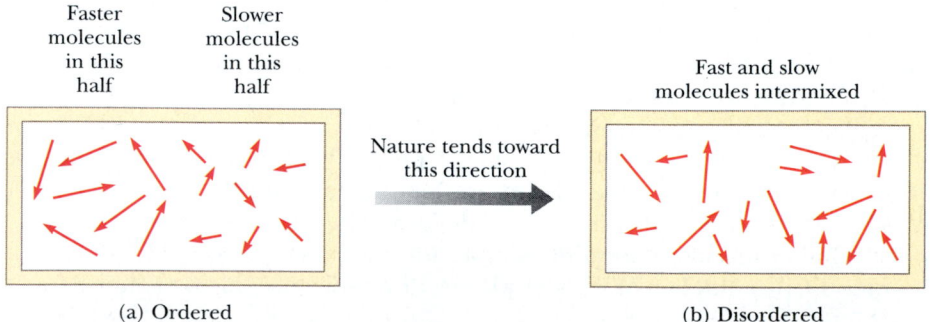

Faster molecules in this half   Slower molecules in this half

Nature tends toward this direction

Fast and slow molecules intermixed

(a) Ordered   (b) Disordered

Figure 22.18 A container of gas in two equally probable states of molecular motion. (a) An ordered arrangement, which is one of a few and therefore a collectively unlikely set. (b) A disordered arrangement, which is one of many and therefore a collectively likely set.

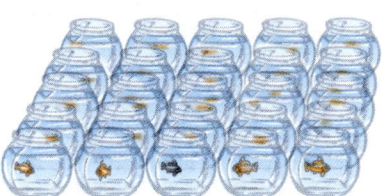

Figure 22.19 By tossing a coin into a jar, the carnival-goer can win the fish in the jar. It is more likely that the coin will land in a jar containing a goldfish than in the one containing the black fish.

ing molecules in Figure 22.18b (a disordered microstate). You might expect the ordered microstate to be very unlikely because random motions tend to mix the slow- and fast-moving molecules uniformly. Yet *individually* each of these microstates is equally probable. However, there are far more disordered microstates than ordered microstates, and so a macrostate corresponding to a large number of equivalent disordered microstates is much more probable than a macrostate corresponding to a small number of equivalent ordered microstates.

Figure 22.19 shows a real-world example of this concept. There are two possible macrostates for the carnival game—winning a goldfish and winning a black fish. Because only one jar in the array of jars contains a black fish, only one possible microstate corresponds to the macrostate of winning a black fish. A large number of microstates are described by the coin's falling into a jar containing a goldfish. Thus, for the macrostate of winning a goldfish, there are many equivalent microstates. As a result, the probability of winning a goldfish is much greater than the probability of winning a black fish. If there are 24 goldfish and 1 black fish, the probability of winning the black fish is 1 in 25. This assumes that all microstates have the same probability, a situation that may not be quite true for the situation shown in Figure 22.19. If you are an accurate coin tosser and you are aiming for the edge of the array of jars, then the probability of the coin's landing in a jar near the edge is likely to be greater than the probability of its landing in a jar near the center.

Let us consider a similar type of probability problem for 100 molecules in a container. At any given moment, the probability of one molecule's being in the left part of the container shown in Figure 22.20a as a result of random motion is $\frac{1}{2}$. If there are two molecules, as shown in Figure 22.20b, the probability of both being in the left part is $(\frac{1}{2})^2$ or 1 in 4. If there are three molecules (Fig. 22.20c), the probability of all of them being in the left portion at the same moment is $(\frac{1}{2})^3$, or 1 in 8. For 100 independently moving molecules, the probability that the 50 fastest ones will be found in the left part at any moment is $(\frac{1}{2})^{50}$. Likewise, the probability that the remaining 50 slower molecules will be found in the right part at any moment is $(\frac{1}{2})^{50}$. Therefore, the probability of finding this fast-slow separation as a result of random motion is the product $(\frac{1}{2})^{50}(\frac{1}{2})^{50} = (\frac{1}{2})^{100}$, which corresponds to about 1 in 10^{30}. When this calculation is extrapolated from 100 molecules to the number in 1 mol of gas (6.02×10^{23}), the ordered arrangement is found to be *extremely* improbable!

QuickLab

Roll a pair of dice 100 times and record the total number of spots appearing on the dice for each throw. Which total comes up most frequently? Is this expected?

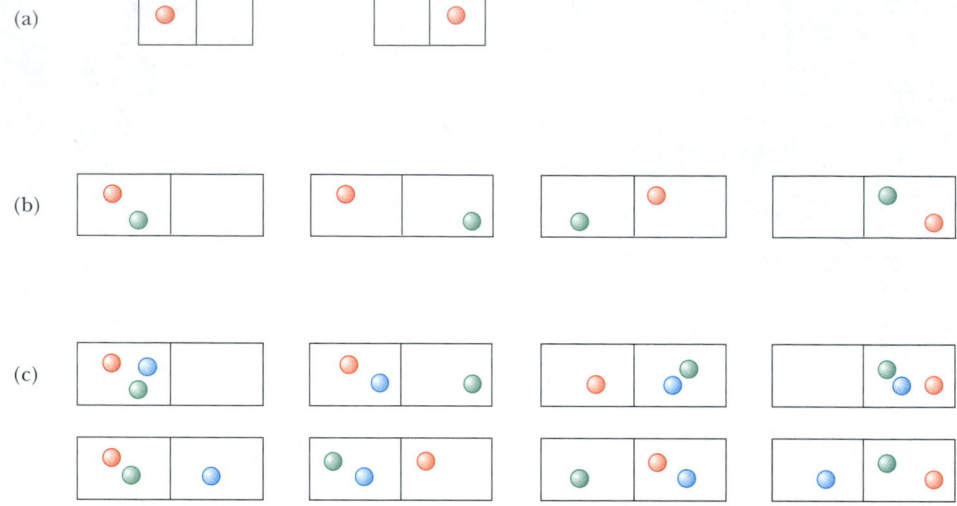

Figure 22.20 (a) One molecule in a two-sided container has a 1-in-2 chance of being on the left side. (b) Two molecules have a 1-in-4 chance of being on the left side at the same time. (c) Three molecules have a 1-in-8 chance of being on the left side at the same time.

EXAMPLE 22.10   Adiabatic Free Expansion—One Last Time

Let us verify that the macroscopic and microscopic approaches to the calculation of entropy lead to the same conclusion for the adiabatic free expansion of an ideal gas. Suppose that 1 mol of gas expands to four times its initial volume. As we have seen for this process, the initial and final temperatures are the same. (a) Using a macroscopic approach, calculate the entropy change for the gas. (b) Using statistical considerations, calculate the change in entropy for the gas and show that it agrees with the answer you obtained in part (a).

Solution   (a) Using Equation 22.13, we have

$$\Delta S = nR \ln\left(\frac{V_f}{V_i}\right) = (1) R \ln\left(\frac{4V_i}{V_i}\right) = \boxed{R \ln 4}$$

(b) The number of microstates available to a single molecule in the initial volume V_i is $w_i = V_i/V_m$. For 1 mol (N_A molecules), the number of available microstates is

$$W_i = w_i{}^{N_A} = \left(\frac{V_i}{V_m}\right)^{N_A}$$

The number of microstates for all N_A molecules in the final volume $V_f = 4V_i$ is

$$W_f = \left(\frac{V_f}{V_m}\right)^{N_A} = \left(\frac{4V_i}{V_m}\right)^{N_A}$$

Thus, the ratio of the number of final microstates to initial microstates is

$$\frac{W_f}{W_i} = 4^{N_A}$$

Using Equation 22.18, we obtain

$$\Delta S = k_B \ln W_f - k_B \ln W_i = k_B \ln\left(\frac{W_f}{W_i}\right)$$

$$= k_B \ln(4^{N_A}) = N_A k_B \ln 4 = \boxed{R \ln 4}$$

The answer is the same as that for part (a), which dealt with macroscopic parameters.

CONCEPTUAL EXAMPLE 22.11   Let's Play Marbles!

Suppose you have a bag of 100 marbles. Fifty of the marbles are red, and 50 are green. You are allowed to draw four marbles from the bag according to the following rules: Draw one marble, record its color, and return it to the bag. Then draw another marble. Continue this process until you have drawn and returned four marbles. What are the possible

macrostates for this set of events? What is the most likely macrostate? What is the least likely macrostate?

Solution   Because each marble is returned to the bag before the next one is drawn, the probability of drawing a red marble is always the same as the probability of drawing a green one. All the possible microstates and macrostates are shown in Table 22.1. As this table indicates, there is only one way to draw four red marbles, and so there is only one microstate. However, there are four possible microstates that correspond to the macrostate of one green marble and three red marbles; six microstates that correspond to two green marbles and two red marbles; four microstates that corre-spond to three green marbles and one red marble; and one microstate that corresponds to four green marbles. The most likely macrostate—two red marbles and two green marbles—corresponds to the most disordered microstates. The least likely macrostates—four red marbles or four green mar-bles—correspond to the most ordered microstates.

TABLE 22.1   Possible Results of Drawing Four Marbles from a Bag

Macrostate	Possible Microstates	Total Number of Microstates
All R	RRRR	1
1G, 3R	RRRG, RRGR, RGRR, GRRR	4
2G, 2R	RRGG, RGRG, GRRG, RGGR, GRGR, GGRR	6
3G, 1R	GGGR, GGRG, GRGG, RGGG	4
All G	GGGG	1

SUMMARY

A **heat engine** is a device that converts internal energy to other useful forms of energy. The net work done by a heat engine in carrying a working substance through a cyclic process ($\Delta E_{int} = 0$) is

$$W = Q_h - Q_c \qquad \textbf{(22.1)}$$

where Q_h is the energy absorbed from a hot reservoir and Q_c is the energy ex-pelled to a cold reservoir.

The **thermal efficiency** e of a heat engine is

$$e = \frac{W}{Q_h} = 1 - \frac{Q_c}{Q_h} \qquad \textbf{(22.2)}$$

The **second law of thermodynamics** can be stated in the following two ways:

- It is impossible to construct a heat engine that, operating in a cycle, produces no effect other than the absorption of energy from a reservoir and the perfor-mance of an equal amount of work (the Kelvin–Planck statement).
- It is impossible to construct a cyclic machine whose sole effect is the continuous transfer of energy from one object to another object at a higher temperature without the input of energy by work (the Clausius statement).

In a **reversible** process, the system can be returned to its initial conditions along the same path shown on a PV diagram, and every point along this path is an equilibrium state. A process that does not satisfy these requirements is **irre-versible**. **Carnot's theorem** states that no real heat engine operating (irre-versibly) between the temperatures T_c and T_h can be more efficient than an en-gine operating reversibly in a Carnot cycle between the same two temperatures.

The **thermal efficiency** of a heat engine operating in the Carnot cycle is

$$e_C = 1 - \frac{T_c}{T_h} \qquad \textbf{(22.4)}$$

You should be able to use this equation (or an equivalent form involving a ratio of heats) to determine the maximum possible efficiency of any heat engine.

The second law of thermodynamics states that when real (irreversible) processes occur, the degree of disorder in the system plus the surroundings increases. When a process occurs in an isolated system, the state of the system becomes more disordered. The measure of disorder in a system is called **entropy** *S*. Thus, another way in which the second law can be stated is

- The entropy of the Universe increases in all real processes.

The **change in entropy** *dS* of a system during a process between two infinitesimally separated equilibrium states is

$$dS = \frac{dQ_r}{T} \tag{22.8}$$

where dQ_r is the energy transfer by heat for a reversible process that connects the initial and final states. The change in entropy of a system during an arbitrary process between an initial state and a final state is

$$\Delta S = \int_i^f \frac{dQ_r}{T} \tag{22.9}$$

The value of ΔS for the system is the same for all paths connecting the initial and final states. The change in entropy for a system undergoing any reversible, cyclic process is zero, and when such a process occurs, the entropy of the Universe remains constant.

From a microscopic viewpoint, entropy is defined as

$$S \equiv k_B \ln W \tag{22.18}$$

where k_B is Boltzmann's constant and W is the number of microstates available to the system for the existing macrostate. Because of the statistical tendency of systems to proceed toward states of greater probability and greater disorder, all natural processes are irreversible, and entropy increases. Thus, entropy is a measure of microscopic disorder.

QUESTIONS

1. Is it possible to convert internal energy to mechanical energy? Describe a process in which such a conversion occurs.

2. What are some factors that affect the efficiency of automobile engines?

3. In practical heat engines, which are we able to control more: the temperature of the hot reservoir, or the temperature of the cold reservoir? Explain.

4. A steam-driven turbine is one major component of an electric power plant. Why is it advantageous to have the temperature of the steam as high as possible?

5. Is it possible to construct a heat engine that creates no thermal pollution? What does this tell us about environmental considerations for an industrialized society?

6. Discuss three common examples of natural processes that involve an increase in entropy. Be sure to account for all parts of each system under consideration.

7. Discuss the change in entropy of a gas that expands (a) at constant temperature and (b) adiabatically.

8. In solar ponds constructed in Israel, the Sun's energy is concentrated near the bottom of a salty pond. With the proper layering of salt in the water, convection is prevented, and temperatures of 100°C may be reached. Can you estimate the maximum efficiency with which useful energy can be extracted from the pond?

9. The vortex tube (Fig. Q22.9) is a T-shaped device that takes in compressed air at 20 atm and 20°C and gives off air at −20°C from one flared end and air at 60°C from the other flared end. Does the operation of this device vi-

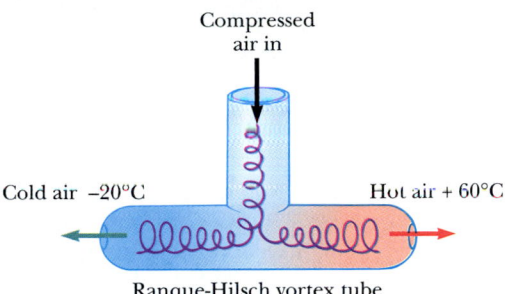

Compressed
air in

Cold air –20°C   Hot air + 60°C

Ranque-Hilsch vortex tube

Figure Q22.9

olate the second law of thermodynamics? If not, explain why not.

10. Why does your automobile burn more gas in winter than in summer?

11. Can a heat pump have a coefficient of performance (COP) less than unity? Explain.

12. Give some examples of irreversible processes that occur in nature.

13. Give an example of a process in nature that is nearly reversible.

14. A thermodynamic process occurs in which the entropy of a system changes by -8.0 J/K. According to the second law of thermodynamics, what can you conclude about the entropy change of the environment?

15. If a supersaturated sugar solution is allowed to evaporate slowly, sugar crystals form in the container. Hence, sugar molecules go from a disordered form (in solution) to a highly ordered crystalline form. Does this process violate the second law of thermodynamics? Explain.

16. How could you increase the entropy of 1 mol of a metal that is at room temperature? How could you decrease its entropy?

17. A heat pump is to be installed in a region where the average outdoor temperature in the winter months is $-20°C$. In view of this, why would it be advisable to place the outdoor compressor unit deep in the ground? Why are heat pumps not commonly used for heating in cold climates?

18. Suppose your roommate is "Mr. Clean" and tidies up your messy room after a big party. That is, your roommate is increasing order in the room. Does this represent a violation of the second law of thermodynamics?

19. Discuss the entropy changes that occur when you (a) bake a loaf of bread and (b) consume the bread.

20. The device shown in Figure Q22.20, which is called a thermoelectric converter, uses a series of semiconductor cells to convert internal energy to electrical energy. In the photograph on the left, both legs of the device are at the same temperature and no electrical energy is produced. However, when one leg is at a higher temperature than the other, as shown in the photograph on the right, electrical energy is produced as the device extracts energy from the hot reservoir and drives a small electric motor. (a) Why does the temperature differential produce electrical energy in this demonstration? (b) In what sense does this intriguing experiment demonstrate the second law of thermodynamics?

21. A classmate tells you that it is just as likely for all the air molecules in the room you are both in to be concentrated in one corner (with the rest of the room being a vacuum) as it is for the air molecules to be distributed uniformly about the room in their current state. Is this true? Why doesn't the situation he describes actually happen?

Figure Q22.20   *(Courtesy of PASCO Scientific Company)*

PROBLEMS

1, **2**, **3** = straightforward, intermediate, challenging   ☐ = full solution available in the *Student Solutions Manual and Study Guide*
WEB = solution posted at **http://www.saunderscollege.com/physics/**   💻 = Computer useful in solving problem   📲 = Interactive Physics
☐ = paired numerical/symbolic problems

Section 22.1   Heat Engines and the Second Law of Thermodynamics
Section 22.2   Reversible and Irreversible Processes

1. A heat engine absorbs 360 J of energy and performs 25.0 J of work in each cycle. Find (a) the efficiency of the engine and (b) the energy expelled to the cold reservoir in each cycle.

2. The energy absorbed by an engine is three times greater than the work it performs. (a) What is its thermal efficiency? (b) What fraction of the energy absorbed is expelled to the cold reservoir?

3. A particular engine has a power output of 5.00 kW and an efficiency of 25.0%. Assuming that the engine expels 8 000 J of energy in each cycle, find (a) the energy absorbed in each cycle and (b) the time for each cycle.

4. A heat engine performs 200 J of work in each cycle and has an efficiency of 30.0%. For each cycle, how much energy is (a) absorbed and (b) expelled?

5. An ideal gas is compressed to half its original volume while its temperature is held constant. (a) If 1 000 J of energy is removed from the gas during the compression, how much work is done on the gas? (b) What is the change in the internal energy of the gas during the compression?

6. Suppose that a heat engine is connected to two energy reservoirs, one a pool of molten aluminum (660°C) and the other a block of solid mercury (− 38.9°C). The engine runs by freezing 1.00 g of aluminum and melting 15.0 g of mercury during each cycle. The heat of fusion of aluminum is 3.97×10^5 J/kg; the heat of fusion of mercury is 1.18×10^4 J/kg. What is the efficiency of this engine?

Section 22.3   The Carnot Engine

7. One of the most efficient engines ever built (actual efficiency 42.0%) operates between 430°C and 1 870°C. (a) What is its maximum theoretical efficiency? (b) How much power does the engine deliver if it absorbs 1.40×10^5 J of energy each second from the hot reservoir?

8. A heat engine operating between 80.0°C and 200°C achieves 20.0% of the maximum possible efficiency. What energy input will enable the engine to perform 10.0 kJ of work?

9. A Carnot engine has a power output of 150 kW. The engine operates between two reservoirs at 20.0°C and 500°C. (a) How much energy does it absorb per hour? (b) How much energy is lost per hour in its exhaust?

10. A steam engine is operated in a cold climate where the exhaust temperature is 0°C. (a) Calculate the theoretical maximum efficiency of the engine, using an intake steam temperature of 100°C. (b) If superheated steam at 200°C were used instead, what would be the maximum possible efficiency?

WEB 11. An ideal gas is taken through a Carnot cycle. The isothermal expansion occurs at 250°C, and the isothermal compression takes place at 50.0°C. Assuming that the gas absorbs 1 200 J of energy from the hot reservoir during the isothermal expansion, find (a) the energy expelled to the cold reservoir in each cycle and (b) the net work done by the gas in each cycle.

12. The exhaust temperature of a Carnot heat engine is 300°C. What is the intake temperature if the efficiency of the engine is 30.0%?

13. A power plant operates at 32.0% efficiency during the summer when the sea water for cooling is at 20.0°C. The plant uses 350°C steam to drive turbines. Assuming that the plant's efficiency changes in the same proportion as the ideal efficiency, what would be the plant's efficiency in the winter, when the sea water is at 10.0°C?

14. Argon enters a turbine at a rate of 80.0 kg/min, a temperature of 800°C, and a pressure of 1.50 MPa. It expands adiabatically as it pushes on the turbine blades and exits at a pressure of 300 kPa. (a) Calculate its temperature at the time of exit. (b) Calculate the (maximum) power output of the turning turbine. (c) The turbine is one component of a model closed-cycle gas turbine engine. Calculate the maximum efficiency of the engine.

15. A power plant that would make use of the temperature gradient in the ocean has been proposed. The system is to operate between 5.00°C (water temperature at a depth of about 1 km) and 20.0°C (surface water temperature). (a) What is the maximum efficiency of such a system? (b) If the power output of the plant is 75.0 MW, how much energy is absorbed per hour? (c) In view of your answer to part (a), do you think such a system is worthwhile (considering that there is no charge for fuel)?

16. A 20.0%-efficient real engine is used to speed up a train from rest to 5.00 m/s. It is known that an ideal (Carnot) engine having the same cold and hot reservoirs would accelerate the same train from rest to a speed of 6.50 m/s using the same amount of fuel. Assuming that the engines use air at 300 K as a cold reservoir, find the temperature of the steam serving as the hot reservoir.

17. A firebox is at 750 K, and the ambient temperature is 300 K. The efficiency of a Carnot engine doing 150 J of work as it transports energy between these constant-temperature baths is 60.0%. The Carnot engine must absorb energy 150 J/0.600 = 250 J from the hot reser-

voir and release 100 J of energy into the environment. To follow Carnot's reasoning, suppose that some other heat engine S could have an efficiency of 70.0%. (a) Find the energy input and energy output of engine S as it does 150 J of work. (b) Let engine S operate as in part (a) and run the Carnot engine in reverse. Find the total energy the firebox puts out as both engines operate together and the total energy absorbed by the environment. Show that the Clausius statement of the second law of thermodynamics is violated. (c) Find the energy input and work output of engine S as it exhausts 100 J of energy. (d) Let engine S operate as in (c) and contribute 150 J of its work output to running the Carnot engine in reverse. Find the total energy that the firebox puts out as both engines operate together, the total work output, and the total energy absorbed by the environment. Show that the Kelvin–Planck statement of the second law is violated. Thus, our assumption about the efficiency of engine S must be false. (e) Let the engines operate together through one cycle as in part (d). Find the change in entropy of the Universe. Show that the entropy statement of the second law is violated.

18. At point A in a Carnot cycle, 2.34 mol of a monatomic ideal gas has a pressure of 1 400 kPa, a volume of 10.0 L, and a temperature of 720 K. It expands isothermally to point B, and then expands adiabatically to point C, where its volume is 24.0 L. An isothermal compression brings it to point D, where its new volume is 15.0 L. An adiabatic process returns the gas to point A. (a) Determine all the unknown pressures, volumes, and temperatures as you fill in the following table:

	P	V	T
A	1 400 kPa	10.0 L	720 K
B			
C		24.0 L	
D		15.0 L	

(b) Find the energy added by heat, the work done, and the change in internal energy for each of the following steps: $A \rightarrow B$, $B \rightarrow C$, $C \rightarrow D$, and $D \rightarrow A$. (c) Show that $W_{net}/Q_{in} = 1 - T_C/T_A$, the Carnot efficiency.

Section 22.4 Gasoline and Diesel Engines

19. In a cylinder of an automobile engine just after combustion, the gas is confined to a volume of 50.0 cm^3 and has an initial pressure of 3.00×10^6 Pa. The piston moves outward to a final volume of 300 cm^3, and the gas expands without energy loss by heat. (a) If $\gamma = 1.40$ for the gas, what is the final pressure? (b) How much work is done by the gas in expanding?

20. A gasoline engine has a compression ratio of 6.00 and uses a gas for which $\gamma = 1.40$. (a) What is the efficiency of the engine if it operates in an idealized Otto cycle?

(b) If the actual efficiency is 15.0%, what fraction of the fuel is wasted as a result of friction and energy losses by heat that could by avoided in a reversible engine? (Assume complete combustion of the air–fuel mixture.)

21. A 1.60-L gasoline engine with a compression ratio of 6.20 has a power output of 102 hp. Assuming that the engine operates in an idealized Otto cycle, find the energy absorbed and exhausted each second. Assume that the fuel–air mixture behaves like an ideal gas, with $\gamma = 1.40$.

22. The compression ratio of an Otto cycle, as shown in Figure 22.12, is $V_A/V_B = 8.00$. At the beginning A of the compression process, 500 cm^3 of gas is at 100 kPa and 20.0°C. At the beginning of the adiabatic expansion, the temperature is $T_C = 750$°C. Model the working fluid as an ideal gas, with $E_{int} = nC_V T = 2.50 nRT$ and $\gamma = 1.40$. (a) Fill in the following table to track the states of the gas:

	T (K)	P (kPa)	V (cm^3)	E_{int}
A	293	100	500	
B				
C	1 023			
D				
A				

(b) Fill in the following table to track the processes:

	Q	W	ΔE_{int}
$A \rightarrow B$			
$B \rightarrow C$			
$C \rightarrow D$			
$D \rightarrow A$			
$ABCDA$			

(c) Identify the energy input Q_h, the energy exhaust Q_c, and the net output work W. (d) Calculate the thermal efficiency. (e) Find the number of revolutions per minute that the crankshaft must complete for a one-cylinder engine to have an output power of 1.00 kW = 1.34 hp. (*Hint:* The thermodynamic cycle involves four piston strokes.)

Section 22.5 Heat Pumps and Refrigerators

23. What is the coefficient of performance of a refrigerator that operates with Carnot efficiency between the temperatures -3.00°C and $+27.0$°C?

24. What is the maximum possible coefficient of performance of a heat pump that brings energy from outdoors at -3.00°C into a 22.0°C house? (*Hint:* The heat pump does work W, which is also available to warm up the house.)

25. An ideal refrigerator or ideal heat pump is equivalent to a Carnot engine running in reverse. That is, energy Q_c is absorbed from a cold reservoir, and energy Q_h is rejected to a hot reservoir. (a) Show that the work that must be supplied to run the refrigerator or heat pump is

$$W = \frac{T_h - T_c}{T_c} Q_c$$

(b) Show that the coefficient of performance (COP) of the ideal refrigerator is

$$\text{COP} = \frac{T_c}{T_h - T_c}$$

26. A heat pump (Fig. P22.26) is essentially a heat engine run backward. It extracts energy from colder air outside and deposits it in a warmer room. Suppose that the ratio of the actual energy entering the room to the work done by the device's motor is 10.0% of the theoretical maximum ratio. Determine the energy entering the room per joule of work done by the motor when the inside temperature is 20.0°C and the outside temperature is − 5.00°C.

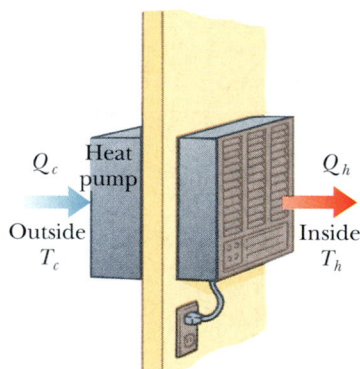

Figure P22.26

WEB **27.** How much work does an ideal Carnot refrigerator require to remove 1.00 J of energy from helium at 4.00 K and reject this energy to a room-temperature (293-K) environment?

28. How much work does an ideal Carnot refrigerator require to remove energy Q from helium at T_c and reject this energy to a room-temperature environment at T_h?

29. A refrigerator has a coefficient of performance equal to 5.00. Assuming that the refrigerator absorbs 120 J of energy from a cold reservoir in each cycle, find (a) the work required in each cycle and (b) the energy expelled to the hot reservoir.

30. A refrigerator maintains a temperature of 0°C in the cold compartment with a room temperature of 25.0°C. It removes energy from the cold compartment at the rate 8 000 kJ/h. (a) What minimum power is required

to operate the refrigerator? (b) At what rate does the refrigerator exhaust energy into the room?

Section 22.6   Entropy

31. An ice tray contains 500 g of water at 0°C. Calculate the change in entropy of the water as it freezes slowly and completely at 0°C.

32. At a pressure of 1 atm, liquid helium boils at 4.20 K. The latent heat of vaporization is 20.5 kJ/kg. Determine the entropy change (per kilogram) of the helium resulting from vaporization.

33. Calculate the change in entropy of 250 g of water heated slowly from 20.0°C to 80.0°C. (*Hint:* Note that $dQ = mc\,dT$.)

34. An airtight freezer holds 2.50 mol of air at 25.0°C and 1.00 atm. The air is then cooled to − 18.0°C. (a) What is the change in entropy of the air if the volume is held constant? (b) What would the change be if the pressure were maintained at 1 atm during the cooling?

Section 22.7   Entropy Changes in Irreversible Processes

35. The temperature at the surface of the Sun is approximately 5 700 K, and the temperature at the surface of the Earth is approximately 290 K. What entropy change occurs when 1 000 J of energy is transferred by radiation from the Sun to the Earth?

36. A 1.00-kg iron horseshoe is taken from a furnace at 900°C and dropped into 4.00 kg of water at 10.0°C. Assuming that no energy is lost by heat to the surroundings, determine the total entropy change of the system (horseshoe and water).

WEB **37.** A 1 500-kg car is moving at 20.0 m/s. The driver brakes to a stop. The brakes cool off to the temperature of the surrounding air, which is nearly constant at 20.0°C. What is the total entropy change?

38. How fast are you personally making the entropy of the Universe increase right now? Make an order-of-magnitude estimate, stating what quantities you take as data and the values you measure or estimate for them.

39. One mole of H_2 gas is contained in the left-hand side of the container shown in Figure P22.39, which has equal volumes left and right. The right-hand side is evacuated. When the valve is opened, the gas streams into the right-hand side. What is the final entropy change of the gas? Does the temperature of the gas change?

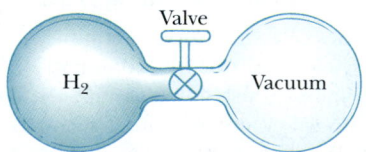

Figure P22.39

40. A rigid tank of small mass contains 40.0 g of argon, initially at 200°C and 100 kPa. The tank is placed into a reservoir at 0°C and is allowed to cool to thermal equilibrium. Calculate (a) the volume of the tank, (b) the change in internal energy of the argon, (c) the energy transferred by heat, (d) the change in entropy of the argon, and (e) the change in entropy of the constant-temperature bath.

41. A 2.00-L container has a center partition that divides it into two equal parts, as shown in Figure P22.41. The left-hand side contains H_2 gas, and the right-hand side contains O_2 gas. Both gases are at room temperature and at atmospheric pressure. The partition is removed, and the gases are allowed to mix. What is the entropy increase of the system?

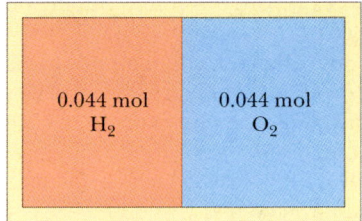

| 0.044 mol | 0.044 mol |
| H_2 | O_2 |

Figure P22.41

42. A 100 000-kg iceberg at $-5.00°C$ breaks away from the polar ice shelf and floats away into the ocean, at $5.00°C$. What is the final change in the entropy of the system after the iceberg has completely melted? (The specific heat of ice is 2010 J/kg·°C.)

43. One mole of an ideal monatomic gas, initially at a pressure of 1.00 atm and a volume of 0.025 0 m³, is heated to a final state with a pressure of 2.00 atm and a volume of 0.040 0 m³. Determine the change in entropy of the gas for this process.

44. One mole of a diatomic ideal gas, initially having pressure P and volume V, expands so as to have pressure $2P$ and volume $2V$. Determine the entropy change of the gas in the process.

(Optional)
Section 22.8   Entropy on a Microscopic Scale

45. If you toss two dice, what is the total number of ways in which you can obtain (a) a 12 and (b) a 7?

46. Prepare a table like Table 22.1 for the following occurrence. You toss four coins into the air simultaneously and then record the results of your tosses in terms of the numbers of heads and tails that result. For example, HHTH and HTHH are two possible ways in which three heads and one tail can be achieved. (a) On the basis of your table, what is the most probable result of a toss? In terms of entropy, (b) what is the most ordered state, and (c) what is the most disordered?

47. Repeat the procedure used to construct Table 22.1 (a) for the case in which you draw three marbles from your bag rather than four and (b) for the case in which you draw five rather than four.

ADDITIONAL PROBLEMS

48. Every second at Niagara Falls, some 5 000 m³ of water falls a distance of 50.0 m (Fig. P22.48). What is the increase in entropy per second due to the falling water? (Assume that the mass of the surroundings is so great that its temperature and that of the water stay nearly constant at 20.0°C. Suppose that a negligible amount of water evaporates.)

Figure P22.48   Niagara Falls.   (*Jan Kopec/Tony Stone Images*)

49. If a 35.0%-efficient Carnot heat engine is run in reverse so that it functions as a refrigerator, what would be the engine's (that is, the refrigerator's) coefficient of performance (COP)?

50. How much work does an ideal Carnot refrigerator use to change 0.500 kg of tap water at 10.0°C into ice at $-20.0°C$? Assume that the freezer compartment is held at $-20.0°C$ and that the refrigerator exhausts energy into a room at 20.0°C.

WEB 51. A house loses energy through the exterior walls and roof at a rate of 5 000 J/s = 5.00 kW when the interior temperature is 22.0°C and the outside temperature is $-5.00°C$. Calculate the electric power required to maintain the interior temperature at 22.0°C for the following two cases: (a) The electric power is used in electric resistance heaters (which convert all of the electricity supplied into internal energy). (b) The electric power is used to drive an electric motor that operates the compressor of a heat pump (which has a coefficient of performance [COP] equal to 60.0% of the Carnot-cycle value).

52. A heat engine operates between two reservoirs at $T_2 =$ 600 K and $T_1 =$ 350 K. It absorbs 1 000 J of energy from the higher-temperature reservoir and performs 250 J of work. Find (a) the entropy change of the Universe ΔS_U for this process and (b) the work W that could have been done by an ideal Carnot engine operating between these two reservoirs. (c) Show that the difference between the work done in parts (a) and (b) is $T_1 \Delta S_U$.

WEB 53. Figure P22.53 represents n mol of an ideal monatomic gas being taken through a cycle that consists of two isothermal processes at temperatures $3T_i$ and T_i and two constant-volume processes. For each cycle, determine,

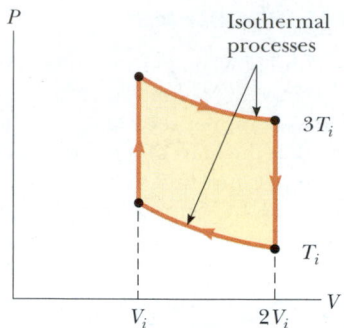

Figure P22.53

in terms of n, R, and T_i, (a) the net energy transferred by heat to the gas and (b) the efficiency of an engine operating in this cycle.

54. A refrigerator has a coefficient of performance (COP) of 3.00. The ice tray compartment is at $-20.0°C$, and the room temperature is $22.0°C$. The refrigerator can convert 30.0 g of water at $22.0°C$ to 30.0 g of ice at $-20.0°C$ each minute. What input power is required? Give your answer in watts.

55. An ideal (Carnot) freezer in a kitchen has a constant temperature of 260 K, while the air in the kitchen has a constant temperature of 300 K. Suppose that the insulation for the freezer is not perfect, such that some energy flows into the freezer at a rate of 0.150 W. Determine the average power that the freezer's motor needs to maintain the constant temperature in the freezer.

56. An electric power plant has an overall efficiency of 15.0%. The plant is to deliver 150 MW of power to a city, and its turbines use coal as the fuel. The burning coal produces steam, which drives the turbines. The steam is then condensed to water at $25.0°C$ as it passes through cooling coils in contact with river water. (a) How many metric tons of coal does the plant consume each day (1 metric ton = 10^3 kg)? (b) What is the total cost of the fuel per year if the delivered price is $8.00/metric ton? (c) If the river water is delivered at $20.0°C$, at what minimum rate must it flow over the cooling coils in order that its temperature not exceed $25.0°C$? (*Note:* The heat of combustion of coal is 33.0 kJ/g.)

57. A power plant, having a Carnot efficiency, produces 1 000 MW of electrical power from turbines that take in steam at 500 K and reject water at 300 K into a flowing river. Assuming that the water downstream is 6.00 K warmer due to the output of the power plant, determine the flow rate of the river.

58. A power plant, having a Carnot efficiency, produces electric power $\mathcal{P}$ from turbines that take in energy from steam at temperature T_h and discharge energy at temperature T_c through a heat exchanger into a flowing river. Assuming that the water downstream is warmer by ΔT due to the output of the power plant, determine the flow rate of the river.

59. An athlete whose mass is 70.0 kg drinks 16 oz (453.6 g) of refrigerated water. The water is at a temperature of $35.0°F$. (a) Neglecting the temperature change of her body that results from the water intake (that is, the body is regarded as a reservoir that is always at $98.6°F$), find the entropy increase of the entire system. (b) Assume that the entire body is cooled by the drink and that the average specific heat of a human is equal to the specific heat of liquid water. Neglecting any other energy transfers by heat and any metabolic energy release, find the athlete's temperature after she drinks the cold water, given an initial body temperature of $98.6°F$. Under *these* assumptions, what is the entropy increase of the entire system? Compare this result with the one you obtained in part (a).

60. One mole of an ideal monatomic gas is taken through the cycle shown in Figure P22.60. The process $A \rightarrow B$ is a reversible isothermal expansion. Calculate (a) the net work done by the gas, (b) the energy added to the gas, (c) the energy expelled by the gas, and (d) the efficiency of the cycle.

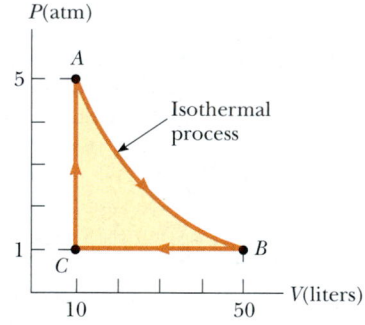

Figure P22.60

61. Calculate the increase in entropy of the Universe when you add 20.0 g of $5.00°C$ cream to 200 g of $60.0°C$ coffee. Assume that the specific heats of cream and coffee are both 4.20 J/g·°C.

62. In 1993 the federal government instituted a requirement that all room air conditioners sold in the United States must have an energy efficiency ratio (EER) of 10 or higher. The EER is defined as the ratio of the cooling capacity of the air conditioner, measured in Btu/h, to its electrical power requirement in watts. (a) Convert the EER of 10.0 to dimensionless form, using the conversion 1 Btu = 1 055 J. (b) What is the appropriate name for this dimensionless quantity? (c) In the 1970s it was common to find room air conditioners with EERs of 5 or lower. Compare the operating costs for 10 000-Btu/h air conditioners with EERs of 5.00 and 10.0 if each air conditioner were to operate for 1 500 h during the summer in a city where electricity costs 10.0¢ per kilowatt-hour.

63. One mole of a monatomic ideal gas is taken through the cycle shown in Figure P22.63. At point A, the pressure, volume, and temperature are P_i, V_i, and T_i, respectively. In terms of R and T_i, find (a) the total energy entering the system by heat per cycle, (b) the total energy leaving the system by heat per cycle, (c) the efficiency of an engine operating in this cycle, and (d) the efficiency of an engine operating in a Carnot cycle between the same temperature extremes.

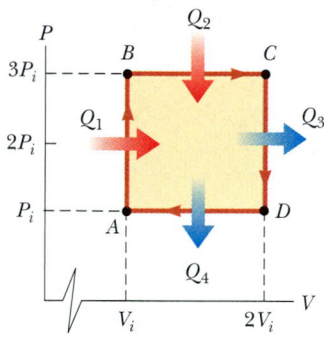

Figure P22.63

64. One mole of an ideal gas expands isothermally. (a) If the gas doubles its volume, show that the work of expansion is $W = RT \ln 2$. (b) Because the internal energy E_{int} of an ideal gas depends solely on its temperature, no change in E_{int} occurs during the expansion. It follows from the first law that the heat input to the gas during the expansion is equal to the energy output by work. Why does this conversion *not* violate the second law?

65. A system consisting of n mol of an ideal gas undergoes a reversible, *isobaric* process from a volume V_i to a volume $3V_i$. Calculate the change in entropy of the gas. (*Hint:* Imagine that the system goes from the initial state to the final state first along an isotherm and then along an adiabatic path—no change in entropy occurs along the adiabatic path.)

66. Suppose you are working in a patent office, and an inventor comes to you with the claim that her heat engine, which employs water as a working substance, has a

thermodynamic efficiency of 0.61. She explains that it operates between energy reservoirs at 4°C and 0°C. It is a very complicated device, with many pistons, gears, and pulleys, and the cycle involves freezing and melting. Does her claim that $e = 0.61$ warrant serious consideration? Explain.

67. An idealized diesel engine operates in a cycle known as the *air-standard diesel cycle*, as shown in Figure 22.13. Fuel is sprayed into the cylinder at the point of maximum compression B. Combustion occurs during the expansion $B \rightarrow C$, which is approximated as an isobaric process. Show that the efficiency of an engine operating in this idealized diesel cycle is

$$e = 1 - \frac{1}{\gamma}\left(\frac{T_D - T_A}{T_C - T_B}\right)$$

68. One mole of an ideal gas ($\gamma = 1.40$) is carried through the Carnot cycle described in Figure 22.10. At point A, the pressure is 25.0 atm and the temperature is 600 K. At point C, the pressure is 1.00 atm and the temperature is 400 K. (a) Determine the pressures and volumes at points A, B, C, and D. (b) Calculate the net work done per cycle. (c) Determine the efficiency of an engine operating in this cycle.

69. A typical human has a mass of 70.0 kg and produces about 2 000 kcal (2.00×10^6 cal) of metabolic energy per day. (a) Find the rate of metabolic energy production in watts and in calories per hour. (b) If none of the metabolic energy were transferred out of the body, and the specific heat of the human body is 1.00 cal/g · °C, what is the rate at which body temperature would rise? Give your answer in degrees Celsius per hour and in degrees Fahrenheit per hour.

70. Suppose that 1.00 kg of water at 10.0°C is mixed with 1.00 kg of water at 30.0°C at constant pressure. When the mixture has reached equilibrium, (a) what is the final temperature? (b) Take $c_P = 4.19$ kJ/kg · K for water. Show that the entropy of the system increases by

$$\Delta S = 4.19 \ln\left[\left(\frac{293}{283}\right)\left(\frac{293}{303}\right)\right] \text{kJ/K}$$

(c) Verify numerically that $\Delta S > 0$. (d) Is the mixing an irreversible process?

ANSWERS TO QUICK QUIZZES

22.1 The cost of heating your home decreases to 25% of the original cost. With electric heating, you receive the same amount of energy for heating your home as enters it by electricity. The COP of 4 for the heat pump means that you are receiving four times as much energy as the energy entering by electricity. With four times as much energy per unit of energy from electricity, you need only one-fourth as much electricity.

22.2 (b) Because the process is reversible and adiabatic, $Q_r = 0$; therefore, $\Delta S = 0$.

22.3 False. The second law states that the entropy of the *Universe* increases in real processes. Although the organization of molecules into ordered leaves and branches represents a decrease in entropy *of the tree*, this organization takes place because of a number of processes in which the tree interacts with its surroundings. If we include the entropy changes associated with all these processes, the entropy change of the Universe during the growth of a tree is still positive.

part 4

Electricity and Magnetism

We now study the branch of physics concerned with electrical and magnetic phenomena. The laws of electricity and magnetism have a central role in the operation of such devices as radios, televisions, electric motors, computers, high-energy accelerators, and other electronic devices. More fundamentally, the interatomic and intermolecular forces responsible for the formation of solids and liquids are electric in origin. Furthermore, such forces as the pushes and pulls between objects and the elastic force in a spring arise from electric forces at the atomic level.

Evidence in Chinese documents suggests that magnetism was known about as early as 2000 B.C. The ancient Greeks observed electrical and magnetic phenomena possibly as early as 700 B.C. They found that a piece of amber, when rubbed, becomes electrified and attracts pieces of straw or feathers. The Greeks knew about magnetic forces from obser-

vations that the naturally occurring stone *magnetite* (Fe_3O_4) is attracted to iron. (The word *electric* comes from *elecktron,* the Greek word for "amber." The word *magnetic* comes from *Magnesia,* the name of the district of Greece where magnetite was first found.)

In 1600, the Englishman William Gilbert discovered that electrification is not limited to amber but rather is a general phenomenon. In the years following this discovery, scientists electrified a variety of objects, including chickens and people! Experiments by Charles Coulomb in 1785 confirmed the inverse-square law for electric forces.

It was not until the early part of the 19th century that scientists established that electricity and magnetism are related phenomena. In 1819, Hans Oersted discovered that a compass needle is deflected when placed near a circuit carrying an electric current. In 1831, Michael Faraday and, almost simultaneously,

Joseph Henry showed that when a wire is moved near a magnet (or, equivalently, when a magnet is moved near a wire), an electric current is established in the wire. In 1873, James Clerk Maxwell used these observations and other experimental facts as a basis for formulating the laws of electromagnetism as we know them today. (*Electromagnetism* is a name given to the combined fields of electricity and magnetism.) Shortly thereafter (around 1888), Heinrich Hertz verified Maxwell's predictions by producing electromagnetic waves in the laboratory. This achievement led to such practical developments as radio and television.

Maxwell's contributions to the field of electromagnetism were especially significant because the laws he formulated are basic to *all* forms of electromagnetic phenomena. His work is as important as Newton's work on the laws of motion and the theory of gravitation.

◀ Paul and Lindamarie Ambrose/FPG International

chapter

23

Electric Fields

*T*he electromagnetic force between charged particles is one of the fundamental forces of nature. We begin this chapter by describing some of the basic properties of electric forces. We then discuss Coulomb's law, which is the fundamental law governing the force between any two charged particles. Next, we introduce the concept of an electric field associated with a charge distribution and describe its effect on other charged particles. We then show how to use Coulomb's law to calculate the electric field for a given charge distribution. We conclude the chapter with a discussion of the motion of a charged particle in a uniform electric field.

23.1 PROPERTIES OF ELECTRIC CHARGES

11.2   A number of simple experiments demonstrate the existence of electric forces and charges. For example, after running a comb through your hair on a dry day, you will find that the comb attracts bits of paper. The attractive force is often strong enough to suspend the paper. The same effect occurs when materials such as glass or rubber are rubbed with silk or fur.

Another simple experiment is to rub an inflated balloon with wool. The balloon then adheres to a wall, often for hours. When materials behave in this way, they are said to be *electrified,* or to have become **electrically charged.** You can easily electrify your body by vigorously rubbing your shoes on a wool rug. The electric charge on your body can be felt and removed by lightly touching (and startling) a friend. Under the right conditions, you will see a spark when you touch, and both of you will feel a slight tingle. (Experiments such as these work best on a dry day because an excessive amount of moisture in the air can cause any charge you build up to "leak" from your body to the Earth.)

In a series of simple experiments, it is found that there are two kinds of electric charges, which were given the names **positive** and **negative** by Benjamin Franklin (1706–1790). To verify that this is true, consider a hard rubber rod that has been rubbed with fur and then suspended by a nonmetallic thread, as shown in Figure 23.1. When a glass rod that has been rubbed with silk is brought near the rubber rod, the two attract each other (Fig. 23.1a). On the other hand, if two charged rubber rods (or two charged glass rods) are brought near each other, as shown in Figure 23.1b, the two repel each other. This observation shows that the rubber and glass are in two different states of electrification. On the basis of these observations, we conclude that **like charges repel one another and unlike charges attract one another.**

Using the convention suggested by Franklin, the electric charge on the glass rod is called positive and that on the rubber rod is called negative. Therefore, any charged object attracted to a charged rubber rod (or repelled by a charged glass rod) must have a positive charge, and any charged object repelled by a charged rubber rod (or attracted to a charged glass rod) must have a negative charge.

Attractive electric forces are responsible for the behavior of a wide variety of commercial products. For example, the plastic in many contact lenses, *etafilcon,* is made up of molecules that electrically attract the protein molecules in human tears. These protein molecules are absorbed and held by the plastic so that the lens ends up being primarily composed of the wearer's tears. Because of this, the wearer's eye does not treat the lens as a foreign object, and it can be worn comfortably. Many cosmetics also take advantage of electric forces by incorporating materials that are electrically attracted to skin or hair, causing the pigments or other chemicals to stay put once they are applied.

QuickLab

Rub an inflated balloon against your hair and then hold the balloon near a thin stream of water running from a faucet. What happens? (A rubbed plastic pen or comb will also work.)

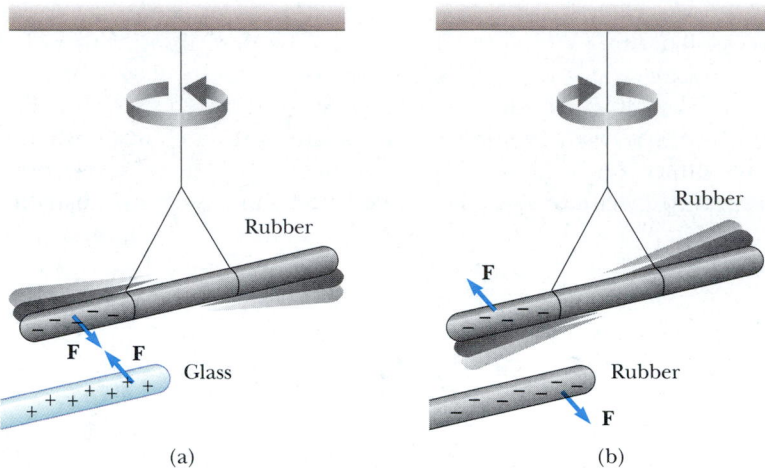

Figure 23.1   (a) A negatively charged rubber rod suspended by a thread is attracted to a positively charged glass rod. (b) A negatively charged rubber rod is repelled by another negatively charged rubber rod.

Charge is conserved

Another important aspect of Franklin's model of electricity is the implication that **electric charge is always conserved.** That is, when one object is rubbed against another, charge is not created in the process. The electrified state is due to a *transfer* of charge from one object to the other. One object gains some amount of negative charge while the other gains an equal amount of positive charge. For example, when a glass rod is rubbed with silk, the silk obtains a negative charge that is equal in magnitude to the positive charge on the glass rod. We now know from our understanding of atomic structure that negatively charged electrons are transferred from the glass to the silk in the rubbing process. Similarly, when rubber is rubbed with fur, electrons are transferred from the fur to the rubber, giving the rubber a net negative charge and the fur a net positive charge. This process is consistent with the fact that neutral, uncharged matter contains as many positive charges (protons within atomic nuclei) as negative charges (electrons).

Quick Quiz 23.1

If you rub an inflated balloon against your hair, the two materials attract each other, as shown in Figure 23.2. Is the amount of charge present in the balloon and your hair after rubbing (a) less than, (b) the same as, or (c) more than the amount of charge present before rubbing?

Figure 23.2   Rubbing a balloon against your hair on a dry day causes the balloon and your hair to become charged.   *(Charles D. Winters).*

Charge is quantized

In 1909, Robert Millikan (1868–1953) discovered that electric charge always occurs as some integral multiple of a fundamental amount of charge e. In modern terms, the electric charge q is said to be **quantized,** where q is the standard symbol used for charge. That is, electric charge exists as discrete "packets," and we can write $q = Ne$, where N is some integer. Other experiments in the same period showed that the electron has a charge $-e$ and the proton has a charge of equal magnitude but opposite sign $+e$. Some particles, such as the neutron, have no charge. A neutral atom must contain as many protons as electrons.

Because charge is a conserved quantity, the net charge in a closed region remains the same. If charged particles are created in some process, they are always created in pairs whose members have equal-magnitude charges of opposite sign.

From our discussion thus far, we conclude that electric charge has the following important properties:

- Two kinds of charges occur in nature, with the property that unlike charges attract one another and like charges repel one another.
- Charge is conserved.
- Charge is quantized.

Properties of electric charge

23.2 INSULATORS AND CONDUCTORS

It is convenient to classify substances in terms of their ability to conduct electric charge:

> Electrical **conductors** are materials in which electric charges move freely, whereas electrical **insulators** are materials in which electric charges cannot move freely.

Materials such as glass, rubber, and wood fall into the category of electrical insulators. When such materials are charged by rubbing, only the area rubbed becomes charged, and the charge is unable to move to other regions of the material.

In contrast, materials such as copper, aluminum, and silver are good electrical conductors. When such materials are charged in some small region, the charge readily distributes itself over the entire surface of the material. If you hold a copper rod in your hand and rub it with wool or fur, it will not attract a small piece of paper. This might suggest that a metal cannot be charged. However, if you attach a wooden handle to the rod and then hold it by that handle as you rub the rod, the rod will remain charged and attract the piece of paper. The explanation for this is as follows: Without the insulating wood, the electric charges produced by rubbing readily move from the copper through your body and into the Earth. The insulating wooden handle prevents the flow of charge into your hand.

Metals are good conductors

Semiconductors are a third class of materials, and their electrical properties are somewhere between those of insulators and those of conductors. Silicon and germanium are well-known examples of semiconductors commonly used in the fabrication of a variety of electronic devices, such as transistors and light-emitting diodes. The electrical properties of semiconductors can be changed over many orders of magnitude by the addition of controlled amounts of certain atoms to the materials.

When a conductor is connected to the Earth by means of a conducting wire or pipe, it is said to be **grounded.** The Earth can then be considered an infinite "sink" to which electric charges can easily migrate. With this in mind, we can understand how to charge a conductor by a process known as **induction.**

To understand induction, consider a neutral (uncharged) conducting sphere insulated from ground, as shown in Figure 23.3a. When a negatively charged rubber rod is brought near the sphere, the region of the sphere nearest the rod obtains an excess of positive charge while the region farthest from the rod obtains an equal excess of negative charge, as shown in Figure 23.3b. (That is, electrons in the region nearest the rod migrate to the opposite side of the sphere. This occurs even if the rod never actually touches the sphere.) If the same experiment is performed with a conducting wire connected from the sphere to ground (Fig. 23.3c), some of the electrons in the conductor are so strongly repelled by the presence of

Charging by induction

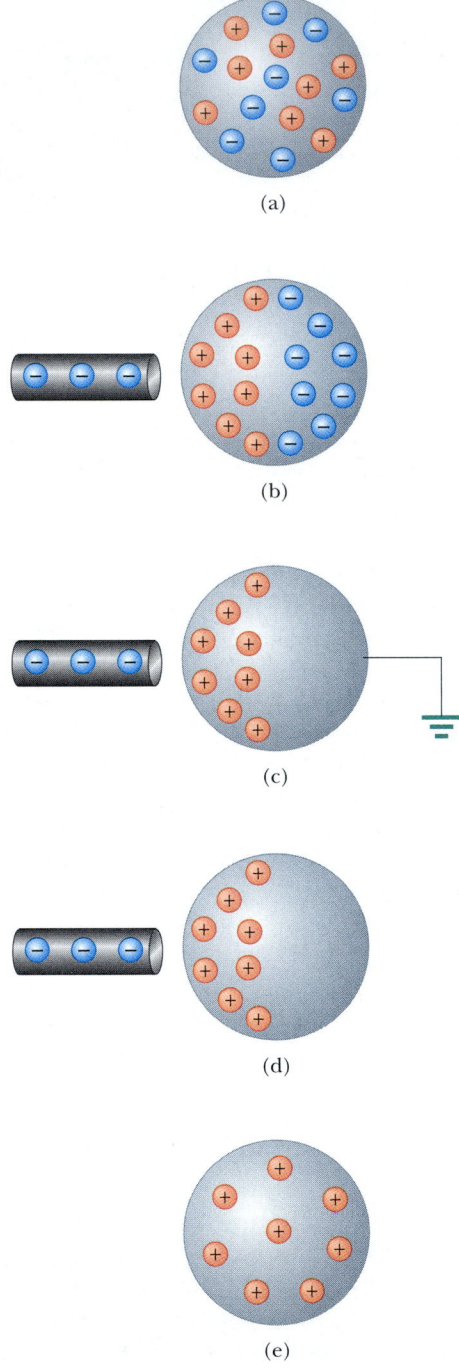

Figure 23.3   Charging a metallic object by *induction* (that is, the two objects never touch each other). (a) A neutral metallic sphere, with equal numbers of positive and negative charges. (b) The charge on the neutral sphere is redistributed when a charged rubber rod is placed near the sphere. (c) When the sphere is grounded, some of its electrons leave through the ground wire. (d) When the ground connection is removed, the sphere has excess positive charge that is nonuniformly distributed. (e) When the rod is removed, the excess positive charge becomes uniformly distributed over the surface of the sphere.

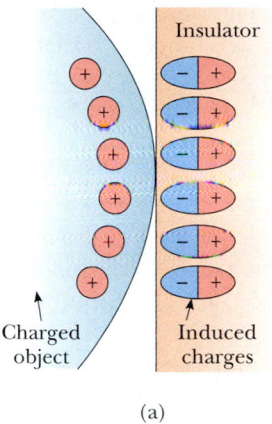

(a)   (b)

Figure 23.4   (a) The charged object on the left induces charges on the surface of an insulator. (b) A charged comb attracts bits of paper because charges are displaced in the paper. *(© 1968 Fundamental Photographs)*

the negative charge in the rod that they move out of the sphere through the ground wire and into the Earth. If the wire to ground is then removed (Fig. 23.3d), the conducting sphere contains an excess of *induced* positive charge. When the rubber rod is removed from the vicinity of the sphere (Fig. 23.3e), this induced positive charge remains on the ungrounded sphere. Note that the charge remaining on the sphere is uniformly distributed over its surface because of the repulsive forces among the like charges. Also note that the rubber rod loses none of its negative charge during this process.

Charging an object by induction requires no contact with the body inducing the charge. This is in contrast to charging an object by rubbing (that is, by *conduction*), which does require contact between the two objects.

A process similar to induction in conductors takes place in insulators. In most neutral molecules, the center of positive charge coincides with the center of negative charge. However, in the presence of a charged object, these centers inside each molecule in an insulator may shift slightly, resulting in more positive charge on one side of the molecule than on the other. This realignment of charge within individual molecules produces an induced charge on the surface of the insulator, as shown in Figure 23.4. Knowing about induction in insulators, you should be able to explain why a comb that has been rubbed through hair attracts bits of electrically neutral paper and why a balloon that has been rubbed against your clothing is able to stick to an electrically neutral wall.

Quick Quiz 23.2

Object A is attracted to object B. If object B is known to be positively charged, what can we say about object A? (a) It is positively charged. (b) It is negatively charged. (c) It is electrically neutral. (d) Not enough information to answer.

23.3 **COULOMB'S LAW**

Charles Coulomb (1736–1806) measured the magnitudes of the electric forces between charged objects using the torsion balance, which he invented (Fig. 23.5).

11.4

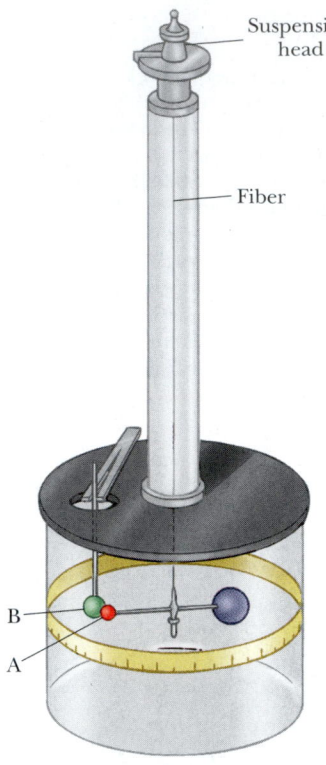

Suspension head

Fiber

B

A

Figure 23.5 Coulomb's torsion balance, used to establish the inverse-square law for the electric force between two charges.

Coulomb confirmed that the electric force between two small charged spheres is proportional to the inverse square of their separation distance r—that is, $F_e \propto 1/r^2$. The operating principle of the torsion balance is the same as that of the apparatus used by Cavendish to measure the gravitational constant (see Section 14.2), with the electrically neutral spheres replaced by charged ones. The electric force between charged spheres A and B in Figure 23.5 causes the spheres to either attract or repel each other, and the resulting motion causes the suspended fiber to twist. Because the restoring torque of the twisted fiber is proportional to the angle through which the fiber rotates, a measurement of this angle provides a quantitative measure of the electric force of attraction or repulsion. Once the spheres are charged by rubbing, the electric force between them is very large compared with the gravitational attraction, and so the gravitational force can be neglected.

Coulomb's experiments showed that the **electric force** between two stationary charged particles

- is inversely proportional to the square of the separation r between the particles and directed along the line joining them;
- is proportional to the product of the charges q_1 and q_2 on the two particles;
- is attractive if the charges are of opposite sign and repulsive if the charges have the same sign.

From these observations, we can express **Coulomb's law** as an equation giving the magnitude of the electric force (sometimes called the *Coulomb force*) between two point charges:

$$F_e = k_e \frac{|q_1||q_2|}{r^2} \tag{23.1}$$

where k_e is a constant called the **Coulomb constant.** In his experiments, Coulomb was able to show that the value of the exponent of r was 2 to within an uncertainty of a few percent. Modern experiments have shown that the exponent is 2 to within an uncertainty of a few parts in 10^{16}.

The value of the Coulomb constant depends on the choice of units. The SI unit of charge is the **coulomb** (C). The Coulomb constant k_e in SI units has the value

$$k_e = 8.987\,5 \times 10^9\ \text{N}\cdot\text{m}^2/\text{C}^2$$

Coulomb constant

This constant is also written in the form

$$k_e = \frac{1}{4\pi\epsilon_0}$$

where the constant ϵ_0 (lowercase Greek epsilon) is known as the *permittivity of free space* and has the value $8.854\,2 \times 10^{-12}\ \text{C}^2/\text{N}\cdot\text{m}^2$.

The smallest unit of charge known in nature is the charge on an electron or proton,[1] which has an absolute value of

Charge on an electron or proton

$$|e| = 1.602\,19 \times 10^{-19}\ \text{C}$$

Therefore, 1 C of charge is approximately equal to the charge of 6.24×10^{18} electrons or protons. This number is very small when compared with the number of

[1] No unit of charge smaller than e has been detected as a free charge; however, recent theories propose the existence of particles called *quarks* having charges $e/3$ and $2e/3$. Although there is considerable experimental evidence for such particles inside nuclear matter, *free* quarks have never been detected. We discuss other properties of quarks in Chapter 46 of the extended version of this text.

TABLE 23.1 **Charge and Mass of the Electron, Proton, and Neutron**

Particle	Charge (C)	Mass (kg)
Electron (e)	$-1.602\ 191\ 7 \times 10^{-19}$	$9.109\ 5 \times 10^{-31}$
Proton (p)	$+1.602\ 191\ 7 \times 10^{-19}$	$1.672\ 61 \times 10^{-27}$
Neutron (n)	0	$1.674\ 92 \times 10^{-27}$

free electrons[2] in 1 cm^3 of copper, which is of the order of 10^{23}. Still, 1 C is a substantial amount of charge. In typical experiments in which a rubber or glass rod is charged by friction, a net charge of the order of 10^{-6} C is obtained. In other words, only a very small fraction of the total available charge is transferred between the rod and the rubbing material.

The charges and masses of the electron, proton, and neutron are given in Table 23.1.

EXAMPLE 23.1 ▶ **The Hydrogen Atom**

The electron and proton of a hydrogen atom are separated (on the average) by a distance of approximately 5.3×10^{-11} m. Find the magnitudes of the electric force and the gravitational force between the two particles.

Solution From Coulomb's law, we find that the attractive electric force has the magnitude

$$F_e = k_e \frac{|e|^2}{r^2} = \left(8.99 \times 10^9\ \frac{\text{N} \cdot \text{m}^2}{\text{C}^2}\right) \frac{(1.60 \times 10^{-19}\ \text{C})^2}{(5.3 \times 10^{-11}\ \text{m})^2}$$

$$= \boxed{8.2 \times 10^{-8}\ \text{N}}$$

Using Newton's law of gravitation and Table 23.1 for the particle masses, we find that the gravitational force has the magnitude

$$F_g = G \frac{m_e m_p}{r^2}$$

$$= \left(6.7 \times 10^{-11}\ \frac{\text{N} \cdot \text{m}^2}{\text{kg}^2}\right)$$

$$\times \frac{(9.11 \times 10^{-31}\ \text{kg})(1.67 \times 10^{-27}\ \text{kg})}{(5.3 \times 10^{-11}\ \text{m})^2}$$

$$= \boxed{3.6 \times 10^{-47}\ \text{N}}$$

The ratio $F_e/F_g \approx 2 \times 10^{39}$. Thus, the gravitational force between charged atomic particles is negligible when compared with the electric force. Note the similarity of form of Newton's law of gravitation and Coulomb's law of electric forces. Other than magnitude, what is a fundamental difference between the two forces?

When dealing with Coulomb's law, you must remember that force is a vector quantity and must be treated accordingly. Thus, the law expressed in vector form for the electric force exerted by a charge q_1 on a second charge q_2, written $\mathbf{F}_{12}$, is

$$\mathbf{F}_{12} = k_e \frac{q_1 q_2}{r^2} \hat{\mathbf{r}} \tag{23.2}$$

where $\hat{\mathbf{r}}$ is a unit vector directed from q_1 to q_2, as shown in Figure 23.6a. Because the electric force obeys Newton's third law, the electric force exerted by q_2 on q_1 is

---

[2] A metal atom, such as copper, contains one or more outer electrons, which are weakly bound to the nucleus. When many atoms combine to form a metal, the so-called *free electrons* are these outer electrons, which are not bound to any one atom. These electrons move about the metal in a manner similar to that of gas molecules moving in a container.

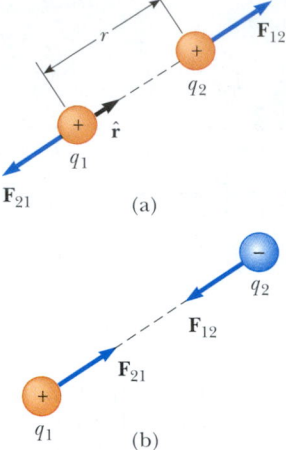

Figure 23.6 Two point charges separated by a distance r exert a force on each other that is given by Coulomb's law. The force $\mathbf{F}_{21}$ exerted by q_2 on q_1 is equal in magnitude and opposite in direction to the force $\mathbf{F}_{12}$ exerted by q_1 on q_2. (a) When the charges are of the same sign, the force is repulsive. (b) When the charges are of opposite signs, the force is attractive.

equal in magnitude to the force exerted by q_1 on q_2 and in the opposite direction; that is, $\mathbf{F}_{21} = -\mathbf{F}_{12}$. Finally, from Equation 23.2, we see that if q_1 and q_2 have the same sign, as in Figure 23.6a, the product $q_1 q_2$ is positive and the force is repulsive. If q_1 and q_2 are of opposite sign, as shown in Figure 23.6b, the product $q_1 q_2$ is negative and the force is attractive. Noting the sign of the product $q_1 q_2$ is an easy way of determining the direction of forces acting on the charges.

Quick Quiz 23.3

Object A has a charge of $+2\ \mu C$, and object B has a charge of $+6\ \mu C$. Which statement is true?

(a) $\mathbf{F}_{AB} = -3\mathbf{F}_{BA}$.   (b) $\mathbf{F}_{AB} = -\mathbf{F}_{BA}$.   (c) $3\mathbf{F}_{AB} = -\mathbf{F}_{BA}$.

When more than two charges are present, the force between any pair of them is given by Equation 23.2. Therefore, the resultant force on any one of them equals the vector sum of the forces exerted by the various individual charges. For example, if four charges are present, then the resultant force exerted by particles 2, 3, and 4 on particle 1 is

$$\mathbf{F}_1 = \mathbf{F}_{21} + \mathbf{F}_{31} + \mathbf{F}_{41}$$

EXAMPLE 23.2   Find the Resultant Force

Consider three point charges located at the corners of a right triangle as shown in Figure 23.7, where $q_1 = q_3 = 5.0\ \mu C$, $q_2 = -2.0\ \mu C$, and $a = 0.10$ m. Find the resultant force exerted on q_3.

Solution   First, note the direction of the individual forces exerted by q_1 and q_2 on q_3. The force F_{23} exerted by q_2 on q_3 is attractive because q_2 and q_3 have opposite signs. The force F_{13} exerted by q_1 on q_3 is repulsive because both charges are positive.

The magnitude of $\mathbf{F}_{23}$ is

$$F_{23} = k_e \frac{|q_2||q_3|}{a^2}$$

$$= \left(8.99 \times 10^9\ \frac{\text{N}\cdot\text{m}^2}{\text{C}^2}\right) \frac{(2.0 \times 10^{-6}\ \text{C})(5.0 \times 10^{-6}\ \text{C})}{(0.10\ \text{m})^2}$$

$$= 9.0\ \text{N}$$

Note that because q_3 and q_2 have opposite signs, $\mathbf{F}_{23}$ is to the left, as shown in Figure 23.7.

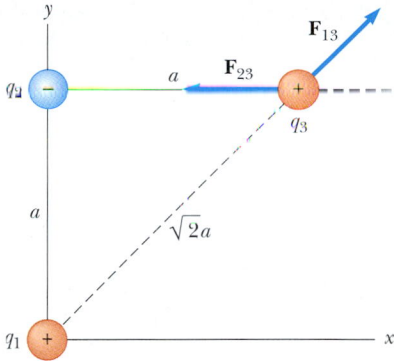

Figure 23.7   The force exerted by q_1 on q_3 is $\mathbf{F}_{13}$. The force exerted by q_2 on q_3 is $\mathbf{F}_{23}$. The resultant force $\mathbf{F}_3$ exerted on q_3 is the vector sum $\mathbf{F}_{13} + \mathbf{F}_{23}$.

The magnitude of the force exerted by q_1 on q_3 is

$$F_{13} = k_e \frac{|q_1||q_3|}{(\sqrt{2}a)^2}$$

$$= \left(8.99 \times 10^9 \; \frac{\text{N} \cdot \text{m}^2}{\text{C}^2}\right) \frac{(5.0 \times 10^{-6}\,\text{C})(5.0 \times 10^{-6}\,\text{C})}{2(0.10\,\text{m})^2}$$

$$= 11\,\text{N}$$

The force $\mathbf{F}_{13}$ is repulsive and makes an angle of 45° with the x axis. Therefore, the x and y components of $\mathbf{F}_{13}$ are equal, with magnitude given by $F_{13} \cos 45° = 7.9\,\text{N}$.

The force $\mathbf{F}_{23}$ is in the negative x direction. Hence, the x and y components of the resultant force acting on q_3 are

$$F_{3x} = F_{13x} + F_{23} = 7.9\,\text{N} - 9.0\,\text{N} = -1.1\,\text{N}$$

$$F_{3y} = F_{13y} = 7.9\,\text{N}$$

We can also express the resultant force acting on q_3 in unit-vector form as

$$\mathbf{F}_3 = \boxed{(-1.1\mathbf{i} + 7.9\mathbf{j})\,\text{N}}$$

Exercise   Find the magnitude and direction of the resultant force $\mathbf{F}_3$.

Answer   8.0 N at an angle of 98° with the x axis.

EXAMPLE 23.3 ▶ Where Is the Resultant Force Zero?

Three point charges lie along the x axis as shown in Figure 23.8. The positive charge $q_1 = 15.0\;\mu\text{C}$ is at $x = 2.00\,\text{m}$, the positive charge $q_2 = 6.00\;\mu\text{C}$ is at the origin, and the resultant force acting on q_3 is zero. What is the x coordinate of q_3?

Solution   Because q_3 is negative and q_1 and q_2 are positive, the forces $\mathbf{F}_{13}$ and $\mathbf{F}_{23}$ are both attractive, as indicated in Figure 23.8. From Coulomb's law, $\mathbf{F}_{13}$ and $\mathbf{F}_{23}$ have magnitudes

$$F_{13} = k_e \frac{|q_1||q_3|}{(2.00 - x)^2} \qquad F_{23} = k_e \frac{|q_2||q_3|}{x^2}$$

For the resultant force on q_3 to be zero, $\mathbf{F}_{23}$ must be equal in magnitude and opposite in direction to $\mathbf{F}_{13}$, or

$$k_e \frac{|q_2||q_3|}{x^2} = k_e \frac{|q_1||q_3|}{(2.00 - x)^2}$$

Noting that k_e and q_3 are common to both sides and so can be dropped, we solve for x and find that

$$(2.00 - x)^2 |q_2| = x^2 |q_1|$$

$$(4.00 - 4.00x + x^2)(6.00 \times 10^{-6}\,\text{C}) = x^2(15.0 \times 10^{-6}\,\text{C})$$

Solving this quadratic equation for x, we find that

$$\boxed{x = 0.775\,\text{m}.}$$ Why is the negative root not acceptable?

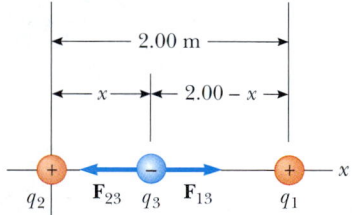

Figure 23.8   Three point charges are placed along the x axis. If the net force acting on q_3 is zero, then the force $\mathbf{F}_{13}$ exerted by q_1 on q_3 must be equal in magnitude and opposite in direction to the force $\mathbf{F}_{23}$ exerted by q_2 on q_3.

EXAMPLE 23.4 ▶ Find the Charge on the Spheres

Two identical small charged spheres, each having a mass of $3.0 \times 10^{-2}\,\text{kg}$, hang in equilibrium as shown in Figure 23.9a. The length of each string is 0.15 m, and the angle θ is 5.0°. Find the magnitude of the charge on each sphere.

Solution   From the right triangle shown in Figure 23.9a,

we see that $\sin\theta = a/L$. Therefore,

$$a = L \sin\theta = (0.15\,\text{m})\sin 5.0° = 0.013\,\text{m}$$

The separation of the spheres is $2a = 0.026\,\text{m}$.

The forces acting on the left sphere are shown in Figure 23.9b. Because the sphere is in equilibrium, the forces in the horizontal and vertical directions must separately add up to zero:

(1) $\sum F_x = T \sin \theta - F_e = 0$

(2) $\sum F_y = T \cos \theta - mg = 0$

From Equation (2), we see that $T = mg / \cos \theta$; thus, T can be

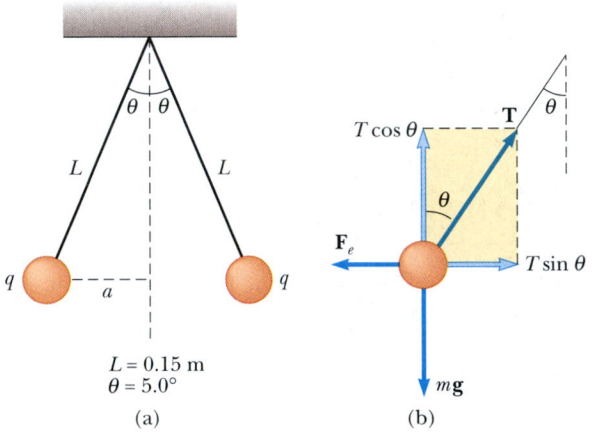

$L = 0.15 \text{ m}$
$\theta = 5.0°$

(a)   (b)

Figure 23.9   (a) Two identical spheres, each carrying the same charge q, suspended in equilibrium. (b) The free-body diagram for the sphere on the left.

eliminated from Equation (1) if we make this substitution. This gives a value for the magnitude of the electric force F_e:

(3) $F_e = mg \tan \theta$

$= (3.0 \times 10^{-2} \text{ kg})(9.80 \text{ m/s}^2) \tan 5.0°$

$= 2.6 \times 10^{-2} \text{ N}$

From Coulomb's law (Eq. 23.1), the magnitude of the electric force is

$$F_e = k_e \frac{|q|^2}{r^2}$$

where $r = 2a = 0.026$ m and $|q|$ is the magnitude of the charge on each sphere. (Note that the term $|q|^2$ arises here because the charge is the same on both spheres.) This equation can be solved for $|q|^2$ to give

$$|q|^2 = \frac{F_e r^2}{k_e} = \frac{(2.6 \times 10^{-2} \text{ N})(0.026 \text{ m})^2}{8.99 \times 10^9 \text{ N} \cdot \text{m}^2/\text{C}^2}$$

$$|q| = \boxed{4.4 \times 10^{-8} \text{ C}}$$

Exercise   If the charge on the spheres were negative, how many electrons would have to be added to them to yield a net charge of -4.4×10^{-8} C?

Answer 2.7×10^{11} electrons.

QuickLab

For this experiment you need two 20-cm strips of transparent tape (mass of each $\approx$ 65 mg). Fold about 1 cm of tape over at one end of each strip to create a handle. Press both pieces of tape side by side onto a table top, rubbing your finger back and forth across the strips. Quickly pull the strips off the surface so that they become charged. Hold the tape handles together and the strips will repel each other, forming an inverted "V" shape. Measure the angle between the pieces, and estimate the excess charge on each strip. Assume that the charges act as if they were located at the center of mass of each strip.

23.4   THE ELECTRIC FIELD

Two field forces have been introduced into our discussions so far—the gravitational force and the electric force. As pointed out earlier, field forces can act through space, producing an effect even when no physical contact between the objects occurs. The gravitational field **g** at a point in space was defined in Section 14.6 to be equal to the gravitational force $\mathbf{F}_g$ acting on a test particle of mass m divided by that mass: $\mathbf{g} \equiv \mathbf{F}_g / m$. A similar approach to electric forces was developed by Michael Faraday and is of such practical value that we shall devote much attention to it in the next several chapters. In this approach, an **electric field** is said to exist in the region of space around a charged object. When another charged object enters this electric field, an electric force acts on it. As an example, consider Figure 23.10, which shows a small positive test charge q_0 placed near a second object carrying a much greater positive charge Q. We define the strength (in other words, the magnitude) of the electric field at the location of the test charge to be the electric force *per unit charge*, or to be more specific

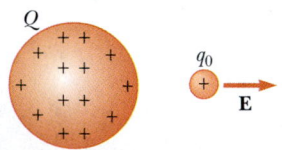

Figure 23.10   A small positive test charge q_0 placed near an object carrying a much larger positive charge Q experiences an electric field **E** directed as shown.

the electric field E at a point in space is defined as the electric force $\mathbf{F}_e$ acting on a positive test charge q_0 placed at that point divided by the magnitude of the test charge:

$$\mathbf{E} \equiv \frac{\mathbf{F}_e}{q_0} \qquad \textbf{(23.3)}$$

Definition of electric field

Note that **E** is the field produced by some charge *external* to the test charge—it is not the field produced by the test charge itself. Also, note that the existence of an electric field is a property of its source. For example, every electron comes with its own electric field.

The vector **E** has the SI units of newtons per coulomb (N/C), and, as Figure 23.10 shows, its direction is the direction of the force a positive test charge experiences when placed in the field. We say that **an electric field exists at a point if a test charge at rest at that point experiences an electric force.** Once the magnitude and direction of the electric field are known at some point, the electric force exerted on *any* charged particle placed at that point can be calculated from

This dramatic photograph captures a lightning bolt striking a tree near some rural homes. *(© Johnny Autery)*

TABLE 23.2	Typical Electric Field Values
Source	*E* **(N/C)**
Fluorescent lighting tube	10
Atmosphere (fair weather)	100
Balloon rubbed on hair	1 000
Atmosphere (under thundercloud)	10 000
Photocopier	100 000
Spark in air	>3 000 000
Near electron in hydrogen atom	5×10^{11}

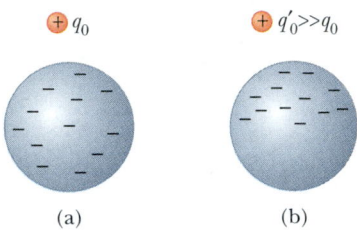

Figure 23.11   (a) For a small enough test charge q_0, the charge distribution on the sphere is undisturbed. (b) When the test charge q_0' is greater, the charge distribution on the sphere is disturbed as the result of the proximity of q_0'.

Equation 23.3. Furthermore, the electric field is said to exist at some point (even empty space) **regardless of whether a test charge is located at that point.** (This is analogous to the gravitational field set up by any object, which is said to exist at a given point regardless of whether some other object is present at that point to "feel" the field.) The electric field magnitudes for various field sources are given in Table 23.2.

When using Equation 23.3, we must assume that the test charge q_0 is small enough that it does not disturb the charge distribution responsible for the electric field. If a vanishingly small test charge q_0 is placed near a uniformly charged metallic sphere, as shown in Figure 23.11a, the charge on the metallic sphere, which produces the electric field, remains uniformly distributed. If the test charge is great enough ($q_0' \gg q_0$), as shown in Figure 23.11b, the charge on the metallic sphere is redistributed and the ratio of the force to the test charge is different: ($F_e'/q_0' \neq F_e/q_0$). That is, because of this redistribution of charge on the metallic sphere, the electric field it sets up is different from the field it sets up in the presence of the much smaller q_0.

To determine the direction of an electric field, consider a point charge q located a distance r from a test charge q_0 located at a point P, as shown in Figure 23.12. According to Coulomb's law, the force exerted by q on the test charge is

$$\mathbf{F}_e = k_e \frac{qq_0}{r^2}\,\hat{\mathbf{r}}$$

where $\hat{\mathbf{r}}$ is a unit vector directed from q toward q_0. Because the electric field at P, the position of the test charge, is defined by $\mathbf{E} = \mathbf{F}_e/q_0$, we find that at P, the electric field created by q is

$$\mathbf{E} = k_e \frac{q}{r^2}\,\hat{\mathbf{r}} \tag{23.4}$$

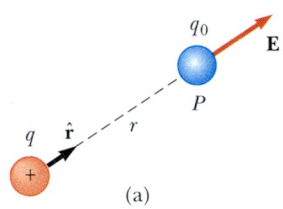

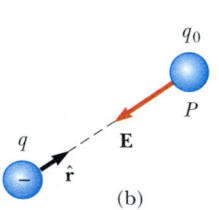

Figure 23.12   A test charge q_0 at point P is a distance r from a point charge q. (a) If q is positive, then the electric field at P points radially outward from q. (b) If q is negative, then the electric field at P points radially inward toward q.

If q is positive, as it is in Figure 23.12a, the electric field is directed radially outward from it. If q is negative, as it is in Figure 23.12b, the field is directed toward it.

To calculate the electric field at a point P due to a group of point charges, we first calculate the electric field vectors at P individually using Equation 23.4 and then add them vectorially. In other words,

> at any point P, the total electric field due to a group of charges equals the vector sum of the electric fields of the individual charges.

This superposition principle applied to fields follows directly from the superposition property of electric forces. Thus, the electric field of a group of charges can

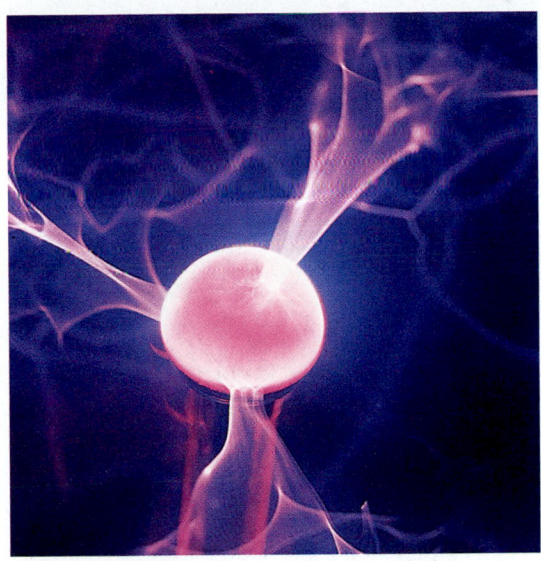

This metallic sphere is charged by a generator so that it carries a net electric charge. The high concentration of charge on the sphere creates a strong electric field around the sphere. The charges then leak through the gas surrounding the sphere, producing a pink glow. *(E. R. Degginer/H. Armstrong Roberts)*

be expressed as

$$\mathbf{E} = k_e \sum_i \frac{q_i}{r_i^2} \hat{\mathbf{r}}_i \qquad \textbf{(23.5)}$$

where r_i is the distance from the ith charge q_i to the point P (the location of the test charge) and $\hat{\mathbf{r}}_i$ is a unit vector directed from q_i toward P.

Quick Quiz 23.4

A charge of $+3\ \mu C$ is at a point P where the electric field is directed to the right and has a magnitude of 4×10^6 N/C. If the charge is replaced with a -3-μC charge, what happens to the electric field at P?

EXAMPLE 23.5 Electric Field Due to Two Charges

A charge $q_1 = 7.0\ \mu C$ is located at the origin, and a second charge $q_2 = -5.0\ \mu C$ is located on the x axis, 0.30 m from the origin (Fig. 23.13). Find the electric field at the point P, which has coordinates (0, 0.40) m.

Solution First, let us find the magnitude of the electric field at P due to each charge. The fields $\mathbf{E}_1$ due to the 7.0-μC charge and $\mathbf{E}_2$ due to the -5.0-μC charge are shown in Figure 23.13. Their magnitudes are

$$E_1 = k_e \frac{|q_1|}{r_1^2} = \left(8.99 \times 10^9\ \frac{\text{N·m}^2}{\text{C}^2}\right) \frac{(7.0 \times 10^{-6}\ \text{C})}{(0.40\ \text{m})^2}$$

$$= 3.9 \times 10^5\ \text{N/C}$$

$$E_2 = k_e \frac{|q_2|}{r_2^2} = \left(8.99 \times 10^9\ \frac{\text{N·m}^2}{\text{C}^2}\right) \frac{(5.0 \times 10^{-6}\ \text{C})}{(0.50\ \text{m})^2}$$

$$= 1.8 \times 10^5\ \text{N/C}$$

The vector $\mathbf{E}_1$ has only a y component. The vector $\mathbf{E}_2$ has an x component given by $E_2 \cos\theta = \frac{3}{5}E_2$ and a negative y component given by $-E_2 \sin\theta = -\frac{4}{5}E_2$. Hence, we can express the vectors as

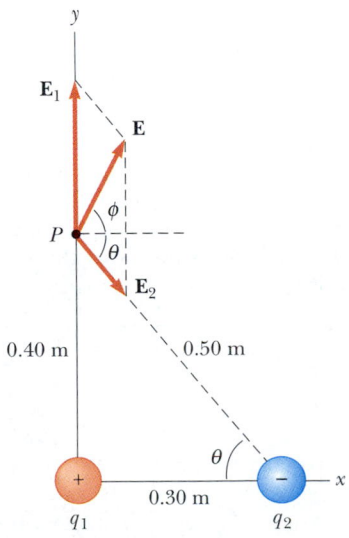

Figure 23.13 The total electric field $\mathbf{E}$ at P equals the vector sum $\mathbf{E}_1 + \mathbf{E}_2$, where $\mathbf{E}_1$ is the field due to the positive charge q_1 and $\mathbf{E}_2$ is the field due to the negative charge q_2.

$$\mathbf{E}_1 = 3.9 \times 10^5 \mathbf{j} \text{ N/C}$$

$$\mathbf{E}_2 = (1.1 \times 10^5 \mathbf{i} - 1.4 \times 10^5 \mathbf{j}) \text{ N/C}$$

The resultant field **E** at *P* is the superposition of $\mathbf{E}_1$ and $\mathbf{E}_2$:

$$\mathbf{E} = \mathbf{E}_1 + \mathbf{E}_2 = \boxed{(1.1 \times 10^5 \mathbf{i} + 2.5 \times 10^5 \mathbf{j}) \text{ N/C}}$$

From this result, we find that **E** has a magnitude of 2.7×10^5 N/C and makes an angle ϕ of 66° with the positive *x* axis.

Exercise   Find the electric force exerted on a charge of 2.0×10^{-8} C located at *P*.

Answer 5.4×10^{-3} N in the same direction as **E**.

EXAMPLE 23.6   Electric Field of a Dipole

An **electric dipole** is defined as a positive charge *q* and a negative charge $-q$ separated by some distance. For the dipole shown in Figure 23.14, find the electric field **E** at *P* due to the charges, where *P* is a distance $y \gg a$ from the origin.

Solution   At *P*, the fields $\mathbf{E}_1$ and $\mathbf{E}_2$ due to the two charges are equal in magnitude because *P* is equidistant from the charges. The total field is $\mathbf{E} = \mathbf{E}_1 + \mathbf{E}_2$, where

$$E_1 = E_2 = k_e \frac{q}{r^2} = k_e \frac{q}{y^2 + a^2}$$

The *y* components of $\mathbf{E}_1$ and $\mathbf{E}_2$ cancel each other, and the *x* components are equal because they are both along the *x* axis. Therefore, **E** is parallel to the *x* axis and has a magnitude equal to $2E_1 \cos \theta$. From Figure 23.14 we see that $\cos \theta = a/r = a/(y^2 + a^2)^{1/2}$. Therefore,

$$E = 2E_1 \cos \theta = 2k_e \frac{q}{(y^2 + a^2)} \frac{a}{(y^2 + a^2)^{1/2}}$$

$$= k_e \frac{2qa}{(y^2 + a^2)^{3/2}}$$

Because $y \gg a$, we can neglect a^2 and write

$$\boxed{E \approx k_e \frac{2qa}{y^3}}$$

Thus, we see that, at distances far from a dipole but along the perpendicular bisector of the line joining the two charges, the magnitude of the electric field created by the dipole varies as $1/r^3$, whereas the more slowly varying field of a point charge varies as $1/r^2$ (see Eq. 23.4). This is because at distant points, the fields of the two charges of equal magnitude and opposite sign almost cancel each other. The $1/r^3$

variation in *E* for the dipole also is obtained for a distant point along the *x* axis (see Problem 21) and for any general distant point.

The electric dipole is a good model of many molecules, such as hydrochloric acid (HCl). As we shall see in later chapters, neutral atoms and molecules behave as dipoles when placed in an external electric field. Furthermore, many molecules, such as HCl, are permanent dipoles. The effect of such dipoles on the behavior of materials subjected to electric fields is discussed in Chapter 26.

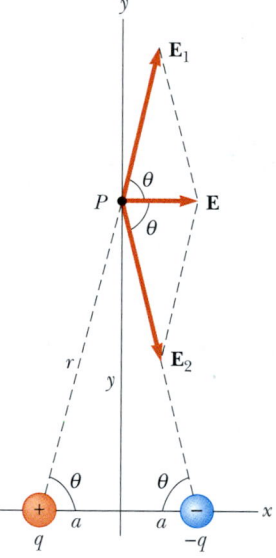

Figure 23.14   The total electric field **E** at *P* due to two charges of equal magnitude and opposite sign (an electric dipole) equals the vector sum $\mathbf{E}_1 + \mathbf{E}_2$. The field $\mathbf{E}_1$ is due to the positive charge *q*, and $\mathbf{E}_2$ is the field due to the negative charge $-q$.

23.5 ELECTRIC FIELD OF A CONTINUOUS CHARGE DISTRIBUTION

Very often the distances between charges in a group of charges are much smaller than the distance from the group to some point of interest (for example, a point where the electric field is to be calculated). In such situations, the system of charges is smeared out, or *continuous*. That is, the system of closely spaced charges is equivalent to a total charge that is continuously distributed along some line, over some surface, or throughout some volume.

To evaluate the electric field created by a continuous charge distribution, we use the following procedure: First, we divide the charge distribution into small elements, each of which contains a small charge Δq, as shown in Figure 23.15. Next, we use Equation 23.4 to calculate the electric field due to one of these elements at a point P. Finally, we evaluate the total field at P due to the charge distribution by summing the contributions of all the charge elements (that is, by applying the superposition principle).

The electric field at P due to one element carrying charge Δq is

$$\Delta \mathbf{E} = k_e \frac{\Delta q}{r^2} \hat{\mathbf{r}}$$

where r is the distance from the element to point P and $\hat{\mathbf{r}}$ is a unit vector directed from the charge element toward P. The total electric field at P due to all elements in the charge distribution is approximately

$$\mathbf{E} \approx k_e \sum_i \frac{\Delta q_i}{r_i^2} \hat{\mathbf{r}}_i$$

where the index i refers to the ith element in the distribution. Because the charge distribution is approximately continuous, the total field at P in the limit $\Delta q_i \to 0$ is

$$\mathbf{E} = k_e \lim_{\Delta q_i \to 0} \sum_i \frac{\Delta q_i}{r_i^2} \hat{\mathbf{r}}_i = k_e \int \frac{dq}{r^2} \hat{\mathbf{r}} \qquad \textbf{(23.6)}$$

where the integration is over the entire charge distribution. This is a vector operation and must be treated appropriately.

We illustrate this type of calculation with several examples, in which we assume the charge is uniformly distributed on a line, on a surface, or throughout a volume. When performing such calculations, it is convenient to use the concept of a charge density along with the following notations:

- If a charge Q is uniformly distributed throughout a volume V, the **volume charge density** ρ is defined by

$$\rho \equiv \frac{Q}{V}$$

where ρ has units of coulombs per cubic meter (C/m^3).

- If a charge Q is uniformly distributed on a surface of area A, the **surface charge density** σ (lowercase Greek sigma) is defined by

$$\sigma \equiv \frac{Q}{A}$$

where σ has units of coulombs per square meter (C/m^2).

- If a charge Q is uniformly distributed along a line of length ℓ, the **linear charge density** λ is defined by

$$\lambda \equiv \frac{Q}{\ell}$$

where λ has units of coulombs per meter (C/m).

A continuous charge distribution

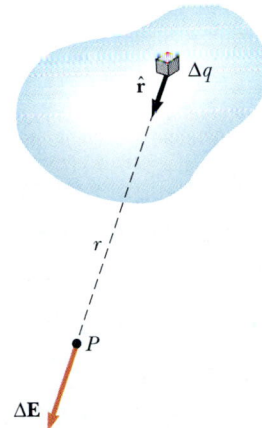

Figure 23.15 The electric field at P due to a continuous charge distribution is the vector sum of the fields $\Delta \mathbf{E}$ due to all the elements Δq of the charge distribution.

Electric field of a continuous charge distribution

Volume charge density

Surface charge density

Linear charge density

• If the charge is nonuniformly distributed over a volume, surface, or line, we have to express the charge densities as

$$\rho = \frac{dQ}{dV} \qquad \sigma = \frac{dQ}{dA} \qquad \lambda = \frac{dQ}{d\ell}$$

where dQ is the amount of charge in a small volume, surface, or length element.

EXAMPLE 23.7   The Electric Field Due to a Charged Rod

A rod of length ℓ has a uniform positive charge per unit length λ and a total charge Q. Calculate the electric field at a point P that is located along the long axis of the rod and a distance a from one end (Fig. 23.16).

Solution Let us assume that the rod is lying along the x axis, that dx is the length of one small segment, and that dq is the charge on that segment. Because the rod has a charge per unit length λ, the charge dq on the small segment is $dq = \lambda\,dx$.

The field $d\mathbf{E}$ due to this segment at P is in the negative x direction (because the source of the field carries a positive charge Q), and its magnitude is

$$dE = k_e \frac{dq}{x^2} = k_e\lambda\,\frac{dx}{x^2}$$

Because every other element also produces a field in the negative x direction, the problem of summing their contributions is particularly simple in this case. The total field at P due to all segments of the rod, which are at different distances from P, is given by Equation 23.6, which in this case becomes[3]

$$E = \int_a^{\ell+a} k_e\lambda\,\frac{dx}{x^2}$$

where the limits on the integral extend from one end of the rod ($x = a$) to the other ($x = \ell + a$). The constants k_e and λ can be removed from the integral to yield

$$E = k_e\lambda \int_a^{\ell+a} \frac{dx}{x^2} = k_e\lambda\left[-\frac{1}{x}\right]_a^{\ell+a}$$

$$= k_e\lambda\left(\frac{1}{a} - \frac{1}{\ell+a}\right) = \boxed{\frac{k_eQ}{a(\ell+a)}}$$

where we have used the fact that the total charge $Q = \lambda\ell$.

If P is far from the rod ($a \gg \ell$), then the ℓ in the denominator can be neglected, and $E \approx k_eQ/a^2$. This is just the form you would expect for a point charge. Therefore, at large values of a/ℓ, the charge distribution appears to be a point charge of magnitude Q. The use of the limiting technique ($a/\ell \to \infty$) often is a good method for checking a theoretical formula.

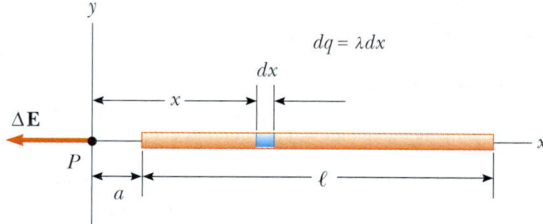

Figure 23.16 The electric field at P due to a uniformly charged rod lying along the x axis. The magnitude of the field at P due to the segment of charge dq is $k_e\,dq/x^2$. The total field at P is the vector sum over all segments of the rod.

EXAMPLE 23.8   The Electric Field of a Uniform Ring of Charge

A ring of radius a carries a uniformly distributed positive total charge Q. Calculate the electric field due to the ring at a point P lying a distance x from its center along the central axis perpendicular to the plane of the ring (Fig. 23.17a).

Solution The magnitude of the electric field at P due to the segment of charge dq is

$$dE = k_e \frac{dq}{r^2}$$

This field has an x component $dE_x = dE\cos\theta$ along the axis and a component $dE_\perp$ perpendicular to the axis. As we see in Figure 23.17b, however, the resultant field at P must lie along the x axis because the perpendicular components of all the

---

[3] It is important that you understand how to carry out integrations such as this. First, express the charge element dq in terms of the other variables in the integral (in this example, there is one variable, x, and so we made the change $dq = \lambda\,dx$). The integral must be over scalar quantities; therefore, you must express the electric field in terms of components, if necessary. (In this example the field has only an x component, so we do not bother with this detail.) Then, reduce your expression to an integral over a single variable (or to multiple integrals, each over a single variable). In examples that have spherical or cylindrical symmetry, the single variable will be a radial coordinate.

various charge segments sum to zero. That is, the perpendicular component of the field created by any charge element is canceled by the perpendicular component created by an element on the opposite side of the ring. Because $r = (x^2 + a^2)^{1/2}$ and $\cos\theta = x/r$, we find that

$$dE_x = dE\cos\theta = \left(k_e \frac{dq}{r^2}\right)\frac{x}{r} = \frac{k_e x}{(x^2 + a^2)^{3/2}}\, dq$$

All segments of the ring make the same contribution to the field at P because they are all equidistant from this point. Thus, we can integrate to obtain the total field at P:

$$E_x = \int \frac{k_e x}{(x^2 + a^2)^{3/2}}\, dq = \frac{k_e x}{(x^2 + a^2)^{3/2}} \int dq$$

$$= \frac{k_e x}{(x^2 + a^2)^{3/2}}\, Q$$

This result shows that the field is zero at $x = 0$. Does this finding surprise you?

Exercise Show that at great distances from the ring ($x \gg a$) the electric field along the axis shown in Figure 23.17 approaches that of a point charge of magnitude Q.

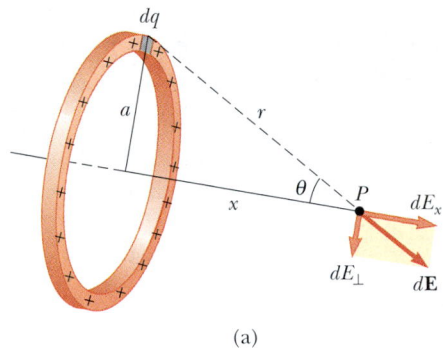

(a)

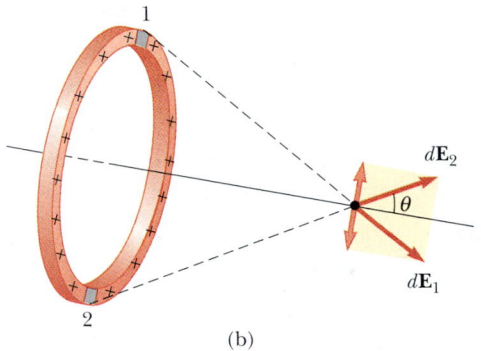

(b)

Figure 23.17 A uniformly charged ring of radius a. (a) The field at P on the x axis due to an element of charge dq. (b) The total electric field at P is along the x axis. The perpendicular component of the field at P due to segment 1 is canceled by the perpendicular component due to segment 2.

EXAMPLE 23.9 ▶ The Electric Field of a Uniformly Charged Disk

A disk of radius R has a uniform surface charge density σ. Calculate the electric field at a point P that lies along the central perpendicular axis of the disk and a distance x from the center of the disk (Fig. 23.18).

Solution If we consider the disk as a set of concentric rings, we can use our result from Example 23.8—which gives the field created by a ring of radius r—and sum the contributions of all rings making up the disk. By symmetry, the field at an axial point must be along the central axis.

The ring of radius r and width dr shown in Figure 23.18 has a surface area equal to $2\pi r\, dr$. The charge dq on this ring is equal to the area of the ring multiplied by the surface charge density: $dq = 2\pi\sigma r\, dr$. Using this result in the equation given for E_x in Example 23.8 (with a replaced by r), we have for the field due to the ring

$$dE = \frac{k_e x}{(x^2 + r^2)^{3/2}} (2\pi\sigma r\, dr)$$

To obtain the total field at P, we integrate this expression over the limits $r = 0$ to $r = R$, noting that x is a constant. This gives

$$E = k_e x\pi\sigma \int_0^R \frac{2r\, dr}{(x^2 + r^2)^{3/2}}$$

$$= k_e x\pi\sigma \int_0^R (x^2 + r^2)^{-3/2}\, d(r^2)$$

$$= k_e x\pi\sigma \left[\frac{(x^2 + r^2)^{-1/2}}{-1/2}\right]_0^R$$

$$= 2\pi k_e \sigma \left(\frac{x}{|x|} - \frac{x}{(x^2 + R^2)^{1/2}}\right)$$

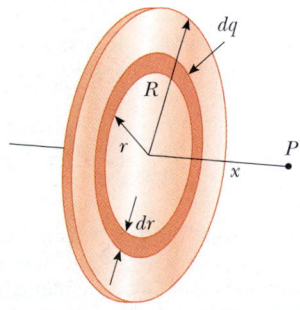

Figure 23.18 A uniformly charged disk of radius R. The electric field at an axial point P is directed along the central axis, perpendicular to the plane of the disk.

This result is valid for all values of *x*. We can calculate the field close to the disk along the axis by assuming that $R \gg x$; thus, the expression in parentheses reduces to unity:

$$E \approx 2\pi k_e \sigma = \frac{\sigma}{2\epsilon_0}$$

where $\epsilon_0 = 1/(4\pi k_e)$ is the permittivity of free space. As we shall find in the next chapter, we obtain the same result for the field created by a uniformly charged infinite sheet.

23.6 ELECTRIC FIELD LINES

A convenient way of visualizing electric field patterns is to draw lines that follow the same direction as the electric field vector at any point. These lines, called **electric field lines,** are related to the electric field in any region of space in the following manner:

- The electric field vector **E** is tangent to the electric field line at each point.
- The number of lines per unit area through a surface perpendicular to the lines is proportional to the magnitude of the electric field in that region. Thus, *E* is great when the field lines are close together and small when they are far apart.

These properties are illustrated in Figure 23.19. The density of lines through surface A is greater than the density of lines through surface B. Therefore, the electric field is more intense on surface A than on surface B. Furthermore, the fact that the lines at different locations point in different directions indicates that the field is nonuniform.

Representative electric field lines for the field due to a single positive point charge are shown in Figure 23.20a. Note that in this two-dimensional drawing we show only the field lines that lie in the plane containing the point charge. The lines are actually directed radially outward from the charge in all directions; thus, instead of the flat "wheel" of lines shown, you should picture an entire sphere of lines. Because a positive test charge placed in this field would be repelled by the positive point charge, the lines are directed radially away from the positive point

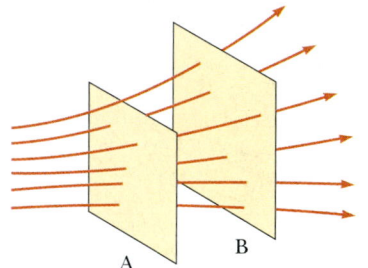

Figure 23.19 Electric field lines penetrating two surfaces. The magnitude of the field is greater on surface A than on surface B.

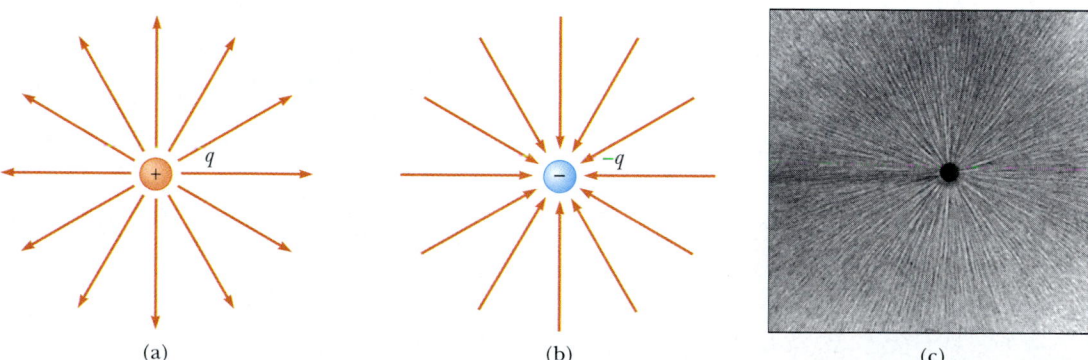

Figure 23.20 The electric field lines for a point charge. (a) For a positive point charge, the lines are directed radially outward. (b) For a negative point charge, the lines are directed radially inward. Note that the figures show only those field lines that lie in the plane containing the charge. (c) The dark areas are small pieces of thread suspended in oil, which align with the electric field produced by a small charged conductor at the center. *(c, Courtesy of Harold M. Waage, Princeton University)*

charge. The electric field lines representing the field due to a single negative point charge are directed toward the charge (Fig. 23.20b). In either case, the lines are along the radial direction and extend all the way to infinity. Note that the lines become closer together as they approach the charge; this indicates that the strength of the field increases as we move toward the source charge.

The rules for drawing electric field lines are as follows:

- The lines must begin on a positive charge and terminate on a negative charge.
- The number of lines drawn leaving a positive charge or approaching a negative charge is proportional to the magnitude of the charge.
- No two field lines can cross.

Rules for drawing electric field lines

Is this visualization of the electric field in terms of field lines consistent with Equation 23.4, the expression we obtained for E using Coulomb's law? To answer this question, consider an imaginary spherical surface of radius r concentric with a point charge. From symmetry, we see that the magnitude of the electric field is the same everywhere on the surface of the sphere. The number of lines N that emerge from the charge is equal to the number that penetrate the spherical surface. Hence, the number of lines per unit area on the sphere is $N/4\pi r^2$ (where the surface area of the sphere is $4\pi r^2$). Because E is proportional to the number of lines per unit area, we see that E varies as $1/r^2$; this finding is consistent with Equation 23.4.

As we have seen, we use electric field lines to qualitatively describe the electric field. One problem with this model is that we always draw a finite number of lines from (or to) each charge. Thus, it appears as if the field acts only in certain directions; this is not true. Instead, the field is *continuous*—that is, it exists at every point. Another problem associated with this model is the danger of gaining the wrong impression from a two-dimensional drawing of field lines being used to describe a three-dimensional situation. Be aware of these shortcomings every time you either draw or look at a diagram showing electric field lines.

We choose the number of field lines starting from any positively charged object to be $C'q$ and the number of lines ending on any negatively charged object to be $C'|q|$, where C' is an arbitrary proportionality constant. Once C' is chosen, the number of lines is fixed. For example, if object 1 has charge Q_1 and object 2 has charge Q_2, then the ratio of number of lines is $N_2/N_1 = Q_2/Q_1$.

The electric field lines for two point charges of equal magnitude but opposite signs (an electric dipole) are shown in Figure 23.21. Because the charges are of equal magnitude, the number of lines that begin at the positive charge must equal the number that terminate at the negative charge. At points very near the charges, the lines are nearly radial. The high density of lines between the charges indicates a region of strong electric field.

Figure 23.22 shows the electric field lines in the vicinity of two equal positive point charges. Again, the lines are nearly radial at points close to either charge, and the same number of lines emerge from each charge because the charges are equal in magnitude. At great distances from the charges, the field is approximately equal to that of a single point charge of magnitude $2q$.

Finally, in Figure 23.23 we sketch the electric field lines associated with a positive charge $+2q$ and a negative charge $-q$. In this case, the number of lines leaving $+2q$ is twice the number terminating at $-q$. Hence, only half of the lines that leave the positive charge reach the negative charge. The remaining half terminate

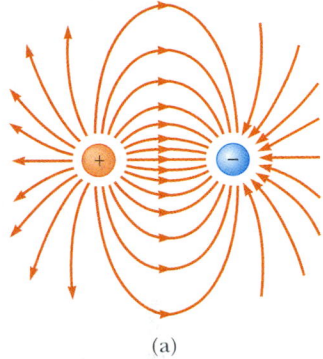

(a)

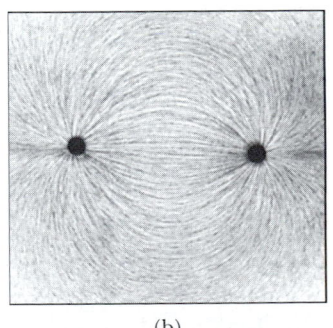

(b)

Figure 23.21 (a) The electric field lines for two point charges of equal magnitude and opposite sign (an electric dipole). The number of lines leaving the positive charge equals the number terminating at the negative charge. (b) The dark lines are small pieces of thread suspended in oil, which align with the electric field of a dipole. *(b, Courtesy of Harold M. Waage, Princeton University)*

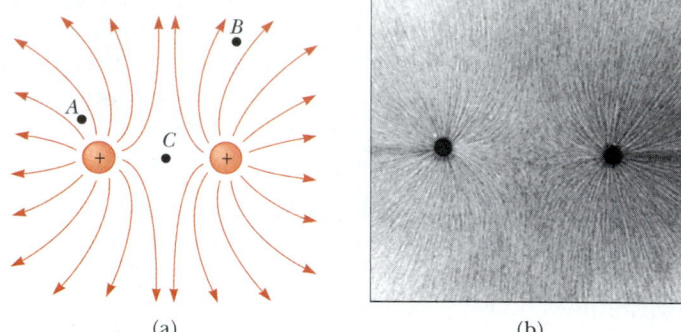

Figure 23.22   (a) The electric field lines for two positive point charges. (The locations *A*, *B*, and *C* are discussed in Quick Quiz 23.5.) (b) Pieces of thread suspended in oil, which align with the electric field created by two equal-magnitude positive charges. *(Photo courtesy of Harold M. Waage, Princeton University)*

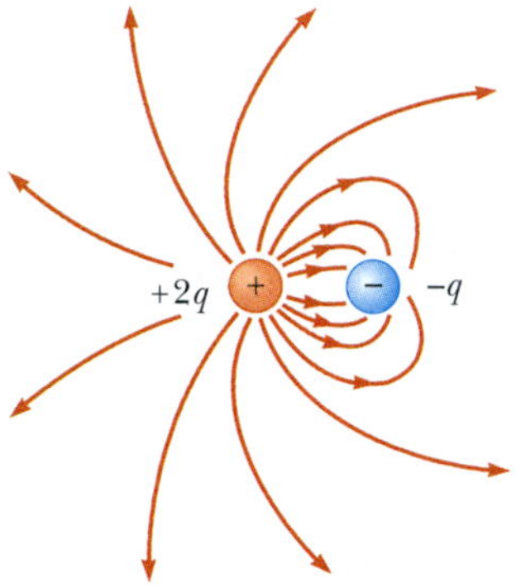

on a negative charge we assume to be at infinity. At distances that are much greater than the charge separation, the electric field lines are equivalent to those of a single charge $+q$.

Quick Quiz 23.5

Rank the magnitude of the electric field at points *A*, *B*, and *C* shown in Figure 23.22a (greatest magnitude first).

Figure 23.23   The electric field lines for a point charge $+2q$ and a second point charge $-q$. Note that two lines leave $+2q$ for every one that terminates on $-q$.

23.7 ▸ MOTION OF CHARGED PARTICLES IN A UNIFORM ELECTRIC FIELD

When a particle of charge q and mass m is placed in an electric field **E**, the electric force exerted on the charge is q**E**. If this is the only force exerted on the particle, it must be the net force and so must cause the particle to accelerate. In this case, Newton's second law applied to the particle gives

$$\mathbf{F}_e = q\mathbf{E} = m\mathbf{a}$$

The acceleration of the particle is therefore

$$\mathbf{a} = \frac{q\mathbf{E}}{m} \tag{23.7}$$

If **E** is uniform (that is, constant in magnitude and direction), then the acceleration is constant. If the particle has a positive charge, then its acceleration is in the direction of the electric field. If the particle has a negative charge, then its acceleration is in the direction opposite the electric field.

EXAMPLE 23.10 ▸   **An Accelerating Positive Charge**

A positive point charge q of mass m is released from rest in a uniform electric field **E** directed along the *x* axis, as shown in Figure 23.24. Describe its motion.

Solution   The acceleration is constant and is given by $q\mathbf{E}/m$. The motion is simple linear motion along the *x* axis. Therefore, we can apply the equations of kinematics in one dimension (see Chapter 2):

$$x_f = x_i + v_{xi}t + \tfrac{1}{2}a_x t^2$$

$$v_{xf} = v_{xi} + a_x t$$

$$v_{xf}^2 = v_{xi}^2 + 2a_x(x_f - x_i)$$

Taking $x_i = 0$ and $v_{xi} = 0$, we have

$$x_f = \tfrac{1}{2}a_x t^2 = \frac{qE}{2m}t^2$$

$$v_{xf} = a_x t = \frac{qE}{m}t$$

$$v_{xf}^2 = 2a_x x_f = \left(\frac{2qE}{m}\right)x_f$$

The kinetic energy of the charge after it has moved a distance $x = x_f - x_i$ is

$$K = \tfrac{1}{2}mv^2 = \tfrac{1}{2}m\left(\frac{2qE}{m}\right)x = qEx$$

We can also obtain this result from the work–kinetic energy theorem because the work done by the electric force is $F_e x = qEx$ and $W = \Delta K$.

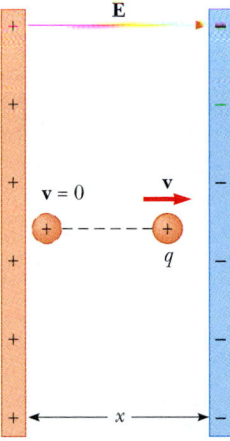

Figure 23.24 A positive point charge q in a uniform electric field **E** undergoes constant acceleration in the direction of the field.

The electric field in the region between two oppositely charged flat metallic plates is approximately uniform (Fig. 23.25). Suppose an electron of charge $-e$ is projected horizontally into this field with an initial velocity $v_i\mathbf{i}$. Because the electric field **E** in Figure 23.25 is in the positive y direction, the acceleration of the electron is in the negative y direction. That is,

$$\mathbf{a} = -\frac{eE}{m}\,\mathbf{j} \tag{23.8}$$

Because the acceleration is constant, we can apply the equations of kinematics in two dimensions (see Chapter 4) with $v_{xi} = v_i$ and $v_{yi} = 0$. After the electron has been in the electric field for a time t, the components of its velocity are

$$v_x = v_i = \text{constant} \tag{23.9}$$

$$v_y = a_y t = -\frac{eE}{m}\,t \tag{23.10}$$

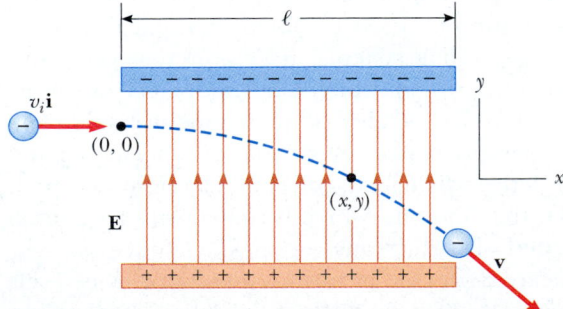

Figure 23.25 An electron is projected horizontally into a uniform electric field produced by two charged plates. The electron undergoes a downward acceleration (opposite **E**), and its motion is parabolic while it is between the plates.

Its coordinates after a time t in the field are

$$x = v_i t \qquad (23.11)$$

$$y = \tfrac{1}{2} a_y t^2 = -\tfrac{1}{2} \frac{eE}{m} t^2 \qquad (23.12)$$

Substituting the value $t = x/v_i$ from Equation 23.11 into Equation 23.12, we see that y is proportional to x^2. Hence, the trajectory is a parabola. After the electron leaves the field, it continues to move in a straight line in the direction of **v** in Figure 23.25, obeying Newton's first law, with a speed $v > v_i$.

Note that we have neglected the gravitational force acting on the electron. This is a good approximation when we are dealing with atomic particles. For an electric field of 10^4 N/C, the ratio of the magnitude of the electric force eE to the magnitude of the gravitational force mg is of the order of 10^{14} for an electron and of the order of 10^{11} for a proton.

EXAMPLE 23.11 ▶ An Accelerated Electron

An electron enters the region of a uniform electric field as shown in Figure 23.25, with $v_i = 3.00 \times 10^6$ m/s and $E = 200$ N/C. The horizontal length of the plates is $\ell = 0.100$ m. (a) Find the acceleration of the electron while it is in the electric field.

Solution The charge on the electron has an absolute value of 1.60×10^{-19} C, and $m = 9.11 \times 10^{-31}$ kg. Therefore, Equation 23.8 gives

$$\mathbf{a} = -\frac{eE}{m}\mathbf{j} = -\frac{(1.60 \times 10^{-19} \text{ C})(200 \text{ N/C})}{9.11 \times 10^{-31} \text{ kg}}\mathbf{j}$$

$$= \boxed{-3.51 \times 10^{13}\,\mathbf{j} \text{ m/s}^2}$$

(b) Find the time it takes the electron to travel through the field.

Solution The horizontal distance across the field is $\ell = 0.100$ m. Using Equation 23.11 with $x = \ell$, we find that the time spent in the electric field is

$$t = \frac{\ell}{v_i} = \frac{0.100 \text{ m}}{3.00 \times 10^6 \text{ m/s}} = \boxed{3.33 \times 10^{-8} \text{ s}}$$

(c) What is the vertical displacement y of the electron while it is in the field?

Solution Using Equation 23.12 and the results from parts (a) and (b), we find that

$$y = \tfrac{1}{2} a_y t^2 = \tfrac{1}{2}(-3.51 \times 10^{13} \text{ m/s}^2)(3.33 \times 10^{-8} \text{ s})^2$$

$$= -0.019\,5 \text{ m} = \boxed{-1.95 \text{ cm}}$$

If the separation between the plates is less than this, the electron will strike the positive plate.

Exercise Find the speed of the electron as it emerges from the field.

Answer 3.22×10^6 m/s.

The Cathode Ray Tube

The example we just worked describes a portion of a cathode ray tube (CRT). This tube, illustrated in Figure 23.26, is commonly used to obtain a visual display of electronic information in oscilloscopes, radar systems, television receivers, and computer monitors. The CRT is a vacuum tube in which a beam of electrons is accelerated and deflected under the influence of electric or magnetic fields. The electron beam is produced by an assembly called an *electron gun* located in the neck of the tube. These electrons, if left undisturbed, travel in a straight-line path until they strike the front of the CRT, the "screen," which is coated with a material that emits visible light when bombarded with electrons.

In an oscilloscope, the electrons are deflected in various directions by two sets of plates placed at right angles to each other in the neck of the tube. (A television

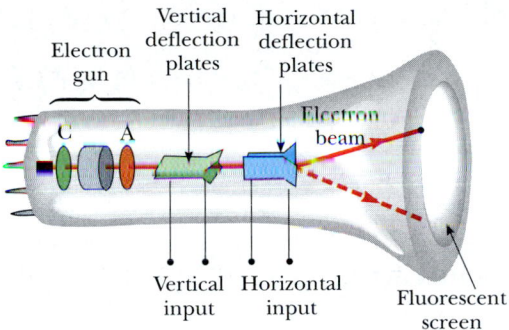

Figure 23.26 Schematic diagram of a cathode ray tube. Electrons leaving the hot cathode C are accelerated to the anode A. In addition to accelerating electrons, the electron gun is also used to focus the beam of electrons, and the plates deflect the beam.

CRT steers the beam with a magnetic field, as discussed in Chapter 29.) An external electric circuit is used to control the amount of charge present on the plates. The placing of positive charge on one horizontal plate and negative charge on the other creates an electric field between the plates and allows the beam to be steered from side to side. The vertical deflection plates act in the same way, except that changing the charge on them deflects the beam vertically.

SUMMARY

Electric charges have the following important properties:

- Unlike charges attract one another, and like charges repel one another.
- Charge is conserved.
- Charge is quantized—that is, it exists in discrete packets that are some integral multiple of the electronic charge.

Conductors are materials in which charges move freely. **Insulators** are materials in which charges do not move freely.

Coulomb's law states that the electric force exerted by a charge q_1 on a second charge q_2 is

$$\mathbf{F}_{12} = k_e \frac{q_1 q_2}{r^2} \,\hat{\mathbf{r}} \tag{23.2}$$

where r is the distance between the two charges and $\hat{\mathbf{r}}$ is a unit vector directed from q_1 to q_2. The constant k_e, called the Coulomb constant, has the value $k_e = 8.99 \times 10^9 \text{ N} \cdot \text{m}^2/\text{C}^2$.

The smallest unit of charge known to exist in nature is the charge on an electron or proton, $|e| = 1.602\ 19 \times 10^{-19} \text{ C}$.

The electric field $\mathbf{E}$ at some point in space is defined as the electric force $\mathbf{F}_e$ that acts on a small positive test charge placed at that point divided by the magnitude of the test charge q_0:

$$\mathbf{E} \equiv \frac{\mathbf{F}_e}{q_0} \tag{23.3}$$

At a distance r from a point charge q, the electric field due to the charge is given by

$$\mathbf{E} = k_e \frac{q}{r^2} \,\hat{\mathbf{r}} \tag{23.4}$$

where $\hat{\mathbf{r}}$ is a unit vector directed from the charge to the point in question. The electric field is directed radially outward from a positive charge and radially inward toward a negative charge.

The electric field due to a group of point charges can be obtained by using the superposition principle. That is, the total electric field at some point equals the vector sum of the electric fields of all the charges:

$$\mathbf{E} = k_e \sum_i \frac{q_i}{r_i^2} \hat{\mathbf{r}}_i \qquad (23.5)$$

The electric field at some point of a continuous charge distribution is

$$\mathbf{E} = k_e \int \frac{dq}{r^2} \hat{\mathbf{r}} \qquad (23.6)$$

where dq is the charge on one element of the charge distribution and r is the distance from the element to the point in question.

Electric field lines describe an electric field in any region of space. The number of lines per unit area through a surface perpendicular to the lines is proportional to the magnitude of $\mathbf{E}$ in that region.

A charged particle of mass m and charge q moving in an electric field $\mathbf{E}$ has an acceleration

$$\mathbf{a} = \frac{q\mathbf{E}}{m} \qquad (23.7)$$

Problem-Solving Hints

Finding the Electric Field

- **Units:**   In calculations using the Coulomb constant k_e $(=1/4\pi\epsilon_0)$, charges must be expressed in coulombs and distances in meters.
- **Calculating the electric field of point charges:**   To find the total electric field at a given point, first calculate the electric field at the point due to each individual charge. The resultant field at the point is the vector sum of the fields due to the individual charges.
- **Continuous charge distributions:**   When you are confronted with problems that involve a continuous distribution of charge, the vector sums for evaluating the total electric field at some point must be replaced by vector integrals. Divide the charge distribution into infinitesimal pieces, and calculate the vector sum by integrating over the entire charge distribution. You should review Examples 23.7 through 23.9.
- **Symmetry:**   With both distributions of point charges and continuous charge distributions, take advantage of any symmetry in the system to simplify your calculations.

QUESTIONS

1. Sparks are often observed (or heard) on a dry day when clothes are removed in the dark. Explain.

2. Explain from an atomic viewpoint why charge is usually transferred by electrons.

3. A balloon is negatively charged by rubbing and then clings to a wall. Does this mean that the wall is positively charged? Why does the balloon eventually fall?

4. A light, uncharged metallic sphere suspended from a thread is attracted to a charged rubber rod. After touching the rod, the sphere is repelled by the rod. Explain.

5. Explain what is meant by the term "a neutral atom."

6. Why do some clothes cling together and to your body after they are removed from a dryer?

7. A large metallic sphere insulated from ground is charged with an electrostatic generator while a person standing on an insulating stool holds the sphere. Why is it safe to do this? Why wouldn't it be safe for another person to touch the sphere after it has been charged?

8. What are the similarities and differences between Newton's law of gravitation, $F_g = Gm_1m_2/r^2$, and Coulomb's law, $F_e = k_e q_1 q_2/r^2$?

9. Assume that someone proposes a theory that states that people are bound to the Earth by electric forces rather than by gravity. How could you prove this theory wrong?

10. How would you experimentally distinguish an electric field from a gravitational field?

11. Would life be different if the electron were positively charged and the proton were negatively charged? Does the choice of signs have any bearing on physical and chemical interactions? Explain.

12. When defining the electric field, why is it necessary to specify that the magnitude of the test charge be very small (that is, why is it necessary to take the limit of $\mathbf{F}_e/q$ as $q \rightarrow 0$)?

13. Two charged conducting spheres, each of radius a, are separated by a distance $r > 2a$. Is the force on either sphere given by Coulomb's law? Explain. (*Hint:* Refer to Chapter 14 on gravitation.)

14. When is it valid to approximate a charge distribution by a point charge?

15. Is it possible for an electric field to exist in empty space? Explain.

16. Explain why electric field lines never cross. (*Hint:* **E** must have a unique direction at all points.)

17. A free electron and free proton are placed in an identical electric field. Compare the electric forces on each particle. Compare their accelerations.

18. Explain what happens to the magnitude of the electric field of a point charge as r approaches zero.

19. A negative charge is placed in a region of space where the electric field is directed vertically upward. What is the direction of the electric force experienced by this charge?

20. A charge $4q$ is a distance r from a charge $-q$. Compare the number of electric field lines leaving the charge $4q$ with the number entering the charge $-q$.

21. In Figure 23.23, where do the extra lines leaving the charge $+2q$ end?

22. Consider two equal point charges separated by some distance d. At what point (other than ∞) would a third test charge experience no net force?

23. A negative point charge $-q$ is placed at the point P near the positively charged ring shown in Figure 23.17. If $x \ll a$, describe the motion of the point charge if it is released from rest.

24. Explain the differences between linear, surface, and volume charge densities, and give examples of when each would be used.

25. If the electron in Figure 23.25 is projected into the electric field with an arbitrary velocity $\mathbf{v}_i$ (at an angle to **E**), will its trajectory still be parabolic? Explain.

26. It has been reported that in some instances people near where a lightning bolt strikes the Earth have had their clothes thrown off. Explain why this might happen.

27. Why should a ground wire be connected to the metallic support rod for a television antenna?

28. A light strip of aluminum foil is draped over a wooden rod. When a rod carrying a positive charge is brought close to the foil, the two parts of the foil stand apart. Why? What kind of charge is on the foil?

29. Why is it more difficult to charge an object by rubbing on a humid day than on a dry day?

PROBLEMS

1, **2**, 3 = straightforward, intermediate, challenging ☐ = full solution available in the *Student Solutions Manual and Study Guide*
WEB = solution posted at **http://www.saunderscollege.com/physics/** 🖥 = Computer useful in solving problem 🌐 = Interactive Physics
▭ = paired numerical/symbolic problems

Section 23.1 Properties of Electric Charges
Section 23.2 Insulators and Conductors
Section 23.3 Coulomb's Law

1. (a) Calculate the number of electrons in a small, electrically neutral silver pin that has a mass of 10.0 g. Silver has 47 electrons per atom, and its molar mass is 107.87 g/mol. (b) Electrons are added to the pin until the net negative charge is 1.00 mC. How many electrons are added for every 10^9 electrons already present?

2. (a) Two protons in a molecule are separated by a distance of 3.80×10^{-10} m. Find the electric force exerted by one proton on the other. (b) How does the magnitude of this force compare with the magnitude of the gravitational force between the two protons? (c) What must be the charge-to-mass ratio of a particle if the magnitude of the gravitational force between two of these particles equals the magnitude of the electric force between them?

WEB 3. Richard Feynman once said that if two persons stood at arm's length from each other and each person had 1% more electrons than protons, the force of repulsion between them would be enough to lift a "weight" equal to that of the entire Earth. Carry out an order-of-magnitude calculation to substantiate this assertion.

4. Two small silver spheres, each with a mass of 10.0 g, are separated by 1.00 m. Calculate the fraction of the elec-

trons in one sphere that must be transferred to the other to produce an attractive force of 1.00×10^4 N (about 1 ton) between the spheres. (The number of electrons per atom of silver is 47, and the number of atoms per gram is Avogadro's number divided by the molar mass of silver, 107.87 g/mol.)

5. Suppose that 1.00 g of hydrogen is separated into electrons and protons. Suppose also that the protons are placed at the Earth's north pole and the electrons are placed at the south pole. What is the resulting compressional force on the Earth?

6. Two identical conducting small spheres are placed with their centers 0.300 m apart. One is given a charge of 12.0 nC, and the other is given a charge of -18.0 nC. (a) Find the electric force exerted on one sphere by the other. (b) The spheres are connected by a conducting wire. Find the electric force between the two after equilibrium has occurred.

7. Three point charges are located at the corners of an equilateral triangle, as shown in Figure P23.7. Calculate the net electric force on the 7.00-μC charge.

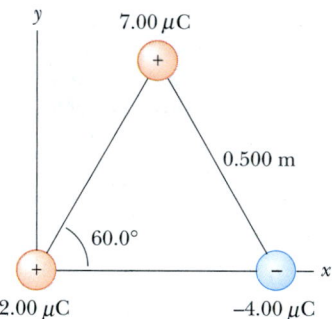

7.00 μC

0.500 m

60.0°

2.00 μC -4.00 μC

Figure P23.7   Problems 7 and 15.

8. Two small beads having positive charges $3q$ and q are fixed at the opposite ends of a horizontal insulating rod extending from the origin to the point $x = d$. As shown in Figure P23.8, a third small charged bead is free to slide on the rod. At what position is the third bead in equilibrium? Can it be in stable equilibrium?

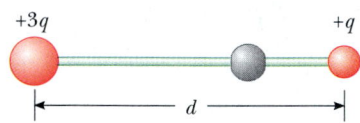

$+3q$ $+q$

d

Figure P23.8

9. **Review Problem.**   In the Bohr theory of the hydrogen atom, an electron moves in a circular orbit about a proton, where the radius of the orbit is 0.529×10^{-10} m. (a) Find the electric force between the two. (b) If this force causes the centripetal acceleration of the electron, what is the speed of the electron?

10. **Review Problem.**   Two identical point charges each having charge $+q$ are fixed in space and separated by a distance d. A third point charge $-Q$ of mass m is free to move and lies initially at rest on a perpendicular bisector of the two fixed charges a distance x from the midpoint of the two fixed charges (Fig. P23.10). (a) Show that if x is small compared with d, the motion of $-Q$ is simple harmonic along the perpendicular bisector. Determine the period of that motion. (b) How fast will the charge $-Q$ be moving when it is at the midpoint between the two fixed charges, if initially it is released at a distance $a \ll d$ from the midpoint?

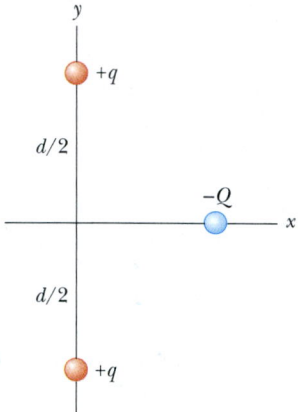

y

$+q$

$d/2$

$-Q$

x

$d/2$

$+q$

Figure P23.10

Section 23.4   The Electric Field

11. What are the magnitude and direction of the electric field that will balance the weight of (a) an electron and (b) a proton? (Use the data in Table 23.1.)

12. An object having a net charge of 24.0 μC is placed in a uniform electric field of 610 N/C that is directed vertically. What is the mass of this object if it "floats" in the field?

13. In Figure P23.13, determine the point (other than infinity) at which the electric field is zero.

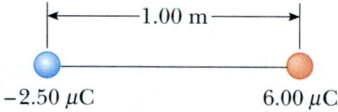

$\leftarrow$——1.00 m——$\rightarrow$

-2.50 μC   6.00 μC

Figure P23.13

14. An airplane is flying through a thundercloud at a height of 2 000 m. (This is a very dangerous thing to do because of updrafts, turbulence, and the possibility of electric discharge.) If there are charge concentrations of $+40.0$ C at a height of 3 000 m within the cloud and of -40.0 C at a height of 1 000 m, what is the electric field E at the aircraft?

15. Three charges are at the corners of an equilateral triangle, as shown in Figure P23.7. (a) Calculate the electric field at the position of the 2.00-μC charge due to the 7.00-μC and $-$4.00-μC charges. (b) Use your answer to part (a) to determine the force on the 2.00-μC charge.

16. Three point charges are arranged as shown in Figure P23.16. (a) Find the vector electric field that the 6.00-nC and $-$3.00-nC charges together create at the origin. (b) Find the vector force on the 5.00-nC charge.

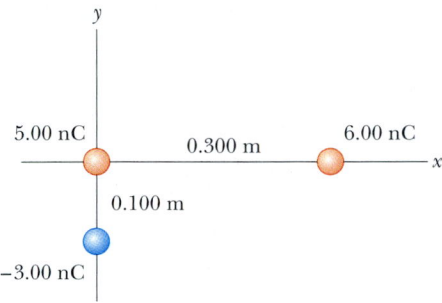

Figure P23.16

WEB 17. Three equal positive charges q are at the corners of an equilateral triangle of side a, as shown in Figure P23.17. (a) Assume that the three charges together create an electric field. Find the location of a point (other than ∞) where the electric field is zero. (*Hint:* Sketch the field lines in the plane of the charges.) (b) What are the magnitude and direction of the electric field at P due to the two charges at the base?

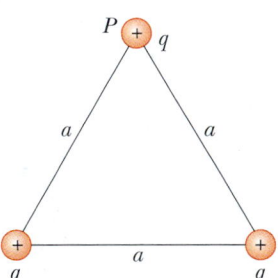

Figure P23.17

18. Two 2.00-μC point charges are located on the x axis. One is at $x = 1.00$ m, and the other is at $x = -1.00$ m. (a) Determine the electric field on the y axis at $y = 0.500$ m. (b) Calculate the electric force on a $-$3.00-μC charge placed on the y axis at $y = 0.500$ m.

19. Four point charges are at the corners of a square of side a, as shown in Figure P23.19. (a) Determine the magnitude and direction of the electric field at the location of charge q. (b) What is the resultant force on q?

20. A point particle having charge q is located at point (x_0, y_0) in the xy plane. Show that the x and y compo-

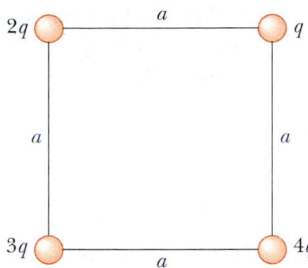

Figure P23.19

nents of the electric field at point (x, y) due to this charge q are

$$E_x = \frac{k_e q(x - x_0)}{[(x - x_0)^2 + (y - y_0)^2]^{3/2}}$$

$$E_y = \frac{k_e q(y - y_0)}{[(x - x_0)^2 + (y - y_0)^2]^{3/2}}$$

21. Consider the electric dipole shown in Figure P23.21. Show that the electric field at a *distant* point along the x axis is $E_x \cong 4k_e qa/x^3$.

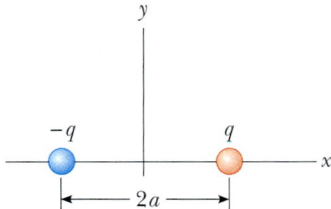

Figure P23.21

22. Consider n equal positive point charges each of magnitude Q/n placed symmetrically around a circle of radius R. (a) Calculate the magnitude of the electric field E at a point a distance x on the line passing through the center of the circle and perpendicular to the plane of the circle. (b) Explain why this result is identical to the one obtained in Example 23.8.

23. Consider an infinite number of identical charges (each of charge q) placed along the x axis at distances a, $2a$, $3a$, $4a$, . . . from the origin. What is the electric field at the origin due to this distribution? *Hint:* Use the fact that

$$1 + \frac{1}{2^2} + \frac{1}{3^2} + \frac{1}{4^2} + \cdots = \frac{\pi^2}{6}$$

Section 23.5 Electric Field of a Continuous Charge Distribution

24. A rod 14.0 cm long is uniformly charged and has a total charge of $-$22.0 μC. Determine the magnitude and direction of the electric field along the axis of the rod at a point 36.0 cm from its center.

25. A continuous line of charge lies along the x axis, extending from $x = +x_0$ to positive infinity. The line carries a uniform linear charge density λ_0. What are the magnitude and direction of the electric field at the origin?

26. A line of charge starts at $x = +x_0$ and extends to positive infinity. If the linear charge density is $\lambda = \lambda_0 x_0/x$, determine the electric field at the origin.

27. A uniformly charged ring of radius 10.0 cm has a total charge of 75.0 μC. Find the electric field on the axis of the ring at (a) 1.00 cm, (b) 5.00 cm, (c) 30.0 cm, and (d) 100 cm from the center of the ring.

28. Show that the maximum field strength E_{max} along the axis of a uniformly charged ring occurs at $x = a/\sqrt{2}$ (see Fig. 23.17) and has the value $Q/(6\sqrt{3}\pi\epsilon_0 a^2)$.

29. A uniformly charged disk of radius 35.0 cm carries a charge density of 7.90×10^{-3} C/m^2. Calculate the electric field on the axis of the disk at (a) 5.00 cm, (b) 10.0 cm, (c) 50.0 cm, and (d) 200 cm from the center of the disk.

30. Example 23.9 derives the exact expression for the electric field at a point on the axis of a uniformly charged disk. Consider a disk of radius $R = 3.00$ cm having a uniformly distributed charge of $+5.20$ μC. (a) Using the result of Example 23.9, compute the electric field at a point on the axis and 3.00 mm from the center. Compare this answer with the field computed from the near-field approximation $E = \sigma/2\epsilon_0$. (b) Using the result of Example 23.9, compute the electric field at a point on the axis and 30.0 cm from the center of the disk. Compare this result with the electric field obtained by treating the disk as a $+5.20$-μC point charge at a distance of 30.0 cm.

31. The electric field along the axis of a uniformly charged disk of radius R and total charge Q was calculated in Example 23.9. Show that the electric field at distances x that are great compared with R approaches that of a point charge $Q = \sigma\pi R^2$. (*Hint:* First show that $x/(x^2 + R^2)^{1/2} = (1 + R^2/x^2)^{-1/2}$, and use the binomial expansion $(1 + \delta)^n \approx 1 + n\delta$ when $\delta \ll 1$.)

32. A piece of Styrofoam having a mass m carries a net charge of $-q$ and floats above the center of a very large horizontal sheet of plastic that has a uniform charge density on its surface. What is the charge per unit area on the plastic sheet?

WEB **33.** A uniformly charged insulating rod of length 14.0 cm is bent into the shape of a semicircle, as shown in Figure P23.33. The rod has a total charge of -7.50 μC. Find the magnitude and direction of the electric field at O, the center of the semicircle.

34. (a) Consider a uniformly charged right circular cylindrical shell having total charge Q, radius R, and height h. Determine the electric field at a point a distance d from the right side of the cylinder, as shown in Figure P23.34. (*Hint:* Use the result of Example 23.8 and treat the cylinder as a collection of ring charges.) (b) Consider now a solid cylinder with the same dimensions and

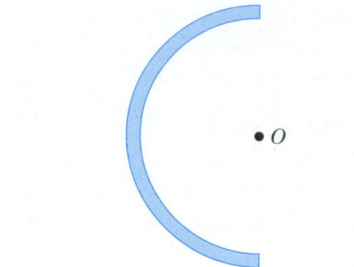

Figure P23.33

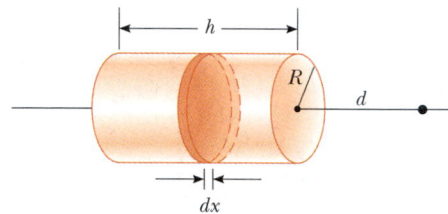

Figure P23.34

carrying the same charge, which is uniformly distributed through its volume. Use the result of Example 23.9 to find the field it creates at the same point.

35. A thin rod of length ℓ and uniform charge per unit length λ lies along the x axis, as shown in Figure P23.35. (a) Show that the electric field at P, a distance y from the rod, along the perpendicular bisector has no x component and is given by $E = 2k_e\lambda \sin \theta_0/y$. (b) Using your result to part (a), show that the field of a rod of infinite length is $E = 2k_e\lambda/y$. (*Hint:* First calculate the field at P due to an element of length dx, which has a charge $\lambda \, dx$. Then change variables from x to θ, using the facts that $x = y \tan \theta$ and $dx = y \sec^2 \theta \, d\theta$, and integrate over θ.)

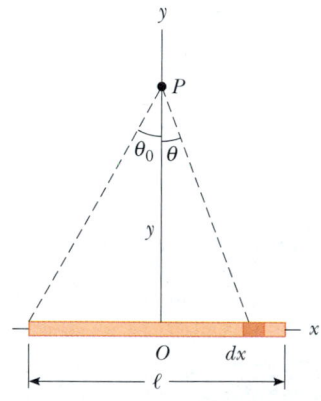

Figure P23.35

36. Three solid plastic cylinders all have a radius of 2.50 cm and a length of 6.00 cm. One (a) carries charge with uniform density 15.0 nC/m² everywhere on its surface. Another (b) carries charge with the same uniform density on its curved lateral surface only. The third (c) carries charge with uniform density 500 nC/m³ throughout the plastic. Find the charge of each cylinder.

37. Eight solid plastic cubes, each 3.00 cm on each edge, are glued together to form each one of the objects (i, ii, iii, and iv) shown in Figure P23.37. (a) If each object carries charge with a uniform density of 400 nC/m³ throughout its volume, what is the charge of each object? (b) If each object is given charge with a uniform density of 15.0 nC/m² everywhere on its exposed surface, what is the charge on each object? (c) If charge is placed only on the edges where perpendicular surfaces meet, with a uniform density of 80.0 pC/m, what is the charge of each object?

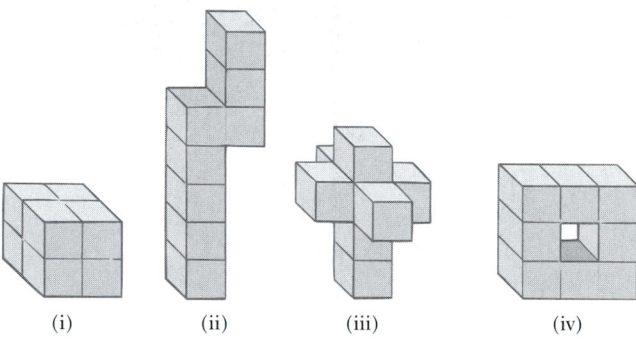

(i)   (ii)   (iii)   (iv)

Figure P23.37

Section 23.6   Electric Field Lines

38. A positively charged disk has a uniform charge per unit area as described in Example 23.9. Sketch the electric field lines in a plane perpendicular to the plane of the disk passing through its center.

39. A negatively charged rod of finite length has a uniform charge per unit length. Sketch the electric field lines in a plane containing the rod.

40. Figure P23.40 shows the electric field lines for two point charges separated by a small distance. (a) Determine the ratio q_1/q_2. (b) What are the signs of q_1 and q_2?

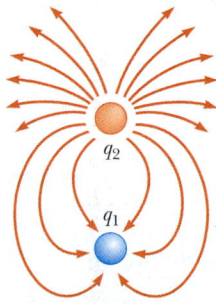

Figure P23.40

Section 23.7   Motion of Charged Particles in a Uniform Electric Field

41. An electron and a proton are each placed at rest in an electric field of 520 N/C. Calculate the speed of each particle 48.0 ns after being released.

42. A proton is projected in the positive x direction into a region of uniform electric field $\mathbf{E} = -6.00 \times 10^5 \mathbf{i}$ N/C. The proton travels 7.00 cm before coming to rest. Determine (a) the acceleration of the proton, (b) its initial speed, and (c) the time it takes the proton to come to rest.

43. A proton accelerates from rest in a uniform electric field of 640 N/C. At some later time, its speed has reached 1.20×10^6 m/s (nonrelativistic, since v is much less than the speed of light). (a) Find the acceleration of the proton. (b) How long does it take the proton to reach this speed? (c) How far has it moved in this time? (d) What is its kinetic energy at this time?

44. The electrons in a particle beam each have a kinetic energy of 1.60×10^{-17} J. What are the magnitude and direction of the electric field that stops these electrons in a distance of 10.0 cm?

WEB 45. The electrons in a particle beam each have a kinetic energy K. What are the magnitude and direction of the electric field that stops these electrons in a distance d?

46. A positively charged bead having a mass of 1.00 g falls from rest in a vacuum from a height of 5.00 m in a uniform vertical electric field with a magnitude of 1.00×10^4 N/C. The bead hits the ground at a speed of 21.0 m/s. Determine (a) the direction of the electric field (up or down) and (b) the charge on the bead.

47. A proton moves at 4.50×10^5 m/s in the horizontal direction. It enters a uniform vertical electric field with a magnitude of 9.60×10^3 N/C. Ignoring any gravitational effects, find (a) the time it takes the proton to travel 5.00 cm horizontally, (b) its vertical displacement after it has traveled 5.00 cm horizontally, and (c) the horizontal and vertical components of its velocity after it has traveled 5.00 cm horizontally.

48. An electron is projected at an angle of 30.0° above the horizontal at a speed of 8.20×10^5 m/s in a region where the electric field is $\mathbf{E} = 390\mathbf{j}$ N/C. Neglecting the effects of gravity, find (a) the time it takes the electron to return to its initial height, (b) the maximum height it reaches, and (c) its horizontal displacement when it reaches its maximum height.

49. Protons are projected with an initial speed $v_i = 9.55 \times 10^3$ m/s into a region where a uniform electric field $\mathbf{E} = (-720\mathbf{j})$ N/C is present, as shown in Figure P23.49. The protons are to hit a target that lies at a horizontal distance of 1.27 mm from the point where the protons are launched. Find (a) the two projection angles θ that result in a hit and (b) the total time of flight for each trajectory.

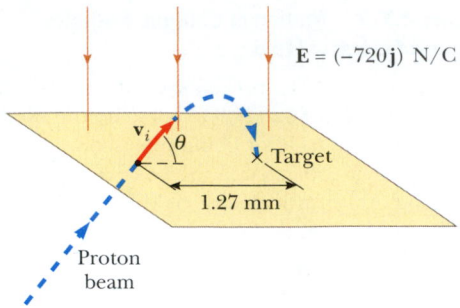

Figure P23.49

ADDITIONAL PROBLEMS

50. Three point charges are aligned along the x axis as shown in Figure P23.50. Find the electric field at (a) the position (2.00, 0) and (b) the position (0, 2.00).

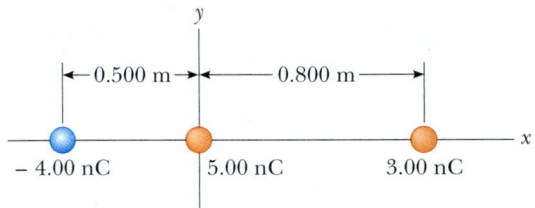

Figure P23.50

51. A uniform electric field of magnitude 640 N/C exists between two parallel plates that are 4.00 cm apart. A proton is released from the positive plate at the same instant that an electron is released from the negative plate. (a) Determine the distance from the positive plate at which the two pass each other. (Ignore the electrical attraction between the proton and electron.) (b) Repeat part (a) for a sodium ion (Na^+) and a chlorine ion (Cl^-).

52. A small, 2.00-g plastic ball is suspended by a 20.0-cm-long string in a uniform electric field, as shown in Figure P23.52. If the ball is in equilibrium when the string

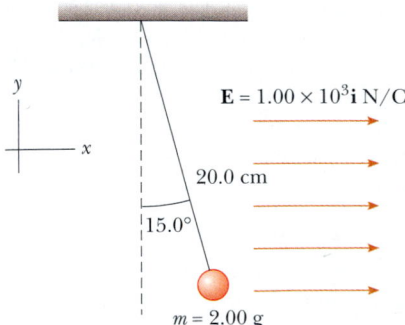

Figure P23.52

makes a 15.0° angle with the vertical, what is the net charge on the ball?

WEB **53.** A charged cork ball of mass 1.00 g is suspended on a light string in the presence of a uniform electric field, as shown in Figure P23.53. When $\mathbf{E} = (3.00\mathbf{i} + 5.00\mathbf{j}) \times 10^5$ N/C, the ball is in equilibrium at $\theta = 37.0°$. Find (a) the charge on the ball and (b) the tension in the string.

54. A charged cork ball of mass m is suspended on a light string in the presence of a uniform electric field, as shown in Figure P23.53. When $\mathbf{E} = (A\mathbf{i} + B\mathbf{j})$ N/C, where A and B are positive numbers, the ball is in equilibrium at the angle θ. Find (a) the charge on the ball and (b) the tension in the string.

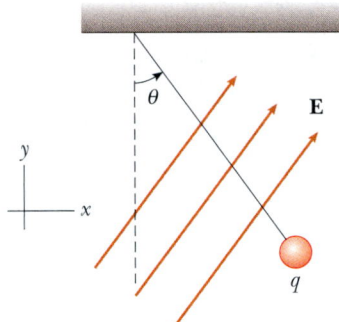

Figure P23.53   Problems 53 and 54.

55. Four identical point charges ($q = +10.0 \ \mu$C) are located on the corners of a rectangle, as shown in Figure P23.55. The dimensions of the rectangle are $L = 60.0$ cm and $W = 15.0$ cm. Calculate the magnitude and direction of the net electric force exerted on the charge at the lower left corner by the other three charges.

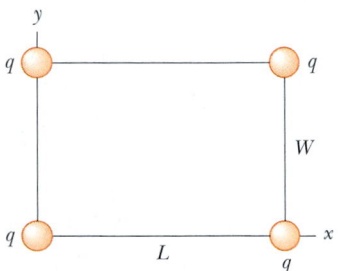

Figure P23.55

56. Three identical small Styrofoam balls ($m = 2.00$ g) are suspended from a fixed point by three nonconducting threads, each with a length of 50.0 cm and with negligible mass. At equilibrium the three balls form an equilateral triangle with sides of 30.0 cm. What is the common charge q carried by each ball?

57. Two identical metallic blocks resting on a frictionless horizontal surface are connected by a light metallic spring having the spring constant $k = 100$ N/m and an unstretched length of 0.300 m, as shown in Figure P23.57a. A total charge of Q is slowly placed on the system, causing the spring to stretch to an equilibrium length of 0.400 m, as shown in Figure P23.57b. Determine the value of Q, assuming that all the charge resides on the blocks and that the blocks are like point charges.

58. Two identical metallic blocks resting on a frictionless horizontal surface are connected by a light metallic spring having a spring constant k and an unstretched length L_i, as shown in Figure P23.57a. A total charge of Q is slowly placed on the system, causing the spring to stretch to an equilibrium length L, as shown in Figure P23.57b. Determine the value of Q, assuming that all the charge resides on the blocks and that the blocks are like point charges.

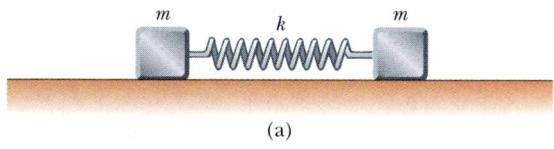

(a)

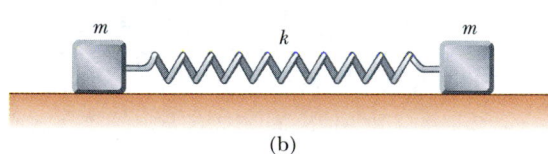

(b)

Figure P23.57   Problems 57 and 58.

59. Identical thin rods of length $2a$ carry equal charges, $+Q$, uniformly distributed along their lengths. The rods lie along the x axis with their centers separated by a distance of $b > 2a$ (Fig. P23.59). Show that the magnitude of the force exerted by the left rod on the right one is given by

$$F = \left(\frac{k_e Q^2}{4a^2} \right) \ln \left(\frac{b^2}{b^2 - 4a^2} \right)$$

60. A particle is said to be nonrelativistic as long as its speed is less than one-tenth the speed of light, or less than 3.00×10^7 m/s. (a) How long will an electron remain nonrelativistic if it starts from rest in a region of an electric field of 1.00 N/C? (b) How long will a proton remain nonrelativistic in the same electric field? (c) Electric fields are commonly much larger than

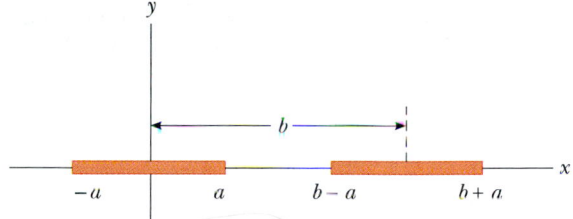

Figure P23.59

1 N/C. Will the charged particle remain nonrelativistic for a shorter or a longer time in a much larger electric field?

61. A line of positive charge is formed into a semicircle of radius $R = 60.0$ cm, as shown in Figure P23.61. The charge per unit length along the semicircle is described by the expression $\lambda = \lambda_0 \cos \theta$. The total charge on the semicircle is 12.0 μC. Calculate the total force on a charge of 3.00 μC placed at the center of curvature.

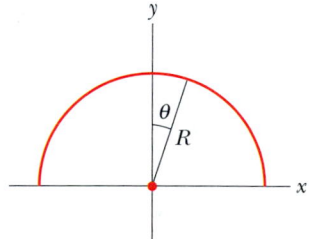

Figure P23.61

62. Two small spheres, each of mass 2.00 g, are suspended by light strings 10.0 cm in length (Fig. P23.62). A uniform electric field is applied in the x direction. The spheres have charges equal to -5.00×10^{-8} C and $+5.00 \times 10^{-8}$ C. Determine the electric field that enables the spheres to be in equilibrium at an angle of $\theta = 10.0°$.

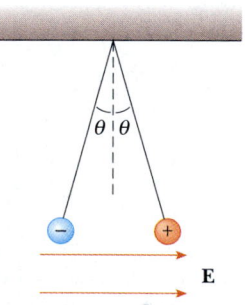

Figure P23.62

63. Two small spheres of mass m are suspended from strings of length ℓ that are connected at a common point. One sphere has charge Q; the other has charge $2Q$. Assume that the angles θ_1 and θ_2 that the strings make with the vertical are small. (a) How are θ_1 and θ_2 related? (b) Show that the distance r between the spheres is

$$r \cong \left(\frac{4k_eQ^2\ell}{mg} \right)^{1/3}$$

64. Three charges of equal magnitude q are fixed in position at the vertices of an equilateral triangle (Fig. P23.64). A fourth charge Q is free to move along the positive x axis under the influence of the forces exerted by the three fixed charges. Find a value for s for which Q is in equilibrium. You will need to solve a transcendental equation.

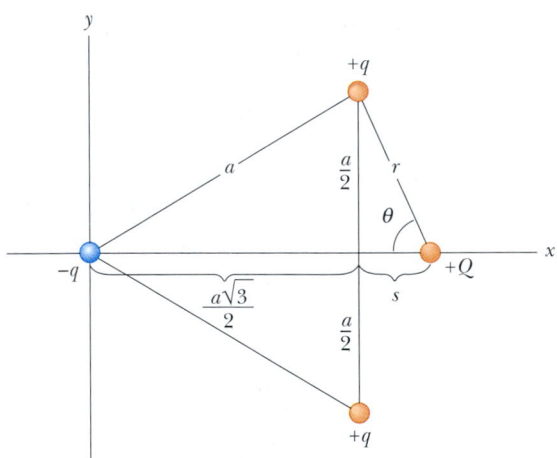

Figure P23.64

65. Review Problem. Four identical point charges, each having charge $+q$, are fixed at the corners of a square of side L. A fifth point charge $-Q$ lies a distance z along the line perpendicular to the plane of the square and passing through the center of the square (Fig. P23.65). (a) Show that the force exerted on $-Q$ by the other four charges is

$$\mathbf{F} = -\frac{4k_eqQz}{\left(z^2 + \dfrac{L^2}{2} \right)^{3/2}}\, \mathbf{k}$$

Note that this force is directed toward the center of the square whether z is positive ($-Q$ above the square) or negative ($-Q$ below the square). (b) If z is small compared with L, the above expression reduces to $\mathbf{F} \approx -(\text{constant})\, z\mathbf{k}$. Why does this imply that the motion of $-Q$ is simple harmonic, and what would be the period of this motion if the mass of $-Q$ were m?

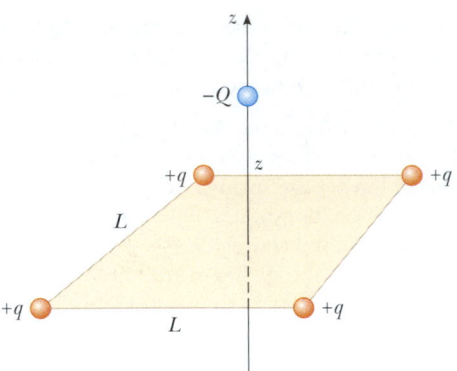

Figure P23.65

66. Review Problem. A 1.00-g cork ball with a charge of 2.00 μC is suspended vertically on a 0.500-m-long light string in the presence of a uniform, downward-directed electric field of magnitude $E = 1.00 \times 10^5$ N/C. If the ball is displaced slightly from the vertical, it oscillates like a simple pendulum. (a) Determine the period of this oscillation. (b) Should gravity be included in the calculation for part (a)? Explain.

67. Three charges of equal magnitude q reside at the corners of an equilateral triangle of side length a (Fig. P23.67). (a) Find the magnitude and direction of the electric field at point P, midway between the negative charges, in terms of k_e, q, and a. (b) Where must a $-4q$ charge be placed so that any charge located at P experiences no net electric force? In part (b), let P be the origin and let the distance between the $+q$ charge and P be 1.00 m.

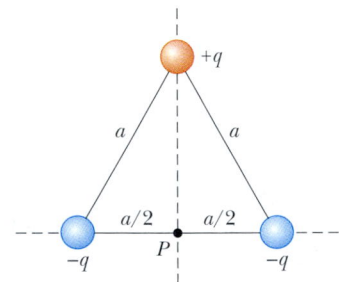

Figure P23.67

68. Two identical beads each have a mass m and charge q. When placed in a hemispherical bowl of radius R with frictionless, nonconducting walls, the beads move, and at equilibrium they are a distance R apart (Fig. P23.68). Determine the charge on each bead.

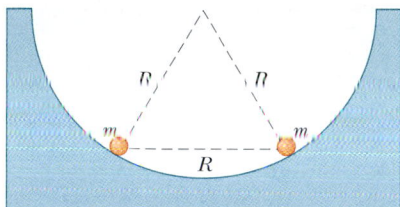

Figure P23.68

69. Eight point charges, each of magnitude q, are located on the corners of a cube of side s, as shown in Figure P23.69. (a) Determine the x, y, and z components of the resultant force exerted on the charge located at point A by the other charges. (b) What are the magnitude and direction of this resultant force?

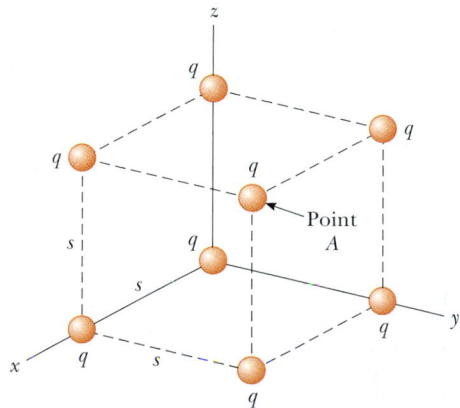

Figure P23.69   Problems 69 and 70.

70. Consider the charge distribution shown in Figure P23.69. (a) Show that the magnitude of the electric field at the center of any face of the cube has a value of $2.18k_e q/s^2$. (b) What is the direction of the electric field at the center of the top face of the cube?

71. A line of charge with a uniform density of 35.0 nC/m lies along the line $y = -15.0$ cm, between the points with coordinates $x = 0$ and $x = 40.0$ cm. Find the electric field it creates at the origin.

72. Three point charges q, $-2q$, and q are located along the x axis, as shown in Figure P23.72. Show that the electric field at P ($y \gg a$) along the y axis is

$$\mathbf{E} = -k_e \frac{3qa^2}{y^4} \mathbf{j}$$

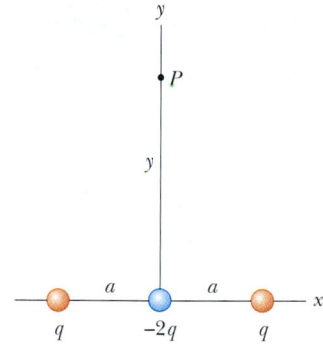

Figure P23.72

This charge distribution, which is essentially that of two electric dipoles, is called an *electric quadrupole*. Note that $\mathbf{E}$ varies as r^{-4} for the quadrupole, compared with variations of r^{-3} for the dipole and r^{-2} for the monopole (a single charge).

73. **Review Problem.**   A negatively charged particle $-q$ is placed at the center of a uniformly charged ring, where the ring has a total positive charge Q, as shown in Example 23.8. The particle, confined to move along the x axis, is displaced a *small* distance x along the axis (where $x \ll a$) and released. Show that the particle oscillates with simple harmonic motion with a frequency

$$f = \frac{1}{2\pi} \left(\frac{k_e qQ}{ma^3} \right)^{1/2}$$

74. **Review Problem.**   An electric dipole in a uniform electric field is displaced slightly from its equilibrium position, as shown in Figure P23.74, where θ is small. The moment of inertia of the dipole is I. If the dipole is released from this position, show that its angular orientation exhibits simple harmonic motion with a frequency

$$f = \frac{1}{2\pi} \sqrt{\frac{2qaE}{I}}$$

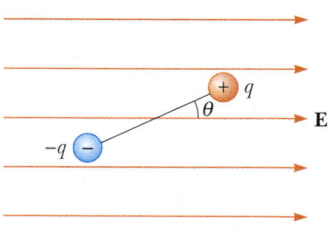

Figure P23.74

Answers to Quick Quizzes

23.1 (b). The amount of charge present after rubbing is the same as that before; it is just distributed differently.

23.2 (d). Object A might be negatively charged, but it also might be electrically neutral with an induced charge separation, as shown in the following figure:

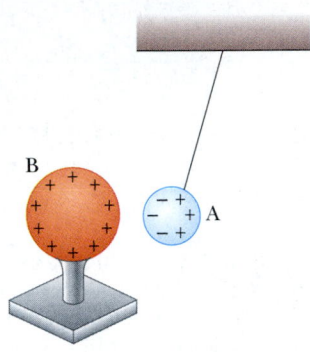

23.3 (b). From Newton's third law, the electric force exerted by object B on object A is equal in magnitude to the force exerted by object A on object B and in the opposite direction—that is, $\mathbf{F}_{AB} = -\mathbf{F}_{BA}$.

23.4 Nothing, if we assume that the source charge producing the field is not disturbed by our actions. Remember that the electric field is created not by the $+3\text{-}\mu\text{C}$ charge or by the $-3\text{-}\mu\text{C}$ charge but by the source charge (unseen in this case).

23.5 *A*, *B*, and *C*. The field is greatest at point *A* because this is where the field lines are closest together. The absence of lines at point *C* indicates that the electric field there is zero.

chapter

24

Gauss's Law

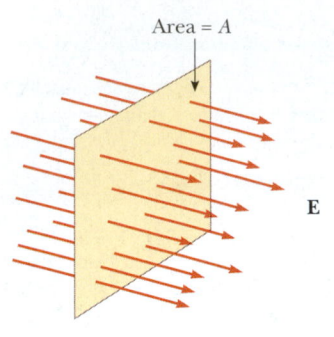

Area = A

E

Figure 24.1 Field lines representing a uniform electric field penetrating a plane of area *A* perpendicular to the field. The electric flux Φ_E through this area is equal to *EA*.

In the preceding chapter we showed how to use Coulomb's law to calculate the electric field generated by a given charge distribution. In this chapter, we describe *Gauss's law* and an alternative procedure for calculating electric fields. The law is based on the fact that the fundamental electrostatic force between point charges exhibits an inverse-square behavior. Although a consequence of Coulomb's law, Gauss's law is more convenient for calculating the electric fields of highly symmetric charge distributions and makes possible useful qualitative reasoning when we are dealing with complicated problems.

24.1 ELECTRIC FLUX

11.6 The concept of electric field lines is described qualitatively in Chapter 23. We now use the concept of electric flux to treat electric field lines in a more quantitative way.

Consider an electric field that is uniform in both magnitude and direction, as shown in Figure 24.1. The field lines penetrate a rectangular surface of area *A*, which is perpendicular to the field. Recall from Section 23.6 that the number of lines per unit area (in other words, the *line density*) is proportional to the magnitude of the electric field. Therefore, the total number of lines penetrating the surface is proportional to the product *EA*. This product of the magnitude of the electric field *E* and surface area *A* perpendicular to the field is called the **electric flux** Φ_E (uppercase Greek phi):

$$\Phi_E = EA \tag{24.1}$$

From the SI units of *E* and *A*, we see that Φ_E has units of newton–meters squared per coulomb ($\text{N} \cdot \text{m}^2/\text{C}$). **Electric flux is proportional to the number of electric field lines penetrating some surface.**

EXAMPLE 24.1   Flux Through a Sphere

What is the electric flux through a sphere that has a radius of 1.00 m and carries a charge of + 1.00 µC at its center?

Solution The magnitude of the electric field 1.00 m from this charge is given by Equation 23.4,

$$E = k_e \frac{q}{r^2} = (8.99 \times 10^9 \, \text{N} \cdot \text{m}^2/\text{C}^2) \frac{1.00 \times 10^{-6} \, \text{C}}{(1.00 \, \text{m})^2}$$

$$= 8.99 \times 10^3 \, \text{N/C}$$

The field points radially outward and is therefore everywhere

perpendicular to the surface of the sphere. The flux through the sphere (whose surface area $A = 4\pi r^2 = 12.6 \, \text{m}^2$) is thus

$$\Phi_E = EA = (8.99 \times 10^3 \, \text{N/C})(12.6 \, \text{m}^2)$$

$$= 1.13 \times 10^5 \, \text{N} \cdot \text{m}^2/\text{C}$$

Exercise What would be the (a) electric field and (b) flux through the sphere if it had a radius of 0.500 m?

Answer (a) $3.60 \times 10^4 \, \text{N/C}$; (b) $1.13 \times 10^5 \, \text{N} \cdot \text{m}^2/\text{C}$.

If the surface under consideration is not perpendicular to the field, the flux through it must be less than that given by Equation 24.1. We can understand this by considering Figure 24.2, in which the normal to the surface of area *A* is at an angle θ to the uniform electric field. Note that the number of lines that cross this area *A* is equal to the number that cross the area *A'*, which is a projection of area *A* aligned perpendicular to the field. From Figure 24.2 we see that the two areas are related by $A' = A \cos \theta$. Because the flux through *A* equals the flux through *A'*, we

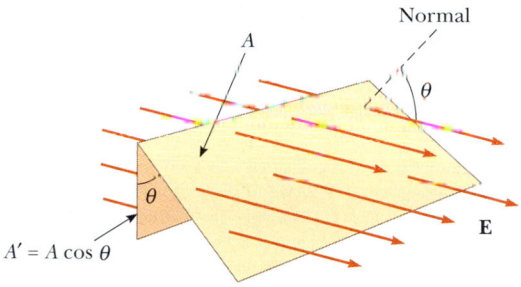

Figure 24.2 Field lines representing a uniform electric field penetrating an area A that is at an angle θ to the field. Because the number of lines that go through the area A' is the same as the number that go through A, the flux through A' is equal to the flux through A and is given by $\Phi_E = EA \cos \theta$.

$A' = A \cos \theta$

QuickLab

Shine a desk lamp onto a playing card and notice how the size of the shadow on your desk depends on the orientation of the card with respect to the beam of light. Could a formula like Equation 24.2 be used to describe how much light was being blocked by the card?

conclude that the flux through A is

$$\Phi_E = EA' = EA \cos \theta \qquad (24.2)$$

From this result, we see that the flux through a surface of fixed area A has a maximum value EA when the surface is perpendicular to the field (in other words, when the normal to the surface is parallel to the field, that is, $\theta = 0°$ in Figure 24.2); the flux is zero when the surface is parallel to the field (in other words, when the normal to the surface is perpendicular to the field, that is, $\theta = 90°$).

We assumed a uniform electric field in the preceding discussion. In more general situations, the electric field may vary over a surface. Therefore, our definition of flux given by Equation 24.2 has meaning only over a small element of area. Consider a general surface divided up into a large number of small elements, each of area ΔA. The variation in the electric field over one element can be neglected if the element is sufficiently small. It is convenient to define a vector $\Delta \mathbf{A}_i$ whose magnitude represents the area of the ith element of the surface and whose direction is *defined to be perpendicular* to the surface element, as shown in Figure 24.3. The electric flux $\Delta \Phi_E$ through this element is

$$\Delta \Phi_E = E_i \Delta A_i \cos \theta = \mathbf{E}_i \cdot \Delta \mathbf{A}_i$$

where we have used the definition of the scalar product of two vectors ($\mathbf{A} \cdot \mathbf{B} = AB \cos \theta$). By summing the contributions of all elements, we obtain the total flux through the surface.[1] If we let the area of each element approach zero, then the number of elements approaches infinity and the sum is replaced by an integral. Therefore, the general definition of electric flux is

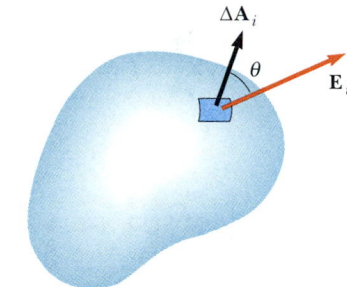

Figure 24.3 A small element of surface area ΔA_i. The electric field makes an angle θ with the vector $\Delta \mathbf{A}_i$, defined as being normal to the surface element, and the flux through the element is equal to $E_i \Delta A_i \cos \theta$.

$$\Phi_E = \lim_{\Delta A_i \to 0} \sum \mathbf{E}_i \cdot \Delta \mathbf{A}_i = \int_{\text{surface}} \mathbf{E} \cdot d\mathbf{A} \qquad (24.3)$$

Definition of electric flux

Equation 24.3 is a *surface integral*, which means it must be evaluated over the surface in question. In general, the value of Φ_E depends both on the field pattern and on the surface.

We are often interested in evaluating the flux through a *closed surface*, which is defined as one that divides space into an inside and an outside region, so that one cannot move from one region to the other without crossing the surface. The surface of a sphere, for example, is a closed surface.

Consider the closed surface in Figure 24.4. The vectors $\Delta \mathbf{A}_i$ point in different directions for the various surface elements, but at each point they are normal to

[1] It is important to note that drawings with field lines have their inaccuracies because a small area element (depending on its location) may happen to have too many or too few field lines penetrating it. We stress that the basic definition of electric flux is $\int \mathbf{E} \cdot d\mathbf{A}$. The use of lines is only an aid for visualizing the concept.

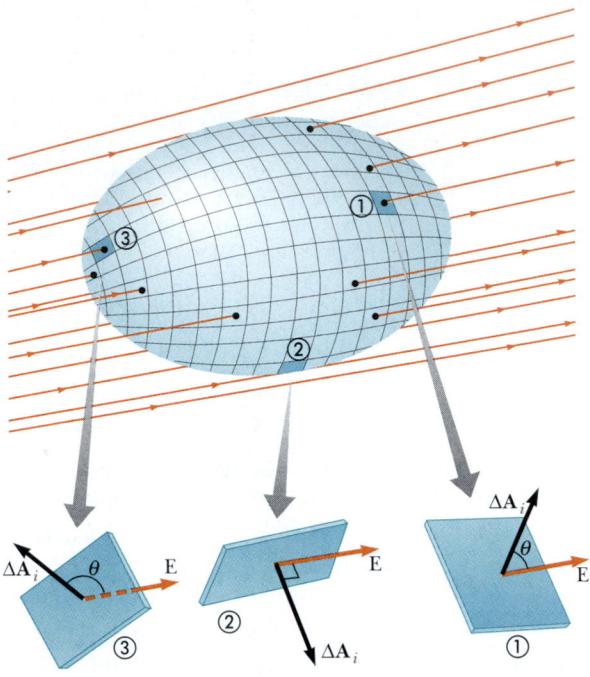

Figure 24.4   A closed surface in an electric field. The area vectors $\Delta\mathbf{A}_i$ are, by convention, normal to the surface and point outward. The flux through an area element can be positive (element ①), zero (element ②), or negative (element ③).

Karl Friedrich Gauss German mathematician and astronomer (1777 – 1855)

the surface and, by convention, always point outward. At the element labeled ①, the field lines are crossing the surface from the inside to the outside and $\theta < 90°$; hence, the flux $\Delta\Phi_E = \mathbf{E} \cdot \Delta\mathbf{A}_i$ through this element is positive. For element ②, the field lines graze the surface (perpendicular to the vector $\Delta\mathbf{A}_i$); thus, $\theta = 90°$ and the flux is zero. For elements such as ③, where the field lines are crossing the surface from outside to inside, $180° > \theta > 90°$ and the flux is negative because $\cos\theta$ is negative. The *net* flux through the surface is proportional to the net number of lines leaving the surface, where the net number means *the number leaving the surface minus the number entering the surface*. If more lines are leaving than entering, the net flux is positive. If more lines are entering than leaving, the net flux is negative. Using the symbol $\oint$ to represent an integral over a closed surface, we can write the net flux Φ_E through a closed surface as

$$\Phi_E = \oint \mathbf{E} \cdot d\mathbf{A} = \oint E_n \, dA \tag{24.4}$$

where E_n represents the component of the electric field normal to the surface. Evaluating the net flux through a closed surface can be very cumbersome. However, if the field is normal to the surface at each point and constant in magnitude, the calculation is straightforward, as it was in Example 24.1. The next example also illustrates this point.

EXAMPLE 24.2   Flux Through a Cube

Consider a uniform electric field $\mathbf{E}$ oriented in the x direction. Find the net electric flux through the surface of a cube of edges ℓ, oriented as shown in Figure 24.5.

Solution   The net flux is the sum of the fluxes through all faces of the cube. First, note that the flux through four of the

faces (③, ④, and the unnumbered ones) is zero because $\mathbf{E}$ is perpendicular to $d\mathbf{A}$ on these faces.

The net flux through faces ① and ② is

$$\Phi_E = \int_1 \mathbf{E} \cdot d\mathbf{A} + \int_2 \mathbf{E} \cdot d\mathbf{A}$$

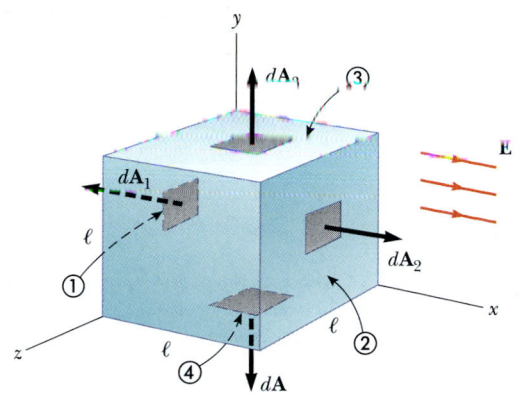

Figure 24.5 A closed surface in the shape of a cube in a uniform electric field oriented parallel to the x axis. The net flux through the closed surface is zero. Side ④ is the bottom of the cube, and side ① is opposite side ②.

For ①, **E** is constant and directed inward but $d\mathbf{A}$ is directed outward ($\theta = 180°$); thus, the flux through this face is

$$\int_1 \mathbf{E} \cdot d\mathbf{A} = \int_1 E(\cos 180°)\, dA = -E \int_1 dA = -EA = -E\ell^2$$

because the area of each face is $A = \ell^2$.

For ②, **E** is constant and outward and in the same direction as $d\mathbf{A}\,(\theta = 0°)$; hence, the flux through this face is

$$\int_2 \mathbf{E} \cdot d\mathbf{A} = \int_2 E(\cos 0°)\, dA = E \int_2 dA = +EA = E\ell^2$$

Therefore, the net flux over all six faces is

$$\Phi_E = -E\ell^2 + E\ell^2 + 0 + 0 + 0 + 0 = \boxed{0}$$

24.2 GAUSS'S LAW

In this section we describe a general relationship between the net electric flux through a closed surface (often called a *gaussian surface*) and the charge enclosed by the surface. This relationship, known as *Gauss's law,* is of fundamental importance in the study of electric fields.

Let us again consider a positive point charge q located at the center of a sphere of radius r, as shown in Figure 24.6. From Equation 23.4 we know that the magnitude of the electric field everywhere on the surface of the sphere is $E = k_e q/r^2$. As noted in Example 24.1, the field lines are directed radially outward and hence perpendicular to the surface at every point on the surface. That is, at each surface point, **E** is parallel to the vector $\Delta\mathbf{A}_i$ representing a local element of area ΔA_i surrounding the surface point. Therefore,

$$\mathbf{E} \cdot \Delta\mathbf{A}_i = E\,\Delta A_i$$

and from Equation 24.4 we find that the net flux through the gaussian surface is

$$\Phi_E = \oint \mathbf{E} \cdot d\mathbf{A} = \oint E\, dA = E \oint dA$$

where we have moved E outside of the integral because, by symmetry, E is constant over the surface and given by $E = k_e q/r^2$. Furthermore, because the surface is spherical, $\oint dA = A = 4\pi r^2$. Hence, the net flux through the gaussian surface is

$$\Phi_E = \frac{k_e q}{r^2}(4\pi r^2) = 4\pi k_e q$$

Recalling from Section 23.3 that $k_e = 1/(4\pi\epsilon_0)$, we can write this equation in the form

$$\Phi_E = \frac{q}{\epsilon_0} \qquad \text{(24.5)}$$

We can verify that this expression for the net flux gives the same result as Example 24.1: $\Phi_E = (1.00 \times 10^{-6}\ \text{C})/(8.85 \times 10^{-12}\ \text{C}^2/\text{N}\cdot\text{m}^2) = 1.13 \times 10^5\ \text{N}\cdot\text{m}^2/\text{C}$.

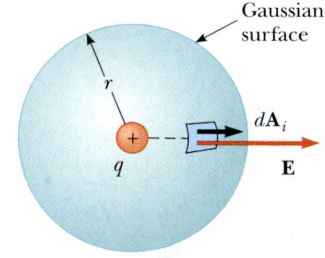

Figure 24.6 A spherical gaussian surface of radius r surrounding a point charge q. When the charge is at the center of the sphere, the electric field is everywhere normal to the surface and constant in magnitude.

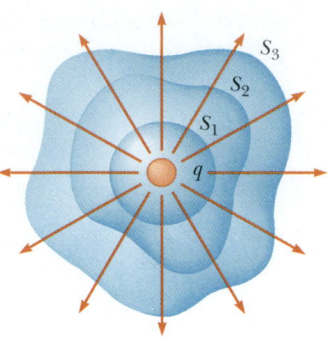

Figure 24.7   Closed surfaces of various shapes surrounding a charge q. The net electric flux is the same through all surfaces.

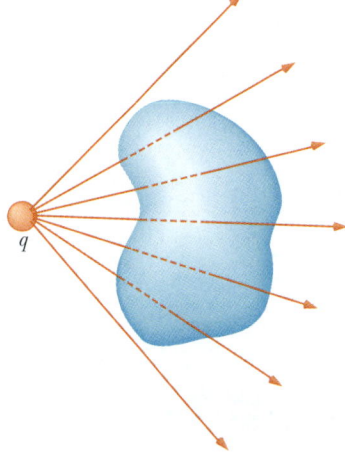

Figure 24.8   A point charge located *outside* a closed surface. The number of lines entering the surface equals the number leaving the surface.

The net electric flux through a closed surface is zero if there is no charge inside

Note from Equation 24.5 that the net flux through the spherical surface is proportional to the charge inside. The flux is independent of the radius r because the area of the spherical surface is proportional to r^2, whereas the electric field is proportional to $1/r^2$. Thus, in the product of area and electric field, the dependence on r cancels.

Now consider several closed surfaces surrounding a charge q, as shown in Figure 24.7. Surface S_1 is spherical, but surfaces S_2 and S_3 are not. From Equation 24.5, the flux that passes through S_1 has the value q/ϵ_0. As we discussed in the previous section, flux is proportional to the number of electric field lines passing through a surface. The construction shown in Figure 24.7 shows that the number of lines through S_1 is equal to the number of lines through the nonspherical surfaces S_2 and S_3. Therefore, we conclude that the net flux through *any* closed surface is independent of the shape of that surface. **The net flux through any closed surface surrounding a point charge q is given by q/ϵ_0.**

Now consider a point charge located *outside* a closed surface of arbitrary shape, as shown in Figure 24.8. As you can see from this construction, any electric field line that enters the surface leaves the surface at another point. The number of electric field lines entering the surface equals the number leaving the surface. Therefore, we conclude that **the net electric flux through a closed surface that surrounds no charge is zero.** If we apply this result to Example 24.2, we can easily see that the net flux through the cube is zero because there is no charge inside the cube.

Quick Quiz 24.1

Suppose that the charge in Example 24.1 is just outside the sphere, 1.01 m from its center. What is the total flux through the sphere?

Let us extend these arguments to two generalized cases: (1) that of many point charges and (2) that of a continuous distribution of charge. We once again use the superposition principle, which states that **the electric field due to many charges is the vector sum of the electric fields produced by the individual charges.** Therefore, we can express the flux through any closed surface as

$$\oint \mathbf{E} \cdot d\mathbf{A} = \oint (\mathbf{E}_1 + \mathbf{E}_2 + \cdots) \cdot d\mathbf{A}$$

where $\mathbf{E}$ is the total electric field at any point on the surface produced by the vector addition of the electric fields at that point due to the individual charges.

Consider the system of charges shown in Figure 24.9. The surface S surrounds only one charge, q_1; hence, the net flux through S is q_1/ϵ_0. The flux through S due to charges q_2 and q_3 outside it is zero because each electric field line that enters S at one point leaves it at another. The surface S' surrounds charges q_2 and q_3; hence, the net flux through it is $(q_2 + q_3)/\epsilon_0$. Finally, the net flux through surface S'' is zero because there is no charge inside this surface. That is, *all* the electric field lines that enter S'' at one point leave at another.

Gauss's law, which is a generalization of what we have just described, states that the net flux through *any* closed surface is

$$\Phi_E = \oint \mathbf{E} \cdot d\mathbf{A} = \frac{q_{\text{in}}}{\epsilon_0} \qquad \textbf{(24.6)}$$

Gauss's law

where q_{in} represents the net charge inside the surface and $\mathbf{E}$ represents the electric field at any point on the surface.

A formal proof of Gauss's law is presented in Section 24.6. When using Equation 24.6, you should note that although the charge q_{in} is the net charge inside the gaussian surface, $\mathbf{E}$ represents the *total electric field*, which includes contributions from charges both inside and outside the surface.

In principle, Gauss's law can be solved for $\mathbf{E}$ to determine the electric field due to a system of charges or a continuous distribution of charge. In practice, however, this type of solution is applicable only in a limited number of highly symmetric situations. As we shall see in the next section, Gauss's law can be used to evaluate the electric field for charge distributions that have spherical, cylindrical, or planar symmetry. If one chooses the gaussian surface surrounding the charge distribution carefully, the integral in Equation 24.6 can be simplified. You should also note that a gaussian surface is a mathematical construction and need not coincide with any real physical surface.

Gauss's law is useful for evaluating E when the charge distribution has high symmetry

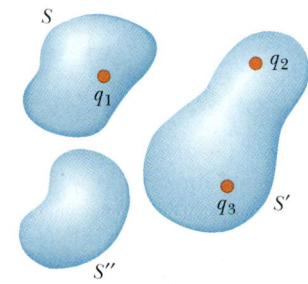

Figure 24.9 The net electric flux through any closed surface depends only on the charge *inside* that surface. The net flux through surface S is q_1/ϵ_0, the net flux through surface S' is $(q_2 + q_3)/\epsilon_0$, and the net flux through surface S'' is zero.

Quick Quiz 24.2

For a gaussian surface through which the net flux is zero, the following four statements *could be true*. Which of the statements *must be true*? (a) There are no charges inside the surface. (b) The net charge inside the surface is zero. (c) The electric field is zero everywhere on the surface. (d) The number of electric field lines entering the surface equals the number leaving the surface.

CONCEPTUAL EXAMPLE 24.3

A spherical gaussian surface surrounds a point charge q. Describe what happens to the total flux through the surface if (a) the charge is tripled, (b) the radius of the sphere is doubled, (c) the surface is changed to a cube, and (d) the charge is moved to another location inside the surface.

Solution (a) The flux through the surface is tripled because flux is proportional to the amount of charge inside the surface.

(b) The flux does not change because all electric field lines from the charge pass through the sphere, regardless of its radius.

(c) The flux does not change when the shape of the gaussian surface changes because all electric field lines from the charge pass through the surface, regardless of its shape.

(d) The flux does not change when the charge is moved to another location inside that surface because Gauss's law refers to the total charge enclosed, regardless of where the charge is located inside the surface.

24.3 ▶ APPLICATION OF GAUSS'S LAW TO CHARGED INSULATORS

As mentioned earlier, Gauss's law is useful in determining electric fields when the charge distribution is characterized by a high degree of symmetry. The following examples demonstrate ways of choosing the gaussian surface over which the surface integral given by Equation 24.6 can be simplified and the electric field determined. In choosing the surface, we should always take advantage of the symmetry of the charge distribution so that we can remove *E* from the integral and solve for it. The goal in this type of calculation is to determine a surface that satisfies one or more of the following conditions:

1. The value of the electric field can be argued by symmetry to be constant over the surface.
2. The dot product in Equation 24.6 can be expressed as a simple algebraic product *E dA* because **E** and *d***A** are parallel.
3. The dot product in Equation 24.6 is zero because **E** and *d***A** are perpendicular.
4. The field can be argued to be zero over the surface.

All four of these conditions are used in examples throughout the remainder of this chapter.

EXAMPLE 24.4 ▶ The Electric Field Due to a Point Charge

Starting with Gauss's law, calculate the electric field due to an isolated point charge *q*.

Solution A single charge represents the simplest possible charge distribution, and we use this familiar case to show how to solve for the electric field with Gauss's law. We choose a spherical gaussian surface of radius *r* centered on the point charge, as shown in Figure 24.10. The electric field due to a positive point charge is directed radially outward by symmetry and is therefore normal to the surface at every point. Thus, as in condition (2), **E** is parallel to *d***A** at each point. Therefore, **E** · *d***A** = *E dA* and Gauss's law gives

$$\Phi_E = \oint \mathbf{E} \cdot d\mathbf{A} = \oint E \, dA = \frac{q}{\epsilon_0}$$

By symmetry, *E* is constant everywhere on the surface, which satisfies condition (1), so it can be removed from the integral. Therefore,

$$\oint E \, dA = E \oint dA = E(4\pi r^2) = \frac{q}{\epsilon_0}$$

where we have used the fact that the surface area of a sphere is $4\pi r^2$. Now, we solve for the electric field:

$$E = \frac{q}{4\pi\epsilon_0 r^2} = \boxed{k_e \frac{q}{r^2}}$$

This is the familiar electric field due to a point charge that we developed from Coulomb's law in Chapter 23.

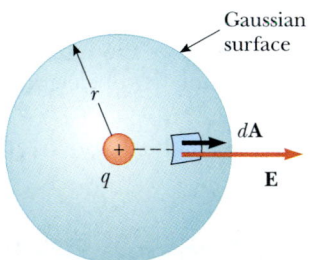

Figure 24.10   The point charge *q* is at the center of the spherical gaussian surface, and **E** is parallel to *d***A** at every point on the surface.

EXAMPLE 24.5 ▶ A Spherically Symmetric Charge Distribution

11.6

An insulating solid sphere of radius *a* has a uniform volume charge density *ρ* and carries a total positive charge *Q* (Fig. 24.11). (a) Calculate the magnitude of the electric field at a point outside the sphere.

Solution Because the charge distribution is spherically symmetric, we again select a spherical gaussian surface of radius *r*, concentric with the sphere, as shown in Figure 24.11a. For this choice, conditions (1) and (2) are satisfied, as they were for the point charge in Example 24.4. Following the line of reasoning given in Example 24.4, we find that

$$E = k_e \frac{Q}{r^2} \qquad \text{(for } r > a\text{)}$$

Note that this result is identical to the one we obtained for a point charge. Therefore, we conclude that, for a uniformly charged sphere, the field in the region external to the sphere is *equivalent* to that of a point charge located at the center of the sphere.

(b) Find the magnitude of the electric field at a point inside the sphere.

Solution In this case we select a spherical gaussian surface having radius $r < a$, concentric with the insulated sphere (Fig. 24.11b). Let us denote the volume of this smaller sphere by V'. To apply Gauss's law in this situation, it is important to recognize that the charge q_{in} within the gaussian surface of volume V' is less than Q. To calculate q_{in}, we use the fact that $q_{in} = \rho V'$:

$$q_{in} = \rho V' = \rho(\tfrac{4}{3}\pi r^3)$$

By symmetry, the magnitude of the electric field is constant everywhere on the spherical gaussian surface and is normal to the surface at each point—both conditions (1) and (2) are satisfied. Therefore, Gauss's law in the region $r < a$ gives

$$\oint E \, dA = E \oint dA = E(4\pi r^2) = \frac{q_{in}}{\epsilon_0}$$

Solving for E gives

$$E = \frac{q_{in}}{4\pi\epsilon_0 r^2} = \frac{\rho \tfrac{4}{3}\pi r^3}{4\pi\epsilon_0 r^2} = \frac{\rho}{3\epsilon_0} r$$

Because $\rho = Q/\tfrac{4}{3}\pi a^3$ by definition and since $k_e = 1/(4\pi\epsilon_0)$, this expression for E can be written as

$$E = \frac{Qr}{4\pi\epsilon_0 a^3} = \frac{k_e Q}{a^3} r \qquad \text{(for } r < a\text{)}$$

Note that this result for E differs from the one we obtained in part (a). It shows that $E \to 0$ as $r \to 0$. Therefore, the result eliminates the problem that would exist at $r = 0$ if E varied as $1/r^2$ inside the sphere as it does outside the sphere. That is, if $E \propto 1/r^2$ for $r < a$, the field would be infinite at $r = 0$, which is physically impossible. Note also that the expressions for parts (a) and (b) match when $r = a$.

A plot of E versus r is shown in Figure 24.12.

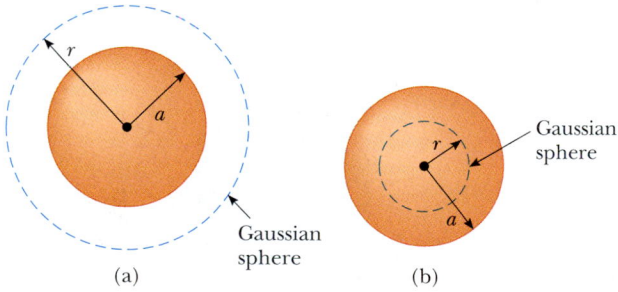

Figure 24.11 A uniformly charged insulating sphere of radius a and total charge Q. (a) The magnitude of the electric field at a point exterior to the sphere is $k_e Q/r^2$. (b) The magnitude of the electric field inside the insulating sphere is due only to the charge *within* the gaussian sphere defined by the dashed circle and is $k_e Qr/a^3$.

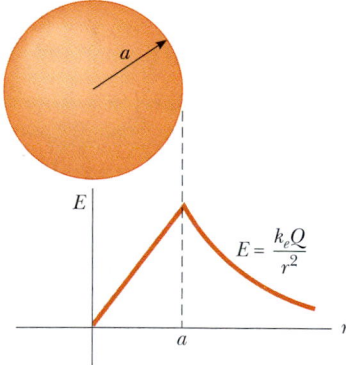

Figure 24.12 A plot of E versus r for a uniformly charged insulating sphere. The electric field inside the sphere ($r < a$) varies linearly with r. The field outside the sphere ($r > a$) is the same as that of a point charge Q located at $r = 0$.

EXAMPLE 24.6 **The Electric Field Due to a Thin Spherical Shell**

A thin spherical shell of radius a has a total charge Q distributed uniformly over its surface (Fig. 24.13a). Find the electric field at points (a) outside and (b) inside the shell.

Solution (a) The calculation for the field outside the shell is identical to that for the solid sphere shown in Example 24.5a. If we construct a spherical gaussian surface of radius $r > a$ concentric with the shell (Fig. 24.13b), the charge inside this surface is Q. Therefore, the field at a point outside

the shell is equivalent to that due to a point charge Q located at the center:

$$E = k_e \frac{Q}{r^2} \qquad \text{(for } r > a\text{)}$$

(b) The electric field inside the spherical shell is zero. This follows from Gauss's law applied to a spherical surface of radius $r < a$ concentric with the shell (Fig. 24.13c). Because of the spherical symmetry of the charge distribution and because the net charge inside the surface is zero—satisfaction of conditions (1) and (2) again—application of Gauss's law shows that $E = 0$ in the region $r < a$.

We obtain the same results using Equation 23.6 and integrating over the charge distribution. This calculation is rather complicated. Gauss's law allows us to determine these results in a much simpler way.

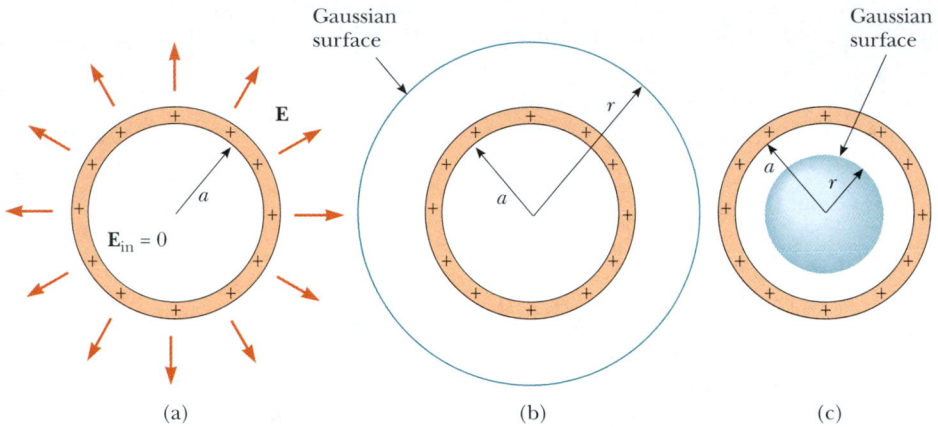

Figure 24.13 (a) The electric field inside a uniformly charged spherical shell is zero. The field outside is the same as that due to a point charge Q located at the center of the shell. (b) Gaussian surface for $r > a$. (c) Gaussian surface for $r < a$.

$\mathcal{E}$XAMPLE 24.7 ▸ A Cylindrically Symmetric Charge Distribution

Find the electric field a distance r from a line of positive charge of infinite length and constant charge per unit length λ (Fig. 24.14a).

Solution The symmetry of the charge distribution requires that **E** be perpendicular to the line charge and directed outward, as shown in Figure 24.14a and b. To reflect the symmetry of the charge distribution, we select a cylindrical gaussian surface of radius r and length ℓ that is coaxial with the line charge. For the curved part of this surface, **E** is constant in magnitude and perpendicular to the surface at each point—satisfaction of conditions (1) and (2). Furthermore, the flux through the ends of the gaussian cylinder is zero because **E** is parallel to these surfaces—the first application we have seen of condition (3).

We take the surface integral in Gauss's law over the entire gaussian surface. Because of the zero value of $\mathbf{E} \cdot d\mathbf{A}$ for the ends of the cylinder, however, we can restrict our attention to only the curved surface of the cylinder.

The total charge inside our gaussian surface is $\lambda\ell$. Applying Gauss's law and conditions (1) and (2), we find that for the curved surface

$$\Phi_E = \oint \mathbf{E} \cdot d\mathbf{A} = E \oint dA = EA = \frac{q_{in}}{\epsilon_0} = \frac{\lambda\ell}{\epsilon_0}$$

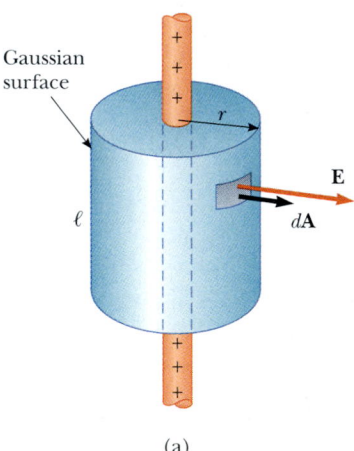

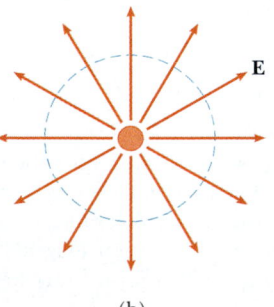

Figure 24.14 (a) An infinite line of charge surrounded by a cylindrical gaussian surface concentric with the line. (b) An end view shows that the electric field at the cylindrical surface is constant in magnitude and perpendicular to the surface.

The area of the curved surface is $A = 2\pi r\ell$; therefore,

$$E(2\pi r\ell) = \frac{\lambda\ell}{\epsilon_0}$$

$$E = \frac{\lambda}{2\pi\epsilon_0 r} = 2k_e\frac{\lambda}{r} \qquad (24.7)$$

Thus, we see that the electric field due to a cylindrically symmetric charge distribution varies as $1/r$, whereas the field external to a spherically symmetric charge distribution varies as $1/r^2$. Equation 24.7 was also derived in Chapter 23 (see Problem 35[b]), by integration of the field of a point charge.

If the line charge in this example were of finite length, the result for E would not be that given by Equation 24.7. A finite line charge does not possess sufficient symmetry for us to make use of Gauss's law. This is because the magnitude of

the electric field is no longer constant over the surface of the gaussian cylinder—the field near the ends of the line would be different from that far from the ends. Thus, condition (1) would not be satisfied in this situation. Furthermore, $\mathbf{E}$ is not perpendicular to the cylindrical surface at all points—the field vectors near the ends would have a component parallel to the line. Thus, condition (2) would not be satisfied. When there is insufficient symmetry in the charge distribution, as in this situation, it is necessary to use Equation 23.6 to calculate $\mathbf{E}$.

For points close to a finite line charge and far from the ends, Equation 24.7 gives a good approximation of the value of the field.

It is left for you to show (see Problem 29) that the electric field inside a uniformly charged rod of finite radius and infinite length is proportional to r.

EXAMPLE 24.8   A Nonconducting Plane of Charge

Find the electric field due to a nonconducting, infinite plane of positive charge with uniform surface charge density σ.

Solution   By symmetry, $\mathbf{E}$ must be perpendicular to the plane and must have the same magnitude at all points equidistant from the plane. The fact that the direction of $\mathbf{E}$ is away from positive charges indicates that the direction of $\mathbf{E}$ on one side of the plane must be opposite its direction on the other side, as shown in Figure 24.15. A gaussian surface that reflects the symmetry is a small cylinder whose axis is perpendicular to the plane and whose ends each have an area A and are equidistant from the plane. Because $\mathbf{E}$ is parallel to the curved surface—and, therefore, perpendicular to $d\mathbf{A}$ everywhere on the surface—condition (3) is satisfied and there is no contribution to the surface integral from this surface. For the flat ends of the cylinder, conditions (1) and (2) are satisfied. The flux through each end of the cylinder is EA; hence, the total flux through the entire gaussian surface is just that through the ends, $\Phi_E = 2EA$.

Noting that the total charge inside the surface is $q_{in} = \sigma A$, we use Gauss's law and find that

$$\Phi_E = 2EA = \frac{q_{in}}{\epsilon_0} = \frac{\sigma A}{\epsilon_0}$$

$$E = \frac{\sigma}{2\epsilon_0} \qquad (24.8)$$

Because the distance from each flat end of the cylinder to the plane does not appear in Equation 24.8, we conclude that $E = \sigma/2\epsilon_0$ at any distance from the plane. That is, the field is uniform everywhere.

An important charge configuration related to this example consists of two parallel planes, one positively charged and the other negatively charged, and each with a surface charge density σ (see Problem 58). In this situation, the electric fields due to the two planes add in the region between the planes, resulting in a field of magnitude σ/ϵ_0, and cancel elsewhere to give a field of zero.

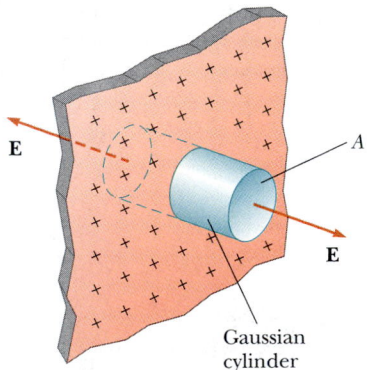

Figure 24.15   A cylindrical gaussian surface penetrating an infinite plane of charge. The flux is EA through each end of the gaussian surface and zero through its curved surface.

CONCEPTUAL EXAMPLE 24.9

Explain why Gauss's law cannot be used to calculate the electric field near an electric dipole, a charged disk, or a triangle with a point charge at each corner.

> **Solution**   The charge distributions of all these configurations do not have sufficient symmetry to make the use of Gauss's law practical. We cannot find a closed surface surrounding any of these distributions that satisfies one or more of conditions (1) through (4) listed at the beginning of this section.

24.4   CONDUCTORS IN ELECTROSTATIC EQUILIBRIUM

As we learned in Section 23.2, a good electrical conductor contains charges (electrons) that are not bound to any atom and therefore are free to move about within the material. When there is no net motion of charge within a conductor, the conductor is in **electrostatic equilibrium.** As we shall see, a conductor in electrostatic equilibrium has the following properties:

1. The electric field is zero everywhere inside the conductor.
2. If an isolated conductor carries a charge, the charge resides on its surface.
3. The electric field just outside a charged conductor is perpendicular to the surface of the conductor and has a magnitude σ/ϵ_0, where σ is the surface charge density at that point.
4. On an irregularly shaped conductor, the surface charge density is greatest at locations where the radius of curvature of the surface is smallest.

Properties of a conductor in electrostatic equilibrium

We verify the first three properties in the discussion that follows. The fourth property is presented here without further discussion so that we have a complete list of properties for conductors in electrostatic equilibrium.

We can understand the first property by considering a conducting slab placed in an external field **E** (Fig. 24.16). We can argue that the electric field inside the conductor *must* be zero under the assumption that we have electrostatic equilibrium. If the field were not zero, free charges in the conductor would accelerate under the action of the field. This motion of electrons, however, would mean that the conductor is not in electrostatic equilibrium. Thus, the existence of electrostatic equilibrium is consistent only with a zero field in the conductor.

Let us investigate how this zero field is accomplished. Before the external field is applied, free electrons are uniformly distributed throughout the conductor. When the external field is applied, the free electrons accelerate to the left in Figure 24.16, causing a plane of negative charge to be present on the left surface. The movement of electrons to the left results in a plane of positive charge on the right surface. These planes of charge create an additional electric field inside the conductor that opposes the external field. As the electrons move, the surface charge density increases until the magnitude of the internal field equals that of the external field, and the net result is a net field of zero inside the conductor. The time it takes a good conductor to reach equilibrium is of the order of 10^{-16} s, which for most purposes can be considered instantaneous.

We can use Gauss's law to verify the second property of a conductor in electrostatic equilibrium. Figure 24.17 shows an arbitrarily shaped conductor. A gaussian surface is drawn inside the conductor and can be as close to the conductor's surface as we wish. As we have just shown, the electric field everywhere inside the conductor is zero when it is in electrostatic equilibrium. Therefore, the electric field must be zero at every point on the gaussian surface, in accordance with condition (4) in Section 24.3. Thus, the net flux through this gaussian surface is zero. From this result and Gauss's law, we conclude that the net charge inside the gaussian sur-

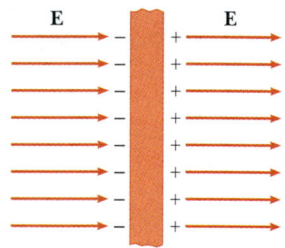

Figure 24.16   A conducting slab in an external electric field **E**. The charges induced on the two surfaces of the slab produce an electric field that opposes the external field, giving a resultant field of zero inside the slab.

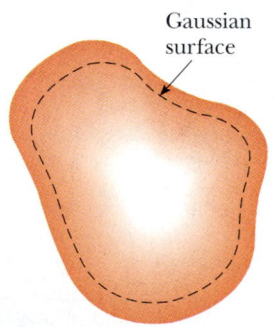

Gaussian surface

Figure 24.17   A conductor of arbitrary shape. The broken line represents a gaussian surface just inside the conductor.

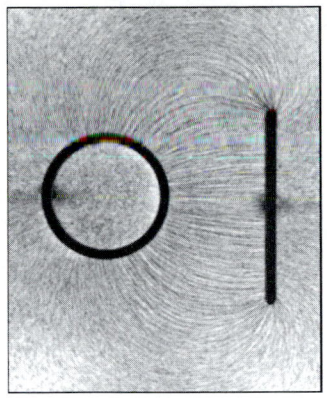

Electric field pattern surrounding a charged conducting plate placed near an oppositely charged conducting cylinder. Small pieces of thread suspended in oil align with the electric field lines. Note that (1) the field lines are perpendicular to both conductors and (2) there are no lines inside the cylinder ($E = 0$). *(Courtesy of Harold M. Waage, Princeton University)*

face is zero. Because there can be no net charge inside the gaussian surface (which is arbitrarily close to the conductor's surface), **any net charge on the conductor must reside on its surface.** Gauss's law does not indicate how this excess charge is distributed on the conductor's surface.

We can also use Gauss's law to verify the third property. We draw a gaussian surface in the shape of a small cylinder whose end faces are parallel to the surface of the conductor (Fig. 24.18). Part of the cylinder is just outside the conductor, and part is inside. The field is normal to the conductor's surface from the condition of electrostatic equilibrium. (If **E** had a component parallel to the conductor's surface, the free charges would move along the surface; in such a case, the conductor would not be in equilibrium.) Thus, we satisfy condition (3) in Section 24.3 for the curved part of the cylindrical gaussian surface—there is no flux through this part of the gaussian surface because **E** is parallel to the surface. There is no flux through the flat face of the cylinder inside the conductor because here **E** = 0—satisfaction of condition (4). Hence, the net flux through the gaussian surface is that through only the flat face outside the conductor, where the field is perpendicular to the gaussian surface. Using conditions (1) and (2) for this face, the flux is EA, where E is the electric field just outside the conductor and A is the area of the cylinder's face. Applying Gauss's law to this surface, we obtain

$$\Phi_E = \oint E\,dA = EA = \frac{q_{in}}{\epsilon_0} = \frac{\sigma A}{\epsilon_0}$$

where we have used the fact that $q_{in} = \sigma A$. Solving for E gives

$$E = \frac{\sigma}{\epsilon_0} \qquad\qquad \textbf{(24.9)}$$

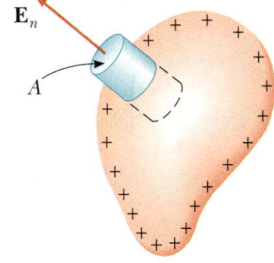

Figure 24.18   A gaussian surface in the shape of a small cylinder is used to calculate the electric field just outside a charged conductor. The flux through the gaussian surface is $E_n A$. Remember that **E** is zero inside the conductor.

> Electric field just outside a charged conductor

EXAMPLE 24.10   A Sphere Inside a Spherical Shell

A solid conducting sphere of radius a carries a net positive charge $2Q$. A conducting spherical shell of inner radius b and outer radius c is concentric with the solid sphere and carries a net charge $-Q$. Using Gauss's law, find the electric field in the regions labeled ①, ②, ③, and ④ in Figure 24.19 and the charge distribution on the shell when the entire system is in electrostatic equilibrium.

Solution   First note that the charge distributions on both the sphere and the shell are characterized by spherical symmetry around their common center. To determine the electric field at various distances r from this center, we construct a spherical gaussian surface for each of the four regions of interest. Such a surface for region ② is shown in Figure 24.19.

To find E inside the solid sphere (region ①), consider a

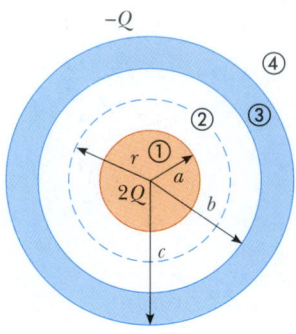

Figure 24.19   A solid conducting sphere of radius a and carrying a charge $2Q$ surrounded by a conducting spherical shell carrying a charge $-Q$.

gaussian surface of radius $r < a$. Because there can be no charge inside a conductor in electrostatic equilibrium, we see that $q_{in} = 0$; thus, on the basis of Gauss's law and symmetry, $E_1 = 0$ for $r < a$.

In region ②—between the surface of the solid sphere and the inner surface of the shell—we construct a spherical gaussian surface of radius r where $a < r < b$ and note that the charge inside this surface is $+2Q$ (the charge on the solid sphere). Because of the spherical symmetry, the electric field

lines must be directed radially outward and be constant in magnitude on the gaussian surface. Following Example 24.4 and using Gauss's law, we find that

$$E_2 A = E_2 (4\pi r^2) = \frac{q_{in}}{\epsilon_0} = \frac{2Q}{\epsilon_0}$$

$$E_2 = \frac{2Q}{4\pi\epsilon_0 r^2} = \boxed{\frac{2k_e Q}{r^2}} \qquad \text{(for } a < r < b\text{)}$$

In region ④, where $r > c$, the spherical gaussian surface we construct surrounds a total charge of $q_{in} = 2Q + (-Q) = Q$. Therefore, application of Gauss's law to this surface gives

$$E_4 = \boxed{\frac{k_e Q}{r^2}} \qquad \text{(for } r > c\text{)}$$

In region ③, the electric field must be zero because the spherical shell is also a conductor in equilibrium. If we construct a gaussian surface of radius r where $b < r < c$, we see that q_{in} must be zero because $E_3 = 0$. From this argument, we conclude that the charge on the inner surface of the spherical shell must be $-2Q$ to cancel the charge $+2Q$ on the solid sphere. Because the net charge on the shell is $-Q$, we conclude that its outer surface must carry a charge $+Q$.

Quick Quiz 24.3

How would the electric flux through a gaussian surface surrounding the shell in Example 24.10 change if the solid sphere were off-center but still inside the shell?

Optional Section

24.5 EXPERIMENTAL VERIFICATION OF GAUSS'S LAW AND COULOMB'S LAW

When a net charge is placed on a conductor, the charge distributes itself on the surface in such a way that the electric field inside the conductor is zero. Gauss's law shows that there can be no net charge inside the conductor in this situation. In this section, we investigate an experimental verification of the absence of this charge.

We have seen that Gauss's law is equivalent to Equation 23.6, the expression for the electric field of a distribution of charge. Because this equation arises from Coulomb's law, we can claim theoretically that Gauss's law and Coulomb's law are equivalent. Hence, it is possible to test the validity of both laws by attempting to detect a net charge inside a conductor or, equivalently, a nonzero electric field inside the conductor. If a nonzero field is detected within the conductor, Gauss's law and Coulomb's law are invalid. Many experiments, including early work by Faraday, Cavendish, and Maxwell, have been performed to detect the field inside a conductor. In all reported cases, no electric field could be detected inside a conductor.

Here is one of the experiments that can be performed.[2] A positively charged metal ball at the end of a silk thread is lowered through a small opening into an uncharged hollow conductor that is insulated from ground (Fig. 24.20a). The positively charged ball induces a negative charge on the inner wall of the hollow conductor, leaving an equal positive charge on the outer wall (Fig. 24.20b). The presence of positive charge on the outer wall is indicated by the deflection of the needle of an electrometer (a device used to measure charge and that measures charge only on the outer surface of the conductor). The ball is then lowered and allowed to touch the inner surface of the hollow conductor (Fig. 24.20c). Charge is transferred between the ball and the inner surface so that neither is charged after contact is made. The needle deflection remains unchanged while this happens, indicating that the charge on the outer surface is unaffected. When the ball is removed, the electrometer reading remains the same (Fig. 24.20d). Furthermore, the ball is found to be uncharged; this verifies that charge was transferred between the ball and the inner surface of the hollow conductor. The overall effect is that the charge that was originally on the ball now appears on the hollow conductor. The fact that the deflection of the needle on the electrometer measuring the charge on the outer surface remained unchanged regardless of what was happening inside the hollow conductor indicates that the net charge on the system always resided on the outer surface of the conductor.

If we now apply another positive charge to the metal ball and place it near the outside of the conductor, it is repelled by the conductor. This demonstrates that $\mathbf{E} \neq 0$ outside the conductor, a finding consistent with the fact that the conductor carries a net charge. If the charged metal ball is now lowered into the interior of the charged hollow conductor, it exhibits no evidence of an electric force. This shows that $\mathbf{E} = 0$ inside the hollow conductor.

This experiment verifies the predictions of Gauss's law and therefore verifies Coulomb's law. The equivalence of Gauss's law and Coulomb's law is due to the inverse-square behavior of the electric force. Thus, we can interpret this experiment as verifying the exponent of 2 in the $1/r^2$ behavior of the electric force. Experiments by Williams, Faller, and Hill in 1971 showed that the exponent of r in Coulomb's law is $(2 + \delta)$, where $\delta = (2.7 \pm 3.1) \times 10^{-16}$!

In the experiment we have described, the charged ball hanging in the hollow conductor would show no deflection even in the case in which an external electric field is applied to the entire system. The field inside the conductor is still zero. This ability of conductors to "block" external electric fields is utilized in many places, from electromagnetic shielding for computer components to thin metal coatings on the glass in airport control towers to keep radar originating outside the tower from disrupting the electronics inside. Cellular telephone users riding trains like the one pictured at the beginning of the chapter have to speak loudly to be heard above the noise of the train. In response to complaints from other passengers, the train companies are considering coating the windows with a thin metallic conductor. This coating, combined with the metal frame of the train car, blocks cellular telephone transmissions into and out of the train.

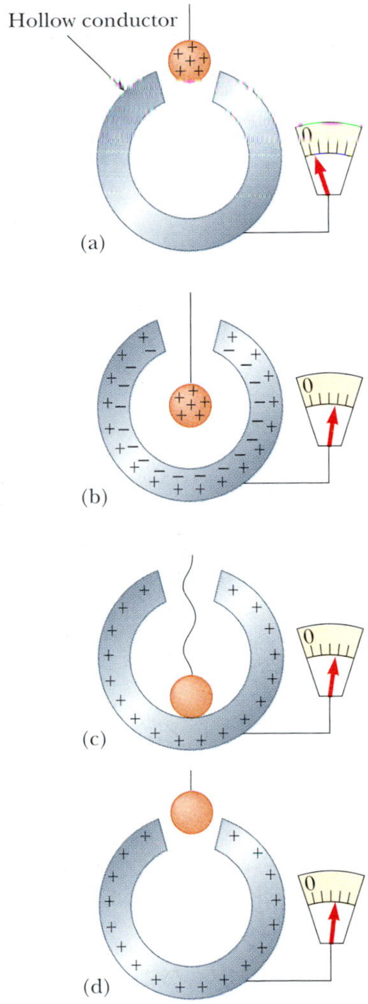

Figure 24.20 An experiment showing that any charge transferred to a conductor resides on its surface in electrostatic equilibrium. The hollow conductor is insulated from ground, and the small metal ball is supported by an insulating thread.

QuickLab

Wrap a radio or cordless telephone in aluminum foil and see if it still works. Does it matter if the foil touches the antenna?

---

[2] The experiment is often referred to as *Faraday's ice-pail experiment* because Faraday, the first to perform it, used an ice pail for the hollow conductor.

Optional Section

24.6 ▶ FORMAL DERIVATION OF GAUSS'S LAW

One way of deriving Gauss's law involves *solid angles*. Consider a spherical surface of radius r containing an area element ΔA. The solid angle $\Delta \Omega$ (uppercase Greek omega) subtended at the center of the sphere by this element is defined to be

$$\Delta \Omega \equiv \frac{\Delta A}{r^2}$$

From this equation, we see that $\Delta \Omega$ has no dimensions because ΔA and r^2 both have dimensions L^2. The dimensionless unit of a solid angle is the **steradian.** (You may want to compare this equation to Equation 10.1b, the definition of the radian.) Because the surface area of a sphere is $4\pi r^2$, the total solid angle subtended by the sphere is

$$\Omega = \frac{4\pi r^2}{r^2} = 4\pi \text{ steradians}$$

Now consider a point charge q surrounded by a closed surface of arbitrary shape (Fig. 24.21). The total electric flux through this surface can be obtained by evaluating $\mathbf{E} \cdot \Delta \mathbf{A}$ for each small area element ΔA and summing over all elements. The flux through each element is

$$\Delta \Phi_E = \mathbf{E} \cdot \Delta \mathbf{A} = E \, \Delta A \cos \theta = k_e q \frac{\Delta A \cos \theta}{r^2}$$

where r is the distance from the charge to the area element, θ is the angle between the electric field $\mathbf{E}$ and $\Delta \mathbf{A}$ for the element, and $E = k_e q / r^2$ for a point charge. In Figure 24.22, we see that the projection of the area element perpendicular to the radius vector is $\Delta A \cos \theta$. Thus, the quantity $\Delta A \cos \theta / r^2$ is equal to the solid angle $\Delta \Omega$ that the surface element ΔA subtends at the charge q. We also see that $\Delta \Omega$ is equal to the solid angle subtended by the area element of a spherical surface of radius r. Because the total solid angle at a point is 4π steradians, the total flux

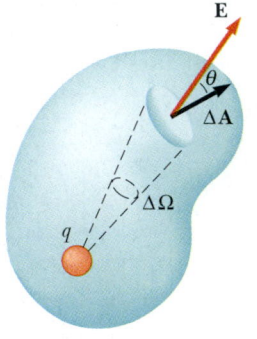

Figure 24.21   A closed surface of arbitrary shape surrounds a point charge q. The net electric flux through the surface is independent of the shape of the surface.

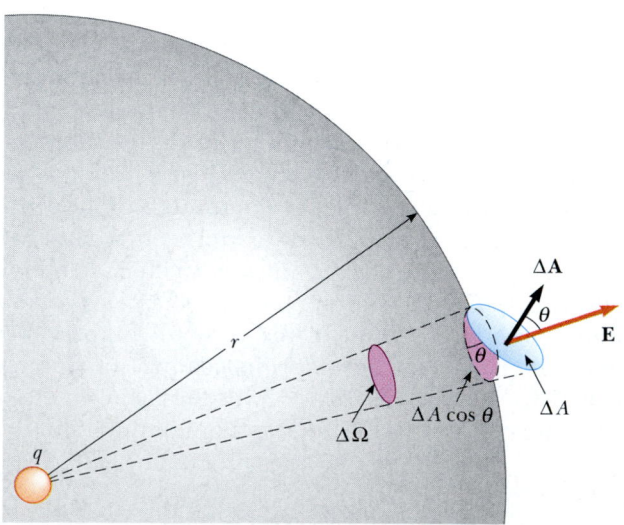

Figure 24.22   The area element ΔA subtends a solid angle $\Delta \Omega = (\Delta A \cos \theta)/r^2$ at the charge q.

through the closed surface is

$$\Phi_E = k_e q \oint \frac{dA \cos \theta}{r^2} = k_e q \oint d\Omega = 4\pi k_e q = \frac{q}{\epsilon_0}$$

Thus we have derived Gauss's law, Equation 24.6. Note that this result is independent of the shape of the closed surface and independent of the position of the charge within the surface.

SUMMARY

Electric flux is proportional to the number of electric field lines that penetrate a surface. If the electric field is uniform and makes an angle θ with the normal to a surface of area A, the electric flux through the surface is

$$\Phi_E = EA \cos \theta \tag{24.2}$$

In general, the electric flux through a surface is

$$\Phi_E = \int_{\text{surface}} \mathbf{E} \cdot d\mathbf{A} \tag{24.3}$$

You need to be able to apply Equations 24.2 and 24.3 in a variety of situations, particularly those in which symmetry simplifies the calculation.

Gauss's law says that the net electric flux Φ_E through any closed gaussian surface is equal to the *net* charge inside the surface divided by ϵ_0:

$$\Phi_E = \oint \mathbf{E} \cdot d\mathbf{A} = \frac{q_{\text{in}}}{\epsilon_0} \tag{24.6}$$

Using Gauss's law, you can calculate the electric field due to various symmetric charge distributions. Table 24.1 lists some typical results.

TABLE 24.1 Typical Electric Field Calculations Using Gauss's Law

Charge Distribution	Electric Field	Location
Insulating sphere of radius R, uniform charge density, and total charge Q	$k_e \dfrac{Q}{r^2}$	$r > R$
	$k_e \dfrac{Q}{R^3} r$	$r < R$
Thin spherical shell of radius R and total charge Q	$k_e \dfrac{Q}{r^2}$	$r > R$
	0	$r < R$
Line charge of infinite length and charge per unit length λ	$2k_e \dfrac{\lambda}{r}$	Outside the line
Nonconducting, infinite charged plane having surface charge density σ	$\dfrac{\sigma}{2\epsilon_0}$	Everywhere outside the plane
Conductor having surface charge density σ	$\dfrac{\sigma}{\epsilon_0}$	Just outside the conductor
	0	Inside the conductor

A conductor in **electrostatic equilibrium** has the following properties:

1. The electric field is zero everywhere inside the conductor.
2. Any net charge on the conductor resides entirely on its surface.
3. The electric field just outside the conductor is perpendicular to its surface and has a magnitude σ/ϵ_0, where σ is the surface charge density at that point.
4. On an irregularly shaped conductor, the surface charge density is greatest where the radius of curvature of the surface is the smallest.

Problem-Solving Hints

Gauss's law, as we have seen, is very powerful in solving problems involving highly symmetric charge distributions. In this chapter, you encountered three kinds of symmetry: planar, cylindrical, and spherical. It is important to review Examples 24.4 through 24.10 and to adhere to the following procedure when using Gauss's law:

- Select a gaussian surface that has a symmetry to match that of the charge distribution and satisfies one or more of the conditions listed in Section 24.3. For point charges or spherically symmetric charge distributions, the gaussian surface should be a sphere centered on the charge as in Examples 24.4, 24.5, 24.6, and 24.10. For uniform line charges or uniformly charged cylinders, your gaussian surface should be a cylindrical surface that is coaxial with the line charge or cylinder as in Example 24.7. For planes of charge, a useful choice is a cylindrical gaussian surface that straddles the plane, as shown in Example 24.8. These choices enable you to simplify the surface integral that appears in Gauss's law and represents the total electric flux through that surface.
- Evaluate the q_{in}/ϵ_0 term in Gauss's law, which amounts to calculating the total electric charge q_{in} inside the gaussian surface. If the charge density is uniform (that is, if λ, σ, or ρ is constant), simply multiply that charge density by the length, area, or volume enclosed by the gaussian surface. If the charge distribution is *nonuniform,* integrate the charge density over the region enclosed by the gaussian surface. For example, if the charge is distributed along a line, integrate the expression $dq = \lambda\,dx$, where dq is the charge on an infinitesimal length element dx. For a plane of charge, integrate $dq = \sigma\,dA$, where dA is an infinitesimal element of area. For a volume of charge, integrate $dq = \rho\,dV$, where dV is an infinitesimal element of volume.
- Once the terms in Gauss's law have been evaluated, solve for the electric field on the gaussian surface if the charge distribution is given in the problem. Conversely, if the electric field is known, calculate the charge distribution that produces the field.

QUESTIONS

1. The Sun is lower in the sky during the winter than it is in the summer. How does this change the flux of sunlight hitting a given area on the surface of the Earth? How does this affect the weather?
2. If the electric field in a region of space is zero, can you conclude no electric charges are in that region? Explain.

3. If more electric field lines are leaving a gaussian surface than entering, what can you conclude about the net charge enclosed by that surface?
4. A uniform electric field exists in a region of space in which there are no charges. What can you conclude about the net electric flux through a gaussian surface placed in this region of space?

5. If the total charge inside a closed surface is known but the distribution of the charge is unspecified, can you use Gauss's law to find the electric field? Explain.

6. Explain why the electric flux through a closed surface with a given enclosed charge is independent of the size or shape of the surface.

7. Consider the electric field due to a nonconducting infinite plane having a uniform charge density. Explain why the electric field does not depend on the distance from the plane in terms of the spacing of the electric field lines.

8. Use Gauss's law to explain why electric field lines must begin or end on electric charges. (*Hint:* Change the size of the gaussian surface.)

9. On the basis of the repulsive nature of the force between like charges and the freedom of motion of charge within the conductor, explain why excess charge on an isolated conductor must reside on its surface.

10. A person is placed in a large, hollow metallic sphere that is insulated from ground. If a large charge is placed on the sphere, will the person be harmed upon touching the inside of the sphere? Explain what will happen if the person also has an initial charge whose sign is opposite that of the charge on the sphere.

11. How would the observations described in Figure 24.20 differ if the hollow conductor were grounded? How would they differ if the small charged ball were an insulator rather than a conductor?

12. What other experiment might be performed on the ball in Figure 24.20 to show that its charge was transferred to the hollow conductor?

13. What would happen to the electrometer reading if the charged ball in Figure 24.20 touched the inner wall of the conductor? the outer wall?

14. You may have heard that one of the safer places to be during a lightning storm is inside a car. Why would this be the case?

15. Two solid spheres, both of radius R, carry identical total charges Q. One sphere is a good conductor, while the other is an insulator. If the charge on the insulating sphere is uniformly distributed throughout its interior volume, how do the electric fields outside these two spheres compare? Are the fields identical inside the two spheres?

PROBLEMS

1, 2, 3 = straightforward, intermediate, challenging ☐ = full solution available in the *Student Solutions Manual and Study Guide*
WEB = solution posted at **http://www.saunderscollege.com/physics/** 💻 = Computer useful in solving problem 🎮 = Interactive Physics
☐ = paired numerical/symbolic problems

Section 24.1   Electric Flux

1. An electric field with a magnitude of 3.50 kN/C is applied along the x axis. Calculate the electric flux through a rectangular plane 0.350 m wide and 0.700 m long if (a) the plane is parallel to the yz plane; (b) the plane is parallel to the xy plane; and (c) the plane contains the y axis, and its normal makes an angle of 40.0° with the x axis.

2. A vertical electric field of magnitude 2.00×10^4 N/C exists above the Earth's surface on a day when a thunderstorm is brewing. A car with a rectangular size of approximately 6.00 m by 3.00 m is traveling along a roadway sloping downward at 10.0°. Determine the electric flux through the bottom of the car.

3. A 40.0-cm-diameter loop is rotated in a uniform electric field until the position of maximum electric flux is found. The flux in this position is measured to be 5.20×10^5 N·m²/C. What is the magnitude of the electric field?

4. A spherical shell is placed in a uniform electric field. Find the total electric flux through the shell.

5. Consider a closed triangular box resting within a horizontal electric field of magnitude $E = 7.80 \times 10^4$ N/C, as shown in Figure P24.5. Calculate the electric flux through (a) the vertical surface, (b) the slanted surface, and (c) the entire surface of the box.

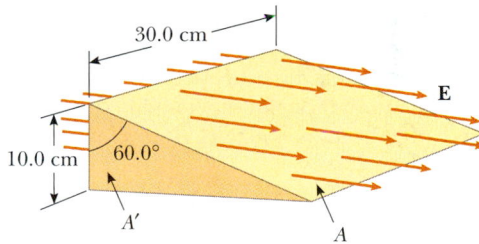

Figure P24.5

6. A uniform electric field $a\mathbf{i} + b\mathbf{j}$ intersects a surface of area A. What is the flux through this area if the surface lies (a) in the yz plane? (b) in the xz plane? (c) in the xy plane?

7. A point charge q is located at the center of a uniform ring having linear charge density λ and radius a, as shown in Figure P24.7. Determine the total electric flux

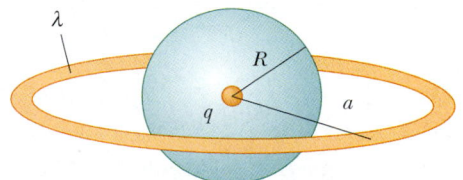

Figure P24.7

through a sphere centered at the point charge and having radius R, where $R < a$.

8. A pyramid with a 6.00-m-square base and height of 4.00 m is placed in a vertical electric field of 52.0 N/C. Calculate the total electric flux through the pyramid's four slanted surfaces.

9. A cone with base radius R and height h is located on a horizontal table. A horizontal uniform field E penetrates the cone, as shown in Figure P24.9. Determine the electric flux that enters the left-hand side of the cone.

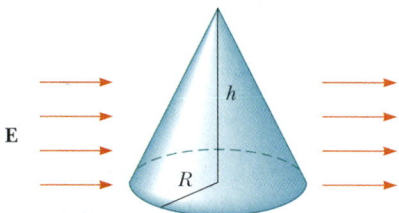

Figure P24.9

Section 24.2   Gauss's Law

10. The electric field everywhere on the surface of a thin spherical shell of radius 0.750 m is measured to be equal to 890 N/C and points radially toward the center of the sphere. (a) What is the net charge within the sphere's surface? (b) What can you conclude about the nature and distribution of the charge inside the spherical shell?

11. The following charges are located inside a submarine: $5.00\ \mu C$, $-9.00\ \mu C$, $27.0\ \mu C$, and $-84.0\ \mu C$. (a) Calculate the net electric flux through the submarine. (b) Is the number of electric field lines leaving the submarine greater than, equal to, or less than the number entering it?

12. Four closed surfaces, S_1 through S_4, together with the charges $-2Q$, Q, and $-Q$ are sketched in Figure P24.12. Find the electric flux through each surface.

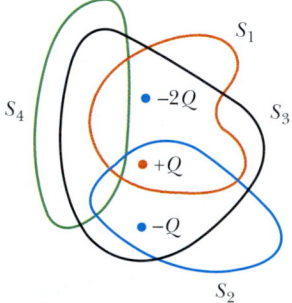

Figure P24.12

13. (a) A point charge q is located a distance d from an infinite plane. Determine the electric flux through the plane due to the point charge. (b) A point charge q is

located a *very small* distance from the center of a *very large* square on the line perpendicular to the square and going through its center. Determine the approximate electric flux through the square due to the point charge. (c) Explain why the answers to parts (a) and (b) are identical.

14. Calculate the total electric flux through the paraboloidal surface due to a constant electric field of magnitude E_0 in the direction shown in Figure P24.14.

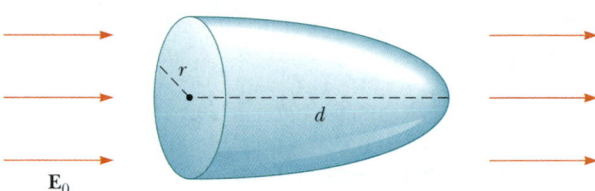

Figure P24.14

15. A point charge Q is located just above the center of the flat face of a hemisphere of radius R, as shown in Figure P24.15. What is the electric flux (a) through the curved surface and (b) through the flat face?

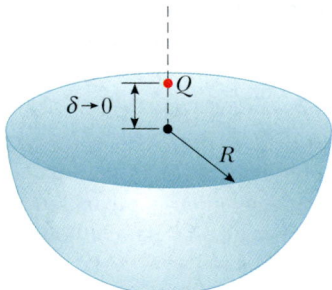

Figure P24.15

16. A point charge of 12.0 μC is placed at the center of a spherical shell of radius 22.0 cm. What is the total electric flux through (a) the surface of the shell and (b) any hemispherical surface of the shell? (c) Do the results depend on the radius? Explain.

17. A point charge of 0.046 2 μC is inside a pyramid. Determine the total electric flux through the surface of the pyramid.

18. An infinitely long line charge having a uniform charge per unit length λ lies a distance d from point O, as shown in Figure P24.18. Determine the total electric flux through the surface of a sphere of radius R centered at O resulting from this line charge. (*Hint:* Consider both cases: when $R < d$, and when $R > d$.)

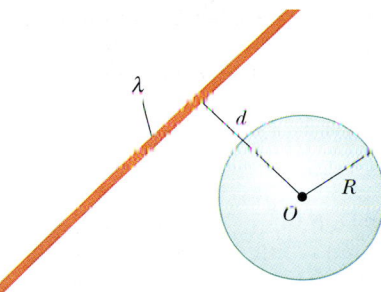

Figure P24.18

19. A point charge $Q = 5.00$ μC is located at the center of a cube of side $L = 0.100$ m. In addition, six other identical point charges having $q = -1.00$ μC are positioned symmetrically around Q, as shown in Figure P24.19. Determine the electric flux through one face of the cube.

20. A point charge Q is located at the center of a cube of side L. In addition, six other identical negative point charges are positioned symmetrically around Q, as shown in Figure P24.19. Determine the electric flux through one face of the cube.

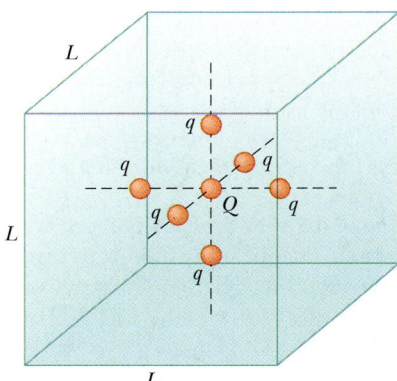

Figure P24.19   Problems 19 and 20.

21. Consider an infinitely long line charge having uniform charge per unit length λ. Determine the total electric flux through a closed right circular cylinder of length L and radius R that is parallel to the line charge, if the distance between the axis of the cylinder and the line charge is d. (*Hint:* Consider both cases: when $R < d$, and when $R > d$.)

22. A 10.0-μC charge located at the origin of a cartesian coordinate system is surrounded by a nonconducting hollow sphere of radius 10.0 cm. A drill with a radius of 1.00 mm is aligned along the z axis, and a hole is drilled in the sphere. Calculate the electric flux through the hole.

23. A charge of 170 μC is at the center of a cube of side 80.0 cm. (a) Find the total flux through each face of the cube. (b) Find the flux through the whole surface of the cube. (c) Would your answers to parts (a) or (b) change if the charge were not at the center? Explain.

24. The total electric flux through a closed surface in the shape of a cylinder is 8.60×10^4 N·m²/C. (a) What is the net charge within the cylinder? (b) From the information given, what can you say about the charge within the cylinder? (c) How would your answers to parts (a) and (b) change if the net flux were -8.60×10^4 N·m²/C?

25. The line ag is a diagonal of a cube (Fig. P24.25). A point charge q is located on the extension of line ag, very close to vertex a of the cube. Determine the electric flux through each of the sides of the cube that meet at the point a.

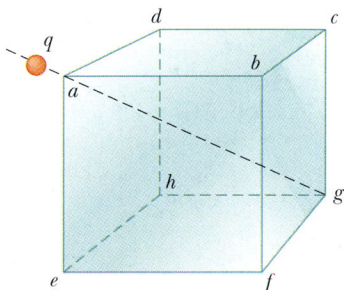

Figure P24.25

Section 24.3   Application of Gauss's Law to Charged Insulators

26. Determine the magnitude of the electric field at the surface of a lead-208 nucleus, which contains 82 protons and 126 neutrons. Assume that the lead nucleus has a volume 208 times that of one proton, and consider a proton to be a sphere of radius 1.20×10^{-15} m.

27. A solid sphere of radius 40.0 cm has a total positive charge of 26.0 μC uniformly distributed throughout its volume. Calculate the magnitude of the electric field (a) 0 cm, (b) 10.0 cm, (c) 40.0 cm, and (d) 60.0 cm from the center of the sphere.

28. A cylindrical shell of radius 7.00 cm and length 240 cm has its charge uniformly distributed on its curved surface. The magnitude of the electric field at a point 19.0 cm radially outward from its axis (measured from the midpoint of the shell) is 36.0 kN/C. Use approximate relationships to find (a) the net charge on the shell and (b) the electric field at a point 4.00 cm from the axis, measured radially outward from the midpoint of the shell.

WEB 29. Consider a long cylindrical charge distribution of radius R with a uniform charge density ρ. Find the electric field at distance r from the axis where $r < R$.

30. A nonconducting wall carries a uniform charge density of 8.60 μC/cm^2. What is the electric field 7.00 cm in front of the wall? Does your result change as the distance from the wall is varied?

31. Consider a thin spherical shell of radius 14.0 cm with a total charge of 32.0 μC distributed uniformly on its surface. Find the electric field (a) 10.0 cm and (b) 20.0 cm from the center of the charge distribution.

32. In nuclear fission, a nucleus of uranium-238, which contains 92 protons, divides into two smaller spheres, each having 46 protons and a radius of 5.90 $\times$ 10^{-15} m. What is the magnitude of the repulsive electric force pushing the two spheres apart?

33. Fill two rubber balloons with air. Suspend both of them from the same point on strings of equal length. Rub each with wool or your hair, so that they hang apart with a noticeable separation between them. Make order-of-magnitude estimates of (a) the force on each, (b) the charge on each, (c) the field each creates at the center of the other, and (d) the total flux of electric field created by each balloon. In your solution, state the quantities you take as data and the values you measure or estimate for them.

34. An insulating sphere is 8.00 cm in diameter and carries a 5.70-μC charge uniformly distributed throughout its interior volume. Calculate the charge enclosed by a concentric spherical surface with radius (a) $r = 2.00$ cm and (b) $r = 6.00$ cm.

35. A uniformly charged, straight filament 7.00 m in length has a total positive charge of 2.00 μC. An uncharged cardboard cylinder 2.00 cm in length and 10.0 cm in radius surrounds the filament at its center, with the filament as the axis of the cylinder. Using reasonable approximations, find (a) the electric field at the surface of the cylinder and (b) the total electric flux through the cylinder.

36. The charge per unit length on a long, straight filament is -90.0 μC/m. Find the electric field (a) 10.0 cm, (b) 20.0 cm, and (c) 100 cm from the filament, where distances are measured perpendicular to the length of the filament.

37. A large flat sheet of charge has a charge per unit area of 9.00 μC/m^2. Find the electric field intensity just above the surface of the sheet, measured from its midpoint.

Section 24.4   Conductors in Electrostatic Equilibrium

38. On a clear, sunny day, a vertical electrical field of about 130 N/C points down over flat ground. What is the surface charge density on the ground for these conditions?

39. A long, straight metal rod has a radius of 5.00 cm and a charge per unit length of 30.0 nC/m. Find the electric field (a) 3.00 cm, (b) 10.0 cm, and (c) 100 cm from the axis of the rod, where distances are measured perpendicular to the rod.

40. A very large, thin, flat plate of aluminum of area A has a total charge Q uniformly distributed over its surfaces. If the same charge is spread uniformly over the *upper* surface of an otherwise identical glass plate, compare the electric fields just above the center of the upper surface of each plate.

41. A square plate of copper with 50.0-cm sides has no net charge and is placed in a region of uniform electric field of 80.0 kN/C directed perpendicularly to the plate. Find (a) the charge density of each face of the plate and (b) the total charge on each face.

42. A hollow conducting sphere is surrounded by a larger concentric, spherical, conducting shell. The inner sphere has a charge $-Q$, and the outer sphere has a charge $3Q$. The charges are in electrostatic equilibrium. Using Gauss's law, find the charges and the electric fields everywhere.

43. Two identical conducting spheres each having a radius of 0.500 cm are connected by a light 2.00-m-long conducting wire. Determine the tension in the wire if 60.0 μC is placed on one of the conductors. (*Hint:* Assume that the surface distribution of charge on each sphere is uniform.)

44. The electric field on the surface of an irregularly shaped conductor varies from 56.0 kN/C to 28.0 kN/C. Calculate the local surface charge density at the point on the surface where the radius of curvature of the surface is (a) greatest and (b) smallest.

45. A long, straight wire is surrounded by a hollow metal cylinder whose axis coincides with that of the wire. The wire has a charge per unit length of λ, and the cylinder has a net charge per unit length of 2λ. From this information, use Gauss's law to find (a) the charge per unit length on the inner and outer surfaces of the cylinder and (b) the electric field outside the cylinder, a distance r from the axis.

46. A conducting spherical shell of radius 15.0 cm carries a net charge of -6.40 μC uniformly distributed on its surface. Find the electric field at points (a) just outside the shell and (b) inside the shell.

WEB 47. A thin conducting plate 50.0 cm on a side lies in the xy plane. If a total charge of 4.00×10^{-8} C is placed on the plate, find (a) the charge density on the plate, (b) the electric field just above the plate, and (c) the electric field just below the plate.

48. A conducting spherical shell having an inner radius of a and an outer radius of b carries a net charge Q. If a point charge q is placed at the center of this shell, determine the surface charge density on (a) the inner surface of the shell and (b) the outer surface of the shell.

49. A solid conducting sphere of radius 2.00 cm has a charge 8.00 μC. A conducting spherical shell of inner radius 4.00 cm and outer radius 5.00 cm is concentric with the solid sphere and has a charge -4.00 μC. Find the electric field at (a) $r = 1.00$ cm, (b) $r = 3.00$ cm, (c) $r = 4.50$ cm, and (d) $r = 7.00$ cm from the center of this charge configuration.

50. A positive point charge is at a distance of $R/2$ from the center of an uncharged thin conducting spherical shell of radius R. Sketch the electric field lines set up by this arrangement both inside and outside the shell.

(Optional)
Section 24.5   Experimental Verification of Gauss's Law and Coulomb's Law

Section 24.6   Formal Derivation of Gauss's Law

51. A sphere of radius R surrounds a point charge Q, located at its center. (a) Show that the electric flux through a circular cap of half-angle θ (Fig. P24.51) is

$$\Phi_E = \frac{Q}{2\epsilon_0}(1 - \cos\theta)$$

What is the flux for (b) $\theta = 90°$ and (c) $\theta = 180°$?

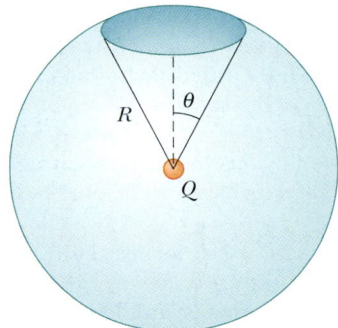

Figure P24.51

ADDITIONAL PROBLEMS

52. A nonuniform electric field is given by the expression $\mathbf{E} = ay\mathbf{i} + bz\mathbf{j} + cx\mathbf{k}$, where a, b, and c are constants. Determine the electric flux through a rectangular surface in the xy plane, extending from $x = 0$ to $x = w$ and from $y = 0$ to $y = h$.

53. A solid insulating sphere of radius a carries a net positive charge $3Q$, uniformly distributed throughout its volume. Concentric with this sphere is a conducting spherical shell with inner radius b and outer radius c, and having a net charge $-Q$, as shown in Figure P24.53. (a) Construct a spherical gaussian surface of radius $r > c$ and find the net charge enclosed by this surface. (b) What is the direction of the electric field at $r > c$? (c) Find the electric field at $r > c$. (d) Find the electric field in the region with radius r where $c > r > b$. (e) Construct a spherical gaussian surface of radius r, where $c > r > b$, and find the net charge enclosed by this surface. (f) Construct a spherical gaussian surface of radius r, where $b > r > a$, and find the net charge enclosed by this surface. (g) Find the electric field in the region $b > r > a$. (h) Construct a spherical gaussian surface of radius $r < a$, and find an expression for the

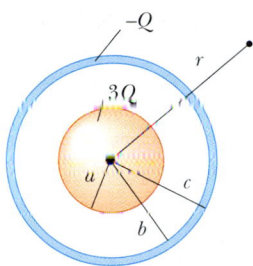

Figure P24.53

net charge enclosed by this surface, as a function of r. Note that the charge inside this surface is less than $3Q$. (i) Find the electric field in the region $r < a$. (j) Determine the charge on the inner surface of the conducting shell. (k) Determine the charge on the outer surface of the conducting shell. (l) Make a plot of the magnitude of the electric field versus r.

54. Consider two identical conducting spheres whose surfaces are separated by a small distance. One sphere is given a large net positive charge, while the other is given a small net positive charge. It is found that the force between them is attractive even though both spheres have net charges of the same sign. Explain how this is possible.

WEB 55. A solid, insulating sphere of radius a has a uniform charge density ρ and a total charge Q. Concentric with this sphere is an uncharged, conducting hollow sphere whose inner and outer radii are b and c, as shown in Figure P24.55. (a) Find the magnitude of the electric field in the regions $r < a$, $a < r < b$, $b < r < c$, and $r > c$. (b) Determine the induced charge per unit area on the inner and outer surfaces of the hollow sphere.

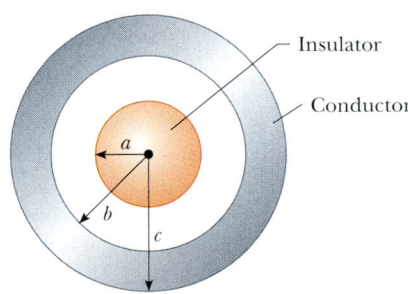

Figure P24.55   Problems 55 and 56.

56. For the configuration shown in Figure P24.55, suppose that $a = 5.00$ cm, $b = 20.0$ cm, and $c = 25.0$ cm. Furthermore, suppose that the electric field at a point 10.0 cm from the center is 3.60×10^3 N/C radially inward, while the electric field at a point 50.0 cm from the center is 2.00×10^2 N/C radially outward. From this information, find (a) the charge on the insulating sphere,

(b) the net charge on the hollow conducting sphere, and (c) the total charge on the inner and outer surfaces of the hollow conducting sphere.

57. An infinitely long cylindrical insulating shell of inner radius a and outer radius b has a uniform volume charge density ρ (C/m³). A line of charge density λ (C/m) is placed along the axis of the shell. Determine the electric field intensity everywhere.

58. Two infinite, nonconducting sheets of charge are parallel to each other, as shown in Figure P24.58. The sheet on the left has a uniform surface charge density σ, and the one on the right has a uniform charge density $-\sigma$. Calculate the value of the electric field at points (a) to the left of, (b) in between, and (c) to the right of the two sheets. (*Hint:* See Example 24.8.)

σ

$-\sigma$

Figure P24.58

WEB 59. Repeat the calculations for Problem 58 when both sheets have *positive* uniform surface charge densities of value σ.

60. A sphere of radius $2a$ is made of a nonconducting material that has a uniform volume charge density ρ. (Assume that the material does not affect the electric field.) A spherical cavity of radius a is now removed from the sphere, as shown in Figure P24.60. Show that the electric field within the cavity is uniform and is given by $E_x = 0$ and $E_y = \rho a/3\epsilon_0$. (*Hint:* The field within the cavity is the superposition of the field due to the original uncut sphere, plus the field due to a sphere

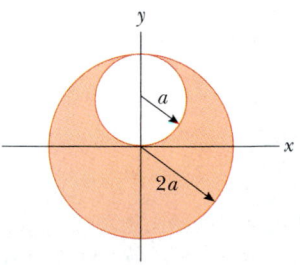

Figure P24.60

the size of the cavity with a uniform negative charge density $-\rho$.)

61. **Review Problem.** An early (incorrect) model of the hydrogen atom, suggested by J. J. Thomson, proposed that a positive cloud of charge $+e$ was uniformly distributed throughout the volume of a sphere of radius R, with the electron an equal-magnitude negative point charge $-e$ at the center. (a) Using Gauss's law, show that the electron would be in equilibrium at the center and, if displaced from the center a distance $r < R$, would experience a restoring force of the form $F = -Kr$, where K is a constant. (b) Show that $K = k_e e^2/R^3$. (c) Find an expression for the frequency f of simple harmonic oscillations that an electron of mass m_e would undergo if displaced a short distance ($< R$) from the center and released. (d) Calculate a numerical value for R that would result in a frequency of electron vibration of 2.47×10^{15} Hz, the frequency of the light in the most intense line in the hydrogen spectrum.

62. A closed surface with dimensions $a = b = 0.400$ m and $c = 0.600$ m is located as shown in Figure P24.62. The electric field throughout the region is nonuniform and given by $\mathbf{E} = (3.0 + 2.0x^2)\,\mathbf{i}$ N/C, where x is in meters. Calculate the net electric flux leaving the closed surface. What net charge is enclosed by the surface?

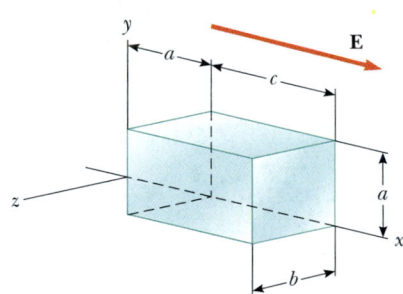

Figure P24.62

63. A solid insulating sphere of radius R has a nonuniform charge density that varies with r according to the expression $\rho = Ar^2$, where A is a constant and $r < R$ is measured from the center of the sphere. (a) Show that the electric field outside ($r > R$) the sphere is $E = AR^5/5\epsilon_0 r^2$. (b) Show that the electric field inside ($r < R$) the sphere is $E = Ar^3/5\epsilon_0$. (*Hint:* Note that the total charge Q on the sphere is equal to the integral of $\rho\, dV$, where r extends from 0 to R; also note that the charge q within a radius $r < R$ is less than Q. To evaluate the integrals, note that the volume element dV for a spherical shell of radius r and thickness dr is equal to $4\pi r^2\, dr$.)

64. A point charge Q is located on the axis of a disk of radius R at a distance b from the plane of the disk (Fig. P24.64). Show that if one fourth of the electric flux from the charge passes through the disk, then $R = \sqrt{3}b$.

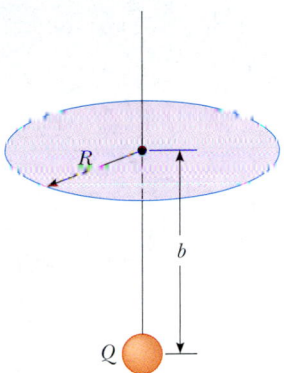

Figure P24.64

a frequency described by the expression

$$f = \frac{1}{2\pi} \sqrt{\frac{\rho e}{m_e \epsilon_0}}$$

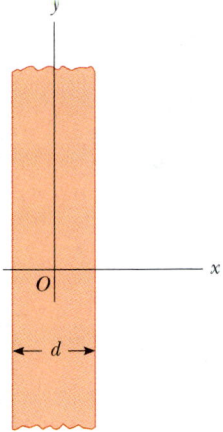

Figure P24.67   Problems 67 and 68.

65. A spherically symmetric charge distribution has a charge density given by $\rho = a/r$, where a is constant. Find the electric field as a function of r. (*Hint:* Note that the charge within a sphere of radius R is equal to the integral of $\rho\, dV$, where r extends from 0 to R. To evaluate the integral, note that the volume element dV for a spherical shell of radius r and thickness dr is equal to $4\pi r^2\, dr$.)

66. An infinitely long insulating cylinder of radius R has a volume charge density that varies with the radius as

$$\rho = \rho_0\!\left(a - \frac{r}{b}\right)$$

where ρ_0, a, and b are positive constants and r is the distance from the axis of the cylinder. Use Gauss's law to determine the magnitude of the electric field at radial distances (a) $r < R$ and (b) $r > R$.

67. **Review Problem.**   A slab of insulating material (infinite in two of its three dimensions) has a uniform positive charge density ρ. An edge view of the slab is shown in Figure P24.67. (a) Show that the magnitude of the electric field a distance x from its center and inside the slab is $E = \rho x/\epsilon_0$. (b) Suppose that an electron of charge $-e$ and mass m_e is placed inside the slab. If it is released from rest at a distance x from the center, show that the electron exhibits simple harmonic motion with

68. A slab of insulating material has a nonuniform positive charge density $\rho = Cx^2$, where x is measured from the center of the slab, as shown in Figure P24.67, and C is a constant. The slab is infinite in the y and z directions. Derive expressions for the electric field in (a) the exterior regions and (b) the interior region of the slab $(-d/2 < x < d/2)$.

69. (a) Using the mathematical similarity between Coulomb's law and Newton's law of universal gravitation, show that Gauss's law for gravitation can be written as

$$\oint \mathbf{g} \cdot d\mathbf{A} = -4\pi G m_{\text{in}}$$

where m_{in} is the mass inside the gaussian surface and $\mathbf{g} = \mathbf{F}_g/m$ represents the gravitational field at any point on the gaussian surface. (b) Determine the gravitational field at a distance r from the center of the Earth where $r < R_E$, assuming that the Earth's mass density is uniform.

ANSWERS TO QUICK QUIZZES

24.1 Zero, because there is no net charge within the surface.

24.2 (b) and (d). Statement (a) is not necessarily true because an equal number of positive and negative charges could be present inside the surface. Statement (c) is not necessarily true, as can be seen from Figure 24.8: A nonzero electric field exists everywhere on the surface, but the charge is not enclosed within the surface; thus, the net flux is zero.

24.3 Any gaussian surface surrounding the system encloses the same amount of charge, regardless of how the components of the system are moved. Thus, the flux through the gaussian surface would be the same as it is when the sphere and shell are concentric.

c h a p t e r

25

Electric Potential

The concept of potential energy was introduced in Chapter 8 in connection with such conservative forces as the force of gravity and the elastic force exerted by a spring. By using the law of conservation of energy, we were able to avoid working directly with forces when solving various problems in mechanics. In this chapter we see that the concept of potential energy is also of great value in the study of electricity. Because the electrostatic force given by Coulomb's law is conservative, electrostatic phenomena can be conveniently described in terms of an electric potential energy. This idea enables us to define a scalar quantity known as *electric potential*. Because the electric potential at any point in an electric field is a scalar function, we can use it to describe electrostatic phenomena more simply than if we were to rely only on the concepts of the electric field and electric forces. In later chapters we shall see that the concept of electric potential is of great practical value.

25.1 POTENTIAL DIFFERENCE AND ELECTRIC POTENTIAL

When a test charge q_0 is placed in an electric field $\mathbf{E}$ created by some other charged object, the electric force acting on the test charge is $q_0\mathbf{E}$. (If the field is produced by more than one charged object, this force acting on the test charge is the vector sum of the individual forces exerted on it by the various other charged objects.) The force $q_0\mathbf{E}$ is conservative because the individual forces described by Coulomb's law are conservative. When the test charge is moved in the field by some external agent, the work done by the field on the charge is equal to the negative of the work done by the external agent causing the displacement. For an infinitesimal displacement $d\mathbf{s}$, the work done by the electric field on the charge is $\mathbf{F} \cdot d\mathbf{s} = q_0\mathbf{E} \cdot d\mathbf{s}$. As this amount of work is done by the field, the potential energy of the charge–field system is decreased by an amount $dU = -q_0\mathbf{E} \cdot d\mathbf{s}$. For a finite displacement of the charge from a point A to a point B, the change in potential energy of the system $\Delta U = U_B - U_A$ is

$$\Delta U = -q_0 \int_A^B \mathbf{E} \cdot d\mathbf{s} \qquad \textbf{(25.1)}$$

Change in potential energy

The integration is performed along the path that q_0 follows as it moves from A to B, and the integral is called either a *path integral* or a *line integral* (the two terms are synonymous). Because the force $q_0\mathbf{E}$ is conservative, **this line integral does not depend on the path taken from A to B.**

Quick Quiz 25.1

If the path between A and B does not make any difference in Equation 25.1, why don't we just use the expression $\Delta U = -q_0 Ed$, where d is the straight-line distance between A and B?

The potential energy per unit charge U/q_0 is independent of the value of q_0 and has a unique value at every point in an electric field. This quantity U/q_0 is called the **electric potential** (or simply the **potential**) V. Thus, the electric potential at any point in an electric field is

$$V = \frac{U}{q_0} \qquad \textbf{(25.2)}$$

The fact that potential energy is a scalar quantity means that electric potential also is a scalar quantity.

The **potential difference** $\Delta V = V_B - V_A$ between any two points A and B in an electric field is defined as the change in potential energy of the system divided by the test charge q_0:

Potential difference

$$\Delta V = \frac{\Delta U}{q_0} = -\int_A^B \mathbf{E} \cdot d\mathbf{s} \qquad (25.3)$$

Potential difference should not be confused with difference in potential energy. The potential difference is proportional to the change in potential energy, and we see from Equation 25.3 that the two are related by $\Delta U = q_0 \Delta V$.

Electric potential is a scalar characteristic of an electric field, independent of the charges that may be placed in the field. However, when we speak of potential energy, we are referring to the charge–field system. Because we are usually interested in knowing the electric potential at the location of a charge and the potential energy resulting from the interaction of the charge with the field, we follow the common convention of speaking of the potential energy as if it belonged to the charge.

Because the change in potential energy of a charge is the negative of the work done by the electric field on the charge (as noted in Equation 25.1), the potential difference ΔV between points A and B equals the work per unit charge that an external agent must perform to move a test charge from A to B without changing the kinetic energy of the test charge.

Just as with potential energy, only *differences* in electric potential are meaningful. To avoid having to work with potential differences, however, we often take the value of the electric potential to be zero at some convenient point in an electric field. This is what we do here: arbitrarily establish the electric potential to be zero at a point that is infinitely remote from the charges producing the field. Having made this choice, we can state that the **electric potential at an arbitrary point in an electric field equals the work required per unit charge to bring a positive test charge from infinity to that point.** Thus, if we take point A in Equation 25.3 to be at infinity, the electric potential at any point P is

$$V_P = -\int_\infty^P \mathbf{E} \cdot d\mathbf{s} \qquad (25.4)$$

In reality, V_P represents the potential difference ΔV between the point P and a point at infinity. (Eq. 25.4 is a special case of Eq. 25.3.)

Because electric potential is a measure of potential energy per unit charge, the SI unit of both electric potential and potential difference is joules per coulomb, which is defined as a **volt** (V):

Definition of volt

$$1 \text{ V} \equiv 1 \frac{\text{J}}{\text{C}}$$

That is, 1 J of work must be done to move a 1-C charge through a potential difference of 1 V.

Equation 25.3 shows that potential difference also has units of electric field times distance. From this, it follows that the SI unit of electric field (N/C) can also be expressed in volts per meter:

$$1 \frac{\text{N}}{\text{C}} = 1 \frac{\text{V}}{\text{m}}$$

A unit of energy commonly used in atomic and nuclear physics is the **electron volt** (eV), which is defined as **the energy an electron (or proton) gains or loses by moving through a potential difference of 1 V.** Because 1 V = 1 J/C and because the fundamental charge is approximately 1.60×10^{-19} C, the electron volt is related to the joule as follows:

$$1 \text{ eV} = 1.60 \times 10^{-19} \text{ C} \cdot \text{V} = 1.60 \times 10^{-19} \text{ J} \qquad \textbf{(25.5)}$$

The electron volt

For instance, an electron in the beam of a typical television picture tube may have a speed of 3.5×10^7 m/s. This corresponds to a kinetic energy of 5.6×10^{-16} J, which is equivalent to 3.5×10^3 eV. Such an electron has to be accelerated from rest through a potential difference of 3.5 kV to reach this speed.

25.2 POTENTIAL DIFFERENCES IN A UNIFORM ELECTRIC FIELD

Equations 25.1 and 25.3 hold in all electric fields, whether uniform or varying, but they can be simplified for a uniform field. First, consider a uniform electric field directed along the negative y axis, as shown in Figure 25.1a. Let us calculate the potential difference between two points A and B separated by a distance d, where d is measured parallel to the field lines. Equation 25.3 gives

$$V_B - V_A = \Delta V = -\int_A^B \mathbf{E} \cdot d\mathbf{s} = -\int_A^B E \cos 0° \, ds = -\int_A^B E \, ds$$

Because E is constant, we can remove it from the integral sign; this gives

$$\Delta V = -E \int_A^B ds = -Ed \qquad \textbf{(25.6)}$$

Potential difference in a uniform electric field

The minus sign indicates that point B is at a lower electric potential than point A; that is, $V_B < V_A$. **Electric field lines always point in the direction of decreasing electric potential,** as shown in Figure 25.1a.

Now suppose that a test charge q_0 moves from A to B. We can calculate the change in its potential energy from Equations 25.3 and 25.6:

$$\Delta U = q_0 \, \Delta V = -q_0 Ed \qquad \textbf{(25.7)}$$

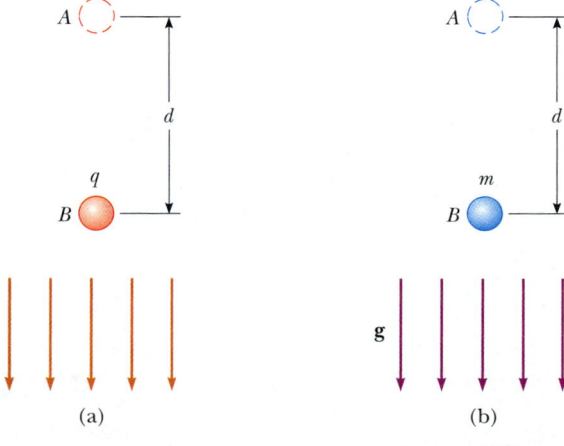

(a)

(b)

Figure 25.1 (a) When the electric field $\mathbf{E}$ is directed downward, point B is at a lower electric potential than point A. A positive test charge that moves from point A to point B loses electric potential energy. (b) A mass m moving downward in the direction of the gravitational field $\mathbf{g}$ loses gravitational potential energy.

QuickLab

It takes an electric field of about 30 000 V/cm to cause a spark in dry air. Shuffle across a rug and reach toward a doorknob. By estimating the length of the spark, determine the electric potential difference between your finger and the doorknob after shuffling your feet but before touching the knob. (If it is very humid on the day you attempt this, it may not work. Why?)

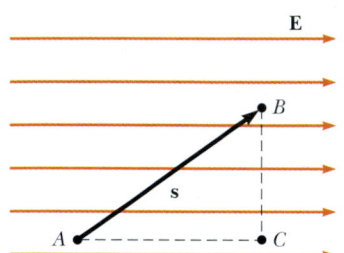

Figure 25.2 A uniform electric field directed along the positive *x* axis. Point *B* is at a lower electric potential than point *A*. Points *B* and *C* are at the *same* electric potential.

11.9

An equipotential surface

From this result, we see that if q_0 is positive, then ΔU is negative. We conclude that **a positive charge loses electric potential energy when it moves in the direction of the electric field.** This means that an electric field does work on a positive charge when the charge moves in the direction of the electric field. (This is analogous to the work done by the gravitational field on a falling mass, as shown in Figure 25.1b.) If a positive test charge is released from rest in this electric field, it experiences an electric force $q_0\mathbf{E}$ in the direction of $\mathbf{E}$ (downward in Fig. 25.1a). Therefore, it accelerates downward, gaining kinetic energy. **As the charged particle gains kinetic energy, it loses an equal amount of potential energy.**

If q_0 is negative, then ΔU is positive and the situation is reversed: **A negative charge gains electric potential energy when it moves in the direction of the electric field.** If a negative charge is released from rest in the field $\mathbf{E}$, it accelerates in a direction opposite the direction of the field.

Now consider the more general case of a charged particle that is free to move between any two points in a uniform electric field directed along the *x* axis, as shown in Figure 25.2. (In this situation, the charge is not being moved by an external agent as before.) If $\mathbf{s}$ represents the displacement vector between points *A* and *B*, Equation 25.3 gives

$$\Delta V = -\int_A^B \mathbf{E} \cdot d\mathbf{s} = -\mathbf{E} \cdot \int_A^B d\mathbf{s} = -\mathbf{E} \cdot \mathbf{s} \tag{25.8}$$

where again we are able to remove $\mathbf{E}$ from the integral because it is constant. The change in potential energy of the charge is

$$\Delta U = q_0 \Delta V = -q_0 \mathbf{E} \cdot \mathbf{s} \tag{25.9}$$

Finally, we conclude from Equation 25.8 that all points in a plane perpendicular to a uniform electric field are at the same electric potential. We can see this in Figure 25.2, where the potential difference $V_B - V_A$ is equal to the potential difference $V_C - V_A$. (Prove this to yourself by working out the dot product $\mathbf{E} \cdot \mathbf{s}$ for $\mathbf{s}_{A \to B}$, where the angle θ between $\mathbf{E}$ and $\mathbf{s}$ is arbitrary as shown in Figure 25.2, and the dot product for $\mathbf{s}_{A \to C}$, where $\theta = 0$.) Therefore, $V_B = V_C$. **The name equipotential surface is given to any surface consisting of a continuous distribution of points having the same electric potential.**

Note that because $\Delta U = q_0 \Delta V$, no work is done in moving a test charge between any two points on an equipotential surface. The equipotential surfaces of a uniform electric field consist of a family of planes that are all perpendicular to the field. Equipotential surfaces for fields with other symmetries are described in later sections.

Quick Quiz 25.2

The labeled points in Figure 25.3 are on a series of equipotential surfaces associated with an electric field. Rank (from greatest to least) the work done by the electric field on a positively charged particle that moves from *A* to *B*; from *B* to *C*; from *C* to *D*; from *D* to *E*.

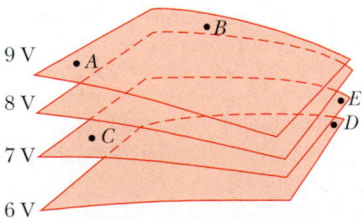

Figure 25.3 Four equipotential surfaces.

EXAMPLE 25.1 ▶ The Electric Field Between Two Parallel Plates of Opposite Charge

A battery produces a specified potential difference between conductors attached to the battery terminals. A 12-V battery is connected between two parallel plates, as shown in Figure 25.4. The separation between the plates is $d = 0.30$ cm, and we assume the electric field between the plates to be uniform.

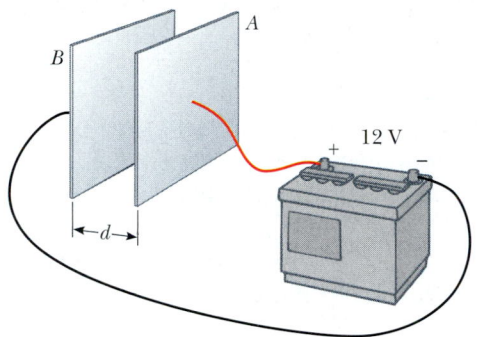

Figure 25.4 A 12-V battery connected to two parallel plates. The electric field between the plates has a magnitude given by the potential difference ΔV divided by the plate separation d.

(This assumption is reasonable if the plate separation is small relative to the plate dimensions and if we do not consider points near the plate edges.) Find the magnitude of the electric field between the plates.

Solution The electric field is directed from the positive plate (A) to the negative one (B), and the positive plate is at a higher electric potential than the negative plate is. The potential difference between the plates must equal the potential difference between the battery terminals. We can understand this by noting that all points on a conductor in equilibrium are at the same electric potential[1]; no potential difference exists between a terminal and any portion of the plate to which it is connected. Therefore, the magnitude of the electric field between the plates is, from Equation 25.6,

$$E = \frac{|V_B - V_A|}{d} = \frac{12 \text{ V}}{0.30 \times 10^{-2} \text{ m}} = 4.0 \times 10^3 \text{ V/m}$$

This configuration, which is shown in Figure 25.4 and called a *parallel-plate capacitor*, is examined in greater detail in Chapter 26.

EXAMPLE 25.2 ▶ Motion of a Proton in a Uniform Electric Field

A proton is released from rest in a uniform electric field that has a magnitude of 8.0×10^4 V/m and is directed along the positive x axis (Fig. 25.5). The proton undergoes a displacement of 0.50 m in the direction of **E**. (a) Find the change in electric potential between points A and B.

Solution Because the proton (which, as you remember, carries a positive charge) moves in the direction of the field, we expect it to move to a position of lower electric potential.

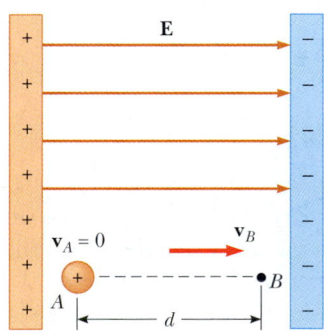

Figure 25.5 A proton accelerates from A to B in the direction of the electric field.

From Equation 25.6, we have

$$\Delta V = -Ed = -(8.0 \times 10^4 \text{ V/m})(0.50 \text{ m})$$

$$= -4.0 \times 10^4 \text{ V}$$

(b) Find the change in potential energy of the proton for this displacement.

Solution

$$\Delta U = q_0 \Delta V = e \Delta V$$

$$= (1.6 \times 10^{-19} \text{ C})(-4.0 \times 10^4 \text{ V})$$

$$= -6.4 \times 10^{-15} \text{ J}$$

The negative sign means the potential energy of the proton decreases as it moves in the direction of the electric field. As the proton accelerates in the direction of the field, it gains kinetic energy and at the same time loses electric potential energy (because energy is conserved).

Exercise Use the concept of conservation of energy to find the speed of the proton at point B.

Answer 2.77×10^6 m/s.

[1] The electric field vanishes within a conductor in electrostatic equilibrium; thus, the path integral $\int \mathbf{E} \cdot d\mathbf{s}$ between any two points in the conductor must be zero. A more complete discussion of this point is given in Section 25.6.

Consider an isolated positive point charge q. Recall that such a charge produces an electric field that is directed radially outward from the charge. To find the electric potential at a point located a distance r from the charge, we begin with the general expression for potential difference:

$$V_B - V_A = -\int_A^B \mathbf{E} \cdot d\mathbf{s}$$

where A and B are the two arbitrary points shown in Figure 25.6. At any field point, the electric field due to the point charge is $\mathbf{E} = k_e q \hat{\mathbf{r}}/r^2$ (Eq. 23.4), where $\hat{\mathbf{r}}$ is a unit vector directed from the charge toward the field point. The quantity $\mathbf{E} \cdot d\mathbf{s}$ can be expressed as

$$\mathbf{E} \cdot d\mathbf{s} = k_e \frac{q}{r^2} \hat{\mathbf{r}} \cdot d\mathbf{s}$$

Because the magnitude of $\hat{\mathbf{r}}$ is 1, the dot product $\hat{\mathbf{r}} \cdot d\mathbf{s} = ds \cos \theta$, where θ is the angle between $\hat{\mathbf{r}}$ and $d\mathbf{s}$. Furthermore, $ds \cos \theta$ is the projection of $d\mathbf{s}$ onto $\mathbf{r}$; thus, $ds \cos \theta = dr$. That is, any displacement $d\mathbf{s}$ along the path from point A to point B produces a change dr in the magnitude of $\mathbf{r}$, the radial distance to the charge creating the field. Making these substitutions, we find that $\mathbf{E} \cdot d\mathbf{s} = (k_e q/r^2)\,dr$; hence, the expression for the potential difference becomes

$$V_B - V_A = -\int E_r \, dr = -k_e q \int_{r_A}^{r_B} \frac{dr}{r^2} = \frac{k_e q}{r}\bigg]_{r_A}^{r_B}$$

$$V_B - V_A = k_e q \left[\frac{1}{r_B} - \frac{1}{r_A} \right] \tag{25.10}$$

The integral of $\mathbf{E} \cdot d\mathbf{s}$ is *independent* of the path between points A and B—as it must be because the electric field of a point charge is conservative. Furthermore, Equation 25.10 expresses the important result that the potential difference between any two points A and B in a field created by a point charge depends only on the radial coordinates r_A and r_B. It is customary to choose the reference of electric potential to be zero at $r_A = \infty$. With this reference, the electric potential created by a point charge at any distance r from the charge is

$$V = k_e \frac{q}{r} \tag{25.11}$$

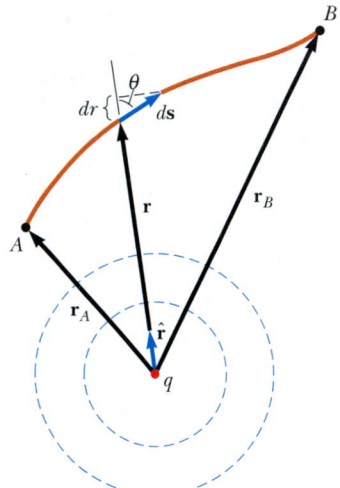

Figure 25.6 The potential difference between points A and B due to a point charge q depends *only* on the initial and final radial coordinates r_A and r_B. The two dashed circles represent cross-sections of spherical equipotential surfaces.

Electric potential created by a point charge

Electric potential is graphed in Figure 25.7 as a function of r, the radial distance from a positive charge in the xy plane. Consider the following analogy to gravitational potential: Imagine trying to roll a marble toward the top of a hill shaped like Figure 25.7a. The gravitational force experienced by the marble is analogous to the repulsive force experienced by a positively charged object as it approaches another positively charged object. Similarly, the electric potential graph of the region surrounding a negative charge is analogous to a "hole" with respect to any approaching positively charged objects. A charged object must be infinitely distant from another charge before the surface is "flat" and has an electric potential of zero.

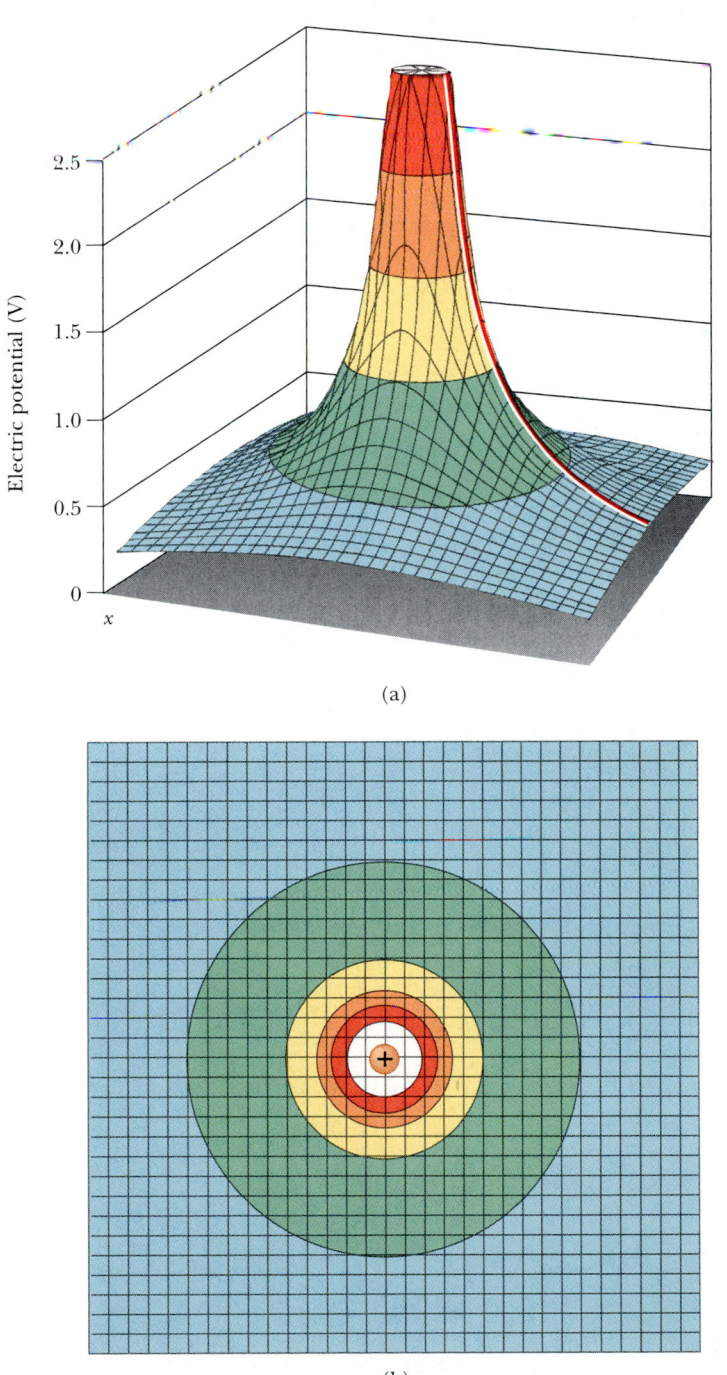

(a)

(b)

Figure 25.7   (a) The electric potential in the plane around a single positive charge is plotted on the vertical axis. (The electric potential function for a negative charge would look like a hole instead of a hill.) The red line shows the $1/r$ nature of the electric potential, as given by Equation 25.11. (b) View looking straight down the vertical axis of the graph in part (a), showing concentric circles where the electric potential is constant. These circles are cross sections of equipotential spheres having the charge at the center.

> **Quick Quiz 25.3** ▸
>
> A spherical balloon contains a positively charged object at its center. As the balloon is inflated to a greater volume while the charged object remains at the center, does the electric potential at the surface of the balloon increase, decrease, or remain the same? How about the magnitude of the electric field? The electric flux?

We obtain the electric potential resulting from two or more point charges by applying the superposition principle. That is, the total electric potential at some point P due to several point charges is the sum of the potentials due to the individual charges. For a group of point charges, we can write the total electric potential at P in the form

Electric potential due to several point charges

$$V = k_e \sum_i \frac{q_i}{r_i} \qquad (25.12)$$

where the potential is again taken to be zero at infinity and r_i is the distance from the point P to the charge q_i. Note that the sum in Equation 25.12 is an algebraic sum of scalars rather than a vector sum (which we use to calculate the electric field of a group of charges). Thus, it is often much easier to evaluate V than to evaluate **E**. The electric potential around a dipole is illustrated in Figure 25.8.

We now consider the potential energy of a system of two charged particles. If V_1 is the electric potential at a point P due to charge q_1, then the work an external agent must do to bring a second charge q_2 from infinity to P without acceleration is $q_2 V_1$. By definition, this work equals the potential energy U of the two-particle system when the particles are separated by a distance r_{12} (Fig. 25.9). Therefore, we can express the potential energy as[2]

Electric potential energy due to two charges

$$U = k_e \frac{q_1 q_2}{r_{12}} \qquad (25.13)$$

Note that if the charges are of the same sign, U is positive. This is consistent with the fact that positive work must be done by an external agent on the system to bring the two charges near one another (because like charges repel). If the charges are of opposite sign, U is negative; this means that negative work must be done against the attractive force between the unlike charges for them to be brought near each other.

If more than two charged particles are in the system, we can obtain the total potential energy by calculating U for every pair of charges and summing the terms algebraically. As an example, the total potential energy of the system of three charges shown in Figure 25.10 is

$$U = k_e \left(\frac{q_1 q_2}{r_{12}} + \frac{q_1 q_3}{r_{13}} + \frac{q_2 q_3}{r_{23}} \right) \qquad (25.14)$$

Physically, we can interpret this as follows: Imagine that q_1 is fixed at the position shown in Figure 25.10 but that q_2 and q_3 are at infinity. The work an external agent must do to bring q_2 from infinity to its position near q_1 is $k_e q_1 q_2 / r_{12}$, which is the first term in Equation 25.14. The last two terms represent the work required to bring q_3 from infinity to its position near q_1 and q_2. (The result is independent of the order in which the charges are transported.)

---

[2] The expression for the electric potential energy of a system made up of two point charges, Equation 25.13, is of the *same* form as the equation for the gravitational potential energy of a system made up of two point masses, $Gm_1 m_2 / r$ (see Chapter 14). The similarity is not surprising in view of the fact that both expressions are derived from an inverse-square force law.

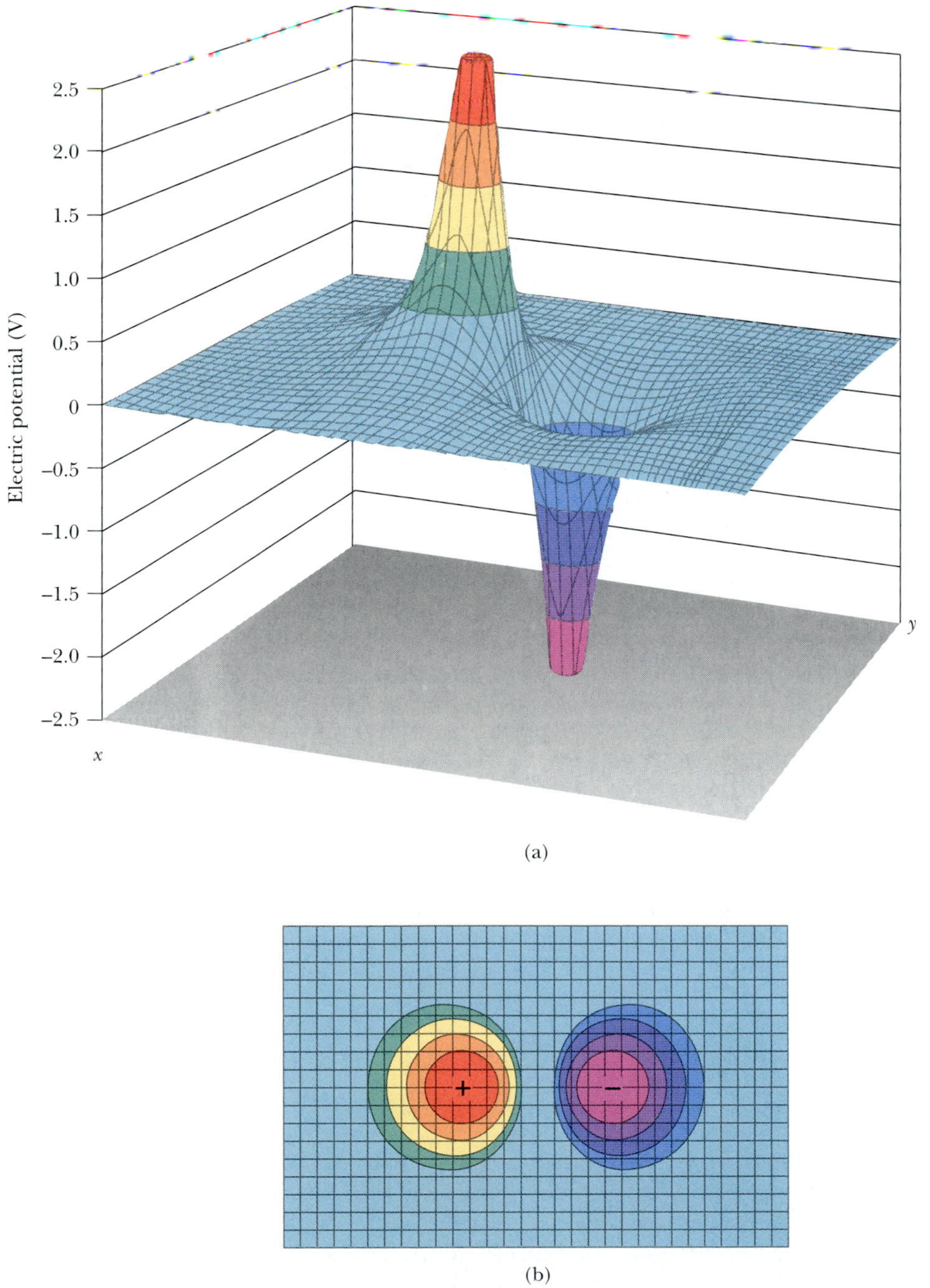

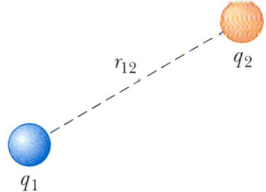

Figure 25. 9 If two point charges are separated by a distance r_{12}, the potential energy of the pair of charges is given by $k_e q_1 q_2 / r_{12}$.

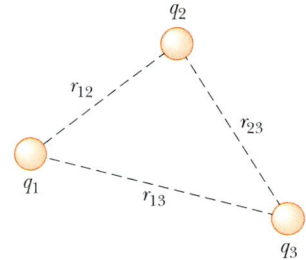

Figure 25.10 Three point charges are fixed at the positions shown. The potential energy of this system of charges is given by Equation 25.14.

(a)

(b)

Figure 25.8 (a) The electric potential in the plane containing a dipole. (b) Top view of the function graphed in part (a).

EXAMPLE 25.3   **The Electric Potential Due to Two Point Charges**

A charge $q_1 = 2.00\ \mu C$ is located at the origin, and a charge $q_2 = -6.00\ \mu C$ is located at $(0, 3.00)$ m, as shown in Figure 25.11a. (a) Find the total electric potential due to these charges at the point P, whose coordinates are $(4.00, 0)$ m.

Solution   For two charges, the sum in Equation 25.12 gives

$$V_P = k_e \left(\frac{q_1}{r_1} + \frac{q_2}{r_2} \right)$$

$$= 8.99 \times 10^9\ \frac{N \cdot m^2}{C^2} \left(\frac{2.00 \times 10^{-6}\ C}{4.00\ m} + \frac{-6.00 \times 10^{-6}\ C}{5.00\ m} \right)$$

$$= \boxed{-6.29 \times 10^3\ V}$$

(b) Find the change in potential energy of a $3.00\text{-}\mu C$ charge as it moves from infinity to point P (Fig. 25.11b).

Solution   When the charge is at infinity, $U_i = 0$, and when the charge is at P, $U_f = q_3 V_P$; therefore,

$$\Delta U = q_3 V_P - 0 = (3.00 \times 10^{-6}\ C)(-6.29 \times 10^3\ V)$$

$$= \boxed{-18.9 \times 10^{-3}\ J}$$

Therefore, because $W = -\Delta U$, positive work would have to be done by an external agent to remove the charge from point P back to infinity.

Exercise   Find the total potential energy of the system illustrated in Figure 25.11b.

Answer $-5.48 \times 10^{-2}\ J$.

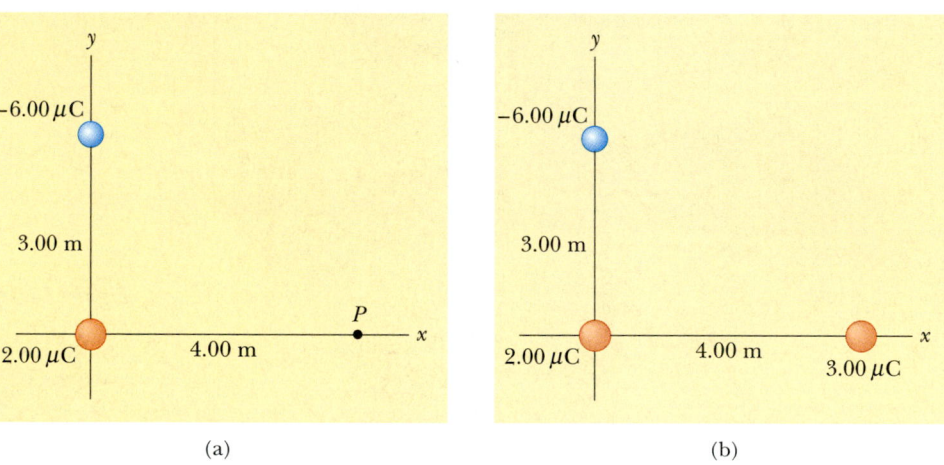

(a)   (b)

Figure 25.11   (a) The electric potential at P due to the two charges is the algebraic sum of the potentials due to the individual charges. (b) What is the potential energy of the three-charge system?

25.4   **OBTAINING THE VALUE OF THE ELECTRIC FIELD FROM THE ELECTRIC POTENTIAL**

The electric field **E** and the electric potential V are related as shown in Equation 25.3. We now show how to calculate the value of the electric field if the electric potential is known in a certain region.

From Equation 25.3 we can express the potential difference dV between two points a distance ds apart as

$$dV = -\mathbf{E} \cdot d\mathbf{s} \tag{25.15}$$

If the electric field has only one component E_x, then $\mathbf{E} \cdot d\mathbf{s} = E_x\ dx$. Therefore, Equation 25.15 becomes $dV = -E_x\ dx$, or

$$E_x = -\frac{dV}{dx} \tag{25.16}$$

That is, the magnitude of the electric field in the direction of some coordinate is equal to the negative of the derivative of the electric potential with respect to that coordinate. Recall from the discussion following Equation 25.8 that the electric potential does not change for any displacement perpendicular to an electric field. This is consistent with the notion, developed in Section 25.2, that equipotential surfaces are perpendicular to the field, as shown in Figure 25.12. A small positive charge placed at rest on an electric field line begins to move along the direction of **E** because that is the direction of the force exerted on the charge by the charge distribution creating the electric field (and hence is the direction of **a**). Because the charge starts with zero velocity, it moves in the direction of the change in velocity—that is, in the direction of **a**. In Figures 25.12a and 25.12b, the charge moves in a straight line because its acceleration vector is always parallel to its velocity vector. The magnitude of **v** increases, but its direction does not change. The situation is different in Figure 25.12c. A positive charge placed at some point near the dipole first moves in a direction parallel to **E** at that point. Because the direction of the electric field is different at different locations, however, the force acting on the charge changes direction, and **a** is no longer parallel to **v**. This causes the moving charge to change direction and speed, but it does not necessarily follow the electric field lines. Recall that it is not the velocity vector but rather the acceleration vector that is proportional to force.

If the charge distribution creating an electric field has spherical symmetry such that the volume charge density depends only on the radial distance r, then the electric field is radial. In this case, $\mathbf{E} \cdot d\mathbf{s} = E_r\, dr$, and thus we can express dV in the form $dV = -E_r\, dr$. Therefore,

$$E_r = -\frac{dV}{dr} \tag{25.17}$$

For example, the electric potential of a point charge is $V = k_e q / r$. Because V is a function of r only, the potential function has spherical symmetry. Applying Equation 25.17, we find that the electric field due to the point charge is $E_r = k_e q / r^2$, a familiar result. Note that the potential changes only in the radial direction, not in

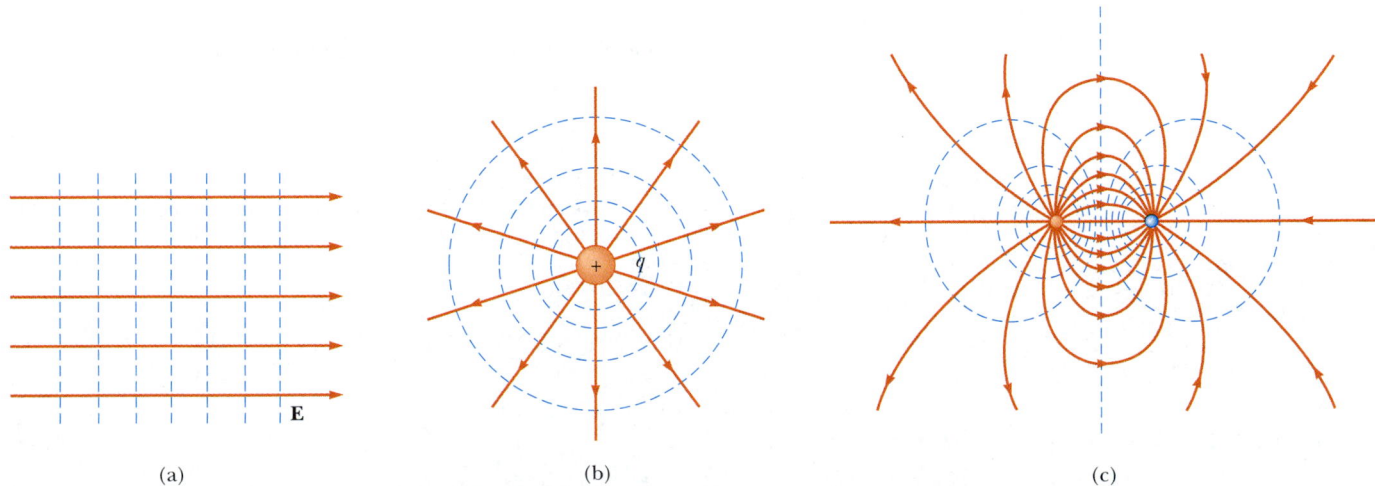

(a)   (b)   (c)

Figure 25.12 Equipotential surfaces (dashed blue lines) and electric field lines (red lines) for (a) a uniform electric field produced by an infinite sheet of charge, (b) a point charge, and (c) an electric dipole. In all cases, the equipotential surfaces are *perpendicular* to the electric field lines at every point. Compare these drawings with Figures 25.2, 25.7b, and 25.8b.

Equipotential surfaces are perpendicular to the electric field lines

any direction perpendicular to r. Thus, V (like E_r) is a function only of r. Again, this is consistent with the idea that **equipotential surfaces are perpendicular to field lines.** In this case the equipotential surfaces are a family of spheres concentric with the spherically symmetric charge distribution (Fig. 25.12b).

The equipotential surfaces for an electric dipole are sketched in Figure 25.12c. When a test charge undergoes a displacement $d\mathbf{s}$ along an equipotential surface, then $dV = 0$ because the potential is constant along an equipotential surface. From Equation 25.15, then, $dV = -\mathbf{E} \cdot d\mathbf{s} = 0$; thus, $\mathbf{E}$ must be perpendicular to the displacement along the equipotential surface. This shows that the equipotential surfaces must *always* be *perpendicular* to the electric field lines.

In general, the electric potential is a function of all three spatial coordinates. If $V(r)$ is given in terms of the cartesian coordinates, the electric field components E_x, E_y, and E_z can readily be found from $V(x, y, z)$ as the partial derivatives[3]

$$E_x = -\frac{\partial V}{\partial x} \qquad E_y = -\frac{\partial V}{\partial y} \qquad E_z = -\frac{\partial V}{\partial z}$$

For example, if $V = 3x^2y + y^2 + yz$, then

$$\frac{\partial V}{\partial x} = \frac{\partial}{\partial x}(3x^2y + y^2 + yz) = \frac{\partial}{\partial x}(3x^2y) = 3y\frac{d}{dx}(x^2) = 6xy$$

EXAMPLE 25.4 ▶ **The Electric Potential Due to a Dipole**

An electric dipole consists of two charges of equal magnitude and opposite sign separated by a distance $2a$, as shown in Figure 25.13. The dipole is along the x axis and is centered at the origin. (a) Calculate the electric potential at point P.

Solution   For point P in Figure 25.13,

$$V = k_e\sum\frac{q_i}{r_i} = k_e\left(\frac{q}{x-a} - \frac{q}{x+a}\right) = \boxed{\frac{2k_eqa}{x^2-a^2}}$$

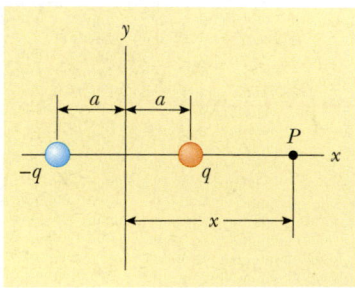

Figure 25.13   An electric dipole located on the x axis.

(How would this result change if point P happened to be located to the left of the negative charge?)

(b) Calculate V and E_x at a point far from the dipole.

Solution   If point P is far from the dipole, such that $x \gg a$, then a^2 can be neglected in the term $x^2 - a^2$, and V becomes

$$V \approx \boxed{\frac{2k_eqa}{x^2}} \qquad (x \gg a)$$

Using Equation 25.16 and this result, we can calculate the electric field at a point far from the dipole:

$$E_x = -\frac{dV}{dx} = \boxed{\frac{4k_eqa}{x^3}} \qquad (x \gg a)$$

(c) Calculate V and E_x if point P is located anywhere between the two charges.

Solution

$$V = k_e\sum\frac{q_i}{r_i} = k_e\left(\frac{q}{a-x} - \frac{q}{x+a}\right) = -\frac{2k_eqx}{x^2-a^2}$$

$$E_x = -\frac{dV}{dx} = -\frac{d}{dx}\left(-\frac{2k_eqx}{x^2-a^2}\right) = 2k_eq\left(\frac{-x^2-a^2}{(x^2-a^2)^2}\right)$$

---

[3] In vector notation, $\mathbf{E}$ is often written

$$\mathbf{E} = -\nabla V = -\left(\mathbf{i}\frac{\partial}{\partial x} + \mathbf{j}\frac{\partial}{\partial y} + \mathbf{k}\frac{\partial}{\partial z}\right)V$$

where ∇ is called the *gradient operator.*

We can check these results by considering the situation at the center of the dipole, where $x = 0$, $V = 0$, and $E_x = -2k_eq/a^2$.

Exercise Verify the electric field result in part (c) by calculating the sum of the individual electric field vectors at the origin due to the two charges.

25.5 ELECTRIC POTENTIAL DUE TO CONTINUOUS CHARGE DISTRIBUTIONS

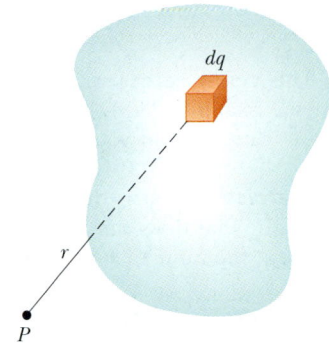

We can calculate the electric potential due to a continuous charge distribution in two ways. If the charge distribution is known, we can start with Equation 25.11 for the electric potential of a point charge. We then consider the potential due to a small charge element dq, treating this element as a point charge (Fig. 25.14). The electric potential dV at some point P due to the charge element dq is

$$dV = k_e \frac{dq}{r} \qquad (25.18)$$

where r is the distance from the charge element to point P. To obtain the total potential at point P, we integrate Equation 25.18 to include contributions from all elements of the charge distribution. Because each element is, in general, a different distance from point P and because k_e is constant, we can express V as

$$V = k_e \int \frac{dq}{r} \qquad (25.19)$$

Figure 25.14 The electric potential at the point P due to a continuous charge distribution can be calculated by dividing the charged body into segments of charge dq and summing the electric potential contributions over all segments.

In effect, we have replaced the sum in Equation 25.12 with an integral. Note that this expression for V uses a particular reference: The electric potential is taken to be zero when point P is infinitely far from the charge distribution.

If the electric field is already known from other considerations, such as Gauss's law, we can calculate the electric potential due to a continuous charge distribution using Equation 25.3. If the charge distribution is highly symmetric, we first evaluate **E** at any point using Gauss's law and then substitute the value obtained into Equation 25.3 to determine the potential difference ΔV between any two points. We then choose the electric potential V to be zero at some convenient point.

We illustrate both methods with several examples.

EXAMPLE 25.5 Electric Potential Due to a Uniformly Charged Ring

(a) Find an expression for the electric potential at a point P located on the perpendicular central axis of a uniformly charged ring of radius a and total charge Q.

Solution Let us orient the ring so that its plane is perpendicular to an x axis and its center is at the origin. We can then take point P to be at a distance x from the center of the ring, as shown in Figure 25.15. The charge element dq is at a distance $\sqrt{x^2 + a^2}$ from point P. Hence, we can express V as

$$V = k_e \int \frac{dq}{r} = k_e \int \frac{dq}{\sqrt{x^2 + a^2}}$$

Because each element dq is at the same distance from point P,

we can remove $\sqrt{x^2 + a^2}$ from the integral, and V reduces to

$$V = \frac{k_e}{\sqrt{x^2 + a^2}} \int dq = \frac{k_eQ}{\sqrt{x^2 + a^2}} \qquad (25.20)$$

The only variable in this expression for V is x. This is not surprising because our calculation is valid only for points along the x axis, where y and z are both zero.

(b) Find an expression for the magnitude of the electric field at point P.

Solution From symmetry, we see that along the x axis **E** can have only an x component. Therefore, we can use Equation 25.16:

$$E_x = -\frac{dV}{dx} = -k_e Q \frac{d}{dx}(x^2 + a^2)^{-1/2}$$

$$= -k_e Q(-\tfrac{1}{2})(x^2 + a^2)^{-3/2}(2x)$$

$$= \frac{k_e Q x}{(x^2 + a^2)^{3/2}} \qquad (25.21)$$

This result agrees with that obtained by direct integration (see Example 23.8). Note that $E_x = 0$ at $x = 0$ (the center of the ring). Could you have guessed this from Coulomb's law?

Exercise   What is the electric potential at the center of the ring? What does the value of the field at the center tell you about the value of V at the center?

Answer $V = k_e Q/a$. Because $E_x = -dV/dx = 0$ at the cen-

ter, V has either a maximum or minimum value; it is, in fact, a maximum.

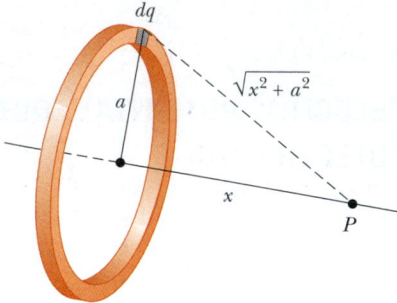

Figure 25.15   A uniformly charged ring of radius a lies in a plane perpendicular to the x axis. All segments dq of the ring are the same distance from any point P lying on the x axis.

EXAMPLE 25.6   Electric Potential Due to a Uniformly Charged Disk

Find (a) the electric potential and (b) the magnitude of the electric field along the perpendicular central axis of a uniformly charged disk of radius a and surface charge density σ.

Solution   (a) Again, we choose the point P to be at a distance x from the center of the disk and take the plane of the disk to be perpendicular to the x axis. We can simplify the problem by dividing the disk into a series of charged rings. The electric potential of each ring is given by Equation 25.20. Consider one such ring of radius r and width dr, as indicated in Figure 25.16. The surface area of the ring is $dA = 2\pi r\, dr$;

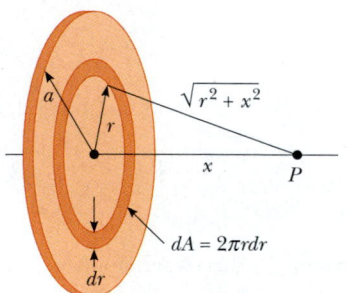

Figure 25.16   A uniformly charged disk of radius a lies in a plane perpendicular to the x axis. The calculation of the electric potential at any point P on the x axis is simplified by dividing the disk into many rings each of area $2\pi r\, dr$.

from the definition of surface charge density (see Section 23.5), we know that the charge on the ring is $dq = \sigma\, dA = \sigma 2\pi r\, dr$. Hence, the potential at the point P due to this ring is

$$dV = \frac{k_e\, dq}{\sqrt{r^2 + x^2}} = \frac{k_e \sigma 2\pi r\, dr}{\sqrt{r^2 + x^2}}$$

To find the *total* electric potential at P, we sum over all rings making up the disk. That is, we integrate dV from $r = 0$ to $r = a$:

$$V = \pi k_e \sigma \int_0^a \frac{2r\, dr}{\sqrt{r^2 + x^2}} = \pi k_e \sigma \int_0^a (r^2 + x^2)^{-1/2}\, 2r\, dr$$

This integral is of the form $u^n\, du$ and has the value $u^{n+1}/(n + 1)$, where $n = -\tfrac{1}{2}$ and $u = r^2 + x^2$. This gives

$$V = 2\pi k_e \sigma[(x^2 + a^2)^{1/2} - x] \qquad (25.22)$$

(b) As in Example 25.5, we can find the electric field at any axial point from

$$E_x = -\frac{dV}{dx} = 2\pi k_e \sigma \left(1 - \frac{x}{\sqrt{x^2 + a^2}}\right) \qquad (25.23)$$

The calculation of V and **E** for an arbitrary point off the axis is more difficult to perform, and we do not treat this situation in this text.

EXAMPLE 25.7 Electric Potential Due to a Finite Line of Charge

A rod of length ℓ located along the x axis has a total charge Q and a uniform linear charge density $\lambda = Q/\ell$. Find the electric potential at a point P located on the y axis a distance a from the origin (Fig. 25.17).

Solution The length element dx has a charge $dq = \lambda\, dx$. Because this element is a distance $r = \sqrt{x^2 + a^2}$ from point P, we can express the potential at point P due to this element as

$$dV = k_e \frac{dq}{r} = k_e \frac{\lambda\, dx}{\sqrt{x^2 + a^2}}$$

To obtain the total potential at P, we integrate this expression over the limits $x = 0$ to $x = \ell$. Noting that k_e and λ are constants, we find that

$$V = k_e \lambda \int_0^\ell \frac{dx}{\sqrt{x^2 + a^2}} = k_e \frac{Q}{\ell} \int_0^\ell \frac{dx}{\sqrt{x^2 + a^2}}$$

This integral has the following value (see Appendix B):

$$\int \frac{dx}{\sqrt{x^2 + a^2}} = \ln(x + \sqrt{x^2 + a^2})$$

Evaluating V, we find that

$$V = \frac{k_e Q}{\ell} \ln\left(\frac{\ell + \sqrt{\ell^2 + a^2}}{a}\right) \tag{25.24}$$

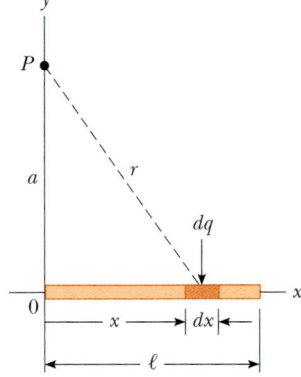

Figure 25.17 A uniform line charge of length ℓ located along the x axis. To calculate the electric potential at P, the line charge is divided into segments each of length dx and each carrying a charge $dq = \lambda\, dx$.

EXAMPLE 25.8 Electric Potential Due to a Uniformly Charged Sphere

An insulating solid sphere of radius R has a uniform positive volume charge density and total charge Q. (a) Find the electric potential at a point outside the sphere, that is, for $r > R$. Take the potential to be zero at $r = \infty$.

Solution In Example 24.5, we found that the magnitude of the electric field outside a uniformly charged sphere of radius R is

$$E_r = k_e \frac{Q}{r^2} \qquad (\text{for } r > R)$$

where the field is directed radially outward when Q is positive. In this case, to obtain the electric potential at an exterior point, such as B in Figure 25.18, we use Equation 25.4 and the expression for E_r given above:

$$V_B = -\int_\infty^r E_r\, dr = -k_e Q \int_\infty^r \frac{dr}{r^2}$$

$$V_B = k_e \frac{Q}{r} \qquad (\text{for } r > R)$$

Note that the result is identical to the expression for the electric potential due to a point charge (Eq. 25.11).

Because the potential must be continuous at $r = R$, we can use this expression to obtain the potential at the surface of the sphere. That is, the potential at a point such as C shown in Figure 25.18 is

$$V_C = k_e \frac{Q}{R} \qquad (\text{for } r = R)$$

(b) Find the potential at a point inside the sphere, that is, for $r < R$.

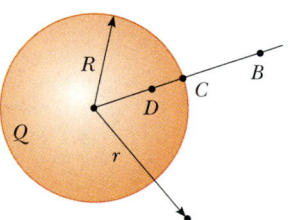

Figure 25.18 A uniformly charged insulating sphere of radius R and total charge Q. The electric potentials at points B and C are equivalent to those produced by a point charge Q located at the center of the sphere, but this is not true for point D.

Solution   In Example 24.5 we found that the electric field inside an insulating uniformly charged sphere is

$$E_r = \frac{k_e Q}{R^3} r \qquad (\text{for } r < R)$$

We can use this result and Equation 25.3 to evaluate the potential difference $V_D - V_C$ at some interior point D:

$$V_D - V_C = -\int_R^r E_r \, dr = -\frac{k_e Q}{R^3} \int_R^r r \, dr = \frac{k_e Q}{2R^3} (R^2 - r^2)$$

Substituting $V_C = k_e Q/R$ into this expression and solving for V_D, we obtain

$$\boxed{V_D = \frac{k_e Q}{2R} \left(3 - \frac{r^2}{R^2} \right)} \qquad (\text{for } r < R) \quad \textbf{(25.25)}$$

At $r = R$, this expression gives a result that agrees with that for the potential at the surface, that is, V_C. A plot of V versus r for this charge distribution is given in Figure 25.19.

Exercise   What are the magnitude of the electric field and the electric potential at the center of the sphere?

Answer $E = 0$; $V_0 = 3k_e Q/2R$.

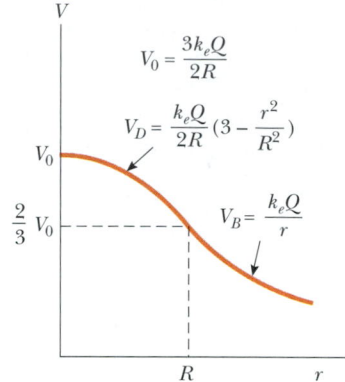

Figure 25.19   A plot of electric potential V versus distance r from the center of a uniformly charged insulating sphere of radius R. The curve for V_D inside the sphere is parabolic and joins smoothly with the curve for V_B outside the sphere, which is a hyperbola. The potential has a maximum value V_0 at the center of the sphere. We could make this graph three dimensional (similar to Figures 25.7a and 25.8a) by spinning it around the vertical axis.

25.6   ELECTRIC POTENTIAL DUE TO A CHARGED CONDUCTOR

In Section 24.4 we found that when a solid conductor in equilibrium carries a net charge, the charge resides on the outer surface of the conductor. Furthermore, we showed that the electric field just outside the conductor is perpendicular to the surface and that the field inside is zero.

We now show that **every point on the surface of a charged conductor in equilibrium is at the same electric potential.** Consider two points A and B on the surface of a charged conductor, as shown in Figure 25.20. Along a surface path connecting these points, **E** is always perpendicular to the displacement $d\mathbf{s}$; there-

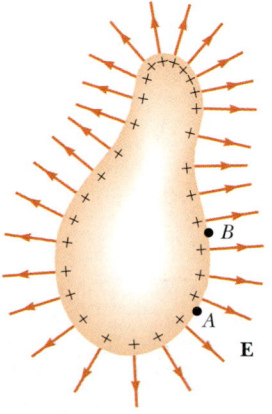

Figure 25.20   An arbitrarily shaped conductor carrying a positive charge. When the conductor is in electrostatic equilibrium, all of the charge resides at the surface, **E** = 0 inside the conductor, and the direction of E just outside the conductor is perpendicular to the surface. The electric potential is constant inside the conductor and is equal to the potential at the surface. Note from the spacing of the plus signs that the surface charge density is nonuniform.

fore $\mathbf{E} \cdot d\mathbf{s} = 0$. Using this result and Equation 25.3, we conclude that the potential difference between A and B is necessarily zero:

$$V_B - V_A = -\int_A^B \mathbf{E} \cdot d\mathbf{s} = 0$$

This result applies to any two points on the surface. Therefore, V is constant everywhere on the surface of a charged conductor in equilibrium. That is,

> the surface of any charged conductor in electrostatic equilibrium is an equipotential surface. Furthermore, because the electric field is zero inside the conductor, we conclude from the relationship $E_r = -dV/dr$ that the electric potential is constant everywhere inside the conductor and equal to its value at the surface.

The surface of a charged conductor is an equipotential surface

Because this is true about the electric potential, no work is required to move a test charge from the interior of a charged conductor to its surface.

Consider a solid metal conducting sphere of radius R and total positive charge Q, as shown in Figure 25.21a. The electric field outside the sphere is $k_e Q/r^2$ and points radially outward. From Example 25.8, we know that the electric potential at the interior and surface of the sphere must be $k_e Q/R$ relative to infinity. The potential outside the sphere is $k_e Q/r$. Figure 25.21b is a plot of the electric potential as a function of r, and Figure 25.21c shows how the electric field varies with r.

When a net charge is placed on a spherical conductor, the surface charge density is uniform, as indicated in Figure 25.21a. However, if the conductor is nonspherical, as in Figure 25.20, the surface charge density is high where the radius of curvature is small and the surface is convex (as noted in Section 24.4), and it is low where the radius of curvature is small and the surface is concave. Because the electric field just outside the conductor is proportional to the surface charge density, we see that **the electric field is large near convex points having small radii of curvature and reaches very high values at sharp points.**

Figure 25.22 shows the electric field lines around two spherical conductors: one carrying a net charge Q, and a larger one carrying zero net charge. In this case, the surface charge density is not uniform on either conductor. The sphere having zero net charge has negative charges induced on its side that faces the

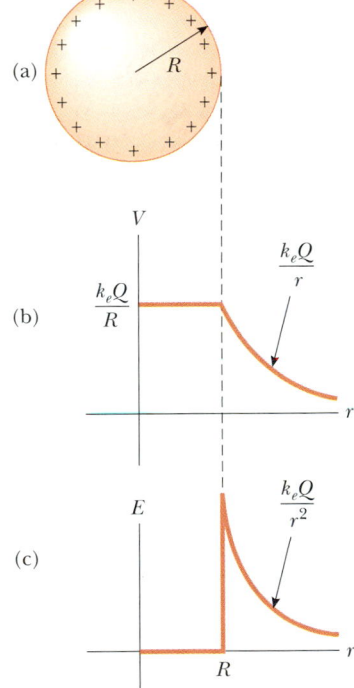

Figure 25.21 (a) The excess charge on a conducting sphere of radius R is uniformly distributed on its surface. (b) Electric potential versus distance r from the center of the charged conducting sphere. (c) Electric field magnitude versus distance r from the center of the charged conducting sphere.

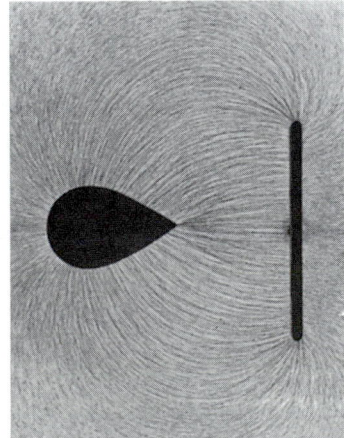

Electric field pattern of a charged conducting plate placed near an oppositely charged pointed conductor. Small pieces of thread suspended in oil align with the electric field lines. The field surrounding the pointed conductor is most intense near the pointed end and at other places where the radius of curvature is small. *(Courtesy of Harold M. Waage, Princeton University)*

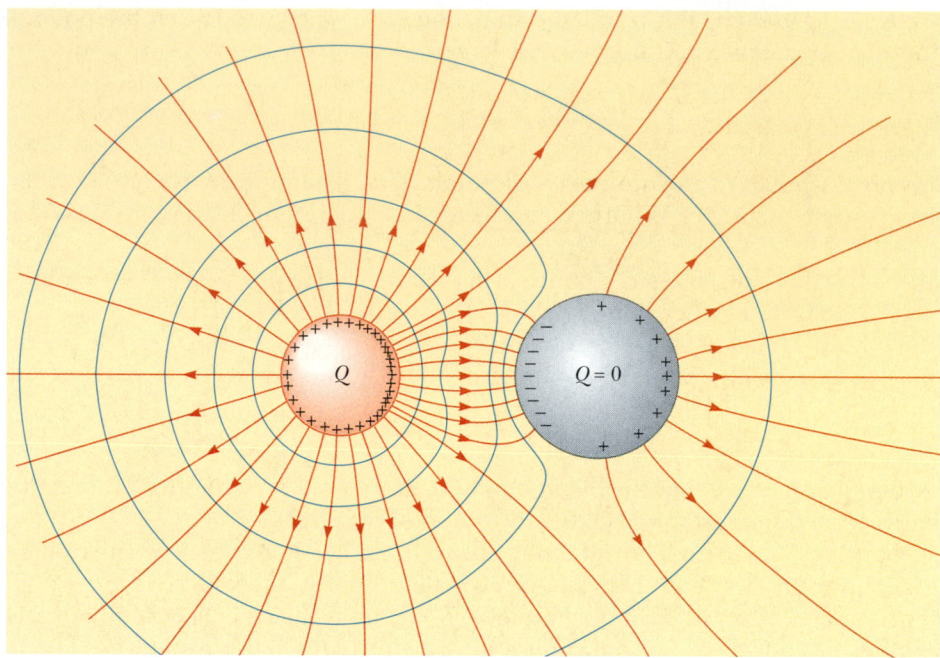

Figure 25.22 The electric field lines (in red) around two spherical conductors. The smaller sphere has a net charge Q, and the larger one has zero net charge. The blue curves are cross-sections of equipotential surfaces.

charged sphere and positive charges induced on its side opposite the charged sphere. The blue curves in the figure represent the cross-sections of the equipotential surfaces for this charge configuration. As usual, the field lines are perpendicular to the conducting surfaces at all points, and the equipotential surfaces are perpendicular to the field lines everywhere. Trying to move a positive charge in the region of these conductors would be like moving a marble on a hill that is flat on top (representing the conductor on the left) and has another flat area partway down the side of the hill (representing the conductor on the right).

EXAMPLE 25.9 ▶ **Two Connected Charged Spheres**

Two spherical conductors of radii r_1 and r_2 are separated by a distance much greater than the radius of either sphere. The spheres are connected by a conducting wire, as shown in Figure 25.23. The charges on the spheres in equilibrium are q_1 and q_2, respectively, and they are uniformly charged. Find the ratio of the magnitudes of the electric fields at the surfaces of the spheres.

Solution Because the spheres are connected by a conducting wire, they must both be at the same electric potential:

$$V = k_e \frac{q_1}{r_1} = k_e \frac{q_2}{r_2}$$

Therefore, the ratio of charges is

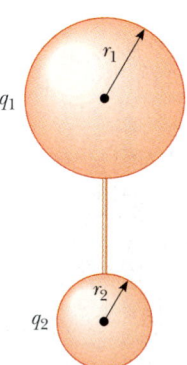

Figure 25.23 Two charged spherical conductors connected by a conducting wire. The spheres are at the *same* electric potential V.

(1) $\dfrac{q_1}{q_2} = \dfrac{r_1}{r_2}$

Because the spheres are very far apart and their surfaces uniformly charged, we can express the magnitude of the electric fields at their surfaces as

$$E_1 = k_e \frac{q_1}{r_1^{\,2}} \quad \text{and} \quad E_2 = k_e \frac{q_2}{r_2^{\,2}}$$

Taking the ratio of these two fields and making use of Equation (1), we find that

$$\frac{E_1}{E_2} = \frac{r_2}{r_1}$$

Hence, the field is more intense in the vicinity of the smaller sphere even though the electric potentials of both spheres are the same.

A Cavity Within a Conductor

Now consider a conductor of arbitrary shape containing a cavity as shown in Figure 25.24. Let us assume that no charges are inside the cavity. **In this case, the electric field inside the cavity must be zero** regardless of the charge distribution on the outside surface of the conductor. Furthermore, the field in the cavity is zero even if an electric field exists outside the conductor.

To prove this point, we use the fact that every point on the conductor is at the same electric potential, and therefore any two points A and B on the surface of the cavity must be at the same potential. Now imagine that a field $\mathbf{E}$ exists in the cavity and evaluate the potential difference $V_B - V_A$ defined by Equation 25.3:

$$V_B - V_A = -\int_A^B \mathbf{E} \cdot d\mathbf{s}$$

If $\mathbf{E}$ is nonzero, we can always find a path between A and B for which $\mathbf{E} \cdot d\mathbf{s}$ is a positive number; thus, the integral must be positive. However, because $V_B - V_A = 0$, the integral of $\mathbf{E} \cdot d\mathbf{s}$ must be zero for all paths between any two points on the conductor, which implies that $\mathbf{E}$ is zero everywhere. This contradiction can be reconciled only if $\mathbf{E}$ is zero inside the cavity. Thus, we conclude that a cavity surrounded by conducting walls is a field-free region as long as no charges are inside the cavity.

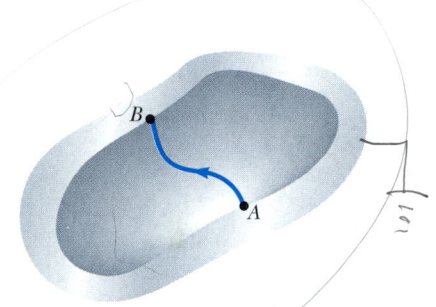

Figure 25.24 A conductor in electrostatic equilibrium containing a cavity. The electric field in the cavity is zero, regardless of the charge on the conductor.

Corona Discharge

A phenomenon known as **corona discharge** is often observed near a conductor such as a high-voltage power line. When the electric field in the vicinity of the conductor is sufficiently strong, electrons are stripped from air molecules. This causes the molecules to be ionized, thereby increasing the air's ability to conduct. The observed glow (or corona discharge) results from the recombination of free electrons with the ionized air molecules. If a conductor has an irregular shape, the electric field can be very high near sharp points or edges of the conductor; consequently, the ionization process and corona discharge are most likely to occur around such points.

Quick Quiz 25.4

(a) Is it possible for the magnitude of the electric field to be zero at a location where the electric potential is not zero? (b) Can the electric potential be zero where the electric field is nonzero?

Optional Section

25.7 ▶ THE MILLIKAN OIL-DROP EXPERIMENT

During the period from 1909 to 1913, Robert Millikan performed a brilliant set of experiments in which he measured *e*, the elementary charge on an electron, and demonstrated the quantized nature of this charge. His apparatus, diagrammed in Figure 25.25, contains two parallel metallic plates. Charged oil droplets from an atomizer are allowed to pass through a small hole in the upper plate. A horizontally directed light beam (not shown in the diagram) is used to illuminate the oil droplets, which are viewed through a telescope whose long axis is at right angles to the light beam. When the droplets are viewed in this manner, they appear as shining stars against a dark background, and the rate at which individual drops fall can be determined.[4]

Let us assume that a single drop having a mass *m* and carrying a charge *q* is being viewed and that its charge is negative. If no electric field is present between the plates, the two forces acting on the charge are the force of gravity *m***g** acting downward and a viscous drag force **F**$_D$ acting upward as indicated in Figure 25.26a. The drag force is proportional to the drop's speed. When the drop reaches its terminal speed *v*, the two forces balance each other ($mg = F_D$).

Now suppose that a battery connected to the plates sets up an electric field between the plates such that the upper plate is at the higher electric potential. In this case, a third force *q***E** acts on the charged drop. Because *q* is negative and **E** is directed downward, this electric force is directed upward, as shown in Figure 25.26b. If this force is sufficiently great, the drop moves upward and the drag force **F**$'_D$ acts downward. When the upward electric force *q***E** balances the sum of the gravitational force and the downward drag force **F**$'_D$, the drop reaches a new terminal speed *v'* in the upward direction.

With the field turned on, a drop moves slowly upward, typically at rates of hundredths of a centimeter per second. The rate of fall in the absence of a field is comparable. Hence, one can follow a single droplet for hours, alternately rising and falling, by simply turning the electric field on and off.

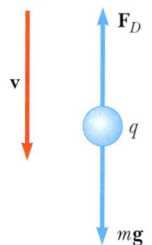

(a) Field off

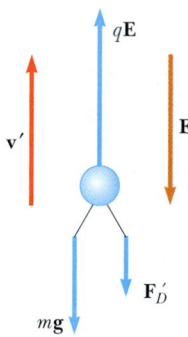

(b) Field on

Figure 25.26 The forces acting on a negatively charged oil droplet in the Millikan experiment.

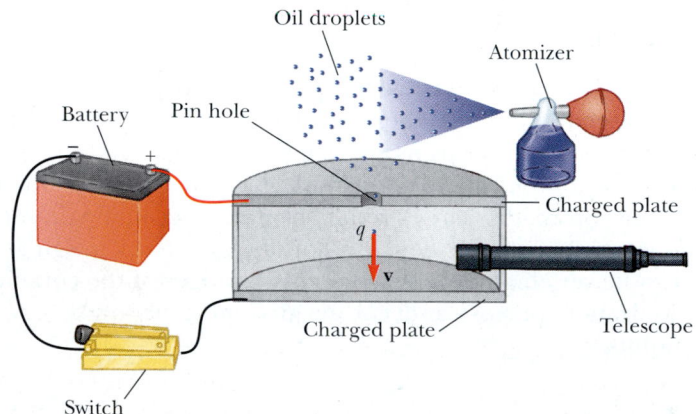

Figure 25.25 Schematic drawing of the Millikan oil-drop apparatus.

[4] At one time, the oil droplets were termed "Millikan's Shining Stars." Perhaps this description has lost its popularity because of the generations of physics students who have experienced hallucinations, near blindness, migraine headaches, and so forth, while repeating Millikan's experiment!

After recording measurements on thousands of droplets, Millikan and his co-workers found that all droplets, to within about 1% precision, had a charge equal to some integer multiple of the elementary charge e:

$$q = ne \qquad n = 0, -1, -2, -3, \ldots$$

where $e = 1.60 \times 10^{-19}$ C. Millikan's experiment yields conclusive evidence that charge is quantized. For this work, he was awarded the Nobel Prize in Physics in 1923.

Optional Section

25.8 APPLICATIONS OF ELECTROSTATICS

The practical application of electrostatics is represented by such devices as lightning rods and electrostatic precipitators and by such processes as xerography and the painting of automobiles. Scientific devices based on the principles of electrostatics include electrostatic generators, the field-ion microscope, and ion-drive rocket engines.

The Van de Graaff Generator

In Section 24.5 we described an experiment that demonstrates a method for transferring charge to a hollow conductor (the Faraday ice-pail experiment). When a charged conductor is placed in contact with the inside of a hollow conductor, all of the charge of the charged conductor is transferred to the hollow conductor. In principle, the charge on the hollow conductor and its electric potential can be increased without limit by repetition of the process.

In 1929 Robert J. Van de Graaff (1901–1967) used this principle to design and build an electrostatic generator. This type of generator is used extensively in nuclear physics research. A schematic representation of the generator is given in Figure 25.27. Charge is delivered continuously to a high-potential electrode by means of a moving belt of insulating material. The high-voltage electrode is a hollow conductor mounted on an insulating column. The belt is charged at point A by means of a corona discharge between comb-like metallic needles and a grounded grid. The needles are maintained at a positive electric potential of typically 10^4 V. The positive charge on the moving belt is transferred to the hollow conductor by a second comb of needles at point B. Because the electric field inside the hollow conductor is negligible, the positive charge on the belt is easily transferred to the conductor regardless of its potential. In practice, it is possible to increase the electric potential of the hollow conductor until electrical discharge occurs through the air. Because the "breakdown" electric field in air is about 3×10^6 V/m, a sphere 1 m in radius can be raised to a maximum potential of 3×10^6 V. The potential can be increased further by increasing the radius of the hollow conductor and by placing the entire system in a container filled with high-pressure gas.

Van de Graaff generators can produce potential differences as large as 20 million volts. Protons accelerated through such large potential differences receive enough energy to initiate nuclear reactions between themselves and various target nuclei. Smaller generators are often seen in science classrooms and museums. If a person insulated from the ground touches the sphere of a Van de Graaff generator, his or her body can be brought to a high electric potential. The hair acquires a net positive charge, and each strand is repelled by all the others. The result is a

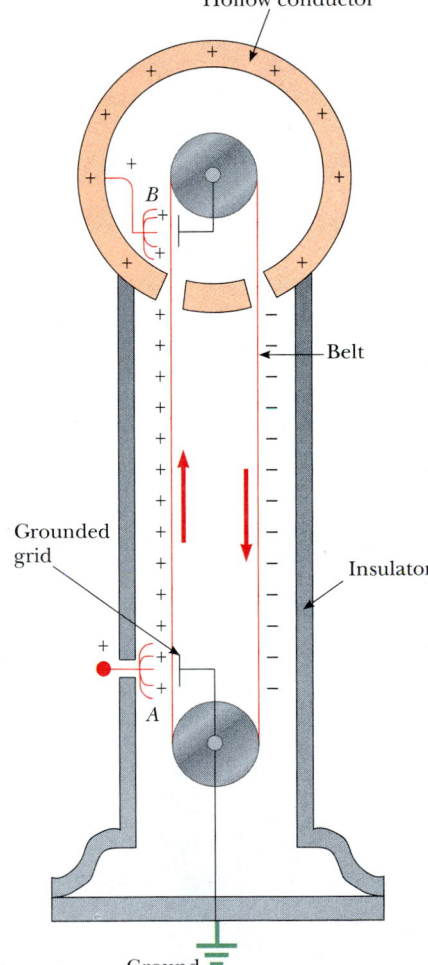

Figure 25.27 Schematic diagram of a Van de Graaff generator. Charge is transferred to the hollow conductor at the top by means of a moving belt. The charge is deposited on the belt at point A and transferred to the hollow conductor at point B.

scene such as that depicted in the photograph at the beginning of this chapter. In addition to being insulated from ground, the person holding the sphere is safe in this demonstration because the total charge on the sphere is very small (on the order of $1\ \mu C$). If this amount of charge accidentally passed from the sphere through the person to ground, the corresponding current would do no harm.

The Electrostatic Precipitator

One important application of electrical discharge in gases is the *electrostatic precipitator*. This device removes particulate matter from combustion gases, thereby reducing air pollution. Precipitators are especially useful in coal-burning power plants and in industrial operations that generate large quantities of smoke. Current systems are able to eliminate more than 99% of the ash from smoke.

Figure 25.28a shows a schematic diagram of an electrostatic precipitator. A high potential difference (typically 40 to 100 kV) is maintained between a wire running down the center of a duct and the walls of the duct, which are grounded. The wire is maintained at a negative electric potential with respect to the walls, so the electric field is directed toward the wire. The values of the field near the wire become high enough to cause a corona discharge around the wire; the discharge ionizes some air molecules to form positive ions, electrons, and such negative ions as O_2^{-}. The air to be cleaned enters the duct and moves near the wire. As the electrons and negative ions created by the discharge are accelerated toward the outer wall by the electric field, the dirt particles in the air become charged by collisions and ion capture. Because most of the charged dirt particles are negative, they too are drawn to the duct walls by the electric field. When the duct is periodically shaken, the particles break loose and are collected at the bottom.

Sprinkle some salt and pepper on an open dish and mix the two together. Now pull a comb through your hair several times and bring the comb to within 1 cm of the salt and pepper. What happens? How is what happens here related to the operation of an electrostatic precipitator?

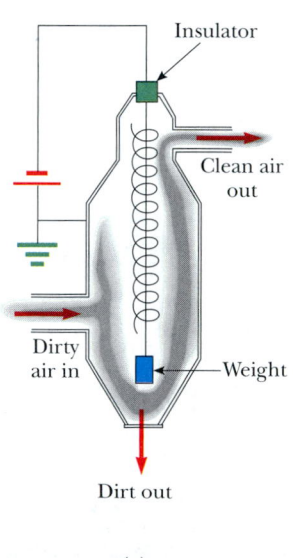

(a)

(b)

(c)

Figure 25.28 (a) Schematic diagram of an electrostatic precipitator. The high negative electric potential maintained on the central coiled wire creates an electrical discharge in the vicinity of the wire. Compare the air pollution when the electrostatic precipitator is (b) operating and (c) turned off. *(b, Rei O'Hara/Black Star/PNI; c, Greig Cranna/Stock, Boston/PNI)*

In addition to reducing the level of particulate matter in the atmosphere (compare Figs. 25.28b and c), the electrostatic precipitator recovers valuable materials in the form of metal oxides.

Xerography and Laser Printers

The basic idea of xerography[5] was developed by Chester Carlson, who was granted a patent for the xerographic process in 1940. The one feature of this process that makes it unique is the use of a photoconductive material to form an image. (A *photoconductor* is a material that is a poor electrical conductor in the dark but that becomes a good electrical conductor when exposed to light.)

The xerographic process is illustrated in Figure 25.29a to d. First, the surface of a plate or drum that has been coated with a thin film of photoconductive material (usually selenium or some compound of selenium) is given a positive electrostatic charge in the dark. An image of the page to be copied is then focused by a lens onto the charged surface. The photoconducting surface becomes conducting only in areas where light strikes it. In these areas, the light produces charge carriers in the photoconductor that move the positive charge off the drum. However, positive

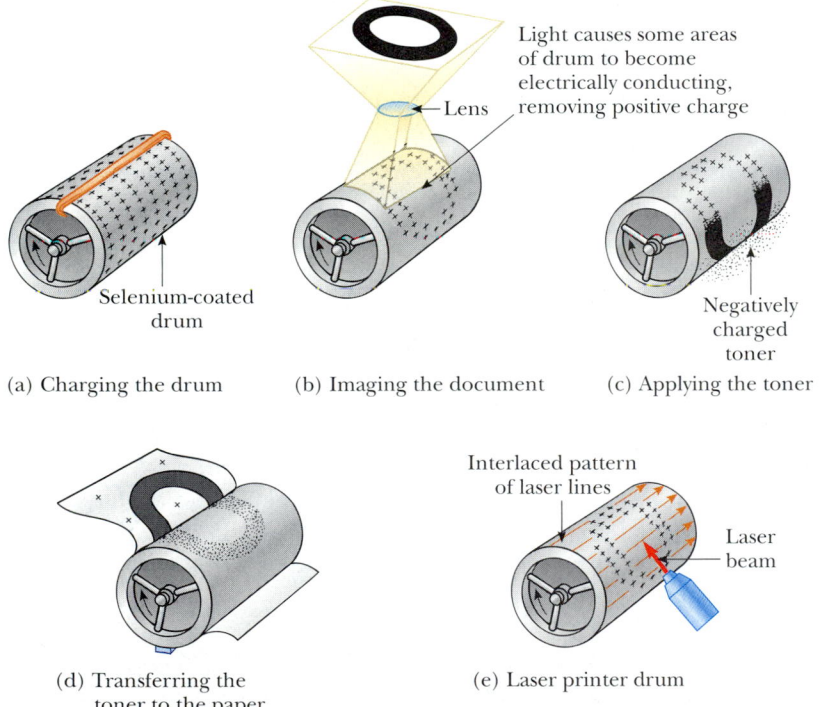

Light causes some areas of drum to become electrically conducting, removing positive charge

Lens

Selenium-coated drum

Negatively charged toner

(a) Charging the drum   (b) Imaging the document   (c) Applying the toner

Interlaced pattern of laser lines

Laser beam

(d) Transferring the toner to the paper   (e) Laser printer drum

Figure 25.29 The xerographic process: (a) The photoconductive surface of the drum is positively charged. (b) Through the use of a light source and lens, an image is formed on the surface in the form of positive charges. (c) The surface containing the image is covered with a negatively charged powder, which adheres only to the image area. (d) A piece of paper is placed over the surface and given a positive charge. This transfers the image to the paper as the negatively charged powder particles migrate to the paper. The paper is then heat-treated to "fix" the powder. (e) A laser printer operates similarly except the image is produced by turning a laser beam on and off as it sweeps across the selenium-coated drum.

[5] The prefix *xero-* is from the Greek word meaning "dry." Note that no liquid ink is used anywhere in xerography.

charges remain on those areas of the photoconductor not exposed to light, leaving a latent image of the object in the form of a positive surface charge distribution.

Next, a negatively charged powder called a *toner* is dusted onto the photoconducting surface. The charged powder adheres only to those areas of the surface that contain the positively charged image. At this point, the image becomes visible. The toner (and hence the image) are then transferred to the surface of a sheet of positively charged paper.

Finally, the toner is "fixed" to the surface of the paper as the toner melts while passing through high-temperature rollers. This results in a permanent copy of the original.

A laser printer (Fig. 25.29e) operates by the same principle, with the exception that a computer-directed laser beam is used to illuminate the photoconductor instead of a lens.

SUMMARY

When a positive test charge q_0 is moved between points A and B in an electric field **E**, the **change in the potential energy** is

$$\Delta U = -q_0 \int_A^B \mathbf{E} \cdot d\mathbf{s} \tag{25.1}$$

The **electric potential** $V = U/q_0$ is a scalar quantity and has units of joules per coulomb (J/C), where $1\,\text{J/C} \equiv 1\,\text{V}$.

The **potential difference** ΔV between points A and B in an electric field **E** is defined as

$$\Delta V = \frac{\Delta U}{q_0} = -\int_A^B \mathbf{E} \cdot d\mathbf{s} \tag{25.3}$$

The potential difference between two points A and B in a uniform electric field **E** is

$$\Delta V = -Ed \tag{25.6}$$

where d is the magnitude of the displacement in the direction parallel to **E**.

An **equipotential surface** is one on which all points are at the same electric potential. Equipotential surfaces are perpendicular to electric field lines.

If we define $V = 0$ at $r_A = \infty$, the electric potential due to a point charge at any distance r from the charge is

$$V = k_e \frac{q}{r} \tag{25.11}$$

We can obtain the electric potential associated with a group of point charges by summing the potentials due to the individual charges.

The **potential energy associated with a pair of point charges** separated by a distance r_{12} is

$$U = k_e \frac{q_1 q_2}{r_{12}} \tag{25.13}$$

This energy represents the work required to bring the charges from an infinite separation to the separation r_{12}. We obtain the potential energy of a distribution of point charges by summing terms like Equation 25.13 over all pairs of particles.

TABLE 25.1   Electric Potential Due to Various Charge Distributions

Charge Distribution	Electric Potential	Location
Uniformly charged ring of radius a	$V = k_e \dfrac{Q}{\sqrt{x^2 + a^2}}$	Along perpendicular central axis of ring, distance x from ring center
Uniformly charged disk of radius a	$V = 2\pi k_e \sigma [(x^2 + a^2)^{1/2} - x]$	Along perpendicular central axis of disk, distance x from disk center
Uniformly charged, *insulating* solid sphere of radius R and total charge Q	$V = k_e \dfrac{Q}{r}$ $V = \dfrac{k_e Q}{2R}\left(3 - \dfrac{r^2}{R^2}\right)$	$r \geq R$ $r < R$
Isolated *conducting* sphere of radius R and total charge Q	$V = k_e \dfrac{Q}{r}$ $V = k_e \dfrac{Q}{R}$	$r > R$ $r \leq R$

If we know the electric potential as a function of coordinates x, y, z, we can obtain the components of the electric field by taking the negative derivative of the electric potential with respect to the coordinates. For example, the x component of the electric field is

$$E_x = -\frac{dV}{dx} \tag{25.16}$$

The **electric potential due to a continuous charge distribution** is

$$V = k_e \int \frac{dq}{r} \tag{25.19}$$

Every point on the surface of a charged conductor in electrostatic equilibrium is at the same electric potential. The potential is constant everywhere inside the conductor and equal to its value at the surface.

Table 25.1 lists electric potentials due to several charge distributions.

Problem-Solving Hints

Calculating Electric Potential

- Remember that electric potential is a scalar quantity, so components need not be considered. Therefore, when using the superposition principle to evaluate the electric potential at a point due to a system of point charges, simply take the algebraic sum of the potentials due to the various charges. However, you must keep track of signs. The potential is positive for positive charges, and it is negative for negative charges.

- Just as with gravitational potential energy in mechanics, only *changes* in electric potential are significant; hence, the point where you choose the potential to be zero is arbitrary. When dealing with point charges or a charge distribution of finite size, we usually define $V = 0$ to be at a point infinitely far from the charges.

- You can evaluate the electric potential at some point P due to a continuous distribution of charge by dividing the charge distribution into infinitesimal elements of charge dq located at a distance r from P. Then, treat one charge element as a point charge, such that the potential at P due to the element is $dV = k_e dq / r$. Obtain the total potential at P by integrating dV over the entire charge distribution. In performing the integration for most problems, you must express dq and r in terms of a single variable. To simplify the integration, consider the geometry involved in the problem carefully. Review Examples 25.5 through 25.7 for guidance.

- Another method that you can use to obtain the electric potential due to a finite continuous charge distribution is to start with the definition of potential difference given by Equation 25.3. If you know or can easily obtain **E** (from Gauss's law), then you can evaluate the line integral of $\mathbf{E} \cdot d\mathbf{s}$. An example of this method is given in Example 25.8.

- Once you know the electric potential at a point, you can obtain the electric field at that point by remembering that the electric field component in a specified direction is equal to the negative of the derivative of the electric potential in that direction. Example 25.4 illustrates this procedure.

QUESTIONS

1. Distinguish between electric potential and electric potential energy.

2. A negative charge moves in the direction of a uniform electric field. Does the potential energy of the charge increase or decrease? Does it move to a position of higher or lower potential?

3. Give a physical explanation of the fact that the potential energy of a pair of like charges is positive whereas the potential energy of a pair of unlike charges is negative.

4. A uniform electric field is parallel to the x axis. In what direction can a charge be displaced in this field without any external work being done on the charge?

5. Explain why equipotential surfaces are always perpendicular to electric field lines.

6. Describe the equipotential surfaces for (a) an infinite line of charge and (b) a uniformly charged sphere.

7. Explain why, under static conditions, all points in a conductor must be at the same electric potential.

8. The electric field inside a hollow, uniformly charged sphere is zero. Does this imply that the potential is zero inside the sphere? Explain.

9. The potential of a point charge is defined to be zero at an infinite distance. Why can we not define the potential of an infinite line of charge to be zero at $r = \infty$?

10. Two charged conducting spheres of different radii are connected by a conducting wire, as shown in Figure 25.23. Which sphere has the greater charge density?

11. What determines the maximum potential to which the dome of a Van de Graaff generator can be raised?

12. Explain the origin of the glow sometimes observed around the cables of a high-voltage power line.

13. Why is it important to avoid sharp edges or points on conductors used in high-voltage equipment?

14. How would you shield an electronic circuit or laboratory from stray electric fields? Why does this work?

15. Why is it relatively safe to stay in an automobile with a metal body during a severe thunderstorm?

16. Walking across a carpet and then touching someone can result in a shock. Explain why this occurs.

PROBLEMS

1, 2, 3 = straightforward, intermediate, challenging   ☐ = full solution available in the *Student Solutions Manual and Study Guide*
WEB = solution posted at **http://www.saunderscollege.com/physics/**   💻 – Computer useful in solving problem   ⬛ = Interactive Physics
☐ = paired numerical/symbolic problems

Section 25.1   Potential Difference and Electric Potential

1. How much work is done (by a battery, generator, or some other source of electrical energy) in moving Avogadro's number of electrons from an initial point where the electric potential is 9.00 V to a point where the potential is − 5.00 V? (The potential in each case is measured relative to a common reference point.)

2. An ion accelerated through a potential difference of 115 V experiences an increase in kinetic energy of 7.37×10^{-17} J. Calculate the charge on the ion.

3. (a) Calculate the speed of a proton that is accelerated from rest through a potential difference of 120 V. (b) Calculate the speed of an electron that is accelerated through the same potential difference.

4. **Review Problem.**   Through what potential difference would an electron need to be accelerated for it to achieve a speed of 40.0% of the speed of light, starting from rest? The speed of light is $c = 3.00 \times 10^8$ m/s; review Section 7.7.

5. What potential difference is needed to stop an electron having an initial speed of 4.20×10^5 m/s?

Section 25.2   Potential Differences in a Uniform Electric Field

6. A uniform electric field of magnitude 250 V/m is directed in the positive x direction. A + 12.0-μC charge moves from the origin to the point (x, y) = (20.0 cm, 50.0 cm). (a) What was the change in the potential energy of this charge? (b) Through what potential difference did the charge move?

7. The difference in potential between the accelerating plates of a TV set is about 25 000 V. If the distance between these plates is 1.50 cm, find the magnitude of the uniform electric field in this region.

8. Suppose an electron is released from rest in a uniform electric field whose magnitude is 5.90×10^3 V/m. (a) Through what potential difference will it have passed after moving 1.00 cm? (b) How fast will the electron be moving after it has traveled 1.00 cm?

WEB 9. An electron moving parallel to the x axis has an initial speed of 3.70×10^6 m/s at the origin. Its speed is reduced to 1.40×10^5 m/s at the point x = 2.00 cm. Calculate the potential difference between the origin and that point. Which point is at the higher potential?

10. A uniform electric field of magnitude 325 V/m is directed in the *negative* y direction as shown in Figure P25.10. The coordinates of point A are (− 0.200, − 0.300) m, and those of point B are (0.400, 0.500) m. Calculate the potential difference $V_B − V_A$, using the blue path.

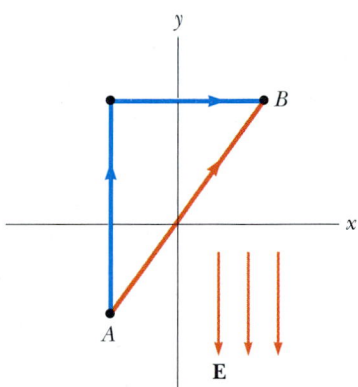

Figure P25.10

11. A 4.00-kg block carrying a charge $Q = 50.0 \ \mu$C is connected to a spring for which k = 100 N/m. The block lies on a frictionless horizontal track, and the system is immersed in a uniform electric field of magnitude $E = 5.00 \times 10^5$ V/m, directed as shown in Figure P25.11. If the block is released from rest when the spring is unstretched (at x = 0), (a) by what maximum amount does the spring expand? (b) What is the equilibrium position of the block? (c) Show that the block's motion is simple harmonic, and determine its period. (d) Repeat part (a) if the coefficient of kinetic friction between block and surface is 0.200.

12. A block having mass m and charge Q is connected to a spring having constant k. The block lies on a frictionless horizontal track, and the system is immersed in a uniform electric field of magnitude E, directed as shown in Figure P25.11. If the block is released from rest when the spring is unstretched (at x = 0), (a) by what maximum amount does the spring expand? (b) What is the equilibrium position of the block? (c) Show that the block's motion is simple harmonic, and determine its period. (d) Repeat part (a) if the coefficient of kinetic friction between block and surface is μ_k.

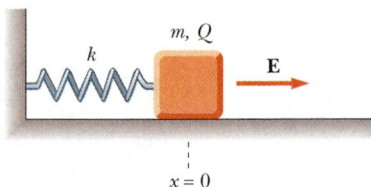

Figure P25.11   Problems 11 and 12.

13. On planet Tehar, the acceleration due to gravity is the same as that on Earth but there is also a strong downward electric field with the field being uniform close to the planet's surface. A 2.00-kg ball having a charge of 5.00 μC is thrown upward at a speed of 20.1 m/s and it hits the ground after an interval of 4.10 s. What is the potential difference between the starting point and the top point of the trajectory?

14. An insulating rod having linear charge density $\lambda =$ 40.0 μC/m and linear mass density $\mu = 0.100$ kg/m is released from rest in a uniform electric field $E =$ 100 V/m directed perpendicular to the rod (Fig. P25.14). (a) Determine the speed of the rod after it has traveled 2.00 m. (b) How does your answer to part (a) change if the electric field is not perpendicular to the rod? Explain.

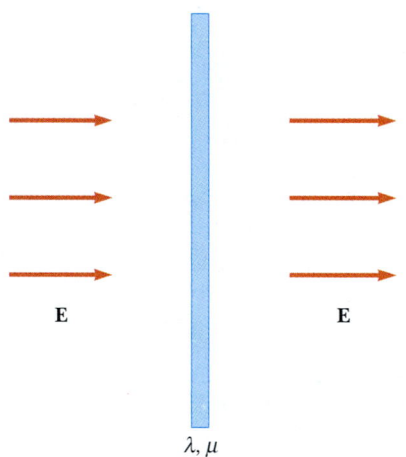

E   E

λ, μ

Figure P25.14

15. A particle having charge $q = +2.00$ μC and mass $m =$ 0.010 0 kg is connected to a string that is $L = 1.50$ m long and is tied to the pivot point P in Figure P25.15. The particle, string, and pivot point all lie on a horizontal table. The particle is released from rest when the

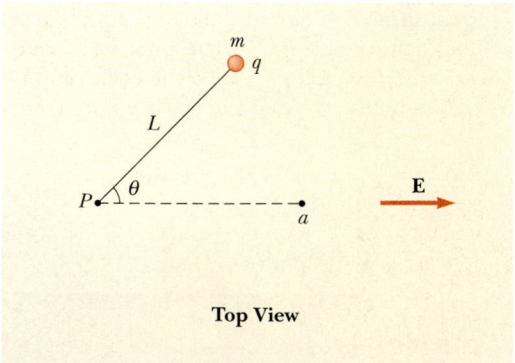

Top View

Figure P25.15

string makes an angle $\theta = 60.0°$ with a uniform electric field of magnitude $E = 300$ V/m. Determine the speed of the particle when the string is parallel to the electric field (point a in Fig. P25.15).

Section 25.3   Electric Potential and Potential Energy Due to Point Charges

Note: Unless stated otherwise, assume a reference level of potential $V = 0$ at $r = \infty$.

16. (a) Find the potential at a distance of 1.00 cm from a proton. (b) What is the potential difference between two points that are 1.00 cm and 2.00 cm from a proton? (c) Repeat parts (a) and (b) for an electron.

17. Given two 2.00-μC charges, as shown in Figure P25.17, and a positive test charge $q = 1.28 \times 10^{-18}$ C at the origin, (a) what is the net force exerted on q by the two 2.00-μC charges? (b) What is the electric field at the origin due to the two 2.00-μC charges? (c) What is the electric potential at the origin due to the two 2.00-μC charges?

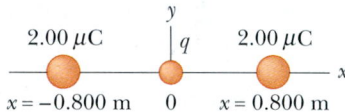

2.00 μC q 2.00 μC

$x = -0.800$ m   0 $x = 0.800$ m

Figure P25.17

18. A charge $+ q$ is at the origin. A charge $- 2q$ is at $x =$ 2.00 m on the x axis. For what finite value(s) of x is (a) the electric field zero? (b) the electric potential zero?

19. The Bohr model of the hydrogen atom states that the single electron can exist only in certain allowed orbits around the proton. The radius of each Bohr orbit is $r = n^2 (0.052\ 9$ nm) where $n = 1, 2, 3, \ldots$. Calculate the electric potential energy of a hydrogen atom when the electron is in the (a) first allowed orbit, $n = 1$; (b) second allowed orbit, $n = 2$; and (c) when the electron has escaped from the atom ($r = \infty$). Express your answers in electron volts.

20. Two point charges $Q_1 = +5.00$ nC and $Q_2 = -3.00$ nC are separated by 35.0 cm. (a) What is the potential energy of the pair? What is the significance of the algebraic sign of your answer? (b) What is the electric potential at a point midway between the charges?

21. The three charges in Figure P25.21 are at the vertices of an isosceles triangle. Calculate the electric potential at the midpoint of the base, taking $q = 7.00$ μC.

22. *Compare this problem with Problem 55 in Chapter 23.* Four identical point charges ($q = +10.0$ μC) are located on the corners of a rectangle, as shown in Figure P23.55. The dimensions of the rectangle are $L = 60.0$ cm and $W = 15.0$ cm. Calculate the electric potential energy of the charge at the lower left corner due to the other three charges.

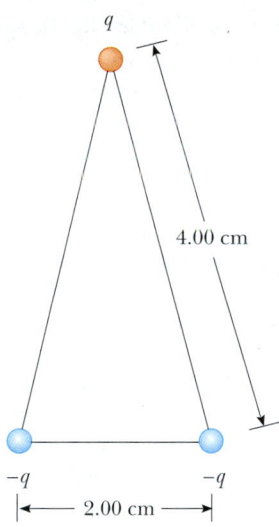

Figure P25.21

WEB **23.** Show that the amount of work required to assemble four identical point charges of magnitude Q at the corners of a square of side s is $5.41 k_e Q^2 / s$.

24. *Compare this problem with Problem 18 in Chapter 23.* Two point charges each of magnitude $2.00 \mu C$ are located on the x axis. One is at $x = 1.00$ m, and the other is at $x = -1.00$ m. (a) Determine the electric potential on the y axis at $y = 0.500$ m. (b) Calculate the electric potential energy of a third charge, of $-3.00 \mu C$, placed on the y axis at $y = 0.500$ m.

25. *Compare this problem with Problem 22 in Chapter 23.* Five equal negative point charges $-q$ are placed symmetrically around a circle of radius R. Calculate the electric potential at the center of the circle.

26. *Compare this problem with Problem 17 in Chapter 23.* Three equal positive charges q are at the corners of an equilateral triangle of side a, as shown in Figure P23.17. (a) At what point, if any, in the plane of the charges is the electric potential zero? (b) What is the electric potential at the point P due to the two charges at the base of the triangle?

27. Review Problem. Two insulating spheres having radii 0.300 cm and 0.500 cm, masses 0.100 kg and 0.700 kg, and charges $-2.00 \mu C$ and $3.00 \mu C$ are released from rest when their centers are separated by 1.00 m. (a) How fast will each be moving when they collide? (*Hint:* Consider conservation of energy and linear momentum.) (b) If the spheres were conductors would the speeds be larger or smaller than those calculated in part (a)? Explain.

28. Review Problem. Two insulating spheres having radii r_1 and r_2, masses m_1 and m_2, and charges $-q_1$ and q_2 are released from rest when their centers are separated by a distance d. (a) How fast is each moving when they

collide? (*Hint:* Consider conservation of energy and conservation of linear momentum.) (b) If the spheres were conductors, would the speeds be greater or less than those calculated in part (a)?

29. A small spherical object carries a charge of 8.00 nC. At what distance from the center of the object is the potential equal to 100 V? 50.0 V? 25.0 V? Is the spacing of the equipotentials proportional to the change in potential?

30. Two point charges of equal magnitude are located along the y axis equal distances above and below the x axis, as shown in Figure P25.30. (a) Plot a graph of the potential at points along the x axis over the interval $-3a < x < 3a$. You should plot the potential in units of $k_e Q / a$. (b) Let the charge located at $-a$ be negative and plot the potential along the y axis over the interval $-4a < y < 4a$.

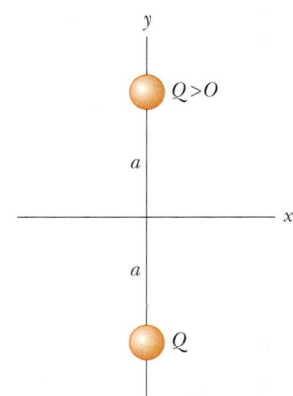

Figure P25.30

31. In Rutherford's famous scattering experiments that led to the planetary model of the atom, alpha particles (charge $+2e$, mass $= 6.64 \times 10^{-27}$ kg) were fired at a gold nucleus (charge $+79e$). An alpha particle, initially very far from the gold nucleus, is fired with a velocity of 2.00×10^7 m/s directly toward the center of the nucleus. How close does the alpha particle get to this center before turning around? Assume the gold nucleus remains stationary.

32. An electron starts from rest 3.00 cm from the center of a uniformly charged insulating sphere of radius 2.00 cm and total charge 1.00 nC. What is the speed of the electron when it reaches the surface of the sphere?

33. Calculate the energy required to assemble the array of charges shown in Figure P25.33, where $a = 0.200$ m, $b = 0.400$ m, and $q = 6.00 \mu C$.

34. Four identical particles each have charge q and mass m. They are released from rest at the vertices of a square of side L. How fast is each charge moving when their distance from the center of the square doubles?

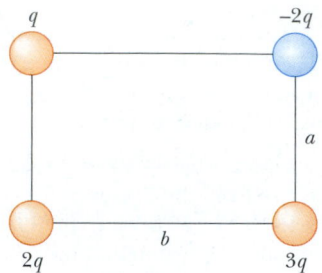

Figure P25.33

35. How much work is required to assemble eight identical point charges, each of magnitude q, at the corners of a cube of side s?

Section 25.4   Obtaining the Value of the Electric Field from the Electric Potential

36. The potential in a region between $x = 0$ and $x = 6.00$ m is $V = a + bx$ where $a = 10.0$ V and $b = -7.00$ V/m. Determine (a) the potential at $x = 0$, 3.00 m, and 6.00 m and (b) the magnitude and direction of the electric field at $x = 0$, 3.00 m, and 6.00 m.

WEB **37.** Over a certain region of space, the electric potential is $V = 5x - 3x^2y + 2yz^2$. Find the expressions for the x, y, and z components of the electric field over this region. What is the magnitude of the field at the point P, which has coordinates $(1, 0, -2)$ m?

38. The electric potential inside a charged spherical conductor of radius R is given by $V = k_e Q/R$ and outside the conductor is given by $V = k_e Q/r$. Using $E_r = -dV/dr$, derive the electric field (a) inside and (b) outside this charge distribution.

39. It is shown in Example 25.7 that the potential at a point P a distance a above one end of a uniformly charged rod of length ℓ lying along the x axis is

$$V = \frac{k_e Q}{\ell} \ln\left(\frac{\ell + \sqrt{\ell^2 + a^2}}{a}\right)$$

Use this result to derive an expression for the y component of the electric field at P. (*Hint:* Replace a with y.)

40. When an uncharged conducting sphere of radius a is placed at the origin of an xyz coordinate system that lies in an initially uniform electric field $\mathbf{E} = E_0\mathbf{k}$, the resulting electric potential is

$$V(x, y, z) = V_0 - E_0 z + \frac{E_0 a^3 z}{(x^2 + y^2 + z^2)^{3/2}}$$

for points outside the sphere, where V_0 is the (constant) electric potential on the conductor. Use this equation to determine the x, y, and z components of the resulting electric field.

Section 25.5   Electric Potential Due to Continuous Charge Distributions

41. Consider a ring of radius R with the total charge Q spread uniformly over its perimeter. What is the potential difference between the point at the center of the ring and a point on its axis a distance $2R$ from the center?

42. *Compare this problem with Problem 33 in Chapter 23.* A uniformly charged insulating rod of length 14.0 cm is bent into the shape of a semicircle, as shown in Figure P23.33. If the rod has a total charge of -7.50 μC, find the electric potential at O, the center of the semicircle.

43. A rod of length L (Fig. P25.43) lies along the x axis with its left end at the origin and has a nonuniform charge density $\lambda = \alpha x$ (where α is a positive constant). (a) What are the units of α? (b) Calculate the electric potential at A.

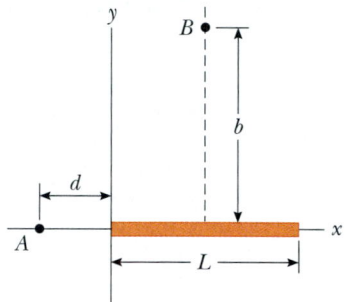

Figure P25.43   Problems 43 and 44.

44. For the arrangement described in the previous problem, calculate the electric potential at point B that lies on the perpendicular bisector of the rod a distance b above the x axis.

45. Calculate the electric potential at point P on the axis of the annulus shown in Figure P25.45, which has a uniform charge density σ.

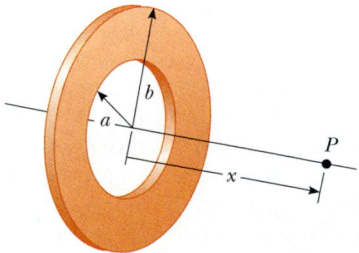

Figure P25.45

46. A wire of finite length that has a uniform linear charge density λ is bent into the shape shown in Figure P25.46. Find the electric potential at point O.

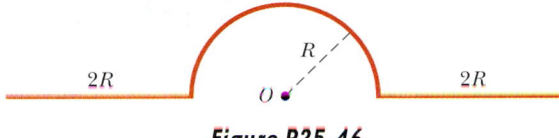

Figure P25.16

Section 25.6   Electric Potential Due to a Charged Conductor

47. How many electrons should be removed from an initially uncharged spherical conductor of radius 0.300 m to produce a potential of 7.50 kV at the surface?

48. Two charged spherical conductors are connected by a long conducting wire, and a charge of 20.0 μC is placed on the combination. (a) If one sphere has a radius of 4.00 cm and the other has a radius of 6.00 cm, what is the electric field near the surface of each sphere? (b) What is the electric potential of each sphere?

WEB 49. A spherical conductor has a radius of 14.0 cm and charge of 26.0 μC. Calculate the electric field and the electric potential at (a) $r = 10.0$ cm, (b) $r = 20.0$ cm, and (c) $r = 14.0$ cm from the center.

50. Two concentric spherical conducting shells of radii $a = 0.400$ m and $b = 0.500$ m are connected by a thin wire, as shown in Figure P25.50. If a total charge $Q = 10.0$ μC is placed on the system, how much charge settles on each sphere?

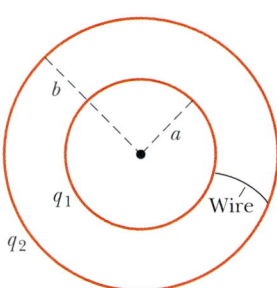

Figure P25.50

(Optional)
Section 25.7   The Millikan Oil-Drop Experiment

(Optional)
Section 25.8   Applications of Electrostatics

51. Consider a Van de Graaff generator with a 30.0-cm-diameter dome operating in dry air. (a) What is the maximum potential of the dome? (b) What is the maximum charge on the dome?

52. The spherical dome of a Van de Graaff generator can be raised to a maximum potential of 600 kV; then additional charge leaks off in sparks, by producing breakdown of the surrounding dry air. Determine (a) the charge on the dome and (b) the radius of the dome.

ADDITIONAL PROBLEMS

53. The liquid-drop model of the nucleus suggests that high-energy oscillations of certain nuclei can split the nucleus into two unequal fragments plus a few neutrons. The fragments acquire kinetic energy from their mutual Coulomb repulsion. Calculate the electric potential energy (in electron volts) of two spherical fragments from a uranium nucleus having the following charges and radii: $38e$ and 5.50×10^{-15} m; $54e$ and 6.20×10^{-15} m. Assume that the charge is distributed uniformly throughout the volume of each spherical fragment and that their surfaces are initially in contact at rest. (The electrons surrounding the nucleus can be neglected.)

54. On a dry winter day you scuff your leather-soled shoes across a carpet and get a shock when you extend the tip of one finger toward a metal doorknob. In a dark room you see a spark perhaps 5 mm long. Make order-of-magnitude estimates of (a) your electric potential and (b) the charge on your body before you touch the doorknob. Explain your reasoning.

55. The charge distribution shown in Figure P25.55 is referred to as a linear quadrupole. (a) Show that the potential at a point on the x axis where $x > a$ is

$$V = \frac{2k_eQa^2}{x^3 - xa^2}$$

(b) Show that the expression obtained in part (a) when $x \gg a$ reduces to

$$V = \frac{2k_eQa^2}{x^3}$$

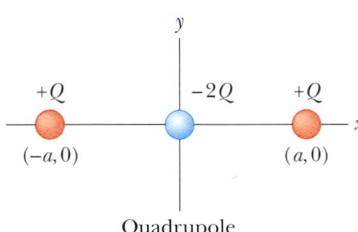

Quadrupole

Figure P25.55

56. (a) Use the exact result from Problem 55 to find the electric field at any point along the axis of the linear quadrupole for $x > a$. (b) Evaluate E at $x = 3a$ if $a = 2.00$ mm and $Q = 3.00$ μC.

57. At a certain distance from a point charge, the magnitude of the electric field is 500 V/m and the electric potential is -3.00 kV. (a) What is the distance to the charge? (b) What is the magnitude of the charge?

58. An electron is released from rest on the axis of a uniform positively charged ring, 0.100 m from the ring's

center. If the linear charge density of the ring is $+0.100 \ \mu C/m$ and the radius of the ring is 0.200 m, how fast will the electron be moving when it reaches the center of the ring?

59. (a) Consider a uniformly charged cylindrical shell having total charge Q, radius R, and height h. Determine the electrostatic potential at a point a distance d from the right side of the cylinder, as shown in Figure P25.59. (*Hint:* Use the result of Example 25.5 by treating the cylinder as a collection of ring charges.) (b) Use the result of Example 25.6 to solve the same problem for a solid cylinder.

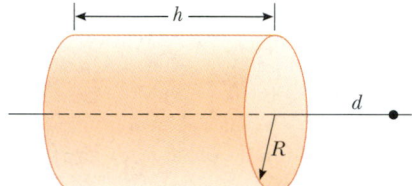

Figure P25.59

60. Two parallel plates having charges of equal magnitude but opposite sign are separated by 12.0 cm. Each plate has a surface charge density of 36.0 nC/m^2. A proton is released from rest at the positive plate. Determine (a) the potential difference between the plates, (b) the energy of the proton when it reaches the negative plate, (c) the speed of the proton just before it strikes the negative plate, (d) the acceleration of the proton, and (e) the force on the proton. (f) From the force, find the magnitude of the electric field and show that it is equal to that found from the charge densities on the plates.

61. Calculate the work that must be done to charge a spherical shell of radius R to a total charge Q.

62. A Geiger–Müller counter is a radiation detector that essentially consists of a hollow cylinder (the cathode) of inner radius r_a and a coaxial cylindrical wire (the anode) of radius r_b (Fig. P25.62). The charge per unit length on the anode is λ, while the charge per unit length on the cathode is $-\lambda$. (a) Show that the magnitude of the potential difference between the wire and the cylinder in the sensitive region of the detector is

$$\Delta V = 2k_e \lambda \ln\left(\frac{r_a}{r_b}\right)$$

(b) Show that the magnitude of the electric field over that region is given by

$$\Delta E = \frac{V}{\ln(r_a/r_b)} \left(\frac{1}{r}\right)$$

where r is the distance from the center of the anode to the point where the field is to be calculated.

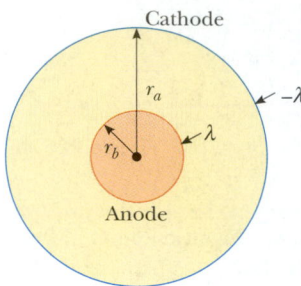

Figure P25.62

WEB 63. From Gauss's law, the electric field set up by a uniform line of charge is

$$\mathbf{E} = \left(\frac{\lambda}{2\pi\epsilon_0 r}\right)\hat{\mathbf{r}}$$

where $\hat{\mathbf{r}}$ is a unit vector pointing radially away from the line and λ is the charge per unit length along the line. Derive an expression for the potential difference between $r = r_1$ and $r = r_2$.

64. A point charge q is located at $x = -R$, and a point charge $-2q$ is located at the origin. Prove that the equipotential surface that has zero potential is a sphere centered at $(-4R/3, 0, 0)$ and having a radius $r = 2R/3$.

65. Consider two thin, conducting, spherical shells as shown in cross-section in Figure P25.65. The inner shell has a radius $r_1 = 15.0$ cm and a charge of 10.0 nC. The outer shell has a radius $r_2 = 30.0$ cm and a charge of -15.0 nC. Find (a) the electric field $\mathbf{E}$ and (b) the electric potential V in regions A, B, and C, with $V = 0$ at $r = \infty$.

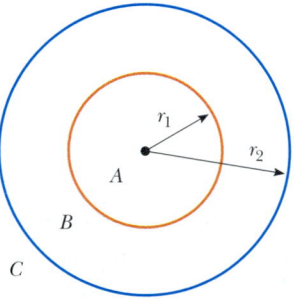

Figure P25.65

66. The x axis is the symmetry axis of a uniformly charged ring of radius R and charge Q (Fig. P25.66). A point charge Q of mass M is located at the center of the ring. When it is displaced slightly, the point charge acceler-

ates along the x axis to infinity. Show that the ultimate speed of the point charge is

$$v = \left(\frac{2k_eQ^2}{MR}\right)^{1/2}$$

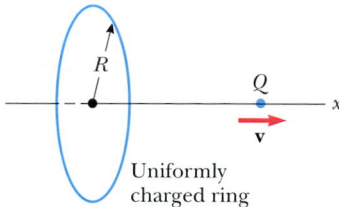

Figure P25.66

67. An infinite sheet of charge that has a surface charge density of 25.0 nC/m² lies in the yz plane, passes through the origin, and is at a potential of 1.00 kV at the point $y = 0$, $z = 0$. A long wire having a linear charge density of 80.0 nC/m lies parallel to the y axis and intersects the x axis at $x = 3.00$ m. (a) Determine, as a function of x, the potential along the x axis between wire and sheet. (b) What is the potential energy of a 2.00-nC charge placed at $x = 0.800$ m?

68. The thin, uniformly charged rod shown in Figure P25.68 has a linear charge density λ. Find an expression for the electric potential at P.

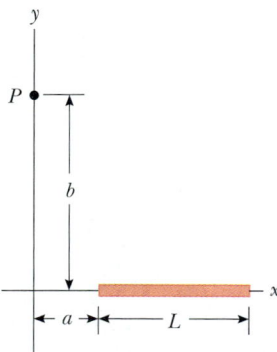

Figure P25.68

69. A dipole is located along the y axis as shown in Figure P25.69. (a) At a point P, which is far from the dipole $(r \gg a)$, the electric potential is

$$V = k_e \frac{p \cos \theta}{r^2}$$

where $p = 2qa$. Calculate the radial component E_r and the perpendicular component E_θ of the associated electric field. Note that $E_\theta = -(1/r)(\partial V/\partial \theta)$. Do these results seem reasonable for $\theta = 90°$ and $0°$? for $r = 0$?

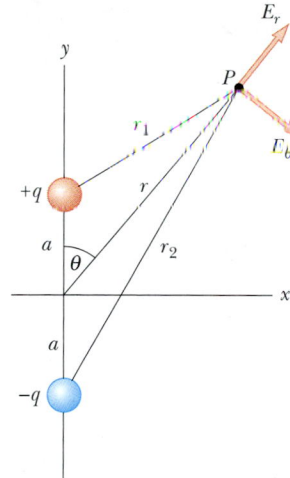

Figure P25.69

(b) For the dipole arrangement shown, express V in terms of cartesian coordinates using $r = (x^2 + y^2)^{1/2}$ and

$$\cos \theta = \frac{y}{(x^2 + y^2)^{1/2}}$$

Using these results and taking $r \gg a$, calculate the field components E_x and E_y.

70. Figure P25.70 shows several equipotential lines each labeled by its potential in volts. The distance between the lines of the square grid represents 1.00 cm. (a) Is the magnitude of the field bigger at A or at B? Why? (b) What is **E** at B? (c) Represent what the field looks like by drawing at least eight field lines.

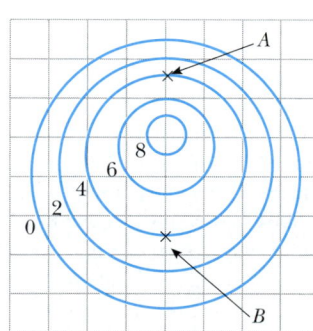

Figure P25.70

71. A disk of radius R has a nonuniform surface charge density $\sigma = Cr$, where C is a constant and r is measured from the center of the disk (Fig. P25.71). Find (by direct integration) the potential at P.

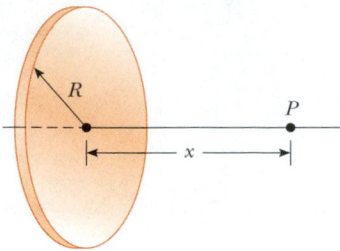

Figure P25.71

72. A solid sphere of radius R has a uniform charge density ρ and total charge Q. Derive an expression for its total

electric potential energy. (*Hint:* Imagine that the sphere is constructed by adding successive layers of concentric shells of charge $dq = (4\pi r^2\, dr)\rho$ and use $dU = V\, dq$.)

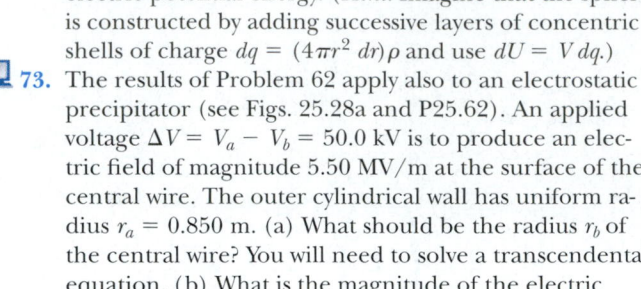 **73.** The results of Problem 62 apply also to an electrostatic precipitator (see Figs. 25.28a and P25.62). An applied voltage $\Delta V = V_a - V_b = 50.0$ kV is to produce an electric field of magnitude 5.50 MV/m at the surface of the central wire. The outer cylindrical wall has uniform radius $r_a = 0.850$ m. (a) What should be the radius r_b of the central wire? You will need to solve a transcendental equation. (b) What is the magnitude of the electric field at the outer wall?

ANSWERS TO QUICK QUIZZES

25.1 We do if the electric field is uniform. (This is precisely what we do in the next section.) In general, however, an electric field changes from one place to another.

25.2 $B \rightarrow C$, $C \rightarrow D$, $A \rightarrow B$, $D \rightarrow E$. Moving from B to C decreases the electric potential by 2 V, so the electric field performs 2 J of work on each coulomb of charge that moves. Moving from C to D decreases the electric potential by 1 V, so 1 J of work is done by the field. It takes no work to move the charge from A to B because the electric potential does not change. Moving from D to E increases the electric potential by 1 V, and thus the field does -1 J of work, just as raising a mass to a higher elevation causes the gravitational field to do negative work on the mass.

25.3 The electric potential decreases in inverse proportion to the radius (see Eq. 25.11). The electric field magnitude decreases as the reciprocal of the radius squared (see Eq. 23.4). Because the surface area increases as r^2 while the electric field magnitude decreases as $1/r^2$, the electric flux through the surface remains constant (see Eq. 24.1).

25.4 (a) Yes. Consider four equal charges placed at the corners of a square. The electric potential graph for this situation is shown in the figure. At the center of the square, the electric field is zero because the individual fields from the four charges cancel, but the potential is not zero. This is also the situation inside a charged conductor. (b) Yes again. In Figure 25.8, for instance, the

electric potential is zero at the center of the dipole, but the magnitude of the field at that point is not zero. (The two charges in a dipole are by definition of opposite sign; thus, the electric field lines created by the two charges extend from the positive to the negative charge and do not cancel anywhere.) This is the situation we presented in Example 25.4c, in which the equations we obtained give $V = 0$ and $E_x \neq 0$.

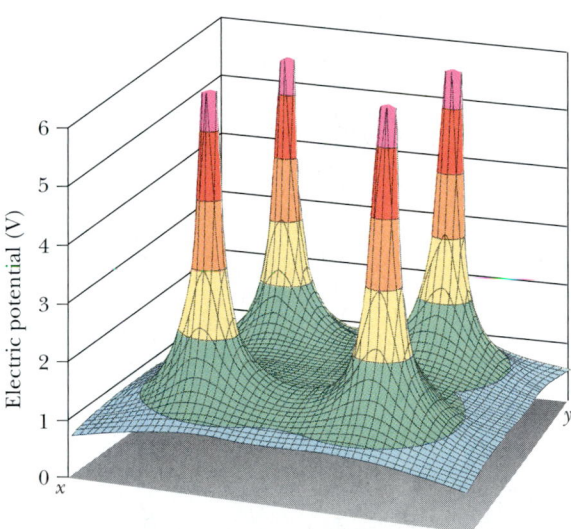

DANGER

HAZARDOUS VOLTAGE INSIDE. DO NOT OPEN.
GEFÄHRLICHE SPANNUNG. ABDECKUNG NICHT ÖFFNEN.
TENSION DANGEREUSE À L'INTÉRIEUR. NE PAS OUVRIR.
VOLTAJE PELIGROSO EN EL INTERIOR. NO ABRA.
TENSIONE PERICOLOSA ALL'INTERNO. NON APRIRE.
FARLIG ELEKTRISK SPÆNDING INDENI, LUK IKKE OP.
HIERBINNEN GENAARLIJK VOLTAGE. NIET OPENMAKEN.
SISÄPUOLELLA VAARALLINEN JÄNNITE. ÄLÄ AVAA.
FARLIG SPENNING. MÅ IKKE ÅPNES.
NÃO ABRA. VOLTAGEM PERIGOSA NO INTERIOR.
FARLIG SPÄNNING INNUTI. ÖPPNAS EJ.

101-7931

chapter

Capacitance and Dielectrics

In this chapter, we discuss *capacitors*—devices that store electric charge. Capacitors are commonly used in a variety of electric circuits. For instance, they are used to tune the frequency of radio receivers, as filters in power supplies, to eliminate sparking in automobile ignition systems, and as energy-storing devices in electronic flash units.

A capacitor consists of two conductors separated by an insulator. We shall see that the capacitance of a given capacitor depends on its geometry and on the material—called a *dielectric*—that separates the conductors.

26.1 DEFINITION OF CAPACITANCE

13.5 Consider two conductors carrying charges of equal magnitude but of opposite sign, as shown in Figure 26.1. Such a combination of two conductors is called a **capacitor.** The conductors are called *plates*. A potential difference ΔV exists between the conductors due to the presence of the charges. Because the unit of potential difference is the volt, a potential difference is often called a **voltage.** We shall use this term to describe the potential difference across a circuit element or between two points in space.

What determines how much charge is on the plates of a capacitor for a given voltage? In other words, what is the *capacity* of the device for storing charge at a particular value of ΔV? Experiments show that the quantity of charge Q on a capacitor[1] is linearly proportional to the potential difference between the conductors; that is, $Q \propto \Delta V$. The proportionality constant depends on the shape and separation of the conductors.[2] We can write this relationship as $Q = C\Delta V$ if we define capacitance as follows:

Definition of capacitance

> The **capacitance** C of a capacitor is the ratio of the magnitude of the charge on either conductor to the magnitude of the potential difference between them:
>
> $$C \equiv \frac{Q}{\Delta V} \qquad (26.1)$$

Note that by definition *capacitance is always a positive quantity*. Furthermore, the potential difference ΔV is always expressed in Equation 26.1 as a positive quantity. Because the potential difference increases linearly with the stored charge, the ratio $Q/\Delta V$ is constant for a given capacitor. Therefore, capacitance is a measure of a capacitor's ability to store charge and electric potential energy.

From Equation 26.1, we see that capacitance has SI units of coulombs per volt. The SI unit of capacitance is the **farad** (F), which was named in honor of Michael Faraday:

$$1 \text{ F} = 1 \text{ C/V}$$

The farad is a very large unit of capacitance. In practice, typical devices have capacitances ranging from microfarads (10^{-6} F) to picofarads (10^{-12} F). For practical purposes, capacitors often are labeled "mF" for microfarads and "mmF" for micromicrofarads or, equivalently, "pF" for picofarads.

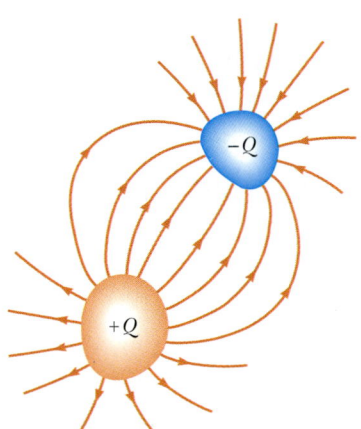

Figure 26.1 A capacitor consists of two conductors carrying charges of equal magnitude but opposite sign.

---

[1] Although the total charge on the capacitor is zero (because there is as much excess positive charge on one conductor as there is excess negative charge on the other), it is common practice to refer to the magnitude of the charge on either conductor as "the charge on the capacitor."

[2] The proportionality between ΔV and Q can be proved from Coulomb's law or by experiment.

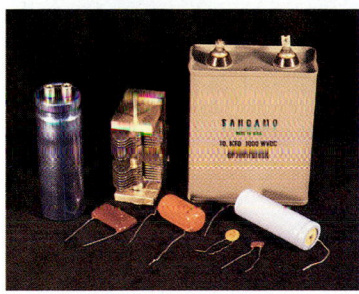

A collection of capacitors used in a variety of applications. *(Henry Leap and Jim Lehman)*

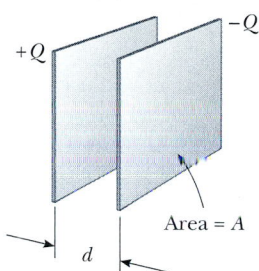

Figure 26.2   A parallel-plate capacitor consists of two parallel conducting plates, each of area A, separated by a distance d. When the capacitor is charged, the plates carry equal amounts of charge. One plate carries positive charge, and the other carries negative charge.

Let us consider a capacitor formed from a pair of parallel plates, as shown in Figure 26.2. Each plate is connected to one terminal of a battery (not shown in Fig. 26.2), which acts as a source of potential difference. If the capacitor is initially uncharged, the battery establishes an electric field in the connecting wires when the connections are made. Let us focus on the plate connected to the negative terminal of the battery. The electric field applies a force on electrons in the wire just outside this plate; this force causes the electrons to move onto the plate. This movement continues until the plate, the wire, and the terminal are all at the same electric potential. Once this equilibrium point is attained, a potential difference no longer exists between the terminal and the plate, and as a result no electric field is present in the wire, and the movement of electrons stops. The plate now carries a negative charge. A similar process occurs at the other capacitor plate, with electrons moving from the plate to the wire, leaving the plate positively charged. In this final configuration, the potential difference across the capacitor plates is the same as that between the terminals of the battery.

Suppose that we have a capacitor rated at 4 pF. This rating means that the capacitor can store 4 pC of charge for each volt of potential difference between the two conductors. If a 9-V battery is connected across this capacitor, one of the conductors ends up with a net charge of -36 pC and the other ends up with a net charge of $+36$ pC.

26.2   CALCULATING CAPACITANCE

We can calculate the capacitance of a pair of oppositely charged conductors in the following manner: We assume a charge of magnitude Q, and we calculate the potential difference using the techniques described in the preceding chapter. We then use the expression $C = Q/\Delta V$ to evaluate the capacitance. As we might expect, we can perform this calculation relatively easily if the geometry of the capacitor is simple.

We can calculate the capacitance of an isolated spherical conductor of radius R and charge Q if we assume that the second conductor making up the capacitor is a concentric hollow sphere of infinite radius. The electric potential of the sphere of radius R is simply $k_e Q/R$, and setting $V = 0$ at infinity as usual, we have

$$C = \frac{Q}{\Delta V} = \frac{Q}{k_e Q/R} = \frac{R}{k_e} = 4\pi\epsilon_0 R \tag{26.2}$$

This expression shows that the capacitance of an isolated charged sphere is proportional to its radius and is independent of both the charge on the sphere and the potential difference.

QuickLab

Roll some socks into balls and stuff them into a shoebox. What determines how many socks fit in the box? Relate how hard you push on the socks to ΔV for a capacitor. How does the size of the box influence its "sock capacity"?

The capacitance of a pair of conductors depends on the geometry of the conductors. Let us illustrate this with three familiar geometries, namely, parallel plates, concentric cylinders, and concentric spheres. In these examples, we assume that the charged conductors are separated by a vacuum. The effect of a dielectric material placed between the conductors is treated in Section 26.5.

Parallel-Plate Capacitors

Two parallel metallic plates of equal area A are separated by a distance d, as shown in Figure 26.2. One plate carries a charge Q, and the other carries a charge $-Q$. Let us consider how the geometry of these conductors influences the capacity of the combination to store charge. Recall that charges of like sign repel one another. As a capacitor is being charged by a battery, electrons flow into the negative plate and out of the positive plate. If the capacitor plates are large, the accumulated charges are able to distribute themselves over a substantial area, and the amount of charge that can be stored on a plate for a given potential difference increases as the plate area is increased. Thus, we expect the capacitance to be proportional to the plate area A.

Now let us consider the region that separates the plates. If the battery has a constant potential difference between its terminals, then the electric field between the plates must increase as d is decreased. Let us imagine that we move the plates closer together and consider the situation before any charges have had a chance to move in response to this change. Because no charges have moved, the electric field between the plates has the same value but extends over a shorter distance. Thus, the magnitude of the potential difference between the plates $\Delta V = Ed$ (Eq. 25.6) is now smaller. The difference between this new capacitor voltage and the terminal voltage of the battery now exists as a potential difference across the wires connecting the battery to the capacitor. This potential difference results in an electric field in the wires that drives more charge onto the plates, increasing the potential difference between the plates. When the potential difference between the plates again matches that of the battery, the potential difference across the wires falls back to zero, and the flow of charge stops. Thus, moving the plates closer together causes the charge on the capacitor to increase. If d is increased, the charge decreases. As a result, we expect the device's capacitance to be inversely proportional to d.

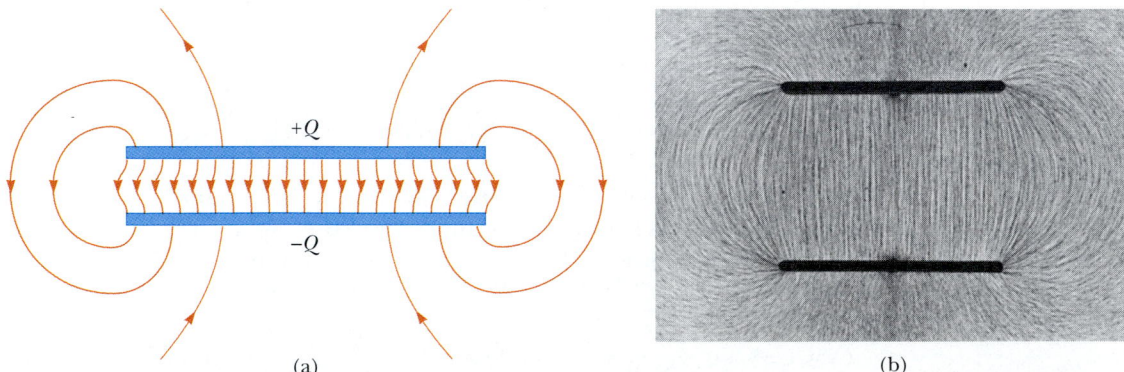

(a)   (b)

Figure 26.3   (a) The electric field between the plates of a parallel-plate capacitor is uniform near the center but nonuniform near the edges. (b) Electric field pattern of two oppositely charged conducting parallel plates. Small pieces of thread on an oil surface align with the electric field. *(b, Courtesy of Harold M. Waage, Princeton University)*

We can verify these physical arguments with the following derivation. The surface charge density on either plate is $\sigma = Q/A$. If the plates are very close together (in comparison with their length and width), we can assume that the electric field is uniform between the plates and is zero elsewhere. According to the last paragraph of Example 24.8, the value of the electric field between the plates is

$$E = \frac{\sigma}{\epsilon_0} = \frac{Q}{\epsilon_0 A}$$

Because the field between the plates is uniform, the magnitude of the potential difference between the plates equals Ed (see Eq. 25.6); therefore,

$$\Delta V = Ed = \frac{Qd}{\epsilon_0 A}$$

Substituting this result into Equation 26.1, we find that the capacitance is

$$C = \frac{Q}{\Delta V} = \frac{Q}{Qd/\epsilon_0 A}$$

$$C = \frac{\epsilon_0 A}{d} \tag{26.3}$$

That is, **the capacitance of a parallel-plate capacitor is proportional to the area of its plates and inversely proportional to the plate separation,** just as we expect from our conceptual argument.

A careful inspection of the electric field lines for a parallel-plate capacitor reveals that the field is uniform in the central region between the plates, as shown in Figure 26.3a. However, the field is nonuniform at the edges of the plates. Figure 26.3b is a photograph of the electric field pattern of a parallel-plate capacitor. Note the nonuniform nature of the electric field at the ends of the plates. Such end effects can be neglected if the plate separation is small compared with the length of the plates.

Quick Quiz 26.1

Many computer keyboard buttons are constructed of capacitors, as shown in Figure 26.4. When a key is pushed down, the soft insulator between the movable plate and the fixed plate is compressed. When the key is pressed, the capacitance (a) increases, (b) decreases, or (c) changes in a way that we cannot determine because the complicated electric circuit connected to the keyboard button may cause a change in ΔV.

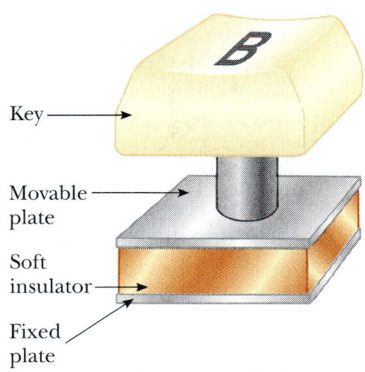

Key
Movable plate
Soft insulator
Fixed plate

Figure 26.4 One type of computer keyboard button.

EXAMPLE 26.1 **Parallel-Plate Capacitor**

A parallel-plate capacitor has an area $A = 2.00 \times 10^{-4} \text{ m}^2$ and a plate separation $d = 1.00$ mm. Find its capacitance.

Solution From Equation 26.3, we find that

$$C = \epsilon_0 \frac{A}{d} = (8.85 \times 10^{-12} \text{ C}^2/\text{N} \cdot \text{m}^2) \left(\frac{2.00 \times 10^{-4} \text{ m}^2}{1.00 \times 10^{-3} \text{ m}} \right)$$

$$= 1.77 \times 10^{-12} \text{ F} = \boxed{1.77 \text{ pF}}$$

Exercise What is the capacitance for a plate separation of 3.00 mm?

Answer 0.590 pF.

Cylindrical and Spherical Capacitors

From the definition of capacitance, we can, in principle, find the capacitance of any geometric arrangement of conductors. The following examples demonstrate the use of this definition to calculate the capacitance of the other familiar geometries that we mentioned: cylinders and spheres.

EXAMPLE 26.2 ▶ The Cylindrical Capacitor

A solid cylindrical conductor of radius a and charge Q is coaxial with a cylindrical shell of negligible thickness, radius $b > a$, and charge $-Q$ (Fig. 26.5a). Find the capacitance of this cylindrical capacitor if its length is ℓ.

Solution   It is difficult to apply physical arguments to this configuration, although we can reasonably expect the capacitance to be proportional to the cylinder length ℓ for the same reason that parallel-plate capacitance is proportional to plate area: Stored charges have more room in which to be distributed. If we assume that ℓ is much greater than a and b, we can neglect end effects. In this case, the electric field is perpendicular to the long axis of the cylinders and is confined to the region between them (Fig. 26.5b). We must first calculate the potential difference between the two cylinders, which is given in general by

$$V_b - V_a = -\int_a^b \mathbf{E} \cdot d\mathbf{s}$$

where $\mathbf{E}$ is the electric field in the region $a < r < b$. In Chapter 24, we showed using Gauss's law that the magnitude of the electric field of a cylindrical charge distribution having linear charge density λ is $E_r = 2k_e\lambda/r$ (Eq. 24.7). The same result applies here because, according to Gauss's law, the charge on the outer cylinder does not contribute to the electric field inside it. Using this result and noting from Figure 26.5b that $\mathbf{E}$ is along r, we find that

$$V_b - V_a = -\int_a^b E_r\, dr = -2k_e\lambda \int_a^b \frac{dr}{r} = -2k_e\lambda \ln\!\left(\frac{b}{a}\right)$$

Substituting this result into Equation 26.1 and using the fact that $\lambda = Q/\ell$, we obtain

$$C = \frac{Q}{\Delta V} = \frac{Q}{\dfrac{2k_e Q}{\ell}\ln\!\left(\dfrac{b}{a}\right)} = \boxed{\dfrac{\ell}{2k_e\ln\!\left(\dfrac{b}{a}\right)}} \qquad \textbf{(26.4)}$$

where ΔV is the magnitude of the potential difference, given

by $\Delta V = |V_b - V_a| = 2k_e\lambda \ln\ (b/a)$, a positive quantity. As predicted, the capacitance is proportional to the length of the cylinders. As we might expect, the capacitance also depends on the radii of the two cylindrical conductors. From Equation 26.4, we see that the capacitance per unit length of a combination of concentric cylindrical conductors is

$$\frac{C}{\ell} = \frac{1}{2k_e\ln\!\left(\dfrac{b}{a}\right)} \qquad \textbf{(26.5)}$$

An example of this type of geometric arrangement is a *coaxial cable*, which consists of two concentric cylindrical conductors separated by an insulator. The cable carries electrical signals in the inner and outer conductors. Such a geometry is especially useful for shielding the signals from any possible external influences.

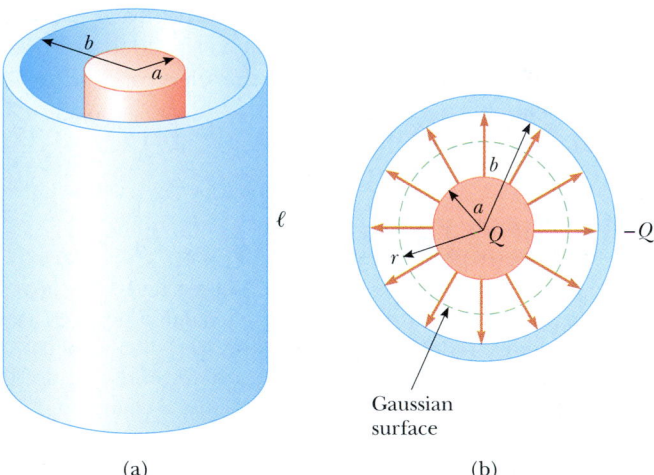

(a)   (b)

Figure 26.5   (a) A cylindrical capacitor consists of a solid cylindrical conductor of radius a and length ℓ surrounded by a coaxial cylindrical shell of radius b. (b) End view. The dashed line represents the end of the cylindrical gaussian surface of radius r and length ℓ.

EXAMPLE 26.3 ▶ The Spherical Capacitor

A spherical capacitor consists of a spherical conducting shell of radius b and charge $-Q$ concentric with a smaller conducting sphere of radius a and charge Q (Fig. 26.6). Find the capacitance of this device.

Solution   As we showed in Chapter 24, the field outside a spherically symmetric charge distribution is radial and given by the expression $k_e Q/r^2$. In this case, this result applies to the field between the spheres ($a < r < b$). From

Gauss's law we see that only the inner sphere contributes to this field. Thus, the potential difference between the spheres is

$$V_b - V_a = -\int_a^b E_r\,dr = -k_e Q \int_a^b \frac{dr}{r^2} = k_e Q \left[\frac{1}{r} \right]_a^b$$

$$= k_e Q \left(\frac{1}{b} - \frac{1}{a} \right)$$

The magnitude of the potential difference is

$$\Delta V = |V_b - V_a| = k_e Q \frac{(b-a)}{ab}$$

Substituting this value for ΔV into Equation 26.1, we obtain

$$C = \frac{Q}{\Delta V} = \frac{ab}{k_e(b-a)} \qquad \textbf{(26.6)}$$

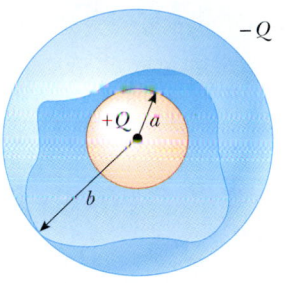

Figure 26.6 A spherical capacitor consists of an inner sphere of radius a surrounded by a concentric spherical shell of radius b. The electric field between the spheres is directed radially outward when the inner sphere is positively charged.

Exercise Show that as the radius b of the outer sphere approaches infinity, the capacitance approaches the value $a/k_e = 4\pi\epsilon_0 a$.

Quick Quiz 26.2

What is the magnitude of the electric field in the region outside the spherical capacitor described in Example 26.3?

26.3 COMBINATIONS OF CAPACITORS

Two or more capacitors often are combined in electric circuits. We can calculate the equivalent capacitance of certain combinations using methods described in this section. The circuit symbols for capacitors and batteries, as well as the color codes used for them in this text, are given in Figure 26.7. The symbol for the capacitor reflects the geometry of the most common model for a capacitor—a pair of parallel plates. The positive terminal of the battery is at the higher potential and is represented in the circuit symbol by the longer vertical line.

Parallel Combination

Two capacitors connected as shown in Figure 26.8a are known as a *parallel combination* of capacitors. Figure 26.8b shows a circuit diagram for this combination of capacitors. The left plates of the capacitors are connected by a conducting wire to the positive terminal of the battery and are therefore both at the same electric potential as the positive terminal. Likewise, the right plates are connected to the negative terminal and are therefore both at the same potential as the negative terminal. Thus, **the individual potential differences across capacitors connected in parallel are all the same and are equal to the potential difference applied across the combination.**

In a circuit such as that shown in Figure 26.8, the voltage applied across the combination is the terminal voltage of the battery. Situations can occur in which

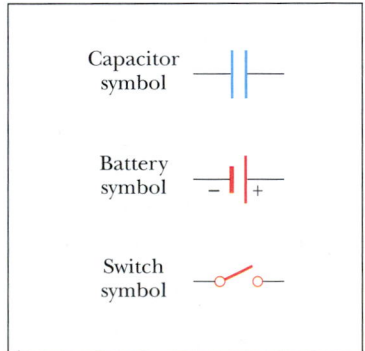

Figure 26.7 Circuit symbols for capacitors, batteries, and switches. Note that capacitors are in blue and batteries and switches are in red.

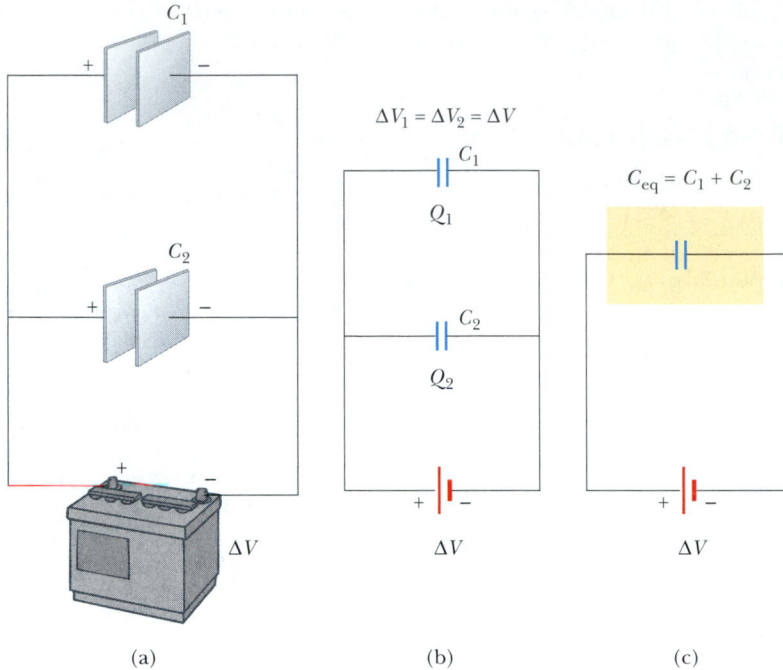

$\Delta V_1 = \Delta V_2 = \Delta V$

$C_{eq} = C_1 + C_2$

(a) (b) (c)

Figure 26.8 (a) A parallel combination of two capacitors in an electric circuit in which the potential difference across the battery terminals is ΔV. (b) The circuit diagram for the parallel combination. (c) The equivalent capacitance is $C_{eq} = C_1 + C_2$.

the parallel combination is in a circuit with other circuit elements; in such situations, we must determine the potential difference across the combination by analyzing the entire circuit.

When the capacitors are first connected in the circuit shown in Figure 26.8, electrons are transferred between the wires and the plates; this transfer leaves the left plates positively charged and the right plates negatively charged. The energy source for this charge transfer is the internal chemical energy stored in the battery, which is converted to electric potential energy associated with the charge separation. The flow of charge ceases when the voltage across the capacitors is equal to that across the battery terminals. The capacitors reach their maximum charge when the flow of charge ceases. Let us call the maximum charges on the two capacitors Q_1 and Q_2. The *total charge* Q stored by the two capacitors is

$$Q = Q_1 + Q_2 \qquad \textbf{(26.7)}$$

That is, **the total charge on capacitors connected in parallel is the sum of the charges on the individual capacitors.** Because the voltages across the capacitors are the same, the charges that they carry are

$$Q_1 = C_1\,\Delta V \qquad Q_2 = C_2\,\Delta V$$

Suppose that we wish to replace these two capacitors by one *equivalent capacitor* having a capacitance C_{eq}, as shown in Figure 26.8c. The effect this equivalent capacitor has on the circuit must be exactly the same as the effect of the combination of the two individual capacitors. That is, the equivalent capacitor must store Q units of charge when connected to the battery. We can see from Figure 26.8c that the voltage across the equivalent capacitor also is ΔV because the equivalent capacitor is connected directly across the battery terminals. Thus, for the equivalent capacitor,

$$Q = C_{eq} \Delta V$$

Substituting these three relationships for charge into Equation 26.7, we have

$$C_{eq} \Delta V = C_1 \Delta V + C_2 \Delta V$$

$$C_{eq} = C_1 + C_2 \qquad \binom{\text{parallel}}{\text{combination}}$$

If we extend this treatment to three or more capacitors connected in parallel, we find the equivalent capacitance to be

$$C_{eq} = C_1 + C_2 + C_3 + \cdots \qquad \text{(parallel combination)} \qquad \textbf{(26.8)}$$

Thus, **the equivalent capacitance of a parallel combination of capacitors is greater than any of the individual capacitances.** This makes sense because we are essentially combining the areas of all the capacitor plates when we connect them with conducting wire.

Series Combination

Two capacitors connected as shown in Figure 26.9a are known as a *series combination* of capacitors. The left plate of capacitor 1 and the right plate of capacitor 2 are connected to the terminals of a battery. The other two plates are connected to each other and to nothing else; hence, they form an isolated conductor that is initially uncharged and must continue to have zero net charge. To analyze this combination, let us begin by considering the uncharged capacitors and follow what happens just after a battery is connected to the circuit. When the battery is con-

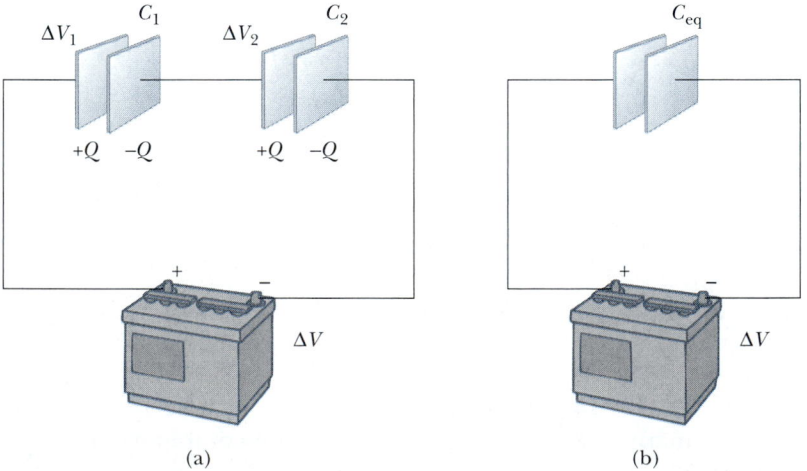

Figure 26.9 (a) A series combination of two capacitors. The charges on the two capacitors are the same. (b) The capacitors replaced by a single equivalent capacitor. The equivalent capacitance can be calculated from the relationship

$$\frac{1}{C_{eq}} = \frac{1}{C_1} + \frac{1}{C_2}$$

nected, electrons are transferred out of the left plate of C_1 and into the right plate of C_2. As this negative charge accumulates on the right plate of C_2, an equivalent amount of negative charge is forced off the left plate of C_2, and this left plate therefore has an excess positive charge. The negative charge leaving the left plate of C_2 travels through the connecting wire and accumulates on the right plate of C_1. As a result, all the right plates end up with a charge $-Q$, and all the left plates end up with a charge $+Q$. Thus, **the charges on capacitors connected in series are the same.**

From Figure 26.9a, we see that the voltage ΔV across the battery terminals is split between the two capacitors:

$$\Delta V = \Delta V_1 + \Delta V_2 \tag{26.9}$$

where ΔV_1 and ΔV_2 are the potential differences across capacitors C_1 and C_2, respectively. In general, **the total potential difference across any number of capacitors connected in series is the sum of the potential differences across the individual capacitors.**

Suppose that an equivalent capacitor has the same effect on the circuit as the series combination. After it is fully charged, the equivalent capacitor must have a charge of $-Q$ on its right plate and a charge of $+Q$ on its left plate. Applying the definition of capacitance to the circuit in Figure 26.9b, we have

$$\Delta V = \frac{Q}{C_{eq}}$$

Because we can apply the expression $Q = C\,\Delta V$ to each capacitor shown in Figure 26.9a, the potential difference across each is

$$\Delta V_1 = \frac{Q}{C_1} \qquad \Delta V_2 = \frac{Q}{C_2}$$

Substituting these expressions into Equation 26.9 and noting that $\Delta V = Q/C_{eq}$, we have

$$\frac{Q}{C_{eq}} = \frac{Q}{C_1} + \frac{Q}{C_2}$$

Canceling Q, we arrive at the relationship

$$\frac{1}{C_{eq}} = \frac{1}{C_1} + \frac{1}{C_2} \qquad \left(\begin{array}{l}\text{series}\\\text{combination}\end{array}\right)$$

When this analysis is applied to three or more capacitors connected in series, the relationship for the equivalent capacitance is

$$\frac{1}{C_{eq}} = \frac{1}{C_1} + \frac{1}{C_2} + \frac{1}{C_3} + \cdots \qquad \left(\begin{array}{l}\text{series}\\\text{combination}\end{array}\right) \tag{26.10}$$

This demonstrates that **the equivalent capacitance of a series combination is always less than any individual capacitance in the combination.**

EXAMPLE 26.4   Equivalent Capacitance

Find the equivalent capacitance between a and b for the combination of capacitors shown in Figure 26.10a. All capacitances are in microfarads.

Solution   Using Equations 26.8 and 26.10, we reduce the combination step by step as indicated in the figure. The 1.0-μF and 3.0-μF capacitors are in parallel and combine according to the expression $C_{eq} = C_1 + C_2 = 4.0 \ \mu F$. The 2.0-$\mu F$ and 6.0-μF capacitors also are in parallel and have an equivalent capacitance of 8.0 μF. Thus, the upper branch in Figure 26.10b consists of two 4.0-μF capacitors in series, which combine as follows.

$$\frac{1}{C_{eq}} = \frac{1}{C_1} + \frac{1}{C_2} = \frac{1}{4.0 \ \mu F} + \frac{1}{4.0 \ \mu F} = \frac{1}{2.0 \ \mu F}$$

$$C_{eq} = \frac{1}{1/2.0 \ \mu F} = 2.0 \ \mu F$$

The lower branch in Figure 26.10b consists of two 8.0-μF capacitors in series, which combine to yield an equivalent capacitance of 4.0 μF. Finally, the 2.0-μF and 4.0-μF capacitors in Figure 26.10c are in parallel and thus have an equivalent capacitance of 6.0 μF.

Exercise Consider three capacitors having capacitances of 3.0 μF, 6.0 μF, and 12 μF. Find their equivalent capacitance when they are connected (a) in parallel and (b) in series.

Answer (a) 21 μF; (b) 1.7 μF.

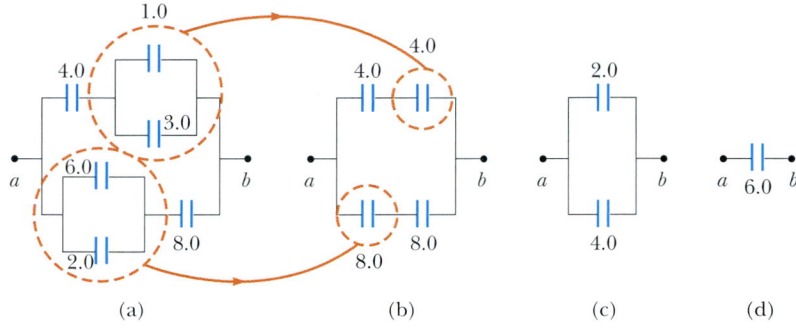

(a)   (b)   (c)   (d)

Figure 26.10 To find the equivalent capacitance of the capacitors in part (a), we reduce the various combinations in steps as indicated in parts (b), (c), and (d), using the series and parallel rules described in the text.

26.4 ENERGY STORED IN A CHARGED CAPACITOR

Almost everyone who works with electronic equipment has at some time verified that a capacitor can store energy. If the plates of a charged capacitor are connected by a conductor, such as a wire, charge moves between the plates and the connecting wire until the capacitor is uncharged. The discharge can often be observed as a visible spark. If you should accidentally touch the opposite plates of a charged capacitor, your fingers act as a pathway for discharge, and the result is an electric shock. The degree of shock you receive depends on the capacitance and on the voltage applied to the capacitor. Such a shock could be fatal if high voltages are present, such as in the power supply of a television set. Because the charges can be stored in a capacitor even when the set is turned off, unplugging the television does not make it safe to open the case and touch the components inside.

Consider a parallel-plate capacitor that is initially uncharged, such that the initial potential difference across the plates is zero. Now imagine that the capacitor is connected to a battery and develops a maximum charge Q. (We assume that the capacitor is charged slowly so that the problem can be considered as an electrostatic system.) When the capacitor is connected to the battery, electrons in the wire just outside the plate connected to the negative terminal move into the plate to give it a negative charge. Electrons in the plate connected to the positive terminal move out of the plate into the wire to give the plate a positive charge. Thus, charges move only a small distance in the wires.

To calculate the energy of the capacitor, we shall assume a different process—one that does not actually occur but gives the same final result. We can make this

QuickLab

Here's how to find out whether your calculator has a capacitor to protect values or programs during battery changes: Store a number in your calculator's memory, remove the calculator battery for a moment, and then quickly replace it. Was the number that you stored preserved while the battery was out of the calculator? (You may want to write down any critical numbers or programs that are stored in the calculator before trying this!)

assumption because the energy in the final configuration does not depend on the actual charge-transfer process. We imagine that we reach in and grab a small amount of positive charge on the plate connected to the negative terminal and apply a force that causes this positive charge to move over to the plate connected to the positive terminal. Thus, we do work on the charge as we transfer it from one plate to the other. At first, no work is required to transfer a small amount of charge dq from one plate to the other.[3] However, once this charge has been transferred, a small potential difference exists between the plates. Therefore, work must be done to move additional charge through this potential difference. As more and more charge is transferred from one plate to the other, the potential difference increases in proportion, and more work is required.

Suppose that q is the charge on the capacitor at some instant during the charging process. At the same instant, the potential difference across the capacitor is $\Delta V = q/C$. From Section 25.2, we know that the work necessary to transfer an increment of charge dq from the plate carrying charge $-q$ to the plate carrying charge q (which is at the higher electric potential) is

$$dW = \Delta V \, dq = \frac{q}{C} \, dq$$

This is illustrated in Figure 26.11. The total work required to charge the capacitor from $q = 0$ to some final charge $q = Q$ is

$$W = \int_0^Q \frac{q}{C} \, dq = \frac{1}{C} \int_0^Q q \, dq = \frac{Q^2}{2C}$$

The work done in charging the capacitor appears as electric potential energy U stored in the capacitor. Therefore, we can express the potential energy stored in a charged capacitor in the following forms:

Energy stored in a charged capacitor

$$U = \frac{Q^2}{2C} = \tfrac{1}{2} Q \, \Delta V = \tfrac{1}{2} C (\Delta V)^2 \qquad \textbf{(26.11)}$$

This result applies to any capacitor, regardless of its geometry. We see that for a given capacitance, the stored energy increases as the charge increases and as the potential difference increases. In practice, there is a limit to the maximum energy

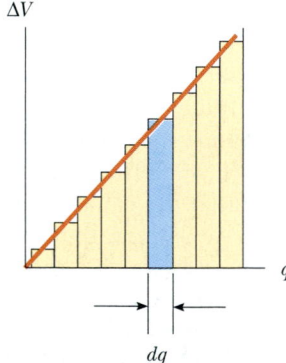

Figure 26.11   A plot of potential difference versus charge for a capacitor is a straight line having a slope $1/C$. The work required to move charge dq through the potential difference ΔV across the capacitor plates is given by the area of the shaded rectangle. The total work required to charge the capacitor to a final charge Q is the triangular area under the straight line, $W = \tfrac{1}{2} Q \, \Delta V$. (Don't forget that 1 V = 1 J/C; hence, the unit for the area is the joule.)

[3] We shall use lowercase q for the varying charge on the capacitor while it is charging, to distinguish it from uppercase Q, which is the total charge on the capacitor after it is completely charged.

(or charge) that can be stored because, at a sufficiently great value of ΔV, discharge ultimately occurs between the plates. For this reason, capacitors are usually labeled with a maximum operating voltage.

Quick Quiz 26.3

You have three capacitors and a battery. How should you combine the capacitors and the battery in one circuit so that the capacitors will store the maximum possible energy?

We can consider the energy stored in a capacitor as being stored in the electric field created between the plates as the capacitor is charged. This description is reasonable in view of the fact that the electric field is proportional to the charge on the capacitor. For a parallel-plate capacitor, the potential difference is related to the electric field through the relationship $\Delta V = Ed$. Furthermore, its capacitance is $C = \epsilon_0 A / d$ (Eq. 26.3). Substituting these expressions into Equation 26.11, we obtain

$$U = \frac{1}{2}\frac{\epsilon_0 A}{d}(E^2 d^2) = \frac{1}{2}(\epsilon_0 A d)E^2 \qquad \textbf{(26.12)}$$

Energy stored in a parallel-plate capacitor

Because the volume V (volume, not voltage!) occupied by the electric field is Ad, the *energy per unit volume* $u_E = U/V = U/Ad$, known as the *energy density*, is

$$u_E = \tfrac{1}{2}\epsilon_0 E^2 \qquad \textbf{(26.13)}$$

Energy density in an electric field

Although Equation 26.13 was derived for a parallel-plate capacitor, the expression is generally valid. That is, the **energy density in any electric field is proportional to the square of the magnitude of the electric field at a given point.**

This bank of capacitors stores electrical energy for use in the particle accelerator at FermiLab, located outside Chicago. Because the electric utility company cannot provide a large enough burst of energy to operate the equipment, these capacitors are slowly charged up, and then the energy is rapidly "dumped" into the accelerator. In this sense, the setup is much like a fire-protection water tank on top of a building. The tank collects water and stores it for situations in which a lot of water is needed in a short time. (*FermiLab Visual Media Services*)

EXAMPLE 26.5 ▶ Rewiring Two Charged Capacitors

Two capacitors C_1 and C_2 (where $C_1 > C_2$) are charged to the same initial potential difference ΔV_i, but with opposite polarity. The charged capacitors are removed from the battery, and their plates are connected as shown in Figure 26.12a. The switches S_1 and S_2 are then closed, as shown in Figure 26.12b. (a) Find the final potential difference ΔV_f between a and b after the switches are closed.

Solution   Let us identify the left-hand plates of the capacitors as an isolated system because they are not connected to the right-hand plates by conductors. The charges on the left-hand plates before the switches are closed are

$$Q_{1i} = C_1 \,\Delta V_i \qquad \text{and} \qquad Q_{2i} = -C_2 \,\Delta V_i$$

The negative sign for Q_{2i} is necessary because the charge on the left plate of capacitor C_2 is negative. The total charge Q in the system is

$$(1) \qquad Q = Q_{1i} + Q_{2i} = (C_1 - C_2)\,\Delta V_i$$

After the switches are closed, the total charge in the system remains the same:

$$(2) \qquad Q = Q_{1f} + Q_{2f}$$

The charges redistribute until the entire system is at the same potential ΔV_f. Thus, the final potential difference across C_1 must be the same as the final potential difference across C_2. To satisfy this requirement, the charges on the capacitors after the switches are closed are

$$Q_{1f} = C_1 \,\Delta V_f \qquad \text{and} \qquad Q_{2f} = C_2 \,\Delta V_f$$

Dividing the first equation by the second, we have

$$\frac{Q_{1f}}{Q_{2f}} = \frac{C_1 \,\Delta V_f}{C_2 \,\Delta V_f} = \frac{C_1}{C_2}$$

$$(3) \qquad Q_{1f} = \frac{C_1}{C_2}\, Q_{2f}$$

Combining Equations (2) and (3), we obtain

$$Q = Q_{1f} + Q_{2f} = \frac{C_1}{C_2}\, Q_{2f} + Q_{2f} = Q_{2f}\left(1 + \frac{C_1}{C_2}\right)$$

$$Q_{2f} = Q\left(\frac{C_2}{C_1 + C_2}\right)$$

Using Equation (3) to find Q_{1f} in terms of Q, we have

$$Q_{1f} = \frac{C_1}{C_2}\, Q_{2f} = \frac{C_1}{C_2}\, Q\left(\frac{C_2}{C_1 + C_2}\right) = Q\left(\frac{C_1}{C_1 + C_2}\right)$$

Finally, using Equation 26.1 to find the voltage across each capacitor, we find that

$$\Delta V_{1f} = \frac{Q_{1f}}{C_1} = \frac{Q\left(\dfrac{C_1}{C_1 + C_2}\right)}{C_1} = \frac{Q}{C_1 + C_2}$$

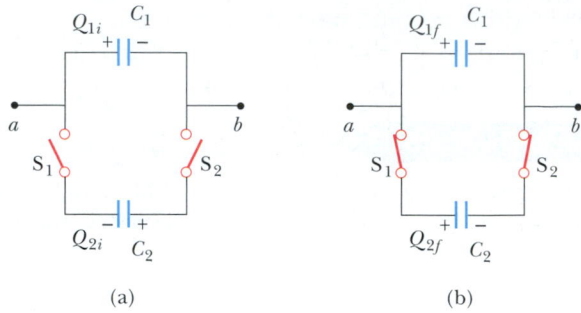

Figure 26.12

$$\Delta V_{2f} = \frac{Q_{2f}}{C_2} = \frac{Q\left(\dfrac{C_2}{C_1 + C_2}\right)}{C_2} = \frac{Q}{C_1 + C_2}$$

As noted earlier, $\Delta V_{1f} = \Delta V_{2f} = \Delta V_f$.

To express ΔV_f in terms of the given quantities C_1, C_2, and ΔV_i, we substitute the value of Q from Equation (1) to obtain

$$\Delta V_f = \left(\frac{C_1 - C_2}{C_1 + C_2}\right)\Delta V_i$$

(b) Find the total energy stored in the capacitors before and after the switches are closed and the ratio of the final energy to the initial energy.

Solution   Before the switches are closed, the total energy stored in the capacitors is

$$U_i = \tfrac{1}{2}C_1(\Delta V_i)^2 + \tfrac{1}{2}C_2(\Delta V_i)^2 = \tfrac{1}{2}(C_1 + C_2)(\Delta V_i)^2$$

After the switches are closed, the total energy stored in the capacitors is

$$U_f = \tfrac{1}{2}C_1(\Delta V_f)^2 + \tfrac{1}{2}C_2(\Delta V_f)^2 = \tfrac{1}{2}(C_1 + C_2)(\Delta V_f)^2$$

$$= \frac{1}{2}(C_1 + C_2)\left(\frac{Q}{C_1 + C_2}\right)^2 = \frac{1}{2}\frac{Q^2}{C_1 + C_2}$$

Using Equation (1), we can express this as

$$U_f = \frac{1}{2}\frac{Q^2}{(C_1 + C_2)} = \frac{1}{2}\frac{(C_1 - C_2)^2(\Delta V_i)^2}{(C_1 + C_2)}$$

Therefore, the ratio of the final energy stored to the initial energy stored is

$$\frac{U_f}{U_i} = \frac{\dfrac{1}{2}\dfrac{(C_1 - C_2)^2(\Delta V_i)^2}{(C_1 + C_2)}}{\dfrac{1}{2}(C_1 + C_2)(\Delta V_i)^2} = \left(\frac{C_1 - C_2}{C_1 + C_2}\right)^2$$

This ratio is less than unity, indicating that the final energy is less than the initial energy. At first, you might think that the law of energy conservation has been violated, but this is not the case. The "missing" energy is radiated away in the form of electromagnetic waves, as we shall see in Chapter 34.

Quick Quiz 26.4

You charge a parallel-plate capacitor, remove it from the battery, and prevent the wires connected to the plates from touching each other. When you pull the plates apart, do the following quantities increase, decrease, or stay the same? (a) C; (b) Q; (c) E between the plates; (d) ΔV; (e) energy stored in the capacitor.

Quick Quiz 26.5

Repeat Quick Quiz 26.4, but this time answer the questions for the situation in which the battery remains connected to the capacitor while you pull the plates apart.

One device in which capacitors have an important role is the *defibrillator* (Fig. 26.13). Up to 360 J is stored in the electric field of a large capacitor in a defibrillator when it is fully charged. The defibrillator can deliver all this energy to a patient in about 2 ms. (This is roughly equivalent to 3 000 times the power output of a 60-W lightbulb!) The sudden electric shock stops the fibrillation (random contractions) of the heart that often accompany heart attacks and helps to restore the correct rhythm.

A camera's flash unit also uses a capacitor, although the total amount of energy stored is much less than that stored in a defibrillator. After the flash unit's capacitor is charged, tripping the camera's shutter causes the stored energy to be sent through a special lightbulb that briefly illuminates the subject being photographed.

web
To learn more about defibrillators, visit
www.physiocontrol.com

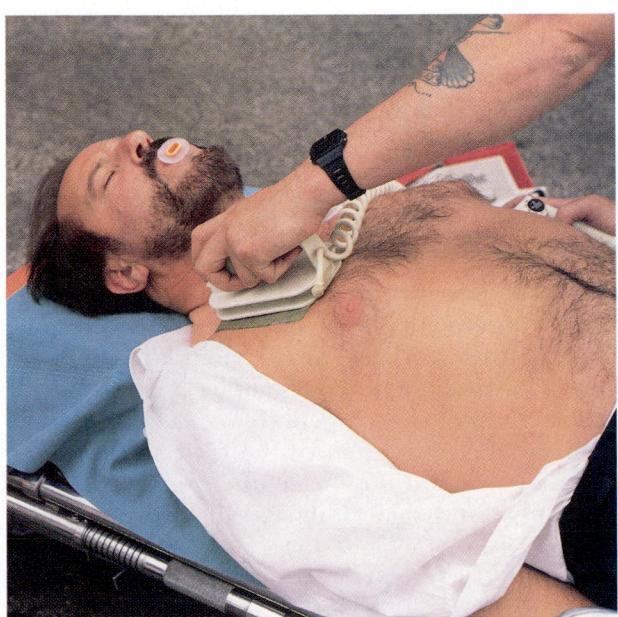

Figure 26.13 In a hospital or at an emergency scene, you might see a patient being revived with a defibrillator. The defibrillator's paddles are applied to the patient's chest, and an electric shock is sent through the chest cavity. The aim of this technique is to restore the heart's normal rhythm pattern. *(Adam Hart-Davis/Science Photo Library/Photo Researchers, Inc.)*

26.5 > CAPACITORS WITH DIELECTRICS

A **dielectric** is a nonconducting material, such as rubber, glass, or waxed paper. When a dielectric is inserted between the plates of a capacitor, the capacitance increases. If the dielectric completely fills the space between the plates, the capacitance increases by a dimensionless factor κ, which is called the **dielectric constant.** The dielectric constant is a property of a material and varies from one material to another. In this section, we analyze this change in capacitance in terms of electrical parameters such as electric charge, electric field, and potential difference; in Section 26.7, we shall discuss the microscopic origin of these changes.

We can perform the following experiment to illustrate the effect of a dielectric in a capacitor: Consider a parallel-plate capacitor that without a dielectric has a charge Q_0 and a capacitance C_0. The potential difference across the capacitor is $\Delta V_0 = Q_0/C_0$. Figure 26.14a illustrates this situation. The potential difference is measured by a *voltmeter*, which we shall study in greater detail in Chapter 28. Note that no battery is shown in the figure; also, we must assume that no charge can flow through an ideal voltmeter, as we shall learn in Section 28.5. Hence, there is no path by which charge can flow and alter the charge on the capacitor. If a dielectric is now inserted between the plates, as shown in Figure 26.14b, the voltmeter indicates that the voltage between the plates decreases to a value ΔV. The voltages with and without the dielectric are related by the factor κ as follows:

$$\Delta V = \frac{\Delta V_0}{\kappa}$$

Because $\Delta V < \Delta V_0$, we see that $\kappa > 1$.

Because the charge Q_0 on the capacitor does not change, we conclude that the capacitance must change to the value

$$C = \frac{Q_0}{\Delta V} = \frac{Q_0}{\Delta V_0/\kappa} = \kappa \frac{Q_0}{\Delta V_0}$$

The capacitance of a filled capacitor is greater than that of an empty one by a factor κ.

$$C = \kappa C_0 \tag{26.14}$$

That is, the capacitance *increases* by the factor κ when the dielectric completely fills the region between the plates.[4] For a parallel-plate capacitor, where $C_0 = \epsilon_0 A/d$ (Eq. 26.3), we can express the capacitance when the capacitor is filled with a dielectric as

$$C = \kappa \frac{\epsilon_0 A}{d} \tag{26.15}$$

From Equations 26.3 and 26.15, it would appear that we could make the capacitance very large by decreasing d, the distance between the plates. In practice, the lowest value of d is limited by the electric discharge that could occur through the dielectric medium separating the plates. For any given separation d, the maximum voltage that can be applied to a capacitor without causing a discharge depends on the **dielectric strength** (maximum electric field) of the dielectric. If the magnitude of the electric field in the dielectric exceeds the dielectric strength, then the insulating properties break down and the dielectric begins to conduct. Insulating materials have values of κ greater than unity and dielectric strengths

[4] If the dielectric is introduced while the potential difference is being maintained constant by a battery, the charge increases to a value $Q = \kappa Q_0$. The additional charge is supplied by the battery, and the capacitance again increases by the factor κ.

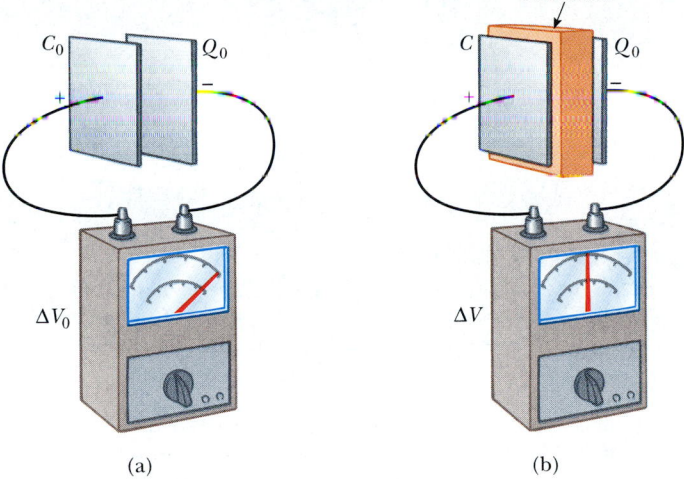

Figure 26.14 A charged capacitor (a) before and (b) after insertion of a dielectric between the plates. The charge on the plates remains unchanged, but the potential difference decreases from ΔV_0 to $\Delta V = \Delta V_0/\kappa$. Thus, the capacitance *increases* from C_0 to κC_0.

greater than that of air, as Table 26.1 indicates. Thus, we see that a dielectric provides the following advantages:

- Increase in capacitance
- Increase in maximum operating voltage
- Possible mechanical support between the plates, which allows the plates to be close together without touching, thereby decreasing d and increasing C

TABLE 26.1	Dielectric Constants and Dielectric Strengths of Various Materials at Room Temperature	
Material	**Dielectric Constant κ**	**Dielectric Strength[a] (V/m)**
Air (dry)	1.000 59	3×10^6
Bakelite	4.9	24×10^6
Fused quartz	3.78	8×10^6
Neoprene rubber	6.7	12×10^6
Nylon	3.4	14×10^6
Paper	3.7	16×10^6
Polystyrene	2.56	24×10^6
Polyvinyl chloride	3.4	40×10^6
Porcelain	6	12×10^6
Pyrex glass	5.6	14×10^6
Silicone oil	2.5	15×10^6
Strontium titanate	233	8×10^6
Teflon	2.1	60×10^6
Vacuum	1.000 00	—
Water	80	—

[a] The dielectric strength equals the maximum electric field that can exist in a dielectric without electrical breakdown. Note that these values depend strongly on the presence of impurities and flaws in the materials.

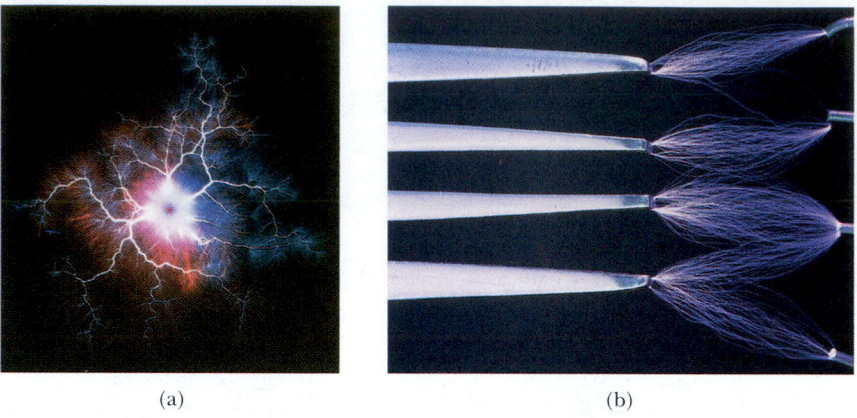

(a)   (b)

(a) Kirlian photograph created by dropping a steel ball into a high-energy electric field. Kirlian photography is also known as *electrophotography*. (b) Sparks from static electricity discharge between a fork and four electrodes. Many sparks were used to create this image because only one spark forms for a given discharge. Note that the bottom prong discharges to both electrodes at the bottom right. The light of each spark is created by the excitation of gas atoms along its path. *(a, Henry Dakin/Science Photo Library; b, Adam Hart-Davis/Science Photo Library)*

Types of Capacitors

Commercial capacitors are often made from metallic foil interlaced with thin sheets of either paraffin-impregnated paper or Mylar as the dielectric material. These alternate layers of metallic foil and dielectric are rolled into a cylinder to form a small package (Fig. 26.15a). High-voltage capacitors commonly consist of a number of interwoven metallic plates immersed in silicone oil (Fig. 26.15b). Small capacitors are often constructed from ceramic materials. Variable capacitors (typically 10 to 500 pF) usually consist of two interwoven sets of metallic plates, one fixed and the other movable, and contain air as the dielectric.

Often, an *electrolytic capacitor* is used to store large amounts of charge at relatively low voltages. This device, shown in Figure 26.15c, consists of a metallic foil in contact with an *electrolyte*—a solution that conducts electricity by virtue of the motion of ions contained in the solution. When a voltage is applied between the foil and the electrolyte, a thin layer of metal oxide (an insulator) is formed on the foil,

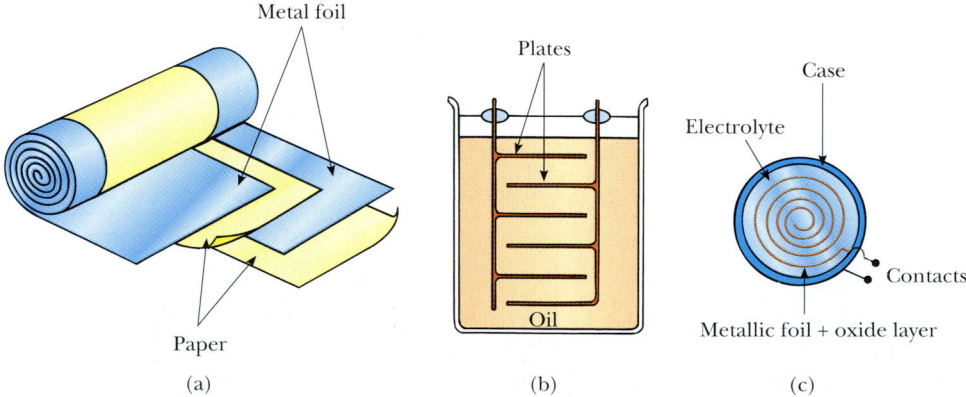

Figure 26.15 Three commercial capacitor designs. (a) A tubular capacitor, whose plates are separated by paper and then rolled into a cylinder. (b) A high-voltage capacitor consisting of many parallel plates separated by insulating oil. (c) An electrolytic capacitor.

and this layer serves as the dielectric. Very large values of capacitance can be obtained in an electrolytic capacitor because the dielectric layer is very thin, and thus the plate separation is very small.

Electrolytic capacitors are not reversible as are many other capacitors—they have a polarity, which is indicated by positive and negative signs marked on the device. When electrolytic capacitors are used in circuits, the polarity must be aligned properly. If the polarity of the applied voltage is opposite that which is intended, the oxide layer is removed and the capacitor conducts electricity instead of storing charge.

Quick Quiz 26.6

If you have ever tried to hang a picture, you know it can be difficult to locate a wooden stud in which to anchor your nail or screw. A carpenter's stud-finder is basically a capacitor with its plates arranged side by side instead of facing one another, as shown in Figure 26.16. When the device is moved over a stud, does the capacitance increase or decrease?

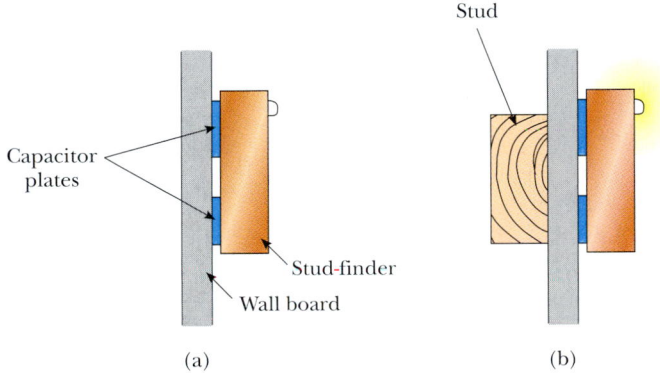

Figure 26.16   A stud-finder. (a) The materials between the plates of the capacitor are the wallboard and air. (b) When the capacitor moves across a stud in the wall, the materials between the plates are the wallboard and the wood. The change in the dielectric constant causes a signal light to illuminate.

EXAMPLE 26.6   **A Paper-Filled Capacitor**

A parallel-plate capacitor has plates of dimensions 2.0 cm by 3.0 cm separated by a 1.0-mm thickness of paper. (a) Find its capacitance.

Solution   Because $\kappa = 3.7$ for paper (see Table 26.1), we have

$$C = \kappa \frac{\epsilon_0 A}{d} = 3.7(8.85 \times 10^{-12} \ C^2/N \cdot m^2)\left(\frac{6.0 \times 10^{-4} \ m^2}{1.0 \times 10^{-3} \ m}\right)$$

$$= 20 \times 10^{-12} \ F = \boxed{20 \ pF}$$

(b) What is the maximum charge that can be placed on the capacitor?

Solution   From Table 26.1 we see that the dielectric strength of paper is 16×10^6 V/m. Because the thickness of the paper is 1.0 mm, the maximum voltage that can be applied before breakdown is

$$\Delta V_{max} = E_{max}d = (16 \times 10^6 \ V/m)(1.0 \times 10^{-3} \ m)$$

$$= 16 \times 10^3 \ V$$

Hence, the maximum charge is

$$Q_{max} = C\Delta V_{max} = (20 \times 10^{-12} \ F)(16 \times 10^3 \ V) = \boxed{0.32 \ \mu C}$$

Exercise   What is the maximum energy that can be stored in the capacitor?

Answer 2.6×10^{-3} J.

EXAMPLE 26.7   Energy Stored Before and After

A parallel-plate capacitor is charged with a battery to a charge Q_0, as shown in Figure 26.17a. The battery is then removed, and a slab of material that has a dielectric constant κ is inserted between the plates, as shown in Figure 26.17b. Find the energy stored in the capacitor before and after the dielectric is inserted.

Solution   The energy stored in the absence of the dielectric is (see Eq. 26.11):

$$U_0 = \frac{Q_0{}^2}{2C_0}$$

After the battery is removed and the dielectric inserted, the *charge on the capacitor remains the same*. Hence, the energy stored in the presence of the dielectric is

$$U = \frac{Q_0{}^2}{2C}$$

But the capacitance in the presence of the dielectric is $C = \kappa C_0$, so U becomes

$$U = \frac{Q_0{}^2}{2\kappa C_0} = \frac{U_0}{\kappa}$$

Because $\kappa > 1$, the final energy is less than the initial energy. We can account for the "missing" energy by noting that the dielectric, when inserted, gets pulled into the device (see the following discussion and Figure 26.18). An external agent must do negative work to keep the dielectric from accelerating. This work is simply the difference $U - U_0$. (Alternatively, the positive work done by the system on the external agent is $U_0 - U$.)

Exercise   Suppose that the capacitance in the absence of a dielectric is 8.50 pF and that the capacitor is charged to a potential difference of 12.0 V. If the battery is disconnected and a slab of polystyrene is inserted between the plates, what is $U_0 - U$?

Answer   373 pJ.

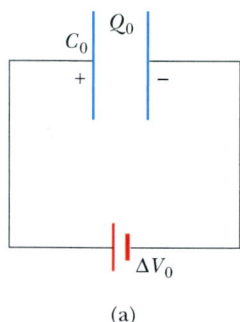

(a)

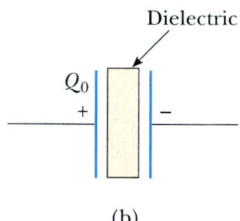

(b)

Figure 26.17

As we have seen, the energy of a capacitor not connected to a battery is lowered when a dielectric is inserted between the plates; this means that negative work is done on the dielectric by the external agent inserting the dielectric into the capacitor. This, in turn, implies that a force that draws it into the capacitor must be acting on the dielectric. This force originates from the nonuniform nature of the electric field of the capacitor near its edges, as indicated in Figure 26.18. The horizontal component of this *fringe field* acts on the induced charges on the surface of the dielectric, producing a net horizontal force directed into the space between the capacitor plates.

Quick Quiz 26.7

A fully charged parallel-plate capacitor remains connected to a battery while you slide a dielectric between the plates. Do the following quantities increase, decrease, or stay the same? (a) C; (b) Q; (c) E between the plates; (d) ΔV; (e) energy stored in the capacitor.

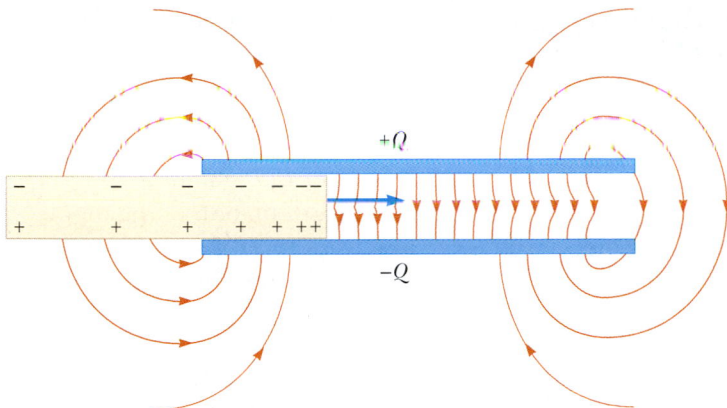

Figure 26.18   The nonuniform electric field near the edges of a parallel-plate capacitor causes a dielectric to be pulled into the capacitor. Note that the field acts on the induced surface charges on the dielectric, which are nonuniformly distributed.

Optional Section

26.6 ⟩ ELECTRIC DIPOLE IN AN ELECTRIC FIELD

We have discussed the effect on the capacitance of placing a dielectric between the plates of a capacitor. In Section 26.7, we shall describe the microscopic origin of this effect. Before we can do so, however, we need to expand upon the discussion of the electric dipole that we began in Section 23.4 (see Example 23.6). The electric dipole consists of two charges of equal magnitude but opposite sign separated by a distance $2a$, as shown in Figure 26.19. The **electric dipole moment** of this configuration is defined as the vector **p** directed from $-q$ to $+q$ along the line joining the charges and having magnitude $2aq$:

$$p \equiv 2aq \tag{26.16}$$

Now suppose that an electric dipole is placed in a uniform electric field **E**, as shown in Figure 26.20. We identify **E** as the field *external* to the dipole, distinguishing it from the field *due to* the dipole, which we discussed in Section 23.4. The field **E** is established by some other charge distribution, and we place the dipole into this field. Let us imagine that the dipole moment makes an angle θ with the field.

The electric forces acting on the two charges are equal in magnitude but opposite in direction as shown in Figure 26.20 (each has a magnitude $F = qE$). Thus, the net force on the dipole is zero. However, the two forces produce a net torque on the dipole; as a result, the dipole rotates in the direction that brings the dipole moment vector into greater alignment with the field. The torque due to the force on the positive charge about an axis through O in Figure 26.20 is $Fa \sin \theta$, where $a \sin \theta$ is the moment arm of F about O. This force tends to produce a clockwise rotation. The torque about O on the negative charge also is $Fa \sin \theta$; here again, the force tends to produce a clockwise rotation. Thus, the net torque about O is

$$\tau = 2Fa \sin \theta$$

Because $F = qE$ and $p = 2aq$, we can express τ as

$$\tau = 2aqE \sin \theta = pE \sin \theta \tag{26.17}$$

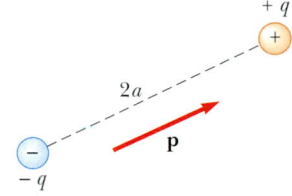

Figure 26.19   An electric dipole consists of two charges of equal magnitude but opposite sign separated by a distance of $2a$. The electric dipole moment **p** is directed from $-q$ to $+q$.

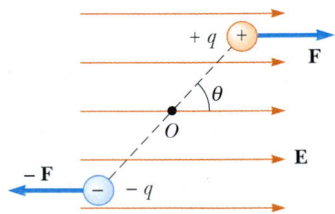

Figure 26.20   An electric dipole in a uniform external electric field. The dipole moment **p** is at an angle θ to the field, causing the dipole to experience a torque.

It is convenient to express the torque in vector form as the cross product of the vectors **p** and **E**:

$$\boldsymbol{\tau} = \mathbf{p} \times \mathbf{E} \tag{26.18}$$

We can determine the potential energy of the system of an electric dipole in an external electric field as a function of the orientation of the dipole with respect to the field. To do this, we recognize that work must be done by an external agent to rotate the dipole through an angle so as to cause the dipole moment vector to become less aligned with the field. The work done is then stored as potential energy in the system of the dipole and the external field. The work dW required to rotate the dipole through an angle $d\theta$ is $dW = \tau\, d\theta$ (Eq. 10.22). Because $\tau = pE \sin\theta$ and because the work is transformed into potential energy U, we find that, for a rotation from θ_i to θ_f, the change in potential energy is

$$U_f - U_i = \int_{\theta_i}^{\theta_f} \tau\, d\theta = \int_{\theta_i}^{\theta_f} pE \sin\theta\, d\theta = pE \int_{\theta_i}^{\theta_f} \sin\theta\, d\theta$$

$$= pE\left[-\cos\theta\right]_{\theta_i}^{\theta_f} = pE(\cos\theta_i - \cos\theta_f)$$

The term that contains $\cos\theta_i$ is a constant that depends on the initial orientation of the dipole. It is convenient for us to choose $\theta_i = 90°$, so that $\cos\theta_i = \cos 90° = 0$. Furthermore, let us choose $U_i = 0$ at $\theta_i = 90°$ as our reference of potential energy. Hence, we can express a general value of $U = U_f$ as

$$U = -pE \cos\theta \tag{26.19}$$

We can write this expression for the potential energy of a dipole in an electric field as the dot product of the vectors **p** and **E**:

$$U = -\mathbf{p} \cdot \mathbf{E} \tag{26.20}$$

To develop a conceptual understanding of Equation 26.19, let us compare this expression with the expression for the potential energy of an object in the gravitational field of the Earth, $U = mgh$ (see Chapter 8). The gravitational expression includes a parameter associated with the object we place in the field—its mass m. Likewise, Equation 26.19 includes a parameter of the object in the electric field—its dipole moment p. The gravitational expression includes the magnitude of the gravitational field g. Similarly, Equation 26.19 includes the magnitude of the electric field E. So far, these two contributions to the potential energy expressions appear analogous. However, the final contribution is somewhat different in the two cases. In the gravitational expression, the potential energy depends on how high we lift the object, measured by h. In Equation 26.19, the potential energy depends on the angle θ through which we rotate the dipole. In both cases, we are making a change in the system. In the gravitational case, the change involves moving an object in a *translational* sense, whereas in the electrical case, the change involves moving an object in a *rotational* sense. In both cases, however, once the change is made, the system tends to return to the original configuration when the object is released: the object of mass m falls back to the ground, and the dipole begins to rotate back toward the configuration in which it was aligned with the field. Thus, apart from the type of motion, the expressions for potential energy in these two cases are similar.

Molecules are said to be *polarized* when a separation exists between the average position of the negative charges and the average position of the positive charges in the molecule. In some molecules, such as water, this condition is always present—such molecules are called **polar molecules.** Molecules that do not possess a permanent polarization are called **nonpolar molecules.**

We can understand the permanent polarization of water by inspecting the geometry of the water molecule. In the water molecule, the oxygen atom is bonded to the hydrogen atoms such that an angle of 105° is formed between the two bonds (Fig. 26.21). The center of the negative charge distribution is near the oxygen atom, and the center of the positive charge distribution lies at a point midway along the line joining the hydrogen atoms (the point labeled × in Fig. 26.21). We can model the water molecule and other polar molecules as dipoles because the average positions of the positive and negative charges act as point charges. As a result, we can apply our discussion of dipoles to the behavior of polar molecules.

Microwave ovens take advantage of the polar nature of the water molecule. When in operation, microwave ovens generate a rapidly changing electric field that causes the polar molecules to swing back and forth, absorbing energy from the field in the process. Because the jostling molecules collide with each other, the energy they absorb from the field is converted to internal energy, which corresponds to an increase in temperature of the food.

Another household scenario in which the dipole structure of water is exploited is washing with soap and water. Grease and oil are made up of nonpolar molecules, which are generally not attracted to water. Plain water is not very useful for removing this type of grime. Soap contains long molecules called *surfactants*. In a long molecule, the polarity characteristics of one end of the molecule can be different from those at the other end. In a surfactant molecule, one end acts like a nonpolar molecule and the other acts like a polar molecule. The nonpolar end can attach to a grease or oil molecule, and the polar end can attach to a water molecule. Thus, the soap serves as a chain, linking the dirt and water molecules together. When the water is rinsed away, the grease and oil go with it.

A symmetric molecule (Fig. 26.22a) has no permanent polarization, but polarization can be induced by placing the molecule in an electric field. A field directed to the left, as shown in Figure 26.22b, would cause the center of the positive charge distribution to shift to the left from its initial position and the center of the negative charge distribution to shift to the right. This *induced polarization* is the effect that predominates in most materials used as dielectrics in capacitors.

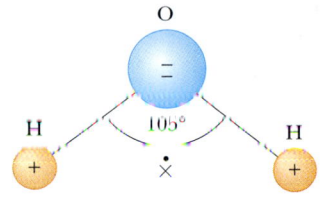

Figure 26.21   The water molecule, H_2O, has a permanent polarization resulting from its bent geometry. The center of the positive charge distribution is at the point ×.

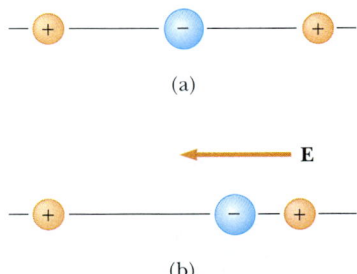

Figure 26.22   (a) A symmetric molecule has no permanent polarization. (b) An external electric field induces a polarization in the molecule.

EXAMPLE 26.8 ▸ The H_2O Molecule

The water (H_2O) molecule has an electric dipole moment of 6.3×10^{-30} C·m. A sample contains 10^{21} water molecules, with the dipole moments all oriented in the direction of an electric field of magnitude 2.5×10^5 N/C. How much work is required to rotate the dipoles from this orientation ($\theta = 0°$) to one in which all the dipole moments are perpendicular to the field ($\theta = 90°$)?

Solution   The work required to rotate one molecule 90° is equal to the difference in potential energy between the 90° orientation and the 0° orientation. Using Equation 26.19, we

obtain

$$W = U_{90} - U_0 = (-pE \cos 90°) - (-pE \cos 0°)$$
$$= pE = (6.3 \times 10^{-30} \text{ C·m})(2.5 \times 10^5 \text{ N/C})$$
$$= 1.6 \times 10^{-24} \text{ J}$$

Because there are 10^{21} molecules in the sample, the *total* work required is

$$W_{\text{total}} = (10^{21})(1.6 \times 10^{-24} \text{ J}) = \boxed{1.6 \times 10^{-3} \text{ J}}$$

Optional Section

26.7 AN ATOMIC DESCRIPTION OF DIELECTRICS

In Section 26.5 we found that the potential difference ΔV_0 between the plates of a capacitor is reduced to $\Delta V_0/\kappa$ when a dielectric is introduced. Because the potential difference between the plates equals the product of the electric field and the separation d, the electric field is also reduced. Thus, if $\mathbf{E}_0$ is the electric field without the dielectric, the field in the presence of a dielectric is

$$\mathbf{E} = \frac{\mathbf{E}_0}{\kappa} \qquad\qquad \text{(26.21)}$$

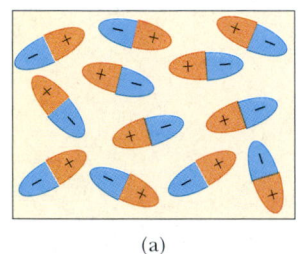

(a)

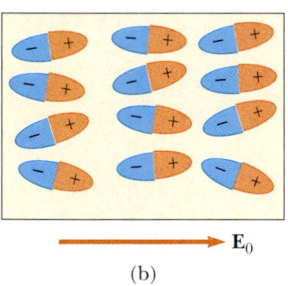

$\mathbf{E}_0$

(b)

Figure 26.23 (a) Polar molecules are randomly oriented in the absence of an external electric field. (b) When an external field is applied, the molecules partially align with the field.

Let us first consider a dielectric made up of polar molecules placed in the electric field between the plates of a capacitor. The dipoles (that is, the polar molecules making up the dielectric) are randomly oriented in the absence of an electric field, as shown in Figure 26.23a. When an external field $\mathbf{E}_0$ due to charges on the capacitor plates is applied, a torque is exerted on the dipoles, causing them to partially align with the field, as shown in Figure 26.23b. We can now describe the dielectric as being polarized. The degree of alignment of the molecules with the electric field depends on temperature and on the magnitude of the field. In general, the alignment increases with decreasing temperature and with increasing electric field.

If the molecules of the dielectric are nonpolar, then the electric field due to the plates produces some charge separation and an *induced dipole moment*. These induced dipole moments tend to align with the external field, and the dielectric is polarized. Thus, we can polarize a dielectric with an external field regardless of whether the molecules are polar or nonpolar.

With these ideas in mind, consider a slab of dielectric material placed between the plates of a capacitor so that it is in a uniform electric field $\mathbf{E}_0$, as shown in Figure 26.24a. The electric field due to the plates is directed to the right and polarizes the dielectric. The net effect on the dielectric is the formation of an *induced positive surface charge density* σ_{ind} on the right face and an equal negative surface charge density $-\sigma_{ind}$ on the left face, as shown in Figure 26.24b. These induced surface charges on the dielectric give rise to an induced electric field $\mathbf{E}_{ind}$ in the direction opposite the external field $\mathbf{E}_0$. Therefore, the net electric field $\mathbf{E}$ in the

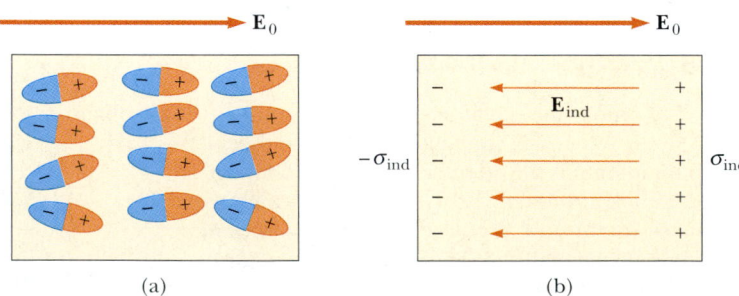

(a)   (b)

Figure 26.24 (a) When a dielectric is polarized, the dipole moments of the molecules in the dielectric are partially aligned with the external field $\mathbf{E}_0$. (b) This polarization causes an induced negative surface charge on one side of the dielectric and an equal induced positive surface charge on the opposite side. This separation of charge results in a reduction in the net electric field within the dielectric.

dielectric has a magnitude

$$E = E_0 - E_{ind} \qquad \textbf{(26.22)}$$

In the parallel-plate capacitor shown in Figure 26.25, the external field E_0 is related to the charge density σ on the plates through the relationship $E_0 = \sigma/\epsilon_0$. The induced electric field in the dielectric is related to the induced charge density σ_{ind} through the relationship $E_{ind} = \sigma_{ind}/\epsilon_0$. Because $E = E_0/\kappa = \sigma/\kappa\epsilon_0$, substitution into Equation 26.22 gives

$$\frac{\sigma}{\kappa\epsilon_0} = \frac{\sigma}{\epsilon_0} - \frac{\sigma_{ind}}{\epsilon_0}$$

$$\sigma_{ind} = \left(\frac{\kappa - 1}{\kappa}\right)\sigma \qquad \textbf{(26.23)}$$

Because $\kappa > 1$, this expression shows that the charge density σ_{ind} induced on the dielectric is less than the charge density σ on the plates. For instance, if $\kappa = 3$, we see that the induced charge density is two-thirds the charge density on the plates. If no dielectric is present, then $\kappa = 1$ and $\sigma_{ind} = 0$ as expected. However, if the dielectric is replaced by an electrical conductor, for which $E = 0$, then Equation 26.22 indicates that $E_0 = E_{ind}$; this corresponds to $\sigma_{ind} = \sigma$. That is, the surface charge induced on the conductor is equal in magnitude but opposite in sign to that on the plates, resulting in a net electric field of zero in the conductor.

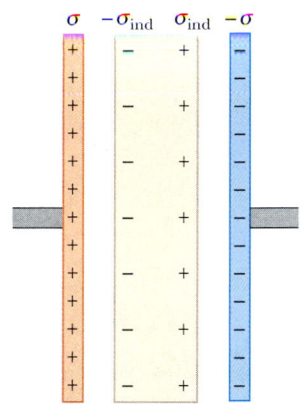

Figure 26.25 Induced charge on a dielectric placed between the plates of a charged capacitor. Note that the induced charge density on the dielectric is *less* than the charge density on the plates.

EXAMPLE 26.9 Effect of a Metallic Slab

A parallel-plate capacitor has a plate separation d and plate area A. An uncharged metallic slab of thickness a is inserted midway between the plates. (a) Find the capacitance of the device.

Solution We can solve this problem by noting that any charge that appears on one plate of the capacitor must induce a charge of equal magnitude but opposite sign on the near side of the slab, as shown in Figure 26.26a. Consequently, the net charge on the slab remains zero, and the electric field inside the slab is zero. Hence, the capacitor is equivalent to two capacitors in series, each having a plate separation $(d - a)/2$, as shown in Figure 26.26b.

Using the rule for adding two capacitors in series (Eq. 26.10), we obtain

$$\frac{1}{C} = \frac{1}{C_1} + \frac{1}{C_2} = \frac{1}{\dfrac{\epsilon_0 A}{(d - a)/2}} + \frac{1}{\dfrac{\epsilon_0 A}{(d - a)/2}}$$

$$C = \frac{\epsilon_0 A}{d - a}$$

Note that C approaches infinity as a approaches d. Why?

(b) Show that the capacitance is unaffected if the metallic slab is infinitesimally thin.

Solution In the result for part (a), we let $a \to 0$:

$$C = \lim_{a \to 0} \frac{\epsilon_0 A}{d - a} = \frac{\epsilon_0 A}{d}$$

which is the original capacitance.

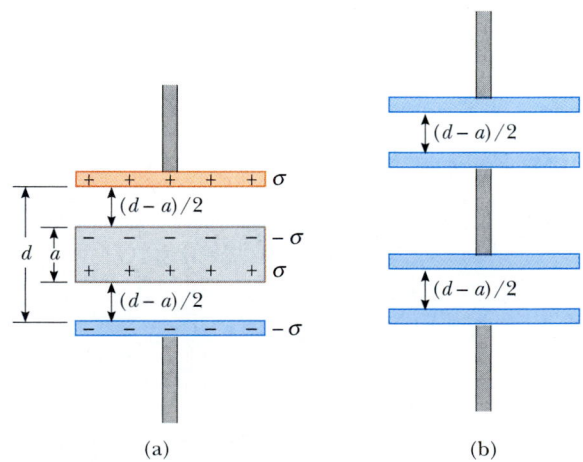

(a)   (b)

Figure 26.26 (a) A parallel-plate capacitor of plate separation d partially filled with a metallic slab of thickness a. (b) The equivalent circuit of the device in part (a) consists of two capacitors in series, each having a plate separation $(d - a)/2$.

(c) Show that the answer to part (a) does not depend on where the slab is inserted.

Solution Let us imagine that the slab in Figure 26.26a is moved upward so that the distance between the upper edge of the slab and the upper plate is b. Then, the distance between the lower edge of the slab and the lower plate is $d - b - a$. As in part (a), we find the total capacitance of the series combination:

$$\frac{1}{C} = \frac{1}{C_1} + \frac{1}{C_2} = \frac{1}{\dfrac{\epsilon_0 A}{b}} + \frac{1}{\dfrac{\epsilon_0 A}{d - b - a}}$$

$$= \frac{b}{\epsilon_0 A} + \frac{d - b - a}{\epsilon_0 A} = \frac{d - a}{\epsilon_0 A}$$

$$C = \frac{\epsilon_0 A}{d - a}$$

This is the same result as in part (a). It is independent of the value of b, so it does not matter where the slab is located.

EXAMPLE 26.10   A Partially Filled Capacitor

A parallel-plate capacitor with a plate separation d has a capacitance C_0 in the absence of a dielectric. What is the capacitance when a slab of dielectric material of dielectric constant κ and thickness $\frac{1}{3}d$ is inserted between the plates (Fig. 26.27a)?

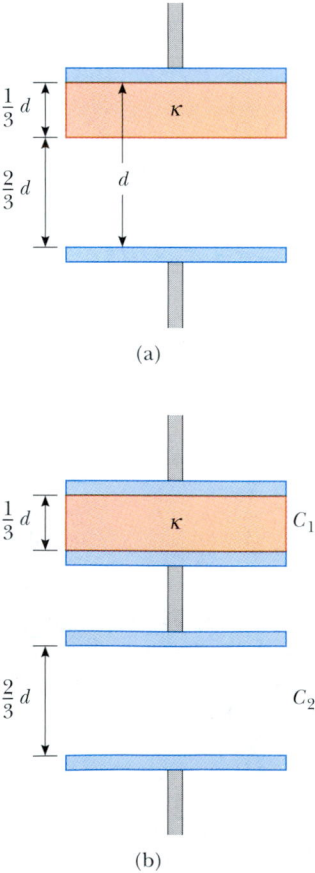

(a)

(b)

Figure 26.27 (a) A parallel-plate capacitor of plate separation d partially filled with a dielectric of thickness $d/3$. (b) The equivalent circuit of the capacitor consists of two capacitors connected in series.

Solution In Example 26.9, we found that we could insert a metallic slab between the plates of a capacitor and consider the combination as two capacitors in series. The resulting capacitance was independent of the location of the slab. Furthermore, if the thickness of the slab approaches zero, then the capacitance of the system approaches the capacitance when the slab is absent. From this, we conclude that we can insert an infinitesimally thin metallic slab anywhere between the plates of a capacitor without affecting the capacitance. Thus, let us imagine sliding an infinitesimally thin metallic slab along the bottom face of the dielectric shown in Figure 26.27a. We can then consider this system to be the series combination of the two capacitors shown in Figure 26.27b: one having a plate separation $d/3$ and filled with a dielectric, and the other having a plate separation $2d/3$ and air between its plates.

From Equations 26.15 and 26.3, the two capacitances are

$$C_1 = \frac{\kappa \epsilon_0 A}{d/3} \quad \text{and} \quad C_2 = \frac{\epsilon_0 A}{2d/3}$$

Using Equation 26.10 for two capacitors combined in series, we have

$$\frac{1}{C} = \frac{1}{C_1} + \frac{1}{C_2} = \frac{d/3}{\kappa \epsilon_0 A} + \frac{2d/3}{\epsilon_0 A}$$

$$= \frac{d}{3\epsilon_0 A}\left(\frac{1}{\kappa} + 2\right) = \frac{d}{3\epsilon_0 A}\left(\frac{1 + 2\kappa}{\kappa}\right)$$

$$C = \left(\frac{3\kappa}{2\kappa + 1}\right)\frac{\epsilon_0 A}{d}$$

Because the capacitance without the dielectric is $C_0 = \epsilon_0 A/d$, we see that

$$C = \left(\frac{3\kappa}{2\kappa + 1}\right)C_0$$

SUMMARY

A **capacitor** consists of two conductors carrying charges of equal magnitude but opposite sign. The **capacitance** C of any capacitor is the ratio of the charge Q on either conductor to the potential difference ΔV between them:

$$C \equiv \frac{Q}{\Delta V} \tag{26.1}$$

This relationship can be used in situations in which any two of the three variables are known. It is important to remember that this ratio is constant for a given configuration of conductors because the capacitance depends only on the geometry of the conductors and not on an external source of charge or potential difference.

The SI unit of capacitance is coulombs per volt, or the **farad** (F), and $1\ \text{F} = 1\ \text{C/V}$.

Capacitance expressions for various geometries are summarized in Table 26.2.

If two or more capacitors are connected in parallel, then the potential difference is the same across all of them. The equivalent capacitance of a parallel combination of capacitors is

$$C_{\text{eq}} = C_1 + C_2 + C_3 + \cdots \tag{26.8}$$

If two or more capacitors are connected in series, the charge is the same on all of them, and the equivalent capacitance of the series combination is given by

$$\frac{1}{C_{\text{eq}}} = \frac{1}{C_1} + \frac{1}{C_2} + \frac{1}{C_3} + \cdots \tag{26.10}$$

These two equations enable you to simplify many electric circuits by replacing multiple capacitors with a single equivalent capacitance.

Work is required to charge a capacitor because the charging process is equivalent to the transfer of charges from one conductor at a lower electric potential to another conductor at a higher potential. The work done in charging the capacitor to a charge Q equals the electric potential energy U stored in the capacitor, where

$$U = \frac{Q^2}{2C} = \tfrac{1}{2} Q \Delta V = \tfrac{1}{2} C (\Delta V)^2 \tag{26.11}$$

TABLE 26.2 Capacitance and Geometry		
Geometry	**Capacitance**	**Equation**
Isolated charged sphere of radius R (second charged conductor assumed at infinity)	$C = 4\pi\epsilon_0 R$	26.2
Parallel-plate capacitor of plate area A and plate separation d	$C = \epsilon_0 \dfrac{A}{d}$	26.3
Cylindrical capacitor of length ℓ and inner and outer radii a and b, respectively	$C = \dfrac{\ell}{2k_e \ln\left(\dfrac{b}{a}\right)}$	26.4
Spherical capacitor with inner and outer radii a and b, respectively	$C = \dfrac{ab}{k_e (b - a)}$	26.6

When a dielectric material is inserted between the plates of a capacitor, the capacitance increases by a dimensionless factor κ, called the **dielectric constant:**

$$C = \kappa C_0 \qquad (26.14)$$

where C_0 is the capacitance in the absence of the dielectric. The increase in capacitance is due to a decrease in the magnitude of the electric field in the presence of the dielectric and to a corresponding decrease in the potential difference between the plates—if we assume that the charging battery is removed from the circuit before the dielectric is inserted. The decrease in the magnitude of $\mathbf{E}$ arises from an internal electric field produced by aligned dipoles in the dielectric. This internal field produced by the dipoles opposes the applied field due to the capacitor plates, and the result is a reduction in the net electric field.

The **electric dipole moment p** of an electric dipole has a magnitude

$$p \equiv 2aq \qquad (26.16)$$

The direction of the electric dipole moment vector is from the negative charge toward the positive charge.

The torque acting on an electric dipole in a uniform electric field $\mathbf{E}$ is

$$\tau = \mathbf{p} \times \mathbf{E} \qquad (26.18)$$

The potential energy of an electric dipole in a uniform external electric field $\mathbf{E}$ is

$$U = -\mathbf{p} \cdot \mathbf{E} \qquad (26.20)$$

Problem-Solving Hints

Capacitors

- Be careful with units. When you calculate capacitance in farads, make sure that distances are expressed in meters and that you use the SI value of ϵ_0. When checking consistency of units, remember that the unit for electric fields can be either N/C or V/m.
- When two or more capacitors are connected in parallel, the potential difference across each is the same. The charge on each capacitor is proportional to its capacitance; hence, the capacitances can be added directly to give the equivalent capacitance of the parallel combination. The equivalent capacitance is always larger than the individual capacitances.
- When two or more capacitors are connected in series, they carry the same charge, and the sum of the potential differences equals the total potential difference applied to the combination. The sum of the reciprocals of the capacitances equals the reciprocal of the equivalent capacitance, which is always less than the capacitance of the smallest individual capacitor.
- A dielectric increases the capacitance of a capacitor by a factor κ (the dielectric constant) over its capacitance when air is between the plates.
- For problems in which a battery is being connected or disconnected, note whether modifications to the capacitor are made while it is connected to the battery or after it has been disconnected. If the capacitor remains connected to the battery, the voltage across the capacitor remains unchanged (equal to the battery voltage), and the charge is proportional to the capaci-

tance, although it may be modified (for instance, by the insertion of a dielectric). If you disconnect the capacitor from the battery before making any modifications to the capacitor, then its charge remains fixed. In this case, as you vary the capacitance, the voltage across the plates changes according to the expression $\Delta V = Q/C$.

QUESTIONS

1. If you were asked to design a capacitor in a situation for which small size and large capacitance were required, what factors would be important in your design?

2. The plates of a capacitor are connected to a battery. What happens to the charge on the plates if the connecting wires are removed from the battery? What happens to the charge if the wires are removed from the battery and connected to each other?

3. A farad is a very large unit of capacitance. Calculate the length of one side of a square, air-filled capacitor that has a plate separation of 1 m. Assume that it has a capacitance of 1 F.

4. A pair of capacitors are connected in parallel, while an identical pair are connected in series. Which pair would be more dangerous to handle after being connected to the same voltage source? Explain.

5. If you are given three different capacitors C_1, C_2, C_3, how many different combinations of capacitance can you produce?

6. What advantage might there be in using two identical capacitors in parallel connected in series with another identical parallel pair rather than a single capacitor?

7. Is it always possible to reduce a combination of capacitors to one equivalent capacitor with the rules we have developed? Explain.

8. Because the net charge in a capacitor is always zero, what does a capacitor store?

9. Because the charges on the plates of a parallel-plate capacitor are of opposite sign, they attract each other. Hence, it would take positive work to increase the plate separation. What happens to the external work done in this process?

10. Explain why the work needed to move a charge Q through a potential difference ΔV is $W = Q\Delta V$, whereas the energy stored in a charged capacitor is $U = \frac{1}{2}Q\Delta V$. Where does the $\frac{1}{2}$ factor come from?

11. If the potential difference across a capacitor is doubled, by what factor does the stored energy change?

12. Why is it dangerous to touch the terminals of a high-voltage capacitor even after the applied voltage has been turned off? What can be done to make the capacitor safe to handle after the voltage source has been removed?

13. Describe how you can increase the maximum operating voltage of a parallel-plate capacitor for a fixed plate separation.

14. An air-filled capacitor is charged, disconnected from the power supply, and, finally, connected to a voltmeter. Explain how and why the voltage reading changes when a dielectric is inserted between the plates of the capacitor.

15. Using the polar molecule description of a dielectric, explain how a dielectric affects the electric field inside a capacitor.

16. Explain why a dielectric increases the maximum operating voltage of a capacitor even though the physical size of the capacitor does not change.

17. What is the difference between dielectric strength and the dielectric constant?

18. Explain why a water molecule is permanently polarized. What type of molecule has no permanent polarization?

19. If a dielectric-filled capacitor is heated, how does its capacitance change? (Neglect thermal expansion and assume that the dipole orientations are temperature dependent.)

PROBLEMS

1, 2, 3 = straightforward, intermediate, challenging ☐ = full solution available in the *Student Solutions Manual and Study Guide*
WEB = solution posted at **http://www.saunderscollege.com/physics/** 🖥 = Computer useful in solving problem 📀 = Interactive Physics
☐ = paired numerical/symbolic problems

Section 26.1 Definition of Capacitance

1. (a) How much charge is on each plate of a 4.00-μF capacitor when it is connected to a 12.0-V battery? (b) If this same capacitor is connected to a 1.50-V battery, what charge is stored?

2. Two conductors having net charges of +10.0 μC and −10.0 μC have a potential difference of 10.0 V. Determine (a) the capacitance of the system and (b) the potential difference between the two conductors if the charges on each are increased to +100 μC and −100 μC.

Section 26.2   Calculating Capacitance

3. An isolated charged conducting sphere of radius 12.0 cm creates an electric field of 4.90×10^4 N/C at a distance 21.0 cm from its center. (a) What is its surface charge density? (b) What is its capacitance?

4. (a) If a drop of liquid has capacitance 1.00 pF, what is its radius? (b) If another drop has radius 2.00 mm, what is its capacitance? (c) What is the charge on the smaller drop if its potential is 100 V?

5. Two conducting spheres with diameters of 0.400 m and 1.00 m are separated by a distance that is large compared with the diameters. The spheres are connected by a thin wire and are charged to 7.00 μC. (a) How is this total charge shared between the spheres? (Neglect any charge on the wire.) (b) What is the potential of the system of spheres when the reference potential is taken to be $V = 0$ at $r = \infty$?

6. Regarding the Earth and a cloud layer 800 m above the Earth as the "plates" of a capacitor, calculate the capacitance if the cloud layer has an area of 1.00 km². Assume that the air between the cloud and the ground is pure and dry. Assume that charge builds up on the cloud and on the ground until a uniform electric field with a magnitude of 3.00×10^6 N/C throughout the space between them makes the air break down and conduct electricity as a lightning bolt. What is the maximum charge the cloud can hold?

WEB 7. An air-filled capacitor consists of two parallel plates, each with an area of 7.60 cm², separated by a distance of 1.80 mm. If a 20.0-V potential difference is applied to these plates, calculate (a) the electric field between the plates, (b) the surface charge density, (c) the capacitance, and (d) the charge on each plate.

8. A 1-megabit computer memory chip contains many 60.0-fF capacitors. Each capacitor has a plate area of 21.0×10^{-12} m². Determine the plate separation of such a capacitor (assume a parallel-plate configuration). The characteristic atomic diameter is 10^{-10} m = 0.100 nm. Express the plate separation in nanometers.

9. When a potential difference of 150 V is applied to the plates of a parallel-plate capacitor, the plates carry a surface charge density of 30.0 nC/cm². What is the spacing between the plates?

10. A variable air capacitor used in tuning circuits is made of N semicircular plates each of radius R and positioned a distance d from each other. As shown in Figure P26.10, a second identical set of plates is enmeshed with its plates halfway between those of the first set. The second set can rotate as a unit. Determine the capacitance as a function of the angle of rotation θ, where $\theta = 0$ corresponds to the maximum capacitance.

WEB 11. A 50.0-m length of coaxial cable has an inner conductor that has a diameter of 2.58 mm and carries a charge of 8.10 μC. The surrounding conductor has an inner diameter of 7.27 mm and a charge of $-$8.10 μC. (a) What is the capacitance of this cable? (b) What is

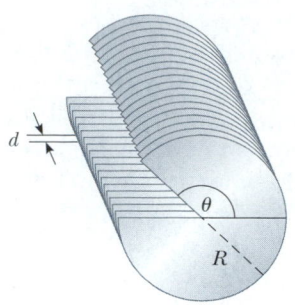

Figure P26.10

the potential difference between the two conductors? Assume the region between the conductors is air.

12. A 20.0-μF spherical capacitor is composed of two metallic spheres, one having a radius twice as large as the other. If the region between the spheres is a vacuum, determine the volume of this region.

13. A small object with a mass of 350 mg carries a charge of 30.0 nC and is suspended by a thread between the vertical plates of a parallel-plate capacitor. The plates are separated by 4.00 cm. If the thread makes an angle of 15.0° with the vertical, what is the potential difference between the plates?

14. A small object of mass m carries a charge q and is suspended by a thread between the vertical plates of a parallel-plate capacitor. The plate separation is d. If the thread makes an angle θ with the vertical, what is the potential difference between the plates?

15. An air-filled spherical capacitor is constructed with inner and outer shell radii of 7.00 and 14.0 cm, respectively. (a) Calculate the capacitance of the device. (b) What potential difference between the spheres results in a charge of 4.00 μC on the capacitor?

16. Find the capacitance of the Earth. (*Hint:* The outer conductor of the "spherical capacitor" may be considered as a conducting sphere at infinity where V approaches zero.)

Section 26.3   Combinations of Capacitors

17. Two capacitors $C_1 = 5.00$ μF and $C_2 = 12.0$ μF are connected in parallel, and the resulting combination is connected to a 9.00-V battery. (a) What is the value of the equivalent capacitance of the combination? What are (b) the potential difference across each capacitor and (c) the charge stored on each capacitor?

18. The two capacitors of Problem 17 are now connected in series and to a 9.00-V battery. Find (a) the value of the equivalent capacitance of the combination, (b) the voltage across each capacitor, and (c) the charge on each capacitor.

19. Two capacitors when connected in parallel give an equivalent capacitance of 9.00 pF and an equivalent ca-

pacitance of 2.00 pF when connected in series. What is the capacitance of each capacitor?

20. Two capacitors when connected in parallel give an equivalent capacitance of C_p and an equivalent capacitance of C_s when connected in series. What is the capacitance of each capacitor?

WEB **21.** Four capacitors are connected as shown in Figure P26.21. (a) Find the equivalent capacitance between points a and b. (b) Calculate the charge on each capacitor if $\Delta V_{ab} = 15.0$ V.

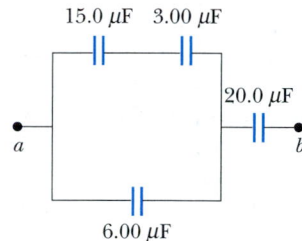

15.0 μF   3.00 μF
20.0 μF
a
b
6.00 μF

Figure P26.21

22. Evaluate the equivalent capacitance of the configuration shown in Figure P26.22. All the capacitors are identical, and each has capacitance C.

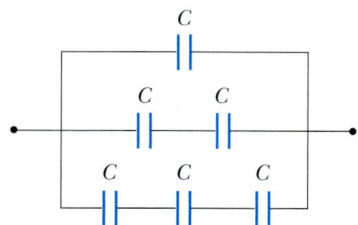

C
C C
C C C

Figure P26.22

23. Consider the circuit shown in Figure P26.23, where $C_1 = 6.00$ μF, $C_2 = 3.00$ μF, and $\Delta V = 20.0$ V. Capacitor C_1 is first charged by the closing of switch S_1. Switch S_1 is then opened, and the charged capacitor is connected to the uncharged capacitor by the closing of S_2. Calculate the initial charge acquired by C_1 and the final charge on each.

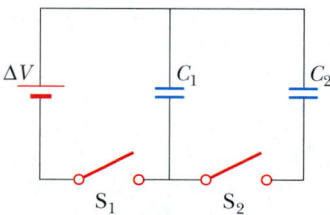

ΔV C_1 C_2
S_1 S_2

Figure P26.23

24. According to its design specification, the timer circuit delaying the closing of an elevator door is to have a capacitance of 32.0 μF between two points A and B. (a) When one circuit is being constructed, the inexpensive capacitor installed between these two points is found to have capacitance 34.8 μF. To meet the specification, one additional capacitor can be placed between the two points. Should it be in series or in parallel with the 34.8-μF capacitor? What should be its capacitance? (b) The next circuit comes down the assembly line with capacitance 29.8 μF between A and B. What additional capacitor should be installed in series or in parallel in that circuit, to meet the specification?

25. The circuit in Figure P26.25 consists of two identical parallel metallic plates connected by identical metallic springs to a 100-V battery. With the switch open, the plates are uncharged, are separated by a distance $d = 8.00$ mm, and have a capacitance $C = 2.00$ μF. When the switch is closed, the distance between the plates decreases by a factor of 0.500. (a) How much charge collects on each plate and (b) what is the spring constant for each spring? (*Hint:* Use the result of Problem 35.)

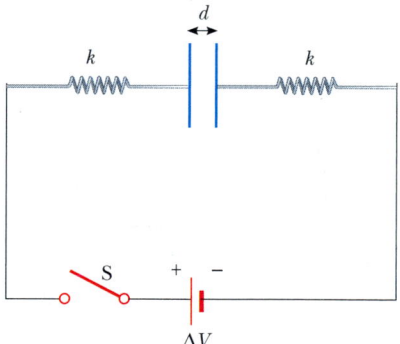

d
k k
S $+$ $-$
ΔV

Figure P26.25

26. Figure P26.26 shows six concentric conducting spheres, A, B, C, D, E, and F having radii R, $2R$, $3R$, $4R$, $5R$, and $6R$, respectively. Spheres B and C are connected by a conducting wire, as are spheres D and E. Determine the equivalent capacitance of this system.

27. A group of identical capacitors is connected first in series and then in parallel. The combined capacitance in parallel is 100 times larger than for the series connection. How many capacitors are in the group?

28. Find the equivalent capacitance between points a and b for the group of capacitors connected as shown in Figure P26.28 if $C_1 = 5.00$ μF, $C_2 = 10.0$ μF, and $C_3 = 2.00$ μF.

29. For the network described in the previous problem if the potential difference between points a and b is 60.0 V, what charge is stored on C_3?

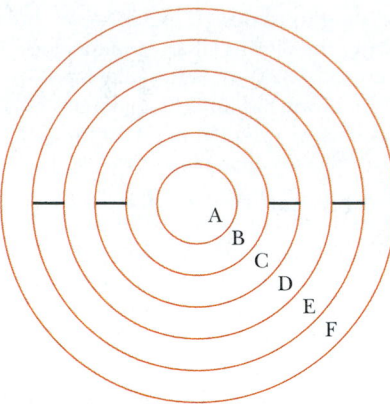

Figure P26.26

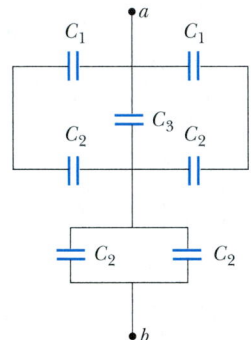

Figure P26.28   Problems 28 and 29.

30. Find the equivalent capacitance between points a and b in the combination of capacitors shown in Figure P26.30.

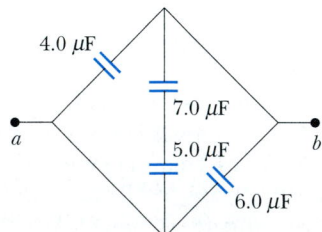

Figure P26.30

Section 26.4   Energy Stored in a Charged Capacitor

31. (a) A 3.00-μF capacitor is connected to a 12.0-V battery. How much energy is stored in the capacitor? (b) If the capacitor had been connected to a 6.00-V battery, how much energy would have been stored?

32. Two capacitors $C_1 = 25.0\ \mu$F and $C_2 = 5.00\ \mu$F are connected in parallel and charged with a 100-V power supply. (a) Draw a circuit diagram and calculate the total

energy stored in the two capacitors. (b) What potential difference would be required across the same two capacitors connected in series so that the combination stores the same energy as in part (a)? Draw a circuit diagram of this circuit.

33. A parallel-plate capacitor is charged and then disconnected from a battery. By what fraction does the stored energy change (increase or decrease) when the plate separation is doubled?

34. A uniform electric field $E = 3\ 000$ V/m exists within a certain region. What volume of space contains an energy equal to 1.00×10^{-7} J? Express your answer in cubic meters and in liters.

WEB **35.** A parallel-plate capacitor has a charge Q and plates of area A. Show that the force exerted on each plate by the other is $F = Q^2/2\epsilon_0 A$. (*Hint:* Let $C = \epsilon_0 A/x$ for an arbitrary plate separation x; then require that the work done in separating the two charged plates be $W = \int F\ dx$.)

36. Plate a of a parallel-plate, air-filled capacitor is connected to a spring having force constant k, and plate b is fixed. They rest on a table top as shown (top view) in Figure P26.36. If a charge $+Q$ is placed on plate a and a charge $-Q$ is placed on plate b, by how much does the spring expand?

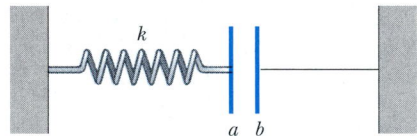

Figure P26.36

37. **Review Problem.** A certain storm cloud has a potential difference of 1.00×10^8 V relative to a tree. If, during a lightning storm, 50.0 C of charge is transferred through this potential difference and 1.00% of the energy is absorbed by the tree, how much water (sap in the tree) initially at 30.0°C can be boiled away? Water has a specific heat of 4 186 J/kg $\cdot$ °C, a boiling point of 100°C, and a heat of vaporization of 2.26×10^6 J/kg.

38. Show that the energy associated with a conducting sphere of radius R and charge Q surrounded by a vacuum is $U = k_e Q^2/2R$.

39. Einstein said that energy is associated with mass according to the famous relationship $E = mc^2$. Estimate the radius of an electron, assuming that its charge is distributed uniformly over the surface of a sphere of radius R and that the mass–energy of the electron is equal to the total energy stored in the resulting nonzero electric field between R and infinity. (See Problem 38. Experimentally, an electron nevertheless appears to be a point particle. The electric field close to the electron must be described by quantum electrodynamics, rather than the classical electrodynamics that we study.)

Section 26.5   Capacitors with Dielectrics

40. Find the capacitance of a parallel-plate capacitor that uses Bakelite as a dielectric, if each of the plates has an area of 5.00 cm^2 and the plate separation is 2.00 mm.

41. Determine (a) the capacitance and (b) the maximum voltage that can be applied to a Teflon-filled parallel-plate capacitor having a plate area of 1.75 cm^2 and plate separation of 0.040 0 mm.

42. (a) How much charge can be placed on a capacitor with air between the plates before it breaks down, if the area of each of the plates is 5.00 cm^2? (b) Find the maximum charge if polystyrene is used between the plates instead of air.

43. A commercial capacitor is constructed as shown in Figure 26.15a. This particular capacitor is rolled from two strips of aluminum separated by two strips of paraffin-coated paper. Each strip of foil and paper is 7.00 cm wide. The foil is 0.004 00 mm thick, and the paper is 0.025 0 mm thick and has a dielectric constant of 3.70. What length should the strips be if a capacitance of 9.50×10^{-8} F is desired? (Use the parallel-plate formula.)

44. The supermarket sells rolls of aluminum foil, plastic wrap, and waxed paper. Describe a capacitor made from supermarket materials. Compute order-of-magnitude estimates for its capacitance and its breakdown voltage.

45. A capacitor that has air between its plates is connected across a potential difference of 12.0 V and stores 48.0 μC of charge. It is then disconnected from the source while still charged. (a) Find the capacitance of the capacitor. (b) A piece of Teflon is inserted between the plates. Find its new capacitance. (c) Find the voltage and charge now on the capacitor.

46. A parallel-plate capacitor in air has a plate separation of 1.50 cm and a plate area of 25.0 cm^2. The plates are charged to a potential difference of 250 V and disconnected from the source. The capacitor is then immersed in distilled water. Determine (a) the charge on the plates before and after immersion, (b) the capacitance and voltage after immersion, and (c) the change in energy of the capacitor. Neglect the conductance of the liquid.

47. A conducting spherical shell has inner radius a and outer radius c. The space between these two surfaces is filled with a dielectric for which the dielectric constant is κ_1 between a and b, and κ_2 between b and c (Fig. P26.47). Determine the capacitance of this system.

48. A wafer of titanium dioxide ($\kappa = 173$) has an area of 1.00 cm^2 and a thickness of 0.100 mm. Aluminum is evaporated on the parallel faces to form a parallel-plate capacitor. (a) Calculate the capacitance. (b) When the capacitor is charged with a 12.0-V battery, what is the magnitude of charge delivered to each plate? (c) For the situation in part (b), what are the free and induced surface charge densities? (d) What is the magnitude E of the electric field?

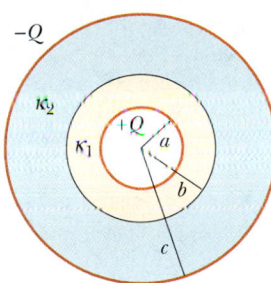

Figure P26.47

49. Each capacitor in the combination shown in Figure P26.49 has a breakdown voltage of 15.0 V. What is the breakdown voltage of the combination?

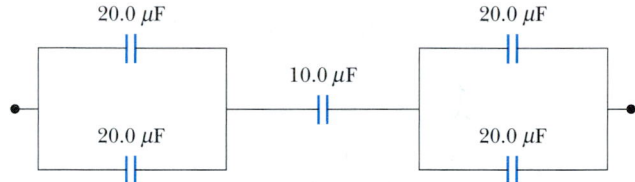

Figure P26.49

(Optional)
Section 26.6   Electric Dipole in an Electric Field

50. A small rigid object carries positive and negative 3.50-nC charges. It is oriented so that the positive charge is at the point $(-1.20$ mm, 1.10 mm$)$ and the negative charge is at the point $(1.40$ mm, -1.30 mm$)$. (a) Find the electric dipole moment of the object. The object is placed in an electric field $\mathbf{E} = (7\ 800\mathbf{i} - 4\ 900\mathbf{j})$ N/C. (b) Find the torque acting on the object. (c) Find the potential energy of the object in this orientation. (d) If the orientation of the object can change, find the difference between its maximum and its minimum potential energies.

51. A small object with electric dipole moment $\mathbf{p}$ is placed in a nonuniform electric field $\mathbf{E} = E(x)\ \mathbf{i}$. That is, the field is in the x direction, and its magnitude depends on the coordinate x. Let θ represent the angle between the dipole moment and the x direction. (a) Prove that the dipole experiences a net force $F = p(dE/dx) \cos\theta$ in the direction toward which the field increases. (b) Consider the field created by a spherical balloon centered at the origin. The balloon has a radius of 15.0 cm and carries a charge of 2.00 μC. Evaluate dE/dx at the point $(16$ cm, 0, $0)$. Assume that a water droplet at this point has an induced dipole moment of $(6.30\mathbf{i})$ nC$\cdot$m. Find the force on it.

(Optional)
Section 26.7   An Atomic Description of Dielectrics

52. A detector of radiation called a Geiger–Muller counter consists of a closed, hollow, conducting cylinder with a fine wire along its axis. Suppose that the internal diameter of the cylinder is 2.50 cm and that the wire along the axis has a diameter of 0.200 mm. If the dielectric strength of the gas between the central wire and the cylinder is 1.20×10^6 V/m, calculate the maximum voltage that can be applied between the wire and the cylinder before breakdown occurs in the gas.

53. The general form of Gauss's law describes how a charge creates an electric field in a material, as well as in a vacuum. It is

$$\oint \mathbf{E} \cdot d\mathbf{A} = \frac{q}{\epsilon}$$

where $\epsilon = \kappa \epsilon_0$ is the permittivity of the material. (a) A sheet with charge Q uniformly distributed over its area A is surrounded by a dielectric. Show that the sheet creates a uniform electric field with magnitude $E = Q/2A\epsilon$ at nearby points. (b) Two large sheets of area A carrying opposite charges of equal magnitude Q are a small distance d apart. Show that they create a uniform electric field of magnitude $E = Q/A\epsilon$ between them. (c) Assume that the negative plate is at zero potential. Show that the positive plate is at a potential $Qd/A\epsilon$. (d) Show that the capacitance of the pair of plates is $A\epsilon/d = \kappa A\epsilon_0/d$.

ADDITIONAL PROBLEMS

54. For the system of capacitors shown in Figure P26.54, find (a) the equivalent capacitance of the system, (b) the potential difference across each capacitor, (c) the charge on each capacitor, and (d) the total energy stored by the group.

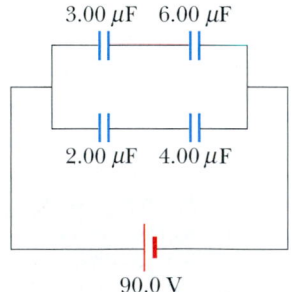

3.00 μF   6.00 μF

2.00 μF   4.00 μF

90.0 V

Figure P26.54

55. Consider two *long*, parallel, and oppositely charged wires of radius d with their centers separated by a distance D. Assuming the charge is distributed uniformly on the surface of each wire, show that the capacitance per unit length of this pair of wires is

$$\frac{C}{\ell} = \frac{\pi \epsilon_0}{\ln\left(\dfrac{D - d}{d}\right)}$$

56. A 2.00-nF parallel-plate capacitor is charged to an initial potential difference $\Delta V_i = 100$ V and then isolated. The dielectric material between the plates is mica ($\kappa = 5.00$). (a) How much work is required to withdraw the mica sheet? (b) What is the potential difference of the capacitor after the mica is withdrawn?

WEB **57.** A parallel-plate capacitor is constructed using a dielectric material whose dielectric constant is 3.00 and whose dielectric strength is 2.00×10^8 V/m. The desired capacitance is 0.250 μF, and the capacitor must withstand a maximum potential difference of 4 000 V. Find the minimum area of the capacitor plates.

58. A parallel-plate capacitor is constructed using three dielectric materials, as shown in Figure P26.58. You may assume that $\ell \gg d$. (a) Find an expression for the capacitance of the device in terms of the plate area A and d, κ_1, κ_2, and κ_3. (b) Calculate the capacitance using the values $A = 1.00$ cm^2, $d = 2.00$ mm, $\kappa_1 = 4.90$, $\kappa_2 = 5.60$, and $\kappa_3 = 2.10$.

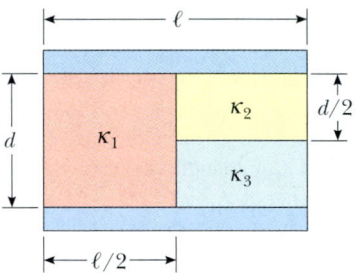

Figure P26.58

59. A conducting slab of thickness d and area A is inserted into the space between the plates of a parallel-plate capacitor with spacing s and surface area A, as shown in Figure P26.59. The slab is not necessarily halfway between the capacitor plates. What is the capacitance of the system?

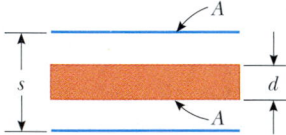

Figure P26.59

60. (a) Two spheres have radii a and b and their centers are a distance d apart. Show that the capacitance of this system is

$$C \approx \frac{4\pi\epsilon_0}{\dfrac{1}{a} + \dfrac{1}{b} - \dfrac{2}{d}}$$

provided that d is large compared with a and b. (*Hint:* Because the spheres are far apart, assume that the charge on one sphere does not perturb the charge distribution on the other sphere. Thus, the potential of each sphere is expressed as that of a symmetric charge distribution, $V = k_e Q/r$, and the total potential at each sphere is the sum of the potentials due to each sphere. (b) Show that as d approaches infinity the above result reduces to that of two isolated spheres in series.

61. When a certain air-filled parallel-plate capacitor is connected across a battery, it acquires a charge (on each plate) of q_0. While the battery connection is maintained, a dielectric slab is inserted into and fills the region between the plates. This results in the accumulation of an *additional* charge q on each plate. What is the dielectric constant of the slab?

62. A capacitor is constructed from two square plates of sides ℓ and separation d. A material of dielectric constant κ is inserted a distance x into the capacitor, as shown in Figure P26.62. (a) Find the equivalent capacitance of the device. (b) Calculate the energy stored in the capacitor if the potential difference is ΔV. (c) Find the direction and magnitude of the force exerted on the dielectric, assuming a constant potential difference ΔV. Neglect friction. (d) Obtain a numerical value for the force assuming that $\ell = 5.00$ cm, $\Delta V = 2\,000$ V, $d = 2.00$ mm, and the dielectric is glass ($\kappa = 4.50$). (*Hint:* The system can be considered as two capacitors connected in *parallel.*)

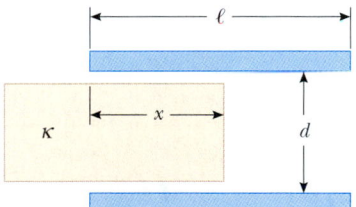

Figure P26.62   Problems 62 and 63.

63. A capacitor is constructed from two square plates of sides ℓ and separation d, as suggested in Figure P26.62. You may assume that d is much less than ℓ. The plates carry charges $+Q_0$ and $-Q_0$. A block of metal has a width ℓ, a length ℓ, and a thickness slightly less than d. It is inserted a distance x into the capacitor. The charges on the plates are not disturbed as the block slides in. In a static situation, a metal prevents an electric field from penetrating it. The metal can be thought of as a perfect dielectric, with $\kappa \to \infty$. (a) Calculate the stored energy as a function of x. (b) Find the direction and magnitude of the force that acts on the metallic block. (c) The area of the advancing front face of the block is essentially equal to ℓd. Considering the force on the block as acting on this face, find the stress (force per area) on it. (d) For comparison, express the energy density in the electric field between the capacitor plates in terms of Q_0, ℓ, d, and ϵ_0.

64. When considering the energy supply for an automobile, the energy per unit mass of the energy source is an important parameter. Using the following data, compare the energy per unit mass (J/kg) for gasoline, lead–acid batteries, and capacitors. (The ampere A will be introduced in Chapter 27 and is the SI unit of electric current. 1 A = 1 C/s.)

Gasoline: 126 000 Btu/gal; density = 670 kg/m^3
Lead–acid battery: 12.0 V; 100 A·h; mass = 16.0 kg
Capacitor: potential difference at full charge = 12.0 V; capacitance = 0.100 F; mass = 0.100 kg

65. An isolated capacitor of unknown capacitance has been charged to a potential difference of 100 V. When the charged capacitor is then connected in parallel to an uncharged 10.0-μF capacitor, the voltage across the combination is 30.0 V. Calculate the unknown capacitance.

66. A certain electronic circuit calls for a capacitor having a capacitance of 1.20 pF and a breakdown potential of 1 000 V. If you have a supply of 6.00-pF capacitors, each having a breakdown potential of 200 V, how could you meet this circuit requirement?

67. In the arrangement shown in Figure P26.67, a potential difference ΔV is applied, and C_1 is adjusted so that the voltmeter between points b and d reads zero. This "balance" occurs when $C_1 = 4.00$ μF. If $C_3 = 9.00$ μF and $C_4 = 12.0$ μF, calculate the value of C_2.

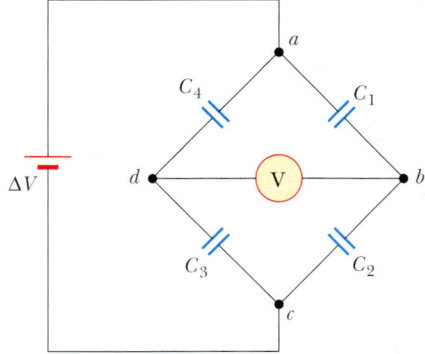

Figure P26.67

68. It is possible to obtain large potential differences by first charging a group of capacitors connected in parallel and then activating a switch arrangement that in effect disconnects the capacitors from the charging source and from each other and reconnects them in a series arrangement. The group of charged capacitors is then discharged in series. What is the maximum potential difference that can be obtained in this manner by using ten capacitors each of 500 μF and a charging source of 800 V?

69. A parallel-plate capacitor of plate separation d is charged to a potential difference ΔV_0. A dielectric slab

of thickness d and dielectric constant κ is introduced between the plates *while the battery remains connected to the plates.* (a) Show that the ratio of energy stored after the dielectric is introduced to the energy stored in the empty capacitor is $U/U_0 = \kappa$. Give a physical explanation for this increase in stored energy. (b) What happens to the charge on the capacitor? (Note that this situation is not the same as Example 26.7, in which the battery was removed from the circuit before the dielectric was introduced.)

70. A parallel-plate capacitor with plates of area A and plate separation d has the region between the plates filled with two dielectric materials as in Figure P26.70. Assume that $d \ll L$ and that $d \ll W$. (a) Determine the capacitance and (b) show that when $\kappa_1 = \kappa_2 = \kappa$ your result becomes the same as that for a capacitor containing a single dielectric, $C = \kappa \epsilon_0 A / d$.

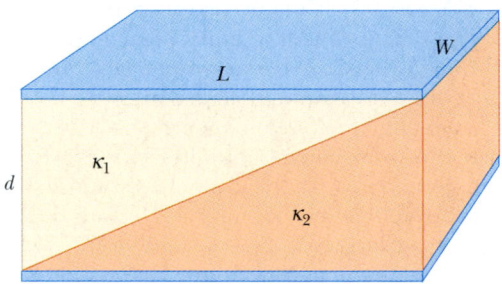

Figure P26.70

71. A vertical parallel-plate capacitor is half filled with a dielectric for which the dielectric constant is 2.00 (Fig. P26.71a). When this capacitor is positioned horizontally, what fraction of it should be filled with the same dielectric (Fig. P26.71b) so that the two capacitors have equal capacitance?

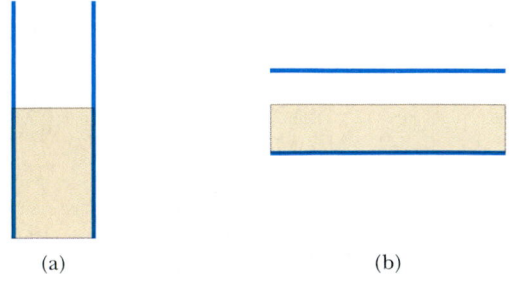

(a)   (b)

Figure P26.71

72. Capacitors $C_1 = 6.00 \ \mu\text{F}$ and $C_2 = 2.00 \ \mu\text{F}$ are charged as a parallel combination across a 250-V battery. The ca-

pacitors are disconnected from the battery and from each other. They are then connected positive plate to negative plate and negative plate to positive plate. Calculate the resulting charge on each capacitor.

73. The inner conductor of a coaxial cable has a radius of 0.800 mm, and the outer conductor's inside radius is 3.00 mm. The space between the conductors is filled with polyethylene, which has a dielectric constant of 2.30 and a dielectric strength of 18.0×10^6 V/m. What is the maximum potential difference that this cable can withstand?

74. You are optimizing coaxial cable design for a major manufacturer. Show that for a given outer conductor radius b, maximum potential difference capability is attained when the radius of the inner conductor is $a = b/e$ where e is the base of natural logarithms.

75. Calculate the equivalent capacitance between the points a and b in Figure P26.75. Note that this is not a simple series or parallel combination. (*Hint:* Assume a potential difference ΔV between points a and b. Write expressions for ΔV_{ab} in terms of the charges and capacitances for the various possible pathways from a to b, and require conservation of charge for those capacitor plates that are connected to each other.)

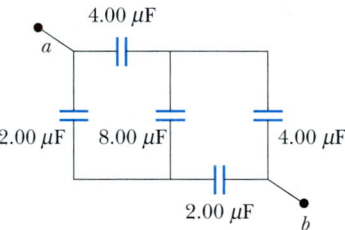

Figure P26.75

76. Determine the effective capacitance of the combination shown in Figure P26.76. (*Hint:* Consider the symmetry involved!)

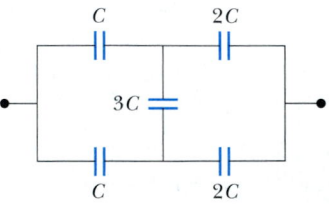

Figure P26.76

ANSWERS TO QUICK QUIZZES

26.1 (a) because the plate separation is decreased. Capacitance depends only on how a capacitor is constructed and not on the external circuit.

26.2 Zero. If you construct a spherical gaussian surface outside and concentric with the capacitor, the net charge inside the surface is zero. Applying Gauss's law to this configuration, we find that $E = 0$ at points outside the capacitor.

26.3 For a given voltage, the energy stored in a capacitor is proportional to C: $U = C(\Delta V)^2/2$. Thus, you want to maximize the equivalent capacitance. You do this by connecting the three capacitors in parallel, so that the capacitances add.

26.4 (a) C decreases (Eq. 26.3). (b) Q stays the same because there is no place for the charge to flow. (c) E remains constant (see Eq. 24.8 and the paragraph following it). (d) ΔV increases because $\Delta V = Q/C$, Q is constant (part b), and C decreases (part a). (e) The energy stored in the capacitor is proportional to both Q and ΔV (Eq. 26.11) and thus increases. The additional energy comes from the work you do in pulling the two plates apart.

26.5 (a) C decreases (Eq. 26.3). (b) Q decreases. The battery supplies a constant potential difference ΔV; thus, charge must flow out of the capacitor if $C = Q/\Delta V$ is to de-

crease. (c) E decreases because the charge density on the plates decreases. (d) ΔV remains constant because of the presence of the battery. (e) The energy stored in the capacitor decreases (Eq. 26.11).

26.6 It increases. The dielectric constant of wood (and of all other insulating materials, for that matter) is greater than 1; therefore, the capacitance increases (Eq. 26.14). This increase is sensed by the stud-finder's special circuitry, which causes an indicator on the device to light up.

26.7 (a) C increases (Eq. 26.14). (b) Q increases. Because the battery maintains a constant ΔV, Q must increase if C ($= Q/\Delta V$) increases. (c) E between the plates remains constant because $\Delta V = Ed$ and neither ΔV nor d changes. The electric field due to the charges on the plates increases because more charge has flowed onto the plates. The induced surface charges on the dielectric create a field that opposes the increase in the field caused by the greater number of charges on the plates. (d) The battery maintains a constant ΔV. (e) The energy stored in the capacitor increases (Eq. 26.11). You would have to push the dielectric into the capacitor, just as you would have to do positive work to raise a mass and increase its gravitational potential energy.

chapter

27

Current and Resistance

Chapter Outline

Thus far our treatment of electrical phenomena has been confined to the study of charges at rest, or *electrostatics*. We now consider situations involving electric charges in motion. We use the term *electric current*, or simply *current*, to describe the rate of flow of charge through some region of space. Most practical applications of electricity deal with electric currents. For example, the battery in a flashlight supplies current to the filament of the bulb when the switch is turned on. A variety of home appliances operate on alternating current. In these common situations, the charges flow through a conductor, such as a copper wire. It also is possible for currents to exist outside a conductor. For instance, a beam of electrons in a television picture tube constitutes a current.

This chapter begins with the definitions of current and current density. A microscopic description of current is given, and some of the factors that contribute to the resistance to the flow of charge in conductors are discussed. A classical model is used to describe electrical conduction in metals, and some of the limitations of this model are cited.

27.1 ELECTRIC CURRENT

It is instructive to draw an analogy between water flow and current. In many localities it is common practice to install low-flow showerheads in homes as a water-conservation measure. We quantify the flow of water from these and similar devices by specifying the amount of water that emerges during a given time interval, which is often measured in liters per minute. On a grander scale, we can characterize a river current by describing the rate at which the water flows past a particular location. For example, the flow over the brink at Niagara Falls is maintained at rates between 1 400 m^3/s and 2 800 m^3/s.

Now consider a system of electric charges in motion. Whenever there is a net flow of charge through some region, a **current** is said to exist. To define current more precisely, suppose that the charges are moving perpendicular to a surface of area A, as shown in Figure 27.1. (This area could be the cross-sectional area of a wire, for example.) **The current is the rate at which charge flows through this surface.** If ΔQ is the amount of charge that passes through this area in a time interval Δt, the **average current** I_{av} is equal to the charge that passes through A per unit time:

$$I_{av} = \frac{\Delta Q}{\Delta t} \qquad \textbf{(27.1)}$$

If the rate at which charge flows varies in time, then the current varies in time; we define the **instantaneous current** I as the differential limit of average current:

$$I \equiv \frac{dQ}{dt} \qquad \textbf{(27.2)}$$

The SI unit of current is the **ampere** (A):

$$1\,A = \frac{1\,C}{1\,s} \qquad \textbf{(27.3)}$$

That is, 1 A of current is equivalent to 1 C of charge passing through the surface area in 1 s.

The charges passing through the surface in Figure 27.1 can be positive or negative, or both. **It is conventional to assign to the current the same direction as the flow of positive charge.** In electrical conductors, such as copper or alu-

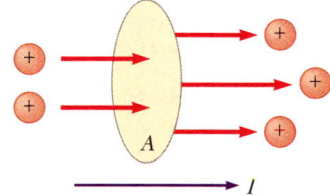

Figure 27.1 Charges in motion through an area A. The time rate at which charge flows through the area is defined as the current I. The direction of the current is the direction in which positive charges flow when free to do so.

Electric current

The direction of the current

minum, the current is due to the motion of negatively charged electrons. Therefore, when we speak of current in an ordinary conductor, **the direction of the current is opposite the direction of flow of electrons.** However, if we are considering a beam of positively charged protons in an accelerator, the current is in the direction of motion of the protons. In some cases—such as those involving gases and electrolytes, for instance—the current is the result of the flow of both positive and negative charges.

If the ends of a conducting wire are connected to form a loop, all points on the loop are at the same electric potential, and hence the electric field is zero within and at the surface of the conductor. Because the electric field is zero, there is no net transport of charge through the wire, and therefore there is no current. The current in the conductor is zero even if the conductor has an excess of charge on it. However, if the ends of the conducting wire are connected to a battery, all points on the loop are not at the same potential. The battery sets up a potential difference between the ends of the loop, creating an electric field within the wire. The electric field exerts forces on the conduction electrons in the wire, causing them to move around the loop and thus creating a current.

It is common to refer to a moving charge (positive or negative) as a mobile **charge carrier.** For example, the mobile charge carriers in a metal are electrons.

Microscopic Model of Current

We can relate current to the motion of the charge carriers by describing a microscopic model of conduction in a metal. Consider the current in a conductor of cross-sectional area A (Fig. 27.2). The volume of a section of the conductor of length Δx (the gray region shown in Fig. 27.2) is $A\,\Delta x$. If n represents the number of mobile charge carriers per unit volume (in other words, the charge carrier density), the number of carriers in the gray section is $nA\,\Delta x$. Therefore, the charge ΔQ in this section is

$$\Delta Q = \text{number of carriers in section} \times \text{charge per carrier} = (nA\,\Delta x)q$$

where q is the charge on each carrier. If the carriers move with a speed v_d, the distance they move in a time Δt is $\Delta x = v_d\,\Delta t$. Therefore, we can write ΔQ in the form

$$\Delta Q = (nAv_d\,\Delta t)q$$

If we divide both sides of this equation by Δt, we see that the average current in the conductor is

$$I_{\text{av}} = \frac{\Delta Q}{\Delta t} = nqv_d A \tag{27.4}$$

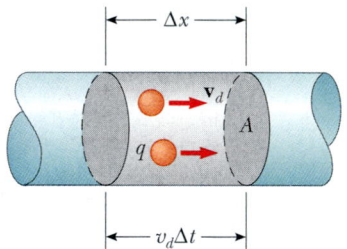

Δx

$\mathbf{v}_d$

A

q

$v_d\Delta t$

Figure 27.2 A section of a uniform conductor of cross-sectional area A. The mobile charge carriers move with a speed v_d, and the distance they travel in a time Δt is $\Delta x = v_d\,\Delta t$. The number of carriers in the section of length Δx is $nAv_d\,\Delta t$, where n is the number of carriers per unit volume.

Average current in a conductor

The speed of the charge carriers v_d is an average speed called the **drift speed.** To understand the meaning of drift speed, consider a conductor in which the charge carriers are free electrons. If the conductor is isolated—that is, the potential difference across it is zero—then these electrons undergo random motion that is analogous to the motion of gas molecules. As we discussed earlier, when a potential difference is applied across the conductor (for example, by means of a battery), an electric field is set up in the conductor; this field exerts an electric force on the electrons, producing a current. However, the electrons do not move in straight lines along the conductor. Instead, they collide repeatedly with the metal atoms, and their resultant motion is complicated and zigzag (Fig. 27.3). Despite the collisions, the electrons move slowly along the conductor (in a direction opposite that of $\mathbf{E}$) at the drift velocity $\mathbf{v}_d$.

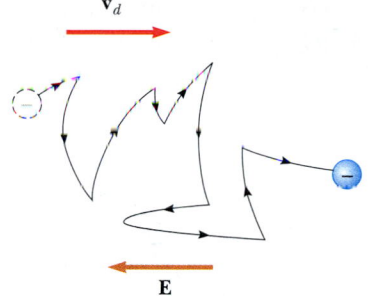

Figure 27.3   A schematic representation of the zigzag motion of an electron in a conductor. The changes in direction are the result of collisions between the electron and atoms in the conductor. Note that the net motion of the electron is opposite the direction of the electric field. Each section of the zigzag path is a parabolic segment.

We can think of the atom–electron collisions in a conductor as an effective internal friction (or drag force) similar to that experienced by the molecules of a liquid flowing through a pipe stuffed with steel wool. The energy transferred from the electrons to the metal atoms during collision causes an increase in the vibrational energy of the atoms and a corresponding increase in the temperature of the conductor.

Quick Quiz 27.1

Consider positive and negative charges moving horizontally through the four regions shown in Figure 27.4. Rank the current in these four regions, from lowest to highest.

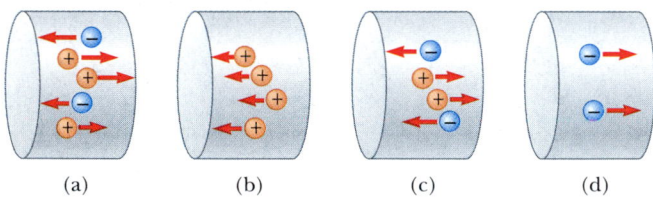

(a)   (b)   (c)   (d)   **Figure 27.4**

EXAMPLE 27.1   **Drift Speed in a Copper Wire**

The 12-gauge copper wire in a typical residential building has a cross-sectional area of 3.31×10^{-6} m². If it carries a current of 10.0 A, what is the drift speed of the electrons? Assume that each copper atom contributes one free electron to the current. The density of copper is 8.95 g/cm³.

Solution   From the periodic table of the elements in Appendix C, we find that the molar mass of copper is 63.5 g/mol. Recall that 1 mol of any substance contains Avogadro's number of atoms (6.02×10^{23}). Knowing the density of copper, we can calculate the volume occupied by 63.5 g ($=1$ mol) of copper:

$$V = \frac{m}{\rho} = \frac{63.5 \text{ g}}{8.95 \text{ g/cm}^3} = 7.09 \text{ cm}^3$$

Because each copper atom contributes one free electron to the current, we have

$$n = \frac{6.02 \times 10^{23} \text{ electrons}}{7.09 \text{ cm}^3} (1.00 \times 10^6 \text{ cm}^3/\text{m}^3)$$

$$= 8.49 \times 10^{28} \text{ electrons/m}^3$$

From Equation 27.4, we find that the drift speed is

$$v_d = \frac{I}{nqA}$$

where q is the absolute value of the charge on each electron. Thus,

$$v_d = \frac{I}{nqA}$$

$$= \frac{10.0 \text{ C/s}}{(8.49 \times 10^{28} \text{ m}^{-3})(1.60 \times 10^{-19} \text{ C})(3.31 \times 10^{-6} \text{ m}^2)}$$

$$= 2.22 \times 10^{-4} \text{ m/s}$$

Exercise   If a copper wire carries a current of 80.0 mA, how many electrons flow past a given cross-section of the wire in 10.0 min?

Answer 3.0×10^{20} electrons.

Example 27.1 shows that typical drift speeds are very low. For instance, electrons traveling with a speed of 2.46×10^{-4} m/s would take about 68 min to travel 1 m! In view of this, you might wonder why a light turns on almost instantaneously when a switch is thrown. In a conductor, the electric field that drives the free electrons travels through the conductor with a speed close to that of light. Thus, when you flip on a light switch, the message for the electrons to start moving through the wire (the electric field) reaches them at a speed on the order of 10^8 m/s.

27.2 ▶ RESISTANCE AND OHM'S LAW

13.3 In Chapter 24 we found that no electric field can exist inside a conductor. However, this statement is true *only* if the conductor is in static equilibrium. The purpose of this section is to describe what happens when the charges in the conductor are allowed to move.

Charges moving in a conductor produce a current under the action of an electric field, which is maintained by the connection of a battery across the conductor. An electric field can exist in the conductor because the charges in this situation are in motion—that is, this is a *nonelectrostatic* situation.

Consider a conductor of cross-sectional area A carrying a current I. The **current density** J in the conductor is defined as the current per unit area. Because the current $I = nqv_dA$, the current density is

$$J \equiv \frac{I}{A} = nqv_d \qquad (27.5)$$

where J has SI units of A/m^2. This expression is valid only if the current density is uniform and only if the surface of cross-sectional area A is perpendicular to the direction of the current. In general, the current density is a vector quantity:

Current density

$$\mathbf{J} = nq\mathbf{v}_d \qquad (27.6)$$

From this equation, we see that current density, like current, is in the direction of charge motion for positive charge carriers and opposite the direction of motion for negative charge carriers.

A current density $\mathbf{J}$ and an electric field $\mathbf{E}$ are established in a conductor whenever a potential difference is maintained across the conductor. If the potential difference is constant, then the current also is constant. In some materials, the current density is proportional to the electric field:

Ohm's law

$$\mathbf{J} = \sigma \mathbf{E} \qquad (27.7)$$

where the constant of proportionality σ is called the **conductivity** of the conductor.[1] Materials that obey Equation 27.7 are said to follow **Ohm's law,** named after Georg Simon Ohm (1787–1854). More specifically, Ohm's law states that

> for many materials (including most metals), the ratio of the current density to the electric field is a constant σ that is independent of the electric field producing the current.

Materials that obey Ohm's law and hence demonstrate this simple relationship between $\mathbf{E}$ and $\mathbf{J}$ are said to be *ohmic*. Experimentally, it is found that not all materials have this property, however, and materials that do not obey Ohm's law are said to

---

[1] Do not confuse conductivity σ with surface charge density, for which the same symbol is used.

be *nonohmic*. Ohm's law is not a fundamental law of nature but rather an empirical relationship valid only for certain materials.

Quick Quiz 27.2

Suppose that a current-carrying ohmic metal wire has a cross-sectional area that gradually becomes smaller from one end of the wire to the other. How do drift velocity, current density, and electric field vary along the wire? Note that the current must have the same value everywhere in the wire so that charge does not accumulate at any one point.

We can obtain a form of Ohm's law useful in practical applications by considering a segment of straight wire of uniform cross-sectional area A and length ℓ, as shown in Figure 27.5. A potential difference $\Delta V = V_b - V_a$ is maintained across the wire, creating in the wire an electric field and a current. If the field is assumed to be uniform, the potential difference is related to the field through the relationship[2]

$$\Delta V = E\ell$$

Therefore, we can express the magnitude of the current density in the wire as

$$J = \sigma E = \sigma \frac{\Delta V}{\ell}$$

Because $J = I/A$, we can write the potential difference as

$$\Delta V = \frac{\ell}{\sigma} J = \left(\frac{\ell}{\sigma A}\right) I$$

The quantity $\ell/\sigma A$ is called the **resistance** R of the conductor. We can define the resistance as the ratio of the potential difference across a conductor to the current through the conductor:

$$R \equiv \frac{\ell}{\sigma A} \equiv \frac{\Delta V}{I} \qquad \textbf{(27.8)}$$

Resistance of a conductor

From this result we see that resistance has SI units of volts per ampere. One volt per ampere is defined to be 1 **ohm** (Ω):

$$1\ \Omega \equiv \frac{1\ \text{V}}{1\ \text{A}} \qquad \textbf{(27.9)}$$

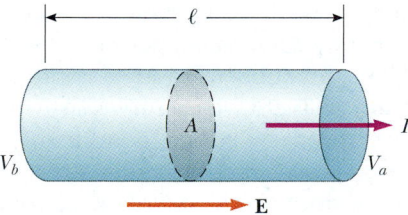

Figure 27.5 A uniform conductor of length ℓ and cross-sectional area A. A potential difference $\Delta V = V_b - V_a$ maintained across the conductor sets up an electric field **E**, and this field produces a current I that is proportional to the potential difference.

---

[2] This result follows from the definition of potential difference:

$$V_b - V_a = -\int_a^b \mathbf{E} \cdot d\mathbf{s} = E \int_0^\ell dx = E\ell$$

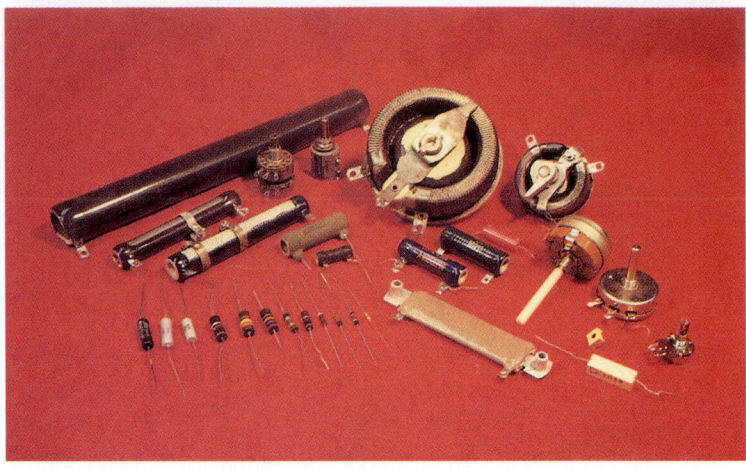

An assortment of resistors used in electric circuits. *(Henry Leap and Jim Lehman)*

This expression shows that if a potential difference of 1 V across a conductor causes a current of 1 A, the resistance of the conductor is 1 Ω. For example, if an electrical appliance connected to a 120-V source of potential difference carries a current of 6 A, its resistance is 20 Ω.

Equation 27.8 solved for potential difference ($\Delta V = I\ell/\sigma A$) explains part of the chapter-opening puzzler: How can a bird perch on a high-voltage power line without being electrocuted? Even though the potential difference between the ground and the wire might be hundreds of thousands of volts, that between the bird's feet (which is what determines how much current flows through the bird) is very small.

The inverse of conductivity is **resistivity**[3] ρ:

Resistivity

$$\rho \equiv \frac{1}{\sigma}$$

(27.10)

where ρ has the units ohm-meters ($\Omega \cdot$ m). We can use this definition and Equation 27.8 to express the resistance of a uniform block of material as

Resistance of a uniform conductor

$$R = \rho \frac{\ell}{A}$$

(27.11)

Every ohmic material has a characteristic resistivity that depends on the properties of the material and on temperature. Additionally, as you can see from Equation 27.11, the resistance of a sample depends on geometry as well as on resistivity. Table 27.1 gives the resistivities of a variety of materials at 20°C. Note the enormous range, from very low values for good conductors such as copper and silver, to very high values for good insulators such as glass and rubber. An ideal conductor would have zero resistivity, and an ideal insulator would have infinite resistivity.

Equation 27.11 shows that the resistance of a given cylindrical conductor is proportional to its length and inversely proportional to its cross-sectional area. If the length of a wire is doubled, then its resistance doubles. If its cross-sectional area is doubled, then its resistance decreases by one half. The situation is analogous to the flow of a liquid through a pipe. As the pipe's length is increased, the

---

[3] Do not confuse resistivity with mass density or charge density, for which the same symbol is used.

TABLE 27.1	**Resistivities and Temperature Coefficients of Resistivity for Various Materials**	
Material	**Resistivity[a] ($\Omega \cdot$ m)**	**Temperature Coefficient $\alpha[(°C)^{-1}]$**
Silver	1.59×10^{-8}	3.8×10^{-3}
Copper	1.7×10^{-8}	3.9×10^{-3}
Gold	2.44×10^{-8}	3.4×10^{-3}
Aluminum	2.82×10^{-8}	3.9×10^{-3}
Tungsten	5.6×10^{-8}	4.5×10^{-3}
Iron	10×10^{-8}	5.0×10^{-3}
Platinum	11×10^{-8}	3.92×10^{-3}
Lead	22×10^{-8}	3.9×10^{-3}
Nichrome[b]	1.50×10^{-6}	0.4×10^{-3}
Carbon	3.5×10^{-5}	-0.5×10^{-3}
Germanium	0.46	-48×10^{-3}
Silicon	640	-75×10^{-3}
Glass	10^{10} to 10^{14}	
Hard rubber	$\approx 10^{13}$	
Sulfur	10^{15}	
Quartz (fused)	75×10^{16}	

[a] All values at 20°C.

[b] A nickel–chromium alloy commonly used in heating elements.

resistance to flow increases. As the pipe's cross-sectional area is increased, more liquid crosses a given cross-section of the pipe per unit time. Thus, more liquid flows for the same pressure differential applied to the pipe, and the resistance to flow decreases.

Most electric circuits use devices called **resistors** to control the current level in the various parts of the circuit. Two common types of resistors are the *composition resistor*, which contains carbon, and the *wire-wound resistor*, which consists of a coil of wire. Resistors' values in ohms are normally indicated by color-coding, as shown in Figure 27.6 and Table 27.2.

Ohmic materials have a linear current–potential difference relationship over a broad range of applied potential differences (Fig. 27.7a). The slope of the *I*-versus-ΔV curve in the linear region yields a value for $1/R$. Nonohmic materials

Figure 27.6   The colored bands on a resistor represent a code for determining resistance. The first two colors give the first two digits in the resistance value. The third color represents the power of ten for the multiplier of the resistance value. The last color is the tolerance of the resistance value. As an example, the four colors on the circled resistors are red ($= 2$), black ($= 0$), orange ($= 10^3$), and gold ($= 5\%$), and so the resistance value is $20 \times 10^3 \ \Omega = 20 \ k\Omega$ with a tolerance value of $5\% = 1 \ k\Omega$. (The values for the colors are from Table 27.2.) *(SuperStock)*

TABLE 27.2	Color Coding for Resistors		
Color	**Number**	**Multiplier**	**Tolerance**
Black	0	1	
Brown	1	10^1	
Red	2	10^2	
Orange	3	10^3	
Yellow	4	10^4	
Green	5	10^5	
Blue	6	10^6	
Violet	7	10^7	
Gray	8	10^8	
White	9	10^9	
Gold		10^{-1}	5%
Silver		10^{-2}	10%
Colorless			20%

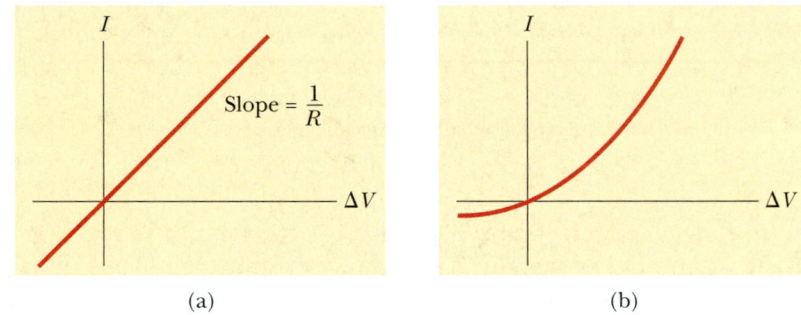

Figure 27.7   (a) The current–potential difference curve for an ohmic material. The curve is linear, and the slope is equal to the inverse of the resistance of the conductor. (b) A nonlinear current–potential difference curve for a semiconducting diode. This device does not obey Ohm's law.

have a nonlinear current–potential difference relationship. One common semi-conducting device that has nonlinear I-versus-ΔV characteristics is the *junction diode* (Fig. 27.7b). The resistance of this device is low for currents in one direction (positive ΔV) and high for currents in the reverse direction (negative ΔV). In fact, most modern electronic devices, such as transistors, have nonlinear current–potential difference relationships; their proper operation depends on the particular way in which they violate Ohm's law.

Quick Quiz 27.3

What does the slope of the curved line in Figure 27.7b represent?

Quick Quiz 27.4

Your boss asks you to design an automobile battery jumper cable that has a low resistance. In view of Equation 27.11, what factors would you consider in your design?

EXAMPLE 27.2 ▸ The Resistance of a Conductor

Calculate the resistance of an aluminum cylinder that is 10.0 cm long and has a cross-sectional area of 2.00×10^{-4} m². Repeat the calculation for a cylinder of the same dimensions and made of glass having a resistivity of 3.0×10^{10} Ω·m.

Solution   From Equation 27.11 and Table 27.1, we can calculate the resistance of the aluminum cylinder as follows:

$$R = \rho \frac{\ell}{A} = (2.82 \times 10^{-8}\ \Omega \cdot m) \left(\frac{0.100\ m}{2.00 \times 10^{-4}\ m^2} \right)$$

$$= \boxed{1.41 \times 10^{-5}\ \Omega}$$

Similarly, for glass we find that

$$R = \rho \frac{\ell}{A} = (3.0 \times 10^{10}\ \Omega \cdot m) \left(\frac{0.100\ m}{2.00 \times 10^{-4}\ m^2} \right)$$

$$= \boxed{1.5 \times 10^{13}\ \Omega}$$

As you might guess from the large difference in resistivi-

ties, the resistance of identically shaped cylinders of aluminum and glass differ widely. The resistance of the glass cylinder is 18 orders of magnitude greater than that of the aluminum cylinder.

Electrical insulators on telephone poles are often made of glass because of its low electrical conductivity. (*J.H. Robinson/Photo Researchers, Inc.*)

EXAMPLE 27.3 ▸ The Resistance of Nichrome Wire

(a) Calculate the resistance per unit length of a 22-gauge Nichrome wire, which has a radius of 0.321 mm.

Solution   The cross-sectional area of this wire is

$$A = \pi r^2 = \pi (0.321 \times 10^{-3}\ m)^2 = 3.24 \times 10^{-7}\ m^2$$

The resistivity of Nichrome is 1.5×10^{-6} Ω·m (see Table 27.1). Thus, we can use Equation 27.11 to find the resistance per unit length:

$$\frac{R}{\ell} = \frac{\rho}{A} = \frac{1.5 \times 10^{-6}\ \Omega \cdot m}{3.24 \times 10^{-7}\ m^2} = \boxed{4.6\ \Omega/m}$$

(b) If a potential difference of 10 V is maintained across a 1.0-m length of the Nichrome wire, what is the current in the wire?

Solution   Because a 1.0-m length of this wire has a resistance of 4.6 Ω, Equation 27.8 gives

$$I = \frac{\Delta V}{R} = \frac{10\ V}{4.6\ \Omega} = \boxed{2.2\ A}$$

Note from Table 27.1 that the resistivity of Nichrome wire is about 100 times that of copper. A copper wire of the same radius would have a resistance per unit length of only 0.052 Ω/m. A 1.0-m length of copper wire of the same radius would carry the same current (2.2 A) with an applied potential difference of only 0.11 V.

Because of its high resistivity and its resistance to oxidation, Nichrome is often used for heating elements in toasters, irons, and electric heaters.

Exercise   What is the resistance of a 6.0-m length of 22-gauge Nichrome wire? How much current does the wire carry when connected to a 120-V source of potential difference?

Answer   28 Ω; 4.3 A.

Exercise   Calculate the current density and electric field in the wire when it carries a current of 2.2 A.

Answer 6.8×10^6 A/m²; 10 N/C.

EXAMPLE 27.4 ▸ The Radial Resistance of a Coaxial Cable

Coaxial cables are used extensively for cable television and other electronic applications. A coaxial cable consists of two cylindrical conductors. The gap between the conductors is completely filled with silicon, as shown in Figure 27.8a, and current leakage through the silicon is unwanted. (The cable is designed to conduct current along its length.) The radius of the inner conductor is $a = 0.500$ cm, the radius of the outer one is $b = 1.75$ cm, and the length of the cable is $L = 15.0$ cm. Calculate the resistance of the silicon between the two conductors.

Solution   In this type of problem, we must divide the object whose resistance we are calculating into concentric elements of infinitesimal thickness dr (Fig. 27.8b). We start by using the differential form of Equation 27.11, replacing ℓ with r for the distance variable: $dR = \rho\, dr/A$, where dR is the resistance of an element of silicon of thickness dr and surface area A. In this example, we take as our representative concentric element a hollow silicon cylinder of radius r, thickness dr, and length L, as shown in Figure 27.8. Any current that passes from the inner conductor to the outer one must pass radially through this concentric element, and the area through which this current passes is $A = 2\pi rL$. (This is the curved surface area—circumference multiplied by length—of our hollow silicon cylinder of thickness dr.) Hence, we can write the resistance of our hollow cylinder of silicon as

$$dR = \frac{\rho}{2\pi rL}\, dr$$

Because we wish to know the total resistance across the entire thickness of the silicon, we must integrate this expression from $r = a$ to $r = b$:

$$R = \int_a^b dR = \frac{\rho}{2\pi L} \int_a^b \frac{dr}{r} = \frac{\rho}{2\pi L} \ln\left(\frac{b}{a}\right)$$

Substituting in the values given, and using $\rho = 640\ \Omega \cdot m$ for silicon, we obtain

$$R = \frac{640\ \Omega \cdot m}{2\pi(0.150\ m)} \ln\left(\frac{1.75\ cm}{0.500\ cm}\right) = \boxed{851\ \Omega}$$

Exercise   If a potential difference of 12.0 V is applied between the inner and outer conductors, what is the value of the total current that passes between them?

Answer   14.1 mA.

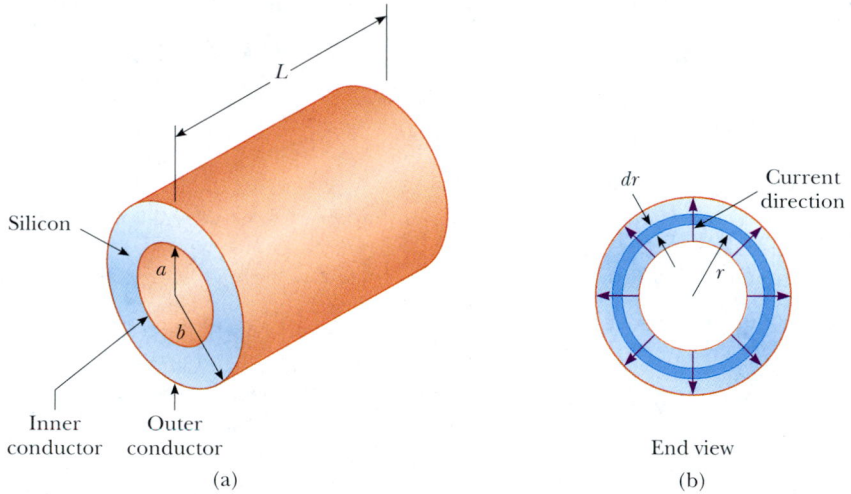

Figure 27.8   A coaxial cable. (a) Silicon fills the gap between the two conductors. (b) End view, showing current leakage.

Labels in figure: L, Silicon, a, b, Inner conductor, Outer conductor, (a), dr, Current direction, r, End view, (b)

27.3 A MODEL FOR ELECTRICAL CONDUCTION

In this section we describe a classical model of electrical conduction in metals that was first proposed by Paul Drude in 1900. This model leads to Ohm's law and shows that resistivity can be related to the motion of electrons in metals. Although the Drude model described here does have limitations, it nevertheless introduces concepts that are still applied in more elaborate treatments.

Consider a conductor as a regular array of atoms plus a collection of free electrons, which are sometimes called *conduction* electrons. The conduction electrons, although bound to their respective atoms when the atoms are not part of a solid, gain mobility when the free atoms condense into a solid. In the absence of an electric field, the conduction electrons move in random directions through the conductor with average speeds of the order of 10^6 m/s. The situation is similar to the motion of gas molecules confined in a vessel. In fact, some scientists refer to conduction electrons in a metal as an *electron gas*. There is no current through the conductor in the absence of an electric field because the drift velocity of the free electrons is zero. That is, on the average, just as many electrons move in one direction as in the opposite direction, and so there is no net flow of charge.

This situation changes when an electric field is applied. Now, in addition to undergoing the random motion just described, the free electrons drift slowly in a direction opposite that of the electric field, with an average drift speed v_d that is much smaller (typically 10^{-4} m/s) than their average speed between collisions (typically 10^6 m/s).

Figure 27.9 provides a crude description of the motion of free electrons in a conductor. In the absence of an electric field, there is no net displacement after many collisions (Fig. 27.9a). An electric field **E** modifies the random motion and causes the electrons to drift in a direction opposite that of **E** (Fig. 27.9b). The slight curvature in the paths shown in Figure 27.9b results from the acceleration of the electrons between collisions, which is caused by the applied field.

In our model, we assume that the motion of an electron after a collision is independent of its motion before the collision. We also assume that the excess energy acquired by the electrons in the electric field is lost to the atoms of the conductor when the electrons and atoms collide. The energy given up to the atoms increases their vibrational energy, and this causes the temperature of the conductor to increase. The temperature increase of a conductor due to resistance is utilized in electric toasters and other familiar appliances.

We are now in a position to derive an expression for the drift velocity. When a free electron of mass m_e and charge $q\ (= -e)$ is subjected to an electric field **E**, it experiences a force $\mathbf{F} = q\mathbf{E}$. Because $\Sigma\mathbf{F} = m_e\mathbf{a}$, we conclude that the acceleration of the electron is

$$\mathbf{a} = \frac{q\mathbf{E}}{m_e} \qquad (27.12)$$

This acceleration, which occurs for only a short time between collisions, enables the electron to acquire a small drift velocity. If t is the time since the last collision and $\mathbf{v}_i$ is the electron's initial velocity the instant after that collision, then the velocity of the electron after a time t is

$$\mathbf{v}_f = \mathbf{v}_i + \mathbf{a}t = \mathbf{v}_i + \frac{q\mathbf{E}}{m_e}t \qquad (27.13)$$

We now take the average value of **v** over all possible times t and all possible values of $\mathbf{v}_i$. If we assume that the initial velocities are randomly distributed over all possible values, we see that the average value of $\mathbf{v}_i$ is zero. The term $(q\mathbf{E}/m_e)t$ is the velocity added by the field during one trip between atoms. If the electron starts with zero velocity, then the average value of the second term of Equation 27.13 is $(q\mathbf{E}/m_e)\tau$, where τ is the *average time interval between successive collisions*. Because the average value of $\mathbf{v}_f$ is equal to the drift velocity,[4] we have

$$\mathbf{v}_f = \mathbf{v}_d = \frac{q\mathbf{E}}{m_e}\tau \qquad (27.14)$$

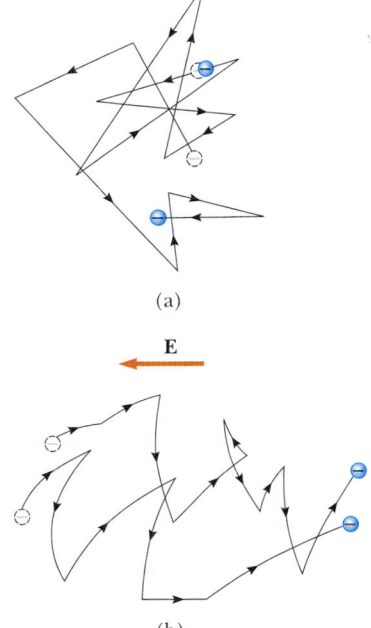

(a)

E

(b)

Figure 27.9 (a) A schematic diagram of the random motion of two charge carriers in a conductor in the absence of an electric field. The drift velocity is zero. (b) The motion of the charge carriers in a conductor in the presence of an electric field. Note that the random motion is modified by the field, and the charge carriers have a drift velocity.

Drift velocity

---

[4] Because the collision process is random, each collision event is *independent* of what happened earlier. This is analogous to the random process of throwing a die. The probability of rolling a particular number on one throw is independent of the result of the previous throw. On average, the particular number comes up every sixth throw, starting at any arbitrary time.

We can relate this expression for drift velocity to the current in the conductor. Substituting Equation 27.14 into Equation 27.6, we find that the magnitude of the current density is

Current density

$$J = nqv_d = \frac{nq^2E}{m_e}\tau \tag{27.15}$$

where n is the number of charge carriers per unit volume. Comparing this expression with Ohm's law, $J = \sigma E$, we obtain the following relationships for conductivity and resistivity:

Conductivity

$$\sigma = \frac{nq^2\tau}{m_e} \tag{27.16}$$

Resistivity

$$\rho = \frac{1}{\sigma} = \frac{m_e}{nq^2\tau} \tag{27.17}$$

According to this classical model, conductivity and resistivity do not depend on the strength of the electric field. This feature is characteristic of a conductor obeying Ohm's law.

The average time between collisions τ is related to the average distance between collisions ℓ (that is, the *mean free path;* see Section 21.7) and the average speed $\bar{v}$ through the expression

$$\tau = \frac{\ell}{\bar{v}} \tag{27.18}$$

EXAMPLE 27.5   Electron Collisions in a Wire

(a) Using the data and results from Example 27.1 and the classical model of electron conduction, estimate the average time between collisions for electrons in household copper wiring.

Solution   From Equation 27.17, we see that

$$\tau = \frac{m_e}{nq^2\rho}$$

where $\rho = 1.7 \times 10^{-8}\ \Omega \cdot m$ for copper and the carrier density is $n = 8.49 \times 10^{28}$ electrons/m^3 for the wire described in Example 27.1. Substitution of these values into the expression above gives

$$\tau = \frac{(9.11 \times 10^{-31}\ \text{kg})}{(8.49 \times 10^{28}\ \text{m}^{-3})(1.6 \times 10^{-19}\ \text{C})^2(1.7 \times 10^{-8}\ \Omega \cdot \text{m})}$$

$$= \boxed{2.5 \times 10^{-14}\ \text{s}}$$

(b) Assuming that the average speed for free electrons in copper is 1.6×10^6 m/s and using the result from part (a), calculate the mean free path for electrons in copper.

Solution

$$\ell = \bar{v}\tau = (1.6 \times 10^6\ \text{m/s})(2.5 \times 10^{-14}\ \text{s})$$

$$= \boxed{4.0 \times 10^{-8}\ \text{m}}$$

which is equivalent to 40 nm (compared with atomic spacings of about 0.2 nm). Thus, although the time between collisions is very short, an electron in the wire travels about 200 atomic spacings between collisions.

Although this classical model of conduction is consistent with Ohm's law, it is not satisfactory for explaining some important phenomena. For example, classical values for $\bar{v}$ calculated on the basis of an ideal-gas model (see Section 21.6) are smaller than the true values by about a factor of ten. Furthermore, if we substitute $\ell/\bar{v}$ for τ in Equation 27.17 and rearrange terms so that $\bar{v}$ appears in the numerator, we find that the resistivity ρ is proportional to $\bar{v}$. According to the ideal-gas model, $\bar{v}$ is proportional to $\sqrt{T}$; hence, it should also be true that $\rho \propto \sqrt{T}$. This is in disagreement with the fact that, for pure metals, resistivity depends linearly on temperature. We are able to account for the linear dependence only by using a quantum mechanical model, which we now describe briefly.

According to quantum mechanics, electrons have wave-like properties. If the array of atoms in a conductor is regularly spaced (that is, it is periodic), then the wave-like character of the electrons enables them to move freely through the conductor, and a collision with an atom is unlikely. For an idealized conductor, no collisions would occur, the mean free path would be infinite, and the resistivity would be zero. Electron waves are scattered only if the atomic arrangement is irregular (not periodic) as a result of, for example, structural defects or impurities. At low temperatures, the resistivity of metals is dominated by scattering caused by collisions between electrons and defects or impurities. At high temperatures, the resistivity is dominated by scattering caused by collisions between electrons and atoms of the conductor, which are continuously displaced from the regularly spaced array as a result of thermal agitation. The thermal motion of the atoms causes the structure to be irregular (compared with an atomic array at rest), thereby reducing the electron's mean free path.

27.4 RESISTANCE AND TEMPERATURE

Over a limited temperature range, the resistivity of a metal varies approximately linearly with temperature according to the expression

$$\rho = \rho_0[1 + \alpha(T - T_0)] \tag{27.19}$$

Variation of ρ with temperature

where ρ is the resistivity at some temperature T (in degrees Celsius), ρ_0 is the resistivity at some reference temperature T_0 (usually taken to be 20°C), and α is the **temperature coefficient of resistivity.** From Equation 27.19, we see that the temperature coefficient of resistivity can be expressed as

$$\alpha = \frac{1}{\rho_0} \frac{\Delta\rho}{\Delta T} \tag{27.20}$$

Temperature coefficient of resistivity

where $\Delta\rho = \rho - \rho_0$ is the change in resistivity in the temperature interval $\Delta T = T - T_0$.

The temperature coefficients of resistivity for various materials are given in Table 27.1. Note that the unit for α is degrees Celsius^{-1} [(°C)$^{-1}$]. Because resistance is proportional to resistivity (Eq. 27.11), we can write the variation of resistance as

$$R = R_0[1 + \alpha(T - T_0)] \tag{27.21}$$

Use of this property enables us to make precise temperature measurements, as shown in the following example.

EXAMPLE 27.6 A Platinum Resistance Thermometer

A resistance thermometer, which measures temperature by measuring the change in resistance of a conductor, is made from platinum and has a resistance of 50.0 Ω at 20.0°C. When immersed in a vessel containing melting indium, its resistance increases to 76.8 Ω. Calculate the melting point of the indium.

Solution Solving Equation 27.21 for ΔT and using the α value for platinum given in Table 27.1, we obtain

$$\Delta T = \frac{R - R_0}{\alpha R_0} = \frac{76.8\ \Omega - 50.0\ \Omega}{[3.92 \times 10^{-3}\ (°C)^{-1}](50.0\ \Omega)} = 137°C$$

Because $T_0 = 20.0°C$, we find that T, the temperature of the melting indium sample, is   157°C.

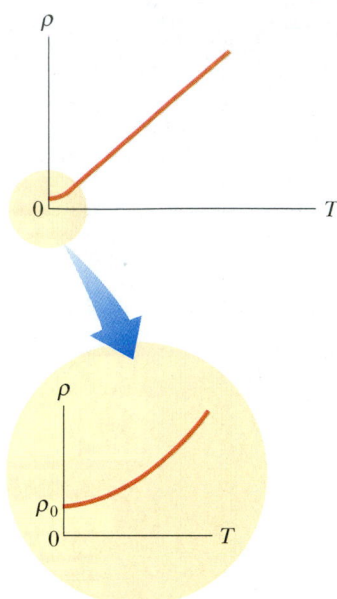

Figure 27.10 Resistivity versus temperature for a metal such as copper. The curve is linear over a wide range of temperatures, and ρ increases with increasing temperature. As *T* approaches absolute zero (inset), the resistivity approaches a finite value ρ_0.

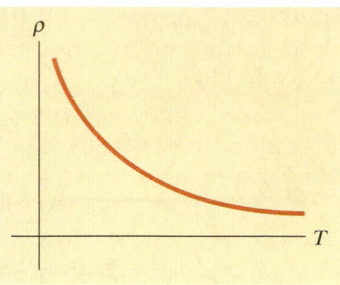

Figure 27.11 Resistivity versus temperature for a pure semiconductor, such as silicon or germanium.

For metals like copper, resistivity is nearly proportional to temperature, as shown in Figure 27.10. However, a nonlinear region always exists at very low temperatures, and the resistivity usually approaches some finite value as the temperature nears absolute zero. This residual resistivity near absolute zero is caused primarily by the collision of electrons with impurities and imperfections in the metal. In contrast, high-temperature resistivity (the linear region) is predominantly characterized by collisions between electrons and metal atoms.

Notice that three of the α values in Table 27.1 are negative; this indicates that the resistivity of these materials decreases with increasing temperature (Fig. 27.11). This behavior is due to an increase in the density of charge carriers at higher temperatures.

Because the charge carriers in a semiconductor are often associated with impurity atoms, the resistivity of these materials is very sensitive to the type and concentration of such impurities. We shall return to the study of semiconductors in Chapter 43 of the extended version of this text.

Quick Quiz 27.5

When does a lightbulb carry more current—just after it is turned on and the glow of the metal filament is increasing, or after it has been on for a few milliseconds and the glow is steady?

Optional Section

27.5 ▶ **SUPERCONDUCTORS**

There is a class of metals and compounds whose resistance decreases to zero when they are below a certain temperature T_c, known as the *critical temperature*. These materials are known as **superconductors.** The resistance–temperature graph for a superconductor follows that of a normal metal at temperatures above T_c (Fig. 27.12). When the temperature is at or below T_c, the resistivity drops suddenly to zero. This phenomenon was discovered in 1911 by the Dutch physicist Heike Kamerlingh-Onnes (1853–1926) as he worked with mercury, which is a superconductor below 4.2 K. Recent measurements have shown that the resistivities of superconductors below their T_c values are less than $4 \times 10^{-25} \ \Omega \cdot m$—around 10^{17} times smaller than the resistivity of copper and in practice considered to be zero.

Today thousands of superconductors are known, and as Figure 27.13 illustrates, the critical temperatures of recently discovered superconductors are substantially higher than initially thought possible. Two kinds of superconductors are recognized. The more recently identified ones, such as $YBa_2Cu_3O_7$, are essentially ceramics with high critical temperatures, whereas superconducting materials such

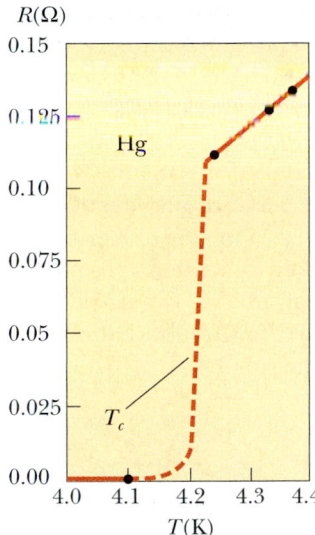

Figure 27.12 Resistance versus temperature for a sample of mercury (Hg). The graph follows that of a normal metal above the critical temperature T_c. The resistance drops to zero at T_c, which is 4.2 K for mercury.

A small permanent magnet levitated above a disk of the superconductor $YBa_2Cu_3O_7$, which is at 77 K. *(Courtesy of IBM Research Laboratory)*

as those observed by Kamerlingh-Onnes are metals. If a room-temperature superconductor is ever identified, its impact on technology could be tremendous.

The value of T_c is sensitive to chemical composition, pressure, and molecular structure. It is interesting to note that copper, silver, and gold, which are excellent conductors, do not exhibit superconductivity.

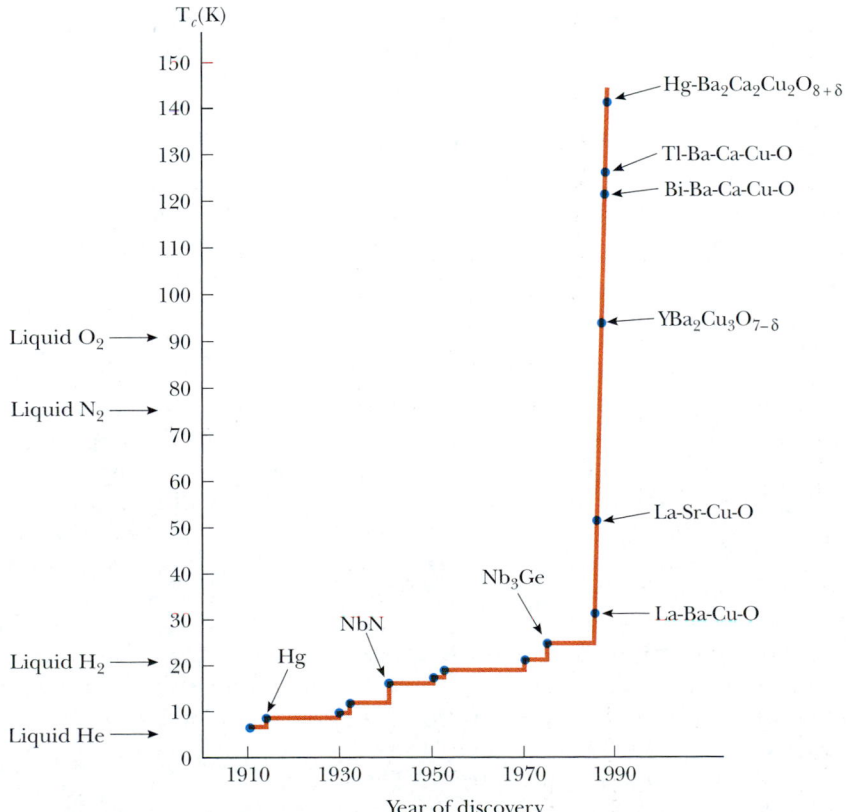

Figure 27.13 Evolution of the superconducting critical temperature since the discovery of the phenomenon.

One of the truly remarkable features of superconductors is that once a current is set up in them, it persists *without any applied potential difference* (because $R = 0$). Steady currents have been observed to persist in superconducting loops for several years with no apparent decay!

An important and useful application of superconductivity is in the development of superconducting magnets, in which the magnitudes of the magnetic field are about ten times greater than those produced by the best normal electromagnets. Such superconducting magnets are being considered as a means of storing energy. Superconducting magnets are currently used in medical magnetic resonance imaging (MRI) units, which produce high-quality images of internal organs without the need for excessive exposure of patients to x-rays or other harmful radiation.

For further information on superconductivity, see Section 43.8.

27.6 · ELECTRICAL ENERGY AND POWER

If a battery is used to establish an electric current in a conductor, the chemical energy stored in the battery is continuously transformed into kinetic energy of the charge carriers. In the conductor, this kinetic energy is quickly lost as a result of collisions between the charge carriers and the atoms making up the conductor, and this leads to an increase in the temperature of the conductor. In other words, the chemical energy stored in the battery is continuously transformed to internal energy associated with the temperature of the conductor.

Consider a simple circuit consisting of a battery whose terminals are connected to a resistor, as shown in Figure 27.14. (Resistors are designated by the symbol ⎓⎓⎓⎓.) Now imagine following a positive quantity of charge ΔQ that is moving clockwise around the circuit from point *a* through the battery and resistor back to point *a*. Points *a* and *d* are *grounded* (ground is designated by the symbol ⏚); that is, we take the electric potential at these two points to be zero. As the charge moves from *a* to *b* through the battery, its electric potential energy *U increases* by an amount $\Delta V \Delta Q$ (where ΔV is the potential difference between *b* and *a*), while the chemical potential energy in the battery *decreases* by the same amount. (Recall from Eq. 25.9 that $\Delta U = q \Delta V$.) However, as the charge moves from *c* to *d* through the resistor, it *loses* this electric potential energy as it collides with atoms in the resistor, thereby producing internal energy. If we neglect the resistance of the connecting wires, no loss in energy occurs for paths *bc* and *da*. When the charge arrives at point *a*, it must have the same electric potential energy (zero) that it had at the start.[5] Note that because charge cannot build up at any point, the current is the same everywhere in the circuit.

The rate at which the charge ΔQ loses potential energy in going through the resistor is

$$\frac{\Delta U}{\Delta t} = \frac{\Delta Q}{\Delta t} \Delta V = I \Delta V$$

where *I* is the current in the circuit. In contrast, the charge regains this energy when it passes through the battery. Because the rate at which the charge loses energy equals the power $\mathcal{P}$ delivered to the resistor (which appears as internal energy), we have

$$\mathcal{P} = I \Delta V \qquad \qquad \textbf{(27.22)}$$

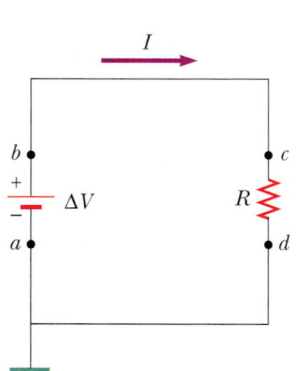

Figure 27.14 A circuit consisting of a resistor of resistance *R* and a battery having a potential difference ΔV across its terminals. Positive charge flows in the clockwise direction. Points *a* and *d* are grounded.

Power

---

[5] Note that once the current reaches its steady-state value, there is *no* change in the kinetic energy of the charge carriers creating the current.

In this case, the power is supplied to a resistor by a battery. However, we can use Equation 27.22 to determine the power transferred to *any* device carrying a current I and having a potential difference ΔV between its terminals.

Using Equation 27.22 and the fact that $\Delta V = IR$ for a resistor, we can express the power delivered to the resistor in the alternative forms

$$\mathscr{P} = I^2 R = \frac{(\Delta V)^2}{R} \qquad \text{(27.23)}$$

Power delivered to a resistor

When I is expressed in amperes, ΔV in volts, and R in ohms, the SI unit of power is the watt, as it was in Chapter 7 in our discussion of mechanical power. The power lost as internal energy in a conductor of resistance R is called *joule heating*[6]; this transformation is also often referred to as an $I^2 R$ loss.

A battery, a device that supplies electrical energy, is called either a *source of electromotive force* or, more commonly, an *emf source*. The concept of emf is discussed in greater detail in Chapter 28. (The phrase *electromotive force* is an unfortunate choice because it describes not a force but rather a potential difference in volts.) **When the internal resistance of the battery is neglected, the potential difference between points *a* and *b* in Figure 27.14 is equal to the emf $\mathcal{E}$ of the battery**—that is, $\Delta V = V_b - V_a = \mathcal{E}$. This being true, we can state that the current in the circuit is $I = \Delta V/R = \mathcal{E}/R$. Because $\Delta V = \mathcal{E}$, the power supplied by the emf source can be expressed as $\mathscr{P} = I\mathcal{E}$, which equals the power delivered to the resistor, $I^2 R$.

When transporting electrical energy through power lines, such as those shown in Figure 27.15, utility companies seek to minimize the power transformed to internal energy in the lines and maximize the energy delivered to the consumer. Because $\mathscr{P} = I\Delta V$, the same amount of power can be transported either at high currents and low potential differences or at low currents and high potential differences. Utility companies choose to transport electrical energy at low currents and high potential differences primarily for economic reasons. Copper wire is very expensive, and so it is cheaper to use high-resistance wire (that is, wire having a small cross-sectional area; see Eq. 27.11). Thus, in the expression for the power delivered to a resistor, $\mathscr{P} = I^2 R$, the resistance of the wire is fixed at a relatively high value for economic considerations. The $I^2 R$ loss can be reduced by keeping the current I as low as possible. In some instances, power is transported at potential differences as great as 765 kV. Once the electricity reaches your city, the potential difference is usually reduced to 4 kV by a device called a *transformer*. Another transformer drops the potential difference to 240 V before the electricity finally reaches your home. Of course, each time the potential difference decreases, the current increases by the same factor, and the power remains the same. We shall discuss transformers in greater detail in Chapter 33.

Figure 27.15   Power companies transfer electrical energy at high potential differences. *(Comstock)*

Quick Quiz 27.6

The same potential difference is applied to the two lightbulbs shown in Figure 27.16. Which one of the following statements is true?
(a) The 30-W bulb carries the greater current and has the higher resistance.
(b) The 30-W bulb carries the greater current, but the 60-W bulb has the higher resistance.

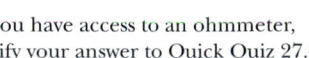

QuickLab

If you have access to an ohmmeter, verify your answer to Quick Quiz 27.6 by testing the resistance of a few lightbulbs.

___

[6] It is called *joule heating* even though the process of heat does not occur. This is another example of incorrect usage of the word *heat* that has become entrenched in our language.

Figure 27.16   These light-bulbs operate at their rated power only when they are connected to a 120-V source. *(George Semple)*

(c) The 30-W bulb has the higher resistance, but the 60-W bulb carries the greater current.
(d) The 60-W bulb carries the greater current and has the higher resistance.

QuickLab

From the labels on household appliances such as hair dryers, televisions, and stereos, estimate the annual cost of operating them.

Quick Quiz 27.7

For the two lightbulbs shown in Figure 27.17, rank the current values at points *a* through *f*, from greatest to least.

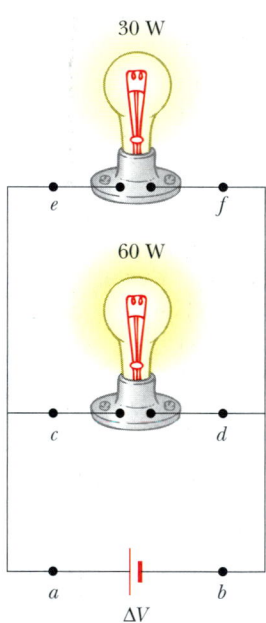

Figure 27.17   Two lightbulbs connected across the same potential difference. The bulbs operate at their rated power only if they are connected to a 120-V battery.

EXAMPLE 27.7   Power in an Electric Heater

An electric heater is constructed by applying a potential difference of 120 V to a Nichrome wire that has a total resistance of 8.00 Ω. Find the current carried by the wire and the power rating of the heater.

Solution   Because $\Delta V = IR$, we have

$$I = \frac{\Delta V}{R} = \frac{120 \text{ V}}{8.00 \ \Omega} = \boxed{15.0 \text{ A}}$$

We can find the power rating using the expression $\mathcal{P} = I^2 R$:

$$\mathcal{P} = I^2 R = (15.0 \text{ A})^2 (8.00 \ \Omega) = \boxed{1.80 \text{ kW}}$$

If we doubled the applied potential difference, the current would double but the power would quadruple because $\mathcal{P} = (\Delta V)^2 / R$.

EXAMPLE 27.8 ▶ The Cost of Making Dinner

Estimate the cost of cooking a turkey for 4 h in an oven that operates continuously at 20.0 A and 240 V

Solution   The power used by the oven is

$$\mathcal{P} = I\,\Delta V = (20.0 \text{ A})(240 \text{ V}) = 4\,800 \text{ W} = 4.80 \text{ kW}$$

Because the energy consumed equals power × time, the amount of energy for which you must pay is

$$\text{Energy} = \mathcal{P}t = (4.80 \text{ kW})(4 \text{ h}) = 19.2 \text{ kWh}$$

If the energy is purchased at an estimated price of 8.00¢ per kilowatt hour, the cost is

$$\text{Cost} = (19.2 \text{ kWh})(\$0.080/\text{kWh}) = \boxed{\$1.54}$$

Demands on our dwindling energy supplies have made it necessary for us to be aware of the energy requirements of our electrical devices. Every electrical appliance carries a label that contains the information you need to calculate the appliance's power requirements. In many cases, the power consumption in watts is stated directly, as it is on a lightbulb. In other cases, the amount of current used by the device and the potential difference at which it operates are given. This information and Equation 27.22 are sufficient for calculating the operating cost of any electrical device.

Exercise   What does it cost to operate a 100-W lightbulb for 24 h if the power company charges \$0.08/kWh?

Answer   \$0.19.

EXAMPLE 27.9 ▶ Current in an Electron Beam

In a certain particle accelerator, electrons emerge with an energy of 40.0 MeV (1 MeV = 1.60×10^{-13} J). The electrons emerge not in a steady stream but rather in pulses at the rate of 250 pulses/s. This corresponds to a time between pulses of 4.00 ms (Fig. 27.18). Each pulse has a duration of 200 ns, and the electrons in the pulse constitute a current of 250 mA. The current is zero between pulses. (a) How many electrons are delivered by the accelerator per pulse?

Solution   We use Equation 27.2 in the form $dQ = I\,dt$ and integrate to find the charge per pulse. While the pulse is on, the current is constant; thus,

$$Q_{\text{pulse}} = I\int dt = I\,\Delta t = (250 \times 10^{-3} \text{ A})(200 \times 10^{-9} \text{ s})$$

$$= 5.00 \times 10^{-8} \text{ C}$$

Dividing this quantity of charge per pulse by the electronic charge gives the number of electrons per pulse:

$$\text{Electrons per pulse} = \frac{5.00 \times 10^{-8} \text{ C/pulse}}{1.60 \times 10^{-19} \text{ C/electron}}$$

$$= \boxed{3.13 \times 10^{11} \text{ electrons/pulse}}$$

(b) What is the average current per pulse delivered by the accelerator?

Solution   Average current is given by Equation 27.1, $I_{\text{av}} = \Delta Q/\Delta t$. Because the time interval between pulses is 4.00 ms, and because we know the charge per pulse from part (a), we obtain

$$I_{\text{av}} = \frac{Q_{\text{pulse}}}{\Delta t} = \frac{5.00 \times 10^{-8} \text{ C}}{4.00 \times 10^{-3} \text{ s}} = \boxed{12.5 \ \mu\text{A}}$$

This represents only 0.005% of the peak current, which is 250 mA.

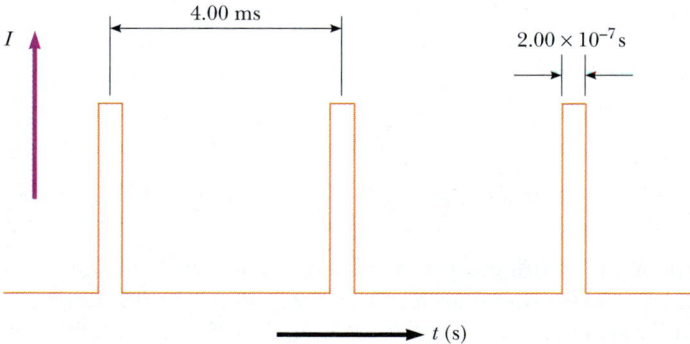

Figure 27.18   Current versus time for a pulsed beam of electrons.

(c) What is the maximum power delivered by the electron beam?

Solution   By definition, power is energy delivered per unit time. Thus, the maximum power is equal to the energy delivered by a pulse divided by the pulse duration:

$$\mathcal{P} = \frac{E}{\Delta t}$$

$$= \frac{(3.13 \times 10^{11} \text{ electrons/pulse})(40.0 \text{ MeV/electron})}{2.00 \times 10^{-7} \text{ s/pulse}}$$

$$= (6.26 \times 10^{19} \text{ MeV/s})(1.60 \times 10^{-13} \text{ J/MeV})$$

$$= 1.00 \times 10^7 \text{ W} = \boxed{10.0 \text{ MW}}$$

We could also compute this power directly. We assume that each electron had zero energy before being accelerated. Thus, by definition, each electron must have gone through a potential difference of 40.0 MV to acquire a final energy of 40.0 MeV. Hence, we have

$$\mathcal{P} = I \Delta V = (250 \times 10^{-3} \text{ A})(40.0 \times 10^6 \text{ V}) = \boxed{10.0 \text{ MW}}$$

Summary

The **electric current** I in a conductor is defined as

$$I \equiv \frac{dQ}{dt} \tag{27.2}$$

where dQ is the charge that passes through a cross-section of the conductor in a time dt. The SI unit of current is the **ampere** (A), where 1 A = 1 C/s.

The average current in a conductor is related to the motion of the charge carriers through the relationship

$$I_{av} = nqv_d A \tag{27.4}$$

where n is the density of charge carriers, q is the charge on each carrier, v_d is the drift speed, and A is the cross-sectional area of the conductor.

The magnitude of the **current density** J in a conductor is the current per unit area:

$$J \equiv \frac{I}{A} = nqv_d \tag{27.5}$$

The current density in a conductor is proportional to the electric field according to the expression

$$\mathbf{J} = \sigma \mathbf{E} \tag{27.7}$$

The proportionality constant σ is called the **conductivity** of the material of which the conductor is made. The inverse of σ is known as **resistivity** ρ ($\rho = 1/\sigma$). Equation 27.7 is known as **Ohm's law,** and a material is said to obey this law if the ratio of its current density $\mathbf{J}$ to its applied electric field $\mathbf{E}$ is a constant that is independent of the applied field.

The **resistance** R of a conductor is defined either in terms of the length of the conductor or in terms of the potential difference across it:

$$R \equiv \frac{\ell}{\sigma A} \equiv \frac{\Delta V}{I} \tag{27.8}$$

where ℓ is the length of the conductor, σ is the conductivity of the material of which it is made, A is its cross-sectional area, ΔV is the potential difference across it, and I is the current it carries.

The SI unit of resistance is volts per ampere, which is defined to be 1 **ohm** (Ω); that is, $1\ \Omega = 1$ V/A. If the resistance is independent of the applied potential difference, the conductor obeys Ohm's law.

In a classical model of electrical conduction in metals, the electrons are treated as molecules of a gas. In the absence of an electric field, the average velocity of the electrons is zero. When an electric field is applied, the electrons move (on the average) with a **drift velocity** $\mathbf{v}_d$ that is opposite the electric field and given by the expression

$$\mathbf{v}_d = \frac{q\mathbf{E}}{m_e}\tau \qquad \textbf{(27.14)}$$

where τ is the average time between electron–atom collisions, m_e is the mass of the electron, and q is its charge. According to this model, the resistivity of the metal is

$$\rho = \frac{m_e}{nq^2\tau} \qquad \textbf{(27.17)}$$

where n is the number of free electrons per unit volume.

The resistivity of a conductor varies approximately linearly with temperature according to the expression

$$\rho = \rho_0[1 + \alpha(T - T_0)] \qquad \textbf{(27.19)}$$

where α is the **temperature coefficient of resistivity** and ρ_0 is the resistivity at some reference temperature T_0.

If a potential difference ΔV is maintained across a resistor, the **power,** or rate at which energy is supplied to the resistor, is

$$\mathcal{P} = I\,\Delta V \qquad \textbf{(27.22)}$$

Because the potential difference across a resistor is given by $\Delta V = IR$, we can express the power delivered to a resistor in the form

$$\mathcal{P} = I^2R = \frac{(\Delta V)^2}{R} \qquad \textbf{(27.23)}$$

The electrical energy supplied to a resistor appears in the form of internal energy in the resistor.

QUESTIONS

1. Newspaper articles often contain statements such as "10 000 volts of electricity surged through the victim's body." What is wrong with this statement?

2. What is the difference between resistance and resistivity?

3. Two wires A and B of circular cross-section are made of the same metal and have equal lengths, but the resistance of wire A is three times greater than that of wire B. What is the ratio of their cross-sectional areas? How do their radii compare?

4. What is required in order to maintain a steady current in a conductor?

5. Do all conductors obey Ohm's law? Give examples to justify your answer.

6. When the voltage across a certain conductor is doubled, the current is observed to increase by a factor of three. What can you conclude about the conductor?

7. In the water analogy of an electric circuit, what corresponds to the power supply, resistor, charge, and potential difference?

8. Why might a "good" electrical conductor also be a "good" thermal conductor?

9. On the basis of the atomic theory of matter, explain why the resistance of a material should increase as its temperature increases.

10. How does the resistance for copper and silicon change with temperature? Why are the behaviors of these two materials different?

11. Explain how a current can persist in a superconductor in the absence of any applied voltage.

12. What single experimental requirement makes superconducting devices expensive to operate? In principle, can this limitation be overcome?

13. What would happen to the drift velocity of the electrons in a wire and to the current in the wire if the electrons could move freely without resistance through the wire?

14. If charges flow very slowly through a metal, why does it not require several hours for a light to turn on when you throw a switch?

15. In a conductor, the electric field that drives the electrons through the conductor propagates with a speed that is almost the same as the speed of light, even though the drift velocity of the electrons is very small. Explain how these can both be true. Does a given electron move from one end of the conductor to the other?

16. Two conductors of the same length and radius are connected across the same potential difference. One conductor has twice the resistance of the other. To which conductor is more power delivered?

17. Car batteries are often rated in ampere-hours. Does this designate the amount of current, power, energy, or charge that can be drawn from the battery?

18. If you were to design an electric heater using Nichrome wire as the heating element, what parameters of the wire could you vary to meet a specific power output, such as 1 000 W?

19. Consider the following typical monthly utility rate structure: $2.00 for the first 16 kWh, 8.00¢/kWh for the next 34 kWh, 6.50¢/kWh for the next 50 kWh, 5.00¢/kWh for the next 100 kWh, 4.00¢/kWh for the next 200 kWh, and 3.50¢/kWh for all kilowatt-hours in excess of 400 kWh. On the basis of these rates, determine the amount charged for 327 kWh.

PROBLEMS

1, 2, 3 = straightforward, intermediate, challenging   ☐ = full solution available in the *Student Solutions Manual and Study Guide*
WEB = solution posted at **http://www.saunderscollege.com/physics/**   🖥 = Computer useful in solving problem   🎮 = Interactive Physics
☐ = paired numerical/symbolic problems

Section 27.1   Electric Current

1. In a particular cathode ray tube, the measured beam current is 30.0 μA. How many electrons strike the tube screen every 40.0 s?

2. A teapot with a surface area of 700 cm^2 is to be silver plated. It is attached to the negative electrode of an electrolytic cell containing silver nitrate ($Ag^+NO_3^-$). If the cell is powered by a 12.0-V battery and has a resistance of 1.80 Ω, how long does it take for a 0.133-mm layer of silver to build up on the teapot? (The density of silver is 10.5 $\times$ 10^3 kg/m^3.)

WEB **3.** Suppose that the current through a conductor decreases exponentially with time according to the expression $I(t) = I_0 e^{-t/\tau}$, where I_0 is the initial current (at $t = 0$) and τ is a constant having dimensions of time. Consider a fixed observation point within the conductor. (a) How much charge passes this point between $t = 0$ and $t = \tau$? (b) How much charge passes this point between $t = 0$ and $t = 10\tau$? (c) How much charge passes this point between $t = 0$ and $t = \infty$?

4. In the Bohr model of the hydrogen atom, an electron in the lowest energy state follows a circular path at a distance of 5.29 $\times$ 10^{-11} m from the proton. (a) Show that the speed of the electron is 2.19 $\times$ 10^6 m/s. (b) What is the effective current associated with this orbiting electron?

5. A small sphere that carries a charge of 8.00 nC is whirled in a circle at the end of an insulating string. The angular frequency of rotation is 100π rad/s. What average current does this rotating charge represent?

6. A small sphere that carries a charge q is whirled in a circle at the end of an insulating string. The angular frequency of rotation is ω. What average current does this rotating charge represent?

7. The quantity of charge q (in coulombs) passing through a surface of area 2.00 cm^2 varies with time according to the equation $q = 4.00t^3 + 5.00t + 6.00$, where t is in seconds. (a) What is the instantaneous current through the surface at $t = 1.00$ s? (b) What is the value of the current density?

8. An electric current is given by the expression $I(t) = 100 \sin(120\pi t)$, where I is in amperes and t is in seconds. What is the total charge carried by the current from $t = 0$ to $t = 1/240$ s?

9. Figure P27.9 represents a section of a circular conductor of nonuniform diameter carrying a current of 5.00 A. The radius of cross-section A_1 is 0.400 cm. (a) What is the magnitude of the current density across A_1? (b) If the current density across A_2 is one-fourth the value across A_1, what is the radius of the conductor at A_2?

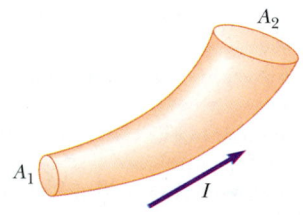

Figure P27.9

10. A Van de Graaff generator produces a beam of 2.00-MeV *deuterons*, which are heavy hydrogen nuclei containing a proton and a neutron. (a) If the beam current is 10.0 μA, how far apart are the deuterons? (b) Is their electrostatic repulsion a factor in beam stability? Explain.

11. The electron beam emerging from a certain high-energy electron accelerator has a circular cross-section of radius 1.00 mm. (a) If the beam current is 8.00 μA, what is the current density in the beam, assuming that it is uniform throughout? (b) The speed of the electrons is so close to the speed of light that their speed can be taken as $c = 3.00 \times 10^8$ m/s with negligible error. Find the electron density in the beam. (c) How long does it take for Avogadro's number of electrons to emerge from the accelerator?

12. An aluminum wire having a cross-sectional area of 4.00×10^{-6} m^2 carries a current of 5.00 A. Find the drift speed of the electrons in the wire. The density of aluminum is 2.70 g/cm^3. (Assume that one electron is supplied by each atom.)

Section 27.2 Resistance and Ohm's Law

13. A lightbulb has a resistance of 240 Ω when operating at a voltage of 120 V. What is the current through the lightbulb?

14. A resistor is constructed of a carbon rod that has a uniform cross-sectional area of 5.00 mm^2. When a potential difference of 15.0 V is applied across the ends of the rod, there is a current of 4.00×10^{-3} A in the rod. Find (a) the resistance of the rod and (b) the rod's length.

WEB 15. A 0.900-V potential difference is maintained across a 1.50-m length of tungsten wire that has a cross-sectional area of 0.600 mm^2. What is the current in the wire?

16. A conductor of uniform radius 1.20 cm carries a current of 3.00 A produced by an electric field of 120 V/m. What is the resistivity of the material?

17. Suppose that you wish to fabricate a uniform wire out of 1.00 g of copper. If the wire is to have a resistance of $R = 0.500$ Ω, and if all of the copper is to be used, what will be (a) the length and (b) the diameter of this wire?

18. (a) Make an order-of-magnitude estimate of the resistance between the ends of a rubber band. (b) Make an order-of-magnitude estimate of the resistance between the 'heads' and 'tails' sides of a penny. In each case, state what quantities you take as data and the values you measure or estimate for them. (c) What would be the order of magnitude of the current that each carries if it were connected across a 120-V power supply? (WARNING! Do not try this at home!)

19. A solid cube of silver (density = 10.5 g/cm^3) has a mass of 90.0 g. (a) What is the resistance between opposite faces of the cube? (b) If there is one conduction electron for each silver atom, what is the average drift speed of electrons when a potential difference of 1.00×10^{-5} V is applied to opposite faces? (The

atomic number of silver is 47, and its molar mass is 107.87 g/mol.)

20. A metal wire of resistance R is cut into three equal pieces that are then connected side by side to form a new wire whose length is equal to one-third the original length. What is the resistance of this new wire?

21. A wire with a resistance R is lengthened to 1.25 times its original length by being pulled through a small hole. Find the resistance of the wire after it has been stretched.

22. Aluminum and copper wires of equal length are found to have the same resistance. What is the ratio of their radii?

23. A current density of 6.00×10^{-13} A/m^2 exists in the atmosphere where the electric field (due to charged thunderclouds in the vicinity) is 100 V/m. Calculate the electrical conductivity of the Earth's atmosphere in this region.

24. The rod in Figure P27.24 (not drawn to scale) is made of two materials. Both have a square cross section of 3.00 mm on a side. The first material has a resistivity of 4.00×10^{-3} $\Omega \cdot$ m and is 25.0 cm long, while the second material has a resistivity of 6.00×10^{-3} $\Omega \cdot$ m and is 40.0 cm long. What is the resistance between the ends of the rod?

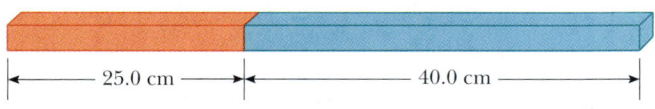

Figure P27.24

Section 27.3 A Model for Electrical Conduction

WEB 25. If the drift velocity of free electrons in a copper wire is 7.84×10^{-4} m/s, what is the electric field in the conductor?

26. If the current carried by a conductor is doubled, what happens to the (a) charge carrier density? (b) current density? (c) electron drift velocity? (d) average time between collisions?

27. Use data from Example 27.1 to calculate the collision mean free path of electrons in copper, assuming that the average thermal speed of conduction electrons is 8.60×10^5 m/s.

Section 27.4 Resistance and Temperature

28. While taking photographs in Death Valley on a day when the temperature is 58.0°C, Bill Hiker finds that a certain voltage applied to a copper wire produces a current of 1.000 A. Bill then travels to Antarctica and applies the same voltage to the same wire. What current does he register there if the temperature is −88.0°C? Assume that no change occurs in the wire's shape and size.

29. A certain lightbulb has a tungsten filament with a resistance of 19.0 Ω when cold and of 140 Ω when hot. Assuming that Equation 27.21 can be used over the large

temperature range involved here, find the temperature of the filament when hot. (Assume an initial temperature of 20.0°C.)

30. A carbon wire and a Nichrome wire are connected in series. If the combination has a resistance of 10.0 kΩ at 0°C, what is the resistance of each wire at 0°C such that the resistance of the combination does not change with temperature? (Note that the equivalent resistance of two resistors in series is the sum of their resistances.)

31. An aluminum wire with a diameter of 0.100 mm has a uniform electric field with a magnitude of 0.200 V/m imposed along its entire length. The temperature of the wire is 50.0°C. Assume one free electron per atom. (a) Using the information given in Table 27.1, determine the resistivity. (b) What is the current density in the wire? (c) What is the total current in the wire? (d) What is the drift speed of the conduction electrons? (e) What potential difference must exist between the ends of a 2.00-m length of the wire if the stated electric field is to be produced?

32. **Review Problem.** An aluminum rod has a resistance of 1.234 Ω at 20.0°C. Calculate the resistance of the rod at 120°C by accounting for the changes in both the resistivity and the dimensions of the rod.

33. What is the fractional change in the resistance of an iron filament when its temperature changes from 25.0°C to 50.0°C?

34. The resistance of a platinum wire is to be calibrated for low-temperature measurements. A platinum wire with a resistance of 1.00 Ω at 20.0°C is immersed in liquid nitrogen at 77 K ($-$196°C). If the temperature response of the platinum wire is linear, what is the expected resistance of the platinum wire at $-$196°C? ($\alpha_{platinum} = 3.92 \times 10^{-3}/°C$)

35. The temperature of a tungsten sample is raised while a copper sample is maintained at 20°C. At what temperature will the resistivity of the tungsten sample be four times that of the copper sample?

36. A segment of Nichrome wire is initially at 20.0°C. Using the data from Table 27.1, calculate the temperature to which the wire must be heated if its resistance is to be doubled.

Section 27.6 Electrical Energy and Power

37. A toaster is rated at 600 W when connected to a 120-V source. What current does the toaster carry, and what is its resistance?

38. In a hydroelectric installation, a turbine delivers 1 500 hp to a generator, which in turn converts 80.0% of the mechanical energy into electrical energy. Under these conditions, what current does the generator deliver at a terminal potential difference of 2 000 V?

WEB 39. **Review Problem.** What is the required resistance of an immersion heater that increases the temperature of 1.50 kg of water from 10.0°C to 50.0°C in 10.0 min while operating at 110 V?

40. **Review Problem.** What is the required resistance of an immersion heater that increases the temperature of a mass m of liquid water from T_1 to T_2 in a time t while operating at a voltage ΔV?

41. Suppose that a voltage surge produces 140 V for a moment. By what percentage does the power output of a 120-V, 100-W lightbulb increase? (Assume that its resistance does not change.)

42. A 500-W heating coil designed to operate from 110 V is made of Nichrome wire 0.500 mm in diameter. (a) Assuming that the resistivity of the Nichrome remains constant at its 20.0°C value, find the length of wire used. (b) Now consider the variation of resistivity with temperature. What power does the coil of part (a) actually deliver when it is heated to 1 200°C?

43. A coil of Nichrome wire is 25.0 m long. The wire has a diameter of 0.400 mm and is at 20.0°C. If it carries a current of 0.500 A, what are (a) the magnitude of the electric field in the wire and (b) the power delivered to it? (c) If the temperature is increased to 340°C and the potential difference across the wire remains constant, what is the power delivered?

44. Batteries are rated in terms of ampere-hours (A·h): For example, a battery that can produce a current of 2.00 A for 3.00 h is rated at 6.00 A·h. (a) What is the total energy, in kilowatt-hours, stored in a 12.0-V battery rated at 55.0 A·h? (b) At a rate of $0.060 0 per kilowatt-hour, what is the value of the electricity produced by this battery?

45. A 10.0-V battery is connected to a 120-Ω resistor. Neglecting the internal resistance of the battery, calculate the power delivered to the resistor.

46. It is estimated that each person in the United States (population = 270 million) has one electric clock, and that each clock uses energy at a rate of 2.50 W. To supply this energy, about how many metric tons of coal are burned per hour in coal-fired electricity generating plants that are, on average, 25.0% efficient? (The heat of combustion for coal is 33.0 MJ/kg.)

47. Compute the cost per day of operating a lamp that draws 1.70 A from a 110-V line if the cost of electrical energy is $0.060 0/kWh.

48. **Review Problem.** The heating element of a coffeemaker operates at 120 V and carries a current of 2.00 A. Assuming that all of the energy transferred from the heating element is absorbed by the water, calculate how long it takes to heat 0.500 kg of water from room temperature (23.0°C) to the boiling point.

49. A certain toaster has a heating element made of Nichrome resistance wire. When the toaster is first connected to a 120-V source of potential difference (and the wire is at a temperature of 20.0°C), the initial current is 1.80 A. However, the current begins to decrease as the resistive element warms up. When the toaster has reached its final operating temperature, the current has dropped to 1.53 A. (a) Find the power the toaster con-

sumes when it is at its operating temperature. (b) What is the final temperature of the heating element?

50. To heat a room having ceilings 8.0 ft high, about 10.0 W of electric power are required per square foot. At a cost of $0.080 0/kWh, how much does it cost per day to use electricity to heat a room measuring 10.0 ft × 15.0 ft?

51. Estimate the cost of one person's routine use of a hair dryer for 1 yr. If you do not use a blow dryer yourself, observe or interview someone who does. State the quantities you estimate and their values.

ADDITIONAL PROBLEMS

52. One lightbulb is marked "25 W 120 V," and another "100 W 120 V"; this means that each bulb converts its respective power when plugged into a constant 120-V potential difference. (a) Find the resistance of each bulb. (b) How long does it take for 1.00 C to pass through the dim bulb? How is this charge different at the time of its exit compared with the time of its entry? (c) How long does it take for 1.00 J to pass through the dim bulb? How is this energy different at the time of its exit compared with the time of its entry? (d) Find the cost of running the dim bulb continuously for 30.0 days if the electric company sells its product at $0.070 0 per kWh. What product *does* the electric company sell? What is its price for one SI unit of this quantity?

53. A high-voltage transmission line with a diameter of 2.00 cm and a length of 200 km carries a steady current of 1 000 A. If the conductor is copper wire with a free charge density of 8.00×10^{28} electrons/m^3, how long does it take one electron to travel the full length of the cable?

54. A high-voltage transmission line carries 1 000 A starting at 700 kV for a distance of 100 mi. If the resistance in the wire is 0.500 Ω/mi, what is the power loss due to resistive losses?

55. A more general definition of the temperature coefficient of resistivity is

$$\alpha = \frac{1}{\rho} \frac{d\rho}{dT}$$

where ρ is the resistivity at temperature T. (a) Assuming that α is constant, show that

$$\rho = \rho_0 e^{\alpha(T - T_0)}$$

where ρ_0 is the resistivity at temperature T_0. (b) Using the series expansion ($e^x \cong 1 + x$ for $x \ll 1$), show that the resistivity is given approximately by the expression $\rho = \rho_0[1 + \alpha(T - T_0)]$ for $\alpha(T - T_0) \ll 1$.

56. A copper cable is to be designed to carry a current of 300 A with a power loss of only 2.00 W/m. What is the required radius of the copper cable?

WEB 57. An experiment is conducted to measure the electrical resistivity of Nichrome in the form of wires with different lengths and cross-sectional areas. For one set of

measurements, a student uses 30-gauge wire, which has a cross-sectional area of 7.30×10^{-8} m^2. The student measures the potential difference across the wire and the current in the wire with a voltmeter and ammeter, respectively. For each of the measurements given in the table taken on wires of three different lengths, calculate the resistance of the wires and the corresponding values of the resistivity. What is the average value of the resistivity, and how does this value compare with the value given in Table 27.1?

L (m)	ΔV (V)	I (A)	R (Ω)	ρ (Ω·m)
0.540	5.22	0.500		
1.028	5.82	0.276		
1.543	5.94	0.187		

58. An electric utility company supplies a customer's house from the main power lines (120 V) with two copper wires, each of which is 50.0 m long and has a resistance of 0.108 Ω per 300 m. (a) Find the voltage at the customer's house for a load current of 110 A. For this load current, find (b) the power that the customer is receiving and (c) the power lost in the copper wires.

59. A straight cylindrical wire lying along the x axis has a length of 0.500 m and a diameter of 0.200 mm. It is made of a material described by Ohm's law with a resistivity of $\rho = 4.00 \times 10^{-8}$ Ω·m. Assume that a potential of 4.00 V is maintained at $x = 0$, and that $V = 0$ at $x = 0.500$ m. Find (a) the electric field **E** in the wire, (b) the resistance of the wire, (c) the electric current in the wire, and (d) the current density **J** in the wire. Express vectors in vector notation. (e) Show that $\mathbf{E} = \rho\mathbf{J}$.

60. A straight cylindrical wire lying along the x axis has a length L and a diameter d. It is made of a material described by Ohm's law with a resistivity ρ. Assume that a potential V is maintained at $x = 0$, and that $V = 0$ at $x = L$. In terms of L, d, V, ρ, and physical constants, derive expressions for (a) the electric field in the wire, (b) the resistance of the wire, (c) the electric current in the wire, and (d) the current density in the wire. Express vectors in vector notation. (e) Show that $\mathbf{E} = \rho\mathbf{J}$.

61. The potential difference across the filament of a lamp is maintained at a constant level while equilibrium temperature is being reached. It is observed that the steady-state current in the lamp is only one tenth of the current drawn by the lamp when it is first turned on. If the temperature coefficient of resistivity for the lamp at 20.0°C is 0.004 50 (°C)$^{-1}$, and if the resistance increases linearly with increasing temperature, what is the final operating temperature of the filament?

62. The current in a resistor decreases by 3.00 A when the potential difference applied across the resistor decreases from 12.0 V to 6.00 V. Find the resistance of the resistor.

63. An electric car is designed to run off a bank of 12.0-V batteries with a total energy storage of 2.00×10^7 J. (a) If the electric motor draws 8.00 kW, what is the current delivered to the motor? (b) If the electric motor draws 8.00 kW as the car moves at a steady speed of 20.0 m/s, how far will the car travel before it is "out of juice"?

64. Review Problem. When a straight wire is heated, its resistance is given by the expression $R = R_0[1 + \alpha(T - T_0)]$ according to Equation 27.21, where α is the temperature coefficient of resistivity. (a) Show that a more precise result, one that accounts for the fact that the length and area of the wire change when heated, is

$$R = \frac{R_0[1 + \alpha(T - T_0)][1 + \alpha'(T - T_0)]}{[1 + 2\alpha'(T - T_0)]}$$

where α' is the coefficient of linear expansion (see Chapter 19). (b) Compare these two results for a 2.00-m-long copper wire of radius 0.100 mm, first at 20.0°C and then heated to 100.0°C.

65. The temperature coefficients of resistivity in Table 27.1 were determined at a temperature of 20°C. What would they be at 0°C? (*Hint:* The temperature coefficient of resistivity at 20°C satisfies the expression $\rho = \rho_0[1 + \alpha(T - T_0)]$, where ρ_0 is the resistivity of the material at $T_0 = 20°C$. The temperature coefficient of resistivity α' at 0°C must satisfy the expression $\rho = \rho_0'[1 + \alpha'T]$, where ρ_0' is the resistivity of the material at 0°C.)

66. A resistor is constructed by shaping a material of resistivity ρ into a hollow cylinder of length L and with inner and outer radii r_a and r_b, respectively (Fig. P27.66). In use, the application of a potential difference between the ends of the cylinder produces a current parallel to the axis. (a) Find a general expression for the resistance of such a device in terms of L, ρ, r_a, and r_b. (b) Obtain a numerical value for R when $L = 4.00$ cm, $r_a = 0.500$ cm, $r_b = 1.20$ cm, and $\rho = 3.50 \times 10^5 \ \Omega \cdot$m. (c) Now suppose that the potential difference is applied between the inner and outer surfaces so that the resulting current flows radially outward. Find a general expression for the resistance of the device in terms of L, ρ,

r_a, and r_b. (d) Calculate the value of R, using the parameter values given in part (b).

67. In a certain stereo system, each speaker has a resistance of 4.00 Ω. The system is rated at 60.0 W in each channel, and each speaker circuit includes a fuse rated at 4.00 A. Is this system adequately protected against overload? Explain your reasoning.

68. A close analogy exists between the flow of energy due to a temperature difference (see Section 20.7) and the flow of electric charge due to a potential difference. The energy dQ and the electric charge dq are both transported by free electrons in the conducting material. Consequently, a good electrical conductor is usually a good thermal conductor as well. Consider a thin conducting slab of thickness dx, area A, and electrical conductivity σ, with a potential difference dV between opposite faces. Show that the current $I = dq/dt$ is given by the equation on the left:

Charge conduction	Analogous thermal conduction (Eq. 20.14)
$\dfrac{dq}{dt} = \sigma A \left| \dfrac{dV}{dx} \right|$	$\dfrac{dQ}{dt} = kA \left| \dfrac{dT}{dx} \right|$

In the analogous thermal conduction equation on the right, the rate of energy flow dQ/dt (in SI units of joules per second) is due to a temperature gradient dT/dx in a material of thermal conductivity k. State analogous rules relating the direction of the electric current to the change in potential and relating the direction of energy flow to the change in temperature.

69. Material with uniform resistivity ρ is formed into a wedge, as shown in Figure P27.69. Show that the resistance between face A and face B of this wedge is

$$R = \rho \frac{L}{w(y_2 - y_1)} \ln\left(\frac{y_2}{y_1}\right)$$

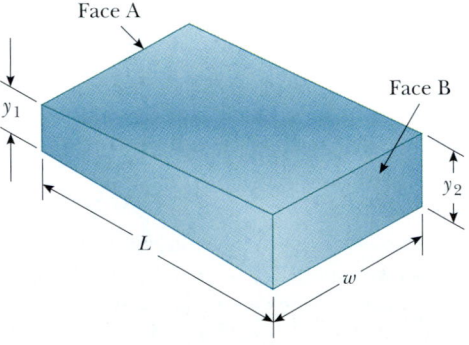

Face A

Face B

y_1

y_2

L

w

Figure P27.69

70. A material of resistivity ρ is formed into the shape of a truncated cone of altitude h, as shown in Figure P27.70.

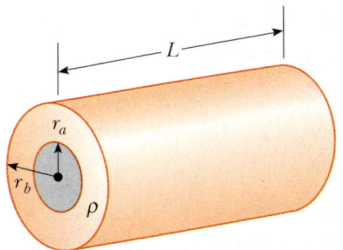

L

r_a

r_b

ρ

Figure P27.66

The bottom end has a radius b, and the top end has a radius a. Assuming that the current is distributed uniformly over any particular cross-section of the cone so that the current density is not a function of radial position (although it does vary with position along the axis

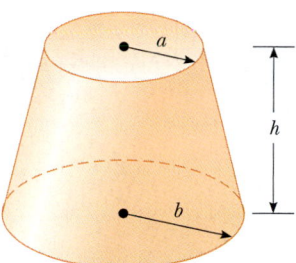

Figure P27.70

of the cone), show that the resistance between the two ends is given by the expression

$$R = \frac{\rho}{\pi}\left(\frac{h}{ab}\right)$$

71. The current–voltage characteristic curve for a semiconductor diode as a function of temperature T is given by the equation

$$I = I_0(e^{e\Delta V/k_B T} - 1)$$

Here, the first symbol e represents the base of the natural logarithm. The second e is the charge on the electron. The k_B is Boltzmann's constant, and T is the absolute temperature. Set up a spreadsheet to calculate I and $R = (\Delta V)/I$ for $\Delta V = 0.400$ V to 0.600 V in increments of 0.005 V. Assume that $I_0 = 1.00$ nA. Plot R versus ΔV for $T = 280$ K, 300 K, and 320 K.

ANSWERS TO QUICK QUIZZES

27.1 d, b = c, a. The current in part (d) is equivalent to two positive charges moving to the left. Parts (b) and (c) each represent four positive charges moving in the same direction because negative charges moving to the left are equivalent to positive charges moving to the right. The current in part (a) is equivalent to five positive charges moving to the right.

27.2 Every portion of the wire carries the same current even though the wire constricts. As the cross-sectional area decreases, the drift velocity must increase in order for the constant current to be maintained, in accordance with Equation 27.4. Equations 27.5 and 27.6 indicate that the current density also increases. An increasing electric field must be causing the increasing current density, as indicated by Equation 27.7. If you were to draw this situation, you would show the electric field lines being compressed into the smaller area, indicating increasing magnitude of the electric field.

27.3 $1/R$. The curvature of the line indicates that the device is nonohmic (that is, its resistance varies with potential difference). Being the definition of resistance, Equation 27.8 still applies, giving different values for R at different points on the curve.

27.4 The cable should be as short as possible but still able to reach from one vehicle to another (small ℓ), it should be quite thick (large A), and it should be made of a material with a low resistivity ρ. Referring to Table 27.1, you should probably choose copper or aluminum because

the only two materials in the table that have lower ρ values—silver and gold—are prohibitively expensive for your purposes.

27.5 Just after it is turned on. When the filament is at room temperature, its resistance is low, and hence the current is relatively large ($I = \Delta V/R$). As the filament warms up, its resistance increases, and the current decreases. Older lightbulbs often fail just as they are turned on because this large initial current "spike" produces rapid temperature increase and stress on the filament.

27.6 (c). Because the potential difference ΔV is the same across the two bulbs and because the power delivered to a conductor is $\mathcal{P} = I\Delta V$, the 60-W bulb, with its higher power rating, must carry the greater current. The 30-W bulb has the higher resistance because it draws less current at the same potential difference.

27.7 $I_a = I_b > I_c = I_d > I_e = I_f$. The current I_a leaves the positive terminal of the battery and then splits to flow through the two bulbs; thus, $I_a = I_c + I_e$. From Quick Quiz 27.6, we know that the current in the 60-W bulb is greater than that in the 30-W bulb. (Note that all the current does not follow the "path of least resistance," which in this case is through the 60-W bulb.) Because charge does not build up in the bulbs, we know that all the charge flowing into a bulb from the left must flow out on the right; consequently, $I_c = I_d$ and $I_e = I_f$. The two currents leaving the bulbs recombine to form the current back into the battery, $I_f + I_d = I_b$.

chapter

28

Direct Current Circuits

his chapter is concerned with the analysis of some simple electric circuits that contain batteries, resistors, and capacitors in various combinations. The analysis of these circuits is simplified by the use of two rules known as *Kirchhoff's rules*, which follow from the laws of conservation of energy and conservation of electric charge. Most of the circuits analyzed are assumed to be in *steady state*, which means that the currents are constant in magnitude and direction. In Section 28.4 we discuss circuits in which the current varies with time. Finally, we describe a variety of common electrical devices and techniques for measuring current, potential difference, resistance, and emf.

28.1 ELECTROMOTIVE FORCE

In Section 27.6 we found that a constant current can be maintained in a closed circuit through the use of a source of *emf*, which is a device (such as a battery or generator) that produces an electric field and thus may cause charges to move around a circuit. One can think of a source of emf as a "charge pump." When an electric potential difference exists between two points, the source moves charges "uphill" from the lower potential to the higher. The emf $\mathcal{E}$ describes the work done per unit charge, and hence the SI unit of emf is the volt.

Consider the circuit shown in Figure 28.1, consisting of a battery connected to a resistor. We assume that the connecting wires have no resistance. The positive terminal of the battery is at a higher potential than the negative terminal. If we neglect the internal resistance of the battery, the potential difference across it (called the *terminal voltage*) equals its emf. However, because a real battery always has some internal resistance r, the terminal voltage is not equal to the emf for a battery in a circuit in which there is a current. To understand why this is so, consider the circuit diagram in Figure 28.2a, where the battery of Figure 28.1 is represented by the dashed rectangle containing an emf $\mathcal{E}$ in series with an internal resistance r. Now imagine moving through the battery clockwise from a to b and measuring the electric potential at various locations. As we pass from the negative terminal to the positive terminal, the potential *increases* by an amount $\mathcal{E}$. However, as we move through the resistance r, the potential *decreases* by an amount Ir, where I is the current in the circuit. Thus, the terminal voltage of the battery $\Delta V = V_b - V_a$ is[1]

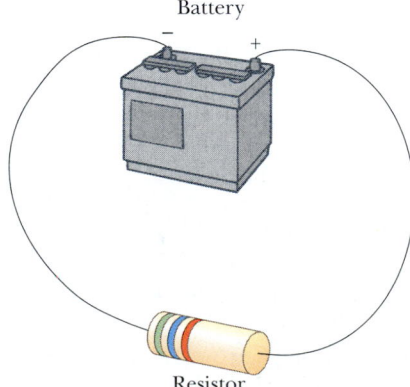

Battery

Resistor

Figure 28.1   A circuit consisting of a resistor connected to the terminals of a battery.

[1] The terminal voltage in this case is less than the emf by an amount Ir. In some situations, the terminal voltage may *exceed* the emf by an amount Ir. This happens when the direction of the current is *opposite* that of the emf, as in the case of charging a battery with another source of emf.

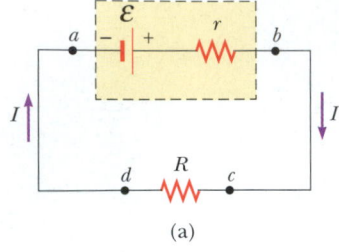

(a)

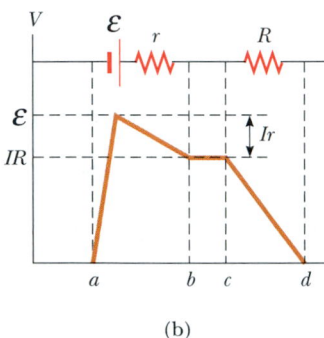

(b)

Figure 28.2 (a) Circuit diagram of a source of emf $\mathcal{E}$ (in this case, a battery), of internal resistance r, connected to an external resistor of resistance R. (b) Graphical representation showing how the electric potential changes as the circuit in part (a) is traversed clockwise.

$$\Delta V = \mathcal{E} - Ir \qquad (28.1)$$

From this expression, note that $\mathcal{E}$ is equivalent to the **open-circuit voltage**—that is, the *terminal voltage when the current is zero.* The emf is the voltage labeled on a battery—for example, the emf of a D cell is 1.5 V. The actual potential difference between the terminals of the battery depends on the current through the battery, as described by Equation 28.1.

Figure 28.2b is a graphical representation of the changes in electric potential as the circuit is traversed in the clockwise direction. By inspecting Figure 28.2a, we see that the terminal voltage ΔV must equal the potential difference across the external resistance R, often called the **load resistance.** The load resistor might be a simple resistive circuit element, as in Figure 28.1, or it could be the resistance of some electrical device (such as a toaster, an electric heater, or a lightbulb) connected to the battery (or, in the case of household devices, to the wall outlet). The resistor represents a *load* on the battery because the battery must supply energy to operate the device. The potential difference across the load resistance is $\Delta V = IR$. Combining this expression with Equation 28.1, we see that

$$\mathcal{E} = IR + Ir \qquad (28.2)$$

Solving for the current gives

$$I = \frac{\mathcal{E}}{R + r} \qquad (28.3)$$

This equation shows that the current in this simple circuit depends on both the load resistance R external to the battery and the internal resistance r. If R is much greater than r, as it is in many real-world circuits, we can neglect r.

If we multiply Equation 28.2 by the current I, we obtain

$$I\mathcal{E} = I^2R + I^2r \qquad (28.4)$$

This equation indicates that, because power $\mathcal{P} = I\Delta V$ (see Eq. 27.22), the total power output $I\mathcal{E}$ of the battery is delivered to the external load resistance in the amount I^2R and to the internal resistance in the amount I^2r. Again, if $r \ll R$, then most of the power delivered by the battery is transferred to the load resistance.

EXAMPLE 28.1   **Terminal Voltage of a Battery**

A battery has an emf of 12.0 V and an internal resistance of 0.05 Ω. Its terminals are connected to a load resistance of 3.00 Ω. (a) Find the current in the circuit and the terminal voltage of the battery.

Solution Using first Equation 28.3 and then Equation 28.1, we obtain

$$I = \frac{\mathcal{E}}{R + r} = \frac{12.0\ \text{V}}{3.05\ \Omega} = \boxed{3.93\ \text{A}}$$

$$\Delta V = \mathcal{E} - Ir = 12.0\ \text{V} - (3.93\ \text{A})(0.05\ \Omega) = \boxed{11.8\ \text{V}}$$

To check this result, we can calculate the voltage across the load resistance R:

$$\Delta V = IR = (3.93\ \text{A})(3.00\ \Omega) = 11.8\ \text{V}$$

(b) Calculate the power delivered to the load resistor, the power delivered to the internal resistance of the battery, and the power delivered by the battery.

Solution The power delivered to the load resistor is

$$\mathcal{P}_R = I^2R = (3.93\ \text{A})^2(3.00\ \Omega) = \boxed{46.3\ \text{W}}$$

The power delivered to the internal resistance is

$$\mathcal{P}_r = I^2r = (3.93\ \text{A})^2(0.05\ \Omega) = \boxed{0.772\ \text{W}}$$

Hence, the power delivered by the battery is the sum of these quantities, or 47.1 W. You should check this result, using the expression $\mathcal{P} = I\mathcal{E}$.

EXAMPLE 28.2 ▶ Matching the Load

Show that the maximum power delivered to the load resistance R in Figure 28.2a occurs when the load resistance matches the internal resistance—that is, when $R = r$.

Solution The power delivered to the load resistance is equal to I^2R, where I is given by Equation 28.3:

$$\mathcal{P} = I^2R = \frac{\mathcal{E}^2R}{(R + r)^2}$$

When $\mathcal{P}$ is plotted versus R as in Figure 28.3, we find that $\mathcal{P}$ reaches a maximum value of $\mathcal{E}^2/4r$ at $R = r$. We can also prove this by differentiating $\mathcal{P}$ with respect to R, setting the result equal to zero, and solving for R. The details are left as a problem for you to solve (Problem 57).

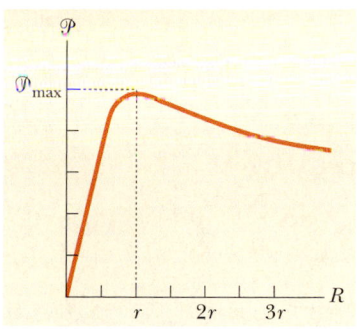

Figure 28.3 Graph of the power $\mathcal{P}$ delivered by a battery to a load resistor of resistance R as a function of R. The power delivered to the resistor is a maximum when the load resistance equals the internal resistance of the battery.

28.2 ▶ RESISTORS IN SERIES AND IN PARALLEL

Suppose that you and your friends are at a crowded basketball game in a sports arena and decide to leave early. You have two choices: (1) your whole group can exit through a single door and walk down a long hallway containing several concession stands, each surrounded by a large crowd of people waiting to buy food or souvenirs; or (2) each member of your group can exit through a separate door in the main hall of the arena, where each will have to push his or her way through a single group of people standing by the door. In which scenario will less time be required for your group to leave the arena?

It should be clear that your group will be able to leave faster through the separate doors than down the hallway where each of you has to push through several groups of people. We could describe the groups of people in the hallway as acting in *series*, because each of you must push your way through all of the groups. The groups of people around the doors in the arena can be described as acting in *parallel*. Each member of your group must push through only one group of people, and each member pushes through a *different* group of people. This simple analogy will help us understand the behavior of currents in electric circuits containing more than one resistor.

When two or more resistors are connected together as are the lightbulbs in Figure 28.4a, they are said to be in *series*. Figure 28.4b is the circuit diagram for the lightbulbs, which are shown as resistors, and the battery. In a series connection, all the charges moving through one resistor must also pass through the second resistor. (This is analogous to all members of your group pushing through the crowds in the single hallway of the sports arena.) Otherwise, charge would accumulate between the resistors. Thus,

for a series combination of resistors, the currents in the two resistors are the same because any charge that passes through R_1 must also pass through R_2.

The potential difference applied across the series combination of resistors will divide between the resistors. In Figure 28.4b, because the voltage drop[2] from a to b

---

[2] The term *voltage drop* is synonymous with a decrease in electric potential across a resistor and is used often by individuals working with electric circuits.

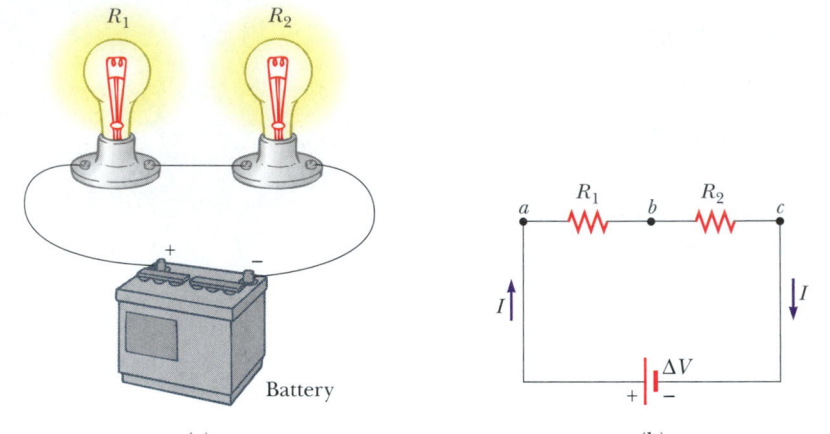

(a)   (b)   (c)

Figure 28.4 (a) A series connection of two resistors R_1 and R_2. The current in R_1 is the same as that in R_2. (b) Circuit diagram for the two-resistor circuit. (c) The resistors replaced with a single resistor having an equivalent resistance $R_{eq} = R_1 + R_2$.

equals IR_1 and the voltage drop from b to c equals IR_2, the voltage drop from a to c is

$$\Delta V = IR_1 + IR_2 = I(R_1 + R_2)$$

Therefore, we can replace the two resistors in series with a single resistor having an *equivalent resistance R_{eq}*, where

$$R_{eq} = R_1 + R_2 \qquad (28.5)$$

The resistance R_{eq} is equivalent to the series combination $R_1 + R_2$ in the sense that the circuit current is unchanged when R_{eq} replaces $R_1 + R_2$.

The equivalent resistance of three or more resistors connected in series is

$$R_{eq} = R_1 + R_2 + R_3 + \cdots \qquad (28.6)$$

This relationship indicates that **the equivalent resistance of a series connection of resistors is always greater than any individual resistance.**

Quick Quiz 28.1

If a piece of wire is used to connect points b and c in Figure 28.4b, does the brightness of bulb R_1 increase, decrease, or stay the same? What happens to the brightness of bulb R_2?

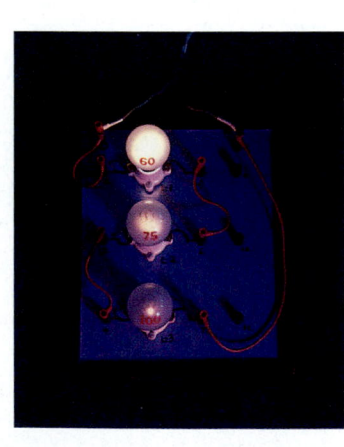

A series connection of three light-bulbs, all rated at 120 V but having power ratings of 60 W, 75 W, and 200 W. Why are the intensities of the bulbs different? Which bulb has the greatest resistance? How would their relative intensities differ if they were connected in parallel? *(Henry Leap and Jim Lehman)*

Now consider two resistors connected in *parallel*, as shown in Figure 28.5. When the current I reaches point a in Figure 28.5b, called a *junction*, it splits into two parts, with I_1 going through R_1 and I_2 going through R_2. A **junction** is any point in a circuit where a current can split (just as your group might split up and leave the arena through several doors, as described earlier.) This split results in less current in each individual resistor than the current leaving the battery. Because charge must be conserved, the current I that enters point a must equal the total current leaving that point:

$$I = I_1 + I_2$$

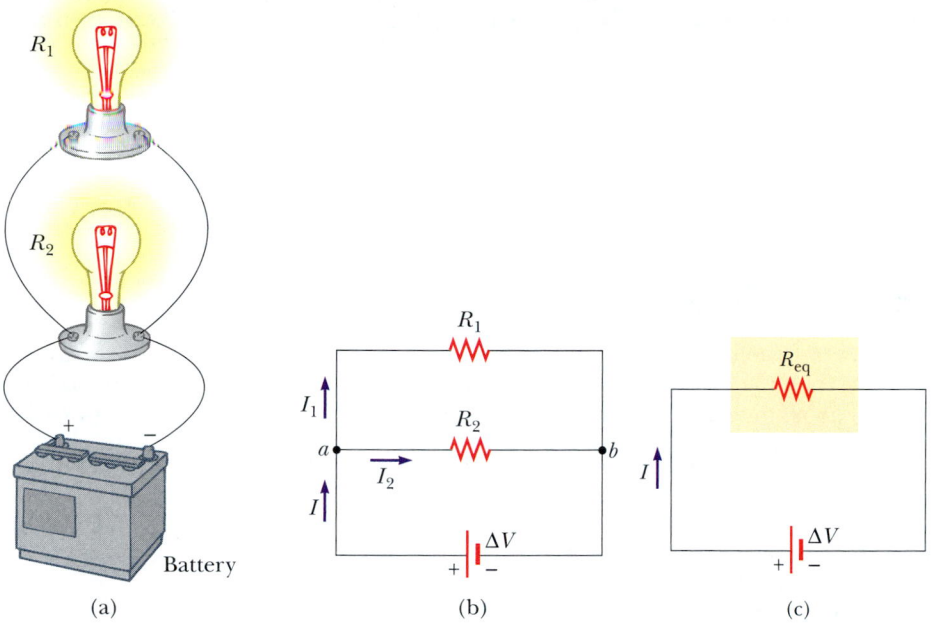

Figure 28.5   (a) A parallel connection of two resistors R_1 and R_2. The potential difference across R_1 is the same as that across R_2. (b) Circuit diagram for the two-resistor circuit. (c) The resistors replaced with a single resistor having an equivalent resistance $R_{eq} = (R_1^{-1} + R_2^{-1})^{-1}$.

As can be seen from Figure 28.5, both resistors are connected directly across the terminals of the battery. Thus,

> when resistors are connected in parallel, the potential differences across them are the same.

Because the potential differences across the resistors are the same, the expression $\Delta V = IR$ gives

$$I = I_1 + I_2 = \frac{\Delta V}{R_1} + \frac{\Delta V}{R_2} = \Delta V \left(\frac{1}{R_1} + \frac{1}{R_2} \right) = \frac{\Delta V}{R_{eq}}$$

From this result, we see that the equivalent resistance of two resistors in parallel is given by

$$\frac{1}{R_{eq}} = \frac{1}{R_1} + \frac{1}{R_2} \tag{28.7}$$

or

$$R_{eq} = \frac{1}{\dfrac{1}{R_1} + \dfrac{1}{R_2}}$$

An extension of this analysis to three or more resistors in parallel gives

$$\frac{1}{R_{eq}} = \frac{1}{R_1} + \frac{1}{R_2} + \frac{1}{R_3} + \cdots \tag{28.8}$$

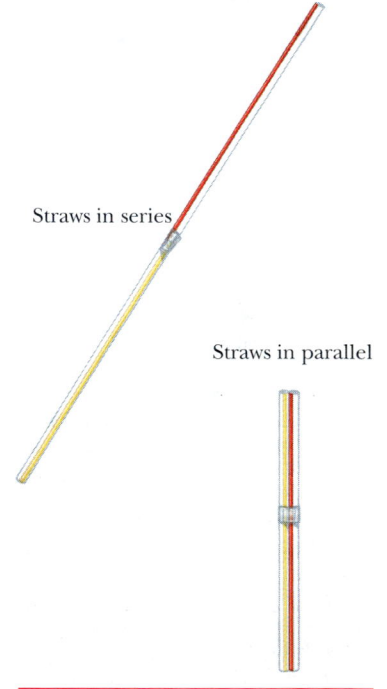

QuickLab

Tape one pair of drinking straws end to end, and tape a second pair side by side. Which pair is easier to blow through? What would happen if you were comparing three straws taped end to end with three taped side by side?

Straws in series

Straws in parallel

The equivalent resistance of several resistors in parallel

Three lightbulbs having power ratings of 25 W, 75 W, and 150 W, connected in parallel to a voltage source of about 100 V. All bulbs are rated at the same voltage. Why do the intensities differ? Which bulb draws the most current? Which has the least resistance? *(Henry Leap and Jim Lehman)*

We can see from this expression that **the equivalent resistance of two or more resistors connected in parallel is always less than the least resistance in the group.**

Household circuits are always wired such that the appliances are connected in parallel. Each device operates independently of the others so that if one is switched off, the others remain on. In addition, the devices operate on the same voltage.

Quick Quiz 28.2

Assume that the battery of Figure 28.1 has zero internal resistance. If we add a second resistor in series with the first, does the current in the battery increase, decrease, or stay the same? How about the potential difference across the battery terminals? Would your answers change if the second resistor were connected in parallel to the first one?

Quick Quiz 28.3

Are automobile headlights wired in series or in parallel? How can you tell?

EXAMPLE 28.3   Find the Equivalent Resistance

Four resistors are connected as shown in Figure 28.6a. (a) Find the equivalent resistance between points a and c.

Solution   The combination of resistors can be reduced in steps, as shown in Figure 28.6. The 8.0-Ω and 4.0-Ω resistors are in series; thus, the equivalent resistance between a and b is 12 Ω (see Eq. 28.5). The 6.0-Ω and 3.0-Ω resistors are in parallel, so from Equation 28.7 we find that the equivalent resistance from b to c is 2.0 Ω. Hence, the equivalent resistance

from a to c is   14 Ω.

(b) What is the current in each resistor if a potential difference of 42 V is maintained between a and c?

Solution   The currents in the 8.0-Ω and 4.0-Ω resistors are the same because they are in series. In addition, this is the same as the current that would exist in the 14-Ω equivalent resistor subject to the 42-V potential difference. Therefore, using Equation 27.8 ($R = \Delta V/I$) and the results from part (a), we obtain

$$I = \frac{\Delta V_{ac}}{R_{eq}} = \frac{42 \text{ V}}{14 \text{ }\Omega} = 3.0 \text{ A}$$

This is the current in the 8.0-Ω and 4.0-Ω resistors. When this 3.0-A current enters the junction at b, however, it splits, with part passing through the 6.0-Ω resistor (I_1) and part through the 3.0-Ω resistor (I_2). Because the potential difference is ΔV_{bc} across each of these resistors (since they are in parallel), we see that $(6.0 \text{ }\Omega)I_1 = (3.0 \text{ }\Omega)I_2$, or $I_2 = 2I_1$. Using this result and the fact that $I_1 + I_2 = 3.0$ A, we find that $I_1 = 1.0$ A and

$I_2 = 2.0$ A. We could have guessed this at the start by noting that the current through the 3.0-Ω resistor has to be twice that through the 6.0-Ω resistor, in view of their relative resistances and the fact that the same voltage is applied to each of them.

As a final check of our results, note that $\Delta V_{bc} = (6.0 \text{ }\Omega)I_1 = (3.0 \text{ }\Omega)I_2 = 6.0$ V   and $\Delta V_{ab} = (12 \text{ }\Omega)I = 36$ V;   therefore, $\Delta V_{ac} = \Delta V_{ab} + \Delta V_{bc} = 42$ V, as it must.

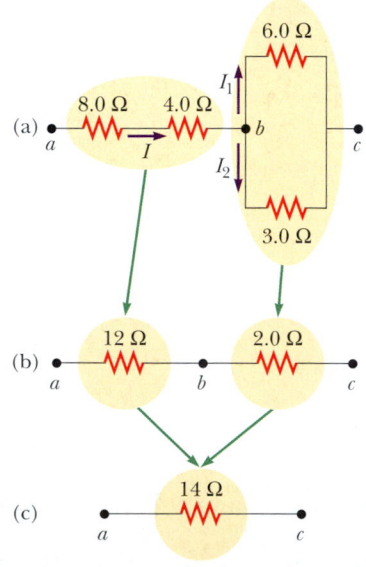

Figure 28.6

EXAMPLE 28.4   Three Resistors in Parallel

Three resistors are connected in parallel as shown in Figure 28.7. A potential difference of 18 V is maintained between points a and b. (a) Find the current in each resistor.

Solution   The resistors are in parallel, and so the potential difference across each must be 18 V. Applying the relationship $\Delta V = IR$ to each resistor gives

$$I_1 = \frac{\Delta V}{R_1} = \frac{18 \text{ V}}{3.0 \text{ }\Omega} = \boxed{6.0 \text{ A}}$$

$$I_2 = \frac{\Delta V}{R_2} = \frac{18 \text{ V}}{6.0 \text{ }\Omega} = \boxed{3.0 \text{ A}}$$

$$I_3 = \frac{\Delta V}{R_3} = \frac{18 \text{ V}}{9.0 \text{ }\Omega} = \boxed{2.0 \text{ A}}$$

(b) Calculate the power delivered to each resistor and the total power delivered to the combination of resistors.

Solution   We apply the relationship $\mathcal{P} = (\Delta V)^2/R$ to each resistor and obtain

$$\mathcal{P}_1 = \frac{\Delta V^2}{R_1} = \frac{(18 \text{ V})^2}{3.0 \text{ }\Omega} = \boxed{110 \text{ W}}$$

$$\mathcal{P}_2 = \frac{\Delta V^2}{R_2} = \frac{(18 \text{ V})^2}{6.0 \text{ }\Omega} = \boxed{54 \text{ W}}$$

$$\mathcal{P}_3 = \frac{\Delta V^2}{R_3} = \frac{(18 \text{ V})^2}{9.0 \text{ }\Omega} = \boxed{36 \text{ W}}$$

This shows that the smallest resistor receives the most power. Summing the three quantities gives a total power of 200 W.

(c) Calculate the equivalent resistance of the circuit.

Solution   We can use Equation 28.8 to find R_{eq}:

$$\frac{1}{R_{eq}} = \frac{1}{3.0 \text{ }\Omega} + \frac{1}{6.0 \text{ }\Omega} + \frac{1}{9.0 \text{ }\Omega}$$

$$= \frac{6}{18 \text{ }\Omega} + \frac{3}{18 \text{ }\Omega} + \frac{2}{18 \text{ }\Omega} = \frac{11}{18 \text{ }\Omega}$$

$$R_{eq} = \frac{18 \text{ }\Omega}{11} = \boxed{1.6 \text{ }\Omega}$$

Exercise   Use R_{eq} to calculate the total power delivered by the battery.

Answer   200 W.

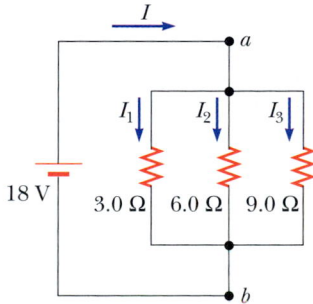

Figure 28.7   Three resistors connected in parallel. The voltage across each resistor is 18 V.

EXAMPLE 28.5   Finding R_{eq} by Symmetry Arguments

Consider five resistors connected as shown in Figure 28.8a. Find the equivalent resistance between points a and b.

Solution   In this type of problem, it is convenient to assume a current entering junction a and then apply symmetry

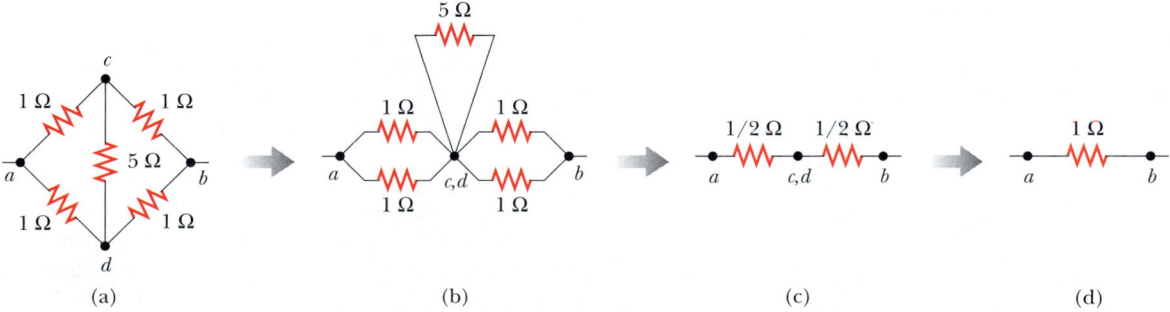

(a)   (b)   (c)   (d)

Figure 28.8   Because of the symmetry in this circuit, the 5-Ω resistor does not contribute to the resistance between points a and b and therefore can be disregarded when we calculate the equivalent resistance.

arguments. Because of the symmetry in the circuit (all 1-Ω resistors in the outside loop), the currents in branches *ac* and *ad* must be equal; hence, the electric potentials at points *c* and *d* must be equal. This means that $\Delta V_{cd} = 0$ and, as a result, points *c* and *d* may be connected together without affecting the circuit, as in Figure 28.8b. Thus, the 5-Ω resistor may be removed from the circuit and the remaining circuit then reduced as in Figures 28.8c and d. From this reduction we see that the equivalent resistance of the combination is 1 Ω. Note that the result is 1 Ω regardless of the value of the resistor connected between *c* and *d*.

CONCEPTUAL EXAMPLE 28.6   Operation of a Three-Way Lightbulb

Figure 28.9 illustrates how a three-way lightbulb is constructed to provide three levels of light intensity. The socket of the lamp is equipped with a three-way switch for selecting different light intensities. The bulb contains two filaments. When the lamp is connected to a 120-V source, one filament receives 100 W of power, and the other receives 75 W. Explain how the two filaments are used to provide three different light intensities.

Solution   The three light intensities are made possible by applying the 120 V to one filament alone, to the other filament alone, or to the two filaments in parallel. When switch S_1 is closed and switch S_2 is opened, current passes only through the 75-W filament. When switch S_1 is open and switch S_2 is closed, current passes only through the 100-W filament. When both switches are closed, current passes through both filaments, and the total power is 175 W.

If the filaments were connected in series and one of them were to break, no current could pass through the bulb, and the bulb would give no illumination, regardless of the switch position. However, with the filaments connected in parallel, if one of them (for example, the 75-W filament) breaks, the bulb will still operate in two of the switch positions as current passes through the other (100-W) filament.

Exercise   Determine the resistances of the two filaments and their parallel equivalent resistance.

Answer   144 Ω, 192 Ω, 82.3 Ω.

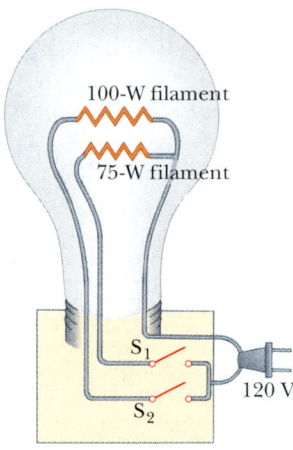

Figure 28.9   A three-way lightbulb.

APPLICATION   Strings of Lights

Strings of lights are used for many ornamental purposes, such as decorating Christmas trees. Over the years, both parallel and series connections have been used for multilight strings powered by 120 V.[3] Series-wired bulbs are safer than parallel-wired bulbs for indoor Christmas-tree use because series-wired bulbs operate with less light per bulb and at a lower temperature. However, if the filament of a single bulb fails (or if the bulb is removed from its socket), all the lights on the string are extinguished. The popularity of series-wired light strings diminished because troubleshooting a failed bulb was a tedious, time-consuming chore that involved trial-and-error substitution of a good bulb in each socket along the string until the defective bulb was found.

In a parallel-wired string, each bulb operates at 120 V. By design, the bulbs are brighter and hotter than those on a series-wired string. As a result, these bulbs are inherently more dangerous (more likely to start a fire, for instance), but if one bulb in a parallel-wired string fails or is removed, the rest of the bulbs continue to glow. (A 25-bulb string of 4-W bulbs results in a power of 100 W; the total power becomes substantial when several strings are used.)

A new design was developed for so-called "miniature" lights wired in series, to prevent the failure of one bulb from extinguishing the entire string. The solution is to create a connection (called a jumper) across the filament after it fails. (If an alternate connection existed across the filament before

---

[3] These and other household devices, such as the three-way lightbulb in Conceptual Example 28.6 and the kitchen appliances shown in this chapter's Puzzler, actually operate on alternating current (ac), to be introduced in Chapter 33.

it failed, each bulb would represent a parallel circuit; in this circuit, the current would flow through the alternate connection, forming a short circuit, and the bulb would not glow.) When the filament breaks in one of these miniature light-bulbs, 120 V appears across the bulb because no current is present in the bulb and therefore no drop in potential occurs across the other bulbs. Inside the lightbulb, a small loop covered by an insulating material is wrapped around the filament leads. An arc burns the insulation and connects the filament leads when 120 V appears across the bulb—that is, when the filament fails. This "short" now completes the circuit through the bulb even though the filament is no longer active (Fig. 28.10).

Suppose that all the bulbs in a 50-bulb miniature-light string are operating. A 2.4-V potential drop occurs across each bulb because the bulbs are in series. The power input to this style of bulb is 0.34 W, so the total power supplied to the string is only 17 W. We calculate the filament resistance at the operating temperature to be $(2.4 \text{ V})^2 / (0.34 \text{ W}) = 17 \, \Omega$. When the bulb fails, the resistance across its terminals is reduced to zero because of the alternate jumper connection mentioned in the preceding paragraph. All the other bulbs not only stay on but glow more brightly because the total resistance of the string is reduced and consequently the current in each bulb increases.

Let us assume that the operating resistance of a bulb remains at 17 Ω even though its temperature rises as a result of the increased current. If one bulb fails, the potential drop across each of the remaining bulbs increases to 2.45 V, the current increases from 0.142 A to 0.145 A, and the power increases to 0.354 W. As more lights fail, the current keeps rising, the filament of each bulb operates at a higher temperature, and the lifetime of the bulb is reduced. It is therefore a good idea to check for failed (nonglowing) bulbs in such a series-wired string and replace them as soon as possible, in order to maximize the lifetimes of all the bulbs.

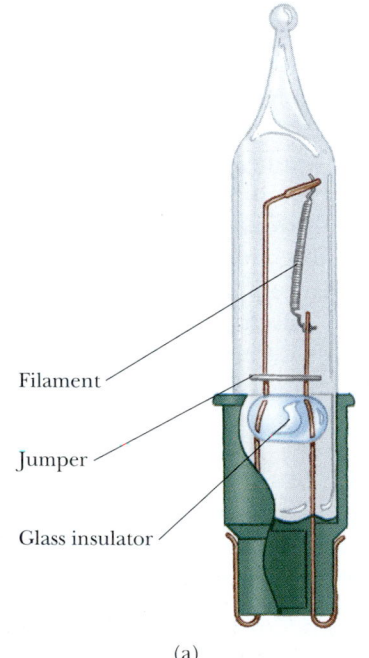

Filament

Jumper

Glass insulator

(a)

(b)

Figure 28.10 (a) Schematic diagram of a modern "miniature" holiday lightbulb, with a jumper connection to provide a current path if the filament breaks. (b) A Christmas-tree lightbulb. *(George Semple)*

28.3 KIRCHHOFF'S RULES

As we saw in the preceding section, we can analyze simple circuits using the expression $\Delta V = IR$ and the rules for series and parallel combinations of resistors. Very often, however, it is not possible to reduce a circuit to a single loop. The procedure for analyzing more complex circuits is greatly simplified if we use two principles called **Kirchhoff's rules:**

1. The sum of the currents entering any junction in a circuit must equal the sum of the currents leaving that junction:

$$\sum I_{\text{in}} = \sum I_{\text{out}} \qquad \text{(28.9)}$$

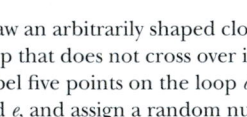

QuickLab

Draw an arbitrarily shaped closed loop that does not cross over itself. Label five points on the loop a, b, c, d, and e, and assign a random number to each point. Now start at a and work your way around the loop, calculating the difference between each pair of adjacent numbers. Some of these differences will be positive, and some will be negative. Add the differences together, making sure you accurately keep track of the algebraic signs. What is the sum of the differences all the way around the loop?

2. The sum of the potential differences across all elements around any closed circuit loop must be zero:

$$\sum_{\substack{\text{closed} \\ \text{loop}}} \Delta V = 0 \qquad (28.10)$$

Kirchhoff's first rule is a statement of conservation of electric charge. All current that enters a given point in a circuit must leave that point because charge cannot build up at a point. If we apply this rule to the junction shown in Figure 28.11a, we obtain

$$I_1 = I_2 + I_3$$

Figure 28.11b represents a mechanical analog of this situation, in which water flows through a branched pipe having no leaks. The flow rate into the pipe equals the total flow rate out of the two branches on the right.

Kirchhoff's second rule follows from the law of conservation of energy. Let us imagine moving a charge around the loop. When the charge returns to the starting point, the charge–circuit system must have the same energy as when the charge started from it. The sum of the increases in energy in some circuit elements must equal the sum of the decreases in energy in other elements. The potential energy decreases whenever the charge moves through a potential drop $-IR$ across a resistor or whenever it moves in the reverse direction through a source of emf. The potential energy increases whenever the charge passes through a battery from the negative terminal to the positive terminal. Kirchhoff's second rule applies only for circuits in which an electric potential is defined at each point; this criterion may not be satisfied if changing electromagnetic fields are present, as we shall see in Chapter 31.

In justifying our claim that Kirchhoff's second rule is a statement of conservation of energy, we imagined carrying a charge around a loop. When applying this rule, we imagine *traveling* around the loop and consider changes in *electric potential*, rather than the changes in *potential energy* described in the previous paragraph. You should note the following sign conventions when using the second rule:

- Because charges move from the high-potential end of a resistor to the low-potential end, if a resistor is traversed in the direction of the current, the change in potential ΔV across the resistor is $-IR$ (Fig. 28.12a).
- If a resistor is traversed in the direction *opposite* the current, the change in potential ΔV across the resistor is $+IR$ (Fig. 28.12b).
- If a source of emf (assumed to have zero internal resistance) is traversed in the direction of the emf (from $-$ to $+$), the change in potential ΔV is $+\mathcal{E}$ (Fig. 28.12c). The emf of the battery increases the electric potential as we move through it in this direction.
- If a source of emf (assumed to have zero internal resistance) is traversed in the direction opposite the emf (from $+$ to $-$), the change in potential ΔV is $-\mathcal{E}$ (Fig. 28.12d). In this case the emf of the battery reduces the electric potential as we move through it.

Limitations exist on the numbers of times you can usefully apply Kirchhoff's rules in analyzing a given circuit. You can use the junction rule as often as you need, so long as each time you write an equation you include in it a current that has not been used in a preceding junction-rule equation. In general, the number of times you can use the junction rule is one fewer than the number of junction

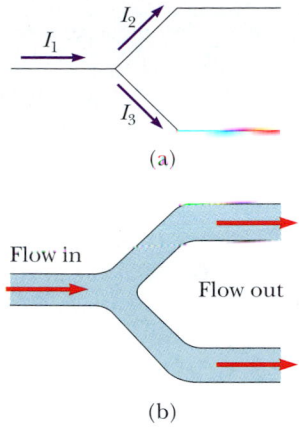

(a)

Flow in

Flow out

(b)

Figure 28.11 (a) Kirchhoff's junction rule. Conservation of charge requires that all current entering a junction must leave that junction. Therefore, $I_1 = I_2 + I_3$. (b) A mechanical analog of the junction rule: the amount of water flowing out of the branches on the right must equal the amount flowing into the single branch on the left.

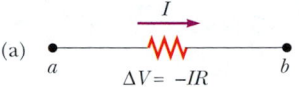
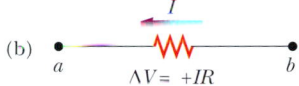
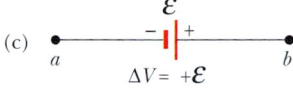
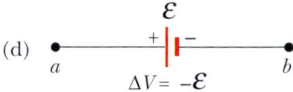

Figure 28.12 Rules for determining the potential changes across a resistor and a battery. (The battery is assumed to have no internal resistance.) Each circuit element is traversed from left to right.

points in the circuit. You can apply the loop rule as often as needed, so long as a new circuit element (resistor or battery) or a new current appears in each new equation. In general, **in order to solve a particular circuit problem, the number of independent equations you need to obtain from the two rules equals the number of unknown currents.**

Complex networks containing many loops and junctions generate great numbers of independent linear equations and a correspondingly great number of unknowns. Such situations can be handled formally through the use of matrix algebra. Computer programs can also be written to solve for the unknowns.

The following examples illustrate how to use Kirchhoff's rules. In all cases, it is assumed that the circuits have reached steady-state conditions—that is, the currents in the various branches are constant. Any capacitor **acts as an open circuit;** that is, the current in the branch containing the capacitor is zero under steady-state conditions.

Problem-Solving Hints

Kirchhoff's Rules

- Draw a circuit diagram, and label all the known and unknown quantities. You must assign a *direction* to the current in each branch of the circuit. Do not be alarmed if you guess the direction of a current incorrectly; your result will be negative, but *its magnitude will be correct.* Although the assignment of current directions is arbitrary, you must adhere rigorously to the assigned directions when applying Kirchhoff's rules.
- Apply the junction rule to any junctions in the circuit that provide new relationships among the various currents.

- Apply the loop rule to as many loops in the circuit as are needed to solve for the unknowns. To apply this rule, you must correctly identify the change in potential as you imagine crossing each element in traversing the closed loop (either clockwise or counterclockwise). Watch out for errors in sign!
- Solve the equations simultaneously for the unknown quantities.

EXAMPLE 28.7   A Single-Loop Circuit

A single-loop circuit contains two resistors and two batteries, as shown in Figure 28.13. (Neglect the internal resistances of the batteries.) (a) Find the current in the circuit.

Solution We do not need Kirchhoff's rules to analyze this simple circuit, but let us use them anyway just to see how they are applied. There are no junctions in this single-loop circuit; thus, the current is the same in all elements. Let us assume that the current is clockwise, as shown in Figure 28.13. Traversing the circuit in the clockwise direction, starting at a, we see that $a \rightarrow b$ represents a potential change of $+\mathcal{E}_1$, $b \rightarrow c$ represents a potential change of $-IR_1$, $c \rightarrow d$ represents a potential change of $-\mathcal{E}_2$, and $d \rightarrow a$ represents a potential change of $-IR_2$. Applying Kirchhoff's loop rule gives

$$\sum \Delta V = 0$$
$$\mathcal{E}_1 - IR_1 - \mathcal{E}_2 - IR_2 = 0$$

Solving for I and using the values given in Figure 28.13, we obtain

$$I = \frac{\mathcal{E}_1 - \mathcal{E}_2}{R_1 + R_2} = \frac{6.0\text{ V} - 12\text{ V}}{8.0\ \Omega + 10\ \Omega} = \boxed{-0.33\text{ A}}$$

The negative sign for I indicates that the direction of the current is opposite the assumed direction.

(b) What power is delivered to each resistor? What power is delivered by the 12-V battery?

Solution

$$\mathcal{P}_1 = I^2 R_1 = (0.33\text{ A})^2 (8.0\ \Omega) = \boxed{0.87\text{ W}}$$

$$\mathcal{P}_2 = I^2 R_2 = (0.33\text{ A})^2 (10\ \Omega) = \boxed{1.1\text{ W}}$$

Hence, the total power delivered to the resistors is $\mathcal{P}_1 + \mathcal{P}_2 = 2.0$ W.

The 12-V battery delivers power $I\mathcal{E}_2 = 4.0$ W. Half of this power is delivered to the two resistors, as we just calculated. The other half is delivered to the 6-V battery, which is being charged by the 12-V battery. If we had included the internal resistances of the batteries in our analysis, some of the power would appear as internal energy in the batteries; as a result, we would have found that less power was being delivered to the 6-V battery.

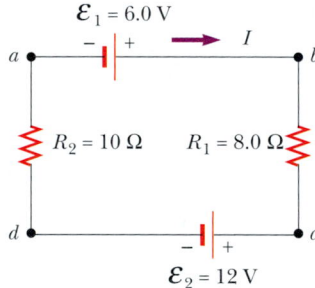

Figure 28.13 A series circuit containing two batteries and two resistors, where the polarities of the batteries are in opposition.

EXAMPLE 28.8   Applying Kirchhoff's Rules

Find the currents I_1, I_2, and I_3 in the circuit shown in Figure 28.14.

Solution Notice that we cannot reduce this circuit to a simpler form by means of the rules of adding resistances in series and in parallel. We must use Kirchhoff's rules to analyze this circuit. We arbitrarily choose the directions of the currents as labeled in Figure 28.14. Applying Kirchhoff's junction rule to junction c gives

$$(1) \qquad I_1 + I_2 = I_3$$

We now have one equation with three unknowns—I_1, I_2, and I_3. There are three loops in the circuit—*abcda*, *befcb*, and *aefda*. We therefore need only two loop equations to determine the unknown currents. (The third loop equation would give no new information.) Applying Kirchhoff's loop rule to loops *abcda* and *befcb* and traversing these loops clockwise, we obtain the expressions

$$(2) \quad abcda \quad 10\text{ V} - (6\ \Omega)I_1 - (2\ \Omega)I_3 = 0$$

$$(3) \quad befcb \quad -14\text{ V} + (6\ \Omega)I_1 - 10\text{ V} - (4\ \Omega)I_2 = 0$$

Note that in loop *befcb* we obtain a positive value when traversing the 6-Ω resistor because our direction of travel is opposite the assumed direction of I_1.

Expressions (1), (2), and (3) represent three independent equations with three unknowns. Substituting Equation (1) into Equation (2) gives

$$10\text{ V} - (6\text{ }\Omega)I_1 - (2\text{ }\Omega)(I_1 + I_2) = 0$$

(4) $\qquad 10\text{ V} = (8\text{ }\Omega)I_1 + (2\text{ }\Omega)I_2$

Dividing each term in Equation (3) by 2 and rearranging gives

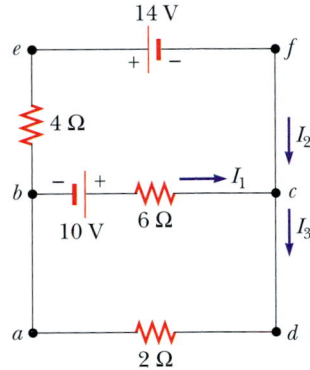

Figure 28.14 A circuit containing three loops.

(5) $\qquad -12\text{ V} = -(3\text{ }\Omega)I_1 + (2\text{ }\Omega)I_2$

Subtracting Equation (5) from Equation (4) eliminates I_2, giving

$$22\text{ V} = (11\text{ }\Omega)I_1$$

$$I_1 = \boxed{2\text{ A}}$$

Using this value of I_1 in Equation (5) gives a value for I_2:

$$(2\text{ }\Omega)I_2 = (3\text{ }\Omega)I_1 - 12\text{ V} = (3\text{ }\Omega)(2\text{ A}) - 12\text{ V} = -6\text{ V}$$

$$I_2 = \boxed{-3\text{ A}}$$

Finally,

$$I_3 = I_1 + I_2 = \boxed{-1\text{ A}}$$

The fact that I_2 and I_3 are both negative indicates only that the currents are opposite the direction we chose for them. However, the numerical values are correct. What would have happened had we left the current directions as labeled in Figure 28.14 but traversed the loops in the opposite direction?

Exercise Find the potential difference between points *b* and *c*.

Answer 2 V.

EXAMPLE 28.9 A Multiloop Circuit

(a) Under steady-state conditions, find the unknown currents I_1, I_2, and I_3 in the multiloop circuit shown in Figure 28.15.

Solution First note that because the capacitor represents an open circuit, there is no current between *g* and *b* along path *ghab* under steady-state conditions. Therefore, when the charges associated with I_1 reach point *g*, they all go through the 8.00-V battery to point *b*; hence, $I_{gb} = I_1$. Labeling the currents as shown in Figure 28.15 and applying Equation 28.9 to junction *c*, we obtain

(1) $\quad I_1 + I_2 = I_3$

Equation 28.10 applied to loops *defcd* and *cfgbc*, traversed clockwise, gives

(2) *defcd* $\quad 4.00\text{ V} - (3.00\text{ }\Omega)I_2 - (5.00\text{ }\Omega)I_3 = 0$

(3) *cfgbc* $\quad (3.00\text{ }\Omega)I_2 - (5.00\text{ }\Omega)I_1 + 8.00\text{ V} = 0$

From Equation (1) we see that $I_1 = I_3 - I_2$, which, when substituted into Equation (3), gives

(4) $\quad (8.00\text{ }\Omega)I_2 - (5.00\text{ }\Omega)I_3 + 8.00\text{ V} = 0$

Subtracting Equation (4) from Equation (2), we eliminate I_3 and find that

$$I_2 = -\frac{4.00\text{ V}}{11.0\text{ }\Omega} = \boxed{-0.364\text{ A}}$$

Because our value for I_2 is negative, we conclude that the direction of I_2 is from *c* to *f* through the 3.00-Ω resistor. Despite

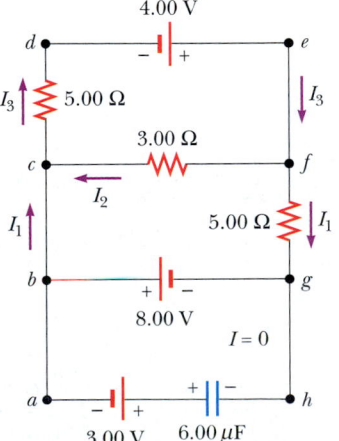

Figure 28.15 A multiloop circuit. Kirchhoff's loop rule can be applied to *any* closed loop, including the one containing the capacitor.

this interpretation of the direction, however, we must continue to use this negative value for I_2 in subsequent calculations because our equations were established with our original choice of direction.

Using $I_2 = -0.364$ A in Equations (3) and (1) gives

$$I_1 = \boxed{1.38 \text{ A}} \qquad I_3 = \boxed{1.02 \text{ A}}$$

(b) What is the charge on the capacitor?

Solution   We can apply Kirchhoff's loop rule to loop *bghab* (or any other loop that contains the capacitor) to find the potential difference ΔV_{cap} across the capacitor. We enter this potential difference in the equation without reference to a sign convention because the charge on the capacitor depends only on the magnitude of the potential difference. Moving clockwise around this loop, we obtain

$$-8.00 \text{ V} + \Delta V_{cap} - 3.00 \text{ V} = 0$$

$$\Delta V_{cap} = 11.0 \text{ V}$$

Because $Q = C \Delta V_{cap}$ (see Eq. 26.1), the charge on the capacitor is

$$Q = (6.00 \text{ } \mu\text{F})(11.0 \text{ V}) = \boxed{66.0 \text{ } \mu\text{C}}$$

Why is the left side of the capacitor positively charged?

Exercise   Find the voltage across the capacitor by traversing any other loop.

Answer   11.0 V.

Exercise   Reverse the direction of the 3.00-V battery and answer parts (a) and (b) again.

Answer   (a) $I_1 = 1.38$ A, $I_2 = -0.364$ A, $I_3 = 1.02$ A; (b) 30 μC.

28.4 · RC CIRCUITS

So far we have been analyzing steady-state circuits, in which the current is constant. In circuits containing capacitors, the current may vary in time. A circuit containing a series combination of a resistor and a capacitor is called an **RC circuit.**

Charging a Capacitor

Let us assume that the capacitor in Figure 28.16 is initially uncharged. There is no current while switch S is open (Fig. 28.16b). If the switch is closed at $t = 0$, however, charge begins to flow, setting up a current in the circuit, and the capacitor begins to charge.[4] Note that during charging, charges do not jump across the capacitor plates because the gap between the plates represents an open circuit. Instead, charge is transferred between each plate and its connecting wire due to the electric field established in the wires by the battery, until the capacitor is fully charged. As the plates become charged, the potential difference across the capacitor increases. The value of the maximum charge depends on the voltage of the battery. Once the maximum charge is reached, the current in the circuit is zero because the potential difference across the capacitor matches that supplied by the battery.

To analyze this circuit quantitatively, let us apply Kirchhoff's loop rule to the circuit after the switch is closed. Traversing the loop clockwise gives

$$\mathcal{E} - \frac{q}{C} - IR = 0 \tag{28.11}$$

where q/C is the potential difference across the capacitor and IR is the potential

---

[4] In previous discussions of capacitors, we assumed a steady-state situation, in which no current was present in any branch of the circuit containing a capacitor. Now we are considering the case *before* the steady-state condition is realized; in this situation, charges are moving and a current exists in the wires connected to the capacitor.

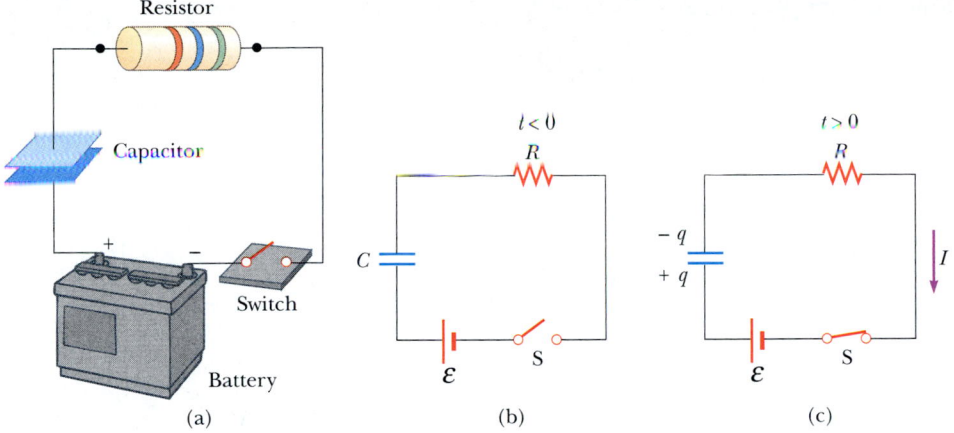

Figure 28.16 (a) A capacitor in series with a resistor, switch, and battery. (b) Circuit diagram representing this system at time $t < 0$, before the switch is closed. (c) Circuit diagram at time $t > 0$, after the switch has been closed.

difference across the resistor. We have used the sign conventions discussed earlier for the signs on $\mathcal{E}$ and IR. For the capacitor, notice that we are traveling in the direction from the positive plate to the negative plate; this represents a decrease in potential. Thus, we use a negative sign for this voltage in Equation 28.11. Note that q and I are *instantaneous* values that depend on time (as opposed to steady-state values) as the capacitor is being charged.

We can use Equation 28.11 to find the initial current in the circuit and the maximum charge on the capacitor. At the instant the switch is closed ($t = 0$), the charge on the capacitor is zero, and from Equation 28.11 we find that the initial current in the circuit I_0 is a maximum and is equal to

$$I_0 = \frac{\mathcal{E}}{R} \qquad \text{(current at } t = 0)\qquad \textbf{(28.12)}$$

Maximum current

At this time, the potential difference from the battery terminals appears entirely across the resistor. Later, when the capacitor is charged to its maximum value Q, charges cease to flow, the current in the circuit is zero, and the potential difference from the battery terminals appears entirely across the capacitor. Substituting $I = 0$ into Equation 28.11 gives the charge on the capacitor at this time:

$$Q = C\mathcal{E} \qquad \text{(maximum charge)} \qquad \textbf{(28.13)}$$

Maximum charge on the capacitor

To determine analytical expressions for the time dependence of the charge and current, we must solve Equation 28.11—a single equation containing two variables, q and I. The current in all parts of the series circuit must be the same. Thus, the current in the resistance R must be the same as the current flowing out of and into the capacitor plates. This current is equal to the time rate of change of the charge on the capacitor plates. Thus, we substitute $I = dq/dt$ into Equation 28.11 and rearrange the equation:

$$\frac{dq}{dt} = \frac{\mathcal{E}}{R} - \frac{q}{RC}$$

To find an expression for q, we first combine the terms on the right-hand side:

$$\frac{dq}{dt} = \frac{C\mathcal{E}}{RC} - \frac{q}{RC} = -\frac{q - C\mathcal{E}}{RC}$$

Now we multiply by dt and divide by $q - C\mathcal{E}$ to obtain

$$\frac{dq}{q - C\mathcal{E}} = -\frac{1}{RC}\,dt$$

Integrating this expression, using the fact that $q = 0$ at $t = 0$, we obtain

$$\int_0^q \frac{dq}{q - C\mathcal{E}} = -\frac{1}{RC}\int_0^t dt$$

$$\ln\left(\frac{q - C\mathcal{E}}{-C\mathcal{E}}\right) = -\frac{t}{RC}$$

From the definition of the natural logarithm, we can write this expression as

Charge versus time for a capacitor being charged

$$q(t) = C\mathcal{E}\,(1 - e^{-t/RC}) = Q(1 - e^{-t/RC}) \qquad \textbf{(28.14)}$$

where e is the base of the natural logarithm and we have made the substitution $C\mathcal{E} = Q$ from Equation 28.13.

We can find an expression for the charging current by differentiating Equation 28.14 with respect to time. Using $I = dq/dt$, we find that

Current versus time for a charging capacitor

$$I(t) = \frac{\mathcal{E}}{R}\,e^{-t/RC} \qquad \textbf{(28.15)}$$

Plots of capacitor charge and circuit current versus time are shown in Figure 28.17. Note that the charge is zero at $t = 0$ and approaches the maximum value $C\mathcal{E}$ as $t \to \infty$. The current has its maximum value $I_0 = \mathcal{E}/R$ at $t = 0$ and decays exponentially to zero as $t \to \infty$. The quantity RC, which appears in the exponents of Equations 28.14 and 28.15, is called the **time constant** τ of the circuit. It represents the time it takes the current to decrease to $1/e$ of its initial value; that is, in a time τ, $I = e^{-1}I_0 = 0.368I_0$. In a time 2τ, $I = e^{-2}I_0 = 0.135I_0$, and so forth. Likewise, in a time τ, the charge increases from zero to $C\mathcal{E}\,(1 - e^{-1}) = 0.632C\mathcal{E}$.

The following dimensional analysis shows that τ has the units of time:

$$[\tau] = [RC] = \left[\frac{\Delta V}{I} \times \frac{Q}{\Delta V}\right] = \left[\frac{Q}{Q/\Delta t}\right] = [\Delta t] = \text{T}$$

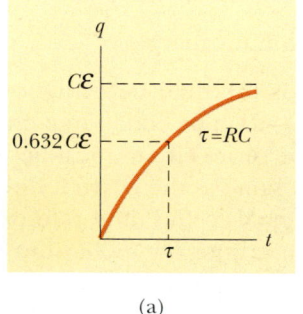
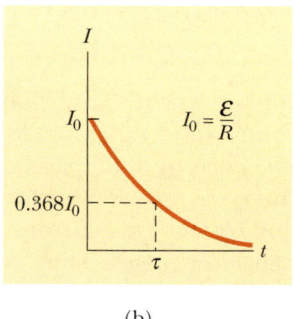

(a)   (b)

Figure 28.17 (a) Plot of capacitor charge versus time for the circuit shown in Figure 28.16. After a time interval equal to one time constant τ has passed, the charge is 63.2% of the maximum value $C\mathcal{E}$. The charge approaches its maximum value as t approaches infinity. (b) Plot of current versus time for the circuit shown in Figure 28.16. The current has its maximum value $I_0 = \mathcal{E}/R$ at $t = 0$ and decays to zero exponentially as t approaches infinity. After a time interval equal to one time constant τ has passed, the current is 36.8% of its initial value.

Because $\tau = RC$ has units of time, the combination t/RC is dimensionless, as it must be in order to be an exponent of e in Equations 28.14 and 28.15.

The energy output of the battery as the capacitor is fully charged is $Q\mathcal{E} = C\mathcal{E}^2$. After the capacitor is fully charged, the energy stored in the capacitor is $\frac{1}{2}Q\mathcal{E} = \frac{1}{2}C\mathcal{E}^2$, which is just half the energy output of the battery. It is left as a problem (Problem 60) to show that the remaining half of the energy supplied by the battery appears as internal energy in the resistor.

Discharging a Capacitor

Now let us consider the circuit shown in Figure 28.18, which consists of a capacitor carrying an initial charge Q, a resistor, and a switch. The *initial* charge Q is not the same as the *maximum* charge Q in the previous discussion, unless the discharge occurs after the capacitor is fully charged (as described earlier). When the switch is open, a potential difference Q/C exists across the capacitor and there is zero potential difference across the resistor because $I = 0$. If the switch is closed at $t = 0$, the capacitor begins to discharge through the resistor. At some time t during the discharge, the current in the circuit is I and the charge on the capacitor is q (Fig. 28.18b). The circuit in Figure 28.18 is the same as the circuit in Figure 28.16 except for the absence of the battery. Thus, we eliminate the emf $\mathcal{E}$ from Equation 28.11 to obtain the appropriate loop equation for the circuit in Figure 28.18:

$$-\frac{q}{C} - IR = 0 \tag{28.16}$$

When we substitute $I = dq/dt$ into this expression, it becomes

$$-R\frac{dq}{dt} = \frac{q}{C}$$

$$\frac{dq}{q} = -\frac{1}{RC}\,dt$$

Integrating this expression, using the fact that $q = Q$ at $t = 0$, gives

$$\int_Q^q \frac{dq}{q} = -\frac{1}{RC}\int_0^t dt$$

$$\ln\left(\frac{q}{Q}\right) = -\frac{t}{RC}$$

$$\boxed{q(t) = Qe^{-t/RC}} \tag{28.17}$$

Differentiating this expression with respect to time gives the instantaneous current as a function of time:

$$\boxed{I(t) = \frac{dq}{dt} = \frac{d}{dt}\left(Qe^{-t/RC}\right) = -\frac{Q}{RC}e^{-t/RC}} \tag{28.18}$$

where $Q/RC = I_0$ is the initial current. The negative sign indicates that the current direction now that the capacitor is discharging is opposite the current direction when the capacitor was being charged. (Compare the current directions in Figs. 28.16c and 28.18b.) We see that both the charge on the capacitor and the current decay exponentially at a rate characterized by the time constant $\tau = RC$.

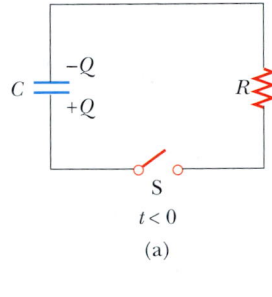

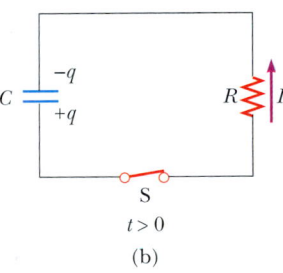

Figure 28.18 (a) A charged capacitor connected to a resistor and a switch, which is open at $t < 0$. (b) After the switch is closed, a current that decreases in magnitude with time is set up in the direction shown, and the charge on the capacitor decreases exponentially with time.

Charge versus time for a discharging capacitor

Current versus time for a discharging capacitor

CONCEPTUAL EXAMPLE 28.10   Intermittent Windshield Wipers

Many automobiles are equipped with windshield wipers that can operate intermittently during a light rainfall. How does the operation of such wipers depend on the charging and discharging of a capacitor?

Solution   The wipers are part of an *RC* circuit whose time constant can be varied by selecting different values of *R*

through a multiposition switch. As it increases with time, the voltage across the capacitor reaches a point at which it triggers the wipers and discharges, ready to begin another charging cycle. The time interval between the individual sweeps of the wipers is determined by the value of the time constant.

EXAMPLE 28.11   Charging a Capacitor in an *RC* Circuit

An uncharged capacitor and a resistor are connected in series to a battery, as shown in Figure 28.19. If $\mathcal{E} = 12.0$ V, $C = 5.00\ \mu$F, and $R = 8.00 \times 10^5\ \Omega$, find the time constant of the circuit, the maximum charge on the capacitor, the maximum current in the circuit, and the charge and current as functions of time.

Solution   The time constant of the circuit is $\tau = RC = (8.00 \times 10^5\ \Omega)(5.00 \times 10^{-6}\ \mathrm{F}) = 4.00$ s.   The maximum charge on the capacitor is $Q = C\mathcal{E} = (5.00\ \mu\mathrm{F})(12.0\ \mathrm{V}) = 60.0\ \mu$C.   The maximum current in the circuit is $I_0 = \mathcal{E}/R = (12.0\ \mathrm{V})/(8.00 \times 10^5\ \Omega) = 15.0\ \mu$A. Using these values and Equations 28.14 and 28.15, we find that

$$q(t) = \boxed{(60.0\ \mu\mathrm{C})(1 - e^{-t/4.00\ \mathrm{s}})}$$

$$I(t) = \boxed{(15.0\ \mu\mathrm{A})\, e^{-t/4.00\ \mathrm{s}}}$$

Graphs of these functions are provided in Figure 28.20.

Exercise   Calculate the charge on the capacitor and the current in the circuit after one time constant has elapsed.

Answer   37.9 μC, 5.52 μA.

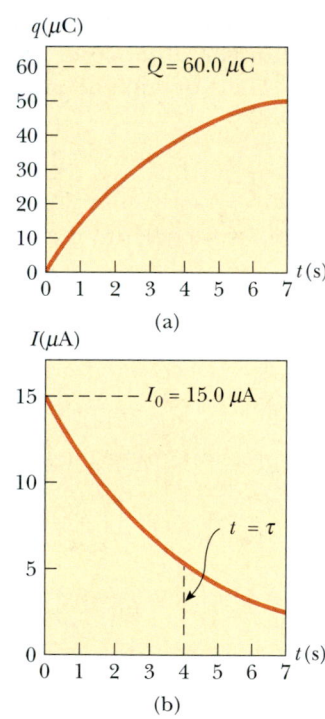

Figure 28.20   Plots of (a) charge versus time and (b) current versus time for the *RC* circuit shown in Figure 28.19, with $\mathcal{E} = 12.0$ V, $R = 8.00 \times 10^5\ \Omega$, and $C = 5.00\ \mu$F.

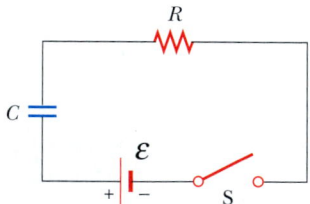

Figure 28.19   The switch of this series *RC* circuit, open for times $t < 0$, is closed at $t = 0$.

EXAMPLE 28.12   Discharging a Capacitor in an *RC* Circuit

Consider a capacitor of capacitance C that is being discharged through a resistor of resistance R, as shown in Figure 28.18. (a) After how many time constants is the charge on the capacitor one-fourth its initial value?

Solution   The charge on the capacitor varies with time according to Equation 28.17, $q(t) = Qe^{-t/RC}$. To find the time it takes q to drop to one-fourth its initial value, we substitute $q(t) = Q/4$ into this expression and solve for t:

$$\frac{Q}{4} = Qe^{-t/RC}$$

$$\frac{1}{4} = e^{-t/RC}$$

Taking logarithms of both sides, we find

$$-\ln 4 = -\frac{t}{RC}$$

$$t = RC(\ln 4) = 1.39RC = \boxed{1.39\tau}$$

(b) The energy stored in the capacitor decreases with time as the capacitor discharges. After how many time constants is this stored energy one-fourth its initial value?

Solution   Using Equations 26.11 ($U = Q^2/2C$) and 28.17, we can express the energy stored in the capacitor at any time t as

$$U = \frac{q^2}{2C} = \frac{(Qe^{-t/RC})^2}{2C} = \frac{Q^2}{2C}e^{-2t/RC} = U_0 e^{-2t/RC}$$

where $U_0 = Q^2/2C$ is the initial energy stored in the capacitor. As in part (a), we now set $U = U_0/4$ and solve for t:

$$\frac{U_0}{4} = U_0 e^{-2t/RC}$$

$$\frac{1}{4} = e^{-2t/RC}$$

Again, taking logarithms of both sides and solving for t gives

$$t = \tfrac{1}{2}RC(\ln 4) = 0.693RC = \boxed{0.693\tau}$$

Exercise   After how many time constants is the current in the circuit one-half its initial value?

Answer $0.693RC = 0.693\tau$.

EXAMPLE 28.13   ▶   **Energy Delivered to a Resistor**

A 5.00-μF capacitor is charged to a potential difference of 800 V and then discharged through a 25.0-kΩ resistor. How much energy is delivered to the resistor in the time it takes to fully discharge the capacitor?

Solution   We shall solve this problem in two ways. The first way is to note that the initial energy in the circuit equals the energy stored in the capacitor, $C\mathcal{E}^2/2$ (see Eq. 26.11). Once the capacitor is fully discharged, the energy stored in it is zero. Because energy is conserved, the initial energy stored in the capacitor is transformed into internal energy in the resistor. Using the given values of C and $\mathcal{E}$, we find

$$\text{Energy} = \tfrac{1}{2}C\mathcal{E}^2 = \tfrac{1}{2}(5.00 \times 10^{-6}\,\text{F})(800\,\text{V})^2 = \boxed{1.60\,\text{J}}$$

The second way, which is more difficult but perhaps more instructive, is to note that as the capacitor discharges through the resistor, the rate at which energy is delivered to the resistor is given by I^2R, where I is the instantaneous current given by Equation 28.18. Because power is defined as the time rate of change of energy, we conclude that the energy delivered to the resistor must equal the time integral of $I^2 R\,dt$:

$$\text{Energy} = \int_0^{\infty} I^2 R\,dt = \int_0^{\infty} (I_0 e^{-t/RC})^2\,R\,dt$$

To evaluate this integral, we note that the initial current I_0 is equal to $\mathcal{E}/R$ and that all parameters except t are constant. Thus, we find

$$(1) \qquad \text{Energy} = \frac{\mathcal{E}^2}{R}\int_0^{\infty} e^{-2t/RC}\,dt$$

This integral has a value of $RC/2$; hence, we find

$$\text{Energy} = \tfrac{1}{2}C\mathcal{E}^2$$

which agrees with the result we obtained using the simpler approach, as it must. Note that we can use this second approach to find the total energy delivered to the resistor at *any* time after the switch is closed by simply replacing the upper limit in the integral with that specific value of t.

Exercise   Show that the integral in Equation (1) has the value $RC/2$.

Optional Section

28.5 ▶ ELECTRICAL INSTRUMENTS

The Ammeter

A device that measures current is called an **ammeter.** The current to be measured must pass directly through the ammeter, so the ammeter must be connected in se-

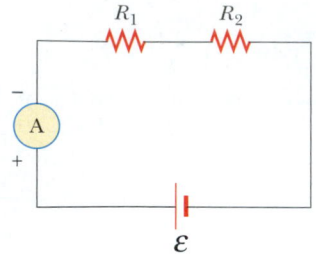

Figure 28.21 Current can be measured with an ammeter connected in series with the resistor and battery of a circuit. An ideal ammeter has zero resistance.

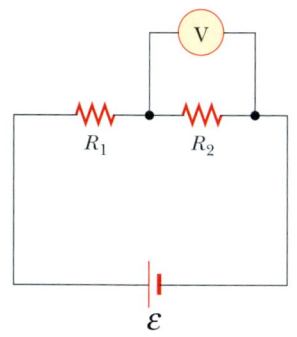

Figure 28.22 The potential difference across a resistor can be measured with a voltmeter connected in parallel with the resistor. An ideal voltmeter has infinite resistance.

ries with other elements in the circuit, as shown in Figure 28.21. When using an ammeter to measure direct currents, you must be sure to connect it so that current enters the instrument at the positive terminal and exits at the negative terminal.

Ideally, an ammeter should have zero resistance so that the current being measured is not altered. In the circuit shown in Figure 28.21, this condition requires that the resistance of the ammeter be much less than $R_1 + R_2$. Because any ammeter always has some internal resistance, the presence of the ammeter in the circuit slightly reduces the current from the value it would have in the meter's absence.

The Voltmeter

A device that measures potential difference is called a **voltmeter.** The potential difference between any two points in a circuit can be measured by attaching the terminals of the voltmeter between these points without breaking the circuit, as shown in Figure 28.22. The potential difference across resistor R_2 is measured by connecting the voltmeter in parallel with R_2. Again, it is necessary to observe the polarity of the instrument. The positive terminal of the voltmeter must be connected to the end of the resistor that is at the higher potential, and the negative terminal to the end of the resistor at the lower potential.

An ideal voltmeter has infinite resistance so that no current passes through it. In Figure 28.22, this condition requires that the voltmeter have a resistance much greater than R_2. In practice, if this condition is not met, corrections should be made for the known resistance of the voltmeter.

The Galvanometer

The **galvanometer** is the main component in analog ammeters and voltmeters. Figure 28.23a illustrates the essential features of a common type called the *D'Arsonval galvanometer.* It consists of a coil of wire mounted so that it is free to rotate on a pivot in a magnetic field provided by a permanent magnet. The basic op-

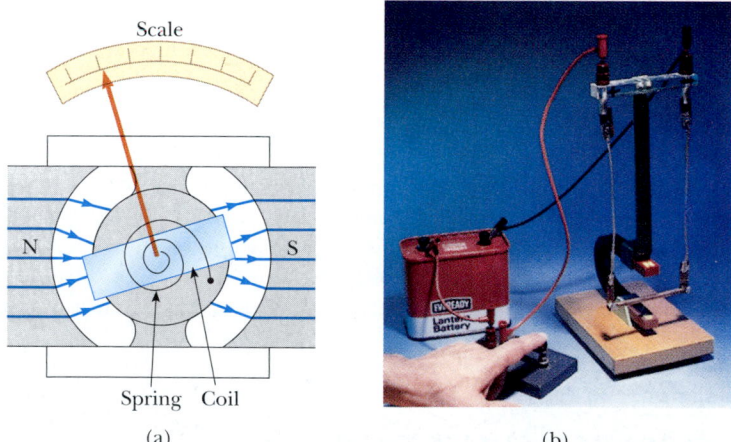

(a)   (b)

Figure 28.23 (a) The principal components of a D'Arsonval galvanometer. When the coil situated in a magnetic field carries a current, the magnetic torque causes the coil to twist. The angle through which the coil rotates is proportional to the current in the coil because of the counteracting torque of the spring. (b) A large-scale model of a galvanometer movement. Why does the coil rotate about the vertical axis after the switch is closed? *(Courtesy of Jim Lehman)*

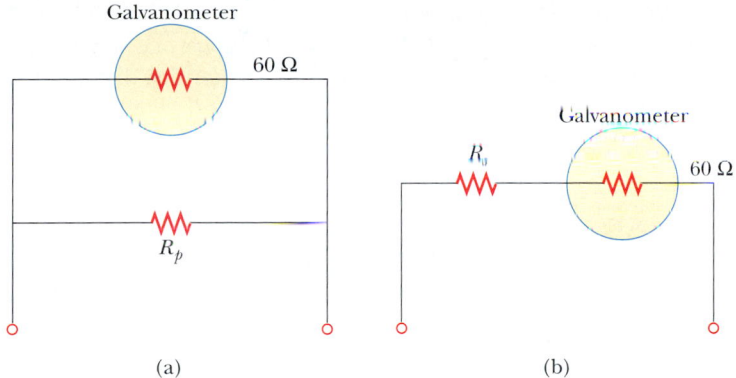

Figure 28.24 (a) When a galvanometer is to be used as an ammeter, a shunt resistor R_p is connected in parallel with the galvanometer. (b) When the galvanometer is used as a voltmeter, a resistor R_s is connected in series with the galvanometer.

eration of the galvanometer makes use of the fact that a torque acts on a current loop in the presence of a magnetic field (Chapter 29). The torque experienced by the coil is proportional to the current through it: the larger the current, the greater the torque and the more the coil rotates before the spring tightens enough to stop the rotation. Hence, the deflection of a needle attached to the coil is proportional to the current. Once the instrument is properly calibrated, it can be used in conjunction with other circuit elements to measure either currents or potential differences.

A typical off-the-shelf galvanometer is often not suitable for use as an ammeter, primarily because it has a resistance of about 60 Ω. An ammeter resistance this great considerably alters the current in a circuit. You can understand this by considering the following example: The current in a simple series circuit containing a 3-V battery and a 3-Ω resistor is 1 A. If you insert a 60-Ω galvanometer in this circuit to measure the current, the total resistance becomes 63 Ω and the current is reduced to 0.048 A!

A second factor that limits the use of a galvanometer as an ammeter is the fact that a typical galvanometer gives a full-scale deflection for currents of the order of 1 mA or less. Consequently, such a galvanometer cannot be used directly to measure currents greater than this value. However, it can be converted to a useful ammeter by placing a shunt resistor R_p in parallel with the galvanometer, as shown in Figure 28.24a. The value of R_p must be much less than the galvanometer resistance so that most of the current to be measured passes through the shunt resistor.

A galvanometer can also be used as a voltmeter by adding an external resistor R_s in series with it, as shown in Figure 28.24b. In this case, the external resistor must have a value much greater than the resistance of the galvanometer to ensure that the galvanometer does not significantly alter the voltage being measured.

The Wheatstone Bridge

An unknown resistance value can be accurately measured using a circuit known as a **Wheatstone bridge** (Fig. 28.25). This circuit consists of the unknown resistance R_x; three known resistances R_1, R_2, and R_3 (where R_1 is a calibrated variable resistor); a galvanometer; and a battery. The known resistor R_1 is varied until the galvanometer reading is zero—that is, until there is no current from a to b. Under this condition the bridge is said to be balanced. Because the electric potential at

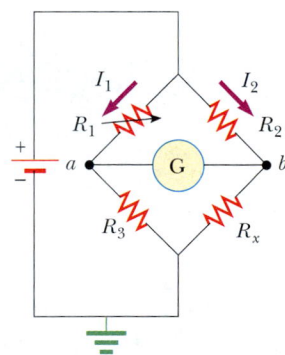

Figure 28.25 Circuit diagram for a Wheatstone bridge, an instrument used to measure an unknown resistance R_x in terms of known resistances R_1, R_2, and R_3. When the bridge is balanced, no current is present in the galvanometer. The arrow superimposed on the circuit symbol for resistor R_1 indicates that the value of this resistor can be varied by the person operating the bridge.

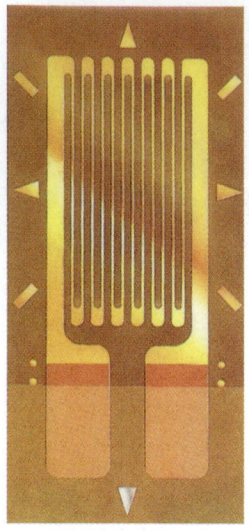

The strain gauge, a device used for experimental stress analysis, consists of a thin coiled wire bonded to a flexible plastic backing. The gauge measures stresses by detecting changes in the resistance of the coil as the strip bends. Resistance measurements are made with this device as one element of a Wheatstone bridge. Strain gauges are commonly used in modern electronic balances to measure the masses of objects.

Figure 28.26 Voltages, currents, and resistances are frequently measured with digital multimeters like this one. *(Henry Leap and Jim Lehman)*

point *a* must equal the potential at point *b* when the bridge is balanced, the potential difference across R_1 must equal the potential difference across R_2. Likewise, the potential difference across R_3 must equal the potential difference across R_x. From these considerations we see that

$$(1) \qquad I_1 R_1 = I_2 R_2$$

$$(2) \qquad I_1 R_3 = I_2 R_x$$

Dividing Equation (1) by Equation (2) eliminates the currents, and solving for R_x, we find that

$$R_x = \frac{R_2 R_3}{R_1} \qquad \textbf{(28.19)}$$

A number of similar devices also operate on the principle of null measurement (that is, adjustment of one circuit element to make the galvanometer read zero). One example is the capacitance bridge used to measure unknown capacitances. These devices do not require calibrated meters and can be used with any voltage source.

Wheatstone bridges are not useful for resistances above $10^5 \, \Omega$, but modern electronic instruments can measure resistances as high as $10^{12} \, \Omega$. Such instruments have an extremely high resistance between their input terminals. For example, input resistances of $10^{10} \, \Omega$ are common in most digital multimeters, which are devices that are used to measure voltage, current, and resistance (Fig. 28.26).

The Potentiometer

A **potentiometer** is a circuit that is used to measure an unknown emf $\mathcal{E}_x$ by comparison with a known emf. In Figure 28.27, point *d* represents a sliding contact that is used to vary the resistance (and hence the potential difference) between points *a* and *d*. The other required components are a galvanometer, a battery of known emf $\mathcal{E}_0$, and a battery of unknown emf $\mathcal{E}_x$.

With the currents in the directions shown in Figure 28.27, we see from Kirchhoff's junction rule that the current in the resistor R_x is $I - I_x$, where I is the current in the left branch (through the battery of emf $\mathcal{E}_0$) and I_x is the current in the right branch. Kirchhoff's loop rule applied to loop *abcda* traversed clockwise gives

$$-\mathcal{E}_x + (I - I_x)R_x = 0$$

Because current I_x passes through it, the galvanometer displays a nonzero reading. The sliding contact at *d* is now adjusted until the galvanometer reads zero (indicating a balanced circuit and that the potentiometer is another null-measurement device). Under this condition, the current in the galvanometer is zero, and the potential difference between *a* and *d* must equal the unknown emf $\mathcal{E}_x$:

$$\mathcal{E}_x = IR_x$$

Next, the battery of unknown emf is replaced by a standard battery of known emf $\mathcal{E}_s$, and the procedure is repeated. If R_s is the resistance between *a* and *d* when balance is achieved this time, then

$$\mathcal{E}_s = IR_s$$

where it is assumed that I remains the same. Combining this expression with the preceding one, we see that

$$\mathcal{E}_x = \frac{R_x}{R_s} \mathcal{E}_s \qquad \textbf{(28.20)}$$

If the resistor is a wire of resistivity ρ, its resistance can be varied by using the sliding contact to vary the length L, indicating how much of the wire is part of the circuit. With the substitutions $R_s = \rho L_s/A$ and $R_x = \rho L_x/A$, Equation 28.20 becomes

$$\mathcal{E}_x = \frac{L_x}{L_s}\mathcal{E}_s \qquad \text{(28.21)}$$

where L_x is the resistor length when the battery of unknown emf $\mathcal{E}_x$ is in the circuit and L_s is the resistor length when the standard battery is in the circuit.

The sliding-wire circuit of Figure 28.27 without the unknown emf and the galvanometer is sometimes called a *voltage divider*. This circuit makes it possible to tap into any desired smaller portion of the emf $\mathcal{E}_0$ by adjusting the length of the resistor.

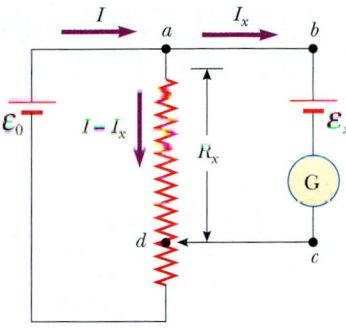

Figure 28.27 Circuit diagram for a potentiometer. The circuit is used to measure an unknown emf $\mathcal{E}_x$.

Optional Section

28.6 HOUSEHOLD WIRING AND ELECTRICAL SAFETY

Household circuits represent a practical application of some of the ideas presented in this chapter. In our world of electrical appliances, it is useful to understand the power requirements and limitations of conventional electrical systems and the safety measures that prevent accidents.

In a conventional installation, the utility company distributes electric power to individual homes by means of a pair of wires, with each home connected in parallel to these wires. One wire is called the *live wire*,[5] as illustrated in Figure 28.28, and the other is called the *neutral wire*. The potential difference between these two wires is about 120 V. This voltage alternates in time, with the neutral wire connected to ground and the potential of the live wire oscillating relative to ground. Much of what we have learned so far for the constant-emf situation (direct current) can also be applied to the alternating current that power companies supply to businesses and households. (Alternating voltage and current are discussed in Chapter 33.)

A meter is connected in series with the live wire entering the house to record the household's usage of electricity. After the meter, the wire splits so that there are several separate circuits in parallel distributed throughout the house. Each circuit contains a circuit breaker (or, in older installations, a fuse). The wire and circuit breaker for each circuit are carefully selected to meet the current demands for that circuit. If a circuit is to carry currents as large as 30 A, a heavy wire and an appropriate circuit breaker must be selected to handle this current. A circuit used to power only lamps and small appliances often requires only 15 A. Each circuit has its own circuit breaker to accommodate various load conditions.

As an example, consider a circuit in which a toaster oven, a microwave oven, and a coffee maker are connected (corresponding to R_1, R_2, and R_3 in Figure 28.28 and as shown in the chapter-opening photograph). We can calculate the current drawn by each appliance by using the expression $\mathcal{P} = I\,\Delta V$. The toaster oven, rated at 1 000 W, draws a current of 1 000 W/120 V = 8.33 A. The microwave oven, rated at 1 300 W, draws 10.8 A, and the coffee maker, rated at 800 W, draws 6.67 A. If the three appliances are operated simultaneously, they draw a total cur-

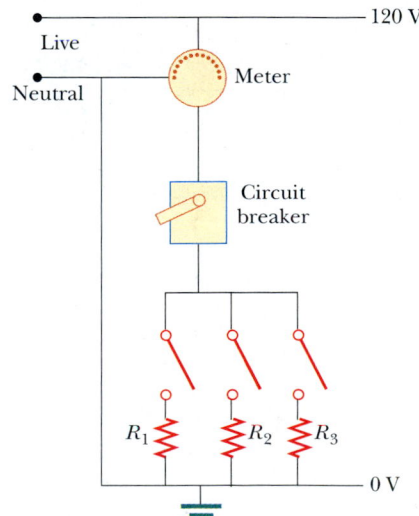

Figure 28.28 Wiring diagram for a household circuit. The resistances represent appliances or other electrical devices that operate with an applied voltage of 120 V.

---

[5] *Live wire* is a common expression for a conductor whose electric potential is above or below ground potential.

Figure 28.29 A power connection for a 240-V appliance. *(George Semple)*

Figure 28.30 A three-pronged power cord for a 120-V appliance. *(George Semple)*

rent of 25.8 A. Therefore, the circuit should be wired to handle at least this much current. If the rating of the circuit breaker protecting the circuit is too small—say, 20 A—the breaker will be tripped when the third appliance is turned on, preventing all three appliances from operating. To avoid this situation, the toaster oven and coffee maker can be operated on one 20-A circuit and the microwave oven on a separate 20-A circuit.

Many heavy-duty appliances, such as electric ranges and clothes dryers, require 240 V for their operation. The power company supplies this voltage by providing a third wire that is 120 V below ground potential (Fig. 28.29). The potential difference between this live wire and the other live wire (which is 120 V above ground potential) is 240 V. An appliance that operates from a 240-V line requires half the current of one operating from a 120-V line; therefore, smaller wires can be used in the higher-voltage circuit without overheating.

Electrical Safety

When the live wire of an electrical outlet is connected directly to ground, the circuit is completed and a short-circuit condition exists. A *short circuit* occurs when almost zero resistance exists between two points at different potentials; this results in a very large current. When this happens accidentally, a properly operating circuit breaker opens the circuit and no damage is done. However, a person in contact with ground can be electrocuted by touching the live wire of a frayed cord or other exposed conductor. An exceptionally good (although very dangerous) ground contact is made when the person either touches a water pipe (normally at ground potential) or stands on the ground with wet feet. The latter situation represents a good ground because normal, nondistilled water is a conductor because it contains a large number of ions associated with impurities. This situation should be avoided at all cost.

Electric shock can result in fatal burns, or it can cause the muscles of vital organs, such as the heart, to malfunction. The degree of damage to the body depends on the magnitude of the current, the length of time it acts, the part of the body touched by the live wire, and the part of the body through which the current passes. Currents of 5 mA or less cause a sensation of shock but ordinarily do little or no damage. If the current is larger than about 10 mA, the muscles contract and the person may be unable to release the live wire. If a current of about 100 mA passes through the body for only a few seconds, the result can be fatal. Such a large current paralyzes the respiratory muscles and prevents breathing. In some cases, currents of about 1 A through the body can produce serious (and sometimes fatal) burns. In practice, no contact with live wires is regarded as safe whenever the voltage is greater than 24 V.

Many 120-V outlets are designed to accept a three-pronged power cord such as the one shown in Figure 28.30. (This feature is required in all new electrical installations.) One of these prongs is the live wire at a nominal potential of 120 V. The second, called the "neutral," is nominally at 0 V and carries current to ground. The third, round prong is a safety ground wire that normally carries no current but is both grounded and connected directly to the casing of the appliance. If the live wire is accidentally shorted to the casing (which can occur if the wire insulation wears off), most of the current takes the low-resistance path through the appliance to ground. In contrast, if the casing of the appliance is not properly grounded and a short occurs, anyone in contact with the appliance experiences an electric shock because the body provides a low-resistance path to ground.

Special power outlets called *ground-fault interrupters* (GFIs) are now being used in kitchens, bathrooms, basements, exterior outlets, and other hazardous areas of new homes. These devices are designed to protect persons from electric shock by sensing small currents (≈ 5 mA) leaking to ground. (The principle of their operation is described in Chapter 31.) When an excessive leakage current is detected, the current is shut off in less than 1 ms.

Is a circuit breaker wired in series or in parallel with the device it is protecting?

SUMMARY

The **emf** of a battery is equal to the voltage across its terminals when the current is zero. That is, the emf is equivalent to the **open-circuit voltage** of the battery.

The **equivalent resistance** of a set of resistors connected in **series** is

$$R_{eq} = R_1 + R_2 + R_3 + \cdots \tag{28.6}$$

The **equivalent resistance** of a set of resistors connected in **parallel** is

$$\frac{1}{R_{eq}} = \frac{1}{R_1} + \frac{1}{R_2} + \frac{1}{R_3} + \cdots \tag{28.8}$$

If it is possible to combine resistors into series or parallel equivalents, the preceding two equations make it easy to determine how the resistors influence the rest of the circuit.

Circuits involving more than one loop are conveniently analyzed with the use of **Kirchhoff's rules:**

1. The sum of the currents entering any junction in an electric circuit must equal the sum of the currents leaving that junction:

$$\sum I_{in} = \sum I_{out} \tag{28.9}$$

2. The sum of the potential differences across all elements around any circuit loop must be zero:

$$\sum_{\substack{closed \\ loop}} \Delta V = 0 \tag{28.10}$$

The first rule is a statement of conservation of charge; the second is equivalent to a statement of conservation of energy.

When a resistor is traversed in the direction of the current, the change in potential ΔV across the resistor is $-IR$. When a resistor is traversed in the direction opposite the current, $\Delta V = +IR$. When a source of emf is traversed in the direction of the emf (negative terminal to positive terminal), the change in potential is $+\mathcal{E}$. When a source of emf is traversed opposite the emf (positive to negative), the change in potential is $-\mathcal{E}$. The use of these rules together with Equations 28.9 and 28.10 allows you to analyze electric circuits.

If a capacitor is charged with a battery through a resistor of resistance R, the charge on the capacitor and the current in the circuit vary in time according to the expressions

$$q(t) = Q(1 - e^{-t/RC}) \qquad (28.14)$$

$$I(t) = \frac{\mathcal{E}}{R} e^{-t/RC} \qquad (28.15)$$

where $Q = C\mathcal{E}$ is the maximum charge on the capacitor. The product RC is called the **time constant** τ of the circuit. If a charged capacitor is discharged through a resistor of resistance R, the charge and current decrease exponentially in time according to the expressions

$$q(t) = Qe^{-t/RC} \qquad (28.17)$$

$$I(t) = -\frac{Q}{RC} e^{-t/RC} \qquad (28.18)$$

where Q is the initial charge on the capacitor and $Q/RC = I_0$ is the initial current in the circuit. Equations 28.14, 28.15, 28.17, and 28.18 permit you to analyze the current and potential differences in an RC circuit and the charge stored in the circuit's capacitor.

QUESTIONS

1. Explain the difference between load resistance in a circuit and internal resistance in a battery.
2. Under what condition does the potential difference across the terminals of a battery equal its emf? Can the terminal voltage ever exceed the emf? Explain.
3. Is the direction of current through a battery always from the negative terminal to the positive one? Explain.
4. How would you connect resistors so that the equivalent resistance is greater than the greatest individual resistance? Give an example involving three resistors.
5. How would you connect resistors so that the equivalent resistance is less than the least individual resistance? Give an example involving three resistors.
6. Given three lightbulbs and a battery, sketch as many different electric circuits as you can.
7. Which of the following are the same for each resistor in a series connection—potential difference, current, power?
8. Which of the following are the same for each resistor in a parallel connection—potential difference, current, power?
9. What advantage might there be in using two identical resistors in parallel connected in series with another identical parallel pair, rather than just using a single resistor?
10. An incandescent lamp connected to a 120-V source with a short extension cord provides more illumination than the same lamp connected to the same source with a very long extension cord. Explain why.
11. When can the potential difference across a resistor be positive?
12. In Figure 28.15, suppose the wire between points g and h is replaced by a 10-Ω resistor. Explain why this change does not affect the currents calculated in Example 28.9.

13. Describe what happens to the lightbulb shown in Figure Q28.13 after the switch is closed. Assume that the capacitor has a large capacitance and is initially uncharged, and assume that the light illuminates when connected directly across the battery terminals.

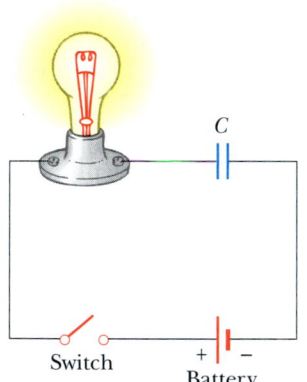

Figure Q28.13

14. What are the internal resistances of an ideal ammeter? of an ideal voltmeter? Do real meters ever attain these ideals?
15. Although the internal resistances of all sources of emf were neglected in the treatment of the potentiometer (Section 28.5), it is really not necessary to make this assumption. Explain why internal resistances play no role in the measurement of $\mathcal{E}_x$.

16. Why is it dangerous to turn on a light when you are in the bathtub?

17. Suppose you fall from a building, and on your way down you grab a high-voltage wire. Assuming that you are hanging from the wire, will you be electrocuted? If the wire then breaks, should you continue to hold onto an end of the wire as you fall?

18. What advantage does 120-V operation offer over 240 V? What are its disadvantages compared with 240 V?

19. When electricians work with potentially live wires, they often use the backs of their hands or fingers to move the wires. Why do you suppose they employ this technique?

20. What procedure would you use to try to save a person who is "frozen" to a live high-voltage wire without endangering your own life?

21. If it is the current through the body that determines the seriousness of a shock, why do we see warnings of high *voltage* rather than high *current* near electrical equipment?

22. Suppose you are flying a kite when it strikes a high-voltage wire. What factors determine how great a shock you receive?

23. A series circuit consists of three identical lamps that are connected to a battery as shown in Figure Q28.23. When switch S is closed, what happens (a) to the intensities of lamps A and B, (b) to the intensity of lamp C, (c) to the current in the circuit, and (d) to the voltage across the three lamps? (e) Does the power delivered to the circuit increase, decrease, or remain the same?

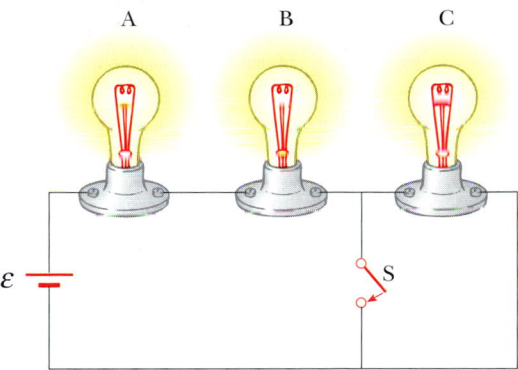

Figure Q28.23

24. If your car's headlights are on when you start the ignition, why do they dim while the car is starting?

25. A ski resort consists of a few chair lifts and several interconnected downhill runs on the side of a mountain, with a lodge at the bottom. The lifts are analogous to batteries, and the runs are analogous to resistors. Describe how two runs can be in series. Describe how three runs can be in parallel. Sketch a junction of one lift and two runs. State Kirchhoff's junction rule for ski resorts. One of the skiers, who happens to be carrying an altimeter, stops to warm up her toes each time she passes the lodge. State Kirchhoff's loop rule for altitude.

PROBLEMS

1, 2, 3 = straightforward, intermediate, challenging ☐ = full solution available in the *Student Solutions Manual and Study Guide*
WEB = solution posted at **http://www.saunderscollege.com/physics/** 🖥 = Computer useful in solving problem 🖱 = Interactive Physics
☐ = paired numerical/symbolic problems

Section 28.1 Electromotive Force

WEB 1. A battery has an emf of 15.0 V. The terminal voltage of the battery is 11.6 V when it is delivering 20.0 W of power to an external load resistor R. (a) What is the value of R? (b) What is the internal resistance of the battery?

2. (a) What is the current in a 5.60-Ω resistor connected to a battery that has a 0.200-Ω internal resistance if the terminal voltage of the battery is 10.0 V? (b) What is the emf of the battery?

3. Two 1.50-V batteries—with their positive terminals in the same direction—are inserted in series into the barrel of a flashlight. One battery has an internal resistance of 0.255 Ω, the other an internal resistance of 0.153 Ω. When the switch is closed, a current of 600 mA occurs in the lamp. (a) What is the lamp's resistance? (b) What percentage of the power from the batteries appears in the batteries themselves, as represented by an increase in temperature?

4. An automobile battery has an emf of 12.6 V and an internal resistance of 0.080 0 Ω. The headlights have a total resistance of 5.00 Ω (assumed constant). What is the potential difference across the headlight bulbs (a) when they are the only load on the battery and (b) when the starter motor, which takes an additional 35.0 A from the battery, is operated?

Section 28.2 Resistors in Series and in Parallel

5. The current in a loop circuit that has a resistance of R_1 is 2.00 A. The current is reduced to 1.60 A when an additional resistor $R_2 = 3.00$ Ω is added in series with R_1. What is the value of R_1?

6. (a) Find the equivalent resistance between points a and b in Figure P28.6. (b) Calculate the current in each resistor if a potential difference of 34.0 V is applied between points a and b.

7. A television repairman needs a 100-Ω resistor to repair a malfunctioning set. He is temporarily out of resistors

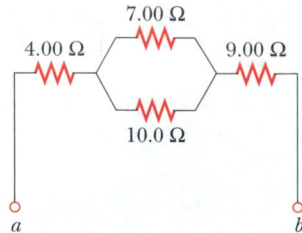

Figure P28.6

of this value. All he has in his toolbox are a 500-Ω resistor and two 250-Ω resistors. How can he obtain the desired resistance using the resistors he has on hand?

8. A lightbulb marked "75 W [at] 120 V" is screwed into a socket at one end of a long extension cord in which each of the two conductors has a resistance of 0.800 Ω. The other end of the extension cord is plugged into a 120-V outlet. Draw a circuit diagram, and find the actual power delivered to the bulb in this circuit.

WEB 9. Consider the circuit shown in Figure P28.9. Find (a) the current in the 20.0-Ω resistor and (b) the potential difference between points a and b.

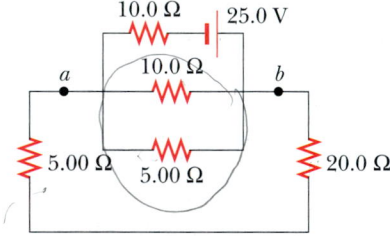

Figure P28.9

10. Four copper wires of equal length are connected in series. Their cross-sectional areas are 1.00 cm², 2.00 cm², 3.00 cm², and 5.00 cm². If a voltage of 120 V is applied to the arrangement, what is the voltage across the 2.00-cm² wire?

11. Three 100-Ω resistors are connected as shown in Figure P28.11. The maximum power that can safely be delivered to any one resistor is 25.0 W. (a) What is the maximum voltage that can be applied to the terminals a and b? (b) For the voltage determined in part (a), what is

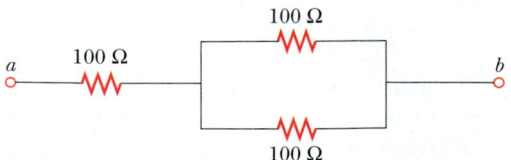

Figure P28.11

the power delivered to each resistor? What is the total power delivered?

12. Using only three resistors—2.00 Ω, 3.00 Ω, and 4.00 Ω—find 17 resistance values that can be obtained with various combinations of one or more resistors. Tabulate the combinations in order of increasing resistance.

13. The current in a circuit is tripled by connecting a 500-Ω resistor in parallel with the resistance of the circuit. Determine the resistance of the circuit in the absence of the 500-Ω resistor.

14. The power delivered to the top part of the circuit shown in Figure P28.14 does not depend on whether the switch is opened or closed. If $R = 1.00\ \Omega$, what is R'? Neglect the internal resistance of the voltage source.

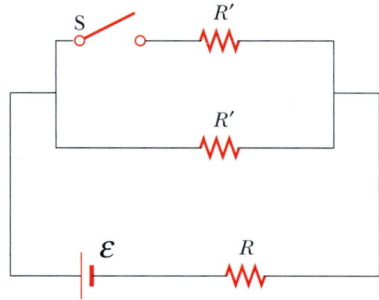

Figure P28.14

15. Calculate the power delivered to each resistor in the circuit shown in Figure P28.15.

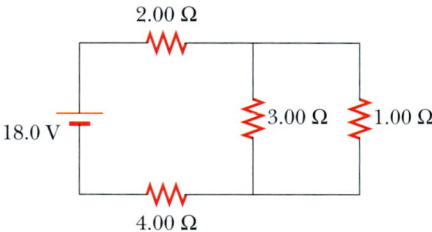

Figure P28.15

16. Two resistors connected in series have an equivalent resistance of 690 Ω. When they are connected in parallel, their equivalent resistance is 150 Ω. Find the resistance of each resistor.

17. In Figures 28.4 and 28.5, let $R_1 = 11.0\ \Omega$, let $R_2 = 22.0\ \Omega$, and let the battery have a terminal voltage of 33.0 V. (a) In the parallel circuit shown in Figure 28.5, which resistor uses more power? (b) Verify that the sum of the power (I^2R) used by each resistor equals the power supplied by the battery ($I\Delta V$). (c) In the series circuit, which resistor uses more power? (d) Verify that the sum of the power (I^2R) used by each resistor equals

the power supplied by the battery ($\mathcal{P} = I\Delta V$).

(e) Which circuit configuration uses more power?

Section 28.3 Kirchhoff's Rules

Note: The currents are not necessarily in the direction shown for some circuits

18. The ammeter shown in Figure P28.18 reads 2.00 A. Find I_1, I_2, and $\mathcal{E}$.

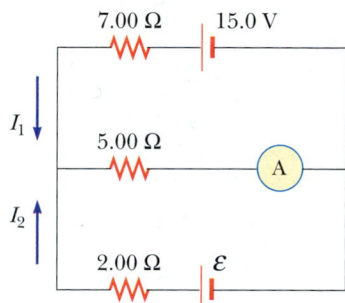

Figure P28.18

WEB 19. Determine the current in each branch of the circuit shown in Figure P28.19.

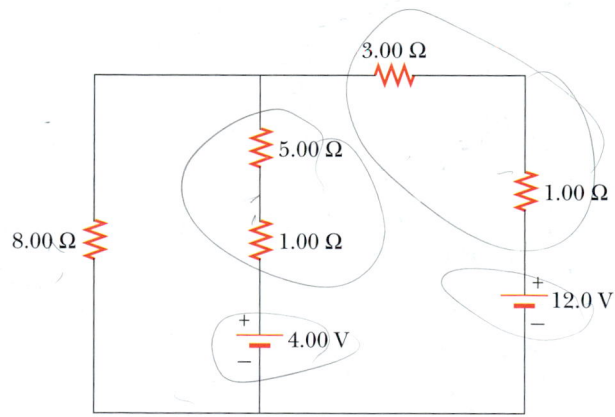

Figure P28.19 Problems 19, 20, and 21.

20. In Figure P28.19, show how to add just enough ammeters to measure every different current that is flowing. Show how to add just enough voltmeters to measure the potential difference across each resistor and across each battery.

21. The circuit considered in Problem 19 and shown in Figure P28.19 is connected for 2.00 min. (a) Find the energy supplied by each battery. (b) Find the energy delivered to each resistor. (c) Find the total amount of energy converted from chemical energy in the battery to internal energy in the circuit resistance.

22. (a) Using Kirchhoff's rules, find the current in each resistor shown in Figure P28.22 and (b) find the potential difference between points *c* and *f*. Which point is at the higher potential?

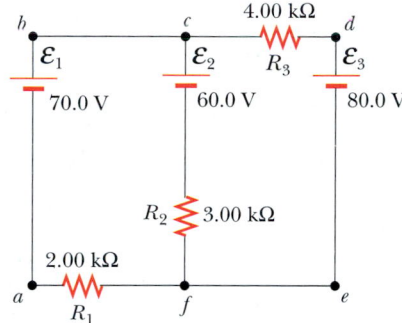

Figure P28.22

23. If $R = 1.00$ kΩ and $\mathcal{E} = 250$ V in Figure P28.23, determine the direction and magnitude of the current in the horizontal wire between *a* and *e*.

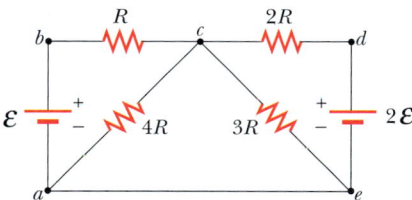

Figure P28.23

24. In the circuit of Figure P28.24, determine the current in each resistor and the voltage across the 200-Ω resistor.

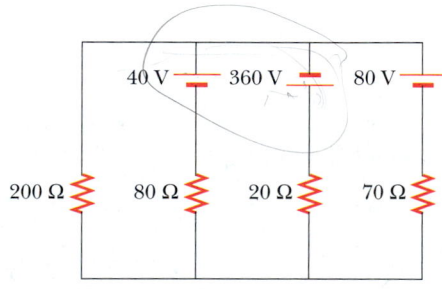

Figure P28.24

25. A dead battery is charged by connecting it to the live battery of another car with jumper cables (Fig. P28.25). Determine the current in the starter and in the dead battery.

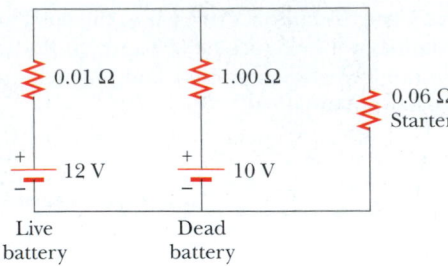

Live   Dead
battery   battery

Figure P28.25

26. For the network shown in Figure P28.26, show that the resistance $R_{ab} = \frac{27}{17}$ Ω.

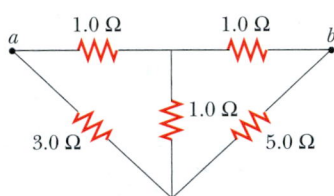

Figure P28.26

27. For the circuit shown in Figure P28.27, calculate (a) the current in the 2.00-Ω resistor and (b) the potential difference between points a and b.

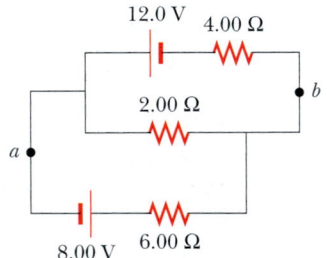

Figure P28.27

28. Calculate the power delivered to each of the resistors shown in Figure P28.28.

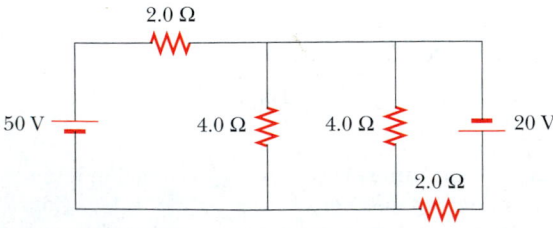

Figure P28.28

Section 28.4   RC Circuits

WEB 29. Consider a series RC circuit (see Fig. 28.16) for which $R = 1.00$ MΩ, $C = 5.00$ μF, and $\mathcal{E} = 30.0$ V. Find (a) the time constant of the circuit and (b) the maximum charge on the capacitor after the switch is closed. (c) If the switch is closed at $t = 0$, find the current in the resistor 10.0 s later.

30. A 2.00-nF capacitor with an initial charge of 5.10 μC is discharged through a 1.30-kΩ resistor. (a) Calculate the current through the resistor 9.00 μs after the resistor is connected across the terminals of the capacitor. (b) What charge remains on the capacitor after 8.00 μs? (c) What is the maximum current in the resistor?

31. A fully charged capacitor stores energy U_0. How much energy remains when its charge has decreased to half its original value?

32. In the circuit of Figure P28.32, switch S has been open for a long time. It is then suddenly closed. Determine the time constant (a) before the switch is closed and (b) after the switch is closed. (c) If the switch is closed at $t = 0$, determine the current through it as a function of time.

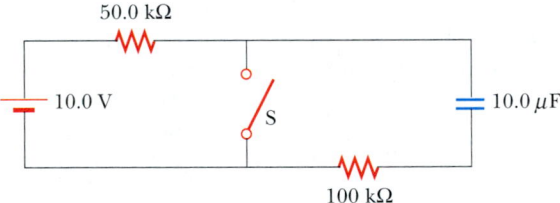

Figure P28.32

33. The circuit shown in Figure P28.33 has been connected for a long time. (a) What is the voltage across the capacitor? (b) If the battery is disconnected, how long does it take the capacitor to discharge to one-tenth its initial voltage?

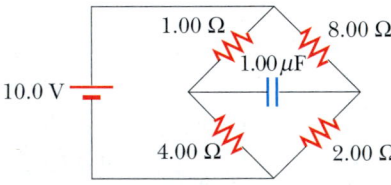

Figure P28.33

34. A 4.00-MΩ resistor and a 3.00-μF capacitor are connected in series with a 12.0-V power supply. (a) What is the time constant for the circuit? (b) Express the current in the circuit and the charge on the capacitor as functions of time.

35. Dielectric materials used in the manufacture of capacitors are characterized by conductivities that are small but not zero. Therefore, a charged capacitor slowly loses its charge by "leaking" across the dielectric. If a certain 3.60-μF capacitor leaks charge such that the potential difference decreases to half its initial value in 4.00 s, what is the equivalent resistance of the dielectric?

36. Dielectric materials used in the manufacture of capacitors are characterized by conductivities that are small but not zero. Therefore, a charged capacitor slowly loses its charge by "leaking" across the dielectric. If a capacitor having capacitance C leaks charge such that the potential difference decreases to half its initial value in a time t, what is the equivalent resistance of the dielectric?

37. A capacitor in an RC circuit is charged to 60.0% of its maximum value in 0.900 s. What is the time constant of the circuit?

(Optional)
Section 28.5   Electrical Instruments

38. A typical galvanometer, which requires a current of 1.50 mA for full-scale deflection and has a resistance of 75.0 Ω, can be used to measure currents of much greater values. A relatively small shunt resistor is wired in parallel with the galvanometer (refer to Fig. 28.24a) so that an operator can measure large currents without causing damage to the galvanometer. Most of the current then flows through the shunt resistor. Calculate the value of the shunt resistor that enables the galvanometer to be used to measure a current of 1.00 A at full-scale deflection. (*Hint:* Use Kirchhoff's rules.)

39. The galvanometer described in the preceding problem can be used to measure voltages. In this case a large resistor is wired in series with the galvanometer in a way similar to that shown in Figure 28.24b. This arrangement, in effect, limits the current that flows through the galvanometer when large voltages are applied. Most of the potential drop occurs across the resistor placed in series. Calculate the value of the resistor that enables the galvanometer to measure an applied voltage of 25.0 V at full-scale deflection.

40. A galvanometer with a full-scale sensitivity of 1.00 mA requires a 900-Ω series resistor to make a voltmeter reading full scale when 1.00 V is measured across the terminals. What series resistor is required to make the same galvanometer into a 50.0-V (full-scale) voltmeter?

41. Assume that a galvanometer has an internal resistance of 60.0 Ω and requires a current of 0.500 mA to produce full-scale deflection. What resistance must be connected in parallel with the galvanometer if the combination is to serve as an ammeter that has a full-scale deflection for a current of 0.100 A?

42. A Wheatstone bridge of the type shown in Figure 28.25 is used to make a precise measurement of the resistance of a wire connector. If $R_3 = 1.00$ kΩ and the bridge is balanced by adjusting R_1 such that $R_1 = 2.50R_2$, what is R_x?

43. Consider the case in which the Wheatstone bridge shown in Figure 28.25 is unbalanced. Calculate the current through the galvanometer when $R_x = R_3 = 7.00\ \Omega$, $R_2 = 21.0\ \Omega$, and $R_1 = 14.0\ \Omega$. Assume that the voltage across the bridge is 70.0 V, and neglect the galvanometer's resistance.

44. **Review Problem.**   A Wheatstone bridge can be used to measure the strain ($\Delta L / L_i$) of a wire (see Section 12.4), where L_i is the length before stretching, L is the length after stretching, and $\Delta L = L - L_i$. Let $\alpha = \Delta L / L_i$. Show that the resistance is $R = R_i(1 + 2\alpha + \alpha^2)$ for any length, where $R_i = \rho L_i / A_i$. Assume that the resistivity and volume of the wire stay constant.

45. Consider the potentiometer circuit shown in Figure 28.27. If a standard battery with an emf of 1.018 6 V is used in the circuit and the resistance between a and d is 36.0 Ω, the galvanometer reads zero. If the standard battery is replaced by an unknown emf, the galvanometer reads zero when the resistance is adjusted to 48.0 Ω. What is the value of the emf?

46. *Meter loading.* Work this problem to five-digit precision. Refer to Figure P28.46. (a) When a 180.00-Ω resistor is put across a battery with an emf of 6.000 0 V and an internal resistance of 20.000 Ω, what current flows in the resistor? What will be the potential difference across it? (b) Suppose now that an ammeter with a resistance of 0.500 00 Ω and a voltmeter with a resistance of

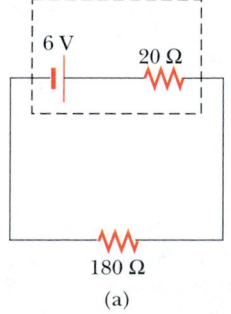
(a)

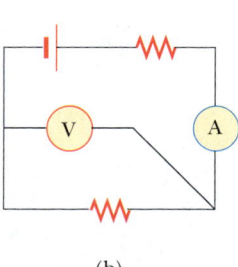
(b)

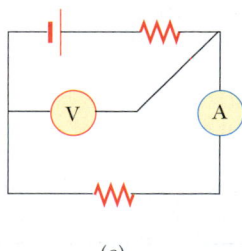
(c)

Figure P28.46

20 000 Ω are added to the circuit, as shown in Figure P28.46b. Find the reading of each. (c) One terminal of one wire is moved, as shown in Figure P28.46c. Find the new meter readings.

(Optional)
Section 28.6   Household Wiring and Electrical Safety

WEB **47.** An electric heater is rated at 1 500 W, a toaster at 750 W, and an electric grill at 1 000 W. The three appliances are connected to a common 120-V circuit. (a) How much current does each draw? (b) Is a 25.0-A circuit breaker sufficient in this situation? Explain your answer.

48. An 8.00-ft extension cord has two 18-gauge copper wires, each with a diameter of 1.024 mm. What is the I^2R loss in this cord when it carries a current of (a) 1.00 A? (b) 10.0 A?

49. Sometimes aluminum wiring has been used instead of copper for economic reasons. According to the National Electrical Code, the maximum allowable current for 12-gauge copper wire with rubber insulation is 20 A. What should be the maximum allowable current in a 12-gauge aluminum wire if it is to have the same I^2R loss per unit length as the copper wire?

50. Turn on your desk lamp. Pick up the cord with your thumb and index finger spanning its width. (a) Compute an order-of-magnitude estimate for the current that flows through your hand. You may assume that at a typical instant the conductor inside the lamp cord next to your thumb is at potential $\sim 10^2$ V and that the conductor next to your index finger is at ground potential (0 V). The resistance of your hand depends strongly on the thickness and moisture content of the outer layers of your skin. Assume that the resistance of your hand between fingertip and thumb tip is $\sim 10^4$ Ω. You may model the cord as having rubber insulation. State the other quantities you measure or estimate and their values. Explain your reasoning. (b) Suppose that your body is isolated from any other charges or currents. In order-of-magnitude terms, describe the potential of your thumb where it contacts the cord and the potential of your finger where it touches the cord.

ADDITIONAL PROBLEMS

51. Four 1.50-V AA batteries in series are used to power a transistor radio. If the batteries can provide a total charge of 240 C, how long will they last if the radio has a resistance of 200 Ω?

52. A battery has an emf of 9.20 V and an internal resistance of 1.20 Ω. (a) What resistance across the battery will extract from it a power of 12.8 W? (b) a power of 21.2 W?

53. Calculate the potential difference between points a and b in Figure P28.53, and identify which point is at the higher potential.

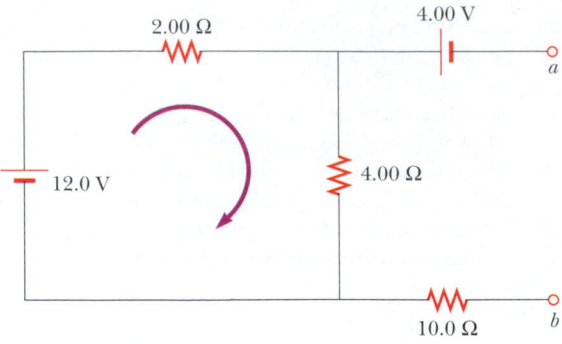

Figure P28.53

54. A 10.0-μF capacitor is charged by a 10.0-V battery through a resistance R. The capacitor reaches a potential difference of 4.00 V at a time 3.00 s after charging begins. Find R.

55. When two unknown resistors are connected in series with a battery, 225 W is delivered to the combination with a total current of 5.00 A. For the same total current, 50.0 W is delivered when the resistors are connected in parallel. Determine the values of the two resistors.

56. When two unknown resistors are connected in series with a battery, a total power $\mathcal{P}_s$ is delivered to the combination with a total current of I. For the same total current, a total power $\mathcal{P}_p$ is delivered when the resistors are connected in parallel. Determine the values of the two resistors.

57. A battery has an emf $\mathcal{E}$ and internal resistance r. A variable resistor R is connected across the terminals of the battery. Determine the value of R such that (a) the potential difference across the terminals is a maximum, (b) the current in the circuit is a maximum, (c) the power delivered to the resistor is a maximum.

58. A power supply has an open-circuit voltage of 40.0 V and an internal resistance of 2.00 Ω. It is used to charge two storage batteries connected in series, each having an emf of 6.00 V and internal resistance of 0.300 Ω. If the charging current is to be 4.00 A, (a) what additional resistance should be added in series? (b) Find the power delivered to the internal resistance of the supply, the I^2R loss in the batteries, and the power delivered to the added series resistance. (c) At what rate is the chemical energy in the batteries increasing?

59. The value of a resistor R is to be determined using the ammeter-voltmeter setup shown in Figure P28.59. The ammeter has a resistance of 0.500 Ω, and the voltmeter has a resistance of 20 000 Ω. Within what range of actual values of R will the measured values be correct, to within 5.00%, if the measurement is made using (a) the circuit shown in Figure P28.59a? (b) the circuit shown in Figure P28.59b?

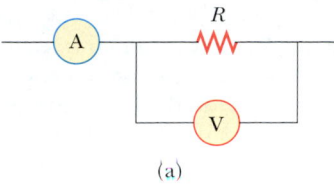

(a)

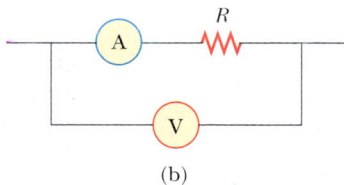

(b)

Figure P28.59

60. A battery is used to charge a capacitor through a resistor, as shown in Figure 28.16. Show that half the energy supplied by the battery appears as internal energy in the resistor and that half is stored in the capacitor.

61. The values of the components in a simple series RC circuit containing a switch (Fig. 28.16) are $C = 1.00 \ \mu F$, $R = 2.00 \times 10^6 \ \Omega$, and $\mathcal{E} = 10.0$ V. At the instant 10.0 s after the switch is closed, calculate (a) the charge on the capacitor, (b) the current in the resistor, (c) the rate at which energy is being stored in the capacitor, and (d) the rate at which energy is being delivered by the battery.

62. The switch in Figure P28.62a closes when $V_c > 2V/3$ and opens when $V_c < V/3$. The voltmeter reads a voltage as plotted in Figure P28.62b. What is the period T of the waveform in terms of R_A, R_B, and C?

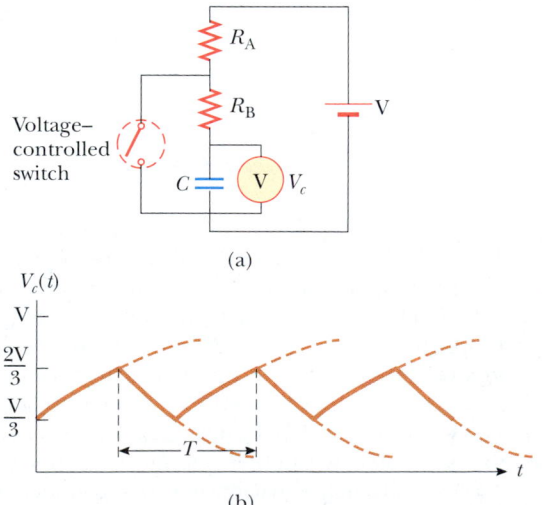

(a)

(b)

Figure P28.62

63. Three 60.0-W, 120-V lightbulbs are connected across a 120-V power source, as shown in Figure P28.63. Find (a) the total power delivered to the three bulbs and (b) the voltage across each. Assume that the resistance of each bulb conforms to Ohm's law (even though in reality the resistance increases markedly with current).

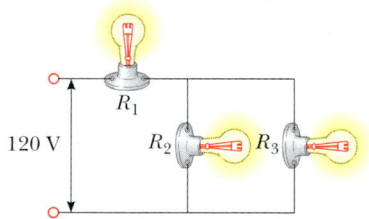

Figure P28.63

64. Design a multirange voltmeter capable of full-scale deflection for 20.0 V, 50.0 V, and 100 V. Assume that the meter movement is a galvanometer that has a resistance of 60.0 Ω and gives a full-scale deflection for a current of 1.00 mA.

65. Design a multirange ammeter capable of full-scale deflection for 25.0 mA, 50.0 mA, and 100 mA. Assume that the meter movement is a galvanometer that has a resistance of 25.0 Ω and gives a full-scale deflection for 1.00 mA.

66. A particular galvanometer serves as a 2.00-V full-scale voltmeter when a 2 500-Ω resistor is connected in series with it. It serves as a 0.500-A full-scale ammeter when a 0.220-Ω resistor is connected in parallel with it. Determine the internal resistance of the galvanometer and the current required to produce full-scale deflection.

67. In Figure P28.67, suppose that the switch has been closed for a length of time sufficiently long for the capacitor to become fully charged. (a) Find the steady-state current in each resistor. (b) Find the charge Q on the capacitor. (c) The switch is opened at $t = 0$. Write an equation for the current I_{R_2} in R_2 as a function of time, and (d) find the time that it takes for the charge on the capacitor to fall to one-fifth its initial value.

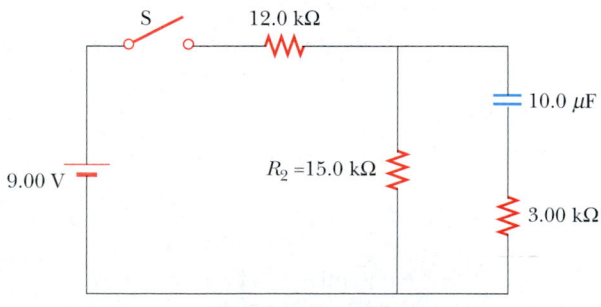

Figure P28.67

68. The circuit shown in Figure P28.68 is set up in the laboratory to measure an unknown capacitance C with the use of a voltmeter of resistance $R = 10.0\ \text{M}\Omega$ and a battery whose emf is 6.19 V. The data given in the table below are the measured voltages across the capacitor as a function of time, where $t = 0$ represents the time at which the switch is opened. (a) Construct a graph of $\ln(\mathcal{E}/\Delta V)$ versus t, and perform a linear least-squares fit to the data. (b) From the slope of your graph, obtain a value for the time constant of the circuit and a value for the capacitance.

ΔV (V)	t (s)	$\ln(\mathcal{E}/\Delta V)$
6.19	0	
5.55	4.87	
4.93	11.1	
4.34	19.4	
3.72	30.8	
3.09	46.6	
2.47	67.3	
1.83	102.2	

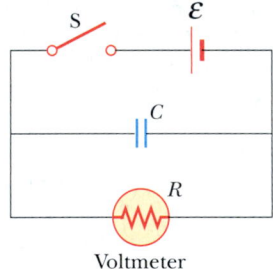

Figure P28.68

69. (a) Using symmetry arguments, show that the current through any resistor in the configuration of Figure P28.69 is either $I/3$ or $I/6$. All resistors have the same resistance r. (b) Show that the equivalent resistance between points a and b is $(5/6)r$.

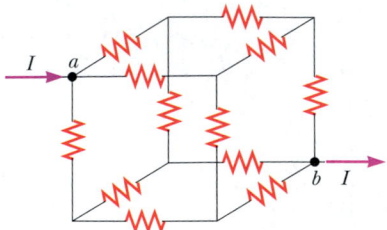

Figure P28.69

70. The student engineer of a campus radio station wishes to verify the effectiveness of the lightning rod on the antenna mast (Fig. P28.70). The unknown resistance R_x is between points C and E. Point E is a true ground but is inaccessible for direct measurement since this stratum is several meters below the Earth's surface. Two identical rods are driven into the ground at A and B, introducing an unknown resistance R_y. The procedure is as follows. Measure resistance R_1 between points A and B, then connect A and B with a heavy conducting wire and measure resistance R_2 between points A and C. (a) Derive a formula for R_x in terms of the observable resistances R_1 and R_2. (b) A satisfactory ground resistance would be $R_x < 2.00\ \Omega$. Is the grounding of the station adequate if measurements give $R_1 = 13.0\ \Omega$ and $R_2 = 6.00\ \Omega$?

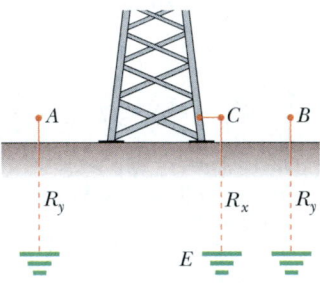

Figure P28.70

71. Three 2.00-Ω resistors are connected as shown in Figure P28.71. Each can withstand a maximum power of 32.0 W without becoming excessively hot. Determine the maximum power that can be delivered to the combination of resistors.

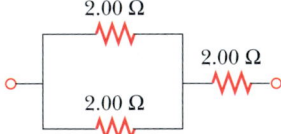

Figure P28.71

72. The circuit in Figure P28.72 contains two resistors, $R_1 = 2.00\ \text{k}\Omega$ and $R_2 = 3.00\ \text{k}\Omega$, and two capacitors, $C_1 = 2.00\ \mu\text{F}$ and $C_2 = 3.00\ \mu\text{F}$, connected to a battery with emf $\mathcal{E} = 120$ V. If no charges exist on the capacitors before switch S is closed, determine the charges q_1 and q_2 on capacitors C_1 and C_2, respectively, after the switch is closed. (*Hint:* First reconstruct the circuit so that it becomes a simple RC circuit containing a single resistor and single capacitor in series, connected to the battery, and then determine the total charge q stored in the equivalent circuit.)

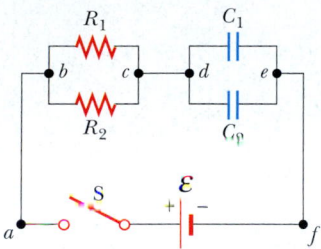

Figure P28.72

73. Assume that you have a battery of emf $\mathcal{E}$ and three identical lightbulbs, each having constant resistance R. What is the total power from the battery if the bulbs are connected (a) in series? (b) in parallel? (c) For which connection do the bulbs shine the brightest?

ANSWERS TO QUICK QUIZZES

28.1 Bulb R_1 becomes brighter. Connecting b to c "shorts out" bulb R_2 and changes the total resistance of the circuit from $R_1 + R_2$ to just R_1. Because the resistance has decreased (and the potential difference supplied by the battery does not change), the current through the battery increases. This means that the current through bulb R_1 increases, and bulb R_1 glows more brightly. Bulb R_2 goes out because the new piece of wire provides an almost resistance-free path for the current; hence, essentially zero current exists in bulb R_2.

28.2 Adding another series resistor increases the total resistance of the circuit and thus reduces the current in the battery. The potential difference across the battery terminals would increase because the reduced current results in a smaller voltage decrease across the internal resistance.

If the second resistor were connected in parallel, the total resistance of the circuit would decrease, and an increase in current through the battery would result. The potential difference across the terminals would decrease because the increased current results in a greater voltage decrease across the internal resistance.

28.3 They must be in parallel because if one burns out, the other continues to operate. If they were in series, one failed headlamp would interrupt the current throughout the entire circuit, including the other headlamp.

28.4 Because the circuit breaker trips and opens the circuit when the current in that circuit exceeds a certain preset value. The circuit breaker must be in series to sense the appropriate current (see Fig. 28.28).

chapter

29

Magnetic Fields

Chapter Outline

Many historians of science believe that the compass, which uses a magnetic needle, was used in China as early as the 13th century B.C., its invention being of Arabic or Indian origin. The early Greeks knew about magnetism as early as 800 B.C. They discovered that the stone magnetite (Fe_3O_4) attracts pieces of iron. Legend ascribes the name *magnetite* to the shepherd Magnes, the nails of whose shoes and the tip of whose staff stuck fast to chunks of magnetite while he pastured his flocks.

In 1269 a Frenchman named Pierre de Maricourt mapped out the directions taken by a needle placed at various points on the surface of a spherical natural magnet. He found that the directions formed lines that encircled the sphere and passed through two points diametrically opposite each other, which he called the *poles* of the magnet. Subsequent experiments showed that every magnet, regardless of its shape, has two poles, called *north* and *south* poles, that exert forces on other magnetic poles just as electric charges exert forces on one another. That is, like poles repel each other, and unlike poles attract each other.

The poles received their names because of the way a magnet behaves in the presence of the Earth's magnetic field. If a bar magnet is suspended from its midpoint and can swing freely in a horizontal plane, it will rotate until its north pole points to the Earth's geographic North Pole and its south pole points to the Earth's geographic South Pole.[1] (The same idea is used in the construction of a simple compass.)

In 1600 William Gilbert (1540–1603) extended de Maricourt's experiments to a variety of materials. Using the fact that a compass needle orients in preferred directions, he suggested that the Earth itself is a large permanent magnet. In 1750 experimenters used a torsion balance to show that magnetic poles exert attractive or repulsive forces on each other and that these forces vary as the inverse square of the distance between interacting poles. Although the force between two magnetic poles is similar to the force between two electric charges, there is an important difference. Electric charges can be isolated (witness the electron and proton), whereas **a single magnetic pole has never been isolated.** That is, **magnetic poles are always found in pairs.** All attempts thus far to detect an isolated magnetic pole have been unsuccessful. No matter how many times a permanent magnet is cut in two, each piece always has a north and a south pole. (There is some theoretical basis for speculating that magnetic *monopoles*—isolated north or south poles—may exist in nature, and attempts to detect them currently make up an active experimental field of investigation.)

The relationship between magnetism and electricity was discovered in 1819 when, during a lecture demonstration, the Danish scientist Hans Christian Oersted found that an electric current in a wire deflected a nearby compass needle.[2] Shortly thereafter, André Ampère (1775–1836) formulated quantitative laws for calculating the magnetic force exerted by one current-carrying electrical conductor on another. He also suggested that on the atomic level, electric current loops are responsible for *all* magnetic phenomena.

In the 1820s, further connections between electricity and magnetism were demonstrated by Faraday and independently by Joseph Henry (1797–1878). They

An electromagnet is used to move tons of scrap metal. (*Jeffrey Sylvester/FPG International*)

Hans Christian Oersted
Danish physicist (1777–1851)
(*North Wind Picture Archives*)

---

[1] Note that the Earth's geographic North Pole is magnetically a south pole, whereas its geographic South Pole is magnetically a north pole. Because *opposite* magnetic poles attract each other, the pole on a magnet that is attracted to the Earth's geographic North Pole is the magnet's *north* pole and the pole attracted to the Earth's geographic South Pole is the magnet's *south* pole.

[2] The same discovery was reported in 1802 by an Italian jurist, Gian Dominico Romognosi, but was overlooked, probably because it was published in the newspaper *Gazetta de Trentino* rather than in a scholarly journal.

showed that an electric current can be produced in a circuit either by moving a magnet near the circuit or by changing the current in a nearby circuit. These observations demonstrate that a changing magnetic field creates an electric field. Years later, theoretical work by Maxwell showed that the reverse is also true: A changing electric field creates a magnetic field.

A similarity between electric and magnetic effects has provided methods of making permanent magnets. In Chapter 23 we learned that when rubber and wool are rubbed together, both become charged—one positively and the other negatively. In an analogous fashion, one can magnetize an unmagnetized piece of iron by stroking it with a magnet. Magnetism can also be induced in iron (and other materials) by other means. For example, if a piece of unmagnetized iron is placed near (but not touching) a strong magnet, the unmagnetized piece eventually becomes magnetized.

This chapter examines the forces that act on moving charges and on current-carrying wires in the presence of a magnetic field. The source of the magnetic field itself is described in Chapter 30.

29.1 ▶ THE MAGNETIC FIELD

In our study of electricity, we described the interactions between charged objects in terms of electric fields. Recall that an electric field surrounds any stationary or moving electric charge. In addition to an electric field, the region of space surrounding any *moving* electric charge also contains a magnetic field, as we shall see in Chapter 30. A magnetic field also surrounds any magnetic substance.

Historically, the symbol **B** has been used to represent a magnetic field, and this is the notation we use in this text. The direction of the magnetic field **B** at any location is the direction in which a compass needle points at that location. Figure 29.1 shows how the magnetic field of a bar magnet can be traced with the aid of a compass. Note that the magnetic field lines outside the magnet point away from north poles and toward south poles. One can display magnetic field patterns of a bar magnet using small iron filings, as shown in Figure 29.2.

We can define a magnetic field **B** at some point in space in terms of the magnetic force $\mathbf{F}_B$ that the field exerts on a test object, for which we use a charged particle moving with a velocity **v**. For the time being, let us assume that no electric or gravitational fields are present at the location of the test object. Experiments on various charged particles moving in a magnetic field give the following results:

- The magnitude F_B of the magnetic force exerted on the particle is proportional to the charge q and to the speed v of the particle.

These refrigerator magnets are similar to a series of very short bar magnets placed end to end. If you slide the back of one refrigerator magnet in a circular path across the back of another one, you can feel a vibration as the two series of north and south poles move across each other. *(George Semple)*

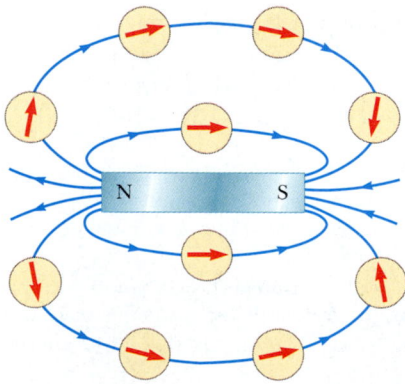

Figure 29.1   Compass needles can be used to trace the magnetic field lines of a bar magnet.

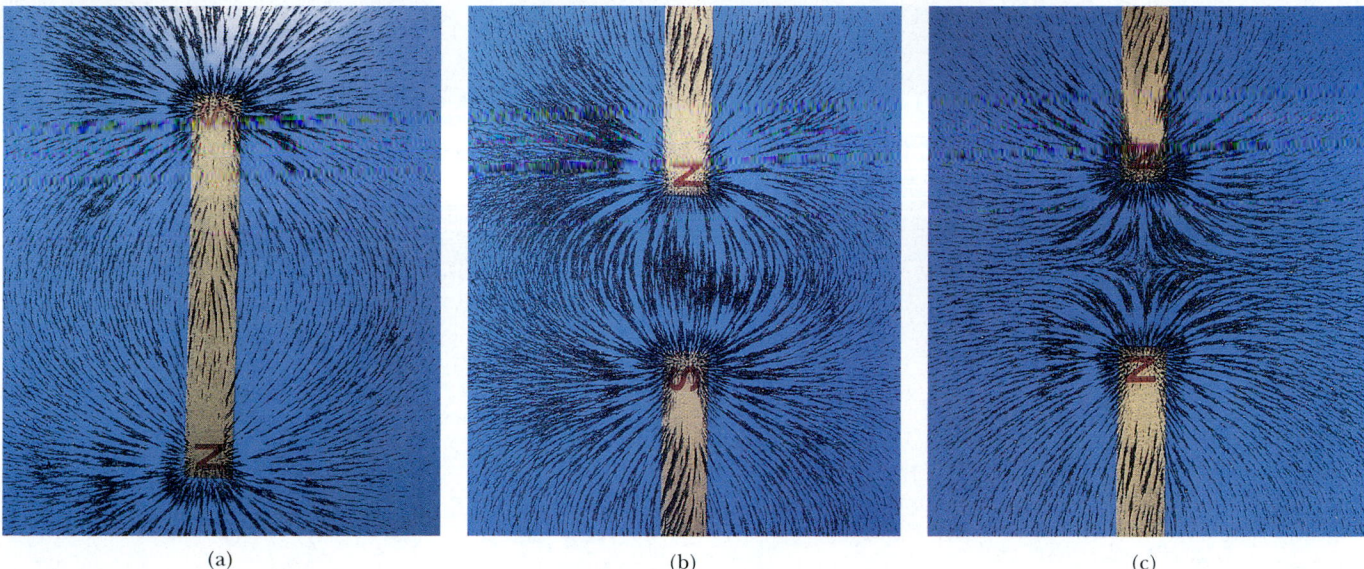

Figure 29.2 (a) Magnetic field pattern surrounding a bar magnet as displayed with iron filings. (b) Magnetic field pattern between *unlike* poles of two bar magnets. (c) Magnetic field pattern between *like* poles of two bar magnets. *(Henry Leap and Jim Lehman)*

- The magnitude and direction of $\mathbf{F}_B$ depend on the velocity of the particle and on the magnitude and direction of the magnetic field $\mathbf{B}$.
- When a charged particle moves parallel to the magnetic field vector, the magnetic force acting on the particle is zero.
- When the particle's velocity vector makes any angle $\theta \neq 0$ with the magnetic field, the magnetic force acts in a direction perpendicular to both $\mathbf{v}$ and $\mathbf{B}$; that is, $\mathbf{F}_B$ is perpendicular to the plane formed by $\mathbf{v}$ and $\mathbf{B}$ (Fig. 29.3a).

> Properties of the magnetic force on a charge moving in a magnetic field $\mathbf{B}$

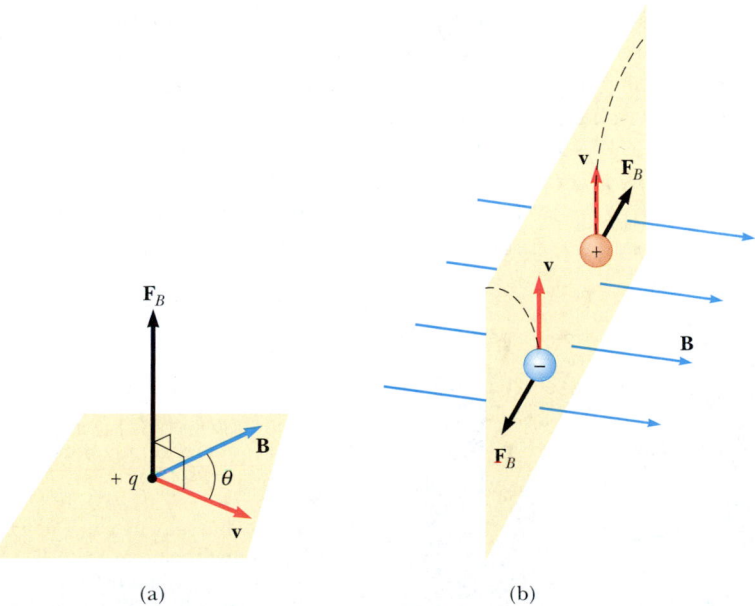

Figure 29.3 The direction of the magnetic force $\mathbf{F}_B$ acting on a charged particle moving with a velocity $\mathbf{v}$ in the presence of a magnetic field $\mathbf{B}$. (a) The magnetic force is perpendicular to both $\mathbf{v}$ and $\mathbf{B}$. (b) Oppositely directed magnetic forces $\mathbf{F}_B$ are exerted on two oppositely charged particles moving at the same velocity in a magnetic field.

The blue-white arc in this photograph indicates the circular path followed by an electron beam moving in a magnetic field. The vessel contains gas at very low pressure, and the beam is made visible as the electrons collide with the gas atoms, which then emit visible light. The magnetic field is produced by two coils (not shown). The apparatus can be used to measure the ratio e/m_e for the electron. *(Courtesy of Central Scientific Company)*

- The magnetic force exerted on a positive charge is in the direction opposite the direction of the magnetic force exerted on a negative charge moving in the same direction (Fig. 29.3b).
- The magnitude of the magnetic force exerted on the moving particle is proportional to sin θ, where θ is the angle the particle's velocity vector makes with the direction of **B**.

We can summarize these observations by writing the magnetic force in the form

$$\mathbf{F}_B = q\mathbf{v} \times \mathbf{B} \qquad (29.1)$$

where the direction of $\mathbf{F}_B$ is in the direction of $\mathbf{v} \times \mathbf{B}$ if q is positive, which by definition of the cross product (see Section 11.2) is perpendicular to both **v** and **B**. We can regard this equation as an operational definition of the magnetic field at some point in space. That is, the magnetic field is defined in terms of the force acting on a moving charged particle.

Figure 29.4 reviews the right-hand rule for determining the direction of the cross product $\mathbf{v} \times \mathbf{B}$. You point the four fingers of your right hand along the direction of **v** with the palm facing **B** and curl them toward **B**. The extended thumb, which is at a right angle to the fingers, points in the direction of $\mathbf{v} \times \mathbf{B}$. Because

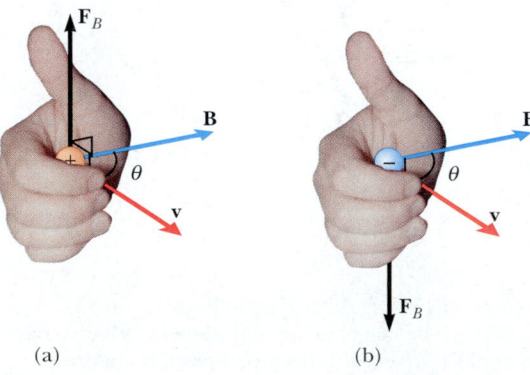

(a)   (b)

Figure 29.4 The right-hand rule for determining the direction of the magnetic force $\mathbf{F}_B = q\mathbf{v} \times \mathbf{B}$ acting on a particle with charge q moving with a velocity **v** in a magnetic field **B**. The direction of $\mathbf{v} \times \mathbf{B}$ is the direction in which the thumb points. (a) If q is positive, $\mathbf{F}_B$ is upward. (b) If q is negative, $\mathbf{F}_B$ is downward, antiparallel to the direction in which the thumb points.

$\mathbf{F}_B = q\mathbf{v} \times \mathbf{B}$, $\mathbf{F}_B$ is in the direction of $\mathbf{v} \times \mathbf{B}$ if q is positive (Fig. 29.4a) and opposite the direction of $\mathbf{v} \times \mathbf{B}$ if q is negative (Fig. 29.4b). (If you need more help understanding the cross product, you should review pages 333 to 334, including Fig. 11.8.)

The magnitude of the magnetic force is

$$F_B = |q|vB \sin \theta \tag{29.2}$$

Magnitude of the magnetic force on a charged particle moving in a magnetic field

where θ is the smaller angle between $\mathbf{v}$ and $\mathbf{B}$. From this expression, we see that F is zero when $\mathbf{v}$ is parallel or antiparallel to $\mathbf{B}$ ($\theta = 0$ or $180°$) and maximum ($F_{B,\,max} = |q|vB$) when $\mathbf{v}$ is perpendicular to $\mathbf{B}$ ($\theta = 90°$).

Quick Quiz 29.1

What is the maximum work that a constant magnetic field $\mathbf{B}$ can perform on a charge q moving through the field with velocity $\mathbf{v}$?

There are several important differences between electric and magnetic forces:

Differences between electric and magnetic forces

- The electric force acts in the direction of the electric field, whereas the magnetic force acts perpendicular to the magnetic field.
- The electric force acts on a charged particle regardless of whether the particle is moving, whereas the magnetic force acts on a charged particle only when the particle is in motion.
- The electric force does work in displacing a charged particle, whereas the magnetic force associated with a steady magnetic field does no work when a particle is displaced.

From the last statement and on the basis of the work–kinetic energy theorem, we conclude that the kinetic energy of a charged particle moving through a magnetic field cannot be altered by the magnetic field alone. In other words,

when a charged particle moves with a velocity $\mathbf{v}$ through a magnetic field, the field can alter the direction of the velocity vector but cannot change the speed or kinetic energy of the particle.

A magnetic field cannot change the speed of a particle

From Equation 29.2, we see that the SI unit of magnetic field is the newton per coulomb-meter per second, which is called the **tesla** (T):

$$1\ \mathrm{T} = \frac{\mathrm{N}}{\mathrm{C} \cdot \mathrm{m/s}}$$

Because a coulomb per second is defined to be an ampere, we see that

$$1\ \mathrm{T} = 1\ \frac{\mathrm{N}}{\mathrm{A} \cdot \mathrm{m}}$$

A non-SI magnetic-field unit in common use, called the *gauss* (G), is related to the tesla through the conversion $1\ \mathrm{T} = 10^4\ \mathrm{G}$. Table 29.1 shows some typical values of magnetic fields.

Quick Quiz 29.2

The north-pole end of a bar magnet is held near a positively charged piece of plastic. Is the plastic attracted, repelled, or unaffected by the magnet?

TABLE 29.1	Some Approximate Magnetic Field Magnitudes
Source of Field	**Field Magnitude (T)**
Strong superconducting laboratory magnet	30
Strong conventional laboratory magnet	2
Medical MRI unit	1.5
Bar magnet	10^{-2}
Surface of the Sun	10^{-2}
Surface of the Earth	0.5×10^{-4}
Inside human brain (due to nerve impulses)	10^{-13}

EXAMPLE 29.1　An Electron Moving in a Magnetic Field

An electron in a television picture tube moves toward the front of the tube with a speed of 8.0×10^6 m/s along the x axis (Fig. 29.5). Surrounding the neck of the tube are coils of wire that create a magnetic field of magnitude 0.025 T, directed at an angle of 60° to the x axis and lying in the xy plane. Calculate the magnetic force on and acceleration of the electron.

Solution　Using Equation 29.2, we can find the magnitude of the magnetic force:

$$F_B = |q|vB \sin \theta$$
$$= (1.6 \times 10^{-19}\,\text{C})(8.0 \times 10^6\,\text{m/s})(0.025\,\text{T})(\sin 60°)$$
$$= \boxed{2.8 \times 10^{-14}\,\text{N}}$$

Because $\mathbf{v} \times \mathbf{B}$ is in the positive z direction (from the right-hand rule) and the charge is negative, $\mathbf{F}_B$ is in the negative z direction.

The mass of the electron is 9.11×10^{-31} kg, and so its acceleration is

$$a = \frac{F_B}{m_e} = \frac{2.8 \times 10^{-14}\,\text{N}}{9.11 \times 10^{-31}\,\text{kg}} = \boxed{3.1 \times 10^{16}\,\text{m/s}^2}$$

in the negative z direction.

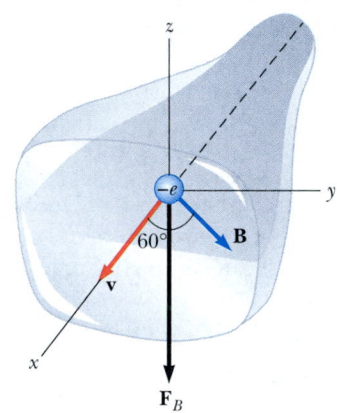

Figure 29.5　The magnetic force $\mathbf{F}_B$ acting on the electron is in the negative z direction when $\mathbf{v}$ and $\mathbf{B}$ lie in the xy plane.

29.2　MAGNETIC FORCE ACTING ON A CURRENT-CARRYING CONDUCTOR

12.3　If a magnetic force is exerted on a single charged particle when the particle moves through a magnetic field, it should not surprise you that a current-carrying wire also experiences a force when placed in a magnetic field. This follows from the fact that the current is a collection of many charged particles in motion; hence, the resultant force exerted by the field on the wire is the vector sum of the individual forces exerted on all the charged particles making up the current. The force exerted on the particles is transmitted to the wire when the particles collide with the atoms making up the wire.

Before we continue our discussion, some explanation of the notation used in this book is in order. To indicate the direction of **B** in illustrations, we sometimes present perspective views, such as those in Figures 29.5, 29.6a, and 29.7. In flat il-

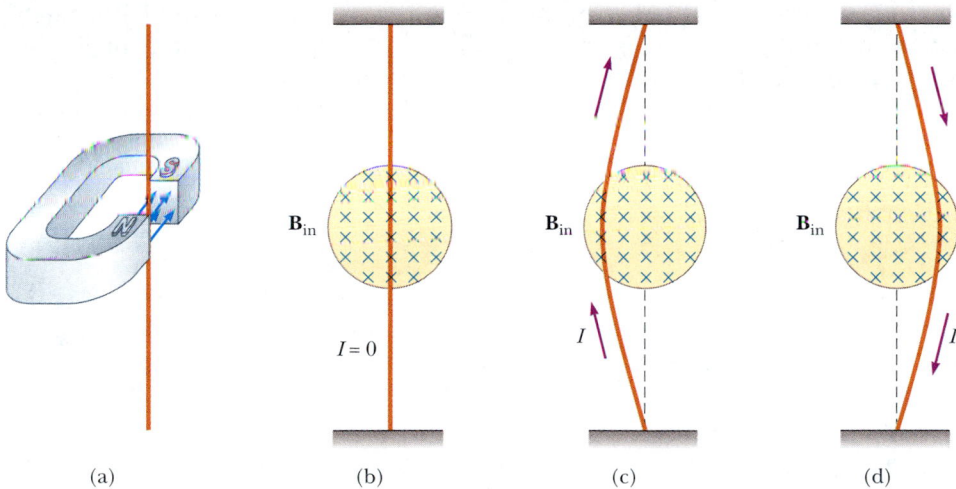

Figure 29.6   (a) A wire suspended vertically between the poles of a magnet. (b) The setup shown in part (a) as seen looking at the south pole of the magnet, so that the magnetic field (blue crosses) is directed into the page. When there is no current in the wire, it remains vertical. (c) When the current is upward, the wire deflects to the left. (d) When the current is downward, the wire deflects to the right.

lustrations, such as in Figure 29.6b to d, we depict a magnetic field directed into the page with blue crosses, which represent the tails of arrows shot perpendicularly and away from you. In this case, we call the field $\mathbf{B}_{in}$, where the subscript "in" indicates "into the page." If $\mathbf{B}$ is perpendicular and directed out of the page, we use a series of blue dots, which represent the tips of arrows coming toward you (see Fig. P29.56). In this case, we call the field $\mathbf{B}_{out}$. If $\mathbf{B}$ lies in the plane of the page, we use a series of blue field lines with arrowheads, as shown in Figure 29.7.

One can demonstrate the magnetic force acting on a current-carrying conductor by hanging a wire between the poles of a magnet, as shown in Figure 29.6a. For ease in visualization, part of the horseshoe magnet in part (a) is removed to show the end face of the south pole in parts (b), (c), and (d) of Figure 29.6. The magnetic field is directed into the page and covers the region within the shaded circles. When the current in the wire is zero, the wire remains vertical, as shown in Figure 29.6b. However, when a current directed upward flows in the wire, as shown in Figure 29.6c, the wire deflects to the left. If we reverse the current, as shown in Figure 29.6d, the wire deflects to the right.

Let us quantify this discussion by considering a straight segment of wire of length L and cross-sectional area A, carrying a current I in a uniform magnetic field $\mathbf{B}$, as shown in Figure 29.7. The magnetic force exerted on a charge q moving with a drift velocity $\mathbf{v}_d$ is $q\mathbf{v}_d \times \mathbf{B}$. To find the total force acting on the wire, we multiply the force $q\mathbf{v}_d \times \mathbf{B}$ exerted on one charge by the number of charges in the segment. Because the volume of the segment is AL, the number of charges in the segment is nAL, where n is the number of charges per unit volume. Hence, the total magnetic force on the wire of length L is

$$\mathbf{F}_B = (q\mathbf{v}_d \times \mathbf{B})\, nAL$$

We can write this expression in a more convenient form by noting that, from Equation 27.4, the current in the wire is $I = nqv_dA$. Therefore,

$$\mathbf{F}_B = I\mathbf{L} \times \mathbf{B} \qquad\qquad \text{(29.3)}$$

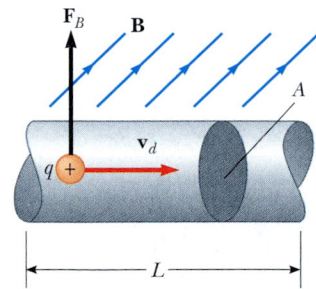

Figure 29.7   A segment of a current-carrying wire located in a magnetic field $\mathbf{B}$. The magnetic force exerted on each charge making up the current is $q\mathbf{v}_d \times \mathbf{B}$, and the net force on the segment of length L is $I\mathbf{L} \times \mathbf{B}$.

Force on a segment of a wire in a uniform magnetic field

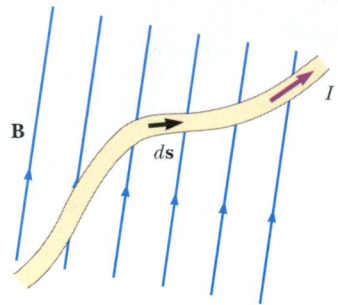

Figure 29.8 A wire segment of arbitrary shape carrying a current I in a magnetic field **B** experiences a magnetic force. The force on any segment $d\mathbf{s}$ is $I\, d\mathbf{s} \times \mathbf{B}$ and is directed out of the page. You should use the right-hand rule to confirm this force direction.

where **L** is a vector that points in the direction of the current I and has a magnitude equal to the length L of the segment. Note that this expression applies only to a straight segment of wire in a uniform magnetic field.

Now let us consider an arbitrarily shaped wire segment of uniform cross-section in a magnetic field, as shown in Figure 29.8. It follows from Equation 29.3 that the magnetic force exerted on a small segment of vector length $d\mathbf{s}$ in the presence of a field **B** is

$$d\mathbf{F}_B = I\, d\mathbf{s} \times \mathbf{B} \qquad (29.4)$$

where $d\mathbf{F}_B$ is directed out of the page for the directions assumed in Figure 29.8. We can consider Equation 29.4 as an alternative definition of **B**. That is, we can define the magnetic field **B** in terms of a measurable force exerted on a current element, where the force is a maximum when **B** is perpendicular to the element and zero when **B** is parallel to the element.

To calculate the total force $\mathbf{F}_B$ acting on the wire shown in Figure 29.8, we integrate Equation 29.4 over the length of the wire:

$$\mathbf{F}_B = I \int_a^b d\mathbf{s} \times \mathbf{B} \qquad (29.5)$$

where a and b represent the end points of the wire. When this integration is carried out, the magnitude of the magnetic field and the direction the field makes with the vector $d\mathbf{s}$ (in other words, with the orientation of the element) may differ at different points.

Now let us consider two special cases involving Equation 29.5. In both cases, the magnetic field is taken to be constant in magnitude and direction.

Case 1 A curved wire carries a current I and is located in a uniform magnetic field **B**, as shown in Figure 29.9a. Because the field is uniform, we can take **B** outside the integral in Equation 29.5, and we obtain

$$\mathbf{F}_B = I\left(\int_a^b d\mathbf{s}\right) \times \mathbf{B} \qquad (29.6)$$

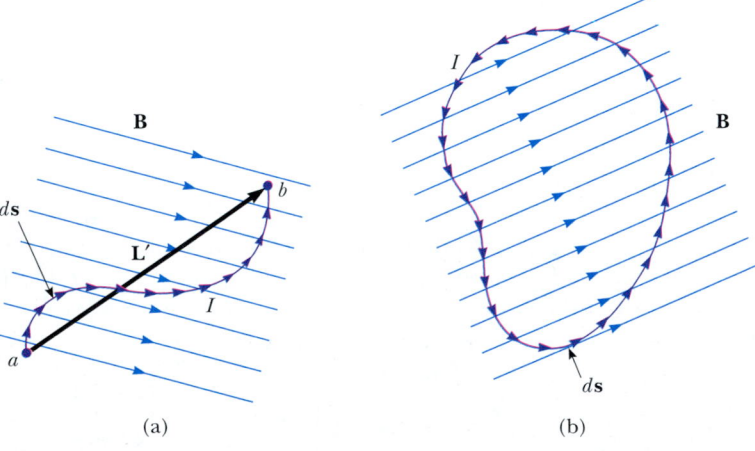

(a)   (b)

Figure 29.9 (a) A curved wire carrying a current I in a uniform magnetic field. The total magnetic force acting on the wire is equivalent to the force on a straight wire of length L' running between the ends of the curved wire. (b) A current-carrying loop of arbitrary shape in a uniform magnetic field. The net magnetic force on the loop is zero.

But the quantity $\int_a^b d\mathbf{s}$ represents the *vector sum* of all the length elements from a to b. From the law of vector addition, the sum equals the vector $\mathbf{L}'$, directed from a to b. Therefore, Equation 29.6 reduces to

$$\mathbf{F}_B = I\mathbf{L}' \times \mathbf{B} \tag{29.7}$$

Case 2 An arbitrarily shaped closed loop carrying a current I is placed in a uniform magnetic field, as shown in Figure 29.9b. We can again express the force acting on the loop in the form of Equation 29.6, but this time we must take the vector sum of the length elements $d\mathbf{s}$ over the entire loop:

$$\mathbf{F}_B = I\left(\oint d\mathbf{s}\right) \times \mathbf{B}$$

Because the set of length elements forms a closed polygon, the vector sum must be zero. This follows from the graphical procedure for adding vectors by the polygon method. Because $\oint d\mathbf{s} = 0$, we conclude that $\mathbf{F}_B = 0$:

> The net magnetic force acting on any closed current loop in a uniform magnetic field is zero.

EXAMPLE 29.2   Force on a Semicircular Conductor

A wire bent into a semicircle of radius R forms a closed circuit and carries a current I. The wire lies in the xy plane, and a uniform magnetic field is directed along the positive y axis, as shown in Figure 29.10. Find the magnitude and direction of the magnetic force acting on the straight portion of the wire and on the curved portion.

Solution The force $\mathbf{F}_1$ acting on the straight portion has a magnitude $F_1 = ILB = 2IRB$ because $L = 2R$ and the wire is oriented perpendicular to $\mathbf{B}$. The direction of $\mathbf{F}_1$ is out of the page because $\mathbf{L} \times \mathbf{B}$ is along the positive z axis. (That is, $\mathbf{L}$ is to the right, in the direction of the current; thus, according to the rule of cross products, $\mathbf{L} \times \mathbf{B}$ is out of the page in Fig. 29.10.)

To find the force $\mathbf{F}_2$ acting on the curved part, we first write an expression for the force $d\mathbf{F}_2$ on the length element $d\mathbf{s}$ shown in Figure 29.10. If θ is the angle between $\mathbf{B}$ and $d\mathbf{s}$, then the magnitude of $d\mathbf{F}_2$ is

$$dF_2 = I\,|\,d\mathbf{s} \times \mathbf{B}\,| = IB\sin\theta\,ds$$

To integrate this expression, we must express ds in terms of θ. Because $s = R\theta$, we have $ds = R\,d\theta$, and we can make this substitution for dF_2:

$$dF_2 = IRB\sin\theta\,d\theta$$

To obtain the total force F_2 acting on the curved portion, we can integrate this expression to account for contributions from all elements $d\mathbf{s}$. Note that the direction of the force on every element is the same: into the page (because $d\mathbf{s} \times \mathbf{B}$ is into the page). Therefore, the resultant force $\mathbf{F}_2$ on the

curved wire must also be into the page. Integrating our expression for dF_2 over the limits $\theta = 0$ to $\theta = \pi$ (that is, the entire semicircle) gives

$$F_2 = IRB\int_0^\pi \sin\theta\,d\theta = IRB\Big[-\cos\theta\Big]_0^\pi$$

$$= -IRB(\cos\pi - \cos 0) = -IRB(-1-1) = 2IRB$$

Because $\mathbf{F}_2$, with a magnitude of $2IRB$, is directed into the page and because $\mathbf{F}_1$, with a magnitude of $2IRB$, is directed out of the page, the net force on the closed loop is zero. This result is consistent with Case 2 described earlier.

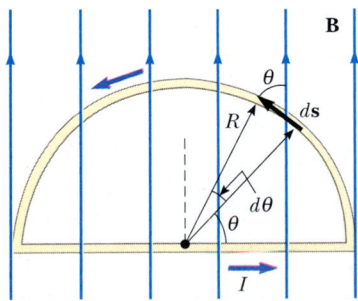

Figure 29.10 The net force acting on a closed current loop in a uniform magnetic field is zero. In the setup shown here, the force on the straight portion of the loop is $2IRB$ and directed out of the page, and the force on the curved portion is $2IRB$ directed into the page.

The four wires shown in Figure 29.11 all carry the same current from point A to point B through the same magnetic field. Rank the wires according to the magnitude of the magnetic force exerted on them, from greatest to least.

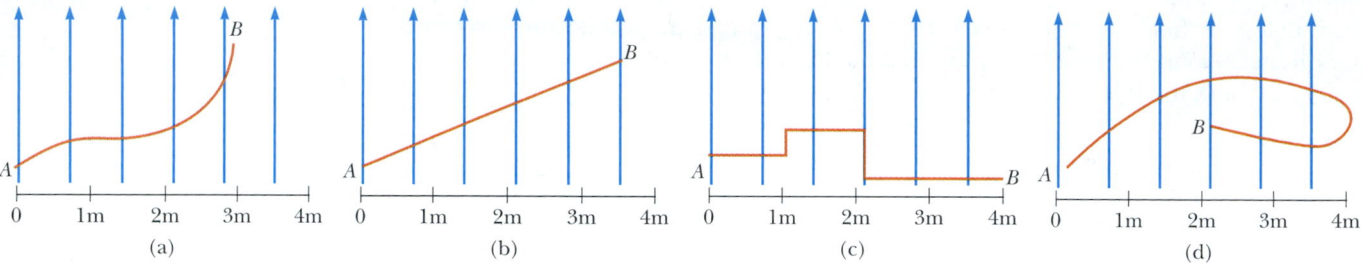

Figure 29.11   Which wire experiences the greatest magnetic force?

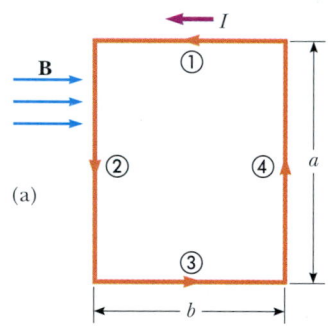

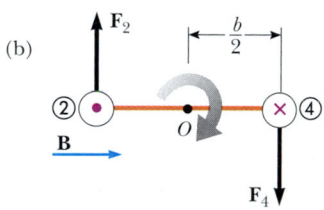

Figure 29. 12   (a) Overhead view of a rectangular current loop in a uniform magnetic field. No forces are acting on sides ① and ③ because these sides are parallel to **B**. Forces are acting on sides ② and ④, however. (b) Edge view of the loop sighting down sides ② and ④ shows that the forces **F**$_2$ and **F**$_4$ exerted on these sides create a torque that tends to twist the loop clockwise. The purple dot in the left circle represents current in wire ② coming toward you; the purple cross in the right circle represents current in wire ④ moving away from you.

29.3 ▸ TORQUE ON A CURRENT LOOP IN A UNIFORM MAGNETIC FIELD

In the previous section, we showed how a force is exerted on a current-carrying conductor placed in a magnetic field. With this as a starting point, we now show that a torque is exerted on any current loop placed in a magnetic field. The results of this analysis will be of great value when we discuss motors in Chapter 31.

Consider a rectangular loop carrying a current I in the presence of a uniform magnetic field directed parallel to the plane of the loop, as shown in Figure 29.12a. No magnetic forces act on sides ① and ③ because these wires are parallel to the field; hence, **L × B** = 0 for these sides. However, magnetic forces do act on sides ② and ④ because these sides are oriented perpendicular to the field. The magnitude of these forces is, from Equation 29.3,

$$F_2 = F_4 = IaB$$

The direction of **F**$_2$, the force exerted on wire ② is out of the page in the view shown in Figure 29.12a, and that of **F**$_4$, the force exerted on wire ④, is into the page in the same view. If we view the loop from side ③ and sight along sides ② and ④, we see the view shown in Figure 29.12b, and the two forces **F**$_2$ and **F**$_4$ are directed as shown. Note that the two forces point in opposite directions but are *not* directed along the same line of action. If the loop is pivoted so that it can rotate about point O, these two forces produce about O a torque that rotates the loop clockwise. The magnitude of this torque τ_{max} is

$$\tau_{max} = F_2 \frac{b}{2} + F_4 \frac{b}{2} = (IaB) \frac{b}{2} + (IaB) \frac{b}{2} = IabB$$

where the moment arm about O is $b/2$ for each force. Because the area enclosed by the loop is $A = ab$, we can express the maximum torque as

$$\tau_{max} = IAB \qquad\qquad (29.8)$$

Remember that this maximum-torque result is valid only when the magnetic field is parallel to the plane of the loop. The sense of the rotation is clockwise when viewed from side ③, as indicated in Figure 29.12b. If the current direction were reversed, the force directions would also reverse, and the rotational tendency would be counterclockwise.

Now let us suppose that the uniform magnetic field makes an angle $\theta < 90°$ with a line perpendicular to the plane of the loop, as shown in Figure 29.13a. For convenience, we assume that **B** is perpendicular to sides ① and ③. In this case, the magnetic forces $\mathbf{F}_2$ and $\mathbf{F}_4$ exerted on sides ② and ④ cancel each other and produce no torque because they pass through a common origin. However, the forces acting on sides ① and ③, $\mathbf{F}_1$ and $\mathbf{F}_3$, form a couple and hence produce a torque about *any point*. Referring to the end view shown in Figure 29.13b, we note that the moment arm of $\mathbf{F}_1$ about the point O is equal to $(a/2) \sin \theta$. Likewise, the moment arm of $\mathbf{F}_3$ about O is also $(a/2) \sin \theta$. Because $F_1 = F_3 = IbB$, the net torque about O has the magnitude

$$\tau = F_1 \frac{a}{2} \sin \theta + F_3 \frac{a}{2} \sin \theta$$

$$= IbB\left(\frac{a}{2} \sin \theta\right) + IbB\left(\frac{a}{2} \sin \theta\right) = IabB \sin \theta$$

$$= IAB \sin \theta$$

where $A = ab$ is the area of the loop. This result shows that the torque has its maximum value IAB when the field is perpendicular to the normal to the plane of the loop ($\theta = 90°$), as we saw when discussing Figure 29.12, and that it is zero when the field is parallel to the normal to the plane of the loop ($\theta = 0$). As we see in Figure 29.13, the loop tends to rotate in the direction of decreasing values of θ (that is, such that the area vector **A** rotates toward the direction of the magnetic field).

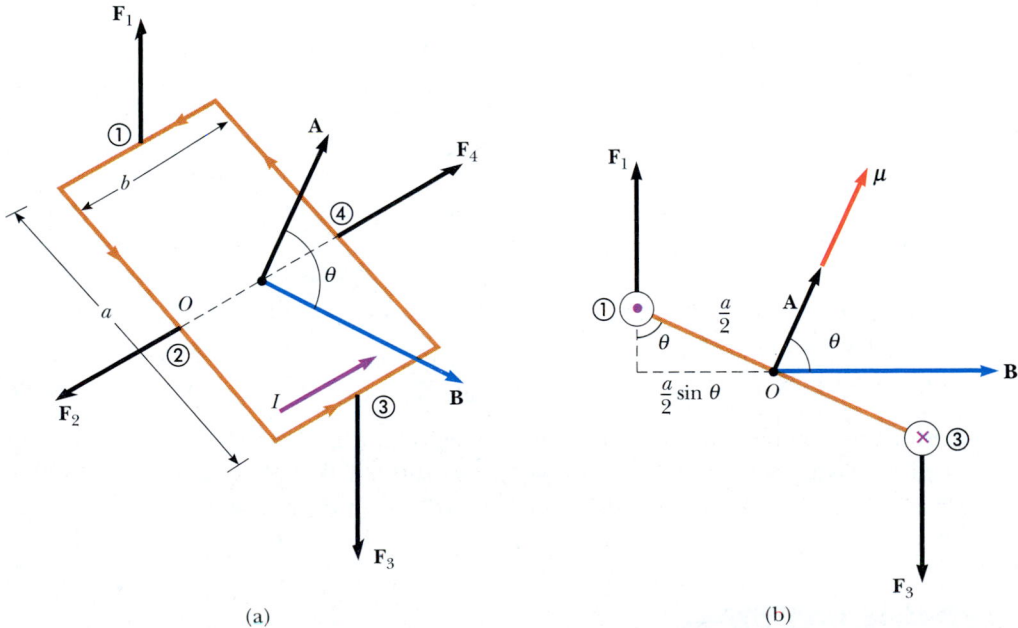

(a)   (b)

Figure 29.13   (a) A rectangular current loop in a uniform magnetic field. The area vector **A** perpendicular to the plane of the loop makes an angle θ with the field. The magnetic forces exerted on sides ② and ④ cancel, but the forces exerted on sides ① and ③ create a torque on the loop. (b) Edge view of the loop sighting down sides ① and ③.

Quick Quiz 29.4

Describe the forces on the rectangular current loop shown in Figure 29.13 if the magnetic field is directed as shown but increases in magnitude going from left to right.

A convenient expression for the torque exerted on a loop placed in a uniform magnetic field **B** is

Torque on a current loop

$$\boldsymbol{\tau} = I\mathbf{A} \times \mathbf{B} \qquad (29.9)$$

where **A**, the vector shown in Figure 29.13, is perpendicular to the plane of the loop and has a magnitude equal to the area of the loop. We determine the direction of **A** using the right-hand rule described in Figure 29.14. When you curl the fingers of your right hand in the direction of the current in the loop, your thumb points in the direction of **A**. The product $I\mathbf{A}$ is defined to be the **magnetic dipole moment $\boldsymbol{\mu}$** (often simply called the "magnetic moment") of the loop:

Magnetic dipole moment of a current loop

$$\boldsymbol{\mu} = I\mathbf{A} \qquad (29.10)$$

The SI unit of magnetic dipole moment is ampere–meter^2 (A·m^2). Using this definition, we can express the torque exerted on a current-carrying loop in a magnetic field **B** as

$$\boldsymbol{\tau} = \boldsymbol{\mu} \times \mathbf{B} \qquad (29.11)$$

Note that this result is analogous to Equation 26.18, $\boldsymbol{\tau} = \mathbf{p} \times \mathbf{E}$, for the torque exerted on an electric dipole in the presence of an electric field **E**, where **p** is the electric dipole moment.

Although we obtained the torque for a particular orientation of **B** with respect to the loop, the equation $\boldsymbol{\tau} = \boldsymbol{\mu} \times \mathbf{B}$ is valid for any orientation. Furthermore, although we derived the torque expression for a rectangular loop, the result is valid for a loop of any shape.

If a coil consists of N turns of wire, each carrying the same current and enclosing the same area, the total magnetic dipole moment of the coil is N times the magnetic dipole moment for one turn. The torque on an N-turn coil is N times that on a one-turn coil. Thus, we write $\boldsymbol{\tau} = N\boldsymbol{\mu}_{\text{loop}} \times \mathbf{B} = \boldsymbol{\mu}_{\text{coil}} \times \mathbf{B}$.

In Section 26.6, we found that the potential energy of an electric dipole in an electric field is given by $U = -\mathbf{p} \cdot \mathbf{E}$. This energy depends on the orientation of the dipole in the electric field. Likewise, the potential energy of a magnetic dipole in a magnetic field depends on the orientation of the dipole in the magnetic field and is given by

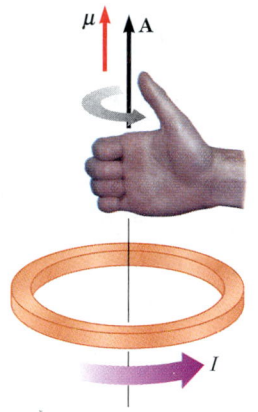

Figure 29.14 Right-hand rule for determining the direction of the vector **A**. The direction of the magnetic moment $\boldsymbol{\mu}$ is the same as the direction of **A**.

$$U = -\boldsymbol{\mu} \cdot \mathbf{B} \qquad (29.12)$$

From this expression, we see that a magnetic dipole has its lowest energy $U_{\text{min}} = -\mu B$ when $\boldsymbol{\mu}$ points in the same direction as **B**. The dipole has its highest energy $U_{\text{max}} = +\mu B$ when $\boldsymbol{\mu}$ points in the direction opposite **B**.

Quick Quiz 29.5

Rank the magnitude of the torques acting on the rectangular loops shown in Figure 29.15, from highest to lowest. All loops are identical and carry the same current.

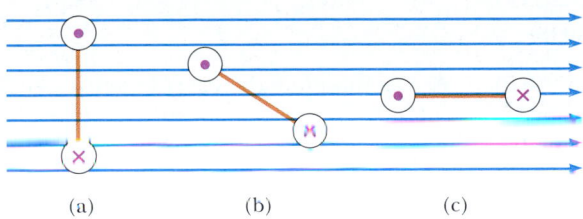

Figure 29.15   Which current loop (seen edge-on) experiences the greatest torque?

EXAMPLE 29.3   The Magnetic Dipole Moment of a Coil

A rectangular coil of dimensions 5.40 cm × 8.50 cm consists of 25 turns of wire and carries a current of 15.0 mA. A 0.350-T magnetic field is applied parallel to the plane of the loop. (a) Calculate the magnitude of its magnetic dipole moment.

Solution   Because the coil has 25 turns, we modify Equation 29.10 to obtain

$$\mu_{coil} = NIA = (25)(15.0 \times 10^{-3}\,\text{A})(0.054\,0\,\text{m})(0.085\,0\,\text{m})$$

$$= \boxed{1.72 \times 10^{-3}\,\text{A}\cdot\text{m}^2}$$

(b) What is the magnitude of the torque acting on the loop?

Solution   Because **B** is perpendicular to μ_{coil}, Equation 29.11 gives

$$\tau = \mu_{coil}B = (1.72 \times 10^{-3}\,\text{A}\cdot\text{m}^2)(0.350\,\text{T})$$

$$= \boxed{6.02 \times 10^{-4}\,\text{N}\cdot\text{m}}$$

Exercise   Show that the units $\text{A}\cdot\text{m}^2\cdot\text{T}$ reduce to the torque units $\text{N}\cdot\text{m}$.

Exercise   Calculate the magnitude of the torque on the coil when the field makes an angle of (a) 60° and (b) 0° with **μ**.

Answer   (a) $5.21 \times 10^{-4}\,\text{N}\cdot\text{m}$; (b) zero.

web

For more information on torquers, visit the Web sites of some of the companies that supply these devices to NASA:

http://www.smad.com   or
http://www.itaco.com

EXAMPLE 29.4   Satellite Attitude Control

Many satellites use coils called *torquers* to adjust their orientation. These devices interact with the Earth's magnetic field to create a torque on the spacecraft in the x, y, or z direction. The major advantage of this type of attitude-control system is that it uses solar-generated electricity and so does not consume any thruster fuel.

If a typical device has a magnetic dipole moment of 250 A · m², what is the maximum torque applied to a satellite when its torquer is turned on at an altitude where the magnitude of the Earth's magnetic field is 3.0×10^{-5} T?

Solution   We once again apply Equation 29.11, recognizing that the maximum torque is obtained when the magnetic dipole moment of the torquer is perpendicular to the Earth's magnetic field:

$$\tau_{max} = \mu B = (250\,\text{A}\cdot\text{m}^2)(3.0 \times 10^{-5}\,\text{T})$$

$$= \boxed{7.5 \times 10^{-3}\,\text{N}\cdot\text{m}}$$

Exercise   If the torquer requires 1.3 W of power at a potential difference of 28 V, how much current does it draw when it operates?

Answer   46 mA.

EXAMPLE 29.5   The D'Arsonval Galvanometer

An end view of a D'Arsonval galvanometer (see Section 28.5) is shown in Figure 29.16. When the turns of wire making up the coil carry a current, the magnetic field created by the magnet exerts on the coil a torque that turns it (along with its attached pointer) against the spring. Let us show that the angle of deflection of the pointer is directly proportional to the current in the coil.

Solution   We can use Equation 29.11 to find the torque τ_m the magnetic field exerts on the coil. If we assume that the magnetic field through the coil is perpendicular to the normal to the plane of the coil, Equation 29.11 becomes

$$\tau_m = \mu B$$

(This is a reasonable assumption because the circular cross section of the magnet ensures radial magnetic field lines.) This magnetic torque is opposed by the torque due to the spring, which is given by the rotational version of Hooke's law, $\tau_s = -\kappa\varphi$, where κ is the torsional spring constant and φ is the angle through which the spring turns. Because the coil does not have an angular acceleration when the pointer is at rest, the sum of these torques must be zero:

$$(1) \qquad \tau_m + \tau_s = \mu B - \kappa\varphi = 0$$

Equation 29.10 allows us to relate the magnetic moment of the N turns of wire to the current through them:

$$\mu = NIA$$

We can substitute this expression for μ in Equation (1) to obtain

$$(NIA)B - \kappa\varphi = 0$$

$$\varphi = \frac{NAB}{\kappa} I$$

Thus, the angle of deflection of the pointer is directly proportional to the current in the loop. The factor NAB/κ tells us that deflection also depends on the design of the meter.

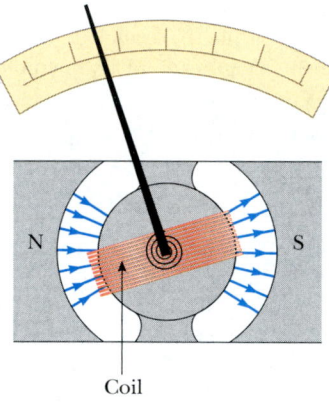

Figure 29.16   End view of a moving-coil galvanometer.

Coil

29.4   MOTION OF A CHARGED PARTICLE IN A UNIFORM MAGNETIC FIELD

12.3

In Section 29.1 we found that the magnetic force acting on a charged particle moving in a magnetic field is perpendicular to the velocity of the particle and that consequently the work done on the particle by the magnetic force is zero. Let us now consider the special case of a positively charged particle moving in a uniform magnetic field with the initial velocity vector of the particle perpendicular to the field. Let us assume that the direction of the magnetic field is into the page. Figure 29.17 shows that the particle moves in a circle in a plane perpendicular to the magnetic field.

The particle moves in this way because the magnetic force $\mathbf{F}_B$ is at right angles to $\mathbf{v}$ and $\mathbf{B}$ and has a constant magnitude qvB. As the force deflects the particle, the directions of $\mathbf{v}$ and $\mathbf{F}_B$ change continuously, as Figure 29.17 shows. Because $\mathbf{F}_B$ always points toward the center of the circle, **it changes only the direction of v and not its magnitude.** As Figure 29.17 illustrates, the rotation is counterclockwise for a positive charge. If q were negative, the rotation would be clockwise. We can use Equation 6.1 to equate this magnetic force to the radial force required to

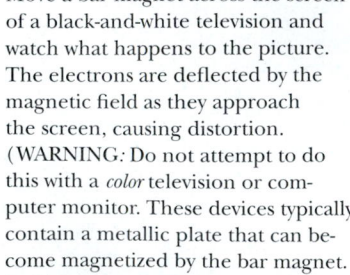

QuickLab

Move a bar magnet across the screen of a black-and-white television and watch what happens to the picture. The electrons are deflected by the magnetic field as they approach the screen, causing distortion. (WARNING: Do not attempt to do this with a *color* television or computer monitor. These devices typically contain a metallic plate that can become magnetized by the bar magnet. If this happens, a repair shop will need to "degauss" the screen.)

keep the charge moving in a circle:

$$\sum F = ma_r$$

$$F_B = qvB = \frac{mv^2}{r}$$

$$r = \frac{mv}{qB} \tag{29.13}$$

That is, the radius of the path is proportional to the linear momentum mv of the particle and inversely proportional to the magnitude of the charge on the particle and to the magnitude of the magnetic field. The angular speed of the particle (from Eq. 10.10) is

$$\omega = \frac{v}{r} = \frac{qB}{m} \tag{29.14}$$

The period of the motion (the time that the particle takes to complete one revolution) is equal to the circumference of the circle divided by the linear speed of the particle:

$$T = \frac{2\pi r}{v} = \frac{2\pi}{\omega} = \frac{2\pi m}{qB} \tag{29.15}$$

These results show that the angular speed of the particle and the period of the circular motion do not depend on the linear speed of the particle or on the radius of the orbit. The angular speed ω is often referred to as the **cyclotron frequency** because charged particles circulate at this angular speed in the type of accelerator called a *cyclotron*, which is discussed in Section 29.5.

If a charged particle moves in a uniform magnetic field with its velocity at some arbitrary angle with respect to **B**, its path is a helix. For example, if the field is directed in the x direction, as shown in Figure 29.18, there is no component of force in the x direction. As a result, $a_x = 0$, and the x component of velocity remains constant. However, the magnetic force $q\mathbf{v} \times \mathbf{B}$ causes the components v_y and v_z to change in time, and the resulting motion is a helix whose axis is parallel to the magnetic field. The projection of the path onto the yz plane (viewed along the x axis) is a circle. (The projections of the path onto the xy and xz planes are sinusoids!) Equations 29.13 to 29.15 still apply provided that v is replaced by $v_\perp = \sqrt{v_y^2 + v_z^2}$.

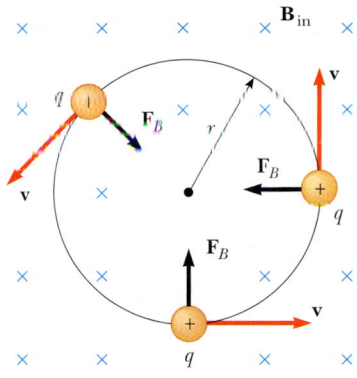

Figure 29.17 When the velocity of a charged particle is perpendicular to a uniform magnetic field, the particle moves in a circular path in a plane perpendicular to **B**. The magnetic force $\mathbf{F}_B$ acting on the charge is always directed toward the center of the circle.

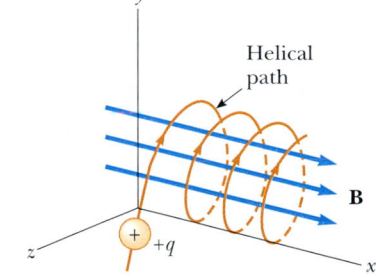

Figure 29.18 A charged particle having a velocity vector that has a component parallel to a uniform magnetic field moves in a helical path.

EXAMPLE 29.6   A Proton Moving Perpendicular to a Uniform Magnetic Field

A proton is moving in a circular orbit of radius 14 cm in a uniform 0.35-T magnetic field perpendicular to the velocity of the proton. Find the linear speed of the proton.

Solution   From Equation 29.13, we have

$$v = \frac{qBr}{m_p} = \frac{(1.60 \times 10^{-19}\,\text{C})(0.35\,\text{T})(14 \times 10^{-2}\,\text{m})}{1.67 \times 10^{-27}\,\text{kg}}$$

$$= \boxed{4.7 \times 10^6\,\text{m/s}}$$

Exercise   If an electron moves in a direction perpendicular to the same magnetic field with this same linear speed, what is the radius of its circular orbit?

Answer 7.6×10^{-5} m.

EXAMPLE 29.7   Bending an Electron Beam

In an experiment designed to measure the magnitude of a uniform magnetic field, electrons are accelerated from rest through a potential difference of 350 V. The electrons travel along a curved path because of the magnetic force exerted on them, and the radius of the path is measured to be 7.5 cm. (Fig. 29.19 shows such a curved beam of electrons.) If the magnetic field is perpendicular to the beam, (a) what is the magnitude of the field?

Solution   First we must calculate the speed of the electrons. We can use the fact that the increase in their kinetic energy must equal the decrease in their potential energy $|e|\Delta V$ (because of conservation of energy). Then we can use Equation 29.13 to find the magnitude of the magnetic field. Because $K_i = 0$ and $K_f = m_e v^2/2$, we have

$$\tfrac{1}{2}m_e v^2 = |e|\Delta V$$

$$v = \sqrt{\frac{2|e|\Delta V}{m_e}} = \sqrt{\frac{2(1.60 \times 10^{-19}\,\text{C})(350\,\text{V})}{9.11 \times 10^{-31}\,\text{kg}}}$$

$$= 1.11 \times 10^7\,\text{m/s}$$

$$B = \frac{m_e v}{|e|r} = \frac{(9.11 \times 10^{-31}\,\text{kg})(1.11 \times 10^7\,\text{m/s})}{(1.60 \times 10^{-19}\,\text{C})(0.075\,\text{m})}$$

$$= \boxed{8.4 \times 10^{-4}\,\text{T}}$$

(b) What is the angular speed of the electrons?

Solution   Using Equation 29.14, we find that

$$\omega = \frac{v}{r} = \frac{1.11 \times 10^7\,\text{m/s}}{0.075\,\text{m}} = \boxed{1.5 \times 10^8\,\text{rad/s}}$$

Exercise   What is the period of revolution of the electrons?

Answer   43 ns.

Figure 29.19   The bending of an electron beam in a magnetic field.   *(Henry Leap and Jim Lehman)*

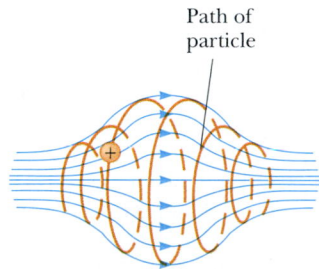

Figure 29.20   A charged particle moving in a nonuniform magnetic field (a magnetic bottle) spirals about the field (red path) and oscillates between the end points. The magnetic force exerted on the particle near either end of the bottle has a component that causes the particle to spiral back toward the center.

When charged particles move in a nonuniform magnetic field, the motion is complex. For example, in a magnetic field that is strong at the ends and weak in the middle, such as that shown in Figure 29.20, the particles can oscillate back and forth between the end points. A charged particle starting at one end spirals along the field lines until it reaches the other end, where it reverses its path and spirals back. This configuration is known as a *magnetic bottle* because charged particles can be trapped within it. The magnetic bottle has been used to confine a *plasma*, a gas consisting of ions and electrons. Such a plasma-confinement scheme could fulfill a crucial role in the control of nuclear fusion, a process that could supply us with an almost endless source of energy. Unfortunately, the magnetic bottle has its problems. If a large number of particles are trapped, collisions between them cause the particles to eventually leak from the system.

The Van Allen radiation belts consist of charged particles (mostly electrons and protons) surrounding the Earth in doughnut-shaped regions (Fig. 29.21). The particles, trapped by the Earth's nonuniform magnetic field, spiral around the field lines from pole to pole, covering the distance in just a few seconds. These particles originate mainly from the Sun, but some come from stars and other heavenly objects. For this reason, the particles are called *cosmic rays*. Most cosmic rays are deflected by the Earth's magnetic field and never reach the atmosphere. However, some of the particles become trapped; it is these particles that make up the Van Allen belts. When the particles are located over the poles, they sometimes collide with atoms in the atmosphere, causing the atoms to emit visible light. Such collisions are the origin of the beautiful Aurora Borealis, or Northern Lights, in the northern hemisphere and the Aurora Australis in the southern hemisphere.

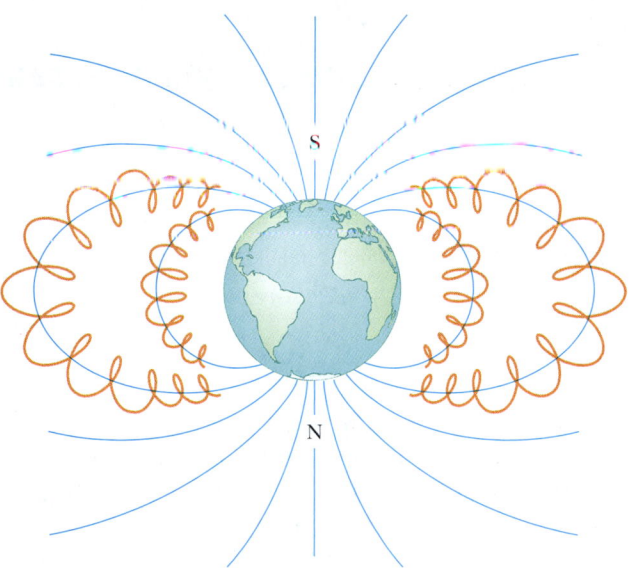

Figure 29.21   The Van Allen belts are made up of charged particles trapped by the Earth's nonuniform magnetic field. The magnetic field lines are in blue and the particle paths in red.

Auroras are usually confined to the polar regions because it is here that the Van Allen belts are nearest the Earth's surface. Occasionally, though, solar activity causes larger numbers of charged particles to enter the belts and significantly distort the normal magnetic field lines associated with the Earth. In these situations an aurora can sometimes be seen at lower latitudes.

12.1 & 12.11

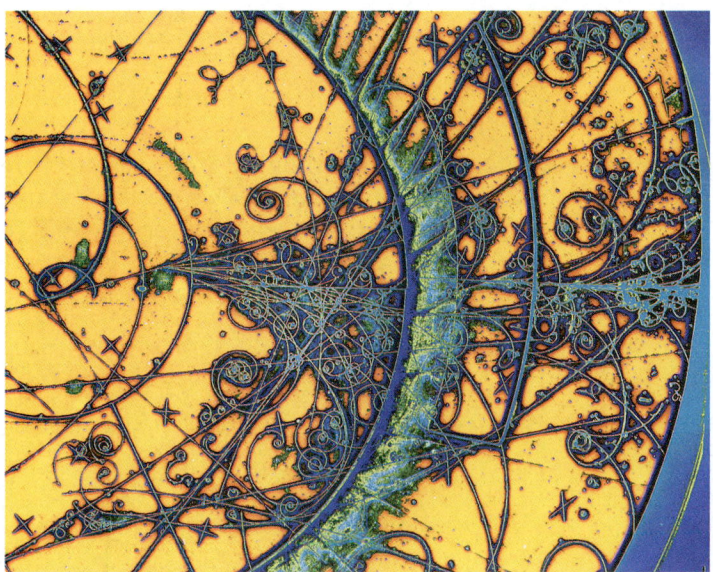

This color-enhanced photograph, taken at CERN, the particle physics laboratory outside Geneva, Switzerland, shows a collection of tracks left by subatomic particles in a bubble chamber. A bubble chamber is a container filled with liquid hydrogen that is superheated, that is, momentarily raised above its normal boiling point by a sudden drop in pressure in the container. Any charged particle passing through the liquid in this state leaves behind a trail of tiny bubbles as the liquid boils in its wake. These bubbles are seen as fine tracks, showing the characteristic paths of different types of particles. The paths are curved because there is an intense applied magnetic field. The tightly wound spiral tracks are due to electrons and positrons.   *(Patrice Loiez, CERN/SPL/Photo Researchers, Inc.)*

Optional Section

29.5 ▶ APPLICATIONS INVOLVING CHARGED PARTICLES MOVING IN A MAGNETIC FIELD

A charge moving with a velocity $\mathbf{v}$ in the presence of both an electric field $\mathbf{E}$ and a magnetic field $\mathbf{B}$ experiences both an electric force $q\mathbf{E}$ and a magnetic force $q\mathbf{v} \times \mathbf{B}$. The total force (called the Lorentz force) acting on the charge is

Lorentz force

$$\sum \mathbf{F} = q\mathbf{E} + q\mathbf{v} \times \mathbf{B} \qquad \text{(29.16)}$$

Velocity Selector

In many experiments involving moving charged particles, it is important that the particles all move with essentially the same velocity. This can be achieved by applying a combination of an electric field and a magnetic field oriented as shown in Figure 29.22. A uniform electric field is directed vertically downward (in the plane of the page in Fig. 29.22a), and a uniform magnetic field is applied in the direction perpendicular to the electric field (into the page in Fig. 29.22a). For q positive, the magnetic force $q\mathbf{v} \times \mathbf{B}$ is upward and the electric force $q\mathbf{E}$ is downward. When the magnitudes of the two fields are chosen so that $qE = qvB$, the particle moves in a straight horizontal line through the region of the fields. From the expression $qE = qvB$, we find that

$$v = \frac{E}{B} \qquad \text{(29.17)}$$

Only those particles having speed v pass undeflected through the mutually perpendicular electric and magnetic fields. The magnetic force exerted on particles moving at speeds greater than this is stronger than the electric force, and the particles are deflected upward. Those moving at speeds less than this are deflected downward.

The Mass Spectrometer

A **mass spectrometer** separates ions according to their mass-to-charge ratio. In one version of this device, known as the *Bainbridge mass spectrometer,* a beam of ions first passes through a velocity selector and then enters a second uniform magnetic field $\mathbf{B}_0$ that has the same direction as the magnetic field in the selector (Fig. 29.23). Upon entering the second magnetic field, the ions move in a semicircle of

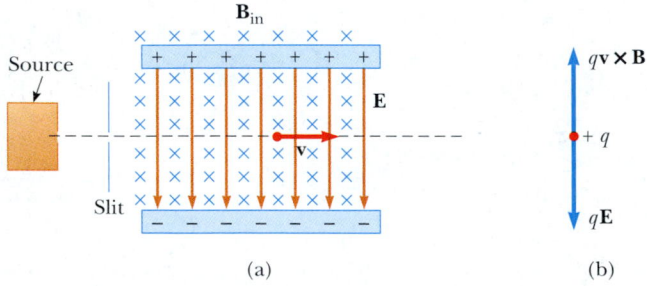

(a)   (b)

Figure 29.22 (a) A velocity selector. When a positively charged particle is in the presence of a magnetic field directed into the page and an electric field directed downward, it experiences a downward electric force $q\mathbf{E}$ and an upward magnetic force $q\mathbf{v} \times \mathbf{B}$. (b) When these forces balance, the particle moves in a horizontal line through the fields.

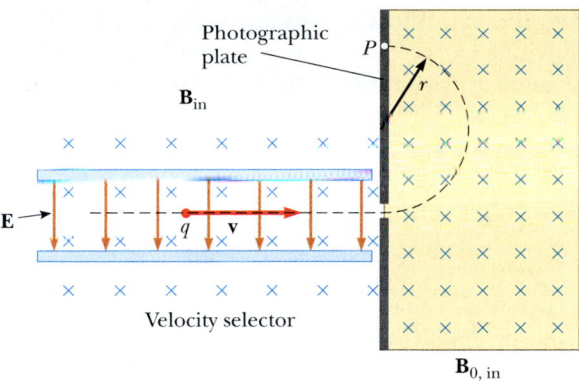

Figure 29.23 A mass spectrometer. Positively charged particles are sent first through a velocity selector and then into a region where the magnetic field $\mathbf{B}_0$ causes the particles to move in a semicircular path and strike a photographic film at P.

radius r before striking a photographic plate at P. If the ions are positively charged, the beam deflects upward, as Figure 29.23 shows. If the ions are negatively charged, the beam would deflect downward. From Equation 29.13, we can express the ratio m/q as

$$\frac{m}{q} = \frac{rB_0}{v}$$

Using Equation 29.17, we find that

$$\frac{m}{q} = \frac{rB_0B}{E} \tag{29.18}$$

Therefore, we can determine m/q by measuring the radius of curvature and knowing the field magnitudes B, B_0, and E. In practice, one usually measures the masses of various isotopes of a given ion, with the ions all carrying the same charge q. In this way, the mass ratios can be determined even if q is unknown.

A variation of this technique was used by J. J. Thomson (1856–1940) in 1897 to measure the ratio e/m_e for electrons. Figure 29.24a shows the basic apparatus he

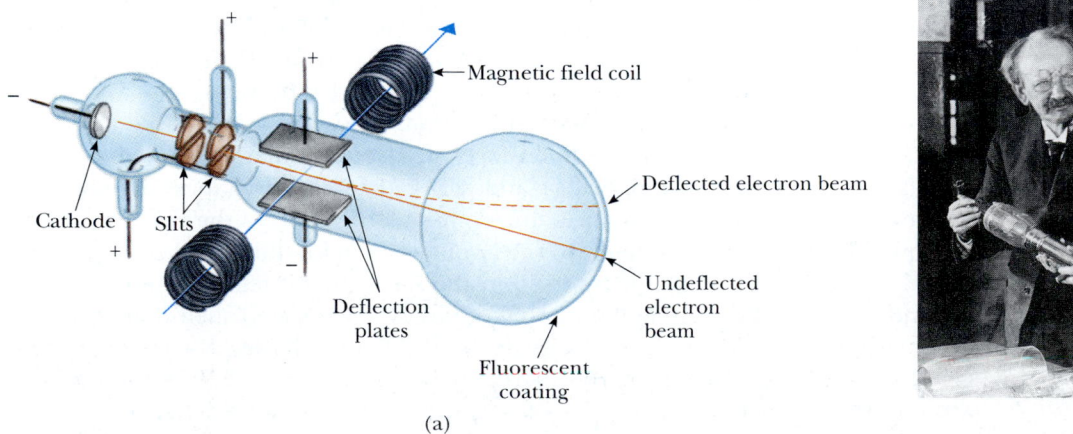

(a)   (b)

Figure 29.24 (a) Thomson's apparatus for measuring e/m_e. Electrons are accelerated from the cathode, pass through two slits, and are deflected by both an electric field and a magnetic field (directed perpendicular to the electric field). The beam of electrons then strikes a fluorescent screen. (b) J. J. Thomson *(left)* in the Cavendish Laboratory, University of Cambridge. It is interesting to note that the man on the right, Frank Baldwin Jewett, is a distant relative of John W. Jewett, Jr., contributing author of this text. *(Bell Telephone Labs/Courtesy of Emilio Segrè Visual Archives)*

used. Electrons are accelerated from the cathode and pass through two slits. They then drift into a region of perpendicular electric and magnetic fields. The magnitudes of the two fields are first adjusted to produce an undeflected beam. When the magnetic field is turned off, the electric field produces a measurable beam deflection that is recorded on the fluorescent screen. From the size of the deflection and the measured values of E and B, the charge-to-mass ratio can be determined. The results of this crucial experiment represent the discovery of the electron as a fundamental particle of nature.

Quick Quiz 29.6

When a photographic plate from a mass spectrometer like the one shown in Figure 29.23 is developed, the three patterns shown in Figure 29.25 are observed. Rank the particles that caused the patterns by speed and m/q ratio.

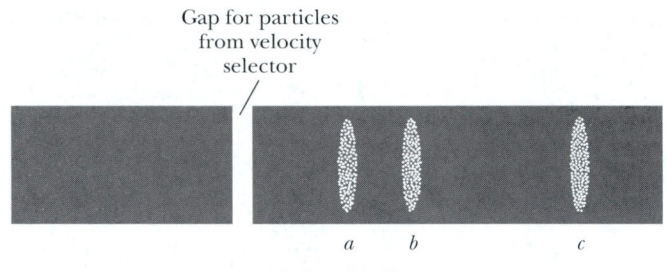

Gap for particles
from velocity
selector

a   *b*   *c*

Figure 29.25

The Cyclotron

A **cyclotron** can accelerate charged particles to very high speeds. Both electric and magnetic forces have a key role. The energetic particles produced are used to bombard atomic nuclei and thereby produce nuclear reactions of interest to researchers. A number of hospitals use cyclotron facilities to produce radioactive substances for diagnosis and treatment.

A schematic drawing of a cyclotron is shown in Figure 29.26. The charges move inside two semicircular containers D_1 and D_2, referred to as *dees*. A high-frequency alternating potential difference is applied to the dees, and a uniform magnetic field is directed perpendicular to them. A positive ion released at P near the center of the magnet in one dee moves in a semicircular path (indicated by the dashed red line in the drawing) and arrives back at the gap in a time $T/2$, where T is the time needed to make one complete trip around the two dees, given by Equation 29.15. The frequency of the applied potential difference is adjusted so that the polarity of the dees is reversed in the same time it takes the ion to travel around one dee. If the applied potential difference is adjusted such that D_2 is at a lower electric potential than D_1 by an amount ΔV, the ion accelerates across the gap to D_2 and its kinetic energy increases by an amount $q\Delta V$. It then moves around D_2 in a semicircular path of greater radius (because its speed has increased). After a time $T/2$, it again arrives at the gap between the dees. By this time, the polarity across the dees is again reversed, and the ion is given another "kick" across the gap. The motion continues so that for each half-circle trip around one dee, the ion gains additional kinetic energy equal to $q\,\Delta V$. When the radius of its path is nearly that of the dees, the energetic ion leaves the system through the exit slit. It is important to note that the operation of the cyclotron is

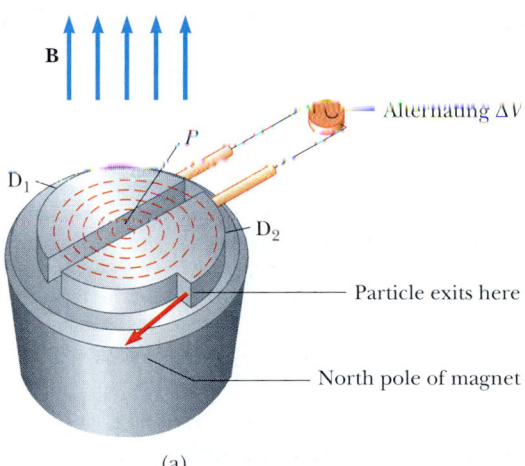

(a)   (b)

Figure 29.26 (a) A cyclotron consists of an ion source at *P*, two dees D$_1$ and D$_2$ across which an alternating potential difference is applied, and a uniform magnetic field. (The south pole of the magnet is not shown.) The red dashed curved lines represent the path of the particles. (b) The first cyclotron, invented by E.O. Lawrence and M.S. Livingston in 1934. *(Courtesy of Lawrence Berkeley Laboratory/University of California)*

based on the fact that *T* is independent of the speed of the ion and of the radius of the circular path.

We can obtain an expression for the kinetic energy of the ion when it exits the cyclotron in terms of the radius *R* of the dees. From Equation 29.13 we know that $v = qBR/m$. Hence, the kinetic energy is

$$K = \tfrac{1}{2}mv^2 = \frac{q^2B^2R^2}{2m} \qquad \textbf{(29.19)}$$

When the energy of the ions in a cyclotron exceeds about 20 MeV, relativistic effects come into play. (Such effects are discussed in Chapter 39.) We observe that *T* increases and that the moving ions do not remain in phase with the applied potential difference. Some accelerators overcome this problem by modifying the period of the applied potential difference so that it remains in phase with the moving ions.

web

More information on these accelerators is available at
http://www.fnal.gov or
http://www.CERN.ch
The CERN site also discusses the creation of the World Wide Web there by physicists in the mid-1990s.

Optional Section

29.6 THE HALL EFFECT

When a current-carrying conductor is placed in a magnetic field, a potential difference is generated in a direction perpendicular to both the current and the magnetic field. This phenomenon, first observed by Edwin Hall (1855–1938) in 1879, is known as the *Hall effect*. It arises from the deflection of charge carriers to one side of the conductor as a result of the magnetic force they experience. The Hall effect gives information regarding the sign of the charge carriers and their density; it can also be used to measure the magnitude of magnetic fields.

The arrangement for observing the Hall effect consists of a flat conductor carrying a current *I* in the *x* direction, as shown in Figure 29.27. A uniform magnetic field **B** is applied in the *y* direction. If the charge carriers are electrons moving in the negative *x* direction with a drift velocity **v**$_d$, they experience an upward mag-

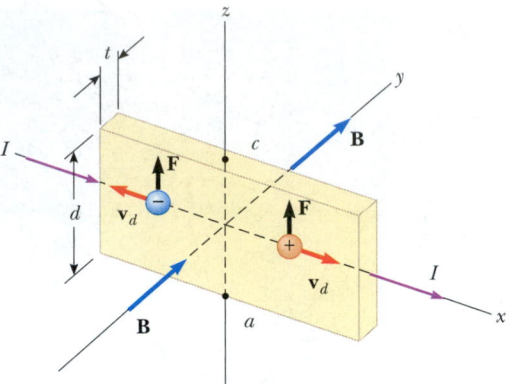

Figure 29.27 To observe the Hall effect, a magnetic field is applied to a current-carrying conductor. When I is in the x direction and **B** in the y direction, both positive and negative charge carriers are deflected upward in the magnetic field. The Hall voltage is measured between points a and c.

netic force $\mathbf{F}_B = q\mathbf{v}_d \times \mathbf{B}$, are deflected upward, and accumulate at the upper edge of the flat conductor, leaving an excess of positive charge at the lower edge (Fig. 29.28a). This accumulation of charge at the edges increases until the electric force resulting from the charge separation balances the magnetic force acting on the carriers. When this equilibrium condition is reached, the electrons are no longer deflected upward. A sensitive voltmeter or potentiometer connected across the sample, as shown in Figure 29.28, can measure the potential difference—known as the **Hall voltage** ΔV_{H}—generated across the conductor.

If the charge carriers are positive and hence move in the positive x direction, as shown in Figures 29.27 and 29.28b, they also experience an upward magnetic force $q\mathbf{v}_d \times \mathbf{B}$. This produces a buildup of positive charge on the upper edge and leaves an excess of negative charge on the lower edge. Hence, the sign of the Hall voltage generated in the sample is opposite the sign of the Hall voltage resulting from the deflection of electrons. The sign of the charge carriers can therefore be determined from a measurement of the polarity of the Hall voltage.

In deriving an expression for the Hall voltage, we first note that the magnetic force exerted on the carriers has magnitude $qv_d B$. In equilibrium, this force is balanced by the electric force qE_{H}, where E_{H} is the magnitude of the electric field due to the charge separation (sometimes referred to as the *Hall field*). Therefore,

$$qv_d B = qE_{\mathrm{H}}$$

$$E_{\mathrm{H}} = v_d B$$

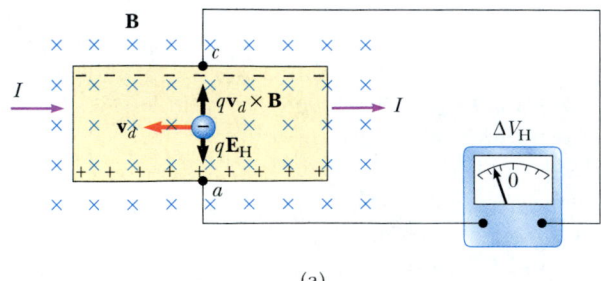

(a)

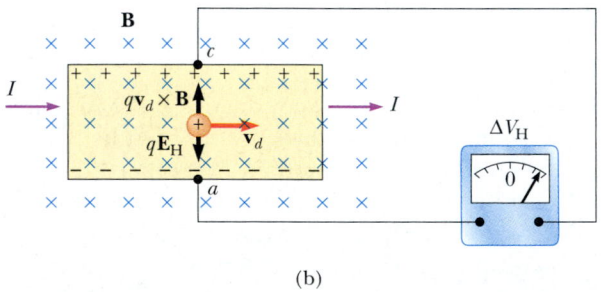

(b)

Figure 29.28 (a) When the charge carriers in a Hall effect apparatus are negative, the upper edge of the conductor becomes negatively charged, and c is at a lower electric potential than a. (b) When the charge carriers are positive, the upper edge becomes positively charged, and c is at a higher potential than a. In either case, the charge carriers are no longer deflected when the edges become fully charged, that is, when there is a balance between the electrostatic force qE_{H} and the magnetic deflection force qvB.

If d is the width of the conductor, the Hall voltage is

$$\Delta V_H = E_H d = v_d B d \qquad \text{(29.20)}$$

Thus, the measured Hall voltage gives a value for the drift speed of the charge carriers if d and B are known.

We can obtain the charge carrier density n by measuring the current in the sample. From Equation 27.4, we can express the drift speed as

$$v_d = \frac{I}{nqA} \qquad \text{(29.21)}$$

where A is the cross-sectional area of the conductor. Substituting Equation 29.21 into Equation 29.20, we obtain

$$\Delta V_H = \frac{IBd}{nqA} \qquad \text{(29.22)}$$

Because $A = td$, where t is the thickness of the conductor, we can also express Equation 29.22 as

$$\Delta V_H = \frac{IB}{nqt} = \frac{R_H IB}{t} \qquad \text{(29.23)}$$

The Hall voltage

where $R_H = 1/nq$ is the **Hall coefficient.** This relationship shows that a properly calibrated conductor can be used to measure the magnitude of an unknown magnetic field.

Because all quantities in Equation 29.23 other than nq can be measured, a value for the Hall coefficient is readily obtainable. The sign and magnitude of R_H give the sign of the charge carriers and their number density. In most metals, the charge carriers are electrons, and the charge carrier density determined from Hall-effect measurements is in good agreement with calculated values for such metals as lithium (Li), sodium (Na), copper (Cu), and silver (Ag), whose atoms each give up one electron to act as a current carrier. In this case, n is approximately equal to the number of conducting electrons per unit volume. However, this classical model is not valid for metals such as iron (Fe), bismuth (Bi), and cadmium (Cd) or for semiconductors. These discrepancies can be explained only by using a model based on the quantum nature of solids.

EXAMPLE 29.8 ▸ The Hall Effect for Copper

A rectangular copper strip 1.5 cm wide and 0.10 cm thick carries a current of 5.0 A. Find the Hall voltage for a 1.2-T magnetic field applied in a direction perpendicular to the strip.

Solution If we assume that one electron per atom is available for conduction, we can take the charge carrier density to be $n = 8.49 \times 10^{28}$ electrons/m^3 (see Example 27.1). Substituting this value and the given data into Equation 29.23 gives

$$\Delta V_H = \frac{IB}{nqt}$$

$$= \frac{(5.0 \text{ A})(1.2 \text{ T})}{(8.49 \times 10^{28} \text{ m}^{-3})(1.6 \times 10^{-19} \text{ C})(0.001\,0 \text{ m})}$$

$$\Delta V_H = \boxed{0.44 \ \mu\text{V}}$$

Such an extremely small Hall voltage is expected in good conductors. (Note that the width of the conductor is not needed in this calculation.)

In semiconductors, n is much smaller than it is in metals that contribute one electron per atom to the current; hence, the Hall voltage is usually greater because it varies as the inverse of n. Currents of the order of 0.1 mA are generally used for such materials. Consider a piece of silicon that has the same dimensions as the copper strip in this example and whose value for $n = 1.0 \times 10^{20}$ electrons/m^3. Taking $B = 1.2$ T and $I = 0.10$ mA, we find that $\Delta V_H = 7.5$ mV. A potential difference of this magnitude is readily measured.

SUMMARY

The magnetic force that acts on a charge q moving with a velocity $\mathbf{v}$ in a magnetic field $\mathbf{B}$ is

$$\mathbf{F}_B = q\mathbf{v} \times \mathbf{B} \tag{29.1}$$

The direction of this magnetic force is perpendicular both to the velocity of the particle and to the magnetic field. The magnitude of this force is

$$F_B = |q|vB \sin \theta \tag{29.2}$$

where θ is the smaller angle between $\mathbf{v}$ and $\mathbf{B}$. The SI unit of $\mathbf{B}$ is the **tesla** (T), where $1\ T = 1\ N/A \cdot m$.

When a charged particle moves in a magnetic field, the work done by the magnetic force on the particle is zero because the displacement is always perpendicular to the direction of the force. The magnetic field can alter the direction of the particle's velocity vector, but it cannot change its speed.

If a straight conductor of length L carries a current I, the force exerted on that conductor when it is placed in a uniform magnetic field $\mathbf{B}$ is

$$\mathbf{F}_B = I\mathbf{L} \times \mathbf{B} \tag{29.3}$$

where the direction of $\mathbf{L}$ is in the direction of the current and $|\mathbf{L}| = L$.

If an arbitrarily shaped wire carrying a current I is placed in a magnetic field, the magnetic force exerted on a very small segment $d\mathbf{s}$ is

$$d\mathbf{F}_B = I\,d\mathbf{s} \times \mathbf{B} \tag{29.4}$$

To determine the total magnetic force on the wire, one must integrate Equation 29.4, keeping in mind that both $\mathbf{B}$ and $d\mathbf{s}$ may vary at each point. Integration gives for the force exerted on a current-carrying conductor of arbitrary shape in a uniform magnetic field

$$\mathbf{F}_B = I\mathbf{L}' \times \mathbf{B} \tag{29.7}$$

where $\mathbf{L}'$ is a vector directed from one end of the conductor to the opposite end. Because integration of Equation 29.4 for a closed loop yields a zero result, the net magnetic force on any closed loop carrying a current in a uniform magnetic field is zero.

The **magnetic dipole moment** $\boldsymbol{\mu}$ of a loop carrying a current I is

$$\boldsymbol{\mu} = I\mathbf{A} \tag{29.10}$$

where the area vector $\mathbf{A}$ is perpendicular to the plane of the loop and $|\mathbf{A}|$ is equal to the area of the loop. The SI unit of $\boldsymbol{\mu}$ is $A \cdot m^2$.

The torque $\boldsymbol{\tau}$ on a current loop placed in a uniform magnetic field $\mathbf{B}$ is

$$\boldsymbol{\tau} = \boldsymbol{\mu} \times \mathbf{B} \tag{29.11}$$

and the potential energy of a magnetic dipole in a magnetic field is

$$U = -\boldsymbol{\mu} \cdot \mathbf{B} \tag{29.12}$$

If a charged particle moves in a uniform magnetic field so that its initial velocity is perpendicular to the field, the particle moves in a circle, the plane of which is perpendicular to the magnetic field. The radius of the circular path is

$$r = \frac{mv}{qB} \tag{29.13}$$

where m is the mass of the particle and q is its charge. The angular speed of the charged particle is

$$\omega = \frac{qB}{m} \qquad\qquad \textbf{(29.14)}$$

QUESTIONS

1. At a given instant, a proton moves in the positive x direction in a region where a magnetic field is directed in the negative z direction. What is the direction of the magnetic force? Does the proton continue to move in the positive x direction? Explain.

2. Two charged particles are projected into a region where a magnetic field is directed perpendicular to their velocities. If the charges are deflected in opposite directions, what can be said about them?

3. If a charged particle moves in a straight line through some region of space, can one say that the magnetic field in that region is zero?

4. Suppose an electron is chasing a proton up this page when suddenly a magnetic field directed perpendicular into the page is turned on. What happens to the particles?

5. How can the motion of a moving charged particle be used to distinguish between a magnetic field and an electric field? Give a specific example to justify your argument.

6. List several similarities and differences between electric and magnetic forces.

7. Justify the following statement: "It is impossible for a constant (in other words, a time-independent) magnetic field to alter the speed of a charged particle."

8. In view of the preceding statement, what is the role of a magnetic field in a cyclotron?

9. A current-carrying conductor experiences no magnetic force when placed in a certain manner in a uniform magnetic field. Explain.

10. Is it possible to orient a current loop in a uniform magnetic field such that the loop does not tend to rotate? Explain.

11. How can a current loop be used to determine the presence of a magnetic field in a given region of space?

12. What is the net force acting on a compass needle in a uniform magnetic field?

13. What type of magnetic field is required to exert a resultant force on a magnetic dipole? What is the direction of the resultant force?

14. A proton moving horizontally enters a region where a uniform magnetic field is directed perpendicular to the proton's velocity, as shown in Figure Q29.14. Describe the subsequent motion of the proton. How would an electron behave under the same circumstances?

15. In a magnetic bottle, what causes the direction of the velocity of the confined charged particles to reverse? (*Hint:* Find the direction of the magnetic force acting on the particles in a region where the field lines converge.)

16. In the cyclotron, why do particles of different velocities take the same amount of time to complete one half-circle trip around one dee?

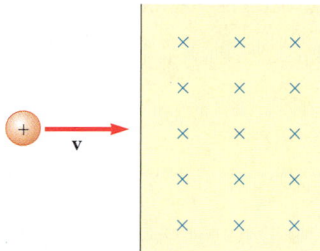

Figure Q29.14

17. The *bubble chamber* is a device used for observing tracks of particles that pass through the chamber, which is immersed in a magnetic field. If some of the tracks are spirals and others are straight lines, what can you say about the particles?

18. Can a constant magnetic field set into motion an electron initially at rest? Explain your answer.

19. You are designing a magnetic probe that uses the Hall effect to measure magnetic fields. Assume that you are restricted to using a given material and that you have already made the probe as thin as possible. What, if anything, can be done to increase the Hall voltage produced for a given magnetic field?

20. The electron beam shown in Figure Q29.20 is projected to the right. The beam deflects downward in the presence of a magnetic field produced by a pair of current-carrying coils. (a) What is the direction of the magnetic field? (b) What would happen to the beam if the current in the coils were reversed?

Figure Q29.20   (*Courtesy of Central Scientific Company*)

PROBLEMS

1, 2, 3 = straightforward, intermediate, challenging   ☐ = full solution available in the *Student Solutions Manual and Study Guide*
WEB = solution posted at **http://www.saunderscollege.com/physics/**   💻 = Computer useful in solving problem   🎮 = Interactive Physics
☐ = paired numerical/symbolic problems

Section 29.1   The Magnetic Field

WEB **1.** Determine the initial direction of the deflection of charged particles as they enter the magnetic fields, as shown in Figure P29.1.

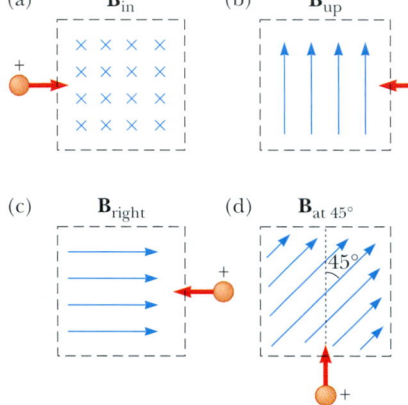

Figure P29.1

2. Consider an electron near the Earth's equator. In which direction does it tend to deflect if its velocity is directed (a) downward, (b) northward, (c) westward, or (d) southeastward?

3. An electron moving along the positive x axis perpendicular to a magnetic field experiences a magnetic deflection in the negative y direction. What is the direction of the magnetic field?

4. A proton travels with a speed of 3.00×10^6 m/s at an angle of 37.0° with the direction of a magnetic field of 0.300 T in the $+y$ direction. What are (a) the magnitude of the magnetic force on the proton and (b) its acceleration?

5. A proton moves in a direction perpendicular to a uniform magnetic field **B** at 1.00×10^7 m/s and experiences an acceleration of 2.00×10^{13} m/s² in the $+x$ direction when its velocity is in the $+z$ direction. Determine the magnitude and direction of the field.

6. An electron is accelerated through 2 400 V from rest and then enters a region where there is a uniform 1.70-T magnetic field. What are (a) the maximum and (b) the minimum values of the magnetic force this charge can experience?

7. At the equator, near the surface of the Earth, the magnetic field is approximately 50.0 μT northward, and the electric field is about 100 N/C downward in fair weather. Find the gravitational, electric, and magnetic forces on an electron with an instantaneous velocity of

6.00×10^6 m/s directed to the east in this environment.

8. A 30.0-g metal ball having net charge $Q = 5.00$ μC is thrown out of a window horizontally at a speed $v = 20.0$ m/s. The window is at a height $h = 20.0$ m above the ground. A uniform horizontal magnetic field of magnitude $B = 0.010$ 0 T is perpendicular to the plane of the ball's trajectory. Find the magnetic force acting on the ball just before it hits the ground.

9. A proton moving at 4.00×10^6 m/s through a magnetic field of 1.70 T experiences a magnetic force of magnitude 8.20×10^{-13} N. What is the angle between the proton's velocity and the field?

10. An electron has a velocity of 1.20 km/s (in the positive x direction) and an acceleration of 2.00×10^{12} m/s² (in the positive z direction) in uniform electric and magnetic fields. If the electric field has a magnitude of 20.0 N/C (in the positive z direction), what can you determine about the magnetic field in the region? What can you not determine?

11. A proton moves with a velocity of $\mathbf{v} = (2\mathbf{i} - 4\mathbf{j} + \mathbf{k})$ m/s in a region in which the magnetic field is $\mathbf{B} = (\mathbf{i} + 2\mathbf{j} - 3\mathbf{k})$ T. What is the magnitude of the magnetic force this charge experiences?

12. An electron is projected into a uniform magnetic field $B = (1.40\mathbf{i} + 2.10\mathbf{j})$ T. Find the vector expression for the force on the electron when its velocity is $\mathbf{v} = 3.70 \times 10^5 \mathbf{j}$ m/s.

Section 29.2   Magnetic Force Acting on a Current-Carrying Conductor

WEB **13.** A wire having a mass per unit length of 0.500 g/cm carries a 2.00-A current horizontally to the south. What are the direction and magnitude of the minimum magnetic field needed to lift this wire vertically upward?

14. A wire carries a steady current of 2.40 A. A straight section of the wire is 0.750 m long and lies along the x axis within a uniform magnetic field of magnitude $B = 1.60$ T in the positive z direction. If the current is in the $+x$ direction, what is the magnetic force on the section of wire?

15. A wire 2.80 m in length carries a current of 5.00 A in a region where a uniform magnetic field has a magnitude of 0.390 T. Calculate the magnitude of the magnetic force on the wire if the angle between the magnetic field and the current is (a) 60.0°, (b) 90.0°, (c) 120°.

16. A conductor suspended by two flexible wires as shown in Figure P29.16 has a mass per unit length of 0.040 0 kg/m. What current must exist in the conductor for the tension in the supporting wires to be zero when the magnetic

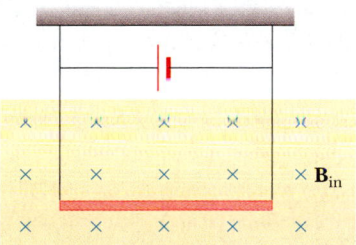

Figure P29.16

field is 3.60 T into the page? What is the required direction for the current?

17. Imagine a very long, uniform wire with a linear mass density of 1.00 g/m that encircles the Earth at its magnetic equator. Suppose that the planet's magnetic field is 50.0 μT horizontally north throughout this region. What are the magnitude and direction of the current in the wire that keep it levitated just above the ground?

18. In Figure P29.18, the cube is 40.0 cm on each edge. Four straight segments of wire — *ab, bc, cd,* and *da* — form a closed loop that carries a current $I = 5.00$ A, in the direction shown. A uniform magnetic field of magnitude $B = 0.020\ 0$ T is in the positive y direction. Determine the magnitude and direction of the magnetic force on each segment.

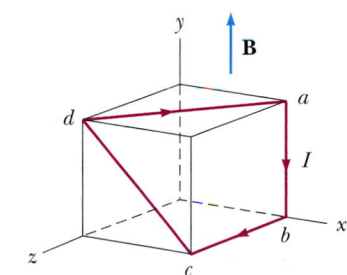

Figure P29.18

19. **Review Problem.** A rod with a mass of 0.720 kg and a radius of 6.00 cm rests on two parallel rails (Fig. P29.19) that are $d = 12.0$ cm apart and $L = 45.0$ cm long. The rod carries a current of $I = 48.0$ A (in the direction shown) and rolls along the rails without slipping. If it starts from rest, what is the speed of the rod as it leaves the rails if a uniform magnetic field of magnitude 0.240 T is directed perpendicular to the rod and the rails?

20. **Review Problem.** A rod of mass m and radius R rests on two parallel rails (Fig. P29.19) that are a distance d apart and have a length L. The rod carries a current I (in the direction shown) and rolls along the rails without slipping. If it starts from rest, what is the speed of the rod as it leaves the rails if a uniform magnetic field **B** is directed perpendicular to the rod and the rails?

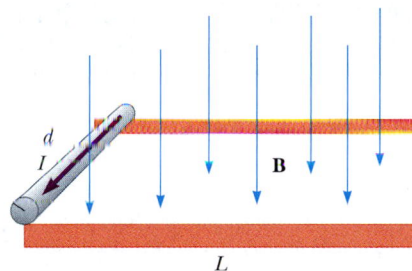

Figure P29.19 Problems 19 and 20.

WEB 21. *A nonuniform magnetic field exerts a net force on a magnetic dipole.* A strong magnet is placed under a horizontal conducting ring of radius r that carries current I, as shown in Figure P29.21. If the magnetic field **B** makes an angle θ with the vertical at the ring's location, what are the magnitude and direction of the resultant force on the ring?

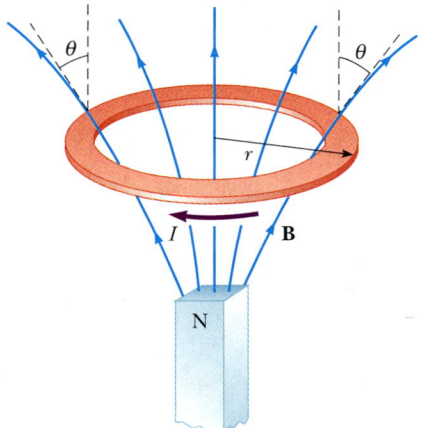

Figure P29.21

22. Assume that in Atlanta, Georgia, the Earth's magnetic field is 52.0 μT northward at 60.0° below the horizontal. A tube in a neon sign carries a current of 35.0 mA between two diagonally opposite corners of a shop window, which lies in a north–south vertical plane. The current enters the tube at the bottom south corner of the window. It exits at the opposite corner, which is 1.40 m farther north and 0.850 m higher up. Between these two points, the glowing tube spells out DONUTS. Use the theorem proved as "Case 1" in the text to determine the total vector magnetic force on the tube.

Section 29.3 **Torque on a Current Loop in a Uniform Magnetic Field**

23. A current of 17.0 mA is maintained in a single circular loop with a circumference of 2.00 m. A magnetic field

of 0.800 T is directed parallel to the plane of the loop.
(a) Calculate the magnetic moment of the loop.
(b) What is the magnitude of the torque exerted on the loop by the magnetic field?

24. A small bar magnet is suspended in a uniform 0.250-T magnetic field. The maximum torque experienced by the bar magnet is 4.60×10^{-3} N·m. Calculate the magnetic moment of the bar magnet.

WEB 25. A rectangular loop consists of $N = 100$ closely wrapped turns and has dimensions $a = 0.400$ m and $b = 0.300$ m. The loop is hinged along the y axis, and its plane makes an angle $\theta = 30.0°$ with the x axis (Fig. P29.25). What is the magnitude of the torque exerted on the loop by a uniform magnetic field $B = 0.800$ T directed along the x axis when the current is $I = 1.20$ A in the direction shown? What is the expected direction of rotation of the loop?

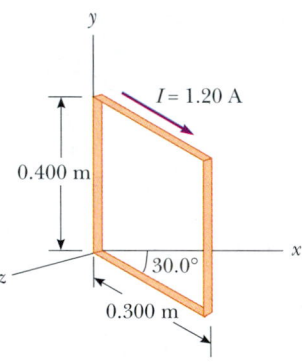

Figure P29.25

26. A long piece of wire of mass 0.100 kg and total length of 4.00 m is used to make a square coil with a side of 0.100 m. The coil is hinged along a horizontal side, carries a 3.40-A current, and is placed in a vertical magnetic field with a magnitude of 0.010 0 T. (a) Determine the angle that the plane of the coil makes with the vertical when the coil is in equilibrium. (b) Find the torque acting on the coil due to the magnetic force at equilibrium.

27. A 40.0-cm length of wire carries a current of 20.0 A. It is bent into a loop and placed with its normal perpendicular to a magnetic field with a strength of 0.520 T. What is the torque on the loop if it is bent into (a) an equilateral triangle, (b) a square, (c) a circle? (d) Which torque is greatest?

28. A current loop with dipole moment $\boldsymbol{\mu}$ is placed in a uniform magnetic field **B**. Prove that its potential energy is $U = -\boldsymbol{\mu} \cdot \mathbf{B}$. You may imitate the discussion of the potential energy of an electric dipole in an electric field given in Chapter 26.

29. The needle of a magnetic compass has a magnetic moment of 9.70 mA·m². At its location, the Earth's magnetic field is 55.0 μT north at 48.0° below the horizontal. (a) Identify the orientations at which the compass needle has minimum potential energy and maximum potential energy. (b) How much work must be done on the needle for it to move from the former to the latter orientation?

30. A wire is formed into a circle having a diameter of 10.0 cm and is placed in a uniform magnetic field of 3.00 mT. A current of 5.00 A passes through the wire. Find (a) the maximum torque on the wire and (b) the range of potential energy of the wire in the field for different orientations of the circle.

Section 29.4   Motion of a Charged Particle in a Uniform Magnetic Field

31. The magnetic field of the Earth at a certain location is directed vertically downward and has a magnitude of 50.0 μT. A proton is moving horizontally toward the west in this field with a speed of 6.20×10^6 m/s. (a) What are the direction and magnitude of the magnetic force that the field exerts on this charge? (b) What is the radius of the circular arc followed by this proton?

32. A singly charged positive ion has a mass of 3.20×10^{-26} kg. After being accelerated from rest through a potential difference of 833 V, the ion enters a magnetic field of 0.920 T along a direction perpendicular to the direction of the field. Calculate the radius of the path of the ion in the field.

33. **Review Problem.**   One electron collides elastically with a second electron initially at rest. After the collision, the radii of their trajectories are 1.00 cm and 2.40 cm. The trajectories are perpendicular to a uniform magnetic field of magnitude 0.044 0 T. Determine the energy (in keV) of the incident electron.

34. A proton moving in a circular path perpendicular to a constant magnetic field takes 1.00 μs to complete one revolution. Determine the magnitude of the magnetic field.

35. A proton (charge $+e$, mass m_p), a deuteron (charge $+e$, mass $2m_p$), and an alpha particle (charge $+2e$, mass $4m_p$) are accelerated through a common potential difference ΔV. The particles enter a uniform magnetic field **B** with a velocity in a direction perpendicular to **B**. The proton moves in a circular path of radius r_p. Determine the values of the radii of the circular orbits for the deuteron r_d and the alpha particle r_α in terms of r_p.

36. **Review Problem.**   An electron moves in a circular path perpendicular to a constant magnetic field with a magnitude of 1.00 mT. If the angular momentum of the electron about the center of the circle is 4.00×10^{-25} J·s, determine (a) the radius of the circular path and (b) the speed of the electron.

37. Calculate the cyclotron frequency of a proton in a magnetic field with a magnitude of 5.20 T.

38. A singly charged ion of mass m is accelerated from rest by a potential difference ΔV. It is then deflected by a uniform magnetic field (perpendicular to the ion's velocity) into a semicircle of radius R. Now a doubly

charged ion of mass m' is accelerated through the same potential difference and deflected by the same magnetic field into a semicircle of radius $R' = 2R$. What is the ratio of the ions' masses?

39. A cosmic-ray proton in interstellar space has an energy of 10.0 MeV and executes a circular orbit having a radius equal to that of Mercury's orbit around the Sun $(5.80 \times 10^{10}$ m$)$. What is the magnetic field in that region of space?

40. A singly charged positive ion moving at 4.60×10^5 m/s leaves a circular track of radius 7.94 mm along a direction perpendicular to the 1.80-T magnetic field of a bubble chamber. Compute the mass (in atomic mass units) of this ion, and identify it from that value.

(Optional)
Section 29.5   Applications Involving Charged Particles Moving in a Magnetic Field

41. A velocity selector consists of magnetic and electric fields described by the expressions $\mathbf{E} = E\mathbf{k}$ and $\mathbf{B} = B\mathbf{j}$. If $B = 0.015\ 0$ T, find the value of E such that a 750-eV electron moving along the positive x axis is undeflected.

42. (a) Singly charged uranium-238 ions are accelerated through a potential difference of 2.00 kV and enter a uniform magnetic field of 1.20 T directed perpendicular to their velocities. Determine the radius of their circular path. (b) Repeat for uranium-235 ions. How does the ratio of these path radii depend on the accelerating voltage and the magnetic field strength?

43. Consider the mass spectrometer shown schematically in Figure 29.23. The electric field between the plates of the velocity selector is 2 500 V/m, and the magnetic field in both the velocity selector and the deflection chamber has a magnitude of 0.035 0 T. Calculate the radius of the path for a singly charged ion having a mass $m = 2.18 \times 10^{-26}$ kg.

44. What is the required radius of a cyclotron designed to accelerate protons to energies of 34.0 MeV using a magnetic field of 5.20 T?

45. A cyclotron designed to accelerate protons has a magnetic field with a magnitude of 0.450 T over a region of radius 1.20 m. What are (a) the cyclotron frequency and (b) the maximum speed acquired by the protons?

46. At the Fermilab accelerator in Batavia, Illinois, protons having momentum 4.80×10^{-16} kg·m/s are held in a circular orbit of radius 1.00 km by an upward magnetic field. What is the magnitude of this field?

WEB 47. The picture tube in a television uses magnetic deflection coils rather than electric deflection plates. Suppose an electron beam is accelerated through a 50.0-kV potential difference and then travels through a region of uniform magnetic field 1.00 cm wide. The screen is located 10.0 cm from the center of the coils and is 50.0 cm wide. When the field is turned off, the electron beam hits the center of the screen. What field magnitude is necessary to deflect the beam to the side of the screen? Ignore relativistic corrections.

(Optional)
Section 29.6   The Hall Effect

48. A flat ribbon of silver having a thickness $t = 0.200$ mm is used in a Hall-effect measurement of a uniform magnetic field perpendicular to the ribbon, as shown in Figure P29.48. The Hall coefficient for silver is $R_H = 0.840 \times 10^{-10}$ m³/C. (a) What is the density of charge carriers in silver? (b) If a current $I = 20.0$ A produces a Hall voltage $\Delta V_H = 15.0\ \mu$V, what is the magnitude of the applied magnetic field?

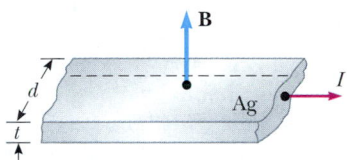

Figure P29.48

49. A section of conductor 0.400 cm thick is used in a Hall-effect measurement. A Hall voltage of 35.0 μV is measured for a current of 21.0 A in a magnetic field of 1.80 T. Calculate the Hall coefficient for the conductor.

50. A flat copper ribbon 0.330 mm thick carries a steady current of 50.0 A and is located in a uniform 1.30-T magnetic field directed perpendicular to the plane of the ribbon. If a Hall voltage of 9.60 μV is measured across the ribbon, what is the charge density of the free electrons? What effective number of free electrons per atom does this result indicate?

51. In an experiment designed to measure the Earth's magnetic field using the Hall effect, a copper bar 0.500 cm thick is positioned along an east–west direction. If a current of 8.00 A in the conductor results in a Hall voltage of 5.10 pV, what is the magnitude of the Earth's magnetic field? (Assume that $n = 8.48 \times 10^{28}$ electrons/m³ and that the plane of the bar is rotated to be perpendicular to the direction of $\mathbf{B}$.)

52. A Hall-effect probe operates with a 120-mA current. When the probe is placed in a uniform magnetic field with a magnitude of 0.080 0 T, it produces a Hall voltage of 0.700 μV. (a) When it is measuring an unknown magnetic field, the Hall voltage is 0.330 μV. What is the unknown magnitude of the field? (b) If the thickness of the probe in the direction of $\mathbf{B}$ is 2.00 mm, find the charge-carrier density (each of charge e).

ADDITIONAL PROBLEMS

53. An electron enters a region of magnetic field of magnitude 0.100 T, traveling perpendicular to the linear boundary of the region. The direction of the field is perpendicular to the velocity of the electron. (a) Determine the time it takes for the electron to leave the "field-filled" region, noting that its path is a semicircle. (b) Find the kinetic energy of the electron if the maximum depth of penetration in the field is 2.00 cm.

54. A 0.200-kg metal rod carrying a current of 10.0 A glides on two horizontal rails 0.500 m apart. What vertical magnetic field is required to keep the rod moving at a constant speed if the coefficient of kinetic friction between the rod and rails is 0.100?

55. Sodium melts at 99°C. Liquid sodium, an excellent thermal conductor, is used in some nuclear reactors to cool the reactor core. The liquid sodium is moved through pipes by pumps that exploit the force on a moving charge in a magnetic field. The principle is as follows: Assume that the liquid metal is in an electrically insulating pipe having a rectangular cross-section of width w and height h. A uniform magnetic field perpendicular to the pipe affects a section of length L (Fig. P29.55). An electric current directed perpendicular to the pipe and to the magnetic field produces a current density J in the liquid sodium. (a) Explain why this arrangement produces on the liquid a force that is directed along the length of the pipe. (b) Show that the section of liquid in the magnetic field experiences a pressure increase JLB.

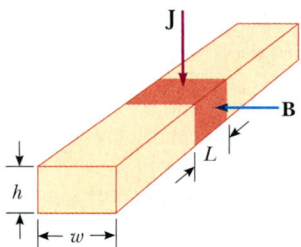

Figure P29.55

56. Protons having a kinetic energy of 5.00 MeV are moving in the positive x direction and enter a magnetic field $\mathbf{B} = (0.050\ 0\ \mathbf{k})$ T directed out of the plane of the page and extending from $x = 0$ to $x = 1.00$ m, as shown in Figure P29.56. (a) Calculate the y component of the protons' momentum as they leave the magnetic field. (b) Find the angle α between the initial velocity vector of the proton beam and the velocity vector after the beam emerges from the field. (*Hint:* Neglect relativistic effects and note that 1 eV = 1.60×10^{-19} J.)

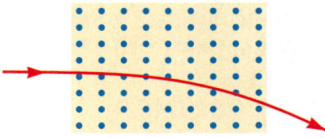

Figure P29.56

57. (a) A proton moving in the $+x$ direction with velocity $\mathbf{v} = v_i\mathbf{i}$ experiences a magnetic force $\mathbf{F} = F_i\mathbf{j}$. Explain what you can and cannot infer about $\mathbf{B}$ from this information. (b) In terms of F_i, what would be the force on a proton in the same field moving with velocity

$\mathbf{v} = -v_i\mathbf{i}$? (c) What would be the force on an electron in the same field moving with velocity $\mathbf{v} = v_i\mathbf{i}$?

58. Review Problem. A wire having a linear mass density of 1.00 g/cm is placed on a horizontal surface that has a coefficient of friction of 0.200. The wire carries a current of 1.50 A toward the east and slides horizontally to the north. What are the magnitude and direction of the smallest magnetic field that enables the wire to move in this fashion?

59. A positive charge $q = 3.20 \times 10^{-19}$ C moves with a velocity $\mathbf{v} = (2\mathbf{i} + 3\mathbf{j} - 1\mathbf{k})$ m/s through a region where both a uniform magnetic field and a uniform electric field exist. (a) What is the total force on the moving charge (in unit–vector notation) if $\mathbf{B} = (2\mathbf{i} + 4\mathbf{j} + 1\mathbf{k})$ T and $\mathbf{E} = (4\mathbf{i} - 1\mathbf{j} - 2\mathbf{k})$ V/m? (b) What angle does the force vector make with the positive x axis?

60. A cosmic-ray proton traveling at half the speed of light is heading directly toward the center of the Earth in the plane of the Earth's equator. Will it hit the Earth? Assume that the Earth's magnetic field is uniform over the planet's equatorial plane with a magnitude of 50.0 μT, extending out 1.30×10^7 m from the surface of the Earth. Assume that the field is zero at greater distances. Calculate the radius of curvature of the proton's path in the magnetic field. Ignore relativistic effects.

61. The circuit in Figure P29.61 consists of wires at the top and bottom and identical metal springs as the left and right sides. The wire at the bottom has a mass of 10.0 g and is 5.00 cm long. The springs stretch 0.500 cm under the weight of the wire, and the circuit has a total resistance of 12.0 Ω. When a magnetic field is turned on, directed out of the page, the springs stretch an additional 0.300 cm. What is the magnitude of the magnetic field? (The upper portion of the circuit is fixed.)

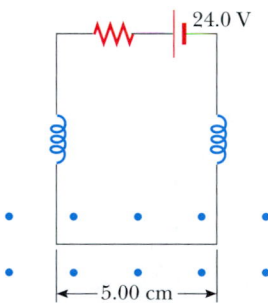

Figure P29.61

62. A hand-held electric mixer contains an electric motor. Model the motor as a single flat compact circular coil carrying electric current in a region where a magnetic field is produced by an external permanent magnet. You need consider only one instant in the operation of the motor. (We will consider motors again in Chapter 31.) The coil moves because the magnetic field exerts torque on the coil, as described in Section 29.3. Make order-of-magnitude estimates of the magnetic field, the torque on the coil, the current in it, its area, and the number of turns in the coil, so that they are related according to Equation 29.11. Note that the input power to the motor is electric, given by $\mathcal{P} = I\,\Delta V$, and the useful output power is mechanical, given by $\mathcal{P} = \tau\omega$.

63. A metal rod having a mass per unit length of 0.010 0 kg/m carries a current of $I = 5.00$ A. The rod hangs from two wires in a uniform vertical magnetic field, as shown in Figure P29.63. If the wires make an angle $\theta = 45.0°$ with the vertical when in equilibrium, determine the magnitude of the magnetic field.

64. A metal rod having a mass per unit length μ carries a current I. The rod hangs from two wires in a uniform vertical magnetic field, as shown in Figure P29.63. If the wires make an angle θ with the vertical when in equilibrium, determine the magnitude of the magnetic field.

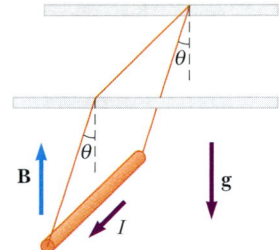

Figure P29.63   Problems 63 and 64.

65. A cyclotron is sometimes used for carbon dating, which we consider in Section 44.6. Carbon-14 and carbon-12 ions are obtained from a sample of the material to be dated and accelerated in the cyclotron. If the cyclotron has a magnetic field of magnitude 2.40 T, what is the difference in cyclotron frequencies for the two ions?

66. A uniform magnetic field of magnitude 0.150 T is directed along the positive x axis. A positron moving at 5.00×10^6 m/s enters the field along a direction that makes an angle of 85.0° with the x axis (Fig. P29.66).

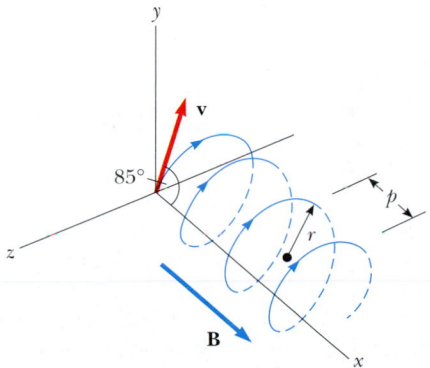

Figure P29.66

The motion of the particle is expected to be a helix, as described in Section 29.4. Calculate (a) the pitch p and (b) the radius r of the trajectory.

67. Consider an electron orbiting a proton and maintained in a fixed circular path of radius $R = 5.29 \times 10^{-11}$ m by the Coulomb force. Treating the orbiting charge as a current loop, calculate the resulting torque when the system is in a magnetic field of 0.400 T directed perpendicular to the magnetic moment of the electron.

68. A singly charged ion completes five revolutions in a uniform magnetic field of magnitude 5.00×10^{-2} T in 1.50 ms. Calculate the mass of the ion in kilograms.

69. A proton moving in the plane of the page has a kinetic energy of 6.00 MeV. It enters a magnetic field of magnitude $B = 1.00$ T directed into the page, moving at an angle of $\theta = 45.0°$ with the straight linear boundary of the field, as shown in Figure P29.69. (a) Find the distance x from the point of entry to where the proton leaves the field. (b) Determine the angle θ' between the boundary and the proton's velocity vector as it leaves the field.

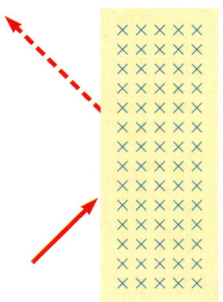

Figure P29.69

70. Table P29.70 shows measurements of a Hall voltage and corresponding magnetic field for a probe used to measure magnetic fields. (a) Plot these data, and deduce a relationship between the two variables. (b) If the mea-

TABLE P29.70

$\Delta V_H(\mu V)$	$B(T)$
0	0.00
11	0.10
19	0.20
28	0.30
42	0.40
50	0.50
61	0.60
68	0.70
79	0.80
90	0.90
102	1.00

surements were taken with a current of 0.200 A and the sample is made from a material having a charge-carrier density of $1.00 \times 10^{26}/m^3$, what is the thickness of the sample?

71. A heart surgeon monitors the flow rate of blood through an artery using an electromagnetic flowmeter (Fig. P29.71). Electrodes A and B make contact with the outer surface of the blood vessel, which has interior diameter 3.00 mm. (a) For a magnetic field magnitude of 0.040 0 T, an emf of 160 μV appears between the electrodes. Calculate the speed of the blood. (b) Verify

that electrode A is positive, as shown. Does the sign of the emf depend on whether the mobile ions in the blood are predominantly positively or negatively charged? Explain.

72. As illustrated in Figure P29.72, a particle of mass m having positive charge q is initially traveling upward with velocity $v\mathbf{j}$. At the origin of coordinates it enters a region between $y = 0$ and $y = h$ containing a uniform magnetic field $B\mathbf{k}$ directed perpendicular out of the page. (a) What is the critical value of v such that the particle just reaches $y = h$? Describe the path of the particle under this condition, and predict its final velocity. (b) Specify the path of the particle and its final velocity if v is less than the critical value. (c) Specify the path of the particle and its final velocity if v is greater than the critical value.

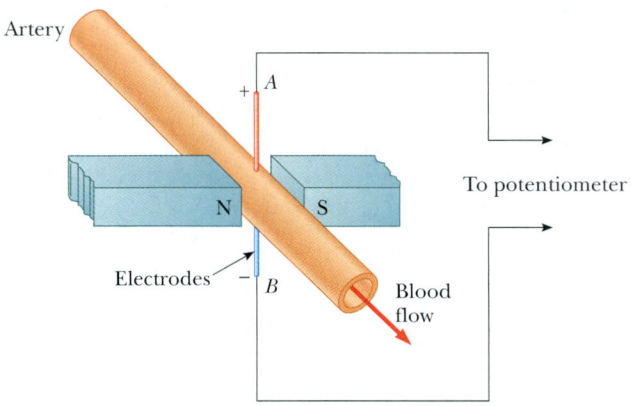

Figure P29.71

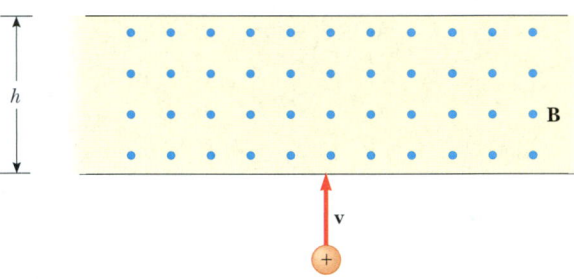

Figure P29.72

ANSWERS TO QUICK QUIZZES

29.1 Zero. Because the magnetic force exerted by the field on the charge is always perpendicular to the velocity of the charge, the field can never do any work on the charge: $W = \mathbf{F}_B \cdot d\mathbf{s} = (\mathbf{F}_B \cdot \mathbf{v}) dt = 0$. Work requires a component of force along the direction of motion.

29.2 Unaffected. The magnetic force exerted by a magnetic field on a charge is proportional to the charge's velocity relative to the field. If the charge is stationary, as in this situation, there is no magnetic force.

29.3 (c), (b), (a), (d). As Example 29.2 shows, we need to be concerned only with the "effective length" of wire perpendicular to the magnetic field or, stated another way, the length of the "magnetic field shadow" cast by the wire. For (c), 4 m of wire is perpendicular to the field. The short vertical pieces experience no magnetic force because their currents are parallel to the field. When the wire in (b) is broken into many short vertical and horizontal segments alternately parallel and perpendicular to the field, we find a total of 3.5 m of horizontal segments perpendicular to the field and therefore experiencing a force. Next comes (a), with 3 m of wire effectively perpendicular to the field. Only 2 m of the wire in (d) experiences a force. The portion carrying current from 2 m to 4 m does experience a force di-

rected out of the page, but this force is canceled by an oppositely directed force acting on the current as it moves from 4 m to 2 m.

29.4 Because it is in the region of the stronger magnetic field, side ③ experiences a greater force than side ①: $F_3 > F_1$. Therefore, in addition to the torque resulting from the two forces, a net force is exerted downward on the loop.

29.5 (c), (b), (a). Because all loops enclose the same area and carry the same current, the magnitude of $\boldsymbol{\mu}$ is the same for all. For (c), $\boldsymbol{\mu}$ points upward and is perpendicular to the magnetic field and $\boldsymbol{\tau} = \boldsymbol{\mu}\mathbf{B}$. This is the maximum torque possible. The next largest cross product of $\boldsymbol{\mu}$ and $\mathbf{B}$ is for (b), in which $\boldsymbol{\mu}$ points toward the upper right (as illustrated in Fig. 29.13b). Finally, $\boldsymbol{\mu}$ for the loop in (a) points along the direction of $\mathbf{B}$; thus, the torque is zero.

29.6 The velocity selector ensures that all three types of particles have the same speed. We cannot determine individual masses or charges, but we can rank the particles by m/q ratio. Equation 29.18 indicates that those particles traveling through the circle of greatest radius have the greatest m/q ratio. Thus, the m/q ranking, from greatest to least value, is c, b, a.

P U Z Z L E R

All three of these commonplace items use magnetism to store information. The cassette can store more than an hour of music, the floppy disk can hold the equivalent of hundreds of pages of information, and many hours of television programming can be recorded on the videotape. How do these devices work? *(George Semple)*

c h a p t e r

30

Sources of the Magnetic Field

In the preceding chapter, we discussed the magnetic force exerted on a charged particle moving in a magnetic field. To complete the description of the magnetic interaction, this chapter deals with the origin of the magnetic field—moving charges. We begin by showing how to use the law of Biot and Savart to calculate the magnetic field produced at some point in space by a small current element. Using this formalism and the principle of superposition, we then calculate the total magnetic field due to various current distributions. Next, we show how to determine the force between two current-carrying conductors, which leads to the definition of the ampere. We also introduce Ampère's law, which is useful in calculating the magnetic field of a highly symmetric configuration carrying a steady current.

This chapter is also concerned with the complex processes that occur in magnetic materials. All magnetic effects in matter can be explained on the basis of atomic magnetic moments, which arise both from the orbital motion of the electrons and from an intrinsic property of the electrons known as spin.

30.1 ▸ THE BIOT–SAVART LAW

Shortly after Oersted's discovery in 1819 that a compass needle is deflected by a current-carrying conductor, Jean-Baptiste Biot (1774–1862) and Félix Savart (1791–1841) performed quantitative experiments on the force exerted by an electric current on a nearby magnet. From their experimental results, Biot and Savart arrived at a mathematical expression that gives the magnetic field at some point in space in terms of the current that produces the field. That expression is based on the following experimental observations for the magnetic field $d\mathbf{B}$ at a point P associated with a length element $d\mathbf{s}$ of a wire carrying a steady current I (Fig. 30.1):

Properties of the magnetic field created by an electric current

- The vector $d\mathbf{B}$ is perpendicular both to $d\mathbf{s}$ (which points in the direction of the current) and to the unit vector $\hat{\mathbf{r}}$ directed from $d\mathbf{s}$ to P.
- The magnitude of $d\mathbf{B}$ is inversely proportional to r^2, where r is the distance from $d\mathbf{s}$ to P.
- The magnitude of $d\mathbf{B}$ is proportional to the current and to the magnitude ds of the length element $d\mathbf{s}$.
- The magnitude of $d\mathbf{B}$ is proportional to $\sin\theta$, where θ is the angle between the vectors $d\mathbf{s}$ and $\hat{\mathbf{r}}$.

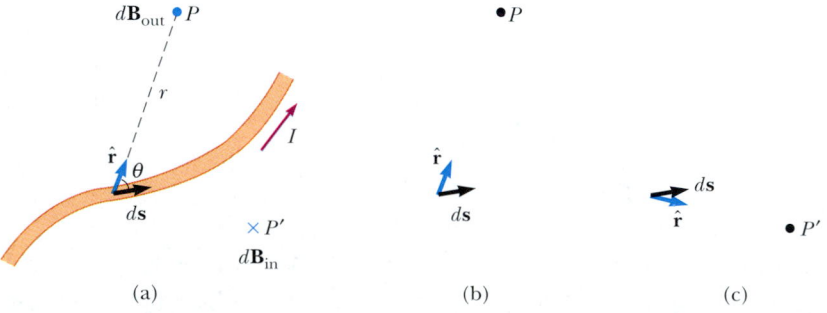

Figure 30.1   (a) The magnetic field $d\mathbf{B}$ at point P due to the current I through a length element $d\mathbf{s}$ is given by the Biot–Savart law. The direction of the field is out of the page at P and into the page at P'. (b) The cross product $d\mathbf{s} \times \hat{\mathbf{r}}$ points out of the page when $\hat{\mathbf{r}}$ points toward P. (c) The cross product $d\mathbf{s} \times \hat{\mathbf{r}}$ points into the page when $\hat{\mathbf{r}}$ points toward P'.

These observations are summarized in the mathematical formula known today as the **Biot–Savart law:**

$$d\mathbf{B} = \frac{\mu_0}{4\pi} \frac{I\, d\mathbf{s} \times \hat{\mathbf{r}}}{r^2} \tag{30.1}$$

where μ_0 is a constant called the **permeability of free space:**

$$\mu_0 = 4\pi \times 10^{-7}\ \text{T} \cdot \text{m/A} \tag{30.2}$$

Permeability of free space

It is important to note that the field $d\mathbf{B}$ in Equation 30.1 is the field created by the current in only a small length element $d\mathbf{s}$ of the conductor. To find the total magnetic field $\mathbf{B}$ created at some point by a current of finite size, we must sum up contributions from all current elements $I\, d\mathbf{s}$ that make up the current. That is, we must evaluate $\mathbf{B}$ by integrating Equation 30.1:

$$\mathbf{B} = \frac{\mu_0 I}{4\pi} \int \frac{d\mathbf{s} \times \hat{\mathbf{r}}}{r^2} \tag{30.3}$$

where the integral is taken over the entire current distribution. This expression must be handled with special care because the integrand is a cross product and therefore a vector quantity. We shall see one case of such an integration in Example 30.1.

Although we developed the Biot–Savart law for a current-carrying wire, it is also valid for a current consisting of charges flowing through space, such as the electron beam in a television set. In that case, $d\mathbf{s}$ represents the length of a small segment of space in which the charges flow.

Interesting similarities exist between the Biot–Savart law for magnetism and Coulomb's law for electrostatics. The current element produces a magnetic field, whereas a point charge produces an electric field. Furthermore, the magnitude of the magnetic field varies as the inverse square of the distance from the current element, as does the electric field due to a point charge. However, the directions of the two fields are quite different. The electric field created by a point charge is radial, but the magnetic field created by a current element is perpendicular to both the length element $d\mathbf{s}$ and the unit vector $\hat{\mathbf{r}}$, as described by the cross product in Equation 30.1. Hence, if the conductor lies in the plane of the page, as shown in Figure 30.1, $d\mathbf{B}$ points out of the page at P and into the page at P'.

Another difference between electric and magnetic fields is related to the source of the field. An electric field is established by an isolated electric charge. The Biot–Savart law gives the magnetic field of an isolated current element at some point, but such an isolated current element cannot exist the way an isolated electric charge can. A current element *must* be part of an extended current distribution because we must have a complete circuit in order for charges to flow. Thus, the Biot–Savart law is only the first step in a calculation of a magnetic field; it must be followed by an integration over the current distribution.

In the examples that follow, it is important to recognize that **the magnetic field determined in these calculations is the field created by a current-carrying conductor.** This field is not to be confused with any additional fields that may be present outside the conductor due to other sources, such as a bar magnet placed nearby.

EXAMPLE 30.1 ▶ Magnetic Field Surrounding a Thin, Straight Conductor

Consider a thin, straight wire carrying a constant current I and placed along the x axis as shown in Figure 30.2. Determine the magnitude and direction of the magnetic field at point P due to this current.

Solution From the Biot–Savart law, we expect that the magnitude of the field is proportional to the current in the wire and decreases as the distance a from the wire to point P increases. We start by considering a length element $d\mathbf{s}$ located a distance r from P. The direction of the magnetic field at point P due to the current in this element is out of the page because $d\mathbf{s} \times \hat{\mathbf{r}}$ is out of the page. In fact, since *all* of the current elements $I\,d\mathbf{s}$ lie in the plane of the page, they all produce a magnetic field directed out of the page at point P. Thus, we have the direction of the magnetic field at point P, and we need only find the magnitude.

Taking the origin at O and letting point P be along the positive y axis, with $\mathbf{k}$ being a unit vector pointing out of the page, we see that

$$d\mathbf{s} \times \hat{\mathbf{r}} = \mathbf{k}\,|\,d\mathbf{s} \times \hat{\mathbf{r}}\,| = \mathbf{k}(dx\,\sin\theta)$$

where, from Chapter 3, $|\,d\mathbf{s} \times \hat{\mathbf{r}}\,|$ represents the magnitude of $d\mathbf{s} \times \hat{\mathbf{r}}$. Because $\hat{\mathbf{r}}$ is a unit vector, the unit of the cross product is simply the unit of $d\mathbf{s}$, which is length. Substitution into Equation 30.1 gives

$$d\mathbf{B} = (dB)\mathbf{k} = \frac{\mu_0 I}{4\pi}\frac{dx\,\sin\theta}{r^2}\,\mathbf{k}$$

Because all current elements produce a magnetic field in the $\mathbf{k}$ direction, let us restrict our attention to the magnitude of the field due to one current element, which is

$$(1) \qquad dB = \frac{\mu_0 I}{4\pi}\frac{dx\,\sin\theta}{r^2}$$

To integrate this expression, we must relate the variables θ, x, and r. One approach is to express x and r in terms of θ. From the geometry in Figure 30.2a, we have

$$(2) \qquad r = \frac{a}{\sin\theta} = a\csc\theta$$

Because $\tan\theta = a/(-x)$ from the right triangle in Figure 30.2a (the negative sign is necessary because $d\mathbf{s}$ is located at a negative value of x), we have

$$x = -a\cot\theta$$

Taking the derivative of this expression gives

$$(3) \qquad dx = a\csc^2\theta\,d\theta$$

Substitution of Equations (2) and (3) into Equation (1) gives

$$(4) \qquad dB = \frac{\mu_0 I}{4\pi}\frac{a\csc^2\theta\,\sin\theta\,d\theta}{a^2\csc^2\theta} = \frac{\mu_0 I}{4\pi a}\sin\theta\,d\theta$$

an expression in which the only variable is θ. We can now obtain the magnitude of the magnetic field at point P by integrating Equation (4) over all elements, subtending angles ranging from θ_1 to θ_2 as defined in Figure 30.2b:

$$B = \frac{\mu_0 I}{4\pi a}\int_{\theta_1}^{\theta_2}\sin\theta\,d\theta = \frac{\mu_0 I}{4\pi a}(\cos\theta_1 - \cos\theta_2) \quad \textbf{(30.4)}$$

We can use this result to find the magnetic field of any straight current-carrying wire if we know the geometry and hence the angles θ_1 and θ_2. Consider the special case of an infinitely long, straight wire. If we let the wire in Figure 30.2b become infinitely long, we see that $\theta_1 = 0$ and $\theta_2 = \pi$ for length elements ranging between positions $x = -\infty$ and $x = +\infty$. Because $(\cos\theta_1 - \cos\theta_2) = (\cos 0 - \cos\pi) = 2$, Equation 30.4 becomes

$$B = \frac{\mu_0 I}{2\pi a} \qquad \textbf{(30.5)}$$

Equations 30.4 and 30.5 both show that the magnitude of

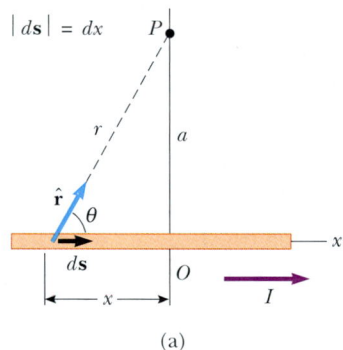

(a)

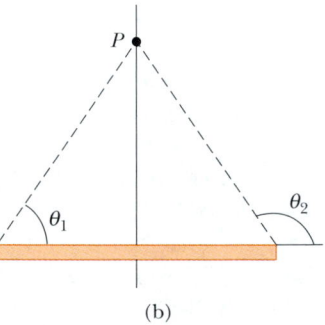

(b)

Figure 30.2 (a) A thin, straight wire carrying a current I. The magnetic field at point P due to the current in each element $d\mathbf{s}$ of the wire is out of the page, so the net field at point P is also out of the page. (b) The angles θ_1 and θ_2, used for determining the net field. When the wire is infinitely long, $\theta_1 = 0$ and $\theta_2 = 180°$.

the magnetic field is proportional to the current and decreases with increasing distance from the wire, as we expected. Notice that Equation 30.5 has the same mathematical form as the expression for the magnitude of the electric field due to a long charged wire (see Eq. 24.7).

Exercise   Calculate the magnitude of the magnetic field 4.0 cm from an infinitely long, straight wire carrying a current of 5.0 A.

Answer 2.5×10^{-5} T.

The result of Example 30.1 is important because a current in the form of a long, straight wire occurs often. Figure 30.3 is a three-dimensional view of the magnetic field surrounding a long, straight current-carrying wire. Because of the symmetry of the wire, the magnetic field lines are circles concentric with the wire and lie in planes perpendicular to the wire. The magnitude of **B** is constant on any circle of radius a and is given by Equation 30.5. A convenient rule for determining the direction of **B** is to grasp the wire with the right hand, positioning the thumb along the direction of the current. The four fingers wrap in the direction of the magnetic field.

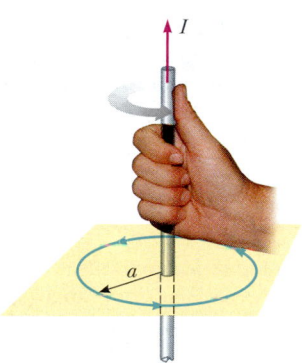

Figure 30.3   The right-hand rule for determining the direction of the magnetic field surrounding a long, straight wire carrying a current. Note that the magnetic field lines form circles around the wire.

EXAMPLE 30.2 ▶ Magnetic Field Due to a Curved Wire Segment

Calculate the magnetic field at point O for the current-carrying wire segment shown in Figure 30.4. The wire consists of two straight portions and a circular arc of radius R, which subtends an angle θ. The arrowheads on the wire indicate the direction of the current.

Solution   The magnetic field at O due to the current in the straight segments AA' and CC' is zero because $d\mathbf{s}$ is parallel to $\hat{\mathbf{r}}$ along these paths; this means that $d\mathbf{s} \times \hat{\mathbf{r}} = 0$. Each length element $d\mathbf{s}$ along path AC is at the same distance R from O, and the current in each contributes a field element $d\mathbf{B}$ directed into the page at O. Furthermore, at every point on AC, $d\mathbf{s}$ is perpendicular to $\hat{\mathbf{r}}$; hence, $|d\mathbf{s} \times \hat{\mathbf{r}}| = ds$. Using this information and Equation 30.1, we can find the magnitude of the field at O due to the current in an element of length ds:

$$dB = \frac{\mu_0 I}{4\pi} \frac{ds}{R^2}$$

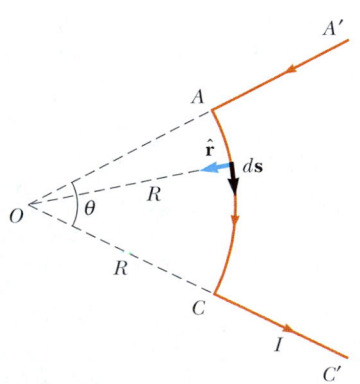

Figure 30.4   The magnetic field at O due to the current in the curved segment AC is into the page. The contribution to the field at O due to the current in the two straight segments is zero.

Because I and R are constants, we can easily integrate this expression over the curved path AC:

$$B = \frac{\mu_0 I}{4\pi R^2} \int ds = \frac{\mu_0 I}{4\pi R^2} s = \boxed{\frac{\mu_0 I}{4\pi R} \theta} \qquad \textbf{(30.6)}$$

where we have used the fact that $s = R\theta$ with θ measured in radians. The direction of $\mathbf{B}$ is into the page at O because $d\mathbf{s} \times \hat{\mathbf{r}}$ is into the page for every length element.

Exercise A circular wire loop of radius R carries a current I. What is the magnitude of the magnetic field at its center?

Answer $\mu_0 I / 2R$.

EXAMPLE 30.3 ▸ Magnetic Field on the Axis of a Circular Current Loop

Consider a circular wire loop of radius R located in the yz plane and carrying a steady current I, as shown in Figure 30.5. Calculate the magnetic field at an axial point P a distance x from the center of the loop.

Solution In this situation, note that every length element $d\mathbf{s}$ is perpendicular to the vector $\hat{\mathbf{r}}$ at the location of the element. Thus, for any element, $d\mathbf{s} \times \hat{\mathbf{r}} = (ds)(1) \sin 90° = ds$. Furthermore, all length elements around the loop are at the same distance r from P, where $r^2 = x^2 + R^2$. Hence, the magnitude of $d\mathbf{B}$ due to the current in any length element $d\mathbf{s}$ is

$$dB = \frac{\mu_0 I}{4\pi} \frac{|d\mathbf{s} \times \hat{\mathbf{r}}|}{r^2} = \frac{\mu_0 I}{4\pi} \frac{ds}{(x^2 + R^2)}$$

The direction of $d\mathbf{B}$ is perpendicular to the plane formed by $\hat{\mathbf{r}}$ and $d\mathbf{s}$, as shown in Figure 30.5. We can resolve this vector into a component dB_x along the x axis and a component dB_y perpendicular to the x axis. When the components dB_y are summed over all elements around the loop, the resultant component is zero. That is, by symmetry the current in any element on one side of the loop sets up a perpendicular component of $d\mathbf{B}$ that cancels the perpendicular component set up by the current through the element diametrically opposite it. Therefore, *the resultant field at* P *must be along the* x *axis* and we can find it by integrating the components $dB_x = dB \cos\theta$. That is, $\mathbf{B} = B_x \mathbf{i}$, where

$$B_x = \oint dB \cos\theta = \frac{\mu_0 I}{4\pi} \oint \frac{ds \cos\theta}{x^2 + R^2}$$

and we must take the integral over the entire loop. Because θ, x, and R are constants for all elements of the loop and because $\cos\theta = R/(x^2 + R^2)^{1/2}$, we obtain

$$B_x = \frac{\mu_0 I R}{4\pi(x^2 + R^2)^{3/2}} \oint ds = \boxed{\frac{\mu_0 I R^2}{2(x^2 + R^2)^{3/2}}} \qquad \textbf{(30.7)}$$

where we have used the fact that $\oint ds = 2\pi R$ (the circumference of the loop).

To find the magnetic field at the center of the loop, we set $x = 0$ in Equation 30.7. At this special point, therefore,

$$B = \frac{\mu_0 I}{2R} \qquad (\text{at } x = 0) \qquad \textbf{(30.8)}$$

which is consistent with the result of the exercise in Example 30.2.

It is also interesting to determine the behavior of the magnetic field far from the loop—that is, when x is much greater than R. In this case, we can neglect the term R^2 in the denominator of Equation 30.7 and obtain

$$B \approx \frac{\mu_0 I R^2}{2x^3} \qquad (\text{for } x \gg R) \qquad \textbf{(30.9)}$$

Because the magnitude of the magnetic moment μ of the loop is defined as the product of current and loop area (see Eq. 29.10)—$\mu = I(\pi R^2)$ for our circular loop—we can express Equation 30.9 as

$$B \approx \frac{\mu_0}{2\pi} \frac{\mu}{x^3} \qquad \textbf{(30.10)}$$

This result is similar in form to the expression for the electric field due to an electric dipole, $E = k_e(2qa/y^3)$ (see Example

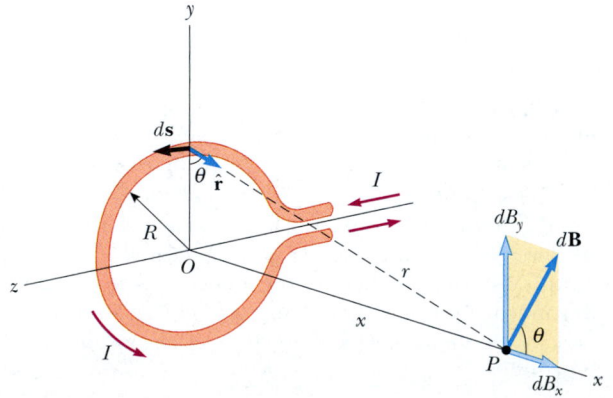

Figure 30.5 Geometry for calculating the magnetic field at a point P lying on the axis of a current loop. By symmetry, the total field $\mathbf{B}$ is along this axis.

23.6), where $2qa = p$ is the electric dipole moment as defined in Equation 26.16.

The pattern of the magnetic field lines for a circular current loop is shown in Figure 30.6a. For clarity, the lines are drawn for only one plane—one that contains the axis of the loop. Note that the field-line pattern is axially symmetric and looks like the pattern around a bar magnet, shown in Figure 30.6c.

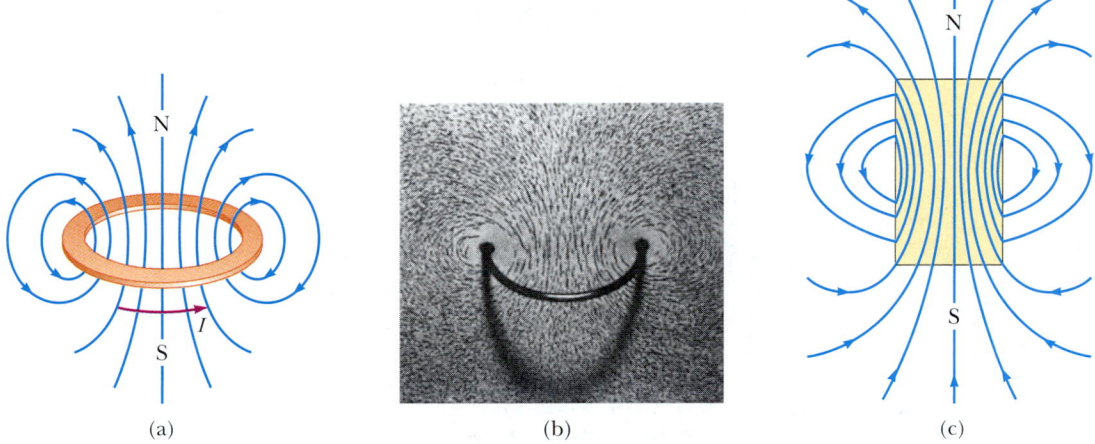

Figure 30.6   (a) Magnetic field lines surrounding a current loop. (b) Magnetic field lines surrounding a current loop, displayed with iron filings *(Education Development Center, Newton, MA).*   (c) Magnetic field lines surrounding a bar magnet. Note the similarity between this line pattern and that of a current loop.

30.2 ▶ THE MAGNETIC FORCE BETWEEN TWO PARALLEL CONDUCTORS

In Chapter 29 we described the magnetic force that acts on a current-carrying conductor placed in an external magnetic field. Because a current in a conductor sets up its own magnetic field, it is easy to understand that two current-carrying conductors exert magnetic forces on each other. As we shall see, such forces can be used as the basis for defining the ampere and the coulomb.

Consider two long, straight, parallel wires separated by a distance a and carrying currents I_1 and I_2 in the same direction, as illustrated in Figure 30.7. We can determine the force exerted on one wire due to the magnetic field set up by the other wire. Wire 2, which carries a current I_2, creates a magnetic field $\mathbf{B}_2$ at the location of wire 1. The direction of $\mathbf{B}_2$ is perpendicular to wire 1, as shown in Figure 30.7. According to Equation 29.3, the magnetic force on a length ℓ of wire 1 is $\mathbf{F}_1 = I_1\boldsymbol{\ell} \times \mathbf{B}_2$. Because $\boldsymbol{\ell}$ is perpendicular to $\mathbf{B}_2$ in this situation, the magnitude of $\mathbf{F}_1$ is $F_1 = I_1\ell B_2$. Because the magnitude of $\mathbf{B}_2$ is given by Equation 30.5, we see that

$$F_1 = I_1\ell B_2 = I_1\ell\left(\frac{\mu_0 I_2}{2\pi a}\right) = \frac{\mu_0 I_1 I_2}{2\pi a}\,\ell \tag{30.11}$$

The direction of $\mathbf{F}_1$ is toward wire 2 because $\boldsymbol{\ell} \times \mathbf{B}_2$ is in that direction. If the field set up at wire 2 by wire 1 is calculated, the force $\mathbf{F}_2$ acting on wire 2 is found to be equal in magnitude and opposite in direction to $\mathbf{F}_1$. This is what we expect be-

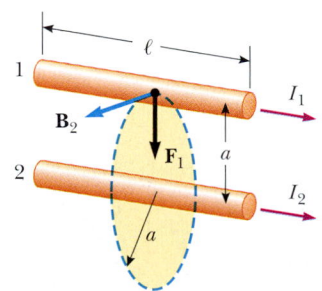

Figure 30.7   Two parallel wires that each carry a steady current exert a force on each other. The field $\mathbf{B}_2$ due to the current in wire 2 exerts a force of magnitude $F_1 = I_1\ell B_2$ on wire 1. The force is attractive if the currents are parallel (as shown) and repulsive if the currents are antiparallel.

cause Newton's third law must be obeyed.[1] When the currents are in opposite directions (that is, when one of the currents is reversed in Fig. 30.7), the forces are reversed and the wires repel each other. Hence, we find that **parallel conductors carrying currents in the same direction attract each other, and parallel conductors carrying currents in opposite directions repel each other.**

Because the magnitudes of the forces are the same on both wires, we denote the magnitude of the magnetic force between the wires as simply F_B. We can rewrite this magnitude in terms of the force per unit length:

$$\frac{F_B}{\ell} = \frac{\mu_0 I_1 I_2}{2\pi a}$$

(30.12)

The force between two parallel wires is used to define the **ampere** as follows:

Definition of the ampere

> When the magnitude of the force per unit length between two long, parallel wires that carry identical currents and are separated by 1 m is 2×10^{-7} N/m, the current in each wire is defined to be 1 A.

The value 2×10^{-7} N/m is obtained from Equation 30.12 with $I_1 = I_2 = 1$ A and $a = 1$ m. Because this definition is based on a force, a mechanical measurement can be used to standardize the ampere. For instance, the National Institute of Standards and Technology uses an instrument called a *current balance* for primary current measurements. The results are then used to standardize other, more conventional instruments, such as ammeters.

The SI unit of charge, the **coulomb,** is defined in terms of the ampere:

Definition of the coulomb

> When a conductor carries a steady current of 1 A, the quantity of charge that flows through a cross-section of the conductor in 1 s is 1 C.

In deriving Equations 30.11 and 30.12, we assumed that both wires are long compared with their separation distance. In fact, only one wire needs to be long. The equations accurately describe the forces exerted on each other by a long wire and a straight parallel wire of limited length ℓ.

Quick Quiz 30.1

For $I_1 = 2$ A and $I_2 = 6$ A in Figure 30.7, which is true: (a) $F_1 = 3F_2$, (b) $F_1 = F_2/3$, or (c) $F_1 = F_2$?

Quick Quiz 30.2

A loose spiral spring is hung from the ceiling, and a large current is sent through it. Do the coils move closer together or farther apart?

---

[1] Although the total force exerted on wire 1 is equal in magnitude and opposite in direction to the total force exerted on wire 2, Newton's third law does not apply when one considers two small elements of the wires that are not exactly opposite each other. This apparent violation of Newton's third law and of the law of conservation of momentum is described in more advanced treatments on electricity and magnetism.

30.3 ▶ AMPÈRE'S LAW

Oersted's 1819 discovery about deflected compass needles demonstrates that a current-carrying conductor produces a magnetic field. Figure 30.8a shows how this effect can be demonstrated in the classroom. Several compass needles are placed in a horizontal plane near a long vertical wire. When no current is present in the wire, all the needles point in the same direction (that of the Earth's magnetic field), as expected. When the wire carries a strong, steady current, the needles all deflect in a direction tangent to the circle, as shown in Figure 30.8b. These observations demonstrate that the direction of the magnetic field produced by the current in the wire is consistent with the right-hand rule described in Figure 30.3. When the current is reversed, the needles in Figure 30.8b also reverse.

Because the compass needles point in the direction of **B**, we conclude that the lines of **B** form circles around the wire, as discussed in the preceding section. By symmetry, the magnitude of **B** is the same everywhere on a circular path centered on the wire and lying in a plane perpendicular to the wire. By varying the current and distance a from the wire, we find that B is proportional to the current and inversely proportional to the distance from the wire, as Equation 30.5 describes.

Now let us evaluate the product $\mathbf{B} \cdot d\mathbf{s}$ for a small length element $d\mathbf{s}$ on the circular path defined by the compass needles, and sum the products for all elements over the closed circular path. Along this path, the vectors $d\mathbf{s}$ and **B** are parallel at each point (see Fig. 30.8b), so $\mathbf{B} \cdot d\mathbf{s} = B\,ds$. Furthermore, the magnitude of **B** is constant on this circle and is given by Equation 30.5. Therefore, the sum of the products $B\,ds$ over the closed path, which is equivalent to the line integral of $\mathbf{B} \cdot d\mathbf{s}$, is

$$\oint \mathbf{B} \cdot d\mathbf{s} = B \oint ds = \frac{\mu_0 I}{2\pi r}\,(2\pi r) = \mu_0 I$$

where $\oint ds = 2\pi r$ is the circumference of the circular path. Although this result was calculated for the special case of a circular path surrounding a wire, it holds

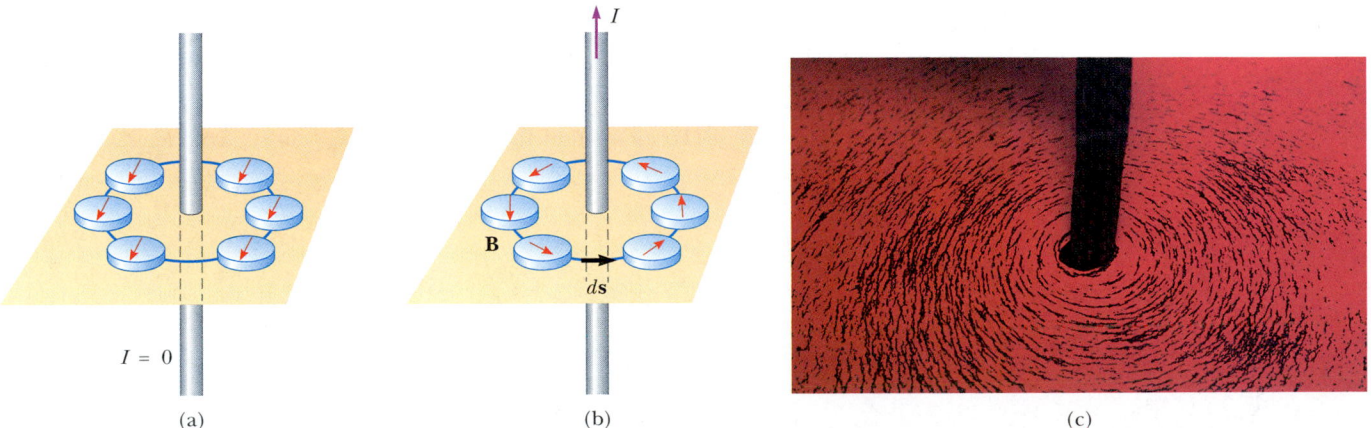

Figure 30.8 (a) When no current is present in the wire, all compass needles point in the same direction (toward the Earth's north pole). (b) When the wire carries a strong current, the compass needles deflect in a direction tangent to the circle, which is the direction of the magnetic field created by the current. (c) Circular magnetic field lines surrounding a current-carrying conductor, displayed with iron filings. *(Henry Leap and Jim Lehman)*

for a closed path of *any* shape surrounding a *current* that remains constant in time. The general case, known as **Ampère's law,** can be stated as follows:

> The line integral of **B** · *d***s** around any closed path equals $\mu_0 I$, where I is the total continuous current passing through any surface bounded by the closed path.
>
> $$\oint \mathbf{B} \cdot d\mathbf{s} = \mu_0 I \qquad \textbf{(30.13)}$$

Ampère's law

Ampère's law describes the creation of magnetic fields by all constant current configurations, but at our mathematical level it is useful only for calculating the magnetic field of current configurations having a high degree of symmetry. Its use is similar to that of Gauss's law in calculating electric fields for highly symmetric charge distributions.

Quick Quiz 30.3

Rank the magnitudes of $\oint \mathbf{B} \cdot d\mathbf{s}$ for the closed paths in Figure 30.9, from least to greatest.

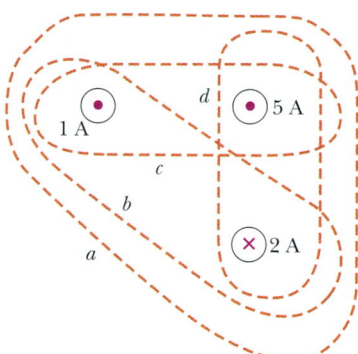

Figure 30.9   Four closed paths around three current-carrying wires.

Quick Quiz 30.4

Rank the magnitudes of $\oint \mathbf{B} \cdot d\mathbf{s}$ for the closed paths in Figure 30.10, from least to greatest.

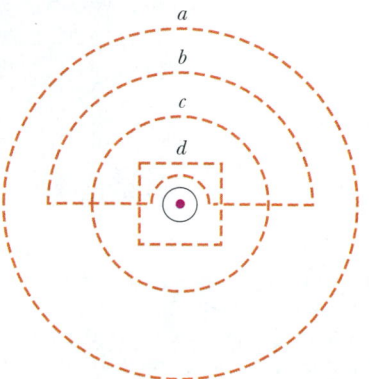

Figure 30.10   Several closed paths near a single current-carrying wire.

EXAMPLE 30.4 ▶ The Magnetic Field Created by a Long Current-Carrying Wire

A long, straight wire of radius R carries a steady current I_0 that is uniformly distributed through the cross-section of the wire (Fig. 30.11). Calculate the magnetic field a distance r from the center of the wire in the regions $r \geq R$ and $r < R$.

Solution For the $r \geq R$ case, we should get the same result we obtained in Example 30.1, in which we applied the Biot–Savart law to the same situation. Let us choose for our path of integration circle 1 in Figure 30.11. From symmetry, **B** must be constant in magnitude and parallel to $d\mathbf{s}$ at every point on this circle. Because the total current passing through the plane of the circle is I_0, Ampère's law gives

$$\oint \mathbf{B} \cdot d\mathbf{s} = B \oint ds = B(2\pi r) = \mu_0 I_0$$

$$B = \frac{\mu_0 I_0}{2\pi r} \qquad \text{(for } r \geq R\text{)} \qquad \textbf{(30.14)}$$

which is identical in form to Equation 30.5. Note how much easier it is to use Ampère's law than to use the Biot–Savart law. This is often the case in highly symmetric situations.

Now consider the interior of the wire, where $r < R$. Here the current I passing through the plane of circle 2 is less than the total current I_0. Because the current is uniform over the cross-section of the wire, the fraction of the current enclosed

by circle 2 must equal the ratio of the area πr^2 enclosed by circle 2 to the cross-sectional area πR^2 of the wire:[2]

$$\frac{I}{I_0} = \frac{\pi r^2}{\pi R^2}$$

$$I = \frac{r^2}{R^2} I_0$$

Following the same procedure as for circle 1, we apply Ampère's law to circle 2:

$$\oint \mathbf{B} \cdot d\mathbf{s} = B(2\pi r) = \mu_0 I = \mu_0 \left(\frac{r^2}{R^2} I_0 \right)$$

$$B = \left(\frac{\mu_0 I_0}{2\pi R^2} \right) r \qquad \text{(for } r < R\text{)} \qquad \textbf{(30.15)}$$

This result is similar in form to the expression for the electric field inside a uniformly charged sphere (see Example 24.5). The magnitude of the magnetic field versus r for this configuration is plotted in Figure 30.12. Note that inside the wire, $B \rightarrow 0$ as $r \rightarrow 0$. Note also that Equations 30.14 and 30.15 give the same value of the magnetic field at $r = R$, demonstrating that the magnetic field is continuous at the surface of the wire.

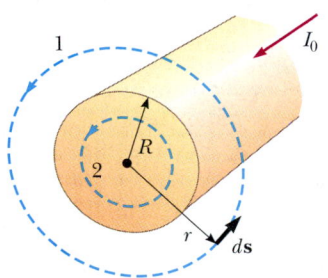

Figure 30.11 A long, straight wire of radius R carrying a steady current I_0 uniformly distributed across the cross-section of the wire. The magnetic field at any point can be calculated from Ampère's law using a circular path of radius r, concentric with the wire.

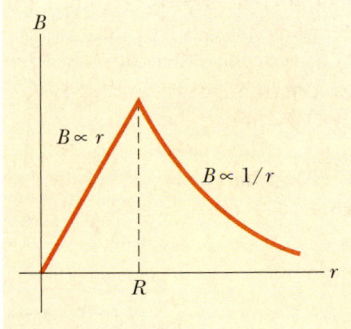

Figure 30.12 Magnitude of the magnetic field versus r for the wire shown in Figure 30.11. The field is proportional to r inside the wire and varies as $1/r$ outside the wire.

EXAMPLE 30.5 ▶ The Magnetic Field Created by a Toroid

A device called a *toroid* (Fig. 30.13) is often used to create an almost uniform magnetic field in some enclosed area. The device consists of a conducting wire wrapped around a ring (a *torus*) made of a nonconducting material. For a toroid hav-

ing N closely spaced turns of wire, calculate the magnetic field in the region occupied by the torus, a distance r from the center.

[2] Another way to look at this problem is to see that the current enclosed by circle 2 must equal the product of the current density $J = I_0/\pi R^2$ and the area πr^2 of this circle.

Solution   To calculate this field, we must evaluate $\oint \mathbf{B} \cdot d\mathbf{s}$ over the circle of radius r in Figure 30.13. By symmetry, we see that the magnitude of the field is constant on this circle and tangent to it, so $\mathbf{B} \cdot d\mathbf{s} = B\, ds$. Furthermore, note that

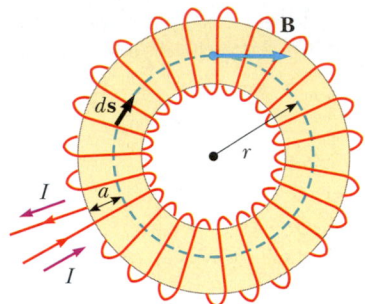

Figure 30.13   A toroid consisting of many turns of wire. If the turns are closely spaced, the magnetic field in the interior of the torus (the gold-shaded region) is tangent to the dashed circle and varies as $1/r$. The field outside the toroid is zero. The dimension a is the cross-sectional radius of the torus.

the circular closed path surrounds N loops of wire, each of which carries a current I. Therefore, the right side of Equation 30.13 is $\mu_0 NI$ in this case.

Ampère's law applied to the circle gives

$$\oint \mathbf{B} \cdot d\mathbf{s} = B \oint ds = B(2\pi r) = \mu_0 NI$$

$$B = \frac{\mu_0 NI}{2\pi r} \qquad (30.16)$$

This result shows that B varies as $1/r$ and hence is nonuniform in the region occupied by the torus. However, if r is very large compared with the cross-sectional radius of the torus, then the field is approximately uniform inside the torus.

For an ideal toroid, in which the turns are closely spaced, the external magnetic field is zero. This can be seen by noting that the net current passing through any circular path lying outside the toroid (including the region of the "hole in the doughnut") is zero. Therefore, from Ampère's law we find that $B = 0$ in the regions exterior to the torus.

EXAMPLE 30.6 ▶ Magnetic Field Created by an Infinite Current Sheet

So far we have imagined currents through wires of small cross-section. Let us now consider an example in which a current exists in an extended object. A thin, infinitely large sheet lying in the yz plane carries a current of linear current density $\mathbf{J}_s$. The current is in the y direction, and J_s represents the current per unit length measured along the z axis. Find the magnetic field near the sheet.

Solution   This situation brings to mind similar calculations involving Gauss's law (see Example 24.8). You may recall that

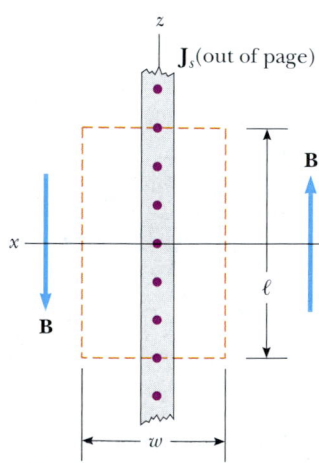

Figure 30.14   End view of an infinite current sheet lying in the yz plane, where the current is in the y direction (out of the page). This view shows the direction of $\mathbf{B}$ on both sides of the sheet.

the electric field due to an infinite sheet of charge does not depend on distance from the sheet. Thus, we might expect a similar result here for the magnetic field.

To evaluate the line integral in Ampère's law, let us take a rectangular path through the sheet, as shown in Figure 30.14. The rectangle has dimensions ℓ and w, with the sides of length ℓ parallel to the sheet surface. The net current passing through the plane of the rectangle is $J_s\ell$. We apply Ampère's law over the rectangle and note that the two sides of length w do not contribute to the line integral because the component of $\mathbf{B}$ along the direction of these paths is zero. By symmetry, we can argue that the magnetic field is constant over the sides of length ℓ because every point on the infinitely large sheet is equivalent, and hence the field should not vary from point to point. The only choices of field direction that are reasonable for the symmetry are perpendicular or parallel to the sheet, and a perpendicular field would pass *through* the current, which is inconsistent with the Biot–Savart law. Assuming a field that is constant in magnitude and parallel to the plane of the sheet, we obtain

$$\oint \mathbf{B} \cdot d\mathbf{s} = \mu_0 I = \mu_0 J_s \ell$$

$$2B\ell = \mu_0 J_s \ell$$

$$B = \mu_0 \frac{J_s}{2}$$

This result shows that *the magnetic field is independent of distance from the current sheet,* as we suspected.

EXAMPLE 30.7 ▶ **The Magnetic Force on a Current Segment**

Wire 1 in Figure 30.15 is oriented along the y axis and carries a steady current I_1. A rectangular loop located to the right of the wire and in the xy plane carries a current I_2. Find the magnetic force exerted by wire 1 on the top wire of length b in the loop, labeled "Wire 2" in the figure.

Solution You may be tempted to use Equation 30.12 to obtain the force exerted on a small segment of length dx of wire 2. However, this equation applies only to two *parallel* wires and cannot be used here. The correct approach is to

consider the force exerted by wire 1 on a small segment $d\mathbf{s}$ of wire 2 by using Equation 29.4. This force is given by $d\mathbf{F}_B = I\,d\mathbf{s} \times \mathbf{B}$, where $I = I_2$ and $\mathbf{B}$ is the magnetic field created by the current in wire 1 at the position of $d\mathbf{s}$. From Ampère's law, the field at a distance x from wire 1 (see Eq. 30.14) is

$$\mathbf{B} = \frac{\mu_0 I_1}{2\pi x}(-\mathbf{k})$$

where the unit vector $-\mathbf{k}$ is used to indicate that the field at $d\mathbf{s}$ points into the page. Because wire 2 is along the x axis, $d\mathbf{s} = dx\,\mathbf{i}$, and we find that

$$d\mathbf{F}_B = \frac{\mu_0 I_1 I_2}{2\pi x}[\mathbf{i} \times (-\mathbf{k})]\,dx = \frac{\mu_0 I_1 I_2}{2\pi}\frac{dx}{x}\mathbf{j}$$

Integrating over the limits $x = a$ to $x = a + b$ gives

$$\mathbf{F}_B = \frac{\mu_0 I_1 I_2}{2\pi}\ln x\Big]_a^{a+b}\mathbf{j} = \frac{\mu_0 I_1 I_2}{2\pi}\ln\left(1 + \frac{b}{a}\right)\mathbf{j}$$

The force points in the positive y direction, as indicated by the unit vector $\mathbf{j}$ and as shown in Figure 30.15.

Exercise What are the magnitude and direction of the force exerted on the bottom wire of length b?

Answer The force has the same magnitude as the force on wire 2 but is directed downward.

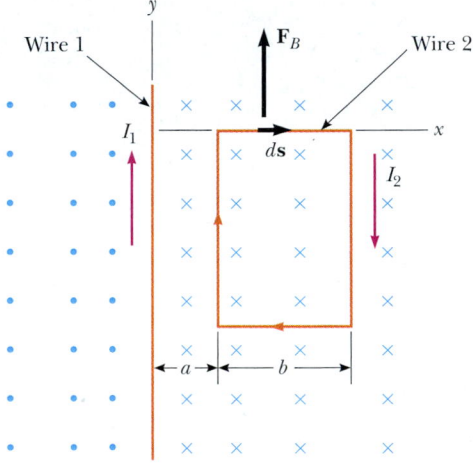

Figure 30.15

Quick Quiz 30.5 ▶

Is a net force acting on the current loop in Example 30.7? A net torque?

30.4 ▶ **THE MAGNETIC FIELD OF A SOLENOID**

A **solenoid** is a long wire wound in the form of a helix. With this configuration, a reasonably uniform magnetic field can be produced in the space surrounded by the turns of wire—which we shall call the *interior* of the solenoid—when the solenoid carries a current. When the turns are closely spaced, each can be approximated as a circular loop, and the net magnetic field is the vector sum of the fields resulting from all the turns.

Figure 30.16 shows the magnetic field lines surrounding a loosely wound solenoid. Note that the field lines in the interior are nearly parallel to one another, are uniformly distributed, and are close together, indicating that the field in this space is uniform and strong. The field lines between current elements on two adjacent turns tend to cancel each other because the field vectors from the two elements are in opposite directions. The field at exterior points such as P is weak because the field due to current elements on the right-hand portion of a turn tends to cancel the field due to current elements on the left-hand portion.

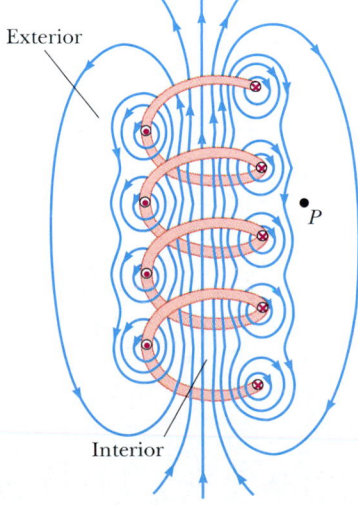

Figure 30.16 The magnetic field lines for a loosely wound solenoid.

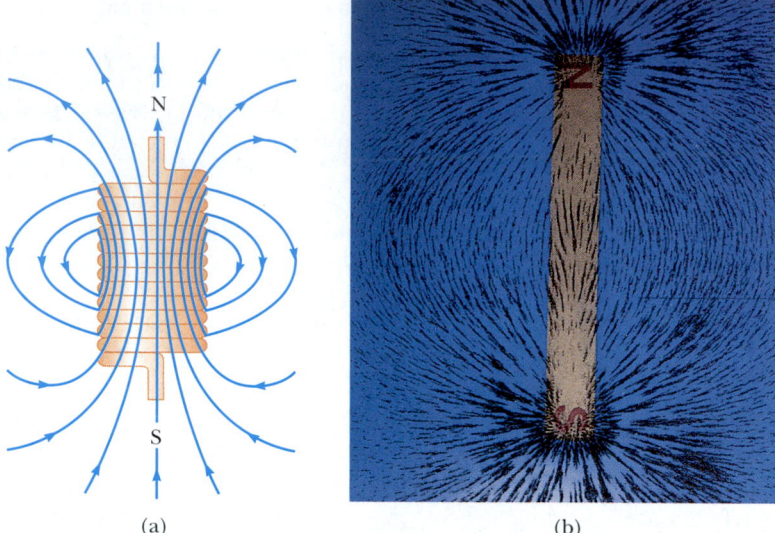

(a)   (b)

Figure 30.17   (a) Magnetic field lines for a tightly wound solenoid of finite length, carrying a steady current. The field in the interior space is nearly uniform and strong. Note that the field lines resemble those of a bar magnet, meaning that the solenoid effectively has north and south poles. (b) The magnetic field pattern of a bar magnet, displayed with small iron filings on a sheet of paper. *(Henry Leap and Jim Lehman)*

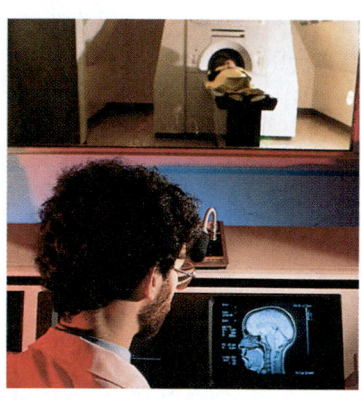

A technician studies the scan of a patient's head. The scan was obtained using a medical diagnostic technique known as magnetic resonance imaging (MRI). This instrument makes use of strong magnetic fields produced by superconducting solenoids. *(Hank Morgan/Science Source)*

If the turns are closely spaced and the solenoid is of finite length, the magnetic field lines are as shown in Figure 30.17a. This field line distribution is similar to that surrounding a bar magnet (see Fig. 30.17b). Hence, one end of the solenoid behaves like the north pole of a magnet, and the opposite end behaves like the south pole. As the length of the solenoid increases, the interior field becomes more uniform and the exterior field becomes weaker. An *ideal solenoid* is approached when the turns are closely spaced and the length is much greater than the radius of the turns. In this case, the external field is zero, and the interior field is uniform over a great volume.

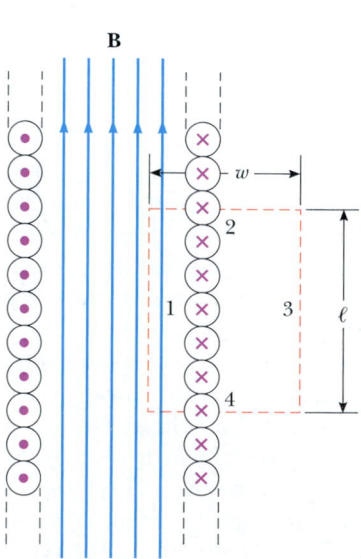

Figure 30.18   Cross-sectional view of an ideal solenoid, where the interior magnetic field is uniform and the exterior field is zero. Ampère's law applied to the red dashed path can be used to calculate the magnitude of the interior field.

We can use Ampère's law to obtain an expression for the interior magnetic field in an ideal solenoid. Figure 30.18 shows a longitudinal cross-section of part of such a solenoid carrying a current I. Because the solenoid is ideal, $\mathbf{B}$ in the interior space is uniform and parallel to the axis, and $\mathbf{B}$ in the exterior space is zero. Consider the rectangular path of length ℓ and width w shown in Figure 30.18. We can apply Ampère's law to this path by evaluating the integral of $\mathbf{B} \cdot d\mathbf{s}$ over each side of the rectangle. The contribution along side 3 is zero because $B = 0$ in this region. The contributions from sides 2 and 4 are both zero because $\mathbf{B}$ is perpendicular to $d\mathbf{s}$ along these paths. Side 1 gives a contribution $B\ell$ to the integral because along this path $\mathbf{B}$ is uniform and parallel to $d\mathbf{s}$. The integral over the closed rectangular path is therefore

$$\oint \mathbf{B} \cdot d\mathbf{s} = \int_{\text{path 1}} \mathbf{B} \cdot d\mathbf{s} = B \int_{\text{path 1}} ds = B\ell$$

QuickLab

Wrap a few turns of wire around a compass, essentially putting the compass inside a solenoid. Hold the ends of the wire to the two terminals of a flashlight battery. What happens to the compass? Is the effect as strong when the compass is outside the turns of wire?

The right side of Ampère's law involves the total current passing through the area bounded by the path of integration. In this case, the total current through the rectangular path equals the current through each turn multiplied by the number of turns. If N is the number of turns in the length ℓ, the total current through the rectangle is NI. Therefore, Ampère's law applied to this path gives

$$\oint \mathbf{B} \cdot d\mathbf{s} = B\ell = \mu_0 NI$$

$$B = \mu_0 \frac{N}{\ell} I = \mu_0 n I \qquad \textbf{(30.17)}$$

Magnetic field inside a solenoid

where $n = N/\ell$ is the number of turns per unit length.

We also could obtain this result by reconsidering the magnetic field of a toroid (see Example 30.5). If the radius r of the torus in Figure 30.13 containing N turns is much greater than the toroid's cross-sectional radius a, a short section of the toroid approximates a solenoid for which $n = N/2\pi r$. In this limit, Equation 30.16 agrees with Equation 30.17.

Equation 30.17 is valid only for points near the center (that is, far from the ends) of a very long solenoid. As you might expect, the field near each end is smaller than the value given by Equation 30.17. At the very end of a long solenoid, the magnitude of the field is one-half the magnitude at the center.

web

For a more detailed discussion of the magnetic field along the axis of a solenoid, visit **www.saunderscollege.com/physics/**

30.5 MAGNETIC FLUX

The flux associated with a magnetic field is defined in a manner similar to that used to define electric flux (see Eq. 24.3). Consider an element of area dA on an arbitrarily shaped surface, as shown in Figure 30.19. If the magnetic field at this element is $\mathbf{B}$, the magnetic flux through the element is $\mathbf{B} \cdot d\mathbf{A}$, where $d\mathbf{A}$ is a vector that is perpendicular to the surface and has a magnitude equal to the area dA. Hence, the total magnetic flux Φ_B through the surface is

$$\Phi_B \equiv \int \mathbf{B} \cdot d\mathbf{A} \qquad \textbf{(30.18)}$$

Definition of magnetic flux

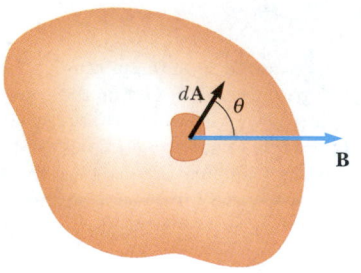

Figure 30.19 The magnetic flux through an area element dA is $\mathbf{B} \cdot d\mathbf{A} = BdA \cos \theta$, where $d\mathbf{A}$ is a vector perpendicular to the surface.

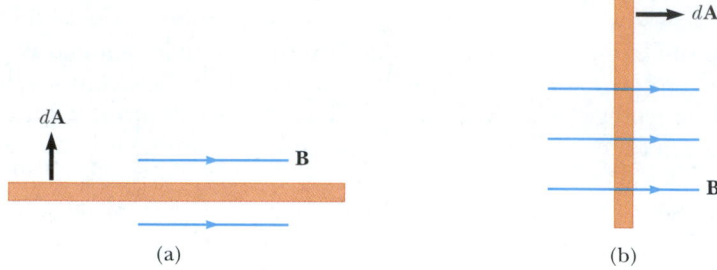

(a)   (b)

Figure 30.20 Magnetic flux through a plane lying in a magnetic field. (a) The flux through the plane is zero when the magnetic field is parallel to the plane surface. (b) The flux through the plane is a maximum when the magnetic field is perpendicular to the plane.

Consider the special case of a plane of area A in a uniform field $\mathbf{B}$ that makes an angle θ with $d\mathbf{A}$. The magnetic flux through the plane in this case is

$$\Phi_B = BA \cos \theta \tag{30.19}$$

If the magnetic field is parallel to the plane, as in Figure 30.20a, then $\theta = 90°$ and the flux is zero. If the field is perpendicular to the plane, as in Figure 30.20b, then $\theta = 0$ and the flux is BA (the maximum value).

The unit of flux is the $\text{T} \cdot \text{m}^2$, which is defined as a *weber* (Wb); 1 Wb = 1 $\text{T} \cdot \text{m}^2$.

EXAMPLE 30.8 ▶ Magnetic Flux Through a Rectangular Loop

A rectangular loop of width a and length b is located near a long wire carrying a current I (Fig. 30.21). The distance between the wire and the closest side of the loop is c. The wire is parallel to the long side of the loop. Find the total magnetic flux through the loop due to the current in the wire.

Solution From Equation 30.14, we know that the magnitude of the magnetic field created by the wire at a distance r from the wire is

$$B = \frac{\mu_0 I}{2\pi r}$$

The factor $1/r$ indicates that the field varies over the loop, and Figure 30.21 shows that the field is directed into the page. Because $\mathbf{B}$ is parallel to $d\mathbf{A}$ at any point within the loop, the magnetic flux through an area element dA is

$$\Phi_B = \int B \, dA = \int \frac{\mu_0 I}{2\pi r} \, dA$$

(Because B is not uniform but depends on r, it cannot be removed from the integral.)

To integrate, we first express the area element (the tan region in Fig. 30.21) as $dA = b \, dr$. Because r is now the only variable in the integral, we have

$$\Phi_B = \frac{\mu_0 I b}{2\pi} \int_c^{a+c} \frac{dr}{r} = \frac{\mu_0 I b}{2\pi} \ln r \Big]_c^{a+c}$$

$$= \frac{\mu_0 I b}{2\pi} \ln\left(\frac{a+c}{c}\right) = \frac{\mu_0 I b}{2\pi} \ln\left(1 + \frac{a}{c}\right)$$

Exercise Apply the series expansion formula for $\ln(1 + x)$ (see Appendix B.5) to this equation to show that it gives a reasonable result when the loop is far from the wire relative to the loop dimensions (in other words, when $c \gg a$).

Answer $\Phi_B \to 0$.

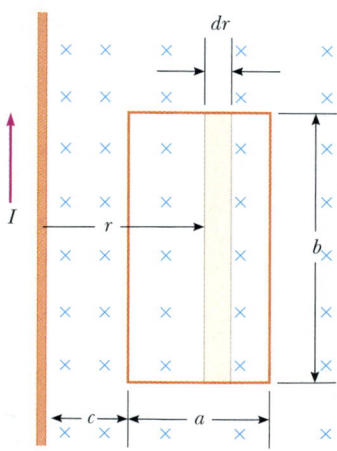

Figure 30.21 The magnetic field due to the wire carrying a current I is not uniform over the rectangular loop.

30.6 ▷ GAUSS'S LAW IN MAGNETISM

In Chapter 24 we found that the electric flux through a closed surface surrounding a net charge is proportional to that charge (Gauss's law). In other words, the number of electric field lines leaving the surface depends only on the net charge within it. This property is based on the fact that electric field lines originate and terminate on electric charges.

The situation is quite different for magnetic fields, which are continuous and form closed loops. In other words, magnetic field lines do not begin or end at any point—as illustrated by the magnetic field lines of the bar magnet in Figure 30.22. Note that for any closed surface, such as the one outlined by the dashed red line in Figure 30.22, the number of lines entering the surface equals the number leaving the surface; thus, the net magnetic flux is zero. In contrast, for a closed surface surrounding one charge of an electric dipole (Fig. 30.23), the net electric flux is not zero.

Gauss's law in magnetism states that

> the net magnetic flux through any closed surface is always zero:
>
> $$\oint \mathbf{B} \cdot d\mathbf{A} = 0 \qquad\qquad \textbf{(30.20)}$$

Gauss's law for magnetism

This statement is based on the experimental fact, mentioned in the opening of Chapter 29, that **isolated magnetic poles (monopoles) have never been detected and perhaps do not exist.** Nonetheless, scientists continue the search be-

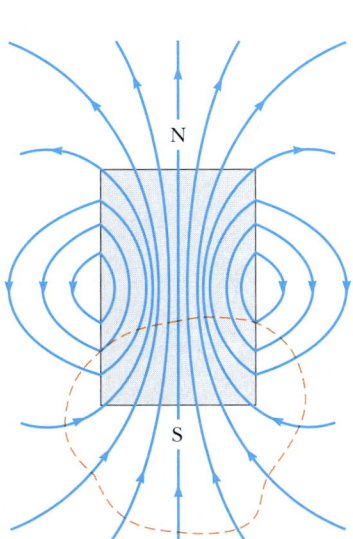

Figure 30.22 The magnetic field lines of a bar magnet form closed loops. Note that the net magnetic flux through the closed surface (dashed red line) surrounding one of the poles (or any other closed surface) is zero.

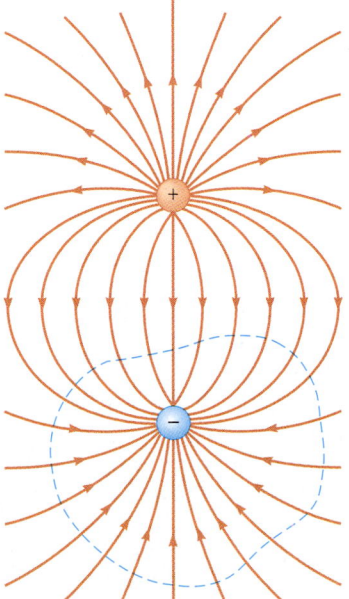

Figure 30.23 The electric field lines surrounding an electric dipole begin on the positive charge and terminate on the negative charge. The electric flux through a closed surface surrounding one of the charges is not zero.

cause certain theories that are otherwise successful in explaining fundamental physical behavior suggest the possible existence of monopoles.

30.7 ▸ DISPLACEMENT CURRENT AND THE GENERAL FORM OF AMPÈRE'S LAW

We have seen that charges in motion produce magnetic fields. When a current-carrying conductor has high symmetry, we can use Ampère's law to calculate the magnetic field it creates. In Equation 30.13, $\oint \mathbf{B} \cdot d\mathbf{s} = \mu_0 I$, the line integral is over any closed path through which the conduction current passes, and the conduction current is defined by the expression $I = dq/dt$. (In this section we use the term *conduction current* to refer to the current carried by the wire, to distinguish it from a new type of current that we shall introduce shortly.) We now show that **Ampère's law in this form is valid only if the electric field is constant in time.** Maxwell recognized this limitation and modified Ampère's law to include time-varying electric fields.

We can understand the problem by considering a capacitor that is being charged as illustrated in Figure 30.24. When a conduction current is present, the charge on the positive plate changes but *no conduction current passes across the gap between the plates*. Now consider the two surfaces S_1 and S_2 in Figure 30.24, bounded by the same path P. Ampère's law states that $\oint \mathbf{B} \cdot d\mathbf{s}$ around this path must equal $\mu_0 I$, where I is the total current through any surface bounded by the path P.

When the path P is considered as bounding S_1, $\oint \mathbf{B} \cdot d\mathbf{s}$ is $\mu_0 I$ because the conduction current passes through S_1. When the path is considered as bounding S_2, however, $\oint \mathbf{B} \cdot d\mathbf{s} = 0$ because no conduction current passes through S_2. Thus, we arrive at a contradictory situation that arises from the discontinuity of the current! Maxwell solved this problem by postulating an additional term on the right side of Equation 30.13, which includes a factor called the **displacement current** I_d, defined as[3]

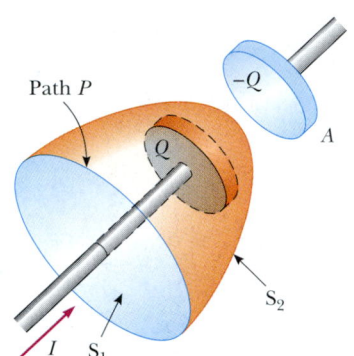

Figure 30.24   Two surfaces S_1 and S_2 near the plate of a capacitor are bounded by the same path P. The conduction current in the wire passes only through S_1. This leads to a contradiction in Ampère's law that is resolved only if one postulates a displacement current through S_2.

| Displacement current |

$$I_d \equiv \epsilon_0 \frac{d\Phi_E}{dt} \qquad \text{(30.21)}$$

where ϵ_0 is the permittivity of free space (see Section 23.3) and $\Phi_E = \int \mathbf{E} \cdot d\mathbf{A}$ is the electric flux (see Eq. 24.3).

As the capacitor is being charged (or discharged), the changing electric field between the plates may be considered equivalent to a current that acts as a continuation of the conduction current in the wire. When the expression for the displacement current given by Equation 30.21 is added to the conduction current on the right side of Ampère's law, the difficulty represented in Figure 30.24 is resolved. No matter which surface bounded by the path P is chosen, either conduction current or displacement current passes through it. With this new term I_d, we can express the general form of Ampère's law (sometimes called the **Ampère–Maxwell law**) as[4]

| Ampère–Maxwell law |

$$\oint \mathbf{B} \cdot d\mathbf{s} = \mu_0 (I + I_d) = \mu_0 I + \mu_0 \epsilon_0 \frac{d\Phi_E}{dt} \qquad \text{(30.22)}$$

---

[3] *Displacement* in this context does not have the meaning it does in Chapter 2. Despite the inaccurate implications, the word is historically entrenched in the language of physics, so we continue to use it.

[4] Strictly speaking, this expression is valid only in a vacuum. If a magnetic material is present, one must change μ_0 and ϵ_0 on the right-hand side of Equation 30.22 to the permeability μ_m and permittivity ϵ characteristic of the material. Alternatively, one may include a magnetizing current I_m on the righthand side of Equation 30.22 to make Ampère's law fully general. On a microscopic scale, I_m is as real as I.

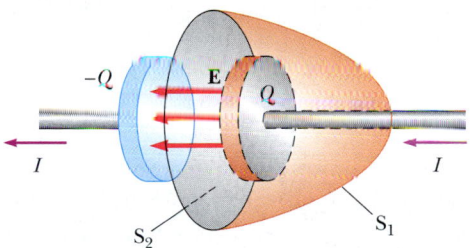

Figure 30.25 Because it exists only in the wires attached to the capacitor plates, the conduction current $I = dQ/dt$ passes through S_1 but not through S_2. Only the displacement current $I_d = \epsilon_0 \, d\Phi_E/dt$ passes through S_2. The two currents must be equal for continuity.

We can understand the meaning of this expression by referring to Figure 30.25. The electric flux through surface S_2 is $\Phi_E = \int \mathbf{E} \cdot d\mathbf{A} = EA$, where A is the area of the capacitor plates and E is the magnitude of the uniform electric field between the plates. If Q is the charge on the plates at any instant, then $E = Q/\epsilon_0 A$ (see Section 26.2). Therefore, the electric flux through S_2 is simply

$$\Phi_E = EA = \frac{Q}{\epsilon_0}$$

Hence, the displacement current through S_2 is

$$I_d = \epsilon_0 \frac{d\Phi_E}{dt} = \frac{dQ}{dt} \tag{30.23}$$

That is, the displacement current through S_2 is precisely equal to the conduction current I through S_1!

By considering surface S_2, we can identify the displacement current as the source of the magnetic field on the surface boundary. The displacement current has its physical origin in the time-varying electric field. The central point of this formalism, then, is that

> magnetic fields are produced both by conduction currents and by time-varying electric fields.

This result was a remarkable example of theoretical work by Maxwell, and it contributed to major advances in the understanding of electromagnetism.

Quick Quiz 30.6

What is the displacement current for a fully charged 3-μF capacitor?

EXAMPLE 30.9   **Displacement Current in a Capacitor**

A sinusoidally varying voltage is applied across an 8.00-μF capacitor. The frequency of the voltage is 3.00 kHz, and the voltage amplitude is 30.0 V. Find the displacement current between the plates of the capacitor.

Solution The angular frequency of the source, from Equation 13.6, is $\omega = 2\pi f = 2\pi(3.00 \times 10^3 \text{ Hz}) = 1.88 \times 10^4 \text{ s}^{-1}$. Hence, the voltage across the capacitor in terms of t is

$$\Delta V = \Delta V_{max} \sin \omega t = (30.0 \text{ V}) \sin(1.88 \times 10^4 t)$$

We can use Equation 30.23 and the fact that the charge on

the capacitor is $Q = C\Delta V$ to find the displacement current:

$$I_d = \frac{dQ}{dt} = \frac{d}{dt}(C\Delta V) = C\frac{d}{dt}(\Delta V)$$

$$= (8.00 \times 10^{-6} \text{ F}) \frac{d}{dt}[(30.0 \text{ V}) \sin(1.88 \times 10^4 t)]$$

$$= (4.52 \text{ A}) \cos(1.88 \times 10^4 t)$$

The displacement current varies sinusoidally with time and has a maximum value of 4.52 A.

Optional Section

30.8 ▸ MAGNETISM IN MATTER

The magnetic field produced by a current in a coil of wire gives us a hint as to what causes certain materials to exhibit strong magnetic properties. Earlier we found that a coil like the one shown in Figure 30.17 has a north pole and a south pole. In general, *any* current loop has a magnetic field and thus has a magnetic dipole moment, including the atomic-level current loops described in some models of the atom. Thus, the magnetic moments in a magnetized substance may be described as arising from these atomic-level current loops. For the Bohr model of the atom, these current loops are associated with the movement of electrons around the nucleus in circular orbits. In addition, a magnetic moment is intrinsic to electrons, protons, neutrons, and other particles; it arises from a property called *spin*.

The Magnetic Moments of Atoms

It is instructive to begin our discussion with a classical model of the atom in which electrons move in circular orbits around the much more massive nucleus. In this model, an orbiting electron constitutes a tiny current loop (because it is a moving charge), and the magnetic moment of the electron is associated with this orbital motion. Although this model has many deficiencies, its predictions are in good agreement with the correct theory, which is expressed in terms of quantum physics.

Consider an electron moving with constant speed v in a circular orbit of radius r about the nucleus, as shown in Figure 30.26. Because the electron travels a distance of $2\pi r$ (the circumference of the circle) in a time T, its orbital speed is $v = 2\pi r/T$. The current I associated with this orbiting electron is its charge e divided by T. Using $T = 2\pi/\omega$ and $\omega = v/r$, we have

$$I = \frac{e}{T} = \frac{e\omega}{2\pi} = \frac{ev}{2\pi r}$$

The magnetic moment associated with this current loop is $\mu = IA$, where $A = \pi r^2$ is the area enclosed by the orbit. Therefore,

$$\mu = IA = \left(\frac{ev}{2\pi r}\right)\pi r^2 = \tfrac{1}{2}evr \tag{30.24}$$

Because the magnitude of the orbital angular momentum of the electron is $L = m_e vr$ (Eq. 11.16 with $\phi = 90°$), the magnetic moment can be written as

$$\mu = \left(\frac{e}{2m_e}\right)L \tag{30.25}$$

This result demonstrates that **the magnetic moment of the electron is proportional to its orbital angular momentum.** Note that because the electron is negatively charged, the vectors $\boldsymbol{\mu}$ and $\mathbf{L}$ point in opposite directions. Both vectors are perpendicular to the plane of the orbit, as indicated in Figure 30.26.

A fundamental outcome of quantum physics is that orbital angular momentum is quantized and is equal to multiples of $\hbar = h/2\pi = 1.05 \times 10^{-34}\,\text{J}\cdot\text{s}$, where h is Planck's constant. The smallest nonzero value of the electron's magnetic moment resulting from its orbital motion is

$$\mu = \sqrt{2}\,\frac{e}{2m_e}\hbar \tag{30.26}$$

We shall see in Chapter 42 how expressions such as Equation 30.26 arise.

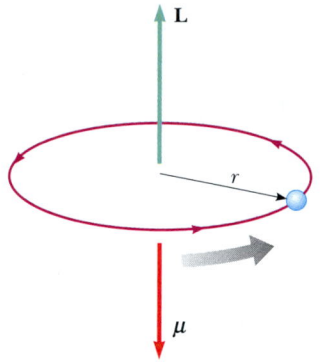

Figure 30.26 An electron moving in a circular orbit of radius r has an angular momentum $\mathbf{L}$ in one direction and a magnetic moment $\boldsymbol{\mu}$ in the opposite direction.

Orbital magnetic moment

Angular momentum is quantized

Because all substances contain electrons, you may wonder why not all substances are magnetic. The main reason is that in most substances, the magnetic moment of one electron in an atom is canceled by that of another electron orbiting in the opposite direction. The net result is that, for most materials, **the magnetic effect produced by the orbital motion of the electrons is either zero or very small.**

In addition to its orbital magnetic moment, an electron has an intrinsic property called **spin** that also contributes to its magnetic moment. In this regard, the electron can be viewed as spinning about its axis while it orbits the nucleus, as shown in Figure 30.27. (Warning: This classical description should not be taken literally because spin arises from relativistic dynamics that must be incorporated into a quantum-mechanical analysis.) The magnitude of the angular momentum S associated with spin is of the same order of magnitude as the angular momentum L due to the orbital motion. The magnitude of the spin angular momentum predicted by quantum theory is

$$S = \frac{\sqrt{3}}{2}\hbar$$

The magnetic moment characteristically associated with the spin of an electron has the value

$$\mu_{\text{spin}} = \frac{e\hbar}{2m_e} \qquad \textbf{(30.27)}$$

This combination of constants is called the **Bohr magneton:**

$$\mu_{\text{B}} = \frac{e\hbar}{2m_e} = 9.27 \times 10^{-24}\,\text{J/T} \qquad \textbf{(30.28)}$$

Thus, atomic magnetic moments can be expressed as multiples of the Bohr magneton. (Note that $1\,\text{J/T} = 1\,\text{A}\cdot\text{m}^2$.)

In atoms containing many electrons, the electrons usually pair up with their spins opposite each other; thus, the spin magnetic moments cancel. However, atoms containing an odd number of electrons must have at least one unpaired electron and therefore some spin magnetic moment. The total magnetic moment of an atom is the vector sum of the orbital and spin magnetic moments, and a few examples are given in Table 30.1. Note that helium and neon have zero moments because their individual spin and orbital moments cancel.

The nucleus of an atom also has a magnetic moment associated with its constituent protons and neutrons. However, the magnetic moment of a proton or neutron is much smaller than that of an electron and can usually be neglected. We can understand this by inspecting Equation 30.28 and replacing the mass of the electron with the mass of a proton or a neutron. Because the masses of the proton and neutron are much greater than that of the electron, their magnetic moments are on the order of 10^3 times smaller than that of the electron.

Magnetization Vector and Magnetic Field Strength

The magnetic state of a substance is described by a quantity called the **magnetization vector M. The magnitude of this vector is defined as the magnetic moment per unit volume of the substance.** As you might expect, the total magnetic field **B** at a point within a substance depends on both the applied (external) field $\mathbf{B}_0$ and the magnetization of the substance.

To understand the problems involved in measuring the total magnetic field **B** in such situations, consider this: Scientists use small probes that utilize the Hall ef-

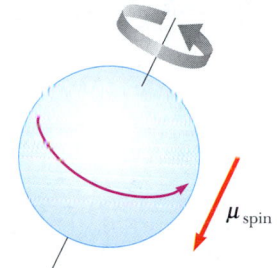

Figure 30.27 Classical model of a spinning electron. This model gives an incorrect magnitude for the magnetic moment, incorrect quantum numbers, and too many degrees of freedom.

Spin angular momentum

Bohr magneton

TABLE 30.1

Magnetic Moments of Some Atoms and Ions

Atom or Ion	Magnetic Moment (10^{-24} J/T)
H	9.27
He	0
Ne	0
Ce^{3+}	19.8
Yb^{3+}	37.1

Magnetization vector **M**

fect (see Section 29.6) to measure magnetic fields. What would such a probe read if it were positioned inside the solenoid mentioned in the QuickLab on page 951 when you inserted the compass? Because the compass is a magnetic material, the probe would measure a total magnetic field **B** that is the sum of the solenoid (external) field $\mathbf{B}_0$ and the (magnetization) field $\mathbf{B}_m$ due to the compass. This tells us that we need a way to distinguish between magnetic fields originating from currents and those originating from magnetic materials. Consider a region in which a magnetic field $\mathbf{B}_0$ is produced by a current-carrying conductor. If we now fill that region with a magnetic substance, the total magnetic field **B** in the region is $\mathbf{B} = \mathbf{B}_0 + \mathbf{B}_m$, where $\mathbf{B}_m$ is the field produced by the magnetic substance. We can express this contribution in terms of the magnetization vector of the substance as $\mathbf{B}_m = \mu_0 \mathbf{M}$; hence, the total magnetic field in the region becomes

$$\mathbf{B} = \mathbf{B}_0 + \mu_0 \mathbf{M} \tag{30.29}$$

When analyzing magnetic fields that arise from magnetization, it is convenient to introduce a field quantity, called the **magnetic field strength H** within the substance. The magnetic field strength represents the effect of the conduction currents in wires on a substance. To emphasize the distinction between the field strength **H** and the field **B**, the latter is often called the *magnetic flux density* or the *magnetic induction*. The magnetic field strength is a vector defined by the relationship $\mathbf{H} = \mathbf{B}_0 / \mu_0 = (\mathbf{B}/\mu_0) - \mathbf{M}$. Thus, Equation 30.29 can be written

$$\mathbf{B} = \mu_0 (\mathbf{H} + \mathbf{M}) \tag{30.30}$$

The quantities **H** and **M** have the same units. In SI units, because **M** is magnetic moment per unit volume, the units are $(\text{ampere})(\text{meter})^2/(\text{meter})^3$, or amperes per meter.

To better understand these expressions, consider the torus region of a toroid that carries a current I. If this region is a vacuum, $\mathbf{M} = 0$ (because no magnetic material is present), the total magnetic field is that arising from the current alone, and $\mathbf{B} = \mathbf{B}_0 = \mu_0 \mathbf{H}$. Because $B_0 = \mu_0 nI$ in the torus region, where n is the number of turns per unit length of the toroid, $H = B_0 / \mu_0 = \mu_0 nI / \mu_0$, or

$$H = nI \tag{30.31}$$

In this case, the magnetic field B in the torus region is due only to the current in the windings of the toroid.

If the torus is now made of some substance and the current I is kept constant, **H** in the torus region remains unchanged (because it depends on the current only) and has magnitude nI. The total field **B**, however, is different from that when the torus region was a vacuum. From Equation 30.30, we see that part of **B** arises from the term $\mu_0 \mathbf{H}$ associated with the current in the toroid, and part arises from the term $\mu_0 \mathbf{M}$ due to the magnetization of the substance of which the torus is made.

Classification of Magnetic Substances

Substances can be classified as belonging to one of three categories, depending on their magnetic properties. **Paramagnetic** and **ferromagnetic** materials are those made of atoms that have permanent magnetic moments. **Diamagnetic** materials are those made of atoms that do not have permanent magnetic moments.

For paramagnetic and diamagnetic substances, the magnetization vector **M** is proportional to the magnetic field strength **H**. For these substances placed in an external magnetic field, we can write

$$\mathbf{M} = \chi \mathbf{H} \tag{30.32}$$

Oxygen, a paramagnetic substance, is attracted to a magnetic field. The liquid oxygen in this photograph is suspended between the poles of the magnet. *(Leon Lewandowski)*

TABLE 30.2	Magnetic Susceptibilities of Some Paramagnetic and Diamagnetic Substances at 300 K			
Paramagnetic Substance	χ	**Diamagnetic Substance**	χ	
Aluminum	2.3×10^{-5}	Bismuth	-1.66×10^{-5}	
Calcium	1.9×10^{-5}	Copper	-9.8×10^{-6}	
Chromium	2.7×10^{-4}	Diamond	-2.2×10^{-5}	
Lithium	2.1×10^{-5}	Gold	-3.6×10^{-5}	
Magnesium	1.2×10^{-5}	Lead	-1.7×10^{-5}	
Niobium	2.6×10^{-4}	Mercury	-2.9×10^{-5}	
Oxygen	2.1×10^{-6}	Nitrogen	-5.0×10^{-9}	
Platinum	2.9×10^{-4}	Silver	-2.6×10^{-5}	
Tungsten	6.8×10^{-5}	Silicon	-4.2×10^{-6}	

where χ (Greek letter chi) is a dimensionless factor called the **magnetic susceptibility.** For paramagnetic substances, χ is positive and **M** is in the same direction as **H**. For diamagnetic substances, χ is negative and **M** is opposite **H**. (It is important to note that this linear relationship between **M** and **H** does not apply to ferromagnetic substances.) The susceptibilities of some substances are given in Table 30.2.

Magnetic susceptibility χ

Substituting Equation 30.32 for **M** into Equation 30.30 gives

$$\mathbf{B} = \mu_0(\mathbf{H} + \mathbf{M}) = \mu_0(\mathbf{H} + \chi\mathbf{H}) = \mu_0(1 + \chi)\mathbf{H}$$

or

$$\mathbf{B} = \mu_m\mathbf{H} \qquad (30.33)$$

where the constant μ_m is called the **magnetic permeability** of the substance and is related to the susceptibility by

$$\mu_m = \mu_0(1 + \chi) \qquad (30.34)$$

Magnetic permeability μ_m

Substances may be classified in terms of how their magnetic permeability μ_m compares with μ_0 (the permeability of free space), as follows:

$$\text{Paramagnetic} \qquad \mu_m > \mu_0$$

$$\text{Diamagnetic} \qquad \mu_m < \mu_0$$

Because χ is very small for paramagnetic and diamagnetic substances (see Table 30.2), μ_m is nearly equal to μ_0 for these substances. For ferromagnetic substances, however, μ_m is typically several thousand times greater than μ_0 (meaning that χ is very great for ferromagnetic substances).

Although Equation 30.33 provides a simple relationship between **B** and **H**, we must interpret it with care when dealing with ferromagnetic substances. As mentioned earlier, **M** is not a linear function of **H** for ferromagnetic substances. This is because the value of μ_m is not only a characteristic of the ferromagnetic substance but also depends on the previous state of the substance and on the process it underwent as it moved from its previous state to its present one. We shall investigate this more deeply after the following example.

EXAMPLE 30.10   **An Iron-Filled Toroid**

A toroid wound with 60.0 turns/m of wire carries a current of 5.00 A. The torus is iron, which has a magnetic permeability of $\mu_m = 5\,000\mu_0$ under the given conditions. Find H and B inside the iron.

Solution   Using Equations 30.31 and 30.33, we obtain

$$H = nI = \left(60.0\ \frac{\text{turns}}{\text{m}}\right)(5.00\ \text{A}) = \boxed{300\ \frac{\text{A·turns}}{\text{m}}}$$

$$B = \mu_m H = 5\,000\mu_0 H$$

$$= 5\,000\left(4\pi \times 10^{-7}\ \frac{\text{T·m}}{\text{A}}\right)\left(300\ \frac{\text{A·turns}}{\text{m}}\right) = \boxed{1.88\ \text{T}}$$

This value of B is 5 000 times the value in the absence of iron!

Exercise   Determine the magnitude of the magnetization vector inside the iron torus.

Answer $M = 1.5 \times 10^6\ \text{A/m}$.

Quick Quiz 30.7

A current in a solenoid having air in the interior creates a magnetic field $\mathbf{B} = \mu_0\mathbf{H}$. Describe qualitatively what happens to the magnitude of $\mathbf{B}$ as (a) aluminum, (b) copper, and (c) iron are placed in the interior.

Ferromagnetism

A small number of crystalline substances in which the atoms have permanent magnetic moments exhibit strong magnetic effects called **ferromagnetism.** Some examples of ferromagnetic substances are iron, cobalt, nickel, gadolinium, and dysprosium. These substances contain atomic magnetic moments that tend to align parallel to each other even in a weak external magnetic field. Once the moments are aligned, the substance remains magnetized after the external field is removed. This permanent alignment is due to a strong coupling between neighboring moments, a coupling that can be understood only in quantum-mechanical terms.

All ferromagnetic materials are made up of microscopic regions called **domains,** regions within which all magnetic moments are aligned. These domains have volumes of about 10^{-12} to 10^{-8} m³ and contain 10^{17} to 10^{21} atoms. The boundaries between the various domains having different orientations are called **domain walls.** In an unmagnetized sample, the domains are randomly oriented so that the net magnetic moment is zero, as shown in Figure 30.28a. When the sample is placed in an external magnetic field, the magnetic moments of the atoms tend to align with the field, which results in a magnetized sample, as in Figure 30.28b. Observations show that domains initially oriented along the external field grow larger at the expense of the less favorably oriented domains. When the external field is removed, the sample may retain a net magnetization in the direction of the original field. At ordinary temperatures, thermal agitation is not sufficient to disrupt this preferred orientation of magnetic moments.

A typical experimental arrangement that is used to measure the magnetic properties of a ferromagnetic material consists of a torus made of the material wound with N turns of wire, as shown in Figure 30.29, where the windings are represented in black and are referred to as the *primary coil.* This apparatus is sometimes referred to as a **Rowland ring.** A *secondary coil* (the red wires in Fig. 30.29) connected to a galvanometer is used to measure the total magnetic flux through the torus. The magnetic field **B** in the torus is measured by increasing the current in the toroid from zero to I. As the current changes, the magnetic flux through

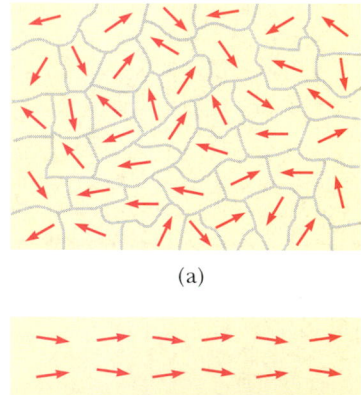

(a)

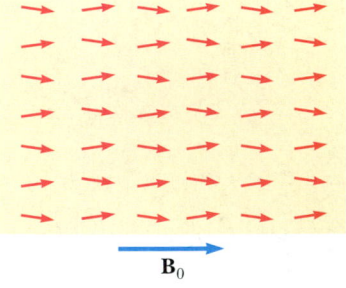

$\mathbf{B}_0$
(b)

Figure 30.28   (a) Random orientation of atomic magnetic moments in an unmagnetized substance. (b) When an external field $\mathbf{B}_0$ is applied, the atomic magnetic moments tend to align with the field, giving the sample a net magnetization vector **M**.

the secondary coil changes by an amount BA, where A is the cross-sectional area of the toroid. As we shall find in Chapter 31, because of this changing flux, an emf that is proportional to the rate of change in magnetic flux is induced in the secondary coil. If the galvanometer is properly calibrated, a value for $\mathbf{B}$ corresponding to any value of the current in the primary coil can be obtained. The magnetic field $\mathbf{B}$ is measured first in the absence of the torus and then with the torus in place. The magnetic properties of the torus material are then obtained from a comparison of the two measurements.

Now consider a torus made of unmagnetized iron. If the current in the primary coil is increased from zero to some value I, the magnitude of the magnetic field strength H increases linearly with I according to the expression $H = nI$. Furthermore, the magnitude of the total field B also increases with increasing current, as shown by the curve from point O to point a in Figure 30.30. At point O, the domains in the iron are randomly oriented, corresponding to $B_m = 0$. As the increasing current in the primary coil causes the external field $\mathbf{B}_0$ to increase, the domains become more aligned until all of them are nearly aligned at point a. At this point the iron core is approaching *saturation*, which is the condition in which all domains in the iron are aligned.

Next, suppose that the current is reduced to zero, and the external field is consequently eliminated. The B versus H curve, called a **magnetization curve,** now follows the path ab in Figure 30.30. Note that at point b, $\mathbf{B}$ is not zero even though the external field is $\mathbf{B}_0 = 0$. The reason is that the iron is now magnetized due to the alignment of a large number of its domains (that is, $\mathbf{B} = \mathbf{B}_m$). At this point, the iron is said to have a *remanent* magnetization.

If the current in the primary coil is reversed so that the direction of the external magnetic field is reversed, the domains reorient until the sample is again unmagnetized at point c, where $B = 0$. An increase in the reverse current causes the iron to be magnetized in the opposite direction, approaching saturation at point d in Figure 30.30. A similar sequence of events occurs as the current is reduced to zero and then increased in the original (positive) direction. In this case the magnetization curve follows the path def. If the current is increased sufficiently, the magnetization curve returns to point a, where the sample again has its maximum magnetization.

The effect just described, called **magnetic hysteresis,** shows that the magnetization of a ferromagnetic substance depends on the history of the substance as well as on the magnitude of the applied field. (The word *hysteresis* means "lagging behind.") It is often said that a ferromagnetic substance has a "memory" because it remains magnetized after the external field is removed. The closed loop in Figure 30.30 is referred to as a hysteresis loop. Its shape and size depend on the proper-

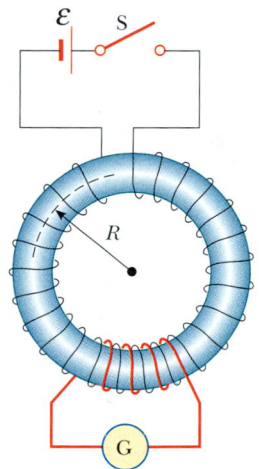

Figure 30.29 A toroidal winding arrangement used to measure the magnetic properties of a material. The torus is made of the material under study, and the circuit containing the galvanometer measures the magnetic flux.

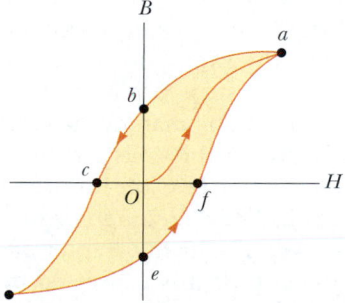

Figure 30.30 Magnetization curve for a ferromagnetic material.

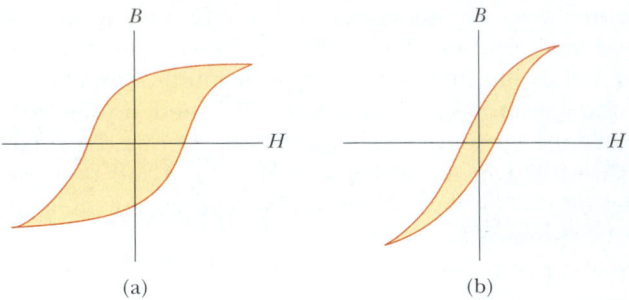

Figure 30.31   Hysteresis loops for (a) a hard ferromagnetic material and (b) a soft ferromagnetic material.

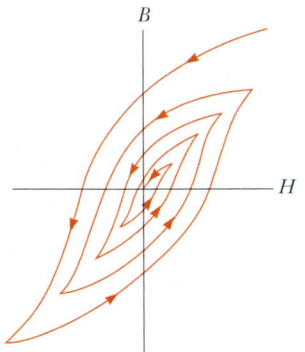

Figure 30.32   Demagnetizing a ferromagnetic material by carrying it through successive hysteresis loops.

ties of the ferromagnetic substance and on the strength of the maximum applied field. The hysteresis loop for "hard" ferromagnetic materials is characteristically wide like the one shown in Figure 30.31a, corresponding to a large remanent magnetization. Such materials cannot be easily demagnetized by an external field. "Soft" ferromagnetic materials, such as iron, have a very narrow hysteresis loop and a small remanent magnetization (Fig. 30.31b.) Such materials are easily magnetized and demagnetized. An ideal soft ferromagnet would exhibit no hysteresis and hence would have no remanent magnetization. A ferromagnetic substance can be demagnetized by being carried through successive hysteresis loops, due to a decreasing applied magnetic field, as shown in Figure 30.32.

Quick Quiz 30.8

Which material would make a better permanent magnet, one whose hysteresis loop looks like Figure 30.31a or one whose loop looks like Figure 30.31b?

The magnetization curve is useful for another reason: **The area enclosed by the magnetization curve represents the work required to take the material through the hysteresis cycle.** The energy acquired by the material in the magnetization process originates from the source of the external field—that is, the emf in the circuit of the toroidal coil. When the magnetization cycle is repeated, dissipative processes within the material due to realignment of the domains result in a transformation of magnetic energy into internal energy, which is evidenced by an increase in the temperature of the substance. For this reason, devices subjected to alternating fields (such as ac adapters for cell phones, power tools, and so on) use cores made of soft ferromagnetic substances, which have narrow hysteresis loops and correspondingly little energy loss per cycle.

   Magnetic computer disks store information by alternating the direction of **B** for portions of a thin layer of ferromagnetic material. Floppy disks have the layer on a circular sheet of plastic. Hard disks have several rigid platters with magnetic coatings on each side. Audio tapes and videotapes work the same way as floppy disks except that the ferromagnetic material is on a very long strip of plastic. Tiny coils of wire in a recording head are placed close to the magnetic material (which is moving rapidly past the head). Varying the current through the coils creates a magnetic field that magnetizes the recording material. To retrieve the information, the magnetized material is moved past a playback coil. The changing magnetism of the material induces a current in the coil, as we shall discuss in Chapter 32. This current is then amplified by audio or video equipment, or it is processed by computer circuitry.

Paramagnetism

Paramagnetic substances have a small but positive magnetic susceptibility ($0 < \chi \ll 1$) resulting from the presence of atoms (or ions) that have permanent magnetic moments. These moments interact only weakly with each other and are randomly oriented in the absence of an external magnetic field. When a paramagnetic substance is placed in an external magnetic field, its atomic moments tend to line up with the field. However, this alignment process must compete with thermal motion, which tends to randomize the magnetic moment orientations.

Pierre Curie (1859–1906) and others since him have found experimentally that, under a wide range of conditions, the magnetization of a paramagnetic substance is proportional to the applied magnetic field and inversely proportional to the absolute temperature:

$$M = C\frac{B_0}{T} \tag{30.35}$$

This relationship is known as **Curie's law** after its discoverer, and the constant C is called **Curie's constant.** The law shows that when $B_0 = 0$, the magnetization is zero, corresponding to a random orientation of magnetic moments. As the ratio of magnetic field to temperature becomes great, the magnetization approaches its saturation value, corresponding to a complete alignment of its moments, and Equation 30.35 is no longer valid.

When the temperature of a ferromagnetic substance reaches or exceeds a critical temperature called the **Curie temperature,** the substance loses its residual magnetization and becomes paramagnetic (Fig. 30.33). Below the Curie temperature, the magnetic moments are aligned and the substance is ferromagnetic. Above the Curie temperature, the thermal agitation is great enough to cause a random orientation of the moments, and the substance becomes paramagnetic. Curie temperatures for several ferromagnetic substances are given in Table 30.3.

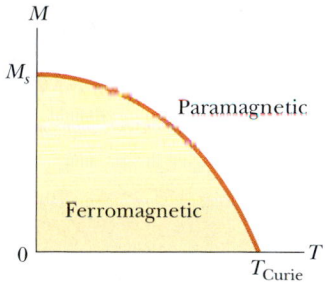

Figure 30.33 Magnetization versus absolute temperature for a ferromagnetic substance. The magnetic moments are aligned below the Curie temperature T_{Curie}, where the substance is ferromagnetic. The substance becomes paramagnetic (magnetic moments unaligned) above T_{Curie}.

TABLE 30.3
Curie Temperatures for Several Ferromagnetic Substances

Substance	T_{curie} (K)
Iron	1 043
Cobalt	1 394
Nickel	631
Gadolinium	317
Fe_2O_3	893

Diamagnetism

When an external magnetic field is applied to a diamagnetic substance, a weak magnetic moment is induced in the direction opposite the applied field. This causes diamagnetic substances to be weakly repelled by a magnet. Although diamagnetism is present in all matter, its effects are much smaller than those of paramagnetism or ferromagnetism, and are evident only when those other effects do not exist.

We can attain some understanding of diamagnetism by considering a classical model of two atomic electrons orbiting the nucleus in opposite directions but with the same speed. The electrons remain in their circular orbits because of the attractive electrostatic force exerted by the positively charged nucleus. Because the magnetic moments of the two electrons are equal in magnitude and opposite in direction, they cancel each other, and the magnetic moment of the atom is zero. When an external magnetic field is applied, the electrons experience an additional force $q\mathbf{v} \times \mathbf{B}$. This added force combines with the electrostatic force to increase the orbital speed of the electron whose magnetic moment is antiparallel to the field and to decrease the speed of the electron whose magnetic moment is parallel to the field. As a result, the two magnetic moments of the electrons no longer cancel, and the substance acquires a net magnetic moment that is opposite the applied field.

web
Visit **www.exploratorium.edu/snacks/ diamagnetism_www/index.html** for an experiment showing that grapes are repelled by magnets!

Figure 30.34 A small permanent magnet levitated above a disk of the superconductor $YBa_2Cu_3O_7$ cooled to liquid nitrogen temperature (77 K). *(U.S. Department of Energy/Science Source/Photo Researchers, Inc.)*

web

For a more detailed description of the unusual properties of superconductors, visit **www.saunderscollege.com/physics/**

As you recall from Chapter 27, a superconductor is a substance in which the electrical resistance is zero below some critical temperature. Certain types of superconductors also exhibit perfect diamagnetism in the superconducting state. As a result, an applied magnetic field is expelled by the superconductor so that the field is zero in its interior. This phenomenon of flux expulsion is known as the **Meissner effect.** If a permanent magnet is brought near a superconductor, the two objects repel each other. This is illustrated in Figure 30.34, which shows a small permanent magnet levitated above a superconductor maintained at 77 K.

EXAMPLE 30.11   Saturation Magnetization

Estimate the saturation magnetization in a long cylinder of iron, assuming one unpaired electron spin per atom.

Solution The saturation magnetization is obtained when all the magnetic moments in the sample are aligned. If the sample contains n atoms per unit volume, then the saturation magnetization M_s has the value

$$M_s = n\mu$$

where μ is the magnetic moment per atom. Because the molar mass of iron is 55 g/mol and its density is 7.9 g/cm^3, the value of n for iron is 8.6×10^{28} atoms/m^3. Assuming that

each atom contributes one Bohr magneton (due to one unpaired spin) to the magnetic moment, we obtain

$$M_s = \left(8.6 \times 10^{28} \, \frac{\text{atoms}}{\text{m}^3}\right)\left(9.27 \times 10^{-24} \, \frac{\text{A} \cdot \text{m}^2}{\text{atom}}\right)$$

$$= 8.0 \times 10^5 \, \text{A/m}$$

This is about one-half the experimentally determined saturation magnetization for iron, which indicates that actually two unpaired electron spins are present per atom.

Optional Section

30.9 THE MAGNETIC FIELD OF THE EARTH

When we speak of a compass magnet having a north pole and a south pole, we should say more properly that it has a "north-seeking" pole and a "south-seeking" pole. By this we mean that one pole of the magnet seeks, or points to, the north geographic pole of the Earth. Because the north pole of a magnet is attracted toward the north geographic pole of the Earth, we conclude that **the Earth's south magnetic pole is located near the north geographic pole, and the Earth's north magnetic pole is located near the south geographic pole.** In fact, the configuration of the Earth's magnetic field, pictured in Figure 30.35, is very much like the one that would be achieved by burying a gigantic bar magnet deep in the interior of the Earth.

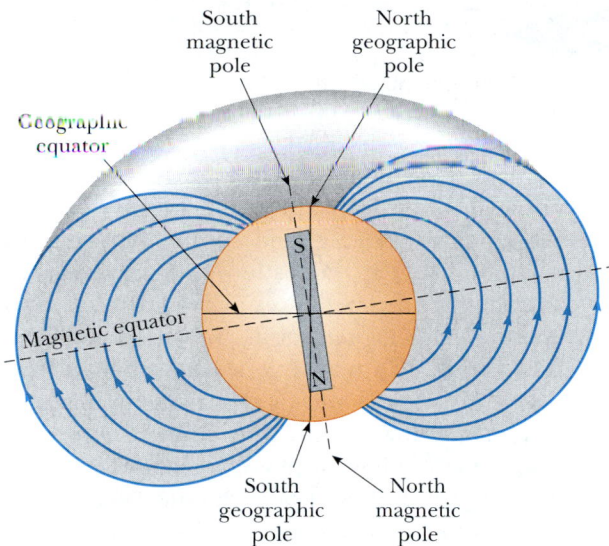

Figure 30.35 The Earth's magnetic field lines. Note that a south magnetic pole is near the north geographic pole, and a north magnetic pole is near the south geographic pole.

QuickLab

A gold ring is very weakly repelled by a magnet. To see this, suspend a 14- or 18-karat gold ring on a long loop of thread, as shown in (a). Gently tap the ring and estimate its period of oscillation. Now bring the ring to rest, letting it hang for a few moments so that you can verify that it is not moving. Quickly bring a very strong magnet to within a few millimeters of the ring, taking care not to bump it, as shown in (b). Now pull the magnet away. Repeat this action many times, matching the oscillation period you estimated earlier. This is just like pushing a child on a swing. A small force applied at the resonant frequency results in a large-amplitude oscillation. If you have a platinum ring, you will be able to see a similar effect except that platinum is weakly attracted to a magnet because it is paramagnetic.

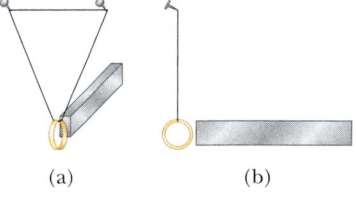

(a)   (b)

If a compass needle is suspended in bearings that allow it to rotate in the vertical plane as well as in the horizontal plane, the needle is horizontal with respect to the Earth's surface only near the equator. As the compass is moved northward, the needle rotates so that it points more and more toward the surface of the Earth. Finally, at a point near Hudson Bay in Canada, the north pole of the needle points directly downward. This site, first found in 1832, is considered to be the location of the south magnetic pole of the Earth. It is approximately 1 300 mi from the Earth's geographic North Pole, and its exact position varies slowly with time. Similarly, the north magnetic pole of the Earth is about 1 200 mi away from the Earth's geographic South Pole.

Because of this distance between the north geographic and south magnetic poles, it is only approximately correct to say that a compass needle points north. The difference between true north, defined as the geographic North Pole, and north indicated by a compass varies from point to point on the Earth, and the difference is referred to as *magnetic declination*. For example, along a line through Florida and the Great Lakes, a compass indicates true north, whereas in Washington state, it aligns 25° east of true north.

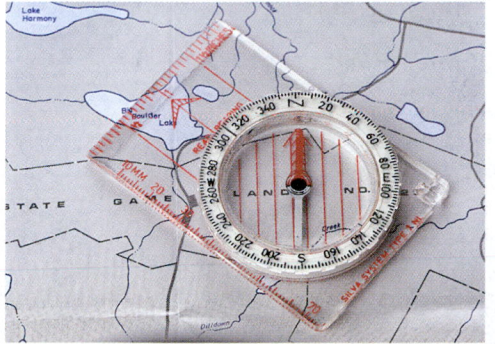

The north end of a compass needle points to the *south* magnetic pole of the Earth. The "north" compass direction varies from true geographic north depending on the magnetic declination at that point on the Earth's surface. *(George Semple)*

Quick Quiz 30.9

If we wanted to cancel the Earth's magnetic field by running an enormous current loop around the equator, which way would the current have to flow: east to west or west to east?

Although the magnetic field pattern of the Earth is similar to the one that would be set up by a bar magnet deep within the Earth, it is easy to understand why the source of the Earth's magnetic field cannot be large masses of permanently magnetized material. The Earth does have large deposits of iron ore deep beneath its surface, but the high temperatures in the Earth's core prevent the iron from retaining any permanent magnetization. Scientists consider it more likely that the true source of the Earth's magnetic field is charge-carrying convection currents in the Earth's core. Charged ions or electrons circulating in the liquid interior could produce a magnetic field just as a current loop does. There is also strong evidence that the magnitude of a planet's magnetic field is related to the planet's rate of rotation. For example, Jupiter rotates faster than the Earth, and space probes indicate that Jupiter's magnetic field is stronger than ours. Venus, on the other hand, rotates more slowly than the Earth, and its magnetic field is found to be weaker. Investigation into the cause of the Earth's magnetism is ongoing.

There is an interesting sidelight concerning the Earth's magnetic field. It has been found that the direction of the field has been reversed several times during the last million years. Evidence for this is provided by basalt, a type of rock that contains iron and that forms from material spewed forth by volcanic activity on the ocean floor. As the lava cools, it solidifies and retains a picture of the Earth's magnetic field direction. The rocks are dated by other means to provide a timeline for these periodic reversals of the magnetic field.

SUMMARY

The **Biot–Savart law** says that the magnetic field $d\mathbf{B}$ at a point P due to a length element $d\mathbf{s}$ that carries a steady current I is

$$d\mathbf{B} = \frac{\mu_0}{4\pi} \frac{I\,d\mathbf{s} \times \hat{\mathbf{r}}}{r^2} \tag{30.1}$$

where $\mu_0 = 4\pi \times 10^{-7}\,\text{T·m/A}$ is the **permeability of free space,** r is the distance from the element to the point P, and $\hat{\mathbf{r}}$ is a unit vector pointing from $d\mathbf{s}$ to point P. We find the total field at P by integrating this expression over the entire current distribution.

The magnetic field at a distance a from a long, straight wire carrying an electric current I is

$$B = \frac{\mu_0 I}{2\pi a} \tag{30.5}$$

The field lines are circles concentric with the wire.

The magnetic force per unit length between two parallel wires separated by a distance a and carrying currents I_1 and I_2 has a magnitude

$$\frac{F_B}{\ell} = \frac{\mu_0 I_1 I_2}{2\pi a} \tag{30.12}$$

The force is attractive if the currents are in the same direction and repulsive if they are in opposite directions.

Ampère's law says that the line integral of $\mathbf{B} \cdot d\mathbf{s}$ around any closed path equals $\mu_0 I$, where I is the total steady current passing through any surface bounded by the closed path:

$$\oint \mathbf{B} \cdot d\mathbf{s} = \mu_0 I \tag{30.13}$$

Using Ampère's law, one finds that the fields inside a toroid and solenoid are

$$B = \frac{\mu_0 N I}{2\pi r} \qquad \text{(toroid)} \tag{30.16}$$

$$B = \mu_0 \frac{N}{\ell} I = \mu_0 n I \qquad \text{(solenoid)} \tag{30.17}$$

where N is the total number of turns.

The **magnetic flux Φ_B** through a surface is defined by the surface integral

$$\Phi_B \equiv \int \mathbf{B} \cdot d\mathbf{A} \tag{30.18}$$

Gauss's law of magnetism states that the net magnetic flux through any closed surface is zero.

The general form of Ampère's law, which is also called the **Ampère-Maxwell law,** is

$$\oint \mathbf{B} \cdot d\mathbf{s} = \mu_0 I + \mu_0 \epsilon_0 \frac{d\Phi_E}{dt} \tag{30.22}$$

This law describes the fact that magnetic fields are produced both by conduction currents and by changing electric fields.

QUESTIONS

1. Is the magnetic field created by a current loop uniform? Explain.

2. A current in a conductor produces a magnetic field that can be calculated using the Biot–Savart law. Because current is defined as the rate of flow of charge, what can you conclude about the magnetic field produced by stationary charges? What about that produced by moving charges?

3. Two parallel wires carry currents in opposite directions. Describe the nature of the magnetic field created by the two wires at points (a) between the wires and (b) outside the wires, in a plane containing them.

4. Explain why two parallel wires carrying currents in opposite directions repel each other.

5. When an electric circuit is being assembled, a common practice is to twist together two wires carrying equal currents in opposite directions. Why does this technique reduce stray magnetic fields?

6. Is Ampère's law valid for all closed paths surrounding a conductor? Why is it not useful for calculating $\mathbf{B}$ for all such paths?

7. Compare Ampère's law with the Biot–Savart law. Which is more generally useful for calculating $\mathbf{B}$ for a current-carrying conductor?

8. Is the magnetic field inside a toroid uniform? Explain.

9. Describe the similarities between Ampère's law in magnetism and Gauss's law in electrostatics.

10. A hollow copper tube carries a current along its length. Why does $\mathbf{B} = 0$ inside the tube? Is $\mathbf{B}$ nonzero outside the tube?

11. Why is $\mathbf{B}$ nonzero outside a solenoid? Why does $\mathbf{B} = 0$ outside a toroid? (Remember that the lines of $\mathbf{B}$ must form closed paths.)

12. Describe the change in the magnetic field in the interior of a solenoid carrying a steady current I (a) if the length of the solenoid is doubled but the number of turns remains the same and (b) if the number of turns is doubled but the length remains the same.

13. A flat conducting loop is positioned in a uniform magnetic field directed along the x axis. For what orientation of the loop is the flux through it a maximum? A minimum?

14. What new concept does Maxwell's general form of Ampère's law include?

15. Many loops of wire are wrapped around a nail and then connected to a battery. Identify the source of $\mathbf{M}$, of $\mathbf{H}$, and of $\mathbf{B}$.

16. A magnet attracts a piece of iron. The iron can then attract another piece of iron. On the basis of domain alignment, explain what happens in each piece of iron.

17. You are stranded on a planet that does not have a magnetic field, with no test equipment. You have two bars of iron in your possession; one is magnetized, and one is not. How can you determine which is which?

18. Why does hitting a magnet with a hammer cause the magnetism to be reduced?

19. Is a nail attracted to either pole of a magnet? Explain what is happening inside the nail when it is placed near the magnet.

20. A Hindu ruler once suggested that he be entombed in a magnetic coffin with the polarity arranged so that he would be forever suspended between heaven and Earth. Is such magnetic levitation possible? Discuss.

21. Why does $\mathbf{M} = 0$ in a vacuum? What is the relationship between $\mathbf{B}$ and $\mathbf{H}$ in a vacuum?

22. Explain why some atoms have permanent magnetic moments and others do not.

23. What factors contribute to the total magnetic moment of an atom?

24. Why is the magnetic susceptibility of a diamagnetic substance negative?

25. Why can the effect of diamagnetism be neglected in a paramagnetic substance?

26. Explain the significance of the Curie temperature for a ferromagnetic substance.

27. Discuss the differences among ferromagnetic, paramagnetic, and diamagnetic substances.

28. What is the difference between hard and soft ferromagnetic materials?

29. Should the surface of a computer disk be made from a hard or a soft ferromagnetic substance?

30. Explain why it is desirable to use hard ferromagnetic materials to make permanent magnets.

31. Would you expect the tape from a tape recorder to be attracted to a magnet? (Try it, but not with a recording you wish to save.)

32. Given only a strong magnet and a screwdriver, how would you first magnetize and then demagnetize the screwdriver?

33. Figure Q30.33 shows two permanent magnets, each having a hole through its center. Note that the upper magnet is levitated above the lower one. (a) How does this occur? (b) What purpose does the pencil serve? (c) What can you say about the poles of the magnets on the basis of this observation? (d) What do you suppose would happen if the upper magnet were inverted?

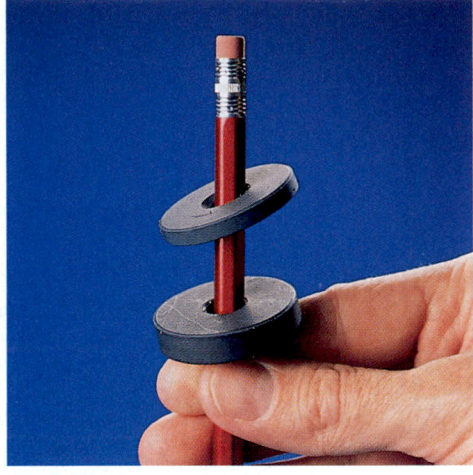

Figure Q30.33 Magnetic levitation using two ceramic magnets. *(Courtesy of Central Scientific Company)*

PROBLEMS

1, 2, 3 = straightforward, intermediate, challenging ☐ = full solution available in the *Student Solutions Manual and Study Guide*
WEB = solution posted at **http://www.saunderscollege.com/physics/** 🖥 = Computer useful in solving problem 🅿 = Interactive Physics
☐ = paired numerical/symbolic problems

Section 30.1 The Biot–Savart Law

1. In Niels Bohr's 1913 model of the hydrogen atom, an electron circles the proton at a distance of 5.29×10^{-11} m with a speed of 2.19×10^{6} m/s. Compute the magnitude of the magnetic field that this motion produces at the location of the proton.

2. A current path shaped as shown in Figure P30.2 produces a magnetic field at P, the center of the arc. If the arc subtends an angle of 30.0° and the radius of the arc is 0.600 m, what are the magnitude and direction of the field produced at P if the current is 3.00 A?

3. (a) A conductor in the shape of a square of edge length $\ell = 0.400$ m carries a current $I = 10.0$ A (Fig. P30.3). Calculate the magnitude and direction of the magnetic

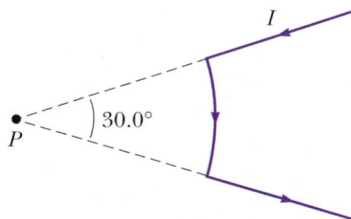

Figure P30.2

field at the center of the square. (b) If this conductor is formed into a single circular turn and carries the same current, what is the value of the magnetic field at the center?

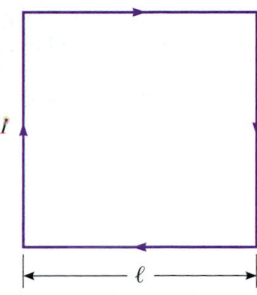

Figure P30.3

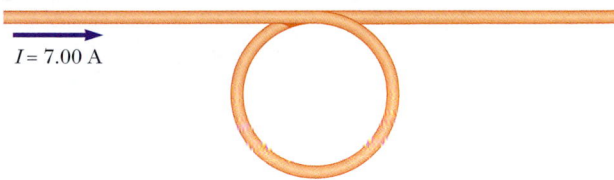

Figure P30.7   Problems 7 and 8.

4. Calculate the magnitude of the magnetic field at a point 100 cm from a long, thin conductor carrying a current of 1.00 A.

WEB 5. Determine the magnetic field at a point P located a distance x from the corner of an infinitely long wire bent at a right angle, as shown in Figure P30.5. The wire carries a steady current I.

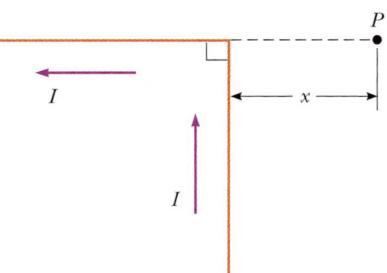

Figure P30.5

6. A wire carrying a current of 5.00 A is to be formed into a circular loop of one turn. If the required value of the magnetic field at the center of the loop is 10.0 μT, what is the required radius?

7. A conductor consists of a circular loop of radius $R = 0.100$ m and two straight, long sections, as shown in Figure P30.7. The wire lies in the plane of the paper and carries a current of $I = 7.00$ A. Determine the magnitude and direction of the magnetic field at the center of the loop.

8. A conductor consists of a circular loop of radius R and two straight, long sections, as shown in Figure P30.7. The wire lies in the plane of the paper and carries a current I. Determine the magnitude and direction of the magnetic field at the center of the loop.

9. The segment of wire in Figure P30.9 carries a current of $I = 5.00$ A, where the radius of the circular arc is $R = 3.00$ cm. Determine the magnitude and direction of the magnetic field at the origin.

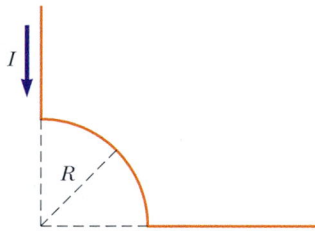

Figure P30.9

10. Consider a flat, circular current loop of radius R carrying current I. Choose the x axis to be along the axis of the loop, with the origin at the center of the loop. Graph the ratio of the magnitude of the magnetic field at coordinate x to that at the origin, for $x = 0$ to $x = 5R$. It may be helpful to use a programmable calculator or a computer to solve this problem.

11. Consider the current-carrying loop shown in Figure P30.11, formed of radial lines and segments of circles whose centers are at point P. Find the magnitude and direction of **B** at P.

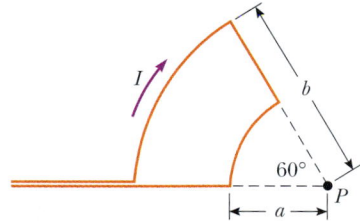

Figure P30.11

12. Determine the magnetic field (in terms of I, a, and d) at the origin due to the current loop shown in Figure P30.12.

13. The loop in Figure P30.13 carries a current I. Determine the magnetic field at point A in terms of I, R, and L.

14. Three long, parallel conductors carry currents of $I = 2.00$ A. Figure P30.14 is an end view of the conductors, with each current coming out of the page. If $a = 1.00$ cm, determine the magnitude and direction of the magnetic field at points A, B, and C.

15. Two long, parallel conductors carry currents $I_1 = 3.00$ A and $I_2 = 3.00$ A, both directed into the page in

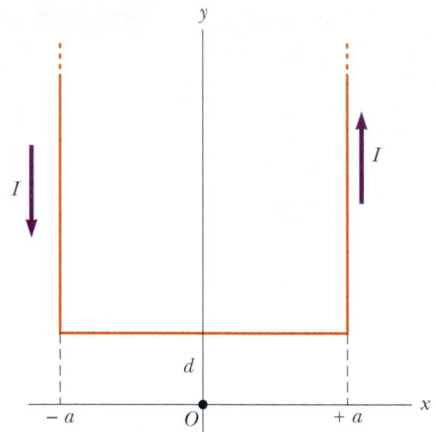

Figure P30.12

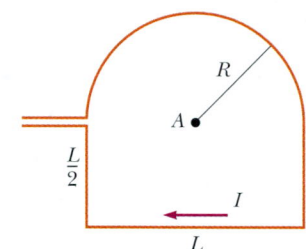

Figure P30.13

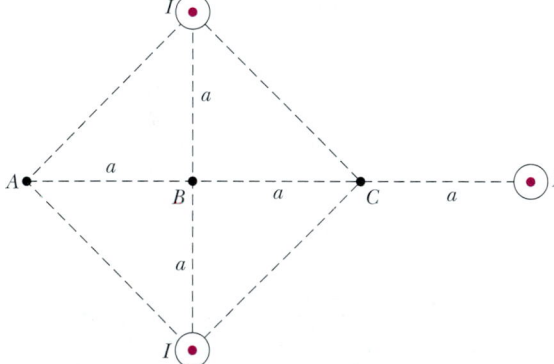

Figure P30.14

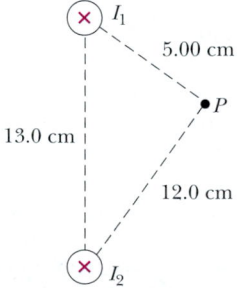

Figure P30.15

Figure P30.15. Determine the magnitude and direction of the resultant magnetic field at P.

Section 30.2   The Magnetic Force Between Two Parallel Conductors

16. Two long, parallel conductors separated by 10.0 cm carry currents in the same direction. The first wire carries current $I_1 = 5.00$ A, and the second carries $I_2 = 8.00$ A. (a) What is the magnitude of the magnetic field created by I_1 and acting on I_2? (b) What is the force per unit length exerted on I_2 by I_1? (c) What is the magnitude of the magnetic field created by I_2 at the location of I_1? (d) What is the force per unit length exerted by I_2 on I_1?

17. In Figure P30.17, the current in the long, straight wire is $I_1 = 5.00$ A, and the wire lies in the plane of the rectangular loop, which carries 10.0 A. The dimensions are $c = 0.100$ m, $a = 0.150$ m, and $\ell = 0.450$ m. Find the magnitude and direction of the net force exerted on the loop by the magnetic field created by the wire.

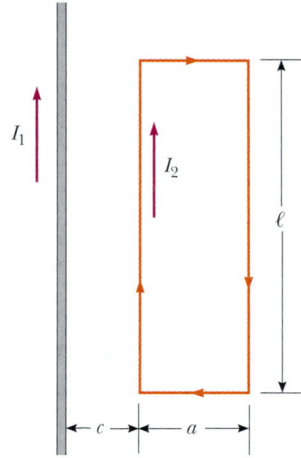

Figure P30.17

18. The unit of magnetic flux is named for Wilhelm Weber. The practical-size unit of magnetic field is named for Johann Karl Friedrich Gauss. Both were scientists at Göttingen, Germany. In addition to their individual accomplishments, they built a telegraph together in 1833. It consisted of a battery and switch that were positioned at one end of a transmission line 3 km long and operated an electromagnet at the other end. (Andre Ampère suggested electrical signaling in 1821; Samuel Morse built a telegraph line between Baltimore and Washington in 1844.) Suppose that Weber and Gauss's transmission line was as diagrammed in Figure P30.18. Two long, parallel wires, each having a mass per unit length of 40.0 g/m, are supported in a horizontal plane by strings 6.00 cm long. When both wires carry the same current I, the wires repel each other so that the angle θ

between the supporting strings is 16.0°. (a) Are the currents in the same direction or in opposite directions? (b) Find the magnitude of the current.

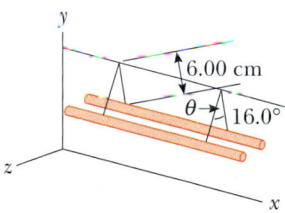

Figure P30.18

Section 30.3 Ampère's Law

WEB **19.** Four long, parallel conductors carry equal currents of $I = 5.00$ A. Figure P30.19 is an end view of the conductors. The direction of the current is into the page at points A and B (indicated by the crosses) and out of the page at C and D (indicated by the dots). Calculate the magnitude and direction of the magnetic field at point P, located at the center of the square with an edge length of 0.200 m.

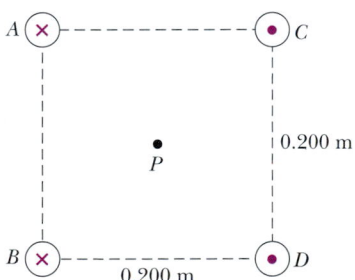

Figure P30.19

20. A long, straight wire lies on a horizontal table and carries a current of 1.20 μA. In a vacuum, a proton moves parallel to the wire (opposite the current) with a constant velocity of 2.30×10^4 m/s at a distance d above the wire. Determine the value of d. You may ignore the magnetic field due to the Earth.

21. Figure P30.21 is a cross-sectional view of a coaxial cable. The center conductor is surrounded by a rubber layer, which is surrounded by an outer conductor, which is surrounded by another rubber layer. In a particular application, the current in the inner conductor is 1.00 A out of the page, and the current in the outer conductor is 3.00 A into the page. Determine the magnitude and direction of the magnetic field at points a and b.

22. The magnetic field 40.0 cm away from a long, straight wire carrying current 2.00 A is 1.00 μT. (a) At what distance is it 0.100 μT? (b) At one instant, the two conductors in a long household extension cord carry equal

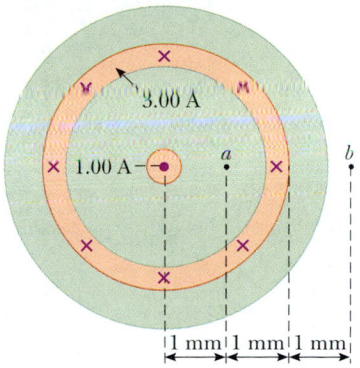

Figure P30.21

2.00-A currents in opposite directions. The two wires are 3.00 mm apart. Find the magnetic field 40.0 cm away from the middle of the straight cord, in the plane of the two wires. (c) At what distance is it one-tenth as large? (d) The center wire in a coaxial cable carries current 2.00 A in one direction, and the sheath around it carries current 2.00 A in the opposite direction. What magnetic field does the cable create at points outside?

23. The magnetic coils of a tokamak fusion reactor are in the shape of a toroid having an inner radius of 0.700 m and an outer radius of 1.30 m. If the toroid has 900 turns of large-diameter wire, each of which carries a current of 14.0 kA, find the magnitude of the magnetic field inside the toroid (a) along the inner radius and (b) along the outer radius.

24. A cylindrical conductor of radius $R = 2.50$ cm carries a current of $I = 2.50$ A along its length; this current is uniformly distributed throughout the cross-section of the conductor. (a) Calculate the magnetic field midway along the radius of the wire (that is, at $r = R/2$). (b) Find the distance beyond the surface of the conductor at which the magnitude of the magnetic field has the same value as the magnitude of the field at $r = R/2$.

WEB **25.** A packed bundle of 100 long, straight, insulated wires forms a cylinder of radius $R = 0.500$ cm. (a) If each wire carries 2.00 A, what are the magnitude and direction of the magnetic force per unit length acting on a wire located 0.200 cm from the center of the bundle? (b) Would a wire on the outer edge of the bundle experience a force greater or less than the value calculated in part (a)?

26. Niobium metal becomes a superconductor when cooled below 9 K. If superconductivity is destroyed when the surface magnetic field exceeds 0.100 T, determine the maximum current a 2.00-mm-diameter niobium wire can carry and remain superconducting, in the absence of any external magnetic field.

27. A long, cylindrical conductor of radius R carries a current I, as shown in Figure P30.27. The current density J, however, is not uniform over the cross-section of the

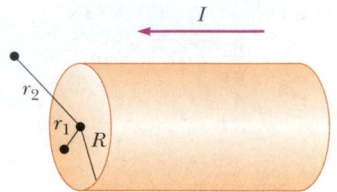

Figure P30.27

conductor but is a function of the radius according to $J = br$, where b is a constant. Find an expression for the magnetic field B (a) at a distance $r_1 < R$ and (b) at a distance $r_2 > R$, measured from the axis.

28. In Figure P30.28, both currents are in the negative x direction. (a) Sketch the magnetic field pattern in the yz plane. (b) At what distance d along the z axis is the magnetic field a maximum?

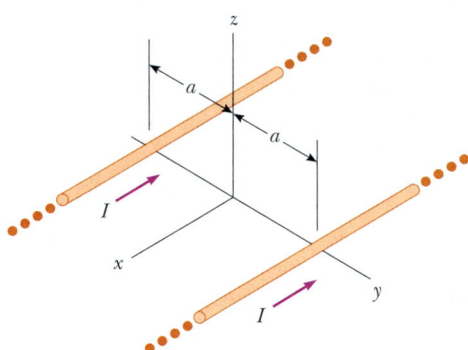

Figure P30.28

Section 30.4   The Magnetic Field of a Solenoid

WEB 29. What current is required in the windings of a long solenoid that has 1 000 turns uniformly distributed over a length of 0.400 m, to produce at the center of the solenoid a magnetic field of magnitude 1.00×10^{-4} T?

30. A superconducting solenoid is meant to generate a magnetic field of 10.0 T. (a) If the solenoid winding has 2 000 turns/m, what current is required? (b) What force per unit length is exerted on the windings by this magnetic field?

31. A solenoid of radius $R = 5.00$ cm is made of a long piece of wire of radius $r = 2.00$ mm, length $\ell = 10.0$ m ($\ell \gg R$) and resistivity $\rho = 1.70 \times 10^{-8}$ $\Omega \cdot$ m. Find the magnetic field at the center of the solenoid if the wire is connected to a battery having an emf $\mathcal{E} = 20.0$ V.

32. A single-turn square loop of wire with an edge length of 2.00 cm carries a clockwise current of 0.200 A. The loop is inside a solenoid, with the plane of the loop perpendicular to the magnetic field of the solenoid. The solenoid has 30 turns/cm and carries a clockwise current of 15.0 A. Find the force on each side of the loop and the torque acting on the loop.

Section 30.5   Magnetic Flux

33. A cube of edge length $\ell = 2.50$ cm is positioned as shown in Figure P30.33. A uniform magnetic field given by $\mathbf{B} = (5.00\mathbf{i} + 4.00\mathbf{j} + 3.00\mathbf{k})$ T exists throughout the region. (a) Calculate the flux through the shaded face. (b) What is the total flux through the six faces?

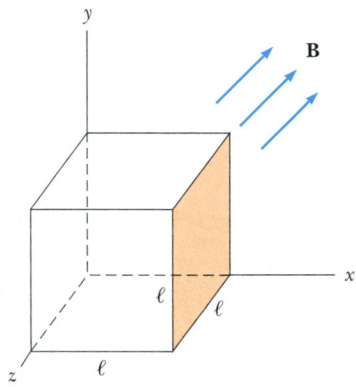

Figure P30.33

34. A solenoid 2.50 cm in diameter and 30.0 cm long has 300 turns and carries 12.0 A. (a) Calculate the flux through the surface of a disk of radius 5.00 cm that is positioned perpendicular to and centered on the axis of the solenoid, as in Figure P30.34a. (b) Figure P30.34b shows an enlarged end view of the same solenoid. Calculate the flux through the blue area, which is defined by an annulus that has an inner radius of 0.400 cm and outer radius of 0.800 cm.

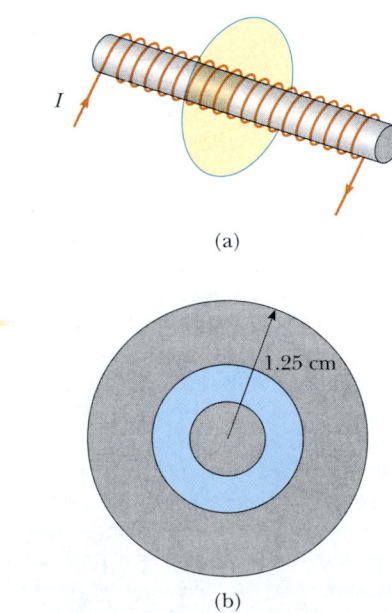

Figure P30.34

35. Consider the hemispherical closed surface in Figure P30.35. If the hemisphere is in a uniform magnetic field that makes an angle θ with the vertical, calculate the magnetic flux (a) through the flat surface S_1 and (b) through the hemispherical surface S_2.

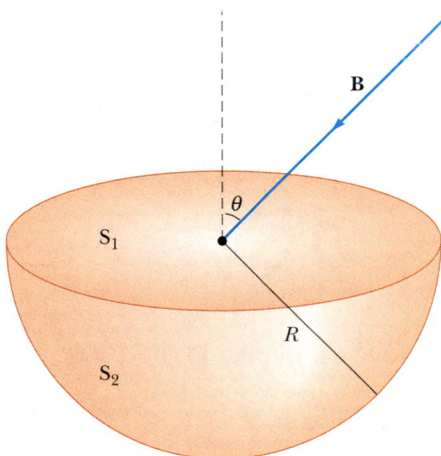

Figure P30.35

Section 30.6 Gauss's Law in Magnetism
Section 30.7 Displacement Current and the General Form of Ampère's Law

36. A 0.200-A current is charging a capacitor that has circular plates 10.0 cm in radius. If the plate separation is 4.00 mm, (a) what is the time rate of increase of electric field between the plates? (b) What is the magnetic field between the plates 5.00 cm from the center?

37. A 0.100-A current is charging a capacitor that has square plates 5.00 cm on each side. If the plate separation is 4.00 mm, find (a) the time rate of change of electric flux between the plates and (b) the displacement current between the plates.

(Optional)
Section 30.8 Magnetism in Matter

38. In Bohr's 1913 model of the hydrogen atom, the electron is in a circular orbit of radius 5.29×10^{-11} m, and its speed is 2.19×10^6 m/s. (a) What is the magnitude of the magnetic moment due to the electron's motion? (b) If the electron orbits counterclockwise in a horizontal circle, what is the direction of this magnetic moment vector?

39. A toroid with a mean radius of 20.0 cm and 630 turns (see Fig. 30.29) is filled with powdered steel whose magnetic susceptibility χ is 100. If the current in the windings is 3.00 A, find B (assumed uniform) inside the toroid.

40. A magnetic field of 1.30 T is to be set up in an iron-core toroid. The toroid has a mean radius of 10.0 cm and magnetic permeability of $5\,000\mu_0$. What current is re-

quired if there are 470 turns of wire in the winding? The thickness of the iron ring is small compared to 10 cm, so the field in the material is nearly uniform.

41. A coil of 500 turns is wound on an iron ring ($\mu_m = 750\mu_0$) with a 20.0-cm mean radius and an 8.00 cm² cross-sectional area. Calculate the magnetic flux Φ_B in this Rowland ring when the current in the coil is 0.500 A.

42. A uniform ring with a radius of 2.00 cm and a total charge of 6.00 μC rotates with a constant angular speed of 4.00 rad/s around an axis perpendicular to the plane of the ring and passing through its center. What is the magnetic moment of the rotating ring?

43. Calculate the magnetic field strength H of a magnetized substance in which the magnetization is 880 kA/m and the magnetic field has a magnitude of 4.40 T.

44. At saturation, the alignment of spins in iron can contribute as much as 2.00 T to the total magnetic field B. If each electron contributes a magnetic moment of 9.27×10^{-24} A·m² (one Bohr magneton), how many electrons per atom contribute to the saturated field of iron? (*Hint:* Iron contains 8.50×10^{28} atoms/m³.)

45. (a) Show that Curie's law can be stated in the following way: The magnetic susceptibility of a paramagnetic substance is inversely proportional to the absolute temperature, according to $\chi = C\mu_0/T$, where C is Curie's constant. (b) Evaluate Curie's constant for chromium.

(Optional)
Section 30.9 The Magnetic Field of the Earth

46. A circular coil of 5 turns and a diameter of 30.0 cm is oriented in a vertical plane with its axis perpendicular to the horizontal component of the Earth's magnetic field. A horizontal compass placed at the center of the coil is made to deflect 45.0° from magnetic north by a current of 0.600 A in the coil. (a) What is the horizontal component of the Earth's magnetic field? (b) The current in the coil is switched off. A "dip needle" is a magnetic compass mounted so that it can rotate in a vertical north-south plane. At this location a dip needle makes an angle of 13.0° from the vertical. What is the total magnitude of the Earth's magnetic field at this location?

47. The magnetic moment of the Earth is approximately 8.00×10^{22} A·m². (a) If this were caused by the complete magnetization of a huge iron deposit, how many unpaired electrons would this correspond to? (b) At two unpaired electrons per iron atom, how many kilograms of iron would this correspond to? (Iron has a density of 7 900 kg/m³ and approximately 8.50×10^{28} atoms/m³.)

ADDITIONAL PROBLEMS

48. A lightning bolt may carry a current of 1.00×10^4 A for a short period of time. What is the resultant magnetic field 100 m from the bolt? Suppose that the bolt extends far above and below the point of observation.

49. The magnitude of the Earth's magnetic field at either pole is approximately 7.00×10^{-5} T. Suppose that the field fades away, before its next reversal. Scouts, sailors, and wire merchants around the world join together in a program to replace the field. One plan is to use a current loop around the equator, without relying on magnetization of any materials inside the Earth. Determine the current that would generate such a field if this plan were carried out. (Take the radius of the Earth as $R_E = 6.37 \times 10^6$ m.)

50. Two parallel conductors carry current in opposite directions, as shown in Figure P30.50. One conductor carries a current of 10.0 A. Point A is at the midpoint between the wires, and point C is a distance $d/2$ to the right of the 10.0-A current. If $d = 18.0$ cm and I is adjusted so that the magnetic field at C is zero, find (a) the value of the current I and (b) the value of the magnetic field at A.

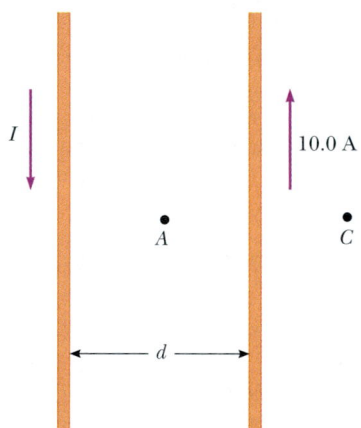

Figure P30.50

51. Suppose you install a compass on the center of the dashboard of a car. Compute an order-of-magnitude estimate for the magnetic field that is produced at this location by the current when you switch on the headlights. How does your estimate compare with the Earth's magnetic field? You may suppose the dashboard is made mostly of plastic.

52. Imagine a long, cylindrical wire of radius R that has a current density $J(r) = J_0(1 - r^2/R^2)$ for $r \le R$ and $J(r) = 0$ for $r > R$, where r is the distance from the axis of the wire. (a) Find the resulting magnetic field inside ($r \le R$) and outside ($r > R$) the wire. (b) Plot the magnitude of the magnetic field as a function of r. (c) Find the location where the magnitude of the magnetic field is a maximum, and the value of that maximum field.

53. A very long, thin strip of metal of width w carries a current I along its length, as shown in Figure P30.53. Find the magnetic field at point P in the diagram. Point P is

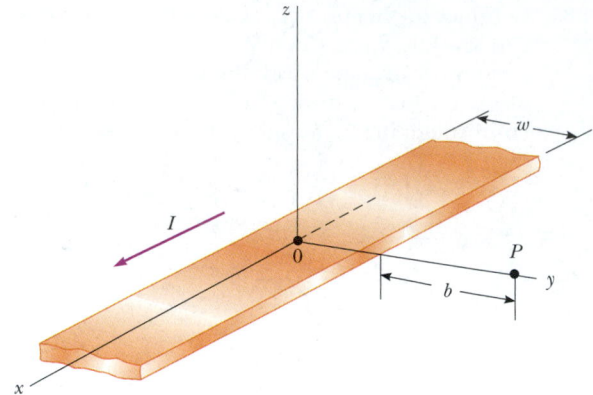

Figure P30.53

in the plane of the strip at a distance b away from the strip.

54. For a research project, a student needs a solenoid that produces an interior magnetic field of 0.030 0 T. She decides to use a current of 1.00 A and a wire 0.500 mm in diameter. She winds the solenoid in layers on an insulating form 1.00 cm in diameter and 10.0 cm long. Determine the number of layers of wire she needs and the total length of the wire.

WEB 55. A nonconducting ring with a radius of 10.0 cm is uniformly charged with a total positive charge of 10.0 μC. The ring rotates at a constant angular speed of 20.0 rad/s about an axis through its center, perpendicular to the plane of the ring. What is the magnitude of the magnetic field on the axis of the ring, 5.00 cm from its center?

56. A nonconducting ring of radius R is uniformly charged with a total positive charge q. The ring rotates at a constant angular speed ω about an axis through its center, perpendicular to the plane of the ring. What is the magnitude of the magnetic field on the axis of the ring a distance $R/2$ from its center?

57. Two circular coils of radius R are each perpendicular to a common axis. The coil centers are a distance R apart, and a steady current I flows in the same direction around each coil, as shown in Figure P30.57. (a) Show that the magnetic field on the axis at a distance x from the center of one coil is

$$B = \frac{\mu_0 I R^2}{2} \left[\frac{1}{(R^2 + x^2)^{3/2}} + \frac{1}{(2R^2 + x^2 - 2Rx)^{3/2}} \right]$$

(b) Show that dB/dx and d^2B/dx^2 are both zero at a point midway between the coils. This means that the magnetic field in the region midway between the coils is uniform. Coils in this configuration are called **Helmholtz coils.**

58. Two identical, flat, circular coils of wire each have 100 turns and a radius of 0.500 m. The coils are arranged as

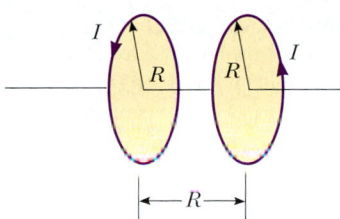

Figure P30.57 Problems 57 and 58.

a set of Helmholtz coils (see Fig. P30.57), parallel and with a separation of 0.500 m. If each coil carries a current of 10.0 A, determine the magnitude of the magnetic field at a point on the common axis of the coils and halfway between them.

59. Two circular loops are parallel, coaxial, and almost in contact, 1.00 mm apart (Fig. P30.59). Each loop is 10.0 cm in radius. The top loop carries a clockwise current of 140 A. The bottom loop carries a counterclockwise current of 140 A. (a) Calculate the magnetic force that the bottom loop exerts on the top loop. (b) The upper loop has a mass of 0.021 0 kg. Calculate its acceleration, assuming that the only forces acting on it are the force in part (a) and its weight. (*Hint:* Think about how one loop looks to a bug perched on the other loop.)

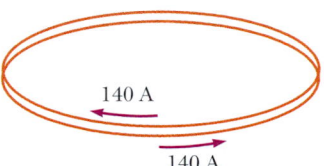

Figure P30.59

60. What objects experience a force in an electric field? Chapter 23 gives the answer: any electric charge, stationary or moving, other than the charge that created the field. What creates an electric field? Any electric charge, stationary or moving, also as discussed in Chapter 23. What objects experience a force in a magnetic field? An electric current or a moving electric charge other than the current or charge that created the field, as discovered in Chapter 29. What creates a magnetic field? An electric current, as you found in Section 30.11, or a moving electric charge, as in this problem. (a) To display how a moving charge creates a magnetic field, consider a charge q moving with velocity $\mathbf{v}$. Define the unit vector $\hat{\mathbf{r}} = \mathbf{r}/r$ to point from the charge to some location. Show that the magnetic field at that location is

$$\mathbf{B} = \frac{\mu_0}{4\pi} \frac{q\mathbf{v} \times \hat{\mathbf{r}}}{r^2}$$

(b) Find the magnitude of the magnetic field 1.00 mm

to the side of a proton moving at 2.00×10^7 m/s. (c) Find the magnetic force on a second proton at this point, moving with the same speed in the opposite direction. (d) Find the electric force on the second proton.

61. *Rail guns* have been suggested for launching projectiles into space without chemical rockets, and for ground-to-air antimissile weapons of war. A tabletop model rail gun (Fig. P30.61) consists of two long parallel horizontal rails 3.50 cm apart, bridged by a bar *BD* of mass 3.00 g. The bar is originally at rest at the midpoint of the rails and is free to slide without friction. When the switch is closed, electric current is very quickly established in the circuit *ABCDEA*. The rails and bar have low electrical resistance, and the current is limited to a constant 24.0 A by the power supply. (a) Find the magnitude of the magnetic field 1.75 cm from a single very long, straight wire carrying current 24.0 A. (b) Find the vector magnetic field at point *C* in the diagram, the midpoint of the bar, immediately after the switch is closed. (*Hint:* Consider what conclusions you can draw from the Biot–Savart law.) (c) At other points along the bar *BD*, the field is in the same direction as at point *C*, but greater in magnitude. Assume that the average effective magnetic field along *BD* is five times larger than the field at *C*. With this assumption, find the vector force on the bar. (d) Find the vector acceleration with which the bar starts to move. (e) Does the bar move with constant acceleration? (f) Find the velocity of the bar after it has traveled 130 cm to the end of the rails.

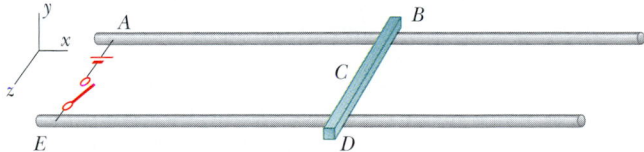

Figure P30.61

62. Two long, parallel conductors carry currents in the same direction, as shown in Figure P30.62. Conductor A carries a current of 150 A and is held firmly in position. Conductor B carries a current I_B and is allowed to slide freely up and down (parallel to A) between a set of nonconducting guides. If the mass per unit length of conductor B is 0.100 g/cm, what value of current I_B will result in equilibrium when the distance between the two conductors is 2.50 cm?

63. Charge is sprayed onto a large nonconducting belt above the left-hand roller in Figure P30.63. The belt carries the charge, with a uniform surface charge density σ, as it moves with a speed v between the rollers as shown. The charge is removed by a wiper at the right-hand roller. Consider a point just above the surface of the moving belt. (a) Find an expression for the magni-

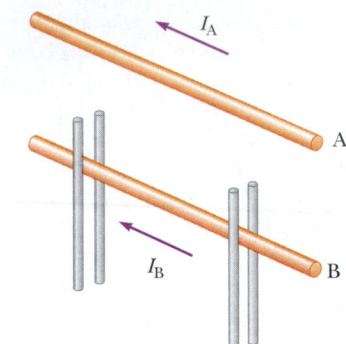

Figure P30.62

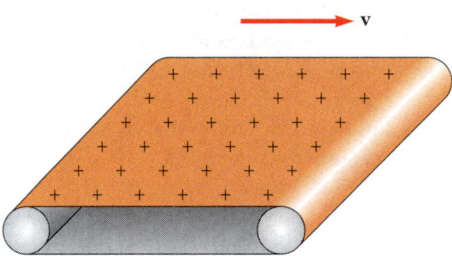

Figure P30.63

tude of the magnetic field **B** at this point. (b) If the belt is positively charged, what is the direction of **B**? (Note that the belt may be considered as an infinite sheet.)

64. A particular paramagnetic substance achieves 10.0% of its saturation magnetization when placed in a magnetic field of 5.00 T at a temperature of 4.00 K. The density of magnetic atoms in the sample is 8.00×10^{27} atoms/m³, and the magnetic moment per atom is 5.00 Bohr magnetons. Calculate the Curie constant for this substance.

65. A bar magnet (mass = 39.4 g, magnetic moment = 7.65 J/T, length = 10.0 cm) is connected to the ceiling by a string. A uniform external magnetic field is applied horizontally, as shown in Figure P30.65. The magnet is in equilibrium, making an angle θ with the horizontal. If $\theta = 5.00°$, determine the magnitude of the applied magnetic field.

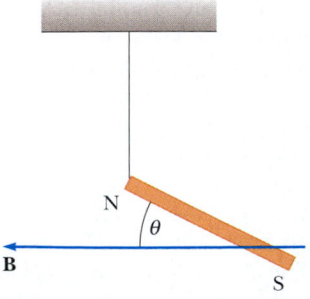

Figure P30.65

66. An infinitely long, straight wire carrying a current I_1 is partially surrounded by a loop, as shown in Figure P30.66. The loop has a length L and a radius R and carries a current I_2. The axis of the loop coincides with the wire. Calculate the force exerted on the loop.

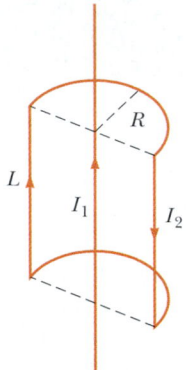

Figure P30.66

67. A wire is bent into the shape shown in Figure P30.67a, and the magnetic field is measured at P_1 when the current in the wire is I. The same wire is then formed into the shape shown in Figure P30.67b, and the magnetic field is measured at point P_2 when the current is again I. If the total length of wire is the same in each case, what is the ratio of B_1/B_2?

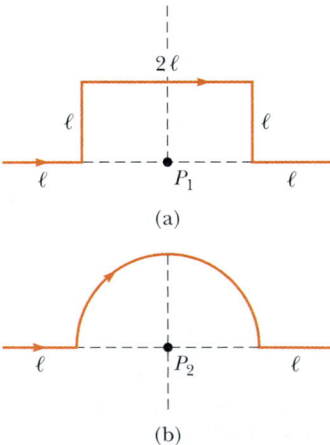

Figure P30.67

68. Measurements of the magnetic field of a large tornado were made at the Geophysical Observatory in Tulsa, Oklahoma, in 1962. If the tornado's field was $B = 15.0$ nT pointing north when the tornado was 9.00 km east of the observatory, what current was carried up or down the funnel of the tornado, modeled as a long straight wire?

69. A wire is formed into a square of edge length L (Fig. P30.69). Show that when the current in the loop is I, the magnetic field at point P, a distance x from the center of the square along its axis, is

$$B = \frac{\mu_0 I L^2}{2\pi(x^2 + L^2/4)\sqrt{x^2 + L^2/2}}$$

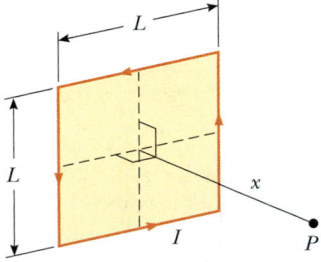

Figure P30.69

70. The force on a magnetic dipole $\boldsymbol{\mu}$ aligned with a nonuniform magnetic field in the x direction is given by $F_x = |\boldsymbol{\mu}| \, dB/dx$. Suppose that two flat loops of wire each have radius R and carry current I. (a) If the loops are arranged coaxially and separated by variable distance x, which is great compared to R, show that the magnetic force between them varies as $1/x^4$. (b) Evaluate the magnitude of this force if $I = 10.0$ A, $R = 0.500$ cm, and $x = 5.00$ cm.

71. A wire carrying a current I is bent into the shape of an exponential spiral $r = e^\theta$ from $\theta = 0$ to $\theta = 2\pi$, as in Figure P30.71. To complete a loop, the ends of the spiral are connected by a straight wire along the x axis. Find the magnitude and direction of **B** at the origin. *Hints:* Use the Biot–Savart law. The angle β between a radial line and its tangent line at any point on the curve $r = f(\theta)$ is related to the function in the following way:

$$\tan \beta = \frac{r}{dr/d\theta}$$

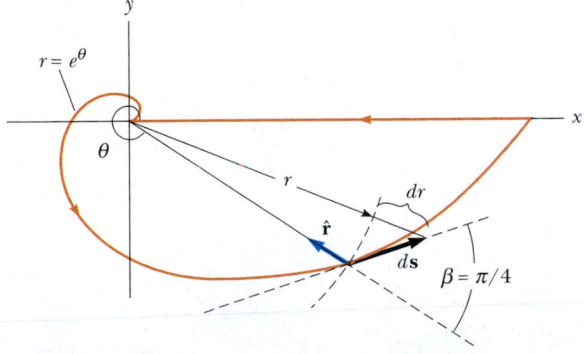

Figure P30.71

Thus, in this case $r = e^\theta$, $\tan \beta = 1$, and $\beta = \pi/4$. Therefore, the angle between $d\mathbf{s}$ and $\hat{\mathbf{r}}$ is $\pi - \beta = 3\pi/4$. Also,

$$ds = \frac{dr}{\sin \pi/4} = \sqrt{2} \, dr$$

72. Table P30.72 contains data taken for a ferromagnetic material. (a) Construct a magnetization curve from the data. Remember that $\mathbf{B} = \mathbf{B}_0 + \mu_0 \mathbf{M}$. (b) Determine the ratio B/B_0 for each pair of values of B and B_0, and construct a graph of B/B_0 versus B_0. (The fraction B/B_0 is called the relative permeability and is a measure of the induced magnetic field.)

TABLE P30.72	
B(T)	B_0 (T)
0.2	4.8×10^{-5}
0.4	7.0×10^{-5}
0.6	8.8×10^{-5}
0.8	1.2×10^{-4}
1.0	1.8×10^{-4}
1.2	3.1×10^{-4}
1.4	8.7×10^{-4}
1.6	3.4×10^{-3}
1.8	1.2×10^{-1}

73. **Review Problem.** A sphere of radius R has a constant volume charge density ρ. Determine the magnetic field at the center of the sphere when it rotates as a rigid body with angular velocity ω about an axis through its center (Fig. P30.73).

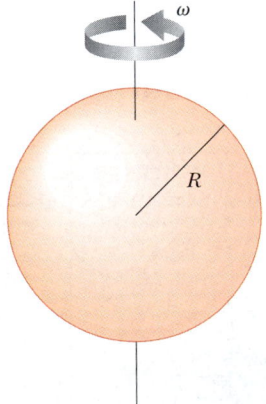

Figure P30.73 Problems 73 and 74.

74. **Review Problem.** A sphere of radius R has a constant volume charge density ρ. Determine the magnetic dipole moment of the sphere when it rotates as a rigid body with angular velocity ω about an axis through its center (see Fig. P30.73).

75. A long, cylindrical conductor of radius a has two cylindrical cavities of diameter a through its entire length, as shown in cross-section in Figure P30.75. A current I is directed out of the page and is uniform through a cross section of the conductor. Find the magnitude and direction of the magnetic field in terms of μ_0, I, r, and a (a) at point P_1 and (b) at point P_2.

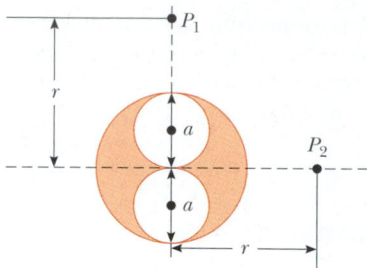

Figure P30.75

ANSWERS TO QUICK QUIZZES

30.1 (c) $F_1 = F_2$ because of Newton's third law. Another way to arrive at this answer is to realize that Equation 30.11 gives the same result whether the multiplication of currents is (2 A)(6 A) or (6 A)(2 A).

30.2 Closer together; the coils act like wires carrying parallel currents and hence attract one another.

30.3 b, d, a, c. Equation 30.13 indicates that the value of the line integral depends only on the net current through each closed path. Path b encloses 1 A, path d encloses 3 A, path a encloses 4 A, and path c encloses 6 A.

30.4 b, then $a = c = d$. Paths a, c, and d all give the same nonzero value $\mu_0 I$ because the size and shape of the paths do not matter. Path b does not enclose the current, and hence its line integral is zero.

30.5 Net force, yes; net torque, no. The forces on the top and bottom of the loop cancel because they are equal in magnitude but opposite in direction. The current in the left side of the loop is parallel to I_1, and hence the force F_L exerted by I_1 on this side is attractive. The current in the right side of the loop is antiparallel to I_1, and hence the force F_R exerted by I_1 on this side of the loop is repulsive. Because the left side is closer to wire 1, $F_L > F_R$ and a net force is directed toward wire 1. Because the

forces on all four sides of the loop lie in the plane of the loop, there is no net torque.

30.6 Zero; no charges flow into a fully charged capacitor, so no change occurs in the amount of charge on the plates, and the electric field between the plates is constant. It is only when the electric field is changing that a displacement current exists.

30.7 (a) Increases slightly; (b) decreases slightly; (c) increases greatly. Equations 30.33 and 30.34 indicate that, when each metal is in place, the total field is $\mathbf{B} = \mu_0(1 + \chi)\mathbf{H}$. Table 30.2 indicates that $\mu_0(1 + \chi)\mathbf{H}$ is slightly greater than $\mu_0\mathbf{H}$ for aluminum and slightly less for copper. For iron, the field can be made thousands of times stronger, as we saw in Example 30.10.

30.8 One whose loop looks like Figure 30.31a because the remanent magnetization at the point corresponding to point b in Figure 30.30 is greater.

30.9 West to east. The lines of the Earth's magnetic field enter the planet in Hudson Bay and emerge from Antarctica; thus, the field lines resulting from the current would have to go in the opposite direction. Compare Figure 30.6a with Figure 30.35.

Calvin and Hobbes   by Bill Watterson

PUZZLER

Before this vending machine will deliver its product, it conducts several tests on the coins being inserted. How can it determine what material the coins are made of without damaging them and without making the customer wait a long time for the results? *(George Semple)*

chapter

31

Faraday's Law

The focus of our studies in electricity and magnetism so far has been the electric fields produced by stationary charges and the magnetic fields produced by moving charges. This chapter deals with electric fields produced by changing magnetic fields.

Experiments conducted by Michael Faraday in England in 1831 and independently by Joseph Henry in the United States that same year showed that an emf can be induced in a circuit by a changing magnetic field. As we shall see, an emf (and therefore a current as well) can be induced in many ways—for instance, by moving a closed loop of wire into a region where a magnetic field exists. The results of these experiments led to a very basic and important law of electromagnetism known as *Faraday's law of induction*. This law states that the magnitude of the emf induced in a circuit equals the time rate of change of the magnetic flux through the circuit.

With the treatment of Faraday's law, we complete our introduction to the fundamental laws of electromagnetism. These laws can be summarized in a set of four equations called *Maxwell's equations*. Together with the *Lorentz force law*, which we discuss briefly, they represent a complete theory for describing the interaction of charged objects. Maxwell's equations relate electric and magnetic fields to each other and to their ultimate source, namely, electric charges.

31.1 ▶ FARADAY'S LAW OF INDUCTION

To see how an emf can be induced by a changing magnetic field, let us consider a loop of wire connected to a galvanometer, as illustrated in Figure 31.1. When a magnet is moved toward the loop, the galvanometer needle deflects in one direction, arbitrarily shown to the right in Figure 31.1a. When the magnet is moved away from the loop, the needle deflects in the opposite direction, as shown in Figure 31.1c. When the magnet is held stationary relative to the loop (Fig. 31.1b), no deflection is observed. Finally, if the magnet is held stationary and the loop is moved either toward or away from it, the needle deflects. From these observations, we conclude that the loop "knows" that the magnet is moving relative to it because it experiences a change in magnetic field strength. Thus, it seems that a relationship exists between current and changing magnetic field.

These results are quite remarkable in view of the fact that **a current is set up even though no batteries are present in the circuit!** We call such a current an *induced current* and say that it is produced by an *induced emf*.

Now let us describe an experiment conducted by Faraday[1] and illustrated in Figure 31.2. A primary coil is connected to a switch and a battery. The coil is wrapped around a ring, and a current in the coil produces a magnetic field when the switch is closed. A secondary coil also is wrapped around the ring and is connected to a galvanometer. No battery is present in the secondary circuit, and the secondary coil is not connected to the primary coil. Any current detected in the secondary circuit must be induced by some external agent.

Initially, you might guess that no current is ever detected in the secondary circuit. However, something quite amazing happens when the switch in the primary

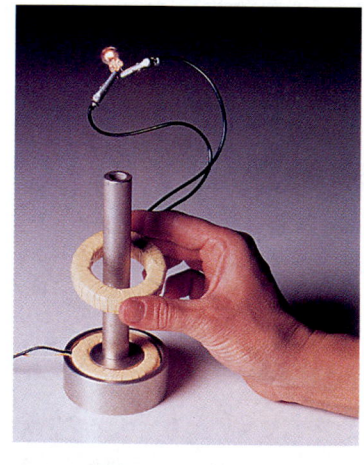

A demonstration of electromagnetic induction. A changing potential difference is applied to the lower coil. An emf is induced in the upper coil as indicated by the illuminated lamp. What happens to the lamp's intensity as the upper coil is moved over the vertical tube?
(Courtesy of Central Scientific Company)

---

[1] A physicist named J. D. Colladon was the first to perform the moving-magnet experiment. To minimize the effect of the changing magnetic field on his galvanometer, he placed the meter in an adjacent room. Thus, as he moved the magnet in the loop, he could not see the meter needle deflecting. By the time he returned next door to read the galvanometer, the needle was back to zero because he had stopped moving the magnet. Unfortunately for Colladon, there must be relative motion between the loop and the magnet for an induced emf and a corresponding induced current to be observed. Thus, physics students learn Faraday's law of induction rather than "Colladon's law of induction."

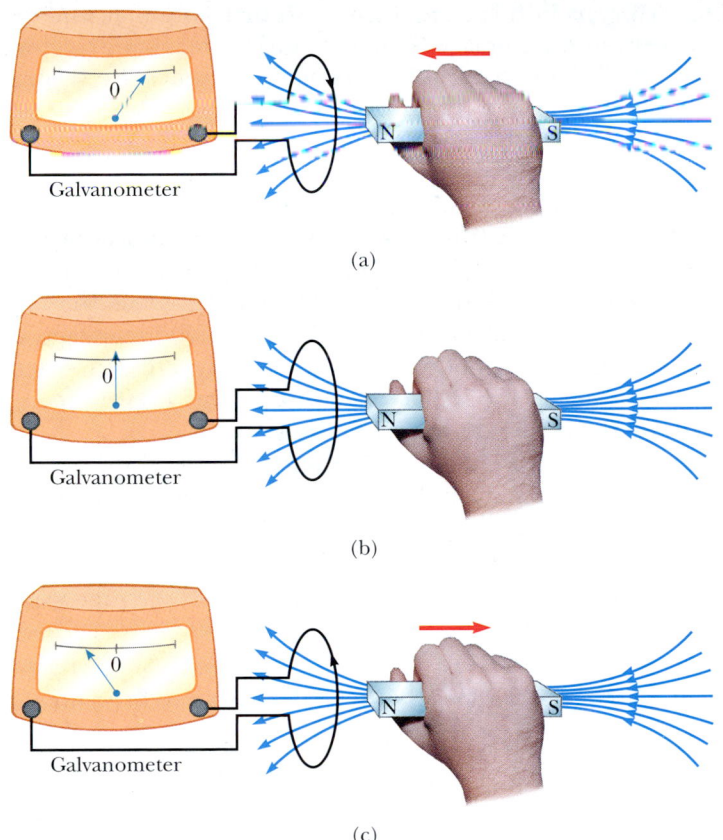

(a)

(b)

(c)

Figure 31.1 (a) When a magnet is moved toward a loop of wire connected to a galvanometer, the galvanometer deflects as shown, indicating that a current is induced in the loop. (b) When the magnet is held stationary, there is no induced current in the loop, even when the magnet is inside the loop. (c) When the magnet is moved away from the loop, the galvanometer deflects in the opposite direction, indicating that the induced current is opposite that shown in part (a). Changing the direction of the magnet's motion changes the direction of the current induced by that motion.

circuit is either suddenly closed or suddenly opened. At the instant the switch is closed, the galvanometer needle deflects in one direction and then returns to zero. At the instant the switch is opened, the needle deflects in the opposite direction and again returns to zero. Finally, the galvanometer reads zero when there is either a steady current or no current in the primary circuit. The key to under-

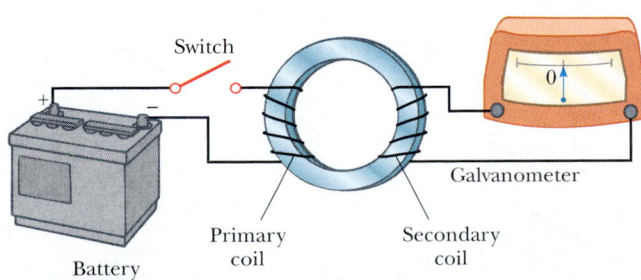

Figure 31.2 Faraday's experiment. When the switch in the primary circuit is closed, the galvanometer in the secondary circuit deflects momentarily. The emf induced in the secondary circuit is caused by the changing magnetic field through the secondary coil.

Michael Faraday (1791–1867)
Faraday, a British physicist and chemist, is often regarded as the greatest experimental scientist of the 1800s. His many contributions to the study of electricity include the invention of the electric motor, electric generator, and transformer, as well as the discovery of electromagnetic induction and the laws of electrolysis. Greatly influenced by religion, he refused to work on the development of poison gas for the British military.
(By kind permission of the President and Council of the Royal Society)

standing what happens in this experiment is to first note that when the switch is closed, the current in the primary circuit produces a magnetic field in the region of the circuit, and it is this magnetic field that penetrates the secondary circuit. Furthermore, when the switch is closed, the magnetic field produced by the current in the primary circuit changes from zero to some value over some finite time, and it is this changing field that induces a current in the secondary circuit.

As a result of these observations, Faraday concluded that **an electric current can be induced in a circuit (the secondary circuit in our setup) by a changing magnetic field.** The induced current exists for only a short time while the magnetic field through the secondary coil is changing. Once the magnetic field reaches a steady value, the current in the secondary coil disappears. In effect, the secondary circuit behaves as though a source of emf were connected to it for a short time. It is customary to say that **an induced emf is produced in the secondary circuit by the changing magnetic field.**

The experiments shown in Figures 31.1 and 31.2 have one thing in common: In each case, an emf is induced in the circuit when the magnetic flux through the circuit changes with time. In general,

> the emf induced in a circuit is directly proportional to the time rate of change of the magnetic flux through the circuit.

This statement, known as **Faraday's law of induction,** can be written

| Faraday's law |

$$\mathcal{E} = -\frac{d\Phi_B}{dt} \tag{31.1}$$

where $\Phi_B = \int \mathbf{B} \cdot d\mathbf{A}$ is the magnetic flux through the circuit (see Section 30.5).

If the circuit is a coil consisting of N loops all of the same area and if Φ_B is the flux through one loop, an emf is induced in every loop; thus, the total induced emf in the coil is given by the expression

$$\mathcal{E} = -N\frac{d\Phi_B}{dt} \tag{31.2}$$

The negative sign in Equations 31.1 and 31.2 is of important physical significance, which we shall discuss in Section 31.3.

Suppose that a loop enclosing an area A lies in a uniform magnetic field $\mathbf{B}$, as shown in Figure 31.3. The magnetic flux through the loop is equal to $BA \cos \theta$;

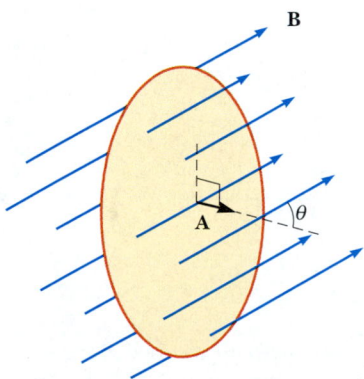

Figure 31.3   A conducting loop that encloses an area A in the presence of a uniform magnetic field $\mathbf{B}$. The angle between $\mathbf{B}$ and the normal to the loop is θ.

hence, the induced emf can be expressed as

$$\mathcal{E} = -\frac{d}{dt}\,(BA\cos\theta) \qquad (31.3)$$

From this expression, we see that an emf can be induced in the circuit in several ways:

- The magnitude of **B** can change with time.
- The area enclosed by the loop can change with time.
- The angle θ between **B** and the normal to the loop can change with time.
- Any combination of the above can occur.

Quick Quiz 31.1

Equation 31.3 can be used to calculate the emf induced when the north pole of a magnet is moved toward a loop of wire, along the axis perpendicular to the plane of the loop passing through its center. What changes are necessary in the equation when the south pole is moved toward the loop?

Some Applications of Faraday's Law

The ground fault interrupter (GFI) is an interesting safety device that protects users of electrical appliances against electric shock. Its operation makes use of Faraday's law. In the GFI shown in Figure 31.4, wire 1 leads from the wall outlet to the appliance to be protected, and wire 2 leads from the appliance back to the wall outlet. An iron ring surrounds the two wires, and a sensing coil is wrapped around part of the ring. Because the currents in the wires are in opposite directions, the net magnetic flux through the sensing coil due to the currents is zero. However, if the return current in wire 2 changes, the net magnetic flux through the sensing coil is no longer zero. (This can happen, for example, if the appliance gets wet, enabling current to leak to ground.) Because household current is alternating (meaning that its direction keeps reversing), the magnetic flux through the sensing coil changes with time, inducing an emf in the coil. This induced emf is used to trigger a circuit breaker, which stops the current before it is able to reach a harmful level.

Another interesting application of Faraday's law is the production of sound in an electric guitar (Fig. 31.5). The coil in this case, called the *pickup coil,* is placed near the vibrating guitar string, which is made of a metal that can be magnetized. A permanent magnet inside the coil magnetizes the portion of the string nearest

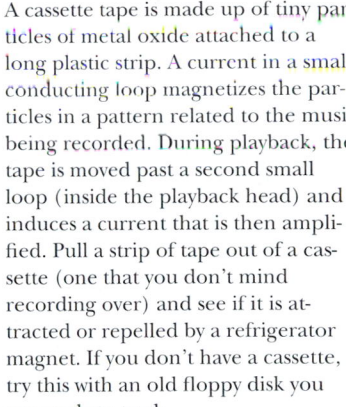

This electric range cooks food on the basis of the principle of induction. An oscillating current is passed through a coil placed below the cooking surface, which is made of a special glass. The current produces an oscillating magnetic field, which induces a current in the cooking utensil. Because the cooking utensil has some electrical resistance, the electrical energy associated with the induced current is transformed to internal energy, causing the utensil and its contents to become hot. *(Courtesy of Corning, Inc.)*

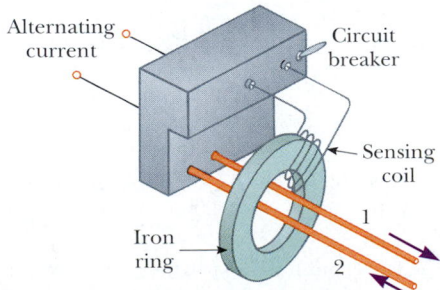

Figure 31.4   Essential components of a ground fault interrupter.

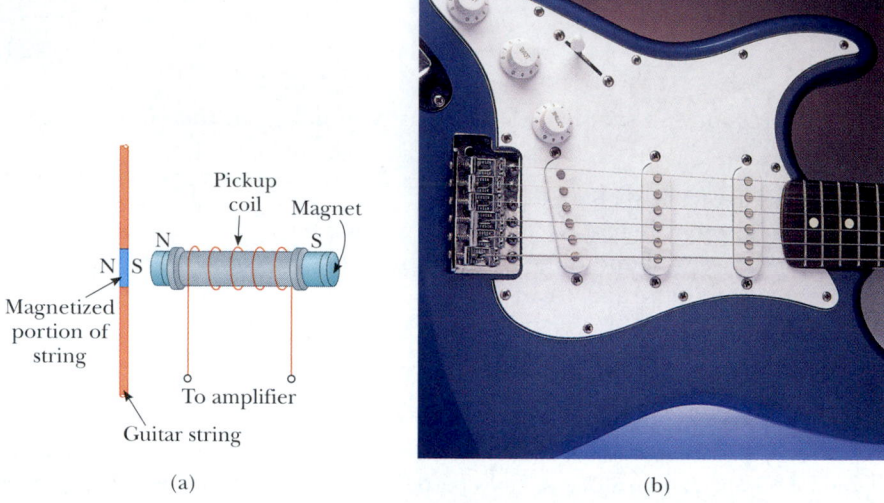

Figure 31.5   (a) In an electric guitar, a vibrating string induces an emf in a pickup coil. (b) The circles beneath the metallic strings of this electric guitar detect the notes being played and send this information through an amplifier and into speakers. (A switch on the guitar allows the musician to select which set of six is used.) How does a guitar "pickup" sense what music is being played? *(b, Charles D. Winters)*

the coil. When the string vibrates at some frequency, its magnetized segment produces a changing magnetic flux through the coil. The changing flux induces an emf in the coil that is fed to an amplifier. The output of the amplifier is sent to the loudspeakers, which produce the sound waves we hear.

$\mathcal{E}$XAMPLE 31.1   One Way to Induce an emf in a Coil

A coil consists of 200 turns of wire having a total resistance of 2.0 Ω. Each turn is a square of side 18 cm, and a uniform magnetic field directed perpendicular to the plane of the coil is turned on. If the field changes linearly from 0 to 0.50 T in 0.80 s, what is the magnitude of the induced emf in the coil while the field is changing?

Solution   The area of one turn of the coil is $(0.18 \text{ m})^2 = 0.032\ 4 \text{ m}^2$. The magnetic flux through the coil at $t = 0$ is zero because $B = 0$ at that time. At $t = 0.80$ s, the magnetic flux through one turn is $\Phi_B = BA = (0.50 \text{ T})(0.032\ 4 \text{ m}^2) = 0.016\ 2 \text{ T} \cdot \text{m}^2$. Therefore, the magnitude of the induced emf

is, from Equation 31.2,

$$|\mathcal{E}| = \frac{N \Delta \Phi_B}{\Delta t} = \frac{200(0.016\ 2 \text{ T} \cdot \text{m}^2 - 0 \text{ T} \cdot \text{m}^2)}{0.80 \text{ s}}$$

$$= 4.1 \text{ T} \cdot \text{m}^2/\text{s} = \boxed{4.1 \text{ V}}$$

You should be able to show that $1 \text{ T} \cdot \text{m}^2/\text{s} = 1 \text{ V}$.

Exercise   What is the magnitude of the induced current in the coil while the field is changing?

Answer   2.0 A.

$\mathcal{E}$XAMPLE 31.2   An Exponentially Decaying B Field

A loop of wire enclosing an area A is placed in a region where the magnetic field is perpendicular to the plane of the loop. The magnitude of **B** varies in time according to the expression $B = B_{max}e^{-at}$, where a is some constant. That is, at $t = 0$ the field is B_{max}, and for $t > 0$, the field decreases exponen-

tially (Fig. 31.6). Find the induced emf in the loop as a function of time.

Solution   Because **B** is perpendicular to the plane of the loop, the magnetic flux through the loop at time $t > 0$ is

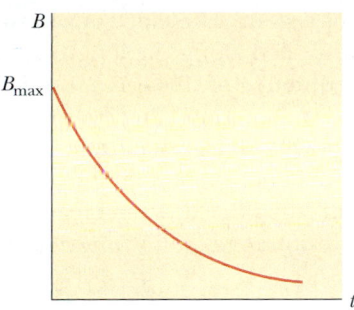

Figure 31.6 Exponential decrease in the magnitude of the magnetic field with time. The induced emf and induced current vary with time in the same way.

$$\Phi_B = BA\cos 0 = AB_{max}e^{-at}$$

Because AB_{max} and a are constants, the induced emf calculated from Equation 31.1 is

$$\mathcal{E} = -\frac{d\Phi_B}{dt} = -AB_{max}\frac{d}{dt}e^{-at} = aAB_{max}e^{-at}$$

This expression indicates that the induced emf decays exponentially in time. Note that the maximum emf occurs at $t = 0$, where $\mathcal{E}_{max} = aAB_{max}$. The plot of $\mathcal{E}$ versus t is similar to the B-versus-t curve shown in Figure 31.6.

CONCEPTUAL EXAMPLE 31.3 ▶ What Is Connected to What?

Two bulbs are connected to opposite sides of a loop of wire, as shown in Figure 31.7. A decreasing magnetic field (confined to the circular area shown in the figure) induces an emf in the loop that causes the two bulbs to light. What happens to the brightness of the bulbs when the switch is closed?

Solution Bulb 1 glows brighter, and bulb 2 goes out. Once the switch is closed, bulb 1 is in the large loop consisting of the wire to which it is attached and the wire connected to the switch. Because the changing magnetic flux is completely enclosed within this loop, a current exists in bulb 1. Bulb 1 now glows brighter than before the switch was closed because it is now the only resistance in the loop. As a result, the current in bulb 1 is greater than when bulb 2 was also in the loop.

Once the switch is closed, bulb 2 is in the loop consisting of the wires attached to it and those connected to the switch. There is no changing magnetic flux through this loop and hence no induced emf.

Exercise What would happen if the switch were in a wire located to the left of bulb 1?

Answer Bulb 1 would go out, and bulb 2 would glow brighter.

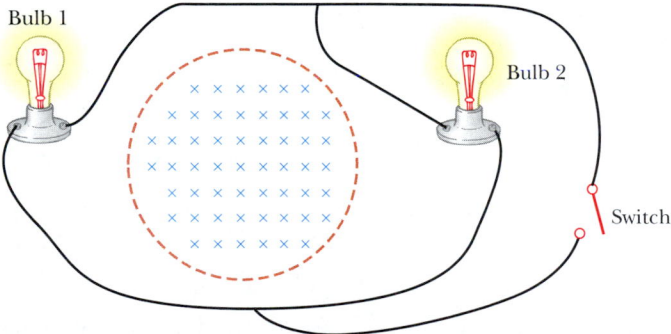

Figure 31.7

31.2 ▶ MOTIONAL EMF

In Examples 31.1 and 31.2, we considered cases in which an emf is induced in a stationary circuit placed in a magnetic field when the field changes with time. In this section we describe what is called **motional emf,** which is the emf induced in a conductor moving through a constant magnetic field.

The straight conductor of length ℓ shown in Figure 31.8 is moving through a uniform magnetic field directed into the page. For simplicity, we assume that the conductor is moving in a direction perpendicular to the field with constant veloc-

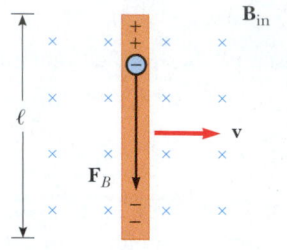

B_{in}

ℓ

F_B

v

Figure 31.8   A straight electrical conductor of length ℓ moving with a velocity **v** through a uniform magnetic field **B** directed perpendicular to **v**. A potential difference $\Delta V = B\ell v$ is maintained between the ends of the conductor.

ity under the influence of some external agent. The electrons in the conductor experience a force $\mathbf{F}_B = q\mathbf{v} \times \mathbf{B}$ that is directed along the length ℓ, perpendicular to both **v** and **B** (Eq. 29.1). Under the influence of this force, the electrons move to the lower end of the conductor and accumulate there, leaving a net positive charge at the upper end. As a result of this charge separation, an electric field is produced inside the conductor. The charges accumulate at both ends until the downward magnetic force qvB is balanced by the upward electric force qE. At this point, electrons stop moving. The condition for equilibrium requires that

$$qE = qvB \qquad \text{or} \qquad E = vB$$

The electric field produced in the conductor (once the electrons stop moving and E is constant) is related to the potential difference across the ends of the conductor according to the relationship $\Delta V = E\ell$ (Eq. 25.6). Thus,

$$\Delta V = E\ell = B\ell v \tag{31.4}$$

where the upper end is at a higher electric potential than the lower end. Thus, **a potential difference is maintained between the ends of the conductor as long as the conductor continues to move through the uniform magnetic field.** If the direction of the motion is reversed, the polarity of the potential difference also is reversed.

A more interesting situation occurs when the moving conductor is part of a closed conducting path. This situation is particularly useful for illustrating how a changing magnetic flux causes an induced current in a closed circuit. Consider a circuit consisting of a conducting bar of length ℓ sliding along two fixed parallel conducting rails, as shown in Figure 31.9a.

For simplicity, we assume that the bar has zero resistance and that the stationary part of the circuit has a resistance R. A uniform and constant magnetic field **B** is applied perpendicular to the plane of the circuit. As the bar is pulled to the right with a velocity **v**, under the influence of an applied force $\mathbf{F}_{app}$, free charges in the bar experience a magnetic force directed along the length of the bar. This force sets up an induced current because the charges are free to move in the closed conducting path. In this case, the rate of change of magnetic flux through the loop and the corresponding induced motional emf across the moving bar are proportional to the change in area of the loop. As we shall see, if the bar is pulled to the right with a constant velocity, the work done by the applied force appears as internal energy in the resistor R (see Section 27.6).

Because the area enclosed by the circuit at any instant is ℓx, where x is the width of the circuit at any instant, the magnetic flux through that area is

$$\Phi_B = B\ell x$$

Using Faraday's law, and noting that x changes with time at a rate $dx/dt = v$, we find that the induced motional emf is

$$\mathcal{E} = -\frac{d\Phi_B}{dt} = -\frac{d}{dt}(B\ell x) = -B\ell\frac{dx}{dt}$$

Motional emf

$$\mathcal{E} = -B\ell v \tag{31.5}$$

Because the resistance of the circuit is R, the magnitude of the induced current is

$$I = \frac{|\mathcal{E}|}{R} = \frac{B\ell v}{R} \tag{31.6}$$

The equivalent circuit diagram for this example is shown in Figure 31.9b.

Let us examine the system using energy considerations. Because no battery is in the circuit, we might wonder about the origin of the induced current and the electrical energy in the system. We can understand the source of this current and energy by noting that the applied force does work on the conducting bar, thereby moving charges through a magnetic field. Their movement through the field causes the charges to move along the bar with some average drift velocity, and hence a current is established. Because energy must be conserved, the work done by the applied force on the bar during some time interval must equal the electrical energy supplied by the induced emf during that same interval. Furthermore, if the bar moves with constant speed, the work done on it must equal the energy delivered to the resistor during this time interval.

As it moves through the uniform magnetic field **B**, the bar experiences a magnetic force **F**$_B$ of magnitude $I\ell B$ (see Section 29.2). The direction of this force is opposite the motion of the bar, to the left in Figure 31.9a. Because the bar moves with constant velocity, the applied force must be equal in magnitude and opposite in direction to the magnetic force, or to the right in Figure 31.9a. (If **F**$_B$ acted in the direction of motion, it would cause the bar to accelerate. Such a situation would violate the principle of conservation of energy.) Using Equation 31.6 and the fact that $F_{\text{app}} = I\ell B$, we find that the power delivered by the applied force is

$$\mathcal{P} = F_{\text{app}}v = (I\ell B)v = \frac{B^2\ell^2v^2}{R} = \frac{\mathcal{E}^2}{R} \qquad (31.7)$$

From Equation 27.23, we see that this power is equal to the rate at which energy is delivered to the resistor I^2R, as we would expect. It is also equal to the power $I\mathcal{E}$ supplied by the motional emf. This example is a clear demonstration of the conversion of mechanical energy first to electrical energy and finally to internal energy in the resistor.

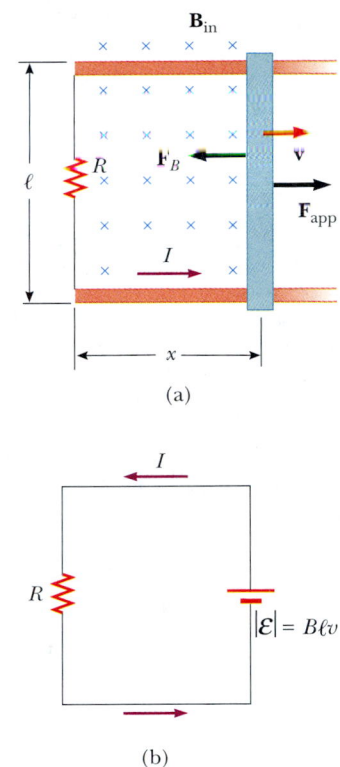

Figure 31.9 (a) A conducting bar sliding with a velocity **v** along two conducting rails under the action of an applied force **F**$_{\text{app}}$. The magnetic force **F**$_B$ opposes the motion, and a counterclockwise current I is induced in the loop. (b) The equivalent circuit diagram for the setup shown in part (a).

Quick Quiz 31.2

As an airplane flies from Los Angeles to Seattle, it passes through the Earth's magnetic field. As a result, a motional emf is developed between the wingtips. Which wingtip is positively charged?

EXAMPLE 31.4   Motional emf Induced in a Rotating Bar

A conducting bar of length ℓ rotates with a constant angular speed ω about a pivot at one end. A uniform magnetic field **B** is directed perpendicular to the plane of rotation, as shown in Figure 31.10. Find the motional emf induced between the ends of the bar.

Solution Consider a segment of the bar of length dr having a velocity **v**. According to Equation 31.5, the magnitude of the emf induced in this segment is

$$d\mathcal{E} = Bv\,dr$$

Because every segment of the bar is moving perpendicular to **B**, an emf $d\mathcal{E}$ of the same form is generated across each. Summing the emfs induced across all segments, which are in series, gives the total emf between the ends of

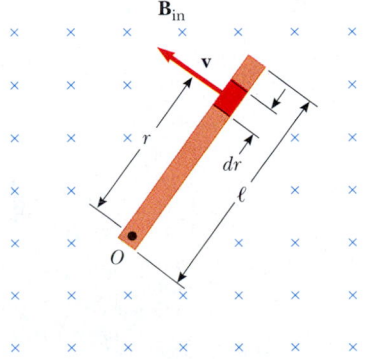

Figure 31.10 A conducting bar rotating around a pivot at one end in a uniform magnetic field that is perpendicular to the plane of rotation. A motional emf is induced across the ends of the bar.

the bar:

$$\mathcal{E} = \int Bv \, dr$$

To integrate this expression, we must note that the linear speed of an element is related to the angular speed ω

through the relationship $v = r\omega$. Therefore, because B and ω are constants, we find that

$$\mathcal{E} = B \int v \, dr = B\omega \int_0^\ell r \, dr = \tfrac{1}{2} B\omega \ell^2$$

EXAMPLE 31.5   Magnetic Force Acting on a Sliding Bar

The conducting bar illustrated in Figure 31.11, of mass m and length ℓ, moves on two frictionless parallel rails in the presence of a uniform magnetic field directed into the page. The bar is given an initial velocity $\mathbf{v}_i$ to the right and is released at $t = 0$. Find the velocity of the bar as a function of time.

Solution   The induced current is counterclockwise, and the magnetic force is $F_B = -I\ell B$, where the negative sign denotes that the force is to the left and retards the motion. This is the only horizontal force acting on the bar, and hence Newton's second law applied to motion in the horizontal direction gives

$$F_x = ma = m\frac{dv}{dt} = -I\ell B$$

From Equation 31.6, we know that $I = B\ell v/R$, and so we can write this expression as

$$m\frac{dv}{dt} = -\frac{B^2\ell^2}{R}v$$

$$\frac{dv}{v} = -\left(\frac{B^2\ell^2}{mR}\right)dt$$

Integrating this equation using the initial condition that $v = v_i$ at $t = 0$, we find that

$$\int_{v_i}^v \frac{dv}{v} = \frac{-B^2\ell^2}{mR}\int_0^t dt$$

$$\ln\left(\frac{v}{v_i}\right) = -\left(\frac{B^2\ell^2}{mR}\right)t = -\frac{t}{\tau}$$

where the constant $\tau = mR/B^2\ell^2$. From this result, we see

that the velocity can be expressed in the exponential form

$$v = v_i e^{-t/\tau}$$

This expression indicates that the velocity of the bar decreases exponentially with time under the action of the magnetic retarding force.

Exercise   Find expressions for the induced current and the magnitude of the induced emf as functions of time for the bar in this example.

Answer $I = \dfrac{B\ell v_i}{R} e^{-t/\tau}$; $\mathcal{E} = B\ell v_i e^{-t/\tau}$. (They both decrease exponentially with time.)

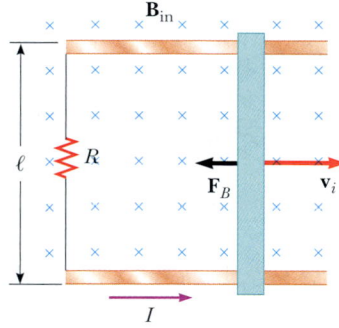

Figure 31.11   A conducting bar of length ℓ sliding on two fixed conducting rails is given an initial velocity $\mathbf{v}_i$ to the right.

31.3   LENZ'S LAW

Faraday's law (Eq. 31.1) indicates that the induced emf and the change in flux have opposite algebraic signs. This has a very real physical interpretation that has come to be known as **Lenz's law**[2]:

[2] Developed by the German physicist Heinrich Lenz (1804–1865).

> The polarity of the induced emf is such that it tends to produce a current that creates a magnetic flux to oppose the change in magnetic flux through the area enclosed by the current loop.

That is, the induced current tends to keep the original magnetic flux through the circuit from changing. As we shall see, this law is a consequence of the law of conservation of energy.

To understand Lenz's law, let us return to the example of a bar moving to the right on two parallel rails in the presence of a uniform magnetic field that we shall refer to as the *external* magnetic field (Fig. 31.12a). As the bar moves to the right, the magnetic flux through the area enclosed by the circuit increases with time because the area increases. Lenz's law states that the induced current must be directed so that the magnetic flux it produces opposes the change in the external magnetic flux. Because the external magnetic flux is increasing into the page, the induced current, if it is to oppose this change, must produce a flux directed out of the page. Hence, the induced current must be directed counterclockwise when the bar moves to the right. (Use the right-hand rule to verify this direction.) If the bar is moving to the left, as shown in Figure 31.12b, the external magnetic flux through the area enclosed by the loop decreases with time. Because the flux is directed into the page, the direction of the induced current must be clockwise if it is to produce a flux that also is directed into the page. In either case, the induced current tends to maintain the original flux through the area enclosed by the current loop.

Let us examine this situation from the viewpoint of energy considerations. Suppose that the bar is given a slight push to the right. In the preceding analysis, we found that this motion sets up a counterclockwise current in the loop. Let us see what happens if we assume that the current is clockwise, such that the direction of the magnetic force exerted on the bar is to the right. This force would accelerate the rod and increase its velocity. This, in turn, would cause the area enclosed by the loop to increase more rapidly; this would result in an increase in the induced current, which would cause an increase in the force, which would produce an increase in the current, and so on. In effect, the system would acquire energy with no additional input of energy. This is clearly inconsistent with all experience and with the law of conservation of energy. Thus, we are forced to conclude that the current must be counterclockwise.

Let us consider another situation, one in which a bar magnet moves toward a stationary metal loop, as shown in Figure 31.13a. As the magnet moves to the right toward the loop, the external magnetic flux through the loop increases with time. To counteract this increase in flux to the right, the induced current produces a flux to the left, as illustrated in Figure 31.13b; hence, the induced current is in the direction shown. Note that the magnetic field lines associated with the induced current oppose the motion of the magnet. Knowing that like magnetic poles repel each other, we conclude that the left face of the current loop is in essence a north pole and that the right face is a south pole.

If the magnet moves to the left, as shown in Figure 31.13c, its flux through the area enclosed by the loop, which is directed to the right, decreases in time. Now the induced current in the loop is in the direction shown in Figure 31.13d because this current direction produces a magnetic flux in the same direction as the external flux. In this case, the left face of the loop is a south pole and the right face is a north pole.

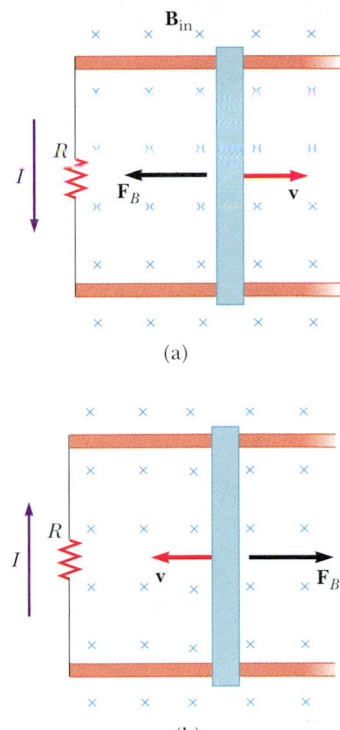

Figure 31.12 (a) As the conducting bar slides on the two fixed conducting rails, the magnetic flux through the area enclosed by the loop increases in time. By Lenz's law, the induced current must be counterclockwise so as to produce a counteracting magnetic flux directed out of the page. (b) When the bar moves to the left, the induced current must be clockwise. Why?

QuickLab

This experiment takes steady hands, a dime, and a strong magnet. After verifying that a dime is not attracted to the magnet, carefully balance the coin on its edge. (This won't work with other coins because they require too much force to topple them.) Hold one pole of the magnet within a millimeter of the face of the dime, but don't bump it. Now very rapidly pull the magnet straight back away from the coin. Which way does the dime tip? Does the coin fall the same way most of the time? Explain what is going on in terms of Lenz's law. You may want to refer to Figure 31.13.

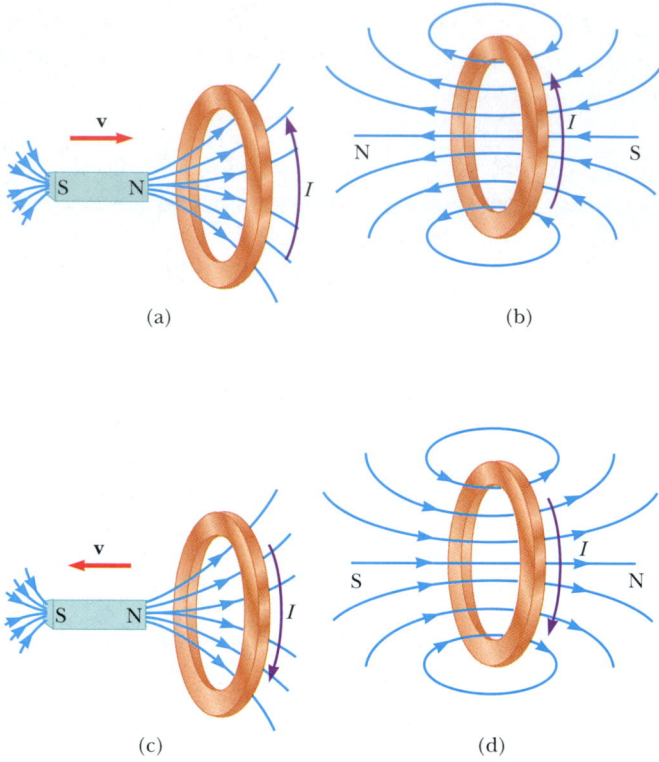

Figure 31.13   (a) When the magnet is moved toward the stationary conducting loop, a current is induced in the direction shown. (b) This induced current produces its own magnetic flux that is directed to the left and so counteracts the increasing external flux to the right. (c) When the magnet is moved away from the stationary conducting loop, a current is induced in the direction shown. (d) This induced current produces a magnetic flux that is directed to the right and so counteracts the decreasing external flux to the right.

Quick Quiz 31.3

Figure 31.14 shows a magnet being moved in the vicinity of a solenoid connected to a galvanometer. The south pole of the magnet is the pole nearest the solenoid, and the gal-

Figure 31.14   When a magnet is moved toward or away from a solenoid attached to a galvanometer, an electric current is induced, indicated by the momentary deflection of the galvanometer needle. *(Richard Megna/Fundamental Photographs)*

vanometer indicates a clockwise (viewed from above) current in the solenoid. Is the person inserting the magnet or pulling it out?

CONCEPTUAL EXAMPLE 31.6   Application of Lenz's Law

A metal ring is placed near a solenoid, as shown in Figure 31.15a. Find the direction of the induced current in the ring (a) at the instant the switch in the circuit containing the solenoid is thrown closed, (b) after the switch has been closed for several seconds, and (c) at the instant the switch is thrown open.

Solution   (a) At the instant the switch is thrown closed, the situation changes from one in which no magnetic flux passes through the ring to one in which flux passes through in the direction shown in Figure 31.15b. To counteract this change in the flux, the current induced in the ring must set up a magnetic field directed from left to right in Figure 31.15b. This requires a current directed as shown.

   (b) After the switch has been closed for several seconds, no change in the magnetic flux through the loop occurs; hence, the induced current in the ring is zero.

   (c) Opening the switch changes the situation from one in which magnetic flux passes through the ring to one in which there is no magnetic flux. The direction of the induced current is as shown in Figure 31.15c because current in this di-

rection produces a magnetic field that is directed right to left and so counteracts the decrease in the field produced by the solenoid.

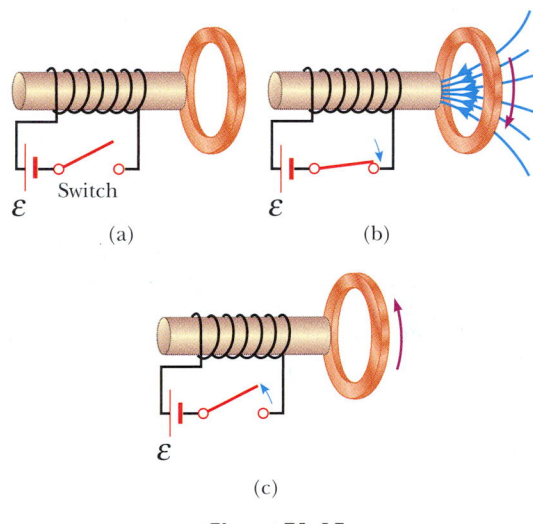

Figure 31.15

CONCEPTUAL EXAMPLE 31.7   A Loop Moving Through a Magnetic Field

A rectangular metallic loop of dimensions ℓ and w and resistance R moves with constant speed v to the right, as shown in Figure 31.16a, passing through a uniform magnetic field **B** directed into the page and extending a distance $3w$ along the x axis. Defining x as the position of the right side of the loop along the x axis, plot as functions of x (a) the magnetic flux through the area enclosed by the loop, (b) the induced motional emf, and (c) the external applied force necessary to counter the magnetic force and keep v constant.

Solution   (a) Figure 31.16b shows the flux through the area enclosed by the loop as a function x. Before the loop enters the field, the flux is zero. As the loop enters the field, the flux increases linearly with position until the left edge of the loop is just inside the field. Finally, the flux through the loop decreases linearly to zero as the loop leaves the field.

   (b) Before the loop enters the field, no motional emf is induced in it because no field is present (Fig. 31.16c). As the right side of the loop enters the field, the magnetic flux directed into the page increases. Hence, according to Lenz's law, the induced current is counterclockwise because it must produce a magnetic field directed out of the page. The motional emf $-B\ell v$ (from Eq. 31.5) arises from the mag-

netic force experienced by charges in the right side of the loop. When the loop is entirely in the field, the change in magnetic flux is zero, and hence the motional emf vanishes. This happens because, once the left side of the loop enters the field, the motional emf induced in it cancels the motional emf present in the right side of the loop. As the right side of the loop leaves the field, the flux inward begins to decrease, a clockwise current is induced, and the induced emf is $B\ell v$. As soon as the left side leaves the field, the emf decreases to zero.

   (c) The external force that must be applied to the loop to maintain this motion is plotted in Figure 31.16d. Before the loop enters the field, no magnetic force acts on it; hence, the applied force must be zero if v is constant. When the right side of the loop enters the field, the applied force necessary to maintain constant speed must be equal in magnitude and opposite in direction to the magnetic force exerted on that side: $F_B = -I\ell B = -B^2\ell^2 v/R$. When the loop is entirely in the field, the flux through the loop is not changing with time. Hence, the net emf induced in the loop is zero, and the current also is zero. Therefore, no external force is needed to maintain the motion. Finally, as the right side leaves the field, the applied force must be equal in magnitude and opposite in direction to the magnetic force acting on the left side of the loop.

From this analysis, we conclude that power is supplied only when the loop is either entering or leaving the field.

Furthermore, this example shows that the motional emf induced in the loop can be zero even when there is motion through the field! A motional emf is induced only when the magnetic flux through the loop *changes in time*.

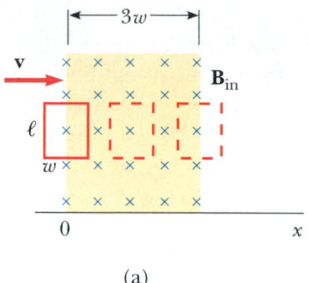

(a)

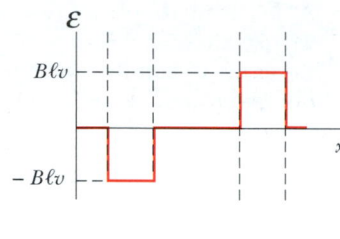

(c)

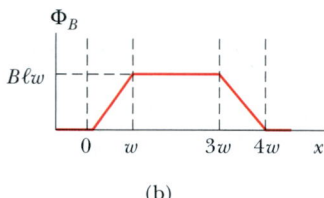

(b)

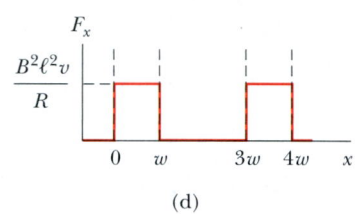

(d)

Figure 31.16 (a) A conducting rectangular loop of width w and length ℓ moving with a velocity **v** through a uniform magnetic field extending a distance $3w$. (b) Magnetic flux through the area enclosed by the loop as a function of loop position. (c) Induced emf as a function of loop position. (d) Applied force required for constant velocity as a function of loop position.

31.4 ▶ INDUCED EMF AND ELECTRIC FIELDS

We have seen that a changing magnetic flux induces an emf and a current in a conducting loop. Therefore, we must conclude that **an electric field is created in the conductor as a result of the changing magnetic flux.** However, this induced electric field has two important properties that distinguish it from the electrostatic field produced by stationary charges: The induced field is nonconservative and time varying.

We can illustrate this point by considering a conducting loop of radius r situated in a uniform magnetic field that is perpendicular to the plane of the loop, as shown in Figure 31.17. If the magnetic field changes with time, then, according to Faraday's law (Eq. 31.1), an emf $\mathcal{E} = -d\Phi_B/dt$ is induced in the loop. The induction of a current in the loop implies the presence of an induced electric field **E**, which must be tangent to the loop because all points on the loop are equivalent. The work done in moving a test charge q once around the loop is equal to $q\mathcal{E}$. Because the electric force acting on the charge is $q\mathbf{E}$, the work done by this force in moving the charge once around the loop is $qE(2\pi r)$, where $2\pi r$ is the circumference of the loop. These two expressions for the work must be equal; therefore, we see that

$$q\mathcal{E} = qE(2\pi r)$$

$$E = \frac{\mathcal{E}}{2\pi r}$$

Using this result, along with Equation 31.1 and the fact that $\Phi_B = BA = \pi r^2 B$ for a

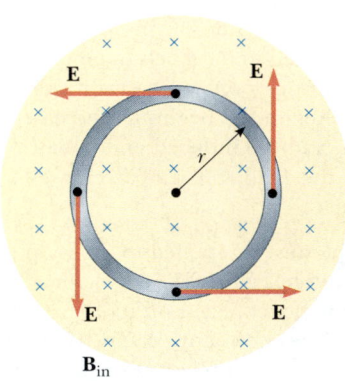

Figure 31.17 A conducting loop of radius r in a uniform magnetic field perpendicular to the plane of the loop. If **B** changes in time, an electric field is induced in a direction tangent to the circumference of the loop.

circular loop, we find that the induced electric field can be expressed as

$$E = -\frac{1}{2\pi r}\frac{d\Phi_B}{dt} = -\frac{r}{2}\frac{dB}{dt}$$

(31.8)

If the time variation of the magnetic field is specified, we can easily calculate the induced electric field from Equation 31.8. The negative sign indicates that the induced electric field opposes the change in the magnetic field.

The emf for any closed path can be expressed as the line integral of $\mathbf{E} \cdot d\mathbf{s}$ over that path: $\mathcal{E} = \oint \mathbf{E} \cdot d\mathbf{s}$. In more general cases, E may not be constant, and the path may not be a circle. Hence, Faraday's law of induction, $\mathcal{E} = -d\Phi_B/dt$, can be written in the general form

$$\oint \mathbf{E} \cdot d\mathbf{s} = -\frac{d\Phi_B}{dt}$$

(31.9)   **Faraday's law in general form**

It is important to recognize that **the induced electric field E in Equation 31.9 is a nonconservative, time-varying field that is generated by a changing magnetic field.** The field $\mathbf{E}$ that satisfies Equation 31.9 cannot possibly be an electrostatic field for the following reason: If the field were electrostatic, and hence conservative, the line integral of $\mathbf{E} \cdot d\mathbf{s}$ over a closed loop would be zero; this would be in contradiction to Equation 31.9.

EXAMPLE 31.8   Electric Field Induced by a Changing Magnetic Field in a Solenoid

A long solenoid of radius R has n turns of wire per unit length and carries a time-varying current that varies sinusoidally as $I = I_{max} \cos \omega t$, where I_{max} is the maximum current and ω is the angular frequency of the alternating current source (Fig. 31.18). (a) Determine the magnitude of the induced electric field outside the solenoid, a distance $r > R$ from its long central axis.

Solution First let us consider an external point and take the path for our line integral to be a circle of radius r centered on the solenoid, as illustrated in Figure 31.18. By sym-

metry we see that the magnitude of $\mathbf{E}$ is constant on this path and that $\mathbf{E}$ is tangent to it. The magnetic flux through the area enclosed by this path is $BA = B\pi R^2$; hence, Equation 31.9 gives

$$\oint \mathbf{E} \cdot d\mathbf{s} = -\frac{d}{dt}(B\pi R^2) = -\pi R^2 \frac{dB}{dt}$$

$$(1) \quad \oint \mathbf{E} \cdot d\mathbf{s} = E(2\pi r) = -\pi R^2 \frac{dB}{dt}$$

The magnetic field inside a long solenoid is given by Equation 30.17, $B = \mu_0 nI$. When we substitute $I = I_{max} \cos \omega t$ into this equation and then substitute the result into Equation (1), we find that

$$E(2\pi r) = -\pi R^2 \mu_0 nI_{max}\frac{d}{dt}(\cos \omega t) = \pi R^2 \mu_0 nI_{max}\omega \sin \omega t$$

$$(2) \quad \boxed{E = \frac{\mu_0 nI_{max}\omega R^2}{2r}\sin \omega t} \quad (\text{for } r > R)$$

Hence, the electric field varies sinusoidally with time and its amplitude falls off as $1/r$ outside the solenoid.

(b) What is the magnitude of the induced electric field inside the solenoid, a distance r from its axis?

Solution For an interior point $(r < R)$, the flux threading an integration loop is given by $B\pi r^2$. Using the same proce-

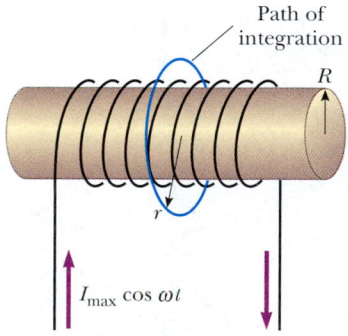

Path of integration

R

r

$I_{max} \cos \omega t$

Figure 31.18 A long solenoid carrying a time-varying current given by $I = I_0 \cos \omega t$. An electric field is induced both inside and outside the solenoid.

dure as in part (a), we find that

$$E(2\pi r) = -\pi r^2 \frac{dB}{dt} = \pi r^2 \mu_0 n I_{max} \omega \sin \omega t$$

(3) $$E = \frac{\mu_0 n I_{max} \omega}{2} r \sin \omega t \qquad (\text{for } r < R)$$

This shows that the amplitude of the electric field induced inside the solenoid by the changing magnetic flux through the solenoid increases linearly with r and varies sinusoidally with time.

Exercise Show that Equations (2) and (3) for the exterior and interior regions of the solenoid match at the boundary, $r = R$.

Exercise Would the electric field be different if the solenoid had an iron core?

Answer Yes, it could be much stronger because the maximum magnetic field (and thus the change in flux) through the solenoid could be thousands of times larger. (See Example 30.10.)

Optional Section

31.5 ▸ GENERATORS AND MOTORS

Electric generators are used to produce electrical energy. To understand how they work, let us consider the **alternating current** (ac) **generator,** a device that converts mechanical energy to electrical energy. In its simplest form, it consists of a loop of wire rotated by some external means in a magnetic field (Fig. 31.19a).

In commercial power plants, the energy required to rotate the loop can be derived from a variety of sources. For example, in a hydroelectric plant, falling water directed against the blades of a turbine produces the rotary motion; in a coal-fired plant, the energy released by burning coal is used to convert water to steam, and this steam is directed against the turbine blades. As a loop rotates in a magnetic field, the magnetic flux through the area enclosed by the loop changes with time; this induces an emf and a current in the loop according to Faraday's law. The ends of the loop are connected to slip rings that rotate with the loop. Connections from these slip rings, which act as output terminals of the generator, to the external circuit are made by stationary brushes in contact with the slip rings.

Turbines turn generators at a hydroelectric power plant. (*Luis Castaneda/The Image Bank*)

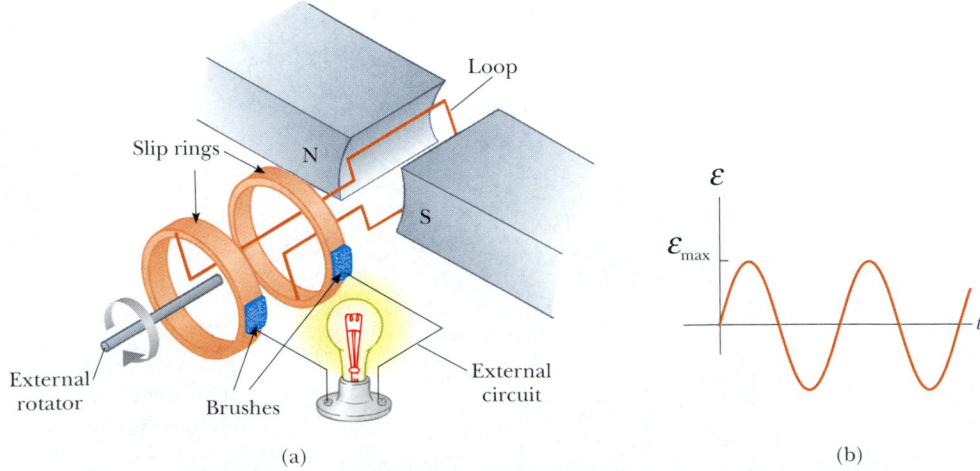

(a)   (b)

Figure 31.19 (a) Schematic diagram of an ac generator. An emf is induced in a loop that rotates in a magnetic field. (b) The alternating emf induced in the loop plotted as a function of time.

Suppose that, instead of a single turn, the loop has N turns (a more practical situation), all of the same area A, and rotates in a magnetic field with a constant angular speed ω. If θ is the angle between the magnetic field and the normal to the plane of the loop, as shown in Figure 31.20, then the magnetic flux through the loop at any time t is

$$\Phi_B = BA \cos \theta = BA \cos \omega t$$

where we have used the relationship $\theta = \omega t$ between angular displacement and angular speed (see Eq. 10.3). (We have set the clock so that $t = 0$ when $\theta = 0$.) Hence, the induced emf in the coil is

$$\mathcal{E} = -N\frac{d\Phi_B}{dt} = -NAB\frac{d}{dt}(\cos \omega t) = NAB\omega \sin \omega t \qquad \textbf{(31.10)}$$

This result shows that the emf varies sinusoidally with time, as was plotted in Figure 31.19b. From Equation 31.10 we see that the maximum emf has the value

$$\mathcal{E}_{max} = NAB\omega \qquad \textbf{(31.11)}$$

which occurs when $\omega t = 90°$ or $270°$. In other words, $\mathcal{E} = \mathcal{E}_{max}$ when the magnetic field is in the plane of the coil and the time rate of change of flux is a maximum. Furthermore, the emf is zero when $\omega t = 0$ or $180°$, that is, when **B** is perpendicular to the plane of the coil and the time rate of change of flux is zero.

The frequency for commercial generators in the United States and Canada is 60 Hz, whereas in some European countries it is 50 Hz. (Recall that $\omega = 2\pi f$, where f is the frequency in hertz.)

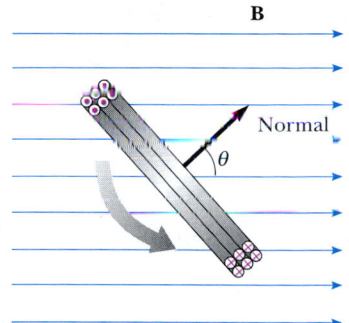

Figure 31.20 A loop enclosing an area A and containing N turns, rotating with constant angular speed ω in a magnetic field. The emf induced in the loop varies sinusoidally in time.

EXAMPLE 31.9 ▶ emf Induced in a Generator

An ac generator consists of 8 turns of wire, each of area $A = 0.090\ 0\ \text{m}^2$, and the total resistance of the wire is $12.0\ \Omega$. The loop rotates in a 0.500-T magnetic field at a constant frequency of 60.0 Hz. (a) Find the maximum induced emf.

Solution First, we note that $\omega = 2\pi f = 2\pi(60.0\ \text{Hz}) = 377\ \text{s}^{-1}$. Thus, Equation 31.11 gives

$$\mathcal{E}_{max} = NAB\omega = 8(0.090\ 0\ \text{m}^2)(0.500\ \text{T})(377\ \text{s}^{-1}) = \boxed{136\ \text{V}}$$

(b) What is the maximum induced current when the output terminals are connected to a low-resistance conductor?

Solution From Equation 27.8 and the results to part (a), we have

$$I_{max} = \frac{\mathcal{E}_{max}}{R} = \frac{136\ \text{V}}{12.0\ \Omega} = \boxed{11.3\ \text{A}}$$

Exercise Determine how the induced emf and induced current vary with time.

Answer $\mathcal{E} = \mathcal{E}_{max} \sin \omega t = (136\ \text{V}) \sin 377t$; $I = I_{max} \sin \omega t = (11.3\ \text{A}) \sin 377t$.

The **direct current** (dc) **generator** is illustrated in Figure 31.21a. Such generators are used, for instance, in older cars to charge the storage batteries used. The components are essentially the same as those of the ac generator except that the contacts to the rotating loop are made using a split ring called a *commutator*.

In this configuration, the output voltage always has the same polarity and pulsates with time, as shown in Figure 31.21b. We can understand the reason for this by noting that the contacts to the split ring reverse their roles every half cycle. At the same time, the polarity of the induced emf reverses; hence, the polarity of the

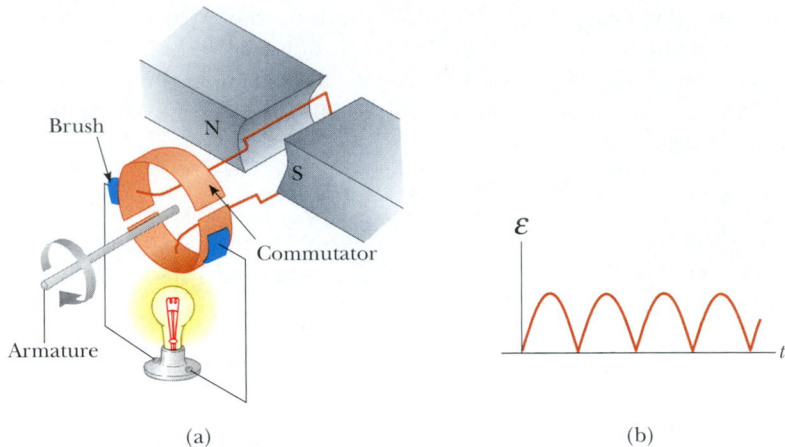

Figure 31.21 (a) Schematic diagram of a dc generator. (b) The magnitude of the emf varies in time but the polarity never changes.

split ring (which is the same as the polarity of the output voltage) remains the same.

A pulsating dc current is not suitable for most applications. To obtain a more steady dc current, commercial dc generators use many coils and commutators distributed so that the sinusoidal pulses from the various coils are out of phase. When these pulses are superimposed, the dc output is almost free of fluctuations.

Motors are devices that convert electrical energy to mechanical energy. Essentially, a motor is a generator operating in reverse. Instead of generating a current by rotating a loop, a current is supplied to the loop by a battery, and the torque acting on the current-carrying loop causes it to rotate.

Useful mechanical work can be done by attaching the rotating armature to some external device. However, as the loop rotates in a magnetic field, the changing magnetic flux induces an emf in the loop; this induced emf always acts to reduce the current in the loop. If this were not the case, Lenz's law would be violated. The back emf increases in magnitude as the rotational speed of the armature increases. (The phrase *back emf* is used to indicate an emf that tends to reduce the supplied current.) Because the voltage available to supply current equals the difference between the supply voltage and the back emf, the current in the rotating coil is limited by the back emf.

When a motor is turned on, there is initially no back emf; thus, the current is very large because it is limited only by the resistance of the coils. As the coils begin to rotate, the induced back emf opposes the applied voltage, and the current in the coils is reduced. If the mechanical load increases, the motor slows down; this causes the back emf to decrease. This reduction in the back emf increases the current in the coils and therefore also increases the power needed from the external voltage source. For this reason, the power requirements for starting a motor and for running it are greater for heavy loads than for light ones. If the motor is allowed to run under no mechanical load, the back emf reduces the current to a value just large enough to overcome energy losses due to internal energy and friction. If a very heavy load jams the motor so that it cannot rotate, the lack of a back emf can lead to dangerously high current in the motor's wire. If the problem is not corrected, a fire could result.

EXAMPLE 31.10   The Induced Current in a Motor

Assume that a motor in which the coils have a total resistance of 10 Ω is supplied by a voltage of 120 V. When the motor is running at its maximum speed, the back emf is 70 V. Find the current in the coils (a) when the motor is turned on and (b) when it has reached maximum speed.

Solution (a) When the motor is turned on, the back emf is zero (because the coils are motionless). Thus, the current in the coils is a maximum and equal to

$$I = \frac{\mathcal{E}}{R} = \frac{120\text{ V}}{10\ \Omega} = \boxed{12\text{ A}}$$

(b) At the maximum speed, the back emf has its maximum value. Thus, the effective supply voltage is that of the external source minus the back emf. Hence, the current is reduced to

$$I = \frac{\mathcal{E} - \mathcal{E}_{back}}{R} = \frac{120\text{ V} - 70\text{ V}}{10\ \Omega} = \frac{50\text{ V}}{10\ \Omega} = \boxed{5.0\text{ A}}$$

Exercise If the current in the motor is 8.0 A at some instant, what is the back emf at this time?

Answer 40 V.

Optional Section

31.6 EDDY CURRENTS

As we have seen, an emf and a current are induced in a circuit by a changing magnetic flux. In the same manner, circulating currents called **eddy currents** are induced in bulk pieces of metal moving through a magnetic field. This can easily be demonstrated by allowing a flat copper or aluminum plate attached at the end of a rigid bar to swing back and forth through a magnetic field (Fig. 31.22). As the plate enters the field, the changing magnetic flux induces an emf in the plate, which in turn causes the free electrons in the plate to move, producing the swirling eddy currents. According to Lenz's law, the direction of the eddy currents must oppose the change that causes them. For this reason, the eddy currents must produce effective magnetic poles on the plate, which are repelled by the poles of the magnet; this gives rise to a repulsive force that opposes the motion of the plate. (If the opposite were true, the plate would accelerate and its energy would

QuickLab

Hang a strong magnet from two strings so that it swings back and forth in a plane. Start it oscillating and determine approximately how much time passes before it stops swinging. Start it oscillating again and quickly bring the flat surface of an aluminum cooking sheet up to within a millimeter of the plane of oscillation, taking care not to touch the magnet. How long does it take the oscillating magnet to stop now?

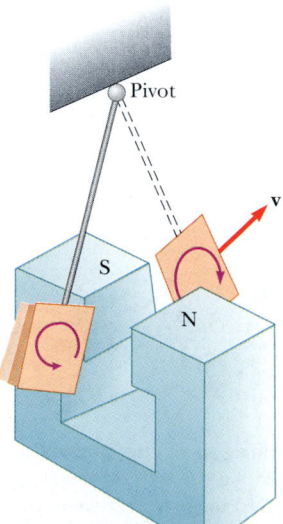

Figure 31.22 Formation of eddy currents in a conducting plate moving through a magnetic field. As the plate enters or leaves the field, the changing magnetic flux induces an emf, which causes eddy currents in the plate.

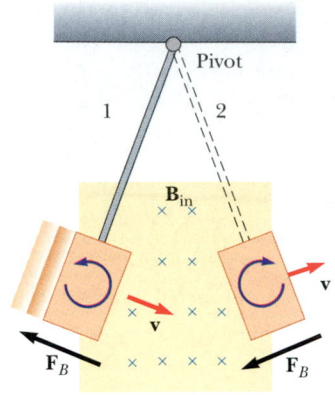

Figure 31.23 As the conducting plate enters the field (position 1), the eddy currents are counterclockwise. As the plate leaves the field (position 2), the currents are clockwise. In either case, the force on the plate is opposite the velocity, and eventually the plate comes to rest.

Figure 31.24 When slots are cut in the conducting plate, the eddy currents are reduced and the plate swings more freely through the magnetic field.

increase after each swing, in violation of the law of conservation of energy.) As you may have noticed while carrying out the QuickLab on page 997, you can "feel" the retarding force by pulling a copper or aluminum sheet through the field of a strong magnet.

As indicated in Figure 31.23, with **B** directed into the page, the induced eddy current is counterclockwise as the swinging plate enters the field at position 1. This is because the external magnetic flux into the page through the plate is increasing, and hence by Lenz's law the induced current must provide a magnetic flux out of the page. The opposite is true as the plate leaves the field at position 2, where the current is clockwise. Because the induced eddy current always produces a magnetic retarding force $\mathbf{F}_B$ when the plate enters or leaves the field, the swinging plate eventually comes to rest.

If slots are cut in the plate, as shown in Figure 31.24, the eddy currents and the corresponding retarding force are greatly reduced. We can understand this by realizing that the cuts in the plate prevent the formation of any large current loops.

The braking systems on many subway and rapid-transit cars make use of electromagnetic induction and eddy currents. An electromagnet attached to the train is positioned near the steel rails. (An electromagnet is essentially a solenoid with an iron core.) The braking action occurs when a large current is passed through the electromagnet. The relative motion of the magnet and rails induces eddy currents in the rails, and the direction of these currents produces a drag force on the moving train. The loss in mechanical energy of the train is transformed to internal energy in the rails and wheels. Because the eddy currents decrease steadily in magnitude as the train slows down, the braking effect is quite smooth. Eddy-current brakes are also used in some mechanical balances and in various machines. Some power tools use eddy currents to stop rapidly spinning blades once the device is turned off.

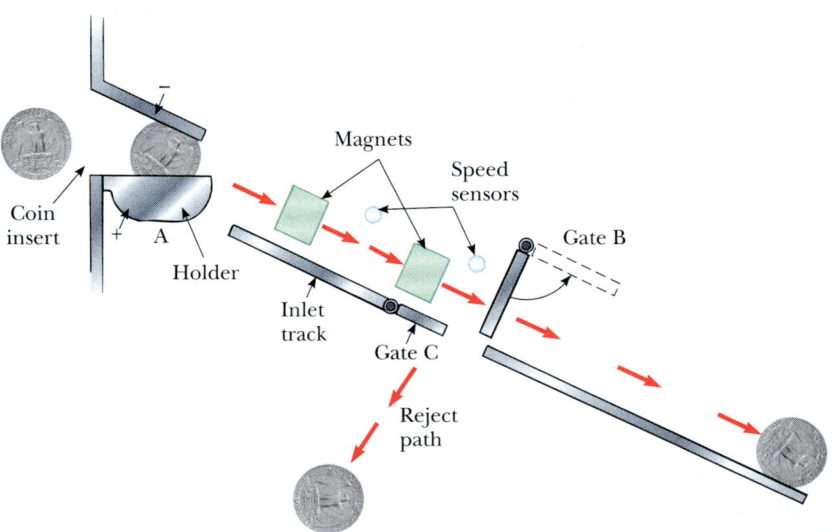

Figure 31.25 As the coin enters the vending machine, a potential difference is applied across the coin at A, and its resistance is measured. If the resistance is acceptable, the holder drops down, releasing the coin and allowing it to roll along the inlet track. Two magnets induce eddy currents in the coin, and magnetic forces control its speed. If the speed sensors indicate that the coin has the correct speed, gate B swings up to allow the coin to be accepted. If the coin is not moving at the correct speed, gate C opens to allow the coin to follow the reject path.

Eddy currents are often undesirable because they represent a transformation of mechanical energy to internal energy. To reduce this energy loss, moving conducting parts are often laminated—that is, they are built up in thin layers separated by a nonconducting material such as lacquer or a metal oxide. This layered structure increases the resistance of the possible paths of the eddy currents and effectively confines the currents to individual layers. Such a laminated structure is used in transformer cores and motors to minimize eddy currents and thereby increase the efficiency of these devices.

Even a task as simple as buying a candy bar from a vending machine involves eddy currents, as shown in Figure 31.25. After entering the slot, a coin is stopped momentarily while its electrical resistance is checked. If its resistance falls within an acceptable range, the coin is allowed to continue down a ramp and through a magnetic field. As it moves through the field, eddy currents are produced in the coin, and magnetic forces slow it down slightly. How much it is slowed down depends on its metallic composition. Sensors measure the coin's speed after it moves past the magnets, and this speed is compared with expected values. If the coin is legal and passes these tests, a gate is opened and the coin is accepted; otherwise, a second gate moves it into the reject path.

31.7 MAXWELL'S WONDERFUL EQUATIONS

We conclude this chapter by presenting four equations that are regarded as the basis of all electrical and magnetic phenomena. These equations, developed by James Clerk Maxwell, are as fundamental to electromagnetic phenomena as Newton's laws are to mechanical phenomena. In fact, the theory that Maxwell developed was more far-reaching than even he imagined because it turned out to be in agreement with the special theory of relativity, as Einstein showed in 1905.

Maxwell's equations represent the laws of electricity and magnetism that we have already discussed, but they have additional important consequences. In Chapter 34 we shall show that these equations predict the existence of electromagnetic waves (traveling patterns of electric and magnetic fields), which travel with a speed $c = 1/\sqrt{\mu_0\epsilon_0} = 3.00 \times 10^8$ m/s, the speed of light. Furthermore, the theory shows that such waves are radiated by accelerating charges.

For simplicity, we present **Maxwell's equations** as applied to free space, that is, in the absence of any dielectric or magnetic material. The four equations are

$$\oint_S \mathbf{E} \cdot d\mathbf{A} = \frac{Q}{\epsilon_0} \qquad\qquad (31.12)$$

Gauss's law

$$\oint_S \mathbf{B} \cdot d\mathbf{A} = 0 \qquad\qquad (31.13)$$

Gauss's law in magnetism

$$\oint \mathbf{E} \cdot d\mathbf{s} = -\frac{d\Phi_B}{dt} \qquad\qquad (31.14)$$

Faraday's law

$$\oint \mathbf{B} \cdot d\mathbf{s} = \mu_0 I + \epsilon_0\mu_0 \frac{d\Phi_E}{dt} \qquad\qquad (31.15)$$

Ampère–Maxwell law

Equation 31.12 is Gauss's law: **The total electric flux through any closed surface equals the net charge inside that surface divided by ϵ_0.** This law relates an electric field to the charge distribution that creates it.

Equation 31.13, which can be considered Gauss's law in magnetism, states that **the net magnetic flux through a closed surface is zero.** That is, the number of magnetic field lines that enter a closed volume must equal the number that leave that volume. This implies that magnetic field lines cannot begin or end at any point. If they did, it would mean that isolated magnetic monopoles existed at those points. The fact that isolated magnetic monopoles have not been observed in nature can be taken as a confirmation of Equation 31.13.

Equation 31.14 is Faraday's law of induction, which describes the creation of an electric field by a changing magnetic flux. This law states that **the emf, which is the line integral of the electric field around any closed path, equals the rate of change of magnetic flux through any surface area bounded by that path.** One consequence of Faraday's law is the current induced in a conducting loop placed in a time-varying magnetic field.

Equation 31.15, usually called the Ampère–Maxwell law, is the generalized form of Ampère's law, which describes the creation of a magnetic field by an electric field and electric currents: **The line integral of the magnetic field around any closed path is the sum of μ_0 times the net current through that path and $\epsilon_0 \mu_0$ times the rate of change of electric flux through any surface bounded by that path.**

Once the electric and magnetic fields are known at some point in space, the force acting on a particle of charge q can be calculated from the expression

| Lorentz force law |

$$\mathbf{F} = q\mathbf{E} + q\mathbf{v} \times \mathbf{B} \tag{31.16}$$

This relationship is called the **Lorentz force law.** (We saw this relationship earlier as Equation 29.16.) Maxwell's equations, together with this force law, completely describe all classical electromagnetic interactions.

It is interesting to note the symmetry of Maxwell's equations. Equations 31.12 and 31.13 are symmetric, apart from the absence of the term for magnetic monopoles in Equation 31.13. Furthermore, Equations 31.14 and 31.15 are symmetric in that the line integrals of $\mathbf{E}$ and $\mathbf{B}$ around a closed path are related to the rate of change of magnetic flux and electric flux, respectively. "Maxwell's wonderful equations," as they were called by John R. Pierce,[3] are of fundamental importance not only to electromagnetism but to all of science. Heinrich Hertz once wrote, "One cannot escape the feeling that these mathematical formulas have an independent existence and an intelligence of their own, that they are wiser than we are, wiser even than their discoverers, that we get more out of them than we put into them."

SUMMARY

Faraday's law of induction states that the emf induced in a circuit is directly proportional to the time rate of change of magnetic flux through the circuit:

$$\mathcal{E} = -\frac{d\Phi_B}{dt} \tag{31.1}$$

where $\Phi_B = \int \mathbf{B} \cdot d\mathbf{A}$ is the magnetic flux.

---

[3] John R. Pierce, *Electrons and Waves*, New York, Doubleday Science Study Series, 1964. Chapter 6 of this interesting book is recommended as supplemental reading.

When a conducting bar of length ℓ moves at a velocity **v** through a magnetic field **B**, where **B** is perpendicular to the bar and to **v**, the **motional emf** induced in the bar is

$$\mathcal{E} = -B\ell v \qquad (31.5)$$

Lenz's law states that the induced current and induced emf in a conductor are in such a direction as to oppose the change that produced them.

A general form of **Faraday's law of induction** is

$$\mathcal{E} = \oint \mathbf{E} \cdot d\mathbf{s} = -\frac{d\Phi_B}{dt} \qquad (31.9)$$

where **E** is the nonconservative, time-varying electric field that is produced by the changing magnetic flux.

When used with the Lorentz force law, $\mathbf{F} = q\mathbf{E} + q\mathbf{v} \times \mathbf{B}$, **Maxwell's equations** describe all electromagnetic phenomena:

$$\oint_S \mathbf{E} \cdot d\mathbf{A} = \frac{Q}{\epsilon_0} \qquad (31.12)$$

$$\oint_S \mathbf{B} \cdot d\mathbf{A} = 0 \qquad (31.13)$$

$$\oint \mathbf{E} \cdot d\mathbf{s} = -\frac{d\Phi_B}{dt} \qquad (31.14)$$

$$\oint \mathbf{B} \cdot d\mathbf{s} = \mu_0 I + \epsilon_0 \mu_0 \frac{d\Phi_E}{dt} \qquad (31.15)$$

The Ampère–Maxwell law (Eq. 31.15) describes how a magnetic field can be produced by both a conduction current and a changing electric flux.

QUESTIONS

1. A loop of wire is placed in a uniform magnetic field. For what orientation of the loop is the magnetic flux a maximum? For what orientation is the flux zero? Draw pictures of these two situations.

2. As the conducting bar shown in Figure Q31.2 moves to the right, an electric field directed downward is set up in the bar. Explain why the electric field would be upward if the bar were to move to the left.

3. As the bar shown in Figure Q31.2 moves in a direction perpendicular to the field, is an applied force required to keep it moving with constant speed? Explain.

4. The bar shown in Figure Q31.4 moves on rails to the right with a velocity **v**, and the uniform, constant magnetic field is directed out of the page. Why is the induced current clockwise? If the bar were moving to the left, what would be the direction of the induced current?

5. Explain why an applied force is necessary to keep the bar shown in Figure Q31.4 moving with a constant speed.

6. A large circular loop of wire lies in the horizontal plane. A bar magnet is dropped through the loop. If the axis of

the magnet remains horizontal as it falls, describe the emf induced in the loop. How is the situation altered if the axis of the magnet remains vertical as it falls?

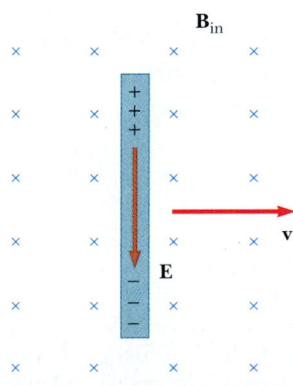

Figure Q31.2 (Questions 2 and 3).

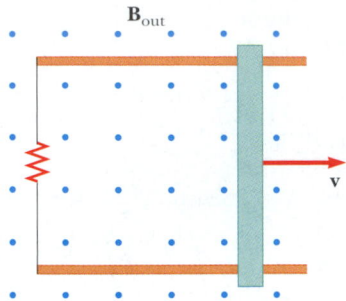

Figure Q31.4   (Questions 4 and 5).

7. When a small magnet is moved toward a solenoid, an emf is induced in the coil. However, if the magnet is moved around inside a toroid, no emf is induced. Explain.

8. Will dropping a magnet down a long copper tube produce a current in the walls of the tube? Explain.

9. How is electrical energy produced in dams (that is, how is the energy of motion of the water converted to alternating current electricity)?

10. In a beam–balance scale, an aluminum plate is sometimes used to slow the oscillations of the beam near equilibrium. The plate is mounted at the end of the beam and moves between the poles of a small horseshoe magnet attached to the frame. Why are the oscillations strongly damped near equilibrium?

11. What happens when the rotational speed of a generator coil is increased?

12. Could a current be induced in a coil by the rotation of a magnet inside the coil? If so, how?

13. When the switch shown in Figure Q31.13a is closed, a cur-

rent is set up in the coil, and the metal ring springs upward (Fig. Q31.13b). Explain this behavior.

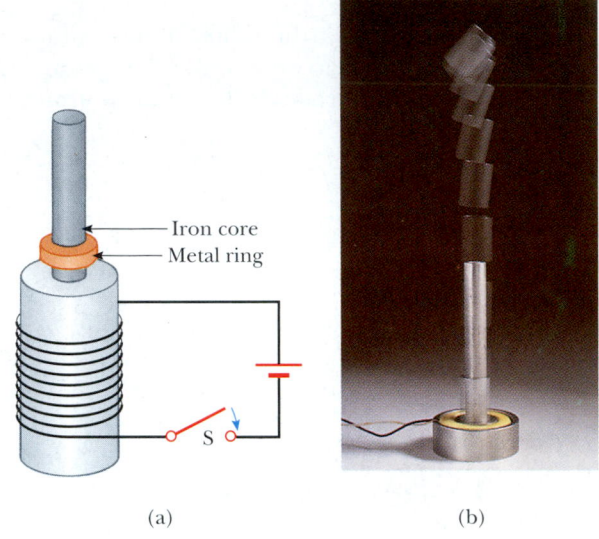

(a)   (b)

Figure Q31.13   (Questions 13 and 14). *(Photo courtesy of Central Scientific Company)*

14. Assume that the battery shown in Figure Q31.13a is replaced by an alternating current source and that the switch is held closed. If held down, the metal ring on top of the solenoid becomes hot. Why?

15. Do Maxwell's equations allow for the existence of magnetic monopoles? Explain.

PROBLEMS

1, 2, 3 = straightforward, intermediate, challenging   ☐ = full solution available in the *Student Solutions Manual and Study Guide*
WEB = solution posted at **http://www.saunderscollege.com/physics/**   🖥 = Computer useful in solving problem   📱 = Interactive Physics
☐ = paired numerical/symbolic problems

Section 31.1   Faraday's Law of Induction
Section 31.2   Motional emf
Section 31.3   Lenz's Law

1. A 50-turn rectangular coil of dimensions 5.00 cm × 10.0 cm is allowed to fall from a position where $B = 0$ to a new position where $B = 0.500$ T and is directed perpendicular to the plane of the coil. Calculate the magnitude of the average emf induced in the coil if the displacement occurs in 0.250 s.

2. A flat loop of wire consisting of a single turn of cross-sectional area 8.00 cm^2 is perpendicular to a magnetic field that increases uniformly in magnitude from 0.500 T to 2.50 T in 1.00 s. What is the resulting induced current if the loop has a resistance of 2.00 Ω?

3. A 25-turn circular coil of wire has a diameter of 1.00 m. It is placed with its axis along the direction of the Earth's magnetic field of 50.0 μT, and then in 0.200 s it is flipped 180°. An average emf of what magnitude is generated in the coil?

4. A rectangular loop of area A is placed in a region where the magnetic field is perpendicular to the plane of the loop. The magnitude of the field is allowed to vary in time according to the expression $B = B_{max}e^{-t/\tau}$, where B_{max} and τ are constants. The field has the constant value B_{max} for $t < 0$. (a) Use Faraday's law to show that the emf induced in the loop is given by

$$\mathcal{E} = (AB_{max}/\tau)e^{-t/\tau}$$

(b) Obtain a numerical value for $\mathcal{E}$ at $t = 4.00$ s when

$A = 0.160$ m^2, $B_{max} = 0.350$ T, and $\tau = 2.00$ s. (c) For the values of A, B_{max}, and τ given in part (b), what is the maximum value of $\mathcal{E}$?

WEB **5.** A strong electromagnet produces a uniform field of 1.60 T over a cross-sectional area of 0.200 m^2. A coil having 200 turns and a total resistance of 20.0 Ω is placed around the electromagnet. The current in the electromagnet is then smoothly decreased until it reaches zero in 20.0 ms. What is the current induced in the coil?

6. A magnetic field of 0.200 T exists within a solenoid of 500 turns and a diameter of 10.0 cm. How rapidly (that is, within what period of time) must the field be reduced to zero if the average induced emf within the coil during this time interval is to be 10.0 kV?

WEB **7.** An aluminum ring with a radius of 5.00 cm and a resistance of 3.00×10^{-4} Ω is placed on top of a long air-core solenoid with 1 000 turns per meter and a radius of 3.00 cm, as shown in Figure P31.7. Assume that the axial component of the field produced by the solenoid over the area of the end of the solenoid is one-half as strong as at the center of the solenoid. Assume that the solenoid produces negligible field outside its cross-sectional area. (a) If the current in the solenoid is increasing at a rate of 270 A/s, what is the induced current in the ring? (b) At the center of the ring, what is the magnetic field produced by the induced current in the ring? (c) What is the direction of this field?

8. An aluminum ring of radius r_1 and resistance R is placed on top of a long air-core solenoid with n turns per meter and smaller radius r_2, as shown in Figure P31.7. Assume that the axial component of the field produced by the solenoid over the area of the end of the solenoid is one-half as strong as at the center of the solenoid. Assume that the solenoid produces negligible field outside its cross-sectional area. (a) If the current in the solenoid is increasing at a rate of $\Delta I/\Delta t$, what is the induced current in the ring? (b) At the center of the ring, what is the magnetic field produced by the induced current in the ring? (c) What is the direction of this field?

9. A loop of wire in the shape of a rectangle of width w and length L and a long, straight wire carrying a current I lie on a tabletop as shown in Figure P31.9. (a) Determine the magnetic flux through the loop due to the current I. (b) Suppose that the current is changing with time according to $I = a + bt$, where a and b are constants. Determine the induced emf in the loop if $b = 10.0$ A/s, $h = 1.00$ cm, $w = 10.0$ cm, and $L = 100$ cm. What is the direction of the induced current in the rectangle?

10. A coil of 15 turns and radius 10.0 cm surrounds a long solenoid of radius 2.00 cm and 1.00×10^3 turns per meter (Fig. P31.10). If the current in the solenoid changes as $I = (5.00$ A$)\sin(120t)$, find the induced emf in the 15-turn coil as a function of time.

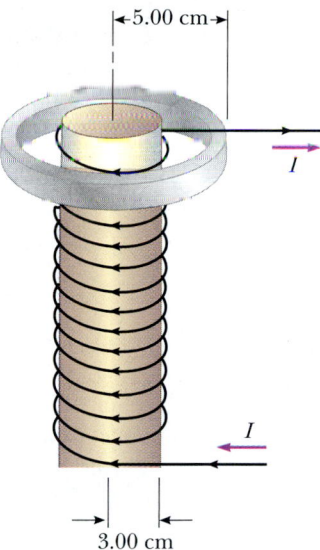

Figure P31.7 Problems 7 and 8.

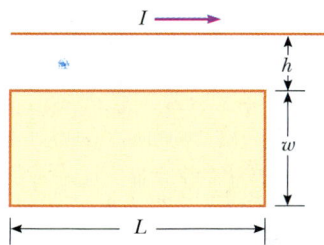

Figure P31.9 Problems 9 and 73.

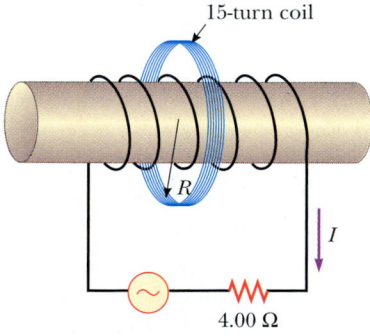

Figure P31.10

11. Find the current through section PQ of length $a = 65.0$ cm shown in Figure P31.11. The circuit is located in a magnetic field whose magnitude varies with time according to the expression $B = (1.00 \times 10^{-3}$ T/s$)t$. Assume that the resistance per length of the wire is 0.100 Ω/m.

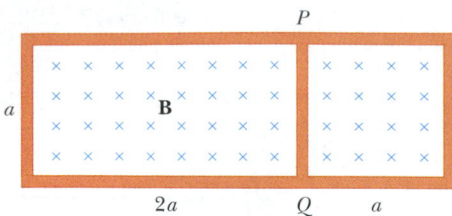

Figure P31.11

12. A 30-turn circular coil of radius 4.00 cm and resistance 1.00 Ω is placed in a magnetic field directed perpendicular to the plane of the coil. The magnitude of the magnetic field varies in time according to the expression $B = 0.010\ 0t + 0.040\ 0t^2$, where t is in seconds and B is in tesla. Calculate the induced emf in the coil at $t = 5.00$ s.

13. A long solenoid has 400 turns per meter and carries a current $I = (30.0\ \text{A})(1 - e^{-1.60t})$. Inside the solenoid and coaxial with it is a coil that has a radius of 6.00 cm and consists of a total of 250 turns of fine wire (Fig. P31.13). What emf is induced in the coil by the changing current?

14. A long solenoid has n turns per meter and carries a current $I = I_{max}(1 - e^{-\alpha t})$. Inside the solenoid and coaxial with it is a coil that has a radius R and consists of a total of N turns of fine wire (see Fig. P31.13). What emf is induced in the coil by the changing current?

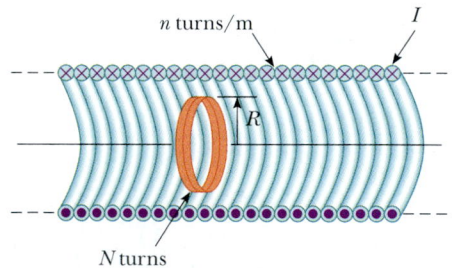

Figure P31.13   Problems 13 and 14.

15. A coil formed by wrapping 50 turns of wire in the shape of a square is positioned in a magnetic field so that the normal to the plane of the coil makes an angle of 30.0° with the direction of the field. When the magnetic field is increased uniformly from 200 μT to 600 μT in 0.400 s, an emf of magnitude 80.0 mV is induced in the coil. What is the total length of the wire?

16. A closed loop of wire is given the shape of a circle with a radius of 0.500 m. It lies in a plane perpendicular to a uniform magnetic field of magnitude 0.400 T. If in 0.100 s the wire loop is reshaped into a square but remains in the same plane, what is the magnitude of the average induced emf in the wire during this time?

17. A toroid having a rectangular cross-section ($a = 2.00$ cm by $b = 3.00$ cm) and inner radius $R = 4.00$ cm consists of 500 turns of wire that carries a current $I = I_{max} \sin \omega t$, with $I_{max} = 50.0$ A and a frequency $f = \omega/2\pi = 60.0$ Hz. A coil that consists of 20 turns of wire links with the toroid, as shown in Figure P31.17. Determine the emf induced in the coil as a function of time.

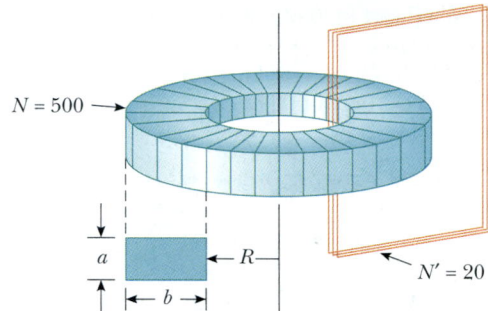

Figure P31.17

18. A single-turn, circular loop of radius R is coaxial with a long solenoid of radius r and length ℓ and having N turns (Fig. P31.18). The variable resistor is changed so that the solenoid current decreases linearly from I_1 to I_2 in an interval Δt. Find the induced emf in the loop.

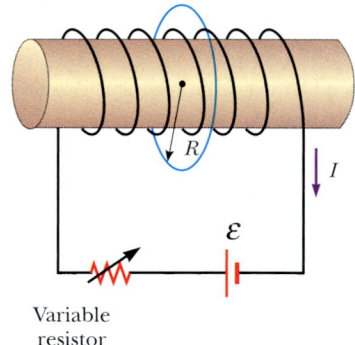

Figure P31.18

19. A circular coil enclosing an area of 100 cm² is made of 200 turns of copper wire, as shown in Figure P31.19. Ini-

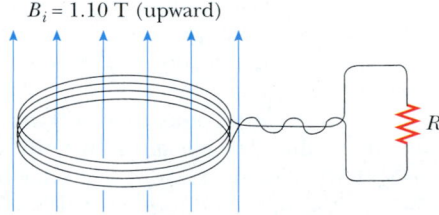

Figure P31.19

tially, a 1.10-T uniform magnetic field points in a perpendicular direction upward through the plane of the coil. The direction of the field then reverses. During the time the field is changing its direction, how much charge flows through the coil if $R = 5.00 \, \Omega$?

20. Consider the arrangement shown in Figure P31.20. Assume that $R = 6.00 \, \Omega$, $\ell = 1.20$ m, and a uniform 2.50-T magnetic field is directed into the page. At what speed should the bar be moved to produce a current of 0.500 A in the resistor?

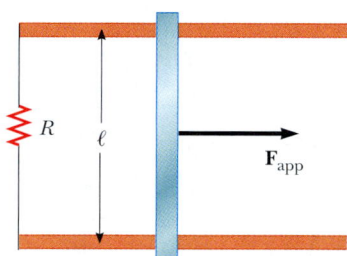

Figure P31.20   Problems 20, 21, and 22.

21. Figure P31.20 shows a top view of a bar that can slide without friction. The resistor is 6.00 Ω and a 2.50-T magnetic field is directed perpendicularly downward, into the paper. Let $\ell = 1.20$ m. (a) Calculate the applied force required to move the bar to the right at a constant speed of 2.00 m/s. (b) At what rate is energy delivered to the resistor?

22. A conducting rod of length ℓ moves on two horizontal, frictionless rails, as shown in Figure P31.20. If a constant force of 1.00 N moves the bar at 2.00 m/s through a magnetic field **B** that is directed into the page, (a) what is the current through an 8.00-Ω resistor R? (b) What is the rate at which energy is delivered to the resistor? (c) What is the mechanical power delivered by the force $\mathbf{F}_{app}$?

23. A Boeing-747 jet with a wing span of 60.0 m is flying horizontally at a speed of 300 m/s over Phoenix, Arizona, at a location where the Earth's magnetic field is 50.0 μT at 58.0° below the horizontal. What voltage is generated between the wingtips?

24. The square loop in Figure P31.24 is made of wires with total series resistance 10.0 Ω. It is placed in a uniform

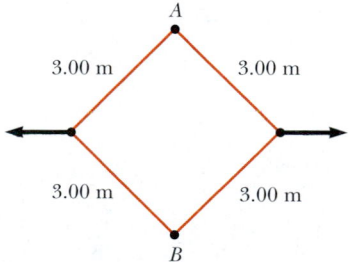

Figure P31.24

0.100-T magnetic field directed perpendicular into the plane of the paper. The loop, which is hinged at each corner, is pulled as shown until the separation between points A and B is 3.00 m. If this process takes 0.100 s, what is the average current generated in the loop? What is the direction of the current?

25. A helicopter has blades with a length of 3.00 m extending outward from a central hub and rotating at 2.00 rev/s. If the vertical component of the Earth's magnetic field is 50.0 μT, what is the emf induced between the blade tip and the center hub?

26. Use Lenz's law to answer the following questions concerning the direction of induced currents: (a) What is the direction of the induced current in resistor R shown in Figure P31.26a when the bar magnet is moved to the left? (b) What is the direction of the current induced in the resistor R right after the switch S in Figure P31.26b is closed? (c) What is the direction of the induced current in R when the current I in Figure P31.26c decreases rapidly to zero? (d) A copper bar is moved to the right while its axis is maintained in a direction perpendicular to a magnetic field, as shown in Figure P31.26d. If the top of the bar becomes positive relative to the bottom, what is the direction of the magnetic field?

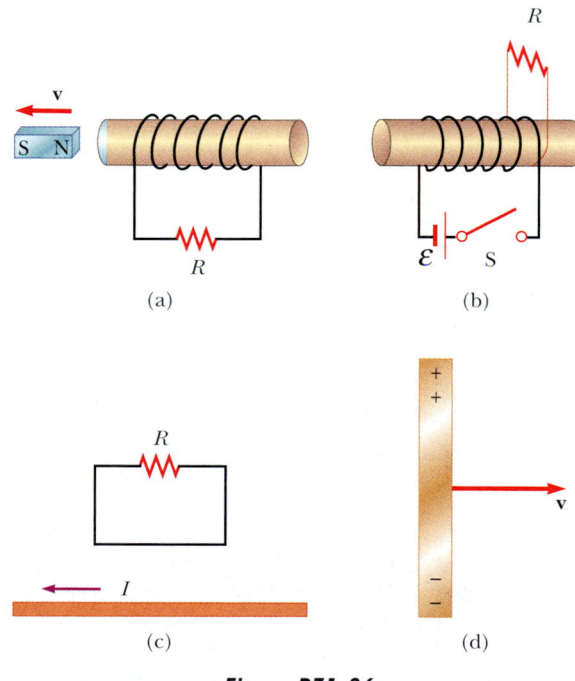

Figure P31.26

27. A rectangular coil with resistance R has N turns, each of length ℓ and width w as shown in Figure P31.27. The coil moves into a uniform magnetic field **B** with a velocity **v**. What are the magnitude and direction of the resultant force on the coil (a) as it enters the magnetic field, (b) as it moves within the field, and (c) as it leaves the field?

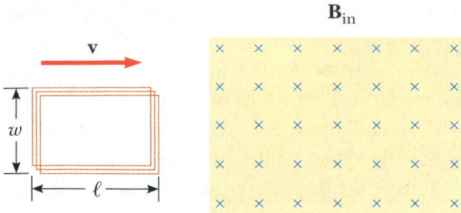

Figure P31.27

28. In 1832 Faraday proposed that the apparatus shown in Figure P31.28 could be used to generate electric current from the water flowing in the Thames River.[4] Two conducting plates of lengths a and widths b are placed facing each other on opposite sides of the river, a distance w apart, and are immersed entirely. The flow velocity of the river is $\mathbf{v}$ and the vertical component of the Earth's magnetic field is B. (a) Show that the current in the load resistor R is

$$I = \frac{abvB}{\rho + abR/w}$$

where ρ is the electrical resistivity of the water. (b) Calculate the short-circuit current ($R = 0$) if $a = 100$ m, $b = 5.00$ m, $v = 3.00$ m/s, $B = 50.0$ μT, and $\rho = 100$ $\Omega \cdot$ m.

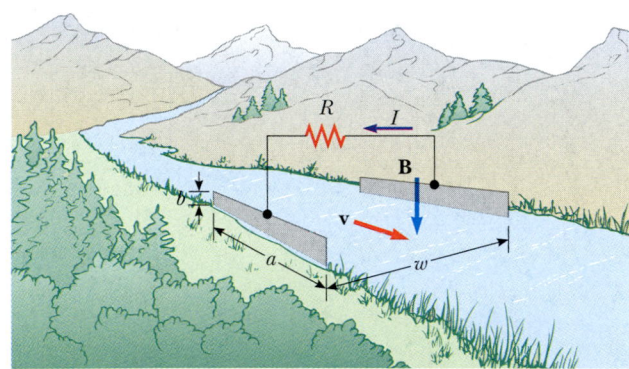

Figure P31.28

29. In Figure P31.29, the bar magnet is moved toward the loop. Is $V_a - V_b$ positive, negative, or zero? Explain.

30. A metal bar spins at a constant rate in the magnetic field of the Earth as in Figure 31.10. The rotation occurs in a region where the component of the Earth's magnetic field perpendicular to the plane of rotation is 3.30×10^{-5} T. If the bar is 1.00 m in length and its angular speed is $5.00\ \pi$ rad/s, what potential difference is developed between its ends?

[4] The idea for this problem and Figure P31.28 is from Oleg D. Jefimenko, *Electricity and Magnetism: An Introduction to the Theory of Electric and Magnetic Fields.* Star City, WV, Electret Scientific Co., 1989.

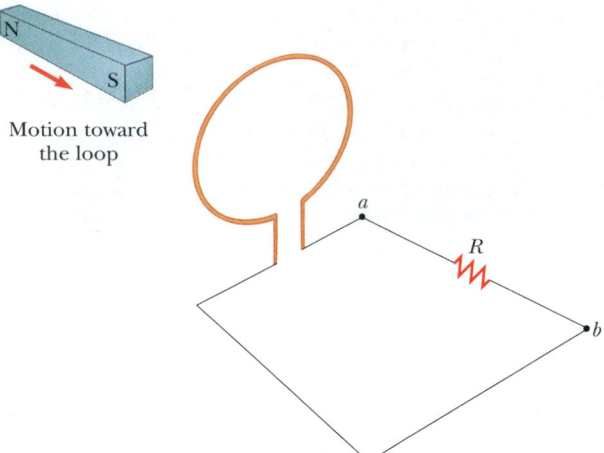

Figure P31.29

31. Two parallel rails with negligible resistance are 10.0 cm apart and are connected by a 5.00-Ω resistor. The circuit also contains two metal rods having resistances of 10.0 Ω and 15.0 Ω sliding along the rails (Fig. P31.31). The rods are pulled away from the resistor at constant speeds 4.00 m/s and 2.00 m/s, respectively. A uniform magnetic field of magnitude 0.010 0 T is applied perpendicular to the plane of the rails. Determine the current in the 5.00-Ω resistor.

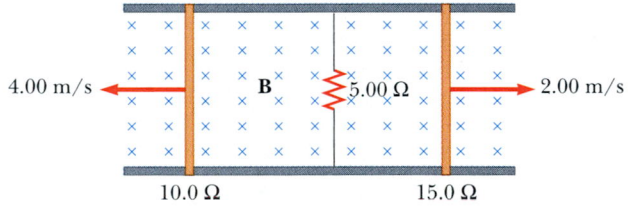

Figure P31.31

Section 31.4 Induced emf and Electric Fields

32. For the situation described in Figure P31.32, the magnetic field changes with time according to the expression $B = (2.00t^3 - 4.00t^2 + 0.800)$ T, and $r_2 = 2R = 5.00$ cm. (a) Calculate the magnitude and direction of

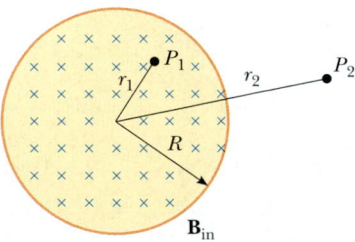

Figure P31.32 Problems 32 and 33.

the force exerted on an electron located at point P_2 when $t = 2.00$ s. (b) At what time is this force equal to zero?

33. A magnetic field directed into the page changes with time according to $B = (0.030\ 0t^2 + 1.40)$ T, where t is in seconds. The field has a circular cross-section of radius $R = 2.50$ cm (see Fig. P31.32). What are the magnitude and direction of the electric field at point P_1 when $t = 3.00$ s and $r_1 = 0.020\ 0$ m?

34. A solenoid has a radius of 2.00 cm and 1 000 turns per meter. Over a certain time interval the current varies with time according to the expression $I = 3e^{0.2t}$, where I is in amperes and t is in seconds. Calculate the electric field 5.00 cm from the axis of the solenoid at $t = 10.0$ s.

35. A long solenoid with 1 000 turns per meter and radius 2.00 cm carries an oscillating current $I = (5.00\text{ A})\sin(100\pi t)$. (a) What is the electric field induced at a radius $r = 1.00$ cm from the axis of the solenoid? (b) What is the direction of this electric field when the current is increasing counterclockwise in the coil?

(Optional)
Section 31.5 Generators and Motors

36. In a 250-turn automobile alternator, the magnetic flux in each turn is $\Phi_B = (2.50 \times 10^{-4}\text{ T}\cdot\text{m}^2)\cos(\omega t)$, where ω is the angular speed of the alternator. The alternator is geared to rotate three times for each engine revolution. When the engine is running at an angular speed of 1 000 rev/min, determine (a) the induced emf in the alternator as a function of time and (b) the maximum emf in the alternator.

WEB 37. A coil of area 0.100 m^2 is rotating at 60.0 rev/s with the axis of rotation perpendicular to a 0.200-T magnetic field. (a) If there are 1 000 turns on the coil, what is the maximum voltage induced in it? (b) What is the orientation of the coil with respect to the magnetic field when the maximum induced voltage occurs?

38. A square coil (20.0 cm × 20.0 cm) that consists of 100 turns of wire rotates about a vertical axis at 1 500 rev/min, as indicated in Figure P31.38. The horizontal component of the Earth's magnetic field at the location of the coil is 2.00×10^{-5} T. Calculate the maximum emf induced in the coil by this field.

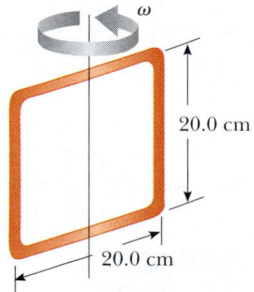

ω

20.0 cm

20.0 cm

Figure P31.38

39. A long solenoid, with its axis along the x axis, consists of 200 turns per meter of wire that carries a steady current of 15.0 A. A coil is formed by wrapping 30 turns of thin wire around a circular frame that has a radius of 8.00 cm. The coil is placed inside the solenoid and mounted on an axis that is a diameter of the coil and coincides with the y axis. The coil is then rotated with an angular speed of 4.00π rad/s. (The plane of the coil is in the yz plane at $t = 0$.) Determine the emf developed in the coil as a function of time.

40. A bar magnet is spun at constant angular speed ω around an axis, as shown in Figure P31.40. A flat rectangular conducting loop surrounds the magnet, and at $t = 0$, the magnet is oriented as shown. Make a qualitative graph of the induced current in the loop as a function of time, plotting counterclockwise currents as positive and clockwise currents as negative.

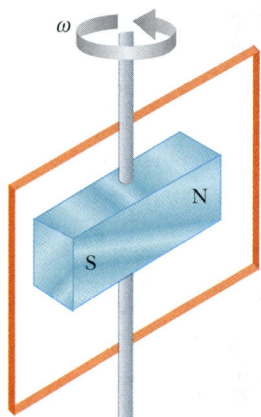

ω

N

S

Figure P31.40

41. (a) What is the maximum torque delivered by an electric motor if it has 80 turns of wire wrapped on a rectangular coil of dimensions 2.50 cm by 4.00 cm? Assume that the motor uses 10.0 A of current and that a uniform 0.800-T magnetic field exists within the motor. (b) If the motor rotates at 3 600 rev/min, what is the peak power produced by the motor?

42. A semicircular conductor of radius $R = 0.250$ m is rotated about the axis AC at a constant rate of 120 rev/min (Fig. P31.42). A uniform magnetic field in all of the lower half of the figure is directed out of the plane of rotation and has a magnitude of 1.30 T. (a) Calculate the maximum value of the emf induced in the conductor. (b) What is the value of the average induced emf for each complete rotation? (c) How would the answers to parts (a) and (b) change if **B** were allowed to extend a distance R above the axis of rotation? Sketch the emf versus time (d) when the field is as drawn in Figure P31.42 and (e) when the field is extended as described in part (c).

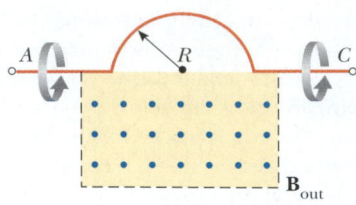

Figure P31.42

43. The rotating loop in an ac generator is a square 10.0 cm on a side. It is rotated at 60.0 Hz in a uniform field of 0.800 T. Calculate (a) the flux through the loop as a function of time, (b) the emf induced in the loop, (c) the current induced in the loop for a loop resistance of 1.00 Ω, (d) the power in the resistance of the loop, and (e) the torque that must be exerted to rotate the loop.

(Optional)
Section 31.6   Eddy Currents

44. A 0.150-kg wire in the shape of a closed rectangle 1.00 m wide and 1.50 m long has a total resistance of 0.750 Ω. The rectangle is allowed to fall through a magnetic field directed perpendicular to the direction of motion of the rectangle (Fig. P31.44). The rectangle accelerates downward as it approaches a terminal speed of 2.00 m/s, with its top not yet in the region of the field. Calculate the magnitude of **B**.

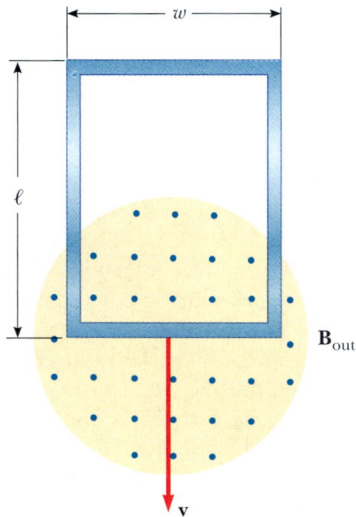

Figure P31.44   Problems 44 and 45.

WEB **45.** A conducting rectangular loop of mass M, resistance R, and dimensions w by ℓ falls from rest into a magnetic field **B** as in Figure P31.44. The loop approaches termi-

nal speed v_t. (a) Show that

$$v_t = \frac{MgR}{B^2 w^2}$$

(b) Why is v_t proportional to R? (c) Why is it inversely proportional to B^2?

46. Figure P31.46 represents an electromagnetic brake that utilizes eddy currents. An electromagnet hangs from a railroad car near one rail. To stop the car, a large steady current is sent through the coils of the electromagnet. The moving electromagnet induces eddy currents in the rails, whose fields oppose the change in the field of the electromagnet. The magnetic fields of the eddy currents exert force on the current in the electromagnet, thereby slowing the car. The direction of the car's motion and the direction of the current in the electromagnet are shown correctly in the picture. Determine which of the eddy currents shown on the rails is correct. Explain your answer.

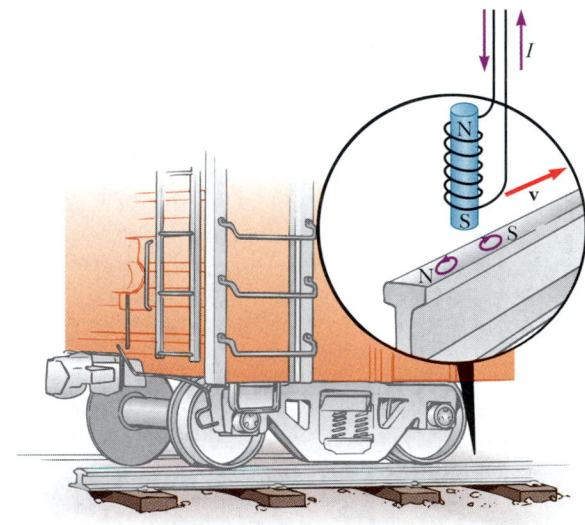

Figure P31.46

Section 31.7   Maxwell's Wonderful Equations

47. A proton moves through a uniform electric field $\mathbf{E} = 50.0\mathbf{j}$ V/m and a uniform magnetic field $\mathbf{B} = (0.200\mathbf{i} + 0.300\mathbf{j} + 0.400\mathbf{k})$ T. Determine the acceleration of the proton when it has a velocity $\mathbf{v} = 200\mathbf{i}$ m/s.

48. An electron moves through a uniform electric field $\mathbf{E} = (2.50\mathbf{i} + 5.00\mathbf{j})$ V/m and a uniform magnetic field $\mathbf{B} = 0.400\mathbf{k}$ T. Determine the acceleration of the electron when it has a velocity $\mathbf{v} = 10.0\mathbf{i}$ m/s.

ADDITIONAL PROBLEMS

49. A steel guitar string vibrates (see Fig. 31.5). The component of the magnetic field perpendicular to the area of a pickup coil nearby is given by

$$B = 50.0 \text{ mT} + (3.20 \text{ mT}) \sin (2\pi 523 \ t/s)$$

The circular pickup coil has 30 turns and radius 2.70 mm. Find the emf induced in the coil as a function of time.

50. Figure P31.50 is a graph of the induced emf versus time for a coil of N turns rotating with angular velocity ω in a uniform magnetic field directed perpendicular to the axis of rotation of the coil. Copy this graph (on a larger scale), and on the same set of axes show the graph of emf versus t (a) if the number of turns in the coil is doubled, (b) if instead the angular velocity is doubled, and (c) if the angular velocity is doubled while the number of turns in the coil is halved.

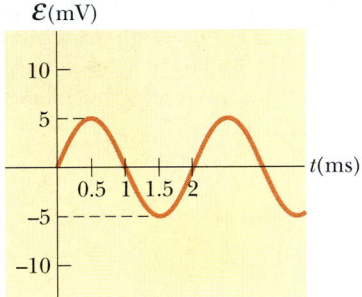

Figure P31.50

51. A technician wearing a brass bracelet enclosing an area of 0.005 00 m² places her hand in a solenoid whose magnetic field is 5.00 T directed perpendicular to the plane of the bracelet. The electrical resistance around the circumference of the bracelet is 0.020 0 Ω. An unexpected power failure causes the field to drop to 1.50 T in a time of 20.0 ms. Find (a) the current induced in the bracelet and (b) the power delivered to the resistance of the bracelet. (*Note:* As this problem implies, you should not wear any metallic objects when working in regions of strong magnetic fields.)

52. Two infinitely long solenoids (seen in cross-section) thread a circuit as shown in Figure P31.52. The magni-

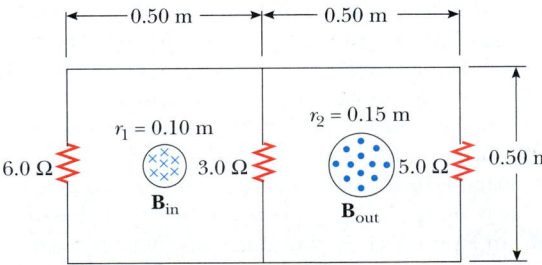

Figure P31.52

tude of **B** inside each is the same and is increasing at the rate of 100 T/s. What is the current in each resistor?

53. A conducting rod of length $\ell = 35.0$ cm is free to slide on two parallel conducting bars, as shown in Figure P31.53. Two resistors $R_1 = 2.00 \ \Omega$ and $R_2 = 5.00 \ \Omega$ are connected across the ends of the bars to form a loop. A constant magnetic field $B = 2.50$ T is directed perpendicular into the page. An external agent pulls the rod to the left with a constant speed of $v = 8.00$ m/s. Find (a) the currents in both resistors, (b) the total power delivered to the resistance of the circuit, and (c) the magnitude of the applied force that is needed to move the rod with this constant velocity.

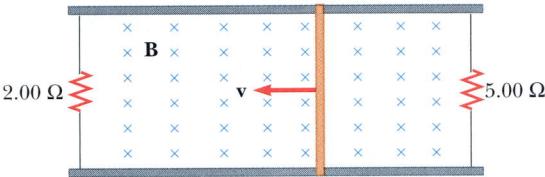

Figure P31.53

54. Suppose you wrap wire onto the core from a roll of cellophane tape to make a coil. Describe how you can use a bar magnet to produce an induced voltage in the coil. What is the order of magnitude of the emf you generate? State the quantities you take as data and their values.

55. A bar of mass m, length d, and resistance R slides without friction on parallel rails, as shown in Figure P31.55. A battery that maintains a constant emf $\mathcal{E}$ is connected between the rails, and a constant magnetic field **B** is directed perpendicular to the plane of the page. If the bar starts from rest, show that at time t it moves with a speed

$$v = \frac{\mathcal{E}}{Bd} \left(1 - e^{-B^2 d^2 t / mR}\right)$$

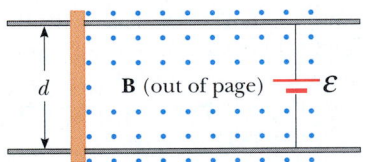

Figure P31.55

56. An automobile has a vertical radio antenna 1.20 m long. The automobile travels at 65.0 km/h on a horizontal road where the Earth's magnetic field is 50.0 μT directed toward the north and downward at an angle of 65.0° below the horizontal. (a) Specify the direction that the automobile should move to generate the maxi-

mum motional emf in the antenna, with the top of the antenna positive relative to the bottom. (b) Calculate the magnitude of this induced emf.

57. The plane of a square loop of wire with edge length $a = 0.200$ m is perpendicular to the Earth's magnetic field at a point where $B = 15.0$ μT, as shown in Figure P31.57. The total resistance of the loop and the wires connecting it to the galvanometer is 0.500 Ω. If the loop is suddenly collapsed by horizontal forces as shown, what total charge passes through the galvanometer?

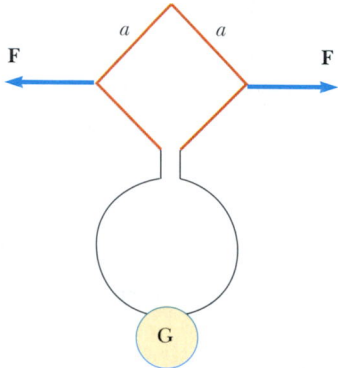

Figure P31.57

58. Magnetic field values are often determined by using a device known as a *search coil*. This technique depends on the measurement of the total charge passing through a coil in a time interval during which the magnetic flux linking the windings changes either because of the motion of the coil or because of a change in the value of B. (a) Show that as the flux through the coil changes from Φ_1 to Φ_2, the charge transferred through the coil will be given by $Q = N(\Phi_2 - \Phi_1)/R$, where R is the resistance of the coil and associated circuitry (galvanometer) and N is the number of turns. (b) As a specific example, calculate B when a 100-turn coil of resistance 200 Ω and cross-sectional area 40.0 cm^2 produces the following results. A total charge of 5.00×10^{-4} C passes through the coil when it is rotated in a uniform field from a position where the plane of the coil is perpendicular to the field to a position where the coil's plane is parallel to the field.

59. In Figure P31.59, the rolling axle, 1.50 m long, is pushed along horizontal rails at a constant speed $v = 3.00$ m/s. A resistor $R = 0.400$ Ω is connected to the rails at points a and b, directly opposite each other. (The wheels make good electrical contact with the rails, and so the axle, rails, and R form a closed-loop circuit. The only significant resistance in the circuit is R.) There is a uniform magnetic field $B = 0.080$ 0 T vertically downward. (a) Find the induced current I in the resistor. (b) What horizontal force F is required to keep the

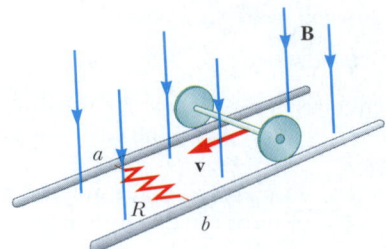

Figure P31.59

axle rolling at constant speed? (c) Which end of the resistor, a or b, is at the higher electric potential? (d) After the axle rolls past the resistor, does the current in R reverse direction? Explain your answer.

60. A conducting rod moves with a constant velocity **v** perpendicular to a long, straight wire carrying a current I as shown in Figure P31.60. Show that the magnitude of the emf generated between the ends of the rod is

$$|\mathcal{E}| = \frac{\mu_0 v I}{2\pi r} \ell$$

In this case, note that the emf decreases with increasing r, as you might expect.

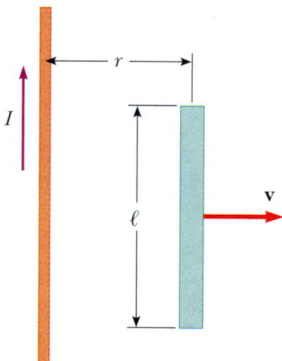

Figure P31.60

61. A circular loop of wire of radius r is in a uniform magnetic field, with the plane of the loop perpendicular to the direction of the field (Fig. P31.61). The magnetic field varies with time according to $B(t) = a + bt$, where a and b are constants. (a) Calculate the magnetic flux through the loop at $t = 0$. (b) Calculate the emf induced in the loop. (c) If the resistance of the loop is R, what is the induced current? (d) At what rate is electrical energy being delivered to the resistance of the loop?

62. In Figure P31.62, a uniform magnetic field decreases at a constant rate $dB/dt = -K$, where K is a positive constant. A circular loop of wire of radius a containing a re-

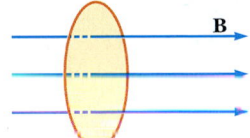

Figure P31.61

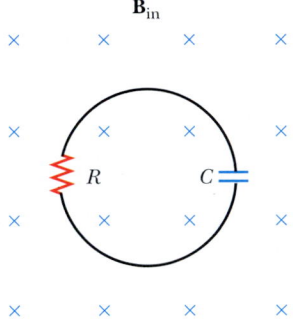

Figure P31.62

sistance R and a capacitance C is placed with its plane normal to the field. (a) Find the charge Q on the capacitor when it is fully charged. (b) Which plate is at the higher potential? (c) Discuss the force that causes the separation of charges.

63. A rectangular coil of 60 turns, dimensions 0.100 m by 0.200 m and total resistance 10.0 Ω, rotates with angular speed 30.0 rad/s about the y axis in a region where a 1.00-T magnetic field is directed along the x axis. The rotation is initiated so that the plane of the coil is perpendicular to the direction of **B** at $t = 0$. Calculate (a) the maximum induced emf in the coil, (b) the maximum rate of change of magnetic flux through the coil, (c) the induced emf at $t = 0.050\ 0$ s, and (d) the torque exerted on the coil by the magnetic field at the instant when the emf is a maximum.

64. A small circular washer of radius 0.500 cm is held directly below a long, straight wire carrying a current of 10.0 A. The washer is located 0.500 m above the top of the table (Fig. P31.64). (a) If the washer is dropped from rest, what is the magnitude of the average induced

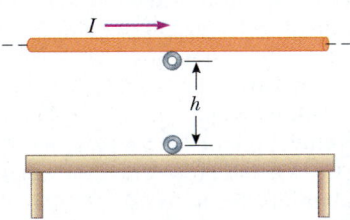

Figure P31.64

emf in the washer from the time it is released to the moment it hits the tabletop? Assume that the magnetic field is nearly constant over the area of the washer and equal to the magnetic field at the center of the washer. (b) What is the direction of the induced current in the washer?

65. To monitor the breathing of a hospital patient, a thin belt is wrapped around the patient's chest. The belt is a 200-turn coil. When the patient inhales, the area encircled by the coil increases by 39.0 cm². The magnitude of the Earth's magnetic field is 50.0 μT and makes an angle of 28.0° with the plane of the coil. If a patient takes 1.80 s to inhale, find the average induced emf in the coil during this time.

66. A conducting rod of length ℓ moves with velocity **v** parallel to a long wire carrying a steady current I. The axis of the rod is maintained perpendicular to the wire with the near end a distance r away, as shown in Figure P31.66. Show that the magnitude of the emf induced in the rod is

$$|\mathcal{E}| = \frac{\mu_0 I}{2\pi} v \ln\left(1 + \frac{\ell}{r}\right)$$

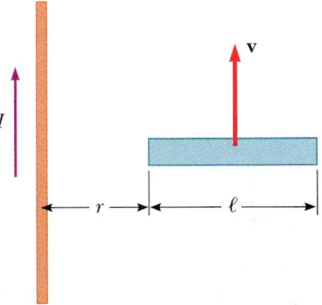

Figure P31.66

67. A rectangular loop of dimensions ℓ and w moves with a constant velocity **v** away from a long wire that carries a current I in the plane of the loop (Fig. P31.67). The to-

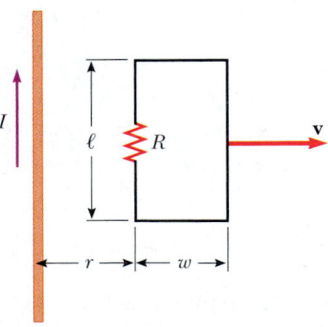

Figure P31.67

tal resistance of the loop is R. Derive an expression that gives the current in the loop at the instant the near side is a distance r from the wire.

68. A horizontal wire is free to slide on the vertical rails of a conducting frame, as shown in Figure P31.68. The wire has mass m and length ℓ, and the resistance of the circuit is R. If a uniform magnetic field is directed perpendicular to the frame, what is the terminal speed of the wire as it falls under the force of gravity?

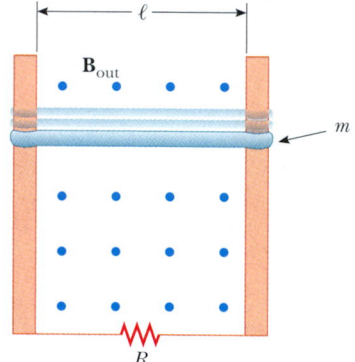

Figure P31.68

69. The magnetic flux threading a metal ring varies with time t according to $\Phi_B = 3(at^3 - bt^2)$ T·m², with $a = 2.00$ s⁻³ and $b = 6.00$ s⁻². The resistance of the ring is 3.00 Ω. Determine the maximum current induced in the ring during the interval from $t = 0$ to $t = 2.00$ s.

70. Review Problem. The bar of mass m shown in Figure P31.70 is pulled horizontally across parallel rails by a massless string that passes over an ideal pulley and is attached to a suspended mass M. The uniform magnetic field has a magnitude B, and the distance between the rails is ℓ. The rails are connected at one end by a load resistor R. Derive an expression that gives the horizon-

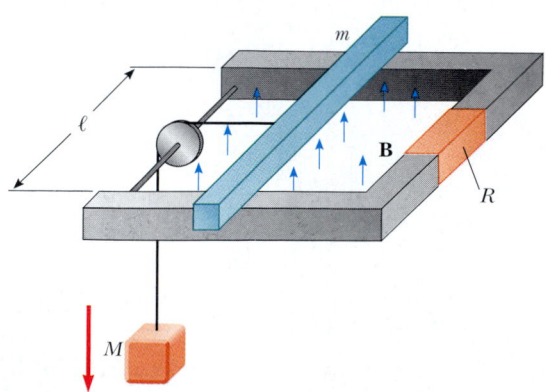

Figure P31.70

tal speed of the bar as a function of time, assuming that the suspended mass is released with the bar at rest at $t = 0$. Assume no friction between rails and bar.

71. A solenoid wound with 2 000 turns/m is supplied with current that varies in time according to $I = 4 \sin(120\pi t)$, where I is in A and t is in s. A small coaxial circular coil of 40 turns and radius $r = 5.00$ cm is located inside the solenoid near its center. (a) Derive an expression that describes the manner in which the emf in the small coil varies in time. (b) At what average rate is energy transformed into internal energy in the small coil if the windings have a total resistance of 8.00 Ω?

72. A wire 30.0 cm long is held parallel to and 80.0 cm above a long wire carrying 200 A and resting on the floor (Fig. P31.72). The 30.0-cm wire is released and falls, remaining parallel with the current-carrying wire as it falls. Assume that the falling wire accelerates at 9.80 m/s² and derive an equation for the emf induced in it. Express your result as a function of the time t after the wire is dropped. What is the induced emf 0.300 s after the wire is released?

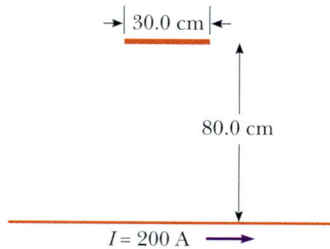

Figure P31.72

WEB 73. A long, straight wire carries a current $I = I_{\max} \sin(\omega t + \phi)$ and lies in the plane of a rectangular coil of N turns of wire, as shown in Figure P31.9. The quantities $I_{\max}$, ω, and ϕ are all constants. Determine the emf induced in the coil by the magnetic field created by the current in the straight wire. Assume $I_{\max} = 50.0$ A, $\omega = 200\pi$ s⁻¹, $N = 100$, $h = w = 5.00$ cm, and $L = 20.0$ cm.

74. A dime is suspended from a thread and hung between the poles of a strong horseshoe magnet as shown in Figure P31.74. The dime rotates at constant angular speed ω about a vertical axis. Letting θ represent the angle between the direction of $\mathbf{B}$ and the normal to the face of the dime, sketch a graph of the torque due to induced currents as a function of θ for $0 < \theta < 2\pi$.

75. The wire shown in Figure P31.75 is bent in the shape of a tent, with $\theta = 60.0°$ and $L = 1.50$ m, and is placed in a uniform magnetic field of magnitude 0.300 T perpendicular to the tabletop. The wire is rigid but hinged at points a and b. If the "tent" is flattened out on the table in 0.100 s, what is the average induced emf in the wire during this time?

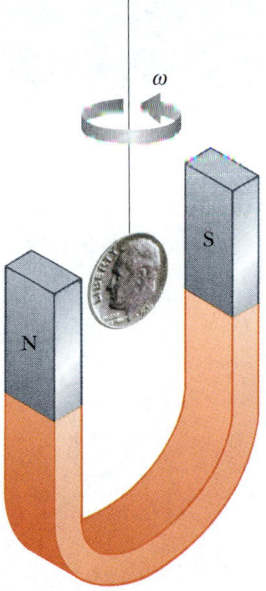

Figure P31.74

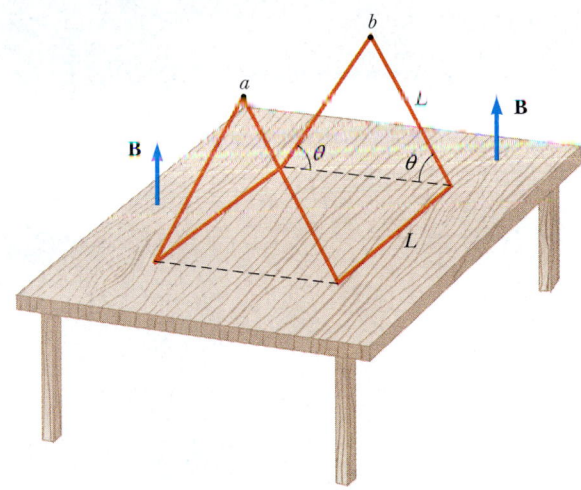

Figure P31.75

ANSWERS TO QUICK QUIZZES

31.1 Because the magnetic field now points in the opposite direction, you must replace θ with $\theta + \pi$. Because $\cos(\theta + \pi) = -\cos\theta$, the sign of the induced emf is reversed.

31.2 The one on the west side of the plane. As we saw in Section 30.9, the Earth's magnetic field has a downward component in the northern hemisphere. As the plane flies north, the right-hand rule illustrated in Figure 29.4 indicates that positive charge experiences a force directed toward the west. Thus, the left wingtip becomes positively charged and the right wingtip negatively charged.

31.3 Inserting. Because the south pole of the magnet is nearest the solenoid, the field lines created by the magnet point upward in Figure 31.14. Because the current induced in the solenoid is clockwise when viewed from above, the magnetic field lines produced by this current point downward in Figure 31.14. If the magnet were being withdrawn, it would create a decreasing upward flux. The induced current would counteract this decrease by producing its own upward flux. This would require a counterclockwise current in the solenoid, contrary to what is observed.

c h a p t e r

32

Inductance

In Chapter 31, we saw that emfs and currents are induced in a circuit when the magnetic flux through the area enclosed by the circuit changes with time. This electromagnetic induction has some practical consequences, which we describe in this chapter. First, we describe an effect known as *self-induction,* in which a time-varying current in a circuit produces in the circuit an induced emf that opposes the emf that initially set up the time-varying current. Self-induction is the basis of the *inductor,* an electrical element that has an important role in circuits that use time-varying currents. We discuss the energy stored in the magnetic field of an inductor and the energy density associated with the magnetic field.

Next, we study how an emf is induced in a circuit as a result of a changing magnetic flux produced by a second circuit; this is the basic principle of *mutual induction.* Finally, we examine the characteristics of circuits that contain inductors, resistors, and capacitors in various combinations.

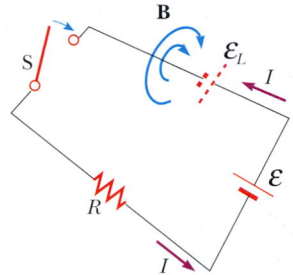

Figure 32.1 After the switch is thrown closed, the current produces a magnetic flux through the area enclosed by the loop. As the current increases toward its equilibrium value, this magnetic flux changes in time and induces an emf in the loop. The battery symbol drawn with dashed lines represents the self-induced emf.

32.1 SELF-INDUCTANCE

In this chapter, we need to distinguish carefully between emfs and currents that are caused by batteries or other sources and those that are induced by changing magnetic fields. We use the adjective *source* (as in the terms *source emf* and *source current*) to describe the parameters associated with a physical source, and we use the adjective *induced* to describe those emfs and currents caused by a changing magnetic field.

Consider a circuit consisting of a switch, a resistor, and a source of emf, as shown in Figure 32.1. When the switch is thrown to its closed position, the source current does not immediately jump from zero to its maximum value $\mathcal{E}/R$. Faraday's law of electromagnetic induction (Eq. 31.1) can be used to describe this effect as follows: As the source current increases with time, the magnetic flux through the circuit loop due to this current also increases with time. This increasing flux creates an induced emf in the circuit. The direction of the induced emf is such that it would cause an induced current in the loop (if a current were not already flowing in the loop), which would establish a magnetic field that would oppose the change in the source magnetic field. Thus, the direction of the induced emf is opposite the direction of the source emf; this results in a gradual rather than instantaneous increase in the source current to its final equilibrium value. This effect is called *self-induction* because the changing flux through the circuit and the resultant induced emf arise from the circuit itself. The emf $\mathcal{E}_L$ set up in this case is called a **self-induced emf.** It is also often called a **back emf.**

As a second example of self-induction, consider Figure 32.2, which shows a coil wound on a cylindrical iron core. (A practical device would have several hun-

Joseph Henry (1797–1878)
Henry, an American physicist, became the first director of the Smithsonian Institution and first president of the Academy of Natural Science. He improved the design of the electromagnet and constructed one of the first motors. He also discovered the phenomenon of self-induction but failed to publish his findings. The unit of inductance, the henry, is named in his honor. *(North Wind Picture Archives)*

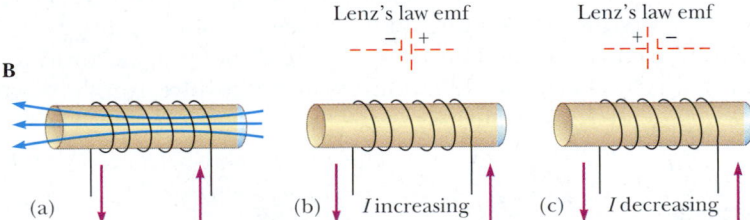

Figure 32.2 (a) A current in the coil produces a magnetic field directed to the left. (b) If the current increases, the increasing magnetic flux creates an induced emf having the polarity shown by the dashed battery. (c) The polarity of the induced emf reverses if the current decreases.

dred turns.) Assume that the source current in the coil either increases or decreases with time. When the source current is in the direction shown, a magnetic field directed from right to left is set up inside the coil, as seen in Figure 32.2a. As the source current changes with time, the magnetic flux through the coil also changes and induces an emf in the coil. From Lenz's law, the polarity of this induced emf must be such that it opposes the change in the magnetic field from the source current. If the source current is increasing, the polarity of the induced emf is as pictured in Figure 32.2b, and if the source current is decreasing, the polarity of the induced emf is as shown in Figure 32.2c.

To obtain a quantitative description of self-induction, we recall from Faraday's law that the induced emf is equal to the negative time rate of change of the magnetic flux. The magnetic flux is proportional to the magnetic field due to the source current, which in turn is proportional to the source current in the circuit. Therefore, **a self-induced emf $\mathcal{E}_L$ is always proportional to the time rate of change of the source current.** For a closely spaced coil of N turns (a toroid or an ideal solenoid) carrying a source current I, we find that

Self-induced emf

$$\mathcal{E}_L = -N\frac{d\Phi_B}{dt} = -L\frac{dI}{dt} \qquad \textbf{(32.1)}$$

where L is a proportionality constant—called the **inductance** of the coil—that depends on the geometry of the circuit and other physical characteristics. From this expression, we see that the inductance of a coil containing N turns is

Inductance of an N-turn coil

$$L = \frac{N\Phi_B}{I} \qquad \textbf{(32.2)}$$

where it is assumed that the same flux passes through each turn. Later, we shall use this equation to calculate the inductance of some special circuit geometries.

From Equation 32.1, we can also write the inductance as the ratio

Inductance

$$L = -\frac{\mathcal{E}_L}{dI/dt} \qquad \textbf{(32.3)}$$

Just as resistance is a measure of the opposition to current ($R = \Delta V/I$), inductance is a measure of the opposition to a *change* in current.

The SI unit of inductance is the **henry** (H), which, as we can see from Equation 32.3, is 1 volt-second per ampere:

$$1\ \mathrm{H} = 1\ \frac{\mathrm{V \cdot s}}{\mathrm{A}}$$

That the inductance of a device depends on its geometry is analogous to the capacitance of a capacitor depending on the geometry of its plates, as we found in Chapter 26. Inductance calculations can be quite difficult to perform for complicated geometries; however, the following examples involve simple situations for which inductances are easily evaluated.

EXAMPLE 32.1   **Inductance of a Solenoid**

Find the inductance of a uniformly wound solenoid having N turns and length ℓ. Assume that ℓ is much longer than the radius of the windings and that the core of the solenoid is air.

Solution   We can assume that the interior magnetic field due to the source current is uniform and given by Equation 30.17:

$$B = \mu_0 n I = \mu_0 \frac{N}{\ell} I$$

where $n = N/\ell$ is the number of turns per unit length. The magnetic flux through each turn is

$$\Phi_B = BA = \mu_0 \frac{NA}{\ell} I$$

where A is the cross-sectional area of the solenoid. Using this expression and Equation 32.2, we find that

$$L = \frac{N\Phi_B}{I} = \frac{\mu_0 N^2 A}{\ell} \qquad \textbf{(32.4)}$$

This result shows that L depends on geometry and is proportional to the square of the number of turns. Because $N = n\ell$, we can also express the result in the form

$$L = \mu_0 \frac{(n\ell)^2}{\ell} A = \mu_0 n^2 A \ell = \mu_0 n^2 V \qquad \textbf{(32.5)}$$

where $V = A\ell$ is the volume of the solenoid.

Exercise What would happen to the inductance if a ferromagnetic material were placed inside the solenoid?

Answer The inductance would increase. For a given current, the magnetic flux is now much greater because of the increase in the field originating from the magnetization of the ferromagnetic material. For example, if the material has a magnetic permeability of $500\mu_0$, the inductance would increase by a factor of 500.

The fact that various materials in the vicinity of a coil can substantially alter the coil's inductance is used to great advantage by traffic engineers. A flat, horizontal coil made of numerous loops of wire is placed in a shallow groove cut into the pavement of the lane approaching an intersection. (See the photograph at the beginning of this chapter.) These loops are attached to circuitry that measures inductance. When an automobile passes over the loops, the change in inductance caused by the large amount of iron passing over the loops is used to control the lights at the intersection.

EXAMPLE 32.2 Calculating Inductance and emf

(a) Calculate the inductance of an air-core solenoid containing 300 turns if the length of the solenoid is 25.0 cm and its cross-sectional area is 4.00 cm².

Solution Using Equation 32.4, we obtain

$$L = \frac{\mu_0 N^2 A}{\ell}$$

$$= (4\pi \times 10^{-7}\,\text{T}\cdot\text{m/A}) \frac{(300)^2(4.00 \times 10^{-4}\,\text{m}^2)}{25.0 \times 10^{-2}\,\text{m}}$$

$$= 1.81 \times 10^{-4}\,\text{T}\cdot\text{m}^2/\text{A} = \boxed{0.181\,\text{mH}}$$

(b) Calculate the self-induced emf in the solenoid if the current through it is decreasing at the rate of 50.0 A/s.

Solution Using Equation 32.1 and given that $dI/dt = -50.0$ A/s, we obtain

$$\mathcal{E}_L = -L\frac{dI}{dt} = -(1.81 \times 10^{-4}\,\text{H})(-50.0\,\text{A/s})$$

$$= \boxed{9.05\,\text{mV}}$$

32.2 RL CIRCUITS

If a circuit contains a coil, such as a solenoid, the self-inductance of the coil prevents the current in the circuit from increasing or decreasing instantaneously. A circuit element that has a large self-inductance is called an **inductor** and has the circuit symbol ⎓⎓⎓. We always assume that the self-inductance of the remainder of a circuit is negligible compared with that of the inductor. Keep in mind, however, that even a circuit without a coil has some self-inductance that can affect the behavior of the circuit.

Because the inductance of the inductor results in a back emf, **an inductor in a circuit opposes changes in the current through that circuit.** If the battery voltage in the circuit is increased so that the current rises, the inductor opposes this change, and the rise is not instantaneous. If the battery voltage is decreased, the presence of the inductor results in a slow drop in the current rather than an immediate drop. Thus, the inductor causes the circuit to be "sluggish" as it reacts to changes in the current.

Quick Quiz 32.1

A switch controls the current in a circuit that has a large inductance. Is a spark more likely to be produced at the switch when the switch is being closed or when it is being opened, or doesn't it matter?

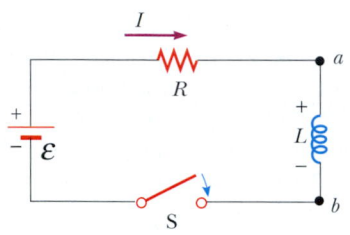

Figure 32.3 A series *RL* circuit. As the current increases toward its maximum value, an emf that opposes the increasing current is induced in the inductor.

Consider the circuit shown in Figure 32.3, in which the battery has negligible internal resistance. This is an **RL circuit** because the elements connected to the battery are a resistor and an inductor. Suppose that the switch S is thrown closed at $t = 0$. The current in the circuit begins to increase, and a back emf that opposes the increasing current is induced in the inductor. The back emf is, from Equation 32.1,

$$\mathcal{E}_L = -L\frac{dI}{dt}$$

Because the current is increasing, dI/dt is positive; thus, $\mathcal{E}_L$ is negative. This negative value reflects the decrease in electric potential that occurs in going from a to b across the inductor, as indicated by the positive and negative signs in Figure 32.3.

With this in mind, we can apply Kirchhoff's loop rule to this circuit, traversing the circuit in the clockwise direction:

$$\mathcal{E} - IR - L\frac{dI}{dt} = 0 \tag{32.6}$$

where IR is the voltage drop across the resistor. (We developed Kirchhoff's rules for circuits with steady currents, but we can apply them to a circuit in which the current is changing if we imagine them to represent the circuit at one *instant* of time.) We must now look for a solution to this differential equation, which is similar to that for the *RC* circuit (see Section 28.4).

A mathematical solution of Equation 32.6 represents the current in the circuit as a function of time. To find this solution, we change variables for convenience, letting $x = \dfrac{\mathcal{E}}{R} - I$, so that $dx = -dI$. With these substitutions, we can write Equation 32.6 as

$$x + \frac{L}{R}\frac{dx}{dt} = 0$$

$$\frac{dx}{x} = -\frac{R}{L}\,dt$$

Integrating this last expression, we have

$$\ln\frac{x}{x_0} = -\frac{R}{L}t$$

where we take the integrating constant to be $-\ln x_0$ and x_0 is the value of x at time $t = 0$. Taking the antilogarithm of this result, we obtain

$$x = x_0 e^{-Rt/L}$$

Because $I = 0$ at $t = 0$, we note from the definition of x that $x_0 = \mathcal{E}/R$. Hence, this last expression is equivalent to

$$\frac{\mathcal{E}}{R} - I = \frac{\mathcal{E}}{R} e^{-Rt/L}$$

$$I - \frac{\mathcal{E}}{R} (1 - e^{-Rt/L})$$

This expression shows the effect of the inductor. The current does not increase instantly to its final equilibrium value when the switch is closed but instead increases according to an exponential function. If we remove the inductance in the circuit, which we can do by letting L approach zero, the exponential term becomes zero and we see that there is no time dependence of the current in this case—the current increases instantaneously to its final equilibrium value in the absence of the inductance.

We can also write this expression as

$$I = \frac{\mathcal{E}}{R} (1 - e^{-t/\tau}) \tag{32.7}$$

where the constant τ is the **time constant** of the RL circuit:

$$\tau = L/R \tag{32.8}$$

Physically, τ is the time it takes the current in the circuit to reach $(1 - e^{-1}) = 0.63$ of its final value $\mathcal{E}/R$. The time constant is a useful parameter for comparing the time responses of various circuits.

Figure 32.4 shows a graph of the current versus time in the RL circuit. Note that the equilibrium value of the current, which occurs as t approaches infinity, is $\mathcal{E}/R$. We can see this by setting dI/dt equal to zero in Equation 32.6 and solving for the current I. (At equilibrium, the change in the current is zero.) Thus, we see that the current initially increases very rapidly and then gradually approaches the equilibrium value $\mathcal{E}/R$ as t approaches infinity.

Let us also investigate the time rate of change of the current in the circuit. Taking the first time derivative of Equation 32.7, we have

$$\frac{dI}{dt} = \frac{\mathcal{E}}{L} e^{-t/\tau} \tag{32.9}$$

From this result, we see that the time rate of change of the current is a maximum (equal to $\mathcal{E}/L$) at $t = 0$ and falls off exponentially to zero as t approaches infinity (Fig. 32.5).

Now let us consider the RL circuit shown in Figure 32.6. The circuit contains two switches that operate such that when one is closed, the other is opened. Suppose that S_1 has been closed for a length of time sufficient to allow the current to reach its equilibrium value $\mathcal{E}/R$. In this situation, the circuit is described completely by the outer loop in Figure 32.6. If S_2 is closed at the instant at which S_1 is opened, the circuit changes so that it is described completely by just the upper loop in Figure 32.6. The lower loop no longer influences the behavior of the circuit. Thus, we have a circuit with no battery ($\mathcal{E} = 0$). If we apply Kirchhoff's loop rule to the upper loop at the instant the switches are thrown, we obtain

$$IR + L\frac{dI}{dt} = 0$$

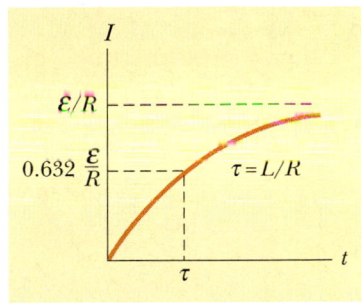

Figure 32.4 Plot of the current versus time for the RL circuit shown in Figure 32.3. The switch is thrown closed at $t = 0$, and the current increases toward its maximum value $\mathcal{E}/R$. The time constant τ is the time it takes I to reach 63% of its maximum value.

Time constant of an RL circuit

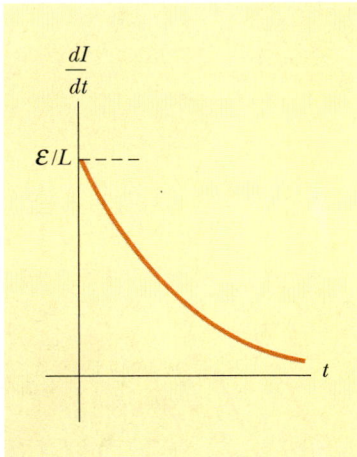

Figure 32.5 Plot of dI/dt versus time for the RL circuit shown in Figure 32.3. The time rate of change of current is a maximum at $t = 0$, which is the instant at which the switch is thrown closed. The rate decreases exponentially with time as I increases toward its maximum value.

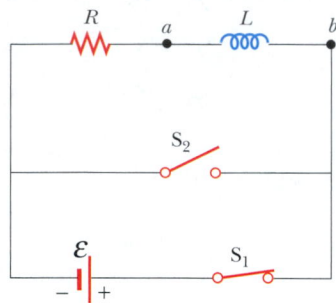

Figure 32.6 An *RL* circuit containing two switches. When S_1 is closed and S_2 open as shown, the battery is in the circuit. At the instant S_2 is closed, S_1 is opened, and the battery is no longer part of the circuit.

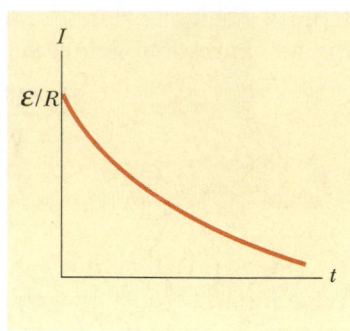

Figure 32.7 Current versus time for the upper loop of the circuit shown in Figure 32.6. For $t < 0$, S_1 is closed and S_2 is open. At $t = 0$, S_2 is closed as S_1 is opened, and the current has its maximum value $\mathcal{E}/R$.

It is left as a problem (Problem 18) to show that the solution of this differential equation is

$$I = \frac{\mathcal{E}}{R}\, e^{-t/\tau} = I_0 e^{-t/\tau} \tag{32.10}$$

where $\mathcal{E}$ is the emf of the battery and $I_0 = \mathcal{E}/R$ is the current at $t = 0$, the instant at which S_2 is closed as S_1 is opened.

    If no inductor were present in the circuit, the current would immediately decrease to zero if the battery were removed. When the inductor is present, it acts to oppose the decrease in the current and to maintain the current. A graph of the current in the circuit versus time (Fig. 32.7) shows that the current is continuously decreasing with time. Note that the slope dI/dt is always negative and has its maximum value at $t = 0$. The negative slope signifies that $\mathcal{E}_L = -L\,(dI/dt)$ is now positive; that is, point a in Figure 32.6 is at a lower electric potential than point b.

Quick Quiz 32.2

Two circuits like the one shown in Figure 32.6 are identical except for the value of L. In circuit A the inductance of the inductor is L_A, and in circuit B it is L_B. Switch S_1 is thrown closed at $t = 0$, while switch S_2 remains open. At $t = 0$, switch S_1 is opened and switch S_2 is closed. The resulting time rates of change for the two currents are as graphed in Figure 32.8. If we assume that the time constant of each circuit is much less than 10 s, which of the following is true? (a) $L_A > L_B$; (b) $L_A < L_B$; (c) not enough information to tell.

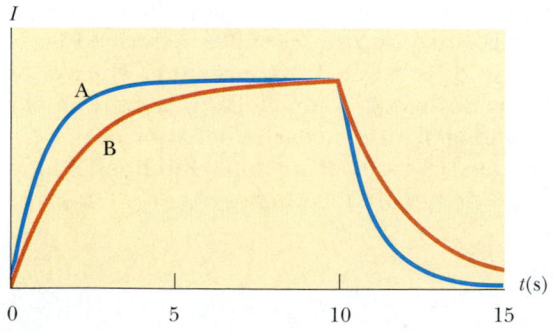

Figure 32.8

EXAMPLE 32.3 ▶ Time Constant of an *RL* Circuit

The switch in Figure 32.9a is thrown closed at $t = 0$. (a) Find the time constant of the circuit.

Solution The time constant is given by Equation 32.8:

$$\tau = \frac{L}{R} = \frac{30.0 \times 10^{-3}\ \text{H}}{6.00\ \Omega} = \boxed{5.00\ \text{ms}}$$

(b) Calculate the current in the circuit at $t = 2.00$ ms.

Solution Using Equation 32.7 for the current as a function of time (with t and τ in milliseconds), we find that at $t = 2.00$ ms

$$I = \frac{\mathcal{E}}{R}\,(1 - e^{-t/\tau}) = \frac{12.0\ \text{V}}{6.00\ \Omega}\,(1 - e^{-0.400}) = \boxed{0.659\ \text{A}}$$

A plot of Equation 32.7 for this circuit is given in Figure 32.9b.

(c) Compare the potential difference across the resistor with that across the inductor.

Solution At the instant the switch is closed, there is no current and thus no potential difference across the resistor. At this instant, the battery voltage appears entirely across the inductor in the form of a back emf of 12.0 V as the inductor tries to maintain the zero-current condition. (The left end of the inductor is at a higher electric potential than the right end.) As time passes, the emf across the inductor decreases and the current through the resistor (and hence the potential difference across it) increases. The sum of the two potential differences at all times is 12.0 V, as shown in Figure 32.10.

Exercise Calculate the current in the circuit and the voltage across the resistor after a time interval equal to one time constant has elapsed.

Answer 1.26 A, 7.56 V.

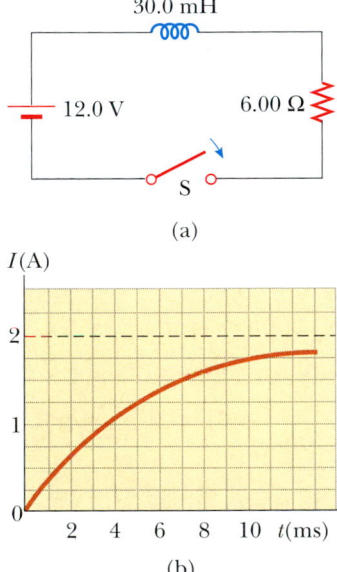

(a)

(b)

Figure 32.9 (a) The switch in this *RL* circuit is thrown closed at $t = 0$. (b) A graph of the current versus time for the circuit in part (a).

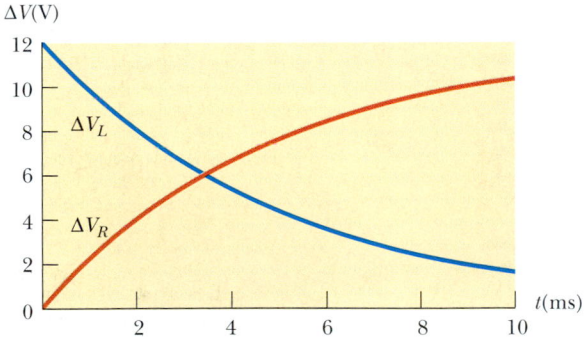

Figure 32.10 The sum of the potential differences across the resistor and inductor in Figure 32.9a is 12.0 V (the battery emf) at all times.

32.3 ▶ ENERGY IN A MAGNETIC FIELD

Because the emf induced in an inductor prevents a battery from establishing an instantaneous current, the battery must do work against the inductor to create a current. Part of the energy supplied by the battery appears as internal energy in the resistor, while the remaining energy is stored in the magnetic field of the inductor. If we multiply each term in Equation 32.6 by I and rearrange the expression, we have

$$I\mathcal{E} = I^2 R + LI\,\frac{dI}{dt} \tag{32.11}$$

This expression indicates that the rate at which energy is supplied by the battery ($I\mathcal{E}$) equals the sum of the rate at which energy is delivered to the resistor, I^2R, and the rate at which energy is stored in the inductor, $LI(dI/dt)$. Thus, Equation 32.11 is simply an expression of energy conservation. If we let U denote the energy stored in the inductor at any time, then we can write the rate dU/dt at which energy is stored as

$$\frac{dU}{dt} = LI\frac{dI}{dt}$$

To find the total energy stored in the inductor, we can rewrite this expression as $dU = LI\,dI$ and integrate:

$$U = \int dU = \int_0^I LI\,dI = L\int_0^I I\,dI$$

Energy stored in an inductor

$$U = \tfrac{1}{2}LI^2 \qquad\qquad (32.12)$$

where L is constant and has been removed from the integral. This expression represents the energy stored in the magnetic field of the inductor when the current is I. Note that this equation is similar in form to Equation 26.11 for the energy stored in the electric field of a capacitor, $U = Q^2/2C$. In either case, we see that energy is required to establish a field.

We can also determine the energy density of a magnetic field. For simplicity, consider a solenoid whose inductance is given by Equation 32.5:

$$L = \mu_0 n^2 A\ell$$

The magnetic field of a solenoid is given by Equation 30.17:

$$B = \mu_0 nI$$

Substituting the expression for L and $I = B/\mu_0 n$ into Equation 32.12 gives

$$U = \tfrac{1}{2}LI^2 = \tfrac{1}{2}\mu_0 n^2 A\ell\left(\frac{B}{\mu_0 n}\right)^2 = \frac{B^2}{2\mu_0}A\ell \qquad\qquad (32.13)$$

Because $A\ell$ is the volume of the solenoid, the energy stored per unit volume in the magnetic field surrounding the inductor is

Magnetic energy density

$$u_B = \frac{U}{A\ell} = \frac{B^2}{2\mu_0} \qquad\qquad (32.14)$$

Although this expression was derived for the special case of a solenoid, it is valid for any region of space in which a magnetic field exists. Note that Equation 32.14 is similar in form to Equation 26.13 for the energy per unit volume stored in an electric field, $u_E = \tfrac{1}{2}\epsilon_0 E^2$. In both cases, the energy density is proportional to the square of the magnitude of the field.

EXAMPLE 32.4   What Happens to the Energy in the Inductor?

Consider once again the RL circuit shown in Figure 32.6, in which switch S_2 is closed at the instant S_1 is opened (at $t = 0$). Recall that the current in the upper loop decays exponentially with time according to the expression $I = I_0 e^{-t/\tau}$, where $I_0 = \mathcal{E}/R$ is the initial current in the circuit and $\tau = L/R$ is the time constant. Show that all the energy initially stored in the magnetic field of the inductor appears as internal energy in the resistor as the current decays to zero.

Solution   The rate dU/dt at which energy is delivered to the resistor (which is the power) is equal to I^2R, where I is the instantaneous current:

$$\frac{dU}{dt} = I^2R = (I_0e^{-Rt/L})^2R = I_0^2Re^{-2Rt/L}$$

To find the total energy delivered to the resistor, we integrate this expression over the limits $t = 0$ to $t \to \infty$ (the upper limit is infinity because it takes an infinite amount of time for the current to reach zero):

$$(1) \qquad U = \int_0^\infty I_0^2Re^{-2Rt/L}dt = I_0^2R\int_0^\infty e^{-2Rt/L}dt$$

The value of the definite integral is $L/2R$ (this is left for the student to show in the exercise at the end of this example), and so U becomes

$$U = I_0^2R\left(\frac{L}{2R}\right) = \frac{1}{2}LI_0^2$$

Note that this is equal to the initial energy stored in the magnetic field of the inductor, given by Equation 32.13, as we set out to prove.

Exercise   Show that the integral on the right-hand side of Equation (1) has the value $L/2R$.

EXAMPLE 32.5   The Coaxial Cable

Coaxial cables are often used to connect electrical devices, such as your stereo system and a loudspeaker. Model a long coaxial cable as consisting of two thin concentric cylindrical conducting shells of radii a and b and length ℓ, as shown in Figure 32.11. The conducting shells carry the same current I in opposite directions. Imagine that the inner conductor carries current to a device and that the outer one acts as a return path carrying the current back to the source. (a) Calculate the self-inductance L of this cable.

Solution   To obtain L, we must know the magnetic flux through any cross-section in the region between the two shells, such as the light blue rectangle in Figure 32.11. Am-

père's law (see Section 30.3) tells us that the magnetic field in the region between the shells is $B = \mu_0I/2\pi r$, where r is measured from the common center of the shells. The magnetic field is zero outside the outer shell ($r > b$) because the net current through the area enclosed by a circular path surrounding the cable is zero, and hence from Ampère's law, $\oint \mathbf{B} \cdot d\mathbf{s} = 0$. The magnetic field is zero inside the inner shell because the shell is hollow and no current is present within a radius $r < a$.

The magnetic field is perpendicular to the light blue rectangle of length ℓ and width $b - a$, the cross-section of interest. Because the magnetic field varies with radial position across this rectangle, we must use calculus to find the total magnetic flux. Dividing this rectangle into strips of width dr, such as the dark blue strip in Figure 32.11, we see that the area of each strip is $\ell\,dr$ and that the flux through each strip is $B\,dA = B\ell\,dr$. Hence, we find the total flux through the entire cross-section by integrating:

$$\Phi_B = \int B\,dA = \int_a^b \frac{\mu_0I}{2\pi r}\,\ell\,dr = \frac{\mu_0I\ell}{2\pi}\int_a^b \frac{dr}{r} = \frac{\mu_0I\ell}{2\pi}\ln\left(\frac{b}{a}\right)$$

Using this result, we find that the self-inductance of the cable is

$$L = \frac{\Phi_B}{I} = \frac{\mu_0\ell}{2\pi}\ln\left(\frac{b}{a}\right)$$

(b) Calculate the total energy stored in the magnetic field of the cable.

Solution   Using Equation 32.12 and the results to part (a) gives

$$U = \tfrac{1}{2}LI^2 = \frac{\mu_0\ell I^2}{4\pi}\ln\left(\frac{b}{a}\right)$$

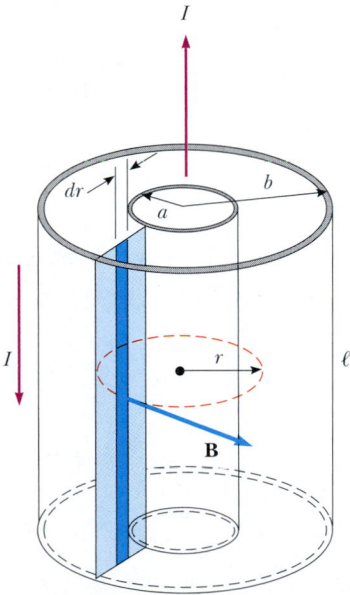

Figure 32.11   Section of a long coaxial cable. The inner and outer conductors carry equal currents in opposite directions.

32.4   MUTUAL INDUCTANCE

Very often, the magnetic flux through the area enclosed by a circuit varies with time because of time-varying currents in nearby circuits. This condition induces an emf through a process known as *mutual induction*, so called because it depends on the interaction of two circuits.

Consider the two closely wound coils of wire shown in cross-sectional view in Figure 32.12. The current I_1 in coil 1, which has N_1 turns, creates magnetic field lines, some of which pass through coil 2, which has N_2 turns. The magnetic flux caused by the current in coil 1 and passing through coil 2 is represented by Φ_{12}. In analogy to Equation 32.2, we define the **mutual inductance** M_{12} of coil 2 with respect to coil 1:

Definition of mutual inductance

$$M_{12} \equiv \frac{N_2 \Phi_{12}}{I_1} \tag{32.15}$$

Quick Quiz 32.3

Referring to Figure 32.12, tell what happens to M_{12} (a) if coil 1 is brought closer to coil 2 and (b) if coil 1 is rotated so that it lies in the plane of the page.

Quick Quiz 32.3 demonstrates that mutual inductance depends on the geometry of both circuits and on their orientation with respect to each other. As the circuit separation distance increases, the mutual inductance decreases because the flux linking the circuits decreases.

If the current I_1 varies with time, we see from Faraday's law and Equation 32.15 that the emf induced by coil 1 in coil 2 is

$$\mathcal{E}_2 = -N_2 \frac{d\Phi_{12}}{dt} = -N_2 \frac{d}{dt}\left(\frac{M_{12}I_1}{N_2}\right) = -M_{12}\frac{dI_1}{dt} \tag{32.16}$$

In the preceding discussion, we assumed that the source current is in coil 1. We can also imagine a source current I_2 in coil 2. The preceding discussion can be repeated to show that there is a mutual inductance M_{21}. If the current I_2 varies with time, the emf induced by coil 2 in coil 1 is

$$\mathcal{E}_1 = -M_{21}\frac{dI_2}{dt} \tag{32.17}$$

In mutual induction, the emf induced in one coil is always proportional to the rate at which the current in the other coil is changing. Although the

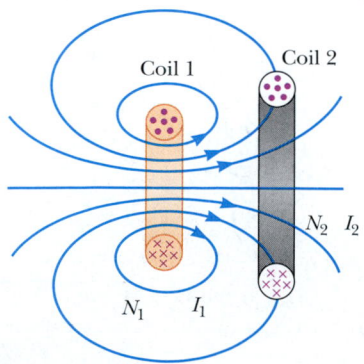

Figure 32.12   A cross-sectional view of two adjacent coils. A current in coil 1 sets up a magnetic flux, part of which passes through coil 2.

proportionality constants M_{12} and M_{21} appear to have different values, it can be shown that they are equal. Thus, with $M_{12} = M_{21} = M$, Equations 32.16 and 32.17 become

$$\mathcal{E}_2 = -M\frac{dI_1}{dt} \quad \text{and} \quad \mathcal{E}_1 = -M\frac{dI_2}{dt}$$

These two equations are similar in form to Equation 32.1 for the self-induced emf $\mathcal{E} = -L(dI/dt)$. The unit of mutual inductance is the henry.

Quick Quiz 32.4

(a) Can you have mutual inductance without self-inductance? (b) How about self-inductance without mutual inductance?

QuickLab

Tune in a relatively weak station on a radio. Now slowly rotate the radio about a vertical axis through its center. What happens to the reception? Can you explain this in terms of the mutual induction of the station's broadcast antenna and your radio's antenna?

EXAMPLE 32.6   "Wireless" Battery Charger

An electric toothbrush has a base designed to hold the toothbrush handle when not in use. As shown in Figure 32.13a, the handle has a cylindrical hole that fits loosely over a matching cylinder on the base. When the handle is placed on the base, a changing current in a solenoid inside the base cylinder induces a current in a coil inside the handle. This induced current charges the battery in the handle.

We can model the base as a solenoid of length ℓ with N_B turns (Fig. 32.13b), carrying a source current I, and having a cross-sectional area A. The handle coil contains N_H turns. Find the mutual inductance of the system.

Solution Because the base solenoid carries a source current I, the magnetic field in its interior is

$$B = \frac{\mu_0 N_B I}{\ell}$$

Because the magnetic flux Φ_{BH} through the handle's coil caused by the magnetic field of the base coil is BA, the mutual inductance is

$$M = \frac{N_H \Phi_{BH}}{I} = \frac{N_H BA}{I} = \boxed{\mu_0 \frac{N_H N_B A}{\ell}}$$

Exercise Calculate the mutual inductance of two solenoids with $N_B = 1\,500$ turns, $A = 1.0 \times 10^{-4}$ m^2, $\ell = 0.02$ m, and $N_H = 800$ turns.

Answer 7.5 mH.

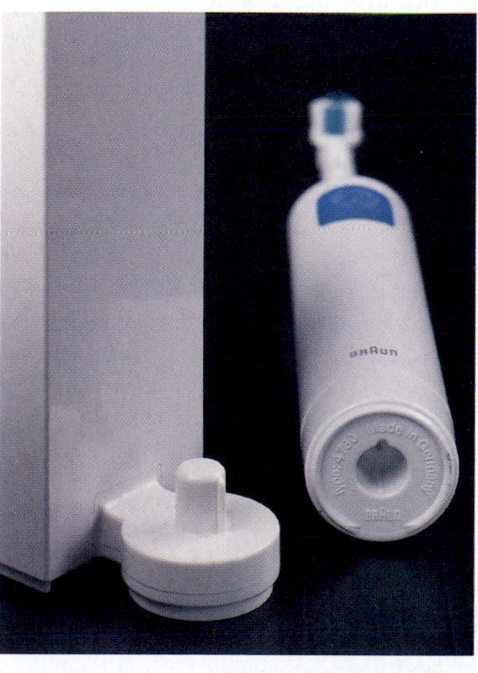

(a)

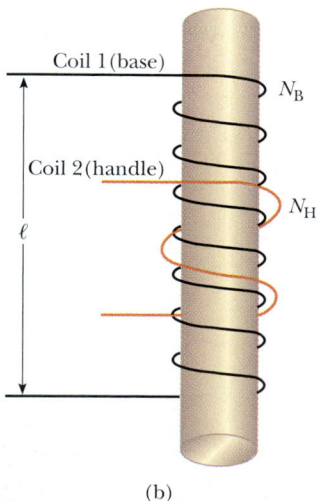

(b)

Figure 32.13   (a) This electric toothbrush uses the mutual induction of solenoids as part of its battery-charging system. (b) A coil of N_H turns wrapped around the center of a solenoid of N_B turns.

32.5 ► OSCILLATIONS IN AN *LC* CIRCUIT

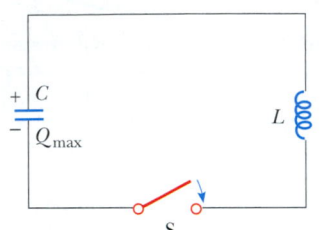

Figure 32.14 A simple *LC* circuit. The capacitor has an initial charge Q_{max}, and the switch is thrown closed at $t = 0$.

When a capacitor is connected to an inductor as illustrated in Figure 32.14, the combination is an **LC circuit.** If the capacitor is initially charged and the switch is then closed, both the current in the circuit and the charge on the capacitor oscillate between maximum positive and negative values. If the resistance of the circuit is zero, no energy is transformed to internal energy. In the following analysis, we neglect the resistance in the circuit. We also assume an idealized situation in which energy is not radiated away from the circuit. We shall discuss this radiation in Chapter 34, but we neglect it for now. With these idealizations—zero resistance and no radiation—the oscillations in the circuit persist indefinitely.

Assume that the capacitor has an initial charge Q_{max} (the maximum charge) and that the switch is thrown closed at $t = 0$. Let us look at what happens from an energy viewpoint.

When the capacitor is fully charged, the energy U in the circuit is stored in the electric field of the capacitor and is equal to $Q_{max}^2/2C$ (Eq. 26.11). At this time, the current in the circuit is zero, and thus no energy is stored in the inductor. After the switch is thrown closed, the rate at which charges leave or enter the capacitor plates (which is also the rate at which the charge on the capacitor changes) is equal to the current in the circuit. As the capacitor begins to discharge after the switch is closed, the energy stored in its electric field decreases. The discharge of the capacitor represents a current in the circuit, and hence some energy is now stored in the magnetic field of the inductor. Thus, energy is transferred from the electric field of the capacitor to the magnetic field of the inductor. When the capacitor is fully discharged, it stores no energy. At this time, the current reaches its maximum value, and all of the energy is stored in the inductor. The current continues in the same direction, decreasing in magnitude, with the capacitor eventually becoming fully charged again but with the polarity of its plates now opposite the initial polarity. This is followed by another discharge until the circuit returns to its original state of maximum charge Q_{max} and the plate polarity shown in Figure 32.14. The energy continues to oscillate between inductor and capacitor.

The oscillations of the *LC* circuit are an electromagnetic analog to the mechanical oscillations of a block–spring system, which we studied in Chapter 13. Much of what we discussed is applicable to *LC* oscillations. For example, we investigated the effect of driving a mechanical oscillator with an external force, which leads to the phenomenon of *resonance*. We observe the same phenomenon in the *LC* circuit. For example, a radio tuner has an *LC* circuit with a natural frequency, which we determine as follows: When the circuit is driven by the electromagnetic oscillations of a radio signal detected by the antenna, the tuner circuit responds with a large amplitude of electrical oscillation only for the station frequency that matches the natural frequency. Thus, only the signal from one station is passed on to the amplifier, even though signals from all stations are driving the circuit at the same time. When you turn the knob on the radio tuner to change the station, you are changing the natural frequency of the circuit so that it will exhibit a resonance response to a different driving frequency.

A graphical description of the energy transfer between the inductor and the capacitor in an *LC* circuit is shown in Figure 32.15. The right side of the figure shows the analogous energy transfer in the oscillating block–spring system studied in Chapter 13. In each case, the situation is shown at intervals of one-fourth the period of oscillation T. The potential energy $\frac{1}{2}kx^2$ stored in a stretched spring is analogous to the electric potential energy $Q_{max}^2/2C$ stored in the capacitor. The kinetic energy $\frac{1}{2}mv^2$ of the moving block is analogous to the magnetic energy $\frac{1}{2}LI^2$

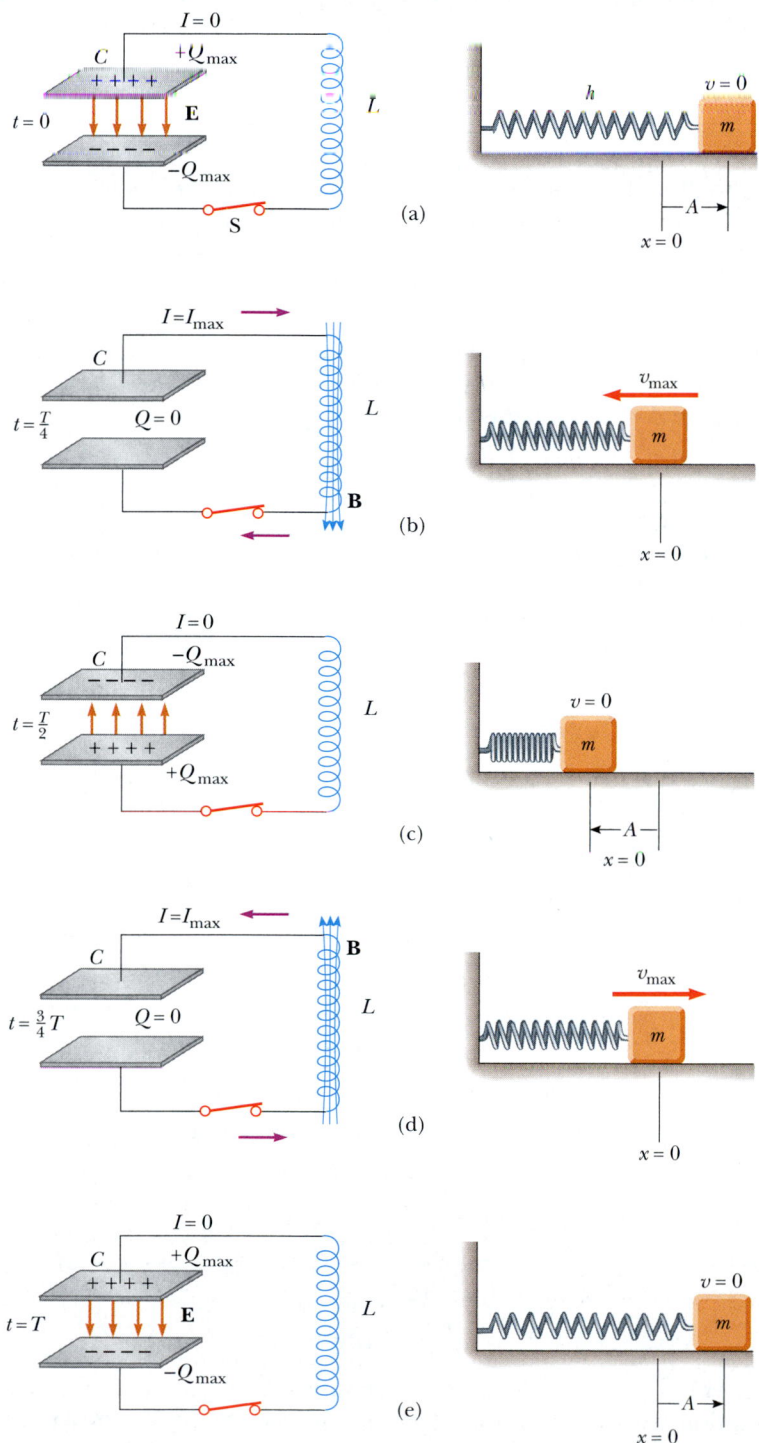

Figure 32.15 Energy transfer in a resistanceless, non-radiating *LC* circuit. The capacitor has a charge Q_{max} at $t = 0$, the instant at which the switch is thrown closed. The mechanical analog of this circuit is a block–spring system.

stored in the inductor, which requires the presence of moving charges. In Figure 32.15a, all of the energy is stored as electric potential energy in the capacitor at $t = 0$. In Figure 32.15b, which is one fourth of a period later, all of the energy is stored as magnetic energy $\frac{1}{2}LI_{max}^2$ in the inductor, where I_{max} is the maximum current in the circuit. In Figure 32.15c, the energy in the LC circuit is stored completely in the capacitor, with the polarity of the plates now opposite what it was in Figure 32.15a. In parts d and e the system returns to the initial configuration over the second half of the cycle. At times other than those shown in the figure, part of the energy is stored in the electric field of the capacitor and part is stored in the magnetic field of the inductor. In the analogous mechanical oscillation, part of the energy is potential energy in the spring and part is kinetic energy of the block.

Let us consider some arbitrary time t after the switch is closed, so that the capacitor has a charge $Q < Q_{max}$ and the current is $I < I_{max}$. At this time, both elements store energy, but the sum of the two energies must equal the total initial energy U stored in the fully charged capacitor at $t = 0$:

Total energy stored in an *LC* circuit

$$U = U_C + U_L = \frac{Q^2}{2C} + \frac{1}{2}LI^2 \tag{32.18}$$

Because we have assumed the circuit resistance to be zero, no energy is transformed to internal energy, and hence *the total energy must remain constant in time.* This means that $dU/dt = 0$. Therefore, by differentiating Equation 32.18 with respect to time while noting that Q and I vary with time, we obtain

The total energy in an ideal *LC* circuit remains constant; $dU/dt = 0$

$$\frac{dU}{dt} = \frac{d}{dt}\left(\frac{Q^2}{2C} + \frac{1}{2}LI^2\right) = \frac{Q}{C}\frac{dQ}{dt} + LI\frac{dI}{dt} = 0 \tag{32.19}$$

We can reduce this to a differential equation in one variable by remembering that the current in the circuit is equal to the rate at which the charge on the capacitor changes: $I = dQ/dt$. From this, it follows that $dI/dt = d^2Q/dt^2$. Substitution of these relationships into Equation 32.19 gives

$$\frac{Q}{C} + L\frac{d^2Q}{dt^2} = 0$$

$$\frac{d^2Q}{dt^2} = -\frac{1}{LC}Q \tag{32.20}$$

We can solve for Q by noting that this expression is of the same form as the analogous Equations 13.16 and 13.17 for a block–spring system:

$$\frac{d^2x}{dt^2} = -\frac{k}{m}x = -\omega^2 x$$

where k is the spring constant, m is the mass of the block, and $\omega = \sqrt{k/m}$. The solution of this equation has the general form

$$x = A\cos(\omega t + \phi)$$

where ω is the angular frequency of the simple harmonic motion, A is the amplitude of motion (the maximum value of x), and ϕ is the phase constant; the values of A and ϕ depend on the initial conditions. Because it is of the same form as the differential equation of the simple harmonic oscillator, we see that Equation 32.20 has the solution

Charge versus time for an ideal *LC* circuit

$$Q = Q_{max}\cos(\omega t + \phi) \tag{32.21}$$

where Q_{max} is the maximum charge of the capacitor and the angular frequency ω is

$$\omega = \frac{1}{\sqrt{LC}} \qquad \textbf{(32.22)}$$

Note that the angular frequency of the oscillations depends solely on the inductance and capacitance of the circuit. This is the *natural frequency* of oscillation of the *LC* circuit.

Because Q varies sinusoidally, the current in the circuit also varies sinusoidally. We can easily show this by differentiating Equation 32.21 with respect to time:

$$I = \frac{dQ}{dt} = -\omega Q_{max} \sin(\omega t + \phi) \qquad \textbf{(32.23)}$$

To determine the value of the phase angle ϕ, we examine the initial conditions, which in our situation require that at $t = 0$, $I = 0$ and $Q = Q_{max}$. Setting $I = 0$ at $t = 0$ in Equation 32.23, we have

$$0 = -\omega Q_{max} \sin \phi$$

which shows that $\phi = 0$. This value for ϕ also is consistent with Equation 32.21 and with the condition that $Q = Q_{max}$ at $t = 0$. Therefore, in our case, the expressions for Q and I are

$$Q = Q_{max} \cos \omega t \qquad \textbf{(32.24)}$$

$$I = -\omega Q_{max} \sin \omega t = -I_{max} \sin \omega t \qquad \textbf{(32.25)}$$

Graphs of Q versus t and I versus t are shown in Figure 32.16. Note that the charge on the capacitor oscillates between the extreme values Q_{max} and $-Q_{max}$, and that the current oscillates between I_{max} and $-I_{max}$. Furthermore, the current is 90° out of phase with the charge. That is, when the charge is a maximum, the current is zero, and when the charge is zero, the current has its maximum value.

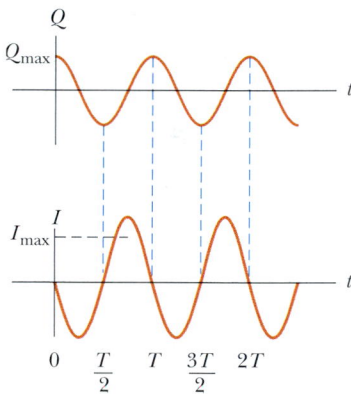

Figure 32.16 Graphs of charge versus time and current versus time for a resistanceless, nonradiating *LC* circuit. Note that Q and I are 90° out of phase with each other.

Quick Quiz 32.5

What is the relationship between the amplitudes of the two curves in Figure 32.16?

Let us return to the energy discussion of the *LC* circuit. Substituting Equations 32.24 and 32.25 in Equation 32.18, we find that the total energy is

$$U = U_C + U_L = \frac{Q_{max}^2}{2C} \cos^2 \omega t + \frac{L I_{max}^2}{2} \sin^2 \omega t \qquad \textbf{(32.26)}$$

This expression contains all of the features described qualitatively at the beginning of this section. It shows that the energy of the *LC* circuit continuously oscillates between energy stored in the electric field of the capacitor and energy stored in the magnetic field of the inductor. When the energy stored in the capacitor has its maximum value $Q_{max}^2/2C$, the energy stored in the inductor is zero. When the energy stored in the inductor has its maximum value $\frac{1}{2} L I_{max}^2$, the energy stored in the capacitor is zero.

Plots of the time variations of U_C and U_L are shown in Figure 32.17. The sum $U_C + U_L$ is a constant and equal to the total energy $Q_{max}^2/2C$ or $L I_{max}^2/2$. Analytical verification of this is straightforward. The amplitudes of the two graphs in Figure 32.17 must be equal because the maximum energy stored in the capacitor

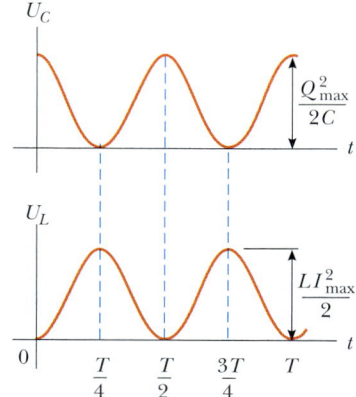

Figure 32.17 Plots of U_C versus t and U_L versus t for a resistanceless, nonradiating *LC* circuit. The sum of the two curves is a constant and equal to the total energy stored in the circuit.

(when $I = 0$) must equal the maximum energy stored in the inductor (when $Q = 0$). This is mathematically expressed as

$$\frac{Q_{max}^2}{2C} = \frac{LI_{max}^2}{2}$$

Using this expression in Equation 32.26 for the total energy gives

$$U = \frac{Q_{max}^2}{2C}(\cos^2 \omega t + \sin^2 \omega t) = \frac{Q_{max}^2}{2C} \qquad \textbf{(32.27)}$$

because $\cos^2 \omega t + \sin^2 \omega t = 1$.

In our idealized situation, the oscillations in the circuit persist indefinitely; however, we remember that the total energy U of the circuit remains constant only if energy transfers and transformations are neglected. In actual circuits, there is always some resistance, and hence energy is transformed to internal energy. We mentioned at the beginning of this section that we are also ignoring radiation from the circuit. In reality, radiation is inevitable in this type of circuit, and the total energy in the circuit continuously decreases as a result of this process.

EXAMPLE 32.7 ▶ An Oscillatory *LC* Circuit

In Figure 32.18, the capacitor is initially charged when switch S_1 is open and S_2 is closed. Switch S_1 is then thrown closed at the same instant that S_2 is opened, so that the capacitor is connected directly across the inductor. (a) Find the frequency of oscillation of the circuit.

Solution Using Equation 32.22 gives for the frequency

$$f = \frac{\omega}{2\pi} = \frac{1}{2\pi\sqrt{LC}}$$

$$= \frac{1}{2\pi[(2.81 \times 10^{-3}\text{ H})(9.00 \times 10^{-12}\text{ F})]^{1/2}}$$

$$= \boxed{1.00 \times 10^6\text{ Hz}}$$

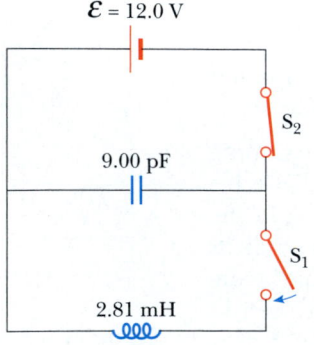

$\mathcal{E} = 12.0\text{ V}$

S_2

9.00 pF

S_1

2.81 mH

Figure 32.18 First the capacitor is fully charged with the switch S_1 open and S_2 closed. Then, S_1 is thrown closed at the same time that S_2 is thrown open.

(b) What are the maximum values of charge on the capacitor and current in the circuit?

Solution The initial charge on the capacitor equals the maximum charge, and because $C = Q/\mathcal{E}$, we have

$$Q_{max} = C\mathcal{E} = (9.00 \times 10^{-12}\text{ F})(12.0\text{ V}) = \boxed{1.08 \times 10^{-10}\text{ C}}$$

From Equation 32.25, we can see how the maximum current is related to the maximum charge:

$$I_{max} = \omega Q_{max} = 2\pi f Q_{max}$$
$$= (2\pi \times 10^6\text{ s}^{-1})(1.08 \times 10^{-10}\text{ C})$$
$$= \boxed{6.79 \times 10^{-4}\text{ A}}$$

(c) Determine the charge and current as functions of time.

Solution Equations 32.24 and 32.25 give the following expressions for the time variation of Q and I:

$$Q = Q_{max}\cos \omega t$$
$$= \boxed{(1.08 \times 10^{-10}\text{ C})\cos[(2\pi \times 10^6\text{ rad/s})t]}$$

$$I = -I_{max}\sin \omega t$$
$$= \boxed{(-6.79 \times 10^{-4}\text{ A})\sin[(2\pi \times 10^6\text{ rad/s})t]}$$

Exercise What is the total energy stored in the circuit?

Answer 6.48×10^{-10} J.

Optional Section

32.6 ▶ THE *RLC* CIRCUIT

We now turn our attention to a more realistic circuit consisting of an inductor, a capacitor, and a resistor connected in series, as shown in Figure 32.19. We let the resistance of the resistor represent all of the resistance in the circuit. We assume that the capacitor has an initial charge Q_{max} before the switch is closed. Once the switch is thrown closed and a current is established, the total energy stored in the capacitor and inductor at any time is given, as before, by Equation 32.18. However, the total energy is no longer constant, as it was in the *LC* circuit, because the resistor causes transformation to internal energy. Because the rate of energy transformation to internal energy within a resistor is I^2R, we have

$$\frac{dU}{dt} = -I^2R$$

where the negative sign signifies that the energy U of the circuit is decreasing in time. Substituting this result into Equation 32.19 gives

$$LI\frac{dI}{dt} + \frac{Q}{C}\frac{dQ}{dt} = -I^2R \tag{32.28}$$

To convert this equation into a form that allows us to compare the electrical oscillations with their mechanical analog, we first use the fact that $I = dQ/dt$ and move all terms to the left-hand side to obtain

$$LI\frac{d^2Q}{dt^2} + \frac{Q}{C}I + I^2R = 0$$

Now we divide through by I:

$$L\frac{d^2Q}{dt^2} + \frac{Q}{C} + IR = 0$$

$$L\frac{d^2Q}{dt^2} + R\frac{dQ}{dt} + \frac{Q}{C} = 0 \tag{32.29}$$

The *RLC* circuit is analogous to the damped harmonic oscillator discussed in Section 13.6 and illustrated in Figure 32.20. The equation of motion for this mechanical system is, from Equation 13.32,

$$m\frac{d^2x}{dt^2} + b\frac{dx}{dt} + kx = 0 \tag{32.30}$$

Comparing Equations 32.29 and 32.30, we see that Q corresponds to the position x of the block at any instant, L to the mass m of the block, R to the damping coefficient b, and C to $1/k$, where k is the force constant of the spring. These and other relationships are listed in Table 32.1.

Because the analytical solution of Equation 32.29 is cumbersome, we give only a qualitative description of the circuit behavior. In the simplest case, when $R = 0$, Equation 32.29 reduces to that of a simple *LC* circuit, as expected, and the charge and the current oscillate sinusoidally in time. This is equivalent to removal of all damping in the mechanical oscillator.

When R is small, a situation analogous to light damping in the mechanical oscillator, the solution of Equation 32.29 is

$$Q = Q_{max}e^{-Rt/2L}\cos\omega_d t \tag{32.31}$$

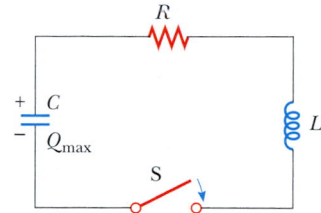

Figure 32.19   A series *RLC* circuit. The capacitor has a charge Q_{max} at $t = 0$, the instant at which the switch is thrown closed.

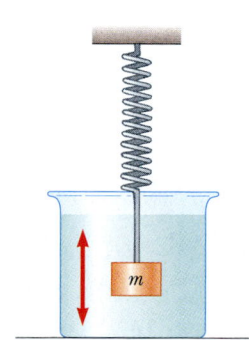

Figure 32.20   A block–spring system moving in a viscous medium with damped harmonic motion is analogous to an *RLC* circuit.

TABLE 32.1   Analogies Between Electrical and Mechanical Systems

Electric Circuit		One-Dimensional Mechanical System
Charge	$Q \leftrightarrow x$	Displacement
Current	$I \leftrightarrow v_x$	Velocity
Potential difference	$\Delta V \leftrightarrow F_x$	Force
Resistance	$R \leftrightarrow b$	Viscous damping coefficient
Capacitance	$C \leftrightarrow 1/k$	(k = spring constant)
Inductance	$L \leftrightarrow m$	Mass
Current = time derivative of charge	$I = \dfrac{dQ}{dt} \leftrightarrow v_x = \dfrac{dx}{dt}$	Velocity = time derivative of position
Rate of change of current = second time derivative of charge	$\dfrac{dI}{dt} = \dfrac{d^2Q}{dt^2} \leftrightarrow a_x = \dfrac{dv_x}{dt} = \dfrac{d^2x}{dt^2}$	Acceleration = second time derivative of position
Energy in inductor	$U_L = \frac{1}{2}LI^2 \leftrightarrow K = \frac{1}{2}mv^2$	Kinetic energy of moving mass
Energy in capacitor	$U_C = \frac{1}{2}\dfrac{Q^2}{C} \leftrightarrow U = \frac{1}{2}kx^2$	Potential energy stored in a spring
Rate of energy loss due to resistance	$I^2R \leftrightarrow bv^2$	Rate of energy loss due to friction
RLC circuit	$L\dfrac{d^2Q}{dt^2} + R\dfrac{dQ}{dt} + \dfrac{Q}{C} = 0 \leftrightarrow m\dfrac{d^2x}{dt^2} + b\dfrac{dx}{dt} + kx = 0$	Damped mass on a spring

where

$$\omega_d = \left[\frac{1}{LC} - \left(\frac{R}{2L} \right)^2 \right]^{1/2} \tag{32.32}$$

is the angular frequency at which the circuit oscillates. That is, the value of the charge on the capacitor undergoes a damped harmonic oscillation in analogy with a mass–spring system moving in a viscous medium. From Equation 32.32, we see that, when $R \ll \sqrt{4L/C}$ (so that the second term in the brackets is much smaller than the first), the frequency ω_d of the damped oscillator is close to that of the undamped oscillator, $1/\sqrt{LC}$. Because $I = dQ/dt$, it follows that the current also undergoes damped harmonic oscillation. A plot of the charge versus time for the damped oscillator is shown in Figure 32.21a. Note that the maximum value of Q decreases after each oscillation, just as the amplitude of a damped block–spring system decreases in time.

Quick Quiz 32.6

Figure 32.21a has two dashed blue lines that form an "envelope" around the curve. What is the equation for the upper dashed line?

When we consider larger values of R, we find that the oscillations damp out more rapidly; in fact, there exists a critical resistance value $R_c = \sqrt{4L/C}$ above which no oscillations occur. A system with $R = R_c$ is said to be *critically damped*. When R exceeds R_c, the system is said to be *overdamped* (Fig. 32.22).

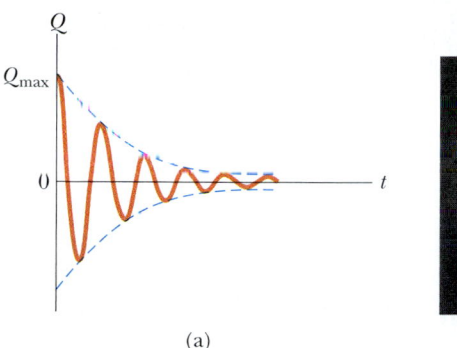

(a)

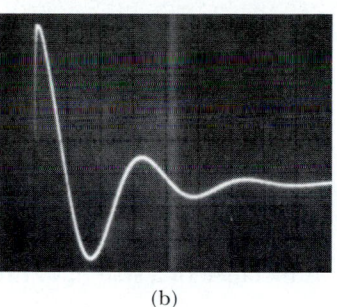

(b)

Figure 32.21 (a) Charge versus time for a damped *RLC* circuit. The charge decays in this way when $R \ll \sqrt{4L/C}$. The *Q*-versus-*t* curve represents a plot of Equation 32.31. (b) Oscilloscope pattern showing the decay in the oscillations of an *RLC* circuit. The parameters used were $R = 75\ \Omega$, $L = 10$ mH, $C = 0.19\ \mu$F, and $f = 300$ Hz. *(Courtesy of J. Rudmin)*

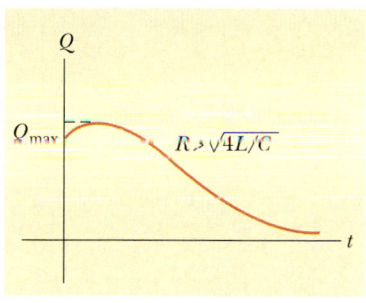

Figure 32.22 Plot of *Q* versus *t* for an overdamped *RLC* circuit, which occurs for values of $R > \sqrt{4L/C}$.

SUMMARY

When the current in a coil changes with time, an emf is induced in the coil according to Faraday's law. The **self-induced emf** is

$$\mathcal{E}_L = -L\frac{dI}{dt} \tag{32.1}$$

where L is the **inductance** of the coil. Inductance is a measure of how much opposition an electrical device offers to a change in current passing through the device. Inductance has the SI unit of **henry** (H), where $1\ \text{H} = 1\ \text{V}\cdot\text{s/A}$.

The inductance of any coil is

$$L = \frac{N\Phi_B}{I} \tag{32.2}$$

where Φ_B is the magnetic flux through the coil and N is the total number of turns. The inductance of a device depends on its geometry. For example, the inductance of an air-core solenoid is

$$L = \frac{\mu_0 N^2 A}{\ell} \tag{32.4}$$

where A is the cross-sectional area, and ℓ is the length of the solenoid.

If a resistor and inductor are connected in series to a battery of emf $\mathcal{E}$, and if a switch in the circuit is thrown closed at $t = 0$, then the current in the circuit varies in time according to the expression

$$I = \frac{\mathcal{E}}{R}(1 - e^{-t/\tau}) \tag{32.7}$$

where $\tau = L/R$ is the time constant of the *RL* circuit. That is, the current increases to an equilibrium value of $\mathcal{E}/R$ after a time that is long compared with τ. If the battery in the circuit is replaced by a resistanceless wire, the current decays exponentially with time according to the expression

$$I = \frac{\mathcal{E}}{R}e^{-t/\tau} \tag{32.10}$$

where $\mathcal{E}/R$ is the initial current in the circuit.

The energy stored in the magnetic field of an inductor carrying a current I is

$$U = \tfrac{1}{2}LI^2 \tag{32.12}$$

This energy is the magnetic counterpart to the energy stored in the electric field of a charged capacitor.

The energy density at a point where the magnetic field is B is

$$u_B = \frac{B^2}{2\mu_0} \tag{32.14}$$

The mutual inductance of a system of two coils is given by

$$M_{12} = \frac{N_2\Phi_{12}}{I_1} = M_{21} = \frac{N_1\Phi_{21}}{I_2} = M \tag{32.15}$$

This mutual inductance allows us to relate the induced emf in a coil to the changing source current in a nearby coil using the relationships

$$\mathcal{E}_2 = -M_{12}\frac{dI_1}{dt} \qquad \text{and} \qquad \mathcal{E}_1 = -M_{21}\frac{dI_2}{dt} \tag{32.16, 32.17}$$

In an LC circuit that has zero resistance and does not radiate electromagnetically (an idealization), the values of the charge on the capacitor and the current in the circuit vary in time according to the expressions

$$Q = Q_{max}\cos(\omega t + \phi) \tag{32.21}$$

$$I = \frac{dQ}{dt} = -\omega Q_{max}\sin(\omega t + \phi) \tag{32.23}$$

where Q_{max} is the maximum charge on the capacitor, ϕ is a phase constant, and ω is the angular frequency of oscillation:

$$\omega = \frac{1}{\sqrt{LC}} \tag{32.22}$$

The energy in an LC circuit continuously transfers between energy stored in the capacitor and energy stored in the inductor. The total energy of the LC circuit at any time t is

$$U = U_C + U_L = \frac{Q_{max}^2}{2C}\cos^2\omega t + \frac{LI_{max}^2}{2}\sin^2\omega t \tag{32.26}$$

At $t = 0$, all of the energy is stored in the electric field of the capacitor ($U = Q_{max}^2/2C$). Eventually, all of this energy is transferred to the inductor ($U = LI_{max}^2/2$). However, the total energy remains constant because energy transformations are neglected in the ideal LC circuit.

QUESTIONS

1. Why is the induced emf that appears in an inductor called a "counter" or "back" emf?

2. The current in a circuit containing a coil, resistor, and battery reaches a constant value. Does the coil have an inductance? Does the coil affect the value of the current?

3. What parameters affect the inductance of a coil? Does the inductance of a coil depend on the current in the coil?

4. How can a long piece of wire be wound on a spool so that the wire has a negligible self-inductance?

5. A long, fine wire is wound as a solenoid with a self-inductance L. If it is connected across the terminals of a battery, how does the maximum current depend on L?

6. For the series RL circuit shown in Figure Q32.6, can the back emf ever be greater than the battery emf? Explain.

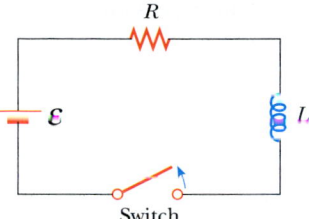

Figure Q32.6

7. Consider this thesis: "Joseph Henry, America's first professional physicist, changed the view of the Universe during a school vacation at the Albany Academy in 1830. Before that time, one could think of the Universe as consisting of just one thing: matter. In Henry's experiment, after a battery is removed from a coil, the energy that keeps the current flowing for a while does not belong to any piece of matter. This energy belongs to the magnetic field surrounding the coil. With Henry's discovery of self-induction, Nature forced us to admit that the Universe consists of fields as well as matter." What in your view constitutes the Universe? Argue for your answer.

8. Discuss the similarities and differences between the energy stored in the electric field of a charged capacitor and the energy stored in the magnetic field of a current-carrying coil.

9. What is the inductance of two inductors connected in series? Does it matter if they are solenoids or toroids?

10. The centers of two circular loops are separated by a fixed distance. For what relative orientation of the loops is their mutual inductance a maximum? a minimum? Explain.

11. Two solenoids are connected in series so that each carries the same current at any instant. Is mutual induction present? Explain.

12. In the LC circuit shown in Figure 32.15, the charge on the capacitor is sometimes zero, even though current is in the circuit. How is this possible?

13. If the resistance of the wires in an LC circuit were not zero, would the oscillations persist? Explain.

14. How can you tell whether an RLC circuit is overdamped or underdamped?

15. What is the significance of critical damping in an RLC circuit?

16. Can an object exert a force on itself? When a coil induces an emf in itself, does it exert a force on itself?

PROBLEMS

1, 2, 3 = straightforward, intermediate, challenging   ☐ = full solution available in the *Student Solutions Manual and Study Guide*
WEB = solution posted at **http://www.saunderscollege.com/physics/**   🖥 = Computer useful in solving problem   🧑 = Interactive Physics
☐ = paired numerical/symbolic problems

Section 32.1   Self-Inductance

1. A coil has an inductance of 3.00 mH, and the current through it changes from 0.200 A to 1.50 A in a time of 0.200 s. Find the magnitude of the average induced emf in the coil during this time.

2. A coiled telephone cord forms a spiral with 70 turns, a diameter of 1.30 cm, and an unstretched length of 60.0 cm. Determine the self-inductance of one conductor in the unstretched cord.

3. A 2.00-H inductor carries a steady current of 0.500 A. When the switch in the circuit is thrown open, the current is effectively zero in 10.0 ms. What is the average induced emf in the inductor during this time?

4. A small air-core solenoid has a length of 4.00 cm and a radius of 0.250 cm. If the inductance is to be 0.060 0 mH, how many turns per centimeter are required?

5. Calculate the magnetic flux through the area enclosed by a 300-turn, 7.20-mH coil when the current in the coil is 10.0 mA.

6. The current in a solenoid is increasing at a rate of 10.0 A/s. The cross-sectional area of the solenoid is $\pi\,\text{cm}^2$, and there are 300 turns on its 15.0-cm length. What is the induced emf opposing the increasing current?

WEB 7. A 10.0-mH inductor carries a current $I = I_{\max} \sin \omega t$, with $I_{\max} = 5.00$ A and $\omega/2\pi = 60.0$ Hz. What is the back emf as a function of time?

8. An emf of 24.0 mV is induced in a 500-turn coil at an instant when the current is 4.00 A and is changing at the rate of 10.0 A/s. What is the magnetic flux through each turn of the coil?

9. An inductor in the form of a solenoid contains 420 turns, is 16.0 cm in length, and has a cross-sectional area of $3.00\,\text{cm}^2$. What uniform rate of decrease of current through the inductor induces an emf of 175 μV?

10. An inductor in the form of a solenoid contains N turns, has length ℓ, and has cross-sectional area A. What uniform rate of decrease of current through the inductor induces an emf $\mathcal{E}$?

11. The current in a 90.0-mH inductor changes with time as $I = t^2 - 6.00t$ (in SI units). Find the magnitude of the induced emf at (a) $t = 1.00$ s and (b) $t = 4.00$ s. (c) At what time is the emf zero?

12. A 40.0-mA current is carried by a uniformly wound air-core solenoid with 450 turns, a 15.0-mm diameter, and 12.0-cm length. Compute (a) the magnetic field inside the solenoid, (b) the magnetic flux through each turn,

and (c) the inductance of the solenoid. (d) Which of these quantities depends on the current?

13. A solenoid has 120 turns uniformly wrapped around a wooden core, which has a diameter of 10.0 mm and a length of 9.00 cm. (a) Calculate the inductance of the solenoid. (b) The wooden core is replaced with a soft iron rod that has the same dimensions but a magnetic permeability $\mu_m = 800\mu_0$. What is the new inductance?

14. A toroid has a major radius R and a minor radius r, and it is tightly wound with N turns of wire, as shown in Figure P32.14. If $R \gg r$, the magnetic field within the region of the torus, of cross-sectional area $A = \pi r^2$, is essentially that of a long solenoid that has been bent into a large circle of radius R. Using the uniform field of a long solenoid, show that the self-inductance of such a toroid is approximately

$$L \cong \mu_0 N^2 A / 2\pi R$$

(An exact expression for the inductance of a toroid with a rectangular cross-section is derived in Problem 64.)

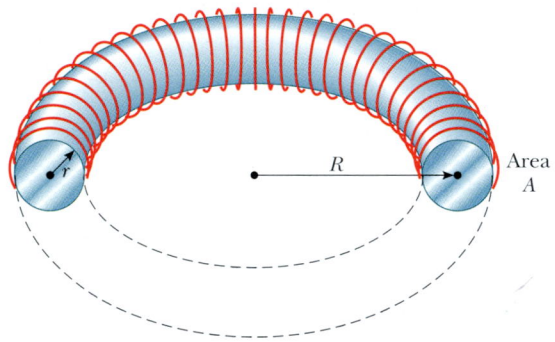

Figure P32.14

15. An emf self-induced in a solenoid of inductance L changes in time as $\mathcal{E} = \mathcal{E}_0 e^{-kt}$. Find the total charge that passes through the solenoid, if the charge is finite.

Section 32.2   RL Circuits

16. Calculate the resistance in an RL circuit in which $L = 2.50$ H and the current increases to 90.0% of its final value in 3.00 s.

17. A 12.0-V battery is connected into a series circuit containing a 10.0-Ω resistor and a 2.00-H inductor. How long will it take the current to reach (a) 50.0% and (b) 90.0% of its final value?

18. Show that $I = I_0 e^{-t/\tau}$ is a solution of the differential equation

$$IR + L\frac{dI}{dt} = 0$$

where $\tau = L/R$ and I_0 is the current at $t = 0$.

19. Consider the circuit in Figure P32.19, taking $\mathcal{E} = 6.00$ V, $L = 8.00$ mH, and $R = 4.00$ Ω. (a) What is

the inductive time constant of the circuit? (b) Calculate the current in the circuit 250 μs after the switch is closed. (c) What is the value of the final steady-state current? (d) How long does it take the current to reach 80.0% of its maximum value?

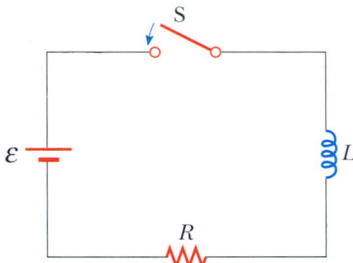

Figure P32.19   Problems 19, 20, 21, and 24.

20. In the circuit shown in Figure P32.19, let $L = 7.00$ H, $R = 9.00$ Ω, and $\mathcal{E} = 120$ V. What is the self-induced emf 0.200 s after the switch is closed?

WEB 21. For the RL circuit shown in Figure P32.19, let $L = 3.00$ H, $R = 8.00$ Ω, and $\mathcal{E} = 36.0$ V. (a) Calculate the ratio of the potential difference across the resistor to that across the inductor when $I = 2.00$ A. (b) Calculate the voltage across the inductor when $I = 4.50$ A.

22. A 12.0-V battery is connected in series with a resistor and an inductor. The circuit has a time constant of 500 μs, and the maximum current is 200 mA. What is the value of the inductance?

23. An inductor that has an inductance of 15.0 H and a resistance of 30.0 Ω is connected across a 100-V battery. What is the rate of increase of the current (a) at $t = 0$ and (b) at $t = 1.50$ s?

24. When the switch in Figure P32.19 is thrown closed, the current takes 3.00 ms to reach 98.0% of its final value. If $R = 10.0$ Ω, what is the inductance?

25. The switch in Figure P32.25 is closed at time $t = 0$. Find the current in the inductor and the current through the switch as functions of time thereafter.

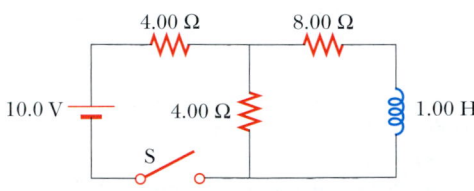

Figure P32.25

26. A series RL circuit with $L = 3.00$ H and a series RC circuit with $C = 3.00$ μF have equal time constants. If the two circuits contain the same resistance R, (a) what is the value of R and (b) what is the time constant?

27. A current pulse is fed to the partial circuit shown in Figure P32.27. The current begins at zero, then becomes 10.0 A between $t = 0$ and $t = 200$ μs, and then is zero once again. Determine the current in the inductor as a function of time.

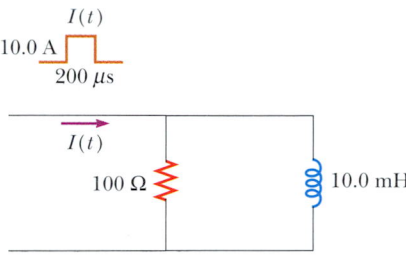

Figure P32.27

28. One application of an RL circuit is the generation of time-varying high voltage from a low-voltage source, as shown in Figure P32.28. (a) What is the current in the circuit a long time after the switch has been in position A? (b) Now the switch is thrown quickly from A to B. Compute the initial voltage across each resistor and the inductor. (c) How much time elapses before the voltage across the inductor drops to 12.0 V?

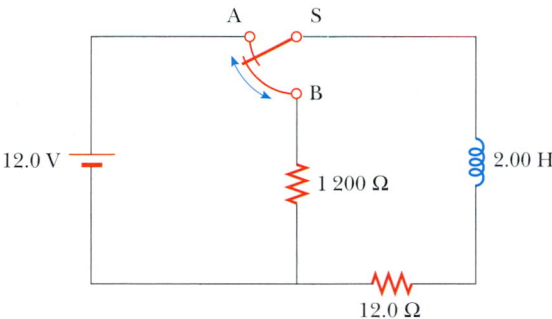

Figure P32.28

WEB 29. A 140-mH inductor and a 4.90-Ω resistor are connected with a switch to a 6.00-V battery, as shown in Figure P32.29. (a) If the switch is thrown to the left (connecting the battery), how much time elapses before the current reaches 220 mA? (b) What is the current in the inductor 10.0 s after the switch is closed? (c) Now the switch is quickly thrown from A to B. How much time elapses before the current falls to 160 mA?

30. Consider two ideal inductors, L_1 and L_2, that have *zero* internal resistance and are far apart, so that their magnetic fields do not influence each other. (a) If these inductors are connected in series, show that they are equivalent to a single ideal inductor having $L_{eq} = L_1 + L_2$. (b) If these same two inductors are connected in parallel, show that they are equivalent to a

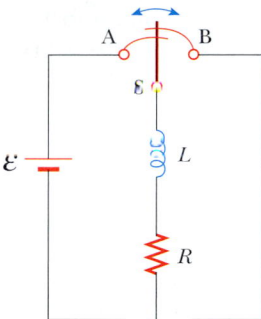

Figure P32.29

single ideal inductor having $1/L_{eq} = 1/L_1 + 1/L_2$. (c) Now consider two inductors L_1 and L_2 that have *nonzero* internal resistances R_1 and R_2, respectively. Assume that they are still far apart so that their magnetic fields do not influence each other. If these inductors are connected in series, show that they are equivalent to a single inductor having $L_{eq} = L_1 + L_2$ and $R_{eq} = R_1 + R_2$. (d) If these same inductors are now connected in parallel, is it necessarily true that they are equivalent to a single ideal inductor having $1/L_{eq} = 1/L_1 + 1/L_2$ and $1/R_{eq} = 1/R_1 + 1/R_2$? Explain your answer.

Section 32.3   Energy in a Magnetic Field

31. Calculate the energy associated with the magnetic field of a 200-turn solenoid in which a current of 1.75 A produces a flux of 3.70×10^{-4} T·m^2 in each turn.

32. The magnetic field inside a superconducting solenoid is 4.50 T. The solenoid has an inner diameter of 6.20 cm and a length of 26.0 cm. Determine (a) the magnetic energy density in the field and (b) the energy stored in the magnetic field within the solenoid.

33. An air-core solenoid with 68 turns is 8.00 cm long and has a diameter of 1.20 cm. How much energy is stored in its magnetic field when it carries a current of 0.770 A?

34. At $t = 0$, an emf of 500 V is applied to a coil that has an inductance of 0.800 H and a resistance of 30.0 Ω. (a) Find the energy stored in the magnetic field when the current reaches half its maximum value. (b) After the emf is connected, how long does it take the current to reach this value?

WEB 35. On a clear day there is a 100-V/m vertical electric field near the Earth's surface. At the same place, the Earth's magnetic field has a magnitude of 0.500×10^{-4} T. Compute the energy densities of the two fields.

36. An RL circuit in which $L = 4.00$ H and $R = 5.00$ Ω is connected to a 22.0-V battery at $t = 0$. (a) What energy is stored in the inductor when the current is 0.500 A? (b) At what rate is energy being stored in the inductor when $I = 1.00$ A? (c) What power is being delivered to the circuit by the battery when $I = 0.500$ A?

37. A 10.0-V battery, a 5.00-Ω resistor, and a 10.0-H inductor are connected in series. After the current in the circuit

has reached its maximum value, calculate (a) the power being supplied by the battery, (b) the power being delivered to the resistor, (c) the power being delivered to in the inductor, and (d) the energy stored in the magnetic field of the inductor.

38. A uniform electric field with a magnitude of 680 kV/m throughout a cylindrical volume results in a total energy of 3.40 μJ. What magnetic field over this same region stores the same total energy?

39. Assume that the magnitude of the magnetic field outside a sphere of radius R is $B = B_0(R/r)^2$, where B_0 is a constant. Determine the total energy stored in the magnetic field outside the sphere and evaluate your result for $B_0 = 5.00 \times 10^{-5}$ T and $R = 6.00 \times 10^6$ m, values appropriate for the Earth's magnetic field.

Section 32.4   Mutual Inductance

40. Two coils are close to each other. The first coil carries a time-varying current given by
$I(t) = (5.00 \text{ A}) e^{-0.025\, 0t} \sin(377t)$. At $t = 0.800$ s, the voltage measured across the second coil is $- 3.20$ V. What is the mutual inductance of the coils?

41. Two coils, held in fixed positions, have a mutual inductance of 100 μH. What is the peak voltage in one when a sinusoidal current given by
$I(t) = (10.0 \text{ A}) \sin(1\,000t)$ flows in the other?

42. An emf of 96.0 mV is induced in the windings of a coil when the current in a nearby coil is increasing at the rate of 1.20 A/s. What is the mutual inductance of the two coils?

43. Two solenoids A and B, spaced close to each other and sharing the same cylindrical axis, have 400 and 700 turns, respectively. A current of 3.50 A in coil A produces an average flux of 300 μT $\cdot$ m^2 through each turn of A and a flux of 90.0 μT $\cdot$ m^2 through each turn of B. (a) Calculate the mutual inductance of the two solenoids. (b) What is the self-inductance of A? (c) What emf is induced in B when the current in A increases at the rate of 0.500 A/s?

44. A 70-turn solenoid is 5.00 cm long and 1.00 cm in diameter and carries a 2.00-A current. A single loop of wire, 3.00 cm in diameter, is held so that the plane of the loop is perpendicular to the long axis of the solenoid, as illustrated in Figure P31.18 (page 1004). What is the mutual inductance of the two if the plane of the loop passes through the solenoid 2.50 cm from one end?

45. Two single-turn circular loops of wire have radii R and r, with $R \gg r$. The loops lie in the same plane and are concentric. (a) Show that the mutual inductance of the pair is $M = \mu_0 \pi r^2/2R$. (*Hint:* Assume that the larger loop carries a current I and compute the resulting flux through the smaller loop.) (b) Evaluate M for $r = 2.00$ cm and $R = 20.0$ cm.

46. On a printed circuit board, a relatively long straight conductor and a conducting rectangular loop lie in the same plane, as shown in Figure P31.9 (page 1003). If

$h = 0.400$ mm, $w = 1.30$ mm, and $L = 2.70$ mm, what is their mutual inductance?

47. Two inductors having self-inductances L_1 and L_2 are connected in parallel, as shown in Figure P32.47a. The mutual inductance between the two inductors is M. Determine the equivalent self-inductance L_{eq} for the system (Fig. P32.47b).

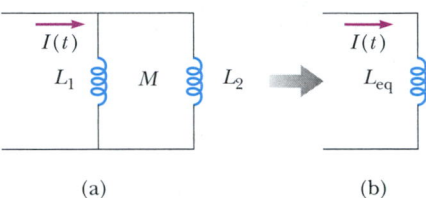

      (a)   (b)

Figure P32.47

Section 32.5   Oscillations in an *LC* Circuit

48. A 1.00-μF capacitor is charged by a 40.0-V power supply. The fully-charged capacitor is then discharged through a 10.0-mH inductor. Find the maximum current in the resulting oscillations.

49. An *LC* circuit consists of a 20.0-mH inductor and a 0.500-μF capacitor. If the maximum instantaneous current is 0.100 A, what is the greatest potential difference across the capacitor?

50. In the circuit shown in Figure P32.50, $\mathcal{E} = 50.0$ V, $R = 250\ \Omega$, and $C = 0.500\ \mu$F. The switch S is closed for a long time, and no voltage is measured across the capacitor. After the switch is opened, the voltage across the capacitor reaches a maximum value of 150 V. What is the inductance L?

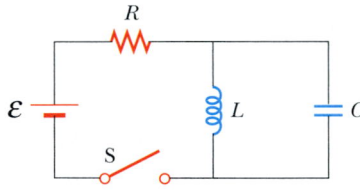

Figure P32.50

51. A fixed inductance $L = 1.05\ \mu$H is used in series with a variable capacitor in the tuning section of a radio. What capacitance tunes the circuit to the signal from a station broadcasting at 6.30 MHz?

52. Calculate the inductance of an *LC* circuit that oscillates at 120 Hz when the capacitance is 8.00 μF.

53. An *LC* circuit like the one shown in Figure 32.14 contains an 82.0-mH inductor and a 17.0-μF capacitor that initially carries a 180-μC charge. The switch is thrown closed at $t = 0$. (a) Find the frequency (in hertz) of the resulting oscillations. At $t = 1.00$ ms, find (b) the charge on the capacitor and (c) the current in the circuit.

54. The switch in Figure P32.54 is connected to point *a* for a long time. After the switch is thrown to point *b*, what are (a) the frequency of oscillation of the *LC* circuit, (b) the maximum charge that appears on the capacitor, (c) the maximum current in the inductor, and (d) the total energy the circuit possesses at $t = 3.00$ s?

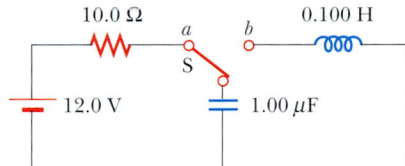

Figure P32.54

WEB 55. An *LC* circuit like that illustrated in Figure 32.14 consists of a 3.30-H inductor and an 840-pF capacitor, initially carrying a 105-μC charge. At $t = 0$ the switch is thrown closed. Compute the following quantities at $t = 2.00$ ms: (a) the energy stored in the capacitor; (b) the energy stored in the inductor; (c) the total energy in the circuit.

(Optional)
Section 32.6   The *RLC* Circuit

56. In Figure 32.19, let $R = 7.60$ Ω, $L = 2.20$ mH, and $C = 1.80$ μF. (a) Calculate the frequency of the damped oscillation of the circuit. (b) What is the critical resistance?

57. Consider an *LC* circuit in which $L = 500$ mH and $C = 0.100$ μF. (a) What is the resonant frequency ω_0? (b) If a resistance of 1.00 kΩ is introduced into this circuit, what is the frequency of the (damped) oscillations? (c) What is the percent difference between the two frequencies?

58. Show that Equation 32.29 in the text is Kirchhoff's loop rule as applied to Figure 32.19.

59. Electrical oscillations are initiated in a series circuit containing a capacitance *C*, inductance *L*, and resistance *R*. (a) If $R \ll \sqrt{4L/C}$ (weak damping), how much time elapses before the amplitude of the current oscillation falls off to 50.0% of its initial value? (b) How long does it take the energy to decrease to 50.0% of its initial value?

ADDITIONAL PROBLEMS

60. Initially, the capacitor in a series *LC* circuit is charged. A switch is closed, allowing the capacitor to discharge, and after time *t* the energy stored in the capacitor is one-fourth its initial value. Determine *L* if *C* is known.

61. A 1.00-mH inductor and a 1.00-μF capacitor are connected in series. The current in the circuit is described by $I = 20.0t$, where *t* is in seconds and *I* is in amperes.

The capacitor initially has no charge. Determine (a) the voltage across the inductor as a function of time, (b) the voltage across the capacitor as a function of time, and (c) the time when the energy stored in the capacitor first exceeds that in the inductor.

62. An inductor having inductance *L* and a capacitor having capacitance *C* are connected in series. The current in the circuit increases linearly in time as described by $I = Kt$. The capacitor is initially uncharged. Determine (a) the voltage across the inductor as a function of time, (b) the voltage across the capacitor as a function of time, and (c) the time when the energy stored in the capacitor first exceeds that in the inductor.

63. A capacitor in a series *LC* circuit has an initial charge *Q* and is being discharged. Find, in terms of *L* and *C*, the flux through each of the *N* turns in the coil, when the charge on the capacitor is $Q/2$.

64. The toroid in Figure P32.64 consists of *N* turns and has a rectangular cross-section. Its inner and outer radii are *a* and *b*, respectively. (a) Show that

$$L = \frac{\mu_0 N^2 h}{2\pi} \ln \frac{b}{a}$$

(b) Using this result, compute the self-inductance of a 500-turn toroid for which $a = 10.0$ cm, $b = 12.0$ cm, and $h = 1.00$ cm. (c) In Problem 14, an approximate formula for the inductance of a toroid with $R \gg r$ was derived. To get a feel for the accuracy of that result, use the expression in Problem 14 to compute the approximate inductance of the toroid described in part (b). Compare the result with the answer to part (b).

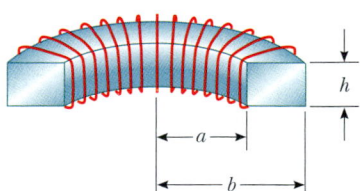

Figure P32.64

65. (a) A flat circular coil does not really produce a uniform magnetic field in the area it encloses, but estimate the self-inductance of a flat circular coil, with radius *R* and *N* turns, by supposing that the field at its center is uniform over its area. (b) A circuit on a laboratory table consists of a 1.5-V battery, a 270-Ω resistor, a switch, and three 30-cm-long cords connecting them. Suppose that the circuit is arranged to be circular. Think of it as a flat coil with one turn. Compute the order of magnitude of its self-inductance and (c) of the time constant describing how fast the current increases when you close the switch.

66. A soft iron rod ($\mu_m = 800 \, \mu_0$) is used as the core of a solenoid. The rod has a diameter of 24.0 mm and is

10.0 cm long. A 10.0-m piece of 22-gauge copper wire (diameter = 0.644 mm) is wrapped around the rod in a single uniform layer, except for a 10.0-cm length at each end, which is to be used for connections. (a) How many turns of this wire can wrap around the rod? (*Hint:* The diameter of the wire adds to the diameter of the rod in determining the circumference of each turn. Also, the wire spirals diagonally along the surface of the rod.) (b) What is the resistance of this inductor? (c) What is its inductance?

67. A wire of nonmagnetic material with radius R carries current uniformly distributed over its cross-section. If the total current carried by the wire is I, show that the magnetic energy per unit length inside the wire is $\mu_0 I^2/16\pi$.

68. An 820-turn wire coil of resistance 24.0 Ω is placed around a 12 500-turn solenoid, 7.00 cm long, as shown in Figure P32.68. Both coil and solenoid have cross-sectional areas of 1.00×10^{-4} m². (a) How long does it take the solenoid current to reach 63.2 percent of its maximum value? Determine (b) the average back emf caused by the self-inductance of the solenoid during this interval, (c) the average rate of change in magnetic flux through the coil during this interval, and (d) the magnitude of the average induced current in the coil.

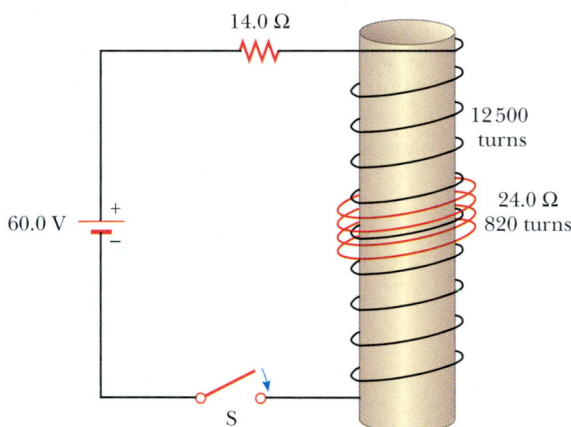

Figure P32.68

69. At $t = 0$, the switch in Figure P32.69 is thrown closed. Using Kirchhoff's laws for the instantaneous currents and voltages in this two-loop circuit, show that the current in the inductor is

$$I(t) = \frac{\mathcal{E}}{R_1}[1 - e^{-(R'/L)t}]$$

where $R' = R_1 R_2/(R_1 + R_2)$.

70. In Figure P32.69, take $\mathcal{E} = 6.00$ V, $R_1 = 5.00$ Ω, and $R_2 = 1.00$ Ω. The inductor has negligible resistance. When the switch is thrown open after having been

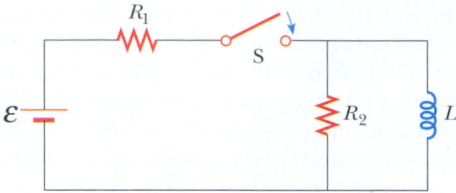

Figure P32.69   Problems 69 and 70.

closed for a long time, the current in the inductor drops to 0.250 A in 0.150 s. What is the inductance of the inductor?

71. In Figure P32.71, the switch is closed for $t < 0$, and steady-state conditions are established. The switch is thrown open at $t = 0$. (a) Find the initial voltage $\mathcal{E}_0$ across L just after $t = 0$. Which end of the coil is at the higher potential: a or b? (b) Make freehand graphs of the currents in R_1 and in R_2 as a function of time, treating the steady-state directions as positive. Show values before and after $t = 0$. (c) How long after $t = 0$ does the current in R_2 have the value 2.00 mA?

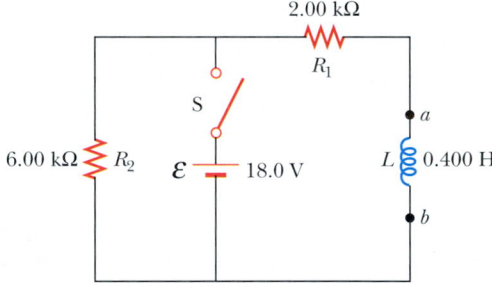

Figure P32.71

72. The switch in Figure P32.72 is thrown closed at $t = 0$. Before the switch is closed, the capacitor is uncharged, and all currents are zero. Determine the currents in L, C, and R and the potential differences across L, C, and R (a) the instant after the switch is closed and (b) long after it is closed.

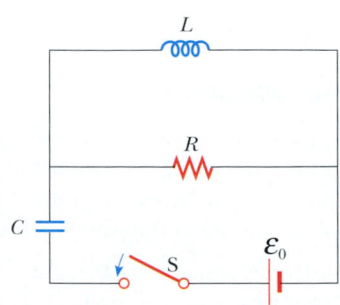

Figure P32.72

73. To prevent damage from arcing in an electric motor, a discharge resistor is sometimes placed in parallel with the armature. If the motor is suddenly unplugged while running, this resistor limits the voltage that appears across the armature coils. Consider a 12.0-V dc motor with an armature that has a resistance of 7.50 Ω and an inductance of 450 mH. Assume that the back emf in the armature coils is 10.0 V when the motor is running at normal speed. (The equivalent circuit for the armature is shown in Fig. P32.73.) Calculate the maximum resistance R that limits the voltage across the armature to 80.0 V when the motor is unplugged.

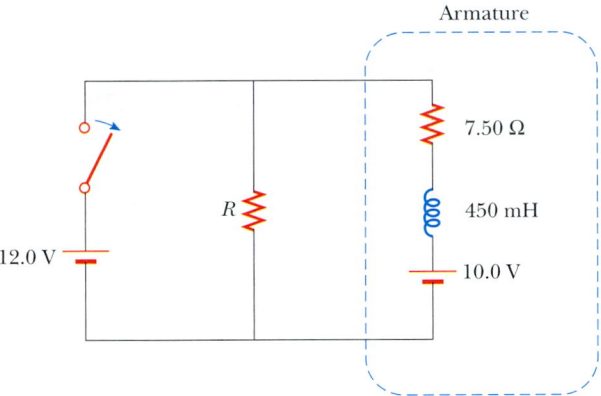

Figure P32.73

74. An air-core solenoid 0.500 m in length contains 1 000 turns and has a cross-sectional area of 1.00 cm². (a) If end effects are neglected, what is the self-inductance? (b) A secondary winding wrapped around the center of the solenoid has 100 turns. What is the mutual inductance? (c) The secondary winding carries a constant current of 1.00 A, and the solenoid is connected to a load of 1.00 kΩ. The constant current is suddenly stopped. How much charge flows through the load resistor?

75. The lead-in wires from a television antenna are often constructed in the form of two parallel wires (Fig. P32.75). (a) Why does this configuration of conductors have an inductance? (b) What constitutes the flux loop for this configuration? (c) Neglecting any magnetic flux inside the wires, show that the inductance of a length x

of this type of lead-in is

$$L = \frac{\mu_0 x}{\pi} \ln\left(\frac{w - a}{a}\right)$$

where a is the radius of the wires and w is their center-to-center separation.

Note: Problems 76 through 79 require the application of ideas from this chapter and earlier chapters to some properties of superconductors, which were introduced in Section 27.5.

76. Review Problem. *The resistance of a superconductor.* In an experiment carried out by S. C. Collins between 1955 and 1958, a current was maintained in a superconducting lead ring for 2.50 yr with no observed loss. If the inductance of the ring was 3.14×10^{-8} H and the sensitivity of the experiment was 1 part in 10^9, what was the maximum resistance of the ring? (*Hint:* Treat this as a decaying current in an RL circuit, and recall that $e^{-x} \cong 1 - x$ for small x.)

77. Review Problem. A novel method of storing electrical energy has been proposed. A huge underground superconducting coil, 1.00 km in diameter, would be fabricated. It would carry a maximum current of 50.0 kA through each winding of a 150-turn Nb₃Sn solenoid. (a) If the inductance of this huge coil were 50.0 H, what would be the total energy stored? (b) What would be the compressive force per meter length acting between two adjacent windings 0.250 m apart?

78. Review Problem. *Superconducting Power Transmission.* The use of superconductors has been proposed for the manufacture of power transmission lines. A single coaxial cable (Fig. P32.78) could carry 1.00×10^3 MW (the output of a large power plant) at 200 kV, dc, over a distance of 1 000 km without loss. An inner wire with a radius of 2.00 cm, made from the superconductor Nb₃Sn, carries the current I in one direction. A surrounding superconducting cylinder, of radius 5.00 cm, would carry the return current I. In such a system, what is the magnetic field (a) at the surface of the inner conductor and (b) at the inner surface of the outer conductor? (c) How much energy would be stored in the space between the conductors in a 1 000-km superconducting line? (d) What is the pressure exerted on the outer conductor?

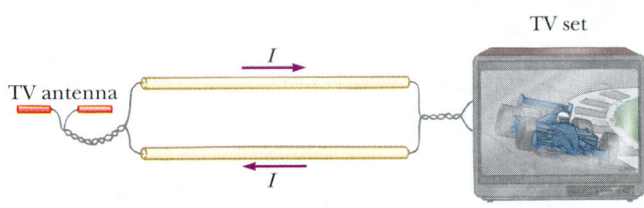

Figure P32.75

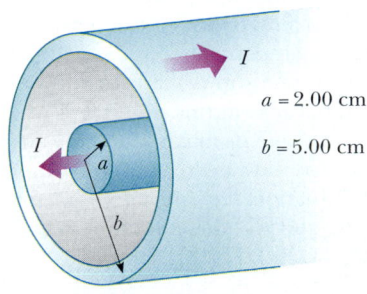

Figure P32.78

79. Review Problem. *The Meissner Effect.* Compare this problem with Problem 63 in Chapter 26 on the force attracting a perfect dielectric into a strong electric field. A fundamental property of a Type I superconducting material is *perfect diamagnetism*, or demonstration of the *Meissner effect*, illustrated in the photograph on page 855 and again in Figure 30.34, and described as follows: The superconducting material has **B** = 0 everywhere inside it. If a sample of the material is placed into an externally produced magnetic field, or if it is cooled to become superconducting while it is in a magnetic field, electric currents appear on the surface of the sample. The currents have precisely the strength and orientation required to make the total magnetic field zero throughout the interior of the sample. The following problem will help you to understand the magnetic force that can then act on the superconducting sample.

Consider a vertical solenoid with a length of 120 cm and a diameter of 2.50 cm consisting of 1 400 turns of copper wire carrying a counterclockwise current of 2.00 A, as shown in Figure P32.79a. (a) Find the magnetic field in the vacuum inside the solenoid. (b) Find the energy density of the magnetic field, and note that the units J/m³ of energy density are the same as the units N/m²(= Pa) of pressure. (c) A superconducting bar 2.20 cm in diameter is inserted partway into the solenoid. Its upper end is far outside the solenoid, where the magnetic field is small. The lower end of the bar is deep inside the solenoid. Identify the direction required for the current on the curved surface of the bar so that the total magnetic field is zero within the bar. The field created by the supercurrents is sketched in Figure P32.79b, and the total field is sketched in Figure

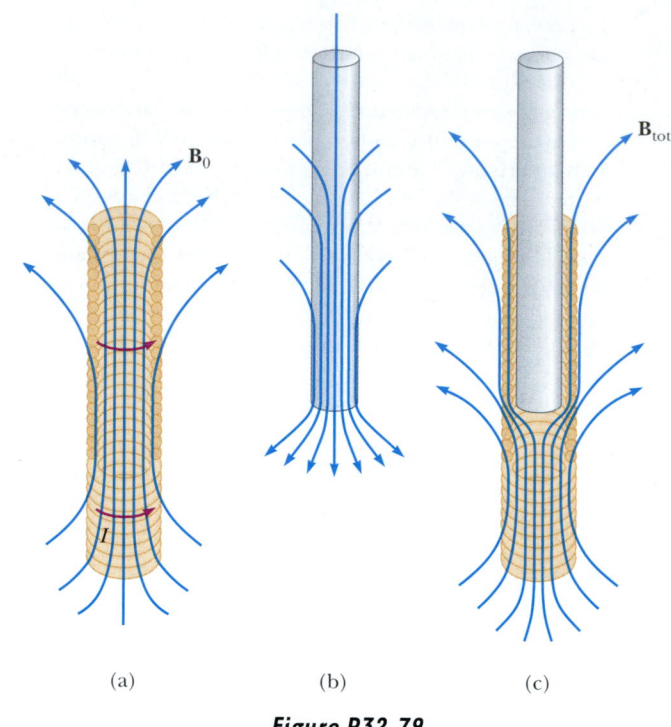

(a)   (b)   (c)

Figure P32.79

P32.79c. (d) The field of the solenoid exerts a force on the current in the superconductor. Identify the direction of the force on the bar. (e) Calculate the magnitude of the force by multiplying the energy density of the solenoid field by the area of the bottom end of the superconducting bar.

ANSWERS TO QUICK QUIZZES

32.1 When it is being opened. When the switch is initially open, there is no current in the circuit; when the switch is then closed, the inductor tends to maintain the no-current condition, and as a result there is very little chance of sparking. When the switch is initially closed, there is current in the circuit; when the switch is then opened, the current decreases. An induced emf is set up across the inductor, and this emf tends to maintain the original current. Sparking can occur as the current bridges the air gap between the poles of the switch.

32.2 (b). Figure 32.8 shows that circuit B has the greater time constant because in this circuit it takes longer for the current to reach its maximum value and then longer for this current to decrease to zero after switch S_2 is closed. Equation 32.8 indicates that, for equal resistances R_A and R_B, the condition $\tau_B > \tau_A$ means that $L_A < L_B$.

32.3 (a) M_{12} increases because the magnetic flux through coil 2 increases. (b) M_{12} decreases because rotation of coil 1 decreases its flux through coil 2.

32.4 (a) No. Mutual inductance requires a system of coils, and each coil has self-inductance. (b) Yes. A single coil has self-inductance but no mutual inductance because it does not interact with any other coils.

32.5 From Equation 32.25, $I_{max} = \omega Q_{max}$. Thus, the amplitude of the I-t graph is ω times the amplitude of the Q-t graph.

32.6 Equation 32.31 without the cosine factor. The dashed lines represent the positive and negative amplitudes (maximum values) for each oscillation period, and it is the $Q = Q_{max}e^{-Rt/2L}$ part of Equation 32.31 that gives the value of the ever-decreasing amplitude.

P U Z Z L E R

Small "black boxes" like this one are commonly used to supply power to electronic devices such as CD players and tape players. Whereas these devices need only about 12 V to operate, wall outlets provide an output of 120 V. What do the black boxes do, and how do they work? *(George Semple)*

c h a p t e r

33

Alternating-Current Circuits

In this chapter we describe alternating-current (ac) circuits. Every time we turn on a television set, a stereo, or any of a multitude of other electrical appliances, we are calling on alternating currents to provide the power to operate them. We begin our study by investigating the characteristics of simple series circuits that contain resistors, inductors, and capacitors and that are driven by a sinusoidal voltage. We shall find that the maximum alternating current in each element is proportional to the maximum alternating voltage across the element. We shall also find that when the applied voltage is sinusoidal, the current in each element is sinusoidal, too, but not necessarily in phase with the applied voltage. We conclude the chapter with two sections concerning transformers, power transmission, and RC filters.

33.1 ▶ AC SOURCES AND PHASORS

An ac circuit consists of circuit elements and a generator that provides the alternating current. As you recall from Section 31.5, the basic principle of the ac generator is a direct consequence of Faraday's law of induction. When a conducting loop is rotated in a magnetic field at constant angular frequency ω, a sinusoidal voltage (emf) is induced in the loop. This instantaneous voltage Δv is

$$\Delta v = \Delta V_{max} \sin \omega t$$

where ΔV_{max} is the maximum output voltage of the ac generator, or the **voltage amplitude.** From Equation 13.6, the angular frequency is

$$\omega = 2\pi f = \frac{2\pi}{T}$$

where f is the frequency of the generator (the voltage source) and T is the period. The generator determines the frequency of the current in any circuit connected to the generator. Because the output voltage of an ac generator varies sinusoidally with time, the voltage is positive during one half of the cycle and negative during the other half. Likewise, the current in any circuit driven by an ac generator is an alternating current that also varies sinusoidally with time. Commercial electric-power plants in the United States use a frequency of 60 Hz, which corresponds to an angular frequency of 377 rad/s.

The primary aim of this chapter can be summarized as follows: If an ac generator is connected to a series circuit containing resistors, inductors, and capacitors, we want to know the amplitude and time characteristics of the alternating current. To simplify our analysis of circuits containing two or more elements, we use graphical constructions called *phasor diagrams*. In these constructions, alternating (sinusoidal) quantities, such as current and voltage, are represented by rotating vectors called **phasors.** The length of the phasor represents the amplitude (maximum value) of the quantity, and the projection of the phasor onto the vertical axis represents the instantaneous value of the quantity. As we shall see, a phasor diagram greatly simplifies matters when we must combine several sinusoidally varying currents or voltages that have different phases.

33.2 ▶ RESISTORS IN AN AC CIRCUIT

Consider a simple ac circuit consisting of a resistor and an ac generator ⎯⊙⎯ , as shown in Figure 33.1. At any instant, the algebraic sum of the voltages around a closed loop in a circuit must be zero (Kirchhoff's loop rule). Therefore, $\Delta v - \Delta v_R = 0$, or[1]

$$\Delta v = \Delta v_R = \Delta V_{max} \sin \omega t \qquad \textbf{(33.1)}$$

where Δv_R is the **instantaneous voltage across the resistor.** Therefore, the instantaneous current in the resistor is

$$i_R = \frac{\Delta v_R}{R} = \frac{\Delta V_{max}}{R} \sin \omega t = I_{max} \sin \omega t \qquad \textbf{(33.2)}$$

where I_{max} is the maximum current:

$$I_{max} = \frac{\Delta V_{max}}{R}$$

Maximum current in a resistor

From Equations 33.1 and 33.2, we see that the instantaneous voltage across the resistor is

$$\Delta v_R = I_{max} R \sin \omega t \qquad \textbf{(33.3)}$$

Let us discuss the current-versus-time curve shown in Figure 33.2a. At point *a*, the current has a maximum value in one direction, arbitrarily called the positive direction. Between points *a* and *b*, the current is decreasing in magnitude but is still in the positive direction. At *b*, the current is momentarily zero; it then begins to increase in the negative direction between points *b* and *c*. At *c*, the current has reached its maximum value in the negative direction.

The current and voltage are in step with each other because they vary identically with time. Because i_R and Δv_R both vary as $\sin \omega t$ and reach their maximum values at the same time, as shown in Figure 33.2a, they are said to be **in phase.** Thus we can say that, for a sinusoidal applied voltage, the current in a resistor is always in phase with the voltage across the resistor.

The current in a resistor is in phase with the voltage

A *phasor diagram* is used to represent current–voltage phase relationships. The lengths of the arrows correspond to ΔV_{max} and I_{max}. The projections of the phasor arrows onto the vertical axis give Δv_R and i_R values. As we showed in Section 13.5, if the phasor arrow is imagined to rotate steadily with angular speed ω, its vertical-axis component oscillates sinusoidally in time. In the case of the single-loop resistive circuit of Figure 33.1, the current and voltage phasors lie along the same line, as in Figure 33.2b, because i_R and Δv_R are in phase.

Note that **the average value of the current over one cycle is zero.** That is, the current is maintained in the positive direction for the same amount of time and at the same magnitude as it is maintained in the negative direction. However, the direction of the current has no effect on the behavior of the resistor. We can understand this by realizing that collisions between electrons and the fixed atoms of the resistor result in an increase in the temperature of the resistor. Although this temperature increase depends on the magnitude of the current, it is independent of the direction of the current.

We can make this discussion quantitative by recalling that the rate at which electrical energy is converted to internal energy in a resistor is the power $\mathcal{P} = i^2 R$, where *i* is the instantaneous current in the resistor. Because this rate is proportional to the square of the current, it makes no difference whether the current is direct or alternating—that is, whether the sign associated with the current is positive or negative. However, the temperature increase produced by an alternating

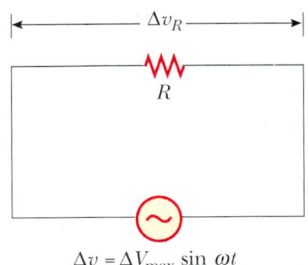

$$\Delta v = \Delta V_{max} \sin \omega t$$

Figure 33.1 A circuit consisting of a resistor of resistance *R* connected to an ac generator, designated by the symbol

.

---

[1] The lowercase symbols v and i are used to indicate the instantaneous values of the voltage and the current.

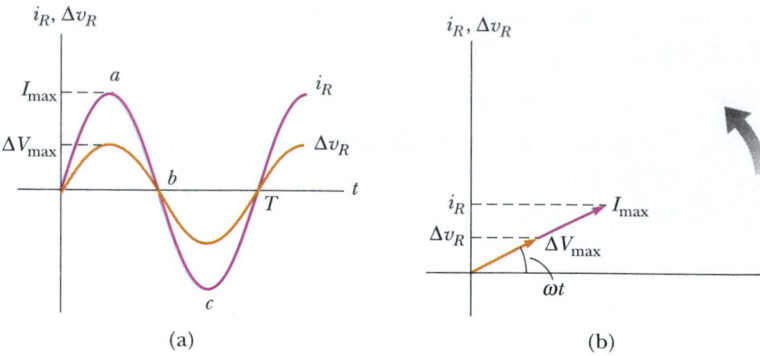

Figure 33.2   (a) Plots of the instantaneous current i_R and instantaneous voltage Δv_R across a resistor as functions of time. The current is in phase with the voltage, which means that the current is zero when the voltage is zero, maximum when the voltage is maximum, and minimum when the voltage is minimum. At time $t = T$, one cycle of the time-varying voltage and current has been completed. (b) Phasor diagram for the resistive circuit showing that the current is in phase with the voltage.

current having a maximum value I_{max} is not the same as that produced by a direct current equal to I_{max}. This is because the alternating current is at this maximum value for only an instant during each cycle (Fig. 33.3a). What is of importance in an ac circuit is an average value of current, referred to as the **rms current.** As we learned in Section 21.1, the notation *rms* stands for *root mean square*, which in this case means the square root of the mean (average) value of the square of the current: $I_{rms} = \sqrt{\overline{i^2}}$. Because i^2 varies as $\sin^2 \omega t$ and because the average value of i^2 is $\frac{1}{2}I_{max}^2$ (see Fig. 33.3b), the rms current is[2]

| rms current |

$$I_{rms} = \frac{I_{max}}{\sqrt{2}} = 0.707 I_{max} \qquad \textbf{(33.4)}$$

This equation states that an alternating current whose maximum value is 2.00 A delivers to a resistor the same power as a direct current that has a value of (0.707) (2.00 A) = 1.41 A. Thus, we can say that the average power delivered to a resistor that carries an alternating current is

| Average power delivered to a resistor |

$$\mathcal{P}_{av} = I_{rms}^2 R$$

[2] That the square root of the average value of i^2 is equal to $I_{max}/\sqrt{2}$ can be shown as follows: The current in the circuit varies with time according to the expression $i = I_{max} \sin \omega t$, so $i^2 = I_{max}^2 \sin^2 \omega t$. Therefore, we can find the average value of i^2 by calculating the average value of $\sin^2 \omega t$. A graph of $\cos^2 \omega t$ versus time is identical to a graph of $\sin^2 \omega t$ versus time, except that the points are shifted on the time axis. Thus, the time average of $\sin^2 \omega t$ is equal to the time average of $\cos^2 \omega t$ when taken over one or more complete cycles. That is,

$$(\sin^2 \omega t)_{av} = (\cos^2 \omega t)_{av}$$

Using this fact and the trigonometric identity $\sin^2 \theta + \cos^2 \theta = 1$, we obtain

$$(\sin^2 \omega t)_{av} + (\cos^2 \omega t)_{av} = 2(\sin^2 \omega t)_{av} = 1$$

$$(\sin^2 \omega t)_{av} = \tfrac{1}{2}$$

When we substitute this result in the expression $i^2 = I_{max}^2 \sin^2 \omega t$, we obtain $(i^2)_{av} = \overline{i^2} = I_{rms}^2 = I_{max}^2/2$, or $I_{rms} = I_{max}/\sqrt{2}$. The factor $1/\sqrt{2}$ is valid only for sinusoidally varying currents. Other waveforms, such as sawtooth variations, have different factors.

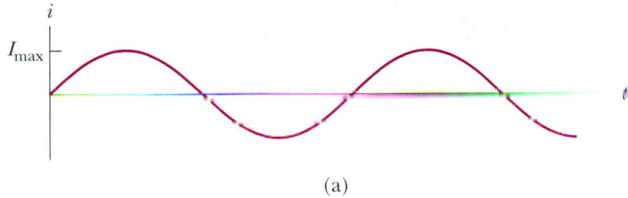

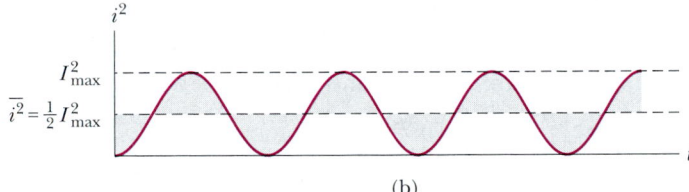

Figure 33.3   (a) Graph of the current in a resistor as a function of time. (b) Graph of the current squared in a resistor as a function of time. Notice that the gray shaded regions *under* the curve and *above* the dashed line for $I_{max}^2/2$ have the same area as the gray shaded regions *above* the curve and *below* the dashed line for $I_{max}^2/2$. Thus, the average value of i^2 is $I_{max}^2/2$.

Alternating voltage also is best discussed in terms of rms voltage, and the relationship is identical to that for current:

$$\Delta V_{rms} = \frac{\Delta V_{max}}{\sqrt{2}} = 0.707\,\Delta V_{max} \qquad \text{(33.5)}$$

rms voltage

When we speak of measuring a 120-V alternating voltage from an electrical outlet, we are referring to an rms voltage of 120 V. A quick calculation using Equation 33.5 shows that such an alternating voltage has a maximum value of about 170 V. One reason we use rms values when discussing alternating currents and voltages in this chapter is that ac ammeters and voltmeters are designed to read rms values. Furthermore, with rms values, many of the equations we use have the same form as their direct-current counterparts.

Quick Quiz 33.1

Which of the following statements might be true for a resistor connected to an ac generator? (a) $\mathcal{P}_{av} = 0$ and $i_{av} = 0$; (b) $\mathcal{P}_{av} = 0$ and $i_{av} > 0$; (c) $\mathcal{P}_{av} > 0$ and $i_{av} = 0$; (d) $\mathcal{P}_{av} > 0$ and $i_{av} > 0$.

EXAMPLE 33.1   What Is the rms Current?

The voltage output of a generator is given by $\Delta v = (200\text{ V})\sin \omega t$. Find the rms current in the circuit when this generator is connected to a 100-Ω resistor.

Solution Comparing this expression for voltage output with the general form $\Delta v = \Delta V_{max}\sin \omega t$, we see that $\Delta V_{max} = 200$ V. Thus, the rms voltage is

$$\Delta V_{rms} = \frac{\Delta V_{max}}{\sqrt{2}} = \frac{200\text{ V}}{\sqrt{2}} = 141\text{ V}$$

Therefore,

$$I_{rms} = \frac{\Delta V_{rms}}{R} = \frac{141\text{ V}}{100\text{ }\Omega} = \boxed{1.41\text{ A}}$$

Exercise   Find the maximum current in the circuit.

Answer   2.00 A.

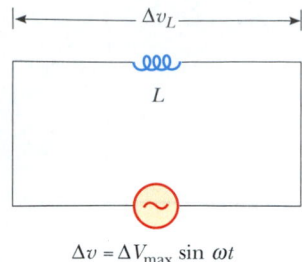

$$\Delta v = \Delta V_{max} \sin \omega t$$

Figure 33.4 A circuit consisting of an inductor of inductance L connected to an ac generator.

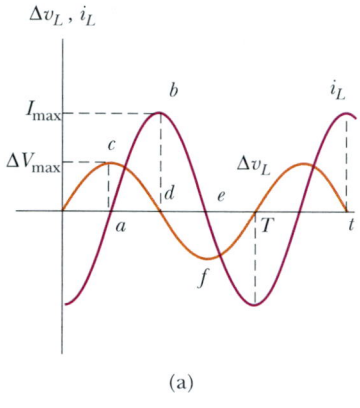

(a)

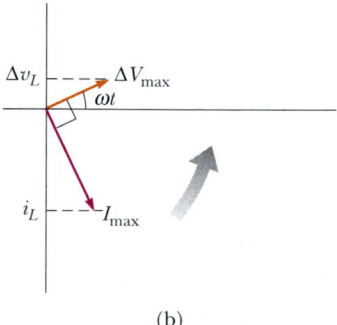

(b)

Figure 33.5 (a) Plots of the instantaneous current i_L and instantaneous voltage Δv_L across an inductor as functions of time. The current lags behind the voltage by 90°. (b) Phasor diagram for the inductive circuit, showing that the current lags behind the voltage by 90°.

The current in an inductor lags the voltage by 90°

33.3 INDUCTORS IN AN AC CIRCUIT

Now consider an ac circuit consisting only of an inductor connected to the terminals of an ac generator, as shown in Figure 33.4. If $\Delta v_L = \mathcal{E}_L = -L(di/dt)$ is the self-induced instantaneous voltage across the inductor (see Eq. 32.1), then Kirchhoff's loop rule applied to this circuit gives $\Delta v + \Delta v_L = 0$, or

$$\Delta v - L \frac{di}{dt} = 0$$

When we substitute $\Delta V_{max} \sin \omega t$ for Δv and rearrange, we obtain

$$L \frac{di}{dt} = \Delta V_{max} \sin \omega t \tag{33.6}$$

Solving this equation for di, we find that

$$di = \frac{\Delta V_{max}}{L} \sin \omega t \, dt$$

Integrating this expression[3] gives the instantaneous current in the inductor as a function of time:

$$i_L = \frac{\Delta V_{max}}{L} \int \sin \omega t \, dt = -\frac{\Delta V_{max}}{\omega L} \cos \omega t \tag{33.7}$$

When we use the trigonometric identity $\cos \omega t = -\sin(\omega t - \pi/2)$, we can express Equation 33.7 as

$$i_L = \frac{\Delta V_{max}}{\omega L} \sin\left(\omega t - \frac{\pi}{2}\right) \tag{33.8}$$

Comparing this result with Equation 33.6, we see that the instantaneous current i_L in the inductor and the instantaneous voltage v_L across the inductor are out of phase by $(\pi/2)$ rad = 90°.

In general, inductors in an ac circuit produce a current that is out of phase with the ac voltage. A plot of voltage and current versus time is provided in Figure 33.5a. At point a, the current begins to increase in the positive direction. At this instant the rate of change of current is at a maximum, and thus the voltage across the inductor is also at a maximum. As the current increases between points a and b, di/dt (the slope of the current curve) gradually decreases until it reaches zero at point b. As a result, the voltage across the inductor is decreasing during this same time interval, as the curve segment between c and d indicates. Immediately after point b, the current begins to decrease, although it still has the same direction it had during the previous quarter cycle (from a to b). As the current decreases to zero (from b to e), a voltage is again induced in the inductor (d to f), but the polarity of this voltage is opposite that of the voltage induced between c and d (because back emfs are always directed to oppose the change in the current). Note that the voltage reaches its maximum value one quarter of a period before the current reaches its maximum value. Thus, we see that

for a sinusoidal applied voltage, the current in an inductor always lags behind the voltage across the inductor by 90° (one-quarter cycle in time).

[3] We neglect the constant of integration here because it depends on the initial conditions, which are not important for this situation.

The phasor diagram for the inductive circuit of Figure 33.4 is shown in Figure 33.5b.

From Equation 33.7 we see that the current in an inductive circuit reaches its maximum value when $\cos \omega t = -1$:

$$I_{max} = \frac{\Delta V_{max}}{\omega L} = \frac{\Delta V_{max}}{X_L} \qquad (33.9)$$

Maximum current in an inductor

where the quantity X_L, called the **inductive reactance,** is

$$X_L = \omega L \qquad (33.10)$$

Inductive reactance

Equation 33.9 indicates that, for a given applied voltage, the maximum current decreases as the inductive reactance increases. The expression for the rms current in an inductor is similar to Equation 33.9, with I_{max} replaced by I_{rms} and ΔV_{max} replaced by ΔV_{rms}.

Inductive reactance, like resistance, has units of ohms. However, unlike resistance, reactance depends on frequency as well as on the characteristics of the inductor. Note that the reactance of an inductor in an ac circuit increases as the frequency of the current increases. This is because at higher frequencies, the instantaneous current must change more rapidly than it does at the lower frequencies; this causes an increase in the maximum induced emf associated with a given maximum current.

Using Equations 33.6 and 33.9, we find that the instantaneous voltage across the inductor is

$$\Delta v_L = -L \frac{di}{dt} = -\Delta V_{max} \sin \omega t = -I_{max} X_L \sin \omega t \qquad (33.11)$$

CONCEPTUAL EXAMPLE 33.2

Figure 33.6 shows a circuit consisting of a series combination of an alternating voltage source, a switch, an inductor, and a lightbulb. The switch is thrown closed, and the circuit is allowed to come to equilibrium so that the lightbulb glows steadily. An iron rod is then inserted into the interior of the inductor. What happens to the brightness of the lightbulb, and why?

Solution   The bulb gets dimmer. As the rod is inserted, the inductance increases because the magnetic field inside the inductor increases. According to Equation 33.10, this increase in L means that the inductive reactance of the inductor also increases. The voltage across the inductor increases while the voltage across the lightbulb decreases. With less

voltage across it, the lightbulb glows more dimly. In theatrical productions of the early 20th century, this method was used to dim the lights in the theater gradually.

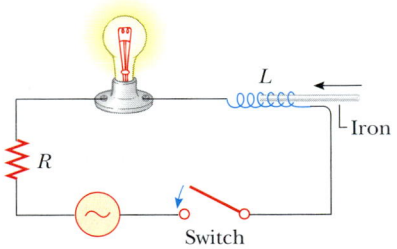

Figure 33.6

EXAMPLE 33.3   A Purely Inductive ac Circuit

In a purely inductive ac circuit (see Fig. 33.4), $L = 25.0$ mH and the rms voltage is 150 V. Calculate the inductive reactance and rms current in the circuit if the frequency is 60.0 Hz.

Solution   Equation 33.10 gives

$$X_L = \omega L = 2\pi f L = 2\pi(60.0 \text{ Hz})(25.0 \times 10^{-3} \text{ H}) = \boxed{9.42 \ \Omega}$$

From a modified version of Equation 33.9, the rms current is

$$I_{rms} = \frac{\Delta V_{L,rms}}{X_L} = \frac{150 \text{ V}}{9.42 \ \Omega} = \boxed{15.9 \text{ A}}$$

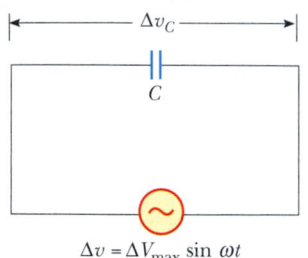

Figure 33.7 A circuit consisting of a capacitor of capacitance C connected to an ac generator.

33.4 ▶ CAPACITORS IN AN AC CIRCUIT

Figure 33.7 shows an ac circuit consisting of a capacitor connected across the terminals of an ac generator. Kirchhoff's loop rule applied to this circuit gives $\Delta v - \Delta v_C = 0$, or

$$\Delta v = \Delta v_C = \Delta V_{max} \sin \omega t \tag{33.12}$$

where Δv_C is the instantaneous voltage across the capacitor. We know from the definition of capacitance that $C = q/\Delta v_C$; hence, Equation 33.12 gives

$$q = C \Delta V_{max} \sin \omega t \tag{33.13}$$

where q is the instantaneous charge on the capacitor. Because $i = dq/dt$, differentiating Equation 33.13 gives the instantaneous current in the circuit:

$$i_C = \frac{dq}{dt} = \omega C \Delta V_{max} \cos \omega t \tag{33.14}$$

Using the trigonometric identity

$$\cos \omega t = \sin\left(\omega t + \frac{\pi}{2}\right)$$

we can express Equation 33.14 in the alternative form

$$i_C = \omega C \Delta V_{max} \sin\left(\omega t + \frac{\pi}{2}\right) \tag{33.15}$$

Comparing this expression with Equation 33.12, we see that the current is $\pi/2$ rad = 90° out of phase with the voltage across the capacitor. A plot of current and voltage versus time (Fig. 33.8a) shows that the current reaches its maximum value one quarter of a cycle sooner than the voltage reaches its maximum value.

Looking more closely, we see that the segment of the current curve from a to b indicates that the current starts out at a relatively high value. We can understand this by recognizing that there is no charge on the capacitor at $t = 0$; as a consequence, nothing in the circuit except the resistance of the wires can hinder the flow of charge at this instant. However, the current decreases as the voltage across the capacitor increases (from c to d on the voltage curve), and the capacitor is charging. When the voltage is at point d, the current reverses and begins to increase in the opposite direction (from b to e on the current curve). During this time, the voltage across the capacitor decreases from d to f because the plates are now losing the charge they accumulated earlier. During the second half of the cycle, the current is initially at its maximum value in the opposite direction (point e) and then decreases as the voltage across the capacitor builds up. The phasor diagram in Figure 33.8b also shows that

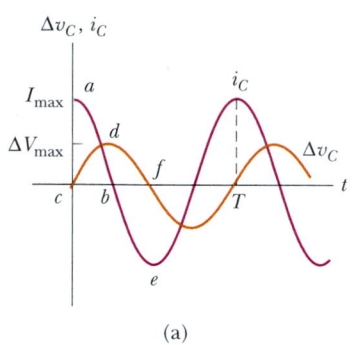

(a)

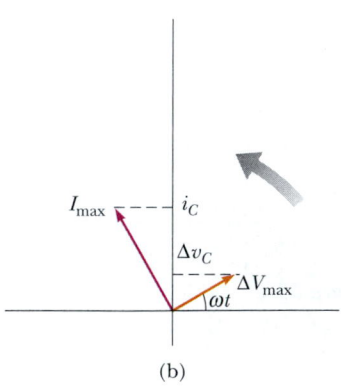

(b)

Figure 33.8 (a) Plots of the instantaneous current i_C and instantaneous voltage Δv_C across a capacitor as functions of time. The voltage lags behind the current by 90°. (b) Phasor diagram for the capacitive circuit, showing that the current leads the voltage by 90°.

for a sinusoidally applied voltage, the current in a capacitor always leads the voltage across the capacitor by 90°.

From Equation 33.14, we see that the current in the circuit reaches its maximum value when $\cos \omega t = 1$:

$$I_{max} = \omega C \Delta V_{max} = \frac{\Delta V_{max}}{X_C} \tag{33.16}$$

where X_C is called the **capacitive reactance:**

$$X_C = \frac{1}{\omega C} \tag{33.17}$$

Capacitive reactance

Note that capacitive reactance also has units of ohms.

The rms current is given by an expression similar to Equation 33.16, with I_{max} replaced by I_{rms} and ΔV_{max} replaced by ΔV_{rms}.

Combining Equations 33.12 and 33.16, we can express the instantaneous voltage across the capacitor as

$$\Delta v_C = \Delta V_{max} \sin \omega t = I_{max} X_C \sin \omega t \tag{33.18}$$

Equations 33.16 and 33.17 indicate that as the frequency of the voltage source increases, the capacitive reactance decreases and therefore the maximum current increases. Again, note that the frequency of the current is determined by the frequency of the voltage source driving the circuit. As the frequency approaches zero, the capacitive reactance approaches infinity, and hence the current approaches zero. This makes sense because the circuit approaches direct-current conditions as ω approaches 0.

EXAMPLE 33.4 ▸ A Purely Capacitive ac Circuit

An 8.00-μF capacitor is connected to the terminals of a 60.0-Hz ac generator whose rms voltage is 150 V. Find the capacitive reactance and the rms current in the circuit.

Solution Using Equation 33.17 and the fact that $\omega = 2\pi f = 377$ s^{-1} gives

$$X_C = \frac{1}{\omega C} = \frac{1}{(377 \text{ s}^{-1})(8.00 \times 10^{-6} \text{ F})} = 332 \text{ }\Omega$$

Hence, from a modified Equation 33.16, the rms current is

$$I_{rms} = \frac{\Delta V_{rms}}{X_C} = \frac{150 \text{ V}}{332 \text{ }\Omega} = 0.452 \text{ A}$$

Exercise If the frequency is doubled, what happens to the capacitive reactance and the current?

Answer X_C is halved, and I_{max} is doubled.

33.5 ▸ THE *RLC* SERIES CIRCUIT

Figure 33.9a shows a circuit that contains a resistor, an inductor, and a capacitor connected in series across an alternating-voltage source. As before, we assume that the applied voltage varies sinusoidally with time. It is convenient to assume that the instantaneous applied voltage is given by

$$\Delta v = \Delta V_{max} \sin \omega t$$

while the current varies as

$$i = I_{max} \sin(\omega t - \phi)$$

where ϕ is the **phase angle** between the current and the applied voltage. Our aim

Phase angle ϕ

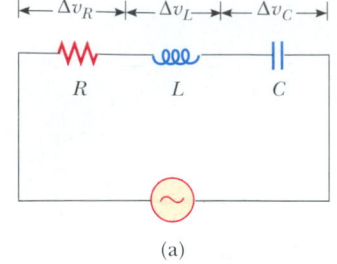

(a)

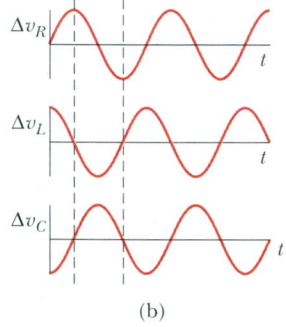

(b)

Figure 33.9   (a) A series circuit consisting of a resistor, an inductor, and a capacitor connected to an ac generator. (b) Phase relationships for instantaneous voltages in the series *RLC* circuit.

is to determine ϕ and I_{max}. Figure 33.9b shows the voltage versus time across each element in the circuit and their phase relationships.

To solve this problem, we must analyze the phasor diagram for this circuit. First, we note that because the elements are in series, the current everywhere in the circuit must be the same at any instant. That is, **the current at all points in a series ac circuit has the same amplitude and phase.** Therefore, as we found in the preceding sections, the voltage across each element has a different amplitude and phase, as summarized in Figure 33.10. In particular, the voltage across the resistor is in phase with the current, the voltage across the inductor leads the current by 90°, and the voltage across the capacitor lags behind the current by 90°. Using these phase relationships, we can express the instantaneous voltages across the three elements as

$$\Delta v_R = I_{max} R \sin \omega t = \Delta V_R \sin \omega t \tag{33.19}$$

$$\Delta v_L = I_{max} X_L \sin\left(\omega t + \frac{\pi}{2}\right) = \Delta V_L \cos \omega t \tag{33.20}$$

$$\Delta v_C = I_{max} X_C \sin\left(\omega t - \frac{\pi}{2}\right) = -\Delta V_C \cos \omega t \tag{33.21}$$

where ΔV_R, ΔV_L, and ΔV_C are the maximum voltage values across the elements:

$$\Delta V_R = I_{max}R \qquad \Delta V_L = I_{max} X_L \qquad \Delta V_C = I_{max}X_C$$

At this point, we could proceed by noting that the instantaneous voltage Δv across the three elements equals the sum

$$\Delta v = \Delta v_R + \Delta v_L + \Delta v_C$$

Quick Quiz 33.2

For the circuit of Figure 33.9a, is the voltage of the ac source equal to (a) the sum of the maximum voltages across the elements, (b) the sum of the instantaneous voltages across the elements, or (c) the sum of the rms voltages across the elements?

Although this analytical approach is correct, it is simpler to obtain the sum by examining the phasor diagram. Because the current at any instant is the same in all

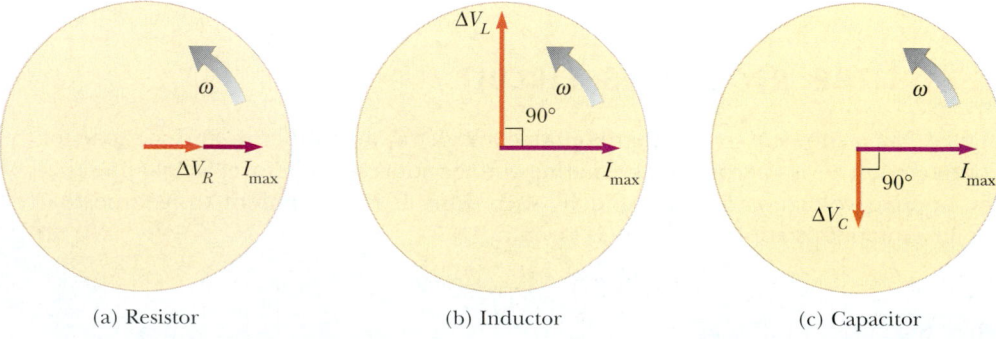

(a) Resistor   (b) Inductor   (c) Capacitor

Figure 33.10   Phase relationships between the voltage and current phasors for (a) a resistor, (b) an inductor, and (c) a capacitor connected in series.

elements, we can obtain a phasor diagram for the circuit. We combine the three phasor pairs shown in Figure 33.10 to obtain Figure 33.11a, in which a single phasor I_{max} is used to represent the current in each element. To obtain the vector sum of the three voltage phasors in Figure 33.11a, we redraw the phasor diagram as in Figure 33.11b. From this diagram, we see that the vector sum of the voltage amplitudes ΔV_R, ΔV_L, and ΔV_C equals a phasor whose length is the maximum applied voltage ΔV_{max}, where the phasor ΔV_{max} makes an angle ϕ with the current phasor I_{max}. Note that the voltage phasors ΔV_L and ΔV_C are in opposite directions along the same line, and hence we can construct the difference phasor $\Delta V_L - \Delta V_C$, which is perpendicular to the phasor ΔV_R. From either one of the right triangles in Figure 33.11b, we see that

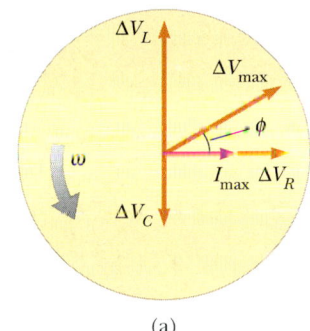

(a)

$$\Delta V_{max} = \sqrt{\Delta V_R{}^2 + (\Delta V_L - \Delta V_C)^2} = \sqrt{(I_{max}R)^2 + (I_{max}X_L - I_{max}X_C)^2}$$

$$\Delta V_{max} = I_{max}\sqrt{R^2 + (X_L - X_C)^2} \tag{33.22}$$

Therefore, we can express the maximum current as

$$I_{max} = \frac{\Delta V_{max}}{\sqrt{R^2 + (X_L - X_C)^2}}$$

The **impedance** Z of the circuit is defined as

$$Z \equiv \sqrt{R^2 + (X_L - X_C)^2} \tag{33.23}$$

where impedance also has units of ohms. Therefore, we can write Equation 33.22 in the form

$$\Delta V_{max} = I_{max}Z \tag{33.24}$$

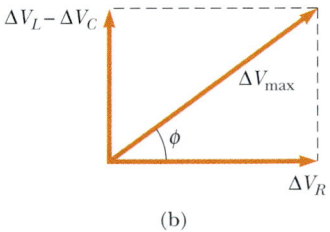

(b)

Figure 33.11 (a) Phasor diagram for the series *RLC* circuit shown in Figure 33.9a. The phasor ΔV_R is in phase with the current phasor I_{max}, the phasor ΔV_L leads I_{max} by 90°, and the phasor ΔV_C lags I_{max} by 90°. The total voltage ΔV_{max} makes an angle ϕ with I_{max}. (b) Simplified version of the phasor diagram shown in (a).

We can regard Equation 33.24 as the ac equivalent of Equation 27.8, which defined *resistance* in a dc circuit as the ratio of the voltage across a conductor to the current in that conductor. Note that the impedance and therefore the current in an ac circuit depend upon the resistance, the inductance, the capacitance, and the frequency (because the reactances are frequency-dependent).

By removing the common factor I_{max} from each phasor in Figure 33.11a, we can construct the *impedance triangle* shown in Figure 33.12. From this phasor diagram we find that the phase angle ϕ between the current and the voltage is

$$\phi = \tan^{-1}\left(\frac{X_L - X_C}{R}\right) \tag{33.25}$$

Also, from Figure 33.12, we see that $\cos\phi = R/Z$. When $X_L > X_C$ (which occurs at high frequencies), the phase angle is positive, signifying that the current lags behind the applied voltage, as in Figure 33.11a. When $X_L < X_C$, the phase angle is negative, signifying that the current leads the applied voltage. When $X_L = X_C$, the phase angle is zero. In this case, the impedance equals the resistance and the current has its maximum value, given by $\Delta V_{max}/R$. The frequency at which this occurs is called the *resonance frequency;* it is described further in Section 33.7.

Table 33.1 gives impedance values and phase angles for various series circuits containing different combinations of elements.

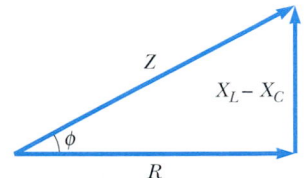

Figure 33.12 An impedance triangle for a series *RLC* circuit gives the relationship $Z = \sqrt{R^2 + (X_L - X_C)^2}$.

TABLE 33.1   **Impedance Values and Phase Angles for Various Circuit-Element Combinations[a]**

Circuit Elements	Impedance Z	Phase Angle ϕ
R (resistor)	R	$0°$
C (capacitor)	X_C	$-90°$
L (inductor)	X_L	$+90°$
R, C	$\sqrt{R^2 + X_C^2}$	Negative, between $-90°$ and $0°$
R, L	$\sqrt{R^2 + X_L^2}$	Positive, between $0°$ and $90°$
R, L, C	$\sqrt{R^2 + (X_L - X_C)^2}$	Negative if $X_C > X_L$ Positive if $X_C < X_L$

[a] In each case, an ac voltage (not shown) is applied across the elements.

Quick Quiz 33.3

Label each part of Figure 33.13 as being $X_L > X_C$, $X_L = X_C$, or $X_L < X_C$.

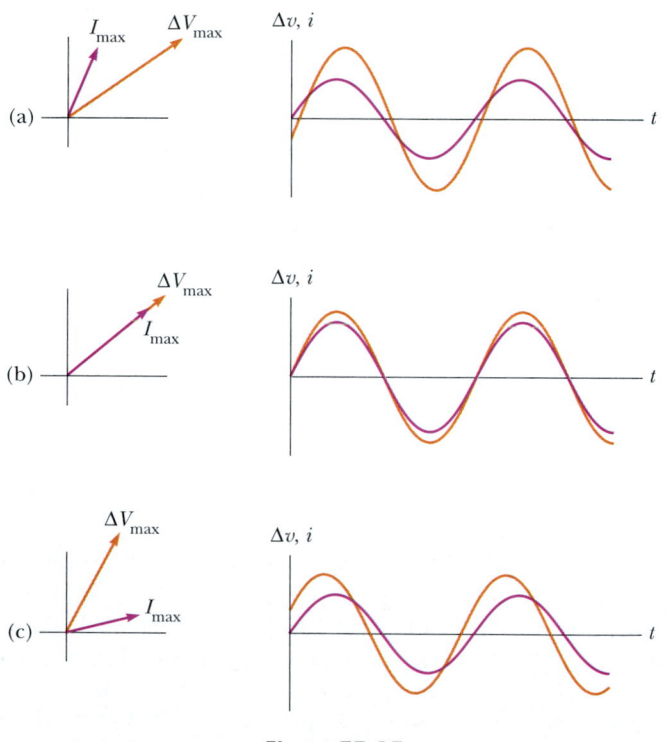

Figure 33.13

EXAMPLE 33.5   Finding *L* from a Phasor Diagram

In a series *RLC* circuit, the applied voltage has a maximum value of 120 V and oscillates at a frequency of 60.0 Hz. The circuit contains an inductor whose inductance can be varied, an 200-Ω resistor, and a 4.00-μF capacitor. What value of *L* should an engineer analyzing the circuit choose such that the voltage across the capacitor lags the applied voltage by 30.0°?

Solution   The phase relationships for the drops in voltage across the elements are shown in Figure 33.14. From the figure we see that the phase angle is $\phi = -60.0°$. This is because the phasors representing I_{max} and ΔV_R are in the same direction (they are in phase). From Equation 33.25, we find that

$$X_L = X_C + R \tan \phi$$

Substituting Equations 33.10 and 33.17 (with $\omega = 2\pi f$) into this expression gives

$$2\pi f L = \frac{1}{2\pi f C} + R \tan \phi$$

$$L = \frac{1}{2\pi f}\left[\frac{1}{2\pi f C} + R \tan \phi\right]$$

Substituting the given values into the equation gives $L =$ 0.84 H.

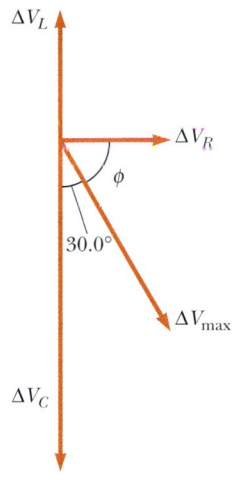

Figure 33.14

EXAMPLE 33.6   Analyzing a Series *RLC* Circuit

A series *RLC* ac circuit has $R = 425 \, \Omega$, $L = 1.25$ H, $C = 3.50 \, \mu$F, $\omega = 377$ s^{-1}, and $\Delta V_{max} = 150$ V. (a) Determine the inductive reactance, the capacitive reactance, and the impedance of the circuit.

Solution   The reactances are $X_L = \omega L =$ 471 Ω and

$X_C = 1/\omega C =$ 758 Ω.   The impedance is

$$Z = \sqrt{R^2 + (X_L - X_C)^2}$$

$$= \sqrt{(425 \, \Omega)^2 + (471 \, \Omega - 758 \, \Omega)^2} = \boxed{513 \, \Omega}$$

(b) Find the maximum current in the circuit.

Solution

$$I_{max} = \frac{V_{max}}{Z} = \frac{150 \text{ V}}{513 \, \Omega} = \boxed{0.292 \text{ A}}$$

(c) Find the phase angle between the current and voltage.

Solution

$$\phi = \tan^{-1}\left(\frac{X_L - X_C}{R}\right) = \tan^{-1}\left(\frac{471 \, \Omega - 758 \, \Omega}{425 \, \Omega}\right)$$

$$= \boxed{-34.0°}$$

Because the circuit is more capacitive than inductive, ϕ is negative and the current leads the applied voltage.

(d) Find both the maximum voltage and the instantaneous voltage across each element.

Solution   The maximum voltages are

$$\Delta V_R = I_{max} R = (0.292 \text{ A})(425 \, \Omega) = \boxed{124 \text{ V}}$$

$$\Delta V_L = I_{max} X_L = (0.292 \text{ A})(471 \, \Omega) = \boxed{138 \text{ V}}$$

$$\Delta V_C = I_{max} X_C = (0.292 \text{ A})(758 \, \Omega) = \boxed{221 \text{ V}}$$

Using Equations 33.19, 33.20, and 33.21, we find that we can write the instantaneous voltages across the three elements as

$$\Delta v_R = \boxed{(124 \text{ V}) \sin 377t}$$

$$\Delta v_L = \boxed{(138 \text{ V}) \cos 377t}$$

$$\Delta v_C = \boxed{(-221 \text{ V}) \cos 377t}$$

Comments   The sum of the maximum voltages across the elements is $\Delta V_R + \Delta V_L + \Delta V_C = 483$ V. Note that this sum is much greater than the maximum voltage of the generator, 150 V. As we saw in Quick Quiz 33.2, the sum of the maximum voltages is a meaningless quantity because when sinusoidally varying quantities are added, *both their amplitudes and their phases* must be taken into account. We know that the maximum voltages across the various elements occur at different times. That is, the voltages must be added in a way that takes account of the different phases. When this is done, Equation 33.22 is satisfied. You should verify this result.

Exercise   Construct a phasor diagram to scale, showing the voltages across the elements and the applied voltage. From your diagram, verify that the phase angle is $-34.0°$.

33.6 ▸ POWER IN AN AC CIRCUIT

No power losses are associated with pure capacitors and pure inductors in an ac circuit. To see why this is true, let us first analyze the power in an ac circuit containing only a generator and a capacitor.

When the current begins to increase in one direction in an ac circuit, charge begins to accumulate on the capacitor, and a voltage drop appears across it. When this voltage drop reaches its maximum value, the energy stored in the capacitor is $\frac{1}{2}C(\Delta V_{max})^2$. However, this energy storage is only momentary. The capacitor is charged and discharged twice during each cycle: Charge is delivered to the capacitor during two quarters of the cycle and is returned to the voltage source during the remaining two quarters. Therefore, **the average power supplied by the source is zero.** In other words, **no power losses occur in a capacitor in an ac circuit.**

Similarly, the voltage source must do work against the back emf of the inductor. When the current reaches its maximum value, the energy stored in the inductor is a maximum and is given by $\frac{1}{2}LI_{max}^2$. When the current begins to decrease in the circuit, this stored energy is returned to the source as the inductor attempts to maintain the current in the circuit.

In Example 28.1 we found that the power delivered by a battery to a dc circuit is equal to the product of the current and the emf of the battery. Likewise, the instantaneous power delivered by an ac generator to a circuit is the product of the generator current and the applied voltage. For the *RLC* circuit shown in Figure 33.9a, we can express the instantaneous power $\mathcal{P}$ as

$$\mathcal{P} = i\,\Delta v = I_{max}\sin(\omega t - \phi)\,\Delta V_{max}\sin\omega t$$

$$= I_{max}\,\Delta V_{max}\sin\omega t\sin(\omega t - \phi) \tag{33.26}$$

Clearly, this result is a complicated function of time and therefore is not very useful from a practical viewpoint. What is generally of interest is the average power over one or more cycles. Such an average can be computed by first using the trigonometric identity $\sin(\omega t - \phi) = \sin\omega t\cos\phi - \cos\omega t\sin\phi$. Substituting this into Equation 33.26 gives

$$\mathcal{P} = I_{max}\,\Delta V_{max}\sin^2\omega t\cos\phi - I_{max}\,\Delta V_{max}\sin\omega t\cos\omega t\sin\phi \tag{33.27}$$

We now take the time average of $\mathcal{P}$ over one or more cycles, noting that I_{max}, ΔV_{max}, ϕ, and ω are all constants. The time average of the first term on the right in Equation 33.27 involves the average value of $\sin^2\omega t$, which is $\frac{1}{2}$ (as shown in footnote 2). The time average of the second term on the right is identically zero because $\sin\omega t\cos\omega t = \frac{1}{2}\sin 2\omega t$, and the average value of $\sin 2\omega t$ is zero. Therefore, we can express the **average power** $\mathcal{P}_{av}$ as

$$\mathcal{P}_{av} = \tfrac{1}{2}I_{max}\,\Delta V_{max}\cos\phi \tag{33.28}$$

It is convenient to express the average power in terms of the rms current and rms voltage defined by Equations 33.4 and 33.5:

$$\mathcal{P}_{av} = I_{rms}\,\Delta V_{rms}\cos\phi \tag{33.29}$$

where the quantity cos ϕ is called the **power factor.** By inspecting Figure 33.11b, we see that the maximum voltage drop across the resistor is given by $\Delta V_R = \Delta V_{max} \cos \phi = I_{max} R$. Using Equation 33.5 and the fact that $\cos \phi = I_{max} R / \Delta V_{max}$, we find that we can express $\mathscr{P}_{av}$ as

$$\mathscr{P}_{av} = I_{rms} \Delta V_{rms} \cos \phi = I_{rms} \left(\frac{\Delta V_{max}}{\sqrt{2}} \right) \frac{I_{max} R}{\Delta V_{max}} = I_{rms} \frac{I_{max} R}{\sqrt{2}}$$

After making the substitution $I_{max} = \sqrt{2} I_{rms}$ from Equation 33.4, we have

$$\mathscr{P}_{av} = I_{rms}^2 R \tag{33.30}$$

Average power delivered to an *RLC* circuit

In words, the **average power delivered by the generator is converted to internal energy in the resistor,** just as in the case of a dc circuit. **No power loss occurs in an ideal inductor or capacitor.** When the load is purely resistive, then $\phi = 0$, $\cos \phi = 1$, and from Equation 33.29 we see that

$$\mathscr{P}_{av} = I_{rms} \Delta V_{rms}$$

Equation 33.29 shows that the power delivered by an ac source to any circuit depends on the phase, and this result has many interesting applications. For example, a factory that uses large motors in machines, generators, or transformers has a large inductive load (because of all the windings). To deliver greater power to such devices in the factory without using excessively high voltages, technicians introduce capacitance in the circuits to shift the phase.

EXAMPLE 33.7   Average Power in an *RLC* Series Circuit

Calculate the average power delivered to the series *RLC* circuit described in Example 33.6.

Solution   First, let us calculate the rms voltage and rms current, using the values of ΔV_{max} and I_{max} from Example 33.6:

$$\Delta V_{rms} = \frac{\Delta V_{max}}{\sqrt{2}} = \frac{150 \text{ V}}{\sqrt{2}} = 106 \text{ V}$$

$$I_{rms} = \frac{I_{max}}{\sqrt{2}} = \frac{0.292 \text{ A}}{\sqrt{2}} = 0.206 \text{ A}$$

Because $\phi = -34.0°$, the power factor, $\cos \phi$, is 0.829; hence, the average power delivered is

$$\mathscr{P}_{av} = I_{rms} \Delta V_{rms} \cos \phi = (0.206 \text{ A})(106 \text{ V})(0.829)$$

$$= 18.1 \text{ W}$$

We can obtain the same result using Equation 33.30.

33.7   RESONANCE IN A SERIES *RLC* CIRCUIT

A series *RLC* circuit is said to be **in resonance** when the current has its maximum value. In general, the rms current can be written

$$I_{rms} = \frac{\Delta V_{rms}}{Z} \tag{33.31}$$

where Z is the impedance. Substituting the expression for Z from Equation 33.23 into 33.31 gives

$$I_{rms} = \frac{\Delta V_{rms}}{\sqrt{R^2 + (X_L - X_C)^2}} \tag{33.32}$$

Because the impedance depends on the frequency of the source, the current in the *RLC* circuit also depends on the frequency. The frequency ω_0 at which $X_L - X_C = 0$ is called the **resonance frequency** of the circuit. To find ω_0, we use the condition $X_L = X_C$, from which we obtain $\omega_0 L = 1/\omega_0 C$, or

Resonance frequency

$$\omega_0 = \frac{1}{\sqrt{LC}}$$ **(33.33)**

Note that this frequency also corresponds to the natural frequency of oscillation of an *LC* circuit (see Section 32.5). Therefore, the current in a series *RLC* circuit reaches its maximum value when the frequency of the applied voltage matches the natural oscillator frequency—which depends only on *L* and *C*. Furthermore, at this frequency the current is in phase with the applied voltage.

Quick Quiz 33.4

What is the impedance of a series *RLC* circuit at resonance? What is the current in the circuit at resonance?

A plot of rms current versus frequency for a series *RLC* circuit is shown in Figure 33.15a. The data assume a constant $\Delta V_{rms} = 5.0$ mV, that $L = 5.0$ μH, and that $C = 2.0$ nF. The three curves correspond to three values of *R*. Note that in each case the current reaches its maximum value at the resonance frequency ω_0. Furthermore, the curves become narrower and taller as the resistance decreases.

By inspecting Equation 33.32, we must conclude that, when $R = 0$, the current becomes infinite at resonance. Although the equation predicts this, real circuits always have some resistance, which limits the value of the current.

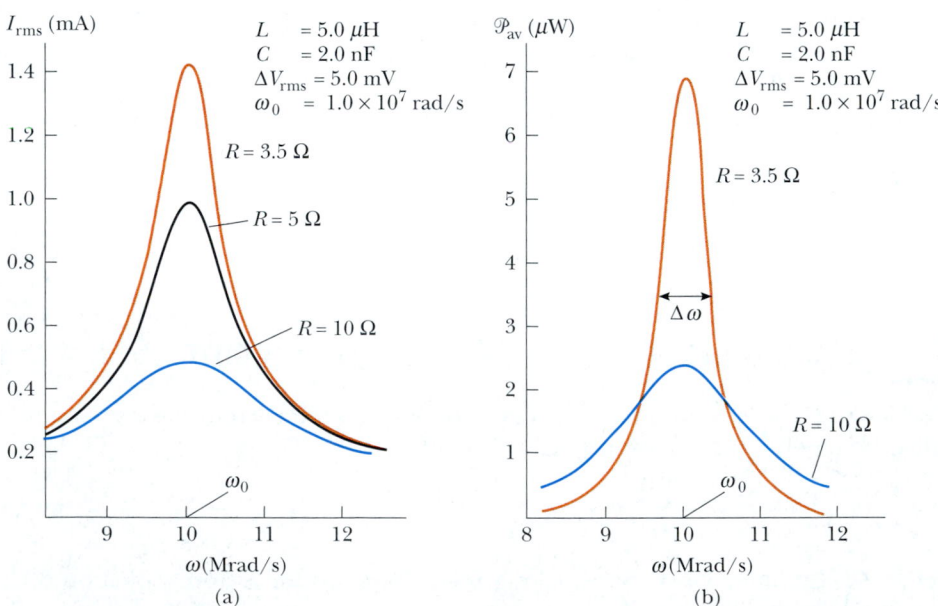

Figure 33.15 (a) The rms current versus frequency for a series *RLC* circuit, for three values of *R*. The current reaches its maximum value at the resonance frequency ω_0. (b) Average power versus frequency for the series *RLC* circuit, for two values of *R*.

It is also interesting to calculate the average power as a function of frequency for a series *RLC* circuit. Using Equations 33.30, 33.31, and 33.23, we find that

$$\mathscr{P}_{av} = I_{rms}^2 R = \frac{(\Delta V_{rms})^2}{Z^2} R = \frac{(\Delta V_{rms})^2 R}{R^2 + (X_L - X_C)^2} \tag{33.34}$$

Because $X_L = \omega L$, $X_C = 1/\omega C$, and $\omega_0^2 = 1/LC$, we can express the term $(X_L - X_C)^2$ as

$$(X_L - X_C)^2 = \left(\omega L - \frac{1}{\omega C}\right)^2 = \frac{L^2}{\omega^2}(\omega^2 - \omega_0^2)^2$$

Using this result in Equation 33.34 gives

$$\mathscr{P}_{av} = \frac{(\Delta V_{rms})^2 \, R\omega^2}{R^2\omega^2 + L^2(\omega^2 - \omega_0^2)^2} \tag{33.35}$$

Average power as a function of frequency in an *RLC* circuit

This expression shows that at resonance, when $\omega = \omega_0$, **the average power is a maximum** and has the value $(\Delta V_{rms})^2/R$. Figure 33.15b is a plot of average power versus frequency for two values of *R* in a series *RLC* circuit. As the resistance is made smaller, the curve becomes sharper in the vicinity of the resonance frequency. This curve sharpness is usually described by a dimensionless parameter known as the **quality factor,** denoted by Q:[4]

$$Q = \frac{\omega_0}{\Delta\omega}$$

Quality factor

where $\Delta\omega$ is the width of the curve measured between the two values of ω for which $\mathscr{P}_{av}$ has half its maximum value, called the *half-power points* (see Fig. 33.15b.) It is left as a problem (Problem 70) to show that the width at the half-power points has the value $\Delta\omega = R/L$, so

$$Q = \frac{\omega_0 L}{R} \tag{33.36}$$

The curves plotted in Figure 33.16 show that a high-*Q* circuit responds to only a very narrow range of frequencies, whereas a low-*Q* circuit can detect a much broader range of frequencies. Typical values of *Q* in electronic circuits range from 10 to 100.

The receiving circuit of a radio is an important application of a resonant circuit. One tunes the radio to a particular station (which transmits a specific electromagnetic wave or signal) by varying a capacitor, which changes the resonant frequency of the receiving circuit. When the resonance frequency of the circuit matches that of the incoming electromagnetic wave, the current in the receiving circuit increases. This signal caused by the incoming wave is then amplified and fed to a speaker. Because many signals are often present over a range of frequencies, it is important to design a high-*Q* circuit to eliminate unwanted signals. In this manner, stations whose frequencies are near but not equal to the resonance frequency give signals at the receiver that are negligibly small relative to the signal that matches the resonance frequency.

QuickLab

Tune a radio to your favorite station. Can you determine what the product of *LC* must be for the radio's tuning circuitry?

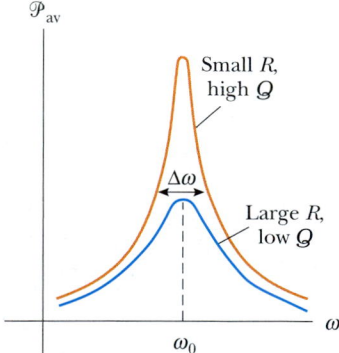

Figure 33.16 Average power versus frequency for a series *RLC* circuit. The width $\Delta\omega$ of each curve is measured between the two points where the power is half its maximum value. The power is a maximum at the resonance frequency ω_0.

---

[4] The quality factor is also defined as the ratio $2\pi E/\Delta E$, where *E* is the energy stored in the oscillating system and ΔE is the energy lost per cycle of oscillation. The quality factor for a mechanical system can also be defined, as noted in Section 13.7.

Quick Quiz 33.5

An airport metal detector (Fig. 33.17) is essentially a resonant circuit. The portal you step through is an inductor (a large loop of conducting wire) that is part of the circuit. The frequency of the circuit is tuned to the resonant frequency of the circuit when there is no metal in the inductor. Any metal on your body increases the effective inductance of the loop and changes the current in it. If you want the detector to be able to detect a small metallic object, should the circuit have a high quality factor or a low one?

Figure 33.17 When you pass through a metal detector, you become part of a resonant circuit. As you step through the detector, the inductance of the circuit changes, and thus the current in the circuit changes. *(Terry Qing/FPG International)*

EXAMPLE 33.8   **A Resonating Series *RLC* Circuit**

Consider a series *RLC* circuit for which $R = 150 \, \Omega$, $L = 20.0$ mH, $\Delta V_{\mathrm{rms}} = 20.0$ V, and $\omega = 5\,000$ s^{-1}. Determine the value of the capacitance for which the current is a maximum.

Solution   The current has its maximum value at the resonance frequency ω_0, which should be made to match the "driving" frequency of $5\,000$ s^{-1}:

$$\omega_0 = 5.00 \times 10^3 \text{ s}^{-1} = \frac{1}{\sqrt{LC}}$$

$$C = \frac{1}{\omega_0^2 L} = \frac{1}{(25.0 \times 10^6 \text{ s}^{-2})(20.0 \times 10^{-3} \text{ H})}$$

$$= \boxed{2.00 \ \mu\text{F}}$$

Exercise   Calculate the maximum value of the rms current in the circuit as the frequency is varied.

Answer   0.133 A.

33.8   ## THE TRANSFORMER AND POWER TRANSMISSION

When electric power is transmitted over great distances, it is economical to use a high voltage and a low current to minimize the I^2R loss in the transmission lines.

Consequently, 350-kV lines are common, and in many areas even higher-voltage (765-kV) lines are under construction. At the receiving end of such lines, the consumer requires power at a low voltage (for safety and for efficiency in design). Therefore, a device is required that can change the alternating voltage and current without causing appreciable changes in the power delivered. The ac transformer is that device.

In its simplest form, the **ac transformer** consists of two coils of wire wound around a core of iron, as illustrated in Figure 33.18. The coil on the left, which is connected to the input alternating voltage source and has N_1 turns, is called the *primary winding* (or the *primary*). The coil on the right, consisting of N_2 turns and connected to a load resistor R, is called the *secondary winding* (or the *secondary*). The purpose of the iron core is to increase the magnetic flux through the coil and to provide a medium in which nearly all the flux through one coil passes through the other coil. Eddy current losses are reduced by using a laminated core. Iron is used as the core material because it is a soft ferromagnetic substance and hence reduces hysteresis losses. Transformation of energy to internal energy in the finite resistance of the coil wires is usually quite small. Typical transformers have power efficiencies from 90% to 99%. In the discussion that follows, we assume an *ideal transformer*, one in which the energy losses in the windings and core are zero.

First, let us consider what happens in the primary circuit when the switch in the secondary circuit is open. If we assume that the resistance of the primary is negligible relative to its inductive reactance, then the primary circuit is equivalent to a simple circuit consisting of an inductor connected to an ac generator. Because the current is 90° out of phase with the voltage, the power factor cos ϕ is zero, and hence the average power delivered from the generator to the primary circuit is zero. Faraday's law states that the voltage ΔV_1 across the primary is

$$\Delta V_1 = -N_1 \frac{d\Phi_B}{dt} \tag{33.37}$$

where Φ_B is the magnetic flux through each turn. If we assume that all magnetic field lines remain within the iron core, the flux through each turn of the primary equals the flux through each turn of the secondary. Hence, the voltage across the secondary is

$$\Delta V_2 = -N_2 \frac{d\Phi_B}{dt} \tag{33.38}$$

Solving Equation 33.37 for $d\Phi_B/dt$ and substituting the result into Equation 33.38, we find that

$$\Delta V_2 = \frac{N_2}{N_1} \Delta V_1 \tag{33.39}$$

When $N_2 > N_1$, the output voltage ΔV_2 exceeds the input voltage ΔV_1. This setup is referred to as a *step-up transformer*. When $N_2 < N_1$, the output voltage is less than the input voltage, and we have a *step-down transformer*.

When the switch in the secondary circuit is thrown closed, a current I_2 is induced in the secondary. If the load in the secondary circuit is a pure resistance, the induced current is in phase with the induced voltage. The power supplied to the secondary circuit must be provided by the ac generator connected to the primary circuit, as shown in Figure 33.19. In an ideal transformer, where there are no losses, the power $I_1 \Delta V_1$ supplied by the generator is equal to the power $I_2 \Delta V_2$ in

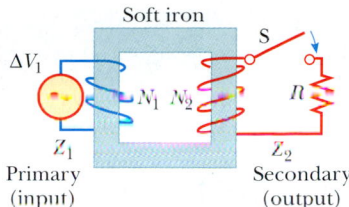

Figure 33.18 An ideal transformer consists of two coils wound on the same iron core. An alternating voltage ΔV_1 is applied to the primary coil, and the output voltage ΔV_2 is across the resistor of resistance R.

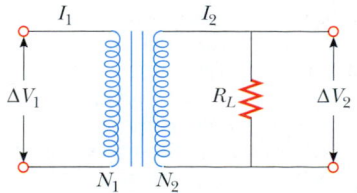

Figure 33.19 Circuit diagram for a transformer.

This cylindrical step-down transformer drops the voltage from 4 000 V to 220 V for delivery to a group of residences. *(George Semple)*

the secondary circuit. That is,

$$I_1 \, \Delta V_1 = I_2 \, \Delta V_2 \tag{33.40}$$

The value of the load resistance R_L determines the value of the secondary current because $I_2 = \Delta V_2/R_L$. Furthermore, the current in the primary is $I_1 = \Delta V_1/R_{eq}$, where

$$R_{eq} = \left(\frac{N_1}{N_2}\right)^2 R_L \tag{33.41}$$

is the equivalent resistance of the load resistance when viewed from the primary side. From this analysis we see that a transformer may be used to match resistances between the primary circuit and the load. In this manner, maximum power transfer can be achieved between a given power source and the load resistance. For example, a transformer connected between the 1-kΩ output of an audio amplifier and an 8-Ω speaker ensures that as much of the audio signal as possible is transferred into the speaker. In stereo terminology, this is called *impedance matching*.

We can now also understand why transformers are useful for transmitting power over long distances. Because the generator voltage is stepped up, the current in the transmission line is reduced, and hence I^2R losses are reduced. In practice, the voltage is stepped up to around 230 000 V at the generating station, stepped down to around 20 000 V at a distributing station, then to 4 000 V for delivery to residential areas, and finally to 120–240 V at the customer's site. The power is supplied by a three-wire cable. In the United States, two of these wires are "hot," with voltages of 120 V with respect to a common ground wire. Home appliances operating on 120 V are connected in parallel between one of the hot wires and ground. Larger appliances, such as electric stoves and clothes dryers, require 240 V. This is obtained across the two hot wires, which are 180° out of phase so that the voltage difference between them is 240 V.

There is a practical upper limit to the voltages that can be used in transmission lines. Excessive voltages could ionize the air surrounding the transmission lines, which could result in a conducting path to ground or to other objects in the vicinity. This, of course, would present a serious hazard to any living creatures. For this reason, a long string of insulators is used to keep high-voltage wires away from their supporting metal towers. Other insulators are used to maintain separation between wires.

Figure 33.20   The primary winding in this transformer is directly attached to the prongs of the plug. The secondary winding is connected to the wire on the right, which runs to an electronic device. Many of these power-supply transformers also convert alternating current to direct current. *(George Semple)*

Many common household electronic devices require low voltages to operate properly. A small transformer that plugs directly into the wall, like the one illustrated in the photograph at the beginning of this chapter, can provide the proper voltage. Figure 33.20 shows the two windings wrapped around a common iron core that is found inside all these little "black boxes." This particular transformer converts the 120-V ac in the wall socket to 12.5-V ac. (Can you determine the ratio of the numbers of turns in the two coils?) Some black boxes also make use of diodes to convert the alternating current to direct current (see Section 33.9).

web

For information on how small transformers and hundreds of other everyday devices operate, visit

http://www.howstuffworks.com

EXAMPLE 33.9   The Economics of ac Power

An electricity-generating station needs to deliver 20 MW of power to a city 1.0 km away. (a) If the resistance of the wires is 2.0 Ω and the electricity costs about 10¢/kWh, estimate what it costs the utility company to send the power to the city for one day. A common voltage for commercial power generators is 22 kV, but a step-up transformer is used to boost the voltage to 230 kV before transmission.

Solution   The power losses in the transmission line are the result of the resistance of the line. We can determine the loss from Equation 27.23, $\mathcal{P} = I^2R$. Because this is an estimate, we can use dc equations and calculate I from Equation 27.22:

$$I = \frac{\mathcal{P}}{\Delta V} = \frac{20 \times 10^6 \text{ W}}{230 \times 10^3 \text{ V}} = 87 \text{ A}$$

Therefore,

$$\mathcal{P} = I^2R = (87 \text{ A})^2(2.0 \text{ Ω}) = 15 \text{ kW}$$

Over the course of a day, the energy loss due to the resistance of the wires is (15 kW)(24 h) = 360 kWh, at a cost of $36.

(b) Repeat the calculation for the situation in which the power plant delivers the electricity at its original voltage of 22 kV.

Solution

$$I = \frac{\mathcal{P}}{\Delta V} = \frac{20 \times 10^6 \text{ W}}{22 \times 10^3 \text{ V}} = 910 \text{ A}$$

$$\mathcal{P} = I^2R = (910 \text{ A})^2(2.0 \text{ Ω}) = 1.7 \times 10^3 \text{ kW}$$

Cost per day = $(1.7 \times 10^3 \text{ kW})(24 \text{ h})(\$0.10/\text{kWh})$

= $4 100

The tremendous savings that are possible through the use of transformers and high-voltage transmission lines, along with the efficiency of using alternating current to operate motors, led to the universal adoption of alternating current instead of direct current for commercial power grids.

Optional Section

33.9   RECTIFIERS AND FILTERS

Portable electronic devices such as radios and compact disc (CD) players are often powered by direct current supplied by batteries. Many devices come with ac–dc converters that provide a readily available alternating-current source if the batteries are low. Such a converter contains a transformer that steps the voltage down from 120 V to typically 9 V and a circuit that converts alternating current to direct current. The process of converting alternating current to direct current is called **rectification,** and the converting device is called a **rectifier.**

The most important element in a rectifier circuit is a **diode,** a circuit element that conducts current in one direction but not the other. Most diodes used in modern electronics are semiconductor devices. The circuit symbol for a diode is ——▶|—— , where the arrow indicates the direction of the current through the diode. A diode has low resistance to current in one direction (the direction of the arrow) and high resistance to current in the opposite direction. We can understand how a diode rectifies a current by considering Figure 33.21a, which shows a diode and a resistor connected to the secondary of a transformer. The transformer reduces the voltage from 120-V ac to the lower voltage that is needed for the device having a resistance R (the load resistance). Because current can pass through the diode in only one direction, the alternating current in the load resistor is reduced to the form shown by the solid curve in Figure 33.21b. The diode conducts current only when the side of the symbol containing the arrowhead has a positive potential relative to the other side. In this situation, the diode acts as a *half-wave rectifier* because current is present in the circuit during only half of each cycle.

When a capacitor is added to the circuit, as shown by the dashed lines and the capacitor symbol in Figure 33.21a, the circuit is a simple dc power supply. The time variation in the current in the load resistor (the dashed curve in Fig. 33.21b) is close to being zero, as determined by the RC time constant of the circuit.

The RC circuit in Figure 33.21a is one example of a **filter circuit,** which is used to smooth out or eliminate a time-varying signal. For example, radios are usually powered by a 60-Hz alternating voltage. After rectification, the voltage still contains a small ac component at 60 Hz (sometimes called *ripple*), which must be filtered. By "filtered," we mean that the 60-Hz ripple must be reduced to a value much less than that of the audio signal to be amplified, because without filtering, the resulting audio signal includes an annoying hum at 60 Hz.

To understand how a filter works, let us consider the simple series RC circuit shown in Figure 33.22a. The input voltage is across the two elements and is represented by $\Delta V_{max} \sin \omega t$. Because we are interested only in maximum values, we can use Equation 33.24, taking $X_L = 0$ and substituting $X_C = 1/\omega C$. This shows that the maximum input voltage is related to the maximum current by

$$\Delta V_{in} = I_{max} Z = I_{max} \sqrt{R^2 + \left(\frac{1}{\omega C}\right)^2}$$

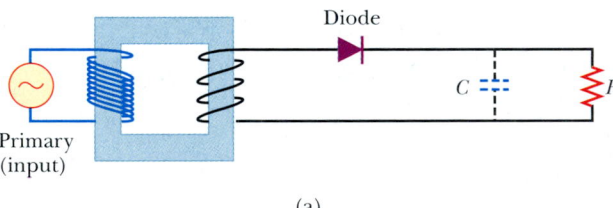

(a)

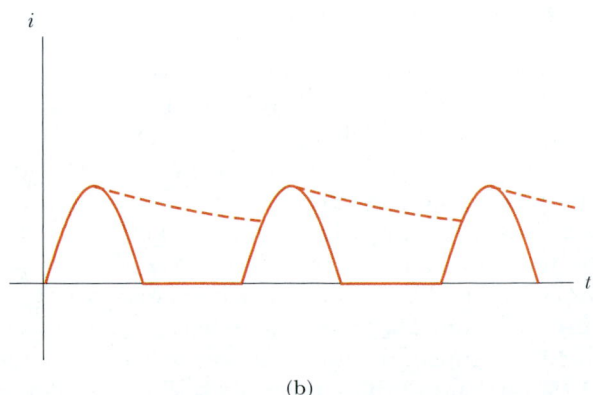

(b)

Figure 33.21   (a) A half-wave rectifier with an optional filter capacitor. (b) Current versus time in the resistor. The solid curve represents the current with no filter capacitor, and the dashed curve is the current when the circuit includes the capacitor.

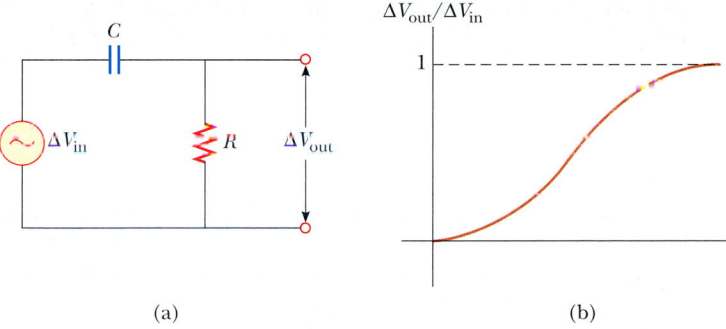

Figure 33.22 (a) A simple RC high-pass filter. (b) Ratio of output voltage to input voltage for an RC high-pass filter as a function of the angular frequency of the circuit.

If the voltage across the resistor is considered to be the output voltage, then the maximum output voltage is

$$\Delta V_{out} = I_{max} R$$

Therefore, the ratio of the output voltage to the input voltage is

$$\frac{\Delta V_{out}}{\Delta V_{in}} = \frac{R}{\sqrt{R^2 + \left(\dfrac{1}{\omega C}\right)^2}} \qquad \textbf{(33.42)}$$

High-pass filter

A plot of this ratio as a function of angular frequency (see Fig. 33.22b) shows that at low frequencies ΔV_{out} is much smaller than ΔV_{in}, whereas at high frequencies the two voltages are equal. Because the circuit preferentially passes signals of higher frequency while blocking low-frequency signals, the circuit is called an RC high-pass filter. Physically, a high-pass filter works because a capacitor "blocks out" direct current and ac current at low frequencies.

Now let us consider the circuit shown in Figure 33.23a, where the output voltage is taken across the capacitor. In this case, the maximum voltage equals the voltage across the capacitor. Because the impedance across the capacitor is $X_C = 1/\omega C$, we have

$$\Delta V_{out} = I_{max} X_C = \frac{I_{max}}{\omega C}$$

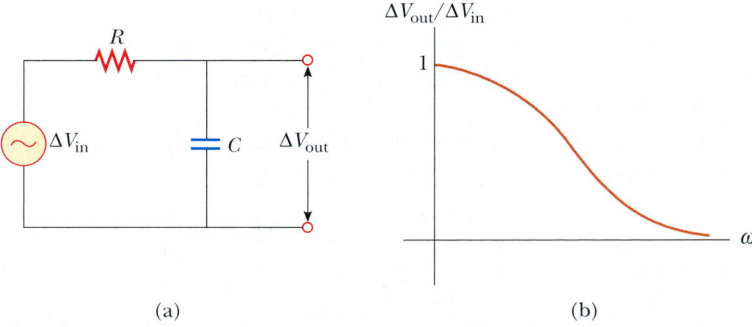

Figure 33.23 (a) A simple RC low-pass filter. (b) Ratio of output voltage to input voltage for an RC low-pass filter as a function of the angular frequency of the circuit.

Therefore, the ratio of the output voltage to the input voltage is

$$\frac{\Delta V_{out}}{\Delta V_{in}} = \frac{1/\omega C}{\sqrt{R^2 + \left(\dfrac{1}{\omega C}\right)^2}} \tag{33.43}$$

This ratio, plotted as a function of ω in Figure 33.23b, shows that in this case the circuit preferentially passes signals of low frequency. Hence, the circuit is called an *RC* low-pass filter.

You may be familiar with crossover networks, which are an important part of the speaker systems for high-fidelity audio systems. These networks utilize low-pass filters to direct low frequencies to a special type of speaker, the "woofer," which is designed to reproduce the low notes accurately. The high frequencies are sent to the "tweeter" speaker.

Quick Quiz 33.6

Suppose you are designing a high-fidelity system containing both large loudspeakers (woofers) and small loudspeakers (tweeters). (a) What circuit element would you place in series with a woofer, which passes low-frequency signals? (b) What circuit element would you place in series with a tweeter, which passes high-frequency signals?

SUMMARY

If an ac circuit consists of a generator and a resistor, the current is in phase with the voltage. That is, the current and voltage reach their maximum values at the same time.

The **rms current** and **rms voltage** in an ac circuit in which the voltages and current vary sinusoidally are given by the expressions

$$I_{rms} = \frac{I_{max}}{\sqrt{2}} = 0.707 I_{max} \tag{33.4}$$

$$\Delta V_{rms} = \frac{\Delta V_{max}}{\sqrt{2}} = 0.707 \Delta V_{max} \tag{33.5}$$

where I_{max} and ΔV_{max} are the maximum values.

If an ac circuit consists of a generator and an inductor, the current lags behind the voltage by 90°. That is, the voltage reaches its maximum value one quarter of a period before the current reaches its maximum value.

If an ac circuit consists of a generator and a capacitor, the current leads the voltage by 90°. That is, the current reaches its maximum value one quarter of a period before the voltage reaches its maximum value.

In ac circuits that contain inductors and capacitors, it is useful to define the **inductive reactance** X_L and the **capacitive reactance** X_C as

$$X_L = \omega L \tag{33.10}$$

$$X_C = \frac{1}{\omega C} \tag{33.17}$$

where ω is the angular frequency of the ac generator. The SI unit of reactance is the ohm.

The **impedance** Z of an RLC series ac circuit, which also has the ohm as its unit, is

$$Z \equiv \sqrt{R^2 + (X_L - X_C)^2} \qquad \text{(33.23)}$$

This expression illustrates that we cannot simply add the resistance and reactances in a circuit. We must account for the fact that the applied voltage and current are out of phase, with the **phase angle** ϕ between the current and voltage being

$$\phi = \tan^{-1}\left(\frac{X_L - X_C}{R}\right) \qquad \text{(33.25)}$$

The sign of ϕ can be positive or negative, depending on whether X_L is greater or less than X_C. The phase angle is zero when $X_L = X_C$.

The **average power** delivered by the generator in an RLC ac circuit is

$$\mathcal{P}_{av} = I_{rms} \Delta V_{rms} \cos \phi \qquad \text{(33.29)}$$

An equivalent expression for the average power is

$$\mathcal{P}_{av} = I_{rms}^2 R \qquad \text{(33.30)}$$

The average power delivered by the generator results in increasing internal energy in the resistor. No power loss occurs in an ideal inductor or capacitor.

The rms current in a series RLC circuit is

$$I_{rms} = \frac{\Delta V_{rms}}{\sqrt{R^2 + (X_L - X_C)^2}} \qquad \text{(33.32)}$$

A series RLC circuit is in resonance when the inductive reactance equals the capacitive reactance. When this condition is met, the current given by Equation 33.32 reaches its maximum value. When $X_L = X_C$ in a circuit, the **resonance frequency** ω_0 of the circuit is

$$\omega_0 = \frac{1}{\sqrt{LC}} \qquad \text{(33.33)}$$

The current in a series RLC circuit reaches its maximum value when the frequency of the generator equals ω_0—that is, when the "driving" frequency matches the resonance frequency.

Transformers allow for easy changes in alternating voltage. Because energy (and therefore power) are conserved, we can write

$$I_1 \Delta V_1 = I_2 \Delta V_2 \qquad \text{(33.40)}$$

to relate the currents and voltages in the primary and secondary windings of a transformer.

QUESTIONS

1. Fluorescent lights flicker on and off 120 times every second. Explain what causes this. Why can't you see it happening?

2. Why does a capacitor act as a short circuit at high frequencies? Why does it act as an open circuit at low frequencies?

3. Explain how the acronyms in the mnemonic "ELI the ICE man" can be used to recall whether current leads voltage or voltage leads current in RLC circuits. (Note that "E" represents voltage.)

4. Why is the sum of the maximum voltages across the elements in a series RLC circuit usually greater than the maximum applied voltage? Doesn't this violate Kirchhoff's second rule?

5. Does the phase angle depend on frequency? What is the phase angle when the inductive reactance equals the capacitive reactance?

6. Energy is delivered to a series RLC circuit by a generator. This energy appears as internal energy in the resistor. What is the source of this energy?

7. Explain why the average power delivered to an *RLC* circuit by the generator depends on the phase between the current and the applied voltage.

8. A particular experiment requires a beam of light of very stable intensity. Why would an ac voltage be unsuitable for powering the light source?

9. Consider a series *RLC* circuit in which *R* is an incandescent lamp, *C* is some fixed capacitor, and *L* is a variable inductance. The source is 120-V ac. Explain why the lamp glows brightly for some values of *L* and does not glow at all for other values.

10. What determines the maximum voltage that can be used on a transmission line?

11. Will a transformer operate if a battery is used for the input voltage across the primary? Explain.

12. How can the average value of a current be zero and yet the square root of the average squared current not be zero?

13. What is the time average of the "square-wave" voltage shown in Figure Q33.13? What is its rms voltage?

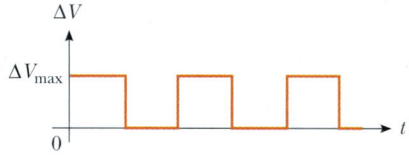

Figure Q33.13

14. Explain how the quality factor is related to the response characteristics of a radio receiver. Which variable most strongly determines the quality factor?

15. Why are the primary and secondary windings of a transformer wrapped on an iron core that passes through both coils?

16. With reference to Figure Q33.16, explain why the capacitor prevents a dc signal from passing between circuits A and B, yet allows an ac signal to pass from circuit A to circuit B. (The circuits are said to be capacitively coupled.)

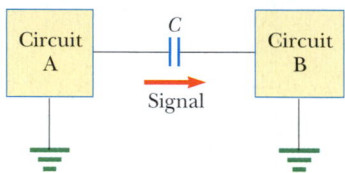

Figure Q33.16

17. With reference to Figure Q33.17, if *C* is made sufficiently large, an ac signal passes from circuit A to ground rather than from circuit A to circuit B. Hence, the capacitor acts as a filter. Explain.

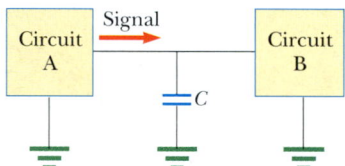

Figure Q33.17

PROBLEMS

1, 2, 3 = straightforward, intermediate, challenging　☐ = full solution available in the *Student Solutions Manual and Study Guide*
WEB = solution posted at **http://www.saunderscollege.com/physics/**　🖥 = Computer useful in solving problem　📱 = Interactive Physics
☐ = paired numerical/symbolic problems

Note: Assume that all ac voltages and currents are sinusoidal unless stated otherwise.

Section 33.1　ac Sources and Phasors

Section 33.2　Resistors in an ac Circuit

1. The rms output voltage of an ac generator is 200 V, and the operating frequency is 100 Hz. Write the equation, giving the output voltage as a function of time.

2. (a) What is the resistance of a lightbulb that uses an average power of 75.0 W when connected to a 60.0-Hz power source having a maximum voltage of 170 V? (b) What is the resistance of a 100-W bulb?

3. An ac power supply produces a maximum voltage $\Delta V_{max} = 100$ V. This power supply is connected to a 24.0-Ω resistor, and the current and resistor voltage are measured with an ideal ac ammeter and voltmeter, as shown in Figure P33.3. What does each meter read?

Note that an ideal ammeter has zero resistance and that an ideal voltmeter has infinite resistance.

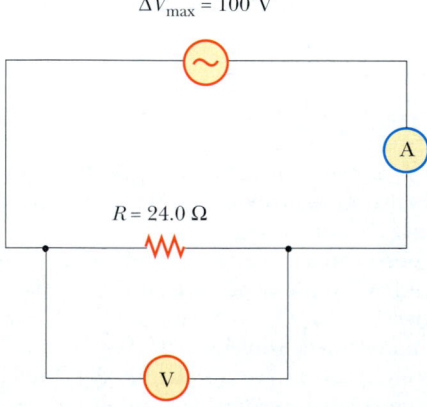

Figure P33.3

4. In the simple ac circuit shown in Figure 33.1, $R = 70.0 \, \Omega$ and $\Delta v = \Delta V_{max} \sin \omega t$. (a) If $\Delta v_R = 0.250 \, \Delta V_{max}$ for the first time at $t = 0.010 \, 0$ s, what is the angular frequency of the generator? (b) What is the next value of t for which $\Delta v_R = 0.250 \Delta V_{max}$?

5. The current in the circuit shown in Figure 33.1 equals 60.0% of the peak current at $t = 7.00$ ms. What is the smallest frequency of the generator that gives this current?

6. Figure P33.6 shows three lamps connected to a 120-V ac (rms) household supply voltage. Lamps 1 and 2 have 150-W bulbs; lamp 3 has a 100-W bulb. Find the rms current and the resistance of each bulb.

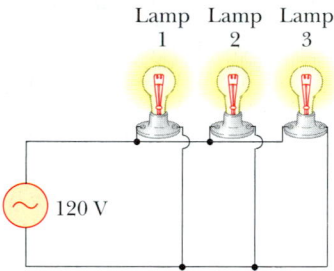

Figure P33.6

7. An audio amplifier, represented by the ac source and resistor in Figure P33.7, delivers to the speaker alternating voltage at audio frequencies. If the source voltage has an amplitude of 15.0 V, $R = 8.20 \, \Omega$, and the speaker is equivalent to a resistance of 10.4 Ω, what time-averaged power is transferred to it?

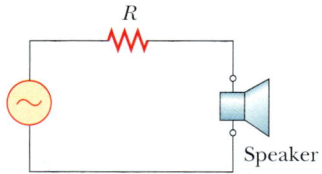

Figure P33.7

Section 33.3 Inductors in an ac Circuit

8. An inductor is connected to a 20.0-Hz power supply that produces a 50.0-V rms voltage. What inductance is needed to keep the instantaneous current in the circuit below 80.0 mA?

9. In a purely inductive ac circuit, such as that shown in Figure 33.4, $\Delta V_{max} = 100$ V. (a) If the maximum current is 7.50 A at 50.0 Hz, what is the inductance L? (b) At what angular frequency ω is the maximum current 2.50 A?

10. An inductor has a 54.0-Ω reactance at 60.0 Hz. What is the maximum current when this inductor is connected to a 50.0-Hz source that produces a 100-V rms voltage?

WEB 11. For the circuit shown in Figure 33.4, $\Delta V_{max} = 80.0$ V, $\omega = 65.0\pi$ rad/s, and $L = 70.0$ mH. Calculate the current in the inductor at $t = 15.5$ ms.

12. A 20.0-mH inductor is connected to a standard outlet ($\Delta V_{rms} = 120$ V, $f = 60.0$ Hz). Determine the energy stored in the inductor at $t = (1/180)$ s, assuming that this energy is zero at $t = 0$.

13. **Review Problem.** Determine the maximum magnetic flux through an inductor connected to a standard outlet ($\Delta V_{rms} = 120$ V, $f = 60.0$ Hz).

Section 33.4 Capacitors in an ac Circuit

14. (a) For what frequencies does a 22.0-μF capacitor have a reactance below 175 Ω? (b) Over this same frequency range, what is the reactance of a 44.0-μF capacitor?

15. What maximum current is delivered by a 2.20-μF capacitor when it is connected across (a) a North American outlet having $\Delta V_{rms} = 120$ V and $f = 60.0$ Hz? (b) a European outlet having $\Delta V_{rms} = 240$ V and $f = 50.0$ Hz?

16. A capacitor C is connected to a power supply that operates at a frequency f and produces an rms voltage ΔV. What is the maximum charge that appears on either of the capacitor plates?

17. What maximum current is delivered by an ac generator with $\Delta V_{max} = 48.0$ V and $f = 90.0$ Hz when it is connected across a 3.70-μF capacitor?

18. A 1.00-mF capacitor is connected to a standard outlet ($\Delta V_{rms} = 120$ V, $f = 60.0$ Hz). Determine the current in the capacitor at $t = (1/180)$ s, assuming that at $t = 0$ the energy stored in the capacitor is zero.

Section 33.5 The *RLC* Series Circuit

19. An inductor ($L = 400$ mH), a capacitor ($C = 4.43 \, \mu$F), and a resistor ($R = 500 \, \Omega$) are connected in series. A 50.0-Hz ac generator produces a peak current of 250 mA in the circuit. (a) Calculate the required peak voltage ΔV_{max}. (b) Determine the phase angle by which the current leads or lags the applied voltage.

20. At what frequency does the inductive reactance of a 57.0-μH inductor equal the capacitive reactance of a 57.0-μF capacitor?

21. A series ac circuit contains the following components: $R = 150 \, \Omega$, $L = 250$ mH, $C = 2.00 \, \mu$F, and a generator with $\Delta V_{max} = 210$ V operating at 50.0 Hz. Calculate the (a) inductive reactance, (b) capacitive reactance, (c) impedance, (d) maximum current, and (e) phase angle between current and generator voltage.

22. A sinusoidal voltage $\Delta v(t) = (40.0 \text{ V}) \sin(100t)$ is applied to a series *RLC* circuit with $L = 160$ mH, $C = 99.0 \, \mu$F, and $R = 68.0 \, \Omega$. (a) What is the impedance of the circuit? (b) What is the maximum current? (c) Determine the numerical values for I_{max}, ω, and ϕ in the equation $i(t) = I_{max} \sin(\omega t - \phi)$.

WEB 23. An *RLC* circuit consists of a 150-Ω resistor, a 21.0-μF capacitor, and a 460-mH inductor, connected in series with a 120-V, 60.0-Hz power supply. (a) What is the phase angle between the current and the applied voltage? (b) Which reaches its maximum earlier, the current or the voltage?

24. A person is working near the secondary of a transformer, as shown in Figure P33.24. The primary voltage is 120 V at 60.0 Hz. The capacitance C_s, which is the stray capacitance between the person's hand and the secondary winding, is 20.0 pF. Assuming that the person has a body resistance to ground $R_b = 50.0$ kΩ, determine the rms voltage across the body. (*Hint:* Redraw the circuit with the secondary of the transformer as a simple ac source.)

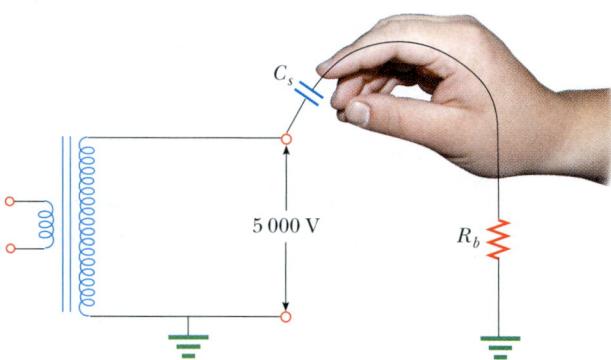

Figure P33.24

25. An ac source with $\Delta V_{max} = 150$ V and $f = 50.0$ Hz is connected between points a and d in Figure P33.25. Calculate the maximum voltages between points (a) a and b, (b) b and c, (c) c and d, and (d) b and d.

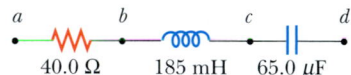

Figure P33.25   Problems 25 and 64.

26. Draw to scale a phasor diagram showing Z, X_L, X_C, and ϕ for an ac series circuit for which $R = 300$ Ω, $C = 11.0$ μF, $L = 0.200$ H, and $f = (500/\pi)$ Hz.

27. A coil of resistance 35.0 Ω and inductance 20.5 H is in series with a capacitor and a 200-V (rms), 100-Hz source. The rms current in the circuit is 4.00 A. (a) Calculate the capacitance in the circuit. (b) What is ΔV_{rms} across the coil?

Section 33.6   Power in an ac Circuit

28. The voltage source in Figure P33.28 has an output $\Delta V_{rms} = 100$ V at $\omega = 1\,000$ rad/s. Determine (a) the current in the circuit and (b) the power supplied by the source. (c) Show that the power delivered to the resistor is equal to the power supplied by the source.

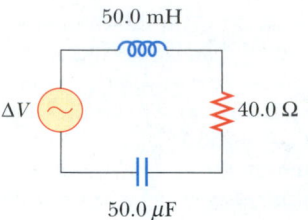

Figure P33.28

WEB **29.** An ac voltage of the form $\Delta v = (100$ V$) \sin(1\,000 t)$ is applied to a series RLC circuit. If $R = 400$ Ω, $C = 5.00$ μF, and $L = 0.500$ H, what is the average power delivered to the circuit?

30. A series RLC circuit has a resistance of 45.0 Ω and an impedance of 75.0 Ω. What average power is delivered to this circuit when $\Delta V_{rms} = 210$ V?

31. In a certain series RLC circuit, $I_{rms} = 9.00$ A, $\Delta V_{rms} = 180$ V, and the current leads the voltage by 37.0°. (a) What is the total resistance of the circuit? (b) What is the reactance of the circuit $(X_L - X_C)$?

32. Suppose you manage a factory that uses many electric motors. The motors create a large inductive load to the electric power line, as well as a resistive load. The electric company builds an extra-heavy distribution line to supply you with a component of current that is 90° out of phase with the voltage, as well as with current in phase with the voltage. The electric company charges you an extra fee for "reactive volt-amps" in addition to the amount you pay for the energy you use. You can avoid the extra fee by installing a capacitor between the power line and your factory. The following problem models this solution.

In an LR circuit, a 120-V (rms), 60.0-Hz source is in series with a 25.0-mH inductor and a 20.0-Ω resistor. What are (a) the rms current and (b) the power factor? (c) What capacitor must be added in series to make the power factor 1? (d) To what value can the supply voltage be reduced if the power supplied is to be the same as that provided before installation of the capacitor?

33. **Review Problem.**   Over a distance of 100 km, power of 100 MW is to be transmitted at 50.0 kV with only 1.00% loss. Copper wire of what diameter should be used for each of the two conductors of the transmission line? Assume that the current density in the conductors is uniform.

34. **Review Problem.**   Suppose power $\mathcal{P}$ is to be transmitted over a distance d at a voltage ΔV, with only 1.00% loss. Copper wire of what diameter should be used for each of the two conductors of the transmission line? Assume that the current density in the conductors is uniform.

35. A diode is a device that allows current to pass in only one direction (the direction indicated by the arrowhead in its circuit-diagram symbol). Find, in terms of ΔV and

R, the average power delivered to the diode circuit shown in Figure P33.35.

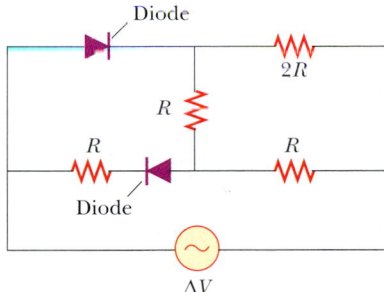

Figure P33.35

Section 33.7 Resonance in a Series *RLC* Circuit

36. The tuning circuit of an AM radio contains an *LC* combination. The inductance is 0.200 mH, and the capacitor is variable, so the circuit can resonate at any frequency between 550 kHz and 1 650 kHz. Find the range of values required for *C*.

37. An *RLC* circuit is used in a radio to tune in to an FM station broadcasting at 99.7 MHz. The resistance in the circuit is 12.0 Ω, and the inductance is 1.40 μH. What capacitance should be used?

38. A series *RLC* circuit has the following values: $L = 20.0$ mH, $C = 100$ nF, $R = 20.0\ \Omega$, and $\Delta V_{max} = 100$ V, with $\Delta v = \Delta V_{max} \sin \omega t$. Find (a) the resonant frequency, (b) the amplitude of the current at the resonant frequency, (c) the *Q* of the circuit, and (d) the amplitude of the voltage across the inductor at resonance.

39. A 10.0-Ω resistor, a 10.0-mH inductor, and a 100-μF capacitor are connected in series to a 50.0-V (rms) source having variable frequency. What is the energy delivered to the circuit during one period if the operating frequency is twice the resonance frequency?

40. A resistor *R*, an inductor *L*, and a capacitor *C* are connected in series to an ac source of rms voltage ΔV and variable frequency. What is the energy delivered to the circuit during one period if the operating frequency is twice the resonance frequency?

41. Compute the quality factor for the circuits described in Problems 22 and 23. Which circuit has the sharper resonance?

Section 33.8 The Transformer and Power Transmission

42. A step-down transformer is used for recharging the batteries of portable devices such as tape players. The turns ratio inside the transformer is 13:1, and it is used with 120-V (rms) household service. If a particular ideal transformer draws 0.350 A from the house outlet, what (a) voltage and (b) current are supplied to a tape player from the transformer? (c) How much power is delivered?

43. A transformer has $N_1 = 350$ turns and $N_2 = 2\ 000$ turns. If the input voltage is $\Delta v(t) = (170\ \text{V})\cos \omega t$, what rms voltage is developed across the secondary coil?

44. A step-up transformer is designed to have an output voltage of 2 200 V (rms) when the primary is connected across a 110-V (rms) source. (a) If there are 80 turns on the primary winding, how many turns are required on the secondary? (b) If a load resistor across the secondary draws a current of 1.50 A, what is the current in the primary under ideal conditions? (c) If the transformer actually has an efficiency of 95.0%, what is the current in the primary when the secondary current is 1.20 A?

45. In the transformer shown in Figure P33.45, the load resistor is 50.0 Ω. The turns ratio $N_1 : N_2$ is 5:2, and the source voltage is 80.0 V (rms). If a voltmeter across the load measures 25.0 V (rms), what is the source resistance R_s?

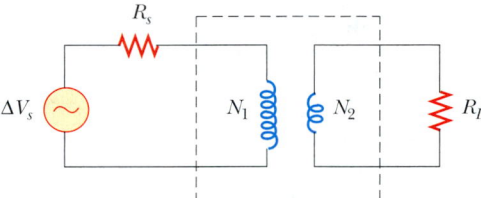

Figure P33.45

46. The secondary voltage of an ignition transformer in a furnace is 10.0 kV. When the primary operates at an rms voltage of 120 V, the primary impedance is 24.0 Ω and the transformer is 90.0% efficient. (a) What turns ratio is required? What are (b) the current in the secondary and (c) the impedance in the secondary?

47. A transmission line that has a resistance per unit length of $4.50 \times 10^{-4}\ \Omega/\text{m}$ is to be used to transmit 5.00 MW over 400 mi (6.44×10^5 m). The output voltage of the generator is 4.50 kV. (a) What is the power loss if a transformer is used to step up the voltage to 500 kV? (b) What fraction of the input power is lost to the line under these circumstances? (c) What difficulties would be encountered on attempting to transmit the 5.00 MW at the generator voltage of 4.50 kV?

(Optional)
Section 33.9 Rectifiers and Filters

48. The *RC* low-pass filter shown in Figure 33.23 has a resistance $R = 90.0\ \Omega$ and a capacitance $C = 8.00$ nF. Calculate the gain ($\Delta V_{out}/\Delta V_{in}$) for input frequencies of (a) 600 Hz and (b) 600 kHz.

WEB 49. The *RC* high-pass filter shown in Figure 33.22 has a resistance $R = 0.500\ \Omega$. (a) What capacitance gives an output signal that has one-half the amplitude of a 300-Hz input signal? (b) What is the gain ($\Delta V_{out}/\Delta V_{in}$) for a 600-Hz signal?

50. The circuit in Figure P33.50 represents a high-pass filter in which the inductor has internal resistance. What is the source frequency if the output voltage ΔV_2 is one-half the input voltage?

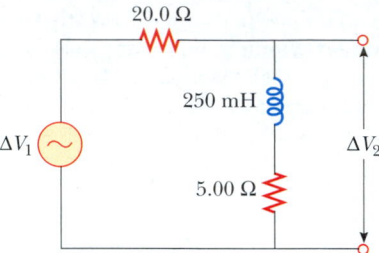

Figure P33.50

51. The resistor in Figure P33.51 represents the midrange speaker in a three-speaker system. Assume that its resistance is constant at 8.00 Ω. The source represents an audio amplifier producing signals of uniform amplitude $\Delta V_{in} = 10.0$ V at all audio frequencies. The inductor and capacitor are to function as a bandpass filter with $\Delta V_{out}/\Delta V_{in} = \frac{1}{2}$ at 200 Hz and at 4 000 Hz. (a) Determine the required values of L and C. (b) Find the maximum value of the gain ratio $\Delta V_{out}/\Delta V_{in}$. (c) Find the frequency f_0 at which the gain ratio has its maximum value. (d) Find the phase shift between ΔV_{in} and ΔV_{out} at 200 Hz, at f_0, and at 4 000 Hz. (e) Find the average power transferred to the speaker at 200 Hz, at f_0, and at 4 000 Hz. (f) Treating the filter as a resonant circuit, find its quality factor.

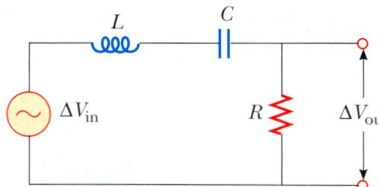

Figure P33.51

52. Show that two successive high-pass filters having the same values of R and C give a combined gain

$$\frac{\Delta V_{out}}{\Delta V_{in}} = \frac{1}{1 + (1/\omega RC)^2}$$

53. Consider a low-pass filter followed by a high-pass filter, as shown in Figure P33.53. If $R = 1\,000$ Ω and $C = 0.050\,0\ \mu$F, determine $\Delta V_{out}/\Delta V_{in}$ for a 2.00-kHz input frequency.

ADDITIONAL PROBLEMS

54. Show that the rms value for the sawtooth voltage shown in Figure P33.54 is $\Delta V_{max}/\sqrt{3}$.

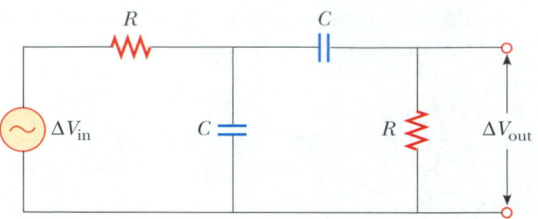

Figure P33.53

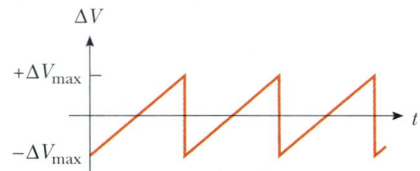

Figure P33.54

WEB **55.** A series RLC circuit consists of an 8.00-Ω resistor, a 5.00-μF capacitor, and a 50.0-mH inductor. A variable frequency source applies an emf of 400 V (rms) across the combination. Determine the power delivered to the circuit when the frequency is equal to one-half the resonance frequency.

56. To determine the inductance of a coil used in a research project, a student first connects the coil to a 12.0-V battery and measures a current of 0.630 A. The student then connects the coil to a 24.0-V (rms), 60.0-Hz generator and measures an rms current of 0.570 A. What is the inductance?

57. In Figure P33.57, find the current delivered by the 45.0-V (rms) power supply (a) when the frequency is very large and (b) when the frequency is very small.

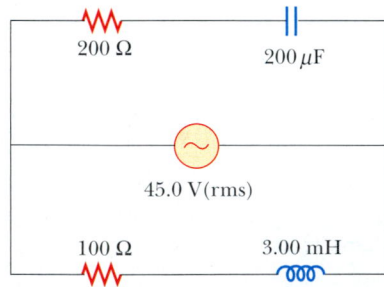

Figure P33.57

58. In the circuit shown in Figure P33.58, assume that all parameters except C are given. (a) Find the current as a function of time. (b) Find the power delivered to the circuit. (c) Find the current as a function of time after *only* switch 1 is opened. (d) After switch 2 is *also* opened, the current and voltage are in phase. Find the capacitance C. (e) Find the impedance of the circuit when both switches are open. (f) Find the maximum energy stored in the capacitor during oscillations. (g) Find the maximum energy stored in the inductor during oscillations. (h) Now the frequency of the voltage source is doubled. Find the phase difference between the current and the voltage. (i) Find the frequency that makes the inductive reactance one-half the capacitive reactance.

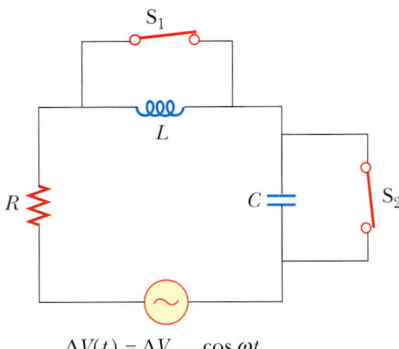

$\Delta V(t) = \Delta V_{max} \cos \omega t$

Figure P33.58

59. As an alternative to the *RC* filters described in Section 33.9, *LC* filters are used as both high- and low-pass filters. However, all real inductors have resistance, as indicated in Figure P33.59, and this resistance must be taken into account. (a) Determine which circuit in Figure P33.59 is the high-pass filter and which is the low-pass filter. (b) Derive the output/input ratio for each

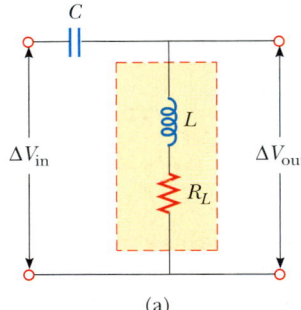

(a)

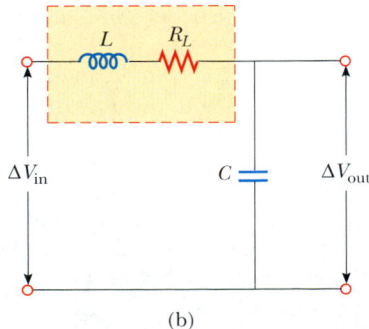

(b)

Figure P33.59

circuit, following the procedure used for the *RC* filters in Section 33.9.

60. An 80.0-Ω resistor and a 200-mH inductor are connected in parallel across a 100-V (rms), 60.0-Hz source. (a) What is the rms current in the resistor? (b) By what angle does the total current lead or lag behind the voltage?

61. Make an order-of-magnitude estimate of the electric current that the electric company delivers to a town from a remote generating station. State the data that you measure or estimate. If you wish, you may consider a suburban bedroom community of 20 000 people.

62. A voltage $\Delta v = (100 \text{ V}) \sin \omega t$ (in SI units) is applied across a series combination of a 2.00-H inductor, a 10.0-μF capacitor, and a 10.0-Ω resistor. (a) Determine the angular frequency ω_0 at which the power delivered to the resistor is a maximum. (b) Calculate the power at that frequency. (c) Determine the two angular frequencies ω_1 and ω_2 at which the power delivered is one-half the maximum value. [The Q of the circuit is approximately $\omega_0/(\omega_2 - \omega_1)$.]

63. Consider a series *RLC* circuit having the following circuit parameters: $R = 200 \ \Omega$, $L = 663$ mH, and $C = 26.5 \ \mu$F. The applied voltage has an amplitude of 50.0 V and a frequency of 60.0 Hz. Find the following: (a) the current I_{max}, including its phase constant ϕ relative to the applied voltage Δv; (b) the voltage ΔV_R across the resistor and its phase relative to the current; (c) the voltage ΔV_C across the capacitor and its phase relative to the current; and (d) the voltage ΔV_L across the inductor and its phase relative to the current.

64. A power supply with $\Delta V_{rms} = 120$ V is connected between points a and d in Figure P33.25. At what frequency will it deliver a power of 250 W?

65. Example 28.2 showed that maximum power is transferred when the internal resistance of a dc source is equal to the resistance of the load. A transformer may be used to provide maximum power transfer between two ac circuits that have different impedances. (a) Show that the ratio of turns N_1/N_2 needed to meet this condition is

$$\frac{N_1}{N_2} = \sqrt{\frac{Z_1}{Z_2}}$$

(b) Suppose you want to use a transformer as an impedance-matching device between an audio amplifier that has an output impedance of 8.00 kΩ and a speaker that has an input impedance of 8.00 Ω. What should your N_1/N_2 ratio be?

66. Figure P33.66a shows a parallel *RLC* circuit, and the corresponding phasor diagram is provided in Figure P33.66b. The instantaneous voltages and rms voltages across the three circuit elements are the same, and each is in phase with the current through the resistor. The currents in *C* and *L* lead or lag behind the current in the resistor, as shown in Figure P33.66b. (a) Show that the rms current delivered by the source is

$$I_{rms} = \Delta V_{rms} \left[\frac{1}{R^2} + \left(\omega C - \frac{1}{\omega L} \right)^2 \right]^{1/2}$$

(b) Show that the phase angle ϕ between ΔV_{rms} and I_{rms} is

$$\tan \phi = R \left(\frac{1}{X_C} - \frac{1}{X_L} \right)$$

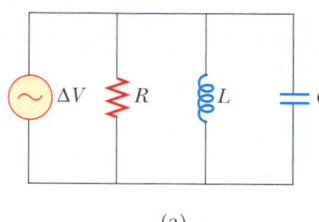

(a)

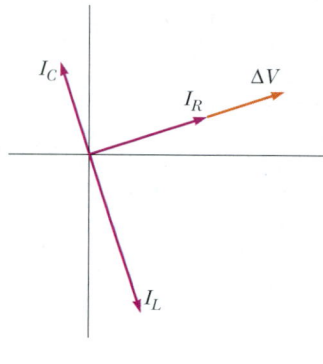

(b)

Figure P33.66

67. An 80.0-Ω resistor, a 200-mH inductor, and a 0.150-μF capacitor are connected *in parallel* across a 120-V (rms) source operating at 374 rad/s. (a) What is the resonant frequency of the circuit? (b) Calculate the rms current in the resistor, the inductor, and the capacitor. (c) What

rms current is delivered by the source? (d) Is the current leading or lagging behind the voltage? By what angle?

68. Consider the phase-shifter circuit shown in Figure P33.68. The input voltage is described by the expression $\Delta v = (10.0 \text{ V}) \sin 200t$ (in SI units). Assuming that $L = 500$ mH, find (a) the value of R such that the output voltage lags behind the input voltage by 30.0° and (b) the amplitude of the output voltage.

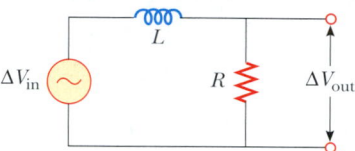

Figure P33.68

69. A series RLC circuit is operating at 2 000 Hz. At this frequency, $X_L = X_C = 1 \ 884 \ \Omega$. The resistance of the circuit is 40.0 Ω. (a) Prepare a table showing the values of X_L, X_C, and Z for $f = 300, 600, 800, 1 \ 000, 1 \ 500, 2 \ 000, 3 \ 000, 4 \ 000, 6 \ 000$, and 10 000 Hz. (b) Plot on the same set of axes X_L, X_C, and Z as functions of $\ln f$.

70. A series RLC circuit in which $R = 1.00 \ \Omega$, $L = 1.00$ mH, and $C = 1.00$ nF is connected to an ac generator delivering 1.00 V (rms). Make a precise graph of the power delivered to the circuit as a function of the frequency, and verify that the full width of the resonance peak at half-maximum is $R/2\pi L$.

71. Suppose the high-pass filter shown in Figure 33.22 has $R = 1 \ 000 \ \Omega$ and $C = 0.050 \ 0 \ \mu$F. (a) At what frequency does $\Delta V_{out} / \Delta V_{in} = \frac{1}{2}$? (b) Plot $\log_{10}(\Delta V_{out} / \Delta V_{in})$ versus $\log_{10}(f)$ over the frequency range from 1 Hz to 1 MHz. (This log–log plot of gain versus frequency is known as a *Bode plot*.)

ANSWERS TO QUICK QUIZZES

33.1 (c) $\mathcal{P}_{av} > 0$ and $i_{av} = 0$. The average power is proportional to the rms current—which, as Figure 33.3 shows, is nonzero even though the average current is zero. Condition (a) is valid only for an open circuit, and conditions (b) and (d) can never be true because $i_{av} = 0$ for ac circuits even though $i_{rms} > 0$.

33.2 (b) Sum of instantaneous voltages across elements. Choices (a) and (c) are incorrect because the unaligned sine curves in Figure 33.9b mean that the voltages are out of phase, so we cannot simply add the maximum (or rms) voltages across the elements. (In other words, $\Delta V \neq \Delta V_R + \Delta V_L + \Delta V_C$ even though it is true that $\Delta v = \Delta v_R + \Delta v_L + \Delta v_C$.)

33.3 (a) $X_L < X_C$. (b) $X_L = X_C$. (c) $X_L > X_C$.

33.4 Equation 33.23 indicates that at resonance (when $X_L = X_C$) the impedance is due strictly to the resistor, $Z = R$. At resonance, the current is given by the expression $I_{rms} = \Delta V_{rms}/R$.

33.5 High. The higher the quality factor, the more sensitive the detector. As you can see from Figure 33.15a, when $Q = \omega_0/\Delta\omega$ is high, as it is in the $R = 3.5 \ \Omega$ case, a slight change in the resonance frequency (as might happen when a small piece of metal passes through the portal) causes a large change in current that can be detected easily.

33.6 (a) An inductor. The current in an inductive circuit decreases with increasing frequency (see Eq. 33.9). Thus, an inductor connected in series with a woofer blocks high-frequency signals and passes low-frequency signals. (b) A capacitor. The current in a capacitive circuit decreases with decreasing frequency (see Eq. 33.16). When a capacitor is connected in series with a tweeter, the capacitor blocks low-frequency signals and passes high-frequency signals.

c h a p t e r

34

Electromagnetic Waves

he waves described in Chapters 16, 17, and 18 are mechanical waves. By definition, the propagation of mechanical disturbances—such as sound waves, water waves, and waves on a string—requires the presence of a medium. This chapter is concerned with the properties of electromagnetic waves, which (unlike mechanical waves) can propagate through empty space.

In Section 31.7 we gave a brief description of Maxwell's equations, which form the theoretical basis of all electromagnetic phenomena. The consequences of Maxwell's equations are far-reaching and dramatic. The Ampère–Maxwell law predicts that a time-varying electric field produces a magnetic field, just as Faraday's law tells us that a time-varying magnetic field produces an electric field. Maxwell's introduction of the concept of displacement current as a new source of a magnetic field provided the final important link between electric and magnetic fields in classical physics.

Astonishingly, Maxwell's equations also predict the existence of electromagnetic waves that propagate through space at the speed of light c. This chapter begins with a discussion of how Heinrich Hertz confirmed Maxwell's prediction when he generated and detected electromagnetic waves in 1887. That discovery has led to many practical communication systems, including radio, television, and radar. On a conceptual level, Maxwell unified the subjects of light and electromagnetism by developing the idea that light is a form of electromagnetic radiation.

Next, we learn how electromagnetic waves are generated by oscillating electric charges. The waves consist of oscillating electric and magnetic fields that are at right angles to each other and to the direction of wave propagation. Thus, electromagnetic waves are transverse waves. Maxwell's prediction of electromagnetic radiation shows that the amplitudes of the electric and magnetic fields in an electromagnetic wave are related by the expression $E = cB$. The waves radiated from the oscillating charges can be detected at great distances. Furthermore, electromagnetic waves carry energy and momentum and hence can exert pressure on a surface.

The chapter concludes with a look at the wide range of frequencies covered by electromagnetic waves. For example, radio waves (frequencies of about 10^7 Hz) are electromagnetic waves produced by oscillating currents in a radio tower's transmitting antenna. Light waves are a high-frequency form of electromagnetic radiation (about 10^{14} Hz) produced by oscillating electrons in atoms.

James Clerk Maxwell Scottish theoretical physicist (1831–1879)
Maxwell developed the electromagnetic theory of light and the kinetic theory of gases, and he explained the nature of color vision and of Saturn's rings. His successful interpretation of the electromagnetic field produced the field equations that bear his name. Formidable mathematical ability combined with great insight enabled Maxwell to lead the way in the study of electromagnetism and kinetic theory. He died of cancer before he was 50. *(North Wind Picture Archives)*

34.1 ▷ MAXWELL'S EQUATIONS AND HERTZ'S DISCOVERIES

In his unified theory of electromagnetism, Maxwell showed that electromagnetic waves are a natural consequence of the fundamental laws expressed in the following four equations (see Section 31.7):

$$\oint_S \mathbf{E} \cdot d\mathbf{A} = \frac{Q}{\epsilon_0} \tag{34.1}$$

$$\oint_S \mathbf{B} \cdot d\mathbf{A} = 0 \tag{34.2}$$

$$\oint \mathbf{E} \cdot d\mathbf{s} = -\frac{d\Phi_B}{dt} \tag{34.3}$$

$$\oint \mathbf{B} \cdot d\mathbf{s} = \mu_0 I + \mu_0 \epsilon_0 \frac{d\Phi_E}{dt} \tag{34.4}$$

As we shall see in the next section, Equations 34.3 and 34.4 can be combined to obtain a wave equation for both the electric field and the magnetic field. In empty space ($Q = 0$, $I = 0$), the solution to these two equations shows that the speed at which electromagnetic waves travel equals the measured speed of light. This result led Maxwell to predict that light waves are a form of electromagnetic radiation.

The experimental apparatus that Hertz used to generate and detect electromagnetic waves is shown schematically in Figure 34.1. An induction coil is connected to a transmitter made up of two spherical electrodes separated by a narrow gap. The coil provides short voltage surges to the electrodes, making one positive and the other negative. A spark is generated between the spheres when the electric field near either electrode surpasses the dielectric strength for air (3×10^6 V/m; see Table 26.1). In a strong electric field, the acceleration of free electrons provides them with enough energy to ionize any molecules they strike. This ionization provides more electrons, which can accelerate and cause further ionizations. As the air in the gap is ionized, it becomes a much better conductor, and the discharge between the electrodes exhibits an oscillatory behavior at a very high frequency. From an electric-circuit viewpoint, this is equivalent to an LC circuit in which the inductance is that of the coil and the capacitance is due to the spherical electrodes.

Because L and C are quite small in Hertz's apparatus, the frequency of oscillation is very high, ≈ 100 MHz. (Recall from Eq. 32.22 that $\omega = 1/\sqrt{LC}$ for an LC circuit.) Electromagnetic waves are radiated at this frequency as a result of the oscillation (and hence acceleration) of free charges in the transmitter circuit. Hertz was able to detect these waves by using a single loop of wire with its own spark gap (the receiver). Such a receiver loop, placed several meters from the transmitter, has its own effective inductance, capacitance, and natural frequency of oscillation. In Hertz's experiment, sparks were induced across the gap of the receiving electrodes when the frequency of the receiver was adjusted to match that of the transmitter. Thus, Hertz demonstrated that the oscillating current induced in the receiver was produced by electromagnetic waves radiated by the transmitter. His experiment is analogous to the mechanical phenomenon in which a tuning fork responds to acoustic vibrations from an identical tuning fork that is oscillating.

Heinrich Rudolf Hertz German physicist (1857–1894)   Hertz made his most important discovery—radio waves—in 1887. After finding that the speed of a radio wave was the same as that of light, he showed that radio waves, like light waves, could be reflected, refracted, and diffracted. Hertz died of blood poisoning at age 36. He made many contributions to science during his short life. The hertz, equal to one complete vibration or cycle per second, is named after him.   *(The Bettmann Archive)*

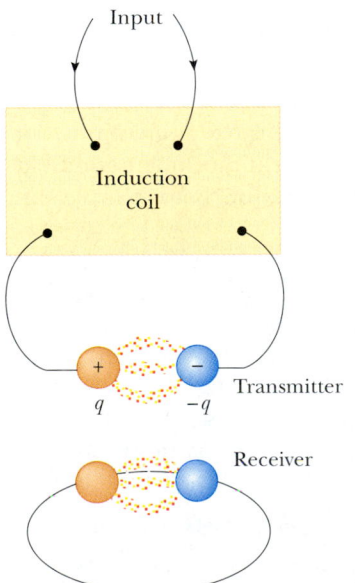

Figure 34.1   Schematic diagram of Hertz's apparatus for generating and detecting electromagnetic waves. The transmitter consists of two spherical electrodes connected to an induction coil, which provides short voltage surges to the spheres, setting up oscillations in the discharge between the electrodes (suggested by the red dots). The receiver is a nearby loop of wire containing a second spark gap.

A large oscillator (bottom) and circular, octagonal, and square receivers used by Heinrich Hertz. *(Photo Deutsches Museum Munich)*

Additionally, Hertz showed in a series of experiments that the radiation generated by his spark-gap device exhibited the wave properties of interference, diffraction, reflection, refraction, and polarization, all of which are properties exhibited by light. Thus, it became evident that the radio-frequency waves Hertz was generating had properties similar to those of light waves and differed only in frequency and wavelength. Perhaps his most convincing experiment was the measurement of the speed of this radiation. Radio-frequency waves of known frequency were reflected from a metal sheet and created a standing-wave interference pattern whose nodal points could be detected. The measured distance between the nodal points enabled determination of the wavelength λ. Using the relationship $v = \lambda f$ (Eq. 16.14), Hertz found that v was close to 3×10^8 m/s, the known speed c of visible light.

34.2 PLANE ELECTROMAGNETIC WAVES

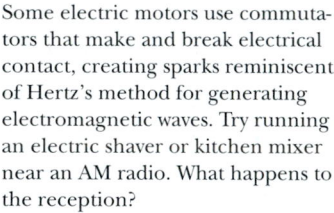

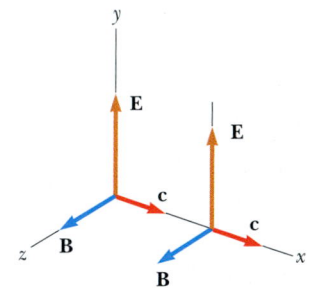

Figure 34.2 An electromagnetic wave traveling at velocity **c** in the positive x direction. The electric field is along the y direction, and the magnetic field is along the z direction. These fields depend only on x and t.

The properties of electromagnetic waves can be deduced from Maxwell's equations. One approach to deriving these properties is to solve the second-order differential equation obtained from Maxwell's third and fourth equations. A rigorous mathematical treatment of that sort is beyond the scope of this text. To circumvent this problem, we assume that the vectors for the electric field and magnetic field in an electromagnetic wave have a specific space–time behavior that is simple but consistent with Maxwell's equations.

To understand the prediction of electromagnetic waves more fully, let us focus our attention on an electromagnetic wave that travels in the x direction (the *direction of propagation*). In this wave, the electric field **E** is in the y direction, and the magnetic field **B** is in the z direction, as shown in Figure 34.2. Waves such as this one, in which the electric and magnetic fields are restricted to being parallel to a pair of perpendicular axes, are said to be **linearly polarized waves.**[1] Furthermore, we assume that at any point P, the magnitudes E and B of the fields depend

---

[1] Waves having other particular patterns of vibration of the electric and magnetic fields include circularly polarized waves. The most general polarization pattern is elliptical.

upon x and t only, and not upon the y or z coordinate. A collection of such waves from individual sources is called a **plane wave.** A surface connecting points of equal phase on all waves, which we call a **wave front,** would be a geometric plane. In comparison, a point source of radiation sends waves out in all directions. A surface connecting points of equal phase is a sphere for this situation, so we call this a **spherical wave.**

We can relate E and B to each other with Equations 34.3 and 34.4. In empty space, where $Q = 0$ and $I = 0$, Equation 34.3 remains unchanged and Equation 34.4 becomes

$$\oint \mathbf{B} \cdot d\mathbf{s} = \mu_0 \epsilon_0 \frac{d\Phi_E}{dt} \tag{34.5}$$

Using Equations 34.3 and 34.5 and the plane-wave assumption, we obtain the following differential equations relating E and B. (We shall derive these equations formally later in this section.) For simplicity, we drop the subscripts on the components E_y and B_z:

$$\frac{\partial E}{\partial x} = -\frac{\partial B}{\partial t} \tag{34.6}$$

$$\frac{\partial B}{\partial x} = -\mu_0 \epsilon_0 \frac{\partial E}{\partial t} \tag{34.7}$$

Note that the derivatives here are partial derivatives. For example, when we evaluate $\partial E/\partial x$, we assume that t is constant. Likewise, when we evaluate $\partial B/\partial t$, x is held constant. Taking the derivative of Equation 34.6 with respect to x and combining the result with Equation 34.7, we obtain

$$\frac{\partial^2 E}{\partial x^2} = -\frac{\partial}{\partial x}\left(\frac{\partial B}{\partial t}\right) = -\frac{\partial}{\partial t}\left(\frac{\partial B}{\partial x}\right) = -\frac{\partial}{\partial t}\left(-\mu_0 \epsilon_0 \frac{\partial E}{\partial t}\right)$$

$$\frac{\partial^2 E}{\partial x^2} = \mu_0 \epsilon_0 \frac{\partial^2 E}{\partial t^2} \tag{34.8}$$

In the same manner, taking the derivative of Equation 34.7 with respect to x and combining it with Equation 34.6, we obtain

$$\frac{\partial^2 B}{\partial x^2} = \mu_0 \epsilon_0 \frac{\partial^2 B}{\partial t^2} \tag{34.9}$$

Equations 34.8 and 34.9 both have the form of the general wave equation[2] with the wave speed v replaced by c, where

$$c = \frac{1}{\sqrt{\mu_0 \epsilon_0}} \tag{34.10}$$

Speed of electromagnetic waves

Taking $\mu_0 = 4\pi \times 10^{-7}$ T·m/A and $\epsilon_0 = 8.854\ 19 \times 10^{-12}$ C^2/N·m^2 in Equation 34.10, we find that $c = 2.997\ 92 \times 10^8$ m/s. Because this speed is precisely the same as the speed of light in empty space, we are led to believe (correctly) that light is an electromagnetic wave.

---

[2] The general wave equation is of the form $(\partial^2 y/\partial x^2) = (1/v^2)(\partial^2 y/\partial t^2)$, where v is the speed of the wave and y is the wave function. The general wave equation was introduced as Equation 16.26, and it would be useful for you to review Section 16.9.

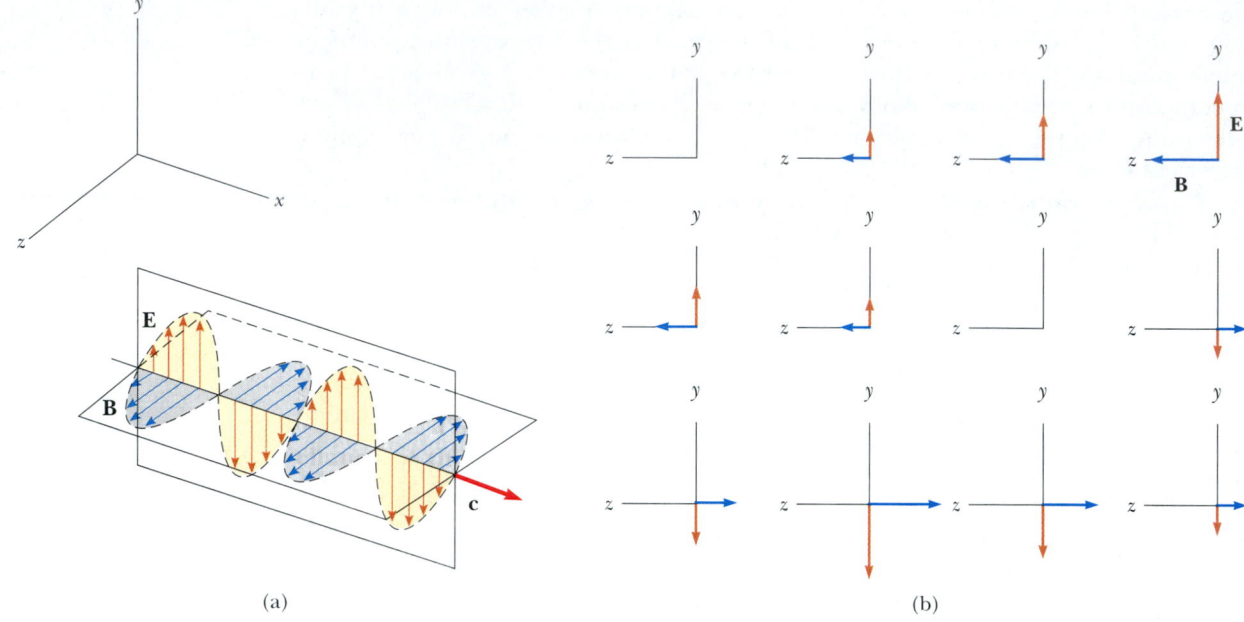

(a)   (b)

Figure 34.3   Representation of a sinusoidal, linearly polarized plane electromagnetic wave moving in the positive x direction with velocity **c**. (a) The wave at some instant. Note the sinusoidal variations of E and B with x. (b) A time sequence illustrating the electric and magnetic field vectors present in the yz plane, as seen by an observer looking in the negative x direction. Note the sinusoidal variations of E and B with t.

The simplest solution to Equations 34.8 and 34.9 is a sinusoidal wave, for which the field amplitudes E and B vary with x and t according to the expressions

Sinusoidal electric and magnetic fields	

$$E = E_{\text{max}} \cos(kx - \omega t) \tag{34.11}$$

$$B = B_{\text{max}} \cos(kx - \omega t) \tag{34.12}$$

where E_{max} and B_{max} are the maximum values of the fields. The angular wave number is the constant $k = 2\pi/\lambda$, where λ is the wavelength. The angular frequency is $\omega = 2\pi f$, where f is the wave frequency. The ratio ω/k equals the speed c:

$$\frac{\omega}{k} = \frac{2\pi f}{2\pi/\lambda} = \lambda f = c$$

We have used Equation 16.14, $v = c = \lambda f$, which relates the speed, frequency, and wavelength of any continuous wave. Figure 34.3a is a pictorial representation, at one instant, of a sinusoidal, linearly polarized plane wave moving in the positive x direction. Figure 34.3b shows how the electric and magnetic field vectors at a fixed location vary with time.

Instructor's Note • Students need to be reminded of the meaning of *phase angle*. The fact that the **E** and **B** vectors are perpendicular can confuse them.

Quick Quiz 34.1

What is the phase difference between B and E in Figure 34.3?

Taking partial derivatives of Equations 34.11 (with respect to x) and 34.12 (with respect to t), we find that

$$\frac{\partial E}{\partial x} = -kE_{max}\sin(kx - \omega t)$$

$$\frac{\partial B}{\partial t} = \omega B_{max}\sin(kx - \omega t)$$

Substituting these results into Equation 34.6, we find that at any instant

$$kE_{max} = \omega B_{max}$$

$$\frac{E_{max}}{B_{max}} = \frac{\omega}{k} = c$$

Using these results together with Equations 34.11 and 34.12, we see that

$$\frac{E_{max}}{B_{max}} = \frac{E}{B} = c \qquad (34.13)$$

That is, **at every instant the ratio of the magnitude of the electric field to the magnitude of the magnetic field in an electromagnetic wave equals the speed of light.**

Finally, note that electromagnetic waves obey the superposition principle (which we discussed in Section 16.4 with respect to mechanical waves) because the differential equations involving E and B are linear equations. For example, we can add two waves with the same frequency simply by adding the magnitudes of the two electric fields algebraically.

Let us summarize the properties of electromagnetic waves as we have described them:

- The solutions of Maxwell's third and fourth equations are wave-like, with both E and B satisfying a wave equation.
- Electromagnetic waves travel through empty space at the speed of light $c = 1/\sqrt{\mu_0\epsilon_0}$.
- The components of the electric and magnetic fields of plane electromagnetic waves are perpendicular to each other and perpendicular to the direction of wave propagation. We can summarize the latter property by saying that electromagnetic waves are transverse waves.
- The magnitudes of **E** and **B** in empty space are related by the expression $E/B = c$.
- Electromagnetic waves obey the principle of superposition.

Properties of electromagnetic waves

EXAMPLE 34.1 An Electromagnetic Wave

A sinusoidal electromagnetic wave of frequency 40.0 MHz travels in free space in the x direction, as shown in Figure 34.4. (a) Determine the wavelength and period of the wave.

Solution Using Equation 16.14 for light waves, $c = \lambda f$, and given that $f = 40.0$ MHz $= 4.00 \times 10^7$ s^{-1}, we have

$$\lambda = \frac{c}{f} = \frac{3.00 \times 10^8 \text{ m/s}}{4.00 \times 10^7 \text{ s}^{-1}} = \boxed{7.50 \text{ m}}$$

The period T of the wave is the inverse of the frequency:

$$T = \frac{1}{f} = \frac{1}{4.00 \times 10^7 \text{ s}^{-1}} = \boxed{2.50 \times 10^{-8} \text{ s}}$$

(b) At some point and at some instant, the electric field has its maximum value of 750 N/C and is along the y axis. Calculate the magnitude and direction of the magnetic field at this position and time.

Solution From Equation 34.13 we see that

$$B_{max} = \frac{E_{max}}{c} = \frac{750 \text{ N/C}}{3.00 \times 10^8 \text{ m/s}} = \boxed{2.50 \times 10^{-6} \text{ T}}$$

Because **E** and **B** must be perpendicular to each other and perpendicular to the direction of wave propagation (x in this case), we conclude that **B** is in the z direction.

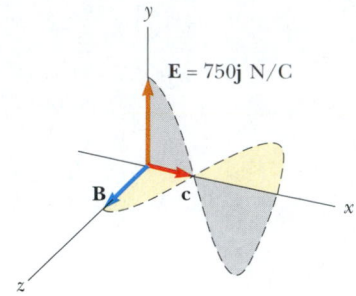

Figure 34.4   At some instant, a plane electromagnetic wave moving in the x direction has a maximum electric field of 750 N/C in the positive y direction. The corresponding magnetic field at that point has a magnitude E/c and is in the z direction.

(c) Write expressions for the space-time variation of the components of the electric and magnetic fields for this wave.

Solution   We can apply Equations 34.11 and 34.12 directly:

$$E = E_{max} \cos(kx - \omega t) = (750 \text{ N/C}) \cos(kx - \omega t)$$

$$B = B_{max} \cos(kx - \omega t) = (2.50 \times 10^{-6} \text{ T}) \cos(kx - \omega t)$$

where

$$\omega = 2\pi f = 2\pi(4.00 \times 10^7 \text{ s}^{-1}) = 2.51 \times 10^8 \text{ rad/s}$$

$$k = \frac{2\pi}{\lambda} = \frac{2\pi}{7.50 \text{ m}} = 0.838 \text{ rad/m}$$

Optional Section

Derivation of Equations 34.6 and 34.7

To derive Equation 34.6, we start with Faraday's law, Equation 34.3:

$$\oint \mathbf{E} \cdot d\mathbf{s} = -\frac{d\Phi_B}{dt}$$

Let us again assume that the electromagnetic wave is traveling in the x direction, with the electric field $\mathbf{E}$ in the positive y direction and the magnetic field $\mathbf{B}$ in the positive z direction.

Consider a rectangle of width dx and height ℓ lying in the xy plane, as shown in Figure 34.5. To apply Equation 34.3, we must first evaluate the line integral of $\mathbf{E} \cdot d\mathbf{s}$ around this rectangle. The contributions from the top and bottom of the rectangle are zero because $\mathbf{E}$ is perpendicular to $d\mathbf{s}$ for these paths. We can express the electric field on the right side of the rectangle as

$$E(x + dx, t) \approx E(x, t) + \frac{dE}{dx}\bigg]_{t \text{ constant}} dx = E(x, t) + \frac{\partial E}{\partial x} dx$$

while the field on the left side is simply $E(x, t)$.[3] Therefore, the line integral over this rectangle is approximately

$$\oint \mathbf{E} \cdot d\mathbf{s} = E(x + dx, t) \cdot \ell - E(x, t) \cdot \ell \approx (\partial E/\partial x) \, dx \cdot \ell \qquad \textbf{(34.14)}$$

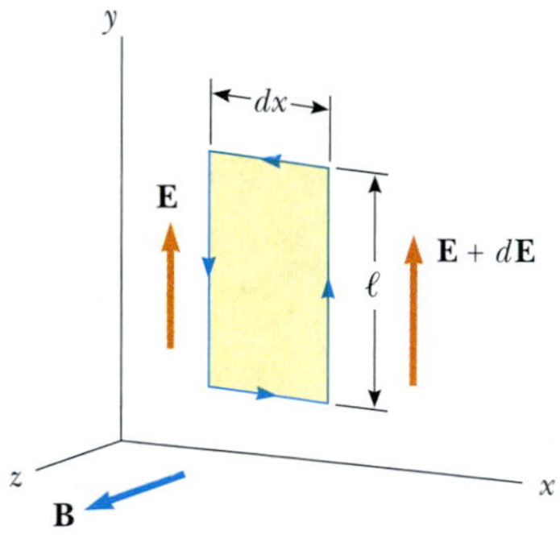

Because the magnetic field is in the z direction, the magnetic flux through the rectangle of area $\ell \, dx$ is approximately $\Phi_B = B\ell \, dx$. (This assumes that dx is very small compared with the wavelength of the wave.) Taking the time derivative of the magnetic flux gives

$$\frac{d\Phi_B}{dt} = \ell \, dx \frac{dB}{dt}\bigg]_{x \text{ constant}} = \ell \, dx \frac{\partial B}{\partial t} \qquad \textbf{(34.15)}$$

Figure 34.5   As a plane wave passes through a rectangular path of width dx lying in the xy plane, the electric field in the y direction varies from $\mathbf{E}$ to $\mathbf{E} + d\mathbf{E}$. This spatial variation in $\mathbf{E}$ gives rise to a time-varying magnetic field along the z direction, according to Equation 34.6.

---

[3] Because dE/dx in this equation is expressed as the change in E with x at a given instant t, dE/dx is equivalent to the partial derivative $\partial E/\partial x$. Likewise, dB/dt means the change in B with time at a particular position x, so in Equation 34.15 we can replace dB/dt with $\partial B/\partial t$.

Substituting Equations 34.14 and 34.15 into Equation 34.3, we obtain

$$\left(\frac{\partial E}{\partial x}\right) dx \cdot \ell = -\ell\, dx\, \frac{\partial B}{\partial t}$$

$$\frac{\partial E}{\partial x} = -\frac{\partial B}{\partial t}$$

This expression is Equation 34.6.

In a similar manner, we can verify Equation 34.7 by starting with Maxwell's fourth equation in empty space (Eq. 34.5). In this case, we evaluate the line integral of $\mathbf{B} \cdot d\mathbf{s}$ around a rectangle lying in the xz plane and having width dx and length ℓ, as shown in Figure 34.6. Noting that the magnitude of the magnetic field changes from $B(x, t)$ to $B(x + dx, t)$ over the width dx, we find the line integral over this rectangle to be approximately

$$\oint \mathbf{B} \cdot d\mathbf{s} = B(x, t) \cdot \ell - B(x + dx, t) \cdot \ell \approx -(\partial B/\partial x)\, dx \cdot \ell \qquad \textbf{(34.16)}$$

The electric flux through the rectangle is $\Phi_E = E\ell\, dx$, which, when differentiated with respect to time, gives

$$\frac{\partial \Phi_E}{\partial t} = \ell\, dx\, \frac{\partial E}{\partial t} \qquad \textbf{(34.17)}$$

Substituting Equations 34.16 and 34.17 into Equation 34.5 gives

$$-(\partial B/\partial x)\, dx \cdot \ell = \mu_0 \epsilon_0 \ell\, dx(\partial E/\partial t)$$

$$\frac{\partial B}{\partial x} = -\mu_0 \epsilon_0 \frac{\partial E}{\partial t}$$

which is Equation 34.7.

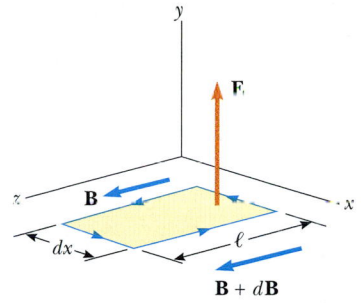

Figure 34.6 As a plane wave passes through a rectangular path of width dx lying in the xz plane, the magnetic field in the z direction varies from $\mathbf{B}$ to $\mathbf{B} + d\mathbf{B}$. This spatial variation in $\mathbf{B}$ gives rise to a time-varying electric field along the y direction, according to Equation 34.7.

34.3 ENERGY CARRIED BY ELECTROMAGNETIC WAVES

Electromagnetic waves carry energy, and as they propagate through space they can transfer energy to objects placed in their path. The rate of flow of energy in an electromagnetic wave is described by a vector $\mathbf{S}$, called the **Poynting vector,** which is defined by the expression

$$\mathbf{S} \equiv \frac{1}{\mu_0} \mathbf{E} \times \mathbf{B} \qquad \textbf{(34.18)}$$

Poynting vector

The magnitude of the Poynting vector represents the rate at which energy flows through a unit surface area perpendicular to the direction of wave propagation. Thus, the magnitude of the Poynting vector represents *power per unit area*. The direction of the vector is along the direction of wave propagation (Fig. 34.7). The SI units of the Poynting vector are $\text{J/s} \cdot \text{m}^2 = \text{W/m}^2$.

As an example, let us evaluate the magnitude of $\mathbf{S}$ for a plane electromagnetic wave where $|\mathbf{E} \times \mathbf{B}| = EB$. In this case,

$$S = \frac{EB}{\mu_0} \qquad \textbf{(34.19)}$$

Magnitude of the Poynting vector for a plane wave

Because $B = E/c$, we can also express this as

$$S = \frac{E^2}{\mu_0 c} = \frac{c}{\mu_0}B^2$$

These equations for S apply at any instant of time and represent the *instantaneous* rate at which energy is passing through a unit area.

What is of greater interest for a sinusoidal plane electromagnetic wave is the time average of S over one or more cycles, which is called the *wave intensity I.* (We discussed the intensity of sound waves in Chapter 17.) When this average is taken, we obtain an expression involving the time average of $\cos^2(kx - \omega t)$, which equals $\frac{1}{2}$. Hence, the average value of S (in other words, the intensity of the wave) is

$$I = S_{av} = \frac{E_{max}B_{max}}{2\mu_0} = \frac{E_{max}^2}{2\mu_0 c} = \frac{c}{2\mu_0}B_{max}^2 \qquad \textbf{(34.20)}$$

Recall that the energy per unit volume, which is the instantaneous energy density u_E associated with an electric field, is given by Equation 26.13,

$$u_E = \tfrac{1}{2}\epsilon_0 E^2$$

and that the instantaneous energy density u_B associated with a magnetic field is given by Equation 32.14:

$$u_B = \frac{B^2}{2\mu_0}$$

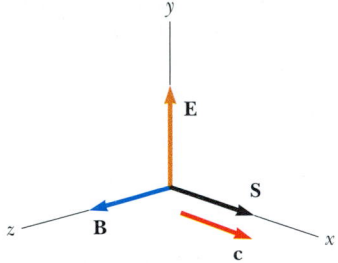

Figure 34.7   The Poynting vector **S** for a plane electromagnetic wave is along the direction of wave propagation.

Because E and B vary with time for an electromagnetic wave, the energy densities also vary with time. When we use the relationships $B = E/c$ and $c = 1/\sqrt{\mu_0\epsilon_0}$, Equation 32.14 becomes

$$u_B = \frac{(E/c)^2}{2\mu_0} = \frac{\mu_0\epsilon_0}{2\mu_0}E^2 = \tfrac{1}{2}\epsilon_0 E^2$$

Comparing this result with the expression for u_E, we see that

$$u_B = u_E = \tfrac{1}{2}\epsilon_0 E^2 = \frac{B^2}{2\mu_0}$$

That is, **for an electromagnetic wave, the instantaneous energy density associated with the magnetic field equals the instantaneous energy density associated with the electric field.** Hence, in a given volume the energy is equally shared by the two fields.

The **total instantaneous energy density** u is equal to the sum of the energy densities associated with the electric and magnetic fields:

$$u = u_E + u_B = \epsilon_0 E^2 = \frac{B^2}{\mu_0}$$

When this total instantaneous energy density is averaged over one or more cycles of an electromagnetic wave, we again obtain a factor of $\frac{1}{2}$. Hence, for any electromagnetic wave, the total average energy per unit volume is

$$u_{av} = \epsilon_0(E^2)_{av} = \tfrac{1}{2}\epsilon_0 E_{max}^2 = \frac{B_{max}^2}{2\mu_0} \qquad \textbf{(34.21)}$$

Comparing this result with Equation 34.20 for the average value of S, we see that

$$I = S_{av} = cu_{av} \qquad \textbf{(34.22)}$$

In other words, **the intensity of an electromagnetic wave equals the average energy density multiplied by the speed of light.**

EXAMPLE 34.2 ▸ Fields on the Page

Estimate the maximum magnitudes of the electric and magnetic fields of the light that is incident on this page because of the visible light coming from your desk lamp. Treat the bulb as a point source of electromagnetic radiation that is about 5% efficient at converting electrical energy to visible light.

Solution Recall from Equation 17.8 that the wave intensity I a distance r from a point source is $I = \mathcal{P}_{av}/4\pi r^2$, where $\mathcal{P}_{av}$ is the average power output of the source and $4\pi r^2$ is the area of a sphere of radius r centered on the source. Because the intensity of an electromagnetic wave is also given by Equation 34.20, we have

$$I = \frac{\mathcal{P}_{av}}{4\pi r^2} = \frac{E_{max}^2}{2\mu_0 c}$$

We must now make some assumptions about numbers to enter in this equation. If we have a 60-W lightbulb, its output at 5% efficiency is approximately 3.0 W in the form of visible light. (The remaining energy transfers out of the bulb by conduction and invisible radiation.) A reasonable distance from the bulb to the page might be 0.30 m. Thus, we have

$$E_{max} = \sqrt{\frac{\mu_0 c \mathcal{P}_{av}}{2\pi r^2}}$$

$$= \sqrt{\frac{(4\pi \times 10^{-7}\,\text{T}\cdot\text{m/A})(3.00 \times 10^8\,\text{m/s})(3.0\,\text{W})}{2\pi(0.30\,\text{m})^2}}$$

$$= \boxed{45\,\text{V/m}}$$

From Equation 34.13,

$$B_{max} = \frac{E_{max}}{c} = \frac{45\,\text{V/m}}{3.00 \times 10^8\,\text{m/s}} = \boxed{1.5 \times 10^{-7}\,\text{T}}$$

This value is two orders of magnitude smaller than the Earth's magnetic field, which, unlike the magnetic field in the light wave from your desk lamp, is not oscillating.

Exercise Estimate the energy density of the light wave just before it strikes this page.

Answer $9.0 \times 10^{-9}\,\text{J/m}^3$.

MOMENTUM AND RADIATION PRESSURE

Electromagnetic waves transport linear momentum as well as energy. It follows that, as this momentum is absorbed by some surface, pressure is exerted on the surface. We shall assume in this discussion that the electromagnetic wave strikes the surface at normal incidence and transports a total energy U to the surface in a time t. Maxwell showed that, if the surface absorbs all the incident energy U in this time (as does a black body, introduced in Chapter 20), the total momentum $\mathbf{p}$ transported to the surface has a magnitude

$$p = \frac{U}{c} \qquad \text{(complete absorption)} \qquad \text{(34.23)}$$

Momentum transported to a perfectly absorbing surface

The pressure exerted on the surface is defined as force per unit area F/A. Let us combine this with Newton's second law:

$$P = \frac{F}{A} = \frac{1}{A}\frac{dp}{dt}$$

If we now replace p, the momentum transported to the surface by light, from Equation 34.23, we have

$$P = \frac{1}{A}\frac{dp}{dt} = \frac{1}{A}\frac{d}{dt}\left(\frac{U}{c}\right) = \frac{1}{c}\frac{(dU/dt)}{A}$$

We recognize $(dU/dt)/A$ as the rate at which energy is arriving at the surface per unit area, which is the magnitude of the Poynting vector. Thus, the radiation pressure P exerted on the perfectly absorbing surface is

$$P = \frac{S}{c} \qquad \textbf{(34.24)}$$

Radiation pressure exerted on a perfectly absorbing surface

Note that Equation 34.24 is an expression for uppercase P, the pressure, while Equation 34.23 is an expression for lowercase p, linear momentum.

If the surface is a perfect reflector (such as a mirror) and incidence is normal, then the momentum transported to the surface in a time t is twice that given by Equation 34.23. That is, the momentum transferred to the surface by the incoming light is $p = U/c$, and that transferred by the reflected light also is $p = U/c$. Therefore,

$$p = \frac{2U}{c} \qquad \text{(complete reflection)} \qquad \textbf{(34.25)}$$

QuickLab

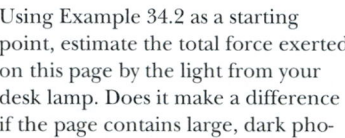

Using Example 34.2 as a starting point, estimate the total force exerted on this page by the light from your desk lamp. Does it make a difference if the page contains large, dark photographs instead of mostly white space?

The momentum delivered to a surface having a reflectivity somewhere between these two extremes has a value between U/c and $2U/c$, depending on the properties of the surface. Finally, the radiation pressure exerted on a perfectly reflecting surface for normal incidence of the wave is[4]

$$P = \frac{2S}{c} \qquad \textbf{(34.26)}$$

Radiation pressure exerted on a perfectly reflecting surface

Although radiation pressures are very small (about 5×10^{-6} N/m^2 for direct sunlight), they have been measured with torsion balances such as the one shown in Figure 34.8. A mirror (a perfect reflector) and a black disk (a perfect absorber) are connected by a horizontal rod suspended from a fine fiber. Normal-incidence light striking the black disk is completely absorbed, so all of the momentum of the light is transferred to the disk. Normal-incidence light striking the mirror is totally reflected, and hence the momentum transferred to the mirror is twice as great as that transferred to the disk. The radiation pressure is determined by measuring the angle through which the horizontal connecting rod rotates. The apparatus must be placed in a high vacuum to eliminate the effects of air currents.

web

Visit **http://pds.jpl.nasa.gov** for more information about missions to the planets. You may also want to read Arthur C. Clarke's 1963 science fiction story *The Wind from the Sun* about a solar yacht race.

NASA is exploring the possibility of *solar sailing* as a low-cost means of sending spacecraft to the planets. Large reflective sheets would be used in much the way canvas sheets are used on earthbound sailboats. In 1973 NASA engineers took ad-

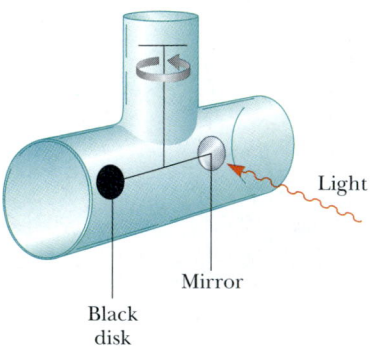

Figure 34.8 An apparatus for measuring the pressure exerted by light. In practice, the system is contained in a high vacuum.

Figure 34.9 Mariner 10 used its solar panels to "sail on sunlight." *(Courtesy of NASA)*

---

[4] For oblique incidence on a perfectly reflecting surface, the momentum transferred is $(2U \cos \theta)/c$ and the pressure is $P = (2S \cos^2 \theta)/c$, where θ is the angle between the normal to the surface and the direction of wave propagation.

vantage of the momentum of the sunlight striking the solar panels of Mariner 10 (Fig. 34.9) to make small course corrections when the spacecraft's fuel supply was running low. (This procedure was carried out when the spacecraft was in the vicinity of the planet Mercury. Would it have worked as well near Pluto?)

CONCEPTUAL EXAMPLE 34.3 ▸ Sweeping the Solar System

A great amount of dust exists in interplanetary space. Although in theory these dust particles can vary in size from molecular size to much larger, very little of the dust in our solar system is smaller than about 0.2 μm. Why?

Solution The dust particles are subject to two significant forces—the gravitational force that draws them toward the Sun and the radiation-pressure force that pushes them away from the Sun. The gravitational force is proportional to the cube of the radius of a spherical dust particle because it is proportional to the mass and therefore to the volume $4\pi r^3/3$ of the particle. The radiation pressure is proportional to the square of the radius because it depends on the planar cross-section of the particle. For large particles, the gravitational force is greater than the force from radiation pressure. For particles having radii less than about 0.2 μm, the radiation-pressure force is greater than the gravitational force, and as a result these particles are swept out of the Solar System.

EXAMPLE 34.4 ▸ Pressure from a Laser Pointer

Many people giving presentations use a laser pointer to direct the attention of the audience. If a 3.0-mW pointer creates a spot that is 2.0 mm in diameter, determine the radiation pressure on a screen that reflects 70% of the light that strikes it. The power 3.0 mW is a time-averaged value.

Solution We certainly do not expect the pressure to be very large. Before we can calculate it, we must determine the Poynting vector of the beam by dividing the time-averaged power delivered via the electromagnetic wave by the cross-sectional area of the beam:

$$S = \frac{\mathcal{P}}{A} = \frac{\mathcal{P}}{\pi r^2} = \frac{3.0 \times 10^{-3}\,\text{W}}{\pi \left(\dfrac{2.0 \times 10^{-3}\,\text{m}}{2}\right)^2} = 955\,\text{W/m}^2$$

This is about the same as the intensity of sunlight at the Earth's surface. (Thus, it is not safe to shine the beam of a laser pointer into a person's eyes; that may more dangerous than looking directly at the Sun.)

Now we can determine the radiation pressure from the laser beam. Equation 34.26 indicates that a completely reflected beam would apply a pressure of $P = 2S/c$. We can model the actual reflection as follows: Imagine that the surface absorbs the beam, resulting in pressure $P = S/c$. Then the surface emits the beam, resulting in additional pressure $P = S/c$. If the surface emits only a fraction f of the beam (so that f is the amount of the incident beam reflected), then the pressure due to the emitted beam is $P = fS/c$. Thus, the total pressure on the surface due to absorption and re-emission (reflection) is

$$P = \frac{S}{c} + f\frac{S}{c} = (1 + f)\,\frac{S}{c}$$

Notice that if $f = 1$, which represents complete reflection, this equation reduces to Equation 34.26. For a beam that is 70% reflected, the pressure is

$$P = (1 + 0.70)\,\frac{955\,\text{W/m}^2}{3.0 \times 10^8\,\text{m/s}} = \boxed{5.4 \times 10^{-6}\,\text{N/m}^2}$$

This is an extremely small value, as expected. (Recall from Section 15.2 that atmospheric pressure is approximately $10^5\,\text{N/m}^2$.)

EXAMPLE 34.5 ▸ Solar Energy

As noted in the preceding example, the Sun delivers about 1 000 W/m^2 of energy to the Earth's surface via electromagnetic radiation. (a) Calculate the total power that is incident on a roof of dimensions 8.00 m × 20.0 m.

Solution The magnitude of the Poynting vector for solar radiation at the surface of the Earth is $S = 1\,000$ W/m^2; this represents the power per unit area, or the light intensity. Assuming that the radiation is incident normal to the roof, we obtain

$$\mathcal{P} = SA = (1\,000\,\text{W/m}^2)(8.00 \times 20.0\,\text{m}^2)$$

$$= \boxed{1.60 \times 10^5\,\text{W}}$$

If all of this power could be converted to electrical energy, it would provide more than enough power for the average home. However, solar energy is not easily harnessed, and the prospects for large-scale conversion are not as bright as may appear from this calculation. For example, the efficiency of conversion from solar to electrical energy is typically 10% for photovoltaic cells. Roof systems for converting solar energy to internal energy are approximately 50% efficient; however, solar energy is associated with other practical problems, such as overcast days, geographic location, and methods of energy storage.

(b) Determine the radiation pressure and the radiation force exerted on the roof, assuming that the roof covering is a perfect absorber.

Solution Using Equation 34.24 with $S = 1\ 000\ \text{W/m}^2$, we find that the radiation pressure is

$$P = \frac{S}{c} = \frac{1\ 000\ \text{W/m}^2}{3.00 \times 10^8\ \text{m/s}} = \boxed{3.33 \times 10^{-6}\ \text{N/m}^2}$$

Because pressure equals force per unit area, this corresponds to a radiation force of

$$F = PA = (3.33 \times 10^{-6}\ \text{N/m}^2)(160\ \text{m}^2) = \boxed{5.33 \times 10^{-4}\ \text{N}}$$

Exercise How much solar energy is incident on the roof in 1 h?

Answer 5.76×10^8 J.

Optional Section

34.5 ▶ RADIATION FROM AN INFINITE CURRENT SHEET

In this section, we describe the electric and magnetic fields radiated by a flat conductor carrying a time-varying current. In the symmetric plane geometry employed here, the mathematics is less complex than that required in lower-symmetry situations.

Consider an infinite conducting sheet lying in the yz plane and carrying a surface current in the y direction, as shown in Figure 34.10. The current is distributed across the z direction such that the current per unit length is $\mathbf{J}_s$. Let us assume that J_s varies sinusoidally with time as

$$J_s = J_{max} \cos \omega t$$

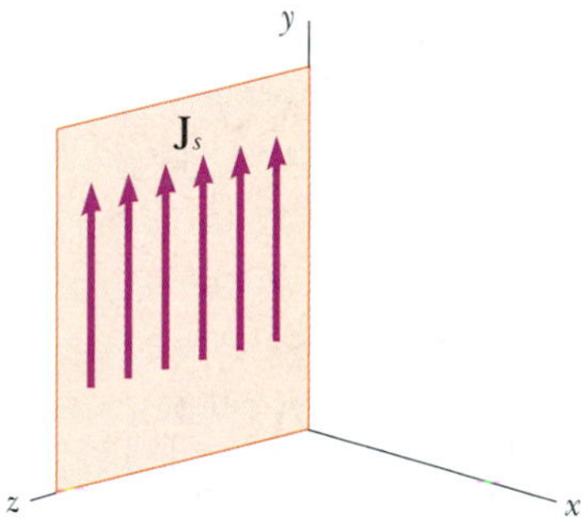

where J_{max} is the amplitude of the current variation and ω is the angular frequency of the variation. A similar problem concerning the case of a steady current was treated in Example 30.6, in which we found that the magnetic field outside the sheet is everywhere parallel to the sheet and lies along the z axis. The magnetic field was found to have a magnitude

$$B_z = \mu_0 \frac{J_s}{2}$$

In the present situation, where J_s varies with time, this equation for B_z is valid only for distances close to the sheet. Substituting the expression for J_s, we have

$$B_z = \frac{\mu_0}{2} J_{max} \cos \omega t \qquad \text{(for small values of x)}$$

To obtain the expression valid for B_z for arbitrary values of x, we can investigate the solution:[5]

$$B_z = \frac{\mu_0 J_{max}}{2} \cos(kx - \omega t) \qquad \textbf{(34.27)}$$

Figure 34.10 A portion of an infinite current sheet lying in the yz plane. The current density is sinusoidal and is given by the expression $J_s = J_{max} \cos \omega t$. The magnetic field is everywhere parallel to the sheet and lies along z.

Radiated magnetic field

---

[5] Note that the solution could also be written in the form $\cos(\omega t - kx)$, which is equivalent to $\cos(kx - \omega t)$. That is, $\cos \theta$ is an even function, which means that $\cos(-\theta) = \cos \theta$.

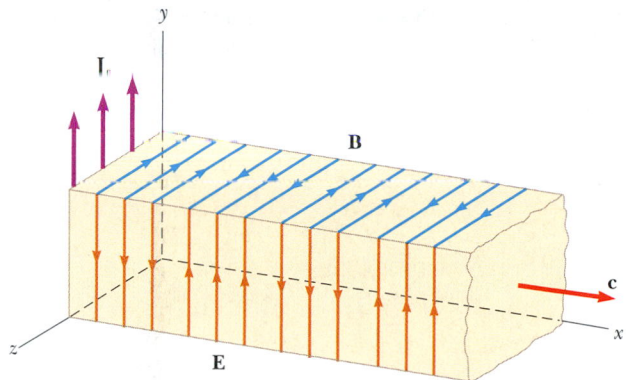

Figure 34.11   Representation of the plane electromagnetic wave radiated by an infinite current sheet lying in the yz plane. The vector **B** is in the z direction, the vector **E** is in the y direction, and the direction of wave motion is along x. Both vector **B** and vector **E** behave according to the expression $\cos(kx - \omega t)$. Compare this drawing with Figure 34.3a.

You should note two things about this solution, which is unique to the geometry under consideration. First, when x is very small, it agrees with our original solution. Second, it satisfies the wave equation as expressed in Equation 34.9. We conclude that the magnetic field lies along the z axis, varies with time, and is characterized by a transverse traveling wave having an angular frequency ω and an angular wave number $k = 2\pi/\lambda$.

We can obtain the electric field radiating from our infinite current sheet by using Equation 34.13:

$$E_y = cB_z = \frac{\mu_0 J_{\max} c}{2} \cos(kx - \omega t) \qquad \textbf{(34.28)}$$

Radiated electric field

That is, the electric field is in the y direction, perpendicular to **B**, and has the same space and time dependencies. These expressions for B_z and E_y show that the radiation field of an infinite current sheet carrying a sinusoidal current is a plane electromagnetic wave propagating with a speed c along the x axis, as shown in Figure 34.11.

We can calculate the Poynting vector for this wave from Equations 34.19, 34.27, and 34.28:

$$S = \frac{EB}{\mu_0} = \frac{\mu_0 J_{\max}^2 c}{4} \cos^2(kx - \omega t) \qquad \textbf{(34.29)}$$

The intensity of the wave, which equals the average value of S, is

$$I = S_{\text{av}} = \frac{\mu_0 J_{\max}^2 c}{8} \qquad \textbf{(34.30)}$$

This intensity represents the power per unit area of the outgoing wave on each side of the sheet. The total rate of energy emitted per unit area of the conductor is $2S_{\text{av}} = \mu_0 J_{\max}^2 c/4$.

EXAMPLE 34.6 ▶ **An Infinite Sheet Carrying a Sinusoidal Current**

An infinite current sheet lying in the yz plane carries a sinusoidal current that has a maximum density of 5.00 A/m. (a) Find the maximum values of the radiated magnetic and electric fields.

Solution   From Equations 34.27 and 34.28, we see that the maximum values of B_z and E_y are

$$B_{\max} = \frac{\mu_0 J_{\max}}{2} \qquad \text{and} \qquad E_{\max} = \frac{\mu_0 J_{\max} c}{2}$$

Using the values $\mu_0 = 4\pi \times 10^{-7}$ T·m/A, $J_{max} = 5.00$ A/m, and $c = 3.00 \times 10^8$ m/s, we get

$$B_{max} = \frac{(4\pi \times 10^{-7} \text{ T·m/A})(5.00 \text{ A/m})}{2}$$

$$= 3.14 \times 10^{-6} \text{ T}$$

$$E_{max} = \frac{(4\pi \times 10^{-7} \text{ T·m/A})(5.00 \text{ A/m})(3.00 \times 10^8 \text{ m/s})}{2}$$

$$= 942 \text{ V/m}$$

(b) What is the average power incident on a flat surface that is parallel to the sheet and has an area of 3.00 m²? (The length and width of this surface are both much greater than the wavelength of the radiation.)

Solution   The intensity, or power per unit area, radiated in each direction by the current sheet is given by Equation 34.30:

$$I = \frac{\mu_0 J_{max}^2 c}{8}$$

$$= \frac{(4\pi \times 10^{-7} \text{ T·m/A})(5.00 \text{ A/m})^2(3.00 \times 10^8 \text{ m/s})}{8}$$

$$= 1.18 \times 10^3 \text{ W/m}^2$$

Multiplying this by the area of the surface, we obtain the incident power:

$$\mathcal{P} = IA = (1.18 \times 10^3 \text{ W/m}^2)(3.00 \text{ m}^2)$$

$$= 3.54 \times 10^3 \text{ W}$$

The result is independent of the distance from the current sheet because we are dealing with a plane wave.

Optional Section

34.6 ▶ PRODUCTION OF ELECTROMAGNETIC WAVES BY AN ANTENNA

Neither stationary charges nor steady currents can produce electromagnetic waves. Whenever the current through a wire changes with time, however, the wire emits electromagnetic radiation. **The fundamental mechanism responsible for this radiation is the acceleration of a charged particle. Whenever a charged particle accelerates, it must radiate energy.**

> Accelerating charges produce electromagnetic radiation

An alternating voltage applied to the wires of an antenna forces an electric charge in the antenna to oscillate. This is a common technique for accelerating charges and is the source of the radio waves emitted by the transmitting antenna of a radio station. Figure 34.12 shows how this is done. Two metal rods are connected to a generator that provides a sinusoidally oscillating voltage. This causes charges to oscillate in the two rods. At $t = 0$, the upper rod is given a maximum positive charge and the bottom rod an equal negative charge, as shown in Figure 34.12a. The electric field near the antenna at this instant is also shown in Figure 34.12a. As the positive and negative charges decrease from their maximum values, the rods become less charged, the field near the rods decreases in strength, and

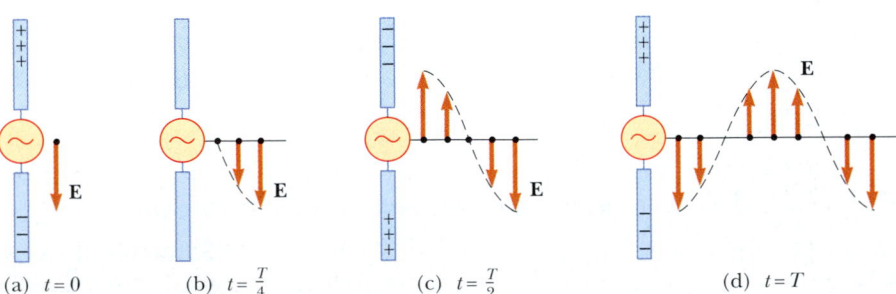

(a) $t = 0$ (b) $t = \dfrac{T}{4}$ (c) $t = \dfrac{T}{2}$ (d) $t = T$

Figure 34.12   The electric field set up by charges oscillating in an antenna. The field moves away from the antenna with the speed of light.

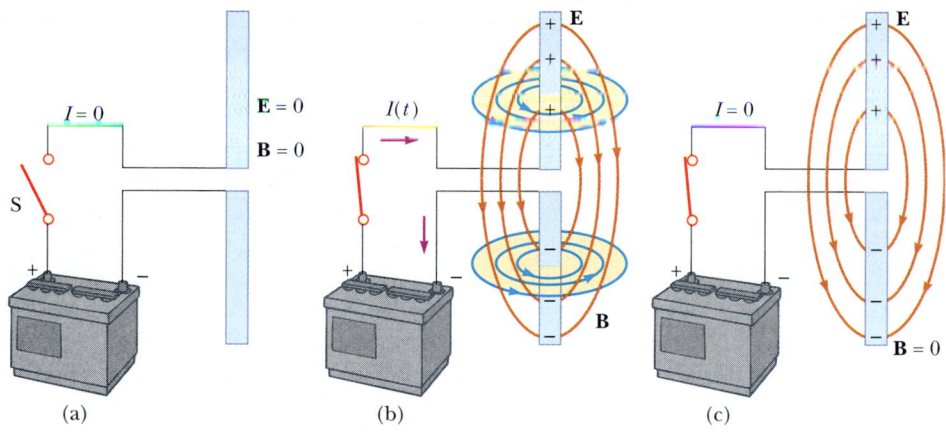

Figure 34.13 A pair of metal rods connected to a battery. (a) When the switch is open and no current exists, the electric and magnetic fields are both zero. (b) Immediately after the switch is closed, the rods are being charged (so a current exists). Because the current is changing, the rods generate changing electric and magnetic fields. (c) When the rods are fully charged, the current is zero, the electric field is a maximum, and the magnetic field is zero.

the downward-directed maximum electric field produced at $t = 0$ moves away from the rod. (A magnetic field oscillating in a direction perpendicular to the plane of the diagram in Fig. 34.12 accompanies the oscillating electric field, but it is not shown for the sake of clarity.) When the charges on the rods are momentarily zero (Fig. 34.12b), the electric field at the rod has dropped to zero. This occurs at a time equal to one quarter of the period of oscillation.

As the generator charges the rods in the opposite sense from that at the beginning, the upper rod soon obtains a maximum negative charge and the lower rod a maximum positive charge (Fig. 34.12c); this results in an electric field near the rod that is directed upward after a time equal to one-half the period of oscillation. The oscillations continue as indicated in Figure 34.12d. The electric field near the antenna oscillates in phase with the charge distribution. That is, the field points down when the upper rod is positive and up when the upper rod is negative. Furthermore, the magnitude of the field at any instant depends on the amount of charge on the rods at that instant.

As the charges continue to oscillate (and accelerate) between the rods, the electric field they set up moves away from the antenna at the speed of light. As you can see from Figure 34.12, one cycle of charge oscillation produces one wavelength in the electric-field pattern.

Next, consider what happens when two conducting rods are connected to the terminals of a battery (Fig. 34.13). Before the switch is closed, the current is zero, so no fields are present (Fig. 34.13a). Just after the switch is closed, positive charge begins to build up on one rod and negative charge on the other (Fig. 34.13b), a situation that corresponds to a time-varying current. The changing charge distribution causes the electric field to change; this in turn produces a magnetic field around the rods.[6] Finally, when the rods are fully charged, the current is zero; hence, no magnetic field exists at that instant (Fig. 34.13c).

Now let us consider the production of electromagnetic waves by a *half-wave antenna*. In this arrangement, two conducting rods are connected to a source of alter-

---

[6] We have neglected the fields caused by the wires leading to the rods. This is a good approximation if the circuit dimensions are much less than the length of the rods.

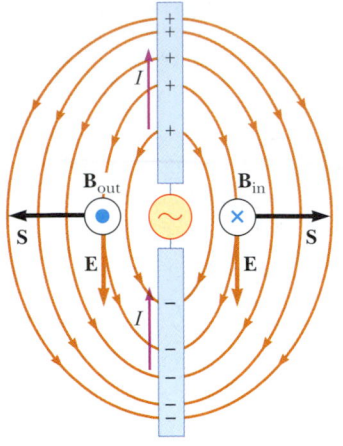

Figure 34.14 A half-wave antenna consists of two metal rods connected to an alternating voltage source. This diagram shows **E** and **B** at an instant when the current is upward. Note that the electric field lines resemble those of a dipole (shown in Fig. 23.21).

nating voltage (such as an *LC* oscillator), as shown in Figure 34.14. The length of each rod is equal to one quarter of the wavelength of the radiation that will be emitted when the oscillator operates at frequency *f*. The oscillator forces charges to accelerate back and forth between the two rods. Figure 34.14 shows the configuration of the electric and magnetic fields at some instant when the current is upward. The electric field lines resemble those of an electric dipole. (As a result, this type of antenna is sometimes called a *dipole antenna*.) Because these charges are continuously oscillating between the two rods, the antenna can be approximated by an oscillating electric dipole. The magnetic field lines form concentric circles around the antenna and are perpendicular to the electric field lines at all points. The magnetic field is zero at all points along the axis of the antenna. Furthermore, **E** and **B** are 90° out of phase in time because the current is zero when the charges at the outer ends of the rods are at a maximum.

At the two points where the magnetic field is shown in Figure 34.14, the Poynting vector **S** is directed radially outward. This indicates that energy is flowing away from the antenna at this instant. At later times, the fields and the Poynting vector change direction as the current alternates. Because **E** and **B** are 90° out of phase at points near the dipole, the net energy flow is zero. From this, we might conclude (incorrectly) that no energy is radiated by the dipole.

However, we find that energy is indeed radiated. Because the dipole fields fall off as $1/r^3$ (as shown in Example 23.6 for the electric field of a static dipole), they are not important at great distances from the antenna. However, at these great distances, something else causes a type of radiation different from that close to the antenna. The source of this radiation is the continuous induction of an electric field by the time-varying magnetic field and the induction of a magnetic field by the time-varying electric field, predicted by Equations 34.3 and 34.4. The electric and magnetic fields produced in this manner are in phase with each other and vary as $1/r$. The result is an outward flow of energy at all times.

The electric field lines produced by a dipole antenna at some instant are shown in Figure 34.15 as they propagate away from the antenna. Note that the intensity and the power radiated are a maximum in a plane that is perpendicular to the antenna and passing through its midpoint. Furthermore, the power radiated is

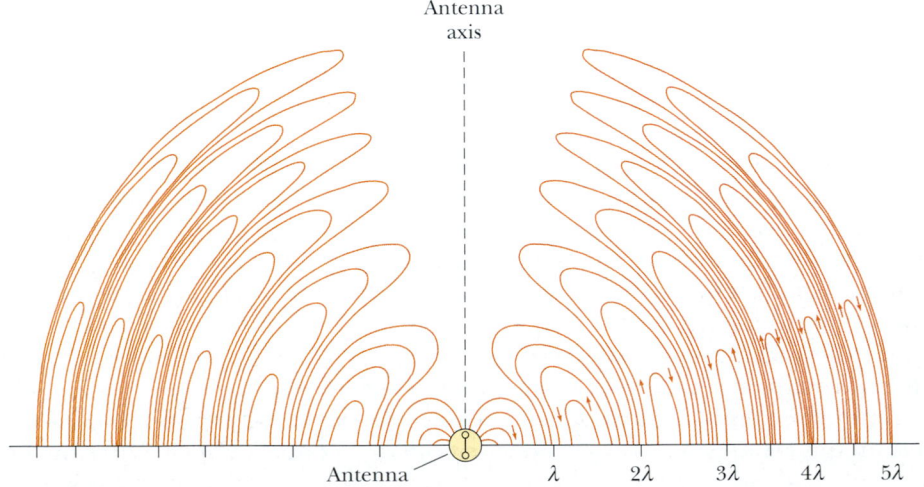

Figure 34.15 Electric field lines surrounding a dipole antenna at a given instant. The radiation fields propagate outward from the antenna with a speed *c*.

zero along the antenna's axis. A mathematical solution to Maxwell's equations for the dipole antenna shows that the intensity of the radiation varies as $(\sin^2\theta)/r^2$, where θ is measured from the axis of the antenna. The angular dependence of the radiation intensity is sketched in Figure 34.16.

Electromagnetic waves can also induce currents in a receiving antenna. The response of a dipole receiving antenna at a given position is a maximum when the antenna axis is parallel to the electric field at that point and zero when the axis is perpendicular to the electric field.

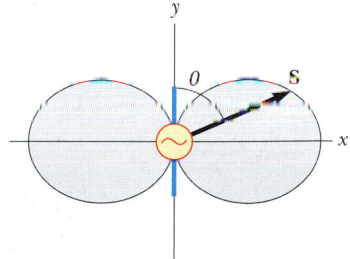

Figure 34.16 Angular dependence of the intensity of radiation produced by an oscillating electric dipole.

Quick Quiz 34.2

If the plane electromagnetic wave in Figure 34.11 represents the signal from a distant radio station, what would be the best orientation for your portable radio antenna — (a) along the x axis, (b) along the y axis, or (c) along the z axis?

QuickLab

Rotate a portable radio (with a telescoping antenna) about a horizontal axis while it is tuned to a weak station. Can you use what you learn from this movement to verify the answer to Quick Quiz 34.2?

34.7 THE SPECTRUM OF ELECTROMAGNETIC WAVES

The various types of electromagnetic waves are listed in Figure 34.17, which shows the **electromagnetic spectrum.** Note the wide ranges of frequencies and wavelengths. No sharp dividing point exists between one type of wave and the next. Remember that **all forms of the various types of radiation are produced by the same phenomenon — accelerating charges.** The names given to the types of waves are simply for convenience in describing the region of the spectrum in which they lie.

Radio waves are the result of charges accelerating through conducting wires. Ranging from more than 10^4 m to about 0.1 m in wavelength, they are generated by such electronic devices as LC oscillators and are used in radio and television communication systems.

Radio waves

Microwaves have wavelengths ranging from approximately 0.3 m to 10^{-4} m and are also generated by electronic devices. Because of their short wavelengths, they are well suited for radar systems and for studying the atomic and molecular properties of matter. Microwave ovens (in which the wavelength of the radiation is $\lambda = 0.122$ m) are an interesting domestic application of these waves. It has been suggested that solar energy could be harnessed by beaming microwaves to the Earth from a solar collector in space.[7]

Microwaves

Infrared waves have wavelengths ranging from 10^{-3} m to the longest wavelength of visible light, 7×10^{-7} m. These waves, produced by molecules and room-temperature objects, are readily absorbed by most materials. The infrared (IR) energy absorbed by a substance appears as internal energy because the energy agitates the atoms of the object, increasing their vibrational or translational motion, which results in a temperature increase. Infrared radiation has practical and scientific applications in many areas, including physical therapy, IR photography, and vibrational spectroscopy.

Infrared waves

Visible light, the most familiar form of electromagnetic waves, is the part of the electromagnetic spectrum that the human eye can detect. Light is produced by the rearrangement of electrons in atoms and molecules. The various wavelengths of visible light, which correspond to different colors, range from red ($\lambda \approx 7 \times 10^{-7}$ m) to violet ($\lambda \approx 4 \times 10^{-7}$ m). The sensitivity of the human eye is a func-

Visible light waves

[7] P. Glaser, "Solar Power from Satellites," *Phys. Today*, February 1977, p. 30.

Satellite-dish television antennas receive television-station signals from satellites in orbit around the Earth. *(Courtesy of Thompson Consumer Electronics)*

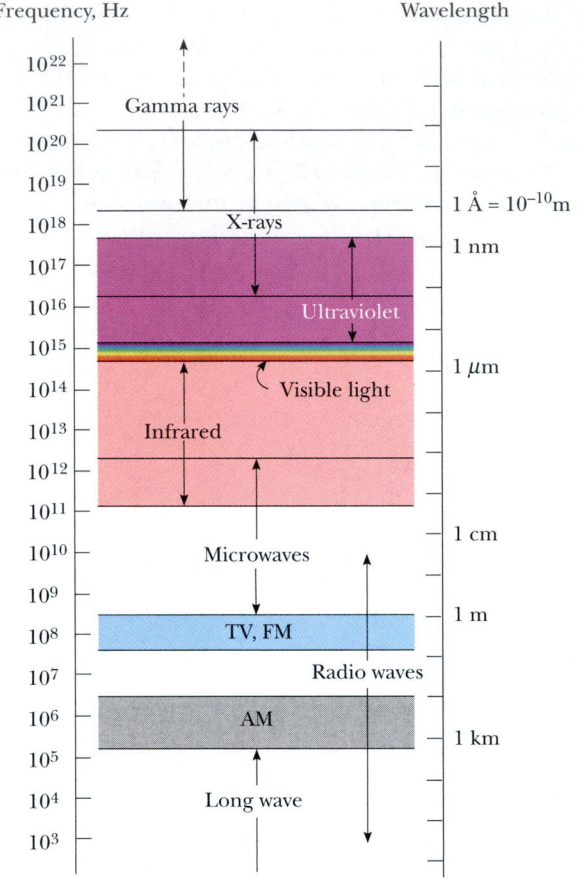

Figure 34.17   The electromagnetic spectrum. Note the overlap between adjacent wave types.

tion of wavelength, being a maximum at a wavelength of about 5.5×10^{-7} m. With this in mind, why do you suppose tennis balls often have a yellow-green color?

Ultraviolet waves

Ultraviolet waves cover wavelengths ranging from approximately 4×10^{-7} m to 6×10^{-10} m. The Sun is an important source of ultraviolet (UV) light, which is the main cause of sunburn. Sunscreen lotions are transparent to visible light but absorb most UV light. The higher a sunscreen's solar protection factor (SPF), the greater the percentage of UV light absorbed. Ultraviolet rays have also been implicated in the formation of cataracts, a clouding of the lens inside the eye. Wearing sunglasses that do not block UV light is worse for your eyes than wearing no sunglasses. The lenses of any sunglasses absorb some visible light, thus causing the wearer's pupils to dilate. If the glasses do not also block UV light, then more damage may be done to the lens of the eye because of the dilated pupils. If you wear no sunglasses at all, your pupils are contracted, you squint, and a lot less UV light enters your eyes. High-quality sunglasses block nearly all the eye-damaging UV light.

Most of the UV light from the Sun is absorbed by ozone (O_3) molecules in the Earth's upper atmosphere, in a layer called the stratosphere. This ozone shield converts lethal high-energy UV radiation to infrared radiation, which in turn warms the stratosphere. Recently, a great deal of controversy has arisen concerning the possible depletion of the protective ozone layer as a result of the chemicals emitted from aerosol spray cans and used as refrigerants.

X-rays have wavelengths in the range from approximately 10^{-8} m to 10^{-12} m. The most common source of x-rays is the deceleration of high energy electrons bombarding a metal target. X-rays are used as a diagnostic tool in medicine and as a treatment for certain forms of cancer. Because x-rays damage or destroy living tissues and organisms, care must be taken to avoid unnecessary exposure or overexposure. X-rays are also used in the study of crystal structure because x-ray wavelengths are comparable to the atomic separation distances in solids (about 0.1 nm).

X-rays

Gamma rays are electromagnetic waves emitted by radioactive nuclei (such as ^{60}Co and ^{137}Cs) and during certain nuclear reactions. High-energy gamma rays are a component of cosmic rays that enter the Earth's atmosphere from space. They have wavelengths ranging from approximately 10^{-10} m to less than 10^{-14} m. They are highly penetrating and produce serious damage when absorbed by living tissues. Consequently, those working near such dangerous radiation must be protected with heavily absorbing materials, such as thick layers of lead.

Gamma rays

Quick Quiz 34.3

The *AM* in *AM* radio stands for *amplitude modulation*, and *FM* stands for *frequency modulation*. (The word *modulate* means "to change.") If our eyes could see the electromagnetic waves from a radio antenna, how could you tell an AM wave from an FM wave?

EXAMPLE 34.7   A Half-Wave Antenna

A half-wave antenna works on the principle that the optimum length of the antenna is one-half the wavelength of the radiation being received. What is the optimum length of a car antenna when it receives a signal of frequency 94.0 MHz?

Solution   Equation 16.14 tells us that the wavelength of

the signal is

$$\lambda = \frac{c}{f} = \frac{3.00 \times 10^8 \text{ m/s}}{9.40 \times 10^7 \text{ Hz}} = 3.19 \text{ m}$$

Thus, to operate most efficiently, the antenna should have a length of $(3.19 \text{ m})/2 = 1.60$ m. For practical reasons, car antennas are usually one-quarter wavelength in size.

SUMMARY

Electromagnetic waves, which are predicted by Maxwell's equations, have the following properties:

- The electric field and the magnetic field each satisfy a wave equation. These two wave equations, which can be obtained from Maxwell's third and fourth equations, are

$$\frac{\partial^2 E}{\partial x^2} = \mu_0 \epsilon_0 \frac{\partial^2 E}{\partial t^2} \qquad \textbf{(34.8)}$$

$$\frac{\partial^2 B}{\partial x^2} = \mu_0 \epsilon_0 \frac{\partial^2 B}{\partial t^2} \qquad \textbf{(34.9)}$$

- The waves travel through a vacuum with the speed of light c, where

$$c = \frac{1}{\sqrt{\mu_0 \epsilon_0}} = 3.00 \times 10^8 \text{ m/s} \qquad \textbf{(34.10)}$$

- The electric and magnetic fields are perpendicular to each other and perpendicular to the direction of wave propagation. (Hence, electromagnetic waves are transverse waves.)
- The instantaneous magnitudes of **E** and **B** in an electromagnetic wave are related by the expression

$$\frac{E}{B} = c \tag{34.13}$$

- The waves carry energy. The rate of flow of energy crossing a unit area is described by the Poynting vector **S**, where

$$\mathbf{S} \equiv \frac{1}{\mu_0} \mathbf{E} \times \mathbf{B} \tag{34.18}$$

- They carry momentum and hence exert pressure on surfaces. If an electromagnetic wave whose Poynting vector is **S** is completely absorbed by a surface upon which it is normally incident, the radiation pressure on that surface is

$$P = \frac{S}{c} \quad \text{(complete absorption)} \tag{34.24}$$

If the surface totally reflects a normally incident wave, the pressure is doubled.

The electric and magnetic fields of a sinusoidal plane electromagnetic wave propagating in the positive x direction can be written

$$E = E_{max} \cos(kx - \omega t) \tag{34.11}$$

$$B = B_{max} \cos(kx - \omega t) \tag{34.12}$$

where ω is the angular frequency of the wave and k is the angular wave number. These equations represent special solutions to the wave equations for E and B. Because $\omega = 2\pi f$ and $k = 2\pi/\lambda$, where f and λ are the frequency and wavelength, respectively, it is found that

$$\frac{\omega}{k} = \lambda f = c$$

The average value of the Poynting vector for a plane electromagnetic wave has a magnitude

$$S_{av} = \frac{E_{max} B_{max}}{2\mu_0} = \frac{E_{max}^2}{2\mu_0 c} = \frac{c}{2\mu_0} B_{max}^2 \tag{34.20}$$

The intensity of a sinusoidal plane electromagnetic wave equals the average value of the Poynting vector taken over one or more cycles.

The electromagnetic spectrum includes waves covering a broad range of wavelengths, from long radio waves at more than 10^4 m to gamma rays at less than 10^{-14} m.

QUESTIONS

1. For a given incident energy of an electromagnetic wave, why is the radiation pressure on a perfectly reflecting surface twice as great as that on a perfectly absorbing surface?

2. Describe the physical significance of the Poynting vector.

3. Do all current-carrying conductors emit electromagnetic waves? Explain.

4. What is the fundamental cause of electromagnetic radiation?

5. Electrical engineers often speak of the radiation resistance of an antenna. What do you suppose they mean by this phrase?

6. If a high-frequency current is passed through a solenoid containing a metallic core, the core warms up by induc-

tion. This process also cooks foods in microwave ovens. Explain why the materials warm up in these situations.

7. Before the advent of cable television and satellite dishes, homeowners either mounted a television antenna on the roof or used "rabbit ears" atop their sets (Fig. Q34.7). Certain orientations of the receiving antenna on a television set gave better reception than others. Furthermore, the best orientation varied from station to station. Explain.

Figure Q34.7 Questions 7, 12, 13, and 14. The V-shaped antenna is the VHF antenna. *(George Semple)*

8. Does a wire connected to the terminals of a battery emit an electromagnetic wave? Explain.

9. If you charge a comb by running it through your hair and then hold the comb next to a bar magnet, do the electric and magnetic fields that are produced constitute an electromagnetic wave?

10. An empty plastic or glass dish is cool to the touch right after it is removed from a microwave oven. How can this be possible? (Assume that your electric bill has been paid.)

11. Often when you touch the indoor antenna on a radio or television receiver, the reception instantly improves. Why?

12. Explain how the (dipole) VHF antenna of a television set works. (See Fig. Q34.7.)

13. Explain how the UHF (loop) antenna of a television set works. (See Fig. Q34.7.)

14. Explain why the voltage induced in a UHF (loop) antenna depends on the frequency of the signal, whereas the voltage in a VHF (dipole) antenna does not. (See Fig. Q34.7.)

15. List as many similarities and differences between sound waves and light waves as you can.

16. What does a radio wave do to the charges in the receiving antenna to provide a signal for your car radio?

17. What determines the height of an AM radio station's broadcast antenna?

18. Some radio transmitters use a "phased array" of antennas. What is their purpose?

19. What happens to the radio reception in an airplane as it flies over the (vertical) dipole antenna of the control tower?

20. When light (or other electromagnetic radiation) travels across a given region, what oscillates?

21. Why should an infrared photograph of a person look different from a photograph of that person taken with visible light?

22. Suppose a creature from another planet had eyes that were sensitive to infrared radiation. Describe what the creature would see if it looked around the room you are now in. That is, what would be bright and what would be dim?

PROBLEMS

1, 2, 3 = straightforward, intermediate, challenging □ = full solution available in the *Student Solutions Manual and Study Guide*
WEB = solution posted at **http://www.saunderscollege.com/physics/** 💻 = Computer useful in solving problem   📓 = Interactive Physics
□ = paired numerical/symbolic problems

Section 34.1 Maxwell's Equations and Hertz's Discoveries

Section 34.2 Plane Electromagnetic Waves

Note: Assume that the medium is vacuum unless specified otherwise.

1. If the North Star, Polaris, were to burn out today, in what year would it disappear from our vision? Take the distance from the Earth to Polaris as 6.44×10^{18} m.

2. The speed of an electromagnetic wave traveling in a transparent nonmagnetic substance is $v = 1/\sqrt{\kappa \mu_0 \epsilon_0}$, where κ is the dielectric constant of the substance. Determine the speed of light in water, which has a dielectric constant at optical frequencies of 1.78.

3. An electromagnetic wave in vacuum has an electric field amplitude of 220 V/m. Calculate the amplitude of the corresponding magnetic field.

4. Calculate the maximum value of the magnetic field of an electromagnetic wave in a medium where the speed of light is two thirds of the speed of light in vacuum and where the electric field amplitude is 7.60 mV/m.

WEB **5.** Figure 34.3a shows a plane electromagnetic sinusoidal wave propagating in what we choose as the x direction. Suppose that the wavelength is 50.0 m, and the electric field vibrates in the xy plane with an amplitude of 22.0 V/m. Calculate (a) the frequency of the wave and (b) the magnitude and direction of **B** when the electric field has its maximum value in the negative y direction. (c) Write an expression for B in the form

$$B = B_{max} \cos(kx - \omega t)$$

with numerical values for B_{max}, k, and ω.

6. Write down expressions for the electric and magnetic fields of a sinusoidal plane electromagnetic wave having a frequency of 3.00 GHz and traveling in the positive x direction. The amplitude of the electric field is 300 V/m.

7. In SI units, the electric field in an electromagnetic wave is described by

$$E_y = 100 \sin(1.00 \times 10^7 x - \omega t)$$

Find (a) the amplitude of the corresponding magnetic field, (b) the wavelength λ, and (c) the frequency f.

8. Verify by substitution that the following equations are solutions to Equations 34.8 and 34.9, respectively:

$$E = E_{max} \cos(kx - \omega t)$$
$$B = B_{max} \cos(kx - \omega t)$$

9. Review Problem. A standing-wave interference pattern is set up by radio waves between two metal sheets 2.00 m apart. This is the shortest distance between the plates that will produce a standing-wave pattern. What is the fundamental frequency?

10. A microwave oven is powered by an electron tube called a magnetron, which generates electromagnetic waves of frequency 2.45 GHz. The microwaves enter the oven and are reflected by the walls. The standing-wave pattern produced in the oven can cook food unevenly, with hot spots in the food at antinodes and cool spots at nodes, so a turntable is often used to rotate the food and distribute the energy. If a microwave oven intended for use with a turntable is instead used with a cooking dish in a fixed position, the antinodes can appear as burn marks on foods such as carrot strips or cheese. The separation distance between the burns is measured to be 6 cm ± 5%. From these data, calculate the speed of the microwaves.

Section 34.3   Energy Carried by Electromagnetic Waves

11. How much electromagnetic energy per cubic meter is contained in sunlight, if the intensity of sunlight at the Earth's surface under a fairly clear sky is 1 000 W/m²?

12. An AM radio station broadcasts isotropically (equally in all directions) with an average power of 4.00 kW. A dipole receiving antenna 65.0 cm long is at a location 4.00 miles from the transmitter. Compute the emf that is induced by this signal between the ends of the receiving antenna.

13. What is the average magnitude of the Poynting vector 5.00 miles from a radio transmitter broadcasting isotropically with an average power of 250 kW?

14. A monochromatic light source emits 100 W of electromagnetic power uniformly in all directions. (a) Calculate the average electric-field energy density 1.00 m from the source. (b) Calculate the average magnetic-field energy density at the same distance from the source. (c) Find the wave intensity at this location.

WEB **15.** A community plans to build a facility to convert solar radiation to electric power. They require 1.00 MW of power, and the system to be installed has an efficiency of 30.0% (that is, 30.0% of the solar energy incident on the surface is converted to electrical energy). What must be the effective area of a perfectly absorbing surface used in such an installation, assuming a constant intensity of 1 000 W/m²?

16. Assuming that the antenna of a 10.0-kW radio station radiates spherical electromagnetic waves, compute the maximum value of the magnetic field 5.00 km from the antenna, and compare this value with the surface magnetic field of the Earth.

WEB **17.** The filament of an incandescent lamp has a 150-Ω resistance and carries a direct current of 1.00 A. The filament is 8.00 cm long and 0.900 mm in radius. (a) Calculate the Poynting vector at the surface of the filament. (b) Find the magnitude of the electric and magnetic fields at the surface of the filament.

18. In a region of free space the electric field at an instant of time is $\mathbf{E} = (80.0\mathbf{i} + 32.0\mathbf{j} - 64.0\mathbf{k})$ N/C and the magnetic field is $\mathbf{B} = (0.200\mathbf{i} + 0.080\,0\mathbf{j} + 0.290\mathbf{k})$ μT. (a) Show that the two fields are perpendicular to each other. (b) Determine the Poynting vector for these fields.

19. A lightbulb filament has a resistance of 110 Ω. The bulb is plugged into a standard 120-V (rms) outlet and emits 1.00% of the electric power delivered to it as electromagnetic radiation of frequency f. Assuming that the bulb is covered with a filter that absorbs all other frequencies, find the amplitude of the magnetic field 1.00 m from the bulb.

20. A certain microwave oven contains a magnetron that has an output of 700 W of microwave power for an electrical input power of 1.40 kW. The microwaves are entirely transferred from the magnetron into the oven chamber through a waveguide, which is a metal tube of rectangular cross-section with a width of 6.83 cm and a height of 3.81 cm. (a) What is the efficiency of the magnetron? (b) Assuming that the food is absorbing all the microwaves produced by the magnetron and that no energy is reflected back into the waveguide, find the direction and magnitude of the Poynting vector, averaged over time, in the waveguide near the entrance to the oven chamber. (c) What is the maximum electric field strength at this point?

21. High-power lasers in factories are used to cut through cloth and metal (Fig. P34.21). One such laser has a beam diameter of 1.00 mm and generates an electric field with an amplitude of 0.700 MV/m at the target. Find (a) the amplitude of the magnetic field produced, (b) the intensity of the laser, and (c) the power delivered by the laser.

Figure P34.21 A laser cutting device mounted on a robot arm is being used to cut through a metallic plate. *(Philippe Plailly/SPL/Photo Researchers)*

22. At what distance from a 100-W electromagnetic-wave point source does $E_{max} = 15.0$ V/m?

23. A 10.0-mW laser has a beam diameter of 1.60 mm. (a) What is the intensity of the light, assuming it is uniform across the circular beam? (b) What is the average energy density of the beam?

24. At one location on the Earth, the rms value of the magnetic field caused by solar radiation is 1.80 μT. From this value, calculate (a) the average electric field due to solar radiation, (b) the average energy density of the solar component of electromagnetic radiation at this location, and (c) the magnitude of the Poynting vector for the Sun's radiation. (d) Compare the value found in part (c) with the value of the solar intensity given in Example 34.5.

Section 34.4 Momentum and Radiation Pressure

25. A radio wave transmits 25.0 W/m² of power per unit area. A flat surface of area A is perpendicular to the direction of propagation of the wave. Calculate the radiation pressure on it if the surface is a perfect absorber.

26. A plane electromagnetic wave of intensity 6.00 W/m² strikes a small pocket mirror, of area 40.0 cm², held perpendicular to the approaching wave. (a) What momen-

tum does the wave transfer to the mirror each second? (b) Find the force that the wave exerts on the mirror.

27. A possible means of space flight is to place a perfectly reflecting aluminized sheet into orbit around the Earth and then use the light from the Sun to push this "solar sail." Suppose a sail of area 6.00×10^5 m² and mass 6 000 kg is placed in orbit facing the Sun. (a) What force is exerted on the sail? (b) What is the sail's acceleration? (c) How long does it take the sail to reach the Moon, 3.84×10^8 m away? Ignore all gravitational effects, assume that the acceleration calculated in part (b) remains constant, and assume a solar intensity of 1 340 W/m².

28. A 100-mW laser beam is reflected back upon itself by a mirror. Calculate the force on the mirror.

WEB **29.** A 15.0-mW helium–neon laser ($\lambda = 632.8$ nm) emits a beam of circular cross-section with a diameter of 2.00 mm. (a) Find the maximum electric field in the beam. (b) What total energy is contained in a 1.00-m length of the beam? (c) Find the momentum carried by a 1.00-m length of the beam.

30. Given that the intensity of solar radiation incident on the upper atmosphere of the Earth is 1 340 W/m², determine (a) the solar radiation incident on Mars, (b) the total power incident on Mars, and (c) the total force acting on the planet. (d) Compare this force to the gravitational attraction between Mars and the Sun (see Table 14.2).

31. A plane electromagnetic wave has an intensity of 750 W/m². A flat rectangular surface of dimensions 50.0 cm × 100 cm is placed perpendicular to the direction of the wave. If the surface absorbs half of the energy and reflects half, calculate (a) the total energy absorbed by the surface in 1.00 min and (b) the momentum absorbed in this time.

(Optional)
Section 34.5 Radiation from an Infinite Current Sheet

32. A large current-carrying sheet emits radiation in each direction (normal to the plane of the sheet) with an intensity of 570 W/m². What maximum value of sinusoidal current density is required?

33. A rectangular surface of dimensions 120 cm × 40.0 cm is parallel to and 4.40 m away from a much larger conducting sheet in which a sinusoidally varying surface current exists that has a maximum value of 10.0 A/m. (a) Calculate the average power that is incident on the smaller sheet. (b) What power per unit area is radiated by the larger sheet?

(Optional)
Section 34.6 Production of Electromagnetic Waves by an Antenna

34. Two hand-held radio transceivers with dipole antennas are separated by a great fixed distance. Assuming that the transmitting antenna is vertical, what fraction of the maximum received power will occur in the receiving antenna when it is inclined from the vertical by (a) 15.0°? (b) 45.0°? (c) 90.0°?

35. Two radio-transmitting antennas are separated by half the broadcast wavelength and are driven in phase with each other. In which directions are (a) the strongest and (b) the weakest signals radiated?

36. Figure 34.14 shows a Hertz antenna (also known as a half-wave antenna, since its length is $\lambda/2$). The antenna is far enough from the ground that reflections do not significantly affect its radiation pattern. Most AM radio stations, however, use a Marconi antenna, which consists of the top half of a Hertz antenna. The lower end of this (quarter-wave) antenna is connected to earth ground, and the ground itself serves as the missing lower half. What are the heights of the Marconi antennas for radio stations broadcasting at (a) 560 kHz and (b) 1 600 kHz?

37. **Review Problem.**   Accelerating charges radiate electromagnetic waves. Calculate the wavelength of radiation produced by a proton in a cyclotron with a radius of 0.500 m and a magnetic field with a magnitude of 0.350 T.

38. **Review Problem.**   Accelerating charges radiate electromagnetic waves. Calculate the wavelength of radiation produced by a proton in a cyclotron of radius R and magnetic field B.

Section 34.7   The Spectrum of Electromagnetic Waves

39. (a) Classify waves with frequencies of 2 Hz, 2 kHz, 2 MHz, 2 GHz, 2 THz, 2 PHz, 2 EHz, 2 ZHz, and 2 YHz on the electromagnetic spectrum. (b) Classify waves with wavelengths of 2 km, 2 m, 2 mm, 2 μm, 2 nm, 2 pm, 2 fm, and 2 am.

40. Compute an order-of-magnitude estimate for the frequency of an electromagnetic wave with a wavelength equal to (a) your height; (b) the thickness of this sheet of paper. How is each wave classified on the electromagnetic spectrum?

41. The human eye is most sensitive to light having a wavelength of 5.50×10^{-7} m, which is in the green–yellow region of the visible electromagnetic spectrum. What is the frequency of this light?

42. Suppose you are located 180 m from a radio transmitter. (a) How many wavelengths are you from the transmitter if the station calls itself 1150 AM? (The AM band frequencies are in kilohertz.) (b) What if this station were 98.1 FM? (The FM band frequencies are in megahertz.)

43. What are the wavelengths of electromagnetic waves in free space that have frequencies of (a) 5.00×10^{19} Hz and (b) 4.00×10^{9} Hz?

44. A radar pulse returns to the receiver after a total travel time of 4.00×10^{-4} s. How far away is the object that reflected the wave?

45. *This just in!* An important news announcement is transmitted by radio waves to people sitting next to their radios, 100 km from the station, and by sound waves to people sitting across the newsroom, 3.00 m from the newscaster. Who receives the news first? Explain. Take the speed of sound in air to be 343 m/s.

46. The U.S. Navy has long proposed the construction of extremely low-frequency (ELF) communication systems. Such waves could penetrate the oceans to reach distant submarines. Calculate the length of a quarter-wavelength antenna for a transmitter generating ELF waves with a frequency of 75.0 Hz. How practical is this?

47. What are the wavelength ranges in (a) the AM radio band (540–1 600 kHz), and (b) the FM radio band (88.0–108 MHz)?

48. There are 12 VHF television channels (Channels 2–13) that lie in the range of frequencies between 54.0 MHz and 216 MHz. Each channel is assigned a width of 6.0 MHz, with the two ranges 72.0–76.0 MHz and 88.0–174 MHz reserved for non-TV purposes. (Channel 2, for example, lies between 54.0 and 60.0 MHz.) Calculate the wavelength ranges for (a) Channel 4, (b) Channel 6, and (c) Channel 8.

ADDITIONAL PROBLEMS

49. Assume that the intensity of solar radiation incident on the cloud tops of Earth is 1 340 W/m². (a) Calculate the total power radiated by the Sun, taking the average Earth–Sun separation to be 1.496×10^{11} m. (b) Determine the maximum values of the electric and magnetic fields at the Earth's location due to solar radiation.

50. The intensity of solar radiation at the top of the Earth's atmosphere is 1 340 W/m³. Assuming that 60% of the incoming solar energy reaches the Earth's surface and assuming that you absorb 50% of the incident energy, make an order-of-magnitude estimate of the amount of solar energy you absorb in a 60-min sunbath.

51. WEB **Review Problem.**   In the absence of cable input or a satellite dish, a television set can use a dipole-receiving antenna for VHF channels and a loop antenna for UHF channels (see Fig. Q34.7). The UHF antenna produces an emf from the changing magnetic flux through the loop. The TV station broadcasts a signal with a frequency f, and the signal has an electric-field amplitude E_{max} and a magnetic-field amplitude B_{max} at the location of the receiving antenna. (a) Using Faraday's law, derive an expression for the amplitude of the emf that appears in a single-turn circular loop antenna with a radius r, which is small compared to the wavelength of the wave. (b) If the electric field in the signal points vertically, what should be the orientation of the loop for best reception?

52. Consider a small, spherical particle of radius r located in space a distance R from the Sun. (a) Show that the ratio F_{rad}/F_{grav} is proportional to $1/r$, where F_{rad} is the

force exerted by solar radiation and F_{grav} is the force of gravitational attraction. (b) The result of part (a) means that, for a sufficiently small value of r the force exerted on the particle by solar radiation exceeds the force of gravitational attraction. Calculate the value of r for which the particle is in equilibrium under the two forces. (Assume that the particle has a perfectly absorbing surface and a mass density of 1.50 g/cm³. Let the particle be located 3.75×10^{11} m from the Sun, and use 214 W/m² as the value of the solar intensity at that point.)

53. A dish antenna with a diameter of 20.0 m receives (at normal incidence) a radio signal from a distant source, as shown in Figure P34.53. The radio signal is a continuous sinusoidal wave with amplitude $E_{max} = 0.200$ μV/m. Assume that the antenna absorbs all the radiation that falls on the dish. (a) What is the amplitude of the magnetic field in this wave? (b) What is the intensity of the radiation received by this antenna? (c) What power is received by the antenna? (d) What force is exerted on the antenna by the radio waves?

Figure P34.53

54. A parallel-plate capacitor has circular plates of radius r separated by distance ℓ. It has been charged to voltage ΔV and is being discharged as current i is drawn from it. Assume that the plate separation ℓ is very small compared to r, so the electric field is essentially constant in the volume between the plates and is zero outside this volume. Note that the displacement current between the capacitor plates creates a magnetic field. (a) Determine the magnitude and direction of the Poynting vector at the cylindrical surface surrounding the electric field volume. (b) Use the value of the Poynting vector and the lateral surface area of the cylinder to find the total power transfer for the capacitor. (c) What are the changes to these results if the direction of the current is reversed, so the capacitor is charging?

55. A section of a very long air-core solenoid, far from either end, forms an inductor with radius r, length ℓ, and

n turns of wire per unit length. At a particular instant, the solenoid current is i and is increasing at the rate di/dt. Ignore the resistance of the wire. (a) Find the magnitude and direction of the Poynting vector over the interior surface of this section of solenoid. (b) Find the rate at which the energy stored in the magnetic field of the inductor is increasing. (c) Express the power in terms of the voltage ΔV across the inductor.

56. A goal of the Russian space program is to illuminate dark northern cities with sunlight reflected to Earth from a 200-m-diameter mirrored surface in orbit. Several smaller prototypes have already been constructed and put into orbit. (a) Assume that sunlight with an intensity of 1 340 W/m² falls on the mirror nearly perpendicularly, and that the atmosphere of the Earth allows 74.6% of the energy of sunlight to pass through it in clear weather. What power is received by a city when the space mirror is reflecting light to it? (b) The plan is for the reflected sunlight to cover a circle with a diameter of 8.00 km. What is the intensity of the light (the average magnitude of the Poynting vector) received by the city? (c) This intensity is what percentage of the vertical component of sunlight at Saint Petersburg in January, when the sun reaches an angle of 7.00° above the horizon at noon?

57. In 1965 Arno Penzias and Robert Wilson discovered the cosmic microwave radiation that was left over from the Big Bang expansion of the Universe. Suppose the energy density of this background radiation is equal to 4.00×10^{-14} J/m³. Determine the corresponding electric-field amplitude.

58. A hand-held cellular telephone operates in the 860- to 900-MHz band and has a power output of 0.600 W from an antenna 10.0 cm long (Fig. P34.58). (a) Find the average magnitude of the Poynting vector 4.00 cm from the antenna, at the location of a typical person's head. Assume that the antenna emits energy with cylindrical wave fronts. (The actual radiation from antennas follows a more complicated pattern, as suggested by Fig. 34.15.) (b) The ANSI/IEEE C95.1-1991 maximum exposure standard is 0.57 mW/cm² for persons living near

Figure P34.58. (©1998 Adam Smith/FPG International)

cellular telephone base stations, who would be continuously exposed to the radiation. Compare the answer to part (a) with this standard.

59. A linearly polarized microwave with a wavelength of 1.50 cm is directed along the positive x axis. The electric field vector has a maximum value of 175 V/m and vibrates in the xy plane. (a) Assume that the magnetic-field component of the wave can be written in the form $B = B_{max} \sin(kx - \omega t)$, and give values for B_{max}, k, and ω. Also, determine in which plane the magnetic-field vector vibrates. (b) Calculate the magnitude of the Poynting vector for this wave. (c) What maximum radiation pressure would this wave exert if it were directed at normal incidence onto a perfectly reflecting sheet? (d) What maximum acceleration would be imparted to a 500-g sheet (perfectly reflecting and at normal incidence) with dimensions of 1.00 m $\times$ 0.750 m?

60. *Review Section 20.7 on thermal radiation.* (a) An elderly couple have installed a solar water heater on the roof of their house (Fig. P34.60). The solar-energy collector consists of a flat closed box with extraordinarily good thermal insulation. Its interior is painted black, and its front face is made of insulating glass. Assume that its emissivity for visible light is 0.900 and its emissivity for infrared light is 0.700. Assume that the noon Sun shines in perpendicular to the glass, with intensity 1 000 W/m², and that no water is then entering or leaving the box. Find the steady-state temperature of the interior of the box. (b) The couple have built an identical box with no water tubes. It lies flat on the ground in front of the house. They use it as a cold frame, where they plant seeds in early spring. If the same noon Sun is at an elevation angle of 50.0°, find the steady-state temperature of the interior of this box, assuming that the ventilation slots are tightly closed.

Figure P34.60 *(©Bill Banaszewski/Visuals Unlimited)*

61. An astronaut, stranded in space 10.0 m from his spacecraft and at rest relative to it, has a mass (including equipment) of 110 kg. Since he has a 100-W light source that forms a directed beam, he decides to use the beam as a photon rocket to propel himself continuously toward the spacecraft. (a) Calculate how long it takes him to reach the spacecraft by this method. (b) Suppose, instead, that he decides to throw the light source away in a direction opposite the spacecraft. If the light source has a mass of 3.00 kg and, after being thrown, moves at 12.0 m/s *relative to the recoiling astronaut*, how long does it take for the astronaut to reach the spacecraft?

62. The Earth reflects approximately 38.0% of the incident sunlight from its clouds and surface. (a) Given that the intensity of solar radiation is 1 340 W/m², what is the radiation pressure on the Earth, in pascals, when the Sun is straight overhead? (b) Compare this to normal atmospheric pressure at the Earth's surface, which is 101 kPa.

63. Lasers have been used to suspend spherical glass beads in the Earth's gravitational field. (a) If a bead has a mass of 1.00 μg and a density of 0.200 g/cm³, determine the radiation intensity needed to support the bead. (b) If the beam has a radius of 0.200 cm, what power is required for this laser?

64. Lasers have been used to suspend spherical glass beads in the Earth's gravitational field. (a) If a bead has a mass m and a density ρ, determine the radiation intensity needed to support the bead. (b) If the beam has a radius r, what power is required for this laser?

65. Review Problem. A 1.00-m-diameter mirror focuses the Sun's rays onto an absorbing plate 2.00 cm in radius, which holds a can containing 1.00 L of water at 20.0°C. (a) If the solar intensity is 1.00 kW/m², what is the intensity on the absorbing plate? (b) What are the maximum magnitudes of the fields **E** and **B**? (c) If 40.0% of the energy is absorbed, how long would it take to bring the water to its boiling point?

66. A microwave source produces pulses of 20.0-GHz radiation, with each pulse lasting 1.00 ns. A parabolic reflector ($R = 6.00$ cm) is used to focus these pulses into a parallel beam of radiation, as shown in Figure P34.66. The average power during each pulse is 25.0 kW. (a) What is the wavelength of these microwaves? (b) What is the total energy contained in each pulse? (c) Compute the average energy density inside each pulse. (d) Determine the amplitude of the electric and magnetic fields in these microwaves. (e) Compute the force exerted on the surface during the 1.00-ns duration of each pulse if the pulsed beam strikes an absorbing surface.

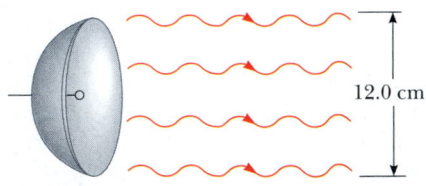

Figure P34.66

67. The electromagnetic power radiated by a nonrelativistic moving point charge q having an acceleration a is

$$\mathcal{P} = \frac{q^2 a^2}{6\pi\epsilon_0 c^3}$$

where ϵ_0 is the permittivity of vacuum (free space) and c is the speed of light in vacuum. (a) Show that the right side of this equation is in watts. (b) If an electron is placed in a constant electric field of 100 N/C, determine the acceleration of the electron and the electromagnetic power radiated by this electron. (c) If a proton is placed in a cyclotron with a radius of 0.500 m and a magnetic field of magnitude 0.350 T, what electromagnetic power is radiated by this proton?

68. A thin tungsten filament with a length of 1.00 m radiates 60.0 W of power in the form of electromagnetic waves. A perfectly absorbing surface, in the form of a hollow cylinder with a radius of 5.00 cm and a length of 1.00 m, is placed concentrically with the filament. Calculate the radiation pressure acting on the cylinder. (Assume that the radiation is emitted in the radial direction, and neglect end effects.)

69. The torsion balance shown in Figure 34.8 is used in an experiment to measure radiation pressure. The suspension fiber exerts an elastic restoring torque. Its torque constant is 1.00×10^{-11} N·m/degree, and the length of the horizontal rod is 6.00 cm. The beam from a 3.00-mW helium–neon laser is incident on the black disk, and the mirror disk is completely shielded. Calculate the angle between the equilibrium positions of the horizontal bar when the beam is switched from "off" to "on."

70. **Review Problem.** The study of Creation suggests a Creator with a remarkable liking for beetles and for small red stars. A red star, typical of the most common kind, radiates electromagnetic waves with a power of 6.00×10^{23} W, which is only 0.159% of the luminosity of the Sun. Consider a spherical planet in a circular orbit around this star. Assume that the emissivity of the planet, as defined in Section 20.7, is equal for infrared and visible light. Assume that the planet has a uniform surface temperature. Identify the projected area over which the planet absorbs starlight, and the radiating area of the planet. If beetles thrive at a temperature of 310 K, what should the radius of the planet's orbit be?

71. A "laser cannon" of a spacecraft has a beam of cross-sectional area A. The maximum electric field in the beam is E. At what rate a will an asteroid accelerate away from the spacecraft if the laser beam strikes the asteroid perpendicularly to its surface, and the surface is nonreflecting? The mass of the asteroid is m. Neglect the acceleration of the spacecraft.

72. A plane electromagnetic wave varies sinusoidally at 90.0 MHz as it travels along the $+x$ direction. The peak value of the electric field is 2.00 mV/m, and it is directed along the $\pm y$ direction. (a) Find the wavelength, the period, and the maximum value of the magnetic field. (b) Write expressions in SI units for the space and time variations of the electric field and of the magnetic field. Include numerical values, and include subscripts to indicate coordinate directions. (c) Find the average power per unit area that this wave propagates through space. (d) Find the average energy density in the radiation (in joules per cubic meter). (e) What radiation pressure would this wave exert upon a perfectly reflecting surface at normal incidence?

ANSWERS TO QUICK QUIZZES

34.1 Zero. Figure 34.3b shows that the **B** and **E** vectors reach their maximum and minimum values at the same time.

34.2 (b) Along the y axis because that is the orientation of the electric field. The electric field moves electrons in the antenna, thus inducing a current that is detected and amplified.

34.3 The AM wave, because its amplitude is changing, would appear to vary in brightness. The FM wave would have changing colors because the color we perceive is related to the frequency of the light.

part 5

Light and Optics

ight is one of the basic "ingredients" of almost all life on the Earth. Plants convert the energy of sunlight to chemical energy through photosynthesis. In addition, light is the principal means by which we are able to transmit and receive information to and from objects around us and throughout the Universe.

The nature and properties of light have been a subject of great interest and speculation since ancient times. The Greeks believed that light consisted of tiny particles (*corpuscles*) that were emitted by a light source and that these particles stimulated the perception of vision upon striking the observer's eye. Newton used this corpuscular theory to explain the reflection and refraction (bending) of light. In 1678, one of Newton's contemporaries, the Dutch scientist Christian Huygens, was able to explain many other

properties of light by proposing that light is a wave. In 1801, Thomas Young showed that light beams can interfere with one another, giving strong support to the wave theory. In 1865, Maxwell developed a brilliant theory that electromagnetic waves travel with the speed of light (see Chapter 34). By this time, the wave theory of light seemed to be firmly established.

However, at the beginning of the 20th century, Max Planck returned to the corpuscular theory of light to explain the radiation emitted by hot objects. Einstein then used the corpuscular theory to explain how electrons are emitted by a metal exposed to light. Today, scientists view light as having a dual nature—that is, light exhibits characteristics of a wave in some situations and characteristics of a particle in other situations.

We shall discuss the particle nature of light in the last part of this text, in which we address modern physics. In Chapters 35 through 38, we concentrate on those aspects of light that are best understood through the wave model. First, we discuss the reflection of light at the boundary between two media and the refraction that occurs as light travels from one medium into another. Then, we use these ideas to study refraction and reflection as light passes through lenses and forms images in mirrors. Next, we describe how the lenses and mirrors used in such instruments as telescopes and microscopes help us view objects not clearly visible to the naked eye. Finally, we study the phenomena of diffraction, polarization, and interference as they apply to light.

◀ Copyright Steve Terrill/The Stock Market.

PUZZLER

The beautiful colors seen inside a well-cut diamond are part of the allure of this gemstone. It now is possible to create artificial gems that sell at a much lower price but have nearly the same sparkle as a diamond. How can you distinguish a cheap imitation from the genuine article by using something you'd find in the kitchen cupboard? *(Charles D. Winters)*

chapter

35

The Nature of Light and the Laws of Geometric Optics

35.1 THE NATURE OF LIGHT

Before the beginning of the 19th century, light was considered to be a stream of particles that either was emitted by the object being viewed or emanated from the eyes of the viewer. Newton, the chief architect of the particle theory of light, held that particles were emitted from a light source and that these particles stimulated the sense of sight upon entering the eye. Using this idea, he was able to explain reflection and refraction.

Most scientists accepted Newton's particle theory. During his lifetime, however, another theory was proposed—one that argued that light might be some sort of wave motion. In 1678, the Dutch physicist and astronomer Christian Huygens showed that a wave theory of light could also explain reflection and refraction. The wave theory did not receive immediate acceptance. It was argued that, if light were some form of wave, it would bend around obstacles; hence, we should be able to see around corners. It is now known that light does indeed bend around the edges of objects. This phenomenon, known as *diffraction*, is not easy to observe because light waves have such short wavelengths. Thus, although Francesco Grimaldi (1618–1663) provided experimental evidence for diffraction in approximately 1660, most scientists rejected the wave theory and adhered to Newton's particle theory for more than a century.

In 1801, Thomas Young (1773–1829) provided the first clear demonstration of the wave nature of light. Young showed that, under appropriate conditions, light rays interfere with each other. Such behavior could not be explained at that time by a particle theory because there was no conceivable way in which two or more particles could come together and cancel one another. Several years later, a French physicist, Augustin Fresnel (1788–1829), performed a number of experiments dealing with interference and diffraction. In 1850, Jean Foucault (1791–1868) provided further evidence of the inadequacy of the particle theory by showing that the speed of light in liquids is less than its speed in air. According to the particle model, the speed of light would be higher in liquids than in air. Additional developments during the 19th century led to the general acceptance of the wave theory of light, the most important resulting from the work of Maxwell, who in 1873 asserted that light was a form of high-frequency electromagnetic wave. As discussed in Chapter 34, Hertz provided experimental confirmation of Maxwell's theory in 1887 by producing and detecting electromagnetic waves. Furthermore, Hertz and other investigators showed that these waves underwent reflection and refraction and exhibited all the other characteristic properties of waves.

Although the wave model and the classical theory of electricity and magnetism were able to explain most known properties of light, they could not explain some subsequent experiments. The most striking of these is the photoelectric effect, also discovered by Hertz: When light strikes a metal surface, electrons are sometimes ejected from the surface. As one example of the difficulties that arose, experiments showed that the kinetic energy of an ejected electron is independent of the light intensity. This finding contradicted the wave theory, which held that a more intense beam of light should add more energy to the electron. An explanation of the photoelectric effect was proposed by Einstein in 1905 in a theory that used the concept of quantization developed by Max Planck (1858–1947) in 1900. The quantization model assumes that the energy of a light wave is present in bundles called *photons;* hence, the energy is said to be quantized. According to Einstein's theory, the energy of a photon is proportional to the frequency of the electromagnetic wave:

$$E = hf \qquad \text{(35.1)} \qquad \text{Energy of a photon}$$

where the constant of proportionality $h = 6.63 \times 10^{-34}$ J·s is Planck's constant (see Section 11.7). It is important to note that this theory retains some features of both the wave theory and the particle theory. The photoelectric effect is the result of energy transfer from a single photon to an electron in the metal, and yet this photon has wave-like characteristics because its energy is determined by the frequency (a wave-like quantity).

In view of these developments, light must be regarded as having a dual nature: **Light exhibits the characteristics of a wave in some situations and the characteristics of a particle in other situations.** Light is light, to be sure. However, the question "Is light a wave or a particle?" is an inappropriate one. Sometimes light acts like a wave, and at other times it acts like a particle. In the next few chapters, we investigate the wave nature of light.

35.2 ▶ MEASUREMENTS OF THE SPEED OF LIGHT

Light travels at such a high speed ($c \approx 3 \times 10^8$ m/s) that early attempts to measure its speed were unsuccessful. Galileo attempted to measure the speed of light by positioning two observers in towers separated by approximately 10 km. Each observer carried a shuttered lantern. One observer would open his lantern first, and then the other would open his lantern at the moment he saw the light from the first lantern. Galileo reasoned that, knowing the transit time of the light beams from one lantern to the other, he could obtain the speed. His results were inconclusive. Today, we realize (as Galileo concluded) that it is impossible to measure the speed of light in this manner because the transit time is so much less than the reaction time of the observers.

We now describe two methods for determining the speed of light.

Roemer's Method

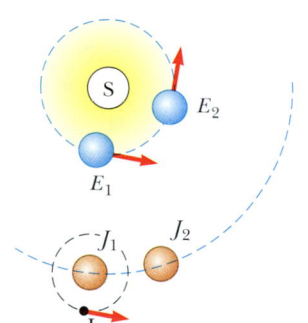

Figure 35.1   Roemer's method for measuring the speed of light. In the time it takes the Earth to travel 90° around the Sun (three months), Jupiter travels only about 7.5° (drawing not to scale).

In 1675, the Danish astronomer Ole Roemer (1644–1710) made the first successful estimate of the speed of light. Roemer's technique involved astronomical observations of one of the moons of Jupiter, Io, which has a period of revolution around Jupiter of approximately 42.5 h. The period of revolution of Jupiter around the Sun is about 12 yr; thus, as the Earth moves through 90° around the Sun, Jupiter revolves through only $(1/12)90° = 7.5°$ (Fig. 35.1).

An observer using the orbital motion of Io as a clock would expect the orbit to have a constant period. However, Roemer, after collecting data for more than a year, observed a systematic variation in Io's period. He found that the periods were longer than average when the Earth was receding from Jupiter and shorter than average when the Earth was approaching Jupiter. If Io had a constant period, Roemer should have seen it become eclipsed by Jupiter at a particular instant and should have been able to predict the time of the next eclipse. However, when he checked the time of the second eclipse as the Earth receded from Jupiter, he found that the eclipse was late. If the interval between his observations was three months, then the delay was approximately 600 s. Roemer attributed this variation in period to the fact that the distance between the Earth and Jupiter changed from one observation to the next. In three months (one quarter of the period of revolution of the Earth around the Sun), the light from Jupiter must travel an additional distance equal to the radius of the Earth's orbit.

Using Roemer's data, Huygens estimated the lower limit for the speed of light to be approximately 2.3×10^8 m/s. This experiment is important historically because it demonstrated that light does have a finite speed and gave an estimate of this speed.

Fizeau's Method

The first successful method for measuring the speed of light by means of purely terrestrial techniques was developed in 1849 by Armand H. L. Fizeau (1819–1896). Figure 35.2 represents a simplified diagram of Fizeau's apparatus. The basic procedure is to measure the total time it takes light to travel from some point to a distant mirror and back. If d is the distance between the light source (considered to be at the location of the wheel) and the mirror and if the transit time for one round trip is t, then the speed of light is $c = 2d/t$.

To measure the transit time, Fizeau used a rotating toothed wheel, which converts a continuous beam of light into a series of light pulses. The rotation of such a wheel controls what an observer at the light source sees. For example, if the pulse traveling toward the mirror and passing the opening at point A in Figure 35.2 should return to the wheel at the instant tooth B had rotated into position to cover the return path, the pulse would not reach the observer. At a greater rate of rotation, the opening at point C could move into position to allow the reflected pulse to reach the observer. Knowing the distance d, the number of teeth in the wheel, and the angular speed of the wheel, Fizeau arrived at a value of $c = 3.1 \times 10^8$ m/s. Similar measurements made by subsequent investigators yielded more precise values for c, approximately $2.997\,9 \times 10^8$ m/s.

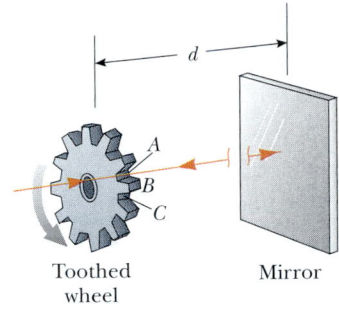

Figure 35.2 Fizeau's method for measuring the speed of light using a rotating toothed wheel. The light source is considered to be at the location of the wheel; thus, the distance d is known.

EXAMPLE 35.1   Measuring the Speed of Light with Fizeau's Wheel

Assume that Fizeau's wheel has 360 teeth and is rotating at 27.5 rev/s when a pulse of light passing through opening A in Figure 35.2 is blocked by tooth B on its return. If the distance to the mirror is 7 500 m, what is the speed of light?

Solution The wheel has 360 teeth, and so it must have 360 openings. Therefore, because the light passes through opening A but is blocked by the tooth immediately adjacent to A, the wheel must rotate through an angle of 1/720 rev in the

time it takes the light pulse to make its round trip. From the definition of angular speed, that time is

$$t = \frac{\theta}{\omega} = \frac{(1/720)\ \text{rev}}{27.5\ \text{rev/s}} = 5.05 \times 10^{-5}\ \text{s}$$

Hence, the speed of light is

$$c = \frac{2d}{t} = \frac{2(7\,500\ \text{m})}{5.05 \times 10^{-5}\ \text{s}} = \boxed{2.97 \times 10^8\ \text{m/s}}$$

35.3 THE RAY APPROXIMATION IN GEOMETRIC OPTICS

The field of **geometric optics** involves the study of the propagation of light, with the assumption that light travels in a fixed direction in a straight line as it passes through a uniform medium and changes its direction when it meets the surface of a different medium or if the optical properties of the medium are nonuniform in either space or time. As we study geometric optics here and in Chapter 36, we use what is called the **ray approximation.** To understand this approximation, first note that the rays of a given wave are straight lines perpendicular to the wave fronts, as discussed in Section 17.4 and illustrated in Figure 35.3 for a plane wave. In the ray approximation, we assume that a wave moving through a medium travels in a straight line in the direction of its rays.

If the wave meets a barrier in which there is a circular opening whose diameter is much larger than the wavelength, as shown in Figure 35.4a, the wave emerg-

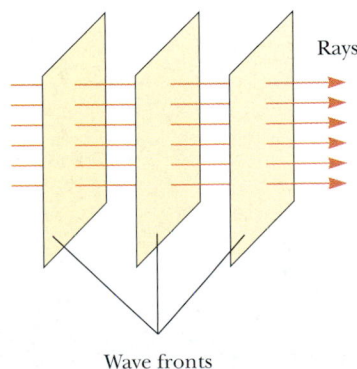

Figure 35.3 A plane wave propagating to the right.

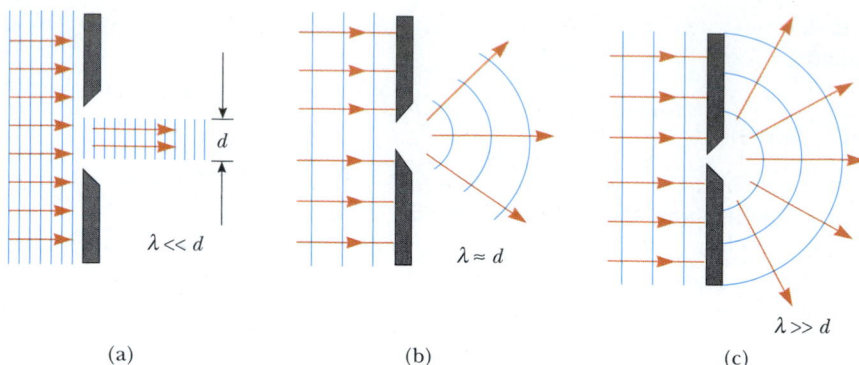

Figure 35.4   A plane wave of wavelength λ is incident on a barrier in which there is an opening of diameter d. (a) When $\lambda \ll d$, the rays continue in a straight-line path, and the ray approximation remains valid. (b) When $\lambda \approx d$, the rays spread out after passing through the opening. (c) When $\lambda \gg d$, the opening behaves as a point source emitting spherical waves.

ing from the opening continues to move in a straight line (apart from some small edge effects); hence, the ray approximation is valid. If the diameter of the opening is of the order of the wavelength, as shown in Figure 35.4b, the waves spread out from the opening in all directions. Finally, if the opening is much smaller than the wavelength, the opening can be approximated as a point source of waves (Fig. 35.4c). Similar effects are seen when waves encounter an opaque object of dimension d. In this case, when $\lambda \ll d$, the object casts a sharp shadow.

The ray approximation and the assumption that $\lambda \ll d$ are used in this chapter and in Chapter 36, both of which deal with geometric optics. This approximation is very good for the study of mirrors, lenses, prisms, and associated optical instruments, such as telescopes, cameras, and eyeglasses.

35.4   REFLECTION

When a light ray traveling in one medium encounters a boundary with another medium, part of the incident light is reflected. Figure 35.5a shows several rays of a beam of light incident on a smooth, mirror-like, reflecting surface. The reflected rays are parallel to each other, as indicated in the figure. The direction of a reflected ray is in the plane perpendicular to the reflecting surface that contains the incident ray. Reflection of light from such a smooth surface is called **specular reflection.** If the reflecting surface is rough, as shown in Figure 35.5b, the surface reflects the rays not as a parallel set but in various directions. Reflection from any rough surface is known as **diffuse reflection.** A surface behaves as a smooth surface as long as the surface variations are much smaller than the wavelength of the incident light.

The difference between these two kinds of reflection explains why it is more difficult to see while driving on a rainy night. If the road is wet, as shown in Figure 35.5c, the smooth surface of the water specularly reflects most of your headlight beams away from your car (and perhaps into the eyes of oncoming drivers). When the road is dry (Fig 35.5d), its rough surface diffusely reflects part of your headlight beam back toward you, allowing you to see the highway clearly. In this book, we concern ourselves only with specular reflection and use the term *reflection* to mean specular reflection.

Consider a light ray traveling in air and incident at an angle on a flat, smooth surface, as shown in Figure 35.6. The incident and reflected rays make angles θ_1

This photograph, taken in Sala-
manca, Spain, shows the reflection of the New Cathedral in the
Tormes River. Are you able to distinguish the cathedral from its
image? *(David Parker/Photo Researchers, Inc.)*

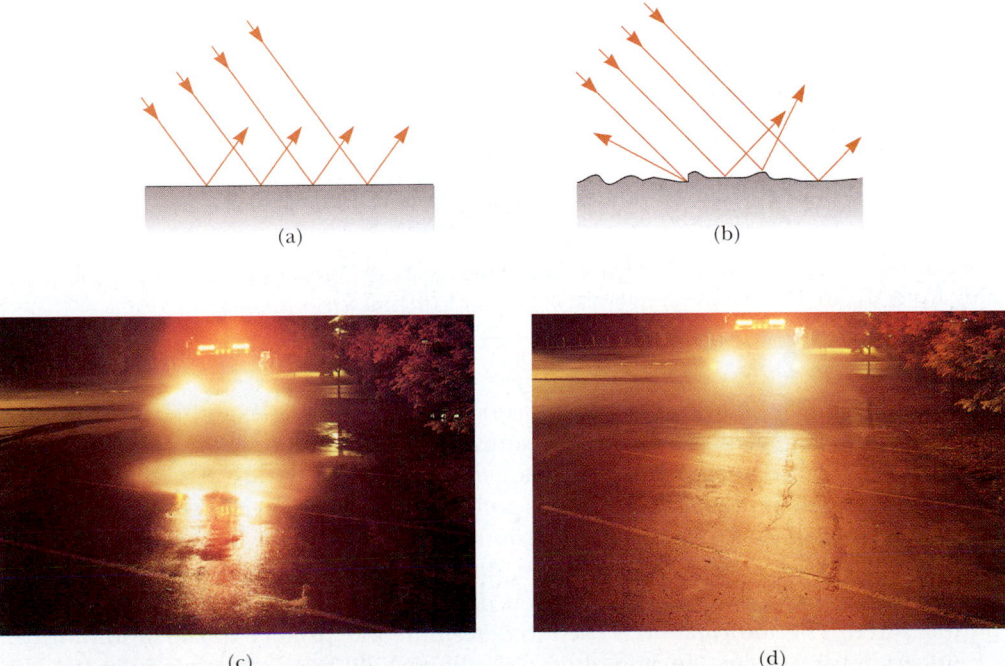

Figure 35.5 Schematic representation of (a) specular reflection, where the reflected rays are
all parallel to each other, and (b) diffuse reflection, where the reflected rays travel in random directions. (c) and (d) Specular and diffuse reflection on the highway. *(Charles D. Winters)*

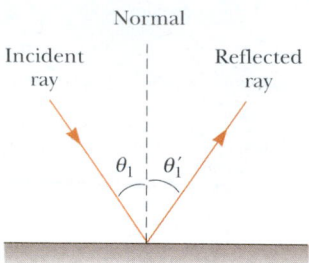

Figure 35.6   According to the law of reflection, $\theta_1 = \theta_1'$. The incident ray, the reflected ray, and the normal all lie in the same plane.

and θ_1', respectively, where the angles are measured from the normal to the rays. (The normal is a line drawn perpendicular to the surface at the point where the incident ray strikes.) Experiments and theory show that **the angle of reflection equals the angle of incidence:**

Law of reflection

$$\theta_1' = \theta_1 \qquad\qquad\qquad (35.2)$$

This relationship is called the **law of reflection.**

EXAMPLE 35.2   The Double-Reflected Light Ray

Two mirrors make an angle of 120° with each other, as illustrated in Figure 35.7. A ray is incident on mirror M_1 at an angle of 65° to the normal. Find the direction of the ray after it is reflected from mirror M_2.

Solution   From the law of reflection, we know that the first reflected ray makes an angle of 65° with the normal. Thus, this ray makes an angle of $90° - 65° = 25°$ with the horizontal.

From the triangle made by the first reflected ray and the two mirrors, we see that the first reflected ray makes an angle of 35° with M_2 (because the sum of the interior angles of any triangle is 180°). Therefore, this ray makes an angle of 55° with the normal to M_2. From the law of reflection, the second reflected ray makes an angle of 55° with the normal to M_2. By comparing the direction of the ray as it is incident on M_1 with its direction after reflecting from M_2, we see that the

ray is reflected through $2(55°) = 120°$, which corresponds to the angle between the mirrors.

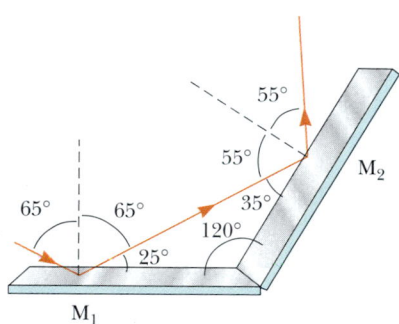

Figure 35.7   Mirrors M_1 and M_2 make an angle of 120° with each other.

If the angle between the two mirrors in the preceding example were 90°, the reflected beam would return to the source parallel to its original path. This phenomenon, called *retroreflection*, has many practical applications. If a third mirror is placed perpendicular to the first two, so that the three form the corner of a cube, retroreflection works in three dimensions. In 1969, a panel of many small reflectors was placed on the Moon by the Apollo 11 astronauts (Fig. 35.8a). A laser beam from the Earth can be reflected directly back on itself and its transit time measured. This information is used to determine the distance to the Moon with an uncertainty of 15 cm. (Imagine how difficult it would be to align a regular flat mirror so that the reflected laser beam would hit a particular location on the Earth!) A more practical application is found in car taillights. Part of the plastic making up the taillight is formed into many tiny cube corners (Fig. 35.8b) so that headlight

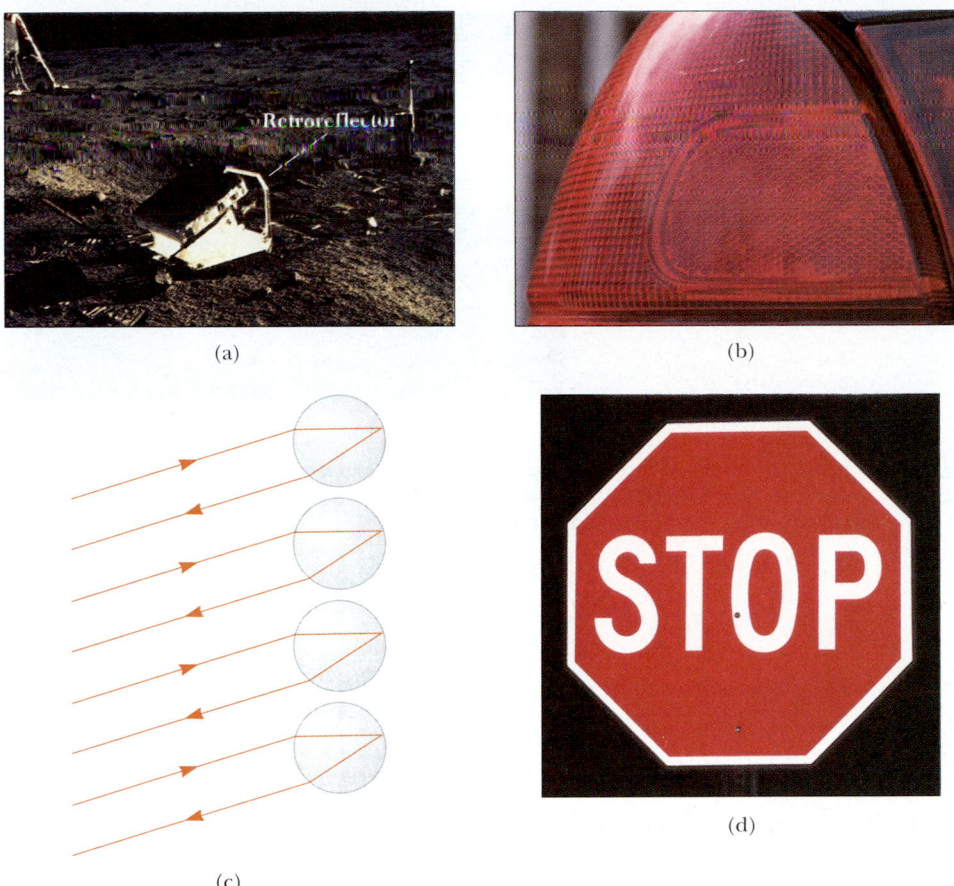

(a)

(b)

(c)

(d)

Figure 35.8 Applications of retroreflection. (a) This panel on the Moon reflects a laser beam directly back to its source on the Earth. (b) An automobile taillight has small retroreflectors that ensure that headlight beams are reflected back toward the car that sent them. (c) A light beam hitting a transparent sphere is retroreflected. (d) This stop sign appears to glow in headlight beams because its surface is covered with a layer of many tiny retroreflecting spheres. What would you see if the sign had a mirror-like surface? *(a, Courtesy of NASA; b and d, George Semple)*

beams from cars approaching from the rear are reflected back to the drivers. Instead of the pyramid-like cube corners, small spherical bumps are sometimes used (Fig. 35.8c). Tiny clear spheres are used in a coating material found on many road signs. The stop sign in Figure 35.8d appears much brighter than it would if it were simply a flat, shiny surface reflecting most of the light hitting it away from the highway.

35.5 REFRACTION

14.4

When a ray of light traveling through a transparent medium encounters a boundary leading into another transparent medium, as shown in Figure 35.9, part of the ray is reflected and part enters the second medium. The part that enters the second medium is bent at the boundary and is said to be **refracted.** The incident ray, the reflected ray, and the refracted ray all lie in the same plane. The **angle of refraction,** θ_2 in Figure 35.9, depends on the properties of the two media and on

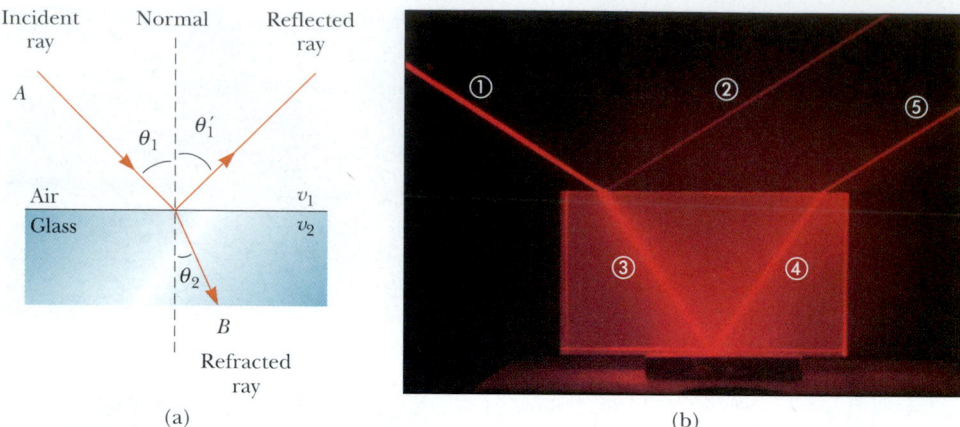

Figure 35.9 (a) A ray obliquely incident on an air–glass interface. The refracted ray is bent toward the normal because $v_2 < v_1$. All rays and the normal lie in the same plane. (b) Light incident on the Lucite block bends both when it enters the block and when it leaves the block. (*b, Henry Leap and Jim Lehman*)

the angle of incidence through the relationship

$$\frac{\sin \theta_2}{\sin \theta_1} = \frac{v_2}{v_1} = \text{constant} \qquad (35.3)$$

where v_1 is the speed of light in the first medium and v_2 is the speed of light in the second medium.

The path of a light ray through a refracting surface is reversible. For example, the ray shown in Figure 35.9a travels from point A to point B. If the ray originated at B, it would travel to the left along line BA to reach point A, and the reflected part would point downward and to the left in the glass.

Quick Quiz 35.1

If beam ① is the incoming beam in Figure 35.9b, which of the other four red lines are reflected beams and which are refracted beams?

From Equation 35.3, we can infer that when light moves from a material in which its speed is high to a material in which its speed is lower, as shown in Figure 35.10a, the angle of refraction θ_2 is less than the angle of incidence θ_1, and the beam is bent toward the normal. If the ray moves from a material in which it moves slowly to a material in which it moves more rapidly, as illustrated in Figure 35.10b, θ_2 is greater than θ_1, and the beam is bent away from the normal.

The behavior of light as it passes from air into another substance and then re-emerges into air is often a source of confusion to students. When light travels in air, its speed is 3×10^8 m/s, but this speed is reduced to approximately 2×10^8 m/s when the light enters a block of glass. When the light re-emerges into air, its speed instantaneously increases to its original value of 3×10^8 m/s. This is far different from what happens, for example, when a bullet is fired through a block of wood. In this case, the speed of the bullet is reduced as it moves through the wood because some of its original energy is used to tear apart the fibers of the

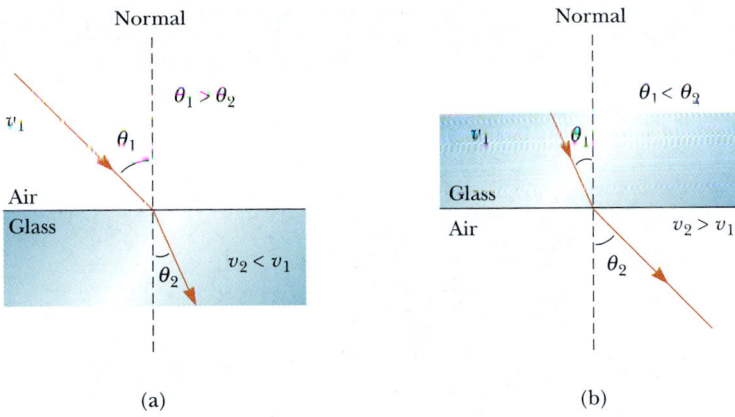

(a)   (b)

Figure 35.10 (a) When the light beam moves from air into glass, the light slows down on entering the glass and its path is bent toward the normal. (b) When the beam moves from glass into air, the light speeds up on entering the air and its path is bent away from the normal.

wood. When the bullet enters the air once again, it emerges at the speed it had just before leaving the block of wood.

To see why light behaves as it does, let us consider Figure 35.11, which represents a beam of light entering a piece of glass from the left. Once inside the glass, the light may encounter an electron bound to an atom, indicated as point A. Let us assume that light is absorbed by the atom; this causes the electron to oscillate (a detail represented by the double-headed vertical arrows). The oscillating electron then acts as an antenna[1] and radiates the beam of light toward an atom at B, where the light is again absorbed. The details of these absorptions and radiations are best explained in terms of quantum mechanics, a subject we study in the extended version of this text. For now, it is sufficient to think of the process as one in which the light passes from one atom to another through the glass. Although light travels from one glass atom to another at 3×10^8 m/s, the absorption and radiation that take place cause the average light speed through the material to slow to 2×10^8 m/s. Once the light emerges into the air, absorption and radiation cease and its speed returns to the original value.

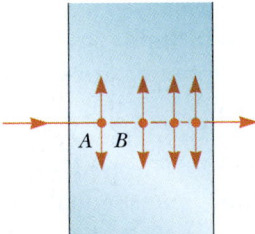

Figure 35.11 Light passing from one atom to another in a medium other than air. The dots are electrons, and the vertical arrows represent their oscillations.

---

[1] When the light frequency is such that it causes the electrons to oscillate near a resonant frequency, the amplitude is so great that atoms of the medium collide with one another and much of the light energy is transformed to internal energy, and thus is absorbed by the medium. The electrons of different materials have different resonant frequencies. This explains why you can see visible light through glass and why the ultraviolet frequencies, which cause sunburn, cannot pass through glass. On the other hand, ultraviolet light can penetrate clouds, whereas visible light cannot. Thus, even though you are not likely to be sunburned through a window, you can get a sunburn if you are outside on a cloudy day.

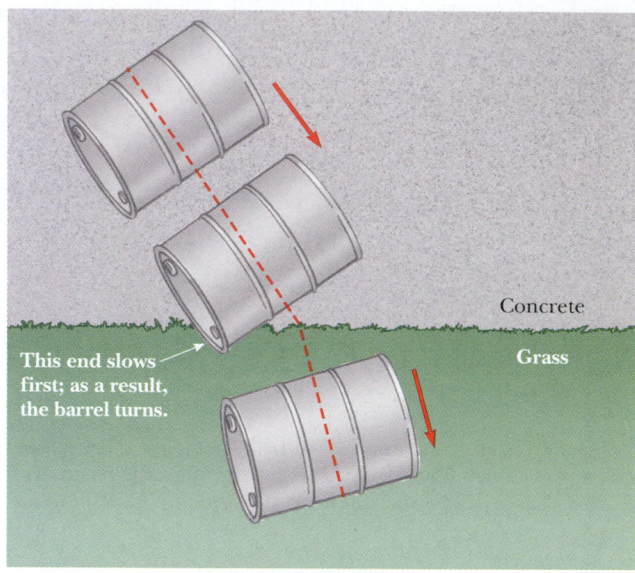

Figure 35.12   Overhead view of a barrel rolling from concrete onto grass.

A mechanical analog of refraction is shown in Figure 35.12. When the left end of the rolling barrel reaches the grass, it slows down, while the right end remains on the concrete and moves at its original speed. This difference in speeds causes the barrel to pivot, and this changes the direction of travel.

Index of Refraction

In general, the speed of light in any material is less than its speed in vacuum. In fact, *light travels at its maximum speed in vacuum*. It is convenient to define the **index of refraction** n of a medium to be the ratio

Index of refraction

$$n \equiv \frac{\text{Speed of light in vacuum}}{\text{Speed of light in a medium}} = \frac{c}{v} \qquad \textbf{(35.4)}$$

From this definition, we see that the index of refraction is a dimensionless number greater than unity because v is always less than c. Furthermore, n is equal to unity for vacuum. The indices of refraction for various substances are listed in Table 35.1.

As light travels from one medium to another, its frequency does not change but its wavelength does. To see why this is so, consider Figure 35.13. Wave fronts pass an observer at point A in medium 1 with a certain frequency and are incident on the boundary between medium 1 and medium 2. The frequency with which the wave fronts pass an observer at point B in medium 2 must equal the frequency at which they pass point A. If this were not the case, then wave fronts would be piling up at the boundary or they would be destroyed or created at the boundary. Because there is no mechanism for this to occur, the frequency must be a constant as a light ray passes from one medium into another. Therefore, because the relationship $v = f\lambda$ (Eq. 16.14) must be valid in both media and because $f_1 = f_2 = f$, we see that

$$v_1 = f\lambda_1 \qquad \text{and} \qquad v_2 = f\lambda_2 \qquad \textbf{(35.5)}$$

Because $v_1 \neq v_2$, it follows that $\lambda_1 \neq \lambda_2$.

QuickLab

Fill a clear drinking glass with water and place a pencil in the water, as shown in the figure below. Observe the pencil from the side at an angle of about 45° to the surface. Why does the pencil appear to be bent at the surface?

TABLE 35.1 Indices of Refraction[a]

Substance	Index of Refraction	Substance	Index of Refraction
Solids at 20°C		**Liquids at 20°C**	
Cubic zirconia	2.20	Benzene	1.501
Diamond (C)	2.419	Carbon disulfide	1.628
Fluorite (CaF_2)	1.434	Carbon tetrachloride	1.461
Fused quartz (SiO_2)	1.458	Ethyl alcohol	1.361
Gallium phosphide	3.50	Glycerin	1.473
Glass, crown	1.52	Water	1.333
Glass, flint	1.66		
Ice (H_2O)	1.309	**Gases at 0°C, 1 atm**	
Polystyrene	1.49	Air	1.000 293
Sodium chloride (NaCl)	1.544	Carbon dioxide	1.000 45

[a]All values are for light having a wavelength of 589 nm in vacuum.

We can obtain a relationship between index of refraction and wavelength by dividing the first Equation 35.5 by the second and then using Equation 35.4:

$$\frac{\lambda_1}{\lambda_2} = \frac{v_1}{v_2} = \frac{c/n_1}{c/n_2} = \frac{n_2}{n_1} \qquad \textbf{(35.6)}$$

This gives

$$\lambda_1 n_1 = \lambda_2 n_2$$

If medium 1 is vacuum, or for all practical purposes air, then $n_1 = 1$. Hence, it follows from Equation 35.6 that the index of refraction of any medium can be expressed as the ratio

$$n = \frac{\lambda}{\lambda_n} \qquad \textbf{(35.7)}$$

where λ is the wavelength of light in vacuum and λ_n is the wavelength in the medium whose index of refraction is n. From Equation 35.7, we see that because $n > 1$, $\lambda_n < \lambda$.

We are now in a position to express Equation 35.3 in an alternative form. If we replace the v_2/v_1 term in Equation 35.3 with n_1/n_2 from Equation 35.6, we obtain

$$n_1 \sin \theta_1 = n_2 \sin \theta_2 \qquad \textbf{(35.8)}$$

Snell's law of refraction

 QuickLab

Tape a coin to the bottom of a large opaque bowl, as shown in figure (a). Stand over the bowl so that you are looking at the coin and then move backwards away from the bowl until you can no longer see the coin over the bowl's rim. Remain at that position, and have a friend fill the bowl with water, as shown in figure (b). You can now see the coin again because the light is refracted at the water–air interface.

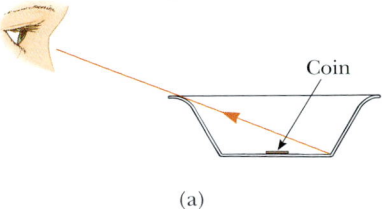

(a)

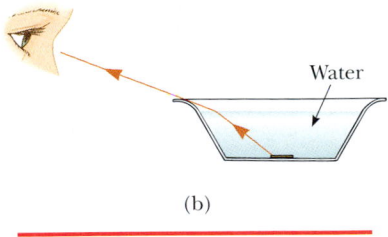

(b)

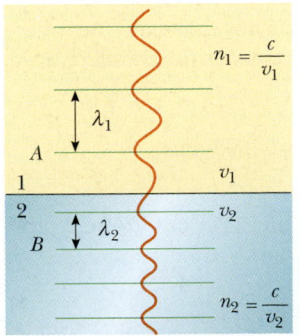

Figure 35.13   As a wave front moves from medium 1 to medium 2, its wavelength changes but its frequency remains constant.

The experimental discovery of this relationship is usually credited to Willebrord Snell (1591–1627) and is therefore known as **Snell's law of refraction.** We shall examine this formula further in Sections 35.6 and 35.9.

EXAMPLE 35.3   **An Index of Refraction Measurement**

A beam of light of wavelength 550 nm traveling in air is incident on a slab of transparent material. The incident beam makes an angle of 40.0° with the normal, and the refracted beam makes an angle of 26.0° with the normal. Find the index of refraction of the material.

Solution   Using Snell's law of refraction (Eq. 35.8) with these data, and taking $n_1 = 1.00$ for air, we have

$$n_1 \sin \theta_1 = n_2 \sin \theta_2$$

$$n_2 = \frac{n_1 \sin \theta_1}{\sin \theta_2} = (1.00)\frac{\sin 40.0°}{\sin 26.0°} = \frac{0.643}{0.438} = \boxed{1.47}$$

From Table 35.1, we see that the material could be fused quartz.

Exercise   What is the wavelength of light in the material?

Answer   374 nm.

EXAMPLE 35.4   **Angle of Refraction for Glass**

A light ray of wavelength 589 nm traveling through air is incident on a smooth, flat slab of crown glass at an angle of 30.0° to the normal, as sketched in Figure 35.14. Find the angle of refraction.

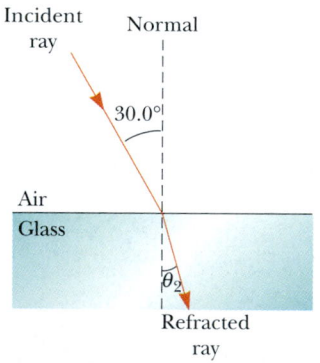

Figure 35.14   Refraction of light by glass.

Solution   We rearrange Snell's law of refraction to obtain

$$\sin \theta_2 = \frac{n_1}{n_2} \sin \theta_1$$

From Table 35.1, we find that $n_1 = 1.00$ for air and that $n_2 = 1.52$ for crown glass. Therefore,

$$\sin \theta_2 = \left(\frac{1.00}{1.52}\right)(\sin 30.0°) = 0.329$$

$$\theta_2 = \sin^{-1}(0.329) = \boxed{19.2°}$$

Because this is less than the incident angle of 30°, the ray is bent toward the normal, as expected. Its change in direction is called the *angle of deviation* and is given by $\delta = |\theta_1 - \theta_2| = 30.0° - 19.2° = 10.8°$.

Exercise   If the light ray moves from inside the glass toward the glass–air interface at an angle of 30.0° to the normal, what is the angle of refraction?

Answer   49.5° away from the normal.

EXAMPLE 35.5   **Laser Light in a Compact Disc**

A laser in a compact disc player generates light that has a wavelength of 780 nm in air. (a) Find the speed of this light once it enters the plastic of a compact disc ($n = 1.55$).

Solution   We expect to find a value less than 3×10^8 m/s because $n > 1$. We can obtain the speed of light in the plastic by using Equation 35.4:

$$v = \frac{c}{n} = \frac{3.00 \times 10^8 \text{ m/s}}{1.55} = \boxed{1.94 \times 10^8 \text{ m/s}}$$

(b) What is the wavelength of this light in the plastic?

Solution   We use Equation 35.7 to calculate the wavelength in plastic, noting that we are given the wavelength in air to be $\lambda = 780$ nm:

$$\lambda_n = \frac{\lambda}{n} = \frac{780 \text{ nm}}{1.55} = \boxed{503 \text{ nm}}$$

Exercise   Find the frequency of the light in air and in the plastic.

Answer 3.85×10^{14} Hz in both cases.

EXAMPLE 35.6 Light Passing Through a Slab

A light beam passes from medium 1 to medium 2, with the latter medium being a thick slab of material whose index of refraction is n_2 (Fig. 35.15). Show that the emerging beam is parallel to the incident beam.

Solution First, let us apply Snell's law of refraction to the upper surface:

$$(1) \qquad \sin \theta_2 = \frac{n_1}{n_2} \sin \theta_1$$

Applying this law to the lower surface gives

$$(2) \qquad \sin \theta_3 = \frac{n_2}{n_1} \sin \theta_2$$

Substituting Equation (1) into Equation (2) gives

$$\sin \theta_3 = \frac{n_2}{n_1} \left(\frac{n_1}{n_2} \sin \theta_1 \right) = \sin \theta_1$$

Therefore, $\theta_3 = \theta_1$, and the slab does not alter the direction of the beam. It does, however, displace the beam parallel to itself.

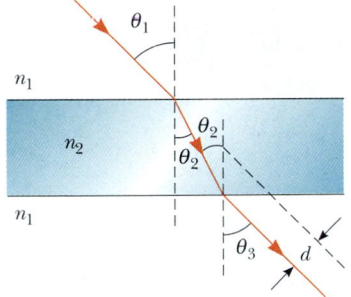

Figure 35.15 When light passes through a flat slab of material, the emerging beam is parallel to the incident beam, and therefore $\theta_1 = \theta_3$. The dashed line drawn parallel to the red ray coming out the bottom of the slab represents the path the light would take if the slab were not there.

35.6 HUYGENS'S PRINCIPLE

In this section, we develop the laws of reflection and refraction by using a geometric method proposed by Huygens in 1678. As noted in Section 35.1, Huygens assumed that light is some form of wave motion rather than a stream of particles. He had no knowledge of the nature of light or of its electromagnetic character. Nevertheless, his simplified wave model is adequate for understanding many practical aspects of the propagation of light.

Huygens's principle is a geometric construction for using knowledge of an earlier wave front to determine the position of a new wave front at some instant. In Huygens's construction,

> all points on a given wave front are taken as point sources for the production of spherical secondary waves, called *wavelets*, which propagate outward through a medium with speeds characteristic of waves in that medium. After some time has elapsed, the new position of the wave front is the surface tangent to the wavelets.

Huygens's principle

First, let us consider a plane wave moving through free space, as shown in Figure 35.16a. At $t = 0$, the wave front is indicated by the plane labeled AA'. In Huygens's construction, each point on this wave front is considered a point source. For clarity, only three points on AA' are shown. With these points as sources for the wavelets, we draw circles, each of radius $c \, \Delta t$, where c is the speed of light in free space and Δt is the time of propagation from one wave front to the next. The surface drawn tangent to these wavelets is the plane BB', which is parallel to AA'. In a similar manner, Figure 35.16b shows Huygens's construction for a spherical wave.

Figure 35.17 shows a convincing demonstration of Huygens's principle. Plane waves coming from far off shore emerge from the openings in the barrier as two-dimensional circular waves propagating outward.

Christian Huygens (1629–1695)
Huygens, a Dutch physicist and astronomer, is best known for his contributions to the fields of optics and dynamics. To Huygens, light was a type of vibratory motion, spreading out and producing the sensation of sight when impinging on the eye. On the basis of this theory, he deduced the laws of reflection and refraction and explained the phenomenon of double refraction. *(Courtesy of Rijksmuseum voor de Geschiedenis der Natuurwetenschappen. Courtesy Niels Bohr Library.)*

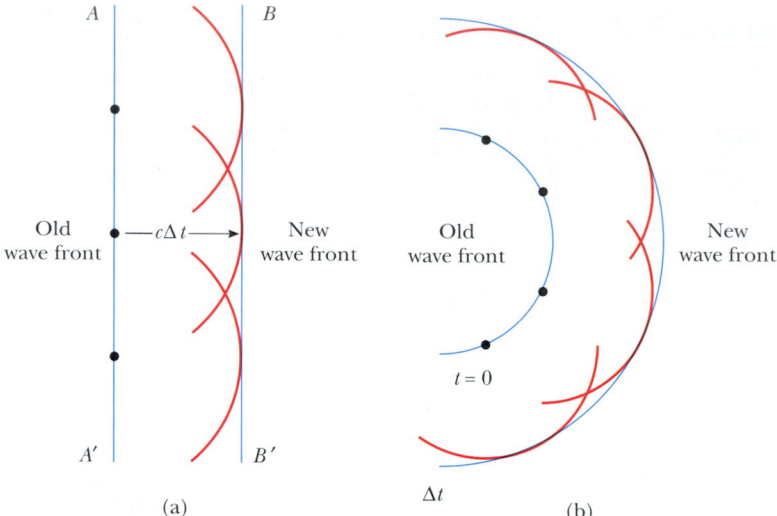

Figure 35.16 Huygens's construction for (a) a plane wave propagating to the right and (b) a spherical wave propagating to the right.

Figure 35.17 This photograph of the beach at Tel Aviv, Israel, shows Huygens wavelets radiating from each opening through a breakwall. Note how the beach has been shaped by the wave action. *(Courtesy of Sabina Zigman/Benjamin Cardozo High School, and by permission of PHYSICS TEACHER, vol. 37, January 1999, p. 55.)*

Huygens's Principle Applied to Reflection and Refraction

The laws of reflection and refraction were stated earlier in this chapter without proof. We now derive these laws, using Huygens's principle.

For the law of reflection, refer to Figure 35.18a. The line AA' represents a wave front of the incident light. As ray 3 travels from A' to C, ray 1 reflects from A and produces a spherical wavelet of radius AD. (Recall that the radius of a Huygens wavelet is $c\,\Delta t$.) Because the two wavelets having radii $A'C$ and AD are in

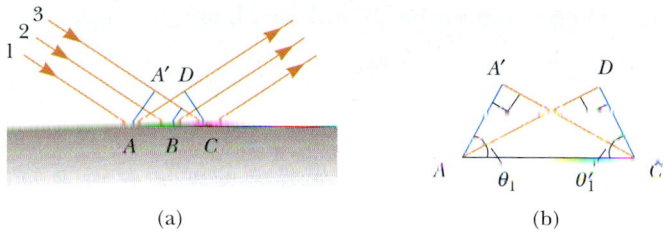

(a)   (b)

Figure 35.18   (a) Huygens's construction for proving the law of reflection. (b) Triangle ADC is congruent to triangle $AA'C$.

the same medium, they have the same speed c; therefore, $A'C = AD$. Meanwhile, the spherical wavelet centered at B has spread only half as far as the one centered at A because ray 2 strikes the surface later than ray 1 does.

From Huygens's principle, we find that the reflected wave front is CD, a line tangent to all the outgoing spherical wavelets. The remainder of our analysis depends on geometry, as summarized in Figure 35.18b. Note that the right triangles ADC and $AA'C$ are congruent because they have the same hypotenuse AC and because $AD = A'C$. From Figure 35.18b, we have

$$\sin \theta_1 = \frac{A'C}{AC} \quad \text{and} \quad \sin \theta_1' = \frac{AD}{AC}$$

Thus,

$$\sin \theta_1 = \sin \theta_1'$$
$$\theta_1 = \theta_1'$$

which is the law of reflection.

Now let us use Huygens's principle and Figure 35.19 to derive Snell's law of refraction. Note that in the time interval Δt, ray 1 moves from A to B and ray 2 moves from A' to C. The radius of the outgoing spherical wavelet centered at A is equal to $v_2 \Delta t$. The distance $A'C$ is equal to $v_1 \Delta t$. Geometric considerations show that angle $A'AC$ equals θ_1 and that angle ACB equals θ_2. From triangles $AA'C$ and ACB, we find that

$$\sin \theta_1 = \frac{v_1 \Delta t}{AC} \quad \text{and} \quad \sin \theta_2 = \frac{v_2 \Delta t}{AC}$$

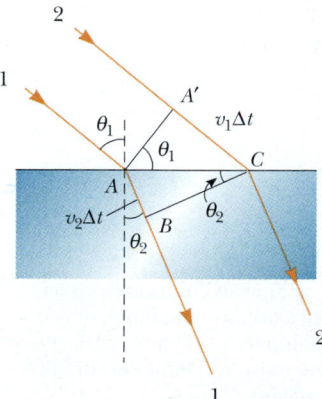

Figure 35.19   Huygens's construction for proving Snell's law of refraction.

If we divide the first equation by the second, we obtain

$$\frac{\sin \theta_1}{\sin \theta_2} = \frac{v_1}{v_2}$$

But from Equation 35.4 we know that $v_1 = c/n_1$ and that $v_2 = c/n_2$. Therefore,

$$\frac{\sin \theta_1}{\sin \theta_2} = \frac{c/n_1}{c/n_2} = \frac{n_2}{n_1}$$

$$n_1 \sin \theta_1 = n_2 \sin \theta_2$$

which is Snell's law of refraction.

35.7 ▷ DISPERSION AND PRISMS

14.4 An important property of the index of refraction is that, for a given material, the index varies with the wavelength of the light passing through the material, as Figure 35.20 shows. This behavior is called **dispersion.** Because n is a function of wavelength, Snell's law of refraction indicates that light of different wavelengths is bent at different angles when incident on a refracting material.

As we see from Figure 35.20, the index of refraction generally decreases with increasing wavelength. This means that blue light bends more than red light does when passing into a refracting material. To understand the effects that dispersion can have on light, let us consider what happens when light strikes a prism, as shown in Figure 35.21. A ray of single-wavelength light incident on the prism from the left emerges refracted from its original direction of travel by an angle δ, called the **angle of deviation.**

Now suppose that a beam of *white light* (a combination of all visible wavelengths) is incident on a prism, as illustrated in Figure 35.22. The rays that emerge spread out in a series of colors known as the **visible spectrum.** These colors, in order of decreasing wavelength, are red, orange, yellow, green, blue, and violet. Clearly, the angle of deviation δ depends on wavelength. Violet light deviates the most, red the least, and the remaining colors in the visible spectrum fall between these extremes. Newton showed that each color has a particular angle of deviation and that the colors can be recombined to form the original white light.

A prism is often used in an instrument known as a **prism spectrometer,** the essential elements of which are shown in Figure 35.23. The instrument is commonly used to study the wavelengths emitted by a light source. Light from the source is sent through a narrow, adjustable slit to produce a parallel, or *collimated*, beam. The light then passes through the prism and is dispersed into a spectrum. The dispersed light is observed through a telescope. The experimenter sees an im-

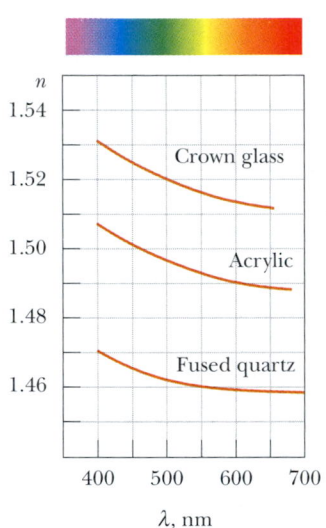

Figure 35.20 Variation of index of refraction with vacuum wavelength for three materials.

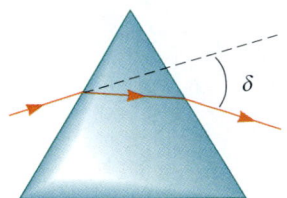

Figure 35.21 A prism refracts a single-wavelength light ray through an angle δ.

Figure 35.22 Different colors are refracted at different angles because the index of refraction of the glass depends on wavelength. Violet light deviates the most; red light deviates the least. *(Courtesy of Bausch & Lomb)*

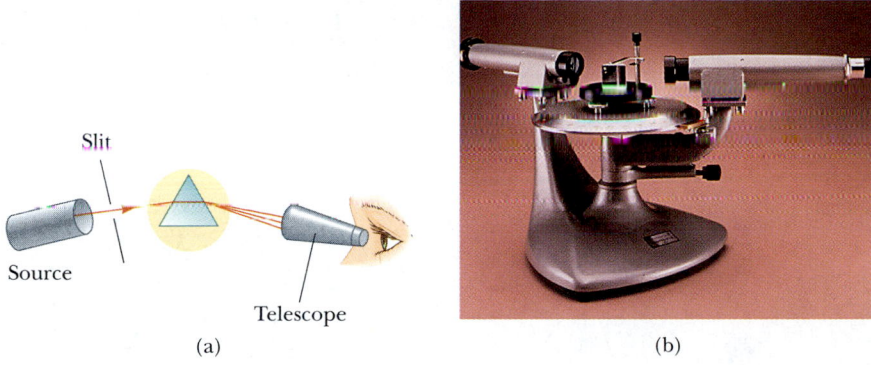

Slit

Source

Telescope

(a)

(b)

Figure 35.23 (a) Diagram of a prism spectrometer. The various colors in the visible spectrum are viewed through a telescope. (b) Photograph of a prism spectrometer. *(Courtesy of Central Scientific Company)*

age of the slit through the eyepiece of the telescope. The telescope can be moved or the prism rotated so that the various images formed by different wavelengths at different angles of deviation can be viewed.

All hot, low-pressure gases emit their own characteristic spectra. Thus, one use of a prism spectrometer is to identify gases. For example, sodium emits two wavelengths, 589.0 nm and 589.6 nm, in the visible spectrum, which appear as two closely spaced yellow lines. Thus, a gas emitting these colors can be identified as having sodium as one of its constituents. Likewise, mercury vapor has its own characteristic spectrum, dominated by lines of four prominent colors—orange, green, blue, and violet—along with some other lines at lower intensity. The particular wavelengths emitted by a gas serve as "fingerprints" of that gas.

The dispersion of light into a spectrum is demonstrated most vividly in nature by the formation of a rainbow, which is often seen by an observer positioned between the Sun and a rain shower. To understand how a rainbow is formed, let us consider Figure 35.24. A ray of sunlight (which is white light) passing overhead strikes a drop of water in the atmosphere and is refracted and reflected as follows: It is first refracted at the front surface of the drop, with the violet light deviating the most and the red light the least. At the back surface of the drop, the light is reflected and returns to the front surface, where it again undergoes refraction as it moves from water into air. The rays leave the drop such that the angle between the incident white light and the most intense returning violet ray is 40° and the angle

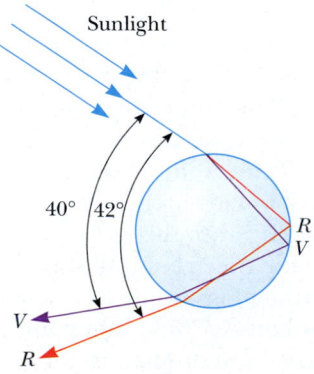

Sunlight

40° 42°

R
V

V

R

Figure 35.24 Refraction of sunlight by a spherical raindrop.

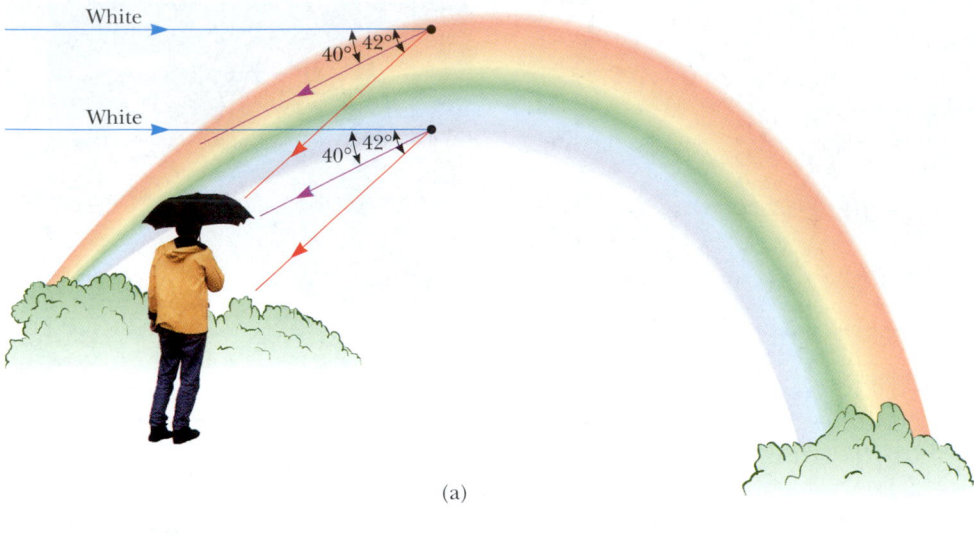

(a)

(b)

Figure 35.25   (a) The formation of a rainbow. (b) A rainbow over Niagara Falls in Ontario, Canada. (*John Edwards/Tony Stone Images*)

between the white light and the most intense returning red ray is 42°. This small angular difference between the returning rays causes us to see a colored bow.

Now let us consider an observer viewing a rainbow, as shown in Figure 35.25a. If a raindrop high in the sky is being observed, the red light returning from the drop can reach the observer because it is deviated the most, but the violet light passes over the observer because it is deviated the least. Hence, the observer sees this drop as being red. Similarly, a drop lower in the sky would direct violet light toward the observer and appears to be violet. (The red light from this drop would strike the ground and not be seen.) The other colors of the spectrum would reach the observer from raindrops lying between these two extreme positions.

It is interesting to observe artists' renderings of rainbows because many times the drawings are incorrect, with violet on the outer part and red on the inside.

Quick Quiz 35.2

Lenses in a camera use refraction to form an image on a film, and you want, as much as possible, all the colors in the light from the object being photographed to be refracted by the same amount. Of the materials shown in Figure 35.20, which would you choose for a camera lens?

EXAMPLE 35.7 ▸ Measuring *n* Using a Prism

Although we do not prove it here, the minimum angle of deviation δ_{min} for a prism occurs when the angle of incidence θ_1 is such that the refracted ray inside the prism makes the same angle α with the normal to the two prism faces,[2] as shown in Figure 35.26. Let us obtain an expression for the index of refraction of the prism material.

Using the geometry shown in Figure 35.26, we find that $\theta_2 = \Phi/2$, where Φ is the apex angle and

$$\theta_1 = \theta_2 + \alpha = \frac{\Phi}{2} + \frac{\delta_{min}}{2} = \frac{\Phi + \delta_{min}}{2}$$

From Snell's law of refraction, with $n_1 = 1$ because medium 1 is air, we have

$$\sin \theta_1 = n \sin \theta_2$$

$$\sin\left(\frac{\Phi + \delta_{min}}{2}\right) = n \sin (\Phi/2)$$

$$n = \frac{\sin\left(\dfrac{\Phi + \delta_{min}}{2}\right)}{\sin (\Phi/2)} \quad \text{(35.9)}$$

Hence, knowing the apex angle Φ of the prism and measuring δ_{min}, we can calculate the index of refraction of the prism material. Furthermore, we can use a hollow prism to determine the values of *n* for various liquids filling the prism.

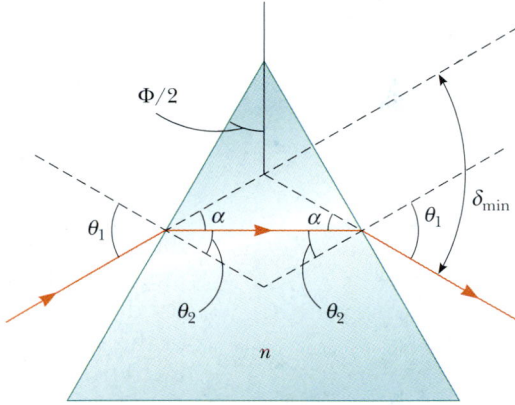

Figure 35.26 A light ray passing through a prism at the minimum angle of deviation δ_{min}.

35.8 ▸ TOTAL INTERNAL REFLECTION

An interesting effect called **total internal reflection** can occur when light attempts to move from a medium having a given index of refraction to one having a lower index of refraction. Consider a light beam traveling in medium 1 and meeting the boundary between medium 1 and medium 2, where n_1 is greater than n_2 (Fig. 35.27a). Various possible directions of the beam are indicated by rays 1 through 5. The refracted rays are bent away from the normal because n_1 is greater than n_2. At some particular angle of incidence θ_c, called the **critical angle,** the refracted light ray moves parallel to the boundary so that $\theta_2 = 90°$ (ray 4 in Fig. 35.27a).

For angles of incidence greater than θ_c, the beam is entirely reflected at the boundary, as shown by ray 5 in Figure 35.27a. This ray is reflected at the boundary as it strikes the perfectly reflecting surface. This ray, and all those like it, obey the

[2] For details, see Chapter 2 of F. A. Jenkins and H. E. White, *Fundamentals of Optics,* New York, McGraw-Hill, 1976.

law of reflection; that is, for these rays, the angle of incidence equals the angle of reflection.

We can use Snell's law of refraction to find the critical angle. When $\theta_1 = \theta_c$, $\theta_2 = 90°$ and Equation 35.8 gives

$$n_1 \sin \theta_c = n_2 \sin 90° = n_2$$

<table>
<tr><td>Critical angle for total internal reflection</td><td>$$\sin \theta_c = \frac{n_2}{n_1}$$ (for $n_1 > n_2$)</td><td>**(35.10)**</td></tr>
</table>

This equation can be used only when n_1 is greater than n_2. That is,

> total internal reflection occurs only when light moves from a medium of a given index of refraction to a medium of lower index of refraction.

If n_1 were less than n_2, Equation 35.10 would give $\sin \theta_c > 1$; this is a meaningless result because the sine of an angle can never be greater than unity.

The critical angle for total internal reflection is small when n_1 is considerably greater than n_2. For example, the critical angle for a diamond ($n = 2.4$) in air is 24°. Any ray inside the diamond that approaches the surface at an angle greater than this is completely reflected back into the stone. This property, combined with proper faceting, causes diamonds to sparkle. The angles of the facets are cut so that light is "caught" inside the stone through multiple internal reflections. These multiple reflections give the light a long path through the medium, and substantial dispersion of colors occurs. By the time the light exits through the top surface of the stone, the rays associated with different colors have been fairly widely separated from one another.

Cubic zirconia also has a high index of refraction and can be made to sparkle very much like a genuine diamond. If a suspect jewel is immersed in corn syrup, the difference in n for the cubic zirconia and that for the syrup is small, and the critical angle is therefore great. This means that more rays escape sooner, and as a result the sparkle completely disappears. A real diamond does not lose all of its sparkle when placed in corn syrup.

Quick Quiz 35.3

If you look closely at Figure 35.28, you will see five rays exiting the bottom of the prism. Why are the two rays at the right much brighter than the other three?

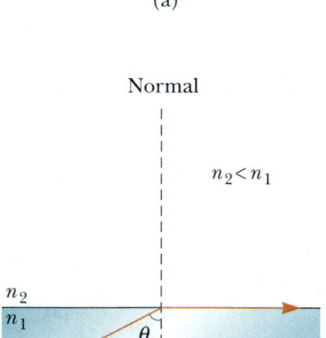

(a)

(b)

Figure 35.27 (a) Rays travel from a medium of index of refraction n_1 into a medium of index of refraction n_2, where $n_2 < n_1$. As the angle of incidence θ_1 increases, the angle of refraction θ_2 increases until θ_2 is 90° (ray 4). For even larger angles of incidence, total internal reflection occurs (ray 5). (b) The angle of incidence producing an angle of refraction equal to 90° is the critical angle, θ_c.

Figure 35.28 Five nonparallel light rays enter a glass prism from the left. The bottom two of these five rays undergo total internal reflection at the right face of the prism. The top three rays are refracted at that face as they leave the prism. (*Courtesy of Henry Leap and Jim Lehman*)

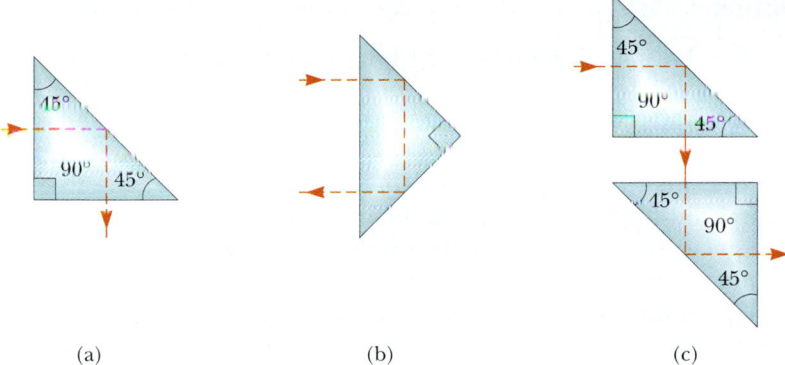

Figure 35.29   Internal reflection in a prism. (a) The direction of travel is changed by 90°. (b) The direction of travel is reversed. (c) Two prisms used as a periscope.

A prism and total internal reflection alter the direction of travel of a light beam. Two possibilities are illustrated in Figure 35.29a and b.

A common application of total internal reflection is in a submarine periscope. In this device, two prisms are arranged as in Figure 35.29c so that an incident beam of light follows the path shown. As a result, one is able to "see around corners" when looking through a periscope.

EXAMPLE 35.8 ▶ A View from the Fish's Eye

(a) Find the critical angle for a water–air boundary (the index of refraction of water is 1.33).

Solution   We can use Figure 35.27 to solve this problem, with the air above the pond having index of refraction n_2 and the water having index of refraction n_1. Applying Equation 35.10, we find that

$$\sin \theta_c = \frac{n_2}{n_1} = \frac{1}{1.33} = 0.752$$

$$\theta_c = \boxed{48.8°}$$

(b) Use this result to predict what a fish sees in a still pond when it looks upward toward the water's surface at angles of 40°, 48.8°, and 60° (Fig. 35.30).

Solution   Because the path of a light ray is reversible, light traveling from medium 2 into medium 1 in Figure 35.27a follows the paths shown, but in the *opposite* direction. A fish looking upward toward the water surface, as shown in Figure 35.30, can see out of the water if it looks toward the surface at an angle less than the critical angle. Thus, when the fish's line of vision makes an angle of 40° with the normal to the surface, light from a cloud, for example, reaches the fish's

eye. At 48.8°, the critical angle for water, the light has to skim along the water's surface before being refracted to the fish's eye; at this angle, the fish can in principle see the whole shore of the pond. At angles greater than the critical angle, the light reaching the fish comes by means of internal reflection at the surface. Thus, at 60°, the fish sees a reflection of the bottom of the pond.

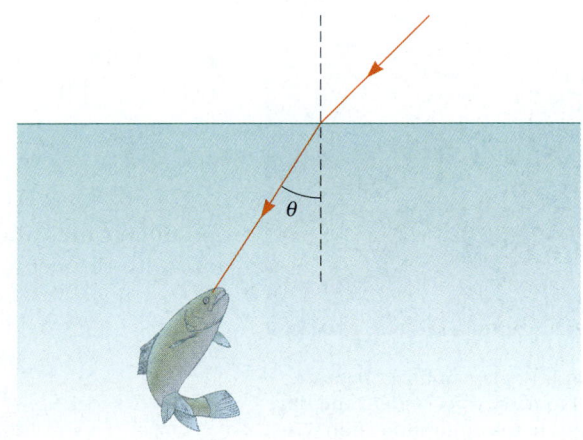

Figure 35.30

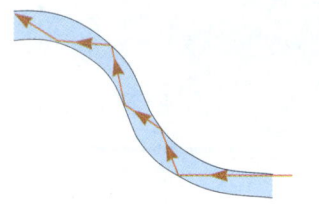

Figure 35.31   Light travels in a curved transparent rod by multiple internal reflections.

Fiber Optics

Another interesting application of total internal reflection is the use of glass or transparent plastic rods to "pipe" light from one place to another. As indicated in Figure 35.31, light is confined to traveling within a rod, even around curves, as the result of successive internal reflections. Such a light pipe is flexible if thin fibers are used rather than thick rods. If a bundle of parallel fibers is used to construct an optical transmission line, images can be transferred from one point to another. This technique is used in a sizable industry known as fiber optics.

Very little light intensity is lost in the fibers as a result of reflections on the sides. Any loss in intensity is due essentially to reflections from the two ends and absorption by the fiber material. Fiber-optic devices are particularly useful for viewing an object at an inaccessible location. For example, physicians often use such devices to examine internal organs of the body or to perform surgery without making large incisions. Fiber-optic cables are replacing copper wiring and coaxial cables for telecommunications because the fibers can carry a much greater volume of telephone calls or other forms of communication than electrical wires can.

Optional Section

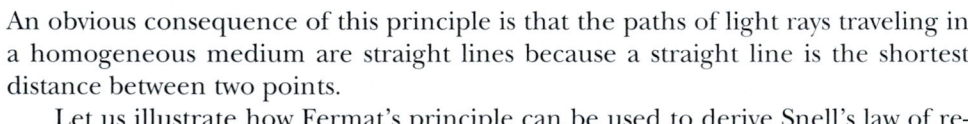

35.9 ▶ FERMAT'S PRINCIPLE

Pierre de Fermat (1601–1665) developed a general principle that can be used to determine light paths. **Fermat's principle** states that

Fermat's principle

> when a light ray travels between any two points, its path is the one that requires the least time.

An obvious consequence of this principle is that the paths of light rays traveling in a homogeneous medium are straight lines because a straight line is the shortest distance between two points.

Let us illustrate how Fermat's principle can be used to derive Snell's law of refraction. Suppose that a light ray is to travel from point P in medium 1 to point Q in medium 2 (Fig. 35.32), where P and Q are at perpendicular distances a and b, respectively, from the interface. The speed of light is c/n_1 in medium 1 and c/n_2 in medium 2. Using the geometry of Figure 35.32, we see that the time it takes the ray to travel from P to Q is

$$t = \frac{r_1}{v_1} + \frac{r_2}{v_2} = \frac{\sqrt{a^2 + x^2}}{c/n_1} + \frac{\sqrt{b^2 + (d-x)^2}}{c/n_2} \qquad (35.11)$$

To obtain the value of x for which t has its minimum value, we take the derivative of t with respect to x and set the derivative equal to zero:

$$\frac{dt}{dx} = \frac{n_1}{c} \frac{d}{dx}\sqrt{a^2 + x^2} + \frac{n_2}{c} \frac{d}{dx}\sqrt{b^2 + (d-x)^2}$$

$$= \frac{n_1}{c}\left(\frac{1}{2}\right)\frac{2x}{(a^2 + x^2)^{1/2}} + \frac{n_2}{c}\left(\frac{1}{2}\right)\frac{2(d-x)(-1)}{[b^2 + (d-x)^2]^{1/2}}$$

$$= \frac{n_1 x}{c(a^2 + x^2)^{1/2}} - \frac{n_2(d-x)}{c[b^2 + (d-x)^2]^{1/2}} = 0$$

Strands of glass optical fibers are used to carry voice, video, and data signals in telecommunication networks. Typical fibers have diameters of 60 μm. *(Dennis O'Clair/ Tony Stone Images)*

or

$$\frac{n_1 x}{(a^2 + x^2)^{1/2}} = \frac{n_2(d - x)}{[b^2 + (d - x)^2]^{1/2}}$$ **(35.12)**

From Figure 35.32,

$$\sin \theta_1 = \frac{x}{(a^2 + x^2)^{1/2}} \qquad \sin \theta_2 = \frac{d - x}{[b^2 + (d - x)^2]^{1/2}}$$

Substituting into Equation 35.12, we find that

$$n_1 \sin \theta_1 = n_2 \sin \theta_2$$

which is Snell's law of refraction.

It is a simple matter to use a similar procedure to derive the law of reflection (see Problem 65).

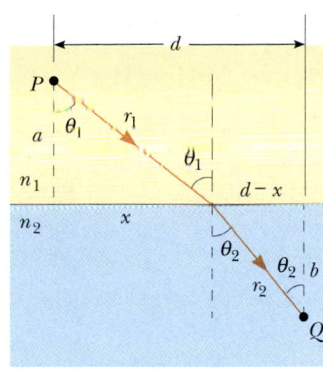

Figure 35.32 Geometry for deriving Snell's law of refraction using Fermat's principle.

CONCEPTUAL EXAMPLE 35.9

Use spreadsheet software or some other tool to create a graph of t versus x based on Equation 35.11.

Solution Consistent with the fact that we are finding the minimum time required for the light to go from point P to point Q, Figure 35.33 shows that a specific value of x results in a minimum time. Changing the values of a, b, d, n_1, and n_2 changes the details of the graph but not the general shape of the function. (When you make up numbers for these parameters so that the spreadsheet software can create a graph, be sure to take $d > x$. You should also select $n_2 > n_1$ so that the situation corresponds to the path shown in Figure 35.32. If $n_1 < n_2$, the path bends the other way.)

This situation is equivalent to the problem of deciding where a lifeguard who can run faster than he can swim should enter the water to help a swimmer in distress. If he enters the water too directly (in other words, at a very small value of θ_1 in Figure 35.32), the distance x is smaller than the value of x that gives the minimum value of the time needed for the guard to get from the starting point on the sand to the swimmer. As a result, he spends too little time running and too much time swimming. The guard's optimum location for entering the water so that he can reach the swimmer in the shortest time is at that interface point that gives the value of x that satisfies Equation 35.12.

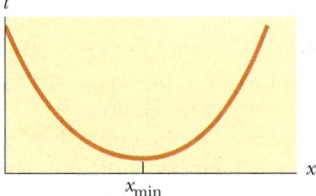

Figure 35.33

SUMMARY

In geometric optics, we use the **ray approximation,** which assumes that a wave travels through a uniform medium in straight lines in the direction of the rays.

The **law of reflection** states that for a light ray traveling in air and incident on a smooth surface, the angle of reflection θ_1' equals the angle of incidence θ_1:

$$\theta_1' = \theta_1$$ **(35.2)**

Light crossing a boundary as it travels from medium 1 to medium 2 is **refracted,** or bent. The angle of refraction θ_2 is defined by the relationship

$$\frac{\sin \theta_2}{\sin \theta_1} = \frac{v_2}{v_1} = \text{constant}$$ **(35.3)**

The **index of refraction** n of a medium is defined by the ratio

$$n \equiv \frac{c}{v} \tag{35.4}$$

where c is the speed of light in a vacuum and v is the speed of light in the medium. In general, n varies with wavelength and is given by

$$n = \frac{\lambda}{\lambda_n} \tag{35.7}$$

where λ is the vacuum wavelength and λ_n is the wavelength in the medium. As light travels from one medium to another, its frequency remains the same.

Snell's law of refraction states that

$$n_1 \sin \theta_1 = n_2 \sin \theta_2 \tag{35.8}$$

where n_1 and n_2 are the indices of refraction in the two media. The incident ray, the reflected ray, the refracted ray, and the normal to the surface all lie in the same plane.

Total internal reflection occurs when light travels from a medium of high index of refraction to one of lower index of refraction. The **critical angle** θ_c for which total reflection occurs at an interface is given by

$$\sin \theta_c = \frac{n_2}{n_1} \qquad (\text{for } n_1 > n_2) \tag{35.10}$$

QUESTIONS

1. Light of wavelength λ is incident on a slit of width d. Under what conditions is the ray approximation valid? Under what circumstances does the slit produce enough diffraction to make the ray approximation invalid?

2. Sound waves have much in common with light waves, including the properties of reflection and refraction. Give examples of these phenomena for sound waves.

3. Does a light ray traveling from one medium into another always bend toward the normal, as shown in Figure 35.9? Explain.

4. As light travels from one medium to another, does the wavelength of the light change? Does the frequency change? Does the speed change? Explain.

5. A laser beam passing through a nonhomogeneous sugar solution follows a curved path. Explain.

6. A laser beam with vacuum wavelength 632.8 nm is incident from air onto a block of Lucite, as shown in Figure 35.9b. Assuming that the line of sight of the photograph is perpendicular to the plane in which the light moves, find the speed, frequency, and wavelength of the light in the Lucite.

7. Suppose blue light were used instead of red light in the experiment shown in Figure 35.9b. Would the refracted beam be bent at a larger or smaller angle?

8. The level of water in a clear, colorless glass is easily observed with the naked eye. The level of liquid helium in a clear glass vessel is extremely difficult to see with the naked eye. Explain.

9. In Example 35.6 we saw that light entering a slab with parallel sides emerges deflected but still parallel to the incoming beam. Our assumption was that the index of refraction of the material did not vary with wavelength. If the slab were made of crown glass (see Fig. 35.20), what would the outgoing beam look like?

10. The F-117A stealth fighter (Fig. Q35.10) is specifically designed to be a *non*-retroreflector of radar. What aspect of its design helps accomplish this? Note that the aircraft's structure consists of many flat angled panels and a flat bottom.

11. Explain why a diamond sparkles more than a glass crystal of the same shape and size.

12. Explain why an oar in the water appears bent.

13. Redesign the periscope of Figure 35.29c so that it can show you where you have been rather than where you are going.

14. Under certain circumstances, sound can be heard over extremely great distances. This frequently happens over a body of water, where the air near the water's surface is cooler than the air higher up. Explain how the refraction of sound waves in such a situation could increase the distance over which sound can be heard.

15. Why do astronomers looking at distant galaxies talk about looking backward in time?

16. A solar eclipse occurs when the Moon passes between the Earth and the Sun. Use a diagram to show why some

Figure Q35.10 *(Courtesy of U.S. Air Force, Langley Air Force Base)*

areas of the Earth see a total eclipse, other areas see a partial eclipse, and most areas see no eclipse.

17. The display windows of some department stores are slanted slightly inward at the bottom. This is to decrease the glare from streetlights or the Sun, which would make it difficult for shoppers to see the display inside. Sketch a light ray reflecting off such a window to show how this technique works.

18. When two colors of light X and Y are sent through a glass prism, X is bent more than Y. Which color travels more slowly in the prism?

19. Why does the arc of a rainbow appear with red on top and violet on the bottom?

20. Under what conditions is a mirage formed? On a hot day, what are we seeing when we observe "water on the road"?

PROBLEMS

1, 2, 3 = straightforward, intermediate, challenging ☐ = full solution available in the *Student Solutions Manual and Study Guide*
WEB = solution posted at **http://www.saunderscollege.com/physics/** 🖥 = Computer useful in solving problem 🔧 = Interactive Physics
☐ = paired numerical/symbolic problems

Section 35.1 The Nature of Light

Section 35.2 Measurements of the Speed of Light

1. The Apollo 11 astronauts set up a highly reflecting panel on the Moon's surface. The speed of light can be found by measuring the time it takes a laser beam to travel from Earth, reflect from the retroreflector, and return to Earth. If this interval is measured to be 2.51 s, what is the measured speed of light? Take the center-to-center distance from the Earth to the Moon to be 3.84×10^8 m, and do not neglect the sizes of the Earth and the Moon.

2. As a result of his observations, Roemer concluded that eclipses of Io by Jupiter were delayed by 22 min during a six-month period as the Earth moved from the point in its orbit where it is closest to Jupiter to the diametrically opposite point where it is farthest from Jupiter. Using 1.50×10^8 km as the average radius of the Earth's orbit around the Sun, calculate the speed of light from these data.

3. In an experiment to measure the speed of light using the apparatus of Fizeau (see Fig. 35.2), the distance between light source and mirror was 11.45 km and the wheel had 720 notches. The experimentally determined value of c was 2.998×10^8 m/s. Calculate the minimum angular speed of the wheel for this experiment.

4. Figure P35.4 shows an apparatus used to measure the speed distribution of gas molecules. It consists of two slotted rotating disks separated by a distance d, with the slots displaced by the angle θ. Suppose that the speed of light is measured by sending a light beam from the left through this apparatus. (a) Show that a light beam will be seen in the detector (that is, will make it through both slots) only if its speed is given by $c = \omega d/\theta$, where ω is the angular speed of the disks and θ is measured in radians. (b) What is the measured speed of light if the distance between the two slotted rotating disks is 2.50 m, the slot in the second disk is displaced 1/60 of 1° from the slot in the first disk, and the disks are rotating at 5 555 rev/s?

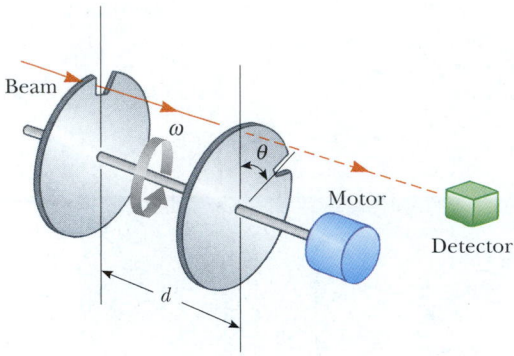

Figure P35.4

Section 35.3   The Ray Approximation in Geometric Optics
Section 35.4   Reflection
Section 35.5   Refraction
Section 35.6   Huygens's Principle

Note: In this section, if an index of refraction value is not given, refer to Table 35.1.

5. A narrow beam of sodium yellow light, with wavelength 589 nm in vacuum, is incident from air onto a smooth water surface at an angle $\theta_1 = 35.0°$. Determine the angle of refraction θ_2 and the wavelength of the light in water.

6. The wavelength of red helium–neon laser light in air is 632.8 nm. (a) What is its frequency? (b) What is its wavelength in glass that has an index of refraction of 1.50? (c) What is its speed in the glass?

7. An underwater scuba diver sees the Sun at an apparent angle of 45.0° from the vertical. What is the actual direction of the Sun?

8. A laser beam is incident at an angle of 30.0° from the vertical onto a solution of corn syrup in water. If the beam is refracted to 19.24° from the vertical, (a) what is the index of refraction of the syrup solution? Suppose that the light is red, with a vacuum wavelength of 632.8 nm. Find its (b) wavelength, (c) frequency, and (d) speed in the solution.

9. Find the speed of light in (a) flint glass, (b) water, and (c) cubic zirconia.

10. A light ray initially in water enters a transparent substance at an angle of incidence of 37.0°, and the transmitted ray is refracted at an angle of 25.0°. Calculate the speed of light in the transparent substance.

WEB 11. A ray of light strikes a flat block of glass ($n = 1.50$) of thickness 2.00 cm at an angle of 30.0° with the normal. Trace the light beam through the glass, and find the angles of incidence and refraction at each surface.

12. Light of wavelength 436 nm in air enters a fishbowl filled with water and then exits through the crown glass wall of the container. What is the wavelength of the light (a) in the water and (b) in the glass?

13. An opaque cylindrical tank with an open top has a diameter of 3.00 m and is completely filled with water. When the setting Sun reaches an angle of 28.0° above the horizon, sunlight ceases to illuminate any part of the bottom of the tank. How deep is the tank?

14. The angle between the two mirrors illustrated in Figure P35.14 is a right angle. The beam of light in the vertical plane *P* strikes mirror 1 as shown. (a) Determine the distance that the reflected light beam travels before striking mirror 2. (b) In what direction does the light beam travel after being reflected from mirror 2?

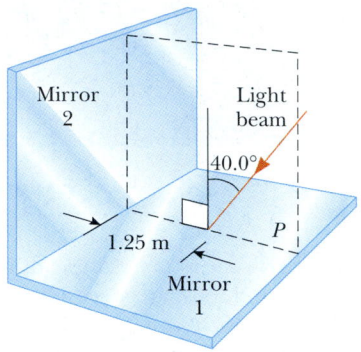

Figure P35.14

15. How many times will the incident beam shown in Figure P35.15 be reflected by each of the parallel mirrors?

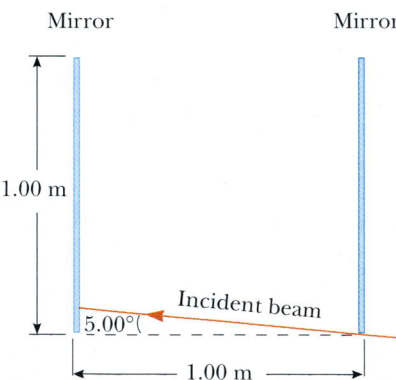

Figure P35.15

16. When the light illustrated in Figure P35.16 passes through the glass block, it is shifted laterally by the distance *d*. If $n = 1.50$, what is the value of *d*?

17. Find the time required for the light to pass through the glass block described in Problem 16.

18. The light beam shown in Figure P35.18 makes an angle of 20.0° with the normal line NN' in the linseed oil. Determine the angles θ and θ'. (The index of refraction for linseed oil is 1.48.)

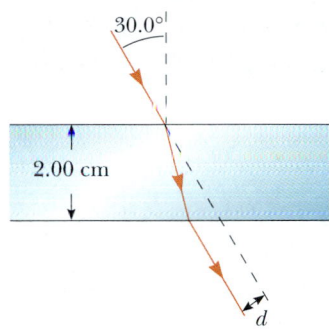

Figure P35.16 Problems 16 and 17.

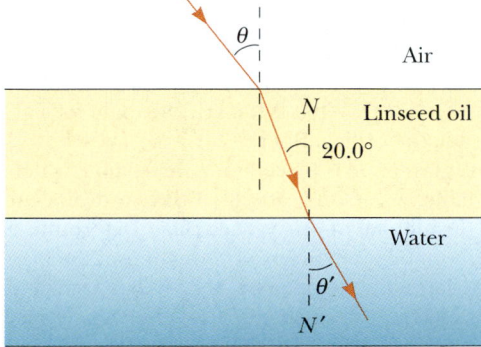

Figure P35.18

19. Two light pulses are emitted simultaneously from a source. Both pulses travel to a detector, but one first passes through 6.20 m of ice. Determine the difference in the pulses' times of arrival at the detector.

20. When you look through a window, by how much time is the light you see delayed by having to go through glass instead of air? Make an order-of-magnitude estimate on the basis of data you specify. By how many wavelengths is it delayed?

21. Light passes from air into flint glass. (a) What angle of incidence must the light have if the component of its velocity perpendicular to the interface is to remain constant? (b) Can the component of velocity parallel to the interface remain constant during refraction?

22. The reflecting surfaces of two intersecting flat mirrors are at an angle of θ ($0° < \theta < 90°$), as shown in Figure

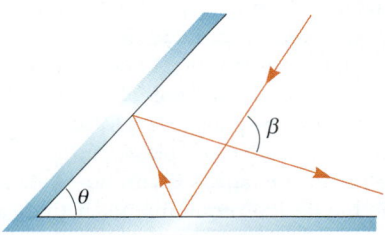

Figure P35.22

P35.22. If a light ray strikes the horizontal mirror, show that the °emerging ray will intersect the incident ray at an angle of $\beta = 180° - 2\theta$.

23. A light ray enters the atmosphere of a planet and descends vertically 20.0 km to the surface. The index of refraction where the light enters the atmosphere is 1.000, and it increases linearly to the surface where it has a value of 1.005. (a) How long does it take the ray to traverse this path? (b) Compare this to the time it takes in the absence of an atmosphere.

24. A light ray enters the atmosphere of a planet and descends vertically to the surface a distance h. The index of refraction where the light enters the atmosphere is 1.000, and it increases linearly to the surface where it has a value of n. (a) How long does it take the ray to traverse this path? (b) Compare this to the time it takes in the absence of an atmosphere.

Section 35.7 Dispersion and Prisms

25. A narrow white light beam is incident on a block of fused quartz at an angle of 30.0°. Find the angular width of the light beam inside the quartz.

26. A ray of light strikes the midpoint of one face of an equiangular glass prism ($n = 1.50$) at an angle of incidence of 30.0°. Trace the path of the light ray through the glass, and find the angles of incidence and refraction at each surface.

27. A prism that has an apex angle of 50.0° is made of cubic zirconia, with $n = 2.20$. What is its angle of minimum deviation?

28. Light with a wavelength of 700 nm is incident on the face of a fused quartz prism at an angle of 75.0° (with respect to the normal to the surface). The apex angle of the prism is 60.0°. Using the value of n from Figure 35.20, calculate the angle (a) of refraction at this first surface, (b) of incidence at the second surface, (c) of refraction at the second surface, and (d) between the incident and emerging rays.

WEB 29. The index of refraction for violet light in silica flint glass is 1.66, and that for red light is 1.62. What is the angular dispersion of visible light passing through a prism of apex angle 60.0° if the angle of incidence is 50.0°? (See Fig. P35.29.)

30. Show that if the apex angle Φ of a prism is small, an approximate value for the angle of minimum deviation is $\delta_{min} = (n - 1)\Phi$.

31. A triangular glass prism with an apex angle of $\Phi = 60.0°$ has an index of refraction $n = 1.50$ (Fig. P35.31). What is the smallest angle of incidence θ_1 for which a light ray can emerge from the other side?

32. A triangular glass prism with an apex angle of Φ has an index of refraction n (Fig. P35.31). What is the smallest angle of incidence θ_1 for which a light ray can emerge from the other side?

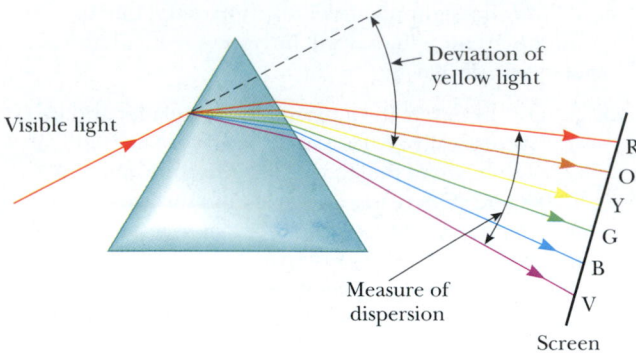

Figure P35.29

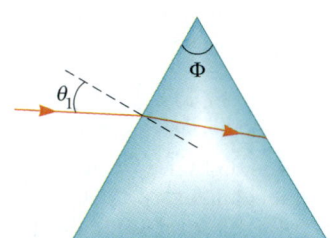

Figure P35.31

33. An experimental apparatus includes a prism made of sodium chloride. The angle of minimum deviation for light of wavelength 589 nm is to be 10.0°. What is the required apex angle of the prism?

34. A triangular glass prism with an apex angle of 60.0° has an index of refraction of 1.50. (a) Show that if its angle of incidence on the first surface is $\theta_1 = 48.6°$, light will pass symmetrically through the prism, as shown in Figure 35.26. (b) Find the angle of deviation δ_{min} for $\theta_1 = 48.6°$. (c) Find the angle of deviation if the angle of incidence on the first surface is 45.6°. (d) Find the angle of deviation if $\theta_1 = 51.6°$.

Section 35.8   Total Internal Reflection

35. For 589-nm light, calculate the critical angle for the following materials surrounded by air: (a) diamond, (b) flint glass, and (c) ice.

36. Repeat Problem 35 for the situation in which the materials are surrounded by water.

37. Consider a common mirage formed by super-heated air just above a roadway. A truck driver whose eyes are 2.00 m above the road, where $n = 1.000\ 3$, looks forward. She perceives the illusion of a patch of water ahead on the road, where her line of sight makes an angle of 1.20° below the horizontal. Find the index of refraction of the air just above the road surface. (*Hint:* Treat this as a problem in total internal reflection.)

38. Determine the maximum angle θ for which the light

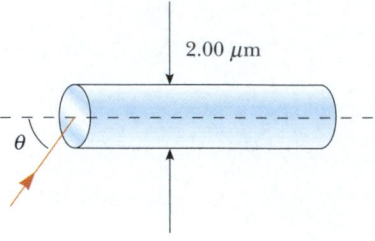

Figure P35.38

rays incident on the end of the pipe shown in Figure P35.38 are subject to total internal reflection along the walls of the pipe. Assume that the pipe has an index of refraction of 1.36 and the outside medium is air.

39. A glass fiber ($n = 1.50$) is submerged in water ($n = 1.33$). What is the critical angle for light to stay inside the optical fiber?

40. A glass cube is placed on a newspaper, which rests on a table. A person reads all of the words the cube covers, through all of one vertical side. Determine the maximum possible index of refraction of the glass.

41. A large Lucite cube ($n = 1.59$) has a small air bubble (a defect in the casting process) below one surface. When a penny (diameter, 1.90 cm) is placed directly over the bubble on the outside of the cube, one cannot see the bubble by looking down into the cube at any angle. However, when a dime (diameter, 1.75 cm) is placed directly over it, one can see the bubble by looking down into the cube. What is the range of the possible depths of the air bubble beneath the surface?

42. A room contains air in which the speed of sound is 343 m/s. The walls of the room are made of concrete, in which the speed of sound is 1 850 m/s. (a) Find the critical angle for total internal reflection of sound at the concrete–air boundary. (b) In which medium must the sound be traveling to undergo total internal reflection? (c) "A bare concrete wall is a highly efficient mirror for sound." Give evidence for or against this statement.

43. In about 1965, engineers at the Toro Company invented a gasoline gauge for small engines, diagrammed in Figure P35.43. The gauge has no moving parts. It consists of a flat slab of transparent plastic fitting vertically into a slot in the cap on the gas tank. None of the plastic has a reflective coating. The plastic projects from the horizontal top down nearly to the bottom of the opaque tank. Its lower edge is cut with facets making angles of 45° with the horizontal. A lawnmower operator looks down from above and sees a boundary between bright and dark on the gauge. The location of the boundary, across the width of the plastic, indicates the quantity of gasoline in the tank. Explain how the gauge works. Explain the design requirements, if any, for the index of refraction of the plastic.

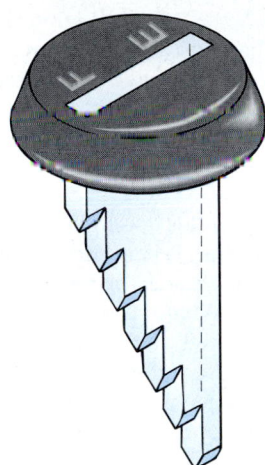

Figure P35.43

(*Optional*)
Section 35.9 Fermat's Principle

44. The shoreline of a lake runs from east to west. A swimmer gets into trouble 20.0 m out from shore and 26.0 m to the east of a lifeguard, whose station is 16.0 m in from the shoreline. The lifeguard takes a negligible amount of time to accelerate. He can run at 7.00 m/s and swim at 1.40 m/s. To reach the swimmer as quickly as possible, in what direction should the lifeguard start running? You will need to solve a transcendental equation numerically.

ADDITIONAL PROBLEMS

45. A narrow beam of light is incident from air onto a glass surface with an index of refraction of 1.56. Find the angle of incidence for which the corresponding angle of refraction is one half the angle of incidence. (*Hint:* You might want to use the trigonometric identity $\sin 2\theta = 2 \sin \theta \cos \theta$.)

46. (a) Consider a horizontal interface between air above and glass with an index of 1.55 below. Draw a light ray incident from the air at an angle of incidence of 30.0°. Determine the angles of the reflected and refracted rays and show them on the diagram. (b) Suppose instead that the light ray is incident from the glass at an angle of incidence of 30.0°. Determine the angles of the reflected and refracted rays and show all three rays on a new diagram. (c) For rays incident from the air onto the air–glass surface, determine and tabulate the angles of reflection and refraction for all the angles of incidence at 10.0° intervals from 0 to 90.0°. (d) Do the same for light rays traveling up to the interface through the glass.

WEB 47. A small underwater pool light is 1.00 m below the surface. The light emerging from the water forms a circle

on the water's surface. What is the diameter of this circle?

48. One technique for measuring the angle of a prism is shown in Figure P35.48. A parallel beam of light is directed on the angle so that the beam reflects from opposite sides. Show that the angular separation of the two beams is given by $B = 2A$.

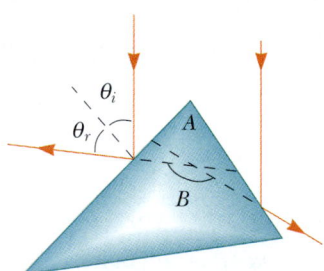

Figure P35.48

49. The walls of a prison cell are perpendicular to the four cardinal compass directions. On the first day of spring, light from the rising Sun enters a rectangular window in the eastern wall. The light traverses 2.37 m horizontally to shine perpendicularly on the wall opposite the window. A young prisoner observes the patch of light moving across this western wall and for the first time forms his own understanding of the rotation of the Earth. (a) With what speed does the illuminated rectangle move? (b) The prisoner holds a small square mirror flat against the wall at one corner of the rectangle of light. The mirror reflects light back to a spot on the eastern wall close beside the window. How fast does the smaller square of light move across that wall? (c) Seen from a latitude of 40.0° north, the rising Sun moves through the sky along a line making a 50.0° angle with the southeastern horizon. In what direction does the rectangular patch of light on the western wall of the prisoner's cell move? (d) In what direction does the smaller square of light on the eastern wall move?

50. The laws of refraction and reflection are the same for sound as for light. The speed of sound in air is 340 m/s, and that of sound in water is 1 510 m/s. If a sound wave approaches a plane water surface at an angle of incidence of 12.0°, what is the angle of refraction?

51. Cold sodium atoms (near absolute zero) in a state called a *Bose–Einstein condensate* can slow the speed of light from its normally high value to a speed approaching that of an automobile in a city. The speed of light in one such medium was recorded as 61.15 km/h. (a) Find the index of refraction of this medium. (b) What is the critical angle for total internal reflection if the condensate is surrounded by vacuum?

52. A narrow beam of white light is incident at 25.0° onto a slab of heavy flint glass 5.00 cm thick. The indices of refraction of the glass at wavelengths of 400 nm and 700 nm are 1.689 and 1.642, respectively. Find the width of the visible beam as it emerges from the slab.

53. A hiker stands on a mountain peak near sunset and observes a rainbow caused by water droplets in the air 8.00 km away. The valley is 2.00 km below the mountain peak and entirely flat. What fraction of the complete circular arc of the rainbow is visible to the hiker? (See Fig. 35.25.)

54. A fish is at a depth d under water. Take the index of refraction of water as 4/3. Show that when the fish is viewed at an angle of refraction θ_1, the apparent depth z of the fish is

$$z = \frac{3d \cos \theta_1}{\sqrt{7 + 9 \cos^2 \theta_1}}$$

WEB 55. A laser beam strikes one end of a slab of material, as shown in Figure P35.55. The index of refraction of the slab is 1.48. Determine the number of internal reflections of the beam before it emerges from the opposite end of the slab.

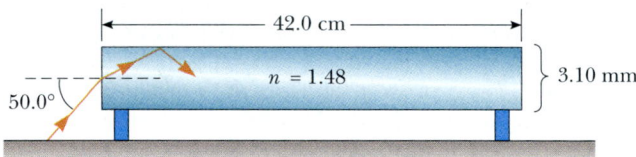

Figure P35.55

56. When light is normally incident on the interface between two transparent optical media, the intensity of the reflected light is given by the expression

$$S_1' = \left(\frac{n_2 - n_1}{n_2 + n_1} \right)^2 S_1$$

In this equation, S_1 represents the average magnitude of the Poynting vector in the incident light (the incident intensity), S_1' is the reflected intensity, and n_1 and n_2 are the refractive indices of the two media. (a) What fraction of the incident intensity is reflected for 589-nm light normally incident on an interface between air and crown glass? (b) In part (a), does it matter whether the light is in the air or in the glass as it strikes the interface? (c) A Bose–Einstein condensate (see Problem 51) has an index of refraction of 1.76×10^7. Find the percent reflection for light falling perpendicularly on its surface. What would the condensate look like?

57. Refer to Problem 56 for a description of the reflected intensity of light normally incident on an interface between two transparent media. (a) When light is normally incident on an interface between vacuum and a transparent medium of index n, show that the intensity S_2 of the transmitted light is given by the expression

$S_2/S_1 = 4n/(n + 1)^2$. (b) Light travels perpendicularly through a diamond slab, surrounded by air, with parallel surfaces of entry and exit. Apply the transmission fraction in part (a) to find the approximate overall transmission through the slab of diamond as a percentage. Ignore light reflected back and forth within the slab.

58. This problem builds upon the results of Problems 56 and 57. Light travels perpendicularly through a diamond slab, surrounded by air, with parallel surfaces of entry and exit. What fraction of the incident intensity is the intensity of the transmitted light? Include the effects of light reflected back and forth inside the slab.

59. The light beam shown in Figure P35.59 strikes surface 2 at the critical angle. Determine the angle of incidence, θ_1.

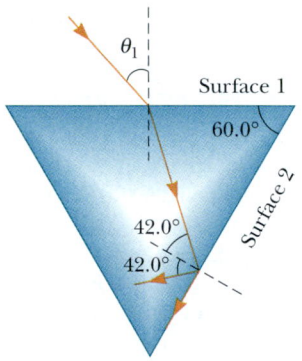

Figure P35.59

60. A 4.00-m-long pole stands vertically in a lake having a depth of 2.00 m. When the Sun is 40.0° above the horizontal, determine the length of the pole's shadow on the bottom of the lake. Take the index of refraction for water to be 1.33.

WEB 61. A light ray of wavelength 589 nm is incident at an angle θ on the top surface of a block of polystyrene, as shown in Figure P35.61. (a) Find the maximum value of θ for which the refracted ray undergoes total internal reflec-

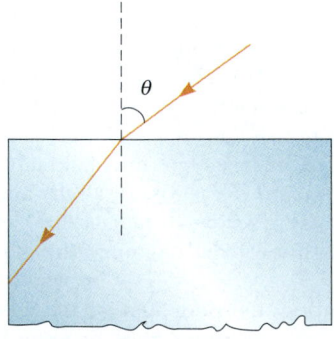

Figure P35.61

tion at the left vertical face of the block. Repeat the calculation for the case in which the polystyrene block is immersed in (b) water and (c) carbon disulfide.

62. A ray of light passes from air into water. For its deviation angle $\delta = |\theta_1 - \theta_2|$ to be 10.0°, what must be its angle of incidence?

63. A shallow glass dish is 4.00 cm wide at the bottom, as shown in Figure P35.63. When an observer's eye is positioned as shown, the observer sees the edge of the bottom of the empty dish. When this dish is filled with water, the observer sees the center of the bottom of the dish. Find the height of the dish.

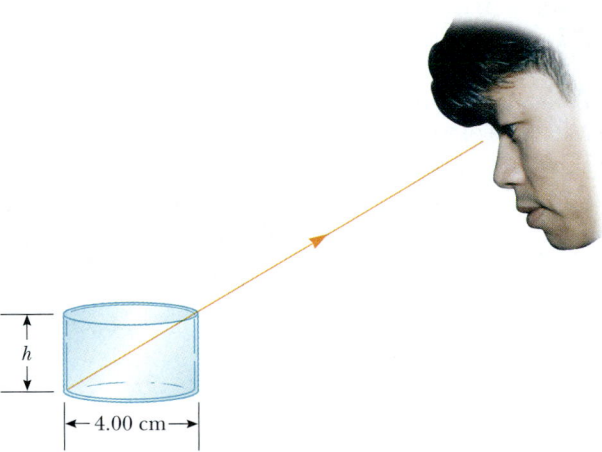

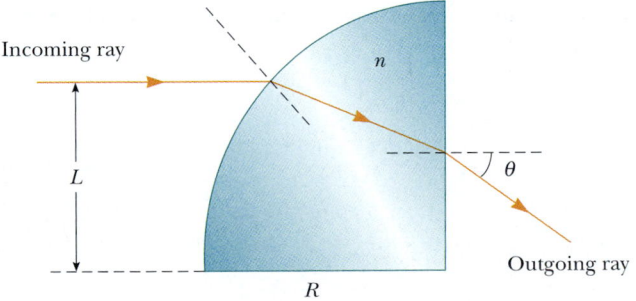

Figure P35.63

64. A material having an index of refraction n is surrounded by a vacuum and is in the shape of a quarter circle of radius R (Fig. P35.64). A light ray parallel to the base of the material is incident from the left at a distance of L above the base and emerges out of the material at the angle θ. Determine an expression for θ.

Incoming ray

n

L

θ

Outgoing ray

R

Figure P35.64

65. Derive the law of reflection (Eq. 35.2) from Fermat's principle of least time. (See the procedure outlined in Section 35.9 for the derivation of the law of refraction from Fermat's principle.)

66. A transparent cylinder of radius $R = 2.00$ m has a mirrored surface on its right half, as shown in Figure P35.66. A light ray traveling in air is incident on the left side of the cylinder. The incident light ray and exiting light ray are parallel and $d = 2.00$ m. Determine the index of refraction of the material.

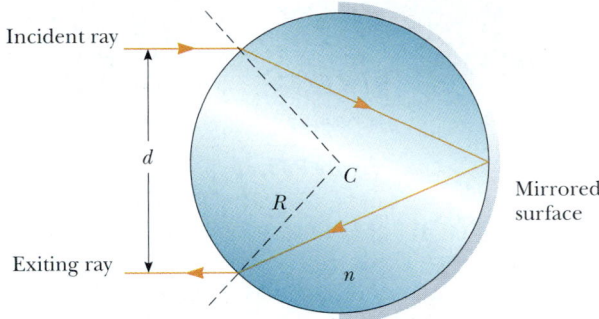

Incident ray

d

C

R

Mirrored surface

Exiting ray

n

Figure P35.66

67. A. H. Pfund's method for measuring the index of refraction of glass is illustrated in Figure P35.67. One face of a slab of thickness t is painted white, and a small hole scraped clear at point P serves as a source of diverging rays when the slab is illuminated from below. Ray PBB' strikes the clear surface at the critical angle and is totally reflected, as are rays such as PCC'. Rays such as PAA' emerge from the clear surface. On the painted surface there appears a dark circle of diameter d, surrounded by an illuminated region, or halo. (a) Derive a formula for n in terms of the measured quantities d and t. (b) What is the diameter of the dark circle if $n = 1.52$ for a slab 0.600 cm thick? (c) If white light is used, the critical angle depends on color caused by dispersion. Is the inner edge of the white halo tinged with red light or violet light? Explain.

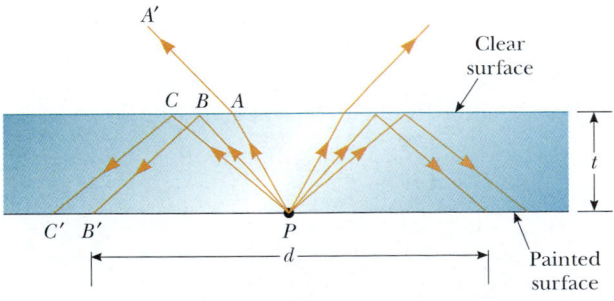

A'

Clear surface

C B A

t

C' B'

P

d

Painted surface

Figure P35.67

68. A light ray traveling in air is incident on one face of a right-angle prism with an index of refraction of $n = 1.50$, as shown in Figure P35.68, and the ray follows the path shown in the figure. If $\theta = 60.0°$ and the base of the prism is mirrored, what is the angle ϕ made by

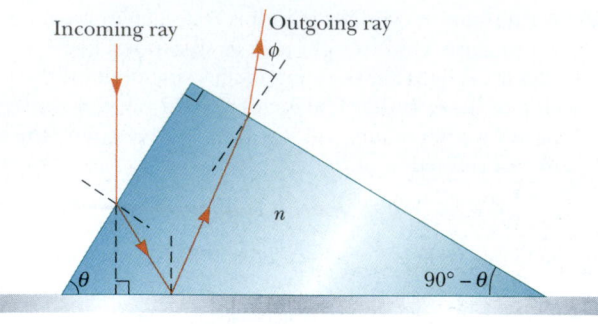

Figure P35.68

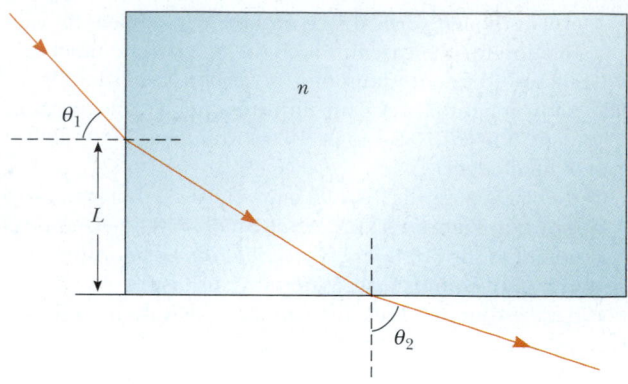

Figure P35.69

the outgoing ray with the normal to the right face of the prism?

69. A light ray enters a rectangular block of plastic at an angle of $\theta_1 = 45.0°$ and emerges at an angle of $\theta_2 = 76.0°$, as shown in Figure P35.69. (a) Determine the index of refraction for the plastic. (b) If the light ray enters the plastic at a point $L = 50.0$ cm from the bottom edge, how long does it take the light ray to travel through the plastic?

70. Students allow a narrow beam of laser light to strike a water surface. They arrange to measure the angle of refraction for selected angles of incidence and record the data shown in the accompanying table. Use the data to verify Snell's law of refraction by plotting the sine of the angle of incidence versus the sine of the angle of refraction. Use the resulting plot to deduce the index of refraction of water.

Angle of Incidence (degrees)	Angle of Refraction (degrees)
10.0	7.5
20.0	15.1
30.0	22.3
40.0	28.7
50.0	35.2
60.0	40.3
70.0	45.3
80.0	47.7

Answers to Quick Quizzes

35.1 Beams ② and ④ are reflected; beams ③ and ⑤ are refracted.

35.2 Fused quartz. An ideal lens would have an index of refraction that does not vary with wavelength so that all colors would be bent through the same angle by the lens. Of the three choices, fused quartz has the least variation in n across the visible spectrum. Thus, it is the best choice for a single-element lens.

35.3 The two rays on the right result from total internal reflection at the right face of the prism. Because all of the light in these rays is reflected (rather than partly refracted), these two rays are brightest. The light from the other three rays is divided into reflected and refracted parts.

PUZZLER

Most car headlights have lines across their faces, like those shown here. Without these lines, the headlights either would not function properly or would be much more likely to break from the jarring of the car on a bumpy road. What is the purpose of the lines? *(George Semple)*

chapter

36

Geometric Optics

This chapter is concerned with the images that result when spherical waves fall on flat and spherical surfaces. We find that images can be formed either by reflection or by refraction and that mirrors and lenses work because of reflection and refraction. We continue to use the ray approximation and to assume that light travels in straight lines. Both of these steps lead to valid predictions in the field called *geometric optics*. In subsequent chapters, we shall concern ourselves with interference and diffraction effects—the objects of study in the field of *wave optics*.

36.1 IMAGES FORMED BY FLAT MIRRORS

We begin by considering the simplest possible mirror, the flat mirror. Consider a point source of light placed at O in Figure 36.1, a distance p in front of a flat mirror. The distance p is called the **object distance.** Light rays leave the source and are reflected from the mirror. Upon reflection, the rays continue to diverge (spread apart), but they appear to the viewer to come from a point I behind the mirror. Point I is called the **image** of the object at O. Regardless of the system under study, we always locate images by extending diverging rays back to a point from which they appear to diverge. **Images are located either at the point from which rays of light *actually* diverge or at the point from which they *appear* to diverge.** Because the rays in Figure 36.1 appear to originate at I, which is a distance q behind the mirror, this is the location of the image. The distance q is called the **image distance.**

Images are classified as real or virtual. **A real image is formed when light rays pass through and diverge from the image point; a virtual image is formed when the light rays do not pass through the image point but appear to diverge from that point.** The image formed by the mirror in Figure 36.1 is virtual. The image of an object seen in a flat mirror is always virtual. Real images can be displayed on a screen (as at a movie), but virtual images cannot be displayed on a screen.

We can use the simple geometric techniques shown in Figure 36.2 to examine the properties of the images formed by flat mirrors. Even though an infinite number of light rays leave each point on the object, we need to follow only two of them to determine where an image is formed. One of those rays starts at P, follows a horizontal path to the mirror, and reflects back on itself. The second ray follows the oblique path PR and reflects as shown, according to the law of reflection. An observer in front of the mirror would trace the two reflected rays back to the point at which they appear to have originated, which is point P' behind the mirror. A continuation of this process for points other than P on the object would result in a virtual image (represented by a yellow arrow) behind the mirror. Because triangles PQR and $P'QR$ are congruent, $PQ = P'Q$. We conclude that **the image formed by an object placed in front of a flat mirror is as far behind the mirror as the object is in front of the mirror.**

Geometry also reveals that the object height h equals the image height h'. Let us define **lateral magnification** M as follows:

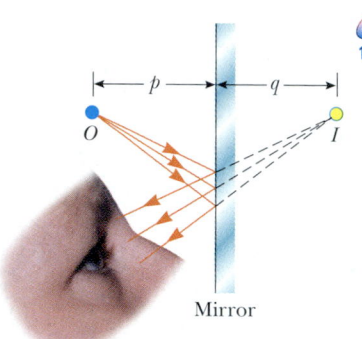

Figure 36.1 An image formed by reflection from a flat mirror. The image point I is located behind the mirror a perpendicular distance q from the mirror (the image distance). Study of Figure 36.2 shows that this image distance is equal to the object distance p.

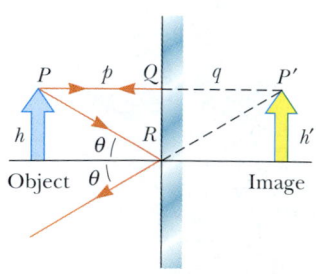

Figure 36.2 A geometric construction that is used to locate the image of an object placed in front of a flat mirror. Because the triangles PQR and $P'QR$ are congruent, $|p| = |q|$ and $h = h'$.

Lateral magnification

$$M \equiv \frac{\text{Image height}}{\text{Object height}} = \frac{h'}{h} \qquad \text{(36.1)}$$

Mt. Hood reflected in Trillium Lake. Why is the image inverted and the same size as the mountain? *(Raymond G. Barnes/Tony Stone Images)*

This is a general definition of the lateral magnification for any type of mirror. For a flat mirror, $M = 1$ because $h' = h$.

Finally, note that a flat mirror produces an image that has an *apparent* left–right reversal. You can see this reversal by standing in front of a mirror and raising your right hand, as shown in Figure 36.3. The image you see raises its left hand. Likewise, your hair appears to be parted on the side opposite your real part, and a mole on your right cheek appears to be on your left cheek.

This reversal is not *actually* a left–right reversal. Imagine, for example, lying on your left side on the floor, with your body parallel to the mirror surface. Now your head is on the left and your feet are on the right. If you shake your feet, the image does not shake its head! If you raise your right hand, however, the image again raises its left hand. Thus, the mirror again appears to produce a left–right reversal but in the up–down direction!

The reversal is actually a *front–back reversal*, caused by the light rays going forward toward the mirror and then reflecting back from it. An interesting exercise is to stand in front of a mirror while holding an overhead transparency in front of you so that you can read the writing on the transparency. You will be able to read the writing on the image of the transparency, also. You may have had a similar experience if you have attached a transparent decal with words on it to the rear window of your car. If the decal can be read from outside the car, you can also read it when looking into your rearview mirror from inside the car.

We conclude that the image that is formed by a flat mirror has the following properties.

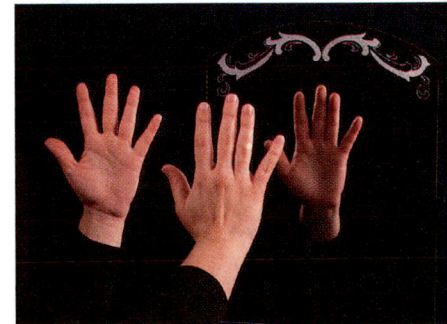

Figure 36.3 The image in the mirror of a person's right hand is reversed front to back. This makes the right hand appear to be a left hand. Notice that the thumb is on the left side of both real hands and on the left side of the image. That the thumb is not on the right side of the image indicates that there is no left-to-right reversal. *(George Semple)*

- The image is as far behind the mirror as the object is in front of the mirror.
- The image is unmagnified, virtual, and upright. (By *upright* we mean that, if the object arrow points upward as in Figure 36.2, so does the image arrow.)
- The image has front–back reversal.

Quick Quiz 36.1

In the overhead view of Figure 36.4, the image of the stone seen by observer 1 is at *C*. Where does observer 2 see the image—at *A*, at *B*, at *C*, at *D*, at *E*, or not at all?

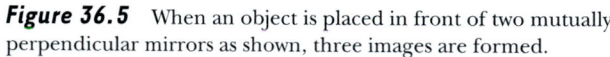

Figure 36.4

CONCEPTUAL EXAMPLE 36.1   Multiple Images Formed by Two Mirrors

Two flat mirrors are at right angles to each other, as illustrated in Figure 36.5, and an object is placed at point *O*. In this situation, multiple images are formed. Locate the positions of these images.

Solution   The image of the object is at I_1 in mirror 1 and at I_2 in mirror 2. In addition, a third image is formed at I_3. This third image is the image of I_1 in mirror 2 or, equivalently, the image of I_2 in mirror 1. That is, the image at I_1 (or I_2) serves as the object for I_3. Note that to form this image at I_3, the rays reflect twice after leaving the object at *O*.

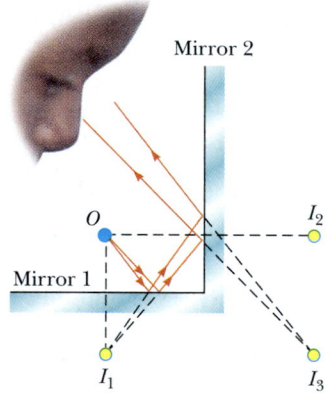

Figure 36.5   When an object is placed in front of two mutually perpendicular mirrors as shown, three images are formed.

CONCEPTUAL EXAMPLE 36.2   The Levitated Professor

The professor in the box shown in Figure 36.6 appears to be balancing himself on a few fingers, with his feet off the floor. He can maintain this position for a long time, and he appears to defy gravity. How was this illusion created?

Solution   This is one of many magicians' optical illusions that make use of a mirror. The box in which the professor stands is a cubical frame that contains a flat vertical mirror positioned in a diagonal plane of the frame. The professor straddles the mirror so that one foot, which you see, is in front of the mirror, and one foot, which you cannot see, is behind the mirror. When he raises the foot in front of the mirror, the reflection of that foot also rises, so he appears to float in air.

Figure 36.6   An optical illusion. *(Courtesy of Henry Leap and Jim Lehman)*

CONCEPTUAL EXAMPLE 36.3 ▷ The Tilting Rearview Mirror

Most rearview mirrors in cars have a day setting and a night setting. The night setting greatly diminishes the intensity of the image in order that lights from trailing vehicles do not blind the driver. How does such a mirror work?

Solution   Figure 36.7 shows a cross-sectional view of a rearview mirror for each setting. The unit consists of a reflective coating on the back of a wedge of glass. In the day setting (Fig. 36.7a), the light from an object behind the car strikes the glass wedge at point 1. Most of the light enters the wedge, refracting as it crosses the front surface, and reflects

from the back surface to return to the front surface, where it is refracted again as it re-enters the air as ray *B* (for *bright*). In addition, a small portion of the light is reflected at the front surface of the glass, as indicated by ray *D* (for *dim*).

This dim reflected light is responsible for the image that is observed when the mirror is in the night setting (Fig. 36.7b). In this case, the wedge is rotated so that the path followed by the bright light (ray *B*) does not lead to the eye. Instead, the dim light reflected from the front surface of the wedge travels to the eye, and the brightness of trailing headlights does not become a hazard.

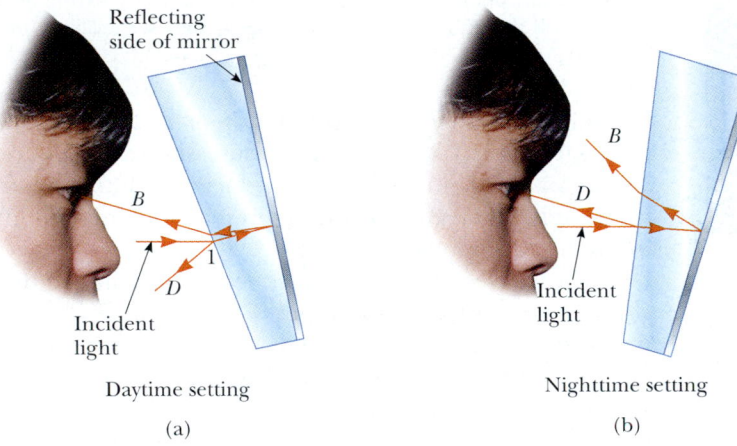

Figure 36.7   Cross-sectional views of a rearview mirror. (a) With the day setting, the silvered back surface of the mirror reflects a bright ray *B* into the driver's eyes. (b) With the night setting, the glass of the unsilvered front surface of the mirror reflects a dim ray *D* into the driver's eyes.

36.2 ▷ IMAGES FORMED BY SPHERICAL MIRRORS

Concave Mirrors

A **spherical mirror,** as its name implies, has the shape of a section of a sphere. This type of mirror focuses incoming parallel rays to a point, as demonstrated by the colored light rays in Figure 36.8. Figure 36.9a shows a cross-section of a spherical mirror, with its surface represented by the solid, curved black line. (The blue band represents the structural support for the mirrored surface, such as a curved piece of glass on which the silvered surface is deposited.) Such a mirror, in which light is reflected from the inner, concave surface, is called a **concave mirror.** The mirror has a radius of curvature R, and its center of curvature is point C. Point V is the center of the spherical section, and a line through C and V is called the **principal axis** of the mirror.

Now consider a point source of light placed at point O in Figure 36.9b, where O is any point on the principal axis to the left of C. Three diverging rays that originate at O are shown. After reflecting from the mirror, these rays converge (come together) at the image point I. They then continue to diverge from I as if an object were there. As a result, we have at point I a real image of the light source at O.

We shall consider in this section only rays that diverge from the object to make a small angle with the principal axis. Such rays are called **paraxial rays.** All such

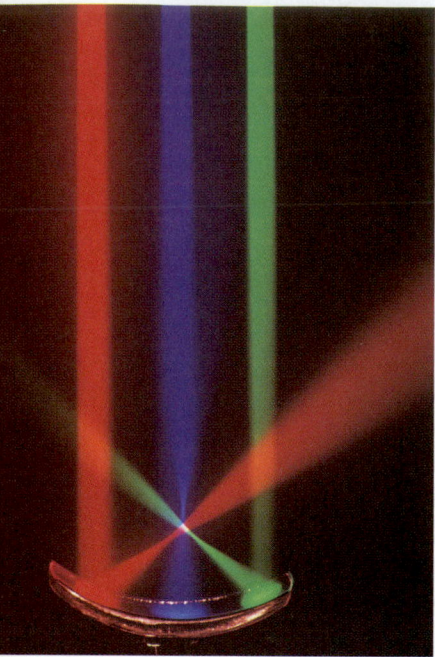

Figure 36.8 Red, blue, and green light rays are reflected by a curved mirror. Note that the focal point where the three colors meet is white. *(Ken Kay/Fundamental Photographs)*

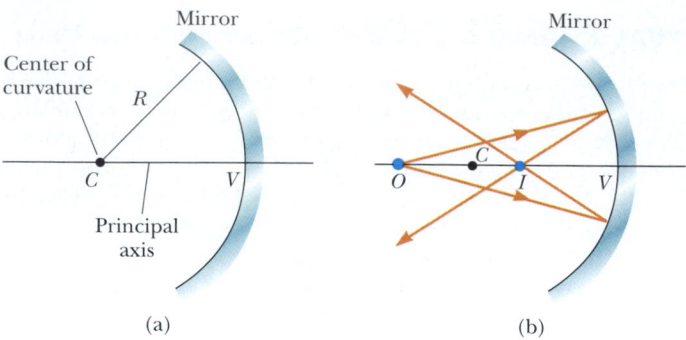

(a)   (b)

Figure 36.9 (a) A concave mirror of radius R. The center of curvature C is located on the principal axis. (b) A point object placed at O in front of a concave spherical mirror of radius R, where O is any point on the principal axis farther than R from the mirror surface, forms a real image at I. If the rays diverge from O at small angles, they all reflect through the same image point.

rays reflect through the image point, as shown in Figure 36.9b. Rays that are far from the principal axis, such as those shown in Figure 36.10, converge to other points on the principal axis, producing a blurred image. This effect, which is called **spherical aberration,** is present to some extent for any spherical mirror and is discussed in Section 36.5.

We can use Figure 36.11 to calculate the image distance q from a knowledge of the object distance p and radius of curvature R. By convention, these distances are measured from point V. Figure 36.11 shows two rays leaving the tip of the object. One of these rays passes through the center of curvature C of the mirror, hitting the mirror perpendicular to the mirror surface and reflecting back on itself. The second ray strikes the mirror at its center (point V) and reflects as shown, obeying the law of reflection. The image of the tip of the arrow is located at the point where these two rays intersect. From the gold right triangle in Figure 36.11, we see that $\tan \theta = h/p$, and from the blue right triangle we see that $\tan \theta = -h'/q$. The negative sign is introduced because the image is inverted, so h' is taken to be negative. Thus, from Equation 36.1 and these results, we find that the magnification of the mirror is

$$M = \frac{h'}{h} = -\frac{q}{p}$$

(36.2)

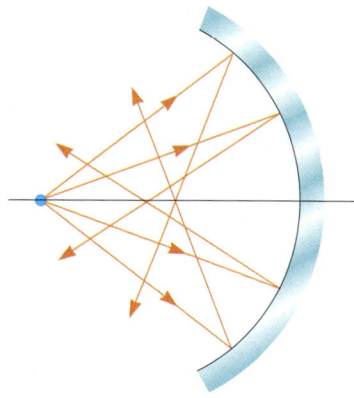

Figure 36.10 Rays diverging from the object at large angles from the principal axis reflect from a spherical concave mirror to intersect the principal axis at different points, resulting in a blurred image. This condition is called *spherical aberration.*

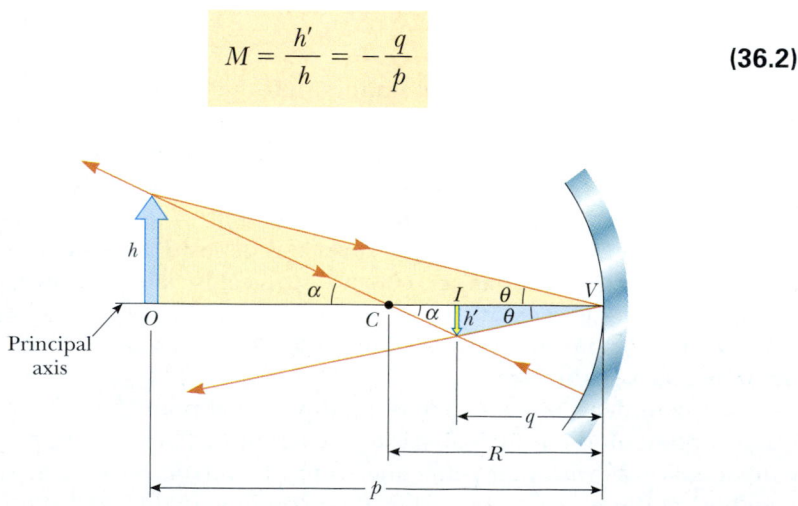

Figure 36.11 The image formed by a spherical concave mirror when the object O lies outside the center of curvature C.

We also note from the two triangles in Figure 36.11 that have α as one angle that

$$\tan \alpha = \frac{h}{p - R} \qquad \text{and} \qquad \tan \alpha = -\frac{h'}{R - q}$$

from which we find that

$$\frac{h'}{h} = -\frac{R - q}{p - R} \tag{36.3}$$

If we compare Equations 36.2 and 36.3, we see that

$$\frac{R - q}{p - R} = \frac{q}{p}$$

Simple algebra reduces this to

$$\frac{1}{p} + \frac{1}{q} = \frac{2}{R} \tag{36.4}$$

Mirror equation in terms of R

This expression is called the **mirror equation.** It is applicable only to paraxial rays.

If the object is very far from the mirror—that is, if p is so much greater than R that p can be said to approach infinity—then $1/p \approx 0$, and we see from Equation 36.4 that $q \approx R/2$. That is, when the object is very far from the mirror, the image point is halfway between the center of curvature and the center point on the mirror, as shown in Figure 36.12a. The incoming rays from the object are essentially parallel in this figure because the source is assumed to be very far from the mirror. We call the image point in this special case the **focal point** F and the image distance the **focal length** f, where

$$f = \frac{R}{2} \tag{36.5}$$

Focal length

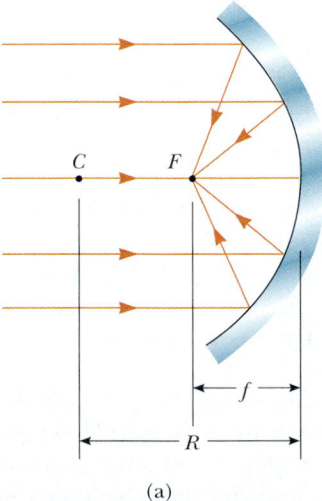

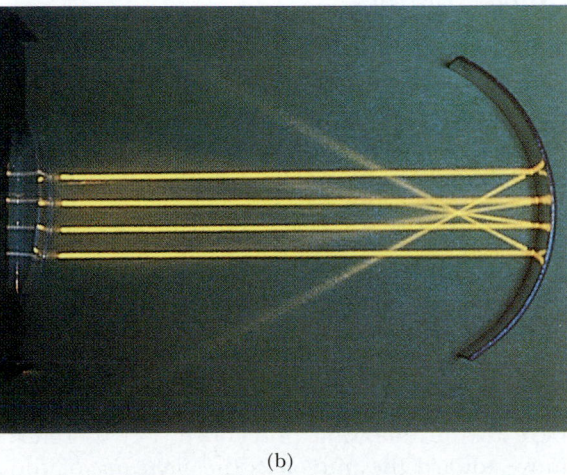

(a)   (b)

Figure 36.12   (a) Light rays from a distant object ($p \approx \infty$) reflect from a concave mirror through the focal point F. In this case, the image distance $q \approx R/2 = f$, where f is the focal length of the mirror. (b) Reflection of parallel rays from a concave mirror. *(Henry Leap and Jim Lehman)*

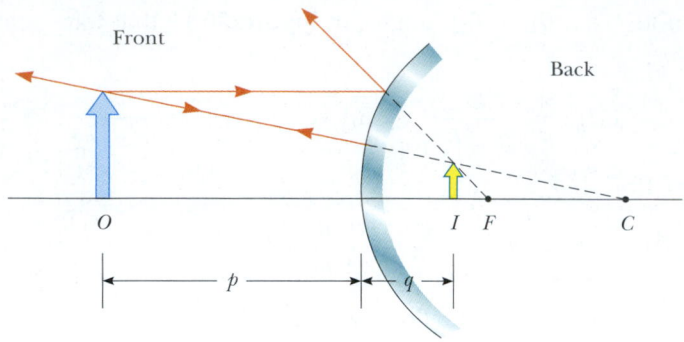

Figure 36.13   Formation of an image by a spherical convex mirror. The image formed by the real object is virtual and upright.

Focal length is a parameter particular to a given mirror and therefore can be used to compare one mirror with another. The mirror equation can be expressed in terms of the focal length:

Mirror equation in terms of f

$$\frac{1}{p} + \frac{1}{q} = \frac{1}{f} \tag{36.6}$$

Notice that the focal length of a mirror depends only on the curvature of the mirror and not on the material from which the mirror is made. This is because the formation of the image results from rays reflected from the surface of the material. We shall find in Section 36.4 that the situation is different for lenses; in that case the light actually passes through the material.

Convex Mirrors

Figure 36.13 shows the formation of an image by a **convex mirror**—that is, one silvered so that light is reflected from the outer, convex surface. This is sometimes called a **diverging mirror** because the rays from any point on an object diverge after reflection as though they were coming from some point behind the mirror. The image in Figure 36.13 is virtual because the reflected rays only appear to originate at the image point, as indicated by the dashed lines. Furthermore, the image is always upright and smaller than the object. This type of mirror is often used in stores to foil shoplifters. A single mirror can be used to survey a large field of view because it forms a smaller image of the interior of the store.

We do not derive any equations for convex spherical mirrors because we can use Equations 36.2, 36.4, and 36.6 for either concave or convex mirrors if we adhere to the following procedure. Let us refer to the region in which light rays move toward the mirror as the *front side* of the mirror, and the other side as the *back side*. For example, in Figures 36.10 and 36.12, the side to the left of the mirrors is the front side, and the side to the right of the mirrors is the back side. Figure 36.14 states the sign conventions for object and image distances, and Table 36.1 summarizes the sign conventions for all quantities.

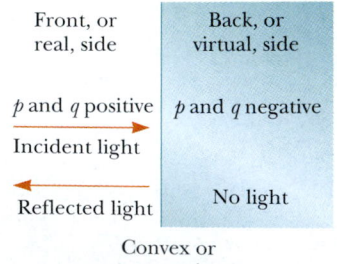

Figure 36.14   Signs of p and q for convex and concave mirrors.

TABLE 36.1 **Sign Conventions for Mirrors**
p is **positive** if object is in **front** of mirror (real object).
p is **negative** if object is in **back** of mirror (virtual object).
q is **positive** if image is in **front** of mirror (real image).
q is **negative** if image is in **back** of mirror (virtual image).
Both f and R are **positive** if center of curvature is in **front** of mirror (concave mirror).
Both f and R are **negative** if center of curvature is in **back** of mirror (convex mirror).
If M is **positive**, image is **upright**.
If M is **negative**, image is **inverted**.

Reflection of parallel lines from a convex cylindrical mirror. The image is virtual, upright, and reduced in size. *(©1990 Richard Megna/Fundamental Photographs)*

Ray Diagrams for Mirrors

The positions and sizes of images formed by mirrors can be conveniently determined with *ray diagrams.* These graphical constructions reveal the nature of the image and can be used to check results calculated from the mirror and magnification equations. To draw a ray diagram, we need to know the position of the object and the locations of the mirror's focal point and center of curvature. We then draw three rays to locate the image, as shown by the examples in Figure 36.15. These rays all start from the same object point and are drawn as follows. We may choose any point on the object; here, we choose the top of the object for simplicity:

- Ray 1 is drawn from the top of the object parallel to the principal axis and is reflected through the focal point *F*.
- Ray 2 is drawn from the top of the object through the focal point and is reflected parallel to the principal axis.
- Ray 3 is drawn from the top of the object through the center of curvature *C* and is reflected back on itself.

The intersection of any two of these rays locates the image. The third ray serves as a check of the construction. The image point obtained in this fashion must always agree with the value of q calculated from the mirror equation.

With concave mirrors, note what happens as the object is moved closer to the mirror. The real, inverted image in Figure 36.15a moves to the left as the object approaches the focal point. When the object is at the focal point, the image is infinitely far to the left. However, when the object lies between the focal point and the mirror surface, as shown in Figure 36.15b, the image is virtual, upright, and enlarged. This latter situation applies in the use of a shaving mirror or a makeup mirror. Your face is closer to the mirror than the focal point, and you see an upright, enlarged image of your face.

In a convex mirror (see Fig. 36.15c), the image of an object is always virtual, upright, and reduced in size. In this case, as the object distance increases, the virtual image decreases in size and approaches the focal point as p approaches infinity. You should construct other diagrams to verify how image position varies with object position.

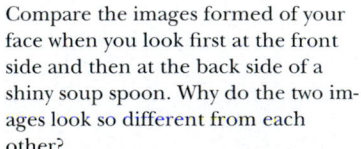

QuickLab

Compare the images formed of your face when you look first at the front side and then at the back side of a shiny soup spoon. Why do the two images look so different from each other?

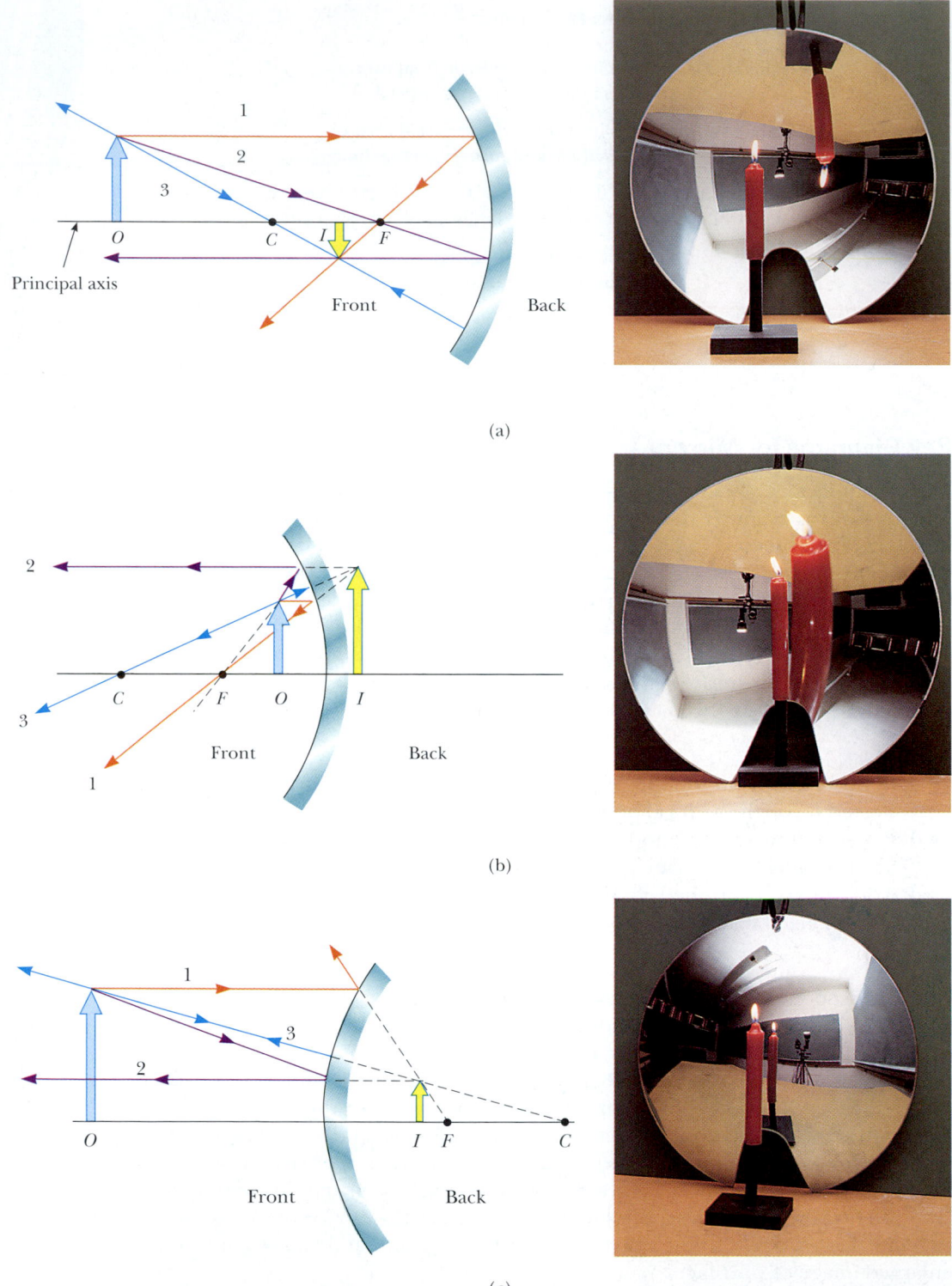

Figure 36.15 Ray diagrams for spherical mirrors, along with corresponding photographs of the images of candles. (a) When the object is located so that the center of curvature lies between the object and a concave mirror surface, the image is real, inverted, and reduced in size. (b) When the object is located between the focal point and a concave mirror surface, the image is virtual, upright, and enlarged. (c) When the object is in front of a convex mirror, the image is virtual, upright, and reduced in size.

EXAMPLE 36.4 ▸ The Image from a Mirror

Assume that a certain spherical mirror has a focal length of +10.0 cm. Locate and describe the image for object distances of (a) 25.0 cm, (b) 10.0 cm, and (c) 5.00 cm.

Solution Because the focal length is positive, we know that this is a concave mirror (see Table 36.1). (a) This situation is analogous to that in Figure 36.15a; hence, we expect the image to be real and closer to the mirror than the object. According to the figure, it should also be inverted and reduced in size. We find the image distance by using the Equation 36.6 form of the mirror equation:

$$\frac{1}{p} + \frac{1}{q} = \frac{1}{f}$$

$$\frac{1}{25.0 \text{ cm}} + \frac{1}{q} = \frac{1}{10.0 \text{ cm}}$$

$$q = \boxed{16.7 \text{ cm}}$$

The magnification is given by Equation 36.2:

$$M = -\frac{q}{p} = -\frac{16.7 \text{ cm}}{25.0 \text{ cm}} = -0.668$$

The fact that the absolute value of M is less than unity tells us that the image is smaller than the object, and the negative sign for M tells us that the image is inverted. Because q is positive, the image is located on the front side of the mirror and is real. Thus, we see that our predictions were correct.

(b) When the object distance is 10.0 cm, the object is located at the focal point. Now we find that

$$\frac{1}{10.0 \text{ cm}} + \frac{1}{q} = \frac{1}{10.0 \text{ cm}}$$

$$q = \boxed{\infty}$$

which means that rays originating from an object positioned at the focal point of a mirror are reflected so that the image is formed at an infinite distance from the mirror; that is, the rays travel parallel to one another after reflection. This is the situation in a flashlight, where the bulb filament is placed at the focal point of a reflector, producing a parallel beam of light.

(c) When the object is at $p = 5.00$ cm, it lies between the focal point and the mirror surface, as shown in Figure 36.15b. Thus, we expect a magnified, virtual, upright image. In this case, the mirror equation gives

$$\frac{1}{5.00 \text{ cm}} + \frac{1}{q} = \frac{1}{10.0 \text{ cm}}$$

$$q = \boxed{-10.0 \text{ cm}}$$

The image is virtual because it is located behind the mirror, as expected. The magnification is

$$M = -\frac{q}{p} = -\left(\frac{-10.0 \text{ cm}}{5.00 \text{ cm}}\right) = 2.00$$

The image is twice as large as the object, and the positive sign for M indicates that the image is upright (see Fig. 36.15b).

Exercise At what object distance is the magnification −1.00?

Answer 20.0 cm.

EXAMPLE 36.5 ▸ The Image from a Convex Mirror

A woman who is 1.5 m tall is located 3.0 m from an anti-shoplifting mirror, as shown in Figure 36.16. The focal length of the mirror is −0.25 m. Find (a) the position of her image and (b) the magnification.

Solution (a) This situation is depicted in Figure 36.15c. We should expect to find an upright, reduced, virtual image. To find the image position, we use Equation 36.6:

$$\frac{1}{p} + \frac{1}{q} = \frac{1}{f} = \frac{1}{-0.25 \text{ m}}$$

$$\frac{1}{q} = \frac{1}{-0.25 \text{ m}} - \frac{1}{3.0 \text{ m}}$$

$$q = \boxed{-0.23 \text{ m}}$$

Figure 36.16 Convex mirrors, often used for security in department stores, provide wide-angle viewing. *(©1990 Paul Silverman/Fundamental Photographs)*

The negative value of q indicates that her image is virtual, or behind the mirror, as shown in Figure 36.15c.

(b) The magnification is

$$M = -\frac{q}{p} = -\left(\frac{-0.23 \text{ m}}{3.0 \text{ m}}\right) = \boxed{0.077}$$

The image is much smaller than the woman, and it is upright because M is positive.

Exercise   Find the height of the image.

Answer   0.12 m.

36.3 IMAGES FORMED BY REFRACTION

In this section we describe how images are formed when light rays are refracted at the boundary between two transparent materials. Consider two transparent media having indices of refraction n_1 and n_2, where the boundary between the two media is a spherical surface of radius R (Fig. 36.17). We assume that the object at O is in the medium for which the index of refraction is n_1, where $n_1 < n_2$. Let us consider the paraxial rays leaving O. As we shall see, all such rays are refracted at the spherical surface and focus at a single point I, the image point.

Figure 36.18 shows a single ray leaving point O and focusing at point I. Snell's law of refraction applied to this refracted ray gives

$$n_1 \sin \theta_1 = n_2 \sin \theta_2$$

Because θ_1 and θ_2 are assumed to be small, we can use the small-angle approximation $\sin \theta \approx \theta$ (angles in radians) and say that

$$n_1 \theta_1 = n_2 \theta_2$$

Now we use the fact that an exterior angle of any triangle equals the sum of the two opposite interior angles. Applying this rule to triangles OPC and PIC in Figure 36.18 gives

$$\theta_1 = \alpha + \beta$$

$$\beta = \theta_2 + \gamma$$

If we combine all three expressions and eliminate θ_1 and θ_2, we find that

$$n_1 \alpha + n_2 \gamma = (n_2 - n_1)\beta \tag{36.7}$$

Looking at Figure 36.18, we see three right triangles that have a common vertical leg of length d. For paraxial rays (unlike the relatively large-angle ray shown in Fig.

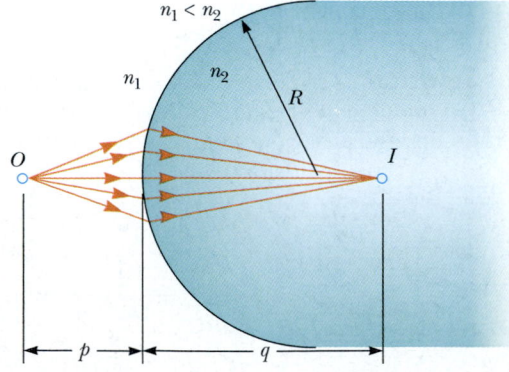

Figure 36.17   An image formed by refraction at a spherical surface. Rays making small angles with the principal axis diverge from a point object at O and are refracted through the image point I.

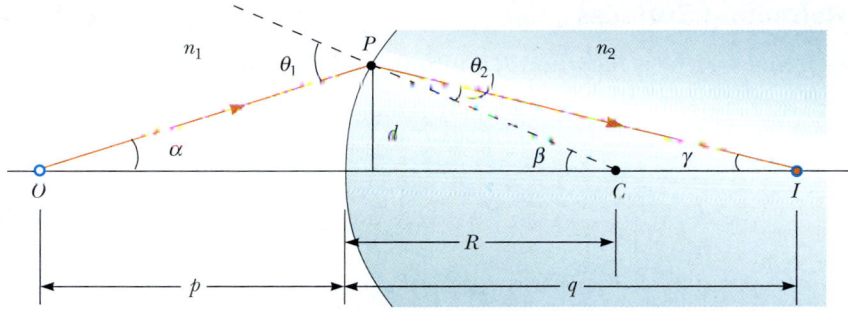

Figure 36.18   Geometry used to derive Equation 36.8.

36.18), the horizontal legs of these triangles are approximately p for the triangle containing angle α, R for the triangle containing angle β, and q for the triangle containing angle γ. In the small-angle approximation, $\tan\theta \approx \theta$, so we can write the approximate relationships from these triangles as follows:

$$\tan\alpha \approx \alpha \approx \frac{d}{p} \qquad \tan\beta \approx \beta \approx \frac{d}{R} \qquad \tan\gamma \approx \gamma \approx \frac{d}{q}$$

We substitute these expressions into Equation 36.7 and divide through by d to get

$$\frac{n_1}{p} + \frac{n_2}{q} = \frac{n_2 - n_1}{R} \qquad\qquad \textbf{(36.8)}$$

For a fixed object distance p, the image distance q is independent of the angle that the ray makes with the axis. This result tells us that all paraxial rays focus at the same point I.

As with mirrors, we must use a sign convention if we are to apply this equation to a variety of cases. We define the side of the surface in which light rays originate as the front side. The other side is called the back side. Real images are formed by refraction in back of the surface, in contrast with mirrors, where real images are formed in front of the reflecting surface. Because of the difference in location of real images, the refraction sign conventions for q and R are opposite the reflection sign conventions. For example, q and R are both positive in Figure 36.18. The sign conventions for spherical refracting surfaces are summarized in Table 36.2.

We derived Equation 36.8 from an assumption that $n_1 < n_2$. This assumption is not necessary, however. Equation 36.8 is valid regardless of which index of refraction is greater.

TABLE 36.2   Sign Conventions for Refracting Surfaces

p is **positive** if object is in **front** of surface (real object).
p is **negative** if object is in **back** of surface (virtual object).

q is **positive** if image is in **back** of surface (real image).
q is **negative** if image is in **front** of surface (virtual image).

R is **positive** if center of curvature is in **back** of convex surface.
R is **negative** if center of curvature is in **front** of concave surface.

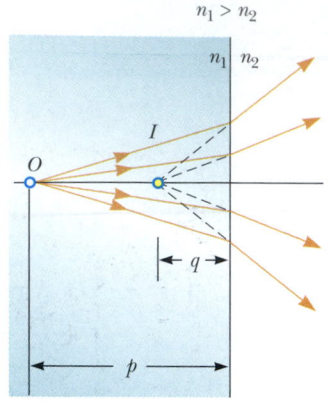

$n_1 > n_2$

Figure 36.19 The image formed by a flat refracting surface is virtual and on the same side of the surface as the object. All rays are assumed to be paraxial.

Flat Refracting Surfaces

If a refracting surface is flat, then R is infinite and Equation 36.8 reduces to

$$\frac{n_1}{p} = -\frac{n_2}{q}$$

$$q = -\frac{n_2}{n_1} p \qquad (36.9)$$

From this expression we see that the sign of q is opposite that of p. Thus, according to Table 36.2, **the image formed by a flat refracting surface is on the same side of the surface as the object.** This is illustrated in Figure 36.19 for the situation in which the object is in the medium of index n_1 and n_1 is greater than n_2. In this case, a virtual image is formed between the object and the surface. If n_1 is less than n_2, the rays in the back side diverge from each other at lesser angles than those in Figure 36.19. As a result, the virtual image is formed to the left of the object.

CONCEPTUAL EXAMPLE 36.6   Let's Go Scuba Diving!

It is well known that objects viewed under water with the naked eye appear blurred and out of focus. However, a scuba diver using a mask has a clear view of underwater objects. (a) Explain how this works, using the facts that the indices of refraction of the cornea, water, and air are 1.376, 1.333, and 1.000 29, respectively.

Solution Because the cornea and water have almost identical indices of refraction, very little refraction occurs when a person under water views objects with the naked eye. In this case, light rays from an object focus behind the retina, resulting in a blurred image. When a mask is used, the air space between the eye and the mask surface provides the normal amount of refraction at the eye–air interface, and the light from the object is focused on the retina.

(b) If a lens prescription is ground into the glass of a mask, should the curved surface be on the inside of the mask, the outside, or both?

Solution If a lens prescription is ground into the glass of the mask so that the wearer can see without eyeglasses, only the inside surface is curved. In this way the prescription is accurate whether the mask is used under water or in air. If the curvature were on the outer surface, the refraction at the outer surface of the glass would change depending on whether air or water were present on the outside of the mask.

EXAMPLE 36.7   Gaze into the Crystal Ball

A dandelion seed ball 4.0 cm in diameter is embedded in the center of a spherical plastic paperweight having a diameter of 6.0 cm (Fig. 36.20a). The index of refraction of the plastic is $n_1 = 1.50$. Find the position of the image of the near edge of the seed ball.

Solution Because $n_1 > n_2$, where $n_2 = 1.00$ is the index of refraction for air, the rays originating from the seed ball are refracted away from the normal at the surface and diverge outward, as shown in Figure 36.20b. Hence, the image is formed inside the paperweight and is virtual. From the given dimensions, we know that the near edge of the seed ball is 1.0 cm beneath the surface of the paperweight. Applying Equation 36.8 and noting from Table 36.2 that R is negative, we obtain

$$\frac{n_1}{p} + \frac{n_2}{q} = \frac{n_2 - n_1}{R}$$

$$\frac{1.50}{1.0 \text{ cm}} + \frac{1}{q} = \frac{1.00 - 1.50}{-3.0 \text{ cm}}$$

$$q = \boxed{-0.75 \text{ cm}}$$

The negative sign for q indicates that the image is in front of the surface—in other words, in the same medium as the object, as shown in Figure 36.20b. Being in the same medium as the object, the image must be virtual (see Table 36.2). The surface of the seed ball appears to be closer to the paperweight surface than it actually is.

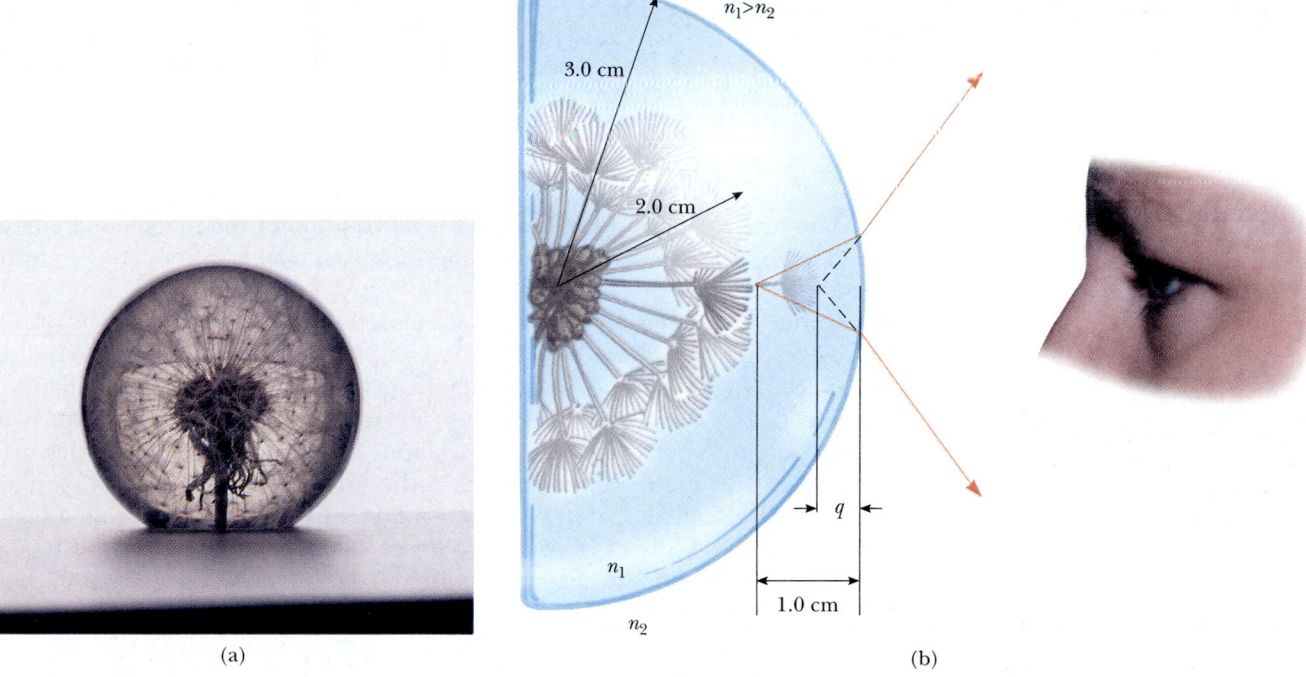

Figure 36.20   (a) An object embedded in a plastic sphere forms a virtual image between the surface of the object and the sphere surface. All rays are assumed paraxial. Because the object is inside the sphere, the front of the refracting surface is the *interior* of the sphere.   (b) Rays from the surface of the object form an image that is still inside the plastic sphere but closer to the plastic surface.   *(George Semple)*

EXAMPLE 36.8   The One That Got Away

A small fish is swimming at a depth d below the surface of a pond (Fig. 36.21). What is the apparent depth of the fish, as viewed from directly overhead?

Solution   Because the refracting surface is flat, R is infinite. Hence, we can use Equation 36.9 to determine the location of the image with $p = d$. Using the indices of refraction given in Figure 36.21, we obtain

$$q = -\frac{n_2}{n_1}\, p = -\frac{1.00}{1.33}\, d = \boxed{-0.752d}$$

Because q is negative, the image is virtual, as indicated by the dashed lines in Figure 36.21. The apparent depth is three-fourths the actual depth.

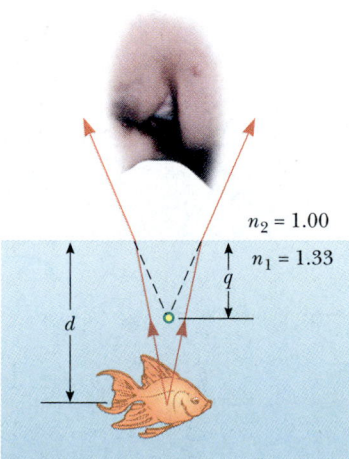

Figure 36.21   The apparent depth q of the fish is less than the true depth d. All rays are assumed to be paraxial.

36.4 ▶ THIN LENSES

Lenses are commonly used to form images by refraction in optical instruments, such as cameras, telescopes, and microscopes. We can use what we just learned about images formed by refracting surfaces to help us locate the image formed by a lens. We recognize that light passing through a lens experiences refraction at two surfaces. The development we shall follow is based on the notion that **the image formed by one refracting surface serves as the object for the second surface.** We shall analyze a thick lens first and then let the thickness of the lens be approximately zero.

Consider a lens having an index of refraction n and two spherical surfaces with radii of curvature R_1 and R_2, as in Figure 36.22. (Note that R_1 is the radius of curvature of the lens surface that the light leaving the object reaches first and that R_2 is the radius of curvature of the other surface of the lens.) An object is placed at point O at a distance p_1 in front of surface 1. If the object were far from surface 1, the light rays from the object that struck the surface would be almost parallel to each other. The refraction at the surface would focus these rays, forming a real image to the right of surface 1 in Figure 36.22 (as in Fig. 36.17). If the object is placed close to surface 1, as shown in Figure 36.22, the rays diverging from the object and striking the surface cover a wide range of angles and are not parallel to each other. In this case, the refraction at the surface is not sufficient to cause the rays to converge on the right side of the surface. They still diverge, although they are closer to parallel than they were before they struck the surface. This results in a virtual image of the object at I_1 to the left of the surface, as shown in Figure 36.22. This image is then used as the object for surface 2, which results in a real image I_2 to the right of the lens.

Let us begin with the virtual image formed by surface 1. Using Equation 36.8 and assuming that $n_1 = 1$ because the lens is surrounded by air, we find that the image I_1 formed by surface 1 satisfies the equation

$$(1) \qquad \frac{1}{p_1} + \frac{n}{q_1} = \frac{n-1}{R_1}$$

where q_1 is a negative number because it represents a virtual image formed on the front side of surface 1.

Now we apply Equation 36.8 to surface 2, taking $n_1 = n$ and $n_2 = 1$. (We make this switch in index because the light rays from I_1 approaching surface 2 are *in the material of the lens,* and this material has index n. We could also imagine removing the object at O, filling all of the space to the left of surface 1 with the mate-

Figure 36.22 To locate the image formed by a lens, we use the virtual image at I_1 formed by surface 1 as the object for the image formed by surface 2. The final image is real and is located at I_2.

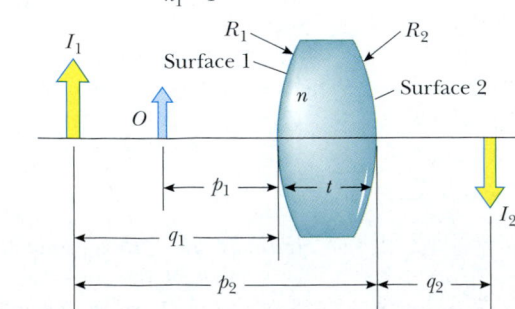

rial of the lens, and placing the object at I_1; the light rays approaching surface 2 would be the same as in the actual situation in Fig. 36.22.) Taking p_2 as the object distance for surface 2 and q_2 as the image distance gives

$$(2) \qquad \frac{n}{p_2} + \frac{1}{q_2} = \frac{1-n}{R_2}$$

We now introduce mathematically the fact that the image formed by the first surface acts as the object for the second surface. We do this by noting from Figure 36.22 that p_2 is the sum of q_1 and t and by setting $p_2 = -q_1 + t$, where t is the thickness of the lens. (Remember that q_1 is a negative number and that p_2 must be positive by our sign convention—thus, we must introduce a negative sign for q_1.) For a *thin* lens (for which the thickness is small compared to the radii of curvature), we can neglect t. In this approximation, we see that $p_2 = -q_1$. Hence, Equation (2) becomes

$$(3) \qquad -\frac{n}{q_1} + \frac{1}{q_2} = \frac{1-n}{R_2}$$

Adding Equations (1) and (3), we find that

$$(4) \qquad \frac{1}{p_1} + \frac{1}{q_2} = (n-1)\left(\frac{1}{R_1} - \frac{1}{R_2}\right)$$

For a thin lens, we can omit the subscripts on p_1 and q_2 in Equation (4) and call the object distance p and the image distance q, as in Figure 36.23. Hence, we can write Equation (4) in the form

$$\frac{1}{p} + \frac{1}{q} = (n-1)\left(\frac{1}{R_1} - \frac{1}{R_2}\right) \qquad \textbf{(36.10)}$$

This expression relates the image distance q of the image formed by a thin lens to the object distance p and to the thin-lens properties (index of refraction and radii of curvature). It is valid only for paraxial rays and only when the lens thickness is much less than R_1 and R_2.

The **focal length** f of a thin lens is the image distance that corresponds to an infinite object distance, just as with mirrors. Letting p approach ∞ and q approach f in Equation 36.10, we see that the inverse of the focal length for a thin lens is

$$\frac{1}{f} = (n-1)\left(\frac{1}{R_1} - \frac{1}{R_2}\right) \qquad \textbf{(36.11)}$$

Lens makers' equation

This relationship is called the **lens makers' equation** because it can be used to determine the values of R_1 and R_2 that are needed for a given index of refraction and a desired focal length f. Conversely, if the index of refraction and the radii of curvature of a lens are given, this equation enables a calculation of the focal length. If the lens is immersed in something other than air, this same equation can be used, with n interpreted as the *ratio* of the index of refraction of the lens material to that of the surrounding fluid.

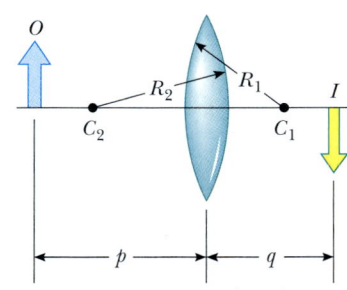

Figure 36.23 Simplified geometry for a thin lens.

Quick Quiz 36.2

What is the focal length of a pane of window glass?

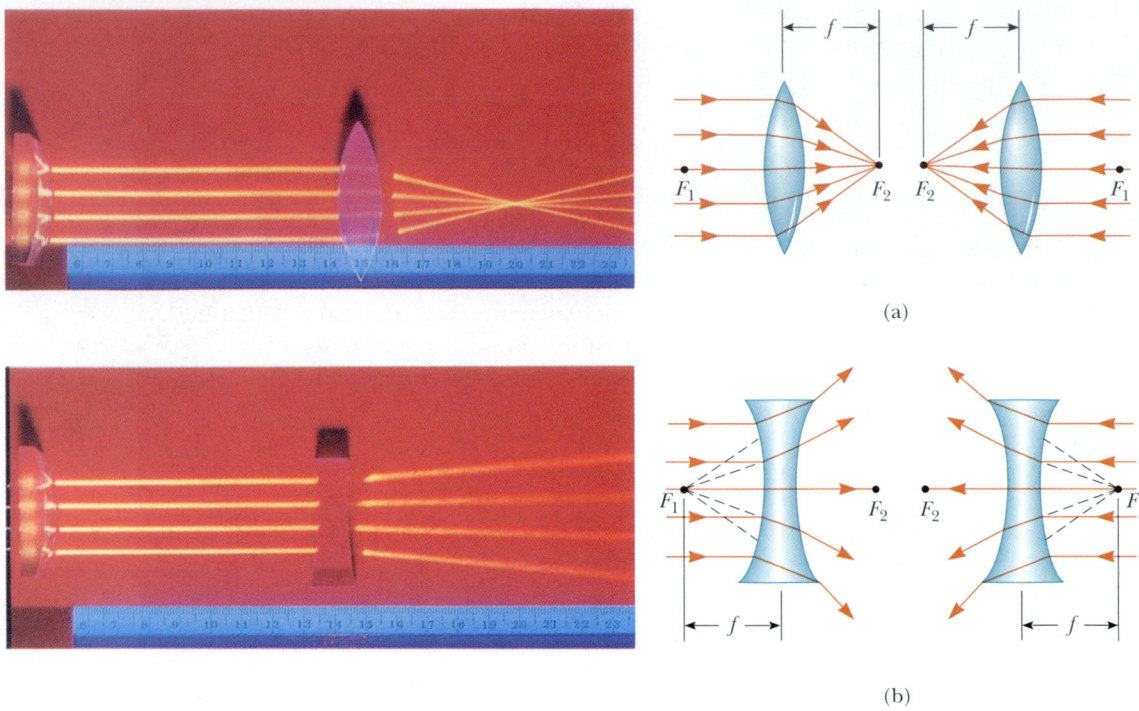

Figure 36.24   (*Left*) Effects of a converging lens (top) and a diverging lens (bottom) on parallel rays. (*Henry Leap and Jim Lehman*)   (*Right*) The object and image focal points of (a) a converging lens and (b) a diverging lens.

Using Equation 36.11, we can write Equation 36.10 in a form identical to Equation 36.6 for mirrors:

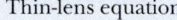
Thin-lens equation

$$\frac{1}{p} + \frac{1}{q} = \frac{1}{f} \tag{36.12}$$

This equation, called the **thin-lens equation,** can be used to relate the image distance and object distance for a thin lens.

Because light can travel in either direction through a lens, each lens has two focal points, one for light rays passing through in one direction and one for rays passing through in the other direction. This is illustrated in Figure 36.24 for a biconvex lens (two convex surfaces, resulting in a converging lens) and a biconcave lens (two concave surfaces, resulting in a diverging lens). Focal point F_1 is sometimes called the *object focal point,* and F_2 is called the *image focal point.*

Figure 36.25 is useful for obtaining the signs of p and q, and Table 36.3 gives the sign conventions for thin lenses. Note that these sign conventions are the same as those for refracting surfaces (see Table 36.2). Applying these rules to a biconvex lens, we see that when $p > f$, the quantities p, q, and R_1 are positive, and R_2 is negative. Therefore, p, q, and f are all positive when a converging lens forms a real image of an object. For a biconcave lens, p and R_2 are positive and q and R_1 are negative, with the result that f is negative.

Various lens shapes are shown in Figure 36.26. Note that a converging lens is thicker at the center than at the edge, whereas a diverging lens is thinner at the center than at the edge.

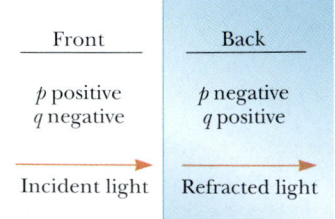

Figure 36.25   A diagram for obtaining the signs of p and q for a thin lens. (This diagram also applies to a refracting surface.)

TABLE 36.3	Sign Conventions for Thin Lenses

p is **positive** if object is in **front** of lens (real object).
p is **negative** if object is in **back** of lens (virtual object).

q is **positive** if image is in **back** of lens (real image).
q is **negative** if image is in **front** of lens (virtual image).

R_1 and R_2 are **positive** if center of curvature is in **back** of lens.
R_1 and R_2 are **negative** if center of curvature is in **front** of lens.

f is **positive** if the lens is **converging.**
f is **negative** if the lens is **diverging.**

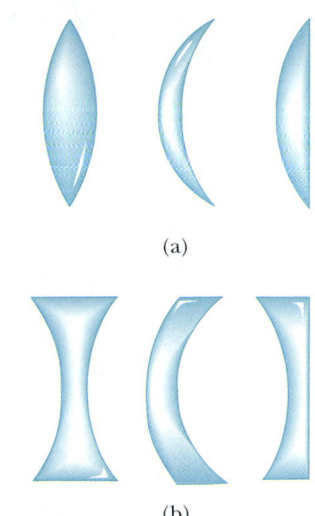

Figure 36.26 Various lens shapes. (a) Biconvex, convex–concave, and plano–convex. These are all converging lenses; they have a positive focal length and are thickest at the middle. (b) Biconcave, convex–concave, and plano–concave. These are all diverging lenses; they have a negative focal length and are thickest at the edges.

Magnification of Images

Consider a thin lens through which light rays from an object pass. As with mirrors (Eq. 36.2), the lateral magnification of the lens is defined as the ratio of the image height h' to the object height h:

$$M = \frac{h'}{h} = -\frac{q}{p}$$

From this expression, it follows that when M is positive, the image is upright and on the same side of the lens as the object. When M is negative, the image is inverted and on the side of the lens opposite the object.

Ray Diagrams for Thin Lenses

Ray diagrams are convenient for locating the images formed by thin lenses or systems of lenses. They also help clarify our sign conventions. Figure 36.27 shows such diagrams for three single-lens situations. To locate the image of a converg-

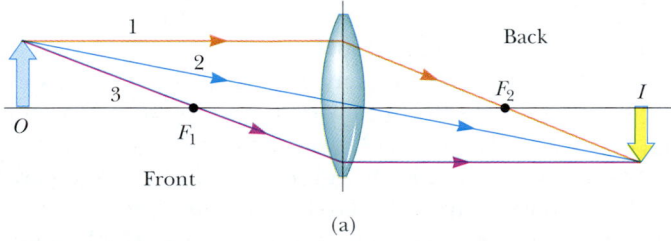

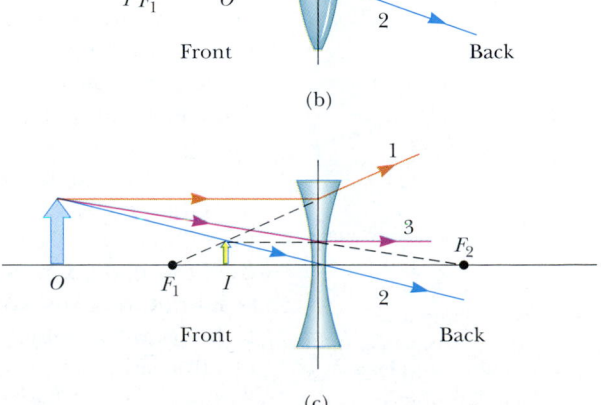

Figure 36.27 Ray diagrams for locating the image formed by a thin lens. (a) When the object is in front of and outside the object focal point F_1 of a converging lens, the image is real, inverted, and on the back side of the lens. (b) When the object is between F_1 and a converging lens, the image is virtual, upright, larger than the object, and on the front side of the lens. (c) When an object is anywhere in front of a diverging lens, the image is virtual, upright, smaller than the object, and on the front side of the lens.

ing lens (Fig. 36.27a and b), the following three rays are drawn from the top of the object:

> • Ray 1 is drawn parallel to the principal axis. After being refracted by the lens, this ray passes through the focal point on the back side of the lens.
> • Ray 2 is drawn through the center of the lens and continues in a straight line.
> • Ray 3 is drawn through that focal point on the front side of the lens (or as if coming from the focal point if $p < f$) and emerges from the lens parallel to the principal axis.

To locate the image of a diverging lens (Fig. 36.27c), the following three rays are drawn from the top of the object:

> • Ray 1 is drawn parallel to the principal axis. After being refracted by the lens, this ray emerges such that it appears to have passed through the focal point on the front side of the lens. (This apparent direction is indicated by the dashed line in Fig. 36.27c.)
> • Ray 2 is drawn through the center of the lens and continues in a straight line.
> • Ray 3 is drawn toward the focal point on the back side of the lens and emerges from the lens parallel to the optic axis.

Quick Quiz 36.3

In Figure 36.27a, the blue object arrow is replaced by one that is much taller than the lens. How many rays from the object will strike the lens?

For the converging lens in Figure 36.27a, where the object is to the left of the object focal point ($p > f_1$), the image is real and inverted. When the object is between the object focal point and the lens ($p < f_1$), as shown in Figure 36.27b, the image is virtual and upright. For a diverging lens (see Fig. 36.27c), the image is always virtual and upright, regardless of where the object is placed. These geometric constructions are reasonably accurate only if the distance between the rays and the principal axis is much less than the radii of the lens surfaces.

It is important to realize that refraction occurs only at the surfaces of the lens. A certain lens design takes advantage of this fact to produce the *Fresnel lens,* a powerful lens without great thickness. Because only the surface curvature is important in the refracting qualities of the lens, material in the middle of a Fresnel lens is removed, as shown in Figure 36.28. Because the edges of the curved segments cause some distortion, Fresnel lenses are usually used only in situations in which image quality is less important than reduction of weight.

The lines that are visible across the faces of most automobile headlights are the edges of these curved segments. A headlight requires a short-focal-length lens to collimate light from the nearby filament into a parallel beam. If it were not for the Fresnel design, the glass would be very thick in the center and quite heavy. The weight of the glass would probably cause the thin edge where the lens is supported to break when subjected to the shocks and vibrations that are typical of travel on rough roads.

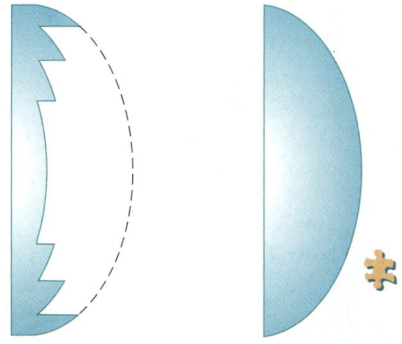

Figure 36.28 The Fresnel lens on the left has the same focal length as the thick lens on the right but is made of much less glass.

Quick Quiz 36.4

If you cover the top half of a lens, which of the following happens to the appearance of the image of an object? (a) The bottom half disappears, (b) the top half disappears; (c) the entire image is visible but has half the intensity; (d) no change occurs; (e) the entire image disappears.

EXAMPLE 36.9   An Image Formed by a Diverging Lens

A diverging lens has a focal length of -20.0 cm. An object 2.00 cm tall is placed 30.0 cm in front of the lens. Locate the image.

Solution   Using the thin-lens equation (Eq. 36.12) with $p = 30.0$ cm and $f = -20.0$ cm, we obtain

$$\frac{1}{30.0 \text{ cm}} + \frac{1}{q} = \frac{1}{-20.0 \text{ cm}}$$

$$q = -12.0 \text{ cm}$$

The negative sign tells us that the image is in front of the lens and virtual, as indicated in Figure 36.27c.

Exercise   Determine both the magnification and the height of the image.

Answer $M = 0.400$; $h' = 0.800$ cm.

EXAMPLE 36.10   An Image Formed by a Converging Lens

A converging lens of focal length 10.0 cm forms an image of each of three objects placed (a) 30.0 cm, (b) 10.0 cm, and (c) 5.00 cm in front of the lens. In each case, find the image distance and describe the image.

Solution   (a) The thin-lens equation can be used again:

$$\frac{1}{p} + \frac{1}{q} = \frac{1}{f}$$

$$\frac{1}{30.0 \text{ cm}} + \frac{1}{q} = \frac{1}{10.0 \text{ cm}}$$

$$q = 15.0 \text{ cm}$$

The positive sign indicates that the image is in back of the lens and real. The magnification is

$$M = -\frac{q}{p} = -\frac{15.0 \text{ cm}}{30.0 \text{ cm}} = -0.500$$

The image is reduced in size by one half, and the negative sign for M means that the image is inverted. The situation is like that pictured in Figure 36.27a.

(b) No calculation is necessary for this case because we know that, when the object is placed at the focal point, the image is formed at infinity. We can readily verify this by substituting $p = 10.0$ cm into the thin-lens equation.

(c) We now move inside the focal point, to an object distance of 5.00 cm:

$$\frac{1}{5.00 \text{ cm}} + \frac{1}{q} = \frac{1}{10.0 \text{ cm}}$$

$$q = -10.0 \text{ cm}$$

$$M = -\frac{q}{p} = -\left(\frac{-10.0 \text{ cm}}{5.00 \text{ cm}}\right) = 2.00$$

The negative image distance indicates that the image is in front of the lens and virtual. The image is enlarged, and the positive sign for M tells us that the image is upright, as shown in Figure 36.27b.

EXAMPLE 36.11   A Lens Under Water

A converging glass lens ($n = 1.52$) has a focal length of 40.0 cm in air. Find its focal length when it is immersed in water, which has an index of refraction of 1.33.

Solution   We can use the lens makers' equation (Eq. 36.11) in both cases, noting that R_1 and R_2 remain the same in air and water:

$$\frac{1}{f_{air}} = (n-1)\left(\frac{1}{R_1} - \frac{1}{R_2}\right)$$

$$\frac{1}{f_{water}} = (n'-1)\left(\frac{1}{R_1} - \frac{1}{R_2}\right)$$

where n' is the ratio of the index of refraction of glass to that of water: $n' = 1.52/1.33 = 1.14$. Dividing the first equation by the second gives

$$\frac{f_{water}}{f_{air}} = \frac{n-1}{n'-1} = \frac{1.52-1}{1.14-1} = 3.71$$

Because $f_{air} = 40.0$ cm, we find that

$$f_{water} = 3.71 f_{air} = 3.71(40.0 \text{ cm}) = \boxed{148 \text{ cm}}$$

The focal length of any glass lens is increased by a factor $(n-1)/(n'-1)$ when the lens is immersed in water.

Combination of Thin Lenses

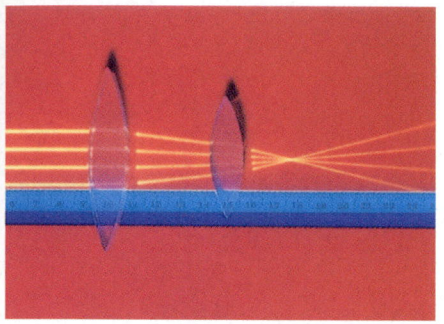

Light from a distant object brought into focus by two converging lenses. *(Henry Leap and Jim Lehman)*

If two thin lenses are used to form an image, the system can be treated in the following manner. First, the image formed by the first lens is located as if the second lens were not present. Then a ray diagram is drawn for the second lens, with the image formed by the first lens now serving as the object for the second lens. The second image formed is the final image of the system. One configuration is particularly straightforward; that is, if the image formed by the first lens lies on the back side of the second lens, then that image is treated as a **virtual object** for the second lens (that is, p is negative). The same procedure can be extended to a system of three or more lenses. The overall magnification of a system of thin lenses equals the product of the magnifications of the separate lenses.

Let us consider the special case of a system of two lenses in contact. Suppose two thin lenses of focal lengths f_1 and f_2 are placed in contact with each other. If p is the object distance for the combination, application of the thin-lens equation (Eq. 36.12) to the first lens gives

$$\frac{1}{p} + \frac{1}{q_1} = \frac{1}{f_1}$$

where q_1 is the image distance for the first lens. Treating this image as the object for the second lens, we see that the object distance for the second lens must be $-q_1$ (negative because the object is virtual). Therefore, for the second lens,

$$\frac{1}{-q_1} + \frac{1}{q} = \frac{1}{f_2}$$

where q is the final image distance from the second lens. Adding these equations eliminates q_1 and gives

$$\frac{1}{p} + \frac{1}{q} = \frac{1}{f_1} + \frac{1}{f_2}$$

Focal length of two thin lenses in contact

$$\frac{1}{f} = \frac{1}{f_1} + \frac{1}{f_2} \tag{36.13}$$

Because the two thin lenses are touching, q is also the distance of the final image from the first lens. Therefore, **two thin lenses in contact with each other are equivalent to a single thin lens having a focal length given by Equation 36.13.**

EXAMPLE 36.12 ▸ Where Is the Final Image?

Even when the conditions just described do not apply, the lens equations yield image position and magnification. For example, two thin converging lenses of focal lengths $f_1 = 10.0$ cm and $f_2 = 20.0$ cm are separated by 20.0 cm, as illustrated in Figure 36.29. An object is placed 15.0 cm to the left of lens 1. Find the position of the final image and the magnification of the system.

Solution First we locate the image formed by lens 1 while ignoring lens 2:

$$\frac{1}{p_1} + \frac{1}{q_1} = \frac{1}{f_1}$$

$$\frac{1}{15.0 \text{ cm}} + \frac{1}{q_1} = \frac{1}{10.0 \text{ cm}}$$

$$q_1 = 30.0 \text{ cm}$$

where q_1 is measured from lens 1. A positive value for q_1 means that this first image is in back of lens 1.

Because q_1 is greater than the separation between the two lenses, this image formed by lens 1 lies 10.0 cm to the right of lens 2. We take this as the object distance for the second lens, so $p_2 = -10.0$ cm, where distances are now measured from lens 2:

$$\frac{1}{p_2} + \frac{1}{q_2} = \frac{1}{f_2}$$

$$\frac{1}{-10.0 \text{ cm}} + \frac{1}{q_2} = \frac{1}{20.0 \text{ cm}}$$

$$q_2 = 6.67 \text{ cm}$$

The final image lies 6.67 cm to the right of lens 2.
The individual magnifications of the images are

$$M_1 = -\frac{q_1}{p_1} = -\frac{30.0 \text{ cm}}{15.0 \text{ cm}} = -2.00$$

$$M_2 = -\frac{q_2}{p_2} = -\frac{6.67 \text{ cm}}{-10.0 \text{ cm}} = 0.667$$

The total magnification M is equal to the product $M_1 M_2 = (-2.00)(0.667) = -1.33$. The final image is real because q_2 is positive. The image is also inverted and enlarged.

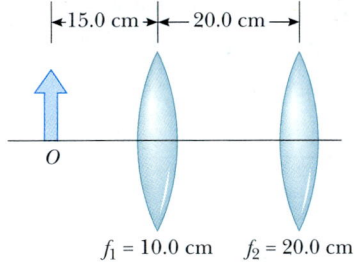

Figure 36.29 A combination of two converging lenses.

CONCEPTUAL EXAMPLE 36.13 ▸ Watch Your p's and q's!

Use a spreadsheet or a similar tool to create two graphs of image distance as a function of object distance—one for a lens for which the focal length is 10 cm and one for a lens for which the focal length is -10 cm.

Solution The graphs are shown in Figure 36.30. In each graph a gap occurs where $p = f$, which we shall discuss. Note the similarity in the shapes—a result of the fact that image and object distances for both lenses are related according to the same equation—the thin-lens equation.

The curve in the upper right portion of the $f = +10$ cm graph corresponds to an object on the *front* side of a lens, which we have drawn as the left side of the lens in our previous diagrams. When the object is at positive infinity, a real image forms at the focal point on the back side (the positive side) of the lens, $q = f$. (The incoming rays are parallel in this case.) As the object gets closer to the lens, the image moves farther from the lens, corresponding to the upward path of the curve. This continues until the object is located at the focal point on the

near side of the lens. At this point, the rays leaving the lens are parallel, making the image infinitely far away. This is described in the graph by the asymptotic approach of the curve to the line $p = f = 10$ cm.

As the object moves inside the focal point, the image becomes virtual and located near $q = -\infty$. We are now following the curve in the lower left portion of Figure 36.30a. As the object moves closer to the lens, the virtual image also moves closer to the lens. As $p \to 0$, the image distance q also approaches 0. Now imagine that we bring the object to the back side of the lens, where $p < 0$. The object is now a virtual object, so it must have been formed by some other lens. For all locations of the virtual object, the image distance is positive and less than the focal length. The final image is real, and its position approaches the focal point as p gets more and more negative.

The $f = -10$ cm graph shows that a distant real object forms an image at the focal point on the front side of the lens. As the object approaches the lens, the image remains virtual and moves closer to the lens. But as we continue toward the left end of the p axis, the object becomes virtual. As the position of this virtual object approaches the focal point, the image recedes toward infinity. As we pass the focal point, the image shifts from a location at positive infinity to one at negative infinity. Finally, as the virtual object continues moving away from the lens, the image is virtual, starts moving in from negative infinity, and approaches the focal point.

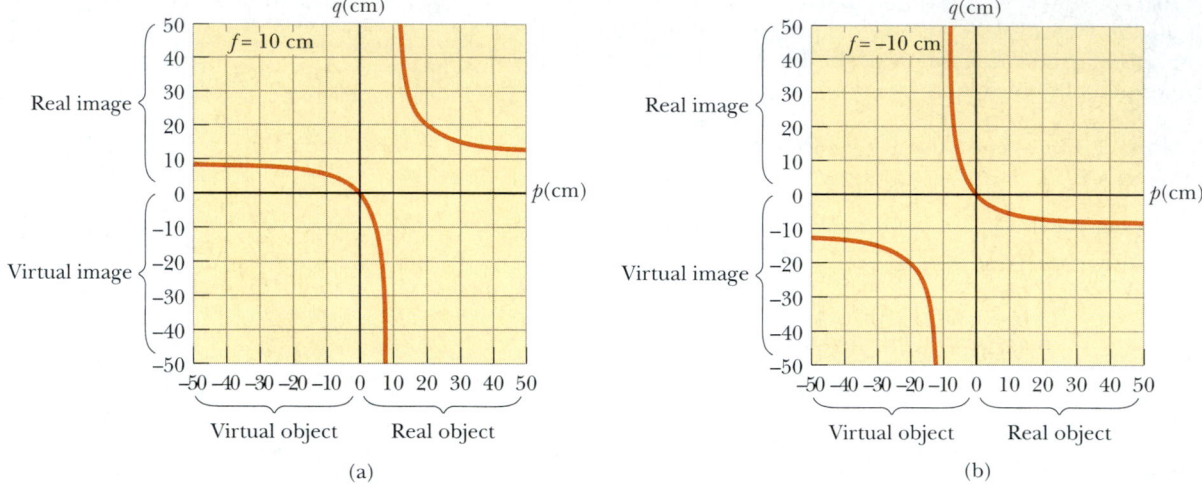

Figure 36.30   (a) Image position as a function of object position for a lens having a focal length of $+10$ cm. (b) Image position as a function of object position for a lens having a focal length of -10 cm.

Optional Section

36.5 ▸ LENS ABERRATIONS

One problem with lenses is imperfect images. The theory of mirrors and lenses that we have been using assumes that rays make small angles with the principal axis and that the lenses are thin. In this simple model, all rays leaving a point source focus at a single point, producing a sharp image. Clearly, this is not always true. When the approximations used in this theory do not hold, imperfect images are formed.

A precise analysis of image formation requires tracing each ray, using Snell's law at each refracting surface and the law of reflection at each reflecting surface. This procedure shows that the rays from a point object do not focus at a single point, with the result that the image is blurred. The departures of actual (imperfect) images from the ideal predicted by theory are called **aberrations.**

Spherical Aberrations

Spherical aberrations occur because the focal points of rays far from the principal axis of a spherical lens (or mirror) are different from the focal points of rays of the same wavelength passing near the axis. Figure 36.31 illustrates spherical aberration for parallel rays passing through a converging lens. Rays passing through points near the center of the lens are imaged farther from the lens than rays passing through points near the edges.

Many cameras have an adjustable aperture to control light intensity and reduce spherical aberration. (An aperture is an opening that controls the amount of light passing through the lens.) Sharper images are produced as the aperture size is reduced because with a small aperture only the central portion of the lens is exposed to the light; as a result, a greater percentage of the rays are paraxial. At the

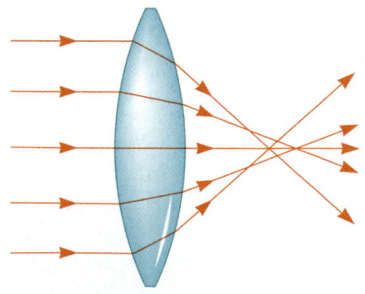

Figure 36.31   Spherical aberration caused by a converging lens. Does a diverging lens cause spherical aberration?

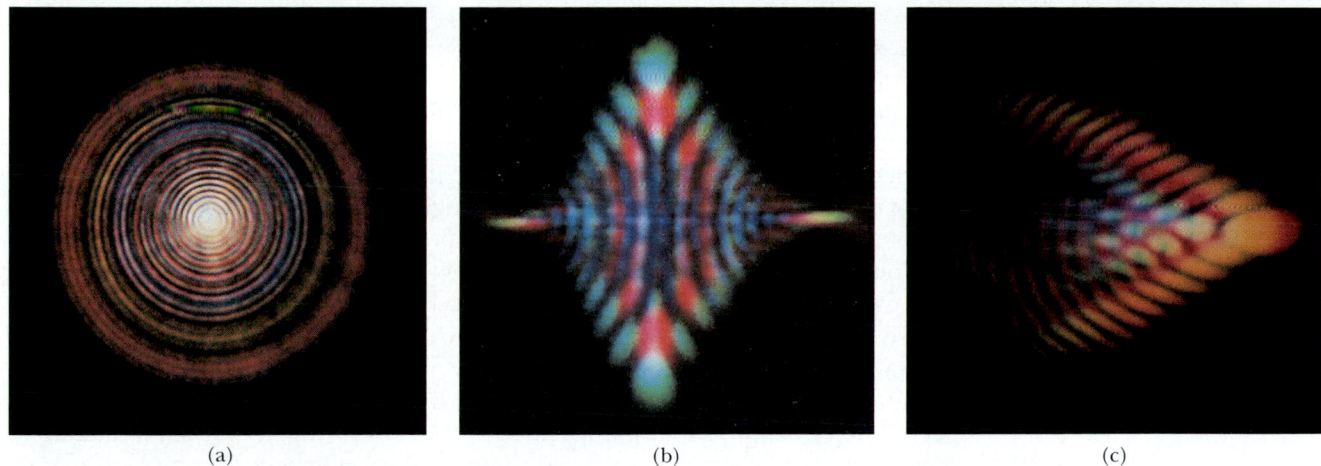

Lens aberrations. (a) *Spherical aberration* occurs when light passing through the lens at different distances from the principal axis is focused at different points. (b) *Astigmatism* occurs for objects not located on the principal axis of the lens. (c) *Coma* occurs as light passing through the lens far from the principal axis and light passing near the center of the lens focus at different parts of the focal plane. *(Photographs by Norman Goldberg)*

same time, however, less light passes through the lens. To compensate for this lower light intensity, a longer exposure time is used.

In the case of mirrors used for very distant objects, spherical aberration can be minimized through the use of a parabolic reflecting surface rather than a spherical surface. Parabolic surfaces are not used often, however, because those with high-quality optics are very expensive to make. Parallel light rays incident on a parabolic surface focus at a common point, regardless of their distance from the principal axis. Parabolic reflecting surfaces are used in many astronomical telescopes to enhance image quality.

Chromatic Aberrations

The fact that different wavelengths of light refracted by a lens focus at different points gives rise to chromatic aberrations. In Chapter 35 we described how the index of refraction of a material varies with wavelength. For instance, when white light passes through a lens, violet rays are refracted more than red rays (Fig. 36.32). From this we see that the focal length is greater for red light than for violet light. Other wavelengths (not shown in Fig. 36.32) have focal points intermediate between those of red and violet.

Chromatic aberration for a diverging lens also results in a shorter focal length for violet light than for red light, but on the front side of the lens. Chromatic aberration can be greatly reduced by combining a converging lens made of one type of glass and a diverging lens made of another type of glass.

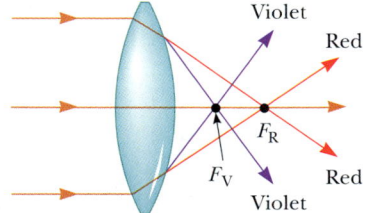

Figure 36.32 Chromatic aberration caused by a converging lens. Rays of different wavelengths focus at different points.

Optional Section

36.6 THE CAMERA

The photographic **camera** is a simple optical instrument whose essential features are shown in Figure 36.33. It consists of a light-tight box, a converging lens that produces a real image, and a film behind the lens to receive the image. One focuses the camera by varying the distance between lens and film. This is accom-

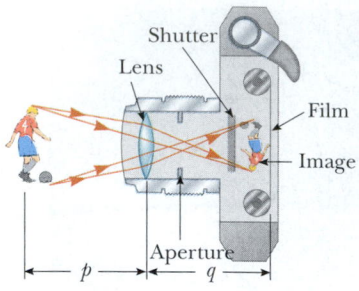

Figure 36.33 Cross-sectional view of a simple camera. Note that in reality, $p \gg q$.

plished with an adjustable bellows in older-style cameras and with some other mechanical arrangement in modern cameras. For proper focusing—which is necessary for the formation of sharp images—the lens-to-film distance depends on the object distance as well as on the focal length of the lens.

The shutter, positioned behind the lens, is a mechanical device that is opened for selected time intervals, called *exposure times*. One can photograph moving objects by using short exposure times, or photograph dark scenes (with low light levels) by using long exposure times. If this adjustment were not available, it would be impossible to take stop-action photographs. For example, a rapidly moving vehicle could move enough in the time that the shutter was open to produce a blurred image. Another major cause of blurred images is the movement of the camera while the shutter is open. To prevent such movement, either short exposure times or a tripod should be used, even for stationary objects. Typical shutter speeds (that is, exposure times) are 1/30, 1/60, 1/125, and 1/250 s. For handheld cameras, the use of slower speeds can result in blurred images (due to movement), but the use of faster speeds reduces the gathered light intensity. In practice, stationary objects are normally shot with an intermediate shutter speed of 1/60 s.

More expensive cameras have an aperture of adjustable diameter to further control the intensity of the light reaching the film. As noted earlier, when an aperture of small diameter is used, only light from the central portion of the lens reaches the film; in this way spherical aberration is reduced.

The intensity I of the light reaching the film is proportional to the area of the lens. Because this area is proportional to the square of the diameter D, we conclude that I is also proportional to D^2. Light intensity is a measure of the rate at which energy is received by the film per unit area of the image. Because the area of the image is proportional to q^2, and $q \approx f$ (when $p \gg f$, so p can be approximated as infinite), we conclude that the intensity is also proportional to $1/f^2$, and thus $I \propto D^2/f^2$. The brightness of the image formed on the film depends on the light intensity, so we see that the image brightness depends on both the focal length and the diameter of the lens.

The ratio f/D is called the **f-number** of a lens:

$$f\text{-number} \equiv \frac{f}{D} \tag{36.14}$$

Hence, the intensity of light incident on the film can be expressed as

$$I \propto \frac{1}{(f/D)^2} \propto \frac{1}{(f\text{-number})^2} \tag{36.15}$$

The *f*-number is often given as a description of the lens "speed." The lower the *f*-number, the wider the aperture and the higher the rate at which energy from the light exposes the film—thus, a lens with a low *f*-number is a "fast" lens. The conventional notation for an *f*-number is "$f/$" followed by the actual number. For example, "$f/4$" means an *f*-number of 4—it *does not* mean to divide f by 4! Extremely fast lenses, which have *f*-numbers as low as approximately $f/1.2$, are expensive because it is very difficult to keep aberrations acceptably small with light rays passing through a large area of the lens. Camera lens systems (that is, combinations of lenses with adjustable apertures) are often marked with multiple *f*-numbers, usually $f/2.8$, $f/4$, $f/5.6$, $f/8$, $f/11$, and $f/16$. Any one of these settings can be selected by adjusting the aperture, which changes the value of D. Increasing the setting from one *f*-number to the next higher value (for example, from $f/2.8$ to $f/4$) decreases the area of the aperture by a factor of two. The lowest *f*-number setting on a camera lens corresponds to a wide-open aperture and the use of the maximum possible lens area.

Simple cameras usually have a fixed focal length and a fixed aperture size, with an *f*-number of about *f*/11. This high value for the *f*-number allows for a large **depth of field,** meaning that objects at a wide range of distances from the lens form reasonably sharp images on the film. In other words, the camera does not have to be focused.

EXAMPLE 36.14 Finding the Correct Exposure Time

The lens of a certain 35-mm camera (where 35 mm is the width of the film strip) has a focal length of 55 mm and a speed (an *f*-number) of *f*/1.8. The correct exposure time for this speed under certain conditions is known to be (1/500) s. (a) Determine the diameter of the lens.

Solution   From Equation 36.14, we find that

$$D = \frac{f}{f\text{-number}} = \frac{55 \text{ mm}}{1.8} = \boxed{31 \text{ mm}}$$

(b) Calculate the correct exposure time if the *f*-number is changed to *f*/4 under the same lighting conditions.

Solution   The total light energy hitting the film is proportional to the product of the intensity and the exposure time. If I is the light intensity reaching the film, then in a time t

the energy per unit area received by the film is proportional to It. Comparing the two situations, we require that $I_1 t_1 = I_2 t_2$, where t_1 is the correct exposure time for *f*/1.8 and t_2 is the correct exposure time for *f*/4. Using this result together with Equation 36.15, we find that

$$\frac{t_1}{(f_1\text{-number})^2} = \frac{t_2}{(f_2\text{-number})^2}$$

$$t_2 = \left(\frac{f_2\text{-number}}{f_1\text{-number}}\right)^2 t_1$$

$$= \left(\frac{4}{1.8}\right)^2 \left(\frac{1}{500} \text{ s}\right) \approx \boxed{\frac{1}{100} \text{ s}}$$

As the aperture size is reduced, exposure time must increase.

Optional Section

36.7 THE EYE

Like a camera, a normal eye focuses light and produces a sharp image. However, the mechanisms by which the eye controls the amount of light admitted and adjusts to produce correctly focused images are far more complex, intricate, and effective than those in even the most sophisticated camera. In all respects, the eye is a physiological wonder.

Figure 36.34 shows the essential parts of the human eye. Light entering the eye passes through a transparent structure called the *cornea,* behind which are a clear liquid (the *aqueous humor*), a variable aperture (the *pupil,* which is an opening in the *iris*), and the *crystalline lens.* Most of the refraction occurs at the outer surface of the eye, where the cornea is covered with a film of tears. Relatively little refraction occurs in the crystalline lens because the aqueous humor in contact with the lens has an average index of refraction close to that of the lens. The iris, which is the colored portion of the eye, is a muscular diaphragm that controls pupil size. The iris regulates the amount of light entering the eye by dilating the pupil in low-light conditions and contracting the pupil in high-light conditions. The *f*-number range of the eye is from about *f*/2.8 to *f*/16.

The cornea–lens system focuses light onto the back surface of the eye, the *retina*, which consists of millions of sensitive receptors called *rods* and *cones.* When stimulated by light, these receptors send impulses via the optic nerve to the brain,

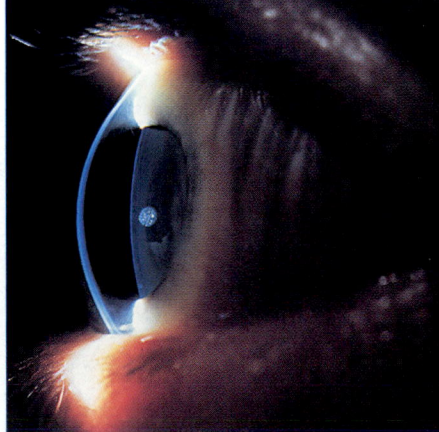

Close-up photograph of the cornea of the human eye. *(From Lennart Nilsson, in collaboration with Jan Lindberg, Behold Man: A Photographic Journey of Discovery Inside the Body, Boston, Little, Brown & Co., 1974.)*

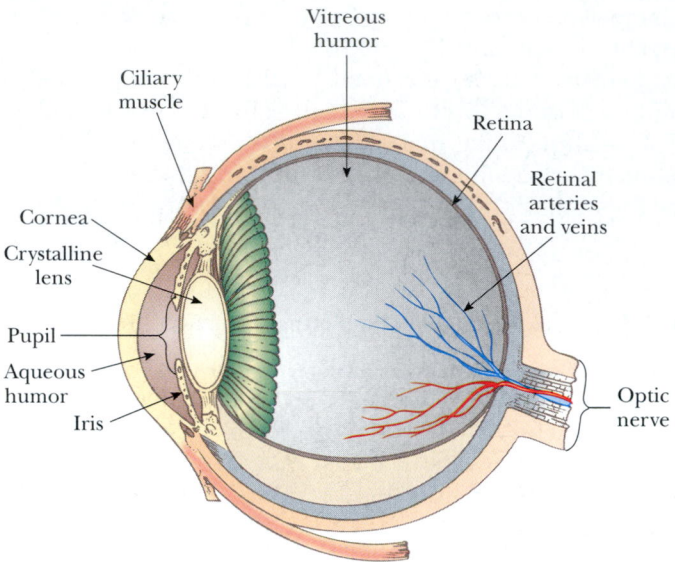

Figure 36.34   Essential parts of the eye.

where an image is perceived. By this process, a distinct image of an object is observed when the image falls on the retina.

The eye focuses on an object by varying the shape of the pliable crystalline lens through an amazing process called **accommodation.** An important component of accommodation is the *ciliary muscle,* which is situated in a circle around the rim of the lens. Thin filaments, called *zonules,* run from this muscle to the edge of the lens. When the eye is focused on a distant object, the ciliary muscle is relaxed, tightening the zonules that attach the muscle to the edge of the lens. The force of the zonules causes the lens to flatten, increasing its focal length. For an object distance of infinity, the focal length of the eye is equal to the fixed distance between lens and retina, about 1.7 cm. The eye focuses on nearby objects by tensing the ciliary muscle, which relaxes the zonules. This action allows the lens to bulge a bit, and its focal length decreases, resulting in the image being focused on the retina. All these lens adjustments take place so swiftly that we are not even aware of the change. In this respect, even the finest electronic camera is a toy compared with the eye.

Accommodation is limited in that objects that are very close to the eye produce blurred images. The **near point** is the closest distance for which the lens can accommodate to focus light on the retina. This distance usually increases with age and has an average value of 25 cm. Typically, at age 10 the near point of the eye is about 18 cm. It increases to about 25 cm at age 20, to 50 cm at age 40, and to 500 cm or greater at age 60. The **far point** of the eye represents the greatest distance for which the lens of the relaxed eye can focus light on the retina. A person with normal vision can see very distant objects, such as the Moon, and thus has a far point near infinity.

Recall that the light leaving the mirror in Figure 36.8 becomes white where it comes together but then diverges into separate colors again. Because nothing but air exists at the point where the rays cross (and hence nothing exists to cause the colors to separate again), seeing white light as a result of a combination of colors must be a visual illusion. In fact, this is the case. Only three types of color-sensitive

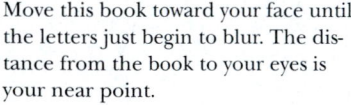

QuickLab

Move this book toward your face until the letters just begin to blur. The distance from the book to your eyes is your near point.

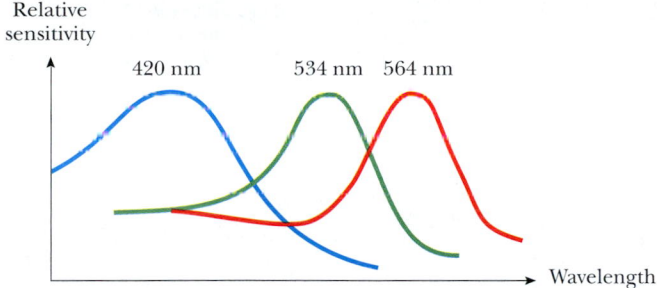

Relative
sensitivity

420 nm   534 nm   564 nm

Wavelength

Figure 36.35   Approximate color sensitivity of the three types of cones in the retina.

cells are present in the retina; they are called red, green, and blue cones because of the peaks of the color ranges to which they respond (Fig. 36.35). If the red and green cones are stimulated simultaneously (as would be the case if yellow light were shining on them), the brain interprets what we see as yellow. If all three types of cones are stimulated by the separate colors red, blue, and green, as in Figure 36.8, we see white. If all three types of cones are stimulated by light that contains *all* colors, such as sunlight, we again see white light.

Color televisions take advantage of this visual illusion by having only red, green, and blue dots on the screen. With specific combinations of brightness in these three primary colors, our eyes can be made to see any color in the rainbow. Thus, the yellow lemon you see in a television commercial is not really yellow, it is red and green! The paper on which this page is printed is made of tiny, matted, translucent fibers that scatter light in all directions; the resultant mixture of colors appears white to the eye. Snow, clouds, and white hair are not really white. In fact, there is no such thing as a white pigment. The appearance of these things is a consequence of the scattering of light containing all colors, which we interpret as white.

Conditions of the Eye

When the eye suffers a mismatch between the focusing range of the lens–cornea system and the length of the eye, with the result that light rays reach the retina before they converge to form an image, as shown in Figure 36.36a, the condition is known as **farsightedness** (or *hyperopia*). A farsighted person can usually see faraway objects clearly but not nearby objects. Although the near point of a normal eye is approximately 25 cm, the near point of a farsighted person is much farther away. The eye of a farsighted person tries to focus by accommodation—that is, by shortening its focal length. This works for distant objects, but because the focal length of the farsighted eye is greater than normal, the light from nearby objects cannot be brought to a sharp focus before it reaches the retina, and it thus causes a blurred image. The refracting power in the cornea and lens is insufficient to focus the light from all but distant objects satisfactorily. The condition can be corrected by placing a converging lens in front of the eye, as shown in Figure 36.36b. The lens refracts the incoming rays more toward the principal axis before entering the eye, allowing them to converge and focus on the retina.

A person with **nearsightedness** (or *myopia*), another mismatch condition, can focus on nearby objects but not on faraway objects. In the case of *axial myopia*, the nearsightedness is caused by the lens being too far from the retina. In *refractive my-*

QuickLab

Pour a pile of salt or sugar into your palm. Compare its white appearance with the transparency of a single grain.

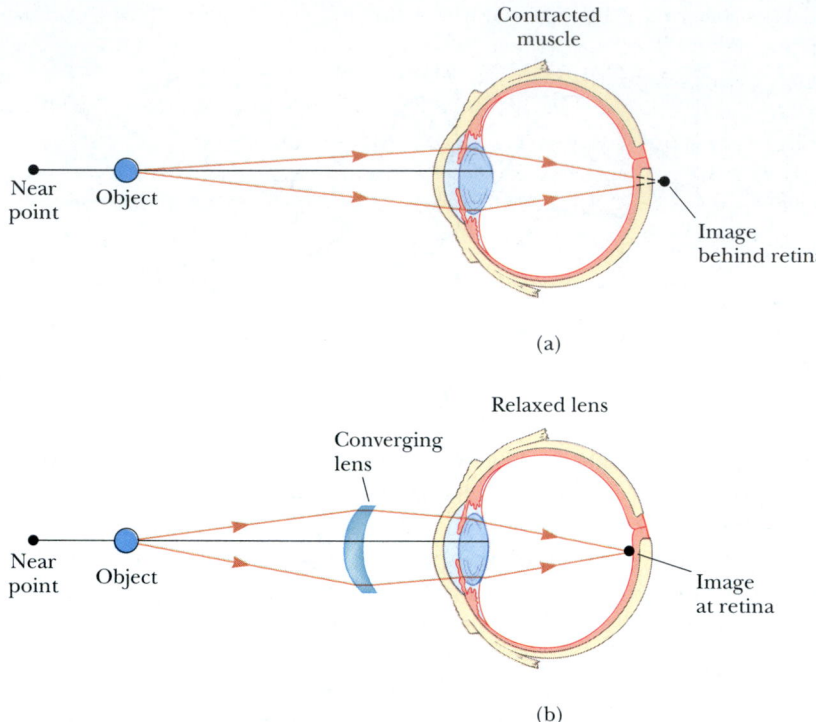

Figure 36.36 (a) When a farsighted eye looks at an object located between the near point and the eye, the image point is behind the retina, resulting in blurred vision. The eye muscle contracts to try to bring the object into focus. (b) Farsightedness is corrected with a converging lens.

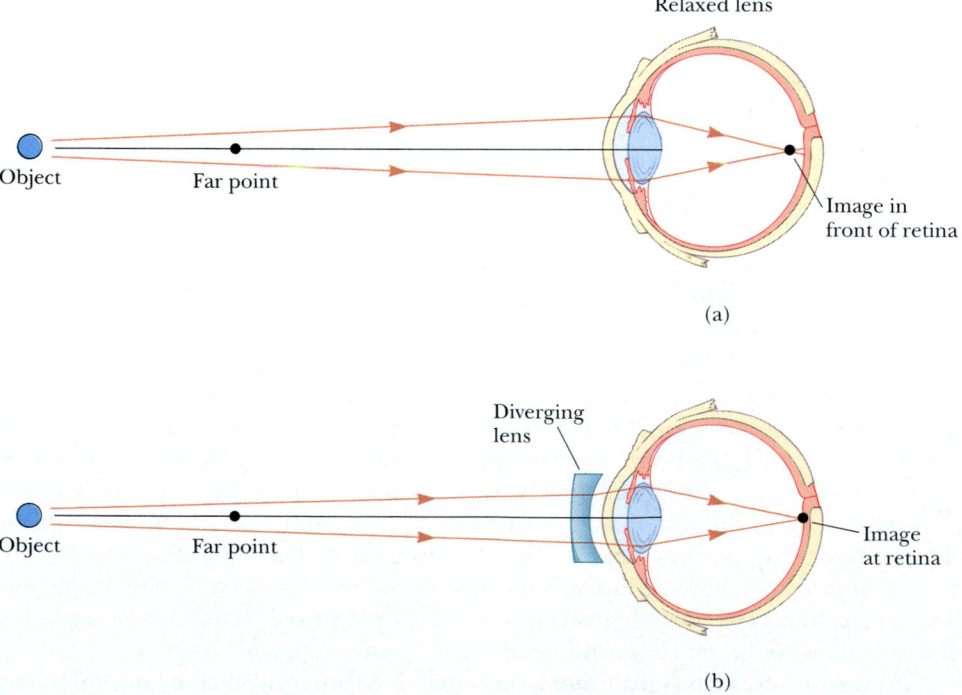

Figure 36.37 (a) When a nearsighted eye looks at an object that lies beyond the eye's far point, the image is formed in front of the retina, resulting in blurred vision. (b) Nearsightedness is corrected with a diverging lens.

opia, the lens–cornea system is too powerful for the length of the eye. The far point of the nearsighted eye is not infinity and may be less than 1 m. The maximum focal length of the nearsighted eye is insufficient to produce a sharp image on the retina, and rays from a distant object converge to a focus in front of the retina. They then continue past that point, diverging before they finally reach the retina and causing blurred vision (Fig. 36.37a). Nearsightedness can be corrected with a diverging lens, as shown in Figure 36.37b. The lens refracts the rays away from the principal axis before they enter the eye, allowing them to focus on the retina.

Quick Quiz 36.5

Which glasses in Figure 36.38 correct nearsightedness and which correct farsightedness?

(a)   (b)

Figure 36.38 *(George Semple)*

Beginning in middle age, most people lose some of their accommodation ability as the ciliary muscle weakens and the lens hardens. Unlike farsightedness, which is a mismatch between focusing power and eye length, **presbyopia** (literally, "old-age vision") is due to a reduction in accommodation ability. The cornea and lens do not have sufficient focusing power to bring nearby objects into focus on the retina. The symptoms are the same as those of farsightedness, and the condition can be corrected with converging lenses.

In the eye defect known as **astigmatism,** light from a point source produces a line image on the retina. This condition arises when either the cornea or the lens or both are not perfectly symmetric. Astigmatism can be corrected with lenses that have different curvatures in two mutually perpendicular directions.

Optometrists and ophthalmologists usually prescribe lenses[1] measured in **diopters:**

> The **power** P of a lens in diopters equals the inverse of the focal length in meters: $P = 1/f$.

For example, a converging lens of focal length $+20$ cm has a power of $+5.0$ diopters, and a diverging lens of focal length -40 cm has a power of -2.5 diopters.

EXAMPLE 36.15 ▶ **A Case of Nearsightedness**

A particular nearsighted person is unable to see objects clearly when they are beyond 2.5 m away (the far point of this particular eye). What should the focal length be in a lens prescribed to correct this problem?

Solution The purpose of the lens in this instance is to "move" an object from infinity to a distance where it can be seen clearly. This is accomplished by having the lens produce an image at the far point. From the thin-lens equation, we have

$$\frac{1}{p} + \frac{1}{q} = \frac{1}{\infty} + \frac{1}{-2.5 \text{ m}} = \frac{1}{f}$$

$$f = \boxed{-2.5 \text{ m}}$$

Why did we use a negative sign for the image distance? As you should have suspected, the lens must be a diverging lens (one with a negative focal length) to correct nearsightedness.

Exercise What is the power of this lens?

Answer -0.40 diopter.

Optional Section

36.8 ▶ THE SIMPLE MAGNIFIER

The simple magnifier consists of a single converging lens. As the name implies, this device increases the apparent size of an object.

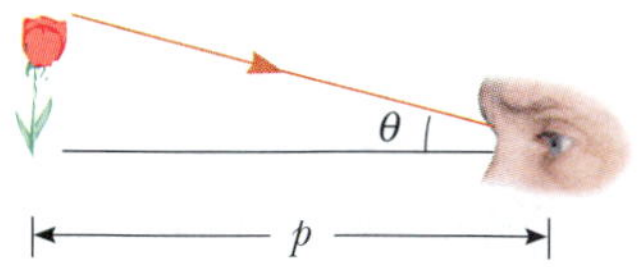

Suppose an object is viewed at some distance p from the eye, as illustrated in Figure 36.39. The size of the image formed at the retina depends on the angle θ subtended by the object at the eye. As the object moves closer to the eye, θ increases and a larger image is observed. However, an average normal eye cannot focus on an object closer than about 25 cm, the near point (Fig. 36.40a). Therefore, θ is maximum at the near point.

To further increase the apparent angular size of an object, a converging lens can be placed in front of the eye as in Figure 36.40b, with the object located at point O, just inside the focal point of the lens. At this location, the lens forms a virtual, upright, enlarged image. We define **angular magnification** m as the ratio of the angle subtended by an object with a lens in use (angle θ in Fig. 36.40b) to the angle subtended by the object placed at the near point with no lens in use (angle

Figure 36.39 The size of the image formed on the retina depends on the angle θ subtended at the eye.

---

[1] The word *lens* comes from *lentil,* the name of an Italian legume. (You may have eaten lentil soup.) Early eyeglasses were called "glass lentils" because the biconvex shape of their lenses resembled the shape of a lentil. The first lenses for farsightedness and presbyopia appeared around 1280; concave eyeglasses for correcting nearsightedness did not appear for more than 100 years after that.

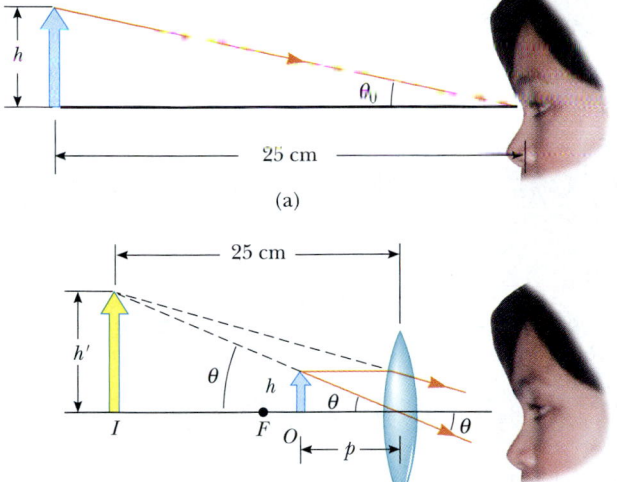

(a)

(b)

Figure 36.40 (a) An object placed at the near point of the eye ($p = 25$ cm) subtends an angle $\theta_0 \approx h/25$ at the eye. (b) An object placed near the focal point of a converging lens produces a magnified image that subtends an angle $\theta \approx h'/25$ at the eye.

θ_0 in Fig. 36.40a):

$$m \equiv \frac{\theta}{\theta_0} \tag{36.16}$$

Angular magnification with the object at the near point

The angular magnification is a maximum when the image is at the near point of the eye—that is, when $q = -25$ cm. The object distance corresponding to this image distance can be calculated from the thin-lens equation:

$$\frac{1}{p} + \frac{1}{-25 \text{ cm}} = \frac{1}{f}$$

$$p = \frac{25f}{25 + f}$$

where f is the focal length of the magnifier in centimeters. If we make the small-angle approximations

$$\tan \theta_0 \approx \theta_0 \approx \frac{h}{25} \quad \text{and} \quad \tan \theta \approx \theta \approx \frac{h}{p} \tag{36.17}$$

Equation 36.16 becomes

$$m_{\text{max}} = \frac{\theta}{\theta_0} = \frac{h/p}{h/25} = \frac{25}{p} = \frac{25}{25f/(25 + f)}$$

$$m_{\text{max}} = 1 + \frac{25 \text{ cm}}{f} \tag{36.18}$$

Although the eye can focus on an image formed anywhere between the near point and infinity, it is most relaxed when the image is at infinity. For the image formed by the magnifying lens to appear at infinity, the object has to be at the focal point of the lens. In this case, Equations 36.17 become

$$\theta_0 \approx \frac{h}{25} \quad \text{and} \quad \theta \approx \frac{h}{f}$$

and the magnification is

$$m_{min} = \frac{\theta}{\theta_0} = \frac{25 \text{ cm}}{f} \qquad (36.19)$$

With a single lens, it is possible to obtain angular magnifications up to about 4 without serious aberrations. Magnifications up to about 20 can be achieved by using one or two additional lenses to correct for aberrations.

EXAMPLE 36.16   **Maximum Magnification of a Lens**

What is the maximum magnification that is possible with a lens having a focal length of 10 cm, and what is the magnification of this lens when the eye is relaxed?

Solution   The maximum magnification occurs when the image is located at the near point of the eye. Under these circumstances, Equation 36.18 gives

$$m_{max} = 1 + \frac{25 \text{ cm}}{f} = 1 + \frac{25 \text{ cm}}{10 \text{ cm}} = \boxed{3.5}$$

When the eye is relaxed, the image is at infinity. In this case, we use Equation 36.19:

$$m_{min} = \frac{25 \text{ cm}}{f} = \frac{25 \text{ cm}}{10 \text{ cm}} = \boxed{2.5}$$

Optional Section

36.9   **THE COMPOUND MICROSCOPE**

A simple magnifier provides only limited assistance in inspecting minute details of an object. Greater magnification can be achieved by combining two lenses in a device called a **compound microscope,** a schematic diagram of which is shown in Figure 36.41a. It consists of one lens, the *objective,* that has a very short focal length

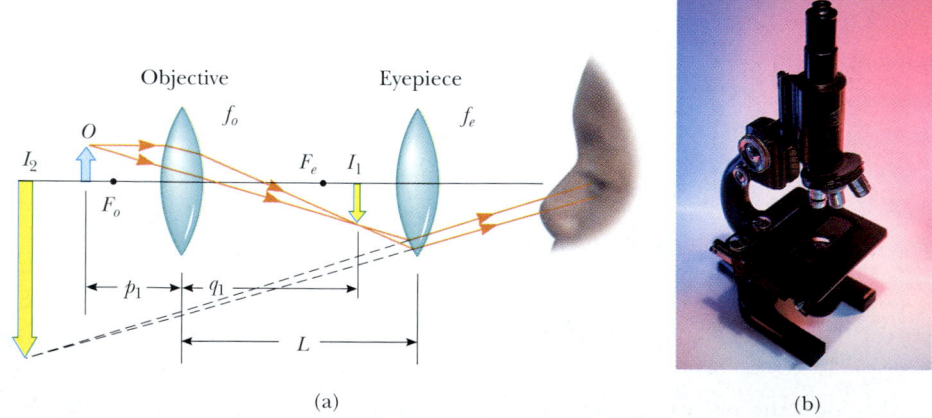

(a)   (b)

Figure 36.41   (a) Diagram of a compound microscope, which consists of an objective lens and an eyepiece lens. (b) A compound microscope. The three-objective turret allows the user to choose from several powers of magnification. Combinations of eyepieces with different focal lengths and different objectives can produce a wide range of magnifications. *(Henry Leap and Jim Lehman)*

$f_o < 1$ cm and a second lens, the *eyepiece,* that has a focal length f_e of a few centimeters. The two lenses are separated by a distance L that is much greater than either f_o or f_e. The object, which is placed just outside the focal point of the objective, forms a real, inverted image at I_1, and this image is located at or close to the focal point of the eyepiece. The eyepiece, which serves as a simple magnifier, produces at I_2 a virtual, inverted image of I_1. The lateral magnification M_1 of the first image is $-q_1/p_1$. Note from Figure 36.41a that q_1 is approximately equal to L and that the object is very close to the focal point of the objective: $p_1 \approx f_o$. Thus, the lateral magnification by the objective is

$$M_1 \approx -\frac{L}{f_o}$$

The angular magnification by the eyepiece for an object (corresponding to the image at I_1) placed at the focal point of the eyepiece is, from Equation 36.19,

$$m_e = \frac{25 \text{ cm}}{f_e}$$

The overall magnification of the compound microscope is defined as the product of the lateral and angular magnifications:

$$M = M_1 m_e = -\frac{L}{f_o}\left(\frac{25 \text{ cm}}{f_e}\right) \tag{36.20}$$

The negative sign indicates that the image is inverted.

The microscope has extended human vision to the point where we can view previously unknown details of incredibly small objects. The capabilities of this instrument have steadily increased with improved techniques for precision grinding of lenses. An often-asked question about microscopes is: "If one were extremely patient and careful, would it be possible to construct a microscope that would enable the human eye to see an atom?" The answer is no, as long as light is used to illuminate the object. The reason is that, for an object under an optical microscope (one that uses visible light) to be seen, the object must be at least as large as a wavelength of light. Because the diameter of any atom is many times smaller than the wavelengths of visible light, the mysteries of the atom must be probed using other types of "microscopes."

The ability to use other types of waves to "see" objects also depends on wavelength. We can illustrate this with water waves in a bathtub. Suppose you vibrate your hand in the water until waves having a wavelength of about 15 cm are moving along the surface. If you hold a small object, such as a toothpick, so that it lies in the path of the waves, it does not appreciably disturb the waves; they continue along their path "oblivious" to it. Now suppose you hold a larger object, such as a toy sailboat, in the path of the 15-cm waves. In this case, the waves are considerably disturbed by the object. Because the toothpick was smaller than the wavelength of the waves, the waves did not "see" it (the intensity of the scattered waves was low). Because it is about the same size as the wavelength of the waves, however, the boat creates a disturbance. In other words, the object acts as the source of scattered waves that appear to come from it.

Light waves behave in this same general way. The ability of an optical microscope to view an object depends on the size of the object relative to the wavelength of the light used to observe it. Hence, we can never observe atoms with an optical microscope[2] because their dimensions are small (≈ 0.1 nm) relative to the wavelength of the light (≈ 500 nm).

Optional Section

36.10 THE TELESCOPE

Two fundamentally different types of **telescopes** exist; both are designed to aid in viewing distant objects, such as the planets in our Solar System. The **refracting telescope** uses a combination of lenses to form an image, and the **reflecting telescope** uses a curved mirror and a lens.

The lens combination shown in Figure 36.42a is that of a refracting telescope. Like the compound microscope, this telescope has an objective and an eyepiece. The two lenses are arranged so that the objective forms a real, inverted image of the distant object very near the focal point of the eyepiece. Because the object is essentially at infinity, this point at which I_1 forms is the focal point of the objective. Hence, the two lenses are separated by a distance $f_o + f_e$, which corresponds to the length of the telescope tube. The eyepiece then forms, at I_2, an enlarged, inverted image of the image at I_1.

The angular magnification of the telescope is given by θ/θ_o, where θ_o is the angle subtended by the object at the objective and θ is the angle subtended by the final image at the viewer's eye. Consider Figure 36.42a, in which the object is a very great distance to the left of the figure. The angle θ_o (to the *left* of the objective) subtended by the object at the objective is the same as the angle (to the *right* of the objective) subtended by the first image at the objective. Thus,

$$\tan \theta_o \approx \theta_o \approx -\frac{h'}{f_o}$$

where the negative sign indicates that the image is inverted.

The angle θ subtended by the final image at the eye is the same as the angle that a ray coming from the tip of I_1 and traveling parallel to the principal axis makes with the principal axis after it passes through the lens. Thus,

$$\tan \theta \approx \theta \approx \frac{h'}{f_e}$$

We have not used a negative sign in this equation because the final image is not inverted; the object creating this final image I_2 is I_1, and both it and I_2 point in the same direction. To see why the adjacent side of the triangle containing angle θ is f_e and not $2f_e$, note that we must use only the bent length of the refracted ray. Hence, the angular magnification of the telescope can be expressed as

$$m = \frac{\theta}{\theta_o} = \frac{h'/f_e}{-h'/f_o} = -\frac{f_o}{f_e} \tag{36.21}$$

and we see that the angular magnification of a telescope equals the ratio of the objective focal length to the eyepiece focal length. The negative sign indicates that the image is inverted.

Quick Quiz 36.6

Why isn't the lateral magnification given by Equation 36.1 a useful concept for telescopes?

---

[2] Single-molecule near-field optic studies are routinely performed with visible light having wavelengths of about 500 nm. The technique uses very small apertures to produce images having resolution as small as 10 nm.

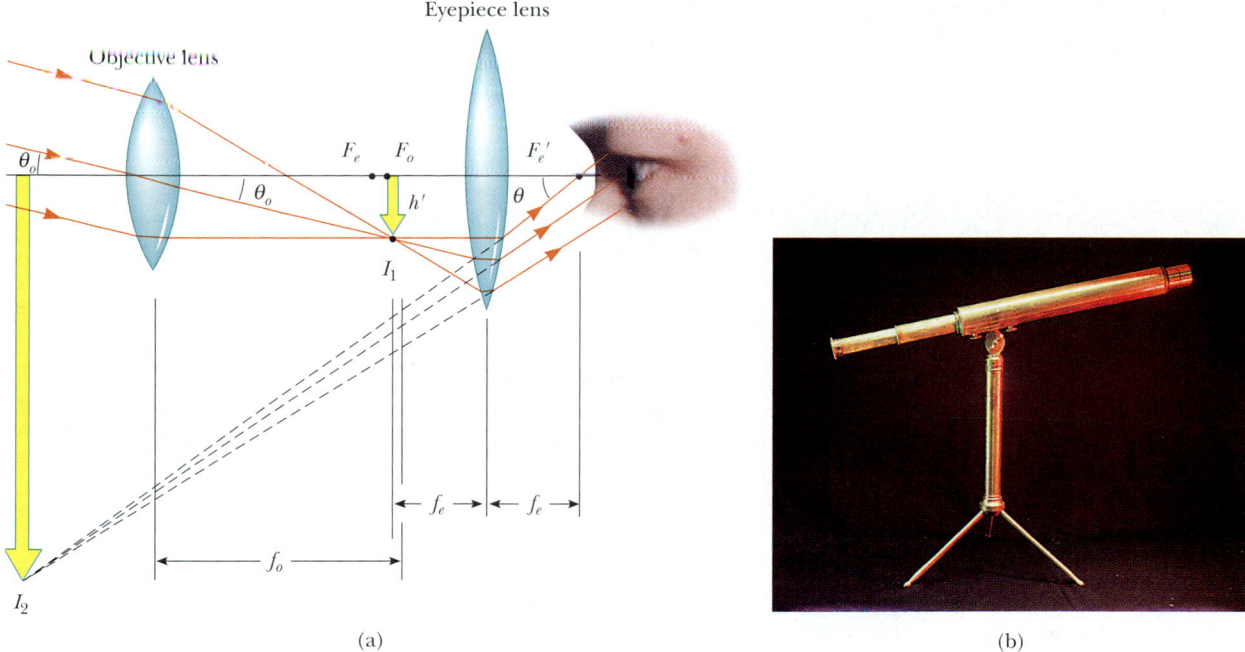

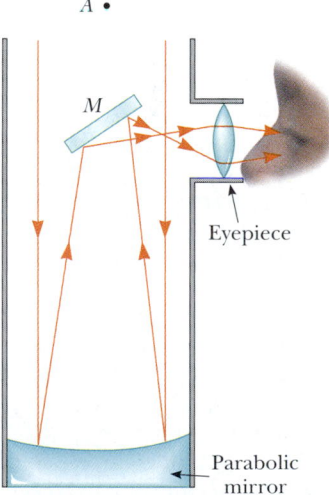

Figure 36.42   (a) Lens arrangement in a refracting telescope, with the object at infinity. (b) A refracting telescope. *(Henry Leap and Jim Lehman)*

When we look through a telescope at such relatively nearby objects as the Moon and the planets, magnification is important. However, stars are so far away that they always appear as small points of light no matter how great the magnification. A large research telescope that is used to study very distant objects must have a great diameter to gather as much light as possible. It is difficult and expensive to manufacture large lenses for refracting telescopes. Another difficulty with large lenses is that their weight leads to sagging, which is an additional source of aberration. These problems can be partially overcome by replacing the objective with a concave mirror, which results in a reflecting telescope. Because light is reflected from the mirror and does not pass through a lens, the mirror can have rigid supports on the back side. Such supports eliminate the problem of sagging.

Figure 36.43 shows the design for a typical reflecting telescope. Incoming light rays pass down the barrel of the telescope and are reflected by a parabolic mirror at the base. These rays converge toward point A in the figure, where an image would be formed. However, before this image is formed, a small, flat mirror M reflects the light toward an opening in the side of the tube that passes into an eyepiece. This particular design is said to have a Newtonian focus because Newton developed it. Note that in the reflecting telescope the light never passes through glass (except through the small eyepiece). As a result, problems associated with chromatic aberration are virtually eliminated.

The largest reflecting telescopes in the world are at the Keck Observatory on Mauna Kea, Hawaii. The site includes two telescopes with diameters of 10 m, each containing 36 hexagonally shaped, computer-controlled mirrors that work together to form a large reflecting surface. In contrast, the largest refracting telescope in the world, at the Yerkes Observatory in Williams Bay, Wisconsin, has a diameter of only 1 m.

Figure 36.43   A Newtonian-focus reflecting telescope.

web

For more information on the Keck telescopes, visit
http://www2.keck.hawaii.edu:3636/

SUMMARY

The **lateral magnification** M of a mirror or lens is defined as the ratio of the image height h' to the object height h:

$$M = \frac{h'}{h} \tag{36.1}$$

In the paraxial ray approximation, the object distance p and image distance q for a spherical mirror of radius R are related by the **mirror equation:**

$$\frac{1}{p} + \frac{1}{q} = \frac{2}{R} = \frac{1}{f} \tag{36.4, 36.6}$$

where $f = R/2$ is the **focal length** of the mirror.

An image can be formed by refraction from a spherical surface of radius R. The object and image distances for refraction from such a surface are related by

$$\frac{n_1}{p} + \frac{n_2}{q} = \frac{n_2 - n_1}{R} \tag{36.8}$$

where the light is incident in the medium for which the index of refraction is n_1 and is refracted in the medium for which the index of refraction is n_2.

The inverse of the **focal length** f of a thin lens surrounded by air is given by the **lens makers' equation:**

$$\frac{1}{f} = (n - 1)\left(\frac{1}{R_1} - \frac{1}{R_2}\right) \tag{36.11}$$

Converging lenses have positive focal lengths, and **diverging lenses** have negative focal lengths.

For a thin lens, and in the paraxial ray approximation, the object and image distances are related by the **thin-lens equation:**

$$\frac{1}{p} + \frac{1}{q} = \frac{1}{f} \tag{36.12}$$

QUESTIONS

1. What is wrong with the caption of the cartoon shown in Figure Q36.1?
2. Using a simple ray diagram, such as the one shown in Figure 36.2, show that a flat mirror whose top is at eye level need not be as long as you are for you to see your entire body in it.
3. Consider a concave spherical mirror with a real object. Is the image always inverted? Is the image always real? Give conditions for your answers.
4. Repeat the preceding question for a convex spherical mirror.
5. Why does a clear stream of water, such as a creek, always appear to be shallower than it actually is? By how much is its depth apparently reduced?
6. Consider the image formed by a thin converging lens. Under what conditions is the image (a) inverted, (b) upright, (c) real, (d) virtual, (e) larger than the object, and (f) smaller than the object?
7. Repeat Question 6 for a thin diverging lens.
8. Use the lens makers' equation to verify the sign of the focal length of each of the lenses in Figure 36.26.

"Most mirrors reverse left and right. This one reverses top and bottom."

Figure Q36.1

9. If a cylinder of solid glass or clear plastic is placed above the words LEAD OXIDE and viewed from the side as shown in Figure Q36.9, the LEAD appears inverted but the OXIDE does not. Explain.

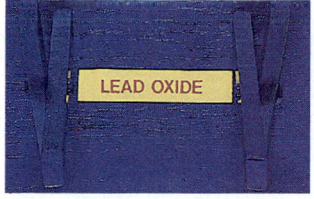 

Figure Q36.9 *(Henry Leap and Jim Lehman)*

10. If the camera "sees" a movie actor's reflection in a mirror, what does the actor see in the mirror?

11. Explain why a mirror cannot give rise to chromatic aberration.

12. Why do some automobile mirrors have printed on them the statement "Objects in mirror are closer than they appear"? (See Fig. Q36.12.)

Figure Q36.12 *(THE FAR SIDE ©1985 FARWORKS, INC. Used by permission. All rights reserved.)*

13. Why do some emergency vehicles have the symbol ƎƆИA⅃UＢＭA written on the front?

14. Explain why a fish in a spherical goldfish bowl appears larger than it really is.

15. Lenses used in eyeglasses, whether converging or diverging, are always designed such that the middle of the lens curves away from the eye, like the center lenses of Figure 36.26a and b. Why?

16. A mirage is formed when the air gets gradually cooler with increasing altitude. What might happen if the air grew gradually warmer with altitude? This often happens over bodies of water or snow-covered ground; the effect is called *looming*.

17. Consider a spherical concave mirror, with an object positioned to the left of the mirror beyond the focal point. Using ray diagrams, show that the image moves to the left as the object approaches the focal point.

18. In a Jules Verne novel, a piece of ice is shaped into a magnifying lens to focus sunlight to start a fire. Is this possible?

19. The *f*-number of a camera is the focal length of the lens divided by its aperture (or diameter). How can the *f*-number of the lens be changed? How does changing this number affect the required exposure time?

20. A solar furnace can be constructed through the use of a concave mirror to reflect and focus sunlight into a furnace enclosure. What factors in the design of the reflecting mirror would guarantee very high temperatures?

21. One method for determining the position of an image, either real or virtual, is by means of *parallax*. If a finger or another object is placed at the position of the image, as shown in Figure Q36.21, and the finger and the image are viewed simultaneously (the image is viewed through the lens if it is virtual), the finger and image have the same parallax; that is, if the image is viewed from different positions, it will appear to move along with the finger. Use this method to locate the image formed by a lens. Explain why the method works.

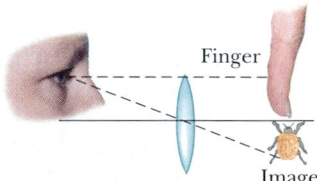

Finger

Image

Figure Q36.21

22. Figure Q36.22 shows a lithograph by M. C. Escher titled *Hand with Reflection Sphere (Self-Portrait in Spherical Mirror).* Escher had this to say about the work: "The picture shows a spherical mirror, resting on a left hand. But as a print is the reverse of the original drawing on stone, it was my right hand that you see depicted. (Being left-handed, I needed my left hand to make the drawing.) Such a globe reflection collects almost one's whole surroundings in one disk-shaped image. The whole room, four walls, the

floor, and the ceiling, everything, albeit distorted, is compressed into that one small circle. Your own head, or more exactly the point between your eyes, is the absolute center. No matter how you turn or twist yourself, you can't get out of that central point. You are immovably the focus, the unshakable core, of your world." Comment on the accuracy of Escher's description.

23. You can make a corner reflector by placing three flat mirrors in the corner of a room where the ceiling meets the walls. Show that no matter where you are in the room, you can see yourself reflected in the mirrors—upside down.

Figure Q36.22 *(M. C. Escher/Cordon Art–Baarn–Holland. All rights reserved.)*

PROBLEMS

1, 2, 3 = straightforward, intermediate, challenging   ☐ = full solution available in the *Student Solutions Manual and Study Guide*
WEB = solution posted at **http://www.saunderscollege.com/physics/**   🖥 = Computer useful in solving problem   🔲 = Interactive Physics
☐ = paired numerical/symbolic problems

Section 36.1   Images Formed by Flat Mirrors

1. Does your bathroom mirror show you older or younger than you actually are? Compute an order-of-magnitude estimate for the age difference, based on data that you specify.

2. In a church choir loft, two parallel walls are 5.30 m apart. The singers stand against the north wall. The organist faces the south wall, sitting 0.800 m away from it. To enable her to see the choir, a flat mirror 0.600 m wide is mounted on the south wall, straight in front of her. What width of the north wall can she see? *Hint:* Draw a top-view diagram to justify your answer.

3. Determine the minimum height of a vertical flat mirror in which a person 5′10″ in height can see his or her full image. (A ray diagram would be helpful.)

4. Two flat mirrors have their reflecting surfaces facing each other, with an edge of one mirror in contact with an edge of the other, so that the angle between the mirrors is α. When an object is placed between the mirrors, a number of images are formed. In general, if the angle α is such that $n\alpha = 360°$, where n is an integer, the number of images formed is $n - 1$. Graphically, find all the image positions for the case $n = 6$ when a point object is between the mirrors (but not on the angle bisector).

5. A person walks into a room with two flat mirrors on opposite walls, which produce multiple images. When the person is 5.00 ft from the mirror on the left wall and 10.0 ft from the mirror on the right wall, find the distances from that person to the first three images seen in the mirror on the left.

Section 36.2   Images Formed by Spherical Mirrors

6. A concave spherical mirror has a radius of curvature of 20.0 cm. Find the location of the image for object distances of (a) 40.0 cm, (b) 20.0 cm, and (c) 10.0 cm. For each case, state whether the image is real or virtual and upright or inverted, and find the magnification.

7. At an intersection of hospital hallways, a convex mirror is mounted high on a wall to help people avoid collisions. The mirror has a radius of curvature of 0.550 m. Locate and describe the image of a patient 10.0 m from the mirror. Determine the magnification.

8. A large church has a niche in one wall. On the floor plan it appears as a semicircular indentation of radius 2.50 m. A worshiper stands on the center line of the niche, 2.00 m out from its deepest point, and whispers a prayer. Where is the sound concentrated after reflection from the back wall of the niche?

WEB **9.** A spherical convex mirror has a radius of curvature of 40.0 cm. Determine the position of the virtual image and the magnification (a) for an object distance of 30.0 cm and (b) for an object distance of 60.0 cm. (c) Are the images upright or inverted?

10. The height of the real image formed by a concave mirror is four times the object height when the object is 30.0 cm in front of the mirror. (a) What is the radius of curvature of the mirror? (b) Use a ray diagram to locate this image.

11. A concave mirror has a radius of curvature of 60.0 cm. Calculate the image position and magnification of an object placed in front of the mirror (a) at a distance of 90.0 cm and (b) at a distance of 20.0 cm. (c) In each case, draw ray diagrams to obtain the image characteristics.

12. A concave mirror has a focal length of 40.0 cm. Determine the object position for which the resulting image is upright and four times the size of the object.

13. A spherical mirror is to be used to form, on a screen 5.00 m from the object, an image five times the size of the object. (a) Describe the type of mirror required. (b) Where should the mirror be positioned relative to the object?

14. A rectangle 10.0 cm × 20.0 cm is placed so that its right edge is 40.0 cm to the left of a concave spherical mirror, as in Figure P36.14. The radius of curvature of the mirror is 20.0 cm. (a) Draw the image formed by this mirror. (b) What is the area of the image?

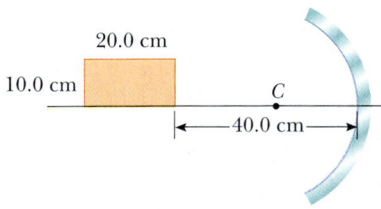

Figure P36.14

15. A dedicated sports-car enthusiast polishes the inside and outside surfaces of a hubcap that is a section of a sphere. When she looks into one side of the hubcap, she sees an image of her face 30.0 cm in back of the hubcap. She then flips the hubcap over and sees another image of her face 10.0 cm in back of the hubcap. (a) How far is her face from the hubcap? (b) What is the radius of curvature of the hubcap?

16. An object is 15.0 cm from the surface of a reflective spherical Christmas-tree ornament 6.00 cm in diameter. What are the magnification and position of the image?

17. A ball is dropped from rest 3.00 m directly above the vertex of a concave mirror that has a radius of 1.00 m and lies in a horizontal plane. (a) Describe the motion of the ball's image in the mirror. (b) At what time do the ball and its image coincide?

Section 36.3   Images Formed by Refraction

18. A flint-glass plate ($n = 1.66$) rests on the bottom of an aquarium tank. The plate is 8.00 cm thick (vertical dimension) and covered with water ($n = 1.33$) to a depth of 12.0 cm. Calculate the apparent thickness of the plate as viewed from above the water. (Assume nearly normal incidence.)

19. A cubical block of ice 50.0 cm on a side is placed on a level floor over a speck of dust. Find the location of the image of the speck if the index of refraction of ice is 1.309.

20. A simple model of the human eye ignores its lens entirely. Most of what the eye does to light happens at the transparent cornea. Assume that this outer surface has a 6.00-mm radius of curvature, and assume that the eyeball contains just one fluid with an index of refraction of 1.40. Prove that a very distant object will be imaged on the retina, 21.0 mm behind the cornea. Describe the image.

21. A glass sphere ($n = 1.50$) with a radius of 15.0 cm has a tiny air bubble 5.00 cm above its center. The sphere is viewed looking down along the extended radius containing the bubble. What is the apparent depth of the bubble below the surface of the sphere?

22. A transparent sphere of unknown composition is observed to form an image of the Sun on the surface of the sphere opposite the Sun. What is the refractive index of the sphere material?

23. One end of a long glass rod ($n = 1.50$) is formed into a convex surface of radius 6.00 cm. An object is positioned in air along the axis of the rod. Find the image positions corresponding to object distances of (a) 20.0 cm, (b) 10.0 cm, and (c) 3.00 cm from the end of the rod.

24. A goldfish is swimming at 2.00 cm/s toward the front wall of a rectangular aquarium. What is the apparent speed of the fish as measured by an observer looking in from outside the front wall of the tank? The index of refraction of water is 1.33.

25. A goldfish is swimming inside a spherical plastic bowl of water, with an index of refraction of 1.33. If the goldfish is 10.0 cm from the wall of the 15.0-cm-radius bowl, where does it appear to an observer outside the bowl?

Section 36.4   Thin Lenses

26. A contact lens is made of plastic with an index of refraction of 1.50. The lens has an outer radius of curvature of + 2.00 cm and an inner radius of curvature of + 2.50 cm. What is the focal length of the lens?

WEB **27.** The left face of a biconvex lens has a radius of curvature of magnitude 12.0 cm, and the right face has a radius of curvature of magnitude 18.0 cm. The index of refraction of the glass is 1.44. (a) Calculate the focal length of the lens. (b) Calculate the focal length if the radii of curvature of the two faces are interchanged.

28. A converging lens has a focal length of 20.0 cm. Locate the image for object distances of (a) 40.0 cm, (b) 20.0 cm, and (c) 10.0 cm. For each case, state whether the image is real or virtual and upright or inverted. Find the magnification in each case.

29. A thin lens has a focal length of 25.0 cm. Locate and describe the image when the object is placed (a) 26.0 cm and (b) 24.0 cm in front of the lens.

30. An object positioned 32.0 cm in front of a lens forms an image on a screen 8.00 cm behind the lens. (a) Find the focal length of the lens. (b) Determine the magnification. (c) Is the lens converging or diverging?

WEB **31.** The nickel's image in Figure P36.31 has twice the diameter of the nickel and is 2.84 cm from the lens. Determine the focal length of the lens.

Figure P36.31

32. A magnifying glass is a converging lens of focal length 15.0 cm. At what distance from a postage stamp should you hold this lens to get a magnification of $+2.00$?

33. A transparent photographic slide is placed in front of a converging lens with a focal length of 2.44 cm. The lens forms an image of the slide 12.9 cm from the slide. How far is the lens from the slide if the image is (a) real? (b) virtual?

34. A person looks at a gem with a jeweler's loupe—a converging lens that has a focal length of 12.5 cm. The loupe forms a virtual image 30.0 cm from the lens. (a) Determine the magnification. Is the image upright or inverted? (b) Construct a ray diagram for this arrangement.

35. Suppose an object has thickness dp so that it extends from object distance p to $p + dp$. Prove that the thickness dq of its image is given by $(-q^2/p^2)\,dp$, so the longitudinal magnification $dq/dp = -M^2$, where M is the lateral magnification.

36. The projection lens in a certain slide projector is a single thin lens. A slide 24.0 mm high is to be projected so that its image fills a screen 1.80 m high. The slide-to-screen distance is 3.00 m. (a) Determine the focal length of the projection lens. (b) How far from the slide should the lens of the projector be placed to form the image on the screen?

37. An object is positioned 20.0 cm to the left of a diverging lens with focal length $f = -32.0$ cm. Determine (a) the

location and (b) the magnification of the image. (c) Construct a ray diagram for this arrangement.

38. Figure P36.38 shows a thin glass ($n = 1.50$) converging lens for which the radii of curvature are $R_1 = 15.0$ cm and $R_2 = -12.0$ cm. To the left of the lens is a cube with a face area of 100 cm^2. The base of the cube is on the axis of the lens, and the right face is 20.0 cm to the left of the lens. (a) Determine the focal length of the lens. (b) Draw the image of the square face formed by the lens. What type of geometric figure is this? (c) Determine the area of the image.

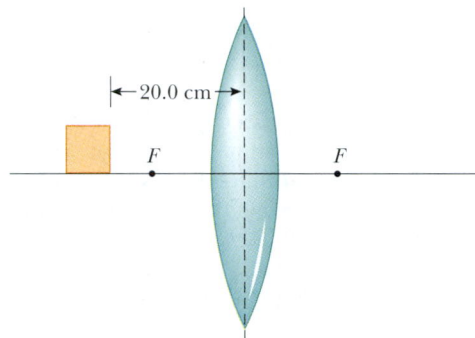

Figure P36.38

39. An object is 5.00 m to the left of a flat screen. A converging lens for which the focal length is $f = 0.800$ m is placed between object and screen. (a) Show that two lens positions exist that form images on the screen, and determine how far these positions are from the object. (b) How do the two images differ from each other?

40. An object is at a distance d to the left of a flat screen. A converging lens with focal length $f < d/4$ is placed between object and screen. (a) Show that two lens positions exist that form an image on the screen, and determine how far these positions are from the object. (b) How do the two images differ from each other?

41. Figure 36.33 diagrams a cross-section of a camera. It has a single lens with a focal length of 65.0 mm, which is to form an image on the film at the back of the camera. Suppose the position of the lens has been adjusted to focus the image of a distant object. How far and in what direction must the lens be moved to form a sharp image of an object that is 2.00 m away?

(Optional)
Section 36.5   Lens Aberrations

42. The magnitudes of the radii of curvature are 32.5 cm and 42.5 cm for the two faces of a biconcave lens. The glass has index 1.53 for violet light and 1.51 for red light. For a very distant object, locate and describe (a) the image formed by violet light and (b) the image formed by red light.

43. Two rays traveling parallel to the principal axis strike a large plano–convex lens having a refractive index of 1.60 (Fig. P36.43). If the convex face is spherical, a ray near the edge does not pass through the focal point (spherical aberration occurs). If this face has a radius of curvature of 20.0 cm and the two rays are $h_1 = 0.500$ cm and $h_2 = 12.0$ cm from the principal axis, find the difference in the positions where they cross the principal axis.

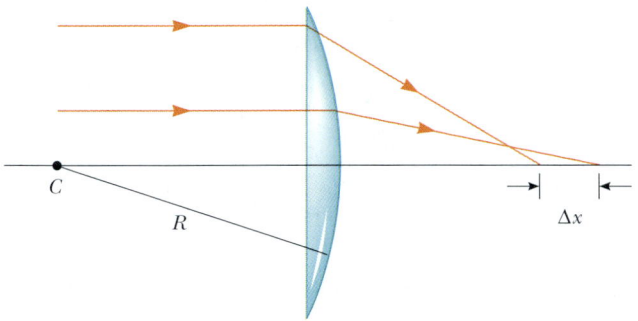

Figure P36.43

(*Optional*)
Section 36.7 The Eye

44. The accommodation limits for Nearsighted Nick's eyes are 18.0 cm and 80.0 cm. When he wears his glasses, he can see faraway objects clearly. At what minimum distance can he see objects clearly?

45. A nearsighted person cannot see objects clearly beyond 25.0 cm (her far point). If she has no astigmatism and contact lenses are prescribed for her, what power and type of lens are required to correct her vision?

46. A person sees clearly when he wears eyeglasses that have a power of − 4.00 diopters and sit 2.00 cm in front of his eyes. If he wants to switch to contact lenses, which are placed directly on the eyes, what lens power should be prescribed?

(*Optional*)
Section 36.8 The Simple Magnifier
Section 36.9 The Compound Microscope
Section 36.10 The Telescope

47. A philatelist examines the printing detail on a stamp, using a biconvex lens with a focal length of 10.0 cm as a simple magnifier. The lens is held close to the eye, and the lens-to-object distance is adjusted so that the virtual image is formed at the normal near point (25.0 cm). Calculate the magnification.

48. A lens that has a focal length of 5.00 cm is used as a magnifying glass. (a) Where should the object be placed to obtain maximum magnification? (b) What is the magnification?

49. The distance between the eyepiece and the objective lens in a certain compound microscope is 23.0 cm. The focal length of the eyepiece is 2.50 cm, and that of the objective is 0.400 cm. What is the overall magnification of the microscope?

50. The desired overall magnification of a compound microscope is 140×. The objective alone produces a lateral magnification of 12.0×. Determine the required focal length of the eyepiece.

51. The Yerkes refracting telescope has a 1.00-m-diameter objective lens with a focal length of 20.0 m. Assume that it is used with an eyepiece that has a focal length of 2.50 cm. (a) Determine the magnification of the planet Mars as seen through this telescope. (b) Are the Martian polar caps seen right side up or upside down?

52. Astronomers often take photographs with the objective lens or the mirror of a telescope alone, without an eyepiece. (a) Show that the image size h' for this telescope is given by $h' = fh/(f - p)$, where h is the object size, f is the objective focal length, and p is the object distance. (b) Simplify the expression in part (a) for the case in which the object distance is much greater than objective focal length. (c) The "wingspan" of the International Space Station is 108.6 m, the overall width of its solar-panel configuration. Find the width of the image formed by a telescope objective of focal length 4.00 m when the station is orbiting at an altitude of 407 km.

53. Galileo devised a simple terrestrial telescope that produces an upright image. It consists of a converging objective lens and a diverging eyepiece at opposite ends of the telescope tube. For distant objects, the tube length is the objective focal length less the absolute value of the eyepiece focal length. (a) Does the user of the telescope see a real or virtual image? (b) Where is the final image? (c) If a telescope is to be constructed with a tube 10.0 cm long and a magnification of 3.00, what are the focal lengths of the objective and eyepiece?

54. A certain telescope has an objective mirror with an aperture diameter of 200 mm and a focal length of 2 000 mm. It captures the image of a nebula on photographic film at its prime focus with an exposure time of 1.50 min. To produce the same light energy per unit area on the film, what is the required exposure time to photograph the same nebula with a smaller telescope, which has an objective lens with an aperture diameter of 60.0 mm and a focal length of 900 mm?

ADDITIONAL PROBLEMS

55. The distance between an object and its upright image is 20.0 cm. If the magnification is 0.500, what is the focal length of the lens that is being used to form the image?

56. The distance between an object and its upright image is d. If the magnification is M, what is the focal length of the lens that is being used to form the image?

57. The lens and mirror in Figure P36.57 have focal lengths of + 80.0 cm and − 50.0 cm, respectively. An object is

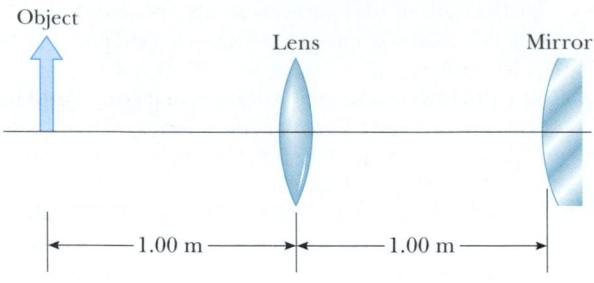

Figure P36.57

placed 1.00 m to the left of the lens, as shown. Locate the final image, which is formed by light that has gone through the lens twice. State whether the image is upright or inverted, and determine the overall magnification.

58. Your friend needs glasses with diverging lenses of focal length − 65.0 cm for both eyes. You tell him he looks good when he does not squint, but he is worried about how thick the lenses will be. If the radius of curvature of the first surface is $R_1 = 50.0$ cm and the high-index plastic has a refractive index of 1.66, (a) find the required radius of curvature of the second surface. (b) Assume that the lens is ground from a disk 4.00 cm in diameter and 0.100 cm thick at the center. Find the thickness of the plastic at the edge of the lens, measured parallel to the axis. *Hint:* Draw a large cross-sectional diagram.

59. The object in Figure P36.59 is midway between the lens and the mirror. The mirror's radius of curvature is 20.0 cm, and the lens has a focal length of −16.7 cm. Considering only the light that leaves the object and travels first toward the mirror, locate the final image formed by this system. Is this image real or virtual? Is it upright or inverted? What is the overall magnification?

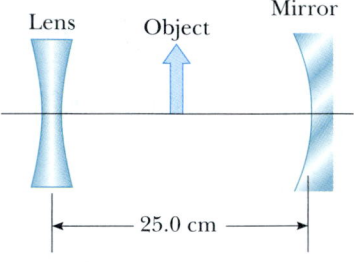

Figure P36.59

60. An object placed 10.0 cm from a concave spherical mirror produces a real image 8.00 cm from the mirror. If the object is moved to a new position 20.0 cm from the

mirror, what is the position of the image? Is the latter image real or virtual?

WEB 61. A parallel beam of light enters a glass hemisphere perpendicular to the flat face, as shown in Figure P36.61. The radius is $R = 6.00$ cm, and the index of refraction is $n = 1.560$. Determine the point at which the beam is focused. (Assume paraxial rays.)

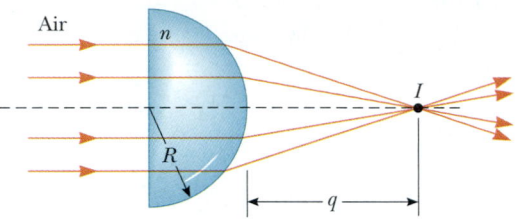

Figure P36.61

62. **Review Problem.** A spherical lightbulb with a diameter of 3.20 cm radiates light equally in all directions, with a power of 4.50 W. (a) Find the light intensity at the surface of the bulb. (b) Find the light intensity 7.20 m from the center of the bulb. (c) At this 7.20-m distance, a lens is set up with its axis pointing toward the bulb. The lens has a circular face with a diameter of 15.0 cm and a focal length of 35.0 cm. Find the diameter of the image of the bulb. (d) Find the light intensity at the image.

63. An object is placed 12.0 cm to the left of a diverging lens with a focal length of − 6.00 cm. A converging lens with a focal length of 12.0 cm is placed a distance d to the right of the diverging lens. Find the distance d that corresponds to a final image at infinity. Draw a ray diagram for this case.

64. Assume that the intensity of sunlight is 1.00 kW/m^2 at a particular location. A highly reflecting concave mirror is to be pointed toward the Sun to produce a power of at least 350 W at the image. (a) Find the required radius R_a of the circular face area of the mirror. (b) Now suppose the light intensity is to be at least 120 kW/m^2 at the image. Find the required relationship between R_a and the radius of curvature R of the mirror. The disk of the Sun subtends an angle of 0.533° at the Earth.

WEB 65. The disk of the Sun subtends an angle of 0.533° at the Earth. What are the position and diameter of the solar image formed by a concave spherical mirror with a radius of curvature of 3.00 m?

66. Figure P36.66 shows a thin converging lens for which the radii are $R_1 = 9.00$ cm and $R_2 = − 11.0$ cm. The lens is in front of a concave spherical mirror of radius $R = 8.00$ cm. (a) If its focal points F_1 and F_2 are 5.00 cm from the vertex of the lens, determine its index of refraction. (b) If the lens and mirror are 20.0 cm apart and an object is placed 8.00 cm to the left of the

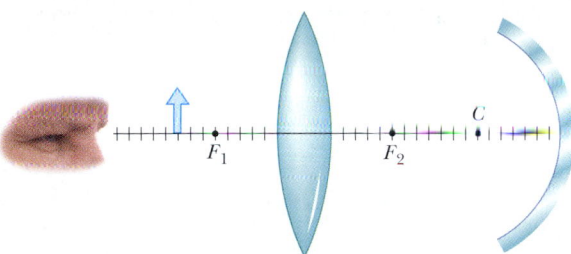

Figure P36.66

lens, determine the position of the final image and its magnification as seen by the eye in the figure. (c) Is the final image inverted or upright? Explain.

67. In a darkened room, a burning candle is placed 1.50 m from a white wall. A lens is placed between candle and wall at a location that causes a larger, inverted image to form on the wall. When the lens is moved 90.0 cm toward the wall, another image of the candle is formed. Find (a) the two object distances that produce the specified images and (b) the focal length of the lens. (c) Characterize the second image.

68. A thin lens of focal length f lies on a horizontal front-surfaced flat mirror. How far above the lens should an object be held if its image is to coincide with the object?

69. A compound microscope has an objective of focal length 0.300 cm and an eyepiece of focal length 2.50 cm. If an object is 3.40 mm from the objective, what is the magnification? (*Hint:* Use the lens equation for the objective.)

70. Two converging lenses with focal lengths of 10.0 cm and 20.0 cm are positioned 50.0 cm apart, as shown in Figure P36.70. The final image is to be located between the lenses, at the position indicated. (a) How far to the left

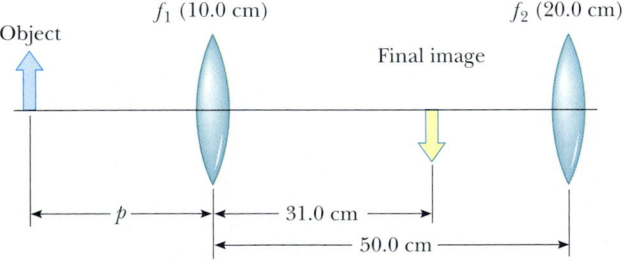

Figure P36.70

of the first lens should the object be? (b) What is the overall magnification? (c) Is the final image upright or inverted?

71. A cataract-impaired lens in an eye may be surgically removed and replaced by a manufactured lens. The focal length required for the new lens is determined by the lens-to-retina distance, which is measured by a sonar-

like device, and by the requirement that the implant provide for correct distant vision. (a) If the distance from lens to retina is 22.4 mm, calculate the power of the implanted lens in diopters. (b) Since no accommodation occurs and the implant allows for correct distant vision, a corrective lens for close work or reading must be used. Assume a reading distance of 33.0 cm, and calculate the power of the lens in the reading glasses.

72. A floating strawberry illusion consists of two parabolic mirrors, each with a focal length of 7.50 cm, facing each other so that their centers are 7.50 cm apart (Fig. P36.72). If a strawberry is placed on the lower mirror, an image of the strawberry is formed at the small opening at the center of the top mirror. Show that the final image is formed at that location, and describe its characteristics. (*Note:* A very startling effect is to shine a flashlight beam on these images. Even at a glancing angle, the incoming light beam is seemingly reflected off the images! Do you understand why?)

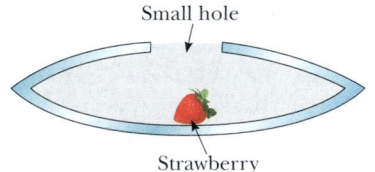

Figure P36.72   *(© Michael Levin/Opti-Gone Associates)*

73. An object 2.00 cm high is placed 40.0 cm to the left of a converging lens with a focal length of 30.0 cm. A diverging lens with a focal length of −20.0 cm is placed 110 cm to the right of the converging lens. (a) Determine the final position and magnification of the final image. (b) Is the image upright or inverted? (c) Repeat parts (a) and (b) for the case in which the second lens is a converging lens with a focal length of +20.0 cm.

ANSWERS TO QUICK QUIZZES

36.1 At *C*. A ray traced from the stone to the mirror and then to observer 2 looks like this:

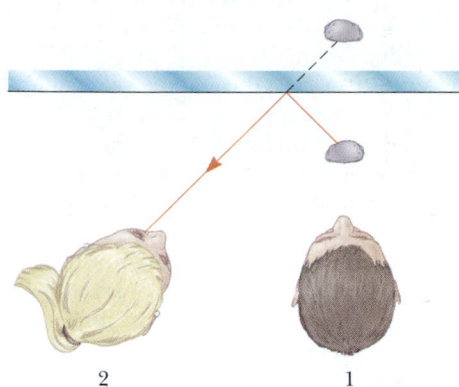

Figure QQA36.1

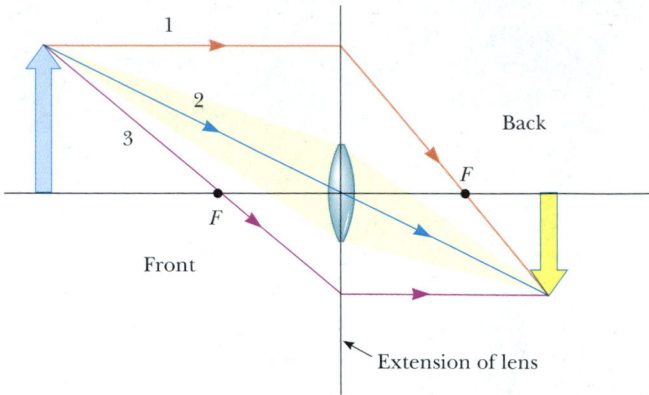

Figure QQA36.2

36.2 The focal length is infinite. Because the flat surfaces of the pane have infinite radii of curvature, Equation 36.11 indicates that the focal length is also infinite. Parallel rays striking the pane focus at infinity, which means that they remain parallel after passing through the glass.

36.3 An infinite number. In general, an infinite number of rays leave each point of any object and travel outward in all directions. (The three principal rays that we use to locate an image make up a selected subset of the infinite number of rays.) When an object is taller than a lens, we merely extend the plane containing the lens, as shown in Figure QQA36.2.

36.4 (c) The entire image is visible but has half the intensity. Each point on the object is a source of rays that travel in all directions. Thus, light from all parts of the object goes through all parts of the lens and forms an image. If you block part of the lens, you are blocking some of the

rays, but the remaining ones still come from all parts of the object.

36.5 The eyeglasses on the left are diverging lenses, which correct for nearsightedness. If you look carefully at the edge of the person's face through the lens, you will see that everything viewed through these glasses is reduced in size. The eyeglasses on the right are converging lenses, which correct for farsightedness. These lenses make everything that is viewed through them look larger.

36.6 The lateral magnification of a telescope is not well defined. For viewing with the eye relaxed, the user may slightly adjust the position of the eyepiece to place the final image I_2 in Figure 36.42a at infinity. Then, its height and its lateral magnification also are infinite. The angular magnification of a telescope as we define it is the factor by which the telescope increases in the diameter—on the retina of the viewer's eye—of the real image of an extended object.

PUZZLER

The brilliant colors seen in peacock feathers are not caused by pigments in the feathers. If they are not produced by pigments, how *are* those beautiful colors created? *(Terry Qing/FPG International)*

c h a p t e r

37

Interference of Light Waves

I n the preceding chapter on geometric optics, we used light rays to examine what happens when light passes through a lens or reflects from a mirror. Here in Chapter 37 and in the next chapter, we are concerned with *wave optics,* the study of interference, diffraction, and polarization of light. These phenomena cannot be adequately explained with the ray optics used in Chapter 36. We now learn how treating light as waves rather than as rays leads to a satisfying description of such phenomena.

37.1   CONDITIONS FOR INTERFERENCE

In Chapter 18, we found that the adding together of two mechanical waves can be constructive or destructive. In constructive interference, the amplitude of the resultant wave is greater than that of either individual wave, whereas in destructive interference, the resultant amplitude is less than that of either individual wave. Light waves also interfere with each other. Fundamentally, all interference associated with light waves arises when the electromagnetic fields that constitute the individual waves combine.

If two lightbulbs are placed side by side, no interference effects are observed because the light waves from one bulb are emitted independently of those from the other bulb. The emissions from the two lightbulbs do not maintain a constant phase relationship with each other over time. Light waves from an ordinary source such as a lightbulb undergo random changes about once every 10^{-8} s. Therefore, the conditions for constructive interference, destructive interference, or some intermediate state last for lengths of time of the order of 10^{-8} s. Because the eye cannot follow such short-term changes, no interference effects are observed. (In 1993 interference from two separate light sources was photographed in an extremely fast exposure. Nonetheless, we do not ordinarily see interference effects because of the rapidly changing phase relationship between the light waves.) Such light sources are said to be **incoherent.**

Interference effects in light waves are not easy to observe because of the short wavelengths involved (from 4×10^{-7} m to 7×10^{-7} m). For sustained interference in light waves to be observed, the following conditions must be met:

- The sources must be **coherent** — that is, they must maintain a constant phase with respect to each other.
- The sources should be **monochromatic** — that is, of a single wavelength.

Conditions for interference

We now describe the characteristics of coherent sources. As we saw when we studied mechanical waves, two sources (producing two traveling waves) are needed to create interference. In order to produce a stable interference pattern, **the individual waves must maintain a constant phase relationship with one another.** As an example, the sound waves emitted by two side-by-side loudspeakers driven by a single amplifier can interfere with each other because the two speakers are coherent — that is, they respond to the amplifier in the same way at the same time.

A common method for producing two coherent light sources is to use one monochromatic source to illuminate a barrier containing two small openings (usually in the shape of slits). The light emerging from the two slits is coherent because a single source produces the original light beam and the two slits serve only to separate the original beam into two parts (which, after all, is what was done to the sound signal from the side-by-side loudspeakers). Any random change in the light emitted by the source occurs in both beams at the same time, and as a result interference effects can be observed when the light from the two slits arrives at a viewing screen.

37.2 YOUNG'S DOUBLE-SLIT EXPERIMENT

Interference in light waves from two sources was first demonstrated by Thomas Young in 1801. A schematic diagram of the apparatus that Young used is shown in Figure 37.1a. Light is incident on a first barrier in which there is a slit S_0. The waves emerging from this slit arrive at a second barrier that contains two parallel slits S_1 and S_2. These two slits serve as a pair of coherent light sources because waves emerging from them originate from the same wave front and therefore maintain a constant phase relationship. The light from S_1 and S_2 produces on a viewing screen a visible pattern of bright and dark parallel bands called **fringes** (Fig. 37.1b). When the light from S_1 and that from S_2 both arrive at a point on the screen such that constructive interference occurs at that location, a bright fringe appears. When the light from the two slits combines destructively at any location on the screen, a dark fringe results. Figure 37.2 is a photograph of an interference pattern produced by two coherent vibrating sources in a water tank.

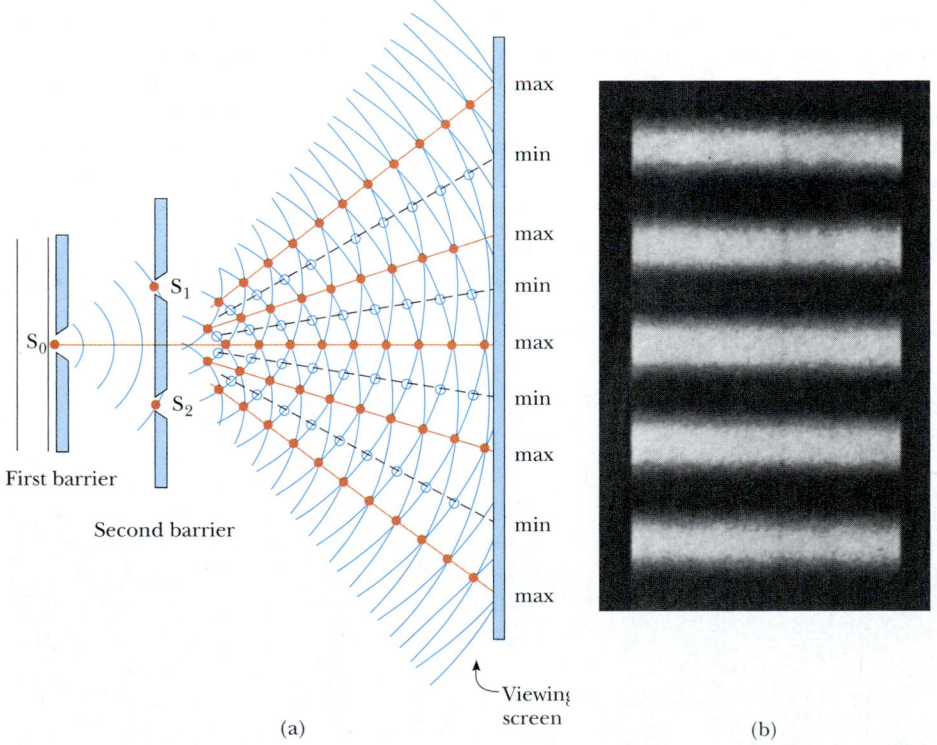

(a)

(b)

Figure 37.1 (a) Schematic diagram of Young's double-slit experiment. Slits S_1 and S_2 behave as coherent sources of light waves that produce an interference pattern on the viewing screen (drawing not to scale). (b) An enlargement of the center of a fringe pattern formed on the viewing screen with many slits could look like this.

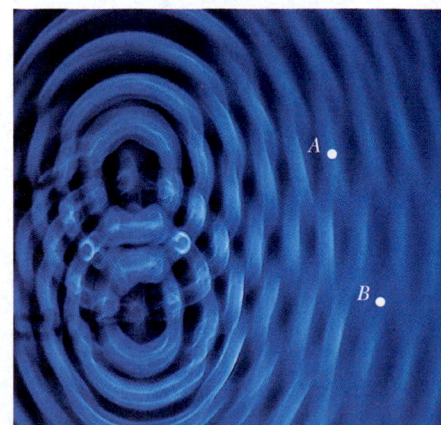

Figure 37.2 An interference pattern involving water waves is produced by two vibrating sources at the water's surface. The pattern is analogous to that observed in Young's double-slit experiment. Note the regions of constructive (*A*) and destructive (*B*) interference. *(Richard Megna/Fundamental Photographs)*

Quick Quiz 37.1

If you were to blow smoke into the space between the second barrier and the viewing screen of Figure 37.1a, what would you see?

QuickLab

Look through the fabric of an umbrella at a distant streetlight. Can you explain what you see? (The fringe pattern in Figure 37.1b is from rectangular slits. The fabric of the umbrella creates a two-dimensional set of square holes.)

Quick Quiz 37.2

Figure 37.2 is an overhead view of a shallow water tank. If you wanted to use a small ruler to measure the water's depth, would this be easier to do at location *A* or at location *B*?

Figure 37.3 shows some of the ways in which two waves can combine at the screen. In Figure 37.3a, the two waves, which leave the two slits in phase, strike the screen at the central point *P*. Because both waves travel the same distance, they arrive at *P* in phase. As a result, constructive interference occurs at this location, and a bright fringe is observed. In Figure 37.3b, the two waves also start in phase, but in this case the upper wave has to travel one wavelength farther than the lower wave to reach point *Q*. Because the upper wave falls behind the lower one by exactly one wavelength, they still arrive in phase at *Q*, and so a second bright fringe appears at this location. At point *R* in Figure 37.3c, however, midway between points *P* and *Q*, the upper wave has fallen half a wavelength behind the lower wave. This means that a trough of the lower wave overlaps a crest of the upper wave; this gives rise to destructive interference at point *R*. For this reason, a dark fringe is observed at this location.

We can describe Young's experiment quantitatively with the help of Figure 37.4. The viewing screen is located a perpendicular distance *L* from the double-slitted barrier. S_1 and S_2 are separated by a distance *d*, and the source is monochromatic. To reach any arbitrary point *P*, a wave from the lower slit travels farther than a wave from the upper slit by a distance $d \sin \theta$. This distance is called the **path difference** δ (lowercase Greek delta). If we assume that r_1 and r_2 are parallel, which is approximately true because *L* is much greater than *d*, then δ is given by

Path difference

$$\delta = r_2 - r_1 = d \sin \theta \qquad \textbf{(37.1)}$$

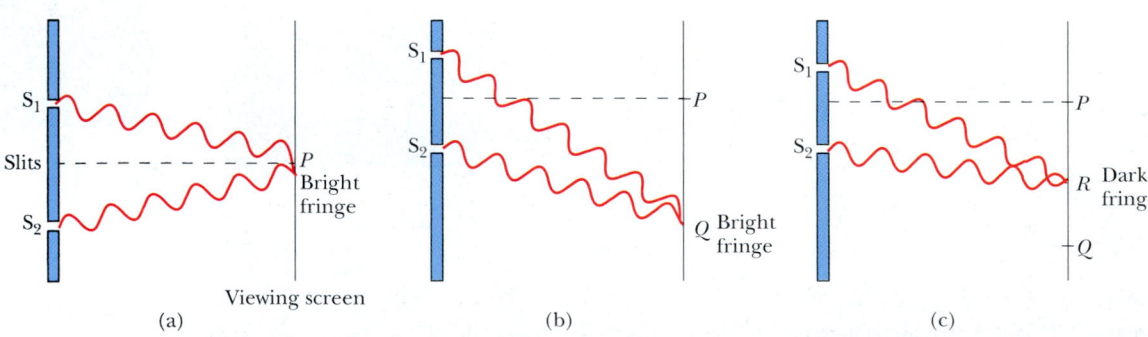

Figure 37.3   (a) Constructive interference occurs at point *P* when the waves combine. (b) Constructive interference also occurs at point *Q*. (c) Destructive interference occurs at *R* when the two waves combine because the upper wave falls half a wavelength behind the lower wave (all figures not to scale).

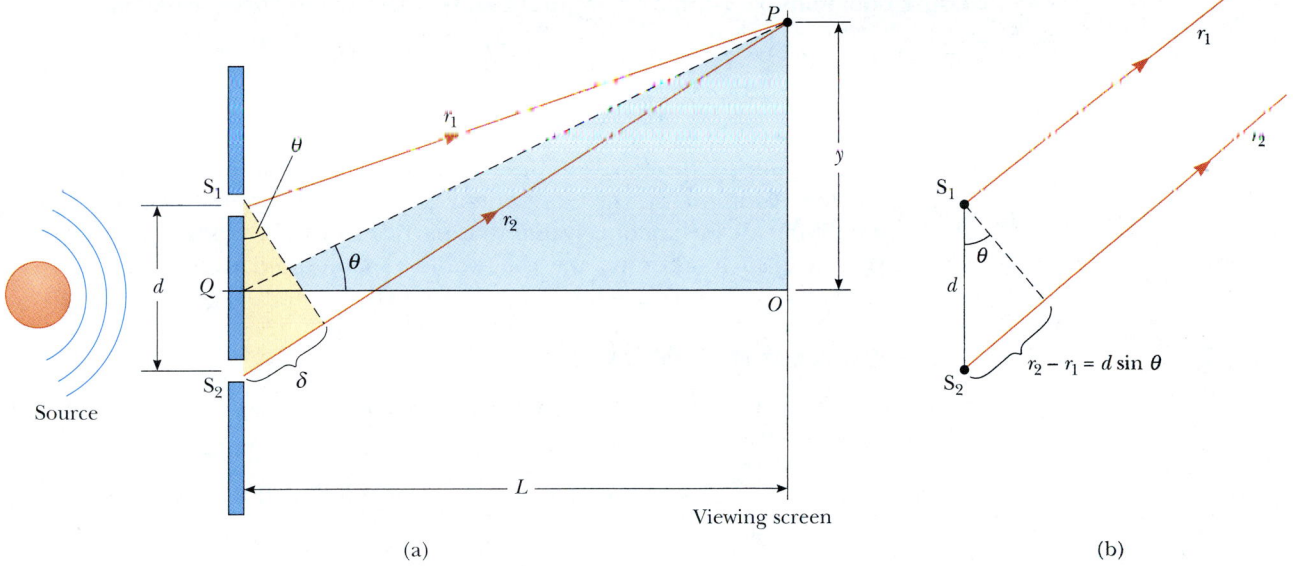

Figure 37.4 (a) Geometric construction for describing Young's double-slit experiment (not to scale). (b) When we assume that r_1 is parallel to r_2, the path difference between the two rays is $r_2 - r_1 = d \sin \theta$. For this approximation to be valid, it is essential that $L \gg d$.

The value of δ determines whether the two waves are in phase when they arrive at point P. If δ is either zero or some integer multiple of the wavelength, then the two waves are in phase at point P and constructive interference results. Therefore, the condition for bright fringes, or **constructive interference,** at point P is

$$\delta = d \sin \theta = m\lambda \qquad m = 0, \pm 1, \pm 2, \ldots \qquad \textbf{(37.2)}$$

Conditions for constructive interference

The number m is called the **order number.** The central bright fringe at $\theta = 0$ ($m = 0$) is called the *zeroth-order maximum*. The first maximum on either side, where $m = \pm 1$, is called the *first-order maximum*, and so forth.

When δ is an odd multiple of $\lambda/2$, the two waves arriving at point P are 180° out of phase and give rise to destructive interference. Therefore, the condition for dark fringes, or **destructive interference,** at point P is

$$d \sin \theta = (m + \tfrac{1}{2})\lambda \qquad m = 0, \pm 1, \pm 2, \ldots \qquad \textbf{(37.3)}$$

Conditions for destructive interference

It is useful to obtain expressions for the positions of the bright and dark fringes measured vertically from O to P. In addition to our assumption that $L \gg d$, we assume that $d \gg \lambda$. These can be valid assumptions because in practice L is often of the order of 1 m, d a fraction of a millimeter, and λ a fraction of a micrometer for visible light. Under these conditions, θ is small; thus, we can use the approximation $\sin \theta \approx \tan \theta$. Then, from triangle OPQ in Figure 37.4, we see that

$$y = L \tan \theta \approx L \sin \theta \qquad \textbf{(37.4)}$$

Solving Equation 37.2 for $\sin \theta$ and substituting the result into Equation 37.4, we see that the positions of the bright fringes measured from O are given by the expression

$$y_{bright} = \frac{\lambda L}{d} m \qquad \textbf{(37.5)}$$

Using Equations 37.3 and 37.4, we find that the dark fringes are located at

$$y_{\text{dark}} = \frac{\lambda L}{d} \left(m + \tfrac{1}{2}\right) \tag{37.6}$$

As we demonstrate in Example 37.1, Young's double-slit experiment provides a method for measuring the wavelength of light. In fact, Young used this technique to do just that. Additionally, the experiment gave the wave model of light a great deal of credibility. It was inconceivable that particles of light coming through the slits could cancel each other in a way that would explain the dark fringes.

EXAMPLE 37.1   Measuring the Wavelength of a Light Source

A viewing screen is separated from a double-slit source by 1.2 m. The distance between the two slits is 0.030 mm. The second-order bright fringe ($m = 2$) is 4.5 cm from the center line. (a) Determine the wavelength of the light.

Solution   We can use Equation 37.5, with $m = 2$, $y_2 = 4.5 \times 10^{-2}$ m, $L = 1.2$ m, and $d = 3.0 \times 10^{-5}$ m:

$$\lambda = \frac{d y_2}{m L} = \frac{(3.0 \times 10^{-5} \text{ m})(4.5 \times 10^{-2} \text{ m})}{2(1.2 \text{ m})}$$

$$= 5.6 \times 10^{-7} \text{ m} = \boxed{560 \text{ nm}}$$

(b) Calculate the distance between adjacent bright fringes.

Solution   From Equation 37.5 and the results of part (a), we obtain

$$y_{m+1} - y_m = \frac{\lambda L(m + 1)}{d} - \frac{\lambda L m}{d}$$

$$= \frac{\lambda L}{d} = \frac{(5.6 \times 10^{-7} \text{ m})(1.2 \text{ m})}{3.0 \times 10^{-5} \text{ m}}$$

$$= 2.2 \times 10^{-2} \text{ m} = \boxed{2.2 \text{ cm}}$$

Note that the spacing between all fringes is equal.

EXAMPLE 37.2   Separating Double-Slit Fringes of Two Wavelengths

A light source emits visible light of two wavelengths: $\lambda = 430$ nm and $\lambda' = 510$ nm. The source is used in a double-slit interference experiment in which $L = 1.5$ m and $d = 0.025$ mm. Find the separation distance between the third-order bright fringes.

Solution   Using Equation 37.5, with $m = 3$, we find that the fringe positions corresponding to these two wavelengths are

$$y_3 = \frac{\lambda L}{d} m = 3 \frac{\lambda L}{d} = 7.74 \times 10^{-2} \text{ m}$$

$$y_3' = \frac{\lambda' L}{d} m = 3 \frac{\lambda' L}{d} = 9.18 \times 10^{-2} \text{ m}$$

Hence, the separation distance between the two fringes is

$$\Delta y = y_3' - y_3 = 9.18 \times 10^{-2} \text{ m} - 7.74 \times 10^{-2} \text{ m}$$

$$= 1.4 \times 10^{-2} \text{ m} = \boxed{1.4 \text{ cm}}$$

37.3   INTENSITY DISTRIBUTION OF THE DOUBLE-SLIT INTERFERENCE PATTERN

Note that the edges of the bright fringes in Figure 37.1b are fuzzy. So far we have discussed the locations of only the centers of the bright and dark fringes on a distant screen. We now direct our attention to the intensity of the light at other points between the positions of maximum constructive and destructive interference. In other words, we now calculate the distribution of light intensity associated with the double-slit interference pattern.

Again, suppose that the two slits represent coherent sources of sinusoidal waves such that the two waves from the slits have the same angular frequency ω and a constant phase difference ϕ. The total magnitude of the electric field at point P on the screen in Figure 37.5 is the vector superposition of the two waves. Assuming that the two waves have the same amplitude E_0, we can write the magnitude of the electric field at point P due to each wave separately as

$$E_1 = E_0 \sin \omega t \quad \text{and} \quad E_2 = E_0 \sin(\omega t + \phi) \tag{37.7}$$

Although the waves are in phase at the slits, *their phase difference ϕ at point P depends on the path difference* $\delta = r_2 - r_1 = d \sin \theta$. Because a path difference of λ (constructive interference) corresponds to a phase difference of 2π rad, we obtain the ratio

$$\frac{\delta}{\lambda} = \frac{\phi}{2\pi}$$

$$\phi = \frac{2\pi}{\lambda}\, \delta = \frac{2\pi}{\lambda}\, d \sin \theta \tag{37.8}$$

This equation tells us precisely how the phase difference ϕ depends on the angle θ in Figure 37.4.

Using the superposition principle and Equation 37.7, we can obtain the magnitude of the resultant electric field at point P:

$$E_P = E_1 + E_2 = E_0[\sin \omega t + \sin(\omega t + \phi)] \tag{37.9}$$

To simplify this expression, we use the trigonometric identity

$$\sin A + \sin B = 2 \sin\left(\frac{A + B}{2}\right) \cos\left(\frac{A - B}{2}\right)$$

Taking $A = \omega t + \phi$ and $B = \omega t$, we can write Equation 37.9 in the form

$$E_P = 2E_0 \cos\left(\frac{\phi}{2}\right) \sin\left(\omega t + \frac{\phi}{2}\right) \tag{37.10}$$

This result indicates that the electric field at point P has the same frequency ω as the light at the slits, but that the amplitude of the field is multiplied by the factor $2 \cos(\phi/2)$. To check the consistency of this result, note that if $\phi = 0, 2\pi, 4\pi, \ldots,$ then the electric field at point P is $2E_0$, corresponding to the condition for constructive interference. These values of ϕ are consistent with Equation 37.2 for constructive interference. Likewise, if $\phi = \pi, 3\pi, 5\pi, \ldots,$ then the magnitude of the electric field at point P is zero; this is consistent with Equation 37.3 for destructive interference.

Finally, to obtain an expression for the light intensity at point P, recall from Section 34.3 that *the intensity of a wave is proportional to the square of the resultant electric field magnitude at that point* (Eq. 34.20). Using Equation 37.10, we can therefore express the light intensity at point P as

$$I \propto E_P^2 = 4E_0^2 \cos^2\left(\frac{\phi}{2}\right) \sin^2\left(\omega t + \frac{\phi}{2}\right)$$

Most light-detecting instruments measure time-averaged light intensity, and the time-averaged value of $\sin^2(\omega t + \phi/2)$ over one cycle is $\frac{1}{2}$. Therefore, we can write the average light intensity at point P as

$$I = I_{max} \cos^2\left(\frac{\phi}{2}\right) \tag{37.11}$$

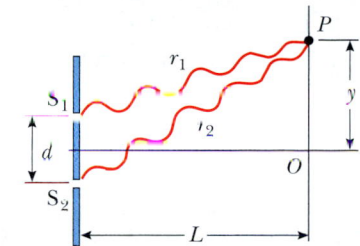

Figure 37.5   Construction for analyzing the double-slit interference pattern. A bright fringe, or intensity maximum, is observed at O.

Phase difference

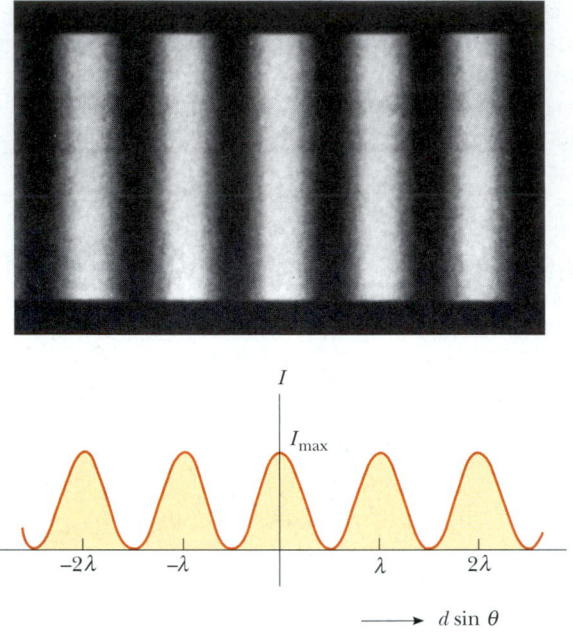

Figure 37.6 Light intensity versus $d \sin \theta$ for a double-slit interference pattern when the screen is far from the slits $(L \gg d)$. *(Photograph from M. Cagnet, M. Francon, and J.C. Thierr, Atlas of Optical Phenomena, Berlin, Springer-Verlag, 1962.)*

where I_{max} is the maximum intensity on the screen and the expression represents the time average. Substituting the value for ϕ given by Equation 37.8 into this expression, we find that

$$I = I_{max} \cos^2\left(\frac{\pi d \sin \theta}{\lambda}\right) \qquad \text{(37.12)}$$

Alternatively, because $\sin \theta \approx y/L$ for small values of θ in Figure 37.4, we can write Equation 37.12 in the form

$$I = I_{max} \cos^2\left(\frac{\pi d}{\lambda L} y\right) \qquad \text{(37.13)}$$

Constructive interference, which produces light intensity maxima, occurs when the quantity $\pi dy/\lambda L$ is an integral multiple of π, corresponding to $y = (\lambda L/d)m$. This is consistent with Equation 37.5.

A plot of light intensity versus $d \sin \theta$ is given in Figure 37.6. Note that the interference pattern consists of equally spaced fringes of equal intensity. Remember, however, that this result is valid only if the slit-to-screen distance L is much greater than the slit separation, and only for small values of θ.

We have seen that the interference phenomena arising from two sources depend on the relative phase of the waves at a given point. Furthermore, the phase difference at a given point depends on the path difference between the two waves. **The resultant light intensity at a point is proportional to the square of the resultant electric field at that point.** That is, the light intensity is proportional to $(E_1 + E_2)^2$. It would be incorrect to calculate the light intensity by adding the intensities of the individual waves. This procedure would give $E_1^2 + E_2^2$, which of course is not the same as $(E_1 + E_2)^2$. Note, however, that $(E_1 + E_2)^2$ has the same *average* value as $E_1^2 + E_2^2$ when the time average is taken over all values of the phase difference between E_1 and E_2. Hence, the law of conservation of energy is not violated.

37.4 PHASOR ADDITION OF WAVES

In the preceding section, we combined two waves algebraically to obtain the resultant wave amplitude at some point on a screen. Unfortunately, this analytical procedure becomes cumbersome when we must add several wave amplitudes. Because we shall eventually be interested in combining a large number of waves, we now describe a graphical procedure for this purpose.

Let us again consider a sinusoidal wave whose electric field component is given by

$$E_1 = E_0 \sin \omega t$$

where E_0 is the wave amplitude and ω is the angular frequency. This wave can be represented graphically by a phasor of magnitude E_0 rotating about the origin counterclockwise with an angular frequency ω, as shown in Figure 37.7a. Note that the phasor makes an angle ωt with the horizontal axis. The projection of the phasor on the vertical axis represents E_1, the magnitude of the wave disturbance at some time t. Hence, as the phasor rotates in a circle, the projection E_1 oscillates along the vertical axis about the origin.

Now consider a second sinusoidal wave whose electric field component is given by

$$E_2 = E_0 \sin(\omega t + \phi)$$

This wave has the same amplitude and frequency as E_1, but its phase is ϕ with respect to E_1. The phasor representing E_2 is shown in Figure 37.7b. We can obtain the resultant wave, which is the sum of E_1 and E_2, graphically by redrawing the phasors as shown in Figure 37.7c, in which the tail of the second phasor is placed at the tip of the first. As with vector addition, the resultant phasor $\mathbf{E}_R$ runs from the tail of the first phasor to the tip of the second. Furthermore, $\mathbf{E}_R$ rotates along with the two individual phasors at the same angular frequency ω. The projection of $\mathbf{E}_R$ along the vertical axis equals the sum of the projections of the two other phasors: $E_P = E_1 + E_2$.

It is convenient to construct the phasors at $t = 0$ as in Figure 37.8. From the geometry of one of the right triangles, we see that

$$\cos \alpha = \frac{E_R/2}{E_0}$$

which gives

$$E_R = 2E_0 \cos \alpha$$

Because the sum of the two opposite interior angles equals the exterior angle ϕ, we see that $\alpha = \phi/2$; thus,

$$E_R = 2E_0 \cos\left(\frac{\phi}{2}\right)$$

Hence, the projection of the phasor $\mathbf{E}_R$ along the vertical axis at any time t is

$$E_P = E_R \sin\left(\omega t + \frac{\phi}{2}\right) = 2E_0 \cos(\phi/2) \sin\left(\omega t + \frac{\phi}{2}\right)$$

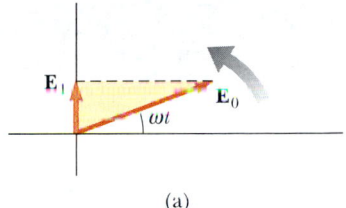

(a)

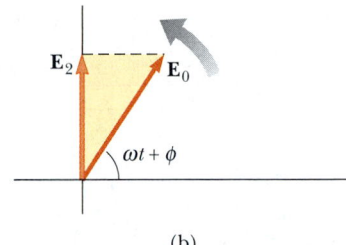

(b)

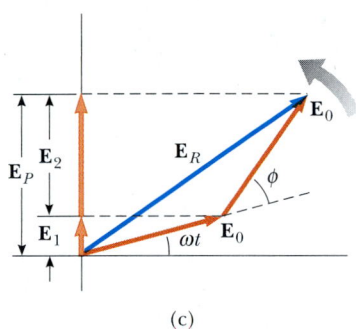

(c)

Figure 37.7 (a) Phasor diagram for the wave disturbance $E_1 = E_0 \sin \omega t$. The phasor is a vector of length E_0 rotating counterclockwise. (b) Phasor diagram for the wave $E_2 = E_0 \sin(\omega t + \phi)$. (c) The disturbance $\mathbf{E}_R$ is the resultant phasor formed from the phasors of parts (a) and (b).

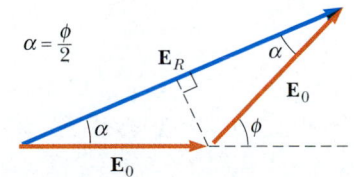

Figure 37.8 A reconstruction of the resultant phasor $\mathbf{E}_R$. From the geometry, note that $\alpha = \phi/2$.

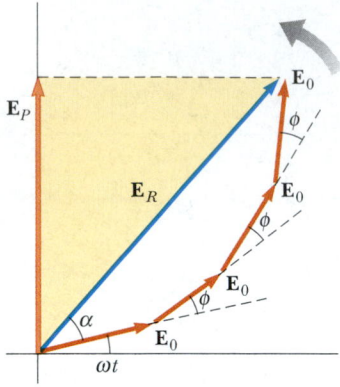

Figure 37.9 The phasor $\mathbf{E}_R$ is the resultant of four phasors of equal amplitude E_0. The phase of $\mathbf{E}_R$ with respect to the first phasor is α.

This is consistent with the result obtained algebraically, Equation 37.10. The resultant phasor has an amplitude $2E_0 \cos(\phi/2)$ and makes an angle $\phi/2$ with the first phasor. Furthermore, the average light intensity at point P, which varies as E_P^2, is proportional to $\cos^2(\phi/2)$, as described in Equation 37.11.

We can now describe how to obtain the resultant of several waves that have the same frequency:

- Represent the waves by phasors, as shown in Figure 37.9, remembering to maintain the proper phase relationship between one phasor and the next.
- The resultant phasor $\mathbf{E}_R$ is the vector sum of the individual phasors. At each instant, the projection of $\mathbf{E}_R$ along the vertical axis represents the time variation of the resultant wave. The phase angle α of the resultant wave is the angle between $\mathbf{E}_R$ and the first phasor. From Figure 37.9, drawn for four phasors, we see that the phasor of the resultant wave is given by the expression $E_P = E_R \sin(\omega t + \alpha)$.

Phasor Diagrams for Two Coherent Sources

As an example of the phasor method, consider the interference pattern produced by two coherent sources. Figure 37.10 represents the phasor diagrams for various values of the phase difference ϕ and the corresponding values of the path difference δ, which are obtained from Equation 37.8. The light intensity at a point is a maximum when $\mathbf{E}_R$ is a maximum; this occurs at $\phi = 0, 2\pi, 4\pi, \ldots$. The light intensity at some point is zero when $\mathbf{E}_R$ is zero; this occurs at $\phi = \pi, 3\pi, 5\pi, \ldots$. These results are in complete agreement with the analytical procedure described in the preceding section.

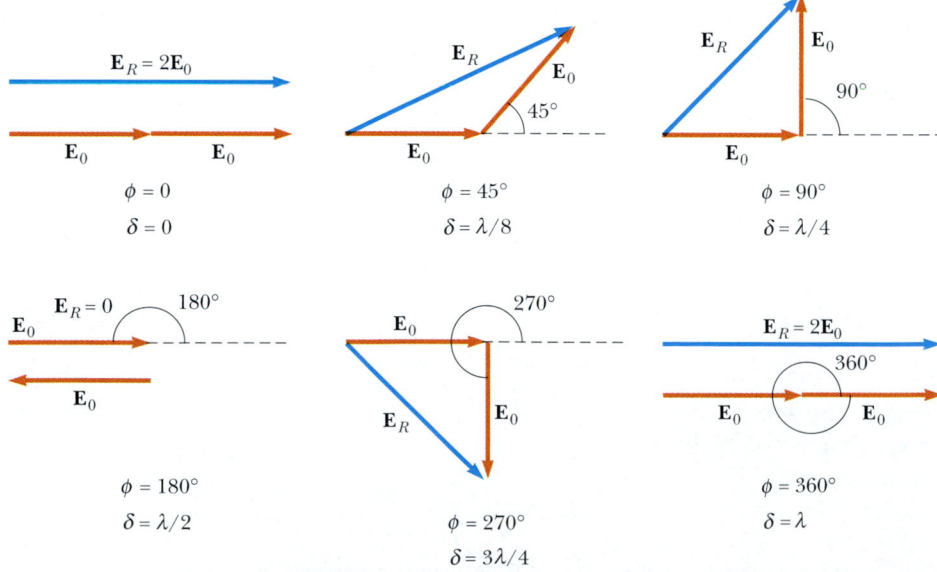

Figure 37.10 Phasor diagrams for a double-slit interference pattern. The resultant phasor $\mathbf{E}_R$ is a maximum when $\phi = 0, 2\pi, 4\pi, \ldots$ and is zero when $\phi = \pi, 3\pi, 5\pi, \ldots$.

Three-Slit Interference Pattern

Using phasor diagrams, let us analyze the interference pattern caused by three equally spaced slits. We can express the electric field components at a point P on the screen caused by waves from the individual slits as

$$E_1 = E_0 \sin \omega t$$

$$E_2 = E_0 \sin(\omega t + \phi)$$

$$E_3 = E_0 \sin(\omega t + 2\phi)$$

where ϕ is the phase difference between waves from adjacent slits. We can obtain the resultant magnitude of the electric field at point P from the phasor diagram in Figure 37.11.

The phasor diagrams for various values of ϕ are shown in Figure 37.12. Note that the resultant magnitude of the electric field at P has a maximum value of $3E_0$, a condition that occurs when $\phi = 0, \pm 2\pi, \pm 4\pi, \ldots$. These points are called *primary maxima*. Such primary maxima occur whenever the three phasors are aligned as shown in Figure 37.12a. We also find secondary maxima of amplitude E_0 occurring between the primary maxima at points where $\phi = \pm \pi, \pm 3\pi, \ldots$. For these points, the wave from one slit exactly cancels that from another slit (Fig. 37.12d). This means that only light from the third slit contributes to the resultant, which consequently has a total amplitude of E_0. Total destructive interference occurs whenever the three phasors form a closed triangle, as shown in Figure 37.12c. These points where $E_R = 0$ correspond to $\phi = \pm 2\pi/3, \pm 4\pi/3, \ldots$. You should be able to construct other phasor diagrams for values of ϕ greater than π.

Figure 37.13 shows multiple-slit interference patterns for a number of configurations. For three slits, note that the primary maxima are nine times more intense than the secondary maxima as measured by the height of the curve. This is because the intensity varies as E_R^2. For N slits, the intensity of the primary maxima is N^2 times greater than that due to a single slit. As the number of slits increases, the primary maxima increase in intensity and become narrower, while the secondary maxima decrease in intensity relative to the primary maxima. Figure 37.13 also shows that as the number of slits increases, the number of secondary maxima also increases. In fact, the number of secondary maxima is always $N - 2$, where N is the number of slits.

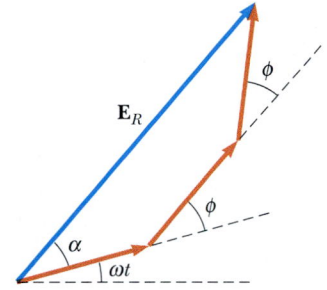

Figure 37.11 Phasor diagram for three equally spaced slits.

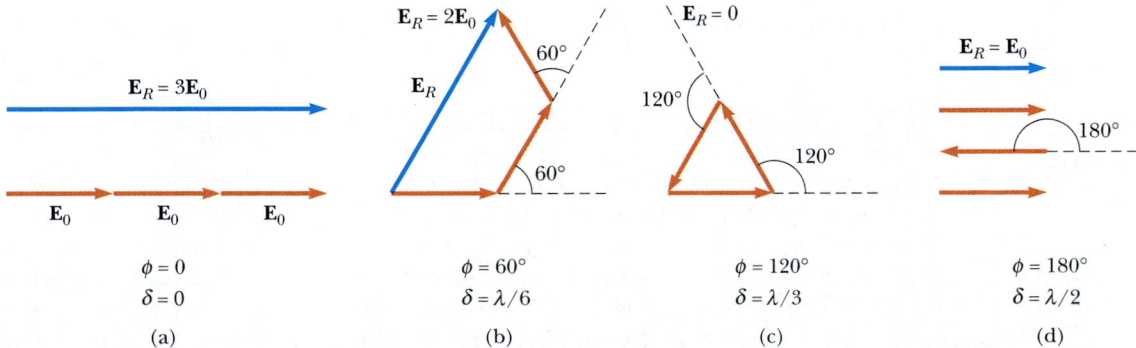

Figure 37.12 Phasor diagrams for three equally spaced slits at various values of ϕ. Note from (a) that there are primary maxima of amplitude $3E_0$ and from (d) that there are secondary maxima of amplitude E_0.

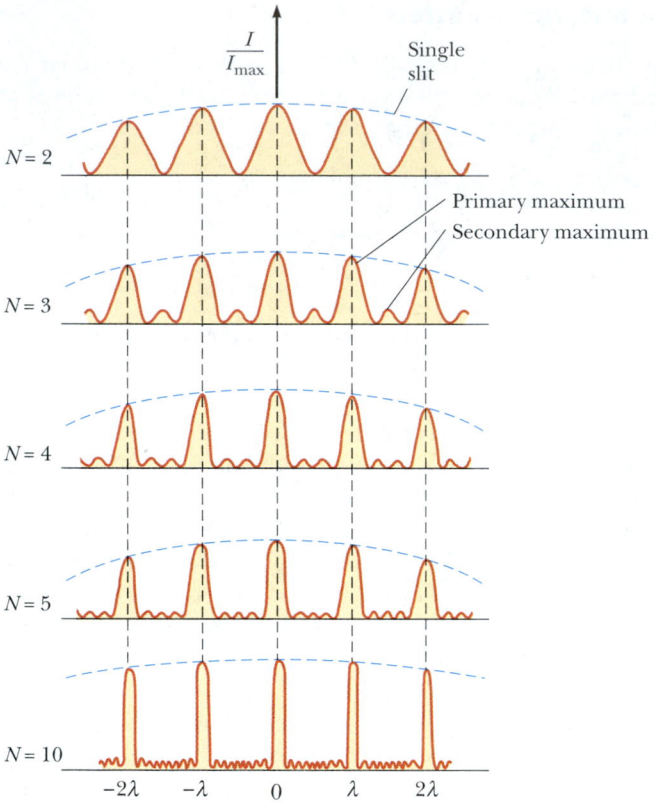

Figure 37.13   Multiple-slit interference patterns. As *N*, the number of slits, is increased, the primary maxima (the tallest peaks in each graph) become narrower but remain fixed in position, and the number of secondary maxima increases. For any value of *N*, the decrease in intensity in maxima to the left and right of the central maximum, indicated by the blue dashed arcs, is due to diffraction, which is discussed in Chapter 38.

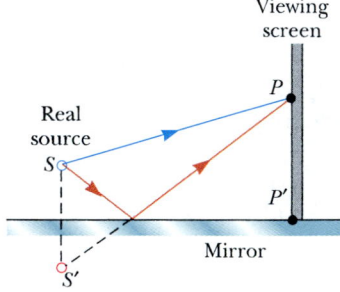

Figure 37.14   Lloyd's mirror. An interference pattern is produced at point *P* on the screen as a result of the combination of the direct ray (blue) and the reflected ray (red). The reflected ray undergoes a phase change of 180°.

Quick Quiz 37.3

Using Figure 37.13 as a model, sketch the interference pattern from six slits.

37.5 CHANGE OF PHASE DUE TO REFLECTION

Young's method for producing two coherent light sources involves illuminating a pair of slits with a single source. Another simple, yet ingenious, arrangement for producing an interference pattern with a single light source is known as *Lloyd's mirror* (Fig. 37.14). A light source is placed at point *S* close to a mirror, and a viewing screen is positioned some distance away at right angles to the mirror. Light waves can reach point *P* on the screen either by the direct path *SP* or by the path involving reflection from the mirror. The reflected ray can be treated as a ray originating from a virtual source at point *S'*. As a result, we can think of this arrangement as a double-slit source with the distance between points S and S' comparable to length d in Figure 37.4. Hence, at observation points far from the source $(L \gg d)$, we expect waves from points S and S' to form an interference pattern just like the one we see from two real coherent sources. An interference pattern is indeed observed. However, the positions of the dark and bright fringes are reversed relative to the pattern created by two real coherent sources (Young's experiment). This is because the coherent sources at points S and S' differ in phase by 180°, a phase change produced by reflection.

To illustrate this further, consider point P', the point where the mirror intersects the screen. This point is equidistant from points S and S'. If path difference alone were responsible for the phase difference, we would see a bright fringe at point P' (because the path difference is zero for this point), corresponding to the central bright fringe of the two-slit interference pattern. Instead, we observe a dark fringe at point P' because of the 180° phase change produced by reflection. In general,

> an electromagnetic wave undergoes a phase change of 180° upon reflection from a medium that has a higher index of refraction than the one in which the wave is traveling.

It is useful to draw an analogy between reflected light waves and the reflections of a transverse wave pulse on a stretched string (see Section 16.6). The reflected pulse on a string undergoes a phase change of 180° when reflected from the boundary of a denser medium, but no phase change occurs when the pulse is reflected from the boundary of a less dense medium. Similarly, an electromagnetic wave undergoes a 180° phase change when reflected from a boundary leading to an optically denser medium, but no phase change occurs when the wave is reflected from a boundary leading to a less dense medium. These rules, summarized in Figure 37.15, can be deduced from Maxwell's equations, but the treatment is beyond the scope of this text.

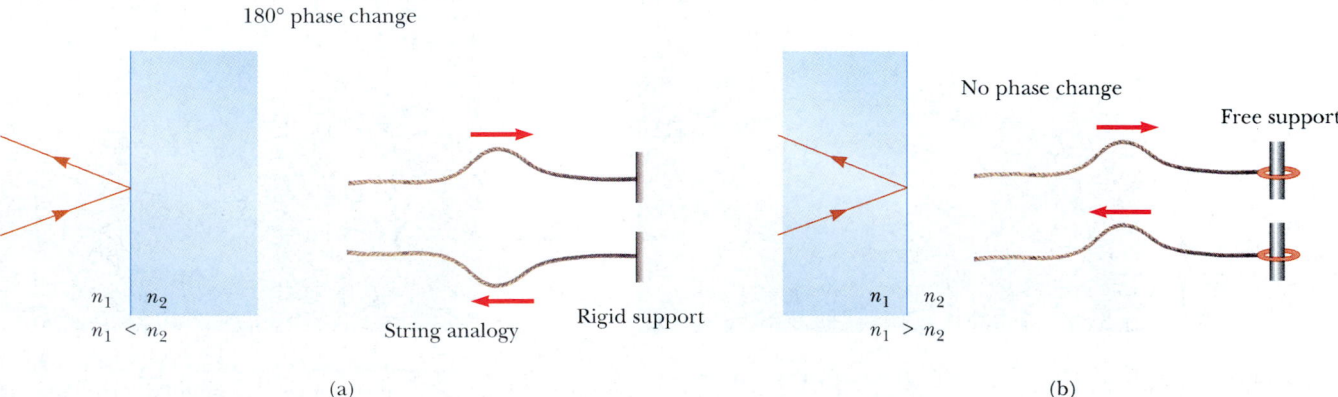

Figure 37.15 (a) For $n_1 < n_2$, a light ray traveling in medium 1 when reflected from the surface of medium 2 undergoes a 180° phase change. The same thing happens with a reflected pulse traveling along a string fixed at one end. (b) For $n_1 > n_2$, a light ray traveling in medium 1 undergoes no phase change when reflected from the surface of medium 2. The same is true of a reflected wave pulse on a string whose supported end is free to move.

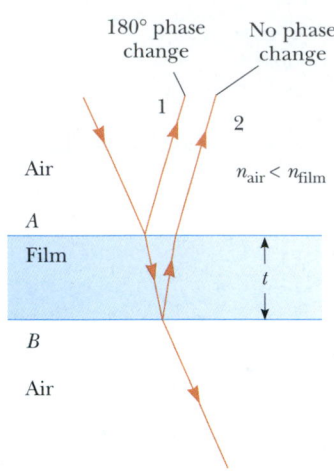

Figure 37.16 Interference in light reflected from a thin film is due to a combination of rays reflected from the upper and lower surfaces of the film.

37.6 › INTERFERENCE IN THIN FILMS

Interference effects are commonly observed in thin films, such as thin layers of oil on water or the thin surface of a soap bubble. The varied colors observed when white light is incident on such films result from the interference of waves reflected from the two surfaces of the film.

Consider a film of uniform thickness t and index of refraction n, as shown in Figure 37.16. Let us assume that the light rays traveling in air are nearly normal to the two surfaces of the film. To determine whether the reflected rays interfere constructively or destructively, we first note the following facts:

- A wave traveling from a medium of index of refraction n_1 toward a medium of index of refraction n_2 undergoes a 180° phase change upon reflection when $n_2 > n_1$ and undergoes no phase change if $n_2 < n_1$.
- The wavelength of light λ_n in a medium whose refraction index is n (see Section 35.5) is

$$\lambda_n = \frac{\lambda}{n} \qquad (37.14)$$

where λ is the wavelength of the light in free space.

Let us apply these rules to the film of Figure 37.16, where $n_{film} > n_{air}$. Reflected ray 1, which is reflected from the upper surface (A), undergoes a phase change of 180° with respect to the incident wave. Reflected ray 2, which is reflected from the lower film surface (B), undergoes no phase change because it is reflected from a medium (air) that has a lower index of refraction. Therefore, ray 1 is 180° out of phase with ray 2, which is equivalent to a path difference of $\lambda_n/2$.

Interference in soap bubbles. The colors are due to interference between light rays reflected from the front and back surfaces of the thin film of soap making up the bubble. The color depends on the thickness of the film, ranging from black where the film is thinnest to red where it is thickest. *(Dr. Jeremy Burgess/ Science Photo Library)*

The brilliant colors in a peacock's feathers are due to interference. The multilayer structure of the feathers causes constructive interference for certain colors, such as blue and green. The colors change as you view a peacock's feathers from different angles. Iridescent colors of butterflies and hummingbirds are the result of similar interference effects. *(©Diane Schiumo 1988/Fundamental Photographs)*

However, we must also consider that ray 2 travels an extra distance $2t$ before the waves recombine in the air above surface A. If $2t = \lambda_n/2$, then rays 1 and 2 recombine in phase, and the result is constructive interference. In general, the condition for constructive interference in such situations is

$$2t = (m + \tfrac{1}{2})\lambda_n \qquad m = 0, 1, 2, \ldots \qquad (37.15)$$

This condition takes into account two factors: (1) the difference in path length for the two rays (the term $m\lambda_n$) and (2) the 180° phase change upon reflection (the term $\lambda_n/2$). Because $\lambda_n = \lambda/n$, we can write Equation 37.15 as

$$2nt = (m + \tfrac{1}{2})\lambda \qquad m = 0, 1, 2, \ldots \qquad (37.16)$$

If the extra distance $2t$ traveled by ray 2 corresponds to a multiple of λ_n, then the two waves combine out of phase, and the result is destructive interference. The general equation for destructive interference is

$$2nt = m\lambda \qquad m = 0, 1, 2, \ldots \qquad (37.17)$$

The foregoing conditions for constructive and destructive interference are valid when the medium above the top surface of the film is the same as the medium below the bottom surface. The medium surrounding the film may have a refractive index less than or greater than that of the film. In either case, the rays reflected from the two surfaces are out of phase by 180°. If the film is placed between two different media, one with $n < n_{\text{film}}$ and the other with $n > n_{\text{film}}$, then the conditions for constructive and destructive interference are reversed. In this case, either there is a phase change of 180° for both ray 1 reflecting from surface A and ray 2 reflecting from surface B, or there is no phase change for either ray; hence, the net change in relative phase due to the reflections is zero.

> Conditions for constructive interference in thin films

> Conditions for destructive interference in thin films

Quick Quiz 37.4

In Figure 37.17, where does the oil film thickness vary the least?

Newton's Rings

Another method for observing interference in light waves is to place a plano-convex lens on top of a flat glass surface, as shown in Figure 37.18a. With this arrangement, the air film between the glass surfaces varies in thickness from zero at the point of contact to some value t at point P. If the radius of curvature R of the lens is much greater than the distance r, and if the system is viewed from above using light of a single wavelength λ, a pattern of light and dark rings is observed, as shown in Figure 37.18b. These circular fringes, discovered by Newton, are called **Newton's rings**.

The interference effect is due to the combination of ray 1, reflected from the flat plate, with ray 2, reflected from the curved surface of the lens. Ray 1 undergoes a phase change of 180° upon reflection (because it is reflected from a medium of higher refractive index), whereas ray 2 undergoes no phase change (because it is reflected from a medium of lower refractive index). Hence, the conditions for constructive and destructive interference are given by Equations 37.16 and 37.17, respectively, with $n = 1$ because the film is air.

The contact point at O is dark, as seen in Figure 37.18b, because ray 1 undergoes a 180° phase change upon external reflection (from the flat surface); in con-

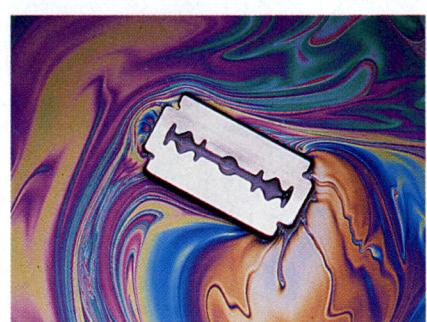

Figure 37.17 A thin film of oil floating on water displays interference, as shown by the pattern of colors produced when white light is incident on the film. Variations in film thickness produce the interesting color pattern. The razor blade gives one an idea of the size of the colored bands. *(Peter Aprahamian/Science Photo Library)*

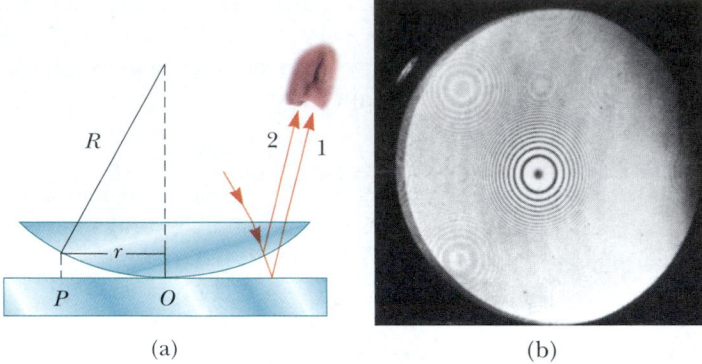

(a)   (b)

Figure 37.18   (a) The combination of rays reflected from the flat plate and the curved lens surface gives rise to an interference pattern known as Newton's rings. (b) Photograph of Newton's rings. *(Courtesy of Bausch and Lomb Optical Company)*

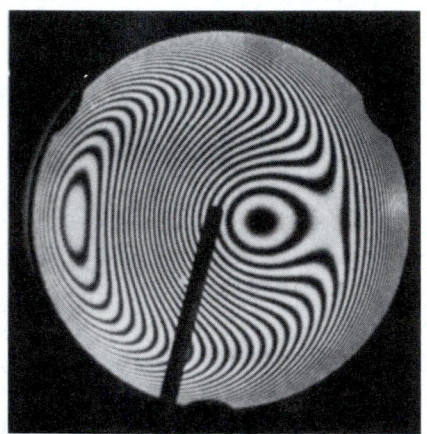

Figure 37.19   This asymmetrical interference pattern indicates imperfections in the lens of a Newton's-rings apparatus. *(From Physical Science Study Committee,* College Physics, *Lexington, MA, Heath, 1968.)*

trast, ray 2 undergoes no phase change upon internal reflection (from the curved surface).

Using the geometry shown in Figure 37.18a, we can obtain expressions for the radii of the bright and dark bands in terms of the radius of curvature R and wavelength λ. For example, the dark rings have radii given by the expression $r \approx \sqrt{m\lambda R/n}$. The details are left as a problem for you to solve (see Problem 67). We can obtain the wavelength of the light causing the interference pattern by measuring the radii of the rings, provided R is known. Conversely, we can use a known wavelength to obtain R.

One important use of Newton's rings is in the testing of optical lenses. A circular pattern like that pictured in Figure 37.18b is obtained only when the lens is ground to a perfectly symmetric curvature. Variations from such symmetry might produce a pattern like that shown in Figure 37.19. These variations indicate how the lens must be reground and repolished to remove the imperfections.

Problem-Solving Hints

Thin-Film Interference

You should keep the following ideas in mind when you work thin-film interference problems:

• Identify the thin film causing the interference.
• The type of interference that occurs is determined by the phase relationship between the portion of the wave reflected at the upper surface of the film and the portion reflected at the lower surface.
• Phase differences between the two portions of the wave have two causes: (1) differences in the distances traveled by the two portions and (2) phase changes that may occur upon reflection.
• When the distance traveled and phase changes upon reflection are both taken into account, the interference is constructive if the equivalent path difference between the two waves is an integral multiple of λ, and it is destructive if the path difference is $\lambda/2$, $3\lambda/2$, $5\lambda/2$, and so forth.

QuickLab

Observe the colors appearing to swirl on the surface of a soap bubble. What do you see just before a bubble bursts? Why?

EXAMPLE 37.3 ▸ Interference in a Soap Film

Calculate the minimum thickness of a soap-bubble film ($n = 1.33$) that results in constructive interference in the reflected light if the film is illuminated with light whose wavelength in free space is $\lambda = 600$ nm.

Solution The minimum film thickness for constructive interference in the reflected light corresponds to $m = 0$ in Equation 37.16. This gives $2nt = \lambda/2$, or

$$t = \frac{\lambda}{4n} = \frac{600 \text{ nm}}{4(1.33)} = \boxed{113 \text{ nm}}$$

Exercise What other film thicknesses produce constructive interference?

Answer 338 nm, 564 nm, 789 nm, and so on.

EXAMPLE 37.4 ▸ Nonreflective Coatings for Solar Cells

Solar cells—devices that generate electricity when exposed to sunlight—are often coated with a transparent, thin film of silicon monoxide (SiO, $n = 1.45$) to minimize reflective losses from the surface. Suppose that a silicon solar cell ($n = 3.5$) is coated with a thin film of silicon monoxide for this purpose (Fig. 37.20). Determine the minimum film thickness that produces the least reflection at a wavelength of 550 nm, near the center of the visible spectrum.

Solution The reflected light is a minimum when rays 1 and 2 in Figure 37.20 meet the condition of destructive interference. Note that both rays undergo a 180° phase change upon reflection—ray 1 from the upper SiO surface and ray 2 from the lower SiO surface. The net change in phase due to reflection is therefore zero, and the condition for a reflection minimum requires a path difference of $\lambda_n/2$. Hence,

$2t = \lambda/2n$, and the required thickness is

$$t = \frac{\lambda}{4n} = \frac{550 \text{ nm}}{4(1.45)} = \boxed{94.8 \text{ nm}}$$

A typical uncoated solar cell has reflective losses as high as 30%; a SiO coating can reduce this value to about 10%. This significant decrease in reflective losses increases the cell's efficiency because less reflection means that more sunlight enters the silicon to create charge carriers in the cell. No coating can ever be made perfectly nonreflecting because the required thickness is wavelength-dependent and the incident light covers a wide range of wavelengths.

Glass lenses used in cameras and other optical instruments are usually coated with a transparent thin film to reduce or eliminate unwanted reflection and enhance the transmission of light through the lenses.

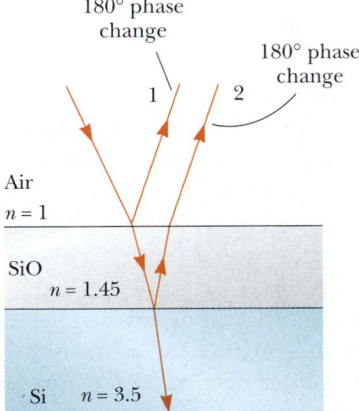

Figure 37.20 Reflective losses from a silicon solar cell are minimized by coating the surface of the cell with a thin film of silicon monoxide.

This camera lens has several coatings (of different thicknesses) that minimize reflection of light waves having wavelengths near the center of the visible spectrum. As a result, the little light that is reflected by the lens has a greater proportion of the far ends of the spectrum and appears reddish-violet. *(Kristen Brochmann/Fundamental Photographs)*

EXAMPLE 37.5   **Interference in a Wedge-Shaped Film**

A thin, wedge-shaped film of refractive index n is illuminated with monochromatic light of wavelength λ, as illustrated in Figure 37.21a. Describe the interference pattern observed for this case.

Solution   The interference pattern, because it is created by a thin film of variable thickness surrounded by air, is a series of alternating bright and dark parallel fringes. A dark fringe corresponding to destructive interference appears at point O, the apex, because here the upper reflected ray undergoes a 180° phase change while the lower one undergoes no phase change.

According to Equation 37.17, other dark minima appear when $2nt = m\lambda$; thus, $t_1 = \lambda/2n$, $t_2 = \lambda/n$, $t_3 = 3\lambda/2n$, and so on. Similarly, the bright maxima appear at locations where

the thickness satisfies Equation 37.16, $2nt = (m + \frac{1}{2})\lambda$, corresponding to thicknesses of $\lambda/4n$, $3\lambda/4n$, $5\lambda/4n$, and so on.

If white light is used, bands of different colors are observed at different points, corresponding to the different wavelengths of light (see Fig. 37.21b). This is why we see different colors in soap bubbles.

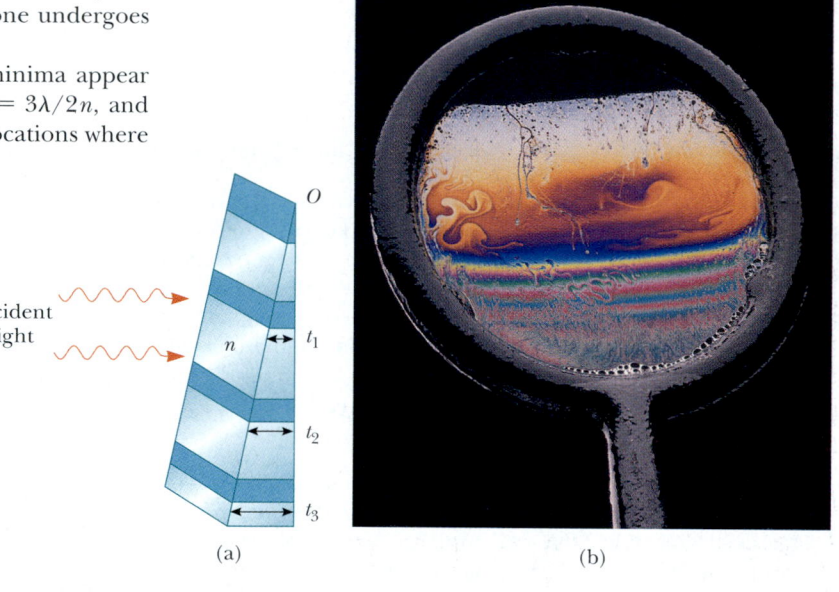

Figure 37.21   (a) Interference bands in reflected light can be observed by illuminating a wedge-shaped film with monochromatic light. The darker areas correspond to regions where rays tend to cancel each other because of interference effects. (b) Interference in a vertical film of variable thickness. The top of the film appears darkest where the film is thinnest. *(Richard Megna/Fundamental Photographs)*

(a)   (b)

Optional Section

37.7   THE MICHELSON INTERFEROMETER

The **interferometer,** invented by the American physicist A. A. Michelson (1852–1931), splits a light beam into two parts and then recombines the parts to form an interference pattern. The device can be used to measure wavelengths or other lengths with great precision.

A schematic diagram of the interferometer is shown in Figure 37.22. A ray of light from a monochromatic source is split into two rays by mirror M, which is inclined at 45° to the incident light beam. Mirror M, called a *beam splitter,* transmits half the light incident on it and reflects the rest. One ray is reflected from M vertically upward toward mirror M_1, and the second ray is transmitted horizontally through M toward mirror M_2. Hence, the two rays travel separate paths L_1 and L_2. After reflecting from M_1 and M_2, the two rays eventually recombine at M to produce an interference pattern, which can be viewed through a telescope. The glass plate P, equal in thickness to mirror M, is placed in the path of the horizontal ray to ensure that the two returning rays travel the same thickness of glass.

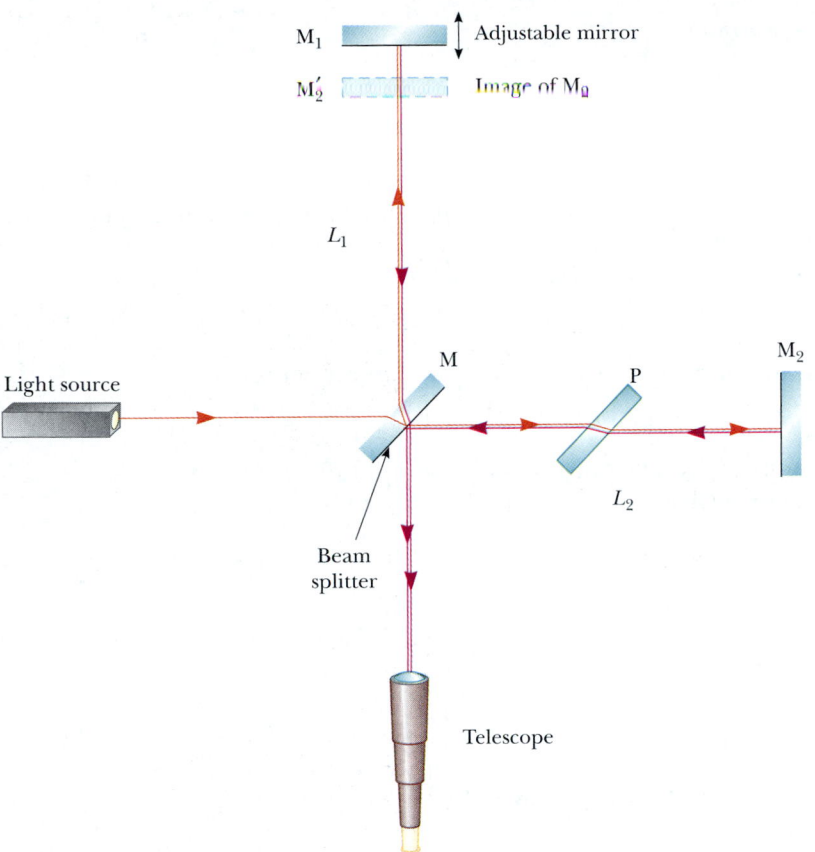

Figure 37.22 Diagram of the Michelson interferometer. A single ray of light is split into two rays by mirror M, which is called a beam splitter. The path difference between the two rays is varied with the adjustable mirror M_1. As M_1 is moved toward M, an interference pattern moves across the field of view.

The interference condition for the two rays is determined by their path length differences. When the two rays are viewed as shown, the image of M_2 produced by the mirror M is at M_2', which is nearly parallel to M_1. (Because M_1 and M_2 are not exactly perpendicular to each other, the image M_2' is at a slight angle to M_1.) Hence, the space between M_2' and M_1 is the equivalent of a wedge-shaped air film. The effective thickness of the air film is varied by moving mirror M_1 parallel to itself with a finely threaded screw adjustment. Under these conditions, the interference pattern is a series of bright and dark parallel fringes as described in Example 37.5. As M_1 is moved, the fringe pattern shifts. For example, if a dark fringe appears in the field of view (corresponding to destructive interference) and M_1 is then moved a distance $\lambda/4$ toward M, the path difference changes by $\lambda/2$ (twice the separation between M_1 and M_2'). What was a dark fringe now becomes a bright fringe. As M_1 is moved an additional distance $\lambda/4$ toward M, the bright fringe becomes a dark fringe. Thus, the fringe pattern shifts by one-half fringe each time M_1 is moved a distance $\lambda/4$. The wavelength of light is then measured by counting the number of fringe shifts for a given displacement of M_1. If the wavelength is accurately known (as with a laser beam), mirror displacements can be measured to within a fraction of the wavelength.

SUMMARY

Interference in light waves occurs whenever two or more waves overlap at a given point. A sustained interference pattern is observed if (1) the sources are coherent and (2) the sources have identical wavelengths.

In Young's double-slit experiment, two slits S_1 and S_2 separated by a distance d are illuminated by a single-wavelength light source. An interference pattern consisting of bright and dark fringes is observed on a viewing screen. The condition for bright fringes (**constructive interference**) is

$$d \sin \theta = m\lambda \qquad m = 0, \pm 1, \pm 2, \ldots \qquad \textbf{(37.2)}$$

The condition for dark fringes (**destructive interference**) is

$$d \sin \theta = (m + \tfrac{1}{2})\lambda \qquad m = 0, \pm 1, \pm 2, \ldots \qquad \textbf{(37.3)}$$

The number m is called the **order number** of the fringe.

The **intensity** at a point in the double-slit interference pattern is

$$I = I_{max} \cos^2\left(\frac{\pi d \sin \theta}{\lambda}\right) \qquad \textbf{(37.12)}$$

where I_{max} is the maximum intensity on the screen and the expression represents the time average.

A wave traveling from a medium of index of refraction n_1 toward a medium of index of refraction n_2 undergoes a 180° phase change upon reflection when $n_2 > n_1$ and undergoes no phase change when $n_2 < n_1$.

The condition for constructive interference in a film of thickness t and refractive index n surrounded by air is

$$2nt = (m + \tfrac{1}{2})\lambda \qquad m = 0, 1, 2, \ldots \qquad \textbf{(37.16)}$$

where λ is the wavelength of the light in free space.

Similarly, the condition for destructive interference in a thin film is

$$2nt = m\lambda \qquad m = 0, 1, 2, \ldots \qquad \textbf{(37.17)}$$

QUESTIONS

1. What is the necessary condition on the path length difference between two waves that interfere (a) constructively and (b) destructively?

2. Explain why two flashlights held close together do not produce an interference pattern on a distant screen.

3. If Young's double-slit experiment were performed under water, how would the observed interference pattern be affected?

4. In Young's double-slit experiment, why do we use monochromatic light? If white light is used, how would the pattern change?

5. Consider a dark fringe in an interference pattern, at which almost no light is arriving. Light from both slits is arriving at this point, but the waves are canceling. Where does the energy go?

6. An oil film on water appears brightest at the outer regions, where it is thinnest. From this information, what can you say about the index of refraction of oil relative to that of water?

7. In our discussion of thin-film interference, we looked at light *reflecting* from a thin film. Consider one light ray, the direct ray, that transmits through the film without reflecting. Consider a second ray, the reflected ray, that transmits through the first surface, reflects from the second, reflects again from the first, and then transmits out into the air, parallel to the direct ray. For normal incidence, how thick must the film be, in terms of the wavelength of light, for the outgoing rays to interfere destructively? Is it the same thickness as for reflected destructive interference?

8. Suppose that you are watching television connected to an antenna rather than a cable system. If an airplane flies near your location, you may notice wavering ghost images in the television picture. What might cause this?

9. If we are to observe interference in a thin film, why must the film not be very thick (on the order of a few wavelengths)?

10. A lens with outer radius of curvature R and index of re-

fraction n rests on a flat glass plate, and the combination is illuminated with white light from above. Is there a dark spot or a light spot at the center of the lens? What does it mean if the observed rings are noncircular?

11. Why is the lens on a high-quality camera coated with a thin film?

12. Why is it so much easier to perform interference experiments with a laser than with an ordinary light source?

PROBLEMS

1, 2, 3 = straightforward, intermediate, challenging   ☐ = full solution available in the *Student Solutions Manual and Study Guide*
WEB = solution posted at **http://www.saunderscollege.com/physics/**   🖥 = Computer useful in solving problem   🔲 = Interactive Physics
☐ = paired numerical/symbolic problems

Section 37.1   Conditions for Interference
Section 37.2   Young's Double-Slit Experiment

1. A laser beam ($\lambda = 632.8$ nm) is incident on two slits 0.200 mm apart. How far apart are the bright interference fringes on a screen 5.00 m away from the slits?

2. A Young's interference experiment is performed with monochromatic light. The separation between the slits is 0.500 mm, and the interference pattern on a screen 3.30 m away shows the first maximum 3.40 mm from the center of the pattern. What is the wavelength?

WEB 3. Two radio antennas separated by 300 m as shown in Figure P37.3 simultaneously broadcast identical signals at the same wavelength. A radio in a car traveling due north receives the signals. (a) If the car is at the position of the second maximum, what is the wavelength of the signals? (b) How much farther must the car travel to encounter the next minimum in reception? (*Note:* Do not use the small-angle approximation in this problem.)

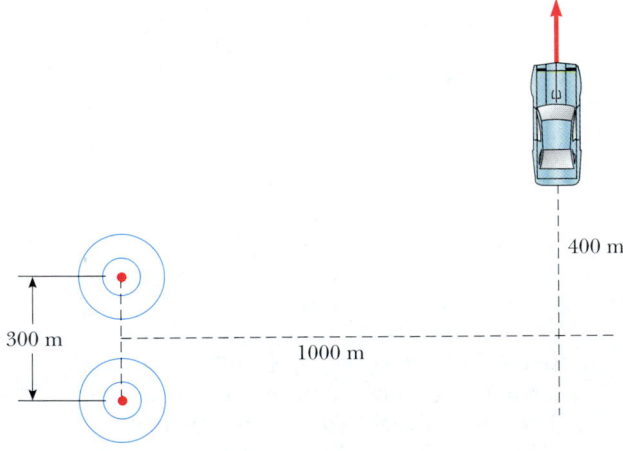

Figure P37.3

4. In a location where the speed of sound is 354 m/s, a 2 000-Hz sound wave impinges on two slits 30.0 cm apart. (a) At what angle is the first maximum located? (b) If the sound wave is replaced by 3.00-cm microwaves, what slit separation gives the same angle for the first maximum? (c) If the slit separation is 1.00 μm, what frequency of light gives the same first maximum angle?

WEB 5. Young's double-slit experiment is performed with 589-nm light and a slit-to-screen distance of 2.00 m. The tenth interference minimum is observed 7.26 mm from the central maximum. Determine the spacing of the slits.

6. The two speakers of a boom box are 35.0 cm apart. A single oscillator makes the speakers vibrate in phase at a frequency of 2.00 kHz. At what angles, measured from the perpendicular bisector of the line joining the speakers, would a distant observer hear maximum sound intensity? minimum sound intensity? (Take the speed of sound as 340 m/s.)

7. A pair of narrow, parallel slits separated by 0.250 mm are illuminated by green light ($\lambda = 546.1$ nm). The interference pattern is observed on a screen 1.20 m away from the plane of the slits. Calculate the distance (a) from the central maximum to the first bright region on either side of the central maximum and (b) between the first and second dark bands.

8. Light with a wavelength of 442 nm passes through a double-slit system that has a slit separation $d = 0.400$ mm. Determine how far away a screen must be placed so that a dark fringe appears directly opposite both slits, with just one bright fringe between them.

9. A riverside warehouse has two open doors, as illustrated in Figure P37.9. Its walls are lined with sound-absorbing material. A boat on the river sounds its horn. To person A, the sound is loud and clear. To person B, the sound is barely audible. The principal wavelength of the sound waves is 3.00 m. Assuming that person B is at the position of the first minimum, determine the distance between the doors, center to center.

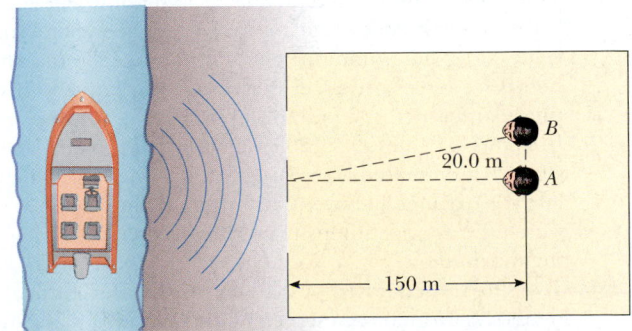

Figure P37.9

10. Two slits are separated by 0.320 mm. A beam of 500-nm light strikes the slits, producing an interference pattern. Determine the number of maxima observed in the angular range $-30.0° < \theta < 30.0°$.

11. In Figure 37.4 let $L = 1.20$ m and $d = 0.120$ mm, and assume that the slit system is illuminated with monochromatic 500-nm light. Calculate the phase difference between the two wavefronts arriving at point P when (a) $\theta = 0.500°$ and (b) $y = 5.00$ mm. (c) What is the value of θ for which the phase difference is 0.333 rad? (d) What is the value of θ for which the path difference is $\lambda/4$?

12. Coherent light rays of wavelength λ strike a pair of slits separated by distance d at an angle of θ_1, as shown in Figure P37.12. If an interference maximum is formed at an angle of θ_2 a great distance from the slits, show that $d(\sin \theta_2 - \sin \theta_1) = m\lambda$, where m is an integer.

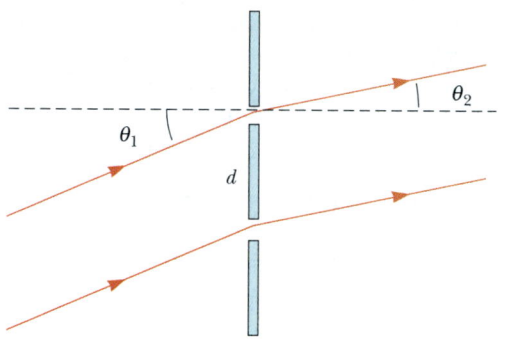

Figure P37.12

13. In the double-slit arrangement of Figure 37.4, $d = 0.150$ mm, $L = 140$ cm, $\lambda = 643$ nm, and $y = 1.80$ cm. (a) What is the path difference δ for the rays from the two slits arriving at point P? (b) Express this path difference in terms of λ. (c) Does point P correspond to a maximum, a minimum, or an intermediate condition?

Section 37.3   Intensity Distribution of the Double-Slit Interference Pattern

14. The intensity on the screen at a certain point in a double-slit interference pattern is 64.0% of the maximum value. (a) What minimum phase difference (in radians) between sources produces this result? (b) Express this phase difference as a path difference for 486.1-nm light.

WEB 15. In Figure 37.4, let $L = 120$ cm and $d = 0.250$ cm. The slits are illuminated with coherent 600-nm light. Calculate the distance y above the central maximum for which the average intensity on the screen is 75.0% of the maximum.

16. Two slits are separated by 0.180 mm. An interference pattern is formed on a screen 80.0 cm away by 656.3-nm light. Calculate the fraction of the maximum intensity 0.600 cm above the central maximum.

17. Two narrow parallel slits separated by 0.850 mm are illuminated by 600-nm light, and the viewing screen is 2.80 m away from the slits. (a) What is the phase difference between the two interfering waves on a screen at a point 2.50 mm from the central bright fringe? (b) What is the ratio of the intensity at this point to the intensity at the center of a bright fringe?

18. Monochromatic coherent light of amplitude E_0 and angular frequency ω passes through three parallel slits each separated by a distance d from its neighbor. (a) Show that the time-averaged intensity as a function of the angle θ is

$$I(\theta) = I_{max}\left[1 + 2\cos\left(\frac{2\pi d \sin \theta}{\lambda}\right)\right]^2$$

(b) Determine the ratio of the intensities of the primary and secondary maxima.

Section 37.4   Phasor Addition of Waves

19. Marie Cornu invented phasors in about 1880. This problem helps you to see their utility. Find the amplitude and phase constant of the sum of two waves represented by the expressions

$$E_1 = (12.0 \text{ kN/C}) \sin(15x - 4.5t)$$

and

$$E_2 = (12.0 \text{ kN/C}) \sin(15x - 4.5t + 70°)$$

(a) by using a trigonometric identity (see Appendix B) and (b) by representing the waves by phasors. (c) Find the amplitude and phase constant of the sum of the three waves represented by

$$E_1 = (12.0 \text{ kN/C}) \sin(15x - 4.5t + 70°)$$

$$E_2 = (15.5 \text{ kN/C}) \sin(15x - 4.5t - 80°)$$

and

$$E_3 = (17.0 \text{ kN/C}) \sin(15x - 4.5t + 160°)$$

20. The electric fields from three coherent sources are described by $E_1 = E_0 \sin \omega t$, $E_2 = E_0 \sin(\omega t + \phi)$, and $E_3 = E_0 \sin(\omega t + 2\phi)$. Let the resultant field be represented by $E_P = E_R \sin(\omega t + \alpha)$. Use phasors to find E_R and α when (a) $\phi = 20.0°$, (b) $\phi = 60.0°$, and (c) $\phi = 120°$. (d) Repeat when $\phi = (3\pi/2)$ rad.

WEB 21. Determine the resultant of the two waves $E_1 = 6.0 \sin(100 \pi t)$ and $E_2 = 8.0 \sin(100 \pi t + \pi/2)$.

22. Suppose that the slit openings in a Young's double-slit experiment have different sizes so that the electric fields and the intensities from each slit are different. If $E_1 = E_{01} \sin(\omega t)$ and $E_2 = E_{02} \sin(\omega t + \phi)$, show that the resultant electric field is $E = E_0 \sin(\omega t + \theta)$, where

$$E_0 = \sqrt{E_{01}^2 + E_{02}^2 + 2E_{01} E_{02} \cos \phi}$$

and

$$\sin \theta = \frac{E_{02} \sin \phi}{E_0}$$

23. Use phasors to find the resultant (magnitude and phase angle) of two fields represented by $E_1 = 12 \sin \omega t$ and $E_2 = 18 \sin(\omega t + 60°)$. (Note that in this case the amplitudes of the two fields are unequal.)

24. Two coherent waves are described by the expressions

$$E_1 = E_0 \sin\left(\frac{2\pi x_1}{\lambda} - 2\pi ft + \frac{\pi}{6}\right)$$

$$E_2 = E_0 \sin\left(\frac{2\pi x_2}{\lambda} - 2\pi ft + \frac{\pi}{8}\right)$$

Determine the relationship between x_1 and x_2 that produces constructive interference when the two waves are superposed.

25. When illuminated, four equally spaced parallel slits act as multiple coherent sources, each differing in phase from the adjacent one by an angle ϕ. Use a phasor diagram to determine the smallest value of ϕ for which the resultant of the four waves (assumed to be of equal amplitude) is zero.

26. Sketch a phasor diagram to illustrate the resultant of $E_1 = E_{01} \sin \omega t$ and $E_2 = E_{02} \sin(\omega t + \phi)$, where $E_{02} = 1.50E_{01}$ and $\pi/6 \le \phi \le \pi/3$. Use the sketch and the law of cosines to show that, for two coherent waves, the resultant intensity can be written in the form

$I_R = I_1 + I_2 + 2\sqrt{I_1 I_2} \cos \phi$.

27. Consider N coherent sources described by $E_1 = E_0 \sin(\omega t + \phi)$, $E_2 = E_0 \sin(\omega t + 2\phi)$, $E_3 = E_0 \sin(\omega t + 3\phi)$, . . . , $E_N = E_0 \sin(\omega t + N\phi)$. Find the minimum value of ϕ for which $E_R = E_1 + E_2 + E_3 + \ldots + E_N$ is zero.

Section 37.5   Change of Phase Due to Reflection

Section 37.6   Interference in Thin Films

28. A soap bubble ($n = 1.33$) is floating in air. If the thickness of the bubble wall is 115 nm, what is the wavelength of the light that is most strongly reflected?

29. An oil film ($n = 1.45$) floating on water is illuminated by white light at normal incidence. The film is 280 nm thick. Find (a) the dominant observed color in the reflected light and (b) the dominant color in the transmitted light. Explain your reasoning.

30. A thin film of oil ($n = 1.25$) is located on a smooth, wet pavement. When viewed perpendicular to the pavement, the film appears to be predominantly red (640 nm) and has no blue color (512 nm). How thick is the oil film?

31. A possible means for making an airplane invisible to radar is to coat the plane with an antireflective polymer. If radar waves have a wavelength of 3.00 cm and the index of refraction of the polymer is $n = 1.50$, how thick would you make the coating?

32. A material having an index of refraction of 1.30 is used to coat a piece of glass ($n = 1.50$). What should be the minimum thickness of this film if it is to minimize reflection of 500-nm light?

33. A film of MgF_2 ($n = 1.38$) having a thickness of 1.00×10^{-5} cm is used to coat a camera lens. Are any wavelengths in the visible spectrum intensified in the reflected light?

34. Astronomers observe the chromosphere of the Sun with a filter that passes the red hydrogen spectral line of wavelength 656.3 nm, called the H_α line. The filter consists of a transparent dielectric of thickness d held between two partially aluminized glass plates. The filter is held at a constant temperature. (a) Find the minimum value of d that produces maximum transmission of perpendicular H_α light, if the dielectric has an index of refraction of 1.378. (b) Assume that the temperature of the filter increases above its normal value and that its index of refraction does not change significantly. What happens to the transmitted wavelength? (c) The dielectric will also pass what near-visible wavelength? One of the glass plates is colored red to absorb this light.

35. A beam of 580-nm light passes through two closely spaced glass plates, as shown in Figure P37.35. For what minimum nonzero value of the plate separation d is the transmitted light bright?

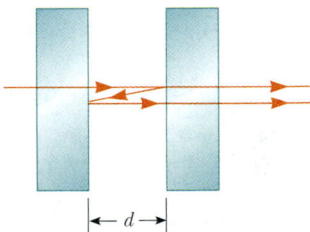

Figure P37.35

36. When a liquid is introduced into the air space between the lens and the plate in a Newton's-rings apparatus, the diameter of the tenth ring changes from 1.50 to 1.31 cm. Find the index of refraction of the liquid.

WEB 37. An air wedge is formed between two glass plates separated at one edge by a very fine wire, as shown in Figure P37.37. When the wedge is illuminated from above by 600-nm light, 30 dark fringes are observed. Calculate the radius of the wire.

Figure P37.37

38. Two rectangular flat glass plates ($n = 1.52$) are in contact along one end and separated along the other end by a sheet of paper 4.00×10^{-3} cm thick (see Fig. P37.37). The top plate is illuminated by monochromatic light ($\lambda = 546.1$ nm). Calculate the number of dark parallel bands crossing the top plate (include the dark band at zero thickness along the edge of contact between the two plates).

39. Two glass plates 10.0 cm long are in contact at one end and separated at the other end by a thread 0.050 0 mm in diameter. Light containing the two wavelengths 400 nm and 600 nm is incident perpendicularly. At what distance from the contact point is the next dark fringe?

(Optional)
Section 37.7   The Michelson Interferometer

40. Light of wavelength 550.5 nm is used to calibrate a Michelson interferometer, and mirror M_1 is moved 0.180 mm. How many dark fringes are counted?

41. Mirror M_1 in Figure 37.22 is displaced a distance ΔL. During this displacement, 250 fringe reversals (formation of successive dark or bright bands) are counted. The light being used has a wavelength of 632.8 nm. Calculate the displacement ΔL.

42. Monochromatic light is beamed into a Michelson interferometer. The movable mirror is displaced 0.382 mm; this causes the interferometer pattern to reproduce itself 1 700 times. Determine the wavelength and the color of the light.

43. One leg of a Michelson interferometer contains an evacuated cylinder 3.00 cm long having glass plates on each end. A gas is slowly leaked into the cylinder until a pressure of 1 atm is reached. If 35 bright fringes pass on the screen when light of wavelength 633 nm is used, what is the index of refraction of the gas?

44. One leg of a Michelson interferometer contains an evacuated cylinder of length L having glass plates on each end. A gas is slowly leaked into the cylinder until a pressure of 1 atm is reached. If N bright fringes pass on the screen when light of wavelength λ is used, what is the index of refraction of the gas?

ADDITIONAL PROBLEMS

45. One radio transmitter A operating at 60.0 MHz is 10.0 m from another similar transmitter B that is 180° out of phase with transmitter A. How far must an observer move from transmitter A toward transmitter B along the line connecting A and B to reach the nearest point where the two beams are in phase?

46. Raise your hand and hold it flat. Think of the space between your index finger and your middle finger as one slit, and think of the space between middle finger and ring finger as a second slit. (a) Consider the interference resulting from sending coherent visible light perpendicularly through this pair of openings. Compute an order-of-magnitude estimate for the angle between adja-

cent zones of constructive interference. (b) To make the angles in the interference pattern easy to measure with a plastic protractor, you should use an electromagnetic wave with frequency of what order of magnitude? How is this wave classified on the electromagnetic spectrum?

47. In a Young's double-slit experiment using light of wavelength λ, a thin piece of Plexiglas having index of refraction n covers one of the slits. If the center point on the screen is a dark spot instead of a bright spot, what is the minimum thickness of the Plexiglas?

48. Review Problem.   A flat piece of glass is held stationary and horizontal above the flat top end of a 10.0-cm-long vertical metal rod that has its lower end rigidly fixed. The thin film of air between the rod and glass is observed to be bright by reflected light when it is illuminated by light of wavelength 500 nm. As the temperature is slowly increased by 25.0°C, the film changes from bright to dark and back to bright 200 times. What is the coefficient of linear expansion of the metal?

49. A certain crude oil has an index of refraction of 1.25. A ship dumps 1.00 m³ of this oil into the ocean, and the oil spreads into a thin uniform slick. If the film produces a first-order maximum of light of wavelength 500 nm normally incident on it, how much surface area of the ocean does the oil slick cover? Assume that the index of refraction of the ocean water is 1.34.

50. Interference effects are produced at point P on a screen as a result of direct rays from a 500-nm source and reflected rays off the mirror, as shown in Figure P37.50. If the source is 100 m to the left of the screen and 1.00 cm above the mirror, find the distance y (in millimeters) to the first dark band above the mirror.

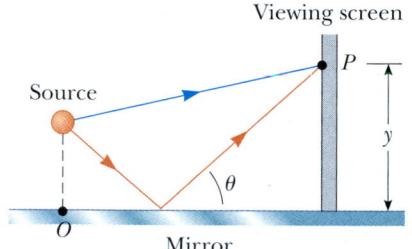

Figure P37.50

51. Astronomers observed a 60.0-MHz radio source both directly and by reflection from the sea. If the receiving dish is 20.0 m above sea level, what is the angle of the radio source above the horizon at first maximum?

52. The waves from a radio station can reach a home receiver by two paths. One is a straight-line path from transmitter to home, a distance of 30.0 km. The second path is by reflection from the ionosphere (a layer of ionized air molecules high in the atmosphere). Assume that this reflection takes place at a point midway between the receiver and the transmitter. The wavelength broadcast by the radio station is 350 m. Find the minimum height of the ionospheric layer that produces destructive interference between the direct and reflected beams. (Assume that no phase changes occur on reflection.)

53. Measurements are made of the intensity distribution in a Young's interference pattern (see Fig. 37.6). At a particular value of y, it is found that $I/I_{max} = 0.810$ when 600-nm light is used. What wavelength of light should be used if the relative intensity at the same location is to be reduced to 64.0%?

54. In a Young's interference experiment, the two slits are separated by 0.150 mm, and the incident light includes light of wavelengths $\lambda_1 = 540$ nm and $\lambda_2 = 450$ nm. The overlapping interference patterns are formed on a screen 1.40 m from the slits. Calculate the minimum distance from the center of the screen to the point where a bright line of the λ_1 light coincides with a bright line of the λ_2 light.

55. An air wedge is formed between two glass plates in contact along one edge and slightly separated at the opposite edge. When the plates are illuminated with monochromatic light from above, the reflected light has 85 dark fringes. Calculate the number of dark fringes that would appear if water ($n = 1.33$) were to replace the air between the plates.

56. Our discussion of the techniques for determining constructive and destructive interference by reflection from a thin film in air has been confined to rays striking the film at nearly normal incidence. Assume that a ray is incident at an angle of 30.0° (relative to the normal) on a film with an index of refraction of 1.38. Calculate the minimum thickness for constructive interference if the light is sodium light with a wavelength of 590 nm.

57. The condition for constructive interference by reflection from a thin film in air as developed in Section 37.6 assumes nearly normal incidence. Show that if the light is incident on the film at a nonzero angle ϕ_1 (relative to the normal), then the condition for constructive interference is $2nt \cos \theta_2 = (m + \frac{1}{2})\lambda$, where θ_2 is the angle of refraction.

58. (a) Both sides of a uniform film that has index of refraction n and thickness d are in contact with air. For normal incidence of light, an intensity minimum is observed in the reflected light at λ_2, and an intensity maximum is observed at λ_1, where $\lambda_1 > \lambda_2$. If no intensity minima are observed between λ_1 and λ_2, show that the integer m in Equations 37.16 and 37.17 is given by $m = \lambda_1/2(\lambda_1 - \lambda_2)$. (b) Determine the thickness of the film if $n = 1.40$, $\lambda_1 = 500$ nm, and $\lambda_2 = 370$ nm.

59. Figure P37.59 shows a radio wave transmitter and a receiver separated by a distance d and located a distance h above the ground. The receiver can receive signals both directly from the transmitter and indirectly from signals that reflect off the ground. Assume that the ground is level between the transmitter and receiver and that a 180° phase shift occurs upon reflection. Determine the longest wavelengths that interfere (a) constructively and (b) destructively.

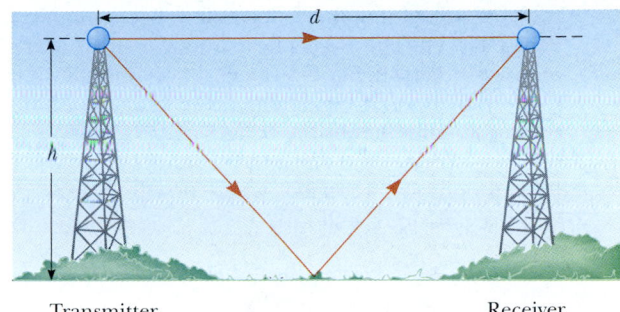

Transmitter   Receiver

Figure P37.59

60. Consider the double-slit arrangement shown in Figure P37.60, where the separation d is 0.300 mm and the distance L is 1.00 m. A sheet of transparent plastic ($n = 1.50$) 0.050 0 mm thick (about the thickness of this page) is placed over the upper slit. As a result, the central maximum of the interference pattern moves upward a distance y'. Find y'.

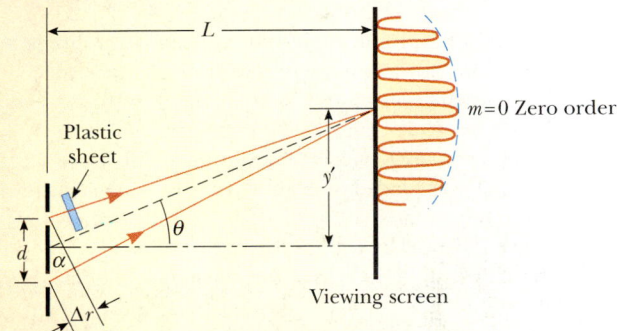

Figure P37.60   Problems 60 and 61.

61. Consider the double-slit arrangement shown in Figure P37.60, where the slit separation is d and the slit to screen distance is L. A sheet of transparent plastic having an index of refraction n and thickness t is placed over the upper slit. As a result, the central maximum of the interference pattern moves upward a distance y'. Find y'.

62. Waves broadcast by a 1 500-kHz radio station arrive at a home receiver by two paths. One is a direct path, and the other is from reflection off an airplane directly above the receiver. The airplane is approximately 100 m above the receiver, and the direct distance from station to home is 20.0 km. What is the precise height of the airplane if destructive interference is occurring? (Assume that no phase change occurs on reflection.)

63. In a Newton's-rings experiment, a plano-convex glass ($n = 1.52$) lens having a diameter of 10.0 cm is placed on a flat plate, as shown in Figure 37.18a. When 650-nm light is incident normally, 55 bright rings are observed, with the last ring right on the edge of the lens. (a) What is the radius of curvature of the convex surface of the lens? (b) What is the focal length of the lens?

64. A piece of transparent material having an index of refraction n is cut into the shape of a wedge, as shown in Figure P37.64. The angle of the wedge is small, and monochromatic light of wavelength λ is normally incident from above. If the height of the wedge is h and the width is ℓ, show that bright fringes occur at the positions $x = \lambda\ell(m + \frac{1}{2})/2hn$ and that dark fringes occur at the positions $x = \lambda\ell m/2hn$, where $m = 0, 1, 2, \ldots$ and x is measured as shown.

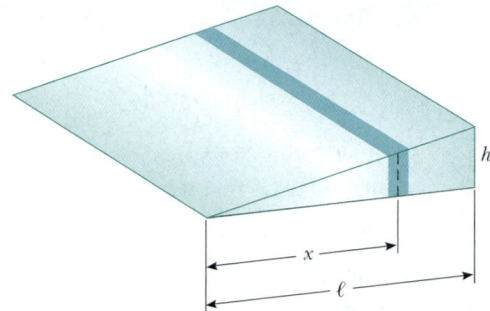

Figure P37.64

65. Use phasor addition to find the resultant amplitude and phase constant when the following three harmonic functions are combined: $E_1 = \sin(\omega t + \pi/6)$, $E_2 = 3.0 \sin(\omega t + 7\pi/2)$, $E_3 = 6.0 \sin(\omega t + 4\pi/3)$.

66. A plano-convex lens having a radius of curvature of $r = 4.00$ m is placed on a concave reflecting surface whose radius of curvature is $R = 12.0$ m, as shown in Figure P37.66. Determine the radius of the 100th bright ring if 500-nm light is incident normal to the flat surface of the lens.

67. A plano-convex lens has index of refraction n. The curved side of the lens has radius of curvature R and rests on a flat glass surface of the same index of refraction, with a film of index n_{film} between them. The lens is illuminated from above by light of wavelength λ. Show that the dark Newton's rings have radii given approximately by

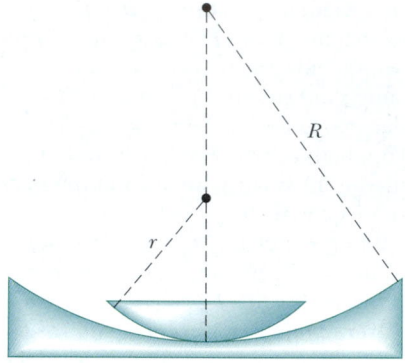

Figure P37.66

$$r \cong \sqrt{m\lambda R/n_{\text{film}}}$$

where m is an integer and r is much less than R.

68. A soap film ($n = 1.33$) is contained within a rectangular wire frame. The frame is held vertically so that the film drains downward and becomes thicker at the bottom than at the top, where the thickness is essentially zero. The film is viewed in white light with near-normal incidence, and the first violet ($\lambda = 420$ nm) interference band is observed 3.00 cm from the top edge of the film. (a) Locate the first red ($\lambda = 680$ nm) interference band. (b) Determine the film thickness at the positions of the violet and red bands. (c) What is the wedge angle of the film?

69. Interference fringes are produced using Lloyd's mirror and a 606-nm source, as shown in Figure 37.14. Fringes 1.20 mm apart are formed on a screen 2.00 m from the real source S. Find the vertical distance h of the source above the reflecting surface.

70. Slit 1 of a double slit is wider than slit 2, so that the light from slit 1 has an amplitude 3.00 times that of the light from slit 2. Show that Equation 37.11 is replaced by the equation $I = (4I_{\text{max}}/9)(1 + 3 \cos^2 \phi/2)$ for this situation.

ANSWERS TO QUICK QUIZZES

37.1 Bands of light along the orange lines interspersed with dark bands running along the dashed black lines.

37.2 At location B. At A, which is on a line of constructive interference, the water surface undulates so much that you probably could not determine the depth. Because B is on a line of destructive interference, the water level does not change, and you should be able to read the ruler easily.

37.3 The graph is shown in Figure QQA37.1. The width of the primary maxima is slightly narrower than the $N = 5$ primary width but wider than the $N = 10$ primary width. Because $N = 6$, the secondary maxima are $\frac{1}{36}$ as intense as the primary maxima.

37.4 The greater the variation in thickness, the narrower the bands of color (like the lines on a topographic map).

The widest bands are the gold ones along the left edge of the photograph and at the bottom right corner of the razor blade. Thus, the thickness of the oil film changes most slowly with position in these areas.

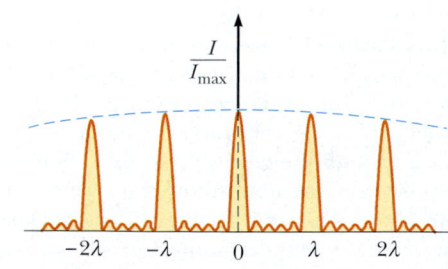

Figure QQA37.1

P U Z Z L E R

At sunset, the sky is ablaze with brilliant reds, pinks, and oranges. Yet, we wouldn't be able to see this sunset were it not for the fact that someone else is simultaneously seeing a blue sky. What causes the beautiful colors of a sunset, and why must the sky be blue somewhere else for us to enjoy one? (© W. A. Banaszewski/Visuals Unlimited)

c h a p t e r

38

Diffraction and Polarization

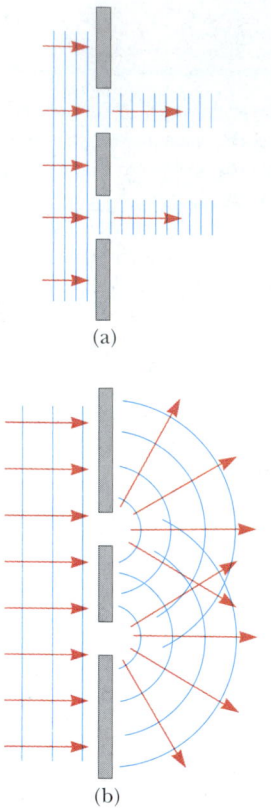

Figure 38.1 (a) If light waves did not spread out after passing through the slits, no interference would occur. (b) The light waves from the two slits overlap as they spread out, filling what we expect to be shadowed regions with light and producing interference fringes.

When light waves pass through a small aperture, an interference pattern is observed rather than a sharp spot of light. This behavior indicates that light, once it has passed through the aperture, spreads beyond the narrow path defined by the aperture into regions that would be in shadow if light traveled in straight lines. Other waves, such as sound waves and water waves, also have this property of spreading when passing through apertures or by sharp edges. This phenomenon, known as diffraction, can be described only with a wave model for light.

In Chapter 34, we learned that electromagnetic waves are transverse. That is, the electric and magnetic field vectors are perpendicular to the direction of wave propagation. In this chapter, we see that under certain conditions these transverse waves can be polarized in various ways.

38.1 INTRODUCTION TO DIFFRACTION

In Section 37.2 we learned that an interference pattern is observed on a viewing screen when two slits are illuminated by a single-wavelength light source. If the light traveled only in its original direction after passing through the slits, as shown in Figure 38.1a, the waves would not overlap and no interference pattern would be seen. Instead, Huygens's principle requires that the waves spread out from the slits as shown in Figure 38.1b. In other words, the light deviates from a straight-line path and enters the region that would otherwise be shadowed. As noted in Section 35.1, this divergence of light from its initial line of travel is called **diffraction.**

In general, diffraction occurs when waves pass through small openings, around obstacles, or past sharp edges, as shown in Figure 38.2. When an opaque object is placed between a point source of light and a screen, no sharp boundary exists on the screen between a shadowed region and an illuminated region. The illuminated region above the shadow of the object contains alternating light and dark fringes. Such a display is called a **diffraction pattern.**

Figure 38.3 shows a diffraction pattern associated with the shadow of a penny. A bright spot occurs at the center, and circular fringes extend outward from the shadow's edge. We can explain the central bright spot only by using the wave the-

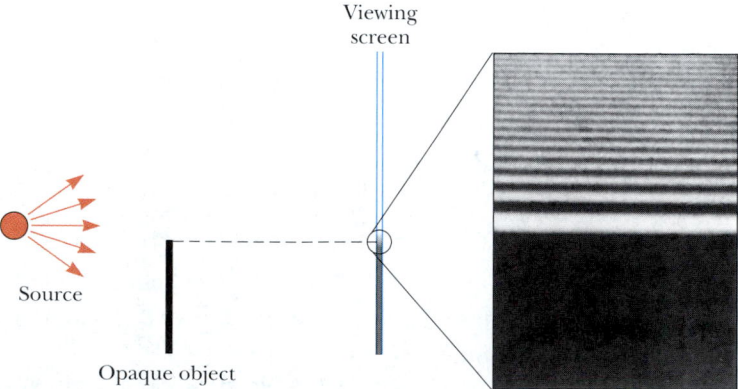

Figure 38.2 Light from a small source passes by the edge of an opaque object. We might expect no light to appear on the screen below the position of the edge of the object. In reality, light bends around the top edge of the object and enters this region. Because of these effects, a diffraction pattern consisting of bright and dark fringes appears in the region above the edge of the object.

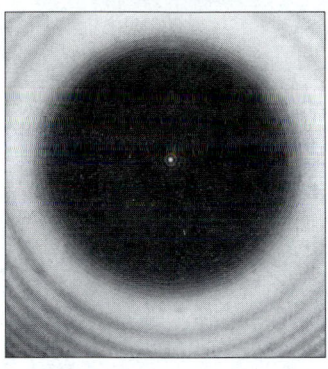

Figure 38.3 Diffraction pattern created by the illumination of a penny, with the penny positioned midway between screen and light source. *(P.M. Rinard, Am. J. Phys. 44:70, 1976.)*

ory of light, which predicts constructive interference at this point. From the viewpoint of geometric optics (in which light is viewed as rays traveling in straight lines), we expect the center of the shadow to be dark because that part of the viewing screen is completely shielded by the penny.

It is interesting to point out an historical incident that occurred shortly before the central bright spot was first observed. One of the supporters of geometric optics, Simeon Poisson, argued that if Augustin Fresnel's wave theory of light were valid, then a central bright spot should be observed in the shadow of a circular object illuminated by a point source of light. To Poisson's astonishment, the spot was observed by Dominique Arago shortly thereafter. Thus, Poisson's prediction reinforced the wave theory rather than disproving it.

In this chapter we restrict our attention to **Fraunhofer diffraction,** which occurs, for example, when all the rays passing through a narrow slit are approximately parallel to one another. This can be achieved experimentally either by placing the screen far from the opening used to create the diffraction or by using a converging lens to focus the rays once they pass through the opening, as shown in Figure 38.4a. A bright fringe is observed along the axis at $\theta = 0$, with alternating dark and bright fringes occurring on either side of the central bright one. Figure 38.4b is a photograph of a single-slit Fraunhofer diffraction pattern.

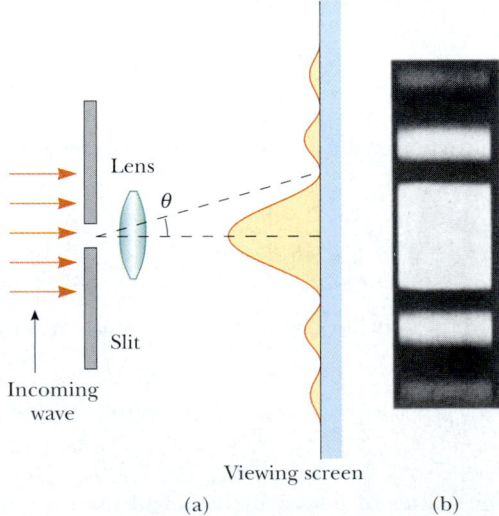

(a)   (b)

Figure 38.4 (a) Fraunhofer diffraction pattern of a single slit. The pattern consists of a central bright fringe flanked by much weaker maxima alternating with dark fringes (drawing not to scale). (b) Photograph of a single-slit Fraunhofer diffraction pattern. *(From M. Cagnet, M. Francon, and J. C. Thierr,* Atlas of Optical Phenomena, *Berlin, Springer-Verlag, 1962, Plate 18.)*

38.2 ▶ DIFFRACTION FROM NARROW SLITS

Until now, we have assumed that slits are point sources of light. In this section, we abandon that assumption and see how the finite width of slits is the basis for understanding Fraunhofer diffraction.

We can deduce some important features of this phenomenon by examining waves coming from various portions of the slit, as shown in Figure 38.5. According to Huygens's principle, **each portion of the slit acts as a source of light waves.** Hence, light from one portion of the slit can interfere with light from another portion, and the resultant light intensity on a viewing screen depends on the direction θ.

To analyze the diffraction pattern, it is convenient to divide the slit into two halves, as shown in Figure 38.5. Keeping in mind that all the waves are in phase as they leave the slit, consider rays 1 and 3. As these two rays travel toward a viewing screen far to the right of the figure, ray 1 travels farther than ray 3 by an amount equal to the path difference $(a/2) \sin \theta$, where a is the width of the slit. Similarly, the path difference between rays 2 and 4 is also $(a/2) \sin \theta$. If this path difference is exactly half a wavelength (corresponding to a phase difference of 180°), then the two waves cancel each other and destructive interference results. This is true for any two rays that originate at points separated by half the slit width because the phase difference between two such points is 180°. Therefore, waves from the upper half of the slit interfere destructively with waves from the lower half when

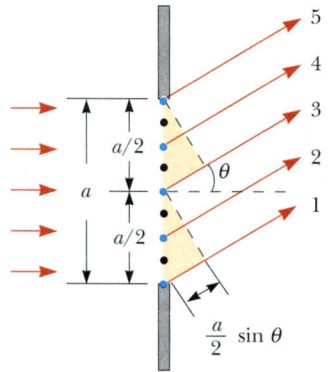

Figure 38.5 Diffraction of light by a narrow slit of width a. Each portion of the slit acts as a point source of light waves. The path difference between rays 1 and 3 or between rays 2 and 4 is $(a/2)\sin \theta$ (drawing not to scale).

$$\frac{a}{2} \sin \theta = \frac{\lambda}{2}$$

or when

$$\sin \theta = \frac{\lambda}{a}$$

If we divide the slit into four equal parts and use similar reasoning, we find that the viewing screen is also dark when

$$\sin \theta = \frac{2\lambda}{a}$$

Likewise, we can divide the slit into six equal parts and show that darkness occurs on the screen when

$$\sin \theta = \frac{3\lambda}{a}$$

Therefore, the general condition for destructive interference is

Condition for destructive interference

$$\boxed{\sin \theta = m\frac{\lambda}{a}} \qquad m = \pm 1, \pm 2, \pm 3, \ldots \qquad \textbf{(38.1)}$$

This equation gives the values of θ for which the diffraction pattern has zero light intensity—that is, when a dark fringe is formed. However, it tells us nothing about the variation in light intensity along the screen. The general features of the intensity distribution are shown in Figure 38.6. A broad central bright fringe is ob-

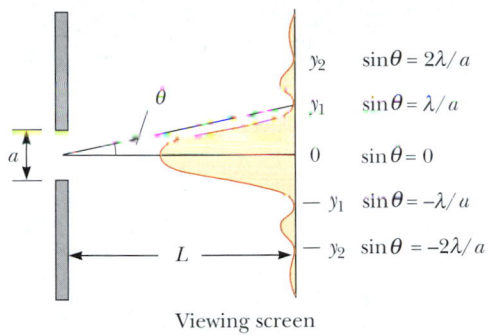

$y_2 \qquad \sin\theta = 2\lambda/a$

$y_1 \qquad \sin\theta = \lambda/a$

$0 \qquad \sin\theta = 0$

$-y_1 \qquad \sin\theta = -\lambda/a$

$-y_2 \qquad \sin\theta = -2\lambda/a$

Viewing screen

Figure 38.6 Intensity distribution for a Fraunhofer diffraction pattern from a single slit of width a. The positions of two minima on each side of the central maximum are labeled (drawing not to scale).

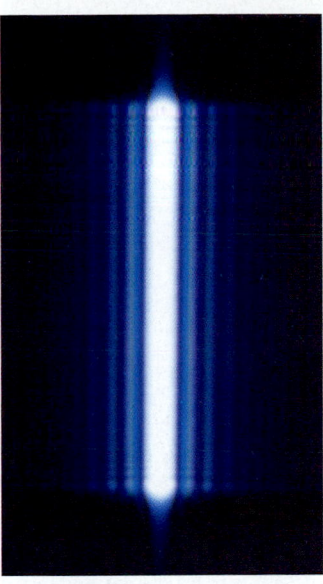

The diffraction pattern that appears on a screen when light passes through a narrow vertical slit. The pattern consists of a broad central bright fringe and a series of less intense and narrower side bright fringes.

served; this fringe is flanked by much weaker bright fringes alternating with dark fringes. The various dark fringes occur at the values of θ that satisfy Equation 38.1. Each bright-fringe peak lies approximately halfway between its bordering dark-fringe minima. Note that the central bright maximum is twice as wide as the secondary maxima.

Quick Quiz 38.1

If the door to an adjoining room is slightly ajar, why is it that you can hear sounds from the room but cannot see much of what is happening in the room?

EXAMPLE 38.1 Where Are the Dark Fringes?

Light of wavelength 580 nm is incident on a slit having a width of 0.300 mm. The viewing screen is 2.00 m from the slit. Find the positions of the first dark fringes and the width of the central bright fringe.

Solution The two dark fringes that flank the central bright fringe correspond to $m = \pm 1$ in Equation 38.1. Hence, we find that

$$\sin\theta = \pm\frac{\lambda}{a} = \pm\frac{5.80 \times 10^{-7} \text{ m}}{0.300 \times 10^{-3} \text{ m}} = \pm 1.93 \times 10^{-3}$$

From the triangle in Figure 38.6, note that $\tan\theta = y_1/L$. Because θ is very small, we can use the approximation $\sin\theta \approx \tan\theta$; thus, $\sin\theta \approx y_1/L$. Therefore, the positions of the first minima measured from the central axis are given by

$$y_1 \approx L\sin\theta = \pm L\frac{\lambda}{a} = \boxed{\pm 3.87 \times 10^{-3} \text{ m}}$$

The positive and negative signs correspond to the dark fringes on either side of the central bright fringe. Hence, the width of the central bright fringe is equal to $2|y_1| = 7.74 \times 10^{-3}$ m $= 7.74$ mm. Note that this value is much greater than the width of the slit. However, as the slit width is increased, the diffraction pattern narrows, corresponding to smaller values of θ. In fact, for large values of a, the various maxima and minima are so closely spaced that only a large central bright area resembling the geometric image of the slit is observed. This is of great importance in the design of lenses used in telescopes, microscopes, and other optical instruments.

Exercise Determine the width of the first-order ($m = 1$) bright fringe.

Answer 3.87 mm.

Intensity of Single-Slit Diffraction Patterns

We can use phasors to determine the light intensity distribution for a single-slit diffraction pattern. Imagine a slit divided into a large number of small zones, each of width Δy as shown in Figure 38.7. Each zone acts as a source of coherent radiation,

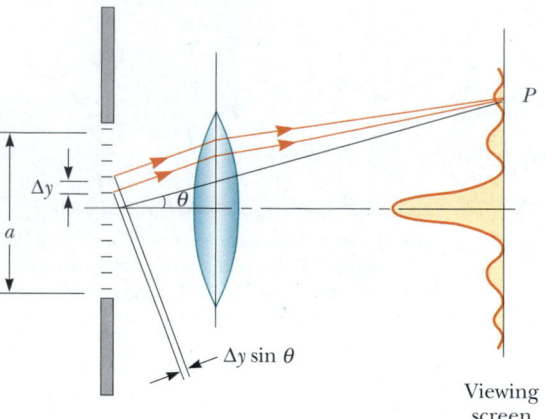

Figure 38.7 Fraunhofer diffraction by a single slit. The light intensity at point P is the resultant of all the incremental electric field magnitudes from zones of width Δy.

and each contributes an incremental electric field of magnitude ΔE at some point P on the screen. We obtain the total electric field magnitude E at point P by summing the contributions from all the zones. The light intensity at point P is proportional to the square of the magnitude of the electric field (see Section 37.3).

The incremental electric field magnitudes between adjacent zones are out of phase with one another by an amount $\Delta\beta$, where the phase difference $\Delta\beta$ is related to the path difference $\Delta y \sin\theta$ between adjacent zones by the expression

$$\Delta\beta = \frac{2\pi}{\lambda}\Delta y \sin\theta \qquad (38.2)$$

To find the magnitude of the total electric field on the screen at any angle θ, we sum the incremental magnitudes ΔE due to each zone. For small values of θ, we can assume that all the ΔE values are the same. It is convenient to use phasor diagrams for various angles, as shown in Figure 38.8. When $\theta = 0$, all phasors are aligned as shown in Figure 38.8a because all the waves from the various zones are in phase. In this case, the total electric field at the center of the screen is $E_0 = N\Delta E$, where N is the number of zones. The resultant magnitude E_R at some small angle θ is shown in Figure 38.8b, where each phasor differs in phase from an adjacent one by an amount $\Delta\beta$. In this case, E_R is the vector sum of the incremental

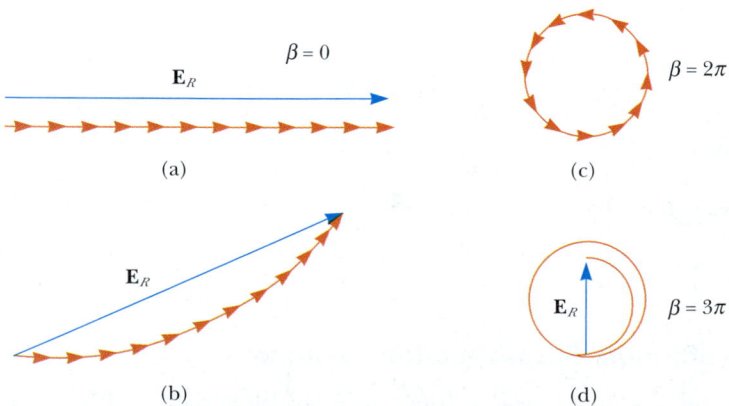

Figure 38.8 Phasor diagrams for obtaining the various maxima and minima of a single-slit diffraction pattern.

magnitudes and hence is given by the length of the chord. Therefore, $E_R < E_0$. The total phase difference β between waves from the top and bottom portions of the slit is

$$\beta = N\,\Delta\beta = \frac{2\pi}{\lambda}\,N\,\Delta y \sin\theta = \frac{2\pi}{\lambda}\,a\sin\theta \qquad \textbf{(38.3)}$$

where $a = N\,\Delta y$ is the width of the slit.

As θ increases, the chain of phasors eventually forms the closed path shown in Figure 38.8c. At this point, the vector sum is zero, and so $E_R = 0$, corresponding to the first minimum on the screen. Noting that $\beta = N\,\Delta\beta = 2\pi$ in this situation, we see from Equation 38.3 that

$$2\pi = \frac{2\pi}{\lambda}\,a\sin\theta$$

$$\sin\theta = \frac{\lambda}{a}$$

That is, the first minimum in the diffraction pattern occurs where $\sin\theta = \lambda/a$; this is in agreement with Equation 38.1.

At greater values of θ, the spiral chain of phasors tightens. For example, Figure 38.8d represents the situation corresponding to the second maximum, which occurs when $\beta = 360° + 180° = 540°$ (3π rad). The second minimum (two complete circles, not shown) corresponds to $\beta = 720°$ (4π rad), which satisfies the condition $\sin\theta = 2\lambda/a$.

We can obtain the total electric field magnitude E_R and light intensity I at any point P on the screen in Figure 38.7 by considering the limiting case in which Δy becomes infinitesimal (dy) and N approaches ∞. In this limit, the phasor chains in Figure 38.8 become the red curve of Figure 38.9. The arc length of the curve is E_0 because it is the sum of the magnitudes of the phasors (which is the total electric field magnitude at the center of the screen). From this figure, we see that at some angle θ, the resultant electric field magnitude E_R on the screen is equal to the chord length. From the triangle containing the angle $\beta/2$, we see that

$$\sin\frac{\beta}{2} = \frac{E_R/2}{R}$$

where R is the radius of curvature. But the arc length E_0 is equal to the product $R\beta$, where β is measured in radians. Combining this information with the previous expression gives

$$E_R = 2R\sin\frac{\beta}{2} = 2\left(\frac{E_0}{\beta}\right)\sin\frac{\beta}{2} = E_0\left[\frac{\sin(\beta/2)}{\beta/2}\right]$$

Because the resultant light intensity I at point P on the screen is proportional to the square of the magnitude E_R, we find that

$$I = I_{max}\left[\frac{\sin(\beta/2)}{\beta/2}\right]^2 \qquad \textbf{(38.4)}$$

where I_{max} is the intensity at $\theta = 0$ (the central maximum). Substituting the expression for β (Eq. 38.3) into Equation 38.4, we have

$$I = I_{max}\left[\frac{\sin(\pi a\sin\theta/\lambda)}{\pi a\sin\theta/\lambda}\right]^2 \qquad \textbf{(38.5)}$$

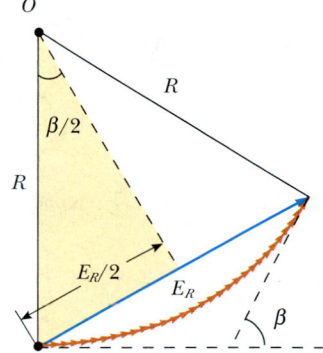

Figure 38.9 Phasor diagram for a large number of coherent sources. All the ends of the phasors lie on the circular red arc of radius R. The resultant electric field magnitude E_R equals the length of the chord.

Intensity of a single-slit Fraunhofer diffraction pattern

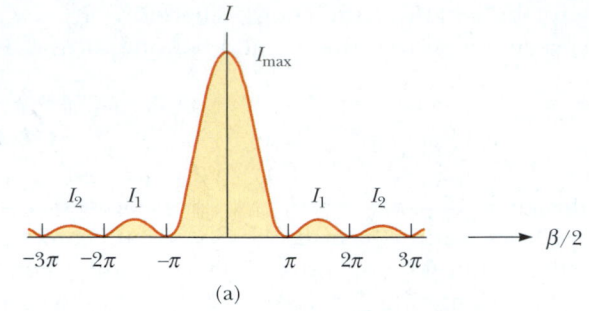

(a)

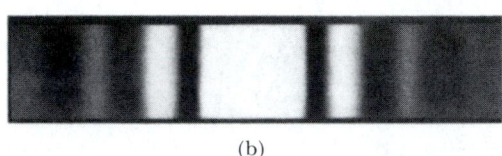

(b)

Figure 38.10   (a) A plot of light intensity I versus $\beta/2$ for the single-slit Fraunhofer diffraction pattern. (b) Photograph of a single-slit Fraunhofer diffraction pattern. *(From M. Cagnet, M. Francon, and J. C. Thierr, Atlas of Optical Phenomena, Berlin, Springer-Verlag, 1962, Plate 18.)*

From this result, we see that minima occur when

$$\frac{\pi a \sin \theta}{\lambda} = m\pi$$

or

Condition for intensity minima

$$\sin \theta = m\frac{\lambda}{a} \qquad m = \pm 1, \pm 2, \pm 3, \ldots$$

in agreement with Equation 38.1.

Figure 38.10a represents a plot of Equation 38.5, and Figure 38.10b is a photograph of a single-slit Fraunhofer diffraction pattern. Note that most of the light intensity is concentrated in the central bright fringe.

EXAMPLE 38.2 ▸ Relative Intensities of the Maxima

Find the ratio of the intensities of the secondary maxima to the intensity of the central maximum for the single-slit Fraunhofer diffraction pattern.

Solution   To a good approximation, the secondary maxima lie midway between the zero points. From Figure 38.10a, we see that this corresponds to $\beta/2$ values of $3\pi/2$, $5\pi/2$, $7\pi/2$, ... Substituting these values into Equation 38.4 gives for the first two ratios

$$\frac{I_1}{I_{max}} = \left[\frac{\sin(3\pi/2)}{(3\pi/2)}\right]^2 = \frac{1}{9\pi^2/4} = \boxed{0.045}$$

$$\frac{I_2}{I_{max}} = \left[\frac{\sin(5\pi/2)}{5\pi/2}\right]^2 = \frac{1}{25\pi^2/4} = \boxed{0.016}$$

That is, the first secondary maxima (the ones adjacent to the central maximum) have an intensity of 4.5% that of the central maximum, and the next secondary maxima have an intensity of 1.6% that of the central maximum.

Exercise   Determine the intensity, relative to the central maximum, of the secondary maxima corresponding to $m = \pm 3$.

Answer   0.008 3.

Intensity of Two-Slit Diffraction Patterns

When more than one slit is present, we must consider not only diffraction due to the individual slits but also the interference of the waves coming from different slits. You may have noticed the curved dashed line in Figure 37.13, which indicates a decrease in intensity of the interference maxima as θ increases. This decrease is due to diffraction. To determine the effects of both interference and diffraction, we simply combine Equation 37.12 and Equation 38.5:

$$I = I_{max} \cos^2\left(\frac{\pi d \sin\theta}{\lambda}\right)\left[\frac{\sin(\pi a \sin\theta/\lambda)}{\pi a \sin\theta/\lambda}\right]^2 \tag{38.6}$$

Although this formula looks complicated, it merely represents the diffraction pattern (the factor in brackets) acting as an "envelope" for a two-slit interference pattern (the cosine-squared factor), as shown in Figure 38.11.

Equation 37.2 indicates the conditions for interference maxima as $d \sin\theta = m\lambda$, where d is the distance between the two slits. Equation 38.1 specifies that the first diffraction minimum occurs when $a \sin\theta = \lambda$, where a is the slit width. Dividing Equation 37.2 by Equation 38.1 (with $m = 1$) allows us to determine which interference maximum coincides with the first diffraction minimum:

$$\frac{d \sin\theta}{a \sin\theta} = \frac{m\lambda}{\lambda}$$

$$\frac{d}{a} = m \tag{38.7}$$

In Figure 38.11, $d/a = 18\ \mu m/3.0\ \mu m = 6$. Thus, the sixth interference maximum (if we count the central maximum as $m = 0$) is aligned with the first diffraction minimum and cannot be seen.

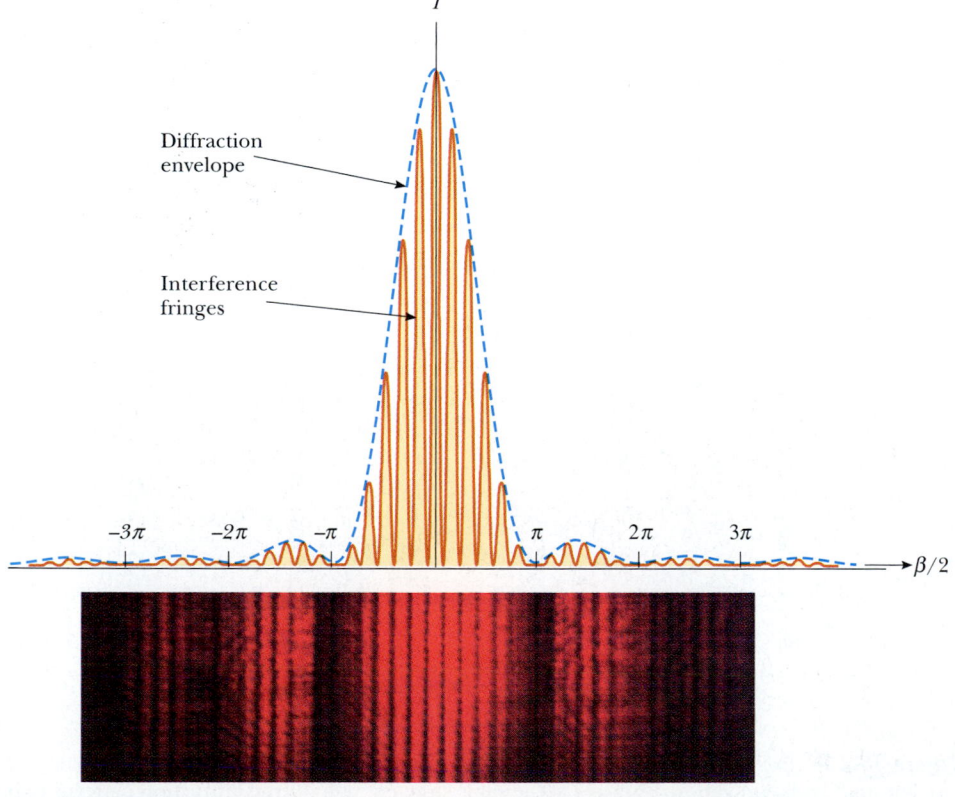

Figure 38.11 The combined effects of diffraction and interference. This is the pattern produced when 650-nm light waves pass through two 3.0-μm slits that are 18 μm apart. Notice how the diffraction pattern acts as an "envelope" and controls the intensity of the regularly spaced interference maxima. *(Photograph courtesy of Central Scientific Company)*

Quick Quiz 38.2

Using Figure 38.11 as a starting point, make a sketch of the combined diffraction and interference pattern for 650-nm light waves striking two 3.0-μm slits located 9.0 μm apart.

38.3 ▶ RESOLUTION OF SINGLE-SLIT AND CIRCULAR APERTURES

The ability of optical systems to distinguish between closely spaced objects is limited because of the wave nature of light. To understand this difficulty, let us consider Figure 38.12, which shows two light sources far from a narrow slit of width a. The sources can be considered as two noncoherent point sources S_1 and S_2—for example, they could be two distant stars. If no diffraction occurred, two distinct bright spots (or images) would be observed on the viewing screen. However, because of diffraction, each source is imaged as a bright central region flanked by weaker bright and dark fringes. What is observed on the screen is the sum of two diffraction patterns: one from S_1, and the other from S_2.

If the two sources are far enough apart to keep their central maxima from overlapping, as shown in Figure 38.12a, their images can be distinguished and are said to be *resolved*. If the sources are close together, however, as shown in Figure 38.12b, the two central maxima overlap, and the images are not resolved. In determining whether two images are resolved, the following condition is often used:

> When the central maximum of one image falls on the first minimum of the other image, the images are said to be just resolved. This limiting condition of resolution is known as **Rayleigh's criterion.**

Figure 38.13 shows diffraction patterns for three situations. When the objects are far apart, their images are well resolved (Fig. 38.13a). When the angular sepa-

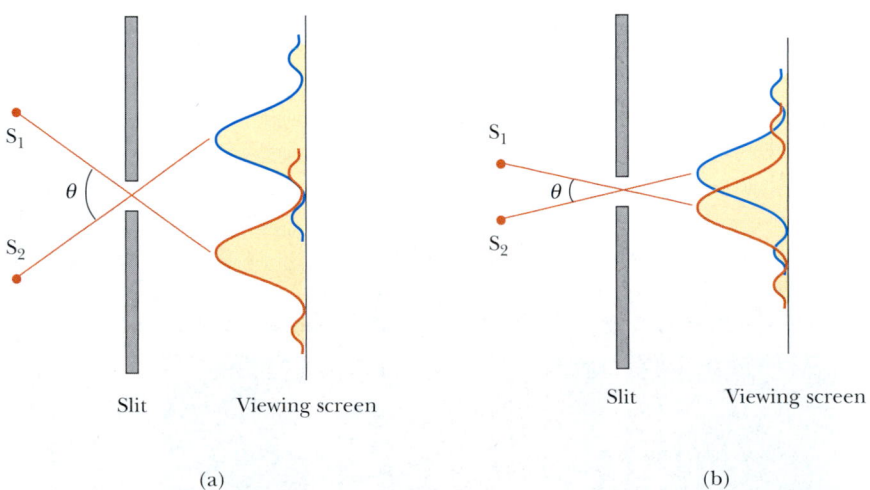

Slit Viewing screen	Slit Viewing screen
(a)	(b)

Figure 38.12 Two point sources far from a narrow slit each produce a diffraction pattern. (a) The angle subtended by the sources at the slit is large enough for the diffraction patterns to be distinguishable. (b) The angle subtended by the sources is so small that their diffraction patterns overlap, and the images are not well resolved. (Note that the angles are greatly exaggerated. The drawing is not to scale.)

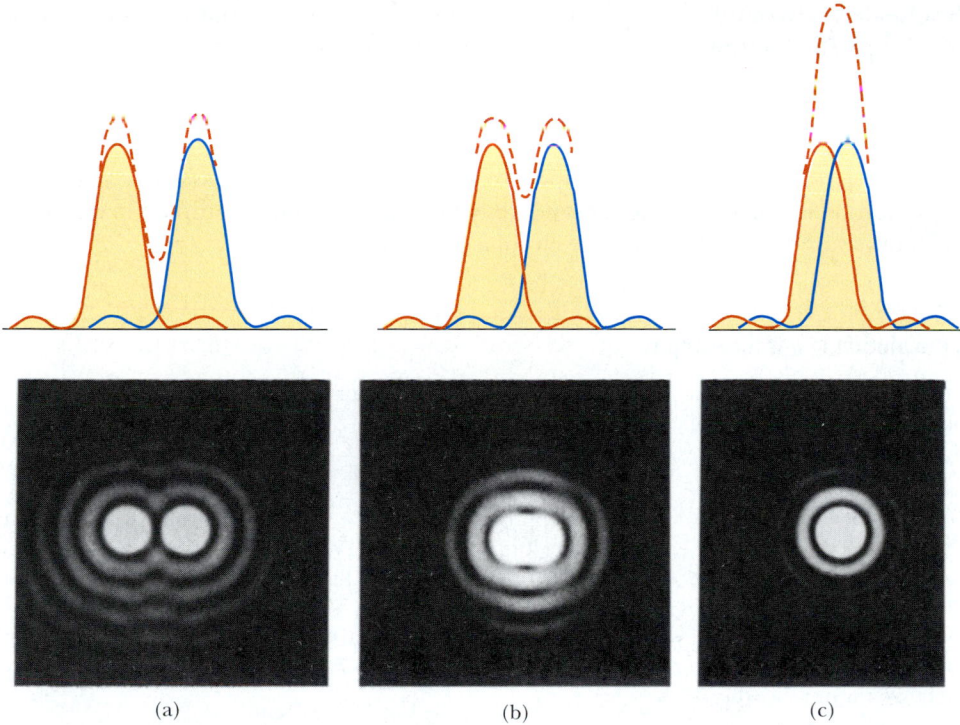

(a)   (b)   (c)

Figure 38.13 Individual diffraction patterns of two point sources (solid curves) and the resultant patterns (dashed curves) for various angular separations of the sources. In each case, the dashed curve is the sum of the two solid curves. (a) The sources are far apart, and the patterns are well resolved. (b) The sources are closer together such that the angular separation just satisfies Rayleigh's criterion, and the patterns are just resolved. (c) The sources are so close together that the patterns are not resolved. *(From M. Cagnet, M. Francon, and J. C. Thierr,* Atlas of Optical Phenomena, *Berlin, Springer-Verlag, 1962, Plate 16.)*

ration of the objects satisfies Rayleigh's criterion (Fig. 38.13b), the images are just resolved. Finally, when the objects are close together, the images are not resolved (Fig. 38.13c).

From Rayleigh's criterion, we can determine the minimum angular separation θ_{min} subtended by the sources at the slit for which the images are just resolved. Equation 38.1 indicates that the first minimum in a single-slit diffraction pattern occurs at the angle for which

$$\sin \theta = \frac{\lambda}{a}$$

where a is the width of the slit. According to Rayleigh's criterion, this expression gives the smallest angular separation for which the two images are resolved. Because $\lambda \ll a$ in most situations, $\sin \theta$ is small, and we can use the approximation $\sin \theta \approx \theta$. Therefore, the limiting angle of resolution for a slit of width a is

$$\theta_{min} = \frac{\lambda}{a} \qquad \textbf{(38.8)}$$

where θ_{min} is expressed in radians. Hence, the angle subtended by the two sources at the slit must be greater than λ / a if the images are to be resolved.

Many optical systems use circular apertures rather than slits. The diffraction pattern of a circular aperture, shown in Figure 38.14, consists of a central circular

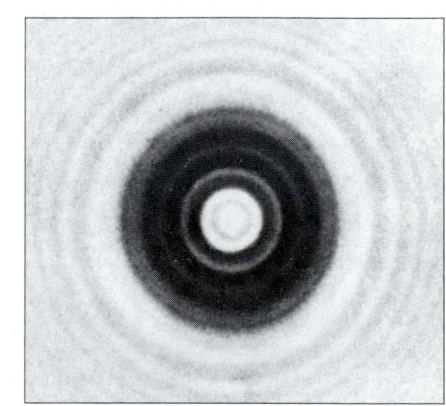

Figure 38.14 The diffraction pattern of a circular aperture consists of a central bright disk surrounded by concentric bright and dark rings. *(From M. Cagnet, M. Francon, and J. C. Thierr,* Atlas of Optical Phenomena, *Berlin, Springer-Verlag, 1962, Plate 34.)*

bright disk surrounded by progressively fainter bright and dark rings. Analysis shows that the limiting angle of resolution of the circular aperture is

$$\theta_{min} = 1.22 \frac{\lambda}{D} \qquad (38.9)$$

Limiting angle of resolution for a circular aperture

where D is the diameter of the aperture. Note that this expression is similar to Equation 38.8 except for the factor 1.22, which arises from a complex mathematical analysis of diffraction from the circular aperture.

EXAMPLE 38.3 ▶ Limiting Resolution of a Microscope

Light of wavelength 589 nm is used to view an object under a microscope. If the aperture of the objective has a diameter of 0.900 cm, (a) what is the limiting angle of resolution?

Solution (a) Using Equation 38.9, we find that the limiting angle of resolution is

$$\theta_{min} = 1.22 \left(\frac{589 \times 10^{-9} \text{ m}}{0.900 \times 10^{-2} \text{ m}} \right) = \boxed{7.98 \times 10^{-5} \text{ rad}}$$

This means that any two points on the object subtending an angle smaller than this at the objective cannot be distinguished in the image.

(b) If it were possible to use visible light of any wavelength, what would be the maximum limit of resolution for this microscope?

Solution To obtain the smallest limiting angle, we have to use the shortest wavelength available in the visible spectrum.

Violet light (400 nm) gives a limiting angle of resolution of

$$\theta_{min} = 1.22 \left(\frac{400 \times 10^{-9} \text{ m}}{0.900 \times 10^{-2} \text{ m}} \right) = \boxed{5.42 \times 10^{-5} \text{ rad}}$$

(c) Suppose that water ($n = 1.33$) fills the space between the object and the objective. What effect does this have on resolving power when 589-nm light is used?

Solution We find the wavelength of the 589-nm light in the water using Equation 35.7:

$$\lambda_{water} = \frac{\lambda_{air}}{n_{water}} = \frac{589 \text{ nm}}{1.33} = 443 \text{ nm}$$

The limiting angle of resolution at this wavelength is now smaller than that calculated in part (a):

$$\theta_{min} = 1.22 \left(\frac{443 \times 10^{-9} \text{ m}}{0.900 \times 10^{-2} \text{ m}} \right) = \boxed{6.00 \times 10^{-5} \text{ rad}}$$

EXAMPLE 38.4 ▶ Resolution of a Telescope

The Hale telescope at Mount Palomar has a diameter of 200 in. What is its limiting angle of resolution for 600-nm light?

Solution Because $D = 200$ in. $= 5.08$ m and $\lambda = 6.00 \times 10^{-7}$ m, Equation 38.9 gives

$$\theta_{min} = 1.22 \frac{\lambda}{D} = 1.22 \left(\frac{6.00 \times 10^{-7} \text{ m}}{5.08 \text{ m}} \right)$$

$$= \boxed{1.44 \times 10^{-7} \text{ rad} \approx 0.03 \text{ s of arc}}$$

Any two stars that subtend an angle greater than or equal to this value are resolved (if atmospheric conditions are ideal).

The Hale telescope can never reach its diffraction limit because the limiting angle of resolution is always set by at-

mospheric blurring. This seeing limit is usually about 1 s of arc and is never smaller than about 0.1 s of arc. (This is one of the reasons for the superiority of photographs from the Hubble Space Telescope, which views celestial objects from an orbital position above the atmosphere.)

Exercise The large radio telescope at Arecibo, Puerto Rico, has a diameter of 305 m and is designed to detect 0.75-m radio waves. Calculate the minimum angle of resolution for this telescope and compare your answer with that for the Hale telescope.

Answer 3.0×10^{-3} rad (10 min of arc), more than 10 000 times larger (that is, *worse*) than the Hale minimum.

EXAMPLE 38.5 ▶ Resolution of the Eye

Estimate the limiting angle of resolution for the human eye, assuming its resolution is limited only by diffraction.

Solution Let us choose a wavelength of 500 nm, near the center of the visible spectrum. Although pupil diameter varies from person to person, we estimate a diameter of 2 mm. We use Equation 38.9, taking $\lambda = 500$ nm and $D = 2$ mm:

$$\theta_{min} = 1.22\,\frac{\lambda}{D} = 1.22\left(\frac{5.00 \times 10^{-7}\ \text{m}}{2 \times 10^{-3}\ \text{m}}\right)$$

$$\approx 3 \times 10^{-4}\ \text{rad} \approx 1\ \text{min of arc}$$

We can use this result to determine the minimum separation distance d between two point sources that the eye can distinguish if they are a distance L from the observer (Fig. 38.15). Because θ_{min} is small, we see that

$$\sin\theta_{min} \approx \theta_{min} \approx \frac{d}{L}$$

$$d = L\theta_{min}$$

For example, if the point sources are 25 cm from the eye (the near point), then

$$d = (25\ \text{cm})(3 \times 10^{-4}\ \text{rad}) = 8 \times 10^{-3}\ \text{cm}$$

This is approximately equal to the thickness of a human hair.

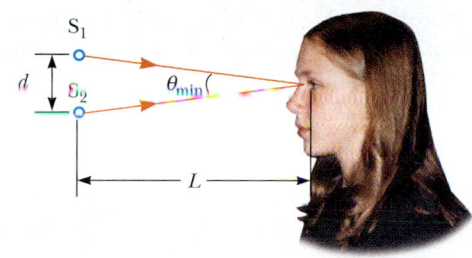

Figure 38.15 Two point sources separated by a distance d as observed by the eye.

Exercise Suppose that the pupil is dilated to a diameter of 5.0 mm and that two point sources 3.0 m away are being viewed. How far apart must the sources be if the eye is to resolve them?

Answer 0.037 cm.

APPLICATION ▶ Loudspeaker Design

The three-way speaker system shown in Figure 38.16 contains a woofer, a midrange speaker, and a tweeter. The small-diameter tweeter is for high frequencies, and the large-diameter woofer is for low frequencies. The midrange speaker, of intermediate diameter, is used for the frequency band above the high-frequency cutoff of the woofer and below the low-frequency cutoff of the tweeter. Circuits known as crossover networks include low-pass, midrange, and high-pass filters that direct the electrical signal to the appropriate speaker. The effective aperture size of a speaker is approximately its diameter. Because the wavelengths of sound waves are comparable to the typical sizes of the speakers, diffraction effects determine the angular radiation pattern. To be most useful, a speaker should radiate sound over a broad range of angles so that the listener does not have to stand at a particular spot in the room to hear maximum sound intensity. On the basis of the angular radiation pattern, let us investigate the frequency range for which a 6-in. (0.15-m) midrange speaker is most useful.

The speed of sound in air is 344 m/s, and for a circular aperture, diffraction effects become important when $\lambda = 1.22D$, where D is the speaker diameter. Therefore, we would expect this speaker to radiate non-uniformly for all frequencies above

$$\frac{344\ \text{m/s}}{1.22(0.15\ \text{m})} = 1\,900\ \text{Hz}$$

Suppose our design specifies that the midrange speaker operates between 500 Hz (the high-frequency woofer cutoff) and 2 000 Hz. Measurements of the dispersion of radiated

Figure 38.16 An audio speaker system for high-fidelity sound reproduction. The tweeter is at the top, the midrange speaker is in the middle, and the woofer is at the bottom. (*International Stock Photography*)

sound at a suitably great distance from the speaker yield the angular profiles of sound intensity shown in Figure 38.17. In examining these plots, we see that the dispersion pattern for a 500-Hz sound is fairly uniform. This angular range is suffi-

ciently great for us to say that this midrange speaker satisfies the design criterion. The intensity of a 2 000-Hz sound decreases to about half its maximum value about 30° from the centerline.

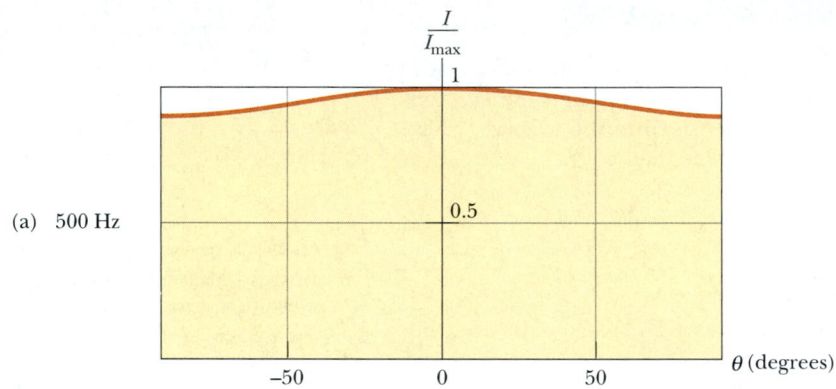

(a)   500 Hz

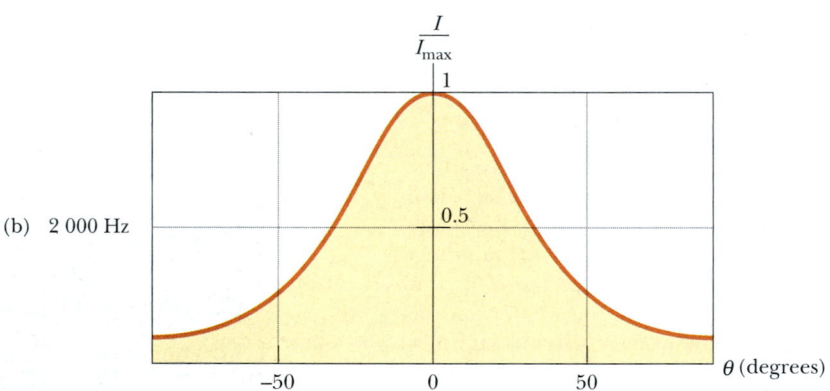

(b)   2 000 Hz

Figure 38.17   Angular dispersion of sound intensity *I* for a midrange speaker at (a) 500 Hz and (b) 2 000 Hz.

38.4 ▶ THE DIFFRACTION GRATING

The **diffraction grating,** a useful device for analyzing light sources, consists of a large number of equally spaced parallel slits. A *transmission grating* can be made by cutting parallel lines on a glass plate with a precision ruling machine. The spaces between the lines are transparent to the light and hence act as separate slits. A *reflection grating* can be made by cutting parallel lines on the surface of a reflective material. The reflection of light from the spaces between the lines is specular, and the reflection from the lines cut into the material is diffuse. Thus, the spaces between the lines act as parallel sources of reflected light, like the slits in a transmission grating. Gratings that have many lines very close to each other can have very small slit spacings. For example, a grating ruled with 5 000 lines/cm has a slit spacing $d = (1/5\ 000)$ cm $= 2.00 \times 10^{-4}$ cm.

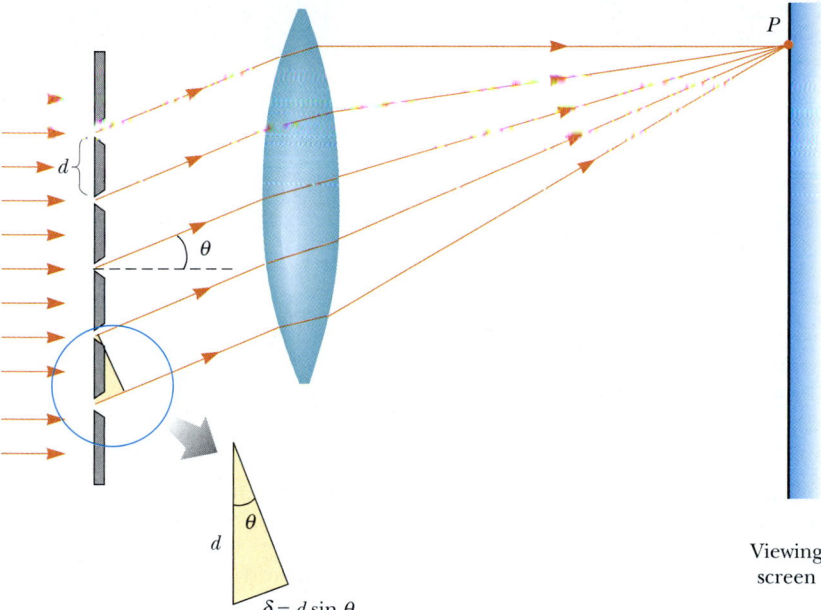

Figure 38.18   Side view of a diffraction grating. The slit separation is d, and the path difference between adjacent slits is $d \sin \theta$.

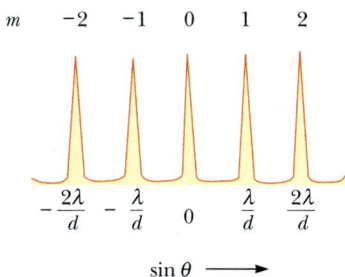

Figure 38.19   Intensity versus $\sin \theta$ for a diffraction grating. The zeroth-, first-, and second-order maxima are shown.

A section of a diffraction grating is illustrated in Figure 38.18. A plane wave is incident from the left, normal to the plane of the grating. A converging lens brings the rays together at point P. The pattern observed on the screen is the result of the combined effects of interference and diffraction. Each slit produces diffraction, and the diffracted beams interfere with one another to produce the final pattern.

The waves from all slits are in phase as they leave the slits. However, for some arbitrary direction θ measured from the horizontal, the waves must travel different path lengths before reaching point P. From Figure 38.18, note that the path difference δ between rays from any two adjacent slits is equal to $d \sin \theta$. If this path difference equals one wavelength or some integral multiple of a wavelength, then waves from all slits are in phase at point P and a bright fringe is observed. Therefore, the condition for maxima in the interference pattern at the angle θ is

$$d \sin \theta = m\lambda \qquad m = 0, 1, 2, 3, \ldots \qquad (38.10)$$

◀ Condition for interference maxima for a grating

We can use this expression to calculate the wavelength if we know the grating spacing and the angle θ. If the incident radiation contains several wavelengths, the mth-order maximum for each wavelength occurs at a specific angle. All wavelengths are seen at $\theta = 0$, corresponding to $m = 0$, the zeroth-order maximum. The first-order maximum ($m = 1$) is observed at an angle that satisfies the relationship $\sin \theta = \lambda/d$; the second-order maximum ($m = 2$) is observed at a larger angle θ, and so on.

The intensity distribution for a diffraction grating obtained with the use of a monochromatic source is shown in Figure 38.19. Note the sharpness of the principal maxima and the broadness of the dark areas. This is in contrast to the broad bright fringes characteristic of the two-slit interference pattern (see Fig. 37.6). Because the principal maxima are so sharp, they are very much brighter than two-slit

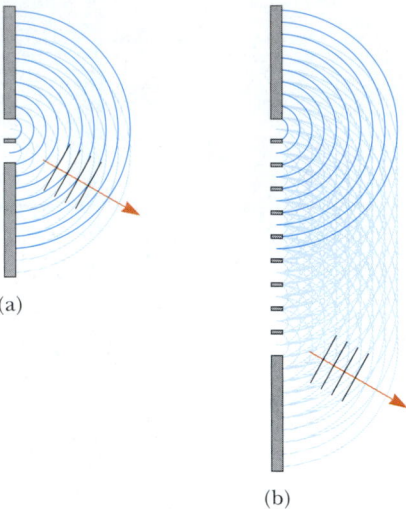

(a)

(b)

Figure 38.20   (a) Addition of two wave fronts from two slits. (b) Addition of ten wave fronts from ten slits. The resultant wave is much stronger in part (b) than in part (a).

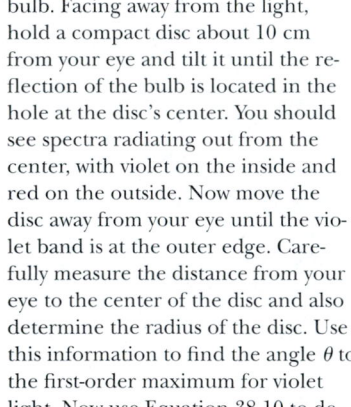

QuickLab

Stand a couple of meters from a light-bulb. Facing away from the light, hold a compact disc about 10 cm from your eye and tilt it until the reflection of the bulb is located in the hole at the disc's center. You should see spectra radiating out from the center, with violet on the inside and red on the outside. Now move the disc away from your eye until the violet band is at the outer edge. Carefully measure the distance from your eye to the center of the disc and also determine the radius of the disc. Use this information to find the angle θ to the first-order maximum for violet light. Now use Equation 38.10 to determine the spacing between the grooves on the disc. The industry standard is 1.6 μm. How close did you come?

interference maxima. The reason for this is illustrated in Figure 38.20, in which the combination of multiple wave fronts for a ten-slit grating is compared with the wave fronts for a two-slit system. Actual gratings have thousands of times more slits, and therefore the maxima are even stronger.

A schematic drawing of a simple apparatus used to measure angles in a diffraction pattern is shown in Figure 38.21. This apparatus is a diffraction grating spectrometer. The light to be analyzed passes through a slit, and a collimated beam of light is incident on the grating. The diffracted light leaves the grating at angles that satisfy Equation 38.10, and a telescope is used to view the image of the slit. The wavelength can be determined by measuring the precise angles at which the images of the slit appear for the various orders.

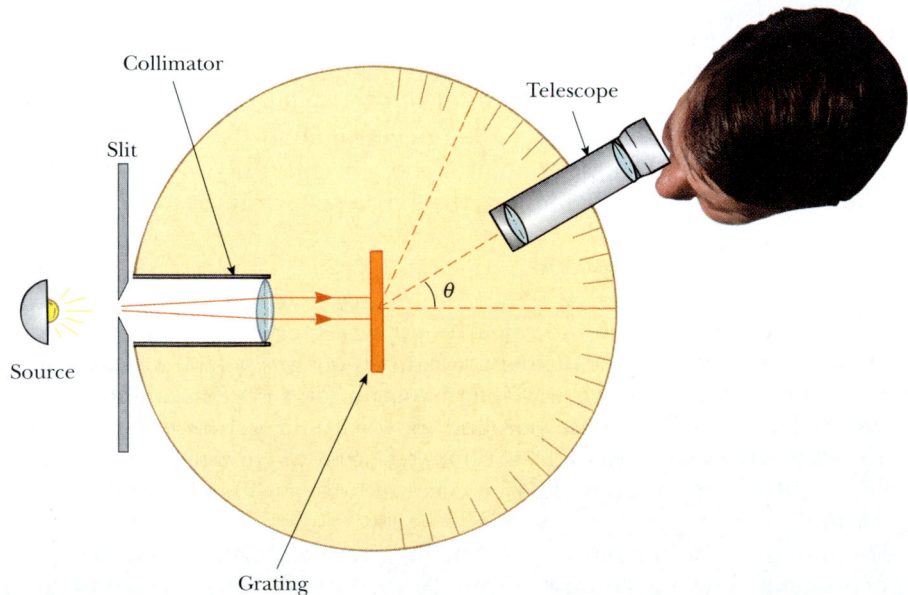

Figure 38.21   Diagram of a diffraction grating spectrometer. The collimated beam incident on the grating is diffracted into the various orders at the angles θ that satisfy the equation $d \sin \theta = m\lambda$, where $m = 0, 1, 2, \ldots$.

CONCEPTUAL EXAMPLE 38.6 ▶ A Compact Disc Is a Diffraction Grating

Light reflected from the surface of a compact disc is multi-colored, as shown in Figure 38.22. The colors and their intensities depend on the orientation of the disc relative to the eye and relative to the light source. Explain how this works.

Solution The surface of a compact disc has a spiral grooved track (with adjacent grooves having a separation on the order of 1 μm). Thus, the surface acts as a reflection grating. The light reflecting from the regions between these closely spaced grooves interferes constructively only in certain directions that depend on the wavelength and on the direction of the incident light. Any one section of the disc serves as a diffraction grating for white light, sending different colors in different directions. The different colors you see when viewing one section change as the light source, the disc, or you move to change the angles of incidence or diffraction.

Figure 38.22 A compact disc observed under white light. The colors observed in the reflected light and their intensities depend on the orientation of the disc relative to the eye and relative to the light source. *(Kristen Brochmann/Fundamental Photographs)*

EXAMPLE 38.7 ▶ The Orders of a Diffraction Grating

Monochromatic light from a helium-neon laser ($\lambda = 632.8$ nm) is incident normally on a diffraction grating containing 6 000 lines per centimeter. Find the angles at which the first-order, second-order, and third-order maxima are observed.

Solution First, we must calculate the slit separation, which is equal to the inverse of the number of lines per centimeter:

$$d = \frac{1}{6\,000} \text{ cm} = 1.667 \times 10^{-4} \text{ cm} = 1\,667 \text{ nm}$$

For the first-order maximum ($m = 1$), we obtain

$$\sin \theta_1 = \frac{\lambda}{d} = \frac{632.8 \text{ nm}}{1\,667 \text{ nm}} = 0.379\,6$$

$$\theta_1 = \boxed{22.31°}$$

For the second-order maximum ($m = 2$), we find

$$\sin \theta_2 = \frac{2\lambda}{d} = \frac{2(632.8 \text{ nm})}{1\,667 \text{ nm}} = 0.759\,2$$

$$\theta_2 = \boxed{49.39°}$$

For $m = 3$, we find that $\sin \theta_3 = 1.139$. Because $\sin \theta$ cannot exceed unity, this does not represent a realistic solution. Hence, only zeroth-, first-, and second-order maxima are observed for this situation.

Resolving Power of the Diffraction Grating

The diffraction grating is most useful for measuring wavelengths accurately. Like the prism, the diffraction grating can be used to disperse a spectrum into its wavelength components. Of the two devices, the grating is the more precise if one wants to distinguish two closely spaced wavelengths.

For two nearly equal wavelengths λ_1 and λ_2 between which a diffraction grating can just barely distinguish, the **resolving power** R of the grating is defined as

$$R = \frac{\lambda}{\lambda_2 - \lambda_1} = \frac{\lambda}{\Delta\lambda} \tag{38.11}$$

Resolving power

where $\lambda = (\lambda_1 + \lambda_2)/2$ and $\Delta\lambda = \lambda_2 - \lambda_1$. Thus, a grating that has a high resolving power can distinguish small differences in wavelength. If N lines of the grating are illuminated, it can be shown that the resolving power in the mth-order diffraction is

$$R = Nm \qquad (38.12)$$

Thus, resolving power increases with increasing order number and with increasing number of illuminated slits.

Note that $R = 0$ for $m = 0$; this signifies that all wavelengths are indistinguishable for the zeroth-order maximum. However, consider the second-order diffraction pattern ($m = 2$) of a grating that has 5 000 rulings illuminated by the light source. The resolving power of such a grating in second order is $R = 5\,000 \times 2 = 10\,000$. Therefore, for a mean wavelength of, for example, 600 nm, the minimum wavelength separation between two spectral lines that can be just resolved is $\Delta\lambda = \lambda/R = 6.00 \times 10^{-2}$ nm. For the third-order principal maximum, $R = 15\,000$ and $\Delta\lambda = 4.00 \times 10^{-2}$ nm, and so on.

One of the most interesting applications of diffraction is holography, which is used to create three-dimensional images found practically everywhere, from credit cards to postage stamps. The production of these special diffracting films is discussed in Chapter 42 of the extended version of this text.

EXAMPLE 38.8   Resolving Sodium Spectral Lines

When an element is raised to a very high temperature, the atoms emit radiation having discrete wavelengths. The set of wavelengths for a given element is called its *atomic spectrum*. Two strong components in the atomic spectrum of sodium have wavelengths of 589.00 nm and 589.59 nm. (a) What must be the resolving power of a grating if these wavelengths are to be distinguished?

Solution

$$R = \frac{\lambda}{\Delta\lambda} = \frac{589.30 \text{ nm}}{589.59 \text{ nm} - 589.00 \text{ nm}} = \frac{589.30}{0.59} = \boxed{999}$$

(b) To resolve these lines in the second-order spectrum, how many lines of the grating must be illuminated?

Solution   From Equation 38.12 and the results to part (a), we find that

$$N = \frac{R}{m} = \frac{999}{2} = \boxed{500 \text{ lines}}$$

Optional Section

38.5   DIFFRACTION OF X-RAYS BY CRYSTALS

In principle, the wavelength of any electromagnetic wave can be determined if a grating of the proper spacing (of the order of λ) is available. X-rays, discovered by Wilhelm Roentgen (1845–1923) in 1895, are electromagnetic waves of very short wavelength (of the order of 0.1 nm). It would be impossible to construct a grating having such a small spacing by the cutting process described at the beginning of Section 38.4. However, the atomic spacing in a solid is known to be about 0.1 nm. In 1913, Max von Laue (1879–1960) suggested that the regular array of atoms in a crystal could act as a three-dimensional diffraction grating for x-rays. Subsequent experiments confirmed this prediction. The diffraction patterns are complex because of the three-dimensional nature of the crystal. Nevertheless, x-ray diffraction has proved to be an invaluable technique for elucidating crystalline structures and for understanding the structure of matter.[1]

Figure 38.23 is one experimental arrangement for observing x-ray diffraction from a crystal. A collimated beam of x-rays is incident on a crystal. The diffracted beams are very intense in certain directions, corresponding to constructive interference from waves reflected from layers of atoms in the crystal. The diffracted beams can be detected by a photographic film, and they form an array of spots known as a *Laue pattern*. One can deduce the crystalline structure by analyzing the positions and intensities of the various spots in the pattern.

The arrangement of atoms in a crystal of sodium chloride (NaCl) is shown in Figure 38.24. Each unit cell (the geometric solid that repeats throughout the crystal) is a cube having an edge length a. A careful examination of the NaCl structure shows that the ions lie in discrete planes (the shaded areas in Fig. 38.24). Now suppose that an incident x-ray beam makes an angle θ with one of the planes, as shown in Figure 38.25. The beam can be reflected from both the upper plane and the lower one. However, the beam reflected from the lower plane travels farther than the beam reflected from the upper plane. The effective path difference is $2d \sin \theta$. The two beams reinforce each other (constructive interference) when this path difference equals some integer multiple of λ. The same is true for reflection from the entire family of parallel planes. Hence, the condition for constructive interference (maxima in the reflected beam) is

$$2d \sin \theta = m\lambda \qquad m = 1, 2, 3, \ldots \qquad \textbf{(38.13)}$$

This condition is known as **Bragg's law,** after W. L. Bragg (1890–1971), who first derived the relationship. If the wavelength and diffraction angle are measured, Equation 38.13 can be used to calculate the spacing between atomic planes.

Quick Quiz 38.3

When you receive a chest x-ray at a hospital, the rays pass through a series of parallel ribs in your chest. Do the ribs act as a diffraction grating for x-rays?

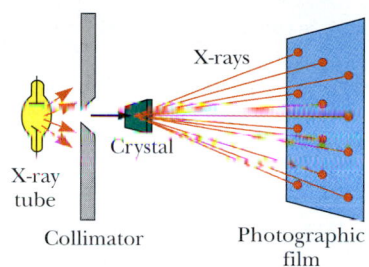

Figure 38.23 Schematic diagram of the technique used to observe the diffraction of x-rays by a crystal. The array of spots formed on the film is called a Laue pattern.

Bragg's law

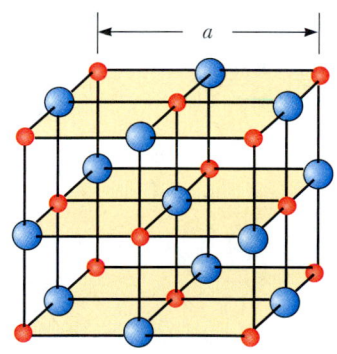

Figure 38.24 Crystalline structure of sodium chloride (NaCl). The blue spheres represent Cl^- ions, and the red spheres represent Na^+ ions. The length of the cube edge is $a = 0.562\ 737$ nm.

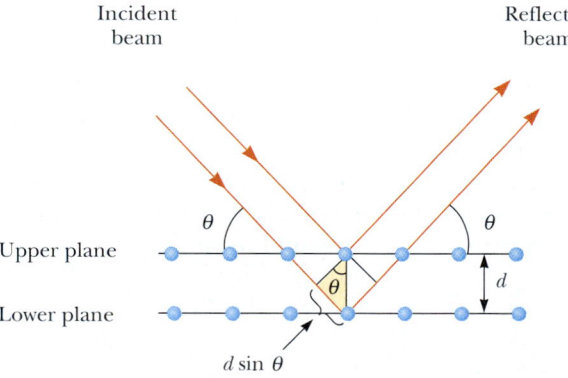

Incident beam — Reflected beam

Upper plane

Lower plane

$d \sin \theta$

Figure 38.25 A two-dimensional description of the reflection of an x-ray beam from two parallel crystalline planes separated by a distance d. The beam reflected from the lower plane travels farther than the one reflected from the upper plane by a distance $2d \sin \theta$.

[1] For more details on this subject, see Sir Lawrence Bragg, "X-Ray Crystallography," *Sci. Am.* 219:58–70, 1968.

38.6 ▶ POLARIZATION OF LIGHT WAVES

In Chapter 34 we described the transverse nature of light and all other electromagnetic waves. Polarization is firm evidence of this transverse nature.

An ordinary beam of light consists of a large number of waves emitted by the atoms of the light source. Each atom produces a wave having some particular orientation of the electric field vector **E**, corresponding to the direction of atomic vibration. The *direction of polarization* of each individual wave is defined to be the direction in which the electric field is vibrating. In Figure 38.26, this direction happens to lie along the *y* axis. However, an individual electromagnetic wave could have its **E** vector in the *yz* plane, making any possible angle with the *y* axis. Because all directions of vibration from a wave source are possible, the resultant electromagnetic wave is a superposition of waves vibrating in many different directions. The result is an **unpolarized** light beam, represented in Figure 38.27a. The direction of wave propagation in this figure is perpendicular to the page. The arrows show a few possible directions of the electric field vectors for the individual waves making up the resultant beam. At any given point and at some instant of time, all these individual electric field vectors add to give one resultant electric field vector.

As noted in Section 34.2, a wave is said to be **linearly polarized** if the resultant electric field **E** vibrates in the same direction *at all times* at a particular point, as shown in Figure 38.27b. (Sometimes, such a wave is described as *plane-polarized*, or simply *polarized*.) The plane formed by **E** and the direction of propagation is called the *plane of polarization* of the wave. If the wave in Figure 38.26 represented the resultant of all individual waves, the plane of polarization is the *xy* plane.

It is possible to obtain a linearly polarized beam from an unpolarized beam by removing all waves from the beam except those whose electric field vectors oscillate in a single plane. We now discuss four processes for producing polarized light from unpolarized light.

Polarization by Selective Absorption

The most common technique for producing polarized light is to use a material that transmits waves whose electric fields vibrate in a plane parallel to a certain direction and that absorbs waves whose electric fields vibrate in all other directions.

In 1938, E. H. Land (1909–1991) discovered a material, which he called *polaroid*, that polarizes light through selective absorption by oriented molecules. This material is fabricated in thin sheets of long-chain hydrocarbons. The sheets are stretched during manufacture so that the long-chain molecules align. After a sheet is dipped into a solution containing iodine, the molecules become good electrical conductors. However, conduction takes place primarily along the hydrocarbon chains because electrons can move easily only along the chains. As a result, the

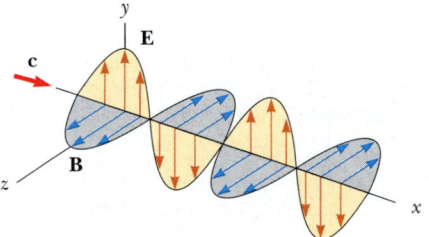

Figure 38.26 Schematic diagram of an electromagnetic wave propagating at velocity **c** in the *x* direction. The electric field vibrates in the *xy* plane, and the magnetic field vibrates in the *xz* plane.

molecules readily absorb light whose electric field vector is parallel to their length and allow light through whose electric field vector is perpendicular to their length.

It is common to refer to the direction perpendicular to the molecular chains as the *transmission axis*. In an ideal polarizer, all light with **E** parallel to the transmission axis is transmitted, and all light with **E** perpendicular to the transmission axis is absorbed.

Figure 38.28 represents an unpolarized light beam incident on a first polarizing sheet, called the *polarizer*. Because the transmission axis is oriented vertically in the figure, the light transmitted through this sheet is polarized vertically. A second polarizing sheet, called the *analyzer*, intercepts the beam. In Figure 38.28, the analyzer transmission axis is set at an angle θ to the polarizer axis. We call the electric field vector of the transmitted beam $\mathbf{E}_0$. The component of $\mathbf{E}_0$ perpendicular to the analyzer axis is completely absorbed. The component of $\mathbf{E}_0$ parallel to the analyzer axis, which is allowed through by the analyzer, is $E_0 \cos \theta$. Because the intensity of the transmitted beam varies as the square of its magnitude, we conclude that the intensity of the (polarized) beam transmitted through the analyzer varies as

$$I = I_{max} \cos^2 \theta \qquad (38.14)$$

where I_{max} is the intensity of the polarized beam incident on the analyzer. This expression, known as **Malus's law,**[2] applies to any two polarizing materials whose transmission axes are at an angle θ to each other. From this expression, note that the intensity of the transmitted beam is maximum when the transmission axes are parallel ($\theta = 0$ or $180°$) and that it is zero (complete absorption by the analyzer) when the transmission axes are perpendicular to each other. This variation in transmitted intensity through a pair of polarizing sheets is illustrated in Figure 38.29. Because the average value of $\cos^2 \theta$ is $\frac{1}{2}$, the intensity of the light passed through an ideal polarizer is one-half the intensity of unpolarized light.

Polarization by Reflection

When an unpolarized light beam is reflected from a surface, the reflected light may be completely polarized, partially polarized, or unpolarized, depending on the angle of incidence. If the angle of incidence is $0°$, the reflected beam is unpolarized. For other angles of incidence, the reflected light is polarized to some ex-

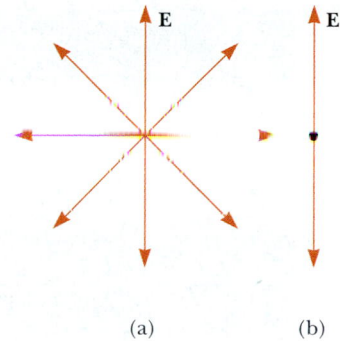

Figure 38.27 (a) An unpolarized light beam viewed along the direction of propagation (perpendicular to the page). The transverse electric field can vibrate in any direction in the plane of the page with equal probability. (b) A linearly polarized light beam with the electric field vibrating in the vertical direction.

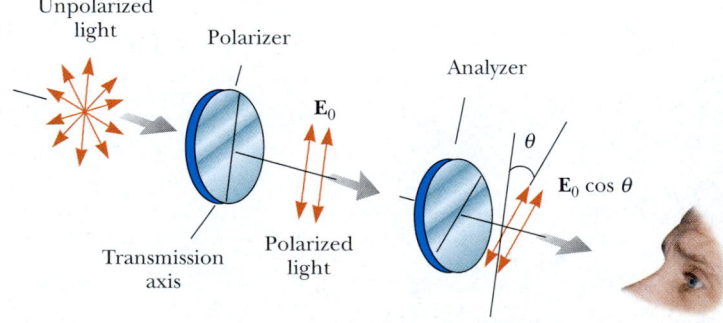

Figure 38.28 Two polarizing sheets whose transmission axes make an angle θ with each other. Only a fraction of the polarized light incident on the analyzer is transmitted through it.

[2] Named after its discoverer, E. L. Malus (1775–1812). Malus discovered that reflected light was polarized by viewing it through a calcite ($CaCO_3$) crystal.

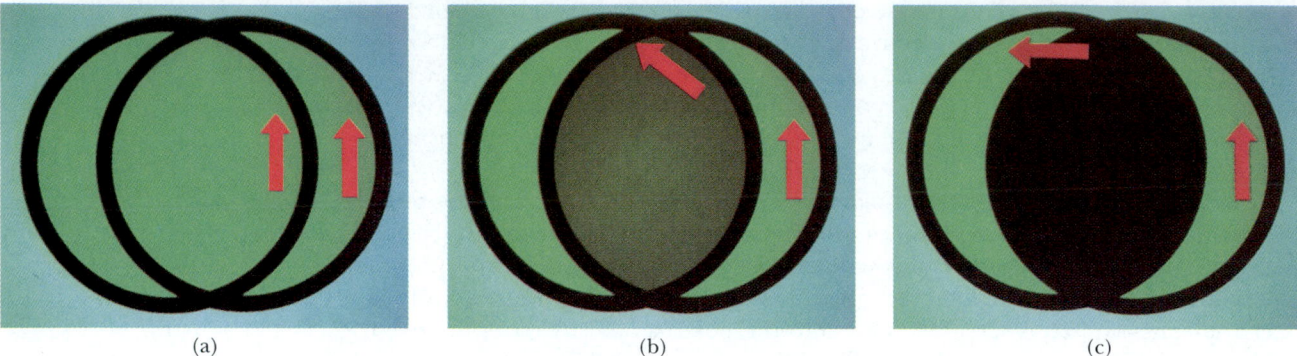

(a)   (b)   (c)

Figure 38.29 The intensity of light transmitted through two polarizers depends on the relative orientation of their transmission axes. (a) The transmitted light has maximum intensity when the transmission axes are aligned with each other. (b) The transmitted light has lesser intensity when the transmission axes are at an angle of 45° with each other. (c) The transmitted light intensity is a minimum when the transmission axes are at right angles to each other. *(Henry Leap and Jim Lehman)*

tent, and for one particular angle of incidence, the reflected light is completely polarized. Let us now investigate reflection at that special angle.

Suppose that an unpolarized light beam is incident on a surface, as shown in Figure 38.30a. Each individual electric field vector can be resolved into two components: one parallel to the surface (and perpendicular to the page in Fig. 38.30, represented by the dots), and the other (represented by the red arrows) perpendicular both to the first component and to the direction of propagation. Thus, the polarization of the entire beam can be described by two electric field components in these directions. It is found that the parallel component reflects more strongly than the perpendicular component, and this results in a partially polarized reflected beam. Furthermore, the refracted beam is also partially polarized.

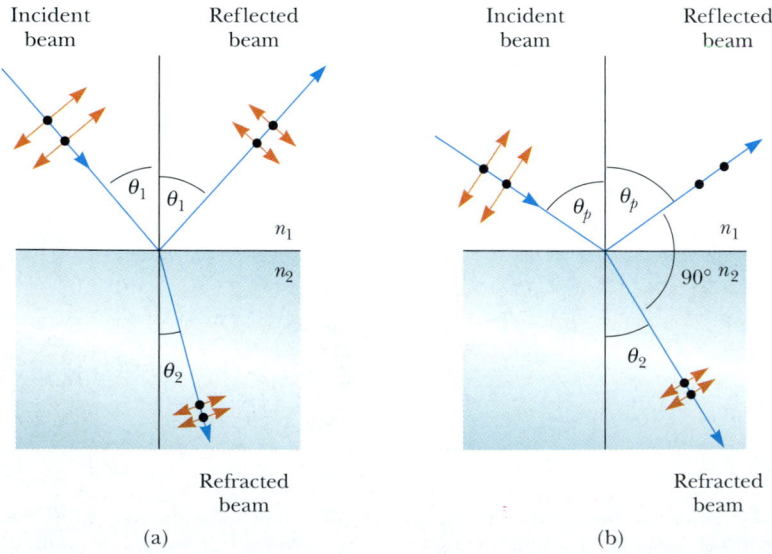

(a)   (b)

Figure 38.30 (a) When unpolarized light is incident on a reflecting surface, the reflected and refracted beams are partially polarized. (b) The reflected beam is completely polarized when the angle of incidence equals the polarizing angle θ_p, which satisfies the equation $n = \tan \theta_p$.

Now suppose that the angle of incidence θ_1 is varied until the angle between the reflected and refracted beams is 90°, as shown in Figure 38.30b. At this particular angle of incidence, the reflected beam is completely polarized (with its electric field vector parallel to the surface), and the refracted beam is still only partially polarized. The angle of incidence at which this polarization occurs is called the **polarizing angle** θ_p.

We can obtain an expression relating the polarizing angle to the index of refraction of the reflecting substance by using Figure 38.30b. From this figure, we see that $\theta_p + 90° + \theta_2 = 180°$; thus, $\theta_2 = 90° - \theta_p$. Using Snell's law of refraction (Eq. 35.8) and taking $n_1 = 1.00$ for air and $n_2 = n$, we have

$$n = \frac{\sin \theta_1}{\sin \theta_2} = \frac{\sin \theta_p}{\sin \theta_2}$$

Because $\sin \theta_2 = \sin(90° - \theta_p) = \cos \theta_p$, we can write this expression for n as $n = \sin \theta_p / \cos \theta_p$, which means that

$$n = \tan \theta_p \tag{38.15}$$

This expression is called **Brewster's law,** and the polarizing angle θ_p is sometimes called **Brewster's angle,** after its discoverer, David Brewster (1781–1868). Because n varies with wavelength for a given substance, Brewster's angle is also a function of wavelength.

Polarization by reflection is a common phenomenon. Sunlight reflected from water, glass, and snow is partially polarized. If the surface is horizontal, the electric field vector of the reflected light has a strong horizontal component. Sunglasses made of polarizing material reduce the glare of reflected light. The transmission axes of the lenses are oriented vertically so that they absorb the strong horizontal component of the reflected light. If you rotate sunglasses 90°, they will not be as effective at blocking the glare from shiny horizontal surfaces.

QuickLab

Devise a way to use a protractor, desklamp, and polarizing sunglasses to measure Brewster's angle for the glass in a window. From this, determine the index of refraction of the glass. Compare your results with the values given in Table 35.1.

Polarization by Double Refraction

Solids can be classified on the basis of internal structure. Those in which the atoms are arranged in a specific order are called *crystalline;* the NaCl structure of Figure 38.24 is just one example of a crystalline solid. Those solids in which the atoms are distributed randomly are called *amorphous.* When light travels through an amorphous material, such as glass, it travels with a speed that is the same in all directions. That is, glass has a single index of refraction. In certain crystalline materials, however, such as calcite and quartz, the speed of light is not the same in all directions. Such materials are characterized by two indices of refraction. Hence, they are often referred to as **double-refracting** or **birefringent** materials.

Upon entering a calcite crystal, unpolarized light splits into two plane-polarized rays that travel with different velocities, corresponding to two angles of refraction, as shown in Figure 38.31. The two rays are polarized in two mutually perpendicular directions, as indicated by the dots and arrows. One ray, called the **ordinary (O) ray,** is characterized by an index of refraction n_O that is the same in all directions. This means that if one could place a point source of light inside the crystal, as shown in Figure 38.32, the ordinary waves would spread out from the source as spheres.

The second plane-polarized ray, called the **extraordinary (E) ray,** travels with different speeds in different directions and hence is characterized by an index of refraction n_E that varies with the direction of propagation. The point source in Fig-

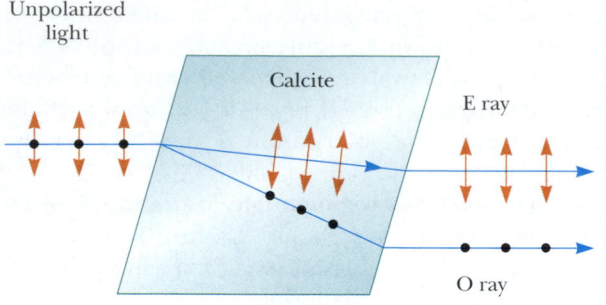

Figure 38.31 Unpolarized light incident on a calcite crystal splits into an ordinary (O) ray and an extraordinary (E) ray. These two rays are polarized in mutually perpendicular directions (drawing not to scale).

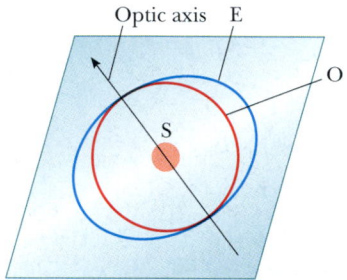

Figure 38.32 A point source S inside a double-refracting crystal produces a spherical wave front corresponding to the ordinary ray and an elliptical wave front corresponding to the extraordinary ray. The two waves propagate with the same velocity along the optic axis.

ure 38.32 sends out an extraordinary wave having wave fronts that are elliptical in cross-section. Note from Figure 38.32 that there is one direction, called the **optic axis,** along which the ordinary and extraordinary rays have the same speed, corresponding to the direction for which $n_O = n_E$. The difference in speed for the two rays is a maximum in the direction perpendicular to the optic axis. For example, in calcite, $n_O = 1.658$ at a wavelength of 589.3 nm, and n_E varies from 1.658 along the optic axis to 1.486 perpendicular to the optic axis. Values for n_O and n_E for various double-refracting crystals are given in Table 38.1.

If we place a piece of calcite on a sheet of paper and then look through the crystal at any writing on the paper, we see two images, as shown in Figure 38.33. As can be seen from Figure 38.31, these two images correspond to one formed by the ordinary ray and one formed by the extraordinary ray. If the two images are viewed through a sheet of rotating polarizing glass, they alternately appear and disappear because the ordinary and extraordinary rays are plane-polarized along mutually perpendicular directions.

Polarization by Scattering

When light is incident on any material, the electrons in the material can absorb and reradiate part of the light. Such absorption and reradiation of light by electrons in the gas molecules that make up air is what causes sunlight reaching an observer on the Earth to be partially polarized. You can observe this effect—called **scattering**—by looking directly up at the sky through a pair of sunglasses whose lenses are made of polarizing material. Less light passes through at certain orientations of the lenses than at others.

Figure 38.34 illustrates how sunlight becomes polarized when it is scattered. An unpolarized beam of sunlight traveling in the horizontal direction (parallel to

Figure 38.33 A calcite crystal produces a double image because it is a birefringent (double-refracting) material. *(Henry Leap and Jim Lehman)*

TABLE 38.1	Indices of Refraction for Some Double-Refracting Crystals at a Wavelength of 589.3 nm		
Crystal	n_O	n_E	n_O/n_E
Calcite ($CaCO_3$)	1.658	1.486	1.116
Quartz (SiO_2)	1.544	1.553	0.994
Sodium nitrate ($NaNO_3$)	1.587	1.336	1.188
Sodium sulfite ($NaSO_3$)	1.565	1.515	1.033
Zinc chloride ($ZnCl_2$)	1.687	1.713	0.985
Zinc sulfide (ZnS)	2.356	2.378	0.991

the ground) strikes a molecule of one of the gases that make up air, setting the electrons of the molecule into vibration. These vibrating charges act like the vibrating charges in an antenna. The horizontal component of the electric field vector in the incident wave results in a horizontal component of the vibration of the charges, and the vertical component of the vector results in a vertical component of vibration. If the observer in Figure 38.34 is looking straight up (perpendicular to the original direction of propagation of the light), the vertical oscillations of the charges send no radiation toward the observer. Thus, the observer sees light that is completely polarized in the horizontal direction, as indicated by the red arrows. If the observer looks in other directions, the light is partially polarized in the horizontal direction.

Some phenomena involving the scattering of light in the atmosphere can be understood as follows. When light of various wavelengths λ is incident on gas molecules of diameter d, where $d \ll \lambda$, the relative intensity of the scattered light varies as $1/\lambda^4$. The condition $d \ll \lambda$ is satisfied for scattering from oxygen (O_2) and nitrogen (N_2) molecules in the atmosphere, whose diameters are about 0.2 nm. Hence, short wavelengths (blue light) are scattered more efficiently than long wavelengths (red light). Therefore, when sunlight is scattered by gas molecules in the air, the short-wavelength radiation (blue) is scattered more intensely than the long-wavelength radiation (red).

When you look up into the sky in a direction that is not toward the Sun, you see the scattered light, which is predominantly blue; hence, you see a blue sky. If you look toward the west at sunset (or toward the east at sunrise), you are looking in a direction toward the Sun and are seeing light that has passed through a large distance of air. Most of the blue light has been scattered by the air between you and the Sun. The light that survives this trip through the air to you has had much of its blue component scattered and is thus heavily weighted toward the red end of the spectrum; as a result, you see the red and orange colors of sunset. However, a blue sky is seen by someone to your west for whom it is still a quarter hour before sunset.

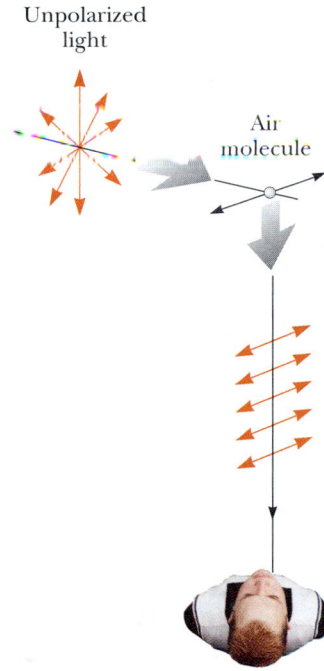

Figure 38.34 The scattering of unpolarized sunlight by air molecules. The scattered light traveling perpendicular to the incident light is plane-polarized because the vertical vibrations of the charges in the air molecule send no light in this direction.

Optical Activity

Many important applications of polarized light involve materials that display **optical activity.** A material is said to be optically active if it rotates the plane of polarization of any light transmitted through the material. The angle through which the light is rotated by a specific material depends on the length of the path through the material and on concentration if the material is in solution. One optically active material is a solution of the common sugar dextrose. A standard method for determining the concentration of sugar solutions is to measure the rotation produced by a fixed length of the solution.

Molecular asymmetry determines whether a material is optically active. For example, some proteins are optically active because of their spiral shape. Other materials, such as glass and plastic, become optically active when stressed. Suppose that an unstressed piece of plastic is placed between a polarizer and an analyzer so that light passes from polarizer to plastic to analyzer. When the plastic is unstressed and the analyzer axis is perpendicular to the polarizer axis, none of the polarized light passes through the analyzer. In other words, the unstressed plastic has no effect on the light passing through it. If the plastic is stressed, however, the regions of greatest stress rotate the polarized light through the largest angles. Hence, a series of bright and dark bands is observed in the transmitted light, with the bright bands corresponding to regions of greatest stress.

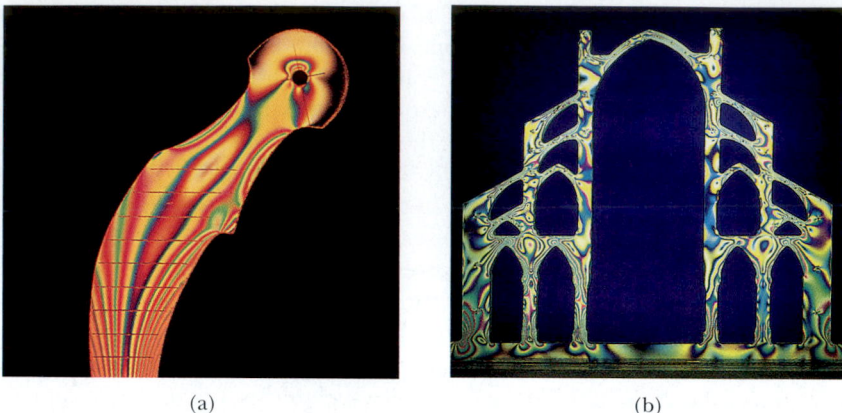

(a)   (b)

Figure 38.35   (a) Strain distribution in a plastic model of a hip replacement used in a medical research laboratory. The pattern is produced when the plastic model is viewed between a polarizer and analyzer oriented perpendicular to each other. *(Sepp Seitz 1981)*   (b) A plastic model of an arch structure under load conditions observed between perpendicular polarizers. Such patterns are useful in the optimum design of architectural components. *(Peter Aprahamian/Science Photo Library)*

Engineers often use this technique, called *optical stress analysis,* in designing structures ranging from bridges to small tools. They build a plastic model and analyze it under different load conditions to determine regions of potential weakness and failure under stress. Some examples of a plastic model under stress are shown in Figure 38.35.

The liquid crystal displays found in most calculators have their optical activity changed by the application of electric potential across different parts of the display. Try using a pair of polarizing sunglasses to investigate the polarization used in the display of your calculator.

SUMMARY

Diffraction is the deviation of light from a straight-line path when the light passes through an aperture or around an obstacle.

The **Fraunhofer diffraction pattern** produced by a single slit of width a on a distant screen consists of a central bright fringe and alternating bright and dark fringes of much lower intensities. The angles θ at which the diffraction pattern has zero intensity, corresponding to destructive interference, are given by

$$\sin \theta = m \frac{\lambda}{a} \qquad m = \pm 1, \pm 2, \pm 3, \ldots \tag{38.1}$$

How the intensity I of a single-slit diffraction pattern varies with angle θ is given by the expression

$$I = I_{max}\left[\frac{\sin (\beta/2)}{\beta/2} \right]^2 \tag{38.4}$$

where $\beta = (2\pi a \sin \theta)/\lambda$ and I_{max} is the intensity at $\theta = 0$.

Rayleigh's criterion, which is a limiting condition of resolution, states that two images formed by an aperture are just distinguishable if the central maximum of the diffraction pattern for one image falls on the first minimum of the diffraction pattern for the other image. The limiting angle of resolution for a slit of width a is $\theta_{min} = \lambda/a$, and the limiting angle of resolution for a circular aperture of diameter D is $\theta_{min} = 1.22\lambda/D$.

A **diffraction grating** consists of a large number of equally spaced, identical slits. The condition for intensity maxima in the interference pattern of a diffraction grating for normal incidence is

$$d \sin\theta = m\lambda \qquad m = 0, 1, 2, 3, \ldots \qquad \textbf{(38.10)}$$

where d is the spacing between adjacent slits and m is the order number of the diffraction pattern. The resolving power of a diffraction grating in the mth order of the diffraction pattern is

$$R = Nm \qquad \textbf{(38.12)}$$

where N is the number of lines in the grating that are illuminated.

When polarized light of intensity I_0 is emitted by a polarizer and then incident on an analyzer, the light transmitted through the analyzer has an intensity equal to $I_{max} \cos^2\theta$, where θ is the angle between the polarizer and analyzer transmission axes.

In general, reflected light is partially polarized. However, reflected light is completely polarized when the angle of incidence is such that the angle between the reflected and refracted beams is 90°. This angle of incidence, called the **polarizing angle** θ_p, satisfies **Brewster's law:**

$$n = \tan\theta_p \qquad \textbf{(38.15)}$$

where n is the index of refraction of the reflecting medium.

QUESTIONS

1. Why can you hear around corners but not see around them?

2. Observe the shadow of your book when it is held a few inches above a table while illuminated by a lamp several feet above it. Why is the shadow somewhat fuzzy at the edges?

3. Knowing that radio waves travel at the speed of light and that a typical AM radio frequency is 1 000 kHz while an FM radio frequency might be 100 MHz, estimate the wavelengths of typical AM and FM radio signals. Use this information to explain why FM radio stations often fade out when you drive through a short tunnel or underpass but AM radio stations do not.

4. Describe the change in width of the central maximum of the single-slit diffraction pattern as the width of the slit is made narrower.

5. Assuming that the headlights of a car are point sources, estimate the maximum observer-to-car distance at which the headlights are distinguishable from each other.

6. A laser beam is incident at a shallow angle on a machinist's ruler that has a finely calibrated scale. The engraved rulings on the scale give rise to a diffraction pattern on a screen. Discuss how you can use this arrangement to obtain a measure of the wavelength of the laser light.

7. Certain sunglasses use a polarizing material to reduce the intensity of light reflected from shiny surfaces. What orientation of polarization should the material have to be most effective?

8. During the "day" on the Moon (that is, when the Sun is visible), you see a black sky and the stars are clearly visible. During the day on the Earth, you see a blue sky and no stars. Account for this difference.

9. You can make the path of a light beam visible by placing dust in the air (perhaps by shaking a blackboard eraser in the path of the light beam). Explain why you can see the beam under these circumstances.

10. Is light from the sky polarized? Why is it that clouds seen through Polaroid glasses stand out in bold contrast to the sky?

11. If a coin is glued to a glass sheet and the arrangement is held in front of a laser beam, the projected shadow has diffraction rings around its edge and a bright spot in the center. How is this possible?

12. If a fine wire is stretched across the path of a laser beam, is it possible to produce a diffraction pattern?

13. How could the index of refraction of a flat piece of dark obsidian glass be determined?

PROBLEMS

1, 2, 3 = straightforward, intermediate, challenging   ☐ = full solution available in the *Student Solutions Manual and Study Guide*
WEB = solution posted at **http://www.saunderscollege.com/physics/**   🖥 = Computer useful in solving problem   🧩 = Interactive Physics
☐ = paired numerical/symbolic problems

Section 38.1   Introduction to Diffraction

Section 38.2   Diffraction from Narrow Slits

1. Helium-neon laser light ($\lambda = 632.8$ nm) is sent through a 0.300-mm-wide single slit. What is the width of the central maximum on a screen 1.00 m from the slit?

2. A beam of green light is diffracted by a slit with a width of 0.550 mm. The diffraction pattern forms on a wall 2.06 m beyond the slit. The distance between the positions of zero intensity on both sides of the central bright fringe is 4.10 mm. Calculate the wavelength of the laser light.

WEB 3. A screen is placed 50.0 cm from a single slit, which is illuminated with 690-nm light. If the distance between the first and third minima in the diffraction pattern is 3.00 mm, what is the width of the slit?

4. Coherent microwaves of wavelength 5.00 cm enter a long, narrow window in a building otherwise essentially opaque to the microwaves. If the window is 36.0 cm wide, what is the distance from the central maximum to the first-order minimum along a wall 6.50 m from the window?

5. Sound with a frequency of 650 Hz from a distant source passes through a doorway 1.10 m wide in a sound-absorbing wall. Find the number and approximate directions of the diffraction-maximum beams radiated into the space beyond.

6. Light with a wavelength of 587.5 nm illuminates a single slit 0.750 mm in width. (a) At what distance from the slit should a screen be located if the first minimum in the diffraction pattern is to be 0.850 mm from the center of the screen? (b) What is the width of the central maximum?

7. A diffraction pattern is formed on a screen 120 cm away from a 0.400-mm-wide slit. Monochromatic 546.1-nm light is used. Calculate the fractional intensity I/I_0 at a point on the screen 4.10 mm from the center of the principal maximum.

8. The second-order bright fringe in a single-slit diffraction pattern is 1.40 mm from the center of the central maximum. The screen is 80.0 cm from a slit of width 0.800 mm. Assuming that the incident light is monochromatic, calculate the light's approximate wavelength.

9. If the light in Figure 38.5 strikes the single slit at an angle β from the perpendicular direction, show that Equation 38.1, the condition for destructive interference, must be modified to read

$$\sin \theta = m\left(\frac{\lambda}{a}\right) - \sin \beta$$

10. Coherent light with a wavelength of 501.5 nm is sent through two parallel slits in a large flat wall. Each slit is 0.700 μm wide, and the slits' centers are 2.80 μm apart. The light falls on a semicylindrical screen, with its axis at the midline between the slits. (a) Predict the direction of each interference maximum on the screen, as an angle away from the bisector of the line joining the slits. (b) Describe the pattern of light on the screen, specifying the number of bright fringes and the location of each. (c) Find the intensity of light on the screen at the center of each bright fringe, expressed as a fraction of the light intensity I_0 at the center of the pattern.

Section 38.3   Resolution of Single-Slit and Circular Apertures

11. The pupil of a cat's eye narrows to a vertical slit of width 0.500 mm in daylight. What is the angular resolution for horizontally separated mice? Assume that the average wavelength of the light is 500 nm.

12. Find the radius of a star image formed on the retina of the eye if the aperture diameter (the pupil) at night is 0.700 cm and the length of the eye is 3.00 cm. Assume that the representative wavelength of starlight in the eye is 500 nm.

WEB 13. A helium-neon laser emits light that has a wavelength of 632.8 nm. The circular aperture through which the beam emerges has a diameter of 0.500 cm. Estimate the diameter of the beam 10.0 km from the laser.

14. On the night of April 18, 1775, a signal was to be sent from the steeple of Old North Church in Boston to Paul Revere, who was 1.80 mi away: "One if by land, two if by sea." At what minimum separation did the sexton have to set the lanterns for Revere to receive the correct message? Assume that Revere's pupils had a diameter of 4.00 mm at night and that the lantern light had a predominant wavelength of 580 nm.

15. The Impressionist painter Georges Seurat created paintings with an enormous number of dots of pure pigment, each of which was approximately 2.00 mm in diameter. The idea was to locate colors such as red and green next to each other to form a scintillating canvas (Fig. P38.15). Outside what distance would one be unable to discern individual dots on the canvas? (Assume that $\lambda = 500$ nm and that the pupil diameter is 4.00 mm.)

16. A binary star system in the constellation Orion has an angular interstellar separation of 1.00×10^{-5} rad. If $\lambda = 500$ nm, what is the smallest diameter a telescope must have to just resolve the two stars?

Figure P38.15   *Sunday Afternoon on the Isle of La Grande Jatte,* by Georges Seurat. *(SuperStock)*

17. A child is standing at the edge of a straight highway watching her grandparents' car driving away at 20.0 m/s. The air is perfectly clear and steady, and after 10.0 min the car's two taillights appear to merge into one. Assuming the diameter of the child's pupils is 5.00 mm, estimate the width of the car.

18. Suppose that you are standing on a straight highway and watching a car moving away from you at a speed v. The air is perfectly clear and steady, and after a time t the taillights appear to merge into one. Assuming the diameter of your pupil is d, estimate the width of the car.

19. A circular radar antenna on a Coast Guard ship has a diameter of 2.10 m and radiates at a frequency of 15.0 GHz. Two small boats are located 9.00 km away from the ship. How close together could the boats be and still be detected as two objects?

20. If we were to send a ruby laser beam ($\lambda = 694.3$ nm) outward from the barrel of a 2.70-m-diameter telescope, what would be the diameter of the big red spot when the beam hit the Moon 384 000 km away? (Neglect atmospheric dispersion.)

21. The angular resolution of a radio telescope is to be 0.100° when the incident waves have a wavelength of 3.00 mm. What minimum diameter is required for the telescope's receiving dish?

22. When Mars is nearest the Earth, the distance separating the two planets is 88.6×10^6 km. Mars is viewed through a telescope whose mirror has a diameter of 30.0 cm. (a) If the wavelength of the light is 590 nm, what is the angular resolution of the telescope? (b) What is the smallest distance that can be resolved between two points on Mars?

Section 38.4   The Diffraction Grating

Note: In the following problems, assume that the light is incident normally on the gratings.

23. White light is spread out into its spectral components by a diffraction grating. If the grating has 2 000 lines per centimeter, at what angle does red light of wavelength 640 nm appear in first order?

24. Light from an argon laser strikes a diffraction grating that has 5 310 lines per centimeter. The central and first-order principal maxima are separated by 0.488 m on a wall 1.72 m from the grating. Determine the wavelength of the laser light.

WEB **25.** The hydrogen spectrum has a red line at 656 nm and a violet line at 434 nm. What is the angular separation between two spectral lines obtained with a diffraction grating that has 4 500 lines per centimeter?

26. A helium–neon laser ($\lambda = 632.8$ nm) is used to calibrate a diffraction grating. If the first-order maximum occurs at 20.5°, what is the spacing between adjacent grooves in the grating?

27. Three discrete spectral lines occur at angles of 10.09°, 13.71°, and 14.77° in the first-order spectrum of a grating spectroscope. (a) If the grating has 3 660 slits per centimeter, what are the wavelengths of the light? (b) At what angles are these lines found in the second-order spectrum?

28. A diffraction grating has 800 rulings per millimeter. A beam of light containing wavelengths from 500 to 700 nm hits the grating. Do the spectra of different orders overlap? Explain.

WEB **29.** A diffraction grating with a width of 4.00 cm has been ruled with 3 000 grooves per centimeter. (a) What is the resolving power of this grating in the first three orders? (b) If two monochromatic waves incident on this grating have a mean wavelength of 400 nm, what is their wavelength separation if they are just resolved in the third order?

30. Show that, whenever white light is passed through a diffraction grating of any spacing size, the violet end of the continuous visible spectrum in third order always overlaps the red light at the other end of the second-order spectrum.

31. A source emits 531.62-nm and 531.81-nm light. (a) What minimum number of lines is required for a grating that resolves the two wavelengths in the first-order spectrum? (b) Determine the slit spacing for a grating 1.32 cm wide that has the required minimum number of lines.

32. Two wavelengths λ and $\lambda + \Delta\lambda$ (with $\Delta\lambda \ll \lambda$) are incident on a diffraction grating. Show that the angular separation between the spectral lines in the mth order spectrum is

$$\Delta\theta = \frac{\Delta\lambda}{\sqrt{(d/m)^2 - \lambda^2}}$$

where d is the slit spacing and m is the order number.

33. A grating with 250 lines per millimeter is used with an incandescent light source. Assume that the visible spectrum ranges in wavelength from 400 to 700 nm. In how many orders can one see (a) the entire visible spectrum and (b) the short-wavelength region?

34. A diffraction grating has 4 200 rulings per centimeter. On a screen 2.00 m from the grating, it is found that for a particular order m, the maxima corresponding to two closely spaced wavelengths of sodium (589.0 nm and 589.6 nm) are separated by 1.59 mm. Determine the value of m.

(Optional)
Section 38.5 Diffraction of X-Rays by Crystals

35. Potassium iodide (KI) has the same crystalline structure as NaCl, with $d = 0.353$ nm. A monochromatic x-ray beam shows a diffraction maximum when the grazing angle is 7.60°. Calculate the x-ray wavelength. (Assume first order.)

36. A wavelength of 0.129 nm characterizes K_α x-rays from zinc. When a beam of these x-rays is incident on the surface of a crystal whose structure is similar to that of NaCl, a first-order maximum is observed at 8.15°. Calculate the interplanar spacing on the basis of this information.

WEB 37. If the interplanar spacing of NaCl is 0.281 nm, what is the predicted angle at which 0.140-nm x-rays are diffracted in a first-order maximum?

38. The first-order diffraction maximum is observed at 12.6° for a crystal in which the interplanar spacing is 0.240 nm. How many other orders can be observed?

39. Monochromatic x-rays of the K_α line from a nickel target ($\lambda = 0.166$ nm) are incident on a potassium chloride (KCl) crystal surface. The interplanar distance in KCl is 0.314 nm. At what angle (relative to the surface) should the beam be directed for a second-order maximum to be observed?

40. In water of uniform depth, a wide pier is supported on pilings in several parallel rows 2.80 m apart. Ocean waves of uniform wavelength roll in, moving in a direction that makes an angle of 80.0° with the rows of posts. Find the three longest wavelengths of waves that will be strongly reflected by the pilings.

Section 38.6 Polarization of Light Waves

41. Unpolarized light passes through two polaroid sheets. The axis of the first is vertical, and that of the second is at 30.0° to the vertical. What fraction of the initial light is transmitted?

42. Three polarizing disks whose planes are parallel are centered on a common axis. The direction of the transmission axis in each case is shown in Figure P38.42 relative to the common vertical direction. A plane-polarized beam of light with E_0 parallel to the vertical reference direction is incident from the left on the first disk with an intensity of $I_i = 10.0$ units (arbitrary). Calculate the transmitted intensity I_f when (a) $\theta_1 = 20.0°$, $\theta_2 = 40.0°$, and $\theta_3 = 60.0°$; (b) $\theta_1 = 0°$, $\theta_2 = 30.0°$, and $\theta_3 = 60.0°$.

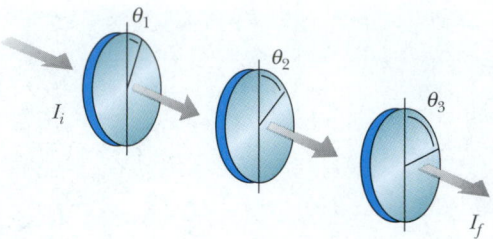

Figure P38.42 Problems 42 and 48.

43. Plane-polarized light is incident on a single polarizing disk with the direction of E_0 parallel to the direction of the transmission axis. Through what angle should the disk be rotated so that the intensity in the transmitted beam is reduced by a factor of (a) 3.00, (b) 5.00, (c) 10.0?

44. The angle of incidence of a light beam onto a reflecting surface is continuously variable. The reflected ray is found to be completely polarized when the angle of incidence is 48.0°. What is the index of refraction of the reflecting material?

45. The critical angle for total internal reflection for sapphire surrounded by air is 34.4°. Calculate the polarizing angle for sapphire.

46. For a particular transparent medium surrounded by air, show that the critical angle for total internal reflection and the polarizing angle are related by the expression $\cot \theta_p = \sin \theta_c$.

47. How far above the horizon is the Moon when its image reflected in calm water is completely polarized? ($n_{water} = 1.33$.)

ADDITIONAL PROBLEMS

48. In Figure P38.42, suppose that the transmission axes of the left and right polarizing disks are perpendicular to each other. Also, let the center disk be rotated on the common axis with an angular speed ω. Show that if unpolarized light is incident on the left disk with an intensity I_{max}, the intensity of the beam emerging from the right disk is

$$I = \frac{1}{16} I_{max}(1 - \cos 4\omega t)$$

This means that the intensity of the emerging beam is modulated at a rate that is four times the rate of rotation of the center disk. [*Hint:* Use the trigonometric identities $\cos^2 \theta = (1 + \cos 2\theta)/2$ and $\sin^2 \theta = (1 - \cos 2\theta)/2$, and recall that $\theta = \omega t$.]

49. You want to rotate the plane of polarization of a polarized light beam by 45.0° with a maximum intensity reduction of 10.0%. (a) How many sheets of perfect polarizers do you need to achieve your goal? (b) What is the angle between adjacent polarizers?

50. Figure P38.50 shows a megaphone in use. Construct a theoretical description of how a megaphone works. You may assume that the sound of your voice radiates just through the opening of your mouth. Most of the information in speech is carried not in a signal at the fundamental frequency, but rather in noises and in harmonics, with frequencies of a few thousand hertz. Does your theory allow any prediction that is simple to test?

Figure P38.50 *(Susan Allen Sigmon/Allsport USA)*

51. Light from a helium-neon laser ($\lambda = 632.8$ nm) is incident on a single slit. What is the maximum width for which no diffraction minima are observed?

52. What are the approximate dimensions of the smallest object on Earth that astronauts can resolve by eye when they are orbiting 250 km above the Earth? Assume that $\lambda = 500$ nm and that a pupil's diameter is 5.00 mm.

53. Review Problem. A beam of 541-nm light is incident on a diffraction grating that has 400 lines per millimeter. (a) Determine the angle of the second-order ray. (b) If the entire apparatus is immersed in water, what is the new second-order angle of diffraction? (c) Show that the two diffracted rays of parts (a) and (b) are related through the law of refraction.

54. The Very Large Array is a set of 27 radio telescope dishes in Caton and Socorro Counties, New Mexico (Fig. P38.54). The antennas can be moved apart on railroad tracks, and their combined signals give the resolving power of a synthetic aperture 36.0 km in diameter. (a) If the detectors are tuned to a frequency of 1.40 GHz, what is the angular resolution of the VLA? (b) Clouds of hydrogen radiate at this frequency. What must be the separation distance for two clouds at the center of the galaxy, 26 000 lightyears away, if they are to be resolved? (c) As the telescope looks up, a circling hawk looks down. For comparison, find the angular resolution of the hawk's eye. Assume that it is most sensitive to green light having a wavelength of 500 nm and that it has a pupil with a diameter of 12.0 mm. (d) A mouse is on the ground 30.0 m below. By what distance must the mouse's whiskers be separated for the hawk to resolve them?

Figure P38.54 A rancher in New Mexico rides past one of the 27 radio telescopes that make up the Very Large Array (VLA). *(© Danny Lehman)*

55. Grote Reber was a pioneer in radio astronomy. He constructed a radio telescope with a 10.0-m diameter receiving dish. What was the telescope's angular resolution for 2.00-m radio waves?

56. A 750-nm light beam hits the flat surface of a certain liquid, and the beam is split into a reflected ray and a refracted ray. If the reflected ray is completely polarized at 36.0°, what is the wavelength of the refracted ray?

57. Light of wavelength 500 nm is incident normally on a diffraction grating. If the third-order maximum of the diffraction pattern is observed at 32.0°, (a) what is the number of rulings per centimeter for the grating? (b) Determine the total number of primary maxima that can be observed in this situation.

58. Light strikes a water surface at the polarizing angle. The part of the beam refracted into the water strikes a submerged glass slab (index of refraction, 1.50), as shown in Figure P38.58. If the light reflected from the upper surface of the slab is completely polarized, what is the angle between the water surface and the glass slab?

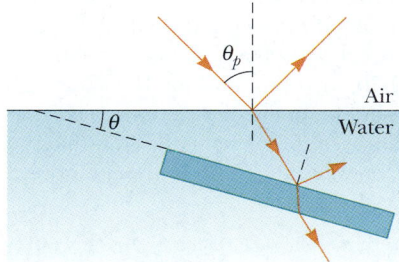

Figure P38.58

59. An American standard television picture is composed of about 485 horizontal lines of varying light intensity. Assume that your ability to resolve the lines is limited only by the Rayleigh criterion and that the pupils of your eyes are 5.00 mm in diameter. Calculate the ratio of minimum viewing distance to the vertical dimension of the picture such that you will not be able to resolve the lines. Assume that the average wavelength of the light coming from the screen is 550 nm.

60. (a) If light traveling in a medium for which the index of refraction is n_1 is incident at an angle θ on the surface of a medium of index n_2 so that the angle between the reflected and refracted rays is β, show that

$$\tan \theta = \frac{n_2 \sin \beta}{n_1 - n_2 \cos \beta}$$

[*Hint:* Use the identity $\sin(A + B) = \sin A \cos B + \cos A \sin B$.] (b) Show that this expression for $\tan \theta$ reduces to Brewster's law when $\beta = 90°$, $n_1 = 1$, and $n_2 = n$.

61. Suppose that the single slit in Figure 38.6 is 6.00 cm wide and in front of a microwave source operating at 7.50 GHz. (a) Calculate the angle subtended by the first minimum in the diffraction pattern. (b) What is the relative intensity I/I_{max} at $\theta = 15.0°$? (c) Consider the case when there are two such sources, separated laterally by 20.0 cm, behind the slit. What must the maximum distance between the plane of the sources and the slit be if the diffraction patterns are to be resolved? (In this case, the approximation $\sin \theta \approx \tan \theta$ is not valid because of the relatively small value of a/λ.)

62. Two polarizing sheets are placed together with their transmission axes crossed so that no light is transmitted. A third sheet is inserted between them with its transmission axis at an angle of 45.0° with respect to each of the other axes. Find the fraction of incident unpolarized light intensity transmitted by the three-sheet combination. (Assume that each polarizing sheet is ideal.)

63. Figure P38.63a is a three-dimensional sketch of a birefringent crystal. The dotted lines illustrate how a thin parallel-faced slab of material could be cut from the larger specimen with the optic axis of the crystal parallel to the faces of the plate. A section cut from the crystal in this manner is known as a *retardation plate*. When a beam of light is incident on the plate perpendicular to the direction of the optic axis, as shown in Figure P38.63b, the O ray and the E ray travel along a single straight line but with different speeds. (a) Letting the thickness of the plate be d, show that the phase difference between the O ray and the E ray is

$$\theta = \frac{2\pi d}{\lambda} (n_O - n_E)$$

where λ is the wavelength in air. (b) If in a particular case the incident light has a wavelength of 550 nm, what is the minimum value of d for a quartz plate for which $\theta = \pi/2$? Such a plate is called a *quarter-wave plate*. (Use values of n_O and n_E from Table 38.1.)

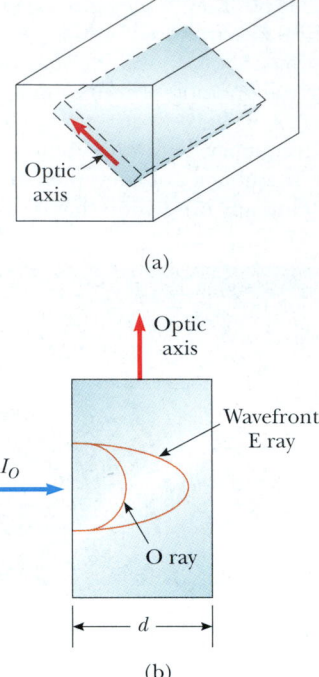

(a)

(b)

Figure P38.63

64. Derive Equation 38.12 for the resolving power of a grating, $R = Nm$, where N is the number of lines illuminated and m is the order in the diffraction pattern. Remember that Rayleigh's criterion (see Section 38.3) states that two wavelengths will be resolved when the principal maximum for one falls on the first minimum for the other.

65. Light of wavelength 632.8 nm illuminates a single slit, and a diffraction pattern is formed on a screen 1.00 m from the slit. Using the data in the table on the following page, plot relative intensity versus distance. Choose an appropriate value for the slit width a, and on the same graph used for the experimental data, plot the theoretical expression for the relative intensity

$$\frac{I}{I_{max}} = \frac{\sin^2(\beta/2)}{(\beta/2)^2}$$

What value of a gives the best fit of theory and experiment?

66. How much diffraction spreading does a light beam undergo? One quantitative answer is the *full width at half maximum* of the central maximum of the Fraunhofer diffraction pattern of a single slit. You can evaluate this angle of spreading in this problem and in the next. (a) In Equation 38.4, define $\beta/2 = \phi$ and show that, at the point where $I = 0.5 I_{max}$, we must have $\sin \phi = \phi/\sqrt{2}$. (b) Let $y_1 = \sin \phi$ and $y_2 = \phi/\sqrt{2}$. Plot y_1 and y_2 on the same set of axes over a range from $\phi = 1$ rad to $\phi = \pi/2$ rad. Determine ϕ from the point of intersection of the two curves. (c) Then show that, if the

Relative Intensity	Distance from Center of Central Maximum (mm)
1.00	0
0.95	0.8
0.80	1.6
0.60	2.4
0.39	3.2
0.21	4.0
0.079	4.8
0.014	5.6
0.003	6.5
0.015	7.3
0.036	8.1
0.047	8.9
0.043	9.7
0.029	10.5
0.013	11.3
0.002	12.1
0.000 3	12.9
0.005	13.7
0.012	14.5
0.016	15.3
0.015	16.1
0.010	16.9
0.004 4	17.7
0.000 6	18.5
0.000 3	19.3
0.003	20.2

fraction λ/a is not large, the angular full width at half maximum of the central diffraction maximum is $\Delta\theta = 0.886\lambda/a$.

67. Another method to solve the equation $\phi = \sqrt{2}\sin\phi$ in Problem 66 is to use a calculator, guess a first value of ϕ, see if it fits, and continue to update your estimate until the equation balances. How many steps (iterations) does this take?

68. In the diffraction pattern of a single slit, described by the equation

$$I_\theta = I_{max}\left[\frac{\sin(\beta/2)}{\beta/2}\right]^2$$

with $\beta = (2\pi a \sin\theta)/\lambda$, the central maximum is at $\beta = 0$ and the side maxima are *approximately* at $\beta/2 = (m + \frac{1}{2})\pi$ for $m = 1, 2, 3, \ldots$. Determine more precisely (a) the location of the first side maximum, where $m = 1$, and (b) the location of the second side maximum. Observe in Figure 38.10a that the graph of intensity versus $\beta/2$ has a horizontal tangent at maxima and also at minima. You will need to solve a transcendental equation.

69. A *pinhole camera* has a small circular aperture of diameter D. Light from distant objects passes through the aperture into an otherwise dark box, falling upon a screen located a distance L away. If D is too large, the display on the screen will be fuzzy because a bright point in the field of view will send light onto a circle of diameter slightly larger than D. On the other hand, if D is too small, diffraction will blur the display on the screen. The screen shows a reasonably sharp image if the diameter of the central disk of the diffraction pattern, specified by Equation 38.9, is equal to D at the screen. (a) Show that for monochromatic light with plane wave fronts and $L \gg D$, the condition for a sharp view is fulfilled if $D^2 = 2.44\,\lambda L$. (b) Find the optimum pinhole diameter if 500-nm light is projected onto a screen 15.0 cm away.

ANSWERS TO QUICK QUIZZES

38.1 The space between the slightly open door and the door-frame acts as a single slit. Sound waves have wavelengths that are approximately the same size as the opening and so are diffracted and spread throughout the room you are in. Because light wavelengths are much smaller than the slit width, they are virtually undiffracted. As a result, you must have a direct line of sight to detect the light waves.

38.2 The situation is like that depicted in Figure 38.11 except that now the slits are only half as far apart. The diffraction pattern is the same, but the interference pattern is stretched out because d is smaller. Because $d/a = 3$, the third interference maximum coincides with the first diffraction minimum. Your sketch should look like the figure to the right.

38.3 Yes, but no diffraction effects are observed because the separation distance between adjacent ribs is so much greater than the wavelength of the x-rays.

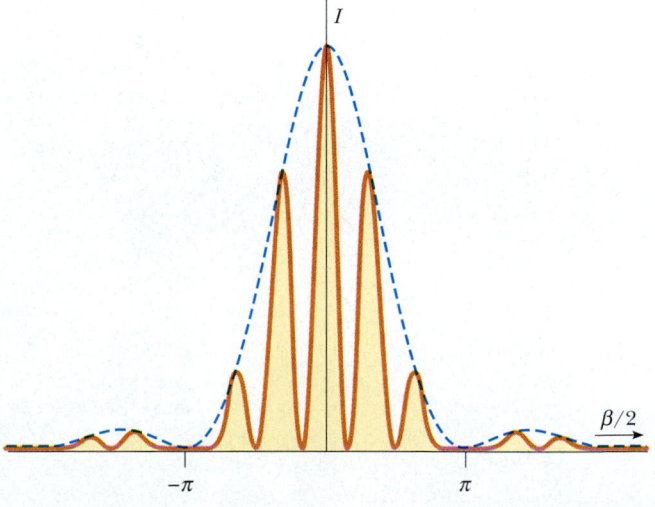

Approximate size of
Earth for comparison

Modern Physics

At the end of the 19th century, many scientists believed that they had learned most of what there was to know about physics. Newton's laws of motion and his theory of universal gravitation, Maxwell's theoretical work in unifying electricity and magnetism, and the laws of thermodynamics and kinetic theory were highly successful in explaining a variety of phenomena.

As the 19th century turned to the 20th, however, a major revolution shook the world of physics. In 1900 Planck provided the basic ideas that led to the formulation of the quantum theory, and in 1905 Einstein formulated his brilliant special theory of relativity. The excitement of the times is captured in Einstein's own words: "It was a marvelous time to be alive." Both ideas were to have a profound effect on our understanding of nature. Within a few decades, these two theories inspired new developments and theories in the fields of atomic physics, nuclear physics, and condensed-matter physics.

In Chapter 39 we introduce the special theory of relativity. The theory provides us with a new and deeper view of physical laws. Although the concepts underlying this theory often violate our common sense, the theory correctly predicts the results of experiments involving speeds near the speed of light. In the extended version of this textbook, *Physics for Scientists and Engineers with Modern Physics,* we cover the basic concepts of quantum mechanics and their application to atomic and molecular physics, and we introduce solid-state physics, nuclear physics, particle physics, and cosmology.

You should keep in mind that, although the physics developed during the 20th century has led to a multitude of important technological achievements, the story is still incomplete. Discoveries will continue to evolve during our lifetimes, and many of these discoveries will deepen or refine our understanding of nature and the world around us. It is still a "marvelous time to be alive."

◄ Courtesy of the SOHO-EIT Consortium.

c h a p t e r

39

Relativity

Most of our everyday experiences and observations have to do with objects that move at speeds much less than the speed of light. Newtonian mechanics was formulated to describe the motion of such objects, and this formalism is still very successful in describing a wide range of phenomena that occur at low speeds. It fails, however, when applied to particles whose speeds approach that of light.

Experimentally, the predictions of Newtonian theory can be tested at high speeds by accelerating electrons or other charged particles through a large electric potential difference. For example, it is possible to accelerate an electron to a speed of $0.99c$ (where c is the speed of light) by using a potential difference of several million volts. According to Newtonian mechanics, if the potential difference is increased by a factor of 4, the electron's kinetic energy is four times greater and its speed should double to $1.98c$. However, experiments show that the speed of the electron—as well as the speed of any other particle in the Universe—always remains less than the speed of light, regardless of the size of the accelerating voltage. Because it places no upper limit on speed, Newtonian mechanics is contrary to modern experimental results and is clearly a limited theory.

In 1905, at the age of only 26, Einstein published his special theory of relativity. Regarding the theory, Einstein wrote:

> The relativity theory arose from necessity, from serious and deep contradictions in the old theory from which there seemed no escape. The strength of the new theory lies in the consistency and simplicity with which it solves all these difficulties . . . . [1]

Although Einstein made many other important contributions to science, the special theory of relativity alone represents one of the greatest intellectual achievements of all time. With this theory, experimental observations can be correctly predicted over the range of speeds from $v = 0$ to speeds approaching the speed of light. At low speeds, Einstein's theory reduces to Newtonian mechanics as a limiting situation. It is important to recognize that Einstein was working on electromagnetism when he developed the special theory of relativity. He was convinced that Maxwell's equations were correct, and in order to reconcile them with one of his postulates, he was forced into the bizarre notion of assuming that space and time are not absolute.

This chapter gives an introduction to the special theory of relativity, with emphasis on some of its consequences. The special theory covers phenomena such as the slowing down of clocks and the contraction of lengths in moving reference frames as measured by a stationary observer. We also discuss the relativistic forms of momentum and energy, as well as some consequences of the famous mass–energy formula, $E = mc^2$.

In addition to its well-known and essential role in theoretical physics, the special theory of relativity has practical applications, including the design of nuclear power plants and modern global positioning system (GPS) units. These devices do not work if designed in accordance with nonrelativistic principles.

We shall have occasion to use relativity in some subsequent chapters of the extended version of this text, most often presenting only the outcome of relativistic effects.

---

[1] A. Einstein and L. Infeld, *The Evolution of Physics,* New York, Simon and Schuster, 1961.

39.1 ▷ THE PRINCIPLE OF GALILEAN RELATIVITY

To describe a physical event, it is necessary to establish a frame of reference. You should recall from Chapter 5 that Newton's laws are valid in all inertial frames of reference. Because an inertial frame is defined as one in which Newton's first law is valid, we can say that **an inertial frame of reference is one in which an object is observed to have no acceleration when no forces act on it.** Furthermore, any system moving with constant velocity with respect to an inertial system must also be an inertial system.

There is no preferred inertial reference frame. This means that the results of an experiment performed in a vehicle moving with uniform velocity will be identical to the results of the same experiment performed in a stationary vehicle. The formal statement of this result is called the **principle of Galilean relativity:**

> The laws of mechanics must be the same in all inertial frames of reference.

Let us consider an observation that illustrates the equivalence of the laws of mechanics in different inertial frames. A pickup truck moves with a constant velocity, as shown in Figure 39.1a. If a passenger in the truck throws a ball straight up, and if air effects are neglected, the passenger observes that the ball moves in a vertical path. The motion of the ball appears to be precisely the same as if the ball were thrown by a person at rest on the Earth. The law of gravity and the equations of motion under constant acceleration are obeyed whether the truck is at rest or in uniform motion.

Now consider the same situation viewed by an observer at rest on the Earth. This stationary observer sees the path of the ball as a parabola, as illustrated in Figure 39.1b. Furthermore, according to this observer, the ball has a horizontal component of velocity equal to the velocity of the truck. Although the two observers disagree on certain aspects of the situation, they agree on the validity of Newton's laws and on such classical principles as conservation of energy and conservation of linear momentum. This agreement implies that no mechanical experiment can detect any difference between the two inertial frames. The only thing that can be detected is the relative motion of one frame with respect to the other. That is, the notion of absolute motion through space is meaningless, as is the notion of a preferred reference frame.

Inertial frame of reference

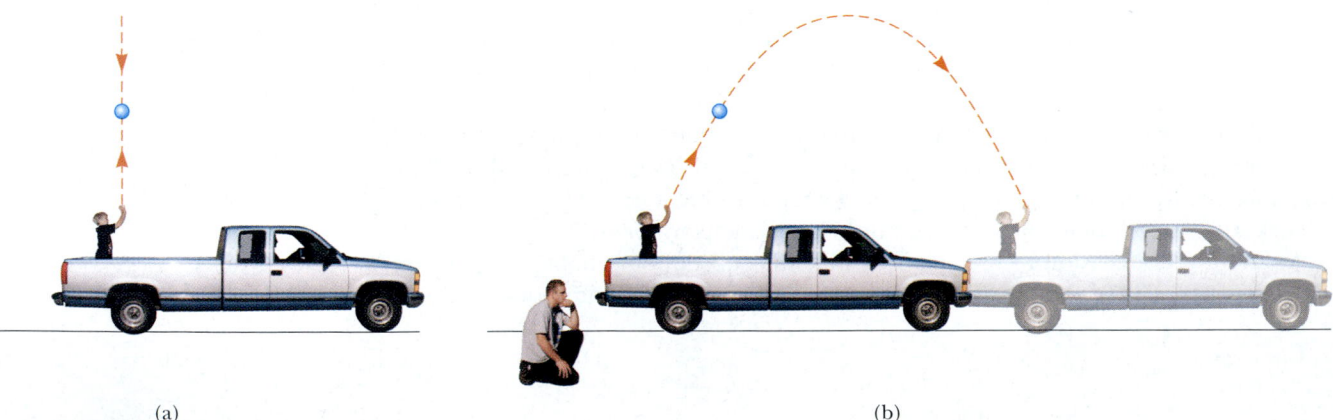

(a)   (b)

Figure 39.1   (a) The observer in the truck sees the ball move in a vertical path when thrown upward. (b) The Earth observer sees the path of the ball as a parabola.

Quick Quiz 39.1

Which observer in Figure 39.1 is right about the ball's path?

Suppose that some physical phenomenon, which we call an *event*, occurs in an inertial system. The event's location and time of occurrence can be specified by the four coordinates (x, y, z, t). We would like to be able to transform these coordinates from one inertial system to another one moving with uniform relative velocity.

Consider two inertial systems S and S' (Fig. 39.2). The system S' moves with a constant velocity **v** along the xx' axes, where **v** is measured relative to S. We assume that an event occurs at the point P and that the origins of S and S' coincide at $t = 0$. An observer in S describes the event with space–time coordinates (x, y, z, t), whereas an observer in S' uses the coordinates (x', y', z', t') to describe the same event. As we see from Figure 39.2, the relationships between these various coordinates can be written

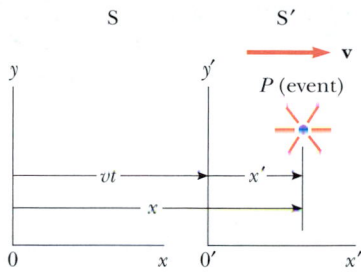

Figure 39.2 An event occurs at a point P. The event is seen by two observers in inertial frames S and S', where S' moves with a velocity **v** relative to S.

$$x' = x - vt$$
$$y' = y$$
$$z' = z \qquad\qquad \textbf{(39.1)}$$
$$t' = t$$

Galilean space–time transformation equations

These equations are the **Galilean space–time transformation equations.** Note that time is assumed to be the same in both inertial systems. That is, within the framework of classical mechanics, all clocks run at the same rate, regardless of their velocity, so that the time at which an event occurs for an observer in S is the same as the time for the same event in S'. Consequently, the time interval between two successive events should be the same for both observers. Although this assumption may seem obvious, it turns out to be incorrect in situations where v is comparable to the speed of light.

Now suppose that a particle moves a distance dx in a time interval dt as measured by an observer in S. It follows from Equations 39.1 that the corresponding distance dx' measured by an observer in S' is $dx' = dx - v\,dt$, where frame S' is moving with speed v relative to frame S. Because $dt = dt'$, we find that

$$\frac{dx'}{dt} = \frac{dx}{dt} - v$$

or

$$u'_x = u_x - v \qquad\qquad \textbf{(39.2)}$$

Galilean velocity transformation equation

where u_x and u'_x are the x components of the velocity relative to S and S', respectively. (We use the symbol **u** for particle velocity rather than **v**, which is used for the relative velocity of two reference frames.) This is the **Galilean velocity transformation equation.** It is used in everyday observations and is consistent with our intuitive notion of time and space. As we shall soon see, however, it leads to serious contradictions when applied to electromagnetic waves.

Quick Quiz 39.2

Applying the Galilean velocity transformation equation, determine how fast (relative to the Earth) a baseball pitcher with a 90-mi/h fastball can throw a ball while standing in a boxcar moving at 110 mi/h.

The Speed of Light

It is quite natural to ask whether the principle of Galilean relativity also applies to electricity, magnetism, and optics. Experiments indicate that the answer is no. Recall from Chapter 34 that Maxwell showed that the speed of light in free space is $c = 3.00 \times 10^8$ m/s. Physicists of the late 1800s thought that light waves moved through a medium called the *ether* and that the speed of light was c only in a special, absolute frame at rest with respect to the ether. The Galilean velocity transformation equation was expected to hold in any frame moving at speed v relative to the absolute ether frame.

Because the existence of a preferred, absolute ether frame would show that light was similar to other classical waves and that Newtonian ideas of an absolute frame were true, considerable importance was attached to establishing the existence of the ether frame. Prior to the late 1800s, experiments involving light traveling in media moving at the highest laboratory speeds attainable at that time were not capable of detecting changes as small as $c \pm v$. Starting in about 1880, scientists decided to use the Earth as the moving frame in an attempt to improve their chances of detecting these small changes in the speed of light.

As observers fixed on the Earth, we can say that we are stationary and that the absolute ether frame containing the medium for light propagation moves past us with speed v. Determining the speed of light under these circumstances is just like determining the speed of an aircraft traveling in a moving air current, or wind; consequently, we speak of an "ether wind" blowing through our apparatus fixed to the Earth.

A direct method for detecting an ether wind would use an apparatus fixed to the Earth to measure the wind's influence on the speed of light. If v is the speed of the ether relative to the Earth, then the speed of light should have its maximum value, $c + v$, when propagating downwind, as shown in Figure 39.3a. Likewise, the speed of light should have its minimum value, $c - v$, when propagating upwind, as shown in Figure 39.3b, and an intermediate value, $(c^2 - v^2)^{1/2}$, in the direction perpendicular to the ether wind, as shown in Figure 39.3c. If the Sun is assumed to be at rest in the ether, then the velocity of the ether wind would be equal to the orbital velocity of the Earth around the Sun, which has a magnitude of approximately 3×10^4 m/s. Because $c = 3 \times 10^8$ m/s, it should be possible to detect a change in speed of about 1 part in 10^4 for measurements in the upwind or downwind directions. However, as we shall see in the next section, all attempts to detect such changes and establish the existence of the ether wind (and hence the absolute frame) proved futile! (You may want to return to Problem 40 in Chapter 4 to see a situation in which the Galilean velocity transformation equation does hold.)

If it is assumed that the laws of electricity and magnetism are the same in all inertial frames, a paradox concerning the speed of light immediately arises. We can understand this by recognizing that Maxwell's equations seem to imply that the speed of light always has the fixed value 3.00×10^8 m/s in all inertial frames, a result in direct contradiction to what is expected based on the Galilean velocity transformation equation. According to Galilean relativity, the speed of light should not be the same in all inertial frames.

For example, suppose a light pulse is sent out by an observer S′ standing in a boxcar moving with a velocity **v** relative to a stationary observer standing alongside the track (Fig. 39.4). The light pulse has a speed c relative to S′. According to Galilean relativity, the pulse speed relative to S should be $c + v$. This is in contradiction to Einstein's special theory of relativity, which, as we shall see, postulates that the speed of the pulse is the same for all observers.

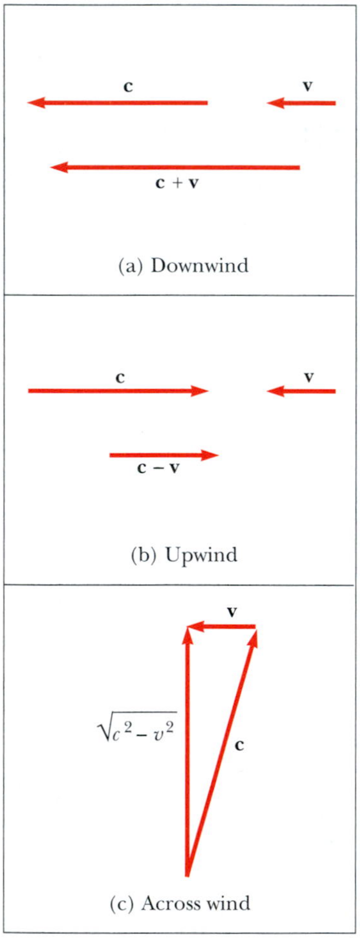

Figure 39.3 If the velocity of the ether wind relative to the Earth is **v** and the velocity of light relative to the ether is **c**, then the speed of light relative to the Earth is (a) $c + v$ in the downwind direction, (b) $c - v$ in the upwind direction, and (c) $(c^2 - v^2)^{1/2}$ in the direction perpendicular to the wind.

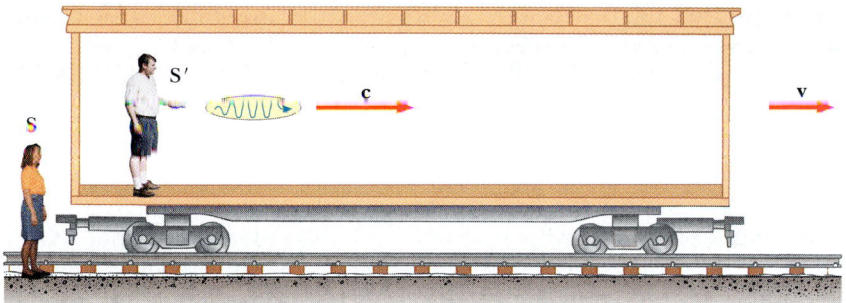

Figure 39.4 A pulse of light is sent out by a person in a moving boxcar. According to Galilean relativity, the speed of the pulse should be $c + v$ relative to a stationary observer.

To resolve this contradiction in theories, we must conclude that either (1) the laws of electricity and magnetism are not the same in all inertial frames or (2) the Galilean velocity transformation equation is incorrect. If we assume the first alternative, then a preferred reference frame in which the speed of light has the value c must exist and the measured speed must be greater or less than this value in any other reference frame, in accordance with the Galilean velocity transformation equation. If we assume the second alternative, then we are forced to abandon the notions of absolute time and absolute length that form the basis of the Galilean space–time transformation equations.

39.2 THE MICHELSON–MORLEY EXPERIMENT

The most famous experiment designed to detect small changes in the speed of light was first performed in 1881 by Albert A. Michelson (see Section 37.7) and later repeated under various conditions by Michelson and Edward W. Morley (1838–1923). We state at the outset that the outcome of the experiment contradicted the ether hypothesis.

The experiment was designed to determine the velocity of the Earth relative to that of the hypothetical ether. The experimental tool used was the Michelson interferometer, which was discussed in Section 37.7 and is shown again in Figure 39.5. Arm 2 is aligned along the direction of the Earth's motion through space. The Earth moving through the ether at speed v is equivalent to the ether flowing past the Earth in the opposite direction with speed v. This ether wind blowing in the direction opposite the direction of Earth's motion should cause the speed of light measured in the Earth frame to be $c - v$ as the light approaches mirror M_2 and $c + v$ after reflection, where c is the speed of light in the ether frame.

The two beams reflected from M_1 and M_2 recombine, and an interference pattern consisting of alternating dark and bright fringes is formed. The interference pattern was observed while the interferometer was rotated through an angle of 90°. This rotation supposedly would change the speed of the ether wind along the arms of the interferometer. The rotation should have caused the fringe pattern to shift slightly but measurably, but measurements failed to show any change in the interference pattern! The Michelson–Morley experiment was repeated at different times of the year when the ether wind was expected to change direction

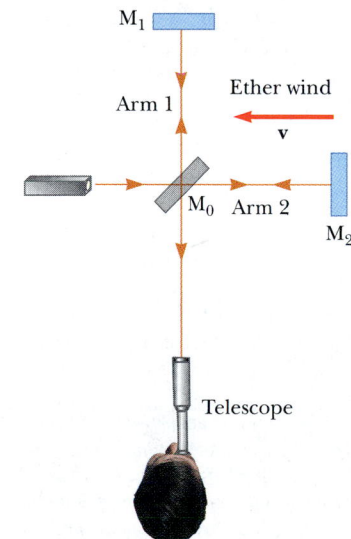

Figure 39.5 According to the ether wind theory, the speed of light should be $c - v$ as the beam approaches mirror M_2 and $c + v$ after reflection.

and magnitude, but the results were always the same: **no fringe shift of the magnitude required was ever observed.**[2]

The negative results of the Michelson–Morley experiment not only contradicted the ether hypothesis but also showed that it was impossible to measure the absolute velocity of the Earth with respect to the ether frame. However, as we shall see in the next section, Einstein offered a postulate for his special theory of relativity that places quite a different interpretation on these null results. In later years, when more was known about the nature of light, the idea of an ether that permeates all of space was relegated to the ash heap of worn-out concepts. **Light is now understood to be an electromagnetic wave, which requires no medium for its propagation.** As a result, the idea of an ether in which these waves could travel became unnecessary.

Albert Einstein (1879–1955)
Einstein, one of the greatest physicists of all times, was born in Ulm, Germany. In 1905, at the age of 26, he published four scientific papers that revolutionized physics. Two of these papers were concerned with what is now considered his most important contribution: the special theory of relativity.

In 1916, Einstein published his work on the general theory of relativity. The most dramatic prediction of this theory is the degree to which light is deflected by a gravitational field. Measurements made by astronomers on bright stars in the vicinity of the eclipsed Sun in 1919 confirmed Einstein's prediction, and as a result Einstein became a world celebrity.

Einstein was deeply disturbed by the development of quantum mechanics in the 1920s despite his own role as a scientific revolutionary. In particular, he could never accept the probabilistic view of events in nature that is a central feature of quantum theory. The last few decades of his life were devoted to an unsuccessful search for a unified theory that would combine gravitation and electromagnetism. *(AIP Niels Bohr Library)*

Optional Section

Details of the Michelson–Morley Experiment

To understand the outcome of the Michelson–Morley experiment, let us assume that the two arms of the interferometer in Figure 39.5 are of equal length L. We shall analyze the situation as if there were an ether wind, because that is what Michelson and Morley expected to find. As noted above, the speed of the light beam along arm 2 should be $c - v$ as the beam approaches M_2 and $c + v$ after the beam is reflected. Thus, the time of travel to the right is $L/(c - v)$, and the time of travel to the left is $L/(c + v)$. The total time needed for the round trip along arm 2 is

$$t_1 = \frac{L}{c + v} + \frac{L}{c - v} = \frac{2Lc}{c^2 - v^2} = \frac{2L}{c}\left(1 - \frac{v^2}{c^2}\right)^{-1}$$

Now consider the light beam traveling along arm 1, perpendicular to the ether wind. Because the speed of the beam relative to the Earth is $(c^2 - v^2)^{1/2}$ in this case (see Fig. 39.3), the time of travel for each half of the trip is $L/(c^2 - v^2)^{1/2}$, and the total time of travel for the round trip is

$$t_2 = \frac{2L}{(c^2 - v^2)^{1/2}} = \frac{2L}{c}\left(1 - \frac{v^2}{c^2}\right)^{-1/2}$$

Thus, the time difference between the horizontal round trip (arm 2) and the vertical round trip (arm 1) is

$$\Delta t = t_1 - t_2 = \frac{2L}{c}\left[\left(1 - \frac{v^2}{c^2}\right)^{-1} - \left(1 - \frac{v^2}{c^2}\right)^{-1/2}\right]$$

Because $v^2/c^2 \ll 1$, we can simplify this expression by using the following binomial expansion after dropping all terms higher than second order:

$$(1 - x)^n \approx 1 - nx \qquad \text{for } x \ll 1$$

In our case, $x = v^2/c^2$, and we find that

$$\Delta t = t_1 - t_2 \approx \frac{Lv^2}{c^3} \tag{39.3}$$

This time difference between the two instants at which the reflected beams arrive at the viewing telescope gives rise to a phase difference between the beams,

---

[2] From an Earth observer's point of view, changes in the Earth's speed and direction of motion in the course of a year are viewed as ether wind shifts. Even if the speed of the Earth with respect to the ether were zero at some time, six months later the speed of the Earth would be 60 km/s with respect to the ether, and as a result a fringe shift should be noticed. No shift has ever been observed, however.

producing an interference pattern when they combine at the position of the telescope. A shift in the interference pattern should be detected when the interferometer is rotated through 90° in a horizontal plane, so that the two beams exchange roles. This results in a time difference twice that given by Equation 39.3. Thus, the path difference that corresponds to this time difference is

$$\Delta d = c(2\,\Delta t) = \frac{2Lv^2}{c^2}$$

Because a change in path length of one wavelength corresponds to a shift of one fringe, the corresponding fringe shift is equal to this path difference divided by the wavelength of the light:

$$\text{Shift} = \frac{2Lv^2}{\lambda c^2} \tag{39.4}$$

In the experiments by Michelson and Morley, each light beam was reflected by mirrors many times to give an effective path length L of approximately 11 m. Using this value and taking v to be equal to 3.0×10^4 m/s, the speed of the Earth around the Sun, we obtain a path difference of

$$\Delta d = \frac{2(11\text{ m})(3.0 \times 10^4\text{ m/s})^2}{(3.0 \times 10^8\text{ m/s})^2} = 2.2 \times 10^{-7}\text{ m}$$

This extra travel distance should produce a noticeable shift in the fringe pattern. Specifically, using 500-nm light, we expect a fringe shift for rotation through 90° of

$$\text{Shift} = \frac{\Delta d}{\lambda} = \frac{2.2 \times 10^{-7}\text{ m}}{5.0 \times 10^{-7}\text{ m}} \approx 0.44$$

The instrument used by Michelson and Morley could detect shifts as small as 0.01 fringe. However, **it detected no shift whatsoever in the fringe pattern.** Since then, the experiment has been repeated many times by different scientists under a wide variety of conditions, and no fringe shift has ever been detected. Thus, it was concluded that the motion of the Earth with respect to the postulated ether cannot be detected.

Many efforts were made to explain the null results of the Michelson–Morley experiment and to save the ether frame concept and the Galilean velocity transformation equation for light. All proposals resulting from these efforts have been shown to be wrong. No experiment in the history of physics received such valiant efforts to explain the absence of an expected result as did the Michelson–Morley experiment. The stage was set for Einstein, who solved the problem in 1905 with his special theory of relativity.

39.3 ▸ EINSTEIN'S PRINCIPLE OF RELATIVITY

In the previous section we noted the impossibility of measuring the speed of the ether with respect to the Earth and the failure of the Galilean velocity transformation equation in the case of light. Einstein proposed a theory that boldly removed these difficulties and at the same time completely altered our notion of space and time.[3] He based his special theory of relativity on two postulates:

[3] A. Einstein, "On the Electrodynamics of Moving Bodies," *Ann. Physik* 17:891, 1905. For an English translation of this article and other publications by Einstein, see the book by H. Lorentz, A. Einstein, H. Minkowski, and H. Weyl, *The Principle of Relativity,* Dover, 1958.

1. **The principle of relativity:** The laws of physics must be the same in all inertial reference frames.
2. **The constancy of the speed of light:** The speed of light in vacuum has the same value, $c = 3.00 \times 10^8$ m/s, in all inertial frames, regardless of the velocity of the observer or the velocity of the source emitting the light.

The first postulate asserts that *all* the laws of physics—those dealing with mechanics, electricity and magnetism, optics, thermodynamics, and so on—are the same in all reference frames moving with constant velocity relative to one another. This postulate is a sweeping generalization of the principle of Galilean relativity, which refers only to the laws of mechanics. From an experimental point of view, Einstein's principle of relativity means that any kind of experiment (measuring the speed of light, for example) performed in a laboratory at rest must give the same result when performed in a laboratory moving at a constant velocity past the first one. Hence, no preferred inertial reference frame exists, and it is impossible to detect absolute motion.

Note that postulate 2 is required by postulate 1: If the speed of light were not the same in all inertial frames, measurements of different speeds would make it possible to distinguish between inertial frames; as a result, a preferred, absolute frame could be identified, in contradiction to postulate 1.

Although the Michelson–Morley experiment was performed before Einstein published his work on relativity, it is not clear whether or not Einstein was aware of the details of the experiment. Nonetheless, the null result of the experiment can be readily understood within the framework of Einstein's theory. According to his principle of relativity, the premises of the Michelson–Morley experiment were incorrect. In the process of trying to explain the expected results, we stated that when light traveled against the ether wind its speed was $c - v$, in accordance with the Galilean velocity transformation equation. However, if the state of motion of the observer or of the source has no influence on the value found for the speed of light, one always measures the value to be c. Likewise, the light makes the return trip after reflection from the mirror at speed c, not at speed $c + v$. Thus, the motion of the Earth does not influence the fringe pattern observed in the Michelson–Morley experiment, and a null result should be expected.

If we accept Einstein's theory of relativity, we must conclude that relative motion is unimportant when measuring the speed of light. At the same time, we shall see that we must alter our common-sense notion of space and time and be prepared for some bizarre consequences. It may help as you read the pages ahead to keep in mind that our common-sense ideas are based on a lifetime of everyday experiences and not on observations of objects moving at hundreds of thousands of kilometers per second.

39.4 CONSEQUENCES OF THE SPECIAL THEORY OF RELATIVITY

Before we discuss the consequences of Einstein's special theory of relativity, we must first understand how an observer located in an inertial reference frame describes an event. As mentioned earlier, an event is an occurrence describable by three space coordinates and one time coordinate. Different observers in different inertial frames usually describe the same event with different coordinates.

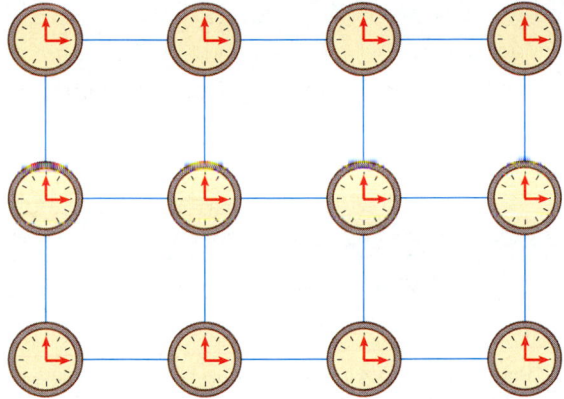

Figure 39.6 In studying relativity, we use a reference frame consisting of a coordinate grid and a set of synchronized clocks.

The reference frame used to describe an event consists of a coordinate grid and a set of synchronized clocks located at the grid intersections, as shown in Figure 39.6 in two dimensions. The clocks can be synchronized in many ways with the help of light signals. For example, suppose an observer is located at the origin with a master clock and sends out a pulse of light at $t = 0$. The pulse takes a time r/c to reach a clock located a distance r from the origin. Hence, this clock is synchronized with the master clock if this clock reads r/c at the instant the pulse reaches it. This procedure of synchronization assumes that the speed of light has the same value in all directions and in all inertial frames. Furthermore, the procedure concerns an event recorded by an observer in a specific inertial reference frame. An observer in some other inertial frame would assign different space–time coordinates to events being observed by using another coordinate grid and another array of clocks.

As we examine some of the consequences of relativity in the remainder of this section, we restrict our discussion to the concepts of simultaneity, time, and length, all three of which are quite different in relativistic mechanics from what they are in Newtonian mechanics. For example, in relativistic mechanics the distance between two points and the time interval between two events depend on the frame of reference in which they are measured. That is, **in relativistic mechanics there is no such thing as absolute length or absolute time.** Furthermore, **events at different locations that are observed to occur simultaneously in one frame are not observed to be simultaneous in another frame moving uniformly past the first.**

Simultaneity and the Relativity of Time

A basic premise of Newtonian mechanics is that a universal time scale exists that is the same for all observers. In fact, Newton wrote that "Absolute, true, and mathematical time, of itself, and from its own nature, flows equably without relation to anything external." Thus, Newton and his followers simply took simultaneity for granted. In his special theory of relativity, Einstein abandoned this assumption.

Einstein devised the following thought experiment to illustrate this point. A boxcar moves with uniform velocity, and two lightning bolts strike its ends, as illustrated in Figure 39.7a, leaving marks on the boxcar and on the ground. The marks on the boxcar are labeled A' and B', and those on the ground are labeled A and B. An observer O' moving with the boxcar is midway between A' and B', and a ground observer O is midway between A and B. The events recorded by the observers are the striking of the boxcar by the two lightning bolts.

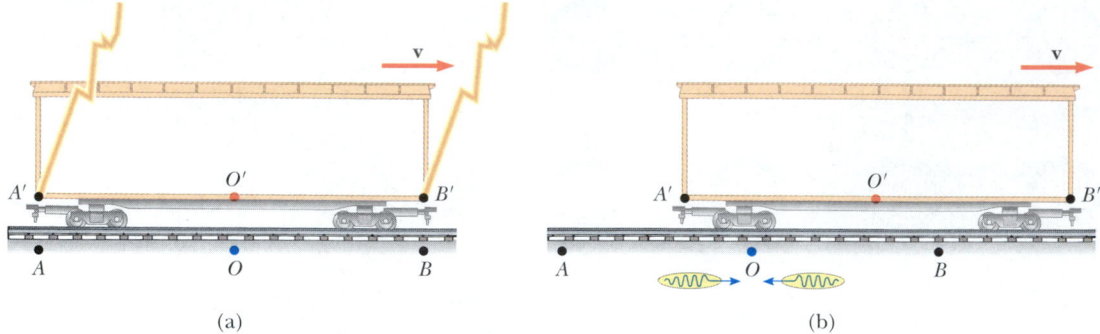

Figure 39.7   (a) Two lightning bolts strike the ends of a moving boxcar. (b) The events appear to be simultaneous to the stationary observer O, standing midway between A and B. The events do not appear to be simultaneous to observer O', who claims that the front of the car is struck before the rear. Note that in (b) the leftward-traveling light signal has already passed O' but the rightward-traveling signal has not yet reached O'.

The light signals recording the instant at which the two bolts strike reach observer O at the same time, as indicated in Figure 39.7b. This observer realizes that the signals have traveled at the same speed over equal distances, and so rightly concludes that the events at A and B occurred simultaneously. Now consider the same events as viewed by observer O'. By the time the signals have reached observer O, observer O' has moved as indicated in Figure 39.7b. Thus, the signal from B' has already swept past O', but the signal from A' has not yet reached O'. In other words, O' sees the signal from B' before seeing the signal from A'. According to Einstein, *the two observers must find that light travels at the same speed.* Therefore, observer O' concludes that the lightning strikes the front of the boxcar before it strikes the back.

This thought experiment clearly demonstrates that the two events that appear to be simultaneous to observer O do not appear to be simultaneous to observer O'. In other words,

> two events that are simultaneous in one reference frame are in general not simultaneous in a second frame moving relative to the first. That is, simultaneity is not an absolute concept but rather one that depends on the state of motion of the observer.

Quick Quiz 39.3

Which observer in Figure 39.7 is correct?

The central point of relativity is this: Any inertial frame of reference can be used to describe events and do physics. **There is no preferred inertial frame of reference.** However, observers in different inertial frames always measure different time intervals with their clocks and different distances with their meter sticks. Nevertheless, all observers agree on the forms of the laws of physics in their respective frames because these laws must be the same for all observers in uniform motion. For example, the relationship $F = ma$ in a frame S has the same form $F' = ma'$ in a frame S' that is moving at constant velocity relative to frame S. It is the alteration of time and space that allows the laws of physics (including Maxwell's equations) to be the same for all observers in uniform motion.

Time Dilation

We can illustrate the fact that observers in different inertial frames always measure different time intervals between a pair of events by considering a vehicle moving to the right with a speed v, as shown in Figure 39.8a. A mirror is fixed to the ceiling of the vehicle, and observer O' at rest in this system holds a laser a distance d below the mirror. At some instant, the laser emits a pulse of light directed toward the mirror (event 1), and at some later time after reflecting from the mirror, the pulse arrives back at the laser (event 2). Observer O' carries a clock C' and uses it to measure the time interval Δt_p between these two events. (The subscript p stands for *proper*, as we shall see in a moment.) Because the light pulse has a speed c, the time it takes the pulse to travel from O' to the mirror and back to O' is

$$\Delta t_p = \frac{\text{Distance traveled}}{\text{Speed}} = \frac{2d}{c} \qquad \textbf{(39.5)}$$

This time interval Δt_p measured by O' requires only a single clock C' located at the same place as the laser in this frame.

Now consider the same pair of events as viewed by observer O in a second frame, as shown in Figure 39.8b. According to this observer, the mirror and laser are moving to the right with a speed v, and as a result the sequence of events appears entirely different. By the time the light from the laser reaches the mirror, the mirror has moved to the right a distance $v\,\Delta t/2$, where Δt is the time it takes the light to travel from O' to the mirror and back to O' as measured by O. In other words, O concludes that, because of the motion of the vehicle, if the light is to hit the mirror, it must leave the laser at an angle with respect to the vertical direction. Comparing Figure 39.8a and b, we see that the light must travel farther in (b) than in (a). (Note that neither observer "knows" that he is moving. Each is at rest in his own inertial frame.)

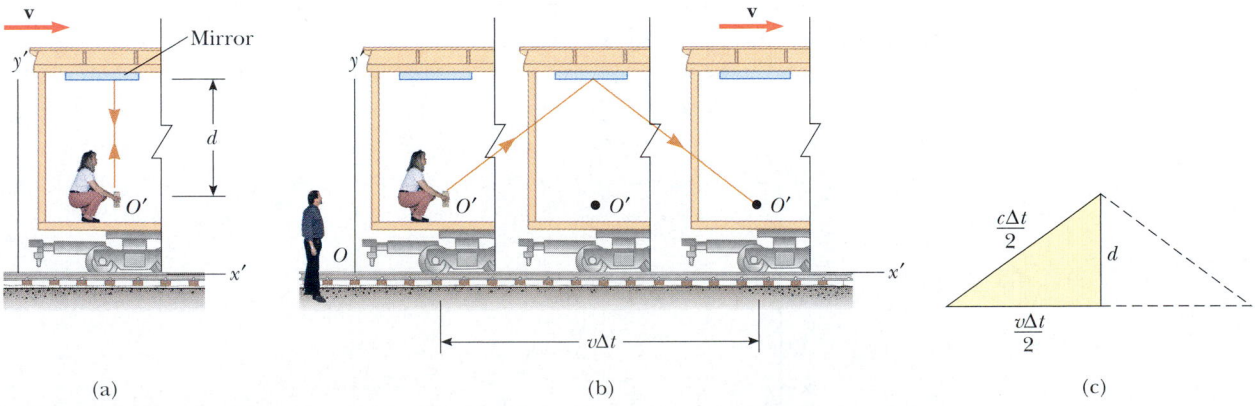

    (a)   (b)   (c)

Figure 39.8 (a) A mirror is fixed to a moving vehicle, and a light pulse is sent out by observer O' at rest in the vehicle. (b) Relative to a stationary observer O standing alongside the vehicle, the mirror and O' move with a speed v. Note that what observer O measures for the distance the pulse travels is greater than $2d$. (c) The right triangle for calculating the relationship between Δt and Δt_p.

According to the second postulate of the special theory of relativity, both observers must measure c for the speed of light. Because the light travels farther in the frame of O, it follows that the time interval Δt measured by O is longer than the time interval Δt_p measured by O'. To obtain a relationship between these two time intervals, it is convenient to use the right triangle shown in Figure 39.8c. The Pythagorean theorem gives

$$\left(\frac{c\,\Delta t}{2}\right)^2 = \left(\frac{v\,\Delta t}{2}\right)^2 + d^2$$

Solving for Δt gives

$$\Delta t = \frac{2d}{\sqrt{c^2 - v^2}} = \frac{2d}{c\sqrt{1 - \dfrac{v^2}{c^2}}} \tag{39.6}$$

Because $\Delta t_p = 2d/c$, we can express this result as

Time dilation

$$\Delta t = \frac{\Delta t_p}{\sqrt{1 - \dfrac{v^2}{c^2}}} = \gamma\,\Delta t_p \tag{39.7}$$

where

$$\gamma = (1 - v^2/c^2)^{-1/2} \tag{39.8}$$

Because γ is always greater than unity, this result says that **the time interval Δt measured by an observer moving with respect to a clock is longer than the time interval Δt_p measured by an observer at rest with respect to the clock.** (That is, $\Delta t > \Delta t_p$.) This effect is known as **time dilation.** Figure 39.9 shows that as the velocity approaches the speed of light, γ increases dramatically. Note that for speeds less than one tenth the speed of light, γ is very nearly equal to unity.

The time interval Δt_p in Equations 39.5 and 39.7 is called the **proper time.** (In German, Einstein used the term *Eigenzeit*, which means "own-time.") In general, **proper time is the time interval between two events measured by an observer who sees the events occur at the same point in space.** Proper time is always the time measured with a single clock (clock C' in our case) at rest in the frame in which the events take place.

If a clock is moving with respect to you, it appears to fall behind (tick more slowly than) the clocks it is passing in the grid of synchronized clocks in your reference frame. Because the time interval $\gamma(2d/c)$, the interval between ticks of a moving clock, is observed to be longer than $2d/c$, the time interval between ticks of an identi-

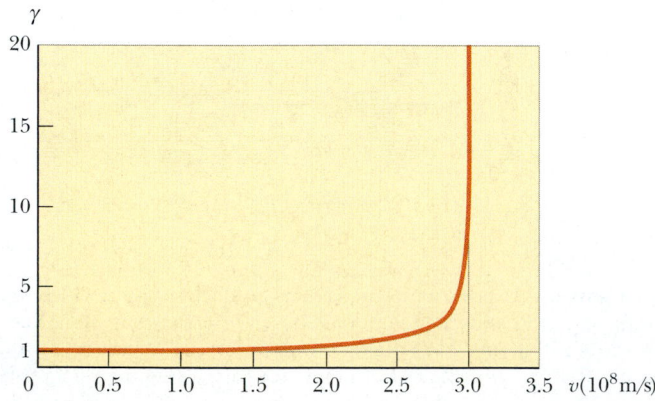

Figure 39.9 Graph of γ versus v. As the velocity approaches the speed of light, γ increases rapidly.

cal clock in your reference frame, it is often said that a moving clock runs more slowly than a clock in your reference frame by a factor γ. This is true for mechanical clocks as well as for the light clock just described. We can generalize this result by stating that all physical processes, including chemical and biological ones, slow down relative to a stationary clock when those processes occur in a moving frame. For example, the heartbeat of an astronaut moving through space would keep time with a clock inside the spaceship. Both the astronaut's clock and heartbeat would be slowed down relative to a stationary clock back on the Earth (although the astronaut would have no sensation of life slowing down in the spaceship).

Quick Quiz 39.4

A rocket has a clock built into its control panel. Use Figure 39.9 to determine approximately how fast the rocket must be moving before its clock appears to an Earth-bound observer to be ticking at one fifth the rate of a clock on the wall at Mission Control. What does an astronaut in the rocket observe?

Bizarre as it may seem, time dilation is a verifiable phenomenon. An experiment reported by Hafele and Keating provided direct evidence of time dilation.[4] Time intervals measured with four cesium atomic clocks in jet flight were compared with time intervals measured by Earth-based reference atomic clocks. In order to compare these results with theory, many factors had to be considered, including periods of acceleration and deceleration relative to the Earth, variations in direction of travel, and the fact that the gravitational field experienced by the flying clocks was weaker than that experienced by the Earth-based clock. The results were in good agreement with the predictions of the special theory of relativity and can be explained in terms of the relative motion between the Earth and the jet aircraft. In their paper, Hafele and Keating stated that "Relative to the atomic time scale of the U.S. Naval Observatory, the flying clocks lost 59 ± 10 ns during the eastward trip and gained 273 ± 7 ns during the westward trip . . . . These results provide an unambiguous empirical resolution of the famous clock paradox with macroscopic clocks."

Another interesting example of time dilation involves the observation of *muons*, unstable elementary particles that have a charge equal to that of the electron and a mass 207 times that of the electron. Muons can be produced by the collision of cosmic radiation with atoms high in the atmosphere. These particles have a lifetime of $2.2~\mu s$ when measured in a reference frame in which they are at rest or moving slowly. If we take $2.2~\mu s$ as the average lifetime of a muon and assume that its speed is close to the speed of light, we find that these particles travel only approximately 600 m before they decay (Fig. 39.10a). Hence, they cannot reach the Earth from the upper atmosphere where they are produced. However, experiments show that a large number of muons do reach the Earth. The phenomenon of time dilation explains this effect. Relative to an observer on the Earth, the muons have a lifetime equal to $\gamma\tau_p$, where $\tau_p = 2.2~\mu s$ is the lifetime in the frame traveling with the muons or the proper lifetime. For example, for a muon speed of $v = 0.99c$, $\gamma \approx 7.1$ and $\gamma\tau_p \approx 16~\mu s$. Hence, the average distance traveled as measured by an observer on the Earth is $\gamma v\tau_p \approx 4\,800$ m, as indicated in Figure 39.10b.

In 1976, at the laboratory of the European Council for Nuclear Research (CERN) in Geneva, muons injected into a large storage ring reached speeds of ap-

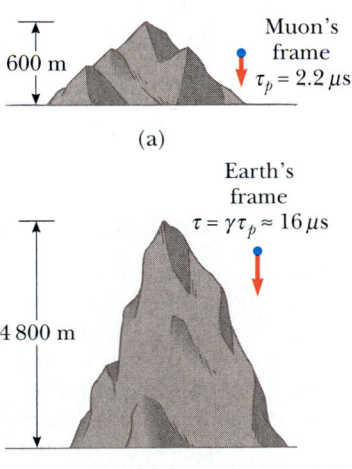

Figure 39.10 (a) Muons moving with a speed of $0.99c$ travel approximately 600 m as measured in the reference frame of the muons, where their lifetime is about $2.2~\mu s$. (b) The muons travel approximately $4\,800$ m as measured by an observer on the Earth.

---

[4] J. C. Hafele and R. E. Keating, "Around the World Atomic Clocks: Relativistic Time Gains Observed," *Science,* 177:168, 1972.

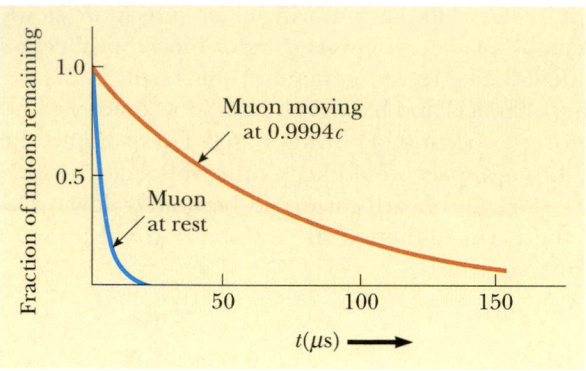

Figure 39.11 Decay curves for muons at rest and for muons traveling at a speed of 0.9994c.

proximately 0.9994c. Electrons produced by the decaying muons were detected by counters around the ring, enabling scientists to measure the decay rate and hence the muon lifetime. The lifetime of the moving muons was measured to be approximately 30 times as long as that of the stationary muon (Fig. 39.11), in agreement with the prediction of relativity to within two parts in a thousand.

EXAMPLE 39.1 ▸ What Is the Period of the Pendulum?

The period of a pendulum is measured to be 3.0 s in the reference frame of the pendulum. What is the period when measured by an observer moving at a speed of 0.95c relative to the pendulum?

Solution Instead of the observer moving at 0.95c, we can take the equivalent point of view that the observer is at rest and the pendulum is moving at 0.95c past the stationary observer. Hence, the pendulum is an example of a moving clock.

The proper time is $\Delta t_p = 3.0$ s. Because a moving clock

runs more slowly than a stationary clock by a factor γ, Equation 39.7 gives

$$\Delta t = \gamma \Delta t_p = \frac{1}{\sqrt{1 - \dfrac{(0.95c)^2}{c^2}}} \Delta t_p = \frac{1}{\sqrt{1 - 0.902}} \Delta t_p$$

$$= (3.2)(3.0 \text{ s}) = \boxed{9.6 \text{ s}}$$

That is, a moving pendulum takes longer to complete a period than a pendulum at rest does.

EXAMPLE 39.2 ▸ How Long Was Your Trip?

Suppose you are driving your car on a business trip and are traveling at 30 m/s. Your boss, who is waiting at your destination, expects the trip to take 5.0 h. When you arrive late, your excuse is that your car clock registered the passage of 5.0 h but that you were driving fast and so your clock ran more slowly than your boss's clock. If your car clock actually did indicate a 5.0-h trip, how much time passed on your boss's clock, which was at rest on the Earth?

Solution We begin by calculating γ from Equation 39.8:

$$\gamma = \frac{1}{\sqrt{1 - \dfrac{v^2}{c^2}}} = \frac{1}{\sqrt{1 - \dfrac{(3 \times 10^1 \text{ m/s})^2}{(3 \times 10^8 \text{ m/s})^2}}} = \frac{1}{\sqrt{1 - 10^{-14}}}$$

If you try to determine this value on your calculator, you will probably get $\gamma = 1$. However, if we perform a binomial expansion, we can more precisely determine the value as

$$\gamma = (1 - 10^{-14})^{-1/2} \approx 1 + \tfrac{1}{2}(10^{-14}) = 1 + 5.0 \times 10^{-15}$$

This result indicates that at typical automobile speeds, γ is not much different from 1.

Applying Equation 39.7, we find Δt, the time interval measured by your boss, to be

$$\Delta t = \gamma \Delta t_p = (1 + 5.0 \times 10^{-15})(5.0 \text{ h})$$

$$= 5.0 \text{ h} + 2.5 \times 10^{-14} \text{ h} = \boxed{5.0 \text{ h} + 0.09 \text{ ns}}$$

Your boss's clock would be only 0.09 ns ahead of your car clock. You might want to try another excuse!

The Twins Paradox

An intriguing consequence of time dilation is the so-called *twins paradox* (Fig. 39.12). Consider an experiment involving a set of twins named Speedo and Goslo. When they are 20 yr old, Speedo, the more adventuresome of the two, sets out on an epic journey to Planet X, located 20 ly from the Earth. Furthermore, his spaceship is capable of reaching a speed of 0.95*c* relative to the inertial frame of his twin brother back home. After reaching Planet X, Speedo becomes homesick and immediately returns to the Earth at the same speed 0.95*c*. Upon his return, Speedo is shocked to discover that Goslo has aged 42 yr and is now 62 yr old. Speedo, on the other hand, has aged only 13 yr.

At this point, it is fair to raise the following question—which twin is the traveler and which is really younger as a result of this experiment? From Goslo's frame of reference, he was at rest while his brother traveled at a high speed. But from Speedo's perspective, it is he who was at rest while Goslo was on the high-speed space journey. According to Speedo, he himself remained stationary while Goslo and the Earth raced away from him on a 6.5-yr journey and then headed back for another 6.5 yr. This leads to an apparent contradiction. Which twin has developed signs of excess aging?

To resolve this apparent paradox, recall that the special theory of relativity deals with inertial frames of reference moving relative to each other at uniform speed. However, the trip in our current problem is not symmetrical. Speedo, the space traveler, must experience a series of accelerations during his journey. As a result, his speed is not always uniform, and consequently he is not in an inertial frame. He cannot be regarded as always being at rest while Goslo is in uniform motion because to do so would be an incorrect application of the special theory of relativity. Therefore, there is no paradox. During each passing year noted by Goslo, slightly less than 4 months elapsed for Speedo.

The conclusion that Speedo is in a noninertial frame is inescapable. Each twin observes the other as accelerating, but it is Speedo that actually undergoes dynamical acceleration due to the real forces acting on him. The time required to accelerate and decelerate Speedo's spaceship may be made very small by using large rockets, so that Speedo can claim that he spends most of his time traveling to Planet X at 0.95*c* in an inertial frame. However, Speedo must slow down, reverse his motion, and return to the Earth in an altogether different inertial frame. At the very best,

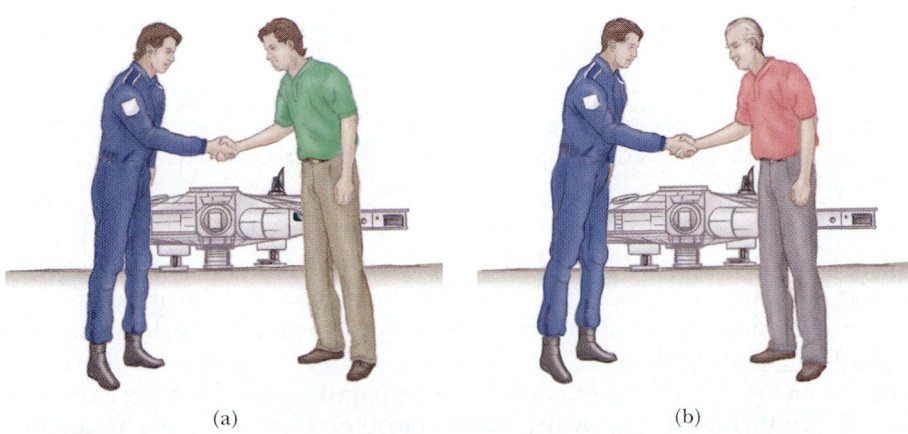

(a)   (b)

Figure 39.12   (a) As one twin leaves his brother on the Earth, both are the same age.
(b) When Speedo returns from his journey to Planet X, he is younger than his twin Goslo.

Speedo is in two different inertial frames during his journey. Only Goslo, who is in a single inertial frame, can apply the simple time-dilation formula to Speedo's trip. Thus, Goslo finds that instead of aging 42 yr, Speedo ages only $(1 - v^2/c^2)^{1/2}(42 \text{ yr}) = 13 \text{ yr}$. Conversely, Speedo spends 6.5 yr traveling to Planet X and 6.5 yr returning, for a total travel time of 13 yr, in agreement with our earlier statement.

Quick Quiz 39.5

Suppose astronauts are paid according to the amount of time they spend traveling in space. After a long voyage traveling at a speed approaching c, would a crew rather be paid according to an Earth-based clock or their spaceship's clock?

Length Contraction

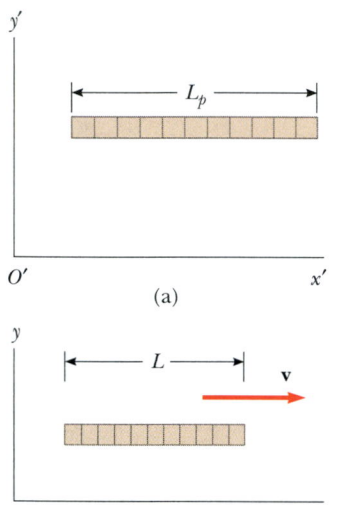

Figure 39.13　(a) A stick measured by an observer in a frame attached to the stick (that is, both have the same velocity) has its proper length L_p. (b) The stick measured by an observer in a frame in which the stick has a velocity **v** relative to the frame is shorter than its proper length L_p by a factor $(1 - v^2/c^2)^{1/2}$.

Length contraction

The measured distance between two points also depends on the frame of reference. **The proper length L_p of an object is the length measured by someone at rest relative to the object.** The length of an object measured by someone in a reference frame that is moving with respect to the object is always less than the proper length. This effect is known as **length contraction.**

Consider a spaceship traveling with a speed v from one star to another. There are two observers: one on the Earth and the other in the spaceship. The observer at rest on the Earth (and also assumed to be at rest with respect to the two stars) measures the distance between the stars to be the proper length L_p. According to this observer, the time it takes the spaceship to complete the voyage is $\Delta t = L_p/v$. Because of time dilation, the space traveler measures a smaller time of travel by the spaceship clock: $\Delta t_p = \Delta t/\gamma$. The space traveler claims to be at rest and sees the destination star moving toward the spaceship with speed v. Because the space traveler reaches the star in the time Δt_p, he or she concludes that the distance L between the stars is shorter than L_p. This distance measured by the space traveler is

$$L = v\,\Delta t_p = v\frac{\Delta t}{\gamma}$$

Because $L_p = v\,\Delta t$, we see that

$$L = \frac{L_p}{\gamma} = L_p\left(1 - \frac{v^2}{c^2}\right)^{1/2} \tag{39.9}$$

where $(1 - v^2/c^2)^{1/2}$ is a factor less than unity. This result may be interpreted as follows:

> If an object has a proper length L_p when it is at rest, then when it moves with speed v in a direction parallel to its length, it contracts to the length $L = L_p(1 - v^2/c^2)^{1/2} = L_p/\gamma$.

For example, suppose that a stick moves past a stationary Earth observer with speed v, as shown in Figure 39.13. The length of the stick as measured by an observer in a frame attached to the stick is the proper length L_p shown in Figure 39.13a. The length of the stick L measured by the Earth observer is shorter than L_p by the factor $(1 - v^2/c^2)^{1/2}$. Furthermore, length contraction is a symmetrical effect: If the stick is at rest on the Earth, an observer in a moving frame would measure its length to be shorter by the same factor $(1 - v^2/c^2)^{1/2}$. Note that **length contraction takes place only along the direction of motion.**

It is important to emphasize that proper length and proper time are measured in different reference frames. As an example of this point, let us return to the decaying muons moving at speeds close to the speed of light. An observer in the muon reference frame measures the proper lifetime (that is, the time interval τ_p), whereas an Earth-based observer measures a dilated lifetime. However, the Earth-based observer measures the proper height (the length L_p) of the mountain in Figure 39.10b. In the muon reference frame, this height is less than L_p, as the figure shows. Thus, in the muon frame, length contraction occurs but time dilation does not. In the Earth-based reference frame, time dilation occurs but length contraction does not. Thus, when calculations on the muon are performed in both frames, the effect of "offsetting penalties" is seen, and the outcome of the experiment in one frame is the same as the outcome in the other frame!

EXAMPLE 39.3   **The Contraction of a Spaceship**

A spaceship is measured to be 120.0 m long and 20.0 m in diameter while at rest relative to an observer. If this spaceship now flies by the observer with a speed of $0.99c$, what length and diameter does the observer measure?

The diameter measured by the observer is still 20.0 m because the diameter is a dimension perpendicular to the motion and length contraction occurs only along the direction of motion.

Solution   From Equation 39.9, the length measured by the observer is

$$L = L_p \sqrt{1 - \frac{v^2}{c^2}} = (120.0 \text{ m})\sqrt{1 - \frac{(0.99c)^2}{c^2}} = \boxed{17 \text{ m}}$$

Exercise   If the ship moves past the observer with a speed of $0.100\ 0c$, what length does the observer measure?

Answer   119.4 m.

EXAMPLE 39.4   **How Long Was Your Car?**

In Example 39.2, you were driving at 30 m/s and claimed that your clock was running more slowly than your boss's stationary clock. Although your statement was true, the time dilation was negligible. If your car is 4.3 m long when it is parked, how much shorter does it appear to a stationary roadside observer as you drive by at 30 m/s?

Solution   The observer sees the horizontal length of the car to be contracted to a length

$$L = L_p \sqrt{1 - \frac{v^2}{c^2}} \approx L_p\left(1 - \frac{1}{2}\frac{v^2}{c^2}\right)$$

where we have again used the binomial expansion for the factor $\sqrt{1 - \frac{v^2}{c^2}}$. The roadside observer sees the car's length as having changed by an amount $L_p - L$:

$$L_p - L \approx \frac{L_p}{2}\left(\frac{v^2}{c^2}\right) = \left(\frac{4.3 \text{ m}}{2}\right)\left(\frac{3.0 \times 10^1 \text{ m/s}}{3.0 \times 10^8 \text{ m/s}}\right)^2$$

$$= \boxed{2.2 \times 10^{-14} \text{ m}}$$

This is much smaller than the diameter of an atom!

EXAMPLE 39.5   **A Voyage to Sirius**

An astronaut takes a trip to Sirius, which is located a distance of 8 lightyears from the Earth. (Note that 1 lightyear (ly) is the distance light travels through free space in 1 yr.) The astronaut measures the time of the one-way journey to be 6 yr. If the spaceship moves at a constant speed of $0.8c$, how can the 8-ly distance be reconciled with the 6-yr trip time measured by the astronaut?

Solution   The 8 ly represents the proper length from the Earth to Sirius measured by an observer seeing both bodies

nearly at rest. The astronaut sees Sirius approaching her at $0.8c$ but also sees the distance contracted to

$$\frac{8 \text{ ly}}{\gamma} = (8 \text{ ly})\sqrt{1 - \frac{v^2}{c^2}} = (8 \text{ ly})\sqrt{1 - \frac{(0.8c)^2}{c^2}} = 5 \text{ ly}$$

Thus, the travel time measured on her clock is

$$t = \frac{d}{v} = \frac{5 \text{ ly}}{0.8c} = 6 \text{ yr}$$

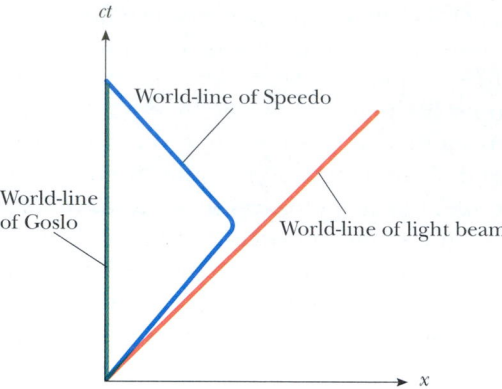

Figure 39.14   The twins paradox on a space–time graph. The twin who stays on the Earth has a world-line along the t axis. The path of the traveling twin through space–time is represented by a world-line that changes direction.

Space–Time Graphs

It is sometimes helpful to make a *space–time graph,* in which time is the ordinate and displacement is the abscissa. The twins paradox is displayed in such a graph in Figure 39.14. A path through space–time is called a **world-line.** At the origin, the world-lines of Speedo and Goslo coincide because the twins are in the same location at the same time. After Speedo leaves on his trip, his world-line diverges from that of his brother. At their reunion, the two world-lines again come together. Note that Goslo's world-line is vertical, indicating no displacement from his original location. Also note that it would be impossible for Speedo to have a world-line that crossed the path of a light beam that left the Earth when he did. To do so would require him to have a speed greater than c.

World-lines for light beams are diagonal lines on space–time graphs, typically drawn at 45° to the right or left of vertical, depending on whether the light beam is traveling in the direction of increasing or decreasing x. These two world-lines means that all possible future events for Goslo and Speedo lie within two 45° lines extending from the origin. Either twin's presence at an event outside this "light cone" would require that twin to move at a speed greater than c, which, as we shall see in Section 39.5, is not possible. Also, the only past events that Goslo and Speedo could have experienced occurred within two similar 45° world-lines that approach the origin from below the x axis.

Quick Quiz 39.6

How is acceleration indicated on a space–time graph?

The Relativistic Doppler Effect

Another important consequence of time dilation is the shift in frequency found for light emitted by atoms in motion as opposed to light emitted by atoms at rest. This phenomenon, known as the Doppler effect, was introduced in Chapter 17 as it pertains to sound waves. In the case of sound, the motion of the source with respect to the medium of propagation can be distinguished from the motion of the observer with respect to the medium. Light waves must be analyzed differently, however, because they require no medium of propagation, and no method exists for distinguishing the motion of a light source from the motion of the observer.

If a light source and an observer approach each other with a relative speed v, the frequency f_{obs} measured by the observer is

$$f_{obs} = \frac{\sqrt{1 + v/c}}{\sqrt{1 - v/c}} f_{source} \qquad (39.10)$$

where f_{source} is the frequency of the source measured in its rest frame. Note that this relativistic Doppler shift formula, unlike the Doppler shift formula for sound, depends only on the relative speed v of the source and observer and holds for relative speeds as great as c. As you might expect, the formula predicts that $f_{obs} > f_{source}$ when the source and observer approach each other. We obtain the expression for the case in which the source and observer recede from each other by replacing v with $-v$ in Equation 39.10.

The most spectacular and dramatic use of the relativistic Doppler effect is the measurement of shifts in the frequency of light emitted by a moving astronomical object such as a galaxy. Spectral lines normally found in the extreme violet region for galaxies at rest with respect to the Earth are shifted by about 100 nm toward the red end of the spectrum for distant galaxies—indicating that these galaxies are *receding* from us. The American astronomer Edwin Hubble (1889–1953) performed extensive measurements of this *red shift* to confirm that most galaxies are moving away from us, indicating that the Universe is expanding.

39.5 > THE LORENTZ TRANSFORMATION EQUATIONS

We have seen that the Galilean transformation equations are not valid when v approaches the speed of light. In this section, we state the correct transformation equations that apply for all speeds in the range $0 \leq v < c$.

Suppose that an event that occurs at some point P is reported by two observers, one at rest in a frame S and the other in a frame S′ that is moving to the right with speed v, as in Figure 39.15. The observer in S reports the event with space–time coordinates (x, y, z, t), and the observer in S′ reports the same event using the coordinates (x', y', z', t'). We would like to find a relationship between these coordinates that is valid for all speeds.

The equations that are valid from $v = 0$ to $v = c$ and enable us to transform coordinates from S to S′ are the **Lorentz transformation equations**:

$$x' = \gamma(x - vt)$$
$$y' = y$$
$$z' = z \qquad (39.11)$$
$$t' = \gamma\left(t - \frac{v}{c^2} x\right)$$

Lorentz transformation equations for S → S′

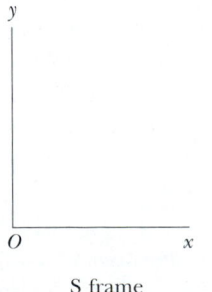

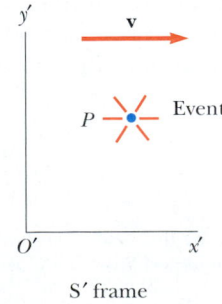

S frame

S′ frame

Figure 39.15 An event that occurs at some point P is observed by two persons, one at rest in the S frame and the other in the S′ frame, which is moving to the right with a speed v.

These transformation equations were developed by Hendrik A. Lorentz (1853–1928) in 1890 in connection with electromagnetism. However, it was Einstein who recognized their physical significance and took the bold step of interpreting them within the framework of the special theory of relativity.

Note the difference between the Galilean and Lorentz time equations. In the Galilean case, $t = t'$, but in the Lorentz case the value for t' assigned to an event by an observer O' standing at the origin of the S$'$ frame in Figure 39.15 depends both on the time t and on the coordinate x as measured by an observer O standing in the S frame. This is consistent with the notion that an event is characterized by four space–time coordinates (x, y, z, t). In other words, in relativity, space and time are not separate concepts but rather are closely interwoven with each other.

If we wish to transform coordinates in the S$'$ frame to coordinates in the S frame, we simply replace v by $-v$ and interchange the primed and unprimed coordinates in Equations 39.11:

$$x = \gamma(x' + vt')$$
$$y = y'$$
$$z = z'$$

(39.12)

$$t = \gamma\left(t' + \frac{v}{c^2}x'\right)$$

When $v \ll c$, the Lorentz transformation equations should reduce to the Galilean equations. To verify this, note that as v approaches zero, $v/c \ll 1$ and $v^2/c^2 \ll 1$; thus, $\gamma = 1$, and Equations 39.11 reduce to the Galilean space–time transformation equations:

$$x' = x - vt \qquad y' = y \qquad z' = z \qquad t' = t$$

In many situations, we would like to know the difference in coordinates between two events or the time interval between two events as seen by observers O and O'. We can accomplish this by writing the Lorentz equations in a form suitable for describing pairs of events. From Equations 39.11 and 39.12, we can express the differences between the four variables x, x', t, and t' in the form

$$\left.\begin{array}{l} \Delta x' = \gamma(\Delta x - v\,\Delta t) \\[2mm] \Delta t' = \gamma\left(\Delta t - \dfrac{v}{c^2}\,\Delta x\right) \end{array}\right\} \text{S} \to \text{S}' \qquad \textbf{(39.13)}$$

$$\left.\begin{array}{l} \Delta x = \gamma(\Delta x' + v\,\Delta t') \\[2mm] \Delta t = \gamma\left(\Delta t' + \dfrac{v}{c^2}\,\Delta x'\right) \end{array}\right\} \text{S}' \to \text{S} \qquad \textbf{(39.14)}$$

where $\Delta x' = x_2' - x_1'$ and $\Delta t' = t_2' - t_1'$ are the differences measured by observer O' and $\Delta x = x_2 - x_1$ and $\Delta t = t_2 - t_1$ are the differences measured by observer O. (We have not included the expressions for relating the y and z coordinates because they are unaffected by motion along the x direction.[5])

---

[5] Although relative motion of the two frames along the x axis does not change the y and z coordinates of an object, it does change the y and z velocity components of an object moving in either frame, as we shall soon see.

EXAMPLE 39.6 ▶ Simultaneity and Time Dilation Revisited

Use the Lorentz transformation equations in difference form to show that (a) simultaneity is not an absolute concept and that (b) moving clocks run more slowly than stationary clocks.

Solution (a) Suppose that two events are simultaneous according to a moving observer O', such that $\Delta t' = 0$. From the expression for Δt given in Equation 39.14, we see that in this case the time interval Δt measured by a stationary observer O is $\Delta t = \gamma v\, \Delta x'/c^2$. That is, the time interval for the same two events as measured by O is nonzero, and so the events do not appear to be simultaneous to O.

(b) Suppose that observer O' finds that two events occur at the same place ($\Delta x' = 0$) but at different times ($\Delta t' \neq 0$). In this situation, the expression for Δt given in Equation 39.14 becomes $\Delta t = \gamma \Delta t'$. This is the equation for time dilation found earlier (Eq. 39.7), where $\Delta t' = \Delta t_p$ is the proper time measured by a clock located in the moving frame of observer O'.

Exercise Use the Lorentz transformation equations in difference form to confirm that $L = L_p/\gamma$ (Eq. 39.9).

Derivation of the Lorentz Velocity Transformation Equation

Once again S is our stationary frame of reference, and S′ is our frame moving at a speed v relative to S. Suppose that an object has a speed u'_x measured in the S′ frame, where

$$u'_x = \frac{dx'}{dt'} \tag{39.15}$$

Using Equation 39.11, we have

$$dx' = \gamma(dx - v\, dt)$$

$$dt' = \gamma\left(dt - \frac{v}{c^2}\, dx\right)$$

Substituting these values into Equation 39.15 gives

$$u'_x = \frac{dx'}{dt'} = \frac{dx - v\, dt}{dt - \dfrac{v}{c^2}\, dx} = \frac{\dfrac{dx}{dt} - v}{1 - \dfrac{v}{c^2}\dfrac{dx}{dt}}$$

But dx/dt is just the velocity component u_x of the object measured by an observer in S, and so this expression becomes

$$u'_x = \frac{u_x - v}{1 - \dfrac{u_x v}{c^2}} \tag{39.16}$$

◀ Lorentz velocity transformation equation for S → S′

If the object has velocity components along the y and z axes, the components as measured by an observer in S′ are

$$u'_y = \frac{u_y}{\gamma\left(1 - \dfrac{u_x v}{c^2}\right)} \quad \text{and} \quad u'_z = \frac{u_z}{\gamma\left(1 - \dfrac{u_x v}{c^2}\right)} \tag{39.17}$$

Note that u'_y and u'_z do not contain the parameter v in the numerator because the relative velocity is along the x axis.

The speed of light is the speed limit of the Universe. It is the maximum possible speed for energy transfer and for information transfer. Any object with mass must move at a lower speed.

When u_x and v are both much smaller than c (the nonrelativistic case), the denominator of Equation 39.16 approaches unity, and so $u'_x \approx u_x - v$, which is the Galilean velocity transformation equation. In the other extreme, when $u_x = c$, Equation 39.16 becomes

$$u'_x = \frac{c - v}{1 - \dfrac{cv}{c^2}} = \frac{c\left(1 - \dfrac{v}{c}\right)}{1 - \dfrac{v}{c}} = c$$

From this result, we see that an object moving with a speed c relative to an observer in S also has a speed c relative to an observer in S′—independent of the relative motion of S and S′. Note that this conclusion is consistent with Einstein's second postulate—that the speed of light must be c relative to all inertial reference frames. Furthermore, the speed of an object can never exceed c. That is, the speed of light is the ultimate speed. We return to this point later when we consider the energy of a particle.

To obtain u_x in terms of u'_x, we replace v by $-v$ in Equation 39.16 and interchange the roles of u_x and u'_x:

Lorentz velocity transformation equations for S′ → S

$$u_x = \frac{u'_x + v}{1 + \dfrac{u'_x v}{c^2}} \tag{39.18}$$

EXAMPLE 39.7   **Relative Velocity of Spaceships**

Two spaceships A and B are moving in opposite directions, as shown in Figure 39.16. An observer on the Earth measures the speed of ship A to be $0.750c$ and the speed of ship B to be $0.850c$. Find the velocity of ship B as observed by the crew on ship A.

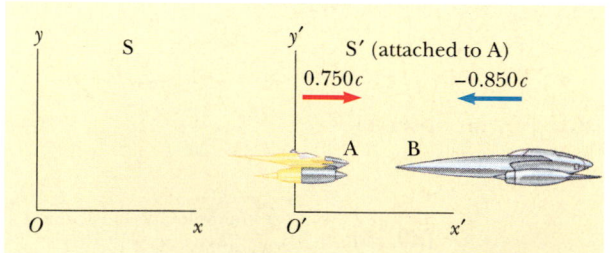

Figure 39.16   Two spaceships A and B move in opposite directions. The speed of B relative to A is *less* than c and is obtained from the relativistic velocity transformation equation.

Solution   We can solve this problem by taking the S′ frame as being attached to ship A, so that $v = 0.750c$ relative to the Earth (the S frame). We can consider ship B as moving with a velocity $u_x = -0.850c$ relative to the Earth. Hence, we can obtain the velocity of ship B relative to ship A by using Equation 39.16:

$$u'_x = \frac{u_x - v}{1 - \dfrac{u_x v}{c^2}} = \frac{-0.850c - 0.750c}{1 - \dfrac{(-0.850c)(0.750c)}{c^2}} = \boxed{-0.977c}$$

The negative sign indicates that ship B is moving in the negative x direction as observed by the crew on ship A. Note that the speed is less than c. That is, a body whose speed is less than c in one frame of reference must have a speed less than c in any other frame. (If the Galilean velocity transformation equation were used in this example, we would find that $u'_x = u_x - v = -0.850c - 0.750c = -1.60c$, which is impossible. The Galilean transformation equation does not work in relativistic situations.)

EXAMPLE 39.8   **The Speeding Motorcycle**

Imagine a motorcycle moving with a speed $0.80c$ past a stationary observer, as shown in Figure 39.17. If the rider tosses a ball in the forward direction with a speed of $0.70c$ relative to himself, what is the speed of the ball relative to the stationary observer?

Solution The speed of the motorcycle relative to the stationary observer is $v = 0.80c$. The speed of the ball in the frame of reference of the motorcyclist is $u_x' = 0.70c$. Therefore, the speed u_x of the ball relative to the stationary observer is

$$u_x = \frac{u_x' + v}{1 + \dfrac{u_x' v}{c^2}} = \frac{0.70c + 0.80c}{1 + \dfrac{(0.70c)(0.80c)}{c^2}} = \boxed{0.90c}$$

Exercise Suppose that the motorcyclist turns on the headlight so that a beam of light moves away from him with a speed c in the forward direction. What does the stationary observer measure for the speed of the light?

Answer c.

Figure 39.17 A motorcyclist moves past a stationary observer with a speed of $0.80c$ and throws a ball in the direction of motion with a speed of $0.70c$ relative to himself.

EXAMPLE 39.9   Relativistic Leaders of the Pack

Two motorcycle pack leaders named David and Emily are racing at relativistic speeds along perpendicular paths, as shown in Figure 39.18. How fast does Emily recede as seen by David over his right shoulder?

Solution Figure 39.18 represents the situation as seen by a police officer at rest in frame S, who observes the following:

$$\text{David:} \qquad u_x = 0.75c \qquad u_y = 0$$

$$\text{Emily:} \qquad u_x = 0 \qquad u_y = -0.90c$$

To calculate Emily's speed of recession as seen by David, we take S′ to move along with David and then calculate u_x' and u_y' for Emily using Equations 39.16 and 39.17:

$$u_x' = \frac{u_x - v}{1 - \dfrac{u_x v}{c^2}} = \frac{0 - 0.75c}{1 - \dfrac{(0)(0.75c)}{c^2}} = -0.75c$$

$$u_y' = \frac{u_y}{\gamma\left(1 - \dfrac{u_x v}{c^2}\right)} = \frac{\sqrt{1 - \dfrac{(0.75c)^2}{c^2}}\,(-0.90c)}{\left(1 - \dfrac{(0)(0.75c)}{c^2}\right)} = -0.60c$$

Thus, the speed of Emily as observed by David is

$$u' = \sqrt{(u_x')^2 + (u_y')^2} = \sqrt{(-0.75c)^2 + (-0.60c)^2} = \boxed{0.96c}$$

Note that this speed is less than c, as required by the special theory of relativity.

Exercise Use the Galilean velocity transformation equation to calculate the classical speed of recession for Emily as observed by David.

Answer $1.2c$.

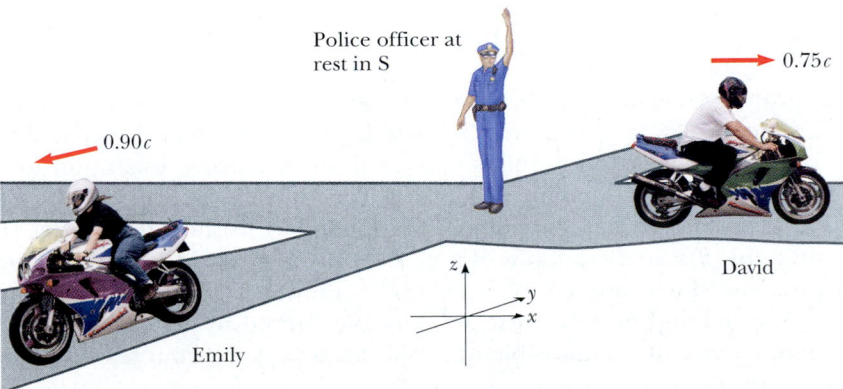

Police officer at rest in S

$0.90c$

$0.75c$

Emily

David

Figure 39.18 David moves to the east with a speed $0.75c$ relative to the police officer, and Emily travels south at a speed $0.90c$ relative to the officer.

39.6 ▷ RELATIVISTIC LINEAR MOMENTUM AND THE RELATIVISTIC FORM OF NEWTON'S LAWS

We have seen that in order to describe properly the motion of particles within the framework of the special theory of relativity, we must replace the Galilean transformation equations by the Lorentz transformation equations. Because the laws of physics must remain unchanged under the Lorentz transformation, we must generalize Newton's laws and the definitions of linear momentum and energy to conform to the Lorentz transformation equations and the principle of relativity. These generalized definitions should reduce to the classical (nonrelativistic) definitions for $v \ll c$.

First, recall that the law of conservation of linear momentum states that when two isolated objects collide, their combined total momentum remains constant. Suppose that the collision is described in a reference frame S in which linear momentum is conserved. If we calculate the velocities in a second reference frame S′ using the Lorentz velocity transformation equation and the classical definition of linear momentum, $\mathbf{p} = m\mathbf{u}$ (where $\mathbf{u}$ is the velocity of either object), we find that linear momentum is *not* conserved in S′. However, because the laws of physics are the same in all inertial frames, linear momentum must be conserved in all frames. In view of this condition and assuming that the Lorentz velocity transformation equation is correct, we must modify the definition of linear momentum to satisfy the following conditions:

- Linear momentum $\mathbf{p}$ must be conserved in all collisions.
- The relativistic value calculated for $\mathbf{p}$ must approach the classical value $m\mathbf{u}$ as $\mathbf{u}$ approaches zero.

For any particle, the correct relativistic equation for linear momentum that satisfies these conditions is

Definition of relativistic linear momentum

$$\mathbf{p} \equiv \frac{m\mathbf{u}}{\sqrt{1 - \dfrac{u^2}{c^2}}} = \gamma m\mathbf{u} \qquad (39.19)$$

where $\mathbf{u}$ is the velocity of the particle and m is the mass of the particle. When u is much less than c, $\gamma = (1 - u^2/c^2)^{-1/2}$ approaches unity and $\mathbf{p}$ approaches $m\mathbf{u}$. Therefore, the relativistic equation for $\mathbf{p}$ does indeed reduce to the classical expression when u is much smaller than c.

The relativistic force $\mathbf{F}$ acting on a particle whose linear momentum is $\mathbf{p}$ is defined as

$$\mathbf{F} \equiv \frac{d\mathbf{p}}{dt} \qquad (39.20)$$

where $\mathbf{p}$ is given by Equation 39.19. This expression, which is the relativistic form of Newton's second law, is reasonable because it preserves classical mechanics in the limit of low velocities and requires conservation of linear momentum for an isolated system ($\mathbf{F} = 0$) both relativistically and classically.

It is left as an end-of-chapter problem (Problem 63) to show that under relativistic conditions, the acceleration $\mathbf{a}$ of a particle decreases under the action of a constant force, in which case $a \propto (1 - u^2/c^2)^{3/2}$. From this formula, note that as the particle's speed approaches c, the acceleration caused by any finite force approaches zero. Hence, it is impossible to accelerate a particle from rest to a speed $u \geq c$.

EXAMPLE 39.10 **Linear Momentum of an Electron**

An electron, which has a mass of 9.11×10^{-31} kg, moves with a speed of $0.750c$. Find its relativistic momentum and compare this value with the momentum calculated from the classical expression.

Solution Using Equation 39.19 with $u = 0.750c$, we have

$$p = \frac{m_e u}{\sqrt{1 - \dfrac{u^2}{c^2}}}$$

$$= \frac{(9.11 \times 10^{-31}\ \text{kg})(0.750 \times 3.00 \times 10^8\ \text{m/s})}{\sqrt{1 - \dfrac{(0.750c)^2}{c^2}}}$$

$$= 3.10 \times 10^{-22}\ \text{kg}\cdot\text{m/s}$$

The (incorrect) classical expression gives

$$p_{\text{classical}} = m_e u = 2.05 \times 10^{-22}\ \text{kg}\cdot\text{m/s}$$

Hence, the correct relativistic result is 50% greater than the classical result!

39.7 RELATIVISTIC ENERGY

We have seen that the definition of linear momentum and the laws of motion require generalization to make them compatible with the principle of relativity. This implies that the definition of kinetic energy must also be modified.

To derive the relativistic form of the work–kinetic energy theorem, let us first use the definition of relativistic force, Equation 39.20, to determine the work done on a particle by a force F:

$$W = \int_{x_1}^{x_2} F\, dx = \int_{x_1}^{x_2} \frac{dp}{dt}\, dx \tag{39.21}$$

for force and motion both directed along the x axis. In order to perform this integration and find the work done on the particle and the relativistic kinetic energy as a function of u, we first evaluate dp/dt:

$$\frac{dp}{dt} = \frac{d}{dt}\frac{mu}{\sqrt{1 - \dfrac{u^2}{c^2}}} = \frac{m(du/dt)}{\left(1 - \dfrac{u^2}{c^2}\right)^{3/2}}$$

Substituting this expression for dp/dt and $dx = u\, dt$ into Equation 39.21 gives

$$W = \int_0^t \frac{m(du/dt)\, u\, dt}{\left(1 - \dfrac{u^2}{c^2}\right)^{3/2}} = m \int_0^u \frac{u}{\left(1 - \dfrac{u^2}{c^2}\right)^{3/2}}\, du$$

where we use the limits 0 and u in the rightmost integral because we have assumed that the particle is accelerated from rest to some final speed u. Evaluating the integral, we find that

$$W = \frac{mc^2}{\sqrt{1 - \dfrac{u^2}{c^2}}} - mc^2 \tag{39.22}$$

Recall from Chapter 7 that the work done by a force acting on a particle equals the change in kinetic energy of the particle. Because of our assumption that the initial speed of the particle is zero, we know that the initial kinetic energy is zero. We therefore conclude that the work W is equivalent to the relativistic kinetic energy K:

$$K = \frac{mc^2}{\sqrt{1 - \dfrac{u^2}{c^2}}} - mc^2 = \gamma mc^2 - mc^2 \qquad \textbf{(39.23)}$$

This equation is routinely confirmed by experiments using high-energy particle accelerators.

At low speeds, where $u/c \ll 1$, Equation 39.23 should reduce to the classical expression $K = \frac{1}{2}mu^2$. We can check this by using the binomial expansion $(1 - x^2)^{-1/2} \approx 1 + \frac{1}{2}x^2 + \dots$ for $x \ll 1$, where the higher-order powers of x are neglected in the expansion. In our case, $x = u/c$, so that

$$\frac{1}{\sqrt{1 - \dfrac{u^2}{c^2}}} = \left(1 - \frac{u^2}{c^2}\right)^{-1/2} \approx 1 + \frac{1}{2}\frac{u^2}{c^2}$$

Substituting this into Equation 39.23 gives

$$K \approx mc^2\left(1 + \frac{1}{2}\frac{u^2}{c^2}\right) - mc^2 = \frac{1}{2}mu^2$$

which is the classical expression for kinetic energy. A graph comparing the relativistic and nonrelativistic expressions is given in Figure 39.19. In the relativistic case, the particle speed never exceeds c, regardless of the kinetic energy. The two curves are in good agreement when $u \ll c$.

The constant term mc^2 in Equation 39.23, which is independent of the speed of the particle, is called the **rest energy** E_R of the particle (see Section 8.9). The term γmc^2, which does depend on the particle speed, is therefore the sum of the kinetic and rest energies. We define γmc^2 to be the **total energy** E:

Total energy = kinetic energy + rest energy

$$E = \gamma mc^2 = K + mc^2 \qquad \textbf{(39.24)}$$

or

$$E = \frac{mc^2}{\sqrt{1 - \dfrac{u^2}{c^2}}} \qquad \textbf{(39.25)}$$

This is Einstein's famous equation about mass–energy equivalence.

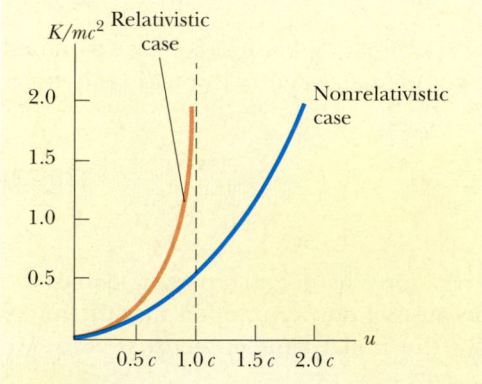

Figure 39.19   A graph comparing relativistic and nonrelativistic kinetic energy. The energies are plotted as a function of speed. In the relativistic case, u is always less than c.

The relationship $E = K + mc^2$ shows that **mass is a form of energy,** where c^2 in the rest energy term is just a constant conversion factor. This expression also shows that a small mass corresponds to an enormous amount of energy, a concept fundamental to nuclear and elementary-particle physics.

In many situations, the linear momentum or energy of a particle is measured rather than its speed. It is therefore useful to have an expression relating the total energy E to the relativistic linear momentum p. This is accomplished by using the expressions $E = \gamma mc^2$ and $p = \gamma mu$. By squaring these equations and subtracting, we can eliminate u (Problem 39). The result, after some algebra, is[6]

$$E^2 = p^2 c^2 + (mc^2)^2 \qquad \textbf{(39.26)}$$

Energy–momentum relationship

When the particle is at rest, $p = 0$ and so $E = E_R = mc^2$. For particles that have zero mass, such as photons, we set $m = 0$ in Equation 39.26 and see that

$$E = pc \qquad \textbf{(39.27)}$$

This equation is an exact expression relating total energy and linear momentum for photons, which always travel at the speed of light.

Finally, note that because the mass m of a particle is independent of its motion, m must have the same value in all reference frames. For this reason, m is often called the **invariant mass.** On the other hand, because the total energy and linear momentum of a particle both depend on velocity, these quantities depend on the reference frame in which they are measured.

Because m is a constant, we conclude from Equation 39.26 that the quantity $E^2 - p^2 c^2$ must have the same value in all reference frames. That is, $E^2 - p^2 c^2$ is invariant under a Lorentz transformation. (Equations 39.26 and 39.27 do not make provision for potential energy.)

When we are dealing with subatomic particles, it is convenient to express their energy in electron volts because the particles are usually given this energy by acceleration through a potential difference. The conversion factor, as you recall from Equation 25.5, is

$$1.00 \text{ eV} = 1.602 \times 10^{-19} \text{ J}$$

For example, the mass of an electron is 9.109×10^{-31} kg. Hence, the rest energy of the electron is

$$m_e c^2 = (9.109 \times 10^{-31} \text{ kg})(2.9979 \times 10^8 \text{ m/s})^2 = 8.187 \times 10^{-14} \text{ J}$$
$$= (8.187 \times 10^{-14} \text{ J})(1 \text{ eV}/1.602 \times 10^{-19} \text{ J}) = 0.511 \text{ MeV}$$

EXAMPLE 39.11 **The Energy of a Speedy Electron**

An electron in a television picture tube typically moves with a speed $u = 0.250c$. Find its total energy and kinetic energy in electron volts.

Solution Using the fact that the rest energy of the electron is 0.511 MeV together with Equation 39.25, we have

$$E = \frac{m_e c^2}{\sqrt{1 - \dfrac{u^2}{c^2}}} = \frac{0.511 \text{ MeV}}{\sqrt{1 - \dfrac{(0.250c)^2}{c^2}}}$$

$$= 1.03(0.511 \text{ MeV}) = \boxed{0.528 \text{ MeV}}$$

This is 3% greater than the rest energy.

We obtain the kinetic energy by subtracting the rest energy from the total energy:

$$K = E - m_e c^2 = 0.528 \text{ MeV} - 0.511 \text{ MeV} = \boxed{0.017 \text{ MeV}}$$

---

[6] One way to remember this relationship is to draw a right triangle having a hypotenuse of length E and legs of lengths pc and mc^2.

EXAMPLE 39.12   **The Energy of a Speedy Proton**

(a) Find the rest energy of a proton in electron volts.

Solution

$$E_R = m_p c^2 = (1.67 \times 10^{-27} \text{ kg})(3.00 \times 10^8 \text{ m/s})^2$$
$$= (1.50 \times 10^{-10} \text{ J})(1.00 \text{ eV}/1.60 \times 10^{-19} \text{ J})$$

$$= \boxed{938 \text{ MeV}}$$

(b) If the total energy of a proton is three times its rest energy, with what speed is the proton moving?

Solution   Equation 39.25 gives

$$E = 3 m_p c^2 = \frac{m_p c^2}{\sqrt{1 - \dfrac{u^2}{c^2}}}$$

$$3 = \frac{1}{\sqrt{1 - \dfrac{u^2}{c^2}}}$$

Solving for u gives

$$1 - \frac{u^2}{c^2} = \frac{1}{9}$$

$$\frac{u^2}{c^2} = \frac{8}{9}$$

$$u = \frac{\sqrt{8}}{3} c = \boxed{2.83 \times 10^8 \text{ m/s}}$$

(c) Determine the kinetic energy of the proton in electron volts.

Solution   From Equation 39.24,

$$K = E - m_p c^2 = 3 m_p c^2 - m_p c^2 = 2 m_p c^2$$

Because $m_p c^2 = 938$ MeV, $K = \boxed{1\,880 \text{ MeV}}$

(d) What is the proton's momentum?

Solution   We can use Equation 39.26 to calculate the momentum with $E = 3 m_p c^2$:

$$E^2 = p^2 c^2 + (m_p c^2)^2 = (3 m_p c^2)^2$$

$$p^2 c^2 = 9(m_p c^2)^2 - (m_p c^2)^2 = 8(m_p c^2)^2$$

$$p = \sqrt{8}\,\frac{m_p c^2}{c} = \sqrt{8}\,\frac{(938 \text{ MeV})}{c} = \boxed{2\,650 \text{ MeV}/c}$$

The unit of momentum is written MeV/c for convenience.

39.8   **EQUIVALENCE OF MASS AND ENERGY**

To understand the equivalence of mass and energy, consider the following thought experiment proposed by Einstein in developing his famous equation $E = mc^2$. Imagine an isolated box of mass M_{box} and length L initially at rest, as shown in Figure 39.20a. Suppose that a pulse of light is emitted from the left side of the box, as depicted in Figure 39.20b. From Equation 39.27, we know that light of energy E carries linear momentum $p = E/c$. Hence, if momentum is to be conserved, the box must recoil to the left with a speed v. If it is assumed that the box is very massive, the recoil speed is much less than the speed of light, and conservation of momentum gives $M_{\text{box}} v = E/c$, or

$$v = \frac{E}{M_{\text{box}} c}$$

The time it takes the light pulse to move the length of the box is approximately $\Delta t = L/c$. In this time interval, the box moves a small distance Δx to the left, where

$$\Delta x = v \Delta t = \frac{EL}{M_{\text{box}} c^2}$$

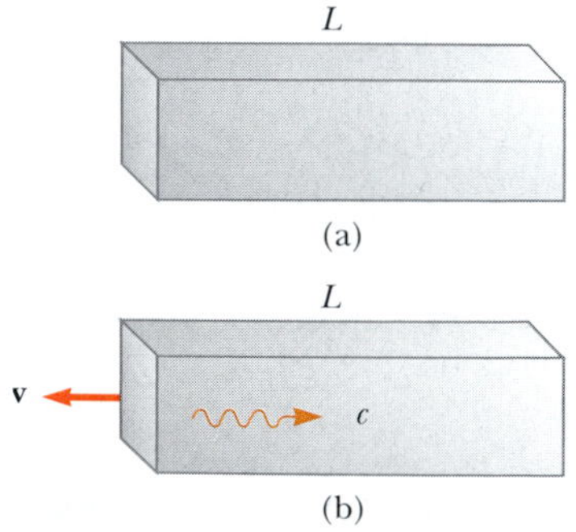

Figure 39.20   (a) A box of length L at rest. (b) When a light pulse directed to the right is emitted at the left end of the box, the box recoils to the left until the pulse strikes the right end.

The light then strikes the right end of the box and transfers its momentum to the box, causing the box to stop. With the box in its new position, its center of mass appears to have moved to the left. However, its center of mass cannot have moved because the box is an isolated system. Einstein resolved this perplexing situation by assuming that in addition to energy and momentum, light also carries mass. If M_{pulse} is the effective mass carried by the pulse of light and if the center of mass of the system (box plus pulse of light) is to remain fixed, then

$$M_{pulse}L = M_{box}\Delta x$$

Solving for M_{pulse}, and using the previous expression for Δx, we obtain

$$M_{pulse} = \frac{M_{box}\Delta x}{L} = \frac{M_{box}}{L}\frac{EL}{M_{box}c^2} = \frac{E}{c^2}$$

or

$$E = M_{pulse}c^2$$

Thus, Einstein reached the profound conclusion that "if a body gives off the energy E in the form of radiation, its mass diminishes by E/c^2, . . ."

Although we derived the relationship $E = mc^2$ for light energy, the equivalence of mass and energy is universal. Equation 39.24, $E = \gamma mc^2$, which represents the total energy of any particle, suggests that even when a particle is at rest ($\gamma = 1$) it still possesses enormous energy because it has mass. Probably the clearest experimental proof of the equivalence of mass and energy occurs in nuclear and elementary particle interactions, where large amounts of energy are released and the energy release is accompanied by a decrease in mass. Because energy and mass are related, we see that the laws of conservation of energy and conservation of mass are one and the same. Simply put, this law states that

the energy of a system of particles before interaction must equal the energy of the system after interaction, where energy of the *i*th particle is given by the expression

$$E_i = \frac{m_i c^2}{\sqrt{1 - \dfrac{u_i^2}{c^2}}} = \gamma m_i c^2$$

Conversion of mass–energy

The release of enormous quantities of energy from subatomic particles, accompanied by changes in their masses, is the basis of all nuclear reactions. In a conventional nuclear reactor, a uranium nucleus undergoes *fission*, a reaction that creates several lighter fragments having considerable kinetic energy. The combined mass of all the fragments is less than the mass of the parent uranium nucleus by an amount Δm. The corresponding energy Δmc^2 associated with this mass difference is exactly equal to the total kinetic energy of the fragments. This kinetic energy raises the temperature of water in the reactor, converting it to steam for the generation of electric power.

In the nuclear reaction called *fusion*, two atomic nuclei combine to form a single nucleus. The fusion reaction in which two deuterium nuclei fuse to form a helium nucleus is of major importance in current research and the development of controlled-fusion reactors. The decrease in mass that results from the creation of one helium nucleus from two deuterium nuclei is $\Delta m = 4.25 \times 10^{-29}$ kg. Hence, the corresponding excess energy that results from one fusion reaction is $\Delta mc^2 =$

3.83×10^{-12} J $= 23.9$ MeV. To appreciate the magnitude of this result, note that if 1 g of deuterium is converted to helium, the energy released is about 10^{12} J! At the current cost of electrical energy, this quantity of energy would be worth about $70 000.

CONCEPTUAL EXAMPLE 39.13

Because mass is a measure of energy, can we conclude that the mass of a compressed spring is greater than the mass of the same spring when it is not compressed?

Solution Recall that when a spring of force constant k is compressed (or stretched) from its equilibrium position a distance x, it stores elastic potential energy $U = kx^2/2$. Ac-

cording to the special theory of relativity, any change in the total energy of a system is equivalent to a change in the mass of the system. Therefore, the mass of a compressed (or stretched) spring is greater than the mass of the spring in its equilibrium position by an amount U/c^2.

EXAMPLE 39.14   Binding Energy of the Deuteron

A deuteron, which is the nucleus of a deuterium atom, contains one proton and one neutron and has a mass of 2.013 553 u. This total deuteron mass is not equal to the sum of the masses of the proton and neutron. Calculate the mass difference and determine its energy equivalence, which is called the *binding energy* of the nucleus.

Solution Using atomic mass units (u), we have

$$m_p = \text{mass of proton} = 1.007\ 276\ \text{u}$$

$$m_n = \text{mass of neutron} = 1.008\ 665\ \text{u}$$

$$m_p + m_n = 2.015\ 941\ \text{u}$$

The mass difference Δm is therefore 0.002 388 u. By defini-

tion, 1 u $= 1.66 \times 10^{-27}$ kg, and therefore

$$\Delta m = 0.002\ 388\ \text{u} = \boxed{3.96 \times 10^{-30}\ \text{kg}}$$

Using $E = \Delta mc^2$, we find that the binding energy is

$$E = \Delta mc^2 = (3.96 \times 10^{-30}\ \text{kg})(3.00 \times 10^8\ \text{m/s})^2$$

$$= 3.56 \times 10^{-13}\ \text{J} = \boxed{2.23\ \text{MeV}}$$

Therefore, the minimum energy required to separate the proton from the neutron of the deuterium nucleus (the binding energy) is 2.23 MeV.

39.9   RELATIVITY AND ELECTROMAGNETISM

Consider two frames of reference S and S' that are in relative motion, and assume that a single charge q is at rest in the S' frame of reference. According to an observer in this frame, an electric field surrounds the charge. However, an observer in frame S says that the charge is in motion and therefore measures both an electric field and a magnetic field. The magnetic field measured by the observer in frame S is created by the moving charge, which constitutes an electric current. In other words, electric and magnetic fields are viewed differently in frames of reference that are moving relative to each other. We now describe one situation that shows how an electric field in one frame of reference is viewed as a magnetic field in another frame of reference.

A positive test charge q is moving parallel to a current-carrying wire with velocity **v** relative to the wire in frame S, as shown in Figure 39.21a. We assume that the net charge on the wire is zero and that the electrons in the wire also move with velocity **v** in a straight line. The leftward current in the wire produces a magnetic field that forms circles around the wire and is directed into the page at the loca-

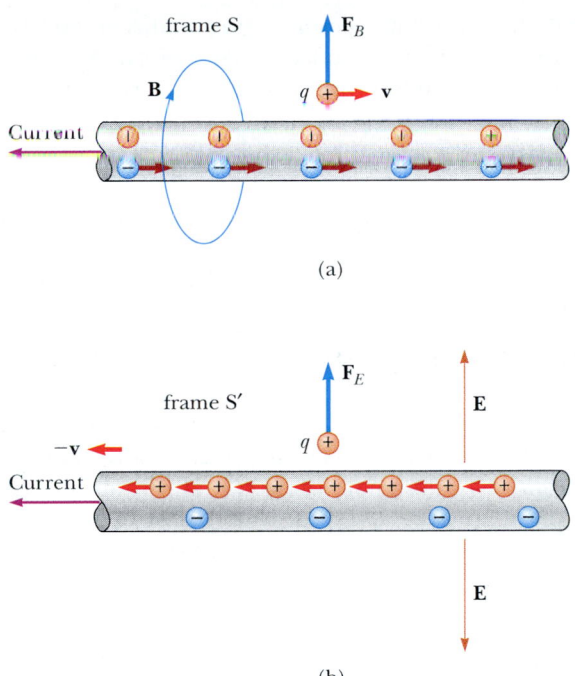

Figure 39.21 (a) In frame S, the positive charge q moves to the right with a velocity **v**, and the current-carrying wire is stationary. A magnetic field **B** surrounds the wire, and charge experiences a *magnetic* force directed away from the wire. (b) In frame S′, the wire moves to the left with a velocity −**v**, and the charge q is stationary. The wire creates an electric field **E**, and the charge experiences an *electric* force directed away from the wire.

tion of the moving test charge. Therefore, a magnetic force $\mathbf{F}_B = q\mathbf{v} \times \mathbf{B}$ directed away from the wire is exerted on the test charge. However, no electric force acts on the test charge because the net charge on the wire is zero when viewed in this frame.

Now consider the same situation as viewed from frame S′, where the test charge is at rest (Figure 39.21b). In this frame, the positive charges in the wire move to the left, the electrons in the wire are at rest, and the wire still carries a current. Because the test charge is not moving in this frame, $\mathbf{F}_B = q\mathbf{v} \times \mathbf{B} = 0$; there is no magnetic force exerted on the test charge when viewed in this frame. However, if a force is exerted on the test charge in frame S′, the frame of the wire, as described earlier, a force must be exerted on it in any other frame. What is the origin of this force in frame S, the frame of the test charge?

The answer to this question is provided by the special theory of relativity. When the situation is viewed in frame S, as in Figure 39.21a, the positive charges are at rest and the electrons in the wire move to the right with a velocity **v**. Because of length contraction, the electrons appear to be closer together than their proper separation. Because there is no net charge on the wire this contracted separation must equal the separation between the stationary positive charges. The situation is quite different when viewed in frame S′, shown in Figure 39.21b. In this frame, the positive charges appear closer together because of length contraction, and the electrons in the wire are at rest with a separation that is greater than that viewed in frame S. Therefore, there is a net positive charge on the wire when viewed in frame S′. This net positive charge produces an electric field pointing away from the wire toward the test charge, and so the test charge experiences an electric force directed away from the wire. Thus, what was viewed as a magnetic field (and a corresponding magnetic force) in the frame of the wire transforms into an electric field (and a corresponding electric force) in the frame of the test charge.

Optional Section

39.10   THE GENERAL THEORY OF RELATIVITY

Up to this point, we have sidestepped a curious puzzle. Mass has two seemingly different properties: a *gravitational attraction* for other masses and an *inertial* property that resists acceleration. To designate these two attributes, we use the subscripts g and i and write

$$\text{Gravitational property} \qquad F_g = m_g g$$

$$\text{Inertial property} \qquad \Sigma F = m_i a$$

The value for the gravitational constant G was chosen to make the magnitudes of m_g and m_i numerically equal. Regardless of how G is chosen, however, the strict proportionality of m_g and m_i has been established experimentally to an extremely high degree: a few parts in 10^{12}. Thus, it appears that gravitational mass and inertial mass may indeed be exactly proportional.

But why? They seem to involve two entirely different concepts: a force of mutual gravitational attraction between two masses, and the resistance of a single mass to being accelerated. This question, which puzzled Newton and many other physicists over the years, was answered when Einstein published his theory of gravitation, known as his *general theory of relativity*, in 1916. Because it is a mathematically complex theory, we offer merely a hint of its elegance and insight.

In Einstein's view, the remarkable coincidence that m_g and m_i seemed to be proportional to each other was evidence of an intimate and basic connection between the two concepts. He pointed out that no mechanical experiment (such as dropping a mass) could distinguish between the two situations illustrated in Figure 39.22a and b. In each case, the dropped briefcase undergoes a downward acceleration g relative to the floor.

Einstein carried this idea further and proposed that *no* experiment, mechanical or otherwise, could distinguish between the two cases. This extension to include all phenomena (not just mechanical ones) has interesting consequences. For example, suppose that a light pulse is sent horizontally across the elevator, as depicted in Figure 39.22c. During the time it takes the light to make the trip, the right wall of the elevator has accelerated upward. This causes the light to arrive at a location lower on the wall than the spot it would have hit if the elevator were not accelerating. Thus, in the frame of the elevator, the trajectory of the light pulse bends downward as the elevator accelerates upward to meet it. Because the accelerating elevator cannot be distinguished from a nonaccelerating one located in a gravitational field, Einstein proposed that a beam of light *should also be bent downward by a gravitational field*. Experiments have verified the effect, although the bending is small. A laser aimed at the horizon falls less than 1 cm after traveling 6 000 km. (No such bending is predicted in Newton's theory of gravitation.)

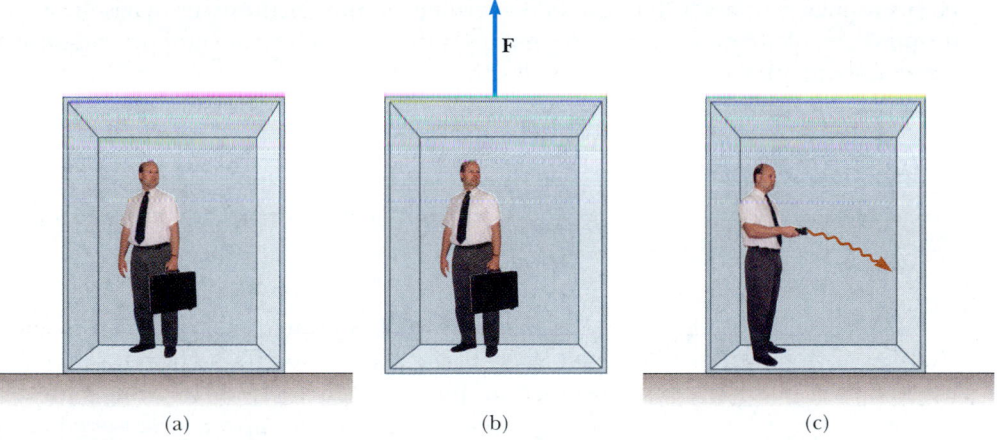

Figure 39.22   (a) The observer is at rest in a uniform gravitational field **g**. (b) The observer is in a region where gravity is negligible, but the frame of reference is accelerated by an external force **F** that produces an acceleration **g**. According to Einstein, the frames of reference in parts (a) and (b) are equivalent in every way. No local experiment can distinguish any difference between the two frames. (c) If parts (a) and (b) are truly equivalent, as Einstein proposed, then a ray of light should bend in a gravitational field.

The two postulates of Einstein's **general theory of relativity** are

- All the laws of nature have the same form for observers in any frame of reference, whether accelerated or not.
- In the vicinity of any point, a gravitational field is equivalent to an accelerated frame of reference in the absence of gravitational effects. (This is the *principle of equivalence.*)

The second postulate implies that gravitational mass and inertial mass are completely equivalent, not just proportional. What were thought to be two different types of mass are actually identical.

One interesting effect predicted by the general theory is that time scales are altered by gravity. A clock in the presence of gravity runs more slowly than one located where gravity is negligible. Consequently, the frequencies of radiation emitted by atoms in the presence of a strong gravitational field are *red-shifted* to lower frequencies when compared with the same emissions in the presence of a weak field. This gravitational red shift has been detected in spectral lines emitted by

This Global Positioning System (GPS) unit incorporates relativistically corrected time calculations in its analysis of signals it receives from orbiting satellites. These corrections allow the unit to determine its position on the Earth's surface to within a few meters. If the corrections were not made, the location error would be about 1 km. *(Courtesy of Trimble Navigation Limited)*

atoms in massive stars. It has also been verified on the Earth by comparison of the frequencies of gamma rays (a high-energy form of electromagnetic radiation) emitted from nuclei separated vertically by about 20 m.

Two identical clocks are in the same house, one upstairs in a bedroom and the other downstairs in the kitchen. Which clock runs more slowly?

The second postulate suggests that a gravitational field may be "transformed away" at any point if we choose an appropriate accelerated frame of reference—a freely falling one. Einstein developed an ingenious method of describing the acceleration necessary to make the gravitational field "disappear." He specified a concept, the *curvature of space–time,* that describes the gravitational effect at every point. In fact, the curvature of space–time completely replaces Newton's gravitational theory. According to Einstein, there is no such thing as a gravitational force. Rather, the presence of a mass causes a curvature of space–time in the vicinity of the mass, and this curvature dictates the space–time path that all freely moving objects must follow. In 1979, John Wheeler summarized Einstein's general theory of relativity in a single sentence: "Space tells matter how to move and matter tells space how to curve."

Consider two travelers on the surface of the Earth walking directly toward the North Pole but from different starting locations. Even though both say they are walking due north, and thus should be on parallel paths, they see themselves getting closer and closer together, as if they were somehow attracted to each other. The curvature of the Earth causes this effect. In a similar way, what we are used to thinking of as the gravitational attraction between two masses is, in Einstein's view, two masses curving space–time and as a result moving toward each other, much like two bowling balls on a mattress rolling together.

One prediction of the general theory of relativity is that a light ray passing near the Sun should be deflected into the curved space–time created by the Sun's mass. This prediction was confirmed when astronomers detected the bending of starlight near the Sun during a total solar eclipse that occurred shortly after World War I (Fig. 39.23). When this discovery was announced, Einstein became an international celebrity.

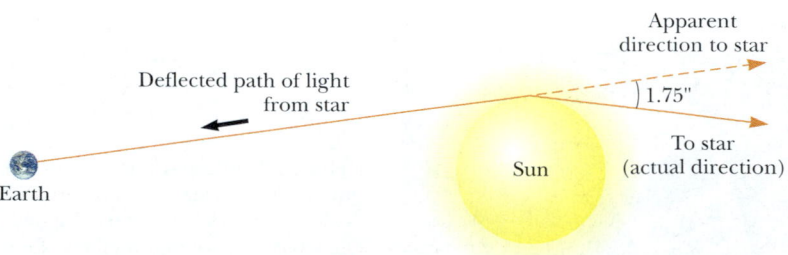

Figure 39.23   Deflection of starlight passing near the Sun. Because of this effect, the Sun or some other remote object can act as a *gravitational lens.* In his general theory of relativity, Einstein calculated that starlight just grazing the Sun's surface should be deflected by an angle of 1.75″.

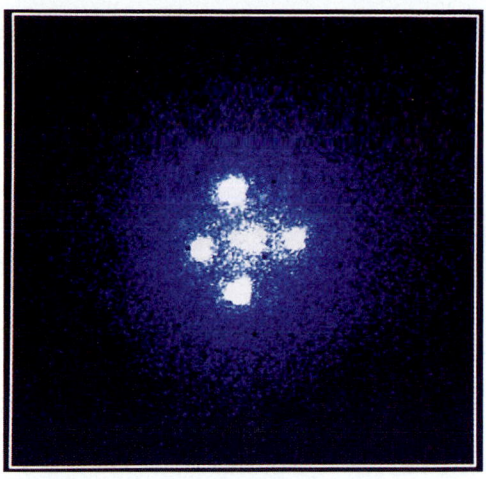

Einstein's cross. The four bright spots are images of the same galaxy that have been bent around a massive object located between the galaxy and the Earth. The massive object acts like a lens, causing the rays of light that were diverging from the distant galaxy to converge on the Earth. (If the intervening massive object had a uniform mass distribution, we would see a bright ring instead of four spots.) *(Courtesy of NASA)*

If the concentration of mass becomes very great, as is believed to occur when a large star exhausts its nuclear fuel and collapses to a very small volume, a **black hole** may form. Here, the curvature of space–time is so extreme that, within a certain distance from the center of the black hole, all matter and light become trapped.

SUMMARY

The two basic postulates of the special theory of relativity are

- The laws of physics must be the same in all inertial reference frames.
- The speed of light in vacuum has the same value, $c = 3.00 \times 10^8$ m/s, in all inertial frames, regardless of the velocity of the observer or the velocity of the source emitting the light.

Three consequences of the special theory of relativity are

- Events that are simultaneous for one observer are not simultaneous for another observer who is in motion relative to the first.
- Clocks in motion relative to an observer appear to be slowed down by a factor $\gamma = (1 - v^2/c^2)^{-1/2}$. This phenomenon is known as **time dilation.**
- The length of objects in motion appears to be contracted in the direction of motion by a factor $1/\gamma = (1 - v^2/c^2)^{1/2}$. This phenomenon is known as **length contraction.**

To satisfy the postulates of special relativity, the Galilean transformation equations must be replaced by the **Lorentz transformation equations:**

$$x' = \gamma(x - vt)$$
$$y' = y$$
$$z' = z \qquad \qquad \textbf{(39.11)}$$
$$t' = \gamma\left(t - \frac{v}{c^2}x\right)$$

where $\gamma = (1 - v^2/c^2)^{-1/2}$.

The relativistic form of the **velocity transformation equation** is

$$u'_x = \frac{u_x - v}{1 - \dfrac{u_x v}{c^2}} \qquad (39.16)$$

where u_x is the speed of an object as measured in the S frame and u'_x is its speed measured in the S′ frame.

The relativistic expression for the **linear momentum** of a particle moving with a velocity **u** is

$$\mathbf{p} \equiv \frac{m\mathbf{u}}{\sqrt{1 - \dfrac{u^2}{c^2}}} = \gamma m \mathbf{u} \qquad (39.19)$$

The relativistic expression for the **kinetic energy** of a particle is

$$K = \gamma mc^2 - mc^2 \qquad (39.23)$$

where mc^2 is called the **rest energy** of the particle.

The total energy E of a particle is related to the mass m of the particle through the famous **mass–energy** equivalence expression:

$$E = \frac{mc^2}{\sqrt{1 - \dfrac{u^2}{c^2}}} \qquad (39.25)$$

The relativistic linear momentum is related to the total energy through the equation

$$E^2 = p^2 c^2 + (mc^2)^2 \qquad (39.26)$$

QUESTIONS

1. What two speed measurements do two observers in relative motion always agree on?

2. A spaceship in the shape of a sphere moves past an observer on the Earth with a speed $0.5c$. What shape does the observer see as the spaceship moves past?

3. An astronaut moves away from the Earth at a speed close to the speed of light. If an observer on Earth measures the astronaut's dimensions and pulse rate, what changes (if any) would the observer measure? Would the astronaut measure any changes about himself?

4. Two identical clocks are synchronized. One is then put in orbit directed eastward around the Earth while the other remains on Earth. Which clock runs slower? When the moving clock returns to Earth, are the two still synchronized?

5. Two lasers situated on a moving spacecraft are triggered simultaneously. An observer on the spacecraft claims to see the pulses of light simultaneously. What condition is necessary so that a second observer agrees?

6. When we say that a moving clock runs more slowly than a stationary one, does this imply that there is something physically unusual about the moving clock?

7. List some ways our day-to-day lives would change if the speed of light were only 50 m/s.

8. Give a physical argument that shows that it is impossible to accelerate an object of mass m to the speed of light, even if it has a continuous force acting on it.

9. It is said that Einstein, in his teenage years, asked the question, "What would I see in a mirror if I carried it in my hands and ran at the speed of light?" How would you answer this question?

10. Some distant star-like objects, called *quasars*, are receding from us at half the speed of light (or greater). What is the speed of the light we receive from these quasars?

11. How is it possible that photons of light, which have zero mass, have momentum?

12. With regard to reference frames, how does general relativity differ from special relativity?

13. Describe how the results of Example 39.7 would change if, instead of fast spaceships, two ordinary cars were approaching each other at highway speeds.

14. Two objects are identical except that one is hotter than the other. Compare how they respond to identical forces.

PROBLEMS

1, 2, 3 = straightforward, intermediate, challenging   ☐ = full solution available in the *Student Solutions Manual and Study Guide*
WEB = solution posted at http://www.saunderscollege.com/physics/   💻 = Computer useful in solving problem   📕 = Interactive Physics
☐ = paired numerical/symbolic problems

Section 39.1   The Principle of Galilean Relativity

1. A 2 000-kg car moving at 20.0 m/s collides and locks together with a 1 500-kg car at rest at a stop sign. Show that momentum is conserved in a reference frame moving at 10.0 m/s in the direction of the moving car.

2. A ball is thrown at 20.0 m/s inside a boxcar moving along the tracks at 40.0 m/s. What is the speed of the ball relative to the ground if the ball is thrown (a) forward? (b) backward? (c) out the side door?

3. In a laboratory frame of reference, an observer notes that Newton's second law is valid. Show that it is also valid for an observer moving at a constant speed, small compared with the speed of light, relative to the laboratory frame.

4. Show that Newton's second law is *not* valid in a reference frame moving past the laboratory frame of Problem 3 with a constant acceleration.

Section 39.2   The Michelson–Morley Experiment
Section 39.3   Einstein's Principle of Relativity
Section 39.4   Consequences of the Special Theory of Relativity

5. How fast must a meter stick be moving if its length is observed to shrink to 0.500 m?

6. At what speed does a clock have to move if it is to be seen to run at a rate that is one-half the rate of a clock at rest?

7. An astronaut is traveling in a space vehicle that has a speed of $0.500c$ relative to the Earth. The astronaut measures his pulse rate at 75.0 beats per minute. Signals generated by the astronaut's pulse are radioed to Earth when the vehicle is moving in a direction perpendicular to a line that connects the vehicle with an observer on the Earth. What pulse rate does the Earth observer measure? What would be the pulse rate if the speed of the space vehicle were increased to $0.990c$?

8. The proper length of one spaceship is three times that of another. The two spaceships are traveling in the same direction and, while both are passing overhead, an Earth observer measures the two spaceships to have the same length. If the slower spaceship is moving with a speed of $0.350c$, determine the speed of the faster spaceship.

9. An atomic clock moves at 1 000 km/h for 1 h as measured by an identical clock on Earth. How many nanoseconds slow will the moving clock be at the end of the 1-h interval?

10. If astronauts could travel at $v = 0.950c$, we on Earth would say it takes $(4.20/0.950) = 4.42$ yr to reach Alpha Centauri, 4.20 ly away. The astronauts disagree. (a) How much time passes on the astronauts' clocks? (b) What distance to Alpha Centauri do the astronauts measure?

WEB 11. A spaceship with a proper length of 300 m takes $0.750\ \mu s$ to pass an Earth observer. Determine the speed of this spaceship as measured by the Earth observer.

12. A spaceship of proper length L_p takes time t to pass an Earth observer. Determine the speed of this spaceship as measured by the Earth observer.

13. A muon formed high in the Earth's atmosphere travels at speed $v = 0.990c$ for a distance of 4.60 km before it decays into an electron, a neutrino, and an antineutrino ($\mu^- \rightarrow e^- + \nu + \bar{\nu}$). (a) How long does the muon live, as measured in its reference frame? (b) How far does the muon travel, as measured in its frame?

14. **Review Problem.**   In 1962, when Mercury astronaut Scott Carpenter orbited the Earth 22 times, the press stated that for each orbit he aged 2 millionths of a second less than he would have had he remained on Earth. (a) Assuming that he was 160 km above the Earth in a circular orbit, determine the time difference between someone on Earth and the orbiting astronaut for the 22 orbits. You will need to use the approximation $\sqrt{1-x} \approx 1 - x/2$ for small x. (b) Did the press report accurate information? Explain.

15. The pion has an average lifetime of 26.0 ns when at rest. In order for it to travel 10.0 m, how fast must it move?

16. For what value of v does $\gamma = 1.01$? Observe that for speeds less than this value, time dilation and length contraction are less-than-one-percent effects.

17. A friend passes by you in a spaceship traveling at a high speed. He tells you that his ship is 20.0 m long and that the identically constructed ship you are sitting in is 19.0 m long. According to your observations, (a) how long is your ship, (b) how long is your friend's ship, and (c) what is the speed of your friend's ship?

18. An interstellar space probe is launched from Earth. After a brief period of acceleration it moves with a constant velocity, 70.0% of the speed of light. Its nuclear-powered batteries supply the energy to keep its data transmitter active continuously. The batteries have a lifetime of 15.0 yr as measured in a rest frame. (a) How long do the batteries on the space probe last as measured by Mission Control on Earth? (b) How far is the probe from Earth when its batteries fail, as measured by Mission Control? (c) How far is the probe from Earth when its batteries fail, as measured by its built-in trip odometer? (d) For what total time after launch are data

received from the probe by Mission Control? Note that radio waves travel at the speed of light and fill the space between the probe and Earth at the time of battery failure.

19. **Review Problem.** An alien civilization occupies a brown dwarf, nearly stationary relative to the Sun, several lightyears away. The extraterrestrials have come to love original broadcasts of *The Ed Sullivan Show*, on our television channel 2, at carrier frequency 57.0 MHz. Their line of sight to us is in the plane of the Earth's orbit. Find the difference between the highest and lowest frequencies they receive due to the Earth's orbital motion around the Sun.

20. Police radar detects the speed of a car (Fig. P39.20) as follows: Microwaves of a precisely known frequency are broadcast toward the car. The moving car reflects the microwaves with a Doppler shift. The reflected waves are received and combined with an attenuated version of the transmitted wave. Beats occur between the two microwave signals. The beat frequency is measured. (a) For an electromagnetic wave reflected back to its source from a mirror approaching at speed v, show that the reflected wave has frequency

$$f = f_{source} \frac{c + v}{c - v}$$

where f_{source} is the source frequency. (b) When v is much less than c, the beat frequency is much less than the transmitted frequency. In this case, use the approximation $f + f_{source} \cong 2f_{source}$ and show that the beat frequency can be written as $f_b = 2v/\lambda$. (c) What beat frequency is measured for a car speed of 30.0 m/s if the microwaves have frequency 10.0 GHz? (d) If the beat frequency measurement is accurate to ± 5 Hz, how accurate is the velocity measurement?

21. *The red shift.* A light source recedes from an observer with a speed v_{source}, which is small compared with c. (a) Show that the fractional shift in the measured wavelength is given by the approximate expression

$$\frac{\Delta\lambda}{\lambda} \approx \frac{v_{source}}{c}$$

This phenomenon is known as the red shift because the visible light is shifted toward the red. (b) Spectroscopic measurements of light at $\lambda = 397$ nm coming from a galaxy in Ursa Major reveal a red shift of 20.0 nm. What is the recessional speed of the galaxy?

Section 39.5   The Lorentz Transformation Equations

22. A spaceship travels at $0.750c$ relative to Earth. If the spaceship fires a small rocket in the forward direction, how fast (relative to the ship) must it be fired for it to travel at $0.950c$ relative to Earth?

WEB **23.** Two jets of material from the center of a radio galaxy fly away in opposite directions. Both jets move at $0.750c$ relative to the galaxy. Determine the speed of one jet relative to that of the other.

24. A moving rod is observed to have a length of 2.00 m, and to be oriented at an angle of 30.0° with respect to the direction of motion (Fig. P39.24). The rod has a speed of $0.995c$. (a) What is the proper length of the rod? (b) What is the orientation angle in the proper frame?

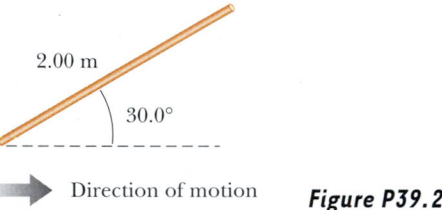

2.00 m

30.0°

⟶ Direction of motion   *Figure P39.24*

25. A Klingon space ship moves away from the Earth at a speed of $0.800c$ (Fig. P39.25). The starship *Enterprise* pursues at a speed of $0.900c$ relative to the Earth. Observers on Earth see the *Enterprise* overtaking the Klingon ship at a relative speed of $0.100c$. With what speed is the *Enterprise* overtaking the Klingon ship as seen by the crew of the *Enterprise*?

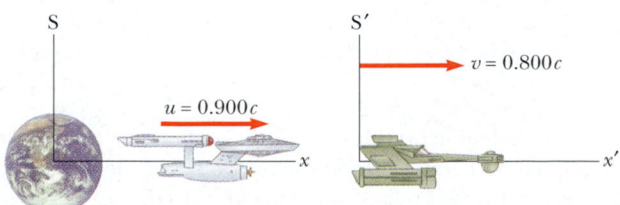

S   S′

                                               $v = 0.800c$

$u = 0.900c$

x   x′

Figure P39.25

Figure P39.20 (*Trent Steffler/David R. Frazier Photolibrary*)

26. A red light flashes at position $x_R = 3.00$ m and time $t_R = 1.00 \times 10^{-9}$ s, and a blue light flashes at $x_B = 5.00$ m and $t_B = 9.00 \times 10^{-9}$ s (all values are measured in the S reference frame). Reference frame S' has its origin at the same point as S at $t = t' = 0$; frame S' moves constantly to the right. Both flashes are observed to occur at the same place in S'. (a) Find the relative velocity between S and S'. (b) Find the location of the two flashes in frame S'. (c) At what time does the red flash occur in the S' frame?

Section 39.6 Relativistic Linear Momentum and the Relativistic Form of Newton's Laws

27. Calculate the momentum of an electron moving with a speed of (a) $0.010\,0c$, (b) $0.500c$, (c) $0.900c$.

28. The nonrelativistic expression for the momentum of a particle, $p = mu$, can be used if $u \ll c$. For what speed does the use of this formula yield an error in the momentum of (a) 1.00 percent and (b) 10.0 percent?

29. A golf ball travels with a speed of 90.0 m/s. By what fraction does its relativistic momentum p differ from its classical value mu? That is, find the ratio $(p - mu)/mu$.

30. Show that the speed of an object having momentum p and mass m is

$$ u = \frac{c}{\sqrt{1 + (mc/p)^2}} $$

WEB 31. An unstable particle at rest breaks into two fragments of *unequal* mass. The mass of the lighter fragment is 2.50×10^{-28} kg, and that of the heavier fragment is 1.67×10^{-27} kg. If the lighter fragment has a speed of $0.893c$ after the breakup, what is the speed of the heavier fragment?

Section 39.7 Relativistic Energy

32. Determine the energy required to accelerate an electron (a) from $0.500c$ to $0.900c$ and (b) from $0.900c$ to $0.990c$.

33. Find the momentum of a proton in MeV/c units if its total energy is twice its rest energy.

34. Show that, for any object moving at less than one-tenth the speed of light, the relativistic kinetic energy agrees with the result of the classical equation $K = mu^2/2$ to within less than 1%. Thus, for most purposes, the classical equation is good enough to describe these objects, whose motion we call *nonrelativistic*.

WEB 35. A proton moves at $0.950c$. Calculate its (a) rest energy, (b) total energy, and (c) kinetic energy.

36. An electron has a kinetic energy five times greater than its rest energy. Find (a) its total energy and (b) its speed.

37. A cube of steel has a volume of 1.00 cm^3 and a mass of 8.00 g when at rest on the Earth. If this cube is now given a speed $u = 0.900c$, what is its density as measured by a stationary observer? Note that relativistic density is E_R/c^2V.

38. An unstable particle with a mass of 3.34×10^{-27} kg is initially at rest. The particle decays into two fragments that fly off with velocities of $0.987c$ and $-0.868c$. Find the masses of the fragments. (*Hint*: Conserve both mass–energy and momentum.)

39. Show that the energy–momentum relationship $E^2 = p^2c^2 + (mc^2)^2$ follows from the expressions $E = \gamma mc^2$ and $p = \gamma mu$.

40. A proton in a high-energy accelerator is given a kinetic energy of 50.0 GeV. Determine (a) its momentum and (b) its speed.

41. In a typical color television picture tube, the electrons are accelerated through a potential difference of 25 000 V. (a) What speed do the electrons have when they strike the screen? (b) What is their kinetic energy in joules?

42. Electrons are accelerated to an energy of 20.0 GeV in the 3.00-km-long Stanford Linear Accelerator. (a) What is the γ factor for the electrons? (b) What is their speed? (c) How long does the accelerator appear to them?

43. A pion at rest ($m_\pi = 270 m_e$) decays to a muon ($m_\mu = 206 m_e$) and an antineutrino ($m_{\bar{\nu}} \approx 0$). The reaction is written $\pi^- \rightarrow \mu^- + \bar{\nu}$. Find the kinetic energy of the muon and the antineutrino in electron volts. (*Hint*: Relativistic momentum is conserved.)

Section 39.8 Equivalence of Mass and Energy

44. Make an order-of-magnitude estimate of the ratio of mass increase to the original mass of a flag as you run it up a flagpole. In your solution explain what quantities you take as data and the values you estimate or measure for them.

45. When 1.00 g of hydrogen combines with 8.00 g of oxygen, 9.00 g of water is formed. During this chemical reaction, 2.86×10^5 J of energy is released. How much mass do the constituents of this reaction lose? Is the loss of mass likely to be detectable?

46. A spaceship of mass 1.00×10^6 kg is to be accelerated to $0.600c$. (a) How much energy does this require? (b) How many kilograms of matter would it take to provide this much energy?

47. In a nuclear power plant the fuel rods last 3 yr before they are replaced. If a plant with rated thermal power 1.00 GW operates at 80.0% capacity for the 3 yr, what is the loss of mass of the fuel?

48. A ^{57}Fe nucleus at rest emits a 14.0-keV photon. Use the conservation of energy and momentum to deduce the kinetic energy of the recoiling nucleus in electron volts. (Use $Mc^2 = 8.60 \times 10^{-9}$ J for the final state of the ^{57}Fe nucleus.)

49. The power output of the Sun is 3.77×10^{26} W. How much mass is converted to energy in the Sun each second?

50. A gamma ray (a high-energy photon of light) can produce an electron (e^-) and a positron (e^+) when

it enters the electric field of a heavy nucleus: $\gamma \rightarrow e^+ + e^-$. What minimum γ-ray energy is required to accomplish this task? (*Hint:* The masses of the electron and the positron are equal.)

Section 39.9   Relativity and Electromagnetism

51. As measured by observers in a reference frame S, a particle having charge q moves with velocity $\mathbf{v}$ in a magnetic field $\mathbf{B}$ and an electric field $\mathbf{E}$. The resulting force on the particle is then measured to be $\mathbf{F} = q(\mathbf{E} + \mathbf{v} \times \mathbf{B})$. Another observer moves along with the charged particle and also measures its charge to be q but measures the electric field to be $\mathbf{E}'$. If both observers are to measure the same force $\mathbf{F}$, show that $\mathbf{E}' = \mathbf{E} + \mathbf{v} \times \mathbf{B}$.

ADDITIONAL PROBLEMS

52. An electron has a speed of $0.750c$. Find the speed of a proton that has (a) the same kinetic energy as the electron; (b) the same momentum as the electron.

WEB 53. The cosmic rays of highest energy are protons, which have kinetic energy on the order of 10^{13} MeV. (a) How long would it take a proton of this energy to travel across the Milky Way galaxy, having a diameter of $\sim 10^5$ ly, as measured in the proton's frame? (b) From the point of view of the proton, how many kilometers across is the galaxy?

54. A spaceship moves away from the Earth at $0.500c$ and fires a shuttle craft in the forward direction at $0.500c$ relative to the ship. The pilot of the shuttle craft launches a probe at forward speed $0.500c$ relative to the shuttle craft. Determine (a) the speed of the shuttle craft relative to the Earth and (b) the speed of the probe relative to the Earth.

55. The net nuclear fusion reaction inside the Sun can be written as $4^1\text{H} \rightarrow {}^4\text{He} + \Delta E$. If the rest energy of each hydrogen atom is 938.78 MeV and the rest energy of the helium-4 atom is 3 728.4 MeV, what is the percentage of the starting mass that is released as energy?

56. An astronaut wishes to visit the Andromeda galaxy (2.00 million lightyears away), making a one-way trip that will take 30.0 yr in the spaceship's frame of reference. If his speed is constant, how fast must he travel relative to the Earth?

57. An alien spaceship traveling at $0.600c$ toward the Earth launches a landing craft with an advance guard of purchasing agents. The lander travels in the same direction with a velocity $0.800c$ relative to the spaceship. As observed on the Earth, the spaceship is 0.200 ly from the Earth when the lander is launched. (a) With what velocity is the lander observed to be approaching by observers on the Earth? (b) What is the distance to the Earth at the time of lander launch, as observed by the aliens? (c) How long does it take the lander to reach the Earth as observed by the aliens on the mother ship? (d) If the lander has a mass of 4.00×10^5 kg, what is its

kinetic energy as observed in the Earth reference frame?

58. A physics professor on the Earth gives an exam to her students, who are on a rocket ship traveling at speed v relative to the Earth. The moment the ship passes the professor, she signals the start of the exam. She wishes her students to have time T_0 (rocket time) to complete the exam. Show that she should wait a time (Earth time) of

$$T = T_0 \sqrt{\frac{1 - v/c}{1 + v/c}}$$

before sending a light signal telling them to stop. (*Hint:* Remember that it takes some time for the second light signal to travel from the professor to the students.)

59. Spaceship I, which contains students taking a physics exam, approaches the Earth with a speed of $0.600c$ (relative to the Earth), while spaceship II, which contains professors proctoring the exam, moves at $0.280c$ (relative to the Earth) directly toward the students. If the professors stop the exam after 50.0 min have passed on their clock, how long does the exam last as measured by (a) the students? (b) an observer on the Earth?

60. Energy reaches the upper atmosphere of the Earth from the Sun at the rate of 1.79×10^{17} W. If all of this energy were absorbed by the Earth and not re-emitted, how much would the mass of the Earth increase in 1 yr?

61. A supertrain (proper length, 100 m) travels at a speed of $0.950c$ as it passes through a tunnel (proper length, 50.0 m). As seen by a trackside observer, is the train ever completely within the tunnel? If so, with how much space to spare?

62. Imagine that the entire Sun collapses to a sphere of radius R_g such that the work required to remove a small mass m from the surface would be equal to its rest energy mc^2. This radius is called the *gravitational radius* for the Sun. Find R_g. (It is believed that the ultimate fate of very massive stars is to collapse beyond their gravitational radii into black holes.)

63. A charged particle moves along a straight line in a uniform electric field $\mathbf{E}$ with a speed of u. If the motion and the electric field are both in the x direction, (a) show that the acceleration of the charge q in the x direction is given by

$$a = \frac{du}{dt} = \frac{qE}{m}\left(1 - \frac{u^2}{c^2}\right)^{3/2}$$

(b) Discuss the significance of the dependence of the acceleration on the speed. (c) If the particle starts from rest at $x = 0$ at $t = 0$, how would you proceed to find the speed of the particle and its position after a time t has elapsed?

64. (a) Show that the Doppler shift $\Delta\lambda$ in the wavelength of light is described by the expression

$$\frac{\Delta\lambda}{\lambda} + 1 = \sqrt{\frac{c - v}{c + v}}$$

where λ is the source wavelength and v is the speed of relative approach between source and observer. (b) How fast would a motorist have to be going for a red light to appear green? Take 650 nm as a typical wavelength for red light, and one of 550 nm as typical for green.

65. A rocket moves toward a mirror at $0.800c$ relative to the reference frame S in Figure P39.65. The mirror is stationary relative to S. A light pulse emitted by the rocket travels toward the mirror and is reflected back to the rocket. The front of the rocket is 1.80×10^{12} m from the mirror (as measured by observers in S) at the moment the light pulse leaves the rocket. What is the total travel time of the pulse as measured by observers in (a) the S frame and (b) the front of the rocket?

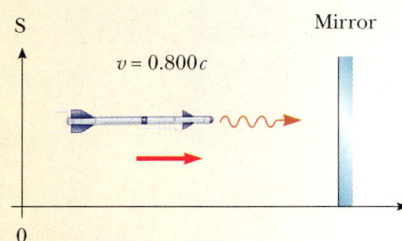

Figure P39.65   Problems 65 and 66.

66. An observer in a rocket moves toward a mirror at speed v relative to the reference frame labeled by S in Figure P39.65. The mirror is stationary with respect to S. A light pulse emitted by the rocket travels toward the mirror and is reflected back to the rocket. The front of the rocket is a distance d from the mirror (as measured by observers in S) at the moment the light pulse leaves the rocket. What is the total travel time of the pulse as measured by observers in (a) the S frame and (b) the front of the rocket?

67. Ted and Mary are playing a game of catch in frame S′, which is moving at $0.600c$, while Jim in frame S watches the action (Fig. P39.67). Ted throws the ball to Mary at $0.800c$ (according to Ted) and their separation (measured in S′) is 1.80×10^{12} m. (a) According to Mary,

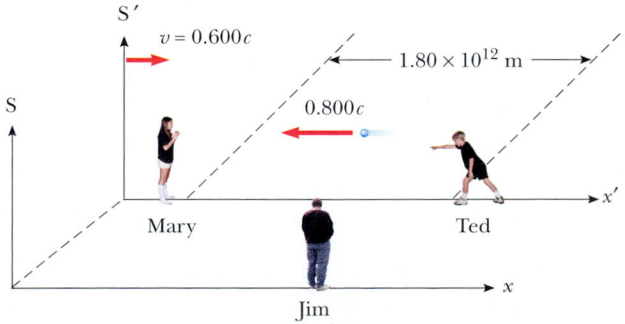

Figure P39.67

how fast is the ball moving? (b) According to Mary, how long does it take the ball to reach her? (c) According to Jim, how far apart are Ted and Mary, and how fast is the ball moving? (d) According to Jim, how long does it take the ball to reach Mary?

68. A rod of length L_0 moving with a speed v along the horizontal direction makes an angle θ_0 with respect to the x' axis. (a) Show that the length of the rod as measured by a stationary observer is $L = L_0[1 - (v^2/c^2)\cos^2\theta_0]^{1/2}$. (b) Show that the angle that the rod makes with the x axis is given by $\tan\theta = \gamma\tan\theta_0$. These results show that the rod is both contracted and rotated. (Take the lower end of the rod to be at the origin of the primed coordinate system.)

69. Consider two inertial reference frames S and S′, where S′ is moving to the right with a constant speed of $0.600c$ as measured by an observer in S. A stick of proper length 1.00 m moves to the left toward the origins of both S and S′, and the length of the stick is 50.0 cm as measured by an observer in S′. (a) Determine the speed of the stick as measured by observers in S and S′. (b) What is the length of the stick as measured by an observer in S?

70. Suppose our Sun is about to explode. In an effort to escape, we depart in a spaceship at $v = 0.800c$ and head toward the star Tau Ceti, 12.0 ly away. When we reach the midpoint of our journey from the Earth, we see our Sun explode and, unfortunately, at the same instant we see Tau Ceti explode as well. (a) In the spaceship's frame of reference, should we conclude that the two explosions occurred simultaneously? If not, which occurred first? (b) In a frame of reference in which the Sun and Tau Ceti are at rest, did they explode simultaneously? If not, which exploded first?

71. The light emitted by a galaxy shows a continuous distribution of wavelengths because the galaxy is composed of billions of different stars and other thermal emitters. Nevertheless, some narrow gaps occur in the continuous spectrum where light has been absorbed by cooler gases in the outer photospheres of normal stars. In particular, ionized calcium atoms at rest produce strong absorption at a wavelength of 394 nm. For a galaxy in the constellation Hydra, 2 billion lightyears away, this absorption line is shifted to 475 nm. How fast is the galaxy moving away from the Earth? (*Note:* The assumption that the recession speed is small compared with c, as made in Problem 21, is not a good approximation here.)

72. Prepare a graph of the relativistic kinetic energy and the classical kinetic energy, both as a function of speed, for an object with a mass of your choice. At what speed does the classical kinetic energy underestimate the relativistic value by 1 percent? By 5 percent? By 50 percent?

73. The total volume of water in the oceans is approximately 1.40×10^9 km³. The density of sea water is 1 030 kg/m³, and the specific heat of the water is 4 186 J/(kg · °C). Find the increase in mass of the oceans produced by an increase in temperature of 10.0°C.

ANSWERS TO QUICK QUIZZES

39.1 They both are because they can report only what they see. They agree that the person in the truck throws the ball up and then catches it a bit later.

39.2 It depends on the direction of the throw. Taking the direction in which the train is traveling as the positive x direction, use the values $u_x' = +90$ mi/h and $v = +110$ mi/h, with u_x in Equation 39.2 being the value you are looking for. If the pitcher throws the ball in the same direction as the train, a person at rest on the Earth sees the ball moving at 110 mi/h + 90 mi/h = 200 mi/h. If the pitcher throws in the opposite direction, the person on the Earth sees the ball moving in the same direction as the train but at only 110 mi/h − 90 mi/h = 20 mi/h.

39.3 Both are correct. Although the two observers reach different conclusions, each is correct in her or his own reference frame because the concept of simultaneity is not absolute.

39.4 About 2.9×10^8 m/s, because this is the speed at which $\gamma = 5$. For every 5 s ticking by on the Mission Control clock, the Earth-bound observer (with a powerful telescope!) sees the rocket clock ticking off 1 s. The astronaut sees her own clock operating at a normal rate. To her, Mission Control is moving away from her at a speed of 2.9×10^8 m/s, and she sees the Mission Control clock as running slow. Strange stuff, this relativity!

39.5 If their on-duty time is based on clocks that remain on the Earth, they will have larger paychecks. Less time will have passed for the astronauts in their frame of reference than for their employer back on the Earth.

39.6 By a curved line. This can be seen in the middle of Speedo's world-line in Figure 39.14, where he turns around and begins his trip home.

39.7 The downstairs clock runs more slowly because it is closer to the Earth and hence experiences a stronger gravitational field than the upstairs clock does.

PUZZLER

This street in Las Vegas is filled with "neon" lights of various bright colors. How do these lights work, and what determines the color of a particular tube of light? *(Dembinsky Photo Associates)*

chapter

40

Introduction to Quantum Physics

In the preceding chapter, we discussed the fact that Newtonian mechanics must be replaced by Einstein's special theory of relativity when we are dealing with particle speeds comparable to the speed of light. As the 20th century progressed, many experimental and theoretical problems were resolved by the special theory. There were many other problems, however, for which classical physics could not provide a theoretical answer. Attempts to apply the laws of classical physics to explain the behavior of matter on the atomic scale were consistently unsuccessful. For example, blackbody radiation, the photoelectric effect, and the emission of sharp spectral lines from atoms could not be explained within the framework of classical physics.

As physicists sought new ways to solve these puzzles, another revolution took place in physics between 1900 and 1930. A new theory called *quantum mechanics* was highly successful in explaining the behavior of atoms, molecules, and nuclei. Like the special theory of relativity, the quantum theory requires a modification of our ideas concerning the physical world.

The basic ideas of quantum theory were introduced by Max Planck, but most of the subsequent mathematical developments and interpretations were made by a number of other distinguished physicists, including Einstein, Bohr, Schrödinger, de Broglie, Heisenberg, Born, and Dirac. Despite the great success of the quantum theory, Einstein frequently played the role of its critic, especially with regard to the manner in which the theory was interpreted. In particular, Einstein did not accept Heisenberg's interpretation of the uncertainty principle, which states that it is impossible to obtain a precise simultaneous measurement of the position and the velocity of a particle. According to this principle, the best we can do is predict the *probability* of the future of a system, contrary to the deterministic view held by Einstein.[1]

Because an extensive study of quantum theory is beyond the scope of this book, this chapter is simply an introduction to its underlying principles.

40.1 ▷ BLACKBODY RADIATION AND PLANCK'S HYPOTHESIS

An object at any temperature emits radiation sometimes referred to as **thermal radiation,** which we discussed in Section 20.7. The characteristics of this radiation depend on the temperature and properties of the object. At low temperatures, the wavelengths of thermal radiation are mainly in the infrared region of the electromagnetic spectrum, and hence the radiation is not observed by the eye. As the temperature of the object increases, the object eventually begins to glow red—in other words, enough visible radiation is emitted so that the object appears to glow. At sufficiently high temperatures, the object appears to be white, as in the glow of the hot tungsten filament of a lightbulb. Careful study shows that, as the temperature of an object increases, the thermal radiation it emits consists of a continuous distribution of wavelengths from the infrared, visible, and ultraviolet portions of the spectrum.

From a classical viewpoint, thermal radiation originates from accelerated charged particles in the atoms near the surface of the object; those charged particles emit radiation much as small antennas do. The thermally agitated particles can have a distribution of accelerations, which accounts for the continuous spec-

---

[1] Einstein's views on the probabilistic nature of quantum theory are expressed in his statement, "God does not play dice with the Universe."

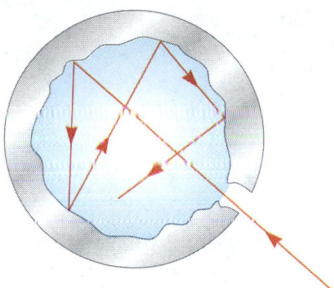

Figure 40.1   The opening to the cavity inside a hollow object is a good approximation of a black body. Light entering the small opening strikes the far wall, where some is absorbed and some is reflected at a random angle. The light continues to be reflected, and at each reflection a portion of the light is absorbed by the cavity walls. After many reflections, essentially all of the incident energy has been absorbed.

trum of radiation emitted by the object. By the end of the 19th century, however, it became apparent that the classical theory of thermal radiation was inadequate. The basic problem was in understanding the observed distribution of wavelengths in the radiation emitted by a black body. As we saw in Section 20.7, a black body is an ideal system that absorbs all radiation incident on it. A good approximation of a black body is a hole leading to the inside of a hollow object, as shown in Figure 40.1. The nature of the radiation emitted through the hole leading to the cavity depends only on the temperature of the cavity walls and not on the material of which the walls are made. The spaces between lumps of hot charcoal (Fig. 40.2) emit light that is very much like blackbody radiation.

Figure 40.3 shows how the energy of blackbody radiation varies with temperature and wavelength. As the temperature of the black body increases, two distinct behaviors are observed. The first effect is that the peak of the distribution shifts to shorter wavelengths. This is why the object described at the beginning of this section changes from not appearing to glow (peak in the infrared) to glowing red (peak in the near infrared with some visible at the red end of the spectrum) to glowing white (peak in the visible). This shift is found to obey the following rela-

QuickLab

Use a black marker or pieces of black electrical tape to make a very dark area on the outside of a shoe box. Poke a hole in the center of the dark area with a pencil. Now put a lid on the box and compare the blackness of the hole with the blackness of the surrounding dark area. The hole acts like a black body.

Figure 40.2   The glow emanating from the spaces between these hot charcoal briquettes is, to a close approximation, blackbody radiation. The color of the light depends only upon the temperature of the briquettes. *(Corbis)*

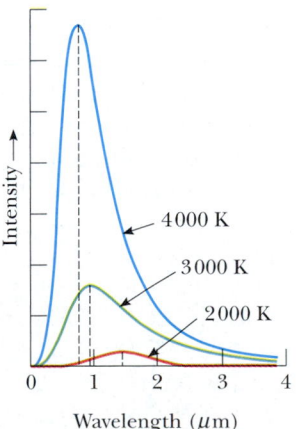

Figure 40.3   Intensity of blackbody radiation versus wavelength at three temperatures. Note that the amount of radiation emitted (the area under a curve) increases with increasing temperature.

tionship, called **Wien's displacement law:**

$$\lambda_{max} T = 2.898 \times 10^{-3}\ \mathrm{m \cdot K} \qquad \textbf{(40.1)}$$

where λ_{max} is the wavelength at which the curve peaks and T is the absolute temperature of the object emitting the radiation. The wavelength at the peak of the curve is inversely proportional to the absolute temperature; that is, as the temperature increases, the peak is "displaced" to shorter wavelengths.

The second effect is that the total amount of energy the object emits increases with temperature. This is described by Stefan's law, given in Equation 20.18, which we wrote in the form $\mathcal{P} = \sigma A e T^4$. Recalling that $I = \mathcal{P}/A$ is the intensity of radiation at the surface of the object and that $e = 1$ for a black body, we can write Stefan's law in the form $I = \sigma T^4$.

To describe the distribution of energy from a black body, it is useful to define $I(\lambda, T)\, d\lambda$ to be the power per unit area emitted in the wavelength interval $d\lambda$. The result of a calculation based on a classical model of blackbody radiation known as the **Rayleigh–Jeans law** is

Rayleigh–Jeans law

$$I(\lambda, T) = \frac{2\pi c k_B T}{\lambda^4} \qquad \textbf{(40.2)}$$

where k_B is Boltzmann's constant. In this classical model of blackbody radiation, the atoms in the cavity walls are treated as a set of oscillators that emit electromagnetic waves at all wavelengths. This model leads to an average energy per oscillator that is proportional to T.

An experimental plot of the blackbody radiation spectrum is shown in Figure 40.4, together with the theoretical prediction of the Rayleigh–Jeans law. At long wavelengths, the Rayleigh–Jeans law is in reasonable agreement with experimental data, but at short wavelengths major disagreement is apparent. We can understand the disagreement by noting that as λ approaches zero, the function $I(\lambda, T)$ given by Equation 40.2 approaches infinity. Hence, not only should short wavelengths predominate in a blackbody spectrum, but also the energy emitted by any black body should become infinite in the limit of zero wavelength. In contrast to this prediction, the experimental data plotted in Figure 40.4 show that as λ approaches zero, $I(\lambda, T)$ also approaches zero. This mismatch of theory and experiment was so disconcerting that scientists called it the *ultraviolet catastrophe*. (This name is a misnomer because the "catastrophe"—infinite energy—occurs as the wavelength approaches zero, not the ultraviolet wavelengths.)

Another discrepancy between theory and experiment concerns the total power emitted by the black body. Experimentally, the total power per unit area

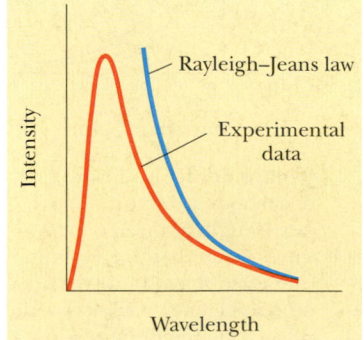

Figure 40.4   Comparison of experimental results and the curve predicted by the Rayleigh–Jeans law for the distribution of blackbody radiation.

given by $\int_0^\infty I(\lambda, T)\,d\lambda$ remains finite even though the Rayleigh–Jeans law (Eq. 40.2) says it should diverge to infinity.

In 1900, Planck derived a formula for blackbody radiation that was in complete agreement with experiment at all wavelengths. Planck's analysis led to the red curve in Figure 40.4. The function he proposed is

$$I(\lambda, T) = \frac{2\pi hc^2}{\lambda^5(e^{hc/\lambda k_B T} - 1)}$$

(40.3)

This function includes a parameter h, which Planck adjusted so that his curve matched the experimental data at all wavelengths. The value of this parameter is found to be independent of the material of which the black body is made and independent of the temperature. Rather than a variable parameter, it is a fundamental constant of nature. The value of h, Planck's constant, which we saw first in Chapter 11 and again in Chapter 35, is

$$h = 6.626 \times 10^{-34}\,\text{J} \cdot \text{s}$$

(40.4)

At long wavelengths, Equation 40.3 reduces to the Rayleigh–Jeans expression, Equation 40.2, and at short wavelengths it predicts an exponential decrease in $I(\lambda, T)$ with decreasing wavelength, in agreement with experimental results.

In his theory, Planck made two bold and controversial assumptions concerning the nature of the oscillating molecules at the surface of the black body:

1. The molecules can have only *discrete* values of energy E_n, given by

$$E_n = nhf$$

(40.5)

where n is a positive integer called a **quantum number** and f is the natural frequency of oscillation of the molecules. This is quite different from the classical model of the harmonic oscillator, in which the energy of identical oscillators is related to the amplitude of the motion and unrelated to the frequency. Because the energy of a molecule can have only discrete values given by Equation 40.5, we say the energy is *quantized*. Each discrete energy value represents a different *quantum state* for the molecule, with each value of n representing a specific quantum state. When the molecule is in the $n = 1$ quantum state, its energy is hf; when it is in the $n = 2$ quantum state, its energy is $2hf$; and so on.

2. The molecules emit or absorb energy in discrete packets that later came to be called **photons.** The molecules emit or absorb these photons by "jumping" from one quantum state to another. If the jump is downward from one state to an adjacent lower state, Equation 40.5 shows that the amount of energy radi-

Max Planck (1858–1947)
Planck introduced the concept of "quantum of action" (Planck's constant, h) in an attempt to explain the spectral distribution of blackbody radiation, which laid the foundations for quantum theory. In 1918 he was awarded the Nobel Prize for this discovery of the quantized nature of energy. *(Photo courtesy of AIP Niels Bohr Library, W. F. Meggers Collection)*

Quantization of energy

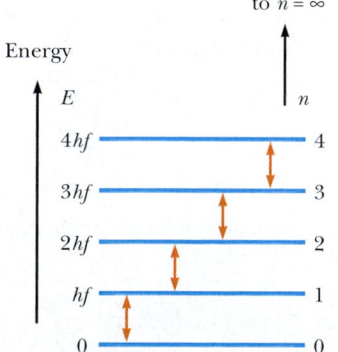

Figure 40.5 Allowed energy levels for a molecule that oscillates at its natural frequency f. Allowed transitions are indicated by the double-headed arrows.

ated by the molecule in a single photon equals *hf.* Hence, the energy of one photon corresponding to the energy difference between two adjacent quantum states is

Energy of a photon

$$E = hf \qquad \textbf{(40.6)}$$

A molecule emits or absorbs energy only when it changes quantum states. If it remains in one quantum state, no energy is emitted or absorbed. Figure 40.5 shows the quantized energy levels and transitions between adjacent states.

The key point in Planck's theory is the radical assumption of quantized energy states. This development marked the birth of the quantum theory. When Planck presented his theory, most scientists (including Planck!) did not consider the quantum concept to be realistic. Hence, he and others continued to search for a more rational explanation of blackbody radiation. However, subsequent developments showed that a theory based on the quantum concept (rather than on classical concepts) had to be used to explain many other phenomena at the atomic level.

Quick Quiz 40.1

Which is more likely to cause sunburn (because more energy is absorbed by skin cells): (a) infrared light, (b) visible light, or (c) ultraviolet light?

EXAMPLE 40.1   Thermal Radiation from Different Objects

Find the peak wavelength of the radiation emitted by each of the following: (a) the human body when the skin temperature is 35°C.

Solution   From Wien's displacement law (Eq. 40.1), we have $\lambda_{max}T = 2.898 \times 10^{-3}$ m·K. Solving for λ_{max}, noting that 35°C corresponds to an absolute temperature of 308 K, we have

$$\lambda_{max} = \frac{2.898 \times 10^{-3}\ \text{m·K}}{308\ \text{K}} = \boxed{9.4\ \mu\text{m}}$$

This radiation is in the infrared region of the spectrum and is invisible to the human eye. Some animals (pit vipers, for instance) are able to detect radiation of this wavelength and therefore can locate warm-blooded prey even in the dark.

(b) The tungsten filament of a lightbulb, which operates at 2 000 K.

Solution   Following the same procedure as in part (a), we find

$$\lambda_{max} = \frac{2.898 \times 10^{-3}\ \text{m·K}}{2\ 000\ \text{K}} = \boxed{1.4\ \mu\text{m}}$$

This also is in the infrared, meaning that most of the energy emitted by a lightbulb is not visible to us.

(c) The Sun, which has a surface temperature of about 5 800 K.

Solution   Again following the same procedure, we have

$$\lambda_{max} = \frac{2.898 \times 10^{-3}\ \text{m·K}}{5\ 800\ \text{K}} = \boxed{0.50\ \mu\text{m}}$$

This is near the center of the visible spectrum, about the color of a yellow-green tennis ball. Because it is the most prevalent color in sunlight, our eyes have evolved to be most sensitive to light of approximately this wavelength.

EXAMPLE 40.2   The Quantized Oscillator

A 2.0-kg block is attached to a massless spring that has a force constant of $k = 25$ N/m. The spring is stretched 0.40 m from its equilibrium position and released. (a) Find the total energy of the system and the frequency of oscillation according to classical calculations.

Solution   The total energy of a simple harmonic oscillator having an amplitude A is $\frac{1}{2}kA^2$ (Eq. 13.22). Therefore,

$$E = \tfrac{1}{2}kA^2 = \tfrac{1}{2}(25\ \text{N/m})(0.40\ \text{m})^2 = \boxed{2.0\ \text{J}}$$

The frequency of oscillation is (Eq. 13.19)

$$f = \frac{1}{2\pi}\sqrt{\frac{k}{m}} = \frac{1}{2\pi}\sqrt{\frac{25 \text{ N/m}}{2.0 \text{ kg}}} = \boxed{0.56 \text{ Hz}}$$

(b) Assuming that the energy is quantized, find the quantum number n for the system.

Solution   If the energy is quantized, we have

$$E_n = nhf = n(6.626 \times 10^{-34}\text{ J}\cdot\text{s})(0.56\text{ Hz}) = 2.0\text{ J}$$

$$n = \boxed{5.4 \times 10^{33}}$$

(c) How much energy is carried away in a one-quantum change?

Solution   From Equation 40.6,

$$E = hf = (6.626 \times 10^{-34}\text{ J}\cdot\text{s})(0.56\text{ Hz}) = \boxed{3.7 \times 10^{-34}\text{ J}}$$

This energy carried away by a one-quantum change is such a small fraction of the total energy of the oscillator that we cannot detect it. Thus, even though the energy of a spring–block system is quantized and does indeed decrease by small quantum jumps, our senses perceive the decrease as continuous. Quantum effects become important and measurable only on the submicroscopic level of atoms and molecules.

40.2 THE PHOTOELECTRIC EFFECT

In the latter part of the 19th century, experiments showed that light incident on certain metal surfaces caused electrons to be emitted from the surfaces. This phenomenon, which we first met in Section 35.1, is known as the **photoelectric effect,** and the emitted electrons are called **photoelectrons.**

Figure 40.6 is a diagram of an apparatus in which the photoelectric effect can occur. An evacuated glass or quartz tube contains a metallic plate E connected to the negative terminal of a battery and another metallic plate C that is connected to the positive terminal of the battery. When the tube is kept in the dark, the ammeter reads zero, indicating no current in the circuit. However, when plate E is illuminated by light having a wavelength shorter than some particular wavelength that depends on the metal used to make plate E, a current is detected by the ammeter, indicating a flow of charges across the gap between plates E and C. This current arises from photoelectrons emitted from the negative plate (the emitter) and collected at the positive plate (the collector).

Figure 40.7 is a plot of photoelectric current versus potential difference ΔV between plates E and C for two light intensities. At large values of ΔV, the current reaches a maximum value. In addition, the current increases as the intensity of the incident light increases, as you might expect. Finally, when ΔV is negative—that is, when the battery in the circuit is reversed to make plate E positive and plate C negative—the current drops to a very low value because most of the emitted photoelectrons are repelled by the now negative plate C. In this situation, only those

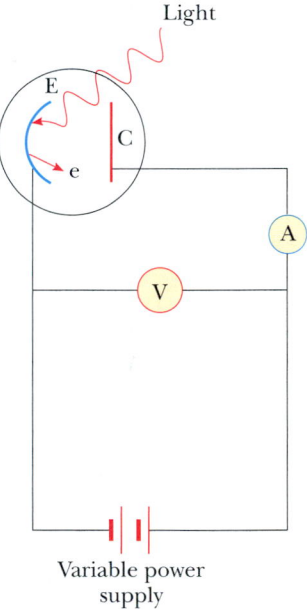

Figure 40.6   A circuit diagram for observing the photoelectric effect. When light strikes the plate E (the emitter), photoelectrons are ejected from the plate. Electrons moving from plate E to plate C (the collector) constitute a current in the circuit.

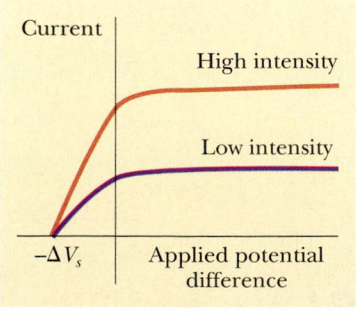

Figure 40.7   Photoelectric current versus applied potential difference for two light intensities. The current increases with intensity but reaches a saturation level for large values of ΔV. At voltages equal to or more negative than $-\Delta V_s$, the stopping potential, the current is zero.

photoelectrons having a kinetic energy greater than the magnitude of $e\,\Delta V$ reach plate C, where e is the charge on the electron.

When ΔV is equal to or more negative than $-\Delta V_s$, the **stopping potential,** no photoelectrons reach C and the current is zero. The stopping potential is independent of the radiation intensity. The maximum kinetic energy of the photoelectrons is related to the stopping potential through the relationship

$$K_{\max} = e\,\Delta V_s \tag{40.7}$$

Several features of the photoelectric effect could not be explained by classical physics or by the wave theory of light:

- No photoelectrons are emitted if the frequency of the incident light falls below some **cutoff frequency** f_c, which is characteristic of the material being illuminated. This is inconsistent with wave theory, which predicts that the photoelectric effect should occur at any frequency, provided the light intensity is sufficiently high.
- The maximum kinetic energy of the photoelectrons is independent of light intensity. According to wave theory, light of higher intensity should carry more energy into the metal per unit time and therefore eject photoelectrons having higher kinetic energies.
- The maximum kinetic energy of the photoelectrons increases with increasing light frequency. The wave theory predicts no relationship between photoelectron energy and incident light frequency.
- Photoelectrons are emitted from the surface almost instantaneously (less than 10^{-9} s after the surface is illuminated), even at low light intensities. Classically, we expect the photoelectrons to require some time to absorb the incident radiation before they acquire enough kinetic energy to escape from the metal.

A successful explanation of the photoelectric effect was given by Einstein in 1905, the same year he published his special theory of relativity. As part of a general paper on electromagnetic radiation, for which he received the Nobel Prize in 1921, Einstein extended Planck's concept of quantization to electromagnetic waves. He assumed that light (or any other electromagnetic wave) of frequency f can be considered a stream of photons. Each photon has an energy E, given by Equation 40.6, $E = hf$. A suggestive image of several photons, not to be taken too literally, is shown in Figure 40.8.

In Einstein's model, a photon is so localized that it gives *all* its energy hf to a single electron in the metal. According to Einstein, the maximum kinetic energy for these liberated photoelectrons is

Photoelectric effect equation

$$K_{\max} = hf - \phi \tag{40.8}$$

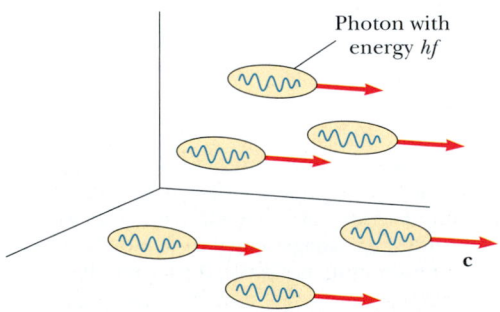

Photon with energy hf

c

Figure 40.8 A representation of photons. Each photon has a discrete energy hf.

where ϕ is called the **work function** of the metal. The work function represents the minimum energy with which an electron is bound in the metal and is on the order of a few electron volts. Table 40.1 lists work functions for various metals.

With the photon theory of light, we can explain the previously mentioned features of the photoelectric effect that we cannot understand using concepts of classical physics:

- That the effect is not observed below a cutoff frequency follows from the fact that the energy of the photon must be greater than or equal to ϕ. If the energy of the incoming photon does not satisfy this condition, the electrons are never ejected from the surface, regardless of the light intensity.
- That K_{max} is independent of light intensity can be understood by means of the following argument: If the light intensity is doubled, the number of photons is doubled, which doubles the number of photoelectrons emitted. However, their maximum kinetic energy, which equals $hf - \phi$, depends only on the light frequency and the work function, not on the light intensity.
- That K_{max} increases with increasing frequency is easily understood with Equation 40.8.
- That photoelectrons are emitted almost instantaneously is consistent with the particle theory of light, in which the incident energy arrives at the surface in small packets and there is a one-to-one interaction between photons and photoelectrons. In this interaction the photon's energy is imparted to an electron that then has enough energy to leave the metal. This is in contrast to the wave theory, in which the incident energy is distributed uniformly over a large area of the metal surface.

Experimental observation of a linear relationship between K_{max} and f would be a final confirmation of Einstein's theory. Indeed, such a linear relationship is observed, as sketched in Figure 40.9. The intercept on the horizontal axis gives the cutoff frequency below which no photoelectrons are emitted, regardless of light intensity. The frequency is related to the work function through the relationship $f_c = \phi / h$. The cutoff frequency corresponds to a **cutoff wavelength** of

$$\lambda_c = \frac{c}{f_c} = \frac{c}{\phi / h} = \frac{hc}{\phi} \tag{40.9}$$

where we have used Equation 16.14 and c is the speed of light. Wavelengths greater than λ_c incident on a material having a work function ϕ do not result in the emission of photoelectrons.

One of the first practical uses of the photoelectric effect was as the detector in the light meter of a camera. Light reflected from the object to be photographed struck a photoelectric surface in the meter, causing it to emit photoelectrons that then passed through a sensitive ammeter. The magnitude of the current in the ammeter depended on the light intensity.

The phototube, another early application of the photoelectric effect, acts much like a switch in an electric circuit. It produces a current in the circuit when light of sufficiently high frequency falls on a metallic plate in the phototube but produces no current in the dark. Phototubes were used in burglar alarms and in the detection of the soundtrack on motion picture film. Modern semiconductor devices have now replaced older devices based on the photoelectric effect.

TABLE 40.1
Work Functions of Selected Metals

Metal	ϕ (eV)
Na	2.46
Al	4.08
Cu	4.70
Zn	4.31
Ag	4.73
Pt	6.35
Pb	4.14
Fe	4.50

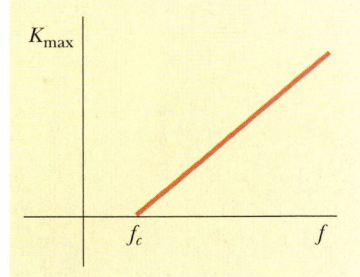

Figure 40.9 A plot of K_{max} of photoelectrons as a function of incident-light frequency in a typical photoelectric effect experiment. Photons having a frequency less than f_c do not have sufficient energy to eject an electron from the metal.

Quick Quiz 40.2

What does the slope of the line in Figure 40.9 represent? What does the y intercept represent? How would such a series of these graphs for different metals compare with one another?

Quick Quiz 40.3

Make a sketch of how classical physicists expected Figure 40.9 to look.

EXAMPLE 40.3   The Photoelectric Effect for Sodium

A sodium surface is illuminated with light having a wavelength of 300 nm. The work function for sodium metal is 2.46 eV. Find (a) the maximum kinetic energy of the ejected photoelectrons and (b) the cutoff wavelength for sodium.

Solution   (a) The energy of each photon in the illuminating light beam is

$$E = hf = \frac{hc}{\lambda} = \frac{(6.626 \times 10^{-34}\,\text{J}\cdot\text{s})(3.00 \times 10^{8}\,\text{m/s})}{300 \times 10^{-9}\,\text{m}}$$

$$= 6.63 \times 10^{-19}\,\text{J} = \frac{6.63 \times 10^{-19}\,\text{J}}{1.60 \times 10^{-19}\,\text{J/eV}} = 4.14\,\text{eV}$$

Using Equation 40.8 gives

$$K_{\text{max}} = hf - \phi = 4.14\,\text{eV} - 2.46\,\text{eV} = \boxed{1.68\,\text{eV}}$$

(b) We can calculate the cutoff wavelength from Equation 40.9 after we convert ϕ from electron volts to joules:

$$\phi = (2.46\,\text{eV})(1.60 \times 10^{-19}\,\text{J/eV}) = 3.94 \times 10^{-19}\,\text{J}$$

$$\lambda_c = \frac{hc}{\phi} = \frac{(6.626 \times 10^{-34}\,\text{J}\cdot\text{s})(3.00 \times 10^{8}\,\text{m/s})}{3.94 \times 10^{-19}\,\text{J}}$$

$$= 5.05 \times 10^{-7}\,\text{m} = \boxed{505\,\text{nm}}$$

This wavelength is in the yellow-green region of the visible spectrum.

Exercise   Calculate the maximum speed of the photoelectrons under the conditions described in this example.

Answer 7.68×10^{5} m/s.

40.3   THE COMPTON EFFECT

In 1919, Einstein concluded that a photon of energy E travels in a single direction (unlike a spherical wave) and carries a momentum equal to $E/c = hf/c$. In his own words, "If a bundle of radiation causes a molecule to emit or absorb an energy packet hf, then momentum of quantity hf/c is transferred to the molecule, directed along the line of the bundle for absorption and opposite the bundle for emission." In 1923, Arthur Holly Compton (1892–1962) and Peter Debye (1884–1966) independently carried Einstein's idea of photon momentum further.

Prior to 1922, Compton and his co-workers had accumulated evidence showing that the classical wave theory of light failed to explain the scattering of x-rays from electrons. According to classical theory, electromagnetic waves of frequency f_0 incident on electrons should have two effects, as shown in Figure 40.10a: (1) Radiation pressure (see Section 34.4) should cause the electrons to accelerate in the direction of propagation of the waves, and (2) the oscillating electric field of the incident radiation should set the electrons into oscillation at the apparent frequency f', where f' is the frequency in the frame of the moving electrons. This apparent frequency f' is different from the frequency f_0 of the incident radiation because of the Doppler effect (see Section 17.5): Each electron first absorbs as a moving particle and then reradiates as a moving particle, thereby exhibiting two Doppler shifts in the frequency of radiation.

Because different electrons will move at different speeds after the interaction, depending on the amount of energy absorbed from the electromagnetic waves, the scattered wave frequency at a given angle should show a distribution of Doppler-shifted values. Contrary to this prediction, Compton's experiments showed that, at a given angle, only *one* frequency of radiation was observed. Compton and his co-workers realized that they could explain these experiments by treat-

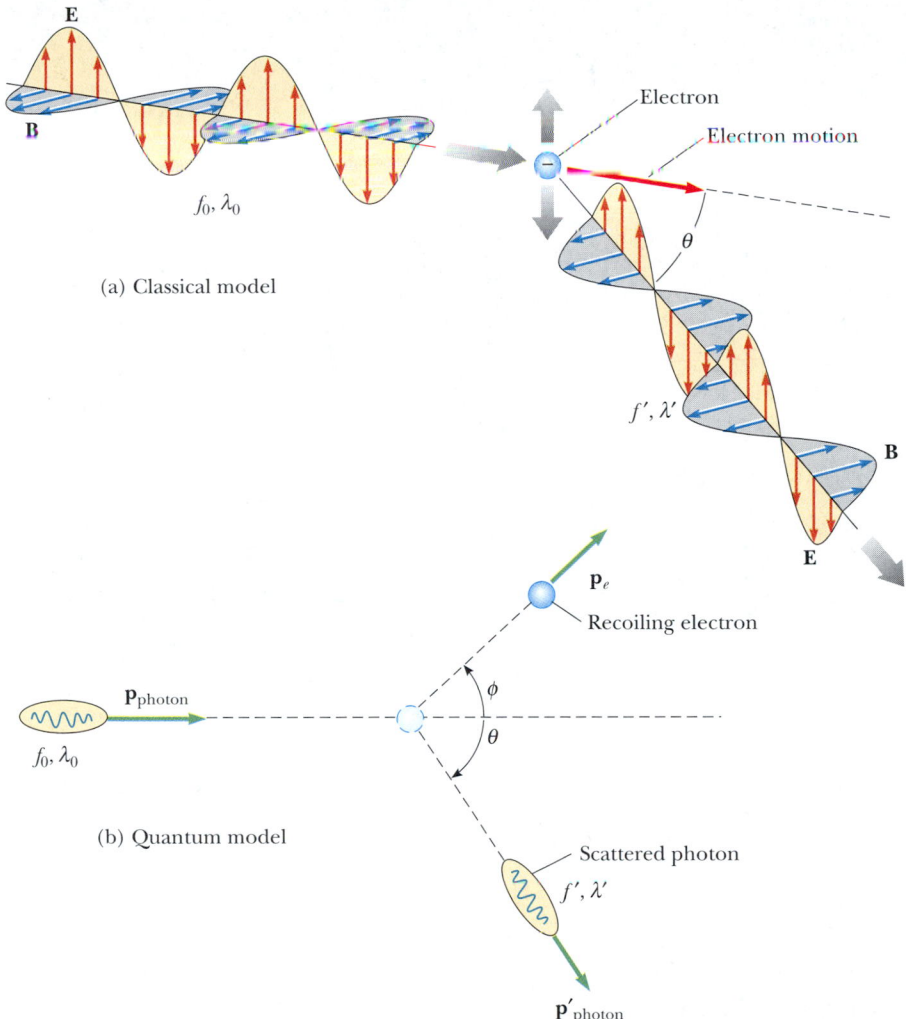

(a) Classical model

(b) Quantum model

Figure 40.10 X-ray scattering from an electron: (a) the classical model; (b) the quantum model.

ing photons not as waves but rather as point-like particles having energy hf and momentum hf/c and by assuming that the energy and momentum of any colliding photon–electron pair are conserved. Compton was adopting a particle model for something that was well known as a wave, and today this scattering phenomenon is known as the **Compton effect.** Figure 40.10b shows the quantum picture of the exchange of momentum and energy between an individual x-ray photon and an electron.

The second difference between the classical and quantum models is also shown in Figure 40.10b. In the classical model, the electron is pushed along the direction of propagation of the incident x-ray by radiation pressure. In the quantum model, the electron is scattered through an angle ϕ with respect to this direction, as if this were a billiard-ball type collision. (The symbol ϕ used here is not to be confused with work function, which was discussed in the previous section.)

Figure 40.11a is a schematic diagram of the apparatus used by Compton. The x-rays, scattered from a graphite target, were analyzed with a rotating crystal spectrometer, and the intensity was measured with an ionization chamber that generated a current proportional to the intensity. The incident beam consisted of mono-

Arthur Holly Compton
(1892–1962)

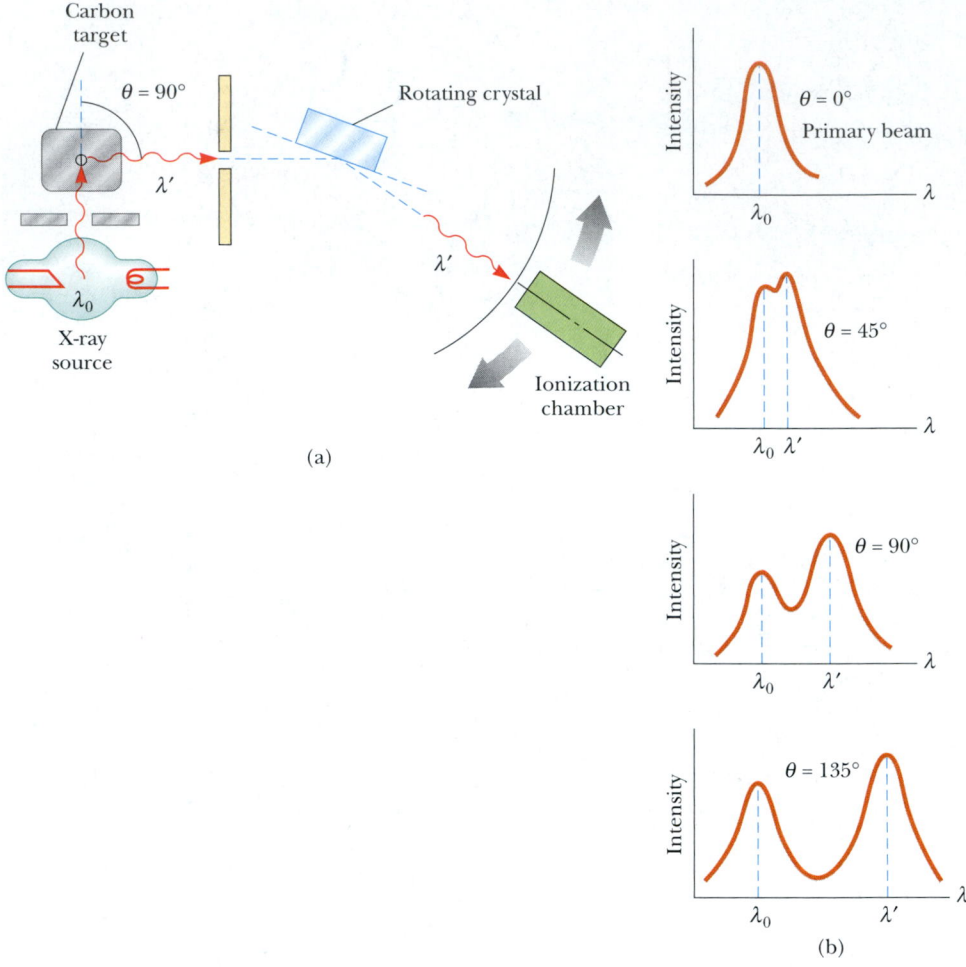

Figure 40.11   (a) Schematic diagram of Compton's apparatus. The wavelength was measured with a rotating crystal spectrometer using graphite (carbon) as the target. (b) Scattered x-ray intensity versus wavelength for Compton scattering at $\theta = 0°$, 45°, 90°, and 135°.

chromatic x-rays of wavelength $\lambda_0 = 0.071$ nm. The experimental intensity-versus-wavelength plots observed by Compton for four scattering angles (corresponding to θ in Fig. 40.10) are shown in Figure 40.11b. The graphs for the three nonzero angles show two peaks, one at λ_0 and one at $\lambda' > \lambda_0$. The shifted peak at λ' is caused by the scattering of x-rays from free electrons, and it was predicted by Compton to depend on scattering angle as

Compton shift equation

$$\lambda' - \lambda_0 = \frac{h}{m_e c}(1 - \cos\theta) \qquad (40.10)$$

where m_e is the mass of the electron. This expression is known as the **Compton shift equation,** and the factor $h/m_e c$ is called the **Compton wavelength** λ_C of the electron. It has a currently accepted value of

Compton wavelength

$$\lambda_C = \frac{h}{m_e c} = 0.002\ 43 \text{ nm}$$

The unshifted peak at λ_0 in Figure 40.11b is caused by x-rays scattered from electrons tightly bound to the target atoms. This unshifted peak also is predicted by Equation 40.10 if the electron mass is replaced with the mass of a carbon atom, which is about 23 000 times the mass of the electron. Thus, there is a wavelength shift for scattering from an electron bound to an atom, but it is so small that it was undetectable in Compton's experiment.

Compton's measurements were in excellent agreement with the predictions of Equation 40.10. It is fair to say that these results were the first that really convinced many physicists of the fundamental validity of quantum theory!

Quick Quiz 40.4

Note that for any given scattering angle θ, Equation 40.10 gives the same value for the Compton wavelength shift for any wavelength. Keeping this in mind, explain why the experiment is normally done with x-rays rather than visible light.

Derivation of the Compton Shift Equation

We can derive the Compton shift equation by assuming that the photon behaves like a particle and collides elastically with a free electron initially at rest, as shown in Figure 40.12a. In this model, the photon is treated as a particle having energy $E = hf = hc/\lambda$ and mass zero. In the scattering process, the total energy and total linear momentum of the system must be conserved. Applying the principle of conservation of energy to this process gives

$$\frac{hc}{\lambda_0} = \frac{hc}{\lambda'} + K_e$$

where hc/λ_0 is the energy of the incident photon, hc/λ' is the energy of the scat-

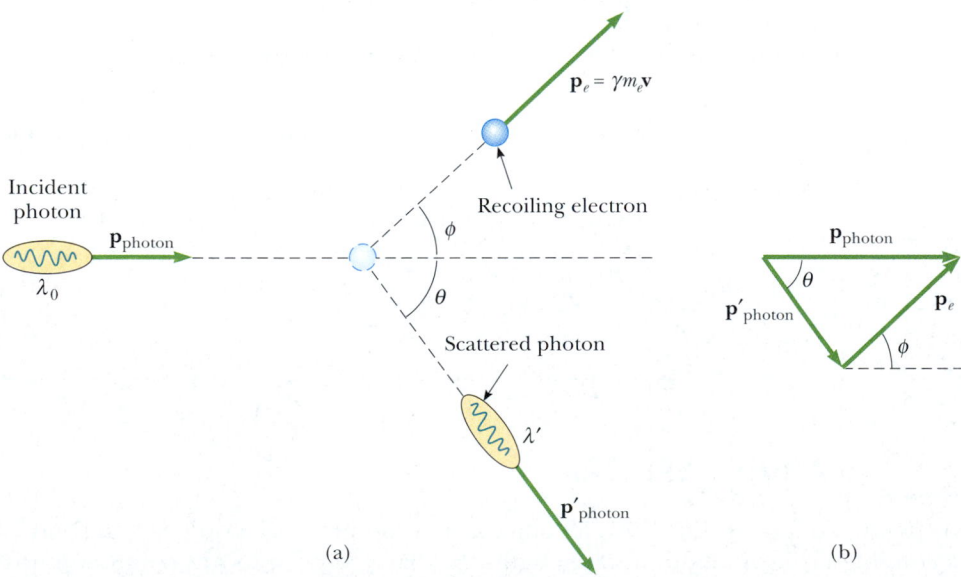

(a)   (b)

Figure 40.12 (a) Compton scattering of a photon by an electron. The scattered photon has less energy (longer wavelength) than the incident photon. (b) Momentum vectors for Compton scattering.

tered photon, and K_e is the kinetic energy of the recoiling electron. Because the electron may recoil at speeds comparable to the speed of light, we must use the relativistic expression $K_e = \gamma m_e c^2 - m_e c^2$ (Eq. 39.23). Therefore,

$$\frac{hc}{\lambda_0} = \frac{hc}{\lambda'} + \gamma m_e c^2 - m_e c^2 \tag{40.11}$$

where $\gamma = 1/\sqrt{1 - v^2/c^2}$.

Next, we apply the law of conservation of momentum to this collision, noting that both the x and y components of momentum are conserved. Equation 39.27 shows that the momentum of a photon has a magnitude $p = E/c$, and we know from Equation 40.6 that $E = hf$. Therefore, $p = hf/c$. Substituting λf for c (Eq. 16.14) in this expression gives us $p = h/\lambda$. Because the relativistic expression for the momentum of the recoiling electron is $p_e = \gamma m_e v$ (Eq. 39.19), we obtain the following expressions for the x and y components of linear momentum, where the angles are as described in Figure 40.12b:

$$x \text{ component:} \qquad \frac{h}{\lambda_0} = \frac{h}{\lambda'} \cos\theta + \gamma m_e v \cos\phi \tag{40.12}$$

$$y \text{ component:} \qquad 0 = \frac{h}{\lambda'} \sin\theta - \gamma m_e v \sin\phi \tag{40.13}$$

By eliminating v and ϕ from Equations 40.11 to 40.13, we obtain a single expression that relates the remaining three variables (λ', λ_0, and θ). After some algebra (see Problem 68), we obtain the Compton shift equation:

$$\Delta\lambda = \lambda' - \lambda_0 = \frac{h}{m_e c}(1 - \cos\theta)$$

EXAMPLE 40.4 ▸ **Compton Scattering at 45°**

X-rays of wavelength $\lambda_0 = 0.200$ nm are scattered from a block of material. The scattered x-rays are observed at an angle of 45.0° to the incident beam. Calculate their wavelength.

Solution The shift in wavelength of the scattered x-rays is given by Equation 40.10:

$$\Delta\lambda = \lambda' - \lambda_0 = \frac{h}{m_e c}(1 - \cos\theta)$$

$$= \frac{6.626 \times 10^{-34}\,\text{J·s}}{(9.11 \times 10^{-31}\,\text{kg})(3.00 \times 10^8\,\text{m/s})}(1 - \cos 45.0°)$$

$$= 7.10 \times 10^{-13}\,\text{m} = 0.000\ 710\ \text{nm}$$

Hence, the wavelength of the scattered x-ray at this angle is

$$\lambda' = \Delta\lambda + \lambda_0 = \boxed{0.200\ 710\ \text{nm}}$$

Exercise Find the fraction of energy lost by the photon in this collision.

Answer $\Delta E/E = 0.003\ 54$.

40.4 ▸ ATOMIC SPECTRA

As pointed out in Section 40.1, all objects emit thermal radiation characterized by a continuous distribution of wavelengths. In sharp contrast to this continuous distribution spectrum is the discrete **line spectrum** emitted by a low-pressure gas subject to an electric discharge. (Electric discharge occurs when the gas is subjected to a potential difference that creates an electric field larger than the dielec-

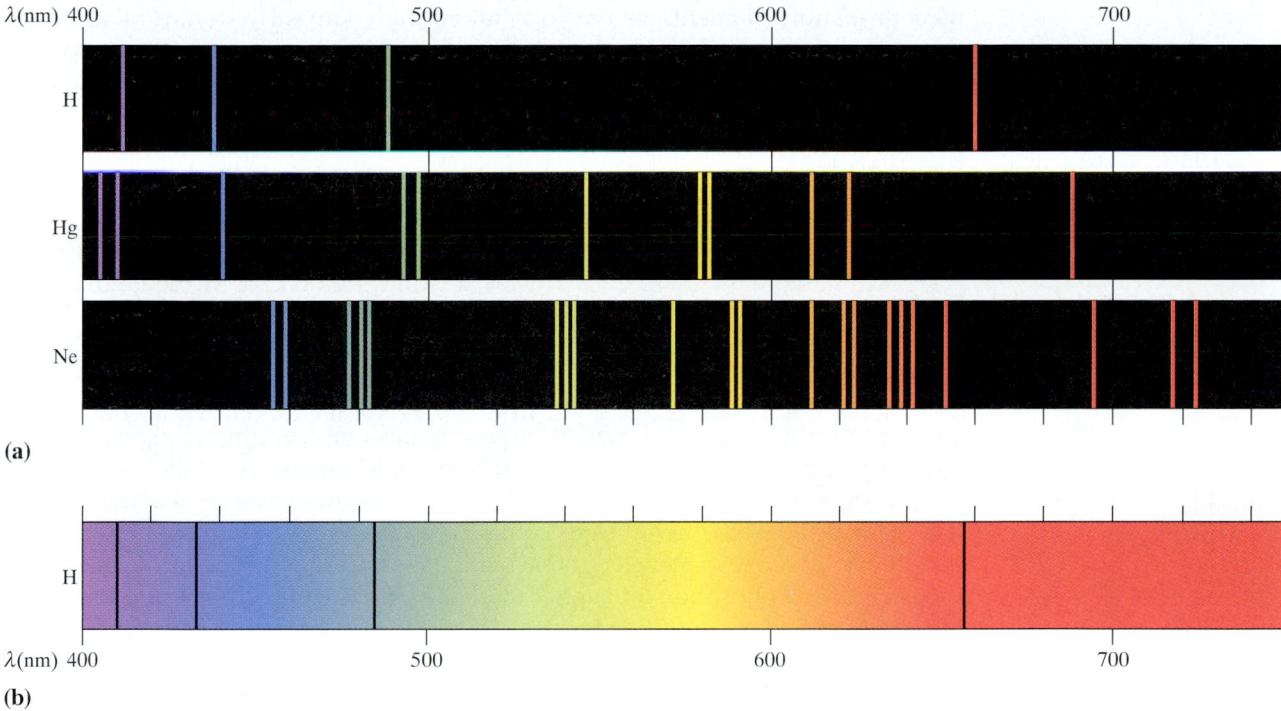

Figure 40.13 (a) Emission line spectra for hydrogen, mercury, and neon. (b) The absorption spectrum for hydrogen. Note that the dark absorption lines occur at the same wavelengths as the hydrogen emission lines in (a). *(K. W. Whitten, R. E. Davis, and M. L. Peck, General Chemistry, 6th ed., Philadelphia, Saunders College Publishing, 2000.)*

tric strength of the gas.) Observation and analysis of this emitted light is called **emission spectroscopy.**

When the light from a gas discharge is examined with a spectroscope, it is found to consist of a few bright lines of color on a generally dark background. (The *lines* are due to the collimation of the light through a slit.) This discrete line spectrum contrasts sharply with the continuous rainbow of colors seen when a glowing solid is viewed through a spectroscope. Furthermore, as you can see from Figure 40.13a, the wavelengths contained in a given line spectrum are characteristic of the element emitting the light. The simplest line spectrum is that for atomic hydrogen, and we describe this spectrum in detail. Other atoms exhibit completely different line spectra. Because no two elements have the same line spectrum, this phenomenon represents a practical and sensitive technique for identifying the elements present in unknown samples.

Another form of spectroscopy very useful in analyzing substances is **absorption spectroscopy.** An absorption spectrum is obtained by passing light from a continuous source through a gas or dilute solution of the element being analyzed. The absorption spectrum consists of a series of dark lines superimposed on the continuous spectrum of the light source, as shown in Figure 40.13b for atomic hydrogen. In general, not all of the lines present in the emission spectrum of an element are present in the element's absorption spectrum.

The absorption spectrum of an element has many practical applications. For example, the continuous spectrum of radiation emitted by the Sun must pass through the cooler gases of the solar atmosphere and through the Earth's atmosphere. The various absorption lines observed in the solar spectrum have been used to identify elements in the solar atmosphere. In early studies of the solar spectrum, experimenters found some lines that did not correspond to any known element. A new element had been discovered! The new element was named helium, after the Greek word for Sun, *helios.* Helium was subsequently isolated from subterranean gas on the Earth.

Scientists are able to examine the light from stars other than our Sun in this fashion, but elements other than those present on the Earth have never been detected. Absorption spectroscopy has also been a useful technique in analyzing heavy-metal contamination of the food chain. For example, the first determination of high levels of mercury in tuna was made with the use of atomic absorption spectroscopy.

The discrete emissions of light from gas discharges are used in "neon" signs, as seen in the opening photograph of this chapter. Neon, the first gas used in these types of signs and the gas after which they are named, emits strongly in the red region. As a result, a glass tube filled with neon gas emits bright red light when an applied voltage causes a continuous discharge. Early signs used different gases to provide different colors, although the brightness of these signs was generally very low. Many present-day "neon" signs contain mercury vapor, which emits strongly in the ultraviolet range of the electromagnetic spectrum. The inside of the glass tube is coated with a phosphor, a material that emits a particular color when it absorbs ultraviolet radiation from the mercury. The color of the light from the tube is due to the particular phosphor chosen. A fluorescent light operates in the same manner, with a white-emitting phosphor coating the inside of the glass tube.

From 1860 to 1885, scientists accumulated a great deal of data on atomic emissions using spectroscopic measurements. In 1885, a Swiss school teacher, Johann Jacob Balmer (1825–1898), found an empirical equation that correctly predicted the wavelengths of four visible emission lines of hydrogen: H_α(red), H_β(green), H_γ(blue), and H_δ(violet). Figure 40.14 shows these and other lines (in the ultraviolet) in the emission spectrum of hydrogen. The complete set of lines is called the **Balmer series.** The four visible lines occur at the wavelengths 656.3 nm, 486.1 nm, 434.1 nm, and 410.2 nm. The wavelengths of these lines can be described by the following equation, which is a modification of Balmer's original equation made by Johannes Rydberg (1854–1919):

$$\frac{1}{\lambda} = R_H\left(\frac{1}{2^2} - \frac{1}{n^2}\right) \tag{40.14}$$

where n may have integral values of 3, 4, 5, . . . and R_H is a constant now called the **Rydberg constant.** If the wavelength is in meters, R_H has the value $1.097\,373\,2 \times 10^7$ m^{-1}. The line in the Balmer series at 656.3 nm corresponds to $n = 3$ in Equation 40.14; the line at 486.1 nm corresponds to $n = 4$, and so on. The measured spectral lines agree with this empirical formula to within 0.1%.

Other lines in the spectrum of hydrogen were found following Balmer's discovery. These spectra are called the Lyman, Paschen, and Brackett series after their discoverers. The wavelengths of the lines in these series can be calculated through the use of the following empirical formulas:

$$\frac{1}{\lambda} = R_H\left(1 - \frac{1}{n^2}\right) \qquad n = 2, 3, 4, \ldots \tag{40.15}$$

$$\frac{1}{\lambda} = R_H\left(\frac{1}{3^2} - \frac{1}{n^2}\right) \qquad n = 4, 5, 6, \ldots \tag{40.16}$$

$$\frac{1}{\lambda} = R_H\left(\frac{1}{4^2} - \frac{1}{n^2}\right) \qquad n = 5, 6, 7, \ldots \tag{40.17}$$

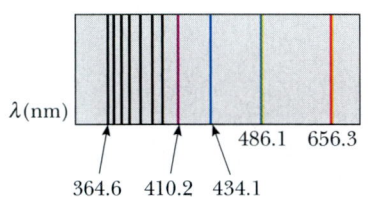

λ(nm)

486.1   656.3

364.6   410.2   434.1

Figure 40.14 The Balmer series of spectral lines for atomic hydrogen. The line labeled 364.6 is the shortest-wavelength line and is in the ultraviolet region of the electromagnetic spectrum. The other labeled lines are in the visible region.

Balmer series

Rydberg constant

Lyman series

Paschen series

Brackett series

All of these equations were purely empirical; this means that no theoretical basis existed for them. They simply worked. In the next section, we discuss the remarkable achievement of a theory for the hydrogen atom that provided a theoretical basis for these equations.

40.5 BOHR'S QUANTUM MODEL OF THE ATOM

At the beginning of the 20th century, scientists were perplexed by the failure of classical physics to explain the characteristics of atomic spectra. Why did atoms of a given element exhibit only certain spectral lines? Furthermore, why did the atoms absorb only those wavelengths that they emitted? In 1913, Niels Bohr provided an explanation of atomic spectra that includes some features of the currently accepted theory. Bohr's theory contained a combination of ideas from Planck's original quantum theory, Einstein's photon theory of light, early models of the atom, and Newtonian mechanics. Using the simplest atom, hydrogen, Bohr described a model of what he thought must be the atom's structure. His model of the hydrogen atom contains some classical features, as well as some revolutionary postulates that could not be justified within the framework of classical physics.

The basic ideas of the Bohr theory as it applies to the hydrogen atom are as follows:

1. The electron moves in circular orbits around the proton under the influence of the Coulomb force of attraction, as shown in Figure 40.15.
2. Only certain electron orbits are stable. These stable orbits are ones in which the electron does not emit energy in the form of radiation. Hence, the total energy of the atom remains constant, and classical mechanics can be used to describe the electron's motion. Note that this representation is completely different from the classical model of an electron in a circular orbit. According to classical physics, the centripetally accelerated electron should continuously emit radiation, losing energy and eventually spiraling into the nucleus.
3. Radiation is emitted by the atom when the electron "jumps" from a more energetic initial orbit to a lower-energy orbit. This jump cannot be visualized or treated classically. In particular, the frequency f of the photon emitted in the jump is related to the change in the atom's energy and *is independent of the frequency of the electron's orbital motion*. The frequency of the emitted radiation is found from the energy-conservation expression

$$E_i - E_f = hf \qquad \textbf{(40.18)}$$

where E_i is the energy of the initial state, E_f is the energy of the final state, and $E_i > E_f$.
4. The size of the allowed electron orbits is determined by a condition imposed on the electron's orbital angular momentum: The allowed orbits are those for which the electron's orbital angular momentum about the nucleus is an integral multiple of $\hbar = h/2\pi$:

$$m_e v r = n\hbar \qquad n = 1, 2, 3, \ldots \qquad \textbf{(40.19)}$$

Using these four assumptions, we can calculate the allowed energy levels and emission wavelengths of the hydrogen atom. We can find the electric potential energy of the system shown in Figure 40.15 from Equation 25.13, $U = k_e q_1 q_2 / r = -k_e e^2 / r$, where k_e is the Coulomb constant and the negative sign arises from the charge $-e$ on the electron. Thus, the total energy of the atom,

Assumptions of the Bohr theory

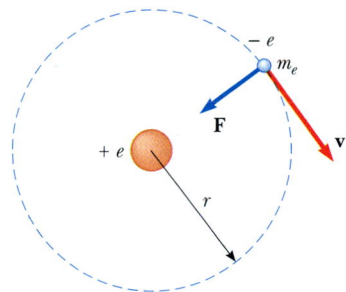

Figure 40.15 Diagram representing Bohr's model of the hydrogen atom, in which the orbiting electron is allowed to be only in specific orbits of discrete radii.

which contains both kinetic and potential energy terms, is

$$E = K + U = \tfrac{1}{2} m_e v^2 - k_e \frac{e^2}{r} \tag{40.20}$$

Applying Newton's second law to this system, we see that the Coulomb attractive force $k_e e^2/r^2$ exerted on the electron must equal the mass times the centripetal acceleration ($a = v^2/r$) of the electron:

$$\frac{k_e e^2}{r^2} = \frac{m_e v^2}{r}$$

From this expression, we see that the kinetic energy of the electron is

$$K = \frac{m_e v^2}{2} = \frac{k_e e^2}{2r} \tag{40.21}$$

Substituting this value of K into Equation 40.20, we find that the total energy of the atom is

$$E = -\frac{k_e e^2}{2r} \tag{40.22}$$

Note that the total energy is negative, indicating a bound electron–proton system. This means that energy in the amount of $k_e e^2/2r$ must be added to the atom to remove the electron and make the total energy of the system zero.

We can obtain an expression for r, the radius of the allowed orbits, by solving Equations 40.19 and 40.21 for v and equating the results:

$$v^2 = \frac{n^2 \hbar^2}{m_e^2 r^2} = \frac{k_e e^2}{m_e r}$$

$$r_n = \frac{n^2 \hbar^2}{m_e k_e e^2} \qquad n = 1, 2, 3, \ldots \tag{40.23}$$

This equation shows that the radii have discrete values—they are quantized. The result is based on the *assumption* that the electron can exist only in certain allowed orbits determined by the integer n.

The orbit with the smallest radius, called the **Bohr radius** a_0, corresponds to $n = 1$ and has the value

$$a_0 = \frac{\hbar^2}{m_e k_e e^2} = 0.052\ 9 \text{ nm} \tag{40.24}$$

We obtain a general expression for the radius of any orbit in the hydrogen atom by substituting Equation 40.24 into Equation 40.23:

Radii of Bohr orbits in hydrogen

$$r_n = n^2 a_0 = n^2 (0.052\ 9 \text{ nm}) \tag{40.25}$$

Bohr's theory gave a value of the right order of magnitude for the radius of a hydrogen atom from first principles rather than from any empirical assumption about orbit size. This result was a striking triumph for Bohr's theory. The first three Bohr orbits are shown to scale in Figure 40.16.

The quantization of orbit radii immediately leads to energy quantization. We can see this by substituting $r_n = n^2 a_0$ into Equation 40.22, obtaining for the allowed energy levels

Allowed energies of the hydrogen atom

$$E_n = -\frac{k_e e^2}{2a_0} \left(\frac{1}{n^2} \right) \qquad n = 1, 2, 3, \ldots \tag{40.26}$$

Inserting numerical values into this expression, we have

$$E_n = -\frac{13.606}{n^2} \text{ eV} \qquad n = 1, 2, 3, \ldots \qquad \textbf{(40.27)}$$

Only energies satisfying this equation (called **energy levels**) are permitted. The lowest allowed energy level, called the **ground state,** has $n = 1$ and energy $E_1 = -13.606$ eV. The next energy level, the **first excited state,** has $n = 2$ and energy $E_2 = E_1/2^2 = -3.401$ eV. Figure 40.17 is an energy level diagram showing the energies of these discrete energy states and the corresponding quantum numbers n. The uppermost level, corresponding to $n = \infty$ (or $r = \infty$) and $E = 0$, represents the state for which the electron is removed from the atom. The minimum energy required to ionize the atom (that is, to completely remove an electron in the ground state from the proton's influence) is called the **ionization energy.** As can be seen from Figure 40.17, the ionization energy for hydrogen in the ground state, based on Bohr's calculation, is 13.6 eV. This constituted another major achievement for the Bohr theory because the ionization energy for hydrogen had already been measured to be 13.6 eV.

Equations 40.18 and 40.26 can be used to calculate the frequency of the photon emitted when the electron jumps from an outer orbit to an inner orbit:

$$f = \frac{E_i - E_f}{h} = \frac{k_e e^2}{2a_0 h} \left(\frac{1}{n_f^2} - \frac{1}{n_i^2} \right) \qquad \textbf{(40.28)}$$

Because the quantity measured experimentally is wavelength, it is convenient to use $c = f\lambda$ to convert frequency to wavelength:

$$\frac{1}{\lambda} = \frac{f}{c} = \frac{k_e e^2}{2a_0 hc} \left(\frac{1}{n_f^2} - \frac{1}{n_i^2} \right) \qquad \textbf{(40.29)}$$

The remarkable fact is that this expression, which is purely theoretical, is identical to the general form of the empirical relationships discovered by Balmer and Rydberg and given by Equations 40.14 to 40.17,

$$\frac{1}{\lambda} = R_H \left(\frac{1}{n_f^2} - \frac{1}{n_i^2} \right) \qquad \textbf{(40.30)}$$

provided the constant $k_e e^2/2a_0 hc$ is equal to the experimentally determined Rydberg constant $R_H = 1.097\,373\,2 \times 10^7 \text{ m}^{-1}$. After Bohr demonstrated that these two quantities agree to within approximately 1%, this work was soon recognized as the crowning achievement of his new theory of quantum mechanics. Furthermore, Bohr showed that all of the spectral series for hydrogen have a natural interpretation in his theory. Figure 40.17 shows these spectral series as transitions between energy levels.

Bohr immediately extended his model for hydrogen to other elements in which all but one electron had been removed. Ionized elements such as He^+, Li^{2+}, and Be^{3+} were suspected to exist in hot stellar atmospheres, where atomic collisions frequently have enough energy to completely remove one or more atomic electrons. Bohr showed that many mysterious lines observed in the spectra of the Sun and several other stars could not be due to hydrogen but were correctly predicted by his theory if attributed to singly ionized helium. In general, to describe a single electron orbiting a fixed nucleus of charge $+Ze$, where Z is the atomic number of the element (see Section 1.2), Bohr's theory

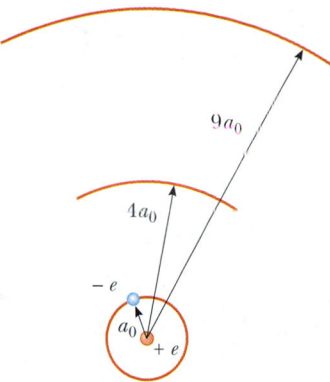

Figure 40.16 The first three circular orbits predicted by the Bohr model of the hydrogen atom.

Frequency of a photon emitted from hydrogen

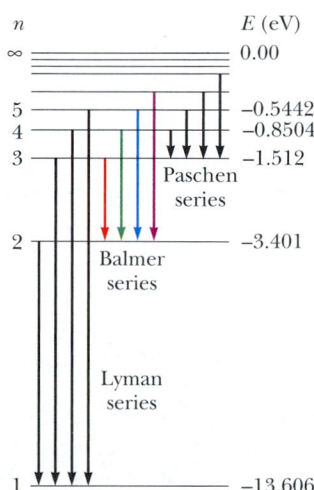

Figure 40.17 An energy level diagram for hydrogen. The discrete allowed energies are plotted on the vertical axis. Nothing is plotted on the horizontal axis, but the horizontal extent of the diagram is made large enough to show allowed transitions. Quantum numbers are given on the left and energies (in electron volts) on the right.

gives

$$r_n = (n^2) \frac{a_0}{Z} \tag{40.31}$$

$$E_n = -\frac{k_e e^2}{2a_0} \left(\frac{Z^2}{n^2} \right) \qquad n = 1, 2, 3, \ldots \tag{40.32}$$

EXAMPLE 40.5 ▶ Spectral Lines from the Star ξ-Puppis

Some mysterious lines observed in 1896 in the emission spectrum of the star ξ-Puppis (ξ is the Greek letter xi) fit the empirical formula

$$\frac{1}{\lambda} = R_H \left(\frac{1}{(n_f/2)^2} - \frac{1}{(n_i/2)^2} \right)$$

Show that these lines can be explained by the Bohr theory as originating from He$^+$.

Solution   The ion He$^+$ has $Z = 2$. Thus, the allowed energy levels are given by Equation 40.32 as

$$E_n = -\frac{k_e e^2}{2a_0} \left(\frac{4}{n^2} \right)$$

Using Equation 40.28, we find

$$f = \frac{E_i - E_f}{h} = \frac{k_e e^2}{2a_0 h} \left(\frac{4}{n_f^2} - \frac{4}{n_i^2} \right)$$

$$= \frac{k_e e^2}{2a_0 h} \left(\frac{1}{(n_f/2)^2} - \frac{1}{(n_i/2)^2} \right)$$

$$\frac{1}{\lambda} = \frac{f}{c} = \frac{k_e e^2}{2a_0 hc} \left(\frac{1}{(n_f/2)^2} - \frac{1}{(n_i/2)^2} \right)$$

This is the desired solution when we recognize that $R_H \equiv k_e e^2 / 2a_0 hc$ (see text discussion immediately following Eq. 40.30).

EXAMPLE 40.6 ▶ Electronic Transitions in Hydrogen

(a) The electron in a hydrogen atom makes a transition from the $n = 2$ energy state to the ground state ($n = 1$). Find the wavelength and frequency of the emitted photon.

Solution   We can use Equation 40.30 directly to obtain λ, with $n_i = 2$ and $n_f = 1$:

$$\frac{1}{\lambda} = R_H \left(\frac{1}{n_f^2} - \frac{1}{n_i^2} \right)$$

$$= R_H \left(\frac{1}{1^2} - \frac{1}{2^2} \right) = \frac{3R_H}{4}$$

$$\lambda = \frac{4}{3R_H} = \frac{4}{3(1.097 \times 10^7 \text{ m}^{-1})}$$

$$= 1.215 \times 10^{-7} \text{ m} = \boxed{121.5 \text{ nm}} \quad \text{(ultraviolet)}$$

Because $c = f\lambda$, the frequency of the photon is

$$f = \frac{c}{\lambda} = \frac{3.00 \times 10^8 \text{ m/s}}{1.215 \times 10^{-7} \text{ m}} = \boxed{2.47 \times 10^{15} \text{ Hz}}$$

(b) In interstellar space, highly excited hydrogen atoms called Rydberg atoms have been observed. Find the wavelength to which radio astronomers must tune to detect signals from electrons dropping from the $n = 273$ level to $n = 272$.

Solution   We can again use Equation 40.30, this time with $n_i = 273$ and $n_f = 272$:

$$\frac{1}{\lambda} = R_H \left(\frac{1}{n_f^2} - \frac{1}{n_i^2} \right)$$

$$\lambda = \boxed{0.922 \text{ m}}$$

(c) What is the radius of the electron orbit for a Rydberg atom for which $n = 272$?

Solution   Using Equation 40.25, we find

$$r_{272} = (272)^2 (0.052\ 9 \text{ nm}) = \boxed{3.91 \ \mu\text{m}}$$

This is large enough that the atom is on the verge of becoming macroscopic!

EXAMPLE 40.7   The Balmer Series for Hydrogen

The Balmer series for the hydrogen atom corresponds to electronic transitions that terminate in the $n = 2$ state, as shown in Figure 40.18. (a) Find the longest-wavelength photon emitted in this series and determine its energy.

Solution   The longest-wavelength (lowest-energy) photon in the Balmer series results from the transition from $n = 3$ to $n = 2$. This is the lowest-energy photon in this series because it involves the smallest possible energy change. Equation 40.30 gives

$$\frac{1}{\lambda} = R_H \left(\frac{1}{n_f^2} - \frac{1}{n_i^2} \right)$$

$$\frac{1}{\lambda_{max}} = R_H \left(\frac{1}{2^2} - \frac{1}{3^2} \right) = \frac{5}{36} R_H$$

$$\lambda_{max} = \frac{36}{5R_H} = \frac{36}{5(1.097 \times 10^7 \text{ m}^{-1})}$$

$$= \boxed{656.3 \text{ nm}} \qquad (\text{red})$$

The energy of this photon is

$$E_{photon} = hf = \frac{hc}{\lambda_{max}}$$

$$= \frac{(6.626 \times 10^{-34} \text{ J} \cdot \text{s})(2.998 \times 10^8 \text{ m/s})}{656.3 \times 10^{-9} \text{ m}}$$

$$= 3.03 \times 10^{-19} \text{ J} = \boxed{1.89 \text{ eV}}$$

We could also obtain the energy by using the expression $hf = E_3 - E_2$, where E_2 and E_3 can be calculated from Equation 40.26.

(b) Find the shortest-wavelength photon emitted in the Balmer series.

Solution   The shortest-wavelength photon in the Balmer series is emitted when the electron makes a transition from $n = \infty$ to $n = 2$. Therefore,

$$\frac{1}{\lambda_{min}} = R_H \left(\frac{1}{2^2} - \frac{1}{\infty} \right) = \frac{R_H}{4}$$

$$\lambda_{min} = \frac{4}{R_H} = \frac{4}{1.097 \times 10^7 \text{ m}^{-1}} = \boxed{364.6 \text{ nm}}$$

This wavelength is in the ultraviolet region and corresponds to the series limit.

Exercise   Find the energy of this shortest-wavelength photon.

Answer   3.40 eV.

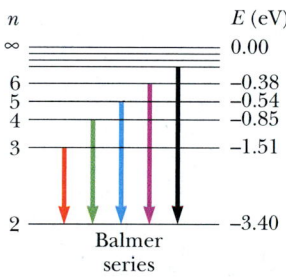

Figure 40.18   Transitions responsible for the Balmer series for the hydrogen atom. All transitions terminate at the $n = 2$ energy level. Energy levels are not drawn to scale.

Bohr's Correspondence Principle

In our study of relativity, we found that Newtonian mechanics is a special case of relativistic mechanics and is usable only when v is much less than c. Similarly, **quantum physics is in agreement with classical physics where the difference between quantized levels becomes vanishingly small.** This principle, first set forth by Bohr, is called the **correspondence principle.**

For example, consider an electron orbiting the hydrogen atom with $n > 10\,000$. For such large values of n, the energy differences between adjacent levels approach zero, and therefore the levels are nearly continuous. Consequently, the classical model is reasonably accurate in describing the system for large values of n. According to the classical picture, the frequency of the light emitted by the atom is equal to the frequency of revolution of the electron in its orbit about the nucleus. Calculations show that for $n > 10\,000$, this frequency is different from that predicted by quantum mechanics by less than 0.015%.

40.6 ▶ PHOTONS AND ELECTROMAGNETIC WAVES

Phenomena such as the photoelectric effect and the Compton effect offer ironclad evidence that when light (or other forms of electromagnetic radiation) and matter interact, the light behaves as if it were composed of particles having energy hf and momentum h/λ. An obvious question at this point is, "How can light be considered a photon (in other words, a particle) when we know it is a wave?" On the one hand, we describe light in terms of photons having energy and momentum. On the other hand, we recognize that light and other electromagnetic waves exhibit interference and diffraction effects, which are consistent only with a wave interpretation.

Which model is correct? Is light a wave or a particle? The answer depends on the phenomenon being observed. Some experiments can be explained either better or solely with the photon model, whereas others are explained better or solely with the wave model. The end result is that **we use both models and admit that the true nature of light is not describable in terms of any single classical picture.** However, you should recognize that the same light beam that can eject photoelectrons from a metal (meaning that the beam consists of photons) can also be diffracted by a grating (meaning that the beam is a wave). In other words, **the particle model and the wave model of light complement each other.**

The success of the particle model of light in explaining the photoelectric effect and the Compton effect raises many other questions. If light is a particle, what is the meaning of its "frequency" and "wavelength," and which of these two properties determines its energy and momentum? Is light *simultaneously* a wave and a particle? Although photons have no rest energy (a nonobservable quantity because a photon cannot be at rest!), is there a simple expression for the *effective mass* of a moving photon? If photons have effective mass, do they experience gravitational attraction? What is the spatial extent of a photon, and how does an electron absorb or scatter one photon? Although some of these questions are answerable, others are difficult to answer because our experiences from the everyday macroscopic world are far different from the behavior of microscopic particles. Many of these questions stem from classical analogies such as colliding billiard balls and water waves breaking on a shore. Quantum mechanics gives light a more fluid and flexible nature by incorporating both the particle model and the wave model as necessary and complementary. Hence,

> light has a dual nature: It exhibits both wave and particle characteristics.

To understand why photons are compatible with electromagnetic waves, consider 2.5-MHz radio waves as an example. The energy of a photon having this frequency is only about 10^{-8} eV, too small to allow the photon to be detected. A sensitive radio receiver might require as many as 10^{10} of these photons to produce a detectable signal. Such a large number of photons would appear, on the average, as a continuous wave. With so many photons reaching the detector every second, it is unlikely that any graininess would appear in the detected signal. That is, with 2.5-MHz waves, one would not be able to detect the individual photons striking the antenna.

Now consider what happens as we go to higher frequencies. In the visible region, it is possible to observe both the particle characteristics and the wave characteristics of light. As we mentioned earlier, a beam of visible light shows interference phenomena (thus, it is a wave) and at the same time can produce photoelectrons (thus, it is a particle). At even higher frequencies, the momentum and energy of the photon increase. Consequently, the particle nature of light becomes more evident than its wave nature. For example, absorption of an x-ray photon is easily detected as a single event but wave effects are very difficult to observe.

40.7 ▶ THE WAVE PROPERTIES OF PARTICLES

Students introduced to the dual nature of light often find the concept difficult to accept. In the world around us, we are accustomed to regarding such things as baseballs solely as particles and such things as sound waves solely as forms of wave motion. Every large-scale observation can be interpreted by considering either a wave explanation or a particle explanation, but in the world of photons and electrons, such distinctions are not as sharply drawn. Even more disconcerting is the fact that, under certain conditions, the things we unambiguously call "particles" exhibit wave characteristics!

In 1923, in his doctoral dissertation, Louis de Broglie postulated that **because photons have both wave and particle characteristics, perhaps all forms of matter have both properties.** This was a highly revolutionary idea with no experimental confirmation at that time. According to de Broglie, electrons, just like light, have a dual particle-wave nature. Accompanying every electron is a wave (not an electromagnetic wave!). He explained the source of this assertion in his 1929 Nobel Prize acceptance speech:

Louis de Broglie (1892–1987)
A French physicist, de Broglie was awarded the Nobel Prize in 1929 for his prediction of the wave nature of electrons. *(AIP Niels Bohr Library)*

> On the one hand the quantum theory of light cannot be considered satisfactory since it defines the energy of a light corpuscle by the equation $E = hf$ containing the frequency f. Now a purely corpuscular theory contains nothing that enables us to define a frequency; for this reason alone, therefore, we are compelled, in the case of light, to introduce the idea of a corpuscle and that of periodicity simultaneously. On the other hand, determination of the stable motion of electrons in the atom introduces integers, and up to this point the only phenomena involving integers in physics were those of interference and of normal modes of vibration. This fact suggested to me the idea that electrons too could not be considered simply as corpuscles, but that periodicity must be assigned to them also.

In Section 39.7, we found that the relationship between the energy and the linear momentum of a photon, which has a rest energy of zero, is $p = E/c$. We also know that the energy of a photon is $E = hf = hc/\lambda$. Thus, the momentum of a photon can be expressed as

$$p = \frac{E}{c} = \frac{hc}{c\lambda} = \frac{h}{\lambda}$$

From this equation we see that the photon wavelength can be specified by its momentum: $\lambda = h/p$. De Broglie suggested that material particles of momentum p have a characteristic wavelength $\lambda = h/p$. Because the momentum of a particle of mass m and speed v is $p = mv$, the **de Broglie wavelength** of that particle is[2]

$$\lambda = \frac{h}{p} = \frac{h}{mv} \qquad\qquad \textbf{(40.33)}$$

---

[2] The de Broglie wavelength for a particle moving at *any* speed v is $\lambda = h/\gamma mv$, where $\gamma = (1 + v^2/c^2)^{-1/2}$.

Furthermore, in analogy with photons, de Broglie postulated that the frequencies of **matter waves** (that is, waves associated with particles having nonzero rest energy) obey the Einstein relationship $E = hf$, where E is the total energy of the particle, so that

$$f = \frac{E}{h} \qquad (40.34)$$

The dual nature of matter is apparent in these two equations because each contains both particle concepts (mv and E) and wave concepts (λ and f). The fact that these relationships are established experimentally for photons makes the de Broglie hypothesis that much easier to accept.

The Davisson–Germer Experiment

De Broglie's proposal in 1923 that matter exhibits both wave and particle properties was regarded as pure speculation. If particles such as electrons had wave properties, then under the correct conditions they should exhibit diffraction effects. Only three years later, C. J. Davisson (1881–1958) and L. H. Germer (1896–1971) of the United States succeeded in measuring the wavelength of electrons. Their important discovery provided the first experimental confirmation of the matter waves proposed by de Broglie.

Interestingly, the intent of the initial Davisson–Germer experiment was not to confirm the de Broglie hypothesis. In fact, their discovery was made by accident (as is often the case). The experiment involved the scattering of low-energy electrons (about 54 eV) from a nickel target in a vacuum. During one experiment, the nickel surface was badly oxidized because of an accidental break in the vacuum system. After the target was heated in a flowing stream of hydrogen to remove the oxide coating, electrons scattered by it exhibited intensity maxima and minima at specific angles. The experimenters finally realized that the nickel had formed large crystalline regions upon heating and that the regularly spaced planes of atoms in these regions served as a diffraction grating for electron matter waves.

Shortly thereafter, Davisson and Germer performed more extensive diffraction measurements on electrons scattered from single-crystal targets. Their results showed conclusively the wave nature of electrons and confirmed the de Broglie relationship $p = h/\lambda$. In the same year, G. P. Thomson (1892–1975) of Scotland also observed electron diffraction patterns by passing electrons through very thin gold foils. Diffraction patterns have since been observed for helium atoms, hydrogen atoms, and neutrons. Hence, the universal nature of matter waves has been established in various ways.

The problem of understanding the dual nature of matter and radiation is conceptually difficult because the two models seem to contradict each other. This problem as it applies to light was discussed earlier. Bohr helped to resolve this problem in his **principle of complementarity,** which states that the **wave and particle models of either matter or radiation complement each other.** Neither model can be used exclusively to describe matter or radiation adequately. Because humans can only generate mental images based on their experiences from the everyday world (baseballs, water waves, and so forth), we use both descriptions in a complementary manner to explain any given set of data from the quantum world.

EXAMPLE 40.8 The Wavelength of an Electron

Calculate the de Broglie wavelength for an electron ($m = 9.11 \times 10^{-31}$ kg) moving at 1.00×10^7 m/s.

Solution Equation 40.33 gives

$$\lambda = \frac{h}{mv} = \frac{6.63 \times 10^{-34} \, \text{J} \cdot \text{s}}{(9.11 \times 10^{-31} \, \text{kg})(1.00 \times 10^7 \, \text{m/s})}$$

$$= \boxed{7.28 \times 10^{-11} \, \text{m}}$$

Exercise Find the de Broglie wavelength of a stone of mass 50 g thrown with a speed of 40 m/s.

Answer 3.3×10^{-34} m.

EXAMPLE 40.9 An Accelerated Charged Particle

A particle of charge q and mass m has been accelerated from rest through a potential difference ΔV. Find an expression for its de Broglie wavelength.

Solution When a charged particle is accelerated from rest through a potential difference ΔV, the gain in kinetic energy $\frac{1}{2}mv^2$ must equal the loss in potential energy $q\Delta V$:

$$\tfrac{1}{2}mv^2 = q\,\Delta V$$

Because $p = mv$, we can express this equation in the form

$$\frac{p^2}{2m} = q\,\Delta V$$

$$p = \sqrt{2mq\,\Delta V}$$

Substituting this expression for p into Equation 40.33 gives

$$\lambda = \frac{h}{p} = \boxed{\frac{h}{\sqrt{2mq\,\Delta V}}}$$

Exercise Calculate the de Broglie wavelength of an electron accelerated through a potential difference of 50 V.

Answer 0.174 nm.

SUMMARY

The characteristics of blackbody radiation cannot be explained using classical concepts. Planck introduced the quantum concept when he assumed that the atomic oscillators responsible for this radiation exist only in discrete energy states. Radiation is emitted in single quantized packets whenever an oscillator makes a transition between discrete energy states.

The **photoelectric effect** is a process whereby electrons are ejected from a metal surface when light is incident on that surface. Einstein provided a successful explanation of this effect by extending Planck's quantum hypothesis to electromagnetic radiation. In this model, light is viewed as a stream of light particles, or **photons,** each having energy $E = hf$, where f is the frequency and h is Planck's constant. The maximum kinetic energy of the ejected photoelectron is

$$K_{\max} = hf - \phi \tag{40.8}$$

where ϕ is the **work function** of the metal.

X-rays are scattered at various angles by electrons in a target. In such a scattering event, a shift in wavelength is observed for the scattered x-rays, and the phenomenon is known as the **Compton effect.** Classical physics does not explain this effect. If the x-ray is treated as a photon, conservation of energy and linear momentum applied to the photon–electron collisions yields for the Compton shift:

$$\lambda' - \lambda_0 = \frac{h}{m_e c} (1 - \cos\theta) \tag{40.10}$$

where m_e is the mass of the electron, c is the speed of light, and θ is the scattering angle.

The Bohr model of the atom is successful in describing the spectra of atomic hydrogen and hydrogen-like ions. One of the basic assumptions of the model is that the electron can exist only in discrete orbits such that the angular momentum mvr is an integral multiple of $h/2\pi = \hbar$. When we assume circular orbits and a simple Coulomb attraction between electron and proton, the energies of the quantum states for hydrogen are calculated to be

$$E_n = -\frac{k_e e^2}{2a_0} \left(\frac{1}{n^2} \right) \qquad n = 1, 2, 3, \ldots \tag{40.26}$$

where k_e is the Coulomb constant, e is the electronic charge, n is an integer called the **quantum number,** and $a_0 = 0.052\,9$ nm is the **Bohr radius.**

If the electron in a hydrogen atom makes a transition from an orbit whose quantum number is n_i to one whose quantum number is n_f, where $n_f < n_i$, a photon is emitted by the atom, and the frequency of this photon is

$$f = \frac{k_e e^2}{2a_0 h} \left(\frac{1}{n_f{}^2} - \frac{1}{n_i{}^2} \right) \tag{40.28}$$

Light has a dual nature in that it has both wave and particle characteristics. Some experiments can be explained either better or solely by the particle model, whereas others can be explained either better or solely by the wave model.

Every object of mass m and momentum $p = mv$ has wave properties, with a wavelength given by the de Broglie relationship:

$$\lambda = \frac{h}{p} = \frac{h}{mv} \tag{40.33}$$

Questions

1. What assumptions were made by Planck in dealing with the problem of blackbody radiation? Discuss the consequences of these assumptions.

2. The classical model of blackbody radiation given by the Rayleigh–Jeans law has two major flaws. Identify them and explain how Planck's law deals with them.

3. If the photoelectric effect is observed for one metal, can you conclude that the effect will also be observed for another metal under the same conditions? Explain.

4. In the photoelectric effect, explain why the stopping potential depends on the frequency of light but not on the intensity.

5. Suppose the photoelectric effect occurs in a gaseous target rather than a solid plate. Will photoelectrons be produced at all frequencies of the incident photon? Explain.

6. How does the Compton effect differ from the photoelectric effect?

7. What assumptions did Compton make in dealing with the scattering of a photon from an electron?

8. The Bohr theory of the hydrogen atom is based upon several assumptions. Discuss these assumptions and their significance. Do any of them contradict classical physics?

9. Suppose that the electron in the hydrogen atom obeyed classical mechanics rather than quantum mechanics. Why should such a "hypothetical" atom emit a continuous spectrum rather than the observed line spectrum?

10. Can the electron in the ground state of hydrogen absorb a photon of energy (a) less than 13.6 eV and (b) greater than 13.6 eV?

11. Why would the spectral lines of diatomic hydrogen be different from those of monatomic hydrogen?

12. Explain why, in the Bohr model, the total energy of the atom is negative.

13. An x-ray photon is scattered by an electron. What happens to the frequency of the scattered photon relative to that of the incident photon?

14. Why does the existence of a cutoff frequency in the photoelectric effect favor a particle theory for light rather than a wave theory?

15. A student claims that he is going to eject electrons from a piece of metal by placing a radio transmitter antenna ad-jacent to the metal and sending a strong AM radio signal into the antenna. The work function of a metal is typically a few electron volts. Will this work?

16. All objects radiate energy. Why, then, are we not able to see all objects in a dark room?

17. Which has more energy, a photon of ultraviolet radiation or a photon of yellow light?

18. Why was the Davisson–Germer experiment involving the diffraction of electrons so important?

PROBLEMS

1, 2, 3 = straightforward, intermediate, challenging ☐ = full solution available in the *Student Solutions Manual and Study Guide*
WEB = solution posted at **http://www.saunderscollege.com/physics/** 💻 = Computer useful in solving problem 🧲 = Interactive Physics
☐ = paired numerical/symbolic problems

Section 40.1 Blackbody Radiation and Planck's Hypothesis

1. The human eye is most sensitive to 560-nm light. What is the temperature of a black body that would radiate most intensely at this wavelength?

2. (a) Lightning produces a maximum air temperature on the order of 10^4 K, whereas (b) a nuclear explosion produces a temperature on the order of 10^7 K. Use Wien's displacement law to find the order of magnitude of the wavelength of the thermally produced photons radiated with greatest intensity by each of these sources. Name the part of the electromagnetic spectrum where you would expect each to radiate most strongly.

3. (a) Assuming that the tungsten filament of a lightbulb is a black body, determine its peak wavelength if its temperature is 2 900 K. (b) Why does your answer to part (a) suggest that more energy from a lightbulb goes into infrared radiation than into visible light?

4. A black body at 7 500 K consists of an opening of diameter 0.050 0 mm, looking into an oven. Find the number of photons per second escaping the hole and having wavelengths between 500 nm and 501 nm.

5. Consider a black body of surface area 20.0 cm² and temperature 5 000 K. (a) How much power does it radiate? (b) At what wavelength does it radiate most intensely? Find the spectral power per wavelength at (c) this wavelength and at wavelengths of (d) 1.00 nm (an x- or γ-ray), (e) 5.00 nm (ultraviolet light or an x-ray), (f) 400 nm (at the boundary between UV and visible light), (g) 700 nm (at the boundary between visible and infrared light), (h) 1.00 mm (infrared light or a microwave), and (i) 10.0 cm (a microwave or radio wave). (j) About how much power does the object radiate as visible light?

6. The radius of our Sun is 6.96×10^8 m, and its total power output is 3.77×10^{26} W. (a) Assuming that the Sun's surface emits as a black body, calculate its surface temperature. (b) Using the result of part (a), find λ_{max} for the Sun.

WEB 7. Calculate the energy, in electron volts, of a photon whose frequency is (a) 620 THz, (b) 3.10 GHz, (c) 46.0 MHz. (d) Determine the corresponding wavelengths for these photons and state the classification of each on the electromagnetic spectrum.

8. A sodium-vapor lamp has a power output of 10.0 W. Using 589.3 nm as the average wavelength of this source, calculate the number of photons emitted per second.

9. An FM radio transmitter has a power output of 150 kW and operates at a frequency of 99.7 MHz. How many photons per second does the transmitter emit?

10. The average threshold of dark-adapted (scotopic) vision is 4.00×10^{-11} W/m² at a central wavelength of 500 nm. If light having this intensity and wavelength enters the eye and the pupil is open to its maximum diameter of 8.50 mm, how many photons per second enter the eye?

11. A simple pendulum has a length of 1.00 m and a mass of 1.00 kg. If the amplitude of oscillations of the pendulum is 3.00 cm, estimate the quantum number for the pendulum.

12. **Review Problem.** A star moving away from the Earth at $0.280c$ emits radiation that we measure to be most intense at the wavelength 500 nm. Determine the surface temperature of this star.

13. Show that at short wavelengths or low temperatures, Planck's radiation law (Eq. 40.3) predicts an exponential decrease in $I(\lambda, T)$ given by *Wien's radiation law*:

$$I(\lambda, T) = \frac{2\pi hc^2}{\lambda^5} e^{-hc/\lambda k_B T}$$

14. Show that at long wavelengths, Planck's radiation law (Eq. 40.3) reduces to the Rayleigh–Jeans law (Eq. 40.2).

Section 40.2 The Photoelectric Effect

15. Molybdenum has a work function of 4.20 eV. (a) Find the cutoff wavelength and cutoff frequency for the pho-

toelectric effect. (b) Calculate the stopping potential if the incident light has a wavelength of 180 nm.

16. Electrons are ejected from a metal surface with speeds ranging up to 4.60×10^5 m/s when light with a wavelength of $\lambda = 625$ nm is used. (a) What is the work function of the surface? (b) What is the cutoff frequency for this surface?

17. Lithium, beryllium, and mercury have work functions of 2.30 eV, 3.90 eV, and 4.50 eV, respectively. If 400-nm light is incident on each of these metals, determine (a) which metals exhibit the photoelectric effect and (b) the maximum kinetic energy for the photoelectrons in each case.

18. A student studying the photoelectric effect from two different metals records the following information: (i) the stopping potential for photoelectrons released from metal 1 is 1.48 V larger than that for metal 2, and (ii) the cutoff frequency for metal 1 is 40.0% smaller than that for metal 2. Determine the work function for each metal.

19. Two light sources are used in a photoelectric experiment to determine the work function for a particular metal surface. When green light from a mercury lamp ($\lambda = 546.1$ nm) is used, a stopping potential of 0.376 V reduces the photocurrent to zero. (a) Based on this measurement, what is the work function for this metal? (b) What stopping potential would be observed when using the yellow light from a helium discharge tube ($\lambda = 587.5$ nm)?

20. When 445-nm light strikes a certain metal surface, the stopping potential is 70.0% of that which results when 410-nm light strikes the same metal surface. Based on this information and the following table of work functions, identify the metal involved in the experiment.

Metal	Work Function (eV)
Cesium	1.90
Potassium	2.23
Silver	4.73
Tungsten	4.58

21. From the scattering of sunlight, Thomson calculated the classical radius of the electron as having a value of 2.82×10^{-15} m. If sunlight with an intensity of 500 W/m^2 falls on a disk with this radius, calculate the time required to accumulate 1.00 eV of energy. Assume that light is a classical wave and that the light striking the disk is completely absorbed. How does your result compare with the observation that photoelectrons are emitted promptly (within 10^{-9} s)?

22. **Review Problem.** An isolated copper sphere of radius 5.00 cm, initially uncharged, is illuminated by ultraviolet light of wavelength 200 nm. What charge will the photoelectric effect induce on the sphere? The work function for copper is 4.70 eV.

23. **Review Problem.** A light source emitting radiation at 7.00×10^{14} Hz is incapable of ejecting photoelectrons from a certain metal. In an attempt to use this source to eject photoelectrons from the metal, the source is given a velocity toward the metal. (a) Explain how this procedure produces photoelectrons. (b) When the speed of the light source is equal to $0.280c$, photoelectrons just begin to be ejected from the metal. What is the work function of the metal? (c) When the speed of the light source is increased to $0.900c$, determine the maximum kinetic energy of the photoelectrons.

Section 40.3   The Compton Effect

24. Calculate the energy and momentum of a photon of wavelength 700 nm.

25. X-rays having an energy of 300 keV undergo Compton scattering from a target. If the scattered rays are detected at 37.0° relative to the incident rays, find (a) the Compton shift at this angle, (b) the energy of the scattered x-ray, and (c) the energy of the recoiling electron.

26. A 0.110-nm photon collides with a stationary electron. After the collision, the electron moves forward and the photon recoils backwards. Find the momentum and kinetic energy of the electron.

WEB 27. A 0.001 60-nm photon scatters from a free electron. For what (photon) scattering angle does the recoiling electron have kinetic energy equal to the energy of the scattered photon?

28. In a Compton scattering experiment, a photon is scattered through an angle of 90.0°, and the electron is scattered through an angle of 20.0°. Determine the wavelength of the scattered photon.

29. A 0.880-MeV photon is scattered by a free electron initially at rest such that the scattering angle of the scattered electron equals that of the scattered photon ($\theta = \phi$ in Fig. 40.10b). (a) Determine the angles θ and ϕ. (b) Determine the energy and momentum of the scattered photon. (c) Determine the kinetic energy and momentum of the scattered electron.

30. A photon having energy E_0 is scattered by a free electron initially at rest such that the scattering angle of the scattered electron equals that of the scattered photon ($\theta = \phi$ in Fig. 40.10b). (a) Determine the angles θ and ϕ. (b) Determine the energy and momentum of the scattered photon. (c) Determine the kinetic energy and momentum of the scattered electron.

31. A 0.700-MeV photon scatters off a free electron such that the scattering angle of the photon is twice the scattering angle of the electron (Fig. P40.31). (a) Determine the scattering angle for the electron and (b) the final speed of the electron.

32. A photon having wavelength λ scatters off a free electron at A (Fig. P40.32), producing a second photon having wavelength λ'. This photon then scatters off another free electron at B, producing a third photon

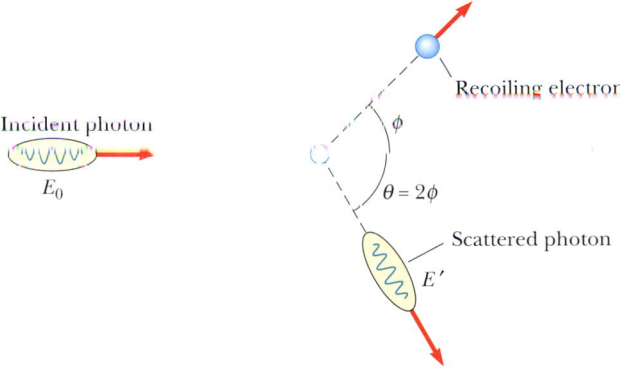

Figure P40.31

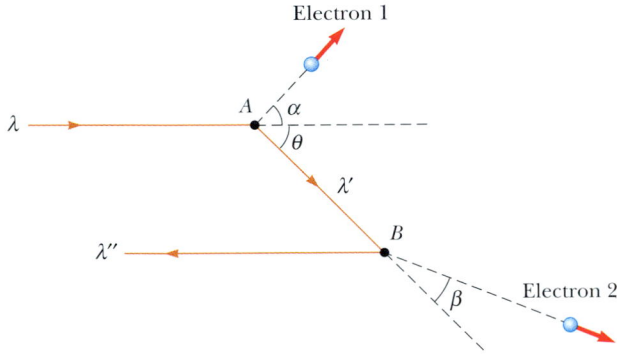

Figure P40.32

having wavelength λ'' and moving in a direction directly opposite the original photon as shown in Figure P40.32. Determine the numerical value of $\Delta\lambda = \lambda'' - \lambda$.

33. After a 0.800-nm x-ray photon scatters from a free electron, the electron recoils at 1.40×10^6 m/s. (a) What was the Compton shift in the photon's wavelength? (b) Through what angle was the photon scattered?

34. Find the maximum fractional energy loss for a 0.511-MeV gamma ray that is Compton scattered from a free (a) electron and (b) proton.

Section 40.4 Atomic Spectra

35. Show that the wavelengths for the Balmer series satisfy the equation

$$\lambda = \frac{364.5 n^2}{n^2 - 4} \text{ nm} \qquad \text{where } n = 3, 4, 5, \ldots$$

36. (a) Suppose that the Rydberg constant were given by $R_H = 2.00 \times 10^7$ m^{-1}. In what part of the electromagnetic spectrum would the Balmer series lie? (b) Repeat for $R_H = 0.500 \times 10^7$ m^{-1}.

37. (a) What value of n is associated with the 94.96-nm line in the Lyman hydrogen series? (b) Could this wavelength be associated with the Paschen or Brackett series?

38. (a) Compute the shortest wavelength in each of these hydrogen spectral series: Lyman, Balmer, Paschen, and Brackett. (b) Compute the energy (in electron volts) of the highest-energy photon produced in each series.

39. Liquid oxygen has a bluish color, meaning that it preferentially absorbs light toward the red end of the visible spectrum. Although the oxygen molecule (O_2) does not strongly absorb visible radiation, it does absorb strongly at 1 269 nm, which is in the infrared region of the spectrum. Research has shown that it is possible for two colliding O_2 molecules to absorb a single photon, sharing its energy equally. The transition that both molecules undergo is the same transition that results when they absorb 1 269-nm radiation. What is the wavelength of the single photon that causes this double transition? What is the color of this radiation?

Section 40.5 Bohr's Quantum Model of the Atom

40. For a hydrogen atom in its ground state, use the Bohr model to compute (a) the orbital speed of the electron, (b) the kinetic energy of the electron, and (c) the electric potential energy of the atom.

WEB 41. A hydrogen atom is in its first excited state ($n = 2$). Using the Bohr theory of the atom, calculate (a) the radius of the orbit, (b) the linear momentum of the electron, (c) the angular momentum of the electron, (d) the kinetic energy, (e) the potential energy, and (f) the total energy.

42. Four possible transitions for a hydrogen atom are as follows:

    (A) $n_i = 2; n_f = 5$ (B) $n_i = 5; n_f = 3$

    (C) $n_i = 7; n_f = 4$ (D) $n_i = 4; n_f = 7$

(a) Which transition emits the shortest wavelength photon? (b) In which transition does the atom gain the most energy? (c) In which transition(s) does the atom lose energy?

43. A photon is emitted as a hydrogen atom undergoes a transition from the $n = 6$ state to the $n = 2$ state. Calculate (a) the energy, (b) the wavelength, and (c) the frequency of the emitted photon.

44. How much energy is required to ionize hydrogen (a) when it is in the ground state? (b) when it is in the state for which $n = 3$?

45. Show that the speed of the electron in the nth Bohr orbit in hydrogen is given by

$$v_n = \frac{k_e e^2}{n\hbar}$$

46. (a) Calculate the angular momentum of the Moon due to its orbital motion about the Earth. In your calculation, use 3.84×10^8 m as the average Earth–Moon distance and 2.36×10^6 s as the period of the Moon in its orbit. (b) Determine the corresponding quantum number if the Moon's angular momentum is given by the

Bohr assumption $mvr = n\hbar$. (c) By what fraction would the Earth–Moon distance have to be increased to increase the quantum number by 1?

47. A monochromatic beam of light is absorbed by a collection of ground-state hydrogen atoms in such a way that six different wavelengths are observed when the hydrogen relaxes back to the ground state. What is the wavelength of the incident beam?

48. Two hydrogen atoms collide head-on and end up with zero kinetic energy. Each then emits a 121.6-nm photon ($n = 2$ to $n = 1$ transition). At what speed were the atoms moving before the collision?

49. (a) Construct an energy level diagram for the He^+ ion, for which $Z = 2$. (b) What is the ionization energy for He^+?

50. What is the radius of the first Bohr orbit in (a) He^+, (b) Li^{2+}, and (c) Be^{3+}?

51. A particle of charge q and mass m, moving with a constant speed v and perpendicular to a constant magnetic field B, follows a circular path. If the angular momentum about the center of this circle is quantized so that $mvr = n\hbar$, show that the allowed radii for the particle are

$$r_n = \sqrt{\frac{n\hbar}{qB}}$$

where $n = 1, 2, 3, \ldots$.

52. An electron is in the nth Bohr orbit of the hydrogen atom. (a) Show that the period of the electron is $T = t_0 n^3$, and determine the numerical value of t_0. (b) On the average, an electron remains in the $n = 2$ orbit for about 10 μs before it jumps down to the $n = 1$ (ground-state) orbit. How many revolutions does the electron make before it jumps to the ground state? (c) If one revolution of the electron is defined as an "electron year" (analogous to an Earth year being one revolution of the Earth around the Sun), does the electron in the $n = 2$ orbit "live" very long? Explain.

Section 40.6 Photons and Electromagnetic Waves
Section 40.7 The Wave Properties of Particles

53. Calculate the de Broglie wavelength for a proton moving with a speed of 1.00×10^6 m/s.

54. Calculate the de Broglie wavelength for an electron that has kinetic energy (a) 50.0 eV and (b) 50.0 keV.

55. (a) An electron has kinetic energy 3.00 eV. Find its wavelength. (b) A photon has energy 3.00 eV. Find its wavelength.

56. In the Davisson–Germer experiment, 54.0-eV electrons were diffracted from a nickel lattice. If the first maximum in the diffraction pattern was observed at $\phi = 50.0°$ (Fig. P40.56), what was the lattice spacing a?

WEB 57. The nucleus of an atom is on the order of 10^{-14} m in diameter. For an electron to be confined to a nucleus, its de Broglie wavelength would have to be of this order of magnitude or smaller. (a) What would be the kinetic energy of an electron confined to this region? (b) On

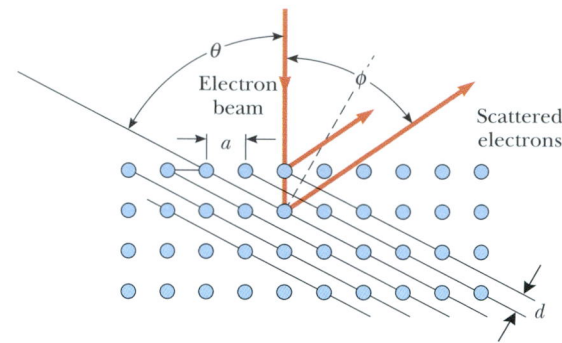

Figure P40.56

the basis of this result, would you expect to find an electron in a nucleus? Explain.

58. Robert Hofstadter won the 1961 Nobel Prize in physics for his pioneering work in scattering 20-GeV electrons from nuclei. (a) What is the γ-factor for a 20.0-GeV electron, where $\gamma = (1 - v^2/c^2)^{-1/2}$? (b) What is the momentum of the electron in kg·m/s? (c) What is the wavelength of a 20.0-GeV electron and how does it compare with the size of a nucleus?

59. (a) Show that the frequency f and wavelength λ of a freely moving particle are related by the expression

$$\left(\frac{f}{c}\right)^2 = \frac{1}{\lambda^2} + \frac{1}{\lambda_C^2}$$

where $\lambda_C = h/mc$ is the Compton wavelength of the particle. (b) Is it ever possible for a particle having nonzero mass to have the same wavelength *and* frequency as a photon? Explain.

60. After learning about de Broglie's hypothesis that particles of momentum p have wave characteristics with wavelength $\lambda = h/p$, an 80.0-kg student has grown concerned about being diffracted when passing through a 75.0-cm-wide doorway. Assume that significant diffraction occurs when the width of the diffraction aperture is less than 10.0 times the wavelength of the wave being diffracted. (a) Determine the maximum speed at which the student can pass through the doorway in order to be significantly diffracted. (b) With that speed, how long will it take the student to pass through the doorway if it is 15.0 cm thick? Compare your result to the currently accepted age of the Universe, which is 4×10^{17} s. (c) Should this student worry about being diffracted?

61. What is the speed of an electron if its de Broglie wavelength equals its Compton wavelength? (*Hint:* If you get an answer of c, see Problem 71.)

ADDITIONAL PROBLEMS

62. Figure P40.62 shows the stopping potential versus incident photon frequency for the photoelectric effect for

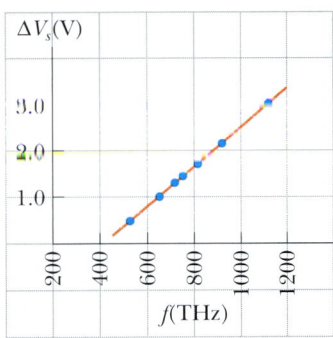

Figure P40.62

sodium. Use the graph to find (a) the work function, (b) the ratio h/e, and (c) the cutoff wavelength. (Data taken from R. A. Millikan, *Phys. Rev.* 7:362, 1916.)

63. Photons of wavelength 450 nm are incident on a metal. The most energetic electrons ejected from the metal are bent into a circular arc of radius 20.0 cm by a magnetic field with a magnitude of 2.00×10^{-5} T. What is the work function of the metal?

64. Photons of wavelength λ are incident on a metal. The most energetic electrons ejected from the metal are bent into a circular arc of radius R by a magnetic field having a magnitude B. What is the work function of the metal?

WEB 65. The table below shows data obtained in a photoelectric experiment. (a) Using these data, make a graph similar to Figure 40.9 that plots as a straight line. From the graph, determine (b) an experimental value for Planck's constant (in joule–seconds) and (c) the work function (in electron volts) for the surface. (Two significant figures for each answer are sufficient.)

Wavelength (nm)	Maximum Kinetic Energy of Photoelectrons (eV)
588	0.67
505	0.98
445	1.35
399	1.63

66. A 200-MeV photon is scattered at 40.0° by a free proton initially at rest. (a) Find the energy (in MeV) of the scattered photon. (b) What kinetic energy (in MeV) does the proton acquire?

67. Positronium is a hydrogen-like atom consisting of a positron (a positively charged electron) and an electron revolving around each other. Using the Bohr model, find the allowed radii (relative to the center of mass of the two particles) and the allowed energies of the system.

68. Derive the formula for the Compton shift (Eq. 40.10) from Equations 40.11, 40.12, and 40.13.

69. *An example of the correspondence principle.* Use Bohr's model of the hydrogen atom to show that when the electron moves from the state n to the state $n - 1$, the frequency of the emitted light is

$$f = \frac{2\pi^2 m_e k_e^2 e^4}{h^3} \left[\frac{2n - 1}{(n-1)^2 n^2} \right]$$

Show that as $n \rightarrow \infty$, this expression varies as $1/n^3$ and reduces to the classical frequency one expects the atom to emit. (*Hint:* To calculate the classical frequency, note that the frequency of revolution is $v/2\pi r$, where r is given by Eq. 40.25.)

70. Show that a photon cannot transfer all of its energy to a free electron. (*Hint:* Note that energy and momentum must be conserved.)

71. Show that the speed of a particle having de Broglie wavelength λ and Compton wavelength $\lambda_C = h/(mc)$ is

$$v = \frac{c}{\sqrt{1 + (\lambda/\lambda_C)^2}}$$

72. The Lyman series for a (new?) one-electron atom is observed in the light from a distant galaxy. The wavelengths of the first four lines and the short-wavelength limit of this series are given by the energy-level diagram in Figure P40.72. Based on this information, calculate (a) the energies of the ground state and first four excited states for this one-electron atom and (b) the wavelengths of the first three lines and the short-wavelength limit in the Balmer series for this atom. (c) Show that the wavelengths of the first four lines and the short-wavelength limit of the Lyman series for the hydrogen atom are all 60.0% of the wavelengths for the Lyman se-

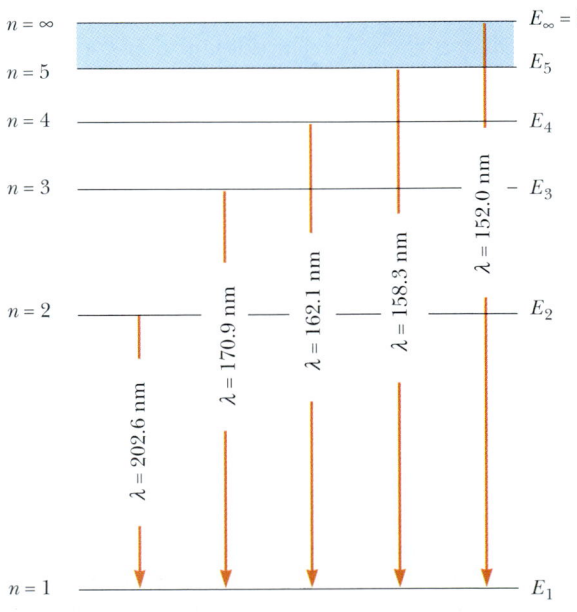

Figure P40.72

ries in the one-electron atom described in part (b). (d) Based on this observation, explain why this atom could be hydrogen.

73. The total power per unit area radiated by a black body at a temperature T is the area under the $I(\lambda, T)$-versus-λ curve, as shown in Figure 40.3. (a) Show that this power per unit area is

$$\int_0^\infty I(\lambda, T)\,d\lambda = \sigma T^4$$

where $I(\lambda, T)$ is given by Planck's radiation law and σ is a constant independent of T. This result is known as the Stefan–Boltzmann law (see Section 20.7). To carry out the integration, you should make the change of variable $x = hc/\lambda k_B T$ and use the fact that

$$\int_0^\infty \frac{x^3\,dx}{e^x - 1} = \frac{\pi^4}{15}$$

(b) Show that the Stefan–Boltzmann constant σ has the value

$$\sigma = \frac{2\pi^5 k_B^4}{15 c^2 h^3} = 5.67 \times 10^{-8}\ \text{W/m}^2 \cdot \text{K}^4$$

74. Derive Wien's displacement law from Planck's law. Proceed as follows: In Figure 40.3 note that the wavelength at which a black body radiates with greatest intensity is the wavelength for which the graph of $I(\lambda, T)$ versus λ has a horizontal tangent. From Equation 40.3 evaluate the derivative $dI/d\lambda$. Set it equal to zero. Solve the resulting transcendental equation numerically to prove $hc/\lambda_{max} k_B T = 4.965 \ldots$, or $\lambda_{max} T = hc/4.965 k_B$. Evaluate the constant as precisely as possible and compare it with Wien's experimental value.

75. A photon of initial energy E_0 undergoes Compton scattering at an angle θ from a free electron (mass m_e) initially at rest. Using relativistic equations for energy and momentum conservation, derive the following relationship for the final energy E' of the scattered photon:

$$E' = E_0[1 + (E_0/m_e c^2)(1 - \cos\theta)]^{-1}$$

76. As we learned in Section 39.4, a muon has a charge of $-e$ and a mass equal to 207 times the mass of an electron. Muonic lead is formed when a lead nucleus captures a muon. According to the Bohr theory, what are the radius and energy of the ground state of muonic lead?

77. An electron initially at rest recoils from a head-on collision with a photon. Show that the kinetic energy acquired by the electron is $2hfa/(1 + 2a)$, where a is the ratio of the photon's initial energy to the rest energy of the electron.

78. The spectral distribution function $I(\lambda, T)$ for an ideal black body at absolute temperature T is shown in Figure P40.78. (a) Show that the percentage of the total power radiated per unit area in the range $0 \le \lambda \le \lambda_{max}$

is

$$\frac{A}{A + B} = 1 - \frac{15}{\pi^4} \int_0^{4.965} \frac{x^3}{e^x - 1}\,dx$$

independent of the value of T. (b) Using numerical integration, show that this ratio is approximately $1/4$.

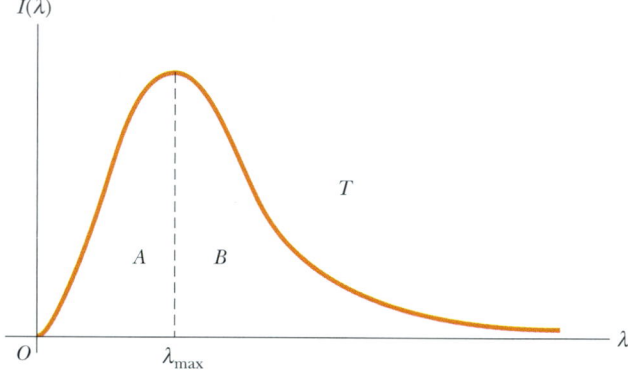

Figure P40.78

79. Show that the ratio of the Compton wavelength λ_C to the de Broglie wavelength $\lambda = h/p$ for a relativistic electron is

$$\frac{\lambda_C}{\lambda} = \left[\left(\frac{E}{m_e c^2}\right)^2 - 1\right]^{1/2}$$

where E is the total energy of the electron and m_e is its mass.

80. The neutron has a mass of 1.67×10^{-27} kg. Neutrons emitted in nuclear reactions can be slowed down via collisions with matter. They are referred to as thermal neutrons once they come into thermal equilibrium with their surroundings. The average kinetic energy $(3k_B T/2)$ of a thermal neutron is approximately 0.04 eV. Calculate the de Broglie wavelength of a neutron with a kinetic energy of 0.040 0 eV. How does it compare with the characteristic atomic spacing in a crystal? Would you expect thermal neutrons to exhibit diffraction effects when scattered by a crystal?

81. A photon with wavelength λ_0 moves toward a free electron that is moving with speed u in the same direction

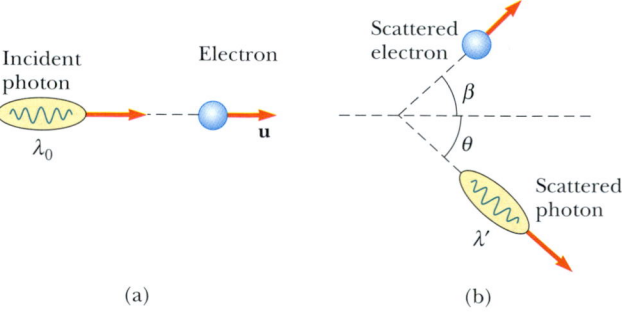

(a)   (b)

Figure P40.81

as the photon (Fig. P40.81a). If the photon scatters through an angle θ (Fig. P40.81b), show that the wavelength of the scattered photon is

$$\lambda' = \lambda_0 \left(\frac{1 - (u/c)\cos\theta}{1 - u/c} \right) + \frac{h}{m_e c} \sqrt{\frac{1 + u/c}{1 - u/c}} \, (1 - \cos\theta)$$

ANSWERS TO QUICK QUIZZES

40.1 (c). Ultraviolet light has the highest frequencies of the three, and hence each photon delivers more energy to a skin cell. (This explains why you can become sunburned on a cloudy day: Clouds block visible light but not much ultraviolet light. You usually do not become sunburned through window glass, even though you can feel the warmth due to the Sun's infrared rays, because the glass blocks ultraviolet light.)

40.2 Comparing Equation 40.8 with the slope–intercept form of the equation for a straight line, $y = mx + b$, we see that the slope in Figure 40.9 is Planck's constant h and that the y intercept is $-\phi$, the negative of the work function. If a different metal were used, the slope would remain the same but the work function would be different. Thus, data for different metals appear as parallel lines on the graph.

40.3 Classical physics predicts that light of sufficient intensity causes photoelectron emission, independent of frequency and certainly without a cutoff frequency. Also, the greater the intensity, the greater the maximum kinetic energy, with some time delay in emission at low intensities. Thus, the classical expectation (which did not match experiment) yields a graph that looks like this:

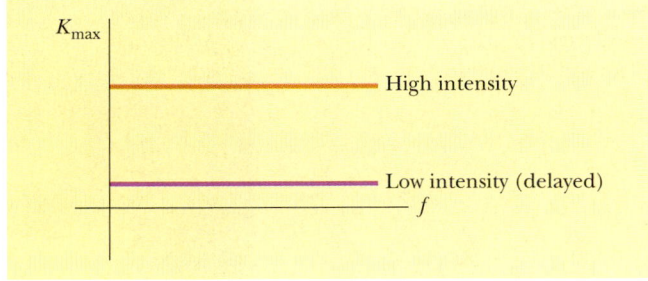

40.4 The fractional change in wavelength $\Delta\lambda / \lambda_0$ is greater (and thus easier to measure) for smaller wavelengths, and x-rays have much smaller wavelengths than visible light.

ROSE IS ROSE reprinted by permission of United Feature Syndicate, Inc.

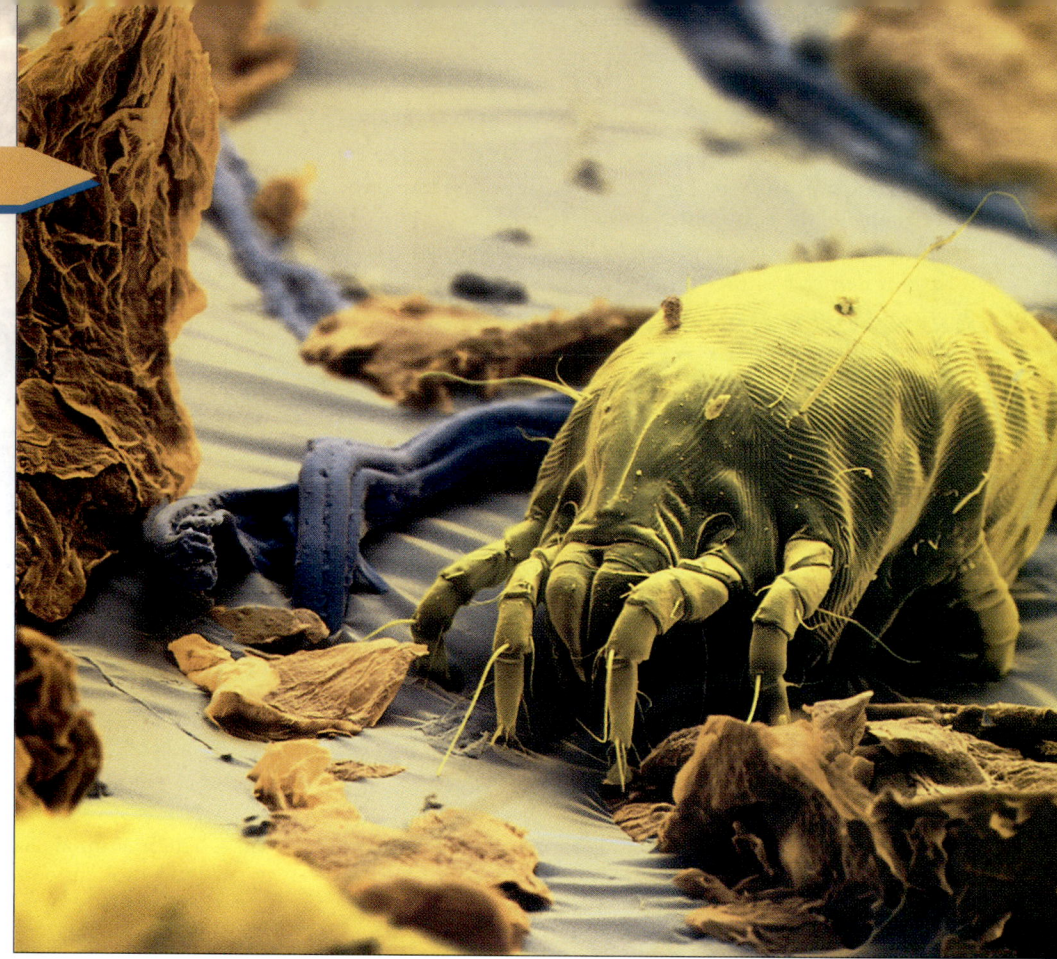

c h a p t e r

Quantum Mechanics

he Bohr model of the hydrogen atom, which we presented in Chapter 40, has severe limitations. It depicts the electron as moving on the circumference of a flat circle, but scattering experiments show that the electron fills a sphere around the nucleus, with an exponentially decreasing probability of being found at greater and greater distances from the nucleus. The Bohr model does not take account of the wave motion of the electron. Bohr supposed that the minimum angular momentum of the electron was $\hbar$; in fact, it is zero. Also, the model cannot be extended to explain the absorption and emission spectra of complex atoms, nor does it predict such details as variations in spectral line intensities and the splittings observed in certain spectral lines under controlled laboratory conditions. Finally, it does not enable us to understand how atoms interact with each other and how such interactions affect the observed physical and chemical properties of matter.

In this chapter we introduce *quantum mechanics,* an extremely successful theory for explaining atomic structure. This theory, developed from 1925 to 1926 by Erwin Schrödinger, Werner Heisenberg, and others, addresses the limitations of the Bohr model and enables us to understand a host of phenomena involving atoms, molecules, nuclei, and solids. Basically, we shall be studying the equation of motion of matter waves, as well as some of the basic features of quantum mechanics and their application to simple one-dimensional systems. For example, we shall treat the problem of a particle confined to a potential well having infinitely high barriers.

41.1 ▶ THE DOUBLE-SLIT EXPERIMENT REVISITED

As we saw in Chapter 40, the concept of wave–particle duality in modern physics is very difficult to understand. One way to crystallize our ideas about this duality is to consider the diffraction of electrons passing through a double slit. This experiment shows the impossibility of *simultaneously* measuring wave and particle properties and embodies all the bizarre consequences of quantum mechanics.

Consider a beam of electrons all having the same energy and all incident on a double-slit barrier, as shown in Figure 41.1, where the slit widths are much less than the slit separation D. An electron detector is positioned far from the slits at a distance much greater than D. **If the detector detects electrons at different positions for a sufficiently long period of time, one finds an interference pattern representing the number of electrons arriving at any position along the detector line.** Such an interference pattern cannot occur if electrons behave as classical particles, and so we must infer that the electrons are behaving as waves. If the experiment is carried out at lower beam intensities over a long period of time, the interference pattern is still observed. At first, one observes only individual blips that are like photon "bullets" hitting in an apparently random pattern, but after long exposure a pattern of blips is observed. This is illustrated in the computer-simulated patterns in Figure 41.2. Note that the interference pattern becomes clearer as the number of electrons reaching the detector increases.

If a single electron produces in-phase waves as it reaches one of the slits, standard wave theory can be used to find the angular separation θ between the central probability maximum and its neighboring minimum. The minimum occurs when the path-length difference between paths A and B in Figure 41.1 is half a wavelength, or

$$D \sin \theta = \frac{\lambda}{2}$$

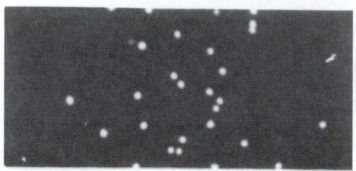

(a) After 28 electrons

(b) After 1000 electrons

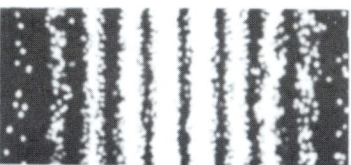

(c) After 10000 electrons

(d) Two-slit electron pattern

Figure 41.2 (a), (b), (c) Computer-simulated interference patterns for a beam of electrons incident on a double slit. *(From E. R. Huggins,* Physics I, *New York, W. A. Benjamin, 1968)* (d) Photograph of a double-slit interference pattern produced by electrons. *(From C. Jönsson,* Zeitschrift für Physik *161:454, 1961; used with permission.)*

Wave function ψ

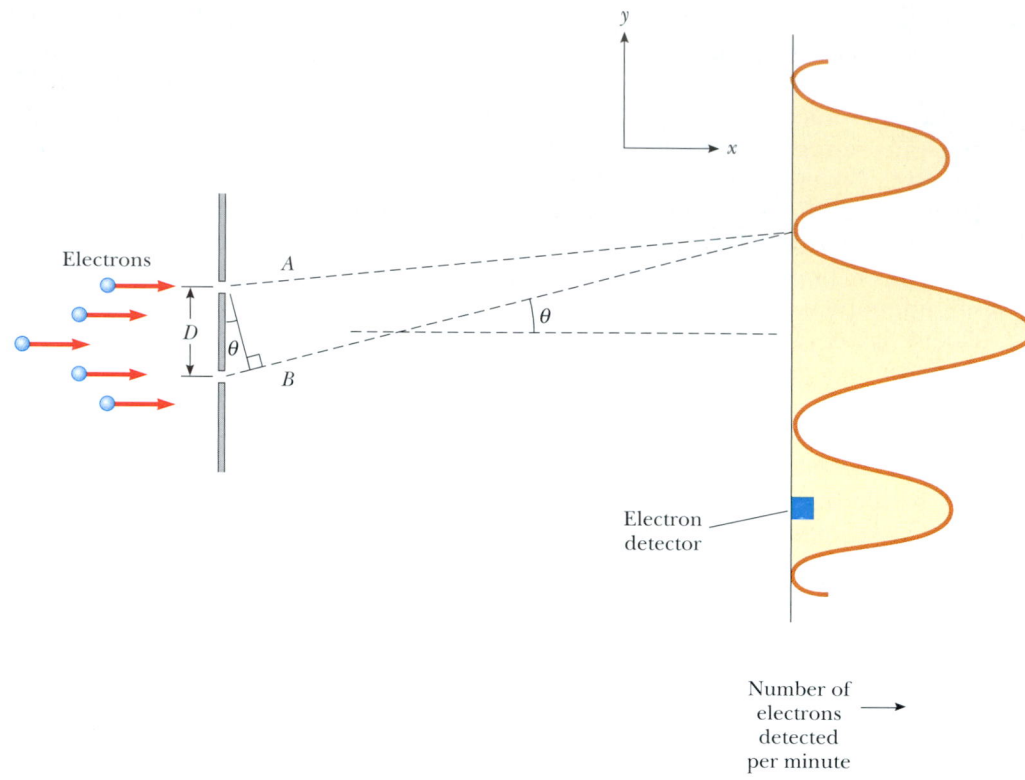

Figure 41.1 Electron diffraction. The slit separation D is much greater than the individual slit widths and much less than the distance between slits and detector.

Because the electron's de Broglie wavelength is given by $\lambda = h/p_x$, we see that, for small θ,

$$\sin \theta \approx \theta = \frac{h}{2p_x D}$$

Thus, the dual nature of electrons is clearly shown in this experiment: **Although the electrons are detected as particles at a localized spot at some instant of time, the probability of arrival at that spot is determined by the intensity of two interfering matter waves.**

In quantum mechanics, matter waves are described by the complex-valued **wave function** ψ. The absolute square $|\psi|^2 = \psi^*\psi$, where ψ^* is the complex conjugate of ψ, gives the probability of finding a particle at a given point at some instant. The wave function contains all the information that can be known about the particle.

Let us use the notion of the wave function to investigate some other unusual results from the double-slit experiment. If one slit is covered during the experiment, the result is a symmetric curve that peaks around the center of the open slit, much like the pattern formed by bullets shot through a hole in armor plate. The two overlapping blue curves in the center of Figure 41.3 are plots of electrons detected per minute with only one slit open. These curves are expressed as $|\psi_1|^2 = \psi_1^*\psi_1$ and $|\psi_2|^2 = \psi_2^*\psi_2$, where ψ_1 and ψ_2 represent the electron passing through slit 1 and slit 2, respectively.

If an experiment is performed with slit 2 blocked for the first half of the experiment and then slit 1 blocked during the remaining time, the accumulated pattern of electrons detected per minute, shown by the single blue curve on the right in Figure 41.3, is completely different from the pattern obtained with both slits open (red curve). In the single-slit curve, a maximum probability of arrival no longer occurs at $\theta = 0$. In fact, **the interference pattern has been lost, and the accumulated result is simply the sum of the individual results.** Because the electron must pass through either slit 1 or slit 2, it is just as localized and indivisible at the slits as it is when measured at the detector. Thus, the blue pattern on the right in Figure 41.3 must represent the sum of those electrons that come through slit 1, $|\psi_1|^2$, and those that come through slit 2, $|\psi_2|^2$.

When both slits are open, it is tempting to assume that the electron goes through either slit 1 or slit 2 and that the counts per minute are again given by $|\psi_1|^2 + |\psi_2|^2$. However, the experimental results, indicated by the red interference pattern in Figure 41.3, contradict this assumption. Thus, our assumption that the electron is localized and goes through only one slit when both slits are open must be wrong (a painful conclusion!). Somehow the wave property of the electron has a presence at both slits.

To find the probability of detecting the electron at a particular point at the detector when both slits are open, we may say that the electron is in a *superposition state*, given by

$$\psi = \psi_1 + \psi_2$$

Thus, the probability of detecting the electron at the detector is $|\psi_1 + \psi_2|^2$, and not $|\psi_1|^2 + |\psi_2|^2$. Because in general matter waves that start out in phase at the slits travel different distances to the detector, ψ_1 and ψ_2 have a relative phase difference ϕ at the detector. Using a phasor diagram (Fig. 41.4) to find $|\psi_1 + \psi_2|^2$ immediately yields

$$|\psi|^2 = |\psi_1 + \psi_2|^2 = |\psi_1|^2 + |\psi_2|^2 + 2|\psi_1||\psi_2| \cos \phi$$

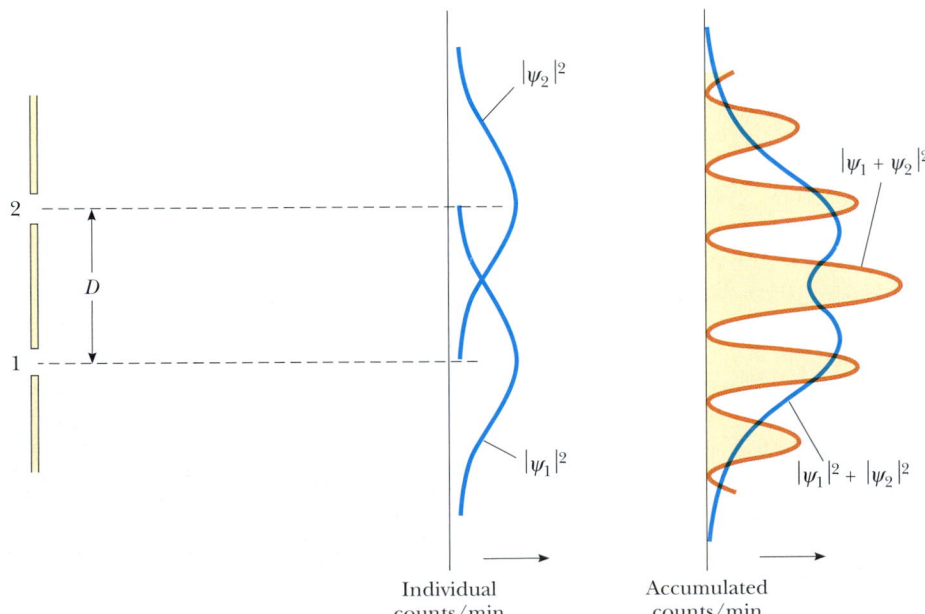

Figure 41.3 The two blue curves in the middle represent the patterns of the individual slits with the upper or lower slit closed. The single blue curve on the right represents the accumulated pattern of counts per minute when each slit is closed half the time. The red curve represents the diffraction pattern with both slits open at the same time.

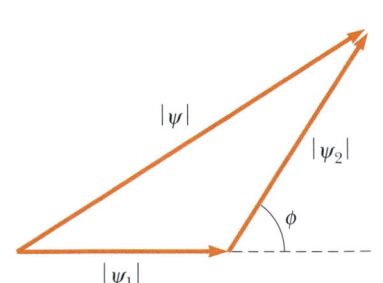

Figure 41.4 Phasor diagram to represent the addition of two complex quantities ψ_1 and ψ_2.

where $|\psi_1|^2$ is the probability of detection with slit 1 open and slit 2 closed, and $|\psi_2|^2$ is the probability of detection with slit 2 open and slit 1 closed. The term $2|\psi_1||\psi_2|\cos\phi$ in this expression is the interference term, which arises from the relative phase ϕ of the waves, in analogy with the phasor addition used in wave optics (see Chapter 37).

To interpret these results, we are forced to conclude that *an electron's wave property interacts with both slits simultaneously*. If we attempt to determine experimentally which slit the electron goes through, the simple act of measurement destroys the interference pattern. Thus, it is impossible to make such a determination. In effect, we can say only that **the electron passes through both slits!** The same arguments apply to photons.

Quick Quiz 41.1

Describe the signal from an electron detector as it is moved laterally far in front of three slits from which electrons with the same energy are being diffracted.

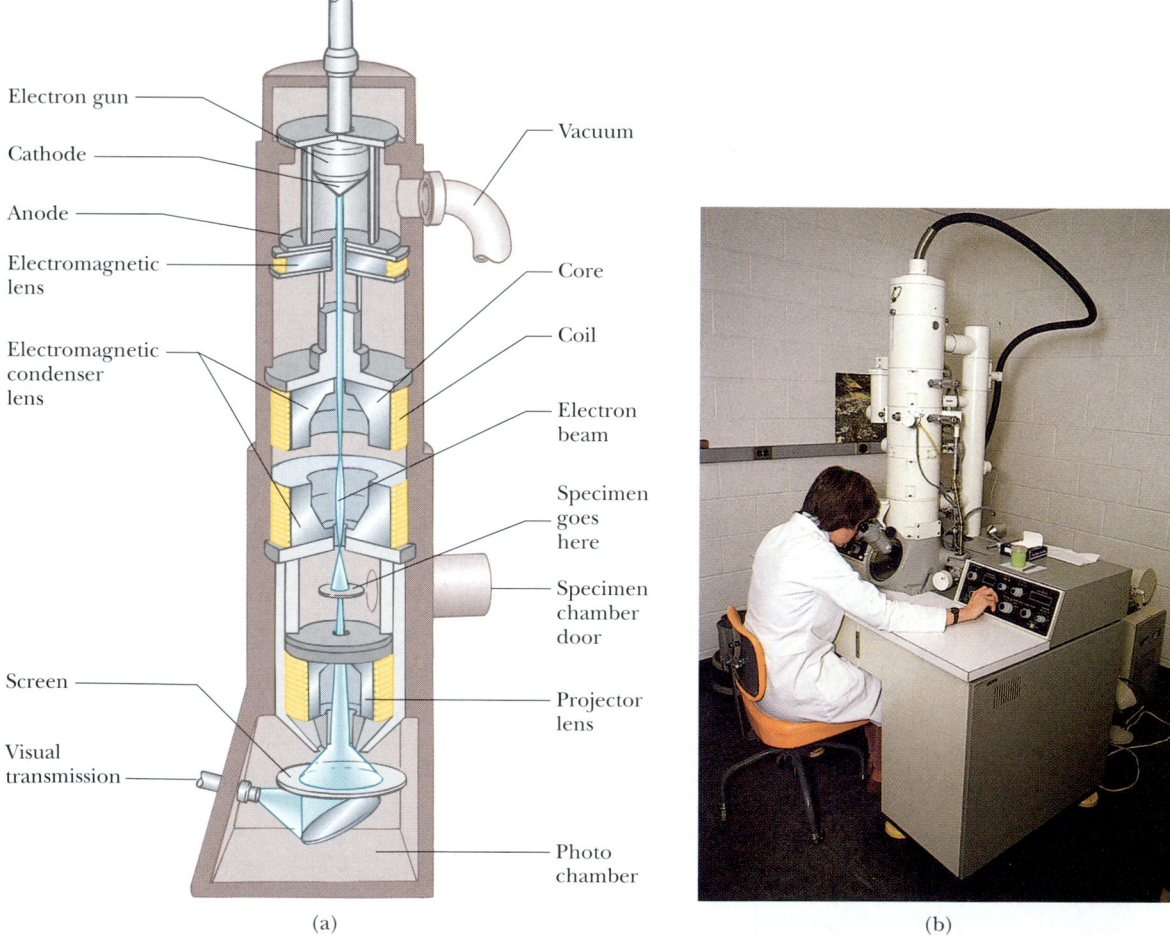

Electron gun

Cathode

Anode

Electromagnetic lens

Electromagnetic condenser lens

Screen

Visual transmission

Vacuum

Core

Coil

Electron beam

Specimen goes here

Specimen chamber door

Projector lens

Photo chamber

(a)

(b)

Figure 41.5 (a) Diagram of a transmission electron microscope for viewing a thinly sectioned sample. The "lenses" that control the electron beam are magnetic deflection coils. (b) An electron microscope. *(W. Ormerod/Visuals Unlimited)*

The Electron Microscope

A practical device that relies on the wave characteristics of electrons is the **electron microscope.** A *transmission* electron microscope, used for viewing flat, very thin samples, is shown in Figure 41.5. In many respects it is similar to an optical microscope, but the electron microscope has a much greater resolving power because it can accelerate electrons to very high kinetic energies, giving them very short wavelengths. No microscope can resolve details that are significantly smaller than the wavelength of the radiation used to illuminate the object. Typically, the wavelengths of electrons are about 100 times shorter than those of the visible light used in optical microscopes. As a result, an electron microscope with ideal lenses would be able to distinguish details about 100 times smaller than those distinguished by an optical microscope. (Radiation of the same wavelength as the electrons in an electron microscope is in the x-ray region of the spectrum.)

The electron beam in an electron microscope is controlled by electrostatic or magnetic deflection, which acts on the electrons to focus the beam to an image. Rather than examining the image through an eyepiece as in an optical microscope, the viewer looks at an image formed on a fluorescent screen. (The viewing screen must be fluorescent because otherwise the image produced would not be visible.)

A photograph taken by a *scanning* electron microscope, which operates in a somewhat different manner to reveal surface details of a three-dimensional sample, is shown at the beginning of the chapter.

41.2 THE UNCERTAINTY PRINCIPLE

If you were to measure the position and speed of a particle at any instant, you would always be faced with experimental uncertainties in your measurements. According to classical mechanics, no fundamental barrier to an ultimate refinement of the apparatus or experimental procedures exists. In other words, it is possible, in principle, to make such measurements with arbitrarily small uncertainty. Quantum theory predicts, however, that such a barrier does exist. In 1927, Werner Heisenberg (1901–1976) introduced this notion, which is now known as the **Heisenberg uncertainty principle:**

> If a measurement of position is made with precision Δx and a simultaneous measurement of linear momentum is made with precision Δp_x, then the product of the two uncertainties can never be smaller than $\hbar/2$:

$$\Delta x \, \Delta p_x \geq \frac{\hbar}{2} \tag{41.1}$$

◀ Heisenberg uncertainty principle

where $\hbar = h/2\pi$. In other words, **it is physically impossible to measure simultaneously the exact position and exact linear momentum of a particle.** If Δx is very small, then Δp_x is large, and vice versa. Heisenberg was careful to point out that the inescapable uncertainties Δx and Δp_x do not arise from imperfections in measuring instruments. Rather, they arise from the quantum structure of matter—from effects such as the unpredictable recoil of an electron when struck by a photon or the diffraction of light or electrons passing through a small opening.

To understand the uncertainty principle, consider the following thought experiment introduced by Heisenberg: Suppose you wish to measure the position and linear momentum of an electron as accurately as possible. You might be able to do this by viewing the electron with a powerful light microscope. For you to see the electron and thus determine its location, at least one photon of light must bounce off the electron, as shown in Figure 41.6a, and then pass through the microscope into your eye, as shown in Figure 41.6b. When it strikes the electron, however, the photon transfers some unknown amount of its momentum to the electron. Thus, in the process of your locating the electron very accurately—that is, making Δx very small by using light with a short wavelength (and consequently a high momentum)—the very light that enables you to succeed changes the electron's momentum to some undeterminable extent (making Δp_x very great).

Let us analyze the collision by first noting that the incoming photon has momentum h/λ. As a result of the collision, the photon transfers part or all of its momentum along the x axis to the electron. Thus, the *uncertainty* in the electron's momentum after the collision is as great as the momentum of the incoming photon: $\Delta p_x = h/\lambda$. Furthermore, because the photon also has wave properties, we expect to be able to determine its position to within one wavelength of the light being used to view it, so $\Delta x = \lambda$. Multiplying these two uncertainties gives

$$\Delta x \, \Delta p_x = \lambda \left(\frac{h}{\lambda} \right) = h$$

This value h represents the minimum in the products of the uncertainties. Because the uncertainty can always be greater than this minimum, we have

$$\Delta x \, \Delta p_x \geq h$$

Apart from the numerical factor $1/4\pi$ introduced by Heisenberg's more precise analysis, this result agrees with Equation 41.1.

Quick Quiz 41.2

To determine the location of an electron, we can send it through a narrow slit. The narrower the slit, the more precisely we know the electron's location. Why does this fact not provide an escape from the limitations of the Heisenberg uncertainty principle?

The Heisenberg uncertainty principle enables us to better understand the dual wave–particle nature of light and matter. We have seen that the wave description of whatever entity we are studying is quite different from the particle description. Therefore, if an experiment (such as the photoelectric effect) is designed to reveal the particle character of, say, an electron, the electron's wave character becomes less apparent. If an experiment (such as diffraction from a crystal) is designed to measure the electron's wave properties, its particle character becomes less apparent.

Another uncertainty relationship sets a limit on the accuracy with which the energy of a system ΔE can be measured in a finite time interval Δt:

$$\Delta E \, \Delta t \geq \frac{\hbar}{2} \tag{41.2}$$

This relationship is plausible if a frequency measurement of any wave is considered. For example, consider measuring the frequency of a 1 000-Hz electromagnetic wave. If our frequency-measuring device has a fixed sensitivity of ± 1 cycle, in 1 s we measure a frequency of $(1\,000 \pm 1)$ cycles/1 s, but in 2 s we measure a frequency of $(2\,000 \pm 1)$ cycles/2 s. Thus, the uncertainty in frequency Δf is inversely proportional to Δt, the time interval during which the measurement is

Werner Heisenberg German theoretical physicist (1901–1976)
Heisenberg made many significant contributions to physics, including his famous uncertainty principle, for which he received a Nobel Prize in 1932; the development of an abstract model of quantum mechanics called matrix mechanics; the prediction of two forms of molecular hydrogen; and theoretical models of the nucleus.
(Courtesy of the University of Hamburg)

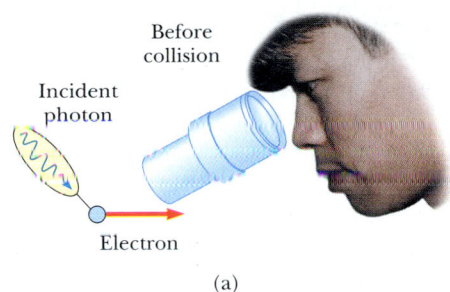

Before collision

Incident photon

Electron

(a)

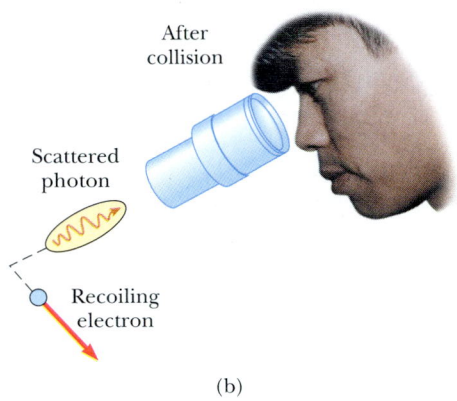

After collision

Scattered photon

Recoiling electron

(b)

Figure 41.6 A thought experiment for viewing an electron with a powerful optical microscope. (a) The electron is moving to the right before colliding with the photon. (b) The electron recoils (its momentum changes) as a result of the collision with the photon.

made. This relationship may be stated as

$$\Delta f \, \Delta t \approx 1$$

Because all quantum systems are wave-like and can be described by the relationship $E = hf$, we may substitute $\Delta f = \Delta E / h$ into the preceding expression to obtain

$$\Delta E \, \Delta t \approx h$$

in basic agreement with Equation 41.2, apart from the factor of $1/4\pi$.

We conclude this section with examples of the types of calculations that can be performed with the uncertainty principle. These back-of-the-envelope calculations are surprising for their simplicity and for their essential description of quantum systems of which the details are unknown.

CONCEPTUAL EXAMPLE 41.1   Is the Bohr Model Realistic?

According to the Bohr model of the hydrogen atom, the electron in the ground state moves in a circular orbit of radius 0.529×10^{-10} m. In view of the Heisenberg uncertainty principle, is this model realistic?

Solution According to the uncertainty principle, the product $\Delta p_r \, \Delta r \geq \hbar/2$, where Δp_r is the uncertainty in the linear momentum of the electron in the radial direction. Let us calculate this uncertainty. The model specifies the radius of the circular orbit very precisely. When we quote the radius to three significant digits, we imply that the uncertainty in the radial position is at most $\Delta r \approx 0.000\,5 \times 10^{-10}$ m. The corresponding uncertainty in momentum of the electron in the radial direction is at least

$$\Delta p_r \approx \frac{\hbar}{2\,\Delta r} = \frac{1.05 \times 10^{-34}\,\text{J}\cdot\text{s}}{2(0.000\,5 \times 10^{-10}\,\text{m})} \approx 1 \times 10^{-21}\,\text{kg}\cdot\text{m/s}$$

The corresponding uncertainty in the radial speed of the electron (using a nonrelativistic calculation) is

$$\Delta v_r \approx \frac{\Delta p_r}{m_e} = \frac{1 \times 10^{-21}\,\text{kg}\cdot\text{m/s}}{9.1 \times 10^{-31}\,\text{kg}} \approx 1 \times 10^{9}\,\text{m/s}$$

A relativistic calculation would also give a large uncertainty in the speed. Because the uncertainty of the radial speed is on the order of ten times the speed of light, we must conclude that the Bohr model is not a reasonable description of the hydrogen atom!

EXAMPLE 41.2 Locating an Electron

The speed of an electron is measured to be 5.00×10^3 m/s to an accuracy of 0.003 00%. Find the minimum uncertainty in determining the position of this electron.

Solution The momentum of the electron is

$$p_x = mv = (9.11 \times 10^{-31} \text{ kg})(5.00 \times 10^3 \text{ m/s})$$
$$= 4.56 \times 10^{-27} \text{ kg·m/s}$$

The uncertainty in p_x is 0.003 00% of this value:

$$\Delta p_x = (0.000\ 030\ 0)(4.56 \times 10^{-27} \text{ kg·m/s})$$
$$= 1.37 \times 10^{-31} \text{ kg·m/s}$$

We can now calculate the minimum uncertainty in position by using this value of Δp_x and Equation 41.1:

$$\Delta x\, \Delta p_x \geq \frac{\hbar}{2}$$

$$\Delta x \geq \frac{\hbar}{2\Delta p_x} = \frac{1.05 \times 10^{-34} \text{ J·s}}{2(1.37 \times 10^{-31} \text{ kg·m/s})} = \boxed{0.383 \text{ mm}}$$

EXAMPLE 41.3 The Width of Spectral Lines

Although an excited atom can radiate at any time from $t = 0$ to $t = \infty$, the average time after excitation at which a group of atoms radiates is called the **lifetime** τ. (a) If $\tau = 1.0 \times 10^{-8}$ s, use the uncertainty principle to compute the line width Δf produced by this finite lifetime.

Solution We use $\Delta E\, \Delta t \geq \hbar/2$, where $\Delta E = h\,\Delta f$ and $\Delta t = 1.0 \times 10^{-8}$ s is the average time available to measure the excited state. Thus, the minimum value of Δf is

$$\Delta f = \frac{1}{4\pi(1.0 \times 10^{-8} \text{ s})} = \boxed{8.0 \times 10^6 \text{ Hz}}$$

Note that ΔE is the uncertainty in the energy of the excited atom. It is also the uncertainty in the energy of the photon emitted by an atom in this state. (Note that in Bohr's theory, spectral lines should have vanishingly small line widths because the energy levels are precise.)

(b) If the wavelength of the spectral line involved in this process is 500 nm, what is the fractional broadening $\Delta f/f$?

Solution First we find the frequency f of this line:

$$f = \frac{c}{\lambda} = \frac{3.00 \times 10^8 \text{ m/s}}{500 \times 10^{-9} \text{ m}} = 6.00 \times 10^{14} \text{ Hz}$$

Hence,

$$\frac{\Delta f}{f} = \frac{8.0 \times 10^6 \text{ Hz}}{6.00 \times 10^{14} \text{ Hz}} = \boxed{1.3 \times 10^{-8}}$$

This narrow natural line width can be seen with a sensitive interferometer. Usually, however, temperature and pressure effects overshadow the natural line width and broaden the line through mechanisms associated with the Doppler effect and collisions.

41.3 PROBABILITY DENSITY

Chapters 34, 37, and 40 revealed various aspects of light, and we can give a profound summary of the nature of light as follows: A photon is a quantum particle that has zero mass and transports energy and momentum as it moves as a wave of electric and magnetic fields. Its equation of motion is the wave equation for electromagnetic waves:

$$\frac{\partial^2 E}{\partial x^2} = \mu_0 \epsilon_0 \frac{\partial^2 E}{\partial t^2}$$

for the electric field, and a similar equation for the magnetic field. The intensity of the wave is proportional to the square of the electric field and is measured as the rate of photon bombardment at a detector.

Our purpose in the current chapter is to give an analogous account of any material particle (one having nonzero mass). As noted in Section 41.1, the probability of finding a matter particle at a given point at some instant of time is given by $|\psi|^2$, the absolute square of a complex-valued wave function ψ. This wave function contains all the information that can be known about the particle. This interpretation of matter waves was first suggested by Max Born (1882–1970) in 1928. In 1926

Erwin Schrödinger (1887–1961) proposed a wave equation that describes how matter waves change in space and time. (The analogous propagation of electromagnetic waves is governed by Maxwell's equations.) The *Schrödinger equation* represents a key element in the theory of quantum mechanics.

A question arises quite naturally from the statement that matter has both a wave nature and a particle nature: If we are describing a particle, what do we conceive is waving? In the cases of waves on strings, water waves, and sound waves, the wave is represented by some quantity that varies with time and position. In a similar manner, the wave function ψ for matter waves depends on both the positions of all the particles in a system and on time, and therefore is often written $\psi(x, y, z, t)$. If ψ is known for a particle, then the particular properties of that particle can be described. In fact, the fundamental problem of quantum mechanics is this: Given the wave function at some instant, find the wave function at some later time t.

In Section 40.7 we found that the de Broglie equation relates the momentum of a particle to its wavelength through the relationship $p = h/\lambda$. If a free particle has a precisely known momentum, its wave function is a sinusoidal wave of wavelength $\lambda = h/p$, and the particle has equal probability of being at any point along the x axis. The wave function for such a free particle moving along the x axis can be written as

$$\psi(x) = A \sin\left(\frac{2\pi x}{\lambda}\right) = A \sin(kx) \tag{41.3}$$

where $k = 2\pi/\lambda$ is the angular wave number and A is a constant amplitude. As we mentioned earlier, the wave function is generally a function of both position and time. Equation 41.3 represents the part of the wave function that depends on position only. For this reason, we can view $\psi(x)$ as a snapshot of the wave function at a given instant, as shown in Figure 41.7a. The wave function for a particle whose wavelength is not precisely defined is shown in Figure 41.7b. Because the wavelength is not precisely defined, it follows that the linear momentum is known only approximately. That is, if the momentum of the particle were measured, the result would have any value over some range, determined by the spread in wavelength. The greater the uncertainty in the momentum, the more the particle is localized. This is reflected in an increased probability density at the location of the particle.

Although we cannot measure ψ, we saw in Section 41.1 that we can measure $|\psi|^2$, a quantity that describes the probability of finding the particle at a particular location and at a certain time. To be more specific, if ψ represents a single particle, then $|\psi(x)|^2$—called the **probability density**—is the probability per unit volume that the particle will be found within an infinitesimal volume containing the point x. This interpretation, first suggested by Born in 1928, can also be stated in the following manner: If dV is a small volume element surrounding some point, then the probability of finding the particle in that volume element is $|\psi|^2 dV$. In this chapter we deal only with one-dimensional systems, in which the particle must be located along the x axis; thus, we replace dV with dx. In this case, the probability $P(x) dx$ that the particle will be found in the infinitesimal interval dx around the point x is

$$P(x) \, dx = |\psi|^2 \, dx$$

◀ Probability density $|\psi|^2$

Because the particle must be somewhere along the x axis, the sum of the probabilities over all values of x must be 1:

$$\int_{-\infty}^{\infty} |\psi|^2 \, dx = 1 \tag{41.4}$$

◀ Normalization condition on ψ

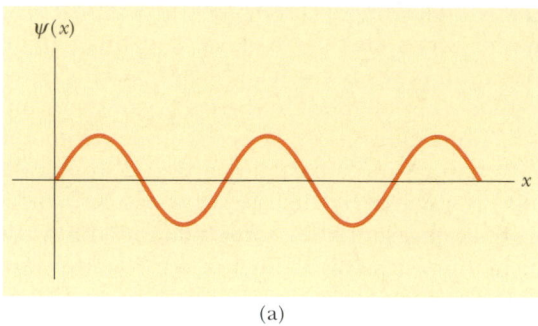

(a)

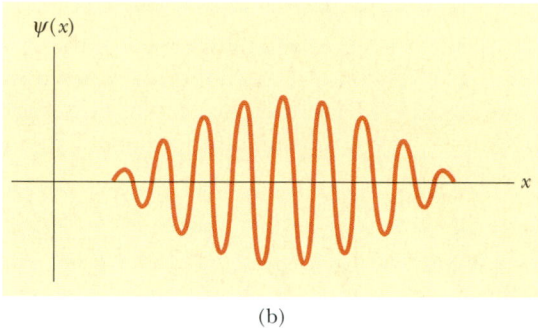

Figure 41.7 (a) Wave function for a particle whose wavelength is precisely known. (b) Wave function for a particle whose wavelength is not precisely known and hence whose momentum is known only over some range of values.

(b)

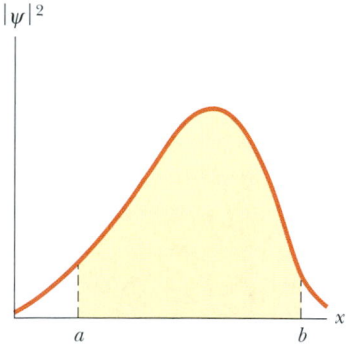

Figure 41.8 The probability of a particle's being in the interval $a \leq x \leq b$ is the area under the curve from a to b.

Any wave function satisfying Equation 41.4 is said to be normalized and to fulfill the **normalization condition.** Normalization is simply a statement that the particle exists at some point at all times. Therefore, although it is not possible to specify the position of a particle with complete certainty, it is possible, through $|\psi|^2$, to specify the probability of observing it at a given location. Furthermore, *the probability of finding the particle in the interval $a \leq x \leq b$ is*

$$P_{ab} = \int_a^b |\psi|^2 \, dx \qquad \textbf{(41.5)}$$

The probability P_{ab} is the area under the curve of probability density versus x between $x = a$ and $x = b$ in Figure 41.8.

Experimentally, a finite probability of finding a particle at some point and at some instant always exists, so the value of the probability must lie between the limits 0 and 1. For example, if the probability density is 0.3 m^{-1} at some point, the probability of finding the particle in some small interval Δx centered on that point is $0.3 \, \Delta x$.

The wave function ψ satisfies a wave equation, just as the electric field associated with an electromagnetic wave satisfies a wave equation that follows from Maxwell's equations. The wave equation satisfied by ψ, which is the Schrödinger equation, cannot be derived from any more fundamental laws, but ψ can be computed from it. Although ψ itself cannot be measured, all measurable quantities of a particle, such as its energy and linear momentum, can be derived from a knowledge of ψ. For example, once the wave function for a particle is known, it is possible to calculate the average position x of the particle, after many experimental trials. This average position is called the **expectation value** of x and is defined by the equation

$$\langle x \rangle \equiv \int_{-\infty}^{\infty} x \, |\psi|^2 \, dx \tag{41.6}$$

(Brackets $\langle \; \rangle$ denote expectation values.) This expression implies that the particle is in a definite state, so the probability density is time-independent. Note that the expectation value is equivalent to the average value of x that would be obtained if we were dealing with a large number of particles in the same state. Furthermore, we can find the expectation value of any function $f(x)$ by using Equation 41.6 with x replaced by $f(x)$.

41.4 A PARTICLE IN A BOX

From a classical viewpoint, if a particle is confined to moving parallel to an x axis and to bouncing back and forth between two impenetrable walls (Fig. 41.9), its motion is easy to describe. If the speed of the particle is v, then the magnitude of its linear momentum (mv) remains constant, as does its kinetic energy. Furthermore, classical physics places no restrictions on the values of its momentum and energy. The quantum-mechanical approach to this problem is quite different and requires that we find the appropriate wave function consistent with the given conditions.

Before we address this problem, it is instructive to review the classical situation of standing waves on a stretched string (see Sections 18.2 and 18.3). If a string of length L is fixed at both ends, standing waves set up in the string must have nodes at the ends, as shown in Figure 41.10, because the wave function must vanish at the boundaries. Standing waves exist only when the length L of the string is some integral multiple of half-wavelengths. That is, we require that

$$L = n \frac{\lambda}{2}$$

or

$$\lambda = \frac{2L}{n} \qquad n = 1, 2, 3, \ldots$$

This result shows that *the wavelength for a standing wave on a string is quantized.*

As we saw in Section 18.2, each point on a standing wave oscillates with simple harmonic motion. Furthermore, all points oscillate with the same frequency, but the amplitude y of the simple harmonic motion of any particle in the medium differs from one point to the next and depends on how far a given point is from one end. We found that the position-dependent part of the wave function for a standing wave is

$$y(x) = A \sin(kx) \tag{41.7}$$

where A is the maximum amplitude of the wave and $k = 2\pi/\lambda$. Because $\lambda = 2L/n$, we see that

$$k = \frac{2\pi}{\lambda} = \frac{2\pi}{2L/n} = n\frac{\pi}{L}$$

Substituting this result into Equation 41.7 gives

$$y(x) = A \sin\left(\frac{n\pi x}{L}\right)$$

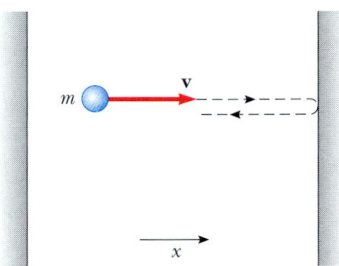

Figure 41.9 A particle of mass m and velocity **v** confined to moving parallel to the x axis and bouncing between two impenetrable walls.

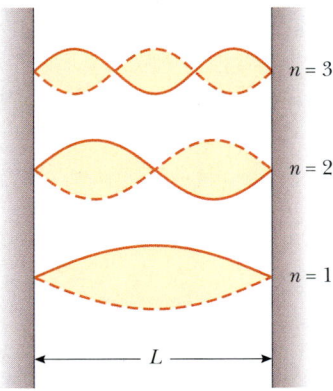

Figure 41.10 Standing waves set up in a stretched string of length L.

From this expression, we see that the wave function for a standing wave on a string meets the required boundary conditions—namely, that for all values of n, $y = 0$ at $x = 0$ and at $x = L$. The wave functions for $n = 1$, 2, and 3 are plotted in Figure 41.10.

Now let us return to the quantum-mechanical description of a particle in a box. Because the walls are impenetrable, the wave function $\psi(x) = 0$ for $x \leq 0$ and for $x \geq L$, where L is now the distance between the two walls. This means that the particle can never be found outside the box. Furthermore, because the wave function must be continuous everywhere, we require that $\psi(0) = \psi(L) = 0$. Only wave functions that satisfy this condition are allowed. In analogy with standing waves on a string, the allowed wave functions for the particle in the box are sinusoidal and are given by

Allowed wave functions for a particle in a box

$$\psi(x) = A \sin\left(\frac{n\pi x}{L}\right) \qquad n = 1, 2, 3, \ldots \qquad \textbf{(41.8)}$$

where A is the maximum value of the wave function. This expression shows that, for a particle confined to a box and having a well-defined de Broglie wavelength, ψ is represented by a sinusoidal wave. The allowed wavelengths are those for which $L = n\lambda/2$. These allowed states of the system are called **stationary states** because they are standing waves.

Figure 41.11 shows plots of ψ versus x and $|\psi|^2$ versus x for $n = 1$, 2, and 3. As we shall soon see, these states correspond to the three lowest allowed energies for the particle. For $n = 1$, the probability of finding the particle is greatest at $x = L/2$—this is the *most probable position* for a particle in this state. For $n = 2$, $|\psi|^2$ is a maximum at $x = L/4$ and again at $x = 3L/4$; this means that both points are equally likely places for a particle in this state to be found.

There are also points within the box at which it is impossible to find the particle. For $n = 2$, $|\psi|^2$ is zero at the midpoint, $x = L/2$; for $n = 3$, $|\psi|^2 = 0$ at $x = L/3$ and $x = 2L/3$; and so on. But how does our particle get from one place

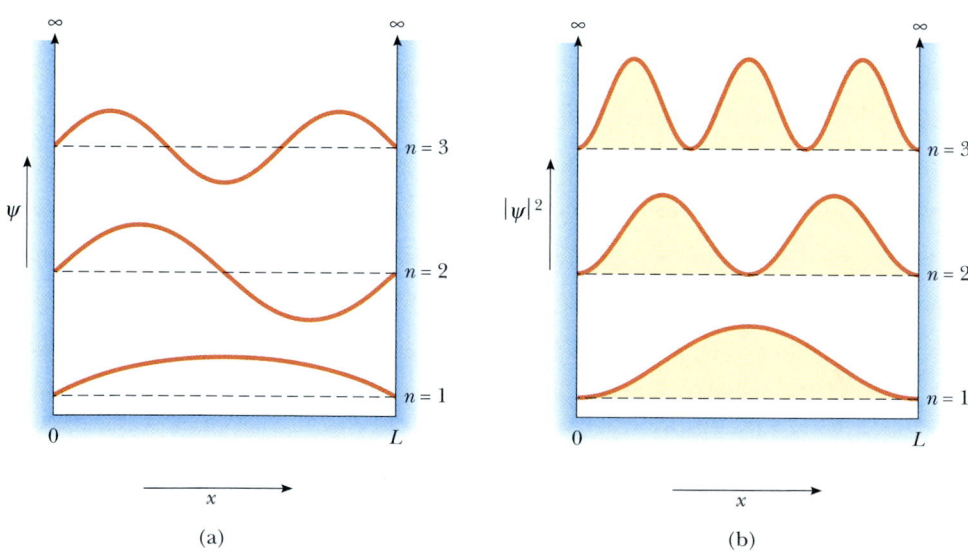

(a)   (b)

Figure 41.11   The first three allowed stationary states for a particle confined to a one-dimensional box. (a) The wave functions ψ for $n = 1$, 2, and 3. (b) The probability densities $|\psi|^2$ for $n = 1$, 2, and 3.

to another when no probability exists for its ever being at points between? This is one of the bizarre consequences of quantum mechanics—we must give up our notion that a particle moves from one point to another by occupying all intervening positions. In quantum mechanics, objects are not viewed as particles but as more complicated things having both particle *and* wave attributes.

Quick Quiz 41.3

Redraw Figure 41.11b, the probability of finding a particle at a particular location in a box, on the basis of classical mechanics instead of quantum mechanics.

Quick Quiz 41.4

(a) Make a sketch like Figure 41.11b for $n = 20$. Imagine placing within the confines of the box a detector that samples the probability of finding a particle within some narrow limits Δx. What would the detector measure as n approached infinity?

Because the wavelengths of the particle are restricted by the condition $\lambda = 2L/n$, the magnitude of the linear momentum is restricted to the values

$$p = \frac{h}{\lambda} = \frac{h}{2L/n} = \frac{nh}{2L}$$

The potential energy is constant inside the box, and it is convenient to set it at $U = 0$. Therefore, the total energy of the particle is equal to its kinetic energy. Using $p = mv$, we find that the allowed values of the energy are

$$E_n = \frac{1}{2}mv^2 = \frac{p^2}{2m} = \frac{(nh/2L)^2}{2m}$$

$$E_n = \left(\frac{h^2}{8mL^2}\right)n^2 \qquad n = 1, 2, 3, \ldots \qquad \textbf{(41.9)}$$

As we see from this expression, *the energy of the particle is quantized,* as we would expect. The lowest allowed energy corresponds to $n = 1$, for which $E_1 = h^2/8mL^2$. Because $E_n = n^2 E_1$, the excited states corresponding to $n = 2, 3, 4, \ldots$ have energies given by $4E_1, 9E_1, 16E_1, \ldots$. Figure 41.12 is an energy level diagram describing the positions of the allowed states. Note that the state $n = 0$ is not allowed. This means that, according to quantum mechanics, the particle can never be at rest. The least energy the particle can have, corresponding to $n = 1$, is called the **zero-point energy.** This result clearly contradicts the classical viewpoint, in which $E = 0$ is an acceptable state. In our quantum-mechanical analysis, only nonzero positive values of E are allowed because the total energy E equals the kinetic energy and the potential energy is zero.

Energy levels are of special importance for the following reason. If the particle is electrically charged, it can emit a photon when it drops from an excited state, such as E_3, to a lower-lying state, such as E_2. It can also absorb a photon whose energy matches the difference in energy between two allowed states. For example, if the photon frequency is f, the particle jumps from state E_1 to state E_2 if $hf = E_2 - E_1$. As noted in Chapter 40, photon emission and absorption can be observed by spectroscopy, and spectral wavelengths are a direct measurement of such energy differences.

Allowed energies for a particle in a box

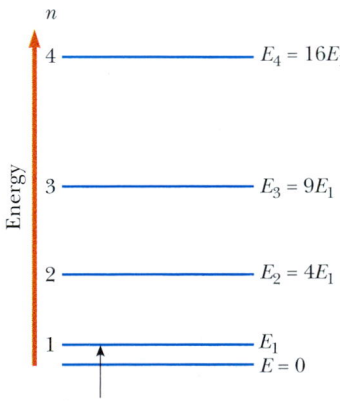

Figure 41.12 Energy level diagram for a particle confined to a one-dimensional box of width L. The lowest allowed energy is $E_1 = h^2/8mL^2$.

EXAMPLE 41.4   A Bound Electron

An electron is confined between two impenetrable walls 0.200 nm apart. Determine the energy levels for the states $n = 1, 2,$ and 3.

Solution   For the state $n = 1$, Equation 41.9 gives

$$E_1 = \frac{h^2}{8mL^2} = \frac{(6.63 \times 10^{-34} \, \text{J} \cdot \text{s})^2}{8(9.11 \times 10^{-31} \, \text{kg})(2.00 \times 10^{-10} \, \text{m})^2}$$

$$= 1.51 \times 10^{-18} \, \text{J} = \boxed{9.42 \, \text{eV}}$$

For $n = 2$ and $n = 3$, $E_2 = 4E_1 = 37.7 \, \text{eV}$ and $E_3 = 9E_1 = 84.8 \, \text{eV}$. Although this is a rather primitive model, it can be used to describe an electron trapped in a vacant crystal site.

EXAMPLE 41.5   Energy Quantization for a Macroscopic Object

A 1.00-mg object is confined to moving between two rigid walls separated by 1.00 cm. Calculate the minimum speed of the object.

Solution   The minimum speed corresponds to the state for which $n = 1$. Using Equation 41.9 with $n = 1$ gives the zero-point energy:

$$E_1 = \frac{h^2}{8mL^2} = \frac{(6.63 \times 10^{-34} \, \text{J} \cdot \text{s})^2}{8(1.00 \times 10^{-6} \, \text{kg})(1.00 \times 10^{-2} \, \text{m})^2}$$

$$= 5.49 \times 10^{-58} \, \text{J}$$

Because $E = K = \frac{1}{2}mv^2$, we have

$$\tfrac{1}{2}mv^2 = 5.49 \times 10^{-58} \, \text{J}$$

$$v = \left[\frac{2(5.49 \times 10^{-58} \, \text{J})}{1.00 \times 10^{-6} \, \text{kg}}\right]^{1/2} = \boxed{3.31 \times 10^{-26} \, \text{m/s}}$$

This speed is so small that the object can be considered to be at rest, which is what one would expect for the minimum speed of a macroscopic object.

Exercise   If the speed of the particle is 3.00 cm/s, find its energy and the value of n that corresponds to this energy.

Answer $4.50 \times 10^{-10} \, \text{J}$; $n = 9.05 \times 10^{23}$. (Note that for values of n this great, we would never be able to distinguish the quantized nature of the energy levels because the difference between the $n = 9.05 \times 10^{23}$ and $n + 1 = 9.05 \times 10^{23} + 1$ levels is so small.)

EXAMPLE 41.6   Model of an Atom

An atom can be viewed as several electrons moving around a positively charged nucleus, where the electrons are subject mainly to the electrical attraction of the nucleus. (This attraction is partially "screened" by the inner-core electrons and is therefore diminished.) Figure 41.13 represents the potential energy of the electron as a function of r. (a) Use the simple model of a particle in a box to *estimate* the energy (in electron volts) required to raise an electron from the state $n = 1$ to the state $n = 2$, assuming that the atom has a radius of 0.100 nm.

Solution   Using Equation 41.9 and taking the length L of the box to be 0.200 nm (the diameter of the atom) and $m = 9.11 \times 10^{-31}$ kg, we find that, as in Example 41.4,

$$E_n = \left(\frac{h^2}{8mL^2}\right)n^2$$

$$= \frac{(6.63 \times 10^{-34} \, \text{J} \cdot \text{s})^2}{8(9.11 \times 10^{-31} \, \text{kg})(2.00 \times 10^{-10} \, \text{m})^2}\, n^2$$

$$= (1.51 \times 10^{-18})n^2 \, \text{J} = 9.42n^2 \, \text{eV}$$

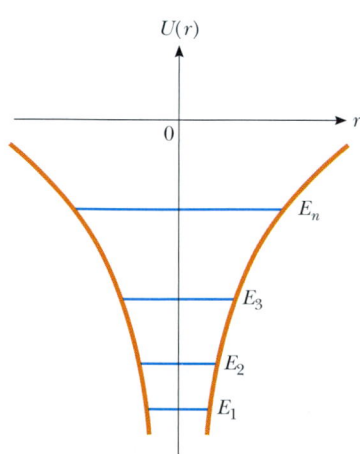

Figure 41.13   Model of potential energy versus r for an atom.

Hence, the energy difference between the states $n = 1$ and $n = 2$ is

$$\Delta E = E_2 - E_1 = 9.42(2)^0 \text{ eV} - 9.42(1)^0 \text{ eV} = \boxed{28.3 \text{ eV}}$$

(b) Calculate the wavelength of the photon that would cause this transition.

Solution Using the fact that $\Delta E = hc/\lambda$, we obtain

$$\lambda = \frac{hc}{\Delta E} = \frac{(6.63 \times 10^{-34} \text{ J} \cdot \text{s})(3.00 \times 10^8 \text{ m/s})}{(28.3 \text{ eV} \times 1.60 \times 10^{-19} \text{ J/eV})}$$

$$= 4.39 \times 10^{-8} \text{ m} = \boxed{43.9 \text{ nm}}$$

This wavelength is in the far ultraviolet region, and it is interesting to note that the result is roughly correct. Although this oversimplified model gives a good estimate for transitions between lowest-lying levels of the atom, the estimate becomes progressively worse for higher-energy transitions.

41.5 THE SCHRÖDINGER EQUATION

As we mentioned earlier, the wave function for de Broglie waves must satisfy an equation developed by Schrödinger. One of the methods of quantum mechanics is to determine a solution to this equation, which in turn yields the allowed wave functions and energy levels of the system under consideration. Proper manipulation of the wave functions enables calculation of all measurable features of the system.

In Section 16.9 we derived Equation 16.26, the general form of the wave equation for waves traveling along the x axis:

$$\frac{\partial^2 y}{\partial x^2} = \frac{1}{v^2} \frac{\partial^2 y}{\partial t^2} \tag{41.10}$$

where v is the wave speed and the variable y depends on x and t. Matter waves are more complicated and do not obey this wave equation.

The wave function for a particle confined to one dimension is

$$\Psi(x, t) = \psi(x)\, e^{-i\omega t} \tag{41.11}$$

where ω is the angular frequency of the matter wave and $\Psi(x, t)$ represents the full time-dependent and space-dependent wave function. In our investigations, we shall need to focus only on $\psi(x)$, the spatial part of the wave function. This function satisfies the equation

$$\frac{d^2\psi}{dx^2} = -\frac{2m}{\hbar^2}(E - U)\psi \tag{41.12}$$

This is the famous **Schrödinger equation** as it applies to a particle confined to moving along the x axis. Because this equation is independent of time, it is commonly referred to as the *time-independent Schrödinger equation*. (We shall not discuss the time-dependent Schrödinger equation in this text.)

In principle, if the potential energy $U(x)$ is known for the system, we can solve Equation 41.12 and obtain the wave functions and energies for the allowed states. Because U may vary with position, it may be necessary to solve the equation in pieces. In the process, the wave functions for the different regions must join smoothly at the boundaries. In the language of mathematics, we require that $\psi(x)$ be *continuous*. Furthermore, for $\psi(x)$ to obey the normalization condition (see text following Equation 41.4), we require that $\psi(x)$ approach zero as x approaches

Erwin Schrödinger Austrian theoretical physicist (1887–1961)
Schrödinger is best known as the creator of quantum mechanics. He also produced important papers in the fields of statistical mechanics, color vision, and general relativity. Schrödinger did much to hasten the universal acceptance of quantum theory by demonstrating the mathematical equivalence between his quantum mechanics and the more abstract matrix mechanics developed by Heisenberg.

Time-independent Schrödinger equation

Required conditions for $\psi(x)$

$\pm \infty$. Finally, $\psi(x)$ must be *single-valued*, and $d\psi/dx$ must also be continuous for finite values of $U(x)$.

The task of solving the Schrödinger equation may be very difficult, depending on the form of the potential energy function. As it turns out, the Schrödinger equation has been extremely successful in explaining the behavior of atomic and nuclear systems, whereas classical physics has failed to explain this behavior. Furthermore, when quantum mechanics is applied to macroscopic objects, the results agree with classical physics, as required by the correspondence principle.

The Particle in a Box Revisited

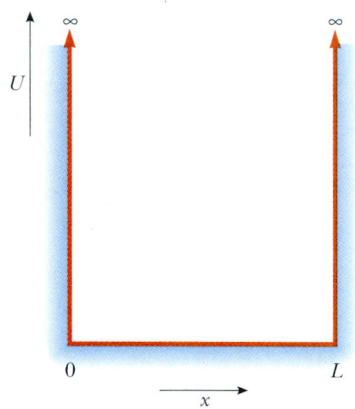

Figure 41.14 A one-dimensional box of width L and infinitely high walls.

Let us solve the Schrödinger equation for our particle in a one-dimensional box of width L (Fig. 41.14). The walls are infinitely high, corresponding to $U(x) = \infty$ for $x = 0$ and $x = L$. The potential energy is constant within the box, and it is again convenient to choose $U = 0$ as its value. Hence, in the region $0 < x < L$, we can express the Schrödinger equation in the form

$$\frac{d^2\psi}{dx^2} = -\frac{2mE}{\hbar^2}\,\psi = -k^2\psi \qquad \textbf{(41.13)}$$

where

$$k = \frac{\sqrt{2mE}}{\hbar}$$

Because the walls are infinitely high, the particle cannot exist outside the box. The particle is permanently bound in the box and cannot be found outside the interval $0 < x < L$. Consequently, $\psi(x)$ must be zero outside the box and at the walls. The solution of Equation 41.13 that meets the boundary conditions $\psi(x) = 0$ at $x = 0$ and $x = L$ is

$$\psi(x) = A\sin(kx) \qquad \textbf{(41.14)}$$

This can easily be verified by substitution into Equation 41.13. Note that the first boundary condition, $\psi(0) = 0$, is satisfied by Equation 41.14 because $\sin 0 = 0$. The second boundary condition, $\psi(L) = 0$, is satisfied only if kL is an integer multiple of π—that is, if $kL = n\pi$, where n is an integer. Because $k = \sqrt{2mE}/\hbar$, we see that

$$kL = \frac{\sqrt{2mE}}{\hbar}L = n\pi$$

Solving for the allowed energies E gives

$$E_n = \left(\frac{h^2}{8mL^2}\right)n^2$$

Likewise, the allowed wave functions are given by

$$\psi_n(x) = A\sin\left(\frac{n\pi x}{L}\right)$$

These results agree with those obtained in the previous section (Eqs. 41.8 and 41.9). It is left as a problem (Problem 25) to show that the normalization constant A for this solution is equal to $(2/L)^{1/2}$.

Optional Section

41.6 ▸ A PARTICLE IN A WELL OF FINITE HEIGHT

Consider a particle having a potential energy that is zero in the region $0 < x < L$—which we call a *potential well*—and has a finite value U outside this region, as in Figure 41.15. If the total energy E of the particle is less than U, classically the particle is permanently bound in the potential well. However, according to quantum mechanics, a finite probability exists that the particle can be found outside the well even if $E < U$. That is, the wave function ψ is generally nonzero outside the well—in regions I and III in Figure 41.15—so the probability density $|\psi|^2$ is also nonzero in these regions.

In region II, where $U = 0$, the allowed wave functions are again sinusoidal because they represent solutions of Equation 41.13. However, the boundary conditions no longer require that ψ be zero at the walls, as was the case with infinitely high walls.

The Schrödinger equation for regions I and III may be written

$$\frac{d^2\psi}{dx^2} = \frac{2m(U - E)}{\hbar^2}\psi \qquad (41.15)$$

Because $U > E$, the coefficient on the right-hand side is necessarily positive. Therefore, we can express Equation 41.15 in the form

$$\frac{d^2\psi}{dx^2} = C^2\psi \qquad (41.16)$$

where $C^2 = 2m(U - E)/\hbar^2$ is a positive constant in regions I and III. As you can verify by substitution, the general solution of Equation 41.16 is

$$\psi = Ae^{Cx} + Be^{-Cx}$$

where A and B are constants.

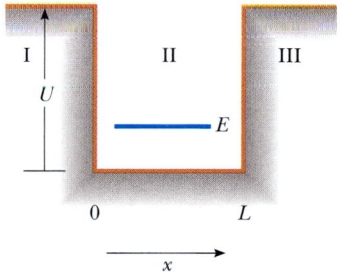

Figure 41.15 Potential-energy diagram of a well of finite height U and width L. The total energy E of the particle is less than U.

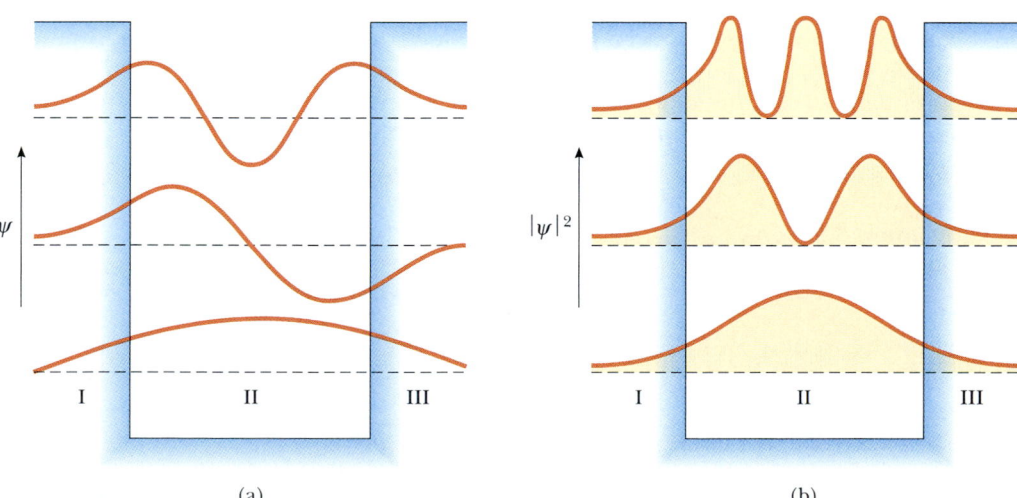

(a)   (b)

Figure 41.16   (a) Wave functions ψ and (b) probability densities $|\psi|^2$ for the lowest three energy states for a particle in a potential well of finite height.

We can use this general solution as a starting point for determining the appropriate solution for regions I and III. The function we choose for our solution must remain finite over the entire region under consideration. In region I, where $x < 0$, we must rule out the term Be^{-Cx}. In other words, we must require that $B = 0$ in region I to avoid an infinite value for ψ for large negative values of x. Likewise, in region III, where $x > L$, we must rule out the term Ae^{Cx}; this is accomplished by taking $A = 0$ in this region. This choice avoids an infinite value for ψ for large positive x values. Hence, the solutions in regions I and III are

$$\psi_{\mathrm{I}} = Ae^{Cx} \qquad \text{for } x < 0$$

$$\psi_{\mathrm{III}} = Be^{-Cx} \qquad \text{for } x > L$$

In region II the wave function is sinusoidal and has the general form

$$\psi_{\mathrm{II}}(x) = F\sin(kx) + G\cos(kx)$$

where F and G are constants.

These results show that the wave functions outside the potential well (where classical physics forbids the presence of the particle) decay exponentially with distance. At great negative x values, ψ_{I} approaches zero exponentially; at great positive x values, ψ_{III} approaches zero exponentially. These functions, together with the sinusoidal solution in region II, are shown in Figure 41.16a for the first three energy states. In evaluating the complete wave function, we require that

$$\psi_{\mathrm{I}} = \psi_{\mathrm{II}} \qquad \text{and} \qquad \frac{d\psi_{\mathrm{I}}}{dx} = \frac{d\psi_{\mathrm{II}}}{dx} \qquad \text{at } x = 0$$

$$\psi_{\mathrm{II}} = \psi_{\mathrm{III}} \qquad \text{and} \qquad \frac{d\psi_{\mathrm{II}}}{dx} = \frac{d\psi_{\mathrm{III}}}{dx} \qquad \text{at } x = L$$

Figure 41.16b plots the probability densities for these states. Note that in each case the inside and outside wave functions join smoothly at the boundaries of the potential well. These boundary conditions and plots follow from the Schrödinger equation.

Figure 41.16a shows that the wave functions ψ are not equal to zero at the walls of the potential well and in the exterior regions. Therefore, the probability densities $|\psi|^2$ are nonzero at these points. The fact that ψ is nonzero at the walls increases the de Broglie wavelength in region II (compare the case of a particle in a potential well of infinite depth; see Fig. 41.11), and this, in turn, lowers the energy and linear momentum of the particle.

Optional Section

41.7 ⟩ TUNNELING THROUGH A BARRIER

A very interesting and peculiar phenomenon occurs when a particle strikes a barrier of finite height and width. Consider a particle of energy E incident on a rectangular barrier of height U and width L, where $E < U$ (Fig. 41.17). Classically, the particle is reflected because it does not have sufficient energy to cross or even penetrate the barrier. Thus, regions II and III are classically *forbidden* to the particle.

According to quantum mechanics, however, **all regions are accessible to the particle, regardless of its energy,** because the amplitude of the de Broglie matter wave associated with the particle is nonzero everywhere. A typical waveform for this case, illustrated in Figure 41.17, shows the wave penetrating into the barrier and beyond. The wave functions are sinusoidal to the left (region I) and right (region III) of the barrier and join smoothly with an exponentially decaying function within the barrier (region II).

Because the probability of locating the particle is proportional to $|\psi|^2$, which is nonzero within the barrier and beyond, we conclude that the particle may be found in region III. This barrier penetration is in complete contradiction to classical physics. The possibility of finding the particle on the far side of the barrier is called **tunneling** or **barrier penetration.** A detailed analysis shows that if tunneling takes place, the barrier must be sufficiently narrow that the time of passage Δt is very short. Then the uncertainty in energy $\Delta E \leq \hbar/2\Delta t$ is so great that we cannot actually attribute negative kinetic energy to the tunneling particle.

Hard as it is to believe, there is a finite (although *extremely* small) possibility that a marble placed inside a shoebox will suddenly appear outside the box (see Problem 47)! However, you would probably have to wait through several life spans of the Universe to observe this.

The probability of tunneling can be described with a transmission coefficient T and a reflection coefficient R. The transmission coefficient is the probability that the particle passes through the barrier, and the reflection coefficient is the probability that the particle is reflected by the barrier. Because the incident particle is either reflected or transmitted, we must require that $T + R = 1$. An approximate expression for T when $T \ll 1$ (a very high or very wide barrier) is

$$T \approx e^{-2CL} \qquad\qquad \textbf{(41.17)}$$

where

$$C = \frac{\sqrt{2m(U - E)}}{\hbar} \qquad\qquad \textbf{(41.18)}$$

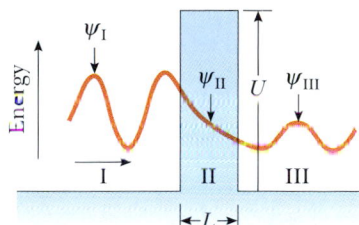

Figure 41.17 Wave function ψ for a particle incident from the left on a barrier of height U. The wave function is sinusoidal in regions I and III but exponentially decaying in region II.

EXAMPLE 41.7   Transmission Coefficient for an Electron

A 30-eV electron is incident on a barrier whose cross-section is a square of height 40 eV. What is the probability that the electron will tunnel through the barrier if its thickness is (a) 1.0 nm? (b) 0.10 nm?

Solution   (a) In this situation, the quantity $U - E$ has the value

$$U - E = (40 \text{ eV} - 30 \text{ eV}) = 10 \text{ eV} = 1.6 \times 10^{-18} \text{ J}$$

Using Equation 41.18, we find that

$$2CL = 2\,\frac{\sqrt{2(9.11 \times 10^{-31} \text{ kg})(1.6 \times 10^{-18} \text{ J})}}{1.054 \times 10^{-34} \text{ J} \cdot \text{s}}$$

$$\times\,(1.0 \times 10^{-9} \text{ m}) = 32.4$$

Thus, the probability of tunneling through the barrier is

$$T \approx e^{-2CL} = e^{-32.4} = \boxed{8.5 \times 10^{-15}}$$

The electron has about 1 chance in 10^{14} of tunneling through the 1.0-nm thickness.

(b) For $L = 0.10$ nm, $2CL = 3.24$, and

$$T \approx e^{-2CL} = e^{-3.24} = \boxed{0.039}$$

Now the electron has a high probability (a 4% chance) of penetrating the barrier. Thus, reducing the thickness of the barrier by only one order of magnitude increases the probability of tunneling by about 12 orders of magnitude!

Some Applications of Tunneling

As we have seen, tunneling is a quantum phenomenon, a manifestation of the wave nature of matter. Many examples for which tunneling is very important exist in nature, on the atomic and nuclear scales.

- **Tunnel diode**   The tunnel diode is a semiconductor device consisting of two oppositely charged regions separated by a very narrow electrically neutral region. The electric current (basically, the rate of tunneling) can be controlled over a wide range by varying the potential difference across the charged regions, which is equivalent to changing the height of the barrier.

- **Josephson junction**   The Josephson junction consists of two superconductors separated by a thin insulating oxide layer, 1 to 2 nm thick. Under appropriate conditions, electrons in the superconductors travel as pairs and tunnel from one superconductor to the other through the oxide layer. Several effects have been observed in this type of junction. For example, a direct current is observed across the junction *in the absence of electric or magnetic fields.* The current is proportional to sin ϕ, where ϕ is the phase difference between the wave functions in the two superconductors. When a potential difference ΔV is applied across the junction, the current oscillates with a frequency $f = 2e\,\Delta V/h$, where e is the charge on the electron.

- **Alpha decay**   One form of radioactive decay is the emission of alpha particles (the nuclei of helium atoms) by unstable heavy nuclei. To escape from the nucleus, an alpha particle must penetrate a barrier that arises from the combination of the attractive nuclear force and the electrical repulsion between the positively charged alpha particle and the rest of the (positively charged) nucleus. Occasionally an alpha particle tunnels through the barrier, which explains the basic mechanism for this type of decay and the great variations in the mean lifetimes of various radioactive nuclei.

- **Solar energy**   According to classical physics, positively charged hydrogen ions in the Sun cannot overcome their mutual repulsion and penetrate the barrier caused by electrical repulsion. Quantum-mechanically, however, the ions are able to tunnel through the barrier and fuse together to form helium. This is the basic reaction that powers the Sun and, indirectly, almost everything else in the Solar System.

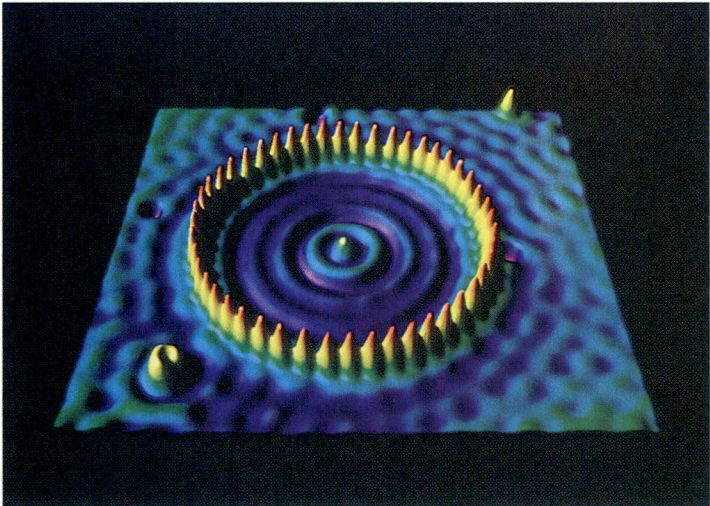

Figure 41.18   A quantum corral consisting of a ring of 48 iron atoms located on a copper surface. The diameter of the ring is 143 nm, and the photograph was obtained using a low-temperature scanning tunneling microscope. Corrals and other structures can confine electron waves. The study of such structures will play an important role in determining the future of small electronic devices. *(IBM Corporation Research Division)*

- **Quantum traps** Scientists are beginning to experiment with *quantum dots* that trap a single electron and *quantum corrals* made up of a small number of atoms, as shown in Figure 41.18. Such tiny traps may eventually be put to use in electronic devices.
- **Scanning tunneling microscopes,** discussed in Section 41.8.

Optional Section

41.8 ▶ THE SCANNING TUNNELING MICROSCOPE[1]

One of the basic phenomena of quantum mechanics—tunneling—is at the heart of a very practical device, the scanning tunneling microscope (STM), which enables us to obtain highly detailed images of surfaces at resolutions comparable to the size of a *single atom*. Figures 41.18 (iron atoms on copper) and 41.19 (the surface of a piece of graphite) show what the STM can do. Note the high quality of the images and the discernible rings of carbon atoms in Figure 41.19. What makes this image so remarkable is that its resolution is about 0.2 nm. For an optical microscope, the resolution is limited by the wavelength of the light used to make the image. Thus, an optical microscope has a resolution no better than 200 nm, about half the wavelength of visible light, and so could never show the detail displayed in Figure 41.19. An ideal electron microscope (Section 41.1) could have a resolution of 0.2 nm by using electron waves of this wavelength, given by the de Broglie formula $\lambda = h/p$. The linear momentum p of an electron required to give this wavelength is 10 000 eV/c, corresponding to an electron speed of 2% of the speed of light. Electrons traveling at this speed would penetrate into the interior of the graphite in Figure 41.19 and thus could not give us information about individual surface atoms.

The STM achieves its very fine resolution by using the basic idea shown in Figure 41.20. An electrically conducting probe with a very sharp tip is brought near the surface to be studied. The empty space between tip and surface represents the "barrier" we have been discussing, and the tip and surface are the two walls of the "potential well." Because electrons obey quantum rules rather than Newtonian rules, they can "tunnel" across the barrier of empty space. If a voltage is applied between surface and tip, electrons in the atoms of the surface material can be made to tunnel preferentially from surface to tip to produce a tunneling current. In this way the tip samples the distribution of electrons just above the surface.

Because of the nature of tunneling, the STM is very sensitive to the distance z from tip to surface—in other words, to the thickness of the barrier (see Example 41.7). The reason is that in the empty space between tip and surface, the electron wave function falls off exponentially (see Fig. 41.17, region II) with a decay length of order 0.1 nm; that is, the wave function decreases by $1/e$ over that distance. For distances $z > 1$ nm (that is, beyond a few atomic diameters), essentially no tunneling takes place. This exponential behavior causes the current of electrons tunneling from surface to tip to depend very strongly on z. This sensitivity is the basis of the operation of the STM: By monitoring the tunneling current as the tip is scanned over the surface, scientists obtain a sensitive measure of the topography of the electron distribution on the surface. The result of this scan is used to make images like that in Figure 41.19. In this way the STM can measure the height of surface features to within 0.001 nm, approximately 1/100 of an atomic diameter!

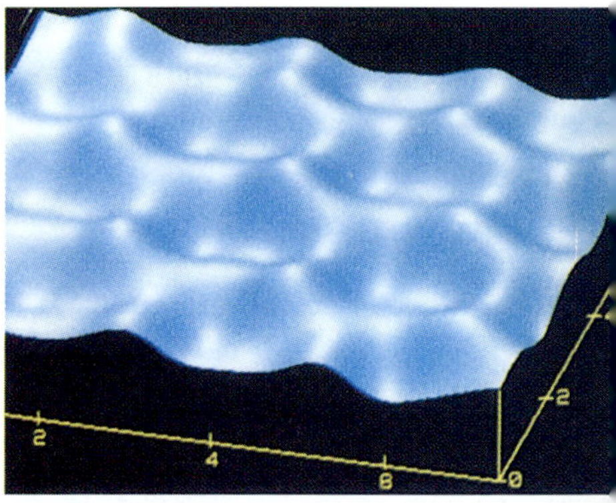

Figure 41.19 The surface of graphite as "viewed" with a scanning tunneling microscope. This type of microscope enables scientists to see details with a lateral resolution of about 0.2 nm and a vertical resolution of 0.001 nm. The contours seen here represent the ring-like arrangement of individual carbon atoms on the crystal surface.

---

[1] This section was written by Roger A. Freedman and Paul K. Hansma, University of California—Santa Barbara.

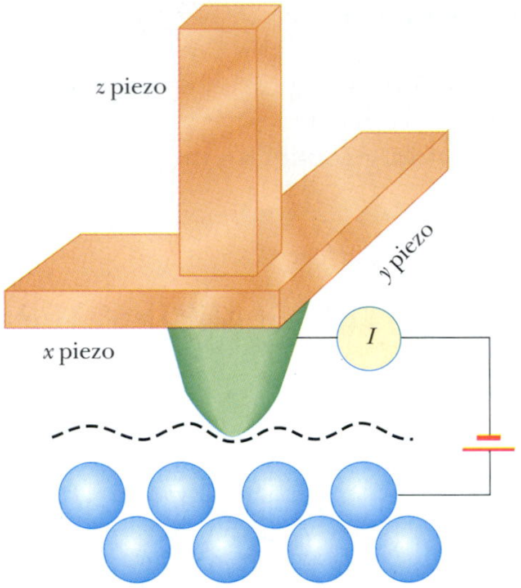

Figure 41.20   Schematic view of an STM. The tip, shown as a rounded cone, is mounted on a piezoelectric *xyz* scanner. A scan of the tip over the sample can reveal contours of the surface down to the atomic level. An STM image is composed of a series of scans displaced laterally from one another. *(Based on a drawing from P. K. Hansma, V. B. Elings, O. Marti, and C. Bracker, Science 242:209, 1988. Copyright 1988 by the AAAS.)*

You can see just how sensitive the STM is by examining Figure 41.19. Of the six carbon atoms in each ring, three appear lower than the other three. In fact, all six atoms are at the same level, but all have slightly different electron distributions. The three atoms that appear lower are bonded to other carbon atoms directly beneath them in the underlying atomic layer; as a result, their electron distributions, which are responsible for the bonding, extend downward beneath the surface. The atoms in the surface layer that appear higher do not lie directly over subsurface atoms and hence are not bonded to any underlying atoms. For these higher-appearing atoms, the electron distribution extends upward into the space above the surface. This extra electron density is what makes these electrons appear higher in Figure 41.19, because what the STM maps is the topography of the electron distribution.

The STM has, however, one serious limitation: It depends on the electrical conductivity of the sample and the tip. Unfortunately, most materials are not electrically conductive at their surfaces. Even metals, which are usually excellent electrical conductors, are covered with nonconductive oxides. A newer microscope, the atomic force microscope (AFM), overcomes this limitation. It measures the electric force acting between a tip and the sample, rather than an electric current. This force, which typically is a result of the exclusion principle, depends very strongly on the tip–sample separation distance, just as the electron tunneling current does in an STM. Thus, the AFM has comparable sensitivity for measuring topography and has become widely used for technological applications.

Perhaps the most remarkable thing about the STM is that its operation is based on a quantum-mechanical phenomenon—tunneling—that was well understood in the 1920s, even though the first STM was not built until the 1980s. What other applications of quantum mechanics may yet be waiting to be discovered?

Optional Section

41.9 ▶ THE SIMPLE HARMONIC OSCILLATOR

Finally, let us consider the problem of a particle that is subject to a linear restoring force $F = -kx$, where x is the magnitude of the displacement of the particle from equilibrium ($x = 0$) and k is the force constant. (This is an important situation to understand because the forces between atoms in a solid can be approximated by this type of interaction.) The classical motion of a particle subject to such a force is simple harmonic motion, which was discussed in Chapter 13. The potential energy of the system is, from Equation 13.21,

$$U = \tfrac{1}{2}kx^2 = \tfrac{1}{2}m\omega^2 x^2$$

where the angular frequency of vibration is $\omega = \sqrt{k/m}$. Classically, if the particle is displaced from its equilibrium position and released, it oscillates between the points $x = -A$ and $x = A$, where A is the amplitude of the motion. Furthermore, its total energy E is, from Equation 13.22,

$$E = K + U = \tfrac{1}{2}kA^2 = \tfrac{1}{2}m\omega^2 A^2$$

In the classical model, any value of E is allowed, including $E = 0$, which is the total energy when the particle is at rest at $x = 0$.

The Schrödinger equation for this problem is obtained by substituting $U = \tfrac{1}{2}m\omega^2 x^2$ into Equation 41.12:

$$\frac{d^2\psi}{dx^2} = -\left[\left(\frac{2mE}{\hbar^2}\right) - \left(\frac{m\omega}{\hbar}\right)^2 x^2\right]\psi \qquad \textbf{(41.19)}$$

The mathematical technique for solving this equation is beyond the level of this text. However, it is instructive to guess at a solution. We take as our guess the following wave function:

$$\psi = Be^{-Cx^2} \qquad \textbf{(41.20)}$$

Substituting this function into Equation 41.19, we find that it is a satisfactory solution to the Schrödinger equation, provided that

$$C = \frac{m\omega}{2\hbar} \qquad \text{and} \qquad E = \tfrac{1}{2}\hbar\omega$$

It turns out that the solution we have guessed corresponds to the ground state of the system, which has an energy $\tfrac{1}{2}\hbar\omega$, the zero-point energy of the system. Because $C = m\omega/2\hbar$, it follows from Equation 41.20 that the wave function for this state is

$$\psi = Be^{-(m\omega/2\hbar)x^2} \qquad \textbf{(41.21)}$$

Wave function for the ground state of a simple harmonic oscillator

This is only one solution to Equation 41.19. The remaining solutions that describe the excited states are more complicated, but all solutions include the exponential factor e^{-Cx^2}.

The energy levels of a harmonic oscillator are quantized, as we would expect when we use quantum mechanics to analyze the situation. The energy of the state for which the quantum number is n is

$$E_n = (n + \tfrac{1}{2})\hbar\omega \qquad n = 0, 1, 2, \ldots$$

Allowed energies for a simple harmonic oscillator

The state $n = 0$ corresponds to the ground state, where $E_0 = \tfrac{1}{2}\hbar\omega$; the state $n = 1$ corresponds to the first excited state, where $E_1 = \tfrac{3}{2}\hbar\omega$; and so on. The energy level diagram for this system is shown in Figure 41.21. Note that the separations between adjacent levels are equal and are given by

$$\Delta E = \hbar\omega \qquad \textbf{(41.22)}$$

The red curves in Figure 41.22 indicate the probability densities $|\psi|^2$ for the first three states of a simple harmonic oscillator. The blue curves represent the classical probability densities that correspond to the same energy and are provided

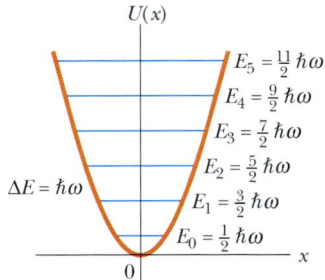

Figure 41.21 Energy level diagram for a simple harmonic oscillator. The levels are equally spaced, with separation $\hbar\omega$. The zero-point energy is $E_0 = \tfrac{1}{2}\hbar\omega$.

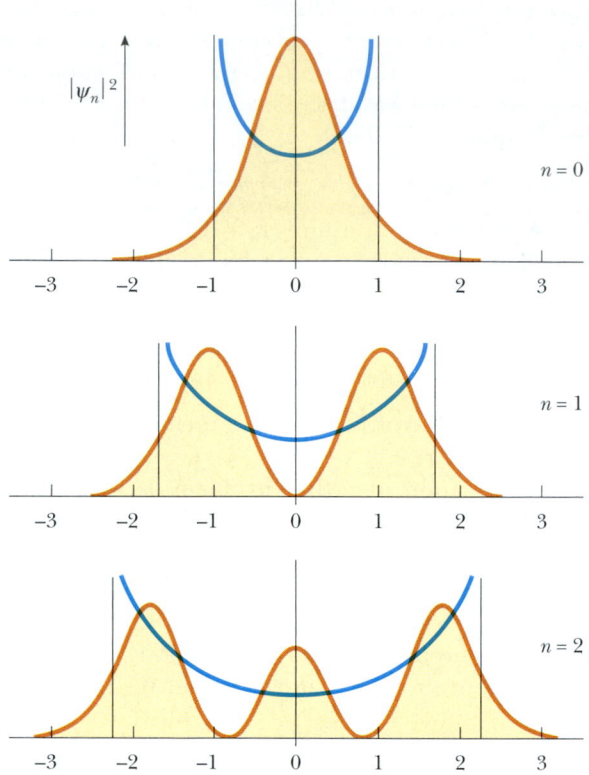

Figure 41.22 The red curves represent probability densities $|\psi|^2$ for the first three states of a simple harmonic oscillator. The blue curves represent classical probability densities corresponding to the same energies. *(From C.W. Sherwin,* Introduction to Quantum Mechanics, *New York, Holt, Rinehart and Winston, 1959. Used with permission.)*

for comparison. Note that as *n* increases, the agreement between the classical and quantum-mechanical results improves, as expected.

Quick Quiz 41.5

(a) Why do the classical probability density curves in Figure 41.22 bend up at the ends? (b) What do you expect the quantum-mechanical probability density curves to look like at very large values of *n*?

SUMMARY

The **Heisenberg uncertainty principle** states that if a measurement of position is made with precision Δx and a simultaneous measurement of linear momentum is made with precision Δp_x, the product of the two uncertainties can never be less than $\hbar/2$:

$$\Delta x \, \Delta p_x \geq \frac{\hbar}{2} \tag{41.1}$$

In quantum mechanics, de Broglie matter waves are represented by a wave function $\psi(x, y, z, t)$. The probability per unit volume (or probability density) that a particle will be found at a point is $|\psi|^2$. If the particle is confined to moving along the x axis, then the probability that it is located in an interval dx is $|\psi|^2\, dx$. Furthermore, the sum of all these probabilities over all values of x must be 1:

$$\int_{-\infty}^{\infty} |\psi|^2\, dx = 1 \qquad \textbf{(41.4)}$$

This is called the **normalization condition.** The measured position x of the particle, averaged over many trials, is called the **expectation value** of x and is defined by

$$\langle x \rangle \equiv \int_{-\infty}^{\infty} x\,|\psi|^2\, dx \qquad \textbf{(41.6)}$$

If a particle of mass m is confined to moving in a one-dimensional box of width L whose walls are impenetrable, we require that ψ be zero at the walls and outside the box. The allowed wave functions for the particle are given by

$$\psi(x) = A \sin\left(\frac{n\pi x}{L}\right) \qquad n = 1, 2, 3, \ldots \qquad \textbf{(41.8)}$$

where A is the maximum value of ψ. The particle has a well-defined wavelength λ with values such that $L = n\lambda/2$. These allowed states are called **stationary states** of the system. The energies of a particle in a box are quantized and are given by

$$E_n = \left(\frac{h^2}{8mL^2}\right) n^2 \qquad n = 1, 2, 3, \ldots \qquad \textbf{(41.9)}$$

The wave function must satisfy the **Schrödinger equation.** The time-independent Schrödinger equation for a particle confined to moving along the x axis is

$$\frac{d^2\psi}{dx^2} = -\frac{2m}{\hbar^2}(E - U)\psi \qquad \textbf{(41.12)}$$

where E is the total energy of the system and U is the potential energy.

The approach of quantum mechanics is to solve Equation 41.12 for ψ and E, given the potential energy $U(x)$ for the system. In doing so, we must place restrictions on $\psi(x)$: (1) $\psi(x)$ must be continuous, (2) $\psi(x)$ must approach zero as x approaches $\pm\infty$, (3) $\psi(x)$ must be single-valued, and (4) $d\psi/dx$ must be continuous for all finite values of $U(x)$.

QUESTIONS

1. Is an electron a particle or a wave? Support your answer by citing some experimental results.

2. An electron and a proton are accelerated from rest through the same potential difference. Which particle has the greater wavelength?

3. If matter has a wave nature, why is this wave-like characteristic not observable in our daily experiences?

4. In what way does Bohr's model of the hydrogen atom violate the uncertainty principle?

5. Why is it impossible to measure simultaneously, with infinite accuracy, the position and speed of a particle?

6. In describing electrons passing through a slit and arriving at a screen, the physicist Richard Feynman said that "electrons arrive in lumps, like particles, but the probability of arrival of these lumps is determined as the intensity of the waves would be. It is in this sense that the electron behaves sometimes like a particle and sometimes like a wave." Elaborate on this point in your own words. (For a further discussion of this point, see R. Feynman, *The Character of Physical Law*, Cambridge, MA, MIT Press, 1980, Chapter 6.)

7. For a particle in a box, the probability density at certain points is zero, as seen in Figure 41.11b. Does this imply that the particle cannot move across these points? Explain.

8. Discuss the relationship between zero-point energy and the uncertainty principle.

9. As a particle of energy E is reflected from a potential barrier of height U, where $E < U$, how does the amplitude of the reflected wave change as the barrier height is reduced?

10. A philosopher once said that "it is necessary for the very existence of science that the same conditions always produce the same results." In view of what has been discussed in this chapter, present an argument showing that this statement is false. How might the statement be reworded to make it true?

11. In quantum mechanics it is possible for the energy E of a particle to be less than the potential energy, but classically this is not possible. Explain.

12. Consider two square wells of the same width, one with finite walls and the other with infinite walls. Compare the energy and momentum of a particle trapped in the finite well with the energy and momentum of an identical particle in the infinite well.

13. Why is it impossible for the lowest energy state of a harmonic oscillator to be zero?

14. Why is an electron microscope more suitable than an optical microscope for "seeing" objects less than 1 μm in size?

15. What is the Schrödinger equation? How is it useful in describing atomic phenomena?

PROBLEMS

1, 2, 3 = straightforward, intermediate, challenging ☐ = full solution available in the *Student Solutions Manual and Study Guide*
WEB = solution posted at **http://www.saunderscollege.com/physics/** 💻 = Computer useful in solving problem 🖳 = Interactive Physics
☐ = paired numerical/symbolic problems

Section 41.1 The Double-Slit Experiment Revisited

WEB **1.** Neutrons traveling at 0.400 m/s are directed through a double slit having a 1.00-mm separation. An array of detectors is placed 10.0 m from the slit. (a) What is the de Broglie wavelength of the neutrons? (b) How far off axis is the first zero-intensity point on the detector array? (c) When a neutron reaches a detector, can we say which slit the neutron passed through? Explain.

2. A modified oscilloscope is used to perform an electron interference experiment. Electrons are incident on a pair of narrow slits 0.060 0 μm apart. The bright bands in the interference pattern are separated by 0.400 mm on a screen 20.0 cm from the slits. Determine the potential difference through which the electrons were accelerated to give this pattern.

3. (a) Show that the wavelength of a neutron is

$$\lambda = \frac{2.86 \times 10^{-11}}{\sqrt{K_n}} \ \text{m}$$

where K_n is the kinetic energy of the neutron in electron volts. (b) What is the wavelength of a 1.00-keV neutron?

4. The distance between adjacent atoms in crystals is on the order of 0.1 nm. The use of electrons in diffraction studies of crystals requires that the de Broglie wavelength of the electrons be of the same order as the distance between atoms of the crystals. What is the order of magnitude of the minimum required energy (in electron volts) of electrons to be used for this purpose?

5. The resolving power of a microscope depends on the wavelength used. If one wished to "see" an atom, a resolution of approximately 1.00×10^{-11} m would be required. (a) If electrons are used (in an electron microscope), what minimum kinetic energy is required for the electrons? (b) If photons are used, what minimum photon energy is needed to obtain the required resolution?

6. Electrons are accelerated through 40 000 V in an electron microscope. What, ideally, is the smallest observable distance between objects seen with this microscope?

7. A beam of electrons with a kinetic energy of 1.00 MeV is incident normally on an array of atoms in rows separated by 0.250 nm. If the array acts like a flat diffraction grating, in what direction can we expect the electrons in the fifth order to be moving?

Section 41.2 The Uncertainty Principle

8. Suppose Fuzzy, a quantum-mechanical duck, lives in a world in which $h = 2\pi$ J·s. Fuzzy has a mass of 2.00 kg and initially is known to be within a pond 1.00 m wide. (a) What is the minimum uncertainty in his speed? (b) Assuming that this uncertainty in speed prevails for 5.00 s, determine the uncertainty in position after that time.

WEB **9.** An electron ($m_e = 9.11 \times 10^{-31}$ kg) and a bullet ($m = 0.020\ 0$ kg) each have a speed of 500 m/s, accurate to within 0.010 0%. Within what limits could we determine the positions of the objects?

10. An air rifle is used to shoot 1.00-g particles at 100 m/s through a hole of diameter 2.00 mm. How far from the rifle must an observer be to see the beam spread by 1.00 cm because of the uncertainty principle? Compare this answer with the diameter of the visible Universe (2×10^{26} m).

11. A light source is used to determine the location of an electron in an atom to a precision of 0.050 0 nm. What is the minimum possible uncertainty in the speed of the electron?

12. Use the uncertainty principle to show that if an electron were confined inside an atomic nucleus of diameter 2×10^{-15} m, it would have to be moving relativistically, while a proton confined to the same nucleus could be moving nonrelativistically.

13. A woman on a ladder drops small pellets toward a point target on the floor. (a) Show that, according to the uncertainty principle, the average miss distance must be at least

$$\Delta x_f = (2/\hbar m)^{1/2}(2/Hg)^{1/4}$$

where H is the initial height of each pellet above the floor and m is the mass of each pellet. Assume that the spread in impact points is given by $\Delta x_f = \Delta x_i + (\Delta v_x)t$. (b) If $H = 2.00$ m and $m = 0.500$ g, what is Δx_f?

Section 41.3   Probability Density

14. The wave function for a particle is

$$\psi(x) = \sqrt{\frac{a}{\pi(x^2 + a^2)}}$$

for $a > 0$ and $-\infty < x < +\infty$. Determine the probability that the particle is located somewhere between $x = -a$ and $x = +a$.

WEB 15. A free electron has a wave function

$$\psi(x) = A \sin(5.00 \times 10^{10} x)$$

where x is in meters. Find (a) the de Broglie wavelength, (b) the linear momentum, and (c) the kinetic energy in electron volts.

Section 41.4   A Particle in a Box

16. An electron that has an energy of approximately 6 eV moves between rigid walls 1.00 nm apart. Find (a) the quantum number n for the energy state that the electron occupies, and (b) the precise energy of the electron.

WEB 17. An electron is contained in a one-dimensional box of width 0.100 nm. (a) Draw an energy-level diagram for the electron for levels up to $n = 4$. (b) Find the wavelengths of all photons that can be emitted by the electron in making transitions from the $n = 4$ state to the $n = 1$ state (by all spontaneous paths).

18. An electron is confined to a one-dimensional region in which its ground-state ($n = 1$) energy is 2.00 eV. (a) What is the width of the region? (b) How much energy is required to promote the electron to its first excited state?

19. A ruby laser emits 694.3-nm light. Assuming that this light is due to a transition of an electron in a box from the $n = 2$ state to the $n = 1$ state, find the width of the box.

20. A laser emits light of wavelength λ. Assuming that this light is due to a transition of an electron in a box from the $n = 2$ state to the $n = 1$ state, find the width of the box.

21. The nuclear potential energy that binds protons and neutrons in a nucleus is often approximated by a square

well. Imagine a proton confined in an infinitely high square well of width 10.0 fm, a typical nuclear diameter. Calculate the wavelength and energy associated with the photon emitted when the proton moves from the $n = 2$ state to the ground state. In what region of the electromagnetic spectrum does this wavelength belong?

22. An alpha particle in a nucleus can be modeled as a particle moving in a box of width 1.00×10^{-14} m (the approximate diameter of a nucleus). Using this model, estimate the energy and momentum of an alpha particle in its lowest energy state ($m_\alpha = 6.64 \times 10^{-27}$ kg).

23. Use the particle-in-a-box model to calculate the first three energy levels of a neutron trapped in a nucleus of diameter 20.0 fm. Do the energy-level differences have a realistic order of magnitude?

24. A particle in an infinitely deep square well has a wave function that is given by

$$\psi_2(x) = \sqrt{\frac{2}{L}} \sin\left(\frac{2\pi x}{L}\right)$$

for $0 \leq x \leq L$ and is zero otherwise. (a) Determine the expectation value of x. (b) Determine the probability of finding the particle near $L/2$, by calculating the probability that the particle lies in the range $0.490L \leq x \leq 0.510L$. (c) Determine the probability of finding the particle near $L/4$, by calculating the probability that the particle lies in the range $0.240L \leq x \leq 0.260L$. (d) Argue that no contradiction exists between the result of part (a) and the results of parts (b) and (c).

25. The wave function for a particle confined to moving in a one-dimensional box is

$$\psi(x) = A \sin\left(\frac{n\pi x}{L}\right)$$

Use the normalization condition on ψ to show that

$$A = \sqrt{\frac{2}{L}}$$

Hint: Because the box width is L, the wave function is zero for $x < 0$ and for $x > L$, so the normalization condition (Eq. 41.4) reduces to

$$\int_0^L |\psi|^2 \, dx = 1$$

26. The wave function of an electron is

$$\psi_2(x) = \sqrt{\frac{2}{L}} \sin\left(\frac{2\pi x}{L}\right)$$

Calculate the probability of finding the electron between $x = 0$ and $x = L/4$.

27. An electron in an infinitely deep square well has a wave function that is given by

$$\psi_2(x) = \sqrt{\frac{2}{L}} \sin\left(\frac{2\pi x}{L}\right)$$

for $0 \le x \le L$ and is zero otherwise. What are the most probable positions of the electron?

28. A particle in an infinite square well has a wave function that is given by

$$\psi_1(x) = \sqrt{\frac{2}{L}} \sin\left(\frac{\pi x}{L}\right)$$

for $0 \le x \le L$ and is zero otherwise. (a) Determine the probability of finding the particle between $x = 0$ and $x = L/3$. (b) Use the result of this calculation and symmetry arguments to find the probability of finding the particle between $x = L/3$ and $x = 2L/3$. Do not reevaluate the integral. (c) Compare the result of part (a) with the classical probability.

29. A proton is confined to moving in a one-dimensional box of width 0.200 nm. (a) Find the lowest possible energy of the proton. (b) What is the lowest possible energy of an electron confined to the same box? (c) How do you account for the great difference in your results for (a) and (b)?

30. Consider a particle moving in a one-dimensional box for which the walls are at $x = -L/2$ and $x = L/2$. (a) Write the wave functions and probability densities for $n = 1$, $n = 2$, and $n = 3$. (b) Sketch the wave functions and probability densities. (*Hint:* Make an analogy with the case of a particle in a box for which the walls are at $x = 0$ and $x = L$.)

Section 41.5   The Schrödinger Equation

31. Show that the wave function $\psi = Ae^{i(kx-\omega t)}$ is a solution to the Schrödinger equation (Eq. 41.12), where $k = 2\pi/\lambda$ and $U = 0$.

32. The wave function of a particle is given by

$$\psi(x) = A\cos(kx) + B\sin(kx)$$

where A, B, and k are constants. Show that ψ is a solution of the Schrödinger equation (Eq. 41.12), assuming that the particle is free ($U = 0$), and find the corresponding energy E of the particle.

33. A particle of mass m moves in a potential well of width $2L$. Its potential energy is infinite for $x < -L$ and for $x > +L$. Inside the region $-L < x < L$, its potential energy is given by

$$U(x) = \frac{-\hbar^2 x^2}{mL^2(L^2 - x^2)}$$

In addition, the particle is in a stationary state described by the wave function $\psi(x) = A(1 - x^2/L^2)$ for $-L < x < +L$, and by $\psi(x) = 0$ elsewhere. (a) Determine the energy of the particle in terms of $\hbar$, m, and L. (*Hint:* Use the Schrödinger equation, Eq. 41.12.) (b) Show that $A = (15/16L)^{1/2}$. (c) Determine the probability that the particle is located between $x = -L/3$ and $x = +L/3$.

34. In a region of space, a particle with zero total energy has a wave function

$$\psi(x) = Axe^{-x^2/L^2}$$

(a) Find the potential energy U as a function of x.
(b) Make a sketch of $U(x)$ versus x.

(Optional)
Section 41.6   A Particle in a Well of Finite Height

35. Suppose a particle is trapped in its ground state in a box that has infinitely high walls (see Fig. 41.11). Now suppose the left-hand wall is suddenly lowered to a finite height and width. (a) Qualitatively sketch the wave function for the particle a short time later. (b) If the box has a width L, what is the wavelength of the wave that penetrates the left-hand wall?

36. Sketch the wave function $\psi(x)$ and the probability density $|\psi(x)|^2$ for the $n = 4$ state of a particle in a finite potential well. (See Fig. 41.16.)

(Optional)
Section 41.7   Tunneling Through a Barrier

37. An electron with kinetic energy $E = 5.00$ eV is incident on a barrier with thickness $L = 0.200$ nm and height $U = 10.0$ eV (Fig. P41.37). What is the probability that the electron (a) will tunnel through the barrier? (b) will be reflected?

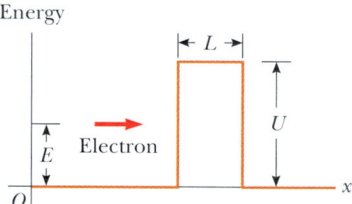

Figure P41.37   Problems 37 and 38.

38. An electron having total energy $E = 4.50$ eV approaches a rectangular energy barrier with $U = 5.00$ eV and $L = 950$ pm, as in Figure P41.37. Classically, the electron cannot pass through the barrier because $E < U$. However, quantum-mechanically a finite probability of tunneling exists. Calculate this probability, which is the transmission coefficient.

39. In Problem 38, by how much would the width L of the potential barrier have to be increased for the chance of an incident 4.50-eV electron tunneling through the barrier to be one in a million?

(Optional)
Section 41.8   The Scanning Tunneling Microscope

40. A scanning tunneling microscope (STM) can precisely determine the depths of surface features because the current through its tip is very sensitive to differences in the width of the gap between the tip and the sample surface. Assume that in this direction the electron wave function falls off exponentially with a decay length of 0.100 nm—that is, with $C = 10.0/\text{nm}$. Determine the ratio of the current when the STM tip is 0.500 nm above a surface feature to the current when the tip is 0.515 nm above the surface.

41. The design criterion for a typical scanning tunneling microscope specifies that it must be able to detect, on the sample below its tip, surface features that differ in height by only 0.002 00 nm. What percentage change in electron transmission must the electronics of the STM be able to detect, to achieve this resolution? Assume that the electron transmission coefficient is e^{-2CL}, with $C = 10.0/\text{nm}$.

(Optional)

Section 41.9   The Simple Harmonic Oscillator

42. Show that Equation 41.21 is a solution of Equation 41.19 with energy $E = \frac{1}{2}\hbar\omega$.

43. A one-dimensional harmonic oscillator wave function is

$$\psi = Axe^{-bx^2}$$

(a) Show that ψ satisfies Equation 41.19. (b) Find b and the total energy E. (c) Is this a ground state or a first excited state?

44. A quantum simple harmonic oscillator consists of an electron bound by a restoring force proportional to its displacement from a certain equilibrium point. The proportionality constant is 8.99 N/m. What is the longest wavelength of light that can excite the oscillator?

45. (a) Normalize the wave function for the ground state of a simple harmonic oscillator. That is, apply Equation 41.4 to Equation 41.21 and find the required value for the coefficient B, in terms of m, ω, and constants. (b) Determine the probability of finding the oscillator in a narrow interval $-\delta/2 < x < \delta/2$ around its equilibrium position.

46. The total energy of a particle moving with simple harmonic motion along the x axis is

$$E = \frac{p_x^2}{2m} + \frac{kx^2}{2}$$

where p_x is the momentum of the particle and k is the spring constant. (a) Using the uncertainty principle, show that this expression can also be written

$$E \ge \frac{p_x^2}{2m} + \frac{k\hbar^2}{8p_x^2}$$

(b) Show that the minimum energy of the harmonic oscillator is

$$E_{\min} = K + U = \frac{1}{4}\hbar\sqrt{\frac{k}{m}} + \frac{\hbar\omega}{4} = \frac{\hbar\omega}{2}$$

ADDITIONAL PROBLEMS

47. Keeping a constant speed of 0.8 m/s, a marble rolls back and forth across a shoebox. Make an order-of-magnitude estimate of the probability of its escaping through the wall of the box by quantum tunneling. State the quantities you take as data and the values you measure or estimate for them.

48. A particle of mass 2.00×10^{-28} kg is confined to a one-dimensional box of width 1.00×10^{-10} m. For $n = 1$, what are (a) the particle's wavelength, (b) its momentum, and (c) its ground-state energy?

WEB 49. An electron is represented by the time-independent wave function

$$\psi(x) = \begin{cases} Ae^{-\alpha x} & \text{for } x > 0 \\ Ae^{+\alpha x} & \text{for } x < 0 \end{cases}$$

(a) Sketch the wave function as a function of x. (b) Sketch the probability that the electron is found between x and $x + dx$. (c) Argue that this can be a physically reasonable wave function. (d) Normalize the wave function. (e) Determine the probability of finding the electron somewhere in the range

$$x_1 = -\frac{1}{2\alpha} \quad \text{to} \quad x_2 = \frac{1}{2\alpha}$$

50. Particles incident from the left are confronted with a step in potential energy as shown in Figure P41.50. The step has a height U, and the particles have energy $E > U$. Classically, we would expect all of the particles to continue on, although with reduced speed. According to quantum mechanics, a fraction of the particles are reflected at the barrier. (a) Prove that the reflection coefficient R for this case is

$$R = \frac{(k_1 - k_2)^2}{(k_1 + k_2)^2}$$

where $k_1 = 2\pi/\lambda_1$ and $k_2 = 2\pi/\lambda_2$ are the angular wavenumbers for the incident and transmitted particles. Proceed as follows: Show that the wave function $\psi_1 = A\cos k_1 x + B\cos(-k_1 x)$ satisfies the Schrödinger equation in region 1, where $x < 0$. Here $A\cos k_1 x$ represents the incident beam, and $B\cos(-k_1 x)$ represents the reflected particles. Show that $\psi_2 = C\cos k_2 x$ satisfies the Schrödinger equation in region 2, for $x > 0$.

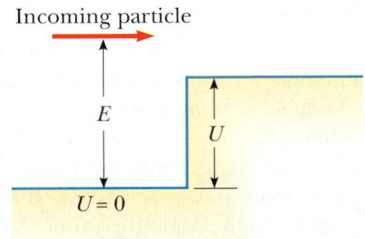

Figure P41.50 Problems 50 and 51.

Impose the boundary conditions $\psi_1 = \psi_2$ and $d\psi_1/dx = d\psi_2/dx$ at $x = 0$, to find the relationship between B and A. Then evaluate $R = B^2/A^2$. (b) A particle that has kinetic energy $E = 7.00$ eV is incident from a region where the potential energy is zero onto one in which $U = 5.00$ eV. Find its probability of being reflected and its probability of being transmitted.

51. Particles incident from the left are confronted with a step in potential energy as shown in Figure P41.50. The step has a height U, and the particles have energy $E = 2U$. Classically, all the particles would pass into the region of higher potential energy at the right. However, according to quantum mechanics, a fraction of the particles are reflected at the barrier. Use the result of Problem 50 to determine the fraction of the incident particles that are reflected. (This situation is analogous to the partial reflection and transmission of light striking an interface between two different media.)

52. An electron is trapped at a defect in a crystal. (A defect is an imperfection in the otherwise orderly arrangement of atoms.) The defect may be modeled as a one-dimensional, rigid-walled box of width 1.00 nm. (a) Sketch the wave functions and probability densities for the $n = 1$ and $n = 2$ states. (b) For the $n = 1$ state, calculate the probability of finding the electron between $x_1 = 0.150$ nm and $x_2 = 0.350$ nm, where $x = 0$ is the left side of the box. (c) Repeat part (b) for the $n = 2$ state. (d) Calculate the energies, in electron volts, of the $n = 1$ and $n = 2$ states. *Hint:* For parts (b) and (c), use Equation 41.5 and note that

$$\int \sin^2 ax \, dx = \tfrac{1}{2}x - \frac{1}{4a}\sin 2ax$$

53. Johnny Jumper's favorite trick is to step out of his 16th-story window and fall 50.0 m into a pool. A news reporter takes a picture of 75.0-kg Johnny just before he makes a splash, using an exposure time of 5.00 ms. Find (a) Johnny's de Broglie wavelength at this moment, (b) the uncertainty of his kinetic energy measurement during such a period of time, and (c) the percent error caused by such an uncertainty.

54. A π^0 meson is an unstable particle produced in high-energy particle collisions. Its rest energy is about 135 MeV, and it exists for an average lifetime of only 8.70×10^{-17} s before decaying into two gamma rays. Using the uncertainty principle, estimate the fractional uncertainty $\Delta m/m$ in its mass determination.

55. An atom in an excited state 1.80 eV above the ground state remains in that excited state 2.00 μs before moving to the ground state. Find the (a) frequency and (b) wavelength of the emitted photon. (c) Find the approximate uncertainty in energy of the photon.

56. An atom in an excited state E above the ground state remains in that excited state for a time T before moving to the ground state. Find the (a) frequency and (b) wavelength of the emitted photon. (c) Find the approximate uncertainty in energy of the photon.

57. For a particle described by a wave function $\psi(x)$, the expectation value of a physical quantity $f(x)$ associated with the particle is defined by

$$\langle f(x) \rangle \equiv \int_{-\infty}^{\infty} f(x)|\psi|^2 \, dx$$

For a particle in a one-dimensional box extending from $x = 0$ to $x = L$, show that

$$\langle x^2 \rangle = \frac{L^2}{3} - \frac{L^2}{2n^2\pi^2}$$

58. A particle is described by the wave function

$$\psi(x) = \begin{cases} A\cos\left(\dfrac{2\pi x}{L}\right) & \text{for } -\dfrac{L}{4} \le x \le \dfrac{L}{4} \\ 0 & \text{for other values of } x \end{cases}$$

(a) Determine the normalization constant A. (b) What is the probability that the particle will be found between $x = 0$ and $x = L/8$ if its position is measured? (*Hint:* Use Eq. 41.5.)

59. A particle has a wave function

$$\psi(x) = \begin{cases} \sqrt{\dfrac{2}{a}}\, e^{-x/a} & \text{for } x > 0 \\ 0 & \text{for } x < 0 \end{cases}$$

(a) Find and sketch the probability density. (b) Find the probability that the particle will be at any point where $x < 0$. (c) Show that ψ is normalized, and then find the probability that the particle will be found between $x = 0$ and $x = a$.

60. A certain electron microscope accelerates electrons to an energy of 65.0 keV. (a) Find the wavelength of these electrons. (b) If one can resolve two points separated by at least 50.0 wavelengths, what is the smallest separation (or the minimum-sized object) that can be resolved with this microscope?

61. An electron of momentum p is at a distance r from a stationary proton. The electron has kinetic energy $K = p^2/2m_e$, potential energy $U = -k_e e^2/r$, and total energy $E = K + U$. If the electron is bound to the proton to form a hydrogen atom, its average position is at the proton, but the uncertainty in its position is approximately equal to the radius r of its orbit. The electron's average vector momentum is zero, but its average squared momentum is approximately equal to the squared uncertainty in its momentum, as given by the uncertainty principle. Treating the atom as a one-dimensional system, (a) estimate the uncertainty in the electron's momentum in terms of r. (b) Estimate the electron's kinetic, potential, and total energies in terms of r. (c) The actual value of r is the one that *minimizes the total energy,* resulting in a stable atom. Find that value of r and the resulting total energy. Compare your answer with the predictions of the Bohr theory.

62. A particle of mass m is placed in a one-dimensional box of width L. Assume that the box is so small that the par-

ticle's motion is *relativistic*, so $E = p^2/2m$ is not valid. (a) Derive an expression for the energy levels of the particle. (b) If the particle is an electron in a box of width $L = 1.00 \times 10^{-12}$ m, find its lowest possible kinetic energy. By what percentage is the nonrelativistic formula in error? (*Hint:* See Eq. 39.26.)

63. Consider a "crystal" consisting of two nuclei and two electrons as shown in Figure P41.63. (a) Taking into account all the pairs of interactions, find the potential energy of the system as a function of d. (b) Assuming that the electrons are restricted to a one-dimensional box of width $3d$, find the minimum kinetic energy of the two electrons. (c) Find the value of d for which the total energy is a minimum. (d) Compare this value of d with the spacing of atoms in lithium, which has a density of 0.530 g/cm^3 and an atomic mass of 7 u. (This type of calculation can be used to estimate the densities of crystals and certain stars.)

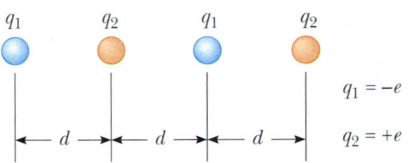

Figure P41.63

64. *The simple harmonic oscillator in an excited state.* The wave function

$$\psi(x) = Bxe^{-(m\omega/2\hbar)x^2}$$

is also a solution to the simple-harmonic-oscillator problem. (a) Find the energy of this state. (b) At what position are you least likely to find the particle? (c) At what positions are you most likely to find the particle? (d) Determine the value of B required to normalize the wave function. (e) Determine the classical probability of finding the particle in an interval of small width δ centered at the position $x = 2(\hbar/m\omega)^{1/2}$. (f) What is the actual probability of finding the particle in this interval?

65. *Normalization of wave functions:* (a) Find the normalization constant A for a wave function made up of the two lowest states of a particle in a box:

$$\psi(x) = A\left[\sin\left(\frac{\pi x}{L}\right) + 4 \sin\left(\frac{2\pi x}{L}\right)\right]$$

(b) A particle is described in the space $-a \le x \le a$ by the wave function

$$\psi(x) = A\cos\left(\frac{\pi x}{2a}\right) + B\sin\left(\frac{\pi x}{a}\right)$$

Determine the relationship between the values of A and B that are required for normalization. (*Hint:* Use the identity $\sin 2\theta = 2 \sin\theta \cos\theta$.)

66. A two-slit electron diffraction experiment is done with slits of *unequal* widths. The number of electrons reach-

ing the screen per second when only slit 1 is open is 25.0 times the number of electrons reaching the screen per second when only slit 2 is open. When both slits are open, an interference pattern results in which the destructive interference is not complete. Find the ratio of the probability that an electron will arrive at an interference maximum to the probability that an electron will arrive at an adjacent interference minimum. (*Hint:* Use the superposition principle.)

67. **Review Problem.** Consider an extension of Young's double-slit experiment, performed with photons. Think of Figure 41.3 as a top view, with the reader looking down on the apparatus. The viewing screen can be a large flat array of *charge-coupled detectors*. Each cell in the array registers individual photons with high efficiency, so we can see where individual photons strike the screen in real time. We cover slit 1 with a polarizer with its transmission axis horizontal, and slit 2 with a polarizer with vertical transmission axis. Any one photon is either absorbed by a polarizing filter or allowed to pass through. The photons that come through a polarizer have their electric field oscillating in the plane defined by their direction of motion and the filter axis. Now we place another large square sheet of polarizing material just in front of the screen. For experimental trial 1, we make the transmission axis of this third polarizer horizontal. This choice, in effect, blocks slit 2. After many photons have been sent through the apparatus, their distribution on the viewing screen is shown by the blue curve $|\psi_1|^2$ in Figure 41.3. For trial 2, we turn the polarizer at the screen to make its transmission axis vertical. Then the screen receives photons only by way of slit 2, and their distribution is shown as $|\psi_2|^2$. For trial 3, we temporarily remove the third sheet of polarizing material. Then the interference pattern shown by the red curve $|\psi_1 + \psi_2|^2$ appears. (a) Is the light that is arriving at the screen to form the interference pattern polarized? Explain your answer. (b) Next, in trial 4 we replace the large square of polarizing material in front of the screen and set its transmission axis to 45°, halfway between horizontal and vertical. What appears on the screen? (c) Suppose we repeat all of trials 1 through 4 with very low light intensity, so that only one photon is present in the apparatus at a time. What are the results now? (d) We go back to high light intensity for convenience, and in trial 5 make the large square of polarizer turn slowly and steadily about a rotation axis through its center and perpendicular to its area. What appears on the screen? (e) At last, we go back to very low light intensity and replace the large square sheet of polarizing plastic with a flat layer of liquid crystal, to which we can apply an electric field in either a horizontal or a vertical direction. With the applied field we can very rapidly switch the liquid crystal so that it transmits only photons with horizontal electric field, acts as a polarizer with a vertical transmission axis, or transmits all photons with high efficiency. We keep track of photons as they are emitted individually by the source. For each photon we

wait until it has passed through the pair of slits. Then we quickly choose the setting of the liquid crystal, and make that photon encounter a horizontal polarizer, a vertical polarizer, or no polarizer before it arrives at the detector array. We can alternate among the conditions

we set up earlier in trials 1, 2, and 3. We keep track of our settings of the liquid crystal and sort out how photons behave under the different conditions, to end up with full sets of data for all three of those trials. What are the results?

ANSWERS TO QUICK QUIZZES

41.1 The diffraction pattern looks like the pattern for light waves going through three slits, which is shown in Figure 37.13.

41.2 If the slit is wide, as in part (a) of the figure at the bottom of the page, we cannot precisely know an individual electron's horizontal position within the electron beam. If we squeeze the beam through a narrower slit, as in part (b) of the figure, we do decrease Δx, but the increasing effects of diffraction mean that we have increased the uncertainty in p_x. (Depending on its momentum, an electron can appear anywhere in a wide horizontal area of the viewing screen.)

41.3 Classically, we expect the particle to bounce back and forth between the two walls at constant speed. Thus, we are as likely to find it at the left side of the box as at the middle, the right side, or anywhere else inside the box. Our graph of probability density versus x would therefore be a horizontal line, with a total area under the line of unity.

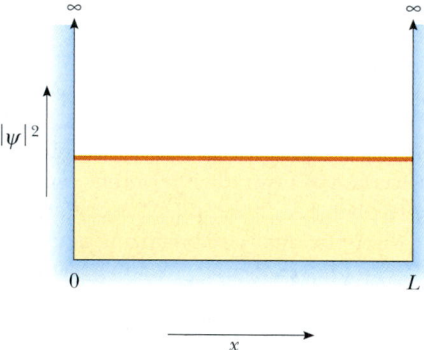

41.4 Figure 41.11b drawn for $n = 20$ would have 20 peaks quite close together. As the value of n increases, the peaks become progressively closer to one another, and

the detector is more likely to detect several peaks and valleys at once. As $n \rightarrow \infty$, the detector reports the average value of many cycles of the oscillating wave function. It gives this same average value anywhere inside the box, matching the classical answer given in Quick Quiz 41.3.

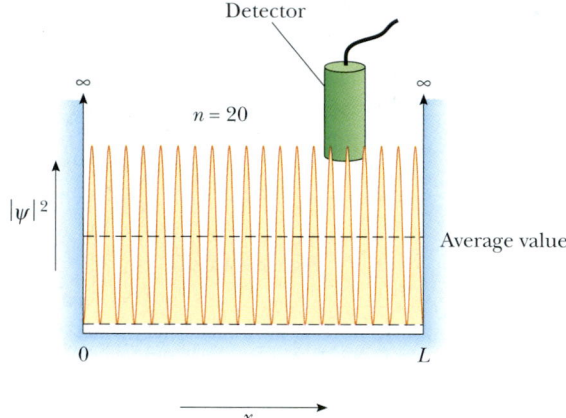

41.5 (a) The upward-curving parts of the classical graphs represent higher probability density values, which means the harmonic oscillator spends more time near the maximum displacement points. A block oscillating at the end of a vertical spring, for example, moves more slowly when the spring is near its fully stretched and fully compressed positions. Thus, a person taking a quick glance is more likely to see the block near one of these two points of maximum excursion from equilibrium. (b) As n increases, if we average over the unresolvable peaks and valleys in the probability density, the quantum-mechanical predictions become progressively closer to classical predictions; for great enough n, the two curves are indistinguishable from each other.

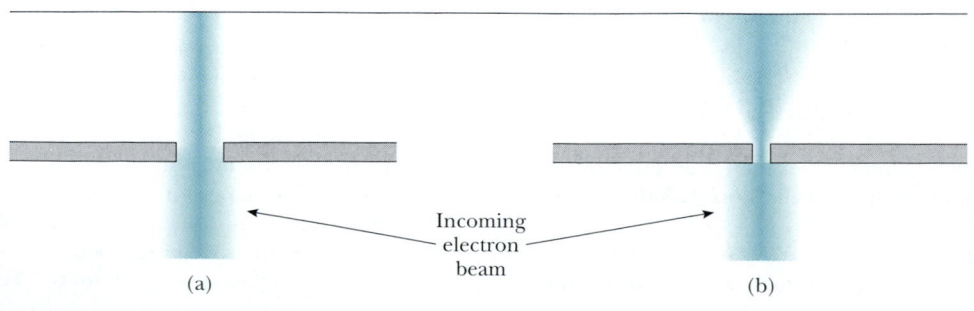

c h a p t e r

42

Atomic Physics

In Chapter 41 we introduced some of the basic concepts and techniques used in quantum mechanics, along with their applications to various one-dimensional systems. This chapter applies quantum mechanics to the real world of atomic structure, and a large portion of the chapter is quantum mechanics applied to the study of the hydrogen atom. Understanding the hydrogen atom, the simplest atomic system, is important for several reasons:

- The hydrogen atom is the only atomic system that can be solved exactly.
- Much of what is learned about the hydrogen atom, with its single electron, can be extended to such single-electron ions as He^+ and Li^{2+}.
- The hydrogen atom is an ideal system for performing precise tests of theory against experiment and for improving our overall understanding of atomic structure.
- The quantum numbers that are used to characterize the allowed states of hydrogen can also be used to describe the allowed states of more complex atoms, and such description enables us to understand the periodic table of the elements. This understanding is one of the greatest triumphs of quantum mechanics.
- The basic ideas about atomic structure must be well understood before we attempt to deal with the complexities of molecular structures and the electronic structure of solids.

The full mathematical solution of the Schrödinger equation applied to the hydrogen atom gives a complete and beautiful description of the atom's properties. However, because the mathematical procedures that are involved are beyond the scope of this text, the details are omitted. The solutions for some states of hydrogen are discussed, together with the quantum numbers used to characterize various allowed stationary states. We also discuss the physical significance of the quantum numbers and the effect of a magnetic field on certain quantum states.

A new physical idea, the *exclusion principle,* is presented in this chapter. This principle is extremely important for understanding the properties of multielectron atoms and the arrangement of elements in the periodic table. In fact, the implications of the exclusion principle are almost as far-reaching as those of the Schrödinger equation.

Finally, we apply our knowledge of atomic structure to describe the mechanisms involved in the production of x-rays and in the operation of a laser.

42.1   EARLY MODELS OF THE ATOM

The model of the atom in the days of Newton was a tiny, hard, indestructible sphere. Although this model provided a good basis for the kinetic theory of gases, new models had to be devised when experiments revealed the electrical nature of atoms. J. J. Thomson suggested a model that describes the atom as a volume of positive charge with electrons embedded throughout the volume, much like the seeds in a watermelon or raisins in thick pudding (Fig. 42.1).

In 1911, Ernest Rutherford (1871–1937) and his students Hans Geiger and Ernest Marsden performed a critical experiment that showed that Thomson's model could not be correct. In this experiment, a beam of positively charged alpha particles (helium nuclei) was projected into a thin metallic foil, such as the target in Figure 42.2a. Most of the particles passed through the foil as if it were empty space. But some of the results of the experiment were astounding: Many of the particles deflected from their original direction of travel were scattered through *large* angles. Some were even deflected backward, reversing their

Electron

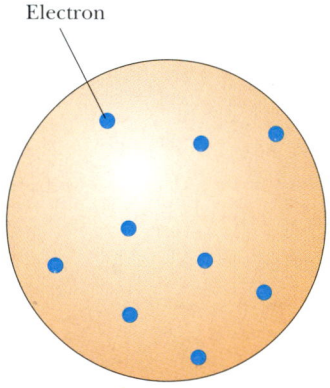

Figure 42.1   Thomson's model of the atom: negatively charged electrons in a volume of continuous positive charge.

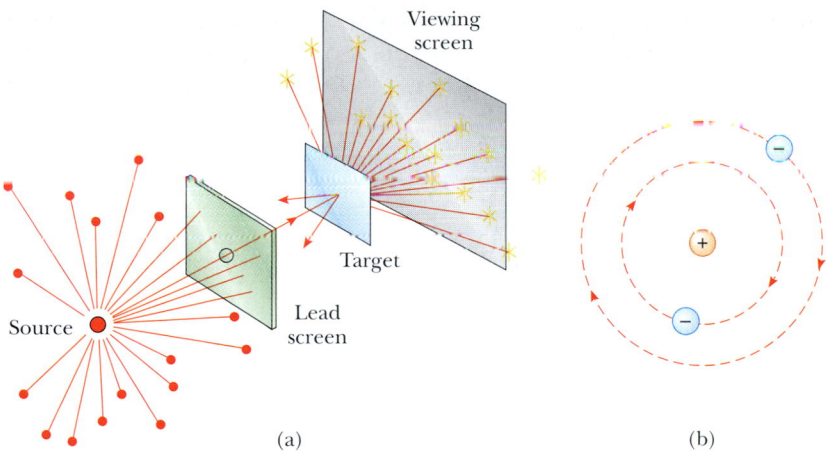

(a) (b)

Figure 42.2 (a) Rutherford's technique for observing the scattering of alpha particles from a thin foil target. The source is a naturally occurring radioactive substance, such as radium. (b) Rutherford's planetary model of the atom.

direction of travel! When Geiger informed Rutherford that some alpha particles were scattered backward, Rutherford wrote, "It was quite the most incredible event that has ever happened to me in my life. It was almost as incredible as if you fired a 15-inch shell at a piece of tissue paper and it came back and hit you."

Such large deflections were not expected on the basis of Thomson's model. According to that model, the positive charge of an atom in the foil is spread out over such a great volume (the entire atom) that a positively charged alpha particle would never come close enough to a positive charge strong enough to cause any large-angle deflections. Rutherford explained his astonishing results by developing a new atomic model, one that assumed that the positive charge in the atom was concentrated in a region that was small relative to the size of the atom. He called this concentration of positive charge the **nucleus** of the atom. Any electrons belonging to the atom were assumed to be in the relatively large volume outside the nucleus. To explain why these electrons were not pulled into the nucleus, Rutherford modeled them as moving in orbits around the positively charged nucleus in the same manner as the planets orbit the Sun (Fig. 42.2b).

Two basic difficulties exist with Rutherford's planetary model. As we saw in Chapter 40, an atom emits certain characteristic frequencies of electromagnetic radiation and no others; the Rutherford model cannot explain this phenomenon. A second difficulty is that Rutherford's electrons are undergoing a centripetal acceleration. According to Maxwell's theory of electromagnetism, centripetally accelerated charges revolving with frequency f should radiate electromagnetic waves of frequency f. Unfortunately, this classical model leads to disaster when applied to the atom. As the electron radiates energy, the radius of its orbit steadily decreases and its frequency of revolution increases. This would lead to an ever-increasing frequency of emitted radiation and an ultimate collapse of the atom as the electron plunges into the nucleus (Fig. 42.3).

Now the stage was set for Bohr. To circumvent the erroneous deductions of electrons falling into the nucleus and a continuous emission from atoms, Bohr postulated that classical radiation theory did not hold for atomic-sized systems. By applying Planck's ideas of quantized energy levels to orbiting atomic electrons, he overcame the problem of a classical electron that continuously loses energy. Thus, Bohr postulated that atoms are generally confined to stable, nonradiating energy

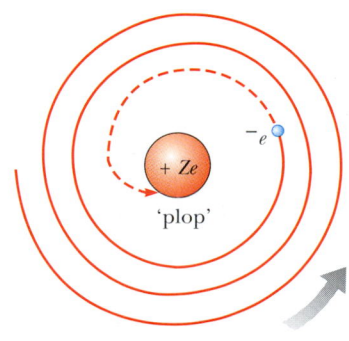

Figure 42.3 The classical model of the nuclear atom.

levels, each level representing a stationary state (see Section 40.5). Furthermore, he applied Einstein's concept of the photon to arrive at an expression for the frequency of the light that is emitted when an electron jumps from one stationary state to another.

One of the first indications that the Bohr theory needed to be modified arose when improved spectroscopic techniques were used to examine the spectral lines of hydrogen. It was found that many of the lines in the Balmer and other series were not single lines at all. Instead, each was a group of lines spaced very close together. An additional difficulty arose when it was observed that, in some situations, certain single spectral lines were split into three closely spaced lines when the atoms were placed in a strong magnetic field. Efforts to explain these deviations from the Bohr model led to improvements in the theory. One of the changes introduced into the original theory was the postulate that the electron could spin on its axis. In addition, Arnold Sommerfeld (1868–1951) improved the Bohr theory by introducing the theory of relativity into the analysis of the electron's motion.

42.2 ▷ THE HYDROGEN ATOM REVISITED

In Chapter 40 we described how the Bohr model views the electron as a particle orbiting around the nucleus in nonradiating, quantized energy levels. The de Broglie model assigns a wavelike nature to the electron, a model that allows some deeper understanding of the hydrogen atom. However, that model does not address all objections to the Bohr model and introduces some of its own difficulties. Fortunately, those difficulties are removed when the methods of quantum mechanics are used to describe atoms.

The potential energy function for the hydrogen atom is

$$U(r) = -k_e \frac{e^2}{r} \qquad \textbf{(42.1)}$$

where $k_e = 8.99 \times 10^9 \, \text{N} \cdot \text{m}^2/\text{C}^2$ is the Coulomb constant, and r is the radial distance from the proton (situated at $r = 0$) to the electron. Figure 42.4 is a plot of this function versus r/a_0, where a_0 is the Bohr radius, 0.052 9 nm (see Eq. 40.24).

The formal procedure for solving the problem of the hydrogen atom is to substitute $U(r)$ into the Schrödinger equation and find appropriate solutions to the equation, as we did for the particle in a box in Chapter 41. The present problem is more complicated, however, because it is three-dimensional and because U depends on the radial coordinate r. We do not attempt to carry out these solutions. Rather, we simply describe their properties and some of their implications with regard to atomic structure.

According to quantum mechanics, the energies of the allowed states for the hydrogen atom are

$$E_n = -\left(\frac{k_e e^2}{2a_0}\right)\frac{1}{n^2} = -\frac{13.606}{n^2} \text{ eV} \qquad n = 1, 2, 3, \ldots \qquad \textbf{(42.2)}$$

This result is in exact agreement with that obtained in the Bohr theory.

In this solution to the Schrödinger equation, three quantum numbers all having integer values are needed for each stationary state, corresponding to three independent degrees of freedom for the electron: **principal quantum number** n, **orbital quantum number** ℓ, and **orbital magnetic quantum number** m_ℓ. A

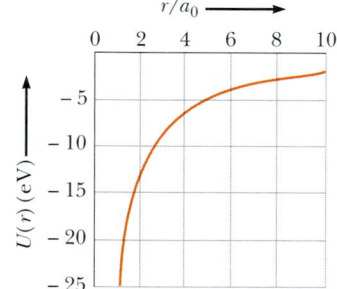

Figure 42.4 Potential energy $U(r)$ versus the ratio r/a_0 for the hydrogen atom. The constant a_0 is the Bohr radius, and r is the electron–proton separation distance.

Allowed energies for the hydrogen atom

TABLE 42.1 Three Quantum Numbers for the Hydrogen Atom

Quantum Number	Name	Allowed Values	Number of Allowed States
n	Principal quantum number	$1, 2, 3, \ldots$	Any number
ℓ	Orbital quantum number	$0, 1, 2, \ldots, n-1$	n
m_ℓ	Orbital magnetic quantum number	$-\ell, -\ell+1, \ldots,$ $0, \ldots, \ell-1, \ell$	$2\ell+1$

fourth quantum number, resulting from a relativistic treatment of the hydrogen atom, is discussed in Section 42.3.

Certain important relationships exist among these three quantum numbers, as well as certain restrictions on their values:

> The values of n can range from 1 to ∞.
> The values of ℓ can range from 0 to $n-1$.
> The values of m_ℓ can range from $-\ell$ to ℓ.

Restrictions on the values of quantum numbers

For example, if $n = 1$, only $\ell = 0$ and $m_\ell = 0$ are permitted. If $n = 2$, ℓ may be 0 or 1; if $\ell = 0$, then $m_\ell = 0$; but if $\ell = 1$, then m_ℓ may be 1, 0, or -1. Table 42.1 summarizes the rules for determining the allowed values of ℓ and m_ℓ for a given n.

For historical reasons, **all states having the same principal quantum number are said to form a shell.** Shells are identified by the letters K, L, M, . . . , which designate the states for which $n = 1, 2, 3, \ldots$. Likewise, **all states having the same values of n and ℓ are said to form a subshell.** The letters[1] s, p, d, f, g, h, . . . are used to designate the subshells for which $\ell = 0, 1, 2, 3, \ldots$. For example, the state designated by $3p$ has the quantum numbers $n = 3$ and $\ell = 1$, and the $2s$ state has the quantum numbers $n = 2$ and $\ell = 0$. These notations are summarized in Table 42.2.

States that violate the rules given in Table 42.1 do not exist. For instance, the $2d$ state, which would have $n = 2$ and $\ell = 2$, cannot exist because the highest allowed value of ℓ is $n - 1$, which in this case is 1. Thus, for $n = 2$, $2s$ and $2p$ are allowed states but $2d$, $2f$, . . . are not. For $n = 3$, the allowed subshells are $3s$, $3p$, and $3d$.

TABLE 42.2 Atomic Shell and Subshell Notations

n	Shell Symbol	ℓ	Subshell Symbol
1	K	0	s
2	L	1	p
3	M	2	d
4	N	3	f
5	O	4	g
6	P	5	h
. . .		. . .	

[1] The first four of these letters come from early classifications of spectral lines: sharp, principal, diffuse, and fundamental. The remaining letters are in alphabetical order.

EXAMPLE 42.1 The *n* = 2 Level of Hydrogen

For a hydrogen atom, determine the number of allowed states corresponding to the principal quantum number *n* = 2, and calculate the energies of these states.

Solution When *n* = 2, ℓ can be 0 or 1. If ℓ = 0, the only value that m_ℓ can have is 0; for ℓ = 1, m_ℓ can be − 1, 0, or 1. Hence, we have one state, designated as the 2*s* state, that is associated with the quantum numbers *n* = 2, ℓ = 0, and m_ℓ = 0, and three states, designated as 2*p* states, for which the quantum numbers are *n* = 2, ℓ = 1, m_ℓ = − 1; *n* = 2, ℓ = 1, m_ℓ = 0; and *n* = 2, ℓ = 1, m_ℓ = 1.

Because all four of these states have the same principal quantum number *n* = 2, they all have the same energy, according to Equation 42.2:

$$E_2 = - \frac{13.606 \text{ eV}}{2^2} = \boxed{-3.401 \text{ eV}}$$

Exercise How many possible states exist for the *n* = 3 level of hydrogen? For the *n* = 4 level?

Answer 9; 16.

42.3 THE SPIN MAGNETIC QUANTUM NUMBER

In Example 42.1, we found four quantum states corresponding to *n* = 2. In reality, however, eight such states occur. The additional four states can be explained by requiring a fourth quantum number for each state: **spin magnetic quantum number** m_s.

The need for this new quantum number came about because of an unusual feature that was noted in the spectra of certain gases, such as sodium vapor. Close examination of one prominent line in the emission spectrum of sodium reveals that the line is, in fact, two closely spaced lines called a *doublet*. The wavelengths of these lines occur in the yellow region of the electromagnetic spectrum at 589.0 nm and 589.6 nm. In 1925, when this doublet was first observed, atomic theory could not explain it. To resolve this dilemma, Samuel Goudsmit (1902–1978) and George Uhlenbeck (1900–1988), following a suggestion made by the Austrian physicist Wolfgang Pauli (1900–1958), proposed the spin quantum number.

To describe this new quantum number, it is convenient (but technically incorrect) to think of the electron as spinning about its axis as it orbits the nucleus, as described in Section 30.8. Only two directions exist for the electron spin, as illustrated in Figure 42.5. If the direction of spin is as shown in Figure 42.5a, the electron is said to have *spin up*. If the direction of spin is reversed as in Figure 42.5b, the electron is said to have *spin down*. In the presence of an external magnetic field, the energy of the electron is slightly different for the two spin directions, and this energy difference accounts for the sodium doublet. The quantum numbers associated with the spin of the electron are $m_s = \frac{1}{2}$ for the spin-up state and $m_s = -\frac{1}{2}$ for the spin-down state.

The classical description of electron spin—as resulting from a spinning electron—is incorrect because quantum mechanics tells us that a rotational degree of freedom would require too many quantum numbers, and more recent theory indicates that the electron is a point particle, without spatial extent. Thus, the electron cannot be considered to be spinning as pictured in Figure 42.5. Despite this conceptual difficulty, all experimental evidence supports the idea that an electron does have some intrinsic property that can be described by the spin magnetic quantum number. Sommerfeld and Paul Dirac (1902–1984) showed that this fourth quantum number originates in the relativistic properties of the electron.

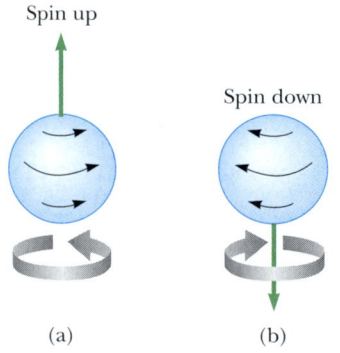

Spin up

Spin down

(a)   (b)

Figure 42.5 The spin of an electron can be either (a) up or (b) down relative to an external magnetic field.

EXAMPLE 42.2   ▶   **Putting Some Spin on Hydrogen**

For a hydrogen atom, determine the quantum numbers associated with the possible states that correspond to the principal quantum number $n = 2$.

Solution   With the addition of the spin quantum number, we have the possibilities given in the accompanying table.

Exercise   Show that for $n = 3$, there are 18 possible states.

n	ℓ	m_ℓ	m_s	Subshell	Shell	Number of Electrons in Subshell
2	0	0	$\frac{1}{2}$	$2s$	L	2
2	0	0	$-\frac{1}{2}$			
2	1	1	$\frac{1}{2}$			
2	1	1	$-\frac{1}{2}$			
2	1	0	$\frac{1}{2}$	$2p$	L	6
2	1	0	$-\frac{1}{2}$			
2	1	-1	$\frac{1}{2}$			
2	1	-1	$-\frac{1}{2}$			

42.4   ▶   THE WAVE FUNCTIONS FOR HYDROGEN

If we neglect electron spin for the present, the potential energy of the hydrogen atom depends only on the radial distance r between nucleus and electron. We therefore expect that some of the allowed states for this atom can be represented by wave functions that depend only on r. This indeed is the case. The simplest wave function for hydrogen is the one that describes the 1s state and is designated $\psi_{1s}(r)$:

$$\psi_{1s}(r) = \frac{1}{\sqrt{\pi a_0{}^3}} e^{-r/a_0} \tag{42.3}$$

Wave function for hydrogen in its ground state

where a_0 is the Bohr radius. Note that ψ_{1s} approaches zero as r approaches ∞ and is normalized as presented (see Eq. 41.4). Furthermore, because ψ_{1s} depends only on r, it is *spherically symmetric*. This, in fact, is true for all s states.

Recall that the probability of finding the electron in any region is equal to an integral of the probability density $|\psi|^2$ over the region. The probability density for the 1s state is

$$|\psi_{1s}|^2 = \left(\frac{1}{\pi a_0{}^3}\right) e^{-2r/a_0} \tag{42.4}$$

and the actual probability of finding the electron in a volume element dV is $|\psi|^2 \, dV$. It is convenient to define the *radial probability density function P(r)* as the probability per unit radial length of finding the electron in a spherical shell of radius r and thickness dr. Thus, $P(r) \, dr$ is the probability of finding the electron in this shell. The volume dV of such an infinitesimally thin shell equals its surface area $4\pi r^2$ multiplied by the shell thickness dr (Fig. 42.6), so we can write this probability as

$$P(r) \, dr = |\psi|^2 \, dV = |\psi|^2 \, 4\pi r^2 \, dr$$

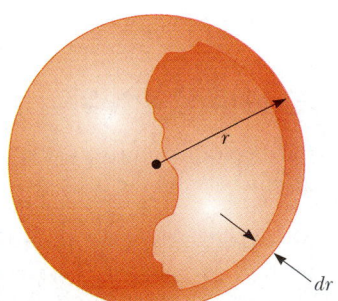

Figure 42.6   A spherical shell of radius r and thickness dr has a volume equal to $4\pi r^2 \, dr$.

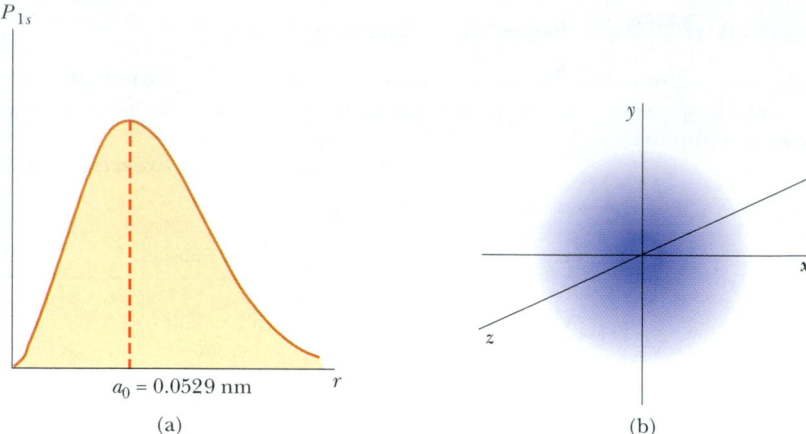

Figure 42.7 (a) The probability of finding the electron as a function of distance from the nucleus for the hydrogen atom in the $1s$ (ground) state. Note that the probability has its maximum value when r equals the Bohr radius a_0. (b) The spherical electronic charge distribution for the hydrogen atom in its $1s$ state.

Thus, the radial probability density function is

$$P(r) = 4\pi r^2 |\psi|^2 \tag{42.5}$$

Substituting Equation 42.4 into Equation 42.5 gives the radial probability density function for the hydrogen atom in its ground state:

Radial probability density for the
$1s$ state of hydrogen

$$P_{1s}(r) = \left(\frac{4r^2}{a_0^3}\right) e^{-2r/a_0} \tag{42.6}$$

A plot of the function $P_{1s}(r)$ versus r is presented in Figure 42.7a. The peak of the curve corresponds to the most probable value of r for this particular state. The spherical symmetry of the radial probability density function is shown in Figure 42.7b.

Quick Quiz 42.1

Sketch a cross-section of the three-dimensional charge distribution shown in Figure 42.7b, imagining the sphere as being "sliced" in the xy plane.

EXAMPLE 42.3   The Ground State of Hydrogen

Calculate the most probable value of r for an electron in the ground state of the hydrogen atom.

Solution The most probable value of r corresponds to the peak of the plot of $P(r)$ versus r. Because the slope of the curve at this point is zero, we can evaluate the most probable value of r by setting $dP/dr = 0$ and solving for r. Using Equation 42.6, we obtain

$$\frac{dP}{dr} = \frac{d}{dr}\left[\left(\frac{4r^2}{a_0^3}\right)e^{-2r/a_0}\right] = 0$$

Carrying out the derivative operation and simplifying the expression, we obtain

$$e^{-2r/a_0}\frac{d}{dr}(r^2) + r^2\frac{d}{dr}(e^{-2r/a_0}) = 0$$

$$2re^{-2r/a_0} + r^2(-2/a_0)e^{-2r/a_0} = 0$$

$$(1) \qquad 2r[1 - (r/a_0)]e^{-2r/a_0} = 0$$

This expression is satisfied if

$$1 - \frac{r}{a_0} = 0$$

$$r = a_0$$

The most probable value of r is the Bohr radius! This result and Equation 42.2 are interesting connections between the Bohr theory and the more sophisticated quantum theory.

Equation (1) is also satisfied at $r = 0$. This is a point of *minimum* probability, which is equal to zero, as seen in Figure 42.7a.

EXAMPLE 42.4 ▶ Probabilities for the Electron in Hydrogen

Calculate the probability that the electron in the ground state of hydrogen will be found outside the first Bohr radius.

Solution The probability is found by integrating the radial probability density function for this state $P_{1s}(r)$ from the Bohr radius a_0 to ∞. Using Equation 42.6, we obtain

$$P = \int_{a_0}^{\infty} P_{1s}(r) \, dr = \frac{4}{a_0^3} \int_{a_0}^{\infty} r^2 e^{-2r/a_0} \, dr$$

We can put the integral in dimensionless form by changing variables from r to $z = 2r/a_0$. Noting that $z = 2$ when $r = a_0$ and that $dr = (a_0/2) \, dz$, we obtain

$$P = \frac{1}{2} \int_{2}^{\infty} z^2 e^{-z} \, dz = -\frac{1}{2} (z^2 + 2z + 2) e^{-z} \Big|_{2}^{\infty}$$

$$P = 5e^{-2} = 0.677 \text{ or } 67.7\%$$

Example 42.3 shows that, for the ground state of hydrogen, the most probable value of r equals the Bohr radius a_0. It turns out that the average value of r for the ground state of hydrogen is $\frac{3}{2}a_0$, which is 50% greater than the most probable value (see Problem 49). The reason the average value is so much greater is the asymmetry in the radial probability density function (Fig. 42.7a), which has more area to the right of the peak. According to quantum mechanics, the atom has no sharply defined boundary. Therefore, the probability distribution for the electron can be viewed as a diffuse region of space, commonly referred to as an *electron cloud*.

The next-simplest wave function for the hydrogen atom is the one corresponding to the 2s state ($n = 2$, $\ell = 0$). The normalized wave function for this state is

$$\psi_{2s}(r) = \frac{1}{4\sqrt{2\pi}} \left(\frac{1}{a_0}\right)^{3/2} \left(2 - \frac{r}{a_0}\right) e^{-r/2a_0} \qquad \textbf{(42.7)}$$

Wave function for hydrogen in the 2s state

Again we see that ψ_{2s} depends only on r and is spherically symmetric. The energy corresponding to this state is $E_2 = -(13.606/4)$ eV $= -3.401$ eV. This energy level represents the first excited state of hydrogen. Plots of the radial probability density function for this state and several others are shown in Figure 42.8. The plot for the 2s state has two peaks. In this case, the most probable value corresponds to that value of r that has the highest value of $P(\approx 5a_0)$. An electron in the 2s state would be much farther from the nucleus (on the average) than an electron in the 1s state. The average value of r is even greater for the 3d, 3p, and 4d states.

As we have mentioned, all s states have spherically symmetric wave functions. The other states are not spherically symmetric. For example, the three wave functions corresponding to the states for which $n = 2$, $\ell = 1$ ($m_\ell = 1$, 0, or -1) can be expressed as appropriate linear combinations of the three p states. Although quantum mechanics limits our knowledge of angular momentum to the projection along any one axis at a time, these p states may be described mathematically as linear combinations of mutually perpendicular functions p_x, p_y, and p_z, as repre-

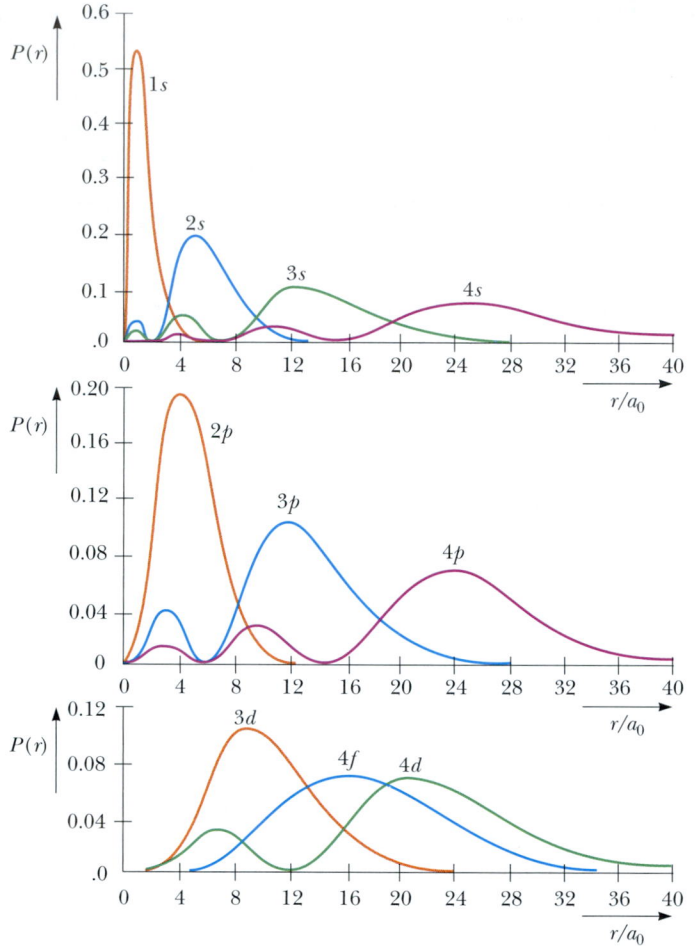

Figure 42.8   The radial probability density function versus r/a_0 for several states of the hydrogen atom.   *(From E. U. Condon and G. H. Shortley,* The Theory of Atomic Spectra, *Cambridge, England, Cambridge University Press, 1953. Used with permission.)*

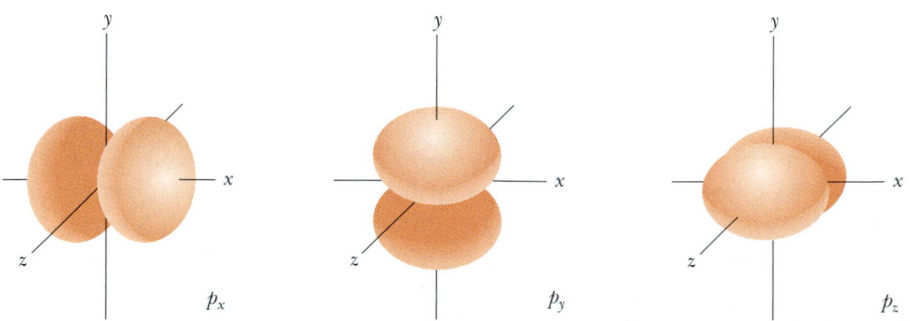

Figure 42.9   Angular dependence of the charge distribution for an electron in a p state. The three charge distributions p_x, p_y, and p_z have the same structure and differ only in their orientations in space.

sented in Figure 42.9, where only the angular dependence of these functions is shown. Note that the three clouds have identical structure but differ in their orientations with respect to the x, y, and z axes. The nonspherical wave functions for these states are

$$\psi_{2p_x} = xF(r)$$
$$\psi_{2p_y} = yF(r) \qquad \textbf{(42.8)}$$
$$\psi_{2p_z} = zF(r)$$

Wave functions for the $2p$ state

where $F(r)$ is some exponential function of r. Wave functions with a highly directional character such as these are convenient for describing chemical bonding, the formation of molecules, and chemical properties.

42.5 THE OTHER QUANTUM NUMBERS

The energy of a particular state in the hydrogen atom depends on the principal quantum number n. Now let us see what the other three quantum numbers contribute to our atomic model.

The Orbital Quantum Number ℓ

If a particle moves in a circle of radius r, the magnitude of its angular momentum relative to the center of the circle is $L = mvr$. The direction of **L** is perpendicular to the plane of the circle and is given by a right-hand rule.[2] According to classical physics, L can have any value. However, the Bohr model of hydrogen postulates that the magnitude of the angular momentum of the electron is restricted to multiples of $\hbar$; that is, $mvr = n\hbar$. This model must be modified because it predicts (incorrectly) that the ground state of hydrogen ($n = 1$) has one unit of angular momentum. Furthermore, if L is taken to be zero in the Bohr model, we are forced to accept a picture of the electron as a particle oscillating along a straight line through the nucleus, a physically unacceptable situation.

These difficulties are resolved with the quantum-mechanical model of the atom. According to quantum mechanics, an atom in a state whose principal quantum number is n can take on the following *discrete* values of the magnitude of the orbital angular momentum:

$$L = \sqrt{\ell(\ell + 1)}\,\hbar \qquad \ell = 0, 1, 2, \ldots, n - 1 \qquad \textbf{(42.9)}$$

Allowed values of L

Because ℓ is restricted to the values $\ell = 0, 1, 2, \ldots, n - 1$, we see that $L = 0$ (corresponding to $\ell = 0$) is an acceptable value of the magnitude of the angular momentum. The fact that L can be zero in this model serves to point out the inherent difficulties in any attempt to describe results based on quantum mechanics in terms of a purely particle-like (classical) model. In the quantum-mechanical interpretation, the electron cloud for the $L = 0$ state is spherically symmetric and has no fundamental axis of revolution.

---

[2] See Sections 11.3 and 11.4 for details on angular momentum and a review of this material.

EXAMPLE 42.5   Calculating *L* for a *p* State

Calculate the magnitude of the orbital angular momentum of an electron in a *p* state of hydrogen.

Solution   Because we know that $\hbar = 1.054 \times 10^{-34}$ J·s, we can use Equation 42.9 to calculate *L*. With $\ell = 1$ for a *p* state, we have

$$L = \sqrt{1(1 + 1)}\,\hbar = \sqrt{2}\hbar = \boxed{1.49 \times 10^{-34}\ \text{J·s}}$$

This number is extremely small relative to, say, the orbital angular momentum of the Earth orbiting the Sun, which is approximately 2.7×10^{40} J·s. The quantum number that describes *L* for macroscopic objects, such as the Earth, is so large that the separation between adjacent states cannot be measured. Once again, the correspondence principle is upheld.

The Orbital Magnetic Quantum Number m_ℓ

Because angular momentum is a vector, its direction must be specified. Recall from Chapter 30 that an orbiting electron can be considered an effective current loop with a corresponding magnetic moment. Such a moment placed in a magnetic field **B** interacts with the field. Suppose a weak magnetic field applied along the *z* axis defines a direction in space. According to quantum mechanics, L^2 and L_z (the projection of **L** along the *z* axis) can have only discrete values. The orbital magnetic quantum number m_ℓ specifies the allowed values of the *z* component of the orbital angular momentum according to the expression

Allowed values of L_z

$$L_z = m_\ell \hbar \tag{42.10}$$

Space quantization

The quantization of the direction of **L** with respect to an external magnetic field is often referred to as **space quantization.**

Let us look at the possible orientations of **L** for a given value of ℓ. Recall that m_ℓ can have values ranging from $-\ell$ to ℓ. If $\ell = 0$, then $m_\ell = 0$ and $L_z = 0$. If $\ell = 1$, then the possible values of m_ℓ are -1, 0, and 1; hence, L_z may be $-\hbar$, 0, or $\hbar$. If $\ell = 2$, then m_ℓ can be -2, -1, 0, 1, or 2, corresponding to L_z values of $-2\hbar$, $-\hbar$, 0, $\hbar$, $2\hbar$, and so on.

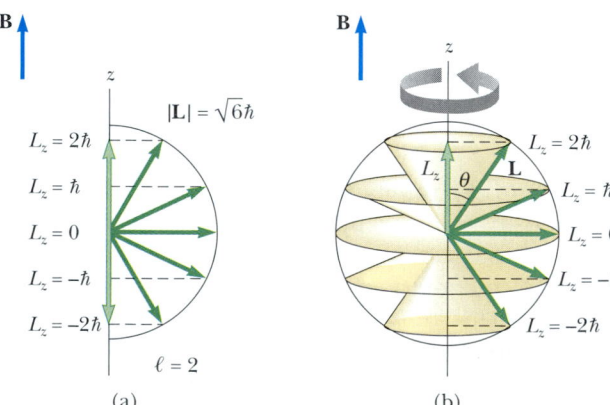

(a)   (b)

Figure 42.10   (a) The allowed projections of the orbital angular momentum **L** for the case $\ell = 2$. (b) The orbital angular momentum vector **L** lies on the surface of a cone and precesses about the *z* axis when a magnetic field **B** is applied in the *z* direction.

Figure 42.10a shows a vector model that describes space quantization for $\ell = 2$. Note that **L** can never be aligned parallel or antiparallel to **B** because L_z must be less than the total angular momentum L. For L_z to be zero, **L** must be perpendicular to **B**. From a three dimensional viewpoint, **L** must lie on the surface of a cone that makes an angle θ with the z axis, as shown in Figure 42.10b. From the figure, we see that θ is also quantized and that its values are specified through the relationship

$$\cos\theta = \frac{L_z}{|\mathbf{L}|} = \frac{m_\ell}{\sqrt{\ell(\ell+1)}} \qquad (42.11)$$

Note that m_ℓ is never greater than ℓ, and therefore θ can never be zero. (Classically, θ can have any value.)

Because of the uncertainty principle, **L** does not point in a specific direction. We can imagine it to trace out a cone in space. If **L** were known exactly, then all three components L_x, L_y, and L_z would be specified. For the moment, let us assume that this is the case and suppose that the electron moves in the xy plane, so **L** is in the z direction and the z component of its linear momentum $p_z = 0$. This means that p_z is precisely known, which is in violation of the uncertainty principle, $\Delta p_z \Delta z \geq \hbar/2$. In reality, only the magnitude of **L** and one component (say, L_z) can have definite values. In other words, quantum mechanics allows us to specify L and L_z but not L_x and L_y. Because the direction of **L** is constantly changing as we imagine it to precess about the z axis, the average values of L_x and L_y are zero and L_z maintains a fixed value of $m_\ell \hbar$.

The additional energy levels provided by the orbital magnetic quantum number explain the *Zeeman effect,* in which spectral lines are observed to split when a magnetic field is present, as shown in Figure 42.11.

 θ is quantized

QuickLab

Spin a top or a gyroscope rapidly, and watch as its axis of rotation slowly precesses about a vertical line passing through the point of support. This models the precession of the angular-momentum vector as shown in Figure 42.10b. In the photograph, Wolfgang Pauli and Niels Bohr are seeing this for themselves. *(Courtesy of AIP Niels Bohr Library, Margarethe Bohr Collection)*

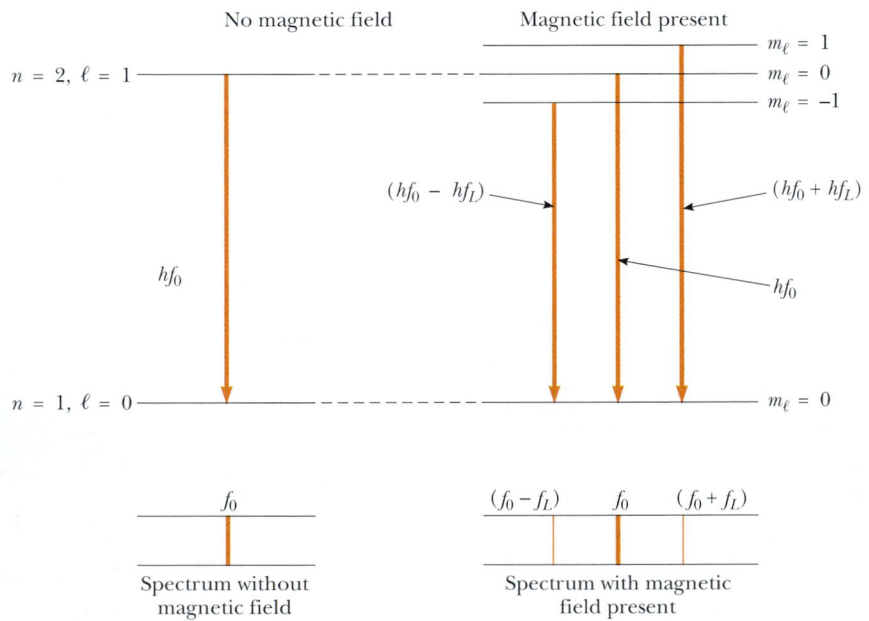

Figure 42.11 Split energy levels for the ground and first excited states of a hydrogen atom immersed in a magnetic field **B**. An atom in one of the excited states decays to the ground state with the emission of a photon, giving rise to emission lines at f_0, $f_0 + f_L$, and $f_0 - f_L$. This is the Zeeman effect. When **B** = 0, only the line at f_0 is observed.

EXAMPLE 42.6   Space Quantization for Hydrogen

Consider the hydrogen atom in the $\ell = 3$ state. Calculate the magnitude of **L** and the allowed values of L_z and θ.

Solution   We can calculate the magnitude of the orbital angular momentum using Equation 42.9:

$$L = \sqrt{\ell(\ell + 1)}\,\hbar = \sqrt{3(3 + 1)}\,\hbar = \boxed{2\sqrt{3}\,\hbar}$$

The allowed values of L_z can be calculated using $L_z = m_\ell \hbar$ with $m_\ell = -3, -2, -1, 0, 1, 2,$ and 3:

$$L_z = -3\hbar, -2\hbar, -\hbar, 0, \hbar, 2\hbar, \text{ and } 3\hbar$$

Finally, we calculate the allowed values of θ using Equation 42.11:

$$\cos \theta = \frac{m_\ell}{2\sqrt{3}}$$

Substituting the allowed values of m_ℓ gives

$$\cos \theta = \pm 0.866, \pm 0.577, \pm 0.289, \text{ and } 0$$

$$\theta = 30.0°, 54.8°, 73.2°, 90.0°, 107°, 125°, \text{ and } 150°$$

Quick Quiz 42.2

Make two drawings like Figure 42.10a, one for $\ell = 1$ and the other for $\ell = 3$.

The Spin Magnetic Quantum Number m_s

In 1921, Otto Stern (1888–1969) and Walter Gerlach (1889–1979) performed an experiment that demonstrated space quantization. However, their results were not in quantitative agreement with the theory that existed at that time. In their experiment, a beam of silver atoms sent through a nonuniform magnetic field was split into two components (Fig. 42.12). They repeated the experiment using other atoms, and in each case the beam split into two or more components. The classical argument is as follows: If the z direction is chosen to be the direction of the maximum nonuniformity of **B**, the net magnetic force on the atoms is along the z axis and is proportional to the component of the magnetic moment μ of the atom in the z direction. (See Quick Quiz 29.4 for a review of the cause of this.) Classically, μ can have any orientation, so the deflected beam should be spread out continuously. According to quantum mechanics, however, the deflected beam has several

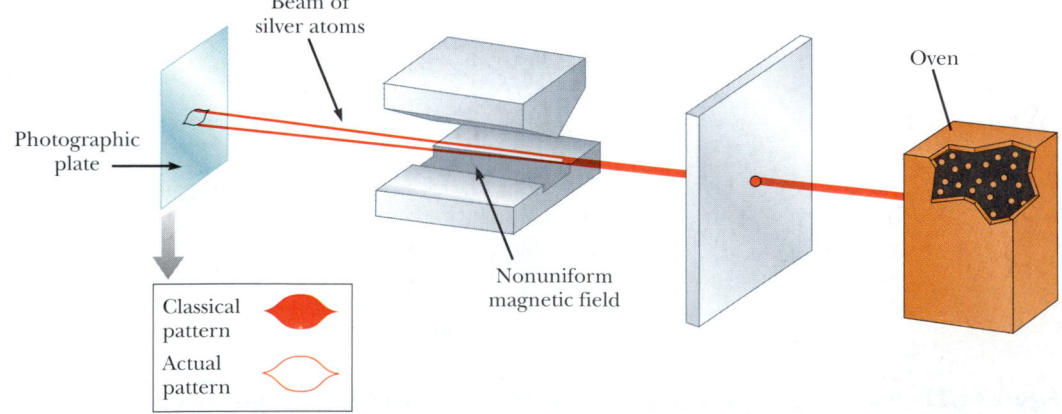

Figure 42.12   The apparatus used by Stern and Gerlach to verify space quantization. A beam of silver atoms is split in two by a nonuniform magnetic field.

components, and the number of components determines the number of possible values of μ_z. Hence, because the Stern–Gerlach experiment showed split beams, space quantization was at least qualitatively verified.

For the moment, let us assume that the magnetic moment μ of the atom is due to the orbital angular momentum. Because μ_z is proportional to m_ℓ, the number of possible values of μ_z is $2\ell + 1$. Furthermore, because ℓ is an integer, the number of values of μ_z is always odd. This prediction is clearly not consistent with Stern and Gerlach's observation of two components (an *even* number) in the deflected beam of silver atoms. Hence, we are forced to conclude that either quantum mechanics is incorrect or the model is in need of refinement.

In 1927, Phipps and Taylor repeated the Stern–Gerlach experiment using a beam of hydrogen atoms. That experiment was important because it dealt with an atom containing a single electron in its ground state, for which the quantum theory makes reliable predictions. Recall that $\ell = 0$ for hydrogen in its ground state, so $m_\ell = 0$. Hence, we would not expect the beam to be deflected by the field because the magnetic moment μ of the atom is zero. However, the beam in the Phipps–Taylor experiment was again split into two components. On the basis of that result, we can come to only one conclusion: Something other than the orbital motion is contributing to the magnetic moment.

As we learned earlier, Goudsmit and Uhlenbeck had proposed that the electron has an intrinsic angular momentum apart from its orbital angular momentum. From a classical viewpoint, this intrinsic angular momentum is attributed to the spinning of the charged electron about its own axis and hence is called electron spin.[3] In other words, the total angular momentum of the electron in a particular electronic state contains both an orbital contribution $\mathbf{L}$ and a spin contribution $\mathbf{S}$. The Phipps–Taylor result confirmed the hypothesis of Goudsmit and Uhlenbeck.

Quick Quiz 42.3

Explain why classical theory predicts the result labeled "classical pattern" in Figure 42.12.

In 1929, Dirac used the relativistic form of the total energy of a system to solve the relativistic wave equation for the electron in a potential well. His analysis confirmed the fundamental nature of electron spin. (Spin, like mass and charge, is an intrinsic property of a particle, independent of its surroundings.) Furthermore, the analysis showed that electron spin can be described by a single quantum number s, whose value can be only $\frac{1}{2}$. The spin angular momentum of the electron *never changes*. This notion contradicts classical laws, which dictate that a rotating charge slows down in the presence of an applied magnetic field because of the Faraday emf that accompanies the changing field. Furthermore, if the electron were viewed as a spinning ball of charge subject to classical laws, parts of it near its surface would be rotating with speeds exceeding the speed of light. Thus, the classical picture must not be pressed too far; ultimately, the spinning electron is a quantum entity defying any simple classical description.

The magnitude of the **spin angular momentum S** for the electron is

$$S = \sqrt{s(s + 1)}\,\hbar = \frac{\sqrt{3}}{2}\hbar \qquad (42.12)$$

Spin angular momentum of an electron

---

[3] Physicists often use the word *spin* when referring to spin angular momentum. For example, it is common to make the statement "The electron has a spin of $\frac{1}{2}$."

Like orbital angular momentum **L**, spin angular momentum **S** is quantized in space, as described in Figure 42.13. It can have two orientations relative to an external magnetic field, specified by the spin magnetic quantum number $m_s = \pm\frac{1}{2}$. The z component of spin angular momentum is

$$S_z = m_s\hbar = \pm\frac{1}{2}\hbar \tag{42.13}$$

The two values $\pm\hbar/2$ for S_z correspond to the two possible orientations for **S** shown in Figure 42.13. The value $m_s = +\frac{1}{2}$ refers to the spin-up case, and the value $m_s = -\frac{1}{2}$ refers to the spin-down case.

The spin magnetic moment $\boldsymbol{\mu}_{\text{spin}}$ of the electron is related to its spin angular momentum **S** by the expression

$$\boldsymbol{\mu}_{\text{spin}} = -\frac{e}{m_e}\mathbf{S} \tag{42.14}$$

where e is the electronic charge and m_e is the mass of the electron. Because $S_z = \pm\frac{1}{2}\hbar$, the z component of the spin magnetic moment can have the values

$$\mu_{\text{spin},\,z} = \pm\frac{e\hbar}{2m_e} \tag{42.15}$$

As we learned in Section 30.8, the quantity $e\hbar/2m_e$ is the Bohr magneton $\mu_{\text{B}} = 9.27 \times 10^{-24}\,\text{J/T}$. Note that the ratio of magnetic moment to angular momentum is twice as great for spin angular momentum (Eq. 42.14) as it is for orbital angular momentum (Eq. 30.25). The factor of 2 is explained in a relativistic treatment first carried out by Dirac.

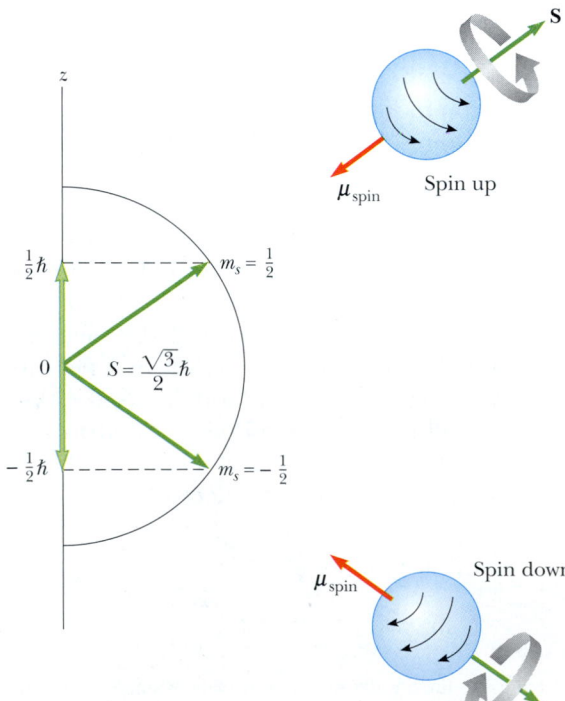

Figure 42.13 Spin angular momentum **S** exhibits space quantization. This figure shows the two allowed orientations of the spin angular momentum vector **S** and the spin magnetic moment $\boldsymbol{\mu}_{\text{spin}}$ for a spin $\frac{1}{2}$ particle, such as the electron.

Today physicists explain the Stern–Gerlach experiment as follows. The observed magnetic moments for both silver and hydrogen are due to spin angular momentum only, with no contribution from orbital angular momentum. A single-electron atom such as hydrogen has its electron spin quantized in the magnetic field in such a way that the z component of spin angular momentum is either $\frac{1}{2}\hbar$ or $-\frac{1}{2}\hbar$, corresponding to $m_s = \pm\frac{1}{2}$. Electrons with spin $+\frac{1}{2}$ are deflected downward, and those with spin $-\frac{1}{2}$ are deflected upward.

The Stern–Gerlach experiment provided two important results. First, it verified the concept of space quantization. Second, it showed that spin angular momentum exists—even though this property was not recognized until four years after the experiments were performed.

42.6 THE EXCLUSION PRINCIPLE AND THE PERIODIC TABLE

Earlier we found that the state of a hydrogen atom is specified by four quantum numbers: n, ℓ, m_ℓ, and m_s. As it turns out, the state of an electron in any other atom may also be specified by this same set of quantum numbers. In fact, these four quantum numbers can be used to describe all the electronic states of an atom regardless of the number of electrons in its structure.

An obvious question that arises here is: "How many electrons can be in a particular quantum state?" Pauli answered this important question in 1925, in a statement known as the **exclusion principle:**

> No two electrons in the same atom can ever be in the same quantum state; therefore, no two electrons in the same atom can have the same set of quantum numbers.

Exclusion principle

If this principle were not valid, every electron in an atom would end up in the lowest possible energy state of the atom, and the chemical behavior of the elements would be grossly modified. Nature as we know it would not exist!

In reality, we can view the electronic structure of complex atoms as a succession of filled levels increasing in energy. As a general rule, the order of filling of an atom's subshells is as follows. Once a subshell is filled, the next electron goes into the lowest-energy vacant subshell. We can understand this behavior by recognizing that if the atom were not in the lowest energy state available to it, it would radiate energy until it reached this state.

Before we discuss the electronic configuration of various elements, it is convenient to define an *orbital* as the state of an electron characterized by the quantum numbers n, ℓ, and m_ℓ. From the exclusion principle, we see that **only two electrons can be present in any orbital.** One of these electrons has a spin magnetic quantum number $m_s = +\frac{1}{2}$, and the other has $m_s = -\frac{1}{2}$. Because each orbital is limited to two electrons, the number of electrons that can occupy the various shells is also limited.

Table 42.3 shows the number of allowed quantum states for an atom for which $n = 3$. The arrows pointing upward indicate an atom in which the electron is described by $m_s = \frac{1}{2}$, and those pointing downward indicate that $m_s = -\frac{1}{2}$. The $n = 1$ shell can accommodate only two electrons because $m_\ell = 0$ means that only one orbital is allowed. (The three quantum numbers describing this orbital are $n = 1$, $\ell = 0$, and $m_\ell = 0$.) The $n = 2$ shell has two subshells, one for $\ell = 0$ and

TABLE 42.3   **Allowed Quantum States for an Atom Having $n = 3$**

n	1	2				3								
ℓ	0	0	1			0	1			2				
m_ℓ	0	0	1	0	-1	0	1	0	-1	2	1	0	-1	-2
m_s	↑↓	↑↓	↑↓	↑↓	↑↓	↑↓	↑↓	↑↓	↑↓	↑↓	↑↓	↑↓	↑↓	↑↓

Wolfgang Pauli Austrian theoretical physicist (1900–1958) An extremely talented theoretician who made important contributions in many areas of modern physics, Pauli gained public recognition at the age of 21 with a masterful review article on relativity that is still considered one of the finest and most comprehensive introductions to the subject. His other major contributions were the discovery of the exclusion principle, the explanation of the connection between particle spin and statistics, and theories of relativistic quantum electrodynamics, the neutrino hypothesis, and the hypothesis of nuclear spin.
(CERN, courtesy of AIP Emilio Segre Visual Archive)

one for $\ell = 1$. The $\ell = 0$ subshell is limited to two electrons because $m_\ell = 0$. The $\ell = 1$ subshell has three allowed orbitals, corresponding to $m_\ell = 1$, 0, and -1. Because each orbital can accommodate two electrons, the $\ell = 1$ subshell can hold six electrons. Thus, the $n = 2$ shell can contain eight electrons. The $n = 3$ shell has three subshells ($\ell = 0, 1, 2$) and nine orbitals and can accommodate up to 18 electrons. In general, each shell can accommodate up to $2n^2$ electrons.

The exclusion principle can be illustrated by an examination of the electronic arrangement in a few of the lighter atoms. First, recall from Section 1.2 that the atomic number Z of any element is the number of protons in the nucleus of an atom of that element. Hydrogen ($Z = 1$) has only one electron—which, in the ground state of the atom, can be described by either of two sets of quantum numbers: 1, 0, 0, $\frac{1}{2}$ or 1, 0, 0, $-\frac{1}{2}$. This electronic configuration is often written $1s^1$. The notation $1s$ refers to a state for which $n = 1$ and $\ell = 0$, and the superscript indicates that one electron is present in the s subshell.

Neutral helium ($Z = 2$) has two electrons. In the ground state, their quantum numbers are 1, 0, 0, $\frac{1}{2}$, and 1, 0, 0, $-\frac{1}{2}$. No other possible combinations of quantum numbers exist for this level, and we say that the K shell is filled. This electronic configuration is written $1s^2$.

Neutral lithium ($Z = 3$) has three electrons. In the ground state, two of these are in the $1s$ subshell. The third is in the $2s$ subshell because this subshell is slightly lower in energy than the $2p$ subshell.[4] Hence, the electronic configuration for lithium is $1s^2 2s^1$.

The electronic configurations of lithium and the next several elements are provided in Figure 42.14. The electronic configuration of beryllium ($Z = 4$), with its four electrons, is $1s^2 2s^2$, and boron ($Z = 5$) has a configuration of $1s^2 2s^2 2p^1$. The $2p$ electron in boron may be described by any of six equally probable sets of quantum numbers. In Figure 42.14, we show this electron in the leftmost $2p$ box with spin up, but it is equally likely to be in any $2p$ box with spin either up or down.

Carbon ($Z = 6$) has six electrons, giving rise to a question concerning how to assign the two $2p$ electrons. Do they go into the same orbital with paired spins (↑ ↓), or do they occupy different orbitals with unpaired spins (↑ ↑)? Experimental data show that the most stable configuration (that is, the one that is energetically preferred) is the latter, in which the spins are unpaired. Hence, the two $2p$ electrons in carbon and the three $2p$ electrons in nitrogen ($Z = 7$) have unpaired spins, as Figure 42.14 shows. The general rule that governs such situations, called **Hund's rule,** states that

Hund's rule

> when an atom has orbitals of equal energy, the order in which they are filled by electrons is such that a maximum number of electrons have unpaired spins.

---

[4] To a first approximation, energy depends only on the quantum number n, as we have discussed. Because of the effect of the electronic charge shielding the nuclear charge in multielectron atoms, however, energy depends on ℓ also. We shall discuss these shielding effects in Section 42.7.

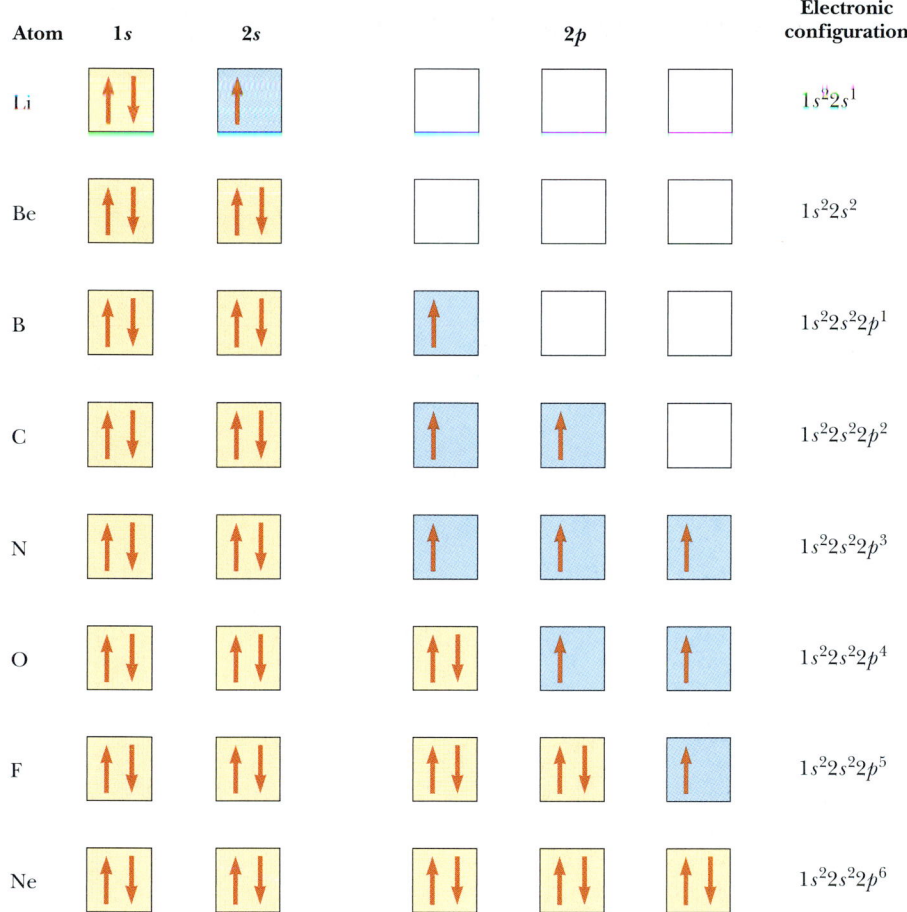

Atom	1s	2s	2p			Electronic configuration
Li	↑↓	↑				$1s^2 2s^1$
Be	↑↓	↑↓				$1s^2 2s^2$
B	↑↓	↑↓	↑			$1s^2 2s^2 2p^1$
C	↑↓	↑↓	↑	↑		$1s^2 2s^2 2p^2$
N	↑↓	↑↓	↑	↑	↑	$1s^2 2s^2 2p^3$
O	↑↓	↑↓	↑↓	↑	↑	$1s^2 2s^2 2p^4$
F	↑↓	↑↓	↑↓	↑↓	↑	$1s^2 2s^2 2p^5$
Ne	↑↓	↑↓	↑↓	↑↓	↑↓	$1s^2 2s^2 2p^6$

Figure 42.14 The filling of electronic states must obey both the exclusion principle and Hund's rule.

Some exceptions to this rule occur in elements having subshells that are close to being filled or half-filled.

A complete list of electronic configurations is provided in Table 42.4. In 1871, without any understanding of quantum mechanics that we now have, the Russian chemist Dmitri Mendeleev (1834–1907) made an early attempt at finding some order among the elements. He was trying to organize the elements for the table of contents of a book he was writing. He arranged the atoms in a table similar to that shown in Appendix C, according to their atomic masses and chemical similarities. The first table Mendeleev proposed contained many blank spaces, and he boldly stated that the gaps were there only because the elements had not yet been discovered. By noting the columns in which some missing elements should be located, he was able to make rough predictions about their chemical properties. Within 20 years of this announcement, most of these elements were indeed discovered.

The elements in the **periodic table** are arranged so that all those in a column have similar chemical properties. For example, consider the elements in the last column, which are all gases at room temperature: He (helium), Ne (neon), Ar (argon), Kr (krypton), Xe (xenon), and Rn (radon). The outstanding characteristic of all these elements is that they do not normally take part in chemical reactions—

TABLE 42.4   Electronic Configuration of the Elements

Atomic Number Z	Symbol	Ground-State Configuration	Ionization Energy (eV)
1	H	$1s^1$	13.595
2	He	$1s^2$	24.581
3	Li	[He] $2s^1$	5.39
4	Be	$2s^2$	9.320
5	B	$2s^2 2p^1$	8.296
6	C	$2s^2 2p^2$	11.256
7	N	$2s^2 2p^3$	14.545
8	O	$2s^2 2p^4$	13.614
9	F	$2s^2 2p^5$	17.418
10	Ne	$2s^2 2p^6$	21.559
11	Na	[Ne] $3s^1$	5.138
12	Mg	$3s^2$	7.644
13	Al	$3s^2 3p^1$	5.984
14	Si	$3s^2 3p^2$	8.149
15	P	$3s^2 3p^3$	10.484
16	S	$3s^2 3p^4$	10.357
17	Cl	$3s^2 3p^5$	13.01
18	Ar	$3s^2 3p^6$	15.755
19	K	[Ar] $4s^1$	4.339
20	Ca	$4s^2$	6.111
21	Sc	$3d^1 4s^2$	6.54
22	Ti	$3d^2 4s^2$	6.83
23	V	$3d^3 4s^2$	6.74
24	Cr	$3d^5 4s^1$	6.76
25	Mn	$3d^5 4s^2$	7.432
26	Fe	$3d^6 4s^2$	7.87
27	Co	$3d^7 4s^2$	7.86
28	Ni	$3d^8 4s^2$	7.633
29	Cu	$3d^{10} 4s^1$	7.724
30	Zn	$3d^{10} 4s^2$	9.391
31	Ga	$3d^{10} 4s^2 4p^1$	6.00
32	Ge	$3d^{10} 4s^2 4p^2$	7.88
33	As	$3d^{10} 4s^2 4p^3$	9.81
34	Se	$3d^{10} 4s^2 4p^4$	9.75
35	Br	$3d^{10} 4s^2 4p^5$	11.84
36	Kr	$3d^{10} 4s^2 4p^6$	13.996
37	Rb	[Kr] $5s^1$	4.176
38	Sr	$5s^2$	5.692
39	Y	$4d^1 5s^2$	6.377
40	Zr	$4d^2 5s^2$	
41	Nb	$4d^4 5s^1$	6.881
42	Mo	$4d^5 5s^1$	7.10
43	Tc	$4d^5 5s^2$	7.228
44	Ru	$4d^7 5s^1$	7.365

Note: The bracket notation is used as a shorthand method to avoid repetition in indicating inner-shell electrons. Thus, [He] represents $1s^2$, [Ne] represents $1s^2 2s^2 2p^6$, [Ar] represents $1s^2 2s^2 2p^6 3s^2 3p^6$, and so on.

TABLE 42.4   *Continued*

Atomic Number Z	Symbol	Ground-State Configuration	Ionization Energy (eV)
45	Rh	$4d^85s^1$	7.461
46	Pd	$4d^{10}$	8.33
47	Ag	$4d^{10}5s^1$	7.574
48	Cd	$4d^{10}5s^2$	8.991
49	In	$4d^{10}5s^25p^1$	
50	Sn	$4d^{10}5s^25p^2$	7.342
51	Sb	$4d^{10}5s^25p^3$	8.639
52	Te	$4d^{10}5s^25p^4$	9.01
53	I	$4d^{10}5s^25p^5$	10.454
54	Xe	$4d^{10}5s^25p^6$	12.127
55	Cs	[Xe] $6s^1$	3.893
56	Ba	$6s^2$	5.210
57	La	$5d^16s^2$	5.61
58	Ce	$4f^15d^16s^2$	6.54
59	Pr	$4f^36s^2$	5.48
60	Nd	$4f^46s^2$	5.51
61	Pm	$4f^56s^2$	
62	Fm	$4f^66s^2$	5.6
63	Eu	$4f^76s^2$	5.67
64	Gd	$4f^75d^16s^2$	6.16
65	Tb	$4f^96s^2$	6.74
66	Dy	$4f^{10}6s^2$	
67	Ho	$4f^{11}6s^2$	
68	Er	$4f^{12}6s^2$	
69	Tm	$4f^{13}6s^2$	
70	Yb	$4f^{14}6s^2$	6.22
71	Lu	$4f^{14}5d^16s^2$	6.15
72	Hf	$4f^{14}5d^26s^2$	7.0
73	Ta	$4f^{14}5d^36s^2$	7.88
74	W	$4f^{14}5d^46s^2$	7.98
75	Re	$4f^{14}5d^56s^2$	7.87
76	Os	$4f^{14}5d^66s^2$	8.7
77	Ir	$4f^{14}5d^76s^2$	9.2
78	Pt	$4f^{14}5d^86s^2$	8.88
79	Au	[Xe, $4f^{14}5d^{10}$] $6s^1$	9.22
80	Hg	$6s^2$	10.434
81	Tl	$6s^26p^1$	6.106
82	Pb	$6s^26p^2$	7.415
83	Bi	$6s^26p^3$	7.287
84	Po	$6s^26p^4$	8.43
85	At	$6s^26p^5$	
86	Rn	$6s^26p^6$	10.745
87	Fr	[Rn] $7s^1$	
88	Ra	$7s^2$	5.277
89	Ac	$6d^17s^2$	6.9
90	Th	$6d^27s^2$	
91	Pa	$5f^26d^17s^2$	

continued

TABLE 42.4   Continued

Atomic Number Z	Symbol	Ground-State Configuration	Ionization Energy (eV)
92	U	$5f^3 6d^1 7s^2$	4.0
93	Np	$5f^4 6d^1 7s^2$	
94	Pu	$5f^6 7s^2$	
95	Am	$5f^7 7s^2$	
96	Cm	$5f^7 6d^1 7s^2$	
97	Bk	$5f^8 6d^1 7s^2$	
98	Cf	$5f^{10} 6d^0 7s^2$	
99	Es	$5f^{11} 6d^0 7s^2$	
100	Fm	$5f^{12} 6d^0 7s^2$	
101	Md	$5f^{13} 6d^0 7s^2$	
102	No	$5f^{14} 6d^0 7s^2$	
103	Lr	$5f^{14} 6d^1 7s^2$	
104	Rf	$5f^{14} 6d^2 7s^2$	

that is, they do not join with other atoms to form molecules. They are therefore called *inert gases.*

We can partially understand this behavior by looking at the electronic configurations in Table 42.4. The chemical behavior of an element depends on the outermost shell that contains electrons. Shells inside the outermost one are filled and do not contribute to chemical behavior. The electronic configuration for helium is $1s^2$—the $n = 1$ shell (which is the outermost shell because it is the only shell) is filled. Additionally, the energy of the atom in this configuration is considerably lower than the energy for the configuration in which an electron is in the next available level, the $2s$ subshell. Next, look at the electronic configuration for neon, $1s^2 2s^2 2p^6$. Again, the outermost shell ($n = 2$ in this case) is filled, and a wide gap in energy occurs between the filled $2p$ subshell and the next available one, the $3s$ subshell. Argon has the configuration $1s^2 2s^2 2p^6 3s^2 3p^6$. Here, it is only the $3p$ subshell that is filled, but again a wide gap in energy occurs between the filled $3p$ subshell and the next available one, the $3d$ subshell. We could continue this procedure through all the inert gases; the pattern remains the same. An inert gas is formed when either a shell or a subshell is filled and a large gap in energy occurs between the filled shell or subshell and the next highest available one.

42.7 ATOMIC SPECTRA

In Chapter 40 we briefly discussed the origin of the visible spectral lines for the hydrogen atom and for hydrogen-like ions. Recall that an atom emits electromagnetic radiation if the atom in an excited state makes a transition to a lower energy state. The set of wavelengths that is observed when a specific atom undergoes such processes is called the **emission spectrum** for that atom. Likewise, atoms having electrons in the ground-state configuration can absorb electromagnetic radiation at specific wavelengths, giving rise to an **absorption spectrum.** Such spectra can be used to identify elements.

The energy level diagram for hydrogen is shown in Figure 42.15. The various diagonal lines represent allowed transitions between stationary states. Whenever an atom makes a transition from a higher energy state to a lower one, a photon of

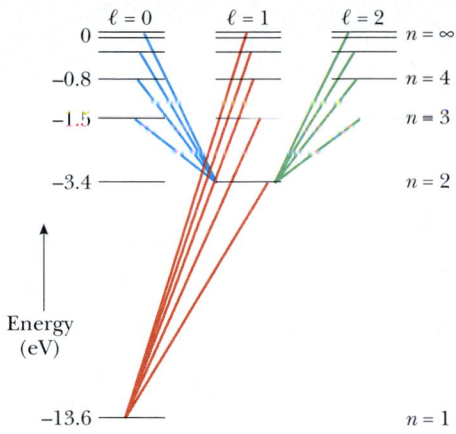

Figure 42.15 Some allowed electronic transitions for hydrogen, represented by the colored lines. These transitions must obey the selection rule $\Delta \ell = \pm 1$.

light is emitted. The frequency of this photon is $f = \Delta E / h$, where ΔE is the energy difference between the two states and h is Planck's constant. The **selection rules** for the *allowed transitions* are

$$\Delta \ell = \pm 1 \qquad \text{and} \qquad \Delta m_\ell = 0, \pm 1 \qquad \textbf{(42.16)}$$

Selection rules for allowed atomic transitions

Transitions that do not obey these selection rules are said to be *forbidden*. (Such transitions can occur, but their probability is low relative to the probability of allowed transitions.)

Because the orbital angular momentum of an atom changes when a photon is emitted or absorbed (that is, as a result of a transition between states) and because angular momentum must be conserved, we conclude that **the photon involved in the process must carry angular momentum.** In fact, the photon has an angular momentum equivalent to that of a particle having a spin of 1. Hence, a photon has energy, linear momentum, and angular momentum.

The photon carries angular momentum

Recall from Equation 40.32 that the allowed energies for one-electron atoms, such as hydrogen and He$^+$, are

$$E_n = -\frac{k_e e^2}{2a_0} \left(\frac{Z^2}{n^2} \right) = -\frac{13.6 Z^2}{n^2} \text{ eV} \qquad \textbf{(42.17)}$$

Allowed energies for one-electron atoms

For multielectron atoms, the positive nuclear charge Ze is largely shielded by the negative charge of the inner-shell electrons. Hence, the outer electrons interact with a net charge that is much smaller than the nuclear charge. The expression for the allowed energies for multielectron atoms has the same form as Equation 42.17 with Z replaced by an effective atomic number Z_{eff}:

$$E_n = -\frac{13.6 Z_{\text{eff}}^2}{n^2} \text{ eV} \qquad \textbf{(42.18)}$$

Allowed energies for multielectron atoms

where Z_{eff} depends on n and ℓ.

It is interesting to plot ionization energy (see Section 40.5) versus atomic number Z, as in Figure 42.16. Note the pattern of $\Delta Z = 2, 8, 8, 18, 18, 32$ for the various peaks. This pattern follows from the exclusion principle and helps explain why the elements repeat their chemical properties in groups. For example, the peaks at $Z = 2, 10, 18,$ and 36 correspond to the elements helium, neon, argon, and krypton, which all have filled outermost shells. These elements have relatively high ionization energies and similar chemical behavior.

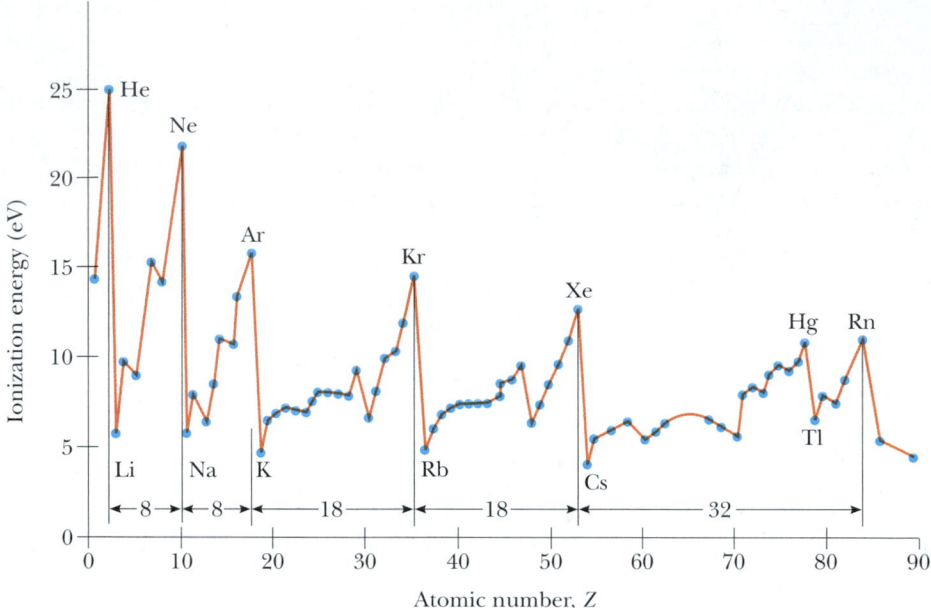

Figure 42.16   Ionization energy of the elements versus atomic number Z.   *(Adapted from J. Orear,* Physics, *New York, Macmillan, 1979.)*

X-Ray Spectra

X-rays are emitted when high-energy electrons or any other charged particles bombard a metal target. The x-ray spectrum typically consists of a broad continuous band containing a series of sharp lines, as shown in Figure 42.17. The continuous spectrum is the result of collisions between incoming electrons and atoms in the target. The kinetic energy lost by the electrons during the collisions emerges as the energy ($E = hf$) of the x-ray photons radiated from the target. The sharp lines superimposed on the continuous spectrum are known as **characteristic x-rays** because they are characteristic of the target material. They were discovered in 1908, but their origin remained unexplained until the details of atomic structure, particularly the shell structure of the atom, were discovered.

Characteristic x-ray emission occurs when a bombarding electron that collides with a target atom has sufficient energy to remove an inner-shell electron from the atom. The vacancy created in the shell is filled when an electron from a higher level drops down into it. This transition is accompanied by the emission of a photon whose energy equals the difference in energy between the two levels. Typically, the energy of such transitions is greater than 10 000 eV, and the emitted photons have wavelengths in the range of 0.001 nm to 0.1 nm, in the x-ray region of the electromagnetic spectrum.

Let us assume that the incoming electron has dislodged an atomic electron from the innermost shell—the K shell. If the vacancy is filled by an electron dropping from the next higher shell—the L shell—the photon emitted has an energy corresponding to the K_α characteristic x-ray line on the curve of Figure 42.17. If the vacancy is filled by an electron dropping from the M shell, the K_β line in Figure 42.17 is produced.

Other characteristic x-ray lines are formed when electrons drop from upper levels to vacancies other than those in the K shell. For example, L lines are produced when vacancies in the L shell are filled by electrons dropping from higher

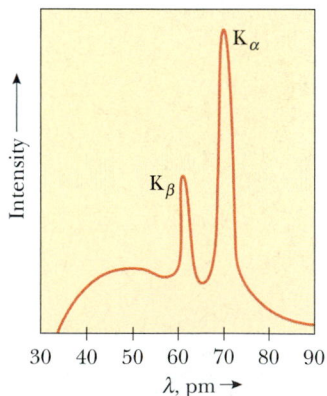

Figure 42.17   The x-ray spectrum of a metal target consists of a broad continuous spectrum containing a number of sharp lines; the lines are due to *characteristic x-rays*. The data shown were obtained when 37-keV electrons bombarded a molybdenum target.

shells. An L_α line is produced as an electron drops from the M shell to the L shell, and an L_β line is produced by a transition from the N shell to the L shell.

Although multielectron atoms cannot be analyzed exactly with either the Bohr model or the Schrödinger equation, we can apply our knowledge of Gauss's law from Chapter 24 to make some surprisingly accurate estimates of expected x-ray energies and wavelengths. Consider an atom of atomic number Z in which one of the two electrons in the K shell has been ejected. Imagine that we draw a gaussian sphere just inside the most probable radius of the L electrons. The electric field at the position of the L electrons is a combination of the fields created by the nucleus, the single K electron, the other L electrons, and the outer electrons. The wave functions of the outer electrons are such that the electrons have a very high probability of being farther from the nucleus than the L electrons are. Thus, they are much more likely to be outside the gaussian surface than inside and, on the average, do not contribute significantly to the electric field at the position of the L electrons. The effective charge inside the gaussian surface is the positive nuclear charge and one negative charge due to the single K electron. If we ignore the interactions between L electrons, a single L electron behaves as if it experiences an electric field due to a charge $(Z - 1)e$ enclosed by the gaussian surface. The nuclear charge is shielded by the electron in the K shell such that Z_{eff} in Equation 42.18 is $Z - 1$. For higher-level shells, the nuclear charge is shielded by electrons in all of the inner shells.

We can now use Equation 42.18 to estimate the energy associated with an electron in the L shell:

$$E_L = -(Z - 1)^2 \frac{13.6}{2^2} \text{ eV}$$

After the atom makes the transition, there are two electrons in the K shell. Using a similar argument for a gaussian surface drawn just inside the most probable radius for the single K electron, we can argue that the energy associated with one of these electrons is approximately that of a one-electron atom with the nuclear charge reduced by the negative charge of the other electron. Thus,

$$E_K = -(Z - 1)^2(13.6 \text{ eV}) \tag{42.19}$$

As Example 42.7 shows, the energy of the atom with an electron in an M shell can be estimated in a similar fashion. Taking the energy difference between the initial and final levels, we can then calculate the energy and wavelength of the emitted photon.

Quick Quiz 42.4

Note in Figure 42.17 that the continuous spectrum stops abruptly at the cutoff wavelength of about 34 pm. Why does a cutoff wavelength occur?

In 1914, Henry G. J. Moseley (1887–1915) plotted the Z values for a number of elements versus $\sqrt{1/\lambda}$, where λ is the wavelength of the K_α line of each element. He found that the plot is a straight line, as in Figure 42.18. This is consistent with rough calculations of the energy levels given by Equation 42.19. From this plot, Moseley determined the Z values of elements that had not yet been discovered and produced a periodic table in excellent agreement with the known chemical properties of the elements. Until that experiment, atomic numbers had been merely placeholders for the elements that appeared in the periodic table, the elements being ordered according to mass.

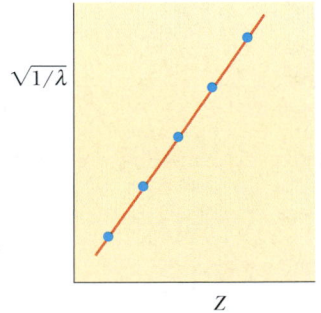

Figure 42.18 A Moseley plot of $\sqrt{1/\lambda}$ versus Z, where λ is the wavelength of the K_α x-ray line of the element of atomic number Z.

EXAMPLE 42.7 The Energy of an X-Ray

Determine the energy of the characteristic x-ray emitted from a tungsten (chemical symbol W) target when an electron drops from the M shell ($n = 3$) to a vacancy in the K shell ($n = 1$).

Solution The atomic number of tungsten is $Z = 74$. Using Equation 42.19, we see that the energy associated with the electron in the K shell is approximately

$$E_{\text{K}} = -(74 - 1)^2(13.6 \text{ eV}) = -72\,500 \text{ eV}$$

An electron in the M shell is subject to an effective nuclear charge that depends on the number of electrons in the $n = 1$ and $n = 2$ states because these electrons shield the M electrons from the nucleus. Because there are eight electrons in the $n = 2$ state and one remaining in the $n = 1$ state, roughly nine electrons shield M electrons from the nucleus, so $Z_{\text{eff}} = Z - 9$. Hence, the energy associated with an electron in the M shell is, from Equation 42.18,

$$E_{\text{M}} = -\frac{13.6 Z_{\text{eff}}^2}{3^2} \text{ eV} = -\frac{13.6(Z - 9)^2}{3^2} \text{ eV}$$

$$= -\frac{(13.6)(74 - 9)^2}{9} \text{ eV} = -6\,380 \text{ eV}$$

Therefore, the emitted x-ray has an energy equal to $E_{\text{M}} - E_{\text{K}} = -6\,380 \text{ eV} - (-72\,500 \text{ eV}) = 66\,100 \text{ eV}$. Despite the approximations we have made in developing Equations 42.18 and 42.19 and the estimation of the effective nuclear charge, this result is in excellent agreement with measurements made on x-rays from tungsten targets.

Exercise Calculate the wavelength of the emitted x-ray for this transition.

Answer 0.018 8 nm.

42.8 ATOMIC TRANSITIONS

We have seen that an atom absorbs and emits electromagnetic radiation only at frequencies that correspond to the energy separation between allowed states. Let us now look at the details of these processes. Consider an atom having the allowed energy levels labeled E_1, E_2, E_3, . . . in Figure 42.19. When radiation is incident on the atom, only those photons whose energy hf matches the energy separation ΔE between two energy levels can be absorbed by the atom. Figure 42.20 is a schematic diagram representing this process, which is called **stimulated absorption** because the photon stimulates the atom to make the upward transition. At ordinary temperatures, most of the atoms in a sample are in the ground state. If a vessel containing many atoms of a gaseous element is illuminated with radiation of all possible photon frequencies (that is, a continuous spectrum), only those photons having energy $E_2 - E_1$, $E_3 - E_1$, $E_4 - E_1$, $E_3 - E_2$, $E_4 - E_2$, and so on are absorbed by the atoms. As a result of this absorption, some of the atoms are raised to allowed higher energy levels, which, as we learned in Section 40.5, are called **excited states.**

Once an atom is in an excited state, some probability exists that the excited atom will jump back to a lower energy level and emit a photon in the process, as in Figure 42.21. This process is known as **spontaneous emission** because it happens randomly, without requiring an event to trigger the transition. Typically, an atom remains in an excited state for only about 10^{-8} s.

When an atom in an excited state returns to the ground state via two or more intermediate steps, the photons emitted during the process are lower in energy than the original photon absorbed by the atom. This process is called *fluorescence*. In a fluorescent light tube, electrons leaving a filament at the end of the tube collide with atoms of mercury vapor present in the tube, causing the mercury atoms to be elevated into excited states. As these atoms make transitions to lower states, they emit ultraviolet photons that strike a coating on the inner surface of the tube. The coating absorbs the photons and emits visible light by means of fluorescence.

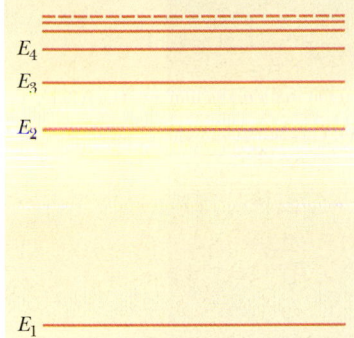

Figure 42.19 Energy level diagram of an atom having various allowed states. The lowest-energy state E_1 is the ground state. All others are excited states.

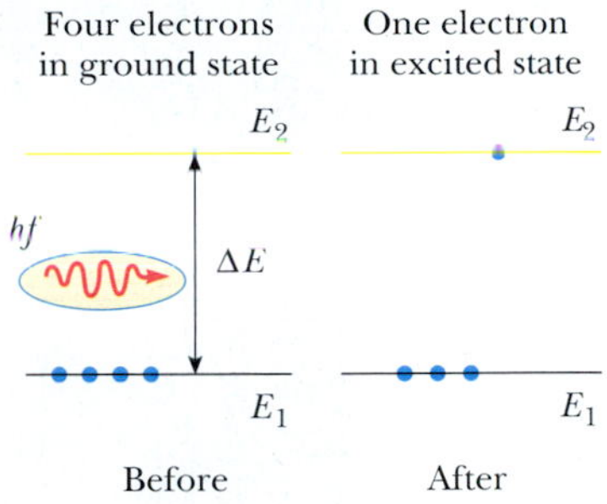

Figure 42.20 Stimulated absorption of a photon. The dots represent electrons. One electron is transferred from the ground state to the excited state when the atom absorbs a photon of energy $hf = E_2 - E_1$.

Phosphorescent materials glow because of a similar process, but the excited atoms may remain in an excited state for periods ranging from a few seconds to several hours. Eventually, the atoms drop to the ground state and while doing so emit visible light. For this reason, phosphorescent materials emit light long after being placed in the dark.

Quick Quiz 42.5

Make a drawing similar to Figure 42.21 for fluorescence.

In addition to spontaneous emission, **stimulated emission** occurs. Suppose an atom is in an excited state E_2, as in Figure 42.22. If the excited state is a *metastable state*—that is, if its lifetime is much longer than the typical 10^{-8} s life-

QuickLab

Place an object that glows in the dark in a drawer for a day. While the room is dark, open the drawer and observe the object. Does it glow? Now expose the object to light from an incandescent lightbulb. Turn the light off, and note the brightness of the glowing object in the dark. Next, expose the object to light from a fluorescent light tube (which emits some ultraviolet light). Turn the light off, and again note the brightness of the glow. Is the object now brighter or dimmer than it was after exposure to the incandescent bulb?

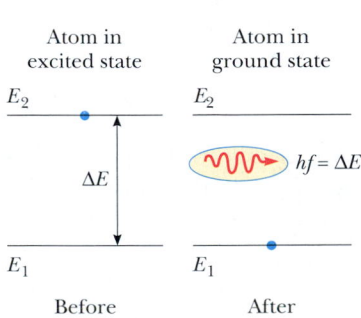

Figure 42.21 Spontaneous emission of a photon by an atom that is initially in the excited state E_2. When the atom falls to the ground state, it emits a photon of energy $hf = E_2 - E_1$.

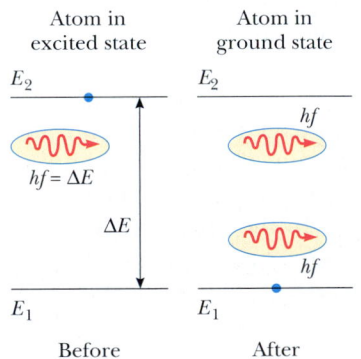

Figure 42.22 Stimulated emission of a photon by an incoming photon of energy hf. Initially, the atom is in the excited state. The incoming photon stimulates the atom to emit a second photon of energy $hf = E_2 - E_1$.

Figure 42.23   A single barium ion (the small dot in the center) glows because it is stimulated by a laser beam (not shown). The surrounding structure is the electromagnetic trap that holds the ion in place. *(Courtesy of David Wineland, National Institute of Standards and Technology)*

time of excited states—then the time interval until spontaneous emission occurs will be relatively long. Let us imagine that in that interval a photon of energy $hf = E_2 - E_1$ is incident on the atom. One possibility is that the photon energy will be sufficient for the photon to ionize the atom. Another possibility is that the interaction between the incoming photon and the atom will cause the atom to return to the ground state and thereby emit a second photon with energy $hf = E_2 - E_1$. In this process the incident photon is not absorbed; thus, after the stimulated emission, two photons with identical energy exist—the incident photon and the emitted photon. The two are in phase—an important consideration in lasers, which we shall discuss in the next section.

In the mid-1980s it became possible to electromagnetically "trap" a single ion (Fig. 42.23) and stimulate it to emit light. This procedure directly confirmed the existence of discrete energy levels in atoms.

Optional Section

42.9 ▶ LASERS AND HOLOGRAPHY

We have described how an incident photon can cause atomic energy transitions either upward (stimulated absorption) or downward (stimulated emission). The two processes are equally probable. When light is incident on a collection of atoms, a

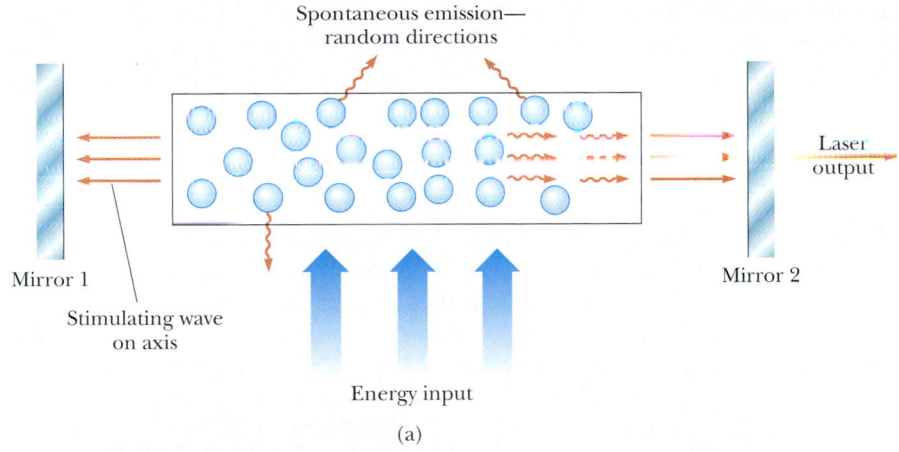

Spontaneous emission—
random directions

Laser
output

Mirror 1

Stimulating wave
on axis

Mirror 2

Energy input

(a)

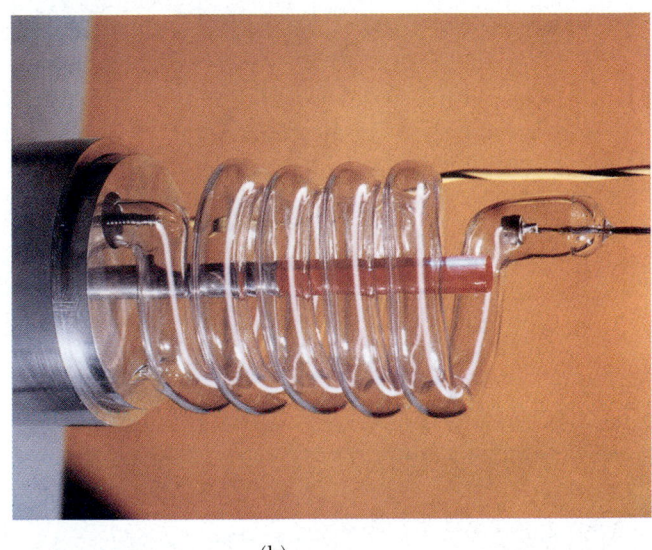

(b)

Figure 42.24 (a) Schematic diagram of a laser design. The tube contains the atoms that are the active medium. An external source of energy (for example, an optical or electrical device) "pumps" the atoms to excited states. The parallel end mirrors confine the photons to the tube, but mirror 2 is slightly transparent. (b) Photograph of the first ruby laser, showing the flash lamp surrounding the ruby rod. *(b, Courtesy of HRL Laboratories LLC, Malibu, CA)*

net absorption of energy usually occurs because, when the system is in thermal equilibrium, many more atoms are in the ground state than in excited states. However, if the situation can be inverted so that more atoms are in an excited state than in the ground state, a net emission of photons can result. Such a condition is called **population inversion.**

This, in fact, is the fundamental principle involved in the operation of a **laser**—an acronym for *light amplification by stimulated emission of radiation.* The amplification corresponds to a buildup of photons in the system as the result of a chain reaction of events. The following three conditions must be satisfied to achieve laser action:

- The system must be in a state of population inversion.
- The excited state of the system must be a metastable state. When this condition is met, stimulated emission occurs before spontaneous emission.

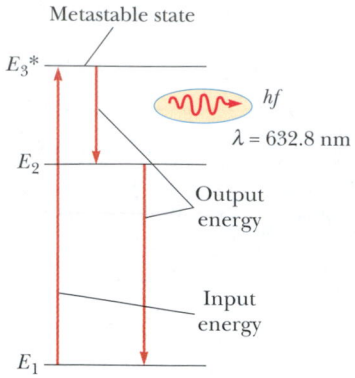

Figure 42.25 Energy level diagram for a neon atom in a helium–neon laser. The atom emits 632.8-nm photons through stimulated emission in the transition $E_3{}^*$–E_2. This is the source of coherent light in the laser.

• The emitted photons must be confined in the system long enough to stimulate further emission from other excited atoms. This confinement is achieved through the use of reflecting mirrors at the ends of the system. One end is made totally reflecting, and the other is slightly transparent to allow part of the laser beam to escape.

In a helium–neon gas laser, a mixture of helium and neon is confined in a sealed glass tube (Fig. 42.24a). An oscillator connected to the tube causes electrons to sweep through it, colliding with the gas atoms and raising them to excited states. As Figure 42.25 shows, neon atoms are excited to state $E_3{}^*$ through this process (the asterisk * indicates a metastable state) and also as a result of collisions with excited helium atoms. Stimulated emission occurs as the neon atoms make a transition to state E_2 and neighboring excited atoms are stimulated. This results in the production of coherent light at a wavelength of 632.8 nm.

Applications

Since the development of the first laser in 1960 (shown in Fig. 42.24b), tremendous growth has occurred in laser technology. Lasers that cover wavelengths in the infrared, visible, and ultraviolet regions are now available. Applications include surgical "welding" of detached retinas, precision surveying and length measurement, precision cutting of metals and other materials (such as the fabric in Figure 42.26), and telephone communication along optical fibers. These and other applications are possible because of the unique characteristics of laser light. In addition to being highly monochromatic, laser light is also highly directional and can be sharply focused to produce regions of extremely intense light energy (with energy densities 10^{12} times that in the flame of a typical cutting torch).

Lasers are used in precision long-range distance measurement (range finding). In recent years it has become important, for astronomical and geophysical

Figure 42.26 This robot carrying laser scissors, which can cut up to 50 layers of fabric at a time, is one of the many applications of laser technology. *(Philippe Plailly/SPL/Photo Researchers, Inc.)*

purposes, to measure as precisely as possible the distances from various points on the surface of the Earth to a point on the Moon's surface. To facilitate this, the Apollo astronauts set up a 0.5-m square of reflector prisms on the Moon, which enables laser pulses directed from an Earth station to be retroreflected to the same station (see Fig. 35.8a). Using the known speed of light and the measured round-trip travel time of a 1-ns pulse, the Earth–Moon distance, 380 000 km, can be determined to a precision of better than 10 cm.

Medical applications use the fact that various laser wavelengths can be absorbed in specific biological tissues. For example, certain laser procedures have greatly reduced blindness in glaucoma and diabetes patients. Glaucoma is a widespread eye condition characterized by a high fluid pressure in the eye, a condition that can lead to destruction of the optic nerve. A simple laser operation (iridectomy) can "burn" open a tiny hole in a clogged membrane, relieving the destructive pressure. A serious side effect of diabetes is neovascularization, the proliferation of weak blood vessels, which often leak blood. When this occurs in the retina, vision deteriorates (diabetic retinopathy) and finally is destroyed. It is now possible to direct the green light from an argon ion laser through the clear eye lens and eye fluid, focus on the retina edges, and photocoagulate the leaky vessels. Even people who have only minor vision defects such as nearsightedness are benefiting from the use of lasers to reshape the cornea, changing its focal length and reducing the need for eyeglasses.

Laser surgery is now an everyday occurrence at hospitals around the world. Infrared light at 10 μm from a carbon dioxide laser can cut through muscle tissue, primarily by vaporizing the water contained in cellular material. Laser power of about 100 W is required in this technique. The advantage of the "laser knife" over conventional methods is that laser radiation cuts tissue and coagulates blood at the same time, leading to a substantial reduction in blood loss. In addition, the technique virtually eliminates cell migration, an important consideration when tumors are being removed.

A laser beam can be trapped in fine glass-fiber light guides (endoscopes) by means of total internal reflection. The light fibers can thus be introduced through natural orifices, conducted around internal organs, and directed to specific interior body locations, eliminating the need for invasive surgery. For example, bleeding in the gastrointestinal tract can be optically cauterized by fiber-optic endoscopes inserted through the mouth.

In biological and medical research, it is often important to isolate and collect unusual cells for study and growth. A laser cell separator exploits the fact that specific cells can be tagged with fluorescent dyes. All cells are then dropped from a tiny charged nozzle and laser-scanned for the dye tag. If triggered by the correct light-emitting tag, a small voltage applied to parallel plates deflects the falling electrically charged cell into a collection beaker. This is an efficient method for extracting the proverbial needle from the haystack.

One of the most unusual and interesting applications of the laser is in the production of three-dimensional images in a process called **holography.** Figure 42.27 shows how a hologram is made. Light from the laser is split into two parts by a half-silvered mirror at B. After passing through lens L_1, which diverges the light rays, one part of the beam reflects off the object to be photographed and strikes an ordinary photographic film. The other part of the beam is diverged by lens L_2, reflects from mirrors M_1 and M_2, and finally strikes the film. The two beams overlap on the film to form an extremely complicated interference pattern. Such an interference pattern can be produced only if the phase relationship of the waves in the two beams is constant throughout the exposure of the film. This condition is met by illuminating the scene with coherent laser radiation. The hologram records not

Figure 42.27 (a) Arrangement for producing a hologram. (b) Photograph of a hologram that uses a cylindrical film. *(Courtesy of Central Scientific Company.)*

only the intensity of the light scattered from the object (as in a conventional photograph) but the phase difference between the beam reflected from the mirrors and the beam scattered from the object. Because of this phase difference, the interference pattern that is formed produces an image having a full three-dimensional perspective.

A hologram is best viewed by allowing coherent light to pass through the developed film as one looks back along the direction from which the beam comes. The light passing through the hologram is diffracted. It emerges in a form identical to the light that left the object while the hologram was being recorded; as a result, viewing the hologram is almost like looking through a window at the object. Figure 42.27b is a photograph of a hologram that was made using a cylindrical film. One sees a three-dimensional image in which the perspective changes as the viewer's head moves.

SUMMARY

Quantum mechanics can be applied to the hydrogen atom by the use of the potential energy function $U(r) = -k_e e^2/r$ in the Schrödinger equation. The solution to this equation yields wave functions for allowed states and allowed energies:

$$E_n = -\left(\frac{k_e e^2}{2a_0}\right)\frac{1}{n^2} = -\frac{13.606}{n^2}\text{ eV} \qquad n = 1, 2, 3, \ldots \qquad \textbf{(42.2)}$$

where n is the **principal quantum number.** The allowed wave functions depend on three quantum numbers: n, ℓ, and m_ℓ, where ℓ is the **orbital quantum number** and m_ℓ is the **orbital magnetic quantum number.** The restrictions on the quantum numbers are

$$n = 1, 2, 3, \ldots$$

$$\ell = 0, 1, 2, \ldots, n - 1$$

$$m_\ell = -\ell, -\ell + 1, \ldots, \ell - 1, \ell$$

All states having the same principal quantum number n form a **shell,** identified by the letters K, L, M, . . . (corresponding to $n = 1, 2, 3, \ldots$). All states having the same values of n and ℓ form a **subshell,** designated by the letters $s, p, d, f, \ldots$ (corresponding to $\ell = 0, 1, 2, 3, \ldots$).

To completely describe a quantum state, it is necessary to include a fourth quantum number m_s, called the **spin magnetic quantum number.** This quantum number can have only two values, $\pm\frac{1}{2}$.

An atom in a state characterized by a specific value of n can have the following values of L, the magnitude of the atom's orbital angular momentum **L**:

$$L = \sqrt{\ell(\ell+1)}\,\hbar \qquad \ell = 0, 1, 2, \ldots, n-1 \tag{42.9}$$

The allowed values of the projection of **L** along the z axis are

$$L_z = m_\ell \hbar \tag{42.10}$$

Only discrete values of L_z are allowed, as determined by the restrictions on m_ℓ. This quantization of L_z is referred to as **space quantization.**

The electron has an intrinsic angular momentum called the **spin angular momentum.** That is, the total angular momentum of an electron in an atom has two contributions, one arising from the spin of the electron (**S**) and one arising from the orbital motion of the electron (**L**). Electron spin can be described by a single quantum number $s = \frac{1}{2}$. The magnitude of the spin angular momentum is

$$S = \frac{\sqrt{3}}{2}\,\hbar \tag{42.12}$$

and the z component of **S** is

$$S_z = m_s \hbar = \pm\tfrac{1}{2}\hbar \tag{42.13}$$

That is, the spin angular momentum is also quantized in space, as specified by the spin magnetic quantum number $m_s = \pm\frac{1}{2}$.

The magnetic moment $\boldsymbol{\mu}_{\text{spin}}$ associated with the spin angular momentum of an electron is

$$\boldsymbol{\mu}_{\text{spin}} = -\frac{e}{m_e}\,\mathbf{S} \tag{42.14}$$

The z component of $\boldsymbol{\mu}_{\text{spin}}$ can have the values

$$\mu_{\text{spin},z} = \pm\frac{e\hbar}{2m_e} \tag{42.15}$$

The **exclusion principle** states that **no two electrons in an atom can be in the same quantum state.** In other words, no two electrons can have the same set of quantum numbers n, ℓ, m_ℓ, and m_s. Using this principle, the electronic configurations of the elements can be determined. This serves as a basis for understanding atomic structure and the chemical properties of the elements.

The allowed electronic transitions between any two levels in an atom are governed by the **selection rules**

$$\Delta\ell = \pm 1 \qquad \text{and} \qquad \Delta m_\ell = 0, \pm 1 \tag{42.16}$$

The x-ray spectrum of a metal target consists of a set of sharp characteristic lines superimposed on a broad continuous spectrum. **Characteristic x-rays** are emitted by atoms when an electron undergoes a transition from an outer shell to a vacancy in an inner shell.

Atomic transitions can be described with three processes: **stimulated absorption,** in which an incoming photon raises the atom to a higher energy state; **spontaneous emission,** in which the atom makes a transition to a lower energy state, emitting a photon; and **stimulated emission,** in which an incident photon causes an excited atom to make a downward transition, emitting a photon identical to the incident one.

QUESTIONS

1. Why are three quantum numbers needed to describe the state of a one-electron atom (neglecting spin)?
2. Compare the Bohr theory and the Schrödinger treatment of the hydrogen atom. Comment on the total energy and the orbital angular momentum.
3. Why is the direction of the orbital angular momentum of an electron opposite that of its magnetic moment?
4. Why is a nonuniform magnetic field used in the Stern–Gerlach experiment?
5. Could the Stern–Gerlach experiment be performed with ions rather than neutral atoms? Explain.
6. Describe some experiments that support the conclusion that the spin magnetic quantum number for electrons can have only the values $\pm\frac{1}{2}$.
7. Discuss some of the consequences of the exclusion principle.
8. Why do lithium, potassium, and sodium exhibit similar chemical properties?
9. Explain why a photon must have a spin of 1.
10. An energy of about 21 eV is required to excite an electron in a helium atom from the $1s$ state to the $2s$ state.

The same transition for the He^+ ion requires approximately twice as much energy. Explain.
11. Does the intensity of light from a laser fall off as $1/r^2$?
12. The absorption or emission spectrum of a gas consists of lines that broaden as the density of gas molecules increases. Why do you suppose this occurs?
13. How is it possible that electrons, whose positions are described by a probability distribution around a nucleus, can exist in states of *definite* energy (e.g., $1s$, $2p$, $3d$, . . .)?
14. It is easy to understand how two electrons (one spin up, one spin down) can fill the $1s$ shell for a helium atom. How is it possible that eight more electrons can fit into the 2s, $2p$ level to complete the $1s^2 2s^2 2p^6$ shell for a neon atom?
15. In 1914, Henry Moseley discovered how to define the atomic number of an element from its characteristic x-ray spectrum. How was this possible? (*Hint:* See Figs. 42.17 and 42.18.)
16. What are the advantages of using monochromatic light to view a holographic image?
17. Why is stimulated emission so important in the operation of a laser?

PROBLEMS

1, **2**, **3** = straightforward, intermediate, challenging   ☐ = full solution available in the *Student Solutions Manual and Study Guide*
WEB = solution posted at **http://www.saunderscollege.com/physics/**   🖥 = Computer useful in solving problem   📰 = Interactive Physics
☐ = paired numerical/symbolic problems

Section 42.1   Early Models of the Atom

1. In the Rutherford scattering experiment, 4.00-MeV alpha particles (^{4}He nuclei containing 2 protons and 2 neutrons) scatter off gold nuclei (containing 79 protons and 118 neutrons). If an alpha particle makes a direct head-on collision with the gold nucleus and scatters backward at 180°, determine (a) the distance of closest approach of the alpha particle to the gold nucleus and (b) the maximum force exerted on the alpha particle. Assume that the gold nucleus remains fixed throughout the entire process.

2. In the Rutherford scattering experiment, alpha particles of energy E (^{4}He nuclei containing 2 protons and 2 neutrons) scatter off a target whose atoms have an atomic number Z. If an alpha particle makes a direct head-on collision with a target nucleus and scatters backward at 180°, determine (a) the distance of closest approach of the alpha particle to the target nucleus and (b) the maximum force exerted on the alpha particle. Assume that the target nucleus remains fixed throughout the entire process.

Section 42.2   The Hydrogen Atom Revisited

3. A photon with energy 2.28 eV is barely capable of causing a photoelectric effect when it strikes a sodium plate. Suppose that the photon is instead absorbed by hydrogen. Find (a) the minimum n for a hydrogen atom that can be ionized by such a photon and (b) the speed of the released electron far from the nucleus.

4. The Balmer series for the hydrogen atom corresponds to electronic transitions that terminate in the state with quantum number $n = 2$, as shown in Figure 40.18. (a) Consider the photon of longest wavelength; determine its energy and wavelength. (b) Consider the spectral line of shortest wavelength; find its photon energy and wavelength.

5. A general expression for the energy levels of one-electron atoms and ions is

$$E_n = -\frac{\mu k_e^2 q_1^2 q_2^2}{2\hbar^2 n^2}$$

where k_e is the Coulomb constant, q_1 and q_2 are the charges of the two particles, and μ is the reduced mass, given by $\mu = m_1 m_2 / (m_1 + m_2)$. In Problem 4 we found that the wavelength for the $n = 3$ to $n = 2$ transition of the hydrogen atom is 656.3 nm (visible red light). What are the wavelengths for this same transition in (a) positronium, which consists of an electron and a positron, and (b) singly ionized helium? (*Note:* A positron is a positively charged electron.)

6. Ordinary hydrogen gas is a mixture of two kinds of atoms (isotopes) containing either one- or two-particle nuclei. These isotopes are hydrogen-1 with a proton nucleus and deuterium with a deuteron nucleus. (A deuteron is one proton and one neutron bound together.) Hydrogen-1 and deuterium have identical chemical properties but can be separated via an ultra-centrifuge or other methods. Their emission spectra show lines of the same colors at very slightly different wavelengths. (a) Use the equation given in Problem 5 to show that the difference in wavelength between the hydrogen and deuterium spectral lines associated with a particular electronic transition is given by

$$\lambda_H - \lambda_D = (1 - \mu_H/\mu_D)\lambda_H$$

(b) Evaluate the wavelength difference for the H_α line of hydrogen, with wavelength 656.3 nm, emitted by an atom making a transition from an $n = 3$ state to an $n = 2$ state.

Section 42.3 The Spin Magnetic Quantum Number

7. List the possible sets of quantum numbers for electrons in (a) the $3d$ subshell and (b) the $3p$ subshell.

Section 42.4 The Wave Functions for Hydrogen

8. Plot the wave function $\psi_{1s}(r)$ (see Eq. 42.3) and the radial probability density function $P_{1s}(r)$ (see Eq. 42.6) for hydrogen. Let r range from 0 to $1.5a_0$, where a_0 is the Bohr radius.

9. The ground-state wave function for the electron in a hydrogen atom is

$$\psi_{1s}(r) = \frac{1}{\sqrt{\pi a_0{}^3}} e^{-r/a_0}$$

where r is the radial coordinate of the electron and a_0 is the Bohr radius. (a) Show that the wave function as given is normalized. (b) Find the probability of locating the electron between $r_1 = a_0/2$ and $r_2 = 3a_0/2$.

10. The wave function for an electron in the $2p$ state of hydrogen is described by the expression

$$\psi_{2p}(r) = \frac{1}{\sqrt{3(2a_0)^{3/2}}} \frac{r}{a_0} e^{-r/2a_0}$$

What is the most likely distance from the nucleus to find an electron in the $2p$ state? (See Fig. 42.8.)

WEB 11. Show that the $1s$ wave function for an electron in hydrogen,

$$\psi_{1s}(r) = \frac{1}{\sqrt{\pi a_0{}^3}} e^{-r/a_0}$$

satisfies the radially symmetric Schrödinger equation,

$$-\frac{\hbar^2}{2m_e}\left(\frac{d^2\psi}{dr^2} + \frac{2}{r}\frac{d\psi}{dr}\right) - \frac{k_e e^2}{r}\psi = E\psi$$

12. During a particular period of time, an electron in the ground state of a hydrogen atom is "observed" 1 000 times at a distance $a_0/2$ from the nucleus. How many times is this electron observed at a distance $2a_0$ from the nucleus during this period of "observation"?

Section 42.5 The Other Quantum Numbers

13. Calculate the angular momentum for an electron in (a) the $4d$ state and (b) the $6f$ state.

14. If an electron has an orbital angular momentum of 4.714×10^{-34} J·s, what is the orbital quantum number for the state of the electron?

15. A hydrogen atom is in its fifth excited state. The atom emits a 1 090-nm wavelength photon. Determine the maximum possible orbital angular momentum of the electron after emission.

16. Find all possible values of L, L_z, and θ for an electron in a $3d$ state of hydrogen.

WEB 17. How many sets of quantum numbers are possible for an electron for which (a) $n = 1$, (b) $n = 2$, (c) $n = 3$, (d) $n = 4$, and (e) $n = 5$? Check your results to show that they agree with the general rule that the number of sets of quantum numbers is equal to $2n^2$.

18. The z component of the electron's spin magnetic moment is given by the Bohr magneton, $\mu_B = e\hbar/2m_e$. Show that the Bohr magneton has the numerical value 9.27×10^{-24} J/T $= 5.79 \times 10^{-5}$ eV/T.

19. (a) Find the mass density of a proton, picturing it as a solid sphere of radius 1.00×10^{-15} m. (b) Consider a classical model of an electron as a solid sphere with the same density as the proton. Find its radius. (c) If this electron possesses spin angular momentum $I\omega = \hbar/2$ because of classical rotation about the z axis, determine the speed of a point on the equator of the electron, and (d) compare this speed to the speed of light.

20. All objects, large and small, behave quantum-mechanically. (a) Estimate the quantum number ℓ for the Earth in its orbit about the Sun. (b) What energy change (in joules) would occur if the Earth made a transition to an adjacent allowed state?

21. Like the electron, the nucleus of an atom has spin angular momentum and a corresponding magnetic moment. The z component of the spin magnetic moment for a nucleus is characterized by the *nuclear magneton* $\mu_n = e\hbar/2m_p$, where m_p is the proton mass. (a) Calculate the value of μ_n in joules per tesla and in electron volts per tesla. (b) Determine the ratio μ_n/μ_B, and comment on your result.

22. An electron is in the N shell. Determine the maximum value of the z component of its angular momentum.

23. The ρ-meson has a charge of $-e$, a spin quantum number of 1, and a mass 1 507 times that of the electron.

Imagine that the electrons in atoms are replaced by ρ-mesons, and list the possible sets of quantum numbers for ρ-mesons in the $3d$ subshell.

Section 42.6 The Exclusion Principle and the Periodic Table

24. (a) Write out the electronic configuration for the ground state of oxygen ($Z = 8$). (b) Write out the values for the set of quantum numbers n, ℓ, m_ℓ, and m_s for each electron in oxygen.

25. Going down the periodic table, which subshell is filled first, the $3d$ or the $4s$ subshell? Which electronic configuration has a lower energy: [Ar]$3d^4 4s^2$ or [Ar]$3d^5 4s^1$? Which has the greater number of unpaired spins? Identify this element, and discuss Hund's rule in this case. (*Note:* The notation [Ar] represents the filled configuration for argon.)

26. Two electrons in the same atom both have $n = 3$ and $\ell = 1$. List the quantum numbers for the possible states of the atom. (b) How many states would be possible if the exclusion principle were inoperative?

27. Consider an atom in its ground state, with its outer electrons completely filling the M shell. (a) Identify the atom. (b) List the number of electrons in each subshell.

28. For a neutral atom of element 110, what would be the probable electronic configuration?

WEB 29. (a) Scanning through Table 42.4 in order of increasing atomic number, note that the electrons fill the subshells in such a way that the subshells with the lowest values of $n + \ell$ are filled first. If two subshells have the same value of $n + \ell$, the one with the lower value of n is filled first. Using these two rules, write the order in which the subshells are filled through $n + \ell = 7$. (b) Predict the chemical valence for the elements that have atomic numbers 15, 47, and 86, and compare your predictions with the actual valences.

30. Devise a table similar to that shown in Figure 42.14 for atoms containing 11 through 19 electrons. Use Hund's rule and educated guesswork.

Section 42.7 Atomic Spectra

31. (a) Determine the possible values of the quantum numbers ℓ and m_ℓ for the He$^+$ ion in the state corresponding to $n = 3$. (b) What is the energy of this state?

32. If you wish to produce 10.0-nm x-rays in the laboratory, what is the minimum voltage you must use in accelerating the electrons?

33. A tungsten target is struck by electrons that have been accelerated from rest through a 40.0-kV potential difference. Find the shortest wavelength of the radiation emitted.

34. In x-ray production, electrons are accelerated through a high voltage ΔV and then decelerated by striking a target. Show that the shortest-wavelength x-ray that can be produced is

$$\lambda_{min} = \frac{1\,240\ \text{nm} \cdot \text{V}}{\Delta V}$$

35. Use the method illustrated in Example 42.7 to calculate the wavelength of the x-ray emitted from a molybdenum target ($Z = 42$) when an electron moves from the L shell ($n = 2$) to the K shell ($n = 1$).

36. The wavelength of characteristic x-rays corresponding to the K$_\beta$ line is 0.152 nm. Determine the material in the target.

37. Electrons are shot into a bismuth target, and x-rays are emitted. Determine (a) the M-to L-shell transitional energy for Bi and (b) the wavelength of the x-ray emitted when an electron falls from the M shell into the L shell.

38. The K series of the discrete spectrum of tungsten contains wavelengths of 0.018 5 nm, 0.020 9 nm, and 0.021 5 nm. The K-shell ionization energy is 69.5 keV. Determine the ionization energies of the L, M, and N shells. Sketch the transitions.

39. When the outermost electron of an alkali atom is excited, it is found that states with the same n but different ℓ have slightly different energies because they penetrate into the central core to different degrees. Low-ℓ orbitals penetrate more, while higher-ℓ orbitals penetrate less. The wavelengths of absorption lines are given approximately by the equation

$$1/\lambda_{n\ell \to n'\ell'} = R_H[(n - \delta_\ell)^{-2} - (n' - \delta_{\ell'})^{-2}]$$

Observe that it is like Equation 40.29, which describes hydrogen, but with the principal quantum numbers n replaced by effective quantum numbers n replaced by effective quantum numbers. Here δ_ℓ is the "quantum defect" associated with orbital quantum number ℓ. The value of δ_ℓ is independent of n. For sodium (Na), $\delta_0 = 1.35$. The longest wavelength for an absorption transition carrying Na from its ground state to a state with higher principal quantum number is 330 nm. (a) For what other value of ℓ can you determine the quantum defect, and (b) what is the value of that defect?

Section 42.8 Atomic Transitions

40. The familiar yellow light from a sodium-vapor street lamp results from the $3p \to 3s$ transition in ^{11}Na. Evaluate the wavelength of this light given that the energy difference $E_{3p} - E_{3s} = 2.10$ eV.

41. Assume that a great number n of identical atoms are in a first excited state. The rate dn/dt at which this population will de-excite is $dn/dt = -Pn$, where P is the quantum transition probability rate. The transition rate is given, in turn, by $P = A + u_f B$, where A is the Einstein coefficient for spontaneous emission and B is the Einstein coefficient for stimulated emission due to the presence of photons with energy density u_f per unit frequency. Einstein showed that these coefficients are related by $A = 16\pi^2 \hbar B/\lambda^3$. For an atomic transition of wavelength 645 nm, what must be the energy density of photons for stimulated emission to be as important as spontaneous emission?

(Optional)

Section 42.9   Lasers and Holography

42. The carbon dioxide laser is one of the most powerful developed. The energy difference between the two laser levels is 0.117 eV. Determine the frequency and wavelength of the radiation emitted by this laser. In what portion of the electromagnetic spectrum is this radiation?

WEB 43. A ruby laser delivers a 10.0-ns pulse of 1.00 MW average power. If the photons have a wavelength of 694.3 nm, how many are contained in the pulse?

44. An important characteristic of a laser is its gain G, specifying the relative enhancement of light-beam intensity over the length L of the laser. When $G = 1.05$, a 5% increase in intensity occurs as the light makes one pass through the laser. The gain is given by

$$G = e^{\sigma(n_u - n_\ell)L}$$

In this equation σ is the atomic-absorption cross-section for the lasing transition, with units of length squared. It is related to the quantum transition probability. The variables n_u and n_ℓ are the number densities (units of length^{-3}) of active atoms in the upper and lower energy states of the laser transition. If $L = 0.500$ m and $\sigma = 1.00 \times 10^{-18}$ m^2 for a particular laser, what number density inversion $n_u - n_\ell$ must be maintained to have a gain of 1.05?

45. The number N of atoms in a particular state is called the population of that state. This number depends on the energy of that state and the temperature. In thermal equilibrium the population of atoms in a state of energy E_n is given by a Boltzmann distribution expression

$$N = N_g e^{-(E_n - E_g)/k_B T}$$

where T is the absolute temperature and N_g is the population of the ground state, of energy E_g. (a) Before the power is switched on, the neon atoms in a laser are in thermal equilibrium at 27.0°C. Find the equilibrium ratio of the populations of the states E_3^* and E_2 shown in Figure 42.25. (b) Find the equilibrium ratio at 4.00 K of the populations of the two states in a ruby laser that can produce a light beam of wavelength 694.3 nm.

46. Lasers operate by a clever artificial production of a "population inversion" between the upper and lower atomic energy states involved in the lasing transition. This means that more atoms occur with electrons in the upper excited state than in the lower one. Consider the ruby laser transition at 694.3 nm. Assume that 2% more atoms occur in the upper state than in the lower. For simplicity, assume that both levels have only one quantum state associated with them. (a) To demonstrate how unnatural such a situation is, find the temperature for which the Boltzmann distribution describes a 2.00% population inversion. (b) Why does such a situation not occur naturally?

ADDITIONAL PROBLEMS

47. An Nd:YAG laser used in eye surgery emits a 3.00-mJ pulse in 1.00 ns, focused to a spot 30.0 μm in diameter on the retina. (a) Find (in SI units) the power per unit area at the retina. (This quantity is called the irradiance.) (b) What energy is delivered to an area of molecular size, taken as a circular area 0.600 nm in diameter?

48. **Review Problem.** (a) How much energy is required to cause an electron in hydrogen to move from the $n = 1$ state to the $n = 2$ state? (b) Suppose the electrons gain this energy through collisions among hydrogen atoms at a high temperature. At what temperature would the average atomic kinetic energy $3k_B T/2$, where k_B is the Boltzmann constant, be great enough to excite the electrons?

49. Show that the average value of r for the 1s state of hydrogen has the value $3a_0/2$. (*Hint:* Use Eq. 42.6.)

50. Find the average (expectation) value of $1/r$ in the 1s state of hydrogen. It is given by

$$\langle 1/r \rangle = \int_{\text{all space}} |\psi|^2 (1/r)\, dV = \int_0^\infty P(r)\,(1/r)\, dr$$

Is the result equal to the inverse of the average value of r?

51. Suppose a hydrogen atom is in the 2s state. Taking $r = a_0$, calculate values for (a) $\psi_{2s}(a_0)$, (b) $|\psi_{2s}(a_0)|^2$, and (c) $P_{2s}(a_0)$. (*Hint:* Use Eq. 42.7.)

52. As noted in a previous chapter, the muon is an elementary particle with the charge of an electron but a mass 207 times greater than that of an electron. Muonium is an "atom" composed of a muon and a proton. Using the formula for the energy levels of hydrogen-like atoms given in Problem 42.5, find the ionization energy of ground-state muonium.

53. A pulsed ruby laser emits light at 694.3 nm. For a 14.0-ps pulse containing 3.00 J of energy, find (a) the physical length of the pulse as it travels through space and (b) the number of photons in it. (c) If the beam has a circular cross-section of 0.600 cm diameter, find the number of photons per cubic millimeter.

54. A pulsed laser emits light of wavelength λ. For a pulse of duration t having energy E, find (a) the physical length of the pulse as it travels through space and (b) the number of photons in it. (c) If the beam has a circular cross section of diameter d, find the number of photons per unit volume.

55. (a) Show that the most probable radial position for an electron in the 2s state of hydrogen is $r = 5.236a_0$. (b) Show that the wave function given by Equation 42.7 is normalized.

56. The force on a magnetic moment μ_z in a nonuniform magnetic field B_z is given by $F_z = \mu_z(dB_z/dz)$. If a beam of silver atoms travels a horizontal distance of 1.00 m through such a field and each atom has a speed of 100 m/s, how strong must be the field gradient dB_z/dz to deflect the beam 1.00 mm?

57. An electron in chromium moves from the $n = 2$ state to the $n = 1$ state without emitting a photon. Instead, the excess energy is transferred to an outer electron (one in the $n = 4$ state), which is then ejected by the atom. (This is called an Auger [pronounced 'ohjay'] process, and the ejected electron is referred to as an Auger electron.) Use the Bohr theory to find the kinetic energy of the Auger electron.

58. Suppose the ionization energy of an atom is 4.10 eV. In the spectrum of this same atom, we observe emission lines with wavelengths 310 nm, 400 nm, and 1 377.8 nm. Use this information to construct the energy-level diagram with the fewest levels. Assume that the higher levels are closer together.

59. All atoms have the same size, to an order of magnitude. (a) To show this, estimate the diameters for aluminum (with molar mass = 27.0 g/mol and density 2.70 g/cm³) and uranium (with molar mass = 238 g/mol and density 18.9 g/cm³). (b) What do the results imply about the wave functions for inner-shell electrons as we progress to higher and higher atomic mass atoms? (*Hint:* The molar volume is approximately $D^3 N_A$, where D is the atomic diameter and N_A is Avogadro's number.)

60. In interstellar space, atomic hydrogen produces the sharp spectral line called the 21-cm radiation, which astronomers find most helpful in detecting clouds of hydrogen between stars. This radiation is useful because interstellar dust that obscures visible wavelengths is transparent to these radio wavelengths. The radiation is not generated by an electronic transition between energy states characterized by n. Instead, in the ground state ($n = 1$), the electron and proton spins may be parallel or antiparallel, with a resultant slight difference in these energy states. (a) Which condition has the higher energy? (b) More precisely, the line has wavelength 21.11 cm. What is the energy difference between the states? (c) The average lifetime in the excited state is about 10^7 yr. Calculate the associated uncertainty in energy of the excited energy level.

61. For hydrogen in the $1s$ state, what is the probability of finding the electron farther than $2.50 a_0$ from the nucleus?

62. For the ground state of hydrogen, what is the probability of finding the electron closer to the nucleus than the Bohr radius?

WEB 63. According to classical physics, a charge e moving with an acceleration a radiates at a rate

$$\frac{dE}{dt} = -\frac{1}{6\pi\epsilon_0}\frac{e^2 a^2}{c^3}$$

(a) Show that an electron in a classical hydrogen atom (see Fig. 42.3) spirals into the nucleus at a rate

$$\frac{dr}{dt} = -\frac{e^4}{12\pi^2\epsilon_0{}^2 r^2 m_e{}^2 c^3}$$

(b) Find the time it takes the electron to reach $r = 0$, starting from $r_0 = 2.00 \times 10^{-10}$ m.

64. In a lithium atom, the electron cloud of the outer electron overlaps with the electron clouds of the two K-shell electrons. A detailed calculation of the effective charge exerting an electric force on another electron may be made using quantum mechanics. For the case of the lithium atom, the effective charge on each inner electron is $-0.85e$. Use this value to find (a) the effective charge on the nucleus as "seen" by the outer valence electron and (b) the ionization energy (compare this with 5.4 eV).

65. **Review Problem.** In the technique known as electron spin resonance (ESR), a sample containing unpaired electrons is placed in a magnetic field. Consider the simplest situation, in which only one electron is present and therefore only two energy states are possible, corresponding to $m_s = \pm\frac{1}{2}$. In ESR, the absorption of a photon causes the electron's spin magnetic moment to flip from a lower energy state to a higher energy state. (The lower energy state corresponds to the case in which the magnetic moment $\boldsymbol{\mu}_{\text{spin}}$ is aligned with the magnetic field, and the higher energy state corresponds to the case in which $\boldsymbol{\mu}_{\text{spin}}$ is aligned opposite the field.) What photon frequency is required to excite an ESR transition in a 0.350-T magnetic field?

66. Figure P42.66 shows the energy-level diagrams of He and Ne. An electrical voltage excites the He atom from its ground state to its excited state of 20.61 eV. The excited He atom collides with an Ne atom in its ground state and excites this atom to the state at 20.66 eV. Lasing action takes place for electronic transitions from E_3^* to E_2 in the Ne atoms. Show that the wavelength of the red He–Ne laser light is approximately 633 nm.

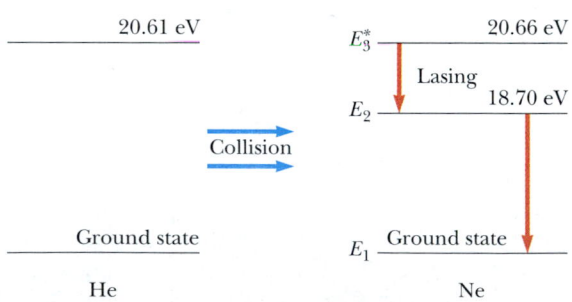

Figure P42.66

67. A dimensionless number that often appears in atomic physics is the fine-structure constant $\alpha = k_e e^2/\hbar c$, where k_e is the Coulomb constant. (a) Obtain a numerical value for $1/\alpha$. (b) In scattering experiments, the electron size is taken to be the classical electron radius, $r_e = k_e e^2/m_e c^2$. In terms of α, what is the ratio of the

Compton wavelength (see Section 40.3), $\lambda_C = h/m_e c$, to the classical electron radius? (c) In terms of α, what is the ratio of the Bohr radius, a_0, to the Compton wavelength? (d) In terms of α, what is the ratio of the Rydberg wavelength, $1/R_H$, to the Bohr radius (see Section 40.5)?

68. Show that the wave function for an electron in the $2s$-state in hydrogen

$$\psi_{2s}(r) = \frac{1}{4\sqrt{2\pi}} \left(\frac{1}{a_0}\right)^{3/2} \left(2 - \frac{r}{a_0}\right) e^{-r/2a_0}$$

satisfies the radially symmetric Schrödinger equation given in Problem 11.

69. A collimated light beam of frequency f passes in the x direction through a sample of a transparent substance with index of refraction n. The length of its path is L. The frequency is tuned to be resonant with a transition between two atomic levels in the substance. The beam can induce stimulated emission from atoms in the upper state, and can be absorbed by atoms in the lower state. The beam intensity is therefore a function $I(x)$ of position. The number of transitions per unit time and per area that the beam will induce over a small distance dx in the material is equal to $BNI(x)\,n\,dx/c$, where B is Einstein's coefficient for the transition (related to the transition probability) and N is the number density (i.e., population density) of atoms of the initial state in the transition. The same equation holds for both stimulated emission and absorption. Show that if I_0 is the intensity of the beam before it enters the material, the intensity of the beam at the other end is

$$I(L) = I_0 e^{-\alpha L}$$

where $\alpha = hf\,B\,\Delta N n/c$, and where ΔN is the difference in number densities between lower and upper states. (*Hint:* Intensity is (energy/time)/area, and photons have energy.)

70. **Review Problem.** The 1997 Nobel Prize in physics was awarded to Steven Chu, Claude Cohen-Tannoudji, and William Phillips for "the development of methods to cool and trap atoms with laser light." One part of their work was the production of a beam of atoms (mass $\sim 10^{-25}$ kg) that move at a speed on the order of 1 km/s, similar to the speed of molecules in air at room temperature. An intense laser light beam tuned to a visible atomic transition (assume 500 nm) is then directed straight into the atomic beam. That is, the atomic beam and light beam are counterpropagating. An atom in the ground state immediately absorbs a photon. Total momentum is conserved in the absorption process. After a lifetime on the order of 10^{-8} s, the excited atom radiates by spontaneous emission. It has an equal probability of emitting a photon in any direction. Thus, the average "recoil" of the atom is zero over many absorption and emission cycles. (a) Estimate the average deceleration of the atomic beam. (b) What is the order of magnitude of the distance over which the atoms in the beam will be brought to a halt?

ANSWERS TO QUICK QUIZZES

42.1 The three-dimensional charge distribution shown in Figure 42.7b is not uniform—it peaks at the Bohr radius. Figure 42.7a represents the probability of finding the electron as a function of distance from the center of the nucleus. Because that probability is a function of r but not of x or y individually, the chance of finding the electron in the xy plane is a maximum at any point for which $r = a_0$. For $r < a_0$, the probability drops rapidly—indicating that we are unlikely to find the electron very close to or inside the nucleus. As r becomes very great, the probability again approaches zero, meaning that the bound electron does not have a significant probability of being far away from the nucleus either. Imagine that you are looking down the z axis of Figure 42.7b, toward the xy plane. The peak area, where we are most likely to find the electron, would appear darkest, and the areas of lower probability would be lighter.

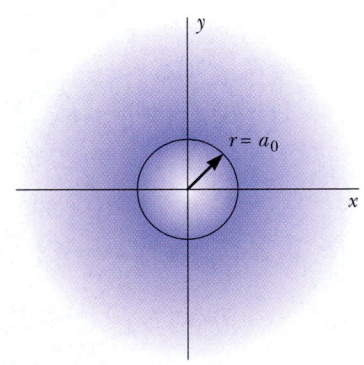

42.2 The $\ell = 3$ drawing is a graphical representation of the results of Example 42.6. Similar calculations yield the magnitude and direction of the angular momentum for the $\ell = 1$ case.

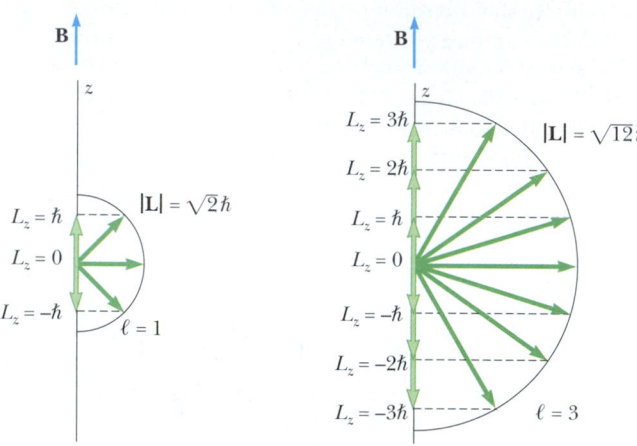

42.3 Classical physics puts no constraints on the angular momentum **L** of the atoms in the beam. Hence the magnetic moment **μ** for an atom can point in any direction. Each atom interacts differently with the nonuniform magnetic field and is deflected accordingly. This random variation in deflection results in a smoothly distributed exposure pattern on the photographic plate, like the classical pattern in Figure 42.12. Of course, this turns out not to be the case experimentally.

42.4 The bombarding electrons have an energy of 37 keV. The cutoff wavelength corresponds to one of these electrons losing all of its kinetic energy in a single collision and having that energy be emitted from the target as a single photon. We can calculate the wavelength of this photon from Equations 16.14 and 40.6:
$\lambda = c/f = hc/E = 34$ pm. Wavelengths shorter than this can appear in the continuous spectrum only if the energy of the bombarding electrons is increased.

42.5

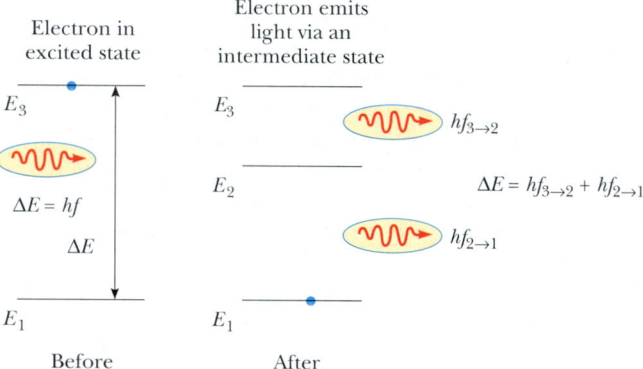

PUZZLER

This wine glass and the silverware behind it are very different, yet the same laws of physics govern the appearance of both. What makes glass transparent and gives metals their shine? *(George Semple)*

c h a p t e r

43

Molecules and Solids

The beautiful symmetry and regularity of crystalline solids have both stimulated and allowed rapid progress in the field of solid state physics in the 20th century. The most random atomic arrangement, that of a gas, was well understood in the 1800s, as we discussed in Chapter 21. In the 1900s great progress was first made in accounting for the properties of the most regular atomic arrangements, those of crystalline solids. Quite recently, our understanding of liquids and amorphous (irregular) solids has advanced. The recent interest in the physics of low-cost amorphous materials has been driven by their use in such devices as solar cells, memory elements, and fiber optic waveguides.

In this chapter, we study the aggregates of atoms known as molecules. We begin by describing the bonding mechanisms in molecules, the various modes of molecular excitation, and the radiation emitted or absorbed by molecules. We then take the next logical step and show how molecules combine to form solids. Then, by examining their electronic distributions, we explain the differences between insulating, conducting, semiconducting, and superconducting materials. The chapter also includes discussions of semiconducting junctions and several semiconductor devices, and it concludes with further treatment of superconductors.

43.1 MOLECULAR BONDS

The energy of a stable molecule is less than the total energy of the separated atoms. The bonding mechanisms in a molecule are primarily due to electric forces between atoms (or ions). When two atoms are separated by an infinite distance, the electric force between them is zero, as is the electric potential energy of the system they constitute. As the atoms are brought closer together, both attractive and repulsive forces act. At very large separation distances, the dominant forces are attractive. At small separation distances, electrostatic forces and the exclusion principle result in a repulsive force, as we shall discuss shortly.

The potential energy of a system of atoms can be positive or negative, depending on the distance between the constituent atoms. As we saw in Example 8.11, the total potential energy of a system of two atoms can be approximated by an expres-

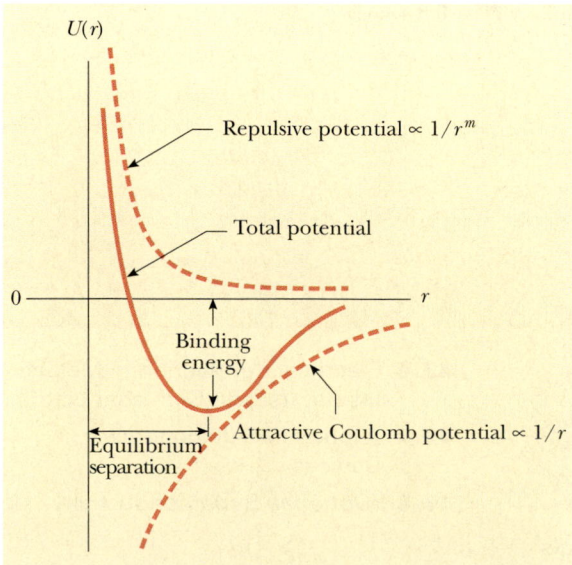

Figure 43.1   Total energy as a function of internuclear separation distance for a system of two atoms.

sion of the form

$$U = -\frac{A}{r^n} + \frac{B}{r^m} \qquad (43.1)$$

where r is the internuclear separation distance, A and B are parameters associated with the attractive and repulsive forces, and n and m are small integers. Total potential energy versus internuclear separation distance for such a two-atom system is graphed in Figure 43.1. At large separation distances, the slope of the curve is positive, corresponding to a net attractive force. When the atoms are close together, the slope is negative, indicating a net repulsive force. At the equilibrium separation distance, the attractive and repulsive forces just balance, the potential energy has its minimum value, and the slope of the curve is zero.

A complete description of the binding mechanisms in molecules is highly complex because binding involves the mutual interactions of many particles. In this section, therefore, we discuss only some simplified models: ionic bonding, covalent bonding, van der Waals bonding, and hydrogen bonding.

Ionic Bonding

When two atoms combine in such a way that one atom gives one or more of its outer electrons to the other atom, the bond formed is called an **ionic bond.** Ionic bonds are fundamentally caused by the Coulomb attraction between oppositely charged ions.

A familiar example of an ionically bonded solid is sodium chloride, NaCl, which is common table salt. Sodium, which has the electronic configuration $1s^2 2s^2 2p^6 3s$, is ionized relatively easily, giving up its $3s$ electron to form a Na$^+$ ion. The energy required to ionize the atom to form Na$^+$ is 5.1 eV. Chlorine, which has the electronic configuration $1s^2 2s^2 2p^5$, is one electron short of the filled-shell structure of argon. Because filled-shell configurations are energetically more favorable than unfilled-shell configurations, the Cl$^-$ ion is more stable than the neutral Cl atom. The energy released when an atom takes on an electron is called the **electron affinity** of the atom. For chlorine, the electron affinity is 3.7 eV. Therefore, the energy required to form Na$^+$ and Cl$^-$ from isolated atoms is $5.1 - 3.7 = 1.4$ eV.

Total energy versus internuclear separation distance for NaCl is graphed in Figure 43.2. The total energy has a minimum value of -4.2 eV at the equilibrium separation distance, which is about 0.24 nm. This means that the energy required to break the Na$^+$–Cl$^-$ bond and form neutral sodium and chlorine atoms, called the **dissociation energy,** is 4.2 eV.

When the two ions are brought to within 0.24 nm of each other, their filled outer shells overlap, and this results in a repulsion between the shells. This repulsion is partly electrostatic in origin and partly the result of the exclusion principle. Because all electrons must obey the exclusion principle, some of them in the overlapping shells are forced into higher energy states, and the system energy increases, as if a repulsive force existed between them.

Quick Quiz 43.1

Figure 43.2 shows the total energy versus the internuclear separation distance for Na$^+$ and Cl$^-$ ions. Once the ions are more than 0.24 nm apart, the energy increases but not without limit. What is the maximum value of the energy for $r > 0.24$ nm?

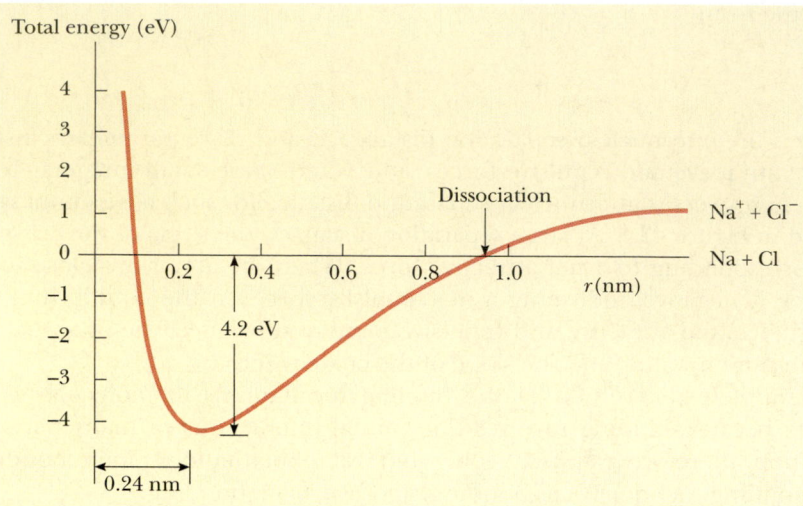

Figure 43.2   Total energy versus internuclear separation distance for Na^+ and Cl^- ions. The energy required to separate the NaCl molecule into neutral atoms of Na and Cl is 4.2 eV.

Covalent Bonding

A **covalent bond** between two atoms is one in which electrons supplied by either one or both atoms are shared. Many diatomic molecules, such as H_2, F_2, and CO, owe their stability to covalent bonds. In the case of the H_2 molecule, the two electrons are equally shared between the nuclei and occupy what is called a *molecular orbital*. The electron density is great in the region between the two nuclei, with the electrons acting as the "glue" holding the nuclei together.

The molecular orbital formed from the *s* orbitals of the two hydrogen atoms in H_2 is shown in Figure 43.3. Because of the exclusion principle, the two electrons in the ground state of H_2 must have antiparallel spins. Also because of the exclusion principle, if a third H atom is brought near the H_2 molecule, the third electron would have to occupy a higher energy level, which is an energetically unfavorable situation. Hence, the H_3 molecule is not stable and does not form.

Stable molecules more complex than H_2, such as H_2O, CO_2, and CH_4, also contain covalent bonds. Consider methane, CH_4, a typical organic molecule shown schematically in the electron-sharing diagram of Figure 43.4a. In this case, one covalent bond is formed between the carbon atom and each hydrogen atom, resulting in a total of four C–H covalent bonds. The geometrical arrangement of the four bonds is shown in Figure 43.4b. The four hydrogen nuclei are at the corners of a regular tetrahedron, and the carbon nucleus is at the center.

Because the outermost molecular orbitals of covalent molecules are full, the interactions between such molecules are quite weak. In fact, many covalent molecules form gases or liquids rather than solids.

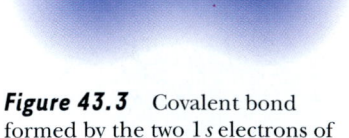

Figure 43.3   Covalent bond formed by the two 1 *s* electrons of the H_2 molecule. The depth of blue color at any location is proportional to the probability of finding an electron in that location.

Van der Waals Bonding

If two molecules are some distance apart, they are attracted to each other by weak electrostatic forces called **van der Waals forces.** Likewise, atoms that do not form ionic or covalent bonds are attracted to each other by van der Waals forces. For this reason, at sufficiently low temperatures at which thermal excitations are negli-

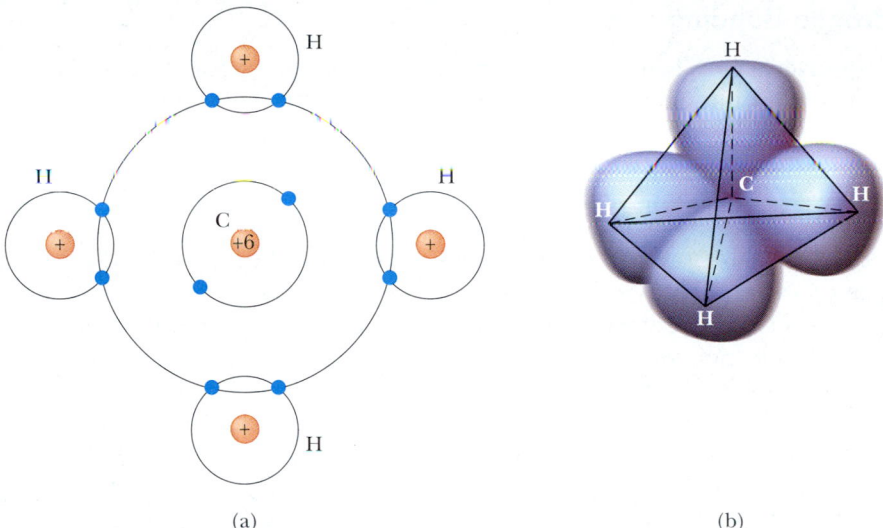

(a)   (b)

Figure 43.4   (a) A highly schematic representation of the four covalent bonds in the CH_4 molecule. (b) The spatial arrangement of these four bonds. The carbon atom is at the center of a tetrahedron having hydrogen atoms at its corners. The electron density is greatest between the nuclei.

gible, gases first condense to liquids and then solidify (with the exception of helium, which does not solidify at atmospheric pressure).

There are three types of van der Waals forces. The first type, called the *dipole–dipole force,* is an interaction between two molecules each having a permanent electric dipole moment—for example, polar molecules such as H_2O have permanent electric dipole moments and attract other polar molecules (Fig. 43.5). In effect, one molecule interacts with the electric field produced by another molecule.

The second type, the *dipole–induced dipole force,* results when a polar molecule having a permanent electric dipole moment induces a dipole moment in a nonpolar molecule.

The third type is called the *dispersion force,* an attractive force that occurs between two nonpolar molecules. In this case, the interaction results from the fact that, although the average dipole moment of a nonpolar molecule is zero, the average of the square of the dipole moment is nonzero because of charge fluctuations. Consequently, two nonpolar molecules near each other tend to be correlated so as to produce an attractive van der Waals force.

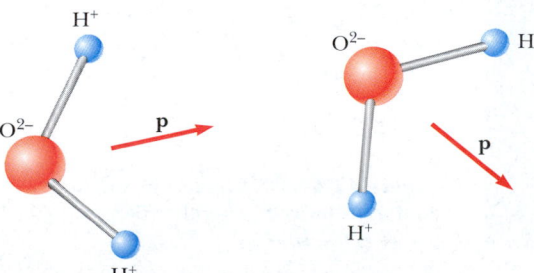

Figure 43.5   Water molecules have a permanent electric dipole moment **p**. The molecules attract each other because the electric field produced by one molecule interacts with and orients the dipole moment of a nearby molecule.

Hydrogen Bonding

Because hydrogen has only one electron, it is expected to form a covalent bond with only one other atom. A hydrogen atom in a given molecule also can form a second type of bond, however, bonding with an atom in another molecule via a **hydrogen bond.** Let us use the water molecule H_2O as an example. In the two covalent bonds in this molecule, the electrons from the hydrogen atoms are more likely to be found near the oxygen atom than near the hydrogen atoms. This leaves essentially bare protons at the positions of the hydrogen atoms. This unshielded positive charge can be attracted to the negative end of another polar molecule. Because the proton is unshielded by electrons, the negative end of the other molecule can come very close to the proton to form a bond that is strong enough to form a solid crystalline structure, such as that of ice. The bonds *within* a water molecule are covalent, but the bonds *between* water molecules in ice are hydrogen bonds. Because hydrogen bonds are relatively weak, ice melts at the low temperature of 0°C.

The hydrogen bond has a binding energy of about 0.1 eV. Although it is relatively weak compared with other chemical bonds, hydrogen bonding is the mechanism responsible for the linking of biological molecules and polymers. For example, in the case of the famous DNA (deoxyribonucleic acid) molecule, which has a double-helix structure (Fig. 43.6), hydrogen bonds formed by the sharing of a proton between two atoms create linkages between the turns of the helix.

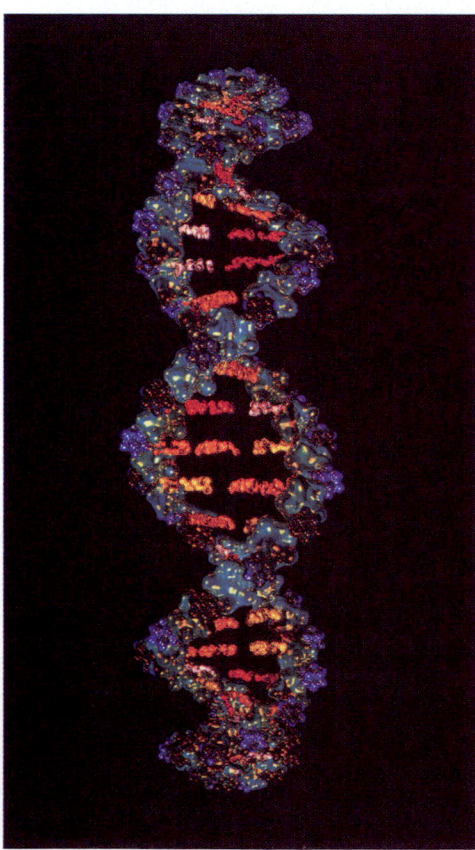

Figure 43.6 DNA molecules are held together by hydrogen bonds. *(Douglas Struthers/Tony Stone Images)*

APPLICATION ▶ Measuring Molecular Bonding Forces with an Atomic Force Microscope (Example 8.11 Revisited)

As noted in Section 41.8, an atomic force microscope (AFM) uses a very sharp tip mounted on a cantilever in close proximity to a surface in order to image surface topography with nanometer resolution. The AFM is similar to the scanning tunneling microscope, except here the tip interacts with the surface to measure force instead of a tunneling current. A variation of the AFM technique allows the measurement of bonding forces between atoms or between linkages in biological molecules or other macromolecules, as illustrated in Figure 43.7. In this figure, the molecule is bonded to a surface that can move vertically with nanometer precision. The force exerted on the tip where another part of the molecule is anchored causes the cantilever to bend. The cantilever can be approximated as a simple spring; thus its deflection is proportional to the force exerted on it (Hooke's law). The AFM can measure forces as minute as piconewtons. In an experiment to measure bonding forces, a specially prepared tip is carefully brought in contact with a surface coated with the molecules of interest. Several molecules might attach to the tip, but careful continued extension (by moving the surface downward) leaves attached to the tip only the longest molecule bridging the surface–tip distance. The force needed to break the next weakest bond is measured by recording cantilever deflection as the tip is retracted.

Consider the force required to break a single covalent bond in a molecule. We can greatly simplify this problem by considering bond breakage for a potential energy function with which we are familiar: the van der Waals interaction described earlier in this section. The potential energy function stated in general form in Equation 43.1 takes the form of the Lennard–Jones equation cited in Example 8.11:

$$U(r) = 4\epsilon\left[\left(\frac{\sigma}{r}\right)^{12} - \left(\frac{\sigma}{r}\right)^{6}\right]$$

This function is plotted in Figure 43.8 for argon, an inert gas that interacts by the van der Waals force, with experimentally determined parameters $\sigma = 0.340$ nm and $\epsilon = 0.0104$ eV. The shape of this potential energy function is generic for many types of bonds (compare with the plot in Fig. 43.2). The position of the energy minimum represents the equilibrium distance of the bond. Repulsive interactions between inner-shell electrons cause a large increase in energy if the atoms are moved closer together, and the interaction energy approaches zero when the atoms are sufficiently far apart. (Note that for very small deviations from the equilibrium distance, the first term in a Taylor series expansion is quadratic, which is the form of the potential energy function for a harmonic oscillator.)

We calculate the equilibrium internuclear distance by finding the position of the minimum in $U(r)$, that is, where $dU(r)/dr = 0$:

$$\frac{dU(r)}{dr} = -\frac{4\epsilon}{\sigma}\left[12\left(\frac{\sigma}{r}\right)^{13} - 6\left(\frac{\sigma}{r}\right)^{7}\right] = 0$$

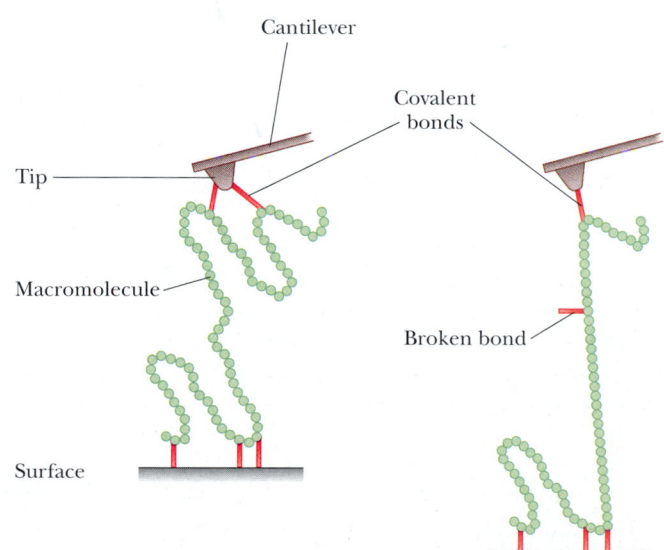

Figure 43.7 Measuring bonding forces with an AFM tip. A single macromolecular chain is covalently attached to the tip at two sites. Increased stretching breaks one bond, producing a sudden change in the force versus separation curve as the separation is increased. *(Adapted from M. Grandbois, M. Beyer, M. Reif, H. Clausen-Schaumann, H. Goub, "How Strong Is a Covalent Bond?" Science 283: 1727–1730, 1999.)*

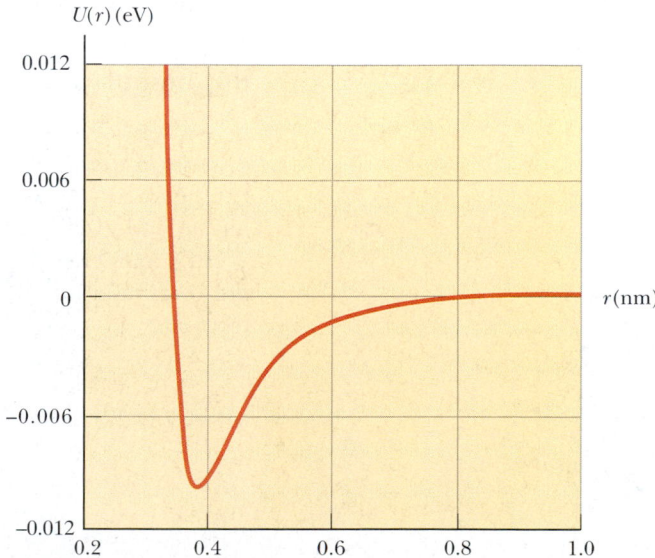

Figure 43.8 Potential energy versus internuclear separation distance for two argon atoms bonded by the van der Waals interaction.

This is equivalent to the position of zero net force. Solving for the equilibrium position r_{eq}, we find it at

$$r_{eq} = 2^{1/6}\,\sigma$$

which for argon is 0.382 nm.

When force is applied by an AFM, the bond ruptures at the point at which the force is a maximum. Because the force is $F = -dU/dr$, the maximum attractive force occurs at $d^2U/dr^2 = 0$. We calculate

$$\frac{d^2U(r)}{dr^2} = \frac{4\epsilon}{\sigma^2}\left[156\left(\frac{\sigma}{r}\right)^{14} - 42\left(\frac{\sigma}{r}\right)^{8}\right] = 0$$

and find that $r_{rupture} = 0.423$ nm, which corresponds to a force of 7.33×10^{-2} eV/nm $= 11.7$ pN. A cantilever with a spring constant of 0.12 N/m would deflect about 0.1 nm at this force. Using similar techniques, Grandbois and coworkers[1] found that the covalent silicon–carbon bond breaks with a force of about 2 nN and that the sulfur–gold bond breaks with a force of about 1.4 nN.

A complete analysis of an AFM experiment would also include the potential energy of the bending cantilever.[2] However, our simplified analysis of bond rupture for a model system suggests how AFM techniques can be applied to more complex and important macromolecular systems.

43.2 THE ENERGY AND SPECTRA OF MOLECULES

As in the case of atoms, we can study the structure and properties of molecules by examining the radiation they emit or absorb. Before we describe these processes, it is important to understand the various ways of exciting a molecule.

Consider a single molecule in the gaseous phase. The energy of the molecule can be divided into four categories: (1) electronic energy, due to the interactions between the molecule's electrons and nuclei; (2) translational energy, due to the motion of the molecule's center of mass through space; (3) rotational energy, due to the rotation of the molecule about its center of mass; and (4) vibrational energy, due to the vibration of the molecule's constituent atoms:

> **Total energy of a molecule**

$$E = E_{el} + E_{trans} + E_{rot} + E_{vib}$$

The electronic energy of a molecule is very complex because it involves the interaction of many charged particles, but various techniques have been developed to approximate its values. Because the translational energy is unrelated to internal structure, this molecular energy is unimportant in interpreting molecular spectra.

Rotational Motion of Molecules

Let us consider the rotation of a molecule around its center of mass, confining our discussion to the diatomic case (Fig. 43.9a) but noting that the same ideas can be extended to polyatomic molecules. A diatomic molecule aligned along an x axis has only two rotational degrees of freedom, corresponding to rotations around the y and z axes. If ω is the angular frequency of rotation around one of these axes, the rotational kinetic energy of the molecule about that axis can be expressed in the form

> **Moment of inertia for a diatomic molecule**

$$E_{rot} = \tfrac{1}{2}I\omega^2 \tag{43.2}$$

where I is the moment of inertia of the molecule, given by

$$I = \left(\frac{m_1 m_2}{m_1 + m_2}\right)r^2 = \mu r^2 \tag{43.3}$$

---

[1] M. Grandbois, M. Beyer, M. Rief, H. Clausen–Schaumann, and H. Gaub, "How Strong Is a Covalent Bond?" *Science* 283:1727–1730, 1999.

[2] For details, see B. Shapiro and H. Qian, "A Quantitative Analysis of Single Protein–Ligand Complex Separation with the Atomic Force Microscope," *Biophys. Chem.* 67:211–219, 1997.

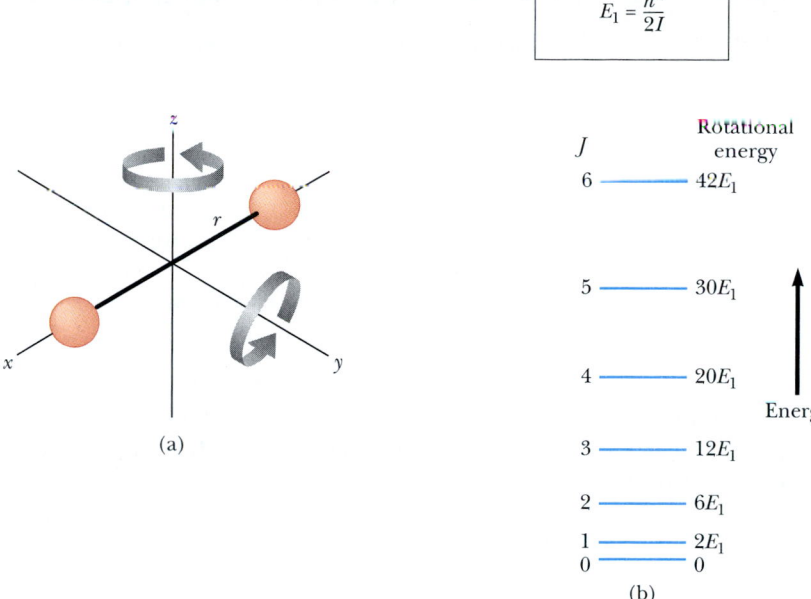

$$E_1 = \frac{\hbar^2}{2I}$$

(a)

J	Rotational energy
6	$42E_1$
5	$30E_1$
4	$20E_1$
3	$12E_1$
2	$6E_1$
1	$2E_1$
0	0

Energy

(b)

Figure 43.9 (a) A diatomic molecule oriented along the x axis has two rotational degrees of freedom, corresponding to rotation around the y and z axes. (b) Allowed rotational energies of a diatomic molecule calculated with Equation 43.6.

where m_1 and m_2 are the masses of the atoms that form the molecule, r is the atomic separation, and μ is the **reduced mass** of the molecule:

$$\mu = \frac{m_1 m_2}{m_1 + m_2} \qquad \textbf{(43.4)}$$

The magnitude of the angular momentum of the molecule is $I\omega$, which classically can have any value. Quantum mechanics, however, restricts angular momentum to the values

$$I\omega = \sqrt{J(J+1)}\,\hbar \qquad J = 0, 1, 2, \ldots \qquad \textbf{(43.5)}$$

where J is an integer called the **rotational quantum number.** Substituting Equation 43.5 into Equation 43.2, we obtain an expression for the allowed values of the rotational kinetic energy:

$$E_{\text{rot}} = \tfrac{1}{2} I\omega^2 = \frac{1}{2I}(I\omega)^2 = \frac{\left(\sqrt{J(J+1)}\,\hbar\right)^2}{2I}$$

$$E_{\text{rot}} = \frac{\hbar^2}{2I} J(J+1) \qquad J = 0, 1, 2, \ldots \qquad \textbf{(43.6)}$$

Thus, we see that **the rotational energy of the molecule is quantized and depends on its moment of inertia.** The allowed rotational energies of a diatomic molecule are plotted in Figure 43.9b.

For most molecules, the transitions between adjacent rotational energy levels result in radiation that lies in the microwave range of frequencies ($f \sim 10^{11}$ Hz). When a molecule absorbs a microwave photon, the molecule jumps from a lower rotational energy level to a higher one. The allowed rotational transitions of linear

TABLE 43.1   Several Rotational Transitions of the CO Molecule

Rotational Transition	Wavelength of Absorbed Photon (m)	Frequency of Absorbed Photon (Hz)
$J = 0 \to J = 1$	2.60×10^{-3}	1.15×10^{11}
$J = 1 \to J = 2$	1.30×10^{-3}	2.30×10^{11}
$J = 2 \to J = 3$	8.77×10^{-4}	3.46×10^{11}
$J = 3 \to J = 4$	6.50×10^{-4}	4.61×10^{11}

From G. M. Barrows, *The Structure of Molecules*, New York, W. A. Benjamin, 1963.

molecules are regulated by the selection rule $\Delta J = \pm 1$. That is, an absorption line in the microwave spectrum of a linear molecule corresponds to an energy separation equal to $E_J - E_{J-1}$. From Equation 43.6, we see that the allowed transitions are given by the condition

$$\Delta E = E_J - E_{J-1} = \frac{\hbar^2}{2I}[J(J+1) - (J-1)J]$$

Separation between adjacent rotational levels

$$\Delta E = \frac{\hbar^2}{I}J = \frac{h^2}{4\pi^2 I}J \tag{43.7}$$

where J is the rotational quantum number of the higher energy state. Because $\Delta E = hf$, where f is the frequency of the absorbed photon, we see that the allowed frequency for the transition $J = 0$ to $J = 1$ is $f_1 = h/4\pi^2 I$. The frequency corresponding to the $J = 1$ to $J = 2$ transition is $2f_1$, and so on. These predictions are in excellent agreement with the observed frequencies.

The wavelengths and frequencies for the microwave absorption spectrum of the carbon monoxide molecule are given in Table 43.1. From these data, we can evaluate the moment of inertia and bond length of the molecule.

EXAMPLE 43.1   Rotation of the CO Molecule

The $J = 0$ to $J = 1$ rotational transition of the CO molecule occurs at 1.15×10^{11} Hz. (a) Use this information to calculate the moment of inertia of the molecule.

Solution   From Equation 43.7, we see that the energy difference between the $J = 0$ and $J = 1$ rotational levels is $h^2/4\pi^2 I$. Equating this ΔE value to the energy of the absorbed photon, we have

$$\Delta E = \frac{h^2}{4\pi^2 I} = hf$$

Solving for I gives

$$I = \frac{h}{4\pi^2 f} = \frac{6.626 \times 10^{-34}\,\text{J}\cdot\text{s}}{4\pi^2(1.15 \times 10^{11}\,\text{s}^{-1})}$$

$$= \boxed{1.46 \times 10^{-46}\,\text{kg}\cdot\text{m}^2}$$

(b) Calculate the bond length of the molecule.

Solution   We can use Equation 43.3 to calculate the bond length, but we first need to know the value for the reduced mass μ of the CO molecule:

$$\mu = \frac{m_1 m_2}{m_1 + m_2} = \frac{(12\,\text{u})(16\,\text{u})}{12\,\text{u} + 16\,\text{u}} = 6.86\,\text{u}$$

$$= (6.86\,\text{u})\left(1.66 \times 10^{-27}\,\frac{\text{kg}}{\text{u}}\right) = 1.14 \times 10^{-26}\,\text{kg}$$

where we have used the fact that $1\,\text{u} = 1.66 \times 10^{-27}$ kg.

Substituting this value and the result of part (a) into Equation 43.3 and solving for r, we obtain

$$r = \sqrt{\frac{I}{\mu}} = \sqrt{\frac{1.46 \times 10^{-46}\,\text{kg}\cdot\text{m}^2}{1.14 \times 10^{-26}\,\text{kg}}}$$

$$= 1.13 \times 10^{-10}\,\text{m} = \boxed{0.113\,\text{nm}}$$

Vibrational Motion of Molecules

A molecule is a flexible structure in which the atoms are bonded together by what can be considered "effective springs" (see Fig. 13.11). If disturbed, the molecule can vibrate and acquire vibrational energy. This vibrational motion and corresponding vibrational energy can be altered if the molecule is exposed to electromagnetic waves of the proper frequency.

Consider the diatomic molecule shown in Figure 43.10a. Its effective spring has a force constant k. A plot of potential energy versus atomic separation for such a molecule is sketched in Figure 43.10b, where r_0 is the equilibrium atomic separation. According to classical mechanics, the frequency of vibration for this system is

$$f = \frac{1}{2\pi}\sqrt{\frac{k}{\mu}} \qquad \textbf{(43.8)}$$

where again μ is the reduced mass given by Equation 43.4.

As we expect, the quantum mechanical solution to this system shows that the energy is quantized, with allowed energies

$$E_{\text{vib}} = (v + \tfrac{1}{2})hf \qquad v = 0, 1, 2, \ldots \qquad \textbf{(43.9)}$$

where v is an integer called the **vibrational quantum number.** If the system is in the lowest vibrational state, for which $v = 0$, its zero-point energy is $\frac{1}{2}hf$. The accompanying vibration—the *zero-point motion*—is always present, even if the molecule is not excited. In the first excited state, $v = 1$ and the vibrational energy is $\frac{3}{2}hf$, and so on.

Substituting Equation 43.8 into Equation 43.9, we obtain the following expression for the vibrational energy:

$$E_{\text{vib}} = (v + \tfrac{1}{2})\frac{h}{2\pi}\sqrt{\frac{k}{\mu}} \qquad v = 0, 1, 2, \ldots \qquad \textbf{(43.10)}$$

◀ Allowed values of vibrational energy

The selection rule for the allowed vibrational transitions is $\Delta v = \pm 1$. From Equation 43.10, we see that the energy difference between any two successive vibra-

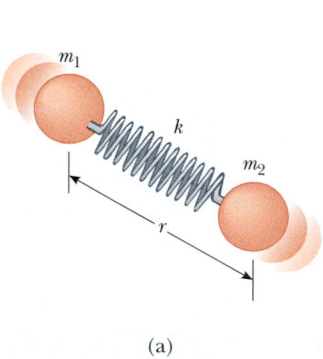

(a)

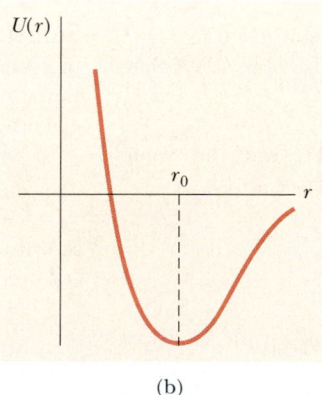

(b)

Figure 43.10 (a) Effective-spring model of a diatomic molecule. The vibration is along the molecular axis. (b) Plot of the potential energy of a diatomic molecule versus atomic separation distance, where r_0 is the equilibrium separation distance of the atoms.

TABLE 43.2 Photon Frequency and Effective-Spring Force Constant for $v = 0$ to $v = 1$ Transition in Some Diatomic Molecules

Molecule	Photon Frequency (Hz)	Force Constant k (N/m)
HF	8.72×10^{13}	970
HCl	8.66×10^{13}	480
HBr	7.68×10^{13}	410
HI	6.69×10^{13}	320
CO	6.42×10^{13}	1 850
NO	5.63×10^{13}	1 530

From G. M. Barrows, *The Structure of Molecules*, New York, W. A. Benjamin, 1963. The k values were calculated from Equation 43.11.

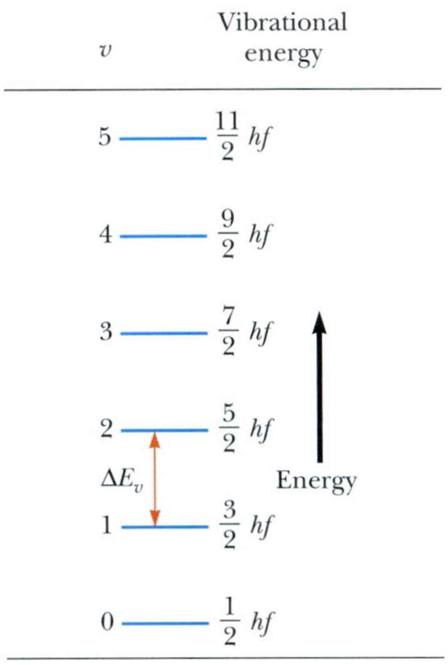

Figure 43.11 Allowed vibrational energies of a diatomic molecule, where f is the frequency of vibration of the molecule, given by Equation 43.8. The spacings between adjacent vibrational levels are equal if the molecule behaves as a harmonic oscillator.

tional levels is

$$\Delta E_{\text{vib}} = \frac{h}{2\pi}\sqrt{\frac{k}{\mu}} = hf \tag{43.11}$$

The vibrational energies of a diatomic molecule are plotted in Figure 43.11. At ordinary temperatures, most molecules have vibrational energies corresponding to the $v = 0$ state because the spacing between vibrational states is much greater than $k_B T$. The molecules are not thermally excited into higher states.

Transitions between vibrational levels are caused by absorption in the infrared region of the spectrum. That is, a molecule jumps from a lower to a higher vibrational energy level by absorbing a photon having a frequency in the infrared range. The photon frequencies corresponding to the $v = 0$ to $v = 1$ transition are listed in Table 43.2 for several diatomic molecules, together with the force constants of the effective springs holding the molecules together. The latter values were calculated by using Equation 43.11. The "stiffness" of a bond can be measured by the size of the effective force constant.

EXAMPLE 43.2   Vibration of the CO Molecule

The frequency of the photon that causes the $v = 0$ to $v = 1$ transition in the CO molecule is 6.42×10^{13} Hz. (a) Calculate the force constant k for this molecule.

Solution We can use Equation 43.11 and the value $\mu = 1.14 \times 10^{-26}$ kg we calculated in Example 43.1b:

$$\frac{h}{2\pi}\sqrt{\frac{k}{\mu}} = hf$$

$$k = 4\pi^2\mu f^2$$
$$= 4\pi^2(1.14 \times 10^{-26}\ \text{kg})(6.42 \times 10^{13}\ \text{s}^{-1})^2$$
$$= \boxed{1.85 \times 10^3\ \text{N/m}}$$

(b) What is the maximum amplitude of vibration for this molecule in the $v = 0$ vibrational state?

Solution The maximum potential energy stored in the molecule is $\frac{1}{2}kA^2$, where A is the amplitude of vibration. Equating this maximum energy to the vibrational energy given by Equation 43.10 with $v = 0$, we have

$$\frac{1}{2}kA^2 = \frac{h}{4\pi}\sqrt{\frac{k}{\mu}}$$

Substituting the value $k = 1.85 \times 10^3$ N/m and the value for μ from part (a), we obtain

$$A^2 = \frac{h}{2\pi k}\sqrt{\frac{k}{\mu}}$$
$$= \frac{6.626 \times 10^{-34}\ \text{J}\cdot\text{s}}{2\pi(1.85 \times 10^3\ \text{N/m})}\sqrt{\frac{1.85 \times 10^3\ \text{N/m}}{1.14 \times 10^{-26}\ \text{kg}}}$$
$$= 2.30 \times 10^{-23}\ \text{m}^2$$

Thus,

$$A = 4.79 \times 10^{-12}\,\text{m} = \boxed{4.79 \times 10^{-3}\,\text{nm}}$$

Comparing this result with the bond length of 0.113 nm we

calculated in Example 43.1b, we see that the amplitude of vibration is about 4% of the bond length. Thus, we see that infrared spectroscopy provides useful information on the elastic properties (bond strengths) of molecules.

Molecular Spectra

In general, a molecule vibrates and rotates simultaneously. To a first approximation, these motions are independent of each other, and so the total energy of the molecule is the sum of Equations 43.6 and 43.9:

$$E = (v + \tfrac{1}{2})hf + \frac{\hbar^2}{2I}J(J+1) \tag{43.12}$$

The energy levels of any molecule can be calculated from this expression, and each level is indexed by the two quantum numbers J and v. From these calculations, an energy level diagram like the one shown in Figure 43.12a can be constructed. For each allowed value of the vibrational quantum number v, there is a complete set of rotational levels corresponding to $J = 0, 1, 2, \ldots$. Note that the energy separation between successive rotational levels is much smaller than the separation between successive vibrational levels. As noted earlier, most molecules

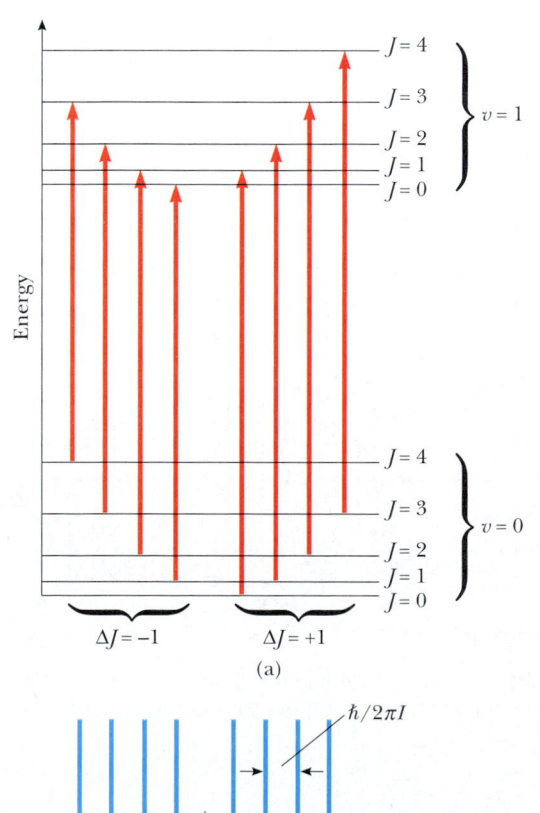

(a)

(b)

Figure 43.12   (a) Absorptive transitions between the $v = 0$ and $v = 1$ vibrational states of a diatomic molecule. The transitions obey the selection rule $\Delta J = \pm 1$ and fall into two sequences, those for $\Delta J = +1$ and those for $\Delta J = -1$. The transition energies are given by Equation 43.12. (b) Expected lines in the absorption spectrum of a molecule. The lines to the right of the center mark correspond to transitions in which J changes by $+1$; the lines to the left of the center mark correspond to transitions for which J changes by -1. These same lines appear in the emission spectrum.

at ordinary temperatures are in the $v = 0$ vibrational state; these molecules can be in various rotational states, as Figure 43.12a shows.

When a molecule absorbs an infrared photon, the vibrational quantum number v increases by one unit while the rotational quantum number J either increases or decreases by one unit, as can be seen in Figure 43.12. Thus, the molecular absorption spectrum consists of two groups of lines: one group to the right of center and satisfying the selection rules $\Delta J = 1$ and $\Delta v = 1$, and the other group to the left of center and satisfying[3] the selection rules $\Delta J = -1$ and $\Delta v = 1$.

The energies of the absorbed photons can be calculated from Equation 43.12:

$$\Delta E = hf + \frac{\hbar^2}{I}\,(J + 1) \qquad J = 0, 1, 2, \ldots \; (\Delta J = +1) \qquad \textbf{(43.13)}$$

$$\Delta E = hf - \frac{\hbar^2}{I}\,J \qquad\qquad J = 1, 2, 3, \ldots \; (\Delta J = -1) \qquad \textbf{(43.14)}$$

where now J is the rotational quantum number of the *initial* state. Equation 43.13 generates the series of equally spaced lines *higher* than the frequency f, whereas Equation 43.14 generates the series *lower* than this frequency. Adjacent lines are separated in frequency by the fundamental unit $\hbar/2\pi I$; we can see this by substituting hf for ΔE in Equation 43.7 and setting $J = 1$. Figure 43.12b shows the expected frequencies in the absorption spectrum of the molecule; these same frequencies appear in the emission spectrum.

The absorption spectrum of the HCl molecule shown in Figure 43.13 follows this pattern very well and reinforces our model. However, one peculiarity is apparent: Each line is split into a doublet. This doubling occurs because two chlorine isotopes (see Section 1.2) were present in the sample used to obtain this spectrum. Because the isotopes have different masses, the two HCl molecules have different values of I.

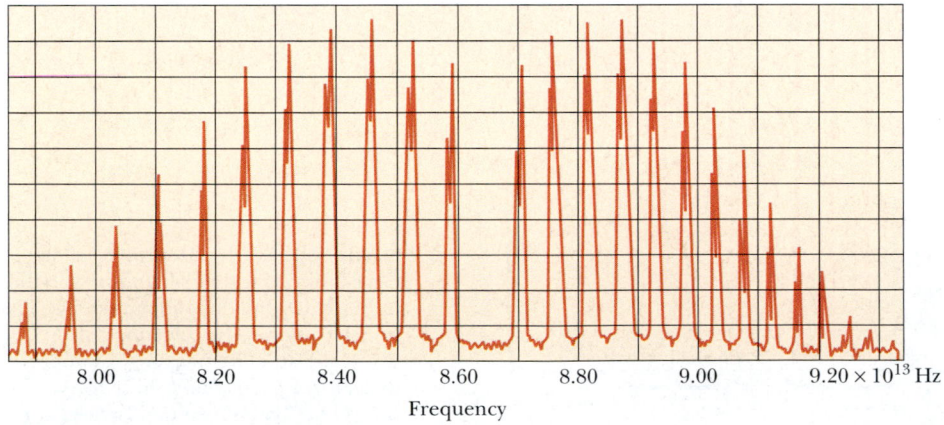

Figure 43.13   Absorption spectrum of the HCl molecule. Each line is split into a doublet because the sample contained two chlorine isotopes that had different masses and therefore different moments of inertia.

---

[3] The selection rule $\Delta J = \pm 1$ implies that the photon causing the transition is a spin-1 particle having a spin quantum number $s = 1$. Hence, this selection rule describes conservation of angular momentum for the system that consists of the molecule and the photon.

Quick Quiz 43.2

Using Figure 43.13, estimate the moment of inertia of an HCl molecule.

For CO_2 molecules, most of the absorption lines are in the infrared portion of the spectrum. Thus, visible light from the Sun is not absorbed by atmospheric CO_2 but instead strikes the Earth's surface, warming it. In turn, the Earth emits infrared radiation. These IR waves are absorbed by the CO_2 in the air instead of radiating out into space. Thus, atmospheric CO_2 acts like a one-way valve for energy from the Sun. The burning of fossil fuels can add more CO_2 to the atmosphere. Many scientists fear that substantial climatic changes might result from an enhanced "greenhouse effect."

43.3 BONDING IN SOLIDS

A crystalline solid consists of a large number of atoms arranged in a regular array, forming a periodic (in other words, repeating) structure. Two of the bonding mechanisms described in Section 43.1—ionic and covalent—are appropriate to use in describing bonds in solids. For example, the ions in the NaCl crystal are ionically bonded, as already noted, and the carbon atoms in the crystal that we call diamond form covalent bonds with one another. The metallic bond described at the end of this section is responsible for the cohesion of copper, silver, sodium, and other solid metals.

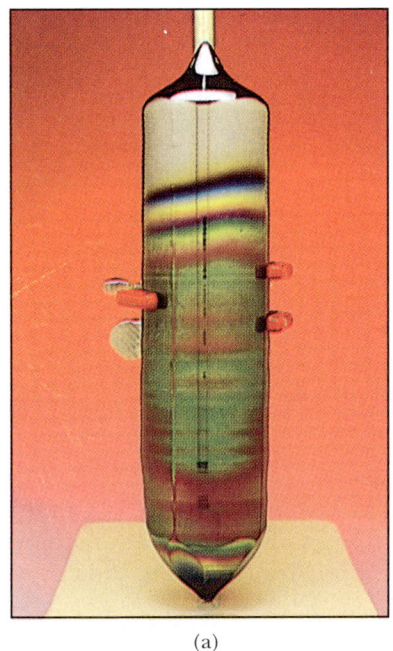

(a)   (b)

Crystalline solids. (a) A cylinder of nearly pure crystalline silicon (Si), approximately 25 cm long. Such crystals are cut into wafers and processed to make various semiconductor devices. (b) Although this crystal is called a Herkimer "diamond," it is natural quartz (SiO_2), one of the most common minerals on the Earth. Quartz crystals are used to make special camera lenses and prisms and in certain electronic applications. *(Charles D. Winters)*

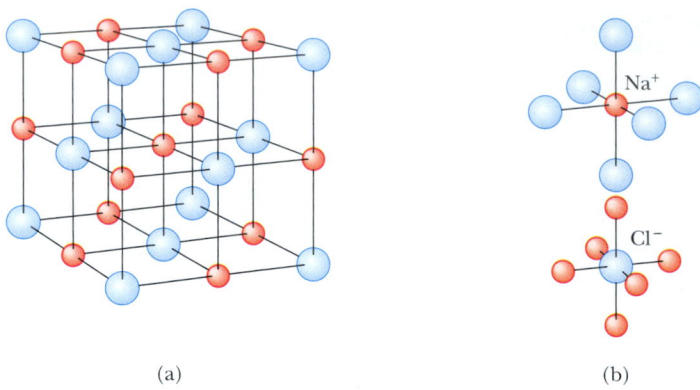

(a)   (b)

Figure 43.14   (a) Crystalline structure of NaCl. (b) Each positive sodium ion (red spheres) is surrounded by six negative chloride ions (blue spheres), and each chloride ion is surrounded by six sodium ions.

Ionic Solids

Many crystals are formed by ionic bonding, in which the dominant interaction between ions is the Coulomb interaction. Consider the NaCl crystal in Figure 43.14. Each Na^+ ion has six nearest-neighbor Cl^- ions, and each Cl^- ion has six nearest-neighbor Na^+ ions. Each Na^+ ion is attracted to its six Cl^- neighbors. The corresponding attractive potential energy is $-6k_e e^2/r$, where k_e is the Coulomb constant and r is the separation distance between each Na^+ and Cl^-. In addition, there are 12 Na^+ ions at a distance of $\sqrt{2}\,r$ from the Na^+, and these 12 positive ions exert weaker repulsive forces on the central Na^+. Furthermore, beyond these 12 Na^+ ions are more Cl^- ions that exert an attractive force, and so on. The net effect of all these interactions is a resultant negative electric potential energy

$$U_{attractive} = -\alpha k_e \frac{e^2}{r} \qquad \textbf{(43.15)}$$

where α is a pure number known as the **Madelung constant.** The value of α depends only on crystalline structure. For example, $\alpha = 1.747\,6$ for the NaCl structure. When the constituent ions of a crystal are brought close together, a repulsive force exists because of electrostatic forces and the exclusion principle, as discussed in Section 43.1. The potential energy term B/r^m in Equation 43.1 accounts for this. Therefore, the total potential energy is

$$U_{total} = -\alpha k_e \frac{e^2}{r} + \frac{B}{r^m} \qquad \textbf{(43.16)}$$

where m in this expression is some small integer.

A plot of total potential energy versus ion separation distance is shown in Figure 43.15. The potential energy has its minimum value U_0 at the equilibrium separation, when $r = r_0$. It is left as a problem for you (Problem 47) to show that

$$U_0 = -\alpha k_e \frac{e^2}{r_0}\left(1 - \frac{1}{m}\right) \qquad \textbf{(43.17)}$$

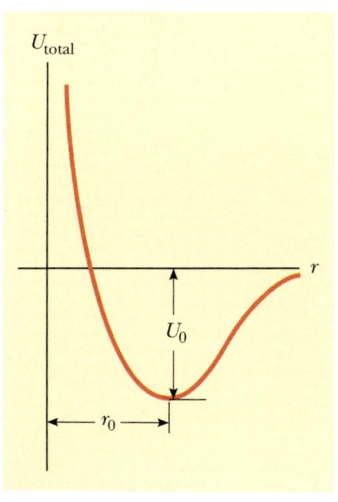

Figure 43.15   Total potential energy versus ion separation distance for an ionic solid, where U_0 is the ionic cohesive energy and r_0 is the equilibrium separation distance between ions.

This minimum energy U_0 is called the **ionic cohesive energy** of the solid, and its absolute value represents the energy required to separate the solid into a collection of isolated positive and negative ions. Its value for NaCl is -7.84 eV per ion pair.

To calculate the **atomic cohesive energy,** which is the binding energy relative to the energy of the neutral atoms, we must add 5.14 eV to the ionic cohesive energy value to account for the transition from Na^+ to Na, and we must subtract 3.61 eV to account for the conversion of Cl^- to Cl. Thus, the atomic cohesive energy of NaCl is

$$-7.84 \text{ eV} + 5.14 \text{ eV} - 3.61 \text{ eV} = -6.31 \text{ eV}$$

Ionic crystals have the following general properties:

- They form relatively stable, hard crystals. (The melting point of NaCl is 801°C.)
- They are poor electrical conductors because they contain no free electrons.
- They have high vaporization temperatures.
- They are transparent to visible radiation but absorb strongly in the infrared region. No visible light is absorbed because the shells formed by the electrons in ionic solids are so tightly bound that visible radiation does not possess sufficient energy to promote electrons to the next allowed shell. Infrared radiation is absorbed strongly because the vibrations of the ions have a natural resonant frequency in the low-energy infrared region.
- Many are quite soluble in polar liquids, such as water. The polar solvent molecules exert an attractive electric force on the charged ions, which breaks the ionic bonds and dissolves the solid.

> Properties of ionic solids

Covalent Solids

Solid carbon, in the form of diamond, is a crystal whose atoms are covalently bonded. Because atomic carbon has the electronic configuration $1s^2 2s^2 2p^2$, it is four electrons short of filling its $n = 2$ shell, which can accommodate eight electrons. Hence, two carbon atoms have a strong attraction for each other, with a cohesive energy of 7.37 eV.

In the diamond structure, each carbon atom is covalently bonded to four other carbon atoms located at four corners of a cube, as shown in Figure 43.16a.

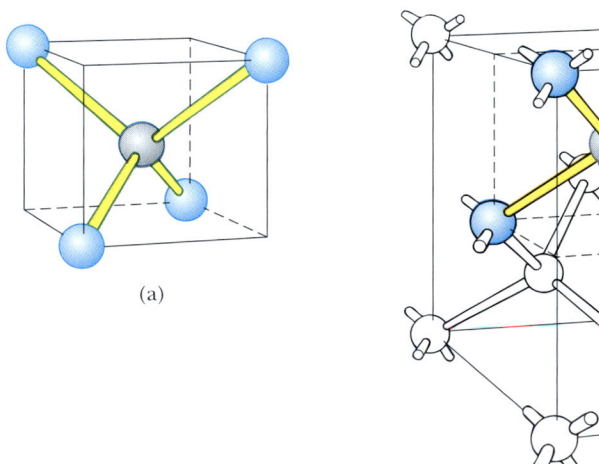

(a)

(b)

Figure 43.16 (a) Each carbon atom in a diamond crystal is covalently bonded to four other carbon atoms and forms a tetrahedral structure. (b) The crystal structure of diamond, showing the tetrahedral bond arrangement.

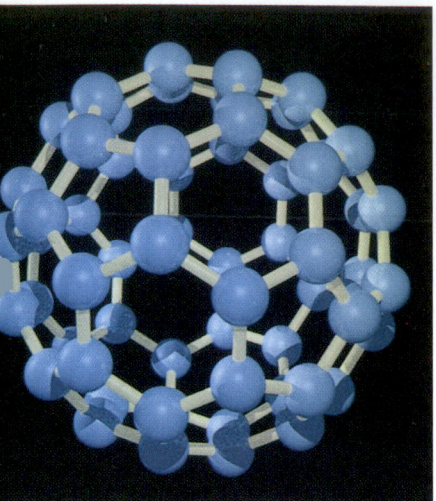

Figure 43.17   Computer rendering of a "buckyball," short for the molecule buckminsterfullerene. These nearly spherical molecular structures that look like soccer balls were named for R. Buckminster Fuller, inventor of the geodesic dome. This form of carbon, C_{60}, was discovered by astrophysicists while investigating the carbon gas that exists between stars. Scientists are actively studying the properties and potential uses of buckminsterfullerene and related molecules. *(Charles D. Winters)*

To form such a configuration of bonds, one $2s$ electron of each atom must be promoted to the $2p$ subshell so that the electronic configuration becomes $1s^2 2s^1 2p^3$, which corresponds to a half-filled p subshell. The promotion of this electron requires an energy input of about 4 eV.

The crystalline structure of diamond is shown in Figure 43.16b. Note that each carbon atom forms covalent bonds with four nearest-neighbor atoms. The basic structure of diamond is called tetrahedral (each carbon atom is at the center of a regular tetrahedron), and the angle between the bonds is 109.5°. Such other crystals as silicon and germanium have a similar structure.

When carbon atoms form a large hollow structure, the compound is called **buckminsterfullerene** after the famous architect who invented the geodesic dome. The unique shape of this molecule (Fig. 43.17) provides a "cage" to hold other atoms or molecules. Related structures, called "buckytubes" because of their long, narrow cylindrical arrangements of carbon atoms, may provide the basis for extremely strong, yet lightweight materials.

The atomic cohesive energies of some covalent solids are given in Table 43.3. The large energies account for the hardness of covalent solids. Diamond is particularly hard and has an extremely high melting point (about 4 000 K). Covalently bonded solids are usually very hard, have high bond energies and high melting points, and are good electrical insulators.

Metallic Solids

Metallic bonds are generally weaker than ionic or covalent bonds. The outer electrons in the atoms of a metal are relatively free to move throughout the material, and the number of such mobile electrons in a metal is large. The metallic structure can be viewed as a "sea" or a "gas" of nearly free electrons surrounding a lattice of positive ions (Fig. 43.18). The binding mechanism in a metal is the attractive force between the positive ions and the electron gas. Metals have a cohesive energy in the range of 1 to 3 eV, which is less than the cohesive energies of ionic or covalent solids.

Light interacts strongly with the free electrons in metals. Hence, visible light is absorbed and re-emitted quite close to the surface of a metal, which accounts for the shiny nature of metal surfaces. (Compare this to the transparency of the glass shown in the photograph at the beginning of this chapter. Visible light does not interact strongly with the electrons of the glass.) In addition to the high electrical conductivity of metals produced by the free electrons, the nondirectional nature

Metal ion

Electron gas

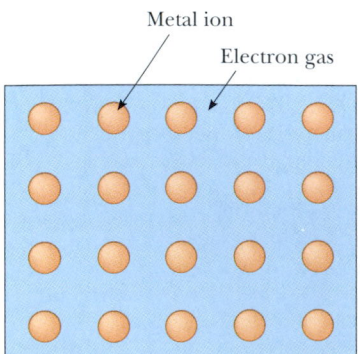

Figure 43.18   Highly schematic diagram of a metal. The blue area represents the electron gas, and the red circles represent the positive metal ions.

TABLE 43.3	Cohesive Energies of Some Covalent Solids
Solid	**Cohesive Energy (eV)**
C (diamond)	7.37
Si	4.63
Ge	3.85
InAs	5.70
SiC	6.15
ZnS	6.32
CuCl	9.24

of the metallic bond allows many different types of metal atoms to be dissolved in a host metal in varying amounts. The resulting *solid solutions*, or *alloys*, may be designed to have particular properties, such as tensile strength, ductility, electrical and thermal conductivity, and resistance to corrosion. Such properties are usually controllable and in many cases predictable.

43.4 BAND THEORY OF SOLIDS

If two identical atoms are very far apart, they do not interact and their electronic energy levels can be considered to be those of isolated atoms. Suppose that the two atoms are sodium, each having a lone 3s electron that has a specific, well-defined energy. As the two sodium atoms are brought closer together, their outer orbits begin to overlap. When the interaction between them is sufficiently strong, two 3s energy levels are formed,[4] as shown in Figure 43.19a.

When a large number of atoms are brought together to form a solid, a similar phenomenon occurs. As the atoms are brought close together, the various isolated-atom energy levels split. This splitting in levels for six atoms in close proximity is shown in Figure 43.19b. In this case, there are six energy levels corresponding to six different combinations of isolated-atom wave functions. Because the range of energy values into which overlapping levels split is not a function of the number of atoms being combined, the six-atom energy levels are more closely spaced than the two-atom levels.

If we extend this argument to the large number of atoms found in solids (of the order of 10^{23} atoms per cm^3), we obtain a large number of levels so closely spaced that they may be regarded as a continuous **band** of energy levels, as shown in Figure 43.19c. In the case of sodium, it is customary to refer to the continuous distribution of allowed energy levels as the 3s band because the band originates from the 3s levels of the individual sodium atoms.

In general, a crystalline solid has a large number of allowed energy bands that arise from the various atomic energy levels. Figure 43.20 shows the allowed energy bands of sodium. Note that energy gaps, called *forbidden energy bands*, occur between the allowed bands.

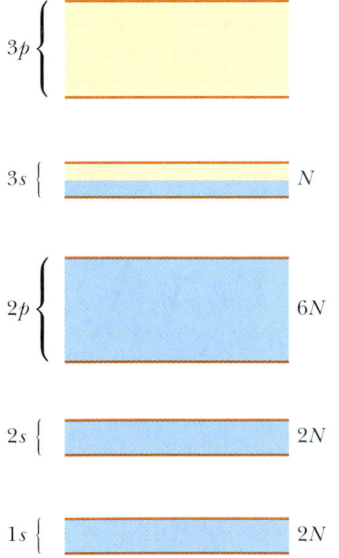

Figure 43.20 Energy bands of sodium. Note the energy gaps (white regions) between the allowed bands; electrons cannot occupy states that lie in these forbidden gaps. Blue represents energy bands occupied by the 11 sodium electrons when the atom is in its ground state. Gold represents energy bands that are empty.

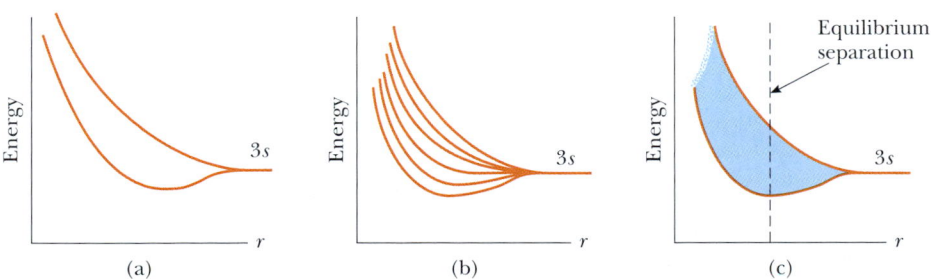

Figure 43.19 (a) Splitting of the 3s levels when two sodium atoms are brought together. (b) Splitting of the 3s levels when six sodium atoms are brought together. (c) Formation of a 3s band when a large number of sodium atoms are assembled to form a solid.

[4] There is a split in energy levels because of the two ways the wave functions of the atoms can combine. If ψ_1 and ψ_2 are the two wave functions, the $\psi_1 + \psi_2$ state results in a high electron density between the two atomic cores, whereas the $\psi_1 - \psi_2$ state has zero probability of finding the electron between the two cores. Because the electron is more tightly bound (has lower energy) when it is between the atoms, the $\psi_1 + \psi_2$ state has slightly lower energy than the $\psi_1 - \psi_2$ state. Thus, there are two slightly separated energy levels for the electron when the two atoms are brought close together.

For N atoms combined in any solid, each energy band contains N energy levels. The $1s$, $2s$, and $2p$ bands of sodium are each full, as indicated by the blue-shaded areas in Figure 43.20. An energy level in which the orbital angular momentum is ℓ can hold $2(2\ell + 1)$ electrons. The factor 2 arises from the two possible electron spin orientations, and the factor $2\ell + 1$ corresponds to the number of possible orientations of the orbital angular momentum. The capacity of each band for a system of N atoms is $2(2\ell + 1)N$ electrons. Hence, the $1s$ and $2s$ bands each contain $2N$ electrons ($\ell = 0$), and the $2p$ band contains $6N$ electrons ($\ell = 1$). Because sodium has only one $3s$ electron and there are a total of N atoms in the solid, the $3s$ band contains only N electrons and is only half full, as indicated by the half-blue, half-gold coloring in Figure 43.20. The $3p$ band, which is above the $3s$ band, is completely empty (all gold in the figure).

In Section 43.6, we shall discuss how the band theory allows us to understand the behavior of conductors, insulators, and semiconductors.

43.5 ▸ FREE-ELECTRON THEORY OF METALS

In Section 27.3 we described a classical theory of electrical conduction in metals that led to Ohm's law. According to this theory, a metal is modeled as a classical gas of conduction electrons moving through a fixed lattice of ion cores. Although this theory predicts the correct functional form of Ohm's law, it does not predict the correct values of electrical and thermal conductivities.

In this section, we discuss the free-electron theory of metals, which remedies the shortcomings of the classical model by taking into account the wave nature of the electrons. In this model, one imagines that the outer-shell electrons are free to move through the metal but are trapped within a cavity formed by the surface of the metal.

Statistical physics can be applied to a collection of particles in an effort to relate microscopic properties to macroscopic properties. In the case of electrons, it is necessary to use *quantum statistics*, with the requirement that each state of the system can be occupied by only one electron. Each state is specified by a set of quantum numbers. The probability that a particular state having energy E is occupied by one of the electrons in a solid is given by

Fermi–Dirac distribution function

$$f(E) = \frac{1}{e^{(E-E_F)/k_B T} + 1} \qquad (43.18)$$

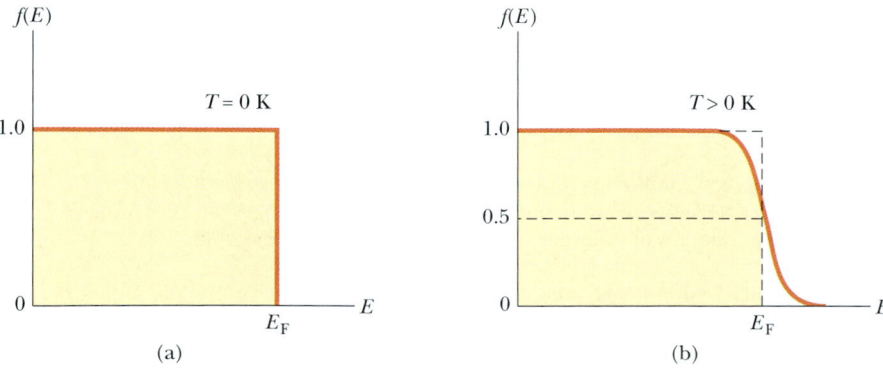

Figure 43.21 Plot of the Fermi–Dirac distribution function $f(E)$ versus energy at (a) $T = 0$ K and (b) $T > 0$ K. The energy E_F is the Fermi energy.

where $f(E)$ is called the **Fermi–Dirac distribution function** and E_F is called the **Fermi energy.** This expression describes how energy is distributed among the electrons. A plot of $f(E)$ versus E at $T = 0$ K is shown in Figure 43.21a. Note that $f(E) = 1$ for $E < E_F$ and $f(E) = 0$ for $E > E_F$. That is, at 0 K all states having energies less than the Fermi energy are occupied, and all states having energies greater than the Fermi energy are vacant. A plot of $f(E)$ versus E at some temperature $T > 0$ K is shown in Figure 43.21b. This curve shows that as T increases, the distribution rounds off slightly, with states between E and $E - k_BT$ losing population and states between E and $E + k_BT$ gaining population. The Fermi energy E_F also depends on temperature, but the dependence is weak in metals.

Quick Quiz 43.3

In Figure 43.21b, what is the physical meaning of the curved part of the plot near E_F?

Quick Quiz 43.4

Where is the Fermi energy level in Figure 43.20?

In Section 41.4 we found that if a particle of mass m is confined to move in a one-dimensional box of length L, the allowed states have quantized energy levels:

$$E_n = \frac{h^2}{8mL^2}\, n^2 = \frac{\hbar^2\pi^2}{2mL^2}\, n^2 \qquad n = 1, 2, 3, \ldots$$

The wave functions for these allowed states are standing waves given by $\psi = A\sin(n\pi x/L)$, which satisfy the boundary condition $\psi = 0$ at $x = 0$ and $x = L$.

Now imagine a piece of metal in the shape of a solid cube of sides L and volume L^3, and let us focus on one electron that is free to move anywhere in this volume. In this model, we require that $\psi(x, y, z) = 0$ at the boundaries. This requirement results in solutions that are standing waves in three dimensions. It can be shown (see Problem 30) that the energy for such an electron is

$$E = \frac{\hbar^2\pi^2}{2m_eL^2}\, (n_x^2 + n_y^2 + n_z^2) \qquad \textbf{(43.19)}$$

where m_e is the mass of the electron and n_x, n_y, and n_z are quantum numbers. Again, the energy levels are quantized, and each is characterized by this set of three quantum numbers (one for each degree of freedom) and the spin quantum number m_s. For example, the ground state, corresponding to $n_x = n_y = n_z = 1$, has an energy equal to $3\hbar^2\pi^2/2m_eL^2$.

If the quantum numbers are treated as continuous variables, the number of allowed states per unit volume that have energies between E and $E + dE$ is

$$g(E)\, dE = CE^{1/2}dE \qquad \textbf{(43.20)}$$

where

$$C = \frac{8\sqrt{2}\,\pi m_e^{3/2}}{h^3} \qquad \textbf{(43.21)}$$

The function $g(E) = CE^{1/2}$ is called the **density-of-states function.**

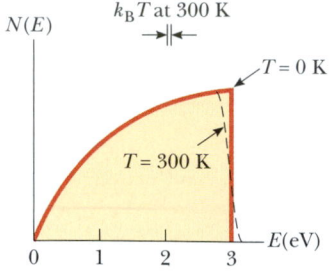

Figure 43.22 Plot of electronic distribution versus energy in a metal at $T = 0$ K (curved and vertical rust lines) and $T = 300$ K (curved rust line and dashed black line). The Fermi energy E_F is 3 eV.

In a metal in thermal equilibrium, the number of electrons per unit volume that have energy between E and $E + dE$ is equal to the product $f(E)g(E)\,dE$:

$$N(E)\,dE = C\frac{E^{1/2}dE}{e^{(E-E_F)/k_B T} + 1} \tag{43.22}$$

A plot of $N(E)$ versus E for two temperatures is given in Figure 43.22.

If n_e is the total number of electrons per unit volume, we require that

$$n_e = \int_0^\infty N(E)\,dE = C\int_0^\infty \frac{E^{1/2}dE}{e^{(E-E_F)/k_B T} + 1} \tag{43.23}$$

We can use this condition to calculate the Fermi energy. At $T = 0$ K, the Fermi distribution function $f(E) = 1$ for $E < E_F$ and $f(E) = 0$ for $E > E_F$. Therefore, at $T = 0$ K, Equation 43.23 becomes

$$n_e = C\int_0^{E_F} E^{1/2}dE = \tfrac{2}{3}C E_F^{3/2} \tag{43.24}$$

Substituting Equation 43.21 into Equation 43.24 gives for the Fermi energy at 0 K

$$E_F(0) = \frac{h^2}{2m_e}\left(\frac{3n_e}{8\pi}\right)^{2/3} \tag{43.25}$$

According to this result, E_F shows a gradual increase with increasing electron concentration. This is expected because the electrons fill the available energy states, two electrons per state, in accordance with the exclusion principle, up to the Fermi energy.

The order of magnitude of the Fermi energy for metals is about 5 eV. Representative values for various metals are given in Table 43.4, together with values for the Fermi speed v_F of the electrons, which is the speed of the electrons when their energy is equal to the Fermi energy. The Fermi speed is defined by the relationship

$$\tfrac{1}{2}m_e v_F^2 \equiv E_F \tag{43.26}$$

Table 43.4 also lists Fermi temperatures T_F, defined by the relationship

$$k_B T_F \equiv E_F \tag{43.27}$$

It is left as a problem for you (Problem 29) to show that the average energy of a free electron in a metal at 0 K is

$$E_{av} = \tfrac{3}{5}E_F \tag{43.28}$$

In summary, we can consider a metal to be a system comprising a very large number of energy levels available to the free electrons. These electrons fill these levels in accordance with the exclusion principle, beginning with $E = 0$ and ending with E_F. At $T = 0$ K, all levels below the Fermi energy are filled and all levels above the Fermi energy are empty. Although the levels are discrete, they are so close together that the electrons have an almost continuous distribution of energy. At 300 K, a very small fraction of the free electrons are excited above the Fermi energy.

TABLE 43.4 Calculated Values of Various Parameters for Metals at 300 K Based on the Free-Electron Theory

Metal	Electron Concentration (m^{-3})	Fermi Energy (eV)	Fermi Speed (m/s)	Fermi Temperature (K)
Li	4.70×10^{28}	4.72	1.29×10^{6}	5.48×10^{4}
Na	2.65×10^{28}	3.23	1.07×10^{6}	3.75×10^{4}
K	1.40×10^{28}	2.12	0.86×10^{6}	2.46×10^{4}
Cu	8.49×10^{28}	7.05	1.57×10^{6}	8.12×10^{4}
Ag	5.85×10^{28}	5.48	1.39×10^{6}	6.36×10^{4}
Au	5.90×10^{28}	5.53	1.39×10^{6}	6.41×10^{4}

EXAMPLE 43.3   The Fermi Energy of Gold

Each atom of gold (Au) contributes one free electron to the metal. Compute (a) the Fermi energy, (b) the Fermi speed, and (c) the Fermi temperature for gold.

Solution   (a) The concentration of free electrons in gold is $5.90 \times 10^{28}\ m^{-3}$ (see Table 43.4). Substitution of this value into Equation 43.25 gives

$$E_F(0) = \frac{h^2}{2m_e}\left(\frac{3n_e}{8\pi}\right)^{2/3}$$

$$= \frac{(6.626 \times 10^{-34}\ \text{J·s})^2}{2(9.11 \times 10^{-31}\ \text{kg})}\left(\frac{3 \times 5.90 \times 10^{28}\ m^{-3}}{8\pi}\right)^{2/3}$$

$$= 8.85 \times 10^{-19}\ \text{J} = \boxed{5.53\ \text{eV}}$$

(b) The Fermi speed is defined by Equation 43.26, $\frac{1}{2}m_e v_F^2 = E_F$. Solving for v_F gives

$$v_F = \left(\frac{2E_F}{m_e}\right)^{1/2} = \left(\frac{2 \times 8.85 \times 10^{-19}\ \text{J}}{9.11 \times 10^{-31}\ \text{kg}}\right)^{1/2}$$

$$= \boxed{1.39 \times 10^{6}\ \text{m/s}}$$

(c) The Fermi temperature is defined by Equation 43.27:

$$T_F = \frac{E_F}{k_B} = \frac{8.85 \times 10^{-19}\ \text{J}}{1.38 \times 10^{-23}\ \text{J/K}} = \boxed{6.41 \times 10^{4}\ \text{K}}$$

Thus, the temperature of a gas of classical particles would have to be raised to approximately 64 000 K to have an average energy per particle equal to the Fermi energy at 0 K!

43.6 ELECTRICAL CONDUCTION IN METALS, INSULATORS, AND SEMICONDUCTORS

Good electrical conductors contain a high density of charge carriers, and the density of charge carriers in insulators is nearly zero. Semiconductors are a class of technologically important materials in which charge-carrier densities are intermediate between those of insulators and those of conductors. In this section, we discuss the mechanisms of conduction in these three classes of materials. The enormous variation in electrical conductivity of these materials can be explained in terms of energy bands.

Metals

If a material is to conduct electricity, the charge carriers in the material must be free to move in response to an applied electric field. Let us consider the electrons in a metal as the charge carriers we shall investigate. The motion of the electrons in response to an electric field represents an increase in energy corresponding to the additional kinetic energy of the moving electrons. Thus, to respond to an electric field, electrons must move upward to a higher energy state on an energy level diagram. For them to be able to do this, energy states must be available above the filled states in the band. If a band is completely filled with electrons, no such states are available, and the electrons cannot respond to the electric field by moving.

In Section 43.4, we described the energy-band configuration for the ground state of metallic sodium. We can obtain a better understanding of how metals act as electrical conductors by considering a half-filled band, such as the $3s$ band of sodium.

Figure 43.23 shows a half-filled band in a metal at $T = 0$ K, where the blue represents levels filled with electrons. Because electrons obey Fermi–Dirac statistics, all levels below the Fermi energy are filled with electrons, and all levels above the Fermi energy are empty. The Fermi energy lies in the middle of the band, as was discussed for sodium in Quick Quiz 43.4. At temperatures slightly greater than 0 K, some electrons are thermally excited to levels above E_F, but overall there is little change from the 0 K case. However, **if a potential difference is applied to the metal, electrons having energies near the Fermi energy require only a small amount of additional energy from the applied field to reach nearby empty energy states above the Fermi energy.** Thus, electrons in a metal experiencing only a small applied field are free to move because there are many empty levels available close to the occupied energy levels. We conclude from this high degree of electron mobility that metals are excellent electrical conductors.

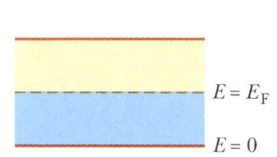

Metal

Figure 43.23 Half-filled band of a metal, an electrical conductor. At $T = 0$ K, the Fermi energy lies in the middle of the band.

Insulators

Now consider the two outermost energy bands of a material, where the lower band is filled with electrons and the higher band is empty at 0 K (Fig. 43.24). It is common to refer to the energy separation between the outermost filled band and the adjacent empty band as the **energy gap** E_g of the material. The energy gap for an insulator is large (≈ 10 eV). The lower, filled band is called the **valence band,** and the upper, empty band is the **conduction band.** The Fermi energy lies somewhere in the energy gap, as shown in Figure 43.24. At 300 K (room temperature), $k_B T = 0.025$ eV, which is much smaller than the energy gap in an insulator. At such temperatures, the Fermi–Dirac distribution predicts very few electrons thermally excited into the conduction band. Thus, although an insulator has many vacant states in its conduction band that can accept electrons, so few electrons occupy these states that the overall electrical conductivity is very small, resulting in a high resistivity for insulators.

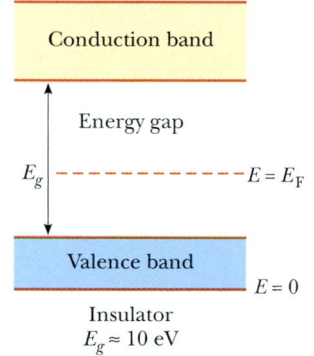

Insulator
$E_g \approx 10$ eV

Figure 43.24 An electrical insulator at $T = 0$ K has a filled valence band and an empty conduction band. The Fermi level lies somewhere between these bands in the region known as the energy gap.

Semiconductors

Semiconductors have the band structure of an insulator and an energy gap on the order of 1 eV. Table 43.5 shows the energy gaps for some representative materials. At $T = 0$ K, all electrons in these materials are in the valence band, and no energy is available to excite them across the energy gap. Thus, semiconductors are poor conductors at very low temperatures. At ordinary temperatures, however, the situation is quite different. For example, the conductivity of silicon at room temperature is about 1.6×10^{-3} $(\Omega \cdot m)^{-1}$.

The band structure of a semiconductor is shown in Figure 43.25. Because the Fermi level is located near the middle of the gap for a semiconductor and because

TABLE 43.5	Energy Gap Values for Some Semiconductors*	
	E_g (eV)	
Crystal	**0 K**	**300 K**
Si	1.17	1.14
Ge	0.744	0.67
InP	1.42	1.35
GaP	2.32	2.26
GaAs	1.52	1.43
CdS	2.582	2.42
CdTe	1.607	1.45
ZnO	3.436	3.2
ZnS	3.91	3.6

*From C. Kittel, *Introduction to Solid State Physics*, 5th ed., New York, John Wiley & Sons, 1976.

E_g is small, appreciable numbers of electrons are thermally excited from the valence band to the conduction band. There are many empty levels in the conduction band; therefore, a small applied potential difference can easily raise the energy of the electrons in the conduction band, resulting in a moderate current. Because electrons' being thermally excited across the narrow gap is more probable at higher temperatures, the conductivity of semiconductors increases rapidly with temperature. This contrasts sharply with the conductivity of metals, which decreases slowly with temperature, as described at the end of Section 27.3.

Charge carriers in a semiconductor can be negative or positive, or both. When an electron moves from the valence band into the conduction band, it leaves behind a vacant site, called a **hole,** in the otherwise filled valence band. This hole (electron-deficient site) appears as a positive charge $+e$ and acts as a charge carrier in the sense that a free electron from a nearby site can transfer into the hole. Whenever an electron does so, it creates a new hole at the site it abandoned. Thus, the net effect can be viewed as the hole migrating through the material in the direction opposite the direction of electron movement.

In a pure crystal containing only one element or one compound, there are equal numbers of conduction electrons and holes. Such combinations of charges are called **electron-hole pairs,** and a pure semiconductor that contains such pairs is called an **intrinsic semiconductor** (Fig. 43.26). In the presence of an external electric field, the holes move in the direction of the field, and the conduction electrons move in the direction opposite the field.

Doped Semiconductors

When impurities are added to a semiconductor, both the band structure of the semiconductor and its resistivity are modified. The process of adding impurities, called **doping,** is important in making devices and semiconductors having well-defined regions of different conductivities. For example, when an atom containing five outer-shell electrons, such as arsenic, is added to a semiconductor, four of the electrons form covalent bonds with atoms of the semiconductor and one is left

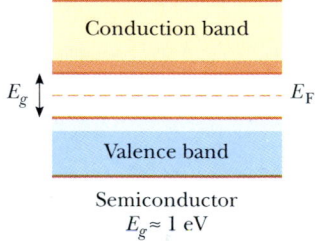

Figure 43.25 Band structure of a semiconductor at ordinary temperatures ($T \approx 300$ K). The energy gap is much smaller than in an insulator, and many electrons occupy states in the conduction band.

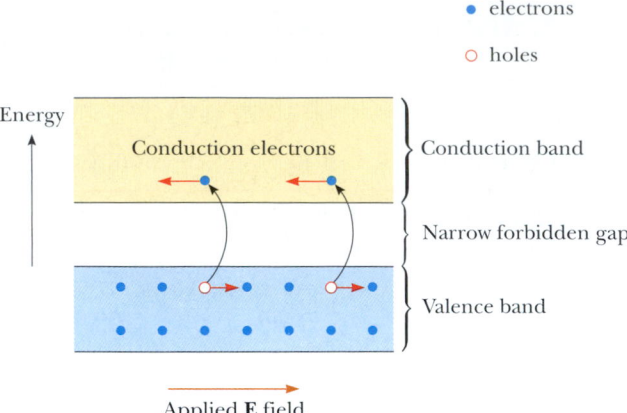

Figure 43.26 Movement of charges (holes and electrons) in an intrinsic semiconductor. The electrons move in the direction opposite the direction of the external electric field, and the holes move in the direction of the field.

over (Fig. 43.27a). This extra electron is nearly free of its parent atom and has an energy level that lies in the energy gap, just below the conduction band (Fig. 43.27b). Such a pentavalent atom in effect donates an electron to the structure and hence is referred to as a **donor atom.** Because the spacing between the energy level of the electron of the donor atom and the bottom of the conduction band is very small (typically, about 0.05 eV), only a small amount of thermal excitation is needed to cause this electron to move into the conduction band. (Recall that the average energy of an electron at room temperature is about $k_\mathrm{B}T \approx 0.025$ eV). Semiconductors doped with donor atoms are called **n-type semiconductors** because the majority of charge carriers are electrons, which are **n**egatively charged.

If a semiconductor is doped with atoms containing three outer-shell electrons, such as indium and aluminum, the three form covalent bonds with neighboring

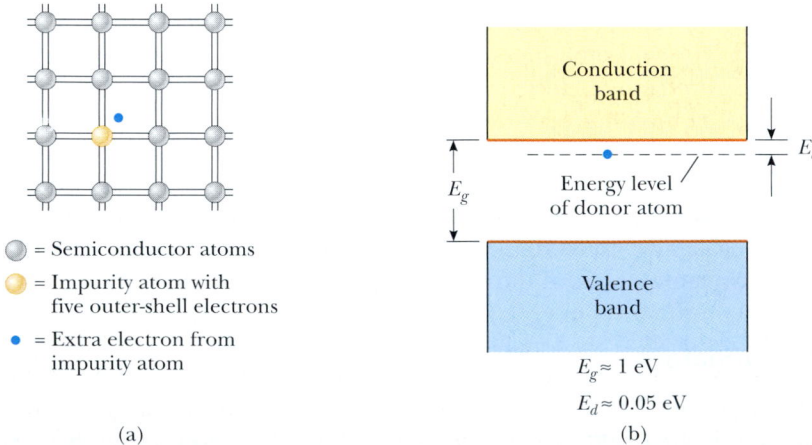

Figure 43.27 (a) Two-dimensional representation of a semiconductor (gray) containing an impurity atom (yellow-orange) that has five outer-shell electrons. Each double line represents a covalent bond. (b) Energy-band diagram for a semiconductor in which the nearly free electron of the impurity atom lies in the forbidden gap, just below the bottom of the conduction band.

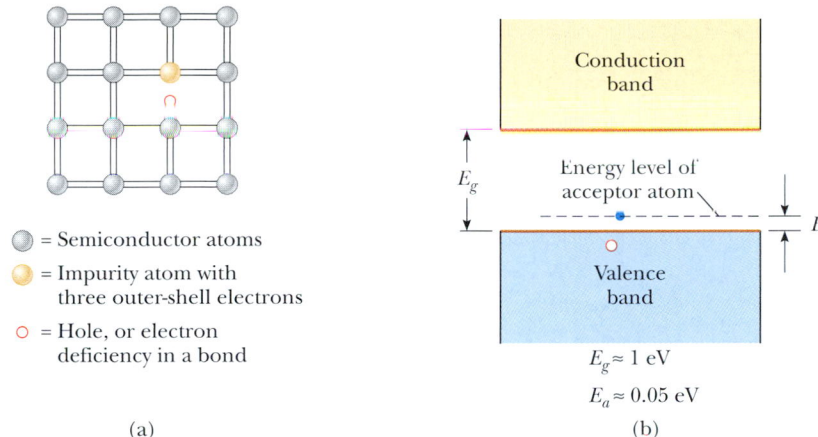

Figure 43.28   (a) Two-dimensional representation of a semiconductor (gray) containing an impurity atom (yellow-orange) having three outer-shell electrons. (b) Energy-band diagram for a semiconductor in which the hole resulting from the trivalent impurity atom lies in the forbidden gap, just above the top of the valence band.

semiconductor atoms, leaving an electron deficiency—a hole—where the fourth bond would be if an impurity-atom electron were available to form it (Fig. 43.28a). The energy level of this hole lies in the energy gap, just above the valence band, as shown in Figure 43.28b. An electron from the valence band has enough energy at room temperature to fill this impurity level, leaving behind a hole in the valence band. Because a trivalent atom in effect accepts an electron from the valence band, such impurities are referred to as **acceptor atoms.** A semiconductor doped with trivalent (acceptor) impurities is known as a ***p*-type semiconductor** because the majority of charge carriers are **p**ositively charged holes.

When conduction in a semiconductor is the result of acceptor or donor impurities, the material is called an **extrinsic semiconductor.** The typical range of doping densities for extrinsic semiconductors is 10^{13} to 10^{19} cm^{-3}, whereas the electron density in a typical semiconductor is roughly 10^{21} cm^{-3}.

Optional Section

43.7 ▸ SEMICONDUCTOR DEVICES

The *p–n* Junction

Now let us consider what happens when a *p*-type semiconductor is joined to an *n*-type semiconductor to form a *p–n* junction. The junction consists of the three distinct regions shown in Figure 43.29a: a *p* region, a depletion region, and an *n* region.

The depletion region, which extends several micrometers to either side of the center of the junction, may be visualized as arising when the two halves of the junction are brought together. The mobile *n*-side donor electrons nearest the junction (deep-blue area in Fig. 43.29a) diffuse to the *p* side, leaving behind immobile positive ions. At the same time, holes from the *p* side nearest the junction diffuse to the *n* side and leave behind a region (brown area in Fig. 43.29a) of fixed negative ions.

The depletion region contains an internal electric field (arising from the charges of the fixed ions) on the order of 10^4 to 10^6 V/cm (see Fig. 43.29b). This

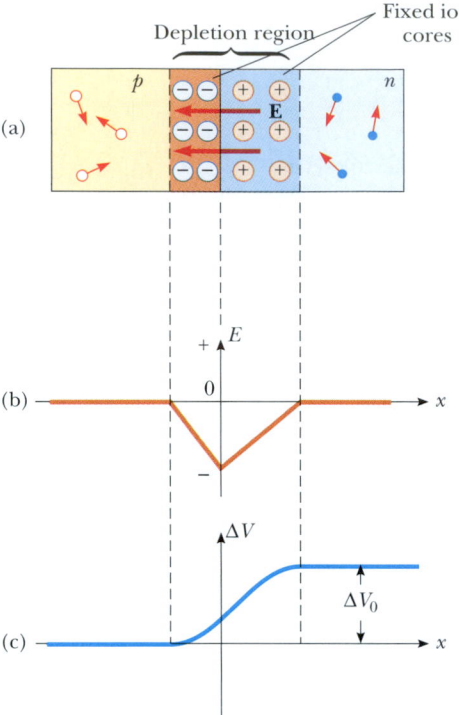

Figure 43.29   (a) Physical arrangement of a p–n junction. (b) Internal electric field versus x for the p–n junction. (c) Internal electric potential difference ΔV versus x for the p–n junction. The potential difference ΔV_0 represents the potential difference across the junction in the absence of an applied electric field.

field sweeps mobile charges out of the depletion region. Thus, the depletion region is so named because it is depleted of mobile charge carriers. This internal electric field creates an internal potential difference ΔV_0 that prevents further diffusion of holes and electrons across the junction and thereby ensures zero current through the junction when no potential difference is applied.

Perhaps the most notable feature of the p–n junction is its ability to pass current in only one direction. Such action is easiest to understand in terms of the potential difference graph shown in Figure 43.29c. If a voltage ΔV is applied to the junction such that the p side is connected to the positive terminal of a voltage source, as shown in Figure 43.30a, the internal potential difference ΔV_0 across the junction is decreased; the decrease results in a current that increases exponentially with increasing forward voltage, or *forward bias*. For *reverse bias* (where the n side of the junction is connected to the positive terminal of a voltage source), the internal potential difference ΔV_0 increases with increasing reverse bias; the increase results in a very small reverse current that quickly reaches a saturation value I_0. The current–voltage relationship for an ideal diode is

$$I = I_0(e^{e\Delta V/k_B T} - 1) \qquad (43.29)$$

where the first e is the base of the natural logarithm, the second e represents the magnitude of the electron charge, k_B is Boltzmann's constant, and T is the temperature in kelvins. Figure 43.30b shows an I–ΔV plot characteristic of a real p–n junction.

Quick Quiz 43.5

Does the p–n junction described in Figure 43.30 obey Ohm's law?

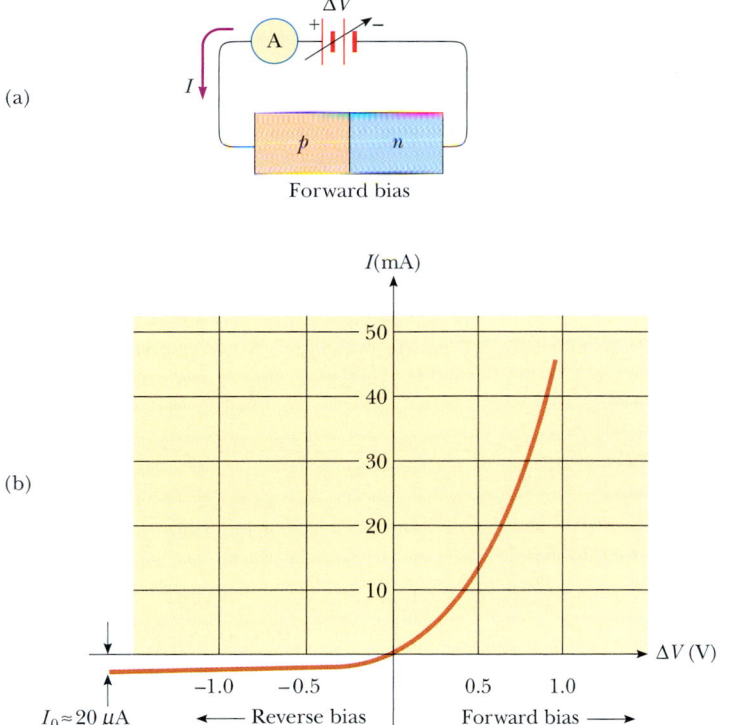

Figure 43.30   (a) Schematic of a $p-n$ junction under forward bias. (b) The characteristic curve for a real $p-n$ junction.

Light-Emitting and Light-Absorbing Diodes

Light emission and absorption in semiconductors is similar to light emission and absorption by gaseous atoms, except that in the discussion of semiconductors discrete atomic energy levels must be replaced by bands. As shown in Figure 43.31a, an electron excited electrically into the conduction band can easily recombine with a hole (especially if the electron is injected into a p region). As this recombination takes place, a photon of energy E_g is emitted. Light-emitting diodes (LEDs) and semiconductor lasers are common examples of devices using this phenomenon.

Figure 43.31   (a) Light emission from a semiconductor. (b) Light absorption by a semiconductor.

The *Sojourner,* seen here cruising on the surface of Mars in 1997, used photovoltaic solar cells to convert sunlight to electricity. (*Courtesy of NASA*)

Conversely, an electron in the valence band may absorb a photon of light and be promoted to the conduction band, leaving a hole behind (Fig. 43.31b). One device that operates on this principle is the photovoltaic solar cell.

EXAMPLE 43.4 ▶ **Where's the Remote?**

Estimate the band gap of the semiconductor in the infrared LED of a typical television remote control.

Solution In Chapter 34 we learned that the wavelength of infrared light ranges from 700 nm to 1 mm. Let us pick a number that is easy to work with, such as 1 000 nm. (This is

not a bad estimate. Remotes typically operate in the range of 880 to 950 nm.)

The energy of a photon is given by $E = hc/\lambda$, and so the energy of the photons from the remote control is about 2.0×10^{-19} J = 1.2 eV. This corresponds to an energy gap E_g of approximately 1.2 eV in the LED's semiconductor.

The Junction Transistor

The invention of the transistor by John Bardeen (1908–1991), Walter Brattain (1902–1987), and William Shockley (1910–1989) in 1948 totally revolutionized the world of electronics. For this work, these three men shared a Nobel Prize in 1956. By 1960, the transistor had replaced the vacuum tube in many electronic applications. The advent of the transistor created a multitrillion dollar industry that produces such popular devices as pocket radios, hand-held calculators, computers, television receivers, and electronic games.

One form of the junction transistor consists of a semiconducting material in which a very narrow *n* region is sandwiched between two *p* regions. This configuration is called a **pnp transistor.** Another configuration is the **npn transistor,** which consists of a *p* region sandwiched between two *n* regions. Because the operation of the two transistors is essentially the same, we describe only the *pnp* transistor. The structure of the *pnp* transistor, together with its circuit symbol, is shown in Figure 43.32. The outer regions are called the **emitter** and the **collector,** and the narrow central region is called the **base.** The configuration contains two junctions: the emitter–base interface and the collector–base interface.

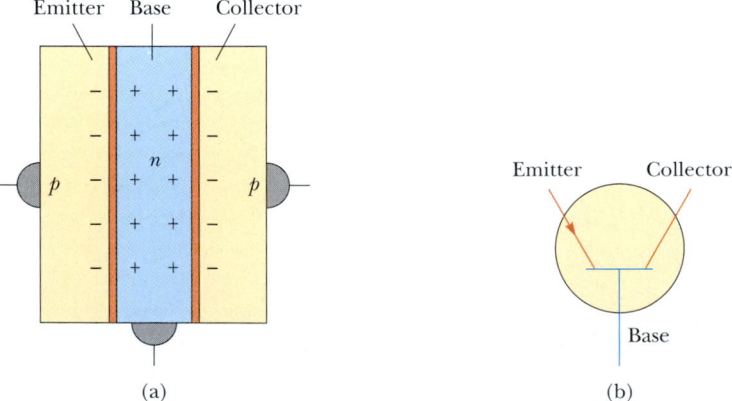

(a) (b)

Figure 43.32 (a) The *pnp* transistor consists of an *n* region (base) sandwiched between two *p* regions (emitter and collector). (b) Circuit symbol for the *pnp* transistor.

Suppose that a voltage is applied to the transistor such that the emitter is at a higher electric potential than the collector. (This is accomplished with the battery labeled ΔV_{ec} in Fig. 43.33.) If we think of the transistor as two $p-n$ junctions back to back, we see that the emitter–base junction is forward biased and that the base–collector junction is reverse-biased. The emitter is heavily doped relative to the base, and as a result nearly all the current consists of holes moving across the emitter–base junction. Most of these holes do not recombine with electrons in the base because it is very narrow. Instead, they are accelerated across the reverse-biased base-collector junction, producing the emitter current I_e shown in Figure 43.33.

Although only a small percentage of holes recombine in the base, those that do limit the emitter current to a small value because positive charge carriers accumulating in the base prevent holes from flowing in. To prevent this current limitation, some of the positive charge on the base must be drawn off; this is accomplished by connecting the base to the battery labeled ΔV_{eb}, as shown in Figure 43.33. Those positive charges that are not swept across the base–collector junction leave the base through this added pathway. **This base current I_b is very small, but a small change in it can significantly change the collector current I_c.** If the transistor is properly biased, the collector (output) current is directly proportional to the base (input) current, and the transistor acts as a current amplifier. This condition may be written

$$I_c = \beta I_b$$

where β, the *current gain* factor, is typically in the range from 10 to 100.

The transistor may be used to amplify a small signal. A small voltage to be amplified is placed in series with the battery ΔV_{eb}. The input signal produces a small variation in the base current, resulting in a large change in the collector current and hence a large change in the voltage across the output resistor.

The Integrated Circuit

Invented independently by Jack Kilby (b. 1923) at Texas Instruments in late 1958 and by Robert Noyce (b. 1927) at Fairchild Camera and Instrument in early 1959, the integrated circuit has been justly called "the most remarkable technology ever to hit mankind." Kilby's first device is shown in Figure 43.34. Integrated circuits have indeed started a "second industrial revolution" and are found at the heart of

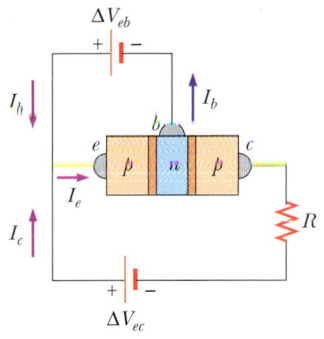

Figure 43.33 A bias voltage ΔV_{eb} applied to the base as shown produces a small base current I_b that is used to control the collector current I_c in a *pnp* transistor.

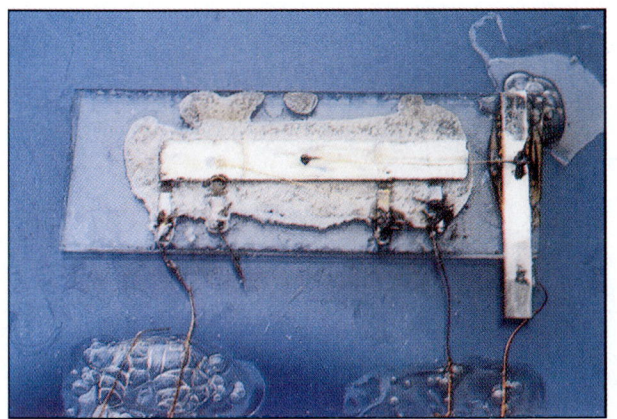

Figure 43.34 Jack Kilby's first integrated circuit, tested on September 12, 1958. *(Courtesy of Texas Instruments, Inc.)*

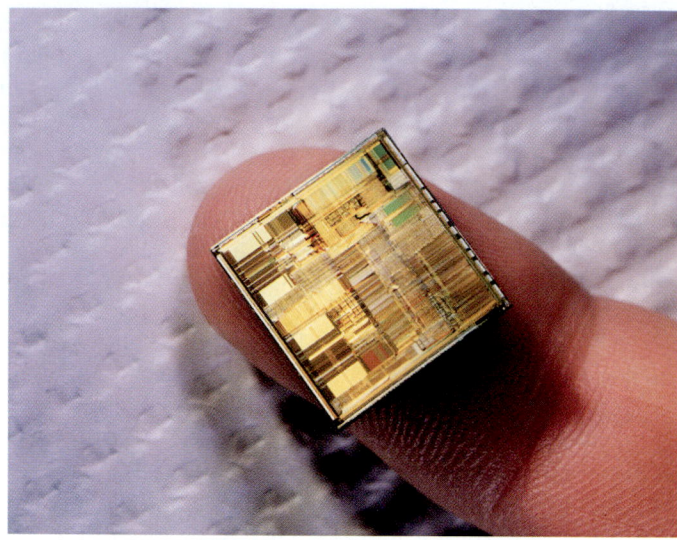

Figure 43.35 Integrated circuits continue to shrink in size and price while simultaneously growing in capability. *(Courtesy of Intel Corporation)*

computers, watches, cameras, automobiles, aircraft, robots, space vehicles, and all sorts of communication and switching networks.

In simplest terms, an **integrated circuit** is a collection of interconnected transistors, diodes, resistors, and capacitors fabricated on a single piece of silicon known as a *chip*. State-of-the-art chips easily contain several million components within a 1-cm^2 area (Fig. 43.35), with the number of components per square inch having doubled every year since the integrated circuit was invented. Figure 43.36 illustrates the dramatic advances made in chip technology in the past 30 years.

Integrated circuits were invented partly to solve the interconnection problem spawned by the transistor. In the era of vacuum tubes, power and size considera-

web

For more information, visit
www.intel.com

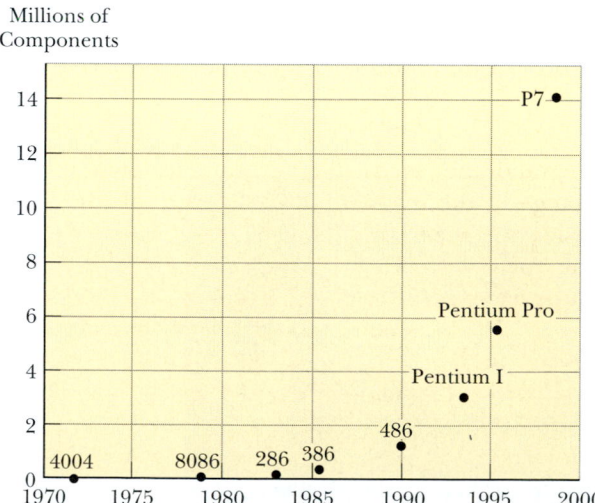

Figure 43.36 This plot illustrates dramatic advances in chip technology: The number of components fitting on a single computer chip versus year of manufacture.

tions of individual components set modest limits on the number of components that could be interconnected in a given circuit. With the advent of the tiny, low-power, highly reliable transistor, design limits on the number of components disappeared and were replaced by the problem of wiring together hundreds of thousands of components. The magnitude of this problem can be appreciated when we consider that second-generation computers (consisting of discrete transistors rather than integrated circuits) contained several hundred thousand components requiring more than a million joints that had to be hand-soldered and tested.

In addition to solving the interconnection problem, integrated circuits possess the advantages of miniaturization and fast response, two attributes critical for high-speed computers. The fast response results from the miniaturization and close packing of components, because the response time of a circuit depends on the time it takes for electrical signals traveling at about 0.3 m/ns to pass from one component to another. This time is reduced by the closely packed components.

Optional Section

43.8 ▸ SUPERCONDUCTIVITY

We learned in Section 27.5 that there is a class of metals and compounds known as **superconductors** whose electrical resistance decreases to virtually zero below a certain temperature T_c called the *critical temperature* (Table 43.6). Let us now look at these amazing materials in greater detail, using what we have just learned about the properties of solids to help us understand the behavior of superconductors.

Let us start by examining the Meissner effect, described in Section 30.8 as the exclusion of magnetic flux from the interior of superconductors. Simple arguments based on the laws of electricity and magnetism can be used to show that the magnetic field inside a superconductor cannot change with time. According to Equation 27.8, $R = \Delta V/I$, and because the potential difference ΔV across a conductor is proportional to the electric field inside the conductor, we see that the electric field is proportional to the resistance of the conductor. Thus, because $R = 0$ for a superconductor at or below its critical temperature, *the electric field in its interior must be zero*. Now recall that Faraday's law of induction can be expressed in the form shown in Equation 31.9:

$$\oint \mathbf{E} \cdot d\mathbf{s} = -\frac{d\Phi_B}{dt} \qquad (43.30)$$

That is, the line integral of the electric field around any closed loop is equal to the negative rate of change in the magnetic flux Φ_B through the loop. Because **E** is zero everywhere inside the superconductor, the integral over any closed path inside the superconductor is zero. Hence, $d\Phi_B/dt = 0$; this tells us that **the magnetic flux in the superconductor cannot change.** From this information, we can conclude that $B \,(=\Phi_B/A)$ must remain constant inside the superconductor.

Before 1933, it was assumed that superconductivity was a manifestation of perfect conductivity. If a perfect conductor is cooled below its critical temperature in the presence of an applied magnetic field, the field should be trapped in the interior of the conductor even after the external field is removed. In addition, the final state of the perfect conductor should depend on which occurs first, the application of the field or the cooling to below T_c. If the field is applied after the material has been cooled, the field should be expelled from the superconductor. If the field is applied before the material is cooled, the field should not be ex-

TABLE 43.6
Critical Temperatures for Various Superconductors

Material	T_c (K)
Zn	0.88
Al	1.19
Sn	3.72
Hg	4.15
Pb	7.18
Nb	9.46
Nb_3Sn	18.05
Nb_3Ge	23.2
$YBa_2Cu_3O_7$	92
Bi-Sr-Ca-Cu-O	105
Tl-Ba-Ca-Cu-O	125

pelled once the material has been cooled. In 1933, however, W. Hans Meissner and Robert Ochsenfeld discovered that, when a metal becomes superconducting in the presence of a weak magnetic field, the field is expelled. Thus, the same final state $\mathbf{B} = 0$ is achieved whether the field is applied before or after the material is cooled below its critical temperature.

The Meissner effect is illustrated in Figure 43.37 for a superconducting material in the shape of a long cylinder. Note that the field penetrates the cylinder when its temperature is greater than T_c (Fig. 43.37a). As the temperature is lowered to below T_c, however, the field lines are spontaneously expelled from the interior of the superconductor (Fig. 43.37b). Thus, a superconductor is more than a perfect conductor (resistivity $\rho = 0$); it is also a perfect diamagnet ($\mathbf{B} = 0$). The property that $\mathbf{B} = 0$ in the interior of a superconductor is as fundamental as the property of zero resistance. If the magnitude of the applied magnetic field exceeds a critical value B_c, defined as the value of B that destroys a material's superconducting properties, the field again penetrates the sample.

Because a superconductor is a perfect diamagnet having a negative magnetic susceptibility, it repels a permanent magnet. In fact, one can perform a demonstration of the Meissner effect by floating a small permanent magnet above a superconductor and achieving magnetic levitation, as seen in Figure 30.34.

You should recall from our study of electricity that a good conductor expels static electric fields by moving charges to its surface. In effect, the surface charges produce an electric field that exactly cancels the externally applied field inside the conductor. In a similar manner, a superconductor expels magnetic fields by forming surface currents. To see why this happens, consider again the superconductor shown in Figure 43.37. Let us assume that the sample is initially at a temperature $T > T_c$, as illustrated in Figure 43.37a, so that the magnetic field penetrates the cylinder. As the cylinder is cooled to a temperature $T < T_c$, the field is expelled, as shown in Figure 43.37b. Surface currents induced on the superconductor's surface produce a magnetic field that exactly cancels the externally applied field inside the superconductor. As you would expect, the surface currents disappear when the external magnetic field is removed.

An important development in physics that elicited much excitement in the scientific community was the discovery of high-temperature copper oxide–based superconductors. The excitement began with a 1986 publication by J. Georg Bed-

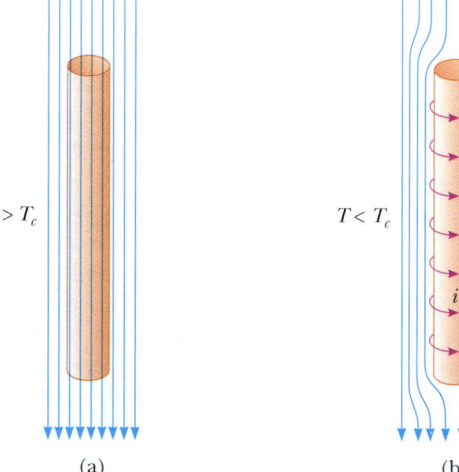

$T > T_c$ $T < T_c$

(a)   (b)

Figure 43.37   A superconductor in the form of a long cylinder in the presence of an external magnetic field. (a) At temperatures above T_c, the field lines penetrate the cylinder because it is in its normal state. (b) When the cylinder is cooled to $T < T_c$ and becomes superconducting, magnetic flux is excluded from its interior by the induction of surface currents.

norz (b. 1950) and K. Alex Müller (b. 1927), scientists at the IBM Zurich Research Laboratory in Switzerland. In their paper,[5] Bednorz and Müller reported strong evidence for superconductivity at 30 K in an oxide of barium, lanthanum, and copper. They were awarded the Nobel Prize for physics in 1987 for their remarkable discovery. Shortly thereafter, a new family of compounds was open for investigation, and research activity in the field of superconductivity proceeded vigorously. In early 1987, groups at the University of Alabama at Huntsville and the University of Houston announced superconductivity at about 92 K in an oxide of yttrium, barium, and copper ($YBa_2Cu_3O_7$). Later that year, teams of scientists from Japan and the United States reported superconductivity at 105 K in an oxide of bismuth, strontium, calcium, and copper. More recently, scientists have reported superconductivity at temperatures as high as 150 K in an oxide containing mercury (see Fig. 27.13). Today, one cannot rule out the possibility of room-temperature superconductivity, and the mechanisms responsible for the behavior of high-temperature superconductors are still under investigation. The search for novel superconducting materials continues both for scientific reasons and because practical applications become more probable and widespread as the critical temperature is raised.

SUMMARY

Two or more atoms combine to form molecules because of a net attractive force between the atoms. The mechanisms responsible for molecular bonding can be classified as follows:

- **Ionic bonds** form primarily because of the Coulomb attraction between oppositely charged ions. Sodium chloride (NaCl) is one example.
- **Covalent bonds** form when the constituent atoms of a molecule share electrons. For example, the two electrons of the H_2 molecule are equally shared between the two nuclei.
- **Van der Waals bonds** are weak electrostatic bonds between atoms that do not form ionic or covalent bonds. These bonds are responsible for the condensation of inert gas atoms and nonpolar molecules into the liquid phase.
- **Hydrogen bonds** form between the center of positive charge in a polar molecule that includes one or more hydrogen atoms and the center of negative charge in another polar molecule.

The energy of a gas molecule consists of contributions from the electronic energy in the bonds and from the translational, rotational, and vibrational motions of the molecule.

The allowed values of the rotational energy of a diatomic molecule are

$$E_{rot} = \frac{\hbar^2}{2I} J(J + 1) \qquad J = 0, 1, 2, \ldots \qquad \textbf{(43.6)}$$

where I is the moment of inertia of the molecule and J is an integer called the **rotational quantum number.** The selection rule for transitions between rotational states is given by $\Delta J = \pm 1$.

The allowed values of the vibrational energy of a diatomic molecule are

$$E_{vib} = (v + \tfrac{1}{2}) \frac{h}{2\pi} \sqrt{\frac{k}{\mu}} \qquad v = 0, 1, 2, \ldots \qquad \textbf{(43.10)}$$

where v is the **vibrational quantum number,** k is the force constant of the "effec-

[5] J. G. Bednorz and K. A. Müller, *Z. Phys. B* 64:189, 1986.

tive spring" bonding the molecule, and μ is the **reduced mass** of the molecule. The selection rule for allowed vibrational transitions is $\Delta v = \pm 1$, and the energy difference between any two adjacent levels is the same regardless of which two levels are involved.

Bonding mechanisms in solids can be classified in a manner similar to the schemes for molecules. For example, the Na^+ and Cl^- ions in NaCl form **ionic bonds,** while the carbon atoms in diamond form **covalent bonds.** The **metallic bond** is characterized by a net attractive force between positive ion cores and the mobile free electrons of a metal.

In a crystalline solid, the energy levels of the system form a set of **bands.** Electrons occupy the lowest-energy states, with no more than one electron per state. Energy gaps are present between the bands of allowed states.

In the **free-electron theory of metals,** the free electrons fill the quantized levels in accordance with the Pauli exclusion principle. The number of states per unit volume available to the conduction electrons having energies between E and $E + dE$ is

$$N(E)\,dE = C\frac{E^{1/2}\,dE}{e^{(E-E_F)/k_B T} + 1} \tag{43.22}$$

where C is a constant and E_F is the **Fermi energy.** At $T = 0$ K, all levels below E_F are filled, all levels above E_F are empty, and

$$E_F(0) = \frac{h^2}{2m_e}\left(\frac{3n_e}{8\pi}\right)^{2/3} \tag{43.25}$$

where n_e is the total number of conduction electrons per unit volume. Only those electrons having energies near E_F can contribute to the electrical conductivity of the metal.

A **semiconductor** is a material having an energy gap of approximately 1 eV and a valence band that is filled at $T = 0$ K. Because of the small energy gap, a significant number of electrons can be thermally excited from the valence band into the conduction band. The band structures and electrical properties of a semiconductor can be modified by the addition of either donor atoms containing five outer-shell electrons (such as arsenic) or acceptor atoms containing three outer-shell electrons (such as indium). A semiconductor **doped** with donor impurity atoms is called an **_n_-type semiconductor,** and one doped with acceptor impurity atoms is called a **_p_-type semiconductor.** The energy levels of these impurity atoms fall within the energy gap of the material.

QUESTIONS

1. Discuss the three major forms of excitation of a molecule (other than translational motion) and the relative energies associated with these three forms.

2. Explain the role of the Pauli exclusion principle in describing the electrical properties of metals.

3. Discuss the properties of a material that determine whether it is a good electrical insulator or a good conductor.

4. Table 43.5 shows that the energy gaps for semiconductors decrease with increasing temperature. What do you suppose accounts for this behavior?

5. The resistivity of metals increases with increasing temperature, whereas the resistivity of an intrinsic semiconductor decreases with increasing temperature. Explain.

6. Discuss the differences in the band structures of metals, insulators, and semiconductors. How does the band-structure model enable you to better understand the electrical properties of these materials?

7. Discuss models for the different types of bonds that form stable molecules.

8. Discuss the electrical, physical, and optical properties of

ionically bonded solids. Compare your expectations with tabulated properties for such solids.

9. Discuss the electrical and physical properties of covalently bonded solids. Compare your expectations with tabulated properties for such solids.

10. Discuss the electrical and physical properties of metals.

11. When a photon is absorbed by a semiconductor, an electron–hole pair is created. Give a physical explanation of this statement, using the energy-band model as the basis for your description.

12. Pentavalent atoms such as arsenic are donor atoms in a semiconductor such as silicon, while trivalent atoms such as indium are acceptors. Inspect the periodic table in Appendix C, and determine what other elements might make good donors or acceptors.

13. What are the essential assumptions made in the free-

electron theory of metals? How does the energy-band model differ from the free-electron theory in describing the properties of metals?

14. How do the vibrational and rotational levels of heavy hydrogen (D_2) molecules compare with those of H_2 molecules?

15. Which is easier to excite in a diatomic molecule, rotational or vibrational motion?

16. The energy of visible light ranges between 1.8 and 3.1 eV. Does this explain why silicon, with an energy gap of 1.1 eV (see Table 43.5), appears opaque, whereas diamond, with an energy gap of 5.5 eV, appears transparent?

17. Why is a *pnp* or *npn* sandwich (whose central region is very thin) essential to transistor operation?

18. How can the analysis of the rotational spectrum of a molecule lead to an estimate of the size of that molecule?

PROBLEMS

1, 2, 3 = straightforward, intermediate, challenging   ☐ = full solution available in the *Student Solutions Manual and Study Guide*

WEB = solution posted at **http://www.saunderscollege.com/physics/**   ▣ = Computer useful in solving problem   ▣ = Interactive Physics

☐ = paired numerical/symbolic problems

Section 43.1   Molecular Bonds

WEB 1. **Review Problem.** A K^+ ion and a Cl^- ion are separated by a distance of 5.00×10^{-10} m. Assuming that the two ions act like point charges, determine (a) the force each ion exerts on the other and (b) the potential energy of attraction in electron volts.

2. Potassium chloride is an ionically bonded molecule, sold as a salt substitute for use in a low-sodium diet. The electron affinity of chlorine is 3.6 eV. An energy input of 0.7 eV is required to form separate K^+ and Cl^- ions from separate K and Cl atoms. What is the ionization energy of K?

3. One description of the potential energy of a diatomic molecule is given by the Lennard–Jones potential,

$$U = \frac{A}{r^{12}} - \frac{B}{r^6}$$

where A and B are constants. Find, in terms of A and B, (a) the value r_0 at which the energy is a minimum and (b) the energy E required to break up a diatomic molecule. (c) Evaluate r_0 in meters and E in electron volts for the H_2 molecule. In your calculations, use the values $A = 0.124 \times 10^{-120}$ eV·m^{12} and $B = 1.488 \times 10^{-60}$ eV·m^6. (*Note:* Although this potential is widely used for modeling, it is known to have serious defects. For example, its behavior at both small and large values of r is greatly in error.)

4. A van der Waals dispersion force between helium atoms produces a very shallow potential well, with a depth on the order of 1 meV. At about what temperature would you expect helium to condense?

Section 43.2   The Energy and Spectra of Molecules

5. The cesium iodide (CsI) molecule has an atomic separation of 0.127 nm. (a) Determine the energy of the lowest excited rotational state and the frequency of the photon absorbed in the $J = 0$ to $J = 1$ transition. (b) What would be the fractional change in this frequency if the estimate of the atomic separation is off by 10%?

6. The CO molecule makes a transition from the $J = 1$ to the $J = 2$ rotational state when it absorbs a photon of frequency 2.30×10^{11} Hz. Find the moment of inertia of this molecule from these data.

WEB 7. A HCl molecule is excited to its first rotational-energy level, corresponding to $J = 1$. If the distance between its nuclei is 0.127 5 nm, what is the angular speed of the molecule about its center of mass?

8. A diatomic molecule consists of two atoms having masses m_1 and m_2 and separated by a distance r. Show that the moment of inertia about an axis through the center of mass of the molecule is given by Equation 43.3, $I = \mu r^2$.

9. (a) Calculate the moment of inertia of a NaCl molecule about its center of mass. The atoms are separated by a distance $r = 0.28$ nm. (b) Calculate the wavelength of radiation emitted if a NaCl molecule undergoes a transition from the $J = 2$ state to the $J = 1$ state.

10. The rotational spectrum of the HCl molecule contains lines with wavelengths of 0.060 4, 0.069 0, 0.080 4, 0.096 4, and 0.120 4 mm. What is the moment of inertia of the molecule?

WEB 11. If the effective force constant of a vibrating HCl molecule is $k = 480$ N/m, find the energy difference between the ground state and the first excited vibrational level.

12. Use the data in Table 43.2 to calculate the minimum amplitude of vibration for (a) the HI molecule and (b) the HF molecule. Which has the weaker bond?

13. The nuclei of the O_2 molecule are separated by 1.2×10^{-10} m. The mass of each oxygen atom in the molecule is 2.66×10^{-26} kg. (a) Determine the rotational energies of an oxygen molecule in electron volts for the levels corresponding to $J = 0$, 1, and 2. (b) The effective force constant k between the atoms in the oxygen molecule is 1 177 N/m. Determine the vibrational energies (in electron volts) corresponding to $v = 0$, 1, and 2.

14. Figure P43.14 is a model of a benzene molecule. All atoms lie in a plane, and the carbon atoms form a regular hexagon, as do the hydrogen atoms. The carbon atoms are 0.110 nm apart center to center. Determine the allowed energies of rotation about an axis perpendicular to the plane of the paper through the center point O. Hydrogen and carbon atoms have masses of 1.67×10^{-27} kg and 1.99×10^{-26} kg, respectively.

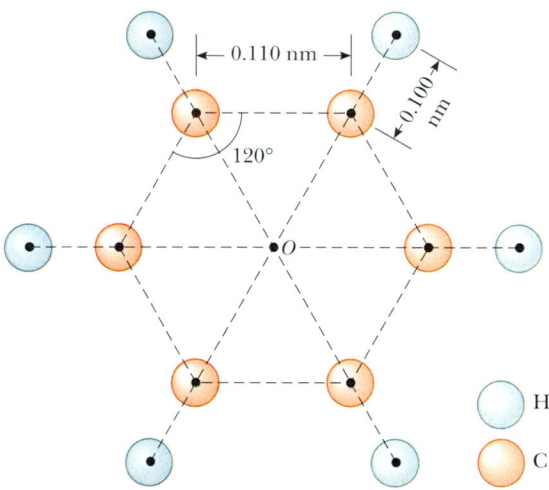

Figure P43.14

15. If the CO molecule were rigid, the rotational transition into what J state would absorb the same wavelength photon as the 0 to 1 vibrational transition? (Use information given in Table 43.2.)

16. Photons of what frequencies can be spontaneously emitted by CO molecules in the state with $v = 1$ and $J = 0$?

17. Most of the mass of an atom is in its nucleus. Model the mass distribution in a diatomic molecule as two spheres, each of radius 2.00×10^{-15} m and mass 1.00×10^{-26} kg, located at points along the x axis as shown in Figure 43.9 and separated by 2.00×10^{-10} m. Rotation about the axis joining the nuclei in the diatomic molecule is ordinarily ignored because the first

excited state would have an energy that is too high to access. To see why, calculate the ratio of the energy of the first excited state for rotation about the x axis, to the energy of the first excited state for rotation about the y axis.

Section 43.3 Bonding in Solids

18. Use a magnifying glass to look at the table salt that comes out of a salt shaker. Compare what you see with Figure 43.14(a). The distance between a sodium ion and a nearest-neighbor chloride ion is 0.261 nm. Make an order-of-magnitude estimate of the number N of atoms in a typical grain of salt. Suppose that you had a number of grains of salt equal to this number N. What would be the volume of this quantity of salt?

19. Use Equation 43.17 to calculate the ionic cohesive energy for NaCl. Take $\alpha = 1.747\,6$, $r_0 = 0.281$ nm, and $m = 8$.

20. The distance between the K^+ and Cl^- ions in a KCl crystal is 0.314 nm. Calculate the distances from one K^+ ion to its nearest-neighbor K^+ ions, to its second-nearest neighbor K^+ ions, and to its third-nearest neighbor K^+ ions.

21. Consider a one-dimensional chain of alternating positive and negative ions. Show that the potential energy associated with an ion in this hypothetical crystal is

$$U(r) = -k_e \, \alpha \, \frac{e^2}{r}$$

where the Madelung constant is $\alpha = 2 \ln 2$ and r is the interionic spacing. [*Hint:* Use the series expansion for $\ln(1 + x)$.]

Section 43.4 Band Theory of Solids
Section 43.5 Free-Electron Theory of Metals

22. Show that Equation 43.25 can be expressed as $E_F = (3.65 \times 10^{-19}) n^{2/3}$ eV where E_F is in electron volts when n is in electrons per cubic meter.

23. The Fermi energy for silver is 5.48 eV. Silver has a density of 10.6×10^3 kg/m^3 and an atomic mass of 108. Use this information to show that silver has one free electron per atom.

24. (a) Find the typical speed of a conduction electron in copper, taking its kinetic energy as equal to the Fermi energy, 7.05 eV. (b) How does this compare with a drift speed of 0.1 mm/s?

25. Sodium is a monovalent metal having a density of 0.971 g/cm^3 and molar mass of 23.0 g/mol. Use this information to calculate (a) the density of charge carriers, (b) the Fermi energy, and (c) the Fermi speed for sodium.

26. When solid silver starts to melt, what is the approximate fraction of the conduction electrons that are thermally excited above the Fermi level?

WEB 27. Calculate the energy of a conduction electron in silver at 800 K if the probability of finding an electron in that state is 0.950. The Fermi energy is 5.48 eV at this temperature.

28. Consider a cube of gold 1.00 mm on an edge. Calculate the approximate number of conduction electrons in this cube whose energies lie in the range 4.000 to 4.025 eV.

29. Show that the average kinetic energy of a conduction electron in a metal at 0 K is $E_{av} = \frac{3}{5} E_F$. (*Hint:* In general, the average kinetic energy is

$$ E_{av} = \frac{1}{n_e} \int EN(E)\, dE $$

where n_e is the density of electrons, $N(E)\, dE$ is given by Equation 43.22, and the integral is over all possible values of the energy.)

30. **Review Problem.** An electron moves in a three-dimensional box of edge length L and volume L^3. If the wave function of the particle is $\psi = A\sin(k_x x)\sin(k_y y)\sin(k_z z)$, show that its energy is given by Equation 43.19:

$$ E = \frac{\hbar^2 \pi^2}{2 m_e L^2} (n_x^2 + n_y^2 + n_z^2) $$

where the quantum numbers (n_x, n_y, n_z) are integers ≥ 1. (*Hint:* The Schrödinger equation in three dimensions may be written

$$ \frac{\partial^2 \psi}{\partial x^2} + \frac{\partial^2 \psi}{\partial y^2} + \frac{\partial^2 \psi}{\partial z^2} = \frac{\hbar^2}{2 m_e} (U - E)\psi $$

To confine the electron inside the box, take $U = 0$ inside and $U = \infty$ outside.)

31. (a) Consider a system of electrons confined to a three-dimensional box. Calculate the ratio of the number of allowed energy levels at 8.50 eV to the number at 7.00 eV. (b) Copper has a Fermi energy of 7.0 eV at 300 K. Calculate the ratio of the number of occupied levels at an energy of 8.50 eV to the number at the Fermi energy. Compare your answer with that obtained in part (a).

Section 43.6   Electrical Conduction in Metals, Insulators, and Semiconductors

32. The energy gap for silicon at 300 K is 1.14 eV. (a) Find the lowest frequency photon that will promote an electron from the valence band to the conduction band. (b) What is the wavelength of this photon?

33. Light from a hydrogen discharge tube is incident on a CdS crystal. Which spectral lines from the Balmer series are absorbed and which are transmitted?

34. A light-emitting diode (LED) made of the semiconductor GaAsP emits red light ($\lambda = 650$ nm). Determine the energy band gap E_g in the semiconductor.

WEB 35. Most solar radiation has a wavelength of 1 μm or less. What energy gap should the material in a solar cell have to absorb this radiation? Is silicon appropriate (see Table 43.5)?

36. Assume you are to build a scientific instrument that is thermally isolated from its surroundings, but such that you can use an external laser to raise the temperature of a target inside it. (It might be a calorimeter, but these design criteria could apply to other devices as well.) Since you know that diamond is transparent and a good thermal insulator, you decide to use a diamond window in the apparatus. Diamond has an energy gap of 5.5 eV between its valence and conduction bands. What is the shortest laser wavelength you can use to warm the sample inside?

(Optional)
Section 43.7   Semiconductor Devices

Note: Problem 71 in Chapter 27 also applies to this section.

37. For what value of the bias voltage ΔV in Equation 43.29 does (a) $I = 9.00\, I_0$? (b) $I = -0.900\, I_0$? Assume $T = 300$ K.

38. The diode shown in Figure 43.30 is connected in series with a battery and a 150-Ω resistor. What battery emf is required for a current of 25.0 mA?

39. You put a diode in a microelectronic circuit to protect the system in case an untrained person installs the battery backwards. In the correct forward-bias situation, the current is 200 mA with a potential difference of 100 mV across the diode at room temperature (300 K). If the battery were reversed, what would be the magnitude of the current through the diode?

(Optional)
Section 43.8   Superconductivity

Note: Problem 26 in Chapter 30 and Problems 76 through 79 in Chapter 32 also apply to this section.

40. A thin rod of superconducting material 2.50 cm long is placed into a 0.540-T magnetic field with its cylindrical axis along the magnetic field lines. (a) Sketch the directions of the applied field and the induced surface current. (b) Find the magnitude of the surface current on the curved surface of the rod.

41. Determine the current generated in a superconducting ring of niobium metal 2.00 cm in diameter if a 0.020 0-T magnetic field in a direction perpendicular to the ring is suddenly decreased to zero. The inductance of the ring is 3.10×10^{-8} H.

42. *A convincing demonstration of zero resistance.* A direct and relatively simple demonstration of zero dc resistance can be carried out using the four-point probe method. The probe shown in Figure P43.42 consists of a disk of $YBa_2Cu_3O_7$ (a high-T_c superconductor) to which four wires are attached by indium solder or some other suitable contact material. Current is maintained through the sample by applying a dc voltage between points a and b, and it is measured with a dc ammeter. The current can be varied with the variable resistance R. The potential difference ΔV_{cd} between c and d is measured

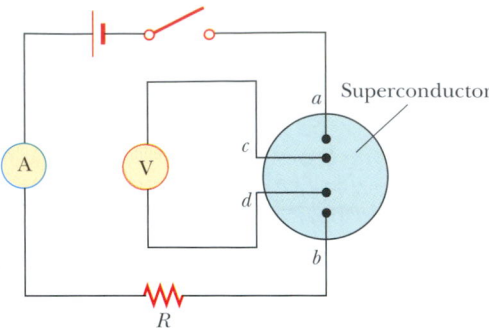

Figure P43.42   Circuit diagram used in the four-point probe measurement of the dc resistance of a sample. A dc digital ammeter is used to measure the current, and the potential difference between c and d is measured with a dc digital voltmeter. Note that there is no voltage source in the inner loop circuit where ΔV_{cd} is measured.

with a digital voltmeter. When the probe is immersed in liquid nitrogen, the sample quickly cools to 77 K, below the critical temperature of the material, 92 K. The current remains approximately constant, but ΔV_{cd} *drops abruptly to zero.* (a) Explain this observation on the basis of what you know about superconductors. (b) The data in Table P43.42 represent actual values of ΔV_{cd} for different values of I taken on the sample at room temperature. Make an I–ΔV plot of the data, and determine whether the sample behaves in a linear manner. From the data obtain a value for the dc resistance of the sample at room temperature. (c) At room temperature it is found that $\Delta V_{cd} = 2.234$ mV for $I = 100.3$ mA, but after

TABLE P43.42	Current versus Potential Difference ΔV_{cd} Measured in a Bulk Ceramic Sample of $YBa_2Cu_3O_7$ at Room Temperature[a]
I (mA)	**ΔV_{cd} (mV)**
57.8	1.356
61.5	1.441
68.3	1.602
76.8	1.802
87.5	2.053
102.2	2.398
123.7	2.904
155	3.61

[a]The current was supplied by a 6-V battery in series with a variable resistor R. The values of R ranged from 10 to 100 Ω. The data are from the author's (RAS) laboratory.

the sample is cooled to 77 K, $\Delta V_{cd} = 0$ and $I = 98.1$ mA. What do you think might cause the slight decrease in current?

ADDITIONAL PROBLEMS

43. As you will learn in Chapter 44, carbon-14 (^{14}C) is an isotope of carbon. It has the same chemical electronic structure of the much more abundant isotope carbon-12 (^{12}C), but it has different nuclear properties. Its mass is 14 u, greater because it has two extra neutrons in its nucleus. Assume that the CO molecular potential is the same for both isotopes of carbon and that the tables and examples in Section 43.2 refer to carbon monoxide with carbon-12 atoms. (a) What is the vibrational frequency of ^{14}CO? (b) What is the moment of inertia of ^{14}CO? (c) What wavelengths of light can be absorbed by ^{14}CO in the ($v = 0$, $J = 10$) state that will cause it to end up in the $v = 1$ level?

44. The effective spring constant associated with bonding in the N_2 molecule is 2 297 N/m. The nitrogen atoms each have a mass of 2.32×10^{-26} kg, and their nuclei are 0.120 nm apart. Assume that the molecule is rigid and in the ground vibrational state. Calculate the J value of the rotational state that has the same energy as the first excited vibrational state.

45. The hydrogen molecule comes apart (dissociates) when it is excited internally by 4.5 eV. Assuming that this molecule behaves like a harmonic oscillator having classical angular frequency $\omega = 8.28 \times 10^{14}$ rad/s, find the highest vibrational quantum number for a state below the 4.5-eV dissociation energy.

46. Under pressure, liquid helium can solidify as each atom bonds with four others, and each bond has an average energy of 1.74×10^{-23} J. Find the latent heat of fusion for helium in joules per gram. (The molar mass of He is 4 g/mol.)

47. Show that the ionic cohesive energy of an ionically bonded solid is given by Equation 43.17. (*Hint:* Start with Equation 43.16, and note that $dU/dr = 0$ at $r = r_0$.)

48. The dissociation energy of ground-state molecular hydrogen is 4.48 eV, whereas it only takes 3.96 eV to dissociate it when it starts in the first excited vibrational state with $J = 0$. Using this information, determine the depth of the H_2 molecular potential-energy function.

49. A particle moves in one-dimensional motion in a region where its potential energy is

$$U(x) = \frac{A}{x^3} - \frac{B}{x}$$

where $A = 0.150$ eV·nm^3 and $B = 3.68$ eV·nm. The general shape of this function is shown in Figure 43.15, where x replaces r. (a) Find the static equilibrium position x_0 of the particle. (b) Determine the depth U_0 of this potential well. (c) In moving along the x axis, what maximum force toward the negative x direction does the particle experience?

50. A particle of mass m moves in one-dimensional motion in a region where its potential energy is

$$U(x) = \frac{A}{x^3} - \frac{B}{x}$$

where A and B are constants with appropriate units. The general shape of this function is shown in Figure 43.15, where x replaces r. (a) Find the static equilibrium position x_0 of the particle in terms of m, A, and B. (b) Determine the depth U_0 of this potential well. (c) In moving along the x axis, what maximum force toward the negative x direction does the particle experience?

51. As an alternative to Equation 43.1, another useful model for the potential energy of a diatomic molecule is the Morse potential

$$U(r) = B[e^{-a(r-r_0)} - 1]^2$$

where B, a, and r_0 are parameters used to adjust the shape of the potential and its depth. (a) What is the equilibrium separation of the nuclei? (b) What is the

depth of the potential well, that is, the difference in energy between the potential's minimum value and its asymptote as r approaches infinity? (c) If μ is the reduced mass of the system of two nuclei, what is the vibrational frequency of the diatomic molecule in its ground state? (Assume that the potential is nearly parabolic about the well minimum.) (d) What amount of energy needs to be supplied to the ground-state molecule to separate the two nuclei to infinity?

52. The Fermi–Dirac distribution function can be written as

$$f(E) = \frac{1}{e^{(E-E_F)/k_B T} + 1} = \frac{1}{e^{(E/E_F - 1)T_F/T} + 1}$$

Write a spreadsheet to calculate and plot $f(E)$ versus E/E_F at a fixed temperature T. Examine the curves obtained for $T = 0.1\,T_F$, $0.2\,T_F$, and $0.5\,T_F$, where $T_F = E_F/k_B$.

53. The Madelung constant may be found by summing an infinite alternating series of terms giving the electrostatic potential energy between an Na^+ ion and its 6 nearest Cl^- neighbors, its 12 next-nearest Na^+ neighbors, and so on (Fig. 43.14a). (a) From this expression, show that the first three terms of the series yield $\alpha = 2.13$ for the NaCl structure. (b) Does this series converge rapidly? Calculate the fourth term as a check.

ANSWERS TO QUICK QUIZZES

43.1. This maximum value is the energy needed to move the sodium and chlorine ions infinitely far apart. This is sometimes called the *activation energy* and, as noted earlier, is 1.4 eV.

43.2. The spacing between adjacent peaks is approximately 0.08×10^{13} Hz. Because these lines are separated in frequency by $\hbar/2\pi I$, the moment of inertia is $1.05 \times 10^{-34}\,\text{J}\cdot\text{s}/(2\pi)\,(0.08 \times 10^{13}\,\text{Hz}) = 2.1 \times 10^{-47}\,\text{kg}\cdot\text{m}^2$, which is not much different from the value for the CO molecule calculated in Example 43.1.

43.3. At any temperature above absolute zero, internal energy $k_B T$ (≈ 0.025 eV near room temperature ≈ 300 K) is available, and this energy causes some of the electrons to have energies greater than E_F. The Fermi–Dirac distribution function $f(E)$ gives the probability of finding

an electron in a particular energy level. In Figure 43.21b, that probability is not quite 1.0 for electrons having initial energies slightly less than E_F because those electrons can absorb some of the available internal energy and now have energies greater than E_F. This results in the nonzero value of $f(E)$ for energies slightly greater than E_F.

43.4. At the blue–gold boundary in the $3s$ band. Some electrons in the $3s$ blue area have enough energy to move into the $3s$ gold area. Thus, the horizontal boundary in Figure 43.20 and the curved part of Figure 43.21b represent the same thing.

43.5. No. If Ohm's law were obeyed, the current I would be directly proportional to the potential difference ΔV across the device (see Eq. 27.8, $I = \Delta V/R$). Instead, the curve in Figure 43.30 has a slope that varies with ΔV.

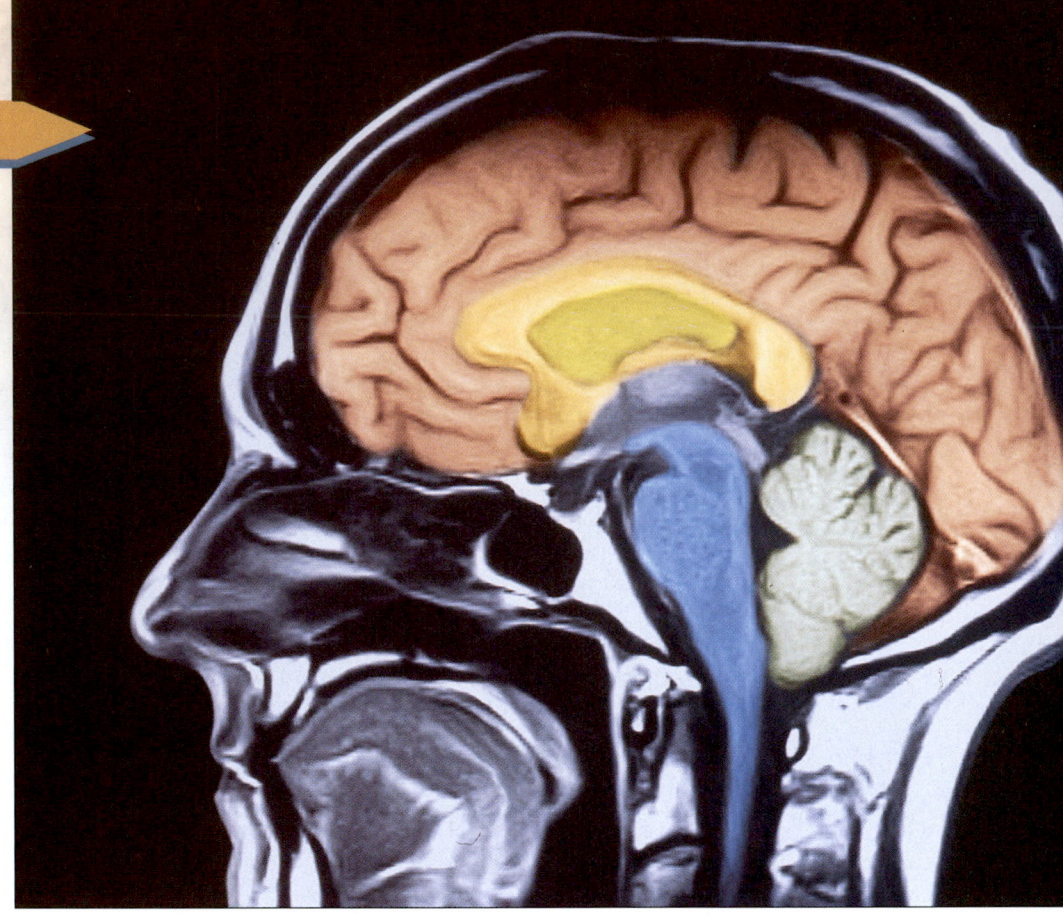

chapter

Nuclear Structure

In 1896, the year that marks the birth of nuclear physics, the French physicist Henri Becquerel (1852–1908) discovered radioactivity in uranium compounds. A great deal of research followed this discovery as scientists attempted to understand the nature of the radiation emitted by radioactive nuclei. Pioneering work by Rutherford showed that the emitted radiation was of three types: alpha, beta, and gamma rays, classified according to the nature of their electric charge and their ability to penetrate matter and ionize air. Later experiments showed that alpha rays are helium nuclei, beta rays are electrons, and gamma rays are high-energy photons.

In 1911, Rutherford, Geiger, and Marsden performed the alpha particle scattering experiments described in Section 42.1. These experiments established that (a) the nucleus of an atom can be regarded as essentially a point mass and point charge and that (b) most of the atomic mass is contained in the nucleus. Subsequent studies revealed the presence of a new type of force, the short-range nuclear force, which is predominant at distances less than approximately 10^{-14} m and is zero for large distances.

Other milestones in the development of nuclear physics include

- The observation of nuclear reactions in 1930 by Cockroft and Walton using artificially accelerated nuclei
- The discovery of the neutron in 1932 by Chadwick and the conclusion that neutrons make up about half of the nucleus
- The discovery of artificial radioactivity in 1933 by Joliot and Irene Curie
- The discovery of nuclear fission in 1938 by Hahn and Strassmann
- The development of the first controlled fission reactor in 1942 by Fermi and his collaborators

In this chapter we discuss the properties and structure of the atomic nucleus. We start by describing the basic properties of nuclei, and this description is followed by a discussion of nuclear forces and binding energy, nuclear models, and the phenomenon of radioactivity. We then discuss nuclear reactions and the various processes by which nuclei decay.

44.1 ▶ SOME PROPERTIES OF NUCLEI

All nuclei are composed of two types of particles: protons and neutrons. The only exception is the ordinary hydrogen nucleus, which is a single proton. In describing the atomic nucleus, we use the following quantities:

- The **atomic number** Z, which equals the number of protons in the nucleus (the atomic number is sometimes called the *charge number*)
- The **neutron number** N, which equals the number of neutrons in the nucleus
- The **mass number** A, which equals the number of **nucleons** (neutrons plus protons) in the nucleus

In representing nuclei, it is convenient to use the symbol $^{A}_{Z}\text{X}$ to show how many protons and neutrons are present, where X represents the chemical symbol of the element. For example, $^{56}_{26}\text{Fe}$ (iron) has mass number 56 and atomic number 26; therefore, it contains 26 protons and 30 neutrons. When no confusion is likely to arise, we omit the subscript Z because the chemical symbol can always be used to determine Z.

The nuclei of all atoms of a particular element contain the same number of protons but often contain different numbers of neutrons. As noted in Section 1.2, nuclei that are related in this way are called **isotopes.**

The isotopes of an element have the same Z value but different N and A values.

The natural abundance of isotopes can differ substantially. For example, $^{11}_{6}C$, $^{12}_{6}C$, $^{13}_{6}C$, and $^{14}_{6}C$ are four isotopes of carbon. The natural abundance of the $^{12}_{6}C$ isotope is approximately 98.9%, whereas that of the $^{13}_{6}C$ isotope is only about 1.1%. Some isotopes, such as $^{11}_{6}C$ and $^{14}_{6}C$, do not occur naturally but can be produced by nuclear reactions in the laboratory or by cosmic rays.

Even the simplest element, hydrogen, has isotopes: $^{1}_{1}H$, the ordinary hydrogen nucleus; $^{2}_{1}H$, deuterium; and $^{3}_{1}H$, tritium.

Charge and Mass

The proton carries a single positive charge, equal in magnitude to the charge e on the electron ($|e| = 1.6 \times 10^{-19}$ C). The neutron is electrically neutral, as its name implies. Because the neutron has no charge, it is difficult to detect.

Nuclear masses can be measured with great precision with the use of a mass spectrometer (see Section 29.5) and by the analysis of nuclear reactions. The proton is approximately 1 836 times as massive as the electron, and the masses of the proton and the neutron are almost equal. In Chapter 1, we defined the atomic mass unit u in such a way that the mass of one atom of the isotope ^{12}C is exactly 12 u, where 1 u = $1.660\ 540 \times 10^{-27}$ kg. According to this definition, the proton and neutron each have a mass of approximately 1 u, and the electron has a mass that is only a small fraction of this value:

$$\text{Mass of proton} = 1.007\ 276\ u$$

$$\text{Mass of neutron} = 1.008\ 665\ u$$

$$\text{Mass of electron} = 0.000\ 548\ 6\ u$$

One might wonder how six protons and six neutrons, each having a mass larger than 1 u, can be combined with six electrons to form a carbon-12 atom having a mass of exactly 12 u. The extra mass of the separated particles appears as binding energy when the particles are combined to form the nucleus. We shall discuss this point in more detail in Section 44.3.

TABLE 44.1   Mass of Selected Particles in Various Units

Particle	Mass		
	kg	**u**	**MeV/c^2**
Proton	$1.672\ 62 \times 10^{-27}$	1.007 276	938.28
Neutron	$1.674\ 93 \times 10^{-27}$	1.008 665	939.57
Electron	$9.109\ 39 \times 10^{-31}$	$5.48\ 579 \times 10^{-4}$	0.510 999
$^{1}_{1}H$ atom	$1.673\ 53 \times 10^{-27}$	1.007 825	938.783
$^{4}_{2}He$ nucleus	$6.644\ 66 \times 10^{-27}$	4.001 506	3 727.38
$^{12}_{6}C$ atom	$1.992\ 65 \times 10^{-27}$	12.000 000	11 177.9

Because the rest energy of a particle is given by $E_R = mc^2$, it is often convenient to express the atomic mass unit in terms of its *rest energy equivalent*. For one atomic mass unit, we have

$$E_R = mc^2 = (1.660\ 540 \times 10^{-27}\ \text{kg})(2.997\ 924\ 58 \times 10^8\ \text{m/s})^2$$
$$= 931.494\ \text{MeV}$$

where we have used the conversion $1\ \text{eV} = 1.602\ 177 \times 10^{-19}\ \text{J}$.

Nuclear physicists often express mass in terms of the unit MeV/c^2, where

$$1\ \text{u} \equiv 931.494\ \text{MeV}/c^2$$

The masses of several nuclei and atoms are given in Table 44.1. The masses and some other properties of selected isotopes are provided in Appendix A.3.

EXAMPLE 44.1   **The Atomic Mass Unit**

Use Avogadro's number to show that $1\ \text{u} = 1.66 \times 10^{-27}\ \text{kg}$.

Solution   From the definition of the mole given in Section 1.3, we know that exactly 12 g ($=1$ mol) of ^{12}C contains Avogadro's number of atoms, where $N_A = 6.02 \times 10^{23}$ atoms/mol. Thus, the mass of one carbon atom is

$$\text{Mass of one } {}^{12}\text{C atom} = \frac{0.012\ \text{kg}}{6.02 \times 10^{23}\ \text{atoms}}$$
$$= 1.99 \times 10^{-26}\ \text{kg}$$

Because one atom of ^{12}C is defined to have a mass of 12.0 u, we find that

$$1\ \text{u} = \frac{1.99 \times 10^{-26}\ \text{kg}}{12.0} = 1.66 \times 10^{-27}\ \text{kg}$$

The Size and Structure of Nuclei

In Rutherford's scattering experiments, positively charged nuclei of helium atoms (alpha particles) were directed at a thin piece of metallic foil. As the alpha particles moved through the foil, they often passed near a metal nucleus. Because of the positive charge on both the incident particles and the nuclei, the particles were deflected from their straight-line paths by the Coulomb repulsive force. Some particles were even deflected straight backward! These particles apparently were moving directly toward a nucleus, on a head-on collision course.

Rutherford used conservation of energy to find an expression for the separation distance d at which an alpha particle approaching a nucleus head-on is turned around by Coulomb repulsion. In such a head-on collision, he reasoned, the kinetic energy of the incoming particle must be converted completely to electric potential energy when the particle stops at the point of closest approach and turns around (Fig. 44.1). If we equate the initial kinetic energy of the alpha particle to the electric potential energy of the system (alpha particle of mass m plus target nucleus of atomic number Z), we have

$$\tfrac{1}{2}mv^2 = k_e \frac{q_1 q_2}{r} = k_e \frac{(2e)(Ze)}{d}$$

Solving for d, the distance of closest approach, we obtain

$$d = \frac{4k_e Ze^2}{mv^2}$$

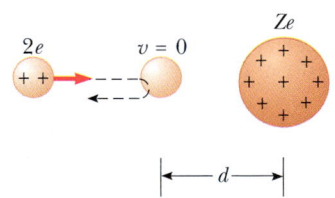

Figure 44.1   An alpha particle on a head-on collision course with a nucleus of charge Ze. Because of the Coulomb repulsion between the like charges, the alpha particle approaches to a distance d from the nucleus, called the distance of closest approach.

From this expression, Rutherford found that the alpha particles approached nuclei to within 3.2×10^{-14} m when the foil was made of gold. Thus, the radius of the gold nucleus must be less than this value. For silver atoms, the distance of closest approach was found to be 2×10^{-14} m. From the results of his scattering experiments, Rutherford concluded that the positive charge in an atom is concentrated in a small sphere, which he called the nucleus, whose radius is no greater than about 10^{-14} m.

Because such small lengths are common in nuclear physics, an often-used convenient length unit is the femtometer (fm), which is sometimes called the **fermi** and is defined as

$$1 \text{ fm} \equiv 10^{-15} \text{ m}$$

In the early 1920s it was known that the nucleus of an atom contains Z protons and has a mass nearly equivalent to that of A protons, where, on the average, $A \approx 2Z$ for lighter nuclei ($Z \le 20$) and $A > 2Z$ for heavier nuclei. To account for the nuclear mass, Rutherford proposed that each nucleus must also contain $A - Z$ neutral particles that he called neutrons. In 1932, the British physicist James Chadwick (1891–1974) discovered the neutron and was awarded the Nobel Prize for this important work.

Since the time of Rutherford's scattering experiments, a multitude of other experiments have shown that most nuclei are approximately spherical and have an average radius given by

Figure 44.2 A nucleus can be modeled as a cluster of tightly packed spheres, where each sphere is a nucleon.

Nuclear radius

$$r = r_0 A^{1/3} \tag{44.1}$$

where r_0 is a constant equal to 1.2×10^{-15} m and A is the mass number. Because the volume of a sphere is proportional to the cube of its radius, it follows from Equation 44.1 that the volume of a nucleus (assumed to be spherical) is directly proportional to A, the total number of nucleons. This proportionality suggests that *all nuclei have nearly the same density*. When nucleons combine to form a nucleus, they combine as though they were tightly packed spheres (Fig. 44.2). This fact has led to an analogy between the nucleus and a drop of liquid, in which the density of the drop is independent of its size. We shall discuss the liquid-drop model of the nucleus in Section 44.4.

EXAMPLE 44.2 **The Volume and Density of a Nucleus**

Find (a) an approximate expression for the mass of a nucleus of mass number A, (b) an expression for the volume of this nucleus in terms of A, and (c) a numerical value for the density of the nucleus.

Solution (a) The mass of the proton is approximately equal to that of the neutron. Thus, if the mass of one of these particles is m, the mass of the nucleus is approximately Am.

(b) Assuming the nucleus is spherical and using Equation 44.1, we find that the volume is

$$V = \tfrac{4}{3}\pi r^3 = \tfrac{4}{3}\pi r_0^3 A$$

(c) The nuclear density is

$$\rho_n = \frac{\text{mass}}{\text{volume}} = \frac{Am}{\tfrac{4}{3}\pi r_0^3 A} = \frac{3m}{4\pi r_0^3}$$

Taking $r_0 = 1.2 \times 10^{-15}$ m and $m = 1.67 \times 10^{-27}$ kg, we find that

$$\rho_n = \frac{3(1.67 \times 10^{-27} \text{ kg})}{4\pi(1.2 \times 10^{-15} \text{ m})^3} = 2.3 \times 10^{17} \text{ kg/m}^3$$

The nuclear density is approximately 2.3×10^{14} times as great as the density of water ($\rho_{\text{water}} = 1.0 \times 10^3$ kg/m^3).

Exercise If the Earth were compressed until it had this density, how large would it be?

Answer A sphere of diameter 370 m!

Nuclear Stability

Because the nucleus is viewed as a closely packed collection of protons and neutrons, you might be surprised that it can exist. Because like charges (the protons) in close proximity exert very large repulsive Coulomb forces on each other, these forces should cause the nucleus to fly apart. However, nuclei are stable because of the **nuclear force,** a very-short-range (about 2 fm) attractive force that acts between all nuclear particles. The protons attract each other by means of the nuclear force, and, at the same time, they repel each other through the Coulomb force. The nuclear force also acts between pairs of neutrons and between neutrons and protons.

There are approximately 400 stable nuclei; hundreds of other nuclei have been observed, but they are unstable. A plot of neutron number N versus atomic number Z for a number of stable nuclei is given in Figure 44.3. The stable nuclei are represented by the blue dots, which lie in a narrow range called the *line of stability.* Note that light nuclei are most stable if they contain an equal number of protons and neutrons; that is, if $N = Z$. Also, note that heavy nuclei are more stable if the number of neutrons exceeds the number of protons—above $Z = 20$, the line

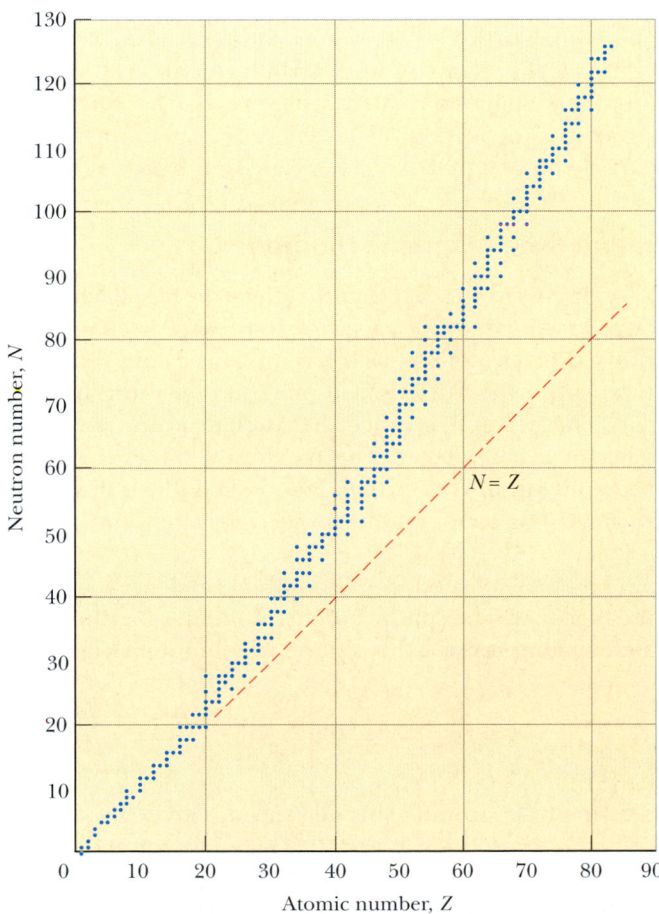

web

For a very detailed, "clickable" version of Figure 44.3, visit Brookhaven National Lab at

www.dne.bnl.gov/CoN/index.html

Figure 44.3   Neutron number N versus atomic number Z for the stable nuclei (blue dots). These nuclei lie in a narrow band called the line of stability. The dashed line corresponds to the condition $N = Z$.

of stability deviates upward from the line representing $N = Z$. This can be understood by recognizing that, as the number of protons increases, the strength of the Coulomb force increases, which tends to break the nucleus apart. As a result, more neutrons are needed to keep the nucleus stable because neutrons experience only the attractive nuclear force. Eventually, the repulsive Coulomb forces between protons cannot be compensated by the addition of more neutrons. This occurs when $Z = 83$, meaning that elements that contain more than 83 protons do not have stable nuclei.

Quick Quiz 44.1

The blue dots in Figure 44.3 form a sequence of vertically oriented groups. What do these groups represent?

It is interesting that most stable nuclei have an even value of A. Furthermore, only eight nuclei have odd values for both Z and N. Certain values of Z and N correspond to nuclei with unusually high stability. These values, called **magic numbers,** are

$$Z \quad \text{or} \quad N = 2, 8, 20, 28, 50, 82 \tag{44.2}$$

For example, the alpha particle (two protons and two neutrons), which has $Z = 2$ and $N = 2$, is very stable. The unusual stability of nuclei having progressively larger magic numbers suggests a shell structure of the nucleus similar to the atomic shell structure. In Section 44.4 we briefly discuss a nuclear model, the independent-particle model, that explains magic numbers.

Nuclear Spin and Spin Magnetic Moment

In Chapter 42, we discussed the fact that the electron has an intrinsic angular momentum, which we called spin. Nuclei also have spin because their component particles—neutrons and protons—each have intrinsic spin $\frac{1}{2}$, as well as orbital angular momentum within the nucleus. The magnitude of the nuclear angular momentum is $\sqrt{I(I + 1)}\,\hbar$, where I is called the **nuclear spin quantum number** and may be an integer or a half-integer. The maximum value of the z component of the spin angular momentum vector is $I\hbar$. Figure 44.4 illustrates the possible orientations of the nuclear spin vector and its projections along the z axis for the case in which $I = \frac{3}{2}$.

Nuclear spin has an associated, corresponding nuclear magnetic moment, similar to that of the electron. The spin magnetic moment of a nucleus is measured in terms of the **nuclear magneton** μ_n, a unit of moment defined as

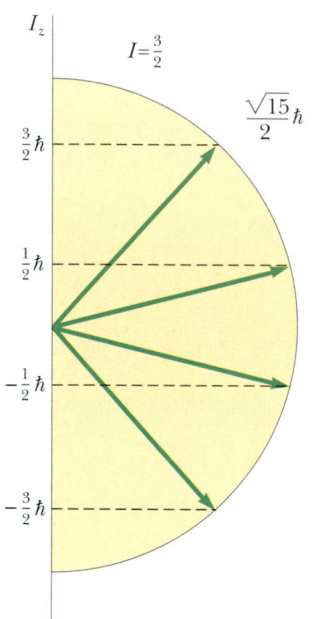

Figure 44.4   Possible orientations of the nuclear spin angular momentum vector and its projections along the z axis for the case $I = \frac{3}{2}$.

Nuclear magneton

$$\mu_n \equiv \frac{e\hbar}{2m_p} = 5.05 \times 10^{-27}\,\text{J/T} \tag{44.3}$$

where m_p is the mass of the proton. This definition is analogous to that of the Bohr magneton μ_B, which corresponds to the spin magnetic moment of a free electron (see Section 42.5). Note that μ_n is smaller than μ_B ($= 9.274 \times 10^{-24}\,\text{J/T}$) by a factor of approximately 2 000 because of the large difference between the proton mass and the electron mass.

The magnetic moment of a free proton is $2.792\,8\mu_n$. Unfortunately, there is no general theory of nuclear magnetism that explains this value. Surprisingly, the uncharged neutron also has a magnetic moment, which has a value of $-1.913\,5\mu_n$. The minus sign indicates that this moment is opposite the spin angular momentum of the neutron. The existence of a magnetic moment for the neutron suggests that it is not fundamental but rather has an underlying structure. We shall explore such structure in Chapter 46.

Quick Quiz 44.2

Which do you expect not to vary substantially between different isotopes of an element? (a) atomic mass, (b) nuclear spin magnetic moment, (c) chemical properties?

44.2 NUCLEAR MAGNETIC RESONANCE AND MAGNETIC RESONANCE IMAGING

Nuclear magnetic moments, as well as electronic magnetic moments, precess when placed in an external magnetic field. The frequency at which they precess, called the **Larmor precessional frequency** ω_p, is directly proportional to the magnitude of the magnetic field. This is described schematically in Figure 44.5a, in which the external magnetic field is along the z axis. For example, the Larmor frequency of a proton in a 1-T magnetic field is 42.577 MHz. The potential energy of a magnetic dipole moment $\boldsymbol{\mu}$ in an external magnetic field $\mathbf{B}$ is given by $-\boldsymbol{\mu} \cdot \mathbf{B}$. When the magnetic moment $\boldsymbol{\mu}$ is lined up with the field as closely as quantum physics allows, the potential energy of the dipole moment in the field has its minimum value $E_{\min}$. When the projection of $\boldsymbol{\mu}$ is as antiparallel to the field as possible, the potential energy has its maximum value $E_{\max}$. In general, there are other energy states between these values that correspond to the quantized directions of the magnetic moment with respect to the field. For a nucleus with spin $\frac{1}{2}$, there are only two allowed states, with energies $E_{\min}$ and $E_{\max}$. These two energy states are shown in Figure 44.5b.

Larmor precessional frequency

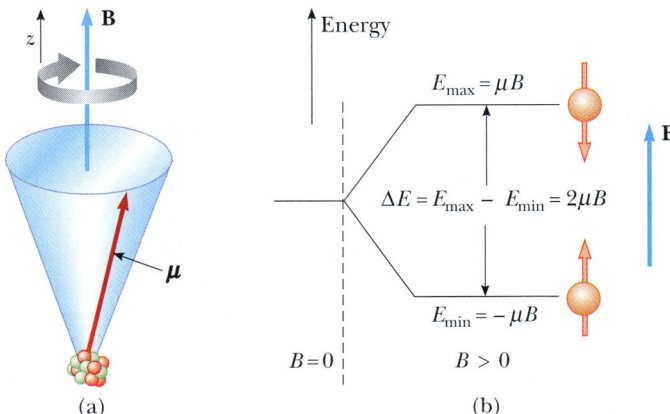

$E_{\max} = \mu B$

$\Delta E = E_{\max} - E_{\min} = 2\mu B$

$E_{\min} = -\mu B$

$B = 0$ $B > 0$

(a)   (b)

Figure 44.5 (a) When a nucleus is placed in an external magnetic field $\mathbf{B}$, the nuclear spin magnetic moment precesses about the magnetic field with a frequency proportional to the magnitude of the field. (b) A nucleus with spin $\frac{1}{2}$ can occupy one of two energy states when placed in an external magnetic field. The lower energy state $E_{\min}$ corresponds to the case where the spin is aligned with the field as much as possible, and the higher energy state $E_{\max}$ corresponds to the case where the spin is opposite the field as much as possible.

Nuclear magnetic resonance

web

For more information on MRI and many other topics dealing with physics, visit **http://physics.miningco.com/msubtext. htm**

It is possible to observe transitions between these two spin states using a technique called **NMR,** for **nuclear magnetic resonance.** A constant magnetic field (**B** in Fig. 44.5a) is introduced to define a z axis. A second, weaker, oscillating magnetic field is then applied perpendicular to **B**. When the frequency of the oscillating field is adjusted to match the Larmor precessional frequency of the nuclei in the sample, a torque that acts on the precessing magnetic moments causes them to "flip" between the two spin states shown in Figure 44.5b. These transitions result in a net absorption of energy by the nuclei, an absorption that can be detected electronically.

A diagram of the apparatus used in nuclear magnetic resonance is illustrated in Figure 44.6. The energy absorbed by the nuclei is supplied by the generator producing the oscillating magnetic field. Nuclear magnetic resonance and a related technique called *electron spin resonance* are extremely important methods for studying nuclear and atomic systems and the ways in which these systems interact with their surroundings.

A widely used medical diagnostic technique called **MRI,** for **magnetic resonance imaging,** is based on nuclear magnetic resonance. Because nearly two thirds of the atoms in the human body are hydrogen (which gives a strong signal), MRI works exceptionally well for viewing internal tissues. The patient is placed inside a large solenoid that supplies a time-constant magnetic field whose magnitude varies spatially across the body. Because of the variation in the field, protons in different parts of the body precess at different frequencies, so the resonance signal can be used to provide information about the positions of the protons. A computer is used to analyze the position information to provide data for constructing a final image. An MRI scan showing incredible detail in internal body structure is shown in Figure 44.7.

The main advantage of MRI over other imaging techniques is that it causes minimal cellular damage. The photons associated with the radio-frequency signals used in MRI have energies of only about 10^{-7} eV. Because molecular bond strengths are much larger (approximately 1 eV), the radio-frequency radiation causes little cellular damage. In comparison, x-rays have energies ranging from 10^4 to 10^6 eV and can cause considerable cellular damage. Thus, despite some individ-

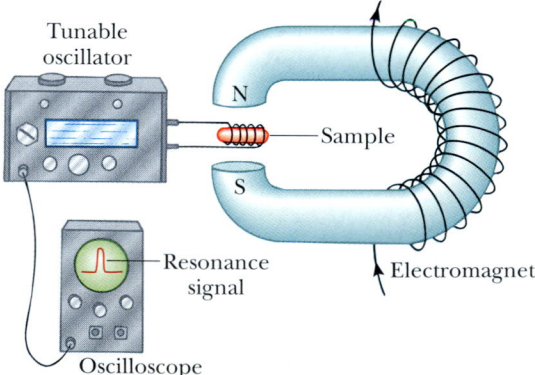

Figure 44.6   Experimental arrangement for nuclear magnetic resonance. The radio-frequency magnetic field created by the coil surrounding the sample and provided by the variable-frequency oscillator must be perpendicular to the constant magnetic field created by the electromagnet. When the nuclei in the sample meet the resonance condition, the nuclei absorb energy from the radio-frequency field of the coil, and this absorption changes the characteristics of the circuit in which the coil is included. Most modern NMR spectrometers use superconducting magnets at fixed field strengths and operate at frequencies of approximately 200 MHz.

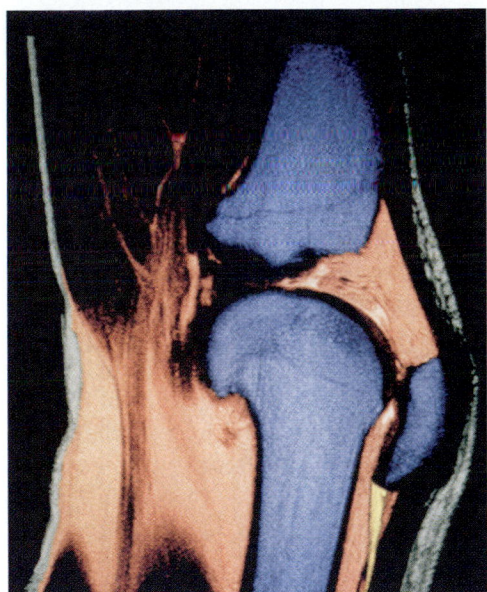

Figure 44.7 An MRI scan of a knee.
(Susie Leavines/Photo Researchers, Inc.)

uals' fears of the word *nuclear* associated with MRI, the radio-frequency radiation involved is overwhelmingly safer than the x-rays that these individuals might accept more readily! A disadvantage of MRI is that the equipment required to conduct the procedure is quite expensive.

The magnetic field produced by the solenoid is sufficient to lift a car, and the radio signal is about the same magnitude as that from a small commercial broadcasting station!

44.3 BINDING ENERGY AND NUCLEAR FORCES

The total mass of a nucleus is always less than the sum of the masses of its individual nucleons. Because mass is a measure of energy, **the total energy of the bound system (the nucleus) is less than the combined energy of the separated nucleons.** As we learned in Example 39.14, this difference in energy is called the **binding energy** of the nucleus and can be thought of as the energy that must be added to a nucleus to break it apart into its components. Therefore, in order to separate a nucleus into protons and neutrons, energy must be delivered to the system.

Conservation of energy and the Einstein mass–energy equivalence relationship show that the binding energy E_b of any nucleus of mass M_A is

$$E_b \text{ (MeV)} = (Zm_p + Nm_n - M_A) \times 931.494 \text{ MeV/u} \qquad (44.4)$$

◄ Binding energy of a nucleus

where m_p is the mass of the proton, m_n is the mass of the neutron, and the masses are all expressed in atomic mass units. In practice, it is often more convenient to use the mass of neutral atoms (nuclear mass plus mass of electrons) in computing binding energy because mass spectrometers generally measure atomic masses.[1]

---

[1] It is possible to use atomic masses rather than nuclear masses because electron masses cancel in the calculations.

A plot of binding energy per nucleon E_b/A as a function of mass number A for various stable nuclei is shown in Figure 44.8. Note that the curve in Figure 44.8 peaks in the vicinity of $A = 60$. That is, nuclei having mass numbers either greater or less than 60 are not as strongly bound as those near the middle of the periodic table. The higher values of binding energy per nucleon near $A = 60$ imply that energy is released when a heavy nucleus splits, or *fissions*, into two lighter nuclei. Energy is released in fission because the nucleons in each product nucleus are more tightly bound to one another than are the nucleons in the original nucleus. The important process of fission and a second important process of *fusion*, in which energy is released as light nuclei combine, are considered in detail in Chapter 45.

Another important feature of Figure 44.8 is that the binding energy per nucleon is approximately constant at around 8 MeV per nucleon for all nuclei with $A > 50$. For these nuclei, the nuclear forces are said to be *saturated*, meaning that, in the closely packed structure shown in Figure 44.2, a particular nucleon can form attractive bonds with only a limited number of other nucleons.

Quick Quiz 44.3

Figure 44.8 shows that, above about $A = 50$, an approximately constant amount of energy is necessary to remove a nucleon from the nucleus. Compare this with Figure 42.16, which shows the widely varying amounts of energy necessary to remove an electron from an atom. Why is there such a difference between the two graphs?

Figure 44.8 provides insight into fundamental questions about the origin of the chemical elements. In the early life of the Universe, there were only hydrogen and helium. Clouds of cosmic gas and dust coalesced under gravitational forces to form stars. As a star ages, it produces heavier elements from the lighter elements contained within it, beginning by fusing hydrogen atoms to form helium. This process continues as the star becomes older, generating atoms having larger and larger atomic numbers, up through the isotope of iron having $A = 56$, which is at the peak of the curve shown in Figure 44.8.

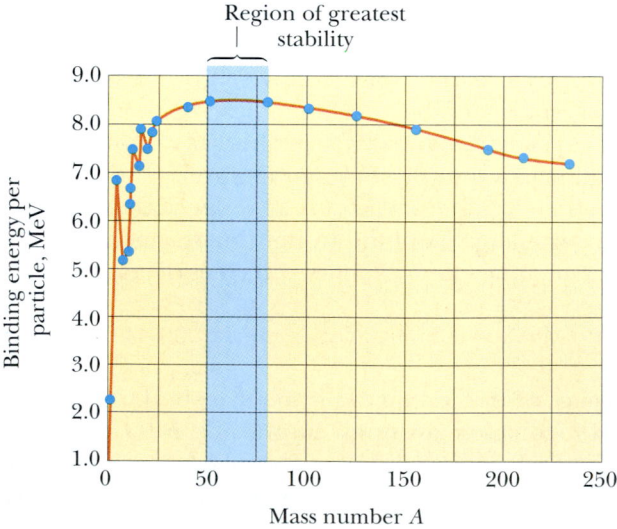

Figure 44.8   Binding energy per nucleon versus mass number for nuclei that lie along the line of stability in Figure 44.3.

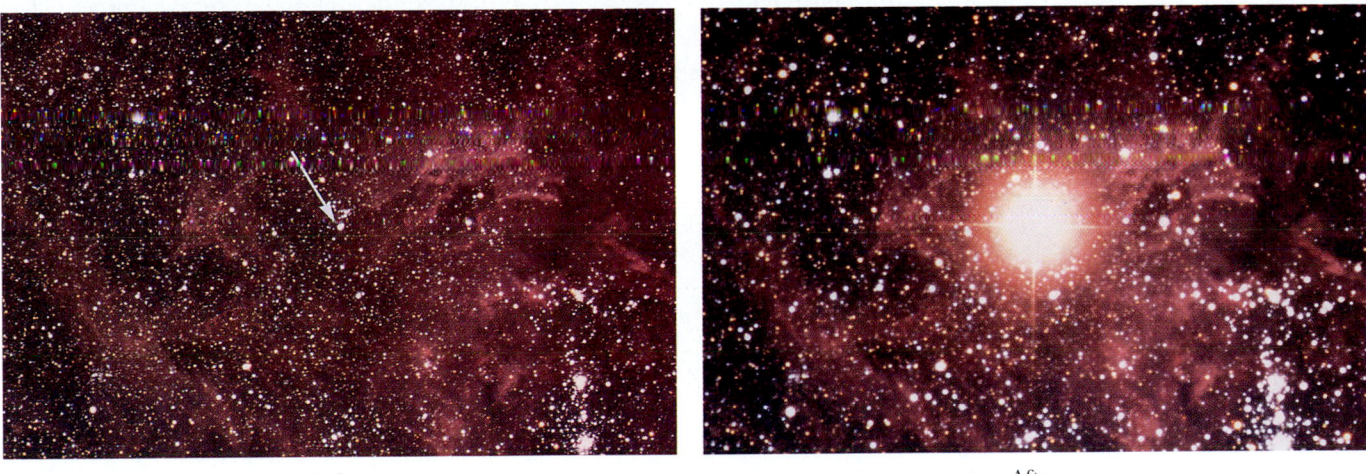

Before   After

Before and after images of the 1987A supernova in the Large Magellanic Cloud. The arrow in the "before" image points to the star that exploded. This was the brightest supernova seen within the past several hundred years. *(Anglo-Australian Telescope Board)*

It takes additional energy to create elements with mass numbers larger than 56 because of their lower binding energies per nucleon. This energy comes from the supernova explosion that occurs at the end of some large stars' lives. Thus, all the heavy atoms in your body were produced from the explosions of ancient stars. You are literally made of stardust!

The general features of the nuclear force responsible for the binding energy of nuclei have been revealed in a wide variety of experiments and are as follows:

- The nuclear force is attractive and is the strongest force in nature.
- It is a short-range force that falls to zero when the separation between nucleons exceeds several fermis. This limited range is evidenced by scattering experiments and shown in the neutron–proton potential energy plot of Figure 44.9a, obtained by scattering neutrons from a target containing hydrogen. The potential energy well is 40 to 50 MeV deep and contains a strong repulsive component that prevents the separation distance between nucleons from being less than about 0.4 fm.
- The magnitude of the nuclear force depends on the relative spin orientations of the nucleons.
- Scattering experiments and other indirect evidence show that the nuclear force is independent of the charge of the interacting nucleons. For this reason, high-speed electrons can be used to probe the properties of nuclei. The charge independence also means that the only difference between neutron–proton (n–p) and proton–proton (p–p) interactions is that the p–p potential energy is a superposition of nuclear and Coulomb interactions, as shown in Figure 44.9b. At separation distances less than 2 fm, p–p and n–p potential energies are nearly identical, but for distances greater than this, the p–p potential energy is positive, with a maximum of about 1 MeV at 4 fm.

Quick Quiz 44.4

Can either curve in Figure 44.9 be vertical at some point?

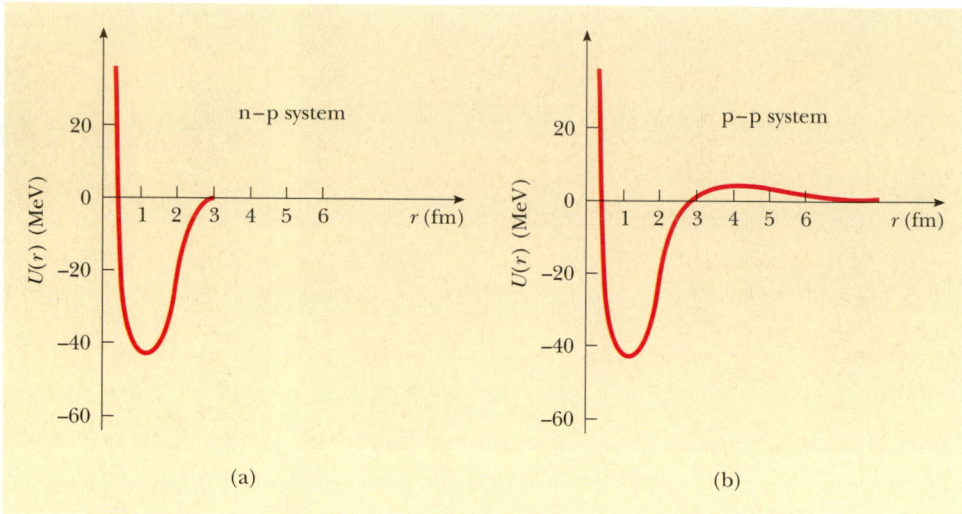

Figure 44.9 (a) Potential energy versus separation distance for a neutron–proton system. (b) Potential energy versus separation distance for a proton–proton system. The difference in the two curves is due to the large Coulomb repulsion in the case of the proton–proton interaction.

44.4 NUCLEAR MODELS

Although the details of the nuclear force are still not well understood, several nuclear models have been proposed, and these are useful in understanding general features of nuclear experimental data and the mechanisms responsible for binding energy. The liquid-drop model accounts for nuclear binding energy, and the independent-particle model accounts for the existence of stable isotopes.

Liquid-Drop Model

In 1936, Bohr proposed treating nucleons like molecules in a drop of liquid. In this **liquid-drop model,** the nucleons interact strongly with one another and undergo frequent collisions as they jiggle around within the nucleus. This jiggling motion is analogous to the thermally agitated motion of molecules in a drop of liquid.

Three major effects influence the binding energy of the nucleus in the liquid-drop model:

- **The volume effect.** Figure 44.8 shows that, for $A > 50$, the binding energy per nucleon is approximately constant, indicating that the nuclear force saturates. This tells us that the binding energy of the nucleus is proportional to A and therefore proportional to the nuclear volume. The contribution to the binding energy of the entire nucleus is $C_1 A$, where C_1 is an adjustable constant.
- **The surface effect.** Because nucleons on the surface of the drop have fewer neighbors than those in the interior, surface nucleons reduce the binding energy by an amount proportional to their number. Because the number of surface nucleons is proportional to the surface area $4\pi r^2$ of the nucleus, and because $r^2 \propto A^{2/3}$ (Eq. 44.1), the surface term can be expressed as $-C_2 A^{2/3}$, where C_2 is a constant.

- **The Coulomb repulsion effect.** Each proton repels every other proton in the nucleus. The corresponding potential energy per pair of interacting protons is $k_e e^2 / r$, where k_e is the Coulomb constant. The total Coulomb energy represents the work required to assemble Z protons from infinity to a sphere of volume V. This energy is proportional to the number of proton pairs $Z(Z - 1)/2$ and inversely proportional to the nuclear radius. Consequently, the reduction in energy that results from the Coulomb effect is $-C_3 Z(Z - 1)/A^{1/3}$.

Another small effect that lowers the binding energy is significant for nuclei having a large excess of neutrons—in other words, heavy nuclei. This effect gives rise to a binding energy term of the form $-C_4 (N - Z)^2 / A$.

Adding these contributions, we get as the total binding energy

$$E_b = C_1 A - C_2 A^{2/3} - C_3 \frac{Z(Z - 1)}{A^{1/3}} - C_4 \frac{(N - Z)^2}{A} \qquad \textbf{(44.5)}$$

Semiempirical binding energy formula

This equation, often referred to as the **semiempirical binding energy formula,** contains four constants that are adjusted to fit the expression to experimental data. For nuclei having $A \geq 15$, the constants have the values

$$C_1 = 15.7 \text{ MeV} \qquad C_2 = 17.8 \text{ MeV}$$

$$C_3 = 0.71 \text{ MeV} \qquad C_4 = 23.6 \text{ MeV}$$

Equation 44.5, together with these constants, fits the known nuclear mass values very well. However, the liquid-drop model does not account for some finer details of nuclear structure, such as stability rules and angular momentum. On the other hand, it does provide a qualitative description of nuclear fission, shown schematically in Figure 44.10.

The Independent-Particle Model

In our second model of the nucleus, the **independent-particle model,** often called the *shell model,* each nucleon is assumed to exist in a shell, similar to an atomic shell for an electron. The nucleons exist in quantized energy states, and there are few collisions between nucleons. Obviously, the assumptions of this model differ greatly from those made in the liquid-drop model.

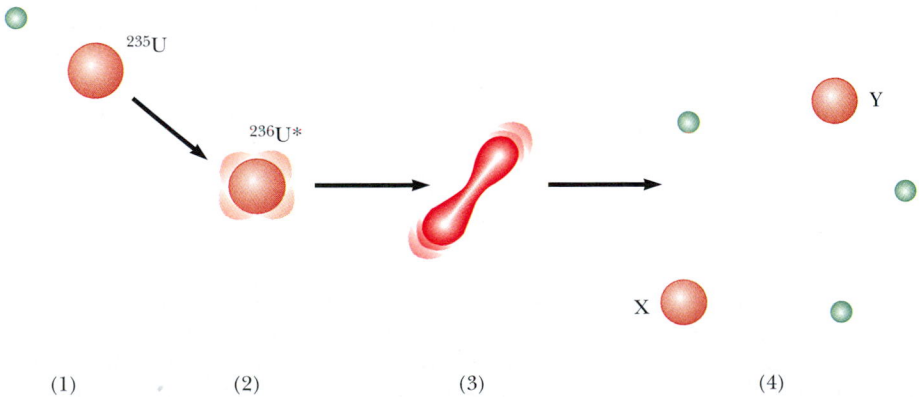

(1)   (2)   (3)   (4)

Figure 44.10 Steps leading to fission according to the liquid-drop model of the nucleus.

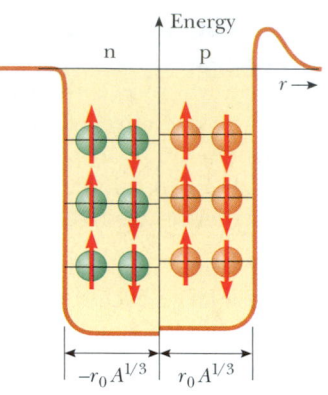

Figure 44.11 A square potential well containing 12 nucleons. The red circles represent protons, and the green circles represent neutrons. The energy levels for the protons are slightly higher than those for the neutrons because of the Coulomb potential experienced by the protons. The difference in the levels increases as Z increases. Note that only two nucleons having opposite spins can occupy a given level, as required by the exclusion principle.

Maria Goeppert-Mayer
(1906–1972) Goeppert-Mayer was born and educated in Germany. She is best known for her development of the shell model (independent-particle model) of the nucleus, published in 1950. A similar model was simultaneously developed by Hans Jensen, a German scientist. Goeppert-Mayer and Jensen were awarded the Nobel Prize in physics in 1963 for their extraordinary work in understanding the structure of the nucleus. *(Courtesy of Louise Barker/AIP Niels Bohr Library)*

The quantized states occupied by the nucleons can be described by a set of quantum numbers. Because both the proton and the neutron have spin $\frac{1}{2}$, the exclusion principle can be applied to describe the allowed states (as we did for electrons in Chapter 42). That is, each state can contain only two protons (or two neutrons) having *opposite* spins (Fig. 44.11). The protons have a set of allowed states, and these states differ from those of the neutrons because the two species move in different potential wells. The proton energy levels are higher than the neutron levels because the protons experience a superposition of Coulomb potential energy and nuclear potential energy, whereas the neutrons experience only a nuclear potential energy.

One factor influencing the observed characteristics of nuclear ground states is *nuclear spin–orbit* effects. Unlike the spin–orbit interaction between the spin of an electron and its orbital motion in an atom, an interaction that is magnetic in origin, the spin–orbit effect for nucleons in a nucleus is due to the nuclear force. It is much stronger than in the atomic case, and it has opposite sign. When these effects are taken into account, the independent-particle model is able to account for the observed magic numbers.

Finally, the independent-particle model helps us understand why nuclei containing an even number of protons and neutrons are more stable than other nuclei. (There are 160 even–even isotopes.) Any particular state is filled when it contains two protons (or two neutrons) having opposite spins. An extra proton or neutron can be added to the nucleus only at the expense of increasing the energy of the nucleus. This increase in energy leads to a nucleus that is less stable than the original nucleus. A careful inspection of the stable nuclei shows that the majority have a special stability when their nucleons combine in pairs, which results in a total angular momentum of zero. This accounts for the large number of high-stability nuclei (those having high binding energies) with the magic numbers given by Equation 44.2.

44.5 RADIOACTIVITY

In 1896, Henri Becquerel accidentally discovered that uranyl potassium sulfate crystals emit an invisible radiation that can darken a photographic plate when the plate is covered to exclude light. After a series of experiments, he concluded that the radiation emitted by the crystals was of a new type, one that requires no external stimulation and was so penetrating that it could darken protected photographic plates and ionize gases. This process of spontaneous emission of radiation by uranium was soon to be called **radioactivity.**

Subsequent experiments by other scientists showed that other substances were more powerfully radioactive. The most significant investigations of this type were conducted by Marie and Pierre Curie. After several years of careful and laborious chemical separation processes on tons of pitchblende, a radioactive ore, the Curies reported the discovery of two previously unknown elements, both radioactive. These were named polonium and radium. Subsequent experiments, including Rutherford's famous work on alpha-particle scattering, suggested that radioactivity is the result of the *decay*, or disintegration, of unstable nuclei.

Three types of radioactive decay occur in a radioactive substance: alpha (α) decay, in which the emitted particles are ^{4}He nuclei; beta (β) decay, in which the emitted particles are either electrons or positrons; and gamma (γ) decay, in which the emitted "rays" are high-energy photons. A **positron** is a particle like the electron in all respects except that the positron has a charge of $+e$ (in other words, the positron is the *antimatter twin* of the electron). The symbol e$^-$ is used to designate an electron, and e$^+$ designates a positron.

It is possible to distinguish among these three forms of radiation by using the scheme described in Figure 44.12. The radiation from a radioactive sample is directed into a region in which there is a magnetic field. The radiation beam splits into three components, two bending in opposite directions and the third experiencing no change in direction. From this simple observation, we can conclude that the radiation of the undeflected beam carries no charge (the gamma ray), the component deflected upward corresponds to positively charged particles (alpha particles), and the component deflected downward corresponds to negatively charged particles (e$^-$). If the beam includes a positron (e$^+$), it is deflected upward like the alpha particle but follows a different trajectory due to its smaller mass.

The three types of radiation have quite different penetrating powers. Alpha particles barely penetrate a sheet of paper, beta particles (electrons and positrons) can penetrate a few millimeters of aluminum, and gamma rays can penetrate several centimeters of lead.

The rate at which a particular decay process occurs in a radioactive sample is proportional to the number of radioactive nuclei present (that is, the number of nuclei that have not yet decayed). If N is the number of radioactive nuclei present

Marie Curie (1867–1934) In 1903 the Polish scientist Marie Curie shared the Nobel Prize in physics with her husband Pierre and with Becquerel for their studies of radioactive substances. In 1911 she was awarded a Nobel Prize in chemistry for the discovery of radium and polonium. She died of leukemia caused by years of exposure to radioactive substances. "I persist in believing that the ideas that then guided us are the only ones which can lead to true social progress. We cannot hope to build a better world without improving the individual. Toward this end, each of us must work toward his own highest development, accepting at the same time his share of responsibility in the general life of humanity." *(FPG International)*

The hands and numbers of this luminous watch contain minute amounts of radium mixed with a phosphorescent material. The radioactive decay of radium causes the watch to glow in the dark. *(©1990 Richard Megna/Fundamental Photographs)*

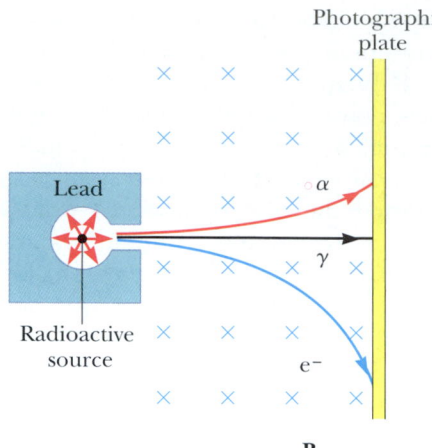

Figure 44.12 The radiation from a radioactive source can be separated into three components by using a magnetic field to deflect the charged particles. The photographic plate at the right records the events. The gamma ray is not deflected by the magnetic field.

at some instant, the rate of change of N is

$$\frac{dN}{dt} = -\lambda N \qquad (44.6)$$

where λ, called the **decay constant,** is the probability of decay per nucleus per second. The minus sign indicates that dN/dt is negative; that is, N decreases in time.

Equation 44.6 can be written in the form

$$\frac{dN}{N} = -\lambda \, dt$$

the solution to which is

Exponential decay

$$N = N_0 e^{-\lambda t} \qquad (44.7)$$

where the constant N_0 represents the number of radioactive nuclei at $t = 0$. Equation 44.7 shows that the number of radioactive nuclei in a sample decreases exponentially with time.

The **decay rate** R, which is the number of decays per second, can be obtained by differentiating Equation 44.7 with respect to time:

Half-life equation

$$R = \left| \frac{dN}{dt} \right| = N_0 \lambda e^{-\lambda t} = R_0 e^{-\lambda t} \qquad (44.8)$$

where $R = \lambda N$ and $R_0 = N_0 \lambda$ is the decay rate at $t = 0$. The decay rate R of a sample is often referred to as its **activity.** Note that both N and R decrease exponentially with time. The plot of N versus t shown in Figure 44.13 illustrates the exponential nature of the decay.

Another parameter useful in characterizing nuclear decay is **half-life** $T_{1/2}$:

The **half-life** of a radioactive substance is the time it takes half of a given number of radioactive nuclei to decay.

Setting $N = N_0/2$ and $t = T_{1/2}$ in Equation 44.7 gives

$$\frac{N_0}{2} = N_0 e^{-\lambda T_{1/2}}$$

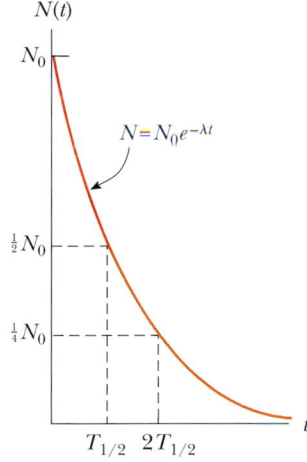

Figure 44.13 Plot of the exponential decay law for radioactive nuclei. The vertical axis represents the number of radioactive nuclei present at any time t, and the horizontal axis is time. The time $T_{1/2}$ is the half-life of the sample.

Canceling the N_0 factors and then taking the reciprocal of both sides, we obtain $e^{\lambda T_{1/2}} = 2$. Taking the natural logarithm of both sides gives

$$T_{1/2} = \frac{\ln 2}{\lambda} = \frac{0.693}{\lambda} \qquad \textbf{(44.9)}$$

This is a convenient expression relating half-life $T_{1/2}$ to decay constant λ. After an elapsed time equal to one half-life, there are $N_0/2$ radioactive nuclei remaining (by definition); after two half-lives, half of these have decayed and $N_0/4$ radioactive nuclei are left; after three half-lives, $N_0/8$ are left, and so on. In general, after n half-lives, the number of radioactive nuclei remaining is $N_0/2^n$.

A frequently used unit of activity is the **curie** (Ci), defined as

$$1 \text{ Ci} \equiv 3.7 \times 10^{10} \text{ decays/s}$$

The curie

This value was originally selected because it is the approximate activity of 1 g of radium. The SI unit of activity is the **becquerel** (Bq):

$$1 \text{ Bq} \equiv 1 \text{ decay/s}$$

The becquerel

Therefore, $1 \text{ Ci} = 3.7 \times 10^{10}$ Bq. The curie is a rather large unit, and the more frequently used activity units are the millicurie and the microcurie.

EXAMPLE 44.3 ▶ **How Many Nuclei Are Left?**

The isotope carbon-14, $^{14}_{6}\text{C}$, is radioactive and has a half-life of 5 730 years. If you start with a sample of 1 000 carbon-14 nuclei, how many will still be around in 22 920 years?

Solution In 5 730 years, half the sample will have decayed, leaving 500 carbon-14 nuclei remaining. In another 5 730 years (for a total elapsed time of 11 460 years), the number will be reduced to 250 nuclei. After another 5 730 years (total time 17 190 years), 125 remain. Finally, after four half-lives (22 920 years), only about 62 remain.

These numbers represent ideal circumstances. Radioactive decay is in reality an averaging process over a very large number of atoms, and the actual outcome depends on statistics. Our original sample in this example contained only 1 000 nuclei, certainly not a very large number. Thus, if we were to count the number remaining after one half-life for this small sample, it probably would not be exactly 500.

EXAMPLE 44.4 ▸ The Activity of Radium

The half-life of the radioactive nucleus radium-226, $^{226}_{88}\text{Ra}$, is 1.6×10^3 yr. (a) What is the decay constant λ of this nucleus?

Solution We can calculate λ using Equation 44.9 and the fact that

$$T_{1/2} = (1.6 \times 10^3 \text{ yr})(3.15 \times 10^7 \text{ s/yr})$$
$$= 5.0 \times 10^{10} \text{ s}$$

Therefore,

$$\lambda = \frac{0.693}{T_{1/2}} = \frac{0.693}{5.0 \times 10^{10} \text{ s}} = \boxed{1.4 \times 10^{-11} \text{ s}^{-1}}$$

Note that this result is also the probability that any single $^{226}_{88}\text{Ra}$ nucleus will decay in a time interval of 1 s.

(b) If a sample contains 3.0×10^{16} $^{226}_{88}\text{Ra}$ nuclei at $t = 0$, determine its activity in curies at this time.

Solution By definition (Eq. 44.8) R_0, the activity at $t = 0$, is λN_0, where N_0 is the number of radioactive nuclei present at $t = 0$. With $N_0 = 3.0 \times 10^{16}$, we have

$$R_0 = \lambda N_0 = (1.4 \times 10^{-11} \text{ s}^{-1})(3.0 \times 10^{16})$$
$$= (4.2 \times 10^5 \text{ decays/s})\left(\frac{1 \text{ Ci}}{3.7 \times 10^{10} \text{ decays/s}}\right)$$
$$= \boxed{11 \ \mu\text{Ci}}$$

(c) What is the activity in becquerels after the sample is 2.0×10^3 yr old?

Solution We use Equation 44.8 and the fact that $t = 2.0 \times 10^3$ yr $= (2.0 \times 10^3 \text{ yr})(3.15 \times 10^7 \text{ s/yr}) = 6.3 \times 10^{10}$ s:

$$R = R_0 e^{-\lambda t}$$
$$= (4.2 \times 10^5 \text{ decays/s})e^{-(1.4 \times 10^{-11} \text{ s}^{-1})(6.3 \times 10^{10} \text{ s})}$$
$$= 1.7 \times 10^5 \text{ decays/s} = \boxed{1.7 \times 10^5 \text{ Bq}}$$

EXAMPLE 44.5 ▸ The Activity of Carbon

A radioactive sample contains 3.50 μg of pure $^{11}_{6}\text{C}$, which has a half-life of 20.4 min. (a) Determine the number of nuclei in the sample at $t = 0$.

Solution The molar mass of $^{11}_{6}\text{C}$ is approximately 11.0 g/mol, and so 11.0 g contains Avogadro's number (6.02×10^{23}) of nuclei. Therefore, 3.50 μg contains N nuclei, where

$$\frac{N}{6.02 \times 10^{23} \text{ nuclei/mol}} = \frac{3.50 \times 10^{-6} \text{ g}}{11.0 \text{ g/mol}}$$

$$N = \boxed{1.92 \times 10^{17} \text{ nuclei}}$$

(b) What is the activity in becquerels of the sample initially and after 8.00 h?

Solution With $T_{1/2} = 20.4$ min $= 1\,224$ s, the decay constant is

$$\lambda = \frac{0.693}{T_{1/2}} = \frac{0.693}{1\,224 \text{ s}} = 5.66 \times 10^{-4} \text{ s}^{-1}$$

Therefore, the initial activity of the sample is

$$R_0 = \lambda N_0 = (5.66 \times 10^{-4} \text{ s}^{-1})(1.92 \times 10^{17})$$
$$= \boxed{1.09 \times 10^{14} \text{ Bq}}$$

We use Equation 44.8 to find the activity at $t = 8.00$ h $= 2.88 \times 10^4$ s:

$$R = R_0 e^{-\lambda t} = (1.09 \times 10^{14} \text{ Bq})e^{-(5.66 \times 10^{-4} \text{ s}^{-1})(2.88 \times 10^4 \text{ s})}$$
$$= \boxed{9.09 \times 10^6 \text{ Bq}}$$

A listing of activity versus time for this situation is given in Table 44.2.

Exercise Calculate the number of radioactive nuclei remaining after 8.00 h.

Answer 1.60×10^{10} nuclei.

TABLE 44.2 Activity Versus Time for the Sample Described in Example 44.5

t (h)	R (Bq)
0	1.09×10^{14}
1	1.41×10^{13}
2	1.84×10^{12}
3	2.39×10^{11}
4	3.12×10^{10}
5	4.06×10^9
6	5.28×10^8
7	6.88×10^7
8	9.09×10^6

EXAMPLE 44.6 **A Radioactive Isotope of Iodine**

A sample of the isotope ^{131}I, which has a half-life of 8.04 days, has an activity of 5.0 mCi at the time of shipment. Upon receipt in a medical laboratory, the activity is 4.2 mCi. How much time has elapsed between the two measurements?

Solution We use Equation 44.8 in the form

$$\frac{R}{R_0} = e^{-\lambda t}$$

Taking the natural logarithm of each side, we obtain

$$\ln\left(\frac{R}{R_0}\right) = -\lambda t$$

$$(1) \qquad t = -\frac{1}{\lambda} \ln\left(\frac{R}{R_0}\right)$$

To find λ, we use Equation 44.9:

$$(2) \qquad \lambda = \frac{0.693}{T_{1/2}} = \frac{0.693}{8.04 \text{ days}}$$

Substituting Equation (2) into Equation (1) gives

$$t = -\left(\frac{8.04 \text{ days}}{0.693}\right) \ln\left(\frac{4.2 \text{ mCi}}{5.0 \text{ mCi}}\right) = \boxed{2.0 \text{ days}}$$

44.6 THE DECAY PROCESSES

As we stated in the preceding section, a radioactive nucleus spontaneously decays by one of three processes: alpha decay, beta decay, or gamma decay. Let us discuss these three processes in more detail.

Alpha Decay

A nucleus emitting an alpha particle ($^4_2$He) loses two protons and two neutrons. Therefore, the atomic number Z decreases by 2, the mass number A decreases by 4, and the neutron number N decreases by 2. The decay can be written

$$^A_Z X \longrightarrow \, ^{A-4}_{Z-2} Y + \, ^4_2 He \qquad \textbf{(44.10)}$$

Alpha decay

where X is called the **parent nucleus** and Y the **daughter nucleus.** As a general rule in any decay equation such as this, (1) the sum of the mass numbers A must be the same on both sides of the equation and (2) the sum of the atomic numbers Z must be the same on both sides of the equation. As examples, ^{238}U and ^{226}Ra are both alpha emitters and decay according to the schemes

$$^{238}_{92} U \longrightarrow \, ^{234}_{90} Th + \, ^4_2 He \qquad \textbf{(44.11)}$$

$$^{226}_{88} Ra \longrightarrow \, ^{222}_{86} Rn + \, ^4_2 He \qquad \textbf{(44.12)}$$

The half-life for ^{238}U decay is 4.47×10^9 years, and that for ^{226}Ra decay is 1.60×10^3 years. The decay of ^{226}Ra is shown in Figure 44.14.

When one element changes into another, as happens in alpha decay, the process is called **spontaneous decay.** In any spontaneous decay, relativistic energy and momentum must be conserved. If we call M_X the mass of the parent nucleus, M_Y the mass of the daughter nucleus, and M_α the mass of the alpha particle, we can define the **disintegration energy** Q as

$$Q = (M_X - M_Y - M_\alpha)c^2 \qquad \textbf{(44.13)}$$

The disintegration energy Q

The energy Q is in joules when the masses are in kilograms and c is the speed of light, 3.00×10^8 m/s. However, when the masses are expressed in the more conve-

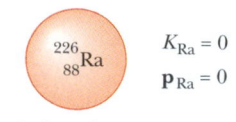

$K_{Ra} = 0$
$\mathbf{p}_{Ra} = 0$

Before decay

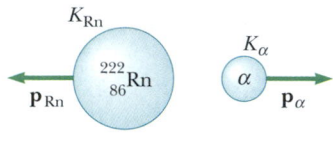

K_{Rn}

K_α

$\mathbf{p}_{Rn}$

$\mathbf{p}_\alpha$

After decay

Figure 44.14   The alpha decay of radium. The radium nucleus is initially at rest. After the decay, the radon nucleus has kinetic energy K_{Rn} and momentum $\mathbf{p}_{Rn}$, and the alpha particle has kinetic energy K_α and momentum $\mathbf{p}_\alpha$.

nient unit u, Q can be calculated in MeV using the expression

$$Q = (M_X - M_Y - M_\alpha) \times 931.494 \text{ MeV/u} \qquad \textbf{(44.14)}$$

The disintegration energy Q appears in the form of kinetic energy in the daughter nucleus and the alpha particle and is sometimes referred to as the Q value of the nuclear reaction. In the case of the ^{226}Ra decay described in Figure 44.14, if the parent nucleus decays at rest, the kinetic energy of the products is 4.87 MeV. Most of this kinetic energy is associated with the alpha particle because this particle is much less massive than the daughter nucleus ^{222}Rn. That is, because momentum must be conserved, the lighter alpha particle recoils with a much higher speed than the daughter nucleus. Generally, less massive particles carry off most of the energy in nuclear decays.

Finally, it is interesting to note that if one assumed that ^{238}U (or any other alpha emitter) decayed by emitting either a proton or a neutron, the mass of the decay products would exceed that of the parent nucleus, corresponding to a negative Q value. A negative Q value indicates that such a proposed decay does not occur spontaneously.

EXAMPLE 44.7   **The Energy Liberated When Radium Decays**

The ^{226}Ra nucleus undergoes alpha decay according to Equation 44.12. Calculate the Q value for this process. Take the masses to be 226.025 402 u for ^{226}Ra, 222.017 571 u for ^{222}Rn, and 4.002 602 u for $^4_2$He, as found in Table A.3.

Solution   We may add 88 electrons to both sides of the reaction 44.12. The differences in electron binding energies are negligible when compared with the Q value for the nuclear decay process. Then, we may use the masses of neutral atoms in Equation 44.14 to see that

$Q = (M_X - M_Y - M_\alpha) \times 931.494 \text{ MeV/u}$

$= (226.025 \ 402 \text{ u} - 222.017 \ 571 \text{ u} - 4.002 \ 602 \text{ u})$
$\quad \times 931.494 \text{ MeV/u}$

$= (0.005 \ 229 \text{ u}) \times (931.494 \text{ MeV/u}) = \boxed{4.87 \text{ MeV}}$

It is left as a problem (Problem 57) to show that the kinetic energy of the alpha particle is about 4.8 MeV, whereas that of the recoiling daughter nucleus is only about 0.1 MeV.

To understand the mechanism of alpha decay, let us imagine a system consisting of (1) the alpha particle, already formed as an entity within the nucleus, and (2) the daughter nucleus that will result when the alpha particle is emitted. Figure 44.15 shows a plot of potential energy versus separation distance r between the alpha particle and the daughter nucleus, where the distance marked R is the range of the nuclear force. The curve represents the combined effects of (1) the Coulomb repulsive energy, which gives the positive peak for $r > R$, and (2) the nuclear attractive force, which causes the curve to be negative for $r < R$. As we saw in Example 44.7, the disintegration energy Q is about 5 MeV, which is the approximate kinetic energy of the alpha particle, represented by the lower dashed line in Figure 44.15.

According to classical physics, the alpha particle is trapped in a potential well. How, then, does it ever escape from the nucleus? The answer to this question was first provided by George Gamow (1904–1968) in 1928 and independently by R. W. Gurney and E. U. Condon in 1929, using quantum mechanics. Briefly, the view of quantum mechanics is that there is always some probability that the particle can tunnel through the barrier (see Section 41.7). Recall that the probability of locating the particle depends on its wave function ψ and that the tunneling probability is measured by $|\psi|^2$. Figure 44.16 is a sketch of the wave function for a particle of

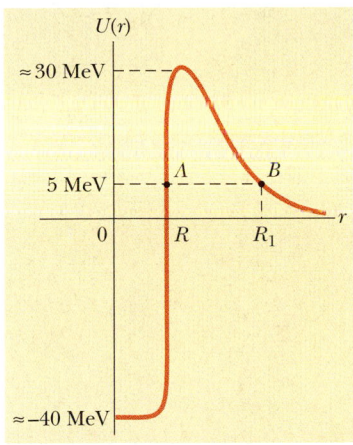

Figure 44.15 Potential energy versus separation distance for a system consisting of an alpha particle and a daughter nucleus. Classically, the energy of the alpha particle is not sufficiently large to overcome the energy barrier, and so the particle should not be able to escape from the nucleus. In reality, the alpha particle does escape by tunneling through the barrier.

energy E meeting a rectangular barrier of finite height, a shape that approximates the nuclear barrier. Note that the wave function exists both inside and outside the barrier. Although the amplitude of the wave function is greatly reduced on the far side of the barrier, its finite value in this region indicates a small but finite probability that the particle can penetrate the barrier. Outside the range of the nuclear force, the function ψ correctly describes the probability that the nucleus will decay. Although the decay probability is constant in time, *the precise moment of decay cannot be predicted.* In general, quantum mechanics implies that the future is indeterminate. (This is in contrast to classical mechanics, where the trajectory of an object can in principle be calculated to an arbitrarily high precision from precise knowledge of its initial coordinates and velocity and of the forces acting on it.) Thus, the fundamental laws of nature are probabilistic and it appears that Einstein was wrong in his famous statement, "God does not roll dice."

Schrödinger's Cat and the Probability of Decay. A radiation detector (see Section 45.6) can be used to show that a radioactive nucleus decays by radiating a particle at a particular moment and in a particular direction. To point out the contrast between this experimental result and its wave function, Erwin Schrödinger imagined a box containing a cat, a radioactive sample, a radiation counter, and a vial of poison. When a nucleus in the sample decays, the counter triggers the administration of lethal poison to the cat. Quantum mechanics correctly predicts the probability of finding the cat dead when the box is opened. However, many questions arise regarding this intriguing thought experiment: Before the box is opened, does the cat have a wave function describing the cat as fractionally dead, with some chance of being alive? Does the act of measurement change the system from a probabilistic state to a definite state? When a particle emitted by a radioactive nucleus is detected at one particular location, does the wave function describing the particle drop to zero instantaneously everywhere else in the Universe? Is there a fundamental difference between a quantum system and a macroscopic system? The answers to such questions are basically unknown.

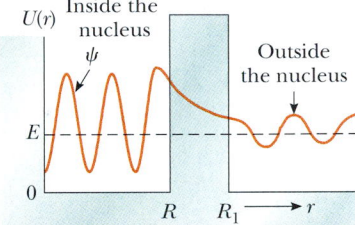

Figure 44.16 The nuclear potential energy is modeled as a rectangular barrier—the blue region extending from R to R_1. The energy of the alpha particle is E, which is less than the height of the barrier. According to quantum mechanics, the alpha particle has some chance of tunneling through the barrier, as indicated by the finite size of the wave function for $r > R_1$.

Quick Quiz 44.5

In alpha decay, the half-life of the decay goes down as the energy of the decay goes up. Why is this?

Artist's rendering of the *Pioneer 10* spacecraft at the end of its useful life, after a 25-year mission. The view is of the Sun and part of the Milky Way galaxy. On the end of the two larger booms are thermoelectric generators that create electricity from the energy given off by ^{238}Pu as the alpha particles it emits collide with surrounding materials. At launch, the generators produced a total power of 160 W. This radioactive power source was necessary because the great distance from the Sun precluded the use of solar panels. *(Courtesy of NASA Ames Home Page)*

 *QuickLab*

Pop open the cover of your smoke detector and see if it describes the radioactive element that is inside. You may also see a warning about not disposing of the detector in the trash. While you have the device open, you should also check the battery.

The Smoke Detector.   A life-saving application of alpha decay is in the household smoke detector, shown in Figure 44.17. Most of the common ones use a radioactive material. The detector consists of an ionization chamber, a sensitive current detector, and an alarm. A weak radioactive source (usually $^{241}_{95}$Am) ionizes the air in the chamber of the detector, creating charged particles. A voltage is maintained between the plates inside the chamber, setting up a small but detectable current in the external circuit. As long as the current is maintained, the alarm is deactivated. However, if smoke drifts into the chamber, the ions become attached to the smoke particles. These heavier particles do not drift as readily as do the lighter ions, which causes a decrease in the detector current. The external circuit senses this decrease in current and sets off the alarm.

Beta Decay

When a radioactive nucleus undergoes beta decay, the daughter nucleus contains the same number of nucleons as the parent nucleus but the atomic number is

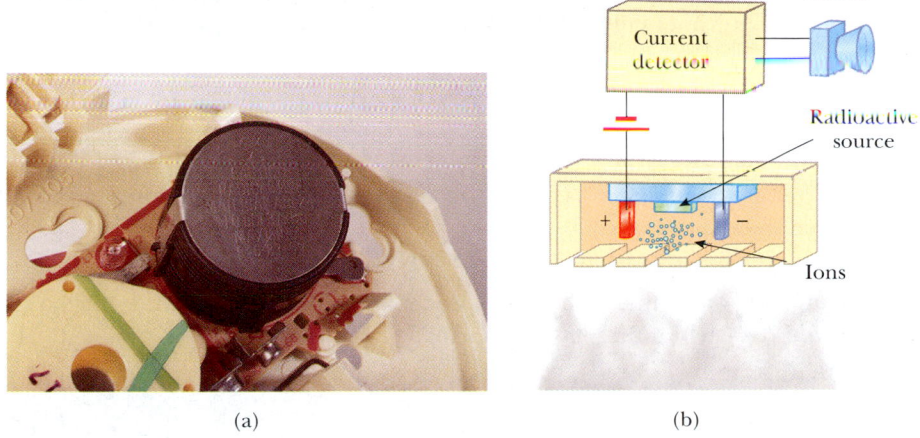

(a) (b)

Figure 44.17 (a) A smoke detector uses alpha decay to determine whether smoke is in the air. (b) Smoke entering the chamber reduces the detected current, causing the alarm to sound. *(a, George Semple)*

changed by 1, which means that the number of protons changes:

$$\ce{^{A}_{Z}X} \rightarrow \ce{^{A}_{Z+1}Y} + \ce{e^-} \qquad \textbf{(44.15)}$$

$$\ce{^{A}_{Z}X} \rightarrow \ce{^{A}_{Z-1}Y} + \ce{e^+} \qquad \textbf{(44.16)}$$

Beta decay

where, as we discussed in Section 44.5, the symbol e^- is used to designate an electron and e^+ designates a positron, with *beta particle* being the general term referring to either. As with alpha decay, the nucleon number and total charge are both conserved in beta decays. From the fact that A does not change but Z does, we conclude that in beta decay, either a neutron changes to a proton (Eq. 44.15) or a proton changes to a neutron (Eq. 44.16). It is also important to note that the electron or positron emitted in these decays is not present beforehand in the nucleus; it is created at the moment of decay from the rest energy of the decaying nucleus.

Two typical beta decay processes are

$$\ce{^{14}_{6}C} \longrightarrow \ce{^{14}_{7}N} + \ce{e^-} \qquad \textbf{(44.17)}$$

$$\ce{^{12}_{7}N} \longrightarrow \ce{^{12}_{6}C} + \ce{e^+} \qquad \textbf{(44.18)}$$

As we shall see later, *beta decay is not described completely by these expressions.* We shall give reasons for this shortly.

Let us consider the energy of the system undergoing beta decay before and after the decay. As with alpha decay, energy must be conserved. Experimentally, it is found that beta particles from a single type of nucleus are emitted over a continuous range of energies (Fig. 44.18). The kinetic energy of the system after the decay is equal to the decrease in mass of the system, that is, the Q value. However, because all decaying nuclei in the sample have the same initial mass, *the Q value must be the same for each decay.* In view of this, why do the emitted particles have the range of kinetic energies shown in Figure 44.18? The law of conservation of energy seems to be violated! And it gets worse: Further analysis of the decay processes described by Equations 44.15 and 44.16 shows that the laws of conservation of both angular momentum (spin) and linear momentum are also violated!

After a great deal of experimental and theoretical study, Pauli in 1930 proposed that a third particle must be present to carry away the "missing" energy and momentum. Fermi later named this particle the **neutrino** (little neutral one) because it had to be electrically neutral and have little or no mass. Although it

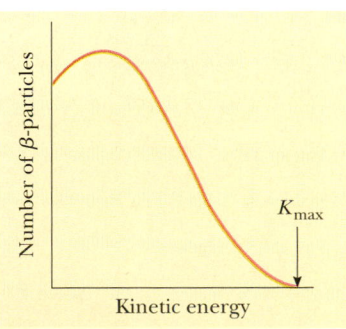

Figure 44.18 A typical beta-decay curve. The maximum kinetic energy observed for the beta particles corresponds to the Q value for the reaction.

eluded detection for many years, the neutrino (symbol ν, Greek letter nu) was finally detected experimentally in 1956. It has the following properties:

Properties of the neutrino

- It has zero electric charge.
- Its mass is either zero (in which case it travels at the speed of light) or very small; there is much recent persuasive experimental evidence that suggests that the neutrino mass is not zero.
- It has a spin of $\frac{1}{2}$, which allows the law of conservation of angular momentum to be satisfied in beta decay.
- It interacts very weakly with matter and is therefore very difficult to detect.

We can now write the beta-decay processes for carbon-14 and nitrogen-12 (Eqs. 44.17 and 44.18) in their correct form:

$$^{14}_{6}\text{C} \longrightarrow\ ^{14}_{7}\text{N} + \text{e}^- + \bar{\nu} \qquad\qquad \textbf{(44.19)}$$

$$^{12}_{7}\text{N} \longrightarrow\ ^{12}_{6}\text{C} + \text{e}^+ + \nu \qquad\qquad \textbf{(44.20)}$$

where the symbol $\bar{\nu}$ represents the **antineutrino,** the antiparticle to the neutrino. We shall discuss antiparticles further in Chapter 46. For now, it suffices to say that **a neutrino is emitted in positron decay and an antineutrino is emitted in electron decay.** As with alpha decay, the decays listed above are analyzed by applying conservation laws, but relativistic expressions must be used for beta particles because their kinetic energy is large (typically 1 MeV) compared with their rest energy of 0.511 MeV.

In Equation 44.19, the number of protons has increased by one and the number of neutrons has decreased by one. We can write the fundamental process of e^- decay in terms of a neutron changing into a proton as follows:

$$\text{n} \longrightarrow \text{p} + \text{e}^- + \bar{\nu} \qquad\qquad \textbf{(44.21)}$$

The electron and the antineutrino are ejected from the nucleus, with the net result that there is one more proton and one fewer neutron, consistent with the changes in Z and $A - Z$. A similar process occurs in e^+ decay, with a proton changing into a neutron, a positron, and a neutrino.

A process that competes with e^+ decay is **electron capture.** This occurs when a parent nucleus captures one of its own orbital electrons and emits a neutrino. The final product after decay is a nucleus whose charge is $Z - 1$:

Electron capture

$$^{A}_{Z}\text{X} + \ ^{0}_{-1}\text{e} \longrightarrow\ ^{A}_{Z-1}\text{Y} + \nu \qquad\qquad \textbf{(44.22)}$$

In most cases, it is a K-shell electron that is captured, and for this reason the process is referred to as **K capture.** One example is the capture of an electron by $^{7}_{4}\text{Be}$:

$$^{7}_{4}\text{Be} + \ ^{0}_{-1}\text{e} \longrightarrow\ ^{7}_{3}\text{Li} + \nu$$

Because the neutrino is very difficult to detect, electron capture is usually observed by the x-rays given off as higher-shell electrons cascade downward to fill the vacancy created in the K shell.

Finally, we specify Q values for the beta-decay processes. The Q values for e^- decay and electron capture are given by $Q = (M_X - M_Y)c^2$, where M_X and M_Y are the masses of neutral atoms. The Q values for e^+ decay are given by $Q = (M_X - M_Y - 2m_e)c^2$. These relationships are useful in determining whether a process is energetically possible.

Quick Quiz 44.6

In beta decay, the kinetic energy of the emitted electron or positron lies somewhere in a relatively large range of possibilities. In alpha decay, the kinetic energy of the emitted alpha particle can have only discrete values. Why is there this difference?

Carbon Dating

The beta decay of ^{14}C (Eq. 44.19) is commonly used to date organic samples. Cosmic rays in the upper atmosphere cause nuclear reactions that create ^{14}C. The ratio of ^{14}C to ^{12}C in the carbon dioxide molecules of our atmosphere has a constant value of approximately 1.3×10^{-12}. The carbon atoms in all living organisms have this same $^{14}C/^{12}C$ ratio because the organisms continuously exchange carbon dioxide with their surroundings. When an organism dies, however, it no longer absorbs ^{14}C from the atmosphere, and so the $^{14}C/^{12}C$ ratio decreases as the ^{14}C decays with a half-life of 5 730 years. It is therefore possible to measure the age of a material by measuring its ^{14}C activity. Using this technique, scientists have been able to identify samples of wood, charcoal, bone, and shell as having lived from 1 000 to 25 000 years ago. This knowledge has helped us reconstruct the history of living organisms—including humans—during this time span.

A particularly interesting example is the dating of the Dead Sea Scrolls. This group of manuscripts was discovered by a shepherd in 1947. Translation showed them to be religious documents, including most of the books of the Old Testament. Because of their historical and religious significance, scholars wanted to know their age. Carbon dating applied to the material in which they were wrapped established their age at approximately 1 950 years.

CONCEPTUAL EXAMPLE 44.8   The Age of Ice Man

In 1991, a German tourist discovered the well-preserved remains of a man, now called the Ice Man, trapped in a glacier in the Italian Alps (Fig. 44.19). Radioactive dating with ^{14}C revealed that this person was alive about 5 300 years ago. Why did scientists date the sample using ^{14}C rather than ^{11}C, which is a beta emitter having a half-life of 20.4 min?

Solution   Because ^{14}C has a half-life of 5 730 years, the fraction of ^{14}C nuclei remaining after one half-life is high enough to allow accurate measurements of changes in the sample's activity. Because ^{11}C has a very short half-life, it is not useful—its activity decreases to a vanishingly small value over the age of the sample, making it impossible to detect.

As a general rule, the isotope chosen to date a sample should have a half-life that is of the same order of magnitude as the age of the sample. If the half-life is much less than the age of the sample, there won't be enough activity left to measure because almost all of the original radioactive nuclei will have decayed away. If the half-life is much greater than the age of the sample, the amount of decay that has taken place since the sample died will be too small to measure. For example, if you have a specimen estimated to have died 50 years ago, neither ^{14}C (5 730 years) nor ^{11}C (20 min) is suitable. If you know your sample contains hydrogen, however, you can measure the activity of ^{3}H (tritium), a beta emitter that has a half-life of 12.3 years.

Figure 44.19   The Ice Man, discovered in 1991 when an Italian glacier melted enough to expose his remains. His possessions, particularly his tools, have shed light on the way people lived in the Bronze Age. Carbon-14 dating was used to determine how long ago this person lived. *(Paul Hanny/Gamma Liaison)*

EXAMPLE 44.9   Radioactive Dating

A 25.0-g piece of charcoal is found in some ruins of an ancient city. The sample shows a ^{14}C activity R of 250 decays/min. How long has the tree this charcoal came from been dead?

Solution   First, let us calculate the decay constant λ for ^{14}C, which has a half-life of 5 730 years.

$$\lambda = \frac{0.693}{T_{1/2}} = \frac{0.693}{(5\ 730\ \text{yr})(3.15 \times 10^7\ \text{s/yr})}$$

$$= 3.84 \times 10^{-12}\ \text{s}^{-1}$$

The number of ^{14}C nuclei can be calculated in two steps. First, the number of ^{12}C nuclei in 25.0 g of carbon is

$$N(^{12}\text{C}) = \frac{6.02 \times 10^{23}\ \text{nuclei/mol}}{12.0\ \text{g/mol}} (25.0\ \text{g})$$

$$= 1.25 \times 10^{24}\ \text{nuclei}$$

Knowing that the ratio of ^{14}C to ^{12}C in the live sample was 1.3×10^{-12}, we see that the number of ^{14}C nuclei in 25.0 g *before* decay was

$$N_0(^{14}\text{C}) = (1.3 \times 10^{-12})(1.25 \times 10^{24}) = 1.6 \times 10^{12}\ \text{nuclei}$$

Hence, the initial activity of the sample was

$$R_0 = N_0\lambda = (1.6 \times 10^{12}\ \text{nuclei})(3.84 \times 10^{-12}\ \text{s}^{-1})$$

$$= 6.14\ \text{decays/s} = 370\ \text{decays/min}$$

We now use Equation 44.8, which relates the activity R at any time t to the initial activity R_0:

$$R = R_0 e^{-\lambda t}$$

$$e^{-\lambda t} = \frac{R}{R_0}$$

Using $R = 250$ decays/min and $R_0 = 370$ decays/min, we calculate t by taking the natural logarithm of both sides of this expression:

$$-\lambda t = \ln\left(\frac{R}{R_0}\right) = \ln\left(\frac{250}{370}\right) = -0.39$$

$$t = \frac{0.39}{\lambda} = \frac{0.39}{3.84 \times 10^{-12}\ \text{s}^{-1}}$$

$$= 1.0 \times 10^{11}\ \text{s} = \boxed{3\ 200\ \text{yr}}$$

Gamma Decay

Very often, a nucleus that undergoes radioactive decay is left in an excited energy state. The nucleus can then undergo a second decay to a lower energy state, perhaps to the ground state, by emitting a high-energy photon:

$$\substack{A\\Z}\text{X}^* \longrightarrow \substack{A\\Z}\text{X} + \gamma \qquad (44.23)$$

where X* indicates a nucleus in an excited state. The typical half-life of an excited nuclear state is 10^{-10} s. Photons emitted in such a de-excitation process are called gamma rays. Such photons have very high energy (1 MeV to 1 GeV) relative to the energy of visible light (about 1 eV). Recall from Sections 40.5 and 42.7 that the energy of a photon emitted or absorbed by an atom equals the difference in energy between the two electronic states involved in the transition. Similarly, a gamma ray photon has an energy hf that equals the energy difference ΔE between two nuclear energy levels. When a nucleus decays by emitting a gamma ray, the only change in the nucleus is that it ends up in a lower energy state.

A nucleus may reach an excited state as the result of a violent collision with another particle. However, it is more common for a nucleus to be in an excited state after it has undergone alpha or beta decay. The following sequence of events represents a typical situation in which gamma decay occurs:

$$\substack{12\\5}\text{B} \longrightarrow \substack{12\\6}\text{C}^* + e^- + \bar{\nu} \qquad (44.24)$$

$$\substack{12\\6}\text{C}^* \longrightarrow \substack{12\\6}\text{C} + \gamma \qquad (44.25)$$

Figure 44.20 shows the decay scheme for ^{12}B, which undergoes beta decay to either of two levels of ^{12}C. It can either (1) decay directly to the ground state of

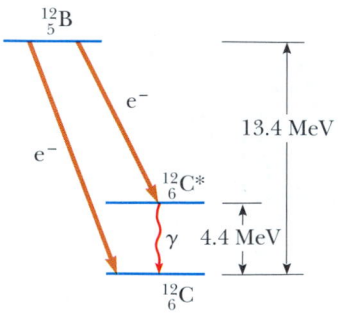

Gamma decay

Figure 44.20   The ^{12}B nucleus undergoes beta decay to either of two levels of ^{12}C: directly to the ground state or to an excited state. Decay to the excited level ^{12}C* is followed by gamma decay to the ground state.

TABLE 44.3	Various Decay Pathways
Alpha decay	$^A_Z X \longrightarrow\ ^{A-4}_{Z-2}Y + ^4_2He$
Beta decay (e^-)	$^A_Z X \longrightarrow\ ^A_{Z+1}Y + e^- + \bar{\nu}$
Beta decay (e^+)	$^A_Z X \longrightarrow\ ^A_{Z-1}Y + e^+ + \nu$
Electron capture	$^A_Z X + e^- \longrightarrow\ ^A_{Z-1}Y + \nu$
Gamma decay	$^A_Z X^* \longrightarrow\ ^A_Z X + \gamma$

^{12}C by emitting a 13.4-MeV electron or (2) undergo beta decay to an excited state of ^{12}C* followed by gamma decay to the ground state. The latter process results in the emission of a 9.0-MeV electron and a 4.4-MeV photon.

The various pathways by which a radioactive nucleus can undergo decay are summarized in Table 44.3.

44.7 NATURAL RADIOACTIVITY

Radioactive nuclei are generally classified into two groups: (1) unstable nuclei found in nature, which give rise to **natural radioactivity,** and (2) unstable nuclei produced in the laboratory through nuclear reactions, which exhibit **artificial radioactivity.**

As Table 44.4 shows, there are three series of naturally occurring radioactive nuclei. Each series starts with a specific long-lived radioactive isotope whose half-life exceeds that of any of its descendants. The three natural series begin with the isotopes ^{238}U, ^{235}U, and ^{232}Th, and the corresponding stable end products are three isotopes of lead: ^{206}Pb, ^{207}Pb, and ^{208}Pb. The fourth series in Table 44.4 begins with ^{237}Np and has as its stable end product ^{209}Bi. The element ^{237}Np is a *transuranic* element (one having an atomic number greater than that of uranium) not found in nature. This element has a half-life of "only" 2.14×10^6 years.

Figure 44.21 shows the successive decays for the ^{232}Th series. Note that ^{232}Th first undergoes alpha decay to ^{228}Ra. Next, ^{228}Ra undergoes two successive beta decays to ^{228}Th. The series continues and finally branches when it reaches ^{212}Bi. At this point, there are two decay possibilities. The end of the decay series is the stable isotope ^{208}Pb. The sequence shown in Figure 44.21 is characterized by a mass-number decrease of either 4 (for alpha decays) or 0 (for beta or gamma decays). The two uranium series are more complex than the ^{232}Th series. Also, there are several naturally occurring radioactive isotopes, such as ^{14}C and ^{40}K, that are not part of any decay series.

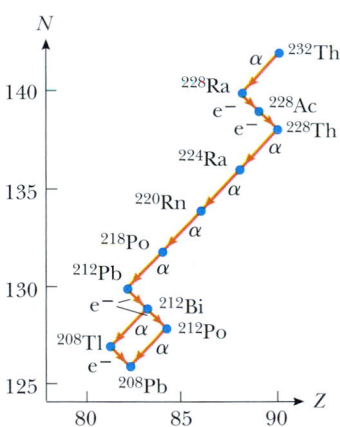

Figure 44.21 Successive decays for the ^{232}Th series.

TABLE 44.4	The Four Radioactive Series		
Series	Starting Isotope	Half-Life (years)	Stable End Product
Uranium ⎫	$^{238}_{92}$U	4.47×10^9	$^{206}_{82}$Pb
Actinium ⎬ Natural	$^{235}_{92}$U	7.04×10^8	$^{207}_{82}$Pb
Thorium ⎭	$^{232}_{90}$Th	1.41×10^{10}	$^{208}_{82}$Pb
Neptunium	$^{237}_{93}$Np	2.14×10^6	$^{209}_{83}$Bi

Because of these radioactive series, our environment is constantly replenished with radioactive elements that would otherwise have disappeared long ago. For example, because the Solar System is approximately 5×10^9 years old, the supply of ^{226}Ra (whose half-life is only 1 600 years) would have been depleted by radioactive decay long ago if it were not for the radioactive series starting with ^{238}U.

44.8 ▶ NUCLEAR REACTIONS

It is possible to change the structure of nuclei by bombarding them with energetic particles. Such collisions, which change the identity of the target nuclei, are called **nuclear reactions.** Rutherford was the first to observe them, in 1919, using naturally occurring radioactive sources for the bombarding particles. Since then, thousands of nuclear reactions have been observed following the development of charged-particle accelerators in the 1930s. With today's advanced technology in particle accelerators and particle detectors, it is possible to achieve particle energies of at least 1 000 GeV = 1 TeV. These high-energy particles are used to create new particles whose properties are helping to solve the mysteries of the nucleus.

Consider a reaction in which a target nucleus X is bombarded by a particle a, resulting in a daughter nucleus Y and a particle b:

Nuclear reaction

$$a + X \longrightarrow Y + b \tag{44.26}$$

Sometimes this reaction is written in the more compact form

$$X(a, b)Y$$

In Section 44.6, the Q value, or disintegration energy, of a radioactive decay was defined as the energy released as a result of the decay process. Likewise, we define the **reaction energy** Q associated with a nuclear reaction as *the total energy released as the result of the reaction:*

Reaction energy Q

$$Q = (M_a + M_X - M_Y - M_b)c^2 \tag{44.27}$$

As an example, consider the reaction ^{7}Li $(p, \alpha)^4$He. The notation p indicates a proton, which is a hydrogen nucleus. Thus, we can write this reaction in the expanded form

$$^1_1\text{H} + {}^7_3\text{Li} \longrightarrow {}^4_2\text{He} + {}^4_2\text{He}$$

Exothermic reaction
Endothermic reaction

The Q value for this reaction is 17.3 MeV. A reaction such as this, for which Q is positive, is called **exothermic.** A reaction for which Q is negative is called **endothermic.** An endothermic reaction does not occur unless the bombarding particle has a kinetic energy greater than Q. The minimum energy necessary for such a reaction to occur is called the **threshold energy.**

Threshold energy

Nuclear reactions must obey the law of conservation of linear momentum. Generally the only force acting on the interacting particles is their mutual force of interaction; that is, there are no external accelerating electric fields present near the colliding particles.

If particles a and b in a nuclear reaction are identical, so that X and Y are also necessarily identical, the reaction is called a **scattering event.** If kinetic energy is conserved as a result of the reaction (that is, if $Q = 0$), it is classified as *elastic scattering.* If kinetic energy is not conserved, $Q \neq 0$ and the reaction is described as *in-*

TABLE 44.5	Q Values for Nuclear Reactions Involving Light Nuclei	
Reaction[a]	**Measured Q-Value (MeV)**	
$^2\text{H}(\text{n}, \gamma)^3\text{H}$	6.257 ± 0.004	
$^2\text{H}(\text{d}, \text{p})^3\text{H}$	4.032 ± 0.004	
$^6\text{Li}(\text{p}, \alpha)^3\text{H}$	4.016 ± 0.005	
$^6\text{Li}(\text{d}, \text{p})^7\text{Li}$	5.020 ± 0.006	
$^7\text{Li}(\text{p}, \text{n})^7\text{Be}$	-1.645 ± 0.001	
$^7\text{Li}(\text{p}, \alpha)^4\text{He}$	17.337 ± 0.007	
$^9\text{Be}(\text{n}, \gamma)^{10}\text{Be}$	6.810 ± 0.006	
$^9\text{Be}(\gamma, \text{n})^8\text{Be}$	-1.666 ± 0.002	
$^9\text{Be}(\text{d}, \text{p})^{10}\text{Be}$	4.585 ± 0.005	
$^9\text{Be}(\text{p}, \alpha)^6\text{Li}$	2.132 ± 0.006	
$^{10}\text{B}(\text{n}, \alpha)^7\text{Li}$	2.793 ± 0.003	
$^{10}\text{B}(\text{p}, \alpha)^7\text{Be}$	1.148 ± 0.003	
$^{12}\text{C}(\text{n}, \gamma)^{13}\text{C}$	4.948 ± 0.004	
$^{13}\text{C}(\text{p}, \text{n})^{13}\text{N}$	-3.003 ± 0.002	
$^{14}\text{N}(\text{n}, \text{p})^{14}\text{C}$	0.627 ± 0.001	
$^{14}\text{N}(\text{n}, \gamma)^{15}\text{N}$	10.833 ± 0.007	
$^{18}\text{O}(\text{p}, \text{n})^{18}\text{F}$	-2.453 ± 0.002	
$^{19}\text{F}(\text{p}, \alpha)^{16}\text{O}$	8.124 ± 0.007	

From C. W. Li, W. Whaling, W. A. Fowler, and C. C. Lauritsen, *Phys. Rev.* 83:512, 1951.

[a] The symbols n, p, d, α, and γ denote the neutron, proton, deuteron, alpha particle, and photon, respectively.

elastic scattering. This terminology is identical to that used in describing collisions between macroscopic objects (Section 9.4).

Measured Q values for a number of nuclear reactions involving light nuclei are given in Table 44.5.

In addition to energy and momentum, the total charge and total number of nucleons must be conserved in any nuclear reaction. For example, consider the reaction $^{19}\text{F}(\text{p}, \alpha)^{16}\text{O}$, which has a Q value of 8.124 MeV. We can show this reaction more completely as

$$^1_1\text{H} + {}^{19}_{9}\text{F} \longrightarrow {}^{16}_{8}\text{O} + {}^4_2\text{He}$$

The total number of nucleons before the reaction $(1 + 19 = 20)$ is equal to the total number after the reaction $(16 + 4 = 20)$. Furthermore, the total charge $(Z = 10)$ is the same before and after the reaction.

SUMMARY

A nucleus is represented by the symbol ^A_ZX, where A is the **mass number** (the total number of nucleons) and Z is the **atomic number** (the total number of protons). The total number of neutrons in a nucleus is the **neutron number** N, where $A = N + Z$. Nuclei having the same Z value but different A and N values are **isotopes** of one another.

Assuming that nuclei are spherical, their radius is given by

$$r = r_0 A^{1/3} \tag{44.1}$$

where $r_0 = 1.2$ fm.

Nuclei are stable because of the **nuclear force** between nucleons. This short-range force dominates the Coulomb repulsive force at distances of less than about 2 fm and is independent of charge. Light nuclei are most stable when the number of protons they contain equals the number of neutrons. Heavy nuclei are most stable when the number of neutrons they contain exceeds the number of protons. The most stable nuclei have Z and N values that are both even.

Nuclei have an intrinsic spin angular momentum of magnitude $\sqrt{I(I+1)}\,\hbar$, where I is the **nuclear spin quantum number.** The magnetic moment of a nucleus is measured in terms of the **nuclear magneton** μ_n, where

$$\mu_n \equiv \frac{e\hbar}{2m_p} = 5.05 \times 10^{-27}\,\text{J/T} \tag{44.3}$$

When a nuclear spin magnetic moment is placed in an external magnetic field, it precesses about the field with a frequency (the **Larmor precessional frequency**) that is proportional to the magnitude of the field.

The difference between the sum of the masses of a group of separate nucleons and the mass of the compound nucleus containing these nucleons, when multiplied by c^2, gives the **binding energy** E_b of the nucleus. We can calculate the binding energy of the nucleus of an atom of mass M_A using the expression

$$E_b(\text{MeV}) = (Zm_p + Nm_n - M_A) \times 931.494\,\text{MeV/u} \tag{44.4}$$

where m_p is the mass of the proton and m_n is the mass of the neutron.

The **liquid-drop model** of nuclear structure treats the nucleons as molecules in a drop of liquid. The three main contributions influencing binding energy are the volume effect, the surface effect, and the Coulomb repulsion effect. Summing such contributions results in the **semiempirical binding energy formula:**

$$E_b = C_1 A - C_2 A^{2/3} - C_3 \frac{Z(Z-1)}{A^{1/3}} - C_4 \frac{(N-Z)^2}{A} \tag{44.5}$$

The **independent-particle model** assumes that each nucleon exists in a shell and can only have discrete energy values. The stability of certain nuclei can be explained with this model.

A radioactive substance decays by **alpha decay, beta decay,** or **gamma decay.** An alpha particle is the ^{4}He nucleus; a beta particle is either an electron (e^-) or a positron (e^+); a gamma particle is a high-energy photon.

If a radioactive material contains N_0 radioactive nuclei at $t=0$, the number N of nuclei remaining after a time t has elapsed is

$$N = N_0 e^{-\lambda t} \tag{44.7}$$

where λ is the **decay constant,** a number equal to the probability per second that a nucleus will decay. The **decay rate,** or **activity,** of a radioactive substance is

$$R = \left|\frac{dN}{dt}\right| = R_0 e^{-\lambda t} \tag{44.8}$$

where $R_0 = N_0 \lambda$ is the activity at $t=0$. The **half-life** $T_{1/2}$ is defined as the time it takes half of a given number of radioactive nuclei to decay, where

$$T_{1/2} = \frac{0.693}{\lambda} \tag{44.9}$$

In alpha decay, a helium nucleus is ejected from the parent nucleus with a definite kinetic energy. A nucleus undergoing beta decay emits either an electron (e^-) and an antineutrino ($\overline{\nu}$) or a positron (e^+) and a neutrino (ν). The electron or positron is ejected with a range of energies. In **electron capture,** the nucleus of an atom absorbs one of its own electrons and emits a neutrino. In gamma decay, a nucleus in an excited state decays to its ground state and emits a gamma ray.

Nuclear reactions can occur when a target nucleus X is bombarded by a particle a, resulting in a daughter nucleus Y and a particle b:

$$a + X \longrightarrow Y + b \qquad \textbf{(44.26)}$$

The energy released in such a reaction, called the **reaction energy** Q, is

$$Q = (M_a + M_X - M_Y - M_b)c^2 \qquad \textbf{(44.27)}$$

QUESTIONS

1. Why are heavy nuclei unstable?
2. The magnetic moment of a proton precesses with a frequency ω_p in the presence of a magnetic field. If the magnetic field magnitude is doubled, what happens to the precessional frequency?
3. Explain why nuclei that are well off the line of stability in Figure 44.3 tend to be unstable.
4. Why do nearly all the naturally occurring isotopes lie above the $N = Z$ line in Figure 44.3?
5. Consider two heavy nuclei X and Y having similar mass numbers. If X has the higher binding energy, which nucleus tends to be more unstable?
6. Discuss the differences between the liquid-drop model and the independent-particle model of the nucleus.
7. How many values of I_z are possible for $I = 5/2$? for $I = 3$?
8. In nuclear magnetic resonance, how does increasing the value of the constant magnetic field change the frequency of the radio-frequency field that excites a particular transition?
9. Would the liquid-drop or independent-particle model be more appropriate to predict the behavior of a nucleus in a fission reaction? Which would be more successful in predicting the magnetic moment of a given nucleus? Which could better explain the γ-ray spectrum of an excited nucleus?
10. If a nucleus has a half-life of 1 year, does this mean it will be completely decayed after 2 years? Explain.
11. What fraction of a radioactive sample has decayed after two half-lives have elapsed?
12. Two samples of the same radioactive nuclide are prepared. Sample A has twice the initial activity of sample B. How does the half-life of A compare with the half-life of B? After each has passed through five half-lives, what is the ratio of their activities?
13. Explain why the half-lives for radioactive nuclei are essentially independent of temperature.

14. The radioactive nucleus $^{226}_{88}\text{Ra}$ has a half-life of approximately 1.6×10^3 years. Being that the Solar System is about 5 billion years old, why do we still find this nucleus in nature?
15. Why is the electron involved in the reaction

$$^{14}_{6}\text{C} \longrightarrow {}^{14}_{7}\text{N} + e^- + \overline{\nu}$$

written as e^-, while the electron involved in the reaction

$$^{7}_{4}\text{Be} + {}^{0}_{-1}e \longrightarrow {}^{7}_{3}\text{Li} + \nu$$

is written as $^{0}_{-1}e$?
16. A free neutron undergoes beta decay with a half-life of about 15 min. Can a free proton undergo a similar decay?
17. Explain how you can carbon date the age of a sample.
18. What is the difference between a neutrino and a photon?
19. Does the Q in Equation 44.27 represent the quantity (final mass $-$ initial mass)c^2, or does it represent the quantity (initial mass $-$ final mass)c^2?
20. Use Equations 44.19 to 44.21 to explain why the neutrino must have a spin of $\frac{1}{2}$.
21. If a nucleus such as ^{226}Ra initially at rest undergoes alpha decay, which has more kinetic energy after the decay, the alpha particle or the daughter nucleus?
22. Can a nucleus emit alpha particles that have different energies? Explain.
23. Explain why many heavy nuclei undergo alpha decay but do not spontaneously emit neutrons or protons.
24. If an alpha particle and an electron have the same kinetic energy, which undergoes the greater deflection when passed through a magnetic field?
25. If film is kept in a wooden box, alpha particles from a radioactive source outside the box cannot expose the film but beta particles can. Explain.
26. Pick any beta decay process and show that the neutrino must have zero charge.

27. Suppose it could be shown that the cosmic ray intensity at the Earth's surface was much greater 10 000 years ago. How would this difference affect what we accept as valid carbon-dated values of the age of ancient samples of once-living matter?

28. Why is carbon dating unable to provide accurate estimates of very old material?

29. Element X has several isotopes. What do these isotopes have in common? How do they differ?

30. Explain the main differences between alpha, beta, and gamma rays.

31. How many protons are there in the nucleus $^{222}_{86}$Rn? How many neutrons? How many orbiting electrons are there in the neutral atom?

PROBLEMS

1, 2, 3 = straightforward, intermediate, challenging ☐ = full solution available in the *Student Solutions Manual and Study Guide*
WEB = solution posted at **http://www.saunderscollege.com/physics/** 🖥 = Computer useful in solving problem 📕 = Interactive Physics
☐ = paired numerical/symbolic problems

Note: Table 44.1 and 44.6 will be useful for many of these problems. A more complete list of atomic masses is given in Table A.3 in Appendix A.

Section 44.1 Some Properties of Nuclei

1. What is the order of magnitude of the number of protons in your body? Of the number of neutrons? Of the number of electrons?

2. Review Problem. Singly ionized carbon is accelerated through 1 000 V and passed into a mass spectrometer to determine the isotopes present (see Chapter 29). The magnitude of the magnetic field in the spectrometer is 0.200 T. (a) Determine the orbit radii for the ^{12}C and the ^{13}C isotopes as they pass through the field. (b) Show that the ratio of radii may be written in the form

$$\frac{r_1}{r_2} = \sqrt{\frac{m_1}{m_2}}$$

and verify that your radii in part (a) agree with this.

3. An α particle ($Z = 2$, mass 6.64×10^{-27} kg) approaches to within 1.00×10^{-14} m of a carbon nucleus ($Z = 6$). What are (a) the maximum Coulomb force on the α particle, (b) the acceleration of the α particle at this point, and (c) the potential energy of the α particle at this point?

4. In a Rutherford scattering experiment, alpha particles having kinetic energy of 7.70 MeV are fired toward a gold nucleus. (a) Use energy conservation to determine the distance of closest approach between the alpha particle and gold nucleus. Assume the nucleus remains at rest. (b) Calculate the de Broglie wavelength for the 7.70-MeV alpha particle and compare it with the distance obtained in part (a). (c) Based on this comparison, why is it proper to treat the alpha particle as a particle and not as a wave in the Rutherford scattering experiment?

5. (a) Use energy methods to calculate the distance of closest approach for a head-on collision between an alpha particle having an initial energy of 0.500 MeV and a gold nucleus (^{197}Au) at rest. (Assume the gold nucleus remains at rest during the collision.) (b) What minimum initial speed must the alpha particle have in order to get as close as 300 fm?

6. How much energy (in MeV units) must an α particle have to reach the surface of a gold nucleus ($Z = 79$, $A = 197$)?

7. Find the radius of (a) a nucleus of $^4_2$He and (b) a nucleus of $^{238}_{92}$U.

8. Find the nucleus that has a radius approximately equal to one-half the radius of uranium $^{238}_{92}$U.

9. A star ending its life with a mass of two times the mass of the Sun is expected to collapse, combining its protons and electrons to form a neutron star. Such a star could be thought of as a gigantic atomic nucleus. If a star of mass $2 \times 1.99 \times 10^{30}$ kg collapsed into neutrons ($m_n = 1.67 \times 10^{-27}$ kg), what would its radius be? (Assume that $r = r_0 A^{1/3}$.)

10. Review Problem. What would be the gravitational force between two golf balls (each with a 4.30-cm diameter), 1.00 meter apart, if they were made of nuclear matter?

11. From Table A.3, identify the stable nuclei that correspond to the magic numbers given by Equation 44.2.

12. For the stable nuclei in Table A.3, identify the number of stable nuclei that are even Z, even N; even Z, odd N; odd Z, even N; and odd Z, odd N.

13. Construct a diagram like that of Figure 44.4 for the case when I equals (a) 5/2 and (b) 4.

Section 44.2 Nuclear Magnetic Resonance and Magnetic Resonance Imaging

14. The Larmor precessional frequency is

$$f = \frac{\Delta E}{h} = \frac{2\mu B}{h}$$

Calculate the radio-wave frequency at which resonance absorption will occur for (a) free neutrons in a magnetic field of 1.00 T, (b) free protons in a magnetic field of 1.00 T, and (c) free protons in the Earth's mag-

TABLE 44.6   **Some Atomic Masses**

Element	Atomic Mass (u)	Element	Atomic Mass (u)
$^4_2$He	4.002 602	$^{27}_{13}$Al	26.981 538
$^7_3$Li	7.016 003	$^{30}_{15}$P	29.978 307
$^9_4$Be	9.012 174	$^{40}_{20}$Ca	39.962 591
$^{10}_{5}$B	10.012 936	$^{42}_{20}$Ca	41.958 618
$^{12}_{6}$C	12.000 000	$^{43}_{20}$Ca	42.958 767
$^{13}_{6}$C	13.003 355	$^{56}_{26}$Fe	55.934 940
$^{14}_{7}$N	14.003 074	$^{64}_{30}$Zn	63.929 144
$^{15}_{7}$N	15.000 108	$^{64}_{29}$Cu	63.929 599
$^{15}_{8}$O	15.003 065	$^{93}_{41}$Nb	92.906 376
$^{17}_{8}$O	16.999 132	$^{197}_{79}$Au	196.966 543
$^{18}_{8}$O	17.999 160	$^{202}_{80}$Hg	201.970 617
$^{18}_{9}$F	18.000 937	$^{216}_{84}$Po	216.001 889
$^{20}_{10}$Ne	19.992 435	$^{220}_{86}$Rn	220.011 369
$^{23}_{11}$Na	22.989 770	$^{234}_{90}$Th	234.043 593
$^{23}_{12}$Mg	22.994 124	$^{238}_{92}$U	238.050 784

netic field at a location where the magnitude of the field is 50.0 μT.

Section 44.3   Binding Energy and Nuclear Forces

15. Calculate the binding energy per nucleon for (a) ^{2}H, (b) ^{4}He, (c) ^{56}Fe, and (d) ^{238}U.

16. The peak of the stability curve occurs at ^{56}Fe. This is why iron is prominent in the spectrum of the Sun and stars. Show that ^{56}Fe has a higher binding energy per nucleon than its neighbors ^{55}Mn and ^{59}Co. Compare your results with Figure 44.8.

WEB 17. Nuclei having the same mass numbers are called *isobars*. The isotope $^{139}_{57}$La is stable. A radioactive isobar $^{139}_{59}$Pr is located below the line of stable nuclei in Figure 44.3 and decays by e^+ emission. Another radioactive isobar of ^{139}La, $^{139}_{55}$Cs, decays by e^- emission and is located above the line of stable nuclei in Figure 44.3. (a) Which of these three isobars has the highest neutron-to-proton ratio? (b) Which has the greatest binding energy per nucleon? (c) Which do you expect to be heavier, ^{139}Pr or ^{139}Cs?

18. Two nuclei having the same mass number are known as *isobars*. Calculate the difference in binding energy per nucleon for the isobars $^{23}_{11}$Na and $^{23}_{12}$Mg. How do you account for the difference?

WEB 19. A pair of nuclei for which $Z_1 = N_2$ and $Z_2 = N_1$ are called mirror isobars (the atomic and neutron numbers are interchanged). Binding energy measurements on these nuclei can be used to obtain evidence of the charge independence of nuclear forces (that is, proton–proton, proton–neutron, and neutron–neutron nuclear forces are equal). Calculate the difference in binding energy for the two mirror isobars $^{15}_{8}$O and $^{15}_{7}$N.

20. The energy required to construct a uniformly charged sphere of total charge Q and radius R is $U = 3k_eQ^2/5R$, where k_e is the Coulomb constant (see Problem 67). Assume that a ^{40}Ca nucleus contains 20 protons uniformly distributed in a spherical volume. (a) How much energy is required to counter the electrostatic repulsion given by the above equation? (*Hint:* First calculate the radius of a ^{40}Ca nucleus.) (b) Calculate the binding energy of ^{40}Ca. (c) Explain what you can conclude from comparing the result of part (b) and that of part (a).

21. Calculate the minimum energy required to remove a neutron from the $^{43}_{20}$Ca nucleus.

Section 44.4   Nuclear Models

22. (a) In the liquid-drop model of nuclear structure, why does the surface-effect term $-C_2A^{2/3}$ have a negative sign? (b) The binding energy of the nucleus increases as the volume-to-surface ratio increases. Calculate this ratio for both spherical and cubical shapes and explain which is more plausible for nuclei.

23. Using the graph in Figure 44.8, estimate how much energy is released when a nucleus of mass number 200 is split into two nuclei each of mass number 100.

24. (a) Use Equation 44.5 to compute the binding energy for $^{56}_{26}$Fe. (b) What percentage is contributed to the binding energy by each of the four terms?

Section 44.5   Radioactivity

25. A radioactive sample contains 1.00×10^{15} atoms and has an activity of 6.00×10^{11} Bq. What is its half-life?

26. The half-life of ^{131}I is 8.04 days. On a certain day, the activity of an iodine-131 sample is 6.40 mCi. What is its activity 40.2 days later?

WEB **27.** A freshly prepared sample of a certain radioactive isotope has an activity of 10.0 mCi. After 4.00 h, its activity is 8.00 mCi. (a) Find the decay constant and half-life. (b) How many atoms of the isotope were contained in the freshly prepared sample? (c) What is the sample's activity 30.0 h after it is prepared?

28. How much time elapses before 90.0% of the radioactivity of a sample of $_{33}^{72}$As disappears, as measured by its activity? The half-life of $_{33}^{72}$As is 26 h.

29. The radioactive isotope ^{198}Au has a half-life of 64.8 h. A sample containing this isotope has an initial activity ($t = 0$) of 40.0 μCi. Calculate the number of nuclei that decay in the time interval between $t_1 = 10.0$ h and $t_2 = 12.0$ h.

30. A radioactive nucleus has half-life $T_{1/2}$. A sample containing these nuclei has initial activity R_0. Calculate the number of nuclei that decay in the time interval between the times t_1 and t_2.

31. Determine the activity of 1.00 g of ^{60}Co. The half-life of ^{60}Co is 5.27 yr.

Section 44.6 The Decay Processes

32. Identify the missing nuclide or particle (X):
(a) $X \rightarrow {}_{28}^{65}Ni + \gamma$
(b) ${}_{84}^{215}Po \rightarrow X + \alpha$
(c) $X \rightarrow {}_{26}^{55}Fe + e^+ + \nu$
(d) ${}_{48}^{109}Cd + X \rightarrow {}_{47}^{109}Ag + \nu$
(e) ${}_{11}^{14}Na + {}_{2}^{4}He \rightarrow X + {}_{8}^{17}O$

33. Find the energy released in the alpha decay

$$_{92}^{238}U \longrightarrow {}_{90}^{234}Th + {}_{2}^{4}He$$

You will find the following mass values useful:

$$M({}_{92}^{238}U) = 238.050\ 784\ u$$

$$M({}_{90}^{234}Th) = 234.043\ 593\ u$$

$$M({}_{2}^{4}He) = 4.002\ 602\ u$$

34. A living specimen in equilibrium with the atmosphere contains one atom of ^{14}C (half-life = 5 730 yr) for every 7.7×10^{11} stable carbon atoms. An archeological sample of wood (cellulose, $C_{12}H_{22}O_{11}$) contains 21.0 mg of carbon. When the sample is placed inside a shielded beta counter with 88.0% counting efficiency, 837 counts are accumulated in one week. Assuming that the cosmic-ray flux and the Earth's atmosphere have not changed appreciably since the sample was formed, find the age of the sample.

35. A ^{3}H nucleus beta decays into ^{3}He by creating an electron and an antineutrino according to the reaction

$$_{1}^{3}H \longrightarrow {}_{2}^{3}He + e^- + \bar{\nu}$$

Use Table A.3 to determine the total energy released in this reaction.

36. Determine which decays can occur spontaneously:
(a) $_{20}^{40}Ca \rightarrow e^+ + {}_{19}^{40}K$

(b) $_{44}^{98}Ru \rightarrow {}_{2}^{4}He + {}_{42}^{94}Mo$
(c) $_{60}^{144}Nd \rightarrow {}_{2}^{4}He + {}_{58}^{140}Ce$

37. The nucleus $_{8}^{15}O$ decays by electron capture. Write (a) the basic nuclear process and (b) the decay process referring to neutral atoms. (c) Determine the energy of the neutrino. Disregard the daughter's recoil.

Section 44.7 Natural Radioactivity

38. A rock sample contains traces of ^{238}U, ^{235}U, ^{232}Th, ^{208}Pb, ^{207}Pb, and ^{206}Pb. Careful analysis shows that the ratio of the amount of ^{238}U to ^{206}Pb is 1.164. (a) Assume that the rock originally contained no lead, and determine the age of the rock. (b) What should be the ratios of ^{235}U to ^{207}Pb and of ^{232}Th to ^{208}Pb so that they would yield the same age for the rock? Neglect the minute amounts of the intermediate decay products in the decay chains. Note that this form of multiple dating gives reliable geological dates.

39. Enter the correct isotope symbol in each open square in Figure P44.39, which shows the sequences of decays starting with uranium-235 and ending with the stable isotope lead-207.

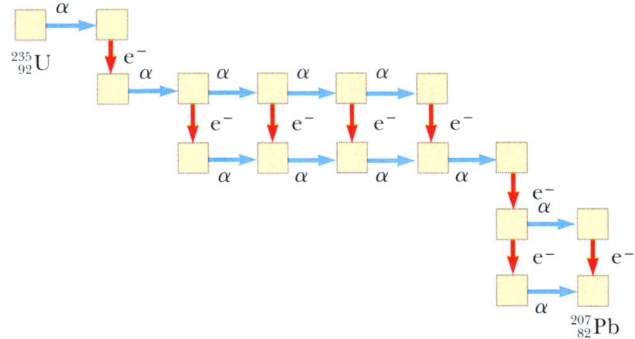

Figure P44.39

40. *Indoor air pollution.* Uranium is naturally present in rock and soil. At one step in its series of radioactive decays, ^{238}U produces the chemically inert gas radon-222, with a half-life of 3.82 days. The radon seeps out of the ground to mix into the atmosphere, typically making open air radioactive with activity 0.3 pCi/L. In homes ^{222}Rn can be a serious pollutant, accumulating to reach much higher activities in enclosed spaces. If the radon radioactivity exceeds 4 pCi/L, the Environmental Protection Agency suggests taking action to reduce it, by reducing infiltration of air from the ground. (a) Convert the activity 4 pCi/L to units of becquerel per cubic meter. (b) How many ^{222}Rn atoms are in one cubic meter of air displaying this activity? (c) What fraction of the mass of the air does the radon constitute?

41. The most common isotope of radon is ^{222}Rn, which has half-life 3.82 days. (a) What fraction of the nuclei that were on Earth one week ago are now undecayed? (b) What fraction of those that existed one year ago? (c) In view of these results, explain why radon remains a problem, contributing significantly to our background radiation exposure.

Section 44.8   Nuclear Reactions

42. The reaction $^{27}_{13}$Al$(\alpha, n)^{30}_{15}$P, achieved in 1934, was the first known in which the product nucleus is radioactive. Calculate the Q value of this reaction.

WEB 43. Natural gold has only one isotope, $^{197}_{79}$Au. If natural gold is irradiated by a flux of slow neutrons, e$^-$ particles are emitted. (a) Write the reaction equation. (b) Calculate the maximum energy of the emitted beta particles. The mass of $^{198}_{80}$Hg is 197.966 743 u.

44. Identify the unknown particles X and X′ in the following nuclear reactions:
    (a) X + $^4_2$He → $^{24}_{12}$Mg + $^1_0$n
    (b) $^{235}_{92}$U + $^1_0$n → $^{90}_{38}$Sr + X + 2$^1_0$n
    (c) 2$^1_1$H → $^2_1$H + X + X′

45. A beam of 6.61-MeV protons is incident on a target of $^{27}_{13}$Al. Those that collide produce the reaction

$$p + {}^{27}_{13}\text{Al} \longrightarrow {}^{27}_{14}\text{Si} + n$$

    ($^{27}_{14}$Si has mass 26.986 721 u.) Neglecting any recoil of the product nucleus, determine the kinetic energy of the emerging neutrons.

46. (a) Suppose $^{10}_5$B is struck by an alpha particle, releasing a proton and a product nucleus in the reaction. What is the product nucleus? (b) An alpha particle and a product nucleus are produced when $^{13}_6$C is struck by a proton. What is the product nucleus?

47. Using the Q values of appropriate reactions from Table 44.5, calculate the masses of ^{8}Be and ^{10}Be in atomic mass units to four decimal places.

48. Determine the Q value associated with the spontaneous fission of ^{236}U into the fragments ^{90}Rb and ^{143}Cs, which have mass 89.914 811 u and 142.927 220 u, respectively. The masses of the other particles involved in the reaction are given in Appendix A.3.

ADDITIONAL PROBLEMS

49. Consider a radioactive sample. Determine the ratio of the number of atoms decaying during the first half of its half-life to the number of atoms decaying during the second half of its half-life.

50. One method of producing neutrons for experimental use is to bombard $^7_3$Li with protons. The neutrons are emitted according to the reaction

$$^1_1\text{H} + {}^7_3\text{Li} \longrightarrow {}^7_4\text{Be} + {}^1_0\text{n}$$

What is the minimum kinetic energy the incident proton must have if this reaction is to occur? You may use the result of Problem 70.

51. A by-product of some fission reactors is the isotope $^{239}_{94}$Pu, an alpha emitter having a half-life of 24 120 years:

$$^{239}_{94}\text{Pu} \longrightarrow {}^{235}_{92}\text{U} + \alpha$$

Consider a sample of 1.00 kg of pure $^{239}_{94}$Pu at $t = 0$. Calculate (a) the number of $^{239}_{94}$Pu nuclei present at $t = 0$ and (b) the initial activity in the sample. (c) How long does the sample have to be stored if a "safe" activity level is 0.100 Bq?

52. (a) The atomic mass of ^{57}Co is 56.936 294 u. Can ^{57}Co decay by e$^+$ emission? Explain. (b) Can ^{14}C decay by e$^-$ emission? Explain. (c) If either answer is yes, what is the range of kinetic energies available for the beta particle?

53. (a) Find the radius of the $^{12}_6$C nucleus. (b) Find the force of repulsion between a proton at the surface of a $^{12}_6$C nucleus and the remaining five protons. (c) How much work (in MeV) has to be done to overcome this electrostatic repulsion to put the last proton into the nucleus? (d) Repeat parts (a), (b), and (c) for $^{238}_{92}$U.

54. The activity of a radioactive sample was measured over 12 h, with the following net count rates:

Time (h)	Counting Rate (counts/min)
1.00	3 100
2.00	2 450
4.00	1 480
6.00	910
8.00	545
10.0	330
12.0	200

(a) Plot the logarithm of counting rate as a function of time. (b) Determine the disintegration constant and half-life of the radioactive nuclei in the sample. (c) What counting rate would you expect for the sample at $t = 0$? (d) Assuming the efficiency of the counting instrument to be 10.0%, calculate the number of radioactive atoms in the sample at $t = 0$.

55. (a) Why is the beta decay p → n + e$^+$ + ν forbidden for a free proton? (b) Why is the same reaction possible if the proton is bound in a nucleus? For example, the following reaction occurs:

$$^{13}_7\text{N} \longrightarrow {}^{13}_6\text{C} + e^+ + \nu$$

(c) How much energy is released in the reaction given in part (b)? [$m(e^+) = 0.000\ 549$ u, $M(^{13}\text{C}) = 13.003\ 355$ u, $M(^{13}\text{N}) = 13.005\ 738$ u]

56. In a piece of rock from the Moon, the ^{87}Rb content is assayed to be 1.82×10^{10} atoms per gram of material, and the ^{87}Sr content is found to be 1.07×10^9 atoms per gram. (a) Calculate the age of the rock. (b) Could the material in the rock actually be much older? What assumption is implicit in using the radioactive dating method? (The relevant decay is ^{87}Rb → ^{87}Sr + e$^-$. The half-life of the decay is 4.75×10^{10} yr.)

WEB **57.** The decay of an unstable nucleus by alpha emission is represented by Equation 44.10. The disintegration energy Q given by Equation 44.13 must be shared by the alpha particle and the daughter nucleus in order to conserve both energy and momentum in the decay process. (a) Show that Q and K_α, the kinetic energy of the alpha particle, are related by the expression

$$Q = K_\alpha \left(1 + \frac{M_\alpha}{M} \right)$$

where M is the mass of the daughter nucleus. (b) Use the result of part (a) to find the energy of the alpha particle emitted in the decay of ^{226}Ra. (See Example 44.7 for the calculation of Q.)

58. The ^{145}Pm nucleus decays by alpha emission. (a) Determine the daughter nucleus. (b) Using the values given in Table A.3, determine the energy released in this decay. (c) What fraction of this energy is carried away by the alpha particle when the recoil of the daughter is taken into account?

59. When, after a reaction or disturbance of any kind, a nucleus is left in an excited state, it can return to its normal (ground) state by emission of a gamma-ray photon (or several photons). This process is illustrated by Equation 44.23. The emitting nucleus must recoil to conserve both energy and momentum. (a) Show that the recoil energy of the nucleus is

$$E_r = \frac{(\Delta E)^2}{2Mc^2}$$

where ΔE is the difference in energy between the excited and ground states of a nucleus of mass M. (b) Calculate the recoil energy of the ^{57}Fe nucleus when it decays by gamma emission from the 14.4-keV excited state. For this calculation, take the mass to be 57 u. (*Hint:* When writing the equation for conservation of energy, use $(Mv)^2/2M$ for the kinetic energy of the recoiling nucleus. Also, assume that $hf \ll Mc^2$ and use the binomial expansion.)

60. After the sudden release of radioactivity from the Chernobyl nuclear reactor accident in 1986, the radioactivity of milk in Poland rose to 2 000 Bq/L due to iodine-131, with half-life 8.04 days. Radioactive iodine is particularly hazardous, because the thyroid gland concentrates iodine. The Chernobyl accident caused a measurable increase in thyroid cancers among children in Belarus. (a) For comparison, find the activity of milk due to potassium. Assume that 1 L of milk contains 2.00 g of potassium, of which 0.011 7% is the isotope ^{40}K that has a half-life 1.28×10^9 yr. (b) After what time would the activity due to iodine fall below that due to potassium?

61. Europeans named a certain direction in the sky as between the horns of Taurus the Bull. On the day they named as July 4, 1054 A.D., a brilliant light appeared there. Europeans left no surviving record of the supernova, which could be seen in daylight for some days. It

faded but remained visible for years, dimming for a time with the 77.1-day half-life of the radioactive cobalt-56 that had been created in the explosion. (a) The remains of the star now form the Crab Nebula. In it, the cobalt-56 has now decreased to what fraction of its original activity? (b) Suppose that an American, of the people called Anasazi, made a charcoal drawing of the supernova. The carbon-14 in the charcoal has now decayed to what fraction of its original activity?

62. A theory of nuclear astrophysics proposes that all the heavy elements, such as uranium, are formed in supernova explosions ending the lives of massive stars. If we assume that at the time of the explosion there were equal amounts of ^{235}U and ^{238}U, how long ago did the star(s) explode that released the elements that formed our Earth? The present ^{235}U/^{238}U ratio is 0.007 25. The half-lives of ^{235}U and ^{238}U are 0.704×10^9 years and 4.47×10^9 years.

63. After determining that the Sun has existed for hundreds of millions of years, but before the discovery of nuclear physics, scientists could not explain why the Sun has continued to burn for such a long time. For example, if it were a coal fire, it would have burned up in about 3 000 years. Assume that the Sun, whose mass is 1.99×10^{30} kg, originally consisted entirely of hydrogen and that its total power output is 3.77×10^{26} W. (a) If the energy-generating mechanism of the Sun is the transforming of hydrogen into helium via the net reaction

$$4{}^1_1\text{H} + 2{}^{\,0}_{-1}\text{e} \longrightarrow {}^4_2\text{He} + 2\nu + \gamma$$

calculate the energy (in joules) given off by this reaction. (b) Determine how many hydrogen atoms constitute the Sun. Take the mass of one hydrogen atom to be 1.67×10^{-27} kg. (c) Assuming that the total power output remains constant, after what time will all the hydrogen be converted into helium, making the Sun die? The actual projected lifetime of the Sun is about 10 billion years, because only the hydrogen in a relatively small core is available as a fuel. Only in the core are temperatures and densities high enough for the fusion reaction to be self-sustaining.

64. (a) One method of producing neutrons for experimental use is bombardment of light nuclei with alpha particles. In one particular arrangement, alpha particles emitted by polonium are incident on beryllium nuclei:

$${}^4_2\text{He} + {}^9_4\text{Be} \longrightarrow {}^{12}_6\text{C} + {}^1_0\text{n}$$

What is the Q value? (b) Neutrons are also often produced by small-particle accelerators. In one design, deuterons accelerated in a Van de Graaff generator bombard other deuterium nuclei:

$${}^2_1\text{H} + {}^2_1\text{H} \longrightarrow {}^3_2\text{He} + {}^1_0\text{n}$$

Is this reaction exothermic or endothermic? Calculate its Q value.

65. Review Problem. Consider the Bohr model of the hydrogen atom, with the electron in the ground state. The magnetic field at the nucleus produced by the orbiting electron has a value of 12.5 T (see Chapter 30, Problem 1). The proton can have its magnetic moment aligned in either of two directions perpendicular to the plane of the electron's orbit. Because of the interaction of the proton's magnetic moment with the electron's magnetic field, there will be a difference in energy between the states with the two different orientations of the proton's magnetic moment. Find that energy difference in eV.

66. Many radioisotopes have important industrial, medical, and research applications. One of these is ^{60}Co, which has a half-life of 5.27 years and decays by the emission of a beta particle (energy 0.31 MeV) and two gamma photons (energies 1.17 MeV and 1.33 MeV). A scientist wishes to prepare a ^{60}Co sealed source that will have an activity of 10.0 Ci after 30.0 months of use. (a) What is the initial mass of ^{60}Co required? (b) At what rate will the source emit energy after 30.0 months?

67. Review Problem. Consider a model of the nucleus in which the positive charge (Ze) is uniformly distributed throughout a sphere of radius R. By integrating the energy density, $\frac{1}{2}\epsilon_0 E^2$, over all space, show that the electrostatic energy may be written

$$U = \frac{3Z^2e^2}{20\pi\epsilon_0 R}$$

68. The ground state of $^{93}_{43}$Tc (molar mass, 92.910 2 g/mol) decays by electron capture and e^+ emission to energy levels of the daughter (molar mass in ground state, 92.906 8 g/mol) at 2.44 MeV, 2.03 MeV, 1.48 MeV, and 1.35 MeV. (a) For which of these levels are electron capture and e^+ decay allowed? (b) Identify the daughter and sketch the decay scheme, assuming all excited states de-excite by direct γ decay to the ground state.

69. Free neutrons have a characteristic half-life of 10.4 min. What fraction of a group of free neutrons with kinetic energy 0.040 0 eV will decay before traveling a distance of 10.0 km?

70. When the nuclear reaction represented by Equation 44.26 is endothermic, the disintegration energy Q is negative. For the reaction to proceed, the incoming particle must have a minimum energy called the thresh-

old energy, E_{th}. Some fraction of the energy of the incident particle is transferred to the compound nucleus to conserve momentum. Therefore, E_{th} must be greater than Q. (a) Show that

$$E_{\text{th}} = -Q\left(1 + \frac{M_a}{M_X}\right)$$

(b) Calculate the threshold energy of the incident alpha particle in the reaction

$$^4_2\text{He} + {}^{14}_7\text{N} \longrightarrow {}^{17}_8\text{O} + {}^1_1\text{H}$$

71. *Student determination of the half-life of ^{137}Ba.* The radioactive barium isotope ^{137}Ba has a relatively short half-life and can be easily extracted from a solution containing radioactive cesium (^{137}Cs). This barium isotope is commonly used in an undergraduate laboratory exercise for demonstrating the radioactive decay law. The data presented in Figure P45.71 were taken by undergraduate students using modest experimental equipment. Determine the half-life for the decay of ^{137}Ba using their data.

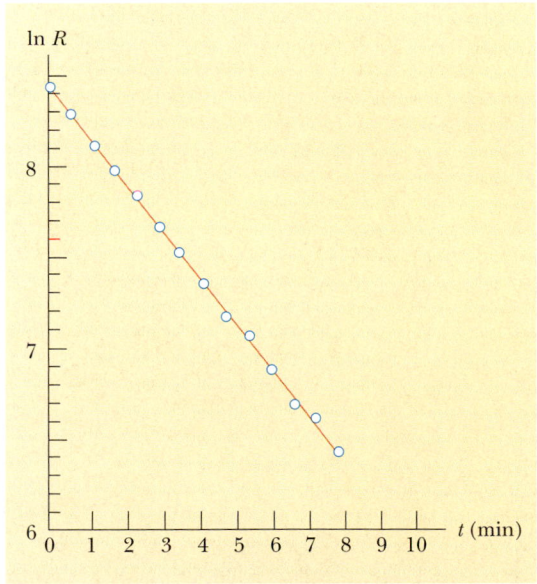

Figure P44.71

ANSWERS TO QUICK QUIZZES

44.1 Within a vertical grouping, the dots represent nuclei all having the same atomic number, so they must all represent the same element. Because the number of neutrons varies within a vertical grouping, each dot must signify an isotope of the element.

44.2 (c). Isotopes are nuclei containing different numbers of neutrons, so the atomic masses are not equal. Spin mag-

netic moments are also different in a family of isotopes because the neutrons have a magnetic moment. Chemical behavior is governed by the electrons. Isotopes of a given element have the same chemical properties because they all contain similar number of electrons.

44.3 Nuclear binding energy per nucleon is approximately constant because the nuclear force is a short-range force. Therefore, a given nucleon interacts only with its nearest neighbors rather than with all other nucleons in the nucleus. No matter how many nucleons are present, pulling one out involves separating it only from its nearest neighbors. The energy to do this is therefore approximately independent of how many nucleons are present.

The electric force binding the electrons in an atom to the nucleus is a long-range force, and so each electron interacts with all the protons in the nucleus. More protons in the nucleus thus means a stronger electron-nucleus attraction. As a result, the energy needed to remove an electron from the atom varies with atomic number.

44.4 No. The negative of the slope of a potential energy–versus–position graph is force. For nucleon separation distances of less than 1 fm, both slopes in Figure 44.9 are negative and so the force is positive, meaning the particles repel each other. From 1 fm to 3 fm the slope is positive, indicating an attractive force (of very short range). If a tangent to the graph were vertical at some point, the force would have to be infinite—a physical impossibility.

44.5 Figure 44.15 shows that the higher the energy of the alpha particle, the thinner the potential barrier. The thinner barrier translates to a higher probability of escape. The higher probability of escape translates to a faster rate of decay, which appears as a shorter half-life.

44.6 In alpha decay, there are only two products—the alpha particle and the daughter nucleus. There are also two conservation principles involved—energy and linear momentum. As a result, the alpha particle must be ejected with a discrete energy to satisfy both conservation principles. There are a small number of discrete energies of the alpha particle, as the daughter nucleus can be left in various excited states, but the allowed energies of the alpha particle are not continuous.

In beta decay, we have the same two conservation principles but three products—the beta particle, the daughter nucleus, and the neutrino. There are many ways that the energy can be divided among the three particles to satisfy the two conservation principles, and as a result the beta particle is emitted over a continuous range of energies.

Nearly everything on the Earth gets its energy from the Sun. Plants use sunlight to make carbohydrates, which are a source of energy for animals. The wind blows because of solar heating of the atmosphere. Automobiles are powered by fossil fuels, which are essentially stored solar energy. All these things get their energy from the Sun, but what is the source of the Sun's energy? *(European Space Agency/Science Photo Library/Photo Researchers, Inc.)*

c h a p t e r

45

Nuclear Fission and Fusion

Chapter Outline

I n this chapter, we are concerned primarily with the two means by which energy can be derived from nuclear reactions: fission, in which a large nucleus splits (or fissions) into two smaller nuclei, and fusion, in which two small nuclei fuse to form a larger one. In either case, there is a release of energy that can be used either destructively (as in bombs) or constructively (as in the production of electric power). We also examine the ways in which radiation interacts with matter and look at several devices used to detect radiation. The chapter concludes with a discussion of some industrial and biological applications of radiation.

45.1 ▶ INTERACTIONS INVOLVING NEUTRONS

To understand nuclear fission and the physics of nuclear reactors, we must first understand how neutrons interact with nuclei. Because of their charge neutrality, neutrons are not subject to Coulomb forces and as a result do not interact electrically with electrons. Therefore, because any piece of matter consists of electrons orbiting tiny atomic nuclei, matter appears quite "open" to free neutrons.

In general, the rate of neutron-induced reactions increases as the neutron kinetic energy decreases. Free neutrons undergo beta decay with a mean lifetime of about 10 min. Once free neutrons enter matter, however, many of them are absorbed by atomic nuclei and are stabilized from decay by the nuclear force of other nucleons.

A **fast neutron** (energy greater than about 1 MeV) traveling through matter undergoes many scattering events with the nuclei. In each event, the neutron gives up some of its kinetic energy to a nucleus. Once the neutron energy is sufficiently low, there is a high probability that the neutron will be captured by a nucleus, an event that is accompanied by the emission of a gamma ray. This **neutron capture** can be written

Neutron capture

$$ {}_0^1 n + {}_Z^A X \longrightarrow {}_Z^{A+1} X^* \longrightarrow {}_Z^{A+1} X + \gamma \qquad \textbf{(45.1)} $$

Once the neutron is captured, the nucleus ${}_Z^{A+1}X^*$ is in an excited state for a very short time before it undergoes gamma decay. Also, the product nucleus ${}_Z^{A+1}X$ is usually radioactive and decays by beta emission.

The neutron-capture rate as neutrons pass through any sample depends on which atoms are contained in the sample and on the energy of the incident neutrons. In addition, the capture rate also depends on the type of atoms in the sample. For some materials and for fast neutrons, elastic collisions dominate. Materials for which this occurs are called **moderators** because they slow down (or moderate) the originally energetic neutrons very effectively. The interaction of neutrons with matter increases with decreasing neutron energy because a slow neutron spends more time in the vicinity of target nuclei. A good moderator should be composed of nuclei that have a low tendency to capture fast neutrons. Moderator nuclei should be of low mass so that more kinetic energy is transferred to them in elastic collisions. For this reason, materials that are abundant in hydrogen, such as paraffin and water, are good moderators for neutrons.

Moderator

Quick Quiz 45.1 ▶

What would be the ideal target particle in a neutron moderator if we were trying to stop the incoming neutrons completely?

Sooner or later, most neutrons bombarding a moderator become **thermal neutrons,** which means they are in thermal equilibrium with the moderator material. Their average kinetic energy at room temperature is, from Equation 21.4,

$$K_{av} = \tfrac{3}{2}k_{B}T \approx \tfrac{3}{2}k_{B}(300 \text{ K}) \approx 0.04 \text{ eV}$$

which corresponds to a neutron root-mean-square speed of about 2 800 m/s. Thermal neutrons have a distribution of speeds, just as the molecules in a container of gas do (see Chapter 21). A high-energy neutron, one whose energy is several MeV, *thermalizes* (that is, reaches K_{av}) in less than 1 ms when incident on a moderator. Thermal neutrons have a very high probability of being captured by the moderator nuclei.

45.2 ▸ NUCLEAR FISSION

As we saw in Section 44.3, nuclear **fission** occurs when a heavy nucleus, such as ^{235}U, splits into two smaller nuclei. In such a reaction, **the combined mass of the daughter nuclei is less than the mass of the parent nucleus,** and the difference in mass is called the **mass defect.** Fission is initiated when a heavy nucleus captures a thermal neutron. Multiplying the mass defect by c^2 gives the numerical value of the released energy. Energy is released because the binding energy per nucleon of the daughter nuclei is about 1 MeV greater than that of the parent nucleus (see Fig. 44.8).

Nuclear fission was first observed in 1938 by Otto Hahn (1879–1968) and Fritz Strassman (b. 1902) following some basic studies by Fermi. After bombarding uranium ($Z = 92$) with neutrons, Hahn and Strassman discovered among the reaction products two medium-mass elements, barium and lanthanum. Shortly thereafter, Lise Meitner (1878–1968) and her nephew Otto Frisch (1904–1979) explained what had happened. The uranium nucleus had split into two nearly equal fragments after absorbing a neutron. Such an occurrence was of considerable interest to physicists attempting to understand the nucleus, but it was to have even more far-reaching consequences. Measurements showed that about 200 MeV of energy was released in each fission event, and this fact was to affect the course of history.

The fission of ^{235}U by thermal neutrons can be represented by the equation

$$^{1}_{0}\text{n} + \,^{235}_{92}\text{U} \longrightarrow \,^{236}_{92}\text{U}^* \longrightarrow \text{X} + \text{Y} + \text{neutrons} \qquad \textbf{(45.2)}$$

where ^{236}U* is an intermediate excited state that lasts only for about 10^{-12} s before splitting into nuclei X and Y, which are called **fission fragments.** In any fission equation, there are many combinations of X and Y that satisfy the requirements of conservation of energy and charge. With uranium, for example, there are about 90 daughter nuclei that can be formed.

Fission also results in the production of several neutrons, typically two or three. On the average, about 2.5 neutrons are released per event. A typical fission reaction for uranium is

$$^{1}_{0}\text{n} + \,^{235}_{92}\text{U} \longrightarrow \,^{141}_{56}\text{Ba} + \,^{92}_{36}\text{Kr} + 3(^{1}_{0}\text{n}) \qquad \textbf{(45.3)}$$

The breakup of the uranium nucleus can be compared to what happens to a drop of water when excess energy is added to it. (Recall the liquid-drop model of the nucleus described in Section 44.4.) Initially, all the atoms in the drop have some energy, but this is not enough to break up the drop. However, if enough energy is added to set the drop into vibration, the drop elongates and compresses

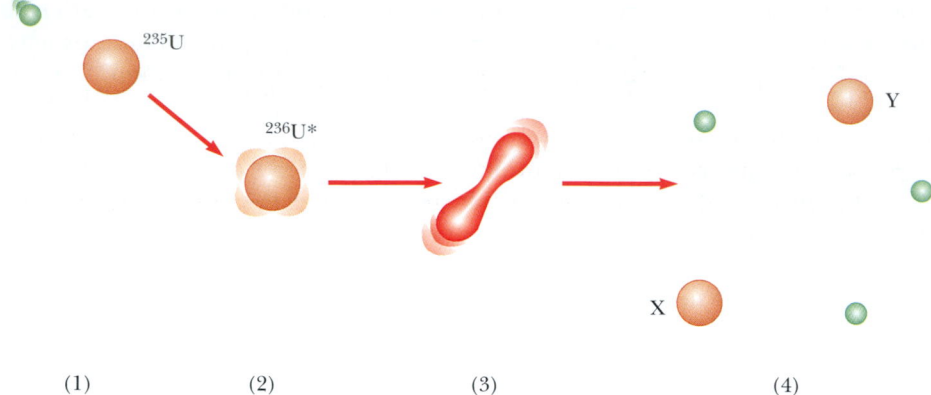

(1)   (2)   (3)   (4)

Figure 45.1   Nuclear fission event as described by the liquid-drop model of the nucleus.

until the amplitude of vibration becomes large enough to cause it to break. In the uranium nucleus, a similar process occurs (Fig. 45.1):

Stage 1.   The ^{235}U nucleus captures a thermal neutron.
Stage 2.   This capture results in the formation of ^{236}U*, and the excess energy of this nucleus causes it to oscillate violently.
Stage 3.   The ^{236}U* nucleus becomes highly distorted, and the force of repulsion between protons in the two halves of the dumbbell shape tends to increase the distortion.
Stage 4.   The nucleus splits into two fragments, emitting several neutrons in the process.

Quick Quiz 45.2

Which of the following is true for stage 3 of a ^{235}U fission event relative to stages 1 and 2? (Refer to Fig. 45.1.) (a) Both the nuclear force and the electrostatic force are smaller. (b) Both forces are greater. (c) The nuclear force is greater and the electrostatic force is smaller. (d) The nuclear force is smaller and the electrostatic force is greater.

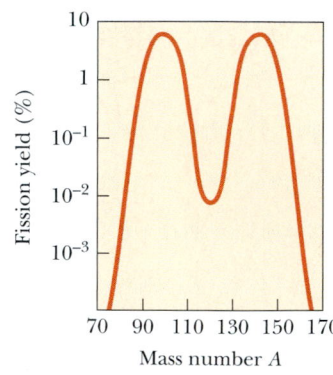

Figure 45.2   Distribution of fission products versus mass number for the fission of ^{235}U bombarded with thermal neutrons. Note that the ordinate scale is logarithmic.

Figure 45.2 is a graph of the distribution of fission products versus mass number A. The most probable products have mass numbers $A \approx 140$ and $A \approx 95$, both of which fall to the left of the stability line shown in Figure 44.3, meaning they contain more neutrons than protons. **These fragments, because they are unstable owing to their excess of neutrons, almost instantaneously release two or three neutrons.** Fragments having values of A other than 140 and 95, but still rich in neutrons, decay to more stable nuclei through a succession of e^- decays, emitting gamma rays in the process.

Let us estimate the disintegration energy Q released in a typical fission process. From Figure 44.8 we see that the binding energy per nucleon is about 7.2 MeV for heavy nuclei ($A \approx 240$) and about 8.2 MeV for nuclei of intermediate mass. This means that the nucleons in fission fragments are more tightly bound and therefore have less mass than the nucleons in a parent nucleus. This decrease in nucleon mass appears as released energy when fission occurs. The amount of energy released is $(8.2 - 7.2)$ MeV per nucleon. Assuming a total of 240 nucleons, we find that the energy released per fission event is

$$Q = (240 \text{ nucleons})\left(8.9\,\frac{\text{MeV}}{\text{nucleon}} - 7.2\,\frac{\text{MeV}}{\text{nucleon}}\right) = 240 \text{ MeV}$$

This is a very large amount of energy relative to the amount released in chemical processes. For example, the energy released in the combustion of one molecule of octane used in gasoline engines is about one-millionth the energy released in a single fission event!

Quick Quiz 45.3

If a heavy nucleus were to fission into just two daughter nuclei, they would be unstable. Why?

Quick Quiz 45.4

Which of the following are possible fission reactions?
(a) ${}^{1}_{0}\text{n} + {}^{235}_{92}\text{U} \rightarrow {}^{140}_{54}\text{Xe} + {}^{94}_{38}\text{Sr} + 2({}^{1}_{0}\text{n})$
(b) ${}^{1}_{0}\text{n} + {}^{235}_{92}\text{U} \rightarrow {}^{132}_{50}\text{Sn} + {}^{101}_{42}\text{Mo} + 3({}^{1}_{0}\text{n})$
(c) ${}^{1}_{0}\text{n} + {}^{239}_{94}\text{Pu} \rightarrow {}^{127}_{53}\text{I} + {}^{93}_{41}\text{Nb} + 3({}^{1}_{0}\text{n})$

EXAMPLE 45.1 ▸ The Energy Released in the Fission of ${}^{235}\text{U}$

Calculate the energy released when 1.00 kg of ${}^{235}\text{U}$ fissions, taking the disintegration energy per event to be $Q = 208$ MeV.

Solution We need to know the number of nuclei in 1.00 kg of uranium. Because $A = 235$, we know that the molar mass of this isotope is 235 g/mol. Therefore, the number of nuclei in our sample is

$$N = \left(\frac{6.02 \times 10^{23} \text{ nuclei/mol}}{235 \text{ g/mol}}\right)(1.00 \times 10^{3} \text{ g})$$

$$= 2.56 \times 10^{24} \text{ nuclei}$$

Hence, the total disintegration energy is

$$E = NQ = (2.56 \times 10^{24} \text{ nuclei})\left(208\,\frac{\text{MeV}}{\text{nucleus}}\right)$$

$$= \boxed{5.32 \times 10^{26} \text{ MeV}}$$

Because 1 MeV is equivalent to 4.45×10^{-20} kWh, we find that $E = 2.37 \times 10^{7}$ kWh. This is enough energy to keep a 100-W lightbulb burning for 30 000 years! If the energy in 1 kg of ${}^{235}\text{U}$ were suddenly released, it would be equivalent to detonating about 20 000 tons of TNT.

45.3 ▸ NUCLEAR REACTORS

In the preceding section, we learned that, when ${}^{235}\text{U}$ fissions, an average of 2.5 neutrons are emitted per event. These neutrons can in turn trigger other nuclei to fission, with the possibility of a chain reaction (Fig. 45.3). Calculations show that if the chain reaction is not controlled (that is, if it does not proceed slowly), it can result in a violent explosion, with the release of an enormous amount of energy. This is the principle behind the first type of nuclear bomb exploded in 1945, an uncontrolled fission reaction. When the reaction is controlled, however, the energy released can be put to less destructive use. In the United States, for example, nearly 20% of the electricity generated each year comes from nuclear power plants, and nuclear power is used extensively in many countries, including France, Japan, and Germany.

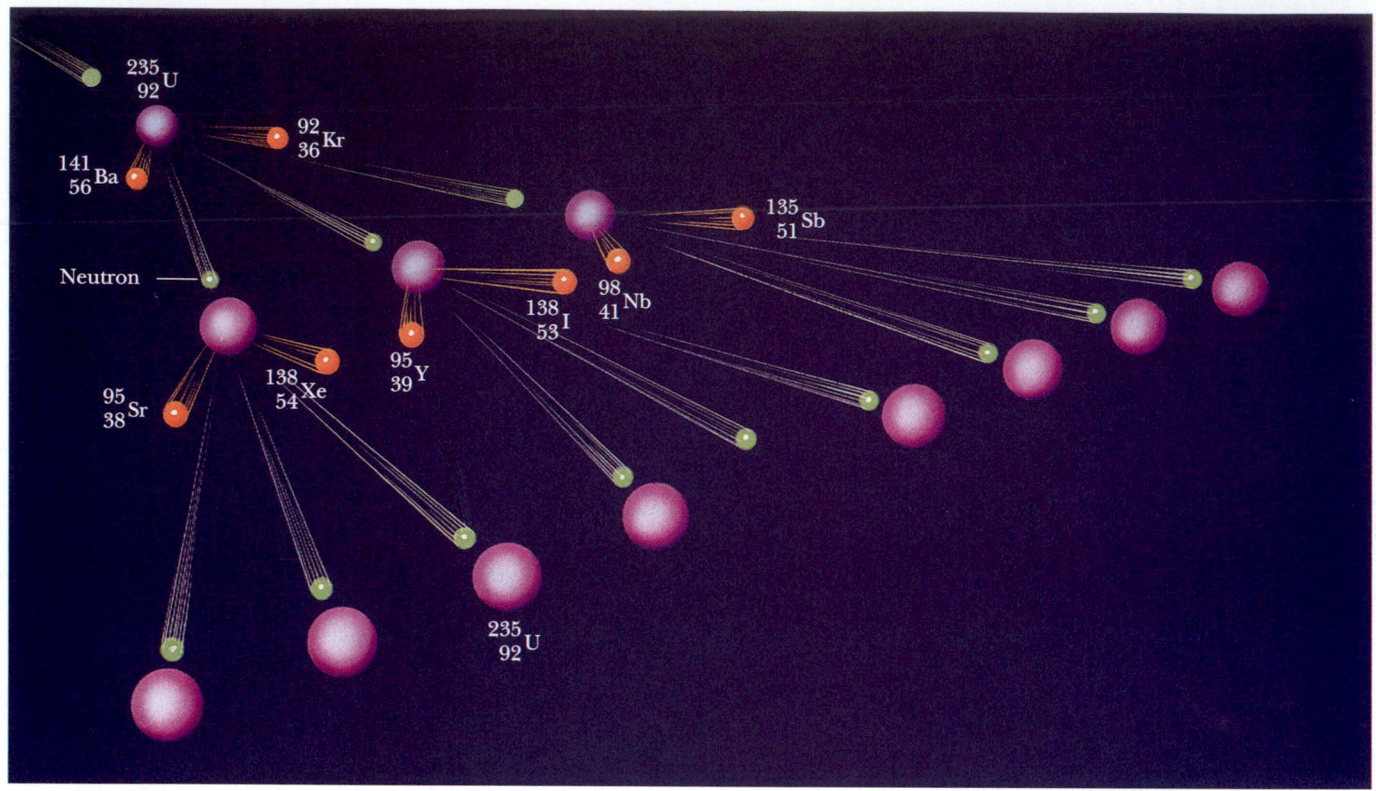

Figure 45.3   A nuclear chain reaction initiated by the capture of a neutron.

A nuclear reactor is a system designed to maintain what is called a **self-sustained chain reaction.** This important process was first achieved in 1942 by Enrico Fermi and his team at the University of Chicago, with naturally occurring uranium as the fuel.[1] Most reactors in operation today also use uranium as fuel. However, naturally occurring uranium contains only about 0.7% of the ^{235}U isotope, with the remaining 99.3% being ^{238}U. This fact is important to the operation of a reactor because ^{238}U almost never fissions. Instead, it tends to absorb neutrons, producing neptunium and plutonium. For this reason, reactor fuels must be artificially *enriched* to contain at least a few percent ^{235}U.

To achieve a self-sustained chain reaction, an average of one neutron emitted in each ^{235}U fission must be captured by another ^{235}U nucleus and cause that nucleus to undergo fission. A useful parameter for describing the level of reactor operation is the **reproduction constant** K, defined as **the average number of neutrons from each fission event that cause another fission event.** As we have seen, K has an average value of 2.5 in the fission of uranium. However, in practice K is less than this because of several factors discussed in the following paragraphs.

A self-sustained chain reaction is achieved when $K = 1$. Under this condition, the reactor is said to be **critical.** When $K < 1$, the reactor is subcritical and the re-

---

[1] Although Fermi's reactor was the first manufactured nuclear reactor, there is evidence that a natural fission reaction may have sustained itself for perhaps hundreds of thousands of years in a deposit of uranium in Gabon, West Africa. See G. Cowan, "A Natural Fission Reactor," *Sci. Am.* 235(5):36, 1976.

action dies out. When $K > 1$, the reactor is supercritical and a runaway reaction occurs. In a nuclear reactor used to furnish power to a utility company, it is necessary to maintain a value of K slightly greater than unity.

In any reactor, a fraction of the neutrons produced in fission leak out of the core before inducing other fission events. If the fraction leaking out is too large, the reactor will not operate. The percentage lost is large if the reactor is very small because leakage is a function of the ratio of surface area to volume. Therefore, a critical feature of the reactor design is an optimal surface area-to-volume ratio.

The neutrons released in fission events are very energetic, having kinetic energies of about 2 MeV. Because the probability of neutron capture increases with decreasing energy, it is necessary to slow these neutrons to thermal energies if they are to be captured and cause other ^{235}U nuclei to fission (see Example 9.8). The energetic neutrons are slowed down by a moderator substance surrounding the fuel.

In the first nuclear reactor ever constructed (Fig. 45.4), Fermi placed bricks of graphite (carbon) between the fuel elements. Carbon nuclei are about 12 times more massive than neutrons, but after several collisions with carbon nuclei, a neutron is slowed sufficiently to increase its likelihood of fission with ^{235}U. In this design, carbon is the moderator; most modern reactors use water as the moderator.

In the process of slowing down, neutrons may be captured by nuclei that do not fission. The most common event of this type is neutron capture by ^{238}U, which constitutes more than 90% of the uranium in the fuel elements. The probability of neutron capture by ^{238}U is very high when the neutrons have high kinetic energies and very low when they have low kinetic energies. Thus, the slowing down of the neutrons by the moderator serves the secondary purpose of making them available for reaction with ^{235}U and decreasing their chances of being captured by ^{238}U.

Enrico Fermi (1901–1954)
Fermi, an Italian physicist, was awarded the Nobel Prize in 1938 for producing transuranic elements by neutron irradiation and for his discovery of nuclear reactions brought about by thermal neutrons. He made many other outstanding contributions to physics, including his theory of beta decay, the free-electron theory of metals, and the development of the world's first fission reactor in 1942. Fermi was truly a gifted theoretical and experimental physicist. He was also well known for his ability to present physics in a clear and exciting manner. "Whatever Nature has in store for mankind, unpleasant as it may be, men must accept, for ignorance is never better than knowledge." *(National Accelerator Laboratory)*

Figure 45.4 Artist's rendition of the world's first nuclear reactor. Because of wartime secrecy, there are few photographs of the completed reactor, which was composed of layers of moderating graphite interspersed with uranium. A self-sustained chain reaction was first achieved on December 2, 1942. Word of the success was telephoned immediately to Washington with this message: "The Italian navigator has landed in the New World and found the natives very friendly." The historic event took place in an improvised laboratory in the racquet court under the stands of the University of Chicago's Stagg Field, and the Italian navigator was Enrico Fermi. *(Courtesy of Chicago Historical Society)*

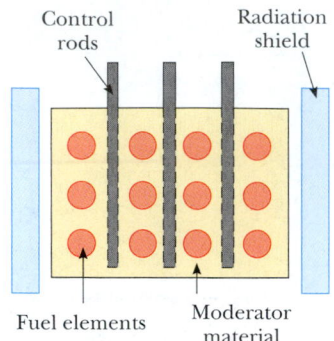

Figure 45.5 Cross-section of a reactor core showing the control rods, fuel elements containing enriched fuel, and moderating material, all surrounded by a radiation shield.

Control of Power Level

It is possible for a reactor to reach the critical stage ($K = 1$) after all the neutron losses described previously are minimized. However, a method of control is needed to maintain a K value near unity. If K rises above this value, the internal energy produced in the reaction could melt the reactor.

The basic design of a nuclear reactor core is shown in Figure 45.5. The fuel elements consist of uranium that has been enriched in the ^{235}U isotope. To control the power level, control rods are inserted into the reactor core. These rods are made of materials, such as cadmium, that are very efficient in absorbing neutrons. By adjusting the number and position of the control rods in the reactor core, the K value can be varied and any power level within the design range of the reactor can be achieved.

Several types of reactor systems convert the kinetic energy of fission fragments to electrical energy. The most common reactor in use in the United States is the pressurized-water reactor (Fig. 45.6), and we shall examine this type because its main parts are common to all reactor designs. Fission events in the reactor core raise the temperature of the water contained in the primary (closed) loop, which is maintained at high pressure to keep the water from boiling. (This water also serves as the moderator.) The hot water is pumped through a heat exchanger, where the internal energy of the water is transferred to the water contained in the secondary loop. The hot water in the secondary loop is converted to steam, which drives a turbine–generator system to create electric power. The water in the secondary loop is isolated from the water in the primary loop to avoid contamination of the secondary water and the steam by radioactive nuclei from the reactor core.

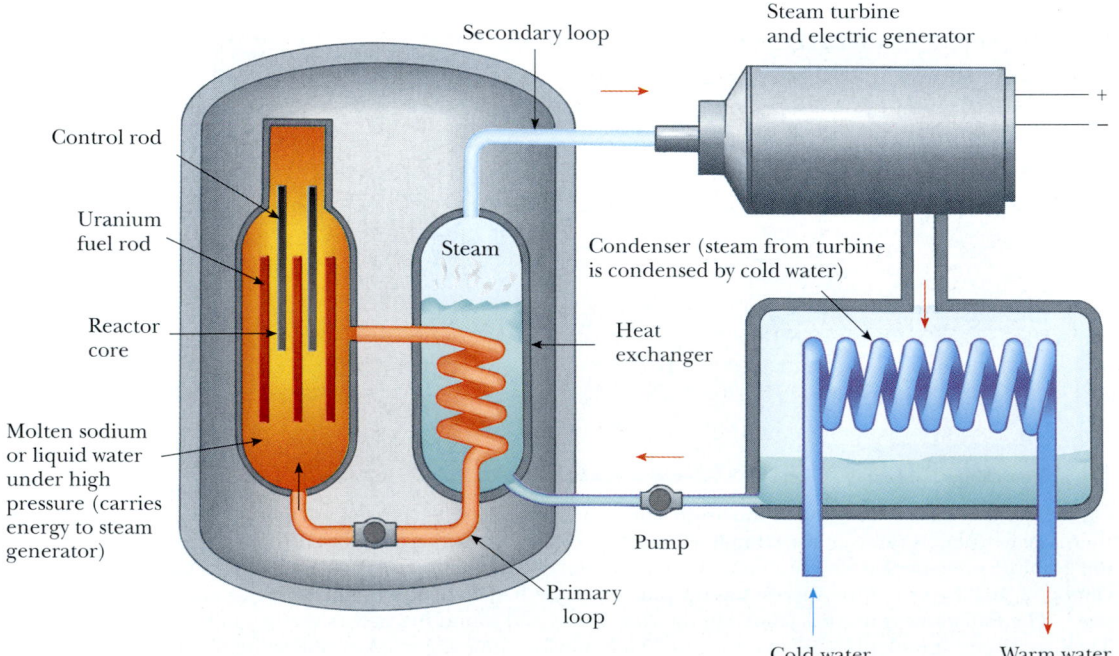

Figure 45.6 Main components of a pressurized-water reactor.

Safety and Waste Disposal

The 1979 near-disaster at a nuclear plant at Three Mile Island in Pennsylvania and the 1986 accident at the Chernobyl reactor in Ukraine rightfully focused attention on reactor safety. The Three Mile Island accident was the result of inadequate control-room instrumentation and poor emergency-response training. There were no injuries or detectable health impacts from the event, even though more than one third of the fuel melted.

This unfortunately was not the case at Chernobyl, where the activity of the materials released immediately after the accident totaled approximately 12×10^{18} Bq and resulted in the evacuation of 116 000 people. At least 237 people suffered from acute radiation sickness and about 800 children later contracted thyroid cancer from the ingestion of radioactive iodine in milk from cows that ate contaminated grass. One conclusion of an international conference studying the Ukraine accident was that "the main causes of the Chernobyl accident were the coincidence of severe deficiencies in the reactor physical design and in the design of the shut-down system and a violation of procedures." Most of these deficiencies have been addressed at plants of similar design in Russia and neighboring countries of the former Soviet Union.

There are no plants of the Chernobyl type in the United States. Many U. S. plants are of the pressurized-water design, as noted earlier.

Commercial reactors achieve safety through careful design and rigid operating protocol, and it is only when these variables are compromised that reactors pose a danger. Radiation exposure and the potential health risks associated with such exposure are controlled by three layers of containment. The fuel and radioactive fission products are contained inside the reactor vessel. Should this vessel rupture, the reactor building acts as a second containment structure to prevent radioactive material from contaminating the environment. Finally, the reactor facilities must be in a remote location to protect the general public from exposure should radiation escape the reactor building.

A continuing concern about nuclear fission reactors is the safe disposal of radioactive material when the reactor core is replaced. This waste material contains long-lived, highly radioactive isotopes and must be stored over long periods of time in such a way that there is no chance of environmental contamination. At present, sealing radioactive wastes in waterproof containers and burying them in deep salt mines seems to be the most promising solution.

Transport of reactor fuel and reactor wastes poses additional safety risks. Accidents during transport of nuclear fuel could expose the public to harmful levels of radiation. The Department of Energy requires stringent crash tests of all containers used to transport nuclear materials. Container manufacturers must demonstrate that their containers will not rupture even in high-speed collisions.

The safety issues associated with nuclear power reactors are complex and often emotional. All sources of energy have associated risks. In each case, the risks must be weighed against the benefits and the availability of the energy source.

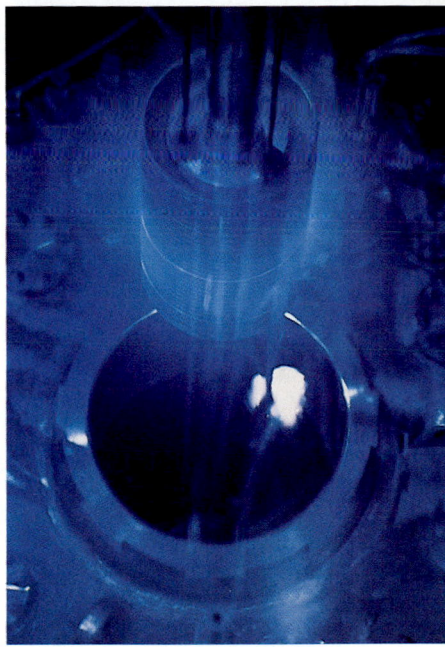

A blue glow from radiation emitted by a fuel element of a reactor at Oak Ridge National Laboratory. The glow results when energetic particles move through the water faster than the speed of light in that medium. *(U.S. Department of Energy/Photo Researchers, Inc.)*

45.4 NUCLEAR FUSION

In Chapter 44 we found that the binding energy for light nuclei ($A < 20$) is much smaller than the binding energy for heavier nuclei. This suggests a process that is the reverse of fission. As we saw in Section 39.8, when two light nuclei combine to form a heavier nucleus, the process is called nuclear **fusion.** Because the mass of the final nucleus is less than the combined masses of the original nuclei, there is a loss of mass accompanied by a release of energy.

Two examples of such energy-liberating fusion reactions are

$$\ce{^1_1H + ^1_1H -> ^2_1H + e^+ + \nu}$$

and

$$\ce{^1_1H + ^2_1H -> ^3_2He + \gamma}$$

This second reaction is followed by either hydrogen–helium fusion or helium–helium fusion:

$$\ce{^1_1H + ^3_2He -> ^4_2He + e^+ + \nu}$$

$$\ce{^3_2He + ^3_2He -> ^4_2He + ^1_1H + ^1_1H}$$

These are the basic reactions in the **proton–proton cycle,** believed to be one of the basic cycles by which energy is generated in the Sun and other stars that contain an abundance of hydrogen. Most of the energy production takes place in the Sun's interior, where the temperature is approximately 1.5×10^7 K. As we shall see later, such high temperatures are required to drive these reactions, and they are therefore called **thermonuclear fusion reactions.** The hydrogen (fusion) bomb, first exploded in 1952, is an example of an uncontrolled thermonuclear fusion reaction. It uses a fission bomb as the "trigger" to create the high temperatures needed for fusion.

Thermonuclear reaction

All of the reactions in the proton–proton cycle are exothermic. An overview of the cycle is that four protons combine to form an alpha particle and two positrons, with the release of 25 MeV of energy.

Fusion Reactions

The enormous amount of energy released in fusion reactions suggests the possibility of harnessing this energy for useful purposes. A great deal of effort is currently under way to develop a sustained and controllable thermonuclear reactor—a fusion power reactor. Controlled fusion is often called the ultimate energy source because of the availability of its fuel source: water. For example, if deuterium were used as the fuel, 0.12 g of it could be extracted from 1 gal of water at a cost of about four cents. Such rates would make the fuel costs of even an inefficient reactor almost insignificant. An additional advantage of fusion reactors is that comparatively few radioactive by-products are formed. For the proton-proton cycle, for instance, the end product is safe, nonradioactive helium. Unfortunately, a thermonuclear reactor that can deliver a net power output spread out over a reasonable time interval is not yet a reality, and many difficulties must be resolved before a successful device is constructed.

The Sun's energy is based, in part, upon a set of reactions in which hydrogen is converted to helium. Unfortunately, the proton–proton interaction is not suitable for use in a fusion reactor because the event requires very high pressures and densities. The process works in the Sun only because of the extremely high density of protons in the Sun's interior.

The reactions that appear most promising for a fusion power reactor involve deuterium ($\ce{^2_1H}$) and tritium ($\ce{^3_1H}$):

$$\ce{^2_1H + ^2_1H -> ^3_2He + ^1_0n} \qquad Q = 3.27 \text{ MeV}$$

$$\ce{^2_1H + ^2_1H -> ^3_1H + ^1_1H} \qquad Q = 4.03 \text{ MeV} \tag{45.4}$$

$$\ce{^2_1H + ^3_1H -> ^4_2He + ^1_0n} \qquad Q = 17.59 \text{ MeV}$$

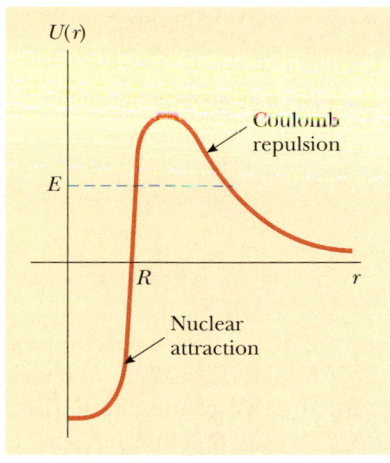

Figure 45.7 Potential energy as a function of separation distance between two deuterons. The Coulomb repulsive force is dominant at long range, and the nuclear force is dominant at short range, where R is of the order of 1 fm. If we neglect tunneling, to undergo fusion the two deuterons require an energy E greater than the height of the barrier.

As noted earlier, deuterium is available in almost unlimited quantities from our lakes and oceans and is very inexpensive to extract. Tritium, however, is radioactive ($T_{1/2} = 12.3$ years) and undergoes beta decay to ^{3}He. For this reason, tritium does not occur naturally to any great extent and must be artificially produced.

One of the major problems in obtaining energy from nuclear fusion is that the Coulomb repulsive force between two charged nuclei must be overcome before they can fuse. Potential energy as a function of the separation distance between two deuterons (deuterium nuclei, each having charge $+ e$) is shown in Figure 45.7. The potential energy is positive in the region $r > R$, where the Coulomb repulsive force dominates ($R \approx 1$ fm), and negative in the region $r < R$, where the nuclear force dominates. The fundamental problem then is to give the two nuclei enough kinetic energy to overcome this repulsive force. This can be accomplished by raising the fuel to extremely high temperatures (to about 10^8 K, far greater than the interior temperature of the Sun). At these high temperatures, the atoms are ionized and the system consists of a collection of electrons and nuclei, commonly referred to as a *plasma*.

> High temperatures are required to overcome the large Coulomb barrier

EXAMPLE 45.2 ▶ The Fusion of Two Deuterons

The separation distance between two deuterons must be about 1.0×10^{-14} m in order for the nuclear force to overcome the repulsive Coulomb force. (a) Calculate the height of the potential barrier due to the repulsive force.

Solution The potential energy associated with two charges separated by a distance r is, from Equation 25.13,

$$U = k_e \frac{q_1 q_2}{r}$$

where k_e is the Coulomb constant. For the case of two deuterons, $q_1 = q_2 = + e$, so that

$$U = k_e \frac{e^2}{r} = \left(8.99 \times 10^9 \ \frac{\text{N} \cdot \text{m}^2}{\text{C}^2}\right) \frac{(1.60 \times 10^{-19} \ \text{C})^2}{1.0 \times 10^{-14} \ \text{m}}$$

$$= 2.3 \times 10^{-14} \ \text{J} = \boxed{0.14 \ \text{MeV}}$$

(b) Estimate the temperature required for a deuteron to overcome the potential barrier, assuming an energy of $\frac{3}{2}k_B T$ per deuteron (where k_B is Boltzmann's constant).

Solution Because the total Coulomb energy of the pair is 0.14 MeV, the Coulomb energy per deuteron is 0.07 MeV = 1.1×10^{-14} J. Setting this energy equal to the average energy per deuteron gives

$$\tfrac{3}{2}k_B T = 1.1 \times 10^{-14} \ \text{J}$$

Solving for T gives

$$T = \frac{2(1.1 \times 10^{-14} \ \text{J})}{3(1.38 \times 10^{-23} \ \text{J/K})} = \boxed{5.3 \times 10^8 \ \text{K}}$$

This calculated temperature is too high because the particles in the plasma have a Maxwellian speed distribution, and therefore some fusion reactions are caused by particles in the high-energy tail of this distribution. Furthermore, even those particles that do not have enough energy to overcome the barrier have some probability of tunneling through. When these effects are taken into account, a temperature of "only" 4×10^8 K appears adequate to fuse the two deuterons.

(c) Find the energy released in the deuterium–deuterium reaction

$$^2_1\text{H} + ^2_1\text{H} \longrightarrow ^3_1\text{H} + ^1_1\text{H}$$

Solution The mass of a single deuterium atom is equal to 2.014 102 u. Thus, the total mass before the reaction is 4.028 204 u. After the reaction, the sum of the masses is 3.016 049 u + 1.007 825 u = 4.023 874 u. The excess mass is 0.004 33 u, equivalent to 4.03 MeV, as was noted in Equation 45.4.

Critical ignition temperature

The temperature at which the power generation rate in any fusion reaction exceeds the loss rate (due to mechanisms such as radiation losses) is called the **critical ignition temperature** T_{ignit}. This temperature for the deuterium–deuterium (D–D) reaction is 4×10^8 K. From the relationship $E \approx k_B T$, this temperature is equivalent to approximately 35 keV. It turns out that the critical ignition temperature for the deuterium–tritium (D–T) reaction is about 4.5×10^7 K, or only 4 keV. A plot of the power $\mathscr{P}_{\text{gen}}$ generated by fusion versus temperature for the two reactions is shown in Figure 45.8. The green straight line represents the power $\mathscr{P}_{\text{lost}}$ lost via the radiation mechanism known as **bremsstrahlung.** In this principal mechanism of energy loss, radiation (primarily x-rays) is emitted as the result of electron–ion collisions within the plasma. The intersections of the $\mathscr{P}_{\text{lost}}$ line with the $\mathscr{P}_{\text{gen}}$ curves give the critical ignition temperatures.

Confinement time

In addition to the high temperature requirements, there are two other critical parameters that determine whether or not a thermonuclear reactor is successful: the **ion density** n and **confinement time** τ, which is the length of time the ions are maintained at $T > T_{\text{ignit}}$. The British physicist J. D. Lawson has shown that the ion density and confinement time must both be large enough to ensure that more fusion energy is released than the amount required to raise the temperature of the plasma. In particular, **Lawson's criterion** states that a net energy output is possible under the following conditions:

Lawson's criterion

$$n\tau \geq 10^{14} \text{ s/cm}^3 \qquad (\text{D–T})$$

$$n\tau \geq 10^{16} \text{ s/cm}^3 \qquad (\text{D–D})$$

(45.5)

A graph of $n\tau$ versus temperature for the D–T and D–D reactions is given in Figure 45.9. The product $n\tau$ is referred to as the **Lawson number** of a reaction.

Lawson's criterion was arrived at by comparing the energy required to heat a given plasma with the energy generated by the fusion process.[2] The energy E_{in} re-

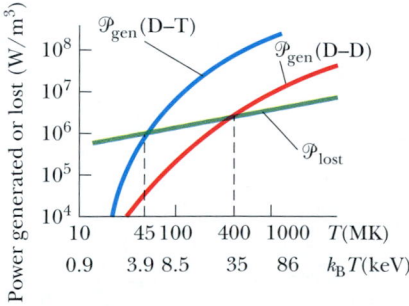

Figure 45.8 Power generated versus temperature for deuterium–deuterium (D–D) and deuterium–tritium (D–T) fusion. The green line represents power lost as a function of temperature. When the generation rate exceeds the loss rate, ignition takes place.

---

[2] Lawson's criterion neglects the energy needed to set up the strong magnetic field used to confine the hot plasma. This energy is expected to be about 20 times greater than the energy required to raise the temperature of the plasma. For this reason, it is necessary either to have a magnetic energy recovery system or to use superconducting magnets.

quired to raise the temperature of the plasma is proportional to the ion density n, which we can express as $E_{in} = C_1 n$, where C_1 is some constant. The energy generated by the fusion process is proportional to $n^2 \tau$, or $E_{gen} = C_2 n^2 \tau$. This may be understood by realizing that the fusion energy released is proportional to both the rate at which interacting ions collide ($\propto n^2$) and the confinement time τ. Net energy is produced when $E_{gen} > E_{in}$. When the constants C_1 and C_2 are calculated for different reactions, the condition that $E_{gen} \geq E_{in}$ leads to Lawson's criterion.

In summary, the three basic requirements of a successful thermonuclear power reactor are

- The plasma temperature must be very high—about 4.5×10^7 K for the D–T reaction and 4×10^8 K for the D–D reaction.
- The ion density must be high. It is necessary to have a high density of interacting nuclei to increase the collision rate between particles.
- The confinement time of the plasma must be long. To meet Lawson's criterion, the product $n\tau$ must be large. For a given value of n, the probability of fusion between two particles increases as τ increases.

Current efforts are aimed at meeting Lawson's criterion at temperatures exceeding T_{ignit}. Although the minimum required plasma densities have been achieved, the problem of confinement time is more difficult. How can a plasma be confined at 10^8 K for 1 s? The two basic techniques under investigation are magnetic confinement and inertial confinement.

Magnetic Confinement

Many fusion-related plasma experiments use **magnetic confinement** to contain the plasma. A toroidal device called a **tokamak,** first developed in Russia, is shown in Figure 45.10a. A combination of two magnetic fields is used to confine and stabilize the plasma: (1) a strong toroidal field produced by the current in the toroidal windings surrounding a donut-shaped vacuum chamber and (2) a weaker "poloidal" field produced by the toroidal current. In addition to confining the plasma, the toroidal current is used to raise its temperature. The resultant helical magnetic field lines spiral around the plasma and keep it from touching the walls of the vacuum chamber. (If the plasma touches the walls, its temperature is reduced and heavy impurities sputtered from the walls "poison" it and lead to large power losses.)

One of the major breakthroughs in magnetic confinement in the 1980s was in the area of auxiliary energy input to reach ignition temperatures. Experiments have shown that injecting a beam of energetic neutral particles into the plasma is a very efficient method of raising it to ignition temperatures (5 to 10 keV). Radio-frequency energy input will probably be needed for reactor-size plasmas.

When it was in operation, the Tokamak Fusion Test Reactor (TFTR, Fig. 45.10b) at Princeton reported central ion temperatures of 510 million degrees Celsius, more than 30 times hotter than the center of the Sun. The $n\tau$ values in the TFTR for the D–T reaction were well above 10^{13} s/cm^3 and close to the value required by Lawson's criterion. In 1991, reaction rates of 6×10^{17} D–T fusions per second were reached in the JET tokamak at Abington, England.

One of the new generations of fusion experiments is the National Spherical Torus Experiment (NSTX) shown in Figure 45.10c. Rather than the donut-shaped plasma of a tokamak, the NSTX produces a spherical plasma that has a hole through its center. The major advantage of the spherical configuration is its ability to confine the plasma at a higher pressure in a given magnetic field. This approach could lead to development of smaller, more economical fusion reactors.

Requirements for a fusion power reactor

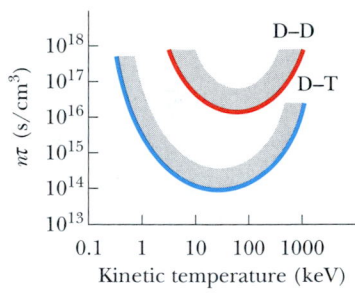

Figure 45.9 The Lawson number $n\tau$ versus temperature for the D–T and D–D fusion reactions. The regions above the colored curves represent favorable conditions for fusion.

web

For more information, visit the Princeton Plasma Physics Laboratory at **www.pppl.gov**

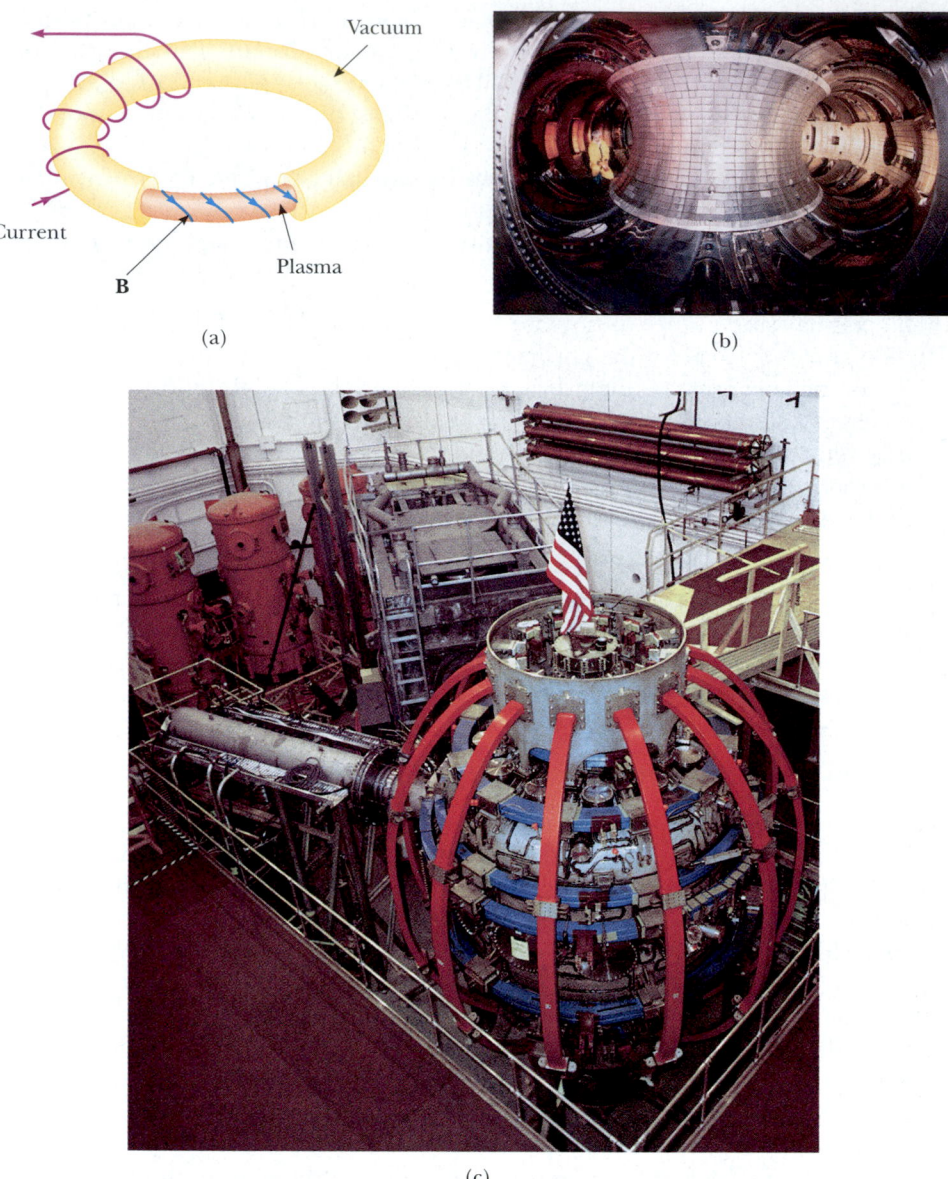

(a)

(b)

(c)

Figure 45.10   (a) Diagram of a tokamak used in the magnetic confinement scheme. (b) Interior view of the recently closed Tokamak Fusion Test Reactor (TFTR) vacuum vessel at the Princeton Plasma Physics Laboratory. *(Courtesy of Princeton Plasma Physics Laboratory)*   (c) The National Spherical Torus Experiment (NSTX) that began operation in March 1999. *(Courtesy of Princeton University)*

An international collaborative effort involving four major fusion programs is currently under way to build a fusion reactor called ITER (International Thermonuclear Experimental Reactor). This facility will address the remaining technological and scientific issues concerning the feasibility of fusion power. The design is completed (Fig. 45.11), and site and construction negotiations are under way. If the planned device works as expected, the Lawson number for ITER will be about six times greater than the current record holder, the JT-60U tokamak in Japan. ITER will produce 1.5 GW of power, and the energy content of the alpha particles

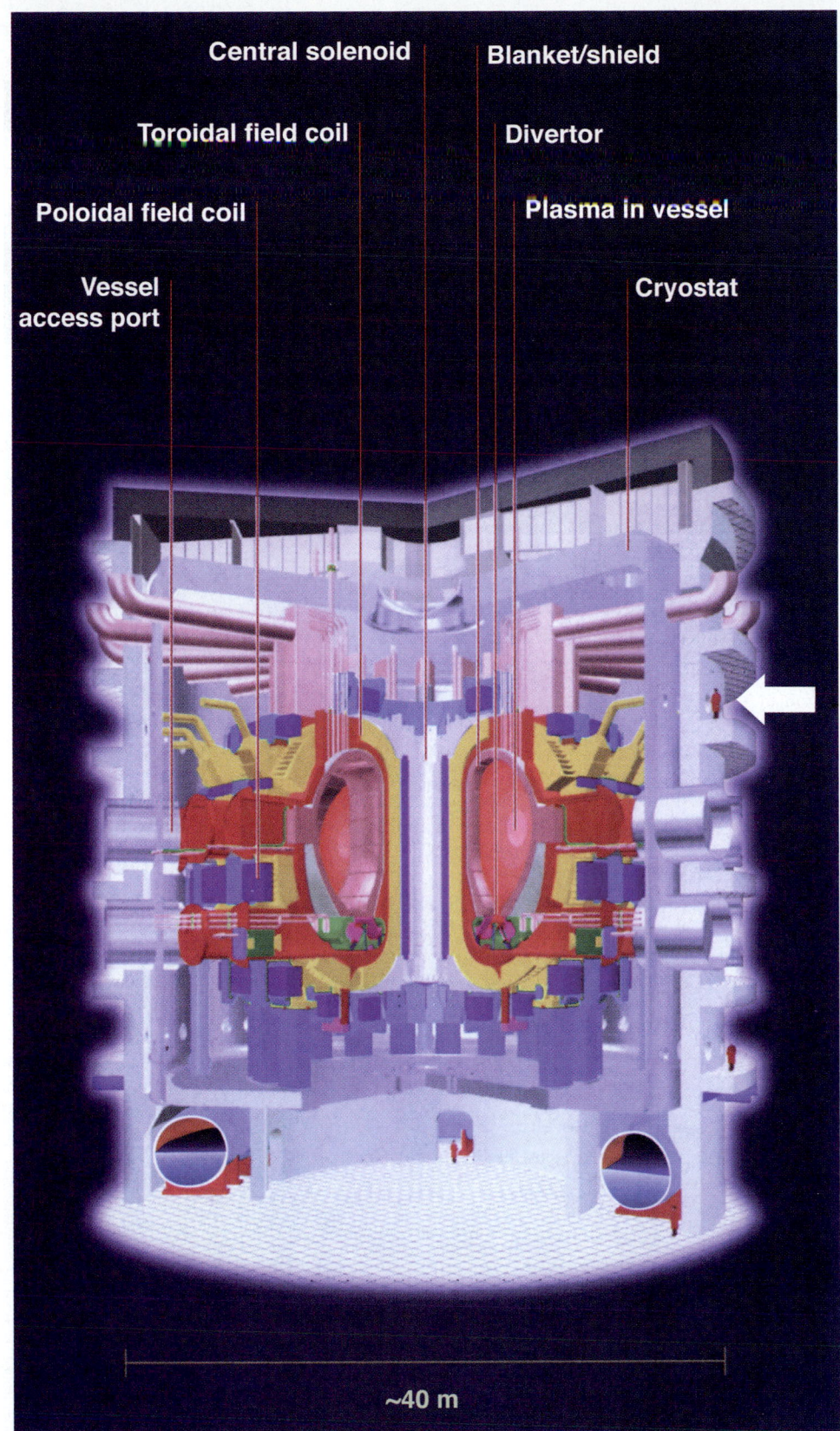

Central solenoid

Toroidal field coil

Poloidal field coil

Vessel access port

Blanket/shield

Divertor

Plasma in vessel

Cryostat

~40 m

Figure 45.11 Cutaway diagram of the ITER (International Thermonuclear Experimental Reactor). Note the size of the reactor relative to that of a person *(arrow)*. *(Courtesy of ITER)*

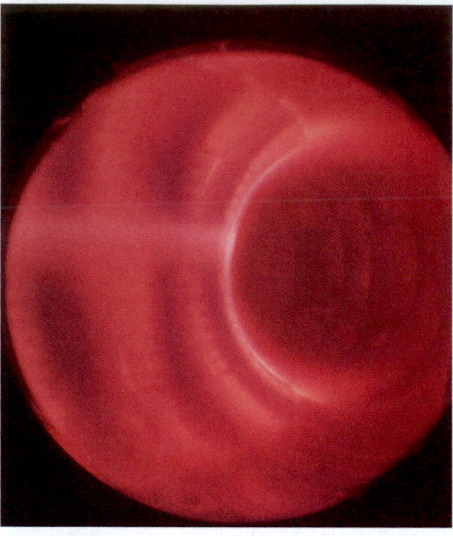

During the operation of the Tokamak Fusion Test Reactor, the plasma discharge was monitored using an optical system that showed the interior of the vacuum vessel. This view of a high-temperature deuterium plasma shows a bright radiation belt in the foreground. *(Courtesy of Princeton Plasma Physics Laboratory)*

inside the reactor will be so intense that they will sustain the fusion reaction, allowing the auxiliary energy sources to be turned off once the reaction is initiated. Such a state of sustained burn is referred to as *ignition*.

EXAMPLE 45.3   Inside a Fusion Reactor

In 1998 the JT-60U tokamak in Japan was operated with a D–T plasma density of 4.8×10^{13} cm^{-3} at a temperature (in energy units) of 16.1 keV. It was able to confine this plasma inside a magnetic field for 1.1 s. (a) Does this meet Lawson's criterion?

Solution   Equation 45.5 says that for a D–T plasma, the Lawson number $n\tau$ must be greater than 10^{14} s/cm^3. For the JT-60U,

$$n\tau = (4.8 \times 10^{13}\ \text{cm}^{-3})(1.1\ \text{s}) = 5.3 \times 10^{13}\ \text{s/cm}^3$$

which is close to meeting Lawson's criterion. In fact, scientists recorded a power gain of 1.25, indicating the reactor was operating slightly past the break-even point and was producing more energy than it required to maintain the plasma.

(b) How does the plasma density compare with the density of atoms in an ideal gas when the gas is at room temperature and pressure?

Solution   The density of atoms in a sample of ideal gas is given by N_A/V_m, where N_A is Avogadro's number and V_m is the molar volume of an ideal gas under standard conditions, 2.24×10^{-2} m^3/mol. Thus, the density of the gas is

$$\frac{N_A}{V_m} = \frac{6.02 \times 10^{23}\ \text{atoms/mol}}{2.24 \times 10^{-2}\ \text{m}^3/\text{mol}} = 2.7 \times 10^{25}\ \text{atoms/m}^3$$

This is more than 500 000 times greater than the plasma density in the reactor.

Inertial Confinement

The second technique for confining a plasma is called **inertial confinement** and makes use of a D–T target that has a very high particle density. In this scheme, the confinement time is very short (typically 10^{-11} to 10^{-9} s), so that, because of their own inertia, the particles do not have a chance to move appreciably from their initial positions. Thus, Lawson's criterion can be satisfied by combining a high particle density with a short confinement time.

Laser fusion is the most common form of inertial confinement. A small D–T pellet, about 1 mm in diameter, is struck simultaneously by several focused, high-

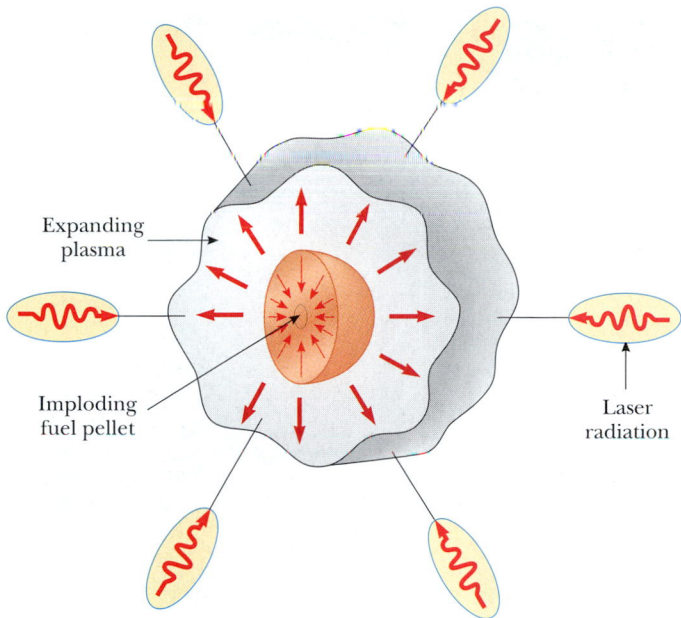

Figure 45.12   In inertial confinement, a D–T fuel pellet fuses when struck by several high-intensity laser beams simultaneously.

intensity laser beams, resulting in a large pulse of input energy that causes the surface of the fuel pellet to evaporate (Fig. 45.12). The escaping particles exert a third-law reaction force on the core of the pellet, resulting in a strong, inwardly moving compressive shock wave. This shock wave increases the pressure and density of the core and produces a corresponding increase in temperature. When the temperature of the core reaches ignition temperature, fusion reactions occur.

Two of the leading laser fusion laboratories in the United States are the Omega facility at the University of Rochester in New York and the Nova facility at Lawrence Livermore National Laboratory in California. The Omega facility focuses 24 laser beams on the target, and the Nova facility employs 10 beams. Figure 45.13a shows the target chamber at Nova, and Figure 45.13b shows the tiny, spherical D–T pellets used. Nova is capable of injecting a power of 2×10^{14} W into a 0.5-mm D–T pellet and has achieved values of $n\tau \approx 5 \times 10^{14}$ s/cm^3 and ion temperatures of 5.0 keV. These values are close to those required for D–T ignition. This steady progress has led the U. S. Department of Energy and other groups to plan a national facility that will involve a laser fusion device with an input energy in the 5–10 MJ range.

Fusion Reactor Design

In the D–T fusion reaction shown in Figure 45.14,

$$^{2}_{1}\text{H} + {}^{3}_{1}\text{H} \longrightarrow {}^{4}_{2}\text{He} + {}^{1}_{0}\text{n} \qquad Q = 17.59 \text{ MeV}$$

the alpha particle carries 20% of the energy and the neutron carries 80%, or about 14 MeV. The alpha particles, because they are charged, are primarily absorbed by the plasma; this causes the plasma's temperature to increase. In contrast, the 14-MeV neutrons, being electrically neutral, pass through the plasma and

(a)

(b)

Figure 45.13   (a) The target chamber of the Nova Laser Facility at Lawrence Livermore Laboratory.   *(Courtesy of University of California Lawrence Livermore National Laboratory and the U. S. Department of Energy)*   (b) Spherical plastic target shells used to contain the D–T fuel, shown clustered on a quarter. The shells have very smooth surfaces and are about 100 nm thick.   *(Courtesy of Los Alamos National Laboratory)*

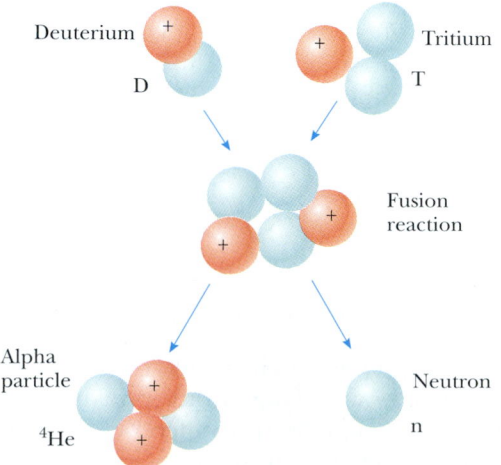

Figure 45.14   Deuterium–tritium fusion. Eighty percent of the energy released is in the 14-MeV neutron.

must be absorbed into a surrounding blanket material, where their large kinetic energy is extracted and used to generate electric power.

One scheme is to use molten lithium metal as the neutron-absorbing material and to circulate the lithium in a closed heat-exchange loop to produce steam and drive turbines as in a conventional power plant. Figure 45.15 shows a diagram of such a reactor. It is estimated that a blanket of lithium about 1 m thick will capture nearly 100% of the neutrons from the fusion of a small D–T pellet.

The capture of neutrons by lithium is described by the reaction

$$\,_{0}^{1}\text{n} + \,_{3}^{6}\text{Li} \longrightarrow \,_{1}^{3}\text{H} + \,_{2}^{4}\text{He}$$

where the kinetic energies of the charged tritium $_{1}^{3}\text{H}$ and alpha particle are converted to internal energy in the molten lithium. An extra advantage of using lithium as the energy-transfer medium is that the tritium produced can be separated from the lithium and returned as fuel to the reactor.

Advantages and Problems of Fusion

If fusion power can ever be harnessed, it will offer several advantages over fission-generated power: (1) low cost and abundance of fuel (deuterium), (2) impossibility of runaway accidents, and (3) lesser radiation hazard. Some of the anticipated problems and disadvantages include (1) scarcity of lithium, (2) limited supply of helium, which is needed for cooling the superconducting magnets used to produce strong confining fields, and (3) structural damage and induced radioactivity caused by neutron bombardment. If such problems and the engineering design factors can be resolved, nuclear fusion will become a feasible source of energy by the middle of the 21st century.

Advantages of fusion

Problem areas and disadvantages of fusion

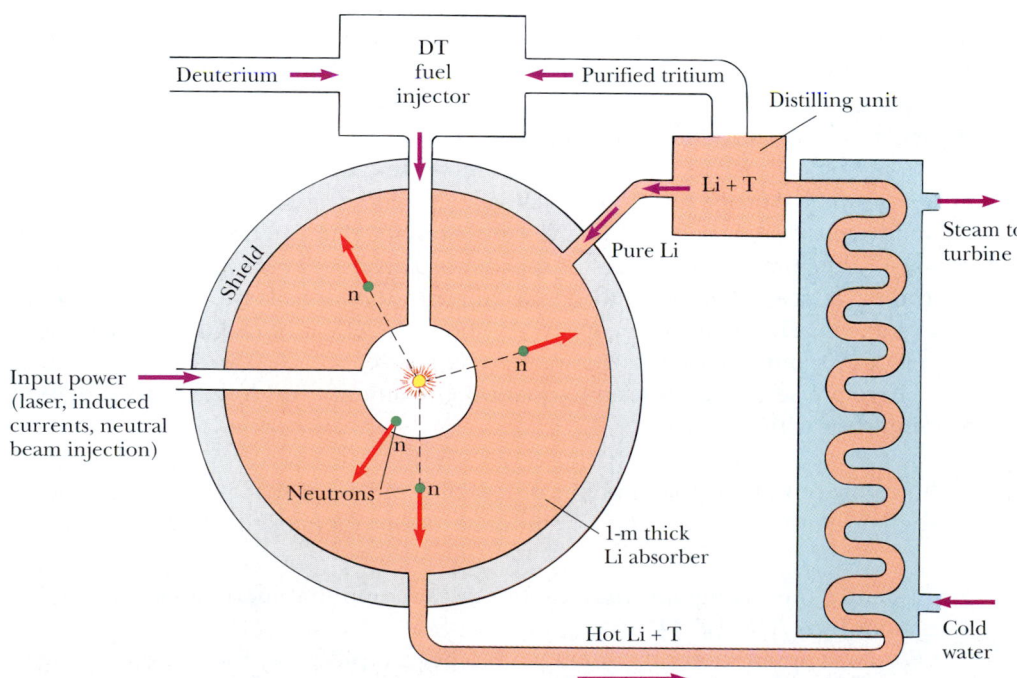

Figure 45.15   Diagram of a fusion reactor.

Optional Section

45.5 ▶ RADIATION DAMAGE IN MATTER

In Chapter 34 we learned that electromagnetic radiation is all around us in the form of radio waves, microwaves, light waves, and so on. In this section, we turn to forms of radiation that can cause severe damage as they pass through matter. These include radiation resulting from radioactive processes and radiation in the form of energetic particles such as neutrons and protons. It is these forms that we refer to here and in the following two sections when we use the word *radiation*.

The degree and type of damage depend upon several factors, including the type and energy of the radiation and the properties of the matter. The metals used in nuclear reactor structures can be severely weakened by high fluxes of energetic neutrons because these high fluxes often lead to metal fatigue. The damage in such situations is in the form of atomic displacements, often resulting in major alterations in the properties of the material. Materials can also be damaged by electromagnetic radiation, such as gamma rays or x-rays, that doesn't displace the atoms in a material but rather strips them of electrons so that they become ions. For example, defects called color centers can be produced in inorganic crystals by irradiating the crystals with x-rays. One extensively studied color center has been identified as an electron trapped in a Cl^- ion vacancy.

Radiation damage in organisms is primarily due to ionization effects in cells. The normal operation of a cell may be disrupted when highly reactive ions are formed as the result of ionizing radiation. For example, hydrogen and the hydroxyl radical OH^- produced from water molecules can induce chemical reactions that may break bonds in proteins and other vital molecules. Furthermore, the ionizing radiation may affect vital molecules directly by removing electrons from their structure. Large doses of radiation are especially dangerous because damage to a great number of molecules in a cell may cause the cell to die. Although the death of a single cell is usually not a problem, the death of many cells may result in irreversible damage to the organism. Cells that divide rapidly, such as those of the digestive tract, reproductive organs, and hair follicles, are especially susceptible. Also, cells that do survive the radiation may become defective. These defective cells can produce more defective cells and lead to cancer.

In biological systems, it is common to separate radiation damage into two categories: somatic damage and genetic damage. *Somatic damage* is that associated with any body cell except the reproductive cells. Somatic damage can lead to cancer or seriously alter the characteristics of specific organisms. *Genetic damage* affects only reproductive cells. Damage to the genes in reproductive cells can lead to defective offspring. Clearly, we must be concerned about the effect of diagnostic treatments such as x-rays and other forms of radiation exposure.

There are several units used to quantify the amount, or dose, of any radiation that interacts with a substance.

The roentgen

> The **roentgen** (R) is that amount of ionizing radiation that produces an electric charge of 3.33×10^{-10} C in 1 cm^3 of air under standard conditions.

Equivalently, the roentgen is that amount of radiation that increases the energy of 1 kg of air by 8.76×10^{-3} J.

For most applications, the roentgen has been replaced by the rad (an acronym for *radiation absorbed dose*):

One **rad** is that amount of radiation that increases the energy of 1 kg of absorbing material by 1×10^{-2} J.

Although the rad is a perfectly good physical unit, it is not the best unit for measuring the degree of biological damage produced by radiation because damage depends not only on the dose but also on the type of the radiation. For example, a given dose of alpha particles causes about ten times more biological damage than an equal dose of x-rays. The **RBE** (relative biological effectiveness) factor for a given type of radiation is **the number of rads of x-radiation or gamma radiation that produces the same biological damage as 1 rad of the radiation being used.** The RBE factors for different types of radiation are given in Table 45.1. The values are only approximate because they vary with particle energy and with the form of the damage. The RBE factor should be considered only a first-approximation guide to the actual effects of radiation.

Finally, the **rem** (radiation equivalent in man) is the product of the dose in rad and the RBE factor:

$$\text{Dose in rem} \equiv \text{dose in rad} \times \text{RBE} \qquad \textbf{(45.6)}$$

According to this definition, 1 rem of any two radiations produces the same amount of biological damage. From Table 45.1, we see that a dose of 1 rad of fast neutrons represents an effective dose of 10 rem, but 1 rad of gamma radiation is equivalent to a dose of only 1 rem.

Low-level radiation from natural sources, such as cosmic rays and radioactive rocks and soil, delivers to each of us a dose of about 0.13 rem/yr; this radiation is called *background radiation*. It is important to note that background radiation varies with geography, with the main factors being altitude (exposure to cosmic rays) and geology (radon gas released by some rock formations, deposits of naturally radioactive minerals).

The upper limit of radiation dose rate recommended by the U.S. government (apart from background radiation) is about 0.5 rem/yr. Many occupations involve much higher radiation exposures, and so an upper limit of 5 rem/yr has been set for combined whole-body exposure. Higher upper limits are permissible for certain parts of the body, such as the hands and the forearms. A dose of 400 to 500 rem results in a mortality rate of about 50% (which means that half the people exposed to this radiation level die). The most dangerous form of exposure is either

TABLE 45.1 **RBE[a] Factors for Several Types of Radiation**

Radiation	RBE Factor
X-rays and gamma rays	1.0
Beta particles	1.0–1.7
Alpha particles	10–20
Thermal neutrons	4–5
Fast neutrons and protons	10
Heavy ions	20

[a]RBE = relative biological effectiveness.

ingestion or inhalation of radioactive isotopes, especially isotopes of those elements the body retains and concentrates, such as ^{90}Sr. In some cases, a dose of 1 000 rem can result from ingesting 1 mCi of radioactive material.

Optional Section

45.6 RADIATION DETECTORS

Various devices have been developed for detecting radiation. These devices are used for a variety of purposes, including medical diagnoses, radioactive dating measurements, measuring background radiation, and measuring the mass, energy, and momentum of particles created in high-energy nuclear reactions.

Ion chamber

In an **ion chamber** (Fig. 45.16), electron–ion pairs are generated as radiation passes through a gas and produce an electrical signal. Two plates in the chamber are connected to a voltage supply and thereby maintained at different electric potentials. The positive plate attracts the electrons, and the negative plate attracts positive ions, causing a current pulse that is proportional to the number of electron–ion pairs produced when a radioactive particle enters the chamber. When an ion chamber is used both to detect the presence of a radioactive particle and to measure its energy, it is called a **proportional counter.**

Geiger counter

The **Geiger counter** (Fig. 45.17) is perhaps the most common form of ion chamber used to detect radioactivity. It can be considered the prototype of all counters that use the ionization of a medium as the basic detection process. It consists of a thin wire electrode aligned along the central axis of a cylindrical metallic tube filled with a gas at low pressure. The wire is maintained at a high positive electric potential (about 10^3 V) relative to the tube. When a high-energy particle resulting, for example, from a radioactive decay enters the tube through a thin

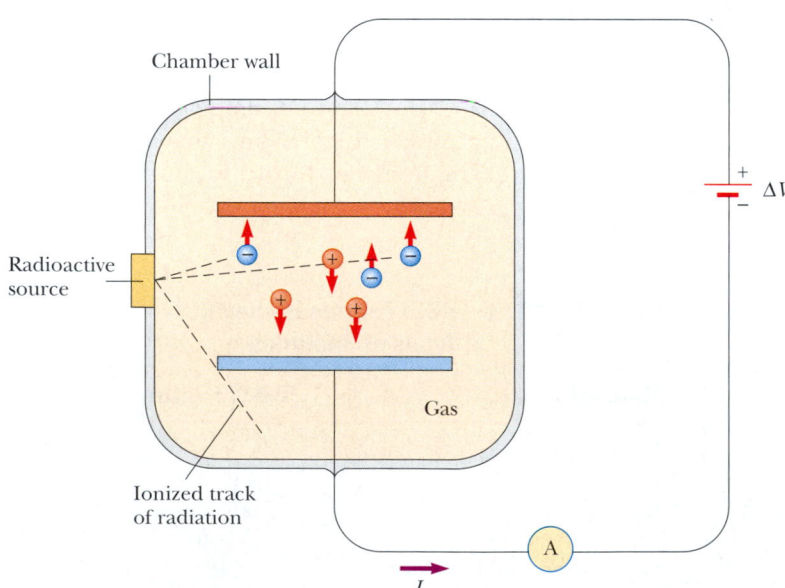

Figure 45.16 Simplified diagram of an ion chamber. The radioactive source creates electrons and positive ions that are collected by the charged plates. The current set up in the external circuit is proportional to a radioactive particle's kinetic energy if the particle stops in the chamber.

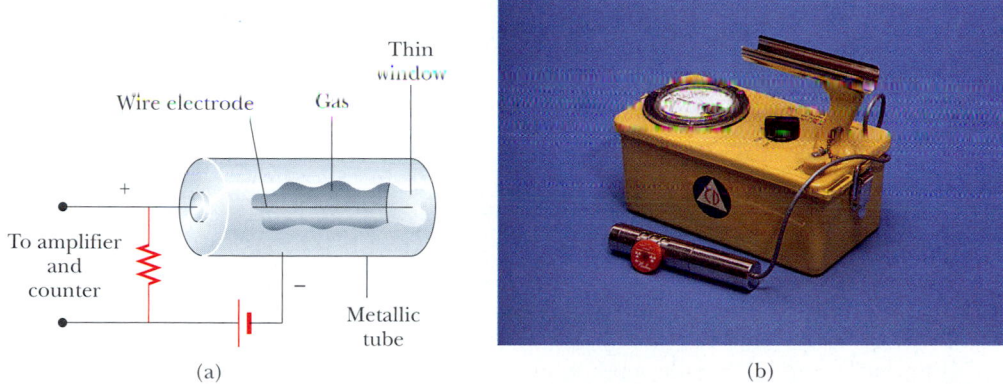

Figure 45.17 (a) Diagram of a Geiger counter. The voltage between the wire electrode and the metallic tube is usually about 1 000 V. (b) A Geiger counter. *(David Rogers)*

window at one end, some of the gas atoms are ionized. The electrons removed from these atoms are attracted toward the wire electrode, and in the process they ionize other atoms in their path. This sequential ionization results in an *avalanche* of electrons that produces a current pulse. After the pulse has been amplified, it can either be used to trigger an electronic counter or be delivered to a loud-speaker that clicks each time a particle is detected. Although a Geiger counter easily detects the presence of a radioactive particle, the energy lost by the particle in the counter is *not* proportional to the current pulse produced. Thus, a Geiger counter cannot be used to measure the energy of a radioactive particle.

A **semiconductor diode detector** is essentially a reverse-bias $p-n$ junction. Recall from Section 43.7 that a $p-n$ junction passes current readily when forward-biased and prohibits a current when reverse-biased. As an energetic particle passes through the junction, electrons are excited into the conduction band and holes are formed in the valence band. The internal electric field sweeps the electrons toward the positive (n) side of the junction and the holes toward the negative (p) side. This movement of electrons and holes creates a pulse of current that is measured with an electronic counter. In a typical device, the duration of the pulse is 10^{-8} s.

A **scintillation counter** usually uses a solid or liquid material whose atoms are easily excited by radiation. The excited atoms then emit visible-light photons when they return to their ground state. Common materials used as scintillators are transparent crystals of sodium iodide and certain plastics. If the scintillator material is attached to one end of a device called a **photomultiplier** (PM) tube as shown in Figure 45.18, the photons emitted by the scintillator can be converted to an electrical signal. The PM tube consists of numerous electrodes, called *dynodes,* whose electric potentials increase in succession along the length of the tube. Between the top of the tube and the scintillator material is a plate called a photocathode. When photons leaving the scintillator hit this plate, electrons are emitted (because of the photoelectric effect). As one of these emitted electrons strikes the first dynode, the electron has sufficient kinetic energy to eject several other electrons from the dynode surface. When these electrons are accelerated to the second dynode, many more electrons are ejected, and thus a multiplication process occurs. The end result is 1 million or more electrons striking the last dynode. Hence, one particle striking the scintillator produces a sizable electrical pulse at the PM output, and this pulse is sent to an electronic counter.

Semiconductor diode detector

Scintillation counter

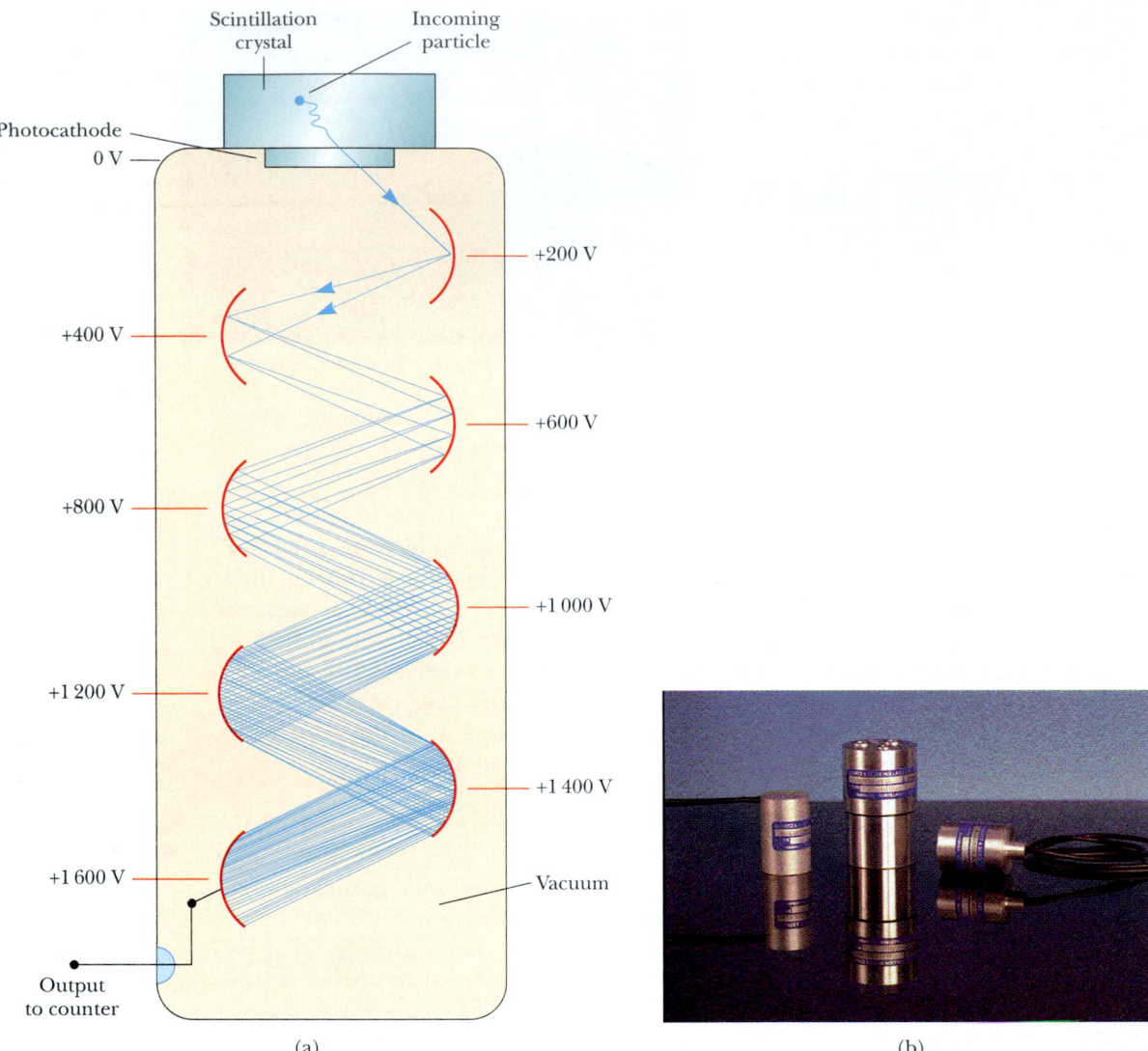

Scintillation crystal

Incoming particle

Photocathode

0 V

+400 V

+800 V

+1 200 V

+1 600 V

+200 V

+600 V

+1 000 V

+1 400 V

Vacuum

Output to counter

(a)

(b)

Figure 45.18 (a) Diagram of a scintillation counter connected to a photomultiplier tube.
(b) The sodium iodide in these scintillation crystals flashes when an energetic particle passes
through, something like the way the atmosphere flashes when a meteor passes through it.

Both the scintillator and the semiconductor diode detector are much more
sensitive than a Geiger counter, mainly because of the higher density of the detecting medium. Both can also be used to measure particle energy if the particle stops
in the detector.

Track detector

Track detectors are various devices used to view the tracks of charged particles directly. High-energy particles produced in particle accelerators may have energies ranging from 10^9 to 10^{12} eV. Thus, they cannot be stopped and cannot have
their energy measured with the small detectors already mentioned. Instead, the
energy and momentum of these energetic particles are found from the curvature
of their path in a magnetic field of known magnitude and direction.

A **photographic emulsion** is the simplest example of a track detector. A
charged particle ionizes the atoms in an emulsion layer. The path of the particle

corresponds to a family of points at which chemical changes have occurred in the emulsion. When the emulsion is developed, the particle's track becomes visible.

A **cloud chamber** contains a gas that has been supercooled to just below its usual condensation point. An energetic radioactive particle passing through ionizes the gas along the particle's path. The ions serve as centers for condensation of the supercooled gas. The track can be seen with the naked eye and can be photographed. A magnetic field can be applied to determine the charges of the radioactive particles, as well as their momentum and energy.

Cloud chamber

A device called a **bubble chamber,** invented in 1952 by D. Glaser, uses a liquid (usually liquid hydrogen) maintained near its boiling point. Ions produced by incoming charged particles leave bubble tracks, which can be photographed (Fig. 45.19). Because the density of the detecting medium in a bubble chamber is much higher than the density of the gas in a cloud chamber, the bubble chamber has a much higher sensitivity.

Bubble chamber

A **spark chamber** is a counting device that consists of an array of conducting parallel plates and is capable of recording a three-dimensional track record. Even-numbered plates are grounded, and odd-numbered plates are maintained at a high electric potential (about 10 kV). The spaces between the plates contain an inert gas at atmospheric pressure. When a charged particle passes through the chamber, gas atoms are ionized, resulting in a current surge and visible sparks along the particle path. These sparks may be photographed or electronically detected and sent to a computer for path reconstruction and determination of particle mass, momentum, and energy.

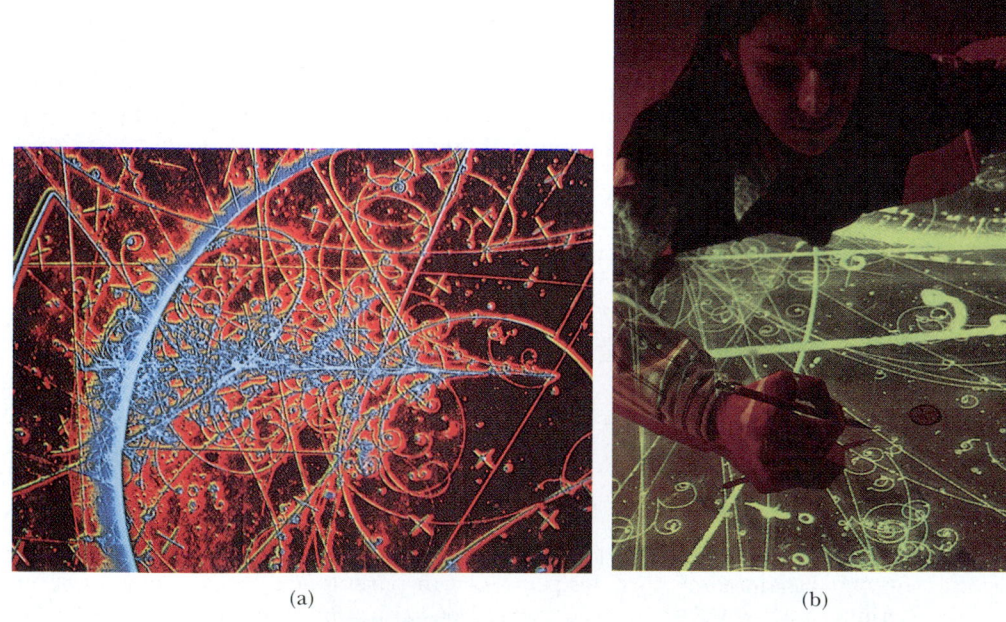

(a)   (b)

Figure 45.19 (a) Artificially colored bubble-chamber photograph showing tracks of particles that have passed through the chamber. *(Photo Researchers, Inc./Science Photo Library)* (b) This research scientist is studying a photograph of particle tracks made in a bubble chamber at Fermilab. The curved tracks are produced by charged particles moving through the chamber in the presence of an applied magnetic field. Negatively charged particles deflect in one direction, while positively charged particles deflect in the opposite direction. *(Dan McCoy/Rainbow)*

Neutron detectors are more difficult to construct than charged-particle detectors because neutrons do not interact electrically with atoms as they pass through matter. Fast neutrons, however, can be detected by filling an ion chamber with hydrogen gas and detecting the resulting ionization by high-speed recoiling protons produced in neutron–proton collisions. Thermal neutrons having energies less than about 1 MeV do not transfer sufficient energy to the protons to be detected in this way; however, they can be detected by using an ion chamber filled with BF_3 gas. In this case, the boron nuclei disintegrate during neutron capture, emitting alpha particles that, because of their charge, are easily detected in the ion chamber.

Optional Section

45.7 ▸ USES OF RADIATION

Nuclear physics applications are extremely widespread in manufacturing, medicine, and biology. Even a brief discussion of all the possibilities would fill an entire book, and to keep such a book up to date would require a number of revisions each year. In this section, we present a few of these applications and the underlying theories supporting them.

Tracing

Radioactive tracers are used to track chemicals participating in various reactions. One of the most valuable uses of radioactive tracers is in medicine. For example, iodine, a nutrient needed by the human body, is obtained largely through the intake of iodized salt and seafood. To evaluate the performance of the thyroid, the patient drinks a very small amount of radioactive sodium iodide containing ^{131}I, an artificially produced isotope of iodine (the natural, nonradioactive isotope is ^{127}I). Two hours later, the amount of iodine in the thyroid gland is determined by measuring the radiation intensity at the neck area. How much or how little ^{131}I is still in the thyroid is a measure of how well that gland is functioning.

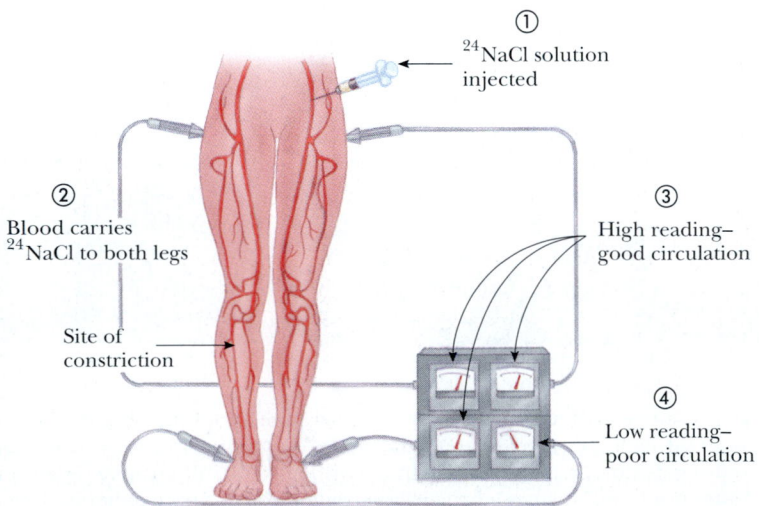

① ^{24}NaCl solution injected

② Blood carries ^{24}NaCl to both legs

③ High reading– good circulation

④ Low reading– poor circulation

Site of constriction

Figure 45.20 A tracer technique for determining the condition of the human circulatory system.

A second medical application is indicated in Figure 45.20. A solution containing radioactive sodium is injected into a vein in the leg, and the time at which the radioisotope arrives at another part of the body is detected with a radiation counter. The elapsed time is a good indication of the presence or absence of constrictions in the circulatory system.

Tracers are also useful in agricultural research. Suppose the best method of fertilizing a plant is to be determined. A certain element in a fertilizer, such as nitrogen, can be *tagged* (identified) with one of its radioactive isotopes. The fertilizer is then sprayed on one group of plants, sprinkled on the ground for a second group, and raked into the soil for a third. A Geiger counter is then used to track the nitrogen through the three groups.

Tracing techniques are as wide ranging as human ingenuity can devise. Present applications range from checking how teeth absorb fluoride to monitoring how cleansers contaminate food-processing equipment to studying deterioration inside an automobile engine. In the latter case, a radioactive material is used in the manufacture of the piston rings, and the oil is checked for radioactivity to determine the amount of wear on the rings.

Materials Analysis

For centuries, a standard method of identifying the elements in a sample of material has been chemical analysis, which involves determining how the material reacts with various chemicals. A second method is spectral analysis, which uses the fact that, when excited, each element emits its own characteristic set of electromagnetic wavelengths. These methods are now supplemented by a third technique, **neutron activation analysis.** Both chemical and spectral methods have the disadvantage that a fairly large sample of the material must be destroyed for the analysis. In addition, extremely small quantities of an element may go undetected by either method. Neutron activation analysis has an advantage over the other two methods in both respects.

When a material is irradiated with neutrons, nuclei in the material absorb the neutrons and are changed to different isotopes, most of which are radioactive. For example, ^{65}Cu absorbs a neutron to become ^{66}Cu, which undergoes beta decay:

$$\ce{^1_0 n + ^{65}_{29}Cu -> ^{66}_{29}Cu -> ^{66}_{30}Zn + e^- + \bar{\nu}}$$

The presence of the copper can be deduced because it is known that ^{66}Cu has a half-life of 5.1 min and decays with the emission of beta particles having maximum energies of 2.63 and 1.59 MeV. Also emitted in the decay of ^{66}Cu is a 1.04-MeV gamma ray. By examining the radiation emitted by a substance after it has been exposed to neutron irradiation, one can detect extremely small amounts of an element in that substance.

Neutron activation analysis is used routinely in a number of industries, for example in commercial aviation for the checking of airline luggage for hidden explosives (Fig. 45.21). The following nonroutine use is of interest. Napoleon died on the island of St. Helena in 1821, supposedly of natural causes. Over the years, suspicion has existed that his death was not all that natural. After his death, his head was shaved and locks of his hair were sold as souvenirs. In 1961, the amount of arsenic in a sample of this hair was measured by neutron activation analysis, and an unusually large quantity of arsenic was found. (Activation analysis is so sensitive that very small pieces of a single hair could be analyzed.) Results showed that the arsenic was fed to him irregularly. In fact, the arsenic concentration pattern corre-

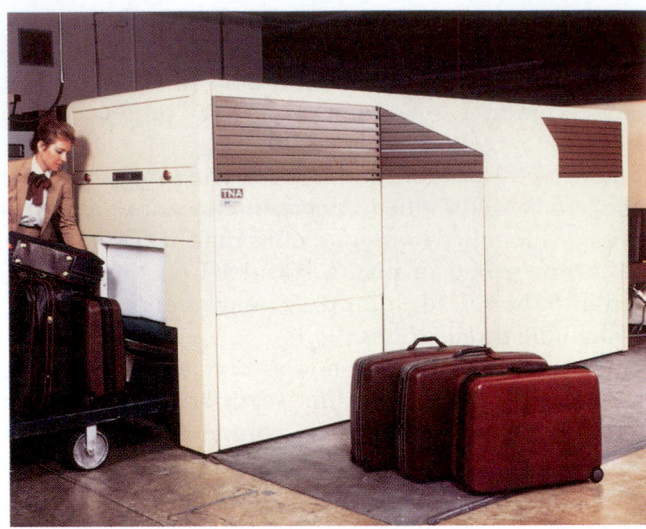

Figure 45.21   This bomb detector irradiates luggage with neutrons. If there are hidden explosives inside, chemicals within the explosive materials become radioactive and can be easily detected. The half-life of the resulting radiation is so short that there is no danger to the security personnel removing the luggage for further inspection. *(Shahn Kermani/Gamma Liaison)*

sponded to the fluctuations in the severity of Napoleon's illness as determined from historical records.

Art historians use neutron activation analysis to detect forgeries. The pigments used in paints have changed throughout history, and old and new pigments react differently to neutron activation. The method can even reveal hidden works of art behind existing paintings.

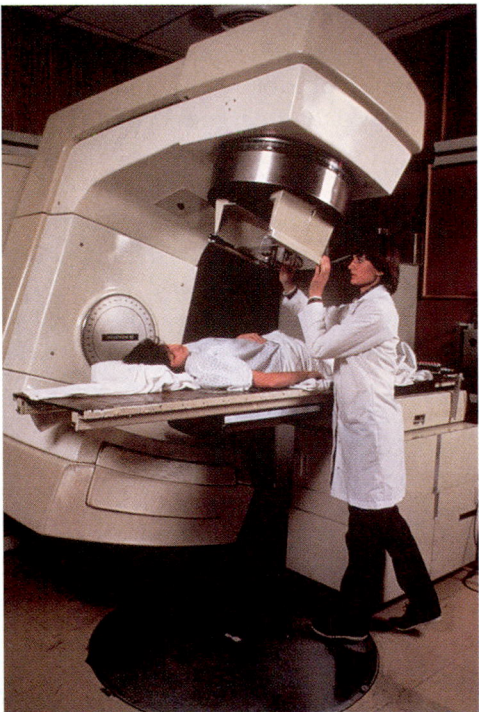

Figure 45.22   This large machine is being set to deliver a dose of radiation from ^{60}Co in an effort to destroy a cancerous tumor. Cancer cells are especially susceptible to this type of therapy because they tend to divide more often than cells of healthy tissue nearby. *(Science/Visuals Unlimited)*

Figure 45.23   The irradiated strawberries on the right have *not* become radioactive. The radiation has killed or incapacitated the mold spores that have spoiled the strawberries on the left. *(Council for Agricultural Science & Technology)*

Another analysis technique that takes advantage of radioactivity is PIXE, **photon-induced x-ray emission.** This process uses x-ray photons to excite the innermost electrons of an atom in the material being analyzed. When the electrons fall back down from their excited states, they give off characteristic spectra that can be evaluated to reveal the types of elements present.

Radiation Therapy

Because radiation causes the most damage to rapidly dividing cells (as discussed in Section 45.5), it is useful in cancer treatment because tumor cells divide extremely rapidly. Several mechanisms can be used to deliver radiation to a tumor. In some cases, a narrow beam of x-rays or radiation from a source such as ^{60}Co is used, as shown in Figure 45.22. In other situations, thin radioactive needles called *seeds* are implanted in the cancerous tissue. The radioactive isotope ^{151}I is used to treat cancer of the thyroid.

Food Preservation

Radiation is finding increasing use as a means of preserving food because exposure to high levels of radiation can destroy or incapacitate bacteria and mold spores (Fig. 45.23). Food preserved this way can be placed in a sealed container (to keep out new spoiling agents) and stored for long periods of time.

SUMMARY

The probability that neutrons are captured as they move through matter generally increases with decreasing neutron energy. A **thermal neutron** is a slow-moving neutron that has a high probability of being captured by a nucleus in a **neutron capture event:**

$$^{1}_{0}n + ^{A}_{Z}X \longrightarrow ^{A+1}_{Z}X + \gamma \tag{45.1}$$

Nuclear fission occurs when a very heavy nucleus, such as ^{235}U, splits into two smaller **fission fragments.** Thermal neutrons can create fission in ^{235}U:

$$^{1}_{0}n + {}^{235}_{92}U \longrightarrow {}^{236}_{92}U^* \longrightarrow X + Y + \text{neutrons} \qquad (45.2)$$

where X and Y are the fission fragments and ^{236}U* is an intermediate excited state. On the average, 2.5 neutrons are released per fission event. The fragments then undergo a series of beta and gamma decays to various stable isotopes. The energy released per fission event is about 200 MeV.

The **reproduction constant** K is the average number of neutrons released from each fission event that cause another event. In a fission reactor, it is necessary to maintain $K \approx 1$. The value of K is affected by such factors as reactor geometry, mean neutron energy, and probability of neutron capture. Neutron energies are regulated with a moderator material that slows down energetic neutrons and therefore increases the probability of neutron capture by other ^{235}U nuclei. The power level of the reactor is adjusted with control rods made of a material that is very efficient in absorbing neutrons.

In **nuclear fusion,** two light nuclei fuse to form a heavier nucleus and release energy. The major obstacle in obtaining useful energy from fusion is the large Coulomb repulsive force between the charged nuclei at small separation distances. Sufficient energy must be supplied to the particles to overcome this Coulomb barrier. The temperature required to produce fusion is of the order of 10^8 K, and at this temperature all matter occurs as a plasma.

In a fusion reactor, the plasma temperature must reach the **critical ignition temperature,** the temperature at which the power generated by the fusion reactions exceeds the power lost in the system. The most promising fusion reaction is the D–T reaction, which has a critical ignition temperature of approximately 4.5×10^7 K. Two critical parameters in fusion reactor design are **ion density** n and **confinement time** τ, the length of time the interacting particles must be maintained at $T > T_{\text{ignit}}$. **Lawson's criterion** states that for the D–T reaction, $n\tau \geq 10^{14}$ s/cm^3.

Questions

1. Explain the function of a moderator in a fission reactor.

2. Why is water a better shield against neutrons than lead or steel?

3. Discuss the advantages and disadvantages of fission reactors from the point of view of safety, pollution, and resources. Make a comparison with power generated from the burning of fossil fuels.

4. Why would a fusion reactor produce less radioactive waste than a fission reactor?

5. Lawson's criterion states that the product of ion density and confinement time must exceed a certain number before a break-even fusion reaction can occur. Why should these two parameters determine the outcome?

6. Why is the temperature required for the D–T fusion less than that needed for the D–D fusion? Estimate the relative importance of Coulomb repulsion and nuclear attraction in each case.

7. What factors make a fusion reaction difficult to achieve?

8. Discuss the similarities and differences between fusion and fission.

9. Discuss the advantages and disadvantages of fusion power from the point of safety, pollution, and resources.

10. Discuss three major problems associated with the development of a controlled fusion reactor.

11. Describe two techniques being pursued in an effort to obtain power from nuclear fusion.

12. If two radioactive samples have the same activity measured in curies, will they necessarily create the same damage to a medium? Explain.

13. Why should a radiologist be extremely cautious about x-ray doses when treating pregnant women?

14. The design of a PM tube might suggest that any number of dynodes may be used to amplify a weak signal. What factors do you suppose would limit the amplification in this device?

15.

And swift, and swift past comprehension
Turn round Earth's beauty and her might.
The heavens blaze in alternation
With deep and chill and miny night.
In mighty currents foams the ocean
Up from the rocks' abyssal base.
With rock and sea torn into motion
In ever-swift celestial race.
And tempests bluster in a contest
From sea to land, from land to sea.
In rage they forge a chain around us
Of deepest meaning, energy.

There flames a lightning disaster
Before the thunder, in its way.
But all Your servants honor, Master,
The gentle order of Your day

Johann Wolfgang von Goethe wrote this song of the archangels in *Faust* a half-century before the law of conservation of energy was recognized. Students often find it convenient to think of a list of several "forms of energy," from kinetic to nuclear. Argue for or against the view that these lines of poetry make an obvious or an oblique reference to every form of energy.

PROBLEMS

1, 2, 3 = straightforward, intermediate, challenging ☐ = full solution available in the *Student Solutions Manual and Study Guide*
WEB = solution posted at **http://www.saunderscollege.com/physics/** 🖥 = Computer useful in solving problem 🖱 = Interactive Physics
☐ = paired numerical/symbolic problems

Section 45.2 Nuclear Fission

1. Find the energy released in the fission reaction

$$^{1}_{0}n + ^{235}_{92}U \longrightarrow ^{98}_{40}Zr + ^{135}_{52}Te + 3(^{1}_{0}n)$$

The atomic masses of the fission products are: $^{98}_{40}Zr$, 97.912 0 u; $^{135}_{52}Te$, 134.908 7 u.

2. Strontium-90 is a particularly dangerous fission product of ^{235}U because it is radioactive and it substitutes for calcium in bones. What other direct fission products would accompany it in the neutron-induced fission of ^{235}U? (*Note:* This reaction may release two, three, or four free neutrons.)

WEB **3.** List the nuclear reactions required to produce ^{233}U from ^{232}Th under fast neutron bombardment.

4. List the nuclear reactions required to produce ^{239}Pu from ^{238}U under fast neutron bombardment.

5. (a) Find the energy released in the following fission reaction:

$$^{1}_{0}n + ^{235}_{92}U \longrightarrow ^{141}_{56}Ba + ^{92}_{36}Kr + 3(^{1}_{0}n)$$

The required masses are

$$M(^{1}_{0}n) = 1.008\ 665\ u$$

$$M(^{235}_{92}U) = 235.043\ 924\ u$$

$$M(^{141}_{56}Ba) = 140.913\ 9\ u$$

$$M(^{92}_{36}Kr) = 91.897\ 3\ u$$

(b) What fraction of the initial mass of the system is given off?

6. A typical nuclear fission power plant produces about 1.00 GW of electric power. Assume that the plant has an overall efficiency of 40.0% and that each fission releases 200 MeV of energy. Calculate the mass of ^{235}U consumed each day.

7. Review Problem. Suppose enriched uranium containing 3.40% of the fissionable isotope $^{235}_{92}U$ is used as fuel for a ship. The water exerts an average frictional drag of 1.00×10^5 N on the ship. How far can the ship travel per kilogram of fuel? Assume that the energy released per fission event is 208 MeV and that the ship's engine has an efficiency of 20.0%.

Section 45.3 Nuclear Reactors

8. To minimize neutron leakage from a reactor, the surface area-to-volume ratio should be a minimum. For a given volume V, calculate this ratio for (a) a sphere, (b) a cube, and (c) a parallelepiped of dimensions $a \times a \times 2a$. (d) Which of these shapes would have minimum leakage? Which would have maximum leakage?

WEB **9.** It has been estimated that there is on the order of 10^9 tons of natural uranium available at concentrations exceeding 100 parts per million, of which 0.7% is ^{235}U. If all the world's energy use (7×10^{12} J/s) were to be supplied by ^{235}U fission, how long would this supply last? (This estimate of uranium supply was taken from K. S. Deffeyes and I. D. MacGregor, *Sci. Am.* 242:66, 1980.)

10. If the reproduction constant (neutron multiplication factor) is 1.000 25 for a chain reaction in a fission reactor and the average time between successive fissions is 1.20 ms, by what factor will the reaction rate increase in 1 min?

11. A large nuclear power reactor produces about 3 000 MW of power in its core. Three months after a reactor is shut down, the core power from radioactive by-products is 10.0 MW. Assuming that each emission delivers 1.00 MeV of energy to the power, find the activity in becquerels three months after the reactor is shut down.

Section 45.4   Nuclear Fusion

12. (a) If a fusion generator were built to create 3.00 GW of power, determine the rate of fuel burning in grams per hour if the D–T reaction is used. (b) Do the same for the D–D reaction assuming that the reaction products are split evenly between (n, ^{3}He) and (p, ^{3}H).

13. Two nuclei having atomic numbers Z_1 and Z_2 approach each other with a total energy E. (a) Suppose they will spontaneously fuse if they approach within a distance of 1.00×10^{-14} m. Find the minimum value of E required to produce fusion, in terms of Z_1 and Z_2. (b) Evaluate the minimum energy for fusion for the D–D and D–T reactions (the first and third reactions in Eq. 45.4).

14. **Review Problem.**   Consider the deuterium-tritium fusion reaction with the tritium nucleus at rest:

$$^2_1\text{H} + {}^3_1\text{H} \longrightarrow {}^4_2\text{He} + {}^1_0\text{n}$$

(a) Suppose that the reactant nuclei will spontaneously fuse if their surfaces touch. From Equation 44.1, determine the required distance of closest approach between their centers. (b) What is the Coulomb potential energy (in eV) at this distance? (c) Suppose the deuteron is fired straight at an originally stationary tritium nucleus with just enough energy to reach the required distance of closest approach. What is the common speed of the deuterium and tritium nuclei as they touch, in terms of the initial deuteron speed v_i? (*Hint:* At this point, the two nuclei have a common speed equal to the center-of-mass speed.) (d) Use energy methods to find the minimum initial deuteron energy required to achieve fusion. (e) Why does the fusion reaction actually occur at much lower deuteron energies than that calculated in part (d)?

WEB 15. To understand why plasma containment is necessary, consider the rate at which an unconfined plasma would be lost. (a) Estimate the rms speed of deuterons in a plasma at 4.00×10^8 K. (b) Estimate the order of magnitude of the time such a plasma would remain in a 10-cm cube if no steps were taken to contain it.

16. Of all the hydrogen in the oceans, 0.030 0% of the mass is deuterium. The oceans have a volume of 317 million mi^3. (a) If all the deuterium in the oceans were fused to $^4_2$He, how many joules of energy would be released? (b) World energy consumption is about 7.00×10^{12} W. If consumption were 100 times greater, how many years would the energy calculated in (a) last?

17. It has been pointed out that fusion reactors are safe from explosion because there is never enough energy in the plasma to do much damage. (a) In 1992, the TFTR reactor achieved an ion temperature of 4.0×10^8 K, an ion density of 2.0×10^{13} cm^{-3}, and a confinement time of 1.4 s. Calculate the amount of energy stored in the plasma of the TFTR reactor. (b) How many kilograms of water could be boiled away by this much energy? (The plasma volume of the TFTR reactor is about 50 m^3.)

18. **Review Problem.**   To confine a stable plasma, the magnetic energy density in the magnetic field (Eq. 32.14) must exceed the pressure $2nk_BT$ of the plasma by a factor of at least 10. In the following, assume a confinement time $\tau = 1.00$ s. (a) Using Lawson's criterion, determine the ion density required for the D–T reaction. (b) From the ignition temperature criterion, determine the required plasma pressure. (c) Determine the magnitude of the magnetic field required to contain the plasma.

19. Find the number of ^{6}Li and the number of ^{7}Li nuclei present in 2.00 kg of lithium. (The natural abundance of ^{6}Li is 7.5%; the remainder is ^{7}Li.)

20. One old prediction for the future was to have a fusion reactor supply energy to dissociate the molecules in garbage into separate atoms and then to ionize the atoms. This material could be put through a giant mass spectrometer, so that trash would be a new source of isotopically pure elements—the mine of the future. Assuming an average atomic mass of 56 and an average charge of 26 (a high estimate, considering all the organic materials), at a beam current of 1.00 MA, how long would it take to process 1.00 metric ton of trash?

(Optional)
Section 45.5   Radiation Damage in Matter

21. **Review Problem.**   A building has become accidentally contaminated with radioactivity. The longest-lived material in the building is strontium-90. ($^{90}_{38}$Sr has an atomic mass 89.907 7 u, and its half-life is 29.1 yr.) If the building initially contained 5.00 kg of this substance uniformly distributed throughout the building (a very unlikely situation) and the safe level is less than 10.0 counts/min, how long will the building be unsafe?

22. **Review Problem.**   A particular radioactive source produces 100 mrad of 2-MeV gamma rays per hour at a distance of 1.00 m. (a) How long could a person stand at this distance before accumulating an intolerable dose of 1 rem? (b) Assuming the radioactive source is a point source, at what distance would a person receive a dose of 10.0 mrad/h?

23. Assume that an x-ray technician takes an average of eight x-rays per day and receives a dose of 5 rem/yr as a result. (a) Estimate the dose in rem per photograph taken. (b) How does the technician's exposure compare with low-level background radiation?

24. When gamma rays are incident on matter the intensity of the gamma rays passing through the material varies with depth x as $I(x) = I_0 e^{-\mu x}$, where μ is the absorption coefficient and I_0 is the intensity of the radiation at the surface of the material. For 0.400-MeV gamma rays in lead, the absorption coefficient is 1.59 cm^{-1}. (a) Determine the "half-thickness" for lead—that is, the thickness of lead that would absorb half the incident gamma rays. (b) What thickness will reduce the radiation by a factor of 10^4?

WEB **25.** A "clever" technician decides to warm some water for his coffee with an x-ray machine. If the machine produces 10.0 rad/s, how long will it take to raise the temperature of a cup of water by 50.0°C?

26. Review Problem. The danger to the body from a high dose of gamma rays is not due to the amount of energy absorbed but occurs because of the ionizing nature of the radiation. To illustrate this, calculate the rise in body temperature that would result if a "lethal" dose of 1 000 rad were absorbed strictly as internal energy. Take the specific heat of living tissue as 4186 J/kg · °C.

27. Technetium-99 is used in certain medical diagnostic procedures. If 1.00×10^{-8} g of ^{99}Tc is injected into a 60.0-kg patient and half of the 0.140-MeV gamma rays are absorbed in the body, determine the total radiation dose received by the patient.

28. Strontium-90 from the testing of atomic bombs can still be found in the atmosphere. Each decay of ^{90}Sr releases 1.1 MeV of energy into the bones of a person who has had strontium replace the calcium. If a 70.0-kg person receives 1.00 ng of ^{90}Sr from contaminated milk, calculate the absorbed dose rate (in J/kg) in one year. Assume the half-life of ^{90}Sr to be 29.1 yr.

(Optional)
Section 45.6   Radiation Detectors

29. In a Geiger tube, the voltage between the electrodes is typically 1.00 kV and the current pulse discharges a 5.00-pF capacitor. (a) What is the energy amplification of this device for a 0.500-MeV electron? (b) How many electrons are avalanched by the initial electron?

30. In a Geiger tube, the voltage between the electrodes is ΔV and the current pulse discharges a capacitor having capacitance C. (a) What is the energy amplification of this device for an electron of energy E? (b) How many electrons are avalanched by the initial electron?

31. In a certain photomultiplier tube there are seven dynodes, having potentials of 100, 200, 300, . . . , 700 V. The average energy required to free an electron from a dynode surface is 10.0 eV. Assume that just one electron is incident and that the tube functions with 100% efficiency. (a) How many electrons are freed at the first dynode? (b) How many electrons are collected by the last dynode? (c) What is the energy available to the counter?

32. (a) Your grandmother recounts to you how, as young children, your father, aunts, and uncles made the screen door slam continually as they ran between the house and the back yard. The time interval between slams varied randomly, but the average slamming rate stayed constant at 38.0/h from dawn to dusk every summer day. If the slamming rate suddenly dropped to zero, the children would have found a nest of baby field mice or gotten into some other mischief requiring adult intervention. How long after the last screen-door slam would a prudent and attentive parent wait before leaving her or his work to see about the children? Explain your reasoning. (b) A student wishes to measure the half life of a radioactive substance, using a small sample. The clicks of her Geiger counter are randomly spaced in time. The counter registers 372 counts during one 5.00-min interval, and 337 counts during the next 5.00 min. The average background rate is 15 counts/min. Find the most probable value for the half-life. (c) Estimate the uncertainty in the half-life determination. Explain your reasoning.

(Optional)
Section 45.7   Uses of Radiation

33. During the manufacture of a steel engine component, radioactive iron (^{59}Fe) is included in the total mass of 0.200 kg. The component is placed in a test engine when the activity due to this isotope is 20.0 μCi. After a 1 000-h test period, oil is removed from the engine and found to contain enough ^{59}Fe to produce 800 disintegrations/min/L of oil. The total volume of oil in the engine is 6.50 L. Calculate the total mass worn from the engine component per hour of operation. (The half-life for ^{59}Fe is 45.1 days.)

34. At some time in your past or future, you may find yourself in a hospital to have a PET scan. The acronym stands for *positron-emission tomography*. In the procedure, a radioactive element that undergoes e^+ decay is introduced into your body. The equipment detects the gamma rays that result from pair annihilation when the emitted positron encounters an electron in your body's tissue. Suppose that you receive an injection of glucose containing on the order of 10^{10} atoms of ^{14}O. Assume that the oxygen is uniformly distributed through 2 L of blood after 5 min. What will be the order of magnitude of the activity of the oxygen atoms in 1 cm^3 of the blood?

35. You want to find out how many ^{65}Cu atoms are in a small sample of material. You bombard it with neutrons to ensure that on the order of 1% of these copper nuclei absorb a neutron. After activation you turn off the neutron flux and then use a highly efficient detector to monitor the gamma radiation that comes out of the sample. Assume that one half of the ^{66}Cu nuclei emit a 1.04-MeV gamma ray in their decay. (The other half of the activated nuclei decay directly to the ground state of ^{66}Ni.) If after 10 min (two half-lives) you have detected 10^4 MeV of photon energy at 1.04 MeV, (a) about how many ^{65}Cu atoms are in the sample? (b) Assume the sample contains natural copper. Refer to the isotopic abundances listed in Table A.3 and estimate the total mass of copper in the sample.

36. When a material of interest is irradiated by neutrons, radioactive atoms are produced continually and some decay according to their characteristic half-lives. (a) If one species of a radioactive nucleus is produced at a constant rate R and its decay is governed by the conventional radioactive decay law, show that the number of radioactive atoms accumulated after an irradiation time t is

$$N = \frac{R}{\lambda}(1 - e^{-\lambda t})$$

(b) What is the maximum number of radioactive atoms that can be produced?

ADDITIONAL PROBLEMS

37. Carbon detonations are powerful nuclear reactions that temporarily tear apart the cores inside massive stars late in their lives. These blasts are produced by carbon fusion, which requires a temperature of about 6×10^8 K to overcome the strong Coulomb repulsion between carbon nuclei. (a) Estimate the repulsive energy barrier to fusion, using the required ignition temperature for carbon fusion. (In other words, what is the average kinetic energy for a carbon nucleus at 6×10^8 K?) (b) Calculate the energy (in MeV) released in each of these "carbon-burning" reactions:

$$^{12}C + {}^{12}C \longrightarrow {}^{20}Ne + {}^4He$$

$$^{12}C + {}^{12}C \longrightarrow {}^{24}Mg + \gamma$$

(c) Calculate the energy (in kWh) given off when 2.00 kg of carbon completely fuses according to the first reaction.

38. The atomic bomb dropped on Hiroshima on August 6, 1945 released 5×10^{13} J of energy (equivalent to that from 12 000 tons of TNT). Estimate (a) the number of $^{235}_{92}U$ nuclei fissioned and (b) the mass of this $^{235}_{92}U$.

39. Compare the fractional mass loss in a typical ^{235}U fission reaction with the fractional mass loss in D–T fusion.

40. **Review Problem.** Consider a nucleus at rest, which then spontaneously splits into two fragments of masses m_1 and m_2. Show that the fraction of the total kinetic energy that is carried by m_1 is

$$\frac{K_1}{K_{tot}} = \frac{m_2}{m_1 + m_2}$$

and the fraction carried by m_2 is

$$\frac{K_2}{K_{tot}} = \frac{m_1}{m_1 + m_2}$$

assuming relativistic corrections can be ignored. (*Note:* If the parent nucleus was moving before the decay, then m_1 and m_2 still divide the kinetic energy as shown, as long as all velocities are measured in the center-of-mass frame of reference, in which the total momentum of the system is zero.)

41. The half-life of tritium is 12.3 yr. If the TFTR fusion reactor contained 50.0 m³ of tritium at a density equal to 2.00×10^{14} ions/cm³, how many curies of tritium were

in the plasma? Compare this value with a fission inventory (the estimated supply of fissionable material) of 4×10^{10} Ci.

42. **Review Problem.** A very slow neutron (with speed approximately equal to zero) can initiate the reaction

$$n + {}^{10}_5B \longrightarrow {}^7_3Li + {}^4_2He$$

If the alpha particle moves away with a speed of 9.30×10^6 m/s, calculate the kinetic energy of the lithium nucleus. Use nonrelativistic formulas.

43. **Review Problem.** A nuclear power plant operates by using the energy released in nuclear fission to convert 20°C water into 400°C steam. How much water could theoretically be converted to steam by the complete fissioning of 1.00 g of ^{235}U at 200 MeV/fission?

44. **Review Problem.** A nuclear power plant operates by using the energy released in nuclear fission to convert liquid water at T_c into steam at T_h. How much water could theoretically be converted to steam by the complete fissioning of a mass m of ^{235}U at 200 MeV/fission?

45. About 1 of every 3 300 water molecules contains one deuterium atom. (a) If all the deuterium nuclei in 1 L of water are fused in pairs according to the D–D reaction $^2H + {}^2H \rightarrow {}^3He + n + 3.27$ MeV, how much energy in joules is liberated? (b) Burning gasoline produces about 3.40×10^7 J/L. Compare the energy obtainable from the fusion of the deuterium in a liter of water with the energy liberated from the burning of a liter of gasoline.

46. The alpha-emitter polonium-210 ($^{210}_{84}Po$) is used in a nuclear energy source on a spacecraft. Determine the initial power output of the energy source if it contains 0.155 kg of ^{210}Po. Assume that the efficiency for conversion of radioactive decay energy to electrical energy is 1.00%.

47. A certain nuclear plant generates 3.065 GW of nuclear power to create 1.000 GW of electric power. Of the waste energy, 3.0% is ejected to the atmosphere and the remainder is passed into a river. A state law requires that the river water be warmed no more than 3.50°C when it is returned to the river. (a) Determine the amount of cooling water necessary (in kg/h and m³/h) to cool the plant. (b) If fission generates 7.80×10^{10} J/g of ^{235}U, determine the rate of fuel burning (in kg/h) of ^{235}U.

48. **Review Problem.** The first nuclear bomb was a fissioning mass of plutonium-239, exploded in the Trinity test, before dawn on July 16, 1945 at Alamogordo, New Mexico. Enrico Fermi was 14 km away, lying on the ground facing away from the bomb. After the whole sky had flashed with unbelievable brightness, Fermi stood up and began dropping bits of paper to the ground. They first fell at his feet in the calm and silent air. As the shock wave passed, about 40 s after the explosion, the paper then in flight jumped about 5 cm away from

ground zero. (a) Assume the shock wave in air propagated equally in all directions without absorption. Find the change in volume of a sphere of radius 14 km as it expands by 5 cm. (b) Find the work $P\,\Delta V$ done by the air in this sphere on the next layer of air farther from the center. (c) Assume the shock wave carried on the order of one tenth of the energy of the explosion. Make an order-of-magnitude estimate of the bomb yield. (d) One ton of exploding trinitrotoluene (TNT) releases energy of 4.2 GJ. What was the order of magnitude of the energy of the Trinity test in equivalent tons of TNT? The dawn revealed the mushroom cloud. Fermi's immediate knowledge of the bomb yield agreed with that determined days later by analysis of elaborate measurements.

49. Natural uranium must be processed to produce uranium enriched in ^{235}U for bombs and power plants. The processing yields a large quantity of nearly pure ^{238}U as a by-product. Because of its high mass density, it is used in armor-piercing artillery shells. (a) Find the edge dimension of a 70.0-kg cube of ^{238}U. (Refer to Table 1.5.) (b) The isotope ^{238}U has a long half-life of 4.47×10^9 yr. As soon as one nucleus decays, it begins a relatively rapid series of 14 steps, which together constitute the net reaction

$$^{238}_{92}\text{U} \longrightarrow 8(^4_2\text{He}) + 6(_{-1}^0\text{e}) + ^{206}_{82}\text{Pb} + 6\bar{\nu} + Q_{net}$$

Find the net decay energy. (Refer to Table A.3.) (c) Argue that a radioactive sample of decay rate R and releasing energy Q per decay has power output $\mathcal{P} = QR$. (d) Consider an artillery shell with a jacket of 70.0 kg of ^{238}U. Find its power output due to the radioactivity of the uranium and its daughters. Assume the shell is old enough that the daughters have reached steady-state amounts. Express the power in joules per year. (e) Assume that a 17-year-old soldier of mass 70.0 kg works in an arsenal where many such artillery shells are stored. If his radiation exposure is limited to 5.00 rem per year, find the rate at which he can absorb the energy of radiation, in joules per year. Assume an average RBE factor of 1.10.

50. A 2.0-MeV neutron is emitted in a fission reactor. If it loses one half its kinetic energy in each collision with a moderator atom, how many collisions must it undergo in order to become a thermal neutron (with energy 0.039 eV)?

WEB 51. Assuming that a deuteron and a triton are at rest when they fuse according to $^2\text{H} + ^3\text{H} \rightarrow ^4\text{He} + n + 17.6$ MeV, determine the kinetic energy acquired by the neutron.

52. A sealed capsule containing the radiopharmaceutical phosphorus-32 ($^{32}_{15}$P), an e^- emitter, is implanted into a patient's tumor. The average kinetic energy of the beta particles is 700 keV. If the initial activity is 5.22 MBq, determine the absorbed dose during a 10.0-day period. Assume the beta particles are completely absorbed in 100 g of tissue. (*Hint:* Find the number of beta particles emitted.)

53. (a) Calculate the energy (in kilowatt-hours) released if 1.00 kg of ^{239}Pu undergoes complete fission and the energy released per fission event is 200 MeV. (b) Calculate the energy (in electron volts) released in the D–T fusion:

$$^2_1\text{H} + ^3_1\text{H} \longrightarrow ^4_2\text{He} + ^1_0\text{n}$$

(c) Calculate the energy (in kilowatt-hours) released if 1.00 kg of deuterium undergoes fusion according to this reaction. (d) Calculate the energy (in kilowatt-hours) released by the combustion of 1.00 kg of coal if each $\text{C} + \text{O}_2 \rightarrow \text{CO}_2$ reaction yields 4.20 eV. (e) List advantages and disadvantages of each of these methods of energy generation.

54. The Sun radiates energy at the rate of 3.77×10^{26} W. Suppose that the net reaction

$$4(^1_1\text{H}) + 2(_{-1}^0\text{e}) \longrightarrow ^4_2\text{He} + 2\nu + \gamma$$

accounts for all the energy released. Calculate the number of protons fused per second.

55. Consider the two nuclear reactions

$$(\text{I}) \qquad \text{A} + \text{B} \longrightarrow \text{C} + \text{E}$$

$$(\text{II}) \qquad \text{C} + \text{D} \longrightarrow \text{F} + \text{G}$$

(a) Show that the net disintegration energy for these two reactions ($Q_{net} = Q_{\text{I}} + Q_{\text{II}}$) is identical to the disintegration energy for the net reaction

$$\text{A} + \text{B} + \text{D} \longrightarrow \text{E} + \text{F} + \text{G}$$

(b) One chain of reactions in the proton–proton cycle in the Sun's interior is

$$^1_1\text{H} + ^1_1\text{H} \longrightarrow ^2_1\text{H} + ^0_1\text{e} + \nu$$

$$^0_1\text{e} + _{-1}^0\text{e} \longrightarrow 2\gamma$$

$$^1_1\text{H} + ^2_1\text{H} \longrightarrow ^3_2\text{He} + \gamma$$

$$^1_1\text{H} + ^3_2\text{He} \longrightarrow ^4_2\text{He} + ^0_1\text{e} + \nu$$

$$^0_1\text{e} + _{-1}^0\text{e} \longrightarrow 2\gamma$$

Based on part (a), what is Q_{net} for this sequence?

56. Suppose the target in a laser fusion reactor is a sphere of solid hydrogen that has a diameter of 1.50×10^{-4} m and a density of 0.200 g/cm^3. Also assume that half of the nuclei are ^{2}H and half are ^{3}H. (a) If 1.00% of a 200-kJ laser pulse is delivered to this sphere, what temperature does the sphere reach? (b) If all of the hydrogen "burns" according to the D–T reaction, how much energy in joules is released?

57. The carbon cycle, first proposed by Hans Bethe in 1939, is another cycle by which energy is released in stars as hydrogen is converted to helium. The carbon cycle requires higher temperatures than the proton–proton cy-

cle. The series of reactions is

$$^{12}C + {}^1H \longrightarrow {}^{13}N + \gamma$$

$$^{13}N \longrightarrow {}^{13}C + e^+ + \nu$$

$$e^+ + e^- \longrightarrow 2\gamma$$

$$^{13}C + {}^1H \longrightarrow {}^{14}N + \gamma$$

$$^{14}N + {}^1H \longrightarrow {}^{15}O + \gamma$$

$$^{15}O \longrightarrow {}^{15}N + e^+ + \nu$$

$$e^+ + e^- \longrightarrow 2\gamma$$

$$^{15}N + {}^1H \longrightarrow {}^{12}C + {}^4He$$

(a) If the proton–proton cycle requires a temperature of 1.5×10^7 K, estimate by proportion the temperature required for the carbon cycle. (b) Calculate the Q value for each step in the carbon cycle and the overall energy released. (c) Do you think the energy carried off by the neutrinos is deposited in the star? Explain.

58. When photons pass through matter, the intensity I of the beam (measured in watts per square meter) decreases exponentially according to

$$I = I_0 e^{-\mu x}$$

where I_0 is the intensity of the incident beam, and I is the intensity of the beam that just passed through a thickness x of material. The constant μ is known as the *linear absorption coefficient,* and its value depends on the absorbing material and the wavelength of the photon beam. This wavelength (or energy) dependence allows us to filter out unwanted wavelengths from a broad-spectrum x-ray beam. (a) Two x-ray beams of wavelengths λ_1 and λ_2 and equal incident intensities pass through the same metallic plate. Show that the ratio of the emergent beam intensities is

$$\frac{I_2}{I_1} = e^{-[\mu_2 - \mu_1]x}$$

(b) Compute the ratio of intensities emerging from an aluminum plate 1.00 mm thick if the incident beam contains equal intensities of 50-pm and 100-pm x-rays. The values of μ for aluminum at these two wavelengths are $\mu_1 = 5.4$ cm^{-1} at 50 pm and $\mu_2 = 41.0$ cm^{-1} at 100 pm. (c) Repeat for an aluminum plate 10.0 mm thick.

ANSWERS TO QUICK QUIZZES

45.1 A proton or another neutron. As we learned in Chapter 9, during an elastic collision, the maximum kinetic energy is transferred when the colliding objects have the same mass (see Example 9.8). Consequently, a neutron loses all of its kinetic energy when it collides head-on with a proton, which has approximately the same mass as the neutron, or with another neutron.

45.2 (a). Both forces decrease with increasing separation. However, because the nuclear (attractive) force is an extremely short-range force, it drops off much more rapidly than the electrostatic (repulsive) force between

protons. Because the attractive force becomes small much faster than the repulsive force, the nucleus fissions.

45.3 According to Figure 44.3, the ratio N/Z increases with increasing Z. As a result, when a heavy nucleus fissions to two lighter nuclei, the lighter nuclei tend to have too many neutrons. This leads to instability, and the nuclei return to the line of stability by further decay processes that reduce the number of neutrons.

45.4 Reactions (a) and (b) because in both cases the Z and A values balance on the two sides of the equations. In reaction (c), $Z_{\text{left}} = Z_{\text{right}}$ but $A_{\text{left}} \neq A_{\text{right}}$.

PUZZLER

Both the circular particle accelerator at Fermilab and the Y-shaped radiotelescope called the Very Large Array are, in a sense, "time machines." These devices allow us to peer back in time and better understand what the Universe was like soon after it was created. How is this possible? What were things like back then? *(Top, courtesy of Fermilab Visual Media Services; bottom, courtesy of NRAO/AUI, photo by Dave Finley)*

c h a p t e r

46

Particle Physics and Cosmology

In this concluding chapter, we examine the various known subatomic particles and the fundamental interactions that govern their behavior. We also discuss the current theory of elementary particles, in which all matter is constructed from only two families of particles, quarks and leptons. Finally, we discuss how clarifications of such models might help scientists understand the birth and evolution of the Universe.

The word *atom* comes from the Greek *atomos,* which means "indivisible." The early Greeks believed that atoms were the indivisible constituents of matter; that is, they regarded them as elementary particles. Experiments in the 1890s and the early part of the 20th century showed that this was not the case, however, and after 1932 physicists viewed all matter as consisting of three constituent particles: electrons, protons, and neutrons. Beginning in the 1940s, many "new" particles were discovered in experiments involving high-energy collisions between known particles. The new particles are characteristically very unstable and have very short half-lives, ranging between 10^{-6} s and 10^{-23} s. So far, more than 300 of them have been catalogued.

Until the 1960s, physicists were bewildered by the great number and variety of subatomic particles that were being discovered. They wondered whether the particles had no systematic relationship connecting them, or whether a pattern was emerging that would provide a better understanding of the elaborate structure in the subatomic world. During the last 40 years, many high-energy particle accelerators have been constructed throughout the world, making it possible to observe collisions of highly energetic particles under controlled laboratory conditions and to see the subatomic world in finer detail. In these years, physicists have tremendously advanced our knowledge of the structure of matter by recognizing that all particles except electrons, photons, and a few others are made of smaller particles called quarks. Protons and neutrons, for example, are not truly elementary but are systems of tightly bound quarks.

46.1 ▶ THE FUNDAMENTAL FORCES IN NATURE

As noted in Section 5.1, all natural phenomena can be described by four fundamental forces acting between particles. In order of decreasing strength, they are the nuclear force, the electromagnetic force, the weak force, and the gravitational force.

The nuclear force, as we mentioned in Chapter 44, represents the glue that holds nucleons together. It is very short-range and is negligible for separations greater than about 10^{-15} m (about the size of the nucleus). The electromagnetic force, which binds atoms and molecules together to form ordinary matter, has about 10^{-2} times the strength of the nuclear force. It is a long-range force that decreases in magnitude as the inverse square of the separation between interacting particles. The weak force is a short-range force that tends to produce instability in certain nuclei. It is responsible for decay processes or the conversion of a neutron into a proton, and its strength is only about 10^{-5} times that of the nuclear force. Finally, the gravitational force is a long-range force that has a strength of only about 10^{-39} times that of the nuclear force. Although this familiar interaction is the force that holds the planets, stars, and galaxies together, its effect on elementary particles is negligible.

In modern physics, interactions between particles are often described in terms of the exchange or continuous emission and absorption of entities called **field particles** or **exchange particles.** In the case of the electromagnetic interaction, for instance, the field particles are photons. In the language of modern physics,

TABLE 46.1 Particle Interactions

Interaction	Relative Strength	Range of Force	Mediating Field Particle
Nuclear	1	Short (≈ 1 fm)	Gluon
Electromagnetic	10^{-2}	∞	Photon
Weak	10^{-5}	Short ($\approx 10^{-3}$ fm)	$W^{\pm}$, Z^0 bosons
Gravitational	10^{-39}	∞	Graviton

the electromagnetic force is said to be *mediated* by photons, and photons are the field particles of the electromagnetic field. Likewise, the nuclear force is mediated by field particles called *gluons* (so called because they "glue" the nucleons together). The weak force is mediated by field particles called W and Z *bosons,* and the gravitational force is mediated by field particles called *gravitons.* These interactions, their ranges, and their relative strengths are summarized in Table 46.1.

46.2 POSITRONS AND OTHER ANTIPARTICLES

In the 1920s, Paul Dirac developed a relativistic quantum-mechanical description of the electron that successfully explained the origin of the electron's spin and its magnetic moment. His theory had one major problem, however: its relativistic wave equation required solutions corresponding to negative energy states, and if negative energy states existed, an electron in a state of positive energy would be expected to make a rapid transition to one of these states, emitting a photon in the process.

Dirac circumvented this difficulty by postulating that all negative energy states are filled. Electrons occupying these negative energy states are said to be in the *Dirac sea* and are not directly observable because the Pauli exclusion principle does not allow them to react to external forces. However, if one of these negative energy states is vacant, leaving a hole in the sea of filled states, the hole can react to external forces and therefore is observable. The way a hole reacts to external forces is similar to the way an electron reacts to the same force except that the hole has a positive charge—it is the *antiparticle* to the electron.

The profound implication of this theory is that *for every particle an antiparticle exists.* The antiparticle for a charged particle has the same mass as the particle but opposite charge. For example, the electron's antiparticle (now called a *positron,* as noted in Section 44.5) has a rest energy of 0.511 MeV and a positive charge of 1.60×10^{-19} C.

Carl Anderson (1905–1991) observed the positron experimentally in 1932, and in 1936 he was awarded a Nobel Prize for his achievement. Anderson discovered the positron while examining tracks created in a cloud chamber by electron-like particles of positive charge (Fig. 46.1). (These early experiments used cosmic rays—mostly energetic protons passing through interstellar space—to initiate high-energy reactions on the order of several GeV.) To discriminate between positive and negative charges, Anderson placed the cloud chamber in a magnetic field, causing moving charges to follow curved paths. He noted that some of the electron-like tracks deflected in a direction corresponding to a positively charged particle.

Paul Adrien Maurice Dirac
British physicist (1902–1984)   Dirac won the Nobel Prize for physics in 1933.   *(Courtesy of AIP Emilio Segré Visual Archives)*

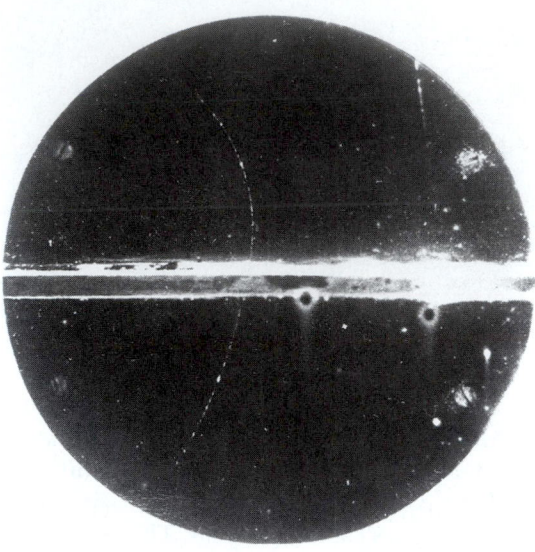

Figure 46.1 The first photograph of a positron track. The particle's track can be seen entering the picture at about the 7 o'clock position and curving upward and to the left. *(Courtesy of Archives, California Institute of Technology.)*

Quick Quiz 46.1

(a) What is the direction of the external magnetic field in Figure 46.1, into or out of the plane of the page? (b) The horizontal line is the edge of a sheet of lead. Why is the curve tighter above the lead than below it?

Since Anderson's discovery, positrons have been observed in a number of experiments. A common source of positrons is **pair production.** In this process, a gamma-ray photon with sufficiently high energy interacts with a nucleus, and an electron–positron pair is created from the photon. (The nucleus is necessary to satisfy the principle of conservation of momentum.) Because the total rest energy of the electron–positron pair is $2m_ec^2 = 1.02$ MeV (where m_e is the mass of the electron), the photon must have at least this much energy to create an electron–positron pair. Thus, electromagnetic energy in the form of a gamma ray is converted to rest energy in accordance with Einstein's famous relationship $E_R = mc^2$. If the gamma-ray photon has energy in excess of the rest energy of the electron and positron, the excess appears as kinetic energy of the two particles. Figure 46.2 shows tracks of electron–positron pairs created by 300-MeV gamma rays striking a lead sheet.

The creation of rest energy from other forms of energy is a general process and occurs in other situations besides pair production. In later sections of this chapter, we shall show how this process can be applied to understand the exchange of field particles between interacting particles.

The reverse process can also occur. Under the proper conditions, an electron and a positron can annihilate each other to produce two gamma-ray photons that have a combined energy of at least 1.02 MeV:

$$e^- + e^+ \longrightarrow 2\gamma$$

Because the initial momentum of the electron–positron system is approximately zero, two gamma rays traveling in opposite directions are necessary in this process to conserve momentum. If all the energy of the system were transformed into one photon, the momentum of the system would be high—momentum would not be conserved. Two photons can move in opposite directions with the result that the

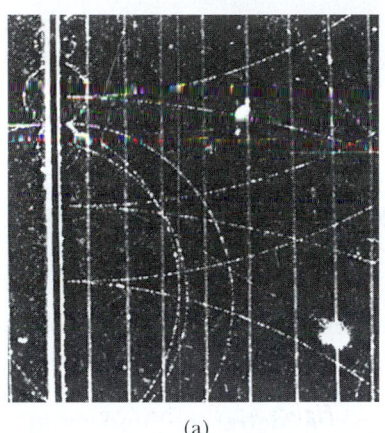
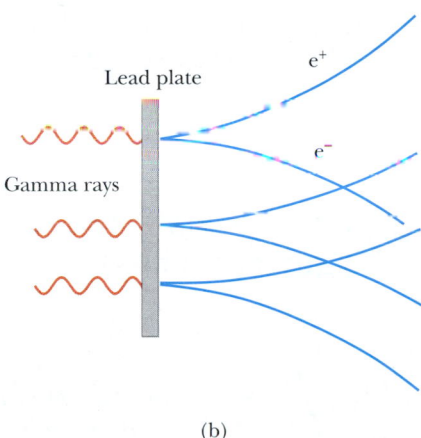

(a) (b)

Figure 46.2 (a) Bubble-chamber tracks of electron–positron pairs produced by 300-MeV gamma rays striking a lead sheet. *(Courtesy Lawrence Berkeley Laboratory, University of California)* (b) The pertinent pair-production events. The positrons deflect upward and the electrons downward because the direction of the applied magnetic field is into the page.

momentum of the electron–positron system remains small and equal to that of the system before annihilation. Very rarely, a proton and an antiproton also annihilate each other to produce two gamma-ray photons.

Practically every known elementary particle has a distinct antiparticle. Among the exceptions are the photon and the neutral pion (π^0). Following the construction of high-energy accelerators in the 1950s, many other antiparticles were revealed. These included the antiproton, discovered by Emilio Segré (1905–1989) and Owen Chamberlain (b. 1920) in 1955, and the antineutron, discovered shortly thereafter.

Electron–positron annihilation is used in the medical diagnostic technique called *positron emission tomography* (PET). The patient is injected with a glucose solution containing a radioactive substance that decays by positron emission, and the material is carried by the blood throughout the body. A positron emitted during a decay event in one of the radioactive nuclei in the glucose solution annihilates with an electron in the surrounding tissue, resulting in two gamma-ray photons emitted in opposite directions. A gamma detector surrounding the patient pinpoints the source of the photons and, with the assistance of a computer, displays an image of the sites at which the glucose accumulates. (Glucose is metabolized rapidly in cancerous tumors and accumulates at those sites, providing a strong signal for a PET detector system.) The images from a PET scan can indicate a wide variety of disorders in the brain, including Alzheimer's disease (Fig. 46.3). In addi-

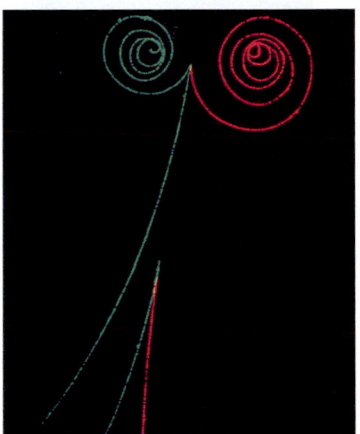

Bubble-chamber photograph of electron (green) and positron (red) tracks produced by energetic gamma rays. The highly curved tracks at the top are due to the electron and positron in an electron–positron pair bending in opposite directions in the magnetic field. *(Lawrence Berkeley Laboratory/Science Photo Library/Photo Researchers, Inc.)*

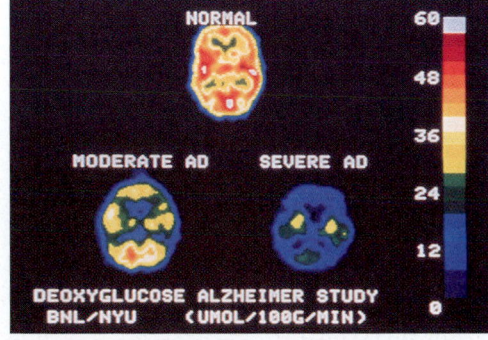

Figure 46.3 PET scans of the brain of a healthy older person and those of patients suffering from Alzheimer's disease. Lighter regions contain higher concentrations of radioactive glucose, indicating higher metabolism rates and therefore increased brain activity. *(Dr. Monty de Leon/New York University Medical Center and National Institute on Aging)*

tion, because glucose metabolizes more rapidly in active areas of the brain, a PET scan can indicate which areas of the brain are involved in the activities in which the patient is engaging at the time of the scan, such as language use, music, and vision.

Hideki Yukawa Japanese physicist (1907–1981) Yukawa was awarded the Nobel Prize in 1949 for predicting the existence of mesons. This photograph of him at work was taken in 1950 in his office at Columbia University. *(UPI/Corbis-Bettman)*

46.3 ▶ MESONS AND THE BEGINNING OF PARTICLE PHYSICS

Physicists in the mid-1930s had a fairly simple view of the structure of matter. The building blocks were the proton, the electron, and the neutron. Three other particles were either known or postulated at the time: the photon, the neutrino, and the positron. Together these six particles were considered the fundamental constituents of matter. With this marvelously simple picture of the world, however, no one was able to answer the following important question: In view of the fact that the protons in any nucleus should strongly repel one another due to their like charges, what is the nature of the force that holds the nucleus together? Scientists recognized that this mysterious force must be much stronger than anything encountered in nature up to that time. This is the nuclear force discussed in Section 44.3 and examined in historical perspective in the following paragraphs.[1]

The first theory to explain the nature of the nuclear force was proposed in 1935 by the Japanese physicist Hideki Yukawa—an effort that earned him a Nobel prize. To understand Yukawa's theory, recall the introduction of field particles in Section 46.1, which stated that each fundamental force is mediated by a field particle exchanged between the interacting particles. Yukawa used this idea to explain the nuclear force, proposing the existence of a new particle whose exchange between nucleons in the nucleus causes the nuclear force. He established that the range of the force is inversely proportional to the mass of this particle and predicted the mass to be about 200 times the mass of the electron. (Yukawa's predicted particle is *not* the gluon mentioned in Section 46.1, which is massless and is today considered to be the field particle for the nuclear force.) Because the new particle would have a mass between that of the electron and that of the proton, it was called a **meson** (from the Greek *meso*, "middle").

In efforts to substantiate Yukawa's predictions, physicists began experimental searches for the meson by studying cosmic rays entering the Earth's atmosphere. In 1937, Carl Anderson and his collaborators discovered a particle of mass 106 MeV/c^2, about 207 times the mass of the electron. This was thought to be Yukawa's meson. However, subsequent experiments showed that the particle interacted very weakly with matter and hence could not be the field particle for the nuclear force. That puzzling situation inspired several theoreticians to propose two mesons having slightly different masses equal to about 200 times that of the electron—one having been discovered by Anderson and the other, still undiscovered, predicted by Yukawa. This idea was confirmed in 1947 with the discovery of the **pi**

[1] The nuclear force that we discussed in Chapter 44 and continue to discuss here was originally called the *strong* force. Once the quark theory (Section 46.9) was established, the phrase *strong force* was reserved for the force between quarks. We shall follow this convention—the strong force occurs between quarks, and the nuclear force occurs between nucleons. The nuclear force is a secondary result of the strong force, as we shall discuss in Section 46.10. Be prepared, however—because of this historical development of the names for these forces, you may find the nuclear force referred to as the strong force in other books.

meson (π), or simply **pion.** The particle discovered by Anderson in 1937, the one initially thought to be Yukawa's meson, is not really a meson. (We shall discuss the requirements for a particle to be a meson in Section 46.4.) Instead, it takes part in the weak and electromagnetic interactions only and is now called the **muon** (μ).

The pion comes in three varieties, corresponding to three charge states: π^+, π^-, and π^0. The π^+ and π^- particles (π^- is the antiparticle of π^+) each have a mass of 139.6 MeV/c^2, and the π^0 mass is 135.0 MeV/c^2. Two muons exist—μ^- and its antiparticle μ^+.

Pions and muons are very unstable particles. For example, the π^-, which has a mean lifetime of 2.6×10^{-8} s, decays to a muon and an antineutrino. The muon, which has a mean lifetime of 2.2 μs, then decays to an electron, a neutrino, and an antineutrino:

$$\pi^- \longrightarrow \mu^- + \bar{\nu} \tag{46.1}$$

$$\mu^- \longrightarrow e^- + \nu + \bar{\nu}$$

Note that for chargeless particles (as well as some charged particles, such as the proton), a bar over the symbol indicates an antiparticle, as in beta decay (see Section 44.6).

The interaction between two particles can be represented in a simple diagram called a **Feynman diagram,** developed by the American physicist Richard P. Feynman. Figure 46.4 is such a diagram for the electromagnetic interaction between two electrons. A Feynman diagram is a qualitative graph of time on the vertical axis versus space on the horizontal axis. It is qualitative in the sense that the actual values of time and space are not important, but the overall appearance of the graph provides a representation of the process. The time evolution of the process can be approximated by starting at the bottom of the diagram and moving your eyes upward.

In the simple case of the electron–electron interaction in Figure 46.4, a photon (the field particle) mediates the electromagnetic force between the electrons. Notice that the entire interaction is represented in the diagram as occurring at a single point in time. Thus, the paths of the electrons appear to undergo a discontinuous change in direction at the moment of interaction. This is different from the *actual* paths, which would be curved due to the continuous exchange of large numbers of field particles. This is another aspect of the qualitative nature of Feynman diagrams.

In the electron–electron interaction, the photon, which transfers energy and momentum from one electron to the other, is called a *virtual photon* because it vanishes during the interaction without having been detected. Virtual photons do not violate the law of conservation of energy because they have a very short lifetime Δt that makes the uncertainty in the energy $\Delta E \approx \hbar/2\,\Delta t$ of the system consisting of two electrons and the photon greater than the photon energy.

Now consider a pion mediating the nuclear force between a proton and a neutron, as in Yukawa's model (Fig. 46.5a). We can reason that the rest energy ΔE_R needed to create a pion of mass m_π is given by Einstein's equation $\Delta E_R = m_\pi c^2$. As with the photon in Figure 46.4, the very existence of the pion would violate the law of conservation of energy if the particle existed for a time greater than $\Delta t \approx \hbar/2\,\Delta E_R$ (from the uncertainty principle), where ΔE_R is the rest energy of the pion and Δt is the time it takes the pion to transfer from one nucleon to the other. Therefore,

$$\Delta t \approx \frac{\hbar}{2\,\Delta E_R} = \frac{\hbar}{2 m_\pi c^2} \tag{46.2}$$

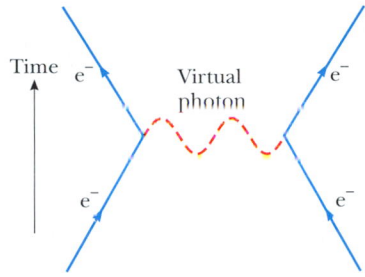

Figure 46.4 Feynman diagram representing a photon mediating the electromagnetic force between two electrons.

Richard Feynman **American physicist (1918–1988)** Feynman with his son, Carl, after winning the Nobel Prize for physics in 1965. The prize was shared by Feynman, Julian Schwinger, and Sin Itiro Tomonaga. *(UPI Telephotos)*

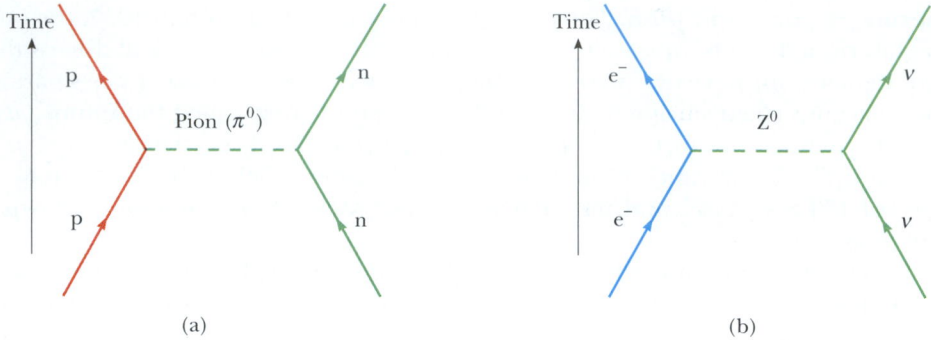

Figure 46.5   (a) Feynman diagram representing a proton and a neutron interacting via the nuclear force with a pion mediating the force. (b) Feynman diagram for an electron and a neutrino interacting via the weak force, with a Z^0 boson mediating the force.

Because the pion cannot travel faster than the speed of light, the maximum distance d it can travel in a time Δt is $c \, \Delta t$. Thus,

$$d = c \, \Delta t \approx \frac{\hbar}{2 m_\pi c} \tag{46.3}$$

From Table 46.1, we know that the range of the nuclear force is approximately 1×10^{-15} m. Using this value for d in Equation 46.3, we estimate the rest energy of the pion to be

$$m_\pi c^2 \approx \frac{\hbar c}{2d} = \frac{(1.05 \times 10^{-34} \, \text{J} \cdot \text{s})(3.00 \times 10^8 \, \text{m/s})}{2(1 \times 10^{-15} \, \text{m})}$$

$$= 1.6 \times 10^{-11} \, \text{J} \approx 100 \, \text{MeV}$$

This is the same order of magnitude as the observed masses of the pions, thus boosting confidence in the field-particle model.

The concept we have just described is quite revolutionary. In effect, it says that a system of two nucleons can change into two nucleons plus a pion, as long as it returns to its original state in a very short time. (Remember that this is the older, historical model in which we assume that the pion is the field particle for the nuclear force; keep in mind that the gluon is the actual field particle in current models, as we shall discuss shortly.) Physicists often say that a nucleon undergoes *fluctuations* as it emits and absorbs pions. As we have seen, these fluctuations are a consequence of a combination of quantum mechanics (through the uncertainty principle) and special relativity (through Einstein's energy–mass relationship $E_R = mc^2$).

This section has dealt with the field particles that were originally proposed to mediate the nuclear force (pions) and those that mediate the electromagnetic force (photons). The graviton, the field particle for the gravitational force, has yet to be observed. The $W^\pm$ and Z^0 particles, which mediate the weak force, were discovered in 1983 by the Italian physicist Carlo Rubbia (b. 1934) and his associates, using a proton–antiproton collider. Rubbia and Simon van der Meer (b. 1925), both at CERN (the European Laboratory for Particle Physics), shared the 1984 Nobel Prize for the discovery of the $W^\pm$ and Z^0 particles and the development of the proton–antiproton collider. Figure 46.5b shows a Feynman diagram for a weak interaction mediated by a Z^0 boson.

What does the infinite range of the electromagnetic and gravitational interactions tell you about the masses of the photon and graviton?

46.4 CLASSIFICATION OF PARTICLES

All particles other than field particles can be classified into two broad categories, hadrons and leptons, according to the interactions in which they take part. Table 46.2 provides a summary of the properties of some of these particles.

Hadrons

Particles that interact through the nuclear force are called **hadrons.** The two classes of hadrons, *mesons* and *baryons*, are distinguished by their masses and spins.

Mesons all have zero or integer spin (0 or 1). As indicated in Section 46.3, the name comes from the expectation that Yukawa's proposed meson mass would lie between the masses of the electron and the proton. Several meson masses do lie in this range, although mesons having masses greater than that of the proton do exist.

All mesons are known to decay finally into electrons, positrons, neutrinos, and photons. The pions are the lightest known mesons; they have masses of about 1.4×10^2 MeV/c^2, and all three pions—π^+, π^-, and π^0—have a spin of 0. (This indicates that the particle discovered by Anderson in 1937, the muon, is not a meson; the muon has spin $\frac{1}{2}$. It belongs in the *lepton* classification, described below.)

Baryons, the second class of hadrons, have masses equal to or greater than the proton mass (the name *baryon* means "heavy" in Greek), and their spin is always a noninteger value ($\frac{1}{2}$ or $\frac{3}{2}$). Protons and neutrons are baryons, as are many other particles. With the exception of the proton, all baryons decay in such a way that the end products include a proton. For example, the baryon called the Ξ hyperon (Greek capital xi) decays to the Λ^0 baryon (Greek capital lambda) in about 10^{-10} s. The Λ^0 then decays to a proton and a π^- in approximately 3×10^{-10} s.

Today it is believed that hadrons are not elementary particles but are composed of more elementary units called quarks, as we shall see in Section 46.9.

Leptons

Leptons (from the Greek *leptos,* meaning "small" or "light") are a group of particles that do not interact by means of the nuclear force. All leptons have spin $\frac{1}{2}$. Whereas hadrons have size and structure, leptons appear to be truly elementary, meaning that they have no structure and are point-like.

Quite unlike the case with hadrons, the number of known leptons is small. Currently, scientists believe that only six leptons exist—the electron, the muon, the tau, and a neutrino associated with each:

$$\begin{pmatrix} e^- \\ \nu_e \end{pmatrix} \quad \begin{pmatrix} \mu^- \\ \nu_\mu \end{pmatrix} \quad \begin{pmatrix} \tau^- \\ \nu_\tau \end{pmatrix}$$

The tau lepton, discovered in 1975, has a mass about twice that of the proton. The neutrino associated with the tau has not yet been observed in the laboratory. Each of the six leptons has an antiparticle.

TABLE 46.2 Some Particles and Their Properties

Category	Particle Name	Symbol	Antiparticle	Mass (MeV/c^2)	B	L_e	L_μ	L_τ	S	Lifetime(s)	Principal Decay Modes[a]
Leptons	Electron	e^-	e^+	0.511	0	+1	0	0	0	Stable	—
	Electron neutrino	ν_e	$\bar{\nu}_e$	$<7\times10^{-6}$	0	+1	0	0	0	Stable	—
	Muon	μ^-	μ^+	105.7	0	0	+1	0	0	2.20×10^{-6}	$e^-\bar{\nu}_e\nu_\mu$
	Muon neutrino	ν_μ	$\bar{\nu}_\mu$	<0.3	0	0	+1	0	0	Stable	—
	Tau	τ^-	τ^+	1784	0	0	0	+1	0	$<4\times10^{-13}$	$\mu^-\bar{\nu}_\mu\nu_\tau,\ e^-\bar{\nu}_e\nu_\tau$
	Tau neutrino	ν_τ	$\bar{\nu}_\tau$	<30	0	0	0	+1	0	Stable	—
Hadrons											
Mesons	Pion	π^+	π^-	139.6	0	0	0	0	0	2.60×10^{-8}	$\mu^+\nu_\mu$
		π^0	Self	135.0	0	0	0	0	0	0.83×10^{-16}	2γ
	Kaon	K^+	K^-	493.7	0	0	0	0	+1	1.24×10^{-8}	$\mu^+\nu_\mu,\ \pi^+\pi^0$
		K^0_S	$\bar{K}^0_S$	497.7	0	0	0	0	+1	0.89×10^{-10}	$\pi^+\pi^-,\ 2\pi^0$
		K^0_L	$\bar{K}^0_L$	497.7	0	0	0	0	+1	5.2×10^{-8}	$\pi^\pm e^\mp\bar{\nu}_e,\ 3\pi^0$ $\pi^\pm\mu^\mp\nu_\mu$
	Eta	η	Self	548.8	0	0	0	0	0	$<10^{-18}$	$2\gamma,\ 3\pi$
		η'	Self	958	0	0	0	0	0	2.2×10^{-21}	$\eta\pi^+\pi^-$
Baryons	Proton	p	$\bar{p}$	938.3	+1	0	0	0	0	Stable	$pe^-\bar{\nu}_e$
	Neutron	n	$\bar{n}$	939.6	+1	0	0	0	0	920	$pe^-\bar{\nu}_e$
	Lambda	Λ^0	$\bar{\Lambda}^0$	1 115.6	+1	0	0	0	−1	2.6×10^{-10}	$p\pi^-,\ n\pi^0$
	Sigma	Σ^+	$\bar{\Sigma}^-$	1 189.4	+1	0	0	0	−1	0.80×10^{-10}	$p\pi^0,\ n\pi^+$
		Σ^0	$\bar{\Sigma}^0$	1 192.5	+1	0	0	0	−1	6×10^{-20}	$\Lambda^0\gamma$
		Σ^-	$\bar{\Sigma}^+$	1 197.3	+1	0	0	0	−1	1.5×10^{-10}	$n\pi^-$
	Xi	Ξ^0	$\bar{\Xi}^0$	1 315	+1	0	0	0	−2	2.9×10^{-10}	$\Lambda^0\pi^0$
		Ξ^-	$\bar{\Xi}^+$	1 321	+1	0	0	0	−2	1.64×10^{-10}	$\Lambda^0\pi^-$
	Omega	Ω^-	Ω^+	1 672	+1	0	0	0	−3	0.82×10^{-10}	$\Xi^0\pi^0,\ \Lambda^0 K^-$

[a] Notations in this column such as $p\pi^-$, $n\pi^0$ mean two possible decay modes. In this case, the two possible decays are $\Lambda^0 \to p + \pi^-$ and $\Lambda^0 \to n + \pi^0$.

Current studies indicate that neutrinos have a small but nonzero mass. If they do have mass, then they cannot travel at the speed of light. Also, so many neutrinos exist that their combined mass may be sufficient to cause all the matter in the Universe to eventually collapse into a single point, which might then explode and create a completely new Universe! We shall discuss this possibility in more detail in Section 46.12.

46.5 CONSERVATION LAWS

In Chapter 44 we learned that conservation laws are important for understanding why certain radioactive decays and nuclear reactions occur and others do not. In general, the laws of conservation of energy, linear momentum, angular momentum, and electric charge provide us with a set of rules that all processes must follow. In the study of elementary particles, a number of additional conservation laws are important. Although the two described here have no theoretical foundation, they are supported by abundant empirical evidence.

Baryon Number

The law of conservation of baryon number tells us that whenever a baryon is created in a decay or nuclear reaction, an antibaryon is also created. This scheme can be quantified by assigning every particle a quantum number, the **baryon number,** as follows: $B = +1$ for all baryons, $B = -1$ for all antibaryons, and $B = 0$ for all other particles. Thus, the **law of conservation of baryon number** states that **whenever a nuclear reaction or decay occurs, the sum of the baryon numbers before the process must equal the sum of the baryon numbers after the process.**

Conservation of baryon number

If baryon number is absolutely conserved, the proton must be absolutely stable. If it were not for the law of conservation of baryon number, the proton could decay to a positron and a neutral pion. However, such a decay has never been observed. At the present, all we can say is that the proton has a half-life of at least 10^{33} years (the estimated age of the Universe is only 10^{10} years). In one recent theory, however, physicists predicted that the proton is unstable. According to this theory, baryon number is not absolutely conserved.

EXAMPLE 46.1 ▶ Checking Baryon Numbers

Use the law of conservation of baryon number to determine whether the following reactions can occur: (a) $p + n \rightarrow p + p + n + \bar{p}$; (b) $p + n \rightarrow p + p + \bar{p}$.

Solution (a) The left side of the equation gives a total baryon number of $1 + 1 = 2$. The right side gives a total baryon number of $1 + 1 + 1 + (-1) = 2$. Thus, baryon number is conserved and the reaction can occur (provided the incoming proton has sufficient energy that energy conservation is satisfied).

(b) The left side of the equation gives a total baryon number of $1 + 1 = 2$. However, the right side gives $1 + 1 + (-1) = 1$. Because baryon number is not conserved, the reaction cannot occur.

EXAMPLE 46.2   Detecting Proton Decay

Measurements taken at the Super Kamiokande neutrino detection facility (Fig. 46.6) indicate that the half-life of protons is at least 10^{33} years. Estimate how long we would have to watch, on average, to observe the decay of a proton in a glass of water.

Solution   Let us estimate that a glass contains about $\frac{1}{4}$ L, or 250 g, of water. The number of molecules of water is

$$\frac{(250\,\text{g})(6.02 \times 10^{23}\,\text{molecules/mol})}{18\,\text{g/mol}} = 8.4 \times 10^{24}\,\text{molecules}$$

Each water molecule contains one proton in each of its two hydrogen atoms plus eight protons in its oxygen atom. Thus, the glass of water contains 8.4×10^{25} protons. The decay constant is given by Equation 44.9:

$$\lambda = \frac{0.693}{T_{1/2}} = \frac{0.693}{10^{33}\,\text{yr}} = 6.9 \times 10^{-34}\,\text{yr}^{-1}$$

This is the probability that any one proton will decay in one year. The probability that any proton in our glass of water will decay in the one-year interval is (Eqs. 44.6 and 44.8)

$$R = (8.4 \times 10^{25})(6.9 \times 10^{-34}\,\text{yr}^{-1}) = 5.8 \times 10^{-8}\,\text{yr}^{-1}$$

So we have to watch our glass of water for $1/R \approx$

17 million years!

Exercise   The Super Kamiokande neutrino facility contains 50 000 tons of water. Estimate the average time between detected proton decays if the half-life is 10^{33} yr.

Answer   Approximately 1 yr.

Figure 46.6   This detector at the Super Kamiokande neutrino facility in Japan is used to study photons and neutrinos. It holds 50 000 tons of highly purified water and 13 000 photomultipliers. The photograph was taken while the detector was being filled. Technicians use a raft to clean the photodetectors before they are submerged. (*Courtesy of KRR [Institute for Cosmic Ray Research], University of Tokyo*)

Lepton Number

Conservation of lepton number

We have three conservation laws involving lepton numbers, one for each variety of lepton. The **law of conservation of electron lepton number** states that whenever a nuclear reaction or decay occurs, **the sum of the electron lepton numbers before the process must equal the sum of the electron lepton numbers after the process.**

The electron and the electron neutrino are assigned an electron lepton number $L_e = +1$, the antileptons e^+ and $\bar{\nu}_e$ are assigned an electron lepton number $L_e = -1$, and all other particles have $L_e = 0$. For example, consider the decay of the neutron:

$$n \longrightarrow p + e^- + \bar{\nu}_e$$

Before the decay, the electron lepton number is $L_e = 0$; after the decay, it is $0 + 1 + (-1) = 0$. Thus, electron lepton number is conserved. (Baryon number must also be conserved, of course, and it is: Before the decay $B = +1$, and after the decay $B = +1 + 0 + 0 = +1$.)

Similarly, when a decay involves muons, the muon lepton number L_μ is conserved. The μ^- and the ν_μ are assigned a muon lepton number $L_\mu = +1$, the antimuon μ^+ and the muon antineutrino $\bar{\nu}_\mu$ are assigned a muon lepton number $L_\mu = -1$, and all other particles have $L_\mu = 0$.

Finally, tau lepton number L_τ is conserved with similar assignments for the tau lepton, its neutrino, and other particles.

EXAMPLE 46.3 ▶ **Checking Lepton Numbers**

Use the law of conservation of lepton numbers to determine which of the following decay schemes can occur: (a) $\mu^- \rightarrow e^- + \bar{\nu}_e + \nu_\mu$; (b) $\pi^+ \rightarrow \mu^+ + \nu_\mu + \nu_e$.

Solution (a) Because this decay involves a muon and an electron, L_μ and L_e must both be conserved. Before the decay, $L_\mu = +1$ and $L_e = 0$. After the decay, $L_\mu = 0 + 0 + 1 = +1$ and $L_e = +1 + (-1) + 0 = 0$. Thus, both numbers are conserved, and on this basis the decay is possible.

(b) Before the decay, $L_\mu = 0$ and $L_e = 0$. After the decay, $L_\mu = -1 + 1 + 0 = 0$, but $L_e = 0 + 0 + 1 = 1$. Thus, the decay is not possible because electron lepton number is not conserved.

Exercise Determine whether the decay $\mu^- \rightarrow e^- + \bar{\nu}_e$ can occur.

Answer No, because L_μ is $+1$ before the decay and 0 after.

Quick Quiz 46.3 ▶

A scientist claims to have observed the decay of an electron into two electron neutrinos. Is this believable?

46.6 ▶ **STRANGE PARTICLES AND STRANGENESS**

Many particles discovered in the 1950s were produced by the interaction of pions with protons and neutrons in the atmosphere. A group of these—the kaon (K), lambda (Λ), and sigma (Σ) particles—exhibited unusual properties both as they were created and as they decayed; hence, they were called *strange particles*.

One unusual property of strange particles is that they are always produced in pairs. For example, when a pion collides with a proton, a highly probable result is the production of two neutral strange particles (Fig. 46.7):

$$\pi^- + p \longrightarrow K^0 + \Lambda^0$$

However, the reaction $\pi^- + p \rightarrow K^0 + n^0$, in which only one of the final particles is strange, never occurs, even though no known conservation laws would be violated and even though the energy of the pion is sufficient to initiate the reaction.

The second peculiar feature of strange particles is that, although they are produced in reactions involving the nuclear interaction at a high rate, they do not decay into particles that interact via the nuclear force at a high rate. Instead, they decay very slowly, which is characteristic of the weak interaction. Their half-lives are in the range 10^{-10} s to 10^{-8} s, whereas most other particles that interact via the nuclear force have lifetimes on the order of 10^{-23} s.

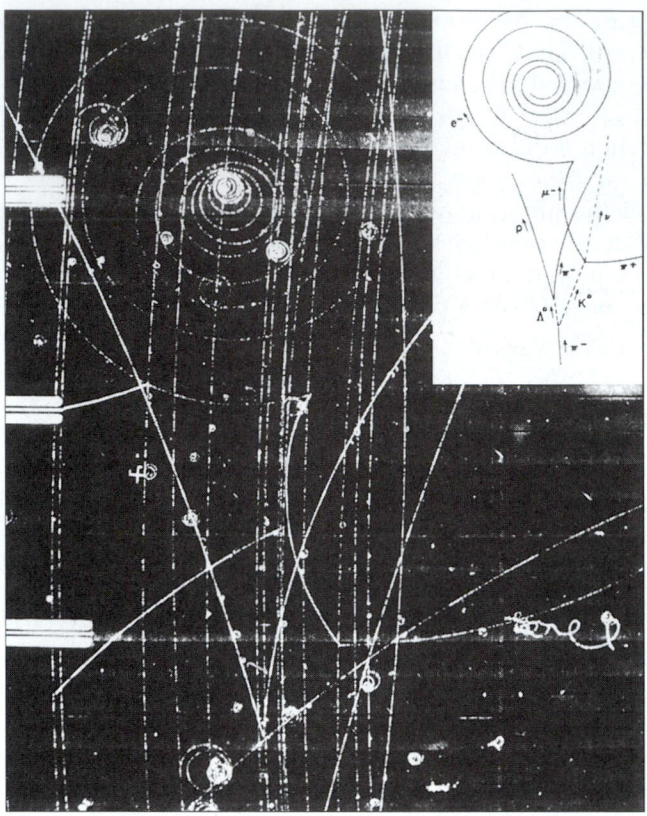

Figure 46.7   This bubble-chamber photograph shows many events, and the inset is a drawing of identified tracks. The strange particles Λ^0 and K^0 are formed at the bottom as a π^- particle interacts with a proton in the reaction $\pi^- + p \rightarrow \Lambda^0 + K^0$. (Note that the neutral particles leave no tracks, as indicated by the dashed lines in the inset.) The Λ^0 then decays in the reaction $\Lambda^0 \rightarrow \pi^- + p$, and the K^0 decays in the reaction $K^0 \rightarrow \pi^+ + \mu^- + \overline{\nu}_\mu$. *(Courtesy Lawrence Berkeley Laboratory, University of California, Photographic Services)*

To explain these unusual properties of strange particles, a new quantum number S, called **strangeness,** was introduced, together with a conservation law. The strangeness numbers for some particles are given in Table 46.2. The production of strange particles in pairs is explained by assigning $S = +1$ to one of the particles, $S = -1$ to the other, and $S = 0$ to all nonstrange particles. The **law of conservation of strangeness** states that **whenever a nuclear reaction or decay occurs, the sum of the strangeness numbers before the process must equal the sum of the strangeness numbers after the process.**

The low decay rate of strange particles can be explained by assuming that the nuclear and electromagnetic interactions obey the law of conservation of strangeness but the weak interaction does not. Because the decay of a strange particle involves the loss of one strange particle, it violates strangeness conservation and hence proceeds slowly via the weak interaction.

> Conservation of strangeness number

EXAMPLE 46.4   **Is Strangeness Conserved?**

(a) Use the law of conservation of strangeness to determine whether the reaction $\pi^0 + n \rightarrow K^+ + \Sigma^-$ occurs.

Solution   From Table 46.2, we see that the initial strangeness is $S = 0 + 0 = 0$. Because the strangeness of the K^+ is

$S = +1$ and the strangeness of the Σ^- is $S = -1$, the strangeness of the final products is $+1 - 1 = 0$. Thus, strangeness is conserved, and the reaction is allowed.

(b) Show that the reaction $\pi^- + p \rightarrow \pi^- + \Sigma^+$ does not conserve strangeness.

Solution Before: $S = 0 + 0 = 0$; after: $S = 0 + (-1) = -1$. Thus, strangeness is not conserved.

Exercise Show that the reaction $p + \pi^- \rightarrow K^0 + \Lambda^0$ obeys the law of conservation of strangeness.

46.7 MAKING PARTICLES AND MEASURING THEIR PROPERTIES

The bewildering array of entries in Table 46.2 leaves one yearning for firm ground. For example, it is natural to wonder about an entry for a particle (Σ^0) that exists for 10^{-20} s and has a mass of 1 192.5 MeV/c^2. How is it possible to detect a particle that exists for only 10^{-20} s? Furthermore, how can its mass be measured when it exists for such a short time? If a standard attribute of a particle is some type of permanence or stability, in what sense is such a fleeting entity a particle? In this section we answer such questions and explain how elementary particles are produced and their properties measured.

Most elementary particles are unstable and are created in nature only rarely, in cosmic ray showers. In the laboratory, however, great numbers of these particles are created in controlled collisions between high-energy particles and a suitable target. The incident particles must have very high energy, and it takes considerable time for electromagnetic fields to accelerate particles to high energies. Thus, stable charged particles such as electrons or protons generally make up the incident beam. In addition, targets must be simple and stable, and the simplest target, hydrogen, serves nicely as both a target (the proton) and a detector.

Figure 46.7 documents a typical event in which a bubble chamber served as both target source and detector. Many parallel tracks of negative pions are visible entering the photograph from the bottom. As the labels in the inset drawing show, one of the pions has hit a stationary proton in the hydrogen and produced two strange particles, Λ^0 and K^0, according to the reaction

$$\pi^- + p \longrightarrow \Lambda^0 + K^0$$

Neither neutral strange particle leaves a track, but their subsequent decay into charged particles can be seen in Figure 46.7. A magnetic field directed into the plane of the page causes the track of each charged particle to curve, and from the measured curvature we can determine the particle's charge and linear momentum. If the mass and momentum of the incident particle are known, we can then usually calculate the product particle's mass, kinetic energy, and speed from the laws of conservation of momentum and energy. Finally, combining a product particle's speed with the length of the track it leaves, we can calculate the particle's lifetime. Figure 46.7 shows that sometimes we can use this lifetime technique even for a neutral particle, which leaves no track. As long as the start and finish of the missing track are known, as well as the particle speed, we can infer the missing track length and so determine the lifetime of the neutral particle.

Resonance Particles

With clever experimental technique and much effort, decay track lengths as short as 10^{-6} m can be measured. This means that lifetimes as short as 10^{-16} s can be measured for high-energy particles traveling at about the speed of light. We arrive at this result by assuming that a decaying particle travels 1 μm at a speed of $0.99c$ in the reference frame of the laboratory, yielding a lifetime of $\Delta t_{\text{lab}} = 1 \times 10^{-6}\,\text{m}/0.99c \approx 3.4 \times 10^{-15}$ s. This is not our final result, however, because we must account for the relativistic effects of time dilation. Because the proper lifetime Δt_p as measured in the decaying particle's reference frame is shorter than the laboratory-frame value Δt_{lab} by a factor of $\sqrt{1 - (v^2/c^2)}$ (see Eq. 39.7), we can calculate the proper lifetime:

$$\Delta t_p = \Delta t_{\text{lab}} \sqrt{1 - \frac{v^2}{c^2}} = (3.4 \times 10^{-15}\,\text{s}) \sqrt{1 - \frac{(0.99c)^2}{c^2}} = 4.8 \times 10^{-16}\,\text{s}$$

Unfortunately, even with Einstein's help, the best answer we can obtain with the track-length method is several orders of magnitude away from lifetimes of 10^{-20} s. How, then, can we detect the presence of particles that exist for time intervals such as 10^{-20} s? As we shall soon see, for such short-lived particles, known as **resonance particles,** all we can do is infer their masses, their lifetimes, and their very existence from data on their decay products.

Let us consider this detection process in detail by examining the case of the resonance particle called the delta plus (Δ^+), which has a mass of 1 232 MeV/c^2 and a lifetime of about 10^{-23} s, even shorter than the most short-lived particle listed in Table 46.2. This particle is produced in the reaction

$$e^- + p \longrightarrow e^- + \Delta^+ \tag{46.4}$$

followed in 10^{-23} s by the decay

$$\Delta^+ \longrightarrow \pi^+ + n \tag{46.5}$$

Because the Δ^+ lifetime is so short, the particle leaves no measurable track in a bubble chamber. It might therefore seem impossible to distinguish the reactions given in Equations 46.4 and 46.5 from the reaction

$$e^- + p \longrightarrow e^- + \pi^+ + n \tag{46.6}$$

in which the reactants of Equation 46.4 decay directly to e^-, π^+, and n with no intermediate step in which a Δ^+ is produced. Distinguishing between these two possibilities is not impossible, however. If a Δ^+ particle exists, it has a distinct rest energy, which must come from the kinetic energy of the incoming particles. If we imagine firing electrons with increasing kinetic energy at protons, eventually we will provide enough energy to the system to create the Δ^+ particle. This is very similar to firing photons of increasing energy at an atom until you fire them with enough energy to excite the atom to a higher quantum state. In fact, the Δ^+ particle is an excited state of the proton, which we can understand via the quark theory discussed in Section 46.9. After the Δ^+ particle is formed, its rest energy becomes the energies of the outgoing pion and neutron. Equation 39.26 can be solved for the rest energy of the Δ^+ particle in terms of its kinetic energy and linear momentum:

$$(m_{\Delta^+}c^2)^2 = E_{\Delta^+}{}^2 - p_{\Delta^+}{}^2 c^2 = E_{\Delta^+}{}^2 - (\mathbf{p}_{\Delta^+})^2 c^2$$

When the Δ^+ particle decays into a pion and a neutron, conservation of energy and momentum requires that

$$E_{\Delta^+} = E_{\pi^+} + E_n \qquad \mathbf{p}_{\Delta^+} = \mathbf{p}_{\pi^+} + \mathbf{p}_n$$

Thus, the rest energy of the Δ^+ particle can be expressed in terms of the energies and momenta of the outgoing particles, which can all be measured in the bubble-

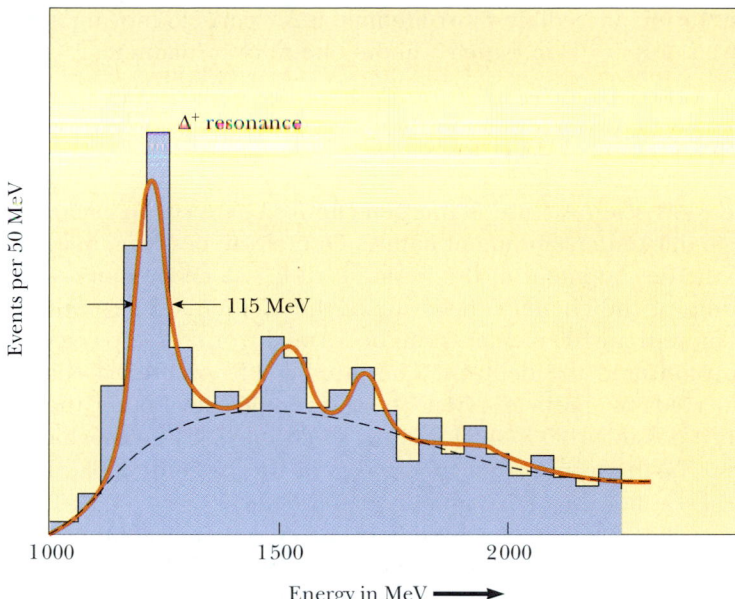

Figure 46.8   Experimental evidence for the existence of the Δ^+ particle. The sharp peak at 1 232 MeV was produced by events in which a Δ^+ formed and promptly decayed to a π^+ and a neutron.

chamber photograph:

$$(m_{\Delta^+}c^2)^2 = (E_{\pi^+} + E_n)^2 - (\mathbf{p}_{\pi^+} + \mathbf{p}_n)^2 c^2$$

Any pions and neutrons that come from the decay of a Δ^+ particle must have energies and momenta that combine in this equation to give the rest energy of the Δ^+ particle. Pions and neutrons coming from the reaction of Equation 46.6 will have a variety of energies and momenta with no particular pattern because the energy of the reactants can divide up in many ways among the three outgoing particles in this reaction. At the energy at which the rest energy of the Δ^+ particle can be created, many reactions occur, as evidenced by the proper combinations of energy and momentum already described.

To show the existence of the Δ^+ particle, we analyze a great number of events in which a π^+ and a neutron are produced. Then the number of events in a given energy range is plotted versus energy. Following this procedure, we obtain a slowly varying curve that has a sharp peak superimposed on it. The peak represents the incident electron energy at which the rest energy of the Δ^+ particle was created, revealing the existence of the particle.

Figure 46.8 is an experimental plot for the Δ^+ particle. The dashed broad curve is produced by direct events in which no Δ^+ was created (see Eq. 46.6). The sharp peak at 1 232 MeV was produced by all the events in which a Δ^+ was formed and decayed to a pion and a neutron. Thus, the rest energy of the Δ^+ particle is 1 232 MeV. Peaks corresponding to two resonance particles with masses greater than that of the Δ^+ particle can also be seen in Figure 46.8.

Graphs such as Figure 46.8 can tell us not only the mass of a short-lived particle but also its lifetime. The width of the resonance peak and the uncertainty relationship $\Delta E\, \Delta t \approx \hbar/2$ are used to infer the lifetime Δt of the particle. The measured width of 115 MeV in Figure 46.8 leads to a lifetime of 0.57×10^{-23} s for the

Δ^+ particle. In this incredibly short lifetime, a Δ^+ particle moving at the limiting speed c travels only 10^{-15} m, which is about one nuclear diameter.

46.8 ▶ FINDING PATTERNS IN THE PARTICLES

One of the tools scientists use is the detection of patterns in data, patterns that contribute to our understanding of nature. One of the best examples of the use of this tool is the development of the periodic table, which provides a fundamental understanding of the chemical behavior of the elements. The periodic table explains how more than 100 elements can be formed from three particles—the electron, the proton, and the neutron. The number of observed particles and resonances observed by particle physicists is even greater than the number of elements. Is it possible that a small number of entities exist from which all of these can be built? Taking a hint from the success of the periodic table, let us explore the historical search for patterns among the particles.

Many classification schemes have been proposed for grouping particles into families. Consider, for instance, the baryons listed in Table 46.2. We can consider these particles as belonging to a group based on the fact that they all have spins of $\frac{1}{2}$. If we plot strangeness versus charge for these baryons using a sloping coordinate system, as in Figure 46.9a, we observe a fascinating pattern: Six of the baryons form a hexagon, and the remaining two are at the hexagon's center.

As a second example, consider the following nine spin-zero mesons listed in Table 46.2: π^+, π^0, π^-, K^+, K^0, K^-, η, η', and the antiparticle $\overline{K}^0$. Figure 46.9b is a plot of strangeness versus charge for this family. Again, a hexagonal pattern emerges. In this case, each particle on the perimeter of the hexagon lies opposite its antiparticle, and the remaining three (which form their own antiparticles) are at the center of the hexagon. These and related symmetric patterns were devel-

Murray Gell-Mann American physicist (b. 1929) Murray Gell-Mann was awarded the Nobel Prize in 1969 for his theoretical studies dealing with subatomic particles.
(Courtesy of Michael R. Dressler)

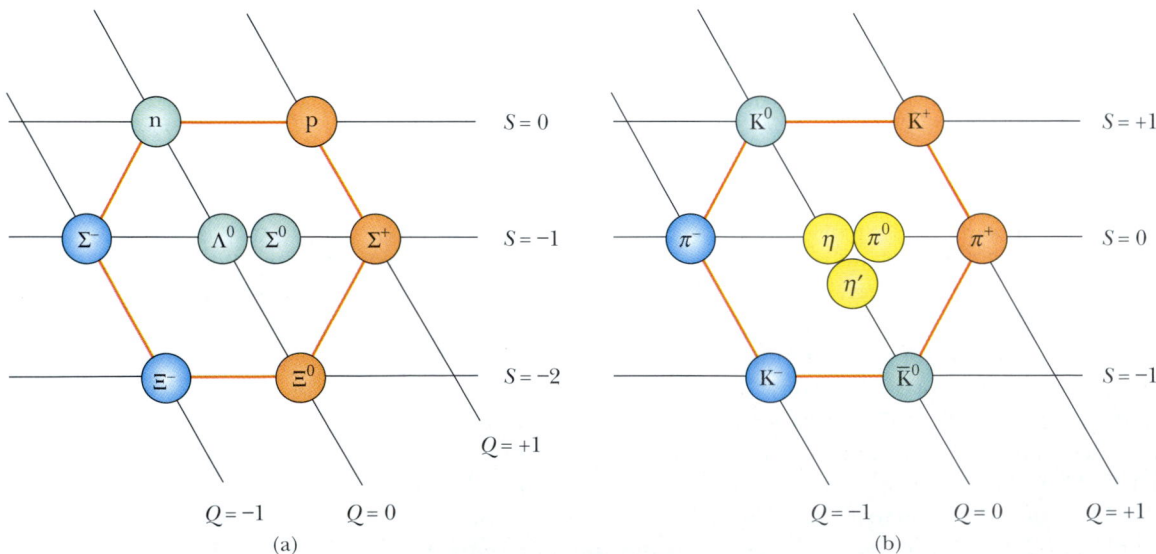

Figure 46.9 (a) The hexagonal eightfold-way pattern for the eight spin-$\frac{1}{2}$ baryons. This strangeness-versus-charge plot uses a sloping axis for charge number Q and a horizontal axis for strangeness S. (b) The eightfold-way pattern for the nine spin-zero mesons.

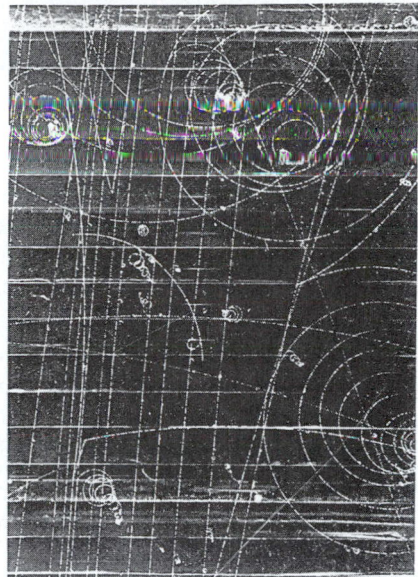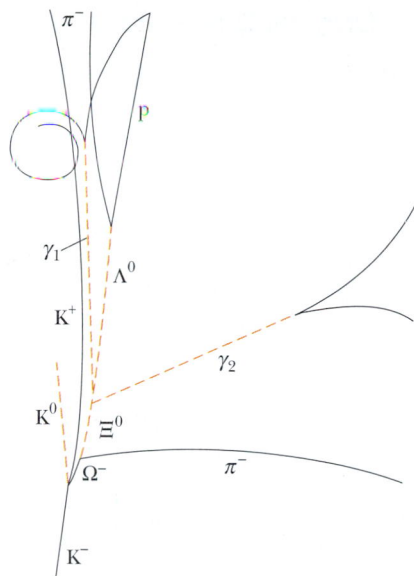

Figure 46.10 Discovery of the Ω^- particle. The K^- particle at the bottom of the photograph collides with a proton to produce the first detected Ω^- particle plus two other particles. *(Courtesy of Brookhaven National Laboratory.)*

oped independently in 1961 by Murray Gell-Mann and Yuval Ne'eman. Gell-Mann called the patterns the **eightfold way,** after the eightfold path to nirvana in Buddhism.

Groups of baryons and mesons can be displayed in many other symmetric patterns within the framework of the eightfold way. For example, the family of spin-$\frac{3}{2}$ baryons contains ten particles arranged in a pattern like that of the pins in a bowling alley. After this pattern was proposed, an empty spot occurred in it, corresponding to a particle that had never been observed. Gell-Mann predicted that the missing particle, which he called the omega minus (Ω^-), should have spin $\frac{3}{2}$, charge -1, strangeness -3, and rest energy $\approx 1\,680$ MeV. Shortly thereafter, in 1964, scientists at the Brookhaven National Laboratory found the missing particle through careful analyses of bubble-chamber photographs (Fig. 46.10) and confirmed all its predicted properties.

The prediction of the particle missing from the eightfold way has much in common with the prediction of missing elements in the periodic table. Whenever a vacancy occurs in an organized pattern of information, experimentalists have a guide for their investigations.

46.9 QUARKS—FINALLY

As we have noted, leptons appear to be truly elementary particles because there are only a few types of them, and they have no measurable size or internal structure. Hadrons, on the other hand, are complex particles having size and structure. The existence of the strangeness-charge patterns of the eightfold way suggests that hadrons have substructure. Furthermore, we know that hundreds of types of hadrons exist and that many of them decay into other hadrons.

The Original Quark Model

In 1963 Gell-Mann and George Zweig independently proposed a model for the substructure of hadrons. According to their model, all hadrons are composite systems of two or three elementary constituents called **quarks.** (Gell-Mann borrowed the word *quark* from the passage "Three quarks for Muster Mark" in James Joyce's *Finnegans Wake.*) The model had three types of quarks, designated by the symbols u, d, and s. These are given the arbitrary names **up, down,** and **strange.** The various types of quarks are called **flavors.** Figure 46.11 is a pictorial representation of the quark compositions of several hadrons.

An unusual property of quarks is that they carry a fractional electronic charge. The u, d, and s quarks have charges of $+2e/3$, $-e/3$, and $-e/3$, respectively. These and other properties of quarks and antiquarks are given in Table 46.3. Notice that quarks have spin $\frac{1}{2}$, which means that all quarks are **fermions,** defined as any particle having half integral spin. As Table 46.3 shows, associated with each quark is an antiquark of opposite charge, baryon number, and strangeness.

The compositions of all hadrons known when Gell-Mann and Zweig presented their model could be completely specified by three simple rules:

- A meson consists of one quark and one antiquark, giving it a baryon number of 0, as required.
- A baryon consists of three quarks.
- An antibaryon consists of three antiquarks.

The theory put forth by Gell-Mann and Zweig is referred to as the *original quark model.*

Quick Quiz 46.4

We have seen a law of lepton number conservation and a law of baryon number conservation. Why is there no law of meson number conservation? (*Hint:* Imagine creating particle-antiparticle pairs from energy, and focus on the creation of a quark–antiquark pair.)

Charm and Other Developments

Although the original quark model was highly successful in classifying particles into families, some discrepancies occurred between its predictions and certain experimental decay rates. Consequently, several physicists proposed a fourth quark flavor in 1967. They argued that if four types of leptons exist (as was thought at the time), then there should also be four flavors of quarks because of an underlying symmetry in nature. The fourth quark, designated c, was assigned a property called **charm.** A *charmed* quark has charge $+2e/3$, just as the up quark does, but its charm distinguishes it from the other three quarks. This introduces a new quantum number C, representing charm. The new quark has charm $C = +1$, its antiquark has charm $C = -1$, and all other quarks have $C = 0$. Charm, like strangeness, is conserved in nuclear and electromagnetic interactions but not in weak interactions.

Evidence that the charmed quark exists began to accumulate in 1974, when a heavy meson called the J/Ψ particle (or simply Ψ, uppercase Greek psi) was discovered independently by two groups, one led by Burton Richter (b. 1931) at the Stanford Linear Accelerator (SLAC), and the other led by Samuel Ting (b. 1936) at the Brookhaven National Laboratory. In 1976 Richter and Ting were awarded a Nobel Prize for this work. The J/Ψ particle does not fit into the three-quark

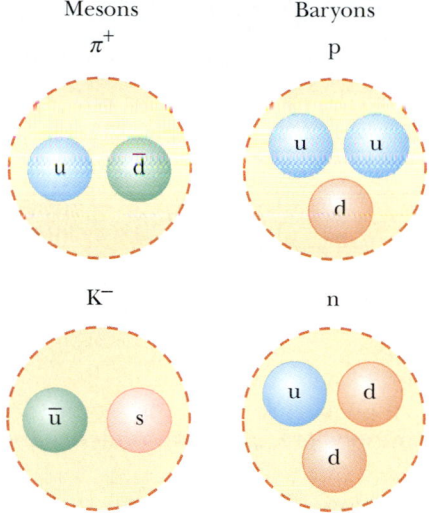

Mesons
π^+

K^-

Baryons
p

n

Figure 46.11 Quark composition of two mesons and two baryons.

model; instead, it has properties of a combination of the proposed charmed quark and its antiquark ($c\bar{c}$). It is much more massive than the other known mesons ($\sim 3\,100$ MeV/c^2), and its lifetime is much longer than the lifetimes of particles that interact via the nuclear force. Soon, related mesons were discovered, corresponding to such quark combinations as $\bar{c}d$ and $c\bar{d}$, all of which have great masses

TABLE 46.3		**Properties of Quarks and Antiquarks**						
				Quarks				
Name	Symbol	Spin	Charge	Baryon Number	Strangeness	Charm	Bottomness	Topness
Up	u	$\frac{1}{2}$	$+\frac{2}{3}e$	$\frac{1}{3}$	0	0	0	0
Down	d	$\frac{1}{2}$	$-\frac{1}{3}e$	$\frac{1}{3}$	0	0	0	0
Strange	s	$\frac{1}{2}$	$-\frac{1}{3}e$	$\frac{1}{3}$	-1	0	0	0
Charmed	c	$\frac{1}{2}$	$+\frac{2}{3}e$	$\frac{1}{3}$	0	$+1$	0	0
Bottom	b	$\frac{1}{2}$	$-\frac{1}{3}e$	$\frac{1}{3}$	0	0	$+1$	0
Top	t	$\frac{1}{2}$	$+\frac{2}{3}e$	$\frac{1}{3}$	0	0	0	$+1$
				Antiquarks				
Name	Symbol	Spin	Charge	Baryon Number	Strangeness	Charm	Bottomness	Topness
Anti-up	$\bar{u}$	$\frac{1}{2}$	$-\frac{2}{3}e$	$-\frac{1}{3}$	0	0	0	0
Anti-down	$\bar{d}$	$\frac{1}{2}$	$+\frac{1}{3}e$	$-\frac{1}{3}$	0	0	0	0
Anti-strange	$\bar{s}$	$\frac{1}{2}$	$+\frac{1}{3}e$	$-\frac{1}{3}$	$+1$	0	0	0
Anti-charmed	$\bar{c}$	$\frac{1}{2}$	$-\frac{2}{3}e$	$-\frac{1}{3}$	0	-1	0	0
Anti-bottom	$\bar{b}$	$\frac{1}{2}$	$+\frac{1}{3}e$	$-\frac{1}{3}$	0	0	-1	0
Anti-top	$\bar{t}$	$\frac{1}{2}$	$-\frac{2}{3}e$	$-\frac{1}{3}$	0	0	0	-1

TABLE 46.4   Quarkᵃ Composition of Mesons

		b	c	s	d	u
Quarks	b	Υ^{-} $(\bar{b}b)$				
	c	B_c^{+} $(\bar{b}c)$	J/Ψ $(\bar{c}c)$			
	s	B_s^{0} $(\bar{b}s)$	D_s^{-} $(\bar{c}s)$	η, η' $(\bar{s}s)$	$\bar{K}^{0}$ $(\bar{d}s)$	K^{-} $(\bar{u}s)$
	d	B^{0} $(\bar{b}d)$	D^{-} $(\bar{c}d)$	K^{0} $(\bar{s}d)$	π^{0}, η, η' $(\bar{d}d)$	π^{-} $(\bar{u}d)$
	u	B^{+} $(\bar{b}u)$	$\bar{D}^{0}$ $(\bar{c}u)$	K^{+} $(\bar{s}u)$	π^{+} $(\bar{d}u)$	π^{0}, η, η' $(\bar{u}u)$

From D. Kestenbaum, "Physicists Find the Last of the Mesons," *Science* 280:35, 1998.

ᵃ The top quark does not form mesons because it decays too quickly.

and long lifetimes. The existence of these new mesons provided firm evidence for the fourth quark flavor.

In 1975, researchers at Stanford University reported strong evidence for the tau (τ) lepton, mass 1 784 MeV/c^2. This was the fifth type of lepton, which led physicists to propose that more flavors of quarks might exist, on the basis of symmetry arguments similar to those leading to the proposal of the charmed quark. These proposals led to more elaborate quark models and the prediction of two new quarks, **top** (t) and **bottom** (b). (Some physicists prefer *truth* and *beauty*.) To distinguish these quarks from the others, quantum numbers called *topness* and *bottomness* (with allowed values $+1, 0, -1$) were assigned to all quarks and antiquarks (see Table 46.3). In 1977, researchers at the Fermi National Laboratory, under the direction of Leon Lederman (b. 1922), reported the discovery of a very massive new meson Υ^{-} (Greek capital upsilon), whose composition is considered to be $b\bar{b}$, providing evidence for the bottom quark. In March 1995, researchers at Fermilab announced the discovery of the top quark (supposedly the last of the quarks that will be found), which has a mass of 173 GeV/c^2.

Table 46.4 lists the quark compositions of mesons formed from the up, down, strange, charmed, and bottom quarks. The meson formed by the combination of the bottom antiquark and the charmed quark ($\bar{b}c$) was the last to be found. It was discovered in 1998.[2] Table 46.5 shows the quark combinations for the baryons

TABLE 46.5   Quark Composition of Several Baryons

Particle	Quark Composition
p	uud
n	udd
Λ^{0}	uds
Σ^{+}	uus
Σ^{0}	uds
Σ^{-}	dds
Ξ^{0}	uss
Ξ^{-}	dss
Ω^{-}	sss

[2] For information about the discovery of the B_c^{+} meson, read D. Kestenbaum, "Physicists Find the Last of the Mesons," *Science* 280:35, 1998.

TABLE 46.6	The Elementary Particles and Their Rest Energies and Charges	
Particle	Rest Energy	Charge
Quarks		
u	360 MeV	$+\frac{2}{3}e$
d	360 MeV	$-\frac{1}{3}e$
c	1 500 MeV	$+\frac{2}{3}e$
s	540 MeV	$-\frac{1}{3}e$
t	173 GeV	$+\frac{2}{3}e$
b	5 GeV	$-\frac{1}{3}e$
Leptons		
e^-	511 keV	$-e$
μ^-	105.7 MeV	$-e$
τ^-	1 784 MeV	$-e$
ν_e	<7 eV	0
ν_μ	<0.3 MeV	0
ν_τ	<30 MeV	0

listed in Table 46.2. Note that only two flavors of quarks, u and d, are contained in all hadrons encountered in ordinary matter (protons and neutrons).

You are probably wondering whether the discoveries of elementary particles will ever end. How many "building blocks" of matter really exist? At the present, physicists believe that the elementary particles in nature are six quarks and six leptons, together with their antiparticles. Table 46.6 lists their rest energies and their charges.

Despite extensive experimental effort, no isolated quark has ever been observed. Physicists now believe that quarks are permanently confined inside ordinary particles because of an exceptionally strong force that prevents them from escaping, called (appropriately) the **strong force**[3] (discussed in Section 46.10). This force increases with separation distance, similar to the force exerted by a stretched spring. One author described its great magnitude as follows:[4]

Quarks are slaves, . . . bound like prisoners of a chain gang. . . . Any locksmith can break the chain between two prisoners, but no locksmith is expert enough to break the gluon chains between quarks. Quarks remain slaves forever.

46.10 ▷ MULTICOLORED QUARKS

Shortly after the concept of quarks was proposed, scientists recognized that certain particles had quark compositions that violated the exclusion principle applied to quarks. In Section 42.6, we applied the exclusion principle to electrons in atoms. The

[3] As a reminder, the original meaning of the term *strong force* was the short-range attractive force between nucleons, which we have called the *nuclear force*. As we shall discuss in Section 46.10, the nuclear force between nucleons is a secondary effect of the strong force between quarks.

[4] H. Fritzsch, *Quarks, the Stuff of Matter.* London, Allen & Lane, 1983.

principle is more general, however, and applies to all particles having half-integral spin ($\frac{1}{2}$, $\frac{3}{2}$, and so on), which we collectively call fermions, as noted in the previous section. Because all quarks are fermions with spin $\frac{1}{2}$, they are expected to follow the exclusion principle. One example of a particle that appears to violate the exclusion principle is the Ω^- (sss) baryon, which contains three strange quarks with parallel spins, giving it a total spin of $\frac{3}{2}$. All three quarks have the same spin quantum number, in violation of the exclusion principle. Other examples of baryons made up of identical quarks having parallel spins are the Δ^{++} (uuu) and the Δ^- (ddd).

To resolve this problem, it was suggested that quarks possess an additional property called **color charge,** not to be confused with the color associated with visible light. This property is similar in many respects to electric charge except that it occurs in three varieties rather than two. The colors assigned to quarks are red, green, and blue, and antiquarks have the colors antired, antigreen, and antiblue. Thus, the colors red, green, and blue serve as the "quantum numbers" for the color of the quark. To satisfy the exclusion principle, the three quarks in any baryon must all have different colors. The three colors "neutralize" to white, as the electric charges + and − neutralize to zero net charge. The quark and antiquark in any meson must be of a color and the corresponding anticolor. The result is that baryons and mesons are always colorless (or white). Thus, the apparent violation of the exclusion principle in the Ω^- baryon is removed because the three quarks in the particle have different colors.

Note that the new property of color increases the number of quarks by a factor of three, since each of the six quarks comes in three colors. Although the concept of color in the quark model was originally conceived to satisfy the exclusion principle, it also provided a better theory for explaining certain experimental results. For example, the modified theory correctly predicts the lifetime of the π^0 meson.

The theory of how quarks interact with each other is called **quantum chromodynamics,** or QCD, to parallel the name *quantum electrodynamics* (the theory of interaction between electric charges). In QCD, each quark is said to carry a color charge, in analogy to electric charge. The strong force between quarks is often called the **color force.** Thus, the terms *strong force* and *color force* are used interchangeably.

In Section 46.1 we stated that the nuclear interaction between hadrons is mediated by massless field particles called **gluons** (analogous to photons for the electromagnetic force). As we shall discuss in greater detail shortly, however, the nuclear force is actually a secondary effect of the strong force between quarks, and so the gluons are actually mediators of the strong force. Let us first investigate this as the primary effect and then investigate how the gluons also mediate the nuclear force. When a quark emits or absorbs a gluon, the quark's color may change. For example, a blue quark that emits a gluon may become a red quark, and a red quark that absorbs this gluon becomes a blue quark.

The color force between quarks is analogous to the electric force between charges: likes repel, and opposites attract. Therefore, two red quarks repel each other, but a red quark is attracted to an antired quark. The attraction between quarks of opposite color to form a meson ($q\bar{q}$) is indicated in Figure 46.12a. Differently colored quarks also attract one another, although with less intensity than the oppositely colored quark and antiquark. For example, a cluster of red, blue, and green quarks all attract one another to form a baryon, as in Figure 46.12b. Thus, every baryon contains three quarks of three different colors.

Although the nuclear force between two colorless hadrons is negligible at large separations, the net strong force between their constituent quarks is not exactly zero at small separations. This residual strong force is the nuclear force that

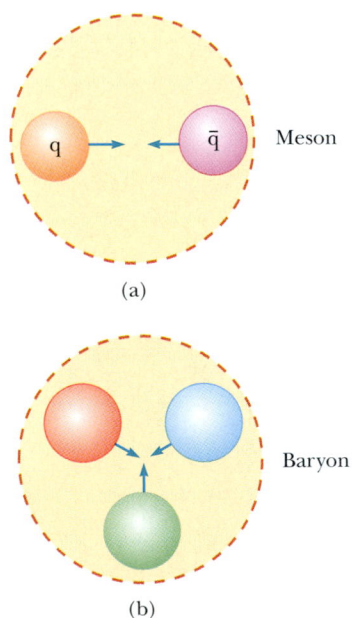

Figure 46.12 (a) A red quark is attracted to an antired quark to form a meson whose quark structure is ($q\bar{q}$). (b) Three quarks of different colors attract one another to form a baryon.

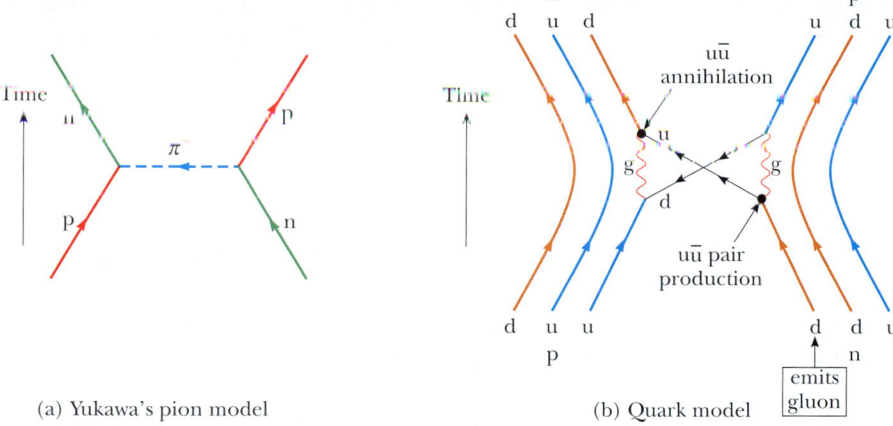

Figure 46.13 (a) A nuclear interaction between a proton and a neutron explained in terms of Yukawa's pion-exchange model. Because the pion carries charge, the proton and neutron switch identities. (b) The same interaction explained in terms of quarks and gluons. Note that the exchanged $\overline{u}d$ quark pair makes a π^- meson.

binds protons and neutrons to form nuclei. It is similar to the force between two electric dipoles. Each dipole is electrically neutral. An electric field surrounds the dipoles, however, because of the separation of the positive and negative charges (see Section 23.6). As a result, an electric force occurs between the dipoles, albeit weaker than the force between single charges.

According to QCD, a more basic explanation of the nuclear force can be given in terms of quarks and gluons. Figure 46.13a shows the nuclear interaction between a neutron and a proton by means of Yukawa's pion, in this case a π^-. This drawing differs from Figure 46.5a, in which the field particle is a π^0; thus, there is no transfer of charge from one nucleon to the other. In Figure 46.13a, the charged pion carries charge from one nucleon to the other, so the nucleons change identities—the proton becomes a neutron and the neutron becomes a proton.

Let us look at the same interaction from the viewpoint of the quark model, shown in Figure 46.13b. In this Feynman diagram, the proton and neutron are represented by their quark constituents. Each quark in the neutron and proton is continuously emitting and absorbing gluons. The energy of a gluon can result in the creation of quark–antiquark pairs. This is similar to the creation of electron–positron pairs in pair production, which we investigated in Section 46.2. When the neutron and proton approach to within 1 fm of each other, these gluons and quarks can be exchanged between the two nucleons, and such exchanges produce the nuclear force. Figure 46.13b depicts one possibility for the process shown in Figure 46.13a. A down quark in the neutron on the right emits a gluon (represented by the wavy line labeled g on the right side). The energy of the gluon is then transformed to create a $u\overline{u}$ pair. The u quark stays within the nucleon (which has now changed to a proton), and the recoiling d quark and the $\overline{u}$ antiquark are transmitted to the proton on the left side of the diagram. Here the $\overline{u}$ annihilates a u quark within the proton (with the creation of a gluon), and the d is captured. Thus, the net effect is to change a u quark to a d quark, and the proton on the left has changed to a neutron.

As the d quark and $\overline{u}$ antiquark in Figure 46.13b transfer between the nucleons, the d and $\overline{u}$ exchange gluons with each other and can be considered to be bound to each other by means of the strong force. If we look back at Table 46.4, we see that this combination is a π^-—Yukawa's field particle! Thus, the quark model of interactions between nucleons is consistent with the pion-exchange model.

46.11 ▶ THE STANDARD MODEL

Scientists now believe that there are three classifications of truly elementary particles: leptons, quarks, and field particles. These three particles are further classified as either **fermions** (quarks and leptons) or **bosons** (field particles). Note that quarks and leptons have spin $\frac{1}{2}$ and hence are fermions, while the field particles have spin 1 or higher and are bosons.

Recall from Section 46.1 that the weak force is believed to be mediated by the W^+, W^-, and Z^0 bosons. These particles are said to have *weak charge,* just as quarks have color charge. Thus, each elementary particle can have mass, electric charge, color charge, and weak charge. Of course, one or more of these could be zero.

In 1979, Sheldon Glashow (b. 1932), Abdus Salam (1926–1996), and Steven Weinberg (b. 1933) won a Nobel Prize for developing a theory that unifies the electromagnetic and weak interactions. This **electroweak theory** postulates that the weak and electromagnetic interactions have the same strength when the particles involved have very high energies. Thus, the two interactions are viewed as different manifestations of a single unifying electroweak interaction. The theory makes many concrete predictions, but perhaps the most spectacular is the prediction of the masses of the W and Z particles at about 82 GeV/c^2 and 93 GeV/c^2, respectively—predictions that are borne out by experiment.

The combination of the electroweak theory and QCD for the strong interaction is referred to in high-energy physics as the **Standard Model.** Although the details of the Standard Model are complex, its essential ingredients can be summarized with the help of Figure 46.14. (The Standard Model does not include the gravitational force at present; however, we include gravity in Figure 46.14 because physicists hope to eventually incorporate this force into a unified theory.) This dia-

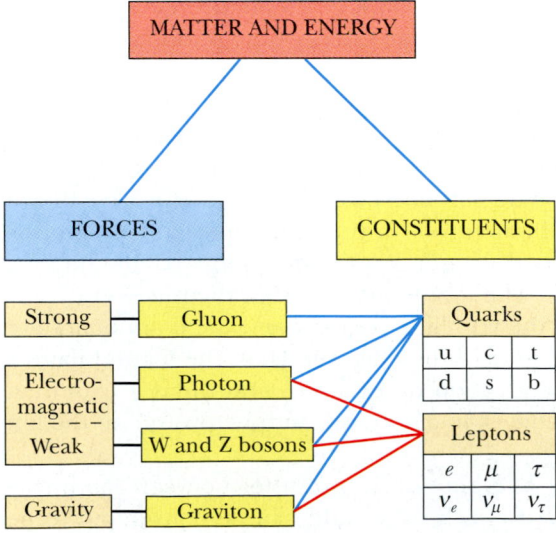

Figure 46.14 The Standard Model of particle physics.

gram shows that quarks participate in all of the fundamental forces and that leptons participate in all except the strong force.

The Standard Model does not answer all questions. A major question that is still unanswered is why, of the two mediators of the electroweak interaction, the photon has no mass but the W and Z bosons do. Because of this mass difference, the electromagnetic and weak forces are quite distinct at low energies but become similar at very high energies, at which the rest energy is negligible relative to the total energy. The behavior as one goes from high to low energies is called *symmetry breaking* because the forces are similar, or symmetric, at high energies but are very different at low energies. The nonzero rest energies of the W and Z bosons raise the question of the origin of particle masses. To resolve this problem, a hypothetical particle called the **Higgs boson,** which provides a mechanism for breaking the electroweak symmetry, has been proposed. The Standard Model modified to include the Higgs mechanism provides a logically consistent explanation of the massive nature of the W and Z bosons. Unfortunately, the Higgs boson has not yet been found, but physicists know that its rest energy should be less than 1 TeV. For us to determine whether the Higgs boson exists, two quarks each having at least 1 TeV of energy must collide, but calculations show that such a collision requires injecting 40 TeV of energy within the volume of a proton.

Scientists are convinced that, because of the limited energy available in conventional accelerators using fixed targets, it is necessary to build colliding-beam accelerators called **colliders.** The concept of colliders is straightforward. Particles that have equal masses and equal kinetic energies, traveling in opposite directions in an accelerator ring, collide head-on to produce the required reaction and form new particles. Because the total momentum of the interacting particles is zero, all of their kinetic energy is available for the reaction. The Large Electron–Positron (LEP) Collider at CERN (Fig. 46.15) and the Stanford Linear Collider collide both electrons and positrons. The Super Proton Synchrotron at CERN accelerates

A technician works on one of the particle detectors at CERN, the European center for particle physics near Geneva, Switzerland. Electrons and positrons accelerated to an energy of 50 GeV collide in a circular tunnel 2 km in circumference, located 100 m underground. *(David Parker/Science Photo Library/Photo Researchers, Inc.)*

Figure 46.15 A view from inside the Large Electron–Positron (LEP) Collider tunnel, which is 27 km in circumference. *(Courtesy of CERN)*

protons and antiprotons to energies of 270 GeV, whereas the world's highest-energy proton accelerator, the Tevatron at the Fermi National Laboratory in Illinois, produces protons at almost 1 000 GeV (1 TeV). The Superconducting Super Collider (SSC), which was being built in Texas, was an accelerator designed to produce 20-TeV protons in a ring 52 mi in circumference. After much debate in Congress and an investment of almost 2 billion dollars, the U.S. Department of Energy canceled the SSC project in October 1993. CERN expects a 2005 completion date for the Large Hadron Collider (LHC), a proton–proton collider that will provide a center-of-mass energy of 14 TeV and enable exploration of Higgs-boson physics. The accelerator will be constructed in the same 27-km circumference tunnel now housing the LEP Collider, and many countries are expected to participate in the project.

Following the success of the electroweak theory, scientists attempted to combine it with QCD in a **grand unification theory** (GUT). In this model, the electroweak force is merged with the strong force to form a grand unified force. One version of the theory considers leptons and quarks as members of the same family that can change into each other by exchanging an appropriate field particle.

web

Visit the Conseil Européen de Recherche Nucléair (now called the European Laboratory for Particle Physics) at **www.CERN.ch**

The World Wide Web was invented at CERN in 1991 as a way for physicists to easily share data.

46.12 ▶ THE COSMIC CONNECTION

In this section we describe one of the most fascinating theories in all of science—the Big Bang theory of the creation of the Universe—and the experimental evidence that supports it. This theory of cosmology states that the Universe had a beginning and, furthermore, that the beginning was so cataclysmic that it is impossible to look back beyond it. According to this theory, the Universe erupted from an infinitely dense singularity about 15 to 20 billion years ago. The first few minutes after the Big Bang saw such extremely high energy that it is believed that all four interactions of physics were unified and all matter was contained in an undifferentiated "quark soup."

The evolution of the four fundamental forces from the Big Bang to the present is shown in Figure 46.16. During the first 10^{-43} s (the ultrahot epoch, $T \sim 10^{32}$ K),

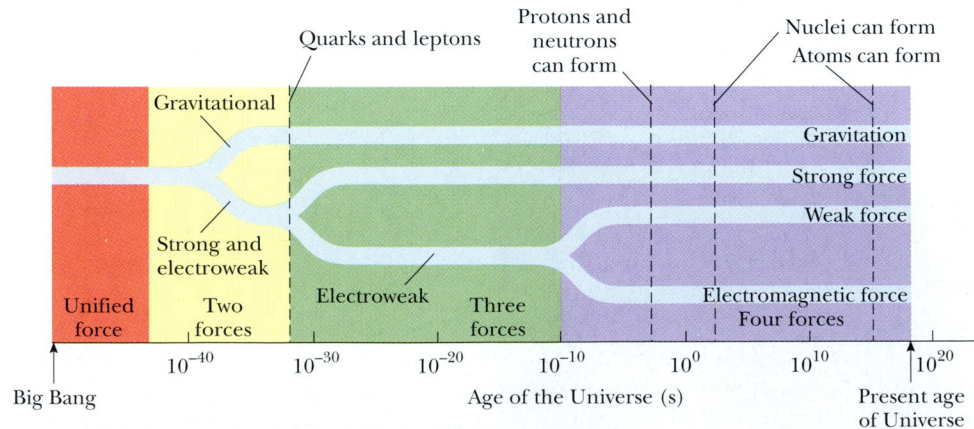

Figure 46.16 A brief history of the Universe from the Big Bang to the present. The four forces became distinguishable during the first nanosecond. Following this, all the quarks combined to form particles that interact via the nuclear force. However, the leptons remained separate and to this day exist as individual, observable particles.

it is presumed that the strong, electroweak, and gravitational forces were joined to form a completely unified force. In the first 10^{-35} s following the Big Bang (the hot epoch, $T \sim 10^{29}$ K), gravity broke free of this unification while the strong and electroweak forces remained as one, described by a grand unification theory. This was a period when particle energies were so great ($> 10^{16}$ GeV) that very massive particles as well as quarks, leptons, and their antiparticles existed. Then, after 10^{-35} s, the Universe rapidly expanded and cooled (the warm epoch, $T \sim 10^{29}$ to 10^{15} K), the strong and electroweak forces parted company, and the grand unification scheme was broken. As the Universe continued to cool, the electroweak force split into the weak force and the electromagnetic force about 10^{-10} s after the Big Bang.

After a few minutes, protons condensed out of the hot soup. For half an hour the Universe underwent thermonuclear detonation, exploding as a hydrogen bomb and producing most of the helium nuclei that exist now. The Universe continued to expand, and its temperature dropped. Until about 700 000 years after the Big Bang, the Universe was dominated by radiation. Energetic radiation prevented matter from forming single hydrogen atoms because collisions would instantly ionize any atoms that happened to form. Photons experienced continuous Compton scattering from the vast numbers of free electrons, resulting in a Universe that was opaque to radiation. By the time the Universe was about 700 000 years old, it had expanded and cooled to about 3 000 K, and protons could bind to electrons to form neutral hydrogen atoms. Because of the quantized energies of the atoms, far more wavelengths of radiation were not absorbed by atoms than were absorbed, and the Universe suddenly became transparent to photons. Radiation no longer dominated the Universe, and clumps of neutral matter steadily grew—first atoms, then molecules, gas clouds, stars, and finally galaxies.

Observation of Radiation from the Primordial Fireball

In 1965, Arno A. Penzias (b. 1933) and Robert W. Wilson (b. 1936) of Bell Laboratories were testing a sensitive microwave receiver and made an amazing discovery. A pesky signal producing a faint background hiss was interfering with their satellite communications experiments. In spite of their valiant efforts, the signal remained. Ultimately, it became clear that they were perceiving microwave background radiation (at a wavelength of 7.35 cm), which represented the leftover "glow" from the Big Bang.

The microwave horn that served as their receiving antenna is shown in Figure 46.17. The intensity of the detected signal remained unchanged as the antenna was pointed in different directions. The fact that the radiation had equal strengths in all directions suggested that the entire Universe was the source of this radiation. Evicting a flock of pigeons from the 20-ft horn and cooling the microwave detector both failed to remove the signal. Through a casual conversation, Penzias and Wilson discovered that a group at Princeton had predicted the residual radiation from the Big Bang and were planning an experiment to attempt to confirm the theory. The excitement in the scientific community was high when Penzias and Wilson announced that they had already observed an excess microwave background compatible with a 3-K blackbody source.

Because Penzias and Wilson made their measurements at a single wavelength, they did not completely confirm the radiation as 3-K blackbody radiation. Subsequent experiments by other groups added intensity data at different wavelengths, as shown in Figure 46.18. The results confirm that the radiation is that of a black body at 2.7 K. This figure is perhaps the most clear-cut evidence for the Big Bang theory. The 1978 Nobel Prize was awarded to Penzias and Wilson for this most important discovery.

Figure 46.17   Robert W. Wilson *(left)* and Arno A. Penzias with the Bell Telephone Laboratories horn-reflector antenna.   *(AT&T Bell Laboratories)*

The discovery of the cosmic background radiation brought with it a problem, however—the radiation was too uniform. Scientists believed that slight fluctuations in this background had to occur in order for such objects as galaxies to form. In 1989, NASA launched a satellite called COBE (KOH-bee), for Cosmic Background Explorer, to study this radiation in greater detail. In 1992, George Smoot

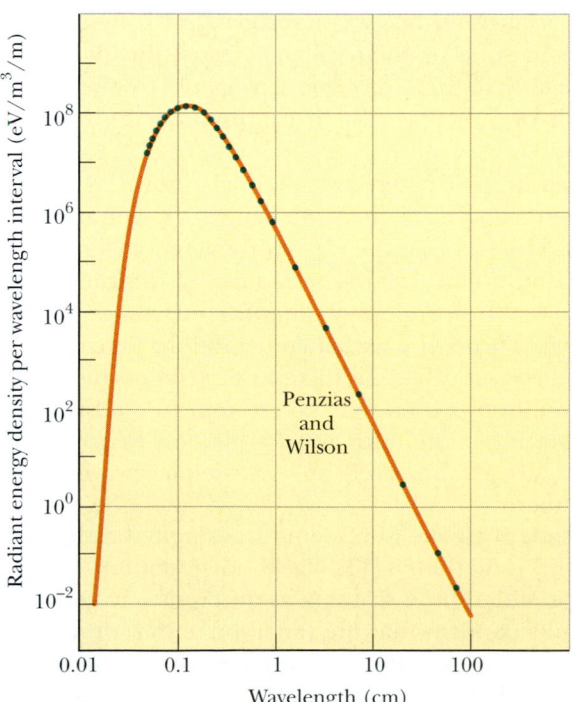

Figure 46.18   Theoretical black-body (red curve) and measured radiation spectra (black points) of the Big Bang. Most of the data were collected from the Cosmic Background Explorer (COBE) satellite. The data of Wilson and Penzias are indicated.

(b. 1945) at the Lawrence Berkeley Laboratory found, on the basis of the data collected, that the background was not perfectly uniform but instead contained irregularities that corresponded to temperature variations of 0.000 3 K. It is these small variations that provided nucleation sites for the formation of the galaxies and other objects we now see in the sky.

Other Evidence for an Expanding Universe

The Big Bang theory of cosmology predicts that the Universe is expanding. Most of the key discoveries supporting the theory of an expanding Universe were made in the 20th century. Vesto Melvin Slipher (1875–1969), an American astronomer, reported in 1912 that most nebulae are receding from the Earth at speeds up to several million miles per hour. Slipher was one of the first scientists to use Doppler shifts (see Section 17.5) in spectral lines to measure velocities.

In the late 1920s, Edwin P. Hubble made the bold assertion that the whole Universe is expanding. From 1928 to 1936, until they reached the limits of the 100-inch telescope, Hubble and Milton Humason (1891–1972) worked at Mount Wilson in California to prove this assertion. The results of that work and of its continuation with the use of a 200-inch telescope in the 1940s showed that the speeds at which galaxies are receding from the Earth increase in direct proportion to their distance R from us (Fig. 46.19). This linear relationship, known as **Hubble's law,** may be written

$$v = HR \qquad \textbf{(46.7)}$$

where H, called the **Hubble parameter,** has the approximate value

$$H \approx 17 \times 10^{-3} \text{ m/s·ly}$$

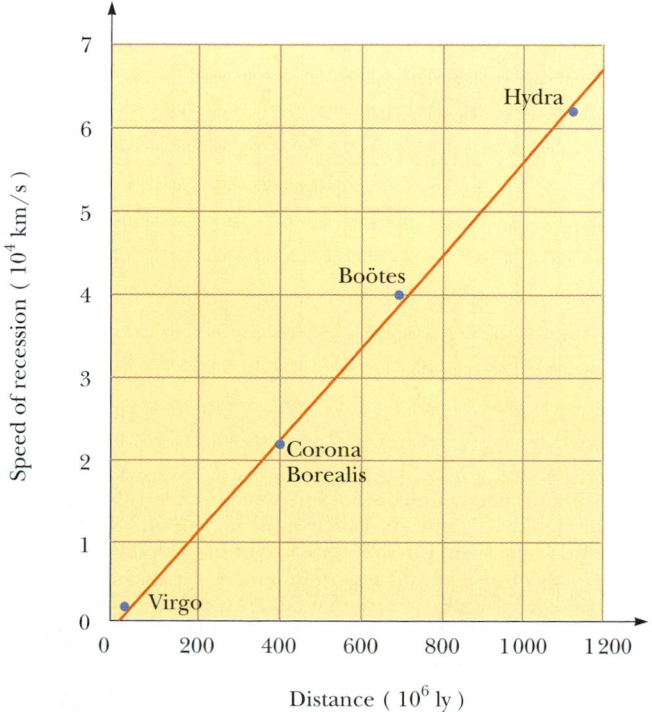

Figure 46.19 Hubble's law: a plot of speed of recession versus distance for four galaxies.

EXAMPLE 46.5   **Recession of a Quasar**

A quasar is an object that appears similar to a star and is very distant from the Earth. Its speed can be determined from Doppler-shift measurements in the light it emits. A certain quasar recedes from the Earth at a speed of $0.55c$. How far away is it?

Solution   We can find the distance through Hubble's law:

$$R = \frac{v}{H} = \frac{(0.55)(3.00 \times 10^8 \text{ m/s})}{17 \times 10^{-3} \text{ m/s} \cdot \text{ly}} = \boxed{9.7 \times 10^9 \text{ ly}}$$

Exercise   Assuming that the quasar has moved at this speed ever since the Big Bang, estimate the age of the Universe.

Answer $t = R/v = 1/H \approx 18$ billion years, which is in reasonable agreement with other calculations.

Will the Universe Expand Forever?

In the 1950s and 1960s, Allan R. Sandage used the 200-inch telescope at Mount Palomar to measure the speeds of galaxies at distances of up to 6 billion lightyears away from the Earth. These measurements showed that these very distant galaxies were moving about 10 000 km/s faster than Hubble's law predicted. According to this result, the Universe must have been expanding more rapidly 1 billion years ago, and consequently we conclude that the expansion rate is slowing[5] (Fig. 46.20). Today, astronomers and physicists are trying to determine the rate of slow-

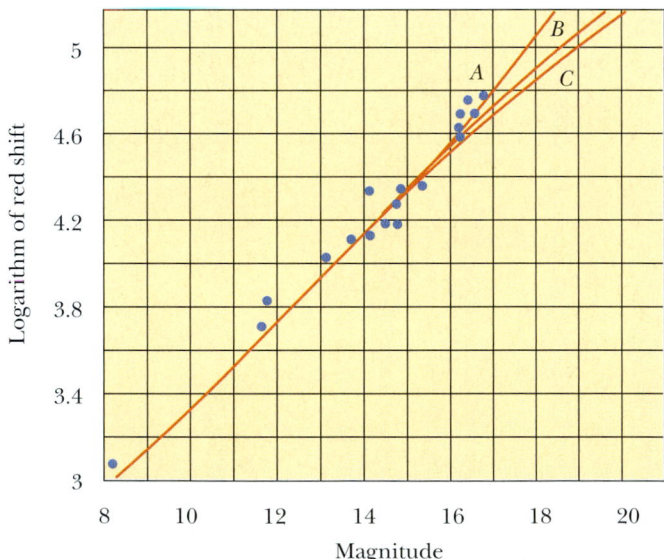

Figure 46.20   Red shift, or speed of recession, versus magnitude (which is related to brightness) of 18 faint galaxy clusters. Significant scatter of the data occurs, so the extrapolation of the curve to the upper right is uncertain. Curve *A* is the trend suggested by the six faintest clusters. Curve *C* corresponds to a Universe having a constant rate of expansion. If more data are taken and the complete set of data indicates a curve that falls between *B* and *C*, the expansion will slow but never stop. If the data fall to the left of *B*, expansion will eventually stop and the Universe will begin to contract.

[5] The data at large distances have large observational uncertainties and may be systematically in error from effects such as abnormal brightness in the most distant visible clusters.

ing. If the average mass density of the Universe is less than some critical value ($\rho_c \approx 3$ atoms/m^3), the galaxies will slow in their outward rush but still escape to infinity. If the average density exceeds the critical value, the expansion will eventually stop and contraction will begin, possibly leading to a superdense state followed by another expansion. In this case, we have an oscillating Universe.

EXAMPLE 46.6   The Critical Density of the Universe

(a) Starting from energy conservation, derive an expression for the critical mass density of the Universe ρ_c in terms of the Hubble parameter H and the universal gravitational constant G.

Solution Figure 46.21 shows a large section of the Universe, contained within a sphere of radius R. The total mass of the galaxies in this volume is M. A galaxy of mass $m \ll M$ that has a speed v at a distance R from the center of the sphere will escape to infinity (at which its speed will approach zero) if the sum of its kinetic energy and the gravitational potential energy of the system—galaxy plus rest of the Universe—is zero at any time. The Universe may be infinite in spatial extent, but Gauss's law implies that only the mass M inside the sphere contributes to the gravitational potential energy of the galaxy:

$$E_{\text{total}} = 0 = K + U = \tfrac{1}{2}mv^2 - \frac{GmM}{R}$$

We substitute for the mass M contained within the sphere the product of the critical density and the volume of the sphere:

$$\tfrac{1}{2}mv^2 = \frac{Gm\frac{4}{3}\pi R^3 \rho_c}{R}$$

Solving for the critical density gives

$$\rho_c = \frac{3v^2}{8\pi G R^2}$$

From Hubble's law, the ratio of v to R is $v/R = H$, so this expression becomes

$$\rho_c = \frac{3H^2}{8\pi G}$$

(b) Estimate a numerical value for the critical density in grams per cubic centimeter.

Solution Using $H = 17 \times 10^{-3}$ m/s·ly, where 1 ly = 9.46×10^{15} m, we find for the critical density

$$\rho_c = \frac{3H^2}{8\pi G} = \frac{3(17 \times 10^{-3}\,\text{m/s·ly})^2}{8\pi(6.67 \times 10^{-11}\,\text{N·m}^2/\text{kg}^2)}\left(\frac{1\,\text{ly}}{9.46 \times 10^{15}\,\text{m}}\right)^2$$

$$= \boxed{6 \times 10^{-27}\,\text{kg/m}^3}$$

Converting this to the requested units, we have $\rho_c = 6 \times 10^{-30}$ g/cm^3. Because the mass of a hydrogen atom is 1.67×10^{-24} g, this value of ρ_c corresponds to 3×10^{-6} hydrogen atoms per cubic centimeter, or 3 atoms per cubic meter.

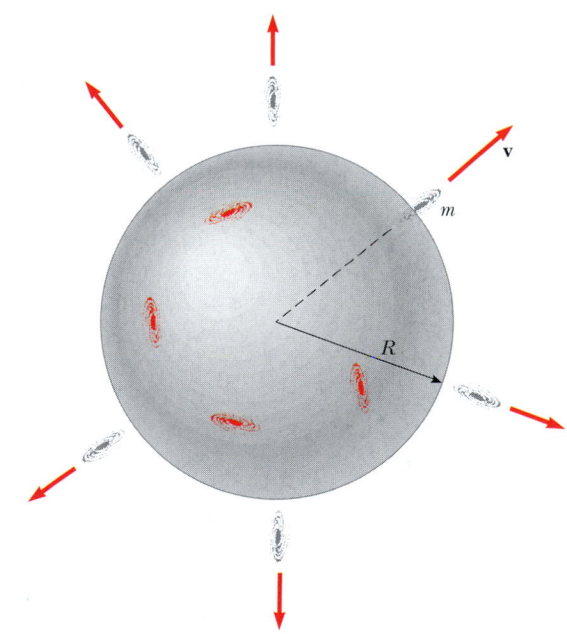

Figure 46.21 The galaxy marked with mass m is escaping from a large cluster of galaxies contained within a spherical volume of radius R. Only the mass within R slows the mass m.

Missing Mass in the Universe?

The luminous matter in galaxies averages out to a Universe density of 5×10^{-33} g/cm^3. The radiation in the Universe has a mass equivalent of approximately 2% of the luminous matter. The total mass of all nonluminous matter (such as interstellar gas and black holes) may be estimated from the speeds of galaxies orbiting each other in a cluster. The higher the galaxy speeds, the more mass in the cluster. Measurements on the Coma cluster of galaxies indicate, surprisingly, that the amount of nonluminous matter is 20 to 30 times the amount of luminous matter present in stars and luminous gas clouds. Yet even this large invisible component of *dark matter*, if extrapolated to the Universe as a whole, leaves the observed mass density a factor of 10 less than ρ_c. The deficit, called *missing mass*, has been the subject of intense theoretical and experimental work, with exotic particles such as axions, photinos, and superstring particles suggested as candidates for the missing mass. Some researchers have made the more mundane proposal that the missing mass is present in neutrinos. In fact, neutrinos are so abundant that a tiny neutrino rest energy on the order of only 20 eV would furnish the missing mass and "close" the Universe. Thus, current experiments designed to measure the rest energy of the neutrino will have an impact on predictions for the future of the Universe.

Although we have some degree of certainty about the beginning of the Universe, we are uncertain about how the story will end. Will the Universe keep on expanding forever, or will it someday collapse and then expand again, perhaps in an endless series of oscillations? Results and answers to these questions remain inconclusive, and the exciting controversy continues.

46.13 PROBLEMS AND PERSPECTIVES

While particle physicists have been exploring the realm of the very small, cosmologists have been exploring cosmic history back to the first microsecond of the Big Bang. Observation of the events that occur when two particles collide in an accelerator is essential for reconstructing the early moments in cosmic history. For this reason, perhaps the key to understanding the early Universe is to first understand the world of elementary particles. Cosmologists and physicists now find that they have many common goals and are joining hands in an attempt to understand the physical world at its most fundamental level.

Our understanding of physics at short distances is far from complete. Particle physics is faced with many questions. Why does so little antimatter exist in the Universe? Is it possible to unify the strong and electroweak theories in a logical and consistent manner? Why do quarks and leptons form three similar but distinct families? Are muons the same as electrons apart from their difference in mass, or do they have other subtle differences that have not been detected? Why are some particles charged and others neutral? Why do quarks carry a fractional charge? What determines the masses of the elementary constituents of matter? Can isolated quarks exist? The questions go on and on. Because of the rapid advances and new discoveries in the field of particle physics, by the time you read this book, some of these questions may be resolved and new ones may emerge.

An important and obvious question that remains is whether leptons and quarks have an underlying structure. If they do, we can envision an infinite number of deeper structure levels. However, if leptons and quarks are indeed the ultimate constituents of matter, as physicists today tend to believe, we should be able to construct a final theory of the structure of matter, just as Einstein dreamed of doing. This theory, whimsically called the Theory of Everything, is a combination of GUT and a quantum theory of gravity. In the view of many physicists, the end of the road is in sight, but how long it will take to reach it is anyone's guess.

SUMMARY

Before quark theory was developed, the four fundamental forces in nature were identified as nuclear, electromagnetic, weak, and gravitational. All the interactions in which these forces take part are mediated by **field particles.** The electromagnetic interaction is mediated by the photon; the weak interaction is mediated by the $W^{\pm}$ and Z^0 bosons; the gravitational interaction is mediated by gravitons; the nuclear interaction is mediated by gluons.

A charged particle and its **antiparticle** have the same mass but opposite charge, and other properties may have opposite values, such as lepton number and baryon number. It is possible to produce particle–antiparticle pairs in nuclear reactions if the available energy is greater than $2mc^2$, where m is the mass of the particle (or antiparticle).

Particles other than field particles are classified as hadrons or leptons. **Hadrons** interact via all four fundamental forces. They have size and structure and are not elementary particles. There are two types—**baryons** and **mesons.** Baryons, which generally are the most massive particles, have nonzero **baryon number** and a spin of $\frac{1}{2}$ or $\frac{3}{2}$. Mesons have baryon number zero and either zero or integral spin.

Leptons have no structure or size and are considered truly elementary. They interact only via the weak, gravitational, and electromagnetic forces. Six types of leptons exist: the electron e^-; the muon μ^-; the tau τ^-; and their neutrinos ν_e, ν_μ, and ν_τ.

In all reactions and decays, quantities such as energy, linear momentum, angular momentum, electric charge, baryon number, and lepton number are strictly conserved. Certain particles have properties called **strangeness** and **charm.** These unusual properties are conserved only in the decays and nuclear reactions that occur via the strong force.

Theorists in elementary particle physics have postulated that all hadrons are composed of smaller units known as **quarks.** Quarks have fractional electric charge and come in six **flavors:** up (u), down (d), strange (s), charmed (c), top (t), and bottom (b). Each baryon contains three quarks, and each meson contains one quark and one antiquark.

According to the theory of **quantum chromodynamics,** quarks have a property called **color,** and the force between quarks is referred to as the **strong force** or the **color force.** The strong force is now considered to be a fundamental force. The nuclear force, which was originally considered to be fundamental, is now understood to be a secondary effect of the strong force, due to gluon exchanges between hadrons.

The electromagnetic and weak forces are now considered to be manifestations of a single force called the **electroweak force.** The combination of quantum chromodynamics and the electroweak theory is called the **Standard Model.**

The background microwave radiation discovered by Penzias and Wilson strongly suggests that the Universe started with a Big Bang 12 to 15 billion years ago. The background radiation is equivalent to that of a black body at 3 K. Various astronomical measurements strongly suggest that the Universe is expanding. According to **Hubble's law,** distant galaxies are receding from the Earth at a speed $v = HR$, where R is the distance from the Earth to the galaxy and H is the **Hubble parameter,** $H \approx 17 \times 10^{-3}$ m/s·ly.

QUESTIONS

1. Name the four fundamental interactions and the field particle that mediates each.
2. Describe the quark model of hadrons, including the properties of quarks.
3. What are the differences between hadrons and leptons?
4. Describe the properties of baryons and mesons and the important differences between them.
5. Particles known as resonances have very short lifetimes, of the order of 10^{-23} s. From this information, would you guess that they are hadrons or leptons? Explain.
6. Kaons all decay into final states that contain no protons or neutrons. What is the baryon number of kaons?
7. The Ξ^0 particle decays by the weak interaction according to the decay mode $\Xi^0 \rightarrow \Lambda^0 + \pi^0$. Would you expect this decay to be fast or slow? Explain.
8. Identify the particle decays listed in Table 46.2 that occur by the weak interaction. Justify your answers.
9. Identify the particle decays listed in Table 46.2 that occur by the electromagnetic interaction. Justify your answers.
10. Two protons in a nucleus interact via the nuclear interaction. Are they also subject to the weak interaction?
11. Discuss the following conservation laws: energy, linear momentum, angular momentum, electric charge, baryon number, lepton number, and strangeness. Are all of these laws based on fundamental properties of nature? Explain.
12. An antibaryon interacts with a meson. Can a baryon be produced in such an interaction? Explain.
13. Describe the essential features of the Standard Model of particle physics.
14. How many quarks are in each of the following: (a) a baryon, (b) an antibaryon, (c) a meson, (d) an antimeson? How do you account for the fact that baryons have half-integral spins while mesons have spins of 0 or 1? (*Hint:* Quarks have spin $\frac{1}{2}$.)
15. In the theory of quantum chromodynamics, quarks come in three colors. How would you justify the statement that "all baryons and mesons are colorless"?
16. Which baryon did Murray Gell-Mann predict in 1961? What is the quark composition of this particle?
17. What is the quark composition of the Ξ^- particle? (See Table 46.5.)
18. The W and Z bosons were first produced at CERN in 1983 (by causing a beam of protons and a beam of antiprotons to meet at high energy). Why was this an important discovery?
19. How did Edwin Hubble (in 1928) determine that the Universe is expanding?
20. **Review Question.** A girl and her grandmother grind corn while the woman tells the girl some illuminating stories. A boy keeps crows away from ripening corn while his grandfather sits in the shade and explains to him the Universe and his place in it. What the children do not understand this summer they will better understand next year. Now you must take the part of the adults. State the most general, most fundamental, most universal truths that you know. If you find yourself repeating someone else's ideas, get the best version of those ideas that you can, and state your source. If there is something you do not understand, make a plan to understand it better within the next year.

PROBLEMS

1, 2, 3 = straightforward, intermediate, challenging   ☐ = full solution available in the *Student Solutions Manual and Study Guide*
WEB = solution posted at **http://www.saunderscollege.com/physics/**   💻 = Computer useful in solving problem   🌐 = Interactive Physics
☐ = paired numerical/symbolic problems

Section 46.1   The Fundamental Forces in Nature
Section 46.2   Positrons and Other Antiparticles

1. A photon produces a proton–antiproton pair according to the reaction $\gamma \rightarrow p + \bar{p}$. What is the minimum possible frequency of the photon? What is its wavelength?
2. Two photons are produced when a proton and an antiproton annihilate each other. What are the minimum frequency and corresponding wavelength of each photon?
3. A photon with an energy $E_\gamma = 2.09$ GeV creates a proton–antiproton pair in which the proton has a ki-

netic energy of 95.0 MeV. What is the kinetic energy of the antiproton? ($m_p c^2 = 938.3$ MeV.)

Section 46.3   Mesons and the Beginning of Particle Physics

4. Occasionally, high-energy muons collide with electrons and produce two neutrinos according to the reaction $\mu^+ + e^- \rightarrow 2\nu$. What kind of neutrinos are these?
5. One of the mediators of the weak interaction is the Z^0 boson, with mass 93 GeV/c^2. Use this information to find the order of magnitude of the range of the weak interaction.

6. A free neutron actually beta-decays by creating a proton, an electron, and an antineutrino according to the reaction $n \rightarrow p + e^- + \bar{\nu}$. Assume, nevertheless, that a free neutron decays by creating a proton and an electron according to the reaction

$$n \longrightarrow p + e^-$$

and assume that the neutron is initially at rest in the laboratory. (a) Determine the energy released in this reaction. (b) Determine the speeds of the proton and electron after the reaction. (Energy and momentum are conserved in the reaction.) (c) Is either of these particles moving at relativistic speeds? Explain.

7. When a high-energy proton or pion traveling near the speed of light collides with a nucleus, it travels an average distance of 3×10^{-15} m before interacting. From this information, find the order of magnitude of the time for the strong interaction to occur.

8. Calculate the range of the force that might be produced by the virtual exchange of a proton.

WEB 9. A neutral pion at rest decays into two photons according to

$$\pi^0 \longrightarrow \gamma + \gamma$$

Find the energy, momentum, and frequency of each photon.

Section 46.4 Classification of Particles

10. Identify the unknown particle on the left side of the following reaction:

$$? + p \longrightarrow n + \mu^+$$

11. Name one possible decay mode (see Table 46.2) for Ω^+, $\overline{K_S^0}$, $\overline{\Lambda^0}$, and $\bar{n}$.

Section 46.5 Conservation Laws

12. Each of the following reactions is forbidden. Determine a conservation law that is violated for each reaction.
    (a) $p + \bar{p} \rightarrow \mu^+ + e^-$
    (b) $\pi^- + p \rightarrow p + \pi^+$
    (c) $p + p \rightarrow p + \pi^+$
    (d) $p + p \rightarrow p + p + n$
    (e) $\gamma + p \rightarrow n + \pi^0$

13. (a) Show that baryon number and charge are conserved in the following reactions of a pion with a proton.

$$\pi^+ + p \longrightarrow K^+ + \Sigma^+ \qquad \textbf{(1)}$$

$$\pi^+ + p \longrightarrow \pi^+ + \Sigma^+ \qquad \textbf{(2)}$$

(b) The first reaction is observed, but the second never occurs. Explain.

14. The first of the following two reactions may occur, but the second cannot. Explain.

$$K_S^0 \longrightarrow \pi^+ + \pi^- \qquad \text{(can occur)}$$

$$\Lambda^0 \longrightarrow \pi^+ + \pi^- \qquad \text{(cannot occur)}$$

WEB 15. The following reactions or decays involve one or more neutrinos. In each case, supply the missing neutrino (ν_e, ν_μ, or ν_τ).
    (a) $\pi^- \rightarrow \mu^- + ?$
    (b) $K^+ \rightarrow \mu^+ + ?$
    (c) $? + p \rightarrow n + e^+$
    (d) $? + n \rightarrow p + e^-$
    (e) $? + n \rightarrow p + \mu^-$
    (f) $\mu^- \rightarrow e^- + ? + ?$

16. A K_S^0 particle at rest decays into a π^+ and a π^-. What will be the speed of each of the pions? The mass of the K_S^0 is 497.7 MeV/c^2, and the mass of each π is 139.6 MeV/c^2.

WEB 17. Determine which of the following reactions can occur. For those that cannot occur, determine the conservation law (or laws) violated.
    (a) $p \rightarrow \pi^+ + \pi^0$
    (b) $p + p \rightarrow p + p + \pi^0$
    (c) $p + p \rightarrow p + \pi^+$
    (d) $\pi^+ \rightarrow \mu^+ + \nu_\mu$
    (e) $n \rightarrow p + e^- + \bar{\nu}_e$
    (f) $\pi^+ \rightarrow \mu^+ + n$

18. (a) Show that the proton-decay reaction

$$p \longrightarrow e^+ + \gamma$$

cannot occur because it violates conservation of baryon number. (b) Imagine that this reaction does occur, and that the proton is initially at rest. Determine the energy and momentum of the positron and photon after the reaction. (*Hint:* Recall that energy and momentum must be conserved in the reaction.) (c) Determine the speed of the positron after the reaction.

19. Determine the type of neutrino or antineutrino involved in each of the following processes.
    (a) $\pi^+ \rightarrow \pi^0 + e^+ + ?$
    (b) $? + p \rightarrow \mu^- + p + \pi^+$
    (c) $\Lambda^0 \rightarrow p + \mu^- + ?$
    (d) $\tau^+ \rightarrow \mu^+ + ? + ?$

Section 46.6 Strange Particles and Strangeness

20. The neutral ρ meson decays by the strong interaction into two pions: $\rho^0 \rightarrow \pi^+ + \pi^-$, half-life 10^{-23} s. The neutral kaon also decays into two pions: $K_S^0 \rightarrow \pi^+ + \pi^-$, half-life 10^{-10} s. How do you explain the difference in half-lives?

21. Determine whether strangeness is conserved in the following decays and reactions.
    (a) $\Lambda^0 \rightarrow p + \pi^-$
    (b) $\pi^- + p \rightarrow \Lambda^0 + K^0$
    (c) $\bar{p} + p \rightarrow \overline{\Lambda^0} + \Lambda^0$
    (d) $\pi^- + p \rightarrow \pi^- + \Sigma^+$
    (e) $\Xi^- \rightarrow \Lambda^0 + \pi^-$
    (f) $\Xi^0 \rightarrow p + \pi^-$

22. For each of the following forbidden decays, determine which conservation law is violated.
    (a) $\mu^- \rightarrow e^- + \gamma$
    (b) $n \rightarrow p + e^- + \nu_e$
    (c) $\Lambda^0 \rightarrow p + \pi^0$
    (d) $p \rightarrow e^+ + \pi^0$
    (e) $\Xi^0 \rightarrow n + \pi^0$

23. Which of the following processes are allowed by the strong interaction, the electromagnetic interaction, the weak interaction, or no interaction at all?
    (a) $\pi^- + p \rightarrow 2\eta$

(b) $K^- + n \rightarrow \Lambda^0 + \pi^-$
(c) $K^- \rightarrow \pi^- + \pi^0$
(d) $\Omega^- \rightarrow \Xi^- + \pi^0$
(e) $\eta \rightarrow 2\gamma$

24. Identify the conserved quantities in the following processes.
    (a) $\Xi^- \rightarrow \Lambda^0 + \mu^- + \nu_\mu$
    (b) $K_S^0 \rightarrow 2\pi^0$
    (c) $K^- + p \rightarrow \Sigma^0 + n$
    (d) $\Sigma^0 \rightarrow \Lambda^0 + \gamma$
    (e) $e^+ + e^- \rightarrow \mu^+ + \mu^-$
    (f) $\bar{p} + n \rightarrow \bar{\Lambda}^0 + \Sigma^-$

25. Fill in the missing particle. Assume that (a) occurs via the strong interaction and that (b) and (c) involve the weak interaction.
    (a) $K^+ + p \rightarrow ? + p$
    (b) $\Omega^- \rightarrow ? + \pi^-$
    (c) $K^+ \rightarrow ? + \mu^+ + \nu_\mu$

Section 46.7 Making Particles and Measuring Their Properties
Section 46.8 Finding Patterns in the Particles
Section 46.9 Quarks—Finally

26. The quark composition of the proton is uud, and that of the neutron is udd. Show that in each case the charge, baryon number, and strangeness of the particle equal, respectively, the sums of these numbers for the quark constituents.

27. (a) Find the number of electrons and the number of each species of quark in 1 L of water. (b) Make an order-of-magnitude estimate of the number of each kind of fundamental matter particle in your body. State your assumptions and the quantities you take as data.

28. The quark compositions of the K^0 and Λ^0 particles are $\bar{s}d$ and uds, respectively. Show that the charge, baryon number, and strangeness of these particles equal, respectively, the sums of these numbers for the quark constituents.

29. Assuming that binding energies can be neglected, find the masses of the u and d quarks from the masses of the proton and neutron.

30. The text stated that the reaction $\pi^- + p \rightarrow K^0 + \Lambda^0$ occurs with high probability, whereas the reaction $\pi^- + p \rightarrow K^0 + n$ never occurs. Analyze these reactions at the quark level. Show that the first reaction conserves the total number of each type of quark, and the second reaction does not.

31. Analyze each reaction in terms of constituent quarks.
    (a) $\pi^- + p \rightarrow K^0 + \Lambda^0$
    (b) $\pi^+ + p \rightarrow K^+ + \Sigma^+$
    (c) $K^- + p \rightarrow K^+ + K^0 + \Omega^-$
    (d) $p + p \rightarrow K^0 + p + \pi^+ + ?$
    In the last reaction, identify the mystery particle.

32. A Σ^0 particle traveling through matter strikes a proton; then a Σ^+ and a gamma ray emerge, as well as a third

particle. Use the quark model of each to determine the identity of the third particle.

33. Identify the particles corresponding to the quark states (a) suu, (b) $\bar{u}d$, (c) $\bar{s}d$, and (d) ssd.

34. What is the electrical charge of the baryons with the quark compositions (a) $\bar{u}\,\bar{u}\,d$ and (b) $\bar{u}\,d\,d$? What are these baryons called?

Section 46.10 Multicolored Quarks
Section 46.11 The Standard Model
Section 46.12 The Cosmic Connection

35. **Review Problem.** Review Section 39.4. Prove that the Doppler shift in wavelength of electromagnetic waves is given by

$$\lambda' = \lambda \sqrt{\frac{1 + v/c}{1 - v/c}}$$

where λ' is the wavelength measured by an observer moving at speed v away from a source radiating waves of wavelength λ.

36. Using Hubble's law (Eq. 46.7), find the wavelength of the 590-nm sodium line emitted from galaxies (a) 2.00×10^6 ly away from Earth, (b) 2.00×10^8 ly away, and (c) 2.00×10^9 ly away. You may use the result of Problem 35.

37. **WEB** A distant quasar is moving away from Earth at such high speed that the blue 434-nm hydrogen line is observed at 650 nm, in the red portion of the spectrum. (a) How fast is the quasar receding? You may use the result of Problem 35. (b) Using Hubble's law, determine the distance from Earth to this quasar.

38. The various spectral lines observed in the light from a distant quasar have longer wavelengths λ'_n than the wavelengths λ_n measured in light from a stationary source. The fractional change in wavelength toward the red is the same for all spectral lines. That is, the redshift parameter Z defined by

$$Z = (\lambda'_n - \lambda_n)/\lambda_n$$

is common to all spectral lines for one object. In terms of Z, determine (a) the speed of recession of the quasar and (b) the distance from Earth to this quasar. Use the result of Problem 35 and Hubble's law.

39. It is mostly your roommate's fault. Nosy astronomers have discovered enough junk and clutter in your dorm room to constitute the missing mass required to close the Universe. After observing your floor, closet, bed, and computer files, they calculate the average density of the observable Universe as $1.20 \, \rho_c$. How many times larger will the Universe become before it begins to collapse? That is, by what factor will the distance between remote galaxies increase in the future?

40. The early Universe was dense with gamma-ray photons of energy $\sim k_B T$ and at such a high temperature that protons and antiprotons were created by the process

$\gamma \rightarrow p + \bar{p}$ as rapidly as they annihilated each other. As the Universe cooled in adiabatic expansion, its temperature fell below a certain value, and proton-pair production became rare. At that time, slightly more protons than antiprotons existed, and essentially all of the protons in the Universe today date from that time. (a) Estimate the order of magnitude of the temperature of the Universe when protons condensed out. (b) Estimate the order of magnitude of the temperature of the Universe when electrons condensed out.

41. **Review Problem.** The cosmic background radiation is blackbody radiation at a temperature of 2.73 K. (a) Determine the wavelength at which this radiation has its maximum intensity. (b) In what part of the electromagnetic spectrum is the peak of the distribution?

Section 46.13 Problems and Perspectives

42. Classical general relativity views the structure of space–time as deterministic and well defined down to arbitrarily small distances. On the other hand, quantum general relativity forbids distances less than the Planck length given by $L = (\hbar G/c^3)^{1/2}$. (a) Calculate the value of the Planck length. The quantum limitation suggests that after the Big Bang, when all of the presently observable section of the Universe was reduced to a point-like singularity, nothing could be observed until that singularity grew larger than the Planck length. Because the size of the singularity grew at the speed of light, we can infer that no observations were possible during the time it took for light to travel the Planck length. (b) Calculate this time, known as the Planck time T, and compare it with the ultrahot epoch mentioned in the text. (c) Does this suggest that we may never know what happened between the time $t = 0$ and the time $t = T$?

ADDITIONAL PROBLEMS

43. The nuclear force can be attributed to the exchange of an elementary particle between protons and neutrons if they are sufficiently close. Take the range of the nuclear force as approximately 1.4×10^{-15} m. (a) Use the uncertainty principle $\Delta E \, \Delta t \geq \hbar/2$ to estimate the mass of the elementary particle if it moves at nearly the speed of light. (b) Using Table 46.2, identify the particle.

44. Name at least one conservation law that prevents each of the following reactions.
    (a) $\pi^- + p \rightarrow \Sigma^+ + \pi^0$
    (b) $\mu^- \rightarrow \pi^- + \nu_e$
    (c) $p \rightarrow \pi^+ + \pi^+ + \pi^-$

WEB 45. The energy flux carried by neutrinos from the Sun is estimated to be on the order of 0.4 W/m^2 at Earth's surface. Estimate the fractional mass loss of the Sun over 10^9 years due to the radiation of neutrinos. (The mass of the Sun is 2×10^{30} kg. The Earth–Sun distance is 1.5×10^{11} m.)

46. Two protons approach each other with 70.4 MeV of kinetic energy and engage in a reaction in which a proton and a positive pion emerge at rest. What third particle, obviously uncharged and therefore difficult to detect, must have been created?

47. **Review Problem.** Supernova 1987A, located about 170 000 ly from Earth, is estimated to have emitted a burst of $\sim 10^{46}$ J of neutrinos. Suppose that the average neutrino energy was 6 MeV and your body presented cross-sectional area 5 000 cm^2. To an order of magnitude, how many of these neutrinos passed through you?

48. A gamma-ray photon strikes a stationary electron. Determine the minimum gamma-ray energy to make this reaction go:

$$\gamma + e^- \longrightarrow e^- + e^- + e^+$$

49. Determine the kinetic energies of the proton and pion resulting from the decay of a Λ^0 at rest:

$$\Lambda^0 \longrightarrow p + \pi^-$$

50. An unstable particle, initially at rest, decays into a proton (rest energy 938.3 MeV) and a negative pion (rest energy 139.5 MeV). A uniform magnetic field of 0.250 T exists perpendicular to the velocities of the created particles. The radius of curvature of each track is found to be 1.33 m. What is the mass of the original unstable particle?

51. A Σ^0 particle at rest decays according to

$$\Sigma^0 \longrightarrow \Lambda^0 + \gamma$$

Find the gamma-ray energy.

52. Two protons approach each other with equal and opposite velocities. What is the minimum kinetic energy of each of the protons if they are to produce a π^+ meson at rest in the following reaction?

$$p + p \longrightarrow p + n + \pi^+$$

53. If a K_S^0 meson at rest decays in 0.900×10^{-10} s, how far will a K_S^0 meson travel if it is moving at $0.960\,c$ through a bubble chamber?

54. A π-meson at rest decays according to $\pi^- \rightarrow \mu^- + \bar{\nu}_\mu$. What energy is carried off by the neutrino? (Assume a massless neutrino that moves off with the speed of light.) $m_\pi c^2 = 139.5$ MeV, $m_\mu c^2 = 105.7$ MeV.

55. **Review Problem.** Use the Boltzmann distribution function $e^{-E/k_B T}$ to calculate the temperature at which 1.00% of a population of photons will have energy greater than 1.00 eV. The energy required to excite an atom is on the order of 1 eV. Thus, as the temperature of the Universe fell below the value you calculate, neutral atoms could form from plasma, and the Universe became transparent. The cosmic background radiation represents our vastly red-shifted view of the opaque fireball of the Big Bang as it was at that time and temperature. The fireball surrounds us; we are embers.

56. What processes are described by the Feynman diagrams in Figure P46.56? What is the exchange particle in each process?

57. Identify the mediators for the two interactions described in the Feynman diagrams shown in Figure P46.57.

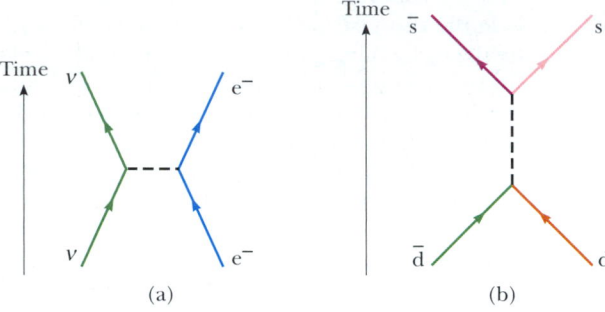

Figure P46.57

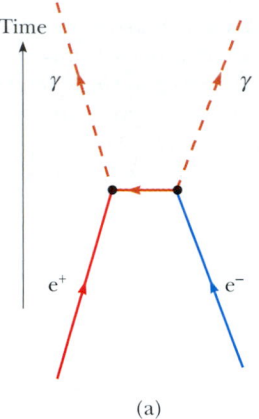

(a)

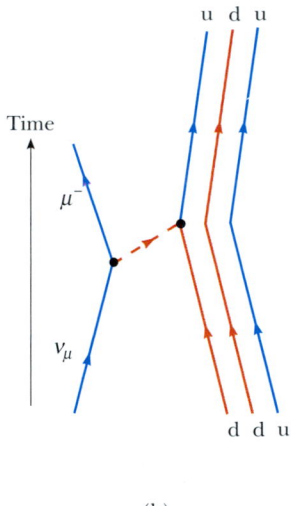

(b)

Figure P46.56

ANSWERS TO QUICK QUIZZES

46.1 (a) Into the plane. The right-hand rule for the positive particle tells you that this is the direction that leads to a force directed toward the center of curvature of the path. (b) The particle must have been slowed (by collisions) during its encounter with the lead, causing it to move in a tighter circular path (see Eq. 29.13, $r = mv/qB$).

46.2 Equation 46.3 indicates that the masses of both of these field particles must be zero; otherwise d would be less than infinity.

46.3 No, because several conservation laws are violated. Electric charge is not conserved because the negative charge on the electron disappears. Electron lepton number is not conserved because an electron with $L_e = 1$ exists before the decay and two neutrinos, each with $L_e = 1$, exist afterward. Angular momentum is not conserved because one spin-$\frac{1}{2}$ particle exists before the decay, and two spin-$\frac{1}{2}$ particles exist afterward.

46.4 We can argue this from the point of view of creating particle–antiparticle pairs from available energy. If energy is converted to rest energy of a lepton–antilepton pair, no net change occurs in lepton number because $L_e = 1$ for the lepton and $L_e = -1$ for the antilepton. Energy can also be transformed to rest energy of a baryon–antibaryon pair. The baryon has $B = +1$ and the antibaryon has $B = -1$, so no net change occurs in baryon number.

Now suppose energy is transformed to rest energy of a quark–antiquark pair. By definition, such a pair is a meson. Thus, a meson has been created from energy—no meson existed before, now one does. Thus, meson number is not conserved. With more energy, we can create more mesons, with no restriction from a conservation law other than the law of conservation of energy.

"Particles, particles, particles."

The Meaning of Success

To earn the respect of intelligent people and to win the affection of children;
To appreciate the beauty in nature and all that surrounds us;
To seek out and nurture the best in others;
To give the gift of yourself to others without the slightest thought of return, for it is in giving that we receive;
To have accomplished a task, whether it be saving a lost soul, healing a sick child, writing a book, or risking your life for a friend;
To have celebrated and laughed with great joy and enthusiasm and sung with exaltation;
To have hope even in times of despair, for as long as you have hope, you have life;
To love and be loved;
To be understood and to understand;
To know that even one life has breathed easier because you have lived;
This is the meaning of success.

RALPH WALDO EMERSON
Modified by Ray Serway, December 1989

TABLE A.1 Conversion Factors

Length

	m	cm	km	in.	ft	mi
1 meter	1	10^2	10^{-3}	39.37	3.281	6.214×10^{-4}
1 centimeter	10^{-2}	1	10^{-5}	0.393 7	3.281×10^{-2}	6.214×10^{-6}
1 kilometer	10^3	10^5	1	3.937×10^4	3.281×10^3	0.621 4
1 inch	2.540×10^{-2}	2.540	2.540×10^{-5}	1	8.333×10^{-2}	1.578×10^{-5}
1 foot	0.304 8	30.48	3.048×10^{-4}	12	1	1.894×10^{-4}
1 mile	1 609	1.609×10^5	1.609	6.336×10^4	5 280	1

Mass

	kg	g	slug	u
1 kilogram	1	10^3	6.852×10^{-2}	6.024×10^{26}
1 gram	10^{-3}	1	6.852×10^{-5}	6.024×10^{23}
1 slug	14.59	1.459×10^4	1	8.789×10^{27}
1 atomic mass unit	1.660×10^{-27}	1.660×10^{-24}	1.137×10^{-28}	1

Note: 1 metric ton = 1 000 kg.

Time

	s	min	h	day	yr
1 second	1	1.667×10^{-2}	2.778×10^{-4}	1.157×10^{-5}	3.169×10^{-8}
1 minute	60	1	1.667×10^{-2}	6.994×10^{-4}	1.901×10^{-6}
1 hour	3 600	60	1	4.167×10^{-2}	1.141×10^{-4}
1 day	8.640×10^4	1 440	24	1	2.738×10^{-5}
1 year	3.156×10^7	5.259×10^5	8.766×10^3	365.2	1

Speed

	m/s	cm/s	ft/s	mi/h
1 meter per second	1	10^2	3.281	2.237
1 centimeter per second	10^{-2}	1	3.281×10^{-2}	2.237×10^{-2}
1 foot per second	0.304 8	30.48	1	0.681 8
1 mile per hour	0.447 0	44.70	1.467	1

Note: 1 mi/min = 60 mi/h = 88 ft/s.

continued

TABLE A.1   Continued

Force

	N	lb
1 newton	1	0.224 8
1 pound	4.448	1

Work, Energy, Heat

	J	ft·lb	eV
1 joule	1	0.737 6	6.242×10^{18}
1 ft·lb	1.356	1	8.464×10^{18}
1 eV	1.602×10^{-19}	1.182×10^{-19}	1
1 cal	4.186	3.087	2.613×10^{19}
1 Btu	1.055×10^3	7.779×10^2	6.585×10^{21}
1 kWh	3.600×10^6	2.655×10^6	2.247×10^{25}

	cal	Btu	kWh
1 joule	0.238 9	9.481×10^{-4}	2.778×10^{-7}
1 ft·lb	0.323 9	1.285×10^{-3}	3.766×10^{-7}
1 eV	3.827×10^{-20}	1.519×10^{-22}	4.450×10^{-26}
1 cal	1	3.968×10^{-3}	1.163×10^{-6}
1 Btu	2.520×10^2	1	2.930×10^{-4}
1 kWh	8.601×10^5	3.413×10^2	1

Pressure

	Pa	atm
1 pascal	1	9.869×10^{-6}
1 atmosphere	1.013×10^5	1
1 centimeter mercury[a]	1.333×10^3	1.316×10^{-2}
1 pound per inch^2	6.895×10^3	6.805×10^{-2}
1 pound per foot^2	47.88	4.725×10^{-4}

	cm Hg	lb/in.^{2}	lb/ft^2
1 newton per meter^2	7.501×10^{-4}	1.450×10^{-4}	2.089×10^{-2}
1 atmosphere	76	14.70	2.116×10^3
1 centimeter mercury[a]	1	0.194 3	27.85
1 pound per inch^2	5.171	1	144
1 pound per foot^2	3.591×10^{-2}	6.944×10^{-3}	1

[a] At 0°C and at a location where the acceleration due to gravity has its "standard" value, 9.806 65 m/s^2.

TABLE A.2   Symbols, Dimensions, and Units of Physical Quantities

Quantity	Common Symbol	Unit[a]	Dimensions[b]	Unit in Terms of Base SI Units
Acceleration	**a**	m/s^2	L/T^2	m/s^2
Amount of substance	n	mole		mol
Angle	θ, ϕ	radian (rad)	1	
Angular acceleration	$\boldsymbol{\alpha}$	rad/s^2	T^{-2}	s^{-2}
Angular frequency	ω	rad/s	T^{-1}	s^{-1}
Angular momentum	**L**	$kg \cdot m^2/s$	ML^2/T	$kg \cdot m^2/s$
Angular velocity	$\boldsymbol{\omega}$	rad/s	T^{-1}	s^{-1}
Area	A	m^2	L^2	m^2
Atomic number	Z			
Capacitance	C	farad (F)	Q^2T^2/ML^2	$A^2 \cdot s^4/kg \cdot m^2$
Charge	q, Q, e	coulomb (C)	Q	$A \cdot s$
Charge density				
Line	λ	C/m	Q/L	$A \cdot s/m$
Surface	σ	C/m^2	Q/L^2	$A \cdot s/m^2$
Volume	ρ	C/m^3	Q/L^3	$A \cdot s/m^3$
Conductivity	σ	$1/\Omega \cdot m$	Q^2T/ML^3	$A^2 \cdot s^3/kg \cdot m^3$
Current	I	AMPERE	Q/T	A
Current density	**J**	A/m^2	Q/T^2	A/m^2
Density	ρ	kg/m^3	M/L^3	kg/m^3
Dielectric constant	κ			
Displacement	**r, s**	METER	L	m
Distance	d, h			
Length	ℓ, L			
Electric dipole moment	**p**	$C \cdot m$	QL	$A \cdot s \cdot m$
Electric field	**E**	V/m	ML/QT^2	$kg \cdot m/A \cdot s^3$
Electric flux	Φ_E	$V \cdot m$	ML^3/QT^2	$kg \cdot m^3/A \cdot s^3$
Electromotive force	ε	volt (V)	ML^2/QT^2	$kg \cdot m^2/A \cdot s^3$
Energy	E, U, K	joule (J)	ML^2/T^2	$kg \cdot m^2/s^2$
Entropy	S	J/K	$ML^2/T^2 \cdot K$	$kg \cdot m^2/s^2 \cdot K$
Force	**F**	newton (N)	ML/T^2	$kg \cdot m/s^2$
Frequency	f	hertz (Hz)	T^{-1}	s^{-1}
Heat	Q	joule (J)	ML^2/T^2	$kg \cdot m^2/s^2$
Inductance	L	henry (H)	ML^2/Q^2	$kg \cdot m^2/A^2 \cdot s^2$
Magnetic dipole moment	$\boldsymbol{\mu}$	$N \cdot m/T$	QL^2/T	$A \cdot m^2$
Magnetic field	**B**	tesla (T) $(= Wb/m^2)$	M/QT	$kg/A \cdot s^2$
Magnetic flux	Φ_B	weber (Wb)	ML^2/QT	$kg \cdot m^2/A \cdot s^2$
Mass	m, M	KILOGRAM	M	kg
Molar specific heat	C	$J/mol \cdot K$		$kg \cdot m^2/s^2 \cdot mol \cdot K$
Moment of inertia	I	$kg \cdot m^2$	ML^2	$kg \cdot m^2$
Momentum	**p**	$kg \cdot m/s$	ML/T	$kg \cdot m/s$
Period	T	s	T	s
Permeability of space	μ_0	$N/A^2 (= H/m)$	ML/Q^2T	$kg \cdot m/A^2 \cdot s^2$
Permittivity of space	ϵ_0	$C^2/N \cdot m^2 (= F/m)$	Q^2T^2/ML^3	$A^2 \cdot s^4/kg \cdot m^3$
Potential	V	volt (V) $(= J/C)$	ML^2/QT^2	$kg \cdot m^2/A \cdot s^3$
Power	$\mathscr{P}$	watt (W) $(= J/s)$	ML^2/T^3	$kg \cdot m^2/s^3$

continued

TABLE A.2 **Continued**

Quantity	Common Symbol	Unit[a]	Dimensions[b]	Unit in Terms of Base SI Units
Pressure	P	pascal (Pa) $= (N/m^2)$	M/LT^2	$kg/m \cdot s^2$
Resistance	R	ohm $(\Omega)(=V/A)$	ML^2/Q^2T	$kg \cdot m^2/A^2 \cdot s^3$
Specific heat	c	$J/kg \cdot K$	$L^2/T^2 \cdot K$	$m^2/s^2 \cdot K$
Speed	v	m/s	L/T	m/s
Temperature	T	KELVIN	K	K
Time	t	SECOND	T	s
Torque	τ	$N \cdot m$	ML^2/T^2	$kg \cdot m^2/s^2$
Volume	V	m^3	L^3	m^3
Wavelength	λ	m	L	m
Work	W	joule $(J)(=N \cdot m)$	ML^2/T^2	$kg \cdot m^2/s^2$

[a] The base SI units are given in uppercase letters.

[b] The symbols M, L, T, and Q denote mass, length, time, and charge, respectively.

TABLE A.3 **Table of Atomic Masses[a]**

Atomic Number Z	Element	Symbol	Chemical Atomic Mass (u)	Mass Number (* Indicates Radioactive) A	Atomic Mass (u)	Percent Abundance	Half-Life (If Radioactive) $T_{1/2}$
0	(Neutron)	n		1*	1.008 665		10.4 min
1	Hydrogen	H	1.007 9	1	1.007 825	99.985	
	Deuterium	D		2	2.014 102	0.015	
	Tritium	T		3*	3.016 049		12.33 yr
2	Helium	He	4.002 60	3	3.016 029	0.000 14	
				4	4.002 602	99.999 86	
				6*	6.018 886		0.81 s
3	Lithium	Li	6.941	6	6.015 121	7.5	
				7	7.016 003	92.5	
				8*	8.022 486		0.84 s
4	Beryllium	Be	9.012 2	7*	7.016 928		53.3 days
				9	9.012 174	100	
				10*	10.013 534		1.5×10^6 yr
5	Boron	B	10.81	10	10.012 936	19.9	
				11	11.009 305	80.1	
				12*	12.014 352		0.020 2 s
6	Carbon	C	12.011	10*	10.016 854		19.3 s
				11*	11.011 433		20.4 min
				12	12.000 000	98.90	
				13	13.003 355	1.10	
				14*	14.003 242		5 730 yr
				15*	15.010 599		2.45 s
7	Nitrogen	N	14.006 7	12*	12.018 613		0.011 0 s
				13*	13.005 738		9.96 min
				14	14.003 074	99.63	
				15	15.000 108	0.37	
				16*	16.006 100		7.13 s
				17*	17.008 450		4.17 s

TABLE A.3 *Continued*

Atomic Number Z	Element	Symbol	Chemical Atomic Mass (u)	Mass Number (* Indicates Radioactive) A	Atomic Mass (u)	Percent Abundance	Half-Life (If Radioactive) $T_{1/2}$
8	Oxygen	O	15.999 4	14*	14.008 595		70.6 s
				15*	15.003 065		122 s
				16	15.994 915	99.761	
				17	16.999 132	0.039	
				18	17.999 160	0.20	
				19*	19.003 577		26.9 s
9	Fluorine	F	18.998 40	17*	17.002 094		64.5 s
				18*	18.000 937		109.8 min
				19	18.998 404	100	
				20*	19.999 982		11.0 s
				21*	20.999.950		4.2 s
10	Neon	Ne	20.180	18*	18.005 710		1.67 s
				19*	19.001 880		17.2 s
				20	19.992 435	90.48	
				21	20.993 841	0.27	
				22	21.991 383	9.25	
				23*	22.994 465		37.2 s
11	Sodium	Na	22.989 87	21*	20.997 650		22.5 s
				22*	21.994 434		2.61 yr
				23	22.989 770	100	
				24*	23.990 961		14.96 h
12	Magnesium	Mg	24.305	23*	22.994 124		11.3 s
				24	23.985 042	78.99	
				25	24.985 838	10.00	
				26	25.982 594	11.01	
				27*	26.984 341		9.46 min
13	Aluminum	Al	26.981 54	26*	25.986 892		7.4×10^5 yr
				27	26.981 538	100	
				28*	27.981 910		2.24 min
14	Silicon	Si	28.086	28	27.976 927	92.23	
				29	28.976 495	4.67	
				30	29.973 770	3.10	
				31*	30.975 362		2.62 h
				32*	31.974 148		172 yr
15	Phosphorus	P	30.973 76	30*	29.978 307		2.50 min
				31	30.973 762	100	
				32*	31.973 908		14.26 days
				33*	32.971 725		25.3 days
16	Sulfur	S	32.066	32	31.972 071	95.02	
				33	32.971 459	0.75	
				34	33.967 867	4.21	
				35*	34.969 033		87.5 days
				36	35.967 081	0.02	
17	Chlorine	Cl	35.453	35	34.968 853	75.77	
				36*	35.968 307		3.0×10^5 yr
				37	36.965 903	24.23	

continued

TABLE A.3   *Continued*

Atomic Number Z	Element	Symbol	Chemical Atomic Mass (u)	Mass Number (* Indicates Radioactive) A	Atomic Mass (u)	Percent Abundance	Half-Life (If Radioactive) $T_{1/2}$
18	Argon	Ar	39.948	36	35.967 547	0.337	
				37*	36.966 776		35.04 days
				38	37.962 732	0.063	
				39*	38.964 314		269 yr
				40	39.962 384	99.600	
				42*	41.963 049		33 yr
19	Potassium	K	39.098 3	39	38.963 708	93.258 1	
				40*	39.964 000	0.011 7	1.28×10^9 yr
				41	40.961 827	6.730 2	
20	Calcium	Ca	40.08	40	39.962 591	96.941	
				41*	40.962 279		1.0×10^5 yr
				42	41.958 618	0.647	
				43	42.958 767	0.135	
				44	43.955 481	2.086	
				46	45.953 687	0.004	
				48	47.952 534	0.187	
21	Scandium	Sc	44.955 9	41*	40.969 250		0.596 s
				45	44.955 911	100	
22	Titanium	Ti	47.88	44*	43.959 691		49 yr
				46	45.952 630	8.0	
				47	46.951 765	7.3	
				48	47.947 947	73.8	
				49	48.947 871	5.5	
				50	49.944 792	5.4	
23	Vanadium	V	50.941 5	48*	47.952 255		15.97 days
				50*	49.947 161	0.25	1.5×10^{17} yr
				51	50.943 962	99.75	
24	Chromium	Cr	51.996	48*	47.954 033		21.6 h
				50	49.946 047	4.345	
				52	51.940 511	83.79	
				53	52.940 652	9.50	
				54	53.938 883	2.365	
25	Manganese	Mn	54.938 05	54*	53.940 361		312.1 days
				55	54.938 048	100	
26	Iron	Fe	55.847	54	53.939 613	5.9	
				55*	54.938 297		2.7 yr
				56	55.934 940	91.72	
				57	56.935 396	2.1	
				58	57.933 278	0.28	
				60*	59.934 078		1.5×10^6 yr
27	Cobalt	Co	58.933 20	59	58.933 198	100	
				60*	59.933 820		5.27 yr
28	Nickel	Ni	58.693	58	57.935 346	68.077	
				59*	58.934 350		7.5×10^4 yr
				60	59.930 789	26.223	
				61	60.931 058	1.140	
				62	61.928 346	3.634	
				63*	62.929 670		100 yr
				64	63.927 967	0.926	

TABLE A.3 Continued

Atomic Number Z	Element	Symbol	Chemical Atomic Mass (u)	Mass Number (* Indicates Radioactive) A	Atomic Mass (u)	Percent Abundance	Half-Life (If Radioactive) $T_{1/2}$
29	Copper	Cu	63.54	63	62.929 599	69.17	
				65	64.927 791	30.83	
30	Zinc	Zn	65.39	64	63.929 144	48.6	
				66	65.926 035	27.9	
				67	66.927 129	4.1	
				68	67.924 845	18.8	
				70	69.925 323	0.6	
31	Gallium	Ga	69.723	69	68.925 580	60.108	
				71	70.924 703	39.892	
32	Germanium	Ge	72.61	70	69.924 250	21.23	
				72	71.922 079	27.66	
				73	72.923 462	7.73	
				74	73.921 177	35.94	
				76	75.921 402	7.44	
33	Arsenic	As	74.921 6	75	74.921 594	100	
34	Selenium	Se	78.96	74	73.922 474	0.89	
				76	75.919 212	9.36	
				77	76.919 913	7.63	
				78	77.917 307	23.78	
				79*	78.918 497		$\leq 6.5 \times 10^4$ yr
				80	79.916 519	49.61	
				82*	81.916 697	8.73	1.4×10^{20} yr
35	Bromine	Br	79.904	79	78.918 336	50.69	
				81	80.916 287	49.31	
36	Krypton	Kr	83.80	78	77.920 400	0.35	
				80	79.916 377	2.25	
				81*	80.916 589		2.1×10^5 yr
				82	81.913 481	11.6	
				83	82.914 136	11.5	
				84	83.911 508	57.0	
				85*	84.912 531		10.76 yr
				86	85.910 615	17.3	
37	Rubidium	Rb	85.468	85	84.911 793	72.17	
				87*	86.909 186	27.83	4.75×10^{10} yr
38	Strontium	Sr	87.62	84	83.913 428	0.56	
				86	85.909 266	9.86	
				87	86.908 883	7.00	
				88	87.905 618	82.58	
				90*	89.907 737		29.1 yr
39	Yttrium	Y	88.905 8	89	88.905 847	100	
40	Zirconium	Zr	91.224	90	89.904 702	51.45	
				91	90.905 643	11.22	
				92	91.905 038	17.15	
				93*	92.906 473		1.5×10^6 yr
				94	93.906 314	17.38	
				96	95.908 274	2.80	

continued

TABLE A.3   *Continued*

Atomic Number Z	Element	Symbol	Chemical Atomic Mass (u)	Mass Number (* Indicates Radioactive) A	Atomic Mass (u)	Percent Abundance	Half-Life (If Radioactive) $T_{1/2}$
41	Niobium	Nb	92.906 4	91*	90.906 988		6.8×10^2 yr
				92*	91.907 191		3.5×10^7 yr
				93	92.906 376	100	
				94*	93.907 280		2×10^4 yr
42	Molybdenum	Mo	95.94	92	91.906 807	14.84	
				93*	92.906 811		3.5×10^3 yr
				94	93.905 085	9.25	
				95	94.905 841	15.92	
				96	95.904 678	16.68	
				97	96.906 020	9.55	
				98	97.905 407	24.13	
				100	99.907 476	9.63	
43	Technetium	Tc		97*	96.906 363		2.6×10^6 yr
				98*	97.907 215		4.2×10^6 yr
				99*	98.906 254		2.1×10^5 yr
44	Ruthenium	Ru	101.07	96	95.907 597	5.54	
				98	97.905 287	1.86	
				99	98.905 939	12.7	
				100	99.904 219	12.6	
				101	100.905 558	17.1	
				102	101.904 348	31.6	
				104	103.905 428	18.6	
45	Rhodium	Rh	102.905 5	103	102.905 502	100	
46	Palladium	Pd	106.42	102	101.905 616	1.02	
				104	103.904 033	11.14	
				105	104.905 082	22.33	
				106	105.903 481	27.33	
				107*	106.905 126		6.5×10^6 yr
				108	107.903 893	26.46	
				110	109.905 158	11.72	
47	Silver	Ag	107.868	107	106.905 091	51.84	
				109	108.904 754	48.16	
48	Cadmium	Cd	112.41	106	105.906 457	1.25	
				108	107.904 183	0.89	
				109*	108.904 984		462 days
				110	109.903 004	12.49	
				111	110.904 182	12.80	
				112	111.902 760	24.13	
				113*	112.904 401	12.22	9.3×10^{15} yr
				114	113.903 359	28.73	
				116	115.904 755	7.49	
49	Indium	In	114.82	113	112.904 060	4.3	
				115*	114.903 876	95.7	4.4×10^{14} yr
50	Tin	Sn	118.71	112	111.904 822	0.97	
				114	113.902 780	0.65	
				115	114.903 345	0.36	
				116	115.901 743	14.53	
				117	116.902 953	7.68	

TABLE A.3   Continued

Atomic Number Z	Element	Symbol	Chemical Atomic Mass (u)	Mass Number (* Indicates Radioactive) A	Atomic Mass (u)	Percent Abundance	Half-Life (If Radioactive) $T_{1/2}$
(50)	(Tin)			118	117.901 605	24.22	
				119	118.903 308	8.58	
				120	119.902 197	32.59	
				121*	120.904 237		55 yr
				122	121.903 439	4.63	
				124	123.905 274	5.79	
51	Antimony	Sb	121.76	121	120.903 820	57.36	
				123	122.904 215	42.64	
				125*	124.905 251		2.7 yr
52	Tellurium	Te	127.60	120	119.904 040	0.095	
				122	121.903 052	2.59	
				123*	122.904 271	0.905	1.3×10^{13} yr
				124	123.902 817	4.79	
				125	124.904 429	7.12	
				126	125.903 309	18.93	
				128*	127.904 463	31.70	$> 8 \times 10^{24}$ yr
				130*	129.906 228	33.87	$\leqslant 1.25 \times 10^{21}$ yr
53	Iodine	I	126.904 5	127	126.904 474	100	
				129*	128.904 984		1.6×10^{7} yr
54	Xenon	Xe	131.29	124	123.905 894	0.10	
				126	125.904 268	0.09	
				128	127.903 531	1.91	
				129	128.904 779	26.4	
				130	129.903 509	4.1	
				131	130.905 069	21.2	
				132	131.904 141	26.9	
				134	133.905 394	10.4	
				136*	135.907 215	8.9	$\geqslant 2.36 \times 10^{21}$ yr
55	Cesium	Cs	132.905 4	133	132.905 436	100	
				134*	133.906 703		2.1 yr
				135*	134.905 891		2×10^{6} yr
				137*	136.907 078		30 yr
56	Barium	Ba	137.33	130	129.906 289	0.106	
				132	131.905 048	0.101	
				133*	132.905 990		10.5 yr
				134	133.904 492	2.42	
				135	134.905 671	6.593	
				136	135.904 559	7.85	
				137	136.905 816	11.23	
				138	137.905 236	71.70	
57	Lanthanum	La	138.905	137*	136.906 462		6×10^{4} yr
				138*	137.907 105	0.090 2	1.05×10^{11} yr
				139	138.906 346	99.909 8	
58	Cerium	Ce	140.12	136	135.907 139	0.19	
				138	137.905 986	0.25	
				140	139.905 434	88.43	
				142*	141.909 241	11.13	$> 5 \times 10^{16}$ yr
59	Praseodymium	Pr	140.907 6	141	140.907 647	100	

continued

TABLE A.3 *Continued*

Atomic Number Z	Element	Symbol	Chemical Atomic Mass (u)	Mass Number (* Indicates Radioactive) A	Atomic Mass (u)	Percent Abundance	Half-Life (If Radioactive) $T_{1/2}$
60	Neodymium	Nd	144.24	142	141.907 718	27.13	
				143	142.909 809	12.18	
				144*	143.910 082	23.80	2.3×10^{15} yr
				145	144.912 568	8.30	
				146	145.913 113	17.19	
				148	147.916 888	5.76	
				150*	149.920 887	5.64	$>1 \times 10^{18}$ yr
61	Promethium	Pm		143*	142.910 928		265 days
				145*	144.912 745		17.7 yr
				146*	145.914 698		5.5 yr
				147*	146.915 134		2.623 yr
62	Samarium	Sm	150.36	144	143.911 996	3.1	
				146*	145.913 043		1.0×10^{8} yr
				147*	146.914 894	15.0	1.06×10^{11} yr
				148*	147.914 819	11.3	7×10^{15} yr
				149*	148.917 180	13.8	$>2 \times 10^{15}$ yr
				150	149.917 273	7.4	
				151*	150.919 928		90 yr
				152	151.919 728	26.7	
				154	153.922 206	22.7	
63	Europium	Eu	151.96	151	150.919 846	47.8	
				152*	151.921 740		13.5 yr
				153	152.921 226	52.2	
				154*	153.922 975		8.59 yr
				155*	154.922 888		4.7 yr
64	Gadolinium	Gd	157.25	148*	147.918 112		75 yr
				150*	149.918 657		1.8×10^{6} yr
				152*	151.919 787	0.20	1.1×10^{14} yr
				154	153.920 862	2.18	
				155	154.922 618	14.80	
				156	155.922 119	20.47	
				157	156.923 957	15.65	
				158	157.924 099	24.84	
				160	159.927 050	21.86	
65	Terbium	Tb	158.925 3	159	158.925 345	100	
66	Dysprosium	Dy	162.50	156	155.924 277	0.06	
				158	157.924 403	0.10	
				160	159.925 193	2.34	
				161	160.926 930	18.9	
				162	161.926 796	25.5	
				163	162.928 729	24.9	
				164	163.929 172	28.2	
67	Holmium	Ho	164.930 3	165	164.930 316	100	
				166*	165.932 282		1.2×10^{3} yr
68	Erbium	Er	167.26	162	161.928 775	0.14	
				164	163.929 198	1.61	
				166	165.930 292	33.6	

TABLE A.3 Continued

Atomic Number Z	Element	Symbol	Chemical Atomic Mass (u)	Mass Number (* Indicates Radioactive) A	Atomic Mass (u)	Percent Abundance	Half-Life (If Radioactive) $T_{1/2}$
(68)	(Erbium)			167	166.932 047	22.95	
				168	167.932 369	27.8	
				170	169.935 462	14.9	
69	Thulium	Tm	168.934 2	169	168.934 213	100	
				171*	170.936 428		1.92 yr
70	Ytterbium	Yb	173.04	168	167.933 897	0.13	
				170	169.934 761	3.05	
				171	170.936 324	14.3	
				172	171.936 380	21.9	
				173	172.938 209	16.12	
				174	173.938 861	31.8	
				176	175.942 564	12.7	
71	Lutecium	Lu	174.967	173*	172.938 930		1.37 yr
				175	174.940 772	97.41	
				176*	175.942 679	2.59	3.78×10^{10} yr
72	Hafnium	Hf	178.49	174*	173.940 042	0.162	2.0×10^{15} yr
				176	175.941 404	5.206	
				177	176.943 218	18.606	
				178	177.943 697	27.297	
				179	178.945 813	13.629	
				180	179.946 547	35.100	
73	Tantalum	Ta	180.947 9	180	179.947 542	0.012	
				181	180.947 993	99.988	
74	Tungsten (Wolfram)	W	183.85	180	179.946 702	0.12	
				182	181.948 202	26.3	
				183	182.950 221	14.28	
				184	183.950 929	30.7	
				186	185.954 358	28.6	
75	Rhenium	Re	186.207	185	184.952 951	37.40	
				187*	186.955 746	62.60	4.4×10^{10} yr
76	Osmium	Os	190.2	184	183.952 486	0.02	
				186*	185.953 834	1.58	2.0×10^{15} yr
				187	186.955 744	1.6	
				188	187.955 832	13.3	
				189	188.958 139	16.1	
				190	189.958 439	26.4	
				192	191.961 468	41.0	
				194*	193.965 172		6.0 yr
77	Iridium	Ir	192.2	191	190.960 585	37.3	
				193	192.962 916	62.7	
78	Platinum	Pt	195.08	190*	189.959 926	0.01	6.5×10^{11} yr
				192	191.961 027	0.79	
				194	193.962 655	32.9	
				195	194.964 765	33.8	
				196	195.964 926	25.3	
				198	197.967 867	7.2	
79	Gold	Au	196.966 5	197	196.966 543	100	

continued

TABLE A.3 *Continued*

Atomic Number Z	Element	Symbol	Chemical Atomic Mass (u)	Mass Number (* Indicates Radioactive) A	Atomic Mass (u)	Percent Abundance	Half-Life (If Radioactive) $T_{1/2}$
80	Mercury	Hg	200.59	196	195.965 806	0.15	
				198	197.966 743	9.97	
				199	198.968 253	16.87	
				200	199.968 299	23.10	
				201	200.970 276	13.10	
				202	201.970 617	29.86	
				204	203.973 466	6.87	
81	Thallium	Tl	204.383	203	202.972 320	29.524	
				204*	203.973 839		3.78 yr
				205	204.974 400	70.476	
		(Ra E″)		206*	205.976 084		4.2 min
		(Ac C″)		207*	206.977 403		4.77 min
		(Th C″)		208*	207.981 992		3.053 min
		(Ra C″)		210*	209.990 057		1.30 min
82	Lead	Pb	207.2	202*	201.972 134		5×10^4 yr
				204*	203.973 020	1.4	$\geqslant 1.4 \times 10^{17}$ yr
				205*	204.974 457		1.5×10^7 yr
				206	205.974 440	24.1	
				207	206.975 871	22.1	
				208	207.976 627	52.4	
		(Ra D)		210*	209.984 163		22.3 yr
		(Ac B)		211*	210.988 734		36.1 min
		(Th B)		212*	211.991 872		10.64 h
		(Ra B)		214*	213.999 798		26.8 min
83	Bismuth	Bi	208.980 3	207*	206.978 444		32.2 yr
				208*	207.979 717		3.7×10^5 yr
				209	208.980 374	100	
		(Ra E)		210*	209.984 096		5.01 days
		(Th C)		211*	210.987 254		2.14 min
				212*	211.991 259		60.6 min
		(Ra C)		214*	213.998 692		19.9 min
				215*	215.001 836		7.4 min
84	Polonium	Po		209*	208.982 405		102 yr
		(Ra F)		210*	209.982 848		138.38 days
		(Ac C′)		211*	210.986 627		0.52 s
		(Th C′)		212*	211.988 842		0.30 μs
		(Ra C′)		214*	213.995 177		164 μs
		(Ac A)		215*	214.999 418		0.001 8 s
		(Th A)		216*	216.001 889		0.145 s
		(Ra A)		218*	218.008 965		3.10 min
85	Astatine	At		215*	214.998 638		$\approx 100 \ \mu$s
				218*	218.008 685		1.6 s
				219*	219.011 294		0.9 min
86	Radon	Rn					
		(An)		219*	219.009 477		3.96 s
		(Tn)		220*	220.011 369		55.6 s
		(Rn)		222*	222.017 571		3.823 days
87	Francium	Fr					
		(Ac K)		223*	223.019 733		22 min

TABLE A.3   Continued

Atomic Number Z	Element	Symbol	Chemical Atomic Mass (u)	Mass Number (* Indicates Radioactive) A	Atomic Mass (u)	Percent Abundance	Half-Life (If Radioactive) $T_{1/2}$
88	Radium	Ra					
		(Ac X)		223*	223.018 499		11.43 days
		(Th X)		224*	224.020 187		3.66 days
		(Ra)		226*	226.025 402		1 600 yr
		(Ms Th$_1$)		228*	228.031 064		5.75 yr
89	Actinium	Ac		227*	227.027 749		21.77 yr
		(Ms Th$_2$)		228*	228.031 015		6.15 h
90	Thorium	Th	232.038 1				
		(Rd Ac)		227*	227.027 701		18.72 days
		(Rd Th)		228*	228.028 716		1.913 yr
				229*	229.031 757		7 300 yr
		(Io)		230*	230.033 127		75.000 yr
		(UY)		231*	231.036 299		25.52 h
		(Th)		232*	232.038 051	100	1.40×10^{10} yr
		(UX$_1$)		234*	234.043 593		24.1 days
91	Protactinium	Pa		231*	231.035 880		32.760 yr
		(Uz)		234*	234.043 300		6.7 h
92	Uranium	U	238.028 9	232*	232.037 131		69 yr
				233*	233.039 630		1.59×10^5 yr
				234*	234.040 946	0.005 5	2.45×10^5 yr
		(Ac U)		235*	235.043 924	0.720	7.04×10^8 yr
				236*	236.045 562		2.34×10^7 yr
		(UI)		238*	238.050 784	99.274 5	4.47×10^9 yr
93	Neptunium	Np		235*	235.044 057		396 days
				236*	236.046 560		1.15×10^5 yr
				237*	237.048 168		2.14×10^6 yr
94	Plutonium	Pu		236*	236.046 033		2.87 yr
				238*	238.049 555		87.7 yr
				239*	239.052 157		2.412×10^4 yr
				240*	240.053 808		6 560 yr
				241*	241.056 846		14.4 yr
				242*	242.058 737		3.73×10^6 yr
				244*	244.064 200		8.1×10^7 yr

[a] The masses in the sixth column are atomic masses, which include the mass of Z electrons. Data are from the National Nuclear Data Center, Brookhaven National Laboratory, prepared by Jagdish K. Tuli, July 1990. The data are based on experimental results reported in *Nuclear Data Sheets* and *Nuclear Physics* and also from *Chart of the Nuclides*, 14th ed. Atomic masses are based on those by A. H. Wapstra, G. Audi, and R. Hoekstra. Isotopic abundances are based on those by N. E. Holden.

APPENDIX B · *Mathematics Review*

These appendices in mathematics are intended as a brief review of operations and methods. Early in this course, you should be totally familiar with basic algebraic techniques, analytic geometry, and trigonometry. The appendices on differential and integral calculus are more detailed and are intended for those students who have difficulty applying calculus concepts to physical situations.

B.1 SCIENTIFIC NOTATION

Many quantities that scientists deal with often have very large or very small values. For example, the speed of light is about 300 000 000 m/s, and the ink required to make the dot over an *i* in this textbook has a mass of about 0.000 000 001 kg. Obviously, it is very cumbersome to read, write, and keep track of numbers such as these. We avoid this problem by using a method dealing with powers of the number 10:

$$10^0 = 1$$
$$10^1 = 10$$
$$10^2 = 10 \times 10 = 100$$
$$10^3 = 10 \times 10 \times 10 = 1000$$
$$10^4 = 10 \times 10 \times 10 \times 10 = 10\,000$$
$$10^5 = 10 \times 10 \times 10 \times 10 \times 10 = 100\,000$$

and so on. The number of zeros corresponds to the power to which 10 is raised, called the **exponent** of 10. For example, the speed of light, 300 000 000 m/s, can be expressed as 3×10^8 m/s.

In this method, some representative numbers smaller than unity are

$$10^{-1} = \frac{1}{10} = 0.1$$

$$10^{-2} = \frac{1}{10 \times 10} = 0.01$$

$$10^{-3} = \frac{1}{10 \times 10 \times 10} = 0.001$$

$$10^{-4} = \frac{1}{10 \times 10 \times 10 \times 10} = 0.000\,1$$

$$10^{-5} = \frac{1}{10 \times 10 \times 10 \times 10 \times 10} = 0.000\,01$$

In these cases, the number of places the decimal point is to the left of the digit 1 equals the value of the (negative) exponent. Numbers expressed as some power of 10 multiplied by another number between 1 and 10 are said to be in **scientific notation.** For example, the scientific notation for 5 943 000 000 is 5.943×10^9 and that for 0.000 083 2 is 8.32×10^{-5}.

When numbers expressed in scientific notation are being multiplied, the following general rule is very useful:

$$10^n \times 10^m = 10^{n+m} \qquad \text{(B.1)}$$

where n and m can be *any* numbers (not necessarily integers). For example, $10^2 \times 10^5 = 10^7$. The rule also applies if one of the exponents is negative: $10^3 \times 10^{-8} = 10^{-5}$.

When dividing numbers expressed in scientific notation, note that

$$\frac{10^n}{10^m} = 10^n \times 10^{-m} = 10^{n-m} \qquad \text{(B.2)}$$

EXERCISES

With help from the above rules, verify the answers to the following:

1. $86\ 400 = 8.64 \times 10^4$
2. $9\ 816\ 762.5 = 9.816\ 762\ 5 \times 10^6$
3. $0.000\ 000\ 039\ 8 = 3.98 \times 10^{-8}$
4. $(4 \times 10^8)(9 \times 10^9) = 3.6 \times 10^{18}$
5. $(3 \times 10^7)(6 \times 10^{-12}) = 1.8 \times 10^{-4}$
6. $\dfrac{75 \times 10^{-11}}{5 \times 10^{-3}} = 1.5 \times 10^{-7}$
7. $\dfrac{(3 \times 10^6)(8 \times 10^{-2})}{(2 \times 10^{17})(6 \times 10^5)} = 2 \times 10^{-18}$

B.2 ▶ ALGEBRA

Some Basic Rules

When algebraic operations are performed, the laws of arithmetic apply. Symbols such as x, y, and z are usually used to represent quantities that are not specified, what are called the **unknowns.**

First, consider the equation

$$8x = 32$$

If we wish to solve for x, we can divide (or multiply) each side of the equation by the same factor without destroying the equality. In this case, if we divide both sides by 8, we have

$$\frac{8x}{8} = \frac{32}{8}$$

$$x = 4$$

Next consider the equation

$$x + 2 = 8$$

In this type of expression, we can add or subtract the same quantity from each side. If we subtract 2 from each side, we get

$$x + 2 - 2 = 8 - 2$$

$$x = 6$$

In general, if $x + a = b$, then $x = b - a$.

Now consider the equation

$$\frac{x}{5} = 9$$

If we multiply each side by 5, we are left with x on the left by itself and 45 on the right:

$$\left(\frac{x}{5}\right)(5) = 9 \times 5$$

$$x = 45$$

In all cases, *whatever operation is performed on the left side of the equality must also be performed on the right side.*

The following rules for multiplying, dividing, adding, and subtracting fractions should be recalled, where a, b, and c are three numbers:

	Rule	**Example**
Multiplying	$\left(\dfrac{a}{b}\right)\left(\dfrac{c}{d}\right) = \dfrac{ac}{bd}$	$\left(\dfrac{2}{3}\right)\left(\dfrac{4}{5}\right) = \dfrac{8}{15}$
Dividing	$\dfrac{(a/b)}{(c/d)} = \dfrac{ad}{bc}$	$\dfrac{2/3}{4/5} = \dfrac{(2)(5)}{(4)(3)} = \dfrac{10}{12}$
Adding	$\dfrac{a}{b} \pm \dfrac{c}{d} = \dfrac{ad \pm bc}{bd}$	$\dfrac{2}{3} - \dfrac{4}{5} = \dfrac{(2)(5) - (4)(3)}{(3)(5)} = -\dfrac{2}{15}$

EXERCISES

In the following exercises, solve for x:

Answers

1. $a = \dfrac{1}{1 + x}$ $x = \dfrac{1 - a}{a}$

2. $3x - 5 = 13$ $x = 6$

3. $ax - 5 = bx + 2$ $x = \dfrac{7}{a - b}$

4. $\dfrac{5}{2x + 6} = \dfrac{3}{4x + 8}$ $x = -\dfrac{11}{7}$

Powers

When powers of a given quantity x are multiplied, the following rule applies:

$$x^n x^m = x^{n+m} \tag{B.3}$$

For example, $x^2 x^4 = x^{2+4} = x^6$.

When dividing the powers of a given quantity, the rule is

$$\frac{x^n}{x^m} = x^{n-m} \tag{B.4}$$

For example, $x^8 / x^2 = x^{8-2} = x^6$.

A power that is a fraction, such as $\frac{1}{3}$, corresponds to a root as follows:

$$x^{1/n} = \sqrt[n]{x} \tag{B.5}$$

For example, $4^{1/3} = \sqrt[3]{4} = 1.5874$. (A scientific calculator is useful for such calculations.)

Finally, any quantity x^n raised to the mth power is

$$(x^n)^m = x^{nm} \tag{B.6}$$

Table B.1 summarizes the rules of exponents.

TABLE B.1
Rules of Exponents

$$x^0 = 1$$
$$x^1 = x$$
$$x^n x^m = x^{n+m}$$
$$x^n / x^m = x^{n-m}$$
$$x^{1/n} = \sqrt[n]{x}$$
$$(x^n)^m = x^{nm}$$

EXERCISES

Verify the following:

1. $3^2 \times 3^3 = 243$
2. $x^5 x^{-8} = x^{-3}$
3. $x^{10} / x^{-5} = x^{15}$
4. $5^{1/3} = 1.709\ 975$ (Use your calculator.)
5. $60^{1/4} = 2.783\ 158$ (Use your calculator.)
6. $(x^4)^3 = x^{12}$

Factoring

Some useful formulas for factoring an equation are

$$ax + ay + az = a(x + y + x) \qquad \text{common factor}$$
$$a^2 + 2ab + b^2 = (a + b)^2 \qquad \text{perfect square}$$
$$a^2 - b^2 = (a + b)(a - b) \qquad \text{differences of squares}$$

Quadratic Equations

The general form of a quadratic equation is

$$ax^2 + bx + c = 0 \tag{B.7}$$

where x is the unknown quantity and a, b, and c are numerical factors referred to as **coefficients** of the equation. This equation has two roots, given by

$$x = \frac{-b \pm \sqrt{b^2 - 4ac}}{2a} \tag{B.8}$$

If $b^2 \geq 4ac$, the roots are real.

EXAMPLE 1

The equation $x^2 + 5x + 4 = 0$ has the following roots corresponding to the two signs of the square-root term:

$$x = \frac{-5 \pm \sqrt{5^2 - (4)(1)(4)}}{2(1)} = \frac{-5 \pm \sqrt{9}}{2} = \frac{-5 \pm 3}{2}$$

$$x_+ = \frac{-5 + 3}{2} = \boxed{-1} \qquad x_- = \frac{-5 - 3}{2} = \boxed{-4}$$

where x_+ refers to the root corresponding to the positive sign and x_- refers to the root corresponding to the negative sign.

EXERCISES

Solve the following quadratic equations:

Answers

1. $x^2 + 2x - 3 = 0$ $x_+ = 1$ $x_- = -3$
2. $2x^2 - 5x + 2 = 0$ $x_+ = 2$ $x_- = \frac{1}{2}$
3. $2x^2 - 4x - 9 = 0$ $x_+ = 1 + \sqrt{22}/2$ $x_- = 1 - \sqrt{22}/2$

Linear Equations

A linear equation has the general form

$$y = mx + b \tag{B.9}$$

where m and b are constants. This equation is referred to as being linear because the graph of y versus x is a straight line, as shown in Figure B.1. The constant b, called the **y-intercept,** represents the value of y at which the straight line intersects the y axis. The constant m is equal to the **slope** of the straight line and is also equal to the tangent of the angle that the line makes with the x axis. If any two points on the straight line are specified by the coordinates (x_1, y_1) and (x_2, y_2), as in Figure B.1, then the slope of the straight line can be expressed as

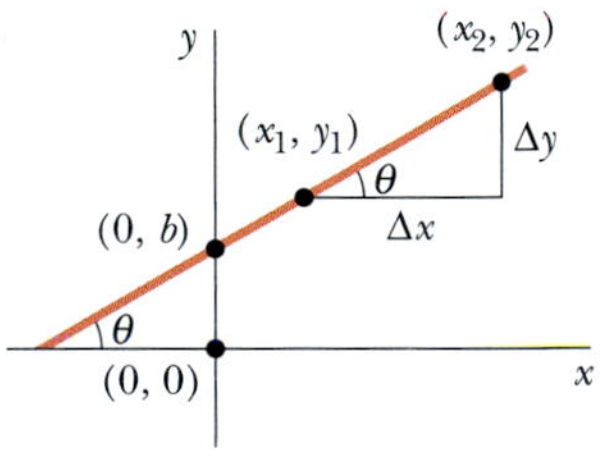

$$\text{Slope} = \frac{y_2 - y_1}{x_2 - x_1} = \frac{\Delta y}{\Delta x} = \tan \theta \tag{B.10}$$

Note that m and b can have either positive or negative values. If $m > 0$, the straight line has a *positive* slope, as in Figure B1. If $m < 0$, the straight line has a *negative* slope. In Figure B.1, both m and b are positive. Three other possible situations are shown in Figure B.2.

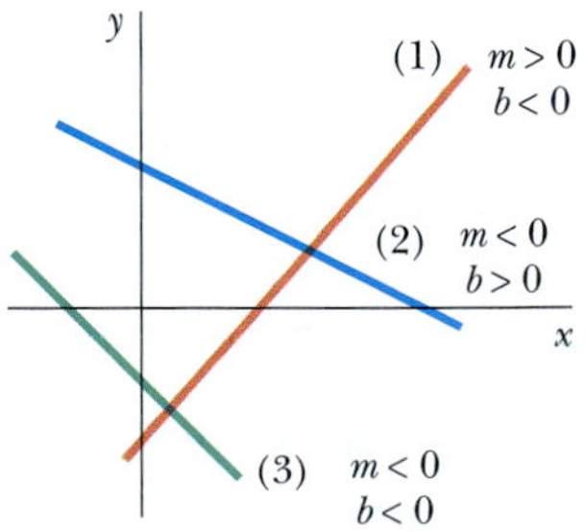

Figure B.1

Figure B.2

EXERCISES

1. Draw graphs of the following straight lines:
   (a) $y = 5x + 3$ (b) $y = -2x + 4$ (c) $y = -3x - 6$
2. Find the slopes of the straight lines described in Exercise 1.

Answers   (a) 5   (b) -2 (c) -3

3. Find the slopes of the straight lines that pass through the following sets of points:

(a) $(0, -4)$ and $(4, 2)$, (b) $(0, 0)$ and $(2, -5)$, and (c) $(-5, 2)$ and $(4, -2)$

Answers   (a) $3/2$ (b) $-5/2$ (c) $-4/9$

Solving Simultaneous Linear Equations

Consider the equation $3x + 5y = 15$, which has two unknowns, x and y. Such an equation does not have a unique solution. For example, note that $(x = 0, y = 3)$, $(x = 5, y = 0)$, and $(x = 2, y = 9/5)$ are all solutions to this equation.

If a problem has two unknowns, a unique solution is possible only if we have *two* equations. In general, if a problem has n unknowns, its solution requires n equations. In order to solve two simultaneous equations involving two unknowns, x and y, we solve one of the equations for x in terms of y and substitute this expression into the other equation.

EXAMPLE 2

Solve the following two simultaneous equations:

(1) $5x + y = -8$

(2) $2x - 2y = 4$

Solution   From (2), $x = y + 2$. Substitution of this into (1) gives

$$5(y + 2) + y = -8$$

$$6y = -18$$

$$y = -3$$

$$x = y + 2 = \boxed{-1}$$

Alternate Solution   Multiply each term in (1) by the factor 2 and add the result to (2):

$$10x + 2y = -16$$

$$\underline{2x - 2y = 4}$$

$$12x = -12$$

$$x = -1$$

$$y = x - 2 = \boxed{-3}$$

Two linear equations containing two unknowns can also be solved by a graphical method. If the straight lines corresponding to the two equations are plotted in a conventional coordinate system, the intersection of the two lines represents the solution. For example, consider the two equations

$$x - y = 2$$

$$x - 2y = -1$$

These are plotted in Figure B.3. The intersection of the two lines has the coordinates $x = 5$, $y = 3$. This represents the solution to the equations. You should check this solution by the analytical technique discussed above.

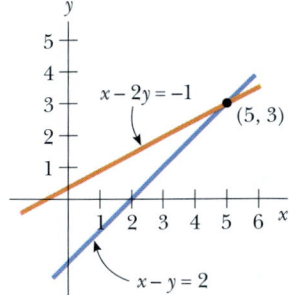

Figure B.3

EXERCISES

Solve the following pairs of simultaneous equations involving two unknowns:

Answers

1. $x + y = 8$ $x = 5, y = 3$
     $x - y = 2$

2. $98 - T = 10a$ $T = 65, a = 3.27$
   $T - 49 = 5a$
3. $6x + 2y = 6$ $x = 2, y = -3$
   $8x - 4y = 28$

Logarithms

Suppose that a quantity x is expressed as a power of some quantity a:

$$x = a^y \tag{B.11}$$

The number a is called the **base** number. The **logarithm** of x with respect to the base a is equal to the exponent to which the base must be raised in order to satisfy the expression $x = a^y$:

$$y = \log_a x \tag{B.12}$$

Conversely, the **antilogarithm** of y is the number x:

$$x = \text{antilog}_a y \tag{B.13}$$

In practice, the two bases most often used are base 10, called the *common* logarithm base, and base $e = 2.718 \ldots$, called Euler's constant or the *natural* logarithm base. When common logarithms are used,

$$y = \log_{10} x \qquad (\text{or } x = 10^y) \tag{B.14}$$

When natural logarithms are used,

$$y = \ln_e x \qquad (\text{or } x = e^y) \tag{B.15}$$

For example, $\log_{10} 52 = 1.716$, so that $\text{antilog}_{10} 1.716 = 10^{1.716} = 52$. Likewise, $\ln_e 52 = 3.951$, so $\text{antiln}_e 3.951 = e^{3.951} = 52$.

In general, note that you can convert between base 10 and base e with the equality

$$\ln_e x = (2.302\ 585) \log_{10} x \tag{B.16}$$

Finally, some useful properties of logarithms are

$$\log(ab) = \log a + \log b$$
$$\log(a/b) = \log a - \log b$$
$$\log(a^n) = n \log a$$
$$\ln e = 1$$
$$\ln e^a = a$$
$$\ln\left(\frac{1}{a}\right) = -\ln a$$

B.3 ▶ GEOMETRY

The **distance** d between two points having coordinates (x_1, y_1) and (x_2, y_2) is

$$d = \sqrt{(x_2 - x_1)^2 + (y_2 - y_1)^2} \tag{B.17}$$

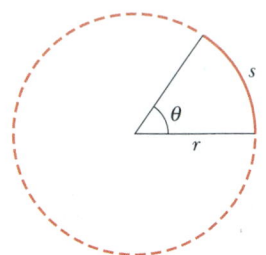

Figure B.4

Radian measure: The arc length s of a circular arc (Fig. B.4) is proportional to the radius r for a fixed value of θ (in radians):

$$s = r\theta$$
$$\theta = \frac{s}{r} \qquad \text{(B.18)}$$

Table B.2 gives the areas and volumes for several geometric shapes used throughout this text:

TABLE B.2   Useful Information for Geometry

Shape	Area or Volume	Shape	Area or Volume
Rectangle	Area $= \ell w$	Sphere	Surface area $= 4\pi r^2$ Volume $= \dfrac{4\pi r^3}{3}$
Circle	Area $= \pi r^2$ (Circumference $= 2\pi r$)	Cylinder	Lateral surface area $= 2\pi r \ell$ Volume $= \pi r^2 \ell$
Triangle	Area $= \frac{1}{2} bh$	Rectangular box	Area $= 2(\ell h + \ell w + hw)$ Volume $= \ell w h$

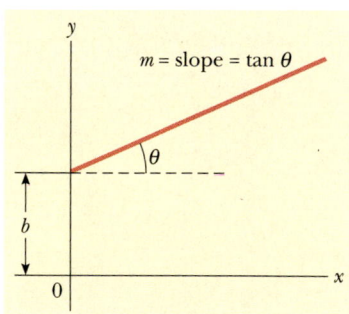

Figure B.5

The equation of a **straight line** (Fig. B.5) is

$$y = mx + b \qquad \text{(B.19)}$$

where b is the y-intercept and m is the slope of the line.

The equation of a **circle** of radius R centered at the origin is

$$x^2 + y^2 = R^2 \qquad \text{(B.20)}$$

The equation of an **ellipse** having the origin at its center (Fig. B.6) is

$$\frac{x^2}{a^2} + \frac{y^2}{b^2} = 1 \qquad \text{(B.21)}$$

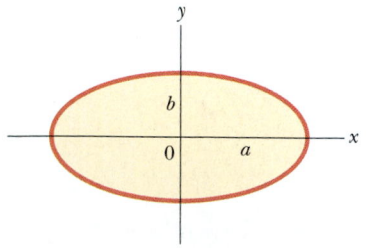

Figure B.6

where a is the length of the semi-major axis (the longer one) and b is the length of the semi-minor axis (the shorter one).

The equation of a **parabola** the vertex of which is at $y = b$ (Fig. B.7) is

$$y = ax^2 + b \qquad \text{(B.22)}$$

The equation of a **rectangular hyperbola** (Fig. B.8) is

$$xy = \text{constant} \qquad \text{(B.23)}$$

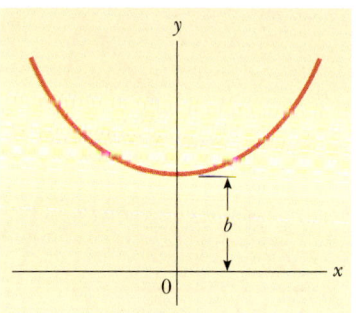

Figure B.7

B.4 ▶ TRIGONOMETRY

That portion of mathematics based on the special properties of the right triangle is called trigonometry. By definition, a right triangle is one containing a 90° angle. Consider the right triangle shown in Figure B.9, where side a is opposite the angle θ, side b is adjacent to the angle θ, and side c is the hypotenuse of the triangle. The three basic trigonometric functions defined by such a triangle are the sine (sin), cosine (cos), and tangent (tan) functions. In terms of the angle θ, these functions are defined by

$$\sin \theta \equiv \frac{\text{side opposite } \theta}{\text{hypotenuse}} = \frac{a}{c} \qquad \text{(B.24)}$$

$$\cos \theta \equiv \frac{\text{side adjacent to } \theta}{\text{hypotenuse}} = \frac{b}{c} \qquad \text{(B.25)}$$

$$\tan \theta \equiv \frac{\text{side opposite } \theta}{\text{side adjacent to } \theta} = \frac{a}{b} \qquad \text{(B.26)}$$

The Pythagorean theorem provides the following relationship between the sides of a right triangle:

$$c^2 = a^2 + b^2 \qquad \text{(B.27)}$$

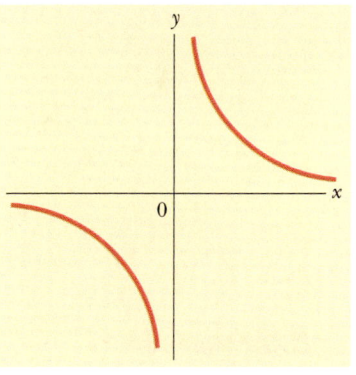

Figure B.8

From the above definitions and the Pythagorean theorem, it follows that

$$\sin^2 \theta + \cos^2 \theta = 1$$

$$\tan \theta = \frac{\sin \theta}{\cos \theta}$$

The cosecant, secant, and cotangent functions are defined by

$$\csc \theta \equiv \frac{1}{\sin \theta} \qquad \sec \theta \equiv \frac{1}{\cos \theta} \qquad \cot \theta \equiv \frac{1}{\tan \theta}$$

The relationships below follow directly from the right triangle shown in Figure B.9:

$$\sin \theta = \cos(90° - \theta)$$

$$\cos \theta = \sin(90° - \theta)$$

$$\cot \theta = \tan(90° - \theta)$$

Some properties of trigonometric functions are

$$\sin(-\theta) = -\sin \theta$$

$$\cos(-\theta) = \cos \theta$$

$$\tan(-\theta) = -\tan \theta$$

The following relationships apply to *any* triangle, as shown in Figure B.10:

$$\alpha + \beta + \gamma = 180°$$

a = opposite side
b = adjacent side
c = hypotenuse

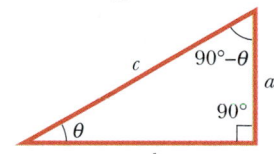

Figure B.9

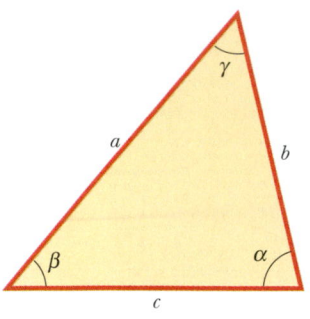

Figure B.10

Law of cosines
$$a^2 = b^2 + c^2 - 2bc\cos\alpha$$
$$b^2 = a^2 + c^2 - 2ac\cos\beta$$
$$c^2 = a^2 + b^2 - 2ab\cos\gamma$$

Law of sines
$$\frac{a}{\sin\alpha} = \frac{b}{\sin\beta} = \frac{c}{\sin\gamma}$$

Table B.3 lists a number of useful trigonometric identities.

TABLE B.3   Some Trigonometric Identities

$\sin^2\theta + \cos^2\theta = 1$	$\csc^2\theta = 1 + \cot^2\theta$
$\sec^2\theta = 1 + \tan^2\theta$	$\sin^2\dfrac{\theta}{2} = \tfrac{1}{2}(1 - \cos\theta)$
$\sin 2\theta = 2\sin\theta\cos\theta$	$\cos^2\dfrac{\theta}{2} = \tfrac{1}{2}(1 + \cos\theta)$
$\cos 2\theta = \cos^2\theta - \sin^2\theta$	$1 - \cos\theta = 2\sin^2\dfrac{\theta}{2}$
$\tan 2\theta = \dfrac{2\tan\theta}{1 - \tan^2\theta}$	$\tan\dfrac{\theta}{2} = \sqrt{\dfrac{1 - \cos\theta}{1 + \cos\theta}}$

$$\sin(A \pm B) = \sin A\cos B \pm \cos A\sin B$$
$$\cos(A \pm B) = \cos A\cos B \mp \sin A\sin B$$
$$\sin A \pm \sin B = 2\sin[\tfrac{1}{2}(A \pm B)]\cos[\tfrac{1}{2}(A \mp B)]$$
$$\cos A + \cos B = 2\cos[\tfrac{1}{2}(A + B)]\cos[\tfrac{1}{2}(A - B)]$$
$$\cos A - \cos B = 2\sin[\tfrac{1}{2}(A + B)]\sin[\tfrac{1}{2}(B - A)]$$

EXAMPLE 3

Consider the right triangle in Figure B.11, in which $a = 2$, $b = 5$, and c is unknown. From the Pythagorean theorem, we have

$$c^2 = a^2 + b^2 = 2^2 + 5^2 = 4 + 25 = 29$$

$$c = \sqrt{29} = \boxed{5.39}$$

To find the angle θ, note that

$$\tan\theta = \frac{a}{b} = \frac{2}{5} = 0.400$$

From a table of functions or from a calculator, we have

$$\theta = \tan^{-1}(0.400) = \boxed{21.8°}$$

where $\tan^{-1}(0.400)$ is the notation for "angle whose tangent is 0.400," sometimes written as arctan (0.400).

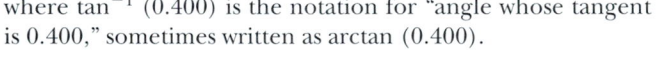
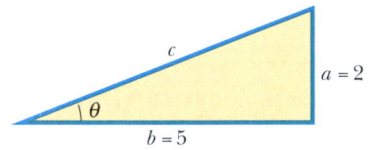

Figure B.11

EXERCISES

1. In Figure B.12, identify (a) the side opposite θ and (b) the side adjacent to ϕ and then find (c) $\cos\theta$, (d) $\sin\phi$, and (e) $\tan\phi$.

Answers   (a) 3, (b) 3, (c) $\frac{4}{5}$, (d) $\frac{4}{5}$, and (e) $\frac{4}{3}$

2. In a certain right triangle, the two sides that are perpendicular to each other are 5 m and 7 m long. What is the length of the third side?

Answer   8.60 m

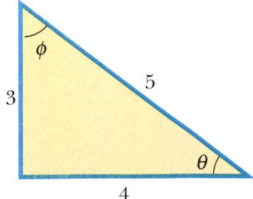

Figure B.12

3. A right triangle has a hypotenuse of length 3 m, and one of its angles is 30°. What is the length of (a) the side opposite the 30° angle and (b) the side adjacent to the 30° angle?

Answers (a) 1.5 m, (b) 2.60 m

B.5 SERIES EXPANSIONS

$$(a + b)^n = a^n + \frac{n}{1!} a^{n-1}b + \frac{n(n-1)}{2!} a^{n-2}b^2 + \cdots$$

$$(1 + x)^n = 1 + nx + \frac{n(n-1)}{2!} x^2 + \cdots$$

$$e^x = 1 + x + \frac{x^2}{2!} + \frac{x^3}{3!} + \cdots$$

$$\ln(1 \pm x) = \pm x - \tfrac{1}{2}x^2 \pm \tfrac{1}{3}x^3 - \cdots$$

$$\left.\begin{array}{l} \sin x = x - \dfrac{x^3}{3!} + \dfrac{x^5}{5!} - \cdots \\[2mm] \cos x = 1 - \dfrac{x^2}{2!} + \dfrac{x^4}{4!} - \cdots \\[2mm] \tan x = x + \dfrac{x^3}{3} + \dfrac{2x^5}{15} + \cdots \quad |x| < \pi/2 \end{array}\right\} x \text{ in radians}$$

For $x \ll 1$, the following approximations can be used[1]:

$$(1 + x)^n \approx 1 + nx \qquad \sin x \approx x$$

$$e^x \approx 1 + x \qquad \cos x \approx 1$$

$$\ln(1 \pm x) \approx \pm x \qquad \tan x \approx x$$

B.6 DIFFERENTIAL CALCULUS

In various branches of science, it is sometimes necessary to use the basic tools of calculus, invented by Newton, to describe physical phenomena. The use of calculus is fundamental in the treatment of various problems in Newtonian mechanics, electricity, and magnetism. In this section, we simply state some basic properties and "rules of thumb" that should be a useful review to the student.

First, a **function** must be specified that relates one variable to another (such as a coordinate as a function of time). Suppose one of the variables is called y (the dependent variable), the other x (the independent variable). We might have a function relationship such as

$$y(x) = ax^3 + bx^2 + cx + d$$

If a, b, c, and d are specified constants, then y can be calculated for any value of x. We usually deal with continuous functions, that is, those for which y varies "smoothly" with x.

[1] The approximations for the functions sin x, cos x, and tan x are for $x \leq 0.1$ rad.

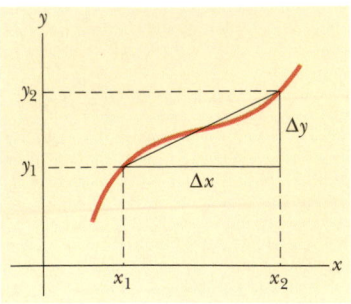

Figure B.13

The **derivative** of y with respect to x is defined as the limit, as Δx approaches zero, of the slopes of chords drawn between two points on the y versus x curve. Mathematically, we write this definition as

$$\frac{dy}{dx} = \lim_{\Delta x \to 0} \frac{\Delta y}{\Delta x} = \lim_{\Delta x \to 0} \frac{y(x + \Delta x) - y(x)}{\Delta x} \tag{B.28}$$

where Δy and Δx are defined as $\Delta x = x_2 - x_1$ and $\Delta y = y_2 - y_1$ (Fig. B.13). It is important to note that dy/dx does not mean dy divided by dx, but is simply a notation of the limiting process of the derivative as defined by Equation B.28.

A useful expression to remember when $y(x) = ax^n$, where a is a *constant* and n is *any* positive or negative number (integer or fraction), is

$$\frac{dy}{dx} = nax^{n-1} \tag{B.29}$$

If $y(x)$ is a polynomial or algebraic function of x, we apply Equation B.29 to *each* term in the polynomial and take $d[\text{constant}]/dx = 0$. In Examples 4 through 7, we evaluate the derivatives of several functions.

EXAMPLE 4

Suppose $y(x)$ (that is, y as a function of x) is given by

$$y(x) = ax^3 + bx + c$$

where a and b are constants. Then it follows that

$$y(x + \Delta x) = a(x + \Delta x)^3$$
$$+ b(x + \Delta x) + c$$

$$y(x + \Delta x) = a(x^3 + 3x^2\Delta x + 3x\Delta x^2 + \Delta x^3)$$
$$+ b(x + \Delta x) + c$$

so

$$\Delta y = y(x + \Delta x) - y(x) = a(3x^2\Delta x + 3x\Delta x^2 + \Delta x^3)$$
$$+ b\Delta x$$

Substituting this into Equation B.28 gives

$$\frac{dy}{dx} = \lim_{\Delta x \to 0} \frac{\Delta y}{\Delta x} = \lim_{\Delta x \to 0} [3ax^2 + 3x\Delta x + \Delta x^2] + b$$

$$\frac{dy}{dx} = \boxed{3ax^2 + b}$$

EXAMPLE 5

$$y(x) = 8x^5 + 4x^3 + 2x + 7$$

Solution Applying Equation B.29 to each term independently, and remembering that d/dx (constant) $= 0$, we have

$$\frac{dy}{dx} = 8(5)x^4 + 4(3)x^2 + 2(1)x^0 + 0$$

$$\frac{dy}{dx} = \boxed{40x^4 + 12x^2 + 2}$$

Special Properties of the Derivative

A. Derivative of the product of two functions If a function $f(x)$ is given by the product of two functions, say, $g(x)$ and $h(x)$, then the derivative of $f(x)$ is defined as

$$\frac{d}{dx} f(x) = \frac{d}{dx}[g(x)h(x)] = g\frac{dh}{dx} + h\frac{dg}{dx} \tag{B.30}$$

B. Derivative of the sum of two functions If a function $f(x)$ is equal to the sum of two functions, then the derivative of the sum is equal to the sum of the derivatives:

$$\frac{d}{dx}f(x) = \frac{d}{dx}[g(x) + h(x)] = \frac{dg}{dx} + \frac{dh}{dx} \qquad \textbf{(B.31)}$$

C. Chain rule of differential calculus If $y = f(x)$ and $x = g(z)$, then dy/dx can be written as the product of two derivatives:

$$\frac{dy}{dz} = \frac{dy}{dx}\frac{dx}{dz} \qquad \textbf{(B.32)}$$

D. The second derivative The second derivative of y with respect to x is defined as the derivative of the function dy/dx (the derivative of the derivative). It is usually written

$$\frac{d^2y}{dx^2} = \frac{d}{dx}\left(\frac{dy}{dx}\right) \qquad \textbf{(B.33)}$$

EXAMPLE 6

Find the derivative of $y(x) = x^3/(x + 1)^2$ with respect to x.

Solution We can rewrite this function as $y(x) = x^3(x + 1)^{-2}$ and apply Equation B.30:

$$\frac{dy}{dx} = (x + 1)^{-2}\frac{d}{dx}(x^3) + x^3\frac{d}{dx}(x + 1)^{-2}$$

$$= (x + 1)^{-2}3x^2 + x^3(-2)(x + 1)^{-3}$$

$$\frac{dy}{dx} = \frac{3x^2}{(x + 1)^2} - \frac{2x^3}{(x + 1)^3}$$

EXAMPLE 7

A useful formula that follows from Equation B.30 is the derivative of the quotient of two functions. Show that

$$\frac{d}{dx}\left[\frac{g(x)}{h(x)}\right] = \frac{h\dfrac{dg}{dx} - g\dfrac{dh}{dx}}{h^2}$$

Solution We can write the quotient as gh^{-1} and then apply Equations B.29 and B.30:

$$\frac{d}{dx}\left(\frac{g}{h}\right) = \frac{d}{dx}(gh^{-1}) = g\frac{d}{dx}(h^{-1}) + h^{-1}\frac{d}{dx}(g)$$

$$= -gh^{-2}\frac{dh}{dx} + h^{-1}\frac{dg}{dx}$$

$$= \frac{h\dfrac{dg}{dx} - g\dfrac{dh}{dx}}{h^2}$$

Some of the more commonly used derivatives of functions are listed in Table B.4.

B.7 INTEGRAL CALCULUS

We think of integration as the inverse of differentiation. As an example, consider the expression

$$f(x) = \frac{dy}{dx} = 3ax^2 + b \qquad \textbf{(B.34)}$$

which was the result of differentiating the function

$$y(x) = ax^3 + bx + c$$

TABLE B.4
Derivatives for Several Functions

$$\frac{d}{dx}(a) = 0$$

$$\frac{d}{dx}(ax^n) = nax^{n-1}$$

$$\frac{d}{dx}(e^{ax}) = ae^{ax}$$

$$\frac{d}{dx}(\sin ax) = a\cos ax$$

$$\frac{d}{dx}(\cos ax) = -a\sin ax$$

$$\frac{d}{dx}(\tan ax) = a\sec^2 ax$$

$$\frac{d}{dx}(\cot ax) = -a\csc^2 ax$$

$$\frac{d}{dx}(\sec x) = \tan x \sec x$$

$$\frac{d}{dx}(\csc x) = -\cot x \csc x$$

$$\frac{d}{dx}(\ln ax) = \frac{1}{x}$$

Note: The letters a and n are constants.

in Example 4. We can write Equation B.34 as $dy = f(x)\,dx = (3ax^2 + b)\,dx$ and obtain $y(x)$ by "summing" over all values of x. Mathematically, we write this inverse operation

$$y(x) = \int f(x)\,dx$$

For the function $f(x)$ given by Equation B.34, we have

$$y(x) = \int (3ax^2 + b)\,dx = ax^3 + bx + c$$

where c is a constant of the integration. This type of integral is called an *indefinite integral* because its value depends on the choice of c.

A general **indefinite integral** $I(x)$ is defined as

$$I(x) = \int f(x)\,dx \qquad \text{(B.35)}$$

where $f(x)$ is called the *integrand* and $f(x) = \dfrac{dI(x)}{dx}$.

For a *general continuous* function $f(x)$, the integral can be described as the area under the curve bounded by $f(x)$ and the x axis, between two specified values of x, say, x_1 and x_2, as in Figure B.14.

The area of the blue element is approximately $f(x_i)\Delta x_i$. If we sum all these area elements from x_1 and x_2 and take the limit of this sum as $\Delta x_i \to 0$, we obtain the *true* area under the curve bounded by $f(x)$ and x, between the limits x_1 and x_2:

$$\text{Area} = \lim_{\Delta x_i \to 0} \sum_i f(x_i)\,\Delta x_i = \int_{x_1}^{x_2} f(x)\,dx \qquad \text{(B.36)}$$

Integrals of the type defined by Equation B.36 are called **definite integrals.**

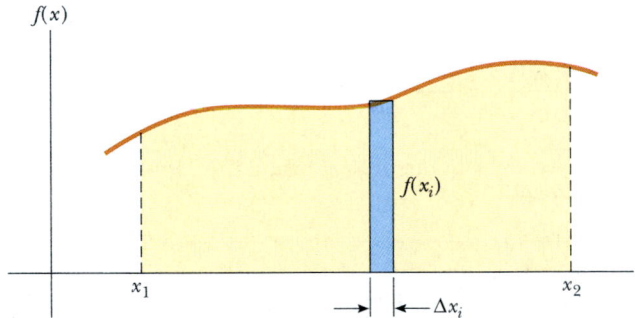

Figure B.14

One common integral that arises in practical situations has the form

$$\int x^n\,dx = \frac{x^{n+1}}{n+1} + c \qquad (n \neq -1) \qquad \text{(B.37)}$$

This result is obvious, being that differentiation of the right-hand side with respect to x gives $f(x) = x^n$ directly. If the limits of the integration are known, this integral becomes a *definite integral* and is written

$$\int_{x_1}^{x_2} x^n\,dx = \frac{x_2^{n+1} - x_1^{n+1}}{n+1} \qquad (n \neq -1) \qquad \text{(B.38)}$$

EXAMPLES

1. $\displaystyle\int_0^a x^2\, dx = \left.\frac{x^3}{3}\right]_0^a = \frac{a^3}{3}$

2. $\displaystyle\int_0^b x^{3/2}\, dx = \left.\frac{x^{5/2}}{5/2}\right]_0^b = \frac{2}{5} b^{5/2}$

3. $\displaystyle\int_3^5 x\, dx = \left.\frac{x^2}{2}\right]_3^5 = \frac{5^2 - 3^2}{2} = 8$

Partial Integration

Sometimes it is useful to apply the method of *partial integration* (also called "integrating by parts") to evaluate certain integrals. The method uses the property that

$$\int u\, dv = uv - \int v\, du \qquad \textbf{(B.39)}$$

where u and v are *carefully* chosen so as to reduce a complex integral to a simpler one. In many cases, several reductions have to be made. Consider the function

$$I(x) = \int x^2 e^x\, dx$$

This can be evaluated by integrating by parts twice. First, if we choose $u = x^2$, $v = e^x$, we get

$$\int x^2 e^x\, dx = \int x^2\, d(e^x) = x^2 e^x - 2\int e^x x\, dx + c_1$$

Now, in the second term, choose $u = x$, $v = e^x$, which gives

$$\int x^2 e^x\, dx = x^2 e^x - 2x e^x + 2\int e^x\, dx + c_1$$

or

$$\int x^2 e^x\, dx = x^2 e^x - 2x e^x + 2 e^x + c_2$$

The Perfect Differential

Another useful method to remember is the use of the *perfect differential,* in which we look for a change of variable such that the differential of the function is the differential of the independent variable appearing in the integrand. For example, consider the integral

$$I(x) = \int \cos^2 x \sin x\, dx$$

This becomes easy to evaluate if we rewrite the differential as $d(\cos x) = -\sin x\, dx$. The integral then becomes

$$\int \cos^2 x \sin x\, dx = -\int \cos^2 x\, d(\cos x)$$

If we now change variables, letting $y = \cos x$, we obtain

$$\int \cos^2 x \sin x\, dx = -\int y^2 dy = -\frac{y^3}{3} + c = -\frac{\cos^3 x}{3} + c$$

Table B.5 lists some useful indefinite integrals. Table B.6 gives Gauss's probability integral and other definite integrals. A more complete list can be found in various handbooks, such as *The Handbook of Chemistry and Physics*, CRC Press.

TABLE B.5   Some Indefinite Integrals (An arbitrary constant should be added to each of these integrals.)

$$\int x^n\, dx = \frac{x^{n+1}}{n+1} \qquad (\text{provided } n \neq -1)$$

$$\int \frac{dx}{x} = \int x^{-1}\, dx = \ln x$$

$$\int \frac{dx}{a+bx} = \frac{1}{b}\ln(a+bx)$$

$$\int \frac{x\, dx}{a+bx} = \frac{x}{b} - \frac{a}{b^2}\ln(a+bx)$$

$$\int \frac{dx}{x(x+a)} = -\frac{1}{a}\ln\frac{x+a}{x}$$

$$\int \frac{dx}{(a+bx)^2} = -\frac{1}{b(a+bx)}$$

$$\int \frac{dx}{a^2+x^2} = \frac{1}{a}\tan^{-1}\frac{x}{a}$$

$$\int \frac{dx}{a^2-x^2} = \frac{1}{2a}\ln\frac{a+x}{a-x} \qquad (a^2-x^2>0)$$

$$\int \frac{dx}{x^2-a^2} = \frac{1}{2a}\ln\frac{x-a}{x+a} \qquad (x^2-a^2>0)$$

$$\int \frac{x\, dx}{a^2\pm x^2} = \pm\tfrac{1}{2}\ln(a^2\pm x^2)$$

$$\int \frac{dx}{\sqrt{a^2-x^2}} = \sin^{-1}\frac{x}{a} = -\cos^{-1}\frac{x}{a} \qquad (a^2-x^2>0)$$

$$\int \frac{dx}{\sqrt{x^2\pm a^2}} = \ln(x+\sqrt{x^2\pm a^2})$$

$$\int \frac{x\, dx}{\sqrt{a^2-x^2}} = -\sqrt{a^2-x^2}$$

$$\int \frac{x\, dx}{\sqrt{x^2\pm a^2}} = \sqrt{x^2\pm a^2}$$

$$\int \sqrt{a^2-x^2}\, dx = \frac{1}{2}\left(x\sqrt{a^2-x^2}+a^2\sin^{-1}\frac{x}{a}\right)$$

$$\int x\sqrt{a^2-x^2}\, dx = -\tfrac{1}{3}(a^2-x^2)^{3/2}$$

$$\int \sqrt{x^2\pm a^2}\, dx = \tfrac{1}{2}[x\sqrt{x^2\pm a^2}\pm a^2\ln(x+\sqrt{x^2\pm a^2})]$$

$$\int x(\sqrt{x^2\pm a^2})\, dx = \tfrac{1}{3}(x^2\pm a^2)^{3/2}$$

$$\int e^{ax}\, dx = \frac{1}{a}e^{ax}$$

$$\int \ln ax\, dx = (x\ln ax) - x$$

$$\int xe^{ax}\, dx = \frac{e^{ax}}{a^2}(ax-1)$$

$$\int \frac{dx}{a+be^{cx}} = \frac{x}{a} - \frac{1}{ac}\ln(a+be^{cx})$$

$$\int \sin ax\, dx = -\frac{1}{a}\cos ax$$

$$\int \cos ax\, dx = \frac{1}{a}\sin ax$$

$$\int \tan ax\, dx = \frac{1}{a}\ln(\cos ax) = \frac{1}{a}\ln(\sec ax)$$

$$\int \cot ax\, dx = \frac{1}{a}\ln(\sin ax)$$

$$\int \sec ax\, dx = \frac{1}{a}\ln(\sec ax+\tan ax) = \frac{1}{a}\ln\left[\tan\left(\frac{ax}{2}+\frac{\pi}{4}\right)\right]$$

$$\int \csc ax\, dx = \frac{1}{a}\ln(\csc ax-\cot ax) = \frac{1}{a}\ln\left(\tan\frac{ax}{2}\right)$$

$$\int \sin^2 ax\, dx = \frac{x}{2} - \frac{\sin 2ax}{4a}$$

$$\int \cos^2 ax\, dx = \frac{x}{2} + \frac{\sin 2ax}{4a}$$

$$\int \frac{dx}{\sin^2 ax} = -\frac{1}{a}\cot ax$$

$$\int \frac{dx}{\cos^2 ax} = \frac{1}{a}\tan ax$$

$$\int \tan^2 ax\, dx = \frac{1}{a}(\tan ax) - x$$

$$\int \cot^2 ax\, dx = -\frac{1}{a}(\cot ax) - x$$

$$\int \sin^{-1} ax\, dx = x(\sin^{-1} ax) + \frac{\sqrt{1-a^2x^2}}{a}$$

$$\int \cos^{-1} ax\, dx = x(\cos^{-1} ax) - \frac{\sqrt{1-a^2x^2}}{a}$$

$$\int \frac{dx}{(x^2+a^2)^{3/2}} = \frac{x}{a^2\sqrt{x^2+a^2}}$$

$$\int \frac{x\, dx}{(x^2+a^2)^{3/2}} = -\frac{1}{\sqrt{x^2+a^2}}$$

TABLE B.6 **Gauss's Probability Integral and Other Definite Integrals**

$$\int_0^\infty x^n e^{-ax} \, dx = \frac{n!}{a^{n+1}}$$

$$I_0 = \int_0^\infty e^{-ax^2} \, dx = \frac{1}{2}\sqrt{\frac{\pi}{a}} \qquad \text{(Gauss's probability integral)}$$

$$I_1 = \int_0^\infty x e^{-ax^2} \, dx = \frac{1}{2a}$$

$$I_2 = \int_0^\infty x^2 e^{-ax^2} \, dx = -\frac{dI_0}{da} = \frac{1}{4}\sqrt{\frac{\pi}{a^3}}$$

$$I_3 = \int_0^\infty x^3 e^{-ax^2} \, dx = -\frac{dI_1}{da} = \frac{1}{2a^2}$$

$$I_4 = \int_0^\infty x^4 e^{-ax^2} \, dx = \frac{d^2 I_0}{da^2} = \frac{3}{8}\sqrt{\frac{\pi}{a^5}}$$

$$I_5 = \int_0^\infty x^5 e^{-ax^2} \, dx = \frac{d^2 I_1}{da^2} = \frac{1}{a^3}$$

$$\cdot$$
$$\cdot$$
$$\cdot$$

$$I_{2n} = (-1)^n \frac{d^n}{da^n} I_0$$

$$I_{2n+1} = (-1)^n \frac{d^n}{da^n} I_1$$

APPENDIX C • *Periodic Table of the Elements*

Symbol — **Ca** 20 — Atomic number
Atomic mass † — 40.08
$4s^2$ — Electron configuration

Group I	Group II	Transition elements						
H 1 1.008 0 $1s^1$								
Li 3 6.94 $2s^1$	**Be** 4 9.012 $2s^2$							
Na 11 22.99 $3s^1$	**Mg** 12 24.31 $3s^2$							
K 19 39.102 $4s^1$	**Ca** 20 40.08 $4s^2$	**Sc** 21 44.96 $3d^1 4s^2$	**Ti** 22 47.90 $3d^2 4s^2$	**V** 23 50.94 $3d^3 4s^2$	**Cr** 24 51.996 $3d^5 4s^1$	**Mn** 25 54.94 $3d^5 4s^2$	**Fe** 26 55.85 $3d^6 4s^2$	**Co** 27 58.93 $3d^7 4s^2$
Rb 37 85.47 $5s^1$	**Sr** 38 87.62 $5s^2$	**Y** 39 88.906 $4d^1 5s^2$	**Zr** 40 91.22 $4d^2 5s^2$	**Nb** 41 92.91 $4d^4 5s^1$	**Mo** 42 95.94 $4d^5 5s^1$	**Tc** 43 (99) $4d^5 5s^2$	**Ru** 44 101.1 $4d^7 5s^1$	**Rh** 45 102.91 $4d^8 5s^1$
Cs 55 132.91 $6s^1$	**Ba** 56 137.34 $6s^2$	57-71*	**Hf** 72 178.49 $5d^2 6s^2$	**Ta** 73 180.95 $5d^3 6s^2$	**W** 74 183.85 $5d^4 6s^2$	**Re** 75 186.2 $5d^5 6s^2$	**Os** 76 190.2 $5d^6 6s^2$	**Ir** 77 192.2 $5d^7 6s^2$
Fr 87 (223) $7s^1$	**Ra** 88 (226) $7s^2$	89-103**	**Rf** 104 (261) $6d^2 7s^2$	**Db** 105 (262) $6d^3 7s^2$	**Sg** 106 (263)	**Bh** 107 (262)	**Hs** 108 (265)	**Mt** 109 (266)

*Lanthanide series	**La** 57 138.91 $5d^1 6s^2$	**Ce** 58 140.12 $5d^1 4f^1 6s^2$	**Pr** 59 140.91 $4f^3 6s^2$	**Nd** 60 144.24 $4f^4 6s^2$	**Pm** 61 (147) $4f^5 6s^2$	**Sm** 62 150.4 $4f^6 6s^2$
**Actinide series	**Ac** 89 (227) $6d^1 7s^2$	**Th** 90 (232) $6d^2 7s^2$	**Pa** 91 (231) $5f^2 6d^1 7s^2$	**U** 92 (238) $5f^3 6d^1 7s^2$	**Np** 93 (239) $5f^4 6d^1 7s^2$	**Pu** 94 (239) $5f^6 6d^0 7s^2$

○ Atomic mass values given are averaged over isotopes in the percentages in which they exist in nature.
† For an unstable element, mass number of the most stable known isotope is given in parentheses.
†† Elements 110, 111, 112, and 114 have not yet been named.
††† For a description of the atomic data, visit **physics.nist.gov/atomic**

	Group III	Group IV	Group V	Group VI	Group VII	Group 0
					H 1 1.008 0 $1s^1$	**He** 2 4.002 6 $1s^2$
	B 5 10.81 $2p^1$	**C** 6 12.011 $2p^2$	**N** 7 14.007 $2p^3$	**O** 8 15.999 $2p^4$	**F** 9 18.998 $2p^5$	**Ne** 10 20.18 $2p^6$
	Al 13 26.98 $3p^1$	**Si** 14 28.09 $3p^2$	**P** 15 30.97 $3p^3$	**S** 16 32.06 $3p^4$	**Cl** 17 35.453 $3p^5$	**Ar** 18 39.948 $3p^6$
Ni 28 58.71 $3d^84s^2$ **Cu** 29 63.54 $3d^{10}4s^1$ **Zn** 30 65.37 $3d^{10}4s^2$	**Ga** 31 69.72 $4p^1$	**Ge** 32 72.59 $4p^2$	**As** 33 74.92 $4p^3$	**Se** 34 78.96 $4p^4$	**Br** 35 79.91 $4p^5$	**Kr** 36 83.80 $4p^6$
Pd 46 106.4 $4d^{10}$ **Ag** 47 107.87 $4d^{10}5s^1$ **Cd** 48 112.40 $4d^{10}5s^2$	**In** 49 114.82 $5p^1$	**Sn** 50 118.69 $5p^2$	**Sb** 51 121.75 $5p^3$	**Te** 52 127.60 $5p^4$	**I** 53 126.90 $5p^5$	**Xe** 54 131.30 $5p^6$
Pt 78 195.09 $5d^96s^1$ **Au** 79 196.97 $5d^{10}6s^1$ **Hg** 80 200.59 $5d^{10}6s^2$	**Tl** 81 204.37 $6p^1$	**Pb** 82 207.2 $6p^2$	**Bi** 83 208.98 $6p^3$	**Po** 84 (210) $6p^4$	**At** 85 (218) $6p^5$	**Rn** 86 (222) $6p^6$
110†† (269) **111††** (272) **112††** (277)		**114††** (289)			

Eu 63 152.0 $4f^76s^2$	**Gd** 64 157.25 $5d^14f^76s^2$	**Tb** 65 158.92 $5d^14f^86s^2$	**Dy** 66 162.50 $4f^{10}6s^2$	**Ho** 67 164.93 $4f^{11}6s^2$	**Er** 68 167.26 $4f^{12}6s^2$	**Tm** 69 168.93 $4f^{13}6s^2$	**Yb** 70 173.04 $4f^{14}6s^2$	**Lu** 71 174.97 $5d^14f^{14}6s^2$
Am 95 (243) $5f^76d^07s^2$	**Cm** 96 (245) $5f^76d^17s^2$	**Bk** 97 (247) $5f^86d^17s^2$	**Cf** 98 (249) $5f^{10}6d^07s^2$	**Es** 99 (254) $5f^{11}6d^07s^2$	**Fm** 100 (253) $5f^{12}6d^07s^2$	**Md** 101 (255) $5f^{13}6d^07s^2$	**No** 102 (255) $6d^07s^2$	**Lr** 103 (257) $6d^17s^2$

APPENDIX D · *SI Units*

TABLE D.1   SI Units

Base Quantity	SI Base Unit Name	Symbol
Length	Meter	m
Mass	Kilogram	kg
Time	Second	s
Electric current	Ampere	A
Temperature	Kelvin	K
Amount of substance	Mole	mol
Luminous intensity	Candela	cd

TABLE D.2   Some Derived SI Units

Quantity	Name	Symbol	Expression in Terms of Base Units	Expression in Terms of Other SI Units
Plane angle	radian	rad	m/m	
Frequency	hertz	Hz	s^{-1}	
Force	newton	N	$kg \cdot m/s^2$	J/m
Pressure	pascal	Pa	$kg/m \cdot s^2$	N/m^2
Energy; work	joule	J	$kg \cdot m^2/s^2$	$N \cdot m$
Power	watt	W	$kg \cdot m^2/s^3$	J/s
Electric charge	coulomb	C	$A \cdot s$	
Electric potential	volt	V	$kg \cdot m^2/A \cdot s^3$	W/A
Capacitance	farad	F	$A^2 \cdot s^4/kg \cdot m^2$	C/V
Electric resistance	ohm	Ω	$kg \cdot m^2/A^2 \cdot s^3$	V/A
Magnetic flux	weber	Wb	$kg \cdot m^2/A \cdot s^2$	$V \cdot s$
Magnetic field intensity	tesla	T	$kg/A \cdot s^2$	
Inductance	henry	H	$kg \cdot m^2/A^2 \cdot s^2$	$T \cdot m^2/A$

All Nobel Prizes in physics are listed (and marked with a P), as well as relevant Nobel Prizes in Chemistry (C). The key dates for some of the scientific work are supplied; they often antedate the prize considerably.

1901 (P) *Wilhelm Roentgen* for discovering x-rays (1895).

1902 (P) *Hendrik A. Lorentz* for predicting the Zeeman effect and *Pieter Zeeman* for discovering the Zeeman effect, the splitting of spectral lines in magnetic fields.

1903 (P) *Antoine-Henri Becquerel* for discovering radioactivity (1896) and *Pierre and Marie Curie* for studying radioactivity.

1904 (P) *Lord Rayleigh* for studying the density of gases and discovering argon.
(C) *William Ramsay* for discovering the inert gas elements helium, neon, xenon, and krypton, and placing them in the periodic table.

1905 (P) *Philipp Lénard* for studying cathode rays, electrons (1898–1899).

1906 (P) *J. J. Thomson* for studying electrical discharge through gases and discovering the electron (1897).

1907 (P) *Albert A. Michelson* for inventing optical instruments and measuring the speed of light (1880s).

1908 (P) *Gabriel Lippmann* for making the first color photographic plate, using interference methods (1891).
(C) *Ernest Rutherford* for discovering that atoms can be broken apart by alpha rays and for studying radioactivity.

1909 (P) *Guglielmo Marconi* and *Carl Ferdinand Braun* for developing wireless telegraphy.

1910 (P) *Johannes D. van der Waals* for studying the equation of state for gases and liquids (1881).

1911 (P) *Wilhelm Wien* for discovering Wien's law giving the peak of a blackbody spectrum (1893).
(C) *Marie Curie* for discovering radium and polonium (1898) and isolating radium.

1912 (P) *Nils Dalén* for inventing automatic gas regulators for lighthouses.

1913 (P) *Heike Kamerlingh Onnes* for the discovery of superconductivity and liquefying helium (1908).

1914 (P) *Max T. F. von Laue* for studying x-rays from their diffraction by crystals, showing that x-rays are electromagnetic waves (1912).
(C) *Theodore W. Richards* for determining the atomic weights of sixty elements, indicating the existence of isotopes.

1915 (P) *William Henry Bragg* and *William Lawrence Bragg*, his son, for studying the diffraction of x-rays in crystals.

1917 (P) *Charles Barkla* for studying atoms by x-ray scattering (1906).

1918 (P) *Max Planck* for discovering energy quanta (1900).

1919 (P) *Johannes Stark,* for discovering the Stark effect, the splitting of spectral lines in electric fields (1913).

1920   (P) *Charles-Édouard Guillaume* for discovering invar, a nickel-steel alloy with low coefficient of expansion.

(C) *Walther Nernst* for studying heat changes in chemical reactions and formulating the third law of thermodynamics (1918).

1921   (P) *Albert Einstein* for explaining the photoelectric effect and for his services to theoretical physics (1905).

(C) *Frederick Soddy* for studying the chemistry of radioactive substances and discovering isotopes (1912).

1922   (P) *Niels Bohr* for his model of the atom and its radiation (1913).

(C) *Francis W. Aston* for using the mass spectrograph to study atomic weights, thus discovering 212 of the 287 naturally occurring isotopes.

1923   (P) *Robert A. Millikan* for measuring the charge on an electron (1911) and for studying the photoelectric effect experimentally (1914).

1924   (P) *Karl M. G. Siegbahn* for his work in x-ray spectroscopy.

1925   (P) *James Franck* and *Gustav Hertz* for discovering the Franck-Hertz effect in electron-atom collisions.

1926   (P) *Jean-Baptiste Perrin* for studying Brownian motion to validate the discontinuous structure of matter and measure the size of atoms.

1927   (P) *Arthur Holly Compton* for discovering the Compton effect on x-rays, their change in wavelength when they collide with matter (1922), and *Charles T. R. Wilson* for inventing the cloud chamber, used to study charged particles (1906).

1928   (P) *Owen W. Richardson* for studying the thermionic effect and electrons emitted by hot metals (1911).

1929   (P) *Louis Victor de Broglie* for discovering the wave nature of electrons (1923).

1930   (P) *Chandrasekhara Venkata Raman* for studying Raman scattering, the scattering of light by atoms and molecules with a change in wavelength (1928).

1932   (P) *Werner Heisenberg* for creating quantum mechanics (1925).

1933   (P) *Erwin Schrödinger* and *Paul A. M. Dirac* for developing wave mechanics (1925) and relativistic quantum mechanics (1927).

(C) *Harold Urey* for discovering heavy hydrogen, deuterium (1931).

1935   (P) *James Chadwick* for discovering the neutron (1932).

(C) *Irène* and *Frédéric Joliot-Curie* for synthesizing new radioactive elements.

1936   (P) *Carl D. Anderson* for discovering the positron in particular and antimatter in general (1932) and *Victor F. Hess* for discovering cosmic rays.

(C) *Peter J. W. Debye* for studying dipole moments and diffraction of x-rays and electrons in gases.

1937   (P) *Clinton Davisson* and *George Thomson* for discovering the diffraction of electrons by crystals, confirming de Broglie's hypothesis (1927).

1938   (P) *Enrico Fermi* for producing the transuranic radioactive elements by neutron irradiation (1934–1937).

1939   (P) *Ernest O. Lawrence* for inventing the cyclotron.

1943   (P) *Otto Stern* for developing molecular-beam studies (1923), and using them to discover the magnetic moment of the proton (1933).

1944   (P) *Isidor I. Rabi* for discovering nuclear magnetic resonance in atomic and molecular beams.

(C) *Otto Hahn* for discovering nuclear fission (1938).

1945   (P) *Wolfgang Pauli* for discovering the exclusion principle (1924).

1946   (P) *Percy W. Bridgman* for studying physics at high pressures.

1947   (P) *Edward V. Appleton* for studying the ionosphere.

1948 (P) *Patrick M. S. Blackett* for studying nuclear physics with cloud-chamber photographs of cosmic-ray interactions.

1949 (P) *Hideki Yukawa* for predicting the existence of mesons (1935).

1950 (P) *Cecil F. Powell* for developing the method of studying cosmic rays with photographic emulsions and discovering new mesons.

1951 (P) *John D. Cockcroft* and *Ernest T. S. Walton* for transmuting nuclei in an accelerator (1932).

(C) *Edwin M. McMillan* for producing neptunium (1940) and *Glenn T. Seaborg* for producing plutonium (1941) and further transuranic elements.

1952 (P) *Felix Bloch* and *Edward Mills Purcell* for discovering nuclear magnetic resonance in liquids and gases (1946).

1953 (P) *Frits Zernike* for inventing the phase-contrast microscope, which uses interference to provide high contrast.

1954 (P) *Max Born* for interpreting the wave function as a probability (1926) and other quantum-mechanical discoveries and *Walther Bothe* for developing the coincidence method to study subatomic particles (1930–1931), producing, in particular, the particle interpreted by Chadwick as the neutron.

1955 (P) *Willis E. Lamb, Jr.*, for discovering the Lamb shift in the hydrogen spectrum (1947) and *Polykarp Kusch* for determining the magnetic moment of the electron (1947).

1956 (P) *John Bardeen, Walter H. Brattain,* and *William Shockley* for inventing the transistor (1956).

1957 (P) *T.-D. Lee* and *C.-N. Yang* for predicting that parity is not conserved in beta decay (1956).

1958 (P) *Pavel A. Čerenkov* for discovering Čerenkov radiation (1935) and *Ilya M. Frank* and *Igor Tamm* for interpreting it (1937).

1959 (P) *Emilio G. Segrè* and *Owen Chamberlain* for discovering the antiproton (1955).

1960 (P) *Donald A. Glaser* for inventing the bubble chamber to study elementary particles (1952).

(C) *Willard Libby* for developing radiocarbon dating (1947).

1961 (P) *Robert Hofstadter* for discovering internal structure in protons and neutrons and *Rudolf L. Mössbauer* for discovering the Mössbauer effect of recoilless gamma-ray emission (1957).

1962 (P) *Lev Davidovich Landau* for studying liquid helium and other condensed matter theoretically.

1963 (P) *Eugene P. Wigner* for applying symmetry principles to elementary-particle theory and *Maria Goeppert Mayer* and *J. Hans D. Jensen* for studying the shell model of nuclei (1947).

1964 (P) *Charles H. Townes, Nikolai G. Basov,* and *Alexandr M. Prokhorov* for developing masers (1951–1952) and lasers.

1965 (P) *Sin-itiro Tomonaga, Julian S. Schwinger,* and *Richard P. Feynman* for developing quantum electrodynamics (1948).

1966 (P) *Alfred Kastler* for his optical methods of studying atomic energy levels.

1967 (P) *Hans Albrecht Bethe* for discovering the routes of energy production in stars (1939).

1968 (P) *Luis W. Alvarez* for discovering resonance states of elementary particles.

1969 (P) *Murray Gell-Mann* for classifying elementary particles (1963).

1970 (P) *Hannes Alfvén* for developing magnetohydrodynamic theory and *Louis Eugène Félix Néel* for discovering antiferromagnetism and ferrimagnetism (1930s).

1971   (P) *Dennis Gabor* for developing holography (1947).

(C) *Gerhard Herzberg* for studying the structure of molecules spectroscopically.

1972   (P) *John Bardeen, Leon N. Cooper,* and *John Robert Schrieffer* for explaining superconductivity (1957).

1973   (P) *Leo Esaki* for discovering tunneling in semiconductors, *Ivar Giaever* for discovering tunneling in superconductors, and *Brian D. Josephson* for predicting the Josephson effect, which involves tunneling of paired electrons (1958–1962).

1974   (P) *Anthony Hewish* for discovering pulsars and *Martin Ryle* for developing radio interferometry.

1975   (P) *Aage N. Bohr, Ben R. Mottelson,* and *James Rainwater* for discovering why some nuclei take asymmetric shapes.

1976   (P) *Burton Richter* and *Samuel C. C. Ting* for discovering the J/psi particle, the first charmed particle (1974).

1977   (P) *John H. Van Vleck, Nevill F. Mott,* and *Philip W. Anderson* for studying solids quantum-mechanically.

(C) *Ilya Prigogine* for extending thermodynamics to show how life could arise in the face of the second law.

1978   (P) *Arno A. Penzias* and *Robert W. Wilson* for discovering the cosmic background radiation (1965) and *Pyotr Kapitsa* for his studies of liquid helium.

1979   (P) *Sheldon L. Glashow, Abdus Salam,* and *Steven Weinberg* for developing the theory that unified the weak and electromagnetic forces (1958–1971).

1980   (P) *Val Fitch* and *James W. Cronin* for discovering CP (charge-parity) violation (1964), which possibly explains the cosmological dominance of matter over antimatter.

1981   (P) *Nicolaas Bloembergen* and *Arthur L. Schawlow* for developing laser spectroscopy and *Kai M. Siegbahn* for developing high-resolution electron spectroscopy (1958).

1982   (P) *Kenneth G. Wilson* for developing a method of constructing theories of phase transitions to analyze critical phenomena.

1983   (P) *William A. Fowler* for theoretical studies of astrophysical nucleosynthesis and *Subramanyan Chandrasekhar* for studying physical processes of importance to stellar structure and evolution, including the prediction of white dwarf stars (1930).

1984   (P) *Carlo Rubbia* for discovering the W and Z particles, verifying the electroweak unification, and *Simon van der Meer,* for developing the method of stochastic cooling of the CERN beam that allowed the discovery (1982–1983).

1985   (P) *Klaus von Klitzing* for the quantized Hall effect, relating to conductivity in the presence of a magnetic field (1980).

1986   (P) *Ernst Ruska* for inventing the electron microscope (1931), and *Gerd Binnig* and *Heinrich Rohrer* for inventing the scanning-tunneling electron microscope (1981).

1987   (P) *J. Georg Bednorz* and *Karl Alex Müller* for the discovery of high temperature superconductivity (1986).

1988   (P) *Leon M. Lederman, Melvin Schwartz,* and *Jack Steinberger* for a collaborative experiment that led to the development of a new tool for studying the weak nuclear force, which affects the radioactive decay of atoms.

1989   (P) *Norman Ramsay* (U.S.) for various techniques in atomic physics; and *Hans Dehmelt* (U.S.) and *Wolfgang Paul* (Germany) for the development of techniques for trapping single charge particles.

1990 (P) *Jerome Friedman, Henry Kendall* (both U.S.), and *Richard Taylor* (Canada) for experiments important to the development of the quark model.

1991 (P) *Pierre-Gilles de Gennes* for discovering that methods developed for studying order phenomena in simple systems can be generalized to more complex forms of matter, in particular to liquid crystals and polymers.

1992 (P) *George Charpak* for developing detectors that trace the paths of evanescent subatomic particles produced in particle accelerators.

1993 (P) *Russell Hulse* and *Joseph Taylor* for discovering evidence of gravitational waves.

1994 (P) *Bertram N. Brockhouse* and *Clifford G. Shull* for pioneering work in neutron scattering.

1995 (P) *Martin L. Perl* and *Frederick Reines* for discovering the tau particle and the neutrino, respectively.

1996 (P) *David M. Lee, Douglas C. Osheroff,* and *Robert C. Richardson* for developing a superfluid using helium-3.

1997 (P) *Steven Chu, Claude Cohen-Tannoudji,* and *William D. Phillips* for developing methods to cool and trap atoms with laser light.

1998 (P) *Robert B. Laughlin, Horst L. Störmer,* and *Daniel C. Tsui* for discovering a new form of quantum fluid with fractionally charged excitations.

Answers to Odd-Numbered Problems

Chapter 1

1. 2.15×10^4 kg/m^3
3. 184 g
5. (a) 7.10 cm^3 (b) 1.18×10^{-29} m^3 (c) 0.228 nm
   (d) 12.7 cm^3, 2.11×10^{-29} m^3, 0.277 nm
7. (a) 4.00 u = 6.64×10^{-24} g (b) 55.9 u =
   9.29×10^{-23} g (c) 207 u = 3.44×10^{-22} g
9. (a) 9.83×10^{-16} g (b) 1.06×10^7 atoms
11. (a) 4.01×10^{25} molecules (b) 3.65×10^4 molecules
13. no
15. (b) only
17. $0.579t$ ft^3/s + $1.19 \times 10^{-9}t^2$ ft^3/s^2
19. 1.39×10^3 m^2
21. (a) 0.071 4 gal/s (b) 2.70×10^{-4} m^3/s (c) 1.03 h
23. 4.05×10^3 m^2
25. 11.4×10^3 kg/m^3
27. 1.19×10^{57} atoms
29. (a) 190 y (b) 2.32×10^4 times
31. 151 μm
33. 1.00×10^{10} lb
35. 3.08×10^4 m^3
37. 5.0 m
39. 2.86 cm
41. $\sim 10^6$ balls
43. $\sim 10^7$ or 10^8 rev
45. $\sim 10^7$ or 10^8 blades
47. $\sim 10^2$ kg; $\sim 10^3$ kg
49. $\sim 10^2$ tuners
51. (a) (346 ± 13) m^2 (b) (66.0 ± 1.3) m
53. $(1.61 \pm 0.17) \times 10^3$ kg/m^3
55. 115.9 m
57. 316 m
59. 4.50 m^2
61. 3.41 m
63. 0.449%
65. (a) 0.529 cm/s (b) 11.5 cm/s
67. 1×10^{10} gal/yr
69. $\sim 10^{11}$ stars

Chapter 2

1. (a) 2.30 m/s (b) 16.1 m/s (c) 11.5 m/s
3. (a) 5 m/s (b) 1.2 m/s (c) -2.5 m/s (d) -3.3 m/s
   (e) 0
5. (a) 3.75 m/s (b) 0

7. (a)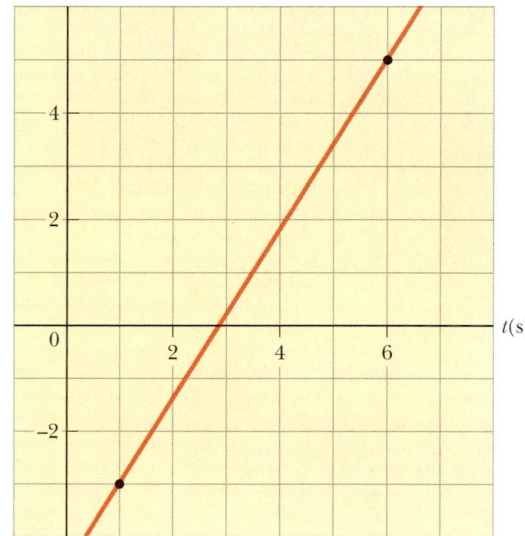

   (b) 1.60 m/s
9. (a) -2.4 m/s (b) -3.8 m/s (c) 4.0 s
11. (a) 5.0 m/s (b) -2.5 m/s (c) 0 (d) 5.0 m/s
13. 1.34×10^4 m/s^2
15. (a)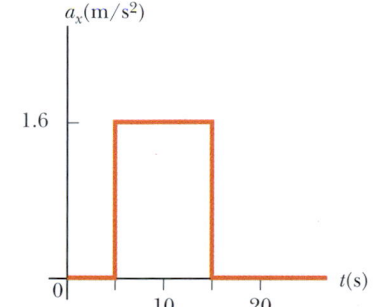

   (b) 1.6 m/s^2 and 0.80 m/s^2
17. (a) 2.00 m (b) -3.00 m/s (c) -2.00 m/s^2
19. (a) 1.3 m/s^2 (b) 2.0 m/s^2 at 3 s (c) at $t = 6$ s and for
    $t > 10$ s (d) -1.5 m/s^2 at 8 s
21. 2.74×10^5 m/s^2, which is 2.79×10^4 g
23. (a) 6.61 m/s (b) -0.448 m/s^2
25. -16.0 cm/s^2
27. (a) 2.56 m (b) -3.00 m/s
29. (a) 8.94 s (b) 89.4 m/s
31. (a) 20.0 s (b) no
33. $x_f - x_i = v_{xf}t - a_x t^2/2$; $v_{xf} = 3.10$ m/s

35. (a) 35.0 s   (b) 15.7 m/s
37. (a) -202 m/s^2 (b) 198 m
39. (a) 3.00 m/s   (b) 6.00 s   (c) -0.300 m/s^2
    (d) 2.05 m/s
41. (a) -4.90 m, -19.6 m, -44.1 m   (b) -9.80 m/s,
    -19.6 m/s, -29.4 m/s
43. (a) 10.0 m/s up   (b) 4.68 m/s down
45. No. In 0.2 s the bill falls out from between David's fin-
    gers.
47. (a) 29.4 m/s   (b) 44.1 m
49. (a) 7.82 m   (b) 0.782 s
51. (a) 1.53 s   (b) 11.5 m   (c) -4.60 m/s, -9.80 m/s^2
53. (a) $a_x = a_{xi} + Jt$, $v_x = v_{xi} + a_{xi}t + \frac{1}{2}Jt^2$,
    $x = x_i + v_{xi}t + \frac{1}{2}a_{xi}t^2 + \frac{1}{6}Jt^3$
55. 0.222 s
57. 0.509 s
59. (a) 41.0 s   (b) 1.73 km   (c) -184 m/s
61. $v_{xi}t + at^2/2$, in agreement with Equation 2.11
63. (a) 5.43 m/s^2 and 3.83 m/s^2 (b) 10.9 m/s and 11.5 m/s
    (c) Maggie by 2.62 m
65. (a) 45.7 s   (b) 574 m   (c) 12.6 m/s   (d) 765 s
67. (a) 2.99 s   (b) -15.4 m/s   (c) 31.3 m/s down and
    34.9 m/s down
69. (a) 5.46 s   (b) 73.0 m   (c) $v_{\text{Stan}} = 22.6$ m/s, $v_{\text{Kathy}} =$
    26.7 m/s
71. (a) See top of next column.
    (b) See top of next column.
73. $0.577v$

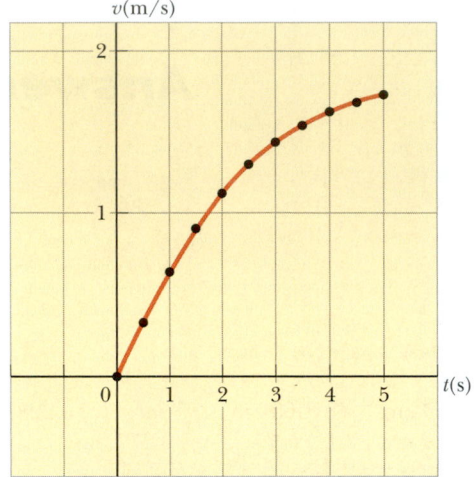

Chapter 2, Problem 71(a)

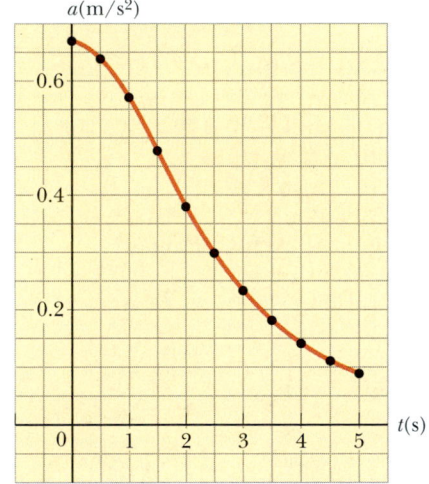

Chapter 2, Problem 71(b)

Chapter 3

1. $(-2.75, -4.76)$ m
3. 1.15; 2.31
5. (a) 2.24 m   (b) 2.24 m at 26.6° from the positive x axis.
7. (a) 484 m   (b) 18.1° north of west
9. 70.0 m
11. (a) approximately 6.1 units at 112°   (b) approximately
    14.8 units at 22°
13. (a) 10.0 m   (b) 15.7 m   (c) 0
15. (a) 5.2 m at 60°   (b) 3.0 m at 330°   (c) 3.0 m at 150°
    (d) 5.2 m at 300°
17. approximately 420 ft at $-3°$
19. 5.83 m at 59.0° to the right of his initial direction
21. 1.31 km north and 2.81 km east
23. (a) 10.4 cm   (b) 35.5°
25. 47.2 units at 122° from the positive x axis.
27. $(-25.0\mathbf{i})$ m $+ (43.3\mathbf{j})$ m
29. 7.21 m at 56.3° from the positive x axis.
31. (a) $2.00\mathbf{i} - 6.00\mathbf{j}$ (b) $4.00\mathbf{i} + 2.00\mathbf{j}$ (c) 6.32   (d) 4.47
    (e) 288°; 26.6° from the positive x axis.
33. (a) $(-11.1\mathbf{i} + 6.40\mathbf{j})$ m   (b) $(1.65\mathbf{i} + 2.86\mathbf{j})$ cm
    (c) $(-18.0\mathbf{i} - 12.6\mathbf{j})$ in.
35. 9.48 m at 166°
37. (a) 185 N at 77.8° from the positive x axis
    (b) $(-39.3\mathbf{i} - 181\mathbf{j})$ N
39. $\mathbf{A} + \mathbf{B} = (2.60\mathbf{i} + 4.50\mathbf{j})$ m

41. 196 cm at $-14.7°$ from the positive x axis.
43. (a) $8.00\mathbf{i} + 12.0\mathbf{j} - 4.00\mathbf{k}$ (b) $2.00\mathbf{i} + 3.00\mathbf{j} - 1.00\mathbf{k}$
    (c) $-24.0\mathbf{i} - 36.0\mathbf{j} + 12.0\mathbf{k}$
45. (a) 5.92 m is the magnitude of $(5.00\mathbf{i} - 1.00\mathbf{j} - 3.00\mathbf{k})$ m
    (b) 19.0 m is the magnitude of $(4.00\mathbf{i} - 11.0\mathbf{j} + 15.0\mathbf{k})$ m
47. 157 km
49. (a) $-3.00\mathbf{i} + 2.00\mathbf{j}$ (b) 3.61 at 146° from the positive
    x axis.   (c) $3.00\mathbf{i} - 6.00\mathbf{j}$
51. (a) $49.5\mathbf{i} + 27.1\mathbf{j}$ (b) 56.4 units at 28.7° from the posi-
    tive x axis.
53. 1.15°
55. (a) 2.00, 1.00, 3.00   (b) 3.74   (c) $\theta_x = 57.7°$, $\theta_y = 74.5°$,
    $\theta_z = 36.7°$
57. 240 m at 237°
59. 390 mi/h at 7.37° north of east
61. $\mathbf{R}_1 = a\mathbf{i} + b\mathbf{j}$; $R_1 = \sqrt{a^2 + b^2}$ (b) $\mathbf{R}_2 = a\mathbf{i} + b\mathbf{j} + c\mathbf{k}$

Chapter 4

1. (a) 4.87 km at 209° from east   (b) 23.3 m/s
   (c) 13.5 m/s at 209°
3. (a) $(18.0t)\mathbf{i} + (4.00t - 4.90t^2)\mathbf{j}$
   (b) $18.0\mathbf{i} + (4.00 - 9.80t)\mathbf{j}$ (c) $-9.80\mathbf{j}$
   (d) $(54.0\mathbf{i} - 32.1\mathbf{j})$ m
   (e) $(18.0\mathbf{i} - 25.4\mathbf{j})$ m/s   (f) $(-9.80\mathbf{j})$ m/s²
5. (a) $(2.00\mathbf{i} + 3.00\mathbf{j})$ m/s²
   (b) $(3.00t + t^2)\mathbf{i}$ m, $(1.50\,t^2 - 2.00t)\mathbf{j}$ m
7. (a) $(0.800\mathbf{i} - 0.300\mathbf{j})$ m/s²   (b) 339°
   (c) $(360\mathbf{i} - 72.7\mathbf{j})$ m, $-15.2°$
9. (a) $(3.34\mathbf{i})$ m/s   (b) $-50.9°$
11. (a) 20.0°   (b) 3.05 s
13. $x = 7.23$ km $y = 1.68$ km
15. 53.1°
17. 22.4° or 89.4°
19. (a) The ball clears by 0.889 m   (b) while descending
21. $d \tan \theta_i - gd^2/(2v_i^2 \cos^2\theta_i)$
23. (a) 0.852 s   (b) 3.29 m/s   (c) 4.03 m/s   (d) 50.8°
    (e) 1.12 s
25. 377 m/s²
27. 10.5 m/s, 219 m/s²
29. (a) 6.00 rev/s   (b) 1.52 km/s²   (c) 1.28 km/s²
31. 1.48 m/s² inward at 29.9° behind the radius
33. (a) 13.0 m/s²   (b) 5.70 m/s   (c) 7.50 m/s²
35. (a)

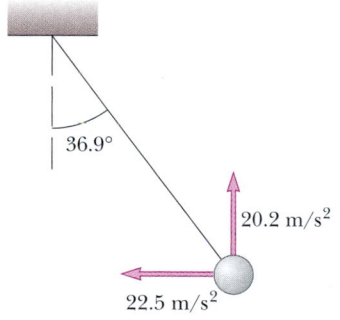

(b) 29.7 m/s²   (c) 6.67 m/s at 36.9° above the
    horizontal
37. 2.02×10^3 s; 21.0% longer
39. 153 km/h at 11.3° north of west
41. (a) 36.9°   (b) 41.6°   (c) 3.00 min
43. 15.3 m
45. $2\,v_i t \cos \theta_i$
47. (b) $45° + \phi/2$; $v_i^2(1 - \sin \phi)/g \cos^2\phi$
49. (a) 41.7 m/s   (b) 3.81 s   (c) $(34.1\mathbf{i} - 13.4\mathbf{j})$ m/s; 36.6 m/s
51. (a) 25.0 m/s² (radial); 9.80 m/s² (tangential)
    (b)

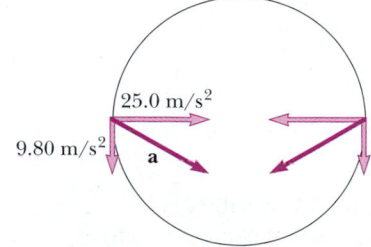

(c) 26.8 m/s² inward at 21.4° below the horizontal
53. 8.94 m/s at $-63.4°$ relative to the positive x axis.
55. 20.0 m
57. (a) 0.600 m   (b) 0.402 m   (c) 1.87 m/s² toward center
    (d) 9.80 m/s² down
59. (a) 6.80 km   (b) 3.00 km vertically above the impact
    point   (c) 66.2°
61. (a) 46.5 m/s   (b) $-77.6°$ (c) 6.34 s
63. (a) 1.53 km   (b) 36.2 s   (c) 4.04 km
65. (a) 20.0 m/s, 5.00 s   (b) $(16.0\mathbf{i} - 27.1\mathbf{j})$ m/s   (c) 6.54 s
    (d) $(24.6\mathbf{i})$ m
67. (a) 43.2 m   (b) $(9.66\mathbf{i} - 25.5\mathbf{j})$ m/s
69. Imagine you are shaking down the mercury in a fever
    thermometer. Starting with your hand at the level of your
    shoulder, move your hand down as fast as you can and
    snap it around an arc at the bottom. ~ 100 m/s² $\approx 10\,g$

Chapter 5

1. (a) 1/3   (b) 0.750 m/s²
3. $(6.00\mathbf{i} + 15.0\mathbf{j})$ N; 16.2 N
5. 312 N
7. (a) $x = vt/2$ (b) $F_g v\mathbf{i}/gt + F_g\mathbf{j}$
9. (a) $(2.50\mathbf{i} + 5.00\mathbf{j})$ N   (b) 5.59 N
11. (a) 3.64×10^{-18} N   (b) 8.93×10^{-30} N is 408 billion
    times smaller.
13. 2.38 kN
15. (a) 5.00 m/s² at 36.9°   (b) 6.08 m/s² at 25.3°
17. (a) $\sim 10^{-22}$ m/s²   (b) $\sim 10^{-23}$ m
19. (a) 0.200 m/s² forward   (b) 10.0 m   (c) 2.00 m/s
21. (a) 15.0 lb up   (b) 5.00 lb up   (c) 0
23. 613 N

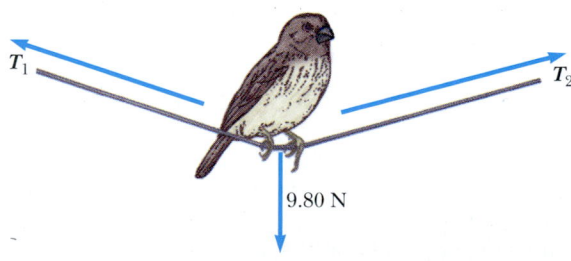

27. (a) 49.0 N   (b) 98.0 N   (c) 24.5 N
29. 8.66 N east
31. 100 N and 204 N
33. 3.73 m
35. $a = F/(m_1 + m_2)$; $T = F m_1/(m_1 + m_2)$
37. (a) $F_x > 19.6$ N   (b) $F_x \le -78.4$ N
    (c) See top of next page.
39. (a) 706 N   (b) 814 N   (c) 706 N   (d) 648 N
41. $\mu_s = 0.306$; $\mu_k = 0.245$
43. (a) 256 m   (b) 42.7 m
45. (a) 1.78 m/s²   (b) 0.368   (c) 9.37 N   (d) 2.67 m/s
47. (a) 0.161   (b) 1.01 m/s²
49. 37.8 N

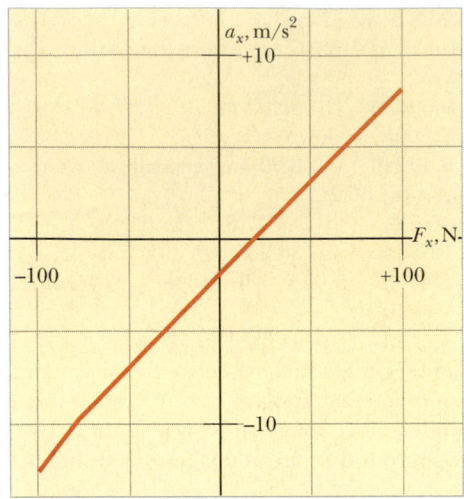

Chapter 5, Problem 37(c)

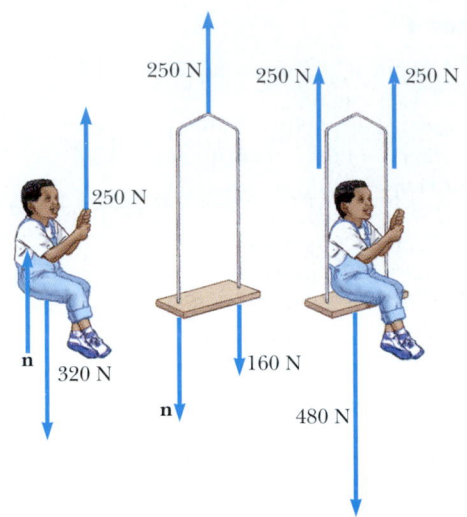

Chapter 5, Problem 55(a)

51. (a)

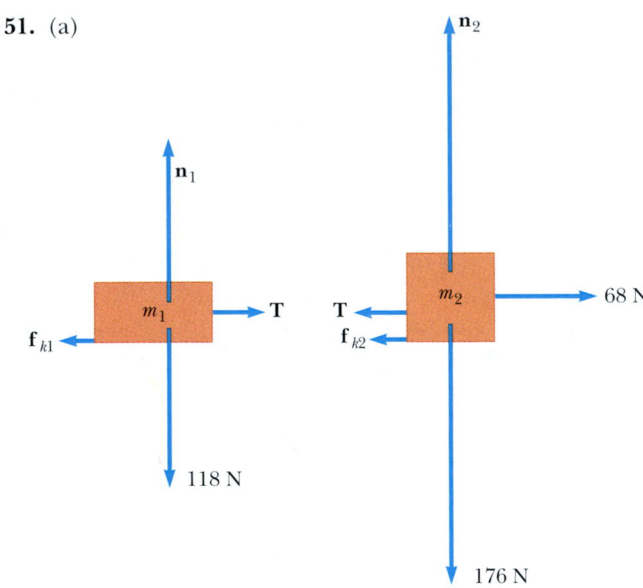

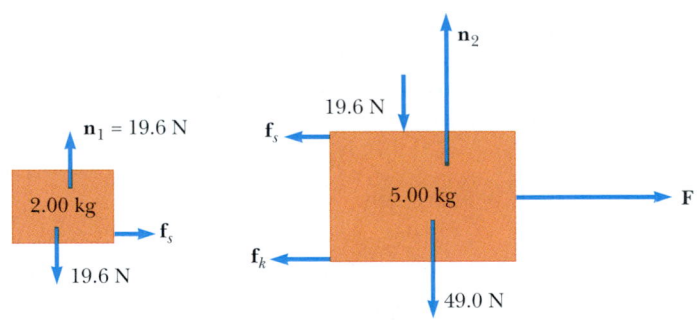

Chapter 5, Problem 67(a)

71. (a)

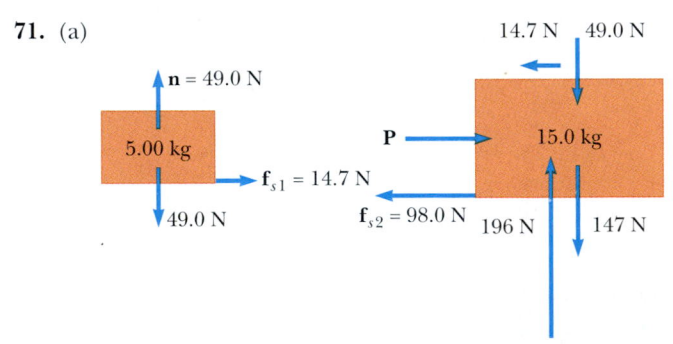

(b) 27.2 N, 1.29 m/s²

53. Any value between 31.7 N and 48.6 N

55. (a) See top of next column.
(b) 0.408 m/s²   (c) 83.3 N

57. 1.18 kN

59. (a) $Mg/2$, $Mg/2$, $Mg/2$, $3Mg/2$, Mg (b) $Mg/2$

61. (b)

θ	0	15.0°	30.0°	45.0°	60.0°
P(N)	40.0	46.4	60.1	94.3	260

63. (a) 19.3°   (b) 4.21 N

65. (a) 2.13 s   (b) 1.67 m

67. (a) See next column.
Static friction between the two blocks accelerates the upper block.   (b) 34.7 N   (c) 0.306

69. $(M + m_1 + m_2)(m_2 g/m_1)$

(b) 113 N   (c) 0.980 m/s² and 1.96 m/s²

73. (a) 0.087 1   (b) 27.4 N

75. (a) 30.7°   (b) 0.843 N

77. (a) 3.34   (b) Either the car would flip over backwards, or the wheels would skid, spinning in place, and the time would increase.

Chapter 6

1. (a) 8.00 m/s   (b) 3.02 N

3. Any speed up to 8.08 m/s

5. 6.22×10^{-12} N

7. (a) 1.52 m/s^2 (b) 1.66 km/s   (c) 6 820 s

9. (a) static friction   (b) 0.085 0

11. $v \leq 14.3$ m/s

13. (a) 68.6 N toward the center of the circle and 784 N up
(b) 0.857 m/s^2

15. No. The jungle lord needs a vine of tensile strength
1.38 kN.

17. (a) 4.81 m/s   (b) 700 N up

19. 3.13 m/s

21. (a) 2.49×10^4 N up   (b) 12.1 m/s

23. (a) 0.822 m/s^2 (b) 37.0 N   (c) 0.0839

25. (a) 17.0°   (b) 5.12 N

27. (a) 491 N   (b) 50.1 kg   (c) 2.00 m/s^2

29. 0.0927°

31. (a) 32.7 s^{-1} (b) 9.80 m/s^2 (c) 4.90 m/s^2

33. 3.01 N

35. (a) 1.47 N·s/m   (b) 2.04×10^{-3} s   (c) 2.94×10^{-2} N

37. (a) 0.0347 s^{-1} (b) 2.50 m/s   (c) $a = -cv$

39. $\sim 10^1$ N

41. (a) 13.7 m/s down

(b)

t (s)	x (m)	v (m/s)
0	0	0
0.2	0	-1.96
0.4	-0.392	-3.88
. . .	. . .	. . .
1.0	-3.77	-8.71
. . . 2.0	-14.4	-12.56
. . . 4.0	-41.0	-13.67

43. (a) 49.5 m/s and 4.95 m/s

(b)

t (s)	y (m)	v (m/s)
0	1 000	0
. . . 1	995	-9.7
. . . 2	980	-18.6
. . . 10	674	-47.7
. . . 10.1	671	-16.7
. . . 12	659	-4.95
. . . 145	0	-4.95

45. (a) 2.33×10^{-4} kg/m   (b) 53 m/s   (c) 42 m/s. The
second trajectory is higher and shorter. In both, the ball
attains maximum height when it has covered about 57% of
its horizontal range, and it attains minimum speed somewhat later. The impact speeds also are both about 30 m/s.

47. (a) $mg - mv^2/R$ (b) $\sqrt{gR}$

49. (a) 2.63 m/s^2 (b) 201 m   (c) 17.7 m/s

51. (a) 9.80 N   (b) 9.80 N   (c) 6.26 m/s

53. (b) 732 N down at the equator and 735 N down at the
poles

59. (a) 1.58 m/s^2 (b) 455 N   (c) 329 N   (d) 397 N upward and 9.15° inward

61. (a) 5.19 m/s   (b) Child + seat:

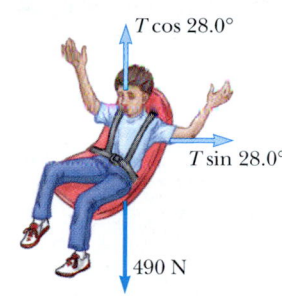

$T = 555$ N

63. (b) 2.54 s; 23.6 rev/min

65. 215 N horizontally inward

67. (a) either 70.4° or 0°   (b) 0°

69. 12.8 N

71. (a)

t (s)	d (m)
0	0
1	4.9
2	18.9
. . . 5	112.6
. . . 10	347.0
11	399.1
. . . 15	611.3
. . . 20	876.5

(b)

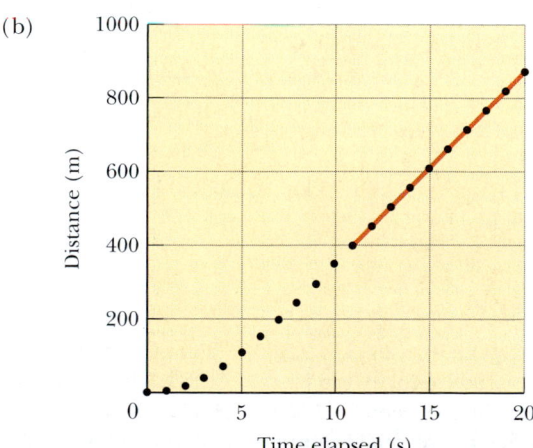

(c) The graph is straight for 11 s $< t <$ 20 s, with slope
53.0 m/s.

Chapter 7

1. 15.0 MJ

3. (a) 32.8 mJ   (b) -32.8 mJ

5. (a) 31.9 J   (b) 0   (c) 0   (d) 31.9 J

7. 4.70 kJ

9. 14.0

11. (a) 16.0 J   (b) 36.9°

13. (a) 11.3°   (b) 156°   (c) 82.3°

15. (a) 24.0 J   (b) -3.00 J   (c) 21.0 J
17. (a) 7.50 J   (b) 15.0 J   (c) 7.50 J   (d) 30.0 J
19. (a) 0.938 cm   (b) 1.25 J
21. 0.299 m/s
23. 12.0 J
25. (b) mgR
27. (a) 1.20 J   (b) 5.00 m/s   (c) 6.30 J
29. (a) 60.0 J   (b) 60.0 J
31. (a) $\sqrt{2W/m}$ (b) W/d
33. (a) 650 J   (b) -588 J   (c) 0   (d) 0   (e) 62.0 J
(f) 1.76 m/s
35. (a) -168 J   (b) -184 J   (c) 500 J   (d) 148 J
(e) 5.64 m/s
37. 2.04 m
39. (a) 22 500 N   (b) 1.33×10^{-4} s
41. (a) 0.791 m/s   (b) 0.531 m/s
43. 875 W
45. 830 N
47. (a) 5 910 W   (b) It is 53.0% of 11 100 W
49. (a) 0.013 5 gal   (b) 73.8   (c) 8.08 kW
51. 5.90 km/L
53. (a) 5.37×10^{-11} J   (b) 1.33×10^{-9} J
55. 90.0 J
59. (a) $(2 + 24t^2 + 72t^4)$ J   (b) $12t$ m/s^2; $48t$ N
(c) $(48t + 288t^3)$ W   (d) 1 250 J
61. -0.047 5 J
63. 878 kN
65. (b) 240 W
67. (a) $\mathbf{F}_1 = (20.5\mathbf{i} + 14.3\mathbf{j})$ N; $\mathbf{F}_2 = (-36.4\mathbf{i} + 21.0\mathbf{j})$ N
(b) $(-15.9\mathbf{i} + 35.3\mathbf{j})$ N   (c) $(-3.18\mathbf{i} + 7.07\mathbf{j})$ m/s^2
(d) $(-5.54\mathbf{i} + 23.7\mathbf{j})$ m/s   (e) $(-2.30\mathbf{i} + 39.3\mathbf{j})$ m
(f) 1 480 J   (g) 1 480 J
69. (a) 4.12 m   (b) 3.35 m
71. 1.68 m/s
73. (a) 14.5 m/s   (b) 1.75 kg   (c) 0.350 kg
75. 0.799 J

Chapter 8

1. (a) 259 kJ, 0, -259 kJ   (b) 0, -259 kJ, -259 kJ
3. (a) -196 J   (b) -196 J   (c) -196 J. The force is conservative.
5. (a) 125 J   (b) 50.0 J   (c) 66.7 J   (d) Nonconservative. The results differ.
7. (a) 40.0 J   (b) -40.0 J   (c) 62.5 J
9. (a) $Ax^2/2 - Bx^3/3$ (b) $\Delta U = 5A/2 - 19B/3$;
$\Delta K = -5A/2 + 19B/3$
11. 0.344 m
13. (a) $v_B = 5.94$ m/s; $v_C = 7.67$ m/s   (b) 147 J
15. $v = (3gR)^{1/2}$, 0.098 0 N down
17. 10.2 m
19. (a) 19.8 m/s   (b) 78.4 J   (c) 1.00
21. (a) 4.43 m/s   (b) 5.00 m
23. (a) 18.5 km, 51.0 km   (b) 10.0 MJ
25. (b) 60.0°
27. 5.49 m/s

29. 2.00 m/s, 2.79 m/s, 3.19 m/s
31. 3.74 m/s
33. (a) -160 J   (b) 73.5 J   (c) 28.8 N   (d) 0.679
35. 489 kJ
37. (a) 1.40 m/s   (b) 4.60 cm after release   (c) 1.79 m/s
39. 1.96 m
41. (A/r^2) away from the other particle
43. (a) $r = 1.5$ mm, stable; 2.3 mm, unstable; 3.2 mm, stable;
$r \to \infty$ neutral   (b) -5.6 J $< E < 1$ J
(c) 0.6 mm $< r <$ 3.6 mm   (d) 2.6 J   (e) 1.5 mm
(f) 4 J
45. (a) $+$ at Ⓑ, $-$ at Ⓓ, 0 at Ⓐ, Ⓒ, and Ⓔ   (b) Ⓒ stable;
Ⓐ and Ⓔ unstable
(c)

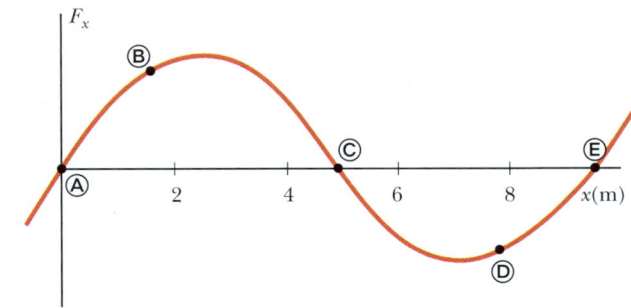

47. (b)

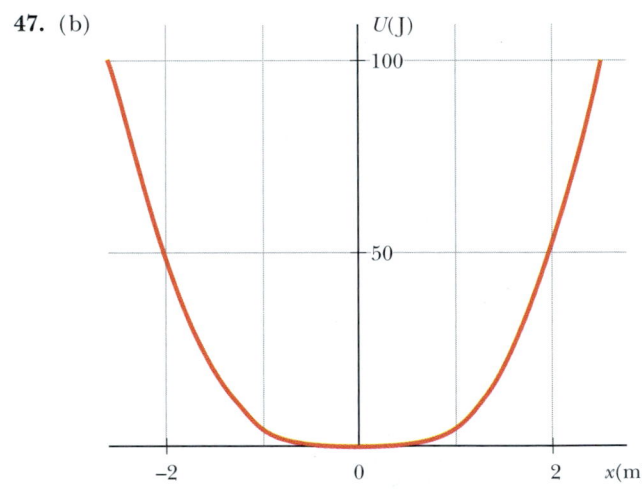

Equilibrium at $x = 0$ (c) $v = \sqrt{0.800\text{J}/m}$
49. (a) 1.50×10^{-10} J   (b) 1.07×10^{-9} J   (c) 9.15×10^{-10} J
51. 48.2° Note that the answer is independent of the pumpkin's mass and of the radius of the dome.
53. (a) 0.225 J   (b) $\Delta E_f = -0.363$ J   (c) No; the normal force changes in a complicated way.
55. $\sim 10^2$ W sustainable power
57. 0.327
59. (a) 23.6 cm   (b) 5.90 m/s^2 up the incline; no.
(c) Gravitational potential energy turns into kinetic energy plus elastic potential energy and then entirely into elastic potential energy.
61. 1.25 m/s

63. (a) 0.400 m   (b) 4.10 m/s   (c) The block stays on the track.

65. (b) 2.06 m/s

67. (b) 1.44 m   (c) 0.400 m   (d) No. A very strong wind pulls the string out horizontally (parallel to the ground). The largest possible equilibrium height is equal to L.

71. (a) 6.15 m/s   (b) 9.87 m/s

73. 0.923 m/s

Chapter 9

1. (a) $(9.00\mathbf{i} - 12.0\mathbf{j})$ kg·m/s   (b) 15.0 kg·m/s at 307°

3. 6.25 cm/s west

5. $\sim 10^{-23}$ m/s

7. (b) $p = \sqrt{2mK}$

9. (a) 13.5 N·s   (b) 9.00 kN   (c) 18.0 kN

11. 260 N normal to the wall

13. 15.0 N in the direction of the initial velocity of the exiting water stream

15. 65.2 m/s

17. 301 m/s

19. (a) $v_{gx} = 1.15$ m/s   (b) $v_{px} = -0.346$ m/s

21. (a) 20.9 m/s east   (b) 8.68 kJ into internal energy

23. (a) 2.50 m/s   (b) 37.5 kJ   (c) Each process is the time-reversal of the other. The same momentum conservation equation describes both.

25. (a) 0.284   (b) 115 fJ and 45.4 fJ

27. 91.2 m/s

29. (a) 2.88 m/s at 32.3° north of east   (b) 783 J into internal energy

31. No; his speed was 41.5 mi/h.

33. 2.50 m/s at $-60.0°$

35. $(3.00\mathbf{i} - 1.20\mathbf{j})$ m/s

37. Orange: $v_i \cos\theta$; yellow: $v_i \sin\theta$

39. (a) $(-9.33\mathbf{i} - 8.33\mathbf{j})$ Mm/s   (b) 439 fJ

41. $\mathbf{r}_{CM} = (11.7\mathbf{i} + 13.3\mathbf{j})$ cm

43. 0.006 73 nm from the oxygen nucleus along the bisector of the angle

45. (a) 15.9 g   (b) 0.153 m

47. 0.700 m

49. (a) $(1.40\mathbf{i} + 2.40\mathbf{j})$ m/s   (b) $(7.00\mathbf{i} + 12.0\mathbf{j})$ kg·m/s

51. (a) 39.0 MN up   (b) 3.20 m/s² up

53. (a) 442 metric tons   (b) 19.2 metric tons

55. (a) $(1.33\mathbf{i})$ m/s   (b) $(-235\mathbf{i})$ N   (c) 0.680 s
(d) $(-160\mathbf{i})$ N·s and $(+160\mathbf{i})$ N·s   (e) 1.81 m
(f) 0.454 m   (g) -427 J   (h) $+107$ J
(i) Equal friction forces act through different distances on person and cart to do different amounts of work on them. The total work on both together, -320 J, becomes $+320$ J of internal energy in this perfectly inelastic collision.

57. 1.39 km/s

59. 240 s

61. 0.980 m

63. (a) 6.81 m/s   (b) 1.00 m

65. $(3Mgx/L)\mathbf{j}$

67. (a) 3.75 kg·m/s²   (b) 3.75 N   (c) 3.75 N   (d) 2.81 J
(e) 1.41 J   (f) Friction between sand and belt converts half of the input work into internal energy.

69. (a) As the child walks to the right, the boat moves to the left and the center of mass remains fixed.   (b) 5.55 m from the pier (c) No, since 6.55 m is less than 7.00 m.

71. (a) 100 m/s   (b) 374 J

73. (a) $\sqrt{2}\,v_i$ for m and $\sqrt{2/3}\,v_i$ for $3m$ (b) 35.3°

75. (a) 3.73 km/s   (b) 153 km

Chapter 10

1. (a) 4.00 rad/s²   (b) 18.0 rad

3. (a) 1 200 rad/s   (b) 25.0 s

5. (a) 5.24 s   (b) 27.4 rad

7. (a) 5.00 rad, 10.0 rad/s, 4.00 rad/s²   (b) 53.0 rad, 22.0 rad/s, 4.00 rad/s²

9. 13.7 rad/s²

11. $\sim 10^7$ rev/y

13. (a) 0.180 rad/s   (b) 8.10 m/s² toward the center of the track

15. (a) 8.00 rad/s   (b) 8.00 m/s, $a_r = -64.0$ m/s², $a_t = 4.00$ m/s²   (c) 9.00 rad

17. (a) 54.3 rev   (b) 12.1 rev/s

19. (a) 126 rad/s   (b) 3.78 m/s   (c) 1.27 km/s²
(d) 20.2 m

21. (a) $-2.73\mathbf{i}$ m $+ 1.24\mathbf{j}$ m   (b) second quadrant, 156°
(c) $-1.85\mathbf{i}$ m/s $- 4.10\mathbf{j}$ m/s
(d) into the third quadrant at 246°

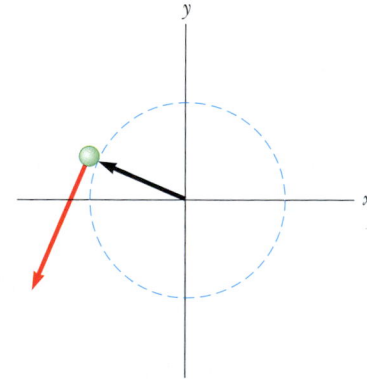

(e) $6.15\mathbf{i}$ m/s² $- 2.78\mathbf{j}$ m/s²
(f) $24.6\mathbf{i}$ N $- 11.1\mathbf{j}$ N

23. (a) 92.0 kg·m², 184 J   (b) 6.00 m/s, 4.00 m/s, 8.00 m/s, 184 J

25. (a) 143 kg·m²   (b) 2.57 kJ

29. 1.28 kg·m²

31. $\sim 10^0 = 1$ kg·m²

33. -3.55 N·m

35. 882 N·m

37. (a) 24.0 N·m   (b) 0.035 6 rad/s²   (c) 1.07 m/s²

39. (a) 0.309 m/s²   (b) 7.67 N and 9.22 N

41. (a) 872 N   (b) 1.40 kN

43. 2.36 m/s

45. (a) 11.4 N, 7.57 m/s^2, 9.53 m/s down   (b) 9.53 m/s

49. (a) $2(Rg/3)^{1/2}$ (b) $4(Rg/3)^{1/2}$ (c) $(Rg)^{1/2}$

51. $\frac{1}{3}\ell$

53. (a) 1.03 s   (b) 10.3 rev

55. (a) 4.00 J   (b) 1.60 s   (c) yes

57. (a) 12.5 rad/s   (b) 128 rad

59. (a) $(3g/L)^{1/2}$ (b) $3g/2L$ (c) $-\frac{3}{2}g\mathbf{i} - \frac{3}{4}g\mathbf{j}$
   (d) $-\frac{3}{2}Mg\mathbf{i} + \frac{1}{4}Mg\mathbf{j}$

61. $\alpha = g(h_2 - h_1)/2\pi R^2$

63. (b) $2gM(\sin\theta - \mu\cos\theta)(m + 2M)^{-1}$

65. 139 m/s

67. 5.80 kg·m^2; the height makes no difference.

69. (a) 2 160 N·m   (b) 439 W

71. (a) 118 N and 156 N   (b) 1.19 kg·m^2

73. (a) $\alpha = -0.176$ rad/s^2 (b) 1.29 rev   (c) 9.26 rev

Chapter 11

1. (a) 500 J   (b) 250 J   (c) 750 J

3. $\frac{7}{10}Mv^2$

5. (a) $\frac{2}{3}g\sin\theta$ for the disk, larger than $\frac{1}{2}g\sin\theta$ for the hoop
   (b) $\frac{1}{3}\tan\theta$

7. 1.21×10^{-4} kg·m^2. The height is unnecessary.

9. $-7.00\mathbf{i} + 16.0\mathbf{j} - 10.0\mathbf{k}$

11. (a) $-17.0\mathbf{k}$ (b) 70.5°

13. (a) 2.00 N·m   (b) $\mathbf{k}$

15. (a) negative z direction   (b) positive z direction

17. 45.0°

19. $(17.5\mathbf{k})$ kg·m^2/s

21. $(60.0\mathbf{k})$ kg·m^2/s

23. $mvR[\cos(vt/R) + 1]\mathbf{k}$

25. (a) zero   (b) $(-mv_i^3\sin^2\theta\cos\theta/2g)\mathbf{k}$
   (c) $(-2mv_i^3\sin^2\theta\cos\theta/g)\mathbf{k}$ (d) The downward force
of gravity exerts a torque in the $-z$ direction.

27. $-m\ell\, gt\cos\theta\,\mathbf{k}$

29. 4.50 kg·m^2/s up

31. (a) 0.433 kg·m^2/s   (b) 1.73 kg·m^2/s

33. (a) $\omega_f = \omega_i I_1/(I_1 + I_2)$ (b) $I_1/(I_1 + I_2)$

35. (a) 1.91 rad/s   (b) 2.53 J, 6.44 J

37. (a) 0.360 rad/s counterclockwise   (b) 99.9 J

39. (a) $mv\ell$ down   (b) $M/(M + m)$

41. (a) $\omega = 2mv_i d/(M + 2m)R^2$ (b) No; some mechanical
energy changes into internal energy.

43. (a) 2.19×10^6 m/s   (b) 2.18×10^{-18} J
   (c) 4.13×10^{16} rad/s

45. $[10Rg(1 - \cos\theta)/7r^2]^{1/2}$

51. (a) 2.70R (b) $F_x = -\frac{20}{7}mg, F_y = -mg$

53. 0.632

55. (a) $v_i r_i/r$ (b) $T = (mv_i^2 r_i^2)r^{-3}$ (c) $\frac{1}{2}mv_i^2(r_i^2/r^2 - 1)$
   (d) 4.50 m/s, 10.1 N, 0.450 J

57. 54.0°

59. (a) 3 750 kg·m^2/s   (b) 1.88 kJ   (c) 3 750 kg·m^2/s
   (d) 10.0 m/s   (e) 7.50 kJ   (f) 5.62 kJ

61. $(M/m)[3ga\,(\sqrt{2} - 1)]^{1/2}$

63. (c) $(8Fd/3M)^{1/2}$

67. (a) 0.800 m/s^2, 0.400 m/s^2 (b) 0.600 N backward on
the plank and forward on the roller, at the top of each
roller; 0.200 N forward on each roller and backward on
the floor, at the bottom of each roller.

Chapter 12

1. 10.0 N up; 6.00 N·m counterclockwise

3. $[(m_1 + m_b)d + m_1\ell/2]/m_2$

5. -0.429 m

7. (3.85 cm, 6.85 cm)

9. $(-1.50$ m, -1.50 m)

11. (a) 859 N   (b) 1 040 N left and upward at 36.9°

13. (a) $f_s = 268$ N, $n = 1\ 300$ N   (b) 0.324

15. (a) 1.04 kN at 60.0°   (b) $(370\mathbf{i} + 900\mathbf{j})$ N

17. 2.94 kN on each rear wheel and 4.41 kN on each front
wheel

19. (a) 29.9 N   (b) 22.2 N

21. (a) 35.5 kN   (b) 11.5 kN   (c) -4.19 kN

23. 88.2 N and 58.8 N

25. 4.90 mm

27. 0.023 8 mm

29. 0.912 mm

31. $\dfrac{8m_1 m_2 g L_i}{\pi d^2 Y(m_1 + m_2)}$

33. (a) 3.14×10^4 N   (b) 6.28×10^4 N

35. 1.80×10^8 N/m^2

37. $n_A = 5.98 \times 10^5$ N, $n_B = 4.80 \times 10^5$ N

39. (a) 0.400 mm   (b) 40.0 kN   (c) 2.00 mm   (d) 2.40 mm
   (e) 48.0 kN

41. (a)

   (b) 69.8 N   (c) 0.877L

43. (a) 160 N right   (b) 13.2 N right   (c) 292 N up
   (d) 192 N

45. (a) $T = F_g(L + d)/\sin\theta(2L + d)$
   (b) $R_x = F_g(L + d)\cot\theta/(2L + d); R_y = F_gL/(2L + d)$

47. 0.789 L

49. 5.08 kN, $R_x = 4.77$ kN, $R_y = 8.26$ kN

51. $T = 2.71$ kN, $R_x = 2.65$ kN

53. (a) $\mu_k = 0.571$; the normal force acts 20.1 cm to the left
of the front edge of the sliding cabinet.   (b) 0.501 m

55. (b) 60.0°
57. (a) $M = (m/2)(2\mu_s \sin\theta - \cos\theta)(\cos\theta - \mu_s \sin\theta)^{-1}$
     (b) $R = (m + M)g(1 + \mu_s^2)^{1/2}$,
     $\Gamma = g[M^2 + \mu_s^2(m + M)^2]^{1/2}$
59. (a) 133 N   (b) $n_A = 429$ N and $n_B = 257$ N
     (c) $R_x = 133$ N and $R_y = -257$ N
61. 66.7 N
65. 1.09 m
67. (a) 4 500 N   (b) 4.50×10^6 N/m²   (c) yes.
69. (a) $P_y = (F_g/L)(d - ah/g)$ (b) 0.306 m
     (c) $\mathbf{P} = (-306\mathbf{i} + 553\mathbf{j})$ N
71. $n_A = n_E = 6.66$ kN; $F_{AB} = 10.4$ kN $= F_{BC} = F_{DC} = F_{DE}$;
     $F_{AC} = 7.94$ kN $= F_{CE}$; $F_{BD} = 15.9$ kN

Chapter 13

1. (a) 1.50 Hz, 0.667 s   (b) 4.00 m   (c) π rad   (d) 2.83 m
3. (a) 20.0 cm   (b) 94.2 cm/s as the particle passes through equilibrium   (c) 17.8 m/s² at the maximum displacement from equilibrium
5. (b) 18.8 cm/s, 0.333 s   (c) 178 cm/s², 0.500 s
     (d) 12.0 cm
7. 0.627 s
9. (a) 40.0 cm/s, 160 cm/s²   (b) 32.0 cm/s, -96.0 cm/s²
     (c) 0.232 s
11. 40.9 N/m
13. (a) 0.750 m   (b) $x = -(0.750 \text{ m}) \sin(2.00 t/s)$
15. 0.628 m/s
17. 2.23 m/s
19. (a) 28.0 mJ   (b) 1.02 m/s   (c) 12.2 mJ   (d) 15.8 mJ
21. (a) 2.61 m/s   (b) 2.38 m/s
23. 2.60 cm and -2.60 cm

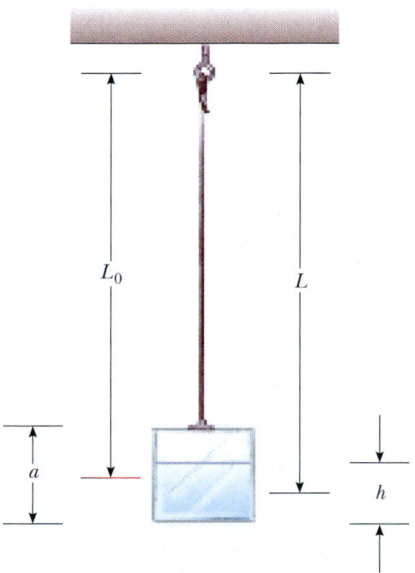

Chapter 13, Problem 57(a)

25. (a) 35.7 m   (b) 29.1 s
27. $\sim 10^0$ s
29. (a) 0.817 m/s   (b) 2.54 rad/s²   (c) 0.634 N
33. 0.944 kg·m²
37. (a) 5.00×10^{-7} kg·m²   (b) 3.16×10^{-4} N·m/rad
39. The x coordinate of the crank pin is $A \cos \omega t$
41. 1.00×10^{-3} s⁻¹
43. (a) 2.95 Hz   (b) 2.85 cm
47. Either 1.31 Hz or 0.641 Hz
49. 6.58 kN/m
51. (a) $2Mg$; $Mg(1 + y/L)$ (b) $T = (4\pi/3)(2L/g)^{1/2}$; 2.68 s
53. 6.62 cm
55. 9.19×10^{13} Hz
57. (a) See bottom of preceding column.
     (b) $\dfrac{dT}{dt} = \dfrac{\pi(dM/dt)}{2\rho a^2 g^{1/2}[L_i + (dM/dt)\,t/2\rho a^2]^{1/2}}$
     (c) $T = 2\pi g^{-1/2}[L_i + (dM/dt)\,t/2\rho a^2]^{1/2}$
59. $f = (2\pi L)^{-1}(gL + kh^2/M)^{1/2}$
61. (a) 3.56 Hz   (b) 2.79 Hz   (c) 2.10 Hz
63. (a) 3.00 s   (b) 14.3 J   (c) 25.5°
65. 0.224 rad/s

Chapter 14

1. $\sim 10^{-7}$ N toward you
3. $\mathbf{g} = (Gm/\ell^2)(\frac{1}{2} + \sqrt{2})$ toward the opposite corner
5. $(-100\mathbf{i} + 59.3\mathbf{j})$ pN
7. (a) 4.39×10^{20} N   (b) 1.99×10^{20} N   (c) 3.55×10^{22} N
9. 0.613 m/s² toward the Earth
11.

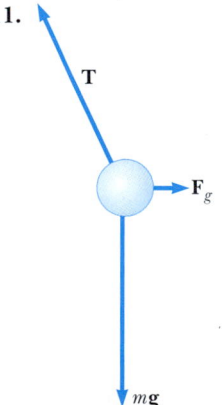

1.000 m $-$ 61.3 nm
15. 12.6×10^{31} kg
17. 1.27
19. 1.90×10^{27} kg
21. 8.92×10^7 m
25. $g = 2MGr(r^2 + a^2)^{-3/2}$ toward the center of mass
27. (a) -4.77×10^9 J   (b) 569 N   (c) 569 N up
29. (a) 1.84×10^9 kg/m³   (b) 3.27×10^6 m/s²
     (c) -2.08×10^{13} J
31. (a) -1.67×10^{-14} J   (b) At the center
33. 1.58×10^{10} J
35. (a) 1.48 h   (b) 7.79 km/s   (c) 6.43×10^9 J

37. 1.66×10^4 m/s
41. 15.6 km/s
43. $GM_E m / 12 R_E$
45. $2 GmM / \pi R^2$ straight up in the picture
47. (a) 7.41×10^{-10} N   (b) 1.04×10^{-8} N
    (c) 5.21×10^{-9} N
49. 2.26×10^{-7}
51. (b) 1.10×10^{32} kg
53. (b) $GMm / 2R$
55. 7.79×10^{14} kg
57. 7.41×10^{-10} N
59. $v_{\text{esc}} = (8 \pi G\rho / 3)^{1/2} R$
61. (a) $v_1 = m_2 (2G/d)^{1/2} (m_1 + m_2)^{-1/2}$
    $v_2 = m_1 (2G/d)^{1/2} (m_1 + m_2)^{-1/2}$
    $v_{\text{rel}} = (2G/d)^{1/2} (m_1 + m_2)^{1/2}$
    (b) $K_1 = 1.07 \times 10^{32}$ J, $K_2 = 2.67 \times 10^{31}$ J
63. (a) $A = M / \pi R^4$ (b) $F = GmM / r^2$ toward the center
    (c) $F = GmMr^2 / R^4$ toward the center
65. 119 km
67. (a) -36.7 MJ   (b) 9.24×10^{10} kg·m²/s
    (c) 5.58 km/s, 10.4 Mm   (d) 8.69 Mm   (e) 134 min
71.

t (s)	x (m)	y (m)	v_x (m/s)	v_y (m/s)
0	0	12 740 000	5 000	0
10	50 000	12 740 000	4 999.9	-24.6
20	99 999	12 739 754	4 999.7	-49.1
30	149 996	12 739 263	4 999.4	$-73.7\ldots$

The object does not hit the Earth; its minimum radius is $1.33 R_E$. Its period is 1.09×10^4 s. A circular orbit would require speed 5.60 km/s.

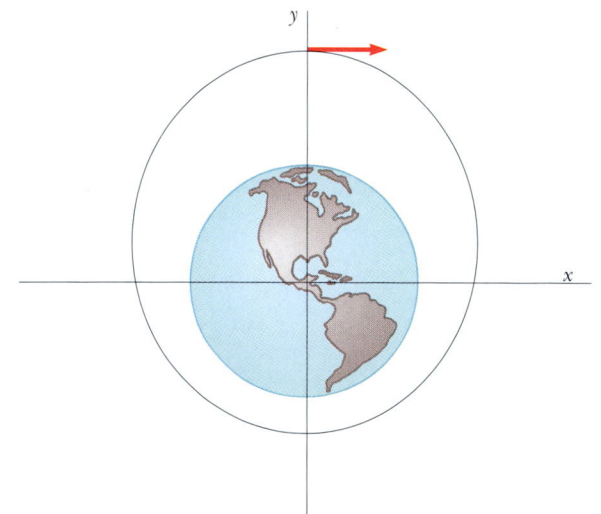

Chapter 15

1. 0.111 kg
3. 6.24 MPa

5. 5.27×10^{18} kg
7. 1.62 m
9. 7.74×10^{-3} m²
11. 271 kN horizontally backward
13. $P_0 + (\rho d/2)(g^2 + a^2)^{1/2}$
15. 0.722 mm
17. 10.5 m; no, some alcohol and water evaporate.
19. 12.6 cm
21. 1.07 m²
23. (a) 9.80 N   (b) 6.17 N
25. (a) 7.00 cm   (b) 2.80 kg
27. $\rho_{\text{oil}} = 1\,250$ kg/m³; $\rho_{\text{sphere}} = 500$ kg/m³
29. 1 430 m³
31. 2.67×10^3 kg
33. (a) 1.06 m/s   (b) 4.24 m/s
35. (a) 17.7 m/s   (b) 1.73 mm
37. 31.6 m/s
39. 68.0 kPa
41. 103 m/s
43. (a) 4.43 m/s   (b) The siphon can be no higher than 10.3 m.
45. $2\sqrt{h(h_0 - h)}$
47. 0.258 N
49. 1.91 m
53. 709 kg/m³
55. top scale 17.3 N; bottom scale 31.7 N
59. 90.04%
61. 4.43 m/s
63. (a) 10.3 m   (b) 0
65. (a) 18.3 mm   (b) 14.3 mm   (c) 8.56 mm
67. (a) 2.65 m/s   (b) 2.31×10^4 Pa
69. (a) 1.25 cm   (b) 13.8 m/s

Chapter 16

1. $y = 6 [(x - 4.5t)^2 + 3]^{-1}$
3. (a) left   (b) 5.00 m/s
5. (a) longitudinal   (b) 665 s
7. (a) 156°   (b) 0.058 4 cm
9. (a) y_1 in $+x$ direction, y_2 in $-x$ direction   (b) 0.750 s
    (c) 1.00 m
11. 30.0 N
13. 1.64 m/s²
15. 13.5 N
17. 586 m/s
19. 32.9 ms
21. 0.329 s
23. (a) See top of next page   (b) 0.125 s
25. 0.319 m
27. 2.40 m/s
29. (a) 0.250 m   (b) 40.0 rad/s   (c) 0.300 rad/m
    (d) 20.9 m   (e) 133 m/s   (f) $+x$
31. (a) $y = (8.00$ cm$) \sin(7.85x + 6\pi t)$
    (b) $y = (8.00$ cm$) \sin(7.85x + 6\pi t - 0.785)$
33. (a) 0.500 Hz, 3.14 rad/s   (b) 3.14 rad/m
    (c) $(0.100$ m$) \sin(3.14x/$m$ - 3.14t/s)$

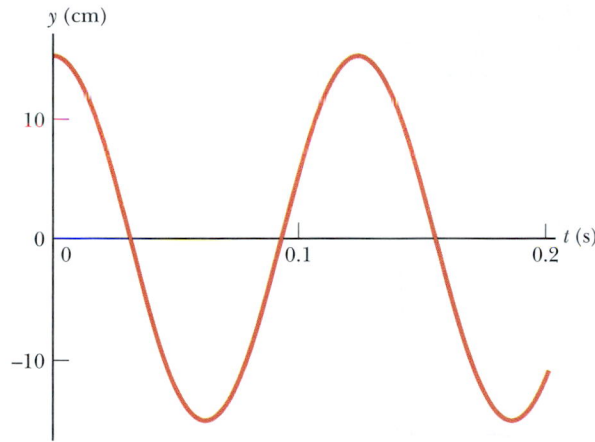

Chapter 16, Problem 23(a)

(d) $(0.100 \text{ m}) \sin(-3.14t/\text{s})$

(e) $(0.100 \text{ m}) \sin(4.71 \text{ rad} - 3.14t/\text{s})$ (f) 0.314 m/s

35. 2.00 cm, 2.98 m, 0.576 Hz, 1.72 m/s

37. (b) 3.18 Hz

41. 55.1 Hz

43. (a) 62.5 m/s   (b) 7.85 m   (c) 7.96 Hz   (d) 21.1 W

45. (a) $A = 40.0$ (b) $A = 7.00$, $B = 0$, $C = 3.00$. One can take the dot product of the given equation with each one of $\mathbf{i}, \mathbf{j},$ and $\mathbf{k}$.   (c) By inspection, $A = 0$, $B = 7.00$ mm, $C = 3.00/\text{m}$, $D = 4.00/\text{s}$, $E = 2.00$. Consider the average value of both sides of the given equation to find A. Then consider the maximum value of both sides to find B. You can evaluate the partial derivative of both sides of the given equation with respect to x and separately with respect to t to obtain equations yielding C and D upon chosen substitutions for x and t. Then substitute $x = 0$ and $t = 0$ to obtain E.

47. It is if $v = (T/\mu)^{1/2}$

49. ~ 1 min

51. (a) $3.33\mathbf{i}$ m/s   (b) -5.48 cm   (c) 0.667 m, 5.00 Hz   (d) 11.0 m/s

53. $(Lm/Mg \sin \theta)^{1/2}$

55. (a) 39.2 N   (b) 0.892 m   (c) 83.6 m/s

57. 14.7 kg

61. (a) $(0.707)2(L/g)^{1/2}$ (b) $L/4$

63. 3.86×10^{-4}

65. (a) $v = (2T_0/\mu_0)^{1/2} = v_0 2^{1/2}$
$v' = (2T_0/3\mu_0)^{1/2} = v_0 (2/3)^{1/2}$
(b) $0.966t_0$

67. 130 m/s, 1.73 km

Chapter 17

1. 5.56 km

3. 7.82 m

5. (a) 27.2 s   (b) 25.7 s; the interval in (a) is longer

7. (a) 153 m/s   (b) 614 m

9. (a) amplitude 2.00 μm, wavelength 40.0 cm, speed 54.6 m/s   (b) -0.433 μm   (c) 1.72 mm/s

11. $\Delta P = (0.2 \text{ Pa}) \sin(62.8x/\text{m} - 2.16 \times 10^4 \, t/\text{s})$

13. (a) 6.52 mm   (b) 20.5 m/s

15. 5.81 m

17. 66.0 dB

19. (a) 3.75 W/m^2 (b) 0.600 W/m^2

21. (a) 1.32×10^{-4} W/m^2 (b) 81.2 dB

23. 65.6 dB

25. (a) 65.0 dB   (b) 67.8 dB   (c) 69.6 dB

27. 1.13 μW

29. (a) 30.0 m   (b) 9.49×10^5 m

31. (a) 332 J   (b) 46.4 dB

33. (a) 75.7-Hz drop   (b) 0.948 m

35. 26.4 m/s

37. 19.3 m

39. (a) 338 Hz   (b) 483 Hz

41. 56.4°

43. (a) 56.3 s   (b) 56.6 km farther along

45. 400 m; 27.5%

47. (a) 23.2 cm   (b) 8.41×10^{-8} m   (c) 1.38 cm

49. (a) 0.515/min   (b) 0.614/min

51. 7.94 km

53. (a) 55.8 m/s   (b) 2 500 Hz

55. Bat is gaining on the insect at the rate of 1.69 m/s.

57. (a)

(b) 0.343 m   (c) 0.303 m   (d) 0.383 m
(e) 1.03 kHz

59. (a) 0.691 m   (b) 691 km

61. 1204.2 Hz

63. (a) 0.948°   (b) 4.40°

65. 1.34×10^4 N

67. 95.5 s

69. (b) 531 Hz

71. (a) 6.45   (b) 0

73. $\sim 10^{11}$ Hz

Chapter 18

1. (a) 9.24 m   (b) 600 Hz

3. 5.66 cm

5. 91.3°

7. (a) 2   (b) 9.28 m and 1.99 m

9. 15.7 m, 31.8 Hz, 500 m/s

11. At 0.089 1 m, 0.303 m, 0.518 m, 0.732 m, 0.947 m, and 1.16 m from one speaker

13. (a) 4.24 cm   (b) 6.00 cm   (c) 6.00 cm   (d) 0.500 cm, 1.50 cm, and 2.50 cm

17. 0.786 Hz, 1.57 Hz, 2.36 Hz, and 3.14 Hz

19. (a) 163 N   (b) 660 Hz

21. 19.976 kHz

23. 31.2 cm from the bridge; 3.84%

25. (a) 350 Hz   (b) 400 kg

27. 0.352 Hz

29. (a) 3.66 m/s   (b) 0.200 Hz

31. (a) 0.357 m   (b) 0.715 m

33. (a) 531 Hz   (b) 42.5 mm

35. around 3 kHz

37. $n(206$ Hz$)$ for $n = 1$ to 9, and $n(84.5$ Hz$)$ for $n = 2$ to 23

39. 239 s

41. 0.502 m and 0.837 m

43. (a) 350 m/s   (b) 1.14 m

45. (a) 19.5 cm   (b) 841 Hz

47. (a) 1.59 kHz   (b) odd-numbered harmonics
(c) 1.11 kHz

49. 5.64 beats/s

51. (a) 1.99 beats/s   (b) 3.38 m/s

53. The second harmonic of E is close to the third harmonic of A, and the fourth harmonic of C# is close to the fifth harmonic of A.

55. (a) 3.33 rad   (b) 283 Hz

57. 3.85 m/s away from the station or 3.77 m/s toward the station

59. 85.7 Hz

61. 31.1 N

63. (a) 59.9 Hz   (b) 20.0 cm

65. (a) 1/2   (b) $[n/(n + 1)]^2\, T$ (c) 9/16

67. 50.0 Hz, 1.70 m

69. (a) $2A \sin(2\pi x/\lambda)\, \cos(2\pi vt/\lambda)$
(b) $2A \sin(\pi x/L)\, \cos(\pi vt/L)$
(c) $2A \sin(2\pi x/L)\, \cos(2\pi vt/L)$
(d) $2A \sin(n\pi x/L)\, \cos(n\pi vt/L)$

Chapter 19

1. (a) $37.0°C = 310$ K   (b) $-20.6°C = 253$ K

3. (a) $-274°C$ (b) 1.27 atm   (c) 1.74 atm

5. (a) $-320°F$ (b) 77.3 K

7. (a) $810°F$ (b) 450 K

9. 3.27 cm

11. (a) 3.005 8 m   (b) 2.998 6 m

13. 55.0°C

15. (a) 0.109 cm²   (b) increase

17. (a) 0.176 mm   (b) 8.78 μm   (c) 0.093 0 cm³

19. (a) 2.52 MN/m²   (b) It will not break.

21. 1.14°C

23. (a) 99.4 cm³   (b) 0.943 cm

25. (a) 3.00 mol   (b) 1.80×10^{24} molecules

27. 1.50×10^{29} molecules

29. 472 K

31. (a) 41.6 mol   (b) 1.20 kg, in agreement with the tabulated density

33. (a) 400 kPa   (b) 449 kPa

35. 2.27 kg

37. 3.67 cm³

39. 4.39 kg

43. (a) 94.97 cm   (b) 95.03 cm

45. 208°C

47. 3.55 cm

49. (a) Expansion makes density drop.   (b) 5×10^{-5} $(°C)^{-1}$

51. (a) $h = nRT/(mg + P_0A)$ (b) 0.661 m

53. $\alpha\,\Delta T$ is much less than 1.

55. (a) 9.49×10^{-5} s   (b) 57.4 s lost

57. (a) $\rho g P_0 V_i(P_0 + \rho g d)^{-1}$ (b) decrease   (c) 10.3 m

61. (a) 5.00 MPa   (b) 9.58×10^{-3}

63. 2.74 m

65. $L_c = 9.17$ cm, $L_s = 14.2$ cm

67. (a) $L_f = L_i e^{\alpha\Delta T}$ (b) 2.00×10^{-4}%; 59.4%

69. (a) 6.17×10^{-3} kg/m   (b) 632 N   (c) 580 N; 192 Hz

Chapter 20

1. $(10.0 + 0.117)°C$

3. 0.234 kJ/kg · °C

5. 29.6°C

7. (a) 0.435 cal/g · °C   (b) beryllium

9. (a) 25.8°C   (b) No

11. 50.7 ks

13. 0.294 g

15. 0.414 kg

17. (a) 0°C   (b) 114 g

19. 59.4°C

21. 1.18 MJ

23. (a) $4P_iV_i$ (b) $T = (P_i/nRV_i)V^2$

25. 466 J

27. 810 J, 506 J, 203 J

29. $Q = -720$ J

31.

	Q	W	ΔE_{int}
BC	$-$	0	$-$
CA	$-$	$-$	$-$
AB	$+$	$+$	$+$

33. (a) 7.50 kJ   (b) 900 K

35. 3.10 kJ; 37.6 kJ

37. (a) 0.041 0 m³   (b) -5.48 kJ   (c) -5.48 kJ

41. 2.40×10^6 cal/s

43. 10.0 kW

45. 51.2°C

47. (a) 0.89 ft² · °F · h/Btu   (b) 1.85 ft² · °F · h/Btu   (c) 2.08

49. (a) $\sim 10^3$ W   (b) decreasing at $\sim 10^{-1}$ K/s

51. 364 K

53. 47.7 g

55. (a) 16.8 L   (b) 0.351 L/s

57. 2.00 kJ/kg · °C

59. 1.87 kJ

61. (a) $4P_iV_i$ (b) $4P_iV_i$ (c) 9.08 kJ

63. 5.31 h

65. 872 g

67. (a) 15.0 mg. Block: $Q = 0$, $W = +5.00$ J, $\Delta E_{int} = 0$, $\Delta K = -5.00$ J; Ice: $Q = 0$, $W = -5.00$ J, $\Delta E_{int} = 5.00$ J, $\Delta K = 0$.

(b) 15.0 mg. Block: $Q = 0$, $W = 0$, $\Delta E_{int} = 5.00$ J, $\Delta K = -5.00$ J; Metal: $Q = 0$, $W = 0$, $\Delta E_{int} = 0$, $\Delta K = 0$. (c) 0.004 04°C. Moving slab. $Q = 0$, $W = +2.50$ J, $\Delta E_{int} = 2.50$ J, $\Delta K = -5.00$ J; Stationary slab: $Q = 0$, $W = -2.50$ J, $\Delta E_{int} = 2.50$ J, $\Delta K = 0$

69. 10.2 h

71. 9.32 kW

Chapter 21

1. 6.64×10^{-27} kg

3. 0.943 N; 1.57 Pa

5. 17.6 kPa

7. 3.32 mol

9. (a) 3.53×10^{23} atoms   (b) 6.07×10^{-21} J
(c) 1.35 km/s

11. (a) 8.76×10^{-21} J for both   (b) 1.62 km/s for helium; 514 m/s for argon

13. 75.0 J

15. (a) 3.46 kJ   (b) 2.45 kJ   (c) 1.01 kJ

17. (a) 118 kJ   (b) 6.03×10^3 kg

19. Between 10^{-2}°C and 10^{-3}°C

21. (a) 316 K   (b) 200 J

23. $9 P_i V_i$

25. (a) 1.39 atm   (b) 366 K, 253 K   (c) 0, 4.66 kJ, -4.66 kJ

27. 227 K

29. (a)

(b) 8.79 L   (c) 900 K   (d) 300 K   (e) 336 J

31. 25.0 kW

33. (a) 9.95 cal/K, 13.9 cal/K   (b) 13.9 cal/K, 17.9 cal/K

35. 2.33×10^{-21} J

37. The ratio of oxygen to nitrogen molecules decreases to 85.5% of its sea-level value.

39. (a) 6.80 m/s   (b) 7.41 m/s   (c) 7.00 m/s

43. 819°C

45. (a) 3.21×10^{12} molecules   (b) 778 km
(c) 6.42×10^{-4} s^{-1}

49. (a) 9.36×10^{-8} m   (b) 9.36×10^{-8} atm   (c) 302 atm

51. (a) 100 kPa, 66.5 L, 400 K, 5.82 kJ, 7.48 kJ, 1.66 kJ

(b) 133 kPa, 49.9 L, 400 K, 5.82 kJ, 5.82 kJ, 0
(c) 120 kPa, 41.6 L, 300 K, 0, -910 J, -910 J
(d) 120 kPa, 43.3 L, 312 K, 722 J, 0, -722 J

55. 0.625

57. (a) Pressure increases as volume decreases.
(d) 0.500 atm^{-1}, 0.300 atm^{-1}

59. 1.09×10^{-3}; 2.69×10^{-2}; 0.529; 1.00; 0.199; 1.01×10^{-41}; 1.25×10^{-1082}

61. (a) Larger-mass molecules settle to the outside.

63. (a) 0.203 mol   (b) $T_B = T_C = 900$ K; $V_C = 15.0$ L

(c, d)	P (atm)	V (L)	T (K)	E_{int} (kJ)
A	1	5	300	0.760
B	3	5	900	2.28
C	1	15	900	2.28

(e) For $A \to B$, lock the piston in place and put the cylinder into an oven at 900 K. For $B \to C$, keep the gas in the oven while gradually letting the gas expand to lift a load on the piston as far as it can. For $C \to A$, move the cylinder from the oven back to the 300-K room and let the gas cool and contract.

(f, g)	Q (kJ)	W (kJ)	ΔE_{int} (kJ)
$A \to B$	1.52	0	1.52
$B \to C$	1.67	1.67	0
$C \to A$	-2.53	-1.01	-1.52
$ABCA$	0.656	0.656	0

65. (a) 3.34×10^{26} molecules   (b) during the 27th day
(c) 2.53×10^6

67. (a) 0.510 m/s   (b) 20 ms

Chapter 22

1. (a) 6.94%   (b) 335 J

3. (a) 10.7 kJ   (b) 0.533 s

5. (a) 1.00 kJ   (b) 0

7. (a) 67.2%   (b) 58.8 kW

9. (a) 869 MJ   (b) 330 MJ

11. (a) 741 J   (b) 459 J

13. 0.330 or 33.0%

15. (a) 5.12%   (b) 5.27 TJ/h   (c) As conventional energy sources become more expensive, or as their true costs are recognized, alternative sources become economically viable.

17. (a) 214 J, 64.3 J
(b) -35.7 J, -35.7 J. The net effect is the transport of energy from the cold to the hot reservoir without expenditure of external work.
(c) 333 J, 233 J
(d) 83.3 J, 83.3 J, 0. The net effect is the expulsion of the energy entering the system by heat, entirely by work, in a cyclic process.
(e) -0.111 J/K. The entropy of the Universe has decreased.

19. (a) 244 kPa   (b) 192 J
21. 146 kW, 70.8 kW
23. 9.00
27. 72.2 J
29. (a) 24.0 J   (b) 144 J
31. -610 J/K
33. 195 J/K
35. 3.27 J/K
37. 1.02 kJ/K
39. 5.76 J/K. Temperature is constant if the gas is ideal.
41. 0.507 J/K
43. 18.4 J/K
45. (a) 1   (b) 6
47. (a)

Macrostate	Possible Microstates	Total Number of Microstates
All R	RRR	1
2R, 1G	RRG, RGR, GRR	3
1R, 2G	GRR, GRG, RGG	3
All G	GGG	1

(b)

Macrostate	Possible Microstates	Total Number of Microstates
All R	RRRRR	1
4R, 1G	RRRRG, RRRGR, RRGRR, RGRRR, GRRRR	5
3R, 2G	RRRGG, RRGRG, RGRRG, GRRRG, RRGGR, RGRGR, GRRGR, RGGRR, GRGRR, GGRRR	10
2R, 3G	GGGRR, GGRGR, GRGGR, RGGGR, GGRRG, GRGRG, RGGRG, GRRGG, RGRGG, RRGGG	10
1R, 4G	GGGGR, GGGRG, GGRGG, GRGGG, RGGGG	5
All G	GGGGG	1

49. 1.86
51. (a) 5.00 kW   (b) 763 W
53. (a) $2nRT_i \ln 2$ (b) 0.273
55. 23.1 mW
57. 5.97×10^4 kg/s
59. (a) 3.19 cal/K   (b) 98.19°F, 2.59 cal/K
61. 1.18 J/K
63. (a) $10.5nRT_i$ (b) $8.50nRT_i$ (c) 0.190   (d) 0.833
65. $nC_P \ln 3$
69. (a) 96.9 W = 8.33×10^4 cal/hr
     (b) 1.19°C/h = 2.14°F/h

Chapter 23

1. (a) 2.62×10^{24} electrons   (b) 2.38 electrons
3. The force is $\sim 10^{26}$ N.
5. 514 kN
7. 0.873 N at 330°
9. (a) 82.2 nN   (b) 2.19 Mm/s
11. (a) 55.8 pN/C down   (b) 102 nN/C up
13. 1.82 m to the left of the negative charge
15. (a) $(18.0\mathbf{i} - 218\mathbf{j})$ kN/C   (b) $(36.0\mathbf{i} - 436\mathbf{j})$ mN
17. (a) The field is zero at the center of the triangle.

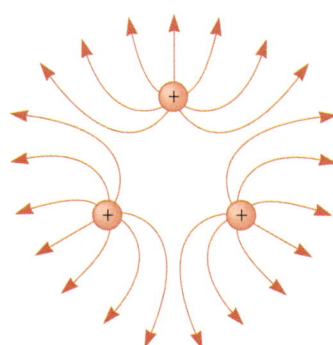

(b) $(1.73\, k_e q/a^2)\mathbf{j}$
19. (a) $5.91 k_e q/a^2$ at 58.8°   (b) $5.91 k_e\, q^2/a^2$ at 58.8°
23. $-(\pi^2 k_e q/6a^2)\,\mathbf{i}$
25. $-(k_e \lambda_0/x_0)\,\mathbf{i}$
27. (a) $6.64\mathbf{i}$ MN/C   (b) $24.1\mathbf{i}$ MN/C   (c) $6.40\mathbf{i}$ MN/C
     (d) $0.664\mathbf{i}$ MN/C, taking the axis of the ring as the x axis
29. (a) 383 MN/C away   (b) 324 MN/C away
     (c) 80.7 MN/C away   (d) 6.68 MN/C away
33. $-21.6\mathbf{i}$ MN/C
37. (a) 86.4 pC for each
     (b) 324 pC, 459 pC, 459 pC, 432 pC
     (c) 57.6 pC, 106 pC, 154 pC, 96.0 pC
39.

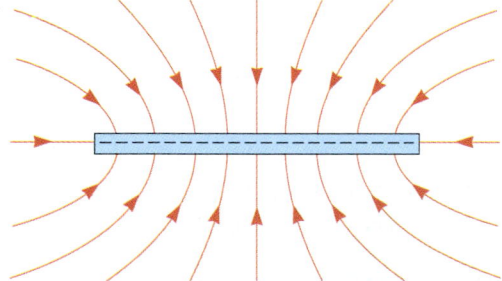

41. 4.39 Mm/s and 2.39 km/s
43. (a) 61.4 Gm/s^2 (b) 19.5 μs   (c) 11.7 m   (d) 1.20 fJ
45. K/ed in the direction of motion
47. (a) 111 ns   (b) 5.67 mm   (c) $(450\mathbf{i} + 102\mathbf{j})$ km/s
49. (a) 36.9°, 53.1°   (b) 167 ns, 221 ns
51. (a) 21.8 μm   (b) 2.43 cm
53. (a) 10.9 nC   (b) 5.43 mN
55. 40.9 N at 263°
57. 26.7 μC
61. $-707\mathbf{j}$ mN

63. (a) $\theta_1 = \theta_2$

65. (b) The object's acceleration is a negative constant times its displacement from equilibrium.
$$T - (\pi 8^{-1/4})(mL^3/k_e qQ)^{1/2}$$

67. (a) $-(4k_e q/3a^2)\mathbf{j}$ (b) $(0, 2.00\text{ m})$

69. (a) $\mathbf{F} = 1.90(k_e q^2/s^2)(\mathbf{i} + \mathbf{j} + \mathbf{k})$
(b) $\mathbf{F} = 3.29(k_e q^2/s^2)$ in the direction away from the diagonally opposite vertex

71. $(-1.36\mathbf{i} + 1.96\mathbf{j})\text{ kN/C}$

Chapter 24

1. (a) $858\text{ N}\cdot\text{m}^2/\text{C}$ (b) 0   (c) $657\text{ N}\cdot\text{m}^2/\text{C}$

3. 4.14 MN/C

5. (a) $-2.34\text{ kN}\cdot\text{m}^2/\text{C}$ (b) $+2.34\text{ kN}\cdot\text{m}^2/\text{C}$ (c) 0

7. q/ϵ_0

9. EhR

11. (a) $-6.89\text{ MN}\cdot\text{m}^2/\text{C}$ (b) The number of lines entering exceeds the number leaving by 2.91 times or more.

13. (a) $q/2\epsilon_0$ (b) $q/2\epsilon_0$ (c) Plane and square look the same to the charge.

15. (a) $+Q/2\epsilon_0$ (b) $-Q/2\epsilon_0$

17. $5.22\text{ kN}\cdot\text{m}^2/\text{C}$

19. $-18.8\text{ kN}\cdot\text{m}^2/\text{C}$

21. 0 when $R < d$ and $\lambda L/\epsilon_0$ when $R > d$

23. (a) $3.20\text{ MN}\cdot\text{m}^2/\text{C}$ (b) $19.2\text{ MN}\cdot\text{m}^2/\text{C}$ (c) The answer to part (a) could change, but the answer to part (b) would stay the same.

25. $-q/24\epsilon_0$

27. (a) 0   (b) 366 kN/C (c) 1.46 MN/C (d) 650 kN/C

29. $\mathbf{E} = \rho r/2\epsilon_0$ away from the axis

31. (a) 0   (b) 7.19 MN/C away from the center

33. (a) $\sim 1\text{ mN}$ (b) $\sim 100\text{ nC}$ (c) $\sim 10\text{ kN/C}$
(d) $\sim 10\text{ kN}\cdot\text{m}^2/\text{C}$

35. (a) 51.4 kN/C outward   (b) $646\text{ N}\cdot\text{m}^2/\text{C}$

37. 508 kN/C up

39. (a) 0   (b) $5\,400\text{ N/C}$ outward   (c) 540 N/C outward

41. (a) $+708\text{ nC/m}^2$ and -708 nC/m^2 (b) $+177\text{ nC}$ and -177 nC

43. 2.00 N

45. (a) $-\lambda, +3\lambda$ (b) $3\lambda/2\pi\epsilon_0 r$ radially outward

47. (a) 80.0 nC/m^2 on each face   (b) $(9.04\mathbf{k})\text{ kN/C}$
(c) $(-9.04\mathbf{k})\text{ kN/C}$

49. (a) 0   (b) 79.9 MN/C radially
(d) 7.35 MN/C radially outwar

51. (b) $Q/2\epsilon_0$ (c) Q/ϵ_0

53. (a) $+2Q$ (b) radially outwar
(d) 0   (e) 0   (f) $3Q$ (g) $3k_e$
(h) $3Qr^3/a^3$ (i) $3k_e Qr/a^3$ ra
(k) $+2Q$ (l) See top of next

55. (a) $\rho r/3\epsilon_0$; $Q/4\pi\epsilon_0 r^2$; 0; $Q/4\pi$
(b) $-Q/4\pi b^2$ and $+Q/4\pi c^2$

57. For $r < a$, $\mathbf{E} = \lambda/2\pi\epsilon_0 r$ radially
$\mathbf{E} = [\lambda + \rho\pi(r^2 - a^2)]/2\pi\epsilon_0 r$
$r > b$, $\mathbf{E} = [\lambda + \rho\pi(b^2 - a^2)]/$

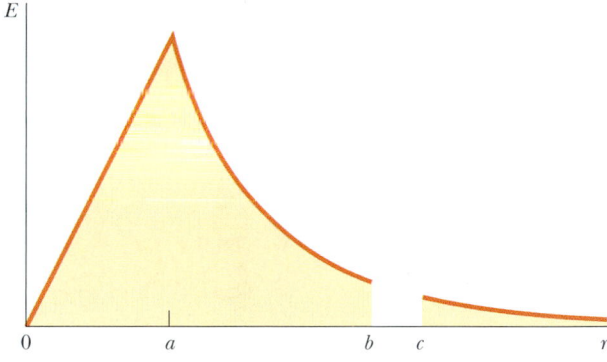

Chapter 24, Problem 53

59. (a) σ/ϵ_0 away from both plates   (b) 0   (c) σ/ϵ_0 away from both plates

61. (c) $f = (1/2\pi)(k_e e^2/m_e R^3)^{1/2}$ (d) 102 pm

65. $\mathbf{E} = a/2\epsilon_0$ radially outward

69. (b) $\mathbf{g} = GM_E r/R_E^3$ radially inward

Chapter 25

1. 1.35 MJ

3. (a) 152 km/s (b) 6.49 Mm/s

5. -0.502 V

7. 1.67 MN/C

9. -38.9 V; the origin

11. (a) 0.500 m (b) 0.250 m (c) 1.26 s (d) 0.343 m

13. 40.2 kV

15. 0.300 m/s

17. (a) $\mathbf{F} = 0$ (b) $\mathbf{E} = 0$ (c) 45.0 kV

19. (a) -27.3 eV (b) -6.81 eV (c) 0

21. -11.0 MV

25. $-5k_e q/R$

27. (a) 10.8 m/s and 1.55 m/s (b) larger

29. 0.720 m, 1.44 m, 2.88 m. No. The radii of the equipotentials are inversely proportional to the potential.

31. 27.4 fm

33. -3.96 J

35. $22.8k_e q^2/s$

37. $\mathbf{E} = (-5 + 6xy)\mathbf{i} + (3x^2 - 2z^2)\mathbf{j} - 4yz\mathbf{k}$; 7.08 N/C

39. $E_y = \dfrac{k_e Q}{\ell y}\left[1 - \dfrac{y^2}{\ell^2 + y^2 + \ell\sqrt{\ell^2 + y^2}}\right]$

41. $-0.553k_e Q/R$

43. (a) C/m^2 (b) $k_e\alpha[L - d\ln(1 + L/d)]$

45. $(\sigma/2\epsilon_0)[(x^2 + b^2)^{1/2} - (x^2 + a^2)^{1/2}]$

47. 1.56×10^{12} electrons removed

49. (a) 0, 1.67 MV (b) 5.85 MN/C away, 1.17 MV
(c) 11.9 MN/C away, 1.67 MV

51. (a) 450 kV (b) $7.50\ \mu\text{C}$

53. 253 MeV

57. (a) 6.00 m (b) $-2.00\ \mu\text{C}$

59. (a) $\dfrac{k_e Q}{h}\ln\left(\dfrac{d + h + \sqrt{(d + h)^2 + R^2}}{d + \sqrt{d^2 + R^2}}\right)$

(b) $\dfrac{k_e Q}{R^2 h}\Big[(d + h)\sqrt{(d + h)^2 + R^2} - d\sqrt{d^2 + R^2}$
$+ R^2 \ln\Big(\dfrac{d + h + \sqrt{(d + h)^2 + R^2}}{d + \sqrt{d^2 + R^2}}\Big) - 2dh - h^2\Big]$

61. $k_e Q^2 / 2R$

63. $V_2 - V_1 = (-\lambda / 2\pi\epsilon_0) \ln(r_2 / r_1)$

65. (a) The fields are radial. $E_A = 0$; $E_B = (89.9 \text{ V/m})/r^2$;
$E_C = (-45 \text{ V/m})/r^2$
(b) $V_A = 150 \text{ V}$; $V_B = -450 \text{ V} + 89.9 \text{ V}/r$; $V_C = -45 \text{ V}/r$

67. (a) $1.00 \text{ kV} - (1.41 \text{ kV/m})\, x - (1.44 \text{ kV}) \ln\Big(1 - \dfrac{x}{3\,\text{m}}\Big)$

(b) $+ 633 \text{ nJ}$

69. (a) $E_r = 2k_e p \cos\theta / r^3$; $E_\theta = k_e p \sin\theta / r^3$; yes; no
(b) $V = k_e p y (x^2 + y^2)^{-3/2}$; $\mathbf{E} = 3k_e p x y\,(x^2 + y^2)^{-5/2}\,\mathbf{i}$
$+ k_e p (2y^2 - x^2)(x^2 + y^2)^{-5/2}\,\mathbf{j}$

71. $V = \pi k_e C\Big[R\sqrt{x^2 + R^2} + x^2 \ln\Big(\dfrac{x}{R + \sqrt{x^2 + R^2}}\Big)\Big]$

73. (a) 1.42 mm (b) 9.20 kV/m

Chapter 26

1. (a) $48.0\ \mu\text{C}$ (b) $6.00\ \mu\text{C}$

3. (a) $1.33\ \mu\text{C/m}^2$ (b) 13.3 pF

5. (a) $5.00\ \mu\text{C}$ on the larger and $2.00\ \mu\text{C}$ on the smaller
sphere   (b) 89.9 kV

7. (a) 11.1 kV/m toward the negative plate
(b) 98.3 nC/m^2 (c) 3.74 pF (d) 74.8 pC

9. $4.42\ \mu\text{m}$

11. (a) 2.68 nF (b) 3.02 kV

13. 1.23 kV

15. (a) 15.6 pF (b) 256 kV

17. (a) $17.0\ \mu\text{F}$ (b) 9.00 V (c) $45.0\ \mu\text{C}$ and $108\ \mu\text{C}$

19. 6.00 pF and 3.00 pF

21. (a) $5.96\ \mu\text{F}$ (b) $89.5\ \mu\text{C}$ on the 20-μF capacitor,
$63.2\ \mu\text{C}$ on the 6-μF capacitor, and $26.3\ \mu\text{C}$ on the 15-μF
and 3-μF capacitors

23. $120\ \mu\text{C}$; $80.0\ \mu\text{C}$ and $40.0\ \mu\text{C}$

25. (a) $400\ \mu\text{C}$ (b) 2.50 kN/m

27. 10

29. $83.6\ \mu\text{C}$

31. (a) $216\ \mu\text{J}$ (b) $54.0\ \mu\text{J}$

33. Stored energy doubles.

37. 9.79 kg

39. 1.40 fm

41. (a) 8.13 nF (b) 2.40 kV

43. 1.04 m

45. (a) $4.00\ \mu\text{F}$ (b) $8.40\ \mu\text{F}$ (c) 5.71 V and $48.0\ \mu\text{C}$

47. $4\pi\kappa_1\kappa_2 abc\epsilon_0 / [\kappa_2 bc - \kappa_1 ab + (\kappa_1 - \kappa_2)ac]$

49. 22.5 V

51. (b) $-8.78 \text{ MN/C} \cdot \text{m}$; $-55.3\mathbf{i} \text{ mN}$

57. 0.188 m^2

59. $\epsilon_0 A / (s - d)$

61. $1 + q/q_0$

63. (a) $Q_0^2 d(\ell - x)/(2\ell^3\epsilon_0)$ (b) $Q_0^2 d/(2\ell^3\epsilon_0)$ to the
right   (c) $Q_0^2/(2\ell^4\epsilon_0)$ (d) $Q_0^2/(2\ell^4\epsilon_0)$

65. $4.29\ \mu\text{F}$

67. $3.00\ \mu\text{F}$

69. (b) $Q/Q_0 = \kappa$

71. $2/3$

73. 19.0 kV

75. $3.00\ \mu\text{F}$

Chapter 27

1. 7.50×10^{15} electrons

3. (a) $0.632\, I_0\tau$ (b) $0.999\,95\, I_0\tau$ (c) $I_0\tau$

5. 400 nA

7. (a) 17.0 A (b) 85.0 kA/m^2

9. (a) 99.5 kA/m^2 (b) 0.800 cm

11. (a) 2.55 A/m^2 (b) $5.31 \times 10^{10} \text{ m}^{-3}$ (c) 1.20×10^{10} s

13. 500 mA

15. 6.43 A

17. (a) 1.82 m (b) $280\ \mu\text{m}$

19. (a) $777 \text{ n}\Omega$ (b) $3.28\ \mu\text{m/s}$

21. $1.56\, R$

23. $6.00 \times 10^{-15}/\Omega \cdot \text{m}$

25. 0.180 V/m

27. 21.2 nm

29. $1.44 \times 10^3 \text{ °C}$

31. (a) $31.5 \text{ n}\Omega \cdot \text{m}$ (b) 6.35 MA/m^2 (c) 49.9 mA
(d) $659\ \mu\text{m/s}$ (e) 0.400 V

33. 0.125

35. 67.6°C

37. 5.00 A; $24.0\ \Omega$

39. $28.9\ \Omega$

41. 36.1%

43. (a) 5.97 V/m (b) 74.6 W (c) 66.1 W

45. 0.833 W

47. 26.9 cents/day

49. (a) 184 W (b) 461°C

51. $\sim \$1$

53. 25.5 yr

57. Experimental resistivity $= 1.47\ \mu\Omega \cdot \text{m} \pm 4\%$,
in agreement with $1.50\ \mu\Omega \cdot \text{m}$

59. (a) $8.00\mathbf{i} \text{ V/m}$ (b) $0.637\ \Omega$ (c) 6.28 A
(d) $200\mathbf{i} \text{ MA/m}^2$

61. $2\,020 \text{°C}$

63. (a) 667 A (b) 50.0 km

65.

Material	$\alpha' = \alpha/(1 - 20\alpha)$
Silver	$4.1 \times 10^{-3}/\text{°C}$
Copper	$4.2 \times 10^{-3}/\text{°C}$
Gold	$3.6 \times 10^{-3}/\text{°C}$
Aluminum	$4.2 \times 10^{-3}/\text{°C}$
Tungsten	$4.9 \times 10^{-3}/\text{°C}$
Iron	$5.6 \times 10^{-3}/\text{°C}$
Platinum	$4.25 \times 10^{-3}/\text{°C}$
Lead	$4.2 \times 10^{-3}/\text{°C}$
Nichrome	$0.4 \times 10^{-3}/\text{°C}$
Carbon	$-0.5 \times 10^{-3}/\text{°C}$
Germanium	$-24 \times 10^{-3}/\text{°C}$
Silicon	$-30 \times 10^{-3}/\text{°C}$

67. No. The fuses should pass no more than 3.87 A.

71. The graphs are as follows:

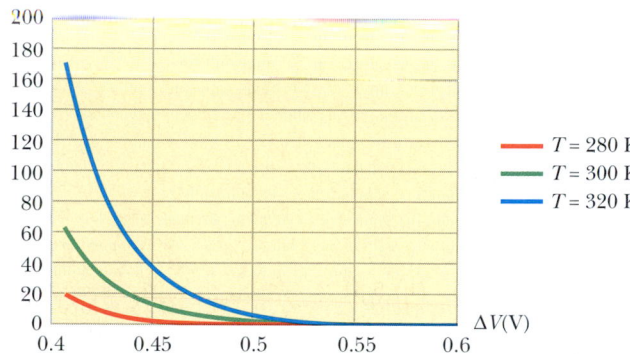

Chapter 28

1. (a) $6.73\ \Omega$ (b) $1.97\ \Omega$

3. (a) $4.59\ \Omega$ (b) 8.16%

5. $12.0\ \Omega$

7. He can put the three resistors in parallel.

9. (a) 227 mA   (b) 5.68 V

11. (a) 75.0 V   (b) 25.0 W, 6.25 W, and 6.25 W; 37.5 W

13. $1.00\ \text{k}\Omega$

15. 14.3 W, 28.5 W, 1.33 W, 4.00 W

17. (a) The $11\text{-}\Omega$ resistor   (b) 148 W $= 148$ W
(c) The $22\text{-}\Omega$ resistor   (d) 33.0 W $= 33.0$ W
(e) The parallel configuration

19. 846 mA down in the $8\text{-}\Omega$ resistor; 462 mA down in the middle branch; 1.31 A up in the right-hand branch

21. (a) -222 J and 1.88 kJ   (b) 687 J, 128 J, 25.6 J, 616 J, 205 J   (c) 1.66 kJ

23. 50.0 mA from a to e

25. Starter, 171 A; battery, 0.283 A

27. (a) 909 mA   (b) -1.82 V

29. (a) 5.00 s   (b) $150\ \mu\text{C}$ (c) $4.06\ \mu\text{A}$

31. $U_0/4$

33. (a) 6.00 V   (b) $8.29\ \mu\text{s}$

35. $1.60\ \text{M}\Omega$

37. 0.982 s

39. $16.6\ \text{k}\Omega$

41. $0.302\ \Omega$

43. 0.588 A

45. 1.36 V

47. (a) Heater, 12.5 A; toaster, 6.25 A; grill, 8.33 A
(b) No; together they would require 27.1 A.

49. 15.5 A

51. 2.22 h

53. 4.00 V, with a at higher potential

55. $6.00\ \Omega$; $3.00\ \Omega$

57. (a) $R \rightarrow \infty$ (b) $R \rightarrow 0$ (c) $R = r$

59. (a) $R \leq 1050\ \Omega$ (b) $R \geq 10.0\ \Omega$

61. (a) $9.93\ \mu\text{C}$ (b) 33.7 nA   (c) 334 nW   (d) 337 nW

63. (a) 40.0 W   (b) 80.0 V, 40.0 V, 40.0 V

65. Place in parallel with the galvanometer a branch consisting of three resistors in series, with contacts between them as follows:

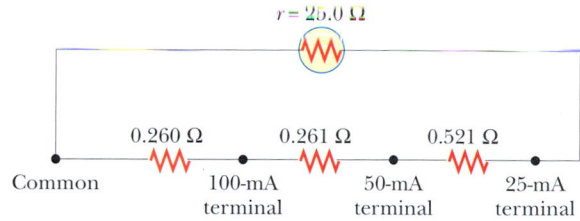

67. (a) 0 in the $3\text{-k}\Omega$ resistor and $333\ \mu\text{A}$ in the others
(b) $50.0\ \mu\text{C}$ (c) $278\ \mu\text{A}e^{-t/180\ \text{ms}}$ (d) 290 ms

71. 48.0 W

73. (a) $\mathcal{E}^2/3R$ (b) $3\mathcal{E}^2/R$ (c) In the parallel connection

Chapter 29

1. (a) up   (b) toward you, out of the plane of the paper
(c) no deflection   (d) into the plane of the paper

3. negative z direction

5. $(-20.9\mathbf{j})$ mT

7. 8.93×10^{-30} N down, 1.60×10^{-17} N up, 4.80×10^{-17} N down

9. $48.8°$ or $131°$

11. 2.34 aN

13. 0.245 T east

15. (a) 4.73 N   (b) 5.46 N   (c) 4.73 N

17. 196 A east

19. 1.07 m/s

21. $2\pi r I B \sin\theta$ up

23. (a) 5.41 mA$\cdot$m^2 (b) 4.33 mN$\cdot$m

25. 9.98 N$\cdot$m; clockwise as seen looking down from above

27. (a) 80.1 mN$\cdot$m   (b) 0.104 N$\cdot$m   (c) 0.132 N$\cdot$m
(d) The circular loop experiences largest torque.

29. (a) minimum: with its north end pointing north at $48.0°$ below the horizontal; maximum: with its north end pointing south at $48.0°$ above the horizontal   (b) $1.07\ \mu\text{J}$

31. (a) 49.7 aN south   (b) 1.29 km

33. 115 keV

35. $r_\alpha = r_d = \sqrt{2}r_p$

37. 4.99×10^8 rad/s

39. 7.88 pT

41. 244 kV/m

43. 0.278 m

45. (a) 4.31×10^7 rad/s   (b) 51.7 Mm/s

47. 70.1 mT

49. 3.70×10^{-9} m^3/C

51. $43.2\ \mu\text{T}$

53. (a) 179 ps   (b) 351 keV

55. (a) The electric current experiences a magnetic force.

57. (a) B_x is indeterminate; B_y is zero; B_z is $-F_i/ev_i$.
(b) $-F_i\mathbf{j}$ (c) $-F_i\mathbf{j}$

59. (a) $(3.52\mathbf{i} - 1.60\mathbf{j})$ aN   (b) $24.4°$

61. 0.588 T

63. 19.6 mT

65. 438 kHz

67. 3.70×10^{-24} N·m

69. (a) 0.501 m   (b) 45.0°

71. (a) 1.33 m/s   (b) No. Positive ions moving toward you in the magnetic field to the right feel an upward magnetic force and migrate upward in the blood vessel. Negative ions moving toward you feel a downward magnetic force and accumulate at the bottom of this section of the vessel. Thus, both species can participate in the generation of the emf.

Chapter 30

1. 12.5 T

3. (a) 28.3 μT into the page   (b) 24.7 μT into the page

5. $\mu_0 I/4\pi x$ into the page

7. 58.0 μT into the page

9. 26.2 μT into the page

11. $\dfrac{\mu_0 I}{12}\left(\dfrac{1}{a} - \dfrac{1}{b}\right)$ out of the page

13. $0.475\mu_0 I/R$ into the page

15. $-13.0\mathbf{j}$ μT

17. $-27.0\mathbf{i}$ μN

19. 20.0 μT toward the bottom of the page

21. 200 μT toward the top of the page; 133 μT toward the bottom of the page

23. (a) 3.60 T   (b) 1.94 T

25. (a) 6.34 mN/m inward   (b) greater

27. (a) $\mu_0 b r_1^2/3$ (b) $\mu_0 b R^3/3r_2$

29. 31.8 mA

31. 464 mT

33. (a) 3.13 mWb   (b) 0

35. (a) $B\pi R^2 \cos\theta$ (b) $B\pi R^2 \cos\theta$

37. (a) 11.3 GV·m/s   (b) 0.100 A

39. 0.191 T

41. 150 μT·m^2

43. 2.62 MA/m

45. (b) 6.45×10^4 K·A/T·m

47. (a) 8.63×10^{45} electrons   (b) 4.01×10^{20} kg

49. 2.00 GA west

51. ~10^{-5} T or 10^{-6} T; on the order of one-tenth as large

53. $\dfrac{\mu_0 I}{2\pi w}\ln\left(1 + \dfrac{w}{b}\right)\mathbf{k}$

55. 143 pT away along the axis

59. (a) 2.46 N up   (b) 107 m/s^2 up

61. (a) 274 μT   (b) $-274\mathbf{j}$ μT   (c) $1.15\mathbf{i}$ mN   (d) $0.384\mathbf{i}$ m/s^2 (e) acceleration is constant.   (f) $0.999\mathbf{i}$ m/s

63. (a) $\mu_0 \sigma v/2$ (b) out of the plane of the page, parallel to the roller axes

65. 28.8 mT

67. $4\sqrt{2}/\pi^2$

71. $\dfrac{\mu_0 I}{4\pi}\left(1 - e^{-2\pi}\right)$ out of the plane of the page

73. $\rho\mu_0\omega R^2/3$

75. (a) $\dfrac{\mu_0 I}{\pi r}\dfrac{(2r^2 - a^2)}{(4r^2 - a^2)}$ to the left

(b) $\dfrac{\mu_0 I}{\pi r}\dfrac{(2r^2 + a^2)}{(4r^2 + a^2)}$ toward the top of the page

Chapter 31

1. 500 mV

3. 9.82 mV

5. 160 A

7. (a) 1.60 A counterclockwise   (b) 20.1 μT   (c) up

9. (a) $(\mu_0 IL/2\pi)\ln(1 + w/h)$ (b) -4.80 μV; current is counterclockwise

11. 283 μA upward

13. $(68.2$ mV$)e^{-1.6t}$ counterclockwise

15. 272 m

17. $(0.422$ V$)\cos\omega t$

19. 0.880 C

21. (a) 3.00 N to the right   (b) 6.00 W

23. 0.763 V with the left-hand wingtip positive

25. 2.83 mV

27. (a) $F = N^2 B^2 w^2 v/R$ to the left   (b) 0   (c) $F = N^2 B^2 w^2 v/R$ to the left

29. negative

31. 145 μA

33. 1.80 mN/C upward and to the left, perpendicular to r_1

35. (a) $(9.87$ mV/m$)\cos(100\pi t)$ (b) clockwise

37. (a) 7.54 kV   (b) The plane of the coil is parallel to **B**.

39. $(28.6$ mV$)\sin(4\pi t)$

41. (a) 0.640 N·m   (b) 241 W

43. (a) $(8.00$ mT·m$^2)\cos(377t)$ (b) $(3.02$ V$)\sin(377t)$ (c) $(3.02$ A$)\sin(377t)$ (d) $(9.10$ W$)\sin^2(377t)$ (e) $(24.1$ mN·m$)\sin^2(377t)$

45. (b) Larger R makes current smaller, so the loop must travel faster to maintain equality of magnetic force and weight. (c) The magnetic force is proportional to the product of the magnetic field and current, while the current is proportional to the magnetic field. If B is halved, the speed must be quadrupled to compensate.

47. $(-2.87\mathbf{j} + 5.75\mathbf{k})$ Gm/s^2

49. -7.22 mV $\cos(2\pi\,523\,t/s)$

51. (a) 43.8 A   (b) 38.3 W

53. (a) 3.50 A and 1.40 A   (b) 34.3 W   (c) 4.29 N

57. 1.20 μC

59. (a) 0.900 A   (b) 0.108 N   (c) b (d) no

61. (a) $a\pi r^2$ (b) $-b\pi r^2$ (c) $-b\pi r^2/R$ (d) $b^2\pi^2 r^4/R$

63. (a) 36.0 V   (b) 600 mWb/s   (c) 35.9 V   (d) 4.32 N·m

65. -10.2 μV

67. $\mu_0 I\ell vw/2\pi Rr(r + w)$

69. 6.00 A

71. (a) $(1.19$ V$)\cos(120\pi t)$ (b) 88.5 mW

73. $(-87.1$ mV$)\cos(200\pi t + \phi)$

75. -6.75 V

Chapter 32

1. 19.5 mV
3. 100 V
5. 240 nT·m²
7. (18.8 V) cos(377t)
9. − 0.421 A/s
11. (a) 360 mV (b) 180 mV (c) 3.00 s
13. (a) 15.8 μH (b) 12.6 mH
15. $\mathcal{E}_0/k^2 L$
17. (a) 0.139 s (b) 0.461 s
19. (a) 2.00 ms (b) 0.176 A (c) 1.50 A (d) 3.22 ms
21. (a) 0.800 (b) 0
23. (a) 6.67 A/s (b) 0.332 A/s
25. (500 mA)(1 − $e^{-10t/s}$), 1.50 A − (0.25 A)$e^{-10t/s}$
27. 0 for t < 0; (10 A)(1 − $e^{-10\,000t}$) for 0 < t < 200 μs;
    (63.9 A)$e^{-10\,000t}$ for t > 200 μs
29. (a) 5.66 ms (b) 1.22 A (c) 58.1 ms
31. 0.064 8 J
33. 2.44 μJ
35. 44.2 nJ/m³ for the **E** field and 995 μJ/m³ for the **B** field
37. (a) 20.0 W (b) 20.0 W (c) 0 (d) 20.0 J
39. $2\pi B_0^2 R^3/\mu_0 = 2.70 \times 10^{18}$ J
41. 1.00 V
43. (a) 18.0 mH (b) 34.3 mH (c) − 9.00 mV
45. (b) 3.95 nH
47. $(L_1 L_2 - M^2)/(L_1 + L_2 - 2M)$
49. 20.0 V
51. 608 pF
53. (a) 135 Hz (b) 119 μC (c) − 114 mA
55. (a) 6.03 J (b) 0.529 J (c) 6.56 J
57. (a) 4.47 krad/s (b) 4.36 krad/s (c) 2.53%
59. 0.693(2L/R) (b) 0.347(2L/R)
61. (a) − 20.0 mV (b) − (10.0 MV/s²)t^2 (c) 63.2 μs
63. $\dfrac{Q}{2N}\sqrt{\dfrac{3L}{C}}$
65. (a) $L \approx (\pi/2)N^2\mu_0 R$ (b) ~100 nH (c) ~1 ns
71. (a) 72.0 V; b
    (b)

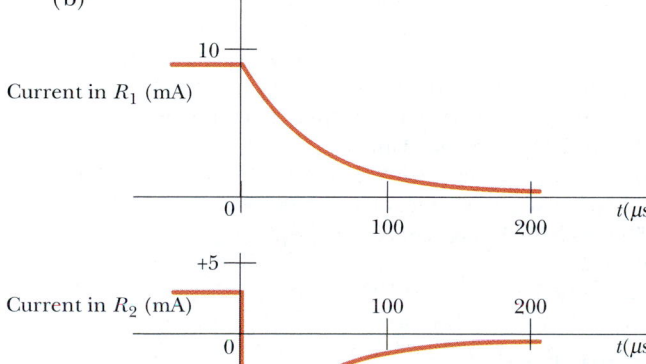

(b)

(c) 75.2 μs
73. 300 Ω
75. (a) It creates a magnetic field. (b) The long narrow rectangular area between the conductors encloses all of the magnetic flux.
77. (a) 62.5 GJ (b) 2 000 N
79. (a) 2.93 mT up (b) 3.42 Pa (c) clockwise (d) up
    (e) 1.30 mN

Chapter 33

1. $\Delta v(t) = (283\text{ V})\sin(628t)$
3. 2.95 A, 70.7 V
5. 14.6 Hz
7. 3.38 W
9. (a) 42.4 mH (b) 942 rad/s
11. 5.60 A
13. 0.450 T·m²
15. (a) 141 mA (b) 235 mA
17. 100 mA
19. (a) 194 V (b) current leads by 49.9°
21. (a) 78.5 Ω (b) 1.59 kΩ (c) 1.52 kΩ (d) 138 mA
    (e) − 84.3°
23. (a) 17.4° (b) voltage leads the current
25. (a) 146 V (b) 213 V (c) 179 V (d) 33.4 V
27. (a) 124 nF (b) 51.5 kV
29. 8.00 W
31. (a) 16.0 Ω (b) − 12.0 Ω
33. 132 mm
35. $11(\Delta V)^2/14R$
37. 1.82 pF
39. 242 mJ
41. 0.591 and 0.987; the circuit in Problem 23
43. 687 V
45. 87.5 Ω
47. (a) 29.0 kW (b) 5.80 × 10⁻³ (c) If the generator were limited to 4 500 V, no more than 17.5 kW could be delivered to the load, never 5 000 kW.
49. (a) 613 μF (b) 0.756
51. (a) 580 μH and 54.6 μF (b) 1 (c) 894 Hz
    (d) ΔV_{out} leads ΔV_{in} by 60.0° at 200 Hz; ΔV_{out} and ΔV_{in} are in phase at 894 Hz; ΔV_{out} lags ΔV_{in} by 60.0° at 4 000 Hz. (e) 1.56 W, 6.25 W, 1.56 W (f) 0.408
53. 0.317
55. 56.7 W
57. (a) 225 mA (b) 450 mA
59. (a) Circuit (a) is a high-pass filter, and circuit (b) is a low-pass filter.
    (b) $\dfrac{\Delta V_{\text{out}}}{\Delta V_{\text{in}}} = \dfrac{(R_L^2 + X_L^2)^{1/2}}{[R_L^2 + (X_L - X_C)^2]^{1/2}}$ for circuit (a)
    $\dfrac{\Delta V_{\text{out}}}{\Delta V_{\text{in}}} = \dfrac{X_C}{[R_L^2 + (X_L - X_C)^2]^{1/2}}$ for circuit (b)
61. ~10² or 10³ A
63. (a) 200 mA; voltage leads by 36.8° (b) 40.0 V; $\phi = 0°$
    (c) 20.0 V; $\phi = -90.0°$ (d) 50.0 V; $\phi = +90.0°$
65. (b) 31.6

67. (a) 919 Hz (b) 1.50 A, 1.60 A, 6.73 mA (c) 2.19 A
(d) current lagging by $\phi = -46.7°$

69. (a)

f (Hz)	X_L (Ω)	X_C (Ω)	Z (Ω)
300	282	12 600	12 300
600	565	6 290	5 720
800	754	4 710	3 960
1 000	942	3 770	2 830
1 500	1 410	2 510	1 100
2 000	1 880	1 880	40.0
3 000	2 830	1 260	1 570
4 000	3 770	942	2 830
6 000	5 660	629	5 030
10 000	9 420	377	9 040

(b)

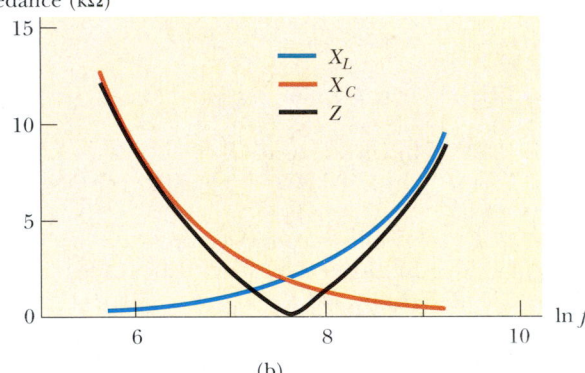

(b)

71. (a) 1.84 kHz
(b)

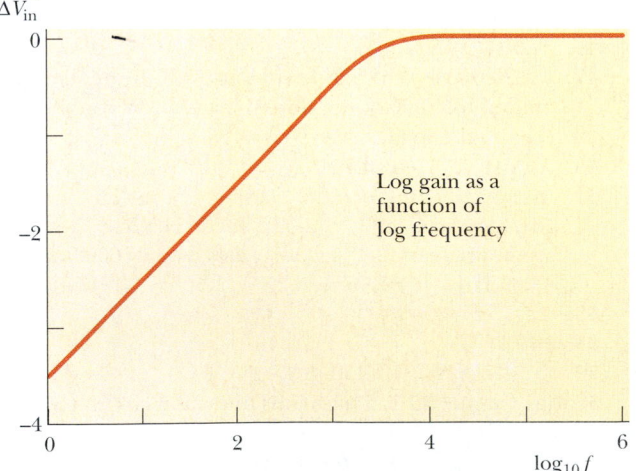

Log gain as a function of log frequency

Chapter 34

1. 2680 A.D.
3. 733 nT
5. (a) 6.00 MHz (b) 73.3 nT $(-\mathbf{k})$
(c) $\mathbf{B} = (-73.3 \text{ nT}) \cos(0.126x - 3.77 \times 10^7 t)\mathbf{k}$

7. (a) 0.333 μT (b) 0.628 μm (c) 477 THz
9. 75.0 MHz
11. 3.33 μJ/m^3
13. 307 μW/m^2
15. 3.33×10^3 m^2
17. (a) 332 kW/m^2 radially inward (b) 1.88 kV/m and 222 μT
19. 29.5 nT
21. (a) 2.33 mT (b) 650 MW/m^2 (c) 510 W
23. (a) 4.97 kW/m^2 (b) 16.6 μJ/m^3
25. 83.3 nPa
27. (a) 5.36 N (b) 8.93×10^{-4} m/s^2 (c) 10.7 days
29. (a) 1.90 kN/C (b) 50.0 pJ (c) 1.67×10^{-19} kg·m/s
31. (a) 11.3 kJ (b) 1.13×10^{-4} kg·m/s
33. (a) 2.26 kW (b) 4.71 kW/m^2
35. (a) away along the perpendicular bisector of the line segment joining the antennas (b) along the extensions of the line segment joining the antennas
37. 56.2 m
39. (a) radio, radio, radio, microwave, infrared, ultraviolet, x-ray, γ-ray, γ-ray; (b) radio, radio, microwave, infrared, ultraviolet or x-ray, x- or γ-ray, γ-ray, γ-ray
41. 545 THz
43. (a) 6.00 pm (b) 7.50 cm
45. The radio audience hears it first, 8.41 ms before the people in the newsroom.
47. (a) 186 m to 556 m (b) 2.78 m to 3.41 m
49. (a) 3.77×10^{26} W (b) 1.01 kV/m and 3.35 μT
51. (a) $2\pi^2 r^2 f B_{\text{max}} \cos\theta$, where θ is the angle between the magnetic field and the normal to the loop (b) The loop should be in the vertical plane that contains the line of sight to the transmitter.
53. (a) 6.67×10^{-16} T (b) 5.31×10^{-17} W/m^2
(c) 1.67×10^{-14} W (d) 5.56×10^{-23} N
55. (a) $\frac{1}{2}\mu_0 n^2 r i \dfrac{di}{dt}$ radially inward (b) $\mu_0 \pi n^2 r^2 \ell i \dfrac{di}{dt}$
(c) $(\Delta V) i$
57. 95.1 mV/m
59. (a) $B_{\text{max}} = 583$ nT, $k = 419$ rad/m, $\omega = 12.6$ Trad/s, the xz plane (b) $S_{\text{av}} = \frac{1}{2}S_{\text{max}} = 40.6$ W/m^2 (c) 271 nPa
(d) 406 nm/s^2
61. (a) 22.6 h (b) 30.6 s
63. (a) 8.32×10^7 W/m^2 (b) 1.05 kW
65. (a) 625 kW/m^2 (b) 21.7 kN/C, 72.4 μT (c) 17.8 min
67. (b) 17.6 Tm/s^2, 1.75×10^{-27} W (c) 1.80×10^{-24} W
69. 3.00×10^{-2} degrees
71. $\epsilon_0 E^2 A/2m$

Chapter 35

1. 299.5 Mm/s
3. 114 rad/s
5. 25.5°, 442 nm
7. 19.5° above the horizon
9. (a) 181 Mm/s (b) 225 Mm/s (c) 136 Mm/s
11. 30.0° and 19.5° at entry; 19.5° and 30.0° at exit

13. 3.39 m

15. six times from mirror 1 and five times from mirror 2

17. 106 ps

19. 0.99 ns

21. (a) 58.9° (b) only if $\theta_1 = 0$

23. (a) 66.8 μs (b) 0.250% longer

25. 0.171°

27. 86.8°

29. 4.61°

31. 27.9°

33. 18.4°

35. (a) 24.4° (b) 37.0° (c) 49.8°

37. 1.000 08

39. 62.4°

41. 1.08 cm $< d <$ 1.17 cm

43. Skylight incident from above travels down the plastic. If the index of refraction of the plastic is greater than 1.41, the rays close in direction to the vertical are totally reflected from the side walls of the slab and from both facets at the bottom of the plastic, where it is not immersed in gasoline. This light returns up inside the plastic and makes it look bright. Where the plastic is immersed in gasoline, total internal reflection is frustrated, and the downward-propagating light passes from the plastic out into the gasoline. Little light is reflected upward, and the gauge looks dark.

45. 77.5°

47. 2.27 m

49. (a) 0.172 mm/s (b) 0.345 mm/s (c) Northward at 50.0° below the horizontal (d) Northward at 50.0° below the horizontal

51. (a) 1.76×10^7 (b) 3.25×10^{-6} degree

53. 62.2%

55. 82 reflections

57. (b) 68.5%

59. 27.5°

61. (a) It always happens. (b) 30.3° (c) It cannot happen.

63. 2.37 cm

67. (a) $n = [1 + (4t/d)^2]^{1/2}$ (b) 2.10 cm (c) violet

69. (a) 1.20 (b) 3.40 ns

Chapter 36

1. $\sim 10^{-9}$ s younger

3. 2'11"

5. 10.0 ft, 30.0 ft, 40.0 ft

7. 0.267 m behind the mirror; virtual, upright, diminished; $M = 0.026\ 7$

9. (a) -12.0 cm; 0.400 (b) -15.0 cm; 0.250 (c) upright

11. (a) $q = 45.0$ cm; $M = -0.500$ (b) $q = -60.0$ cm; $M = 3.00$ (c) Image (a) is real, inverted, and diminished. Image (b) is virtual, upright, and enlarged. The ray diagrams are like Figures 36.15a and 36.15b, respectively.

13. (a) a concave mirror with radius of curvature 2.08 m (b) 1.25 m from the object

15. (a) 15.0 cm (b) 60.0 cm

17. (a) A real image moves from $+0.600$ m to infinity, then a virtual image moves from $-\infty$ to 0. (b) 0.639 s and 0.782 s

19. 38.2 cm below the top surface of the ice

21. 8.57 cm

23. (a) 45.0 cm (b) -90.0 cm (c) -6.00 cm

25. inside the bowl at -9.01 cm

27. (a) 16.4 cm (b) 16.4 cm

29. (a) 650 cm from the lens on the opposite side from the object; real, inverted, enlarged (b) 600 cm from the lens on the same side as the object; virtual, upright, enlarged

31. 2.84 cm

33. (a) either 9.63 cm or 3.27 cm (b) 2.10 cm

37. (a) -12.3 cm, to the left of the lens (b) 0.615 (c)

39. (a) 4.00 m and 1.00 m (b) Whereas both images are real and inverted, the first has magnification -0.250, and the second, -4.00.

41. 2.18 mm away from the film plane

43. 21.3 cm

45. -4.00 diopters; a diverging lens

47. 3.50

49. -575

51. (a) -800 (b) upside down

53. (a) virtual (b) at infinity (c) 15.0 cm, -5.00 cm

55. -40.0 cm

57. 160 cm to the left of the lens; inverted; $M = -0.800$

59. 25.3 cm to the right of the mirror; virtual; upright; enlarged 8.05 times

61. 0.107 m to the right of the vertex of the hemispherical face

63. 8.00 cm

65. 1.50 m in front of the mirror; 1.40 cm (inverted)
67. (a) 30.0 cm and 120 cm   (b) 24.0 cm   (c) real, inverted, diminished, with $M = -0.250$
69. -75.0
71. (a) 44.6 diopters   (b) 3.03 diopters
73. (a) 20.0 cm to the right of the second lens; $M = -6.00$
   (b) inverted   (c) 6.67 cm to the right of the second lens; $M = -2.00$; inverted

Chapter 37

1. 1.58 cm
3. (a) 55.7 m   (b) 124 m
5. 1.54 mm
7. (a) 2.62 mm   (b) 2.62 mm
9. 11.3 m
11. (a) 13.2 rad   (b) 6.28 rad   (c) 0.012 7 degree
   (d) 0.059 7 degree
13. (a) 1.93 μm   (b) 3.00 λ (c) maximum
15. 48.0 μm
17. (a) 7.95 rad   (b) 0.453
19. (a) and (b) 19.7 kN/C at 35.0°   (c) 9.36 kN/C at 169°
21. $10.0 \sin(100\pi t + 0.927)$
23. $26.2 \sin(\omega t + 36.6°)$
25. $\pi/2$
27. $360°/N$
29. (a) green   (b) violet
31. 0.500 cm
33. No reflection maxima in the visible spectrum
35. 290 nm
37. 4.35 μm
39. 1.20 mm
41. 39.6 μm
43. 1.000 369
45. 1.25 m
47. $\lambda/2(n - 1)$
49. 5.00 km^2
51. 3.58°
53. 421 nm
55. 113 dark fringes
59. (a) $2(4h^2 + d^2)^{1/2} - 2d$ (b) $(4h^2 + d^2)^{1/2} - d$
61. $y' = (n - 1)tL/d$
63. (a) 70.6 m   (b) 136 m
65. $7.99 \sin(\omega t + 4.44 \text{ rad})$
69. 0.505 mm

Chapter 38

1. 4.22 mm
3. 0.230 mm
5. Three maxima, at 0° and near 46° to the left and right
7. 0.016 2
11. 1.00 mrad
13. 3.09 m
15. 13.1 m

17. 1.90 m if the predominant wavelength is 650 nm
19. 105 m
21. 2.10 m
23. 7.35°
25. 5.91° in first order, 13.2° in second order, 26.5° in third order
27. (a) 478.7 nm, 647.6 nm, and 696.6 nm   (b) 20.51°, 28.30°, and 30.66°
29. (a) 12 000, 24 000, 36 000   (b) 11.1 pm
31. (a) 2 800 lines   (b) 4.72 μm
33. (a) 5 orders   (b) 10 orders in the short-wavelength region
35. 93.4 pm
37. 14.4°
39. 31.9°
41. 3/8
43. (a) 54.7°   (b) 63.4°   (c) 71.6°
45. 60.5°
47. 36.9° above the horizon
49. (a) 6   (b) 7.50°
51. 632.8 nm
53. (a) 25.6°   (b) 19.0°
55. 0.244 rad = 14.0°
57. (a) 3.53×10^3 lines per centimeter   (b) 11 maxima
59. 15.4
61. (a) 41.8°   (b) 0.593   (c) 0.262 m
63. (b) 15.3 μm
65. $a = 99.5$ μm $\pm 1\%$

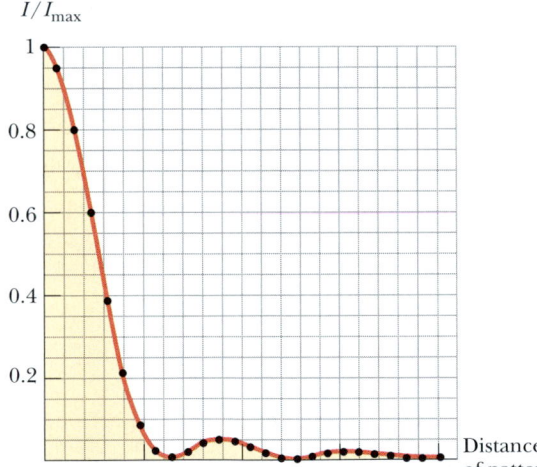

67. $\phi = 1.391\ 557\ 4$ after 17 steps or fewer
69. (b) 0.428 mm

Chapter 39

5. $0.866c$
7. 64.9/min; 10.6/min
9. 1.54 ns
11. $0.800c$

13. (a) 2.18 μs (b) 649 m
15. 0.789c
17. (a) 20.0 m (b) 19.0 m (c) 0.312c
19. 1.13 × 10^4 Hz
21. (b) 0.050 4c
23. 0.960c
25. 0.357c
27. (a) 2.73 × 10^{-24} kg·m/s (b) 1.58 × 10^{-22} kg·m/s
    (c) 5.64 × 10^{-22} kg·m/s
29. 4.50 × 10^{-14}
31. 0.285c
33. 1.63 × 10^3 MeV/c
35. (a) 939 MeV (b) 3.01 GeV (c) 2.07 GeV
37. 18.4 g/cm^3
41. (a) 0.302c (b) 4.00 f J
43. 3.88 MeV and 28.8 MeV
45. 3.18 × 10^{-12} kg, not detectable
47. 0.842 kg
49. 4.19 × 10^9 kg/s
53. (a) a few hundred seconds (b) ~ 10^8 km
55. 0.712%
57. (a) 0.946c (b) 0.160 ly (c) 0.114 yr (d) 7.50 × 10^{22} J
59. (a) 76.0 min (b) 52.1 min
61. Yes, with 18.8 m to spare
63. (b) For u small compared with c, the relativistic accelera-
    tion agrees with the classical expression. As u approaches
    c, the acceleration approaches zero; thus, the object's
    speed can never reach or surpass the speed of light.
    (c) Perform $\int (1 - u^2/c^2)^{-3/2}du = (qE/m)\int dt$ to
    obtain $u = qEct(m^2c^2 + q^2E^2t^2)^{-1/2}$ and then $\int dx =$
    $\int qEct(m^2c^2 + q^2E^2t^2)^{-1/2}dt$ to obtain $x =$
    $(c/qE)[(m^2c^2 + q^2E^2t^2)^{1/2} - mc]$.
65. (a) 6.67 ks (b) 4.00 ks
67. (a) 0.800c (b) 7.50 ks (c) 1.44 Tm, − 0.385c
    (d) 12.5 ks
69. (a) 0.544c, 0.866c (b) 0.833 m
71. 0.185c = 55.4 Mm/s
73. 6.71 × 10^8 kg

Chapter 40

1. 5.18 × 10^3 K
3. (a) 999 nm (b) The infrared region of the spectrum is
   much wider than the visible region, and the spectral distribution function is highest in the infrared.
5. (a) 70.9 kW (b) 580 nm (c) 7.99 × 10^{10} W/m
   (d) 9.42 × 10^{-1226} W/m (e) 1.00 × 10^{-227} W/m
   (f) 5.44 × 10^{10} W/m (g) 7.38 × 10^{10} W/m
   (h) 0.260 W/m (i) 2.60 × 10^{-9} W/m (j) ≈20 kW
7. (a) 2.57 eV (b) 12.8 μeV (c) 191 neV
   (d) 484 nm (visible), 9.68 cm and 6.52 m (radio waves)
9. 2.27 × 10^{30} photons/s
11. 1.34 × 10^{31}
15. (a) 296 nm, 1.01 PHz (b) 2.71 V
17. (a) only lithium (b) 0.808 eV
19. (a) 1.90 eV (b) 0.216 V
21. 148 days; absurdly large
23. (a) The Doppler effect increases the frequency incident
    on the metal. (b) 3.87 eV (c) 8.78 eV
25. (a) 488 fm (b) 268 keV (c) 31.5 keV
27. 70.1°
29. (a) 43.0° (b) 602 keV, 3.21 × 10^{-22} kg·m/s
    (c) 278 keV, 3.21 × 10^{-22} kg·m/s
31. (a) 33.0° (b) 0.785c
33. (a) 2.88 pm (b) 101°
37. (a) 5 (b) no; no
39. 634 nm, red
41. (a) 0.212 nm (b) 9.95 × 10^{-25} kg·m/s
    (c) 2.11 × 10^{-34} kg·m^2/s (d) 3.40 eV (e) − 6.80 eV
    (f) − 3.40 eV
43. (a) 3.03 eV (b) 410 nm (c) 732 THz
47. 97.5 nm
49. (a) $E_n = -54.4$ eV/n^2 for $n = 1, 2, 3, \ldots$
    (b) 54.4 eV
53. 397 fm
55. (a) 0.709 nm (b) 414 nm
57. (a) ~ 100 MeV or more (b) No. With kinetic energy
    much larger than the magnitude of its negative electric
    potential energy, the electron would immediately escape.
59. (b) No. $\lambda^{-2} + \lambda_C^{-2}$ cannot be equal to λ^{-2}.
61. $c/\sqrt{2}$ = 212 Mm/s
63. 1.36 eV
65. (a) See bottom of page.
    (b) 6.4 × 10^{-34} J·s ± 8% (c) 1.4 eV
67. The particles are separated by $r_n = (0.106$ nm$)n^2$, and
    $E_n = -6.80$ eV/n^2, for $n = 1, 2, 3, \ldots$
69. The classical frequency is $4\pi^2 m_e k_e^2 e^4/h^3 n^3$.

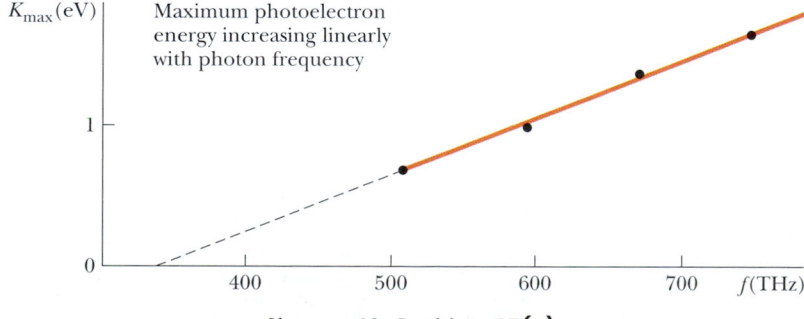

Chapter 40, Problem 65(a)

Chapter 41

1. (a) 993 nm   (b) 4.97 mm   (c) If its detection forms part of an interference pattern, the neutron must have passed through both slits. If we test to see which slit a particular neutron passes through, it will not form part of the interference pattern.
3. (b) 907 fm
5. (a) 15.1 keV   (b) 124 keV
7. At $1.00°$ on both sides of the central maximum
9. Within 1.16 mm for the electron, 5.28×10^{-32} m for the bullet
11. 1.16 Mm/s
13. (b) 0.519 fm
15. (a) 126 pm   (b) 5.27×10^{-24} kg·m/s   (c) 95.5 eV
17. (a) n

```
4 ——————————— 603 eV



3 ——————————— 339 eV


2 ——————————— 151 eV

1 ——————————— 37.7 eV
```

   (b) 2.20 nm, 2.75 nm, 4.12 nm, 4.71 nm, 6.60 nm, 11.0 nm
19. 0.793 nm
21. 202 fm, 6.14 MeV, a gamma ray
23. 0.513 MeV, 2.05 MeV, 4.62 MeV; yes
27. At $L/4$ and at $3L/4$
29. (a) 5.13×10^{-3} eV   (b) 9.41 eV   (c) The electron has much higher energy because it is much less massive.

33. (a) $E = \hbar^2/mL^2$
   (b) Requiring $\int_{-L}^{L} A^2(1 - x^2/L^2)^2\, dx = 1$ gives
   $A = (15/16L)^{1/2}$.   (c) $47/81 = 0.580$

35. (a)

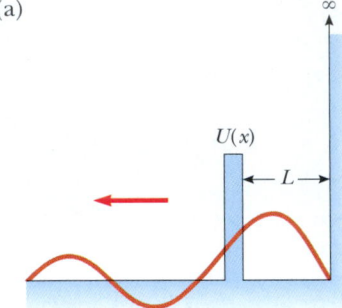

   (b) $2L$
37. (a) 0.010 3   (b) 0.990
39. By 0.959 nm, to 1.91 nm
41. 3.92%
43. (b) $b = m\omega/2\hbar$, $E = 3\hbar\omega/2$ (c) first excited state
45. (a) $B = (m\omega/\pi\hbar)^{1/4}$ (b) $(m\omega/\pi\hbar)^{1/2}\delta$
47. $\sim 10^{-10^{30}}$
49. See bottom of page.
   (c) The wave function is continuous. It shows localization by approaching zero as $x \rightarrow \pm\infty$. It is everywhere finite and can be normalized.   (d) $A = \sqrt{\alpha}$ (e) 0.632
51. 0.029 4
53. (a) 2.82×10^{-37} m   (b) 1.06×10^{-32} J
   (c) $2.87 \times 10^{-35}\%$ or more
55. (a) 434 THz   (b) 691 nm   (c) 165 peV or more
59. (a)

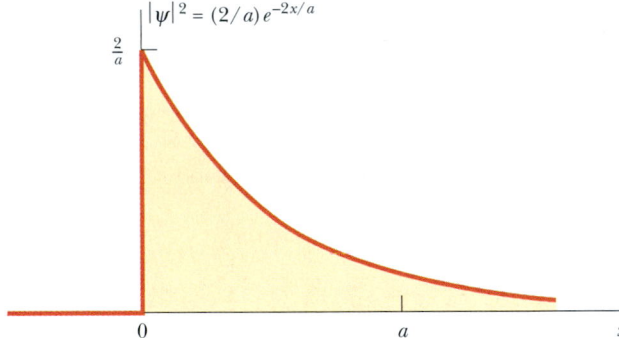

(a)

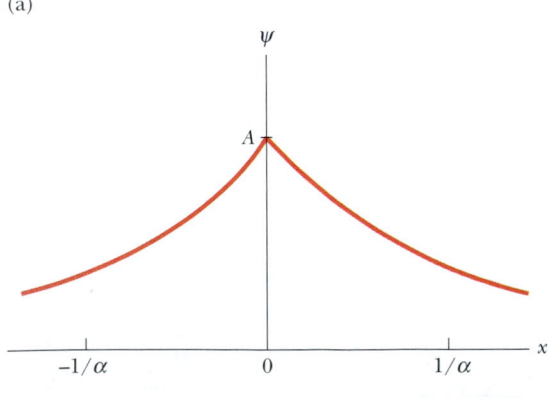

(b)

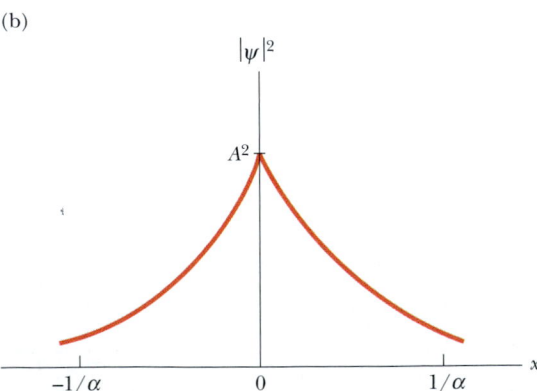

Chapter 41, Problem 49

(b) 0   (c) 0.865

61. (a) $\Delta p \geq \hbar/2r$

(b) Choosing $p \approx \hbar/r$, $E = \hbar^2/2m_e r^2 - k_e e^2/r$

(c) $r = \hbar^2/m_e k_e e^2 = a_0$ and $E = -13.6$ eV, in agreement with the Bohr theory.

63. (a) $-7k_e e^2/3d$ (b) $\hbar^2/36m_e d^2$ (c) 49.9 pm

(d) Li atom spacing is 280 pm, which is 5.62 times greater than answer (c).

65. (a) $A = (2/17L)^{1/2}$ (b) $|A|^2 + |B|^2 = 1/a$

67. (a) The light is unpolarized. It contains both horizontal and vertical electric field oscillations.   (b) The interference pattern appears, but with diminished overall intensity.   (c) The results are the same in each case.   (d) The interference pattern appears and disappears as the polarizer turns, with alternately increasing and decreasing contrast between the bright and dark fringes. The intensity on the screen is precisely zero at the center of a dark fringe four times in each revolution, when the filter axis has turned by 45°, 135°, 225°, and 315° from the vertical.   (e) Looking at the overall light energy arriving at the screen, we see a low-contrast interference pattern. After we sort out the individual photon runs into those for trial 1, those for trial 2, and those for trial 3, we have the original results replicated: The runs for trials 1 and 2 form the two blue graphs in Figure 41.3, and the runs for trial 3 build up the red graph.

Chapter 42

1. (a) 56.8 fm   (b) 11.3 N

3. (a) 3   (b) 520 km/s

5. (a) 1.31 μm   (b) 164 nm

7. (a)

n	ℓ	m_ℓ	m_s
3	2	2	1/2
3	2	2	$-1/2$
3	2	1	1/2
3	2	1	$-1/2$
3	2	0	1/2
3	2	0	$-1/2$
3	2	-1	1/2
3	2	-1	$-1/2$
3	2	-2	1/2
3	2	-2	$-1/2$

(b)

n	ℓ	m_ℓ	m_s
3	1	1	1/2
3	1	1	$-1/2$
3	1	0	1/2
3	1	0	$-1/2$
3	1	-1	1/2
3	1	-1	$-1/2$

9. (b) 0.497

11. It does, with $E = -k_e e^2/2a_0$.

13. (a) $\sqrt{6}\hbar$ (b) $\sqrt{12}\hbar$

15. $\sqrt{6}\hbar$

17. (a) 2   (b) 8   (c) 18   (d) 32   (e) 50

19. (a) 3.99×10^{17} kg/m^3 (b) 81.7 am   (c) 1.77 Tm/s

(d) 5.91×10^3 c

21. (a) 5.05×10^{-27} J/T = 31.6 neV/T   (b) Here μ_n is 1 836 times smaller than μ_B, because a proton is 1 836 times more massive than an electron. The electron has a greater charge-to-mass ratio than any other particle, which gives it a bigger "handle" for a magnetic field to twist.

23. $n = 3$; $\ell = 2$; $m_\ell = -2, -1, 0, 1,$ or 2; $s = 1$; $m_s = -1, 0,$ or 1

25. The $4s$ subshell is filled first. We would expect [Ar]$3d^4 4s^2$ to have lower energy, but [Ar]$3d^5 4s^1$ has more unpaired spins and lower energy, as suggested by Hund's rule. It is the ground-state configuration of chromium.

27. (a) Zn or Cu   (b) $1s^2 2s^2 2p^6 3s^2 3p^6 4s^2 3d^{10} 5s^2$ or $1s^2 2s^2 2p^6 3s^2 3p^6 4s^2 3d^{10} 5s^1$

29. (a) $1s, 2s, 2p, 3s, 3p, 4s, 3d, 4p, 5s, 4d, 5p, 6s, 4f, 5d, 6p, 7s$ (b) Element 15 should have valence $+3$ or -5, and it does. Element 47 should have valence -1, but it has valence $+1$. Element 86 should be inert, and it is.

31. (a) $\ell = 0$ and $m_\ell = 0$; or $\ell = 1$ and $m_\ell = -1, 0,$ or 1; or $\ell = 2$ and $m_\ell = -2, -1, 0, 1,$ or 2 (b) -6.05 eV

33. 0.031 0 nm

35. 0.072 5 nm

37. (a) 14 keV   (b) 89 pm

39. (a) 1   (b) 0.69

41. 6.21×10^{-14} J·s/m^3

43. 3.49×10^{16} photons

45. (a) 1.22×10^{-33} (b) $10^{-2\,253}$

47. (a) 4.24 PW/m^2 (b) 1.20 pJ = 7.50 MeV

51. (a) 1.57×10^{14} m$^{-3/2}$ (b) 2.47×10^{28} m^{-3} (c) 8.69×10^8 m^{-1}

53. (a) 4.20 mm   (b) 1.05×10^{19} photons   (c) 8.82×10^{16}/mm^3

57. 5.39 keV

59. (a) For Al, about 0.255 nm ~ 0.1 nm; for U, about 0.276 nm ~ 0.1 nm.   (b) For an outer electron, the nuclear charge is screened by the interior electrons. For an inner-shell electron, the nuclear charge is unscreened. The distance scale of the wave function, representing the orbital size, is proportional to a_0/Z.

61. 0.125

63. (b) 0.846 ns

65. 9.79 GHz

67. (a) 137   (b) $\lambda_C/r_e = 2\pi/\alpha$ (c) $a_0/\lambda_C = 1/2\pi\alpha$ (d) $1/R_H a_0 = 4\pi/\alpha$

Chapter 43

1. (a) 921 pN toward the other ion   (b) -2.88 eV

3. (a) $(2A/B)^{1/6}$ (b) $B^2/4A$ (c) 74.2 pm, 4.46 eV

5. (a) 40.0 μeV, 9.66 GHz   (b) 20% too large if r is 10% too small

7. 5.69 Trad/s

9. (a) 1.81×10^{-45} kg·m^2 (b) 1.62 cm

11. 0.358 eV

13. (a) 0, 364 μeV, 1.09 meV   (b) 98.2 meV, 295 meV, 491 meV

15. 558

17. 6.25×10^9

19. -7.84 eV

23. An average atom contributes 0.981 electron to the conduction band.

25. (a) 2.54×10^{28} electrons/m^3 (b) 3.15 eV
   (c) 1.05 Mm/s

27. 5.28 eV

31. (a) 1.10   (b) 1.55×10^{-25}; much smaller

33. All of the Balmer lines are absorbed, except for the red line at 656 nm, which is transmitted.

35. 1.24 eV or less; yes

37. (a) 59.5 mV   (b) -59.5 mV

39. 4.19 mA

41. 203 A to produce a magnetic field in the direction of the original field

43. (a) 61.5 THz   (b) 1.59×10^{-46} kg·m^2
   (c) 4.79 μm or 4.96 μm

45. 7

49. (a) 0.350 nm   (b) -7.02 eV   (c) $-1.20\mathbf{i}$ nN

51. (a) r_0 (b) B (c) $(a/\pi)[B/2\mu]^{1/2}$
   (d) $B - (ha/\pi)[B/8\mu]^{1/2}$

53. (b) No. The fourth term is larger than the sum of the first three.

Chapter 44

1. $\sim 10^{28}$, $\sim 10^{28}$, $\sim 10^{28}$

3. (a) 27.6 N   (b) 4.17×10^{27} m/s^2 away from the nucleus
   (c) 1.73 MeV

5. (a) 455 fm   (b) 6.04 Mm/s

7. (a) 1.90 fm   (b) 7.44 fm

9. 16.0 km

11. Z magic: He, O, Ca, Ni, Sn, Pb; N magic: isotopes of H, He, N, O, Cl, K, Ca, V, Cr, Sr, Y, Zr, Xe, Ba, La, Ce, Pr, Nd, Pb, Bi, Po

13.

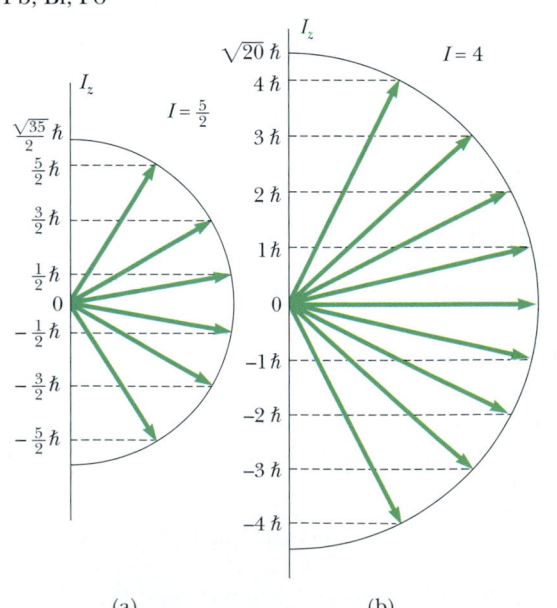

(a)   (b)

15. (a) 1.11 MeV/nucleon   (b) 7.07 MeV/nucleon
   (c) 8.79 MeV/nucleon   (d) 7.57 MeV/nucleon

17. (a) Cs   (b) La   (c) Cs

19. greater for N by 3.54 MeV

21. 7.93 MeV

23. 200 MeV

25. 1.16 ks

27. (a) 1.55×10^{-5}/s, 12.4 h   (b) 2.39×10^{13} atoms
   (c) 1.87 mCi

29. 9.47×10^9 nuclei

31. 41.8 TBq or 4.18×10^{13} decays/s

33. 4.27 MeV

35. 18.6 keV

37. (a) $e^- + p \rightarrow n + \nu$ (b) $^{15}_{8}$O atom $\rightarrow$ $^{15}_{7}$N atom $+ \nu$
   (c) 2.75 MeV

39.

235 [U] $\xrightarrow{\alpha}$ 231 [Th]

$^{235}_{92}$U $\xrightarrow{e^-}$ 227 [Pa] $\xrightarrow{\alpha}$ [Ac] $\xrightarrow{\alpha}$ 223 [Fr] $\xrightarrow{\alpha}$ 219 [At] $\xrightarrow{\alpha}$ 215 [Bi]

[Th] $\xrightarrow{\alpha}$ [Ra] $\xrightarrow{\alpha}$ [Rn] $\xrightarrow{\alpha}$ [Po] $\xrightarrow{\alpha}$ 211 [Pb]

[Bi] $\xrightarrow{\alpha}$ 207 [Tl]

[Po] $\xrightarrow{\alpha}$ [Pb]

$^{207}_{82}$Pb

41. (a) 0.281   (b) 1.65×10^{-29} (c) Radium-226 continuously creates radon

43. (a) $^{197}_{79}$Au $+ ^{1}_{0}$n $\rightarrow ^{198}_{80}$Hg $+ ^{0}_{-1}$e $+ \bar{\nu}$ (b) 7.89 MeV

45. 1 MeV

47. 8.005 3 u, 10.013 5 u

49. $\sqrt{2}$

51. (a) 2.52×10^{24} (b) 2.29 TBq   (c) 1.07 Myr

53. (a) 2.75 fm   (b) 152 N   (c) 2.62 MeV
   (d) 7.44 fm, 379 N, 17.6 MeV

55. (a) Conservation of energy   (b) Electrostatic energy of the nucleus   (c) 1.20 MeV

57. (b) 4.78 MeV

59. (b) 0.001 94 eV

61. (a) $\sim 10^{-1349}$ (b) 0.892

63. (a) 4.28 pJ   (b) 1.19×10^{57} atoms   (c) 107 Gyr

65. 2.20 μeV

69. 0.400%

71. 2.64 min

Chapter 45

1. 192 MeV

3. n $+ ^{232}$Th $\rightarrow ^{233}$Th $\rightarrow ^{233}$Pa $+ e^- + \bar{\nu}$ and
   ^{233}Pa $\rightarrow ^{233}$U $+ e^- + \bar{\nu}$

5. (a) 201 MeV   (b) 0.091 3%

7. 5.80 Mm

9. about 3 000 yr

11. 6.25×10^{19} Bq

13. (a) $2.30 \times 10^{-14} \, Z_1 \, Z_2 \, \mathrm{J}$ (b) 0.144 MeV for both
15. (a) 2.22 Mm/s   (b) $\sim 10^{-7}$ s
17. (a) 1.7×10^7 J   (b) 7.3 kg
19. 1.30×10^{25} ^{6}Li; 1.61×10^{26} ^{7}Li
21. 1.66×10^3 yr
23. (a) 2.5 mrem per x-ray   (b) 5 rem/yr is 38 times 0.13 rem/yr
25. 2.09×10^6 s
27. 1.14 rad
29. (a) 3.12×10^7 (b) 3.12×10^{10} electrons
31. (a) 10   (b) 10^6 (c) 1.00×10^8 eV
33. 4.45×10^{-8} kg/h
35. (a) $\sim 10^6$ (b) $\sim 10^{-15}$ g
37. (a) 8×10^4 eV   (b) 4.62 MeV and 13.9 MeV (c) 1.03×10^7 kWh
39. 0.375% for D–T fusion, which is about four times larger than 0.095 0% for ^{235}U fission.
41. 482 Ci, less than the fission inventory by on the order of 100 million times
43. 2.56×10^4 kg
45. (a) 2.65 GJ   (b) The fusion energy is 78.0 times larger.
47. (a) 4.91×10^8 kg/h $= 4.91 \times 10^5$ m^3/h   (b) 0.141 kg/h
49. (a) 15.5 cm   (b) 51.7 MeV   (c) The number of decays per second is the decay rate R and the energy released in each decay is Q. Then the energy released per time is $\mathscr{P} = QR$.   (d) 227 kJ/yr   (e) 3.18 J/yr
51. 14.1 MeV
53. (a) 2.24×10^7 kWh   (b) 17.6 MeV   (c) 2.34×10^8 kWh (d) 9.36 kWh   (e) Coal is cheap at this moment in human history. We hope that safety and waste disposal problems can be solved so that nuclear energy can be affordable before scarcity drives up the price of fossil fuels.
55. (b) 26.7 MeV
57. (a) 5×10^7 K

(b)

Reaction	Q (MeV)
$^{12}\mathrm{C} + {}^1\mathrm{H} \rightarrow {}^{13}\mathrm{N} + \gamma$	1.94
$^{13}\mathrm{N} \rightarrow {}^{13}\mathrm{C} + e^+ + \nu$	1.20
$e^+ + e^- \rightarrow 2\gamma$	1.02
$^{13}\mathrm{C} + {}^1\mathrm{H} \rightarrow {}^{14}\mathrm{N} + \gamma$	7.55
$^{14}\mathrm{N} + {}^1\mathrm{H} \rightarrow {}^{15}\mathrm{O} + \gamma$	7.30
$^{15}\mathrm{O} \rightarrow {}^{15}\mathrm{N} + e^+ + \nu$	1.73
$e^+ + e^- \rightarrow 2\gamma$	1.02
$^{15}\mathrm{N} + {}^1\mathrm{H} \rightarrow {}^{12}\mathrm{C} + {}^4\mathrm{He}$	4.97
Overall	26.7

(c) Most of the neutrinos leave the star directly after their creation, without interacting with any other particles.

Chapter 46

1. 453 ZHz; 662 am
3. 118 MeV
5. $\sim 10^{-18}$ m
7. $\sim 10^{-23}$ s
9. 67.5 MeV, 67.5 MeV/c, 16.3 ZHz
11. $\Omega^+ \rightarrow \overline{\Lambda}^0 + \mathrm{K}^+$ $\overline{\mathrm{K}}_S^0 \rightarrow \pi^+ + \pi^-$ $\Lambda^0 \rightarrow \overline{\mathrm{p}} + \pi^+$
$\overline{\mathrm{n}} \rightarrow \overline{\mathrm{p}} + e^+ + \nu_e$
13. (b) The second reaction violates strangeness conservation.
15. (a) $\overline{\nu}_\mu$ (b) ν_μ (c) $\overline{\nu}_e$ (d) ν_e (e) ν_μ (f) $\overline{\nu}_e + \nu_\mu$
17. (a), (c), and (f) violate baryon number conservation. (b), (d), and (e) can occur.   (f) violates muon–lepton number conservation
19. (a) ν_e (b) ν_μ (c) $\overline{\nu}_\mu$ (d) $\nu_\mu + \overline{\nu}_\tau$
21. (b) and (c) conserve strangeness.   (a), (d), (e), and (f) violate strangeness conservation.
23. (a) not allowed; violates conservation of baryon number (b) strong interaction   (c) weak interaction (d) weak interaction   (e) electromagnetic interaction
25. (a) K$^+$ (b) Ξ^0 (c) π^0
27. (a) 3.34×10^{26} e$^-$, 9.36×10^{26} u, 8.70×10^{26} d (b) $\sim 10^{28}$ e$^-$, $\sim 10^{29}$ u, $\sim 10^{29}$ d . My strangeness, charm, truth, and beauty are zero.
29. $m_\mathrm{u} = 312$ MeV/c^2 $m_\mathrm{d} = 314$ MeV/c^2
31. (a) The reaction $\overline{\mathrm{u}}\mathrm{d} + \mathrm{uud} \rightarrow \overline{\mathrm{s}}\mathrm{d} + \mathrm{uds}$ has a total of 1 u, 2 d, and 0 s quarks originally and finally.   (b) The reaction $\overline{\mathrm{d}}\mathrm{u} + \mathrm{uud} \rightarrow \overline{\mathrm{s}}\mathrm{u} + \mathrm{uus}$ has a net of 3 u, 0 d, and 0 s before and after.   (c) $\overline{\mathrm{u}}\mathrm{s} + \mathrm{uud} \rightarrow \overline{\mathrm{s}}\mathrm{u} + \overline{\mathrm{s}}\mathrm{d} + \mathrm{sss}$ shows conservation at 1 u, 1 d, and 1 s quark.   (d) The process $\mathrm{uud} + \mathrm{uud} \rightarrow \overline{\mathrm{s}}\mathrm{d} + \mathrm{uud} + \overline{\mathrm{d}}\mathrm{u} + \mathrm{uds}$ nets 4 u, 2 d, 0 s initially and finally; the mystery particle is a Λ^0.
33. (a) Σ^+ (b) π^- (c) K^0 (d) Ξ^-
37. (a) $0.383\,c$ (b) 6.76×10^9 ly
39. 6.00
41. (a) 1.06 mm   (b) microwave
43. (a) ~ 100 MeV/c^2 (b) charged or neutral pion
45. one part in 50 000 000
47. $\sim 10^{14}$
49. 5.35 MeV and 32.3 MeV
51. 74.4 MeV
53. 9.26 cm
55. 2.52×10^3 K
57. (a) Z^0 boson   (b) gluon

Page numbers in *italics* indicate illustrations; page numbers followed by an "n" indicate footnotes, page numbers followed by "t" indicate tables.

Standard Abbreviations and Symbols for Units

Symbol	Unit	Symbol	Unit
A	ampere	K	kelvin
Å	angstrom	kcal	kilocalorie
u	atomic mass unit	kg	kilogram
atm	atmosphere	kmol	kilomole
Btu	British thermal unit	L	liter
C	coulomb	lb	pound
°C	degree Celsius	m	meter
cal	calorie	min	minute
eV	electron volt	mol	mole
°F	degree Fahrenheit	N	newton
F	farad	Pa	pascal
ft	foot	rad	radian
G	gauss	rev	revolution
g	gram	s	second
H	henry	T	tesla
h	hour	V	volt
hp	horsepower	W	watt
Hz	hertz	Wb	weber
in.	inch	Ω	ohm
J	joule		

Mathematical Symbols Used in the Text and Their Meaning

Symbol	Meaning
$=$	is equal to
$\equiv$	is defined as
$\neq$	is not equal to
$\propto$	is proportional to
$\sim$	is on the order of
$>$	is greater than
$<$	is less than
$>>(<<)$	is much greater (less) than
$\approx$	is approximately equal to
Δx	the change in x
$\sum\limits_{i=1}^{N} x_i$	the sum of all quantities x_i from $i = 1$ to $i = N$
$|x|$	the magnitude of x (always a nonnegative quantity)
$\Delta x \rightarrow 0$	Δx approaches zero
$\dfrac{dx}{dt}$	the derivative of x with respect to t
$\dfrac{\partial x}{\partial t}$	the partial derivative of x with respect to t
$\int$	integral

Conversions[a]

Length
1 in. = 2.54 cm (exact)
1 m = 39.37 in. = 3.281 ft
1 ft = 0.304 8 m
12 in. = 1 ft
3 ft = 1 yd
1 yd = 0.914 4 m
1 km = 0.621 mi
1 mi = 1.609 km
1 mi = 5 280 ft
$1 \text{ Å} = 10^{-10}$ m
$1 \mu\text{m} = 1\mu = 10^{-6}$ m $= 10^3$ nm
1 lightyear $= 9.461 \times 10^{15}$ m

Area
$1 \text{ m}^2 = 10^4 \text{ cm}^2 = 10.76 \text{ ft}^2$
$1 \text{ ft}^2 = 0.092\,9 \text{ m}^2 = 144 \text{ in.}^2$
$1 \text{ in.}^2 = 6.452 \text{ cm}^2$

Volume
$1 \text{ m}^3 = 10^6 \text{ cm}^3 = 6.102 \times 10^4 \text{ in.}^3$
$1 \text{ ft}^3 = 1\,728 \text{ in.}^3 = 2.83 \times 10^{-2} \text{ m}^3$
$1 \text{ L} = 1\,000 \text{ cm}^3 = 1.057\,6 \text{ qt} = 0.035\,3 \text{ ft}^3$
$1 \text{ ft}^3 = 7.481 \text{ gal} = 28.32 \text{ L} = 2.832 \times 10^{-2} \text{ m}^3$
$1 \text{ gal} = 3.786 \text{ L} = 231 \text{ in.}^3$

Mass
1 000 kg = 1 t (metric ton)
1 slug = 14.59 kg
$1 \text{ u} = 1.66 \times 10^{-27} \text{ kg} = 931.5 \text{ MeV}/c^2$

Force
1 N = 0.224 8 lb
1 lb = 4.448 N

Velocity
1 mi/h = 1.47 ft/s = 0.447 m/s = 1.61 km/h
1 m/s = 100 cm/s = 3.281 ft/s
1 mi/min = 60 mi/h = 88 ft/s

Acceleration
$1 \text{ m/s}^2 = 3.28 \text{ ft/s}^2 = 100 \text{ cm/s}^2$
$1 \text{ ft/s}^2 = 0.304\,8 \text{ m/s}^2 = 30.48 \text{ cm/s}^2$

Pressure
$1 \text{ bar} = 10^5 \text{ N/m}^2 = 14.50 \text{ lb/in.}^2$
1 atm = 760 mm Hg = 76.0 cm Hg
$1 \text{ atm} = 14.7 \text{ lb/in.}^2 = 1.013 \times 10^5 \text{ N/m}^2$
$1 \text{ Pa} = 1 \text{ N/m}^2 = 1.45 \times 10^{-4} \text{ lb/in.}^2$

Time
$1 \text{ yr} = 365 \text{ days} = 3.16 \times 10^7 \text{ s}$
$1 \text{ day} = 24 \text{ h} = 1.44 \times 10^3 \text{ min} = 8.64 \times 10^4 \text{ s}$

Energy
1 J = 0.738 ft·lb
1 cal = 4.186 J
$1 \text{ Btu} = 252 \text{ cal} = 1.054 \times 10^3 \text{ J}$
$1 \text{ eV} = 1.6 \times 10^{-19} \text{ J}$
$1 \text{ kWh} = 3.60 \times 10^6 \text{ J}$

Power
1 hp = 550 ft·lb/s = 0.746 kW
1 W = 1 J/s = 0.738 ft·lb/s
1 Btu/h = 0.293 W

Some Approximations Useful for Estimation Problems
1 m ≈ 1 yd
1 kg ≈ 2 lb
$1 \text{ N} \approx \frac{1}{4} \text{ lb}$
$1 \text{ L} \approx \frac{1}{4} \text{ gal}$

1 m/s ≈ 2 mi/h
$1 \text{ yr} \approx \pi \times 10^7 \text{ s}$
60 mi/h ≈ 100 ft/s
$1 \text{ km} \approx \frac{1}{2} \text{ mi}$

[a] See Table A.1 of Appendix A for a more complete list.

The Greek Alphabet

Alpha	A	α	Iota	I	ι	Rho	P	ρ
Beta	B	β	Kappa	K	κ	Sigma	Σ	σ
Gamma	Γ	γ	Lambda	Λ	λ	Tau	T	τ
Delta	Δ	δ	Mu	M	μ	Upsilon	Υ	υ
Epsilon	E	ϵ	Nu	N	ν	Phi	Φ	ϕ
Zeta	Z	ζ	Xi	Ξ	ξ	Chi	X	χ
Eta	H	η	Omicron	O	o	Psi	Ψ	ψ
Theta	Θ	θ	Pi	Π	π	Omega	Ω	ω